THE EUROPEAN GARDEN FLORA

The European Garden Flora is the definitive manual for the accurate identification of cultivated ornamental flowering plants. Designed to meet the highest scientific standards, the vocabulary has nevertheless been kept as uncomplicated as possible so that the work is fully accessible to the informed gardener as well as to the professional botanist. This new edition has been thoroughly reorganised and revised, bringing it into line with modern taxonomic knowledge. Although European in name, the Flora covers plants cultivated in most areas of the United States and Canada as well as in non-tropical parts of Asia and Australasia.

Volume III contains accounts of 47 families, including those formerly included in the Leguminosae (Mimosaceae, Caesalpiniaceae, Fabaceae) as well as the large and important Rosaceae. Also included are those families formerly covered by the name Saxifragaceae (Saxifragaceae in the strict sense, Penthoraceae, Grossulariaceae, Parnassiaceae, Hydrangeaceae and Escalloniaceae).

James Cullen has been a professional plant taxonomist for over 50 years, working particularly on the classification and identification of plants in cultivation (especially Rhododendron) at Liverpool and Edinburgh Universities, at the Royal Botanic Garden Edinburgh, and in Cambridge. With the late Dr S. M. Walters, he was the initiator of the first edition of The European Garden Flora and is responsible for two spin-offs, The Orchid Book (1992) and Manual of North European Garden Plants (2001).

Sabina Knees is a taxonomist at the Royal Botanic Garden Edinburgh and although now working on plants of the Middle East, particularly the Flora of the Arabian Peninsula and Socotra, she spent over 20 years working as a horticultural taxonomist for the Royal Horticultural Society and the Royal Botanic Garden Edinburgh and is a founder member of the Horticultural Taxonomy Group (HORTAX). She was editor of The New Plantsman for seven years and worked initially as a research associate and then as a member of the editorial committee on the first edition of The European Garden Flora.

Suzanne Cubey has worked at the Royal Botanic Garden Edinburgh (RBGE) since 1987 originally as a researcher, and then later becoming the Assistant Secretary on the Editorial Board for the first edition of The European Garden Flora. Since 2005 her main role has been as Assistant Herbarium Curator with particular responsibility for the Cultivated Plants, where she curates the cultivated specimens in the RBGE herbarium and manages the vouchering of research material from the living collections.

THE EUROPEAN GARDEN FLORA
FLOWERING PLANTS

*A manual for the identification of plants cultivated
in Europe, both out-of-doors and under glass*

VOLUME III

Angiospermae – Dicotyledons

Second edition

edited by

James Cullen, Sabina G. Knees, H. Suzanne Cubey

assisted by

J.M.H. Shaw, P. Harrold, L. Banfield,
A. Laporte-Bisquit, M.F. Gardner, S. Neale,
G.D. Rowley, N. Zantout & C.D. Brickell

sponsored by

The Stanley Smith (UK) Horticultural Trust
The Royal Botanic Garden Edinburgh
Cambridge University Botanic Garden

CAMBRIDGE
UNIVERSITY PRESS

CAMBRIDGE UNIVERSITY PRESS
Cambridge, New York, Melbourne, Madrid, Cape Town,
Singapore, São Paulo, Delhi, Tokyo, Mexico City

Cambridge University Press
The Edinburgh Building, Cambridge CB2 8RU, UK

Published in the United States of America
by Cambridge University Press, New York

www.cambridge.org
Information on this title: www.cambridge.org/9780521761550

Second edition first published 2011
First edition published 1984–2000

Printed in the United Kingdom at the University Press, Cambridge

A catalogue record for this publication is available from the British Library

Library of Congress Cataloging-in-Publication Data

The European garden flora, flowering plants: a manual for the identification of
plants cultivated in Europe, both out-of-doors and under glass/edited by James
Cullen, Sabina G. Knees, H. Suzanne Cubey; assisted by J.M.H. Shaw ... [et al.]. –
2nd ed.
5 v.; cm.
Includes index.
Contents: Vol. 1. Monocotyledons: Alismataceae to Orchidaceae – Vol. 2.
Dicotyledons: Casuarinaceae to Cruciferae – Vol. 3. Dicotyledons: Resedaceae to
Cyrillaceae – Vol. 4. Dicotyledons: Aquifoliaceae to Hydrophyllaceae – Vol. 5.
Dicotyledons: Boraginaceae to Compositae.
ISBN 978-0-521-76167-3 (set) – ISBN 978-0-521-76147-5 (Vol. 1) – ISBN 978-
0-521-76151-2 (Vol. 2) – ISBN 978-0-521-76155-0 (Vol. 3) – ISBN 978-0-521-
76160-4 (Vol. 4) – ISBN 978-0-521-76164-2 (Vol. 5)
1. Angiosperms–Europe–Identification. 2. Monocotyledons–Europe–Identification.
3. Dicotyledons–Europe–Identification. 4. Plants, Ornamental–Europe–
Identification. I. Cullen, J. (James) II. Knees, Sabina. III. Cubey, H. Suzanne,
1964- IV. Shaw, J. M. H.
QK495.A1E97 2011
635.9094–dc23   2011021496

Volume III ISBN 978-0-521-76155-0 Hardback

CONTENTS

DICOTYLEDONS

MAPS AND FIGURES

CONTRIBUTORS TO THE FIRST EDITION

European Advisers

Mrs E.M. Allen
Professor C.D.K. Cook
Professor H. Ern

Dr H. Heine
Sr A. Pañella
Professor P. Wendelbo

Contributors

J. Akeroyd
J.C.M. Alexander
D. Allison
S. Andrews
G. Argent
K.B. Ashburner
T. Ayres
M-J. Balkwill
P.G. Barnes
P. Berry
S.A. Bird
J.J. Bos
A. Brady
W. Brandenburg
C.D. Brickell
M.I.H. Brooker
N. Brown
S.T. Buczacki
R.K. Brummitt
G.S. Bunting
E.J. Campbell
D.C.G. Cann
S. Carter
D.F. Chamberlain
M. Cheek
J.Y. Clark
W.D. Clayton
J. Compton
B. Conn
A.J. Coombes
C.J. Couper
E.J. Cowley
C.J. Cox
T.B. Croat
J. Cullen
S.J.M. Droop

J. Dransfield
W.W. Eddie
J. Edmondson
U. Eggli
H. Ern
A.R. Ferguson
D. Fräenz
C. Fraile
D. Frodin
J. Fryer
T.W.J. Gadella
A. Galloway
M.F. Gardner
C. Gorman
R. Gornall
Z. Gowler
V. Graham
M.L. Grant
P.S. Green
C. Grey-Wilson
N. Groendijk-Wilders
R.F.L. Hamilton
E.H. Hamlet
C.J. Harrison
E. Haston
F. Hibberd
P. Hoch
J. Howe
G. Hull
D.R. Hunt
J.A. Le Huquet
R. Hyam
B. Hylmö
C. Innes
E. van Jaarsveld
N. Jacobsen

A. James
V. Jamnicky
M. Jebb
S.J.L. Johnstone
L. Jones
P.C. de Jong
P.-M. Jörgensen
S.L. Jury
J. Kendall
C.J. King
S.G. Knees
J.M. Lamond
D.J. Leedy
J.M. Lees
A.C. Leslie
B.E. Leuenberger
D.Z. Li
H. McAllister
B. MacBryde
D. McClintock
F. McIntosh
D. McKean
M.T. Madison
V. Malécot
J. Mann Taylor
J.B. Martinez-Laborde
B. Mathew
V.A. Matthews
H.S. Maxwell
R.D. Meikle
M. Mendum
D. Middleton
R.R. Mill
D.M. Miller
R.J. Mitchell
C.M. Mitchem

B.D. Morley
E.C. Nelson
D. Nicol-Brown
D.H. Nicolson
W.P. Oates
U. Oster
J. Parnell
M. Peña Chocarro
D.M. Percy
F.H. Perring
D. Philcox
D. Pigott
D.A.H. Rae
J.A. Ratter
S.S. Renner
S. Renvoize
A.J. Richards
J.E. Richardson
I.B.K. Richardson
M. de Ridder
E.M. Rix
N.K.B. Robson
G.D. Rowley
A. Rutherford
M.J.P. Scannell
P. Sell
J.M.H. Shaw
A.J. Silverside
N. Sinclair
R.M. Smith
S.A. Spongberg
L. Springate
C.A. Stace
W.T. Stearn
D.C. Stuart
G.J. Swales

D.M. Synnott
N.P. Taylor
M.C. Tebbitt
S.A. Thompson
S. Thornton-Wood
J. Tun-Garrido

T. Upson
R. Vickery
F.T. de Vries
K.S. Walter
S.M. Walters
T. Wangdi

J.D. Wann
M.C. Warwick
M.F. Watson
D.A. Webb
C. Whitefoord
C.M. Whitehouse

A.C. Whiteley
D.O. Wijnands
J.R.I. Wood
P.J.B. Woods
P.F. Yeo

PREFACE TO THE SECOND EDITION

Work on the first edition of this Flora was begun in 1976, and publication of 6 volumes followed reasonably regularly thereafter (vol. II, 1984; vol. I, 1986; vol. III, 1989; vol. IV, 1995; vol. V, 1997 and vol. VI, 2000). During the whole period of production, the work of 175 authors and contributors was supervised by an editorial committee, chaired from 1977 to 1989 by the late Dr S.M. Walters and by myself from 1989 to 2000. The membership of the committee varied, but included (at different times) Dr J.C.M. Alexander, the late Mr A. Brady, Mr C.D. Brickell, Dr J.R. Edmondson, the late Mr P.S. Green, Professor V.H. Heywood, Professor P.-M. Jørgensen, Dr S.L. Jury, Ms S.G. Knees, Dr A. Leslie, the late Mr J. Lewis, Ms V.A. Matthews, Ms H.S. Maxwell, Mrs D.M. Miller, Dr E.C. Nelson, Dr N.K.B. Robson, the late Professor D.A. Webb, the late Professor D.O. Wijnands and the late Dr P.F. Yeo. Over the whole period the project was sponsored by the Royal Horticultural Society of London, the Royal Botanic Garden Edinburgh (where the secretariat was based) and the Stanley Smith Horticultural Trust. Financial and other support was provided by the Cory Foundation of Cambridge University Botanic Garden, the Coke Trust, the Cruden Foundation, the Edinburgh Botanic Garden (Sibbald) Trust, the Humphrey Whitbread Trust, the John S. Cohen Foundation, the Lok Wan Tho Memorial Fund, the Royal Botanic Garden Edinburgh (especially staff time and office space, logistic and IT support), the Royal Horticultural Society, the Stanley Smith Horticultural Trust (the major provider of funds over the whole period), the Will Charitable Trust, the William Adlington Cadbury Charitable Trust and the Wolfson Industrial Fellowship. More detail on the history of the project can be found in an article in *The Plantsman* (**7**: pp. 164–7, 2000).

Following completion of the work on volume VI in 1999, Ms S.G. Knees was appointed to the Stanley Smith Horticultural Fellowship at the Royal Botanic Garden Edinburgh, and for 3 years worked on the correction, updating and improvement of the text (in both paper and computerised form); this work now provides the foundation for the new second edition, which has been edited by myself, Ms S.G. Knees and Ms H.S. Cubey (née Maxwell), assisted by Mr J.M.H. Shaw (Nottingham), Ms L. Banfield (RBG Edinburgh), Mr P. Harrold (RBG Edinburgh), Ms A. Laporte-Bisquit (RBG Edinburgh), Mr M.F. Gardner (RBG Edinburgh), Ms S. Neale (RBG Edinburgh), Mr G.D. Rowley (Reading), Miss N. Zantout (RBG Edinburgh and Bremen) and Mr C.D. Brickell (Pulborough).

We are particularly grateful to the Royal Botanic Garden Edinburgh for allowing Ms Knees and Ms Cubey time from their official duties at the garden to complete the necessary changes to, and organisation of, the text. We are also grateful to all those contributors who have revised the accounts written for the first edition.

The current edition contains many corrections, updates and improvements to the orginal text. The major change is the omission of the ferns and their allies (Pteridophyta) and the Gymnosperms. These groups, considered as cultivated plants, do not fit well with a Flora-type treatment, being mainly cultivars of relatively few genera and species; there are many, more appropriate, guides to them available. Because of this omission, it has been thought sensible to combine the original volumes I & II into a single large volume (I) containing all the Monocotyledons; volumes II–V (equivalent to III, IV, V & VI of edition 1) contain the Dicotyledons. Contributions that have been only lightly revised are presented here under the name of the original author. Those that have been more heavily altered are under the name of the original author plus that of the editor.

Because the original was first conceived during the 1970s, long before the current and continuing storm of changes to family and generic placements caused by the introduction of cladistic and molecular taxonomic methods, the text was organised and families and genera were recognised according to the system presented in H. Melchior's edition (edn 12) of *Syllabus der Pflanzenfamilien* (1964). Use of this system is no longer tenable, but the editors were loath to accept any of the current labile and uncertainly documented systems (e.g. those developed by the Angiosperm Phylogeny Group) as these are more likely to confuse users of this book rather than help them. Instead, for the recognition of genera and families, we have chosen to follow R.K. Brummitt's *Vascular Plants, Families and Genera* (1992), which provides a published and easily available listing of all the families and genera recognised at Kew in the 1990s. Because Brummitt's list is in alphabetical order, the order of families and genera used here is rather arbitrary, but retains essentially that of the genera found in edition 1, thus maintaining some degree of continuity with the original publication.

Many other changes in detail have been made to the text, most of which do not require mention here. However, it is important to note that, though we have tried to list (under the heading 'Literature') taxonomically useful, up-to-date literature for all genera and families, it does not necessarily follow that the text of the Flora will contain any or all of the changes proposed in this literature. Further relevant literature continues to appear on a regular basis and interested readers are referred to *Kew Record*, or the extremely valuable *ePIC* plant information database available on the website of the Royal Botanic Gardens Kew (www.kew.org.uk).

Each volume contains its own index, and a consolidated index to accepted Families and Genera will be found in volume V.

James Cullen
Director, Stanley Smith (UK) Horticultural Trust
Cambridge, April, 2009.

PREFACE TO THE FIRST EDITION

The ideas that led to the production of *The European Garden Flora* derived from two independent sources, and were developed during the late 1960s and early 1970s. One of us (SMW), then Director of the University of Cambridge Botanic Garden and heavily involved in the completion of *Flora Europaea*, had been thinking of how botanic gardens might make some contribution to the problems of horticultural taxonomy and classification. He had indeed discussed these matters with the British amateur botanist David McClintock, who was himself very anxious to further research towards the production of a Flora for gardens. The other (JC), then Assistant Regius Keeper at the Royal Botanic Garden Edinburgh, formerly Assistant Director at Liverpool University Botanic Gardens, Ness, had independently developed similar ideas and had produced some samples to show how such a Flora might be prepared.

These two sources came together at the 1976 Scientific Review Group (of which SMW was a member) of the work of the Royal Botanic Garden Edinburgh. Discussion between us soon revealed our common concerns. The need for such a Flora and the recommendation that the RBG Edinburgh should be actively involved in its production were written into the Review Group report, to become part of the official policy of the Garden.

A meeting of the British botanists and horticulturists who might wish to be involved in such a production was called by us in Autumn 1976 and held at the offices of the Royal Horticultural Society in London. It was agreed that such a Flora should be produced, that its content should be mainly those species, subspecies and (botanical) varieties that were grown in Europe (both outdoors and under glass), that it should be organised from RBG Edinburgh, that it should use a minimum of taxonomic jargon, and that it should seek contributors from all parts of the world. The group at this first meeting constituted itself into the Editorial Committee (which was later strengthened by the inclusion of members from other countries in Europe) to oversee the project, and to organise and administer the grants necessary to see the production through. A twenty-year timescale was envisaged.

For the first two years, the preparation of lists of species that could or should be included in the projected Flora was the main concern. For this purpose, a vacation student (Ms Margaret McDonald, the first of many) was employed by RBG Edinburgh to make an index from all the available nurserymen's catalogues. This early predecessor of *The Plant Finder* became the basis for the content of the Flora. The employment and training of vacation students has been a constant feature of the development of the project.

In 1979 the Editorial Committee approached the Stanley Smith Horticultural Trust for a grant to employ a Research Associate to work full-time at Edinburgh. The application was successful, and in October 1979, Dr Crinan Alexander began work in this capacity. He continued, supported by grants from other donors, until he was succeeded by Sabina G. Knees in 1987. Since 1994 various other workers have been employed on an ad hoc basis; they are individually acknowledged in volumes IV, V and VI.

The first volume of the Flora (volume II) was published in 1984; subsequent volumes followed in 1986, 1989, 1995 and 1997; this, the final volume, will be published in 2000. If we take the originally envisaged twenty-year span as beginning with the first appointment of the Research Associate, then the production has slipped by only a single year.

Full acknowledgement of the contributions of the various authors (now over 200), editors and collaborating institutions has been given in the individual volumes. Here we would like to add some more general acknowledgements. Firstly, to the Royal Botanic Garden Edinburgh, which has fully honoured its commitment to the 1976 Review Group by providing staff-time, secretariat and logistical support (telephones, stationery, computers, etc.) over the whole 24-year period. Similarly, though on a smaller scale, the University of Cambridge Botanic Garden has provided background support. The Royal Horticultural Society has provided funds and facilities over a long period, especially since 1987. Several grant-giving bodies have been associated with the project; again, these are acknowledged in the various volumes, but a special acknowledgment is due to the Stanley Smith Horticultural Trust (now the Stanley Smith (UK) Horticultural Trust) and its first Director, Sir George Taylor FRS, for its continued support, but especially for its faith in the project at the beginning. Finally, all the volumes have been published by Cambridge University Press; we are grateful to them for persisting with the project over such a long period and especially to the two editors who have seen the process through – Martin G. Walters and Dr Maria Murphy.

James Cullen
S. Max Walters
Cambridge, September 1999

xiii

ACKNOWLEDGEMENTS

In addition to the contributors mentioned above, the editors are grateful to the following for editorial support: Bridie Andrews, Jane Bulleid, Christine Couper, Rosanna Colangeli, M.J. Howard, Linda McGee, Valerie Muirhead and Moira Watson. Specialist taxonomic advice was given by Wilhelm Barthlott, David Chamberlain, Ulf Molau and Paddy Woods. Futher horticultural advice was provided by J. Bogner, L. Buchan, R.U. Cranston, P. Cribb, J.D. Donald, R. Kerby, L.A. Lauener, Mrs E. Molesworth-Allan, S. Mayo, A. Paxton, R.L. Shaw, Dr. D. Wills, and Luwasa (Hydroculture) Ltd. Many other individuals provided information over the years and their help is also acknowledged here.

Special thanks are due to Sophie Durlacher, who undertook the arduous task of project managing this second edition.

We are grateful to J.C.M Alexander, Diane Allison, Peter Barnes, Anke Berg, Elaine Campbell, Eleanor Catherine, Mark Coode, Urs Eggli, M. Flanagan, Christine Grey-Wilson, Geoffrey Herklotts, Frances Hibberd, Michael Hickey, Christopher Hogg, Julia Howe, Sarah Johnstone, Christabel King, Rosemary King, Debbie Maizells, Victoria A. Matthews, Suzanne Maxwell, Gillian Meadows, Mary Mendum, Diane Nicol-Brown, Susan Oldfield, Christina Oliver, Donald Pigott, Rodella Purves, Sally Rae, Gordon Rowley, Alison Rutherford, Julian Shaw, Niki Simpson, Rosemary M. Smith, Mark Tebbitt, Margaret Tebbs, Vicky Thomas, Femke de Vries, Maureen Warwick and Rosemary Wise for the preparation of illustrations and also to Edinburgh University Press for allowing us to re-use illustrations from volumes 6 (1978) and 7 (1982) of P.H. Davies (ed.) *Flora of Turkey*.

INTRODUCTION

Amenity horticulture (gardening, landscaping, etc.) touches human life at many points. It is a major leisure activity for very large numbers of people, and is a very important means of improving the environment. The industry that has grown up to support this activity (the nursery trade, landscape architecture and management, public parks, etc.) is a large one, employing a considerable number of people. It is clearly important that the basic material of all this activity, i.e. plants, should be readily identifiable, so that both suppliers and users can have confidence that the material they buy and sell is what it purports to be.

The problems of identifying plants in cultivation are many and various, and derive from several sources, which may be summarised as follows:

(a) Plants in cultivation have originated in all parts of the world, many of them from areas whose wild flora is not well known or documented. Many have been introduced, lost and reintroduced under different names.

(b) Plants in gardens are growing under conditions to which they are not necessarily well adapted, and may therefore show morphological and physiological differences from the original wild stocks.

(c) All plants that become established in cultivation have gone through a process of selection, some of it conscious (selection of the 'best' variants, etc.), some of it unconscious (by methods of cultivation and particularly propagation), so that, again, the populations of a species in cultivation may differ significantly from the wild populations.

(d) Many garden plants have been 'improved' by hybridisation (deliberate or accidental), and so, again, differ from the original stocks.

(e) Finally, and perhaps most importantly, the scientific study of plant classification (taxonomy) has concentrated mainly on wild plants, largely ignoring material in gardens.

Nevertheless, the classification of garden plants has a long and distinguished history. Many of the herbals of pre-Linnaean times (i.e. before 1753) consist partly or largely of descriptions of plants in gardens, and this tradition continued, and perhaps reached its peak, in the late eighteenth and early nineteenth centuries – the period following the publication of Linnaeus' major works, when exploration of the world was at its height. This is the period that saw the founding of *Curtis' Botanical Magazine* (1787) and the publication of J.C. Loudon's *Encyclopaedia of plants* (1829, and many subsequent editions).

The further development of plant taxonomy, from about the middle of the nineteenth century to the present, has seen an increasing divergence between garden and scientific taxonomy, leading on the one hand to such works as the *The Royal Horticultural Society's dictionary of gardening* (1951 and reprints), based on G. Nicholson's *Illustrated dictionary of gardening* (1884–1888), *The new Royal Horticultural Society's dictionary of gardening* (1992), and the very numerous popular, usually illustrated, works on garden plants available today, and, on the other hand, to the Floras, Revisions and Monographs of scientific taxonomy.

Despite this divergence, a number of plant taxonomists realised the importance of the classification and identification of cultivated plants, and produced works of considerable scientific value. Foremost among these stands L.H. Bailey, editor of *The standard cyclopedia of horticulture* (1900, with several subsequent reprints and editions), author of *Manual of cultivated plants* (1924, edn 2, 1949) and founder of the journals *Gentes Herbarium* and *Baileya*. Other important workers in this field are T. Rumpler (*Vilmorin's Blumengärtnerei*, 1879), L. Dippel (*Handbuch der Laubholzkunde*, 1889–93), A. Voss & A. Siebert (*Vilmorin's Blumengärtnerei*, edn 3, 1894–6), C.K. Schneider (*Illustriertes Handbuch der Laubholzkunde*, 1904–12), A. Rehder (*Manual of cultivated trees and shrubs*, 1927, edn 2, 1947), J.W.C. Kirk (*A British garden flora*, 1927), F. Enke (*Parey's Blumengärtnerei*, edn 2, 1958), B.K. Boom (*Flora Cultuurgewassen*, 1959 and proceeding), V.A. Avrorin & M.V. Baranova (*Decorativn'ie Travyanist'ie Rasteniya Dlya Otkritogo Grunta SSSR*, 1977) and R. Mansfeld (*Verzeichnis landwirtschaftlicher and gärtnerischer Kulturpflanzen*, edn 2, 1986).

The present Flora, which, of necessity, is based on original taxonomic studies by many workers, attempts to provide a scientifically accurate and up-to-date means for the identification of plants cultivated for amenity in Europe (i.e. it does not include crops, whether horticultural or agricultural, or garden weeds), and to provide what are currently thought to be their correct names, together with sufficient synonymy to make sense of catalogues and other horticultural works. The needs of informed amateur gardeners have been borne in mind at all stages of the work, and it is hoped that the Flora will meet their needs just as much as it meets the needs of professional taxonomists. The details of the format and use of the Flora are explained in section 2 below (p. xvi).

In writing the work, the Editorial Committee has been fully aware of the difficulties involved. Some of these have been outlined above; others derive from the fact that herbarium material of cultivated plants is scanty and usually poorly annotated, so that material of many species is not available for checking the use of names, or for comparative purposes. Because of these facts, attention has been drawn to numerous problems which cannot be solved but can only be adverted to. The solution of such problems requires much more taxonomic work.

The form in which contributions appear is the responsibility of the Editorial Committee. The vocabulary and the technicalities of plant description are therefore not necessarily those used by the contributors.

1. SELECTION OF SPECIES

The problem of determining which species are in cultivation is complex and difficult, and has no complete and final answer. Many species, for instance, are grown in botanic gardens but not elsewhere; others, particularly orchids, succulents and some alpines, are to be found in the collections of specialists but are not available generally. Yet others have been in cultivation in the past but are now lost, or perhaps linger in a few collections, unrecorded and unpropagated. Further problems arise from the fact that the identification of plants in collections is not always as good as it might be, and some less well-known species probably appear in published lists under the names of other, well-known species (and vice versa).

The Flora attempts to cover all those species that are likely to be found in general collections (i.e. excluding botanic gardens and specialist collections) in Europe, whether they are grown outdoors or under glass. In order to produce a working list of such species, a compilation of all European nursery catalogues available to us was made in 1978 by Margaret McDonald, a vacation student working at the Royal Botanic Garden, Edinburgh. Since then, numerous additions have been made. This list (known as the 'Commercial List'), which includes well over 12,000 specific names, forms the basis of the species included here. Since 1987 the annual production by the Hardy Plant Society (later, the Royal Horticultural Society) of the *Plant Finder* has made the process of ascertaining which plants are on sale very much simpler. Similar publications on the continent of Europe – Erhardt, A. & W., *Pflanzen–Einkaufsführer* (1990), Pereire, A. & Bonduel, P., *Où trouver vos Plantes* (1992) and van der Laar, H.J., *Naamlijst van Houtige Gewassen* (1985–89) – have made the process of scanning catalogues no longer necessary. Another work of great benefit to cultivated plant taxonomy is *Index Hortensis* by Piers Trehane; volume I covering herbaceous perennials was published in 1989; other groups will be covered in subsequent volumes. This work gives the nomenclaturally correct names and synonyms for plants in cultivation and also has extensive lists of cultivars. In addition to the 'Commercial List', *The Plant Finder* and *Index Hortensis*, several works on the flora of gardens have been consulted, and the species covered by them have been carefully considered for inclusion. These works are: *The Royal Horticultural Society's dictionary of gardening*, edn 2 (1956, supplement revised 1969); *The new RHS dictionary of gardening* (1992); Encke, F. (ed.), *Parey's Blumengärtnerei* (1956); Boom, B.K., *Flora der Cultuurgewassen van Nederland* (1959 and proceeding); Bean, W. J., *Trees and shrubs hardy in the British Isles* (edn 8, 1970–81); Krüssmann, G., *Handbuch der Laubgeholze* (edn 2, 1976–78); *Manual of cultivated broad-leaved trees and shrubs* (English edition, translated by M.E. Epp, G.S. Daniels, editor, 1986); Encke, F., Buchheim, G. & Seybold, S. (eds), *Zander's Handworterbuch der Pflanzennamen* (edn 14, 1993), and, since 1986, Mansfeld, R., *Verzeichnis landwirtschaftlicher und gärtnerischer Kulturpflanzen* (edn 2, 1986) and Jelitto L. & Schacht, W., *Die Freiland Schmuckstauden* (1963); Jelitto, L., Schacht, W., & Fessler, A. (eds), (edn 3, 1985), *Hardy herbaceous perennials revised*, English edition (1989); and Brickell, C.D. (ed.), *The Royal Horticultural Society A–Z encyclopedia of garden plants* (2003). Most of the names included in these works are covered by the present Flora, though some have been rejected as referring to plants no longer in general cultivation.

As well as the works cited above, several relating to plants in cultivation in North America have also been consulted: Rehder, A., *Manual of cultivated trees and shrubs* (edn 2, 1947) and *Bibliography of cultivated trees and shrubs* (1949); Bailey, L.H., *Manual of cultivated plants* (edn 2, 1949); *Hortus Third* (edited by the staff of the L.H. Bailey Hortorium, Cornell University, 1976).

The contributors have also drawn on their own experience, as well as that of the family editors, European advisers and other experts, in deciding which species should be included.

Most species have a full entry, being keyed, numbered and described as set out under section 2c below (p. xvii). A few, less commonly cultivated, species (additional species) are not keyed or numbered individually, but are described briefly under the species to which they are most likely to key out in the formal key.

2. USE OF THE FLORA

a. *The taxonomic system followed in the Flora.* Plants are described in this work in a taxonomic order, so that similar genera and species occur close to each other, rendering comparison of descriptions easier than in a work where the entries are alphabetical. The families (and higher groups) follows R. K. Brummitt's *Vascular Plants*: Families and Genera (1992).

The order of the species within each genus has been a matter for the individual author's discretion. In general, however, some established revision of the genus has been followed, or, if no such revision exists, the author's own views on similarity and relationships have governed the order used.

b. *Nomenclature.* The arguments for using Latin names for plants in popular as well as scientific works are often stated and widely accepted, particularly for Floras such as this, which cover an area in which several languages are spoken. Latin names have therefore been used at every taxonomic level. A concise outline of the taxonomic hierarchy and how it is used can be found in C. Jeffrey's *An introduction to plant taxonomy* (1968, edn 2, 1982). Because of the difficulties of providing vernacular names in all the necessary languages (not to say dialects), they have not been included. S. Priszter's *Trees and shrubs of Europe, a dictionary in eight languages* (1983) is a useful source for the vernacular names of woody plants.

Many horticultural reference works omit the authority that should follow every Latin plant name. Knowledge of this authority prevents confusion between specific names that may have been used more than once within the same genus, and makes it possible to find the original description of the species (using *Index Kewensis,* which lists the original references for all Latin names for higher plants published since 1753). In this Flora authorities are therefore given for all names at or below genus level. These are unabbreviated to avoid the obscure contractions that mystify the lay reader, and on occasion, the professional botanist. In most cases we have not thought it necessary to include the initials or qualifying words and letters that often accompany author names, e.g. A. Richard, Reichenbach filius, fil. or f. (the exceptions involve a few, very common surnames).

In scientific taxonomic literature, the authority for a plant name sometimes consists of two names joined together by *ex* or *in*. Such formulae have not been used here: the authority has been shortened in accordance with *The international code of botanical nomenclature*: e. g. *Capparis lasiantha* R. Brown ex de Candolle becomes *Capparis lasiantha* de Candolle; *Viburnum ternatum* Rehder in Sargent becomes *Viburnum ternatum* Rehder. The abbreviations *hort.* and *auct.*, which sometimes stand in place of the authority after Latin names, have not been used in this work as they are often obscure or misleading. The situations described by them can be clearly and unambiguously covered by the terms *invalid, misapplied* or *Anon. Invalid* implies that the name in question has not been validly published in accordance with the Code of Nomenclature, and therefore cannot be accepted. *Misapplied* refers to names which have been applied to the wrong species in gardens or in literature. *Anon.* is used with validly published names for which there is no apparent author.

Gardeners and horticulturists complain bitterly when long-used and well-loved names are replaced by unfamiliar ones. These changes are unavoidable if *The international code of botanical nomenclature* is adhered to. Taxonomic research will doubtless continue to unearth earlier names and will also continue to realign or split up existing groups, as relationships are further investigated. However, the previously accepted names are not lost; in this work they appear as synonyms, given in brackets after the currently accepted name;

they are also included in the index. Dates of publication are not given either for accepted names or for synonyms.

c. *Descriptions and terminology.* Families, genera and species included in the Flora are mostly provided with full-length descriptions. Shorter, diagnostic descriptions are, however, used for genera or species that differ in only a few characters from others already fully described, e.g.:

3. P. vulgaris Linnaeus. Like *P. officinalis* but leaves lanceolate and corolla red …
This implies that the description of *P. vulgaris* is generally similar to that of *P. officinalis* except in the characters mentioned; it should not be assumed that plants of the two species will necessarily look very like each other. Additional species (see p. xvi), subspecies, varieties, formae and cultivars (see p. xviii) are described very briefly and diagnostically.

Unqualified measurements always refer to length (though 'long' is sometimes added in cases where confusion might arise); similarly, two measurements separated by a multiplication sign indicate length and breadth respectively.

The terminology has been simplified as far as is consistent with accuracy. The technical terms that, inevitably, have had to be used are explained in the glossary (p. 567). Technical terms restricted to particular families or genera are defined in the observations following the family or genus description, and are also referred to in the glossary.

d. *Informal keys.* For larger genera and families an informal key is often given; this will not necessarily enable the user to identify precisely every species included, but will provide a guide to the occurrence of the more easily recognised characters. A selection of these characters is given, each of which is followed by the entry-numbers of those species which show that character. In some cases, where only a few species of a genus show a particular character, the alternative states are not specified, e.g.:
   Leaves. Leathery: **18**, **19**.
This means that only species **18** and **19** in the particular genus have leathery leaves; the other species may have leaves of various textures, but they are not leathery.

e. *Formal keys.* For every family containing more than one genus, and for every genus containing more than one full-entry species, a dichotomous key is provided. This form of key, in which a series of decisions must be made between pairs of contrasting characteristics, should lead the user step by step to an entry number followed by the name of a species. The reader should then check this identification with the description of that species; in some cases, other less commonly cultivated species may be mentioned under the description of the full-entry species, when the brief descriptions of these should also be scanned, so that a final identification can be made. A key to all the families of the Dicotyledons to be included in the Flora is provided (p. 1).

f. *Horticultural information.* Notes on the cultural requirements and methods of propagation are generally included in the observations on each genus; more rarely, such information is given in the observations under the family description. These notes are generally brief and very generalised, and merely provide guidance. Reference to general works on gardening is necessary for more detailed information.

g. *Citation of literature.* References to taxonomic books, articles and registration lists are cited for each family and genus, as appropriate. No abbreviations are used in these citations (though very long titles have been shortened). The citation of a particular book or article does not necessarily imply that it has been used in the preparation of the account of the particular genus or family in this work.

h. *Citation of illustrations.* Where possible, references to good illustrations are given for each species, subspecies or variety; the names under which they were originally published (which may be different from those used here) are not normally given. The illustrations may be coloured or black and white, and may be drawings, paintings or photographs. Up to five illustrations per species have been given, and an attempt has been made to choose from widely available, modern works. Where no illustrations are cited, they either do not exist, as far as we know, or those that do are considered to be of doubtful accuracy or of very restricted availability.

There are considerable difficulties with the citation of references to illustrations, particularly with regard to their dates of publication. These difficulties arise from two main sources. 1. Many illustrated plant books of the 19th and early 20th centuries were published in fascicles, as illustrations became available, out of proper order. This means, however, that the dates of publication of individual illustrations are often more or less unknown, and, anyway, these dates do not correspond to the date of the final binding. 2. Many popular illustrated books published since the 1950s have gone through many reprints, each of which bears its own date, so that the same book may well be cited with different dates in different places in this work. We have tried to cite meaningful dates wherever this was possible, but the user should understand that the date given here for a particular volume may not correspond with the date printed in the copy available to him.

In searching for illustrations, use has been made of *Index Londinensis* (1929–31, supplement 1941), R.T. Isaacson's *Flowering plant index of illustration and information* (1979) and an extensive index compiled over the last 30 years at the Royal Botanic Garden, Edinburgh.

Several pages of figures of diagnostic plant parts are included with various groups in the Flora, and should be particularly helpful when plants are being identified by means of the keys. Some of these are original, others have either been redrawn from various sources or are silhouettes of photocopies of leaves.

i. *Geographical distribution.* The wild distribution, as far as it can be ascertained, is given in italics at the end of the description of each species, subspecies or variety. The choice and spelling of place names in general follows *The Times Atlas*, Comprehensive edition (1994 reprint), except:

(1) Well-established English forms of names have been used in preference to less familiar vernacular names, e.g. Crete instead of Kriti, Naples instead of Napoli, Borneo instead of Kalimantan;

(2) New names or spellings will be adopted as soon as they appear in readily available works of reference.

(3) Because of a lack of precise information, certain areas, now politically divided, are referred to under their older names (e.g. 'former Jugoslavia', 'former USSR', etc.).

j. *Hardiness* (see map, p. xix). For every species a hardiness code is given. This gives a tentative indication of the lowest temperatures that the species can withstand:

G2 – needs a heated glasshouse even in south Europe.

G1 – needs a cool glasshouse even in south Europe.

H5 – hardy in favourable areas: withstands 0 to $-5\,°C$ minimum.

H4 – hardy in mild areas: withstands -5 to $-10\,°C$ minimum.

H3 – hardy in cool areas: withstands -10 to $-15\,°C$ minimum.

H2 – hardy almost everywhere: withstands -15 to $-20\,°C$ minimum.

H1 – hardy everywhere: withstands $-20\,°C$ and below.

The map of mean January minima (p. xix) shows the isotherms corresponding to these codes.

k. *Flowering time.* The terms spring, summer, autumn and winter have been used as a guide to flowering times in cultivation in Europe. It is not possible to be more specific when dealing with an area extending from northern Scandinavia to the Mediterranean. In cases where plants do not flower in cultivation, or flower rarely, or whose time of flowering is not recorded, no flowering time is given.

l. *Subspecies, varieties and cultivars.* Subspecies and varieties are included, where appropriate. This is done in various ways, depending on the number of such groups; all are self-explanatory.

No attempt has been made to describe the range of cultivars of any species, either partially or comprehensively. The former is scarcely worth doing, the latter virtually impossible. Reference to individual, commonly grown cultivars is, however, made in various ways:

(1) If a registration list of cultivars exists, it is cited in the 'Literature' paragraph (see section 2g) following the description of the genus.

(2) If a particular cultivar is very widely grown, it may be referred to, either in the description of the species to which it belongs (or most resembles), or in the observations to that species.

(3) If, in a particular species, cultivars are numerous and fall into reasonably distinct groups based on variation in some obvious character, then these groups may be referred to, together with an example of each, in the observations to the species.

m. *Hybrids.* Many hybrids between species (interspecific hybrids) and some between genera (intergeneric hybrids) are in cultivation, and some of them are widely grown. Commonly cultivated interspecific hybrids are, where possible, included as though they were species. Their names, however, include a multiplication sign indicating their hybrid origin; the names of the parents (when known or presumed) are also given. Other hybrids that are less frequently grown are mentioned in the observations to the parent species they most resemble. In some genera, where the number of hybrids is very large, only a small selection of those most commonly grown is mentioned.

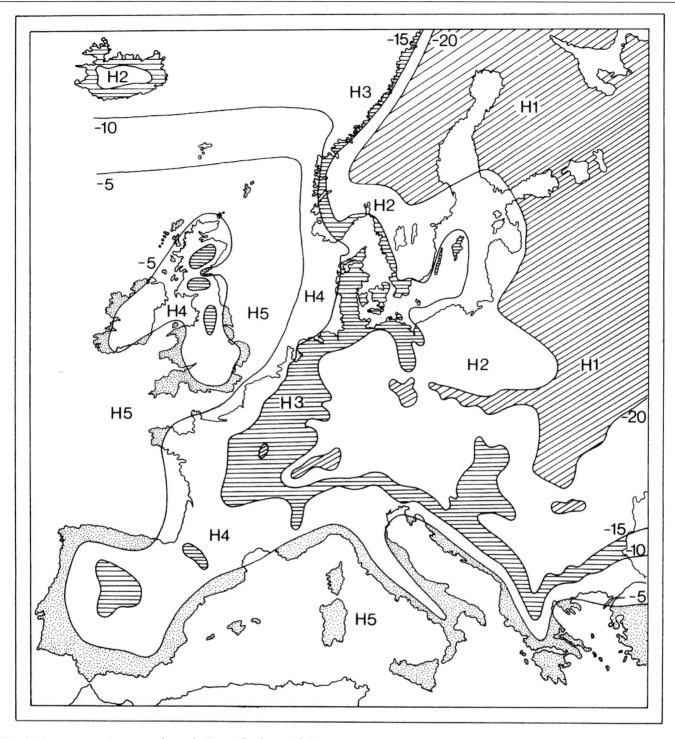

Map 1. Mean minimum January isotherms for Europe (hardiness codes).
(After Krüssmann, *Handbuch der Laubgehölze*, 1960 and *Mitteilungen der
Deutsche Dendrokigische Gesellschaft* **75**: 1983.)

DICOTYLEDONS

KEY TO FAMILIES

KEY TO GROUPS

1a. Perianth of 2 or rarely more whorls, distinguished usually into calyx and corolla, the outermost and inner whorls sharply distinguished by any or all of the following: position, colour, size, texture, shape   2

  b. Perianth of a single whorl or rarely of 2 whorls which are not sharply distinguishable as above (there may be a relatively smooth transition from outer to inner), or completely absent   10

2a. Ovary partly or fully inferior   3

  b. Ovary totally superior   4

3a. Most of the petals free from each other at the base   **Group I**

  b. All petals united into a tube or cup at the base   **Group II**

4a. Corolla made up of petals at least some of which are free from each other at their bases, falling individually except rarely when either attached individually to a ring formed by the united bases of the filaments or joined loosely at the apex   5

  b. All petals united into a tube at the base   9

5a. Ovary of a single carpel with a single style and/or stigma, or made up of several carpels which are entirely free from each other (including their styles)   **Group III**

  b. Ovary of 2 or more carpels which are united to each other at least by their styles, more usually the bodies of the carpels united   6

6a. Stamens more than twice as many as petals   **Group IV**

  b. Stamens up to twice as many as petals   7

7a. Placentation parietal   **Group V**

  b. Placentation axile, apical, basal or free-central   8

8a. Leaves alternate, or reduced to alternate scales   **Group VI**

  b. Leaves opposite or whorled   **Group VII**

9a. Corolla actinomorphic   **Group VIII**

  b. Corolla zygomorphic   **Group IX**

10a. At least the male flowers borne in catkins which are usually deciduous as a whole   **Group X**

  b. Flowers not borne in catkins as above   11

11a. Ovary of a single carpel with a single style and/or stigma, or made up of several carpels which are entirely free from each other (including their styles)   **Group XI**

  b. Ovary of 2 or more carpels which are united to each other at least by their styles, more usually the bodies of the carpels united   12

12a. Stamens borne on the perianth, or ovary inferior   **Group XII**

  b. Stamens free from the perianth, ovary superior   **Group XIII**

Group I

1a. Petals and stamens numerous; plants succulent   2

  b. Petals 10 or fewer, stamens usually fewer than 10; plants usually not succulent   3

2a. Stems succulent, usually with spines; leaves usually absent   **88. Cactaceae**

  b. Leaves succulent; spines usually absent   **81. Aizoaceae**

3a. Anthers opening by terminal pores   **218. Melastomataceae**

  b. Anthers opening by longitudinal slits or by valves   4

4a. Placentation parietal, placentas sometimes intrusive   5

  b. Placentation axile, apical, basal or free-central   9

5a. Leaves with translucent, aromatic glands   **215. Myrtaceae**

  b. Leaves without translucent, aromatic glands   6

6a. Aquatic plants with large, floating, peltate leaves   **109. Nymphaeaceae**

  b. Combination of characters not as above   7

7a. Stamens 8 or more; leaves usually opposite   **142. Hydrangeaceae**

  b. Stamens 4–6; leaves alternate   8

8a. Disc present; leaves usually with gland-tipped teeth   **143. Escalloniaceae**

  b. Disc absent; leaves without gland-tipped teeth   **140. Grossulariaceae**

9a. Placentation free-central; sepals 2   **82. Portulacaceae**

  b. Placentation axile, apical or basal; sepals usually more than 2   10

10a. Stamens as many as and on the same radii as petals; trees or shrubs with simple leaves   **187. Rhamnaceae**

  b. Stamens more numerous than petals or if as many, then not on the same radii as them; plants herbaceous or woody, leaves simple or compound   11

11a. Leaves with translucent, aromatic glands   **215. Myrtaceae**

  b. Leaves without translucent, aromatic glands   12

12a. Style 1   13

  b. Styles 2–numerous   24

13a. Floating aquatic herbs with inflated leaf-stalks   **214. Trapaceae**

  b. Terrestrial herbs, trees or shrubs; leaf-stalks not inflated   14

14a. Inflorescences borne on the surfaces of the leaves (by adnation of the peduncle to the leaf main vein)   **229. Helwingiaceae**

  b. Inflorescences not borne on the leaf surfaces   15

15a. Ovule 1, apical in each cell of the ovary (the ovary may be 1-celled)   16

b. Ovules 2–many in each cell of the ovary (the ovary may be 1-celled)   21

16a. Stamens with swollen, hairy filaments; petals rolled and recurved downwards   **224. Alangiaceae**

b. Stamens without swollen, hairy filaments; petals often borne horizontally, but not as above   17

17a. Ovary with 2 or more cells   **228. Cornaceae**

b. Ovary single-celled   18

18a. Petals 5 (or rarely more), imbricate   19

b. Petals 4, valvate   20

19a. Stigmas 3; leaves evergreen   **227. Griseliniaceae**

b. Stigmas 2; leaves deciduous   **225. Nyssaceae**

20a. Flowers unisexual; petals brownish; leaves evergreen   **230. Aucubaceae**

b. Flowers bisexual; petals various, not brownish; leaves usually deciduous   **228. Cornaceae**

21a. Stamens more than 10; ovary with 8–12 superposed cells; plant a spiny shrub   **216. Punicaceae**

b. Combination of characters not as above   22

22a. Stamens 8–10; plants woody   **219. Combretaceae**

b. Stamens 4–8; plants herbaceous   23

23a. Sap watery; petals 2 or 4; ovary usually 4-celled   **220. Onagraceae**

b. Sap milky; petals 5; ovary 3-celled   **283. Campanulaceae**

24a. Flowers borne in umbels, these sometimes modified, or in superposed whorls; leaves usually compound or much divided   25

b. Flowers not borne in umbels; leaves usually simple, little divided   26

25a. Fruit a schizocarp splitting into 2 mericarps; flowers usually bisexual; petals imbricate in bud and inflexed; usually aromatic herbs without stellate hairs   **233. Umbelliferae**

b. Fruit a berry; flowers often unisexual; petals valvate in bud, not inflexed; plants mostly woody, often with stellate hairs   **232. Araliaceae**

26a. Plants herbaceous   27

b. Plants woody   28

27a. Leaves deeply dissected; stamens usually 8; ovules 1–4, apical   **221. Haloragaceae**

b. Leaves not as above; stamens usually 10; ovules numerous, axile   **139. Saxifragaceae**

28a. Anthers opening by valves; stellate hairs often present   **135. Hamamelidaceae**

b. Anthers opening by slits; stellate hairs absent   29

29a. Leaves opposite, evergreen   **144. Cunoniaceae**

b. Leaves mainly alternate and deciduous, never both evergreen and opposite   **149. Rosaceae**

Group II

Dicotyledons with perianth of 2 distinct whorls (calyx & corolla), ovary partly or fully inferior; petals united to each other at the base.

1a. Leaves whorled, mostly basal, leathery, spiny; inflorescence a spike of many-flowered whorls; calyx 2-lobed   **282. Morinaceae**

b. Combination of characters not as above   2

2a. Inflorescence a head surrounded by an involucre of bracts; ovule always solitary   3

b. Inflorescence and ovules not as above   4

3a. Each flower with a cup-like involucel; anthers not united into a tube around the style   **281. Dipsacaceae**

b. Involucel absent; anthers united into a tube around the style   **287. Compositae**

4a. Stamens 2, united tot he style to form a touch-sensitive column; leaves linear   **286. Stylidiaceae**

b. Combination of characters not as above   5

5a. Leaves alternate or all basal   6

b. Leaves opposite or whorled   15

6a. Anthers opening by pores; fruit a berry or drupe   **237. Ericaceae**

b. Anthers opening by longitudinal slits; fruit various   7

7a. Evergreen trees or shrubs; corolla white, campanulate; ovary half-inferior; placentation free-central, ovules few   **241. Myrsinaceae**

b. Combination of characters not as above   8

8a. Climbers with tendrils and unisexual flowers; stamens 1–5; placentation parietal; fruit berry-like   **212. Cucurbitaceae**

b. Combination of characters not as above   9

9a. Stamens 10–many; plants woody   10

b. Stamens fewer than 6; plants woody or herbaceous   12

10a. Leaves with translucent glands smelling of eucalyptus; corolla completely united, unlobed, falling as a whole   **215. Myrtaceae**

b. Combination of characters not as above   11

11a. Hairs stellate or scale-like; stamens in 1 series, anthers linear   **246. Styracaceae**

b. Hairs absent or not as above; stamens in several series; anthers broad   **247. Symplocaceae**

12a. Stigmas surrounded by a sheath formed from the top of the style   **284. Goodeniaceae**

b. Stigmas not surrounded by a sheath   13

13a. Stamens as many as and on the same radii as the petals   **242. Primulaceae**

b. Stamens not as above   **14**

15a. Stamens 2 or 4, borne on the corolla; sap not milky   **274. Gesneriaceae**

b. Stamens 5, free from the corolla; sap usually milky   **283. Campanulaceae**

15a. Placentation parietal; stamens 2, or 4 and paired   **274. Gesneriaceae**

b. Placentation axile or apical; stamens 1 or more, if 4 then not paired   16

16a. Stamens 1–3; ovary with a single ovule   **280. Valerianaceae**

b. Stamens 4 or 5; ovary with usually 2 or more ovules   17

17a. Leaves divided into 3 leaflets; flowers few, in a head; herbaceous   **279. Adoxaceae**

b. Leaves simple or rarely pinnate; inflorescence various, usually not as above; usually woody   18

18a. Stipules usually borne between the bases of the leaf-stalks and sometimes looking like leaves; ovary usually 2-celled, more rarely 5-celled; corolla usually actinomorphic; fruit capsular, fleshy or schizocarpic   **255. Rubiaceae**

b. Stipules usually absent, when present not as above; ovary usually 3-celled (occasionally 2–5-celled), sometimes only 1 cell fertile; corollas often zygomorphic; fruit a berry or drupe   **278. Caprifoliaceae**

Group III

1a. Ovary apparently consisting of a single carpel, with a single style and/or stigma and a singe cell within, with 1–many ovules   2

b. Ovary consisting of 2 or more carpels which are entirely free from each other, each with its own separate style and stigma   8

2a. Corolla radially symmetric; stamens usually more than 10   3

b. Corolla bilaterally symmetric; stamens usually 10 or fewer   4

3a. Petals valvate; stamens usually much exceeding petals; leaves bipinnate   **151. Mimosaceae**

b. Petals imbricate; stamens not greatly exceeding petals; leaves various, not bipinnate   **149. Rosaceae**

4a. Leaves often pinnate, bipinnate, trifoliolate or palmate, rarely simple or reduced to phyllodes, with stipules   5

b. Leaves often simple, without stipules   6

5a. Upper petal interior (rarely petal 1 or petals absent); seed usually with a straight radical   **152. Caesalpiniaceae**

b. Upper petal exterior; seed usually with an incurved radical   **153. Fabaceae**

6a. Corolla zygomorphic   **104. Ranunculaceae**

b. Corolla actinomorphic   7

7a. Resinous tree or shrub; style set obliquely on the ovary   **170. Anacardiaceae**

b. Non-resinous shrubs or herbs; style not set obliquely on the ovary   **106. Berberidaceae**

8a. Calyx, corolla and stamens perigynous   **149. Rosaceae**

b. Calyx, corolla and stamens hypogynous   9

9a. Aquatic plants with floating or emergent peltate leaves (submerged leaves may be of different shape)   10

b. Terrestrial plants, no leaves peltate   11

10a. Carpels sunk individually in a top-shaped receptacle; sepals 4–5, petals 10–25   **111. Nelumbonaceae**

b. Carpels not sunk in a receptacle; sepals 3, petals 3   **110. Cabombaceae**

11a. Leaves conspicuously succulent   **136. Crassulaceae**

b. Leaves not succulent   12

12a. Plants completely herbaceous   13

b. Plants woody   16

13a. Petals fringed; fruits borne on a common gynophore   **132. Resedaceae**

b. Petals not fringed; gynophore absent   14

14a. Sap milky   **128. Papaveraceae**

b. Sap clear, watery   15

15a. Sepals not all the same size and shape; stamens borne on a nectar-secreting disc   **118. Paeoniaceae**

b. Sepals all similar in shape and size; stamens not borne on a disc, nectar secreted on the petals   **104. Ranunculaceae**

16a. Leaves opposite; each petal keeled inside   **169. Coriariaceae**

b. Leaves alternate; petals not keeled inside   17

17a. Leaves simple, entire or toothed   18

b. Leaves compound or deeply lobed or divided   21

18a. Woody climbers with unisexual flowers; petals 3, 6 or 9   19

b. Shrubs; flowers not as above   20

19a. Stamens united into a fleshy mass; ovules 2–3 per carpel   **95. Schisandraceae**

b. Stamens free; ovules 1 per carpel   **108. Menispermaceae**

20a. Leaves dotted with translucent glands; petals in 2 or more series   **91. Winteraceae**

b. Leaves without translucent glands; petals in a single whorl   **117. Dilleniaceae**

21a. Flowers unisexual; mostly woody climbers, if shrubs, then with blue fruits   **107. Lardizabalaceae**

b. Flowers bisexual; shrubs, fruits never blue   22

22a. Sepals not all the same size and shape; stamens borne on a nectar-secreting disc   **118. Paeoniaceae**

b. Sepals all similar in size and shape; stamens not borne on a disc, nectar secreted on the petals   **104. Ranunculaceae**

Group IV

1a. Herbaceous climber; leaves palmately divided into stalked leaflets; petals 2, stamens 8   **157. Tropaeolaceae**

b. Combination of characters not as above   2

2a. Perianth and stamens hypogynous, borne independently below the superior ovary   3

b. Perianth and stamens perigynous, borne on the edge of a rim or cup which itself is borne below the superior ovary   31

3a. Placentation axile or free-central   4

b. Placentation parietal   20

4a. Placentation free-central; sepals 2   **82. Portulacaceae**

b. Placentation axile; sepals usually more than 2   5

5a. Leaves all basal, tubular, forming insect-trapping pitchers; style peltately dilated   **125. Sarraceniaceae**

b. Leaves not as above; style not peltately dilated   6

6a. Leaves alternate   7

b. Leaves opposite or rarely whorled   19

7a. Anthers opening by terminal pores   8

b. Anthers opening by longitudinal slits   10

8a. Shrubs with simple leaves without stipules, often covered with stellate hairs; stamens inflexed in bud; fruit a berry   **120. Actinidiaceae**

b. Combination of characters not as above   9

9a. Ovary deeply lobed, borne on an enlarged receptacle or gynophore; petals not fringed   **121. Ochnaceae**

b. Ovary not lobed, not borne as above; petals often fringed   **190. Elaeocarpaceae**

10a. Perianth segments of inner whorl (petals) tubular or bifid, nectar-secreting; fruit a group of partly to fully coalescent follicles   **104. Ranunculaceae**

b. Combination of characters not as above   11

11a. Leaves with translucent, aromatic glands   **162. Rutaceae**

b. Leaves without such glands   12

12a. Sap milky; flowers unisexual   **160. Euphorbiaceae**

b. Sap watery; flowers bisexual   13

13a. Succulent herb with spines; bark hard and resinous; stamens 15 in groups of 3 in each of which the central is largest   **156. Geraniaceae**

b. Combination of characters not as above   14

14a. Stipules absent; leaves evergreen   **122. Theaceae**

b. Stipules present; leaves usually deciduous   15

15a. Filaments free; anthers 2-celled   16

b. Filaments united into a tube at least around the ovary, often also around the style; anthers often 1-celled   17

16a. Nectar-secreting disc absent; stamens more than 15; leaves simple   **191. Tiliaceae**

b. Nectar-secreting disc present, conspicuous; stamens 15; leaves dissected   **158. Zygophyllaceae**

17a. Styles divided above, several; stipules often persistent; carpels 5 or more **192. Malvaceae**

b. Style 1, stigma capitate or several; stipules usually deciduous, carpels 2–5 18

18a. Stamens in 2 whorls, those of the outer whorl usually sterile **194. Sterculiaceae**

b. Stamens in several whorls, all fertile **193. Bombacaceae**

19a. Sepals united, falling as a unit; fruit separating into two boat-shaped units **119. Eucryphiaceae**

b. Sepals and fruit not as above **158. Zygophyllaceae**

20a. Aquatic plants with cordate leaves; style and stigmas forming a disc on top of the ovary **109. Nymphaeaceae**

b. Combination of characters not as above 21

21a. Leaves modified into active insect-traps, the 2 halves of the blade fringed and closing rapidly when stimulated **127. Droseraceae**

b. Leaves not as above 22

22a. Leaves opposite 23

b. Leaves alternate 25

23a. Styles numerous; floral parts in 3s **128. Papaveraceae**

b. Styles 1–5; floral parts in 4s or 5s 24

24a. Style 1; stamens not united in bundles; leaves without translucent glands **202. Cistaceae**

b. Styles 3–5, free or variously united below; stamens united in bundles (rarely apparently all free); leaves with translucent or blackish glands **124. Guttiferae**

25a. Small trees with aromatic bark; filaments of the stamens all united **94. Canellaceae**

b. Herbs shrubs or trees, bark not aromatic; filaments free 26

26a. Trees; leaves with stipules; anthers opening by short, pore-like slits 27

b. Herbs or shrubs; leaves usually without stipules; anthers opening by longitudinal slits 28

27a. Anthers horseshoe-shaped; leaves simple, entire **203. Bixaceae**

b. Anthers straight; leaves palmately lobed **204. Cochlospermaceae**

28a. Sepals or rarely 3, quickly deciduous **128. Papaveraceae**

b. Sepals 4–8, persistent in flower 29

29a. Leaves scale-like; styles 5, stigmas 5 **205. Tamaricaceae**

b. Leaves not as above; styles 1, 2, 3 or absent, stigmas 1, 2 or 3 30

30a. Ovary closed at the apex, borne on a stalk (gynophore); none of the petals fringed **130. Capparaceae**

b. Ovary open at the apex, not borne on a stalk; at least some of the petals fringed **132. Resedaceae**

31a. Flowers unisexual; leaf-bases oblique **211. Begoniaceae**

b. Flowers bisexual; leaf-bases not oblique 32

32a. Aquatic plants with cordate leaves **109. Nymphaeaceae**

b. Terrestrial plants; leaves various 33

33a. Carpels 1 or 3, eccentrically placed at the top of, the bottom of, or within the tubular perigynous zone **150. Chrysobalanaceae**

b. Carpels and perigynous zone not as above 34

34a. Stamens united into bundles on the same radii as the petals; staminodes often present; plants usually rough with stinging hairs **209. Loasaceae**

b. Combination of characters not as above 35

35a. Sepals 2, united, falling as a unit as the flower opens; plants herbaceous **128. Papaveraceae**

b. Sepals 4 or 5, usually free, not falling as a unit; mostly trees or shrubs 36

36a. Stamens united into several rings or sheets **217. Lecythidaceae**

b. Stamens not as above 37

37a. Carpels 8–12, superposed **213. Lythraceae**

b. Carpels fewer, side-by-side 38

38a. Leaves with stipules 39

b. Leaves without stipules 40

39a. Leaves alternate; plants woody or herbaceous **149. Rosaceae**

b. Leaves opposite; plants woody **144. Cunoniaceae**

40a. Leaves with translucent, aromatic glands; style 1 **215. Myrtaceae**

b. Leaves without such glands; styles more than 1 **142. Hydrangeaceae**

Group V

1a. Sepals, petals and stamens perigynous, borne on a rim or cup which itself is inserted below the ovary 2

b. Sepals, petals and stamens hypogynous, inserted individually below the ovary 7

2a. Trees; leaves bi- or tripinnate; flowers bilaterally symmetric; stamens 5, of different lengths **133. Moringaceae**

b. Combination of characters not as above 3

3a. Annual aquatic herb; stamens 6 **131. Cruciferae**

b. Combination of characters not as above 4

4a. Flower-stalks slightly united to the leaf-stalks so that the flowers appear to be borne on the latter; petals contorted in bud; carpels 3 **200. Turneraceae**

b. Flower-stalks not united to the leaf-stalks; petals not contorted in bud; carpels usually 2 or 4 5

5a. Stamens 4–6 **143. Escalloniaceae**

b. Stamens 8 or more 6

6a. Ovary surrounded by a disc bearing 10 small staminode-like structures; placentas 5, very intrusive **176. Greyiaceae**

b. Disc absent, without staminodes; placentas 2–4, not intrusive **142. Hydrangeaceae**

7a. Corolla zygomorphic 8

b. Corolla actinomorphic 11

8a. Ovary open at apex; some or all petals fringed **132. Resedaceae**

b. Ovary closed at the apex; no petals fringed 9

9a. Petals and stamens 5; carpels 2 or 3 **198. Violaceae**

b. Petals and stamens 4 or 6; carpels 2 10

10a. Ovary borne on a stalk (gynophore); stamens projecting well beyond the petals **130. Capparaceae**

b. Ovary not borne on a stalk; stamens not projecting beyond petals **129. Fumariaceae**

11a. Petals and stamens numerous **81. Aizoaceae**

b. Petals and fertile stamens each fewer than 10 12

12a. Stamens alternating with much-divided staminodes **141. Parnassiaceae**

b. Stamens not alternating with much-divided staminodes 13

13a. Leaves insect-trapping and -digesting by means of stalked, glandular hairs **127. Droseraceae**

b. Leaves not as above 14

14a. Climbers 15

b. Shrubs or herbaceous plants 16

15a. Plants with tendrils; ovary and stamens borne on a common stalk (androgynophore); corona present **201. Passifloraceae**

  b. Plant without tendrils; ovary and stamens not borne on a common stalk; corona absent **197. Flacourtiaceae**

16a. Petals 4, the outer pair trifid; sepals 2 **129. Fumariaceae**

  b. Petals not as above; sepals 4 or 5   17

17a. Stamens usually 6, 4 longer and 2 shorter, rarely reduced to 2; carpels 2; fruit usually with a secondary septum **131. Cruciferae**

  b. Stamens 4–10, all more or less equal; carpels 2–5; fruit without a secondary septum   18

18a. Petals each with a scale-like appendage at the base of the blade; leaves opposite **206. Frankeniaceae**

  b. Petals without appendages; leaves alternate or all basal   19

19a. Stipules present **198. Violaceae**

  b. Stipules absent   20

20a. Leaves alternate, scale-like **205. Tamaricaceae**

  b. Leaves usually all basal, not scale-like **236. Pyrolaceae**

Group VI

1a. Placentation free-central; ovary of a single cell, at least above   2

  b. Placentation axile, basal or apical; ovary of a single cell of 2 or more cells   3

2a. Shrubs; leaves mostly evergreen with translucent dots or stripes; style 1; sepals never 2 **241. Myrsinaceae**

  b. Combination of characters not as above, sepals usually **282. Portulacaceae**

3a. Stamens (including staminodes) and petals usually of the same number and on the same radii (stamens antepetalous), rarely stamens fewer than petals   4

  b. Stamens not on the same radii as the petals   9

4a. Styles 5, free or shortly joined towards the base; ovule 1, basal, borne on a long, curved funicle **243. Plumbaginaceae**

  b. Combination of characters not as above   5

5a. Fertile stamens 2, staminodes 3; corolla zygomorphic **174. Meliosmaceae**

  b. All stamens (4 or 5) fertile; corolla actinomorphic   6

6a. Sepals, petals and stamens perigynous **187. Rhamnaceae**

  b. Sepals, petals and stamens hypogynous   7

7a. Inflorescences not leaf-opposed; usually trees **180. Corynocarpaceae**

  b. Inflorescences leaf-opposed; climbers with tendrils or rarely shrubs   8

8a. Filaments of stamens free from each other at the base **188. Vitaceae**

  b. Filaments of stamens united to each other at the base **189. Leeaceae**

9a. Anthers opening by clearly defined pores at the apex   10

  b. Anthers opening by longitudinal or horseshoe-shaped slits or by valves   16

10a. Leaves and stems covered in conspicuous glandular hairs on which insects are often trapped   11

  b. Leaves and stems without such hairs   12

11a. Carpels 2; herbs **147. Byblidaceae**

  b. Carpels 3; low shrubs **148. Roridulaceae**

12a. Low shrubs with unisexual flowers; stamens 4, petals 4, some of them often 2–3-lobed **190. Elaeocarpaceae**

  b. Combination of characters not as above   13

13a. Corolla zygomorphic; stamens 8 **168. Polygalaceae**

  b. Corolla actinomorphic; stamens some other number   14

14a. Carpels 3; style divided above into 3 stigmas **235. Clethraceae**

  b. Carpels 4 or more; style undivided or with 4 or more branches   15

15a. Petals about as broad as long, clawed; evergreen herbs or low shrubs; style divided above into 4 or 5 stigmas, rarely unlobed **236. Pyrolaceae**

  b. Petals longe than broad; styles undivided, stigmas 4 or 5 borne in a cup-like sheath **237. Ericaceae**

16a. Corolla zygomorphic   17

  b. Corolla actinomorphic   22

17a. Anthers cohering above the ovary like a cap **177. Balsaminaceae**

  b. Anthers not cohering as above   18

18a. Stamens 8; carpels 3; usually sprawling or climbing plants with peltate or divided leaves **157. Tropaeolaceae**

  b. Characters not as above   19

19a. Leaves with stipules   20

  b. Leaves without stipules   21

20a. Stamens 4, free; stipules borne between the petioles and the stems **175. Melianthaceae**

  b. Stamens 10 or more, filaments united into a tube around the styles; stipules borne laterally to the petioles **156. Geraniaceae**

21a. Plants herbaceous **139. Saxifragaceae**

  b. Plants woody **172. Sapindaceae**

22a. Sepals, petals and stamens perigynous   23

  b. Sepals, petals and stamens hypogynous   26

23a. Style 1, often divided above **181. Celastraceae**

  b. Styles more than 1, often 2 and divergent   24

24a. Fruit an inflated, membranous capsule; leaves compound **182. Staphyleaceae**

  b. Fruit not as above; leaves simple   25

25a. Trees or shrubs; hairs often stellate; anthers usually opening by valves; fruit a few-seeded, woody capsule **135. Hamamelidaceae**

  b. Herbs; hairs simple or absent; fruit a capsule or almost a pair of separate follicles **139. Saxifragaceae**

26a. Petals and stamens both 8 or more; stamens numerous **81. Aizoaceae**

  b. Petals and stamens fewer than 8; stamens usually definite in number   27

27a. Leaves with translucent, aromatic glands **162. Rutaceae**

  b. Leaves without such glands   28

28a. Sap usually milky; flowers unisexual; styles 3, often further divided **160. Euphorbiaceae**

  b. Combination of characters not as above   29

29a. Flower with a well-developed nectar-secreting disc below and around the ovary   30

  b. Disc absent, nectar secreted in other ways   35

30a. Resinous trees or shrubs   31

  b. Herbs, shrubs or trees, not resinous, occasionally aromatic   32

31a. Ovules 2 in each cell of the ovary **165. Burseraceae**

  b. Ovule 1 in each cell of the ovary **170. Anacardiaceae**

32a. Plant herbaceous **183. Stackhousiaceae**

  b. Plant woody   33

33a. Flowers (or at least some of them) functionally unisexual
(i.e. anthers not producing pollen, ovary without ovules)
164. Simaroubaceae
   b. Flowers functionally bisexual   34
34a. Leaves entire or toothed; stamens 4–5, filaments free,
emerging from the disc   **181. Celastraceae**
   b. Leaves usually pinnate; stamens 8–10, filaments united into
a tube, not emerging from the disc   **166. Meliaceae**
35a. Plants herbaceous   36
   b. Plants woody   40
36a. Leaves always simple; ovary 6–10-celled by the
development of 3–5 secondary septa during maturation
159. Linaceae
   b. Leaves lobed or compound; secondary septa absent from the
ovary   37
37a. Leaves without stipules   38
   b. Leaves with stipules   39
38a. Ovary of 3–5 free carpels united only by a common style
154. Limnanthaceae
   b. Ovary of 5 carpels whose bodies are completely united;
styles 5, free   **155. Oxalidaceae**
39a. Anthers 1-celled; leaves soft and mucilaginous; nectar
secreted on the inner surfaces of the sepals
192. Malvaceae
   b. Anthers 2-celled; leaves not soft and mucilaginous; nectar
secreted round the base of the ovary   **156. Geraniaceae**
40a. Filaments of the stamens united below   41
   b. Filaments of stamens completely free from each other   42
41a. Plants succulent, spiny; stamens 8 with woolly filaments;
plants unisexual   **89. Didiereaceae**
   b. Combination of characters not as above
194. Sterculiaceae
42a. Stamens 8–10   43
   b. Stamens 2–6   45
43a. Petals long-clawed, often fringed or toothed; stamens 10;
usually some or all of the sepals with nectar-secreting
appendages on the outside   **167. Malpighiaceae**
   b. Petals neither clawed nor toothed; stamens 8; sepals
without nectar-secreting appendages   44
44a. Leaves pinnate, exstipulate   **172. Sapindaceae**
   b. Leaves simple, toothed, stipulate but stipules soon falling
199. Stachyuraceae
45a. Stamens 2   **248. Oleaceae**
   b. Stamens 3–6   46
46a. Staminodes present in flowers which also contain fertile
stamens   **178. Cyrillaceae**
   b. Staminodes absent from flowers which also contain fertile
stamens   47
47a. Sepals united to each other at the base   48
   b. Sepals entirely free from each other   49
48a. Carpels 3, 1 or 2 of them sterile, the fertile containing 2
apical ovules   **186. Icacinaceae**
   b. Carpels 3 or more, all fertile, each containing 1 or 2 apical
ovules   **179. Aquifoliaceae**
49a. Ovule 1 per cell; petals 3–4   **163. Cneoraceae**
   b. Ovules many per cell; petals 5   **146. Pittosporaceae**

Group VII

1a. Petals and stamens numerous; plants succulent
81. Aizoaceae

   b. Combination of characters not as above   2
2a. Placentation free-central, ovary of a single cell, at least
above   3
   b. Placentation axile, basal or apical, ovary of 1–several cells 4
3a. Sepals usually 2, if more, then petals numerous
82. Portulacaceae
   b. Sepals or calyx-lobes 4 or 5, petals 4 or 5
84. Caryophyllaceae
4a. Corolla zygomorphic   5
   b. Corolla actinomorphic   7
5a. Plants woody; leaves palmate-digitate
173. Hippocastanaceae
   b. Plants herbaceous; leaves various, not palmate-digitate 6
6a. Sepals, petals and stamens hypogynous **156. Geraniaceae**
   b. Sepals, petals and stamens perigynous **213. Lythraceae**
7a. Small hairless annual herb growing in water or on wet
mud; leaves with stipules; seeds pitted **207. Elatinaceae**
   b. Combination of characters not as above   8
8a. Sepals, petals and stamens perigynous   9
   b. Sepals, petals and stamens hypogynous   11
9a. Styles 2 or more; fruit an inflated, bladdery capsule; leaves
trifoliolate or pinnate   **182. Staphyleaceae**
   b. Style 1; fruit various, not as above; leaves simple   10
10a. Perigynous zone prominently ribbed; seeds without arils;
mostly herbs   **213. Lythraceae**
   b. Perigynous zone not ribbed; seeds with arils; shrubs or small
trees   **181. Celastraceae**
11a. Leaves with translucent, aromatic glands **162. Rutaceae**
   b. Leaves without such glands   12
12a. Flower with a well-developed disc, usually nectar-secreting,
below and around the ovary   13
   b. Flower without a disc, nectar secreted in other ways   15
13a. Leaves often palmately lobed; sap sometimes milky; flowers
functionally unisexual; fruit a group of winged samaras;
trees   **171. Aceraceae**
   b. Combination of characters not as above   14
14a. Leaves entire or toothed; stamens 4 or 5, emerging from the
disc; seeds with arils   **181. Celastraceae**
   b. Combination of characters not as above
158. Zygophyllaceae
15a. Plant herbaceous   16
   b. Plant woody   17
16a. Leaves always simple and entire; ovary 6–10-celled by the
development of 3–5 false septa during maturation; fruit a
capsule   **159. Linaceae**
   b. Leaves lobed or compound; ovary without false septa; fruit a
schizocarp   **156. Geraniaceae**
17a. Petals long-clawed, often fringed or toothed; stamens 10;
usually some or all of the sepals with nectar-secreting
appendages outside   **167. Malpighiaceae**
   b. Petals not long-clawed, nor fringed or toothed; stamens 5;
sepals without nectar-secreting appendages outside
146. Pittosoporaceae

Group VIII

1a. Stamens 2, anthers back to back   **248. Oleaceae**
   b. Stamens more than 2, anthers never back to back   2
2a. Carpels several, free; leaves succulent **136. Crassulaceae**
   b. Carpels united, or, if the bodies of the carpels are free, then
the styles united; leaves usually not succulent   3

3a. Corolla papery, translucent, 4-lobed; stamens 4, projecting from the corolla; leaves with parallel veins, often all basal **277. Plantaginaceae**

b. Combination of characters not as above 4

4a. Central flowers of the inflorescence abortive, their bracts forming nectar-secreting pitchers; petals completely united, the corolla falling as a whole as the flower opens **123. Marcgraviaceae**

b. Combination of characters not as above 5

5a. Stamens more than twice as many as corolla-lobes 6

b. Stamens up to twice as many as corolla-lobes 12

6a. Leaves with stipules; filaments of stamens united into a tube around the ovary and style **192. Malvaceae**

b. Leaves without stipules; filaments free 7

7a. Anthers opening by pores **120. Actinidiaceae**

b. Anthers opening by longitudinal slits 8

8a. Leaves with translucent, aromatic glands; calyx cup-like, unlobed **162. Rutaceae**

b. Leaves without such glands; calyx not as above 9

9a. Placentation parietal leaves fleshy **258. Fouquieriaceae**

b. Placentation axile; leaves not fleshy 10

10a. Sap milky; ovules 1 per cell **244. Sapotaceae**

b. Sap not milky; ovules 2 or more per cell 11

11a. Ovules 2 per cell; flowers usually unisexual **245. Ebenaceae**

b. Ovules many per cell; flowers bisexual **122. Theaceae**

12a. Stamens as many as petals and on the same radii as them 13

b. Stamens more or fewer than petals, if as many then not on the same radii as them 20

13a. Tropical trees with milky sap and evergreen leaves **244. Sapotaceae**

b. Tropical or temperate trees, shrubs, herbs or climbers, with watery sap and usually deciduous leaves 14

14a. Placentation axile 15

b. Placentation basal or free-central 16

15a. Climbers with tendrils; stamens free **188. Vitaceae**

b. Upright shrubs without tendrils; stamens with the filaments united below **189. Leeaceae**

16a. Trees or shrubs; fruit a berry or drupe 17

b. Herbs (occasionally woody at the extreme base); fruit a capsule or indehiscent 18

17a. Leaves with translucent glands; anthers opening towards the centre of the flower; staminodes absent **241. Myrsinaceae**

b. Leaves without such glands; anthers opening towards the outside of the flower; staminodes 5 **240. Theophrastaceae**

18a. Sepals 2, free **82. Portulacaceae**

b. Sepals 4 or more, united 19

19a. Corolla persistent and papery in fruit; ovule 1 on a long stalk arising from the base of the ovary **243. Plumbaginaceae**

b. Corolla not persistent and papery in fruit; ovules many, on a free-central placenta **242. Primulaceae**

20a. Flower compressed with 2 planes of symmetry; stamens united in 2 bundles of ½+1+½ **129. Fumariaceae**

b. Combination of characters not as above 21

21a. Leaves bipinnate or replaced by phyllodes; carpel 1; fruit a legume **151. Mimosaceae**

b. Combination of characters not as above 22

22a. Anthers opening by pores (rarely by short, pore-like slits); pollen never in coherent masses 23

b. Anthers opening by longitudinal slits or pollen in coherent masses (pollinia) 24

23a. Stamens free from corolla-tube, often twice as many as corolla-lobes **237. Ericaceae**

b. Stamens borne on the corolla-tube, as many as lobes **266. Solanaceae**

24a. Leaves alternate or all basal; carpels never 2 and free or almost so but united by the common style 25

b. Leaves opposite or rarely alternate, when the carpels are 2 and almost completely free, united by the common style 45

25a. Flowers unisexual; male flowers with a corolla, female flowers without a corolla **160. Euphorbiaceae**

b. Flowers bisexual, all with corollas 26

26a. Plant woody; leaves usually evergreen, often spiny-margined; stigma sessile on top of the ovary **179. Aquifoliaceae**

b. Combination of characters not as above 27

27a. Shrubs with stellate hairs or lepidote scales **246. Styracaceae**

b. Herbs or shrubs, without stellate hairs or lepidote scales 28

28a. Procumbent herbs with milky sap and stamens free from the corolla-tube **283. Campanulaceae**

b. Combination of characters not as above 29

29a. Ovary 5-celled 30

b. Ovary 2–4-celled 32

30a. Placentation parietal; softly wooded tree **208. Caricaceae**

b. Placentation axile; herbs 31

31a. Leaves fleshy; anthers 2-celled; fruit often deeply lobed **265. Nolanaceae**

b. Leaves leathery; anthers 1-celled; fruit a capsule or berry **239. Epacridaceae**

32a. Ovary 3-celled 33

b. Ovary 1-, 2- or 4-celled 36

33a. Trees; stamens free from the corolla-tube **74. Olacaceae**

b. Shrubs, herbs or climbers; stamens borne on the corolla-tube 34

34a. Dwarf, evergreen shrublets; staminodes 5; petals imbricate **234. Diapensiaceae**

b. Herbs or climbers, not evergreen; staminodes absent; petals contorted 35

35a. Climber with tendrils **257. Cobaeaceae**

b. Herbs, without tendrils **256. Polemoniaceae**

36a. (33) Stamens with filaments united into a tube; flowers in heads; stigmas surrounded by a sheath **285. Brunoniaceae**

b. Combination of characters not as above 37

37a. Flowers in spirally coiled cymes, or the calyx with appendages between the lobes; style terminal or arising from between the lobes of the ovary 38

b. Flowers not in spirally coiled cymes, calyx without appendages; style terminal 39

38a. Style terminal; fruit a capsule, usually many-seeded **260. Hydrophyllaceae**

b. Style arising from the depression between the 4 lobes of the ovary; fruit of 4 nutlets or more rarely a 1–4-seeded drupe **261. Boraginaceae**

39a. Placentation parietal   40
  b. Placentation axile   41
40a. Corolla-lobes valvate in bud; leaves simple and cordate or
     peltate, or of 3 leaflets, hairless; aquatic or marsh
     plants   **252. Menyanthaceae**
  b. Corolla lobes imbricate in bud; leaves never as above;
     terrestrial plants   **274. Gesneriaceae**
41a. Ovules 1–2 in each cell of the ovary   42
  b. Ovules 3–many in each cell of the ovary   44
42a. Arching shrubs with small purple flowers in clusters on the
     previous year's wood   **267. Buddlejaceae**
  b. Combination of characters not as above   43
43a. Sepals free; corolla-lobes contorted and infolded in bud;
     twiners, herbs or dwarf shrubs   **259. Convolvulaceae**
  b. Sepals united; corolla-lobes not as above in bud; trees or
     shrubs   **261. Boraginaceae**
44a. Corolla-lobes folded, valvate or contorted in bud; septum of
     the ovary oblique, not in the horizontal plane
                                                         266. Solanaceae
  b. Corolla-lobes variously imbricate but not as above in bud;
     septum of ovary in the horizontal plane
                                                         268. Scrophulariaceae
45a. Trailing, heather-like shrublet   **237. Ericaceae**
  b. Plant not as above   46
46a. Milky sap usually present; fruit usually of 2 almost free
     follicles united by a common style; seeds with silky
     appendages   47
  b. Milky sap absent; fruit a capsule or fleshy, carpels united;
     seeds without silky appendages   48
47a. Pollen granular; corona absent; corolla-lobes valvate in
     bud   **253. Apocynaceae**
  b. Pollen usually in coherent masses (pollinia); corona usually
     present; corolla-lobes valvate or contorted in bud
                                                         254. Asclepiadaceae
48a. Flowers in coiled cymes; usually herbs
                                                         260. Hydrophyllaceae
  b. Flowers not in coiled cymes; herbs or shrubs   49
49a. Placentation parietal; carpels 2   50
  b. Placentation axile; carpels 2, 3 or 5   51
50a. Leaves compound; epicalyx present
                                                         260. Hydrophyllaceae
  b. Leaves simple; epicalyx absent   **251. Gentianaceae**
51a. Stamens fewer than corolla-lobes   **262. Verbenaceae**
  b. Stamens as many as corolla-lobes   52
52a. Carpels 5; shrubs with leaves with spiny margins
                                                         250. Desfontainiaceae
  b. Carpels 2 or 3; herbs or shrubs; leaves not as above   53
53a. Leaves without stipules; carpels 3; corolla-lobes contorted in
     bud; herbs   **256. Polemoniaceae**
  b. Leaves with stipules (often reduced to a ridge between the
     leaf-bases); corolla-lobes variously imbricate or valvate in
     bud; plant usually woody   54
54a. Corolla usually 5-lobed; stellate and/or glandular hairs
     absent   **249. Loganiaceae**
  b. Corolla 4-lobed; stellate and glandular hairs present
                                                         267. Buddlejaceae

Group IX

 1a. Stamens more numerous than the corolla-lobes, or anthers
     opening by pores   2

  b. Stamens as many as corolla-lobes or fewer, anthers not
     opening by pores   6
 2a. Anthers opening by pores; leaves undivided; ovary of 2 or
     more united carpels   3
  b. Anthers opening by longitudinal slits; leaves dissected or
     compound; ovary of a single carpel   5
 3a. The 2 lateral sepals large and petal-like; filaments
     united   **168. Polygalaceae**
  b. No sepals petal-like; filaments free   4
 4a. Shrubs with alternate or apparently whorled leaves;
     stamens 4–27   **237. Ericaceae**
  b. Herbs with opposite leaves; stamens 5
                                                         251. Gentianaceae
 5a. Leaves pinnate or of 3 leaflets; perianth not spurred
                                                         153. Fabaceae
  b. Leaves laciniate; upper petal spurred; upper sepal helmet-
     like or spurred   **104. Ranunculaceae**
 6a. Stamens as many as corolla-lobes; zygomorphy of corolla
     usually weak   7
  b. Stamens fewer than corolla-lobes; zygomorphy of corolla
     pronounced   14
 7a. Stamens on the same radii as the corolla-lobes; placentation
     free-central   **242. Primulaceae**
  b. Stamens on different radii from the corolla-lobes;
     placentation axile   8
 8a. Leaves of 3 leaflets, with translucent, aromatic glands;
     stamens 5, the upper 2 fertile, the lower 3 sterile
                                                         162. Rutaceae
  b. Combination of characters not as above   9
 9a. Ovary of 3 carpels; ovules many   **256. Polemoniaceae**
  b. Ovary of 2 carpels; ovules 4 or many   10
10a. Flowers in coiled cymes; fruit of up to 4 1-seeded nutlets
                                                         261. Boraginaceae
  b. Flowers not in coiled cymes; fruit a many-seeded capsule
                                                                 11
11a. Annual or shortly-lived perennial climber; corolla scarlet at
     first, fading yellow-white   **259. Convolvulaceae**
  b. Combination of characters not as above   12
12a. Corolla-lobes variously imbricate in bud; stamens 2, 4 or 5
     and unequal; leaves usually alternate
                                                         268. Scrophulariaceae
  b. Corolla-lobes contorted in bud; stamens 5, equal   13
13a. Leaves opposite; woody climber   **249. Loganiaceae**
  b. Leaves alternate; annual or perennial herbs
                                                         266. Solanaceae
14a. Placentation axile; ovules 4 or many   15
  b. Placentation parietal, free-central, apical or basal; ovules
     many or 1 or 2   22
15a. Ovules numerous but not in vertical rows in each cell of the
     ovary   16
  b. Ovules 4, or more numerous but then in vertical rows in
     each cell of the ovary   18
16a. Seeds winged; mainly trees, shrubs and climbers with
     opposite, pinnate, digitate or rarely simple leaves
                                                         270. Bignoniaceae
  b. Seeds usually wingless; mainly herbs or shrubs with simple
     leaves   17
17a. Corolla-lobes imbricate in bud; septum of the ovary in the
     horizontal plane; leaves opposite or alternate
                                                         268. Scrophulariaceae

b. Corolla-lobes usually folded, contorted or valvate in bud; septum of ovary oblique, not in the horizontal plane; leaves alternate **266. Solanaceae**

18a. Leaves all alternate, usually with blackish, resinous glands; plants woody **276. Myoporaceae**

b. At least the lower leaves opposite or whorled, none with glands as above; plants herbaceous or woody 19

19a. Fruit a capsule; ovules 4–many, usually in vertical rows in each cell of the ovary 20

b. Fruit not a capsule; ovules 4, side-by-side 21

20a. Leaves all opposite, often prominently marked with cystoliths; flower-stalks without swollen glands at the base; capsule usually opening elastically, seeds usually on hooked stalks **271. Acanthaceae**

b. Upper leaves alternate, cystoliths absent; flower-stalks with swollen glands at the base; capsule not elastic, seeds not on hooked stalks **272. Pedaliaceae**

21a. Style arising from the depression between the 4 lobes of the ovary, or if terminal then corolla with a reduced upper lip; fruit usually of 4 1-seeded nutlets; calyx and corolla often 2-lipped **264. Labiatae**

b. Style terminal; corolla with well-developed upper lip; fruit usually a berry or drupe; calyx often more or less actinomorphic, not 2-lipped **262. Verbenaceae**

22a. Ovules 4–many; fruit a capsule, rarely a berry or drupe 23

b. Ovules 1–2; fruit indehiscent, often dispersed in the persistent calyx 28

23a. Ovary containing 4 ovules side-by-side **262. Verbenaceae**

b. Ovary containing many ovules 24

24a. Placentation free-central; corolla spurred; leaves modified for trapping and digesting insects **275. Lentibulariaceae**

b. Placentation parietal or apical; corolla not spurred, rarely swollen at base; leaves not insectivorous 25

25a. Leaves scale-like, never green; root-parasites **268. Scrophulariaceae**

b. Leaves green, expanded; free-living plants 26

26a. Seeds winged; mainly climbers with opposite, pinnately divided leaves **270. Bignoniaceae**

b. Combination of characters not as above 27

27a. Capsule with a long beak separating into 2 curved horns; plants sticky-velvety **273. Martyniaceae**

b. Capsule without beak or horns; plant velvety or variously hairy or hairless **274. Gesneriaceae**

28a. Flowers in heads surrounded by an involucre of bracts; ovule 1 **269. Globulariaceae**

b. Flowers not in heads, often in spikes; ovules 1 or 2 **268. Scrophulariaceae**

Group X

1a. Stems jointed; leaves reduced to whorls of scales **61. Casuarinaceae**

b. Stems not jointed; leaves not as above 2

2a. Leaves pinnate 3

b. Leaves simple and entire, toothed or lobed (sometimes deeply so) 4

3a. Leaves without stipules; fruit a nut **63. Juglandaceae**

b. Leaves with stipules; fruit a legume **152. Caesalpiniaceae**

4a. Leaves opposite, evergreen, entire; fruit berry-like **231. Garryaceae**

b. Leaves alternate, deciduous or evergreen; fruit not berry-like 5

5a. Ovules many, parietal; seeds many, cottony-hairy; male catkin erect with the stamens projecting between the bracts, or hanging and with fringed bracts **64. Salicaceae**

b. Ovules solitary or few, not parietal; seeds few, not cottony-hairy; male catkins not as above 6

6a. Leaves dotted with aromatic glands **62. Myricaceae**

b. Leaves not dotted with aromatic glands 7

7a. Styles 3, each often branched; fruit splitting into 3 mericarps; seeds with appendages **160. Euphorbiaceae**

b. Styles 1–6, not branched; fruit and seeds not as above 8

8a. Plant with milky sap **70. Moraceae**

b. Plant with clear sap 9

9a. Male catkin simple, i.e. each bract with a single flower attached to it; styles 1 or 3–6 **67. Fagaceae**

b. Male catkin compound, i.e. each bract with 2–3 flowers attached to it; styles 2 10

10a. Nuts small, borne in cone-like catkins; perianth present in male flowers, absent in female, ovary naked **65. Betulaceae**

b. Nuts large, subtended by leaf-like bracts or involucres (cupules); perianth present in female flowers, absent in male; ovary inferior **66. Corylaceae**

Group XI

1a. Ovary apparently of a single carpel 2

b. Ovary of 2 or more free carpels 15

2a. Mostly submerged aquatic herbs with at least the submerged leaves whorled 3

b. Terrestrial plants, sometimes growing in damp places; leaves not whorled 4

3a. Leaves much divided; stamens 10–20, borne beneath the ovary **112. Ceratophyllaceae**

b. Leaves simple, entire; stamen 1, borne on the upper part of the ovary **223. Hippuridaceae**

4a. Leaves with stipules 5

b. Leaves without stipules 9

5a. Rhubarb-like marsh plants with large leaves; stamens 1 or 2 **222. Gunneraceae**

b. Combination of characters not as above 6

6a. Herbs or softly-wooded shrubs, often with stinging hairs; cystoliths present in the leaves; stamens 4 or 5, inflexed in bud, exploding when ripe **72. Urticaceae**

b. Combination of characters not as above 7

7a. Leaves opposite; flowers unisexual **115. Chloranthaceae**

b. Leaves alternate; flowers bisexual 8

8a. Stamens 4; epicalyx present **149. Rosaceae**

b. Stamens 5–7; epicalyx absent **150. Chrysobalanaceae**

9a. Stamens borne on the perianth 10

b. Stamens free from the perianth 12

10a. Trees or shrubs with very hard, leathery leaves; perianth segments free, usually spoon-shaped **73. Proteaceae**

b. Shrubs; leaves deciduous or evergreen but not very hard; perianth-segments united into a tube below 11

11a. Plants covered in lepidote scales; ovule basal **196. Elaeagnaceae**

b. Plants not covered in lepidote scales; ovule apical **195. Thymelaeaceae**

12a. Large evergreen trees or shrubs 13

   b. Herbs or small, deciduous shrubs   14

13a. Plants aromatic; leaves glandular-punctate; anthers opening by valves   **99. Lauraceae**

   b. Plants not aromatic; leaves not glandular-punctate; anthers opening by longitudinal slits   **93. Myristicaceae**

14a. Flowers in racemes; fruit often fleshy; stamens 3–many
               79. Phytolaccaceae

   b. Flowers in cymes; fruit an achene; stamens usually 5
               80. Nyctaginaceae

15a. Trees with bark peeling off in plates; leaves palmately lobed, base of petiole covering the axillary bud; flowers unisexual in hanging, spherical heads   **134. Platanaceae**

   b. Combination of characters not as above   16

16a. Perianth completely absent   17

   b. Perianth present   18

17a. Herbs   **113. Saururaceae**

   b. Small trees or shrubs   **102. Eupteleaceae**

18a. Perianth and stamens perigynous, borne on a rim or cup itself borne below the ovary   19

   b. Perianth and stamens hypogynous, borne independently below the ovary   22

19a. Leaves modified into insect-trapping pitchers
               137. Cephalotaceae

   b. Leaves not modified into pitchers   20

20a. Flowers unisexual; leaves evergreen   **97. Monimiaceae**

   b. Flowers bisexual; leaves deciduous   21

21a. Inner stamens sterile; perianth of many segments; leaves opposite   **98. Calycanthaceae**

   b. Stamens all fertile; perianth of up to 9 segments; leaves usually alternate   **149. Rosaceae**

22a. Leaves with conspicuous stipules which enclose the axillary buds; bark aromatic   **90. Magnoliaceae**

   b. Leaves without stipules; bark usually not aromatic   23

23a. Woody climbers   24

   b. Herbs, shrubs or trees   28

24a. Leaves opposite; flowers bisexual; plant climbing by means of hooked, hardened petioles   **104. Ranunculaceae**

   b. Leaves alternate; flowers unisexual; plant twining   25

25a. Leaves compound; parts of the flower in 3s
               107. Lardizabalaceae

   b. Leaves simple; parts of the flower not usually in 3s   26

26a. Leaves evergreen, leathery, wavy-margined; flowers in dense, cone-like racemes; carpels 5 or more, each 1-seeded   **79. Phytolaccaceae**

   b. Combination of characters not as above   27

27a. Carpels many; seeds not U-shaped   **95. Schisandraceae**

   b. Carpels 3 or 6; seeds usually U-shaped
               108. Menispermaceae

28a. Parts of the flower in 3s; fruits blue **107. Lardizabalaceae**

   b. Combination of characters not as above   29

29a. Perianth-segments 6 or more in 2–3 whorls, sometimes differing a little in size and colour; bark aromatic
               96. Illiciaceae

   b. Combination of characters not as above   30

30a. Trees with rounded, cordate leaves which are opposite on long shoots, alternate on short shoots; flowers axillary, very inconspicuous, unisexual   **103. Cercidiphyllaceae**

   b. Combination of characters not as above   31

31a. Each anther tipped by an enlarged connective; fruit a berry or an aggregate of berries; plants woody **92. Annonaceae**

   b. Anthers not tipped by enlarged connectives; fruit not as above; plants usually herbaceous   **104. Ranunculaceae**

Group XII

1a. Plants aquatic, mostly submerged   2

   b. Plants terrestrial   4

2a. Stamens 8, 4 or 2; leaves deeply divided
               221. Haloragaceae

   b. Stamens 6 or 1; leaves entire or slightly toothed   3

3a. Stamens 6; leaves all basal   **131. Cruciferae**

   b. Stamen 1; leaves opposite

               263. Callitrichaceae

4a. Trees or shrubs   5

   b. Herbs, climbers or parasites   13

5a. Stamens as many as, and on radii alternating with the perianth-segments   **187. Rhamnaceae**

   b. Stamens not as above   6

6a. Stipules present, sometimes falling early   7

   b. Stipules absent   9

7a. Styles 3–6; fruit a nut, surrounded by a scaly cupule
               67. Fagaceae

   b. Styles 2; fruit not as above   8

8a. Leaves alternate; stellate hairs usually present; fruit a woody capsule   **135. Hamamelidaceae**

   b. Leaves opposite; stellate hairs absent; fruit a non-woody capsule   **144. Cunoniaceae**

9a. Ovary superior; leaves opposite; sap sometimes milky; fruit a group of samaras; trees   **171. Aceraceae**

   b. Ovary inferior; combination of other characters not as above   10

10a. Ovary 1-celled, ovule 1, apical, or ovules 1–5, basal   11

   b. Ovary several-celled, or if 1-celled then ovules more than 5, parietal or axile   12

11a. Epigynous zone present above the ovary, bearing the perianth on its rim and stamens on its inner face; ovule 1, apical   **219. Combretaceae**

   b. Epigynous zone absent, perianth and stamens not as above; ovules 1–5, basal   **75. Santalaceae**

12a. Placentation parietal; flowers bisexual, variously arranged but not as below   **142. Hydrangeaceae**

   b. Placentation axile; flowers unisexual in heads consisting of many male flowers surrounding a single female flower, each head subtended by 2 large, white bracts
               226. Davidiaceae

13a. Plants parasitic   14

   b. Plants free-living   15

14a. Two united bracteoles forming a cup-like structure, borne just below the perianth   **76. Loranthaceae**

   b. Bracteoles absent   **77. Viscaceae**

15a. Perianth absent; flowers in spikes **113. Saururaceae**

   b. Perianth present; flowers not usually in spikes   16

16a. Leaf-base oblique; ovary inferior, 3-celled
               211. Begoniaceae

   b. Leaf-base not oblique; ovary not as above   17

17a. Ovary superior   18

   b. Ovary inferior   23

18a. Carpels 3 or rarely 2, ovule 1, basal; perianth persistent in fruit   19

   b. Combination of characters not as above   20

19a. Leaves without stipules; stamens 5   **83. Basellaceae**

b. Leaves with stipules usually united into a sheath (ochrea) around the stem; stamens usually 6–9 **79. Polygonaceae**

20a. Carpels 5 or rarely 6, united only below, free above **138. Penthoraceae**

b. Carpels either 1 or 2–3, fully united 21

21a. Leaves alternate, usually lobed or compound **149. Rosaceae**

b. Leaves opposite 22

22a. Ovule 1, basal; fruit a nut; stipules usually present, often hyaline **85. Illecebraceae**

b. Ovules numerous; fruit a capsule; stipules absent **213. Lythraceae**

23a. Leaves pinnate; ovary open at the apex **210. Datiscaceae**

b. Leaves not pinnate; ovary closed at apex 24

24a. Ovary 6-celled; perianth 3-lobed, or tubular and bilaterally symmetric **116. Aristolochiaceae**

b. Combination of characters not as above 25

25a. Ovules 1–5; seed 1 26

b. Ovules and seeds numerous 27

26a. Perianth-segments thickening in fruit; leaves alternate **86. Chenopodiaceae**

b. Perianth-segments not thickening in fruit; leaves opposite or alternate **75. Santalaceae**

27a. Styles 2; placentation parietal **142. Hydrangeaceae**

b. Style 1; placentation axile **220. Onagraceae**

Group XIII

1a. Aquatic plants, either submerged in part or at least partially covered by flowing water **263. Callitrichaceae**

b. Terrestrial plants 2

2a. Climbers or scramblers, most leaves ending in a tendril-like structure which itself terminates in an insectivorous pitcher **126. Nepenthaceae**

b. Combination of characters not as above 3

3a. Stipules present, sometimes falling early 4

b. Stipules entirely absent 14

4a. Ovary 1-celled, containing a single ovule 5

b. Ovary 1–several-celled, containing 2 or more ovules 9

5a. Styles 2–4, usually 3, free; stipules sheathing the stems **78. Polygonaceae**

b. Style 1, sometimes deeply divided above into 2 stigmas; stipules not as above 6

6a. Ovule basal; herbs or shrubs, flowers never sunk in a fleshy receptacle, leaves never palmately lobed or divided; cystoliths present **72. Urticaceae**

b. Ovule apical; trees, shrubs, herbs or climbers, if herbs then flowers sunk in a fleshy receptacle or leaves palmately lobed or divided; cystoliths absent 7

7a. Herbs or climbers; perianth in male flowers of 5 united segments **71. Cannabaceae**

b. Trees or shrubs; perianth not as above 8

8a. Sap watery; stigma 1; flowers often bisexual; leaves usually oblique at the base **68. Ulmaceae**

b. Sap milky; stigmas 2; flowers usually unisexual; leaves not oblique at the base **70. Moraceae**

9a. Placentation parietal or free-central 10

b. Placentation axile, apical or basal 11

10a. Shrubs, trees or climbers; leaves alternate; placentation parietal **197. Flacourtiaceae**

b. Herbs; leaves usually opposite; placentation free-central **84. Caryophyllaceae**

11a. Leaves large, pinnate; stipules large, palmately veined; irritant hairs present **145. Davidsoniaceae**

b. Combination of characters not as above 12

12a. Sap milky; styles usually 3, often divided; ovules 1–2 per cell **160. Euphorbiaceae**

b. Combination of characters not as above 13

13a. Stellate hairs usually present **194. Sterculiaceae**

b. Stellate hairs absent **113. Saururaceae**

14a. Trees with milky sap; styles 2 **69. Eucommiaceae**

b. Combination of characters not as above 15

15a. Ovary 1-celled, containing a single basal ovule 16

b. Ovary 1–several-celled, containing several ovules 20

16a. Flowers minute, bisexual, usually sunk in a fleshy spike; stigma sessile, usually brush-like **114. Piperaceae**

b. Combination of characters not as above 17

17a. Leaves usually with stipules united into a sheath (ochrea) around the stem; young leaves revolute; stamens 6–9; styles 2–4, usually 3, free **78. Polygonaceae**

b. Combination of characters not as above 18

18a. Leaves usually opposite or whorled; fruit an achene borne in the persistent perianth; style 1, slightly lobed at apex **80. Nyctaginaceae**

b. Combination of characters not as above 19

19a. Filaments united below; perianth usually hyaline and/or papery **87. Amaranthaceae**

b. Filaments free; perianth herbaceous **86. Chenopodiaceae**

20a. Plants woody 21

b. Plants herbaceous 32

21a. Creeping shrublets with heather-like leaves **238. Empetraceae**

b. Trees or upright shrubs; leaves not heather-like 22

22a. Leaves opposite or whorled 23

b. Leaves alternate 26

23a. Leaves whorled; carpels 6–10, each with an individual stigma **101. Trochodendraceae**

b. Leaves opposite; carpels 2 or 3, style usually 1 24

24a. Stamens 2, rarely 3 or 1; leaves usually deciduous, often pinnate **248. Oleaceae**

b. Stamens 4–12; leaves evergreen, simple 25

25a. Stamens 4 **184. Buxaceae**

b. Stamens 8–12 **185. Simmondsiaceae**

26a. Leaves with pellucid, aromatic glands **162. Rutaceae**

b. Leaves without pellucid, aromatic glands 27

27a. Leaves compound or deeply divided **172. Sapindaceae**

b. Leaves simple, entire or slightly lobed or toothed 28

28a. Ovary 2-celled below, 1-celled above; leaves evergreen; flowers unisexual, males with a perianth of 3–8 imbricate segments, females without a perianth **161. Daphniphyllaceae**

b. Combination of characters not as above 29

29a. Resinous trees and shrubs; hypogynous disc usually present; ovary 3–5-celled **170. Anacardiaceae**

b. Non-resinous trees and shrubs; hypogynous disc absent; ovary usually 6- or more-celled 30

30a. Connective of stamens prolonged as appendages; fruit an aggregate of berries **92. Annonaceae**

b. Connective of stamens not appendaged; fruit not as above 31

31a. Perianth-segments, stamens and carpels 4 each; inflorescence catkin-like   **100. Tetracentraceae**

b. Perianth-segments, stamens and carpels each more numerous; inflorescence a raceme   **79. Phytolaccaceae**

32a. Ovary 1-celled, placentation free-central; stamens as many as and on alternating radii with the perianth-segments
          242. Primulaceae

b. Ovary 1- or more-celled,placentation not free-central; stamens not as above   33

33a. Leaves modified into insect-trapping pitchers; style peltate
          125. Sarraceniaceae

b. Leaves not modified into insect-trapping pitchers; style not peltate   34

34a. Placentation parietal; perianth-segments 2, falling quickly   **128. Papaveraceae**

b. Placentation basal or axile; perianth-segments more than 2, not falling quickly   35

35a. Carpels several, united for most of their length, styles not 2 and diverging   **79. Phytolaccaceae**

b. Carpels 2, united only at the base, styles 2, diverging   36

36a. Perianth segments 4; stamens very numerous
          105. Glaucidiaceae

b. Perainth segments 5; stamens 10   **147. Saxifragaceae**

132. RESEDACEAE

Herbs or shrubs. Leaves alternate, simple or divided, stipules minute. Inflorescence a spike or raceme. Flowers unisexual or bisexual, zygomorphic. Calyx, corolla and stamens usually hypogynous. Calyx of 4–8 free sepals. Corolla of 2–8 free petals, usually fringed. Stamens 3–many. Ovary of 2–6 united (rarely free) carpels, open at the top; ovules many, parietal or marginal; styles absent, stigmas surrounding the open apex of the ovary. Fruit a capsule, follicle or berry, usually open and gaping at the apex.

Six genera with 75 species. Two genera are native to Europe, 2 to North America. A few species of *Reseda* are grown as ornamentals or for their scented inflorescences.

Literature: Abdallah, M.S., The Resedaceae, a taxonomical revision of the family, *Mededelingen Landbouwhogeschool Wageningen, Nederland 67* **8**: 1–98 (1967); Abdallah, M.S. & de Wit, C.H.D., The Resedaceae, *Mededelingen Landbouwhogeschool Wageningen, Nederland 78* **14**: 99–416 (1978). Both parts were published together in *Belmontia*, n.s., **8**: 1–416 (1978), and this reference is cited for the illustrations.

1. RESEDA Linnaeus
J.R. Akeroyd & S.G. Knees
Description as family.

A genus of 55 species, with a similar distribution as the family, cultivated largely for their economic uses in perfumery or as dyes. Propagation is usually from seed. Seeds should be sown in their final position, as plants do not transplant readily.

1a. Leaves mostly pinnatifid; capsule distinctly longer than wide 2
  b. Leaves mostly simple, the upper sometimes with 2 or 3 lobes; capsule about as long as wide 3
2a. Petals white; anthers flesh-coloured **2. alba**
  b. Petals and stamens yellow **4. lutea**
3a. Petals 4, yellow; anthers yellow **1. luteola**
  b. Petals 6, yellowish white or greenish; anthers orange-deep red **3. odorata**

1. R. luteola Linnaeus. Illustration: Ross-Craig, Drawings of British plants **4**: pl. 2 (1950); Phillips, Wild flowers of Britain, 75 (1977); *Belmontia*, n.s., **8**: f. 60 (1978); Polunin, Collins guide to wild flowers of Britain and Europe, 196 (1988); Aeschimann et al., Flora Alpina **1**: 613 (2004).

Erect biennial, 50–150 cm; stems little-branched. Leaves 3–12 cm × 4–15 mm, strap-shaped to narrowly oblanceolate, entire, tapered towards base, very shortly stalked. Flowers numerous, in dense racemes to 35 cm. Sepals 4; petals 4, 2–4 mm, the upper 4–8-lobed, shortly stalked, the remainder stalked or not, entire or 4-lobed, yellow. Stamens 20–30; anthers yellow. Capsule 3–4 × 5–6 mm, almost spherical, with 3 or 4 terminal teeth. Seeds *c.* 1 mm, rounded to kidney-shaped, blackish. *Mediterranean area & SW Europe.* H1. Summer.

Cultivated both for ornament and as a source of a yellow dye.

2. R. alba Linnaeus. Illustration: Zohary, Flora Palaestina **1**: pl. 483 (1966); Polunin, Flowers of Europe, pl. 38 (1969); Phillips, Wild flowers of Britain, 97 (1977); *Belmontia*, n.s., **8**: f. 18–20 (1978).

Erect annual to perennial, 30–100 cm, branched above. Leaves 5–10 cm, 1–2-pinnatifid with 5–15 pairs of entire lobes. Flowers in dense conical racemes. Sepals 5 or 6, 3–4 mm. Petals 5 or 6, 3.5–6 mm, triangular, deeply 3-lobed, narrowed into a long stalk, white. Stamens 10–12; anthers flesh-coloured. Capsule 8–16 mm, narrowly obovate or ellipsoid, narrowed at apex, with usually 4 terminal teeth. Seeds *c.* 1 mm, almost kidney-shaped, brown. *Mediterranean area to Middle East.* H1. Summer.

3. R. odorata Linnaeus. Illustration: *Botanical Magazine*, 29 (1787); Hay & Synge, RHS Dictionary of garden plants, 46 (1969); Maire, Flore de l'Afrique du nord **14**: 193 (1976); *Belmontia*, n.s., **8**: f. 67 (1978).

Annual to perennial, 10–50 cm; stems branched. Leaves 5–10 cm × 5–15 mm, oblanceolate to obovate, usually with 1 or 2 lateral lobes. Flowers in loose racemes, fragrant. Sepals 6; petals 6, 4–4.5 mm, yellowish white or greenish, 9–15-lobed. Stamens 20–25, anthers orange to deep red. Capsule 9–10 × 7–11 mm, nodding, almost spherical. Seeds *c.* 1.5 mm, rounded, ridged, greenish yellow. *North Africa, but naturalised in Mediterranean area.* H5. Summer.

Cultivated for ornament and for use in perfumery. 'Goliath' has compact racemes with conspicuous rusty orange anthers; in 'Red Monarch' they are deep red.

4. R. lutea Linnaeus. Illustration: Ross-Craig, Drawings of British plants **4**: pl. 1 (1950); Polunin, Flowers of Europe, pl. 38 (1969); Phillips, Wild flowers of Britain, 37 (1977); *Belmontia*, n.s., **8**: f. 57 (1978).

Somewhat untidy annual to perennial, 30–80 cm; stems branched. Leaves 5–10 cm, 1–2-pinnatifid, with 1–4 pairs of entire lobes. Racemes up to 20 cm, dense; bracts falling early. Sepals 6; petals 6, 3–4 mm, entire or 2- or 3-lobed, yellow. Stamens 12–20; anthers yellow. Capsule 6–12 × 4.5–5.5 mm, sometimes nodding, ovoid-cylindric. Seeds 1.4–1.8 mm, obovoid, blackish. *Mediterranean area & S Europe.* H1. Summer.

133. MORINGACEAE

Trees. Leaves alternate, 2–3-pinnate, exstipulate, though with glands at the base of the stalk. Flowers bisexual, zygomorphic, in panicles. Calyx, corolla and stamens perigynous. Calyx of 5 free sepals. Corolla of 4–5 free petals. Stamens 5+3–5 staminodes, anthers 1-celled. Ovary of 3 united carpels; ovules many, parietal; style 1, long, slender. Fruit a 3-sided capsule.

A family of a single genus of about 12 species occurring in mainly desert areas.

A few of the species are grown as ornamentals.

1. MORINGA Adanson
S.G. Knees
Description as for family.

A genus of 12 species from the drier parts of Africa, Arabia and India. Although easily propagated from seed, the plants die back soon after producing a short stem. The underground storage organ develops quickly and a period of drought should follow. Once established the plant can be treated in a similar way to other caudiciform succulents.

1a. Leaflets 1.5–2.5 cm wide
　　　　　　　　　　　1. ovalifolia
　b. Leaflets 5–12 mm wide　　　　2
2a. Leaflets obtuse, notched　**3. oleifera**
　b. Leaflets acute, mucronate
　　　　　　　　　　　2. drouhardii

1. M. ovalifolia Dinter & A. Berger (*M. ovalifoliolata* Dinter & A. Berger). Illustration: Marloth, Flora of South Africa **1**: 245 (1935); Jacobsen, Handbook of succulent plants **2**: 695 (1960); Flora of southern Africa 185 (1970); Rowley, Caudiciform and pachycaul succulents, 201–2 (1987).

Tree with few branches, 2–6 m, with trunk to 1 m across. Leaves 50–80 × 40–60 cm, twice pinnate; leaflets 2–4 × 1.5–2.5 cm, ovate-narrowly elliptic, mucronate. Flowers radially symmetric; petals 5, white. Fruits 15–30 cm, *c.* 2 cm thick. Seeds 2.5–3.5 cm × 8–15 mm, with 3 membranous wings. *SW Africa.* G1.

2. M. drouhardii Jummelle. Illustration: Koechlin et al., Flore et vegetation de Madagascar, 264 (1974); Flore de Madagascar et Comores, Famille **85**: 35, 39 (1982).

Tree 5–10 m, trunk very swollen towards base. Leaves 20–30 cm, twice or three times pinnate; leaflets 1.5–3 cm × 5–12 mm, oblong-ovate, obtuse, mucronate. Flowers radially symmetric; petals 5, yellowish white. Fruits 30–50 cm, constricted; seed 2–2.5 × 1.8–2 cm, ovoid to triangular, without membranous wings, but with 3 ridges. *SW Madagascar.* G1.

3. M. oleifera Lamarck (*M. pterygosperma* Gaertner). Illustration: Engler Die natürlichen Pflanzenfamilien **3**(2): 243 (1891); Menninger, Flowering trees of the world for tropics and warm climates, pl. 281 (1962); Hortus III, 742 (1976); Flore

de Madagascar et Comores, Famille **85**: 35, 39 (1982).

Small trees, 3–10 m. Leaves 30–60 cm, pinnate; leaflets 9–20 × 5–12 mm, ovate, obtuse, notched. Flowers bilaterally symmetric; petals 4, white. Fruits 20–50 cm, 3-sided; seed 2.4–2.6 cm × 4–7 mm, including membranous wings, 3-sided to spherical. *NW India, but widely naturalised in tropics and subtropics, often planted as street trees.* G1.

---

134. PLATANACEAE

Trees with exfoliating bark, often with stellate hairs. Leaves alternate, palmately lobed, stipulate, leaf-bases covering the axillary buds; ptyxis conduplicate-plicate. Inflorescence a raceme of hanging, spherical heads. Flowers unisexual, actinomorphic. Perianth and stamens perigynous or hypogynous. Perianth of 3–5 free or united segments, sometimes considered as bracts. Stamens 3–4. Ovary of 5–9 carpels; ovules 1 or rarely 2 marginal; styles free. Fruit prickly balls of achenes with persistent styles.

A family of a single genus with about 8 species, occurring in both Europe and North America. Several species and hybrids are cultivated, often as street trees.

1. PLATANUS Linnaeus
A.C. Whiteley
Description as for family.

A genus of about 8 species from North and Central America, south-eastern Europe to Iran and Indo-China. Several species are much planted in cities as shade and street trees, being tolerant of pruning and pollution. All species are sun-loving and do best in warmer areas, preferring a deep loamy soil. Propagation is by seed, hardwood cuttings or layering.

1a. Leaves 3–5-lobed; middle lobe as long as or shorter than wide　　2
　b. Leaves 5–7-lobed; middle lobe longer than wide　　　　　3
2a. Leaves usually 3-lobed; middle lobe shorter than wide; lobes shallowly toothed; fruiting heads solitary
　　　　　　　　　　2. occidentalis
　b. Leaves 3–5-lobed; the middle lobe about as long as wide, set at right angles to the lateral lobes, coarsely toothed at the base; fruiting heads usually 2–4　　　**3. × acerifolia**

3a. Leaves 5–7-lobed; lobes about half as long as blade, coarsely toothed
　　　　　　　　　　　1. orientalis
　b. Leaves 5-lobed, the lobes at least half as long as blade, narrow, entire or slightly toothed　　　　　4
4a. Margins of the leaf-lobes usually entire; fruiting heads 2–4, stalked, not bristly　　　**5. wrightii**
　b. Margins of the leaf-lobes usually slightly toothed; fruiting heads 2–7, stalkless, bristly　　**4. racemosa**

1. P. orientalis Linnaeus. Illustration: Hora, The Oxford encyclopedia of trees of the world, 121 (1981); Mitchell, Field guide to the trees of Britain & northern Europe, pl. 24 (1985); Krüssmann, Manual of cultivated broad-leaved trees & shrubs **2**: 417 (1986).

Tree to 30 m, often branching low and forming a spreading crown, the bark peeling in large plates before becoming rugged with age. Leaves 10–25 cm wide, palmate, slightly shorter than wide, base usually wedge-shaped, occasionally truncate, with usually 5 narrow large lobes and two smaller ones at the base, the larger at least half the length of the blade and with 1–3 large teeth or small lobes each side; leaf-stalks 4–7 cm. Leaves and young shoots at first densely felted with star-shaped hairs, soon falling but persistent on the veins beneath. Fruiting heads 2–6, close together on the stalk, each *c.* 2.5 cm. Achenes conical and often with a hairy surface with persistent styles. *SE Europe.* H3. Spring.

2. P. occidentalis Linnaeus. Illustration: Sargent, Silva of North America **7**: pl. 326, 327 (1895); Hora, The Oxford encyclopedia of trees of the world, 121 (1981); Krüssmann, Manual of cultivated broad-leaved trees & shrubs **2**: 417 (1986).

Tree to 40 m in the wild, with a short trunk and ascending branches, the bark peeling in small plates. Leaves shallowly 3-lobed, 10–18 cm, the lobes acuminate and shallowly toothed, the base truncate or slightly cordate or wedge-shaped, densely hairy at first, persisting only on the veins beneath; leaf-stalks 7–13 cm. Fruiting heads usually solitary, *c.* 2.5 cm, on stalks 7–15 cm. Achenes rounded or truncate, tipped with the short persistent base of the style, hairless on the surface. *E & S North America.* H5. Spring.

Not widely grown, being severely affected by plane anthracnose, *Gnomonia platani*, a fungal disease that kills the young growth.

3. P. × acerifolia (Aiton) Willdenow (*P. orientalis × P. occidentalis; P. × hispanica* Muenchhausen; *P. × hybrida* Brotero). Illustration: Phillips, Trees in Britain, Europe & North America, 46, 165 (1978); Mitchell, Field guide to the trees of Britain and northern Europe, pl. 24 (1985); Krüssmann, Manual of cultivated broad-leaved trees & shrubs, **2**: 416 (1986).

Tree to 35 m with a smooth, tall trunk and rounded head of often rather contorted branches, the bark peeling in large plates. Terminal shoots pendent on older trees. Leaves 10–20 × 12–25 cm, commonly palmate, 3–5-lobed, the central lobe broadly triangular, set at right angles to the lateral lobes, all lobes toothed at the base, sometimes sparsely throughout; base truncate or cordate, occasionally wedge-shaped; leaf-stalks 5–8 cm. Many variations can occur on a single tree. Shoots and leaves covered by a pale brown felt when young, falling quite quickly. Fruiting heads 2–4, *c.* 3 cm, usually separated along the stalk, bristly at first, becoming smoother as persistent styles break off. Achenes conical, with hairs on the surface. *Garden origin.* H3. Spring.

The most widely grown species, frequent in towns and cities, valued for its tolerance of pollution. A number of cultivars is attributable to this hybrid:

'Augustine Henry' has pendent lower branches, a less contorted crown, very freely flaking bark and generally larger leaves, truncate at the base, deeply lobed, regularly toothed, the undersides remaining felted along the veins. Fruiting heads 1–3, not freely produced. Achenes with few surface hairs.

'Pyramidalis' has horizontal lower branches (often removed), ascending upper branches and a loose, broad crown. The bark does not flake freely and soon becomes rugged. Leaves relatively small, mostly 3-lobed, sparsely toothed and rather glossy. Fruiting heads 1 or 2, to 5 cm. A widely planted clone, often mixed with *P. × acerifolia.*

'Mirkovec' is slow-growing and bushy; the young foliage has a bronze tinge.

'Suttneri' has the leaves splashed and streaked with white variegation.

4. P. racemosa Nuttall (*P. californica* Bentham). Illustration: Sargent, Silva of North America, **7**: pl. 328 (1895);

Krüssmann, Manual of cultivated broad-leaved trees & shrubs, **2**: 418 (1986).

Tree 30–40 m in the wild, often branching low down, forming a rounded crown, the bark dark and rugged low down, flaking on the branches. Leaves 3–5-lobed, 17–32 × 15–30 cm, cordate to somewhat wedge-shaped at the base, the lobes at least half the length of the blade, narrowly tapered, sparsely toothed. Young shoots and the undersides of leaves densely downy, the down persisting on veins and leaf-stalks. Fruiting heads 2–7, *c.* 2 cm. Achenes densely hairy, at least when young, acute or rounded; styles persistent. *SW USA.* H5. Spring.

5. P. wrightii S. Watson (*P. racemosa* var. *wrightii* (S. Watson) Benson). Illustration: Sargent, Silva of North America, **7**: pl. 329 (1895).

Tree to 25 m, often branching low down into 2 or 3 trunks, lower branches horizontal, upper erect, bark rugged at base of trunk, becoming scaly upwards, finally smooth. Leaves 15–18 cm, 3–5-lobed, the lobes more than half the length of the blade, narrowly tapered, mostly entire; base cordate to wedge-shaped; leaf stalks 4–8 cm. Young shoots and leaves downy, the down persisting on veins and leaf-stalks. Fruiting heads 2–4, *c.* 2 cm, individually stalked, not bristly. Achenes truncate, hairless, without a persistent style. *SE USA.* H5. Spring.

---

135. HAMAMELIDACEAE

Woody, often with stellate hairs. Leaves usually alternate, simple or lobed, stipulate; ptyxis flat, conduplicate or rarely supervolute. Flowers in spikes, clusters or pairs, unisexual or bisexual, actinomorphic or zygomorphic. Calyx, corolla and stamens perigynous or epigynous; rarely perianth and stamens perigynous or epigynous. Calyx of 4–5 free sepals or united with 4–5 lobes. Corolla of 4–5 (rarely fewer or more) petals. Stamens 4–5 or rarely more, anthers opening by valves. Ovary of 2 united carpels; ovules 1–many per cell, axile; styles 2, free. Fruit a woody capsule.

A family of 28 genera and about 90 species. *Rhodoleia* is sometimes separated off into Rhodoleiaceae.

Literature: Weaver, R.E., The witch hazel family (Hamamelidaceae), *Arnoldia*

36: 69–109 (1976); Wright, D., Hamamelidaceae, a survey of the genera in cultivation, *The Plantsman* **4**: 29–53 (1982).

1a. Leaves deeply palmately divided 2
  b. Leaves simple 3
2a. Leaves deciduous **8. Liquidambar**
  b. Leaves evergreen **4. Exbucklandia**
3a. Leaves at least partly evergreen 4
  b. Leaves quite deciduous 8
4a. Petals 4, white **9. Loropetalum**
  b. Petals absent 5
5a. Flowers in heads 6
  b. Flowers in racemes **3. Distylium**
6a. Flower-heads stalkless 7
  b. Flower-heads with long stalks **12. Rhodoleia**
7a. Leaves evergreen, entire; bracts 5 mm **15. Sycopsis**
  b. Leaves semi-deciduous, shallowly toothed; bracts 10 mm or more **14. × Sycoparrotia**
8a. Bark flaking **10. Parrotia**
  b. Bark not flaking 9
9a. All flowers bisexual 10
  b. Some flowers unisexual 14
10a. Flower-clusters with conspicuous white bracts at base **11. Parrotiopsis**
  b. Flower-clusters or racemes without conspicuous bracts 11
11a. Leaves palmately veined, base cordate **2. Disanthus**
  b. Leaves pinnately veined 12
12a. Flowers in clusters or racemes; petals present 13
  b. Flowers in erect white spikes; petals lacking **6. Fothergilla**
13a. Flowers in hanging yellow racemes; petals ovate **1. Corylopsis**
  b. Flowers in clusters; petals strap-shaped **7. Hamamelis**
14a. Racemes hanging, unisexual **13. Sinowilsonia**
  b. Racemes erect, bisexual in spring, male in autumn **5. Fortunearia**

1. CORYLOPSIS Siebold & Zuccarini
D.O. Wijnands
Deciduous shrubs. Leaves elliptic-ovate to circular, green to glaucous, sharply and finely toothed, veins impressed. Flowers in hanging racemes, bell- to funnel-shaped, yellow, scented.

A genus of 7 species from E Asia.

Literature: Morley, B. & Chao, Jew-Ming, A review of Corylopsis (Hamamelidaceae) *Journal of the Arnold Arboretum*, **58**: 382–415 (1977).

1a. Racemes with 1–5 flowers; leaves to
    5 cm   **3. pauciflora**
  b. Racemes with more than 5 flowers;
    leaves more than 6 cm   2
2a. Sepals apparently absent; nectaries
    truncate   **2. multiflora**
  b. Sepals present; nectaries 2-fid   3
3a. Sepals thin, lanceolate; flowers often
    fewer than 10   **5. spicata**
  b. Sepals fleshy, shortly ovate; flowers
    often more than 10   4
4a. Sepals bluntly triangular, *c.* 1 mm;
    leaves to 8 cm   **1. glabrescens**
  b. Sepals ovate to triangular, *c.* 2 mm;
    leaves to 12 cm   **4. sinensis** var.
                                    sinensis

1. C. glabrescens Franchet & Savatier
(*C. coreana* Uyeki; *C. gotoana* Makino;
C.platypetala misapplied). Illustration:
Taylor's guide to shrubs, 167 (1987);
Brickell (ed.), RHS A–Z encyclopedia of
garden plants, 312 (2003).

Shrub to 6 m. Leaf-stalks 1.5–3 cm,
almost hairless; blades 5–8 × 3–7 cm,
ovate to almost circular, acuminate.
Racemes with 10–20 flowers evenly spaced
along the hairless axis; petals to 8 mm;
anthers yellow or purplish; nectaries bifid,
longer than the sepals. *Japan, Korea.* H3.
Spring.

2. C. multiflora Hance (*C. wilsonii*
Hemsley). Illustration: *Hooker's Icones
Plantarum* **29**: 2819 & 2820 (1909).

Shrub or small tree. Leaf-stalks
1–2.5 cm, hairy, blades 7–15 cm, ovate to
elliptic, acute or tailed at tip. Racemes with
10–20 flowers on a hairy axis; petals
5–6 mm, linear-spathulate; nectaries entire
and truncate. *C China.* H3. Spring.

3. C. pauciflora Siebold & Zuccarini.
Illustration: *Botanical Magazine*, 7736
(1900); Hay & Synge, Dictionary of garden
plants in colour, 192 (1969); Hillier colour
dictionary of trees & shrubs, 81 (1981);
Taylor's guide to trees and shrubs, 166 &
167 (1987).

Spreading small shrub to *c.* 3 m. Leaves
ovate, 3–5 cm. Racemes few-flowered,
2–3 cm, pale yellow; petals to 8 mm,
oblong-ovate; anthers yellow; nectaries
entire and truncate. *Korea, Japan, Taiwan.*
H3. Spring.

Dislikes lime and dry sunny positions.

4. C. sinensis Hemsley var. **sinensis** (*C.
willmottiae* Rehder & Wilson). Illustration:
Hay & Synge, Dictionary of garden plants
in colour, 192 (1969); Everard & Morley,
Wild flowers of the world, pl. 94 (1970);

Journal of the Royal Horticultural Society
102: 250 (1977); Brickell (ed.), RHS A–Z
encyclopedia of garden plants, 312 (2003).

Shrub to 5 m. Leaf-stalks 5–15 mm,
usually silky-hairy; blades 5–12 cm,
obovate to oblong-elliptic, acuminate,
rounded or obliquely cordate at base, hairy
between the veins. Racemes with 10–30
flowers, densely and evenly spaced along
the hairy axis; petals 7–8 mm, almost
circular; anthers yellow; nectaries 2-fid,
longer than the sepals. *China.* H3. Spring.

'Spring Purple' has purple unfolding
leaves.

Var. **calvescens** Rehder & Wilson
(*C. platypetala* Rehder & Wilson; *C.
glaucescens* Handel-Mazzetti; *C. hypoglauca*
Cheng).

Leaf-stalks mostly hairless, occasionally
glandular; blades hairless between the
veins. Petals to 6 mm, circular to kidney-
shaped. *China, Xizang.* H3. Spring.

Forma **veitchiana** (Bean) Morley &
Chao (*C. veitchiana* Bean). Illustration:
Botanical Magazine, 8349 (1910).

Anthers red; leaves hairless beneath.
China (W Hubei). H3. Spring.

5. C. spicata Siebold & Zuccarini.
Illustration: *Botanical Magazine*, 5458
(1864); Hay & Synge, Dictionary of garden
plants in colour, 192 (1969); *Gardening
from 'Which'*, 1987 (March): 70; Hellyer,
Shrubs in colour, 35 (1982).

Shrub to 3 m. Leaf-stalks 1–2.5 cm,
hairy; blades 5–10 cm, ovate to obovate,
shortly acuminate. Racemes with 5–12
flowers clustered toward the end of the
hairy axis; petals pale yellow, 7–9 mm;
anthers reddish purple; nectaries 2-fid,
shorter than the sepals. *Japan.* H3. Spring.

1. DISANTHUS Maximowicz
D.O. Wijnands
Deciduous shrub to 8 m. Leaves alternate,
palmately veined, long-stalked, circular to
ovate, to 10 cm across, entire, hairless,
turning deep red and orange in autumn.
Flowers axillary in short-stalked pairs, dark
purple, 1.5 cm across, faintly scented,
bisexual; petals, sepals, stamens and
staminodes 5, calyx softly hairy with
recurved lobes. Capsule 2-celled with
several seeds in each cell.

A single species native to Japan and
China. Grown mainly for its autumn colour.
Performs best in shade in moist acid soil.

1. D. cercidifolius Maximowicz.
Illustration: *Botanical Magazine*, 8716
(1917); Hay & Synge, Dictionary of garden

plants in colour, 192 (1969); Hillier colour
dictionary of trees & shrubs, 95 (1981);
Gartenpraxis **4**: 43 (1990); Brickell (ed.),
RHS A–Z encyclopedia of garden plants,
382 (2003).

Japan & China. H4. Autumn.

1. DISTYLIUM Siebold & Zuccarini
D.O. Wijnands
Evergreen trees or shrubs. Leaves ovate to
lanceolate, entire or remotely toothed
above the middle.

A genus of 12 species from E Asia and
C America.

Literature: Walker, E.H., A revision of
Distylium and Sycopsis, *Journal of the
Arnold Arboretum* **25**: 319–41 (1944).

1. D. racemosum von Siebold & Zuccarini
(*D. myricoides* misapplied; *Sycopsis tutcheri*
misapplied). Illustration: *Botanical
Magazine*, 9501 (1937); Krüssmann,
Handbuch der Laubgehölze **1**: 480 (1976);
Gartenpraxis **4**: 42 (1990).

Tree to 25 m, in cultivation usually a
shrub; leaves elliptic, acute, 3–7 cm;
racemes to 4 cm, longer in fruit, with
stellate hairs; anthers red. *Japan, Korea,
SE & C China, Ryukyu Islands.* H4.
Spring–summer.

Earlier-flowering plants with narrower,
more pointed leaves and longer racemes
are often grown under the name
D. myricoides Hemsley, a hardier Chinese
species (illustration: Iconographia
cormophytorum Sinicorum **2**: 167, 1987).
Many of these are likely to be
D. racemosum.

4. EXBUCKLANDIA R.W. Brown
D.O. Wijnands
Evergreen trees with alternate leaves;
stipules prominent, paired, enclosing the
axillary buds and inflorescence. Flowers in
heads, unisexual, in groups of 4 sunk in
the floral axis.

A genus of 2 timber-bearing species
related to *Liquidambar*, distributed from
E Himalaya through S China to Sumatra.

1. E. populnea (W. Griffith) R.W. Brown
(*Bucklandia populnea* W. Griffith;
Symingtonia populnea (W. Griffith) van
Steenis). Illustration: *Botanical Magazine*,
6507 (1880); Nakao, Living Himalayan
flowers, t. 194 (1964); Iconographia
cormophytorum Sinicorum **2**: 157 (1987).

Leaves ovate, cordate, 10–15 cm,
palmately veined, leathery, glossy green
above, veins and lower surface reddish;
leaf-stalks 6 cm; stipules obovate, *c.* 3 cm.

Himalaya (Nepal to Bhutan), S China, Malaya, Indonesia (Sumatra).

5. FORTUNEARIA Rehder & Wilson
D.O. Wijnands

Large deciduous shrub to 8 m. Leaves obovate, acuminate, toothed. Flowers greenish, in erect racemes; male flowers *c.* 2.5 cm formed in autumn, bisexual flowers *c.* 5 cm appearing with the leaves; petals 5, strap-shaped.

A single species from China related to *Hamamelis*. Moderately lime-tolerant; not an attractive garden plant, rare in cultivation.

1. F. sinensis Rehder & Wilson. Illustration: Iconographia cormophytorum Sinicorum **2**: 166 (1987).

W & C China. H3. Autumn & spring.

6. FOTHERGILLA Linnaeus
D.O. Wijnands

Deciduous shrubs with alternate, coarsely toothed leaves. Flowers cream, without petals, in terminal spikes or heads like bottle-brushes, slightly before or with the leaves; epigynous zone bell-shaped, with 4–7 lobes; stamens 15–25, white, thick near the apex; ovary 2-celled; fruit a 2-seeded capsule.

A genus of 2 species from south-eastern North America. Propagation is by seed or cuttings, which take long to strike. Requires neutral or lime-free soil.

Literature: Weaver, R.E., The fothergillas, *Arnoldia* **31**: 89–96 (1971).

1a. Leaves 4–6 cm; flowers appearing before the leaves **1. gardenii**
  b. Leaves 6–15 cm; flowers appearing with the leaves **2. major**

1. F. gardenii Linnaeus (*F. alnifolia* Linnaeus filius; *F. carolina* Britton). Illustration: *Botanical Magazine*, 1341 (1811); Hay & Synge, Dictionary of garden plants in colour, 202 (1969); Taylor's guide to shrubs, 198 & 199 (1987); Brickell (ed.), RHS A–Z encyclopedia of garden plants, 453 (2003).

Shrub to *c.* 75 cm. Leaves 4–6 cm, obovate, dull glaucous-green, turning golden yellow in autumn. Flowers cream, before the leaves, in 2–3 cm spikes. *USA (Virginia to Georgia & Alabama).* G1. Spring–summer.

2. F. major (Sims) Loddiges (*F. monticola* Ashe; *F. gardenii* misapplied). Illustration: *Botanical Magazine*, 1342 (1811); *The Plantsman* **4**: 37 (1982); Taylor's guide to

shrubs, 198 & 199 (1987); Phillips & Rix, Shrubs, 151 & 268 (1989).

Shrub to 3 m. Leaves 6–15 cm with stellate hairs, broadly obovate, glaucous green, turning yellow, orange or scarlet in autumn. Flowers with the leaves, in 3–6 cm heads or spikes. *USA (N Carolina to Alabama).* H3. Spring–summer.

Very variable in habit, size of leaves and autumn colour. At least 6 unnamed clones are in cultivation; those with small leaves tend to show the best autumn colour.

7. HAMAMELIS Linnaeus
M. de Ridder

Deciduous shrubs or small trees with stellate hairs. Leaves alternate, pinnately veined, short-stalked, unequal at base, ovate to obovate; margin irregularly toothed, sometimes wavy towards the tip. Flowers bisexual, fragrant, in many, very condensed inflorescences in clusters along the branches. Inflorescence usually with 3 flowers; bracts 1; bracteoles 2, more or less fused. Flowers short-stalked, parts in 4s; calyx persistent, cup-shaped and hairy outside; petals strap-shaped, wrinkled and crisped, 5–25 × 1–2 mm, rolled in bud. Fruit a woody capsule splitting with great power, hurling away the black, shiny seeds.

A genus of 4 species, 2 from E USA and 2 from E Asia, and one hybrid of garden origin (*H. × intermedia* Rehder) between the 2 Asiatic species. Although this hybrid was first observed and noted in the Arnold Arboretum (USA), most cultivars have been selected and marketed in western Europe. These are usually grafted on to rootstocks of *H. virginiana*, which is grown from seed. Propagation by cuttings is also possible. *Hamamelis* needs light shade from the midday sun in dryer and warmer conditions and deep, moist, rich, but not heavy soil.

Literature: de Ridder, M. & Wijnands, D.O., De systematiek van Hamamelis, *Dendroflora* **17**: 6–8 (1981); Grootendorst, H.J., Hamamelis, *Dendroflora* **17**: 9–17 (1981); Lamb, J.G.D. & Nutty, F., The propagation of Hamamelis, *The Plantsman* **6**: 45–8 (1984); Strand, C., Asian witch hazels and their hybrids: a history of Hamamelis in cultivation, *The New Plantsman* **5**(4): 231–45 (1998).

1a. Bracteoles not joined 2
  b. Bracteoles joined or mostly so 3
2a. Petals 5–8 mm, yellow to orange; calyx violet inside; leaves obliquely wedge-shaped at base **4. vernalis**

  b. Petals 10–12 mm, lemon-yellow; calyx yellow-green inside, leaves obliquely cordate at base **5. virginiana**
3a. Bracteoles totally joined; leaves hairless above, with stellate hairs on veins and vein-axils beneath; petals yellow; calyx grey-purple inside **2. japonica**
  b. Bracteoles more or less joined; leaves softly hairy, less so during the growing-season 4
4a. Leaves with stellate hairs on both sides; petals flat, bright yellow; bracteoles joined only at base **3. mollis**
  b. Leaves hairless to sparsely hairy; petals wrinkled and crisped, red to orange or yellow; bracteoles more or less joined **1. × intermedia**

1. H. × intermedia Rehder. Illustration: *Amateur Gardening* 5335, 21 (1987); Phillips & Rix, Shrubs, 14 (1989).

Variable shrub to 4 m, generally intermediate between the parents (*H. mollis* and *H. japonica*) in shape and size of leaf, amount of hair, size of flower etc. *Garden origin.* Winter–spring.

Several clones have arisen, among which are:

'Jelena' (illustration: Brickell (ed.), RHS A–Z encyclopedia of garden plants, 502, 2003): a shrub of vigorous, spreading habit with large, broad, softly hairy leaves; flowers in dense clusters; petals *c.* 2 cm × 2 mm, yellow suffused with rich coppery red, margin yellow. Midwinter.

'Feuer-Zauber': vigorous shrub with strong ascending branches; flowers coppery-red; petals *c.* 1.6 cm × 1.5 mm. Winter–spring.

'Orange Beauty': flowers deep yellow to orange-yellow; petals *c.* 1.6 cm × 1.5 mm. Spring.

2. H. japonica Siebold & Zuccarini (*H. bitchuensis* Makino; *H. megalophylla* Koidzumi). Illustration: Huxley, Deciduous garden trees and shrubs, f. 86 (1979); *Journal of the Royal Horticultural Society* **99**: 22 (1974); Phillips & Rix, Shrubs, 14 (1989).

Variable; commonly a large spreading shrub to 3 m. Leaves obovate, smaller than those of *H. mollis*, becoming hairless and shiny. Flowers small to medium-sized; petals 1–1.5 cm, very narrow, strap-shaped, much twisted and crumpled, yellow with a brown spot at the base, slightly scented. *Japan & Korea.* G3. Spring.

Var. **arborea** (Ottolander) Gumbleton (*H. arborea* Ottolander) is a wide-spreading shrub; flowers like those of the species.

Var. **flavopurpurascens** (Makino) Rehder differs from *H. japonica* in its brownish-red petals, orange-yellow or sulphur yellow towards the tip. *Japan.* Spring.

'Zuccariniana' is a large, distinctly erect shrub, flattening out when older; flowers small, pale sulphur-yellow, with a greenish brown calyx. One of the latest-flowering, usually early spring.

3. H. mollis Oliver. Illustration: *Botanical Magazine*, 7884 (1811); Hay & Synge, *Dictionary of garden plants in colour*, 192 (1969); Everard & Morley, *Wild flowers of the world*, pl. 94 (1970); Phillips & Rix, *Shrubs*, 14 (1989).

Large shrub to 5 m; young branches very downy. Leaves softly hairy, rounded or very broadly obovate, toothed, shortly and abruptly pointed, unequally cordate; leaf-stalk short and downy. Flowers rich golden yellow, very fragrant; petals strap-shaped, *c.* 1.7 cm × 2 mm, not wavy. *China.* Winter–spring.

'Pallida' (illustration: Brickell (ed.), RHS A–Z encyclopedia of garden plants, 502, 2003): leaves unlike those of *H. mollis*, becoming hairless when older; flowers large, strongly scented, sulphur yellow, borne in densely crowded clusters along the naked stems. Winter–early spring.

'Brevipetala': petals only *c.* 1 cm, orange-yellow.

4. H. vernalis Sargent. Illustration: *Botanical Magazine*, 8573 (1914); Everard & Morley, *Wild flowers of the world*, pl. 155 (1970); Taylor's guide to shrubs, 171 (1987).

Upright suckering shrub to 3 m. Leaves obovate to oblong, obtuse, broadly wedge-shaped to truncate at base, 6–12 cm, slightly glaucous beneath, hairless. Flowers very small, 5–8 mm, pale yellow-red. *USA (Missouri, Oklahoma, Louisiana, Alabama).* G1. Winter–spring.

5. H. virginiana Linnaeus (*H. macrophylla* Pursh). Illustration: *Botanical Magazine*, 6684 (1883); *Journal of the Royal Horticultural Society* **94**: f. 41 (1969); Phillips & Rix, *Shrubs*, 256 & 269 (1989).

Large shrub or occasionally a small, broad-crowned tree. Leaves 8–15 cm, turning yellow in autumn. Flowers small to medium-sized, yellow. *E USA.* G1. Autumn.

Often used as a rootstock for the larger-flowered species.

8. LIQUIDAMBAR Linnaeus
D.O. Wijnands
Deciduous unisexual trees with balsamic resin; twigs often with corky ridges. Leaves alternate (opposite in *Acer*, which may look similar), long-stalked, palmately lobed. Flowers without petals, numerous, in spherical heads, male heads in terminal catkins, female heads solitary. Fruiting-heads spherical, woody, with conspicuous persistent styles; capsule splitting along the radial wall. Seed flattened, narrowly winged.

Four widely dispersed species from eastern North America, Turkey and China. They need deep, lime-free soil. *Liquidambar* is sometimes placed with *Altingia* in the separate family *Altingiaceae*.

Literature: Samorodova-Bianki, G., De genere Liquidambar L., *Notulae Systematicae (Leningrad)* **18**: 77–89 (1957); Thomas, J.L., Liquidambar, *Arnoldia* **21**: 59–66 (1961); Santamour, F.S., Interspecific hybridization in Liquidambar, *Forest Science* **18**: 23–6 (1972).

1a. Small tree or shrub with rounded crown; leaf-lobes themselves lobed
                            1. orientalis
  b. Large tree; leaf-lobes entire   2
2a. Leaves with 3 (rarely 5) lobes
                           3. formosana
  b. Leaves with 5 (rarely 7) lobes
                         2. styraciflua

1. L. orientalis Miller. Illustration: Hora, Oxford encyclopaedia of trees of the world, 122 (1981); Brickell (ed.), RHS A–Z encyclopedia of garden plants, 639 (2003).

Slow-growing tree to 10 m with a rounded crown, often a shrub in cultivation. Leaves 5–10 × 6–13 cm, 5-lobed, the lobes themselves lobed. Fruiting-heads 2.5–3 cm, hanging. *W & SW Turkey, Greece (Rhodes).* H4. Spring.

The scented medicinal balsam, liquid storax, is produced by wounding the bark.

2. L. styraciflua Linnaeus. Illustration: Hay & Synge, Dictionary of garden plants in colour, 210 (1969); Perry, Flowers of the world, 136 (1972); Hora, Oxford encyclopaedia of trees of the world, 122 (1981); Brickell (ed.), RHS A–Z encyclopedia of garden plants, 639 (2003).

Tree to 30 m with a pyramidal crown. Leaves 12–15 cm, with 5 (rarely 7) lobes, toothed, shiny green, hairless above and

with tufts of hairs in the axils beneath. *E North America (Connecticut to Florida & Texas), C America.* H3. Spring.

3. L. formosana Hance (*L. acerifolia* Maximowicz; *L. maximowiczii* Miquel). Illustration: Hora, Oxford encyclopaedia of trees of the world, 122 (1981); Iconographia cormophytorum Sinicorum **2**: 159 (1987).

Tree to 10 m. Leaves 8–15 cm wide, with 3 (rarely 5) lobes, toothed, sparsely hairy, matt green. Fruit-clusters *c.* 3 cm across. *S China & Taiwan.* H5. Spring.

Var. **monticola** Rehder & Wilson does not differ significantly.

The Chinese plants known by this name have 3-lobed, almost hairless leaves and are hardier. H4. Spring.

9. LOROPETALUM Reichenbach
D.O. Wijnands
Evergreen shrubs with ovate, slightly asymmetric leaves. Flowers 6–8 in a terminal cluster, petals 4, strap-shaped, white, anthers 4-chambered.

A genus of one or two species from the Himalaya, China and Japan.

1. L. chinense (R. Brown) Oliver. Illustration: de Noailles & Lancaster, Mediterranean plants and gardens, 96 (1977); Krüssmann, Manual of cultivated broad-leaved trees & shrubs **2**: 257 (1986); Taylor's guide to shrubs, 174 & 175 (1987); Phillips & Rix, Shrubs, 15 (1989).

Evergreen spreading shrub to 2 m, with interlacing spreading branches. Leaves broadly ovate to heart-shaped, 2–4 × 1–2 cm, finely toothed, veins deeply impressed. Flowers creamy white, green at the base, similar to *Hamamelis*. Petals narrow, 2 cm × 2.5 mm. *Japan, China, Assam.* H5. Late winter–spring.

10. PARROTIA C.A. Meyer
D.O. Wijnands
Deciduous tree or shrub to 12 m, with flaking bark; branches often hanging. Leaves 6–10 × 4–8 cm, oblong-elliptic to narrowly ovate, wavy and shallowly toothed, glossy green turning orange, yellow, crimson and purple in autumn, short-stalked, with large early-falling stipules. Flowers bisexual, *c.* 1.2 cm across, in dense heads surrounded by large bracts; petals absent; anthers 5–7, red, conspicuous.

A genus of a single species from the SW Caspian area; needs a sunny, dry position;

tolerant of lime. Propagation is by greenwood cuttings.

1. P. persica (de Candolle) C.A. Meyer. Illustration: *Botanical Magazine*, 5744 (1868); Hay & Synge, Dictionary of garden plants in colour, 216 (1969); Perry, Flowers of the world, 136 (1972); Hora, Oxford encyclopaedia of trees of the world, 123 (1981).

N Iran, former USSR (Azerbaijan), along the Caspian Sea. H3. Spring.

'Vanessa' is a tree-forming clone, suitable for street planting.

11. PARROTIOPSIS (Niedenzu) Schneider
D.O. Wijnands
Deciduous tree to 6 m, or more usually an *Alnus*-like shrub. Leaves 3–5 cm, circular, slightly truncate, toothed, turning yellow in autumn. Flowers bisexual, in clusters *c.* 1 cm across, surrounded by 4–6 white bracts (as in *Cornus florida*); petals absent; anthers 15–24.

A genus of a single species from the W Himalaya.

1. P. jacquemontiana (Decaisne) Rehder. Illustration: *Botanical Magazine*, 7501 (1896); *The Plantsman* **4**: 49 (1982); *Gartenpraxis* **4**: 42 (1990).

W Himalaya. H3. Spring & autumn.

12. RHODOLEIA Hooker
D.O. Wijnands
Large evergreen shrub or tree, to 25 m in the wild. Leaves 5–13 × 2–6 cm, lanceolate to broadly ovate, entire, glaucous beneath; stalks 2–4.5 cm. Flowers joined, 5–10 together in hanging, stalked heads *c.* 2 cm across, surrounded by 12–20 reddish brown bracts, petals absent or 1–4, red, anthers 7–11, black. Fruits woody, 2–celled; seeds numerous.

A genus of a single very variable species, distributed from S China to Malaysia; performs best in sheltered woodland.

Literature: Vink, W., Rhodoleia, Flora Malesiana **5**: 371–4 (1957).

1. R. championii Hooker. Illustration: *Botanical Magazine*, 4509 (1850); Vink, Flora Malesiana **5**: 373 (1957); Menninger, Flowering trees of the world, pl. 173 (1962); Walden & Hu, Wild flowers of Hong Kong, pl. 6 (1977).

China (Yunnan, Hong Kong), Burma, Malaya, Indonesia (Sumatra). G5. Summer–spring.

13. SINOWILSONIA Hemsley
D.O. Wijnands
Deciduous bisexual tree to 8 m with stellate hairs. Leaves alternate, elliptic to broadly ovate, to 15 cm, entire. Flowers unisexual, without petals; racemes hanging, male to 6 cm, female to 3 cm. Epigynous zone urn-shaped; styles projecting.

A genus of one species from China.

1. S. henryi Hemsley. Illustration: *Hooker's Icones Plantarum*, pl. 2817 (1909); Iconographia cormophytorum Sinicorum **2**: 166 (1987).

C & W China. H3.

14. × SYCOPARROTIA Endress & Anliker
D.O. Wijnands
Semi-deciduous shrub to 4 m. Similar to *Parrotia* but smaller; stems not flaking. Leaves oblong-elliptic, shallowly spiny-toothed, 5–8 × 2–4 cm, often persistent. Flowers as those of *Parrotia*.

1. S. semidecidua Endress & Anliker. Illustration: Phillips & Rix, Shrubs, 47 (1989).

Garden origin. H3. Spring.

A hybrid between *Parrotia persica* and *Sycopsis sinensis*, originating in Switzerland around 1950.

15. SYCOPSIS Oliver
D.O. Wijnands
Evergreen shrubs or trees. Leaves entire, with pinnate veins, short-stalked. Flowers without petals, male or bisexual, in short racemes or heads, surrounded by softly hairy bracts; male flowers with 8–10 stamens and minute sepals; female flowers with an urn-shaped, 5-lobed flower-tube. Fruit with 2 shining brown seeds.

A genus of about 7 species distributed from NE India to China, Malaysia and New Guinea. Propagation is by cuttings of fairly ripened wood given bottom heat.

1. S. sinensis Oliver. Illustration: *Botanical Magazine*, 655 (1973); Bean, Trees and shrubs hardy in the British Isles, edn 8, **4**: pl. 67 (1980); Hillier colour dictionary of trees & shrubs, 245 (1981); *The Plantsman* **4**: 51 (1982).

Evergreen shrub or small tree to 7 m. Leaves leathery, strongly veined, entire or slightly toothed towards the apex, elliptic to lanceolate. Male flowers more showy than the females; stamens 10, filaments yellow, anthers red. *C & W China.* H3. Spring.

Plants grown under the name *S. tutcheri* Hemsley, a Chinese species not in cultivation, are usually *Distylium racemosum*.

136. CRASSULACEAE

Herbs or shrubs, with succulent leaves. Leaves alternate or opposite, simple, exstipulate. Flowers in cymes, spikes, panicles, racemes or rarely solitary, bisexual, actinomorphic. Calyx, corolla and stamens hypogynous. Calyx of 4–7 or 12–16 free sepals, or united, with 4–5 lobes. Corolla with 4–7 or 12–16 petals, or united with 4–5 corolla-lobes. Stamens 4–5 or 10–12 or rarely more. Ovary of 4–5 or 10–12 free carpels; ovules many, marginal; styles free. Fruit a group of follicles.

There are 30 genera and about 1400 species, all leaf-succulents; almost all of the genera are in cultivation. Thirteen genera are native to Europe, 12 to North America.

The editors are grateful to Mr Gordon D. Rowley for his helpful comments on the revised accounts of this family.

Literature: Eggli, U. (ed.), Illustrated handbook of succulent plants: Crassulaceae (2003).

1a.	Stamens as many as petals	2
b.	Stamens twice as many as petals	4
2a.	Leaves alternate; flowering stems arising from a terminal rosette of leaves	**2. Sinocrassula**
b.	Leaves alternate, opposite or whorled; flowering stems not arising from a terminal rosette of leaves	3
3a.	Leaves opposite; flowers borne in axils of most leaves	**1. Crassula**
b.	Leaves opposite, alternate or whorled; flowers in terminal cymes or corymbs or in axils of upper leaves only	**6. Sedum**
4a.	Leaves opposite	**1. Crassula**
b.	Leaves alternate, in rosettes	5
5a.	Petals with rows of spots	**16. Graptopetalum**
b.	Petals without rows of spots	**2. Sinocrassula**
6a.	Flower-parts in 4 s, with petals united into a tube	**3. Kalanchoë**
b.	Flower-parts in 5 s or more; if 4 (*Rhodiola rosea*) then petals free	7
7a.	Flower-parts 4–5 or, if 6–7, then leaves not in a rosette	8

b. Flower-parts in 6 s (rarely 5 s) or more; leaves in rosettes (some also scattered in annual and biennial plants)　37
8a. Inflorescence terminal; leaves usually not forming a rosette　9
b. Inflorescence lateral; leaves commonly in a rosette　21
9a. Petals free or nearly so　10
b. Petals more or less united　14
10a. Inflorescence equilateral, with branches in all planes　11
b. Inflorescence a one-sided cyme with flowers all in one plane　13
11a. Flowers cup-shaped, with erect petals spreading only at the tips　12
b. Flowers rotate, with petals spreading from low down　**4. Orostachys**
12a. Leaves on non-flowering shoots opposite　**5. Lenophyllum**
b. Leaves on non-flowering shoots not opposite　**6. Sedum**
13a. Perennial stem base covered in adpressed brown scale leaves　**7. Rhodiola**
b. No leaf-scales at bases of stems　**6. Sedum** (including **8. × Sedeveria** in part)
14a. Plants annual　**9. Pistorinia**
b. Plants perennial　15
15a. Sepals small, shorter than the corolla tube　16
b. Sepals conspicuous, as long as or longer than the corolla tube, or if shorter then leaf-like　**10. Villadia**
16a. Leaves alternate　17
b. Leaves opposite　20
17a. Leaves deciduous, gradually replaced by bracts　19
b. Leaves persistent, abruptly replaced by bracts　18
18a. Plant a small shrub　**11. Adromischus**
b. Plant a rosette-forming herb　**18. Rosularia**
19a. Plant dying back to a subterranean tuber　**12. Umbilicus**
b. Plant with persistent, thick, aerial stems　**13. Tylecodon**
20a. Flowers erect, yellow　**14. Chiastophyllum**
b. Flowers nodding, orange, reddish or yellow　**15. Cotyledon**
21a. Petals united into a more or less long bell-shaped tube　22
b. Petals free or only shortly united　24
22a. Flowers vivid magenta　**16. Graptopetalum**
b. Flowers white, yellow or pinkish　23

23a. Stem-forming, with loose rosettes 10 cm or more across; leaves *c.* 1 cm thick　**17. Cremnophila**
b. Stemless or nearly so, with compact rosettes 2–6 cm (rarely to 11 cm) across; leaves to 5 mm thick　**18. Rosularia**
24a. Petals spreading radially from midway　25
b. Petals erect or only slightly spreading above　28
25a. Inflorescence a narrow, equilateral panicle with many short, few-flowered branches　**19. Thompsonella**
b. Inflorescence not as above　26
26a. Petals spotted　**16. Graptopetalum**
b. Petals not spotted　27
27a. Stamens erect　**20. Dudleya**
b. Stamens spreading　**22. Pachyveria**
28a. Petals with a pair of internal basal appendages　**21. Pachyphytum** (including × **Pachyveria** in part)
b. Petals not appendaged　29
29a. Leaves awn-tipped, or scape with large leafy bracts, or petals spotted　**23. × Graptoveria**
b. Leaves not awned, scapes with small scattered bracts only, and petals not spotted　30
30a. Petals uniformly bright yellow　**8. × Sedeveria**
b. Petals not uniformly bright yellow　31
31a. Corolla showy, red, part yellow or rarely green or white, not rolled up in the bud, usually strongly pentagonal; petals thick and fleshy, sharply keeled; sepals more or less spreading; leaves narrow-based, readily detachable　**24. Echeveria**
b. Corolla pallid, rolled up in bud, slightly angled only; petals thin, scarcely keeled; sepals erect or adpressed; leaves broad-based, not readily detachable　**20. Dudleya**
32a. Flower-parts 5–16　33
b. Flower-parts 17–32　**25. Greenovia**
33a. Scales at base of carpels large and petal-like　**26. Monanthes**
b. Scales at base of carpels not petal-like　34
34a. Petals with rows of spots　**16. Graptopetalum**
b. Petals not spotted, or if spotted then spots not in rows　35
35a. Rosettes stemless, tufted (hardy plants)　36
b. Stem-forming plants (tender)　38
36a. Petals united into a short, wide, 6–8-angled tube　**18. Rosularia**

b. Petals free or only shortly united　37
37a. Flower-parts 6–7; flower bell-shaped　**27. Jovibarba**
b. Flower-parts 8–16; flower star-shaped　**28. Sempervivum**
38a. Perennial with terminal leaf rosettes; nectar glands more or less 4-sided　**29. Aeonium**
b. Annual or biennial (rarely perennial) with some scattered leaves; nectar glands 1- or 2-horned　**30. Aichryson**

1. CRASSULA Linnaeus

S.G. Knees, H.S. Maxwell, R. Hyam & G.D. Rowley

Succulent annual or perennial herbs, shrublets or shrubs with cartilaginous or soft-wooded branches; rarely perennating in tubers. Leaves opposite, more or less united, membranous to thickly fleshy, persistent or deciduous. Inflorescence with 1–several dichasia, very rarely a single terminal flower; stem with leaf-like bracts. Flowers spreading or erect. Flower parts usually 5, rarely as few as 3 or as many as 12, with green fleshy sepals; white to pink or rarely yellow or red petals; stamens, nectar scales and free carpels in single whorls; pollen yellow to brown or black. Fruits follicles, splitting open like pea-pods. Seeds ellipsoid, smooth or covered with tubercles.

A genus of about 150 perennial species mainly from southern Africa and a few annuals in Europe, the Americas, Australasia and Africa. Most species and cultivars can be readily propagated from cuttings.

Literature: Higgins, V., Crassulas in cultivation (1949); Tölken, H.R., Crassulaceae, in O.A. Leistner (ed.), Flora of southern Africa **14** (1985); Rowley, G., Crassula, a grower's guide (2003).

1a. Aquatic annuals or perennials, stems trailing, rooting at the nodes　2
b. Succulent annual or perennial herbs or shrubs　3
2a. Stems 2–5 cm, solitary or little-branched; flowers usually stalkless　**53. aquatica**
b. Stems 2–30 cm, branched; flowers stalked　**54. helmsii**
3a. Leaves 2–8 mm, scale-like, closely adpressed to form cord-like, erect stems　4
b. Combination of characters not as above　5

4a. Stems often leafless below; flowers in terminal heads; sepals *c.* 4 mm; anthers brown   **14. ericoides**

b. Stems covered with living or dead leaves; flowers borne in leaf axils; sepals *c.* 1 mm; anthers yellow   **1. muscosa**

5a. Almost stemless perennial, rosettes with only the tips of leaves emerging above soil-level; anthers black; stigmas red   **42. susannae**

b. Combination of characters not as above   6

6a. Leaves in almost stemless basal rosettes   7

b. Stems distinct, leafy throughout   21

7a. Leaves at least twice as long as wide, narrowly triangular to lanceolate   8

b. Leaves about as long as wide, if greater, then broadly ovate to rounded   11

8a. Stems to 25 cm, sometimes producing runners; leaves 2–5 cm   9

b. Stems to 50 cm, without runners; leaves 5–25 cm   10

9a. Flower-stalks hairless, flowers in conical panicles   **19. orbicularis**

b. Flower-stalks hairy, leafy; flowers in flat-topped clusters   **24. setulosa**

10a. Corolla white, pink-tinged or red; anthers dark brown   **26. alba**

b. Corolla yellowish white; anthers yellow   **25. vaginata**

11a. Flowers borne in stalkless clusters among the leaves   **43. mesembrianthemopsis**

b. Flowers borne on clear stems above the leaves   12

12a. Leaves bluish grey, covered with hard whitish, irregularly shaped papillae   **40. tecta**

b. Leaves green or reddish, or if bluish grey, then lacking papillae   13

13a. Leaves crowded together in basal, apparently 2-ranked clusters   **39. alstonii**

b. Leaves overlapping or free in apparently 4-ranked clusters   14

14a. Leaves broadly ovate to rounded, tapered and almost free at base   15

b. Leaves ovate to diamond-shaped, acute, closely overlapping at base   16

15a. Leaves covered with velvety down, light greyish green with a reddish tinge   **51. cotyledonis**

b. Leaves covered with short white hairs, bright green and almost glossy, reddish tinged   **28. capitella**

16a. Leaves distinctly grey-green, margins silky white-hairy margin, very succulent; flowers in leafy, spike-like inflorescences   **29. barbata**

b. Leaves bright green to yellowish green, often tinged red, margins without silky white hairs; flower-stalks with 1 or 2 pairs of leaf-like bracts   17

17a. Calyx-lobes 0.5–1 mm   **27. hemisphaerica**

b. Calyx-lobes 1–3 mm   18

18a. Calyx-lobes 1–1.5 mm   19

b. Calyx-lobes 2–3 mm   20

19a. Leaves 4–7 × 4–6 mm, ovate-elliptic   **22. socialis**

b. Leaves 1–3 × 1–3 cm, obovate to rounded   **21. intermedia**

20a. Stamens with black anthers   **20. montana** subsp. **quadrangularis**

b. Stamens with yellow anthers   **23. exilis**

21a. Leaves at least 4 times as long as wide   22

b. Leaves not more than 2.5 times as long as wide   26

22a. Leaves 1.2–3.5 cm wide   **47. perfoliata**

b. Leaves 1–8 mm wide   23

23a. Leaves 8–30 mm; sepals 1–2.5 mm; petals 3–5 mm   24

b. Leaves 1.8–10 cm; sepals 0.5–2 mm; petals 1–4 mm   25

24a. Stems hairy when young; leaves 1–3 cm; sepals bluntly acute   **15. sarcocaulis**

b. Stems hairless, even when young; leaves 8–15 mm; sepals rounded   **50. subaphylla**

25a. Leaves 1–4 mm wide; sepals 1–2 mm; petals cream   **16. tetragona**

b. Leaves 3–15 mm wide; sepals 0.5–1 mm; petals white tinged pink   **33. macowaniana**

26a. Leaves opposite perfoliate, stem clearly visible between each pair of leaves   27

b. Leaves free, or if united then stem not visible between the leaves   32

27a. Leaves 2–8 cm   **44. grisea**

b. Leaves 3–20 mm   28

28a. Stems less than 10 cm   **4. deltoidea**

b. Stems 10–60 cm   29

29a. Corolla star-shaped; anthers white or purple   **5. pellucida**

b. Corolla tubular; anthers brown, black or yellow   30

30a. Leaves 1.5–2 cm; anthers black or yellow   **32. brevifolia**

b. Leaves 3–20 mm; anthers brown   31

31a. Petals 2–2.5 mm, cream or pale yellow   **30. perforata**

b. Petals 3–4 mm, white, pink or red   **31. rupestris**

32a. Stout-stemmed shrubs with rounded to ovate leaves and stalked terminal inflorescences   33

b. Dwarf succulents with stems covered by leaves, or if shrubby then stems slender; flowers stalked or stalkless   38

33a. Leaves 1.0–2.5 cm; anthers yellow   **8. cordata**

b. Leaves 2–10 cm; anthers purple or black   34

34a. Corolla tubular; anthers black   **49. cultrata**

b. Corolla star-shaped; anthers purple   35

35a. Leaf-stalks 5–20 mm, blades with slightly recurved margins   **9. multicava**

b. Leaves often stalkless, or if stalked then stalk not more than 5 mm   36

36a. Leaves bright glossy green; petals 7–10 mm   **11. ovata**

b. Leaves dull greyish green or whitish green with distinct red margins, not glossy or shiny   37

37a. Leaves stalkless; petals 4–7 mm   **10. lactea**

b. Leaf-stalks to 5 mm, occasionally absent; petals 7–10 mm   **12. arborescens**

38a. Stems obscured by leaves, forming tightly packed, column-like growths   39

b. Stems clearly visible between the leaves   45

39a. Leaves 4–12 mm, ovate to diamond-shaped, closely adpressed to stem, forming a tight square-edged column   **37. pyramidalis**

b. Leaves not as above   40

40a. Leaves at least 3 times as long as wide, narrowly elliptic to triangular   41

b. Leaves as wide as or wider than long, broadly ovate, deeply keeled below   42

41a. Cushion-forming perennial, with leaves clustered at the ends of the branches; petals 5–7 mm   **41. ausensis**

b. Erect perennial or biennial; petals 8–12 mm   **36. alpestris**

42a. Leaves 3–4 × 1–15 mm, adpressed to stem, forming a smooth column to 1.5 cm across **35. barklyi**

b. Leaves 0.5–1.5 × 0.6–2.5 mm, closely adpressed, but not forming smooth columns 43

43a. Flowers borne in tight rounded heads, with stalks obscured by leaves **34. columnaris**

b. Flowers borne on slender stalks 2–8 cm above the leaves 44

44a. Unbranched perennial, occasionally branched from the base; leaves dark greyish green **45. plegmatoides**

b. Well-branched perennial; leaves light blue grey-green, conspicuously dotted **46. deceptor**

45a. Leaf-stalks 5–15 mm **13. nemorosa**

b. Leaves stalkless or very shortly stalked 46

46a. Usually tuberous rooted perennial to 8 cm, with thick woody stems at base; leaves 3–5 mm; flowers concealed among the leaves **3. corallina**

b. Annual or perennial herbs, stems thin, or if woody then flowers stalked and held clear of the leaves 47

47a. Leaves 2–12 mm, stems very slender 48

b. Leaves 1.2–8 cm, stems variable in width 49

48a. Erect annual herb; stems and leaves hairless **2. dichotoma**

b. Creeping perennial; stems rooting at the nodes, stems and leaves very hairy **38. lanuginosa**

49a. Leaves with conspicuous marginal hairs **17. ciliata**

b. Leaf margins hairless but sometimes dotted or toothed 50

50a. Shrubby perennials with stout, well-branched, more or less erect stems 51

b. Slightly woody perennials with weak, often wiry, prostrate or scrambling stems 53

51a. Sepals 1.5–3 mm; anthers yellow **52. nudicaulis**

b. Sepals 3–22 mm; anthers black 52

52a. Sepals 3–3.5 mm; corolla white or cream **48. mesembrianthoides**

b. Sepals 1.5–2.2 cm; corolla bright red or white tinged red **18. coccinea**

53a. Stems usually 4-sided; leaves often cordate at base **6. spathulata**

b. Stems rounded, arising from irregularly shaped tubers; leaves elliptic-ovate, constricted at base **7. sarmentosa**

1. C. muscosa Linnaeus (*C. lycopodioides* Lamarck; *C. ericoides* misapplied). Illustration: Leistner (ed.), Flora of southern Africa **14**: 127 (1985); Graf, Exotica, series 4, edn 12, **1**: 873 (1985); Rowley, Crassula, a grower's guide, 138, 139 (2003).

Scrambling perennial, 10–40 cm, branches woody, hidden by dead and living leaves, not more than 5 mm across. Leaves 2–8 × 1–4 mm, conspicuously 4-ranked, ovate to triangular, flat, pointed or blunt, hairless, leathery, green tinged yellow, grey or brown; leaf sheath 1.5 cm. Flowers solitary or densely clustered in groups of 2–8, malodorous. Sepals *c.* 1 mm, triangular; petals *c.* 2 mm, erect, triangular, keeled, yellow-green to brown; stamens 0.5–1 mm, anthers yellow; ovaries tapered, style less than half ovary length. *Southern Africa*. G1. Summer.

2. C. dichotoma Linnaeus (*Grammanthes gentianoides* (Lamarck) de Candolle; *Crassula gentianoides* Lamarck; *Vauanthes dichotoma* (Linnaeus) O. Kuntze). Illustration: *Botanical Magazine*, 4607 (1851) & 6401 (1878); Leistner (ed.), Flora of southern Africa **14**: 127 (1985); Rowley, Crassula, a grower's guide, 111 (2003).

Thin-stemmed annual, 5–25 cm. Leaves 5–12 × 1–6 mm, elliptic-linear, pointed or blunt, hairless; black hydathodes often present. Flowers tubular, subtended by bracts, borne in terminal cymes. Sepals 4–6 mm, lanceolate-elliptic, blunt; petals 7–20 mm, elliptic-oblanceolate, blunt or pointed, orange-yellow; stamens 5–12 mm, anthers yellow; style half ovary length. *South Africa (Cape Province)*. G1. Spring–summer.

3. C. corallina Thunberg. Illustration: Higgins, Crassulas in cultivation, 36 (1949); Jacobsen, Lexicon of succulent plants, pl. 43 (1974); Rowley, Crassula, a grower's guide, 105 (2003).

Perennial herbs to 8 cm, often with tuberous tap-root; stems sometimes woody below. Leaves 3–5 × 2–5 mm, obovate, stalkless, tapering at tip and base, convex above and below, warty, with a flaking waxy surface. Flowers solitary or clustered, partially obscured by the upper leaves; sepals 1–2 mm, triangular, blunt, grey; petals 2–3.5 mm, obovate-oblong, rounded at tip, pouched below, reflexed above, cream; stamens 1.5–2 mm, anthers yellow; style short or absent. *South Africa, Namibia*. G1.

4. C. deltoidea Thunberg (*C. rhomboidea* N.E. Brown). Illustration: Higgins, Crassulas in cultivation, 40 (1949); Graf, Exotica, series 4, edn 12, **1**: 881 (1985); Rowley, Crassula, a grower's guide, 110 (2003).

Perennial shrub to 10 cm, with fleshy branching stems. Leaves 1–2 cm × 4–15 mm, oblanceolate-ovate or diamond-shaped, fused at base, blunt or pointed, flat above, convex below, grey-green; surface covered in flaking wax. Flowers in rounded terminal clusters; sepals 0.5–2 mm, rounded, hairless; petals 3.5–5 mm, oblanceolate-elliptic, creamy white; stamens 3–4.5 mm, anthers black; styles half carpel length. *SE Africa*. G1.

5. C. pellucida Linnaeus (*C. centauroides* Aiton). Illustration: *Botanical Magazine*, 1765 (1815); Higgins, Crassulas in cultivation, 58 (1949); Rowley, Crassula, a grower's guide, 152 (2003).

Perennials, or rarely annuals, with prostrate stems to 60 cm. Leaves 1–2 cm × 5–12 mm, ovate-elliptic, green, sometimes with brown stripes; margin colourless or red. Flowers solitary or bundled, often hidden by leaves. Sepals 2.5–5 mm, acute, green or colourless; petals 3–5 mm, elliptic, acute, white tinged pink; stamens with white or purple anthers. *South Africa*. G1. Spring–summer.

6. C. spathulata Thunberg (*C. cyclophylla* Schönland & E.G. Baker; *C. latispathulata* Schönland & E.G. Baker). Illustration: Higgins, Crassulas in cultivation, 66 (1949); Rowley, Crassula, a grower's guide, 172 (2003).

Prostrate perennial, sparsely branching, rooting, horizontal stems to 20 cm. Leaves 2–3 × 1.5–2.5 cm, ovate, blunt, base cordate; margin with forward-pointing teeth, often tinged red; leaf-stalk short. Sepals 1–2 mm, triangular-linear, blunt, hairless, tinged red; petals 3.5–5 mm, linear-lanceolate, hooded, white tinged pink; stamens 3–4 mm, anthers purple-pink; styles as long as ovaries. *South Africa (Cape Province)*. G1. Autumn.

7. C. sarmentosa Harvey (*C. ovata* E. Meyer). Illustration: Jacobsen, Lexicon of succulent plants, pl. 48 (1974); Graf, Exotica, series 4, edn 12, **1**: 884 (1985); Rowley, Crassula, a grower's guide, 167 (2003).

Tuberous rooted, scrambling perennial with sparsely branched, climbing or hanging stems to 1 m. Leaves

2–6 × 1–3 cm, ovate-elliptic, pointed, flat, green; hydathodes spread below; margins tinged red, entire or with forward-pointing teeth; leaf-stalk short or absent. Inflorescence terminal, flat or rounded. Sepals 1–3 mm, triangular-linear, pointed; petals 4–8 mm, linear-lanceolate, cream-white, tinged pink, ridged with several projections; stamens 4–6 mm, anthers white, tinged red; styles longer than ovaries. *South Africa (Natal to Cape Province)*. G1. Winter.

8. C. cordata Thunberg. Illustration: Higgins, Crassulas in cultivation, 37 (1949); Graf, Exotica, series 4, edn 12, **1**: 884 (1985); Rowley, Crassula, a grower's guide, 106 (2003).

Perennial to 25 cm, with erect or horizontal, sparsely branched, woody stems. Leaves shortly stalked, 1–2.5 cm × 8–20 mm, ovate, blunt, cordate, grey-green, with red hydathodes scattered above; margin entire, often tinged red. Sepals 1–2 mm, triangular, pointed or blunt; petals 4–5 mm, lanceolate, with a drawn-out point, ridged, light yellowish cream, tinged red; stamens 2–3.5 mm, anthers yellow; style as long as ovaries. *South Africa (Cape Province)*. G1. Early spring.

9. C. multicava Lemaire. Illustration: Everett, New York Botanical Gardens illustrated encyclopedia of horticulture **3**: 904 (1981); Graf, Exotica, series 4, edn 12, **1**: 881 (1985); Brickell (ed.), RHS A–Z encyclopedia of garden plants, 320 (2003); Rowley, Crassula, a grower's guide, 136 (2003).

Prostrate or erect perennial to 40 cm; stems sparsely branched, swollen and rooting at the nodes, woody below. Leaves 2–6.5 × 1.5–4 cm, oblong-ovate or elliptic, tip blunt or notched, hydathodes conspicuous above; margin entire, curved under; leaf-stalk 5–20 mm, sheathed at base. Sepals 4 or 5, triangular, 1–2 mm, ridged; petals 4 or 5, lanceolate, pointed, ridged, 3–6 mm, white or cream, red at tip; stamens 3–5 mm, anthers purple; style as long as ovaries. *South Africa (Cape Province)*. G1. Autumn.

10. C. lactea Solander. Illustration: RHS dictionary of gardening 567 (1974); Jacobsen, Lexicon of succulent plants, pl. 44 (1974); Everett, New York Botanical Gardens illustrated encyclopedia of horticulture **3**: 904 (1981); Rowley, Crassula, a grower's guide, 126 (2003).

Perennial to 20 cm, with occasionally scrambling, thick horizontal stems to 40 cm. Leaves 2.5–7 × 1–3 cm, stalkless, oblanceolate, tapering towards base, convex above and below, dull green; margin entire, horny, white dotted. Sepals 1.5–3 mm, linear, pointed, very fleshy; petals 4–7 mm, lanceolate, yellow-white, pink above, with a projection; stamens 4–5.5 mm, anthers purple; style equal to or exceeding ovary length. *South Africa (Cape Province)*. G1. Summer.

11. C. ovata (Miller) Druce (*Cotyledon ovata* Miller; *Crassula argentea* Thunberg; *C. portulacea* Lamarck; *C. obliqua* Solander; *C. arborescens* misapplied). Illustration: Jacobsen, Lexicon of succulent plants, pl. 45, 46 (1974); Everett, New York Botanical Gardens illustrated encyclopedia of horticulture **3**: 903 (1981); Graf, Tropica, edn 3, 364, 366 (1986); Rowley, Crassula, a grower's guide, 150, 151 (2003).

Perennial shrub 1–2 m, with much-branched stems to 20 cm across. Leaves 2–3 × 1–1.8 cm, elliptic, shiny; margin often red, horny; sometimes stalked, to 4 mm. Flowers in rounded clusters on stems 1–3 cm. Sepals 1–2 mm, triangular; petals 7–10 mm, elliptic-lanceolate, pointed, hooded; stamens with purple anthers. *South Africa (Natal to Cape Province)*. G1. Winter.

Very popular as a house plant grown under the name 'Jade' plant. 'Hummel's Sunset' has leaves that turn red in sunlight.

12. C. arborescens (Miller) Willdenow (*Cotyledon arborescens* Miller). Illustration: *Botanical Magazine*, 384 (1797); Jacobsen, Lexicon of succulent plants, pl. 42 (1974); Graf, Exotica, series 4, edn 12, **1**: 878, 879 (1985); Rowley, Crassula, a grower's guide, 85 (2003).

Shrubby perennial 1–2 m, to 6 cm across at base, with freely branched stems; bark peeling. Leaves 3–7 × 2–4 cm, obovate to round, tip obtuse, base tapered; surface grey bloomed; margin entire, horny, often tinged purple; leaf-stalk absent or to 5 mm. Flowers in rounded heads, with several branches. Sepals 5–7, 1–1.5 mm, triangular, ridged, pointed; petals, 5–7, 7–10 mm, lanceolate, hooded, cream tinged red at tip; stamens 4, 5–6 mm, anthers purple; style half length of ovary. *South Africa (Cape Province)*. G1. Summer.

13. C. nemorosa (Ecklon & Zeyher) Endlicher (*Petrogeton nemorosum* Ecklon & Zeyher; *C. confusa* Schönland & E.G. Baker; *C. coerulescens* Schönland). Illustration: *Cactus and Succulent Journal of Great Britain* **40**: 53 (1978); Rowley, Crassula, a grower's guide, 143 (2003).

Tuberous perennial with prostrate or erect stems 4–15 cm. Leaf blade 4–13 mm across, nearly round or ovate, apex blunt, grey or brown-green; margin entire; leaf-stalk 3–15 mm. Sepals ovate-triangular, pointed or blunt; petals 2–3.5 mm, lanceolate with a long thin point, slightly ridged below, green-yellow; stamens 1.5–2 mm, anthers yellow; styles half ovary length. *South Africa (Western Cape Province)*. G1.

14. C. ericoides Haworth. Illustration: Jacobsen, Lexicon of succulent plants, pl. 44 (1974); Rowley, Crassula, a grower's guide, 113 (2003).

Erect perennial to 12 cm with woody branches. Leaves in 4 series, 3–7 × 1–3 mm, ovate-lanceolate, stalkless, pointed; margins entire. Flowers partly concealed by upper leaves; corolla tubular. Sepals 2–4 mm, linear; petals 3–5 mm, elliptic, pointed, white, ridged on back; stamens 3–4 mm, anthers brown; styles as long as ovaries. *South Africa (Natal & Cape Province)*. G1. Summer.

15. C. sarcocaulis Ecklon & Zeyher. Illustration: Higgins, Crassulas in cultivation, 64 (1949); Graf, Exotica, series 4, edn 12, **1**: 884 (1985); Brickell (ed.), RHS A–Z encyclopedia of garden plants, 320 (2003); Rowley, Crassula, a grower's guide, 166 (2003).

Shrubby perennial with many branched, erect, hairy stems 20–60 cm. Leaves 6–30 × 1–8 mm, lanceolate-elliptic, pointed, stalkless, compressed or almost round in cross-section; margin entire, occasionally hairy. Inflorescence terminal, dense; sepals 1–2 mm, triangular, ridged; petals 3–5 mm, oblanceolate-oblong, ridged but without a projection, white; stamens 2.5–3 mm, anthers brown; styles half ovary length, ovaries kidney-shaped. *Southern Africa*. G1. Late summer.

16. C. tetragona Linnaeus. Illustration: Higgins, Crassulas in cultivation, 70 (1949); Graf, Exotica, series 4, edn 12, **1**: 881 (1985); Graf, Tropica, edn 3, 364 (1986).

Perennial with erect, branched, hairless stems to 1 m. Leaves 1.8–5 cm × 1–4 mm, lanceolate, pointed, stalkless. Sepals 1–2 mm, triangular, blunt; petals 1–3 mm, oblanceolate, elliptic, rounded at tip, ridged on back, but without a projection, cream; stamens 1–2 mm, anthers brown; styles less than one third ovary length. *South Africa (Cape Province)*. G1. Summer.

17. C. ciliata Linnaeus. Illustration: Higgins, Crassulas in cultivation, 30 (1949); Rowley, Crassula, a grower's guide, 98 (2003).

Perennial shrublets to 20 cm, well branched, with persistent old leaves. Leaves 1.5–3 × 0.5–1.2 mm, rounded, oblong to elliptic; margins with green to yellowish hairs. Inflorescence domed on stems 15–25 cm. Sepals 2–2.5 mm, triangular, green-yellow; petals 3.5–4.5 mm, elliptic, cream to pale yellow; stamens with yellow anthers. *South Africa (Cape Province)*. G1. Summer.

18. C. coccinea (Linnaeus) Linnaeus (*Rochea coccinea* Linnaeus). Illustration: *Botanical Magazine*, 495 (1799); Jacobsen, Lexicon of succulent plants, pl. 122 (1974); Graf, Exotica, series 4, edn 12, **1**: 919, 920 (1985); Rowley, Crassula, a grower's guide, 99, 100 (2003).

Shrubby perennial to 60 cm, with few branched stems, well covered with leaves. Leaves 1.2–2.5 cm × 4–15 mm, elliptic to ovate, pointed, green often tinged red; margin with hairs, curved upwards. Flowers stalkless; sepals 1.5–2.2 cm, lanceolate, unequal, pointed; corolla tubular, 3.5–4.5 cm, petals spathulate, recurved, red or white tinged red with a dorsal appendage; stamens 2–3 cm, anthers black; style 1 or 2 times length of ovary. *South Africa (Cape Province)*. G1. Summer.

19. C. orbicularis Linnaeus (*C. rosularis* Haworth). Illustration: Higgins, Crassulas in cultivation, 49 (1949); Jacobsen, Lexicon of succulent plants, pl. 45 (1974); Graf, Exotica, series 4, edn 12, **1**: 880, 883 (1985); Rowley, Crassula, a grower's guide, 148, 149 (2003).

Perennial, 5–25 cm, stems sometimes producing runners. Leaves packed in basal rosettes, 1.5–5 cm × 5–20 mm, elliptic to oblanceolate, pointed; hydathodes often visible on upper surface and margins, but not on lower surface; margin with hairs, sometimes tinged red. Inflorescence with scale-like leaves, flowers stalked. Sepals 1–3 mm, lanceolate-triangular, blunt; petals 2–5 mm, oblanceolate, blunt or pointed, pale yellow-white, tinged pink,

often with a dorsal appendage; stamens 2.5–3.5 mm, anthers yellow; styles half ovary length. *South Africa (Natal)*. G1. Summer.

20. C. montana Thunberg subsp. **quadrangularis** (Schönland) Tölken (*C. quadrangularis* Schönland). Illustration: Higgins, Crassulas in cultivation, 49 (1949); Rowley, Crassula, a grower's guide, 134, 135 (2003).

Cushion-forming, rosetted perennial reaching 4–10 cm. Leaves 4–20 × 4–17 mm, in 4 ranks, obovate to ovate, pointed, adpressed or spreading, greenish brown, dark-spotted; hydathodes red, on margins and over upper surface; margins with hairs. Flowering stems to 10 cm, flowers stalkless, in flat-topped axillary clusters. Sepals 2–3 mm, triangular; petals 3–6 mm, oblong, blunt, reflexed, white, tinged pink, with dorsal projection; stamens 2.5–3 mm, anthers black; style less than a quarter of ovary length. *South Africa (Cape Province)*.

21. C. intermedia Schönland. Illustration: Higgins, Crassulas in cultivation, 31 (1949); Rowley, Crassula, a grower's guide, 124 (2003).

Perennial to 20 cm, like *C. montana* subsp. *quadrangularis* except for flowers with yellow anthers and short reflexed styles. *South Africa (Cape Province)*. G1. Summer.

22. C. socialis Schönland. Illustration: Jacobsen, Lexicon of succulent plants, pl. 48 (1974); Everett, New York Botanical Gardens illustrated encyclopedia of horticulture **3**: 906 (1981); Graf, Exotica, series 4, edn 12, **1**: 882 (1985); Rowley, Crassula, a grower's guide, 171 (2003).

Perennial to 6 cm, like *C. montana* subsp. *quadrangularis* but lower growing, and having pure white flowers with yellow anthers and short, reflexed styles. *South Africa (Cape Province)*. G1. Summer.

23. C. exilis Harvey (*C. bolusii* J.D. Hooker). Illustration: Higgins, Crassulas in cultivation, 35 (1949); Rowley, Crassula, a grower's guide, 114, 115 (2003).

Annual or cushion forming perennial to 10 cm. Leaves 4–4.5 × 1–10 mm, in 4 ranks; blade flat or rounded in cross-section, tip pointed; margin with hairs. Flowers stalked in terminal clusters; sepals 2–2.5 mm, triangular, pointed, margins hairy; petals 4–4.5 mm, oblong-obovate, with a projection, white, tinged pink; stamens 3–3.5 mm, anthers yellow. Fruit

spreading at right angles to flower-stalk. *South Africa (Cape Province)*. G1.

Subsp. **cooperi** (Regel) Tölken (*C. cooperi* Regel; *C. picturata* Boom), has dense spreading cushions and oblanceolate, less fleshy leaves.

A complex group of garden hybrids including *C. justi-corderoyi* misapplied and *C. picturata* hybrids, would also key out here.

24. C. setulosa Harvey (*C. curta* N.E. Brown; *C. milfordiae* Byles). Illustration: Higgins, Crassulas in cultivation, 54 (1949); Rowley, Crassula, a grower's guide, 170 (2003).

Woody-stemmed perennial to 25 cm. Leaves 2–3.5 cm × 1–10 mm, pointed, flat, green tinged red, 4-ranked or densely packed in basal rosettes; hydathodes present on margins and upper surface. Stem-leaves reduced towards top. Inflorescence hairy, bracts triangular. Sepals 1–3 mm, lanceolate-triangular; petals 2.5–4 mm, oblong, white tinged red, with a projection; stamens 2–3 mm, anthers yellow-brown; style one to two-thirds ovary length. *Widespread in southern Africa*. G1. Summer.

25. C. vaginata Ecklon & Zeyher (*C. drakensbergensis* Schönland; *Sedum crassifolium* O. Kuntze). Illustration: Leistner (ed.), Flora of southern Africa **14**: 175 (1985); Rowley, Crassula, a grower's guide, 186 (2003).

Tuberous rooted perennial to 50 cm. Leaves 5–25 cm × 3–35 mm, linear-lanceolate, fused at base, margin with hairs, in basal rosettes. Flowers numerous; sepals 1.5–3.5 mm, triangular, fine-pointed, sometimes toothed; petals 2.5–5 mm, oblong, blunt, hooded, yellowish white, appendage very small; stamens 2.5–3.5 mm, anthers yellow; style one third ovary length. *South Africa to Arabia*. G1. Summer.

26. C. alba Forskål (*C. recurva* N.E. Brown). Illustration: Leistner (ed.), Flora of southern Africa **14**: 175 (1985); Rowley, Crassula, a grower's guide, 82 (2003).

Biennial or perennial to 50 cm. Leaf pairs spirally arranged, leaves mainly in basal rosettes, 6–17 cm × 5–15 mm, linear-lanceolate, stalkless, acute, dark geenish yellow, tinged purple; margin with fine recurved hairs. Flowers numerous; sepals 1–4.5 mm, triangular; petals 3–6 mm, oblong, pointed, hooded and recurved at tip, white or red; stamens 3–5 mm, anthers dark brown. *Southern Africa.* H5.

27. C. hemisphaerica Thunberg (*Purgosea hemisphaerica* (Thunberg) G. Don). Illustration: Jacobsen, Lexicon of succulent plants, pl. 43 (1974); Everett, New York Botanical Gardens illustrated encyclopedia of horticulture **3**: 905 (1981); Graf, Tropica, edn 3, 367 (1986); Rowley, Crassula, a grower's guide, 121 (2003).

Perennial, 5–12 cm, with leaves 4-ranked in few rounded rosettes. Leaves 1–5 cm × 8–25 mm, obovate, bristle-tipped, stalkless; hydathodes on margins and upper surface; margins with hairs. Inflorescence long and tapering, with many small stalkless flowers. Sepals 0.5–1 mm, triangular, hairless; petals 2–3 mm, oblong-lanceolate, pointed or rounded with a dorsal ridge and projection, creamy white; stamens 1.5–2.5 mm, anthers black; style short. *South Africa (Cape Province).* G1. Spring.

28. C. capitella Thunberg. Illustration: Graf, Exotica, series 4, edn 12, **1**: 868, 869 (1985); Rowley, Crassula, a grower's guide, 96, 97 (2003).

Perennial or biennial to 40 cm, stems usually woody. Leaves 1–12 cm × 3–20 mm, the pairs spirally arranged in basal rosettes, usually smallest near the centre of the rosette, ovate to linear-lanceolate, pointed, stalkless, spreading, hairless or hairy, grooved, often red spotted; hydathodes red, scattered over upper surface; margin with fine recurved hairs or papillae. Inflorescence hairless or hairy, spike-like. Sepals 1–4 mm, triangular-lanceolate, pointed, with marginal hairs; petals 2–5 mm, oblong-lanceolate, blunt, hooded, white tinged pink, with a projection on the back; stamens 2–4.5 mm, anthers dark brown; style very short or absent. *South Africa (Cape Province, Lesotho, Orange Free State & Transvaal).* G1.

A complex species with two subspecies in cultivation. Subsp. **nodulosa** (Schönland) Tölken (*C. nodulosa* Schönland; *C. elata* N.E. Brown) has tuberous bases and reflexed leaves, a hairy inflorescence with hairy sepals and short or absent style. Subsp. **thyrsiflora** (Thunberg) Tölken (*C. thyrsiflora* Thunberg; *C. turrita* Thunberg) has hairless leaves, except for the margins, hairless sepals that are occasionally toothed, petals with rounded appendages and style one third to half of ovary length.

29. C. barbata Thunberg. Illustration: Higgins, Crassulas in cultivation, 13 (1949); Graf, Exotica, series 4, edn 12, **1**: 880, 883 (1985); Rowley, Crassula, a grower's guide, 90 (2003).

Annual or biennial to 30 cm. Leaves spreading with erect tips in 4 series and basal rosettes; blade 1–4 × 1–3.5 cm, obovate to nearly round, greyish green; margin with long, tufted hairs. Flowers stalkless in spiked racemes. Sepals 2.5–3 mm, oblong-triangular, blunt; petals 4.5–6 mm, oblong, rounded or pointed with a dorsal projection, white tinged pink; stamens 3–4.5 mm, anthers black, style very short, broad. *South Africa (Cape Province).* H5–G1. Spring.

30. C. perforata Thunberg. Illustration: Jacobsen, Lexicon of succulent plants, pl. 46 (1974); Everett, New York Botanical Gardens illustrated encyclopedia of horticulture **3**: 906 (1981); Graf, Exotica, series 4, edn 12, **1**: 873 (1985); Rowley, Crassula, a grower's guide, 156 (2003).

Shrubby or scrambling perennials with little-branched stems, partially covered in old leaves. Leaves 4–20 × 3–15 mm, ovate, stalkless, united in pairs, pointed or blunt, convex below; hydathodes on upper surface only; margins hairless or with hairs, reddish yellow, horny. Inflorescence with adpressed bracts, flowers stalkless. Sepals 0.5–1 mm, triangular; petals 2–2.5 mm, oblong-elliptic, with small projection; stamens 2 mm, anthers brown; styles one third ovary length. *South Africa (Natal & Cape Province).* G1.

31. C. rupestris Thunberg. Illustration: Jacobsen, Lexicon of succulent plants, pl. 47 (1974); Graf, Exotica, series 4, edn 12, **1**: 873, 880 (1985); Graf, Tropica, edn 3, 366 (1986); Rowley, Crassula, a grower's guide, 162–5 (2003).

Shrubby perennial to 50 cm, with erect, well-branched stems and flaking bark. Leaves 3–15 × 2–13 mm, ovate-lanceolate, stalkless, united in pairs, hairless, brownish red; hydathodes red, spread over upper surface; margins horny, yellow-red. Inflorescence domed; sepals to 1 mm, triangular; petals 3–4 mm, oblong-elliptic, tip rounded, ridged, projection very small, white tinged red; stamens 2–3.5 mm, anthers brown; style half ovary length. *South Africa (Cape Province).* G1.

32. C. brevifolia Harvey. Illustration: Higgins, Crassulas in cultivation, 67 (1949); Graf, Exotica, series 4, edn 12, **1**:

871 (1985); Rowley, Crassula, a grower's guide, 93 (2003).

Shrubby perennial to 50 cm, stems branched. Leaves 1.5–5 cm × 2–6 mm, oblong elliptic, sharply tapered to a blunt tip, spreading and upturned, convex above, convex or keeled below, very fleshy, greyish green, hydathodes on upper surface and margins; margins horny at edge. Inflorescence domed; sepals triangular, blunt; petals 3–5 mm, oblong-elliptic, rounded, ridged with a projection; stamens 2.5–4 mm, anthers black to yellow; styles one third length of ovary. *Namibia & South Africa (Cape Province).* G1.

33. C. macowaniana Schönland & E.G. Baker. Illustration: Leistner (ed.), Flora of southern Africa **14**: 194 (1985); Rowley, Crassula, a grower's guide, 130 (2003).

Perennial shrubs with branched, sometimes prostrate stems and flaking bark. Leaves 2.4–8 cm × 3–15 mm, linear-lanceolate, stalkless, pointed, flat above, convex below, hairless, brownish green. Flowers in domed clusters; sepals 0.5–1 mm, oblong, rounded; petals 2.5–4 mm, oblanceolate, rounded, ridged, white tinged pink, projection very small; stamens 2.5–3 mm, anthers black; style half ovary length; ovaries tapered above. *Namibia, South Africa (Cape Province).* G1. Summer.

34. C. columnaris Thunberg. Illustration: Hunt (ed.), Marshall Cavendish encyclopedia of gardening 382 (1968); Leistner (ed.), Flora of southern Africa **14**: 194 (1985); Graf, Exotica, series 4, edn 12, **1**: 871, 875 (1985); Rowley, Crassula, a grower's guide, 102, 103 (2003).

Biennial or perennial, 3–10 cm, often monocarpic. Leaves 3–12 × 10–25 mm, transversely ovate, blunt, stalkless, grey or brownish green, clasping and dish-shaped with erect hairy margin, densely packed into an erect column 2–3 cm across. Flowers partially hidden by upper leaves, nearly stalkless; sepals 3–4 mm, oblong-elliptic, rounded; petals 7–13 mm, oblong rounded with a small dorsal projection, white or yellowish, tinged red. Stamens 4.5–6 mm, anthers brown to yellow; style very short. *South Africa & Namibia.*

35. C. barklyi N.E. Brown (*C. teres* Marloth). Illustration: *Botanical Magazine*, 8421 (1921); Everett, New York Botanical Gardens illustrated encyclopedia of horticulture **3**: 906 (1981); Graf, Tropica, edn 3, 362 (1986); Rowley, Crassula, a grower's guide, 91 (2003).

Perennial 5–9 cm with horizontal or erect stems branching at base. Leaves 3–4 × 1–15 mm, adpressed to stem to form a smooth column up to 1.5 cm across; stalkless, transversely ovate, convex below, concave above, dish-shaped; margin membranous with erect dense hairs. Flowering-stems almost stalkless, hidden by upper leaves. Sepals 4–5 mm, oblanceolate to oblong, rounded; petals hooded, with a projection, cream; stamens 4.5–5 mm, anthers yellow, style short, wide; ovary conical. *South Africa (Cape Province)*. H5–G1. Winter.

36. C. alpestris Thunberg. Illustration: Higgins, Crassulas in cultivation, 25 (1949); Rowley, Crassula, a grower's guide, 83 (2003).

Perennial, or rarely biennial, erect and sometimes branched. Leaves 1–2 cm × 5–8 mm, triangular, sharply acute, leathery, green-brown. Sepals 2.5–3 mm, rounded with spreading marginal hairs, green-brown; petals 8–12 mm, fused at base, elliptic-oblong, white or cream, often tinged red; stamens with brown anthers. *South Africa (Cape Province)*. G1. Winter–spring.

37. C. pyramidalis Thunberg. Illustration: Jacobsen, Lexicon of succulent plants, pl. 47 (1974); Leistner (ed.), Flora of southern Africa **14**: 194 (1985); Everett, New York Botanical Gardens illustrated encyclopedia of horticulture **3**: 906 (1981); Rowley, Crassula, a grower's guide, 161 (2003).

Perennial with simple or branched, erect or horizontal stems 3–25 cm. Leaves 4–12 × 4–8 mm, ovate, adpressed to stem forming a more or less square column; margins hairless, horny. Inflorescence terminal, with numerous stalkless flowers. Sepals 3–4 mm, lanceolate to spathulate; petals white-cream, 7–11 mm, oblong, hooded, projection transparent; stamens 3.5–4.5 mm, anthers yellow; styles indistinct. *South Africa (Cape Province)*. G1. Spring.

38. C. lanuginosa Harvey. Illustration: Higgins, Crassulas in cultivation, 51 (1949); Rowley, Crassula, a grower's guide, 129 (2003).

Perennial to 15 cm, with horizontal or drooping stems, occasionally rooting at the nodes. Leaves 2–35 × 2–8 mm, obovate to linear-lanceolate, pointed or blunt, convex above and below, hairy; margin with recurved hairs. Flowers terminal, very shortly stalked, clustered; sepals 1.5–2 mm; anthers black; ovaries tapered to lateral stigmas. *South Africa (Cape Province)*. G1. Summer.

39. C. alstonii Marloth. Illustration: Jacobsen, Lexicon of succulent plants, pl. 41 (1974); Leistner (ed.), Flora of southern Africa **14**: 203 (1985); Graf, Exotica, series 4, edn 12, **1**: 872 (1985); Rowley, Crassula, a grower's guide, 84 (2003).

Perennial herb with leaves in 2 rows, forming dense round rosettes 2–5 cm across. Leaf-blade 6–10 mm × 1.2–2 cm, rounded or obovate, tip erect, hairy beneath. Flowers stalkless; sepals triangular, pointed, hairy, margins with hairs; petals 2.5–3 mm, lanceolate-oblong, yellowish cream, pointed; stamens 1.5–2 mm; brownish yellow; stigmas red; ovaries kidney-shaped. *South Africa (Cape Province)*. H3. Autumn.

40. C. tecta Thunberg. Illustration: Higgins, Crassulas in cultivation, 68 (1949); Jacobsen, Lexicon of succulent plants, pl. 48 (1974); Graf, Exotica, series 4, edn 12, **1**: 885 (1985); Rowley, Crassula, a grower's guide, 178, 179 (2003).

Perennial to 20 cm, often branched to form tufts. Leaves 2–3.5 cm × 5–15 mm, oblong-lanceolate, rounded, stalkless, conspicuously white, with papillae, hairy below. Flowers numerous, almost stalkless, in rounded, terminal clusters. Sepals 1.5–2 mm, triangular, hairy; petals 3–4 mm, oblong-lanceolate, blunt with a projection, white-cream; stamens 1.5–2 mm, anthers yellow; stigmas lateral, style very short or absent. *South Africa (Cape Province)*. G1. Summer.

41. C. ausensis P.C. Hutchison (*C. hofmeyeriana* Dinter). Illustration: Higgins, Crassulas in cultivation, 27 (1949); Rowley, Crassula, a grower's guide, 88, 89 (2003).

Perennial with short branches, and tightly grouped, persistent leaves forming dense cushions. Leaves 1.2–3 cm × 4–10 mm, lanceolate to elliptic, sharply acute, flat above, convex below, covered with short hairs, green or brownish. Inflorescence umbel-like, with flowers on stalks 2–4.5 cm. Sepals 2–3 mm, triangular, green, fleshy; petals 5–7 mm, fused at base for *c.* 1 mm, oblong, acute, white-cream, dorsal appendage indistinct;

stamens with brown anthers. *Namibia*. G1. Autumn.

42. C. susannae Rauh & Friedrich. Illustration: Jacobsen, Lexicon of succulent plants, pl. 48 (1974); Graf, Exotica, series 4, edn 12, **1**: 871 (1985); Rowley, Crassula, a grower's guide, 176 (2003).

Perennial 1–4 cm. Leaves 6–10 × 4–8 mm, stalkless, oblong, fleshy, channelled, in 4-ranked rosettes. Flowers few, in terminal racemes. Sepals 1.5–2 mm, triangular-oblong, bluntly pointed, hairy; petals 2.5–3.5 mm, sharply pointed, oblong, ridged, white, projection very small; stamens 2–2.5 mm, anthers black; styles short, wide, stigmas lateral, red. *South Africa (Cape Province)*. G1. Autumn.

43. C. mesembrianthemopsis Dinter. Illustration: Jacobsen, Lexicon of succulent plants, pl. 45 (1974); Graf, Exotica, series 4, edn 12, **1**: 871 (1985); Graf, Tropica, edn 3, 367 (1986); Rowley, Crassula, a grower's guide, 132 (2003).

Similar to *C. susannae*, except: leaves wedge-shaped, triangular in cross-section, not grooved; flowers short-stalked, partially hidden by leaves; sepals to 3 mm, petals 5–7 mm, stamens 3–4 mm, anthers yellow. *Namibia & South Africa (Cape Province)*. G1. Autumn.

44. C. grisea Schönland. Illustration: Higgins, Crassulas in cultivation, 45 (1949); Jacobsen, Lexicon of succulent plants, pl. 43 (1974); Leistner (ed.), Flora of southern Africa **14**: 210 (1985); Rowley, Crassula, a grower's guide, 119 (2003).

Perennial with few erect stems to 20 cm. Leaves persistent, scarcely shrivelling, 2–8 cm × 3–6 mm, linear-lanceolate, grey-green, hairless, hairy or covered with rounded papillae. Flowers in terminal clusters on stalks 2–10 cm. Sepals 1–1.5 mm, oblong-triangular; petals 2–3.5 mm, obtuse, cream fading to brown; stamens with brown anthers. *South Africa (Cape Province) & Namibia*. G1. Summer–autumn.

45. C. plegmatoides Friedrich (*C. deltoidea* Schönland & Baker; *C. arta* in the sense of Higgins & Jacobsen, but not Schönland). Illustration: Jacobsen, Lexicon of succulent plants, pl. 43 (1974); Leistner (ed.), Flora of southern Africa **14**: 210 (1985); Graf, Exotica, series 4, edn 12, **1**: 871 (1985); Rowley, Crassula, a grower's guide, 157 (2003).

Perennial to 15 cm, only branched from base or unbranched. Leaves 5–9 × 7–13 mm, ovate, adpressed to stem to form a 4-angled column. Flower-stalk 3–6 cm; sepals 1.5–2 mm, triangular, fleshy, grey-green; petals 2–3 mm, fused at base, cream fading to brown; stamens with brown anthers. *Namibia & South Africa (Cape Province).* G1. Late summer–early autumn.

46. C. deceptor Schönland & E.G. Baker (*C. deceptrix* Schönland; *C. arta* Schönland). Illustration: Leistner (ed.), Flora of southern Africa **14**: 210 (1985); Graf, Exotica, series 4, edn 12, **1**: 881, 883 (1985); Rowley, Crassula, a grower's guide, 109 (2003).

Perennial to 15 cm, with numerous branches, usually covered with old leaf bases. Leaves 6–15 × 6–15 mm, ovate, stalkless, in 4 ranks, adpressed to stem, forming a square column to 2.5 cm across, hydathodes prominent on both surfaces. Flowers stalkless, spreading; sepals *c.* 1.5 mm, triangular-oblong, blunt-pointed, hairy; petals 2–2.5 mm, elliptic-oblong, pointed or blunt with a dorsal projection, creamish brown; stamens 1–2 mm, anthers brown, stigmas stalkless, red. *South Africa (Cape Province).* G1. Summer.

47. C. perfoliata (Linnaeus) de Candolle (*Rochea falcata* (Wendland) de Candolle var. *acuminata* Ecklon & Zeyher; *R. perfoliata* Linnaeus, in part). Illustration: Jacobsen, Lexicon of succulent plants, pl. 46 (1974); Graf, Exotica, series 4, edn 12, **1**: 873 (1985); Graf, Tropica, edn 3, 366 (1986); Rowley, Crassula, a grower's guide, 154, 155 (2003).

Perennial with erect branches to 1.5 m, covered with coarse papillae. Leaves persistent, 4–12 × 1.2–3.5 cm, lanceolate-triangular, acute, densely covered with papillae, green or greyish, sometimes with purple spots. Inflorescence rounded or flat-topped on stems 3–10 cm. Sepals 1–3 mm, triangular, hairy, fleshy and green; petals 4–7.5 mm, fused at base for *c.* 1 mm, oblong-lanceolate, white, pink or scarlet; stamens with black anthers. *South Africa (Cape Province).* G1.

48. C. mesembryanthoides (Haworth) Dietrich (*Globulea mesembryanthoides* Haworth). Illustration: Higgins, Crassulas in cultivation, 32 (1949); Rowley, Crassula, a grower's guide, 133 (2003).

Perennial, forming shrublet to 40 cm, much-branched with spreading woody branches, old leaves deciduous. Leaves 1–5 cm × 2–3 mm, linear-triangular to linear-elliptic, green-brown, covered with hairs to 1 mm. Flower-stalks 10–30 cm; sepals 3–3.5 mm, linear-triangular, hairy; petals 4.5–6 mm, fused at base for *c.* 1 mm, with elliptic dorsal appendage, white to cream; stamens with black anthers. *South Africa (Cape Province).* G1. Autumn.

49. C. cultrata Linnaeus. Illustration: *Botanical Magazine*, 1940 (1817); Jacobsen, Lexicon of succulent plants, pl. 43 (1974); Graf, Exotica, series 4, edn 12, **1**: 879 (1985); Rowley, Crassula, a grower's guide, 108 (2003).

Erect perennial shrub to 80 cm; stems branched, older bark flaking towards base. Leaves 2.5–10 × 1–3 cm, lanceolate, rounded or blunt at tip, hairless or velvety; margin horny, red. Flowers numerous in loose groups, almost stalkless. Sepals 2–3 mm, triangular-oblong, rounded, hairy; petals 3.5–4.5 mm, fiddle-shaped, with apical projection, cream. Stamens 2.5–3.5 mm; anthers black; style very short or absent. *South Africa (Cape & Natal).*

50. C. subaphylla (Ecklon & Zeyher) Harvey (*Sphaeritis subaphylla* Ecklon & Zeyher). Illustration: Higgins, Crassulas in cultivation, 69 (1949); Leistner (ed.), Flora of southern Africa **14**: 217 (1985); Rowley, Crassula, a grower's guide, 175 (2003).

Perennial shrublet to 80 cm, with wiry woody branches. Leaves 8–15 × 2–3 mm, linear-elliptic to lanceolate, green, grey-green or brown, sometimes with short hairs. Flower-stalks 3–15 cm; sepals 2–2.5, triangular, fleshy, grey-green; petals 3–5 mm, with terminal dorsal appendage; stamens with yellow to brown anthers. *South Africa (Cape Province), Namibia.* G1.

51. C. cotyledonis Thunberg. Illustration: Jacobsen, Lexicon of succulent plants, pl. 42 (1974); Rowley, Crassula, a grower's guide, 107 (2003).

Perennial with basal rosettes; stems woody, to 20 cm. Leaves persistent, 3–6 × 1–2.5 cm, oblong-oblanceolate, covered in coarse recurved hairs and marginal hairs, grey-green to yellowish green. Inflorescences 15–30 cm; flowers many; sepals 2–3 mm, oblong triangular; petals 3–5 mm, cream to pale yellow; stamens with yellow anthers. *South Africa (Cape Province) & Namibia.* G1. Summer.

52. C. nudicaulis Linnaeus. Illustration: Rowley, Crassula, a grower's guide, 144, 145 (2003).

Woody-stemmed perennial with hairy or hairless branches. Leaves in numerous rosettes, 2–9 cm × 4–25 mm, oblong-elliptic, sometimes linear or round, convex or flat above, convex below, hairy or hairless. Flowers in a loose cyme-like inflorescence, sepals 1.5–3 mm, triangular-oblong, hairy or hairless; petals 3–5 mm, fiddle-shaped with a projection, cream; stamens 2.5–3.5 mm, anthers yellow; style sharp; stigmas lateral. *Southern Africa.* G1.

53. C. aquatica (Linnaeus) Schönland (*Tillaea aquatica* Linnaeus). Illustration: Blamey & Grey-Wilson, Illustrated flora of Britain and northern Europe, 164 (1989); Stace, New flora of the British Isles, 379 (1991).

Tiny annual plant of moist habitats. Stems solitary or little-branched, horizontal. Leaves 3–6 mm, linear, fleshy, tips pointed. Flowers borne in leaf axils, solitary, almost stalkless. Sepals 4, triangular, blunt; petals 4, longer than sepals, *c.* 1 mm, white or pale pink, ovate, pointed. N & C Europe, *N Asia & N America.* H1. Summer–autumn.

54. C. helmsii (T. Kirk) Cockayne (*C. recurva* (J.D. Hooker) Ostenfeld; *Tillaea recurva* (J.D. Hooker) Hooker). Illustration: Stace, New flora of the British Isles, 379 (1991); Rowley, Crassula, a grower's guide, 26, 120 (2003).

Perennial aquatic herb with simple or branched erect stems. Leaves 4–20 mm, linear to lanceolate, pointed, joined at the base. Flowers solitary, stalked; sepals 4, oblong, pointed; petals 4, ovate, 1.5–2 mm, white to pink, blunt. *Australia & New Zealand, naturalised in N Europe.* H5. Summer.

Often cultivated in ponds. It should not be introduced into the wild as it often becomes a noxious weed.

2. SINOCRASSULA Berger
S.G. Knees

Succulent perennials, hairless or with minute hairs. Leaves in rosettes, thickened, obtuse or tapering, hair-tipped, lined, blotched or variously marked with reddish brown. Flowering shoots erect, bracts loose; panicles cyme-like or simple with shortly stalked, erect flowers crowded towards apex. Sepals 5; petals 5, whitish with vivid red tips, joined to form spherical or urn-shaped corolla. Stamens 5, in a single whorl.

A genus of 2 species, formerly included in *Crassula* and *Sedum*. Occurring in the Himalaya and western China these plants are almost hardy, but grow best if kept in a cold frame during winter months. Propagation is by leaf cuttings, seed or division.

1a. Leaves 3.5–6 cm, linear-spathulate
           1. indica
  b. Leaves 1.2–2.5 cm, lanceolate
           2. yunnanensis

1. S. indica (Decaisne) Berger (*Crassula indica* Decaisne; *Sedum indicum* (Decaisne) Hamet). Illustration: Jacquemont, Voyages dans l'Indie **4**: 73 (1835); Hooker, Flora of British India **2**: 413(1878).

Stems 10–30 cm, smooth, hairless, creeping and branching. Leaves linear-spathulate, long-tapered, 3.5–6 × 1–1.5 cm, underside convex, greyish green, spotted and lined with reddish brown markings. Flower-stalk 15–28 cm; calyx reddish; petals greenish crimson, *c.* 3 × 1 mm; stamens 2.5–3 mm. *India, Bhutan, Sikkim.* G1. Summer.

Sometimes cultivated under the name *Echeveria maculata.*

2. S. yunnanensis (Franchet) Berger (*Crassula yunnanensis* Franchet; *Sedum indicum* var. *yunnanensis* Hamet). Illustration: Jacobsen, Handbook of succulent plants **2**: 846 (1960); Haage, Cacti & succulents, a practical handbook, 77 (1963); Jacobsen, Lexicon of succulent plants, pl. 133 (1977); Lamb & Lamb, Popular exotic cacti in colour, 159 (1975).

Root-crown short and thick bearing annual stems 3–9 cm. Basal leaves kidney-shaped, 50–70 in dense rosettes; stem-leaves in 3 s, lanceolate, tapered, mucronate; upper side roundish, underside very rounded, 1.2–2.5 cm × 4.5–6 mm, dark bluish green, covered with minute, white and glandular hairs. Inflorescence minutely hairy. *W China (Yunnan).* H5. Summer.

'Cristata', a monstrous form, is often cultivated as a curiosity.

3. KALANCHOË Adanson
L.S. Springate
Annuals to small trees, mostly perennials, erect to prostrate, herbaceous to succulent. Hairs absent or simple, often glandular or branched, sometimes scale-like. Leaves opposite, very rarely alternate or whorled,

entire to twice pinnatisect, usually flattened. Inflorescence terminal, very rarely lateral. Flower-parts in 4 s. Calyx-tube very short to exceeding lobes. Corolla-tube usually much exceeding lobes. Stamens in 2 whorls, partly fused to corolla-tube, often projecting from tube (position of upper whorl of anthers noted in the following descriptions). Plantlets sometimes produced on leaves, leaf-stalks or on inflorescence.

A genus of about 130 species with greatest diversity in Madagascar, also occurring in Africa and southern Asia to the Ryuku Islands and Indonesia and widely naturalised in the tropics. The species flower in winter naturally. Cultivation in low winter light and temperatures in Europe can result in delayed flowering, abnormal flower-head and flower formation and uncharacteristic regrowth. Grown under glass in most porous composts with restricted nutrients. Control of day-length is needed for continuous pot-plant production. Propagation from seed (except for some cultivars), cuttings or plantlets.

Literature: Hamet, R., Monographie de genre Kalanchoë, *Bulletin de L'Herbier Boissier, serie 2,* **7**: 869–900 (1907) & **8**: 17–48 (1908); Boiteau, P. & Mannoni, O., Les plantes grasses de Madagascar. Les Kalanchoë, *Cactus* **12**: 6–10 (1947), **13**: 7–10 (1948), **14**: 23–8 (1948), **15–16**: 37–42 (1949), **17–18**: 57–8 (1949), **19**: 9–14 (1949), **20**: 43–6 (1949), **21**: 69–76 (1949) & **22**: 113–14 (1947–9); Hamet, R. & Marnier-Lapostolle, J., Le genre Kalanchoë au Jardin Botanique 'Les Cedres' (1964); Cufodontis, G., The species of Kalanchoë in Ethiopia and Somalia Republic, *Webbia* **19**: 711–44 (1965); Raadts, E., The genus Kalanchoë (Crassulaceae) in tropical east Africa, *Willdenowia* **8**: 101–57 (1977); van Voorst, A., & Arends, J.C., The origin and chromosome numbers of cultivars of Kalanchoë blossfeldiana von Poellnitz: Their history and evolution, *Euphytica* **31**: 573–84 (1982); Shaw, J.M.H., An investigation of the cultivated Kalanchoë daigremontiana group, with a checklist of Kalanchoë cultivars, *Hanburyana* **3**: 17–79 (2008).

1a. Leaf-hairs 3- or 4-branched, non-glandular, sometimes compressed and scale-like   2

  b. Leaf–hairs simple, often glandular, or absent   8
2a. Leaves alternate   3
  b. Leaves opposite   4
3a. Leaves stalked   **25. rhombopilosa**
  b. Leaves stalkless   **23. tomentosa**
4a. Leaves entire   5
  b. Leaves toothed   7
5a. Leaves with long, soft hairs   **24. eriophylla**
  b. Leaves with dense, scale-like hairs   6
6a. Leaves circular to obovate, very obtuse; corolla-tube less than 6.5 mm   **21. hildebrandtii**
  b. Leaves ovate to lanceolate, more or less acute; corolla-tube more than 6.5 mm   **19. orygalis**
7a. Leaves triangular to hastate, with irregular teeth and lobes   **20. beharensis**
  b. Leaves ovate to obovate, with regular, shallow teeth only   **22. millotii**
8a. Stems climbing, supported by the leaves   9
  b. Stems erect, free-standing or prostrate   10
9a. Adult leaves with irregular pinnate divisions   **9. schizophylla**
  b. All leaves entire to lobed, never with pinnate divisions   **10. beauverdii**
10a. Plant rather glaucous, rarely hairless; leaves triangular, more or less hastate, stalked, non-peltate; margins sinuous with uneven teeth and lobes   **20. beharensis**
  b. Plant clearly different in at least one feature   11
11a. Ovaries and styles divergent   12
  b. Ovaries and at least lower part of styles adpressed   13
12a. Leaves peltate   **1. peltata**
  b. Leaves not peltate   **3. gracilipes × manginii**
13a. Free part of filament longer than part fused to corolla   14
  b. Free part of filament shorter than part fused to corolla   28
14a. Calyx-segments at least as long as tube   15
  b. Calyx-segments shorter than tube   24
15a. No spurs bearing plantlets on leaf margins   16
  b. Spurs bearing plantlets on leaf margins   22
16a. Stems thick (more than 4 mm at base), more or less erect   17
  b. Stems thin, creeping   19

17a. Mature plants more than 40 cm, or flower-heads with more than 30 flowers   **2. miniata**

b. Mature plants less than 40 cm, flower-heads with 3–30 flowers   18

18a. Flower-heads usually 3-branched from base, with 15–30 flowers   **6. 'Wendy'**

b. Flower-heads with single stem, branched at apex, with less than 15 flowers   **7. porphyrocalyx**

19a. Plant glaucous   **26. pumila**

b. Plant green   20

20a. Flowers funnel-shaped, yellow   **5. jongmansii**

b. Flowers urn-shaped, red to purple   21

21a. Leaves obovate to spathulate or oblong-obovate; calyx-segments more than 5 mm; corolla-tube more than 2 cm   **4. manginii**

b. Leaves more or less circular; calyx-segments less than 5 mm; corolla-tube less than 2 cm   **8. uniflora**

22a. Leaves nearly cylindric, entire except for 3–9 teeth at apex   **11. delagoensis**

b. Leaves flat, whole margin toothed   23

23a. Leaves with less than 20 irregular teeth and few spurs bearing plantlets   **12. × houghtonii**

b. Leaves with more than 20 regular teeth, alternating with plantlet-bearing spurs   **13. daigremontiana**

24a. Leaves simple   25

b. Leaves pinnately divided   27

25a. Leaf-blade more than 7.5 cm, with more than 30 teeth   **16. gastonis-bonnieri**

b. Leaf-blade less than 7.5 cm with less than 30 teeth   26

26a. Leaves more or less tapered at base, without auricles, crowded at base of stem   **14. fedtschenkoi**

b. Leaves with very small to prominent auricles, well-spaced at base of stem   **15. laxiflora**

27a. Leaves pinnatifid to pinnatisect; corolla less than 3 cm   **17. prolifera**

b. Leaves divided into 3–5 leaflets; corolla more than 3 cm   **18. pinnata**

28a. Leaves at middle of stem half-cylindric or triangular in section, longer than 5 cm   29

b. All leaves flattened, or shorter than 5 cm   30

29a. Flowers white; leaves entire   **43. bentii**

b. Flowers pink; leaves toothed or lobed   **42. × kewensis**

30a. Corolla-tube more than 2.7 cm   31

b. Corolla-tube less than 2.7 cm   32

31a. Leaves stalked   **40. quartiniana**

b. Leaves stalkless   **41. marmorata**

32a. Inflorescence lateral; stolons long, with a single node, from leaf axils   **27. synsepala**

b. Inflorescence terminal; stolons absent   33

33a. Plant with some hairs   34

b. Plant entirely hairless   38

34a. Some anthers protruding from throat of corolla; flowers pink   **38. petitiana**

b. All anthers inside corolla-tube or flowers red, orange, yellow or whitish   35

35a. Stems sprawling, slender; corolla funnel-shaped   **5. jongmansii**

b. Stems erect, more than 4 mm wide at base; corolla-tube cylindric   36

36a. Some leaves deeply lobed to twice pinnatisect   **32. laciniata**

b. All leaves scalloped to almost entire   37

37a. Plant hairy throughout   **31. lateritia**

b. At least lower part of plant hairless   **30. crenata**

38a. All anthers inside corolla-tube   39

b. Some anthers projecting from throat of corolla   44

39a. Leaves more or less green, stalks slightly to distinctly broadened at base, not easily detached   40

b. Leaves glaucous or stalks not broadened at base, easily detached   42

40a. Corolla-lobes 1–2.5 mm wide, filaments free for less than 1.5 mm; flowers red, orange, yellow or rose   **28. blossfeldiana**

b. Corolla-lobes 2.5–8 mm wide or filaments free for more than 1.5 mm or flowers purplish   41

41a. Leaves evenly scalloped; flowers yellow to red only; plants at least 30 cm   **30. crenata**

b. Leaves with irregular or angular teeth or flowers with pink or purple hue or plants less than 30 cm   **29. blossfeldiana hybrids**

42a. Calyx-lobes to 2.5 mm; corolla twisted around carpels in fruit   **33. rotundifolia**

b. Calyx-lobes usually longer than 2.5 mm; corolla not twisted around carpels in fruit   43

43a. Calyx more or less persistent; leaves usually stalked   **34. glaucescens**

b. Calyx-lobes dropping early and separately; leaves more or less stalkless   **35. flammea**

44a. Corolla-lobes more or less as long as tube   45

b. Corolla-lobes clearly shorter than tube   47

45a. Leaves green   **29. blossfeldiana hybrids**

b. Leaves glaucous   46

46a. Flowers pink   **26. pumila**

b. Flowers yellow   **44. grandiflora**

47a. Calyx-lobes to 2.5 mm   **45. farinacea**

b. Calyx-lobes longer than 2.5 mm   48

48a. Leaves peltate   **39. nyikae**

b. Leaves not peltate   49

49a. Both whorls of anthers projecting from throat of corolla; flowers pink   **38. petitiana**

b. One whorl of anthers projecting from throat of corolla or flowers yellow, orange or red   50

50a. Leaves green   **30. crenata**

b. Leaves glaucous   51

51a. Flowers pink, orange or red; stems not square in section   **35. flammea**

b. Flowers greenish yellow to orange; stems square in section   52

52a. Leaves mostly basal, entire   **37. thyrsiflora**

b. Leaves evenly distributed along stem, toothed   **36. longiflora**

1. K. peltata (Baker) Baillon. Illustration: Jacobsen, Handbook of succulent plants **2**: 657 (1960); Hamet & Marnier-Lapostolle, Le genre Kalanchoë au Jardin Botanique 'Les Cèdres', f. D, 6, 7 (1964); Maire, Flore de l'Afrique du nord **14**: 262 (1976).

Perennial to 2 m, hairless or sparsely covered with simple hairs; stems decumbent. Leaves ovate, unevenly scalloped, peltate, often with reddish spots, 3–12.5 × 2.5–6 cm; stalk thin, 2–10 cm. Inflorescence corymb-like; flowers many, pendent. Corolla rose to red; tube narrow 2.1–2.7 cm; lobes ovate, partly spreading, 3–6 × 4.5–9.5 mm. Anthers and styles projecting from throat of corolla, filaments free up to 6 mm; styles divergent, pressed to side of corolla-tube. *Madagascar*. G2.

2. K. miniata Hilsembach & Bojer (*Bryophyllum miniatum* (Hilsembach & Bojer) Berger). Illustration: Jacobsen, Handbook of succulent plants **2**: 655 (1960), Hamet & Marnier-Lapostolle, Le genre Kalanchoë au Jardin Botanique 'Les Cèdres', f. L, 52, 53 (1964).

Perennial to 80 cm, hairless or inflorescence with simple, glandular hairs. Stems erect or decumbent. Leaves ovate to obovate, usually simple, rarely 3-lobed, usually finely scalloped, rarely entire, 2–13 × 1.2–7.5 cm; base tapered to auriculate; stalk often winged, 7–40 mm, rarely absent. Inflorescence a panicle of several corymb-like clusters; flowers pendent, often partly replaced by bulbils. Corolla bell-shaped, yellow, pink or red; tube 2.3–3.1 cm; lobes recurved, ovate, sometimes purple, 4–6 × 5–7 mm. Anthers and stigmas at throat of corolla; filaments *c.* 5 mm, fused to corolla. *Madagascar.* G2.

3. K. gracilipes (Baker) Baillon × **K. manginii** Hamet & Perrier de la Bâthie. Illustration: Brickell, RHS gardeners' encyclopedia of plants & flowers, 385 (1989).

Similar to *K. manginii*, but differs in all leaves ovate-oblong, with 2–6 distinct teeth; stalk distinct, 4–13 mm. Inflorescence without bulbils. Flower colour similar in bright light, but very pale in shade. Styles divergent. G1–2.

The main cultivar, 'Tessa', differs from its parents in often having 6 or more flowers per cluster.

K. gracilipes is similar, but differs in: hairs entirely absent, calyx-tube more or less as long as lobes, corolla flesh or buff.

4. K. manginii Hamet & Perrier de la Bâthie (*Bryophyllum manginii* (Hamet & Perrier de la Bâthie) Northdurft). Illustration: Jacobsen, Handbook of succulent plants **2**: 653 (1960); Hamet & Marnier-Lapostolle, Le genre Kalanchoë au Jardin Botanique 'Les Cèdres', f. 124, 125 (1964); Jacobsen, Das Sukkulentenlexikon, pl. 41 (1981).

Creeping perennial with sparse, simple, glandular hairs; stems slender; flowering branches 10–40 cm. Leaves obovate to spathulate, entire or rarely 1–3 weak notches near apex, thick, 1.2–3 cm × 6–14 mm; stalk absent or indistinct. Inflorescence open; flowers pendent, usually 1–5, partly replaced by bulbils. Corolla narrowly urceolate, red; tube 2.1–2.8 cm; lobes ovate, 3–7 × 4.5–5.5 mm. Anthers and styles projecting from corolla-tube; filaments fused to corolla for 5–9 mm. *Madagascar.* G1–2.

5. K. jongmansii Hamet & Perrier de la Bâthie. Illustration: *Botanical Magazine*, n.s., 388 (1962); Hamet & Marnier-Lapostolle, Le genre Kalanchoë au Jardin Botanique 'Les Cèdres', f. Q, 43, 44 (1964); *Cactus & Succulent Journal* (US) **55**: 210 (1983).

Sprawling perennial with few simple glandular hairs mainly on the flowering shoots; stems slender; flowering branches ascending, to 30 cm. Leaves linear-elliptic, entire, stalkless, fleshy, 7.5–42 × 2.5–9.5 mm. Flowers few, at end of branches and from upper leaf-axils, more or less erect. Corolla gradually spreading from base, golden yellow; tube 7–22 mm; lobes ovate, 6–9 × 4–6 mm. Anthers and stigmas usually inside corolla-tube; filaments free for up to 6 mm. *Madagascar.* G1–2.

6. K. 'Wendy'; *K. miniata × porphyrocalyx*. Illustration: Brickell, RHS gardeners' encyclopedia of plants & flowers, 384 (1989).

Shrublet to 40 cm, semi-erect, with very few fine simple hairs only in the inflorescences. Stems soon thickening and epidermis separating into plates. Leaves ovate to oblong-lanceolate with uneven, shallow, rounded teeth, 4–7 × 1–2.5 cm, tapered at base; stalk indistinct. Stem of inflorescence usually 3-branched from base with 15–30 pendent flowers. Corolla urn-shaped; tube rose-purple, *c.* 2.3 × 1.3 cm; lobes ovate, yellow, 5 × 6 mm. Anthers and styles just projecting from corolla-tube; filaments fused to corolla for *c.* 5 mm. G2.

Similar to *K. porphyrocalyx*, but coarser and more floriferous when grown in similar conditions. Sometimes self-fertilised, seedlings very variable.

7. K. porphyrocalyx (Baker) Baillon (*Bryophyllum porphyrocalyx* (Baker) Berger). Illustration: *Cactus* (France) **17–18**: 57 (1948); Hamet & Marnier-Lapostolle, Le genre Kalanchoë au Jardin Botanique 'Les Cèdres', f. N, 117, 118 (1964).

Shrublet to 30 cm, semi-erect, with fine glandular hairs in the inflorescence. Stems soon thickening and epidermis separating into plates. Leaves oblong to obovate, with uneven, shallow or deep, rounded teeth, 2.3–5.5 cm × 5–30 mm, tapered at base; stalk indistinct. Flowering-stem single, only branched at apex, usually with 5–15 pendent flowers. Corolla urn-shaped; tube rose to red, 2–3.1 cm × 6–12 mm; lobes ovate, yellow or orange, 3–5.5 × 3–7.5 mm. Anthers and stalks usually

just projecting from corolla-tube. *Madagascar.* G2.

8. K. uniflora (Stapf) Hamet (*Bryophyllum uniflorum* (Stapf) Berger). Illustration: *Botanical Magazine*, 8286 (1909); Hamet & Marnier-Lapostolle, Le genre Kalanchoë au Jardin Botanique 'Les Cèdres', f. G, H, 119 (1964); Hay et al., The dictionary of indoor plants in colour, edn 1, pl. 78 (1974); *Cactus & Succulent Journal* (US) **55**: 201 (1983).

Prostrate perennial, hairless or with fine glandular hairs in the inflorescence; stems slender, rooting. Leaves circular, with few uneven, rounded teeth, both faces convex, 4–15 mm; stalk 1–3 mm. Flowering-stem spreading or pendent, with few pendent flowers. Corolla urn-shaped, red to purple; tube 1–1.9 cm × 6–9 mm across; lobes ovate 3.5–4.5 × 3–6 mm. Anthers and styles just projecting from throat of corolla; filaments fused to corolla to 4 mm. *Madagascar.* G2.

9. K. schizophylla (Baker) Baillon (*Bryophyllum schizophyllum* (Baker) Berger). Illustration: Hamet & Marnier-Lapostolle, Le genre Kalanchoë au Jardin Botanique 'Les Cèdres', f. 65–7 (1964).

Woody climber, supported by hooked leaves, hairless; stem to 8 m. Young leaves ovate, toothed; mature leaves pinnatisect, 13–15 × 4–5 cm; segments more or less linear, with teeth and sometimes 1–3 hooked lobes. Inflorescence large, with few flowers, pinnatisect bracts and many bulbils; flowers dull violet, pendent. Corolla narrowly bell-shaped; tube 1.3–1.7 cm; lobes ovate, 2–3.5 × 2–4 mm. Filaments and styles projecting beyond corolla; filaments fused to corolla for *c.* 5 mm. *Madagascar.* G2.

10. K. beauverdii Hamet (*Bryophyllum beauverdii* (Hamet) Berger; *K. costantinii* Hamet; *K. juelii* Hamet). Illustration: *Cactus* (France) **19**: 12 (1949); Hamet & Marnier-Lapostolle, Le genre Kalanchoë au Jardin Botanique 'Les Cèdres', f. 34, 44, 45 (1964); Jacobsen, Das Sukkulentenlexikon, t. 103 (1970).

Woody climber to 6 m, supported by reflexed leaves, often brownish. Leaves with teeth and plantlets only at apex, stalkless; lower leaves almost cylindric; upper leaves ovate to linear; base rounded to cordate or long-tapered, blade hastate, 2–10 cm × 5–25 mm. Inflorescence a loose panicle of small flower-clusters; flowers usually pendent. Corolla funnel-shaped, greenish to purplish; tube 1.1–1.5 cm; lobes rounded, 1.2–1.7 × 1–1.9 cm. Anthers and styles projecting from corolla-tube; filaments fused to corolla for up to 4 mm. *Madagascar.* G2.

11. K. delagoensis Ecklon & Zeyher (*Bryophyllum delagoense* (Ecklon & Zeyher) Schinz; *B. tubiflorum* Harvey; *K. tubiflora* (Harvey) Hamet). Illustration: *Botanical Magazine*, 9251 (1929); *Journal of the Royal Horticultural Society* **108**: 238 (1983); Leistner (ed.), Flora of southern Africa **14**: 74 (1985); Brickell, RHS gardeners' encyclopedia of plants & flowers, 388 (1989).

Perennial to 1.2 m, hairless, grey-green; stems erect. Leaves opposite on young shoots, in whorls of 3 or sometimes alternate on older shoots, cylindric, stalkless, 1.5–13 cm × 2–6 mm; apex with 3–9 teeth with spurs in between, bearing plantlets. Inflorescence dense, large, corymb-like, flowers pendent. Calyx tube 2–6 mm, lobes 5.5–8 × 3.5–5.5 mm. Corolla pale orange to magenta; tube narrow, 2–2.5 cm; lobes obovate, spreading, 7–12 × 7–10 mm. Anthers and styles just projecting from corolla-tube; filaments fused to corolla for up to 6.5 mm. *Madagascar.* G2.

12. K. × houghtonii D.B.Ward (*K. hybrida* invalid; *K. serrata* misapplied). Illustration: *Cactus and Succulent Journal* **78**: 94 (2005); Graf, Exotica, edn 9, **3**: 883, 685 (1978).

Fertile plants intermediate between *K. delagoensis* and *K. daigremontiana*. Leaves usually opposite, rarely a few alternate, oblong to lanceolate, with sharp teeth and some bulbils, 3–5.5 cm; stalks 1.5–3 cm. Calyx-lobes about as long as tube. G2.

Cultivars include: 'Hybrida' (illustration: Jacobsen, Handbook of succulent plants **2**: f. 860, 1960); 'Fujicho' ('Pink Butterflies', 'Pink Sparkler', Hybrida Variegated) – illustration: Hirose & Yokoi, Variegated plants in color, 154 (1998). This clone can only be propagated by stem cuttings as the bulbils are without chlorophyll.

13. K. daigremontiana Hamet & Perrier de la Bâthie (*Bryophyllum daigremontianum* (Hamet & Perrier de la Bâthie) Berger). Illustration: Hamet & Marnier-Lapostolle, Le genre Kalanchoë au Jardin Botanique 'Les Cèdres', f. 35–8 (1964); Jacobsen, Das Sukkulentenlexikon, t. 103 (1970); Brickell, RHS gardeners' encyclopedia of plants & flowers, 383 (1989); Brickell (ed.),

RHS A–Z encyclopedia of garden plants, 593 (2003).

Erect perennial 30–100 cm, hairless, brownish green. Leaves lanceolate, peltate at least at base of plant, with regular, small, sharp teeth alternate with spurs bearing plantlets, marbled beneath with brown-purple, 6–15 cm × 9–50 mm; stalk 1–3.5 cm. Inflorescence an open panicle with several dense long-stemmed clusters of pendent flowers. Calyx-tube 3–4.5 mm; lobes 3–5.5 × 2–3.5 mm. Corolla greyish violet; tube 1.6–1.9 cm; lobes obovate, partly spreading, 7–8 × 3–4.5 mm. Anthers and styles projecting from corolla-tube; filaments fused to corolla for up to 8 mm. *Madagascar.* G2.

Two other species are frequently grown that will key out here.

K. laetivirens Descoings. Illustration: Rauh, Succulent and xerophytic plants of Madagascar **2**: 318, 319 (1998). Easily distinguished by slightly glaucous green leaves completely without any markings, which turn pink in strong light. Sometimes called *K. daigremontiana* 'Green Form'. *Madagascar.* G2.

K. rosei Raym.-Ham. & H. Perrier (*K. rauhii* invalid). Illustration: Rauh, Succulent and xerophytic plants of Madagascar **2**: 321, 323 (1998). Leaves oblong-lanceolate, marginal teeth not uniform, about 7 pairs. Often seen as the clone 'Lucky Bells'. *Madagascar.* G1.

14. K. fedtschenkoi Hamet & Perrier de la Bâthie (*Bryophyllum fedtschenkoi* (Hamet & Perrier de la Bâthie) Lauzac-Marchal). Illustration: Jacobsen, Handbook of succulent plants **2**: 647 (1960); Hamet & Marnier-Lapostolle, Le genre Kalanchoë au Jardin Botanique 'Les Cèdres', f. 39, 40 (1964); Hortus Third, 329 (1976).

Perennial to 50 cm, hairless, blue-glaucous; stem decumbent with leaves more crowded at base. Leaves obovate to oblong with 2–8 prominent teeth in apical half, 1.2–6 cm × 8–40 mm; base tapered; stalk 1–6 mm. Young flowering-stem bent back at apex, resembling a crosier. Inflorescence a small loose corymb; flowers few, pendent. Calyx tube 1.2–1.3 cm, lobes 6–6.5 × 4.5–5 mm. Corolla bell-shaped, dull red to purple; tube 1.5–2 cm; lobes obovate, c. 6.5 × 4.5 mm. Anthers and styles projecting from corolla-tube; filaments fused for 6–9.5 mm. *Madagascar.* G2.

'Variegata'. Illustration: Brickell, RHS gardeners' encyclopedia of plants & flowers,

383 (1989). Leaves with narrow cream margin.

15. K. laxiflora Baker (*Bryophyllum crenatum* Baker). Illustration: *Botanical Magazine*, 7856 (1902); Hamet & Marnier-Lapostolle, Le genre Kalanchoë au Jardin Botanique 'Les Cèdres', f. 47–9 (1964).

Similar to *K. fedtschenkoi*, but differs in: plant often pale green. Leaves more or less evenly spaced on stem; blade circular or ovate to oblong; teeth blunter; base truncate with very small to large, upturned auricles; stalks 1.2–3.7 cm. *Madagascar.* G2.

16. K. gastonis-bonnieri Hamet & Perrier de la Bâthie (*Bryophyllum gastonis-bonnieri* (Hamet & Perrier de la Bâthie) Lauzac-Marchal). Illustration: Hamet & Marnier-Lapostolle, Le genre Kalanchoë au Jardin Botanique 'Les Cèdres', f. 41, 42 (1964); *Botanical Magazine*, n.s., 451 (2002) – as *Bryophyllum gastonis-bonnieri*.

Mealy perennial to 75 cm; fine glandular hairs only on corolla; sterile shoots forming loose rosettes; flowering stems with about 4 pairs of leaves. Leaves lanceolate, scalloped, with transverse bands of spots, 9–16.5 × 3.7–5.5 cm; stalk broad, indistinct, 3.5–6.3 × 1.2–1.5 cm. Inflorescence corymb-like or paniculate; flowers c. 2–4 cm, pendent. Calyx inflated. Corolla pale, yellowish or reddish; tube narrow, 2.9–3 cm; lobes ovate, partly recurved, 9–11 × 5.5–7.5 mm. Anthers and styles projecting from corolla-tube; filaments fused to corolla for 7–10 mm. *Madagascar.* G2.

17. K. prolifera (Bowie) Hamet (*Bryophyllum proliferum* Bowie). Illustration: *Botanical Magazine*, 5147 (1859); Jacobsen, Handbook of succulent plants **2**: 660 (1960); Hamet & Marnier-Lapostolle, Le genre Kalanchoë au Jardin Botanique 'Les Cèdres', f. 60–2 (1964).

Perennial to 1.5 m, hairless; stems erect. Leaves pinnatifid to pinnatisect, to 45 cm; segments lanceolate to oblong, oblique, scalloped to deeply toothed, 7–15 × 1.5–5 cm; stalks 6–16 cm. Inflorescence an open panicle of small clusters of flowers on long stalks with numerous bulbils; flowers pendent. Calyx inflated. Corolla-tube greenish yellow, 1.8–2.4 cm: lobes ovate, recurved, often pink, 2.5–3.5 × 3–4 mm. Anthers and styles projecting from corolla-tube; filaments fused to corolla for 4.5–8 mm; styles recurved at tip. *Madagascar.* G2.

18. K. pinnata (Lamarck) Persoon (*Bryophyllum pinnatum* (Lamarck) Oken; *B. calycinum* Salisbury). Illustration: *Botanical Magazine*, 1409 (1811); Hamet & Marnier-Lapostolle, Le genre Kalanchoë au Jardin Botanique 'Les Cèdres', f. 57–9 (1964); Flora of tropical east Africa: Crassulaceae, 29 (1977).

Perennial to 2 m, with sparse, fine, glandular hairs only on the corolla; stems erect. Lower leaves simple, upper leaves usually pinnate with 3–5 leaflets and stalk 2.5–7.5 cm; leaflets oblong, scalloped, 5–20 × 3–12 cm. Inflorescence a loose panicle; flowers pendent. Calyx inflated, green or reddish. Corolla-tube 3–4 cm; lobes ovate, partly recurved, reddish. Anthers and styles projecting from throat of corolla; filaments fused for 9–12 mm. *Widespread in tropics.* G2.

19. K. orygalis Baker. Illustration: Jacobsen, Handbook of succulent plants **2**: 657 (1960); Hamet & Marnier-Lapostolle, Le genre Kalanchoë au Jardin Botanique 'Les Cèdres', f. 80, 81 (1964); *Cactus and Succulent Journal* (US) **55**: 204–5 (1983).

Shrublet to 1.5 m, covered in fine, scale-like, star-like, golden hairs; stems erect. Leaves opposite, ovate to lanceolate, entire, 7.5–12 × 5.5–10.5 cm; stalk stout, 5–15 mm. Inflorescence a long, open, panicle, with distant, dense, small flower clusters; flowers more or less erect. Corolla urn-shaped, yellow, tube 8–10 mm, lobes ovate, 3–4 × 3–6 mm. Anthers and stigmas at throat of corolla, filaments free for 1–6 mm. *Madagascar.* G1–2.

20. K. beharensis Drake del Castillo. Illustration: Everard & Morley, Wild flowers of the world, pl. 70 (1970); *Cactus & Succulent Journal* (US) **49**: 269 (1977); Jacobsen, Das Sukkulentenlexikon, t. 103 (1970); Brickell, RHS gardeners' encyclopedia of plants & flowers, 38 (1989).

Much-branched tree eventually to 6 m, often with a single non-flowering stem in cultivation, covered in dense, short, branched, silver or golden hairs, rarely hairless and glaucous. Leaves triangular to hastate, 4.5–35 × 3.5–25 cm; margins wavy, with uneven teeth and lobes; stalks thick, 1.5–13 cm. Inflorescence mostly lateral with an open panicle of small, dense flower clusters; flowers more or less erect. Corolla urn-shaped, yellow-green, tube 6.5–10 mm, lobes obovate, violet-lined, about as long as tube. Anthers and styles

projecting from throat of corolla, filaments free for 4–9.5 mm. *Madagascar.* G1–2.

21. K. hildebrandtii Baillon. Illustration: Jacobsen, Handbook of succulent plants **2**: 650 (1960); Hamet and Marnier-Lapostolle, Le genre Kalanchoë au Jardin Botanique 'Les Cèdres', f. 7, 8 (1964).

Much-branched shrub to 5 m, covered in dense, fine, scale-like, silver, star-like hairs. Leaves circular to obovate, entire, 1.6–4 × 1.3–3.5 cm; stalk 3–7 mm. Inflorescence an open panicle of small, dense flower clusters; flowers more or less erect. Corolla bell-shaped, white or pale green; tube 3.5–5 mm; lobes oblong, 2–3.5 × 1.5–2 mm. Anthers and styles projecting from throat of corolla; filaments free for 1.5–3.5 mm. *Madagascar.* G1–2.

22. K. millotii Hamet & Perrier de la Bâthie. Illustration: Hamet & Marnier-Lapostolle, Le genre Kalanchoë au Jardin Botanique 'Les Cèdres', f. 26–8 (1964); Hay et al., Dictionary of indoor plants in colour, edn 1, pl. 309 (1974); Jacobsen, Das Sukkulentenlexikon, t. 105 (1970).

Much-branched shrublet to 35 cm, covered in dense, short, branched, white hairs. Leaves ovate to obovate, with even, shallow teeth, 3–6 × 2.5–5.5 cm; stalk thick, 5–18 mm. Inflorescence small, dense, corymb-like; flowers erect. Calyx also with simple glandular hairs. Corolla narrowly bell-shaped; tube yellow-green, 9–10.5 mm; lobes oblong, reddish, 3–3.5 mm. Anthers and styles just projecting from throat of corolla; filaments free for 2.5–3 mm. *Madagascar.* G1–2.

23. K. tomentosa Baker, Illustration: Everard & Morley, Wild flowers of the world, pl. 70 (1970); *Cactus & Succulent Journal* (US) **55**: 202 (1983); Brickell, RHS gardeners' encyclopedia of plants & flowers, 387 (1989).

Shrub to 1 m, erect, sparsely branched, covered in stiff, matted, long, silver, branched hairs. Leaves alternate, oblong, stalkless, thick, 2–8 × 1–3 cm; margin entire or with a few teeth towards apex, often brown. Inflorescence a narrow panicle of small, dense flower-clusters; flowers spreading. Corolla bell-shaped, with simple, reddish, glandular hairs und usually branched, non-glandular hairs; tube greenish yellow, 10–12 mm; lobes ovate, usually purplish, 2.5–3.5 × 4–5.5 mm. Anthers and stigmas usually inside corolla-tube; filaments free for 3–6 mm, *Madagascar.* G1.

Several cultivars based on leaf variation are grown.

24. K. eriophylla Hilsembach & Bojer. Illustration: Hamet & Marnier-Lapostolle, Le genre Kalanchoë au Jardin Botanique 'Les Cèdres', f. M, 24, 25 (1964); *Cactus & Succulent Journal* (US) **55**: 204 (1983).

Perennial to 20 cm, covered in soft, matted, long, white branched hairs. Leaves oblong, entire, stalkless, very thick, 1.6–3 cm × 8–15 mm. Inflorescence a tight cluster of 2–7 more or less erect flowers on a slender stem. Corolla bell-shaped, blue-violet; tube 5–6.5 mm; lobes obovate, 4–6 × 3–5.5 mm. Anthers and stigmas more or less at throat of corolla, filaments free for 2.5–3.5 mm. *Madagascar.* G1–2.

25. K. rhombopilosa Mannoni & Boiteau. Illustration: Hamet & Marnier-Lapostolle, Le genre Kalanchoë au Jardin Botanique 'Les Cèdres', f. 126–8 (1964); Graham, Growing succulent plants, 81 (1987).

Slow-growing shrublet to 50 cm; stems and leaves covered in dense, greyish, scale-like, 4-branched hairs; stems erect. Leaves stalked, alternate, obovate to fan-shaped, grey-green usually with red-brown streaks, 2–3 × 1.5–2.5 cm; the apical margin with prominent teeth; sometimes the hairs confined to the apical margin and the leaves copper-coloured; stalk 2–5 mm. Inflorescence a small, open panicle; flowers erect. Corolla urn-shaped, yellow-green, red-lined; tube *c.* 4 mm; lobes ovate, *c.* 2.5 mm. Anthers and styles projecting from throat of corolla; filaments free for 2–3.5 mm. *Madagascar.* G1–2.

26. K. pumila Baker. Illustration: Hay et al., Dictionary of indoor plants in colour, edn 1, pl. 310 (1974); *Cactus & Succulent Journal* (US) **55**: 204 (1983); Blundell, Wild flowers of east Africa, 338 (1987).

Sprawling shrublet to 30 cm, covered in white bloom. Leaves glaucous, obovate, toothed towards apex, 2–3.5 × 1.2–2 cm; base tapered; stalk usually indistinct. Inflorescence a tight cluster of about 5 erect flowers; stem slender, to 6 cm. Corolla narrowly bell-shaped, pink with purple lines; tube 4.5–8.5 mm; lobes obovate, 7.5–10 mm. Anthers and styles projecting from throat of corolla; filaments free for 3–6 mm. *Madagascar.* G1–2.

27. K. synsepala Baker. Illustration: Hamet & Marnier-Lapostolle, Le genre Kalanchoë au Jardin Botanique 'Les

Cèdres', f. 13–17 (1964); *Cactus & Succulent Journal* (US) **55**: 206–7 (1983).

Perennial to 30 cm, hairless or with simple glandular hairs in the inflorescence, bearing long, slender stolons with one internode. Leaves lanceolate to oblanceolate, finely or coarsely toothed or pinnatifid to 15 × 7 cm, tapered; stalk indistinct. Inflorescence on slender, erect, lateral stem; flowers erect in dense clusters. Corolla narrowly bell-shaped; tube 7–12 mm; lobes elliptic, 4.5–7 × 3–4 mm. Anthers and styles projecting from throat of corolla; filaments free for 1–2 mm. *Madagascar.* G2.

28. K. blossfeldiana von Poellnitz (*K. globulifera* var. *coccinea* Perrier de la Bâthie). Illustration: *Botanical Magazine*, 9440 (1936); *Euphytica* **31**: pl. 1 (1982).

Perennial to 30 cm, much-branched, hairless. Leaves ovate to oblong, scalloped, bright green, 2–7.5 × 1.6–4 cm; base rounded; stalk broad, indistinct to 10 mm. Inflorescence of several dense clusters of erect flowers. Calyx-lobes linear, 3–5 × 1–1.5 mm. Corolla-tube cylindric, reddish above, 6–9 × *c.* 1.5 mm; lobes elliptic, bright red, 4–5.5 × 1.5–2.5 mm. Anthers inside corolla-tube; filaments free for up to 1 mm. *Madagascar.* G2.

Cultivars with compact growth and yellow, orange or rose flowers have been selected. Illustration: Hay et al., Dictionary of indoor plants in colour, pl. 308 (1974); *Euphytica* **31**: pl. 1 (1982) – as 'Vulcan geel'. Scarcely cultivated. Plants not closely matching the description above are referred to *K. blossfeldiana* hybrids.

29. K. blossfeldiana hybrids (*K. blossfeldiana* misapplied). Illustration: Brickell (ed.), RHS A–Z encyclopedia of garden plants, 593 (2003).

Perennial to 30 cm, much-branched, hairless. Leaves ovate, deep green, shallowly scalloped, to 12 × 6.5 cm; base rounded or truncate; stalk broad, to 2.3 cm × 8 mm. Inflorescence often one broad, dense, corymb-like cluster of erect flowers. Calyx-lobes lanceolate, 4–5.5 × 2.5 mm. Corolla-tube greenish, 6–7 × 4.5 mm; lobes oblong to 8 × 6.5 mm, in many clear colours from pale yellow and pink to brick-red and purple. Anthers and stigmas at throat of corolla; filaments free for up to 2.5 mm. G2.

Current pot-plant cultivars are described above. They are complex hybrids of inadequately documented origin. Earlier hybrids persist in plant collections and are often confused with other species. Some are intermediate between *K. blossfeldiana* hybrids and various other species, particularly *K. pumila* and *K. flammea*; others more closely resemble *K. blossfeldiana* hybrids and include plants with variegated leaves. Illustration: Hay et al., Dictionary of indoor plants in colour, edn 1, pl. 306–7 (1974); *Euphytica* **31**: pl. 1 (1982) – as 'Chérie'; Brickell, RHS gardeners' encyclopedia of plants & flowers, 385 (1989).

30. K. crenata (Andrews) Haworth (*K. laciniata* misapplied; *K. brasiliensis* Cambessèdes; *K. coccinea* Britten; *K. integra* (Medicus) O. Kuntze, ambiguous). Illustration: Hutchison & Dalziel, Flora of west tropical Africa, edn 2, **1**: 117 (1954); Heywood, Flowering plants of the world, 145 (1978).

Erect perennial to 2 m, hairless or with simple glandular hairs in the flower-head. Leaves ovate to spathulate, usually scalloped, 4–25 × 1.2–15 cm; stalks flattened, to 4 cm. Inflorescence an open panicle; flowers erect. Corolla red to yellow, tube narrow, 8–16 mm; lobes elliptic, spreading, 4–10 × 2.5–5 mm. Anthers and stigmas inside corolla-tube, filaments free for 0.5–3.5 mm. *Natal to Arabia, naturalised in tropical America.* G2.

Asian plants, from Kashmir to the Ryuku Islands and Java, are sometimes distinguished as *K. spathulata* de Candolle, but cannot be readily separated in gardens. The leaves are narrow and sometimes divided into 3 leaflets; the leaf-stalks are sometimes indistinct. Illustration: Li et al., Flora of Taiwan **3**: 14 (1977).

31. K. lateritia Engler (*K. coccinea* Britten var. *subsessilis* Britten; *K. integra* (Medicus) Kuntze var. *subsessilis* (Britten) Cufodontis; *K. velutina* misapplied; *K. zimbabwensis* Rendle). Illustration: *Botanical Magazine*, 787 (1902); Hamet & Marnier-Lapostolle, Le genre Kalanchoë au Jardin Botanique 'Les Cèdres', f. I, 89–91 (1964); Everard & Morley, Wild flowers of the world, pl. 70 (1970).

Erect perennial to 1.5 m, with brownish, simple, usually dense, glandular hairs. Leaves ovate to obovate, more or less scalloped, 4.5–16 × 3–8 cm; base tapered; stalk 5–50 mm, sometimes indistinct. Inflorescence rather dense, paniculate or corymb-like; flowers erect. Corolla red to pale yellow or pink; tube narrow, 8–11 mm; lobes ovate, spreading, 4.5–8 × 1.5–6 mm. Anthers and stigmas inside corolla-tube, filaments free for 1–2 mm. *Kenya to Zimbabwe.* G2.

32. K. laciniata (Linnaeus) de Candolle (*K. schweinfurthii* Penzig).

Erect perennial to 1.2 m, sometimes monocarpic; inflorescence and sometimes upper leaf surface with colourless or brownish, simple, glandular hairs. Leaves simple or pinnate, to 20 cm, usually with 3 or 5 ovate to elliptic, entire or scalloped to pinnatisect leaflets; leaf-stalk 2–5 cm. Inflorescence paniculate or corymb-like, very dense; flowers erect. Corolla greenish white to pale orange; tube narrow, 8–16 mm; lobes ovate, spreading, 3.5–7 × 2–5.5 mm. Anthers and stigmas inside corolla-tube, filaments free for *c.* 1 mm. *Namibia to Ethiopia, S India, Thailand.* G2.

33. K. rotundifolia (Haworth) Haworth (*K. guillauminii* Hamet). Illustration: Hamet & Marnier-Lapostolle, Le genre Kalanchoë au Jardin 'Les Cèdres', f. 108, 109 (1964).

Erect or decumbent perennial, 20–200 cm, hairless, usually glaucous. Leaves ovate to spathulate, circular to linear, entire or scalloped, sometimes deeply 3-lobed, 1–8.5 cm × 5–55 mm; stalk to 2.5 cm. Inflorescence open, paniculate or corymb-like; flowers erect. Calyx-lobes persistent. Corolla pink or orange to deep red; tube narrow, 6–10 mm; lobes elliptic, spreading, 2.5–5 × 1.2 mm, twisted around carpels in fruit. Anthers and stigmas inside corolla-tube, filaments free for *c.* 1 mm. *South Africa to Zimbabwe, Yemen (Socotra).* G2.

34. K. glaucescens Britten (*K. laciniata* misapplied; *K. beniensis* de Wildeman). Illustration: Andrews, Flowering plants of the Anglo-Egyptian Sudan **1**: 77 (1950); Toussaint, Flore du Congo Belge et Ruanda-Urundi **2**: 567 (1951); Blundell, Wild flowers of east Africa, 276 (1987).

Erect or decumbent perennial, 30–120 cm, hairless, glaucous. Leaves ovate to spathulate, with few rounded teeth and sometimes reddish margins and marks on lower surface, 3–10 × 1.2–7 cm; base tapered; stalk 5–25 mm. Inflorescences loose or dense, paniculate or corymb-like; flowers erect. Calyx-lobes persistent. Corolla yellow to red or pink; tube narrow, 5–15 mm; lobes elliptic, spreading, 3–7.5 × 1–2.5 mm. Anthers and stigmas inside corolla-tube, filaments free for *c.* 1 mm. *Sudan to Zaire.* G2.

35. K. flammea Stapf. Illustration: *Botanical Magazine*, 7595 (1898).

Perennial to 40 cm, hairless, glaucous; branches short, thick, erect above. Leaves usually obovate to spathulate, entire, 2.5–9 × 1.5–4 cm, rarely rhombic or scalloped; stalks usually indistinct. Inflorescence dense, corymb-like; flowers erect. Calyx-lobes falling early and separately. Corolla-tube narrow, 8–13 mm; lobes ovate to obovate, spreading, orange or red, rarely rose, never yellow, 4–8 × 4–5 mm. Anthers and stigmas inside corolla-tube, filaments free for *c.* 1 mm. *Somalia*. G2.

36. K. longiflora Schlechter (*K. petitiana* misapplied). Illustration: Hamet & Marnier-Lapostolle, Le genre Kalanchoë au Jardin Botanique 'Les Cèdres', f. 92, 93 (1964); Maire, Flore de l'Afrique du nord **14**: 256 (1976).

Decumbent perennial to 60 cm, hairless, glaucous, sometimes deep red in sun; branches and flower-stems square in section. Leaves obovate or fan-shaped, somewhat concave with few teeth towards apex, 4–8 × 3–8 cm; stalk indistinct, to 1.5 cm. Calyx persistent. Corolla greenish yellow to orange; tube narrow, 1.1–1.7 cm; lobes obovate, spreading, 2–5 × 3–5 mm. Anthers and stigmas just projecting from throat of corolla, filaments free for 0.5–1.5 mm. *Natal*. G2.

37. K. thyrsiflora Harvey. Illustration: *Botanical Magazine*, 7678 (1899); *Flowering Plants of Africa*, 341 (1929); Leistner (ed.), Flora of southern Africa **14**: 70 (1985).

Monocarpic or perennial with basal offsets, to 1.3 m, covered in white bloom; stems square in section. Leaves in a loose basal rosette, oblanceolate, entire, stalkless, 6–14 × 2.5–9 cm, opposite pairs united at base. Inflorescence a long, spike-like panicle to 30 × 8 cm; flowers erect or spreading, strongly scented. Corolla deep yellow, tube narrow, 1.1–2 cm, lobes rounded recurved, 2–5 mm. Anthers and stigmas just projecting from throat of corolla, filaments free for 1–2 mm. *Botswana, Lesotho, South Africa*. G1–2.

38. K. petitiana Richard.

Decumbent perennial to 1.5 m, hairless or with sparse, simple, glandular hairs. Leaves ovate, twice scalloped, 4–16 × 3–11 cm; base tapered to almost cordate with upturned lobes; stalk slender, 1.8–4 cm. Inflorescence an open panicle, flower-clusters dense; flowers erect. Whole

calyx shed in fruit. Corolla pale to deep pink, tube narrow, 1.4–2.4 cm, lobes oblong, recurved, 6–7 × 3.5–4.5 mm. Anthers and styles projecting from throat of corolla. *Ethiopia*. G2.

Very scarce in cultivation, the name is usually applied to *K. longiflora*.

39. K. nyikae Engler (*K. hemsleyana* Cufodontis). Illustration: Hamet & Marnier-Lapostolle, Le genre Kalanchoë au Jardin Botanique 'Les Cèdres', f. 102, 103 (1964).

Erect or decumbent perennial, 40–200 cm, hairless, glaucous. Leaves ovate, more or less entire, peltate with base upturned, 6.5–18 × 5–16 cm; stalk stout, 3–10 cm. Inflorescence dense to open, corymb-like or paniculate; flowers erect. Corolla-tube narrow, 1.7–2.1 cm, lobes ovate or lanceolate, spreading, cream, yellow or pinkish, 8–11 × 3–5.5 mm. Anthers at throat of corolla, filaments free for 0.5–1.5 mm. *Uganda, Kenya, Tanzania*. G2.

40. K. quartiniana Richard.

Perennial to 1 m, hairless, more or less glaucous; stems robust, erect. Leaves obovate to oblong, scalloped, unmarked, 10–18 × 6–10 cm; stalk broad, 3–6 cm. Inflorescence paniculate; flowers erect. Corolla white, tube very narrow, 3–5 cm, lobes obovate, spreading, 1–2 cm × 6–7 mm. Anthers at throat of corolla, filaments free for 0.5–2 mm. *Ethiopia*. G2.

41. K. marmorata Baker (*K. macrantha* Baker; *K. somaliensis* Baker). Illustration: *Botanical Magazine*, 7333 (1894) & 7831 (1902); *Flowering plants of Africa*, 1049 (1947); Jacobsen, Handbook of succulent plants **2**: 654 (1960).

Erect or decumbent perennial to 1.3 m, hairless, glaucous. Leaves obovate, entire, scalloped or sharp-toothed, often with purplish marks on both faces, 6–20 × 3.5–13 cm; base tapered, stalkless. Inflorescence paniculate; flowers erect. Corolla white, rarely with pink or yellow tinge; tube very narrow, 4.5–12 cm; lobes lanceolate to obovate, spreading, 1–2.5 × 6–12 mm. Anthers at throat of corolla, filaments free for 0.5–2 mm. *Sudan to Zaire*. G2.

42. K. × kewensis Thiselton-Dyer; *K. flammea × K. bentii subsp. bentii*. Illustration: *Annals of Botany* **17**: pl. 21–3 (1903); van Laren, Vetplanten, 77 (1932).

Erect or decumbent perennial to 1.2 m, hairless, bronze-green, slightly glaucous.

Leaves cylindric, grooved above, often with several lateral lobes or teeth, 10–30 cm, stalkless; sometimes basal leaves flat, elliptic, toothed. Inflorescence dense, paniculate or corymb-like; flowers erect. Calyx more or less erect. Corolla-tube narrow, *c.* 1.45 cm; lobes ovate, spreading, deep pink, *c.* 8 × 6 mm. Anthers at throat of corolla, filaments free for 0.5–4 mm, degree of fusion irregular. G2.

43. K. bentii J.D. Hooker (*K. teretifolia* misapplied).

Erect perennial to 1.5 m, olive-green, glaucous. Leaves cylindric, thickest in middle, grooved on upper surface, entire, stalkless, 7.5–40 cm × 6–12 mm. Inflorescence paniculate with flat-topped clusters of erect, scented flowers. Calyx spreading. Corolla white; tube narrow, 3–3.5 cm; lobes lanceolate, recurved, 8–16 × 3.5–6.5 mm. Anthers at throat of corolla; filaments free for *c.* 0.5 mm. G2.

Subsp. **bentii**. Illustration: *Botanical Magazine*, 7765 (1901). Hairless. *Yemen*.

Subsp. **somaliensis** Cufodontis. Calyx inside and corolla outside with simple glandular hairs. *Somalia*.

44. K. grandiflora Wight & Arnott. Illustration: *Botanical Magazine*, 5460 (1864); *Addisonia* **22**: 725 (1945).

Erect perennial to 80 cm, hairless, glaucous. Leaves ovate to obovate, weakly scalloped, 4–10 × 3.5–7.5 cm; base tapered; stalk indistinct. Inflorescence paniculate, compact; flowers more or less erect. Calyx spreading. Corolla bright yellow, tube *c.* 1.2 cm × 4–6 mm, lobes obovate, spreading, 1.3–1.7 cm × 5–7 mm. Anthers and styles just projecting from throat of corolla, filaments free for 0.5–2 mm. *S India*. G2.

45. K. farinacea Balfour (*K. scapigera* misapplied). Illustration: *Botanical Magazine*, 7769 (1901); *Flowering plants of Africa*, 1329 (1960); Jacobsen, Das Sukkulentenlexikon, pl. 42 (1970); *British Cactus and Succulent Journal* **4**: 6 (1986).

Perennial to 30 cm, hairless; stems short, erect above. Leaves crowded, obovate, entire with thin grey wax, 2–6 × 1.5–5 cm; base tapering; stalk indistinct. Inflorescence rounded, fairly dense, almost stemless; flowers erect. Corolla red, tube narrow, sometimes yellow towards base, 8–12 mm, lobes ovate, ascending, 4–5.5 × 2–2.5 mm. Anthers and stigmas at throat of corolla, filaments free for 1–2 mm. *Yemen (Socotra)*. G2.

4. OROSTACHYS Fischer
S.G. Knees

Monocarpic perennial herbs with leaves in rosettes. Leaves alternate, linear-elliptic, often with a cartilaginous bristly tip. Flowers in tall, dense spike-like, terminal racemes. Sepals 5, almost equal; petals 5, often joined at the base, spreading, white, yellowish or reddish, often spotted red; stamens 10; styles straight, slender, carpels 5, ovaries stalked. Fruits erect, free.

A genus of about 10 mainly herbaceous species from Europe and temperate Asia, at one time included in *Sedum* but differing in the callus-tipped leaves in dense rosettes and flowers in dense panicles. Plants flower in the second or third year and propagation is usually from seed or offsets. These plants can be grown in the rock garden in a very freely drained soil.

Literature: Ohba, H., Notes towards a monograph of the genus Orostachys (Crassulaceae) (1), *Journal of Japanese Botany* **65**(7): 193–203 (1990).

Leaves. with obvious spiny tip: **1, 2, 3**; lacking spiny tip: **4, 5, 6**.
Flowers. white or pinkish white: **1, 3, 5, 6**; greenish white: **4**; yellowish green: **2**.

la. Leaves with an obvious, hard spiny tip   2
  b. Leaves with hard margins but lacking a spiny tip   4
2a. Flowering-spike 4–10 × 1.5–2 cm
                1. erubescens
  b. Flowering-spike 10–30 × 5–8 cm   3
3a. Petals 3–5 mm, greenish yellow
                 2. spinosa
  b. Petals 6–8 mm, white, tipped red
                 3. chanetii
4a. Leaves distinctly bluish grey; sometimes with cream margins
                 5. iwarengis
  b. Leaves green or yellowish green, never bluish   5
5a. Leaves 2–4 × 1–2 cm   **6. aggregata**
  b. Leaves 1–2 cm × 5–10 mm
                 4. furusei

1. O. erubescens (Maximowicz) Ohwi (*O. japonica* Maximowicz). Illustration: Hayashi, Azegami & Hishiyama, Wild flowers of Japan, 439 (1983); Graf, Tropica, 374 (1978).

Leaves 13–30 × 4–6 mm, narrowly spathulate, fleshy; margin and apex hard, with several teeth, spine 1–3.5 mm. Flowering-stem 6–15 cm, with lanceolate leaves; flowering-spike 4–10 × 1.5–2 cm, stalked; petals 5–7 mm, whitish pink; anthers red at first, darkening to purple. *China, Korea & Japan.* H5–G1. Summer–autumn.

Three variants are recognised and occasionally seen in cultivation: var. **erubescens**, var. **japonica** Maximowicz and var. **polycephala** (Makino) Hara.

2. O. spinosa (Linnaeus) Berger. Illustration: Jacobsen, Handbook of succulent plants **2**: 702 (1960); Graf, Exotica, series 4, edn 12, **1**: 876 (1985); *Gartenpraxis* **8**: 21 (1987).

Tufted perennial with leaves of 2 lengths, in basal rosettes, the centre of which folds up in winter. Leaves 15–25 × 3–5 mm, oblong with spiny white tips 2–4 mm. Flowering-stems 10–30 cm, leafy, with flowers borne in spike-like panicles 5–20 cm. Petal-claws 1–2 mm or absent; petals 3–5 mm, lanceolate, greenish yellow; anthers yellow. Fruits 5–6 mm. *NE & C Asia (Xizang to Mongolia).* H5–G1. Summer.

3. O. chanetii (A. Léveillé) Berger. Illustration: Die natürlichen Pflanzenfamilien **18a**: 464 (1930).

Leaves of 2 lengths, to 2.5 cm × 5 mm, linear, spine-tipped, grey-green. Flowers borne in branched pyramidal panicles 15–30 × 5–8 cm; petals 6–8 mm, white, tipped reddish. *China.* H5–G1. Summer.

4. O. furusei Ohwi.

Sterile stems ascending to 5 cm, with leaves clustered towards apex. Leaves 1–2 cm × 5–10 mm, obovate, wedge-shaped at base. Flowering-stems 5–10 cm, leafy, flowers many. Sepals *c.* 3.5 mm, erect; petals 4.5–5 mm, ovate, greenish white; styles *c.* 1 mm. *Japan.* H5–G1. Summer–autumn.

5. O. iwarengis (Makino) Hara. Illustration: Jacobsen, Lexicon of succulent plants, pl. 112 (1974); Hayashi, Azegami & Hishiyama, Wild flowers of Japan, 439 (1983); *Gartenpraxis* **8**: 8 (1987); *Kakteen und andere Sukkulenten* **41**(10): 222–4 (1990).

Stems erect, to 5 cm, producing many stolons; rosettes persisting over winter. Leaves 3–7 cm × 7–28 mm, oblong to spathulate, fleshy, striking bluish grey. Flowering-stems 5–20 cm, forming narrow cones. Flowers solitary, shortly stalked; sepals 2–3 mm; petals 5–7 mm, white. *Japan, China.* H5–G1. Summer–autumn.

Propagate by decapitating young rosettes and rooting the offsets.

'Fuji', a cultivar with broad cream leaf-margins, is sometimes grown. Illustration: Graf, Tropica, 374 (1978).

6. O. aggregata (Makino) Hara (*Cotyledon aggregata* Makino; *O. malacophylla* (Pallas) Fischer, invalid). Illustration: *Botanical Magazine*, 4098 (1844); Jacobsen, Handbook of succulent plants **2**: 701 (1960); *Journal of Japanese Botany* **65**(7): 197 (1990).

Like *O. iwarengis*, but leaves smaller, 2–4 × 1–2 cm, rounded, blunt, green. *Japan.* H5–G1. Summer–autumn.

5. LENOPHYLLUM Rose
S.G. Knees

Hairless, succulent, perennial herbs. Roots pale brown; stock whitish. Stems erect or spreading, pinkish. Leaves opposite on non-flowering shoots, often crowded together, usually boat-shaped. Flowering-stems terminal with scattered alternate leaves, often branched towards apex. Sepals 5, often very fleshy; petals 5, more or less equalling sepals, erect, yellow, tips recurved; stamens 10, 5 arising directly from petals, 5 from between petals; nectaries oblong, orange or deep yellow; carpels erect, joined at base, green; stigmas white, follicles spreading, brown. Seeds oblong-elliptic, brown.

A genus of 5 or 6 species from northern Mexico and adjacent Texas, closely related to *Echeveria* and with similar cultivation requirements. All can be readily rooted from leaf cuttings or seed.

Literature: The genus Lenophyllum Rose, *Hasseltonia* **2**: 1–19 (1994).

la. Leaves obtuse or rounded at apex   2
  b. Leaves acute   3
2a. Leaves broadest at base
                 3. guttatum
  b. Leaves narrow at base
                 2. weinbergii
3a. Stems 3–5 cm; flowers solitary
                 4. pusillum
  b. Stems 10 cm or more; flowers clustered   4
4a. Leaf pairs crowded together; flowers yellow with reddish tips **1. texanum**
  b. Leaf pairs distant; flowers greenish yellow   **5. acutifolium**

1. L. texanum (J.C. Smith) Rose (*Sedum texanum* J.C. Smith; *Villadia texana* (J.C. Smith) Rose). Illustration: *Annals of the Missouri Botanical Garden* **6**: t. 50 (1895); *Addisonia* **8**: t. 267 (1923); Clausen, Sedum of North America, 586 (1975).

Tufted perennial herb, stems to 10 cm. Leaves 1.5–3.5 cm × 8–18 mm, ovate, elliptic-lanceolate, acute; boat-shaped above, rounded beneath, green when young, turning lavender-green or reddish with age. Flowering-stems erect or spreading, 10–35 cm, pinkish. Sepals acute, 3 mm, oblanceolate-ovate. Petals oblanceolate, oblong, acute, pale primrose-yellow with dark red blotch at tip. Stamens yellow; follicles spreading. All floral parts persistent. *SE USA (Texas), NE Mexico*. G1. Late summer–early winter.

A rather weedy species propagating very easily from leaves, which may fall off at the slightest touch.

2. L. weinbergii Britton. Illustration: *Smithsonian Miscellaneous Collections* **47**: 161 (1904); *Botanical Magazine*, n.s., 750 (1978).

Perennial herb. Leaves 2–4 × 1.8–3 cm, obovate to wedge-shaped, trough-shaped, truncate, more or less concave, obtuse at apex. Flowering-stems to 15 cm, few-flowered with 2 pairs of leaf-like bracts to 1.5 × 1 cm; branches 5, with 1–9 flowers. Sepals obtuse, 4 mm, club-shaped, dull purplish grey or light green, speckled purple. Petals *c.* 7 mm, erect with spreading or recurved tips. Fruits to 7 mm. *NE Mexico (Coahuila)*. G1. Summer.

3. L. guttatum Rose. Illustration: *Smithsonian Miscellaneous Collections* **47**: t. 20 (1904).

Perennial herb. Leaves 1.8–3.5 cm, elliptic-ovate to rhombic, deeply but shallowly grooved above, apex obtuse or rounded, greyish pink or light green, flecked with blackish purple. Flowering-stems to 20 cm; branches 2–6 with 5–12 flowers; flowers almost stalkless. Sepals obtuse, *c.* 6 mm, club-shaped; petals *c.* 5 mm, obtuse, yellowish, eventually drying reddish. *NE Mexico*. G1. Late summer–autumn.

4. L. pusillum Rose. Illustration: Jacobsen, Handbook of succulent plants **2**: 672 (1960).

Perennial herb with matted stems 1–3 cm. Leaves fleshy, 8–15 mm, narrow, acute, upper surface furrowed, lower surface keeled, dirty reddish green; readily falling and rooting. Flowering-stems to 5 cm, with solitary, terminal flowers. Sepals acute; petals 6–7 mm, lemon yellow. *Mexico*. G1. Summer–autumn.

5. L. acutifolium Rose. Illustration: *Smithsonian Miscellaneous Collections* **47**: t. 19 (1904).

Perennial herb, well branched at the base with stems to 10 cm. Leaves arranged in 6–8 distant pairs, lanceolate, acute, upper surface grooved. Flowers numerous, in interrupted spikes, stalkless or very shortly stalked; sepals thick, almost equal, greenish yellow; petals erect below, spreading above, greenish yellow. *Mexico (Nuevo Leon)*.

6. SEDUM Linnaeus.
N. Groendijk-Wilders & L. Springate
Plants hairless, downy or glandular hairy, fleshy, erect or decumbent, sometimes tufted or moss-like. Leaves very variable, opposite, alternate or whorled, entire or rarely toothed. Inflorescence usually terminal, rarely lateral, mostly cyme-like or flowers solitary, white, yellow or rose, rarely red or blue, hermaphrodite, floral parts usually in 5 s, rarely in 4 s, 6 s, 7 s, 8 s or 9s; petals usually free, sometimes fused for almost a third, stamens usually twice as many as petals, rarely equal in number to petals.

A genus of about 280 species, mostly natives of the temperate and colder regions of the northern hemisphere.

Literature: Praeger, R.L., An account of the genus Sedum as found in cultivation, *Journal of the Royal Horticultural Society* **46** (1920–1921); Clausen, R.T., Sedum of the trans-Mexican volcanic belt (1959); Jacobsen, H., Handbook of succulent plants **2** (1960); Clausen, R.T., Sedum of North America north of the Mexican plateau (1975); Köhlein, F., Freilandsukkulenten (1977); 't Hart, H., Biosystematic studies in the Acre-group and the series Rupestria Berger of the genus Sedum L. (Crassulaceae) (1978); Fessler, A., Die Freilandschmuckstauden, edn 3 (1985).

1a.	Plant with somewhat woody stems	2
b.	Plant herbaceous	28
2a.	Leaves opposite, ovate to spherical; flowers yellow	**1. stahlii**
b.	Leaves alternate	3
3a.	Leaves crowded, sometimes seemingly in whorls	4
b.	Leaves not crowded	9
4a.	Flowers light pink to deep scarlet-purple; leaves oblong-lanceolate, thick-fleshy, bluish bloomed	5
b.	Flowers yellow or greenish yellow	6
5a.	Shoots wholly pendent; leaves to 2 cm, slightly bloomed	**2. morganianum**

b.	Shoot-ends curving slightly outwards; leaves to 2.5 cm, densely bloomed	× **Sedeveria 'Harry Butterfield'** (p. 53)
6a.	Flowers orange-yellow; leaves obovate to spathulate, glaucous	**3. palmeri**
b.	Flowers pale yellow or greenish yellow	7
7a.	Plant 15–20 cm; leaves flat, fleshy green	8
b.	Plant to 30 cm; leaves cylindric, very fleshy, grey-green	**4. pachyphyllum**
8a.	Stems erect; leaves without spur beneath	**11. fusiforme**
b.	Stems prostrate; leaves with spur beneath	× **Sedadia amecamecana** (p. 40)
9a.	Flowers greenish yellow or bright orange	10
b.	Flowers white, pink, red or white with red	17
10a.	Stem at first fleshy, later somewhat woody; leaves rounded at apex, pale green, often suffused with red	**5. rubrotinctum**
b.	Stem always woody	11
11a.	Leaves with waxy white, later reddish margin, spathulate, broadly rounded	**6. dendroideum**
b.	Leaves without waxy margin	12
12a.	Leaves glossy or pale green, flat	13
b.	Leaves bluish or grey-green or semi-terete	16
13a.	Flower to 1 cm across; leaves obovate-oblong	**12. nudum**
b.	Flowers more than 1 cm across; leaves obovate to spathulate or oblanceolate to spathulate	14
14a.	Leaves 5–7 cm, oblanceolate to spathulate	**7. praealtum**
b.	Leaves 1.5–4 cm, obovate to spathulate	15
15a.	Leaves 1.5–2 × 1.1–1.3 cm, pale green; sepals 2–3 mm	**8. decumbens**
b.	Leaves 2–4 × 1.2–2 cm, mid-green; sepals 1–1.5 mm	**9. confusum**
16a.	Leaves blue-green, densely grey bloomed	**10. treleasei**
b.	Leaves grey-green, not bloomed	**3. palmeri**
17a.	Leaves reddish blue, oblong-elliptic with rounded apex; flowers white, bell-shaped	**13. craigii**
b.	Leaves green, yellowish or grey	18
18a.	Leaves less than 8 mm	19
b.	Leaves more than 8 mm	20

19a. Leaves obovate or spathulate; nectaries large, red, exposed
14. longipes

b. Leaves ovate to lanceolate; nectaries small, white or pale yellow, concealed **15. moranense**

20a. Leaf-margin toothed or notched 21

b. Leaf-margin entire 22

21a. Leaves ovate, cordate at base (poplar-like), margin coarsely and irregularly toothed
16. populifolium

b. Leaves obovate-oblong, spathulate, margin deeply notched at tip
17. retusum

22a. Inflorescence borne at top of main shoots 23

b. Inflorescence borne on short or long lateral branches 26

23a. Leaves more than 5 mm wide 24

b. Leaves less than 5 mm wide 25

24a. Shrub 50–90 cm; flowers pinkish to dull red, *c.* 1.3 cm across
18. oxypetalum

b. Shrub to 40 cm; flowers greenish white, 1.6–1.8 cm across
19. allantoides

25a. Leaves green, average width 1.5–2.1 mm, average thickness 0.6–1.2 mm; nectaries purple, oblong or spathulate **20. bourgaei**

b. Leaves grey or green, average width 2.3–2.7 mm, average thickness 1.4–2 mm; nectaries white or yellow, kidney-shaped
21. griseum

26a. Leaves dark lustrous green; flower-stalks 1–7 mm **22. lucidum**

b. Leaves dull or yellowish green; flower-stalks 8–20 mm 27

27a. Shrub 7.5–12.5 cm; shoots ascending; inflorescence a panicle
23. adolphii

b. Shrub to 7 cm; shoots prostrate; inflorescence a corymb
24. nussbaumerianum

28a. Perennial with thickened rootstock and slender roots; stems mostly annual with flat leaves; flowers yellow to orange 29

b. Rootstock, roots, stems and leaves formed differently 33

29a. Plant 30–80 cm 30

b. Plant 10–30 cm 31

30a. Plant 30–45 cm, densely hairy; leaves greyish downy, 4–5 × 1–1.2 cm; flowers 8–10 mm across **25. selskianum**

b. Plant to 80 cm, hairless; leaves 5–8 × 1.9–2.1 cm; flowers *c.* 1.2 cm across **26. aizoon**

31a. Stems persistent, creeping, with leaves at tip in winter
27. hybridum

b. Stems annual, not creeping 32

32a. Leaves bright or dark green; flowers 1.5–2 cm across; petals mucronate; plant usually with one terminal, leafy or leafless inflorescence **28. kamtschaticum**

b. Leaves dark green, crowded; flowers 1.2–1.5 cm across; petals acute; plant with many short axillary leafy floriferous branches
29. floriferum

33a. Annual or biennial with flat broad leaves; basal leaves forming a rosette 34

b. Perennial or, when annual or biennial, leaves linear or not forming a rosette 36

34a. Plant hairy, with linear-ovate leaves, green dotted with red; flowers white, often with a few pinkish spots **30. cepaea**

b. Plant downy; flowers rose to crimson red 35

35a. Plant glandular-downy, 5–10 cm; leaves to 1.8 cm, linear to spathulate, obtuse; flowers rose-red, *c.* 9 mm across **31. pilosum**

b. Plant downy, 10–30 cm; leaves 2.5–3 cm, ovate, bluntly acuminate; flowers carmine-red, *c.* 1.3 cm across
32. sempervivoides

36a. Plant annual or biennial (perennial in some variants of *S. hispanicum*) 37

b. Plant perennial 40

37a. Annual with reddish stems, 15–25 cm; leaves flat, margin entire; inflorescence very large, of many 2- to 3-forked branches with flowers in the forks; flowers pale yellow **33. formosanum**

b. Plant 5–15 cm; leaves almost or partly terete; inflorescence not as above; flowers white, pink, rose or blue 38

38a. Flowers many, pale or sometimes sky-blue, 5–6 mm across
34. caeruleum

b. Flowers white to pink, never blue, 6–12 mm across 39

39a. Leaves glandular hairy, 6–12 mm, green; flowers pink, dull rose or white, 6 mm across **35. villosum**

b. Leaves hairless, 1.2–2.5 cm, grey-green, often reddish; flowers pink to white, *c.* 1.2 cm across **36. hispanicum**

40a. Plant with short rootstocks and thick roots 41

b. Plant rootstocks not as above or absent 56

41a. Stems to 30 cm 42

b. Stems more than 30 cm 50

42a. Leaves opposite or in whorls of 3 43

b. Leaves alternate 46

43a. Leaves opposite 44

b. Leaves in whorls of 3 **40. sieboldii**

44a. Plant 5–10 cm; leaves oblong-lanceolate **41. cyaneum**

b. Plant 10–25 cm; leaves circular to obovate 45

45a. Leaves not red-dotted, margin usually entire; flowers pink or pale violet **38. ewersii**

b. Leaves minutely red-dotted, margin with a few blunt teeth; flowers rose-purple, becoming carmine-red
39. cauticola

46a. Flowers to 1 cm across 47

b. Flowers more than 1 cm across 49

47a. Flowers white to pink
45. telephioides

b. Flowers purplish red to purplish pink 48

48a. Leaves 1.2–2.5 cm, grey-green
37. anacampseros

b. Leaves 5–7.5 cm, blue-green
46. telephium

49a. Leaves oblanceolate, entire, grey-green; flowers purplish carmine-rose **41. cyaneum**

b. Leaves linear-lanceolate, toothed towards apex, green; flowers pale pink **42. tatarinowii**

50a. Leaves alternate **46. telephium**

b. Leaves opposite or in whorls 51

51a. Flowers 1–1.3 cm across 52

b. Flowers to 6 mm across 54

52a. Flowers pink; leaves rounded or tapered at base, stalkless, whole plant never purplish
48. spectabile

b. Flowers white to greenish white or rose-white or leaves with distinct short stalks or whole plant purplish 53

53a. Plant 30–60 cm; leaves usually opposite, grey-green, 5–7.5 cm; petals rose-white or greenish white **49. erythrostictum**

b. Plant 50–100 cm; leaves opposite or ternate, dark-green, 5–12.5 cm;

flowers yellowish or greenish white or whole plant purplish
50. maximum

54a. Leaves red-dotted; flowers pale-green **43. verticillatum**

b. Leaves green or green with a red edge; flowers cream to pink 55

55a. Leaves green, base rounded, about as long as internodes
47. pseudospectabile

b. Leaves glaucous, base wedge-shaped, twice as long as internodes
44. spectabile × telephium 'Autumn Joy'

56a. Rootstocks thickening horizontally or contracted; leaves evergreen; flowers mostly white, rarely red or yellow 57

b. Rootstocks absent; plant creeping or erect, flowering shoots monocarpic or sterile 65

57a. Stems 15–30 cm, arising from perennial leafy rosette 58

b. Stems 2–15 cm, annual or lasting up to 18 months; rosette absent, or small and loose 61

58a. Plant hairless 59

b. Plant hairy 60

59a. Leaves oblong-ovate; flowers white with purple midrib and markings towards base, petals ovate
51. glabrum

b. Leaves spathulate; flowers white, petals lanceolate **56. bellum**

60a. Plant hairy; stem-leaves linear or lanceolate, usually less than 10.5 mm across **52. hemsleyanum**

b. Plant minutely hairy; stem-leaves ovate, very fleshy, usually more than 10.5 mm across
53. ebracteatum

61a. Flowers white or tinged red, petals opaque 62

b. Flowers pink, reddish or yellow or petals translucent 63

62a. Flowers bell-shaped, white; petals oblong-ovate, erect below, wide-spreading above **54. wrightii**

b. Flowers star-like; petals lanceolate, spreading **55. cockerellii**

63a. Leaves c. 2.5 cm, green, red at margins and tip **57. versadense**

b. Leaves c. 6 mm, slightly glaucous 64

64a. Plant hairless, fresh green; leaves slightly glaucous; flowers sulphur yellow **58. greggii**

b. Plant glaucous with grey-green leaves; petals whitish or pinkish, translucent **59. alamosanum**

65a. Flowers yellow 66

b. Flowers pink, purple, red, white or cream 91

66a. All leaves opposite or whorled 67

b. Leaves alternate (at least on flowering shoots) 71

67a. Leaves opposite, thick, obovate to spathulate, green or reddish green; flowers c. 1.8 cm across
60. divergens

b. Plant with 3 or more leaves at one level 68

68a. Plant with light green shining tint; leaves 2–3 mm across, narrowly linear, obtuse, in whorls of 3–5, or rarely solitary **61. mexicanum**

b. Plant without green shining tint; leaves mostly ternate 69

69a. Leaves light green, broadest near the apex, obtuse, 5 mm broad; stems, leaves and sepals with sparse fine papillae
62. chauveaudii

b. Leaves pale green, linear or lanceolate, acute or nearly acute at the top; plant without papillae 70

70a. Leaves in whorls of 3, sometimes also opposite, broadly lanceolate, 6 mm broad; flowering-stems short
63. sarmentosum

b. Leaves linear or linear-lanceolate, to 3 mm across; flowering-stems tall **64. lineare**

71a. Leaves spathulate, flat 72

b. Leaves not as above 75

72a. Leaves green, to 1.8 cm; leaves of erect stems elliptic; flowers bright yellow, petals oblong to lanceolate
65. purdyi

b. Leaves tinged red or orange-red when older and 1.2–3 cm; petals lanceolate 73

73a. Leaves shining green, often reddened, very fleshy, 1.2–1.8 cm; petals long-acuminate, ascending; flower-buds c. 1.5 cm
66. oreganum

b. Leaves greyish or bluish green, to 3 cm; petals acute 74

74a. Leaves greyish green or white mealy, often tinged red or purple, white on back; flowers to 1.5 cm across; petals wide-spreading
67. spathulifolium

b. Leaves bluish green, in age orange-red; flowers 6–10 mm across; petals erect below, somewhat spreading above
68. obtusatum subsp. **obtusatum**

75a. Leaves with marginal hairs
69. humifusum

b. Leaves hairless 76

76a. Plant with cypress-like branches; leaves ovate, closely overlapping, c. 1.5 cm **70. muscoideum**

b. Branches not as above 77

77a. Leaves shorter than 8 mm 78

b. Leaves 10 mm or longer 84

78a. Leaves broadest at the base or at the top 79

b. Leaves linear 81

79a. Leaves ovate-triangular to oblong or ovate-rounded, densely overlapping throughout **71. acre**

b. Leaves oblong-elliptic to obovate, overlapping only at shoot tips 80

80a. Stems minutely roughened, red brown, 2 mm thick near tip; follicles erect **72. oaxacanum**

b. Stems smooth, greenish, 1 mm thick near tip; follicles spreading
73. alpestre

81a. Leaves arranged in 6 spiral rows, bright green, 4–7 mm; flowers golden yellow, 5–9 mm across
74. sexangulare

b. Leaves not as above; flowers c. 1.2 cm across 82

82a. Flowers bright yellow, to 1.6 cm across **83. urvillei**

b. Flowers dull yellow, to 1.3 cm across 83

83a. Plant bushy with shrubby growth; leaves green, with papillae, all winter-deciduous except the youngest **75. multiceps**

b. Plant with ascending branches; leaves variable **76. japonicum**

84a. Vegetative rosettes in leaf axils of floral stems **77. stenopetalum**

b. Plant not as above 85

85a. Sterile shoot tips forming propagules covered in dead leaf-bases in summer
78. amplexicaule

b. Fresh leaves persisting through summer at shoot-tips 86

86a. Inflorescence dense, drooping in bud; petals 6–7 87

b. Inflorescence not drooping in bud, loose or flowers with 5 petals 88

87a. Leaves in dense rosette-like cones at tips of sterile shoots
79. forsterianum

b. Leaves evenly spread along sterile shoots, not in cones **80. rupestre**

88a. Plant to 15 cm; leaves to 1.8 cm 89

b. Plant to 60 cm; leaves often greater than 1.8 cm **93. sediforme**

89a. Long prostrate sterile shoots with tufts of leaves at tip; flowers pale yellow, sepals and petals 6 **81. pruinatum**

b. Compact sterile shoots, less than 6 cm; flowers usually bright yellow, sepals and petals 5 90

90a. Leaves oblong-lanceolate, with blunt mucros; flowering-shoot with long stem dividing near apex **82. lanceolatum**

b. Leaves ovate to elliptic, blunt without mucros; flowering-shoot with short stem, soon dividing **83. urvillei**

91a. Flowers red, purple, pink or green and red 92

b. Flowers white or white with tinges of pink, red or green, or greenish red 97

92a. Flowers rose-red outside, whitish inside, *c.* 5 mm across; leaves obovate to spathulate, obtuse, 5-6 mm, spotted with red **84. stevenianum**

b. Flowers white, white tinged red or rose-purple 93

93a. Leaves opposite 94

b. Leaves alternate or ternate 96

94a. Leaves ovate or obovate, obtuse, roundish towards the base, narrowed into a very short leaf-stalk; petals slender, pointed **85. obtusifolium**

b. Leaves not as above; petals acute 95

95a. Leaves obovate, dark green; flowers pink to purple, *c.* 1.3 cm across with semi-erect petals; inflorescence dense and flat **86. spurium**

b. Leaves diamond-shaped to spathulate, bright green; flowers pink with wide-spreading petals, 1.3–1.8 cm across; inflorescence loose and small **87. stoloniferum**

96a. Stems triangular; leaves in whorls of 3, uppermost often alternate, light green; flowers greenish red, 1.3 cm across; petals wide-spreading, later sharply reflexed **88. rhodocarpum**

b. All leaves alternate, green, narrow-linear; flowers rosy purple, 5–14 mm across, 4-parted **89. pulchellum**

97a. Leaves opposite, toothed in upper half **86. spurium**

b. Leaves alternate, in whorls of 3 or 4 or entire 98

98a. Leaves flat, most in whorls of 3 or 4 99

b. Leaves thickened or both alternate and opposite on one plant 100

99a. Leaves in whorls of 3 or opposite, on the flowering-stems alternate, 1.2–2.5 cm, pale green, obovate; flowers *c.* 1.3 cm across **90. ternatum**

b. Leaves in whorls of 4, 3–10 mm, olive-green, oblong to spathulate; flowers 6–10 mm across **91. monregalense**

100a. Inflorescence a raceme; leaves light to bright green **92. magellense**

b. Inflorescense cyme-like; leaves pale green, blue-green, grey-green or dark green 101

101a. Lower leaves with thorn-like points **93. sediforme**

b. Lower leaves lacking thorn-like points 102

102a. Plant partly hairy 103

b. Plant hairless 108

103a. Leaves ovate or diamond-shaped to linear to oblong elliptic or slightly spathulate, base not or scarcely tapered 104

b. Leaves obovate, diamond-shaped to spathulate, distinctly long-tapered towards base 107

104a. Leaves glaucous, pale grey-green or blue-green, ovate, elliptic or oblong to linear, hairy or hairless 105

b. Leaves not glaucous, dull green, ovate to diamond-shaped, elliptic or oblong, downy or hairy 106

105a. Leaves linear or linear-oblong **94. bithynicum**

b. Leaves ovate to elliptic-oblong **95. dasyphyllum**

106a. Petals *c.* 3 mm; much-branched corymb-like inflorescence of more than 25 flowers **96. gypsicola**

b. Petals 5–8.5 mm; open, paniculate inflorescence of less than 25 flowers **97. hirsutum**

107a. Petals fused for 1–1.5 mm; flowers sweetly scented; leaves diamond-shaped to spathulate **98. fragrans**

b. Petals fused for 1.5–2 mm; flowers not scented; leaves oblong to spathulate **Rosularia adenotricha** (p. 65)

108a. Low plant to 2.5 cm with rosette-like, obovate, grey-green leaves; flowers white, intensely scented **99. compactum**

b. Plants more than 2.5 cm, with non-scented flowers 109

109a. Leaves of sterile shoots ovoid 110

b. Leaves long and thin or semi-terete to flat 112

110a. Sepals spurred at base **110. tenellum**

b. Sepals not spurred at base 111

111a. Leaves alternate, 6–10 mm, with uneven covering of waxy scales **100. furfuraceum**

b. Leaves in rows of 4 or 5, 2–6 mm, mealy or smooth and lustrous **101. brevifolium**

112a. Leaves green with red or pink tips or dots 113

b. Leaves green or pale brownish or bluish green, lacking dots 121

113a. Leaves obovate to spathulate; petals fused for 1–3 mm **102. oregonense**

b. Leaves ovate to oblong-elliptic or linear; petals free or fused up to 1 mm 114

114a. Leaves glaucous 115

b. Leaves not glaucous 116

115a. Sterile shoots to 5 cm; leaves ovate to oblong; petals 3–5.5 mm **95. dasyphyllum**

b. Sterile shoots to 15 cm; leaves oblong to linear; petals 4–7 mm **103. diffusum**

116a. Leaves ovate or ovate-elliptic; inflorescence with 1–3 branches 117

b. Leaves more or less linear or cylindric to ovate, or inflorescence corymb-like with more than 3 branches 118

117a. Plant 5–15 cm; leaves 5 mm or more, ovate-oblong, green, tipped red; old leaves persistent with loose silvery bases, mucronate at the top; flowers white; petals acute **104. liebmannianum**

b. Plant 7–8 cm; leaves 4–5 mm, ovate-elliptic, thick, green, often red; flowers white-pink; petals mucronate **105. anglicum**

118a. Plant 3–5 cm 119

b. Plant 5–20 cm 120

119a. Leaves to 6 mm, linear-oblong, nearly cylindric, green, dotted red; inflorescence branches semi-erect, forked, without flowers between the primary and secondary forks; flowers 9–12 mm across, white, with red spots on back **106. albertii**

b. Leaves 6–8 mm, narrowly linear to cylindric, green or red; inflorescence dense, flat; flowers *c.* 6 mm across, white **107. lydium**

120a. Plants 5–10 cm; leaves *c.* 8 mm, linear-oblong, bright green or reddish green; inflorescence branches almost horizontal; flowers white, dotted red on back; follicles spreading **108. gracile**

b. Plants 10–20 cm; leaves 6–12 mm, cylindric to oblong-ovate, dark green, often reddish; inflorescence branches semi-erect; flowers white; follicles erect **109. album**

121a. Leaves 12 mm or longer **122**

b. Leaves of sterile stems *c.* 3 mm **123**

122a. Leaves 1.2–1.8 cm, pale brownish green **111. nevii**

b. Leaves *c.* 2.5 cm, pale or blue-green **112. glaucophyllum**

123a. Leaves to 8 mm; sepals spurred at base; follicles erect **110. tenellum**

b. Leaves to 6 mm; sepals not spurred; follicles spreading **108. gracile**

1. S. stahlii Solms. Illustration: *Botanical Magazine,* 7908 (1903); *Journal of the Royal Horticultural Society* **46**: 222 (1920–1921); Die natürlichen Pflanzenfamilien **18a**: 441 (1930); Clausen, Sedum of the trans-Mexican volcanic belt, f. 49 (1959).

Evergreen, finely downy, prostrate shrub, 10–20 cm. Leaves opposite, 7–14 × 4–8 mm, ovoid to spherical, apex rounded, dark green to brown. Flowers *c.* 1.3 cm, yellow; petals elliptic-lanceolate, short acuminate. *Mexico.* G1. Late summer.

2. S. morganianum Walther. Illustration: Jacobsen, Handbook of succulent plants **2**: 789 (1960); Boom, Flora Cultuurgewassen **3**: kamer- en kasplanten, pl. 130 (1968); Hay & Beckett, Reader's Digest encyclopaedia of garden plants and flowers, edn 2, 658 (1978); Brickell (ed.), RHS A–Z encyclopedia of garden plants, 973 (2003).

Evergreen shrublet with prostrate or creeping branches, to 7.5 cm. Leaves alternate, numerous, oblong-lanceolate, nearly cylindric, acute, *c.* 2 cm × 8 mm, thick-fleshy, pale green, bluish bloomed. Flowers pale pink to deep scarlet purple, *c.* 10 mm across. *Mexico.* G1. Spring.

3. S. palmeri S. Watson (*S. compressum* Rose). Illustration: *Journal of the Royal Horticultural Society* **46**: 233, 235 (1920–1921).

Evergreen hairless shrub, to 22 cm, with ascending branches. Leaves in loose rosettes, oblong to obovate to spathulate, obtuse, sometimes mucronate, 2.5–5 × 1.5–2 cm, usually grey-green. Flowers usually orange-yellow, 7–15 mm across. Petals narrowly lanceolate, acute, spreading. *Mexico.* G1. Winter–early summer.

4. S. pachyphyllum Rose. Illustration: *Contributions from the US National Herbarium* **13**: pl. 58 (1911); *Journal of the Royal Horticultural Society* **46**: 215 (1920–1921); *Gartenschönheit* **8**: 289 (1927).

Evergreen, hairless, erect shrub to 30 cm. Stems branched, somewhat prostrate. Leaves in distinct spiral rows, oblong-lanceolate, cylindric, bluntly rounded, 1.5–4 cm × 8–10 mm, very fleshy, grey-green mostly with a reddish tip. Flowers pale yellow, *c.* 1.5 cm across. Petals oblong-lanceolate. *Mexico.* G1. Winter–spring.

5. S. rubrotinctum R.T. Clausen (*S. guatemalense* misapplied). Illustration: Evans, Handbook of cultivated sedums, pl. 8 (1983).

Small hairless shrublet, 15–30 cm. Branches at first fleshy, later somewhat woody. Leaves alternate, stalkless, cylindric, rounded at apex, 1.3–2 cm × *c.* 5 mm, pale green, often suffused with red. Flowers yellow, 1.3 cm across. *Mexico.* G1. Winter.

'Aurora'. Illustration: Hay & Beckett, Reader's Digest encyclopaedia of garden plants and flowers, edn 2, 659 (1978); Evans, Handbook of cultivated sedums, pl. 8 (1983). Leaves grey-green, tinged rose-red.

6. S. dendroideum de Candolle. Illustration: *Journal of the Royal Horticultural Society* **46**: 208 (1920–1921); *Sedum Society Newsletter* **16**: 5 (1991).

Shrub to 60 cm. Leaves alternate, spathulate to circular, 4.5 × 2 cm, bright green, margin waxy white, later reddish, with a row of red or dark green glandular dots. Flowers yellow, *c.* 1.5 cm across. Petals lanceolate, acute or obtuse. *Mexico, Guatemala.* G1. Spring–summer.

S. × amecamecana Praeger. Illustration: *Journal of the Royal Horticultural Society* **46**: 214 (1920–1921); Clausen, Sedum of the trans-Mexican volcanic belt, 79–80 (1959); Evans, Handbook of cultivated sedums, 136 (1983). Stems prostrate or creeping, with branches 15–20 cm. Leaves narrow elliptic to oblanceolate, with a short spur below, 1.5–2 cm × 5 mm, acute, somewhat recurved, with some scattered along the shoots. Inflorescence terminal, bearing starry, 5-petalled yellow flowers 1.5 cm across. Sepals unequal, green, spreading. *Mexico.* G1–H5. Summer.

S. dendroideum × *S. goldmannii*, previously misreported as *Sedum prealtum* × *Villadia batesii.*

7. S. praealtum de Candolle (*S. dendroideum* de Candolle subsp. *praealtum* (de Candolle) R.T. Clausen). Illustration: *Journal of the Royal Horticultural Society* **46**: 210 (1920–1921).

Evergreen hairless shrub, 30–150 cm. Leaves alternate, oblanceolate to spathulate, bluntly pointed or rounded, 5–7 × 1.5–2 cm, light green throughout, margin without glandular dots. Flowers pale-bright yellow, 1.5–1.8 cm across. Petals narrowly lanceolate, very acute, wide-spreading. *Mexico.* H5 or G1. Spring–summer.

8. S. decumbens R.T. Clausen (*S. confusum* misapplied). Illustration: *Sedum Society Newsletter* **16**: 1, 3 (1991).

Similar to *S. confusum* but differing in its lesser height, smaller leaves, 1.5–2 × 1.1–1.3 cm, pale green; inflorescence branched at apex only, flat-topped, usually 3-branched, flowers *c.* 15; sepals 2–3 mm. *Origin unknown.* G1.

9. S. confusum Hemsley (*S. aoikon* Ulbrich). Illustration: *Journal of the Royal Horticultural Society* **46**: 212 (1920–1921).

Shrub to 30 cm, branches ascending. Leaves alternate, obovate to spathulate, obtuse, thickish, 2–4 × 1.2–2 cm, glossy green throughout, margin with glandular dots. Inflorescence rounded, with 3–5 terminal and lateral branches and *c.* 25 flowers; flowers yellow, 1.2–1.5 cm across. Petals ovate-lanceolate; sepals 1–1.5 mm. *Mexico.* G1. Winter–spring.

10. S. treleasei Rose. Illustration: *Contributions from the US National Herbarium* **13**: 300 (1911); *Journal of the Royal Horticultural Society* **46**: 217 (1920–1921); *Gartenschönheit* **8**: 290 (1927).

Evergreen shrub to 45 cm, with erect or prostrate stems. Leaves alternate, oblong-obovate, obtuse, very fleshy, upper face more or less flattened, 2–4.5 cm × 9–15 mm, blue-green, densely grey-bloomed. Flowers pale-bright yellow, *c.* 1.3 cm across. Petals ovate-lanceolate, acute. *Mexico.* G1. Spring.

11.S. fusiforme Lowe. Illustration: Evans, Handbook of cultivated sedums, 154 (1983).

Small shrub to *c.* 12 cm. Leaves alternate, lanceolate to oblong or elliptic, terete, crowded at tips of branches, to 1.6 cm, grey-green. Flowers greenish yellow, red at the centre, *c.* 1.6 cm across. *Madeira.* G1. Spring & autumn.

12. S. nudum Aiton. Illustration: *Journal of the Royal Horticultural Society* **46**: 252 (1920–1921); Evans, Handbook of cultivated sedums, 274 (1983).

Evergreen shrub with hairless shoots. Leaves alternate, stalkless, obovate-oblong, obtuse, *c.* 10 × 5 mm, thick, pale green. Flowers greenish yellow, to *c.* 1 cm across. Petals linear-lanceolate, acute, wide-spreading. *Madeira.* G1. Spring–summer.

13. S. craigii R.T. Clausen (*Graptopetalum craigii* (Clausen) Clausen). Illustration: Evans, Handbook of cultivated sedums, 146 (1983).

Small shrub with fleshy stems. Leaves alternate, stalkless, oblong-elliptic, apex rounded, 2–5 cm × 9–22 mm, reddish blue. Flowers white, *c.* 6 mm across. Petals erect, narrowed below, apex recurved. *Mexico.* G1. Autumn.

14. S. longipes Rose. Illustration: *Journal of the Royal Horticultural Society* **46**: 203 (1920–1921).

Hairless matted perennial with long arching stems developing from short rosettes. Leaves alternate, obovate or spathulate, fleshy, entire, *c.* 8 × 5 mm, bright green. Flowers reddish, purple or in part whitish, *c.* 7.5 mm across. Petals ovate, obtuse, almost spreading, red in upper part, becoming silvery white near base. Nectaries large, red, prominent. *Mexico.* G1. Winter.

15. S. moranense Humboldt, Bonpland & Kunth. Illustration: *Journal of the Royal Horticultural Society* **46**: 173 (1920–1921); Clausen, Sedum of the trans-Mexican volcanic belt, 260 (1959).

Evergreen hairless small shrub, to 15 cm, much-branched, prostrate, green, later brown and dark red stems, woody below. Leaves alternate, ovate to ovate-lanceolate, obtuse or acute, 3–5 × 2–3 mm, green. Flowers stalkless, white, sometimes pinkish, *c.* 1 cm across. Petals lanceolate, acute or obtuse. *Mexico.* H5–G1. Summer.

Var. **arboreum** (Masters) Praeger is an erect, much-branched, 15–30 cm bush with leaves occasionally in 5 spiral rows.

16. S. populifolium Pallas. Illustration: *Journal of the Royal Horticultural Society* **46**: 148 (1920–1921); Hegi, Illustrierte flora von Mittel-Europa **4**: 527 (1923); *Botanical Magazine*, n.s., 211 (1952); *The Plantsman* **8**: 9 (1986).

Deciduous hairless perennial, woody at base, 20–40 cm, stems greenish or purplish. Leaves alternate, ovate, acute, cordate at base, 1.3–3.5 × 1–2.5 cm, green, coarsely and irregularly toothed. Flowers pink or white, scented, *c.* 1 cm across. Petals lanceolate, acute, spreading. *Former USSR (Siberia).* H1. Summer.

17. S. retusum Hemsley. Illustration: *Journal of the Royal Horticultural Society* **46**: 149 (1920–1921).

Evergreen hairless shrub to 30 cm, bark grey, non-peeling. Leaves alternate, obovate-oblong, spathulate, 1.5–2.5 cm × 6–10 mm, deeply notched at apex. Flowers white with red centre, *c.* 1 cm across. Petals oblong-lanceolate, obtuse, shortly mucronate, erect below, spreading above. *Mexico.* G1. Summer–autumn.

18. S. oxypetalum Humboldt, Bonpland & Kunth. Illustration: *Journal of the Royal Horticultural Society* **46**: 193 (1920–1921).

Hairless shrub, 50–90 cm, bark peeling in papery layers. Leaves alternate, ovate-lanceolate or spathulate, 1.5–5 cm, finely papillose, green, deciduous after flowering. Flowers stellate, pinkish or dull red especially at centre, *c.* 1.3 cm across, scented. Petals lanceolate, acute, wide-spreading from near base. *Mexico.* G1. Summer.

19. S. allantoides Rose. Illustration: *Contributions from the US National Herbarium* **12**: pl. 79 (1909); *Journal of the Royal Horticultural Society* **46**: 154 (1920–1921).

Evergreen, glaucous shrub to 40 cm, branches erect. Leaves alternate, club-shaped, 2–3 cm × 6–25 mm, grey-green, white-grey bloomed. Flowers greenish white, 1.6–1.8 cm across. Petals lanceolate, acute, wide-spreading, often red-spotted near tip. *Mexico.* G1. Summer.

20. S. bourgaei Hemsley. Illustration: *Journal of the Royal Horticultural Society* **46**: 155 (1920–1921); Clausen, Sedum of trans-Mexican volcanic belt, 148 (1959).

Evergreen small shrub, 15–30 cm. Leaves alternate, linear, obtuse, 1–2 cm × *c.* 2 mm, green. Flowers white, usually tipped red, *c.* 1.2 cm across. *C Mexico.* H5–G1. Late summer.

21. S. griseum Praeger (*S. farinosum* misapplied). Illustration: *Journal of the Royal Horticultural Society* **46**: 158 (1920–1921); Clausen, Sedum of the trans-Mexican volcanic belt, 160 (1959).

Small evergreen shrub 15–20 cm, often white-grey bloomed. Leaves alternate, linear or lanceolate-linear, 1–2 cm × *c.* 2.5 mm, green or glaucous-green. Flowers white, 1.2–1.5 cm across. Petals lanceolate. *Mexico.* G1. Winter.

22. S. lucidum R.T. Clausen. Illustration: Clausen, Sedum of the trans-Mexican volcanic belt, 94 (1959): *Kew Magazine* **4**: 75 (1987).

Hairless shrub with prostrate stems, to 45 cm. Leaves alternate, elliptic to oblanceolate or spathulate, stalkless, acute to obtuse, 2.5–5 × 1–1.9 cm, thick, dark green, shining (especially young leaves). Flowers white with musky scent, 9–11 mm across. Petals lanceolate or elliptic-oblong, obtuse or acute. *Mexico.* G1. Late autumn–early spring.

23. S. adolphii Hamet.
Evergreen, hairless shrub, 7.5–12.5 cm, branches ascending. Leaves alternate, upper rather crowded, very fleshy, oblong-oblanceolate, *c.* 3.5 × 5 cm, yellowish green, sometimes with faint reddish margins. Flowers white, sometimes pinkish on back, *c.* 1.8 cm across. Petals ovate-lanceolate, acuminate, wide-spreading. *Mexico.* G1. Spring.

24. S. nussbaumerianum Bitter (*S. adolphii* misapplied). Illustration: Clausen, Sedum of the trans-Mexican volcanic belt, 108 (1959); Boom, Flora Cultuurgewassen **3**: kamer- en kasplanten, pl. 132 (1968).

Hairless shrub with stout spreading branches, *c.* 7 cm, stems reddish brown. Leaves alternate, oblanceolate-elliptic, acute at apex, 1.2–5 cm × 8 mm, yellow-green, edged with yellow, orange or red. Flowers white or pinkish, 1.4–1.6 cm across. Petals linear-lanceolate, acute. *Mexico.* G1. Winter.

25. S. selskianum Regel & Maack (*S. aizoon* Linnaeus subsp. *selskianum* (Regel & Maack) Fröderström). Illustration:

Regel & Maack, Tentamen flora Ussuriensis, **66**: t. 6 (1861); *Gartenflora* **11**: t. 361 (1862); *Journal of the Royal Horticultural Society* **46**: 114 (1920–1921).

Densely hairy perennial herb with annual stems, erect, 30–45 cm. Leaves stalkless, alternate, linear-oblong, crowded, 4–5 × 1.2 cm, dark green, greyish downy, upper half toothed. Flowers yellow, 8–10 mm across. Petals broadly lanceolate, acuminate or mucronate, wide-spreading. *China (Manchuria), former USSR (Amur).* H1. Summer.

26. S. aizoon Linnaeus (*S. maximowiczii* Regel). Illustration: *Journal of the Royal Horticultural Society* **46**: 109 (1920–1921); Hay & Synge, The dictionary of garden plants in colour; pl. 1369 (1969).

Perennial herb, hairless, deciduous, shoots erect, to 80 cm. Leaves alternate, stalkless, ovate-lanceolate to linear-lanceolate, 5–8 × 2 cm, sharply toothed from below the middle. Flowers numerous, yellow to orange, *c.* 1.2 cm across. Petals linear-lanceolate, mucronate, wide-spreading. *Japan, former USSR (Siberia), China.* H1. Summer–autumn.

'Aurantiacum' (illustration: Brickell (ed.), RHS A–Z encyclopedia of garden plants, 971, 2003) has a dark red stem, dark green leaves, dark yellow to deep orange flowers and red fruits.

27. S. hybridum Linnaeus. Illustration: Reichenbach, Icones florae Germanicae et Helveticae **23**: f. 64 (1897); *Journal of the Royal Horticultural Society* **46**: 126 (1920–1921).

Evergreen perennial herb, 10–30 cm, branches persistent, ascending. Leaves alternate, oblanceolate to spathulate, upper part coarsely toothed, lower part entire, 2.5 × 2 cm, bright green, in winter reddish green. Flowers yellow, 1.2–1.4 cm across. Petals mucronate. *Former USSR (Siberia), Mongolia.* H1. Summer.

28. S. kamtschaticum Fischer & Meyer (*S. aizoon* Linnaeus subsp. *kamtschaticum* (Fischer & Meyer) Fröderström). Illustration: *Gartenflora* **32**: 250 (1883); *Journal of the Royal Horticultural Society* **46**: 121 (1920–1921).

Perennial herb, 15–30 cm, branches annual, ascending. Leaves opposite or alternate, stalkless, oblanceolate to spathulate, toothed towards apex, 3–5 × *c.* 1.2 cm, dark green. Flowers orange-yellow, 1.5–2 cm across. Petals lanceolate,

mucronate. *N & C China, Japan, former USSR (Kamchatka).* H1. Late summer.

'Variegatum'. Illustration: Köhlein, Freilandsukkulenten, 161 (1977). Leaves with a creamish white margin.

Var. **ellacombianum** (Praeger) R.T. Clausen (*S. ellacombianum* Praeger). Figure 87(2), p. 43. Illustration: *Journal of the Royal Horticultural Society* **46**: 118 (1920–1921); Köhlein, Freilandsukkulenten, 143 (1977). Deciduous perennial, 15–25 cm. Leaves opposite, obovate to spathulate, 3.5–4 × *c.* 1.8 cm, bright green. Flowers pure yellow, *c.* 1.5 cm across. *Japan.* H3. Summer.

Var. **middendorffianum** (Maximowicz) R.T. Clausen (*S. middendorffianum* Maximowicz; *S. aizoon* Linnaeus subsp. *middendorffianum* (Maximowicz) R.T. Clausen). Illustration: *Journal of the Royal Horticultural Society* **46**: 116 (1920–1921). Plant to 20 cm. Leaves alternate, linear-oblanceolate, *c.* 35 × mm, fresh bright green, toothed near apex. Flowers pale yellow, *c.* 1.5 cm across.

Var. **middendorffianum** forma **diffusum** Praeger (*S. middendorffianum* var. *diffusum* Praeger). Leaves larger, lanceolate to narrowly spathulate, 2.5–5 cm × *c.* 6 mm, sharply toothed in upper part. *N & C Asia, former USSR (E Siberia), China (Manchuria), Mongolia.* H1. Summer.

29. S. floriferum Praeger (*S. kamtschaticum* var. *floriferum* Praeger). Illustration: *Journal of the Royal Horticultural Society* **46**: 123 (1920–1921).

Hairless perennial herb with annual red stems, *c.* 15 cm. Leaves alternate, stalkless, dark green, crowded, obtuse. Flowers yellow, 1.2–1.5 cm across. Petals lanceolate, acute, wide-spreading. *NE China.* H1. Summer.

'Weihenstephaner Gold'. Illustration: Fessler, Die Freilandschmuckstauden, edn 3, 586 (1985); *The Plantsman* **8**: 11 (1986). Forms very compact mats, flowers bright yellow, 1.5–1.8 cm across.

30. S. cepaea Linnaeus. Illustration: *Edwards's Botanical Register*, t. 1391 (1830–1831); Hegi, Illustrierte flora von Mittel-Europa **4**: 520 (1921).

Annual to biennial herb, hairy, to 30 cm. Leaves alternate, opposite or in whorls of 3–4, linear-ovate, 1.2–2.5 cm × 12 mm, dotted with red. Flowers white, often with a few pinkish spots, *c.* 9 mm across. Petals lanceolate, acuminate. *W, C & S Europe, W Turkey.* H2. Early summer.

31. S. pilosum Bieberstein. Illustration: *Botanical Magazine*, 8503 (1913); *Journal of the Royal Horticultural Society* **46**: 284 (1920–1921); *Gartenschönheit* **11**: 199 (1930); Köhlein, Freilandsukkulenten, 143 (1977).

Biennial herb, glandular-downy, 5–10 cm, producing a dense rosette of very numerous leaves in the first year. Leaves linear to spathulate, obtuse, to *c.* 1.8 cm × 2 mm on flowering stems, dark green. Flowers rose-red, *c.* 9 mm across. Petals elliptic-lanceolate, acute, erect below, divergent above. *SW Asia, Caucasus, Iran.* H3. Spring–summer.

32. S. sempervivoides Bieberstein (*S. sempervivum* Sprengel). Illustration: *Botanical Magazine*, 2474 (1824); *Gartenflora* **16**: t. 551 (1876) & **33**: t. 1155 (1884); *Journal of the Royal Horticultural Society* **46**: 282 (1920–1921); Köhlein, Freilandsukkulenten, 143 (1977).

Biennial, 10–30 cm. Leaves in rosettes, alternate on the flowering-stems. Leaves stalkless, obovate to ovate, bluntly acuminate, 2.5–3 × 3–1.8 cm, downy, reddened. Flowers carmine-red, *c.* 1.3 cm across. Petals lanceolate, acute, bright crimson. *SW Asia, Caucasus, N Iran, former USSR (Georgia).* H3. Summer.

33. S. formosanum N.E. Brown. Illustration: *Journal of the Royal Horticultural Society* **46**: 296 (1920–1921).

Annual herb with succulent, reddish stems, 15–25 cm. Leaves alternate, spathulate, obtuse, recurved, *c.* 2.5 × 2 cm, entire. Flowers yellow, *c.* 1.2 cm across. Petals lanceolate, spreading. *S Japan to Taiwan.* H3–4. Summer.

34. S. caeruleum Linnaeus (*S. azureum* Desfontaines). Illustration: *Edwards's Botanical Register*, t. 520 (1820–1821); *Botanical Magazine*, 2224 (1821); *Journal of the Royal Horticultural Society* **46**: 303 (1920–1921); Hay & Synge, The dictionary of garden plants in colour, pl. 3 74 (1969).

Annual hairless herb, 5–10 cm. Leaves alternate, ovate to long-ovate, 9–18 mm, pale green. Flowers pale blue, usually 7-parted, *c.* 6 mm across. Petals oblong, obtuse. *Corsica, Sardinia, Sicily, Algeria.* H5. Summer–autumn.

Often misspelled as *S. coeruleum.*

35. S. villosum Linnaeus. Illustration: Reichenbach, Icones florae Germanicae et Helveticae **23**: t. 52 (1897); *Journal of the Royal Horticultural Society* **46**: 302

Figure 87. Diagnostic details of *Sedum* species. 1, *S. fosterianum* (a, flowering stem; b, habit; c, flower from above; d, fruit). 2, *S. kamtschaticum* var. *ellacombianum* (a, flowering stem; b, habit; c, flower from above; d, fruit). 3, *S. sexangulare* (a, b, habit; c, flower from above; d, fruit). 4, *S. album* 'Murale' (a, b, habit; c, flower from side and above; d, fruit).

(1920–1921); Hegi, Illustrierte flora von Mittel-Europa **4**: pl. 140 (1921); Die natürlichen Pflanzenfamilien **18a**: 441 (1930).

Biennial or sometimes perennial herb with red stems, 7.5–10 cm. Leaves alternate, linear-oblong, obtuse, 6–12 mm, fleshy, glandular hairy, green. Flowers pink, dull rose or white, *c.* 6 mm across. Petals ovate, rather acute. *Scandinavia, across Europe to N Africa, Greenland, Iceland.* H1. Early summer.

36. S. hispanicum Linnaeus (*S. glaucum* Waldstein & Kitaibel). Illustration: Reichenbach, *Iconographia botanica* **9**: 1136 (1831–1833); Journal of the Royal Horticultural Society **46**: f. 178, excluding b (1920–1921); Hegi, Illustrierte flora von Mittel-Europa **4**: 528 (1921).

Annual, biennial or perennial herb, 5–15 cm, hairless or flower-stem with shaggy glandular hairs. Leaves alternate, linear to long-lanceolate, 1.2–2.5 cm, grey-green, often reddish. Flowers pink to white, *c.* 1.2 cm across. Petals lanceolate, acuminate, white often with purplish mid-vein. *Alps, S & SE Europe, Turkey, N Iran.* H3. Summer.

37. S. anacampseros Linnaeus. Illustration: *Botanical Magazine*, 118 (1790); *Journal of the Royal Horticultural Society* **46**: 105 (1920–1921); Hegi, Illustrierte flora von Mittel-Europa **4**: 520 (1921).

Evergreen perennial, 10–25 cm. Leaves alternate, stalkless, ovate to obovate or elliptic, slightly acuminate, 1.2–2.5 × 2–1.8 cm, grey-green. Leaf-margins entire, reddish. Flowers purple, *c.* 6 mm across. Petals oblong-lanceolate, obtuse. *Europe.* H2. Summer.

38. S. ewersii von Ledebour. Illustration: *Gartenflora* **9**: t. 295 (1860); *Journal of the Royal Horticultural Society* **46**: 97 (1920–1921); Polunin & Stainton, Flowers of the Himalaya, pl. 45 (1985).

Deciduous perennial herb, stem ascending, 10–20 cm. Leaves opposite, stalkless, circular to broad-ovate, with cordate base, *c.* 1.8 × 8 cm, bluish green, usually entire. Flowers pink or pale violet, 1–1.2 cm across. Petals oblong-lanceolate, pinkish with darker spots. *W Himalaya, Mongolia.* H1. Summer.

Var. **homophyllum** Praeger (*S. pluricaule* misapplied; *S. cyaneum* misapplied). Illustration: *Journal of the Royal Horticultural Society* **46**: 98

(1920–1921). All parts smaller; leaves obovate, 1.2–1.5 cm × 9 mm, more glaucous than the type. *Only known in cultivation.*

39. S. cauticola Praeger. Illustration: *Journal of the Royal Horticultural Society* **46**: 100 (1920–1921); *Botanical Magazine*, 401 (1962–1963); Hay & Synge, The dictionary of garden plants in colour, pl. 1370 (1969).

Sprawling perennial herb, shoots dark purple, 10–20 cm. Leaves opposite, roundish to spathulate, obovate, *c.* 2.5 × 8 cm, glaucous, minutely dotted with red. Flowers rose-purple, older ones carmine-red, *c.* 1.5 cm across. Petals lanceolate, acute. *Japan.* H3–4. Summer–autumn.

'Ruby Glow' (*S.* 'Robustum'). Illustration: Hay & Synge, The dictionary of garden plants in colour, pl. 1374 (1969) – as *S. spectabile* hybrid 'Ruby Glow'. Perennial herb with ascending stems, 15–25 cm. Leaves as *S. cauticola*, glaucous blue and purplish red. Flowers pinkish red, *c.* 7 mm across. *Garden origin.* H2. Summer–autumn.

A hybrid between *S. cauticola* and *S. telephium.*

40. S. sieboldii Sweet. Illustration: *Botanical Magazine*, 5358 (1863); *Journal of the Royal Horticultural Society* **46**: 102 (1920–1921).

Deciduous, hairless perennial herb with sprawling purplish stems, 15–20 cm. Leaves 3 whorled, stalkless, almost circular, 1.3–2.5 × *c.* 1.9 cm, upper part toothed, blue-green with red edge. Flowers pink, *c.* 1.3 cm across. Petals broadly lanceolate, acute, spreading. *Japan.* H3. Autumn.

'Mediovariegatum'. Illustration: Köhlein, Freilandsukkulenten, 161 (1977). Leaves with large pale yellow central patch. Requires winter protection.

41. S. cyaneum Rudolph. Illustration: *Gartenflora* **27**: t. 972 (1879); *Journal of the Royal Horticultural Society* **46**: 106 (1920–1921).

Deciduous creeping hairless perennial herb, 5–10 cm. Leaves alternate or opposite, oblong-lanceolate, obtuse, entire, 1–2 cm × 6 mm, grey-green. Flowers purplish carmine rose, to 1.3 cm across. Petals ovate-oblong. *Former USSR (E Siberia & Sakhalin).* H1. Summer.

42. S. tatarinowii Maximowicz. Illustration: *Journal of the Royal Horticultural Society* **46**: 103 (1920–1921).

Hairless perennial herb, 10–15 cm, stems almost erect. Leaves alternate, linear-lanceolate, nearly obtuse, 1.3–2.5 cm, fleshy, coarsely toothed towards apex. Flowers pinkish white, *c.* 1.2 cm across. Petals ovate-lanceolate, acute, wide-spreading. *N China, Mongolia.* H1. Summer–autumn.

43. S. verticillatum Linnaeus. Illustration: *Journal of the Royal Horticultural Society* **46**: 95 (1920–1921).

Hairless perennial herb with erect stems, 30–60 cm. Leaves whorled, lower often opposite or in 3s, upper in whorls of 4 or 5, oblong-lanceolate, 5–7.5 × 2.5 cm, green, red dotted. Flowers pale green, *c.* 6 mm across. Petals ovate-lanceolate, acute, wide-spreading. Numerous bulbils produced in the axils of old leaves and bracts. *Japan, former USSR (Kamchatka).* H1. Autumn.

Var. **nipponicum** Praeger. Slender, dwarf plant with opposite leaves.

44. S. spectabile Boreau × **telephium** Linnaeus **'Autumn Joy'** ('Herbstfreude'). Illustration: Hay & Synge, The dictionary of garden plants in colour, pl. 1372 (1969).

Hairless perennial herb with dark glaucous green leaves, 40–60 cm. Flowers cream to light pink, *c.* 5 mm across. Petals semi-erect. Stamens absent. Carpels dark pink, very fleshy, longer than petals, hence dominating the flower colour. *Garden origin.* H1. Summer–autumn.

45. S. telephioides Michaux. Illustration: Britton & Brown, An illustrated flora of the Northern United States **2**: 208 (1897); Clausen, Sedum of North America, 71, 88 (1975).

Hairless perennial, 15–30 cm. Leaves mostly alternate, elliptic to spathulate, obtuse, 1–4.5 × 30 mm, wedge-shaped throughout, entire or remotely toothed. Flowers white to pink, *c.* 1 cm across. Petals lanceolate. *E North America.* H1. Summer–early autumn.

46. S. telephium Linnaeus. Illustration: Reichenbach, Iconographia botanica **8**: 968 (1830); Die natürlichen Pflanzenfamilien **18a**: 441 (1930).

Hairless, erect perennial herb, 20–70 cm. Leaves usually alternate, oblong to ovate-oblong, obtuse, 5–7.5 × 4 cm, irregularly toothed, bluish green. Flowers mostly red-purple, 7–10 mm across. Petals lanceolate, acute, spreading, somewhat recurved. *Europe to Siberia*. H1. Summer.

Subsp. **telephium** (*S. purpureum* (Linnaeus) Schultes; *S. purpurascens* Koch). Illustration: *Journal of the Royal Horticultural Society* **46**: 83 (1920–1921). Leaves alternate, lower leaves wedge-shaped at base, upper leaves truncate at base, fleshy, *c.* 7.5 × 6 cm, dark green. Flowers purple, rarely whitish, *c.* 1 cm across. *Europe, Asia*.

Var. **borderi** Rouy & Camus. Illustration: *The Plantsman* **8**: 6 (1986). Plant with red stems, 25–40 cm. Leaves deeply toothed. Flowers purplish pink, *c.* 7 mm across. *France*.

Subsp. **fabaria** (Koch) Kirschleger. (*S. fabaria* Koch; *S. vulgare* (Haworth) Link). Illustration: *Journal of the Royal Horticultural Society* **46**: 84 (1920–1921). Plant 20–40 cm. Leaves sometimes stalked, lanceolate, dark green. Flowers smaller than in the type. *W & C Europe*.

47. S. pseudospectabile Praeger. Illustration: *Journal of the Royal Horticultural Society* **46**: 91 (1920–1921).

Perennial hairless herb, 30–60 cm. Leaves in 3s or opposite, broadly ovate to obovate, stalkless, base rounded, 3.5–5 × 5–3.5 cm, irregularly toothed, green, often with a red edge. Flowers pink, *c.* 6 mm across. Petals ovate-lanceolate, acute, spreading but not at right angles. *China*. H1. Summer–autumn.

48. S. spectabile Boreau. Illustration: Saunders, Refugium botanicum, **1**: f. 32 (1869); *Gartenflora* **21**: t. 709 (1872); *Journal of the Royal Horticultural Society* **46**: 93 (1920–1921); Hay & Beckett, Reader's Digest encyclopaedia of garden plants and flowers, edn 2, 658 (1978).

Deciduous hairless perennial herb, 30–50 cm. Leaves opposite or 3-whorled, ovate to spathulate, *c.* 7.5 × cm, base entire, toothed above, fleshy, bluish white. Flowers pink, 1–1.3 cm across. Petals lanceolate, acute, spreading. *Korea, China (Manchuria), Japan*. H1. Summer–autumn.

The following cultivars are commonly grown: 'Brilliant', petals bright pink with darker carpels and anthers; 'Carmen', petals darker pink; 'Humile', petals pink as in the species; 'Rosenteller', petals bright pink; 'Septemberglut' (illustration: Köhlein, Freilandsukkulenten, 144, 1977), petals deep pink, the darkest cultivar.

49. S. erythrostictum Miquel (*S. alboroseum* Baker; *S. japonicum* misapplied). Illustration: Saunders, *Refugium botanicum*, f. 33 (1869); *Gartenflora* **21**: t. 709 (1872); *Journal of the Royal Horticultural Society* **46**: 89 (1920–1921).

Glaucous, erect, perennial herb with thickened roots, stem 30–60 cm. Leaves usually opposite, often narrowed towards the base, ovate, 5–7.5 × 5–4 cm, grey-green. Flowers greenish white or rose-white, carpels pink. Petals oblong-lanceolate, acute, wide-spreading. *E Asia*. H1. Summer–early autumn.

'Mediovariegatum'. Illustration: Köhlein, Freilandsukkulenten, 1143 (1977). Leaves yellowish white, margin green; flowers rose.

50. S. maximum Suter (*S. telephium* Linnaeus subsp. *maximum* (Linnaeus) Krocker). Illustration: Reichenbach, Iconographia botanica **8**: f. 969 (1830); *Journal of the Royal Horticultural Society* **46**: 80 (1920–1921); Hegi, Illustrierte flora von Mittel-Europa **4**: pl. 140 (1921).

Hairless, herbaceous perennial with green or red erect stems, 50–100 cm. Leaves opposite or sometimes 3-whorled, stalkless, broadly ovate with a cordate base, obtuse, 5–12.5 × 5 cm, irregularly toothed, dark green. Flowers greenish or greenish white, more rarely pale pink, *c.* 1 cm across. Petals ovate-lanceolate, rather acute. *Europe, Caucasus*. H1. Late summer.

'Atropurpureum'. Illustration: *Botanical Magazine*, n.s., 429 (1962–1963); Hay & Synge, The dictionary of garden plants in colour, pl. 1371 (1969). Leaves and stems deep purple. This cultivar is very variable.

51. S. glabrum (Rose) Praeger. Illustration: *Journal of the Royal Horticultural Society* **46**: 128 (1920–1921).

Evergreen hairless perennial herb to 20 cm. Basal leaves forming a rosette, stem-leaves alternate, oblong-ovate, 3 × 2 cm, pale green. Flowers white, *c.* 1 cm across. Petals ovate, white with purple midribs and markings towards base, recurving apically. *Mexico*. G1. Summer–autumn.

52. S. hemsleyanum Rose. Illustration: Clausen, Sedum of the trans-Mexican volcanic belt, 228 (1959).

Evergreen herb to 30 cm, hairy. Basal leaves forming a rosette, circular; stem-leaves alternate linear or lanceolate, 2–3.5 cm × 10 mm. Flowers white, 8–11 mm across. Petals ovate to

lanceolate, recurving apically. *Mexico*. G1. Autumn–early winter.

53. S. ebracteatum de Candolle. Illustration: *Journal of the Royal Horticultural Society* **46**: 131 (1920–1921); Die natürlichen Pflanzenfamilien **18a**: 441 (1930); Clausen, Sedum of the trans-Mexican volcanic belt, 237 (1959).

Evergreen perennial herb to 30 cm, minutely hairy. Basal leaves forming rosettes, ovate; stem-leaves ovate, 2–2.5 × 1.6 cm. Flowers white, 1–1.3 cm across. Petals ovate, acute, recurved apically. *Mexico*. G1. Autumn.

54. S. wrightii A. Gray. Illustration: *Journal of the Royal Horticultural Society* **46**: 138 (1920–1921); Clausen, Sedum of North America, 203 (1975).

Evergreen hairless perennial herb, 2–15 cm. Leaves crowded into small rosettes, elliptic to ovate or obovate, extremely fleshy, *c.* 10 × mm, pale green with shining papillae. Flowers bell-shaped, white, *c.* 1 cm across. Petals oblong-obovate, obtuse, mucronate, erect below, wide-spreading above. *SW USA*. H4. Summer–autumn.

55. S. cockerellii Britton. Illustration: Clausen, Sedum of North America, 189 (1975).

Evergreen perennial herb to 18 cm, with hairless or papillose shoots. Leaves alternate, on sterile shoots crowded, basal leaves spathulate, acute, *c.* 12 × 6 mm, stem-leaves linear-lanceolate, rather acute, 1.2–2.5 cm × 3–6 mm, green, partly becoming red. Flowers white, *c.* 1.2 cm across. Petals narrowly lanceolate, acute. *USA (California to New Mexico & Texas)*. H3. Late summer.

56. S. bellum Rose (*S. aleurodes* Bitter; *S. farinosum* Rose). Illustration: *Contributions from the US National Herbarium* **13**: pl. 54 (1911); *Journal of the Royal Horticultural Society* **46**: 142 (1920–1921).

Tufted, perennial herb, 7–15 cm. Leaves alternate, in dense rosettes, spathulate, broadest just below the rounded tip, 2–3.5 cm × 8–10 mm, light green, mealy white bloomed. Flowers white, *c.* 1.2 cm across. Petals lanceolate, spreading. *Mexico*. G1. Spring.

57. S. versadense Thompson (*S. chontalense* Alexander). Illustration: *Journal of the Royal Horticultural Society* **46**: 143 (1920–1921); Clausen, Sedum of the trans-Mexican volcanic belt, 222 (1959).

Evergreen, tufted perennial with densely hairy stems, 7–15 cm. Leaves alternate, spathulate or obovate, *c.* 2.5 × 1.2 cm, very fleshy, downy, green, red at tip and edges. Upper leaves in flowering-shoots smaller, more distant, almost acute, hairless. Flowers pale pink, 1.2–1.8 cm across. Petals elliptic, mucronate. *Mexico.* G1. Spring–late summer.

58. S. greggii Hemsley (*S. diversifolium* Rose). Illustration: *Journal of the Royal Horticultural Society* **46**: 145 (1920–1921); Clausen, Sedum of the trans-Mexican volcanic belt, 252 (1959); *Gartenschönheit* **8**: 290 (1927).

Perennial herb, 7–15 cm. Basal leaves in rosettes, papillose. Leaves alternate, elliptic or oblanceolate, *c.* 6 mm or more. Flowers sulphur yellow, *c.* 1 cm across. Petals elliptic or ovate, acute. *Mexico.* G1. Winter–spring.

59. S. alamosanum S. Watson. Illustration: *Journal of the Royal Horticultural Society* **46**: 134 (1920–1921).

Low perennial herb, glaucous, partly papillose; sterile stems *c.* 1 cm; flowering-stems 7–12 cm, dying after flowering, replaced by branches from base. Leaves alternate, densely adpressed, linear-oblong, *c.* 6 mm, grey-green. Flowers few, reddish white, *c.* 1 cm across. Petals broadly lanceolate or elliptic, acute, wide-spreading, whitish or pinkish translucent. *NW Mexico.* G1. Winter–spring.

60. S. divergens S. Watson. Illustration: *Journal of the Royal Horticultural Society* **46**: 220 (1920–1921); Clausen, Sedum of North America, 303 (1975).

Perennial hairless herb, 7–10 cm. Leaves opposite, stalkless, thick, almost spherical, obovate to spathulate, rounded at apex, *c.* 6 × 5 mm, green or reddish green. Flowers yellow, *c.* 1.8 cm across. Petals elliptic-lanceolate, acute or obtuse. *W North America.* H4. Summer.

61. S. mexicanum Britton. Illustration: *Journal of the Royal Horticultural Society* **46**: 231 (1920–1921).

Evergreen hairless perennial herb, stems erect, ascending, to 15 cm. Leaves usually in whorls of 3–5, rarely alternate except at top of flowering-stem. Leaves narrowly linear, obtuse, 6–15 × 2–3 mm, with a light green shining tint. Flowers nearly stalkless, golden yellow, 1–1.3 cm across. Petals elliptic-lanceolate, acute. *Mexico.* G1. Spring–summer.

62. S. chauveaudii Hamet. Illustration: *Journal of the Royal Horticultural Society* **46**: 225 (1920–1921).

Evergreen creeping perennial herb, stems red, leaves and sepals sparsely fine papillose. Leaves oblanceolate, obtuse, 1.2–2 cm × *c.* 5 mm, light green. Flowers yellow, streaked red on back, *c.* 1.5 cm across. *W China (Yunnan).* H4. Summer–autumn.

63. S. sarmentosum Bunge. Illustration: *Journal of the Royal Horticultural Society* **46**: 226 (1920–1921).

Hairless perennial herb with long creeping stems, shoots dying in winter. Leaves usually in 3s, stalkless, elliptic, usually acute, flat, 1.3–2.5 cm × 4–6 mm, pale green. Flowers stalkless, bright yellow, 6–15 mm across. Petals linear-lanceolate, narrowly acute, wide-spreading. *Japan, Korea, N China.* H3. Summer.

64. S. lineare Murray. Illustration: *Journal of the Royal Horticultural Society* **46**: 228 (1920–1921).

Evergreen hairless perennial herb with reddish stems, erect or creeping. Leaves in whorls of 3, linear or linear-lanceolate, nearly acute, rounded on back, *c.* 2.5 cm × 3 mm, pale green. Flowers bright yellow, stalkless, stellate, *c.* 1.6 cm across. Petals narrowly lanceolate, very acute. *Japan.* H5. Spring–summer.

'Robustum'. Illustration: *Journal of the Royal Horticultural Society* **46**: 230 (1920–1921). Plant grey-green, stouter and more branched with paler flowers and broader petals.

'Variegatum'. Leaves with a broad whitish margin.

65. S. purdyi Jepson (*S. spathulifolium* Hooker subsp. *purdyi* Jepson), Illustration: Clausen, Sedum of North America, 469 (1975).

Perennial herb with prostrate or creeping stems, 7–10 cm. Leaves in compact flat rosettes, spathulate, to 2 cm, papillose on margins. Leaves of erect stems elliptic, obtuse, green. Flowers bright yellow to white, *c.* 1.5 cm across. Petals oblong to lanceolate, wide-spreading. *USA (S Oregon, N California).* H5. Spring–early summer.

66. S. oreganum Torrey & A. Gray (*S. obtusatum* misapplied). Illustration: *Journal of the Royal Horticultural Society* **46**: 242 (1920–1921); Clausen, Sedum of North America, 346 (1975).

Evergreen creeping hairless perennial, 5–15 cm. Leaves alternate, wedge-shaped to spathulate, obtuse, 1.2–1.8 × *c.* 1 cm, very fleshy, shining green, often reddened. Flowers yellow, 1.2–1.5 cm across. Petals narrowly lanceolate, long acuminate, ascending. *USA (Alaska to N California).* H3. Summer.

67. S. spathulifolium J.D. Hooker. Illustration: *Journal of the Royal Horticultural Society* **46**: 239 (1920–1921); Clausen, Sedum of North America, 440 (1975); *The Plantsman* **8**: opposite page 1 (1986).

Evergreen hairless perennial herb, 5–15 cm. Leaves in dense flat rosettes, obovate to spathulate, obtuse, *c.* 2.5 × 1 cm, fleshy, greyish green, often tinged red, white on back. Leaves of flowering-shoots alternate, stalkless, oblong, very fleshy. Flowers yellow, to 1.5 cm across. Petals lanceolate, acute, wide-spreading. *N America (British Columbia to California).* H1. Summer.

Var. **majus** Praeger. Illustration: *Journal of the Royal Horticultural Society* **46**: 239 (1920–1921). Rosettes twice as large; leaves longer and broader, more mucronate, green, scarcely glaucous, not suffused with red. The name is misapplied to large glaucous or mealy plants.

'Purpureum' (illustration: Fessler, Die Freilandschmuckstauden, edn 3, 587, 1985) has large rosettes, 3.5–5 cm across, deep purple leaves except when young, when they are white and mealy.

'Cape Blanco' (illustration: Hay & Synge, The dictionary of garden plants in colour, pl. 203 (1969) – as 'Capa Blanca') has rosettes with silver-white mealy leaves.

68. S. obtusatum A. Gray subsp. **obtusatum** (*S. rubroglaucum* Praeger). Illustration: Clausen, Sedum of North America, 359 (1975); Brickell (ed.), RHS A–Z encyclopedia of garden plants, 972 (2003).

Evergreen hairless prostrate perennial, 6–15 cm. Leaves on the sterile stems in loose rosettes, opposite, spathulate to obovate or wedge-shaped, obtuse or sometimes notched, 1.2–2.5 cm × 6–12 mm, entire, bluish green, in age orange-red, leaves on flowering shoots alternate. Flowers bright or pale yellow, 6–10 mm across. Petals lanceolate, erect and partly fused below, somewhat spreading above, less than 3 times sepals. *USA (California).* H3. Summer.

69. S. humifusum Rose. Illustration: *Contributions from the US National Herbarium* **13**: pl. 55 (1911); *Journal of the Royal Horticultural Society* **46**: 244 (1920–1921); *Gartenschönheit* **11**: 199 (1930); Die natürlichen Pflanzenfamilien **18a**: 441 (1930).

Creeping, evergreen perennial herb, stems to 2.5 cm. Leaves obovate, densely overlapping, 2–3 mm, light green, later reddish, with marginal hairs. Flowers solitary, bright yellow. *Mexico*. G1. Spring–summer.

70. S. muscoideum Rose (*S. cupressoides* misapplied). Illustration: *Journal of the Royal Horticultural Society* **46**: 245 (1920–1921).

Evergreen perennial herb, branches like a cypress. Leaves alternate, ovate, *c.* 1.5 mm. Flowers bright yellow. *Mexico*. G1. Summer.

71. S. acre Linnaeus (*S. ukrainae* invalid). Illustration: *Journal of the Royal Horticultural Society* **46**: 246 (1920–1921); Hegi, Illustrierte flora von Mittel-Europa **4**: pl. 140 (1921); Köhlein, Freilandsukkulenten, 144 (1977); Brickell (ed.), RHS A–Z encyclopedia of garden plants, 971 (2003).

Low, creeping, hairless, evergreen, perennial herb, with sterile shoots erect, 2–10 cm. Leaves alternate, crowded, ovoid-triangular to oblong, fleshy, 2–8 mm. Flowers bright yellow, 1–1.5 cm across. Petals lanceolate, acute, wide-spreading. *Europe, W & N Asia, N Africa*. H1. Summer.

'Elegans' has pale silvery shoot-tips and young leaves in spring.

Var. **majus** Masters. Illustration: *Journal of the Royal Horticultural Society* **46**: 247 (1920–1921). Leaves overlapping in 7 rows, pale green. Flowers *c.* 1.5 cm across. Larger and more robust than the type. *Morocco*.'Aureum'. Illustration: Hay & Beckett, Reader's Digest encyclopaedia of garden plants & flowers, edn 2, 657 (1978). Leaves and shoot-tips bright golden yellow in spring; flowers somewhat smaller than in the type.

Var. **krajinae** Domin (*S. krajinae* Domin). Leaves broad triangular-ovate to ovate-roundish, obtuse, 4.5–5 × 3–4 mm, very thick, pale or fresh green. Flowers yellow, 1.2–1.5 cm across. *Slovakia*. H2. Summer.

72. S. oaxacanum Rose. Illustration: *Journal of the Royal Horticultural Society* **46**: 250 (1920–1921); Jacobsen, Handbook of succulent plants **2**: 790 (1960).

Evergreen, creeping, small hairless perennial with reddish stems. Leaves alternate, obovate to oblong-elliptic, obtuse, *c.* 6 × 3 mm, thick, green, often glaucous. Flowers stalkless, yellow, *c.* 7.5 mm across. Petals nearly oblong, very broad below tip. *Mexico*. G1. Spring.

73. S. alpestre Villars. Illustration: Hegi, Illustrierte flora von Mittel-Europa **4**: 536, 539 (1921); 't Hart, Biosystematic studies in the Acre-group and the series Rupestria Berger of the genus Sedum L., 36 (1978).

Dwarf tufted perennial, hairless. Leaves crowded in a rosette, obovate to narrowly oblong, *c.* 6 mm, bright green. Flowers green-yellow, often tinged with red, *c.* 4 mm across, petals ascending. *Mountains of C & S Europe, Turkey*. H3. Summer.

74. S. sexangulare Linnaeus (*S. mite* Gilibert; *S. boloniense* Loiseleur). Figure 87 (3), p. 43. Illustration: *Journal of the Royal Horticultural Society* **46**: 265 (1920–1921); Reichenbach, Icones florae Germanicae et Helveticae **23**: t. 57 (1897).

Evergreen hairless perennial herb with creeping stems, 5–15 cm. Leaves usually arranged in 6 rows, terete, linear-oblong, cylindric, obtuse, 4–7 mm, bright green. Flowers yellow, 5–9 mm across. Petals narrowly lanceolate, acute, wide-spreading. *C Europe to Finland, W France to Romania & Ukraine*. H1. Summer.

75. S. multiceps Cosson & Durieu. Illustration: *Journal of the Royal Horticultural Society* **46**: 264 (1920–1921); Jacobsen, Handbook of succulent plants **2**: f. 1027 (1960); *Gardener's Chronicle* **6**: 204 (1876) & **10**: 717 (1878).

Partly deciduous hairless bushy perennial, 5–15 cm. Leaves alternate, very densely arranged, linear-oblong, obtuse, glaucous, papillose, 6 mm with a flat upper surface. Flowers stalkless, yellow, *c.* 1.3 cm across. Petals oblong-lanceolate, mucronate, wide-spreading. *Algeria*. H5–G1. Summer.

76. S. japonicum von Siebold. Illustration: *Gartenflora* **15**: t. 513 (1866).

Evergreen perennial herb to 15 cm, with ascending branches. Leaves very variable, linear-oblong, *c.* 6 mm. Flowers yellow, *c.* 1.2 cm across. Petals oblong-lanceolate, acuminate. *Japan, China*. H3. Summer.

77. S. stenopetalum Pursh (*S. douglasii* J.D. Hooker; *S. himalense* misapplied).

Illustration: *Gartenflora* **21**: t. 741 (1872); *Journal of the Royal Horticultural Society* **46**: 258 (1920–1921); Clausen, Sedum of North America, 265 (1975).

Evergreen hairless perennial herb, stems 7.5–15 cm, producing vegetative rosettes in the axils of the leaves of the floral stems. Leaves alternate, stalkless, fleshy, linear-lanceolate, acute, 1.2–1.5 cm, bright green often flushed red. Flowers stalkless, pale-bright yellow, 1.2–1.8 cm across. Petals linear-lanceolate, acute. *C & W North America*. H3. Summer.

78. S. amplexicaule de Candolle (*S. tenuifolium* (Sibthorp & Smith) Strobl). Illustration: *Journal of the Royal Horticultural Society* **46**: 280 (1920–1921); 't Hart, Biosystematic studies in the Acre-group and the series Rupestria Berger of the genus Sedum L., 88 (1978).

Hairless perennial herb, 3–15 cm; leaves withering in summer, plant persisting by shoot tips enclosed in the swollen leaf bases, green for the rest of the year. Branches 2.5–10 cm. Leaves alternate, linear, mucronate, membranous at base, 1–2 cm, bluish green. Flowers golden yellow, 1.5–1.8 cm across, floral parts in 5s–8s. Petals oblong-lanceolate. *S Europe, Algeria, Bulgaria, Turkey*. H3. Late spring–early summer.

79. S. forsterianum Smith (*S. elegans* Lejeune; *S. rupestre* misapplied). Figure 87 (1), p. 43. Illustration: Reichenbach, Iconographia botanica **9**: 1135 (1831–1833); *Journal of the Royal Horticultural Society* **46**: 266 (1920–1921); 't Hart, Biosystematic studies in the Acre-group and the series Rupestria Berger of the genus Sedum L., 83 (1978).

Perennial herb. Leaves in dense rosette-like cones at tips of the shoots, linear-elliptic, 1–1.5 cm, green or glaucous. Flowers yellow, *c.* 1.6 cm across. Petals lanceolate-elliptic. *W Europe, Morocco*. H3. Summer.

This species is closely related to *S. rupestre*. Variants with glaucous leaves are sometimes distinguished as subsp. **elegans** (Lejeune) E.F. Warburg.

80. S. rupestre Linnaeus (*S. rupestre* subsp. *reflexum* (Linnaeus) Hegi & Schmid; *S. reflexum* Linnaeus). Illustration: *Journal of the Royal Horticultural Society* **46**: 266, 269 (1920–1921); Hegi, Illustrierte flora von Mittel-Europa **4**: pl. 140 (1921); 't Hart, Biosystematic studies in the Acre-group and the series Rupestria Berger of the genus Sedum L., 100 (1978); *The Plantsman* **8**: 17 (1986).

Evergreen hairless creeping herb, 15–30 cm. Leaves alternate, linear, cylindric, acute, 1.2–1.8 cm, green or grey-green. Flowers pale yellow, 1.3–1.8 cm across. Petals lanceolate-elliptic. Very variable species. *W & C Europe to S Norway & S Sweden, former USSR (W Ukraine), Sicilia.* H1. Summer.

'Cristatum'. Illustration: *Journal of the Royal Horticultural Society* **46**: 271 (1920–1921) – as *S. reflexum* var. *cristatum.* A fasciated variant with flattened stems often *c.* 5 cm broad. In this condition it never flowers.

81. S. pruinatum Brotero. Illustration: *Journal of the Royal Horticultural Society* **46**: 278 (1920–1921); 't Hart, Biosystematic studies in the Acre-group and the series Rupestria Berger of the genus Sedum L., 93 (1978).

Evergreen perennial herb, strongly bloomed, to 10 cm. Leaves alternate, stalkless, linear, acute, to 2 cm, grey-green, upper surface flattened. Flowers pale yellow to straw yellow, 1.5–2 cm across. Petals linear, acute, wide-spreading. *N & C Portugal.* H4. Summer.

82. S. lanceolatum Torrey (*S. stenopetalum* misapplied). Illustration: *Gartenflora* **12**: t. 403 (1862) – as *S. rhodiola* var. *lanceolatum; Journal of the Royal Horticultural Society* **46**: 276 (1920–1921); Clausen, Sedum of North America, 212 (1975).

Small, evergreen, hairless perennial. Leaves alternate, stalkless, linear-lanceolate, obtuse or obtusely mucronate, minutely papillose, 1–1.8 cm × 2–3 mm, bluish green or somewhat purplish. Flowers bright yellow, 1.5–2 cm across. Petals lanceolate, acute or acuminate, widely spreading. *C & W North America.* H3. Early summer.

83. S. urvillei de Candolle (*S. hillebrandtii* Fenzl; *S. sartorianum* Boissier; *S. stribrnyi* Velenovsky). Illustration: *Journal of the Royal Horticultural Society* **46**: 249 (1920–1921); 't Hart, Biosystematic studies in the Acre-group and the series Rupestria Berger of the genus Sedum L., 53 (1978).

Hairless perennial herb to 15 cm. Leaves narrow ovate to linear-oblong, semi-terete or sometimes terete, usually obtuse, grey-green, 4–15 mm. Flowers bright yellow, 0.8–1.6 cm across. Petals lanceolate. *E & C*

Balkans south of Carpathians, SW Asia, SW former USSR, Israel, Lebanon, Syria. H4. Summer.

84. S. stevenianum Rouy & Camus (*S. roseum* von Steven). Illustration: *Journal of the Royal Horticultural Society* **46**: 200 (1920–1921).

Tufted hairless perennial herb, 2–5 cm. Leaves opposite, obovate to spathulate, very obtuse, 5–6 × *c.* 1.5 mm, entire, spotted with red. Flowers rose-red outside, white inside, *c.* 6 mm across. Petals ovate-lanceolate, blunt, not widely spreading. *SW Asia, Caucasus.* H3. Summer.

85. S. obtusifolium C.A. Meyer.

Hairless perennial, 15–25 cm, often glandular hairy towards the apex. Leaves opposite, ovate or obovate, obtuse, roundish towards the base, narrowed into a very short leaf-stalk, 6–30 × 4–20 mm. Flowers pink, reddish or white, *c.* 1 cm across. Petals lanceolate, slender-pointed. *SW Asia, Caucasus, Iran.* H2. Late spring–summer.

86. S. spurium Bieberstein (*S. oppositifolium* Sims; *S. stoloniferum* misapplied). Illustration: *Botanical Magazine*, 2370 (1823); *Journal of the Royal Horticultural Society* **46**: 195 (1920–1921); Hegi, Illustrierte flora von Mittel-Europa **4**: 527 (1921), Hay & Beckett, Reader's Digest encyclopaedia of garden plants & flowers, edn 2, 658 (1978).

Evergreen, somewhat papillose perennial herb with creeping stems, 10–20 cm. Leaves opposite, obovate, 2.5–3 × *c.* 1.8 cm, with teeth in upper half, dark green, with marginal hairs. Flowers pink to purple, white or cream, *c.* 1.3 mm across. Petals lanceolate, acute, almost erect. *N Iran, Caucasus, Armenia.*

The most common cultivars are: 'Album Superbum', with white flowers rarely produced; 'Fuldaglut', leaves dark purplish red, flowers carmine pink; 'Purpurteppich', leaves and flowers dark purplish red; 'Roseum Superbum' (illustration: Köhlein, Freilandsukkulenten, 143, 1977), flowers pink; 'Schorbuser Blut' (illustration: Hay & Synge, The dictionary of garden plants in colour, pl. 204, 1969), leaves with red margin or dark brown leaves, flowers dark purplish pink; 'Variegatum', leaves with creamish pink margin, flowers pink.

87. S. stoloniferum Gmelin (*S. ibericum* von Steven; *S. hybridum* misapplied). Illustration: *Journal of the Royal Horticultural Society* **46**: 197 (1920–1921).

Tufted, hairless or somewhat papillose perennial herb with creeping habit, to 15 cm. Leaves opposite, diamond-shaped to spathulate, obtuse, 1–3 cm × 8–18 mm, bright green, scalloped in upper half, mature leaves without marginal hairs. Flowers pink, 1.3–1.8 cm across. Petals lanceolate, acute, wide-spreading. *N Iran, SW Asia, Caucasus.* H3. Summer.

88. S. rhodocarpum Rose. Illustration: *Contributions from the US National Herbarium* **13**: pl. 59 (1911); *Journal of the Royal Horticultural Society* **46**: 201 (1920–1921).

Evergreen, perennial herb with 3-angled stems. Leaves, in 3s, uppermost leaves alternate, spathulate to circular, *c.* 2 cm, light green, entire. Flowers greenish red, *c.* 1.3 mm across. Petals oblong, acute, wide-spreading, later sharply reflexed. *Mexico.* G1. Winter.

89. S. pulchellum Michaux. Illustration: *Botanical Magazine*, 6223 (1876); *Journal of the Royal Horticultural Society* **46**: 205 (1920–1921); Clausen, Sedum of North America, 143, 153 (1975); Köhlein, Freilandsukkulenten, 143 (1977).

Annual or short-lived perennial, 10–15 cm, stems ascending. Leaves alternate, narrow-linear, cylindric, obtuse, 1.3–2.5 cm × 2–5 mm, green. Flowers stalkless, rosy purple, 5–14 mm across. Petals lanceolate, acute. *USA (Virginia to Georgia, Indiana, Missouri, Texas).* H3. Summer.

90. S. ternatum Michaux. Illustration: *Edwards's Botanical Register*, t. 142 (1816–1817); *Botanical Magazine*, 1977 (1818); *Journal of the Royal Horticultural Society* **46**: 160 (1920–1921); Clausen, Sedum of North America, 93 (1975).

Evergreen, smooth or finely papillose perennial herb, 7–15 cm. Leaves of sterile shoots usually in whorls of 3, obovate to spathulate, apex obtuse usually rounded, 1.2–2.5 cm × 6–12 mm, pale green. Flowers stalkless, white, *c.* 1.3 cm across, parts in 4s. Petals oblong-narrowly lanceolate, acute. *USA (New York & New Jersey to Georgia, Indiana, Michigan, Tennessee).* H3. Summer.

91. S. monregalense Balbis. Illustration: Reichenbach, Iconographia Botanica **3**: 438 (1825); Reichenbach, Icones florae Germanicae et Helveticae **23**: t. 64a (1897); *Journal of the Royal Horticultural Society* **46**: 172 (1920–1921).

Tufted perennial herb, 7–15 cm, minutely hairy. Leaves in whorls of 4, opposite below; oblong to spathulate, obtuse, fleshy, papillose at the apex, 3–10 mm, olive-green. Flowers white, 0.6–1 cm across, glandular hairy. Petals triangular, spreading, white on upper surface, pinkish brown beneath. *SW Alps, Italy, Corsica.* H3. Summer.

92. S. magellense Tenore. Illustration: Tenore, Flora Napolitana **1**: 139, f. 1 (1811–1815); *Journal of the Royal Horticultural Society* **46**: 167 (1920–1921).

Evergreen perennial herb, 7.5–10 cm, tufted with hairless, light green rosettes. Leaves alternate or opposite, 6–10 mm, obovate to elliptic-oblong, obtuse, flat, fleshy, light green. Inflorescence raceme-like, 2.5–5 cm, flowers white or whitish, 6–9 mm across. Petals lanceolate, acute, mucronate, wide-spreading. *Italy, Greece, SW Turkey, NW Africa.* H3. Late spring–early summer.

93. S. sediforme (Jacquin) Pau (*S. altissimum* Poiret; *S. nicaeense* Allioni). Illustration: *Journal of the Royal Horticultural Society* **46**: 272 (1920–1921); Jacobsen, Handbook of succulent plants **2**: 798 (1960); Köhlein, Freilandsukkulenten, 161 (1977); 't Hart, Biosystematic studies in the Acre-group and the series Rupestria Berger of the genus Sedum L., 95 (1978).

Evergreen, hairless perennial herb, 15–60 cm. Leaves alternate, oblong-elliptic to linear-elliptic, acuminate, 1–3 cm × 3–5 mm, grey-green. Flowers greenish white, rarely yellow, *c.* 1.3 cm across. Petals lanceolate-elliptic. *Mediterranean area, C France, Portugal.* H3. Spring–summer.

94. S. bithynicum Boissier (*S. hispanicum* Linnaeus var. *minus* Praeger; *S. lydium* invalid; *S. lydium aureum* invalid; *S. glaucum* invalid). Illustration: *Journal of the Royal Horticultural Society* **46**: f. 178b (1920–1921).

Perennial herb, hairless or flower-stems with shaggy glandular hairs, 2–7.5 cm. Leaves alternate, linear to oblong, nearly cylindric, 4–7 mm, pale green or grey-green. Flowers 9–12 mm across. Petals ovate, acuminate, white often with purplish mid-vein and keel. *Balkans, Turkey.* H3. Summer.

Closely related to the annual *S. hispanicum* Linnaeus from S Europe.

95. S. dasyphyllum Linnaeus (*S. corsicum* Duby). Illustration: *Journal of the Royal*

Horticultural Society **46**: 177 (1920–1921); Hegi, Illustrierte flora von Mittel-Europa **4**: pl. 140 (1921); Fessler, Die Freilandschmuckstauden, edn 3, 528, 585 (1985).

Evergreen, glandular hairy herb, sometimes almost hairless, 2–10 cm. Leaves opposite or alternate, stalkless, ovate to elliptic-oblong, 3–12 mm, blue-green. Flowers white, pinkish on back, *c.* 0.9 cm across. *Europe, N Africa.* H4. Summer.

Var. **suendermannii** Praeger. Illustration: *Journal of the Royal Horticultural Society* **46**: 178 (1920–1921). Larger and with densely overlapping, obovate, densely glandular hairy leaves. Flowers somewhat larger and appearing some weeks later.

96. S. gypsicola Boissier & Reuter. Illustration: *Journal of the Royal Horticultural Society* **46**: 187 (1920–1921). Creeping, evergreen perennial herb, with fine dense grey hairs, 5–10 cm. Leaves overlapping at the apices of the stems in 5 spiral rows, ovate to diamond-shaped, obtuse, 6–8 mm, dull grey-green, often tinged red. Flowers white, *c.* 6 mm across. *Spain, Portugal, Morocco.* H4. Early summer.

97. S. hirsutum Allioni. Illustration: *Journal of the Royal Horticultural Society* **46**: 188 (1920–1921); Valdés et al., Flora vascular de Andalucía occidental **2**: 14 (1987).

Evergreen, perennial herb, 5–10 cm, glandular hairy. Leaves alternate in dense rosettes at ends of shoots, oblong, elliptic or slightly spathulate, obtuse, 6–10 × 2–5 mm. Flowers white to pink. *S France, N Italy, Morocco.* H2. Summer.

Subsp. **baeticum** Rouy (*S. winkleri* (Willkomm) Wolley-Dod). Illustration: Evans, Handbook of cultivated sedums, 313, pl. 15 (1983). Taller and stouter, with larger flowers. *S Spain.*

98. S. fragrans 't Hart (*S. alsinefolium* misapplied). Illustration: *Journal of the Royal Horticultural Society* **46**: 170 (1920–1921); Grey-Wilson & Blamey, Alpine flowers of Britain & Europe, 77 (1979).

Perennial hairy herb. Leaves of sterile shoots in a crowded rosette, diamond-shaped to spathulate, obtuse, 6–15 × *c.* 6 mm, dark green. Flowers white, *c.* 6 mm across. Petals oblong-obovate, abruptly acuminate, erect in lower half,

spreading above. *S France, N Italy.* H4. Summer.

99. S. compactum Rose. Illustration: *Contributions from the US National Herbarium* **13**: pl. 53 (1911); *Journal of the Royal Horticultural Society* **46**: 176 (1920–1921); Gartenschönheit **8**: 290 (1927).

Low perennial herb. Leaves in rosettes, obovate, obtuse, *c.* 3 mm, grey-green. Flowers white, small, urn-shaped, strongly scented. Petals ovate, mucronate. *Mexico.* G1. Summer.

100. S. furfuraceum Moran. Illustration: Evans, Handbook of cultivated sedums, 154 (1983).

Creeping hairless herb. Leaves alternate, crowded, ovoid, obtuse, 6–10 mm, dark green to purplish, with uneven grey scaly covering. Flowers white or pinkish, *c.* 1 cm across. *Mexico.* G1. Early spring.

101. S. brevifolium de Candolle. Illustration: *Journal of the Royal Horticultural Society* **46**: 179 (1920–1921); Valdés et al., Flora vascular de Andalucía occidental **2**: 13 (1987).

Evergreen, hairless tufted herb. Leaves opposite, crowded in 4 rows, ovoid or spherical, 3–4 mm, white-bloomed, flushed with red, sometimes with bloom, purplish or yellow-green; alternate on the flowering-stems. Flowers white, 6–8 mm across. Petals ovate, acute, white with reddish midrib. *SW Europe, Morocco.* H4–5. Summer.

Var. **quinquefarium** Praeger has shoots twice as thick as long, leaves in 5 rows, 5–8 mm, white-bloomed.

102. S. oregonense (Watson) Peck (*S. watsoni* (Britton) Tidestrom; *S. rubroglaucum* misapplied). Illustration: Clausen, Sedum of North America, 411, 413 (1975).

Similar to *S. obtusatum* subsp. *obtusatum,* but differs in its larger size, white or creamish flowers, petals more than 3 times length of sepals. *W USA.* H3. Summer.

103. S. diffusum S. Watson (*S. potosinum* Rose). Illustration: *Journal of the Royal Horticultural Society* **46**: 139 (1920–1921); *Sedum Society Newsletter* **7**: 1, 5 (1988).

Evergreen, hairless perennial, 7–25 cm with prostrate stems. Leaves of sterile shoots alternate, dense, oblong to linear, obtuse, almost terete, 3–12 × 1–3 mm. Flowers *c.* 1.5 cm across. Petals white, sometimes flushed red on back. *Mexico.* H5–G1. Spring–autumn.

104. S. liebmannianum Hemsley. Illustration: Rose, *Contributions from the US National Herbarium* **13**: f. 56 (1911); *Journal of the Royal Horticultural Society* **46**: 175 (1920–1921).

Perennial hairless herb, 5–15 cm, stems ascending, nearly deciduous. Leaves alternate, ovate-oblong, obtuse, overlapping, *c.* 5 mm, green, tipped red, forming a silvery persistent covering when dead. Flowers white, stalkless, *c.* 1 cm across. Petals lanceolate, rather acute. *Mexico & Texas.* G1. Summer.

105. S. anglicum Hudson. Illustration: *Journal of the Royal Horticultural Society* **46**: 181 (1920–1921).

Evergreen, hairless perennial herb, 7–8 cm. Leaves alternate, ovate–elliptic, very thick, 4–5 mm, green, often tinged red. Flowers white–pink, *c.* 1 cm across. Petals lanceolate. *Europe.* H2. Summer.

106. S. albertii Regel. Illustration: *Gartenflora* **19**: t. 1019 (1870).

Small, hairless, evergreen perennial herb, 3–5 cm. Roots thickened, cord–like. Leaves alternate, crowded, linear-oblong, nearly cylindric, obtuse, covered with rough papillae, *c.* 6 mm. Flowers white, 9–12 mm across, stalks absent or *c.* 1 mm. Petals oblong–ovate, mucronate, fused at base. Anthers violet. Follicles erect. *Former USSR (W Siberia, Turkestan).* H1. Late spring–early summer.

107. S. lydium Boissier. Illustration: *Journal of the Royal Horticultural Society* **46**: 189 (1920–1921); Brickell (ed.), RHS A–Z encyclopedia of garden plants, 972 (2003).

Evergreen, hairless perennial herb, 3–5 cm. Leaves alternate, crowded, narrowly linear to cylindric, obtuse, 4–8 × *c.* 1 mm, green or red, minutely papillose at tip. Flowers white, *c.* 6 mm across. Petals lanceolate, acute. Follicles erect. *W Turkey.* H3. Summer.

108. S. gracile C.A. Meyer. Illustration: *Journal of the Royal Horticultural Society* **46**: 190, 192 – as *S. albertii* (1920–1921).

Evergreen, perennial herb, 5–10 cm. Leaves alternate, crowded linear-oblong, nearly cylindric, *c.* 8 mm, bright green, sometimes reddish green. Flowers white, often dotted red on back, 7–8 mm across. Follicles spreading. *N Iran, Caucasus, E Turkey.* H3. Early summer.

109. S. album Linnaeus. Figure 87(4), p. 43. Illustration: Oeder, Flora Danica, t. 66 (1763); Reichenbach, Icones florae Germanicae et Helveticae **23**: t. 55 (1897); *Journal of the Royal Horticultural Society* **46**: 183 (1920–1921); Hegi, Illustrierte flora von Mittel-Europa **4**: pl. 140 (1921).

Evergreen, creeping, almost hairless, perennial herb, branches erect, with some glandular hairs, 10–20 cm. Leaves alternate, cylindric to ovate, 6–12 mm, obtuse, dark green, often reddish. Flowers white, 7–12 mm across. Petals lanceolate to ovate, obtuse. *Europe, N Africa, C Asia.* H3. Summer.

Var. **micranthum** (de Candolle) de Candolle. Illustration: Hegi, Illustrierte flora von Mittel-Europa **4**: 536 (1921). Inflorescence has many small flowers.

The following cultivars are good garden plants: 'Coral Carpet', as ground cover with, in winter, greenish and, in summer, bronze-red leaves; 'Laconicum', with partly green and partly red-brown leaves, petals somewhat broader than in the species; 'Micranthum Chloroticum' is *c.* 5 cm high with bright green leaves, flowers greenish white; 'Murale' has brownish red leaves and pale rose flowers.

110. S. tenellum Bieberstein.

Low, hairless perennial herb. Leaves on sterile stems alternate, crowded, oblong to nearly circular, *c.* 3 mm. Leaves on flowering-stems linear-oblong, cylindric, obtuse, *c.* 6 mm. Flowers white, tinged red, 6–8 mm across. Sepals spurred at base. Follicles erect. *Armenia, N Iran, Caucasus.* H3. Summer.

111. S. nevii A. Gray. Illustration: Clausen, Sedum of North America, 110 (1975).

Evergreen, tufted, hairless herb, 3–10 cm. Sterile shoots forming rather dense rosettes, *c.* 1.8 cm across. Leaves of flowering-shoots alternate, oblanceolate or linear-lanceolate, mostly nearly cylindric, 1.2–1.8 cm × *c.* 4 mm, entire, pale brownish green, ratio of width to thickness less than 1.7, glaucous. Flowers white, to *c.* 1.3 cm across. Petals linear–lanceolate, acute. *USA (Virginia to Alabama).* H1. Summer.

112. S. glaucophyllum R.T. Clausen (*S. nevii* misapplied). Illustration: Clausen, Sedum of North America, 123, 133 (1975).

Perennial herb, 2–10 cm. Leaves of sterile stems in dense rosettes, obovate to spathulate, obtuse or nearly acute, *c.* 1.5 × 1 cm; flowering-stems *c.* 2 cm, leaves oblanceolate, *c.* 2.5 cm × 6 mm, pale or blue-green, ratio of width to thickness more than 2. Flowers stalkless, white, to *c.* 1.2 cm across. Petals lanceolate, acuminate, widely spreading. *E North America.* H3. Late spring–summer.

Some plants with narrow green leaves closely resemble *S. nevii* and are best distinguished by the leaf width to thickness ratio of flowering-shoots. The correct application of *S. beyrichianum* Masters and *S. nevii* var. *beyrichianum* (Masters) Praeger is uncertain; the names are applied to both species in gardens.

7. RHODIOLA Linnaeus

L.S. Springate

Perennial herbs usually with a stout, very short, fleshy, perennial stem, sometimes branched above, often partly, covered in brown triangular scales; branches simple, annual, leafy with terminal flower-heads developing laterally below each terminal bud on the perennial stem. Sometimes perennial stem lengthening, slender, much-branched, forming tufts or shrublets, branches crowned with leaves distinct from those on the annual stems and scales present or absent. Floral parts usually in 4s or 5s. Flowers solitary or few to many in spherical, corymb-like, raceme-like or paniculate heads, bisexual or functionally unisexual, the plants then usually dioecious. Stamens twice as long as petals, absent in female flowers. Carpels present, opposite the petals in female flowers, reduced and opposite the sepals in male flowers.

A genus of 50 species centred on the mountains of E Asia, extending to N America and Europe. The species listed can be grown in pots if protected from winter rain or in raised beds; the coarser species can also be grown in open borders. Flowering may be poor in zones H4 and H5 at low latitudes.

Literature: Praeger, R.L., An account of the genus Sedum as found in cultivation, *Journal of the Royal Horticultural Society* **46** (1920–1921); Clausen, R.T., Sedum of North America north of the Mexican Plateau (1975); Ohba, H., Rhodiola in H. Ohashi, Flora of the eastern Himalaya, 3rd Report, 283–362 (1975); Ohba, H., A revision of the Asiatic species of Sedoideae (Crassulaceae), *Journal of the Faculty of Science of the University of Tokyo, Section 3*, **12**: 337–405 (1980); **13**: 65–119,

121–169 (1981–1982); Evans, R.L., Handbook of cultivated sedums (1983).

Perennial stems. Long and slender: **2, 3**; stout, but lengthening: **13, 14**; compact, stout: **1, 4, 5, 6, 7, 8, 9, 10, 11, 12**.

Dead annual stems. Persistent, dense, in twiggy tufts: **13, 14**.

Flowers. Bisexual: **1, 2, 3, 4, 5, 6, 7**; unisexual: **8, 9, 10, 11, 12, 13, 14**; bell-shaped: **5**.

Petals. White or cream to pale green: **2, 3, 4, 5, 6, 7, 10**; yellow to yellowish green: **8, 10, 12, 14**; pink or reddish to reddish green or brown: **1, 6, 7, 9, 10, 12**; dark red or purplish: **11, 13, (14)**; erect or ascending: **1, 2, 4, 6, 7, 13, 14**; and female flowers: **10, 12**; spreading: **3, 4, 6, 8, 9, 10, 11, 12**.

1a. Perennial stems long, slender, much-branched, forming tufts, or shrublets   2
b. Perennial stem solitary, stout, short or if long then not slender   3
2a. Leaves of perennial stems with distinct stalks more than 4 mm   **2. primuloides**
b. Leaves of perennial stems with indistinct stalks to 3 mm   **3. pachyclados**
3a Perennial stem (caudex) with stalked green leaves; filaments fixed to back of anthers   **1. hobsonii**
b. Perennial stem (caudex) without green leaves (triangular scales sometimes green at first); filaments fixed to base of anthers   4
4a. Inflorescence raceme-like   5
b. Inflorescence corymb-like or head-like   6
5a Leaves linear, to 7 cm; petals 7–8.5 mm   **6. semonovii**
b. Leaves oblong, to 2.5 cm; petals 8–13 mm   **7. rhodantha**
6a. Petals erect, more than 7.5 mm   7
b. Petals spreading or less than 7.5 mm   8
7a. Leaves linear, less than 3 mm wide   **5. dumulosa**
b. Leaves oblong, more than 3 mm wide   **7. rhodantha**
8a. Inflorescence diffuse; leaf length less than twice width   **11. bupleuroides**
b. Inflorescence compact, all flowers touching, or leaf length at least twice width   9
9a. Petals 5–11 mm; flowers bisexual   **4. wallichiana**
b. Petals to 5 mm; flowers unisexual   10

10a. Leaves linear-lanceolate or linear   11
b. Leaves broader   12
11a. Leaves to 1.2 cm   **14. quadrifida**
b. Leaves more than 2.5 cm   **12. kirilowii**
12a. Leaves with papillae above   **13. himalensis**
b. Leaves smooth   13
13a. Inflorescence very condensed; flower-stalks partly fused, to 1 mm   **10. heterodonta**
b. Inflorescence corymb-like; flower-stalks entirely separate, 1–4 mm   14
14a. Petals red; leaves hardly glaucous   **9. integrifolia**
b. Petals yellowish; leaves usually glaucous   **8. rosea**

1. R. hobsonii (Hamet) Fu (*Sedum hobsonii* Hamet). Illustration: Praeger, Account of the genus Sedum in cultivation (1921); Evans, Handbook of cultivated sedums, 75, pl. 1 (1983).

Plant to 12 cm, hairless with two distinct types of leaf. Perennial stem massive, short, crowned by erect, elliptic, sometimes persistent, green leaves 7.5–10 × 3–4 mm, with stalks 7–10 mm. Annual stems spreading, 5–15 cm, with many lanceolate to spathulate, stalkless green leaves, 6–15 × 2–4.5 mm and a cluster of 3–10 flowers. Flowers bisexual, 7–9 mm across, floral parts usually in 5s. Petals elliptic, erect, hooded, rose, 5.5–7.5 × 2–3 mm. Filaments attached to back of anthers. *Bhutan & China (Xizang).* H3. Late summer.

2. R. primuloides (Franchet) Fu (*Sedum primuloides* Franchet). Illustration: *Journal of the Royal Horticultural Society* **46**: 70 (1920–1921); Evans, Handbook of cultivated sedums, 75, pl. 1 (1983).

Plant to 10 cm, hairless. Perennial stem usually long, much-branched, occasionally short, compact; branches sometimes rooting, crowned by rosettes of leaves; leaves bright green, sometimes persistent, circular to oblong to 1 cm × 9 mm, leaf stalks to 9 mm, clasping stem at base. Annual stems to 5 cm, with 1–4 flowers and leaves not clasping. Flowers bisexual, 6–9 mm across; floral parts in 5s. Petals erect, elliptic, white, 5–6.5 × 2.5–3.5 mm. *SW China.* H3. Late summer.

3. R. pachyclados (Aitchison & Hemsley) Ohba (*Sedum pachyclados* Aitchison & Hemsley). Illustration: *Sedum Society Newsletter* **6**: 1 (1988).

Plant to 10 cm, hairless with 2 distinct types of leaf. Perennial stem much-branched, rhizomatous below, forming dense tufts; branches crowned by rosettes of persistent leaves; leaves ovate to obovate with some wavy teeth near the apex, 4–9 × 3–6 mm. Annual stems to 4 cm with 1–10 flowers and leaves obovate, entire, 4.5 cm × 1.5–2 mm. Flowers bisexual; floral parts in 5s. Petals spreading, elliptic, concave, white, 5.5–7 × 2–3 mm. *Afghanistan & Pakistan.* H3. Early summer.

4. R. wallichiana (Hooker) Fu (*Sedum wallichianum* Hooker; *R. asiatica* invalid; *S. crassipes* Hooker & Thomson; *S. crassipes* var. *cholaense* Praeger). Illustration: *Journal of the Royal Horticultural Society* **46**: 59 (1920–1921).

Plant to 30 cm, hairless. Perennial stem massive, short, sometimes with underground stolons. Annual stems erect, with many leaves; leaves linear-lanceolate, upper half with distant teeth, green, 1.2–3 cm × 1–6 mm. Inflorescence dense, corymb-like. Flowers bisexual; floral parts usually in 5s. Petals usually spreading, elliptic, concave, yellowish white to pale greenish, 5–11 × 1.5–2.5 mm. *Kashmir to Sikkim, China (Xizang).* H3. Summer.

Cultivated plants with a few characteristics found in *R. cretinii* (Hamet) Ohba but not in *R. wallichiana* (e.g. calyx-lobes less than 5 mm, petals less than 6.5 mm and stoloniferous stems) are included in *R. wallichiana* here. Illustration: *Journal of the Royal Horticultural Society* **46**: 56, 57 (1920–1921); Evans, Handbook of cultivated sedums, 66, pl. 1 (1983).

5. R. dumulosa (Franchet) Fu (*Sedum dumulosum* Franchet). Illustration: *Journal of the Royal Horticultural Society* **46**: 62 (1920–1921); Evans, Handbook of cultivated sedums, 67, pl. 1 (1983).

Plant to 18 cm, hairless. Perennial stem massive, short. Annual stems erect with very many leaves; leaves linear with lower surface convex, green, 1–2.5 cm × 1.5–2.5 mm. Inflorescence compact, corymb-like with 4–20 flowers. Flowers bisexual, bell-shaped; floral parts usually in 5s. Petals oblong, greenish white to cream, 7.5–12 × 2–2.5 mm, with mucro 0.5–1.5 mm. *Bhutan to Manchuria & N Korea.* H3. Summer.

6. R. semonovii (Regel & Herder) Borissova (*Sedum semonovii* (Regel & Herder) Masters; *Clementsia semonovii* (Regel & Herder) Borissova). Illustration:

Journal of the Royal Horticultural Society **46**: 66 (1920–1921); Evans, Handbook of cultivated sedums, 82, pl. 2 (1983).

Plant to 60 cm, hairless, perennial stem massive, short. Annual stems erect with very many leaves; leaves linear, entire or with remote teeth, green; the middle leaves largest, to 7 cm × 3 mm. Inflorescence raceme-like, dense. Flowers bisexual; floral parts usually in 5s. Calyx-lobes 4–5.5 mm. Petals lanceolate, partly spreading, greenish white to pink, 7–8.5 mm. Carpels as long as petals at flowering. *Central Asia*. H3. Summer.

7. R. rhodantha (Gray) Jacobsen (*S. rhodanthum* Gray; *Clementsia rhodantha* (Gray) Rose). Illustration: *Journal of the Royal Horticultural Society* **46**: 68 (1920–1921); Clausen, Sedum of North America, 476 (1975); Evans, Handbook of cultivated sedums, 78, pl. 1 (1983).

Plant to 40 cm, hairless, perennial stem massive, short. Annual stems erect with many leaves; leaves oblong, entire or with obscure teeth near apex, green, 1.5–2.5 cm × 5–6 mm. Inflorescence raceme-like, sometimes condensed and almost capitate. Flowers bisexual, floral parts in 5s. Calyx-lobes 5–8 mm. Petals erect, elliptic, rose to almost white, 8–13 mm. Carpels usually distinctly shorter than petals at flowering. *USA (Montana to Arizona)*. H3. Summer.

8. R. rosea Linnaeus (*Sedum rosea* (Linnaeus) Scopoli; *S. rhodiola* de Candolle; *R. elongata* (Ledebour) Fischer & Meyer; *R. roanensis* (Britton) Britton). Illustration: *Journal of the Royal Horticultural Society* **46**: 29 (1921); Clausen, Sedum of North America, 518 (1975); Evans, Handbook of cultivated sedums, 79, pl. 1 (1983); Brickell (ed.), RHS A–Z encyclopedia of garden plants, 888 (2003).

Plant 5–30 cm, hairless, dioecious. Perennial stem massive, short. Annual stems few, stout with many leaves, leaves broad-ovate to narrow-oblanceolate, entire or with irregular teeth towards apex, glaucous, 7–40 × 5–17 mm. Inflorescence compact, convex, with many flowers; flower-stalks 1–4 mm; floral parts most often in 4s. All parts yellow or greenish yellow in male flowers, the petals narrow-oblong, concave, spreading, 2.5–3.5 cm × 0.7–1.2 mm. Petals greenish yellow, linear-subulate, 2–2.5 mm in female flowers, the carpels green. *Europe, N Asia, N America, Greenland*. H3. Early summer.

Widespread and very variable; dwarf cultivated plants (illustration: Evans, Handbook of cultivated sedums, pl. 2, 1983) are sometimes incorrectly referred to a race from the Palaeo-Arctic islands as *R. arctica* Borissova, *R. rosea* subsp. *arctica* (Borissova) Löve, *Sedum arcticum* (Borissova) Ronning, or *S. rosea* 'Arcticum' invalid.

9. R. integrifolia Rafinesque (*Sedum integrifolium* (Rafinesque) Nelson; *R. atropurpurea* (Turczaninow) Trautvetter & Meyer; *S. rosea* var. *atropurpureum* (Turczaninow) Praeger). Illustration: *Journal of the Royal Horticultural Society* **46**: f. 5C (1920–1921); Clausen, Sedum of North America, 488, 507 (1975); Evans, Handbook of cultivated sedums, 73, pl. 1 (1983).

Very similar to *R. rosea* in appearance, but differs in having petals usually red, in male flowers spathulate, 2.5–5 × 1.1–1.7 mm, leaves green, not or hardly glaucous and floral parts more often in 5s than in 4s. *Arctic Siberia to Sakhalin & N America*. H4. Early summer.

10. R. heterodonta (Hooker & Thomson) Borissova (*Sedum heterodontum* Hooker & Thomson). Illustration: *Journal of the Royal Horticultural Society* **46**: 35 (1920–1921); Evans, Handbook of cultivated sedums, 69, pl. 1 (1983); Brickell (ed.), RHS A–Z encyclopedia of garden plants, 888 (2003).

Plant to 40 cm, dioecious, hairless. Perennial stem massive, short. Annual stems few, stout, erect, growth continuing after flowering; leaves distant; leaves ovate, often clasping the stem, nearly entire or with few sharp teeth, green or glaucous, 1.2–2.5 × 1–1.5 cm. Flowers numerous, densely packed in a hemispheric head; stalks absent or to 1 mm; floral parts usually in 4s. Male flowers with petals narrowly elliptic, 3.5–4.5 mm; filaments 5.5–9 mm, very prominent. Female flowers with petals linear, 2–2.5 mm. Petals and filaments reddish, yellowish or greenish; anthers red. *Afghanistan & Pamirs to Nepal*. H3. Late spring.

Male clones with whole flower-head appearing brick-red or purplish are most often grown.

11. R. bupleuroides (Hooker & Thomson) Fu (*Sedum bupleuroides* Hooker & Thomson; *S. elongatum* Hooker & Thomson; *S. bhutanense* Praeger). Illustration: *Journal of the Royal Horticultural Society* **46**: 42, 44, 45, 48 (1920–1921); Evans,

Handbook of cultivated sedums, 65, pl. 1 (1983).

Plant 7–90 cm, dioecious, hairless. Perennial stem massive, short. Annual stems few, slender, erect or ascending, with distant leaves; leaves broadly ovate to narrowly elliptic or obovate, cordate to tapered at base, entire or with few shallow teeth, green or somewhat glaucous, 3–95 × 4–45 mm. Flowers usually numerous in loose, open heads; floral parts usually in 5s, dark purplish. Male flowers with petals oblanceolate, concave, widespread, 3–4 mm; stamens 3–4 mm. Female flowers with petals linear, 1.5–3 mm. *Nepal to China (Yunnan)*. H3. Early summer.

12. R. kirilowii (Regel) Maximowicz (*Sedum kirilowii* Regel; *S. longicaule* Praeger; *S. kirilowii* var. *rubrum* Praeger). Illustration: *Journal of the Royal Horticultural Society* **46**: 36, 38, 40 (1920–1921); Evans, Handbook of cultivated sedums, 74, pl. 1 (1983).

Plant to 90 cm, dioecious, hairless. Perennial stem massive, short. Annual stems few, stout, erect, growth continuing after flowering, with very many leaves; leaves linear-lanceolate or linear, entire to sharply toothed throughout, green, 4–9 cm × 3–6 mm. Flowers numerous in dense, corymb-like heads; flower-stalks 3–4.5 mm; floral parts in 5s, yellowish green or brownish red. Male flowers with petals oblanceolate, 3–5 mm, stamens 4–6 mm. Female flowers with linear petals 2.5–3 mm. *Central Asia, N China*. H3. Late spring.

13. R. himalensis (Don) Fu (*Sedum himalense* Don). Illustration: *Journal of the Faculty of Science, University of Tokyo, Section 3*, **13**: 134, 136 (1982).

Plant to 50 cm, dioecious, hairless. Perennial stem massive, ascending, to 40 cm exposed. Annual stems many, erect, reddish, with papillae, persisting when dead, to 30 cm with many leaves; leaves elliptic to oblanceolate, entire or with shallow teeth towards apex, green, upper surface with papillae, 6–20 × 2.5–7 mm. Inflorescence corymb-like, to 50 flowers; male head dense, female loose; the floral parts usually in 5s. Petals ovate-oblong, deep red, 2.5–4.5 × 1.5–2.5 mm. Stamens 0.8–3 mm. *Nepal to China*. H3. Early summer.

Cultivated plants (illustration: Evans, Handbook of cultivated sedums, 70, pl. 1, 1983) differ in some features.

14. R. quadrifida (Pallas) Fischer & Meyer (*Sedum quadrifidum* Pallas).

Plants 3–15 cm, dioecious, hairless. Perennial stem massive, eventually well exposed and branched above. Annual stems many, erect with very many leaves; leaves linear, lower surface convex, entire, 5–12 × 1–1.5 mm. Inflorescence corymb-like with 6–16 flowers, floral parts usually in 4s, mainly yellow. Petals oblong, 2.5–4 × 1.5–2 mm. Stamens as long as petals. Nectaries and fruits deep red. *Arctic Siberia to C Asia & Mongolia*. H3. Summer.

Cultivated plants and those from Pakistan differ in having all flower parts deep red, stems or leaf margins sometimes with papillae and stamens 0.8–2.8 mm, and may be an undescribed species. Cultivated plants incorrectly named *R. fastigiata* (Hooker & Thomson) Fu or *Sedum fastigiatum* Hooker & Thomson (illustration: Evans, Handbook of cultivated sedums, 68, pl. 1, 1983) are rather similar.

8. × SEDEVERIA Walther

G.D. Rowley

Shrubby or stemless perennial, evergreen leaf-succulents. The many, small flowers are borne on lateral inflorescences, and yellow is the dominant colour. They are broadly funnel-shaped with the 5 petals shortly united at the base. Their diverse habit reflects their origin as hybrids between different species of the large genera *Echeveria* and *Sedum*. Only the 2 best known are covered here.

la. Stems pendent, clothed all along with leaves; flowers inverted at the shoot tips
    1. × S. 'Harry Butterfield'
b. Stems prostrate, leafy only towards the ends; flowers borne on erect scapes   **2. × Shummelii**

1. × S. 'Harry Butterfield'; *?Echeveria derenbergii* × *Sedum morganianum* ('Super Burro's Tail'). Illustration: *British Cactus and Succulent Journal* **5**: 11 (1987).

Stems *c.* 5 mm thick, long, pendent, branching from near the base, densely packed throughout with pale green, almost terete, pointed, glaucous, very succulent leaves, 2.5 cm, of a pale whitish green colour. The flowers resemble those of *Sedum morganianum*. G1.

Basket plant with the habit of the 'Burro's Tail', *Sedum morganianum*, p. 40.

2. × S. hummelii Walther; *Echeveria derenbergii* × *Sedum pachyphyllum*

('*S. hummellii*' was a misspelling). Illustration: *Cactus and Succulent Journal* (US) **25**: 20–21 (1953); Jacobsen, Lexicon of succulent plants, t. 124/4 (1974).

Shrublet to 15 cm, recalling *Sedum pachyphyllum*, with prostrate stems 7–8 mm across and densely leafy towards the tips. Leaves narrow elliptic, tapering at both ends, biconvex to almost terete, glaucous pearly green with a sharp red tip, 3.5–5 cm × 8–11 mm. Inflorescence to 15 cm, slender, with scattered bracts, bearing one-sided cymes of *c.* 12 showy bright yellow flowers *c.* 1.5 cm across. Sepals 5, unequal, green, adpressed, 7–8 mm. Petals 1.1–1.2 cm, acuminate. Spring.

A vigorous, almost hardy, free-flowering hybrid, welcomed for its early blooming; × S. 'Alidaea' is apparently the same cultivar. G1–H5.

9. PISTORINIA de Candolle

S.G. Knees

Erect annuals with alternate leaves. Stems hairless below, glandular hairy above. Leaves linear, succulent, stalkless and soon falling. Flowers shortly stalked, numerous in dense corymbs. Sepals 5, petals 5, united into a funnel-shaped corolla, tube long and narrow, lobes spreading or erect. Stamens 10, 5 long, 5 short. Styles slender. Carpels 5. Fruit a many seeded follicle.

A genus of 2 species occurring in the western Mediterranean countries of North Africa and the Iberian peninsula. Propagation is from seed.

1a. Flowers pink, swelling towards base; stems slender   **1. hispanica**
b. Flowers yellow, narrowing towards base; stems robust   **2. brevifolia**

1. P. hispanica (Linnaeus) de Candolle (*Cotyledon hispanica* Linnaeus). Illustration: Polunin & Smythies, Flowers of south-west Europe, pl. 13 (1973); Valdés et al., Flora vascular de Andalucía occidental **2**: 8 (1987).

Stems erect, sometimes branched, 4–15 cm. Leaves 1.2–2.5 cm, oblong, obtuse, covered with glandular hairs. Sepals 1.6–2 mm, narrowly triangular, joined towards the base. Corolla 2–2.5 cm, pinkish purple, swelling towards base. *S Spain, Portugal, NW Africa (Morocco)*. H5. Summer.

2. P. breviflora Boissier. Illustration: Jacobsen, Lexicon of succulent plants, pl.

119 (1974); Valdés et al., Flora vascular de Andalucía occidental **2**: 8 (1987).

Stems erect, rarely branched, 5–12 cm. Leaves almost cylindric, 5–10 mm, acute. Sepals 2–3 mm, linear-lanceolate, free, or joined only at the base. Corolla 1.8–2 cm, yellow, narrowly cone-shaped. *S Spain, NW Africa (Morocco)*. H5. Summer.

10. VILLADIA Rose

S.G. Knees

Succulent, annual or perennial herbs, sometimes woody at base. Stems solitary, trailing or erect. Leaves alternate, cylindric or nearly so. Flowering-stems leafy, terminal. Flowers in racemes or spikes (Section **Villadia**); or in flat cymes of several branches (Section **Altamiranoa**). Sepals 5, equal, erect; petals 5, more or less united; stamens 10; carpels 5, united only at base; styles short and erect.

A genus of 25–30 species of ornamental succulent plants, native of the Americas, from Texas, through Mexico to Peru. Most can be grown outdoors during summer in much of Europe, although they will need glasshouse protection in winter.

Literature: Regnat, H., Die Pflanzenfamilie Crassulaceae: Gattungen der amerikanischen Kontinents, eine Betrachtung: 9 Villadia Rose, *Kakteen und Sukkulenten* **44**: 241–43 (1999).

la. Plant forming short-stemmed mats   2
b. Plant with long, trailing or erect stems   3
2a. Leaves overlapping, closely adpressed to stem; flowers white   **1. imbricata**
b. Leaves spreading; flowers yellow   **2. parva**
3a. Stems not more than 25 cm; leaves 1–1.3 cm; flowers white or pinkish   **3. batesii**
b. Stems usually 35–50 cm; flowers orange, red, yellow, purple, or, if white, leaves only 5–6 mm   4
4a. Leaves 5–6 mm, spreading; flowers white or purplish on stems with 2–5 branches   **4. jurgensii**
b. Leaves 6–25 mm; flowers orange-red or yellowish green in narrow spikes 5
5a. Stems spotted, to 40 cm; flowers orange or reddish, petals hooded   **5. cucullata**
b. Stems unspotted, to 50 cm; flowers yellow or greenish yellow, petals not hooded   **6. guatemalensis**

1. V. imbricata Rose (*Altamiranoa ericoides* (Rose) Jacobs). Illustration: *National Cactus & Succulent Journal* **13**: 76 (1958);

Jacobsen, Handbook of succulent plants **2**: 904 (1960).

Mat-forming perennial with few-branched stems 2–6 cm. Leaves ovate, overlapping, closely adpressed to stem, 5–6 mm, acute, with fine tubercles beneath. Flowers erect, in short spikes; sepals leaf-like; corolla 3–5 mm, white. *Mexico (Oaxaca)*. G1. Summer.

2. V. parva (Hemsley) Jacobsen (*Sedum parvum* Hemsley; *Altamiranoa parva* (Hemsley) Rose).

Freely branched, mat-forming perennial. Leaves 3–6 mm, linear, spreading around stem. Flowers yellow. *Mexico (San Luis Potosí)*. G1.

3. V. batesii (Hemsley) Baehni & Macbride (*Altamiranoa batesii* Hemsley). Illustration: Hemsley, Biologia Centrali-Americana **5**: pl. 19 (1880); Sanchez, Flore del valle de Mexico, pl. 132 (1969).

Hairless perennial with thick roots and several erect or prostrate stems 15–25 cm, arising from base. Leaves 1–1.3 cm × 2–4 mm, linear-lanceolate. Flowering-stems with 2–5 branches. Corolla 2–4 mm wide, white or pinkish. *Mexico*. H5–G1. Summer–autumn.

4. V. jurgensii (Hemsley) Jacobsen (*Cotyledon jurgensii* Hemsley).

Erect or trailing perennial, branching from near the base. Stems woody at base, to 45 cm, with very fine downy hairs. Leaves 5–6 mm, spreading, acute. Flowering-stems with 2–5 branches, each with 2 or 3 flowers. Corolla 5–6 mm wide, white or purplish. *E Mexico*. G1. Summer.

5. V. cucullata Rose.

Tuberous-rooted perennial. Stems spotted, hairless, to 40 cm. Leaves to 2.5 cm, acuminate. Flowers in a narrow spike to 20 cm. Sepals *c.* 2 mm, distinct, green; corolla 2–4 mm, orange or reddish, with finely toothed, hooded petals. *NE Mexico (Coahuila)*. H5–G1. Autumn.

6. V. guatemalensis Rose (*V. laevis* Rose; *Altamiranoa guatemalensis* (Rose) Walther). Illustration: *Contributions to the US National Herbarium* **20**: pl. 81 (1909); *National Cactus & Succulent Journal* **13**: 77 (1958); Jacobsen, Handbook of succulent plants **2**: 904 (1960); Jacobsen, Lexicon of succulent plants, pl. 141 (1974).

Hairless perennial, with freely branching stems to 50 cm. Leaves linear, 6–25 mm, acute. Flowers in a narrow spike; corolla yellow or greenish yellow, *c.* 6 mm wide. *S Mexico, Guatemala*. G1. Autumn.

11. ADROMISCHUS Lemaire
F.T. de Vries

Perennial shrublet with usually prostrate, rarely erect, fleshy or somewhat woody branches. Leaves fleshy, spirally arranged, usually clustered. Inflorescence spike-like, usually *c.* 35 cm, sometimes *c.* 45 cm, with clusters of 1–3 erect or spreading flowers. Calyx with 5 triangular, acute lobes; corolla fused into a tube at the base, with 5 spreading lobes. Stamens 10 in 2 whorls, filaments fused to the lower third of the corolla-tube, anthers either protruding or not. Nectary scales 5. Carpels 5, usually free, narrowed into a short style. Seeds ellipsoid with a constriction at the end, and vertical ridges.

A genus of about 30 species from South Africa and Namibia, closely related to *Cotyledon*. The species with purple and dark green leaf-markings are especially popular, but unfortunately the spots are often lost in cultivation. They grow well in loose, peaty, slightly moist soil, preferably in full sun to ensure a compact plant with highly coloured leaves. If kept warm enough (minimum 15 °C) they can flower nearly all year round, though they can tolerate lower temperatures. Propagation is not difficult, and leaves root easily, forming small, soon very attractive plantlets. Seed and cuttings are also used.

The measurements given are those of the plant without inflorescence.

Literature: Tölken, H.R., Crassulaceae, in O.A. Leistner (ed.), Flora of southern Africa **14**: 37–60 (1985).

1a. Anthers protruding from corolla-tube; corolla-lobes at least as broad as long 2
 b. Anthers not protruding from corolla-tube; corolla-lobes up to 3 times longer than broad 7
2a. Corolla-lobes abruptly narrowed at tip, with wavy, ruffled margins 3
 b. Corolla-lobes gradually tapered at tip, with wavy margins 6
3a. Leaves covered with flaking wax; corolla-lobes 2–2.5 mm, sometimes with a few club-shaped hairs in the throat of the tube
 2. hemisphaericus
 b. Leaves smooth, or with a very slight bloom; corolla-lobes 1–2 mm, without club-shaped hairs 4

4a. Leaves oblong-oblanceolate, more or less flattened
 1. filicaulis subsp. **filicaulis**
 b. Leaves linear-elliptic, rarely linear-lanceolate, terete or almost so 5
5a. Stems erect or prostrate and rarely with fibrous adventitious roots
 1. filicaulis subsp. **filicaulis**
 b. Stems prostrate or decumbent and with stilt roots
 1. filicaulis subsp. **marlothii**
6a. Leaves elliptic to rounded with a marked marginal ridge extending to the base **3. trigynus**
 b. Leaves oblanceolate to obovate, with marginal ridge rarely extending beyond the middle or if so then narrow and not horny
 4. umbraticola
7a. Buds cylindric or slightly angular and spreading; club-shaped hairs present on lower parts of lobes and in throat of corolla–tube 8
 b. Buds distinctly grooved between petals and until flowering adpressed to central axis; club-shaped hairs found mainly in throat of corolla–tube 11
8a. Leaves terete or almost so, apex gradually tapered to both ends, base abruptly wedge-shaped; margin only horny at tip **5. mammillaris**
 b. Leaves flattened, oblanceolate to obovate; horny margin extending right around leaf to the base 9
9a. Leaves evenly and gradually narrowed towards base, oblanceolate, never with spots 10
 b. Leaves abruptly narrowed towards the base **7. maculatus**
10a. Corolla-tube 8–9 mm **6. maximus**
 b. Corolla-tube 10–12 mm
 8. sphenophyllus
11a. Leaves flattened, at least 3 times broader than thick at middle of leaf; leaves with a more or less thick, white bloom 12
 b. Leaves distinctly concave on both surfaces, about as broad, rarely up to twice as broad as thick at middle of leaf; leaves smooth or with a slight white bloom 13
12a. Corolla-lobes 3–5 mm; leaves oblanceolate to round or almost so; leaves with a thick white bloom
 9. leucophyllus
 b. Corolla-lobes 1.5–3 mm; leaves obovate, abruptly narrowed to the base; leaves with a thin white bloom
 12. marianae var. **hallii**

13a. Leaves usually acute, concave towards tip on both surfaces or with a longitudinal groove above; apical gland on each anther raised above pollen-sacs   14

  b. Leaves obtuse, truncate and/or flattened towards tip; apical gland on each anther stalkless   17

14a. Roots tuberous   15

  b. Roots fibrous   16

15a. Leaves smooth, with marginal ridge horny and usually wavy   **12. marianae** var. **hallii**

  b. Leaves warty (sometimes only visible under × 10 lens), with marginal ridge slightly raised but indistinct, not horny and wavy   **12. marianae** var. **immaculatus**

16a. Leaves concave above, marginal ridge horny and raised   **12. marianae** var. **marianae**

  b. Leaves convex above (rarely somewhat concave towards tip), marginal ridge never horny and scarcely raised   **12. marianae** var. **kubusensis**

17a. Club-shaped hairs in throat and on corolla-lobes; stem hairless, occasionally with aerial roots; leaves hairless   **11. cooperi**

  b. Club-shaped hairs usually only in corolla-throat; stem densely covered with brown aerial roots or glandular hairs; leaves usually covered with glandular hairs   18

18a. Stems 4–8 cm, without aerial roots, covered with glandular hairs   **10. cristatus** var. **zeyheri**

  b. Stems 2–4 cm, covered with aerial roots   19

19a. Ridge at tip of leaves narrower than broadest point on leaf; inflorescence with glandular hairs   **10. cristatus** var. **schonlandii**

  b. Ridge at tip of leaf constitutes broadest point of leaf; inflorescence hairless, rarely with a few hairs when young   20

20a. Leaf-blade 1–1.5 times longer than breadth of apical ridge, leaves reversed-triangular, usually with distinct leaf-stalk   **10. cristatus** var. **cristatus**

  b. Leaves 2–5 times longer than breadth of apical ridge, reversed-triangular to club-shaped, hairless or nearly so   **10. cristatus** var. **clavifolius**

1. A. filicaulis (Ecklon & Zeyher) C.A. Smith (*A. kleinioides* C.A. Smith; *Cotyledon mammillaris* J.D. Hooker).

Erect to prostrate, *c.* 35 cm; roots fibrous, often with stiff, adventitious stilt roots. Leaves 2–8 cm × 5–15 mm, lanceolate, elliptic to oblong or oblanceolate, usually terete, sometimes flattened, grey-green to greyish brown, sometimes with darker spots; tip acute or obtuse; base abruptly wedge-shaped. Margin horny, slightly expanded. Buds terete, abruptly narrowed towards tip, spreading. Flowers 1–1.3 cm, yellowish green; corolla-lobes 1–2 mm, pale, often mauve-red on mucro, rough, without club-shaped hairs. Anthers protruding from corolla-tube. *South Africa.* G2.

Subsp. **filicaulis** has branches 7–12 mm across, with peeling bark, leaves elliptic to oblong-oblanceolate, rarely lanceolate, often distinctly flattened above; tip obtuse, rarely acute.

Subsp. **marlothii** (Schönland) Tölken (*A. tricolor* C.A. Smith). Branches 3–8 mm across, with flaking bark, leaves lanceolate, rarely elliptic, usually more or less terete; tip acute.

2. A. hemisphaericus (Linnaeus) Lemaire (*A. rotundifolius* (Haworth) C.A. Smith; *A. bolusii* (Schönland) Berger; *Cotyledon crassifolius* Salisbury). Illustration: *Kakteenkunde*, 17 (1940); Rice & Compton, Wild Flowers of the Cape of Good Hope, pl. 49 (1951); *Cactus & Succulent Journal* **8**: 51 (1953).

Much-branched, low-growing *c.* 30 cm, with fibrous roots. Leaves 1–4.5 × 1–3 cm, oblanceolate to obovate, rarely circular, flattened but often convex on both surfaces, especially beneath, grey-green with often darker spots, and flaking wax; tip obtuse or rounded; base abruptly wedge-shaped. Margin indistinctly horny in upper half, sometimes wavy. Buds terete, abruptly narrowed towards tip, spreading. Corolla-lobes 2–2.5 mm, white or tinged pink, darker around throat and on mucro. Anthers protruding from corolla-tube. *South Africa.* G2.

3. A. trigynus (Burchell) von Poellnitz (*A. rupicolus* C.A. Smith; *A. maculatus* misapplied). Illustration: *Bothalia* **3**: 645 (1939); *National Cactus and Succulent Society Journal* **6**: 35 (1951); Jacobsen, Handbook of succulent plants **1**: 52 (1960); Parey's Blumengärtnerei **1**: 726 (1958).

Short, erect, much-branched plant, with fibrous roots. Leaves 1.5–4 cm × 8–30 mm, elliptic to rounded, rarely oblanceolate, flattened, usually concave above, more or less convex beneath, grey-green, usually with darker spots; tip rounded; base abruptly narrowed. Horny margin extending right around leaf. Buds terete, slightly grooved, gradually tapered towards tip, adpressed at first, later erect. Flowers pale yellowish green, thickly bloomed; corolla-lobes 1.5–2.5 mm, acuminate, off-white to tinged pink, rough and with some club-shaped hairs mainly in throat. Anthers protruding from corolla-tube. *South Africa.* G2.

4. A. umbraticola C.A. Smith. Illustration: *Bothalia* **3**: 646 (1938); Letty, Wildflowers of the Transvaal, 74 (1962); Barkhuizen, Succulents of South Africa, 69 (1978).

More or less branched *c.* 10 cm, with fibrous roots. Leaves 1.5–6.5 cm, oblanceolate, rarely linear-oblanceolate or obovate, flattened, usually convex on both sides, green to grey-green; tip obtuse or rounded; base usually gradually constricted. Margin horny, well developed in the upper half to absent. Buds with longitudinal grooves, gradually tapered towards tip, adpressed at first, later erect. Flowers 1–1.3 cm, pale green, usually tinged pink, bloomed; corolla-lobes 2–2.5 mm, acuminate, off-white to tinged pink, rough and with some club-shaped hairs mainly in throat. Anthers protruding from corolla-tube. *South Africa.* G2.

5. A. mammillaris (Linnaeus filius) Lemaire. Illustration: Rooksby, Desert plant life, 112 (1938); *National Cactus and Succulent Society Journal* **20**: 61 (1948); Jacobsen, Handbook of succulent plants **1**: 48 (1960).

Sparsely branched, creeping, *c.* 15 cm, stems rooting. Leaves 2–5 cm × 5–11 mm, linear-lanceolate, slightly flattened to almost terete, grey-green, sometimes tinged brown; tip tapered gradually; base abruptly wedge-shaped, shortly stalked. Horny margin restricted to tip. Buds terete, erect to spreading or curved towards tip. Flowers 1.1–1.3 cm, grey-green; corolla-lobes 4–5 mm, acute, white often tinged pink, mauve along margins, spreading to recurved, rough; lower lobes and throat with club-shaped hairs. Anthers not protruding from corolla-tube. *South Africa.* G2.

6. A. maximus P.C. Hutchison. Illustration: *Cactus and Succulent Journal* (US) **31**: 131, **132**(1959).

Erect to decumbent, *c.* 20 cm, with fibrous roots. Leaves oblanceolate 6–16 × 2.5–6 cm, flattened, grey-green, spotless; tip usually rounded, rarely obtuse; base wedge-shaped. Margin horny, at least in upper half, sometimes right around leaf. Buds terete, slightly grooved; corolla lobes 2.5–4 mm, acute, white or cream, pale pink tinged, rough; lower lobes and throat with club-shaped hairs. Anthers not protruding. *South Africa.* G2.

7. A. maculatus (Salm-Dyck) Lemaire (*A. mucronatus* Lemaire; *Cotyledon hemisphaerica* Harvey; *A. trigynus* misapplied). Illustration: *Bothalia* **3**: 621 (1938); Barkhuizen, Succulents of South Africa, 68 (1978); *Flowering Plants of Africa* **45**: t. 1776 (1978); Brickell (ed.), RHS A–Z encyclopedia of garden plants, 80 (2003).

Sparsely branched, prostrate, *c.* 15 cm, somewhat woody, with fibrous roots. Leaves 2.5–10 × 1.5–4 cm, obovate or spathulate to oblanceolate, flattened, green to greyish green, with or without purple spots; tip obtuse, often with a mucro or notched; base abruptly wedge-shaped, margin horny all round the leaf. Buds terete, gradually tapered towards tip, spreading. Flowers 8–11 mm, pale yellowish green; corolla-lobes 2.5–5 mm, acute, white or tinged pale pink, mauve along margins, spreading to recurved, rough; lower lobes and throat with club-shaped hairs. Anthers not protruding from corolla-tube. *South Africa.* G2.

8. A. sphenophyllus C.A. Smith (*A. rhombifolius* misapplied). Illustration: *National Cactus and Succulent Journal* (US) **17**: 150, 151 (1945).

Little-branched, to 20 cm, with fibrous roots. Leaves 2.5–6.5 × 1–3.5 cm, oblanceolate, gradually tapered towards base, flattened, grey-green without spots; tip rounded to obtuse, mucronate; base wedge-shaped, margin horny, extending right around the leaf, forming a straight line from broadest point to base. Buds terete, gradually tapered towards tip, spreading. Flowers 9–12 mm, pale green, sometimes tinged red; corolla-lobes 2.5–4 mm, acute, white often tinged pink, with mauve margin, rough; lower lobes and throat with club-shaped hairs. Anthers not protruding from corolla-tube. *South Africa.* G2.

9. A. leucophyllus Uitewaal. Illustration: *Cactus and Succulent Journal* (US) **32**: 136 (1960); *National Cactus and Succulent Society Journal* **9**: 58–9 (1954).

Much-branched, somewhat erect to decumbent *c.* 10 cm, with fibrous roots. Leaves 1.5–4 × 1.5–3 cm, oblanceolate to obovate or circular, flattened, whitish green, with a thick white bloom; tip usually rounded; base abruptly wedge-shaped. Margin distinctly horny, extending right around the leaf, sometimes darker. Buds terete, slightly grooved, gradually tapered towards tip, erect at first, later spreading. Flowers 1.1–1.3 cm; corolla-lobes 3–5 mm, triangular, white with a pink median band, rough and with club-shaped hairs mainly in throat. Anthers not protruding from corolla-tube. *South Africa.* G2.

10. A. cristatus (Haworth) Lemaire. Illustration: Roeder, Sukkulenten, edn 2, **9**: 3 (1931); Jacobsen, Handbook of succulent plants **1**: 48 (1960); Rauh, Die grossartige Welt der Sukkulenten, 69 (1967).

Erect *c.* 8 cm, with fibrous roots at base, and glandular hairs on stem. Leaves 1.5–5 cm × 5–20 mm, reversed-triangular to oblong-elliptic, terete to somewhat flattened, green to grey-green; tip truncate or rounded to more or less broadened and crisped; base wedge-shaped, sometimes stalked. Margin in upper half of the leaf horny, wavy, often darker. Buds terete, slightly grooved, gradually tapered towards tip, erect at first, later spreading. Flowers 1–1.2 cm, grey-green; corolla-lobes 2–3.5 mm, ovate-triangular, white tinged pink, with darker margin, spreading or recurved, rough and with club-shaped hairs mainly in throat. Anthers not protruding from corolla-tube. *South Africa.* G2.

Var. **cristatus** is much-branched, *c.* 4 cm, has red aerial roots and red hairs on the stem, and reversed-triangular leaves, which have much-broadened margins at the tip.

Var. **clavifolius** (Haworth) Tölken (*A. clavifolius* Haworth; *A. poellnitzianus* Werdermann; *Cotyledon cristata* Harvey). Illustration: *National Cactus & Succulent Journal* **24**: 33 (1952); Andersohn, Cacti and succulents, 263 (1983). Much-branched, *c.* 4 cm, covered with aerial roots, and has reversed-triangular to club-shaped leaves, broadened at the tip, narrowed towards base, with a wavy

margin about as broad as leaf. Inflorescence hairless with a waxy bloom.

Var. **schonlandii** (Phillips) Tölken. A brittle, much-branched plant, *c.* 4 cm, with aerial roots and narrowly triangular to oblong-lanceolate leaves with a horny, wavy margin, narrower than the leaf. Inflorescence with glandular hairs.

Var. **zeyheri** (Harvey) Tölken is little-branched, *c.* 8 cm, has no aerial roots, but glandular hairs on the stem, and broadly triangular leaves with much broadened, horny, wavy margins. Inflorescence with glandular hairs.

11. A. cooperi (Baker) Berger (*A. festivus* C.A. Smith; *A. pachylophus* C.A. Smith). Illustration: *Bothalia* **3**: 634 (1938); Barkhuizen, Succulents of Southern Africa, 66 (1978); *Flowering plants of Africa* **45**: t. 1849 (1982).

Much-branched, erect *c.* 10 cm, with thick fibrous roots and sometimes aerial roots. Leaves 2–9 × 1–3 cm, oblanceolate-oblong to spathulate, terete to somewhat flattened, green to grey-green, with or without purple spots; tip obtuse to truncate; base wedge-shaped to stalked. Margin ridged at the tip, wavy and wider than remainder of leaf. Buds terete, slightly grooved, gradually tapered towards tip, erect at first, later spreading. Flowers 9–11 mm, bluish green, glaucous; corolla-lobes 3–4.5 mm, sharply acute, spreading or recurved, pale pink with a thick bloom, wine-red towards margins, rough and with club-shaped hairs mainly in throat. Anthers not protruding from corolla-tube. *South Africa.* G2.

12. A. marianae (Marloth) Berger. Illustration: Rauh, Die grossartige Welt der Sukkulenten, pl. 69 (1967); Barkhuizen, Succulents of South Africa, 69 (1978); Court, Succulent flora of southern Africa, 65 (1981).

Erect, rarely decumbent *c.* 8 cm, with often tuberous roots. Leaves 2–12 cm × 4–25 mm, linear-lanceolate to elliptic, rarely obovate, almost terete with a more or less pronounced central groove above, grey-green to greyish brown, rarely with bloom or spots; tip acute or truncate; base wedge-shaped. Margin at least in places horny. Buds terete, slightly grooved, gradually tapered towards tip, erect at first, later spreading. Flowers 1–1.2 cm, pale pink to white, with a thick bloom; corolla-lobes 2–3 mm, acute, spreading or recurved, white with purple margins, rough and with club-shaped hairs mainly in throat. Anthers not protruding from corolla-tube. *South Africa, SW Namibia.* G2.

Var. marianae. To 10 cm, sparsely branched, clump-forming, with thick old fibrous roots; leaves fairly narrow, oblanceolate to elliptic, rarely lanceolate, concave above, often more or less spotted, with a horny margin, usually distinct from tip to middle of leaf.

Var. hallii (Hutchison) Tölken (*A. casmithianus* von Poellnitz). Illustration: *Cactus and Succulent Journal* (US) **28**: 144–7 (1956). Much-branched with thick stems and thick tuberous roots; leaves 1.5–2.5 cm, obovate to almost circular, rarely oblanceolate, flattened, slightly concave above, rarely faint reddish-spotted, usually bloomed, with a raised, horny, often brown and wavy, margin at the tip.

Var. immaculatus Uitewaal (*Cotyledon herrei* Barker; *A. antidorcadum* von Poellnitz; *A. alveolatus* Hutchison). Illustration: Rauh, Die grossartige Welt der Sukkulenten, pl. 69 (1967); Barkhuizen, Succulents of South Africa, 69 (1978); Court, Succulent flora of southern Africa, 94 (1981). Much-branched, small, with thick stems, often constricted at base; leaves oblanceolate to elliptic, rarely obovate or slightly concave above and below, more or less warty, without purple spots, with a brown or white, raised but not horny margin.

Var. kubusensis (Uitewaal) Tölken (*A. blosianus* P.C. Hutchison; *A. geyeri* P.C. Hutchison). Illustration: *Cactus and Succulent Journal* (US) **29**: 36, 37 (1957) & **32**: 89, 91 (1960). Much-branched, with thick stems, rarely continued in thick fibrous roots; leaves 3–9 cm, oblanceolate, rarely elliptic, somewhat concave on both sides to terete, smooth and rarely purple-spotted, with an often brown, scarcely raised and not horny margin.

12. UMBILICUS de Candolle
S.G. Knees

Perennial, succulent herbs with new shoots arising annually from scaly rhizomes or tubers. Leaves alternate, hairless; stalked at base, becoming stalkless. Flowering-stems terminal, usually solitary, but occasionally branched above into several racemes. Sepals 5, free, about half as long as corolla. Petals 5, united into a bell-shaped corolla; lobes more or less erect. Stamens 10, rarely 5; styles short or absent. Fruit a slender follicle; seeds numerous.

A genus of about 18 species, chiefly in Europe, W Asia, N Africa and the Atlantic Islands. Propagation is usually from seed. Plants are hardy but are often best kept in a cool greenhouse or cold-frame during winter months.

1a. Racemes occupying more than half the total stem; flowers nodding **2**
  b. Racemes occupying less than half the total stem; flowers erect or horizontal **3**
2a. Racemes 15–30 cm; flowers greenish yellow, spotted red **1. rupestris**
  b. Racemes 10–16 cm; flowers yellowish white **2. intermedius**
3a. Flowers horizontal, 4.5–6 mm **3. horizontalis**
  b. Flowers erect, 9–14 mm **4. erectus**

1. U. rupestris (Salisbury) Dandy (*Cotyledon rupestris* Salisbury; *C. tuberosa* (Linnaeus) Halacsy; *Umbilicus pendulinus* de Candolle). Illustration: Polunin, Flowers of Europe, pl. 39 (1969); Phillips, Wild flowers of Britain, 73 (1977); Pignatti, Flora d'Italia **1**: 489 (1982); Blamey & Grey-Wilson, Illustrated flora of Britain and northern Europe, 165 (1989).

Erect, tuberous perennial; tuber rounded, 1–3 cm across, usually covered with a web of fine roots. Stem 20–50 cm. Basal leaves almost circular, peltate, blade shallowly funnel-shaped, 3–10 cm across; margin coarsely scalloped; stalk to 15 cm. Upper stem-leaves becoming progressively smaller and very shortly stalked, not peltate. Flowers pendent, numerous in bracteate, long, cylindric racemes, 15–30 cm. Corolla 5–8 mm, greenish yellow, spotted or streaked with red. Fruit 5–6 mm, a narrowly boat-shaped, membranous follicle; seeds dark brown *c.* 0.5 mm. *Europe, W Asia, N Africa, Azores, Madeira.* H5. Spring–early summer.

2. U. intermedius Boissier. Illustration: Bouloumoy, Flore du Liban et de la Syrie, Atlas, t. 64 (1930).

Like *U. rupestris* but lowest leaves not very succulent. Flowers 6–8 mm, yellowish white, in dense racemes 10–16 cm. *Bulgaria, Cyprus, Turkey, Israel, Egypt.* G1. Spring–summer.

3. U. horizontalis (Gussone) de Candolle (*Cotyledon horizontalis* Gussone). Illustration: Pignatti, Flora d'Italia **1**: 489 (1982); Valdes et al., Flora vascular de Andalucía occidental **2**: 7 (1987).

Like *U. rupestris*, but leaves diminishing abruptly; upper stem covered with scale-like, linear-triangular bracts, *c.* 2 cm × 5 mm or less. Flowers usually horizontal in very congested, unbranched racemes, 10–25 cm. Corolla 4.5–6 mm, yellowish brown. Fruit dark reddish, *c.* 4.5 mm. *Mediterranean area, Azores.* H5–G1. Spring–summer.

4. U. erectus de Candolle (*Cotyledon umbilicus-veneris* Linnaeus; *Cotyledon lutea* Hudson). Illustration: Bouloumoy, Flore du Liban et de la Syrie, Atlas, t. 64 (1930); Pignatti, Flora d'Italia **1**: 489 (1982).

Stout, erect perennial with a tuberous root; stems simple, 30–70 cm. Leaves peltate, 3–7 cm across, stem-leaves ovate-triangular to linear, margins toothed. Flowers more or less erect in simple or, more rarely, branched racemes, 8–25 cm. Corolla tubular 9–14 mm, yellowish. Fruit 5–6 mm. *S Europe, Middle East, NW Africa.* G1. Spring–summer.

13. TYLECODON Tölken
G.D. Rowley

Small to large perennial shrublets with greatly swollen, soft, succulent stem bases and few proportionally thick, sometimes tuberculate branches. Leaves spirally arranged near the stem tips, simple, entire, stalkless, soft and fleshy, early deciduous. Inflorescence a tall, erect scape more or less branched into a loose head of usually erect tubular flowers. Bracts few, narrow, scattered. Sepals 5, green. Petals 5, united into a long tube with recurving lobes. Stamens 10 in 2 whorls, arising from the corolla-tube and usually longer than it. Nectar scales 5, free. Seeds ellipsoid, ridged, constricted at one end. The soft, fleshy stems are poisonous to cattle.

A segregate genus from *Cotyledon*, for which over 30 species have been described, but these variable in both vegetative and floral characters and the number could almost certainly be reduced by a more conservative treatment. The spiralled, deciduous leaves and mostly erect flowers most readily distinguish the genus from *Cotyledon*. They grow in the Cape Province of South Africa, extending into Namibia. The miniatures typically inhabit rock crevices and are difficult to see when dormant and leafless. The taller species (notably *T. paniculatus*) are a conspicuous feature of many landscapes, the massive conical trunk with peeling papery bark resembling that of *Cyphostemma* and the

Mexican tree sedums. All species appeal to devotees of succulents. As extreme xerophytes they require a dry rest when leafless, a porous, well-drained soil and rather sparing watering at any time. Full sun is preferred. In winter the temperature should not be allowed to fall below 4 °C for any length of time. Propagation is mainly by seed, although the thinner-stemmed cuttings of some can be rooted.

Literature: Tölken, H.R., Crassulaceae, in O.A. Leistner (ed.), Flora of southern Africa **14**: 19–35 (1985); Jaarsveld, E.V. & Koutnik, D., Cotyledon and Tylecodon (2004).

1a. Plant covered in the dry, stiff remains of persistent inflorescences like weak thorns　　**5. reticulatus**
 b. Plant not covered in persistent thorny inflorescences　　2
2a. Leaves flat, obovate-spathulate　　**2. paniculatus**
 b. Leaves terete or semi-terete　　3
3a. Stem covered in persistent leaf-bases forming hard tubercles 6 mm or more　　4
 b. Stem without elevated tubercles　　6
4a. Flowers tubular, bright yellow, erect, to 2.5 cm　　**3. papillaris** subsp. **papillaris**
 b. Flowers more or less bell-shaped, not bright yellow, pendent or almost erect, to 1.5 cm　　5
5a. Inflorescence glandular　　**3. papillaris** subsp. **wallichii**
 b. Inflorescence not glandular　　**3. papillaris** subsp. **ecklonianus**
6a. Leaves rarely produced and then short-lived, not in rosettes and confined to a few branches; other branches with minute scales only　　**1. buchholzianus**
 b. All branches at some time bearing leaves in small rosettes　　**4. pearsonii**

1. T. buchholzianus (Schuldt & Stephan) Tölken (*Cotyledon buchholziana* Schuldt & Stephan). Illustration: Tölken, Flowering plants of Africa, t. 1774 (1978); Rauh, Wonderful world of succulents, t. 59/1 (1984); Rowley, Caudiciform and pachycaul succulents, 177 (1987); Jaarsveld & Koutnik, Cotyledon and Tylecodon, 64, 65 (2004).

Compact stem-succulent shrublet made up of numerous, closely packed, erect, fleshy, cylindric, brittle shoots, 6–10 mm across, grey and slightly rough from numerous small, spiralled black leaf-bases. Leaves few and erratically appearing, often at long intervals, scattered, terete, short-lived, greyish green, fusiform, 8–15 × 2–4 mm, ascending. Scape thin and wiry, to 10 cm, more or less persistent, with 1 to few erect red flowers, 1–1.2 cm. Sepals 1.5–2.5 mm. Corolla-tube cylindric. *S Africa (NW Cape Province to Namibia).* G1. Winter.

This curious plant can grow for years without producing leaves, the shoots elongating and bearing minute scales only.

2. T. paniculatus (Linnaeus filius) Tölken (*Cotyledon paniculata* Linnaeus filius; *Cotyledon fascicularis* Aiton). Illustration: *Botanical Magazine,* 5602 (1866); Rauh, Wonderful world of succulents, t. 59/3 & 59/4 (1984); Rowley, Caudiciform and pachycaul succulents, 19, 176, 235 (1987); Jaarsveld & Koutnik, Cotyledon and Tylecodon, 98, 99 (2004).

Stem succulent to 1.6 m, with a swollen soft trunk to 60 cm across at the base, branching above into a few ascending shoots, 1.5–2 cm across, covered in yellowish peeling papery bark. Leaves obovate-spathulate, 6–12 × 3–6 cm, tapered to the base, rounded at the apex, fleshy and brittle, bright green, more or less glandular downy. Scape terminal, in cultivation to 60 cm, branched, with cymes of scattered nodding flowers, 1.2–1.6 cm. Sepals 4–6 mm. Petals dark red with yellowish stripes. *South Africa (W Cape Province), Namibia.* Spring.

3. T. papillaris (Linnaeus) Rowley (*T. cacalioides* (Linnaeus filius) Tölken; *T. wallichii* (Harvey) Tölken; *Cacalia papillaris* Linnaeus; *Cotyledon eckloniana* Harvey). Illustration: Graf, Tropica, 365 (1978); Rauh, Wonderful world of succulents, t. 59/5, 59/7 (1984); Rowley, Caudiciform and pachycaul succulents, 178, 179 (1987); Jaarsveld & Koutnik, Cotyledon and Tylecodon, 146, 147 (2004).

Shrub to 80 cm, with thick succulent branches covered in spirally arranged tubercles formed from the persistent leaf-bases, 6–13 × 3–4 mm, with an obliquely truncate sharp apex. Bark silvery grey. Leaves 5–10 cm × 2–4 mm, linear, terete, hairless, short-lived, pale green, dying back from the tip. Scape terminal, to 60 cm, green parts all covered in glands (glands absent in subsp. **ecklonianus** (Harvey) Rowley), branching near the top, with numerous pendent bright yellow flowers 1.7–2.5 cm (dull greenish yellow and not above 1.3 cm in subsp. **wallichii** (Harvey) Rowley). Sepals 4–12 mm. Corolla-tube cylindric, urn-shaped or cupped. *South Africa (W Cape, Karoo).* G1. Winter.

The natural hybrid with *T. paniculatus,* with tapering stems covered in low tubercles, **T. × fergusoniae** (L. Bolus) Rowley, is also in cultivation.

4. T. pearsonii (Schönland) Tölken (*Cotyledon luteosquamata* von Poellnitz). Illustration: Jacobsen, Handbook of succulent plants **1**: t. 286 (1960); Rowley, Caudiciform and pachycaul succulents, 177, 261 (1987); Jaarsveld & Koutnik, Cotyledon and Tylecodon, 100, 101 (2004).

Small, much-branched, stem-succulent shrublet with finger-like shoots covered in leaf-scars arising from a much swollen base. Leaves terete with a groove down the upperside, 2–4 cm × 2–4 mm, soft and pliable, smooth or finely downy, grey-green. Scape 6–10 cm, the main axis more or less persistent, sparingly branched at the top, bearing a few whitish, red-striped flowers, which are pendent at first but become erect as the buds open. Sepals glandular, 5–6 mm. Corolla-tube urn-shaped, 1.2–1.4 cm. *South Africa (NW Cape) to Namibia.* G1. Winter.

T. pygmaeus (Barker) Tölken (illustration: Jaarsveld & Koutnik, Cotyledon and Tylecodon, 106, 2004), is much reduced, with gnarled stems only a few cm high and *c.* 5 mm across, with papillose leaves 2.5 cm × 6–12 mm.

T. schaeferianus (Dinter) Tölken (*Cotyledon sinus-alexandri* von Poellnitz). Illustration: Rowley, Caudiciform and pachycaul succulents, 177 (1987). Another curious miniature, much in demand with collectors, having smooth club-shaped leaves and relatively large, erect, glandless, pinkish white flowers.

5. T. reticulatus (Linnaeus filius) Tölken (*Cotyledon reticulata* Linnaeus filius). Illustration: Jacobsen, Lexicon of succulent plants, t. 40/2 (1974); Rauh, Wonderful world of succulents, t. 59/2 (1984); Rowley, Caudiciform and pachycaul succulents, 176, 262 (1987); Jaarsveld, & Koutnik, Cotyledon and Tylecodon, 109–11 (2004).

Compact highly stem-succulent shrublet with many stout branches forming a gnarled, roughly hemispherical dome *c.* 30 cm across. Bark greyish, papery, peeling. Leaves almost cylindric, flattened or furrowed above, 2–4.5 cm × *c.* 6 mm, soft and flexible, green downy or smooth. Scape slender, wiry, with a much-branched dichasium, the whole persisting to form a dense tangle of weak, silvery thorns enveloping the plant. Flowers erect, 7–15 mm, yellowish green. Corolla-tube cylindric or urn-shaped. *South Africa (W Cape, Karoo) Namibia*. G1. Winter.

14. CHIASTOPHYLLUM Berger
P.G. Barnes

Fleshy hairless evergreen perennial with prostrate or ascending stems often rooting from the lower nodes. Leaves opposite, broadly elliptic or rounded, scalloped, short-stalked. Inflorescence a loose panicle of arching spike-like branches with small linear bracts and almost stalkless yellow flowers. Calyx-lobes 5, narrowly oblong; petals 5, elliptic, erect, rather fleshy. Stamens 10, filaments attached at base of corolla-lobes. Carpels 5, each with a small scale at the base. Fruiting-head consisting of 5 erect oblong follicles.

A genus of one species. A hardy plant for the rock garden, requiring a moist but well-drained soil and a sunny position. Propagation by cuttings or by division.

1. C. oppositifolium (Ledebour) Berger (*Cotyledon oppositifolia* Ledebour; *Cotyledon simplicifolia* invalid). Illustration: *Botanical Magazine*, 8822 (1919); RHS dictionary of gardening **1**: 459 (1956); Hay & Synge, Dictionary of garden plants in colour, f. 41 (1969); Graf, Exotica, series 4, edn 12, **1**: 905 (1985).

Stems to 25 cm. Leaves 2.5–4 × 1.5–4 cm, stalk 5–10 mm. Flowers 5–7 mm. *Caucasus*. H4. Summer.

A cultivar with variegated foliage has recently been introduced as 'Jim's Pride'.

15. COTYLEDON Linnaeus
G.D. Rowley

Perennial, evergreen, leaf-succulent shrublets with fleshy or later soft woody branches. Leaves simple, entire or scalloped at the apex, thick and soft, opposite in crossed pairs, free, stalkless or nearly so. Inflorescence terminal, an umbrella-like cluster of usually pendent tubular flowers at the top of a long, almost bractless, erect scape; reduced to a few or even a solitary flower in the smallest species. Sepals 5, green. Petals 5 (as distinct from 4 in the otherwise similar *Kalanchoë*), united into a long tube with recurving lobes. Stamens 10 in two whorls, arising from the corolla-tube, and projecting beyond it. Nectary more or less cup-like. Seeds ellipsoid, ridged, constricted at one end.

Linnaeus' genus contained species now classified as *Adromischus*, *Echeveria*, *Tylecodon*, *Umbilicus* and other segregate genera, many of whose species have synonyms under *Cotyledon*. As delimited here, the genus has about 9 species from S Africa extending to the tropics, Ethiopia and Arabia. Most are popular with collectors of succulents and some, notably the different clones of the extremely variable *C. orbiculata*, have long been favoured as conservatory and house plants. They thrive in a porous but nutritious compost with sharp drainage and ample water in summer but only enough in winter to keep the soil from becoming dust dry. In the dormant state some leaf-drop is inevitable. A winter minimum of 4 °C suits them, and most will not suffer from an occasional drop to zero. Propagation is easy from cuttings, and single fallen leaves often root and grow plantlets.

Literature: Tölken, H.R., Crassulaceae, in O.A. Leistner (ed.), Flora of southern Africa **14**: 3–17 (1985); Jaarsveld, E.V. & Koutnik, D., Cotyledon and Tylecodon (2004).

1a. Corolla-lobes bright yellow, spreading, at least twice as long as the tube   **2. campanulata**
  b. Corolla-lobes not bright yellow, recurved, not longer than tube   2
2a. Leaves hairy, scalloped with small round apical teeth   **5. tomentosa**
  b. Leaves hairless, or if hairy then not toothed   3
3a. Corolla-tube bulging into pouches between sepals   **1. barbeyi**
  b. Corolla-tube cylindric or only gradually widened below   4
4a. Corolla-tube 5–16 mm, or, if longer, then bracts several   5
  b. Corolla-tube 2–2.5 cm; bracts 1 or 2 pairs   **3. orbiculata** var. **oblonga**
5a. Flowers solitary or up to 3 **6. woodii**
  b. Flowers clustered in 1–5 dichasia   6
6a. Spreading to procumbent; corolla-tube 5–8 mm   **4. papillaris**
  b. Branching shrublets; corolla-tube 8–16 mm   **3. orbiculata**

1. C. barbeyi Baker (*C. wickensii* Schönland). Illustration: Jacobsen, Handbook of succulent plants **1**: t. 273 (1960); Court, Succulent flora of South Africa, 95 (1981); Leistner (ed.), Flora of southern Africa **14**: f. 1A (1985); Jaarsveld & Koutnik, Cotyledon and Tylecodon, 21, 22 (2004).

Stiffly erect shrubs to 2 m, with numerous thick fleshy branches covered in semi-circular leaf-scars. Leaves very variable in size, shape and thickness from flat and oblanceolate to nearly terete, 6–12 × 2–4.5 cm, smooth or slightly glandular, grey-green. Scape 20–60 cm with a head of more or less pendent flowers 1–3 cm. Sepals 8–10 mm, glandular. Petals orange to deep red, united into a tube that bulges out into nectar-pouches between the 5 sepals; lobes recurved. *Transvaal through E Africa to Arabia*. G1. Spring.

2. C. campanulata Marloth (*C. teretifolia* Thunberg). Illustration: *Botanical Magazine*, 6235 (1876); van Laren, Succulents other than cacti, 72 (1935); Jacobsen, Lexicon of succulent plants, pl. 41 (1974); Jaarsveld & Koutnik, Cotyledon and Tylecodon, 23, 24 (2004).

Sprawling downy shrublets with thick, fleshy, decumbent stems eventually becoming 20 cm or more, branching basally. Leaves nearly terete, slightly wider midway and usually channelled on the upper side, 4–12 cm × 8–15 mm, soft and very succulent, yellowish green, covered in glandular hairs, with a distinctly flattened, obtuse reddish apex. Scape 20–40 cm, with a cluster of more or less pendent, showy, bright yellow flowers about 4 cm across. Sepals 3–5 mm, glandular. Petals united for about a third of their length. *South Africa (E Cape Province)*. G1. Early summer.

3. C. orbiculata Linnaeus (*C. ausana* Dinter; *C. decussata* Sims). Illustration: Graf, Tropica, 361, 363–5 (1978); Rowley, Illustrated encyclopedia of succulents, 123 (1978); Rauh, Wonderful world of succulents, pl. 70 (1984); Jaarsveld & Koutnik, Cotyledon and Tylecodon, 29–31 (2004).

Shrub to 1 m, with thick, brittle branches, exceedingly variable in stature, leaf size and shape. Leaves typically longer than broad, 3.5–8 cm or longer, to 4.5 cm across; margin entire, notched or even incised, sometimes wavy; white- to grey-bloomed, usually hairless, often with a red margin. Scape 25–10 cm, with few, pendent, orange to pinkish red long-tubed flowers. Corolla-tube more or less cylindric, 8–16 mm, with shorter recurved lobes. *South Africa (Cape Province), Namibia & Angola*. G1. Spring.

The commonest species in cultivation since its introduction in 1690, and the most variable, with 54 synonyms according to Tölken (1985). The most popular variants with growers are the miniatures, especially those with scalloped wavy-edged leaves or egg-shaped leaves (var. **oophylla** Dinter and var. **dinteri** Jacobsen).

Var. **oblonga** (Haworth) de Candolle (*C. undulata* Haworth; *C. coruscans* Haworth; *C. whitei* Schönland & E.G. Baker). Illustration: van Laren, Succulents other than cacti, 67 (1935); Jacobsen, Lexicon of succulent plants, pl. 41 (1974); Rauh, Wonderful world of succulents, pl. 70 (1984); Jaarsveld & Koutnik, Cotyledon and Tylecodon, 34, 35 (2004). Vigorous, showy-flowered, decumbent shrublet with branches 1–1.8 cm across, bearing narrow obovate to oblanceolate closely packed leaves. Scape with 1 or less frequently 2 pairs of bracts. Corolla-tube 2–2.5 cm.

4. C. papillaris Linnaeus filius. Illustration: Jaarsveld & Koutnik, Cotyledon and Tylecodon, 37, 38 (2004).

Dwarf, clump-forming, with narrow oblong biconvex acuminate leaves 2–3 cm × 6–10 mm, glaucous grey-green, flushed and margined with red. Scape 10 cm or more, with few, pendent, greenish to yellowish red flowers.

Not to be confused with *Tylecodon papillaris*.

5. C. tomentosa Harvey (*C. ladismithensis* von Poellnitz). Illustration: Graf, Tropica, 363 (1978); Court, Succulent flora of South Africa, 94 (1981); Rauh, Wonderful world of succulents, pl. 70 (1984); Jaarsveld & Koutnik, Cotyledon and Tylecodon, 41–4 (2004).

Compact shrublet to 30 cm or more, with many thin branches. Leaves obovate oblong, tapered to a short stalk at the base, very plump and soft, 1.5–5.5 cm × 8–15 mm, with a more or less scalloped apex with up to 9 teeth, densely felted with hairs. Scape 10–20 cm with around 10 nodding red flowers. Sepals densely felted. Corolla-tube 1.2–1.6 cm with recurved lobes half as long. *South Africa (Cape Province: Ladismith area)*. G1. Spring.

6. C. woodii Schönland & E.G. Baker. Illustration: Jaarsveld & Koutnik, Cotyledon and Tylecodon, 46, 47 (2004).

An attractive, free-flowering, small-growing species with flat, red-edged, obovate, acuminate leaves and scape bearing 1–3 pendent orange to red flowers.

16. GRAPTOPETALUM Rose
S.G. Knees

Evergreen perennial herbs and shrubs. Leaves alternate, mostly in rosettes. Flowers shortly stalked in 1–several alternating cymes, borne in axillary clusters. Sepals 5, almost equal, closely adpressed to corolla; petals 5–7, united below, overlapping in bud, stamens 5–10, outcurved; carpels erect, mostly short-styled.

About 12 species of succulent perennials from the southern United States and Mexico. Closely allied to *Echeveria*, but petals spreading from the middle, with red dots or blotches, often in more or less transverse rows.

Literature: Acevedo Rosas, R., Cameron, K., Sosa, V. & Pell, S., Phylogenetic relationships and morphological patterns in Graptopetalum, *Brittonia* **56**: 185–94 (2004).

1a. Leaves glossy, dark green   2
  b. Leaves covered with a matt greyish or slight purplish bloom   3
2a. Leaves with a bristle-tip to 12 mm; margins with a white wing
                                   1. filiferum
  b. Leaves mucronate, margins reddish
                                     2. bellum
3a. Leaves distinctly club-shaped
                               3. pachyphyllum
  b. Leaves linear to spathulate with a distinct point   4
4a. Plants shrubby with distinct stems   5
  b. Plants herbaceous, often in stemless clusters   7
5a. Shrublet *c.* 15 cm; leaves with distinct purplish bloom, in rosettes of 12–15   **4. amethystinum**
  b. Shrubs 40–200 cm; leaves with a greyish bloom, in rosettes of 15–30   6
6a. Leaves 2–4 cm, with abruptly pointed tips, pale green or reddish
                                  5. fruticosum
  b. Leaves 5–9 cm, rounded at tips, yellowish   **6. grande**
7a. Leaves greyish red; flowers dark red with greenish bands   **7. rusbyi**
  b. Leaves pale lavender, blue-grey or yellowish; flowers white or yellowish with red markings   8
8a. Erect or trailing herb with stems to 30 cm; leaves pale lavender-grey
                               8. paraguayense
  b. Almost stemless herb, producing many stolons; leaves greyish blue
                               9. macdougallii

1. G. filiferum (S. Watson) Whitehead (*Sedum filiferum* S. Watson). Illustration: Rauh, Schöne Kakteen und andere Sukkulenten, 223 (1978); Riha & Subik, Illustrated encyclopedia of cacti and other succulents, 276 (1981).

Stemless perennial with rosettes of 50–300 leaves, to 6 cm across. Leaves 1.2–5 cm × 8–12 mm, spoon-shaped, shiny green with white-winged edges and a bristle-tip to 12 mm. The flowers are borne on stems 5–10 cm, with 2–5 branches, each with 2–5 flowers. Petals 4–7 mm, white, spotted red. *NW Mexico*. G1. Spring–early summer.

2. G. bellum (Moran & Meyran) D.R. Hunt (*Tacitus bellus* Moran & Meyran). Illustration: Rowley, Illustrated encyclopedia of succulents, 128 (1978); *Pacific Horticulture* **39**(4): 36 (1978); *Botanical Magazine*, n.s., 781 (1979); Brickell, RHS gardeners' encyclopedia of plants and flowers, 1398 (1989).

Hairless, succulent perennial herb, with 25–50 leaves in basal rosettes, 3–8 cm across. Leaves 2–3.5 × 1.5–2.8 cm, obtuse, mucronate, dark grey-green, margin reddish brown. Flower-stalks axillary, with 1–3 branches terminating in large, odourless, open flowers to 3.5 cm across. Flower-stalk 1–2 cm, sepals equal, reflexed; petals 6–10 mm across, overlapping in bud, ovate to elliptic, bright crimson-red; stamens 10, filaments red, anthers bright yellow; carpels 5, free to base, 1–1.7 cm; style *c.* 4 mm. *Mexico (Chihuahua)*. G1. Late spring–early summer.

3. G. pachyphyllum Rose. Illustration: *Addisonia* **7**: 247 (1922); Lamb & Lamb, Pocket encyclopedia of cacti and succulents in colour, 261 (1974); Everett, New York Botanical Gardens illustrated encyclopedia of horticulture **5**: 1532 (1981).

Perennial herb with trailing stems to 20 cm, branched at the base. Leaves in rosettes of 20–50, club-shaped, 1.2–1.8 cm, bluish green, tipped red, covered with a greyish bloom. Flowering-stems 2.5–10 cm, with 1–4 branches, each with 2–5 flowers. Petals *c.* 11 mm, creamy yellow with few red spots. *Mexico*. G1. Spring–summer.

4. G. amethystinum (Rose) Walther (*Pachyphytum amethystinum* Rose). Illustration: Everett, New York Botanical Gardens illustrated encyclopedia of horticulture **5**: 1532 (1981); Riha &

Subik, Illustrated encyclopedia of cacti and other succulents, 274 (1981).

Shrublet or perennial herb to *c.* 15 cm. Leaves in loose rosettes of 12–15, 5–7.5 × 2 × 3.5 cm, ovate, to 2 cm thick, bluish grey to green and covered with a purplish bloom. Flowering-stems 15–17 cm, with 3–10 branches, each with 3–6 flowers. Petals *c.* 11 mm, greenish yellow with bands of red dots. *W Mexico.* G1. Spring–early summer.

5. G. fruticosum Moran. Illustration: *Cactus & Succulent Journal* (US) **40**(4): 152–4 (1968).

Shrub to 40 cm, branching from the base, with smooth silvery bark. Leaves 20–30, borne in loose rosettes, 3–7 cm across; spathulate to rhombic, 2–4 cm × 8–20 mm, pale green or reddish, with a slight grey bloom, abruptly pointed at tip. Flowering-stems 12–30 cm, with 5–11, zig-zagged branches, each with 12–50 flowers; petals 7–9 mm, pale yellow with red bands towards tip. *Mexico (Jalisco).* G1. Spring–summer.

6. G. grande Alexander. Illustration: *Cactus & Succulent Journal* (US) **28**: 174–5 (1956).

Large shrub to 2 m, bark grey-green, eventually fissured. Leaves 15–25, rounded, 5–9 × 2–4 cm, and *c.* 5 mm thick, wedge-shaped to spathulate, yellowish with a bluish grey bloom. Flowering-stems to 60 cm, with 8–15 zig-zagged or coiled branches. Flowers to 2.5 cm across; petals *c.* 11 mm, yellow, with red dots; stamens greenish white, anthers red-brown; carpels *c.* 6 mm. *S Mexico.* G1. Winter–spring.

7. G. rusbyi (Greene) Rose (*Cotyledon rusbyi* Greene). Illustration: *Addisonia* **9**: 304 (1924); Lamb & Lamb, Colourful cacti of the American deserts, 101 (1974); Japan Succulent Society, Colour encyclopaedia of succulents, 60 (1981).

Stemless perennial with rosettes of 10–35 leaves, 1.5–5 cm × 3–12 mm, linear to spathulate, the surface covered in rough papillae, greyish red or green. Flowering stems 7–12 cm, with 2 or 3 branches, each with 3–7 flowers. Petals 5–7, *c.* 8 mm, dark red with greenish bands. *N Mexico, USA (Arizona).* G1. Spring–summer.

8. G. paraguayense (N.E. Brown) Walther. Illustration: Graf, Tropica, 374 (1978); Everett, New York Botanical Gardens illustrated encyclopedia of horticulture **5**: 1532 (1981); *Cactus & Succulent Journal* (US) **58**: 48, 49, 53

(1986); Brickell, RHS gardeners' encyclopedia of plants and flowers, 1398 (1989).

Hairless perennial herb to 30 cm. Leaves 2–8 × 1.5–2.5 cm, in loose rosettes of 15–25, flat or hollowed above, keeled beneath, pale lavender-grey, spathulate. Flowers borne on stems to 15 cm with 2–6 branches, each with 3–14 flowers; petals *c.* 1 cm, white, dotted with red. *W Mexico.* G1. Late winter–early spring.

Var. **bernalense** Kimnach & Moran. Illustration: *Cactus & Succulent Journal* (US) **58**(2): 54 (1986), with yellowish leaves not more than 4 cm. Occasionally offered for sale.

9. G. macdougallii Alexander. Illustration: Everett, New York Botanical Gardens illustrated encyclopedia of horticulture **5**: 1533 (1981); Riha & Subik, Illustrated encyclopedia of cacti and other succulents, 275 (1981).

Almost stemless perennial, producing many stolons with short tight rosettes of very blue-grey. Leaves spathulate, 2–4 × 1.5–2 cm, apex abruptly pointed. Flowering-stems to 15 cm, weak, with 1–3 branches, each with 2–5 flowers; petals *c.* 1.3 cm, white to yellowish green with red banding towards the tip. *S Mexico.* G1. Winter–early spring.

17. CREMNOPHILA Rose
S.G. Knees
Hairless, succulent perennial herbs. Stems spreading, trailing or erect. Leaves alternate, fleshy, turgid, arranged in basal rosettes or scattered along stems. Flowering-stems leafy, axillary, deciduous, paniculate, each branch with a solitary flower. Sepals 5, unequal, closely adpressed to corolla; petals 5, united; stamens 10; carpels 5, erect.

A genus of 2 species, both from Mexico, closely allied to *Echeveria*, and sometimes hybridising with it.

Literature: Moran, R., Resurrection of Cremnophila, *Cactus & Succulent Journal* (US) **50**(3): 139–46 (1978).

1a. Leaves grey; petals yellow, spreading **1. nutans**
  b. Leaves green; petals greenish white, erect **2. linguifolia**

1. C. nutans Rose (*Sedum cremnophila* R.T. Clausen; *S. nutans* Rose not Haworth). Illustration: *Addisonia* **1**: t. 25 (1916); *Cactus & Succulent Journal* (US) **50**(3): 139–41 (1978); Graf, Exotica, series 4, edn

12, **1**: 898 (1985); *British Cactus & Succulent Journal* **8**(3): 82–3 (1990).

Stems 7–12 cm. Leaves 20–30, crowded, greyish, rhombic, 2.5–7.5 × 3–3.5 cm, and 1–1.2 cm thick, densely clustered, ascending or decumbent. Panicles *c.* 10 cm, nodding. Petals *c.* 6 mm, bright yellow, spreading, overlapping in bud. *Mexico (Morelos).* H5–G1. Winter.

2. C. linguifolia (Lemaire) Moran (*Echeveria linguifolia* Lemaire). Illustration: *Cactus & Succulent Journal* (US) **50**(3): 142–4 (1978); Everett, New York Botanical Gardens illustrated encyclopaedia of horticultures **3**: 913 (1981).

Stems to 30 cm. Leaves 15–40, crowded or separate, obovate to oblong, spathulate, 2.5–10 × 1.8–5 cm, and 1–1.2 cm thick, obtuse to rounded, green. Flowering-stems 30–45, loose, nodding. Petals *c.* 9 mm, greenish white. *Mexico (Mexico State).* H5–G1. Winter–spring.

18. ROSULARIA (de Candolle) Stapf
U. Eggli
Herbaceous, evergreen, rosette-forming perennials with or without rootstock and/or long tap-root; rosettes single or highly tufted and mat-forming; leaves succulent-juicy, broadly spathulate to oblong, hairless or glandular-hairy; margin entire, glandular-hairy or minutely toothed, some with waxy bloom; inflorescence lateral or terminal, upright or decumbent, often drooping in bud, paniculate with sometimes curled lateral branches; perianth segments 5–9, united for one tenth to three-quarters of their length, tips more or less reflexed; stamens twice as many as petals, carpels free or basally united, upright. Seeds oblong-ellipsoid, 0.5–1.3 mm, dark brownish to ochre, longitudinally striped.

A genus of 25 species including some formerly placed in *Afrovivella* Berger and *Sempervivella* Stapf, mainly from Turkey and adjacent eastern countries, inner Asia, and the Himalaya region. Some (mostly not in cultivation) also occur in Crete and west Mediterranean countries, North Africa and Ethiopia. Although winter-hardy species can be grown in rockeries, most benefit from some protection against excessive rain in winter and should be grown in fast-draining, open compost in shallow pots or pans. Species from the east Mediterranean start their growth cycle in early winter. In species from Turkey, the outer leaves of rosettes dry up during spring and summer, leaving a tightly closed 'bud'. Propagation is usually by offsets or seed. Common pests are aphids on developing inflorescences, and larvae of wine weevil feeding inside enlarged rootstocks or tap-roots. All species of the genus show tremendous variability, which is only partly due to genetic differences. Almost all characters are greatly influenced by the environment, mostly by the availability of water and fertiliser. All subsequent descriptions as well as the key apply to plants grown under 'hard' conditions in order to keep them compact.

Literature: Eggli, U., A monographic study of the genus Rosularia, *Bradleya* **6** (suppl.) 1988.

1a. Rootstock absent, tap-root weakly developed, flowers whitish or yellowish, with or without reddish spots and streaks   2
b. Rootstock present, or mature plants with somewhat to much thickened tap-root   6
2a. Inflorescence lateral   **1. aizoon**
b. Inflorescence terminal   3
3a. Leaves hairless, flowers normally with purplish streaks
          2. serpentinica var. **serpentinica**
b. Leaves glandular-hairy, at least along margin; flowers normally without purplish streaks   4
4a. Outer leaves of rosette flushed bright red   **3. muratdaghensis**
b. All leaves green throughout   5
5a. Inflorescence 10–20 cm
                              4. chrysantha
b. Inflorescence 5–9 cm **5. rechingeri**
6a. Inflorescence terminal, plants normally dying after flowering
                          6. globulariifolia
b. Inflorescence normally lateral, and plants not dying after flowering
                                          7
7a. Inflorescence flowering for more than half of its length, leaves with copious whitish bloom   **7. serrata**
b. Inflorescence flowering for upper third or less   8
8a. Flowers congested into a head-like inflorescence (which may lengthen after flowering)   **8. alpestris**
b. Flowers not congested into a head-like inflorescence   9
9a. Flowers white, saucer-shaped and widely opening, *c.* 1.5 cm across; perianth-segments 6–8; cultivated plants normally with long, thread-like runners   **9. sedoides**

b. Flowers white or various shades of pink, narrowly funnel-shaped; perianth-segments 5; cultivated plants never with thread-like runners   10
10a. Leaves strikingly blue-green or grey-green, sometimes tinged purplish 11
b. Leaves green   12
11a. Leaves blue-green, flat and slightly succulent, margin toothed
          10. sempervivum subsp. **glaucophylla**
b. Leaves grey-green, thickened and very succulent, margin more or less entire
          11. adenotricha subsp. **adenotricha**
12a. Leaves hairless, margin toothed; inflorescence hairless or glandular-hairy
    10. sempervivum subsp. **persica**
b. Leaves glandular-hairy, margin slightly toothed or entire; inflorescence normally glandular-hairy   13
13a. Flowers broadly bell-shaped, 8–12 mm   **12. lineata**
b. Flowers narrowly bell-shaped or tubular, 6–9 mm   14
14a. Rosettes 1.5–3 cm across; leaves 8–18 mm   **13. haussknechtii**
b. Rosettes 3–6 cm across; leaves 1.6–3.2 cm   15
15a. Petals 7–9 mm, pale pink with darker veins
          10. sempervivum subsp. **libanotica**
b. Petals 6–7 mm, pale pink or white
          10. sempervivum subsp. **pestalozzae**

1. R. aizoon (Fenzl) Berger (*R. pallida* (Schott & Kotschy) Stapf; *R. chrysantha* misapplied). Figure 88(3), p. 63. Illustration: Takhtajan, Flora Armenii, **3**: t. 107 (1958); *Bradleya* **6**(suppl.): 42, 49, 64 (1988).

Rosettes globular to slightly flattened-globular, 1.5–3 cm across, pale to fresh green, dry leaves papery and ochre-coloured; leaves glandular-hairy. Inflorescence lateral, 5–9 cm, leafy. Flowers narrowly funnel-shaped (sometimes opening more widely), pale yellowish, rarely with purplish venation. *S, SE, E & NE Turkey, adjacent Armenia.* H3–5.

2. R. serpentinica (Werdermann) Muirhead var. **serpentinica**. Figure 88(6), p. 63. Illustration: *Gartenpraxis* **8**: 8–11

(1987); *Bradleya* **6**(suppl.): 49, 100 (1988).

Rosettes globular to flattened-globular, 1–3 cm across, offsetting and forming compact cushions; leaves hairless (innermost leaves glandular-hairy before a rosette prepares to flower), bluish green, outer leaves attractively flushed purplish, dry leaves ochre and papery. Inflorescence terminal, leafy, glandular-hairy, flowers funnel-shaped, erect, 1.1–1.3 cm, petals white to cream with purplish longitudinal streaks. *SW Turkey.* H4–G1.

A very attractive species, easily grown in porous soil.

Var. **gigantea** Eggli. Figure 89(1), p. 64. Differs in having much larger rosettes (to 11 cm across) and longer inflorescences. The rosettes do not normally produce offsets and the plant has to be propagated by seeds. *SW Turkey.* G1.

3. R. muratdaghensis Kit Tan (*R. platyphylla* misapplied (not *R. platyphylla* (Schrenk) Berger, which is not in cultivation). Figure 89(2), p. 64. Illustration: *Bradleya* **6**(suppl.): 49 (1988) – as sp. A.

Highly tufted with fresh green rosettes, outer leaves flushed bright reddish. Flowering very rarely, flowers as in *R. chrysantha*, to which it is similar. *W Turkey.* H5–G1.

4. R. chrysantha (Boissier) Takhtajan (*R. pallida* misapplied). Figure 89(3), p. 64. Illustration: Jelitto, Schacht & Fessler, Die Freiland Schmuckstauden, edn 3, 545 (1985); *Gartenpraxis* **8**: 8 (1987); *Bradleya* **6**(suppl.): 49, 52 (1988).

Rosettes globular to flattened-globular, 1.5–3 cm across, strongly offsetting, pale green, dry leaves ochre and papery, leaves glandular-hairy. Inflorescence terminal, 10–20 cm, leafy throughout, strongly glandular-hairy, flowers narrowly funnel-shaped, pale greenish cream to pale yellow, with or without purplish venation, or pale dirty olive. *W & SW Turkey.* H3–4.

5. R. rechingeri Jansson (*R. turkestanica* misapplied). Figure 89(4), p. 64. Illustration: *Acta Horti Gotoburgensis* **28**: 185 (1966); Rechinger (ed.), Flora Iranica **72**: t.6 (1970); *Bradleya* **6**(suppl.): 49, 79 (1988).

Very similar to *R. aizoon*, differing in having nearly hairless to slightly glandular leaves, terminal inflorescences and longer flowers. *N Iraq, Iran, Turkey.* H3. Summer.

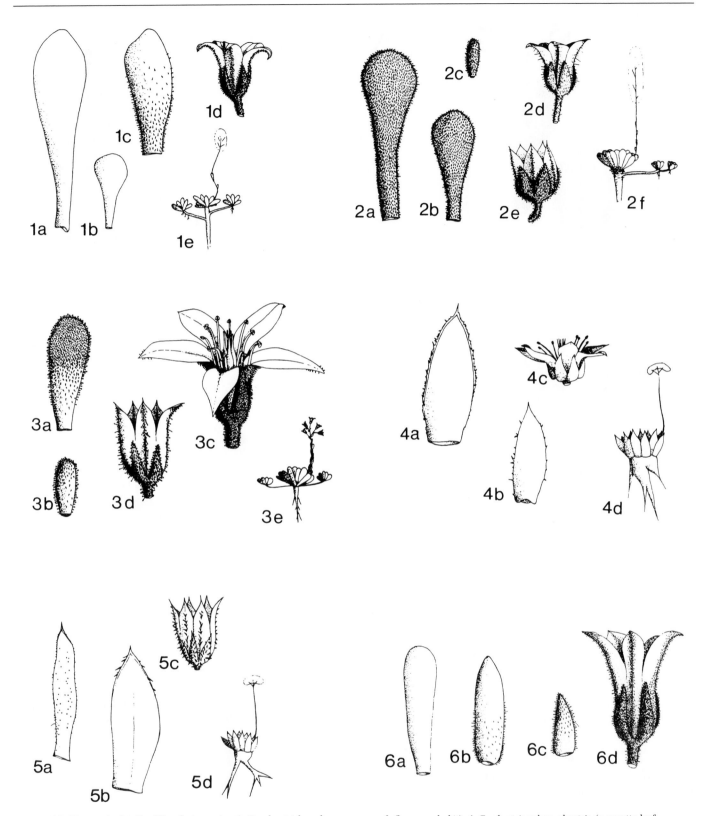

Figure 88. Diagnostic details of *Rosularia* species. 1, *R. adenotricha* subsp. *adenotricha* (a, b, rosette leaf; c, inflorescence leaf; d, flower; e, habit). 2, *R. adenotricha* subsp. *viguieri* (a, b, rosette leaf; c, inflorescence leaf; d, e, flower; f, habit). 3, *R. aizoon* (a, rosette leaf; b, inflorescence leaf; c, d, flower; e, habit). 4, *R. alpestris* subsp. *alpestris* (a, rosette leaf; b, inflorescence leaf; c, flower; d, habit). 5, *R. alpestris* subsp. *marnieri* (a, b, rosette leaf; c, flower; d, habit). 6, *R. serpentinica* var. *serpentinica* (a, rosette leaf; b, c, inflorescence leaves; d, flower).

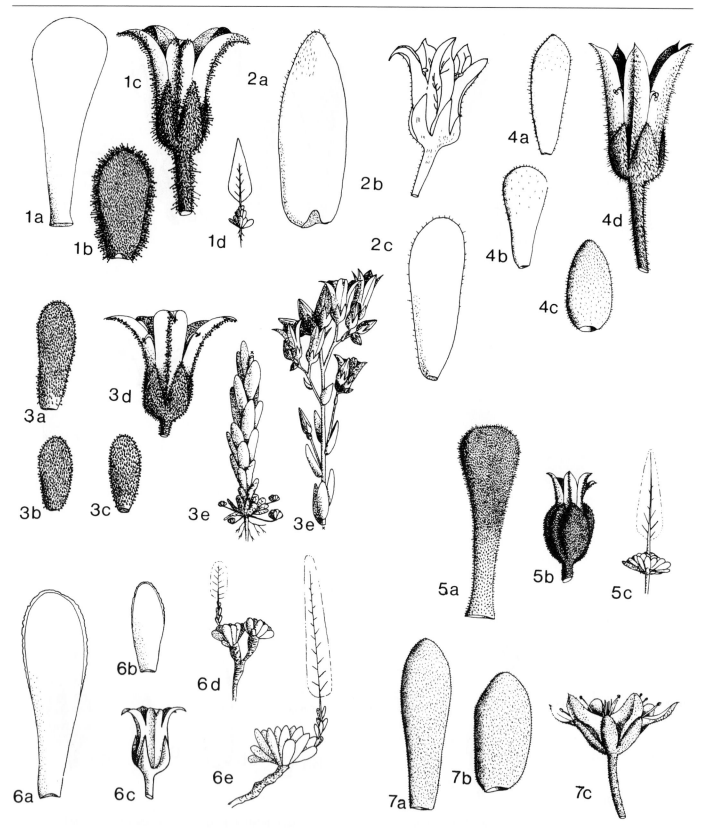

Figure 89. Diagnostic details of *Rosularia* species. 1, *R. serpentinica* var. *gigantea* (a, rosette leaf; b, inflorescence leaf; c, flower; d, habit). 2, R. muratdaghensis (a, inflorescence leaf; b, flower; c, rosette leaf). 3, R. chrysantha (a, rosette leaf; b, inflorescence leaf; c, rosette leaf; d, flower; e, habit). 4, R. rechingeri (a, b, rosette leaf; c, inflorescence leaf; d, flower). 5, R. globulariifolia (a, rosette leaf; b, flower; c, habit). 6, R. serrata (a, rosette leaf; b, inflorescence leaf; c, flower; d, e, habit). 7, R. sedoides (a, rosette leaf; b, inflorescence leaf; c, flower).

Cultivated material under this name is often wrongly identified and represents *R. serpentinica*, but note that *R. turkestanica* (Regel & Winkler) Berger is a synonym of the uncultivated *R. platyphylla*.

6. R. globulariifolia (Fenzl) Berger (*Umbilicus cyprius* Holmboe; *Sedum globulariaefolium* (Fenzl) Hamet: *Rosularia cypria* (Holmboe) Meikle). Figure 89(5), p. 64. Illustration: *Bergens Museums Skrifter* **1**(2): (1914); Hamet, Crassulacearum icones selectae **2**: t. 38 (1956); Meikle, Flora of Cyprus **1**: 646 (1977); *Bradleya* **6** (suppl.): 57, 61 (1988).

Rosettes flattish, open, 4–11 cm across, single or rarely offsetting; leaves flattish, medium green, glandular-hairy and sticky with resinous smell; inflorescence in cultivation nearly always terminal, narrowly pyramidal, 15–40 cm; flowers 5–8 mm, urn-shaped, with large green sepals, petals white or pale pink, more or less erect or slightly reflexed. Spontaneously sets seeds by self-pollination. *E Mediterranean.* H3. Early summer.

In habitat, this species flowers laterally for several years, before it finally produces a terminal inflorescence and then dies. In cultivation, mostly terminal inflorescences are formed, and the plant has to be propagated by seeds.

7. R. serrata (Linnaeus) Berger (*Cotyledon serrata* Linnaeus). Figure 89(6), p. 64. Illustration: Dillenius, Hortus Elthamensis, t. 95 (1732); *Bradleya* **6**(suppl.): 64, 106 (1988).

Rosettes globular or open with more or less erect leaves, 2.5–6 cm; leaves oblong to spathulate with an obscure tip, flattish, grey-green, with copious whitish bloom, glabrous, margins toothed. Inflorescence lateral (very leafy at first and easily mistaken for a rosette), narrowly pyramidal. Flowers small, 6–8 mm, urn-shaped, petals whitish or pale to brownish pink. *SE Aegean Islands, SW coastal Turkey.* G1.

8. R. alpestris (Karelin & Kiriloff) A. Borissova subsp. **alpestris** (*Sempervivella acuminata* (Decaisne) Berger; *Sempervivella mucronata* (Edgeworth) Berger & var. *glabra* Kitamura). Figure 88(4), p. 63. Illustration: *Bradleya* **6**(suppl.): 45, 53 (1988).

Rosettes open-globular, 1.5–6 cm, leaves fresh green, hairless, very lush and succulent, often hard-tipped. Inflorescence 10–20 cm, flowers densely packed at flowering time, later well spaced, opening widely, saucer-shaped, petals white, with or without greenish or purplish venation. *Himalaya, Pamir, China (Xizang), adjacent former USSR.* H3–5.

Subsp. **marnieri** (H. Ohba) Eggli. Figure 88(5), p. 63. Illustration: *Bradleya* **6** (suppl.): 48, 53 (1988). Differs in having very long narrow leaves and smaller, often more greenish, flowers and petals with an irregularly cut margin.

9. R. sedoides (Decaisne) H. Ohba (*Sempervivella sedoides* (Decaisne) Stapf; *Sempervivella alba* (Edgeworth) Stapf; *Rosularia sedoides* var. *alba* (Edgeworth) P. Mitchell). Figure 89(7), p. 64. Illustration: Decaisne, Voyages Inde **4** (Bot.): 74 (1844); Jelitto, Schacht & Fessler, Die Freiland Schmuckstauden, edn 3, 590 (1985); *Succulenta* **66**(1): 20 (1987) – as *Sempervivella mucronata*; *Bradleya* **6**(suppl.): 53, 83 (1988).

Rosettes flattish to flattened-globular, 2–3.5 cm, freely offsetting, with offsets on long, thread-like runners; leaves fresh green, spathulate, glandular hairy. Inflorescence lateral, upright or decumbent, leafy, flowers 1.2–1.8 cm, saucer-shaped, petals white with greenish venation. *N India (Himalaya).* H2–4.

A commonly grown rock garden plant, easily propagated by the copious offsets formed.

10. R. sempervivum (Bieberstein) Berger.

This group of the genus shows a tremendous variability, which makes classification very difficult. Several recognised as species in the past have been subsumed under this species in a recent revision. Intermediates are very common; garden material may moreover be of hybrid origin. Coming from regions with very cold but dry winters, all members of this group are best cultivated in a frost-free greenhouse (G1), some grow adequately under H4–5 conditions. The following subspecies are met with in cultivation:

Subsp. **glaucophylla** Eggli (*R. libanotica* 'Group C' Chamberlain & Muirhead, in Davis (ed.), Flora of Turkey **4**: 219 (1972); *R. spathulata* invalid; *Cotyledon globulariifolia* Baker). Figure 90(1), p. 66. Illustration: Saunders, Refugium botanicum **3**: 201 (1870); *Bradleya* **6** (suppl.): 60, 90 (1988). Striking blue-green flattish rosettes; leaves with toothed margin. Inflorescence hairless or glandular-hairy. Flowers broadly funnel-shaped, 6–9 mm, whitish to pale pink. *S Turkey (Taurus).*

Subsp. **libanotica** (Labillardière) Eggli (*R. libanotica* 'Group E' Chamberlain & Muirhead in Davis (ed.), Flora of Turkey **4**: 219, 1972). Figure 90(2), p. 66. Illustration: Hamet, Crassulacearum icones selectae **3**: 43–4 (1958) – as *Sedum sempervivum*; *Bradleya* **6**(suppl.): 94 (1988). Rosettes medium green, leaves glandular-hairy, flowers broadly funnel-shaped with conspicuously reflexed petal tips, 8–10 mm, pale pink with darker venation. *S Turkey, W Syria, Lebanon, W Jordan, Israel.*

Subsp. **persica** (Boissier) Eggli (*R. persica* (Boissier) Muirhead; *R. radiciflora* Borissova invalid; subsp. *glabra* (Bossier) Muirhead). Figure 90(3), p. 66. Illustration: *Bradleya* **6** (suppl.): 60, 64, 96 (1988). Rosettes flattish, dark green, leaves mostly hairless, sometimes loosely glandular-hairy, margin normally irregularly toothed, often whitish horny. Inflorescence hairless or strongly glandular-hairy, flowers narrowly funnel-shaped, 8–12 mm, clear to dark pink, rarely nearly white. *SW Syria (Hermon), Lebanon, Israel, E Turkey, N & W Iran, NE Iraq, Transcaucasia.*

Very widespread and extremely variable.

Subsp. **pestalozzae** (Boissier) Eggli (*R. pestalozzae* (Boissier) Samuelsson & Fröderström; *R. sempervivum* var. *pestalozzae* (Boissier) Thiébaut, incl. *R. libanotica* 'Group B' Chamberlain & Muirhead in Davis (ed.), Flora of Turkey **4**: 219, 1972). Figure 90(4), p. 66. Illustration: *Bradleya* **6**(suppl.): 60, 99 (1988). Rosettes flattish to flattened-globular, leaves medium green, finely and densely glandular-hairy. Inflorescence glandular-hairy, flowers narrowly funnel-shaped, 6–8 mm, whitish to pale pink. *S Turkey & adjacent Syria.*

11. R. adenotricha (Edgeworth) Jansson (*Sedum adenotrichum* Edgeworth; *Cotyledon tenuicaulis* Aitchison). Figure 88(1), p. 63. Illustration: Saunders, Refugium botanicum **3**: 296 (1871); *Journal of the Linnean Society* **18**: 101 (1882); *Journal of the Royal Horticultural Society* **46**: 164 (1921); *Bradleya* **6**(suppl.): 38, 53 (1988).

Rosettes open, 1.5–2.5 cm, leaves more or less erect, exposed stems slightly thickened, brownish, more or less glossy, often with leaf-scars. Inflorescence 15–20 cm, hairless or glandular-hairy; flowers 6–10 mm, petals white or pale pink

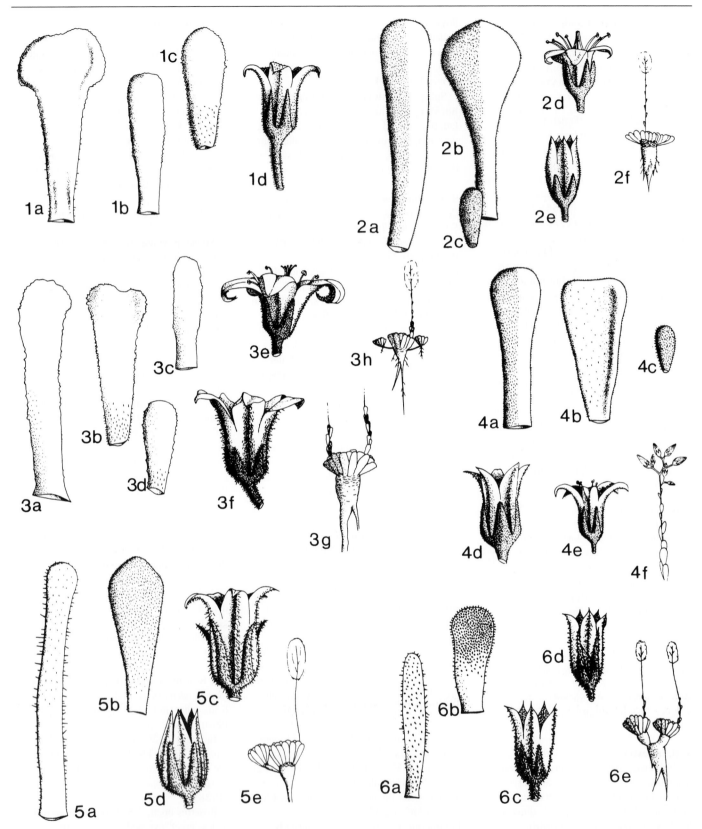

Figure 90. Diagnostic details of *Rosularia* species. 1, *R. sempervivum* subsp. *glaucophylla* (a, b, rosette leaf; c, inflorescence leaf; d, flower). 2, *R. sempervivum* subsp. *libanotica* (a, b, rosette leaf; c, inflorescence leaf; d, e, flower; f, habit). 3, *R. sempervivum* subsp. *persica* (a, b, rosette leaf; c, d, inflorescence leaf; e, f, flower; g, h, habit). 4, *R. sempervivum* subsp. *pestalozzae* (a, b, rosette leaf; c, inflorescence leaf; d, e, flower; f, habit. 5, *R. lineata* a, b, rosette leaf; c, d, flower; e, habit). 6, *R. haussknechtii* (a, b, rosette leaf; c, d, flower; e, habit).

with darker venation. *S Himalaya & Pamir.* H5–G1.

Easily propagated from detached leaves.

Subsp. **viguieri** (Hamet) Jansson (*Sedum viguieri* (Hamet) Fröderström). Figure 88 (2), p. 63. Differs in having medium green, glandular hairy leaves and pure white flowers.

12. R. lineata (Boissier) Berger (*R. setosa* Bywater). Figure 90(5), p. 66. Illustration: Bouloumoy, Nouvelle flore du Liban et de la Syrie **2**: 67 (1970); *Kew Bulletin* **34**(2): 403 (1979); *Bradleya* **6**(suppl.): 57, 69 (1988).

Rosettes flattish, 2–7 cm; leaves oblong to spathulate, with rounded or truncate apex, 2–4 cm, deep green, shortly glandular-hairy, margin glandular-hairy with long bristles. Inflorescence lateral, erect, flowers 8–12 mm, broadly bell-shaped, pale pink with dark pink venation, often foetid. *Lebanon, Syria, Jordan, Israel.* G1.

13. R. haussknechtii (Boissier & Reuter) Berger. Figure 90(6), p. 66. Illustration: *Bradleya* **6**(suppl.): 63, 64 (1988).

Rosettes flattish, 1–3 cm; leaves dull or fresh green, narrowly spathulate, basally narrowed into a long claw, margin entire but with some obvious bristles. Inflorescence lateral, 4–8 cm, flowers narrowly funnel-shaped, bright pink. *SE Turkey.* H4–G1.

Probably not in cultivation; this name has been widely misapplied to many different Turkish species.

19. THOMPSONELLA Britton & Rose
G.D. Rowley
Echeveria-like rosette succulents with a tendency to shed their leaves in winter; stem undeveloped or very short, unbranched. Leaves spreading, thick, soft fleshy, hairless except for marginal hairs, concave above, convex below. Inflorescence lateral, a long dense spike or slender thyrse. Flowers almost stalkless, small, with 5 erect sepals, 5 thin spreading petals with recurved tips, 10 erect stamens and 5 free carpels with slender beak-like tips and narrow bases. Nectar scales thin.

A genus of 2 species in Mexico. Rot-sensitive and difficult to keep in cultivation: water sparingly and keep quite dry in winter when the leaves wither. Propagation by seed; adventitious buds sometimes appear in the inflorescence and may be treated as cuttings.

Literature: Moran, R., *Cactus & Succulent Journal (US)* **41**: 173–5 (1969).

1. T. minutiflora (Rose) Britton & Rose (*Echeveria minutiflora* Rose). Illustrations: *Contributions from the US National Herbarium* **12**: t. 44 (1909); *Cactus & Succulent Journal* (US) **8**: 100–2 (1937) & **41**: 173–5 (1969); Martin & Chapman, Succulents and their cultivation, 160–1 (1977).

Roots thickened; rosette to 17 cm, flat; leaves 10–20, oblanceolate, acute, to 10 × 1.5–2.5 cm, 3–7 mm thick, dark green, sometimes flushed with purple, margin upcurved, more or less curly. Inflorescence to 30 cm, unbranched, with a diminishing series of leaf-like bracts below. Flowers 1–7 in tight cymes, to 1 cm across, dull yellow with red markings recalling those of *Graptopetalum*. *Mexico.* G1. Late summer–autumn.

20. DUDLEYA Britton & Rose
E.H. Hamlet
Perennial herbs with simple or branched rootstocks or small spherical to oblong corms; leaves mainly in basal rosettes, fleshy, flattened to nearly terete, more or less ovate to linear, commonly white, glaucous. Flowering-stems axillary, with stem-leaves much reduced. Flowers in terminal paniculate or cyme-like clusters. Calyx deeply divided into 5 erect, lanceolate-linear to ovate segments. Corolla white to yellow or red, cylindric or bell-shaped; petals united near base, erect or spreading from middle or near tips of petals. Stamens 10, borne on the corolla-tube. Seeds numerous, narrowly ovoid, brown or reddish brown.

A genus of 40 species native to SW USA & NW Mexico. They prefer a well-drained fertile soil in a semi-shaded position and little water in winter. Propagation by division or seed.

Literature: Thiede, J., The genus Dudleya Britton & Rose, its systematics and biology, *Cactus & Succulent Journal* (US) **76**: 224–31 (2004).

Flowering stems. Mealy: **5, 9.** Twisted: **5.**
Petals. White or cream: **4, 6, 8.** Yellow: **1, 2, 3, 6, 7, 9, 10.** Pink or red: **2, 3, 5, 7, 8, 10.**

1a. Leaves linear or narrowly lanceolate 2
 b. Leaves not as above 4
2a. Leaves narrowly lanceolate; flowers yellow or red **10. saxosa**
 b. Leaves linear; flowers not yellow 3
3a. Petals narrowly ovate; leaves acute **8. densiflora**
 b. Petals oblong-lanceolate; leaves acute to almost acuminate **4. edulis**
4a. Rosette leaves 30 or more 5
 b. Rosette leaves fewer than 30 6
5a. Calyx segments triangular; leaves triangular-ovate to oblong-lanceolate **1. candida**
 b. Calyx segments lanceolate; leaves obovate to spathulate **5. pulverulenta**
6a. Flowering-stems mostly white-mealy **9. farinosa**
 b. Flowering-stems not white-mealy 7
7a. At least inner leaves glaucous 9
 b. Leaves not glaucous 8
8a. Leaves oblanceolate to spathulate, 1.5–6 cm **7. cymosa**
 b. Leaves oblong, 5–13 cm **2. cultrata**
9a. Stem to 20 cm 10
 b. Stem rarely more than 5 cm **7. cymosa**
10a. Flowers yellow to red; leaves 5–20 cm **6. caespitosa**
 b. Flowers yellow with red markings; leaves 5–8 cm **3. rigida**

1. D. candida Britton. Illustration: Jacobsen, Lexicon of succulent plants, pl. 54 (1974); *Cactus & Succulent Journal* (US) **46**: 71 (1974); Lamb & Lamb, Illustrated reference on cacti and other succulents **3**: 765 (1963).

Stems 2–6 cm across, branching to form clumps to 80 cm across. Rosettes 7–21 cm across. Leaves 30–70, glaucous or green, 5–11 × 1–3 cm, triangular-ovate to oblong-oblanceolate. Inflorescence to 50 cm, with 3–5 branches. Petals 8.5–13 × 2–3.5 mm, oblong-lanceolate, acute, pale yellow. *Mexico (Baja California).* Spring.

2. D. cultrata Rose. Illustration: *Cactus & Succulent Journal* (US) **40**(4): 170 (1968).

Stems 2–4 cm across, to 20 cm, branching. Rosettes 3–8 cm across. Leaves 20–30, green, 5–13 × 1–1.5 cm, 3–5 mm thick, oblong, pointed. Inflorescence to 40 cm, with about 3 branches. Stem-leaves triangular, acute. Calyx 4.5–6 × 5–6 mm; segments triangular, acute. Petals 10–13 × 3 mm, elliptic, acute, pale yellow. *Mexico (Baja California).* Spring.

3. D. rigida Rose. Illustration: Jacobsen, Lexicon of succulent plants, pl. 55 (1974); *Cactus & Succulent Journal* (US) **59**(5): 188–9 (1987).

Stems erect, to 10 cm or more, 1–3.5 cm across, branching to form clumps to 25 cm across. Rosettes 6–15 cm across. Leaves 10–25, usually slightly glaucous, 5–8 × 2.5–4 cm, 6–10 mm thick, oblong to triangular-ovate, tapering, short-acuminate. Flowering-stems to 50 cm, with 2 or 3 branches. Petals yellow with red markings. *Mexico (Baja California)*. Spring.

4. D. edulis Moran. Illustration: Lamb & Lamb, Illustrated reference on cacti and other succulents **5**: 1430 (1978).

Stems 1.5–4.5 cm across, to 15 cm, branching to form clumps to 40 cm across. Rosettes 5–10 cm across; leaves 15–25, green, 5–20 cm × 4–10 mm, linear, slightly glaucous, acute to almost acuminate. Flowering-stems 15–50 cm, stem-leaves 1–5 cm, turgid, triangular-lanceolate, acute. Inflorescence long, rather open. Flower-stalks to 2 mm; calyx-lobes 2.5–4.5 mm, oblong, acute; petals 7–10 mm, joined for 1–1.5 mm, oblong-lanceolate, acutish, creamy white. *USA (California), Mexico (Baja California)*. Early summer.

5. D. pulverulenta Britton & Rose. Illustration: *Cactus & Succulent Journal* (US) **48**: 4 (1976); Riha & Subik, The illustrated encyclopedia of cacti and other succulents, 254 (1981); Brickell (ed.), RHS A–Z encyclopedia of garden plants, 392 (2003).

Plant mealy-glaucous; stems to 40 cm, 4–9 cm thick. Rosette leaves 30–80, obovate to spathulate, 8–25 × 4–10 cm. Flowering-stems stout, 40–80 cm; branches twisted so that the flowers are more or less on the underside; stem-leaves many, 1–4 cm, broadly ovate, heart-shaped, clasping at base, acute. Flower-stalks slender, 5–30 mm, spreading. Calyx-segments *c.* 4–8 mm, lanceolate, acute, red to glaucous; petals 1.2–1.8 cm, yellow to mostly deep red, joined nearly to the middle. Seeds brown. *USA (California), Mexico (Baja California)*. Late spring–summer.

6. D. caespitosa Britton & Rose. Illustration: *Cactus & Succulent Journal* (US) **26**: 10 (1954).

Stems 1.5–4 cm across, to 20 cm, branching. Rosettes 5–15 cm across; leaves 15–30, 5–20 × 1–5 cm, oblong-oblanceolate, almost shining, only the inner glaucous. Inflorescence to 60 cm, with 3–14 flowers. Calyx 4–6 × 4–8 mm, nearly truncate to tapering at base. Petals

0.8–1.6 cm, white to bright yellow, rarely red. *USA (California)*. Late spring.

7. D. cymosa Britton & Rose (*Cotyledon cymosa* Baker). Illustration: Saunders, Refugium botanicum **1**: f. 68 (1869).

Stems 1–3.5 cm, simple or with few branches. Rosettes usually 6–15 cm across. Leaves 16–25, 1.5–6 × 1–3 cm, mostly evergreen, green to glaucous, usually oblanceolate to spathulate, pointed. Inflorescence with 2 or 3 branches. Calyx 3–6 mm, triangular-ovate, acute. Petals 7–14 mm, yellow to red. *USA (California)*. Late spring–summer.

8. D. densiflora Moran. Illustration: Jacobsen, Lexicon of succulent plants, pl. 55 (1974).

Stems 1–2.5 cm across, to 10 cm, branching. Rosettes 7–25 cm across; leaves 20–40, linear, 6–15 cm × 6–12 mm, 5–8 mm thick, acute, terete above, persistently glaucous. Flowering-stems 15–30 cm, with 3 or more branches; inflorescence dense, more or less rounded. Stem-leaves 1–4 cm, turgid, acute. Flower-stalks 2–5 mm; calyx-lobes 1.5–2.5 mm, triangular-ovate, acute; petals 5–10 mm, narrowly ovate, acute, whitish to pink. *USA (California)*. Summer.

9. D. farinosa Britton & Rose. Illustration: Lamb & Lamb, Illustrated reference on cacti and other succulents **3**: 470 (1963); Jacobsen, Lexicon of succulent plants, pl. 54 (1974); Riha & Subik, The illustrated encyclopedia of cacti and other succulents, 253 (1981).

Stems stout, 1–3 cm thick, often tapered gradually, usually with several rosettes, each with 15–30 leaves. Rosettes 4–10 cm across; leaves 2.5–6 × 1–2.5 cm, 3–6 mm thick, densely white-mealy to green, ovate-oblong, acute, flat on upper surface, lightly rounded beneath. Flowering-stems 10–35 cm, with 3–5 branches, stout, mostly white-mealy; stem-leaves many, 1–2.5 cm, triangular-ovate, concave. Flower-stalks 1–5 mm, stout; calyx 5–8 mm, mostly 5–6 mm wide, rounded at base; lobes deltoid-ovate. Petals 1–1.4 cm, lemon-yellow, oblong, acute. *USA (California, Oregon)*. Summer.

10. D. saxosa Britton & Rose. Illustration: Munz, A flora of southern California, 380 (1974).

Plant pale green or almost glaucous; stems 1–3 cm across. Rosettes 3–10 cm across. Rosette leaves 10–25, narrowly lanceolate, 3–15 cm × 5–25 mm,

1.5–6 mm thick, almost rounded. Flowering-stems 5–40 cm, more or less reddish, with 2 or 3 branches; leaves ovate-lanceolate, slightly clasping. Flower-stalks 1–2 cm; calyx-lobes *c.* 5 mm, lanceolate-ovate, red; petals 0.9–2 cm, oblong-lanceolate, acute, yellow, more or less reddish in age. *USA (California)*. H2–3. Summer.

21. PACHYPHYTUM Link, Klotzsch & Otto
G.D. Rowley
Small, evergreen, hairless, perennial shrublets with thick, fleshy, sparingly branched stems covered in leaf-scars. Leaves simple, entire, stalkless, smooth, thick and highly succulent, spirally arranged in rosettes, mostly white- to bluish-bloomed, often flushed purple. Scape lateral, erect, with small scattered bracts, ending in 1–few pendent or scrolled, one-sided inflorescences, with small tubular flowers in a double row often half-hidden by overlapping bracts, pendent but often becoming erect by subsequent growth of the inflorescence. Sepals 5, equal or unequal, adpressed. Petals 5, shortly united at base, erect with diverging tips, the lower margins inrolled to form 2 scale-like appendages at the base about half as long as the petal. Stamens 10 in 2 whorls. Carpels 5, free.

A genus of around 13 species from Mexico, beloved of succulent collectors for the sculptural effect of the few, thick, chunky leaves with a characteristic waxy bloom, which in *P. oviferum* resemble sugared almonds. They should not be handled, however, as they mark easily. The normal porous but nutritious soil used for most succulents suits *Pachyphytum*, with good drainage; water (preferably from below to avoid stains on the foliage) freely in summer and sparingly in winter. A minimum of 4 °C is adequate, and an occasional drop below zero should do no harm if the plant is dry and resting. Flowering tends to be spread over a long season rather than concentrated into a single month. Most species are adaptable for use as house plants, and add appeal to bowl gardens and floral wreaths. Propagation is by cuttings or single leaves, which throw up one or more plantlets and roots from the leaf-base.

For hybrids with *Echeveria* see under × *Pachyveria*.

Literature: Walther, E., The Genus Pachyphytum, *Cactus and Succulent Journal*

(US) **3**: 9–13 (1931); Poellnitz, K.V., Pachyphytum, *actus Journal (Great Britain)* **5**: 72–5 (1937). (Both with keys to 8–10 species); Thiede, J., Pachyphytum, in U. Eggli (ed.), Illustrated handbook of succulent plants: Crassulaceae (2003).

1a. Petals without basal appendages
       × **Pachyveria scheideckeri**
  b. Petals with 2 scale-like basal
       appendages   2
2a. Leaves yellowish green, never
       glaucous   **6. viride**
  b. Leaves grey, pinkish, dark or bluish
       green, glaucous   3
3a. Leaves flattened; sepals large, at least
       as long as petals   4
  b. Leaves more or less terete;
       sepals smaller, shorter than
       petals   9
4a. Sepals obtuse, widest above the
       middle   5
  b. Sepals acute, widest at or below the
       middle   6
5a. Leaves less than twice as long as
       broad, upperside convex; flowers
       7–15   **5. oviferum**
  b. Leaves more than twice as long as
       broad, upperside flat or concave;
       flowers 12–25   **1. bracteosum**
6a. Inflorescence a recumbent panicle
                      × **Pachyveria sodalis**
  b. Inflorescence erect, raceme-like   7
7a. Flowers on short stalks, crowded   8
  b. Flowers on long stalks, scattered
                      × **Pachyveria mirabilis**
8a. Leaves grey, broadest midway;
       raceme forked
                      × **Pachyveria clavata**
  b. Leaves flushed purple, broadest
       above the middle; racemes mostly
       with 3 branches
              × **Pachyveria pachyphytoides**
9a. Stems sticky near the top; leaves
       broad, 2.5 cm wide or more
                               3. glutinicaule
  b. Stems not sticky; leaves narrower,
       almost as thick as wide   10
10a. Leaves densely packed and faceted
        from mutual pressure   11
  b. Leaves scattered with visible
       internodes, scarcely faceted   12
11a. Leaves 5–6 cm, upperside slightly
        flattened; petal tips recurved
                      × **Pachyveria glauca**
  b. Leaves 2–4 cm, angular terete; petals
       erect   **2. compactum**
12a. Leaves 2–3 times as long as broad;
        petals 1.2–1.5 cm   **4. hookeri**

  b. Leaves to 5 times as long as broad;
       petals 1–1.2 cm
                      × **Pachyveria sobrina**

1. P. bracteosum Link, Klotzsch & Otto. Illustration: *Botanical Magazine*, 4951 (1856); van Laren, Succulents other than cacti, 80 (1935); *Succulenta* **38**: 20 (1959); Rauh, Wonderful world of succulents, t. 77/3 (1984).

Stem 10–30 cm, 1.2–2.5 cm thick, with few thick branches. Leaves obovate, tapering towards the base, apex obtuse or rounded, often mucronate, 7–10 × 2.5 cm, to 1.2 cm thick, bloomed with sometimes a mauve flush. Scape 15–50 cm, unbranched, with oblong or ovate bracts to 2.5 cm. Flowers 12–25; sepals 1.4–2.2 cm; petals deep red, 9 mm. *Mexico (Hidalgo).* G1. Summer–autumn.

2. P. compactum Rose. Illustration: *Succulenta* **38**: 7, 26, 154 (1959); Jacobsen, Lexicon of succulent plants, t. 114/5 (1974); Brickell (ed.), RHS A–Z encyclopedia of garden plants, 758 (2003); Eggli (ed.), Illustrated handbook of succulent plants: Crassulaceae, t. xivd. (2003).

Compact, with thick stems rarely above 10 cm bearing dense rosettes of 30–60 terete, lanceolate, dark to reddish green, bloomed leaves 2–4 cm × 5–10 mm across. Upper leaf surface somewhat flattened or faceted with net-like, paler marginal lines; apex tapered, acute. Scape to 40 cm with about 10 pendent flowers *c.* 8 mm, on stalks to 1.5 cm. Sepals nearly equal, smaller than the petals. Petals orange-red with a bluish green tip. *Mexico (Hidalgo).* G1. Spring.

A cristate clone is commonly grown.

3. P. glutinicaule Moran (*P. brevifolium* misapplied). Illustration: *Cactus and Succulent Journal (US)* **3**: 152 (1932), **4**: 237–8 (1932) & **35**: 37–9 (1963); *Succulenta* **38**: 102–3, 154 (1959); Jacobsen, Lexicon of succulent plants, t. 114/2 (1974); Eggli (ed.), Illustrated handbook of succulent plants: Crassulaceae, t. xxive. (2003).

Stems to 30 cm, decumbent, 7–15 mm thick, sticky when young. Leaves in loose rosettes 7–12 cm across, bluish glaucous, obovate to oblong, mucronate, 3–6 × 2–3.5 cm, and 5–15 mm thick. Scape 15–20 cm, with 6–23 flowers that are pendent at first, erect later. Flower-stalks 4–15 mm; sepals 1–1.6 cm; petals

red, 1.2–1.7 cm. *Mexico (Queretaro, Hidalgo).* G1. Flowers over a long season, mostly winter–spring.

4. P. hookeri (Salm-Dyck) A. Berger (*Cotyledon adunca* Baker; *Pachyphytum uniflorum* Rose). Illustration: Saunders, Refugium botanicum, t. 60 (1869); *Cactus and Succulent Journal* (US) **16**: 159 (1944); *Succulenta* **38**: 132 (1959).

Stems ascending or decumbent when old, little-branched, 9–18 mm thick. Leaves narrow oblong or lanceolate, almost terete, 3–5 cm × 7–17 mm, 5–11 mm thick, obtuse with a minute mucro, green, glaucous. Scape 10–25 cm, with scattered bracts. Flowers 5–18, 1.2–1.5 cm; sepals equal, 5–8 mm; petals *c.* 1.5 cm, yellowish, flushed purple towards the tip. *Mexico (San Luis Potosí).* G1. Mainly spring.

5. P. oviferum Purpus. Illustration: Rowley, Illustrated encyclopedia of succulents, 124 (1978); Rauh, Wonderful world of succulents, t. 77/2 (1984); Brickell (ed.), RHS A–Z encyclopedia of garden plants, 758 (2003); Eggli (ed.), Illustrated handbook of succulent plants: Crassulaceae, t. xxiva. (2003).

Stems short, tending to become prostrate, to 1.3 cm across. Leaves in loose rosettes, obovate, 2.5–5 × 1.8–3 cm, 1–1.6 cm thick, rounded at the edges and somewhat channelled on the upperside, bloomed with often a lavender flush recalling sugared almonds. Scape 8–10 cm, with 7–15 greenish or reddish white flowers. Sepals unequal, 1.5–1.8 cm. Petals *c.* 1 cm. *Mexico (San Luis Potosi).* G1. Winter–spring.

6. P. viride Walther. Illustration: *Cactus and Succulent Journal* (U.S.) **8**: 210–11 (1937); Lamb & Lamb, Illustrated reference on cacti and other succulents **2**: 472 (1959); *Succulenta* **38**: 51 (1959); Jacobsen, Lexicon of succulent plants, t. 114/3 (1974).

Stem squat, 2–3 cm across, heavily leaf-scarred. Leaves crowded at the stem apex, finger-like, spreading and inclined, almost terete, 8–15 × 1.3–2 cm, 1–1.8 cm thick, blunt, yellowish green, never glaucous. Scape 20–30 cm. Flowers 10–22, nodding at first, erect later. Sepals 1.2–2.5 cm; petals shorter, hidden within the calyx, pinkish red with paler margins. *Mexico (Queretaro).* G1. Autumn & spring.

P. fittkaui Moran (illustration: Eggli (ed.), Illustrated handbook of succulent plants: Crassulaceae, t. xxivd, 2003), has shorter, broader, narrowly obovate to oblanceolate leaves. *Mexico.*

22. × PACHYVERIA Haage & Schmidt

G.D. Rowley

More or less shrubby evergreen perennials with very succulent, spirally arranged entire leaves in terminal rosettes as well as more or less scattered along the branches. Leaves oblong to spathulate to nearly cylindrical, upperside sometimes concave, more or less glaucous, bluish green, often flushed purple. Flowering-stems lateral, ending in a one-sided cyme with few or no branches, with scattered bracts like reduced foliage leaves. Sepals adpressed. Corolla bell-shaped, pinkish to orange-red, often darker spotted or blotched. Petals with or without (× *P. scheideckeri*) the scale-like appendages of *Pachyphytum* at the base.

Intergeneric hybrids between species of *Echeveria* and *Pachyphytum*. For key to hybrids and hybrid cultivars, see also under *Pachyphytum.*

Literature: Walther, E., Echeveria hybrids, *Cactus and Succulent Journal (US)* **6**: 53–6 (1934); Keppel, J.C.V., × Pachyveria, *Succulenta* **41**: 6–10, 30–1, 45–6, 88–9 (1962) & **42**: 86–7, 115–17,133–5 (1963).

1a. Petals with a pair of basal
      appendages   2
  b. Petals without basal appendages
                                    5. × scheideckeri
2a. Leaves flattened; sepals large, as long
      as petals or longer   3
  b. Leaves almost terete; sepals small,
      shorter than petals   6
3a. Inflorescence recumbent, paniculate
                                          7. × sodalis
  b. Inflorescence erect, raceme-like   4
4a. Flowers crowded, on short stalks   5
  b. Flowers not crowded; stalks to 6 mm
                                          3. × mirabilis
5a. Leaves grey, broadest midway;
      raceme forked   **1. × clavata**
  b. Leaves flushed purple, broadest
      above the middle; raceme with
      usually 3 branches
                                    4. × pachyphytoides
6a. Leaves crowded and faceted from
      mutual pressure   **2. × glauca**
  b. Leaves loose, not or scarcely faceted
                                          6. × sobrina

1. × **P. clavata** Walther; *Echeveria* sp. × *Pachyphytum bracteosum* (*E. clavifolia* misapplied). Illustration: *Succulenta*

41: 88–9 (1962); Jacobsen, Lexicon of succulent plants, t. 116/1 (1974).

Small shrublets to 40 cm, with few, stout, erect branches with leaves scattered along the upper parts and crowded in a rosette at the ends. Leaves flat, oblong with an obtuse apex, *c.* 10 × 3 cm, grey-glaucous. Inflorescences erect, with 2 scrolled branches packed with numerous small flowers hanging when young, later becoming erect, on short stalks. Sepals broad, as long as corolla. Petals reddish, appendaged. *Garden origin.* G1. Summer.

A cristate variant is also in cultivation.

2. × **P. glauca** Walther; *Echeveria* sp. × *Pachyphytum compactum*. Illustration: *Succulenta* **41**: 45–6 (1962); Jacobsen, Lexicon of succulent plants, t. 116/3 (1974).

Stem undeveloped. Leaves about 40, crowded into a dense rosette, almost terete, slightly flattened on the upperside, 5–6 × 1–1.5 cm, angled from mutual pressure when young, acute, bluish glaucous. Inflorescences 30 cm or more, with scattered small bracts, bearing *c.* 8 pendent flowers. Sepals equal, *c.* 6 mm; petals yellow, with recurving red tips, *c.* 1.2 cm, appendaged at base. *Garden origin.* G1.

'Glossoides' is classified here and develops decumbent stems covered in leaves.

3. × **P. mirabilis** (Deleuil) Walther; *Echeveria scheeri* × *Pachyphytum bracteosum*. Illustration: *Cactus and Succulent Journal* (US) **6**: 54 (1934).

Stem undeveloped. Leaves flat, opaline rose, glaucous, *c.* 60 to a rosette, oblanceolate, acute, *c.* 7 × 2 cm, channelled. Inflorescence simple, erect, to 20 cm, with a single long, loose, hanging, one-sided series of flowers on flower-stalks to 6 mm. Sepals acute, as long as petals. Petals appendaged at the base. *Garden origin.* G1.

4. × **P. pachyphytoides** (Morren) Walther; *Echeveria gibbiflora* × *Pachyphytum bracteosum*. Illustration: *Succulenta* **42**: 86 (1963).

Stem thick, stiffly erect, usually unbranched, 20–40 cm, with spiralled leaf scars, topped by a rosette of flat, spreading, oblanceolate, obtuse or rounded leaves to 12 × 5 cm, bluish glaucous with a purple flush. Inflorescences to 40 cm, with usually 3 scrolled branches bearing numerous, small, crowded, short-stalked flowers.

Sepals longer than corolla. Petals pinkish red, appendaged. *Garden origin.* G1.

Similar to *Pachyphytum bracteosum*, but distinguished by the forked inflorescence, more pointed bracts, sepal shape and pale tips to the petals.

5. × **P. scheideckeri** Walther; *Echeveria secunda* × *Pachyphytum bracteosum*. Illustration: van Laren, Succulents other than cacti, 75 (1935); *Succulenta* **41**: 8, 10 (1962); Jacobsen, Lexicon of succulent plants, t. 116/2 (1974).

Short–stemmed, with *Echeveria*-like rosettes of *c.* 50 oblong or tapered flat leaves 5–7 × 1–2 cm, *c.* 6 mm thick, acute, bluish glaucous and somewhat striped. Inflorescences *c.* 13 cm, with a long pendent tip bearing about 10 starry orange-red flowers. Sepals broad, nearly as long as corolla. Petals *c.* 1.3 cm, without appendages. *Garden origin.* G1.

A fasciated 'Cristata' and variegated 'Albocarinata' are also in cultivation.

6. × **P. sobrina** (Berger) Walther; *Echeveria* sp. × *Pachyphytum hookeri*. Illustration: *Cactus and Succulent Journal* (US) **6**: 54 (1934).

Stem 10–15 cm, to 2 cm thick, with closely packed leaf-rosettes. Leaves *c.* 20, narrow, to 5 times as long as broad, *c.* 5 cm × 9 mm, almost terete, oblanceolate, acute, flattened above, bright green, glaucous. Inflorescences to 25 cm with small scattered bracts, forked at the top. Flowers 6–15, on a pendent branch; flower-stalks 1–1.2 cm. Bracts lanceolate, acute. Sepals shorter than the petals, nearly equal, triangular lanceolate, 6–7 mm. Petals 1–1.2 cm, red, keeled, appendaged at the base. *Garden origin.* G1.

7. × **P. sodalis** (Berger) Walther; *Echeveria* sp. × *Pachyphytum bracteosum*. Illustration: *Cactus and Succulent Journal* (US) **6**: 54 (1934).

Stems short, thick, little-branched, to 3 cm across. Leaves 15–25 in a rosette, flattened, wedge-shaped with a broad blunt tip with a mucro, to 9 × 3.5 cm, ascending, glaucous, bright green flushed purple. Inflorescence recumbent, 40–50 cm, with scattered lanceolate bracts. Branches about 3, pendent, with overlapping bracts. Flower-stalks 6–7 mm. Sepals narrow, acute, unequal, linear-lanceolate, 1–1.5 cm. Petals lanceolate, keeled, finely striped with red, appendaged at base. *Garden origin.* G1.

23. × GRAPTOVERIA Gossot

G.D. Rowley

Shrubby or stemless evergreen perennials with compact or loose rosettes. Leaves entire, succulent, lanceolate to obovate or spathulate, tapered below, acute or mucronate, bluish green, sometimes flushed or blotched purple and more or less glaucous. Inflorescences borne laterally, with usually much-branched cymes and a series of bracts transitional from foliage leaves. Flower-stalks 1–3 cm. Sepals 5, short, adpressed to the corolla tube. Corolla more or less urn-shaped, yellow to pinkish, often red-spotted. Stamens 10, 5 of which recurve between the petals.

Intergeneric hybrids between species of *Echeveria* and *Graptopetalum*. In addition to the hybrids described here, several more have been recently offered in the USA. All are notable for the leaf textures and intense colourings.

Literature: Keppel, J.C.V., An account of the hybrid genus × Graptoveria, *National Cactus and Succulent Society Journal* **35**: 28–31 (1980) & **36**: 13–17 (1981).

1a. Leaves papillose, at least when young   **2. 'Michael Roan'**
  b. Leaves all smooth   2
2a. Leaves to 5 cm, awned, with a purplish bristle tip; stems undeveloped   **3. 'Silver Star'**
  b. Leaves 10 cm or more, not awned; rosettes on stems above ground   **1. 'Fred Ives'**

1. × G. 'Fred Ives'; *Echeveria gibbiflora* × *Graptopetalum paraguayense*. Illustration: *National Cactus and Succulent Society Journal* **36**: 14 (1981).

Vigorous, stem to 3 cm thick, leaf-rosettes 30–40 cm across when the plant is well fed. Leaves hairless, obovate–oblanceolate, *c.* 15 × 6 cm, with the mucro often oblique, bluish green with a purple flush when exposed to sun. Flowering-stem copiously leafy, branching at the top with groups of 5–15 flowers with large persistent bracts. Sepals long, unequal. Petals yellowish. *Garden origin.* G1. Spring.

Cristate and variegated clones are also grown, and being smaller are preferred when space is limited.

2. × G. 'Michael Roan'; *Echeveria setosa* × *Graptopetalum paraguayense*. Illustration: Rowley, Illustrated encyclopedia of succulents, 120, (1978).

Stem very short, *c.* 8 mm thick, branching to form a cluster of compact rosettes *c.* 9 cm across. Leaves blue-green, at first papillose, oblong with an acute apex, convex and keeled below, *c.* 5 × 1 cm. Flowering-stem to 15 cm, almost bractless, branching all round, with numerous small cup-shaped flowers with yellow petals spotted with red. Sepals adpressed. *Garden origin.* G1. Spring.

3. × G. 'Silver Star'; *Echeveria agavoides* × *Graptopetalum filiferum*. Illustration: *Cactus and Succulent Journal* (US) **46**: 135 (1974) & **55**: 261 (1983); Lamb, Neale's photographic reference plates, No. 1787 (undated).

Densely clump-forming with compact almost stemless rosettes 6–10 cm across, made up of numerous, ovate, very fleshy, silvery green leaves, each tapering to a long purplish bristle tip, like a larger version of *Graptopetalum filiferum*. Leaves to 4 × 1.3 cm, lanceolate with the awn 9–11 mm. Flowers on slender lateral inflorescences, intermediate between those of the parents. *Garden origin.* G1. Spring.

24. ECHEVERIA de Candolle

S.G. Knees, H.S. Maxwell, R. Hyam, G.D. Rowley

Succulent, evergreen or occasionally somewhat deciduous, perennial herbs and shrublets. Roots usually fibrous, sometimes thick. Stems usually short, simple, occasionally long or branched. Leaves alternate or spiralled, scattered or in dense rosettes, usually entire and stalkless, fleshy or thin, pointed, not clasping at base, hairless or hairy, glaucous or shiny. Inflorescence lateral, on an erect stem with numerous bracts; flowers in racemes or panicles; sepals 5, more or less equal, fused at base or free, spreading, erect or adpressed; petals 5, red, orange, yellow, white or greenish, fused into a 5-angled tube at base, segments erect or reflexed, thickly keeled; stamens 10, unequal, carpels 5.

A genus of about 150 species from USA, Mexico, Central and South America. Most are easily propagated from leaf or stem cuttings or offsets.

Literature: Ginns, R., Echeverias, The National Succulent Society Handbooks **1** (1968); Walther, E., Echeveria (1972); Carruthers, L. & Ginns, R., Echeveria: A guide to cultivation and identification (1973); Schulz, L. & Kapitany, A.,

Echeveria cultivars (2005); Pilbeam, J. The genus Echeveria (2008).

1a. Leaves conspicuously hairy or with very fine hairs or papillae   2
  b. Leaves hairless, but often covered with a waxy bloom   13
2a. Leaf-hairs very fine (hand lens required), or leaf surface covered with minute papillae   3
  b. Leaf-hairs obvious without the use of a lens   5
3a. Leaves 10–14 cm, with very fine hairs, or sometimes hairless   **1. semivestita**
  b. Leaves 3–7 cm, with fine hairs or pointed papillae   4
4a. Leaves 4–7 cm, surface roughened with minute pointed papillae; corolla 2–2.4 cm   **2. spectabilis**
  b. Leaves not more than 5 cm, surface with fine hairs but not roughened; corolla *c.* 1.6 cm   **3. nodulosa**
5a. Leaf-rosettes stemless, flowers borne in simple or 2-branched, 1-sided racemes   6
  b. Shrublets, often with scattered leaves, or rosettes borne on clear stems; flowers borne all around the stem   7
6a. Leaves hairy over entire surface   **4. setosa**
  b. Leaves hairy only along margins and on keel beneath   **5. ciliata**
7a. Stems often branched, 30–60 cm, rusty brown towards the top, especially when young; leaves in loose terminal clusters   **6. coccinea**
  b. Stems very short or rarely exceeding 30 cm, solitary or branched   8
8a. Leaves thick, turgid; sepals closely adpressed to corolla   9
  b. Leaves usually thin; sepals spreading   10
9a. Leaf-hairs often reddish; sepals clearly joined at the base, less than half as long as corolla   **7. pulvinata**
  b. Leaf-hairs white; sepals almost free at the base, more than half as long as corolla   **8. leucotricha**
10a. Well–branched shrub to 30 cm or occasionally more; flowers few, corolla 3 cm or more   **9. harmsii**
  b. Shrublet 10–20 cm; flowers in racemes, corolla 1.2–2.6 cm   11
11a. Stems to 20 cm; corolla 2.4–2.6 cm   **10. amphoralis**
  b. Stems rarely exceeding 10 cm; corolla 1.2–1.5 cm   12

12a. Leaves to 7 cm, oblanceolate-elliptic;
     flowers borne in compound panicles
                                    11. pilosa
  b. Leaves to 4 cm, obovate-lanceolate;
     flowers borne in simple racemes
                                    12. pringlei
13a. Leaves 15–40 cm   14
  b. Leaves 2–15 cm   18
14a. Leaves 12–25 cm wide
                                    14. gibbiflora
  b. Leaves 5–10 cm wide   15
15a. Leaves 4 times as long as wide
                                    15. acutifolia
  b. Leaves not more than 2–3 times as
     long as wide   16
16a. Leaves rounded and notched at
     apex; flowering-stem to 2 m
                                    16. gigantea
  b. Leaves rounded or pointed with a
     bristle tip; flowering-stem not more
     than 1 m   17
17a. Bracts spurred; corolla 1.2–1.5 cm,
     buff within   **17. fimbriata**
  b. Bracts not spurred; corolla 2–2.5 cm,
     yellow-red within   **18. subrigida**
18a. Leaves less than 1 cm wide   19
  b. Leaves usually 1.5 cm or more
     across   23
19a. Plants stemless, or very short,
     forming crowded rosettes   20
  b. Stems 10–12 cm   21
20a. Sepals to 3 mm, unequal, almost free
                                    19. amoena
  b. Sepals c. 6 mm, equal   **20. bella**
21a. Corolla 1.7–1.9 cm
                                22. macdougallii
  b. Corolla 9–11 mm   22
22a. Leaves 2.5–3 cm   **13. gracilis**
  b. Leaves 3–3.5 cm   **21. johnsonii**
23a. Leaves at least 4 times as long as
     wide   24
  b. Leaves 1–3 times as long as
     wide   32
24a. Leaves not more than 5 cm
                                    23. whitei
  b. Leaves mostly more than 5 cm
     (4–12 cm)   25
25a. Corolla creamy white
                                24. chilonensis
  b. Corolla orange-red, pinkish green or
     yellow   26
26a. Sepals 3–10 mm   27
  b. Sepals 1.1–1.6 cm   30
27a. Corolla 8–10 mm   28
  b. Corolla 1.2–1.4 cm   29
28a. Leaves 5–7 × 1–1.5 cm; corolla
     pinkish orange outside, yellow inside
                                25. carnicolor
  b. Leaves 9–10 × 2–2.5 cm; corolla
     green-yellow   **26. megacalyx**

29a. Stems branched; corolla to 1.4 cm
                                27. sessiliflora
  b. Stems unbranched; corolla not more
     than 1.2 cm   **28. bifida**
30a. Stems tufted; corolla not more than
     10 mm, pinkish green
                           29. sanchez-mejoradae
  b. Stems horizontal or absent; corolla
     1.3–1.5 cm, red, orange or
     yellow   31
31a. Stems horizontal, thin; sepals to 1.6
     cm, linear, pinkish   **30. rosea**
  b. Stems short or absent; sepals to
     1.4 cm, triangular to oblong, bluish
     green or red   **31. strictiflora**
32a. Leaves 10–15 cm   33
  b. Leaves 2–10 cm   39
33a. Stems absent or very short   34
  b. Stems 10–60 cm   36
34a. Stems absent; corolla not more than
     1.2 cm, red at base   **32. paniculata**
  b. Stems very short; corolla 1.5–1.6 cm,
     yellow throughout   35
35a. Sepals not more than 8 mm, unequal
                                33. maculata
  b. Sepals c. 1.3 cm, almost equal
                                    34. lutea
36a. Stems 30–60 cm; leaves green with
     red margins; corolla red at base,
     green above   **35. nuda**
  b. Stems 10–30 cm; leaves green or
     purplish, often glaucous   37
37a. Leaves and stems deep purplish
     green; corolla red **36. atropurpurea**
  b. Leaves greenish; corolla pink-yellow
     or reddish orange   38
38a. Leaf-margins usually wavy; corolla
     pink outside, yellow inside
                                37. crenulata
  b. Leaf-margins wavy or flat; corolla
     red outside, orange inside
                                    38. fulgens
39a. Plants with obvious stems 2–90 cm
                                              40
  b. Plants stemless or with very short
     stems, or tufted   52
40a. Stems 2–10 cm   41
  b. Stems 20–90 cm   46
41a. Stems not more than 2–3 cm
                                    39. moranii
  b. Stems mostly 5–10 cm   42
42a. Stems unbranched   43
  b. Stems branched   44
43a. Leaves 2–5 cm; corolla red
                                    40. affinis
  b. Leaves 5–9 cm; corolla red outside,
     yellow inside   **41. laui**
44a. Stems not more than 6 cm; leaves
     3–3.5 cm; corolla red throughout
                                 42. × pulchella

  b. Stems 6–10 cm; leaves 4–10 cm;
     corolla red outside, pink or yellow
     inside   45
45a. Bracts to 4 cm; corolla pink inside
                                    43. pittieri
  b. Bracts not more than 2.5 cm; corolla
     yellow inside   **44. stolonifera**
46a. Stems 20–30 cm   47
  b. Stems 30–90 cm   48
47a. Leaves 2–7 × 1.5–2 cm; corolla red
     outside, pink inside   **45. australis**
  b. Leaves 8–15 × 4–7 cm; corolla red
     outside, orange inside   **38. fulgens**
48a. Corolla 1.2–1.5 cm, yellow outside,
     orange-yellow inside   **46. bicolor**
  b. Corolla 8–11 mm, red, at least in
     part   49
49a. Sepals not more than 6 mm; corolla
     red throughout, or yellow inside   50
  b. Sepals mostly 8–9 mm; corolla red
     and white or green   51
50a. Shrublet with many branched stems;
     leaf-margin and tips red; bracts to
     2.5 cm; corolla red, orange-yellow on
     margin and inside   **47. multicaulis**
  b. Horizontal or erect herb, stems little-
     branched; bracts to 4 cm; corolla red
     throughout   **48. maxonii**
51a. Leaves 6–13 cm; corolla red at base,
     green above   **35. nuda**
  b. Leaves 2.5–7.5 cm; corolla white at
     base, red above   **49. waltheri**
52a. Leaves 3 times as long as wide   53
  b. Leaves 1–2.5 times as long as wide
                                              56
53a. Leaves 3–6 × 1–2 cm   54
  b. Leaves 6–10 × 2–3 cm   55
54a. Bracts 1–1.2 cm; corolla 1.2–1.5 cm,
     pink outside, yellow-orange inside
                                    50. elegans
  b. Bracts c. 1.5 cm; corolla to 1 cm,
     lemon-yellow   **51. pulidonis**
55a. Corolla c. 1.2 cm, yellow, red at base
                                32. paniculata
  b. Corolla 1.4–1.7 cm, red outside,
     orange inside   **52. schaffneri**
56a. Sepals 1.3–1.6 cm   57
  b. Sepals 2–10 mm   59
57a. Corolla c. 2 cm   **53. runyonii**
  b. Corolla 1.2–1.5 cm   58
58a. Leaves 2–4 × 2–2.5 cm; corolla
     yellow, red on keel and tip
                                54. derenbergii
  b. Leaves 6 cm or more; corolla
     pink outside, orange inside
                                55. cuspidata
59a. Bracts with 2 or 3 spurs at base   60
  b. Bracts entire   61
60a. Bracts with 2 spurs; corolla c. 11
     mm, red-pink   **56. peacockii**

b. Bracts with 3 spurs; corolla
2.5–3 cm, greenish yellow
57. longissima
61a. Sepals 3–6 mm   62
b. Sepals 8–10 mm   68
62a. Leaves 1–1.3 cm across; corolla *c.*
1.6 cm   **58. halbingeri**
b. Leaves 1.4 cm or more across;
corolla 7–12 mm   63
63a. Plants tufted, with numerous
offsetting rosettes   64
b. Rosettes solitary or with few
branches   65
64a. Leaves 1.2–3 cm wide; flowering-
stems solitary, to 30 cm**59. secunda**
b. Leaves 3–4 cm wide; flowering-stems
several, to 50 cm   **60. colorata**
65a. Bracts 5–10 mm; flowering-stem to
50 cm   **61. racemosa**
b. Bracts 1.5–2 cm; flowering-stems
20–40 cm   66
66a. Flowering-stem solitary   67
b. Flowering-stems several
64. obtusifolia
67a. Leaves spotted with red and brown,
in a dense solitary rosette
62. purpusorum
b. Leaves not spotted, rosettes usually
tufted, rarely solitary **63. agavoides**
68a. Bracts 8–15 mm   **65. shaviana**
b. Bracts 2–3 cm   69
69a. Stems very short, branched; bracts 2
cm; corolla orange outside, orange-
yellow inside   **66. humilis**
b. Plant stemless; bracts 3 cm; corolla
red outside, orange inside
67. chihuahuaensis

1. E. semivestita Moran. Illustration:
Walther, Echeveria, 167–9 (1972); Graf,
Exotica, series 4, edn 12, **1**: 886 (1985);
Pilbeam, The genus Echeveria, 248–50
(2008).

Stem short, unbranched. Leaves
10–14 × 1.5–3 cm, lanceolate, covered
with very fine hairs, very rarely hairless,
margins curved upwards and tinged red,
base nearly round in cross-section, borne
in a rosette. Flowering-stem to 55 cm,
bearing a many-flowered panicle; bracts
leaf-like, reduced in upper part; sepals to
1.5 cm, unequal, green-purple, glaucous;
corolla to 1.3 cm, pink-red outside, yellow-
red within. *Mexico*. G1.

2. E. spectabilis Alexander. Illustration:
Walther, Echeveria, 318 (1972); Pilbeam,
The genus Echeveria, 262, 263 (2008).

Shrublet 10–60 cm with much-branched
stems. Leaves 4–7 × 2.5–3 cm, spathulate
to obovate, bristle-tipped, flat, covered with

minute, pointed papillae, green with red
margins, with leaf-stalks, borne in loose
rosettes or scattered. Flowering-stems to
70 cm, several, with *c.* 10-flowered
racemes; bracts *c.* 5 mm, oblong, stalkless;
sepals to 1.8 cm, nearly equal, lanceolate,
pointed, ascending or somewhat spreading;
corolla 2–2.4 cm, red-orange outside,
yellow on margins and within, 5-sided,
tapering to mouth, segments keeled,
pouched at base. *Mexico*. G1. Summer.

3. E. nodulosa (Baker) Otto (*Cotyledon
nodulosa* Baker; *E. discolor* Baker).
Illustration: Walther, Echeveria, 229, 315,
316 (1972); Graf, Exotica, series 4, edn
12, **1**: 895 (1985); Pilbeam, The genus
Echeveria, 186, 187 (2008).

Stems to *c.* 20 cm, branching. Leaves
4–5 × 1–1.5 cm, wedge-shaped to obovate,
thick, concave above, light green, deep red
on margins and keel, surface covered with
fine hairs, but not roughened; borne in
loose rosettes or scattered. Flowers 8–12,
horizontal, in racemes, on several stems to
30 cm; bracts *c.* 3 cm; sepals to 1.5 cm,
nearly equal, linear, pointed, thick,
spreading; corolla to 1.6 cm, red, yellow
within and on margins, hardly tapering to
mouth, segments sharply keeled, thick,
fine-pointed. *Mexico*. G1. Autumn.

4. E. setosa Rose & Purpus. Illustration:
Walther, Echeveria, 18, 240, 399, 400
(1972); Graf, Exotica, series 4, edn 12, **1**:
892 (1985); Graf, Tropica, edn 3, 370
(1986); Pilbeam, The genus Echeveria,
253–6 (2008).

Leaf-rosettes stemless, or very short-
stemmed. Leaves 4–5 × 1.8–2 cm,
numerous, oblanceolate, flat above,
pointed, bristle- tipped, covered with long
hairs throughout. Flowering-stems to
30 cm, several, with racemes of *c.* 10
flowers; bracts *c.* 2.5 cm, oblong, pointed,
thick, ascending; sepals to 10 mm, nearly
equal, triangular to oblong, thick; corolla
1–1.2 cm, yellow with red markings,
yellow within, urn-shaped, 5-sided. *Mexico*.
G1. Summer.

5. E. ciliata Moran. Illustration: Walther,
Echeveria, 240, 401, 402 (1972); Graf,
Tropica, edn 3, 386 (1986); Pilbeam, The
genus Echeveria, 254 (2008) – as *E. setosa*
var. *ciliata*.

Low-growing, hairy perennial with
leaves clustered in basal rosettes; stem to
7 cm. Leaves 3–5 cm, wedge-shaped to
obovate, bristle-tipped, hairy only along the
margins and on keel beneath, 30–70 to a

rosette, leaf-stalk thick. Flowers 4–7 in
simple or one-sided racemes; sepals
c. 4 mm, nearly equal, spreading,
lanceolate to triangular, pointed; corolla
c. 1 cm, yellow-red, tapering to mouth.
Mexico. G1.

6. E. coccinea (Cavanilles) de Candolle
(*Cotyledon coccinea* Cavanilles; *E. pubescens*
Schlechtendal). Illustration: *Botanical
Magazine*, 2572 (1825); Walther,
Echeveria, 389 (1972); Graf, Exotica, series
4, edn 12, **1**: 887 (1985); Pilbeam, The
genus Echeveria, 84–6 (2008).

Densely hairy throughout with
branching stems 30–60 cm, deep rusty
brown when young, later greyish. Leaves
6–8 × 2 cm, oblanceolate, concave above,
convex below, clustered towards ends of
branches, stalk-like at base. Flowers *c.* 25,
borne in spikes to 30 cm; bracts *c.* 3 cm,
leaf-like, soon falling; sepals to 1.3 cm,
unequal, lanceolate, round in cross-section,
pointed; corolla 1–1.2 cm, red outside,
orange-yellow within, 5-sided, cylindric.
Mexico. G1. Autumn–winter.

7. E. pulvinata Rose. Illustration:
Botanical Magazine, 7918 (1903); Walther,
Echeveria, 237, 393 (1972); Brickell, The
RHS gardeners' encyclopedia of plants &
flowers, 385 (1989); Pilbeam, The genus
Echeveria, 221–4 (2008).

Densely hairy throughout, with stems to
20 cm. Leaves 2.5–6.5 × 1.8–3.5 cm,
spathulate to obovate, fine-pointed, thick,
turgid, borne in very loose rosettes, hairs
often reddish. Flowering-stems 20–30 cm,
with racemes of *c.* 15 flowers; bracts round
in cross-section, pointed; sepals 5–9 mm,
nearly equal, clearly joined at base,
adpressed; corolla 1.2–1.9 cm, yellow, keel
red, urn-shaped, 5-angled. *S Mexico*. G1.
Winter–spring.

8. E. leucotricha J.A. Purpus. Illustration:
Walther, Echeveria, 375 (1972); Graf,
Exotica, series 4, edn 12, **1**: 891 (1985);
Graf, Tropica, edn 3, 370 (1986); Pilbeam,
The genus Echeveria, 151–2 (2008).

Hairy shrublet, to 15 cm, with red
branching stems. Leaves 6–8 × 2–2.5 cm,
oblong-lanceolate, blunt, bristle-tipped,
upcurved and red, borne in loose rosettes,
hairs whitish. Flowering-stems to 40 cm,
with 12–15 flowers, spike-like or
paniculate; bracts *c.* 3 cm, obovate-oblong;
sepals 1–1.2 cm, nearly equal, free to base,
adpressed, triangular-linear, pointed;
corolla to 1.8 cm, orange, red on keel,

5-angled, tapering to mouth, segments spreading at tips. *Mexico*. G1. Spring.

9. E. harmsii J.F. Macbride (*Oliverella elegans* Rose; *Oliveranthus elegans* (Rose) Rose; *Echeveria elegans* (Rose) Berger). Illustration: Walther, Echeveria, 224, 410 (1972); Graf, Exotica, series 4, edn 12, **1**: 894, 898 (1985); Graf, Tropica, edn 3, 374 (1986); Pilbeam, The genus Echeveria, 138, 139 (2008).

Shrub, shortly hairy throughout, with branching stems to 30 cm, occasionally more. Leaves 2–5 × 1–1.5 cm, oblanceolate, pointed, light green, margins and tip red, clustered on ends of branches. Flowering-stems many, to 20 cm, few-flowered; bracts *c.* 2 cm, few, pointed; sepals to 1.8 cm, spreading. Corolla 3–3.4 cm, red, yellow on margins and inside, urn-shaped, 5-angled, segments fine-pointed. *Mexico*. Summer.

Many of the hybrids common in cultivation have this as one parent, including 'Set Oliver', 'Victor' and 'Pulv Oliver'. Most would key out here.

10. E. amphoralis Walther. Illustration: *Cactus & Succulent Journal of America* **30**: 149–50 (1958); Walther, Echeveria, 241, 407–8 (1972); Pilbeam, The genus Echeveria, 47 (2008).

Shrublet, hairy throughout with branched stems to 20 cm. Leaves many, 3.5–4 × 2–2.5 cm, wedge-shaped to obovate, bristle-tipped, borne in loose, ill-defined rosettes, leaf-stalk thick. Flowering-stems to 20 cm, several, with few-flowered racemes; bracts 2–3 cm, bristle-tipped, spreading; sepals to 1.4 cm, nearly equal, round in cross-section, pointed, spreading; corolla 2.2–2.6 cm, red at base, yellow above and within. *Mexico*. G1. Summer.

11. E. pilosa J.A. Purpus. Illustration: Jacobsen, Handbook of succulent plants **1**: 382 (1960); Walther, Echeveria, 297 (1972); Brickell (ed.), RHS A–Z encyclopedia of garden plants, 395 (2003); Pilbeam, The genus Echeveria, 281 (2008).

Stem short, not exceeding 10 cm, unbranched, with long hairs throughout. Leaves 6–7 × 1.5–2 cm, oblanceolate to elliptic, tapering to a point, concave above, convex below, *c.* 40 in loose rosettes. Flowers borne in a raceme, paniculate below, stems several, to 30 cm; bracts to 3 cm, many, oblong, pointed; sepals to 1.5 cm, nearly equal, almost round in cross-section, pointed; corolla *c.* 1.2 cm,

orange, keels red, yellow at tip and within, 5-sided, tapering to the mouth. *Mexico*. G1. Spring–summer.

12. E. pringlei (S. Watson) Rose (*Cotyledon pringlei* S. Watson). Illustration: Walther, Echeveria, 241, 403 (1972); Pilbeam, The genus Echeveria, 212, 213 (2008).

Densely hairy throughout, stem to 10 cm and branched. Leaves 3.5–4 × 2–2.5 cm, obovate to lanceolate, pointed, borne in loose rosettes at end of branches or scattered. Cymes 15–25 cm, simple, with 5–18 flowers, becoming horizontal; bracts to 3 cm, leaf-like, reduced in upper part; sepals to 1.4 cm, nearly equal, lanceolate, pointed, spreading; corolla to 1.5 cm, red outside, buff within, urn-shaped, 5-angled. *Mexico*. G1.

13. E. gracilis Walther. Illustration: Walther, Echeveria, 300 (1972); Pilbeam, The genus Echeveria, 128, 129 (2008).

Shrubby at base, stems to 10 cm, horizontal or erect. Leaves 2.5–3 cm × 9–10 mm, *c.* 5 mm thick, oblong to club-shaped, curving upwards, scattered or in loose rosettes. Racemes several, with 10 or more flowers; bracts *c.* 2 cm, many, pointed; sepals *c.* 10 mm, nearly equal, adpressed or widely spreading; corolla *c.* 10 mm, red outside, orange within, 5-angled, tapering slightly, segments thick, pointed. *Mexico*. G1.

14. E. gibbiflora de Candolle (*Cotyledon gibbiflora* (de Candolle) Baker). Illustration: Walther, Echeveria, 223, 226, 227 (1972); Graf, Exotica, series 4, edn 12, **1**: 893 (1985); Graf, Tropica, edn 3, 369 (1986); Pilbeam, The genus Echeveria, 116–19 (2008).

Stem to 30 cm, unbranched, erect, about 5 cm thick. Leaves 35–38 × 10–25 cm, tinged purple, obovate-spathulate, pointed, margins wavy, 12–25 to a rosette. Panicles to 1 m, many-branched with 12 solitary flowers to each branch; bracts 8–10 × 4–5 cm, obovate, bristle-tipped; sepals lavender, to 11 mm, unequal, spreading, lanceolate, pointed; corolla *c.* 1.6 cm, red outside, buff within, cylinder-shaped to bell-shaped, segments slightly spreading at tips. *Mexico*. G1. Autumn–winter.

15. E. acutifolia Lindley. (*E. holwayi* Rose; *Cotyledon acutifolia* (Lindley) Baker; *Cotyledon devensis* N.E. Brown). Illustration: Walther, Echeveria, 214, 215 (1972); Pilbeam, The genus Echeveria, 39 (2008).

Stem to 30 cm, unbranched. Leaves 28–32 × 7–9 cm, obovate to oblong, bristle-tipped or pointed, green tinged red, leaf-stalk long, channelled. Flower-stalks *c.* 1 cm, with 3 or 4 flowers to each panicle; sepals *c.* 6 mm, unequal, ascending or erect; petals 11–12 mm, red, mouth of tube contracted. *S Mexico*. G1. Winter.

16. E. gigantea Rose & Purpus. Illustration: Walther, Echeveria, 198–202 (1972); Graf, Exotica, series 4, edn 12, **1**: 892 (1985); Pilbeam, The genus Echeveria, 121–3 (2008).

Stem to 50 cm, unbranched. Leaves 15–20 × 8–10 cm, grape-green, margins purple, spathulate to obovate, rounded and notched at apex, base of leaf stalk-like, few borne in a loose rosette. Flowers solitary in panicles to 2 m, thick at base; bracts green, margins green, spathulate to oblong, pointed or blunt, numerous, *c.* 1.5 cm; sepals unequal, triangular to oblong-lanceolate, tapering to the tip, spreading; corolla 1.2–1.7 cm, rose-red, not yellow within, 5-sided, not constricted at mouth. *Mexico*. G1. Winter.

17. E. fimbriata C. Thompson. Illustration: Walther, Echeveria, 19, 187 (1972); Pilbeam, The genus Echeveria, 113 (2008).

Stem to 50 cm. Leaves 15–20 × 6–7 cm, almost stalked at base, rounded with a bristle tip, somewhat hairy at first, few, borne in a loose rosette. Flowering-stems 2- or 3-branched, flowers usually solitary; bracts *c.* 4 × 1.8 cm, few, pointed, spurred; sepals light blue to grey-green, to 9 mm, unequal, spreading to erect, pointed. Corolla 1.2–1.5 cm, red-pink outside, buff within, narrowing slightly at mouth, segments thick, spreading at tips. *Mexico*. G1. Winter.

18. E. subrigida (Robinson & Seaton) Rose (*Cotyledon subrigida* Robinson & Seaton; *E. angusta* von Poellnitz; *E. palmeri* Rose; *E. rosei* Nelson & Macbride). Illustration: Jacobsen, Handbook of succulent plants **1**: 386 (1960); Walther, Echeveria, 177, 178 (1972); Graf, Exotica, series 4, edn 12, **1**: 893 (1985); Pilbeam, The genus Echeveria, 271–3 (2008).

Stem to 10 cm, thick. Leaves 15–25 × 5–10 cm, obovate to oblanceolate, pointed, white-glaucous like a *Dudleya*, margins red, upturned and with finely rounded teeth, borne in dense rosettes. Flowering-stems 60–90 cm, 1 or 2 with 6–15 branches, each with panicles of *c.* 15

flowers; bracts few, 3–5 cm; sepals to 2.5 cm, just equal, triangular to lanceolate, ascending, grey-purple; corolla 2–2.5 cm, red, bloomed white outside, yellow-red within, 5-sided, not very constricted at mouth. *Mexico*. G1. Summer.

19. E. amoena L. de Smet (*E. microcalyx* Britton & Rose; *E. pusilla* Berger). Illustration: Pilbeam, The genus Echeveria, 46, 47 (2008).

Mat-forming. Leaves 2–3 cm × 6–8 mm, spathulate to oblanceolate, pointed, round in cross-section, somewhat club-shaped, grey-green with maroon tips, borne in dense rosettes. Flowering-stalks 10–20 cm, erect, with 6–12 nodding flowers; bracts numerous; sepals to 3 mm, unequal almost free, blunt; corolla 8–9 mm, pinkish rose to red, margins yellow, thin, exceeding stamens. *Mexico*. G1. Spring.

20. E. bella Alexander. Illustration: Walther, Echeveria, 357, 358 (1972); Pilbeam, The genus Echeveria, 56, 57 (2008).

Stem very short and branched. Leaves 2–2.8 cm × 3–5 mm, narrowly oblong, pointed, convex above and below, in dense rosettes of 20–30. Flowering-stems *c.* 25 cm, red, with racemes of *c.* 12 flowers; bracts *c.* 2.3 cm × 5 mm, oblong, pointed and spurred; sepals *c.* 6 mm, nearly equal, spreading, almost round, pointed; corolla to 10 mm, tapering to mouth, segments red to yellow spreading at tips and bristled below apex. *S Mexico*. G1. Spring.

21. E. johnsonii Walther. Illustration: Walther, Echeveria, 293 (1972); Carruthers & Ginns, Echeverias: A guide to cultivation & identification, pl. 25 (1973); Pilbeam, The genus Echeveria, 145 (2008).

Stems to *c.* 10 cm, branched. Leaves 3–3.5 cm × 7–9 mm, crowded near end of branches, club-shaped, or oblanceolate, nearly round in cross-section, faintly purple on edges. Flowers *c.* 10, borne in spike-like heads, lateral, to 10 cm; bracts *c.* 2 cm, oblong, narrow, round in cross-section, pointed; corolla 9–11 mm, buff, red on keel, orange-yellow within, 5-sided, straight, segments just spreading at tips. *Ecuador*. G1.

22. E. macdougallii Walther. Illustration: Walther, Echeveria, 303, 304 (1972); Carruthers & Ginns, Echeverias: A guide to cultivation & identification, pl. 32 (1973); Pilbeam, The genus Echeveria, 166 (2008).

Shrublet with stem to 12 cm. Leaves 2–3 cm × 5–10 mm, glaucous, green with red markings, very thick, obovate to club-shaped, nearly round in cross-section, borne in loose rosettes near end of the stems. Flowers few, borne in racemes on stems 10–25 cm; bracts 1.8–2.5 cm, leaf-like; sepals *c.* 10 mm, nearly equal, oblong to elliptic, round in cross-section, blunt; corolla 1.7–1.9 cm, red, yellow on upper edges and within, to 5-sided, straight, tapering to mouth, segments slightly spreading at tips. *Mexico*. G1.

23. E. whitei Rose (*E. buchtienii* von Poellnitz; *E. chilonensis* Walther). Illustration: Walther, Echeveria, 354 (1972); Pilbeam, The genus Echeveria, 301 (2008).

Stem stout, branching. Leaves 3–5 cm × 8–20 mm, spathulate, bristle-tipped, flat above, rounded beneath, glaucous green-brown, *c.* 35 borne to each rosette. Flowering-stalks to 30 cm, sometimes one-sided, branched, with racemes of *c.* 10 flowers; bracts to 1 cm, few, triangular to ovate, pointed; sepals to 6 mm, ascending to erect, linear to lanceolate, pointed; corolla 1–1.5 cm, 5-sided, urn-shaped, red. *Bolivia*. G1. Winter–spring.

24. E. chilonensis (Kuntze) Walther (*E. vanvlietii* von Keppel). Illustration: Pilbeam, The genus Echeveria, 82, 83 (2008).

Stem to 5 cm, 1–2 cm thick. Leaves 4–8 × 1–2 cm, oblanceolate to oblong, with a small point, convex below, grey-green, tinged bronze, 20–25 to each rosette. Flowering-stems 20–50 cm, with 15–40 flowers in each raceme; bracts to 2.5 × 1 cm, ovate; sepals 5–10 mm, unequal, spreading, ascending, triangular to ovate; corolla to 12 mm, creamy white, 5-sided, segments spreading at tips. *Bolivia*. G1. Summer.

25. E. carnicolor (Baker) Morren (*Cotyledon carnicolor* Baker). Illustration: Jacobsen, Handbook of succulent plants **1**: 370 (1960); Walther, Echeveria, 346, 347 (1972); Graf, Exotica, series 4, edn 12, **1**: 894 (1985); Pilbeam, The genus Echeveria, 72 (2008).

Stem short or absent. Leaves 5–7 × 1.5–1.6 cm, in rosettes of about 20, oblanceolate to spathulate, blunt, thick, concave above, convex below, flesh-coloured, papillose. Flowers 6–20, borne in several racemes to 25 cm; bracts *c.* 2 cm,

numerous, easily detached, pointed. Sepals 5–6 mm, lanceolate, nearly equal, almost round in cross-section, pointed; corolla 9–10 mm, pink-orange outside, buff-yellow inside, 5-angled, straight, segments thick, keeled. *E Mexico*. G1. Winter.

26. E. megacalyx Walther. Illustration: Walther, Echeveria, 365, 366 (1972); Pilbeam, The genus Echeveria, 169, 170 (2008).

Stem short, branching. Leaves 9–10 × 2–2.5 cm, thin, oblong to spathulate, many borne in dense rosettes, margin thin, transparent. Flowers solitary, nodding, *c.* 30, in racemes on stems to 45 cm; bracts leaf-like, reduced higher up; sepals 8–10 mm, leaf-like, spreading, nearly equal, pointed; corolla *c.* 8 mm, green-yellow, urn-shaped, tips of segments reflexed. *S Mexico*. G1. Summer–autumn.

27. E. sessiliflora Rose (*E. corallina* Alexander). Illustration: Pilbeam, The genus Echeveria, 251, 252 (2008).

Stem 2–3 cm, branched. Leaves 6–8 × 1.5–1.6 cm, oblanceolate, pointed, glaucous-green, borne in terminal rosettes, margins brown-red. Flowering-stems 40–50 cm, pink, with 20–25 flowers; bracts leaf-like, numerous, reduced in upper part; sepals to 7 mm, unequal, blue-green; corolla to 1.4 cm, conical, segments spreading at tips. *Mexico*. G1.

28. E. bifida Schlechtendal (*E. teretifolia* Kunze). Illustration: Walther, Echeveria, 217, 243, 246 (1972); Pilbeam, The genus Echeveria, 60, 61 (2008).

Stem short, not branched. Leaves tinged red, 3.5–10 × 1.2–2.5 cm, oblanceolate to diamond-shaped, pointed, bristle-tipped, glaucous green. Flowers 20–30, borne in a 2-branched raceme, 25–60 cm; bracts *c.* 3 cm, many, round in cross-section, blunt; sepals to 1 cm, unequal, rigid, rounded, blunt, spreading. Corolla *c.* 1.2 cm, pink-orange outside, yellow within, keeled, petals joined for much of their length, urn-shaped. *Mexico*. G1. Summer.

29. E. sanchez-mejoradae Walther. Illustration: Walther, Echeveria, 109, 110, 213 (1972); Pilbeam, The genus Echeveria, 135–7 (2008) – as *E. halbingeri* var. *sanchez-mejoradae*.

Stems tufted. Leaves *c.* 6 × 1.5 cm, oblanceolate to obovate or linear, tapering towards the base, tip rounded with a small point, borne in a number of rosettes. Flowering-stems to 50 cm, several, unbranched; flowers 10, one-sided; bracts sparse; sepals *c.* 11 mm, triangular to lanceolate, pointed, usually spreading, unequal; corolla 9–10 mm, pink-green, pitcher-shaped. *Mexico.* G1. Spring–summer.

30. E. rosea Lindley (*Cotyledon roseata* Baker). Illustration: Walther, Echeveria, 229, 326 (1972); Pilbeam, The genus Echeveria, 235–7 (2008).

Often epiphytic shrublet, stems little-branched, thin, horizontal. Leaves 5–9 × 1.5–2 cm, oblanceolate to oblong, upturned, more or less flat, pointed, leaf stalk 3-angled, borne in indistinct rosettes. Flowering-stems to 20 cm, flowers borne in spikes or racemes; bracts *c.* 3 mm, oblanceolate and stalked; sepals *c.* 1.6 cm, linear, narrow, spreading and pink-purple; corolla 1.2–1.3 cm, yellow with red tips, straight, segments thin, keeled. *Mexico.* G1. Winter–spring.

31. E. strictiflora A. Gray (*Cotyledon strictiflora* (A. Gray) Baker). Illustration: Walther, Echeveria, 217, 250, 251 (1972); Pilbeam, The genus Echeveria, 266, 267 (2008).

Stem short or absent. Leaves 7–9 × 1.5–2 cm, spreading and ascending, obovate to diamond-shaped, tapering to base and pointed at tip, thin, glaucous, few borne in a loose rosette, margins curled upwards. Flowering-stem to 20 cm, solitary to several, racemes one-sided, occasionally branched with 10–15 flowers; bracts 2–3 cm, many, keeled, pointed; sepals *c.* 1.4 cm, unequal, oblong to triangular, ascending, pointed; corolla 1.4–1.5 cm, red outside, orange-red within, bell or urn-shaped, 5-sided, lobes sharply keeled. *USA (Texas) & Mexico.* G1. Summer.

32. E. paniculata A. Gray (*Cotyledon grayii* Baker; *E. grayii* (Baker) Morren). Illustration: Walther, Echeveria, 371 (1972); Pilbeam, The genus Echeveria, 194, 195 (2008).

Stem absent. Leaves 10–12 × 3–5 cm, *c.* 40 per rosette, spreading, oblanceolate, fine-pointed. Flowers solitary, 2–6 to to each branched panicle, stems to 50 cm; bracts few, lanceolate to ovate; sepals to *c.* 10 cm, unequal, triangular to oblong, pointed, spreading; corolla 1–1.2 cm, yellow, red at base, conical, 5-sided, tapering to mouth. *Mexico.* G1. Summer.

33. E. maculata Rose. Illustration: Walther, Echeveria, 373 (1972); Pilbeam, The genus Echeveria, 195 (2008) – as *E. paniculata* var. *maculata*.

Roots thick, stem short or absent. Leaves 9–15 × 2–3 cm, oblanceolate to obovate, fine-pointed, glaucous, green, up-curved. Flowers solitary, borne in spikes or panicles on stems 30–90 cm; bracts to *c.* 5 cm, nearly round in cross-section, many; sepals to *c.* 8 mm, round in cross-section, pointed, spreading, unequal; corolla 1.5–1.6 cm, narrowly urn-shaped, segments spreading above, yellow. *Mexico.* G1. Spring–summer.

34. E. lutea Rose. Illustration: Walther, Echeveria, 257, 258 (1972); Pilbeam, The genus Echeveria, 257, 258 (2008).

Stems very short. Leaves 11–13 × 3–4 cm, linear to oblanceolate, pointed, margins and tip curved, somewhat keeled, green, tinged purple, borne in crowded rosettes of *c.* 25. Flowering-stems simple or 2-branched with 12–30 solitary flowers; bracts 3–6.5 cm, numerous, round to triangular in cross-section, lanceolate to linear, bristle tip clear; sepals *c.* 1.3 cm, almost equal; corolla to 1.7 cm, yellow, lobes somewhat spreading, 5-sided. *Mexico.* G1. Summer.

35. E. nuda Lindley (*Cotyledon nuda* (Lindley) Baker; *E. navicularis* L. de Smet). Illustration: Walther, Echeveria, 221, 279, 280 (1972); Graf, Exotica, series 4, edn 12, **1**: 894 (1985); Pilbeam, The genus Echeveria, 189 (2008).

Stem 30–60 cm, branched. Leaves 6–13 × 2–5.5 cm, spathulate to obovate, blunt with a bristle tip or pointed, thin, tapering to the base, green, bloomed at first, borne in loose rosettes at ends of branches, margins red. Flowering-stems to 40 cm, with *c.* 20 to each spike; bracts to 2.5 cm, obovate, pointed, thin; sepals to 9 mm, nearly equal, linear to lanceolate, fine-pointed, somewhat ascending; corolla 1–1.1 cm, 5-angled, segments red at base, green above, tip erect or spreading. *Mexico.* G1. Summer.

36. E. atropurpurea (Baker) Morren (*Cotyledon atropurpurea* Baker; *Echeveria sanguinea* misapplied). Illustration: Walther, Echeveria, 232, 329 (1972); Pilbeam, The genus Echeveria, 49, 50 (2008).

Stem 10–15 × 2.5 cm. Leaves 10–12 × 3–5 cm, oblong or spathulate to obovate, dark purple-green, glaucous, forming dense rosettes of *c.* 20 leaves. Flowering-stems 30–60 cm, with 20–25 flowers to each raceme; bracts leaf-like but smaller; sepals *c.* 4 mm, almost equal, lanceolate, spreading to swept back; corolla

to 1.2 cm, red, 5-angled, segments pointed. *Mexico.* G1.

37. E. crenulata Rose. Illustration: Graf, Exotica, series 4, edn 12, **1**: 893 (1985); Graf, Tropica, edn 3, 369, 370 (1986); Pilbeam, The genus Echeveria, 93 (2008).

Stem *c.* 10 cm, branching at base. Leaves 10–12 × 7–9 cm, obovate to diamond-shaped, pointed, with or without a bristle tip, few in a loose rosette, margins wavy or flat, brown. Flowers borne in several panicles to 50 cm; bracts many, *c.* 5 mm, reduced higher up, pointed, sepals to 1 cm, unequal, triangular-oblanceolate, pointed, spreading or swept back; corolla to 1.8 cm, pink outside, yellow within, 5-angled, straight. *Mexico.* G1. Winter.

38. E. fulgens Lamarck (*E. retusa* Lamarck). Illustration: Walther, Echeveria, 160, 161, 162 (1972); Graf, Exotica, series 4, edn 12, **1**: 893 (1985); Pilbeam, The genus Echeveria, 114, 115 (2008).

Stems to 30 cm, occasionally branched. Leaves 8–5 × 4–7 cm, spathulate to obovate, glaucous green, blunt, bristle-tipped, few, borne in rosettes, base stalk-like, margin wavy or flat. Flowering-stems to 90 cm, several, few branched, flowers 20–30; bracts numerous, to 3.5 cm; sepals to 10 mm, unequal, lanceolate to triangular, pointed, green; corolla to 1.5 cm, red outside, orange within, segments slightly spreading at tips. *Mexico.* G1. Winter.

39. E. moranii Walther. (*E. proxima* Walther). Illustration: Walther, Echeveria, 348, 349 (1972); Pilbeam, The genus Echeveria, 176, 177 (2008).

Stems 2–3 cm × 6–8 mm thick. Leaves 4–6 × 2–3 cm, wedge-shaped to obovate, 1 cm thick at centre, concave above, convex below, bristle tipped, papillose when young, green with maroon spots, margins and tip, *c.* 25 to each rosette. Flowering-stems several, 20–50 cm, with *c.* 15 flowers to each raceme; bracts green to blue-grey, *c.* 3 × 1 cm, oblong to elliptic, with transparent spur; sepals *c.* 4 mm, nearly equal, triangular to ovate, pointed, thick; corolla *c.* 1.3 cm, red outside, orange-buff within, 5-sided, conical, segments slightly spreading at tip. *Mexico.* G1. Summer.

40. E. affinis Walther. Illustration: Walther, Echeveria, 73, 74 (1972); Pilbeam, The genus Echeveria, 40, 41 (2008).

Stem to 10 cm, unbranched. Leaves 4.5–5 × 1.8–2.2 cm, oblanceolate to oblong, short-pointed, nearly flat above, convex below, 6–9 mm thick, bright green. Flowering-stems *c.* 30 cm, with 2–5 branches, flowers 15–30; bracts few; sepals erect, nearly equal, triangular ovate, pointed; corolla 9–10 mm, red, bell-shaped, just exceeding stamens. *W Mexico.* G1. Summer.

41. E. laui Moran & Meyran. Illustration: Graf, Tropica, edn 3, 361 (1986); Pilbeam, The genus Echeveria, 148, 149 (2008).

Glaucous perennial, stem to 10 cm, simple. Leaves white, tinged red, 5–9 × 3–4 cm, 6–8 mm thick, flat to concave above, convex below, 30–50 borne in a dense rosette. Flowering-stems to 10 cm, with 9–17 flowers; bracts 1.1–2 cm × 7–14 mm, rounded to obovate, tinged red; sepals to 1.8 cm, ascending, unequal, ovate-elliptic; corolla 1.3–1.6 cm, tapering towards the mouth, red, segments pointed, yellow inside. *Mexico.* G1. Spring.

42. E. × pulchella Berger. Illustration: Graf, Exotica, series 4, edn 12, **1**: 896 (1985).

Stem to 6 cm. Leaves 3–3.5 × 1–1.5 cm, spathulate to oblong, numerous, ascending, thick, concave above, bristle-tipped, green, suffused red in the sun. Flower-stalks 2- or 3-branched, to 20 cm, flowers 10–18; sepals light brown-olive or red, to 4 mm, triangular to ovate, fused for half their length; corolla to 6 mm, red, thin. *Garden origin.* G1. Spring–summer.

A hybrid involving *E. amoena*.

43. E. pittieri Rose. Illustration: Walther, Echeveria, 321 (1972); Graf, Exotica, series 4, edn 12, **1**: 896 (1985); Pilbeam, The genus Echeveria, 209 (2008).

Shrublet, stem branching, to *c.* 10 cm. Leaves 4–10 × 2–3 cm, oblanceolate to elliptic, pointed, stalked, slightly concave above, green-brown at tip, borne in loose rosettes. Flowers in dense spikes with thick erect stems *c.* 20 cm; bracts green-purple, leaf-like, *c.* 4 cm, numerous; sepals to 9 mm thick, pointed; corolla 1.2–1.3 cm, red outside, pink within, bell-shaped, 5-angled, segments spreading at tips. *Guatemala, Nicaragua & Costa Rica.* G1. Winter.

44. E. stolonifera (Baker) Otto (*Cotyledon stolonifera* Baker). Illustration: Walther, Echeveria, 153 (1972); Pilbeam, The genus Echeveria, 265 (2008).

Tuft-forming; stems 5–10 cm, branching, stoloniferous. Leaves 5–8 × 2–4 cm, obovate to spathulate, bristle–tipped, concave above, keeled, glaucous green when young. Flowering-stems 6–10 cm, several; bracts to 2.5 cm, leaf-like, numerous; sepals to 1 cm, spreading, nearly round in cross-section, pointed; corolla *c.* 1.4 cm, red outside, yellow within, segment tips pointed, spreading. *Mexico (possibly a garden hybrid).* G1. Summer.

45. E. australis Rose. Illustration: Walther, Echeveria, 225, 298 (1972); Pilbeam, The genus Echeveria, 51 (2008).

Stems 20–30 cm, little-branched with leaves crowded at their ends, round and smooth. Leaves 2–7 × 1.5–2 cm, wedge-shaped, obovate, rounded, bristle-tipped, thin, keeled, lime-green tinged purple. Flowering-stems to 25 cm, with dense, many-flowered panicles or racemes; bracts *c.* 2.5 cm, many, spreading, oblong-rounded; sepals 8–12 mm, nearly equal, spreading to ascending; corolla 1.1–1.4 cm, 5-angled, segments thick, red outside, pink within. *Costa Rica, Honduras.* G1. Spring–summer.

46. E. bicolor (Humbolt, Bonpland & Kunth) Walther (*Sedum bicolor* Humboldt, Bonpland & Kunth; *E. bracteolata* Link, Klotzsch & Otto; *Cotyledon bracteolata* (Link, Klotzsch & Otto) Baker; *E. subspicata* (Baker) Berger). Illustration: Walther, Echeveria, 336 (1972); Pilbeam, The genus Echeveria, 58, 59 (2008).

Stem to 60 × 2 cm, upright and branched from the base. Leaves 9–10 × 3–3.5 cm, wedge-shaped to ovate, rounded, bristle-tipped, green, stalks angled, borne in ill-defined rosettes. Flowering-stems sturdy, many, with 25 flowers to each raceme; bracts *c.* 5 × 3–5 cm, leaf-like, numerous, becoming reduced towards apex; sepals green, *c.* 1.4 cm, nearly equal, lanceolate and fine-pointed, somewhat keeled; corolla 1.2–1.5 cm, yellow outside, orange-yellow within, 5-sided, tapering to mouth, segments oblong to oval, fine-pointed, keeled. *Venezuela & Colombia.* G1. Winter.

47. E. multicaulis Rose. Illustration: Walther, Echeveria, 228, 312 (1972); Graf, Exotica, series 4, edn 12, **1**: 892 (1985); Pilbeam, The genus Echeveria, 180, 181 (2008).

Shrublet, stems 30–60 cm, much-branched and covered with old leaf-scars. Leaves 3–5 × 2–4 cm, wedge-shaped to obovate, blunt, with a bristle tip, flat, green, appearing in terminal rosettes, margins and tip red. Flowers 6–15, borne in racemes on several stems to 25 cm; bracts to 2.5 cm, uppermost orange, numerous, rounded to obovate, pointed; sepals *c.* 6 mm, red, almost equal, ascending, pointed; corolla *c.* 1 cm, scarlet orange-yellow on margins and within, bell-shaped, twice as wide at mouth as at base, segments keeled. *Mexico.* G1. Winter–spring.

48. E. maxonii Rose (*Cotyledon acutifolia* Hemsley). Illustration: Walther, Echeveria, 295, 296 (1972); Pilbeam, The genus Echeveria, 167, 168 (2008).

Stems 30–80 cm, horizontal or erect and little-branched. Leaves 3–10 × 3–4 cm, spathulate to oblanceolate, blunt, bristle-tipped or pointed, concave above, convex below, papillose, scattered or in loose rosettes. Flowers solitary, *c.* 25 to a raceme, on stems to 60 cm; bracts *c.* 4 cm, many, ovate, pointed; sepals *c.* 5 mm, nearly equal, thick, semicircular in cross-section, elliptic to oblong; corolla *c.* 1 cm, red, somewhat tapering, segments pointed, keeled. *Guatemala.* G1. Winter.

49. E. waltheri Moran & Meyran. Illustration: Pilbeam, The genus Echeveria, 297, 298 (2008).

Stems to 90 cm, horizontal to erect. Leaves 2.5–7.5 × 1.2–2.5 cm, ovate to spathulate, abruptly pointed, margins red. Flowering-stems 30–60 cm, few with many-flowered spikes; sepals to 9 mm, nearly equal, erect; corolla to 9 mm, white below, red above. *Mexico.* G1. Autumn–winter.

50. E. elegans Rose (*E. perelegans* Berger). Illustration: *Botanical Magazine*, 7993 (1905); Walther, Echeveria, 53, 98 (1972); Graf, Tropica, edn 3, 369 (1986); Pilbeam, The genus Echeveria, 105–7 (2008).

Stem short. Leaves 3–6 × 1–2 cm, thick, thinner near the tip, spathulate to oblong, flat above, concave below, bristle-tipped, glaucous white, borne in spherical rosettes, margins translucent. Flowers 5–10, borne in several unbranched stems, 10–15 cm; bracts 1–1.2 cm; sepals *c.* 5 mm, triangular to lanceolate, pointed, unequal, spreading; corolla 1.2–1.5 cm, pink outside, yellow-orange within, spreading at apex. *Mexico.* G1. Spring–summer.

Var. **simulans** (Rose) von Poellnitz (*E. simulans* Rose). Illustration: Walther, Echeveria, 103 (1972); Pilbeam, The genus Echeveria, 259, 260 (2008) – as *E. simulans*. Rosettes flattened, leaves to 7 × 4 cm, white. Inflorescence sometimes 2-branched; sepals to 6 mm; corolla deep pink. *Mexico.* G1. Winter–spring.

51. E. pulidonis Walther. Illustration: Walther, Echeveria, 123 (1972); Carruthers & Ginns, Echeverias: A guide to cultivation & identification, pl. 15 (1973); Pilbeam, The genus Echeveria, 219, 220 (2008).

Leaves borne in a basal rosette, numerous, *c.* 5 × 1.5 cm, ascending, oblong to ovate, tapering to the base, slightly concave above, convex below, bristle-tipped, light green, margins and tips red. Flowers nodding, 10 or more to each one-sided raceme, flowering-stalk to 18 cm; bracts 1.4–1.6 cm; sepals to 6 mm, triangular to ovate, pointed, spreading; corolla 9–10 mm, lemon-yellow, 5-sided, narrowing slightly at mouth. *Mexico.* G1. Spring–summer.

52. E. schaffneri (S. Watson) Rose (*Cotyledon schaffneri* S. Watson). Illustration: Pilbeam, The genus Echeveria, 242, 243 (2008).

Stems short. Leaves 6–10 × *c.* 2 cm, bright green with red margins, oblanceolate to oblong, spreading, with a small abrupt point, borne in dense rosettes. Flowering-stems solitary, 2-branched, to 20 cm, with 8–24 flowers; bracts *c.* 2.5 cm × 7 mm, green, numerous, oblanceolate; sepals to 8 mm, nearly equal, lanceolate, tapering to the tip, spreading; corolla 1.4–1.7 cm, red outside, orange within, 5-angled, not contracted at mouth, segments sharply keeled. *Mexico.* G1. Winter–spring.

53. E. runyonii Walther. Illustration: Graf, Exotica, series 4, edn 12, **1**: 896, 898 (1985); Graf, Tropica, edn 3, 372 (1986); Pilbeam, The genus Echeveria, 240 (2008).

Stems short or absent. Leaves 6–8 × 3–4 cm, wedge-shaped to spathulate, blunt, sometimes shallowly notched, glaucous, borne in open-rosettes. Flowering-stems several, 2-branched, racemes nodding at first; bracts 2–4 cm, many, adpressed, oblanceolate to linear; sepals 1.3–1.5 cm, unequal, spreading; corolla 1.7–2 cm, red-pink, 5-angled, erect. *Mexico.* G1.

Var. **macabeana** Walther differs in having leaves pointed or bristle-tipped.

54. E. derenbergii J.A. Purpus. Illustration: Walther, Echeveria, 273 (1972); Graf, Exotica, series 4, edn 12, **1**: 892, 894 (1985); Graf, Tropica, edn 3, 370 (1986); Pilbeam, The genus Echeveria, 116–19 (2008).

Stem short, much-branched. Leaves 2–4 × 2–2.5 cm, wedge-shaped to obovate, thick, bristle-tipped, pale green, margin and tip red, numerous, borne in dense tuft-forming rosettes. Flowers few, borne in several racemes to 10 cm; bracts *c.* 1.5 cm, pointed, keeled; sepals 7–10 mm, light green, margins red, nearly equal, oblanceolate, fine-pointed, ascending; corolla 1.2–1.5 cm, segments yellow, keel and tip red, somewhat bell-shaped, erect. *Mexico.* G1. Spring–summer.

55. E. cuspidata Rose (*E. parrasensis* Walther). Illustration: Pilbeam, The genus Echeveria, 96, 97 (2008).

Leaves grey-green somewhat glaucous, red-tipped, 6–8 × 3.5–4 cm, oblong-obovate, flat, rather thin, blunt with a small point, numerous, borne in a dense basal rosette. Flowers *c.* 15, borne in several one-sided racemes; bracts to 1.6 cm, fine-pointed; sepals *c.* 8 mm, unequal, lanceolate-oblong, petals *c.* 1.4 cm, tube pink outside, orange within, conical, segment tips just spreading. *Mexico.* G1. Spring–summer.

56. E. peacockii Croucher. Illustration: Graf, Exotica, series 4, edn 12, **1**: 886 (1985); Graf, Tropica, edn 3, 368 (1986); Pilbeam, The genus Echeveria, 200, 201 (2008).

Stem short. Leaves 5–6 × 2–3 cm, obovate–oblanceolate, pointed or bristle-tipped, convex below, flat above, numerous, usually crowded in solitary rosettes. Racemes several, *c.* 30 cm, one-sided; flowers solitary; bracts with 2 spurs at base; corolla *c.* 11 mm, not constricted at mouth, segments keeled, slightly spreading at tip, red-pink. *Mexico.* G1. Summer.

57. E. longissima Walther. Illustration: Walther, Echeveria, 244, 414–16 (1972); Pilbeam, The genus Echeveria, 156, 157 (2008).

Stem short, unbranched. Leaves 6 × 3 cm, wedge-shaped to ovate, bristle-tipped, thick, 20 to each rosette, margins red. Flowering-stems to 20 cm, one-sided, sometimes branched, racemes with 4–12 flowers; bracts *c.* 2 cm, 3-spurred; sepals to 8 mm, urn-shaped, narrow; corolla

2.5–3 cm, green-yellow below, green above. *Mexico.* G1. Summer.

58. E. halbingeri Walther. Illustration: Walther, Echeveria, 121, 122, 231 (1972); Pilbeam, The genus Echeveria, 133, 134 (2008).

Leaves glaucous green, 2–2.5 × 1–1.3 cm, obovate, blunt, with a small point, triangular in cross-section at tip, borne in dense, basal rosettes. Flowering-stems simple, to 12 cm, with 6–9 cm racemes; bracts to 1 cm, triangular in cross-section; sepals to 6 mm, spreading unequal; corolla *c.* 1.6 cm, orange outside, light orange within, urn-shaped, segment tips recurved. *Mexico.* G1. Summer.

59. E. secunda W. Booth (*Cotyledon secunda* Baker; *E. spilota* Kunze). Illustration: *Botanical Magazine*, 8748 (1918); Walther, Echeveria, 130 (1972); Graf, Tropica, edn 3, 368 (1986); Pilbeam, The genus Echeveria, 245–7 (2008).

Stems short and forming numerous offsets. Leaves 2.5–7.5 × 1.2–3 cm, spathulate to wedge-shaped, blunt, bristle-tipped, keeled, glaucous green, margin and tip often red, numerous borne in dense, basal rosettes. Flowering-stem to 30 cm, with 5–15 flowers borne in a simple raceme; sepals spreading, lanceolate, pointed; corolla 7–12 mm, red outside, yellow within, tube narrowing at mouth, lobes swept back. *Mexico.* G1.

Var. **glauca** (Baker) Otto (*E. glauca* Baker). Leaves thin, scarcely keeled, almost flat, to 2 cm across, truncate to rounded.

Var. **pumila** (Schönland) Otto (*E. pumila* Schönland). Leaves to 1.5 cm across, thin and scarcely keeled, almost flat, acute.

60. E. colorata Walther (*E. lindsayana* Walther). Illustration: Graf, Exotica, series 4, edn 12, **1**: 898 (1985); Graf, Tropica, edn 3, 370 (1986); Pilbeam, The genus Echeveria, 87–9 (2008).

Leaves 5–9 × 3–4 cm, oblong to obovate, blunt or bristle-tipped, slightly keeled at tip, light green, tip red, numerous, in dense rosettes at first, later becoming tufted. Flowering-stems several, to 50 cm, tinged pink, 2-branched, 14-flowered; sepals *c.* 3 mm, triangular to ovate, pointed; corolla *c.* 1 cm, pink, yellow to orange at tips and within. *Mexico.* G1. Spring–summer.

61. E. racemosa Schöenland & Chamisso (*E. lurida* Haworth). Illustration: *Botanical Magazine*, 7713 (1900); Walther,

Echeveria, 233, 341 (1972); Pilbeam, The genus Echeveria, 230, 321 (2008).

Stem very short or absent. Leaves 5–8 × 3–3.5 cm, oblanceolate to diamond-shaped, pointed, slightly concave above, borne in dense rosettes of *c.* 15, margins often irregularly cut. Flowering-stems 1 or 2, erect, to 50 cm, thick, racemes with 20–40 flowers; bracts 5–10 mm, obovate, pointed, easily detached; sepals to 6 mm, nearly equal, triangular to ovate, pointed, nearly round in cross-section, spreading; corolla 1–1.2 cm, red-orange outside, orange-buff within, conical or urn-shaped, segments fine-pointed and spreading at tips. *Mexico.* G1. Autumn.

62. E. purpusorum Berger (*Urbinia purpusii* Rose). Illustration: Walther, Echeveria, 124–7 (1972); Carruthers & Ginns, Echeverias: A guide to cultivation & identification, pl. 17 (1973); Graf, Exotica, series 4, edn 12, **1**: 898 (1985); Pilbeam, The genus Echeveria, 225, 226 (2008).

Rosette solitary. Leaves 3–4 × 1.5–2 cm, to 1 cm thick, ovate, tapering to a sharply pointed tip, flat above, convex below, spinach-green, spotted red and brown, borne in dense, solitary rosettes. Flowering-stems *c.* 20 cm, simple with racemes of 6–9 flowers; bracts *c.* 1.5 cm, ovate, pointed; sepals 2–3 mm, pointed, adpressed to petals; corolla *c.* 1.2 cm, pink-red outside, yellow within, fused near base, thick. *Mexico.* G1. Early summer.

63. E. agavoides Lemaire (*Cotyledon agavoides* Baker; *E. obscura* (Rose) von Poellnitz; *Urbinia obscura* Rose; *U. agavoides* (Lemaire) Rose). Illustration: Walther, Echeveria, 82, 83 (1972); Graf, Exotica, series 4, edn 12, **1**: 889 (1985); Graf, Tropica, edn 3, 371 (1986); Pilbeam, The genus Echeveria, 42–4 (2008).

Rosettes tufted or solitary, stem very short. Leaves few, 3–8 × 2.5–3 cm, ovate to triangular, thick, sharply pointed, deep green, waxy, margins transparent. Flowers 10–16, borne on a 2-branched flowering-stalk; bracts few; sepals olive, to 5 mm, spreading; corolla 1–1.2 cm, thin, spreading at tip, pink-orange outside, yellow within. *Mexico.* G1. Spring–summer.

Var. **prolifera** Walther. Leaves numerous, 10–12 × 3–4 cm; petals to 1.6 cm.

Var. **corderoyi** (Baker) von Poellnitz (*Cotyledon corderoyi* Baker; *E. corderoyi* (Baker) Morren; *Urbinia corderoyi* (Baker) Rose). Leaves numerous, ovate, *c.* 6.5 × 3.5 cm; flowering-stems 3-branched; sepals free; petals to 9 mm.

E. × gilva Walther; *E. agavoides × E. elegans.* Illustration: Walther, Echeveria, 116 (1972); Schulz & Kapitany, Echeveria cultivars, 109 (2005). Stems short, branching; leaves 5–8 × 2–2.5 cm, oblong, short-pointed, surface crystalline, margin translucent, up to 30 borne in dense rosettes; flowers borne in several simple racemes to 25 cm; sepals *c.* 5 mm, unequal; corolla *c.* 9 mm, pink below, yellow above. *Garden origin.* G1. Spring.

64. E. obtusifolia Rose. Illustration: Walther, Echeveria, 165, 216 (1972); Pilbeam, The genus Echeveria, 115 (2008) – as *E. fulgens* var. *obtusifolia.*

Stem short. Leaves 4–8 × 2.5–3.5 cm, spathulate to oblanceolate, rounded, bristle-tipped, green with red margins, borne in a rosette. Flowers 9–11, borne on several red, 2-branched stems; bracts numerous, 1.5–2 cm, oblong, pointed; sepals 5–6 mm, unequal, spreading; corolla 1–1.2 cm, orange. *Mexico.* G1. Winter.

65. E. shaviana Walther. Illustration: Walther, Echeveria, 221, 271, 272 (1972); Graf, Exotica, series 4, edn 12, **1**: 889 (1985); Graf, Tropica, edn 3, 369 (1986); Pilbeam, The genus Echeveria, 257, 258 (2008).

Stem short or absent. Leaves *c.* 5 × 1.5–2.5 cm, obovate, tapering to a stalk-like base with a small point, somewhat spathulate, glaucous green, flushed pink, margin wavy, toothed near tip, many borne in crowded rosettes. Flowering-stems to 12 cm, solitary to several, simple, one-sided, nodding with 12–15 flowers; bracts 1–1.5 cm, linear, pointed, spurred; sepals to 9 mm, unequal, linear to lanceolate or triangular, pointed, ascending; corolla 1–1.3 cm, 5-sided, pink, segments keeled, tips slender and spreading. *Mexico.* G1. Summer.

66. E. humilis Rose. Illustration: Pilbeam, The genus Echeveria, 143, 144 (2008).

Stem short and branching. Leaves 4–7 × 2.5 cm, lanceolate to ovate, sharply bristle-tipped, convex below, green-brown, borne in dense, long rosettes. Flowering-stems to 20 cm, simple, occasionally branched; flowers in pendent racemes; bracts *c.* 2 cm, pointed, nearly round in cross-section; sepals to 9 mm, unequal, thick, pointed, ascending; corolla to 1.3 cm, orange outside, orange-yellow within, bell to urn-shaped, segments spreading at tips. *Mexico.* G1. Summer.

67. E. chihuahuaensis von Poellnitz. Illustration: Graf, Exotica, series 4, edn 12, **1**: 887 (1985); Graf, Tropica, edn 3, 361 (1986); Pilbeam, The genus Echeveria, 79–81 (2008).

Stem absent. Leaves 4–6 × 3–4 cm, numerous, obovate to oblong, blunt, short-pointed, thin, white-glaucous green, tip reddish purple. Flowering-stems *c.* 20 cm, simple or branched; bracts 2–3 cm; sepals *c.* 8 mm, unequal, oblong-lanceolate to triangular; corolla *c.* 1.4 cm, red outside, orange within, narrowing at mouth, segments spreading at tips. *Mexico.* G1. Spring–summer.

25. GREENOVIA Webb

J.Y. Clark

Dwarf evergreen perennial succulent herbs. Stem small. Leaves fleshy, spoon-shaped, alternate, stalkless, mostly hairless and glaucous, occasionally glandular-hairy, margins usually entire, occasionally hairy, forming a dense rosette which usually dies after flowering. Flowers yellow, bisexual, borne in a terminal glandular cyme; calyx 16–35-merous; petals and carpels narrow, 16–32, half-fused to the hypogynous disc; stamens twice the number of petals.

A genus of 4 species endemic to the Canary Islands and similar to *Aeonium*, there being a number of putative hybrids between the two genera. *G. aurea* is the most commonly cultivated. The main growing season is from October to March. Watering should only be light in the resting season, perhaps restricted to mist-spraying in most species, although *G. aizoon* prefers to be kept generally moist all year round. Propagation is generally easy from seed and cuttings, although leaves are sometimes difficult to root.

Literature: Praeger, R.L., An account of the Sempervivum group (1932); Bramwell, D. & Bramwell Z., Wild flowers of the Canary Islands (1974).

1a. Leaves densely glandular-hairy, green, remaining as more or less open rosettes in the resting season **1. aizoon**
  b. Leaf surface hairless, glaucous; leaves forming tightly packed rosettes in the resting season 2
2a. Offsets absent; leaf-margin often hairy, especially in small rosettes **2. diplocycla**
  b. Offsets present; leaf margin usually lacking hairs 3

3a. Rosettes less than 8 cm across, rarely
    widely open, with numerous, often
    long-stemmed, offsets; petals and
    carpels 16–22   **3. dodrentalis**
  b. Rosettes 8–20 cm across, opening
    fairly widely in the growing season,
    with a few, usually short-stemmed,
    offsets; petals and carpels 20–35
             4. aurea

1. G. aizoon Bolle. Illustration: Praeger,
An account of the Sempervivum group,
215 (1932); Lamb & Lamb, The illustrated
reference on cacti and other succulents **3**:
840 (1963); Lamb, Neale's photographic
reference plate, 2585a (undated); *Cactus
and Succulent Journal* (US) **39**: 94, 134
(1967).

Much-branched herb forming clumps of
green rosettes, each 4–7 cm across.
Rosettes more or less open during the
summer drought. Stems usually relatively
short, but branches to 15 cm sometimes
formed. Leaves oblong to spathulate,
densely glandular-hairy all over,
3–4 × 1–1.5 cm, apex more or less
truncate, mucronate. Flower-stem
10–13 cm, leafy and glandular-hairy;
flower-head 2–3 × 4–8 cm, with 10–40
flowers; flowers *c.* 1.2 cm across; sepals *c.*
4 mm, 2-fid; petals yellow, 6–7 mm;
stamens 5–6.5 mm; carpels hairy, green,
4–5 mm. *Tenerife.* G1. Late spring.

2. G. diplocycla Webb. Illustration:
Praeger, An account of the Sempervivum
group, 217 (1932); Lamb, Neale's
photographic reference plate, 2586(1),
2586(2) (undated).

Herb, forming a usually solitary rosette,
8–25 cm across, tightly closed during the
summer drought. Stem short. Leaves
spathulate, glaucous, hairless on both
surfaces, margins often sparsely to densely
hairy especially when young,
5–8 × 4–6 cm, apex more or less rounded
to somewhat depressed, sometimes
mucronate. Flower-stem 10–20 cm, leafy
and glandular hairy; flower-head
13–16 × 15–20 cm; flowers *c.* 1.5 cm
across; sepals *c.* 1.7 mm, more or less
hairy; petals 15–25, yellow, 5–6 mm;
stamens 5–5.5 mm; carpels 15–25,
hairy, greenish yellow, *c.* 4 mm. *Palma,
Gomera, Hierro.* G1. Late winter–late
spring.

3. G. dodrentalis (Willdenow) Webb
(*G. gracilis* Bolle). Illustration: Praeger, An
account of the Sempervivum group, 222
(1932); *Cactus and Succulent Journal* (US) **5**:

427 (1933); Jacobsen, Lexicon of succulent
plants, pl. 89.4 (1974).

Much-branched herb forming clumps of
glaucous rosettes, each 3–8 cm across,
tightly closed during the summer drought
and seldom widely open even during the
growing period. Main stem short, offsets
long-stemmed. Leaves spathulate,
glaucous, hairless on both surfaces,
margins finely glandular-hairy only when
young, 2–5 × 1–2 cm, apex more or less
pointed, rounded or somewhat depressed,
sometimes mucronate. Flower-stem
15–25 cm, leafy and glandular-hairy;
flowers *c.* 1.5 cm across; sepals hairy;
petals 16–22, yellow, *c.* 7 mm; stamens
c. 5 mm; carpels 16–22, greenish yellow,
c. 4 mm. *Tenerife.* G1. Late winter–late
spring.

4. G. aurea (C. Smith) Webb & Berthelot.
Illustration: Hunt (ed.), Marshall Cavendish
encyclopedia of gardening, 806 (1968);
Bramwell & Bramwell, Wild flowers of the
Canary Islands, pl. 161 (1974); Jacobsen,
Lexicon of succulent plants, pl. 89.5
(1974); Graf, Tropica, 378 (1984).

Usually branched herb, forming clumps
of glaucous rosettes, each 8–25 cm across,
tightly closed during the summer drought,
fairly open during the growing season.
Main stem short, offsets usually also short-
stemmed but longer than main stem.
Leaves spathulate, glaucous, hairless on
both surfaces, margins sometimes finely
glandular-hairy only when young,
5–10 × 3–6 cm, apex more or less pointed,
rounded or somewhat depressed,
sometimes mucronate. Flower-stem
30–45 cm, leafy and glandular-hairy;
flowers 2–2.5 cm across; sepals hairy;
petals 20–35, yellow, 7–8 mm; stamens *c.*
5 mm; carpels 20–35, greenish yellow, *c.*
5 mm. *Tenerife, Gran Canaria, Gomera,
Hierro.* G1. Spring.

26. MONANTHES Haworth
E.H. Hamlet

Small perennial herbs or shrublets. Stems
simple or branched, erect or spreading.
Leaves alternate, rarely opposite, ovate,
diamond or club-shaped, entire. Inflorescence
a raceme or cyme-like. Flowers greenish,
purplish, or yellowish; calyx saucer-shaped,
with 6–8 parts. Petals linear or lanceolate,
same number as calyx segments. Stamens
twice as many as petals. Nectar-scales at
base of carpels large, petal-like.

A genus of 12 species from Canary
Islands, Salvage Islands and Morocco. In

frost-free climates they can be grown
outdoors. They prefer well-drained sandy
soil. When grown indoors, cool conditions
in winter are necessary. Night
temperatures of 5–10 °C and 10–15 °C by
day are sufficient. Little water is required
in winter and only moderate amounts from
spring to autumn. Propagation is by seed,
stem- and leaf-cuttings.

Literature: Nyffeler, R., A taxonomic
revision of the genus Monanthes Haworth
(Crassulaceae), *Bradleya* **10**: 49–82
(1992).

1a. Stems short, thick, unbranched   2
  b. Stems slender, branched   3
2a. Rosettes loose   **2. brachycaulos**
  b. Rosettes dense   **3. pallens**
3a. Leaves opposite   **1. laxiflora**
  b. Leaves alternate   4
4a. Shrubby, stem erect   **6. anagensis**
  b. Not shrubby, stem prostrate in lower
    part   5
5a. Rosettes more or less spherical;
    flowering-stems with long whitish
    hairs   **4. polyphylla**
  b. Rosettes narrower; flowering-stems
    finely downy   **5. subcrassicaulis**

1. M. laxiflora Bolle. Illustration: Praeger,
Sempervivums, 234, 236 (1932); Kunkel
& Kunkel, Flora de Gran Canaria **3**: 69
(1978); *Bradleya* **10**: 61, 64 (1992).

Perennial with grey stems; branches
rather twisted, at first erect then sprawling.
Leaves 8–10 × 6–8 mm, 3–5 mm thick,
opposite, elliptic to oblanceolate or obovate,
grooved and slightly flattened on face, dark
green, sometimes with silvery mottling.
Flowering-shoots erect, 2–8 cm, with
opposite leaves below; racemes almost bare
with 6–10 flowers. Flowers parts in 5s–8s,
purplish or yellowish. Calyx green, *c.*
3.5 mm with ovate segments. Petals *c.*
4 mm, linear-lanceolate, acute, mostly
greenish with red lines giving a purple
effect. Stamens purplish; anthers pale red.
Canary Islands. Spring.

2. M. brachycaulos (Webb & Berthelot)
Lowe (*M. brachycaulon* Lowe). Illustration:
Praeger, Sempervivums, 228 (1932);
Kunkel & Kunkel, Flora de Gran Canaria **3**:
67 (1978); *Bradleya* **10**: 61, 64 (1992).

Minute perennial; stems erect, short,
thick, bulb-like or cylindric. Leaves forming
a loose rosette, oblanceolate to spathulate,
tapering gradually at base, very fleshy, flat
or concave on face, very convex on back,
1.5–2 cm × 5 mm, 2–2.5 mm thick, green,
mottled with purple, with purple midrib

and stain at base. Flowering-branches lateral, from lower leaf-axils, 4–7 cm, erect, leafy in the middle with small loosely rosulate or crowded hairy leaves, raceme-like in the upper part. Racemes with 5–10 flowers, 2–5 cm, glandular-hairy. Flower parts in 6s or 7s; flowers 8–10 mm across, purplish green or greenish purple. Calyx *c.* 4 mm, green, glandular-hairy with ovate-lanceolate segments. Petals lanceolate or oblanceolate, *c.* 4 × 1 mm. Stamens greenish or purplish; anthers red. *Canary Islands.* Spring.

3. M. pallens Christ. Illustration: Praeger, Sempervivums, 231 (1932); *Bradleya* **10**: 61 (1992).

Perennial with erect, fleshy stems 1–3 cm, cylindric and bare in old plants; leaf-rosettes dense, 1.5–3.5 cm across, slightly concave, borne at ends of stems. Leaves 1–1.8 cm × 3–3.5 mm, 2 mm thick, diamond-shaped to spathulate, thick above, tapering below, green or glaucous, flattish on face, very convex on back, lower part pale, base purple. Flowering-shoots lateral, from outer leaf-axils *c.* 3 cm. Flowers with parts in 6s or 7s; calyx *c.* 4 mm, glandular-hairy, segments oblong-lanceolate, *c.* 3 mm, with red mottling and green papillae. Petals 3.5–4 × 0.75–1 mm, linear-lanceolate, acute, yellowish with red lines. Stamens *c.* 4 mm; filaments reddish, anthers red. *Canary Islands.* Spring.

4. M. polyphylla Haworth. Illustration: Praeger, Sempervivums, 241 (1932); Lamb & Lamb, Illustrated reference on cacti and other succulents **4**: 1150 (1975); Bramwell, Wild flowers of the Canary Islands, pl. 143 (1974); *Gartenpraxis* 1985, 52.

Minute creeping perennial with prostrate, slender stems forming a mat or dense cushion. Rosettes 1–1.5 cm, bluntly conical or ovoid, very dense. Leaves 6–8 × 2–2.5 mm, closely overlapping near the rounded apex, narrowly wedge-shaped to spathulate, fleshy, convex on face and back. Flower-stem arising from centre of rosette, 1–1.8 cm, with 4–8 flowers. Flowers 1–1.2 cm, parts in 6s–8s, green or brownish purple; calyx *c.* 4 mm, hairy; petals 4.5–5 × 0.5–0.8 mm, almost linear, acuminate, hairy on face, back and edges. *Canary Islands.* Spring.

5. M. subcrassicaulis Praeger.
Tufted perennial, not shrubby. Leaf-rosettes dense, *c.* 1 cm across, forming close tufts; leaves overlapping, *c.* 7–10 ×

2.5–3 mm, *c.* 1.5 mm thick, wedge- to club-shaped, shining, dark green, base purple, lower leaves often purplish. Flowering-shoots mostly from the centre of the rosettes, one or several to each rosette, leafless, 3–4 cm, with 1–5 flowers. Flowers with parts in 7s or 8s, purplish. Calyx hairy, striped red, *c.* 3 mm, with lanceolate segments. Petals *c.* 3.5 mm, linear-deltoid, acuminate, hairy. Stamens purplish; anthers purple. *Canary Islands.*

6. M. anagensis Praeger. Illustration: Praeger, Sempervivums, 238 (1932); *Bradleya* **10**: 61 (1992).

Perennial herb or shrublet to 15 cm. Branches grey, twisted; leaves *c.* 2.5 cm × 4 mm, *c.* 3 mm thick, alternate, stalkless, linear, upper surface grooved. Inflorescence raceme-like with 2–6 flowers. Flower parts in 7s, *c.* 1 cm across, greenish yellow. Calyx with triangular segments; petals *c.* 4 mm, triangular-lanceolate, acute, greenish yellow with a reddish vein. Stamens almost equal in length to the petals; filaments reddish; anthers yellow. *Canary Islands.* Spring.

27. JOVIBARBA Opiz
Z.R. Gowler

Like *Sempervivum* but petals 6 or 7 in number, pale yellow, keeled outside, fringed with glandular hairs forming a bell-shaped flower.

A genus of 6 European species, in which no natural hybrids occur.

Literature: Mitchell, P.J., The Sempervivum and Jovibarba handbook (1979); Praeger, R.L., An account of the Sempervivum group (1932).

1a.	Plant without stolons	**1. heuffelii**
b.	Plant with stolons	2
2a.	Rosette-leaves finely hairy on the surface	**2. allionii**
b.	Rosette-leaves hairless on the surface but with hairs along the margins	3
3a.	Rosette-leaves broadest near the middle	**3. arenaria**
b.	Rosette-leaves not broadest near the middle	4
4a.	Rosette-leaves broadest two-thirds of the way up	**4. sobolifera**
b.	Rosette-leaves broadest one third of the way up	5
5a.	Rosette-leaves an intense bluish green colour	**5. preissiana**
b.	Rosette-leaves wholly green or yellow-green with brown tips	**6. hirta**

1. J. heuffelii (Schott) A. & D. Löve (*Sempervivum heuffelii* Schott). Illustration: Praeger, An account of the Sempervivum group, 94 (1932); *Bulletin of the Alpine Garden Society* **6**: 103–4 (1938). Hay & Beckett, Reader's Digest encyclopaedia of garden plants and flowers, 659 (1971); Payne, Plant jewels of the high country, 42 (1972).

Very variable; leaves green, grey, brown, red or purple; rosette-leaves oblong-obovate, gradually narrowing towards the base; margins with stiff white hairs, apex very acute; flowering-stems 8–20 cm, with broad clasping lanceolate leaves; inflorescence dense and flattish; petals 6–7, yellow to white, hairy. *E Carpathians to the Balkan Peninsula.*

This is the only species that does not produce offsets or stolons but increases by the parent rosette splitting into 2 or more equal rosettes. The plant does not die after flowering.

Var. **heuffelii** develops medium-sized rosettes with distinctly hairy leaves.

Var. **glabra** Beck & Szyszylowicz. Rosette-leaves hairless on both sides, obovate and very much contracted to the apex. Petals truncate.

Var. **kopaonikense** (Pančic) P.J. Mitchell (*Sempervivum kopaonikense* Pančic). Hypogynous scales spreading, not erect, causing a bulge at base of calyx and corolla giving the flower a swollen appearance.

Var. **patens** (Grisebach & Schrenk) P.J. Mitchell (*Sempervivum patens* Grisebach & Schrenk). Rosettes minutely hairy.

Cultivars include: 'Aquarius', 'Bermuda', 'Bronze Ingot', 'Cameo', 'Chocoleto', 'Giuseppi Spiny', 'Greenstone', 'Henry Correvon', 'Hystyle', 'Miller's Violet', 'Minuta', 'Pallasii', 'Purple Haze', 'Sundancer', 'Tan', 'Tancredi', 'Torrid Zone', 'Xanthoheuff'.

2. J. allionii (Jordan & Fourreau) D.A. Webb (*Sempervivum allionii* Jordan & Fourreau). Illustration: Praeger, An account of the Sempervivum group, 96 (1932); Pitschman et al, Bilder-Flora der Sudalpen, 105 (1959); Huxley, Mountain flowers, pl. 243 (1967); Payne, Plant jewels of the high country, 20 (1972).

Rosettes 2–3 cm across, almost globular due to incurving of the leaves. Rosette-leaves lanceolate, acuminate, minutely hairy on both surfaces, with longer hairs on margins, yellowish green with a reddish flush on the apex when fully exposed.

Numerous offsets produced on short thin stolons, often stalkless, forming dense matted clumps. Flower-stems slender, 10–15 cm, inflorescence with 2–3, few-flowered, short branches. Sepals minutely hairy; petals greenish white, obviously fringed. *Europe (S Alps, France to the Tyrol).* H4. Early summer.

A compact and easily grown species, which is sometimes slow and reluctant to flower under cultivation.

J. allionii × J. hirta. Illustration: *Houselekes* **16**(1): 19 (1985). Intermediate with incurving rosettes, producing offsets. Rosette-leaves bright green with very few hairs, outer surface of older leaves and tips of new leaves flushed red. Offsets almost globular, on very thin stolons.

J. allionii × J. sobolifera. Leaves deep green with a bright red flush on the leaf backs.

J. × kwediana Mitchell; *J. allionii × J. heuffelii.* Illustration: *Houselekes* **13**(2): 39, 40 (1982).

Rosettes intermediate between the parents, 3.5–4 cm across. Rosette-leaves ovate, concave, mucronate, with scattered hairs on both surfaces, very succulent, 5 mm thick, with marginal hairs. Offsets on very short stolons. Flowering-stem *c.* 12 cm, glandular-hairy. Stem-leaves ovate, *c.* 2 × 1 cm. Inflorescence compact; flowers 1.5–1.75 cm; petals 5–6 erect, fringed, pale yellow.

'Cabaret'; *J. allionii × J. heuffelii* 'Giuseppi Spiny'. Upper half of the incurved rosette leaves flushed red, lower half bright green, rosettes squat and slow-growing.

'Pickwick' has larger and more robust rosettes, longer and more erect, lightish green rosette leaves, flushed red on the upper and outer surfaces. Flowering-stem more than 12 cm.

3. J. arenaria (Koch) Opiz (*Sempervivum arenarium* Koch). Illustration: Reichenbach, Icones florae Germanicae et Helveticae **23**: t. 74 (1897); Hegi, Illustrierte flora von Mittel-Europa **4**: 557, 559 (1921); Praeger, An account of the Sempervivum group, 62 (1932); Huxley, Mountain flowers, pl. 244 (1967).

Rosettes 5–20 mm across, globular, many-leaved. Rosette leaves incurved, lanceolate, broadest near the middle, bright green and flushed with red-brown at apex, hairless except on the leaf-edges. Minute offsets borne on very short (a few mm), slender horizontal stolons from the base of the parent rosette. Flowering-stems

7–12 cm, with erect, lanceolate, acuminate leaves. Inflorescence large for the size of the plant, with 3 forked branches. Petals 6, greenish white, hairy, especially at the tips *c.* 3 times as long as calyx. Flowers are produced infrequently and are seldom seen. *Europe (Alps, S Tirol, Styria & Carinthia).* H4. Summer.

Requires a lime-free soil.

J. × mitchellii Zonneveld; *J. arenaria × J. heuffelii.* Illustration: *Houselekes* **13**(2): 36 (1982). Rosette 2.5–4 cm across. Leaves mucronate, hairless except on margins. Offsets between the lower leaves on short stolons. Flowering-stem *c.* 8 cm, covered in leaves that are lightly hairy on the outside. Flowers bell-shaped, sterile, variously coloured; petals with 2 or 3 long fringes at tip, but nearly absent on keel.

4. J. sobolifera (Sims) Opiz (*Sempervivum soboliferum* Sims). Illustration: *Botanical Magazine*, 1457 (1812); Reichenbach, Icones flora Germanicae et Helveticae **23**: t. 66 (1897); Praeger, An account of the Sempervivum group, 101 (1932); Grey-Wilson & Blamey, Alpine flowers of Britain and Europe, 76 (1979).

Rosettes 2–3 cm across, usually half-closed with incurved leaves. Rosette leaves oblanceolate, *c.* 10 × 5 mm, very fleshy, bright green often with a coppery red flush on the back near the apex, hairless; margins with stiff glandular hairs. Offsets numerous, spherical, borne among the outer and middle leaves on very slender stolons. Flowering-stems 10–20 cm, glandular-hairy, clothed with many, almost-erect leaves. Inflorescence 5–7 cm across, dense and flattish; flowers seldom produced, petals 6, fringed, greenish yellow. *C & E Europe (C Germany to the E Carpathians, N & C Russia).*

'Green Globe'. Rosettes bright green without the red flush.

J. × nixonii Zonneveld. Illustration: *Houselekes* **13**(2): 32, 35 (1982). Rosettes 3–5 cm across. Rosette-leaves abruptly mucronate, hairless except along the margins. Offsets between the lower leaves on very short stolons. Flowering-stem *c.* 8 cm, covered in lightly hairy leaves; flowers sterile. Petals fringed near the top, but not on keel. The plant dies after flowering.

'Jowan'; *J. sobolifera × J. heuffelii.* Rosettes dark almost brownish red; the colour is retained for a considerable proportion of the year. Offsets on very short stolons

around the base of the parent rosette, soon forms a compact cluster.

5. J. preissiana (Domin) Omel'chuk-Myakushko & Chopik (*Sempervivum preissianum* Domin). Illustration: *Botanicheskii Zhurnal* **60**(8): 1184 (1975).

Rosettes 4–5 cm across, stellate; rosette leaves intense bluish green, 1.5–2.5 cm, ovate-lanceolate to long-lanceolate, broadest one third of the way up from the base, gradually pointed at the tip, surface hairless except along margin. Flowering-stems robust, 20–30 cm, softly hairy, with numerous stem-leaves, ovate-lanceolate, gradually pointed, to 8–10 mm across, margins hairy. Inflorescence dense, large and many-flowered. Flower-stalk fleshy and glandular-hairy; flowers almost bell-shaped; petals 6, pale or whitish yellow. *Carpathians.* H3. Summer.

6. J. hirta (Linnaeus) Opiz (*Sempervivum hirtum* Linnaeus). Illustration: Reichenbach, Icones flora Germanicae et Helveticae **23**: t. 73 (1897); Jávorka & Csapody, Iconographia florae Hungaricae, 223 (1930); Praeger, An account of the Sempervivum group, 98–9 (1932); Huxley, Mountain flowers, pl. 241 (1967).

Rosettes 2.5–7 cm across, open and stellate; rosette-leaves broadly lanceolate, broadest one third of the way up, narrowing to an acute apex. Leaves hairless except along margins, wholly green or yellow-green with brown tips, some forms are a bright red-brown when in full exposure; offsets small, very numerous, arising on very thin stolons from the axils of the middle and lower leaves of the parent rosette; flowering-stems 10–20 cm, stout, glandular-hairy; stem-leaves numerous, erect, broadly lanceolate, hairless. Inflorescence dense, flattish or convex, 5–8 cm across with 3 forked branches, each with *c.* 5 flowers; petals 6, erect, pale yellow to greenish white, hairy along the edges. *E Alps, Carpathians, Hungary, NW Balkan Peninsula.* H4. Summer.

Subsp. **hirta.** Rosettes 2.5–5 cm across; leaves 5–6 mm across at or below the middle, without a red apex; margins with glandular hairs. Stem-leaves and sepals hairy on the lower surface, at least on the midribs.

Subsp. **borealis** (Huber) R. Sóo. Rosettes 1–3 cm across; leaves broadest above the middle, often with a red apex; margins with glandular hairs. Surface of sepals hairless.

Subsp. **glabrescens** (Sabransky) R. Sóo & Jávorka. Stem-leaves and sepals hairless except along margins.

Var. **neilreichii** (Schott, Nyman & Kotschy) Konop & Bendak. Rosette leaves 2–3 mm across, broadest in upper third, stem-leaves hairless.

Forma **hildebrandtii** (Schott) Konop & Bendak. Rosette-leaves 8–12 mm across, rosettes grey-green; stem-leaves hairless.

28. SEMPERVIVUM Linnaeus
Z.R. Gowler & M.C. Tebbitt

Succulent perennial herbs with pointed basal leaves in a rosette. Rosettes 1–15 cm across, monocarpic, plant increases by producing young offsets from the leaf axils, usually on runners (stolons). Flower-stems terminal, erect, usually with a covering of leaves. Flowers in a terminal cyme, with 8–16 parts in a star shape, twice as many stamens as petals, equal number of carpels to petals. Petals yellowish, pink, purple or red.

A genus of about 50 species chiefly from Europe but also Turkey, Iran, the Caucasus and Morocco. Sempervivums hybridise easily both in the wild and in cultivation.

Literature: Mitchell, P.J., The Sempervivum & Jovibarba handbook (1973); Praeger, R.L. An account of the Sempervivum group (1932).

1a.	Petals and stamens lacking	
	2. octopodes var. **apetalum**	
b.	Petals and stamens present	2
2a.	Petals yellow or yellow-green	3
b.	Petals not yellow	19
3a.	Petals red, pink or purple at the base	
		4
b.	Petals not coloured at the base	12
4a.	Petals with a green central stripe	
	1. sosnowskii	
b.	Petals without a green central stripe	
		5
5a.	Rosettes not more than 2 cm across	
	2. octopodes	
b.	Rosettes more than 2 cm across	6
6a.	Leaves giving off a strong 'goaty' smell	**3. grandiflorum**
b.	Leaves not smelling	7
7a.	Leaf-tips not coloured	
	4. transcaucasicum	
b.	Leaf-tips coloured	8
8a.	Hairs present on the surfaces of mature rosette-leaves	9
b.	Hairs absent from the surfaces of mature rosette-leaves	11
9a.	Rosettes 3–4 cm across; stamen filaments dark crimson	**6. zeleborii**

b.	Rosettes 4–7 cm across; stamen filaments purple	10
10a.	Leaf-margins with long glandular hairs; flower-stems very leafy	
	5. kindingeri	
b.	Leaf-margins with medium glandular hairs; flower-stems not very leafy	
	4. transcaucasicum	
11a.	Juvenile rosettes glandular-hairy	
	7. armenum	
b.	Juvenile rosettes hairless	**8. wulfenii**
12a.	Petals flushed red near the tip	
	11. ciliosum, see × **praegeri**	
b.	Petals not flushed red near the tip	13
13a.	Rosettes more than 5 cm across	
	9. ruthenicum	
b.	Rosettes 5 cm or less across	14
14a.	Rosettes half open	15
b.	Rosettes open	17
15a.	Rosettes to 2.5 cm across	**10. minus**
b.	Rosettes 3 cm or more across	16
16a.	Hairs on leaf-margin 2–4 mm	
	11. ciliosum	
b.	Hairs on leaf-margin less than 2 mm	
	12. davisii	
17a.	Leaves incurved	18
b.	Leaves not incurved	
	14. glabrifolium	
18a.	Hairs on leaf-margin 2–4 mm	
	11. ciliosum	
b.	Hairs on leaf-margin less than 2 mm	
	13. pittonii	
19a.	Petals white or greenish white	20
b.	Petals pink, purple or red	21
20a.	Petals white with rose-red stripes	
	9. ruthenicum, see	
	× **degenianum**	
b.	Petals greenish white with no stripes	
	15. leucanthum	
21a.	Petals with white lines or white margins	22
b.	Petals without white lines or white margins	34
22a.	Petals with narrow white lines	
	16. erythraeum	
b.	Petals with white margins	23
23a.	Rosette-leaves completely hairless on the surfaces	**17. marmoreum**
b.	Rosette-leaves with some hairs on the surfaces	24
24a.	Rosette-leaves with only a few hairs on the backs	**18. iranicum**
b.	Rosette-leaves with hairs on both sides	25
25a.	Filaments bright lilac, white-striped on the base	
	17. marmoreum var.	
	angustissimum	
b.	Filaments not as above	26
26a.	Leaf-tips coloured	27

b.	Leaf-tips not coloured	31
27a.	Leaf-tips brown	28
b.	Leaf-tips red-purple	30
28a.	Stolons less than 4 cm	**19. giuseppii**
b.	Stolons more than 4 cm	29
29a.	Rosettes flat, many-leaved; stem-leaves ovate to oblong	
	20. ingwersenii	
b.	Rosettes not flat, few-leaved; stem-leaves triangular-lanceolate	
	21. ossetiense	
30a.	Stolons long, to 12 cm	**22. kosaninii**
b.	Stolons very short	
	23. dzhavachischvilii	
31a.	Stolons slender	32
b.	Stolons stout	**24. reginae-amaliae**
32a.	Flower-stems covered in shaggy white hairs	**25. dolomiticum**
b.	Flower-stems not covered in shaggy white hairs	33
33a.	Anthers yellow	
	26. thompsonianum	
b.	Anthers purple	**27. pumilum**
34a.	Surfaces of rosette-leaves hairless	35
b.	Surfaces of rosette-leaves hairy	43
35a.	Rosettes semi-open	36
b.	Rosettes open	38
36a.	Rosette-leaves with brown tips	37
b.	Rosette leaf-tips not coloured	
	28. ballsii	
37a.	Petals with toothed edges	
	29. borissovae	
b.	Petals with smooth edges	
	30. andreanum	
38a.	Leaf-margin hairless, leaves usually grey-green with purple tips	
	31. calcareum	
b.	Leaf-margin with at least a few hairs, leaves not as above	39
39a.	Leaf-tips coloured	40
b.	Leaf-tips not coloured	41
40a.	Flower-stems 5–12 cm	
	32. balcanicum	
b.	Flower-stems 20–40 cm	
	33. tectorum	
41a.	Flower-stem leaves with longitudinal spots	**34. vincentei**
b.	Flower-stem leaves without longitudinal spots	42
42a.	Rosette 5–10 cm across; leaves blue-green	**33. tectorum** var. **glaucum**
b.	Rosette 2.5–3.5 cm across; leaves green	**35. nevadense**
43a.	Leaves covered in long white woolly hairs	**36. arachnoideum**
b.	Leaves not covered in long white woolly hairs	44
44a.	Marginal hairs of unequal lengths	45
b.	Marginal hairs all the same length or absent	47

45a. Rosettes 8 cm or more across
　　　　　　　37. charadzeae
　b. Rosettes less than 6 cm across　　46
46a. Flower-stem *c.* 20 cm
　　　36. arachnoideum, see × **funckii**
　b. Flower-stem *c.* 30 cm　　**38. italicum**
47a. Leaf-tips coloured　　　　　　　48
　b. Leaf-tips not coloured　　　　　51
48a. Leaf-tips dark brown
　　　　　　　39. caucasicum
　b. Leaf-tips red or purple　　　　49
49a. Rosettes 5–6 cm across
　　　　　　　40. cantabricum
　b. Rosettes 4 cm or less across　　50
50a. Flower-stems short
　　　　　　　41. macedonicum
　b. Flower-stems tall (*c.* 33 cm)
　　　　　　　42. altum
51a. Rosettes to 2 cm across　　　　52
　b. Rosettes 4–8 cm across
　　　　　　　43. atlanticum
52a. Leaf-tips with long woolly hairs
　　　36. arachnoideum, see
　　　　　× **barbulatum**
　b. Leaf-tips lacking woolly hairs
　　　　　　　44. montanum

1. S. sosnowskii Ter-Chatschatorova.
Illustration: *Journal of the Sempervivum Society* **10**(1): 26 (1979).

Rosettes 10–12 cm across. Leaves oblong to spathulate, slightly broader in the upper third, hairless except for short hairs along the margins, flushed red at the tips. Offsets numerous, on stout stolons 2–3 cm. Flowering-stem 30–40 cm, covered in short scattered hairs and hairless strap-shaped leaves 6–9 cm. Flowers 3–3.5 cm across. Petals 14–16, yellowish green with a green central stripe and lilac tinge at the base. Petals glandular-hairy; filaments lilac-tinted, glandular-hairy. *Georgia.*

Similar to *S. armenum.*

2. S. octopodes Turrill. Illustration: *Bulletin of the Alpine Garden Society* **6**: 99–100 (1938), **8**: 215 (1940) & **17**: 233 (1949); *Journal of the Sempervivum Society* **9**(3): 85 (1978) & **10**(3): 81 (1979); Payne, Plant jewels of the high country, 50 (1972).

Rosettes 1–2 cm across, incurved or semi-open. Leaves oblanceolate or obovate, slightly obtuse, shortly mucronate, fleshy, densely hairy on both surfaces, glandular marginal hairs getting longer towards the red-brown tips. Offsets on slender, brown stolons to 7 cm. Flower-stem *c.* 9 cm, slender with a compact inflorescence of few flowers. Petals yellow with a pale red spot at the base. Filaments reddish purple,

anthers yellow. *S former Yugoslavia (N Macedonia).* Summer.

Similar to *S. ciliosum* and *S. thompsonianum.* Requires lime-free soil. It is difficult to grow as it dislikes excessive wet or dry at any time of year and is readily attacked by greenfly.

Var. **apetalum** Turrill. Rosettes 2.5–3 cm across with more leaves. Leaves less fleshy and lighter green with very small brown markings on the tips; stolons to 9 cm. Flowers have no petals or stamens but numerous sepals. Easy to grow, withstands winter damp well.

3. S. grandiflorum Haworth
(*S. globiferum* Curtis; *S. globiferum* Gaudin; *S. gaudini* Christ; *S. braunii* Arcangeli; *S. wulfenii* Hoppe subsp. *gaudini* Christ). Illustration: Praeger, An account of the Sempervivum group, 78 (1932); *Bulletin of the Alpine Garden Society* **3**: 273 (1935), **37**: 96 (1969); Nicholson et al., The Oxford book of garden flowers, 91 (1963); Huxley, Mountain flowers, pl. 233 (1967).

Rosettes 2–10 cm across, flat and rather loose. Leaves oblanceolate to wedge-shaped, or almost strap-shaped, green often with a brown tip, densely hairy and quite sticky, giving off a strong, resinous, 'goaty' smell. Offsets nearly spherical on long leafy stolons. Flowers large; petals yellow or greenish yellow tinged purple at the base. *S Switzerland & N Italy.* Summer.

Requires lime-free soil. 'Fasciatum' is a curiously congested form; 'Keston' has yellow-green leaves in spring.

S. × christii Wolf; *S. grandiflorum × S. montanum* (*S. rupicolum* Chenevard & Schmidely). Varies between the parents. Leaves with a small purple tip or wholly green, lightly or densely hairy. Petal colour varies from yellow to purple but is always yellowish towards the tip. *Switzerland to N Italy.*

S. × hayekii Rowley; *S. grandiflorum × S. tectorum.* Rosettes large, sparingly hairy, with marginal hairs on the leaves. Flowers pale yellowish purple, smaller than in *S. grandiflorum* but larger than *S. tectorum.*

4. S. transcaucasicum Muirhead.
Rosettes 4–7 cm across, semi-open. Leaves obovate or oblanceolate, shortly mucronate, densely and finely glandular-hairy on both surfaces, with many marginal hairs, green or yellow-green, upper half tinged pink when exposed. Few offsets produced on stout stolons *c.* 2 cm. Flower-stem 15–18 cm, glandular-hairy, covered in many overlapping glandular

hairy leaves, flushed pink on the outer surfaces. Flowers to 2.5 cm across, consisting of 12–14 petals. Petals greenish yellow, purple at the base, twisted at the tip, densely glandular hairy on the lower surface. Filaments pale purple, the lower half densely hairy, anthers yellow. *Caucasus, Transcaucasus.*

Quite easy to grow but needs some protection from winter damp; seldom flowers. Can be confused with *S. ruthenicum* Schnittspahn & Lehmann and *S. zeleborii* Schott. *S. globiferum* misapplied is a name used to describe all yellow species from the Caucasus and Turkey.

5. S. kindingeri Adamovic. Illustration: Praeger, An account of the Sempervivum group, 83 (1932); Payne, Plant jewels of the high country, 44 (1972); *Bulletin of the Alpine Garden Society* **3**: 280 (1935); *Houslekes* **16**(1): 35 (1985).

Rosettes 4–6 cm across, flat, open. Leaves wedge-shaped to oblong, glandular-hairy, with long marginal hairs, pale yellowish green with a purplish flush at the tips. Offsets on medium-long stolons. Flower-stem to 20 cm, very leafy. Petals 12–14, pale yellow with a pink-red base. Filaments purple, anthers yellow. *Former Yugoslavia (Macedonia), N Greece.*

A rare plant that needs protection in the winter, flowers poorly and develops few offsets.

6. S. zeleborii Schott (*S. ruthenicum* Koch). Illustration: *Journal of the Sempervivum Society* **7**(3): 66 (1976).

Rosettes 3–4 cm across, spherical and compact. Leaves oblong-obovate, shortly mucronate, densely hairy, pale or grey-green with outer leaves flushed pink on exposure, sometimes with a small dark tip. Few offsets produced on short stolons. Flower-stem 10–15 cm. Flowers *c.* 2.5 cm across. Petals 12–14, yellow with a crimson base. Filaments dark crimson, anthers yellow. *SE Bulgaria & S Romania.*

7. S. armenum Boissier & Huet. Illustration: *Bulletin of the Alpine Garden Society* **10**: 236 (1942); *Notes from the Royal Botanic Garden, Edinburgh* **29**: 16, pl. 1A (1969).

Rosettes 4–6 cm across, glaucous. Rosette leaves hairless except along the edges, mucronate, dark purple tips. Few stolons; juvenile rosettes glandular-hairy, hairless on maturity. Flower-stem 10–15 cm, upper leaves hairy. Inflorescence a dense cyme 4–5 cm across.

Flowers with 12–14 pale yellow or greenish petals, purple tinged at the base. Filaments purple, anthers yellow. *Armenia, Turkey.*

A rare species often confused with *S. globiferum* and some variants of *S. tectorum*, considered to be closely related to *S. sosnowskii*. Slow-growing, apt to damp-off in the winter wet.

Var. **insigne** Muirhead. Rosettes 2–3 cm, highly coloured with many offsets. Stem-leaves overlapping. Petals flushed rose-purple violet at base. Filaments violet.

8. S. wulfenii Merten & Koch (*S. globiferum* Wulfen). Illustration: Reichenbach, Icones florae Germanicae et Helveticae **23**: t. 69 (1896–1899); Praeger, An account of the Sempervivum group, 91 (1932); Huxley, Mountain flowers, pl. 232 (1967); Grey-Wilson & Blamey, Alpine flowers of Britain and Europe, 75 (1979).

Rosettes 5–9 cm across, open. Leaves oblong to spathulate, hairless except for the marginal hairs, grey-green with a purple base and darker tips. Few offsets produced on long stolons to 10 cm. Flower-stem 15–25 cm, covered with hairy leaves, which are slightly recurved at the tips. Flowers *c.* 2.5 cm across, funnel-shaped to stellate. Petals 12–15, lemon-yellow with a purple base. Filaments purple, anthers yellow. *C & E Alps.* Summer.

Like *S. tectorum* var. *glaucum* but central leaves closed to form a central bud. Slow-growing, difficult in cultivation, needs winter protection.

9. S. ruthenicum Schnittspahn & Lehmann (*S. globiferum* (Linnaeus) Koch; *S. globiferum* Linnaeus; *S. ruthenicum* Koch). Illustration: *Bulletin of the Alpine Garden Society* **3**: 274, 279; Jelitto & Schacht, Hardy herbaceous perennials **2**: 613 (1990).

Rosettes 5–8 cm across, incurved but more open in summer. Leaves club-shaped, narrowing at the base, shortly glandular-hairy with dense marginal hairs, dark green. Stolons 3–5 cm, stout. Flower-stem 20–30 cm, covered in weakly pointed, oblong leaves with reddish brown tips. Petals yellow with small rounded scales that stand out clearly. Filaments green, anthers yellow. *Romania, Ukraine.*

Quite easy to grow and sometimes confused with *S. zeleborii* Schott (*S. ruthenicum* Koch) and *S. armenum* Boissier & Huet.

S. × degenianum Domokos; *S. banaticum × S. ruthenicum*. Rosettes 8–10 cm across, open. Leaves linear, contracting abruptly towards the tip, broadest in the upper third, dense white hairs on both surfaces, marginal hairs, brown-red. Flower-stem to 30 cm, stem-leaves covered in dense white hairs. Flowers to 2.5 cm across, with 13–14 parts. Petals white with rose-red stripes, glandular-hairy below, hairless above. Filaments densely hairy at the base, anthers pale pink or pale yellow. *E Europe (Banatu to near the lower Danube).* Summer.

10. S. minus Turrill. Illustration: *Bulletin of the Alpine Garden Society* **10**: 235–7 (1942).

Rosettes 1–2.5 cm across, very compact, central leaves closed, outer leaves more open. Leaves oblong-oblanceolate, acute, very short hairs on both surfaces with slightly longer hairs on the margins, dull olive-green, purple at the base, bronze on the outer leaves when exposed. Offsets on very short stolons, almost stemless. Flower-stem 2.5–6.5 cm, slender and with elliptic fleshy leaves. Inflorescence compact with few, large flowers. Petals pale yellow, anthers yellow. *N Turkey.*

The smallest species of *Sempervivum*. Rarely seen in cultivation as it increases slowly and flowers often. May sometimes resemble *S. pumilum*.

11. S. ciliosum Craib (*S. wulfenii* Velenovsky; *S. wulfenii* Hoppe; *S. ciliosum* Pančič; *S. ciliatum* Craib; *S. borisii* Degen & Urumov). Illustration: Praeger, An account of the Sempervivum group, 89 (1932); Payne, Plant jewels of the high country, 32 (1972); *Bulletin of the Alpine Garden Society* **8**: 207 (1940); *Journal of the Sempervivum Society* **7**(1): 19 (1976), **9**(3): 77–8 (1978) & **10**(3): 75, 77 (1979).

Rosettes 3.5–5 cm across, grey-green, flattened spherical, wholly or half-closed, outer leaves tinged red. Leaves oblong-oblanceolate, acute, strongly incurved, hairy, marginal hairs very long, giving the leaves their greyness. Many offsets produced on strong hairy stolons. Yellow flowers on short stems with overlapping leaves. Petals 10–12. *Bulgaria, former Yugoslavia, NW Greece.* Late spring.

Not difficult to grow but susceptible to winter damp.

Var. **borisii** (Degen & Urumov) P. Mitchell, extra densely haired giving a white appearance, stolons short. *Bulgaria.*

Var. **galicicum** A.C. Smith has hairy rosettes becoming plum-red in the summer; stolons extra long, hairy; flowers small, yellow. *Former Yugoslavia (SW Macedonia).*

S. ciliosum × S. marmoreum. Small green rosettes heavily fringed with hair.

S. × praegeri Rowley; *S. ciliosum × S. erythraeum*. Illustration: *Journal of the Sempervivum Society* **7**(2): 28 (1976). Some consider this hybrid to be *S. erythraeum × S. leucanthum*, often sold under the name *S. × praegeri*. Intermediate between the parents. Rosettes like *S. erythraeum*, small and open, but the leaves densely glandular-hairy with both long and short hairs. Leaves in outline like *S. erythraeum*, broader than *S. ciliosum*. Flowers intermediate, petals greenish yellow, flushed with red near the tip, hairy on back and edges. Anthers buff flushed purple. Filaments purplish above, whitish below. *Bulgaria.*

12. S. davisii Muirhead. Illustration: *Notes from the Royal Botanic Garden, Edinburgh* **29**: pl. 3B, 26 (1969).

Rosettes 3–4 cm across, half-open. Leaves oblanceolate or obovate, abruptly mucronate, covered in dense glandular hairs, grey-green sometimes with a brown-red tip. Few offsets produced on stolons 2–3 cm. Flower-stem 10–12 cm, erect, covered in short dense glandular hairs. Numerous stem-leaves, short glandular hairs on both sides, grey-green, with a brown-red apex. Inflorescence compact or expanded, with 20–40 flowers, glandular. Flowers *c.* 2 cm across, 12–14 pale yellow petals. Filaments white, anthers yellow. *Turkey.*

13. S. pittonii Schott, Nyman & Kotschy (*S. braunii* Maly). Illustration: Seboth & Bennett, Alpine plants **2**: t. 34 (1880); Hegi, Illustrierte flora von Mittel-Europa, **4**: 552, 557 (1921); Praeger, An account of the Sempervivum group, 85 (1932); Huxley, Mountain flowers, pl. 234 (1967).

Rosettes 1.5–3 cm across, dense, flattish, many-leaved. Leaves linear-oblanceolate, acute, incurved, glandular-hairy on both sides with glandular marginal hairs, grey-green with a very small purple tip. Stolons 2–3 cm. Flower-stem 2–3 cm, slender with narrow, overlapping, purple-tipped leaves. Inflorescence comparatively small with large flowers 2–2.5 cm across. Petals 9–12, yellow, filaments greenish yellow, anthers yellow. *E Alps.*

Similar to *S. leucanthum*; not difficult to grow but susceptible to winter damp.

14. S. glabrifolium Borissova. Illustration: *Notes from the Royal Botanic Garden, Edinburgh* **29**: pl. 1B, 16 (1969).

Rosettes, *c.* 2.5 cm across. Leaves with marginal hairs, olive-green with heavy purple markings on the upper half, especially towards the outside of the rosette. Numerous swollen offsets produced on short stolons. Petals pale greenish yellow; filaments white, anthers yellow. *E Turkey.*

Very similar to *S. armenum* and *S. sosnowskii*, all have hairless rosette-leaves when mature but are glandular when young. Difficult to grow, will damp-off in summer and winter.

15. S. leucanthum Pančič. Illustration: Praeger, An account of the Sempervivum group, 87 (1932); *Bulletin of the Alpine Garden Society* **3**: 280 (1935); *Journal of the Sempervivum Society* **6**(1): 8, 12 (1975).

Rosettes 2.5–5 cm across, flattish with the inner leaves closed, outer leaves erect. Leaves narrowly wedge-shaped, widest near the tip and finely hairy on both sides, marginal hairs of unequal lengths, pale yellow-green becoming dark green with a dark red-purple tip. Stolons stout, 5–8 cm. Flower-stem tall, slender, bearing a small inflorescence. Petals 11–13 greenish white; filaments white to purple, anthers yellow. *Bulgaria (Rila Mts).* Early summer.

Easy to cultivate but increases slowly and soon dies out as all the rosettes often flower at once. Similar to *S. kindingeri* but rosettes less open with many more leaves, stolons much longer.

16. S. erythraeum Velenovsky (*S. montanum* Velenovsky; *S. cinerascens* Adamovic; *S. leucanthum* Stojanoff & Stojanoff; *S. ballsii* misapplied). Illustration: Praeger, An account of the Sempervivum group, 56 (1932); *Bulletin of the Alpine Garden Society* **3**: 25 (1935) & **8**: 201 (1940); Payne, Plant jewels of the high country, 38 (1972).

Rosettes 2–5 cm across, flat and open. Leaves obovate, spathulate, mucronate, covered in very short white hairs that can often only be seen under a magnifying glass, marginal hairs of unequal length, grey-green, often tinged with purple. Offsets always on short stolons. Flower-stem very hairy, to 20 cm. Flowers *c.* 2 cm wide; petals 11 or 12, purplish red with narrow white lines, with fine white hairs on the lower side. *Bulgaria.* Summer.

17. S. marmoreum Grisebach (*S. montanum* Sibthorp & Smith; *S. schlehanii* Schott; *S. assimile* Schott; *S. blandum* Schott; *S. rubicundum* Schur; *S. tectorum* Boissier; *S. reginae-amaliae* Boissier; *S. montanum* var. *assimile* Stojanoff & Stefanoff; *S. blandum* var. *assimile* Stojanoff & Stefanoff; *S. montanum* var. *blandum* (Schott) misapplied; *S. ornatum* misapplied). Illustration: *Bulletin of the Alpine Garden Society* **6**: 100 (1938) & **8**: 302 (1940); Huxley, Garden perennials and water plants, pl. 251 (1970); *Journal of the Sempervivum Society* **6**(3): 11 (1975).

Rosettes *c.* 6 cm across, flat and open. Leaves obovate to spathulate, broader in the upper part, abruptly mucronate, hairless at maturity except for stout deflexed marginal hairs, green sometimes with darker tips or red on back and base. Offsets on thick stolons *c.* 2 cm. Flower-stem stout, 10–15 cm. Flowers *c.* 2.5 cm wide. Petals 12–13, purple-pink with white margins. *E Europe, Balkans.* Summer.

Var. **angustissimum** Priszter. Rosettes 3–4 cm across. Leaves long and narrow, hairy on both surfaces, top third flushed deep red. Offsets on short stolons. Flower-stem 25–30 cm, finely hairy with few leaves, flushed purple. Inflorescence 7–10 cm across with over 30 flowers. Flowers 1.6–2 cm across. Petals 11–13, bright rose, hairy on margins and lower surface. Filaments bright lilac, white and striped on the base, lower third glandular hairy. *NE Hungary.*

Var. **dinaricum** Becker. Rosettes 1–2.5 cm across. Leaves sharply pointed, dark red-brown at the tips. Inflorescence smaller, flowers with narrow petals. *Bulgaria, former Yugoslavia (Macedonia).*

'Brunneifolium', rosettes compact, uniform brown, becoming red in winter, mature leaves hairless; 'Chocolate' ('Brunneifolium Dark Form'), rosettes slightly less compact than 'Brunneifolium', dark chocolate-coloured; 'Rubrifolium', leaves deep red, tips and margins green; 'Ornatum', rosettes large, ruby-red, tips apple-green, colour only lasts from spring to summer.

S. × versicolor Velenovsky; *S. marmoreum × S. zeleborii.* Illustration: *Bulletin of the Alpine Garden Society* **3**: 273 (1935). Rosettes large. Flowers pale yellow to pale lilac with age, seldom produced. *On roofs in Bulgaria but not found wild.*

18. S. iranicum Bornmüller & Gauba. Rosettes 3–5 cm across, open. Leaves narrowly obovate, mucronate, hairs along the margins and few on the backs. Offsets on very short stolons. Petals rose-pink with narrow white margins; filaments rose, anthers brownish pink. *N Iran.*

Not difficult to grow but susceptible to winter damp.

19. S. giuseppii Wale. Illustration: *Bulletin of the Alpine Garden Society* **9**: 111, 114 (1941); *Journal of the Sempervivum Society* **6**(1): 8 (1975); Brickell (ed.), RHS A–Z encyclopedia of garden plants, 976 (2003).

Rosettes 2.5–3.5 cm across, compact in dense clumps. Leaves ovate, mucronate, pale green with a small brown patch at the apex, densely hairy with long stiff marginal hairs. Numerous offsets on short stolons. Flowers rose-red with narrow white margins. *NW Spain.*

20. S. ingwersenii Wale. Illustration: *Bulletin of the Alpine Garden Society* **10**: 88, 91, 93 (1942).

Rosettes 3–4 cm across, very compact and flattish. Leaves densely but finely hairy and with a small brown apex. Numerous offsets produced on long brownish red stolons. Flowers rarely produced, petals red with narrow white margins. *Caucasus*

Vigorous and easy to cultivate. Like *S. altum* and *S. ossetiense* in habit but with pea-green rosettes and smaller flowers in a more compact inflorescence.

21. S. ossetiense Wale. Illustration: *Bulletin of the Alpine Garden Society* **10**: 100, 102 (1942).

Rosettes *c.* 3 cm across, dense, few-leaved. Leaves oblanceolate to oblong-oblanceolate, very fleshy, covered in short hairs, marginal hairs slightly longer than surface hairs, pea-green with a small brown tip. Offsets on stout stolons 5–8 cm. Flower-stem 8–10 cm. Inflorescence small, few-flowered but individual flowers large. Petals purple with broad white margins; filaments purple, anthers deep red. *Caucasus.*

Difficult to grow, susceptible to winter damp, produces few offsets or flowers. Similar to *S. altum* and *S. ingwersenii*.

22. S. kosaninii Praeger. Illustration: *Bulletin Institut & Jardin Botaniques Belgrade* **1**: 211–12 (1930); Praeger, An account of the Sempervivum group, 54 (1932); *Bulletin of the Alpine Garden Society* **3**: 268 (1935); *Journal of the Sempervivum Society* **7**(2): 43 (1976).

Rosettes 4–8 cm across, open, dense, flattish. Leaves oblanceolate, shortly acuminate, glandular-hairy on both sides, marginal hairs twice as long as other hairs, dark green with a red-purple tip. Offsets on strong leafy stolons to 12 cm. Flower-stem stout, covered in loosely overlapping leaves. Petals reddish purple with white margins, greenish on the back. Filaments purple, anthers light red. *SW former Yugoslavia (Macedonia)*.

23. S. dzhavachischvilii Gurgenidze. Illustration: *Houslekes* **11**: 79, 81 (1980).

Rosettes 3–5 cm. Leaves lanceolate, acuminate, dark green, outside flushed purple, dark purple tips often bent inwards, glandular-hairy on both sides. Offsets with outer leaves longer and increasingly cylindric, on very short stolons. Flower-stem 7–10 cm, stem-leaves covered in unequal length glandular hairs. Flower 2–2.3 cm across. Petals purple-pink with white margins. Filaments dark purple. Anthers pale purple with traces of orange. Rosette-leaves persistent at flowering. *E Caucasus*.

Closely related to *S. pumilum*.

24. S. reginae-amaliae Haláscy. Illustration: *Gartenflora* **83**: 159 (1934); *Bulletin of the Alpine Garden Society* **8**: 62, 201 (1940); Payne, Plant jewels of the high country, 54 (1972); *Houslekes* **14**(2): 55–8 (1983).

Rosettes 2–4 cm across, usually dense-leaved and compact. Leaves spathulate to obovate, acute, thick, evenly hairy on both surfaces with longer marginal hairs, soft green, outer leaves red. Few offsets produced on short stout stolons. Flower-stem 9–12 cm, wider at the base, covered in overlapping ovate leaves, usually flushed rose to red. Petals 10–12, crimson with clearly defined white margins. Filaments crimson, anthers buff sometimes violet. *Greece, S Albania*. Summer.

This name is often wrongly applied to a dark-tipped variant of *Jovibarba heuffelii*. Praeger included it under *S. schlehanii* Schott. Very variable in colour, size and compactness of rosette, but floral characters constant. Easy to grow but slow to increase.

25. S. dolomiticum Facchini. (*S. tectorum* var. *angustifolium* Seybold; *S. lehmanni* Schnittspahn; *S. oligotrichum* Dalla Torre). Illustration: Jávorka & Csapody, Iconographia florae Hungaricae, 222 (1930); Praeger, An account of the Sempervivum group, 51 (1932); Huxley, Mountain flowers, pl. 236 (1967);

Grey-Wilson & Blamey, Alpine flowers of Britain and Europe, 75 (1979).

Rosettes 2–4 cm, erect, semi-open, nearly spherical. Leaves covered in glandular hairs and with long white hairs along the margins, bright green with the outside of older leaves tinged scarlet. Numerous offsets on slender stolons. Flower-stem to 10 cm, thin, covered in shaggy white hairs and purple-tipped leaves. Petals 10–14, rose-red with a dark central stripe and white marginal flecks. *E Alps*. Summer.

Similar to some forms of *S. montanum* but upright leaves denser and more pointed. Not very easy to grow, susceptible to winter damp, flowers only occasionally.

S. dolomiticum × **S. montanum**. A compact growing plant with small deep green rosettes.

26. S. thompsonianum Wale. Illustration: *Bulletin of the Alpine Garden Society* **8**: 208, 212 (1940); *Journal of the Sempervivum Society* **7**(2): 41 (1976).

Rosettes 1.5–2 cm across, many-leaved, nearly spherical, outer leaves erect. Leaves ovate-lanceolate, acute, hairy, marginal hairs of unequal length, *c*. 3 times as long as the surface hairs, yellowish green, outer leaves flushed red. Stolons 5–8 cm, slender and brown. Flower-stems *c*. 8 cm, very slender and covered with few, narrow, lanceolate leaves. Inflorescence few-flowered and compact. Petals deep pink with white margins and yellow tips. Filaments purple, anthers yellow. *S former Yugoslavia (Macedonia)*.

27. S. pumilum Bieberstein (*S. montanum* Eichwald; *S. braunii* Ledebour). Illustration: *Bulletin of the Alpine Garden Society* **3**: 260 (1935) & **10**: 81, 82, 87 (1942); Praeger, An account of the Sempervivum group, 58 (1932); *Houslekes* **11**(2): 41 (1980).

Rosettes 1–2 cm across, spherical. Leaves lanceolate or oblong-lanceolate, acute or shortly acuminate, wholly green, glandular-hairy on both sides, marginal hairs twice as long as surface hairs. Numerous offsets on very slender stolons *c*. 1 cm. Flower-stems 5–8 cm. Inflorescence small with 4–8 relatively large flowers. Petals rosy purple with white margins; filaments purple, anthers red-purple. *Caucasus*.

28. S. ballsii Wale. Illustration: *Kew Bulletin*, pl. 5 (1940); *Bulletin of the Alpine Garden Society* **8**: 205–6 (1940);

Journal of the Sempervivum Society **9**(1): 22 (1978).

Rosettes neat, nearly spherical, densely leaved, *c*. 3 cm across, inner leaves closed with the outer leaves more open and erect, Rosette-leaves obovate, abruptly mucronate, hairless at maturity except for a few marginal hairs on the lower two-thirds, uniform green with a bronze to red tinge on the outermost leaves. Offsets 3–4, produced in spring on stout, basal stolons *c*. 1.5 cm. Flowering-stem to 10 cm, with erect, only slightly overlapping leaves. Inflorescence compact, small, *c*. 4 cm across. Flowers 1.8–2 cm across; petals 12, dull pink; filaments crimson. *NW Greece*. Summer.

29. S. borissovae Wale. Illustration: *Bulletin of the Alpine Garden Society* **10**: 94, 97, 99 (1942); Payne, Plant jewels of the high country, 28 (1972).

Rosettes flat, semi-open, *c*. 3 cm across, in large dense clumps. Rosette-leaves with many marginal hairs but otherwise hairless, green at the base but red-brown in the upper half. Petals rose-red with slightly irregular toothed edges. *Caucasus*.

30. S. andreanum Wale. Illustration: *Bulletin of the Alpine Garden Society* **9**: 105, 117 (1941).

Rosettes 1.5–4 cm across, bright green with brown markings. Outer rosette-leaves erect, tipped with small brown markings. Central leaves in a tight conical bud (like *S. wulfenii*). Offsets produced on short stolons causing the rosettes to grow in tight clumps. Petals pale to bright pink, light red at the base. Anthers yellow. *Spain*. Summer.

31. S. calcareum Jordan (*S. tectorum* var. *calcareum* (Jordan) Cariot & Saint-Lager; *S. racemosum* Jordan & Fourreau; *S. columnare* Jordan & Fourreau; *S. californicum* misapplied, Baker; *S. greenii* Baker). Illustration: *Bulletin of the Alpine Garden Society* **12**: 11 (1944) & **30**: 79 (1962); Grey-Wilson & Blamey, Alpine flowers of Britain and Europe, 75 (1979); *Houslekes* **13**(1): 22 (1982); Jelitto & Schacht, Hardy herbaceous perennials **2**: 611 (1990).

Rosettes large, grey-green, 6 cm or more across. Leaves hairless, tipped with purple on both sides. Flowers pale pink but seldom produced. *French Alps*.

Several cultivars are grown including: 'Benz', a large variant with more than the usual number of leaves; 'Greenii', with smaller neater rosettes; 'Mrs Giuseppi', more compact, leaf-tips deep red; 'Sir William Lawrence', similar to 'Mrs Giuseppi' but rosettes larger and more spherical, leaf-tips red; 'Griggs Surprise' ('Monstrosum') a curious congested form with abnormal growth development that occasionally occurs in gardens, leaves grey-green, round and hollow without an opening at the tip, each leaf ends in a red thorn-like beak, which is often double.

32. S. balcanicum Stojanov.

Rosettes 1.5–2.5 cm across. Leaves oblong-lanceolate, gradually mucronate, green with red tips, hairless except along the margins. Flower-stem 5–12 cm, densely hairy. Stem-leaves oblong to oblong-linear. Inflorescence with few branches. Petals pale lilac, filaments dirty purple, blackish towards apex. Anthers rounded, sulphur-yellow, margins black. *C Balkans*. Summer.

Closely related to *S. erythraeum*.

33. S. tectorum Linnaeus (*S. arvernense* Lecocq & Lomotee; *S. cantalicum* Jordan & Fourreau; *S. lamottei* Boreau). Illustration: Reichenbach, Icones florae Germanicae et Helveticae, **23**: t. 67 (1896–99); Praeger, An account of the Sempervivum group, 5, 66, 71 (1932); Perry, The good gardener's guide, 465 (1974); *Bulletin of the Alpine Garden Society* **47**: 77–8 (1979) & **51**: 70 (1983).

Rosettes 2–8 cm (rarely to 18 cm) across, open. Leaves oblong-lanceolate to obovate, sharply mucronate, hairless except for obvious white marginal hairs, very fleshy, dark green sometimes red-brown or purple, red or white at the base and frequently tipped with a darker colour. Stolons to 4 cm, stout, reddish. Flower-stem 20–40 cm, stout, covered in many white hairs, only the upper stem-leaves hairy on the surfaces. Inflorescence large and flat, flowers 40–100; each *c.* 2.5 cm across. Petals 12–16, dull pink or purple; anthers orange brown. *C Europe to Balkans*. Summer.

The best-known cultivated *Sempervivum*, very variable and will hybridise with many other species.

Var. **tectorum** Linnaeus (*S. murale* Boreau). Rosettes to 18 cm across, flat and open. Leaves obovate-lanceolate, hairless, bright green, whitish at the base; tips well marked with purple-brown. Flower-stem 30–50 cm. Inner whorl of stamens mostly sterile and often replaced by carpels.

Var. **alpinum** Praeger (*S. alpinum* Grisebach & Schenk; *S. boutignyanum* Billot & Grenier; *S. arvernense* forma *boutignyanum* Rouy & Camus; *S. fuscum* Lehmann & Schnittspahn; *S. tectorum* subsp. *alpinum* Wettstein & Hayek). Illustration: Blandford, Garden flowers in colour, 110 (1955). Rosettes 2–6 cm across. Leaves green, always red at the base, sometimes with purple-brown tips. Flower-stem 10–30 cm; inflorescence compact and flattish. *Pyrenees, Alps*.

Var. **glaucum** Praeger (*S. glaucum* Tenore; *S. spectabile* Lehmann & Schnittspahn; *S. tectorum* subsp. *schottii* Hayek; *S. acuminatum* Schott; *S. schottii* Baker). Rosettes 5–10 cm across. Leaves blue-green, white at the base and no markings on the tips. Flower-stem to 60 cm. Variable. *S & E Alps*.

Cultivars include: 'Atropurpureum', 'Boissieri', 'Nigrum', 'Red Flush', 'Royanum', 'Sunset', 'Triste', 'Atroviolaceum'.

S. × calcaratum Baker (*S. tectorum* × *S. calcaratum* misapplied; *S. comollii* misapplied). Rosettes to 15 cm across. Leaves spathulate to oblanceolate, noticeably mucronate, hairless, green-blue to purple with a crimson base. Stolons to 5 cm. Flower-stem *c.* 30 cm. Flowers dull red-purple. Easy to grow and increases quickly.

S. × widderi Lehmann & Schnittspahn; *S. tectorum* × *S. wulfenii* (*S. albidum* Lehmann & Schnittspahn). Intermediate between the parents but many variations. Medium-sized blue-green rosettes. Leaves hairless except for marginal hairs, usually red at the base. Petals yellow, towards the base red or streaked red and yellow. *Switzerland*.

34. S. vincentei Pau.

Rosettes *c.* 2 cm across. Leaves oblong to spathulate, mucronate, hairless except for marginal hairs. Juvenile offsets hairy. Flower-stem to 12 cm, stem-leaves oblong-lanceolate with longitudinal spots. Petals pale red with a purplish base. Filaments purple, hairy. *Spain*.

35. S. nevadense Wale. Illustration: *Bulletin of the Alpine Garden Society* **9**: 107–8 (1941); *Journal of the Sempervivum Society* **8**(3): 57 (1977).

Rosettes 2.5–3.5 cm across, flattish, compact, many-leaved. Leaves obovate, mucronate, fleshy, strongly incurved, hairless except for the short, stout, curved, marginal hairs, green, outer leaves scarlet in winter and pinkish bronze at flowering time. Offsets on short stolons. Flower-stems covered in many fleshy, overlapping leaves.

Inflorescence compact. Petals crimson, filaments dark red, hairless. *Spain*.

36. S. arachnoideum Linnaeus. (*S. sanguineum* Timbal-Lagrave; *Sedum arachnoideum* Krause). Illustration: *Botanical Magazine*, 68 (1942); Praeger, An account of the Sempervivum group, 23, 36 (1932); Reichenbach, Icones florae Germanicae et Helveticae **23**: t. 72 (1896–99); Grey-Wilson & Blamey, Alpine flowers of Britain and Europe, 75 (1979).

Rosettes small, ball-shaped, densely covered with white woolly hairs. Rosette-leaves deep red or shades of green. Offsets stalkless, crowded. Flower-stem 7–12.5 cm. Flowers usually rose-red with 9–12 petals. *Pyrenees, Alps, Apennines, Carpathians*.

A species very variable in rosette size and quantity of hairs. All variants are easy to grow, increase freely and hybridise easily.

Var. **arachnoideum**. Rosettes to 2 cm across, ovoid or spherical, with variable amounts of arachnoid hairs.

'Fasciatum' A curiosity that is fasciated. The fasciation is not very stable.

Var. **glabrescens** Wilkommen (*S. doellianum* Lehmann; *S. heterotrichum* Schott; *S. moggridgei* J.D. Hooker; *S. arachnoideum* var. *doellianum* Jaccard). Rosettes less than 1.5 cm across, flattish with only a small amount of cobweb. Flowers white and produced so freely that it is difficult to culture as it flowers itself to death.

Var. **tomentosum** (Lehmann & Schnittspahn) Hayek (*S. tomentosum* Lehmann & Schnittspahn; *S. webbianum* Lehmann & Schnittspahn; *S. laggeri* Hallier). Rosettes to 4 cm across, large and flattish with a very dense cobweb and an undercolour of dark red in spring.

Common cultivars include: 'Kappa', 'Robin', 'Sultan', 'Rubrum', 'Stansfieldii'.

S. arachnoideum × S. nevadense. Intermediate between the parents. Rosettes green with a flush of red and covered in woolly hair. *Spain (Sierra Nevada)*.

S. × barbulatum Schott; *S. arachnoideum × montanum* (*S. fimbriatum* Schnittspahn & Lehmann; *S. elegans* Lagger; *S. barbatulum* Baker; *S. hausmanii* Nyman; *S. oligotrichum* Baker; *S. × hybridum* Brugger; *S. dolomiticum* Koch; *S. hookeri* misapplied). Illustration: *Chanousia* **1**: 117–18 (1928). Rosettes *c.* 1 cm across. Leaves elliptic-oblong, blunt, finely and densely glandular-hairy on both sides, marginal hairs glandular, longer woolly hairs on the tip. Very small offsets produced on short, slender, bare stolons. Flower-stem to 5 cm, glandular-hairy. Inflorescence with *c.* 6 flowers, glandular-hairy. Flowers with 10 petals, *c.* 2 cm across. Petals purplish rose, glandular-hairy on the back, edges and upper part of face. *European Alps, N Spain.* Summer.

S. × fauconettii Reuter; *S. arachnoideum × S. tectorum (S. flavipilum* Sauter; *S. schnittspahnii* misapplied, not Lagger; *S. villosum* Aiton; *S. × thomayeri* Correvon; *S. thompsonii* Lindsay). Illustration: *Bibliotheca Botanica* **11**(58): t. 6, f. 171 (1902); Payne, Plant jewels of the high country, 38 (1972). Rosettes small and green with a tuft of straggling white hairs at the leaf tips, and on the edges near the tips. Inflorescence similar to *S. arachnoideum,* but flowers not so bright, dull purple-pink. *Europe.*

S. × funckii Koch; (*S. arachnoideum × S. montanum*) *× S. tectorum (S. montanum* Mertens & Koch). Illustration: Hegi, Illustrierte flora von Mittel-Europa, **4**: 552 (1922); Huxley, Mountain flowers, pl. 237 (1967). Rosettes 2.5–4 cm across, compact, flattish, open. Leaves obovate-lanceolate, shortly mucronate, shortly but finely hairy on both surfaces, strong marginal hairs of unequal length, bright green, white at the base, older leaves sometimes with a purplish tinge on the tips. Offsets very numerous on short stems. Flower-stem *c.* 20 cm, with glandular-hairy, erect oblong-ovate leaves. Inflorescence compact and flattish, 6–8 cm across. Flowers dull purple-red. *Alps, Switzerland to Syria.*

S. × morelianum Viviand-Morel; *S. arachnoideum × S. calcareum.* Medium-sized rosettes with tufts of woolly hair on the leaf tips.

S. × roseum Hunter & Nyman; *S. arachnoideum × S. ciliosum (S. fimbriatum* Hegi). Rosettes neat grey-green, leaves hairless, sparingly tipped with woolly hairs.

S. fimbriatum Schott. Rosettes small, compact, spherical. Leaves tinted red and fringed with hairs. Petals yellow with red lines or red with a yellow edge or an intermediate shade. *Austria, Switzerland.*

S. × vaccarii Wilczeck; *S. arachnoideum × S. grandiflorum.* Variable. Rosettes much resembling *S. grandiflorum* but smaller and leaf-tips slightly white-hairy.

37. S. charadzeae Gurgenidze. Illustration: *Houslekes* **15**(2): 37 (1984).

Rosettes 8–12 cm. Leaves spathulate with a sharply narrowing tip, hairs of unequal length on surfaces and margins. Flower-stem 40–50 cm. Inflorescence corymb-like to umbel-like, 3–5 branches, with 60–80 flowers. Flowers dark purple with alternate petals, hairless on the inside of the bases and the outside third hairy. Opposite petals are three-quarters covered with short glandular hairs on both sides. Filaments rose; Anthers pale yellow. *E Georgia.* Summer.

38. S. italicum Ricci.

Rosettes to 5 cm across. Leaves linear, sharply mucronate, covered in short glandular hairs on both surfaces, marginal hairs show an almost regular alternation between long and short. Flower-stem to 30 cm. Petals 11 or 12, purple, glandular-hairy. *C Italy (Apennines).*

Sometimes confused with *S. montanum* Linnaeus.

39. S. caucasicum Boissier (*S. tectorum* Bieberstein; *S. montanum* Meyer; *S. tectorum* subsp. *caucasicum* Berger). Illustration: *Bulletin of the Alpine Garden Society* **10**: 103 (1942).

Rosettes made up of few fleshy leaves, dense, 2–5 cm across. Leaves spathulate, abruptly contracted, mucronate with a dark brown apex, shortly hairy along the edges and sparingly hairy on both sides. Offsets 6 or 7, few-leaved and spherical. Stolons 6–8 cm, thick and reddish nearer the rosette, narrower and greenish further from the rosette, *c.* 8 leaves. Flower-stem 12–20 cm. Petals 14, rose-red with a central stripe. *Caucasus.*

40. S. cantabricum Huber & Sündermann. Illustration: *Feddes Repertorium* **33**: t. 140 (1934); Polunin & Smythies, Flowers of south-west Europe, pl. 9 (1973); *Journal of the Sempervivum Society* **8**(2): 38 (1977); *Houslekes* **11**(1): 7–14 (1980); **15**(2): 33, 35, 36, (1984).

Rosettes 5–6 cm, half open. Leaves hairy on both sides, deep green with dark purple-red tips. Few offsets produced on stout, stiff, leafy stems, stolons to 5 cm. Flower-stems stout, 25–30 cm. Flowers deep crimson with purple filaments. *N Spain.*

Relatively easy to grow but suffers from winter wet.

Subsp. **guadarramense** M.C. Smith. Small rosettes with incurving leaves and purple tips.

Subsp. **urbionense** M.C. Smith. Rosettes light green, hairy, incurving, with pointed leaves tipped red.

S. cantabricum × S. montanum subsp. *stiriacum* Large rosettes strikingly tipped with purple.

41. S. macedonicum Praeger. Illustration: *Bulletin de Institut & Jardin Botaniques Belgrade* **1**: 213 (1930); Praeger, An account of the Sempervivum group, 53 (1932); *Bulletin of the Alpine Garden Society* **8**: 216 (1940).

Rosettes small, similar to *S. montanum* but flatter and more open, often with a reddish tinge, 2–4 cm across, central leaves closed. Leaves broadly oblanceolate, shortly acuminate, densely but minutely hairy, marginal hairs dull green often red near the tip but not purple-tipped. Offsets on long stolons. Flower-stem short and very leafy, bearing a compact inflorescence. Petals dull red-purple; filaments lilac. *SW former Yugoslavia (Macedonia).*

Easy to grow, increases quickly but produces few flowers.

42. S. altum Turrill. Illustration: *Bulletin of the Alpine Garden Society* **10**: 103 (1942).

Rosettes rather loose, 2.5–4 cm across. Leaves oblanceolate, abruptly mucronate-acuminate, light green, glandular-hairy on both surfaces with hairs along the edges. Offsets are produced on stolons 8–12 cm long. Flower-stem *c.* 33 cm, strong, densely glandular-hairy, covered with many linear-lanceolate stem-leaves *c.* 4.5 cm, with glandular hairs around the edges. Inflorescence branching, many-flowered, almost dense, densely glandular-hairy. Flowers 2.6–3 cm across with 12 or 13 pale, red-purple, glandular-hairy petals. Stamens purple, anthers yellow. *Caucasus.* Summer.

On exposure to the sun the tips of the outer leaves may turn red; similar to *S. kosaninii* but flower-stems taller; dislikes winter damp.

43. S. atlanticum Ball (*S. tectorum* var. *atlanticum* J.D. Hooker). Illustration: *Botanical Magazine*, 6055 (1873); Praeger, An account of the Sempervivum group, 62 (1932); *Bulletin of the Alpine Garden Society* **3**: 26 (1935).

Rosettes 4–8 cm across, pale green, flushed red when fully exposed, nearly erect, often asymmetric. Leaves hairy on both sides and abruptly mucronate. Numerous offsets produced on very short stolons forming a regular hump of rosettes. Flowers pink but rarely seen in cultivation. *Morocco (Atlas Mts)*.

'Edward Balls' ('Ball's Form'): larger more symmetrical rosettes, gaining much colour in the summer.

44. S. montanum Linnaeus (*S. flagelliforme* Link). Illustration: Reichenbach, Icones florae Germanicae et Helveticae **23**: t. 68 (1896–99); Praeger, An account of the Sempervivum group, 23, 44 (1932); Huxley, Mountain flowers, pl. 238 (1967); *Bulletin of the Alpine Garden Society* **3**: 286 (1935), **49**: 77, 79 (1981) & **53**: 143 (1985).

Rosettes less than 2 cm across, open, forming a close mat. Leaves oblanceolate, rather acute, finely and densely sticky-hairy on both surfaces, marginal hairs slightly longer, wholly dull green. Many offsets on slender leafy stolons 1–3 cm. Flower-stem 5–8 cm, leafy, few relatively large flowers. Petals 10–15, soft violet-purple. *Pyrenees, Alps, Carpathians, Apeninnes, Corsica.*

A very variable species with a wide distribution. Hybridises easily in nature with *S. arachnoideum, S. grandiflorum, S. nevadense, S.* tectorum and *S. wulfenii,* and with many other species in cultivation.

Var. **braunii** Praeger (*S. wulfenii* Bertolini; *S. stiriacum* var. *braunii* Hayek; *S. montanum* subsp. *stiriacum* var. *braunii* Hayek). Illustration: *Bulletin of the Alpine Garden Society* **3**: 259 (1935). Rosettes *c.* 2.5 cm across, compact, green with nut-brown tips, very finely hairy on both sides. Flowers rose.

Var. **burnatii** Hayek. Illustration: *Bulletin of the Alpine Garden Society* **6**: 259 (1938). Rosettes to 8 cm across, open, light green. Leaves obovate to wedge-shaped. Offsets on long strong stolons. Flower-stem twice as tall as var. *montanum*, flowers lighter purple. Similar to *S. grandiflorum*. *SW Alps & Pyrenees.*

Var. **stiriacum** Hayek (*S. braunii* Koch; *S. funckii* Maly; *S. stiriacum* Hayek). Illustration: *Bulletin of the Alpine Garden Society* **3**: 260 (1935) & **45**: 59 (1977). Rosettes 2–4.5 cm across, more open than var. *montanum*. Leaves oblanceolate with dark red-brown tips in the summer, marginal hairs distinctly longer than the surface hairs. Flowers larger than the type and often darker. *E Alps.*

'Rubrum'. Medium-sized rosettes, mahogany-red in summer.

S. × densum Lehmann & Schnittspahn; probably *S. montanum × S. tectorum.* Small rosettes, leaves finely hairy on both sides, marginal hairs and tips well marked in dark red. Flowers white, anthers rose.

S. × schottii Lehmann & Schnittspahn; *S. montanum × S. tectorum* (*S. verlotti* Lamotte; *S. monticolum* Jordan & Fourreau; *S. modestum* Jordan & Fourreau; *S. parvulum* Jordan & Fourreau; *S. funckii* Jordan & Fourreau; *S. rhaeticum* Brügger; *S. rupestre × candollei* Rouy & Camus). Very variable, generally resembles a small *S. tectorum* but the rosettes are denser and the leaves are glandular-hairy without a purple tip.

29. AEONIUM Webb & Berthelot
S.G. Knees

Succulent, evergreen herbs and shrublets; usually terrestrial, but occasionally epiphytic, monocarpic or perennial. Stems erect or spreading, often woody at base, branched or unbranched; bark fissured or covered with remains of old leaf-bases. Leaves in dense concave or flattish rosettes, alternate, simple, stalked or stalkless, fleshy and succulent, 1–12 mm thick, convex below, flattened or concave above, spathulate, usually hairless, glaucous, sparsely glandular-hairy; margins entire or shallowly toothed, usually hairy. Leaves of vegetative- and flowering-shoots often dissimilar. Inflorescence terminal, subtended by bracts, flowers bisexual; calyx bell-shaped, yellow, green or variegated with red or pink, with 6–16 triangular to lanceolate sepals; corolla-like calyx, but always larger and with petals equalling number of sepals. Stamens twice as many as petals, filaments sometimes widened at base, anthers basifixed, white, yellow or brown. Nectar-producing glands present at base of ovary. Carpels usually equalling sepals in number, rarely fewer, free or loosely attached at their bases; styles erect or spreading, stigma apical; ovaries 1-celled, ovules numerous. Fruit a follicle, seeds numerous, ovate to pear-shaped, ridged, yellowish brown or red-brown.

A genus of 31 species, formerly included in *Sempervivum*, and sometimes found in gardens under that name. Largely endemic to the Canary Islands, Cape Verde Islands and southern Morocco, but with isolated species in eastern Africa and Yemen. Many different growth forms from large shrubby perennials to small stemless biennials are found in the genus and have always attracted horticultural interest.

Propagation is by cuttings or from seed, though the species hybridise freely and consideration must be given to the source of the seed. There are more published names of hybrids than species.

Literature: Praeger, R.L., An account of the Sempervivum group (1932); Liu, H.-Y., Systematics of Aeonium (Crassulaceae), Serial Publications Number 3, National Museum of Natural Science (Taiwan) (1989).

1a. Plants stemless or nearly so 2
  b. Plants with distinct stems at least 25 cm 5
2a. Rosettes usually 4–10 cm across **4. simsii**
  b. Rosettes usually 10–50 cm across 3
3a. Leaves glaucous; stems often with stolons **6. cuneatum**
  b. Leaves green; stolons absent 4
4a. Leaf-margins smooth with hairs 1–2 mm **8. tabuliforme**
  b. Leaf-margins sometimes wavy with hairs 0.6–1 mm **7. canariense**
5a. Stems unbranched 6
  b. Stems branched 9
6a. Leaves yellowish or dark green, sometimes with brownish margins 7
  b. Leaves glaucous, sometimes with reddish margins 8
7a. Bundle-scars visible; leaves oblanceolate to spathulate or oblong; stems to 2.5 m **11. undulatum**
  b. Bundle-scars indistinct; leaves obovate; stems to 60 cm **12. nobile**
8a. Plant 1–2 m; leaves glaucous green, leaving scars 5–8 mm wide **17. urbicum**
  b. Plant 2.5–5 m; leaves very glaucous, often with a pink or purplish tinge and reddish margins, leaving scars 6–18 mm wide **21. hierrense**
9a. Stems with 1–10 lateral branches 10
  b. Stems with 30–100 lateral branches 17
10a. Stems with adventitious roots 11
  b. Stems without adventitious roots 12
11a. Bark with tubercles; leaf-scars slightly raised **18. ciliatum**
  b. Bark with raised netted lines; leaf-scars depressed **15. lancerottense**
12a. Whole plant smelling strongly of balsam **9. balsamiferum**
  b. Plants generally odourless 13
13a. Stem and leaves very sticky, especially when young **13. glutinosum**
  b. Stems and leaves not sticky 14

14a. Leaves dark green with reddish margin, but without any brownish stripes　　**19. percarneum**

b. Leaves sometimes with reddish or purple margins, but usually with brown stripes　　15

15a. Leaf rosettes with raised centre, the young leaves arising erectly　　**14. gorgoneum**

b. Leaf rosettes with flattened centres, the young leaves tightly adpressed to the older ones　　16

16a. Leaf-apex acuminate; marginal hairs 0.3–1 mm　　**10. arboreum**

b. Leaf-apex acute, mucronate; marginal hairs 0.5–2 mm　　**11. undulatum**

17a. Adventitious roots common on lower stems　　18

b. Adventitious roots absent　　20

18a. Leaf-scars not more than 1 mm wide; bark fissured into plates　　**22. castello-paivae**

b. Leaf-scars 2–4 mm wide; bark only slightly fissured, but not into plates　　19

19a. Stem surface covered in netted lines; bark grey or brown; leaf scars c. 2 mm wide　　**16. haworthii**

b. Stem surface unlined; bark pale reddish green or whitish; leaf-scars 3–4 mm wide　　**20. decorum**

20a. Leaf-scars not more than 0.5 mm wide　　**2. goochiae**

b. Leaf-scars 1–1.5 mm wide　　21

21a. Leaf-scars 1.1–1.5 mm wide; plant smelling of balsam　　**1. lindleyi**

b. Leaf-scars c. 1 mm wide; plant usually odourless　　22

22a. Leaf-margin with bead-shaped hairs　　**5. spathulatum**

b. Leaf-margin without bead-shaped hairs　　**3. sedifolium**

1. A lindleyi Webb & Berthelot (*Sempervivum lindleyi* (Webb & Berthelot) Christ; *S. tortuosum* Aiton var. *lindleyi*). Illustration: Praeger, An account of the Sempervivum group, 205 (1932); Graf, Exotica, edn 3, 652 (1963).

Perennial, terrestrial shrublet with c. 50 lateral branches; smelling strongly of balsam. Stems to c. 50 × 3–15 cm, young branches softly hairy, green to brownish, sticky, bark grey or whitish, slightly fissured; leaf-scars 1.1–1.5 mm wide. Leaf-rosettes 4–9 cm across. Leaves 2–4.5 cm × 6–16 mm, obovate to spathulate, yellowish to dark green, sticky. Inflorescence a cyme-like panicle,

2–7 × 3–9 cm, with 15–85 flowers. Sepals 7–9; petals 5–7 × 1.5–2 mm, narrowly elliptic-lanceolate; anthers yellow, nectar-glands yellow; ovaries 2–3 mm, styles 2–3 mm. *Canary Islands (Tenerife)*. H5–G1. Summer–autumn.

2. A. goochiae Webb & Berthelot (*Sempervivum goochiae* Webb & Berthelot). Illustration: Praeger, An account of the Sempervivum group, 207 (1932).

Terrestrial shrublet with c. 30 branched stems to 40 cm; leaf-scars c. 0.5 mm wide, bark pale brown, slightly fissured. Rosettes 3–12 cm across. Leaves 2–6.5 × 1.5–2.5 cm, elliptic to spathulate, narrowing into a stalk-like part c. 0.25 mm wide, pale green or yellowish green, occasionally variegated with red, shortly hairy; margin with short hairs. Inflorescence 2–5 × 3–11 cm, paniculate with 10–45 flowers. Sepals 7 or 8, narrowly triangular, 3–3.5 cm × 0.7–1 mm, apex acute; petals 5–7 × 1–2 mm, very pale yellow or whitish, tinged pink towards centre; stamens 5.5–7 mm; nectar-glands c. 0.6 mm wide, yellowish; ovaries 2–3 mm, styles 2.5–3 mm. *Canary Islands (La Palma)*. H5–G1. Late winter–summer.

3. A. sedifolium (Bolle) Pitard & Proust (*Aichryson sedifolium* Bolle; *Greenovia sedifolium* (Bolle) Christ; *Sempervivum sedifolium* (Bolle) Christ). Illustration: Praeger, An account of the Sempervivum group, 211 (1932); Graf, Exotica, edn 3, 652 (1963); Bramwell & Bramwell, Wild flowers of the Canary Islands, pl. 153 (1974); Jacobsen, Lexicon of succulent plants, pl. 5 (1974).

Perennial, terrestrial shrublet with up to 100 branches. Stems to 40 cm × 1–5 mm, young branches sparsely covered with fine down, dark brown, lustrous, sticky; bark grey or greyish brown, fissured; leaf scars c. 1 mm wide. Leaf-rosettes 1.4–3 cm across, overlapping, becoming almost spherical in very dry conditions. Leaves stalkless, ovate to obovate, green–yellowish green, with pale reddish markings on midrib and towards the apex. Inflorescence 2–7 × 2–5 cm, with 6–15 flowers. Sepals 9–11, elliptic, 2.5–3 mm, variegated with reddish lines; petals 5–7 × 2–2.5 mm, obovate-oblanceolate, yellow, apex often reflexed, toothed, hairless. Stamens 4–5.5 mm; nectar-glands absent; ovaries c. 1.5–2 × 1 mm, styles 2.5–3 mm. *Canary Islands (La Palma)*. G1. Spring–summer.

4. A. simsii (Sweet) Stearn (*A. caespitosum* (Smith) Webb & Berthelot; *Sempervivum ciliatum* Sims; *S. simsii* Sweet; *S. caespitosum* Smith; *S. ciliare* Haworth; *S. barbatum* Hornemann). Illustration: *Botanical Magazine*, 1978 (1818); Praeger, An account of the Sempervivum group, 194 (1932); Graf, Exotica, edn 3, 654 (1963); Bramwell & Bramwell, Wild flowers of the Canary Islands, pl. 158 (1974).

Perennial terrestrial herb; stems tufted, with c. 8 branches, c. 15 cm, forming low hummocks. Rosettes 4–10 cm across, withered leaf remains persisting beneath new growth. Leaves 2–6 cm × 6–20 mm, lanceolate, with red multicellular hairs along margin. Flowers borne on stems 5–30 cm, arising from leaf axils. Sepals 7–9, elliptic; petals 5–6 mm, oblanceolate, yellow; stamens 5–6 mm, anthers yellow. Nectar-glands 0.3–0.4 mm; carpels 1.5–2 mm. *Canary Islands (Gran Canaria)*. G1. Spring–late summer.

Hybridises with *A. canariense*, *A. arboreum*, *A. percarneum*, *A. spathulatum* and *A. undulatum*. Of these the most widely cultivated hybrid is **A. × velutinum** Hill; *A. simsii × A. canariense*.

5. A. spathulatum (Hornemann) Praeger (*Sempervivum spathulatum* Hornemann; *S. lineolare* Haworth; *S. villosum* Lindley; *A. cruentum* Webb & Berthelot; *A. strepsicladum* Webb & Berthelot; *Aichryson pulchellum* C.A. Meyer). Illustration: *Edwards's Botanical Register*, t. 1553 (1832); Praeger, An account of the Sempervivum group, 202 (1932); Bramwell & Bramwell, Wild flowers of the Canary Islands, pl. 157 (1974).

Perennial, terrestrial shrublet. Stems to 60 cm, ascending or almost erect, branches c. 80, green or greyish brown, covered with fine hairs, bark slightly fissured; leaf-scars c. 1 mm wide. Rosettes 1–5 cm across. Leaves 5–25 × 3–9 mm, ovate to spathulate, margin with multicellular hairs to 0.06 mm, and bead-shaped hairs to 1 mm. Inflorescence 3–10 × 3–15 cm, stalks 5–20 cm. Sepals 8–10; petals 8–12 with red stripes; stamens 4–5.5 mm, anthers yellow; nectar-glands absent. Ovaries 2–2.5 cm, styles 1.5–2 mm. *Canary Islands*. G1. Spring–early summer.

6. A. cuneatum Webb & Berthelot (*Sempervivum cuneatum* Webb & Berthelot). Illustration: Praeger, An account of the Sempervivum group, 142 (1932);

Bramwell & Bramwell, Wild flowers of the Canary Islands, pl. 147 (1974).

Perennial terrestrial or epiphytic herb. Stems 1 or 2 erect, very short, often with brown stolons, to 25 cm. Rosettes 15–50 cm across. Leaves 10–25 × 5–8 cm, obovate to oblanceolate, glaucous; margins hairy, with unicellular hairs to 0.4 mm. Inflorescence 18–60 × 12–30 cm. Sepals triangular, 3–4 mm; petals 6.5–7.5 mm, oblanceolate, yellow; stamens 5–6 mm, anthers yellow; nectar-glands obovate *c.* 0.7 mm, greenish. Carpels 3–3.5 mm, styles 3.5–4 mm. *Canary Islands (Tenerife).* G1. Spring–early summer.

Very similar to *A. canariense*, with which it sometimes hybridises.

7. A. canariense (Linnaeus) Webb & Berthelot (*Sempervivum canariense* Linnaeus; *S. latifolium* Salisbury; *A. giganteum* Christ). Illustration: Praeger, An account of the Sempervivum group, 133–5 (1932); Graf, Exotica, edn 3, 655 (1963); Graf, Tropica, 62 (1978); Everett, New York Botanical Garden illustrated encyclopedia of horticulture **1**: 66 (1980).

Perennial terrestrial herb. Stems stout and unbranched to *c.* 5 cm. Rosettes 10–40 cm across. Leaves obovate to oblanceolate, 6–20 × 3–8 cm, green, velvety; margins sometimes wavy with hairs 0.6–1 mm. Inflorescence 15–60 × 12–30 cm. Sepals 6–12; petals 7–10 mm, elliptic to lanceolate; stamens 6–9 mm; styles 2.5–3.5 mm. *Canary Islands.* G1. Spring–summer.

Var. **canariense** from Tenerife, and var. **virgineum** (Christ) H.-Y. Liu (*A. virgineum* Christ) from Gran Canaria are cultivated. Var. *virgineum* is early-flowering and has more cup-shaped rosettes, rarely exceeding 25 cm across, and red-tinged leaves.

8. A. tabuliforme (Haworth) Webb & Berthelot (*Sempervivum tabuliforme* Haworth; *S. complanatum* A. de Candolle; *A. berthelotianum* Bolle; *A. macrolepum* Christ). Illustration: Praeger, An account of the Sempervivum group, 146 (1932); Graf, Exotica, edn 3, 652 (1963); Bramwell & Bramwell, Wild flowers of the Canary Islands, pl. 151 (1974); Jacobsen, Lexicon of succulent plants, pl. 5 (1974).

Biennial or perennial, terrestrial herb; stems solitary, short, erect, occasionally tufted. Rosettes 10–40 cm across, with closely overlapping leaves 4–20 × 2–4 cm, blade obovate to oblanceolate, narrowed towards the base. Flowering-stems 15–30 cm. Sepals 7–9, elliptic,

3–4 × 1.5–2 mm; petals narrowly elliptic 6–7 × 1.5–2 mm, pale yellow; stamens 5–6.5 mm, anthers yellow; nectar-glands 1–1.5 × 0.3–0.5 mm, oblong, whitish. Ovaries 2.5–3.5 × 1.5–2 mm; styles 2.5–3 mm. *Canary Islands (Tenerife).* H5–G1. Spring–summer.

A fasciated variant 'Cristata' is common in cultivation. It hybridises with *A. lindleyi*.

9. A. balsamiferum Webb & Berthelot (*Sempervivum balsamiferum* (Webb & Berthelot) Christ). Illustration: Praeger, An account of the Sempervivum group, 157 (1932).

Perennial terrestrial shrublet; stems with *c.* 8 erect or ascending branches to 1.5 m; rhombic leaf scars 3–9 mm wide. Rosettes 7–18 mm across, generally flattened in the centre. Leaves 3–7 × 1.5–3.5 cm, spathulate, recurved, greyish green, hairless, acute to acuminate. Sepals 7–8; petals 6–8 × 1.2–1.5 mm, lanceolate, yellow, nectar-glands 0.5 mm, wedge-shaped, yellow; ovaries *c.* 3 × 1 mm, hairless; styles *c.* 3.5 mm. *Canary Islands (Lanzarote, Fuerteventura).* G1. Late spring–early summer.

10. A. arboreum (Linnaeus) Webb & Berthelot (*Sempervivum arboreum* Linnaeus; *Sedum arboreum* Linnaeus; *A. manriqueorum* Bolle). Illustration: Praeger, An account of the Sempervivum group, 158, 161 (1932); Graf, Exotica, edn 3, 656 (1963); Bramwell & Bramwell, Wild flowers of the Canary Islands, pl. 152 (1974); Everett, New York Botanical Gardens illustrated encyclopedia of horticulture **1**: 64 (1980).

Perennial, terrestrial shrublet, occasionally epiphytic, with erect branched stems to 2 m × 1–3 cm. Branches *c.* 8; bark light brown or greyish brown; leaf-scars distinct, 3–8 mm wide. Rosettes 10–25 cm across, shrinking to *c.* 5 cm when dry, flattened in centre. Leaves 5–15 × 1.4–5 cm, obovate to oblanceolate, apex acuminate, straight or slightly recurved, occasionally with wavy margins, green or purplish green, or with purple or whitish lines near margin and midrib; margin with curved hairs 0.3–1 mm. Inflorescence dense, conical, ovoid or hemispherical, 7–30 × 7–15 cm. Sepals 9–10, triangular, 2–3.5 × 1–1.4 mm, acuminate; stamens 5–6.5 mm, anthers yellow, nectar glands 0.5–1.2 mm wide. Ovaries 2–3.5 mm, styles 1.5–2.5 mm. Seeds *c.* 0.7 × 0.2 mm. *Canary Islands (Gran Canaria), widely naturalised in the Mediterranean, C & S America and New Zealand.* H5–G1. Flowering for most of the year.

'Albovariegatum': leaf-margins yellowish white. 'Zwartkop': leaves intense blackish purple, unique and highly regarded.

11. A undulatum Webb & Berthelot (*Sempervivum undulatum* Webb & Berthelot). Illustration: Praeger, An account of the Sempervivum group, 155 (1932); Bramwell & Bramwell, Wild flowers of the Canary Islands, pl. 149 (1974); Jacobsen, Lexicon of succulent plants, pl. 5 (1974); Everett, New York Botanical Gardens illustrated encyclopedia of horticulture **1**: 64 (1980).

Perennial shrublet, terrestrial or epiphytic; stems branched or unbranched to 2.5 m, hairless, smooth; bark green, pale brown or grey. Leaf-scars distinct, narrowly transversely rhombic, 3–9 mm wide. Rosettes 10–30 cm across, with flattened centres. Leaves 6–18 × 3–5 cm, oblanceolate or spathulate to oblong, dark green, often with brownish wavy margins; marginal hairs curved, 0.5–2 mm. Inflorescence 12–50 × 12–40 cm, sepals 9–12, obovate, 1.5–1.8 × 0.8–1 mm, hairless, apex slightly notched; petals oblong-lanceolate, 6–8 × 1.2–1.5 mm, yellow, hairless, slightly notched. Stamens 5.5–7 mm, anthers yellow; nectar-glands 0.6–0.8 mm wide. Ovaries *c.* 3 mm, hairless; styles 2 mm. *Canary Islands (Gran Canaria).* G1. Spring.

12. A. nobile (Praeger) Praeger (*Sempervivum nobile* Praeger). Illustration: Praeger, An account of the Sempervivum group, 150, 151 (1932); Bramwell & Bramwell, Wild flowers of the Canary Islands, pl. 159 (1974); Jacobsen, Lexicon of succulent plants, pl. 5 (1974).

Terrestrial monocarpic shrublet with simple, erect stems to 60 cm; bark brown or greyish brown, hairless, rough; leaf-scars narrowly elliptic, 5–15 mm wide. Rosettes 15–60 cm across. Leaves 7–30 × 4–20 cm, thick and leathery, obovate, yellowish green, occasionally with irregular brownish lines, especially on or near midrib, acute; margin with straight lines *c.* 0.5 mm, sometimes hairless when mature. Inflorescence 20–40 × 30–60 cm, broadly dome-shaped; sepals 7–9, 2–3 mm, green, often striped with reddish lines; petals 3–5 mm, lanceolate, purplish red; stamens 3–6 mm, hairless, whitish, tinged red, anthers yellow, glands 4-sided, *c.* 1 mm wide; ovaries 2–3 mm, styles

2 mm. *Canary Islands (La Palma)*. G1.
Spring–summer.

13. A. glutinosum (Alton) Webb &
Berthelot (*Sempervivum glutinosum* Aiton).
Illustration: Praeger, An account of the
Sempervivum group, 144 (1932); *Botanical
Magazine*, 1963 (1818).

Terrestrial shrublet to 1.5 m, with *c.* 5
erect or ascending branches, overtopping
central or oldest stem; bark green, brown
or greyish brown; sticky, hairless; leaf-scars
distinct, 6–12 mm wide. Rosettes
12–22 cm across, glutinous towards centre.
Leaves 7–12 × 3.5–5 cm, obovate,
spathulate, slightly folded, dull pale green,
usually with brown stripes near midrib and
apex; apex acute, margin with scattered
hairs *c.* 0.5 mm. Inflorescence loose,
1.5–4 cm × 2.5–3.5 mm, marked with
reddish stripes. Petals 5–7 × 2–3 mm,
oblong-obovate to oblong-lanceolate,
yellow with red lines; stamens 4 or 5,
hairless, anthers yellow; nectar-glands
0.5–0.8 mm wide, yellow; ovaries
2–2.5 mm, styles 1–3 mm. *Madeira*. G1.
Spring–autumn.

14. A. gorgoneum J.A. Schmidt
(*Sempervivum gorgoneum* J.A. Schmidt).
Illustration: Praeger, An account of the
Sempervivum group, 165 (1932).

Terrestrial shrublet, stems with *c.* 7
branches to 2 m; bark green, yellowish
orange or greyish brown, irregularly
fissured; leaf-scars 4–9 mm wide. Rosettes
9–20 cm across. Leaves 5–10 × 1.5–3 cm,
oblanceolate to spathulate, glaucous green,
often red towards midrib and margin;
marginal hairs *c.* 0.05 mm, straight or
curved. Inflorescence 5–8 × 7–10 cm,
pyramidal. Sepals 8–10, elliptic, 2–3 mm,
often tinged red, acuminate; petals
5–6 × 1–1.5 mm, oblong-lanceolate,
yellowish, with red markings, acuminate
and sometimes recurved. Stamens
5–6.5 mm, anthers yellow; nectar-glands
4-sided, 0.3–0.5 mm wide. Ovaries to
3 mm; styles *c.* 2 mm. Seeds 0.5 × 0.2 mm.
Cape Verde Islands. G1. Autumn–winter.

15. A. lancerottense (Praeger) Praeger
(*Sempervivum lancerottense* Praeger).
Illustration: Praeger, An account of the
Sempervivum group, 191, 192 (1932);
Bramwell & Bramwell, Wild flowers of the
Canary Islands, pl. 155 (1974).

Terrestrial shrublet with *c.* 10 ascending
branches to 60 cm, often with adventitious
roots; bark pale brown or silver-grey; leaf-
scars 3–8 mm wide. Rosettes 10–18 cm

across. Leaves 5–9 × 1.5–4 cm, obovate to
oblanceolate or spathulate, concave, green
to yellowish green, often tinged reddish,
glaucous; margin weakly toothed with
distant hairs. Inflorescence
8–30 × 8–25 cm, dome-shaped. Sepals 7 or
8, triangular, 1.7–3 mm, acuminate,
occasionally pinkish; petals
6–9 × 1–1.5 mm, linear-lanceolate,
whitish, tinged pink, hairless, acuminate.
Stamens 5–8 mm, whitish, anthers yellow
or pinkish; glands 4-sided, 0.5 mm wide.
Ovaries 2.5–3 mm; styles 3–4 mm,
occasionally marked with pink lines. Seeds
c. 0.5 mm. *Canary Islands (Lanzarote)*.
H5–G1. Spring–summer.

16. A. haworthii Webb & Berthelot.
Illustration: Praeger, An account of the
Sempervivum group, 175 (1932); Graf,
Exotica, edn 3, 652 (1963); Brickell (ed.),
RHS A–Z encyclopedia of garden plants, 82
(2003).

Terrestrial shrublet, stems with *c.* 40
branches to 60 cm; bark grey or brown,
smooth or fissured, surface covered with
netted lines, adventitious roots often
present; leaf-scars *c.* 2 mm wide. Rosettes
6–11 cm across. Leaves 3.5–5 × 1.5–3 cm,
obovate, sometimes folded towards the
apex, green, often with glaucous bloom,
apex acute or heart-shaped; margin with
curved hairs 0.4–0.8 mm. Inflorescence
loose, hemispherical, 6–16 cm across;
sepals 7–9, lanceolate; petals pale yellow or
whitish, often with pinkish variegation,
margin finely toothed. Stamens 5.5–8 mm,
filaments white or pinkish, anthers pale
yellow; nectar-glands 4-sided, *c.* 0.8 mm
wide; ovaries 3–5 mm; styles *c.* 3.5 mm.
Seeds *c.* 0.4 × 0.2 mm. *Canary Islands
(Tenerife)*. G1. Spring–summer.

17. A. urbicum (Hornemann) Webb &
Berthelot (*Sempervivum urbicum*
Hornemann). Illustration: Praeger, An
account of the Sempervivum group, 168
(1932); Die Natürlichen Pflanzenfamilien
18a: 431 (1930); Graf, Exotica, edn 3,
655 (1963); Bramwell & Bramwell, Wild
flowers of the Canary Islands, pl. 54
(1974).

Terrestrial shrublet usually with
unbranched stems to 2 m. Bark pale brown
or greyish, hairless, smooth or slightly
fissured; leaf-scars 5–8 mm wide, 1–2 mm
tall. Rosettes 15–32 cm across. Leaves
8–25 × 3–5.5 cm, obovate-oblanceolate,
glaucous green, margin with straight hairs
0.5–1 mm. Inflorescence
15–75 × 10–45 cm, dome-shaped. Sepals

8–10, triangular, 2–3 × 1–1.5 mm,
sometimes reddish variegated, acuminate.
Petals 7–10 × 1.2–2 mm, lanceolate,
whitish except for pink midrib, hairless,
acuminate; stamens 6–10 mm, whitish,
anthers pale yellow or whitish; nectar-
glands 4-sided, *c.* 1 cm wide, whitish.
Ovaries 3.5–4.5 mm, hairless; styles
4–6 mm. Seeds *c.* 0.5 mm. *Canary Islands
(Tenerife, Gran Canaria)*. H5–G1.
Spring–autumn.

18. A. ciliatum Webb & Berthelot
(*Sempervivum ciliatum* Willdenow).
Illustration: Praeger, An account of the
Sempervivum group, 182, 183 (1932);
Santos, Vegetacíon y flora de La Palma,
125 (1983).

Terrestrial shrublet; stems with *c.* 6
branches to 1 m. Bark pale brown or
greyish, hairless, rough and with many
adventitious roots; leaf-scars 3–5 mm wide,
slightly raised. Rosettes 8–20 cm across.
Leaves 4–12 × 2–5 cm, obovate to
oblanceolate or spathulate, often slightly
folded near apex, dark green or yellowish
green, glaucous, apex acute, often
recurved; margin with hairs 0.4–0.8 mm.
Inflorescence 15–40 × 10–35 cm. Sepals
7–9, triangular, 2.5–3 mm, acuminate;
petals 7–10 × 1.2–2 mm, lanceolate,
slightly keeled, whitish; ovaries 2.5–4 mm;
styles 2.5–5 mm. Seeds *c.* 0.5 × 0.2 mm.
Canary Islands (Tenerife). H5–G1.
Spring–summer.

19. A. percarneum (R.P. Murray) Pitard
& Proust (*Sempervivum percarneum* R.P.
Murray). Illustration: Praeger, An account
of the Sempervivum group, 188, 189
(1932).

Terrestrial shrublet; unbranched or
occasionally with up to 8 branches to 1.5
m; bark light brown or greyish, hairless,
smooth or slightly fissured; leaf-scars
3–13 mm wide. Rosettes 8–20 cm across.
Leaves 4.5–10 × 2–4 cm, obovate to
oblanceolate or spathulate, concave, dark
green, glaucous, hairless; margin reddish,
entire or weakly toothed, with curved hairs
0.5–1 mm. Inflorescence
10–30 × 10–25 cm, dome-shaped. Sepals
8–10, triangular, 2–3 mm, green,
acuminate; petals 7–8 mm, lanceolate,
whitish with pink markings near midrib,
acuminate; stamens 5–7 mm, filaments
white or pinkish, anthers yellow or
whitish; glands 0.5–0.8 mm wide, 4-sided,
greenish; ovaries 2–3 mm, pinkish; styles
3–4 mm; seeds *c.* 0.6 × 0.2 mm. *Canary*

Islands (Gran Canaria). H5–G1.
Spring–summer.

20. A. decorum Bolle (*Sempervivum decorum* Bolle). Illustration: Praeger, An account of the Sempervivum group, 180 (1932); Graf, Exotica, edn 3, 654 (1963); Everett, New York Botanical Gardens illustrated encyclopedia of horticulture **1**: 64 (1980).

Terrestrial shrublet, stems with *c.* 50 slender branches to 60 cm; bark slightly fissured, pale reddish green or whitish, hairless, rough with abundant adventitious roots; leaf-scars 3–4 mm wide. Rosettes 5–10 cm across. Leaves 2.5–5 × 1–1.5 cm, obovate-oblanceolate, often slightly folded and recurved near apex, acuminate, glaucous dark green; margin reddish with very short hairs to *c.* 0.5 mm. Inflorescence loose, cylindric, 8–30 × 8–20 cm. Sepals 6–8, triangular, 3–4 mm, green with reddish tinge; petals 7–8 × 2–2.5 mm, lanceolate, whitish with pink coloration towards midrib; stamens 5–7 mm, filaments whitish, anthers pale yellow or whitish; nectar glands *c.* 1 mm wide, whitish; ovaries 2–3.5 mm; styles 3–4 mm. Seeds *c.* 0.5 × 0.2 mm. *Canary Islands (La Gomera).* G1. Spring–summer.

21. A. hierrense (R.P. Murray) Pitard & Proust (*Sempervivum hierrense* R.P. Murray). Illustration: Praeger, An account of the Sempervivum group, 170 (1932).

Terrestrial shrublet, with erect stems to 1.2 m, usually unbranched; bark greyish, hairless, often slightly fissured; leaf-scars 6–18 mm wide, 2.5–5 mm tall. Rosettes 15–60 cm across. Leaves 10–30 × 3–8 cm, obovate to oblanceolate, glaucous green with reddish margin and often pink or purple tinged; margin with hairs 1–2 mm. Inflorescence 15–30 × 12–50 cm, dome-shaped. Sepals 6–9, triangular, 2.5–3 mm; petals 7–9 × 1.5–2 mm, lanceolate, acuminate, whitish; anthers pale yellow; nectar-glands 0.6 mm wide, whitish; ovaries *c.* 3.5 mm; styles *c.* 3.5 mm; seeds *c.* 0.5 × 0.2 mm. *Canary Islands (Hierro, La Palma).* H5–G1. Spring.

22. A. castello-paivae Bolle (*Sempervivum castello-paivae* (Bolle) Christ; *S. paivae* R. Lowe; *A. paivae* (R. Lowe) Lamarck). Illustration: *Botanical Magazine*, 5593 (1866); Praeger, An account of the Sempervivum group, 178 (1932).

Terrestrial, perennial shrublet often with *c.* 60 twisted branched stems to 70 cm; bark pale brown or greyish, fissured into plates; adventitious roots common; leaf-scars *c.* 1 mm wide. Rosettes 3–7 cm across. Leaves 1.5–3.5 cm × 8–20 mm, obovate to spathulate, pale glaucous green or yellowish green, sometimes with reddish lines, hairless; margin with hairs to 0.2 mm. Inflorescence loose, hemispherical, 6–20 × 6–20 cm. Sepals 7–9, triangular, 2.5–3.5 mm, green, acuminate; petals 8–10 × 1–1.5 mm, greenish white. Stamens 5–7 mm, filaments white, anthers white or pale yellow; nectar-glands 0.8–1.5 mm wide, pinkish; ovaries *c.* 3 mm; styles *c.* 3 mm; seeds *c.* 0.5 × 0.2 mm. *Canary Islands (La Gomera).* H5–G1. Spring.

30. AICHRYSON Webb & Berthelot
A.C. Whiteley

Annual to perennial herbs, occasionally shrubs, some species monocarpic. Branching often appearing dichotomous, including that of the inflorescence. Leaves alternate, simple, mostly entire, usually hairy, fleshy. Flowers in terminal panicles, yellow. Calyx cup-shaped with 5–12 fleshy lobes. Petals as many as calyx lobes, almost free. Stamens twice as many as calyx lobes. Carpels as many as calyx lobes, their bases fused with the receptacle. Nectar-glands 2-horned or with several teeth.

A genus of 15 species from Macaronesia and Morocco. They are adapted to survival on rocks and walls. In cultivation they tend to grow much larger than in nature. Several species make interesting pot-plants suitable for a sunny windowsill, requiring the minimum of attention. Propagation by seed or cuttings.

Literature: Praeger, R.L., An Account of the Sempervivum group (1932).

1a. Plants densely clothed with long
    hairs   2
 b. Plants downy or sparsely hairy   3
2a. Leaves 2.5–4 cm, the blade longer
    than the stalk   **4. villosum**
 b. Leaves 3–8 cm, the blade about as
    long as the stalk   **3. laxum**
3a. Congested shrub to 10 cm; leaves to
    1.2 cm × 6 mm   **1. tortuosum**
 b. Loose shrub to 30 cm; leaves to
    4 × 1 cm   **2. × domesticum**

1. A. tortuosum (Aiton) Praeger (*Sempervivum tortuosum* Aiton; *Aeonium tortuosum* Berger). Illustration: Praeger, An account of the Sempervivum group, 105 (1932); *Botanical Magazine*, 296 (1795).

A compact, downy shrub to 10 cm. Stems woody at the base, much-branched, widely diverging, downy at first, becoming hairless. Leaves crowded at the tips of non-flowering-shoots, stalkless, obovate to spathulate, fleshy, to 1.2 cm × 6 mm, and 2–4 mm thick, sticky, sometimes reddish. Flowering-shoots short, with a few widely spaced leaves. Panicles with 2 or 3 branches to 3 cm, few-flowered, stickily hairy. Flowers *c.* 1 cm across, with parts in 8s, on stalks to 7 mm. Calyx *c.* 4 mm, with acute lobes to 3 mm. Petals lanceolate, *c.* 6 mm, the margins finely hairy. Stamens 3–4 mm. Nectar-glands orange, to 1 × 0.5 mm, 2-horned, each horn 2-fid. Carpels 3–4 mm. Styles short. *Canary Islands.* G1. Early summer.

2. A. × domesticum Praeger (*Aeonium domesticum* Berger; *Sempervivum tortuosum* de Candolle). Illustration: Praeger, An account of the Sempervivum group, 107 (1932); Beckett, The RHS encyclopaedia of house plants, 67 (1987).

A sparsely hairy, glandular shrub to 30 cm. Stems woody at base, much-branched, widely diverging. Leaves crowded at the tips of non-flowering-shoots, spathulate to obovate, obtuse, narrowed gradually to the base, to 4 × 1 cm, and 1.5 mm thick, sticky and with a resinous scent. Flowering-branches 10–20 cm, branched up to 3 times, with widely spaced leaves at the base. Flowers 1.3–1.5 cm across, with parts in 7s or 8s; stalks 5–10 mm. Calyx 3–5 mm with acute lobes 2–3 mm. Petals lanceolate, acuminate, 6–7 mm, the tips reflexed. Stamens 4–4.5 mm. Nectar-glands orange or reddish, 0.5–0.75 mm, 2-fid or toothed. Carpels 3–4 mm. Styles recurved. *Garden origin.* G1. Summer.

Probably a hybrid between *A. tortuosum* (Aiton) Praeger and *A. punctatum* (C.A. Smith) Webb & Berthelot. Widely cultivated, especially as the cultivar 'Variegatum', which has leaves margined with creamy white.

3. A. laxum (Haworth) Bramwell (*A. dichotomum* (de Candolle) Webb & Berthelot). Illustration: Praeger, An account of the Sempervivum group, 111 (1932); Bramwell, Wild flowers of the Canary Islands, 58 (1974); Kunkel & Kunkel, Flora de Gran Canaria **3**: 57 (1978); Graf, Exotica, series 4, 860 (1982).

An erect, softly hairy annual or biennial, 10–100 cm. Stems strong, sparingly dichotomously branched, densely covered with hairs to 4 mm. Leaves densely hairy, 3–8 cm, the blade rounded to rhombic, about as long as the stalk, the tip occasionally notched. Inflorescence terminal, large and much-branched. Flowers to 1.5 cm across, with parts in 9s to 12s; stalks 6–12 mm. Calyx shallow, 4–5 mm, with narrow lobes to 3.5 mm. Petals linear to lanceolate, acuminate, 6–7 mm. Stamens *c.* 5 mm. Nectar-glands irregularly 2-horned, orange or yellow, *c.* 0.5 mm. Carpels 3.5–4 mm. Styles recurved. *Canary Islands.* G1. Early summer.

4. A. villosum (Aiton) Webb & Bertholet. Illustration: *Botanical Magazine*, 1809 (1819); Praeger, An account of the Sempervivum group, 117 (1932); Graf, Exotica, series 4, 860 (1982).

A glandular annual to 20 × 30 cm, completely clothed with long hairs. Stem stout at the base, soon branching into long dichotomous shoots. Leaves sometimes crowded, spathulate to diamond-shaped, blade 1.5–2 × 1.5–2 cm, stalk 1–2 cm. Flowers 1.2–1.5 cm across, with parts in 8s (rarely 6s, 7s or 9s); stalks to 1 cm. Calyx 4–5 mm with lanceolate lobes to 3 mm. Petals lanceolate to oval, acute, 5–6 mm. Stamens *c.* 4 mm. Nectar-glands orange, *c.* 0.5 × 0.75 mm, with 4–6 teeth. Carpels 3.5–4.5 mm. Styles recurved. *Azores.* G1. Early summer.

137. CEPHALOTACEAE

Herbs. Leaves alternate, in a rosette, modified into stalked, insectivorous pitchers, exstipulate. Flowers in racemes, bisexual, actinomorphic. Perianth and stamens perigynous. Perianth of 6 free segments. Stamens 12, connectives swollen, glandular. Ovary of 6 free carpels; ovules 1 per carpel, basal or marginal; styles free. Fruit a group of follicles.

A family of a single Australian species superficially resembling *Nepenthes*, although not closely allied. In cultivation it requires a deep pot and a well-drained but moist mixture of peat, leaf-mould and sand; over-watering in winter can damage it. Propagation is by root or leaf cuttings or division.

1. CEPHALOTUS Labillardière
E.C. Nelson
Description as for family.

1. C. follicularis Labillardière. Illustration: *Botanical Magazine*, 3118–19 (1831);

Slack, Carnivorous plants, 88, 206 (1979); Morley & Toelken, Flowering plants in Australia, 141 (1983).

Western Australia. H5–G1. Summer.

A protected plant confined to seasonal swamps.

138. PENTHORACEAE

Perennial herbs. Leaves alternate, exstipulate. Flowers bisexual, actinomorphic, in terminal cymes. Perianth and stamens perigynous. Perianth of 5, somewhat unequal, segments. Stamens 10. Ovary of 5 carpels, united only at the base; styles 5, ovules numerous. Fruit a group of follicles.

A family of a single genus with 2 species, one of which is native to North America. Included within the Saxifragaceae in edition 1.

1. PENTHORUM Linnaeus
M.F. Gardner
Erect perennials with stolons. Leaves alternate, lanceolate with a long, gradually diminishing point at both ends, stalkless, with minute teeth. Flowers in terminal, spirally coiled clusters, yellow-green; sepals 5; petals 5, or absent; stamens 10, carpels 5, united in the lower half; capsules flattened, 5-beaked, upper portion deciduous.

A genus of 1–3 species native to east and south-east Asia and North America. Suitable for planting in boggy waterside margins or even in shallow water.

1. P. sedoides Linnaeus. Illustration: Rickett, Wild flowers of the United States **1**: 201 (1966); Correll & Correll, Aquatic & wetland plants of southwestern United States **2**: 1000 (1975); Voss, Michigan flora, pl. 4 (1985).

Stems hairless to 60 cm, branched above. Leaves not fleshy, pale green becoming bright orange with age. Flowering-stems glandular; petals sometimes absent. Seeds pink with warts, elliptic to ovoid, *c.* 7 mm. *E North America.* H4. Summer.

139. SAXIFRAGACEAE

Herbs. Leaves alternate, often all basal, usually simple, sometimes evergreen, exstipulate or rarely with small stipules;

ptyxis variable. Flowers solitary or in racemes, usually bisexual, actinomorphic or rarely zygomorphic. Calyx, corolla and stamens perigynous or epigynous. Calyx of 4–5 free sepals. Corolla of 4–5 free petals or rarely absent. Stamens 4–5 or 8–10. Ovary of 2 united carpels (rarely more), occasionally almost free; ovules many, axile; styles 2, free, usually divergent. Fruit a capsule.

About 30 genera and 475 species, with genera native to both Europe and North America. Many of the genera are widely cultivated, especially species and hybrids of the largest genus, *Saxifraga*. The genus *Francoa*, with evergreen, usually divided leaves and flowers with parts in 4 s, is sometimes separated off as a family (Francoaceae).

1a. Leaves compound, or simple but deeply divided   2
  b. Leaves simple, entire, toothed, scalloped or shallowly lobed   6
2a. Floral parts, including the ovary, in 4s   **26. Francoa**
  b. Floral parts not in 4s, ovary of 2 or 3 carpels   3
3a. Plants without stem-leaves   **7. Saxifraga**
  b. Plants with at least a few, often small, leaves on the stem   4
4a. Leaves trifoliolate   **15. Tiarella**
  b. Leaves with more than 3 leaflets   5
5a. Leaves palmate, pseudopinnate or pinnate, leaflets leathery, wrinkled; petals usually absent   **2. Rodgersia**
  b. Leaves pinnate, leaflets neither leathery nor wrinkled; petals usually present   **1. Astilbe**
6a. Leaves peltate   7
  b. Leaves not peltate   9
7a. Petals pale yellow   **24. Peltoboykinia**
  b. Petals white or pink   8
8a. Flowers in corymbs; petals pinkish, 4–8 m; stamens 10   **6. Darmera**
  b. Flowers in spike-like cymes; petals creamy white, to 4 mm; stamens 6–8   **3. Astilboides**
9a. Carpels 3   **14. Lithophragma**
  b. Carpels 2   10
10a. Carpels completely free from each other   11
  b. Carpels united at least at the base   12
11a. Leaves oblong to obovate, toothed, leathery   **4. Leptarrhena**
  b. Leaves rounded to heart-shaped, shallowly lobed, not leathery   **10. Jepsonia**

12a. Flowers in bracteate, dichotomous cymes on erect shoots, bracts leaf-like, greenish or yellow **25. Chrysosplenium**

b. Flowers and bracts not as above **13**

13a. Basal leaves stalkless or very shortly stalked **7. Saxifraga**

b. Basal leaves clearly stalked, stalks often long **14**

14a. Flowers with a varying number of stamens in each inflorescence **17. × Heucherella**

b. Flowers in each inflorescence with a constant number of stamens **15**

15a. Stamens 10 or rarely 8 **16**

b. Stamens 3, 5 or 6 **23**

16a. Leaves ovate, coarsely double-toothed and often somewhat lobed **1. Astilbe**

b. Leaves not as above **17**

17a. Capsule with 2 unequal horns (carpels), few-seeded **15. Tiarella**

b. Capsule with 2 equal horns (carpels), many-seeded **18**

18a. Petals divided into linear segments, comb-like **19**

b. Petals entire, not divided into linear segments and comb-like **20**

19a. Petals white or greenish yellow, not reflexed **19. Mitella**

b. Petals greenish white at first, becoming pinkish red, reflexed **20. Tellima**

20a. Petals absent **11. Tanakaea**

b. Petals present **21**

21a. Leaves large, coarse, evergreen, leathery, sheathing at their bases; carpels almost free from each other and the perigynous zone **5. Bergenia**

b. Combination of characters not as above **22**

22a. Styles partially fused above the ovule-bearing parts of the carpels **9. Telesonix**

b. Styules free above the ovule-bearing parts of the carpels **7. Saxifraga**

23a. Stamens 3; petals thread-like, purplish brown **21. Tolmiea**

b. Stamens 5 or 6; petals not as above **24**

24a. Petals deeply divided into linear segments, comb-like **25**

b. Petals entire, not as above **26**

25a. Calyx 1–3 mm; styles less than 1 mm **19. Mitella**

b. Calyx 6–8 mm; styles more than 1 mm **13. Elmera**

26a. Rootstock bearing bulbils **18. Bolandra**

b. Rootstock without bulbils **27**

27a. Plants with horizontal stolons **8. Sullivantia**

b. Plants without horizontal stolons **28**

28a. Ovary 1-celled **16. Heuchera**

b. Ovary 2-celled **29**

29a. Scape leafless; flowers in compact, ebracteate panicles, almost stalkless **23. Mukdenia**

b. Combination of characters not as above **30**

30a. Scapes leafless **22. Bensoniella**

b. Stems leafy (stem-leaves sometimes few) **12. Boykinia**

1. ASTILBE D. Don

P.G. Barnes

Herbaceous perennials with stout rhizomes, sometimes stoloniferous, generally with conspicuous brown hair-like scales. Leaves alternate, both basal and on the stem, pinnately compound or rarely simple; leaflets lanceolate to broadly ovate, strongly toothed, teeth forward-pointing; stipules scarious. Inflorescence a terminal panicle, the branches usually spike-like. Flowers small, white, pink or purplish, sometimes unisexual. Perigynous zone short. Sepals 5; petals 5, linear, narrowly spathulate or rarely absent; stamens 5 or 10; carpels 2, more or less joined at the base, superior; capsule 2-celled, the cells dehiscing inwardly.

A genus of about 14 species, in north-east Asia, the Himalaya and North America. The astilbes are popular plants best suited by moist acid or neutral soils, and a position in full sun. The smaller species and hybrids may be grown in the rock garden or on peat-banks. Propagation is by division in autumn or early spring. Many cultivars are grown, mostly of complex hybrid origin. These show a wide range of flower colour, from whitish pink to lilac and deep red and may have extra floral parts. The name *A. astilboides* Lemoine is of uncertain application: some plants under this name probably belong to *A. japonica*.

Literature: Ievina, S.O. & Lusinya, M.A., Astilby: introdutsiya v Latviiskoi SSR (1975). Descriptions of numerous cultivars of hybrid origin may be found in Thomas, G.S., Perennial garden plants, edn 2 (1982) and in Jelitto, L., Schacht, W. & Fessler, A., Die Freiland Schmuckstauden, edn 3 (1985).

1a. Petals absent; stamens 5 **5. rivularis**

b. Petals present; stamens 10 **2**

2a. Leaves simple **1. simplicifolia**

b. Leaves compound **3**

3a. Petals pink, red, purple or white; inflorescence often with long curled hairs **4**

b. Petals white; inflorescence often with short glandular hairs **6**

4a. Leaflets wedge-shaped at base; sepals red **6. rubra**

b. Leaflets rounded or cordate at base **5**

5a. Inflorescence with erect branches; petals pink **2. chinensis**

b. Inflorescence with spreading branches; petals white **8. koreana**

6a. Leaflets wedge-shaped at base, sharply double-toothed **3. japonica**

b. Leaflets rounded to cordate at base **7**

7a. Plant to 80 cm; leaflets strongly double-toothed; stamens shorter than petals **4. thunbergii**

b. Plant to 1.5 m; leaflets simply toothed; stamens longer than petals **7. grandis**

1. A. simplicifolia Makino. Figure 91(1), p. 97. Illustration: *Gardeners' Chronicle*, 101 (1912); Terasaki, Nippon shokubutsu zufu, f. 697 (1932); Kitamura et al., Coloured illustrations of herbaceous plants of Japan **2**: pl. 35 (1986); Hayashi, Azegami & Hishiyama, Wild flowers of Japan, 424 (1983).

Compact plant with a short rhizome. Stem to 30 cm. Leaves 3–10 × 2–6 cm, simple, ovate, coarsely double-toothed and often somewhat lobed, rather glossy. Inflorescence 10–15 cm, a narrow loose panicle with spreading branches, usually leafless, somewhat glandular-hairy. Flowers short-stalked; petals 2–3 mm, linear, white. Stamens 10, equalling petals; anthers pale yellow. *Japan*. H4. Early summer.

The species is sometimes grown on rock gardens but many plants under the name are hybrids with species with compound leaves.

2. A. chinensis (Maximowicz) Franchet & Savatier. Figure 91(2), p. 97. Illustration: Iconographia cormophytorum Sinicorum **2**: 122 (1972).

Stems to 80 cm, rather hairy above; basal leaves 2–3 times compound or pinnately divided, leaflets 3–8 × 2–4 cm, ovate, acuminate, coarsely toothed. Inflorescence rather narrow with ascending spike-like branches. Flowers almost stalkless, very crowded; petals *c*. 5 mm, linear, rosy-pink; stamens 10, shorter than petals. *NE Asia*. H4. Summer.

More often cultivated are: var. **davidii** Franchet, to 1 m, with purplish pink petals,

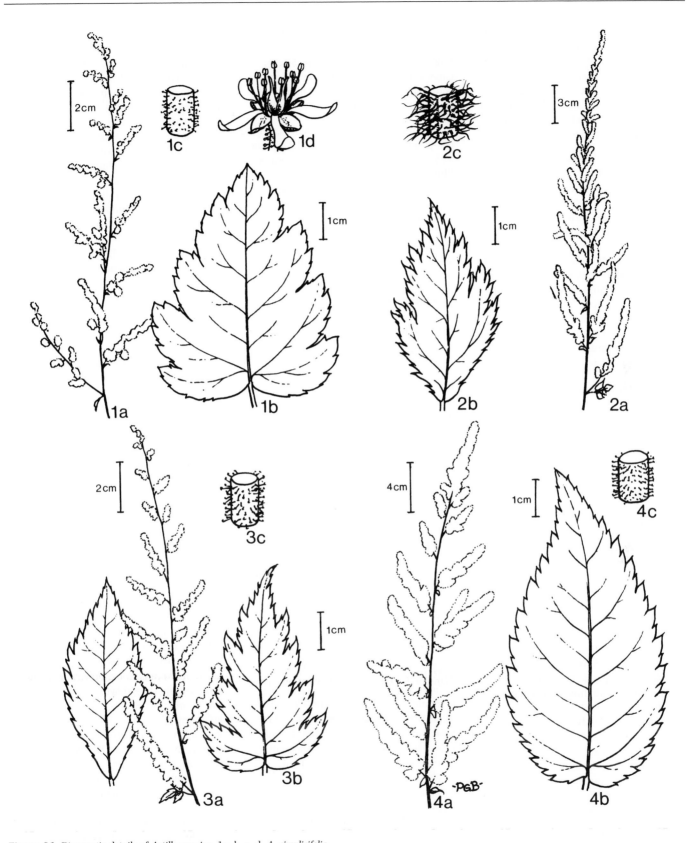

Figure 91. Diagnostic details of *Astilbe* species. 1a, b, c, d, *A. simplicifolia*.
2a, b, c, *A. chinensis*. 3a, c, *A. japonica*, 3b, *A. japonica* subsp. *glaberrima*.
4a, b, c, *A. thunbergii*.

a parent of many garden hybrids, especially those included in **A. × arendsii** Arends; and 'Pumila' (var. *pumila* misapplied), a dwarf variant smaller in all parts, with dense narrow inflorescences usually less than 30 cm, petals purplish pink. The name var. *taquetii* Vilmorin is of uncertain significance but it is said to be a parent of many hybrids.

3. A. japonica (Morren & Decaisne) A. Gray. Figure 91(3a, c), p. 97. Illustration: Step, Favourite flowers of garden and greenhouse **2**: pl. 92 (1897); Makino, Illustrated flora of Nippon, 496 (1942); Satake et al., Wild flowers of Japan **2**: pl. 156 (1985); Kitamura et al., Coloured illustrations of herbaceous plants of Japan **2**: pl. 36 (1986).

Stems 50–90 cm. Leaves compound, leaflets 3–7 × 1–2 cm, lanceolate, acute, sharply double-toothed, base wedge-shaped. Inflorescence with ascending glandular-hairy branches. Petals 3–4 mm, narrowly spathulate, white. Stamens 10, shorter than petals. *Japan.* H4. Summer.

Subsp. **glaberrima** (Nakai) Kitamura (*A. japonica* var. *terrestris* (Nakai) Murata; *A. glaberrima* Nakai var. *saxatilis* (Nakai) Ohba). Figure 91(3b), p. 97. Illustration: *Bulletin of the Alpine Garden Society* **19**: 386 (1951); Satake et al., Wild flowers of Japan **2**: pl. 157 (1985). A dwarf variant, 20–40 cm. Leaflets relatively broader, glossy and incised-toothed. Calyx pink. *Japan (Yakushima).*

4. A. thunbergii (Siebold & Zuccarini) Miquel. Figure 91(4), p. 97. Illustration: Makino, Illustrated flora of Nippon, 496 (1942); Satake et al., Wild flowers of Japan **2**: pl. 156 (1985); Kitamura et al., Coloured illustrations of herbaceous plants of Japan **2**: pl. 36 (1986).

Stem to 80 cm. Basal leaves compound, leaflets 4–12 × 2–5 cm, ovate, long-acuminate, sharply double-toothed, base rounded or cordate. Inflorescence a loose panicle to 30 cm, with dense, fine glandular hairs. Petals 3–4 mm, linear, white. Stamens 10, shorter than petals. *Japan.* H4. Summer.

Var. **formosa** (Nakai) Ohwi has leaflets acute or short acuminate; petals 5–7 mm.

A. × lemoinei Lemoine includes hybrids with **A. astilboides** Lemoine.

5. A. rivularis Hamilton. Figure 92(1), p. 99. Illustration: Collett, Flora Simlensis, 174 (1902); Polunin & Stainton, Flowers of the Himalaya, 466 (1985).

Stems to 1.5 m. Leaves 2 or 3 pinnate, leaflets 5–12 × 3–7 cm, ovate, acuminate, scalloped-toothed, base cordate. Inflorescence a large panicle to 50 cm; calyx-lobes spreading, white or pale pink, petals absent. Stamens 5, *c.* 3 mm. *Himalaya.* H4. Summer.

6. A. rubra Hooker & Thomson. Figure 92 (2), p. 99. Illustration: *Botanical Magazine*, 4959 (1857); Grierson & Long, Flora of Bhutan **1**(3): f. 35 (1983).

Stems to 1 m, often with abundant long brown hairs in inflorescence. Leaves bipinnate or compound, leaflets 4–7 × 2–4 cm, ovate, toothed, acute or obtuse, base wedge-shaped. Petals 4–5 mm, linear, red to purple. Stamens 10, shorter than petals. *Himalaya.* H4. Summer.

7. A. grandis Wilson. Figure 92(3), p. 99. Illustration: *Gardeners' Chronicle* 1905: 426; *Journal of Horticulture* **59**: 331 (1909).

Stems to 1.5 m, with soft hairs and often short glandular ones in the inflorescence. Leaves compound, leaflets 4–12 × 3–7 cm, ovate, acute or shortly acuminate, teeth pointing forwards, cordate or rounded at base. Flowers in a loose panicle of spike-like branches. Petals 4–5 mm, linear, white. Stamens 10, longer than petals. *W China.* H4. Summer.

8. A. koreana Nakai. Figure 92(4), p. 99. Illustration: Hay, Plants for connoisseurs, 18 (1938).

Stems 30–60 cm. Leaves 2- or 3-pinnate; leaflets broadly ovate, teeth pointing forwards, acute or shortly acuminate, base cordate or rounded. Inflorescence a large panicle with spreading hairy branches; flowers pink in bud, petals linear, white. Stamens 10, shorter than petals. *Korea & N China.* H4. Summer.

2. RODGERSIA A. Gray

J. Cullen

Perennial, rhizomatous herbs. Flowering-stems with reduced leaves, most of the foliage arising directly from the rhizome. Leaves long-stalked, pinnate, apparently pinnate or palmate, leaflets toothed, generally hairy or downy beneath. Inflorescence a large, corymb-like, flat-topped cyme with many flowers. Flowers shortly stalked, each with a bract. Sepals 5, greenish, whitish or red, united below. Petals absent. Stamens 5, spreading. Ovary of 2 carpels, which are united in their lower part. Fruit a capsule. Seeds

numerous (often not ripened in cultivation).

A genus of 6 species from E Asia, grown for their bold foliage and large inflorescences. The leaves are pinnate, palmate or apparently pinnate, the last a term referring to leaves, which are divided into a variable number of leaflets staggered along a long or short axis; the lowermost set of leaflets is most commonly 3, and these are borne in a plane opposite to that of all the rest (of which there may be 3, 5 or 7). They are generally easily cultivated in moist, rich soil and are propagated by division of the rhizome or by seed when available.

Literature: Cullen, J., Taxonomic notes on the genus Rodgersia, *Notes from the Royal Botanic Garden, Edinburgh* **34**: 113–23(1975).

1a. At least some of the basal and lower stem-leaves pinnate or apparently pinnate 2
  b. All leaves strictly palmate 3
2a. Leaflets with adpressed bristles above **2. sambucifolia**
  b. Leaflets hairless above **1. pinnata**
3a. Leaflets with 3–5 lobes at the tip; sepals acuminate with straight tips **3. podophylla**
  b. Leaflets not lobed at the tips; sepals ovate, obtuse to rounded, with usually reflexed tips 4
4a. Leaf-teeth hairy on under surface; sepals conspicuously enlarging after flowering **4. aesculifolia**
  b. Leaf-teeth hairless on under surface; sepals not enlarging after flowering **5. henrici**

1. R. pinnata Franchet. Illustration: *Botanical Magazine*, 7892 (1903); *Notes from the Royal Botanic Garden, Edinburgh* **34**: 114 (1975).

Stems to 1 m. Leaves mostly pinnate or apparently pinnate, leaflets leathery, wrinkled, hairless above, erect to spreading and channelled. Sepals white or pink, enlarging or not after flowering, almost as long as to longer than the stamens. Capsule hard, 5–6 mm. *W China (Yunnan, Sichuan).* H4. Summer.

A variable species of which 3 cultivars are available: 'Alba', with white sepals; 'Elegans' with pink sepals; and 'Superba' (illustration: Brickell (ed.) RHS A–Z encyclopedia of garden plants, 908, 2003), a very robust plant with red-suffused leaves and pink sepals, which may perhaps be a hybrid between *R. pinnata* and

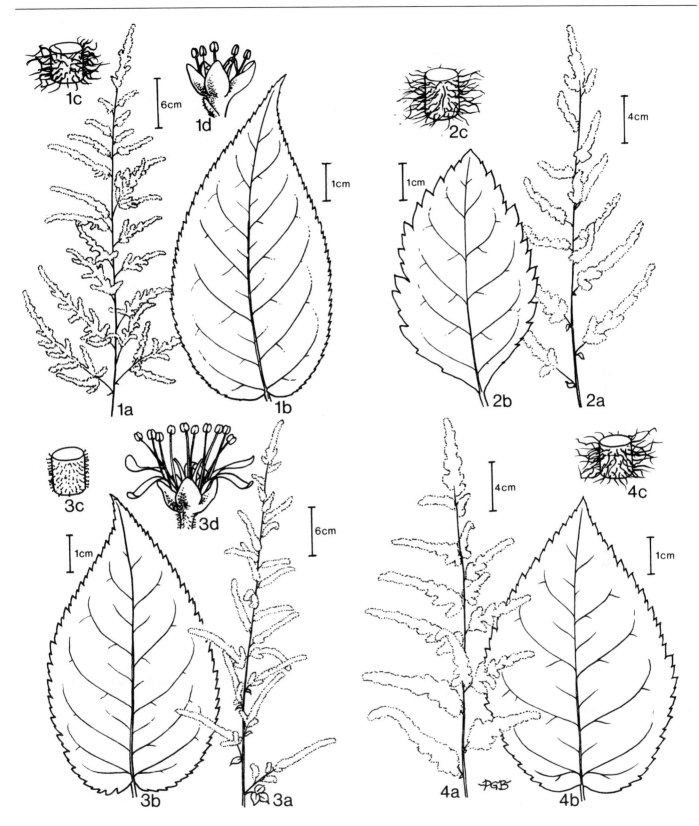

Figure 92. Diagnostic details of *Astilbe* species. 1a, b, c, d, *A. rivularis*.
2a, b, c, *A. rubra*. 3a, b, c, d, *A. grandis*. 4a, b, c, *A. koreana*.

R. aesculifolia. A further presumed hybrid between these species is sometimes grown as *R. purdomii* invalid, a name of no botanical standing.

2. R. sambucifolia Hemsley. Illustration: Parey's Blumengärtnerei **1**: 754 (1958); *Notes from the Royal Botanic Garden, Edinburgh* **34**: 114 (1975); Brickell (ed.), RHS A–Z encyclopedia of garden plants, 908 (2003).

Similar, but smaller in stature, leaves always strictly pinnate, the leaflets with adpressed bristles above. *W China (Yunnan, Sichuan).* H5. Summer.

3. R. podophylla Gray (*R. japonica* Regel). Illustration: *Botanical Magazine*, 6691 (1883); *Journal of the Royal Horticultural Society* **96**(Sept.): front cover (1971); *Notes from the Royal Botanic Garden, Edinburgh* **34**: 114 (1975); Brickell (ed.), RHS A–Z encyclopedia of garden plants, 908 (2003).

Stems to 1 m. Leaves strictly palmate, the leaflets drooping, flat, with 3–5 teeth at the apex. Sepals acuminate at their tips, not enlarging in fruit, considerably exceeded by the stamens. Capsule hard, 4–5 mm. *Japan, Korea.* H3. Summer.

4. R. aesculifolia Batalin. Illustration: Parey's Blumengärtnerei **1**: 753 (1958); *Notes from the Royal Botanic Garden, Edinburgh* **34**: 114 (1975); Brickell (ed.), RHS A–Z encyclopedia of garden plants, 908 (2003).

Stems to 1 m. Leaves strictly palmate, leaflets erect to spreading, channelled, not divided at the apex, hairy on the veins and teeth beneath. Sepals ovate with reflexed tips, white or pink, conspicuously enlarging after flowering. Stamens pink. Capsule hard, 5–7 mm. *N China.* H3. Summer.

5. R. henrici (Franchet) Franchet. Illustration: *Notes from the Royal Botanic Garden, Edinburgh* **34**: 114 (1975).

Very similar to *R. aesculifolia* but leaf-teeth not hairy beneath, sepals not enlarging in fruit. *W China, Burma.* H5. Summer.

3. ASTILBOIDES Engler
J. Cullen
Perennial, brown-hairy rhizomatous herbs to 1.5 m. Leaves mostly basal, long-stalked, blade peltate, almost circular, to 90 cm across, lobed and irregularly toothed. Flowers to 8 mm across, creamy white, in spike-like, terminal cymes to 5 cm; stem-leaves very small. Sepals 4 or 5, triangular,

blunt or notched. Petals 4 or 5, oblong or oblong-lanceolate, somewhat unequal. Stamens 6–8, about as long as the petals. Ovary usually 2-celled (rarely 4-celled), each cell with *c.* 8 ovules; styles 2 (rarely 4), long, with small, capitate stigmas. Capsule usually 2-celled, many-seeded. Seeds pointed at each end.

A genus of a single species from E Asia, easily grown in semi-shade in moist, rich soil. Propagation is by division of the rhizome or by seed.

1. A. tabularis (Hemsley) Engler (*Saxifraga tabularis* Hemsley; *Rodgersia tabularis* (Hemsley) Komarov). Illustration: Hay & Synge, Dictionary of garden plants in colour, pl. 1351 (1969); Huxley, Garden perennials and water plants, pl. 228 (1970); Brickell (ed.), RHS A–Z encyclopedia of garden plants, 154 (2003).

N China, Mongolia, North Korea. H2. Summer.

4. LEPTARRHENA R. Brown
D.M. Miller
Herbs with horizontal, spreading rootstocks. Leaves in basal rosettes, 5–15 cm, oblong to obovate, narrowed to the short, winged stalk, toothed, leathery, dark glossy green above, paler beneath. Inflorescence-stalk 5–30 cm, with 1–3 small leaves clasping the stem. Flowers bisexual in dense, terminal inflorescences. Sepals 5, minute. Petals 5, 2–3 mm, spathulate, persistent, white or tinged with pink. Stamens 10, longer than petals. Stigmas borne on top of the carpels, more or less without styles; carpels 2. Fruit a group of 2 follicles *c.* 8 mm, bright red.

A genus of a single species from North America, which can be grown in semi-shade or sun in a moist but not wet soil. Propagation is by cuttings or division, or by seed in spring.

1. L. amplexifolia (Sternberg) Seringe (*L. pyrolifolia* (D. Don) Seringe). Illustration: Hitchcock et al., Vascular plants of the Pacific northwest **3**: 11 (1971); Rickett, Wild flowers of the United States **5**: pl. 80 (1971); Clark, Wild flowers of British Columbia, 199 (1973).

W North America (Alaska to Oregon). H1. Summer.

5. BERGENIA Moench
P.F. Yeo
Perennial herbs, with stout prostrate rooting stems terminating in leaf-rosettes. Leaves 6–25 × 5–17 cm, usually evergreen;

stalk stout, sheathing at base; blade simple, shallowly toothed or nearly entire, leathery. Flowers numerous, in panicles supported on stout scapes, differing from those of *Saxifraga* in being perigynous and having the carpels nearly free from each other and from the perigynous zone.

A genus of 6–8 species from cool-temperate Asia; 5–6 species are widely cultivated, being used for borders, large rock gardens and wild gardens. Their large tough leaves give ground cover and winter colour and they produce masses of long-lasting but somewhat frost-sensitive flowers in spring. The species are interfertile and individuals are self-incompatible; garden seed, therefore, usually produces hybrids. Indeed, a great many hybrids are now in circulation in gardens. This should be borne in mind when attempting identifications as only a few of them can be described here. The plants are easy to grow and extremely resistant to neglect; however, for good effect propagation every 5 years is recommended. This is done by breaking up the rhizomes and selecting terminal pieces for replanting.

Literature: Borissova, A., De speciebus generis Bergenia Moench Asiae Mediae, *Botanicheskie Materialy* **16**: 97–103 (1954); Pan, J.T., A conspectus of the genus Bergenia Moench, *Acta Phytotaxonomica Sinica* **26**: 120–9 (1988), in Chinese with English summary; Yeo, P. F., Two Bergenia hybrids, *Baileya* **9**: 20–8 (1961); Yeo, P.F., A revision of the genus Bergenia, *Kew Bulletin* **20**: 113–48 (1966); Yeo, P.F., Further observations on Bergenia in cultivation, *Kew Bulletin* **26**: 47–56 (1971); Yeo, P.F., Cultivars of Bergenia in the British Isles, *Baileya* **18**: 96–112 (1972).

1a. Leaf-blade hairless or slightly hairy on margins at base   2
  b. Leaf-blade densely hairy on margins   4
2a. Petals 1.1–2.5 cm; some glands on the flowering branches stalked   **3. purpurascens**
  b. Petals 0.8–1.3 cm; glands on the flowering branches not stalked   3
3a. Leaves ovate, base wedge-shaped or rounded; flowers more or less reflexed; petals elliptic to broadly ovate, tapering into the claw   **1. crassifolia**
  b. Leaves circular, base rounded or cordate, often blistered; flowers mostly erect or upwardly inclined;

petals circular or broadly ovate, abruptly contracted into the claw **2. cordifolia**

4a. Petals at first pink or purple, never white **5**

b. Petals normally white at first, sometimes reddish in age **6**

5a. Flowers reflexed, usually purple, late **3. purpurascens**

b. Flowers not reflexed, pink, redder in age, early **4. × schmidtii**

6a. Leaves circular or broadly ovate, usually deciduous; flowers erect or ascending; petals nearly circular **5. ciliata**

b. Leaves obovate, base wedge-shaped base; petals obovate **6. stracheyi**

1. B. crassifolia (Linnaeus) Fritsch (*Megasea crassifolia* (Linnaeus) Haworth; *B. pacifica* Komarov). Illustration: *Botanical Magazine*, 196 (1792); Step, Favourite flowers of garden and greenhouse **2**: t. 89 (1897); Bloom, Perennials for your garden, 37 (1975) – as *B. cordifolia*.

Plant to 45 cm. Leaf-blade oblong, obovate or broadly ovate, base shallowly cordate, rounded or wedge-shaped, apex rounded or truncate, hairless. Flowering-stem branched near the top with the branches turned to one side, reddish, with stalkless glands. Flowers more or less nodding, bell-shaped. Perigynous zone and calyx together 5–9 mm. Petals 8–15 × 5.5–8 mm, elliptic or obovate, purplish pink. *Former USSR (Siberia & far eastern region).* H1. Spring.

2. B. cordifolia (Haworth) Sternberg (*Megasea cordifolia* Sternberg). Illustration: Perry, Collins guide to border plants, t. 214 (1957); *Kew Bulletin* **20**: 142 (1966).

Plant to 40 cm. Leaf-blade circular, base rounded or cordate, sometimes blistered, hairless. Flowering-stem branched near the top; panicle spherical, becoming umbrella-shaped. Flowering-stem, branches, perigynous zone and sepals flushed with red, with stalkless glands. Flowers very numerous, broadly bell-shaped, becoming erect. Perigynous zone and calyx together 6–8 mm. Petals to 1.2 cm, circular or broadly ovate, with distinct claw, pink. *Former USSR (Siberia).* H1. Spring.

Not regarded as distinct from *B. crassifolia* by Soviet botanists. 'Purpurea' (illustration: Brickell, RHS gardeners' encyclopedia of plants & flowers, 226, 1989), leaves thicker, redder, petals deep purple, more widely overlapping and spreading. Frequently grown.

3. B. purpurascens (J.D. Hooker & Thomson) Engler (*B. delavayi* (Franchet) Engler; *B. beesiana* invalid; *B. yunnanensis* invalid). Illustration: *Botanical Magazine*, 5066 (1858), 117 (1950); Hay & Beckett (eds), Reader's Digest encyclopaedia of garden plants and flowers, edn 2, 81 (1978); Stainton, Flowers of the Himalaya, a supplement, t. 37 (1988).

Plant to 40 cm. Leaf-blade elliptic or ovate-elliptic, base rounded or acute, usually convex and glossy above, reddish, hairless or with a few marginal hairs at base. Panicle compact, with most branches near the top and turned to one side. Flowering-stem, branches, perigynous zone and sepals deep purplish or brownish red with numerous glands, some of them stalked. Flowers few to several, nodding, bell-shaped. Perigynous zone and calyx together 0.8–1.4 cm. Petals 1.5–2.5 × 0.7–0.9 cm, broadly spoon-shaped or obovate, deep purplish red or bright pink. *E Himalaya, N Burma, W China.* H2. Late spring.

'Ballawley' ('Delbees'). Illustration: Bloom, Perennials for your garden, 37 (1975). Leaves large, circular, doubly toothed or scalloped, base cordate, petals to 1.7–1.2 cm, purple.

4. B. × schmidtii (Regel) Silva-Tarouca (*B. ligulata* Engler var. *speciosa* invalid; *B. ornata* Guillaumin; *B. leichtlinii* misapplied; *B. stracheyi* misapplied; *Saxifraga ligulata* var. *speciosa* Verlot). Illustration: *Revue Horticole* **40**: 261 (1868); *Botanical Magazine*, 5967 (1872*)*; *Gartenflora* **27**: t. 946 (1878); *Addisonia* **6**: 29 (1921).

Plant to 40 cm. Leaf-blade usually to 25 cm, obovate to obovate-elliptic, base rounded or shallowly cordate, usually bright green, often drooping, margin hairy. Panicle about as long as its stalk, more or less spherical at first, later becoming hemispherical and rather loose with arching branches. Flowering-stem, branches, perigynous zone and sepals flushed with bright red, with sparse short-stalked or stalkless glands. Flowers at first nodding, soon horizontal to erect, broadly bell-shaped. Perigynous zone and calyx together 6–9 mm. Petals 1.3–1.7 × 0.8–1.3 cm, oblong or broadly ovate, the claw lengthening with age, pink, darker at the claw. *Garden origin.* H2. Winter–spring.

This is the most commonly grown *Bergenia* in Atlantic Europe.

5. B. ciliata (Haworth) Sternberg (*B. ligulata* Engler). Illustration: *Botanical Magazine*, 3406 (1835) & 4915 (1856); *Edwards's Botanical Register* **32**: t. 33 (1846); Polunin & Stainton & Stainton, Flowers of the Himalaya, t. 40 (1985).

Plant to 30 cm. Leaves usually deciduous except for a few especially small ones produced in autumn. Stalks of larger leaves often with the sheath not reaching to within 1 cm of base of blade. Leaf-blade circular to broadly ovate, with rounded or cordate base, margin hairy, surface hairy (forma **ciliata**) or hairless (forma **ligulata** Yeo). Panicle as long as or longer than the stalk, loose, sparingly branched, without stalked glands; branches green to pink. Flowers erect or upwardly inclined, broadly bell-shaped or cup-shaped. Perigynous zone and calyx together 7–12 mm, more or less pink. Petals 1.1–1.8 × 0.7–1.8 cm, broadly obovate or circular, white, becoming red in age. *Himalaya, Assam.* H2. Spring.

6. B. stracheyi (Hooker & Thomson) Engler. Illustration: *Edwards's Botanical Register* **29**: t. 65 (1843); Blatter, Beautiful flowers of Kashmir **1**: 126 (1927); Rau, Illustrations of west Himalayan wild flowering plants, 15 (1964); *Kew Bulletin* **20**: 143 (1966); Polunin & Stainton, Flowers of the Himalaya, t. 40 (1985).

Plant to 25 cm. Leaf-stalk with sheath often reaching on to blade and always to within 1 cm of it. Leaf-blade 6–20 × 3–10 cm, obovate, base more or less wedge-shaped, rarely as much as four-fifths as wide as long, margin hairy, convex above. Panicle to *c.* half as long as stalk, compact, usually with stalked glands; branches usually green. Flowers nodding, bell-shaped. Perigynous zone and calyx together 0.8–1.3 cm, green. Petals 1–1.5 × 0.6–0.8 cm, obovate or spoon-shaped, white, becoming pale pink. *E Afghanistan, former USSR (Tadzhikistan), W Himalaya.* H2. Spring.

B. × spathulata Guillaumin; *B. ciliata × B. stracheyi.* Evergreen but has some leaves with the sheath not reaching to within 1 cm of the base of the blade. It is most commonly represented by 'Gambol', dwarf, with smooth, neat, obovate, sometimes glossy foliage; inflorescence softly hairy.

6. DARMERA Voss
M.F. Gardner
Perennial herb to 65 cm, rhizomes stout, tips clothed by broad stipular leaf-sheaths.

Leaves basal, almost circular, peltate, 5–40 cm across, with 7–15 shallow lobes, each lobe cut and sharply toothed, leaf-stalks 10–150 cm. Flowering-stems 1–20 cm, with long hairs and shorter glandular hairs; flowers borne in many-flowered corymbs, calyx lobes 5, 2.5–3.5 mm; petals white or pale pink, oblong-elliptic to obovate, 4.5–7 mm. Follicles purplish, 6–10 mm, joined with calyx for 1–2 mm.

A genus of a single species native to W North America. A majestic plant for a cool, moist site in dappled shade.

1. D. peltata (Torrey) Voss (*Peltiphyllum peltatum* (Torrey) Engler). Illustration: *Botanical Magazine*, 6074 (1874); RHS dictionary of gardening, 1517 (1951); Hitchcock et al., Flora of the Pacific northwest, 192 (1987); Brickell, RHS gardeners' encyclopedia of plants and flowers, 197 (1989).

British Columbia to N California. H2. Spring.

7. SAXIFRAGA Linnaeus
R.J. Gornall

Perennial, biennial or annual herbs, erect or forming mats or cushions. Leaves alternate (rarely opposite). Flowers solitary or in branched, usually cyme-like inflorescences. Sepals 5. Petals 5. Stamens usually 10, rarely 8, inserted at the junction of the ovary wall and floral-tube. Ovary usually of 2 carpels, with separate styles.

A genus of about 440 species, mostly from N temperate and arctic areas, with a few species as far south as Thailand, Ethiopia and the Andes south to Tierra del Fuego. The richest areas are the mountains of Europe, western North America, the Himalayan-Tibetan area and eastern Asia. Most species require shade from the midday sun, good drainage, a high atmospheric humidity and ample watering in the growing season but less so at other times. Species of woodland or other shady habitats are usually best grown in sandy loam with a mulch of organic material, and a more even supply of water. Species vary as to their preferred pH; this is indicated in the text where appropriate. In cultivation, most lime-loving species grow well in standard 'alpine' compost. Propagation may be readily effected by seed, or, in the case of cushion or mat-forming species, by cuttings consisting of the apical centimetre or two of a leafy

shoot. Full cultivation details may be found in Köhlein (1984) and Harding (1992).

The arrangement followed here is based on Gornall (1987), which recognises 15 sections. By far the most horticulturally important are sections *Porphyrion* (the 'Kabschia' and 'Engleria' saxifrages), *Ligulatae* (the 'silver' or 'encrusted' saxifrages) and *Saxifraga* (which includes the 'mossy' saxifrages). Accounts of the first and last of these, which are by far the best represented in gardens, in addition make use of the ranks of subsection and series where appropriate in an attempt to indicate groups of apparently closely related species. The accounts of the European species are based on the work of Professor D.A. Webb, of Trinity College, Dublin.

In gardens hybrids are commoner than species but only brief accounts of these are given here. Details of the many hybrids and other cultivars in section *Porphyrion* are given in the comprehensive account by Horny et al. (1986); descriptions of many of these hybrids and also of those from sections *Ligulatae* and *Saxifraga* can also be found in Köhlein (1984) and Harding (1992).

Literature: Engler, A. & Irmscher, E., Saxifragaceae-Saxifraga, Das Pflanzenreich **67**: 1–448 (1916) & **69**: 449–709 (1919); Harding, W., Saxifrages. The genus Saxifraga in the wild and in cultivation (1970); Köhlein, F., Saxifrages and related genera (1984); Horny, R. et al., Porophyllum saxifrages (1986); Gornall, R.J., An outline of a revised classification of Saxifraga L., *Botanical Journal of the Linnean Society* **95**: 273–92 (1987); Webb, D.A. & Gornall, R.J., Saxifrages of Europe (1989); Harding, W., Saxifrages. A gardeners' guide to the genus (1992).

Key to groups

1a. Leaves hard or stiff and usually encrusted with at least some lime, without a distinct stalk, margins apparently entire (though sometimes fringed with tooth-like hairs or minute serrations) **Group A**
  b. Leaves soft, fleshy or herbaceous, not leathery, hardly ever encrusted with lime, often with a distinct stalk and a lobed blade 2
2a. Ovary at least one third inferior **Group B**
  b. Ovary superior to less than one third inferior 3

3a. Petals yellow or orange, rarely purple **Group C**
  b. Petals whitish or pink, sometimes with red, green or yellowish spots **Group D**

Group A

1a. Petals bright yellow or orange 2
  b. Petals white, pink, purple, red, pale creamy yellow or greenish 13
2a. Calyx bell- or urn-shaped, enclosing and hiding the petals **79. corymbosa**
  b. Calyx cup- or saucer-shaped, not hiding the petals 3
3a. Larger leaves at least 7 mm across **98. mutata**
  b. Larger leaves to 6 mm across 4
4a. Leaves fleshy, rather soft, usually with only one hydathode **100. aizoides**
  b. Leaves usually stiff and hard, with 3–9 hydathodes 5
5a. Stamens at least as long as petals; apex of leaves with a short point 6
  b. Stamens shorter than petals; apex of leaves with a short point or rounded 12
6a. Leafy shoots forming secondary rosettes of leaves on the stem above the primary basal rosettes 7
  b. Leafy shoots forming primary basal rosettes only 8
7a. Flowering-stem, stalks and perigynous zone hairless **51. subverticillata**
  b. Flowering-stem, stalks and perigynous zone glandular-hairy **52. scleropoda**
8a. Lowest flower-stalks 6–10 mm; leaves soft, recurved at the tip **49. kotschyi**
  b. Lowest flower-stalks absent or to 6 mm; leaves hard, straight 9
9a. Leaves oblong, mucronate; inflorescence narrowly oblong **50. pseudolaevis**
  b. Leaves linear to lanceolate, apiculate; inflorescence spherical to broadly oblong 10
10a. Flower-stalks absent **48. desoulavyi**
  b. Flower-stalks present but short 11
11a. Flowering-stem, flower-stalks and perigynous zone glandular-hairy **46. juniperifolia**
  b. Flowering-stem, flower-stalks and perigynous zone hairless **47. sancta**

12a. Inflorescence usually with not more than 6 flowers; most leaves with a short, straight point; petals notched **45. aretioides**

b. Inflorescence with more than 6 flowers; leaves obtuse or with an incurved point; petals entire **44. ferdinandi-coburgi**

13a. Leaves mostly opposite 14

b. Leaves alternate 20

14a. Margins of the leaf-pairs confluent, base of each leaf without tooth-like hairs 15

b. Margins of the leaf-pairs meeting at an acute angle, base of each leaf with tooth-like hairs 18

15a. Sepals without marginal hairs **84. retusa**

b. Sepals with marginal hairs, at least at the base 16

16a. Leaves nearly circular, rather soft and fleshy **85. biflora**

b. Leaves obovate to oblong, hard and rigid 17

17a. Petals usually pink to purple; leaves at least 5 mm **83. oppositifolia**

b. Petals white; leaves to 5 mm **86. vacillans**

18a. Leaves more than 3 mm **88. georgei**

b. Leaves to 2.5 mm 19

19a. Hairs on flowering-stem with pale red glands **89. alpigena**

b. Hairs on flowering-stem with white glands **87. quadrifaria**

20a. Flowering-stems obsolete or nearly so (less than 1 cm) 21

b. Flowering-stems longer than 1 cm 27

21a. Leaves not markedly thickened or truncate at the tip **76. lowndesii**

b. Leaves thickened and truncate at the tip 22

22a. Leaves with 3–5 equal-sized chalk-glands set in the margin of thickened tip; flowers often more than 1 **61. clivorum**

b. Leaves with 1 chalk-gland (sometimes with a much smaller one on each side of it) set in the margin of thickened tip; flowers usually solitary 23

23a. Tip of leaf fringed with tooth-like hairs **75. hypostoma**

b. Tip of leaf without hairs 24

24a. Sepals broader than long; petals broadly obovate to nearly circular, overlapping 25

b. Sepals at least as long as broad; petals obovate, barely overlapping 26

25a. Leafy shoots forming a loose cushion; petals a bright rose-lilac, entire **76. lowndesii**

b. Leafy shoots forming a dense cushion; petals creamy white, sometimes lobed at the tip **73. lolaensis**

26a. Sepals with 1 chalk-gland, slightly recurved; petals 3.5–4.5 mm across **74. matta-florida**

b. Sepals without chalk-glands, straight; petals 2–3 mm across **72. pulvinaria**

27a. Calyx bell- or urn-shaped, enclosing and hiding the petals 28

b. Calyx cup- or saucer-shaped, not hiding the petals 33

28a. Petals yellow **79. corymbosa**

b. Petals pink or purple 29

29a. Most flowers without stalks; inflorescence not branched 30

b. Most flowers clearly with stalks; inflorescence often branched 32

30a. Leaves linear to linear-oblong, not widened at the tip **82. sempervivum**

b. Leaves obovate to spathulate, widened near the tip 31

31a. Leaves 4–10 × 2–3 mm; inflorescence dull pink, sometimes tinged with green, usually bearing 4–7 flowers **80. porophylla**

b. Leaves 12–35 × 3–8 mm; inflorescence bright red to dark crimson, bearing 10–20 flowers **81. federici-augusti**

32a. Inflorescence sparingly branched, with 1 or 2 flowers on each primary branch **77. media**

b. Inflorescence freely branched, with 3 or 4 flowers on each primary branch **78. stribrnyi**

33a. All leaves less than 1.5 cm 34

b. Larger leaves at least 1.5 cm 61

34a. Leaves recurved in the upper half 35

b. Leaves straight 51

35a. Flowering-stems usually with a solitary flower 36

b. Flowering-stems usually with more than one flower, at least in bud 38

36a. Leaf-tips rounded or almost acute **83. oppositifolia**

b. Leaf-tips mucronate (shortly pointed) 37

37a. Leafy shoots more than 10 cm, forming a loose cushion; leaves at least 5 mm **54. poluniniana**

b. Leafy shoots less than 10 cm, forming a dense cushion; leaves mostly less than 5 mm **53. lilacina**

38a. Leaves glandular-hairy over the whole lower surface and at least half the upper **59. spruneri**

b. Leaves usually hairless on both surfaces, but if hairs are present then they are non-glandular 39

39a. Leaves linear or oblong, 4–5 times as long as broad, acutely pointed 40

b. Leaves obovate or spathulate, rarely linear, less than 4 times as long as broad, mucronate to obtuse 43

40a. Leaves 2–4 mm across 41

b. Leaves to 2 mm across 42

41a. Leaves tapered gradually to a point; petals white or pale pink, rarely crimson **56. scardica**

b. Leaves tapered more abruptly to a point; petals deep rose **63. rhodopetala**

42a. Leaves 10–12 mm; petals 8–10 mm **65. cinerea**

b. Leaves c. 4 mm; petals 4–6 mm **66. caesia**

43a. Leaves with thickened tips in which are set the chalk-glands 44

b. Leaf-tips not thickened 45

44a. Leaves heavily encrusted with lime; petals 5–9 × 3–7 mm **64. stolitzkae**

b. Leaves with only traces of lime; petals 3–6.5 × 1.5–4 mm **62. andersonii**

45a. Petals at least 5 mm 46

b. Petals to 6 mm 50

46a. Leaves usually 5–15 mm, if shorter then petals without a notch at the tip 47

b. Leaves 3–4 mm; petals often with a notch at the tip 49

47a. Flowering-stems at least 2.5 cm 48

b. Flowering-stems less than 2.5 cm **60. alberti**

48a. Leaf-margins fringed with hairs only in the basal half **55. marginata**

b. Leaf-margins fringed with hairs from base to mid-way or more **56. scardica**

49a. Petals obovate, usually with a notch at the tip, fading to deep purple at the base **57. iranica**

b. Petals broadly obovate, without a notch at the tip, fading to pink **58. wendelboi**

50a. Leaves curved outwards from near the base; upper part of flowering stem hairier than lower part **66. caesia**

b. Leaves curved outwards only from near the tip; upper part of flowering-stem less hairy than lower part
67. squarrosa

51a. Tips of leaves tapered to a sharp point   52

   b. Tips of leaves obtuse   57

52a. Leaves to 5 mm   53

   b. Leaves at least 6 mm   54

53a. Leaf-points sharply incurved
69. tombeanensis

   b. Leaf-points straight
68. diapensioides

54a. Flowers solitary, rarely 2 per stem
70. burseriana

   b. Flowers 3–14 per stem   55

55a. Leaves deep green, not clearly encrusted with lime, lanceolate
71. vandellii

   b. Leaves glaucous, usually clearly lime-encrusted, oblong to obovate-elliptic   56

56a. Middle part of leaf-margin entire
55. marginata

   b. Middle part of leaf-margin finely toothed   **56. scardica**

57a. Petals 7–12 mm   58

   b. Petals 3–6 mm   60

58a. Flower-stalks hairless, or very nearly so   **92. cochlearis**

   b. Flower-stalks densely glandular-hairy   59

59a. Leaves usually more than 1.5 mm across, broadest above the middle
55. marginata

   b. Leaves usually less than 1.5 mm across, broadest at or below the middle   **68. diapensioides**

60a. Leaves finely toothed in the upper half; chalk-glands confined to near the margins of the leaves
95. paniculata

   b. Leaves entire in the upper half; chalk-glands scattered over the upper surface of the leaves
97. valdensis

61a. Leaf-rosette solitary, without offsets; plant dying after flowering   62

   b. Leaf-rosettes producing offsets which persist after the flowering rosette dies, thereby forming a clump   63

62a. Leaves glaucous, lime-encrusted, tip almost acute; petals white
90. longifolia

   b. Leaves dull green, not obviously lime-encrusted, mucronate; petals dull pink   **99. florulenta**

63a. Flowering-stem to 10 cm, bearing up to 8 flowers   **55. marginata**

   b. Flowering-stem taller than 10 cm, bearing more than 10 flowers   64

64a. Leaves entire to scalloped; chalk-glands embedded in leaf-margin itself   65

   b. Leaves toothed; chalk-glands next to teeth, on upper leaf-surface   67

65a. Leaves spathulate, obtuse
92. cochlearis

   b. Leaves linear to oblong, if a little expanded near the tip, then almost acute   66

66a. Leaves 2.5–7 mm across; lower branches of inflorescence bearing at least 5 flowers   **91. callosa**

   b. Leaves 2–3 mm across; lower branches of inflorescence bearing not more than 4 flowers   **93. crustata**

67a. Panicle occupying at least half the flowering-stem; lower primary branches bearing at least 13 evenly spaced flowers   **96.cotyledon**

   b. Panicle occupying less than half the flowering-stem; lower primary branches bearing up to 12 flowers crowded at the tips   68

68a. Basal leaves curving downwards near the tip, to make a flat or convex leaf-rosette; primary branches of the inflorescence usually with 4–12 flowers   **94. hostii**

   b. Basal leaves tending to curve upwards, to make a concave or almost hemispherical leaf-rosette; primary branches of the inflorescence usually with 1–3 flowers   **95. paniculata**

Group B

1a. Bulbils present, at or below ground level, in the axils of the basal leaves   2

   b. Bulbils absent from axils of the basal leaves   11

2a. Petals glandular-hairy on the upper surface, at least near the base   3

   b. Petals hairless   6

3a. Bulbils present in the axils of the stem-leaves and bracts
120. bulbifera

   b. Bulbils present only in the axils of the lower leaves   4

4a. Flowering-stem usually branched from the middle or below, bearing a diffuse inflorescence
117. haenseleri

   b. Flowering-stem usually branched only in the top quarter, bearing a compact inflorescence   5

5a. Basal leaves deeply divided into linear-oblong or oblanceolate lobes
118. dichotoma

   b. Basal leaves scalloped or somewhat pinnately lobed, but not deeply so
119. carpetana

6a. Basal leaves deeply divided into 3 primary lobes, which are narrowed or stalked at the base   7

   b. Basal leaves divided up to two-thirds of the way to the base; lobes not narrowed or stalked at the base   8

7a. Petals 12–20 mm   **121. biternata**

   b. Petals 5–8 mm   **122. bourgaeana**

8a. Petals 2–6 mm   9

   b. Petals 7–16 mm   10

9a. Basal leaves semicircular or kidney-shaped, 9 mm or more across; ovary not fully inferior   **114. rivularis**

   b. Basal leaves wedge-shaped, 4–9 mm across; ovary fully inferior
117. haenseleri

10a. Leaves 3-lobed; flowering-stem branched usually from near the base
116. corsica

   b. Leaves scalloped rather than lobed; flowering-stem usually branched from above the middle
115. granulata

11a. Carpels united to above the middle at fruiting stage; flowering stems usually with at least one leaf   12

   b. Carpels separated nearly to the base at fruiting stage; flowering-stems usually leafless   76

12a. Leaves linear-oblong, fleshy; flowers yellow or dark red   **100. aizoides**

   b. Leaves not fleshy; flowers white, cream or greenish yellow   13

13a. Leaves all entire   14

   b. Some leaves toothed, scalloped or lobed   25

14a. Petals conspicuous, longer than sepals, usually pure white   15

   b. Petals inconspicuous, barely longer than sepals, dull in colour   24

15a. Leafy shoots with dormant axillary buds conspicuous at flowering time   16

   b. Leafy shoots without conspicuous, summer-dormant, axillary buds   17

16a. Leaves obovate, obtuse; petals turning pink after pollination
155. erioblasta

   b. Leaves linear-oblong, apiculate; petals staying white after pollination
153. conifera

17a. Leaves hairless except for a few on the margin   18

b. Leaves more or less hairy on the surface 20

18a. Leaves 3–6 mm across; flowering stems with 1 or 2 leaves and 1–3 flowers **159. androsacea**

b. Leaves 1–2 mm across; flowering-stems with 4–7 leaves and 3–9 flowers 19

19a. Leaves and sepals apiculate **164. tenella**

b. Leaves and sepals obtuse **163. glabella**

20a. Petals 2–3 mm; plant annual **165. tridactylites**

b. Petals 4–7 mm; plant perennial 21

21a. Leaves usually lime-encrusted, firm, with a translucent margin; inflorescence with 5–12 flowers **59. spruneri**

b. Leaves not lime-encrusted, soft, without a translucent margin; inflorescence with 1–4 flowers 22

22a. Leafy shoots to 10 cm, crowded into a compact cushion; leaves mostly narrower than 2.5 mm; petals touching **161. muscoides**

b. Leafy shoots shorter and fewer, forming a low mat; leaves mostly wider than 2.5 mm; petals not touching 23

23a. Leaves 3–6 mm across; flowering-stems with 1–3 flowers **159. androsacea**

b. Leaves 6–9 mm across; flowering-stems with 3–7 flowers **160. depressa**

24a. Petals gradually tapered to a narrow base, tip notched **162. presolanensis**

b. Petals not tapered to a narrow base, tip rounded **149. exarata**

25a. Petals greenish, cream or dull yellow 26

b. Petals white, pink or bright red 30

26a. Larger leaves more than 6 mm across, usually wider than long 27

b. Larger leaves to 6 mm across, usually longer than wide 28

27a. Blade of leaves covered in cobweb-like hairs **128. arachnoidea**

b. Blade of leaves hairless, or nearly so **129. paradoxa**

28a. Leaf-segments furrowed on the upper surface 29

b. Leaf-segments flat on the upper surface **149. exarata**

29a. Leaf-segments mucronate, more or less hairless **150. hariotii**

b. Leaf-segments obtuse or almost acute, glandular-hairy **149. exarata**

30a. Mature leaves hairless, though often with stalkless glands 31

b. Mature leaves hairy, at least on the margin or leaf-stalk (hairs may be visible only with a lens) 38

31a. Flowering-stem axillary 32

b. Flowering-stem terminal 35

32a. Ultimate leaf-segments not more than 9

b. Ultimate leaf-segments 9–38 34

33a. Petals 10–11 mm **143. portosanctana**

b. Petals 5–8 mm **141. cuneata**

34a. Leaves usually deeply lobed, antler-like **139. trifurcata**

b. Leaves usually shallowly lobed, not antler-like **142. maderensis**

35a. All leaf-segments obtuse or almost acute 36

b. Some leaf-segments mucronate 37

36a. Leaf-segments usually not more than 2 mm across, sides parallel, upper surface furrowed; petals 3.5–5 mm **136. pentadactylis**

b. Leaf-segments usually more than 2 mm across, sides often curved, upper surface flat; petals 7–14 mm **137. fragilis**

37a. Leaf-segments oblong to elliptic, 1.5–2.5 times as long as wide **138. camposii**

b. Leaf-segments linear, 3–6 times as long as wide **140. canaliculata**

38a. Leafy shoots without conspicuous, dormant axillary buds at flowering time 39

b. Leafy shoots with conspicuous, dormant axillary buds at flowering time 68

39a. Petals prominently notched at the tip, often unequal 40

b. Petals barely notched, if at all, equal 41

40a. Basal leaves divided almost to the base; petals 8–11 mm, touching; biennial **126. petraea**

b. Basal leaves divided not more than halfway to the base; petals 4–8 mm, not touching; perennial **127. berica**

41a. Perennial 42

b. Annual or biennial 65

42a. Petals at least 7 mm 43

b. Petals usually not more than 7 mm 48

43a. Ultimate leaf-segments more than 11 44

b. Ultimate leaf-segments less than 11 46

44a. Flowering-stems 2–5 mm across; aquatic **123. aquatica**

b. Flowering-stems less than 2 mm across; not aquatic 45

45a. Stems forming a compact tuft, herbaceous at the base **124. irrigua**

b. Stems spreading to form a loose cushion, woody at the base **132. geranioides**

46a. Plant with prostrate, non-flowering-shoots bearing at least some entire leaves **151. hypnoides**

b. Plant without prostrate, non-flowering-shoots 47

47a. Petals more than 1 cm **131. pedemontana**

b. Petals not more than 1 cm **144. rosacea**

48a. Leaf-segments furrowed on the upper surface 49

b. Leaf-segments flat on the upper surface 54

49a. Hairs on leaves and calyx with stalks barely longer than the diameter of the gland, giving a finely warty appearance 50

b. Hairs on leaves and calyx with stalks much longer than the diameter of the gland, giving a downy appearance 52

50a. Larger leaves 2–3 cm, in cushions to 30 cm across **133. moncayensis**

b. Larger leaves not more than 15 mm, in cushions rarely more than 10 cm across 51

51a. Stems woody below; leaves dark green; forming a loose cushion; petals 4–5 mm across **135. intricata**

b. Stems scarcely woody below; leaves fresh green; forming a compact cushion; petals to 4 mm across **149. exarata**

52a. Leaves with short, glandular hairs, sometimes sparse; petals about twice as long as wide, not touching **149. exarata**

b. Leaves with a dense cover of long, glandular hairs; petals about 1.5 times as long as wide, usually touching 53

53a. Leaves dark green; rosettes often 2–2.5 cm across; stem-leaves usually lobed **146. pubescens**

b. Leaves fresh green; rosettes not more than 1.5 cm across; stem-leaves usually entire **148. cebennensis**

54a. Stems and leaf-stalks covered with sticky, cobweb-like hairs 5–10 mm **128. arachnoidea**

b. Stems and leaf-stalks with much shorter hairs 55

55a. Leaves with broadly triangular or ovate lobes; leafy shoots short, forming a tuft or mat   56
  b. Leaves with lanceolate, oblong or elliptical lobes; leafy shoots rather long, forming a cushion or rarely a mat   58
56a. Leaves at least 9 mm across with a distinct leaf-stalk   **114. rivularis**
  b. Leaves to 9 mm across without a distinct leaf-stalk   57
57a. Leaves 3–6 mm across; flowering-stems with 1–3 flowers   **159. androsacea**
  b. Leaves 6–9 mm across; flowering-stems with 3–7 flowers   **160. depressa**
58a. Petals more than 5 mm   59
  b. Petals to 5 mm   63
59a. Hairs on the leaves much less than 0.5 mm; foliage with a strong spicy scent   **134. vayredana**
  b. Hairs on the leaves mostly more than 0.5 mm; foliage scarcely scented   60
60a. Leaves with hairs mostly or entirely non-glandular   **144. rosacea**
  b. Leaves with numerous glandular hairs   61
61a. Flowers not more than 10 mm across; sepals and perigynous zone crimson; petals often with red veins   **147. nevadensis**
  b. Flowers at least 10 mm across; sepals and perigynous zone mainly green; petals without red veins   62
62a. Petals 5.5–6.5 mm, usually tinged with green or yellow; leaves mostly 3-lobed   **145. cespitosa**
  b. Petals usually more than 6.5 mm, pure white; some leaves usually with 5 or more lobes   **144. rosacea**
63a. Leafy shoots erect and sometimes very short   64
  b. Leafy shoots prostrate or ascending, usually long   **130. praetermissa**
64a. Sepals and perigynous zone crimson; petals often with red veins; anthers red or orange before shedding pollen   **147. nevadensis**
  b. Sepals and perigynous zone greenish; petals without red veins; anthers yellow before shedding pollen   **149. exarata**
65a. Basal leaves kidney-shaped, with a long, well-defined leaf-stalk   **125. latepetiolata**
  b. Basal leaves oblanceolate, with a short, ill-defined leaf-stalk   66

66a. Flower-stalks to 1 cm in fruit   **166. adscendens**
  b. Flower-stalks 1–2 cm in fruit   67
67a. Petals 2.5–3 mm; annual   **165. tridactylites**
  b. Petals c. 5 mm; biennial   **167. blavii**
68a. Petals 1.1–2 cm   69
  b. Petals 4–10 mm   72
69a. Leaves deeply divided, with 3 primary lobes narrowed to stalk-like bases   **121. biternata**
  b. Leaves palmately lobed, the lobes not narrowed at the base   70
70a. Outer leaves of axillary buds lobed, leaf-like in colour and texture, without cobweb-like hairs   **131. pedemontana**
  b. Outer leaves of axillary buds entire, more or less papery, fringed with cobweb-like hairs   71
71a. Leaves deeply divided into linear-oblong lobes   **154. rigoi**
  b. Leaves divided, half to three-quarters of the way to the base, into lobes that are wedge-shaped at the base   **158. maweana**
72a. Leaf-segments mucronate or apiculate   73
  b. Leaf-segments obtuse to acute   74
73a. Dormant axillary buds without stalks, their outer leaves partly green   **151. hypnoides**
  b. Dormant axillary buds on stalks, their outer leaves papery and translucent   **152. continentalis**
74a. Dormant axillary buds clothed in white, woolly hairs   **155. erioblasta**
  b. Dormant axillary buds hairy, but the hairs not white and woolly   75
75a. Sepals c. 4 × 3 mm; petal-margins recurved; flowering-stem with 1–3 flowers   **156. reuteriana**
  b. Sepals c. 2 × 1.5 mm; petals flat; flowering-stem with at least 4 flowers   **157. globulifera**
76a. Petals purplish crimson at the sides, green elsewhere   **33. hieracifolia**
  b. Petals cream or white   77
77a. Leaves sparingly toothed to almost entire; reddish brown hairs absent from base of stem and leaf-stalks   **32. pensylvanica**
  b. Leaves scalloped to sharply toothed; reddish brown hairs often present at the base of the stem or leaf-stalks   78
78a. Petals with 2 yellow spots near the base, equalling or shorter than the sepals   **29. reflexa**
  b. Petals unspotted, equalling or longer than the sepals   79

79a. Reddish brown hairs present at least in the leaf axils and at the base of the leaf-stalk   80
  b. Reddish brown hairs absent, replaced by chaffy white hairs   82
80a. Leaves evenly and sharply toothed, often densely covered with reddish brown hairs beneath; inflorescence more or less flat-topped   **31. rufidula**
  b. Leaves usually scalloped or coarsely and unevenly toothed; reddish brown hairs mostly confined to the axil and base of the leaf-stalk, sometimes sparingly so; inflorescence pyramidal or or rounded   81
81a. Filaments linear   **26. nivalis**
  b. Filaments club-shaped   **30. occidentalis**
82a. Petals 2–3 mm, less than twice as long as the sepals   **27. tenuis**
  b. Petals 3–6 mm, 2–3 times as long as the sepals   **28. virginiensis**

Group C

1a. Leaves lobed or toothed   2
  b. Leaves entire (but margins may be fringed with stout hairs)   5
2a. Leafy summer-dormant buds in the axils of the stem-leaves   **7. strigosa**
  b. Stem-leaves without leafy summer-dormant buds   3
3a. Sepals reflexed in fruit   **17. sibthorpii**
  b. Sepals upright or spreading in fruit   4
4a. Petals not more than 3 mm, white or pale yellow   **18. hederacea**
  b. Petals more than 4 mm, bright yellow   **16. cymbalaria**
5a. Slender runners produced from the axils of the leaves   6
  b. Runners absent   9
6a. Petals not overlapping or touching   **13. neopropagulifera**
  b. Petals overlapping or touching   7
7a. Leaves shiny; sepals less than one third length of petals   **14. brunonis**
  b. Leaves matt; sepals more than one third length of petals   8
8a. Runners at most only sparingly glandular-hairy   **11. flagellaris**
  b. Runners densely glandular-hairy   **12. mucronulata**
9a. All non-glandular hairs white or cream, not reddish brown   10
  b. Reddish brown non-glandular hairs present on leaves, leaf-stalks, leaf-axils or stems   14
10a. Leaf-margins fringed with stout hairs or bristles   11

b. Leaf-margins without stout hairs or bristles   13

11a. Flowering-stems to 1 cm   **15. eschscholtzii**

  b. Flowering-stems more than 2 cm   12

12a. Flowers stellate, petals widely separated, golden-yellow   **13. neopropagulifera**

  b. Flowers cup-shaped, petals touching or overlapping, pale yellow   **8. punctulata**

13a. Sepals ascending   **9. serpyllifolia**

  b. Sepals reflexed   **10. chrysantha**

14a. Basal leaves soon deciduous; stem-leaves numerous, most more or less equal in size   **6. cardiophylla**

  b. Basal leaves persistent; stem-leaves progressively reduced in size upwards   15

15a. Blade of basal leaves tapering or rounded at the base, the stalk ill-defined, if present at all   16

  b. Blade of basal leaves cordate at the base, with a well-defined stalk   18

16a. Stems and leafy shoots forming a dense cushion   **3. montana**

  b. Stems and leafy shoots solitary or in a loose mat   17

17a. Veins in the sepals remaining separate   **1. hirculus**

  b. Veins in the sepals converging to a point near the tip   **2. hookeri**

18a. Petals yellow   **4. diversifolia**

  b. Petals orange   **5. pardanthina**

Group D

1a. Flowering-stems leafy   2

  b. Flowering-stems leafless   18

2a. Bulbils present in axils of basal leaves   3

  b. Bulbils absent from axils of basal leaves   7

3a. Bulbils replacing at least some of the flowers   4

  b. Bulbils not replacing any flowers   5

4a. Filaments awl-shaped   **113. cernua**

  b. Filaments club-shaped   **43. mertensiana**

5a. Petals more than 7 mm, white   **112. sibirica**

  b. Petals to 5 mm, off-white or pink   6

6a. Filaments club-shaped   **43. mertensiana**

  b. Filaments awl-shaped   **114. rivularis**

7a. Blade of basal leaves heart-shaped or truncate at the base, with a distinct leaf-stalk   8

  b. Blade of basal leaves tapering or parallel-sided at the base, without a well-defined leaf-stalk   11

8a. Bulbils often replacing some flowers; sepals reflexed in fruit; filaments club-shaped   **43. mertensiana**

  b. Bulbils absent; sepals erect or spreading in fruit; filaments awl-shaped   9

9a. Largest leaves at least 2.5 cm across, with at least 15 scalloped or triangular teeth   **110. rotundifolia**

  b. Largest leaves to 2.5 cm across, with 3–11 scalloped or triangular teeth   10

10a. Plant annual; petals white, without spots   **18. hederacea**

  b. Plant perennial; petals white, with purplish-red spots   **111. taygetea**

11a. Capsule streaked or spotted with purple   12

  b. Capsule green to brown   13

12a. Leaves of basal rosettes more than 1 cm; filaments linear   **19. merkii**

  b. Leaves of basal rosettes to 1 cm; filaments club-shaped   **20. tolmiei**

13a. Axillary buds on leafy shoots conspicuous at flowering time   14

  b. Axillary buds on leafy shoots not conspicuous at flowering time   15

14a. Leaves of non-flowering-shoots straight, much longer than subtended axillary bud; leaves of flowering-stem *c.* 1 cm, spreading   **101. aspera**

  b. Leaves of non-flowering-shoots incurved, barely longer than subtended axillary bud; leaves of flowering-stem not more than 5 mm, almost erect   **102. bryoides**

15a. Leaves usually with 3 teeth at the tip   **105. tricuspidata**

  b. Leaves usually entire   16

16a. Flowers cup-shaped   **8. punctulata**

  b. Flowers saucer-shaped   17

17a. Leaves narrowed gradually to a spine-like tip   **103. bronchialis**

  b. Leaves mucronate at the tip   **104. cherlerioides**

18a. Flowers with irregular or bilateral symmetry   19

  b. Flowers with regular symmetry   25

19a. Blade of basal leaves cordate or truncate at the base, with a slender leaf-stalk   20

  b. Blade of basal leaves tapering or parallel-sided at the base, with a broad, often indistinct leaf-stalk   24

20a. Slender runners produced from the axils of the leaves   21

  b. Runners usually absent   22

21a. Leaves green on upper surface   **42. veitchiana**

b. Leaves marked with grey veins on upper surface   **41. stolonifera**

22a. Upper petals linear-lanceolate, without a claw, unspotted; seeds smooth   **38. fortunei**

  b. Upper petals broadly ovate, clawed at the base, spotted yellow or red; seeds warty   23

23a. Leaves palmately divided   **39. cortusifolia**

  b. Leaves scalloped or shallowly lobed   **40. nipponica**

24a. Lower bracts leaf-like; most basal leaves with 5–10 coarse teeth on each side; flowers not replaced by bulbils   **21. clusii**

  b. Lower bracts small; most basal leaves with shallow teeth on each side; at least some flowers often replaced by bulbils   **22. ferruginea**

25a. Leaf-margin with a narrow, translucent border   26

  b. Leaf-margin without a translucent border   29

26a. Petals rarely with red spots; ovary almost white; leaves usually entire in basal third of blade   **106. cuneifolia**

  b. Petals usually with red spots; ovary pink or green; leaves toothed or scalloped almost to the base of the blade   27

27a. Leaf-stalk slender, more or less cylindric; leaves usually hairy on both surfaces   **109. hirsuta**

  b. Leaf-stalk broad, flattened; leaves hairless at least beneath   28

28a. Leaf-stalk with numerous marginal hairs, usually shorter than the scalloped blade   **107. umbrosa**

  b. Leaf-stalk with few marginal hairs, mostly near the base, often at least as long as the boldly toothed blade   **108. spathularis**

29a. Blade of basal leaves circular to kidney-shaped, base heart-shaped; leaf-stalk well defined, slender   30

  b. Blade of basal leaves lanceolate to obovate, base tapered or truncate; leaf-stalk ill-defined or, if present, then broad   33

30a. At least some flowers replaced by bulbils; leaf-stalks with long white hairs   **43. mertensiana**

  b. Inflorescence without bulbils; leaf-stalks usually only sparingly hairy   31

31a. Petals more than 5 mm, shorter than stamens in mature flowers   **37. manschuriensis**

  b. Petals less than 5 mm, longer than stamens   32

32a. Petals broadly elliptic to circular, with 2 green or yellow spots
    35. odontoloma

  b. Petals ovate to oblong, unspotted or with orange spots **34. nelsoniana**

33a. Leaves densely overlapping on leafy shoots 34

  b. Leaves forming a basal rosette 35

34a. Petals *c.* 3 mm across **8. punctulata**

  b. Petals less than 3 mm across
    15. eschscholtzii

35a. Filaments club-shaped 36

  b. Filaments linear 39

36a. Blade of basal leaves more than 10 cm **25. micranthidifolia**

  b. Blade of basal leaves less than 10 cm 37

37a. Axils of the leaves hairless or with white hairs **36. lyallii**

  b. Axils of the leaves with at least a few reddish brown hairs 38

38a. Petals shorter than or equalling sepals, with 2 yellow spots near the base **29. reflexa**

  b. Petals longer than sepals, unspotted
    30. occidentalis

39a. Carpels a dark, blackish crimson
    24. melanocentra

  b. Carpels pinkish, green or brown 40

40a. Petals less than 4 mm; leaves with numerous, sharp, even teeth, often covered with reddish brown hairs
    31. rufidula

  b. Petals 3–8 mm; leaf-margins scalloped or with 3 or 4 blunt teeth on each side, with whitish hairs at least at the base 41

41a. Leaf-margins scalloped or with numerous teeth; ovary at least partly inferior **28. virginiensis**

  b. Leaves with 3 or 4 blunt teeth on each side; ovary superior
    23. stellaris

Section **Ciliatae** Haworth (section *Hirculus* Tausch). Perennials, usually evergreen, some monocarpic; stems solitary or clumped into mats or cushions. Leaves herbaceous, sometimes stiff or leathery, with or without a distinct stalk, margin usually entire. Chalk-glands absent. Flowering-stems usually leafy, bearing 1–several flowers. Petals usually yellow, with swellings near the base. Ovary superior to half inferior; capsule splitting above the middle.

Although several species belonging to this section have been introduced to cultivation, it seems that many have now disappeared from gardens or are at best extremely rare. The following, however, have been more persistent, or have been reintroduced.

1. S. hirculus Linnaeus. Illustration: Ross-Craig, Drawings of British plants **10**: pl. 3 (1957); Keble Martin, The new concise British flora in colour, pl. 32 (1982); Garrard & Streeter, Wild flowers of the British Isles, 91 (1983); Aeschimann et al., Flora alpina **1**: 691 (2004).

Plant with loose rosettes of basal leaves forming a loose mat. Basal leaves 1–3 cm × 3–6 mm, narrowly lanceolate, entire without a distinct leaf-stalk. Reddish brown, non-glandular hairs present on leaf-bases and leaf-axils. Flowering-stem to 3 5 cm, leafy, bearing 1–4 flowers in a loose corymb. Petals 9–16 × 4–6 mm, elliptic to obovate-oblong, bright yellow. Ovary superior. *Circumboreal, Colorado, Romania & Caucasus.* H2. Late summer–autumn.

Several subspecies are recognised but none is common in cultivation. Difficult to grow; needs well-irrigated conditions, such as in a pan of moss.

2. S. hookeri Engler & Irmscher.
Like *S. hirculus* but the veins in the sepals converging to a point near the tip, rather than remaining separate. *E Himalaya, Bhutan, Sikkim, China (S Xizang).* H2. Late summer–autumn.

3. S. montana H. Smith.
Plant forming a cushion. Basal leaf blades 5–15 × 2–4.5 mm, elliptic-lanceolate, tapering to a stalk. Flowering-stems 3–25 cm, leafy, reddish brown, non-glandular hairs densely present especially on upper part. Flowers 1–few. Petals 6–10 × 3–4 mm, elliptic or obovate, yellow. Ovary superior. *China (Yunnan, Sichuan), Bhutan, Sikkim.* ?H2. Summer.

Probably best in a well-drained peaty soil.

4. S. diversifolia Seringe. Illustration: Köhlein, Saxifrages and related genera, 157 (1984).

Plant upright, forming a clump, with a cluster of long-stalked basal leaves with reddish brown, non-glandular hairs; blade 1.5–3 cm, almost as wide, heart-shaped, margin entire. Stem-leaves diminishing in size and with progressively shorter leaf-stalks up the flowering-stem. Flowering-stem 20–30 cm, branched above the middle to form a panicle of numerous dull or golden yellow flowers, each to 2 cm across. Ovary superior. *W China, Kashmir to Bhutan.* H3. Late summer–autumn.

Prefers an acid soil rich in humus in partial shade.

5. S. pardanthina Handel-Mazzetti.
Like *S. diversifolia* but flowering-stems to 30 cm. Petals deep orange, marked with red. *SW China.* H3? Late summer–autumn.

6. S. cardiophylla Franchet.
Plant upright, forming a clump. Basal leaves soon deciduous; stem-leaves 2–2.5 × 1–1.5 cm, broadly ovate, base cordate. Reddish brown, non-glandular hairs at base of leaf-stalks. Inflorescence more or less flat-topped, sparingly branched with only 4–8 orange-yellow flowers. Ovary superior. *Sikkim, China (Sichuan, Yunnan).* H3? Late summer–autumn.

Prefers an acid soil rich in humus in partial shade.

7. S. strigosa Wallich.
Plant upright. Basal leaves soon falling; stem-leaves forming a rosette *c.*one third of the way up the stem; leafy summer dormant buds in the axils of the leaves; leaves toothed at the tip, with long, white, non-glandular hairs adpressed on both surfaces. Flowering-stem bracteate, bearing a solitary (rarely more) flower. Petals 5–6 × 2–3 mm, buttercup-yellow, sometimes white marked with yellow or red. Sepals reflexed. *N India, W China to upper Burma.* H3? Summer–autumn.

Prefers a moist acid soil rich in humus, in partial shade.

8. S. punctulata Engler. Illustration: Harding, Saxifrages, 11 (1992).

Plant forming a dense cushion. Basal leaves 4–6 × 2–3 mm, spathulate, without a distinct stalk. Flowering-stems 2–5 cm, leafy, glandular-hairy, bearing 1–6 flowers. Petals 6–8 × 3–4 mm, white with 2 yellow blotches or pale yellow, and with several red (or black?) spots. *China (Xizang), Nepal, Sikkim.* H2? Late summer–early autumn.

9. S. serpyllifolia Pursh. Illustration: Harding, Saxifrages, 17 (1992).

Plant forming a mat, with non-flowering, procumbent leafy shoots produced from axils of basal or lower stem-leaves. Basal leaves 3–9 mm, spathulate, entire, without a distinct leaf-stalk, hairless except for margins with white, non-glandular hairs. Flowering-stems 2–8 cm, leafy, bearing a solitary flower. Sepals ascending. Petals 4–7 mm, obovate with a short claw, yellow or rarely purple. *Arctic*

Siberia, Japan, W North America. H2.
Summer.

10. S. chrysantha A. Gray

Like *S. serpyllifolia* but leaves almost
hairless; sepals slightly more oblong and
reflexed and petals golden. *USA (Rocky
Mts)*. H2. Summer.

11. S. flagellaris Willdenow. Illustration:
Botanical Magazine, 4261 (1851); Evans,
Alpines '81, 36 (1981); Webb & Gornall,
Saxifrages of Europe, pl. 1 (1989).

Evergreen perennial, with prominent
rosettes of basal leaves, from the axils of
which emerge thread-like runners that
terminate in a leafy bud. Rosette-leaves
7–16 × 2–5 mm, narrowly lanceolate to
elliptic-obovate, margin fringed with stout
hairs. Flowering-stems *c.* 6 cm, leafy,
glandular-hairy, bearing a solitary flower,
or sometimes a loose cyme of 1–4 flowers.
Petals 6–10 mm, obovate, bright yellow.
Ovary superior to almost superior.
*Circumpolar Arctic, W North America (Rocky
Mts), Caucasus & Himalaya to Baikal*. H1.
Summer.

Eight to 10 subspecies can be
recognised, of which the following may be
seen in cultivation.

Subsp. **flagellaris**. Runners without
glandular hairs. Stem-leaves longer than
the internodes. Sepals narrowly triangular.
Ovary superior, with a shallow perigynous
zone. *Caucasus*.

Subsp. **platysepala** (Trautvetter) Porsild.
Runners sparingly glandular-hairy. Stem-
leaves longer than the internodes. Sepals
ovate-triangular. Ovary almost superior,
with a bowl-shaped perigynous zone.
Circumpolar Arctic.

Subsp. **setigera** (Pursh) Tolmatchev.
Runners with glandular hairs. Stem-leaves
shorter than the internodes. Sepals
narrowly triangular. Ovary superior, with
a shallow perigynous zone. *C & E Asia,
W North America*.

Subsp. **crandallii** (Gandoger) Hultén.
Like subsp. *platysepala* but with narrower,
erect petals, and a smaller, funnel-shaped
perigynous zone with white, glandular
hairs in addition to the usual black ones.
W North America.

Best grown in a cool glasshouse in well-
drained medium, dryish in winter but
wetter in the spring/summer growing
season.

12. S. mucronulata Royle. Illustration:
Harding, Saxifrages, 17 (1992) - as *S.
flagellaris* subsp. *sikkimensis*.

Like *S. flagellaris* but runners densely
glandular-hairy; flowers 3–10, funnel-
shaped, petals only twice as long as the
sepals. *E Himalaya, China (S Xizang)*. H2?
Summer.

Cultivation as for *S. flagellaris*. Most
plants in cultivation belong to subsp.
sikkimensis (Hultén) Hara.

13. S. neopropagulifera Hara.
Illustration: Harding, Saxifrages, 18 (1992)

Like *S. flagellaris* but flowering-stems
8–12 cm, bearing solitary, stellate flowers
in which the golden yellow petals do not
overlap. Runners produced in the wild but
not so far seen in cultivation. *Nepal*. H2?
Late spring.

Cultivation as for *S. flagellaris*.

14. S. brunonis Seringe (*S. brunoniana*
Wallich, invalid). Illustration: *Botanical
Magazine*, 8189 (1908); Harding, The
genus Saxifraga, 126 (1970).

Like *S. flagellaris* but stem nearly
hairless, leaves shiny, runners hairless,
flowers more numerous, sepals shorter,
2–2.5 mm, and only one quarter to one
third the length of the petals. *Kashmir to
Bhutan & China (S Xizang)*. H3. Summer.

Cultivation as for *S. flagellaris*.

15. S. eschscholtzii Sternberg.
Illustration: Hultén, Flora of Alaska and
neighbouring territories, 566 (1968).

Plant forming a compact cushion of leafy
shoots. Leaves 1–3 mm, oblong to obovate,
without a distinct leaf-stalk, densely
overlapping, the margin fringed with stout
bristles. Flowering-stems to 1 cm, leafless,
bearing a solitary flower. Petals yellow,
white or pink, soon falling. Dioecious. *E
Siberia to Yukon*. H1. Summer.

Difficult to grow; cultivation in a cool
glasshouse is probably essential.

Section **Cymbalaria** Linnaeus. Annuals or
biennials. Leaves soft or slightly fleshy,
with a distinct stalk, margin lobed, toothed
or nearly entire. Chalk-glands absent.
Flowering-stems leafy, producing flowers
on long axillary flower-stalks. Petals
usually yellow or orange, rarely white,
often with swellings near the base. Ovary
superior, or nearly so; capsule splitting
above the middle.

16. S. cymbalaria Linnaeus. Illustration:
Harding, The genus Saxifraga, 98 (1970);
Köhlein, Saxifrages and related genera,
t. 11 (1984); Webb & Gornall, Saxifrages
of Europe, pl. 2 (1989).

Annual (sometimes biennial?), with
ascending, diffusely branched, leafy stems
10–25 cm. Leaf-stalk to 35 cm, blade
usually *c.* 1 × 1.3 cm but sometimes larger,
kidney-shaped to circular or ovate, more or
less 5–9 lobed, the lobes acute to rounded.
Flowers 2–6 in a loose cyme. Petals
4.5–6 mm, elliptic-oblong with a short
claw, bright yellow towards the tip, duller
orange-yellow at the base, with 2 small
swellings located above the short claw.
Turkey & Caucasus, Romania, Algeria. H3.
Spring–early autumn.

Three varieties are recognised; the one
usually seen in gardens, as a handsome
weed, is var. **huetiana** (Boissier) Engler &
Irmscher. This is distinguished by the
leaves that are sometimes opposite and
have rounded lobes and truncate or
wedge-shaped bases. Prefers a shady place
in the garden.

17. S. sibthorpii Boissier. Illustration:
Webb & Gornall, Saxifrages of Europe, pl. 3
(1989).

Like *S. cymbalaria* but with reflexed,
rather than spreading, sepals, and solitary
flowers on leafless stalks. *Greece & SW
Turkey*. H4. Summer.

Doubtfully in cultivation, most plants
grown under this name are *S. cymbalaria*.

18. S. hederacea Linnaeus.

Like *S. cymbalaria* but petals 2–3 mm,
obovate-elliptic, very pale yellow or white.
E Mediterranean. H4. Spring–early summer.

Section **Merkianae** (Engler & Irmscher)
Gornall. Evergreen perennials, somewhat
woody below, forming mats. Leaves
without a well-defined leaf-stalk; margins
entire or 3-lobed at the tip. Chalk-glands
absent. Flowering-stems more or less leafy,
bearing 1–4 flowers. Sepals reflexed in
fruit. Petals white. Ovary almost superior;
capsule splitting above the middle, streaked
or spotted with purple.

19. S. merkii Fischer.

Leaves of basal rosettes
12–15 × 3–4 mm, obovate or spathulate to
lanceolate, acute at the tip, margin usually
entire but fringed with hairs; leaf-stalk
indistinct. Flowering-stem 3–6 cm,
sparingly leafy, bearing 1–3 flowers at the
top. Petals 6–7 mm, ovate, tapered at the
base to a claw, white. Filaments linear.
Japan. H4. Summer.

20. S. tolmiei Torrey & A. Gray.
Illustration: Hitchcock et al., Vascular

plants of the Pacific northwest, **3**: 57 (1971); Harding, Saxifrages, 16 (1992).

Leaves of basal rosettes 3–10 mm, linear, spathulate or oblanceolate, tapered to an indistinct leaf-stalk, base usually with long hairs; margin entire, cylindric or curved slightly downwards and inwards. Flowering-stems 3–12 cm, leafy, bearing 1–4 flowers in a loose cyme. Petals 3–6 mm, spathulate to broadly oblanceolate, white. Filaments club-shaped. *W North America*. H4. Summer.

Section **Micranthes** (Haworth) D. Don (section *Boraphila* Engler). Perennials, usually evergreen, occasionally deciduous. Stems solitary or few. Leaves usually all basal, herbaceous, or rather fleshy or leathery, with or without a distinct stalk; margin entire, toothed or variously lobed. Chalk-glands absent. Flowering-stems leafless, though sometimes with large, leaf-like bracts, terminating in a loose or dense panicle. Petals usually white, variously spotted or flushed with red or dark purple-black, occasionally greenish. Ovary superior to more than half inferior; capsule splitting to below the middle.

21. S. clusii Gouan. Illustration: Webb & Gornall, Saxifrages of Europe, pl. 5 (1989).

Blades of basal leaves 4–12 × 1.2–4.5 cm, oblanceolate to elliptic-oblong, with 4–10 coarse teeth on each side, tapered to an ill-defined leaf-stalk. Flowering-stems to 40 cm, leafless, branches many, forming a broad, diffuse, bracteate panicle, the terminal flower of which is regular, but the others irregular. Petals 4–7 × 2–3 mm, lanceolate, acute, with a prominent claw, white, unequal, the 3 upper ones longer and with 2 mustard spots at the base. Filaments linear. Ovary superior. *SW Europe*. H4. Summer.

Subsp. **clusii** has sparse, short hairs on the flowering-stem, and no leafy buds replacing the flowers. *Pyrenees, SW France*.

Subsp. **lepismigena** (Planellas) D.A. Webb has numerous long hairs on the flowering-stem, and at least some flowers replaced by leafy buds. *N Spain, N Portugal*. Probably best grown in a cool glasshouse, in acid soil.

Also keying out with this species is **S. michauxii** Britton, which is disconcertingly like *S. clusii* but probably can be distinguished by the larger, coarser teeth of its leaves. *E North America*. H4? Summer–early autumn.

22. S. ferruginea Graham. Illustration: Hitchcock et al., Vascular plants of the Pacific northwest, **3**: 43 (1971).

Blade of basal leaves usually 1.5–5 cm, occasionally more, spathulate to oblanceolate, tapered to a broad, indistinct leaf-stalk, margin sharply but irregularly toothed. Flowering-stems 10–30 cm, panicle diffuse; flowers sometimes replaced by bulbils. Petals 3–6 mm, upper 3 clawed, white with 2 yellow spots, lower 2 elliptic to spathulate, not clawed or spotted. Filaments linear. Ovary superior. *W North America*. H4. Summer.

Best grown in a cool glasshouse.

23. S. stellaris Linnaeus. Illustration: Ross-Craig, Drawings of British plants **10**: pl. 2 (1957). Phillips, Wild flowers of Britain, 189 (1977); Keble Martin, The new concise British flora in colour, pl. 32 (1982); Garrard & Streeter, Wild flowers of the British Isles, 90 (1983).

Leaf-rosettes sometimes forming carpets. Blades of basal leaves 1.2–7 × 0.8–2 cm, oblanceolate to broadly elliptic, with 3–5 blunt, forward-pointing teeth on each margin, tapered at the base to a more or less distinct leaf-stalk. Flowering-stems 5–20 cm, leafless, bearing a loose but rather narrow panicle, bracts smaller than in *S. clusii*. Flowers usually regular. Petals 3–8 × 2–3 mm, lanceolate, with a prominent claw, white with 2 yellow spots near the base. Filaments linear. Ovary superior. *Europe, Greenland & NE Canada*. H2. Summer.

Prefers a moist, acidic soil in a cool glasshouse.

24. S. melanocentra Franchet. Illustration: Harding, Saxifrages, 21 (1992).

Blades of leaves 2–3 × 0.5–2 cm, ovate-oblong with a scalloped or coarsely toothed margin, narrowed to a distinct leaf-stalk. Flowering stems 5–10 cm, bearing 1–5 flowers in a loose cyme. Petals 6–8 × 4–5 mm, obovate, upper surface white with 2 yellow or orange spots at the base, lower surface sometimes becoming purplish. Filaments linear. Ovary superior, carpels a dark, blackish crimson. *SW China, Nepal*. H3? Summer.

Grows best in a cool alpine house.

25. S. micranthidifolia (Haworth) Britton (*S. erosa* Pursh). Illustration: Gleason, Illustrated flora of the north-eastern United States & adjacent Canada **2**: 263 (1952).

Like *S. clusii* in general appearance but usually larger. Leaf–blades to 35 cm, lanceolate to oblanceolate, margin coarsely to rather finely, but irregularly, toothed, tapered to an indistinct leaf-stalk. Axils of basal leaves often with reddish brown, non-glandular hairs. Flowering-stems 30–75 cm, panicle loose. Petals 1.5–3.5 mm (shorter than the stamens), equal, elliptic to spathulate, narrowed at the base to a short claw, with a yellow spot. Filaments club-shaped. Ovary superior. *E North America*. H4? Late spring–summer.

26. S. nivalis Linnaeus. Illustration: Ross-Craig, Drawings of British plants **10**: pl. 1 (1957); Keble Martin, The new concise British flora in colour, pl. 32 (1982); Garrard & Streeter, Wild flowers of the British Isles, 90 (1983).

Blades of basal leaves 1–4 cm, ovate, diamond-shaped or almost circular, tapered to a broad leaf-stalk, usually with some reddish brown, non-glandular hairs on the margin or lower surface; margin scalloped or toothed. Flowering-stem 5–20 cm, more or less leafless, simple or branched near the top in a congested or, occasionally, loose cyme, with conspicuous narrow bracts. Petals 2–3 mm, slightly longer than the sepals, oblong to obovate, white, usually with pink tips. Filaments linear. Ovary half-inferior. *Circumpolar Arctic, southwards in W North America; C Europe & the Altai*. H1. Summer.

Best kept cold and dry in winter (cold-frame) and brought into an alpine house in spring and watered through the summer.

27. S. tenuis (Wahlenberg) Lindman (*S. stricta* Hornemann). Illustration: Ronning, Svalbards Flora, f. 28 (1964).

Like *S. nivalis* but without the reddish brown, non-glandular hairs. It is also usually smaller in all its parts and sometimes has a more open inflorescence. *Distribution uncertain, but probably more strictly arctic than S. nivalis*. H1. Summer.

Cultivation conditions as for *S. nivalis*.

28. S. virginiensis Michaux. Illustration: *Botanical Magazine*, 1664 (1814); Gleason, Illustrated flora of the north-eastern United States & adjacent Canada **2**: 264 (1952).

Leaf-blades to 9 cm, ovate to elliptic, tapered to a short, broad leaf-stalk; margin scalloped to sharply toothed; reddish brown, non-glandular hairs present in the axils. Flowering-stems 6–50 cm, bearing a congested, conical inflorescence. Petals

3–6 mm, mostly 2–3 times as long as sepals, elliptic to spathulate, unspotted. Filaments linear. Ovary less than one third inferior. *E North America.* H3. Spring.

Probably best in a humus-rich soil in partial shade.

29. S. reflexa W.J. Hooker. Illustration: Hultén, *Flora of Alaska and neighbouring territories,* 580 (1968).

Leaf-blades 1–2 cm, circular to ovate, oblanceolate or spathulate, tapered (sometimes abruptly) to a broad leaf-stalk; margin coarsely toothed; reddish brown, non-glandular hairs often present in the axils, short greyish hairs on the upper leaf-surface, lower surface purplish. Flowering-stems 9–60 cm, bearing a loose, rather flat-topped cyme. Petals 2–3 mm, shorter than or equalling the sepals, elliptic-oblong, without a claw, white with 2 yellow spots. Filaments club–shaped. Ovary less than one third inferior. *NW Canada to USA (Alaska).* H3? Summer.

30. S. occidentalis S. Watson. Illustration: Hitchcock et al., *Vascular plants of the Pacific northwest* **3**: 52 (1971).

Like *S. reflexa* but differing chiefly in its unspotted petals (sometimes narrowed to a claw), which are longer than the sepals, and in its linear to only slightly club-shaped filaments. *W North America.* H4. Spring–summer.

31. S. rufidula (Small) Macoun. Illustration: Hitchcock et al., *Vascular plants of the Pacific northwest* **3**: 52 (1971).

Like *S. reflexa* but basal leaves sharply and evenly toothed, often densely covered with reddish brown, non-glandular hairs beneath. Flowering-stems to 15 cm, bracts and calyces with reddish brown, non-glandular hairs. Filaments linear. *W North America.* H4? Spring-summer.

Probably best grown in a cool glasshouse or in a shaded, moist spot in the rock garden.

32. S. pensylvanica Linnaeus. Illustration: Gleason, *Illustrated flora of the north-eastern United States & adjacent Canada* **2**: 265 (1952).

Leaf-blades 10–25 cm, linear to oblanceolate, tapered to a broad, indistinct leaf-stalk; margin sparingly toothed to nearly entire; reddish brown, non-glandular hairs absent. Flowering-stems 25–60 cm, bearing a loose, cylindric inflorescence of flowers clustered at the ends of ascending branches. Petals 2.5–4 mm, linear to elliptic or narrowly obovate. Filaments linear. Ovary less than half inferior, becoming much less so in fruit. *E North America.* H4? Spring-summer.

A distinctive variant, with carpels markedly divergent when ripe, is known as var. **forbesii** (Vasey) Engler & Irmscher. Probably best in a humus-rich soil in partial shade.

Also keying out here is **S. oregana** Howell, a species very much like *S. pensylvanica* and from which there seem to be no consistent differences, apart from in chromosome number and geographical distribution! *W North America.* H4? Spring–summer.

33. S. hieracifolia Willdenow. Illustration: Rasetti, *I fiori delle Alpi,* f. 243 (1980); Webb & Gornall, *Saxifrages of Europe,* pl. 4 (1989); Aeschimann et al., *Flora alpina* **1**: 691 (2004).

Deciduous perennial. Leaf-blades 2–6 × 1–3 cm, ovate-elliptic, tapering to a short, winged leaf-stalk, margin entire or bluntly toothed, reddish brown, non-glandular hairs absent. Flowering-stem to 50 cm, more or less leafless, terminating in a narrow, dense, spike-like panicle, bracts often very conspicuous. Petals 1.5–3 mm, roughly equalling the sepals, but narrower, green with a purplish crimson margin. Filaments linear. Ovary half inferior, less so in fruit. *Circumpolar Arctic, Carpathians, the Altai.* H1. Summer.

Best grown under glass in shade, cold and dry in the winter but with a good supply of water in the summer.

34. S. nelsoniana D. Don (*S. punctata* misapplied). Illustration: Hultén, *Flora of Alaska and neighbouring territories,* 572–4 (1968). Hitchcock et al., *Vascular plants of the Pacific northwest* **3**: 57 (1971).

Leaves in a loose, basal rosette; blade 2–8 × 3–9 cm, kidney-shaped to nearly circular, base cordate, margin regularly scalloped; leaf-stalk to 4 times as long as blade, with non-glandular hairs (if present) whitish. Flowering-stem 10–35 cm, bearing a loose to compact panicle. Petals 2.5–4.5 mm, ovate to oblong, narrowed at the base to a short claw, white or pale pink, occasionally with orange spots. Filaments club-shaped. Ovary superior. *Arctic & subarctic regions from the Urals eastwards to NW North America.* H2. Summer.

Rarely cultivated in Europe; best grown on the edge of a bog-garden.

35. S. odontoloma Piper (*S. arguta* misapplied; *S. odontophylla* invalid). Illustration: Hitchcock et al., *Vascular plants of the Pacific northwest* **3**: 37 (1971).

Like *S. nelsoniana* and often confused with it, differing in its slightly irregular flowers, the larger petals of which are 2–4 mm, broadly elliptic to circular, contracted sharply at the base to a slender claw, with 2 yellow or green spots. *W North America.* H4? Summer.

36. S. lyallii Engler. Illustration: Hitchcock et al., *Vascular plants of the Pacific northwest* **3**: 48 (1971); Hultén, *Flora of Alaska and neighbouring territories,* 578 (1968).

Basal leaf-blades to 8 cm, spathulate to fan-shaped, base contracted to a more or less well-defined leaf-stalk; margin regularly toothed in the upper part. Flowering-stems usually 7–30 cm, bearing a loose, slender, cylindric, few-flowered cyme. Petals 2.5–5 mm, broadly elliptic to almost circular, with a short claw, white, cream or reddish, usually with 2 yellow or green spots. Filaments club-shaped. Ovary superior. *NW North America.* H3? Summer–early autumn.

37. S. manschuriensis (Engler) Komarov. Illustration: *Botanical Magazine,* 8707 (1917); Harding, *The genus Saxifraga,* 97 (1970).

Leaf-blades 4–7 × 6–8 cm, kidney-shaped to nearly circular, base deeply cordate, margin regularly and coarsely toothed or scalloped; leaf-stalk 1.5–2.5 times longer than the blade. Flowering-stem 15–35 cm, branched at the top to form a fairly compact panicle. Flowers with 6–8 petals. Petals 5–6 mm, narrowly oblong-elliptic. Stamens becoming longer than the petals with age, the filaments club-shaped. Ovary superior. *N China (Manchuria), Korea.* H5. Summer.

Best grown in a cool glasshouse in a well-drained soil but with plenty of water in the growing season.

Section **Irregulares** Haworth (section *Diptera* (Borkhausen) Sternberg).

Perennials, deciduous outdoors, evergreen under glass. Leaves usually all basal, herbaceous, or rather fleshy or leathery, margin entire, toothed or variously lobed; leaf-stalks well-defined. Chalk-glands absent. Flowering-stems leafless, though sometimes with large, leaf-like bracts, terminating in a loose panicle.

Flowers with a bilateral or irregular symmetry, the upper 3 or 4 petals smaller than the lower 2 or 1. Petals usually white, variously spotted. Ovary superior to almost superior; capsule splitting only in the upper half.

All the cultivated species of this section thrive best in light woodland, in acidic soils with plenty of humus and a degree of shelter. They do not tolerate much frost and can only be reliably grown outside in H5 areas.

38. S. fortunei J.D. Hooker (*S. cortusifolia* var. *fortunei* (J.D. Hooker) Maximowicz). Illustration: *Botanical Magazine*, 5377 (1863); Bloom, *Alpines for your garden*, 94 (1980).

Leaf-blades 4–6 × 6–10 cm, kidney-shaped to nearly circular, base deeply cordate, margin scalloped; leaf-stalk to 1.5 times as long as the blade, expanded at the base. Flowering-stems to 50 cm, bearing a loose panicle of numerous, irregular flowers. Petals unequal, linear-lanceolate, unspotted, lower 1 or 2 to 2.5 cm, occasionally toothed in the upper part, upper 3 or 4 to 1 cm, entire. Filaments linear, scarcely broadened above. *Japan.* H5. Early autumn.

A variable species, with varieties distinguished mainly on leaf-shape. The cultivars 'Rubrifolia' and 'Wada' are suffused with red throughout, the latter more deeply so.

39. S. cortusifolia Siebold & Zuccarini. Illustration: *Botanical Magazine*, 6680 (1883).

Like *S. fortunei* but flowers smaller, upper, 'short' petals *c.* 4 mm, broadly ovate, spotted with yellow or red, contracted at the base to a claw. *Japan, Korea.* H5. Summer.

40. S. nipponica Makino.

Like *S. fortunei* but with smaller leaves, and flowers usually with 2 long petals rather than 1. *Japan.* H7. Summer.

Dislikes growing in a pot.

41. S. stolonifera Curtis (*S. sarmentosa* Linnaeus filius; *S. cuscutiformis* Loddiges). Illustration: *Botanical Magazine*, 92 (1789).

Leaf-blades 4–9 cm across, kidney-shaped to circular, dark green, hairy and marked with grey veins on the upper surface, but reddish and hairless beneath. Thread-like runners emanate from the axils, root at intervals and generate new plants. Flowering-stems 20–50 cm, bearing a loose panicle of numerous, irregular

flowers. Petals unequal: the upper 3 are 3–4 mm, ovate, and spotted red and yellow near the base; the lower 2 are 1–2 cm, and narrowly elliptic. Filaments club-shaped. *SE China to Japan.* H5–G1. Summer.

There are a few variants, some of which are grown: var. **tricolor** (Lemaire) Maximowicz is characterised by leaves variegated with red and grey. G1. A dwarf variant, often seen under the name *S. cuscutiformis* Loddiges, is hardier than most. H5.

42. S. veitchiana I.B. Balfour.

Like *S. stolonifera* but smaller in stature and leaves not veined with grey. *W China.* H5. Summer.

Section **Heterisia** (Small) A.M. Johnson. Evergreen perennial, with basal leaves in loose rosettes. Leaf-blades with the margins scalloped or toothed; leaf-stalks well-defined. Chalk-glands absent. Flowering-stems more or less leafy, terminating in a loose panicle of regular flowers. Petals white. Ovary superior; capsule splitting to just above the middle.

43. S. mertensiana Bongard. Illustration: Hitchcock et al., *Vascular plants of the Pacific northwest* **3**: 48 (1971).

Leaf-blades 2–8 cm, kidney-shaped to circular; margin shallowly scalloped, lobes toothed or scalloped; leaf-stalk to 5 times as long as blade, often with long hairs. Flowering-stems 15–40 cm, bearing a loose panicle of numerous flowers, some of which are usually replaced by bulbils. Sepals reflexed in fruit. Petals 4–5 mm, oblong-elliptic to obovate, base truncate or with a short claw, white. *W North America.* H4? Spring–summer.

Best cultivated in woodland conditions.

Section **Porphyrion** Tausch. Evergreen perennials, forming cushions or mats. Stems somewhat woody below, producing cylindric leafy shoots or rosettes. Leaves usually hard or leathery, without a distinct stalk, margin entire or finely toothed. Chalk-glands located in pits, usually evident by the conspicuous calcareous encrustation on the leaves. Flowering-stems leafy, produced from persistent leaf-rosettes, bearing flowers singly or in small cymes or racemes. Petals white, pink, purple or yellow. Ovary almost superior to nearly inferior; capsule splitting above the middle.

Recent exploration of the Himalayan region has led to the introduction of several new saxifrages belonging to this

section. In many cases the identity of these plants is not yet known for certain and, regrettably, they have therefore been omitted from the following account.

Hybrids: Since hybrids between species of this section (Nos 44–89) are more widely grown than the parents, it is appropriate to detail them in a check-list, together with the names of some of the more common cultivars. Most have been made in cultivation. In general they are intermediate between the parents in their appearance, but so many selections of each hybrid exist that any general description is not diagnostic and thus of only limited use for purposes of identification. Nevertheless an indication of at least flower colour is given in the notes below. A key to the hybrids is given in Horny et al. (1986), where descriptions of individual cultivars may also be found.

S. × abingdonensis Arundel & Gornall; *S. burseriana* × *S. poluniniana*. Flowers like the first parent and foliage like the second. 'Judith Shackleton'.

S. × anglica Horny, Soják & Webr; *S. aretioides* × *S. lilacina* × *S. media*. Inflorescence of 1–3 flowers with pink to rose-purple petals. 'Beatrix Stanley', 'Cranbourne', 'Winifred'.

S. × anormalis Horny, Soják & Webr; *S. pseudolaevis* × *S. stribrnyi*. Petals carmine-red initially, becoming a drab orange-yellow with reddish streaks later.

S. × apiculata Engler; *S. marginata* × *S. sancta* (*S. × pungens* Sündermann; *S. × malyi* invalid). A vigorous hybrid with pale yellow petals in most of its cultivars. 'Alba' (white petals!), 'Gregor Mendel'.

S. × arco-valleyi Sündermann; *S. lilacina* × *S. marginata*. A slow-growing plant bearing usually solitary flowers with very pale lilac-pink petals in most cultivars. 'Arco'.

S. × baccii Young & Gornall; *S. aretioides* × *S. lilacina* × *S. media* × *S. stolitzkae*. Petals pink-lilac. 'Irene Bacci'.

S. × bertolonii Sündermann; *S. sempervivum* × *S. stribrnyi* (*S. × amabilis* Stapf). A vigorous hybrid, more or less intermediate between the parents, but with the inflorescence often more resembling that of *S. sempervivum*. 'Amabilis'.

S. × biasolettoi Sündermann; *S. federici-angusti* subsp. *grisebachii* × *S. sempervivum*. Similar to subsp. *grisebachii*; inflorescence purple-red to purple-violet. 'Crystalie', 'Feuerkopf', 'Phoenix'.

S. × bilekii Sündermann; *S. ferdinandi-coburgi* × *S. tombeanensis*. Petals a pale yellow. 'Castor', often referred to incorrectly in catalogues as *S. diapensioides lutea* (see under *S. × malbyana*).

S. × boeckeleri Sündermann; *S. ferdinandi-coburgi × S. stribrnyi*. Petals orange-yellow. 'Armida'.

S. × borisii Sündermann; *S. ferdinandi-coburgi × S. marginata*. Petals usually yellow to pale yellow. 'Kyrilli', 'Margarete', 'Pseudo-borisii', 'Vesna'.

S. × boydii Dewar; *S. aretioides × S. burseriana* (*S. aretiastrum* Engler & Irmscher). Flowers 1–3, like those of *S. burseriana* but with yellow petals. 'Aretiastrum', 'Cherrytrees', 'Pilatus', 'Sulphurea'.

S. × boydilacina Horny, Soják & Webr; *S. aretioides × S. burseriana × S. lilacina*. Flowers solitary with pink to pale creamy yellow petals. 'Moonbeam', 'Penelope'.

S. × bursiculata Jenkins; *S. burseriana × S. marginata × S. sancta*. Flowers circular in outline with white petals. 'King Lear'.

S. × byam-groundsii Horny, Soják & Webr; *S. aretioides × S. burseriana × S. marginata*. Flowers usually solitary with pale yellow petals. 'Lenka'.

S. × caroliquarti Lang; *S. albertii × S. lowndesii*. Petals violet-pink. 'Ivana'.

S. × clarkei Sündermann *S. media × S. vandellii*. Flowers in a small panicle, with soft pink petals. 'Sidonia'.

S. × doerfleri Sündermann; *S. federici-angusti* subsp. *grisebachii × S. stribrnyi*. A vigorous hybrid with inflorescences intermediate; flowers pink or purple-red to purple-violet. 'Ignaz Doerfler'.

S. × edithae Sündermann; *S. marginata × S. stribrnyi*. Petals pink or pale pink. 'Bridget'.

S. × elisabethae Sündermann; *S. burseriana × S. sancta* (*S. × godsejfiana* invalid). A very vigorous hybrid bearing flowers usually with yellow petals. 'Boston Spa', 'Carmen', 'Mrs Leng', 'Ochroleuca'.

S. × eudoxiana Sündermann; *S. ferdinandi-coburgi × S. sancta*. Flowers in compact cymes. Petals yellow. 'Gold Dust', 'Haagii'.

S. × fallsvillagensis Horny, Soják & Webr; *S. burseriana × S. marginata × S. tombeanensis*. Petals white. 'Swan'.

S. × finnisiae Horny, Soják & Webr; *S. aizoides × S. aretioides × S. lilacina × S. media*. Petals pale yellow, tinged with orange. 'Parcevalis'.

S. × fleischeri Sündermann; *S. federici-angusti* subsp. *grisebachii × S. corymbosa*. Petals yellowish to red. 'Mephisto'.

S. × fontanae Sündermann; *S. diapensioides × S. ferdinandi-coburgi*. Slow-growing, with yellow petals. 'Amalie'.

S. × geuderi Sündermann; *S. ferdinandi-coburgi × S. aretioides × S. burseriana*. Flowers supposedly solitary, with deep yellow petals. Plants in cultivation under this hybrid binomial do not correspond to the original description, being multiflorous. See Horny et al. (1986) for details. 'Eulenspiegel'.

S. × gloriana Horny, Soják & Webr; *S. lilacina × S. obtusa*. Flowers solitary with pale pinkish lilac petals. The cultivar 'Godiva' is often sold as 'Gloriana'. The whole history of this hybrid is discussed by Horny et al. (1986).

S. × goringana Young & Gornall; *S. aretioides × S. cinerea × S. lilacina × S. media*. Petals deep cerise-pink fading to pale lilac-pink. 'Nancye'.

S. × grata Engler & Irmscher; *S. aretioides × S. ferdinandi-coburgi*. Petals deep yellow. Plants of this hybrid may not be distinguishable from those going under the name *S. × guederi*; see Horny et al. (1986) for details. 'Loeflingii'.

S. × gusmusii Irving & Malby; *S. corymbosa × S. sempervivum*. Flowering-stems drab purple or yellowish, branched or not. Petals yellowish or drab orange. 'Subluteiviridis'.

S. hardingii Horny, Soják & Webr; *S. aretioides × S. burseriana × S. media*. Petals creamy yellow, apricot, orange or pink. 'CM. Prichard', 'Iris Prichard'.

S. × heinrichii Sündermann; *S. aretioides × S. stribrnyi*. Flowers in small cymes, with yellow to orange petals. 'Ernst Heinrich'.

S. × hoerhammeri Engler & Irmscher; *S. federici-angusti* subsp. *grisebachii × S. marginata*. Panicles irregular; petals bright rose. 'Lohengrin'.

S. × hofmannii Sündermann; *S. burseriana × S. sempervivum*. Flowering-stems more like those of *S. sempervivum*, with erect, pink to drab pink petals. 'Bodensee', 'Ferdinand'.

S. × hornibrookii Horny, Soják & Webr; *S. lilacina × S. stribrnyi*. Forms distinctly domed cushions. Flowers 1–5 per stem, according to cultivar and influence of *S. stribrnyi*. Petals spreading, deep wine-red to violet-purple. 'Riverslea'.

S. × ingwersenii Horny, Soják & Webr; *S. lilacina × S. tombeanensis*. Flowering stem with 1–3 flowers with pale pink petals. 'Simplicity'.

S. × irvingii May et al.; *S. burseriana × S. lilacina*. Flowers like those of *S. burseriana*, but with pinkish petals. 'His Majesty', 'Walter Irving', 'Jenkinsiae'.

S. × kayei Horny, Soják & Webr' *S. aretioides × S. burseriana × S. ferdinandi-coburgi × S. sancta*. Petals a very deep yellow. 'Buttercup'.

S. × kellereri Sündermann' *S. burseriana × S. stribrnyi* (*S. × suendermannii* Sündermann *S. × hedwigii* invalid). Flowering-stems with few flowers. Petals spreading, pale to purplish rose. 'Johann Kellerer', 'Kewensis', 'Sundermannii Major'. Plants grown in the British Isles as *S. pseudo-kellereri* invalid are not separable from 'Johann Kellerer'.

S. × laeviformis Horny, Soják & Webr; *S. marginata × S. pseudolaevis*. Flowers in small, compact cymes, with muddy yellow petals. 'Egmont'.

S. × landaueri Horny, Soják & Webr; *S. burseriana × S. marginata × S. stribrnyi*. Petals pale rose. 'Leonore'.

S. × leyboldii Sündermann; *S. marginata × S. vandellii*. Flowers in small cymes, with white petals. 'August Hayek'.

S. × lincolni–fosteri Horny, Soják & Webr; *S. aretioides × S. burseriana × S. diapensioides*. Flowers 1–3, with large, pale yellow petals. 'Salome'.

S. × lismorensis Young & Gornall; *S. aretioides × S. georgei × S. lilacina × S. media*. Petals pink to red. 'Lismore Carmine', 'Lismore Pink'.

S. × luteopurpurea Lapeyrouse; *S. aretioides × S. media* (*S. lapeyrousii* D. Don *S. × ambigua* de Candolle *S. × benthamii* Engler & Irmscher). A natural hybrid found in the Pyrenees. It is at least partly fertile and hence very variable. Leaves and inflorescence are more or less intermediate between those of the parents. Petals yellow to pinkish purple. 'Godroniana'.

S. × malbyana Horny, Soják & Webr; *S. aretioides × S. diapensioides* (*S. diapensioides lutea* invalid). A slow-growing, dense cushion, like *S. diapensioides* in appearance. Flowers in a loose cyme, with pale yellow to cream petals. 'Primulina'.

S. × margoxiana Horny, Soják & Webr; *S. ferdinandi-coburgi × S. marginata × S. sancta*. Flowers stellate with pale yellow petals. 'Parsee'.

S. × mariae-theresiae Sündermann; *S. burseriana × S. federici-angusti* subsp. *grisebachii* (*S. × marie-meresiae* invalid). Petals pink. 'Gaertneri', 'Theresia'.

S. × megaseaeflora May et al.; *S. aretioides × S. burseriana × S. lilacina*

× *S. media*. Flowers usually solitary with petals varying from pale purple to pink or yellow. 'Robin Hood', 'Jupiter'.

S. × millstreamiana Horny, Soják & Webr; *S. burseriana × S. ferdinandi-coburgi × S. tombeanensis*. Petals pale yellow. 'Eliot Hodgkin'.

S. × paulinae Sündermann; *S. burseriana × S. ferdinandi-coburgi* (*S. kolbyi* invalid). Flowers in loose cymes, with yellow petals. 'Franzii', 'Kolbiana'.

S. × petraschii Irving; *S. burseriana × S. tombeanensis* (*S. × assimilis* invalid). Flowers 1–5, with white petals. 'Funkii', 'Hansii', 'Kaspar Maria Sternberg', 'Schelleri'.

S. × poluanglica Bürgel; *S. aretioides × S. lilacina × S. media × S. poluniniana*. Petals rich rose-pink, deepening toward base. 'Redpoll'.

S. × polulacina Bürgel; *S. lilacina × S. poluniniana*. Close to latter parent. Petals clear pink, fading to pale pink with age. 'Kathleen'.

S. × poluteo-purpurea Bürgel; *S. aretioides × S. media × S. poluniniana*. Petals light pink, nectary red. 'Sasava'.

S. × pragensis Horny, Soják & Webr; *S. ferdinandi-coburgi × S. marginata × S. stribrnyi*. Inflorescence racemose. Petals initially yellow, later turning orangeish or salmon-coloured. 'Golden Prague'.

S. × prossenii (Sündermann) Ingwersen; *S. sancta × S. stribrnyi*. Flowering-stems with small cymes of stellate flowers, with bronze to rosy orange petals. 'Prometheus', 'Regina'.

S. × pseudo-kotschyi Sündermann; *S. kotschyi × S. marginata*. Flowering-stems with compact, flat-topped cymes of stellate flowers with yellow petals. 'Denisa'.

S. × rosinae Horny, Soják & Webr; *S. diapensioides × S. marginata*. Forms a cushion of nearly spherical rosettes. Flowering-stems with flat-topped cymes. Petals off-white. 'Rosina Sündermann'.

S. × saleixiana Gaussen & Le Brun; *S. aretioides × S. caesia*. A natural hybrid possibly of this parentage is reported fom the Pyrenees.

S. × salmonica Jenkins; *S. burseriana × S. marginata* (*S. × salomonii* Sündermann, invalid). Flowering-stems with few-flowered cymes. Petals white. 'Friesei', 'Kestoniensis', 'Maria Luisa', 'Obristii', 'Pichleri', 'Pseudo-salomonii', 'Salomonii', 'Schreineri'.

S. × schottii Irving & Malby; *S. corymbosa × S. stribrnyi*. Petals drab yellow or vermillion red. 'Sub-stribrnyi'.

S. × semmleri Horny, Soják & Webr; *S. ferdinandi-coburgi × S. pseudolaevis × S. sancta*. Inflorescence compact and flat-topped with yellow, stellate flowers. 'Martha'.

S. × smithii Horny, Soják & Webr; *S. marginata × S. tombeanensis*. Flowering-stems short, bearing flat-topped cymes of large flowers with white petals. 'Vahlii'.

S. × steinii Sündermann; *S. aretioides × S. tombeanensis*. Flowers several. Petals creamy white. 'Agnes'.

S. × stormonthii Horny, Soják & Webr; *S. desoulavyi × S. sancta*. Flowers in compact cymes, with yellow petals. 'Stella'.

S. × stuartii Sündermann; *S. aretioides × S. media × S. stribrnyi*. Paniculate cyme. Flowers with pale yellow to drab rose-pink petals. 'Lutea', 'Rosea'.

S. × thomasiana Sündermann; *S. stribrnyi × S. tombeanensis*. Flowering-stems tall (6–10 cm) bearing a loose cyme of small flowers with pale pink petals. 'Magdalena'.

S. × tiroliensis Kerner; *S. caesia × S. squarrosa*. Occurs as a natural hybrid in the Dolomites and Julian Alps, associated with the parents.

S. × urumoffii Horny, Soják & Webr; *S. corymbosa × S. ferdinandi-coburgi*. Flowering-stems tall (6–14 cm), bearing a short cyme of small flowers with yellow petals. 'Ivan Urumov'.

S. × webrii Horny; *S. sancta × S. scardica* (*S. sartorii* misapplied). Flowers in flat-topped cymes. Petals yellow. 'Pygmalion'.

S. × wehrhahnii Horny, Soják & Webr; *S. marginata × S. scardica*. Flowers in flat-topped cymes. Petals white. 'Pseudo-scardica'.

S. × wendelacina Horny & Webr; *S. lilacina × S. wendelboi*. Flowering-stem bearing 1 or 2 flowers. Petals pale pink to lilac. 'Wendrush'.

S. × youngiana Harding & Gornall; *S. lilacina × S. marginata × S. stribrnyi*. Petals lilac. 'Lilac Time'.

Subsection **Kabschia** (Engler) Rouy & Camus. Leafy shoots aggregated into cushions or mats. Leaves alternate, the margins with a translucent border. Flowering-stems distinct or virtually lacking. Sepals spreading or semi-erect, shorter than the petals and not concealing them. Petals white, pink, purple or yellow, hairless. Ovary almost superior to inferior.

Series **Aretioideae** (Engler & Irmscher) Gornall. Leaves usually with 3–13 chalk-glands; tips erect or patent.

Flowering-stems distinct, bearing 3–15 flowers. Petals longer than the stamens, yellow. Ovary almost superior to three-quarters inferior.

44. S. ferdinandi-coburgi Kellerer & Sündermann. Illustration: Horny et al., Porophyllum saxifrages, f. 20–2 (1986).

Leafy shoots forming an untidy cushion. Leaves 4–8 × 1–2 mm, oblong-lanceolate, with a short point usually bent upwards; margin entire, except for a few hair-like teeth near the base; chalk-glands producing a conspicuous calcareous encrustation and making the leaves look grey. Flowering stems 3–12 cm, reddish, bearing 3–15 flowers in a flat-topped or a rather long cyme. Petals 5–8 mm, broadly to narrowly obovate, scarcely touching, bright yellow. Ovary half to three-quarters inferior. *E Bulgaria, N Greece.* H2. Spring.

Var. **radoslavoffii** Stojanov (var. *pravislavii* misspelling) and var. **macedonica** Drenov are based on variation in the height of the flowering-stem, in the number of flowers and size of petals; these features, however, are diagnostic of individuals rather than populations, so cultivar status is more appropriate. Plants distributed under the name of the former variety, however, often turn out to be var. **rhodopea** Kellerer & Stojanov, which differs consistently from the type only in its longer rosette-leaves, 6–13 mm.

Prefers an alkaline soil.

45. S. aretioides Lapeyrouse. Illustration: Horny et al., Porophyllum saxifrages, f. 23 (1986); Webb & Gornall, Saxifrages of Europe, pl. 13 (1989).

Similar to *S. ferdinandi-coburgi* but leaves shorter, wider, apex obtuse, or if with a short point, then this held straight and not angled upwards and inwards. Flowering-stems usually with up to 5 flowers, these packed more densely into a more consistently flat-topped cyme. Petals sometimes narrower, tip often notched. *Pyrenees, NW Spain.* H2. Spring.

Prefers an alkaline soil, does best in a cool glasshouse.

Series **Juniperifoliae** (Engler & Irmscher) Gornall. Leaves usually with 3–9 chalk-glands; tips erect or patent, barely if at all recurved. Flowering-stems distinct, bearing 3–19 flowers. Petals shorter than or barely equalling the stamens, not touching, yellow (very rarely white). Ovary three-quarters inferior.

46. S. juniperifolia Adams (*S. juniperina* Bieberstein; *S. pseudosancta* Janka; *S. macedonica* Degen). Illustration: *Botanical Magazine*, n.s., 137 (1951); Köhlein, Saxifrages and related genera, t. 23 (1984); Horny et al., Porophyllum saxifrages, f. 64–6 (1986).

Leafy shoots woody below, columnar, forming a fairly dense cushion. Leaves 9–13 × 1–2.5 mm, linear to linear-lanceolate, tapered to a stiff point; margin minutely toothed towards the base; producing little (if any) calcareous encrustation. Flowering-stems 3–6 cm, usually with many shaggy hairs, bearing 3–8 flowers in a dense, oblong cyme. Petals 5–6 mm, obovate, sometimes narrowly so, yellow. *Bulgaria, NE Turkey, Caucasus.* H2. Spring.

May be more vigorous in partial shade.

47. S. sancta Grisebach (*S. juniperifolia* subsp. *sancta* (Grisebach) D.A. Webb). Illustration: *Botanical Magazine*, n.s., 137 (1951); Köhlein, Saxifrages and related genera, t. 23 (1984); Horny et al., Porophyllum saxifrages, f. 63, 65 (1986).

Like *S. juniperifolia* but leaves usually less than 1 cm, margin with long tooth-like hairs up to near the apex. Flowering-stems hairless; inflorescence more nearly spherical or flat-topped. Stamens barely, if at all, exceeding the petals. *NE Greece, NW Turkey (Mt Ida).* H2. Spring.

Prefers an alkaline soil.

48. S. desoulavyi Oettingen (*S. caucasica* var. *desoulavyi* (Oettingen) Engler & Irmscher). Illustration: Horny et al., Porophyllum saxifrages, f. 67 (1986).

Like *S. sancta* but flowering-stems 1.5–2 cm, sparingly glandular-hairy, producing a spherical head of 3–7 flowers. *Caucasus.* H2. Early spring.

Prefers an alkaline soil and some shade.

49. S. kotschyi Boissier. Illustration: *Botanical Magazine*, 6065 (1873); Horny et al., Porophyllum saxifrages, f. 71 (1986).

Leafy shoots forming a deep cushion. Leaves 4–12 × 1–4 mm, oblong to spathulate, tip mucronate, basal margin with tooth-like hairs; producing an obvious calcareous encrustation. Flowering-stems 1–8 cm, glandular-hairy, bearing a fairly loose, somewhat flat-topped panicle of 4–13 flowers. Petals 3–6 mm, obovate to spathulate, yellow. *C &E Turkey to NW Iran.* H2?. Early spring.

Probably prefers an alkaline soil.

50. S. pseudolaevis Oettingen (*S. laevis* misapplied). Illustration: Horny et al., Porophyllum saxifrages, f. 68 (1986).

Leafy shoots forming a mat. Leaves 4–8 × 2.5–3 mm, oblong-obovate to spathulate, shortly pointed at the tip, shiny, with tooth-like hairs on the margins towards the base. Flowering-stems 2–5 cm, variably glandular-hairy to hairless, bearing a small, oblong cyme of 5–7 flowers. Petals 2.5–4 mm, obovate, tip irregularly lobed, yellow. Styles markedly spreading. *Caucasus.* H2. Early spring.

Prefers an alkaline soil and partial shade.

51. S. subverticillata Boissier. Illustration: Horny et al., Porophyllum saxifrages, f. 73 (1986).

Leafy shoots forming a loose cushion, with secondary leaf-rosettes terminating extended shoots. Leaves 12–20 × 1–2.5 mm, oblong to linear-oblong, sharply pointed; tooth-like hairs on margin at base. Flowering-stems 2.5–4 cm, hairless, bearing a loose panicle of 6–9 flowers. Petals 3–4 mm, narrowly obovate, well separated one from another, pale yellow to cream. Styles unusually long, to 7 mm. *Caucasus.* H2. Early spring.

Var. **colchica** (Albow) Horny & Webr differs in its shorter leaves and slightly longer petals.

Prefers partial shade.

52. S. scleropoda Sommier & Levier. Illustration: Horny et al., Porophyllum saxifrages, f. 72 (1986).

Leafy shoots columnar, woody below, forming a cushion with secondary leaf-rosettes terminating extended shoots. Leaves 3.5–10 × 1–2 mm, lanceolate, oblong or elliptic, tip mucronate, the lower two-thirds with tooth-like hairs on margin, tips recurved. Flowering-stems 2–4 cm, glandular-hairy, bearing a spherical cyme of 12–19 flowers. Petals 2.5–5 mm, narrowly obovate to linear-oblong, well separated, sulphur yellow. *Caucasus.* H2. Spring.

Series **Lilacinac** Gornall. Leaves with 5–9 chalk-glands, leaf-tips usually strongly recurved but not prominently thickened. Flowering-stems usually distinct, bearing solitary flowers. Petals longer than the stamens, usually more or less touching or overlapping, white, pink or purple. Ovary one to three-quarters inferior.

53. S. lilacina Duthie. Illustration: Horny et al., Porophyllum saxifrages, f. 113

(1986); *Bulletin of the Alpine Garden Society*, **58**(3): 272 (1990).

Leafy shoots forming a compact cushion. Leaves 3.5–5.5 × 1–1.5 mm, oblong to spathulate, tip obtuse or shortly pointed, margin recurved, with tooth-like hairs in the lower half. Flowering-stem 1–2 cm. Petals 8–11 mm, obovate, often with wavy margins, tips recurved, deep violet at first but fading later to reveal dark veins. Ovary three-quarters -inferior. *W Pakistan, Kashmir, India.* H2. Spring.

This species is a key parent of many Kabschia cultivars. Prefers an acid soil; does best in a cool glasshouse.

54. S. poluniniana H. Smith. Illustration: Horny et al., Porophyllum saxifrages, f. 107 (1986); Harding, Saxifrages, 54 (1992).

Leafy shoots forming a loose cushion. Leaves 5–6 × 1–2 mm, oblong-spathulate to linear, tip almost acute but with a short point, recurved, moderately thickened, base with tooth-like hairs on margin, calcareous encrustation prominent. Flowering-stems 1.5–2 cm. Petals *c.* 1 cm, obovate, tapering to a slender claw, overlapping, white becoming pink with age (always pink in the sun). Ovary inferior. *Nepal.* H2. Spring.

Prefers an alkaline soil in partial shade.

Series **Marginatae** (Engler & Irmscher) Gornall. Leaves usually with 5–18 chalk-glands, tips usually strongly recurved and often prominently thickened. Flowering-stems distinct, bearing mostly 2–12 flowers. Petals longer than the stamens, usually more or less touching, white, pink or purple. Ovary half to three-quarters inferior.

55. S. marginata Sternberg. Illustration: *Botanical Magazine*, 6702 (1883); Köhlein, Saxifrages and related genera, t. 22 (1984); Horny et al., Porophyllum saxifrages, f. 26–38 (1986); Webb & Gornall, Saxifrages of Europe, f. 16 (1989); Harding, Saxifrages, 71 (1992).

Leaves obovate to narrowly elliptic, 3–13 × 1–5 mm, tip obtuse or mucronate, sometimes recurved; margin entire, but fringed with hairs at least at the base; calcareous encrustation usually obvious. Flowering-stems 3–12 cm, terminating in a small, compact panicle usually of 2–8 flowers. Petals 5–15 mm, obovate, scarcely touching, white but sometimes turning pink with age. *S Italy, Romania, Balkans.* H2. Spring.

The variation in the species is great but difficult to treat owing to its more or less continuous nature. It seems, however, that it is possible to recognise at least two subspecies.

Subsp. **marginata** (*S. rocheliana* Sternberg; *S. coriophylla* Grisebach; *S. boryi* Boissier & Heldreich). Distinguished by its leaves, in which the tips are rounded or obtuse and the marginal hairs are confined to the basal region. A diminutive variant from the Balkans with columnar shoots of small leaves is known as var. **coriophylla** (Grisebach) Engler, and a similar one but with hairless flowering-stems is known as subsp. **bubakii** (Rohlena) Chrtek & Soják.

Subsp. **karadzicensis** (Degen & Kosanin) Chrtek & Soják has small leaves (3–4.5 × 1–2 mm), in which the tips are shortly pointed, bear only one chalk-gland, and the margins are fringed with long hairs from the base almost to the tip; non-glandular hairs also sometimes occur on the leaf surface. *Former Yugoslavia (Karadzice).*

Many plants under this name in cultivation do not appear to match those found in the wild.

It is possible that the enigmatic **S. obtusa** (Sprague) Horny & Weber (*S. scardica* Grisebach var. *obtusa* Sprague; *S. dalmatica* invalid) should be included within *S. marginata* or one of its hybrids. It is possibly sterile and not known for certain in the wild; a full discussion of it and its alleged hybrids is presented by Horny et al. (1986).

Prefers an alkaline soil.

56. S. scardica Grisebach (*S. sartorii* Boissier). Illustration: *Botanical Magazine*, 8243 (1909); Horny et al., Porophyllum saxifrages, f. 45–7 (1986); Webb & Gornall, Saxifrages of Europe, f. 17 (1989).

Leafy shoots crowded (but remaining distinct) into a hard, fairly dense cushion. Leaves 5–15 × 2–4 mm, oblong, acute or with a short point, rather glaucous; margin entire at tip only, middle minutely toothed; calcareous encrustation evident. Flowering-stems 4–12 cm, with a compact cyme of 5–12 flowers. Petals 7–12 mm, obovate, nearly touching, white but sometimes becoming pink or purple with age. *Balkan Peninsula.* H2. Spring.

There is some variation in leaf-size, and those plants with the smallest leaves (*c.* 6 mm) are known as var. **pseudocoriophylla** Engler & Irmscher.

Prefers an alkaline soil in partial shade; probably best in a cool glasshouse.

57. S. iranica Bornmüller. Illustration: Horny et al., Porophyllum saxifrages, f. 74 (1986).

Leafy shoots forming a very compact, hard cushion. Leaves 3–4 × 2.5–4 mm, oblong to broadly obovate, tip obtuse. Flowering-stems 2.5–5 cm, bearing a small cyme of 3–6 flowers. Petals 5–12 mm, obovate, tip usually slightly notched, touching or overlapping, recurved, white becoming deep purple with age mainly at the base. *N Iran (Elburz Mts).* H2. Spring.

Prefers an alkaline soil.

58. S. wendelboi Schönbeck-Temesy. Illustration: Horny et al., Porophyllum saxifrages, f. 76 (1986).

Like *S. iranica*, of which it is probably only a variant, but flowering-stem 3–6 cm; petals more broadly obovate, less recurved, ageing to a pink colour. *Iran.* H2. Spring.

Prefers an alkaline soil.

59. S. spruneri Boissier. Illustration: Horny et al., Porophyllum saxifrages, f. 42–3 (1986); Webb & Gornall, Saxifrages of Europe, f. 18 (1989).

Leafy shoots mostly with dead leaves, but terminating in living rosettes aggregated into a deep cushion. Leaves 4–8 × 2–4 mm, oblong-spathulate to obovate-oblanceolate, obtuse or with a short point, uniquely (in the section) the whole lower surface and at least half the upper glandular-hairy; margin entire. Flowering-stems 3–8 cm, bearing 4–15 flowers in a flat-topped cyme. Petals 4–6 mm, obovate, more or less touching, white. *Balkan Peninsula.* H2. Spring.

Prefers an alkaline soil; best grown in a cool glasshouse.

60. S. alberti Regel & Schmalhausen. Illustration: Horny et al., Porophyllum saxifrages, f. 103 (1986).

Leafy shoots forming a dense, compact cushion. Leaves 5.5–9 × 1–2 mm, linear to oblong, obtuse, margins fringed with tooth-like hairs in the lower half, dull green. Flowering-stems 1.5–2.5 cm, bearing a compact, flat-topped cyme of 4–8 flowers. Petals 6–7 mm, obovate, touching, white. *C Asia (Pamirs, Tien Shan).* H2? Spring.

Probably prefers an acid soil in partial shade.

61. S. clivorum H. Smith. Illustration: Horny et al., Porophyllum saxifrages, f. 119 (1986).

Leafy shoots forming a dense cushion. Leaves 5–6 × 2–2.5 mm, linear-oblong, tip truncate with chalk-glands embedded in margin, margin fringed with a few thick, glandular hairs in the lower half. Flowering-stems *c.* 8 mm, bearing 1–3 flowers. Petals 3–4 mm, obovate, often broadly so, touching, white. *Bhutan, Nepal?* Spring.

62. S. andersonii Engler. Illustration: Horny et al., Porophyllum saxifrages, f. 122 (1986).

Leafy shoots forming a compact (loose in shade) cushion. Leaves 4–10 × 1–4 mm, linear-obovate to obovate, tip obtuse or shortly pointed, thickened into a horseshoe-shaped rim; margin fringed with hairs from the base to about the middle. Flowering-stems 1.5–3.5 cm, bearing 1–5 flowers. Petals 3–6 × 1.5–4 mm, obovate, white ageing to pink, or sometimes pinkish to start with. *Nepal, China (Xizang), Bhutan.* H2? Late spring–summer.

Prefers a well-watered, but well-drained neutral compost. **S. afghanica** Aitchison & Hemsley appears to be similar but is doubtfully in cultivation. *Afghanistan, W Pakistan, Nepal, China (SE Xizang).* H2? Spring.

63. S. rhodopetala H. Smith. Illustration: Horny et al., Porophyllum saxifrages, f. 118 (1986).

Like *S. andersonii* but distinguished by its usually 5–9 flowers with petals a deep rose-pink. *Nepal.* H2? Late spring–summer.

64. S. stolitzkae Engler & Irmscher. Illustration: Evans, Alpines '81, 26 (1981); Horny et al., Porophyllum saxifrages, f. 124 (1986); *Bulletin of the Alpine Garden Society* **56**(1): 55 (1988); Harding, Saxifrages, 71 (1992).

Leafy shoots forming a dense cushion. Leaves 7–9 × 2–3 mm, oblong to slightly spathulate, tip shortly pointed, thickened into a horseshoe-shaped rim, margin fringed with tooth-like hairs in the lower third, glaucous, with a prominent calcareous encrustation. Flowering-stems 3–8 cm, bearing 1–6 flowers. Petals 5–9 mm, obovate to broadly so, tapered to a short claw, margins often wavy, overlapping, white becoming pink with age, especially at the base. *India, Bhutan, Nepal.* H2? Spring.

Prefers an alkaline soil; best in a cool glasshouse.

65. S. cinerea H. Smith. Illustration: Horny et al., Porophyllum saxifrages,

f. 123 (1986); Harding, Saxifrages, 56 (1992).

Leafy shoots forming a low, pad-like cushion. Leaves 10–12 × 1.5–2 mm, linear-oblong, tip acute, base expanded and fringed with a few tooth-like hairs, glaucous-green. Flowering-stems *c.* 8 cm, red, bearing 2–6 flowers in a compact cyme. Petals 8–10 mm, obovate, touching, white. *Nepal.* H2? Spring.

Series **Squarrosae** (Engler & Irmscher) Gornall. Leaves with 5–7 chalk-glands, tips recurved. Flowering-stems distinct, bearing 2–8 flowers. Stamens united at the base by a ribbon of tissue. Petals longer than the stamens, white. Ovary more than three-quarters inferior.

66. S. caesia Linnaeus (*S. recurvifolia* Lapeyrouse). Illustration: Webb & Gornall, Saxifrages of Europe, f. 15 (1989); *Bulletin of the Alpine Garden Society* **58**(3): 273 (1990); Harding, Saxifrages, 36 (1992); Aeschimann et al., Flora alpina **1**: 693 (2004).

Leafy shoots forming a moderately dense cushion. Leaves *c.* 4 × 1–1.5 mm, oblong to spathulate, upper half curving outwards and expanded near the tip, glaucous, with conspicuous calcareous encrustation; margins entire but fringed with hairs at the base. Flowering-stems 4–12 cm, branched near the top to form a small panicle of 2–8 flowers. Petals 4–6 mm, white. *Pyrenees, Alps, Tatra & W Carpathians, W former Yugoslavia.* H2. Late spring.

An alleged hybrid with *S. mutata*, called **S. × forsteri** Stein, is so much like *S. caesia* that the parentage must be doubted. For a hybrid with *S. aizoides* see under that species. Prefers an alkaline soil; best grown in tufa under glass.

67. S. squarrosa Sieber. Illustration: Finkenzeller & Grau, Alpenblumen, 107 (1985); Aeschimann et al., Flora alpina **1**: 693 (2004).

Like *S. caesia* but with narrower, denser leafy shoots aggregated to form a taller, more compact cushion. Leaves oblong (not expanded towards the tip), only the tip curving outwards, less glaucous, with fewer chalk-glands (usually 3 versus 7). Flowering-stem hairier at the base than at the top (the reverse is true of *S. caesia*). *SE Alps.* H2. Spring.

Prefers an alkaline soil; best grown in tufa under glass.

Series **Rigidae** (Engler & Irmscher) Gornall. Leaves with 3–7 chalk-glands, the tips erect or patent, barely, if at all, recurved. Flowering-stems distinct, bearing 1–6 flowers. Petals longer than the stamens, white. Ovary three-quarters inferior.

68. S. diapensioides Bellardi (*S. glauca* Clairville). Illustration: *Bulletin of the Alpine Garden Society* **25**(2): 132 (1957); Finkenzeller & Grau, Alpenblumen, 105 (1985); Horny et al., Porophyllum saxifrages, f. 39 (1986); Aeschimann et al., Flora alpina **1**: 695 (2004).

Leafy shoots forming dense, hard cushions, but remaining more or less distinct, not merging into the mass of foliage. Leaves 3–5 × 1–1.5 mm, oblong, obtuse or with a very short point, glaucous; margin entire; calcareous encrustation prominent. Flowering-stems 3–8 cm, terminating in a small flat-topped cyme of 3–5 flowers. Petals 7–9 mm, white. *W Alps.* H2. Spring.

Prefers an alkaline soil in partial shade.

69. S. tombeanensis Engler. Illustration: Horny et al., Porophyllum saxifrages, f. 41 (1986); Webb & Gornall, Saxifrages of Europe, pl. 10, 11 (1989); Aeschimann et al., Flora alpina **1**: 695 (2004).

Leafy shoots very tightly packed to form a dense, moderately hard cushion. Leaves 2–4.5 × 1–1.5 mm, narrowly oblong to elliptic-lanceolate, with a short point usually bent upwards, darkish green and only slightly glaucous; margin entire; calcareous encrustation slight or not apparent. Flowering-stems 3–7 cm, bearing a small, compact cyme of 1–4 flowers. Petals 8–12 mm, broadly obovate, touching, white. *SE Alps.* H2. Spring.

Prefers an alkaline soil in partial shade.

70. S. burseriana Linnaeus. Illustration: *Botanical Magazine*, 747 (1977); Horny et al., Porophyllum saxifrages, f. 1–19 (1986); Webb & Gornall, Saxifrages of Europe, pl. 12 (1989); Aeschimann et al., Flora alpina **1**: 695 (2004).

Leafy shoots forming a dense mat or cushion. Leaves 6–12 × 1.7–2 mm, mostly pointing upwards, narrowly lanceolate, apex tapered, acuminate, distinctly glaucous, with 5–7 chalk-glands but little evident calcareous encrustation; margin entire, except for a few tooth-like hairs at the base. Flowering-stems 2.5–5 cm, crimson, usually with a solitary flower.

Petals 7–12 mm (sometimes more in cultivars), white. *E Alps.* H2. Spring.

Variation in the size of petals and height of flowering-stems is continuous and difficult to treat. Epithets such as *crenata, major, minor* or *tridentina*, used by gardeners to describe particular clones, are best regarded as cultivars. Prefers an alkaline soil.

71. S. vandellii Sternberg. Illustration: Finkenzeller & Grau, Alpenblumen, 105 (1985); Horny et al., Porophyllum saxifrages, f. 40 (1986); Aeschimann et al., Flora alpina **1**: 695 (2004).

Leafy shoots densely crowded, forming a deep, hard cushion. Leaves 7–9 × 2–3 mm, dark green, narrowly triangular-lanceolate, tapered to a sharp point; margin entire except for a few tooth-like hairs at the base; chalk-glands 5–7, generating only a slight calcareous encrustation, which is soon gone. Flowering-stems 4–6 cm, bearing 3–6 flowers in a compact cyme. Petals 7–9 mm, obovate, touching, white. *Italian Alps.* H2. Spring.

Prefers an alkaline soil in partial shade; does best in a vertical crevice in tufa.

Series **Subsessiliflorae** Gornall. Leaves with a solitary chalk-gland in the margin of the usually thickened, truncated tip, which is not recurved. Flowers stemless or very nearly so, solitary. Petals longer than the stamens, touching or overlapping, white or cream. Ovary half inferior.

72. S. pulvinaria H. Smith (*S. imbricata* Royle, not Lamarck). Illustration: Horny et al., Porophyllum saxifrages, f. 92 (1986).

Leafy shoots forming a low cushion or mat. Leaves 2–4 × 0.5–1.5 mm, oblong, margin fringed with tooth-like hairs from near the tip to the base, lower surface with a prominent keel, green. Petals 4–5 mm, obovate, not recurved, more or less touching, white. *Sino-Himalayan area.* H2? Late spring.

Prefers an acid soil; best in a cool glasshouse.

73. S. lolaensis H. Smith. Illustration: Horny et al., Porophyllum saxifrages, f. 93 (1986).

Leafy shoots forming a dense cushion. Leaves *c.* 3 × 1.5 cm, oblong to elliptic, margin fringed with tooth-like hairs from near the tip to the base, or hairless. Petals *c.* 3.5 mm, broadly obovate to nearly circular, base contracted to a short claw, sometimes irregularly and shallowly lobed, recurved, overlapping, creamy-white. *China*

(SE Xizang), Nepal? H2? Late winter–early spring.

Prefers an acid soil; best grown in a cool glasshouse.

74. S. matta-florida H. Smith. Illustration: Horny et al., Porophyllum saxifrages, f. 89 (1986).

Leafy shoots forming a dense, compact cushion. Leaves 4–6 × 1.5–2 mm, linear to oblong, margin more or less entire except possibly for a few minute teeth. Petals 5–6 mm, broadly obovate, base tapered to a short claw, recurved, overlapping, white. *Bhutan, China (SE Xizang).* H2? Spring.

Prefers an acid soil; best grown in a cool glasshouse.

75. S. hypostoma H. Smith. Illustration: Evans, Alpines '81, 27 (1981); Horny et al., Porophyllum saxifrages, f. 88 (1986).

Leafy shoots forming a dense, compact cushion. Leaves *c.* 4 × 1.5 mm, linear to linear-obovate, tip margin fringed with tooth-like hairs, base entire, green. Petals *c.* 4 mm, broadly obovate, tapered to a short claw, margin wavy, recurved, overlapping, white, set round a glistening green ovary. *Nepal.* H2? Spring.

Prefers an acid soil; best grown in a cool glasshouse.

76. S. lowndesii H. Smith. Illustration: Horny et al., Porophyllum saxifrages, f. 95 (1986); *Bulletin of the Alpine Garden Society* **58**(1): 39 (1990); Harding, Saxifrages, 54 (1992).

Leafy shoots forming a loose cushion. Leaves 5–7 × 2–2.5 mm, linear-obovate to spathulate, soft, obtuse but not markedly thickened at the tip, margin sparingly fringed, base with a few short, glandular-hairs, shiny green. Petals *c.* 7 mm, nearly circular, base tapered to a short claw, overlapping, bright rose-lilac. Ventral walls of the carpels united only at the base (rather than to at least halfway). *Nepal.* H2? Spring.

Subsection **Engleria** (Sündermann) Gornall. Leaves alternate; margin with a translucent border. Flowering-stems distinct, bearing an inflorescence with coloured bracts and sepals. Flowers broadly ellipsoid, with erect sepals, which usually exceed and largely hide the petals. Petals pink to purple, white or yellow, the margins fringed with hairs toward base. Ovary two-thirds or more inferior.

77. S. media Gouan (*S. calyciflora* Lapeyrouse). Illustration: *Botanical Magazine,* 7315 (1893); *Bulletin of the Alpine Garden Society* **23**(1): 39 (1955); Horny et al., Porophyllum saxifrages, f. 48 (1986); Webb & Gornall, Saxifrages of Europe, pl. 15 (1989).

Leafy shoots terminating in prominent rosettes, arranged into a cushion. Leaves 6–17 × 2–4 mm, linear-oblong to oblanceolate, acute, slightly glaucous; margin entire; chalk-glands 7–17, producing a calcareous encrustation of variable density. Flowering-stems 3–12 cm, reddish pink, sometimes with a green tinge. Flowers 2–12, all stalked, in a raceme-like or sparingly branched panicle. Petals 3.5–4 mm, obovate to nearly circular, bright pinkish purple. *Pyrenees.* H2. Late spring.

Prefers an alkaline soil; best grown in a cool glasshouse.

78. S. stribrnyi (Velenovsky) Podpera. Illustration: *Botanical Magazine,* 8946 (1913); Horny et al., Porophyllum saxifrages, f. 55–8 (1986); Webb & Gornall, Saxifrages of Europe, pl. 16 (1989).

Like *S. media* but with larger, more spathulate leaves, and a more freely branched panicle (with 3 or 4 flowers on each primary branch, rather than the 1 or 2 in *S. media*). *Balkans.* H2. Spring.

Prefers an alkaline soil; best grown in a cool glasshouse.

79. S. corymbosa Boissier (*S. luteoviridis* Schott & Kotschy). Illustration: Horny et al., Porophyllum saxifrages, f. 53–4 (1986); Webb & Gornall, Saxifrages of Europe, pl. 18 (1989).

Like *S. media* but flowers in a crowded cyme; petals pale greenish yellow. *SE Europe, Balkans, Turkey.* H2. Spring.

Prefers an alkaline soil.

80. S. porophylla Bertoloni. Illustration: Horny et al., Porophyllum saxifrages, f. 52 (1986); Webb & Gornall, Saxifrages of Europe, f. 20 (1989); Harding, Saxifrages, 35 (1992).

Leafy shoots short, forming distinct rosettes aggregated into a cushion. Leaves 4–10 × 2–3 mm, obovate-spathulate to oblong-oblanceolate, tip obtuse or with a short point, glaucous; margin entire; chalk-glands 5–11, producing a calcareous encrustation of variable density. Flowering-stems 3–8 cm, pink to bright rose-pink. Flowers 4–7, rarely to 12, at least some without stalks, in a slender raceme. Petals *c.* 1.5 mm, obovate, white, sometimes with

a pink stripe, becoming pink with age. *Italy (Apennines).* H2. Late spring.

Prefers an alkaline soil.

81. S. federici-augusti Biasoletto. Illustration: *Botanical Magazine,* 8308 (1910); Horny et al., Porophyllum saxifrages, f. 49–51 (1986); Webb & Gornall, Saxifrages of Europe, pl. 14 (1989); Harding, Saxifrages, 18, 35 (1992).

Like *S. porophylla* but leaves larger, 1.2–3.5 cm × 3–8 mm; inflorescence cherry-red to dark purple-crimson, bearing 10–20 flowers with purplish pink petals. *Balkan Peninsula.* H2. Spring.

Subsp. **federici-augusti** (*S. montenegrina* Engler & Irmscher). Basal-leaves to 1.9 cm, obovate-lanceolate; inflorescence dark purple-crimson, with up to 16 flowers.

Subsp. **grisebachii** (Degen & Dörfler) D.A. Webb (*S. grisebachii* Degen & Dörfler). Basal-leaves to 3.5 cm, usually spathulate; inflorescence crimson to cherry-red, with up to 20 flowers.

Prefers an alkaline soil.

82. S. sempervivum C. Koch. Illustration: Köhlein, Saxifrages and related genera, t. 23 (1984); Horny et al., Porophyllum saxifrages, f. 59–62 (1986).

Like *S. porophylla* but leaves, 5–20 × 1–2 mm, narrower, linear to linear-oblong, apex never widened; flowering-stems 6–20 cm, with 10–20 flowers on well-grown plants; petals reddish purple. *Balkan Peninsula, Turkey.* H2. Spring.

Very narrow-leaved variants have been distinguished as *S. thessalica* Schott or, in gardens, as forma **stenophylla** Boissier. All intermediates occur, however. An albino variant with a pale green inflorescence is in cultivation under the name 'Zita'. Prefers an alkaline soil; best grown in a cool glasshouse.

Subsection **Oppositifoliae** Hayek. Leaves usually opposite, margin without a translucent border. Flowering-stems very short or obsolete. Flowers in small cymes of 1–3, the sepals suberect or spreading, shorter than and not concealing the petals. Petals deep purple, pink or white. Ovary inferior or nearly so.

Series **Oppositifoliae** (Hayek) Gornall. Leafy stem not perfoliate, the leaf-pairs with margins meeting at an acute angle, with tooth-like hairs at the base. Chalk-glands usually 3–7 (rarely solitary). Flowers 1–3 per stem.

83. S. oppositifolia Linnaeus. Illustration: Ross-Craig, Drawings of British plants **10**: pl. 15 (1957); Köhlein, Saxifrages and related genera, t. 8 (1984); Webb & Gornall, Saxifrages of Europe, pl. 17, f. 21 (1989); Harding, Saxifrages, cover, 99 (1992); Aeschimann et al., Flora alpina **1**: 697 (2004).

Stems branched, forming a mat or loose cushion of leafy shoots. Leaves mostly 2–5 × 1.5–2 mm, elliptic or obovate, tip almost acute to rounded, margins entire but with bristly hairs at least near the base; chalk-glands 1–5, producing a calcareous encrustation variable in its intensity. Flowering-stems 1–2 cm, with a solitary flower. Petals mostly 5–12 × 2–7 mm, obovate to elliptic-oblong, pink to deep purple (rarely white), fading to violet. *Circumboreal, extending to the Altai & Himalaya.* H1. Spring–summer.

The species is very variable and has been split into a large number of somewhat ill-defined subspecies and varieties. The following subspecies are the chief sources of garden plants.

Subsp. **oppositifolia**. Leafy shoots forming a mat. Leaves usually oblong to narrowly obovate, hairy up to the tip. Flowering-stems well-developed. Hairs on the sepals usually non-glandular. Petals usually 7–12 mm. *Circumpolar.*

Plants with glandular-hairy sepals have been distinguished as *S. latina* Terraciano (with densely overlapping leaves) or S. murithiana Tissière, but intermediates to subsp. *oppositifolia* always occur.

Subsp. **rudolphiana** (Koch) Engler & Irmscher. Leafy shoots short, dense, forming a tight, flattish cushion. Leaves less than 2 mm long, oblong-obovate, hairy in the basal half only. Flowering-stems very short, nearly obsolete. Sepals with glandular hairs. Petals 5–7 mm. *Alps.*

Subsp. **blepharophylla** (Hayek) Engler & Irmscher. Leafy shoots columnar, forming a compact cushion. Leaves 3–5 mm, broadly obovate, hairy up to the tip. Flowering-stems *c.* 5 mm. Sepals with long, non-glandular hairs. Petals 5–8 mm. *Austrian Alps.*

Subsp. **speciosa** (Dörfler & Hayek) Engler & Irmscher. Leafy shoots forming a moderately compact cushion. Leaves 4–5 mm, broadly obovate to nearly circular, hairy in the basal half. Flowering-stems less than 5 mm. Sepals with non-glandular hairs. Petals 8–12 mm. *Italy (C Apennines).*

Usually prefers a slightly alkaline soil in partial shade.

84. S. retusa Gouan (*S. purpurea* Allioni). Illustration: Webb & Gornall, Saxifrages of Europe, pl. 21 (1989); Aeschimann et al., Flora alpina **1**: 697 & 699 (2004).

Like *S. oppositifolia* but with shiny leaves, often more than one flower per stem, sepals lacking marginal hairs. *Pyrenees, Alps, Tatra, Carpathians & Rila mountains.* H2. Late spring–summer.

Subsp. **retusa**. Flowering-stems to 1.5 cm, bearing 1–3 flowers with hairless perigynous zones.

Subsp. **augustana** (Vaccari) Fournier. Flowering-stems to 5 cm, bearing 3–5 flowers with hairy perigynous zones.

The former subspecies prefers an acid soil, whereas the latter prefers an alkaline one; both do better under glass.

85. S. biflora Allioni (*S. kochii* Bluff, Nees & Schauer; *S. macropetala* A. Kerner). Illustration: Webb & Gornall, Saxifrages of Europe, pl. 20 (1989); Harding, Saxifrages, 100 (1992); Aeschimann et al., Flora alpina **1**: 699 (2004).

Like *S. oppositifolia* but with soft, fleshy, nearly circular leaves; petals widely separated, dull purple or white; prominent yellow nectary disc in the centre of flower. *Alps, NW Greece.* H2. Summer.

Subsp. **biflora**. Leaves with 1 chalk-gland and negligible calcareous encrustation. Flowers usually at least 2 per stem, barely raised above the leaves. Sepals hairy beneath.

Subsp. **epirotica** D.A. Webb. Leaves with 3–5 chalk-glands and prominent calcareous encrustation. Flowers solitary, well above the leaves. Sepals hairless beneath.

Best grown in a cool glasshouse.

86. S. vacillans H. Smith. Illustration: Horny et al., Porophyllum saxifrages, f. 84 (1986).

Leafy shoots forming a fairly compact cushion. Leaves 3.5–5 × 1.5–2 mm, linear, acute, margin fringed with small tooth-like hairs at the base, mostly opposite but alternate on vigorous shoots, with 3–7 chalk-glands. Flowering-stems 1.5–2 cm, bearing 1–2 flowers. Petals 5–6 mm, obovate, margin with irregular, shallow lobes, white. *Bhutan.* Late spring.

Doubtful whether in cultivation.

Series **Tetrameridium** (Engler) Gornall. Leafy stem perfoliate, with margins of the leaf-pairs completely confluent, entire,

lacking basal hairs, usually with a solitary chalk-gland. Flowers solitary.

87. S. quadrifaria Engler & Irmscher. Illustration: Horny et al., Porophyllum saxifrages, f. 79 (1986).

Leafy shoots forming a dense, compact cushion. Leaves 2–2.5 × 1–1.5 mm, ovate, thickened at the somewhat truncated tip, green. Flowers almost stemless. Petals 3–4.5 mm, obovate, more or less touching, white. *China (SE Xizang), Nepal.* H2? Spring.

Prefers an acid soil in partial shade; best grown in a cool glasshouse.

88. S. georgei Anthony. Illustration: Horny et al, Porophyllum saxifrages, f. 81 (1986); Harding, Saxifrages, 100 (1992).

Leafy shoots forming a loose cushion. Leaves 3–3.5 × 2–3 mm, oblong-obovate, tip obtuse, light green. Flowers stemless or nearly so, 4–5-parted. Petals *c.* 5 mm, broadly obovate, base tapered to a short claw, barely touching, white or pale pink. *Nepal, Bhutan, China (Sichuan, NW Yunnan, SE Xizang).* H2? Spring.

Best grown in a cool glasshouse.

89. S. alpigena H. Smith. Illustration: Horny et al., Porophyllum saxifrages, f. 82 (1986).

Like *S. georgei* but leaves 1.5–2.5 × 1–2 mm, ovate to obovate. Flowers 5-parted. Petals *c.* 6 mm, obovate, base tapered to a long claw, white. *Nepal.* H2? Spring.

Prefers an acid soil; best grown in a cool glasshouse.

Section **Ligulatae** Haworth. Evergreen perennials. Stems somewhat woody at the base, producing leaf-rosettes that aggregate to form cushions or mats. Leaves alternate, somewhat fleshy, hard or leathery, usually glaucous, without a distinct stalk, margins entire or finely toothed, often with a translucent border. Chalk-glands located in pits, nearly always evident by the conspicuous calcareous encrustation on the leaves. Flowering-stems leafy, produced from monocarpic leaf-rosettes (sometimes the entire plant is monocarpic), bearing an often large and many-flowered panicle. Petals usually white, often with red spots, rarely pink, yellow or orange. Ovary at least half inferior; capsule splitting above the middle. The rosette bearing the flowering-stem dies after flowering but, in those species which are not monocarpic, offsets arise from the base of the flowering rosette. The species and their hybrids are well known to gardeners as the 'silver' or 'encrusted' saxifrages; hybrids and cultivars are treated in more detail by Köhlein (1984) and Harding (1992).

90. S. longifolia Lapeyrouse. Illustration: Schacht, Rock gardens and their plants, 125 (1963); Köhlein, Saxifrages and related genera, t. 18, 19 (1984); Webb & Gornall, Saxifrages of Europe, f. 22, 33 (1989).

Stem normally unbranched, producing a single leaf-rosette to 15 cm across, with up to 200 living leaves. Leaves 6–11 cm × 4–7 mm, linear, sometimes slightly expanded below the acute tip, margin entire with the numerous chalk-glands set into it and not on the upper surface. Flowering-stem to 60 cm, branched from near the base to form a cylindric to conical panicle with up to 800 flowers, the primary branches bearing 4–10 flowers. Petals *c.* 7 mm, obovate, white, sometimes with crimson spots. Ovary inferior. The species is strictly monocarpic; during flowering the leaves die, and by the time seed is ripe, the plant is dead. *Pyrenees, locally in E Spain, Morocco (High Atlas).* H2. Summer.

Hybridises in the wild with *S. cotyledon*: **S. × superba** Rouy & Camus (*S. × imperialis* invalid; *S. × splendida* invalid); and *S. paniculata*: **S. × lhommei** Coste & Soulié. Garden hybrids with *S. callosa* (*S. × calabrica* invalid) and *S. cochlearis* are grown. That with the former is known as 'Tumbling Waters', that with the latter may include 'Dr Ramsey' and 'Francis Cade', but there is little firm evidence. Prefers an alkaline soil in partial shade; grow in a vertical crevice for full effect.

91. S. callosa Smith (*S. lingulata* Bellardi; *S. lantoscana* Boissier & Reuter). Illustration: *Botanical Magazine*, 8434 (1912); Köhlein, Saxifrages and related genera, t. 14 (1984); Webb & Gornall, Saxifrages of Europe, f. 24 (1989); Aeschimann et al., Flora alpina **1**: 699 (2004).

Leaf-rosettes formed in clumps, offsets usually present, the larger rosettes to 16 cm across. Leaves 4–9 cm × 2.5–7 mm, linear, sometimes expanded near the tip, hard; margin entire except for a few non-glandular hairs near base, numerous chalk-glands set into it and not on the upper surface, calcareous encrustation variable. Flowering-stem 15–40 cm, bearing a many-flowered, narrow panicle which occupies 40–60% of the stem,

primary branches bearing 3–7 flowers. Petals 6–12 mm, obovate to oblanceolate, sometimes with a long claw, white, sometimes with crimson spots near the base. Ovary about three-quarters inferior. *NE Spain, SW Alps, Apennines to S Italy, Sicily & Sardinia.* H2. Late spring–summer.

Subsp. **callosa.** Inflorescence hairless or sparingly glandular-hairy. Plants with leaves linear, scarcely expanded near the tip are known as var. **callosa.** *SW Alps, N Italy.*

Those with leaves oblanceolate to linear, with a more or less diamond-shaped tip are called var. **australis** (Moricand) D.A. Webb. *SW Alps, C & S Italy, Sicily, Sardinia.*

Plants from the western Maritime Alps are commonly found in gardens under the name var. *lantoscana*, but they scarcely differ from var. *australis*.

Subsp. **catalaunica** (Boissier & Reuter) D.A. Webb. Inflorescence fairly densely glandular-hairy. Leaves oblanceolate, fairly short. *NE Spain.*

Hybrids with *S. cotyledon*: **S. × macnabiana** Lindsay (*S. × lindsayana* Engler & Irmscher), *S. cochlearis* (*S. × farreri* invalid) and *S. hostii* (*S. × florairensis* invalid?) have been recorded. The artificial hybrid *S. callosa* × *S. paniculata* probably includes the commonly-grown cultivar 'Kathleen Pinsent'; see *S. longifolia* for a hybrid with that species. Prefers an alkaline soil.

92. S. cochlearis Reichenbach. Illustration: *Botanical Magazine*, 6688 (1883); Köhlein, Saxifrages and related genera, t. 15 (1984); Finkenzeller & Grau, Alpenblumen, 101 (1985); Webb & Gornall, Saxifrages of Europe, f. 25 (1989).

Leaf-rosettes produced in a dense, irregular cushion. Leaves to 45 × 7 mm, usually much less, spathulate to oblanceolate, leathery and fleshy, base often tinged red; margin entire, numerous chalk-glands set into it and not on the upper surface, base with a few hairs. Flowering-stems 5–30 cm, reddish, branched in the upper third to form an open, densely glandular-hairy (at least in the lower half) panicle of 15–25 flowers, rarely to 60. Petals 7–11 mm, oblong-obovate, white, sometimes with crimson spots near the base. Ovary three-quarters inferior. *Maritime Alps, Italy (Portofino Peninsula).* H2. Late spring–summer.

A commonly grown dwarf variant is known by the cultivar name 'Minor'. The cultivar 'Major', also known by the invalid

name *S. cochleata*, may be of hybrid origin.

S. × burnatii Sündermann; *S. cochlearis* × *S. paniculata*. Recorded once from the Maritime Alps; 'Esther' and 'Whitehill' are commonly grown cultivars probably of this parentage. For hybrids with *S. callosa* and *S. longifolia* see under those species. Prefers an alkaline soil.

93. S. crustata Vest (*S. incrustata* invalid; *S. vochinensis* invalid). Illustration: Finkenzeller & Grau, Alpenblumen, 103 (1985); Webb & Gornall, Saxifrages of Europe, f. 26 (1989); Aeschimann et al., Flora alpina **1**: 701 (2004).

Leaf-rosettes 2.5–8 cm across, crowded to form a thick cushion. Leaves usually 1–2.5 cm × 2–3 mm, linear, tip obtuse and scarcely expanded, margin with a few long hairs at the base, entire or very slightly scalloped; chalk-glands numerous, sunk into the margin. Flowering-stems 12–35 cm, glandular-hairy, branched from the middle or above to form a panicle of up to 35 flowers, each primary branch usually with 1–3 flowers aggregated near the tip. Petals 5–6 mm, obovate, white, rarely with red spots. Ovary inferior. *E Alps, N & C former Yugoslavia.* H2. Summer.

Natural hybrids with *S. hostii* (**S. × engleri** Huter; *S. × paradoxa*, invalid) and *S. paniculata* (**S. × pectinata** Schott, Nyman & Kotschy; *S. × fritschiana* invalid; *S. × portae* invalid) are known. Prefers an alkaline soil.

94. S. hostii Tausch (*S. elatior* Mertens & Koch). Illustration: Webb & Gornall, Saxifrages of Europe, f. 27, 28 (1989); Aeschimann et al., Flora alpina **1**: 701 (2004).

Leaf-rosettes 6–18 cm across, forming a loose mat. Leaves 3–10 cm × 4–10 mm, oblong to broadly linear, tip scarcely expanded, more or less glaucous; margins finely toothed, with the chalk-glands situated on the upper surface of the minute marginal teeth, fringed with a few long hairs at the base. Flowering-stems 25–50 cm, glandular-hairy above; branched in the upper half to form a panicle, the primary branches of which usually bear 5–12 flowers. Petals 4–8 mm, elliptic-obovate, white often with purple-red spots. Ovary inferior. *E Alps.* H2. Late spring–summer.

Subsp. **hostii** (*S. hostii* var. *altissima* (Kerner) Engler & Irmscher). Basal leaves to 10 cm, at least 6 mm across, tip obtuse, marginal teeth distinct. Primary branches of panicle with at least 5 flowers. *NE Italy to SE Austria & NW Slovenia.*

Subsp. **rhaetica** (Kerner) Braun-Blanquet. Basal leaves to 5 × 0.7 cm, tapered to an acute tip, marginal teeth obscure. Primary branches of panicle with 3–5 flowers. *N Italy.*

Natural hybrids with *S. paniculata* (**S. × churchillii** Huter) occur; for those with *S. callosa* and *S. crustata* see under those species. Prefers an alkaline soil.

95. S. paniculata Miller (*S. aizoon* Jacquin; *S. recta* Lapeyrouse; *S. zelebori* invalid?). Illustration: Finkenzeller & Grau, Alpenblumen, 103 (1985); Webb & Gornall, Saxifrages of Europe, f. 29 (1989); Aeschimann et al., Flora alpina **1**: 703 (2004).

Leaf-rosettes 1–9 cm across, aggregated into a loose cushion. Leaves mostly 8–35 × 4–5 mm, obovate-oblong to broadly linear, tip obtuse to acute or acuminate, stiff and rather fleshy; margin with forward-pointing teeth, these grading into hairs at the base, chalk-glands situated on the upper surface at the base of each tooth. Leaves tending to curve upwards making the rosette hemispherical rather than flat as in other species. Flowering-stems 6–40 cm, variably glandular-hairy, branched in the upper half to form a narrow panicle, the primary branches of which usually bear 1–3 flowers. Petals 3–6 mm, elliptic-obovate, white, rarely pink or pale yellow, often with reddish purple spots. Ovary three-quarters inferior. *E North America, Greenland, Iceland, S Norway, N Spain through C & S Europe to Caucasus.* H2. Late spring–summer.

A very variable species but one in which most of the variation is continuous. Populations from the Caucasus and neighbouring regions, however, have acuminate leaf-tips and sometimes pink or crimson petals, and these are designated as subsp. **cartilaginea** (Willdenow) D.A. Webb (*S. kolenatiana* Regel; *S. sendtneri* invalid). Populations from the Balkans with a rather similar leaf-shape have been assigned to *S. aizoon* var. *orientalis* Engler. Plants from Monte Baldo, Italy, with neat, crowded rosettes of small leaves (*c.* 5 × 1.5 mm) and reddish flowering-stems are known to gardeners as *S. aizoon* var. *baldensis* Farrer.

There is a host of other variants in gardens to which many names have been given; taken in the context of the variation pattern in the wild, however, it is not practical to give any of them recognition, except possibly as cultivars. Accounts of a representative selection are given by Harding (1992). For hybrids with *S. callosa*, *S. cochlearis*, *S. crustata*, *S. cotyledon*, *S. hostii*, *S. longifolia*, *S. aizoides* and *S. cuneifolia* see under those species.

S. × andrewsii Harvey (possibly a hybrid between *S. paniculata* and *S. spathularis*; *S. × guthrieana* invalid) looks very like a hybrid between sections *Ligulatae* and *Gymnopera*. Its origin is obscure but was probably made accidentally in an Irish garden. It is more or less intermediate between the parents, although the leaves are longer and narrower than might be expected. The species is variable in its preference for acidic or alkaline soils.

96. S. cotyledon Linnaeus (*S. pyramidalis* Lapeyrouse; *S. montavoniensis* Kolb; *S. nepalensis* misapplied; *S. linguaeformis* invalid). Illustration: *Bulletin of the Alpine Garden Society* **22**(1): 50 (1954); Köhlein, Saxifrages and related genera, t. 16 (1984); Webb & Gornall, Saxifrages of Europe, f. 30, 31 (1989); Harding, Saxifrages, 117 (1992).

Principal leaf-rosette 7–12 cm across, usually accompanied by smaller rosettes arising from short axillary runners which soon die, so that the plant never forms large clumps or cushions. Leaves 2–8 cm × 6–20 mm, oblong to oblanceolate or spathulate, margin finely and regularly toothed except at the base where it is fringed with non-glandular hairs, slightly fleshy, not very glaucous; chalk-glands situated in the middle of each marginal tooth on the upper surface, bearing a small amount of calcareous encrustation. Flowering-stems to 70 cm, branched from near the base, or at least from below the middle, to form a pyramidal panicle; primary branch with 8–40 flowers. Petals 7–10 mm, oblanceolate, narrowed at the base to a claw, white, sometimes with red spots or veins. Ovary inferior. *Scandinavia & Iceland, Pyrenees, Alps.* H2. Summer.

Attempts to give taxonomic recognition to populations from the 3 areas listed above cannot be justified, although cultivar status may be appropriate if desired, e.g. 'Norvegica', 'Icelandica' and 'Caterhamensis'.

A natural hybrid with *S. paniculata* (**S. × gaudinii** Brügger; *S. × timbalii* Rouy & Camus; *S. canis-dalmatica* invalid; *S. × speciosa* invalid) is known. For hybrids with *S. callosa* and *S. longifolia* see under those species. The cultivar 'Southside Seedling', whose petals are blotched with crimson, is possibly a hybrid of *S. cotyledon*. Prefers acidic soils, but will grow in moderately calcareous conditions; probably better outside than under glass.

97. S. valdensis de Candolle (*S. compacta* Sternberg; *S. rupestris* Seringe). Illustration: Webb & Gornall, Saxifrages of Europe, f. 32 (1989); Aeschimann et al., Flora alpina **1**: 703 (2004).

Leaf-rosettes 1–3 cm across, aggregated into a dense, hard, mounded cushion. Leaves 3–8 × 2–3 mm, obovate to oblanceolate or spathulate, tip obtuse and somewhat curved downwards, margin entire except for a few hairs fringing the base and lacking a distinct translucent border, markedly glaucous, some chalk-glands set in the margin and others scattered on the upper surface. Flowering-stems 3–11 cm, branched above the middle to form a panicle of 6–12 flowers. Petals 4–5 mm, obovate, white, without spots. Ovary inferior. *SW Alps.* H2. Summer.

Usually prefers alkaline soils.

98. S. mutata Linnaeus. Illustration: Webb & Gornall, Saxifrages of Europe, f. 33 (1989); Harding, Saxifrages, 93 (1992); Aeschimann et al., Flora alpina **1**: 703 (2004).

Leaf-rosettes 5–15 cm across, often solitary, but sometimes with a few offsets. Leaves 2.5–7 cm × 7–15 mm, oblong-oblanceolate, tip obtuse, margin irregularly toothed in the middle section, with a prominent translucent border, base fringed with hairs; somewhat fleshy; chalk-glands situated on the upper surface inside the translucent border, but producing little, if any, calcareous encrustation. Flowering-stem 10–50 cm, bearing a narrow panicle with a very variable number of flowers. Petals 6–8 mm, linear, acute at the tip. Ovary three-quarters inferior. *Alps, S Carpathians, Low Tatra Mts.* H2. Summer–early autumn.

Subsp. **mutata** has the flowering-stem branched from the middle or above. *Alps, Low Tatra Mts.*

Subsp. **demissa** (Schott & Kotschy) D.A. Webb has the flowering-stem branched from or near the base. *S Carpathians.*

For hybrids with *S. caesia* and *S. aizoides* see under those species. Prefers an alkaline soil with a good supply of water.

99. S. florulenta Moretti. Illustration: *Botanical Magazine*, 6102 (1874); Rasetti, I fiori delle Alpi, pl. 263–4 (1980); Heath, Collectors' alpines, their cultivation in frames and alpine houses, 441 (1981); Harding, Saxifrages, 118 (1992).

Monocarpic perennial. Leaf-rosette 5–15 cm across, nearly always solitary. Leaves 3–6 × 0.4–0.7 cm, narrowly spathulate, with a short, spine-like tip; margin with a conspicuous translucent border, entire near the tip, becoming irregularly scalloped below, basal half fringed with tooth-like hairs; chalk-glands near the margin on the upper surface, producing little, if any, calcareous encrustation. Flowering-stem 10–25 cm, branched from near the base to form a dense cylindric panicle of numerous flowers. Petals 5–7 mm, oblanceolate, flesh-pink. Ovary inferior; carpels 3, rather than the 2 found in nearly all other species. *Maritime Alps.* H2. Summer.

Prefers a very well-drained acid soil in partial shade; difficult to grow.

Section **Xanthizoon** Grisebach. Evergreen perennial, forming a mat or cushion. Leaves fleshy, narrow, without a distinct stalk, margin more or less entire. Chalk-glands flush with the surface, not in pits, often without any apparent calcareous encrustation. Flowering-stems leafy, bearing flowers in a loose cyme. Petals yellow, orange or reddish. Ovary half inferior; capsule splitting above the middle.

100. S. aizoides Linnaeus (*S. autumnalis* Jacquin; *S. atrorubens* Bertoloni; *S. crocea* Gaudin). Illustration: Ross-Craig, Drawings of British plants **10**: pl. 14 (1957); Webb & Gornall, Saxifrages of Europe, pl. 22, 24 (1989); Aeschimann et al., Flora alpina **1**: 705 (2004).

Leafy shoots forming a thick mat or loose cushion. Leaves 4–22 × 1.5–4 mm, linear to oblong, tips obtuse, acute or with a short point; medium to darkish green, not glaucous; margin entire or occasionally with 2 short teeth near the tip, usually fringed with forward-pointing tooth-like hairs; chalk-glands usually solitary, near the tip. Flowers solitary or in short, leafy cymes of 2–15. Petals 3–7 mm, usually yellow, often with orange spots, sometimes orange or brick-red. Ovary half inferior, covered on top by a prominent nectary disc. *Arctic-alpine in Europe & North America.* H1. Summer–early autumn.

Var. **atrorubens** (Bertoloni) Sternberg with brick-red petals and deep crimson nectary disc is commonly grown.

The cultivar 'Primulaize' is traditionally supposed to be *S. aizoides* × *umbrosa* 'Primuloides'; **S.** × **larsenii** Sündermann (*S. aizoides* × *S. paniculata*) has been made artificially. Natural hybrids with *S. mutata* (**S.** × **hausmannii** Kerner; *S.* × *regelii* Kerner), *S. caesia* (**S.** × **patens** Gaudin) and *S. squarrosa* are also recorded. The species prefers neutral to alkaline, moist but well-drained soils.

Section **Trachyphyllum** (Gaudin) W.D.J. Koch. Evergreen perennials, forming loose cushions or mats. Leaves narrow, usually lanceolate, without a distinct stalk, stiff, margin entire or 3-lobed at tip, the latter with a short point, fringed with stout, often hooked, hairs. Chalk-glands absent. Axillary buds often prominent but not summer-dormant. Flowering-stems leafy, bearing 1–several flowers in a cyme. Petals white to pale yellow, sometimes with reddish spots or a yellow patch at the base. Ovary superior or nearly so; capsule splitting above the middle.

101. S. aspera Linnaeus (*S. hugueninii* Brügger; *S. etrusca* Pignatti). Illustration: Heathcote, Plants of the Engadine, pl. 91 (1891); Aeschimann et al., Flora alpina **1**: 705 (2004).

Leafy shoots forming an irregular mat, with nearly spherical leaf-rosettes. Leaves of prostrate shoots 5–8 mm, pressed close to the stem at first but spreading later; axils with conspicuous leafy buds at the time of flowering, which are shorter than the subtending leaf. Flowering-stems 7–22 cm, usually bearing 2–7 flowers in an open cyme. Petals 5–7 mm, oblong, base narrowed to a very short claw, white or pale cream, often with a deep yellow patch at the base and reddish spots near the middle. *Pyrenees, Alps, N Apennines.* H2. Summer.

Probably best grown under glass to avoid damage by birds.

102. S. bryoides Linnaeus. Illustration: Webb & Gornall, Saxifrages of Europe, pl. 23, f. 34 (1989); Aeschimann et al., Flora alpina **1**: 705 (2004).

Leafy shoots forming a dense mat or low cushion, with nearly spherical leaf-rosettes. Leaves of prostrate shoots 3.5–8 × 1–1.5 mm, oblong-lanceolate, incurved and remaining so, tip with a short point, surface shiny; axils with leafy buds at time of flowering, which equal or exceed the subtending leaf. Flowering-stems 2–5 cm, bearing a solitary flower. Petals 5–7 mm,

elliptic-oblong to obovate, white with a large patch of deep yellow at the base and some reddish spots near the middle. *High mountains of Europe.* H2. Summer–early autumn.

103. S. bronchialis Linnaeus (*S. spinulosa* Adams). Illustration: Hultén, Flora of Alaska and neighbouring territories, 570 (1968).

Leafy shoots forming a thick mat or low cushion. Leaves usually 5–12 × 1–2.5 mm, narrowly oblong to oblong-lanceolate, tip tapered to a white spine; axils with leafy buds, inconspicuous at flowering. Flowering-stems 5–20 cm, bearing 2–12 flowers in a fairly compact cyme. Petals 3.5–8 mm, oblong-elliptic, yellowish white, variously spotted with yellow or crimson. *Arctic-alpine areas from the Urals eastwards to the Rocky Mts.* H1. Summer.

104. S. cherlerioides D. Don (*S. stelleriana* Merk). Illustration: Hultén, Flora of Alaska and neighbouring territories, 570 (1968).

Like *S. bronchialis* but shoots columnar, leafy; leaves incurved, overlapping, tip mucronate, marginal hairs without glands; flowering-stems 2–6.5 cm. *Aleutian Islands, Kamchatka, E Siberia.* H2. Summer.

105. S. tricuspidata Rottböll. Illustration: Hultén, Flora of Alaska and neighbouring territories, 571 (1968).

Leafy shoots forming a mat or low cushion. Leaves 6–15 × 1.5–6.5 mm, oblanceolate to obovate, tip usually with 3 teeth; axillary buds inconspicuous. Flowering-stems 4–24 cm, bearing an open cyme. Petals 4–7 mm, elliptic, base truncate, white or cream, with yellow, orange and red spots in sequence from the base upwards. *Arctic-alpine North America, Greenland.* H2. Summer.

Section **Gymnopera** D. Don. Evergreen perennials with rather succulent, leathery, stalked leaves in basal rosettes, margins toothed or scalloped. Chalk-glands absent. Flowering-stem leafless, terminating in a many-flowered panicle. Sepals reflexed. Petals white, usually with yellow and reddish pink spots. Ovary superior; capsule splitting above the middle.

106. S. cuneifolia Linnaeus. Illustration: Köhlein, Saxifrages and related genera, t. 9 (1984); Webb & Gornall, Saxifrages of Europe, f. 9, 10 (1989); Aeschimann et al., Flora alpina **1**: 693 (2004).

Stems prostrate, runner-like, producing leaf-rosettes, ultimately forming a mat. Leaf-blade 8–25 × 7–22 mm, usually wedge-shaped, broadly ovate or nearly circular, apex often truncate; hairless; margin toothed, scalloped or entire, with a translucent border. Leaf-stalk flat, 0.4–1.3 times as long as the blade into which it grades, with a very few hairs at the extreme base. Flowering-stem 10–25 cm. Petals 3–5.5 mm, oblong, white, often with a yellow patch at the base and rarely with reddish spots near the middle. *Pyrenees, Alps, Carpathians.* H2. Late spring–summer.

Subsp. **cuneifolia** (*S. cuneifolia* var. *capillipes* Reichenbach; *S. capillaris* invalid) has stems prostrate, rosettes spaced 3–6 cm apart, leaves small (the largest to 2.5 cm, including the stalk), margin almost entire. Flowering-stem slender, nearly hairless, usually bearing up to 10 flowers. *S France (Maritime Alps), N Italy (Tuscany).*

Subsp. **robusta** D.A. Webb usually has leaf-rosettes less than 2 cm apart on the prostrate stems. Largest leaves more than 2.5 cm (including stalk), margin distinctly scallop-toothed. Flowering-stem robust, clearly glandular-hairy, usually bearing more than 10 flowers.

Variegated plants of the species are occasionally found in gardens and are best treated as cultivars, e.g. 'Aureo-maculata' and 'Variegata'.

107. S. umbrosa Linnaeus. Illustration: Keble Martin, The new concise British flora in colour, pl. 32 (1982); Garrard & Streeter, Wild flowers of the British Isles, 91 (1983); Webb & Gornall, Saxifrages of Europe, f. 12 (1989).

Evergreen perennial producing leaf-rosettes from prostrate stems, ultimately forming a low cushion. Leaf-blade 1.5–3 × 1–2 cm, oblong-elliptic, sometimes nearly obovate, usually hairless; margin with 5–10 shallow, forward-pointing scalloped teeth on either side; translucent border 0.2–0.3 mm wide. Leaf-stalk usually one third to half as long as the blade, flat, margin densely glandular hairy. Flowering-stem to 35 cm. Petals *c.* 4 mm, broadly elliptic, tapered to a short claw, white with crimson spots in the middle and 2 yellow ones near the base. *Pyrenees.* H2. Summer.

Var. **hirta** D.A. Webb. Plant dwarf; leaf-blades usually with some hairs. Very similar to those known by the invalid names of *S. umbrosa* var. *primuloides* or *S. umbrosa* var. *minor*, which are sometimes assigned to *S.* × *urbium* (see next

entry); of these 'Clarence Elliott' is popular. Best grown in a somewhat sheltered, shady spot in the garden.

108. S. spathularis Brotero (*S. hibernica* Sternberg; *S. serrata* Sternberg). Illustration: Ross-Craig, Drawings of British plants **10**: pl. 5 (1957); Garrard & Streeter, Wild flowers of the British Isles, 91 (1983); Webb & Gornall, Saxifrages of Europe, pl. 6, f. 13 (1989).

Evergreen perennial producing leaf-rosettes from prostrate, runner-like stems, forming a loosely tufted mat. Leaf-blade 1.5–5 × 1.2–3 cm, circular to oblong-elliptic, hairless; margin with 5–11 teeth on each side (3 or 4 in dwarf plants), teeth usually sharply, but occasionally bluntly, triangular; translucent border *c.* 0.1 mm wide. Leaf-stalk of at least some leaves longer than the blade, flat, with only a few marginal hairs. Flowering-stem to 50 cm. Petals *c.* 5 mm, elliptic, white with several crimson spots near the middle and 2 yellow spots near the base. *SW Europe, Ireland.* H3. Late spring–summer.

S. × **urbium** D.A. Webb (*S. umbrosa* misapplied; *S. spathularis* × *S. umbrosa*). Illustration: Phillips, Wild flowers of Britain, 67 (1977); Stace, New flora of the British Isles, 384 (1991). This garden hybrid is best distinguished from its parents by the leaf-stalk, which is barely longer than the blade, with at least a few hairs on its margins, marginal teeth of the blade projecting more nearly at right angles, and the translucent border of the blade measuring 0.2–0.25 mm in width.

One of the commonest saxifrages in gardens, often called 'London Pride'. Variegated plants are grown under the names 'Aureo-variegata', 'Aureo-punctata' and 'Variegata', are commonly naturalised or persisting in Britain.

S. × **andrewsii** Harvey, a putative hybrid with *S. paniculata*, is grown and is more or less intermediate between the parents. For a hybrid with *S. hirsuta* see under that species.

109. S. hirsuta Linnaeus (*S. geum* misapplied). Illustration: Ross-Craig, Drawings of British plants **10**: pl. 4 (1957); Garrard & Streeter, Wild flowers of the British Isles, 91 (1983); Webb & Gornall, Saxifrages of Europe, f. 14 (1989); Stace, New flora of the British Isles, 384 (1991).

Evergreen perennial producing sprawling, loose rosettes of leaves from prostrate, rhizome-like stems. Leaf-blade

1.5–4 × 1–5 cm, kidney-shaped to broadly elliptic or circular, hairs at least on the upper surface; margin usually 2–13-scalloped or -toothed on either side, translucent border inconspicuous. Leaf-stalk usually 2–3 times as long as the blade, nearly cylindric, hairy all round. Flowering-stem 12–40 cm. Petals 3.5–4 mm, oblong, white but usually with a yellow patch at the base and a few pink spots near the middle. *SW Europe, SW Ireland.* H3. Late spring–summer.

Subsp. **hirsuta** has a leaf-stalk at least twice as long as the blade, which is kidney-shaped to circular, 0.8–1.2 times as long as wide, and whose margin bears at least 6 crenations or teeth on each side. *Range of species.*

Subsp. **paucicrenata** (Gillot) D.A. Webb has a leaf-stalk only to 1.5 times as long as the blade, which is ovate or elliptic-oblong to nearly circular, 1.1–1.8 times as long as wide, and whose margin bears 2–6 crenations on each side. *Pyrenees.*

S. × **geum** Linnaeus; *S. hirsuta* × *S. umbrosa*. Illustration: Stace, New flora of the British Isles, 384 (1991). Distinguished from its parents by the leaf-stalk longer than the oblong blade, with hairs only on the margins of the stalk and only very sparsely scattered over the surface of the blade, margin scalloped. *Pyrenees.*

S. × **polita** (Haworth) Link; *S. hirsuta* × *S. spathularis*. Illustration: Stace, New flora of the British Isles, 384 (1991). Distinguished from its parents by the leaf-stalk longer than the nearly circular blade, hairs scattered sparingly all over the stalk as well as on the surface of the blade, teeth numerous, acutely pointed; translucent border 0.1–0.15 mm. *Ireland, Spain.*

Section **Cotylea** Tausch. Evergreen perennials. Leaves round or kidney-shaped, somewhat fleshy, with distinct stalks, margins scalloped, toothed or slightly lobed. Chalk-glands absent. Flowering-stem leafy, terminating in a many-flowered panicle. Sepals erect or spreading. Petals white, usually with red and yellow spots. Ovary superior; capsule splitting above the middle.

110. S. rotundifolia Linnaeus (*S. repanda* Willdenow; *S. lasiophylla* Schott; *S. olympica* Boissier). Illustration: *Botanical Magazine*, 424 (1798); Webb & Gornall, Saxifrages of Europe, pl. 7, f. 8 (1989); Harding, Saxifrages, 105 (1992); Aeschimann et al., Flora alpina **1**: 693 (2004).

Rhizomatous, producing clumps of leafy shoots with leaves in loose rosettes. Leaf-blade 1.7–4.5 × 3–8.5 cm, kidney-shaped to circular, base cordate, margin scalloped, toothed or palmately lobed. Leaf-stalk 4–18 cm. Flowering-stem 15–100 cm. Petals 6–11 × 2.5–5 mm, narrowly oblong to broadly elliptic, contracted to a short claw, white, usually with crimson-purple spots in the middle, orange to yellow at the base. *C & S Europe, SW Asia.* H2. Late spring–summer.

Subsp. **rotundifolia** has a uniformly narrow leaf-stalk and blade with a narrow, translucent border, usually without fringing hairs. *C & S Europe, SW Asia.* Var. **rotundifolia** has stellate flowers with narrow, spreading petals to 9 mm, lightly spotted. Var. **heucherifolia** (Grisebach & Schenk) Engler has cup-shaped flowers with wide, more or less erect petals to 11 mm, heavily spotted. Var. **apennina** D. A. Webb has large basal leaves, more than 7 cm wide; petals 10.5–11.5 mm, spreading, narrow, heavily spotted.

Subsp. **chrysosplenifolia** (Boissier) D.A. Webb (*S. chrysosplenifolia* Boissier) has a leaf-stalk that widens at the top to grade into the blade, which is nevertheless cordate but lacks a translucent border and is usually fringed with hairs. *Aegean & Balkan Peninsula.* Var. **chrysosplenifolia** has scalloped leaf-margins and its petals lack red spots. Var. **rhodopea** (Velenovsky) D.A. Webb has finely toothed leaf-margins and petals with red spots.

Best naturalised in woodland, except in very cold climates.

111. S. taygetea Boissier & Heldreich.

Like *S. rotundifolia* but of smaller stature, with smaller leaves, 0.5–1.3 × 0.8–2.3 cm, margin with up to 9 scalloped teeth versus at least 13 in *S. rotundifolia*. *Balkan Peninsula.* H2. Summer.

Section **Mesogyne** Sternberg. Usually winter-dormant perennials, with bulbils in the axils of the basal leaves, and sometimes also the stem-leaves, or replacing flowers. Basal leaves thin, with a slender leaf-stalk and a semicircular to kidney-shaped blade, palmately divided into 5–11 lobes. Chalk-glands absent. Flowering-stems leafy; flowers solitary or in a small cyme. Petals white or pink. Ovary superior to one third inferior or more; capsule splitting above the middle. All the species are best kept cold and dry in winter (cold-frame) and brought into an alpine house and watered in spring and summer.

112. S. sibirica Linnaeus. Illustration: Köhlein, Saxifrages and related genera, t. 11 (1984); Webb & Gornall, Saxifrages of Europe, pl. 25 (1989); Harding, Saxifrages, 106 (1992).

Stems solitary or in tufts. Basal leaf-blades 5–20 × 8–30 mm, kidney-shaped, palmately divided into 5–7 broadly ovate, obtuse lobes; stalk long, thin. Flowering-stems 5–18 cm, bearing 2–7 flowers in a compact cyme; stalks to 2.5 cm. Petals 7–14 cm, narrowly obovate, white. Ovary almost superior. *SE Europe, Turkey, Caucasus, C Asia, W Himalaya, S Siberia, China.* H2. Spring–summer.

113. S. cernua Linnaeus. Illustration: Webb & Gornall, Saxifrages of Europe, pl. 27 (1989); Aeschimann et al., Flora alpina 1: 707 (2004).

Stems solitary or in tufts. Basal leaf-blades 5–18 × 9–25 mm, semicircular to kidney-shaped, divided into 3–7, ovate to oblong, almost acute lobes; stalk 2–3 times as long as blade, base sheathing. Flowering-stem 3–30 cm, usually simple but sometimes branched from near the middle, bearing red to blackish purple bulbils in the axils of the stem-leaves; flowers usually solitary at the ends of each stem, or aborted. Petals 7–12 mm, obovate, white. Ovary almost superior. *Circumboreal arctic-alpine.* H1. Summer.

114. S. rivularis Linnaeus. Illustration: Ross-Craig, Drawings of British plants **10**: pl. 9 (1957); Keble Martin, The new concise British flora in colour, pl. 32 (1982); Garrard & Streeter, Wild flowers of the British Isles, 92 (1983); Webb & Gornall, Saxifrages of Europe, pl. 26 (1989).

Stems in tufts or cushions. Basal axillary bulbils germinating before flowering to form slender runners, which produce new plants at the tips. Basal leaf-blades usually 5–12 × 9–17 mm, semicircular or kidney-shaped, divided into 3–7 broadly ovate, obtuse lobes; stalk 2–6 times as long as blade, base sheathing. Flowering-stems 3–15 cm, usually bearing a solitary flower, but sometimes branched from near the middle and then bearing 2–5 flowers on long stalks. Petals 4–5 mm, obovate, white, sometimes tinged with pink. Ovary at least one third inferior. *Circumpolar Arctic, W North America.* H1. Late spring–summer.

Section **Saxifraga.** Usually perennial, occasionally annual or biennial; perennials mostly evergreen, but some summer-dormant perennating by bulbils. Habit varied but often with leafy shoots forming a cushion or mat. Leaves usually rather soft, often lobed or scalloped, usually with a distinct leaf-stalk. Chalk-glands absent. Flowering-stems usually leafy, bearing 1–several flowers in a cyme. Petals usually white, rarely yellowish, pink or red. Ovary half to fully inferior; capsule splitting above the middle.

Subsection **Saxifraga.** Perennials, more or less summer-dormant, forming loose rosettes of basal leaves, with axillary bulbils, or these in the axils of scales on the short stock; bulbils consisting of an outer series of loosely imbricate, papery scales with marginal or surface hairs, surrounding an inner series of fleshy scales. Leaves ovate to kidney-shaped, variously lobed. Flowering-stems terminal. Petals white, rarely tinged or veined with pink. Ovary half or more inferior.

115. S. granulata Linnaeus. Illustration: Ross-Craig, Drawings of British plants **10**: pl. 7 (1957); Phillips, Wild flowers of Britain, 31 (1977); Keble Martin, The new concise British flora in colour, pl. 32 (1982); Garrard & Streeter, Wild flowers of the British Isles, 91 (1983).

Plant forming loose rosettes of stalked basal leaves; bulbils clustered. Basal leaf-blades 6–30 × 8–50 mm, kidney-shaped, base cordate, margin scalloped or toothed, with 5–13 rounded or flat-topped lobes. Leaf-stalk to 5 cm, usually 2–5 times as long as the blade. Flowering-stem usually 10–30 cm, variably branched to form a panicle of 4–30 flowers. Petals 0.7–1.6 cm, obovate to broadly oblanceolate, white, rarely with red veins, hairless. Ovary three-quarters inferior. *Europe (except the SE), N Africa.* H3. Spring–summer.

A variable species, one variant of which, with doubled, sterile flowers, has been grown in gardens since the 17th century and is known as 'Flore Pleno'. For details of the 'mossy' saxifrages, a complex of hybrids between *S. granulata, S. exarata, S. rosacea,* and *S. hypnoides,* see under *S. exarata.* The species is best grown in partial shade in deep soil on the rock garden, or in a meadow.

116. S. corsica (Seringe) Grenier & Godron.

Like a small *S. granulata* but at least some basal leaves deeply 3-lobed. Flowering-stems usually branched, widely spreading, at or from near the base; flower-stalks thread-like, longer than the capsule (unlike *S. granulata* where they mostly equal the capsules). *E Spain, Balearic Islands, Corsica, Sardinia.* H5? Summer.

Subsp. **corsica**. Basal leaves divided to halfway; inflorescence branched at the base. *Corsica, Sardinia.*

Subsp. **cossoniana** (Boissier & Reuter) D.A. Webb. Basal leaves divided to more than halfway; inflorescence branched from above the base. *E Spain.*

Plants from the Balearic Islands (Formentera) are intermediate between the 2 subspecies, but slightly closer to the first.

117. S. haenseleri Boissier & Reuter. Illustration: Webb & Gornall, Saxifrages of Europe, pl. 30 (1989).

Basal leaf-blades 6–10 × 4–9 mm, somewhat wedge-shaped, rather deeply divided into 3–7 oblong, obtuse lobes, tapered to a leaf-stalk to 1.2 cm. Flowering-stem 8–30 cm, branched from the middle or the base, bearing a diffuse panicle of usually 6–12 flowers, but sometimes as many as 40. Petals 5–6 mm, narrowly obovate, white, sometimes with glandular hairs on the upper surface at the base. Ovary inferior. *S Spain.* H5? Spring–early summer.

118. S. dichotoma Willdenow (*S. hervieri* invalid; *S. albarracinensis* Pau). Illustration: Maire, Flore de l'Afrique du Nord **15**: f. 5 (1980).

Basal leaf-blades 5–18 × 7–30 mm, semicircular to fan-shaped, base truncate to wedge-shaped, divided about two-thirds of the way into 3–7 obtuse lobes, which may themselves be lobed, thus generating 5–15 ultimate segments. Leaf-stalk 1.5–4 times as long as the blade. Flowering-stem 6–25 cm, branched above the middle, bearing a narrow cyme of 2–7 flowers. Petals 5–10 mm, narrowly obovate, glandular-hairy, white, often veined or tinged pink. Ovary half to three-quarters inferior. *Spain, Morocco, Algeria.* H5? Spring.

Cultivation difficult.

119. S. carpetana Boissier & Reuter (*S. atlantica* Boissier & Reuter; *S. veronicifolia* Dufour; *S. graeca* Boissier & Heldreich). Illustration: Webb & Gornall, Saxifrages of Europe, pl. 31 (1989).

Basal leaf-blades circular to kidney-shaped, cordate, scalloped, with a long leaf-stalk. Flowering-stems 12–25 cm, branched above the middle to form a compact cyme of 4–13 flowers. Petals 8–12 mm, narrowly obovate, pure white, upper surface glandular-hairy. Ovary three-quarters inferior. *Mediterranean.* H5. Late spring–summer.

Subsp. **carpetana**. Later-formed basal leaf-blades ovate, base wedge-shaped or rounded, with a short leaf-stalk. Sepals *c.* 2 mm. *Iberian Peninsula, N Africa.*

Subsp. **graeca** (Boissier & Heldreich) D.A. Webb. All basal leaf-blades like first-formed. Sepals 2.5–3 mm. *Balkans, Italy, Algeria.*

Subsp. **graeca** can be grown in a pan under glass; subsp. *carpetana* is more difficult.

120. S. bulbifera Linnaeus. Illustration: Aeschimann et al., Flora alpina **1**: 707 (2004).

Basal leaves like those of *S. granulata* or *S. carpetana* but distinguished by the numerous axillary bulbils present on the flowering-stem, and by the compact inflorescence of usually 3–8 flowers. Petals 6–10 mm, obovate-oblong, upper surface glandular-hairy, white. *C & S Europe.* H4. Late spring–summer.

Best grown in a pan, outside and fairly dry in summer but under glass and rather wetter during autumn and winter.

121. S. biternata Boissier. Illustration: Harding, The genus Saxifraga, 20 (1970); *Botanical Magazine*, n.s., 670 (1974).

Stems sprawling, woody below, often with persistent leaf-stalks of dead leaves. Leafy shoots densely glandular-hairy, arranged in a diffuse cushion-like habit. Blade of largest leaves *c.* 4 × 4.5 cm, deeply divided into 3 primary lobes, each narrowed to a stalk-like base (the leaf thus appearing compound), primary lobes themselves 3-lobed, with the secondary lobes narrowed at the base to a short stalk, scalloped or divided again into segments at the tip. In a poorly grown plant the blade may consist only of 3 stalked, scalloped lobes. Leaf-stalks with a sheathing base. Flowering-stems to 10 cm, bearing a loose cyme of 2–6 (occasionally to 15) flowers. Petals 1.2–2 cm, broadly oblanceolate to obovate, white with green veins. Ovary inferior. *S Spain.* H5. Late spring–summer.

Best grown under glass in a poor, alkaline soil.

122. S. bourgaeana Boissier & Reuter.

Like *S. biternata* but with bulbils restricted to lowest leaf-axils, secondary lobes of leaf-blade not narrowed to a stalk, and shorter petals, 5–8 mm. *S Spain.* H5.

Subsection **Triplinervium** (Gaudin) Gornall. Biennials or evergreen perennials, often forming mats or cushions. Leaves obovate to kidney-shaped with the margins variously lobed, or linear to lanceolate with the margins entire. Bulbils absent but summer-dormant leafy buds sometimes present in leaf-axils. Flowering-stems terminal or axillary. Petals white to greenish yellow, sometimes tinged or veined with red. Ovary two-thirds or more inferior.

Series **Aquaticae** (Engler) Pawlowska. Biennials or evergreen perennials, forming mats or clumps. Leaves longer than 4 cm (including the leaf-stalk), with the leaf-stalk longer than the many-lobed blade, robust. Hairs, at least on the leaf-stalks, long and wavy. Petals white. Flowering-stems terminal or axillary. Ovary two-thirds or more inferior.

123. S. aquatica Lapeyrouse (*S. petraea* misapplied). Illustration: Webb & Gornall, Saxifrages of Europe, pl. 51 (1989).

Perennial, forming dense mats to 2 m across. Blade of basal leaves to 2.5 × 3.5 cm, more or less semicircular, divided almost to the base into 3 primary lobes, which are themselves further divided to give 15–27 ovate to narrowly triangular ultimate segments, the latter often overlapping one another. Flowering-stems 25–60 cm, robust, axillary, branched above the middle to form a narrow panicle. Petals 7–9 mm, narrowly obovate, usually white. Ovary three-quarters inferior. *Pyrenees.* H4. Summer.

Requires flowing, neutral or acidic water; probably best by a suitable stream.

124. S. irrigua Bieberstein (*S. ranunculoides* Haworth). Illustration: *Botanical Magazine*, 2207 (1821).

Perennial, forming a tuft of leafy shoots. Blade of basal leaves usually 2.5–3 × 3.5–4 cm, kidney-shaped or semicircular, divided nearly to the base into 3 primary lobes, which are further divided to give 11–35 oblong-lanceolate, almost acute to apiculate, ultimate segments. Flowering-stems 10–20 cm, robust, terminal, bearing in the upper part a flat-topped cyme of 5–12 flowers. Petals 1.2–1.6 cm, oblanceolate, tip almost acute, white. Ovary two-thirds inferior. *Crimea.* H4. Late spring–summer.

Grows well as a large clump in a rock garden.

125. S. latepetiolata Willkomm. Illustration: *Botanical Magazine*, 7056

(1889); Webb & Gornall, Saxifrages of Europe, pl. 56, f. 57 (1989); Harding, Saxifrages, 135 (1992).

Biennial, densely covered with long, sticky, glandular hairs. Basal leaves arranged in a domed 'rosette' to 6 cm high; blade 8–15 mm × 1–2.7 cm, kidney-shaped to semicircular, base more or less cordate, divided to halfway into 5–7 obovate to wedge-shaped, truncate lobes, these further divided in larger leaves into 3 segments; leaf-stalk brittle. Flowering-stem 15–25 cm, robust, terminal, branched from near the base to give a narrow pyramidal panicle of numerous flowers. Petals 7–10 mm, narrowly obovate, white. Ovary inferior. *E Spain.* H5. Late spring–summer.

Probably best grown in a pan in a cool glasshouse.

Series **Arachnoideae** (Engler & Irmscher) Gornall. Biennials or evergreen perennials, with straggling or ascending leafy stems, Leaves usually longer than 4 cm (including the leaf-stalk), usually with leaf-stalk longer than scalloped or lobed blade, thin. Hairs on leaf-stalks long. Flowering-stems terminal. Petals white or pale greenish yellow. Ovary inferior.

126. S. petraea Linnaeus. Illustration: Köhlein, Saxifrages and related genera, t. 11 (1984); Finkenzeller & Grau, Alpenblumen, 115 (1985); Webb & Gornall, Saxifrages of Europe, pl. 52 (1989); Aeschimann et al., Flora alpina **1**: 713 (2004).

Biennial, covered with long, soft, sticky glandular hairs. Blade of basal leaves 1.2–3 × 2.2–3.5 cm, semicircular to diamond-shaped, deeply divided into a narrow central lobe, with 3–5 teeth, and 2 broader lateral lobes, variously toothed or lobed, giving a total of 19–23 ultimate segments. Flowering-stems to 35 cm, brittle, freely branched, bearing small, loose, leafy cymes. Flowers usually slightly irregular, petals to 1.1 cm, tip notched, pure white. *E Alps.* H5. Spring–summer.

Best grown in a pan in alkaline soil in a cool glasshouse.

127. S. berica (Béguinot) D.A. Webb. Illustration: Webb & Gornall, Saxifrages of Europe, f. 58 (1989); Aeschimann et al., Flora alpina **1**: 713 (2004).

Like *S. petraea* but perennial, leaves with a brownish tint; flowers smaller, more asymmetrical; hairs shorter (*c.* 0.5 mm on the flower of *S. berica* versus *c.* 1.5 mm on

S. petraea). *NE Italy (Colli Berici).* H5. Spring–early summer.

Appears to grow best in crevices in tufa under glass.

128. S. arachnoidea Sternberg. Illustration: Finkenzeller & Grau, Alpenblumen, 115 (1985); Webb & Gornall, Saxifrages of Europe, pl. 53 (1989); Aeschimann et al., Flora alpina **1**: 713 (2004).

Perennial, densely covered with wavy, sticky, glandular hairs. Blade of basal leaves usually *c.* 1.2 × 1.4 cm, occasionally to 2 × 3 cm, fan-shaped, circular or elliptic, divided near the tip into 3–5 broadly ovate, obtuse lobes. Flowering-stems 10–20 cm, ascending, brittle, bearing up to 5 flowers in a loose cyme. Petals 2.5–3 mm, oblong, not touching, off-white or cream. Ovary inferior, surmounted by a prominent nectary disc. *N Italy (Giudicarian Alps).* H4? Summer.

Cultivate in a cool glasshouse in a fine, alkaline soil, in shade, with a fairly high humidity.

129. S. paradoxa Sternberg. Illustration: Webb & Gornall, Saxifrages of Europe, f. 59 (1989); Aeschimann et al., Flora alpina **1**: 713 (2004).

Perennial, probably short-lived, almost hairless except for a few long wavy hairs on leaf-stalks and base of the stem. Blade of basal leaves 1.5–2 × 2.5–4 cm, kidney-shaped, base cordate, thin, shiny, with 5–9 shallow, obtuse lobes usually broader than long. Flowering-stems to 20 cm, brittle, barely projecting above the basal leaves, bearing few-flowered cymes. Petals *c.* 1.5 mm, linear, greenish yellow. Ovary inferior, surmounted by a conspicuous nectary-disc. *SE Alps.* H4? Summer.

Difficult to grow; requires deep shade and a moist soil in the growing season.

Series **Axilliflorae** (Willkomm) Pawlowska. Evergreen perennials with prostrate leafy shoots, forming loose mats. Leaves usually less than 1.5 cm long (including the leaf-stalk), divided into 3–5 primary lobes, with the leaf-stalk more or less the same length as the blade. Hairs short. Flowering-stems terminal or axillary. Petals white. Ovary inferior. Capsule narrowly cylindric rather than ovoid or ellipsoid.

130. S. praetermissa D.A. Webb (*S. ajugifolia* misapplied). Illustration: Webb & Gornall, Saxifrages of Europe, f. 46 (1989).

Flowering-stems 10–15 cm, arising from the axils of leaves on prostrate shoots some distance from the upturned tip. *Pyrenees, NW Spain.* H4. Summer.

Sporadically in cultivation.

Series **Ceratophyllae** (Haworth) Pawlowska. Evergreen perennials. Leafy shoots more or less woody below, forming cushions. Leaves usually less than 7 cm (including the leaf-stalk), lobed, often stiff or leathery, the leaf-stalk longer than or equalling the blade; usually with very short, often stalkless, glandular hairs. Flowering-stems terminal or axillary. Petals white. Ovary inferior or nearly so.

131. S. pedemontana Allioni (*S. allionii* Terraciano; *S. pedatifida* misapplied). Illustration: *Botanical Magazine,* n.s., 687 (1975); Finkenzeller & Grau, Alpenblumen, 113 (1975); Webb & Gornall, Saxifrages of Europe, pl. 43–5, f. 48–50 (1989); Harding, Saxifrages, 136 (1992).

Leafy shoots forming a loose cushion. Blade of rosette leaves 8–15 × 9–20 mm, palmately divided into 3–9 elliptic to linear-oblong segments, not furrowed on the upper surface, covered with very short glandular hairs. Flowering-stems 5–18 cm, terminal, branched in the upper half to form a narrow panicle of 2–12 flowers. Petals 9–21 mm, oblanceolate, basal part erect, upper bent outwards, pure white, rarely tinged or veined with red. *Caucasus, Carpathians, Balkan Peninsula, SW Alps, Corsica, Sardinia, Cevennes, Morocco.* H3. Summer–early autumn.

Five subspecies are currently recognised, and a recently discovered dwarf variant from the Caucasus may represent a sixth:

Subsp. **cymosa** Engler. Leaf-blade tapered gradually into a wide leaf-stalk; segments ovate-oblong, obtuse, short, forward-pointing. Flowering-stem to 8 cm. Petals 9–15 × 3–5 mm. *Carpathians, Balkan Peninsula.*

Subsp. **pedemontana**. Leaf-blade tapered gradually to a narrow leaf-stalk; segments oblong, almost acute, short. Flowering-stem to 18 cm. Petals 1.5–2.1 × 6–8 mm, sometimes tinged pink at the base. *SW Alps.*

Subsp. **cervicornis** (Viviani) Engler. Young leaves on non-flowering rosettes incurved; blade contracted to a narrow leaf-stalk; segments narrowly oblong, obtuse, acute or with a short point, long, forward-pointing or divergent. Flowering-stem to 15 cm. Petals 1–1.3 cm × 4–5 mm. *Corsica, Sardinia.*

Subsp. **demnatensis** (Battandier) Maire. Like subsp. *cervicornis* but with larger, more leathery leaves and oblong sepals which are equal to or are longer than the perigynous zone. *Morocco.*

Subsp. **prostii** (Sternberg) D.A. Webb. Leaf-blade contracted to a fairly narrow leaf-stalk; segments broader than in subsp. *cervicornis*, but narrower than in the first 2 subspecies, acute or with a short point. Flowering-stem to 18 cm. Petals 9–12 × 2.5–4 mm. *France (Cevennes).*

Cultivate in acidic soils, either in partial shade on the rock garden or in a cool glasshouse. The last subspecies is best outside.

132. S. geranioides Linnaeus. Illustration: Webb & Gornall, Saxifrages of Europe, f. 51 (1989).

Leafy shoots forming a loose cushion. Leaf-blade *c.* 1.5–2.5 cm, semicircular, deeply divided into 3 primary lobes, these further lobed or toothed, giving a total of 17–25 acute segments, not furrowed on the upper surface, covered with very short glandular hairs. Flowering-stems 15–25 cm, terminal, usually branched from near the middle to give a loose cyme of up to 20 flowers. Petals *c.* 1.2 cm, oblanceolate, white. *Pyrenees.* H4. Summer.

Prefers a moist, acidic soil in partial shade on the rock garden.

133. S. moncayensis D.A. Webb. Illustration: Webb & Gornall, Saxifrages of Europe, pl. 48, f. 52 (1989).

Leafy shoots forming a fairly dense, deep cushion. Leaf-blade 8–11 × 9–15 mm, deeply divided into 3 primary lobes, which are narrowly oblong to oblanceolate, obtuse, furrowed on the upper surface, outer lobes sometimes with a secondary lobe, covered with very short glandular hairs. Flowering-stems 5–10 cm, terminal, branched from the middle to give a loose panicle of 9–35 flowers. Petals 6–7 mm, broadly oblong, narrowed at the erect base, bent outwards in the upper part, not touching. *NE Spain (Sierra de Moncayo).* H4. Late spring–summer.

Prefers partial shade on the rock garden or in a pan in a cool glasshouse.

134. S. vayredana Luizet. Leafy shoots forming a large, fairly dense cushion. Leaf-blade 5–9 × 8–13 mm, deeply divided into 3 primary lobes, these often themselves shortly lobed to give a total of 5–9 acute, spreading segments, not furrowed on the upper surface, covered in very short glandular hairs and smelling very strongly of spice when crushed. Flowering-stems 7–12 cm, terminal, bearing 3–9 flowers in a compact, narrow panicle. Petals 6–7 mm, obovate, pure white, nearly touching. *NE Spain (Sierra de Montseny).* H4. Summer.

Grows well in partial shade on the rock garden or in a pan under glass.

135. S. intricata Lapeyrouse (*S. nervosa* invalid). Illustration: Webb & Gornall, Saxifrages of Europe, pl. 47 (1989).

Like *S. vayredana* but leaf-segments obtuse, forward-pointing, furrowed on the upper surface. Petals 5–6 mm, broadly elliptic to nearly circular, overlapping, white, spreading horizontally. *Pyrenees.* H4. Summer.

Best grown in an acidic soil in a cool glasshouse.

136. S. pentadactylis Lapeyrouse. Leafy shoots forming an open, rather brittle cushion. Leaf-blade divided nearly to the base into 3 primary lobes, the lateral ones further divided into 2 segments, the central one sometimes into 3; segments linear to oblong, obtuse, furrowed on the upper surface; covered with stalkless glands. Flowering-stems 7–17 cm, terminal, bearing 5–50 flowers. Petals 3.5–5 mm, obovate to oblong, usually white. *Pyrenees, N & C Spain.* H4. Summer.

A variable species, rarely found in cultivation despite the frequency with which the name is encountered. Four subspecies are currently recognised, of which subsp. **willkommiana** (Willkomm) Rivas Martinez seems to be the easiest to grow. It is recognised by its leaf-stalk, which equals or is slightly shorter than the blade and is as narrow as the broadest segments, which are widely divergent.

The subspecies differ with respect to their preference for soil pH. The easiest one, subsp. *willkommiana*, likes an acidic soil in a pan in a cool glasshouse.

137. S. fragilis Schrank (*S. corbariensis* Timbal-Lagrave). Illustration: *Botanical Magazine*, n.s., 701 (1975); Webb & Gornall, Saxifrages of Europe, pl. 49, f. 53 (1989).

Leafy shoots forming a loose cushion. Leaf-blade 1–1.7 × 1–3 cm, usually semicircular, divided almost to the base into 3 primary lobes, the lateral ones usually (and the central one always) divided into 2 or 3 segments, usually giving a total of 5–11 obtuse to almost acute, narrowly oblong segments, not furrowed on the upper surface, covered with stalkless glands. Flowering-stems 10–22 cm, terminal, branched in the upper half to form a loose cyme of 5–20 flowers. Petals 7–14 mm, obovate to oblanceolate, tip sometimes notched, pure white. *E Spain, S France.* H4. Spring–summer.

Subsp. **fragilis**, from the northern part of the range, can be distinguished from the more southerly subsp. *valentina* (Willkomm) D.A. Webb by its longer petals (at least 10 mm) and longer stamens (exceeding sepals by 3 mm or more).

Subsp. *fragilis* is the more garden-worthy plant. Best grown in a pan of alkaline soil in a cool glasshouse.

138. S. camposii Boissier & Reuter. Illustration: The plant figured under this name in *Botanical Magazine*, 6640 (1882) is probably *S. corsica* subsp. *cossoniana* or one of its hybrids.

Like *S. fragilis* subsp. *valentina* but with stiffer leaves, at least some leaf-segments mucronate. *S & SE Spain.* H5. Late spring–summer.

139. S. trifurcata Schrader (*S. ceratophylla* Dryander). Illustration: *Botanical Magazine*, 1651 (1814); Webb & Gornall, Saxifrages of Europe, f. 54 (1989).

Leafy shoots forming a fairly open cushion of dark, shiny-green leaves. Leaf-blade to 2 × 3 cm, more or less semicircular, divided three-quarters of the way to the base into 3 primary lobes, these divided further to give 9–17 triangular apiculate segments, not furrowed on the upper surface, many overlapping, the lateral ones strongly recurved, covered with stalkless glands. Flowering-stems to 30 cm, axillary, bearing a loose cyme of 5–15 flowers. Petals 8–11 mm, elliptic-oblong, white. *N Spain.* H3. Late spring–summer.

S. × schraderi Sternberg, a garden hybrid between *S. continentalis* and *S. trifurcata*, approaches *S. trifurcata* in appearance, and is frequent in cultivation under the name *S. trifurcata*.

Grows well in alkaline soil in partial shade on the rock garden.

140. S. canaliculata Engler. Illustration: Webb & Gornall, Saxifrages of Europe, pl. 46, f. 55 (1989).

Leafy shoots forming a loose cushion. Leaf-blade to 1.2 × 1.8 cm, semicircular, deeply divided into 3–11 linear, deeply grooved segments, covered with stalkless glands. Flowering-stems 8–15 cm, terminal, bearing a compact cyme of 5–12 flowers. Petals 8–10 mm, broadly obovate, overlapping, tip often recurving with age, white. *N Spain.* H3. Late spring–summer.

Cultivation requirements as for the previous species.

141. S. cuneata Willdenow. Illustration: Webb & Gornall, Saxifrages of Europe, pl. 50 (1989).

Leafy shoots in loose tufts. Leaf-blade to 2.2 × 2.6 cm, diamond- to fan-shaped, divided to halfway into 3, broadly triangular to ovate, mucronate, primary lobes, which are secondarily 3-lobed on larger leaves, covered with stalkless glands. Flowering-stems 7–30 cm, axillary, branched above the middle to form a narrow panicle of 7–15 flowers. Petals 5–8 mm, obovate, more or less touching, pure white. *N Spain, SW France.* H4. Late spring–summer.

Grows best in a pan of alkaline soil in a cool glasshouse.

142. S. maderensis D. Don. Illustration: *Bocagiana* **33**: 3 (1973).

Leafy shoots forming a loose cushion. Leaf-blade semicircular, kidney-, diamond- or fan-shaped, base cordate or tapered to the leaf-stalk, divided up to halfway into 9–38 obtuse, acute or apiculate segments, covered with stalkless glands, though sometimes with some very short-stalked glandular hairs on the leaf-stalk. Flowering-stems to 10 cm, axillary, bearing up to 13 flowers in a compact to loose cyme. Petals 4–12 mm, white, *Madeira.* H4. Summer.

Var. **maderensis**. Leaf-blades kidney-shaped to semicircular, base cordate to somewhat tapered; segments obtuse to acute. Flowering-stems with up to 13 flowers. Petals 4–10 mm. *Madeira.* Var. **pickeringii** (C. Simon) D.A. Webb & Press. Leaf-blades diamond- to fan-shaped, base strongly tapered; segments acute to apiculate. Flowering-stems with up to 6 flowers. Petals 8–12 mm. *Madeira.* The latter variety performs well in partial shade in alkaline soil on the rock garden; the former is rather better under glass.

143. S. portosanctana Boissier. Illustration: *Bocagiana* **33**: 3 (1973).

Like *S. maderensis* but leaf-blade much longer than broad, base strongly tapered, divided to halfway or more into up to 9 obtuse segments. Petals 1–1.1 cm, *c.* 3

times as long as broad (versus only 2 times in *S. maderensis*). *Porto Santo.* H4. Summer.

Grows well in alkaline soil on the rock garden.

Series **Cespitosae** (Reichenbach) Pawlowska. Evergreen perennials, with leafy shoots forming cushions. Leaves less than 2.5 cm (including the stalk), the blade divided into lobes and usually longer than or equalling the leaf-stalk. Hairs on leaf-stalk not long and wavy. Flowering-stems terminal. Petals white to dull yellow, sometimes tinged or veined with red. Ovary two-thirds or more inferior.

144. S. rosacea Moench (*S. decipiens* Sternberg; *S. palmata* Smith; *S. sternbergii* Willdenow; *S. hirta* misapplied; *S. hibernica* Haworth; *S. quinquifida* Haworth; *S. affinis* D. Don). Illustration: Ross-Craig, Drawings of British plants **10**: pl. 12 (1957); Garrard & Streeter, Wild flowers of the British Isles, 92 (1983); Webb & Gornall, Saxifrages of Europe, f. 44 (1989).

Leafy shoots forming a dense or loose cushion or thin mat. Rosette-leaves 6–25 mm, including the fairly distinct leaf-stalk, which may be longer or shorter than blade; blade divided to halfway or more into 3–5 primary lobes, these sometimes further divided to give a total of up to 11 broadly elliptic to linear-oblong segments, which are obtuse, acute or shortly pointed, but not furrowed on the upper surface. Flowering-stems 4–25 cm, bearing 2–6 flowers in an open cyme. Petals 6–10 mm, obovate, pure white. Ovary two-thirds inferior. *C & NW Europe.* H4. Late spring–summer.

Subsp. **rosacea**. Leaf-segments variable but not shortly pointed, usually broad. Leaf-hairs mostly non-glandular. *C & NW Europe.*

Subsp. **sponhemica** (Gmelin) D.A. Webb. Leaf-segments shortly pointed, narrowly oblong. Leaf-hairs mostly non-glandular. *C Europe.*

Subsp. **hartii** (D.A. Webb) D.A. Webb. Leaf-segments almost acute, broad. Leaf-hairs mostly glandular. *Ireland (Arranmore Island).*

For a commonly grown complex of hybrids with *S. exarata, S. hypnoides* and *S. granulata,* see under *S. exarata.* Grows well on the rock garden and under glass.

145. S. cespitosa Linnaeus (*S. groenlandica* Linnaeus). Illustration: Keble Martin, The new concise British flora in

colour, pl. 32 (1982); Garrard & Streeter, Wild flowers of the British Isles, 92 (1983).

Leafy shoots packed into a dense cushion, rarely a loose mat. Leaves 4–15 mm, including the indistinct leaf-stalk, most divided into 3 oblong-elliptic, obtuse lobes, not furrowed on the upper surface; margin densely glandular-hairy. Flowering-stems 2–10 cm, bearing 1–5 flowers in an open cyme. Petals 5.5–6.5 mm (in European plants), obovate, usually off-white. Ovary two-thirds inferior. *Arctic-circumpolar, W North America.* H1. Summer.

Requires cold and dry winter conditions (cold-frame) and standard alpine house treatment during spring and summer.

146. S. pubescens Pourret (*S. mixta* Lapeyrouse). Illustration: Webb & Gornall, Saxifrages of Europe, pl. 36 (1989).

Leafy shoots forming a loose to compact cushion with rather flat rosettes. Leaves to 1.8 cm, including the broad, indistinct leaf-stalk, but usually much shorter, divided into 5–9 oblong, obtuse lobes, 2–3 times as long as broad, most furrowed on the upper surface, dark green, densely glandular-hairy all over. Flowering-stems 3–10 cm, apparently axillary but actually terminal, bearing a flat-topped cyme of 5–9 flowers. Petals 4–6 mm, broadly obovate to nearly circular, overlapping, white. Ovary nearly inferior. *Pyrenees.* H4. Summer.

Subsp. **pubescens**. Leafy shoots short, loose, with dead leaves soon dropping. Leaf-segments about 3 times as long as broad, rather spreading. Petals pure white. Anthers yellow. Ovary green.

Subsp. **iratiana** (Schultz) Engler & Irmscher. Leafy shoots long, columnar, in a tight cushion, with persistent dead leaves. Leaf-segments about 2 times as long as broad, nearly parallel. Petals sometimes veined with red. Anthers and ovary reddish.

The former subspecies grows well in acidic soil under glass; the latter is more difficult.

147. S. nevadensis Boissier. Illustration: Webb & Gornall, Saxifrages of Europe, pl. 40 (1989).

Like *S. pubescens* but leaves with a more distinct leaf-stalk of 3–4 mm; blade 4–7 × 5–8 mm, divided to halfway into 3–5 oblong, obtuse segments, not furrowed on the upper surface. Petals more consistently red-veined. *S Spain (Sierra Nevada).* H3. Late spring–summer.

Best kept outside in a cold-frame in autumn and winter (protected from the wet) but brought under glass and watered in spring and early summer.

148. S. cebennensis Rouy & Camus (*S. prostiana* (Seringe) Luizet). Illustration: Köhlein, Saxifrages and related genera, t. 5 (1984).

Leafy shoots forming a soft, domed cushion of light green foliage. Leaf-blade *c.* 6 × 5 mm, tapered to a distinct leaf-stalk of *c.* 6 mm, usually divided to halfway into 3–5 obtuse lobes (though some leaves are entire), furrowed on the upper surface, densely glandular-hairy all over. Flowering-stems 5–8 cm, bearing 2 or 3 flowers. Petals 6–8 mm, broadly obovate, overlapping, creamy white. Ovary inferior. *S France.* H4. Late spring–summer.

Best grown under glass in alkaline soil.

149. S. exarata Villars. Illustration: Finkenzeller & Grau, Alpenblumen, 107 (1985); Webb & Gornall, Saxifrages of Europe, pl. 41, 42 (1989); *Bulletin of the Alpine Garden Society* **58**(1): 39 (1990); Aeschimann et al., Flora alpina **1**: 709 & 711 (2004).

Leafy shoots forming a compact cushion. Leaves 4–20 mm, including the fairly distinct leaf-stalk, divided to at least halfway, usually into 3 oblong, obtuse lobes (sometimes entire or with 5–7 lobes), furrowed or flat on the upper surface, variably hairy. Flowering-stems 3–10 cm, bearing a flat-topped cyme of 1–5 flowers. Petals 2.5–6 mm, oblong or narrowly elliptic to broadly obovate, never properly touching, white to greenish yellow, rarely tinged with red. Ovary inferior. *C & S Europe eastwards to the Caucasus.* H3. Late spring–early autumn.

Subsp. **exarata**. Leaf-lobes usually 3–5, spreading, furrowed on the upper surface. Petals 4–6 mm, about 1.5 times as long as broad, broadly obovate, white or pale cream. *Alps, Balkan Peninsula, Turkey, Caucasus.*

Subsp. **pseudoexarata** (Braun-Blanquet) Webb. Leaf-lobes mostly 3, slightly spreading, furrowed on the upper surface. Petals 3.5–4.5 mm, twice as long as broad, oblong to narrowly elliptic, well separated, pale dull yellow. *Alps, Apennines, Balkan Peninsula.*

Subsp. **moschata** (Wulfen) Cavillier (*S. moschata* Wulfen; *S. pyrenaica* Villars; *S. pygmaea* Haworth; *S. rhei* Schott; *S. various* Sieber). Leaves entire or 3-lobed, not furrowed on the upper surface. Petals 3–4 mm, twice as long as broad, oblong to

narrowly elliptic, well separated, yellowish (sometimes tinged with red). *C & S Europe eastwards to the Caucasus, except Apennines & S Balkan Peninsula.*

Subsp. **lamottei** (Luizet) Webb (*S. lamottei* Luizet). Leaves entire to 5-lobed, segments not furrowed on the upper surface, sparingly hairy. Petals 4–5 mm, usually less than twice as long as broad, obovate, nearly touching. *SC France.*

Subsp. **ampullacea** (Tenore) Webb (*S. ampullacea* Tenore). Like subsp. *lamottei* but leaves entire or with 3 very short, parallel lobes; leaf-stalk indistinct. Flowers 1–3. *C Apennines.*

This species is a parent (along with *S. rosacea*, *S. hyphenoides* and others) of the invalidly named *S.* × *arendsii* hybrid complex. Also known as the 'mossy' saxifrages, their precise parentage is unknown, but descriptions of many of the cultivars may be found in Harding (1970, 1992) and Köhlein (1984).

150. S. hariotii Luizet & Soulie. Illustration: Webb & Gornall, Saxifrages of Europe, pl. 37 (1989).

Leafy shoots forming a dense to fairly loose cushion. Leaves 5–9 mm, including the indistinct leaf-stalk, mostly 3-lobed but some entire; lobes oblong and with a short point, furrowed on the upper surface, sparingly glandular-hairy. Flowering-stems 3–7 cm, bearing 3–12 flowers. Petals 3.5–4 mm, oblong, not touching, dull creamy-white, usually with reddish veins. Ovary inferior. *W Pyrenees.* H4. Summer.

Best grown in an alkaline soil under glass.

Series **Gramiferae** (Willkomm) Pawlowska. Evergreen perennials with leafy shoots usually forming mats or cushions. Summer-dormant, leafy buds, with long wavy hairs, present in the leaf-axils of at least some shoots. Leaves lobed or entire, usually less than 4 cm, including leaf-stalk, with the latter longer or shorter than the blade. Flowering stems terminal. Petals white, rarely tinged with pink. Ovary inferior or nearly so.

151. S. hypnoides Linnaeus (*S. leptophylla* D. Don; *S. rupestris* Salisbury). Illustration: Ross-Craig, Drawings of British plants **10**: pl. 10 (1957); Phillips, Wild flowers of Britain, 44 (1977); Keble Martin, The new concise British flora in colour, pl. 32 (1982); Garrard & Streeter, Wild flowers of the British Isles, 92 (1983).

Leafy shoots usually forming a loose mat. Leaves mostly entire and linear-lanceolate on the prostrate shoots, but 3–7-lobed and fan-shaped, with a broad leaf-stalk, in the terminal rosettes; lobes with bristle-like tips, 0.5–0.75 mm; hairs mostly confined to the leaf-stalk. Summer-dormant buds 5–10 × 2–4 mm, sometimes present in the leaf axils of the prostrate shoots; outer bud-leaves with broad, membranous margins and a green centre. Flowering-stems 5–20 cm, branched above to form a loose panicle of 2–7 flowers, the buds of which are nodding. Petals 7–12 mm, elliptic, white, not touching. Ovary three-quarters inferior. *NW Europe.* H3. Spring–summer.

For a complex hybrid with *S. exarata* and others, see under that species. Grows well in partial shade on the rock garden, or under glass.

152. S. continentalis (Engler & Irmscher) D.A. Webb.

Leafy shoots prostrate, forming a compact mat. Stem-leaves entire, linear at top of shoot, to 1–2 cm; shorter and 3-lobed in middle, shortest and 5-lobed at base of shoot. All lobes with a short, bristle-like point. Winter rosette-leaves divided into 3 primary lobes, which are further divided to give a total of 5–13 elliptic to oblong-lanceolate segments. Hairs mostly confined to the leaf-stalk. Summer-dormant buds 7–11 mm, narrowly ellipsoid, on short stalks, always present in at least some leaf axils; outer bud-leaves lanceolate, membranous and translucent except for the midrib. Flowering-stems 8–25 cm, bearing 4–11 flowers. Petals 4–8 mm, elliptic, white, not touching. Ovary three-quarters inferior. *SW Europe.* H5. Spring–summer.

Var. **continentalis**. Leaf-segments very narrow, outer ones usually strongly recurved. *France, NE Spain.*

Var. **cantabrica** (Engler) D.A. Webb. Leaf-segments broad, mostly forward-pointing. *Portugal, W Spain.*

For a garden hybrid with *S. trifurcata* see under that species. Probably usually best in an alkaline soil, either on a rock garden or under glass.

153. S. conifera Cosson & Durieu. Illustration: Webb & Gornall, Saxifrages of Europe, pl. 35 (1989).

Leafy shoots forming a dense mat. Leaves 3–10 mm, linear-lanceolate, entire, shortly pointed, shiny, silvery green; margin fringed with hairs, these long, wavy and non-glandular on the outer leaves of the summer-dormant buds, but short and mainly glandular nearer the top of the shoot. Summer-dormant buds 9–12 × 3–4 mm, spindle-shaped to obconical, on stalks 4–10 mm. Flowering-stems 4–8 cm, bearing 3–7 flowers in a compact cyme. Petals 3–4 mm, narrowly obovate, white. Ovary inferior. *N Spain.* H5. Late spring–summer.

Fairly difficult to grow; an alkaline soil is preferred.

154. S. rigoi Porta. Illustration: Webb & Gornall, Saxifrages of Europe, pl. 34 (1989).

Leafy shoots forming a loose or fairly dense cushion. Leaves to 2.5 cm, including the variable leaf-stalk; blade to 8 × 12 mm, semicircular, deeply 3-lobed, the lateral pair (and sometimes the central one) divided to give 5–7 linear-oblong, almost acute segments; covered in short glandular hairs. Summer-dormant buds 5–8 × 2.5–4 mm, ovoid, oblong or obovoid, on stalks; outer bud-leaves linear, herbaceous. Flowering-stems 7–13 cm, bearing a fairly compact cyme of 2–5 flowers. Petals 1.2–2 cm, oblanceolate, more or less erect. Ovary inferior. *SE Spain, Morocco (Rif Mts).* H5. Late spring–summer.

Probably best grown in an alkaline soil in a cool glasshouse.

155. S. erioblasta Boissier & Reuter. Illustration: Webb & Gornall, Saxifrages of Europe, pl. 39, f. 42 (1989).

Leafy shoots forming a small, compact cushion. Leaves 4–7 mm, oblong-oblanceolate to spathulate, entire or with 3 short, obtuse lobes, leaf-stalk indistinct; covered in short, glandular hairs. Summer-dormant buds 1.5–3 mm across, obovoid, on short stalks; outer bud-leaves similar to the smaller, entire leaves on the leafy shoots, but the inner bud-leaves broader, mostly membranous and translucent. Flowering-stems 4–7 cm, bearing 1–4 flowers in a small cyme. Petals 3.5–5 mm, obovate, nearly touching, white but becoming cherry-pink with age. Ovary inferior. *S Spain.* H5? Late spring–summer.

Probably best grown in an alkaline soil in a cool glasshouse.

156. S. reuteriana Boissier. Illustration: Webb & Gornall, Saxifrages of Europe, pl. 32, f. 40, 41 (1989).

Leafy shoots crowded into a compact, hemispherical cushion. Blade of leaves usually 6–10 mm × 1–1.5 cm, semicircular but divided to about halfway into 3 lobes, the lateral pair usually further divided into 2–4 segments, the central more rarely into 3, to give a total of 5–11 almost acute segments; leaf-stalk much longer than blade, grooved; leaf glandular-hairy. Summer-dormant buds 4–6 mm across, globular to obovoid; outer bud-leaves herbaceous, entire. Flowering-stems 2–5 cm, bearing 1–4 flowers. Petals 8–10 mm, elliptic, margins reflexed near the tip, slightly greenish white. Ovary inferior. *S Spain.* H5? Late spring–summer.

Probably best grown in an alkaline soil in a cool glasshouse.

157. S. globulifera Desfontaines. Illustration: Webb & Gornall, Saxifrages of Europe, pl. 38 (1989).

Leafy shoots forming a cushion. At least some basal leaves spathulate, without a distinct stalk. Blade of most other leaves to 8 × 17 mm, semicircular but divided to about halfway or more into 5 elliptic, acute lobes; leaf-stalk 2–3 times as long as the blade; leaf covered with glandular hairs. Summer-dormant buds 2.5–3.5 mm across, obovoid to nearly spherical, stalked. Flowering-stems 5–15 cm, bearing usually 2–7 flowers. Petals 5–7 mm, narrowly obovate, white. Ovary inferior. *S Spain, Gibraltar, N Africa.* H5? Spring–summer.

A variable species, especially so in N Africa, where several varieties have been described. Probably best grown in an alkaline soil in a cool glasshouse.

158. S. maweana Baker. Illustration: *Botanical Magazine*, 6384 (1878).

Similar to *S. globulifera*. Leafy shoots forming fairly long rosettes of kidney-shaped basal leaves, which are divided into segments. Summer-dormant buds narrowly oblong. Petals 1.2–1.4 cm. H5? Morocco (Rif Mts).

Doubtfully in cultivation.

Subsection **Holophyllae** Engler & Irmscher. Evergreen perennials forming cushions or mats. Leaves entire or with 3 short lobes at the tip. Bulbils and summer-dormant leafy buds absent. Flowering-stems terminal. Petals white or pale greenish yellow. Ovary inferior or nearly so. Most of the species, coming as they do from snow-lie habitats, are very difficult to cultivate and do not persist for long in gardens. Most, consequently, are seen only very infrequently.

159. S. androsacea Linnaeus. Illustration: Rasetti, I fiori delle Alpi, pl. 247 (1980); Aeschimann et al., Flora alpina **1**: 715 (2004).

Leafy shoots axillary, short, erect, forming rosettes of apparently basal leaves, which cluster into tufts or mats. Dead leaves brownish in colour. Leaves 7–30 × 3–6 mm, including the indistinct stalk; blade linear-oblong to narrowly obovate, usually entire but sometimes with 3 almost acute teeth, tip obtuse or almost acute; margin with glandular hairs, some of them to 2 mm. Flowering-stems 2–8 cm, bearing 1–3 flowers. Petals 4–7 mm, oblong or narrowly obovate, not touching, white. *Mountains of Europe, Siberia.* H3. Late spring–summer.

160. S. depressa Sternberg. Illustration: Rasetti, I fiori delle Alpi, pl. 248 (1980); Finkenzeller & Grau, Alpenblumen, 11 (1985); Webb & Gornall, Saxifrages of Europe, pl. 54 (1989); Aeschimann et al., Flora alpina **1**: 715 (2004).

Like *S. androsacea* but leaves 6–9 mm across, shortly 3-lobed, covered with short glandular hairs. Flowering-stems with 3–7 pure white flowers. *Dolomites.* H3. Summer.

161. S. muscoides Allioni (*S. planifolia* Sternberg). Illustration: Webb & Gornall, Saxifrages of Europe, pl. 57 (1989); Aeschimann et al., Flora alpina **1**: 715 (2004).

Leafy shoots axillary, long, columnar, erect, forming soft, deep cushions. Old leaves a greyish colour in the upper half. Leaves usually *c.* 5 × 1.5 mm, including the indistinct stalk, oblong to narrowly elliptic, entire, tip rounded; margin with glandular hairs. Newly dead leaves silvery grey, especially near the tip. Flowering-stems usually 1–5 cm, with 1 or 2 flowers. Petals 3.5–5 mm, obovate, tip obtuse or slightly notched, overlapping, white, cream or pale yellow. *Alps.* H3. Summer.

162. S. presolanensis Engler. Illustration: Finkenzeller & Grau, Alpenblumen, 109 (1985); Webb & Gornall, Saxifrages of Europe, pl. 58 (1989); Aeschimann et al., Flora alpina **1**: 717 (2004).

Like *S. muscoides* but flowering-stems 6–10 cm, bearing 2–8 flowers, in a loose, flat-topped cyme. Petals 3–4 mm, oblong to wedge-shaped, tip usually deeply notched, well-separated, translucent, off-white, variably tinged with pale greenish yellow.

Capsule nearly spherical. *Italian Alps.* H3. Summer.

163. S. glabella Bertoloni. Illustration: Strid, Wild flowers of Mount Olympus, pl. 104 (1980).

Leafy shoots short, ascending or straggling, forming a loose cushion or mat. Dead leaves brownish in colour. Leaves mostly $5–8 \times 1–1.5$ mm, narrowly oblanceolate, obtuse, more or less hairless. Flowering-stems 4–10 cm, leafy, bearing 3–8 flowers in a fairly compact cyme. Petals 2–2.5 mm, broadly obovate, overlapping, white. *C Apennines, Balkan Peninsula.* H4. Summer.

164. S. tenella Wulfen. Illustration: Rasetti, I fiori delle Alpi, pl. 258 (1980); Webb & Gornall, Saxifrages of Europe, pl. 59 (1989); Aeschimann et al., Flora alpina **1**: 719 (2004).

Leafy shoots short, prostrate to ascending, forming a fairly dense mat. Dead leaves a shiny straw or silvery grey colour. Leaves $8–11 \times 1–2$ mm, linear, without a distinct stalk, tip long, pointed, keeled beneath, shiny, straw-coloured or silvery; margin with a narrow, translucent border. Flowering-stem to 15 cm, bearing 3–9 flowers in loose cyme. Petals *c.* 3 mm, obovate, creamy white. *E Alps.* H3. Summer.

Relatively easy to grow either in a shady part of the rock garden or in a cool glasshouse.

Subsection **Tridactylites** (Haworth) Gornall. Annuals or biennials. Leaves entire or with 3–5 short lobes. Bulbils and summer-dormant leafy buds absent. Flowering-stems terminal, bearing a loose, leafy panicle of flowers on long stalks. Petals usually white and notched at the tip. Ovary inferior or nearly so.

165. S. tridactylites Linnaeus. Illustration: Ross-Craig, Drawings of British plants **10**: pl. 6 (1957). Phillips, Wild flowers of Britain, 30 (1977); Keble Martin, The new concise British flora in colour, pl. 32 (1982); Garrard & Streeter, Wild flowers of the British Isles, 91 (1983).

Winter-annual, very variable according to growing conditions, ranging from 20 cm, with a large panicle of more than 50 flowers, to less than 3 cm with a solitary flower. First-formed basal leaves spathulate, later ones oblong, diamond-shaped or semicircular, usually with 3–5 spreading lobes, rarely entire, with or without a stalk, very variable in size but usually *c.* 10×4 mm, including the stalk, not forming a well-defined rosette. Flower-stalks usually 1–2 cm in fruit. Petals 2.5–3 mm, narrowly obovate, tip entire or slightly notched, white. Capsule almost spherical, rounded at the base. *Europe, N Africa, SW Asia, Caucasus.* H3. Spring–summer.

Best grown in a pan in a cool glasshouse; water freely until the first flowers appear, then cut off the water supply and keep in the sun. The leaves should turn red, setting off the small white flowers.

166. S. adscendens Linnaeus. Illustration: Harding, The genus Saxifraga, 123 (1970); Aeschimann et al., Flora alpina **1**: 719 (2004).

Biennial or winter-annual, differing from *S. tridactylites* in its well-developed, persistent rosette of basal leaves; flower-stalks shorter, to 8 mm in fruit; petals longer, 3–5 mm; fruit usually an obovoid capsule tapered at the base. *Scandinavia, mountains of C & S Europe, Turkey, Caucasus, W North America.* H3. Summer.

A variable species, very difficult to cultivate; it seems to need cold winters.

167. S. blavii (Engler) G. Beck. Illustration: Webb & Gornall, Saxifrages of Europe, pl. 60 (1989).

Like *S. adscendens* but with a basal rosette of small leaves that wither at flowering. Leaf-lobes short, forward-pointing. Flower-stalks usually 1–2 cm in fruit. Perigynous zone much smaller than in *S. adscendens*. Petals *c.* 5 mm. Capsule almost spherical, base rounded, broadly ellipsoid. *Balkan Peninsula.* H4? Summer.

Very attractive; probably worth trying in a pan of alkaline soil in a cool glasshouse.

8. SULLIVANTIA Torrey & Gray
DM. Miller
Herbaceous perennials with slender horizontal stolons. Leaves basal and on the stem, alternate, kidney-shaped or rounded, toothed and lobed. Flowers bisexual, radially symmetric in branched inflorescences. Perigynous zone bell-shaped. Sepals 5, erect. Petals 5, clawed. Stamens 5. Carpels 2, ovary 2-celled, stigmas without obvious styles.

A genus of 6 species from C & W USA, represented in cultivation by a single species, which thrives in moist woodland conditions in partial shade. Propagate by division or seed in spring.

1. S. oregana Watson. Illustration: Abrams, Illustrated flora of the Pacific states **2**: f. 2227 (1944); Hitchcock et al., Vascular plants of the Pacific northwest **3**: 61 (1971).

Slender plant to 20 cm. Leaves 2–7 cm across, kidney-shaped, incised to halfway into 7–9 lobes, coarsely toothed, hairless, yellowish green; stalk to 10 cm. Inflorescence to 10 cm, flowers erect. Sepals to 1 mm. Petals to 2 mm, white. Stamens not projecting from the corolla. Flower-stalks reflexed in fruit. *USA (Oregon).* H4. Late spring.

9. TELESONIX Rafinesque
D.M. Miller
Low-growing, herbaceous perennials, with short rhizomes, glandular-hairy. Basal leaves 2–5 cm wide, kidney-shaped, shallowly lobed and scalloped, stalks to 12 cm; stem-leaves smaller. Flowers bisexual, radially symmetric, *c.* 2 cm across, in clusters of 5–25 on reddish stems to 20 cm. Calyx lobes *c.* 4 mm. Petals 5, to 5 mm, ovate to rounded, reddish purple. Stamens 10. Styles 2, carpels 2. Fruit a many-seeded capsule.

A genus of a single species for the alpine house or rock garden that is not always easy to establish in cultivation. It grows best in full sunlight, except in the hottest season, in a very well-drained, gritty soil which, however, should not be allowed to dry out completely. Propagate by seed or cuttings.

1. T. jamesii (Torrey) Rafinesque (*Saxifraga jamesii* Torrey; *Boykinia jamesii* (Torrey) Engelmann). Illustration: Hitchcock et al., Vascular plants of the Pacific northwest **3**: 61 (1971); Rickett, Wild flowers of the United States **6**: pl. 98 (1973).

C to W USA. H2. Summer.

10. JEPSONIA Torrey & Gray
M.F. Gardner
Herbaceous perennial; rootstock corm-like. Leaves mostly basal, 2–6 cm across, rounded to heart-shaped, shallowly lobed, hairy. Flowers borne in terminal cymes, stalks 10–30 cm, slender, glandular-hairy becoming hairless later, perigynous zone bell-shaped, usually purple-veined; sepals 5; petals 5, white, spathulate; stamens 10, shorter than sepals; carpels 2, free. Fruit a pair of follicles, beaked; seeds 4-ridged.

A genus of 1 or 2 species native to N America. Best cultivated in the alpine house, where they require a freely drained compost and semi-shade. Plants should be liberally watered in the autumn and winter, when their growth is active; allow to dry out after flowering.

1. J. parryi (Torrey) Small (*Saxifraga parryi* Torrey). Illustration: *Torrey Bulletin* **23**: pl. 256 (1896); Abrams, Illustrated flora of Pacific states **2**: 353 (1944); *Bulletin of the Alpine Garden Society* **37**: 276 (1969).

California. G1. Autumn–winter.

11. TANAKAEA Franchet & Savatier
P.G. Barnes
Rhizomatous dioecious perennials with creeping rhizomes and slender stolons. Leaves all basal, evergreen, ovate-oblong, acute, base cordate to rounded, with forward-pointing teeth, leathery, stalked. Scapes erect, with many small creamy white flowers in a narrow or broad panicle, with small bracts. Sepals 5, lanceolate; petals absent; stamens 10, longer than sepals. Carpels 2, joined for most of their length, styles free.

A genus of 1 or 2 species in NE Asia. Hardy and easily cultivated in a moist, well-drained soil or in light shade. Propagate by division or by detaching rooted stolons.

1. T. radicans Franchet & Savatier. Illustration: *Botanical Magazine*, 7943 (1904); *Bulletin of the Alpine Garden Society* **30**: 154 (1962); Jelitto, Schacht & Fessler, Die Freiland Schmuckstauden, edn 3, 619 (1985); Kitamura et al., Coloured illustrations of herbaceous plants of Japan **2**: pl. 35 (1986).

Stems to 30 cm. Leaves 3–8 × 2-5 cm, mostly longer than the stalks. Flowers 2–3 mm. *Japan, China.* H4. Spring.

The Chinese plants are sometimes separated as **T. omeiensis** Nakai: this appears to be smaller (to 10 cm), with leaves 1–4 cm on stalks to 5 cm.

12. BOYKINIA Nuttall
D.M. Miller
Herbaceous perennials with short rootstocks, glandular-hairy. Leaves mostly basal (stem-leaves small), kidney-shaped, lobed and toothed; stipules present. Flowers bisexual in loose panicles. Perigynous zone bell-shaped to obconical. Sepals 5. Petals 5, often deciduous. Stamens 5. Carpels 2; styles 2; ovary 2-celled. Fruit a many-seeded capsule.

A genus of about 8 species from N America and E Asia, of which about 3

are in cultivation. They are easily grown in woodland conditions in light shade and fairly moist, humus-rich, preferably lime-free soil. Propagate by seed or by division.

la. Petals more or less equal to sepals **1. rotundifolia**
 b. Petals distinctly longer than sepals 2
2a. Leaf-lobes with sharp bristle-pointed teeth **2. elata**
 b. Leaf-lobes without bristle-pointed teeth **3. aconitifolia**

1. B. rotundifolia Parry. Illustration: Abrams, Illustrated flora of the Pacific states **2**: f. 2230 (1944).

Plant 45–90 cm. Basal leaves 7–15 cm across, rounded, with shallowly rounded, toothed lobes; stipules very small; stalk to 20 cm. Flowers shortly stalked; perigynous zone bell-shaped. Sepals ovate, *c.* 2 mm. Petals obovate, *c.* 2 mm, sometimes unequal, white. Fruits drooping. *W USA (California).* H4. Early summer.

2. B. elata (Nuttall) Greene (*B. occidentalis* Torrey & Gray). Illustration: Abrams, Illustrated flora of the Pacific states **2**: f. 2229 (1944); Hitchcock et al., Vascular plants of the Pacific northwest **3**: 6 (1971); Clark, Wild flowers of the Pacific northwest, 202 (1976).

Plant 30–60 cm with brownish, gland-tipped hairs on the stem. Basal leaves 2–8 cm across, rounded, base cordate, fairly deeply 5–7-lobed, with sharp, bristle-pointed teeth; stalk to 15 cm; stipules brown and scarious or reduced to bristles. Inflorescence with leaf-like bracts. Perigynous zone obconical, especially in fruit. Sepals *c.* 2 mm. Petals 5–6 mm, oblanceolate, white, sometimes tinged with pink. *W North America.* H4. Early summer.

B. major Gray is very similar.

3. B. aconitifolia Nuttall. Illustration: Justice & Bell, Wild flowers of North Carolina, 85 (1968); Gleason, Illustrated flora of the north-eastern United States and adjacent Canada **2**: 267 (1952); Jelitto, Schacht & Fessler, Die Freiland Schmuckstauden, edn 3, 102 (1985).

Plant 30–80 cm. Basal leaves 5–12 cm across, rounded to kidney-shaped, divided to about halfway into 5–9 sharply toothed lobes; stalk to 10 cm (stem-leaves shortly stalked). Inflorescence with leaf-like bracts. Perigynous zone obconical, sticky. Sepals small, lanceolate. Petals 3–5lmm, obovate, white. *E USA.* H4. Summer.

13. ELMERA Rydberg
D.M. Miller
Low-growing herbaceous perennials with slender, horizontal rhizomes. Basal leaves *c.* 2 × 3–5 cm, kidney-shaped, lobes rounded, toothed, somewhat hairy; stalk to 7 cm, glandular-hairy with large, membranous stipules; leaves on flowering-stems 1–4, alternate, smaller. Flowers bisexual, 10–30 in a raceme to 25 cm. Sepals 5, triangular, erect, greenish yellow, *c.* 4 mm. Petals 5, erect, deeply 3–5-fid, yellowish white, 4–6 mm. Stamens 5, shorter than the sepals. Styles 2, thick, carpels 2. Fruit a many-seeded capsule.

A genus of a single species that thrives in a well-drained, fertile soil in full sun to semi-shade. Propagate by division in spring or autumn, or by seed.

1. E. racemosa (Watson) Rydberg (*Heuchera racemosa* Watson; *Tellima racemosa* (Watson) Greene). Illustration: Hitchcock et al., Vascular plants of the Pacific northwest **3**: 11 (1971); Rickett, Wild flowers of the United States **5**: pl. 77 (1971).

NW North America (British Columbia to Washington). H2. Summer.

14. LITHOPHRAGMA (Nuttall) Torrey & Gray
D.M. Miller
Herbaceous perennial with rhizomes bearing bulb-like tubers. Leaves mostly basal, kidney-shaped or rounded, deeply or shallowly lobed, with long stalks. Flowers bisexual, slightly bilaterally symmetric, in few-flowered racemes. Calyx lobes 5. Petals 5, clawed, often deeply divided. Stamens 10. Carpels and styles 3, ovary 1-celled. Fruit a many-seeded capsule.

A genus of 9 species from W North America represented by a single species in cultivation. It is best grown in a fertile soil in a rock garden or woodland, and should be kept relatively dry in summer when it is dormant. Propagate by division, bulbils, or seed.

1. L. parviflora (Hooker) Nuttall (*Tellima parviflora* Hooker). Illustration: Hitchcock et al., Vascular plants of the Pacific northwest **3**: 23 (1971); Rickett, Wild flowers of the United States **5**: pl. 79 (1971).

Plant with glandular hairs in all parts. Basal leaves 1–3 × 1–2.5 cm, divided nearly to the base into 3–5 segments, each deeply 3–5-lobed; stalk to more than 6 cm; stem-leaves 2 or 3, divided into narrower, slightly smaller segments. Raceme with 10 or more flowers, 25–40 cm, often purplish; flower-stalks 2–5 mm, erect. Flowers 1.8–2 cm across, slightly irregular. Sepals 2 mm, acute, green tinged with brown. Petals 8–10 × 5–6 mm, deeply 3–5-lobed, white, veins pink, or pale pink. *W North America*. H3. Spring.

15. TIARELLA Linnaeus
D.M. Miller

Herbaceous perennial with rhizomes. Leaves mostly basal, simple, lobed or made up of 3 leaflets; stipules small; stalks long. Inflorescence a raceme or panicle. Flowers bisexual, small, white, radially symmetric. Sepals 5, coloured. Petals 5, clawed. Stamens 10, protruding, sometimes of unequal size. Styles 2. Fruit a few-seeded capsule with 2 unequal flaps.

A genus of about 7 species of herbaceous plants, 1 from E Asia, the rest from N America. They are easily grown in shaded woodland conditions in a moist, humus-rich soil. Propagate by division in spring or autumn, or by seed sown in a peaty compost.

Literature: Lakela, O., A Monograph of the genus Tiarella Linnaeus in North America, *American Journal of Botany* **24**: 344–51 (1937).

1a.	Inflorescence panicle-like	2
b.	Inflorescence raceme-like	4
2a.	Leaves simple	**6. unifoliata**
b.	Leaves compound	3
3a.	Leaflets shallowly 3-lobed	
		3. trifoliata
b.	Leaflets deeply divided into narrow segments	**4. laciniata**
4a.	Petals subulate to linear; inflorescence open	**5. polyphylla**
b.	Petals narrow but not subulate; inflorescence fairly dense, expanding as fruit develops	5
5a.	Plants producing stolons after one year; petals elliptic to lanceolate	**1. cordifolia**
b.	Plants never producing stolons; petals narrowly lanceolate	**2. wherryi**

1. T. cordifolia Linnaeus. Illustration: *Botanical Magazine*, 1589 (1813); Justice & Bell, Wild flowers of North Carolina, 85 (1968); Brickell (ed.), RHS A–Z encyclopedia of garden plants, 1038 (2003).

Plant with slender, rooting stolons. Basal leaves 5–10 × 3–8 cm, broadly ovate to almost circular, base cordate, 3–5-lobed, unequally toothed, hairy, often marbled bronze; stalks 5–9 cm. Inflorescence 10–30 cm, raceme-like, stalk sometimes with 1 or 2 small stem-leaves. Sepals 2–4 mm, white. Petals 4–8 mm, elliptic to lanceolate, white. Stamens 2–7 mm, all equal; anthers orange. Capsule 4–10 mm. *E North America (Ontario to Alabama)*. H1. Late spring–early summer.

Some plants varying in flower or leaf colour have been given cultivar names.

2. T. wherryi Lakela (*T. cordifolia* var. *collina* Wherry).

Plant without stolons. Basal leaves 7–14 × 6–9 cm, broadly ovate, 3-lobed, unevenly toothed, base cordate, apex acute, hairy, often becoming reddish in autumn; stalks 10–20 cm. Inflorescence raceme-like, 15–35 cm, occasionally with 1–3 small leaves on the stalk. Sepals 1.5–2 mm, white to purple-tinged. Petals 3–5 mm, narrowly lanceolate, white. Stamens 3–5 mm, equal in length, anthers orange. Capsule 5–10 mm. *SE USA*. H3. Late spring–early summer.

3. T. trifoliata Linnaeus. Illustration: Hitchcock et al., Vascular plants of the Pacific northwest **3**: 68 (1971); Rickett, Wild flowers of the United States **6**: pl. 98 (1973); Brickell (ed.), RHS A–Z encyclopedia of garden plants, 1038 (2003).

Plant without stolons. Basal leaves to 9 × 12 cm, made up of 3 leaflets, each leaflet 3–8 × 2–5 cm, diamond-shaped, 3-lobed, unevenly toothed, hairy; leaf-stalk 5–17 cm. Inflorescence narrow panicle-like, open, 15–50 cm, with 2 or 3 leaves. Sepals 1–2 mm, white or sometimes pink-tinged. Petals 2–5 mm, subulate, with twisted tips, white. Stamens 3–5 mm, unequal, anthers cream. Capsule 3–7 mm. *NW North America*. H3. Late spring–early summer.

4. T. laciniata J.D. Hooker.

Very similar to *T. trifoliata*, but has a smaller leaf with leaflets deeply divided into narrow segments. *W North America (Alaska to Oregon)*. H3. Late spring–early summer.

5. T. polyphylla D. Don. Illustration: Ito, Alpine plants in Hokkaido, pl. 397 (1981).

Plant with underground stolons. Basal leaves 2–7 cm, almost circular, cordate, shallowly 5-lobed with unequal teeth, hairy; stalk 2–10 cm. Inflorescence raceme-like, 10–40 cm, open, with 2 or 3 shortly stalked leaves. Sepals 1–2 mm, white. Petals 2–3 mm, subulate to linear, white. Stamens 3–5 mm, slightly unequal, anthers cream. Capsule 7–12 mm. *Japan, China*. H3. Late spring–early summer.

6. T. unifoliata J.D. Hooker.

Like *T. polyphylla*, but differing in its paniculate inflorescence. *W North America (Alaska to Oregon)*. H3. Late spring–early summer.

May be in cultivation.

16. HEUCHERA Linnaeus
D.M. Miller

Herbaceous perennials with semi-woody, scaly, branched rootstocks. Leaves mostly basal, simple, palmately lobed, usually with cordate bases and long stalks. Inflorescences paniculate, sometimes with leafy bracts. Flowers bisexual, usually radially symmetric, occasionally bilaterally symmetric where the perigynous zone is longer on 1 side. Sepals 5. Petals 5 or rarely absent. Stamens 5. Carpels 2, ovary 1-celled, styles 2. Fruit a many-seeded capsule.

A genus of about 55 species from N America, some of which are grown for ground-cover, or for their flowers in rock gardens and borders. They are best grown in a well-drained, fertile soil in full sun or partial shade. Propagate by division in spring or autumn, or by seed (except for the named cultivars). It is advisable to divide and replant large clumps.

Literature: Rosendahl, O.C., Butters, F.K. & Lakela, O., A Monograph of the genus Heuchera, *Minnesota Studies in Plant Science* **2**: 1–180 (1936).

Inflorescence. Diffuse and very open: **4, 5**; spike-like: **1, 2**; compact but not spike-like: **3, 6, 7, 8, 9, 10**.

Flowers. Red or red-tinged: **3, 4, 6, 9, 10**; white: **2, 5, 6**.

Petals. Exceeding calyx: **2, 4, 5, 6, 7, 8, 10**; included in calyx or absent: **1, 3**.

1a.	Petals bright red	**3. sanguinea**
b.	Petals white, pink, greenish or tinged with red, or absent	2
2a.	Stamens shorter than sepals	3
b.	Stamens equalling or exceeding sepals	4
3a.	Petals shorter than sepals or absent, greenish; leaves usually longer than broad	**1. cylindrica**
b.	Petals equalling or exceeding sepals, white; leaves usually broader than long	**2. grossulariifolia**
4a.	Inflorescence wide, open, diffuse	5
b.	Inflorescence narrow to spike-like	6

5a. Leaves broader than long, lobes
   acute; leaf-stalk usually hairless
                               5. glabra
  b. Leaves longer than broad, lobes
   obtuse; leaf-stalk usually with long
   hairs   **4. micrantha**
6a. Flowers bilaterally symmetric   7
  b. Flowers radially symmetric   9
7a. Flowers narrow, pinkish red, or
   white tinged with pink **6. rubescens**
  b. Flowers almost spherical, green, pink
   or white   8
8a. Inflorescence-stalk usually leafless,
   with stiff hairs   **7. richardsonii**
  b. Inflorescence-stalk with 1–3 small
   leaves, hairs not stiff **8. pubescens**
9a. Petals more or less equal to sepals;
   panicle often leafy   **9. americana**
  b. Petals distinctly longer than sepals;
   panicle usually leafless   10
10a. Plant very hairy; stamens more or
   less equalling or just exceeding
   sepals; leaves 3–9 cm across
                            10. pilosissima
  b. Plant slightly hairy; stamens
   obviously exceeding sepals; leaves
   1–5 cm across   **6. rubescens**

1. H. cylindrica Douglas. Illustration:
Edwards's Botanical Register, t. 1924
(1837); Rickett, Wild flowers of the United
States **6**: pl. 99 (1973); Clark, Wild flowers
of British Columbia, 199 (1973).

   Leaves basal, usually 2–7 × 2–6 cm,
broadly ovate to almost circular, deeply
lobed, lobes rounded, margin scalloped,
often hairy; stalk 2–15 cm. Inflorescence-
stalk to 90 cm, leafless, with glandular
hairs. Panicle 3–15 cm, narrow, spike-like.
Flowers 4–7 mm. Sepals 2–4 mm, cream or
tinged with green, Petals 1–2 mm or
absent. Stamens and styles shorter than
sepals. *W North America*. H2. Summer.

   'Greenfinch' (illustration Brickell (ed.),
RHS A–Z encyclopaedia of garden plants,
532, 2003), bears green flowers in tall,
stiff, short-branched panicles.

2. H. grossulariifolia Rydberg.
Illustration: Hitchcock et al., Vascular
plants of the Pacific northwest **3**: 15
(1971); *Bulletin of the Alpine Garden Society*
50: 343 (1982).

   Leaves all basal, 1–6 cm across, circular
to kidney-shaped, broader than long,
deeply 3–5-lobed, coarsely toothed,
sometimes hairless; stalk 1–6 cm.
Inflorescence-stalk to 40 cm, leafless,
sparsely glandular-hairy. Panicle 1–6 cm,
narrow, spike-like. Flowers 3–4 mm. Sepals
1–1.5 mm. Petals 1.5–3 mm, oblanceolate,

white. Stamens and styles shorter than
sepals. *C USA*. H2. Summer.

3. H. sanguinea Engelmann. Illustration:
Botanical Magazine, 6929 (1887); Rickett,
Wild flowers of the United States **4**: pl. 89
(1970); Hay & Beckett (eds), Reader's
Digest encyclopaedia of garden plants and
flowers, edn 4, 340 (1987).

   Basal leaves 2–6 cm across, kidney-
shaped to almost circular, with 5–7
rounded lobes, sharply toothed, glandular-
hairy; stalk 4–12 cm. Flowering-stem to
50 cm, often with 2 or 3 very small leaves.
Panicle 5–15 cm, open; flowers 4–6 in
pendent clusters. Flowers 6–10 mm. Sepals
c. 4 mm, red. Petals, stamens and styles
shorter than sepals. *SW USA*. H3. Summer.

   Many named cultivars have been raised
from this species, including one with white
leaves spotted with green, becoming pink
in winter, known as 'Taff s Joy'.

   H. × brizoides Anon. is presumed to be
a hybrid between *H. sanguinea* and
H. micrantha. It is very similar to, and often
confused with, × *Heucherella tiarelloides* but
differs in the absence of stolons and the
constant presence of 5 stamens. Many
named cultivars are grown.

4. H. micrantha Lindley. Illustration:
Edwards's Botanical Register, t. 1302
(1829); Hitchcock et al., Vascular plants of
the Pacific northwest **3**: 15 (1971); Clark,
Wild flowers of British Columbia, 206
(1973).

   Plants usually with long hairs. Basal
leaves 2–8 × 2–7 cm, kidney-shaped to
oblong, shallowly 5–7-lobed, lobes rounded
with broad teeth; stalk 5–15 cm.
Flowering-stem to 1 m, with or without
small leaves, hairy. Panicle 30 cm or more,
very loose. Flowers 1–3 mm, greenish
white tinged with red. Sepals less than
1 mm. Petals 1–2 mm, very narrow, white.
Stamens and styles projecting. *W North
America*. H2. Summer.

   A plant with reddish purple foliage,
grown as 'Palace Purple' is usually
assigned to this species.

5. H. glabra Willdenow. Illustration:
Hitchcock et al., Vascular plants of the
Pacific northwest **3**: 15 (1971); Rickett,
Wild flowers of the United States **5**: pl. 78
(1971); Clark, Wild flowers of British
Columbia, 207 (1973).

   Plant with few hairs. Basal leaves
2–8 × 3–9 cm, rounded, deeply 5–7-lobed,
lobes acute and with acute teeth; stalk
5–20 cm. Flowering-stem to 70 cm, usually

with 1–3 small leaves; panicle to 25 cm,
open, with long internodes. Flowers
2–4 mm, greenish white. Sepals *c.* 1 mm.
Petals 2–4 mm, very narrow, white.
Stamens and styles projecting. *NW North
America (Oregon to Alaska)*. H1. Summer.

   H. villosa Michaux is very similar,
differing mainly in the more hairy leaf- and
flower-stalks.

6. H. rubescens Torrey. Illustration:
Rickett, Wild flowers of the United States
6: pl. 100 (1973).

   Leaves all basal, 1–5 cm across, broadly
ovate to rounded, deeply 3–7-lobed,
sharply toothed; stalk 2–10 cm.
Inflorescence stalk to 35 cm, leafless;
panicle *c.* 15 cm, narrow. Flowers 3–5 mm,
slightly irregular, narrow, pink or white
tinged with pink. Sepals *c.* 2 mm. Petals
3–4 mm, very narrow, almost white.
Stamens and styles usually exceeding
petals. *W USA*. H2. Summer.

   H. versicolor Greene is very similar,
differing in the position of the stamens, and
is considered by some authors to belong to
this species.

7. H. richardsonii Britton. Illustration:
Hitchcock et al., Vascular plants of the
Pacific northwest **3**: 19 (1971); Rickett,
Wild flowers of the United States **6**: pl. 99
(1973).

   Leaves all basal, 4–13 × 4–11 cm,
rounded, shallowly lobed and toothed,
hairy beneath; stalk 6–20 cm.
Inflorescence-stalk to 80 cm or more,
leafless; panicle to 40 cm, narrowly
cylindric. Flowers 5–10 mm, irregular,
greenish. Sepals 2–3 mm. Petals as long as
or slightly longer than sepals. Stamens
exceeding sepals, styles about equal to
them. *C North America*. H1. Summer.

8. H. pubescens Pursh. Illustration:
Gleason, Illustrated flora of the north-
eastern United States and adjacent Canada
2: 271 (1952).

   Plant with glandular hairs. Basal leaves
4–11 × 3–10 cm, almost circular, with 5–7
triangular lobes, toothed; stalk 10–20 cm.
Flowering-stem to 75 cm, with 1–3 small
leaves; panicle *c.* 25 cm, conical. Flowers
5–10 mm, irregular, greenish tinged with
purple. Sepals 2–3 mm. Petals exceeding
sepals, reddish purple. Stamens and styles
mostly equalling petals, sometimes longer.
E USA. H2. Summer.

9. H. americana Linnaeus (*H. glauca*
Rafinesque). Illustration: Gleason,

Illustrated flora of the north-eastern United States and adjacent Canada **2**: 269 (1952).

Basal leaves 5–14 × 4–12 cm, rounded, 5–9-lobed, toothed, with stiff hairs; stalk 8–25 cm. Flowering-stem to 1 m, often with 1–3 small leaves; panicle 30 cm or more, narrowly cylindric. Flowers 2 –5 mm, regular, green tinged with red. Sepals 1–2 mm. Petals about as long as sepals. Stamens and styles much longer than petals. *E North America*. H2. Summer.

10. H. pilosissima Fischer & Meyer. Illustration: Rickett, Wild flowers of the United States **5**: pl. 78 (1971).

All parts very hairy. Basal leaves 5–9 × 3–9 cm, with 5–7 rounded, toothed lobes; stalk 7–20 cm, with brown hairs. Flowering-stems 20–50 cm, usually leafless but sometimes with 1–3 very small leaves; panicle narrow but loose. Flowers 2–5 mm, spherical, very hairy, tinged red. Sepals 1–2 mm. Petals longer than sepals, oblanceolate, pinkish white. Stamens and styles scarcely projecting. *W USA (California)*. H3. Summer.

17. × HEUCHERELLA Wehrhahn
D.M. Miller
Herbaceous perennials. Leaves mainly basal, ovate or rounded, long-stalked. Flowers small, numerous, in panicles. Sepals 5. Petals 5. Stamens 5–10, often 7–8. Carpels 2, slightly unequal.

Hybrids between species of *Heuchera* and *Tellima*, which can be distinguished from the parents by the variable number of stamens within one inflorescence (5–10 but usually 7 or 8). The plants are easily grown in any good garden soil in sun or partial shade but may take 2 years or more to flower freely. They are sterile, but may be propagated by division or cuttings.

1a. Plant spreading by stolons
1. tiarelloides
  b. Plant forming a compact clump
2. alba

1. × H. tiarelloides (Lemoine) Wehrhahn (*Heuchera tiarelloides* Lemoine). Illustration: *Botanical Magazine*, n.s., 31 (1948); Brickell (ed.), RHS A–Z encyclopedia of garden plants, 533 (2003).

A hybrid between a pink- or red-flowered hybrid *Heuchera* and *Tiarella cordifolia*. Plant spreading by thin, rooting stolons. Basal leaves *c*. 9 cm across, mostly more or less rounded, with 7 shallow lobes with rounded teeth, light green blotched with brownish red when young; leaf-stalk 7–15 cm, hairy, reddish. Inflorescence-stalk to 40 cm, brownish red. Panicle narrow, open. Sepals 2 mm, ovate, spreading, pinkish red. Petals 3 mm, erect, pale pink. Stamens *c*. 3 mm, anthers yellow. *Garden origin*. H3. Late spring–early summer.

2. × H. alba (Lemoine) Steam (*Heuchera tiarelloides* Lemoine var. *alba* Lemoine).

A hybrid between a white-flowered *Heuchera* and *Tiarella wherryi*. Clump-forming, without stolons. Basal leaves *c*. 10 × 9 cm, broadly ovate, 7–9-lobed, sharply toothed, green marked with grey blotches. Inflorescence-stalk to 60 cm, green. Panicle narrow. Sepals *c*. 3 mm, ovate, white. Petals narrow, 3–4 mm, white. Stamens as long as or shorter than petals, anthers brown. *Garden origin*. H3. Late spring–early autumn.

'Bridget Bloom', with light pink flowers, is the variant found most frequently in cultivation.

18. BOLANDRA A. Gray
D.M. Miller
Herbaceous perennials with short rootstocks bearing bulbils. Leaves alternate with large stipules; lower leaves kidney-shaped, palmately veined and long-stalked, the upper leaves smaller, stalkless and with leaf-like stipules. Flowers bisexual, few, in loose panicles with conspicuous leafy bracts. Sepals 5, spreading, linear-lanceolate. Petals 5, linear, erect. Stamens 5, shorter than the petals. Ovary 2-celled; carpels 2. Fruit a capsule containing many seeds.

A genus of 2 species native to western North America, useful for partially shaded areas of rock or wild gardens in moist but not waterlogged soil. Propagate by seed or by division.

1. B. oregana Watson. Illustration: Abrams, Illustrated flora of the Pacific states **2**: f. 2224 (1944); Hitchcock et al., Vascular plants of the Pacific northwest **3**: 6 (1971).

Plant to 60 cm. Basal and lower stem-leaves to 7 cm across, shallowly lobed into 9–12 irregularly and sharply toothed segments; stalk to 15 cm. Panicle 3–7-flowered. Sepals 5–10 mm, purplish. Petals purple, equalling sepals. Stamens with reddish purple filaments. *NW North America*. H4. Late spring–early summer.

19. MITELLA Linnaeus
D.M. Miller
Low-growing herbaceous perennials. Leaves mostly basal, simple, lobed, with cordate bases and long stalks. Inflorescences simple, often 1-sided, occasionally leafy racemes. Flowers numerous, small. Perigynous zone bell-shaped; sepals 5, small. Petals 5, usually deeply cut, green or white. Stamens 5 or 10, alternate or opposite to the petals. Styles 2, short, carpels 2, ovary 1-celled. Capsule containing numerous glossy, black seeds.

A genus of about 20 species native to N America and E Asia, useful as ground cover. All thrive in partial shade in woodland conditions in moist soils, especially if these are rich in organic matter. Propagate by division in spring or autumn, or by seed.

Inflorescence-stalk. Leafy: **2, 4**; leafless: **1, 3, 5, 6, 7, 8**.
Petals. Comb-like: **1, 2, 3, 4, 7, 8**; divided into 3: **5, 6**.
Stamens. 5: **3, 4, 5, 6, 7, 8**; 10: **1, 2**.

1a. Stamens 10	2
b. Stamens 5	3
2a. Inflorescence-stalk usually leafless; petals greenish yellow	**1. nuda**
b. Inflorescence-stalk with 2 opposite leaves; petals white	**2. diphylla**
3a. Stamens on the same radii as the petals	**3. pentandra**
b. Stamens on the same radii as the sepals	4
4a. Inflorescence-stalk with 1–3 leaves	**4. caulescens**
b. Inflorescence-stalk without leaves or occasionally with 1 very small leaf	5
5a. Petals never divided into more than 3 divisions	6
b. Petals comb-like, usually with 5 or more divisions	7
6a. Petals spreading, segments thread-like	**5. stauropetala**
b. Petals erect, segments narrow but not thread-like	**6. trifida**
7a. Leaves 4–8 cm wide, broader than long, slightly hairy	**7. breweri**
b. Leaves 1–4 cm wide, longer than broad, coarsely hairy	**8. ovalis**

1. M. nuda Linnaeus. Illustration: Hitchcock et al., Vascular plants of the Pacific northwest **3**: 25 (1971); Rickett, Wild flowers of the United States **6**: pl. 98 (1973).

Plant with stolons. Basal leaves 1–3 cm, rounded to kidney-shaped, margins scalloped, sparsely hairy. Inflorescence-stalk to 20 cm, rarely with a small leaf at the base; inflorescence 3–12-flowered, flower-stalks 2–6 mm. Sepals 1–2 mm, ovate. Petals to 4 mm, comb-like, greenish yellow. Stamens 10. *N America to E Asia.* H1. Summer.

2. M. diphylla Linnaeus. Illustration: *Edwards's Botanical Register*, t. 166 (1816).

Basal leaves 3–6 cm, broadly ovate, 3–5-lobed, toothed, apex acute to acuminate, hairy. Inflorescence-stalk to 45 cm, with 2 opposite, almost stalkless leaves similar to but smaller than the basal leaves; inflorescence 5–20 flowered, flower-stalks *c.* 2 mm. Sepals 1–2 mm, Petals 2–3 mm, comb-like, white. Stamens 10. *E North America.* H1. Spring–early summer.

3. M. pentandra W.J. Hooker. Illustration: *Botanical Magazine*, 2933 (1829); Hitchcock et al, Vascular plants of the Pacific northwest **3**: 25 (1971); Rickett, Wild flowers of the United States **6**: pl. 99 (1973).

Plant sometimes with stolons. Basal leaves 3–8 × 2–5 cm, ovate, shallowly 5–9-lobed, margins scalloped, sometimes hairy. Inflorescence-stalk to 30 cm; inflorescence with 5–25 flowers, flower-stalks 2–7 mm. Sepals *c.* 2 mm, spreading. Petals 2–3 mm, comb-like, greenish. Stamens 5, on the same radii as the petals. Stigma 2-lobed. *C toW North America.* H2. Spring–early summer.

4. M. caulescens Nuttall. Illustration: Hitchcock et al., Vascular plants of the Pacific northwest **3**: 23 (1971); Rickett, Wild flowers of the United States **6**: pl. 99 (1973).

Plant usually with stolons. Basal leaves 3–7 cm across, rounded, mostly 5-lobed, toothed, sparsely hairy; stem-leaves 1–3, alternate, smaller than basal leaves. Inflorescence-stalk to 40 cm, with up to 25 flowers, opening from top downwards; flower-stalks 2–8 mm. Sepals 1–2 mm, spreading. Petals 3–4 mm, comb-like, greenish purple. Stamens 5, on the same radii as the sepals. Stigma capitate. *C to W North America.* H2. Late spring–early summer.

5. M. stauropetala Piper. Illustration: Hitchcock et al., Vascular plants of the Pacific northwest **3**: 30 (1971).

Basal leaves 2–8 cm across, rounded to broadly ovate, indistinctly lobed, margins scalloped. Inflorescence-stalk to 50 cm, leafless, with 10–30 flowers on one side; flower-stalks *c.* 1 mm. Sepals 2–3 mm, recurved. Petals 2–4 mm, spreading, divided into 3 thread-like segments, greenish white to purple. Stamens 5, on the same radii as the sepals. Stigmas flattened. *C to W North America.* H1. Late spring–early summer.

6. M. trifida Graham. Illustration: Hitchcock et al., Vascular plants of the Pacific northwest **3**: 30 (1971).

Basal leaves 2–6 cm across, rounded to broadly ovate, indistinctly lobed, margin scalloped. Flower-stalk thickened. Sepals 2–3 mm, erect. Petals 2–4 mm, erect, 3-fid, white to purple-tinged. Stamens 5, on the same radii as the sepals. Stigmas flattened. *W North America.* H2. Late spring–early summer.

7. M. breweri A. Gray. Illustration: Hitchcock et al., Vascular plants of the Pacific northwest **3**: 23 (1971); Rickett, Wild flowers of the United States **6**: pl. 99 (1973); Clark, Wild flowers of the Pacific northwest, 215 (1976); Brickell (ed.), RHS A–Z encyclopedia of garden plants, 696 (2003).

Plants with slender rhizomes. Basal leaves 4–8 cm across, rounded, indistinctly lobed and scalloped, sparsely hairy. Inflorescence-stalk to 30 cm, leafless, inflorescence with 20–40 flowers, flower-stalks 1–2 mm. Sepals very small. Petals 1–2 mm, comb-like, greenish yellow. Stamens 5, on the same radii as the sepals. Stigmas bilobed. *C to W North America.* H1. Late spring–early summer.

8. M. ovalis Greene. Illustration: Hitchcock et al., Vascular plants of the Pacific northwest **3**: 25 (1971).

Basal leaves 4–6 × 2–4 cm, ovate, indistinctly lobed, margin scalloped or toothed, coarsely hairy. Inflorescence-stalk to 30 cm, leafless; inflorescence with 20–40 flowers, flower-stalks 1–2 mm, stout. Sepals very small. Petals *c.* 2 mm, dissected into 3–5 thread-like segments, greenish yellow. Stamens 5, on the same radii as the sepals. Stigmas bilobed. *W North America.* H2. Late spring–early summer.

20. TELLIMA R. Brown
D.M. Miller
Herbaceous perennials with thick, short rootstocks. Basal leaves to 10 cm across, kidney-shaped, stiffly hairy, 5–7-lobed, coarsely toothed, light green; stalks 15–25 cm, with long hairs; stem-leaves 2 or 3, smaller, almost stalkless. Flowers bisexual, radially symmetric, 15–30, more or less drooping in a raceme to 30 cm, stalk to 75 cm; flower-stalks to 4 mm. Perigynous zone 8–10 mm, inflated, glandular-hairy. Sepals 5, 3 mm, ovate, pale green. Petals 5, reflexed, 4–6 mm, lanceolate, fringed into linear segments, greenish white turning pinkish red. Stamens 10, filaments very short. Styles 2, divided to halfway, carpels 2. Fruit a many-seeded capsule.

A genus of a single species, which grows well in rock gardens or woodland conditions often naturalising in moist soil rich in organic material. Propagate by seed or division in spring or autumn.

1. T. grandiflora (Pursh) Douglas (*T. odorata* Howell). Illustration: *Edwards's Botanical Register*, t. 1178 (1828); Rickett, Wild flowers of the United States **5**: pl. 82, 83(1971); Brickell (ed.), RHS A–Z encyclopedia of garden plants, 1027 (2003).

W North America (Alaska to California). H3. Late spring–early summer.

The names 'Rubra' and 'Purpurea' have been used for cultivated variants with leaves that become reddish purple, especially in winter.

21. TOLMIEA Torrey & Gray
D.M. Miller
Herbaceous perennials with creeping rhizomes. Basal leaves 5–12 × 4–10 cm, heart-shaped, palmately veined, apex acute, shallowly lobed and toothed, hairy; stalks 5–20 cm, hairy; stem-leaves smaller with shorter stalks. Many leaves produce plantlets at the junction of blade and stalk. Inflorescence a narrow raceme to 60 cm, with 20–50 bisexual, shortly stalked, bilaterally symmetric flowers. Perigynous zone cylindric to funnel-shaped. Sepals 5, 3–4 mm, 2 smaller, ovate, greenish purple. Petals 4, *c.* 6 mm, thread-like, purplish brown. Stamens usually 3, shorter than petals, unequal. Styles 2. Fruit a narrow, many-seeded capsule.

A genus of a single species, which thrives in moist woodland conditions where it often naturalises. It is best grown in a humus-rich soil, and is useful as ground cover; it is also grown as an unusual pot-plant. Propagate by division or planting leaves bearing plantlets in a sandy, peaty soil.

1. T. menziesii (Pursh) Torrey & Gray.
Illustration: Everett, New York Botanical
Gardens illustrated encyclopedia of
horticulture **10**: 3359 (1982); Jelitto,
Schacht & Fessler, Die Freiland
Schmuckstauden, edn 3, 630 (1985).

W North America (S Alaska to California).
H4. Late spring–early summer.

Plants with foliage streaked and splashed
with yellow or white are grown as 'Taff's
Gold'(illustration: Brickell (ed.), RHS A–Z
encyclopedia of garden plants, 1042,
2003), or 'Maculata'.

22. BENSONIELLA Morton
D.M. Miller
Herbaceous perennials with slender, scaly,
branching rhizomes. Leaves all basal,
4–8 cm wide, shallowly 5–7-lobed,
scalloped, base cordate, slightly hairy
beneath; stalk to more than 7 cm. Flowers
bisexual in rather narrow, dense racemes
on stems to 25 cm. Perigynous zone
saucer-shaped, creamy white. Sepals 5,
c. 2 mm, creamy white. Petals 5, 2–3 mm,
linear, white. Stamens 5, anthers pink.
Styles 2. Capsules with many seeds.

A genus of a single species (formerly
known as *Bensonia* Abrams & Bacigalupi),
suitable for growing as ground cover in
moist woodland conditions in semi-shade.
Propagate by division in autumn.

1. B. oregona (Abrams & Bacigalupi)
Morton (*Bensonia oregona* Abrams &
Bacigalupi). Illustration: Abrams,
Illustrated flora of the Pacific states **2**:
f. 2280 (1944).

NW USA (Oregon). H3. Early summer.

23. MUKDENIA Koidzumi
P.G. Barnes
Herbaceous perennials with short scaly
rhizomes. Leaves 1 or 2, palmately lobed,
with forward-pointing teeth, hairless and
slightly fleshy. Scape leafless. Inflorescence
a compact bractless panicle of small,
almost stalkless white flowers. Sepals 5 or
6, oblong, white; petals 5–6. Stamens 5–6,
about equalling the petals; carpels 2, free
to about the middle. Fruit a capsule of 2
locules.

A genus of 2 species in China and
Korea. Easily cultivated in a humus-rich
acid soil that does not dry out. Propagate
by seed or division in early spring.

1. M. rossii (Engler) Koidzumi
(*Aceriphyllum rossii* Engler). Illustration:
Bulletin of the Alpine Garden Society **2**: 286
(1933) – as *Aceriphyllum borisii*;

Iconographia cormophytorum Sinicorum
2: 137 (1972); Jelitto, Schacht & Fessler,
Die Freiland Schmuckstauden, edn 3, 10
(1985); *The Garden* **113**(12): 569 (1988).

Leaves 4–7 × 5–8 cm, circular, palmately
lobed to about halfway, lobes ovate, acute,
with forward-pointing teeth; leaf-stalks
7–9 cm. Scape 20–40 cm, flowers *c.* 6 mm.
N China. H4. Summer.

24. PELTOBOYKINIA (Engelmann)
Hara
D.M. Miller
Large herbaceous perennials with short,
thick rhizomes. Basal leaves few,
10–25 cm, rounded, peltate on long stalks,
palmate with 7–13 shallowly toothed
lobes, almost hairless, glossy green; stem-
leaves 2 or 3, smaller, almost stalkless.
Flowers bisexual, radially symmetric, in
terminal cymes on shoots to 60 cm. Calyx
with 5 erect lobes, each 4–5 mm. Petals 5,
1–1.2 cm × 5 mm, oblanceolate, toothed at
the tip, erect, pale yellow. Stamens 10,
anthers dark brown. Styles 2, *c.* 3 mm.
Capsules 1–1.3 cm.

A genus of 1 or 2 species from Japan,
which are easily grown in the wild garden
in partial shade in humus-rich, moist soils.
Propagate by seed or division in spring.

1. P. tellimoides (Maximowicz) Hara
(*Saxifraga tellimoides* Maximowicz; *Boykinia
tellimoides* (Maximowicz) Engelmann).
Illustration: *Botanical Magazine*, 9002
(1924); Wehrhahn, Die Gartenstauden **2**:
555(1931).

Japan. H4. Spring–early summer.

25. CHRYSOSPLENIUM Linnaeus
P.G. Barnes
Herbaceous perennials with underground
rhizomes or rooting prostrate stems. Leaves
stalked, alternate or opposite, slightly fleshy
and often hairy. Erect stems with leaves
often clustered towards the tips, the lower
leaves much smaller. Flowers in bracted
dichotomous cymes on erect shoots, the
bracts leaf-like, sometimes coloured. Sepals
4 or 5, petals absent. Stamens 8 or 10,
surrounded at the base by a fleshy, 8- or
10-lobed disc. Styles 2, free, carpels 2,
united, more or less inferior, opening along
the inner edge. Seeds numerous, black.

About 60 species in Europe, NE Asia,
N Africa, N America and temperate
S America. Hardy perennials, mostly easily
grown in a moist soil and a sunny position.
A few species are occasionally grown on
peat walls or in damp rock gardens or

stream-sides. Propagate by seed or more
usually by division.

la. Leaves alternate, bracts bright
yellow   **1. davidianum**
b. Leaves opposite, bracts greenish
yellow   **2. oppositifolium**

1. C. davidianum Maximowicz.
Illustration: Iconographia cormophytorum
Sinicorum **2**: 143 (1972); Brickell (ed.),
RHS A–Z encyclopedia of garden plants,
274 (2003).

Mat-forming perennial with stems
5–12 cm. Stem-leaves alternate, the upper
with blades 1.5–5 cm, broadly ovate or
rounded, with sparse coarse hairs. Bracts
5–10 mm, obovate, slightly scalloped,
hairless, bright yellow. Flowers 3–4 mm
across, yellow. *W China.* H4. Spring.

2. C. oppositifolium Linnaeus.
Illustration: Ross-Craig, Drawings of British
plants **10**: pl. 16 (1957); Keble Martin,
Concise British flora in colour, pl. 32
(1965); Phillips, Wild flowers of Britain, 30
(1977); Aeschimann et al., Flora alpina **1**:
721 (2004).

Leafy stems prostrate and rooting;
flowering-stems erect, to 15 cm. Lower
leaves with blades 1–2 cm, circular,
scalloped or entire, mostly longer than the
stalk. Flower-stems with 1–3 pairs of
smaller leaves; bracts greenish yellow.
Flowers 3–4 mm across. *W, C & S Europe.*
H4. Spring.

26. FRANCOA Cavanilles
A.C. Whiteley
More or less evergreen herb. Leaves in a
basal rosette, lyrate, often with winged
leaf-stalk, softly hairy, to 30 × 10 cm.
Flowers white or pink, in dense terminal
racemes on stems to 90 cm, occasionally
branched. Sepals 4, acute, *c.* 5 mm. Petals
4, oblong, white or pink, with or without
darker central markings, *c.* 10 × 3 mm.
Stamens 8. Staminodes simple. Ovary
superior, cylindric. Fruit a cylindric
capsule. Seeds numerous, winged.

A genus of a single, very variable species
from Chile, often split into 4 or 5 species.
Easily grown in most soils in a sunny
position. Also good in pots. Propagate by
seed or occasionally division.

1. F. sonchifolia Cavanilles
(*F. appendiculata* Cavanilles; *F. glabrata* de
Candolle; *F. ramosa* D. Don). Illustration:
Botanical Magazine, 3178 (1832), 3309
(1834), 3824 (1840); *Revue Horticole*,
428–9 (1906).

Chile. H5. Summer.

White-flowered plants are sometimes separated as *F. ramosa* or cultivar 'Alba'; 'Rogerson's Form' has deep pink flowers.

140. GROSSULARIACEAE

Shrubs, often spiny. Leaves alternate, simple, often lobed, stipulate or exstipulate, usually deciduous. Flowers in racemes, unisexual or bisexual, actinomorphic. Calyx, corolla and stamens epigynous. Calyx of 4–5 free sepals. Corolla of 4–5 free petals. Stamens 4–5. Ovary of 2 united carpels; ovules few to many, parietal; styles free or united into a single style lobed at the apex. Fruit a berry.

A single genus with about 150 species, native to both Europe and North America. Many species are cultivated as ornamental shrubs, others for their edible fruit (blackcurrant, redcurrant, gooseberry).

1. RIBES Linnaeus
M.C. Tebbitt

Deciduous or occasionally evergreen, low to medium shrubs; branches with or without bristles and or thorns. Leaves alternate, simple, usually lobed and toothed or scalloped, stalked. Flowers usually bisexual, sometimes unisexual and plants dioecious, usually 5-, rarely 4-parted, in few- to many-flowered racemes. Ovary inferior, styles 2. Fruit a berry, crowned by remains of calyx.

A genus of about 150 species from cool and temperate regions of the northern hemisphere and S America. The majority of species are grown for their flowers or occasionally their attractive foliage; a few species are grown for their edible fruit. All are easily grown in a fertile, well-drained soil in full sun, either in the open or trained against a wall. Plants should be pruned immediately after flowering if straggly, but care should be taken not to cut back shoots too severely as the current year's flowers are borne on the previous year's growth. Propagate by nodal hardwood cuttings in autumn to winter or, in the case of *R. laurifolium*, by semi-ripe cuttings during the summer.

The most frequently encountered species are *R. sanguineum*, *R. odoratum* and *R. uva-crispa*.

1a.	Branches lacking thorns and bristles	2
b.	Branches with thorns and or bristles	53
2a.	Plant evergreen	3
b.	Plant deciduous	6
3a.	Leaves lobed	**56. gayanum**
b.	Leaves not lobed	4
4a.	Leaves 2–4 cm, margin entire or with a few small teeth	**22. viburnifolium**
b.	Leaves 5–10 cm, margin scalloped	5
5a.	Shoots hairy when mature	**70. henryi**
b.	Shoots hairless when mature	**71. laurifolium**
6a.	Flowers consistently in clusters or racemes of 1–3	7
b.	Flowers usually in clusters or racemes of more than 3	11
7a.	Calyx tubular	8
b.	Calyx bell-shaped	9
8a.	Flowers white, greenish to yellowish; styles with long hairs	**18. cereum**
b.	Flowers usually pink; styles usually hairless	**19. inebrians**
9a.	Fruit green	**11 ambiguum**
b.	Fruit red to black	10
10a.	Young branches grey; racemes compact	**45. hirtellum**
b.	Young branches reddish; racemes elongated	**44. grossularioides**
11a.	Plant with flowers only of one sex	12
b.	Plant with bisexual flowers	19
12a.	New leaves appearing late spring; fruit black	13
b.	New leaves appearing early spring; fruit red, yellow or green	14
13a.	Leaves to 2 cm	**65. vilmorinii**
b.	Leaves to 6 cm	**68. luridum**
14a.	Flowers purple-brown or brownish red	15
b.	Flowers yellow-green to green-red	16
15a.	Flowers purple-brown	**67. glaciale**
b.	Flowers brownish red	**66. tenue**
16a.	Shoots glandular hairy	17
b.	Shoots hairless	18
17a.	Leaves usually constantly 3-lobed; racemes 1–5 cm	**62. orientale**
b.	Leaves 3–5-lobed; racemes 3–20 cm	**69. maximowiczii**
18a.	Leaf-lobes obtuse	**63. alpinum**
b.	Leaf-lobes acute	**64. distans**
19a.	Plant prostrate	20
b.	Plant erect, occasionally low-growing	24
20a.	Fruit with glandular hairs or bristles	21
b.	Fruit smooth, lacking hairs and bristles	23
21a.	Leaves 5–7-lobed; leaves unpleasent smelling	**57. glandulosum**
b.	Leaves usually 5-lobed; leaves not unpleasant smelling	22
22a.	Leaf-lobes triangular; racemes compact	**13. coloradensis**
b.	Leaf-lobes ovate; racemes loose	**12. laxiflorum**
23a.	Leaves mostly 3-lobed; fruit red	**5. triste**
b.	Leaves equally 3- and 5-lobed; fruit brownish	**23. procumbens**
24a.	Fruit red	25
b.	Fruit black, blue, purple, brown, yellowish white, green	33
25a.	Flowers with a ring around style-base	26
b.	Flowers without a ring around style-base	27
26a.	Leaf-base deeply cordate with a narrow sinus	**4. rubrum**
b.	Leaf-base shallowly to mediumly cordate, with a wide sinus	**6. warszewiczii**
27a.	Calyx with 5 'warts' inside	**2. mandshuricum**
b.	Calyx lacking 5 'warts' inside	28
28a.	Flowers yellow	29
b.	Flowers greenish, brown, reddish	30
29a.	Branches hairy when young	**1. multiflorum**
b.	Branches hairless when young	**55. fasciculatum**
30a.	Young shoots hairy	**1. multiflorum**
b.	Young shoots hairless	31
31a.	Leaf-base truncate to shallowly cordate, sinus broad; flowers large	**3. spicatum**
b.	Leaf-base usually cordate, sinus medium; flowers small to medium	32
32a.	Sepals rounded, with hairs on margin	**7. petraeum**
b.	Sepals broadly ovate, without hairs on margins	**8. emodense**
33a.	Leaf-lobes 5–7	**20. bracteosum**
b.	Leaf-lobes 3–5	34
34a.	Flowers tubular	35
b.	Flowers cup- to bell-shaped	41
35a.	Racemes to 30 cm	**9. longeracemosum**
b.	Racemes to 15 cm	36
36a.	Leaves with white or grey matted hairs below	37
b.	Leaves hairless or hairy below	38
37a.	Leaves 2–5 cm across, 3-lobed, roughly hairy above	**16. malvaceum**
b.	Leaves 5–10 cm across, 3–5-lobed, softly hairy above	**15. sanguineum**
38a.	Leaves with yellow gland spots on both surfaces	**30. americanum**
b.	Leaves without yellow gland spots on both surfaces	39

39a. Leaf- and flower-stalks with black-
stalked glands   **17. ciliatum**

b. Leaf- and flower-stalks without
black-stalked glands   40

40a. Young shoots hairless or sparsely
hairy   **24. aureum**

b. Young shoots hairy   **25. odoratum**

41a. Racemes to 30 cm
9. longeracemosum

b. Racemes to 15 cm   42

42a. Sepals white on outer surface   43

b. Sepals yellow, green, brown, pink, or
red on outer surface   44

43a. Leaves kidney-shaped, wider than
long   **26. hudsonianum**

b. Leaves circular   47

44a. Branches with prominent yellow
glands; plant very aromatic
27. nigrum

b. Branches without prominent yellow
glands; plant not particularly
aromatic   45

45a. Calyx-cup interior with a slightly
recognisable ring   **6. warszewiczii**

b. Calyx-cup interior without a slightly
recognisable ring   46

46a. Leaf-lobes acute, sharply toothed
28. petiolare

b. Leaf-lobes obtuse, scalloped
14. nevadense

47a. Flowers stalkless   **10. moupinense**

b. Flowers stalked   48

48a. Plant lacking stolons   49

b. Plant stoloniferous   **29. ussuriense**

49a. Leaves to 15 cm across; racemes
erect   **21. japonicum**

b. Leaves to 10 cm across; racemes
pendent   50

50a. Leaves both 3- and 5-lobed on the
same plant   **6. warszewiczii**

b. Leaves mostly 3-lobed   51

51a. Leaf-lobes obtuse
58. magellanicum

b. Leaf-lobes acute   52

52a. Sepals round, with hairs on margin
7. petraeum

b. Sepals broadly ovate, without hairs
on margin   **8. emodense**

53a. Flowers in racemes of 5–20   54

b. Flowers in racemes of 1–5   57

54a. Fruit black   **31. lacustre**

b. Fruit red, yellow or green   55

55a. Branches thornless
69. maximowiczii

b. Branches with at least a few thorns
56

56a. Leaves 3-lobed; racemes erect
60. pulchellum

b. Leaves 5-lobed; racemes pendent
32. montigenum

57a. Plant evergreen; flowers fuchsia-like
33. speciosum

b. Plant deciduous; flowers not fuchsia-
like   58

58a. Fruit smooth, hairless   72

b. Fruit glandular-hairy to bristly   59

59a. Plant with flowers of one sex only
61. giraldii

b. Plant with bisexual flowers   60

60a. Petals rolled inwards   61

b. Petals not rolled inwards   63

61a. Young shoots densely bristly
35. menziesii

b. Young shoots hairy, but not bristly
62

62a. Sepals often only 4; petals pink
34. lobbii

b. Sepals constantly 5; petals white
36. roezlii

63a. Calyx tubular   **47. setosum**

b. Calyx bell- to urn-shaped   64

64a. Petals strap-shaped   65

b. Petals fan-shaped, obovate   66

65a. Branches wavy   **42. stenocarpum**

b. Branches straight   68

66a. Plant with prickles, hairless or hairy
but not bristly   67

b. Plant bristly, without prickles
44. grossularioides

67a. Flowers greenish   **48. uva-crispa**

b. Flowers orange-red   **38. pinetorum**

68a. Leaves hairless to sparsely hairy   69

b. Leaves hairy   70

69a. Ovaries glandular bristly
43. alpestre

b. Ovaries bristly, but only the very
shortest occasionally glandular
37. californicum

70a. Fruit green   **40. burejense**

b. Fruit red or black   71

71a. Sepals white   **47. setosum**

b. Sepals brownish green **50. cynosbati**

72a. Branches grey and thorns large,
hooked   **51. divaricatum**

b. Branches and thorns not as
above   73

73a. Sepals white or yellowish   74

b. Sepals green to red   79

74a. Sepal-lobes yellow **39. quercetorum**

b. Sepal-lobes white   75

75a. Thorns constantly single, thin,
bristles absent   76

b. Thorns 1–3 on a single plant, often
stout, bristles often present   77

76a. Leaf-margin toothed, sparsely hairy
54. curvatum

b. Leaf-margin scalloped, usually
hairless   **53. niveum**

77a. Leaves hairless **49. oxyacanthoides**

b. Leaves finely hairy   78

78a. Prickles usually less than 1 cm, awl-
shaped   **47. setosum**

b. Prickles usually *c.* 1 cm, not awl-
shaped   **46. leptanthum**

79a. Branches occasionally with 5–7
prickles at node   **41. aciculare**

b. Branches with only 1–3 prickles at
node, never 5–7   80

80a. Branches with stout thorns   81

b. Branches without stout thorns,
bristly or prickly only   83

81a. Branches wavy   **42. stenocarpum**

b. Branches straight   82

82a. Leaves slightly 3-lobed
52. rotundifolium

b. Leaves deeply 3–5-lobed
49. oxyacanthoides

83a. Sepal-lobes twice as long as calyx-
cup   **59. diacanthum**

b. Sepal-lobes same length as calyx-cup
45. hirtellum

1. R. multiflorum Kitaibel. Illustration:
Janczewski, Monographie des groseilliers,
274 (1907); Krüssmann, Manual of
cultivated broad-leaved trees & shrubs **3**:
215 (1986).

Erect shrub to 2 m; branches thornless,
ash-grey, hairy when young; buds large.
Leaves *c.* 10 cm, circular, 3–5-lobed,
toothed, grey-white beneath. Flowers bell-
shaped, to 50, in racemes of *c.* 12 cm.
Calyx golden yellow-green. Stamens wide-
spreading. Fruit red. *SE Europe, Balkan
peninsula.* H2. Late spring–early summer.

R. × urceolatum Tausch;
R. multiflorum × R. petraeum. Illustration:
Janczewski, Monographie des groseilliers,
487 (1907). Similar to *R. multiflorum*, but
flowers *c.* 25 in loose racemes, to 12 cm.
Flowers brownish. Fruit red. *Garden origin.*
H1–2. Early summer.

R. × koehneanum Janczewski;
R. multiflorum × R. rubrum. Illustration:
Janczewski, Monographie des groseilliers,
486 (1907). Dome-shaped shrub. Leaves
6.5 cm across, 3–5-lobed. Flowers
brownish, up to 35 in racemes, to 10 cm.
Stamens pink; stamens and style same
length. *Garden origin.* H1–2. Late spring.
The commonly grown garden currant.

2. R. mandshuricum (Maximowicz)
Komarov. Illustration: Janczewski,
Monographie des groseilliers, 275 (1907);
Krüssmann, Manual of cultivated broad-
leaved trees & shrubs **3**: 205 (1986).

Shrub to 2 m, bark usually almost black,
branches spineless. Leaves to 9 × 11 cm,
broadly ovate, 3-lobed, lobes usually acute,
coarsely toothed. Flowers in many-flowered

pendent racemes. Calyx cup-shaped, the inside with 5 'warts', not interconnected by a raised ring. Fruit red. *NE Asia.* Early summer.

3. R. spicatum Robson (*R. rubrum* Linnaeus in part; *R. schlechtendahlii* Lange). Illustration: Krüssmann, Manual of cultivated broad-leaved trees & shrubs **3**: 204 (1986).

Erect shrub to 2 m; branches hairless. Leaves to 10 cm across, circular, 3–5-lobed, base truncate to shallowly cordate, sinus broad. Flowers light green, tinged brown-red, usually in erect racemes; calyx cup-shaped, lacking a ring around style-base. Fruit red. *N Europe, N Asia.* H1 Early summer.

4. R. rubrum Linnaeus (*R. silvestre* (Lamark) Mertens; *R. sativum* (Reichenbach) Symes; *R. vulgare* Lamark). Illustration: Krüssmann, Manual of cultivated broad-leaved trees & shrubs **3**: 204 (1986).

Erect, broad shrub; young shoots slightly hairy, glandular. Leaves to 6 cm across, circular, 3–5-lobed, lobes acute, base deeply cordate, sinus narrow. Flowers greenish to reddish, in pendent to spreading racemes; calyx cup-shaped, style-base with a pentagonal ring. Fruit red. *W Europe.* H1–2. Early summer.

R. × houghtonianum Janczewski (*R.* 'Houghton Castle'). Illustration: Janczewski, Monographie des groseilliers, 479 (1907). Intermediate between the parents. Vigorous shrub; new growth appearing late. Leaves *c.* 6 cm across, hairy, base slightly cordate. Flowers green, tinged brown, 8–18 in racemes, to 5 cm. Fruit red, edible. *Garden origin.* H1. Early summer.

A hybrid between *R. silvestre* × *R. spicatum.*

5. R. triste Pallas (*R. albinervium* Michaux). Illustration: Janczewski, Monographie des groseilliers, 283 (1907); Gleason, Illustrated flora of the north-eastern United States and adjacent Canada **2**: 279 (1952); Krüssmann, Manual of cultivated broad-leaved trees & shrubs **3**: 206 (1986).

Shrub to 50 cm, stems decumbent, spineless. Leaves 6–10 cm across, 3-lobed, lobes toothed, base usually cordate. Flowers in racemes, the later glandular, mostly shorter than the leaves. Flowers reddish. Calyx broadly bell-shaped, sepal-lobes spreading. Fruit *c.* 6 mm across,

smooth, red. *N North America.* H1. Early summer.

6. R. warszewiczii Janczewski. Illustration: Janczewski, Monographie des groseilliers, 285 (1907); Krüssmann, Manual of cultivated broad-leaved trees & shrubs **3**: 204 (1986).

Similar to *R. spicatum.* Erect shrub to 1.5 m. Leaves to 9 cm, rounded, 3–5-lobed, base cordate. Flowers green tinged with red in pendent racemes, the later 5–7 cm. Calyx-cup interior with a slightly recognisable ring around the style-base. Fruits red-black. *E Siberia.* H1. Early summer.

7. R. petraeum Wulfen (*R. petraeum* var. *bullatum* (Otto & Dietrich) Schneider). Illustration: Janczewski, Monographie des groseilliers, 291 (1907); Huxley, Mountain flowers of Europe, 1040 (1986); Krüssmann, Manual of cultivated broad-leaved trees & shrubs **3**: 205 (1986).

Erect shrub to 1.5 m or more; branches grey-brown, hairless. Leaves 7–10 cm, circular, usually 3-lobed, lobes acute, toothed, hairy beneath, base cordate to truncate. Flowers small, bell-shaped, green to reddish, in dense many-flowered racemes. Sepals short, round, hairy on margins. Petals half as long as sepals. Style cone-shaped. Fruit red to purple. *W & C Europe to Siberia.* H1. Early summer.

Var. **altissimum** (Turczaninow) Janczewski. Shrub to 3 m. Leaves to 15 cm across. Flowers pale red, *c.* 20, in racemes, 5–7 cm. *Siberia.*

Var. **atropurpureum** (C.A. Meyer) Schneider. Leaves to 15 cm across, 3-lobed. Flowers purple outside, lighter inside, *c.* 15 in racemes, 2–4 cm. *Siberia.*

Var. **biebersteinii** (Berland) Schneider (*R. caucasicum* Bieberstein; *R. biebersteinii* Berland). Leaves *c.* 12 cm across, usually 5-lobed, hairy beneath. Flowers reddish, racemes to 10 cm. *Caucasus.*

R. × gondouinii Janczewski; *R. petraeum* × *R. silvestre.* Illustration: Janczewski, Monographie des groseilliers, 484 (1907). Intermediate between parents. Young shoots red, hairless, appearing early. Flowers bell-shaped, in almost horizontal racemes. Fruit red. *Garden origin.* H1–2. Early summer.

8. R. emodense Rehder. (*R. himalayense* Decaisne not Royle; *R. meyeri* Schneider not Maximowicz). Illustration: Krüssmann, Manual of cultivated broad-leaved trees & shrubs **3**: 205 (1986).

Similar to *R. petraeum*, but sepals broadly ovate, lacking hairs on margin; petals wedge-shaped, erect. Fruit large, red to black. *Himalayas, China (Yunnan).* H1. Early summer.

9. R. longeracemosum Franchet. Illustration: Janczewski, Monographie des groseilliers, 301 (1907); Krüssmann, Manual of cultivated broad-leaved trees & shrubs **3**: 205 (1986).

Shrub to 3 m; branches thornless, hairless. Leaves to 14 cm, circular, 3–5-lobed, lobes acute, hairless. Flowers tubular to bell-shaped, reddish to greenish, to 15 in loose, pendent racemes to 30 cm; sepals and petals erect. Stamens and style protruding. Fruit black, glossy. *W China (Sichuan, Hubei, Xbang).* H1. Early summer.

10. R. moupinensc Franchet. Illustration: Janczewski, Monographie des groseilliers, 299 (1907); Krüssmann, Manual of cultivated broad-leaved trees & shrubs **3**: 205 (1986).

Erect shrub, 1–2 m; branches thornless, hairless. Leaves variable, to 16 cm across, 3–5-lobed, with scattered stalkless glands on both surfaces. Flowers red or green-red, in racemes 4–12 cm. Petals and sepals erect. Stamens included. Fruit black, glossy. *China.* H1. Early summer.

11. R. ambiguum Maximowicz. Illustration: Janczewski, Monographie des groseilliers, 304 (1907).

Shrub to 60 cm; branches thornless. Leaves 2–5 cm across, kidney-shaped, 3–5-lobed, lobes obtuse, stocky glandular beneath. Flowers 1–2, stalks *c.* 1 cm. Sepals elliptic, greenish. Fruit green, glandular-hairy, translucent. *Japan, China (Sichuan).* H1–2. Early summer.

Commonly found growing on old trees in the wild.

12. R. laxiflorum Pursh (*R. affine* Douglas). Illustration: Janczewski, Monographie des groseilliers, 307 (1907); Krüssmann, Manual of cultivated broad-leaved trees & shrubs **3**: 204 (1986).

Prostrate shrub, branches thornless. Leaves 6–8 cm across, circular, deeply 5-lobed, base cordate, sharply toothed, hairless above, hairy beneath. Flowers 6–12, in loose erect racemes to 8 cm; sepals reddish; petals fan-shaped. *N America (Alaska to N Carolina).* H1. Early summer.

13. R. coloradensis Coville. Illustration: Janczewski, Monographie des groseilliers, 310 (1907); Krüssmann, Manual of cultivated broad-leaved trees & shrubs **3**: 204 (1986).

Low-growing shrub to 50 cm; young shoots hairy. Leaves 5–8 cm; broadly ovate, usually 5-lobed, lobes triangular, hairless above, hairy on veins beneath. Flowers 6–12, in erect, glandular-hairy racemes, to 5 cm. Sepals greenish to reddish; petals fan-shaped, purple. Fruit black. *USA*. H1. Summer.

14. R. nevadense Kellogg. Illustration: Janczewski, Monographie des groseilliers, 316 (1907); Krüssmann, Manual of cultivated broad-leaved trees & shrubs **3**: 203 (1986).

Erect shrub to 1.5 m; branches thornless, usually hairless. Leaves 3–6 cm across, circular, usually 3-lobed, leaves rounded, scalloped. Flowers *c.* 20, in pendent racemes. Sepals pink; petals white, rounded-oblong. Fruit blue, glandular. *USA (California, Sierra Nevada)*. H2–3. Early summer.

15. R. sanguineum Pursh. Illustration: Janczewski, Monographie des groseilliers, 320 (1907); Hay & Beckett, Reader's Digest encyclopedia of garden plants and flowers, 592 (1978); Krüssmann, Manual of cultivated broad-leaved trees & shrubs **3**: 203 (1986); Phillips & Rix, Shrubs, 40 (1989).

Shrub to 4 m; branches spineless, aromatic, glandular. Leaves 5–10 cm across, circular, 3–5-lobed, base cordate, dark green, hairy above, white felted beneath, stalk hairy. Flowers in many-flowered, glandular-hairy, erect or pendent racemes, to 8 cm. Calyx reddish purple, lobes longer than tube. Petals white to reddish purple, half as long as sepals. Fruit to 1 cm across, black, blue-white bloomed. *W America*. H1. Early summer.

The most frequently grown of all the ornamental currants. Many cultivars exist and include: 'Albescens', flowers whitish; 'Atrorubens', compact habit, flowers deep red, small; 'Brocklebankii', slow-growing, flowers yellow; 'Grandiflorum', flowers red, in large racemes; 'Koja', flowers dark red; 'Plenum', slow-growing, flowers double, red; 'Splendens', flowers dark red; and 'Pulborough Scarlet' (illustration: Brickell (ed.), RHS A–Z encyclopedia of garden plants, 906, 2003), flowers deep red with a white centre, in large racemes, one of the best cultivars available today.

R. × fontenayense Janczewski; *R. sanguineum × R. uva-crispa*. Illustration: Janczewski, Monographie des groseilliers,

492 (1907). Intermediate between parents. Shrub to 1 m; branches thornless. Leaves 6–8 cm, circular, 3–5-lobed, hairy beneath. Flowers red, 3–6 in somewhat pendent racemes. Calyx tube short, broader than long. Free-flowering. Fruit purple-black, seldom produced. *Garden origin*. H1. Early summer.

16. R. malvaceum Smith. Illustration: Krüssmann, Manual of cultivated broad-leaved trees & shrubs **3**: 203 (1986).

Very similar to *R. sanguineum* but leaves roughly hairy above and grey felted beneath. Ovary white hairy. *USA (California)*. H2. Early Summer.

R. × bethmontii Janczewski; *R. malvaceum × R. sanguineum*. Habit like *R. sanguineum*; leaves roughly hairy, like *R. malvaceum*. Flowers bright pink; calyx urceolate, style almost twice as long as tube. Pollen sterile. Fruit black; sets fruit only after pollination by another species. *France*. H1–2. Early summer.

17. R. ciliatum Roemer & Schultes. Illustration: Janczewski, Monographie des groseilliers, 329 (1907).

Dome-shaped shrub to 2 m; much branched, branches glandular-hairy when young; leaf- and flower-stalks with black-stalked glands. Leaves 3–5 cm across, 3–5-lobed, coarsely double-toothed, with flat, bristles above, glandular-hairy beneath. Flowers green, outer surface hairy, 6–10 in a pendent raceme. Fruit black, glossy. *Mexico*. H5. Early summer.

18. R. cereum Douglas. Illustration: Janczewski, Monographie des groseilliers, 337 (1907); Krüssmann, Manual of cultivated broad-leaved trees & shrubs **3**: 209 (1986).

Shrub to 1 m, much branched, without spines. Leaves 1–4 cm across, kidney-shaped, 3–5-lobed, toothed. Flowers white, greenish to yellowish, in short, few-flowered racemes. Calyx tubular, sepals longer than petals. Styles with long hairs. Fruit red. *W North America*. H1. Early summer.

19. R. inebrians Lindley (*R. pumilum* Nuttall). Illustration: Janczewski, Monographie des groseilliers, 335 (1907); Krüssmann, Manual of cultivated broad-leaved trees & shrubs **3**: 209 (1986).

Very similar to *R. cereum*, but flowers usually pink, in few-flowered, pendent racemes; ovary usually hairless. Fruit red, glandular. *W North America*. H1. Early summer.

20. R. bracteosum Douglas. Illustration: Janczewski, Monographie des groseilliers, 339 (1907); Krüssmann, Manual of cultivated broad-leaved trees & shrubs **3**: 206 (1986).

Erect shrub to 3 m; young shoots sparsely glandular hairy. Leaves 5–20 cm across, circular, 5–7-lobed, lobes lanceolate, double-toothed. Flowers in erect racemes, to 20 cm, much longer than leaves. Calyx green, tinged purplish red. Petals larger than sepals, white. Fruit spherical, black, white bloomed. *W North America (Alaska to N California)*. H1. Early summer.

R. × fuscescens Janczewski; *R. bracteosum × nigrum*. Habit similar to *R. bracteosum*, but calyx reddish brown, bracts small, linear. Fruit larger. *Garden origin*. H2. Early summer.

21. R. japonicum Maximowicz. Illustration: Janczewski, Monographie des groseilliers, 340 (1907).

Shrub to 2 m; branches thornless. Leaves *c.* 15 cm across, circular, 5-lobed, lobes acute, toothed. Flowers bell-shaped, green or brown, in erect racemes. Fruit black, hairless. *Japan*. H1–2. Early summer.

22. R. viburnifolium A. Gray. Illustration: Janczewski, Monographie des groseilliers, 341 (1907); Krüssmann, Manual of cultivated broad-leaved trees & shrubs **3**: 211 (1986).

Evergreen shrub to 1.5 m; branches thornless. Leaves 2–4 cm, broadly ovate to elliptic, base round, margin entire or with a few small teeth. Flowers pink, in erect racemes, to 2.5 cm. Fruit red. *USA (California)*. H4–5. Early summer.

23. R. procumbens Pallas. Illustration: Janczewski, Monographie des groseilliers, 342 (1907); Krüssmann, Manual of cultivated broad-leaved trees & shrubs **3**: 215 (1986).

Low shrub to 70 cm. Leaves to 8 cm, kidney-shaped, 3–5-lobed. Flowers reddish, in racemes to 4 cm, bracts absent. Fruit brownish, smooth. *E Siberia*. H1. Early summer.

24. R. aureum Pursh (*R. tenuiflorum* Lindley). Illustration: Janczewski, Monographie des groseilliers, 334 (1907); Hay & Beckett, Reader's Digest encyclopedia of garden plants and flowers, 592 (1978); Krüssmann, Manual of cultivated broad-leaved trees & shrubs **3**: 206 (1986).

Erect shrub to 2 m, branches spineless, bark brown. Leaves 3–5 cm, broadly elliptic, 3–5-lobed, coarsely toothed, hairless. Flowers yellow, fragrant, in few- to many-flowered, pendent racemes. Calyx tubular, sepals spreading, but inclining after flowering. Petals commonly becomming reddish. Fruit purple-brown to black. *USA (California), Mexico*. H1. Early summer.

25. R. odoratum Wendland (*R. aureum* misapplied; *R. fragrans* Loddiges). Illustration: Gleason, Illustrated flora of the north-eastern United States and adjacent Canada **2**: 279 (1952); Phillips & Rix, Shrubs, 41 (1989); Brickell (ed.), RHS A–Z encyclopedia of garden plants, 906 (2003).

Similar to and often confused with *R. aureum*; differs in its hairier young shoots. Branches spineless; leaves 3–8 cm across, maple-like, 3–5-lobed, lobes coarsely toothed. Flowers yellow, 5–10, in racemes, cupule 6–10 × c. 1.5 mm, sepals spreading, lobes to half as long as tube. Fruit c. 1 cm across, spherical, black. *C North America*. H1. Early summer.

R. × gordonianum Beaton; *R. petraeum × R. sanguineum*. Illustration: Phillips & Rix, Shrubs, 40 (1989). Shrub to 2.5 m; branches spineless, hairless. Leaves c. 4 cm, circular, 3–5-lobed, shallowly to coarsely toothed, glandular-hairy on both sides. Flowers in long, many-flowered racemes, more erect than in parents; calyx tube c. 3 mm, calyx reddish yellow, or tube red and yellow and lobes red outside, yellow inside. *Garden origin*. H1. Early summer.

26. R. hudsonianum Richards. Illustration: Janczewski, Monographie des groseilliers, 346 (1907); Gleason, Illustrated flora of the north-eastern United States and adjacent Canada **2**: 278 (1952); Krüssmann, Manual of cultivated broad-leaved trees & shrubs **3**: 203 (1986).

Erect shrub to 1.5 m; branches thornless. Leaves to 10 cm across, broadly ovate, more or less hairy, with resin glands, 3–5-lobed, lobes ovate, obtuse to acute, coarsely toothed. Flowers white, in loose erect racemes to 6 cm; calyx cup-shaped, sepals ovate, spreading. Ovary with resin glands. Fruit black, hairless. *N North America*. H1. Early summer.

27. R. nigrum Linnaeus. Illustration: Janczewski, Monographie des groseilliers, 347 (1907); Gleason, Illustrated flora of the north-eastern United States and adjacent Canada **2**: 278 (1952); Krüssmann, Manual of cultivated broad-leaved trees & shrubs **3**: 204 (1986).

Rounded shrub to 2 m; branches with yellowish glands, hairy to hairless when young, yellow, aromatic. Leaves 5–10 cm across, circular, 3–5-lobed, hairy beneath. Flowers bell-shaped, 4–10 cm, in pendent racemes. Calyx greenish outside, reddish white inside. Fruit black, edible. *Europe to C Asia & Himalaya*. H1. Early summer.

'Apiifolium', leaves 3-lobed, very deeply incised and toothed; 'Chlorocarpum', fruit green; 'Coloratura', leaves variegated white; 'Heterophyllum', leaves deeply cleft; 'Marmoratum', leaves deeply lobed, marbled cream; 'Xanthocarpum', fruit yellow to white.

R. × culverwellii MacFarlane (*R. schneideri* Maurer). Habit like *R. nigrum*, leaves like *R. uva-crispa*. Fruit sterile. Resembling a gooseberry without the bristles. *Garden origin*. H2. Early summer.

A hybrid between *R. nigrum* and *R. uva-crispa*.

28. R. petiolare Fischer. Illustration: Krüssmann, Manual of cultivated broad-leaved trees & shrubs **3**: 203 (1986).

Hairless shrub to 1.5 m. Leaves 10–15 cm across, with resin glands beneath. Flowers white, in erect racemes, to 12 cm. Fruit black, not bloomed. *E Siberia, Manchuria, W North America*. H1. Early summer.

29. R. ussuriense Janczewski. Illustration: Janczewski, Monographie des groseilliers, 349 (1907).

Stoloniferous shrub to 1 m; young shoots hairy, with yellow resin-glands, camphor-scented. Flowers yellowish green, 5–9 in racemes. Fruit bluish black. *Manchuria to Korea*. H1. Early summer.

30. R. americanum Miller (*R. floridum* Miller). Illustration: Gleason, Illustrated flora of the north-eastern United States and adjacent Canada **2**: 278 (1952) Krüssmann, Manual of cultivated broad-leaved trees & shrubs **3**: 203 (1986).

Shrub to 1.5 m; branches spineless, grey-brown. Leaves 5–8 cm, circular, usually 3-lobed, lobes acute, sharply toothed, with yellow gland spots on both surfaces (under magnification). Flowers c. 10 in racemes, to slightly longer than leaves, pendent. Calyx tube greenish yellow. Petals small. Fruit c. 6 mm across, black. *N North America*. H1. Summer.

31. R. lacustre (Persoon) Pourret (*R. grossularioides* Michaux; *R. echinatum* Lindley). Illustration: Janczewski, Monographie des groseilliers, 352 (1907); Gleason, Illustrated flora of the north-eastern United States and adjacent Canada **2**: 277 (1952); Krüssmann, Manual of cultivated broad-leaved trees & shrubs **3**: 207 (1986).

Shrub to 1 m; branches slightly pendent, bristly and slightly prickly. Leaves 3–6 cm across, circular, deeply lobed. Flowers greenish red, 4–10 in loose, pendent racemes, to 9 cm. Fruit small, black, densely bristly. *N North America*. H1. Early summer.

Often wrongly named *R. grossularioides* in gardens; the true **R. grossularioides** Maximowicz is a little known, medium-height, bristly to smooth shrub closely related to *R. alpestre* and distinguished from this species by its anthers, which lack glands at their tips.

32. R. montigenum McClatchie (*R. lacustre* var. *molle* A. Gray). Illustration: Janczewski, Monographie des groseilliers, 355 (1907).

Loosely branched shrub, c. 75 cm; branches bristly, with a few thorns. Leaves 1–4 cm across, kidney-shaped, 5-lobed, lobes acute, hairy. Flowers shortly tubular, in short, few-flowered, pendent racemes; calyx brownish green, tube glandular-bristly. Fruit dark red, glandular-bristly. *W North America*. H1. Early summer.

33. R. speciosum Pursh. Illustration: Janczewski, Monographie des groseilliers, 357 (1907); Hay & Beckett, Reader's Digest encyclopedia of garden plants and flowers, 592 (1978); Krüssmann, Manual of cultivated broad-leaved trees & shrubs **3**: 205 (1986).

Evergreen shrub to 4 m; branches grey-brown at first, densely spiny; thorns to 1 cm, in groups of 2 or 3, at nodes. Leaves 1–4 cm, broadly ovate, 3–5-lobed, hairless. Flowers in groups of 3–5, 4-parted, calyx bell-shaped, sepals narrow, red-purple; stamens long protruding. Fruit densely glandular-bristly, red. *USA (California)*. H2–3. Early summer.

One of the most attractive species.

34. R. lobbii Gray (*R. subvestitum* Hooker & Arnott). Illustration: Janczewski, Monographie des groseilliers, 359 (1907); Krüssmann, Manual of cultivated broad-leaved trees & shrubs **3**: 210 (1986).

Shrub to 2 m; branches thorny, hairy when young, thorns in groups of 3, 1–2 cm. Leaves 2–3.5 cm across, circular to heart-shaped, 3–5-lobed, finely hairy above. Flowers large, single or in pairs; sepals deep purple, often only 4. Anthers black, filaments pink. Fruit purple, densely glandular-hairy. *N America (British Columbia to California)*. H2–3. Late spring.

35. R. menziesii Pursh (*R. subvestitum* Hooker & Arnott). Illustration: Janczewski, Monographie des groseilliers, 362 (1907); Krüssmann, Manual of cultivated broad-leaved trees & shrubs **3**: 207 (1986).

Shrub to 2 m. Young shoots densely bristly, with thin spines in groups of 3, 1–2 cm. Leaves 2–4 cm across, 3–5-lobed, hairless to sparsely hairy above, velvety hairy and glandular beneath. Flowers 1 or 2, calyx tubular, lobes 3 times as long as tube, purple. Petals whitish. Stamens long protruding. Fruit densely bristly black. *USA (Oregon to California)*. H5–G1. Early summer.

R. × darwinii F. Koch; *R. menziesii × R. niveum*. Illustration: Krüssmann, Manual of cultivated broad-leaved trees & shrubs **3**: pl. 70 (1986). Vigorous, erect shrub to 3 m. Intermediate between parents. Leaves similar to those of *R. niveum*. Flowers 1–4 in pendent racemes; sepals red, apex whitish, reflexed; petals white. Style to 1.5 cm, reddish. Fruit black, hairless. *Garden origin*. H1–2. Early summer.

36. R. roezlii Regel (*R. amictum* Greene). Illustration: Krüssmann, Manual of cultivated broad-leaved trees & shrubs **3**: 210 (1986).

Erect shrub to 1.5 m; branches with thin spines in groups of 3 at nodes, each to 1.5 cm, young branches hairy but not bristly. Leaves 1.5–2.5 cm, circular, 3–5-lobed, toothed, base cuneate to almost cordate, leaf-stalk short. Flowers in clusters of 1–3; calyx bell-shaped, lobes purple; petals half as long as sepals, white; filaments included, very bristly. Fruit 1–1.5 cm, purple, very bristly. *USA (California)*. H3. Early summer.

37. R. californicum Hooker & Arnott. Illustration: Krüssmann, Manual of cultivated broad-leaved trees & shrubs **3**: 210 (1986).

Similar to *R. menziesii* but branches lacking bristles; leaves hairless or almost so; flowers green or reddish. *USA (California)*. H2. Early summer.

38. R. pinetorum Greene. Illustration: Janczewski, Monographie des groseilliers, 370 (1907); Krüssmann, Manual of cultivated broad-leaved trees & shrubs **3**: 205 (1986).

Shrub to 2 m; branches thorny, but lacking bristles. Leaves 2–3 cm, heart-shaped, deeply 3–5-lobed, hairless above, hairy below. Flowers large, usually solitary, occasionally paired; calyx bell-shaped, orange-red, sepals twice as long as calyx-tube; style hairless. Fruit purple. *USA*. H1–2. Early summer.

39. R. quercetorum Greene

Erect shrub to 1.5 m; branches pendent, thorny, new growth appearing very early. Leaves 1–2 cm, circular, deeply 3–5-lobed. Flowers shortly tubular, yellowish or whitish; sepals spreading; petals short. Ovary and style hairless. Fruit black, hairless. *USA (California)*. H3–4. Early summer.

40. R. burejense F. Schmidt. Illustration: Janczewski, Monographie des groseilliers, 371 (1907).

Shrub to 1 m; branches usually very bristly, with thorns at the nodes, 1 cm. Leaves 2–6 cm across, circular, deeply 3–5-lobed, lobes obtuse, toothed, hairy on both surfaces. Flowers single or in pairs, stalks 3–6 mm; calyx broadly bell-shaped. Stamens protruding, styles hairless. Fruit green, very bristly, edible. *NE Asia, Alaska, N California*. H1. Early summer.

41. R. aciculare Smith. Illustration: Janczewski, Monographie des groseilliers, 373 (1907); Krüssmann, Manual of cultivated broad-leaved trees & shrubs **3**: 205 (1986).

Very similar to *R. burejense* but branches short prickled and bristly, occasionally with 5–7 prickles at the nodes. Flowers pink or light green. Fruit red, green or yellow. *Siberia*. H1. Early summer.

42. R. stenocarpum Maximowicz. Illustration: Janczewski, Monographie des groseilliers, 375 (1907); Krüssmann, Manual of cultivated broad-leaved trees & shrubs **3**: 205 (1986).

Shrub to 2 m; branches very prickly and bristly, thorny. Leaves *c.* 3 cm across, deeply 3–5-lobed. Flowers in groups of 1–3; calyx bell-shaped, reddish; petals white, three-fifths as long as sepals. Fruit greenish to reddish. *NW China*. H1. Early summer.

43. R. alpestre Wallich. Illustration: Janczewski, Monographie des groseilliers, 376 (1907); Polunin & Stainton, Flowers of the Himalaya, 468 (1985); Krüssmann, Manual of cultivated broad-leaved trees & shrubs **3**: 205 (1986).

Vigorous, erect shrub to 3 m; branches very prickly and bristly, reddish when young. Leaves 2–5 cm across, circular, 3–5-lobed, toothed, hairless to slightly hairy. Flowers 1 or 2 on short stalks, small, greenish red and white. Ovary glandular-bristly. Fruit purple-red, glandular-bristly. *Himalaya, W China*. H1. Early summer.

Var. **giganteum** Janczewski. Shrub to 5 m; thorns to 3 cm. Fruit green, lacking bristles. *W China*. H1. Early summer.

44. R. grossularioides Maximowicz. Illustration: Janczewski, Monographie des groseilliers, 377 (1907); Krüssmann, Manual of cultivated broad-leaved trees & shrubs **3**: 205 (1986).

Similar to *R. alpestre* but distinguished by its anthers, which lack glands at their tips. *Japan*. Early summer.

Uncommon in cultivation.

45. R. hirtellum Michaux (*R. oxyacanthoides* J.D. Hooker not Linnaeus; *R. gracile* of Janczewski). Illustration: Gleason, Illustrated flora of north-eastern United States and adjacent Canada **2**: 277 (1952).

Shrub to 1 m; branches thin, grey when young, later dark brown, thornless or with small prickles. Leaves ovate to circular, 3–5-lobed. Flowers 1–3 in short racemes, narrowly bell-shaped; calyx-lobes erect or spreading, greenish or reddish. Fruit purple to black, mostly hairless. *N USA*. Summer.

46. R. leptanthum A. Gray. Illustration: Janczewski, Monographie des groseilliers, 379 (1907); Krüssmann, Manual of cultivated broad-leaved trees & shrubs **3**: 205 (1986).

Erect, dense, spreading shrub usually to 1 m; branches stoutly thorny. Leaves *c.* 2 cm across, circular, 3–5-lobed, lobes obtuse, finely hairy. Flowers numerous, single or in pairs; calyx-base greenish, sepals greenish white, hairy; petals spathulate, white or light reddish. Fruit black. *USA*. H1. Summer.

R. × lydia F. Koch; *R. leptanthum × R. quercetorum*. New growth appearing early. Leaves like *R. leptanthum*; flowers in pairs, light yellow. Fruit small, black. *Garden origin*. H1–2. Late spring.

R. × magdalenae F. Koch; *R. leptanthum × R. uva-crispa*. Very similar to *R. leptanthum*, but flowers pink and white. Fruit black, mostly sterile. *Garden origin.* H1–2. Early summer.

47. R. setosum Lindley (*R. saximontanum* E. Nelson). Illustration: Janczewski, Monographie des groseilliers, 382 (1907); Gleason, Illustrated flora of the north-eastern United States and adjacent Canada **2**: 277 (1952).

Shrub to 1 m; branches with awl-shaped prickles, to 1 cm. Leaves 1–4 cm across, circular, 3–5-lobed, finely hairy. Flowers in groups of 1–3; calyx shortly tubular to bell-shaped, white, sepals half as long as tube. Fruit red to black, slightly bristly to smooth. *NW USA.* H1. Early summer.

48. R. uva-crispa Linnaeus (*R. grossularia* var. *uva-crispa* Smith; *R. grossularia* var. *pubescens* Koch). Illustration: Polunin, Flowers of Europe, pl. **43** (1969); Krüssmann, Manual of cultivated broad-leaved trees & shrubs **3**: 205 (1986).

Broad shrub to 1 m; shoots hairy, prickly. Leaves 2–6 cm across, circular, 3–5-lobed, scalloped, softly hairy beneath. Flowers greenish, 1–3 in racemes; calyx cup-shaped. Ovary downy. Fruit small, yellow or green, hairy. *NE & C Europe.* H1. Early summer.

49. R. oxyacanthoides Linnaeus. Illustration: Janczewski, Monographie des groseilliers, 387 (1907); Krüssmann, Manual of cultivated broad-leaved trees & shrubs **3**: 205 (1986).

Shrub *c.* 1 m; branches bristly, with 1–3 thorns at nodes, each *c.* 1 cm. Leaves 2–4 cm, circular, 3–5-lobed, toothed, hairless. Flowers 1–2, on short stalks. Calyx shortly tubular, lobes longer than tube, greenish white. Fruit purple-red, hairless. *N America.* H1. Early summer.

50. R. cynosbati Linnaeus. Illustration: Janczewski, Monographie des groseilliers, 383 (1907); Gleason, Illustrated flora of the north-eastern United States and adjacent Canada **2**: 276 (1952); Krüssmann, Manual of cultivated broad-leaved trees & shrubs **3**: 207 (1986).

Shrub to 1.5 m; branches pendent, with spines simple or in groups of 3 at nodes, or occasionally thornless. Leaves 3–5 cm across, circular, 3–5-lobed, lobes coarsely toothed. Flowers in groups of 2 or 3; calyx bell-shaped, sepals and petals brownish green, sepals longer than petals. Stamens slightly protruding from calyx. Fruit black,

top half with bristles. *E USA.* H1. Early summer.

51. R. divaricatum Douglas. Illustration: Janczewski, Monographie des groseilliers, 390 (1907); Krüssmann, Manual of cultivated broad-leaved trees & shrubs **3**: 211 (1986).

Shrub to 3 m; branches grey-brown, with stout, hooked thorns to 2 cm. Leaves 2–6 cm across, ovate to circular, 5-lobed, coarsely toothed. Flowers greenish purple, in clusters of 2–4; calyx bell-shaped, sepals slightly longer than petals. Anthers protruding. Fruit dark red to black, white bloomed. *W North America.* H1. Early summer.

R. × succirubrum Zabel; *R. divaricatum × R. niveum.* Illustration: Janczewski, Monographie des groseilliers, 500 (1907). Vigorous, erect shrub; branches hairless, with prickles at nodes, to 2 cm. Leaves 3–5 cm across. Flowers bright pink, 2–4 in pendent racemes. Stamens widely spreading. Fruit black, bloomed. *Garden origin.*

52. R. rotundifolium Michaux (*R. triflorum* Willdenow; *R. gracile* of Pursh not Michaux). Illustration: Janczewski, Monographie des groseilliers, 392 (1907); Gleason, Illustrated flora of the north-eastern United States and adjacent Canada **2**: 276 (1952); Krüssmann, Manual of cultivated broad-leaved trees & shrubs **3**: 205 (1986).

Shrub to 1 m; branches sparsely thorny. Leaves 2–5 cm across, more or less broadly heart-shaped, usually 3-lobed, lobes obtuse, finely hairy. Flowers greenish red, in clusters of 1–3; calyx bell-shaped, sepals twice as long as tube, stamen and style long protruding. Fruit purple. *E & C USA.* H1. Early summer.

Often confused with *R. divaricatum* in cultivation.

53. R. niveum Lindley. Illustration: Janczewski, Monographie des groseilliers, 394 (1907); Krüssmann, Manual of cultivated broad-leaved trees & shrubs **3**: 207 (1986).

Shrub to 3 m; branches with thin solitary thorns, hairless; reddish brown; young growth early. Leaves to 3 cm across, circular, 3–5-lobed, lobes usually rounded, few-toothed, usually hairless. Flowers 1–4, nodding, on thin stalks. Cupule bell-shaped; sepals *c.* 1 cm, white; petals very short, white. Fruit blue-black, hairless. *NW North America.* H1–2. Early summer.

R. × kochii Krüssmann; *R. niveum × R. speciosum.* Slow-growing shrub to 1 m; branches thorny. Flowers 4 or 5 parted; sepals red; petals white to reddish. Fruit brownish, bristly, rarely produced. *Garden origin.* H2–3. Early summer.

R. × robustum Janczewski. Thorns small. Flowers white or pinkish white. Fruit black. *Garden origin.* H1–2. Early summer.

54. R. curvatum Small. Illustration: Janczewski, Monographie des groseilliers, 395 (1907).

Shrub to 1 m, very similar to *R. niveum.* Thorns *c.* 5 mm. Leaves to 3 cm across, usually slightly hairy. Flowers bell-shaped, usually solitary or in pairs; sepals spreading; petals small, white. Ovary glandular. Fruit purple, hairless. *SE North America.* H2–3. Early summer.

55. R. fasciculatum Siebold & Zuccarini. Illustration: Janczewski, Monographie des groseilliers, 396 (1907); Krüssmann, Manual of cultivated broad-leaved trees & shrubs **3**: 209 (1986).

Shrub to 1.5 m, branches spineless, usually hairless when young. Leaves appearing very early and persisting a long time, 4–7 cm, circular to broadly ovate, 3–5-lobed. Flowers yellow, fragrant, in racemes of 2–9; calyx cup-shaped; petals longer than sepals. Fruit red. *Japan, Korea.* H1. Early summer.

56. R. gayanum (Spach) Steudel (*R. villosum* C. Gray not Nuttall). Illustration: Janczewski, Monographie des groseilliers, 427 (1907); Krüssmann, Manual of cultivated broad-leaved trees & shrubs **3**: 202 (1986).

Evergreen shrub, to 1 m; young branches downy. Leaves 3–6 cm, circular, shallowly 3–5-lobed, lobes obtuse, toothed, hairy on both surfaces. Flowers bell-shaped, yellow, fragrant, hairy, in many-flowered racemes, to 6 cm. Fruit black, edible. *Chile.* H4. Early–mid summer.

57. R. glandulosum Weber (*R. prostratum* L'Héretier). Illustration: Janczewski, Monographie des groseilliers, 431 (1907); Gleason, Illustrated flora of the north-eastern United States and adjacent Canada **2**: 278 (1952); Krüssmann, Manual of cultivated broad-leaved trees & shrubs **3**: 204 (1986).

Vigorous prostrate shrub to 40 cm, wide-spreading; young branches sparsely glandular-hairy. Leaves appearing early, 3–8 cm across, circular, 5–7-lobed, unpleasant smelling, hairless above, hairy on veins beneath. Flowers reddish white, 8–12 in ascending racemes, to as long as leaves. Fruit red, glandular-bristly. *N America*. H1. Early summer.

58. R. magellanicum Poiret. Illustration: Janczewski, Monographie des groseilliers, 443 (1907); Moore, Flora of Tierra del Fuego, pl. 6 (1983).

Erect shrub to 2 m; branches spineless. Leaves *c.* 4 cm, circular, usually 3-lobed, sharply toothed, hairless. Flowers in many-flowered racemes; calyx cup-shaped, whitish green; corolla lobes orange-red. Fruit black, edible. *Chile, Argentina*. H4. Summer.

59. R. diacanthum Douglas (*R. saxatile* Pallas). Illustration: Janczewski, Monographie des groseilliers, 390 (1907); Krüssmann, Manual of cultivated broad-leaved trees & shrubs **3**: 207 (1986).

Erect shrub to 2 m; branches with a few short prickles below the nodes. Leaves 2.5–3.5 × 1.5–3 cm, ovate, 3-lobed, lobes toothed. Flowers small, 1–6 in racemes, shorter than the leaves. Calyx greenish yellow, tinged red. Petals red. Fruit small, spherical, red, hairless. *N Asia*. H1. Late spring.

60. R. pulchellum Turczaninow. Illustration: Janczewski, Monographie des groseilliers, 453 (1907).

Prickly shrub to 2 m. Similar to *R. diacanthum* but leaves to 5 cm, more deeply 3-lobed, base wedge-shaped to slightly cordate. Flowers reddish, in racemes to 6 cm. Fruit red. *N China*. H1. Early summer.

Often confused with *R. orientale* var. *heterotrichum*.

61. R. giraldii Janczewski. Illustration: Janczewski, Monographie des groseilliers, 455 (1907); Krüssmann, Manual of cultivated broad-leaved trees & shrubs **3**: 202 (1986).

Spreading shrub to 1 m; branches hairy, bristly and thorny. Leaves *c.* 3.5 cm, circular. Flowers greenish brown, in non-glandular racemes to 7 cm. Fruit red, glandular hairy. *N China (Shaansi)*. H1–2. Early summer.

62. R. orientale Desfontaines (*R. villosum* Wallich; *R. resinosum* Pursh; *R. punctatum* Lindley). Illustration: Janczewski, Monographie des groseilliers, 457 (1907); Polunin & Stainton, Flowers of the Himalaya, pl. 44 (1985); Krüssmann,

Manual of cultivated broad-leaved trees & shrubs **3**: 202 (1986); Stainton, Flowers of the Himalaya, Supplement, pl. 35 (1988).

Shrub to 2 m; branches thornless, red-brown, glandular glutinous, new growth appearing very early. Leaves *c.* 4.5 × 5.5 cm, broadly ovate, usually 3-lobed, hairy beneath, fragrant when crushed. Flowers 5–20, in 1–5 cm, erect racemes. Calyx bell-shaped, green, turning reddish. Fruit scarlet red, glandular-hairy. *Greece to Himalaya & Siberia*. H1. Early summer.

Var. **heterotrichum** (C.A. Meyer) Janczewski. Young branches reddish; flowers reddish; fruit not glandular. *Siberia*.

63. R. alpinum Linnaeus. Illustration: Janczewski, Monographie des groseilliers, 461 (1907); Polunin, Flowers of Europe, pl. 43 (1969); Phillips & Rix, Shrubs, 74 (1989).

Erect shrub to 2 m; much branched, branches spineless, hairless, light grey. Leaves to 5 cm, circular, 3–5-lobed, lobes rounded, toothed, appearing early. Flowers dioecious, in small, erect racemes; calyx cup-shaped, yellowish green; petals longer than sepals; male racemes many-flowered; female racemes few-flowered. Fruit dark red. *Europe to Siberia*. H1. Early summer.

64. R. distans Janczewski (*R. alpinum* var. *mandshuricum* Maximowicz; *R. maximowiczianum* Komarov). Illustration: Krüssmann, Manual of cultivated broad-leaved trees & shrubs **3**: 202 (1986).

Similar to *R. alpinum*; leaves *c.* 5 × 6 cm, base cordate; sepals green. Fruit red, stalk almost absent. *China (Manchuria.)*. H1. Early summer.

65. R. vilmorinii Janczewski. Illustration: Janczewski, Monographie des groseilliers, 462 (1907); Krüssmann, Manual of cultivated broad-leaved trees & shrubs **3**: 202 (1986).

Dioecious shrub; similar to *R. alpinum*; differing in young shoots red, appearing very late. Flowers greenish, 6–12 in pendent racemes, to 6 cm. Fruit black-red. *W China*. H4. Late winter–early spring.

66. R. tenue Janczewski.
Dioecious shrub to 2 m; branches thornless. Flowers brownish red, in racemes. Fruit red. *W Asia*. H4–5. Early summer.

67. R. glaciale Wallich. Illustration: Krüssmann, Manual of cultivated broad-leaved trees & shrubs **3**: 202 (1986);

Stainton, Flowers of the Himalaya, Supplement, pl. 35 (1988).

Dioecious shrub; similar to *R. alpinum*; differing in leaf-lobes acute; flowers appearing soon after leaves; petals purple-brown. Fruit scarlet-red. *China (Yunnan, Hubei, Xizang)*. H1. Late spring.

68. R. luridum J.D. Hooker & Thomson. Illustration: Krüssmann, Manual of cultivated broad-leaved trees & shrubs **3**: 202 (1986).

Dioecious shrub; similar to *R. glaciale*. Branches smooth; new growth appearing later. Leaves to 6 cm, rounded ovate to kidney-shaped, 3–5-lobed. Fruit black. *Himalaya, Sikkim, Nepal, W China (Xizang)*. H1. Early summer.

69. R. maximowiczii Batalin. Illustration: Krüssmann, Manual of cultivated broad-leaved trees & shrubs **3**: 202 (1986).

Dioecious shrub; similar to *R. luridum* but shoots glandular-bristly. Leaves ovate, entire or shallowly 3–5-lobed, coarsely toothed. Flowers red-green, in erect, many-flowered racemes, 3–20 cm. Fruit glandular-hairy or finely bristly, red, yellow or green. *C China (Gansu)*. H1. Early summer.

70. R. henryi Franchet.
Evergreen shrub to 1.2 m, similar to *R. laurifolium* but shoots hairy; leaves to 10 cm, thinner. *C China*. H3–4. Early spring.

71. R. laurifolium Janczewski. Illustration: Phillips & Rix, Shrubs, 41 (1989); Brickell, RHS encyclopedia of plants and flowers, 143 (1986); Brickell (ed.), RHS A–Z encyclopedia of garden plants, 906 (2003).

Evergreen shrub to 1.5 m; young shoots glandular, later hairless. Leaves 5–10 cm, ovate-oblong, scalloped, entire, apex acute, base round, leathery. Flowers greenish yellow, 6–12 in pendent racemes, to 6 cm. Fruit black-red. *W China*. H4. Late winter–early spring.

141. PARNASSIACEAE

Herbs. Leaves mostly basal, simple, exstipulate. Flowers solitary, actinomorphic, bisexual. Calyx, corolla and stamens hypogynous. Calyx of 5 free sepals or 5-lobed. Corolla of 5 free petals. Stamens 5 alternating with 5 multifid staminodes. Ovary of 3–4 united carpels; ovules many,

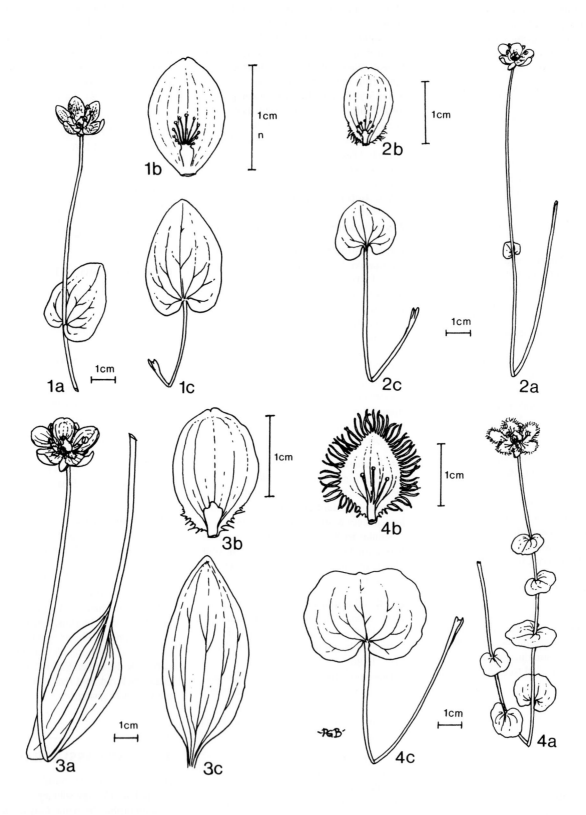

Figure 93. Diagnostic details of *Parnassia* species. 1a, b, c, *P. palustris.*
2a, b, c, *P. fimbriata.* 3a, b, c, *P. nubicola.* 4a, b, c, *P. foliosa.*

parietal; style absent, stigmas more or less sessile. Fruit a capsule.

A family with 2 genera, one native to both Europe and North America, the other to North America. Both are difficult in cultivation and rarely seen. *Lepuropetalon* is sometimes separated off as the Lepuropetalaceae.

1. PARNASSIA Linnaeus
P.G. Barnes

Herbaceous perennials with short rootstocks and alternate, mostly basal leaves. Flowering-stems bearing 1–6 leaves, flowers solitary. Sepals and petals 5, petals white. Stamens 5, alternating with the petals. Staminodes 5, yellowish, nectar-bearing, often fringed, opposite the petals. Ovary superior, ovoid, usually of 4 united carpels; style 1, short; stigmas 4. Fruit an ovoid or obovoid capsule, dehiscing along the middle of each cell.

Estimates of the number of species vary from 15 to 50, in the northern temperate areas. Most are typical of wet acid grassy places and require moist conditions in cultivation. Propagate by division or seed.

1a. Stem-leaves 2–8; petals fringed all round **2. foliosa**
  b. Stem-leaf solitary **2**
2a. Petals not fringed **1. palustris**
  b. Petals fringed in lower half **3**
3a. Leaves ovate; petals 1.2–1.5 cm **4. nubicola**
  b. Leaves rounded or kidney-shaped; petals *c.* 1 cm **3. fimbriata**

1. P. palustris Linnaeus. Figure 93(1), p. 146. Illustration: Ross-Craig, Drawings of British plants **10**: 18 (1957); Keble Martin, Concise British flora in colour, pl. 32 (1965); Kitamura et al., Coloured illustrations of herbaceous plants of Japan **2**: pl. 35 (1986).

Plant 10–30 cm, Basal leaves 1–5 cm, ovate, base cordate, long-stalked. Stem with a single stalkless leaf. Flowers 1.5–3 cm across; petals 8–12 mm, broadly elliptic Staminodes 3–4 mm, spathulate, divided at apex into 7–13 filaments, each tipped with a yellow gland. *Europe & N Asia.* H4. Summer.

2. P. foliosa Hooker & Thomson. Figure 93(4), p. 146. Illustration: Terasaki, Nippon shokubutsu zufu, t. 1440 (1933); Kitamura et al., Coloured illustrations of herbaceous plants of Japan **2**: pl. 35 (1986); Inami, Illustrations of selected

plants from Hiroshima prefecture **4**: 35 (1988).

Plant 15–30 cm. Basal leaves 2–4 × 2–4 cm, circular, base cordate, long-stalked. Stems bearing 2–8 stalkless leaves, 1–2 cm. Petals 1–1.5 cm, ovate, fringed all round the margin. Staminodes divided into 3 gland-tipped filaments. *Japan.* H4. Summer.

The above description is of subsp. **nummularia** (Maximowicz) Kitamura & Murata, which appears to be the one that is generally cultivated. Subspecies **foliosa** from China is larger in all its parts.

3. P. fimbriata K. König. Figure 93(2), p. 146. Illustration: *Bulletin of the Alpine Garden Society* **6**: 118 (1938); Hitchcock et al., Vascular plants of the Pacific northwest **3**: 30 (1971); Clark, Wild flowers of the Pacific northwest, 206, 219 (1976).

Plant 15–30 cm. Basal leaves 2–5 cm, rounded or kidney-shaped, stalk 3–10 cm. Stem-leaf solitary, stalkless. Flower 2–2.5 cm across; petals obovate, *c.* 1 cm, the margins conspicuously fringed near the base. Staminodes obovate, bluntly lobed or fringed with gland-tipped filaments. Capsule *c.* 1 cm. *W North America.* H4. Summer.

Most cultivated plants appear to be var. **hoodiana** Hitchcock, with staminodes bearing 6–10 gland-tipped filaments.

4. P. nubicola Royle. Figure 93(3), p. 146. Illustration: Grierson & Long, Flora of Bhutan **1**(3): f. 36e–h (1983).

Plant 20–30 cm. Basal leaves 3–10 × 2–5 cm, ovate, acute, base cordate, rounded or broadly tapered into a stalk to 12 cm. Stem-leaf solitary, stalkless, the margins hairy towards the base. Sepals 8–10 mm, ovate. Petals obovate, 1.2–1.5 cm × 7–10 mm, usually fringed towards the base. Staminodes oblong, shortly 3-lobed at apex. Capsule 1–1.5 cm. *Himalaya.* H4–5. Summer.

---

142. HYDRANGEACEAE

Herbs or softly wooded shrubs, rarely climbing, many with stellate hairs. Leaves usually opposite, simple, exstipulate. Inflorescences various. Flowers mostly bisexual (sometimes the outer flowers of the inflorescence sterile and with enlarged corollas), actinomorphic (fertile flowers only). Calyx, corolla and stamens hypogynous, perigynous or epigynous.

Calyx of 4–5 free sepals. Corolla of 4–7 free petals. Stamens 4–many. Ovary of 2–7 united carpels, rarely 1-celled; ovules many, axile or parietal; style 1, stigmas head-like or 2–7 and free or almost so. Fruit a capsule or berry.

There are about 17 genera and 170 species, several of the genera native to either Europe or North America. Many are cultivated as ornamental shrubs, especially *Deutzia*, *Hydrangea* and *Philadelphus*. *Philadelphus* and some of its allies are sometimes separated off as the family Philadelphaceae. The whole family was included in the Saxifragaceae in edition 1.

1a. Plants herbaceous **2**
  b. Plants woody shrubs or climbers **3**
2a. Flowering stems with scale-leaves at the base and 2–4 large opposite leaves at the top; stamens numerous **1. Deinanthe**
  b. Leaves several, opposite, borne on the stems; stamens 15 **5. Kirengeshoma**
3a. Plant with both fertile and sterile flowers (occasionally all the flowers sterile in some cultivars) **4**
  b. Plant without sterile flowers **7**
4a. Leaves alternate **9. Cardiandra**
  b. Leaves opposite **5**
5a. Sterile flowers at the tips of the shoots, each with a large, showy, 3–4-lobed calyx; fertile flowers borne lower down the shoots **13. Platycrater**
  b. Sterile flowers borne around the margins of a corymb, more numerous than the fertile in some cultivars **6**
6a. Sterile flowers each reduced to a single coloured bract 2.5 cm or more **7. Schizophragma**
  b. Sterile flowers each reduced to a bilaterally symmetric coloured corolla, not as above **6. Hydrangea**
7a. Plants climbing **8**
  b. Plants upright **9**
8a. Petals 7–10; stamens 20–30; ovary with 7–1- cells **8. Decumaria**
  b. Petals 4 or 5; stamens 8–10; ovary with 4 or 5 cells **14. Pileostegia**
9a. Stamens more than 10 **10**
  b. Stamens 10 or fewer **11**
10a. Evergreen shrub; ovary totally superior **10. Carpenteria**
  b. Deciduous shrub; ovary mostly inferior **4. Philadelphus**
11a. Sepals and petals 4; stamens 8 **12. Fendlera**

   b. Sepals and petals almost always 5;
      stamens 10   12
12a. Ovary mostly superior; leaves with
      thick down beneath   **2. Jamesia**
   b. Ovary inferior; leaves without thick
      down beneath   13
13a. Filaments winged; usually deciduous
      shrubs with a covering of stellate
      hairs   **3. Deutzia**
   b. Filaments not winged; evergreen
      shrubs without stellate hairs
                                 11. Dichroa

1. DEINANTHE Maximowicz
P.G. Barnes
Herbaceous perennials with woody,
creeping rhizomes. Stems erect, with scale-
like leaves at the base and 2–4 large leaves
near the top. Leaves opposite, rather large,
broadly ovate or elliptic, stalked, with
forward-pointing teeth, coarsely hairy.
Inflorescence a hairless corymb, often with
a few small sterile flowers consisting of 3
or 4 sepals. Fertile flowers nodding, with
an inferior, conical ovary and 5 rounded
sepals. Petals 5 or more, rounded,
deciduous; stamens numerous, style 1,
rather prominent with 5 small lobes. Ovary
5-celled, seeds numerous, with a short tail
at each end.
   A genus of 2 species from NE Asia. They
are fairly hardy but require cool, moist and
somewhat shady conditions and an acid
soil. Propagate by division or seed.

   1a. Leaves notched; flowers white, flat
                                   1. bifida
    b. Leaves mostly acuminate; flowers
       blue, cupped   **2. caerulea**

1. D. bifida Maximowicz. Illustration:
Bulletin of the Alpine Garden Society **50**:
223 (1982); Satake et al., Wild flowers of
Japan **2**: pl. 146 (1985); Kitamura et al.,
Coloured illustrations of herbaceous plants
of Japan **2**: f. 278 (1986).
   Stems 20–50 cm. Upper leaves
10–20 × 6–10 cm, elliptic, with forward-
pointing teeth, apex deeply notched.
Flowers *c.* 2 cm across, white. Petals *c.*
1 cm, rounded, spreading widely; anthers
and filaments yellow. *Japan.* H4. Summer.

2. D. caerulea Stapf. Illustration: *Botanical
Magazine*, 8373 (1911); *Bulletin of the
Alpine Garden Society* **50**: 214 (1982);
Thomas, Perennial garden plants, edn 2,
f. 9 (1982); Jelitto, Schacht & Fessler, Die
Freiland Schmuckstauden, edn 3, 188
(1985).
   Stems to 35 cm. Leaves
10–20 × 5–15 cm, broadly ovate, apex

acuminate or occasionally deeply notched.
Flowers 2–3 cm across, pale violet-blue.
Petals 5–8, rounded, 1–1.5 cm, concave.
Filaments and anthers blue. *China.* H4.
Summer.

2. JAMESIA Torrey & Gray
D.M. Miller
Deciduous shrubs to 1 m or sometimes
more, with peeling, brown, papery bark.
Branches with solid pith, downy when
young. Leaves 2–7 × 1.5–5 cm, opposite,
simple, ovate, acute, coarsely toothed,
wrinkled, dull green above, with thick,
grey-white down beneath; stalk downy,
2–15 mm. Leaves on flowering-shoots
similar but smaller. Flowers *c.* 1.5 cm
across, bisexual, slightly fragrant, in erect,
terminal, many-flowered panicles to 5 cm.
Sepals 5, *c.* 3 × 2 mm, ovate to lanceolate,
downy. Petals 5, 8–10 × 5–6 mm, oblong
to obovate, white or pinkish. Stamens 10.
Styles 3–5, united at base. Ovary more or
less superior. Fruit a many-seeded capsule
to 4 mm.
   A genus of a single species from
N America, with orange-red autumn
colour. It grows best in a sunny position in
well-drained, fertile soil. Propagate by seed
or cuttings.

1. J. americana Torrey & Gray.
Illustration: *Botanical Magazine*, 6142
(1875); Krüssmann, Manual of cultivated
broad-leaved trees & shrubs **2**: pl. 71
(1986).
   E North America. H3. Summer.

3. DEUTZIA Thunberg
D.R. McKean
Deciduous or rarely evergreen shrubs with
pith-filled branches and peeling bark. Hairs
often stellate with a varying number of
rays; sometimes a long ray is borne erect,
at right angles to the others. Leaves
opposite, usually shortly stalked, sometimes
toothed, without stipules. Flowers in
racemes, cymes, panicles or corymbs, or
solitary on terminal or axillary shoots.
Calyx-teeth 5. Petals 5, edge-to-edge or
overlapping in bud. Stamens 10, in 2 series
of 5, those of the inner series smaller;
filaments mostly broadly winged, with 2
teeth at the top on either side of the
anther; anthers sometimes shortly stalked
above the filament. Ovary 3- or 4-celled,
inferior. Styles 3 or 4, thickened at apex.
Fruit a capsule.
   A genus of about 60 species, of which
some 50 have at one time or another been

in cultivation (some of them perhaps only
in America). Those included here are all
Asiatic, ranging from the Himalaya to
Japan and the Philippines. They are
notoriously difficult plants to identify
because of the small differences that
separate several of the species and the
great variability of other species. They have
long been favourite shrubs in gardens and
many cultivars have been raised and
distributed (especially by Lemoine's nursery
in the late 19th and early 20th centuries).
A large number of these cultivars and
hybrids still exists; many of them are of
doubtful origin and are particularly difficult
to identify. Cultivation as for *Philadelphus*
(p. 158).
   Literature: Zaikonnikova, I.T., A key to
the species of the genus Deutzia, *Baileya*
19: 133–44 (1975).

 1a. Petals all edge-to-edge in bud   2
  b. Petals overlapping in bud (or partly
     edge-to-edge in the flowers of some
     hybrids)   26
 2a. Flowers in panicles or racemes   3
  b. Flowers solitary or in corymbs or
     cymes   14
 3a. Filaments of outer stamens tapered
     to apex   **11. scabra**
  b. Filaments of all stamens broad,
     toothed at apex   4
 4a. Leaves hairless or almost so beneath
     or with 4–6-rayed hairs   5
  b. Leaves with 8–16-rayed hairs
     beneath   8
 5a. Inflorescence a long panicle or
     raceme, or with very few white
     flowers   6
  b. Inflorescence a broad, loose panicle;
     flowers pinkish or rarely white   7
 6a. Leaves almost hairless beneath
                                 1. gracilis
  b. Leaves with a moderate to dense
     covering of 4- or 5-rayed hairs
     beneath   **7. taiwanensis**
 7a. Calyx-teeth longer than tube
                                  3. × rosea
  b. Calyx-teeth shorter than tube
                                 2. × carnea
 8a. Leaves to 2.5 cm wide and obscurely
     toothed   9
  b. Leaves more than 2.5 cm wide and
     distinctly toothed   10
 9a. Panicles dense, 20–60-flowered
                               4. ningpoensis
  b. Panicles loose, 12–20-flowered
                             6. maximowicziana
10a. Inflorescence a raceme, all stamens
     distinctly toothed   **8. crenata**

b. Inflorescence a panicle, not all stamens distinctly toothed **11**

11a. Calyx-teeth about as long as ovary **9. × magnifica**

b. Calyx-teeth shorter than ovary **12**

12a. Leaves leathery, with dense 14–16-rayed hairs beneath; styles 5 **5. pulchra**

b. Leaves thinner, less densely hairy with hairs with fewer rays; styles less than 5 **13**

13a. Leaves obscurely and finely scalloped; panicle narrow **8. crenata**

b. Leaves with fine forward-pointing teeth; panicle broad and loose **10. schneideriana**

14a. Inflorescence with 1–3 flowers **15**

b. Inflorescence with more than 3 flowers **16**

15a. Leaves white beneath **17. grandiflora**

b. Leaves pale green beneath **16. coreana**

16a. Calyx-teeth shorter than ovary, broadly ovate or triangular; leaves not whitish beneath **13. setchuenensis**

b. Calyx-teeth as long as or longer than ovary; leaves sometimes whitish beneath **17**

17a. Mature leaves on flowering-shoots *c.* 9 × 3 cm **19. longifolia**

b. Mature leaves on flowering-shoots smaller **18**

18a. Petals with reflexed margins **20. reflexa**

b. Petals without reflexed margins **19**

19a. Leaves sharply toothed; calyx reflexed; flowers white **18. vilmoriniae**

b. Leaves finely toothed; calyx not reflexed; flowers pink or white **20**

20a. Inflorescence a dense cyme borne on a very short shoot **21. calycosa**

b. Inflorescence not as above **21**

21a. Leaf underside with stellate hairs, not spreading hairs **22**

b. Leaf underside with stellate hairs **23**

22a. Leaves pale green beneath; flowers *c.* 0.7 cm **22. rehderiana**

b. Leaves white beneath; flowers *c.* 1 cm **14. monbeigii**

23a. Leaves on flowering shoots *c.* 3 × 1.5 cm, with 4- or 5-rayed stellate hairs beneath **23. glomeruliflora**

b. Leaves on flowering-shoots usually larger, stellate hairs with more rays **24**

24a. Leaves with moderately 5–7-rayed stellate hairs beneath (hairs not overlapping) **26. × elegantissima**

b. Leaves with densely (overlapping) 7–12-rayed hairs **25**

25a. Leaf-underside with 7–9-rayed stellate hairs **25. purpurascens**

b. Leaf-underside with 9–17-rayed stellate hairs **24. discolor**

26a. Leaves with very dense 10–13-rayed hairs beneath **15. staminea**

b. Leaves densely shaggy-hairy, moderately stellate-hairy or almost hairless beneath **27**

27a. Leaves densely shaggy-hairy beneath **28**

b. Leaves moderately stellate-hairy or almost hairless beneath **29**

28a. Petals all overlapping in bud; leaves with only spreading hairs beneath **35. mollis**

b. Some petals not overlapping in bud, others overlapping; leaves with both stellate and spreading hairs **36. × wilsonii**

29a. Inflorescence a panicle; some petals overlapping, others edge-to-edge in bud; filaments toothed **30**

b. Inflorescence a corymb; all petals overlapping in bud; filaments toothed or not **33**

30a. Flowers white **31**

b. Flowers pinkish outside **32**

31a. Some of the longer filaments tapered at apex **12. × candida**

b. Longer filaments all toothed **29. × lemoinei**

32a. Calyx-teeth longer than ovary; flowers *c.* 1.5 cm across **28. × maliflora**

b. Calyx-teeth as long as ovary; flowers *c.* 2 cm across **27. × kalmiiflora**

33a. Longer stamens, all tapered to the apex, not toothed **30. parviflora**

b. Longer stamens, all broad, distinctly toothed at apex **34**

34a. Longer stamens, some toothed, others tapered **34. compacta**

b. Longer stamens, all toothed **35**

35a. Lower leaf-surface densely hairy with 5–9-rayed stellate hairs, the surface usually not visible **31. hookeriana**

b. Stellate hairs less dense on lower leaf-surface, the surface visible through the hairs **36**

36a. Longer stamens truncate or shallowly lobed at apex **32. rubens**

b. Longer stamens distinctly toothed at apex, the anthers borne on short stalks **33. corymbosa**

1. D. gracilis Siebold & Zuccarini. Figures 94 (12), p. 150 & 95(17), p. 151. Illustration: Hay & Beckett (eds), Reader's Digest encyclopaedia of garden plants and flowers, 221 (1978); Davis, The gardener's illustrated encyclopaedia of trees & shrubs, 116 (1987); Brickell (ed.), RHS A–Z encyclopedia of garden plants, 366 (2003).

Erect, deciduous shrub, 1–2 m. Leaves 3.5–6(rarely to 10) × 2–4 cm, lanceolate to ovate, with minute, forward-pointing teeth, very thin, bright green, with scattered hairs, 3- or 4-rayed on upper surface and 4- or 5-rayed on the lower. Flowers in racemes or narrow panicles, 40–80 cm; flower-stalks long and slender. Flowers 1–2 cm across, pure white, strongly scented. Calyx-teeth *c.* 1.5 mm, triangular, usually greenish. Ovary *c.* 2 mm. Anthers borne on short stalks above the filament teeth. Styles 5, about the same length as stamens. Capsules *c.* 4 mm across. *Japan.* H3. Spring–early summer.

Plants of this species may need protection from late spring frosts. Var. **nagurae** (Makino) Makino has smaller flowers (*c.* 5 mm across).

D. × candelabrum (Lemoine) Rehder; *D. gracilis × D. scabra*. Illustration: *Revue Horticole*, 175 (1908). Like *D. gracilis* but with larger, many-flowered panicles, coarser leaves and more spreading growth. *Garden origin.* H5. Early summer.

2. D. × carnea (Lemoine) Rehder; *D. × rosea* 'Grandiflora' × *D. scabra*.

Differs from *D. gracilis* by its broad, loose panicles of pink flowers with purplish sepals. *Garden origin.* H3. Spring–early summer.

3. D. × rosea (Lemoine) Rehder; *D. gracilis × D. purpurascens*. Illustration: Kaier, Garden trees and shrubs, pl. 60 (1963); Seabrook, Shrubs for your garden, 63 (1974); Hessayon, Tree and shrub expert, 23 (1983); Krüssmann, Manual of cultivated broad-leaved trees & shrubs **1**: pl. 170, 171 (1984).

Dwarf arching shrub to 1 m. Leaves ovate or oblong to lanceolate, margins with forward-pointing teeth, and with scattered, 4–6-rayed hairs. Flowers in short, broad panicles, carmine outside, white inside. Calyx-teeth lanceolate, longer than the ovary. Styles usually longer than the stamens. *Garden origin.*

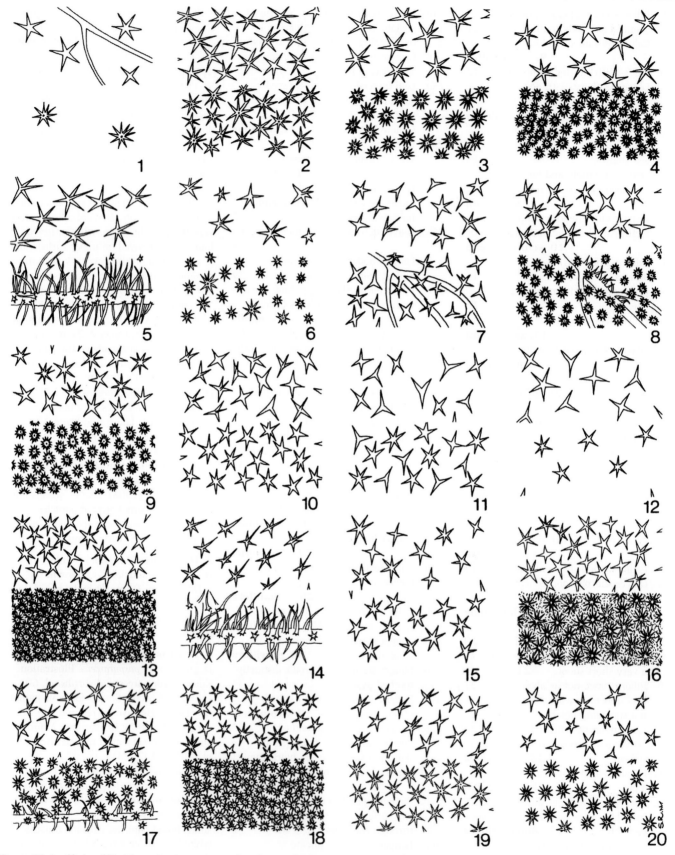

Figure 94. Leaf hairs of *Deutzia* species; top, upper surface; bottom, lower surface (× 30). 1, *D. corymbosa*. 2, *D. purpurascens*. 3, *D. schneideriana*. 4, *D. ningpoensis*. 5, *D. glomeruliflora*. 6, *D. crenata*. 7, *D. scabra*. 8, *D. discolor*. 9, *D. pulchra*. 10, *D. rubens*. 11, *D. setchuenensis*.

12, *D. gracilis*. 13, *D. monbeigii*. 14, *D. mollis*. 15, *D. hookeriana*. 16, *D. grandiflora*. 17, *D. longifolia*. 18, *D. staminea*. 19, *D. parviflora*. 20, *D. calycosa*.

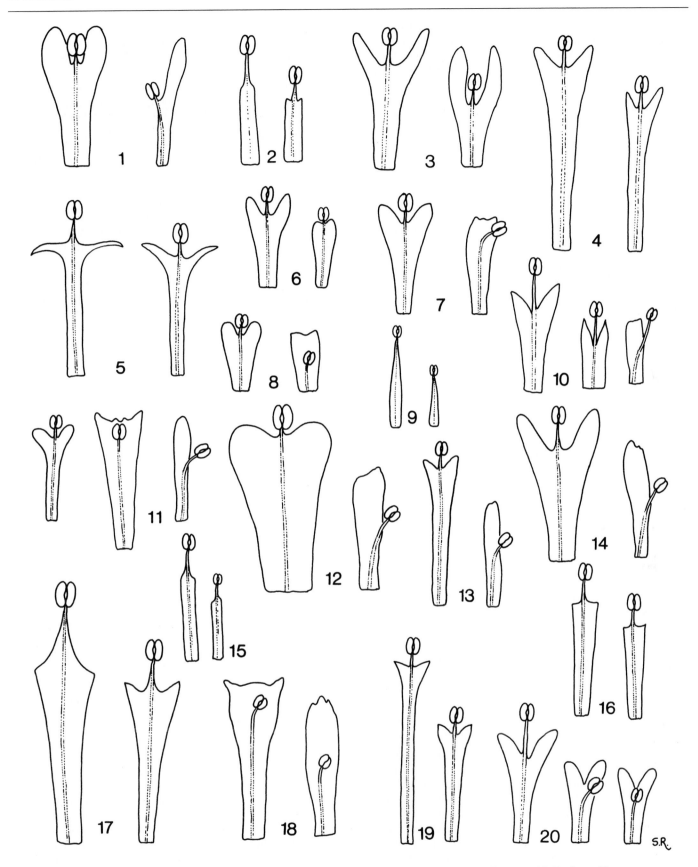

Figure 95. Stamens of *Deutzia* species (× 6.6). 1, *D. calycosa.*
2, *D. parviflora.* 3, *D. longifolia.* 4, *D. crenata.* 5, *D. grandiflora.*
6, *D. staminea.* 7, *D. monbeigii.* 8, *D. hookeriana.* 9, *D. mollis.*

10, *D. corymbosa.* 11, *D. setchuenensis.* 12, *D. glomeruliflora.*
13, *D. schneideriana.* 14, *D. purpurascens.* 15, *D. scabra.* 16, *D. ningpoensis.*
17, *D. gracilis.* 18, *D. rubens.* 19, *D. pulchra.* 20, *D. discolor.*

Several cultivars are grown, including 'Campanulata', with dense panicles of white flowers, 'Carminea' with reddish pink flowers, and 'Eximia' with pale pink flowers, white inside.

4. D. ningpoensis Rehder (*D. chunii* Hu). Figures 94(4), p. 150 & 95(16), p. 151. Illustration: Hu & Chun, Icones plantarum Sinicarum **5**: f. 222 (1937); Krüssmann, Manual of cultivated broad-leaved trees & shrubs **1**: pl. 168, 172 (1984).

Shrub to 2 m. Leaves 3.5–7 × *c.* 2.5 cm, ovate, generally entire, with 5- or 6-rayed hairs above and very dense 12–14-rayed hairs beneath. Inflorescence a narrow panicle to 10 cm. Flowers 5–10 mm across, white or pink, densely crowded. Calyx-teeth *c.* 1 mm, shorter than the ovary (*c.* 2 mm). Stamens indistinctly toothed, the anthers borne on stalks *c.* 0.75 mm above the filaments. *E China (Zhejiang, Anhui)*. H4. Summer.

5. D. pulchra Vidal. Figures 94(9), p. 150 & 95(19), p. 151. Illustration: *Botanical Magazine*, 8962 (1923); Li, Woody flora of Taiwan, f. 89 (1963); Li et al., Flora of Taiwan **3**: f. 468 (1977).

Shrub 2–4 m. Leaves 5–10 × 2–5 cm, lanceolate to narrowly ovate, base tapered or rounded, entire or toothed, thick and leathery, with moderately dense 5–8-rayed (rarely to 13-rayed) hairs above and very dense 14–16-rayed hairs beneath. Flowering-shoot terminal or almost so, bearing a few-flowered panicle to 12 × 7 cm. Flowers upright to reflexed, white. Calyx-teeth *c.* 1 × 2 mm, broadly triangular, shorter than the ovary (to 3.2 mm). Stamens *c.* 8 mm, narrow, toothed. Styles 5, as long as stamens. *Philippines (Luzon) & Taiwan*. H4. Spring.

6. D. maximowicziana Makino (*D. hypoleuca* Maximowicz). Illustration: Kitamura & Okamoto, Coloured illustrations of trees & shrubs of Japan, pl. 202 (1977).

Deciduous shrub to 1.5 m; bark brown, current growth stellate downy. Leaves 3–8 × 1.2–1.8 cm, narrowly oblanceolate, long-pointed, white beneath with dense stellate hairs; moderately stellate-hairy above. Flowers white, in loose, erect 12–20-flowered panicles 5–7 cm. Petals *c.* 7 mm. Longest stamens *c.* 8 mm, slightly toothed, shorter stamens untoothed. Styles 3 or 4, *c.* 1 cm. *Japan*. H4. Spring–early summer.

7. D. taiwanensis (Maximowicz) Schneider (*D. crenata* var. *taiwanensis*). Illustration: Ito, Taiwan shabubutu dyusetu, (Illustration Formosan plants) t. 620 (1927).

Deciduous shrub to 2 m; current growth thinly stellate-hairy. Leaves 5–8 × 2–3 cm, ovate lanceolate, rough with very short stiff hairs, with fine forward-pointing teeth, moderately 3- or 4-rayed stellate hairy above, more densely 4- or 5-rayed hairy beneath. Flowers few in terminal panicles to 6 cm. Petals *c.* 8 × 4 mm, oblong. Calyx-teeth *c.* 1 × 1 mm. Stamens *c.* 5–7 mm indistinctly toothed. Capsule almost spherical, *c.* 3.5 mm across. *Taiwan*. H4. Spring.

Probably rare in cultivation and of doubtful horticultural value because of the small number of insignificant flowers.

8. D. crenata Siebold & Zuccarini. Figures 94(6), p. 150 & 95(4), p. 151. Illustration: Siebold & Zuccarini, Flora of Japan **1**: t. 6 (1835); *Botanical Magazine*, 3838 (1841) – as *D. scabra*; Hu & Chun, Icones plantarum Sinicorum **5**: t. 220 (1937) – as *D. scabra*; Makino, New illustrated flora of Japan, f. 950 (1963) – as *D. sieboldiana*.

Shrub to 2.5 m, with erect branches. Leaves 3–6 × 1.5–3 cm on fertile shoots, much larger on non-flowering-shoots, ovate to ovate-lanceolate, obscurely scalloped, stalked; hairs on upper surface with 4–6 spreading rays and a longer, erect, central ray, those on the lower surface 6–13-rayed. Inflorescence mainly a raceme, more rarely a narrow panicle, 10–15 cm. Flowers 1.5–2 cm across. Calyx-teeth *c.* 1.8 mm, shorter than the ovary (*c.* 2.3 mm), covered with apparently simple as well as stellate hairs. Petals long and narrow, 1–1.5 cm. All stamens toothed. Styles 3 or 4. Capsules *c.* 5 mm across. *Japan, ?SE China*. H3. Spring.

A species with several cultivars and a parent of several hybrids. It has been much confused with *D. scabra*, from which it differs in its stalked leaves subtending the inflorescences and its toothed filaments.

9. D. × magnifica (Lemoine) Rehder (*D. crenata* var. *magnifica* Lemoine). Illustration: Seabrook, Shrubs for your garden, 62 (1974); Hofman, Ornamental shrubs, 77 (1978).

Shrub to 2 m; similar to *D. crenata*, with strong, upright growth and stout branches. Leaves 4–6 cm, ovate-oblong, sharply but finely toothed, rough above and with dense 10–15-rayed hairs beneath. Flowers white,

single or double, in short, dense panicles 4–6 cm. Calyx-teeth as long as ovary. Filaments with large teeth. *Garden origin*. H5. Early summer.

Recent authors have indicated the parentage of this hybrid as *D. scabra* × *D. vilmoriniae*. The original author considered it to be a variety of *D. crenata*, however the parentage given below seems more likely. Several cultivars are available, including: 'Eburnea' with single, white, bell-shaped flowers in loose panicles; 'Latiflora' with single, white flowers to 3 cm across in erect panicles; 'Longiflora', with single flowers with long, narrow petals; and 'Macrothyrsa', a tall plant with many umbel-like panicles along each branch.

10. D. schneideriana Rehder. Figures 94 (3), p. 150 & 95(13), p. 151. Illustration: Hu & Chun, Icones plantarum Sinicarum **5**: f. 221 (1937); Iconographia cormophytorum Sinicorum **2**: t. 1931 (1972).

Shrub 1–2 m. Leaves 9–11 × 3–4 cm, ovate or lanceolate-ovate, margin with fine, forward-pointing teeth, hairs on upper surface with 4 or 5 spreading rays and a longer, erect, central ray, those on lower surface dense, 9–13-rayed, some occasionally with a longer central ray. Inflorescence corymb-like, 9–11 × 4–6 cm. Calyx-teeth *c.* 1 mm, broadly triangular, ovary *c.* 2 mm. Petals narrowly lanceolate, *c.* 1 cm; longer stamens toothed, shorter stamens not toothed, anther borne just below the apex. Styles 3. *China (W Hubei)*. H4. Summer.

Var. **laxiflora** Rehder has looser and broader inflorescences and larger petals (1.2–1.4 cm), and is more often grown.

11. D. scabra Thunberg (*D. sieboldiana* Maximowicz). Figures 94(7), p. 150 & 95 (15), p. 151. Illustration: Makino, New illustrated flora of Japan, f. 951 (1963); Bean, Trees & shrubs hardy in the British Isles, edn 8, **2**: 49 (1973); Hofman, Ornamental shrubs, 78 (1978); Brickell (ed.), RHS A–Z encyclopedia of garden plants, 367 (2003).

Shrub to 2.5 m. Leaves 3–8 × 2–4 cm, broadly ovate, stalked except for those subtending the inflorescences, with coarse, forward-pointing teeth; hairs on the upper surface 3 or 4-rayed, those on the lower 4–6-rayed, some on the veins occasionally with a longer central, erect ray. Panicles broadly pyramidal, loose. Flowers 1–1.5 cm across, white, honey-scented. Calyx-teeth *c.* 1 mm, ovary *c.* 2 mm. Filaments not toothed. Styles usually 3. *Japan.* H4. Summer.

Sometimes confused with *D. crenata.* Several widely cultivated hybrids and cultivars formerly attributed to this species should be referred to *D. crenata.* For *D. × elegantissima* see *D. glomeruliflora.*

12. D. × candida (Lemoine) Rehder (*D. discolor* var. *candida* Lemoine).

Growth upright. Leaves 3.5–5 cm, ovate, with forward-pointing teeth, slightly rough above with short, stiff hairs, with 5–7-rayed hairs beneath. Flowers numerous in panicles, *c.* 2 cm across, white. Calyx-teeth oblong-ovate, about as long as ovary. Petals partly overlapping and partly edge-to-edge in bud. Filaments toothed. Styles 3, shorter than the longer stamens. *Garden origin.* H3. Summer.

A hybrid between *D. × lemoinei* and *D. scabra.*

Var. **compacta** Lemoine is more compact and has smaller flowers. 'Boule de Neige' has a compact habit and denser inflorescences of larger flowers.

13. D. setchuenensis Franchet. Figures 94(11), p. 150 & 95(11), p. 151. Illustration: Iconographia cormophytorum Sinicorum **2**: t. 1926 (1972); Bean, Trees & shrubs hardy in the British Isles, edn 8, **2**: f. 9 (1973); Krüssmann, Manual of cultivated broad-leaved trees & shrubs **1**: f. 316, 317 (1984).

Shrub 1.5–2 m. Leaves *c.* 6 × 2 cm, ovate, usually long acuminate, margin with fine, forward-pointing teeth, upper surface with 4–6-rayed hairs, lower with 3–5-rayed-hairs, some of which have longer central rays. Inflorescence composed of loose corymbs. Flowers to 1 cm across, white, on long and slender stalks. Calyx-teeth *c.* 0.8 mm. Longer stamens toothed, shorter not toothed, with anthers attached below the apex. Ovary *c.* 1 mm. Styles 3, short. *W China.* H5. Summer.

Var. **corymbiflora** (Lemoine) Rehder. Illustration: *Botanical Magazine,* 8255 (1909); Krüssmann, Manual of cultivated broad-leaved trees & shrubs **1**: pl. 169, 170 (1984); Brickell (ed.), RHS A–Z encyclopedia of garden plants, 367 (2003). Leaves 3–11 × 1.5–3 cm; flowers *c.* 1.5 cm across in larger corymbs. This is more commonly found in cultivation than var. *setchuenensis,* and prefers a limey soil.

14. D. monbeigii W.W. Smith. Figures 94(13), p. 150 & 95(7), p. 151. Illustration: *Botanical Magazine,* n.s., 123

(1950); Brickell (ed.), RHS A–Z encyclopedia of garden plants, 367 (2003).

Shrub 1–1.5 m, with slender branches. Leaves 1.5–3 cm × 5–12 mm, ovate-lanceolate, margin with forward-pointing teeth, white beneath with very dense 12–15-rayed hairs, the hairs above less dense, 6–9-rayed. Inflorescence corymb-like or cyme-like, *c.* 6 × 4 cm. Calyx-teeth *c.* 1.5 mm, broad, about the same length as the ovary. Petals *c.* 7 × 3 mm, ovate. Longer stamens *c.* 5 mm, toothed, smaller stamens *c.* 3 mm, with anthers attached below the apex, minutely toothed. Styles 5. *SW China (Yunnan).* H4. Spring–summer.

15. D. staminea Wallich. Figures 94(18), p. 150 & 95(6), p. 151. Illustration: *Edwards's Botanical Register* **33**: pl. 13 & 265 (1847) – as *D. corymbosa;* Schneider, Illustriertes Handbuch der Laubholzkunde **1**: f. 244 (1906).

Small shrub, 1 m (rarely to 2 m). Leaves ovate, apex acuminate, base rounded or wedge-shaped, margin with fine, forward-pointing teeth; upper surface hairy, hairs 9–11-(rarely as few as 7-)rayed, lower surface with dense 10–13-(rarely to 17-) rayed hairs. Inflorescence a panicle or rarely corymb-like, *c.* 5 cm wide. Flowers *c.* 1.5 cm across, white or pink. Calyx-teeth narrowly triangular, 1.5–2 mm, ovary 2–3.5 mm, densely stellate-hairy. Petals boat-shaped, oblong-elliptic, 7–10 × 3–5 mm. Stamens 6–7.5 mm, outer toothed, inner with minute teeth, anthers almost stalkless, attached below the apex. Styles 4, equal in length to the outer stamens. *Himalaya.* H5. Early summer.

Var. **brunoniana** Hooker & Thomson is less hairy and larger flowered.

16. D. coreana H. Léveillé. Illustration: Nakai, Flora sylvatica Koreana **15**: t. 16 (1926); Krüssmann, Manual of cultivated broad-leaved trees & shrubs **1**: f. 316 (1984).

Shrub to 2 m; current growth rough, warty. Leaves ovate to ovate-elliptic, paler beneath, margin finely double-toothed, with teeth pointing forwards, both surfaces moderately covered with 4–6-rayed hairs. Flowers on previous year's growth, solitary, white, in leaf axils. Calyx-tube *c.* 2.5 mm, stellate, teeth *c.* 2 mm, broadly triangular. Both whorls of stamens toothed. Styles 3–4, *c.* 7 mm. *Korea.* H4. Spring.

Probably rare in cultivation.

17. D. grandiflora Bunge. Figures 94(16), p. 150 & 95(5), p. 151. Illustration: Hu & Chun, Icones plantarum Sinicarum **5**: f. 228 (1937); Iconographia cormophytorum Sinicorum **2**: t. 1935 (1972).

Shrub 1.5–2 m, young shoots grey at first with stellate hairs. Leaves on flowering-shoots 2.5–3 × 1–1.5 cm, increasing later to 5 × 2 cm, margins with fine forward-pointing teeth, white beneath with dense, 9- or 10-rayed hairs, above with 5-rayed hairs that have additional central rays. Flowers 1–3 on short leafy shoots, 2.5–3 cm across, white. Calyx-teeth *c.* 3.8 mm, linear-lanceolate. Both sets of filaments toothed. Ovary *c.* 3 mm. Styles 3, longer than stamens. *N China.* H3. Early spring.

This is the earliest flowering species, with the largest flowers; the 1–3-flowered inflorescence is distinctive.

18. D. vilmoriniae Lemoine. Illustration: Krüssmann, Manual of cultivated broad-leaved trees & shrubs **1**: pl. 170 (1984).

Vigorous deciduous shrub to 2 m; current growth rough with stellate hairs. Leaves ovate to oblong-lanceolate, slender pointed, 3–8 × 2–5 cm, sharply toothed, dull green above, grey beneath with stellate hairs; hairs on the midrib appearing simple but with a minute stellate base. Flowers white, *c.* 2.5 cm across; panicles *c.* 8 cm. Calyx-teeth linear lanceolate, reflexed. Filaments toothed. *SW China.* H4. Summer.

19. D. longifolia Franchet. Figures 94(17), p. 150 & 95(3), p. 151. Illustration: *Botanical Magazine,* 8493 (1913); Hu & Chun, Icones plantarum Sinicarum **5**: f. 227 (1937); Iconographia cormophytorum Sinicorum **2**: t. 1933 (1972); Krüssmann, Manual of cultivated broad-leaved trees & shrubs **1**: pl. 168 & f. 316–7 (1984).

Shrub 1.5–2 m, young shoots sparsely hairy with stellate hairs at first, later hairless. Leaves 5–7 × 1–2.5 cm on fertile shoots, lanceolate, base rounded or tapered, apex acuminate, thick and with prominent veins, margins with forward pointing, fine teeth, upper surface dull green, with 5 or 6-rayed hairs, lower surface pale green, with dense 8–12-rayed hairs, hairs on the veins simple. Cymes broad, loose or compact, *c.* 8 × 6 cm. Flowers 2–2.5 cm across; calyx-teeth *c.* 3 mm, narrowly triangular; petals broadly ovate-oblong, white with a purplish pink stripe outside. Outer stamens toothed, inner with a single large tooth and 1 or 2 variable, minute teeth. Ovary *c.* 2.8 mm. *China (Sichuan, Yunnan).* H4. Early summer.

Var. **macropetala** Zaikonnikova. Leaves on fertile shoots 6–11 × 2.5–3 cm; flowers 2.5–3 cm across.

Several cultivars of *D. longifolia* are known, including 'Veitchii' with larger, darker purple flowers and purple flowering-stems, possibly the most handsome *Deutzia*, and 'Elegans' with slender, drooping branches and flowers to 2 cm wide.

D. × hybrida Lemoine; *D. longifolia × D. discolor.* Illustration: *National Horticultural Magazine* **30**: 87 (1951); Hay & Beckett (eds), Reader's Digest encyclopaedia of garden plants and flowers, 221 (1972).

Like *D. longifolia* but leaves 6–10 cm; flowers larger and wider, petals pink with wavy edges; anthers bright yellow. *Garden origin.* H4. Early summer. Usually found as cultivar 'Magician'.

D. × excellens (Lemoine) Rehder; *D. × rosea* 'Grandiflora' × *D. longifolia.* Like *D. longifolia* but leaves 3–6 cm, ovate-oblong, margins with fine, forward-pointing teeth, upper surface rough with short, stiff hairs, lower surface greyish white with 8–12-rayed hairs and simple spreading hairs on the main veins. Flowers in broad, loose corymbs 4–6 cm across, white. *Garden origin.* H4. Summer.

20. D. reflexa Duthie.

Deciduous shrub *c.* 1 m, current growth hairless. Leaves 5–10 × 1–2.5 cm, broadly oblong-lanceolate, long-pointed, faintly toothed, thinly stellate downy above, more densely hairy beneath. Hairs on midrib appearing simple, but with a minute stellate base and a long central arm. Flowers in dense panicles, petals *c.* 8 mm, white, margin reflexed. Stamens toothed. Calyx-teeth narrowly oblong. *C China.* H4. Summer.

21. D. calycosa Rehder. Figures 94(20), p. 150 & 95(1), p. 151.

Shrub to 2 m. Leaves 2–7 × 1–2 cm, ovate, the upper surface with hairs that have 5 or 6 rays and a central ray, lower surface moderately to densely hairy with hairs 9- or 10-(rarely to 15-)rayed. Flowering-shoots with 2 or 3 (rarely to 6) pairs of leaves. Inflorescence a dense cyme, 4–7 cm (rarely to 10 cm), borne on a very short shoot. Flowers *c.* 2 cm across, white with purplish exterior. Calyx-teeth *c.* 3 mm. Ovary *c.* 2 mm. Petals *c.* 9 × 7 mm, broadly

ovate. Inner and outer stamens *c.* 6 mm, outer toothed, inner not toothed, anthers attached just below the apex. *SW China (Yunnan).* H4. Spring.

A recent introduction rare, as yet, in cultivation.

22. D. rehderiana Schneider. Illustration: Krüssmann, Manual of cultivated broad-leaved trees & shrubs **1**: pl. 168, 316 (1984).

Deciduous, dense shrub to 1.5 m; current growth rough with stellate hairs. Bark red-brown, peeling. Leaves 1.5–3 × 1–1.5 cm, ovate, with fine forward-pointing teeth, rough and densely stellate-hairy above, with 4–6-rayed stellate-hairs, paler beneath with 4–8-rayed stellate hairs. Flowers in 3–8-flowered cymes in leaf axils, *c.* 7 mm across, white. Calyx *c.* 4–5 mm, densely stellate, teeth 1.5–2 mm. Styles free *c.* 3 mm. Capsule *c.* 2–3 mm. *W China.* H4. Spring.

23. D. glomeruliflora Franchet. Figures 94(5), p. 150 & 95(12), p. 151. Illustration: Hu & Chun, Icones plantarum Sinicarum **5**: f. 224(1937); Iconographia cormophytorum Sinicorum **2**: t. 1934 (1972); Krüssmann, Manual of cultivated broad-leaved trees & shrubs **1**: pl. 170 (1984).

Shrub 2–3 m, with arching branches. Leaves to 4 × 1.5 cm on flowering-shoots, lanceolate to ovate, acuminate, margin with fine forward-pointing teeth, upper surface with hairs with 3 or 4 rays and a central ray, lower surface densely hairy with 4- or 5-rayed hairs. Cymes numerous, few-flowered, borne on shoots 2.5–5 cm. Flowers white, *c.* 2 cm wide. Calyx-teeth *c.* 4 mm, narrowly lanceolate. Petals 1.3 × 1 cm. Inner and outer stamens *c.* 6 mm, the outer with broad teeth, inner untoothed and with the anthers attached about halfway down each filament. Styles 3–5, *c.* 6 mm. Ovary *c.* 2.5 mm. Capsule *c.* 3.5 × 4 mm. *W China (Sichuan).* H4. Spring.

Possibly no longer commercially available.

24. D. discolor Hemsley. Figures 94(8), p. 150 & 95(20), p. 151. Illustration: Hu & Chun, Icones plantarum Sinicarum **5**: f. 226 (1937); Iconographia cormophytorum Sinicorum **2**: f. 1932 (1972); Krüssmann, Manual of cultivated broad-leaved trees & shrubs **1**: pl. 168, 170 (1984).

Arching shrub 1–2 m. Leaves 4–11 × 1.5–3 cm, narrowly ovate-oblong,

thin, margin with fine, forward-pointing teeth; upper surface with 4- or 5-rayed hairs, a few of them with an additional erect ray, lower surface with dense, 9–12-rayed hairs, some of them, especially on the veins, with an additional central ray, rendering the veins shaggy. Cymes loose. Flowers 1.3–2.5 cm across, white to pink. Calyx-teeth 3–4 mm. Longest stamens *c.* 6 mm, distinctly toothed, inner stamens *c.* 4 mm, less distinctly toothed, anthers borne on stalks above the filaments. Ovary 2–3.5 mm. *China (W Hubei).* H4. Spring.

Several cultivars of this species are available. **D. globosa** Duthie is similar to *D. discolor* but with smaller leaves and cream-coloured flowers. *C China (W Hubei).*

25. D. purpurascens (Henry) Rehder (*D. discolor* Hemsley var. *purpurascens* Henry). Figures 94(2), p. 150 & 95(14), p. 151. Illustration: *Botanical Magazine*, 7708 (1900); *New Flora and Silva* **4**: pl. 103 (1930) & **6**: pl. 106 (1932); Krüssmann, Manual of cultivated broad-leaved trees & shrubs **1**: f. 316 (1984).

Slender arching shrub to 1.5 m. Leaves 4–7 × 1.2–3 cm, ovate to ovate-lanceolate, margin with forward-pointing teeth, upper surface with mainly 5-(rarely 3- or 4-) rayed hairs, lower surface with 7- or 8-rayed hairs, some hairs on both surfaces with erect, central rays. Cymes on short shoots, 4–6 cm. Flowers *c.* 2 cm across, white inside, purplish outside. Calyx-teeth *c.* 4.5 mm, lanceolate. Stamens broadly winged, outer series with broad teeth, inner with minute teeth, anthers attached below apex. Ovary *c.* 2.5 mm. Styles 4. *W China.* H4. Late spring–early summer.

26. D. × elegantissima (Lemoine) Rehder (*D. discolor elegantissima*). Illustration: Hay & Beckett (eds), Reader's Digest encyclopaedia of garden plants and flowers, 221 (1984); Krüssmann, Manual of cultivated broad-leaved trees & shrubs **1**: pl. 172 (1984).

Upright shrub to 1.5 m. Leaves ovate to oblong-ovate, irregularly and sharply toothed, with rather sparse 4–6-rayed hairs beneath. Cymes loose. Flowers *c.* 2 cm across, pink. Inner filaments toothed. *Garden origin.* H4. Early summer.

A hybrid between *D. scabra* and *?D. crenata.* Distinguished from *D. purpurascens* by its broader, more abruptly acuminate leaves, shorter leaf-stalks, and larger stamens, the filaments of the inner series toothed. Due to the confusion between *D. scabra* and *D. crenata, D. scabra* is often recorded as one of the parents of this hybrid. However, the toothing of the inner filaments suggests that *D. crenata* is also involved.

27. D. × kalmiiflora Lemoine; *D. purpurascens × parviflora*. Illustration: Davis, The gardener's illustrated encyclopaedia of trees and shrubs, 117 (1987).

Like *D. purpurascens* but an arching shrub to 1.5 m, less densely hairy. Flowering-shoots 10–30 cm; leaves 3–6 cm, finely toothed. Flowers in upright, umbel-like panicles, each to 2 cm across, deep pink outside, white inside. *Garden origin.* H5. Early summer.

28. D. × maliflora Rehder; *D. × lemoinei × D. purpurascens*.

Like *D. purpurascens* but an upright shrub to 2 m; leaves 2.5–4 cm, ovate-oblong, acuminate, margin with fine, forward-pointing teeth, covered with scattered 5–8-rayed hairs, flowers reddish outside, *c.* 1.5 cm across, in corymbs 3–6 cm broad.

'Boule Rose' is a very floriferous cultivar, a rounded shrub to 1 m with white petals edged with pink.

29. D. × lemoinei Lemoine; *D. gracilis × D. parviflora*. Illustration: *Garden and Forest* **9**: 285 (1896).

Upright shrub to 2 m. Leaves 3–6 cm, elliptic-lanceolate to lanceolate, long-pointed, with sharp, forward-pointing teeth, and 5–8-rayed hairs beneath. Flowers numerous, in panicles or pyramidal corymbs. Calyx-teeth triangular, shorter than ovary. Petals partly overlapping, some edge-to-edge in bud. Filaments toothed. *Garden origin.* H2. Early summer.

Many cultivars have been raised, including 'Avalanche', with dense corymbs on drooping branchlets and 'Boule de Neige', a compact dwarf shrub with large, white flowers.

30. D. parviflora Bunge. Figures 94(19), p. 150 & 95(2), p. 151. Illustration: Schneider, Illustriertes Handbuch der Laubholzkunde **1**: 379, 381 (190406); Csapody & Tóth, Colour atlas of flowering trees and shrubs, 79 (1982).

Shrub to 2 m. Leaves 3–11 × 2–3 cm, ovate, ovate-lanceolate or elliptic, with coarse, forward-pointing teeth, hairs moderately dense above, 5–8-rayed, sparse beneath, 10–12-rayed; simple hairs present along the main veins. Inflorescence broadly corymb-like. Flowers *c.* 1–1.5 cm across, white with orange disc. Calyx-teeth *c.* 1 mm. Stamens tapered at apex or obscurely toothed. Ovary *c.* 2 mm. Styles 3, shorter than the longer stamens. *N China.* H5. Early summer.

Rare in cultivation.

Var. **amurensis** Regel (*D. amurensis* (Regel) Airy-Shaw) is more commonly grown: it is as above but without the simple hairs on the leaf-veins beneath. *N China, Korea.*

D. × myriantha Lemoine; *D. parviflora × setchuenensis*. Illustration: *Gardeners' Chronicle* **52**: 45 (1912). Upright shrub to 1 m; leaves 4–6 cm, oblong-lanceolate, finely toothed, rough on both surfaces, with 5- or 6-rayed hairs beneath. Flowers *c.* 2 cm across, white. Filaments strongly toothed. *Garden origin.* H5. Early summer.

31. D. hookeriana (Schneider) Airy-Shaw (*D. corymbosa* Brown var. *hookeriana* Schneider). Figures 94(15), p. 150 & 95 (8), p. 151.

Shrub, roughly 2 m; like *D. corymbosa* except for the narrower leaves with 3–5-rayed hairs above and dense 5–9-rayed hairs beneath; anthers borne on the side of the filament; flowers sometimes tinged pink. *Himalaya to W China (Yunnan).* H5.

Uncommon in cultivation.

32. D. rubens Rehder (*D. hypoglauca* Rehder). Figures 94(10), p. 150 & 95(18), p. 151. Illustration: *Botanical Magazine*, 9362 (1934); Krüssmann, Manual of cultivated broad-leaved trees & shrubs **1**: pl. 170, 317 (1984).

Erect shrub to 2 m; young shoots stellate-hairy, becoming hairless. Leaves 4–7 × 1.5–3 cm, oblong to ovate-oblong, apex acuminate, base narrowed, or rarely rounded, minutely toothed, thin, with sparse 4-rayed hairs above, with sparse to dense 4–7-rayed hairs beneath. Inflorescence cyme-like. Flowers 8–25 mm across, pink. Calyx-teeth 1–2 mm. Anthers of larger stamens attached at apex of filament or just below it, smaller stamens untoothed but with a small lobe at the apex. Ovary *c.* 2 mm. Styles 3, equalling or shorter than the stamens. *C China (Sichuan, Hubei, Shaanxi).* H4. Early summer.

33. D. corymbosa R. Brown. Figures 94 (1), p. 150 & 95(10), p. 151. Illustration: *Edwards's Botanical Register* **26**: t. 5 (1840); Schneider, Illustriertes Handbuch der Laubholzkunde **1**: 381 (190406).

Shrub to 2 m. Leaves 6–10 × 2.5–5 cm, ovate, apex acuminate, finely to coarsely toothed; hairs not dense, but often denser above than beneath, those above with 6 or 7 short, fat rays, those beneath with 12–14 long, thin rays. Inflorescence corymb-like, borne on a shoot 10–18 cm. Flowers 1–1.5 cm across, white, hawthorn-scented. Calyx-teeth, *c.* 1 × 1.8 mm, Petals broadly obovate. Larger stamens with shallow teeth, smaller stamens mostly lacking teeth, anthers on short stalks 1–2 mm above the filament. Ovary *c.* 2 mm. Styles 3–4, longer than stamens. *W Himalaya.* H5. Early summer.

A frost-tender species. Var. **staurothrix** (Airy-Shaw) Zaikonnikova differs from the above in its large, almost uniformly 4-rayed, cross-shaped hairs.

34. D. compacta Craib. Illustration: *Botanical Magazine*, 8795 (1919); Krüssmann, Manual of cultivated broad-leaved trees & shrubs **1**: f. 316 (1984).

Deciduous shrub to 1.5 m. Young growth stellate-downy at first. Leaves 5–6 × 2–2.5 cm, lanceolate to oblanceolate, apex long-pointed, dull green above, paler beneath. Flowers numerous, in broad corymb-like panicles, white. Petals *c.* 4 × 4, almost circular. *Only known from cultivation, supposed origin China.* H4. Midsummer.

Doubtfully distinct from *D. corymbosa*. 'Lavender Time', with flowers pale purple at first, is a most attractive cultivar.

35. D. mollis Duthie. Figures 94(14), p. 150 & 95(9), p. 151. Illustration: *Botanical Magazine*, 8559 (1914); Bean, Trees & shrubs hardy in the British Isles, edn 8, **2**: 42 (1973); Krüssmann, Manual of cultivated broad-leaved trees & shrubs **l**: pl. 168 & f. 316, 317 (1984).

Vigorous shrub to 1.5–2 m. Leaves large, to 9 × 4 cm, margin coarsely doubly toothed, densely shaggy-hairy beneath, with 4-rayed stellate hairs above, which have conspicuous central rays. Inflorescence paniculate or corymb-like, to 12 × 13 cm, borne on shoots 9–20 cm. Flowers white, *c.* 1 cm across. Calyx-teeth *c.* 1.5 mm. Stamens long, narrow and without teeth. Ovary *c.* 2 mm, densely hairy. *China (W Hubei).* H5. Summer.

36. D. × wilsonii Duthie. Illustration: *Botanical Magazine*, 8083 (1906); Schneider, Illustriertes Handbuch der Laubholzkunde **2**: 933 (1912);

Krüssmann, Manual of cultivated broad-leaved trees & shrubs **2**: pl. 168 (1986).

Vigorous shrub to 2 m; leaves 7–11 cm, apex acuminate, margin with forward-pointing teeth, densely covered with 5–10-rayed stellate hairs, and with spreading hairs on the veins also beneath. Flowers in loose broad corymbs, each *c.* 2 cm across, white. Some stamens toothed, others tapered at the apex. *Garden origin.* H4. Early summer.

4. PHILADELPHUS Linnaeus
D.R. McKean

Shrubs with mainly peeling bark; leaves usually deciduous, opposite, simple. Flowers in racemes, panicles or cymes, or solitary, often strongly scented. Sepals 4. Petals 4. Stamens numerous. Ovary inferior, surmounted by a nectar-secreting disc; carpels 4, united; styles 4, partially or wholly united. Fruit a many-seeded capsule. Seeds usually with tails.

A genus of about 40 species, mainly from E Asia and the Himalaya, N America, S Europe and the Caucasus. They prefer a loamy soil in full sun; pruning should consist of thinning some of the older wood only, as flowers are produced on the previous year's growth. Vigorous shoots should be left unpruned. Propagate by softwood cuttings.

Literature: Koehne, E., Philadelphus, *Gartenflora* **45**: 450–1, 486–8, 500–8, 541–2, 561–3, 596–7, 618–9, 651–2 (1896); Schneider, C.K., Illustriertes Handbuch der Laubholzkunde **1**: 362–74 (190406); Hu, S.Y., The genus Philadelphus, *Journal of the Arnold Arboretum* **35**: 275–333 (1954), **36**: 52–109, 325–68 (1955) & **37**: 15–90 (1956).

1a. Epigynous zone hairless outside 2
  b. Epigynous zone hairy outside 15
2a. Flowers solitary, in 3s or in few-flowered panicles 3
  b. Flowers in distinct, many-flowered racemes or panicles 5
3a. Flowers mainly semi-double or double; bark not peeling; stamens absent or fewer than 20
      16. × cymosus
  b. Flowers single (except in some cultivars of *P. × lemoinei*); bark peeling; stamens more than 20 4
4a. Petals elliptic, acute; stamens 20–35
      17. × falconeri
  b. Petals oblong, obtuse; stamens 60–90 **3. inodorus**

5a. Calyx greenish purple; flowers bell-shaped; flower-buds tinged pink
      15. purpurascens
  b. Calyx green; flowers mainly disc- or cross-shaped; flower-buds white 6
6a. Leaves uniformly long-hairy above and on the veins beneath; stamens 30–40 **14. delavayi**
  b. Leaves with hairs only on the veins and/or vein-angles or hairless or almost so; stamens mainly 25–35 7
7a. Flowers in racemose panicles 8
  b. Flowers in simple racemes 9
8a. Upper leaf-surface hairless or becoming so, lower surface moderately hairy **13. × lemoinei**
  b. Leaves hairy on veins and/or vein-angles only **30. californicus**
9a. Leaves hairless or almost so (rarely slightly hairy beneath in *P. × purpureo-maculatus*) 10
  b. Leaves hairy on the veins on either or both surfaces 11
10a. Sepals each with a small tail at the apex; petals *c.* 1.5 × 1.2 cm; bark grey; stamens 38 **8. intectus**
  b. Sepals not tailed; petals 9–11 × 8 mm; bark dark brown; stamens 25 **10. pekinensis**
11a. Leaves usually shaggy-hairy beneath, becoming hairless above, long-acuminate **9. tomentosus**
  b. Leaves hairy only on veins and/or vein-angles beneath 12
12a. Current growth shaggy at first; leaves sparsely stiff-hairy above; stamens 0–40 **11. brachybotrys**
  b. Current growth hairless or sparsely downy; leaves hairy only on veins and/or vein-angles; stamens to 30 13
13a. Current growth sparsely downy; leaves softly hairy on veins and vein-angles; flowers fragrant
      12. coronarius
  b. Current growth hairless; leaf-veins stiffly hairy; flowers not fragrant 14
14a. Leaves 6–11 cm; flowers disc-shaped
      7. × splendens
  b. Leaves 4–5 cm; (except for some varieties); flowers not as above
      6. lewisii
15a. Epigynous zone hairy outside, the surface visible through the hairs 16
  b. Epigynous zone very densely hairy outside, the surface not visible through the hairs 28
16a. Leaves 1–1.5 cm **27. microphyllus**

  b. Leaves more than 2.5 cm 17
17a. Flowers 3–5, in cymes or corymbs 18
  b. Flowers more numerous in racemose panicles or in distinct racemes 23
18a. Flowers with a pink or purplish centre 19
  b. Flowers without a pink or purplish centre 20
19a. Petals obovate-oblong, *c.* 2 × 1 cm, hairy inside **2. × burkwoodii**
  b. Petals ovate, *c.* 1.1 × 1 cm, not hairy
      5. × purpureo-maculatus
20a. Leaves densely hairy beneath; winter-buds exposed **32. hirsutus**
  b. Leaves moderately hairy beneath; winter-buds hidden beneath the bark 21
21a. Leaves evergreen, with long curved hairs above and beneath
      1. mexicanus
  b. Leaves deciduous, upper surface hairless or soon becoming so 22
22a. Stamens 60–90 **4. floridus**
  b. Stamens 30 or fewer
      28. × polyanthus
23a. Flowers in panicles **31. insignis**
  b. Flowers in racemes 24
24a. Leaves hairy on veins and in vein-angles only **26. satsumi**
  b. Leaves hairy on the surfaces as well as on veins or not hairy 25
25a. Leaves densely long, shaggy-hairy beneath **9. tomentosus**
  b. Leaves hairy only on the veins beneath 26
26a. Hairs on leaf veins sparse, each with a swollen base **24. kansuensis**
  b. Hairs on leaf veins dense, bases of hairs not swollen 27
27a. Disc and style hairless
      22. sericanthus
  b. Disc and style hairy **23. subcanus**
28a. Leaves uniformly long-hairy above and on the veins beneath
      22. sericanthus
  b. Leaves not as above 29
29a. Flowers solitary **29. argyrocalyx**
  b. Flowers in racemes 30
30a. Leaves densely hairy beneath 31
  b. Leaves not densely hairy beneath
      32
31a. New growth hairy **21. incanus**
  b. New growth hairless
      18. pubescens
32a. Current year's growth hairless; flowers usually double **19. × nivalis**
  b. Current year's growth with long shaggy hairs; flowers single or double 33

33a. Flowers single; styles hairy
25. schrenkii
  b. Flowers usually double; styles
     hairless **20. × virginalis**

1. P. mexicanus Schlechtendahl.
Illustration: *Botanical Magazine*, 7600
(1898).

Tender, climbing, evergreen shrub to
5 m; branches long, drooping; bark dark
brown, wrinkled, current growth long-
bristly, winter-buds prominent. Leaves
5–11.5 × 2–5 cm, ovate, adpressed-bristly
on both surfaces, apex long-acuminate,
base rounded to cordate, more or less
entire or with a few tiny teeth. Flowers
3–4 cm across, solitary or in 3s, yellowish
white, rose-scented. *Mexico, Guatemala*. G1.
Summer.

2. P. × burkwoodii Burkwood &
Skipwith.

Dwarf shrub with prominent winter
buds, bark very dark brown in the second
year and eventually peeling; current
growth adpressed stiffly hairy. Leaves
3.5–6.6 × 1.5–3 cm, ovate-elliptic, hairless
above, sparsely hairy beneath. Flowers
5–6 cm across, 1–5, in panicles, cross-
shaped, white with purplish inside,
fragrant. Sepals short-hairy or partially so;
disc and style hairless. *Garden origin*. H4.
Summer.

A hybrid between *P. mexicanus* and
another, unknown, species.

3. P. inodorus Linnaeus. Illustration:
Journal of the Arnold Arboretum **35**: pl. 4
(1954); Krüssmann, Manual of cultivated
broad-leaved trees & shrubs **2**: f. 273
(1985).

Arching shrub to 2–3 m, bark of the
second year chestnut brown, peeling;
current year's growth hairless. Leaves
5–9 × 2–3.5 cm, ovate-elliptic or elliptic,
more or less entire or faintly toothed,
sparsely adpressed hairy or almost hairless
above, hairy on main veins and vein-
angles beneath. Flowers 4–5 cm across, in
cymes of 1, 3 or rarely 9; flower-stalk,
epigynous zone and calyx all hairless.
Stamens 60–90. Style equal to the longest
stamens, hairless, stigmas swollen. Seeds
long-tailed. *SE USA*. H3. Summer.

4. P. floridus Beadle. Illustration:
Schneider, Illustriertes Handbuch der
Laubholzkunde **1**: 366 (190406).

Shrub to 3 m; current year's growth
hairless, brown, second year's growth
chestnut. Leaves 4–10 × 2–6 cm, mostly
ovate-elliptic, apex sharply acuminate, base

rounded or obtuse, almost entire or
inconspicuously and remotely toothed,
evenly adpressed bristly beneath, hairless
above except for a few adpressed bristles on
the veins. Flowers in 3s, rarely solitary or
in racemes, epigynous zone and calyx
shaggy-hairy. Flower 4–5 cm across, disc-
shaped, petals almost circular, *c.* 2.5 cm,
pure white. Stamens 80–90, to 1.5 cm.
Stigma oar-shaped. Seeds long-tailed. *USA
(Georgia)*. H3. Early summer.

Var. **faxonii** Render has smaller flowers
in the shape of a cross, and is sometimes
regarded as a cultivar. *Origin unknown*. H3.
Early summer.

5. P. × purpureo-maculatus Lemoine;
P. × lemoinei × P. coulteri. Illustration:
Botanical Magazine, 8193 (1908).

Shrub to 1.5 m; bark blackish brown,
eventually peeling in the second year;
current growth hairy. Leaves
1–3.5 × 0.6–2.5 cm, broadly ovate, apex
acute, base rounded, entire or almost so,
with a few scattered hairs beneath. Flowers
solitary or in 3s or 5s, fragrant, epigynous
zone with short, rough hairs. Corolla
2.5–3 cm across, disc-shaped, petals almost
circular, white, purplish at base. Stamens
c. 30. *Garden origin*. H3. Summer.

Several cultivars are known, all with the
characteristic pink- or purplish-centred
flowers, flowering late and with a long
flowering season; they include 'Belle Étoile',
'Bicolore', 'Étoile Rose', 'Fantaisie',
'Galathée', 'Nuage Rose', 'Sybille'.

6. P. lewisii Pursh. Illustration: Schneider,
Illustriertes Handbuch der Laubholzkunde
1: 368 (1904–06); Krüssmann, Manual of
cultivated broad-leaved trees & shrubs **2**:
380 (1985).

Erect shrub to 3 m; second year's growth
yellowish to chestnut brown, bark not
peeling but with transverse cracks, current
year's growth hairless except at the nodes.
Leaves 4–5.5 × 2–3.5 cm, apex acute, base
rounded, obtuse or shortly acuminate,
more or less entire or inconspicuously
finely toothed, very sparsely covered with
long, rough hairs on the veins above and
with tufts of hair in the vein-angles
beneath, margin hairy. Flowers 3–4.5 cm
across, in racemes of 5–11, cross-shaped.
Sepals 5–6 × 3 mm, ovate, wide at the
base. Stamens 28–35, the longest being
half the length of the petals. Anthers and
disc hairless. Style shorter than the longest
stamens, hairless, undivided or slightly
divided above. Seeds long-tailed. *W North*

America (British Columbia to California). H3.
Early summer.

Var. **gordonianus** (Lindley) Koehne.
Illustration: McMinn, Manual of Californian
shrubs, 139 (1939); Krüssmann, Manual
of cultivated broad-leaved trees & shrubs **3**:
381 (1986). Leaves more densely hairy
and more strongly toothed, flowers disc-
shaped. *W USA (Washington to California)*.
H3. Summer.

7. P. × splendens Rehder. Illustration:
Journal of the Arnold Arboretum **35**: pl. 3
(1954); Krüssmann, Manual of cultivated
broad-leaved trees & shrubs **2**: f. 271
(1985).

Upright shrub; bark of the second year
dark brown, peeling; current growth
hairless. Leaves 6–11.5 × 2.5–5 cm,
oblong-elliptic, apex acuminate, base
rounded, inconspicuously and finely
toothed or almost entire, hairless or rough,
shaggy hairy on the veins and vein-angles
beneath. Flowers in crowded racemes of
5–9, disc-shaped, *c.* 4 cm across, slightly
scented. Petals rounded, pure white.
Epigynous zone, calyx, disc and style
hairless. Stamens 30. *Garden origin*. H3.
Summer.

Thought to be *P. inodorus* var.
grandiflorus × P. lewisii var. *gordonianus*.

8. P. intectus Beadle (*P. pubescens*
Loiseleur var. *intectus* (Beadle) Moore).

Erect shrub to 5 m; bark silvery not
peeling; current growth hairless. Leaves on
non-flowering-shoots 6–10 × 4–6 cm, ovate
to oblong-elliptic; on flowering-shoots
3–6 × 1.5–3.5 cm; apex acuminate, base
rounded to obtuse, hairless or rarely hairy
beneath, margin with a few forward-
pointing teeth. Flowers *c.* 3 cm across, in
racemes of 5–9, disc-shaped. Epigynous
zone hairless. Sepals ovate, tailed. Stamens
38. Disc and style hairless. Seeds long-
tailed. *SE USA*. H3. Summer.

9. P. tomentosus Royle (*P. coronarius*
Linnaeus var. *tomentosus* (Royle) Hooker &
Thomson). Illustration: Royle, Illustrated
botany of the Himalayan mountains **2**: pl.
46 (1839); Schneider, Illustriertes
Handbuch der Laubholzkunde **1**: 371
(1904–06); Krüssmann, Manual of
cultivated broad-leaved trees & shrubs **3**:
383 (1986).

Shrub to 3 m, similar to *P. coronarius* but
for the downy undersides of the leaves;
second year's bark cinnamon, eventually
peeling off; current growth hairless or
becoming so. Leaves 4–10 × 2–5 cm, ovate

or rarely lanceolate, apex acuminate (often strikingly so), base rounded or obtuse, becoming hairless above, uniformly shaggy hairy beneath or rarely almost hairless. Flowers c. 3 cm across, 5–7 in a raceme, cross-shaped, fragrant. Petals obovate-oblong, cream. Disc and styles hairless, stigmas club-shaped. *N India, Himalaya.* H3. Early summer.

10. P. pekinensis Ruprecht. Illustration: Schneider, Illustriertes Handbuch der Laubholzkunde **1**: 237, 238 (1904–06); *Journal of the Arnold Arboretum* **35**: pl. 3, 4 (1954) & **37**: pl. 5 (1956); Krüssmann, Manual of cultivated broad-leaved trees & shrubs **3**: f. 276 (1986).

Low compact shrub to 2 m; bark of the second year dark brown, peeling; current year's growth hairless. Leaves on non-flowering shoots 6–9 × 2.5–4.6 cm, ovate, those of flowering-shoots 3–7 × 1.5–2.5 cm, apex long-pointed, base rounded or obtuse, toothed, hairless on both surfaces or sometimes with tufts of hairs in the vein-angles nearest the stalk beneath. Racemes with 3–9 yellowish white, fragrant, disc-shaped flowers, 2–3 cm across. Calyx and disc hairless. Sepals ovate, c. 4 mm. Stamens 25. Seeds with short tails. *N & W China.* H3. Early summer.

11. P. brachybotrys (Koehne) Koehne (*P. pekinensis* Ruprecht var. *brachybotrys* Koehne).

Shrub to 3 m; bark of the second year brownish grey, not peeling; current growth shaggy hairy, becoming hairless. Leaves 2–6 × 1–3 cm, ovate, apex shortly acuminate, base rounded, finely toothed or almost entire, sparsely adpressed bristly above and on the veins beneath. Flowers c. 3 cm across, 5–7 in short racemes, cream, disc-shaped. Epigynous zone, calyx, disc and style hairless. Stamens 30–40. Style about as long as stamens, stigma spathulate, inner surface 1 mm, outer 2 mm. Seeds short-tailed. *SE China.* H5. Summer.

There is some doubt as to whether the plants in cultivation under this name are genuine or of hybrid origin.

12. P. coronarius Linnaeus. Illustration: *Journal of the Arnold Arboretum* **35**: pl. 1, 3, 4 (1954); Krüssmann, Manual of cultivated broad-leaved trees & shrubs **3**: f. 274 (1986).

Shrub to 3 m; bark dark brown, slowly peeling in the second year; current growth

sparsely downy, becoming hairless. Leaves 4.5–9 × 2–4.5 cm, ovate, almost hairless but downy on the the major veins and in the vein-angles beneath, margins irregularly and shallowly toothed, apex acuminate, base obtuse or acute. Flowers 2.5–3 cm across, 5–9 in short terminal racemes, creamy white, strongly fragrant. Sepals triangular, acute, hairless. Stamens c. 25. Disc and style hairless. Seeds with long tails. *S Europe (Austria, Italy, Romania), former USSR (Caucasus).* H4. Early summer.

Several cultivars (formerly recognised as varieties or forms) are known, e.g. 'Aureum' (var. *aureus* Anon.) with the young leaves yellowish; 'Deutziflorus' with the petals pointed at the apex; 'Duplex' with double flowers; and 'Variegatus' with white margins to the leaves.

13. P. × lemoinei Lemoine; *P. coronarius* × *P. microphyllus*. Illustration: *Journal of the Arnold Arboretum* **35**: pl. 3 (1954); Krüssmann, Manual of cultivated broad-leaved trees & shrubs **3**: f. 279 (1986); Brickell (ed.), RHS A–Z encyclopedia of garden plants, 800 (2003).

A low compact shrub as wide as high, with peeling bark. Leaves 1.5–2.5 cm × 7–12 mm, ovate, hairless above, sparsely bristly beneath, apex acuminate, base rounded or obtuse, with c. 6 teeth. Flowers usually in 3s, more rarely solitary or in 5s, cross-shaped, c. 3 cm across. Sepals ovate, hairless. Petals notched. Stamens c. 5. *Garden origin.* H4. Summer.

Several cultivars have been raised, e.g. 'Avalanche', 'Candelabre', 'Coup d'Argent', etc.

14. P. delavayi Henry. Illustration: Schneider, Illustriertes Handbuch der Laubholzkunde **1**: 370 (1904–06); *Botanical Magazine*, 9022 (1924); Bean, Trees & shrubs hardy in the British Isles, edn 8, **3**: 130 (1976).

Shrub to 4 m; bark of second year grey-brown, grey or chestnut brown, not peeling; current growth hairless, glaucous. Leaves 2–8 × 2–5 cm on flowering-shoots, much larger on non-flowering-shoots, ovate-lanceolate or ovate-oblong, apex acuminate, base rounded, usually with forward-pointing teeth but sometimes entire; all sparsely bristly above, densely adpressed shaggy-hairy beneath. Flowers 2.5–3.5 cm across, in racemes of 5–9 (rarely more), disc-shaped, pure white, fragrant. Calyx hairless, glaucous, tinged

with purple. Stamens c. 35. Disc and style hairless. *SW China.* H4. Early summer.

15. P. purpurascens (Koehne) Rehder (*P. brachybotrys* var. *purpurascens* Koehne). Illustration: *Botanical Magazine*, 8324 (1910) – as *P. delavayi*; Krüssmann, Manual of cultivated broad-leaved trees & shrubs **3**: pl. 154 (1986).

Shrub to 4 m; bark of second year brown or grey, smooth; current growth hairless. Leaves 1.5–6 cm × 5–30 mm on flowering-shoots, much longer on non-flowering-shoots, ovate to ovate-lanceolate, usually uniformly adpressed bristly above and on the veins beneath, finely toothed. Racemes usually with 5–9 bell-shaped flowers 3–4 cm wide and very fragrant. Calyx green tinged with purple, glaucous. Stamens 25–30. Style and disc hairless. *S China.* H4. Summer.

Var. **venustus** (Koehne) S.Y. Hu (*P. venustus* Koehne) has shaggy hairs on the young growth. *SW China.* H4. Summer.

16. P. × cymosus Rehder (*P. floribundus* Schrader).

Erect shrub to 2.5 m; bark brown, not peeling. Leaves ovate, sparsely toothed, hairy beneath, especially on the veins. Flowers in cymes of 1–5. Sepals and disc hairless. Some stamens may be petal-like. *Garden origin.*

Parentage unknown. Several cultivars are known, e.g. 'Amalthée', 'Bonnierè' (semi-double), 'Bouquet Blanc' (double), 'Conquète' and 'Dresden'.

17. P. × falconeri Nicholson. Illustration: *Journal of the Arnold Arboretum* **35**: pl. 1, 2 (1954); Krüssmann, Manual of cultivated broad-leaved trees & shrubs **3**: pl. 155 (1986).

Shrub to 3 m; bark brown, peeling in the second year; branches slender and pendent, current growth hairless. Leaves 3–6.5 × 1–2.5 cm, ovate or ovate-elliptic, base rounded or obtuse, faintly toothed, veins with adpressed bristles beneath. Flowers to 3 cm across, 3–5 in cymes, abundant, pure white, stellate, petals elliptic, pointed. Styles much longer than stamens, sterile. *Garden origin.* H3. Summer.

A hybrid of unknown parentage, which is often shy to flower.

18. P. pubescens Loiseleur. Illustration: Schneider, Illustriertes Handbuch der Laubholzkunde **1**: 369 (1904–06);

Krüssmann, Manual of cultivated broad-leaved trees & shrubs **3**: f. 272 (1986).

Shrub to 5 m; second year's bark grey, first year's bark not peeling, current growth hairless. Leaves 4–8 × 3–5.5 cm on flowering-shoots, ovate, apex abruptly acuminate, base rounded, remotely toothed or entire, hairless above except for rough, short, stiff hairs on the veins, shaggy bristly hairy beneath. Flowers *c.* 3.5 cm across, 5–11 in racemes, white, scentless. Stamens *c.* 35. Disc and style hairless. Seeds large, with short tails. *SE USA.* H3. Early summer.

P. × monstrosus (Späth) Schelle (*P. inodorus × P. lewisii* var. *gordonianus*) differs only in its less hairy epigynous zone. A hybrid between *P. pubescens* and perhaps *P. inodorus* is also rarely cultivated (*P. × pendulifolius* Carrière).

19. P. × nivalis Jacques.

Arching shrub to 2.5 m; bark dark brown, peeling; current growth hairless. Leaves 5–10 × 2.5–6 cm, ovate or ovate-elliptic, apex acuminate, base rounded or obtuse, faintly toothed, hairless above, uniformly shaggy-hairy beneath. Racemes with 5–7 double flowers. Sepals with dense, long, rough hairs. Corolla 2.5–3.5 cm across, disc-shaped. Disc and style hairless. *Garden origin.* H3. Early summer.

Probably *P. coronarius × P. pubescens*.

20. P. × virginalis Rehder. Illustration: Krüssmann, Manual of cultivated broad-leaved trees & shrubs **3**: pl. 156 (1986).

Stiffly upright shrub to 2.5 m, second year's bark grey, peeling only when old, current growth with shaggy hairs. Leaves 4–7 × 2.5–4.5 cm, ovate, apex shortly acuminate, base rounded, becoming hairless above, uniformly tough-shaggy-hairy beneath. Flowers 4–5 cm across, in racemes, usually double, pure white, very fragrant. Calyx densely hairy. Style hairless. *Garden origin.* H4. Summer–late summer.

Of doubtful origin but with strong characteristics of *P. pubescens*. Several cultivars are grown, including 'Argentine', 'Boule d'Argent' (illustration: Brickell (ed.), RHS A–Z encyclopedia of garden plants, 800, 2003), 'Enchantment', 'Fleur de Neige'.

21. P. incanus Koehne. Illustration: *Journal of the Arnold Arboretum* **35**: pl. 3 (1954).

Erect shrub to 3.5 m; bark grey and smooth to the second year, later peeling; current growth hairy. Leaves 4–8.5 × 2–4 cm on flowering-shoots, to 10 × 6 cm on non-flowering-shoots, ovate-elliptic, apex slender-pointed, base tapered to rounded, sparsely bristly above, adpressed-bristly beneath. Flowers *c.* 2.5 cm across, 7–11 in racemes, white. Calyx and flower-stalk adpressed bristly. Stamens *c.* 34. Seeds very short-tailed. *China (Hubei, Shaanxi).* H4. Late summer.

22. P. sericanthus Koehne. Illustration: *Botanical Magazine*, 8941 (1922); *New Flora and Silva*, pl. 85 (1934); Csapody & Tóth, Colour atlas of flowering trees and shrubs, 79 (1982); Krüssmann, Manual of cultivated broad-leaved trees & shrubs **3**: pl. 154 (1986).

Shrub to 3 m; current growth hairless or soon becoming so, second year's bark grey or grey-brown, slowly peeling. Leaves 4–11 × 1.5–5 cm, ovate-elliptic or elliptic-lanceolate, apex acuminate, base obtuse or rounded, usually coarsely toothed, sparsely adpressed-bristly above and on the veins beneath. Flowers *c.* 2.5 cm across, 7–15 in racemes, pure white, unscented. Calyx and flower-stalk densely adpressed bristly. Disc and style hairless. Seeds short-tailed. *China (Sichuan & Hubei).* H4. Summer.

23. P. subcanus Koehne. Illustration: *New Flora and Silva*, pl. 85 (1934); Csapody & Tóth, Colour atlas of flowering trees and shrubs, 79 (1982).

Erect shrub to 6 m; current growth brown, hairless or soon becoming so, older bark grey-brown, smooth, peeling late. Leaves 4–14 × 1.5–7.5 cm, ovate or ovate-lanceolate, apex acuminate, base rounded or obtuse, obscurely finely toothed on flowering-shoots but with forward-pointing teeth on non-flowering-shoots, all sparsely covered with upright hairs above, shaggy-hairy on the veins beneath. Flowers 5–29 in racemes, which are 2.5–22 cm; flowers 2.5–3 cm across, disc-shaped, pure white, slightly fragrant. Calyx curly-hairy. Petals circular to obovate, long curly-hairy at base. Stamens *c.* 30. Disc downy, lower style hairy. Seeds with very short tails. *W China.* H4. Early summer.

Rare in cultivation; most commonly seen is var. **magdalenae** (Koehne) S.Y. Hu, which is somewhat smaller in all its parts and has the flower-stalks and calyx only slightly downy with curly hairs. *SW China.* H4. Early summer.

24. P. kansuensis (Rehder) S.Y. Hu (*P. pekinensis* Ruprecht var. *kansuensis*

Rehder). Illustration: *Journal of the Arnold Arboretum* **35**: pl. 4 (1954).

Upright shrub to 7 m; current year's growth with curly hairs, eventually becoming hairless, second year's bark grey-brown, peeling. Leaves to 11 × 6.5 cm on non-flowering-shoots, 3–5 × 1–2 cm on flowering-shoots, ovate or ovate-lanceolate, more or less entire or faintly toothed, apex pointed, base obtuse or rounded, uniformly bristly-hairy above, hairs on the veins beneath with swollen bases. Flowers *c.* 2.5 cm across, 5–7 in racemes, disc-shaped, flower-stalks bristly-hairy. Petals oblong-rounded. Stamens *c.* 30, disc bristly at its rim. Style hairless. Seeds short-tailed. *NW China.* H5. Summer.

25. P. schrenkii Ruprecht. Illustration: Schneider, Illustriertes Handbuch der Laubgehölze **1**: 237 (1904–06); Krüssmann, Manual of cultivated broad-leaved trees & shrubs **3**: f. 278 (1986).

Upright shrub to 4 m, second year's bark grey or rarely brown, with transverse cracks, rough hairy at first. Leaves 7–13 × 4–7 cm on non-flowering-shoots, 4.5–7.5 × 1.5–4 cm on flowering-shoots, ovate, occasionally ovate-elliptic, apex acuminate, base acute or obtuse, remotely finely toothed or almost entire, sparsely shaggy-hairy on the main veins beneath, usually hairless above. Flowers 2.5–3.5 cm across, 3–7 in racemes, cross-shaped, very fragrant. Sepals 3–7 mm, ovate. Stamens 25–30. Disc hairless. Style hairy. Seeds short-tailed. *Korea to former USSR (E Siberia).* H3. Summer.

Var. **jackii** Koehne. Illustration: Krüssmann, Manual of cultivated broad-leaved trees & shrubs **3**: f. 278 (1986). Leaves more obviously toothed, the veins hairy beneath. *N China, Korea.* H3. Summer.

26. P. satsumi (Siebold) S.Y. Hu. Illustration: Schneider, Illustriertes Handbuch der Laubholzkunde **1**: 237 (1904–06); Makino, Illustrated flora of Japan, 1461 (1956); Krüssmann, Manual of cultivated broad-leaved trees & shrubs **3**: f. 276 (1986).

Upright shrub to 3 m, second-year twigs brown, bark eventually peeling; current growth becoming hairless. Leaves on non-flowering-shoots 6–9 × 3–5 cm, ovate or broadly elliptic, with coarse, forward-pointing teeth, apex long-acuminate, base obtuse or rounded, those on flowering-shoots 4.5–7 × 1.5–4.5 cm, ovate or ovate-lanceolate, tapered, apex long-acuminate, base obtuse or sometimes rounded; all with sparse bristles or hairless above, with stiff hairs on the veins beneath and in the vein-angles. Flowers c. 3 cm across, 5–7 in a raceme, cross-shaped, slightly fragrant. Petals oblong-obovate. Stamens c. 30. Style hairless, shortly divided at the apex. Seeds with medium tails. *Japan*. H5. Summer.

P. satsumanus Miquel. Similar but with uniformly downy leaves. Seldom found in cultivation. *Japan*. H5. Summer.

27. P. microphyllus Gray. Illustration: Schneider, Illustriertes Handbuch der Laubholzkunde **1**: 234 (1904–06); *Journal of the Arnold Arboretum* **35**: pl. 1, 2, 4 (1954); Csapody & Tóth, Colour atlas of flowering trees and shrubs, 79 (1982); Brickell (ed.), RHS A–Z encyclopedia of garden plants, 801 (2003).

Low, erect, graceful shrub to 1 m; current growth adpressed downy, second year's bark chestnut brown, shiny, soon flaking off. Leaves 1–1.5 cm × 5–7 mm, ovate-elliptic or sometimes lanceolate, entire and with marginal hairs, hairless or becoming so above, softly shaggy hairy beneath, base obtuse, apex acute or obtuse. Flowering-shoots 1.5–4 cm. Flowers c. 3 cm across, 1–2, pure white, very fragrant; cross-shaped. Sepals lanceolate. Stamens c. 32. Seeds with very short tails. *SW USA*. H5. Early summer.

28. P. × polyanthus Rehder.

Erect shrub, bark dark brown, eventually peeling; current year's growth with a few shaggy hairs. Leaves 3.5–5 × 1.5–2.5 cm, ovate, apex acuminate, base rounded or obtuse, entire or with a few sharp teeth, hairless above, sparsely hairy with short, adpressed bristles beneath. Flowers c. 3 cm across, 3–5 in cymes or corymbs, cross-shaped. Epignous zone and calyx downy, sepals tailed at apex. Stamens c. 30. *Garden origin*. H3. Summer.

Thought to be *P. insignis* × *P. × lemoinei*. Various cultivars are grown, including 'Étoile Rose', 'Fantaisie', 'Galathée', 'Nuage Rose', 'Sybille'.

29. P. argyrocalyx Wooton. Illustration: *Journal of the Arnold Arboretum* **35**: pl. 14 (1954); Krüssmann, Manual of cultivated broad-leaved trees & shrubs **3**: f. 273 (1986).

Erect shrub to 2 m; twigs grey-brown, the second-year bark usually intact, current growth with rusty, shaggy hairs. Leaves 1–3.5 cm × 4–15 mm, ovate, ovate-lanceolate or elliptic, apex acute or obtuse,

base obtuse, dark green, hairless or becoming so above, sparsely shaggy-bristly and paler green beneath. Flowers solitary to 3.5 cm across, on short stalks (1–2 mm), cross-shaped, white and slightly fragrant. Calyx densely white woolly. Seeds long-tailed. *USA (New Mexico)*. Summer–late summer.

30. P. californicus Bentham (*P. lewisii* var. *californicus* (Bentham) Torrey). Illustration: Schneider, Illustriertes Handbuch der Laubholzkunde **1**: 234 (190406); Dippel, Handbuch der Laubholzkunde **1**: 181 (1889); *Journal of the Arnold Arboretum* **35**: pl. 2 (1954); Krüssmann, Manual of cultivated broad-leaved trees & shrubs **3**: f. 272 (1986).

Erect shrubs to 3 m; second year's bark dark brown, current year's growth soon becoming hairless. Leaves on non-flowering-shoots 4.5–8 × 3–5 cm, those on flowering-shoots 3–5 × 2–3 cm (rarely to 8 cm), ovate or ovate-elliptic, hairless except for tufts in the vein-angles beneath, apex acute, base acute, obtuse or sometimes rounded, entire or obscurely toothed. Flowers to 2.5 cm across, 3–5 in a panicle, cross-shaped, fragrant. Sepals ovate, hairless. Stamens 25–37. Disc and style hairless. Seed short-tailed. *USA (California)*. H5. Summer.

31. P. insignis Carrière (*P. × insignis* Carrière). Illustration: Schneider, Illustriertes Handbuch der Laubholzkunde **1**: 236 (1904–06); Krüssmann, Manual of cultivated broad-leaved trees & shrubs **3**: f. 277 (1986).

Erect shrub to 4 m; second year's bark grey, (rarely brown) smooth. Leaves 3.5–8 × 1.5–6 cm, ovate or ovate-elliptic, obtuse, apex acute or strongly acuminate, base acute or rounded, more or less entire or faintly toothed, coarsely adpressed-hairy beneath. Flowers 2.5–3.5 cm across, in 3s in panicles. Sepals ovate, usually adpressed hairy. Stamens c. 30. Disc and style hairless. Seeds short-tailed. *W USA (California, Oregon)*. H5. Summer.

Considered by some authors to be a hybrid (*P. californicus* × *P. pubescens*).

32. P. hirsutus Nuttall. Illustration: *Edwards's Botanical Register* **24**: t. 14 (1839); *Botanical Magazine*, 5334 (1862); Schneider, Illustriertes Handbuch der Laubholzkunde **1**: 234 (1904–06).

Low, spreading shrub with slender, slightly twisted arching branches, to 2.5 m; shoots widely divergent, second year's bark

dark brown, peeling, current year's growth with shaggy hairs. Leaves 2.5–7 × 1–5 cm, ovate-elliptic or ovate-lanceolate, apex acuminate, base rounded, sharply toothed, uniformly covered with hairs with swollen bases above and densely shaggy beneath. Flowers c. 2.5 cm across, 1–5 on very short shoots with 1 or 2 pairs of leaves, disc-shaped. Sepals broadly triangular, shaggy hairy. Style and disc hairless. Seeds without tails. *SE USA (North Carolina to Georgia)*. H5. Early summer.

5. KIRENGESHOMA Yatabe
P.G. Barnes
Erect herbaceous perennial, with short, thick rhizomes. Leaves opposite, the lower long-stalked, the upper stalkless, all palmately lobed and coarsely sinuous-toothed. Flowers in terminal and axillary cymes, somewhat nodding on long stalks. Sepals 5, small, petals 5, narrowly ovate, rather thick, pale yellow, not opening widely. Stamens 15, styles usually 3; ovary inferior, 3-celled. Capsule ovoid, 3-celled, seeds flat, with an irregular wing.

A genus of 2 species from NE Asia. Easily cultivated hardy perennials, requiring a moisture retentive but well-drained soil and a position in full sun or part shade. Propagate by division in autumn or early spring. Although seeds are often produced quite freely, germination may be slow and erratic.

1. K. palmata Yatabe. Illustration: *Botanical Magazine*, 7944 (1904); Thomas, Perennial garden plants, f. 13 (1982); Jelitto, Schacht & Fessler, Die Freiland Schmuckstauden, edn 3, 338 (1985); Kitamura et al., Coloured illustrations of herbaceous plants of Japan **2**: pl. 33 (1986).

Stem to 1 m or more, often purplish above. Leaves 10–20 cm, slightly hairy beneath. Flowers 2.5–3.5 cm. *Japan*. H4. Summer.

Plants from Korea have been distinguished as *K. koreana* Nakai. This is said to be taller, with more erect flowers.

6. HYDRANGEA Linnaeus
S.T. Buczacki edited by S.G. Knees
Deciduous or evergreen shrubs, small trees or climbers. Leaves usually rounded-ovate and toothed, opposite or in whorls of 3. Bark often flaking when mature. Fertile flowers bisexual (rarely unisexual), radially symmetric, in panicles or corymbs. Sepals 4 or 5, small, inconspicuous. Petals 4 or 5,

white, blue or pink. Stamens 8 or 10 (rarely more). Ovary inferior, 2–5-celled, containing many ovules. Fruit a 2–5-celled, many-seeded capsule. Many species also bear larger, sterile flowers borne at the outside of the corymb like inflorescences.

A genus of up to 100 species from China, Japan, the Himalaya, the Philippines, Indonesia, N and S America. Many variants have been bred and selected, those with sterile flowers having particular appeal. Many of the popular types of mop-head garden hydrangea are particularly favoured for seaside planting and adopt differing flower colours depending on the relative availability of aluminium ions in the soil. In alkaline soils where aluminium is unavailable, the natural flower colour is pink but a change to blue can be encouraged by supplying the plants with aluminium sulphate, known commercially as 'blueing powder'. Many species are prone to damage from late frosts. Propagate by cuttings or layers.

Literature: McClintock, E., Monograph of the genus Hydrangea, *Proceedings of the Californian Academy of Science* (1957); Haworth-Booth, M., The hydrangeas, edn 5 (1984); Mallet, C., Mallet, R. & Trier, H. van, Hydrangea: species and cultivars **1** (1992); Mallet, C. Hydrangeas: species & cultivars, **2** (1994); Lawson-Hall, T. & Rothera, B., Hydrangeas, a gardeners' guide (1995).

1a. Evergreen, clinging climber   2
 b. Deciduous shrub, tree or climber   3
2a. Flowers in a series of clusters one
    above the other   **16. serratifolia**
 b. Flowers in a single, terminal cluster
                          17. seemanni
3a. Clinging climber   4
 b. Shrub or small tree   5
4a. Leaves coarsely toothed; sterile
    flowers few; stamens usually fewer
    than 15   **8. anomala**
 b. Leaves finely toothed; sterile flowers
    numerous; stamens usually 15 or
    more   **9. petiolaris**
5a. Leaves deeply 5–7-lobed, like those
    of an oak   **2. quercifolia**
 b. Leaves usually toothed, but not
    deeply lobed   6
6a. Leaves small, very coarsely toothed,
    like those of a nettle   **11. hirta**
 b. Leaves fairly finely toothed   7
7a. Flowers in panicles   **12. paniculata**
 b. Flowers in corymbs   8
8a. Corymbs enclosed by c. 6 persistent
    bracts   **3. involucrata**

 b. Corymbs without persistent bracts   9
9a. Leaves hairless on both surfaces,
    often more or less shiny above   10
 b. Leaves at least slightly downy, hairy
    or bristly beneath   11
10a. Sterile flowers white, each 1–1.8 cm
    across   **1. arborescens**
 b. Sterile flowers pink, each 3–5 cm
    across   **14. macrophylla**
11a. Leaves with hairs confined to the
    veins beneath   12
 b. Leaves with hairs or bristles not
    confined to the veins beneath   15
12a. Corymbs with up to 12 pink or blue
    sterile flowers   **15. serrata**
 b. Corymbs with white, white-blue or
    cream sterile flowers   13
13a. Sepals toothed   **10. scandens**
 b. Sepals entire   14
14a. Sterile flowers more than 2 cm
    across   **13. heteromalla**
 b. Sterile flowers less than 2 cm across
                          1. arborescens
15a. Leaves densely softly downy beneath
                                          16
 b. Leaves hairy, bristly or downy
    beneath but not with a dense
    covering   17
16a. Fertile flowers dull white
                          1. arborescens
 b. Fertile flowers white-purple or pink
                          4. aspera
17a. Shoots densely covered with hairs
    and stiff bristles   **7. sargentiana**
 b. Shoots at most finely downy   18
18a. Corymbs to 30 cm across **6. robusta**
 b. Corymbs to 15 cm across   19
19a. Flower-stalks bristly; leaf-stalks 4–17
    cm   **5. longipes**
 b. Flower-stalks downy; leaf-stalks
    1.5–3 cm   **13. heteromalla**

1. H. arborescens Linnaeus. Illustration: *Botanical Magazine*, 437 (1799); Haworth-Booth, The hydrangeas, edn 5, 35 (1984); Davis, The gardener's illustrated encyclopaedia of trees and shrubs, 142 (1987); Taylor's guide to shrubs, 140 (1987).

Fairly loose, open, deciduous shrub, 1–3.5 m. Shoots at first downy, then hairless. Leaves 7.5–17.5 × 5–15 cm, broadly ovate, acuminate, with coarse, forward-pointing teeth, rather shiny, dark green above, paler beneath, hairless or slightly downy beneath on veins and in vein axils:, stalks 2.5–7.5 cm. Corymbs fairly flat, much branched, 5–15 cm across; sterile flowers absent or 1–8; sterile flowers long-stalked, creamy white, each 1–1.8 cm

across; fertile flowers numerous, small, dull white; flower-stalks downy. Capsule 8–10-ribbed. E USA. H2. Summer.

A variable species; the commonest cultivated variant is subsp. **arborescens** 'Grandiflora'; it was originally found wild in Ohio and has a large, cushion-like, heavy head of white sterile flowers only.

Subsp. **discolor** Seringe (*H. cinerea* Small) has tiny warts on the leaves, with sparse downy hairs beneath, and is usually seen as the cultivar 'Sterilis', in which most of the flowers are sterile.

Subsp. **radiata** (Walter) McClintock (*H. radiata* Walter) is a striking plant with darker green leaves and a unique indumentum of thick, downy, white hairs on the leaves beneath.

2. H. quercifolia Bartram (*H. platanifolia* invalid). Illustration: *Botanical Magazine*, 8447 (1912); *American Horticulturist* **54**: 15 (1975); *Practical Gardening* **1987** (August): 37; Davis, The gardener's illustrated encyclopaedia of trees and shrubs, 145 (1987); Brickell (ed.), RHS A–Z encyclopedia of garden plants, 554 (2003).

Fairly loose, rounded, deciduous shrub 1–2.5 m. Shoots thick, stout, at first finely reddish downy, then hairless and flaky. Leaves 7.5–20 × 5–17 cm, broadly ovate-rounded and deeply 5–7-lobed (like those of the oak, *Quercus rubra*: see volume **2**, p. 80 & Figure 51(8)), minutely toothed; stalks 2.5–6 cm. Panicles more or less erect, long-pyramidal, 10–25 cm, with numerous long-stalked, white sterile flowers, which gradually turn purplish, each 2.5–3.5 cm across; fertile flowers numerous, small, white. Flower-stalks loosely hairy. SE USA. H2. Summer.

3. H. involucrata Siebold. Illustration: Kitamura & Okamoto, Coloured illustrations of trees and shrubs of Japan, 197 (1977); Davis, The gardener's illustrated encyclopaedia of trees and shrubs, 143 (1987).

Fairly loose, open, deciduous shrub 1–1.2 m (to 2 m in milder areas). Shoots at first bristly-downy, later hairless. Leaves 7.5–15 × 2.5–6 cm, broadly ovate-oblong, acuminate, finely toothed, bristly, especially above; stalks 6–25 mm. Corymbs irregular, 7.5–12.5 cm across, at first enclosed by c. 6 broadly ovate bracts, which later open outwards and persist, covered with flattened, whitish down. Sterile flowers few, long-stalked, pale blue or faintly pink, each 1.8–2.5 cm across; fertile flowers numerous, small, blue. Flower-stalks slightly downy. *Japan, Taiwan*. H4. Late summer.

Most frequently seen in cultivation as the cultivar 'Hortensis' with more numerous, double, pink-white sterile flowers; it is of Japanese garden origin.

4. H. aspera Don (*H. fulvescens* Rehder; *H. kawakamii* Hayata; *H. rehderiana* Schneider; *H. villosa* Rehder). Illustration: Millar-Gault, The dictionary of trees and shrubs in colour, pl. 237 (1976); Davis, The gardener's illustrated encyclopaedia of trees and shrubs, 142, 146 (1987); *Amateur Gardening* **1987** (Nov 28) 11.

Spreading deciduous shrub or small tree to 4 m. Shoots at first with flattened or spreading hairs, later hairless and often peeling. Leaves 10–25 × 2.5–10 cm, mostly lanceolate to narrowly ovate, acute or acuminate, base rounded or tapered, with spreading or forward-pointing marginal teeth, densely covered with soft down beneath, sparsely hairy above; stalks 2.5–10 cm. Corymbs fairly flattened, to 25 cm across, with few to many white to pale pink or purple, darker-veined, sterile flowers, each with 4 rounded, toothed or entire sepals, each to 2.5 cm across. Fertile flowers small, numerous, white-purple or pink, each with 5 petals, falling early. Flower-stalks hairy. Capsule 2.5–3 mm, hemispherical. *Himalaya, W & C China, Taiwan, Indonesia (Java, Sumatra)*. H5. Summer.

A variable species in respect of overall habit (including some very robust, more or less climbing or scrambling variants), Most familiar in cultivation is the variant previously known as *H. villosa*.

Subsp. **strigosa** (Rehder) McClintock (*H. aspera* var. *macrophylla* Hemsley; *H. strigosa* Rehder; *H. strigosa* var. *macrophylla* (Hemsley) Rehder). Illustration: *Botanical Magazine*, 9324 (1933). As above, but leaves with short, stiff hairs beneath. *China*. H5. Summer–autumn.

5. H. longipes Franchet (*H. aspera* subsp. *robusta* misapplied). Illustration: Schneider, Illustriertes Handbuch der Laubholzkunde **2**: 939 (1912).

Loose, spreading, deciduous shrub, 2–2.5 m. Shoots at first loosely downy, later hairless. Leaves 7.5–17.5 × 3–9 cm, rounded-ovate, abruptly acuminate, base rounded-cordate, sharply toothed, bristly, especially beneath; stalks 4–17 cm. Corymbs fairly flat, 10–15 cm across.

Sterile flowers 8–9, each 1.9–2.4 cm across, white or faintly purple. Fertile flowers numerous, small, white. Flower–stalks bristly. Capsule rounded, hairless, with the calyx at the top. *C C & W China*. H5. Summer–autumn.

6. H. robusta Hooker & Thomson (*H. aspera* subsp. *robusta* (Hooker & Thomson) McClintock; *H. rosthornii* Diels). Illustration: Schneider, Illustriertes Handbuch der Laubholzkunde **2**: 939 (1912).

Loose, spreading deciduous shrub to 4 m, very like *H. longipes* but leaves larger, thicker, densely bristly beneath; corymbs to 30 cm across, with 20 or more large white sterile flowers and blue fertile flowers. *China*. H5. Summer–autumn.

7. H. sargentiana Rehder (*H. aspera* subsp. *sargentiana* (Rehder) McClintock). Illustration: *Botanical Magazine*, 8447 (1933); Bean, Trees & shrubs hardy in the British Isles, edn 8, **2**: 630 (1973); *Journal of the Royal Horticultural Society* **100**: 599 (1975); Davis, The gardener's illustrated encyclopaedia of trees and shrubs, 145 (1987).

Loose, spreading, deciduous shrub to 3 m. Shoots densely covered with small, erect hairs and stiff, translucent bristles. Leaves 10–25 × 5–18 cm, broadly ovate, base rounded, densely covered with short, velvety hairs above, densely bristly beneath; stalks 2.5–11 cm, bristly. Corymbs fairly flattened, 12.5–22.5 cm across, with a few, pink-white sterile flowers confined to the periphery, each to 3 cm across and with 4–5 entire, irregular sepals; fertile flowers numerous, small, pale purple. Flower-stalks hairy. *China (W Hubei)*. H5. Summer.

8. H. anomala D. Don (*H. altissima* Wallich).

Clinging, deciduous climber to 12 m. Shoots hairless or hairy, becoming very rough and peeling. Leaves 7.5–13 × 4–10 cm, ovate, shortly acuminate, base cordate, regularly toothed, hairless except for downy tufts in the vein-axils beneath; stalks 1.5–7.5 cm. Corymbs fairly flat, 15–20 cm across, with few, white, peripheral sterile flowers each 1.5–3.7 cm across and numerous, small, cream fertile flowers. *Himalaya, China*. H4. Early summer.

9. H. petiolaris Siebold & Zuccarini (*H. anomala* subsp. *petiolaris* (Siebold & Zuccarini) McClintock; *H. scandens* Maximowicz). Illustration: *Botanical*

Magazine, 6788 (1884); Bean, Trees & shrubs hardy in the British Isles, edn 8, **2**: 628 (1973); *Practical Gardening* **1986** (Nov): 16; Davis, The gardener's illustrated encyclopaedia of trees and shrubs, 145 (1987).

Clinging, deciduous climber to 20 m. Shoots at first finely hairy or hairless, later rough and peeling. Leaves 3.5–11 × 2.5–8 cm, ovate-rounded, shortly acuminate, base more or less cordate, finely toothed, hairless above, sometimes downy beneath, especially on the veins; stalks 5–40 mm. Corymbs flat, 15–25 cm across, with up to 12 white, peripheral sterile flowers each 2.5–4.5 cm across. Fertile flowers small, numerous, off-white. *Japan, former USSR (Sakhalin), Korea, Taiwan*. H3. Summer.

10. H. scandens (Linnaeus) de Candolle (*H. scandens* Maximowicz; *Viburnum scandens* Linnaeus; *V. virens* Thunberg). Illustration: Haworth-Booth, The hydrangeas, edn 5, 49 (1984).

Spreading or almost pendent shrub to 1 m. Shoots hairless or very finely downy. Leaves 5–9 × 2–4 cm, lanceolate or oblong-ovate, shortly toothed, hairless above, usually finely downy on the veins beneath; stalks *c.* 5 mm. Corymbs fairly flattened, to 7.5 cm across, often abundant, with few white-blue sterile flowers each 1.7–3.8 cm across, with toothed sepals. Fertile flowers numerous, white, with small, clawed petals. *S Japan*. H5. Summer.

Subsp. **chinensis** (Maximowicz) McClintock (*H. chinensis* Maximowicz; *H. davidii* Franchet; *H. umbellata* Rehder) is an altogether larger plant from the E Asian mainland and other areas outside Japan, with tough, woody twigs and more leathery leaves. The name *H. chinensis* covers a number of different variants, including forma **lobbii** Maximowicz from the Philippines with large, white sterile flowers and forma **macrosepala** Hayata from Japan.

11. H. hirta Siebold. Illustration: Schneider, Illustriertes Handbuch der Laubholzkunde **1**: 385, 386 (1904–06).

Very similar to *H. scandens* but plant with nettle-like leaves and smaller flowers. *Japan*. H5. Summer.

Occasionally grown but of little merit as a garden plant.

12. H. paniculata Siebold. Illustration: *Botanical Magazine*, n.s., 301 (1957); Huxley, Deciduous garden trees and shrubs, pl. 92 (1973); Davis, The gardener's illustrated encyclopaedia of trees and shrubs, 145 (1987).

Large deciduous shrub or small tree to 4 m (or, in very favourable conditions, 6 m). Shoots at first downy, then hairless. Leaves 7.5–15 × 3.7–7.5 cm, ovate, acuminate, rounded or sometimes tapered at the base, toothed, sparsely bristly above and on the veins beneath; stalks 1.2–2.5 cm. Panicles pyramidal, 15–20 × 10–13 cm wide at the base, with few white-pink sterile flowers each 1.7–3 cm across; fertile flowers numerous, yellow-white. Flower-stalks downy. *E & S China, Japan, former USSR (Sakhalin).* H3. Summer–autumn.

A variable species including some notable cultivated variants, especially the cultivar 'Grandiflora', in which almost all of the white-pink flowers are sterile, and which, with pruning, can produce panicles to 45 cm. Other widely grown cultivars are: 'Floribunda' with more sterile flowers than the species but fewer than 'Grandiflora'; 'Praecox' an earlier-flowering, more upright plant; 'Tardiva' a poorer but later-flowering variant and 'Unique' (illustration: Brickell (ed.), RHS A–Z encyclopedia of garden plants, 554, 2003), similar to 'Grandiflora' but more vigorous.

13. H. heteromalla D. Don (*H. dumicola* W.W. Smith; *H. hypoglauca* Rehder; *H. khasiana* Hooker & Thomson; *H. mandarinorum* Diels; *H. vestita* Wallich; *H. xanthoneura* Diels). Illustration: Schneider, Illustriertes Handbuch der Laubholzkunde **1**: 385, 389 (1904–06); Millar-Gault, Dictionary of trees and shrubs in colour, pl. 238 (1976); Haworth-Booth, The hydrangeas, edn 5, 57 (1984).

Deciduous shrub to 3 m. Shoots at first with short hairs, later hairless and smooth. Leaves 8.7–20 × 3–14 cm, variable, mostly narrowly ovate, base rounded, wedge-shaped or sometimes cordate, toothed and bristly at the margins, hairless above, downy beneath, at least on the veins; stalks 1.5–3 cm. Corymbs flattened, 15 cm across, with few white sterile flowers, each 2.5–5 cm across. Fertile flowers numerous, small, white. Flower–stalks downy. *Himalayas, W & N China.* H4. Summer.

Variable, especially in the size of the corymbs and colour of the flowers, some being markedly yellow. The cultivar 'Bretschneideri' (*H. pekinensis* invalid) is distinguished by its peeling bark.

14. H. macrophylla (Thunberg) Seringe (*H. hortensis* Siebold; *H. opuloides* (Lamarck) Anon.). Illustration: Siebold & Zuccarini, Flora Japonica, t. 52 (1845); Millar-Gault, Dictionary of trees and shrubs in colour, pl. 239–46 (1976); Davis, The gardener's illustrated encyclopaedia of trees and shrubs, 143, 144 (1987).

Fairly spreading deciduous shrub to 3 m. Shoots more or less hairless. Leaves broadly ovate, acute or acuminate, very coarsely toothed, shining and almost greasy above, smooth beneath; blades 10–20 × 6.5–14 cm, stalks 1.7–5 cm. Corymbs flattened, much branched, with few pink sterile flowers each 3–5 cm across and numerous small blue or pink fertile flowers; flower-stalks hairless. Capsule yellow-brown, erect, 6–8 × 1–3 mm, with 3 apical, diverging, woody styles 1–3 mm. *Japan.* H3. Summer.

The above description is of a maritime plant sometimes called *H. macrophylla* var. *normalis* Wilson (*H. maritima* Haworth-Booth) with the corymb type called 'Lace-cap'. It is considered to be the ancestor of many 'Hortensia' cultivars of both mop-head and lace-cap type. The mop-heads, such as the very common 'Generale Vicomtesse de Vibraye', have spherical corymbs almost wholly composed of sterile flowers. The plant described above is believed to be a true wild species, but it was its variety, grown in Japanese gardens and later called 'Sir Joseph Banks', together with certain old Japanese cultivars of hybrid origin involving *H. macrophylla* and other species, especially 'Otaksa', that were the plants first introduced to the West and from which many modern cultivars derive. 'Sea Foam', which arose as a branch-sport on 'Sir Joseph Banks', is the modern cultivar closest to the wild plant. In most garden variants the flower colour is influenced by the presence and availability of aluminium ions in the soil (see above). Some cultivars have variegated leaves.

Some cultivated lace-caps, such as the familiar 'Bluebird' are, however, derived from *H. serrata* (see below), considered by some authorities as a subspecies of *H. macrophylla*. It is possible that several others, both lace-cap and mop-head, are actually hybrids between these 2 species.

15. H. serrata (Thunberg) Seringe (*H. japonica* Siebold, in part; *H. macrophylla* var. *acuminata* (Siebold & Zuccarini) Makino; *H. macrophylla* subsp. *serrata* (Thunberg) Makino; *H. serrata* forma *acuminata* (Siebold & Zuccarini) Wilson). Illustration: Millar-Gault, Dictionary of trees and shrubs in colour, pl. 247–9 (1976); Hillier's manual of trees & shrubs, edn 5, 152 (1981); Davis, The gardener's illustrated encyclopaedia of trees and shrubs, 146 (1987); *Practical Gardening* **1987**(August): 37 (1987).

Spreading deciduous shrub to 2 m. Shoots at first finely downy, later hairless. Leaves 5–15 × 2.5–7 cm, lanceolate, acuminate, hairless above, veins beneath with short hairs. Corymbs flattened, 5–10 cm across, with up to 12 pink or blue sterile flowers, each 1–1.5 cm across, with 3 entire or variously toothed, variously shaped sepals. Fertile flowers numerous, small, pink or blue. Capsules 1–3 × 0.5–1 cm, erect, yellow-brown, with 3 tiny, apical, diverging woody styles. *Japan & Korea (Quelpaert Island).* H2. Summer.

Lace-cap cultivars derived from *H. serrata* include 'Bluebird' and 'Grayswood' and tend to be hardier than those of *H. macrophylla* origin. Some cultivars never have blue flowers, even in very acid soils. A smaller, compact variant, known as var. **thunbergii** Siebold (sometimes treated as a species, *H. thunbergii* Siebold) is occasionally seen, and differs in its very dark stems and leaves toothed only towards the apex, slightly hairy above.

16. H. serratifolia (Hooker & Arnott) Philippi (*H. integerrima* (Hooker & Arnott) Engler). Illustration: *Botanical Magazine*, n.s., 153 (1951).

Dioecious evergreen, clinging climber, to 30 m in favourable situations. Shoots at first with fine down, later hairless, and bearing aerial roots. Leaves 5–15 × 2.5–7.5 cm, elliptic, acuminate, base tapered, often cordate, almost always entire, leathery, hairless; stalks 5–45 mm. Inflorescences terminal and axillary, to 15 × 9 cm, composed of numerous small corymbs, each at first enclosed by 4 papery bracts. Flowers normally all small, fertile, white, although some variants exist with 1 or few white sterile flowers. *Chile, Argentina.* H4. Summer.

17. H. seemanni Riley.

Similar to *H. serratifolia* but with flowers in a single terminal cluster, sterile flowers sometimes present. *Mexico.* H5–G1.

Occasionally seen in cultivation.

7. SCHIZOPHRAGMA Siebold & Zuccarini
A.C. Whiteley

Deciduous climbing shrubs clinging by aerial roots. Leaves opposite, on long stalks. Flowers white in large terminal cymes, the central flowers small and hermaphrodite, outer ones sterile, reduced to a single large bract on a slender stalk. Fertile flowers with 4 or 5 sepals and petals. Stamens 10. Ovary inferior, united with the calyx tube, 10-celled. Stigma lobed. Fruit a ribbed capsule, dehiscing between the ribs.

A genus of 2 species from E Asia. They do well in most soils, climbing up walls or trees. Propagate by cuttings.

1a. Leaves coarsely toothed; inflorescence 20–25 cm across, bracts 2.5–4 cm **1. hydrangeoides**
  b. Leaves sparingly toothed; inflorescence to 30 cm across, bracts to 9 cm **2. integrifolium**

1. S. hydrangeoides Siebold & Zuccarini. Illustration: *Botanical Magazine*, 8520 (1913); Everett, The New York Botanical Gardens illustrated encyclopedia of horticulture, **9**: 3081 (1982); Krüssmann, Manual of cultivated broad-leaved trees & shrubs, **3**: pl. 106 (1986); Phillips & Rix, Shrubs, 237 (1989).

Stems climbing to 12 m. Leaves broadly ovate, base rounded, heart-shaped or tapering, 10–15 × 6–10 cm, with deep veins and coarse teeth, underside slightly glaucous, with silky hairs. Inflorescence 20–25 cm across, somewhat downy. Fertile flowers slightly scented. Sterile bracts terminal on main branches of cyme, ovate or heart-shaped, 2.5–4 cm, yellowish white. *Japan & Korea*. H4. Summer.

'Roseum' has pinkish bracts.

2. S. integrifolium (Franchet) Oliver (*S. hydrangeoides* var. *integrifolium* Franchet). Illustration: *Botanical Magazine*, 8991 (1923); Hillier colour dictionary of trees & shrubs, 278 (1981); Krussmann, Manual of cultivated broad-leaved trees & shrubs **3**: pl. 106 (1986).

Stems climbing to 10 m. Leaves ovate, acuminate, base heart-shaped or rounded, 7–17 × 4–11 cm, entire or with a few small teeth, hairy on the veins beneath. Inflorescence to 30 cm across. Fertile flowers *c.* 5 mm across. Sterile bracts terminal on main branches of cyme, narrowly ovate, to 9 × 4 cm, white with

darker veins. *S & China & Taiwan*. H4. Summer.

8. DECUMARIA Linnaeus
A.C. Whiteley

Deciduous or evergreen climbing shrubs clinging by aerial roots. Young shoots and buds downy. Leaves opposite, ovate, with or without shallow teeth. Flowers in terminal corymbs or panicles. Petals 7–10, oblong, white. Calyx-teeth 7–10, alternating with petals. Stamens 20–30. Ovary inferior with 7–10 cells. Stigma head-like. Fruit a ribbed capsule dehiscing between the ribs.

A genus of 2 species from SE United States and C China. They do well in most soils in sheltered positions. Propagate by cuttings or occasionally seed.

1a. Deciduous; leaves 7–12 cm **1. barbara**
  b. Evergreen; leaves 3–8 cm **2. sinensis**

1. D. barbara Linnaeus. Illustration: Gleason, New illustrated flora of the northeastern United States and adjacent Canada, **2**: 272 (1952); Krüssmann, Manual of cultivated broad-leaved trees & shrubs **1**: 426 (1984); Phillips & Rix, Shrubs, 237 (1989).

Deciduous to 10 m. Leaves 7–12 × 3–8 cm, slightly downy beneath when young, entire or shallowly toothed towards the tip, leaf-stalk 2.5–5 cm. Flowers *c.* 6 mm across in corymbs 5–8 cm long and wide. Petals 7–10, narrowly oblong, white. Fruit *c.* 8 mm. *SE United States*. H5. Summer.

2. D. sinensis Oliver. Illustration: *Botanical Magazine*, 9429 (1936).

Evergreen to 4 m. Leaves 3–8 × 1–4 cm, hairless, glossy, leaf-stalk 6–20 mm. Flowers *c.* 5 mm across in panicles 4–9 cm long and wide. Petals 7–10, oblong, yellowish white. Stamens prominent. *C China*. H5. Summer.

9. CARDIANDRA Siebold & Zuccarini
M.F. Gardner

Deciduous small shrubs. Leaves alternate, lanceolate, with coarse forward-pointing teeth; stipules absent. Flowers numerous, borne in loose corymbs, outer flowers sterile with 3 petal-like calyx lobes; inner flowers fertile, sepals 4 or 5, triangular-ovate; petals 5, overlapping; stamens numerous; ovary inferior, 3-celled; styles 3. Capsules egg-shaped, crowned by

calyx-limb and styles, opening between the styles, many-seeded, seeds flattened, winged on each side.

A genus of 5 species from China, Japan, and Taiwan.

1. C. alternifolia Siebold & Zuccarini. Illustration: Schneider, Handbuch der Laubholzkunde **1**: 383, 385 (1904–06); Satake et al., Wild flowers of Japan **2**: pl. 146 (1985); Hayashi, Azegami & Hishiyama, Wild flowers of Japan, 432 (1983).

Stems semi-woody, 45–75 cm, downy towards the ends. Leaves 5–15 × 2–2.5 cm, slender-pointed at both ends, with forward-pointing teeth. Flowers borne in terminal, long-stalked corymbs, 7.5–10 cm across, sepals triangular, blunt; petals *c.* 3 mm, white becoming rose-lilac; capsules broadly egg-shaped, *c.* 4 mm. *Japan*. G1. Summer.

10. CARPENTERIA Torrey
D.M. Miller

Evergreen bushy shrubs to 6 m, stems angled, branches pithy. Leaves opposite, to 11 × 2.5 cm, simple, lanceolate, entire, acute, base narrowed, hairless and bright green above, glaucous with short white hairs beneath; stalk to 5 mm. Flowers 5–7 cm across, bisexual, fragrant, in terminal clusters of 3–7. Sepals usually 5, to 10 × 6 mm, ovate, downy. Petals 5, to 2.5 × 2.5 cm, circular, white. Stamens numerous with conspicuous yellow anthers. Style 1, 5–7-lobed. Ovary superior. Fruit a conical capsule containing numerous seeds.

A genus of a single species from western USA, which grows best in a well-drained, sandy soil in a sunny, sheltered position. Pruning is not necessary, but dead or weak shoots should be cut out. It may be grown from seed, but some seedlings may eventually have inferior flowers. Variants with larger flowers, such as 'Ladham's Variety' should be propagated by layering or by cuttings rooted in mist.

1. C. californica Torrey. Illustration: *Botanical Magazine*, 6911 (1886); *Gardeners' Chronicle* **40**: f. 5 (1906); Bean, Trees & shrubs hardy in the British Isles, edn 8, **1**: pl. 28 (1970).

W USA (California). H4. Early summer.

11. DICHROA Loureiro
A.C. Whiteley

Evergreen shrubs. Leaves opposite, simple, usually coarsely toothed. Flowers in terminal panicles, white, blue or pink.

Calyx-lobes 5, rarely 4 or 6. Petals 5, sometimes 6, usually more brightly coloured on inner surfaces. Stamens 10. Ovary inferior, 4-celled. Styles 4. Fruit a small berry. Seeds numerous.

A genus of around 13 species in SE Asia and Indonesia. One species is cultivated. It requires a frost-free climate and will succeed in most soils given ample moisture and slight shade. Propagate by seed or cuttings.

1. D. febrifuga Loureiro (*D. cyanitis* Miquel; *Adamia cyanea* Wallich). Illustration: *Botanical Magazine*, 3046 (1841); Everett, The New York Botanical Gardens illustrated encyclopedia of horticulture **4**: 1061 (1981).

Shrub to 2.5 m. Leaves elliptic, acuminate, tapering gradually to the base, coarsely toothed, 10–20 × 4–9 cm, the veins beneath often reddish. Leaf-stalks 1–2 cm, often reddish. Flowers in panicles to 15 × 20 cm. Petals white to blue or pink, somewhat fleshy, *c.* 5 × 2 mm, boat-shaped. Calyx-lobes triangular, *c.* 1 mm. Stamens blue or violet, *c.* 5 mm. Receptacle green to pink. Fruit spherical, blue, to 5 mm wide. *SE Asia.* H5. Spring–autumn.

12. FENDLERA Engelmann & Gray
D.M. Miller
Deciduous bushy shrubs with ribbed branches. Leaves opposite, ovate to lanceolate, entire, more or less stalkless, 1–3-veined. Flowers bisexual, 1–3, stalked, on short, lateral branches. Sepals 4, small. Petals 4, clawed, ovate with toothed margin. Stamens 8. Styles 4, completely free. Ovary half-inferior. Fruit a many-seeded capsule opening by 4 flaps, surrounded by the persistent calyx.

A genus of 2 or 3 species from SW USA and Mexico. They should be grown in a hot, sunny location in well-drained soil; moist or cool conditions should be avoided. Old flowering branches and weak or crowded shoots should be thinned after flowering. Propagate from seed or by half-ripe cuttings under mist with gentle heat.

1a. Leaves green, more or less hairless or bristly beneath   **1. rupicola**
  b. Leaves with thick, grey-white, densely felted hairs beneath
                  2. wrightii

1. F. rupicola Gray. Illustration: Bailey, Standard cyclopedia of horticulture **2**: f. 1480 (1915).

Shrub to 2 m with downy, grey-yellow young bark. Leaves 2–3 × *c.* 1 cm on vegetative shoots, lanceolate to narrowly oblong, 3-veined, stalkless, rough above, but smaller, linear and clustered on flowering-shoots. Flowers usually solitary, to 3 cm across, fragrant, on short side-branches. Sepals very small. Petals *c.* 1 cm, spreading, white or pink-tinged. Stamens half as long as petals. Capsule more than 1 cm. *SW USA, N Mexico.* H3. Late spring–early summer.

2. F. wrightii (Gray) Heller (*F. rupicola* var. *wrightii* Gray). Illustration: *Botanical Magazine*, 7924 (1903).

Very similar to *F. rupicola*, but leaves and flowers smaller, generally more hairy. *SW USA (Texas to New Mexico), NW Mexico.* H3. Late spring–early summer.

13. PLATYCRATER Siebold & Zuccarini
M.F. Gardner
Deciduous shrub, sometimes prostrate with papery bark. Leaves opposite, 10–15 × 3–7 cm, oblong to broadly lanceolate, sparsely hairy above, hairy below, with forward-pointing teeth, with slender points; leaf-stalks 5–30 cm. Flowers with long slender stalks, of 2 types: sterile ornamental, at the top of cyme branches, 1–3 cm across, calyx large, shallowly 3- or 4-lobed; fertile, less showy, flowers lower down, 2–3 cm across, calyx smaller, deeply 4–lobed to near the base, sepals ovate-lanceolate, petals 4, white, very thick. Stamens numerous; styles 2, persistent; ovary inferior; capsules cone-shaped.

A genus of a single species native to Japan, which prefers a cool, moist site in partial shade.

1. P. arguta Siebold & Zuccarini (*P. serrata* Makino). Illustration: Siebold, Flora Japonica **1**: 27 (1837); Schneider, Illustriertes Handbuch der Laubholzkunde **1**: 383–5 (1904–06); *Journal of Japanese Botany*, **61**: 70 (1986); Hayashi, Woody plants of Japan, 228 (1985).
Japan (Honshu, Shikoku, Kyushu). H5. Summer.

14. PILEOSTEGIA Hooker & Thomson
A.C. Whiteley
Evergreen climbing or prostrate shrubs, clinging by aerial roots. Leaves opposite, simple, usually entire, leathery. Flowers small, white, in terminal panicles. Calyx cup-shaped with 4 or 5 short lobes. Petals 4 or 5, coherent, falling quickly. Stamens 8–10, prominent. Ovary inferior, 4- or

5-celled. Fruit a spherical ribbed capsule, dehiscing between the ribs.

A genus of 3 species from E Asia. One species is cultivated. It does well in most soils, climbing up walls or trees. Propagate by cuttings.

1. P. viburnoides Hooker & Thomson (*Schizophragma viburnoides* (Hooker & Thomson) Stapf). Illustration: *Botanical Magazine*, 9262 (1929); Bean, Trees & shrubs hardy in the British Isles, edn 8, **3**: pl. 31 (1976); Miller-Gault, Dictionary of trees and shrubs in colour, fig. 322 (1976); Phillips & Rix, Shrubs, 232 (1989).

Stems climbing to 6 m. Leaves entire, soon hairless, narrowly oblong to ovate, acute, 6–15 × 2–6 cm, leaf-stalk 5–25 mm. Flowers creamy white, *c.* 9 mm across in panicles 10–15 × 10–15 cm with a central axis and several pairs of dichotomous branches. Stamens 5 mm. *Himalaya, China & Taiwan.* H4. Autumn.

---

143. ESCALLONIACEAE

Trees or shrubs. Leaves mostly alternate, evergreen, exstipulate, usually with gland-tipped teeth. Flowers in racemes, actinomorphic, bisexual. Calyx, corolla and stamens perigynous or epigynous. Calyx with 4–6, usually 5 lobes. Corolla of 4–6 (usually 5) free petals. Stamens 4–6, usually 5. Ovary usually of 2 united carpels; ovules many, parietal; style 1, somewhat bilobed at the apex. Fruit a capsule or berry.

There are about 15 genera with about 70 species. Some species and hybrids of the largest genus, *Escallonia*, are cultivated as ornamental shrubs and hedging plants. The family is often further divided (*Iteaceae*, *Brexiaceae*, etc.), but there is little agreement as to how this should best be done.

1a. Ovary superior   2
  b. Ovary inferior   5
2a. Ovary 1- or 2-celled; flowers in unbranched racemes   3
  b. Ovary 5-celled; flowers in umbels or branched panicles   4
3a. Leaves with large, rounded teeth; flower parts in 6s or 9s; ovary 1-celled   **7. Anopterus**
  b. Leaves finely toothed or with spiny margins; flower parts in 5s; ovary 2-celled   **2. Itea**

4a. Flowers *c.* 8 mm across, in panicles; stamens 5 or 6, not alternating with teeth of the fleshy disc **4. Abrophyllum**
b. Flowers to 3.5 cm across, in umbels; stamens 5, alternating with teeth of the fleshy disc **3. Brexia**
5a. Corolla yellow; leaves with a dense covering of T-shaped hairs beneath; fruit a berry surrounded by the persistent calyx **8. Corokia**
b. Combination of characters not as above 6
6a. Evergreen or deciduous shrubs; leaf-stalks at most 1 cm **6. Escallonia**
b. Evergreen trees, or, if shrubs, then leaf-stalks *c.* 2 cm 7
7a. Leaves 6 cm or more, with glandular-hairy margins **1. Quintinia**
b. Leaves up to 6 cm, margins not glandular-hairy **5. Carpodetus**

1. QUINTINIA de Candolle, A.

M.F. Gardner

Evergreen trees or shrubs. Leaves leathery, alternate, entire or with obscurely forward-pointing teeth with marginal glandular hairs. Flowers borne in many-flowered terminal and axillary racemes or many-flowered panicles; petals 5, white or pale lilac, oblong, spreading or turned-back, overlapping; stamens 5; ovary inferior, 3–5 celled, stigma head-like, 3–5 lobed. Capsules 3–5 ribbed, opening at the top; seeds winged on each side.

A genus of 4 species from C Malaysia, New Guinea, Australia and New Zealand. Successful cultivation is achieved in the cool glasshouse in a loam-enriched soil. Propagate by seed or cuttings.

1a. Petals white; leaf-margins wavy, entire **2. acutifolia**
b. Petals pale lilac; leaf-margins wavy, toothed **1. serrata**

1. Q. serrata A. Cunningham. Illustration: *Hooker's Icones Plantarum* **6**: 558 (1843); Moore & Irwin, The Oxford book of New Zealand plants, 79 (1978); Salmon, The native trees of New Zealand, 196 (1980); Salmon, Field guide to the native trees of New Zealand, 104 (1986).

Tree to *c.* 7 m, shoots clammy. Leaves often blotched, 6–12.5 × 1–2.5 cm, oblong, with coarsely forward-pointing teeth and wavy. Racemes 6–8 cm, flower-stalks *c.* 4 mm; petals 2.5–3 cm, pale lilac.

Capsules 4–5 mm, obovoid. *New Zealand.* G1. Summer.

2. Q. acutifolia Kirk. Illustration: Salmon, The native trees of New Zealand, 194–5 (1980); Salmon, Field guide to the native trees of New Zealand, 104 (1986).

Tree or shrub to 8 m. Leaves yellowish with green veins, 6–16 × 3–5 cm, broadly elliptic-obovate to wedge-shaped, margins wavy, leaf-stalks to 2 cm. Racemes 4–7 cm, flower-stalks *c.* 3 mm; petals 3–3.5 mm, white. Capsules 4–6 mm, obovoid or oblong. *New Zealand.* G1. Summer.

2. ITEA Linnaeus

D.M. Miller

Deciduous or evergreen trees or shrubs. Branches with chambered pith. Leaves alternate, entire or toothed. Flowers small, bisexual, radially symmetric, in terminal or axillary racemes or spikes. Sepals 5, persistent. Petals 5, linear. Stamens 5. Styles 2, united; ovary superior, 2-celled. Fruit a many-seeded capsule.

A genus of about 10 species from N America to E Asia. They grow in most garden soils, unless very dry, in sun or partial shade. Pruning is not essential, except for the removal of dead or crowded branches. Propagate by cuttings taken from ripened wood, or seed if available.

1a. Deciduous shrub **1. virginica**
b. Evergreen shrub 2
2a. Leaves with sharp, spiny, holly-like teeth, ovate to rounded; flowers greenish white in mid- to late summer **2. ilicifolia**
b. Leaves with fewer, fine spines, ovate to lanceolate; flowers dull white in early to mid summer **3. yunnanensis**

1. I. virginica Linnaeus. Illustration: *Botanical Magazine*, 2409 (1823); Bean, Trees & shrubs hardy in the British Isles, edn 8, **2**: pl. 59 (1973); Krüssmann, Manual of cultivated broad-leaved trees & shrubs **2**: pl. 70 (1986).

Deciduous shrub 1–2 m with rather upright habit. Leaves to 10´3 cm, narrow-ovate to ovate, somewhat downy beneath, margins finely toothed; stalk to 1 cm. Flowers fragrant, in dense, cylindric, erect racemes 6–15 cm. Sepals to 2 mm, lanceolate, downy. Petals to 5 mm, narrow, white. Stamens as long as petals. Seeds to 5 mm or more, borne in narrow, downy capsules. *E USA.* H2. Early summer.

2. I. ilicifolia Oliver. Illustration: *Botanical Magazine*, 9090 (1925); Bean, Trees & shrubs hardy in the British Isles, edn 8, **2**: pl. 57 (1973).

Evergreen shrub to 4 m or more. Leaves to 9 × 7 cm, broadly ovate to rounded, apex usually rounded with a short point, margins with stiff spines, dark glossy green above, with small tufts of hairs in the vein-axils beneath; stalk to 1 cm. Flowers in very narrow, pendent racemes to 30 cm. Sepals minute. Petals to 3 mm, narrow, greenish white. *C China.* H5. Mid–late summer.

3. I. yunnanensis Franchet. Illustration: *Iconographia cormophytorum Sinicorum* **2**: 119 (1972).

Evergreen shrub to 3 m. Leaves to 11 × 4 cm, ovate to lanceolate, apex acute, spine-tipped, margins with fine, spiny teeth, dark glossy green above, more or less hairless on both surfaces; stalk to 1.5 cm. Flowers in narrow arching racemes to 18 cm. Petals to 3 mm, very narrow, dull white. Stamens and styles shorter than petals. *W China (Yunnan).* H5. Early–midsummer.

3. BREXIA Thouars

A.C. Whiteley

Evergreen shrub or small tree to 6 m with grey bark. Leaves alternate, oblanceolate to oblong, leathery, obtuse, entire or with a few teeth (especially on young plants), 10–15 × 3–4 cm, with short, stout leaf-stalks. Flowers *c.* 3.5 cm across in umbels of 6–20 on axillary stalks. Calyx of 5 united, rounded lobes. Petals 5, oblong, *c.* 1.2 cm, greenish white. Stamens 5, with fleshy filaments nearly as long as petals, alternating with teeth of the prominent epigynous disc. Ovary superior, conical, 5-celled. Stigma head-like with 5 lobes. Fruit oblong, *c.* 5 cm, woody, indehiscent. Seeds numerous, black.

A genus of a single, somewhat variable species needing a heated greenhouse, but seedlings sometimes used for tropical effect in summer bedding. Propagate by seed or cuttings.

1. B. madagascariensis (Lamarck) Ker–Gawler. Illustration: *Edwards's Botanical Register*, 730 (1823); Nicholson, Illustrated dictionary of gardening **1**: 211 (1884); Nicholson & Mottet, Dictionaire practique d'horticulture et de jardinage, **1**: 413 (1892).

E tropical Africa, Madagascar & Seychelles. G2. Spring–summer.

4. ABROPHYLLUM J.D. Hooker
M.F. Gardner

Shrub or tree. Leaves alternate, ovate to lanceolate, pointed, with forward-pointing teeth. Flowers borne in terminal or axillary panicles, sepals 5 or 6, deciduous; petals 5 or 6, spreading; stamens 5 or 6; ovary superior, 5-celled, stigma stalkless, 5-lobed. Fruit a berry, crowned by a stigma, many-seeded.

A genus of 2 species from E Australia. The cultivated species is an attractive plant suited to a cool, moist, woodland site, in a well-drained soil. Propagate from freshly sown seed or cuttings. Not very widely cultivated.

1. A. ornans J.D. Hooker. Illustration: *Hooker's Icones Plantarum*, pl. 1323 (1880); Jones, Ornamental rainforest plants in Australia, 196 (1986).

Shrub or tree, 3–6 × 2–5 m, young shoots hairy. Leaves 10–22 × 4–10 cm, dark green, toothed in upper part. Flowers white to yellowish, fragrant, each flower *c.* 8 mm across; filaments very short, anthers broadly oblong; Fruit a berry, purplish black, *c.* 8 mm across, spherical, many-seeded. *Australia (New South Wales & Queensland).* G1. Summer.

5. CARPODETUS J.R. Forster & G. Forster
A.C. Whiteley

Shrubs or small trees. Leaves alternate, without stipules. Flowers small, in few-flowered panicles. Calyx-tube united with ovary, bearing 5 or 6 deciduous lobes. Petals and stamens 5 or 6, alternating, inserted at the margin of an epigynous disc. Ovary half-inferior. Stigma capitate. Fruit a spherical, indehiscent capsule with 3–5 cells. Seeds numerous.

A genus of 10 species, 9 in New Guinea and 1 endemic to New Zealand, which is occasionally cultivated. It requires a moist, acid to neutral soil, rich in humus. Propagate by seed or cuttings.

1. C. serratus J.R. Forster & G. Forster. Illustration: Laing & Blackwell, Plants of New Zealand, 187 (1907); Cheeseman, Illustrations of the New Zealand flora **1**: pl. 41 (1914); Salmon, New Zealand flowers & plants in colour, 62 (1967); Irwin, The Oxford book of New Zealand plants, 79 (1978).

Evergreen tree to 10 m. Stems with prominent lenticels, downy when young. Adult leaves elliptic, somewhat leathery,

acute or obtuse with a few small teeth, 4–6 × 2–3 cm. Young leaves 1–3 × 1–2 cm, on zig-zag shoots. Leaf-stalks to 1 cm, downy. Flowers 5–6 mm across, in panicles to 5 × 5 cm. Calyx-lobes *c.* 1 mm. Petals ovate, 3–4 mm, white. Capsule 4–6 × 4–6 mm. *New Zealand.* H5. Summer.

6. ESCALLONIA Mutis
P.G. Barnes & A.C. Whiteley

Evergreen or rarely deciduous shrubs, occasionally small trees; branches, leaves and inflorescences hairy, hairless or with stalked glands. Leaves alternate, usually somewhat leathery, toothed. Flowers in short racemes, panicles or solitary in axils of upper leaves. Calyx-tube short, lobes 5, erect or spreading; petals 5, obovate or circular and spreading or with an erect long narrow claw and spreading circular apical portion, the latter type of flower appearing tubular. Stamens 5, alternating with and about as long as petals. Ovary inferior, 2- or 3-celled; style 1 with a capitate or 2-fid stigma, or rarely 2-fid at apex with 2 distinct stigmas; disk conical or cushion-shaped and surrounding base of style, or flat. Fruit a capsule with numerous seeds.

A genus of 39 species, native to S America, easily cultivated in most soils, in a sunny situation. Many are not very hardy in temperate areas. Resistant to maritime conditions and useful for hedges or as specimen shrubs. Propagate by cuttings.

Literature: Sleumer, H. Die Gattung Escallonia (Saxifragaceae), Verhandelingen der Koninklijke nederlandsche Akademie van Wetenschappen, afdeeling natuurkunde **2**(58) No. 2 (1968).

1a.	Flowers open; petals spreading, not clawed 2
b.	Flowers appearing tubular; petals erect with a long narrow claw, stamens included 5
2a.	Leaves 4–10 cm 3
b.	Leaves less than 4 cm 4
3a.	Flowers in slender racemes **1. pulverulenta**
b.	Flowers in broad panicles **11. bifida**
4a.	Leaves deciduous **4. virgata**
b.	Leaves evergreen **7. leucantha**
5a.	Petals pink-red 6
b.	Petals white, rarely pale pink 8
6a.	Leaves less than twice as long as wide; disc taller than wide **6. rubra**
b.	Leaves more than 2 times longer than wide 7

7a.	Leaves 3–7.5 × 1–2.5 cm; disk almost flat **9. laevis**
b.	Leaves 1.5–3 cm; disk as tall as wide **3. alpina**
8a.	Leaves 1–2.5 cm, branchlets winged **2. rosea**
b.	Leaves usually more than 3 cm, branchlets smooth or angled 9
9a.	Leaves clearly hairy, at least beneath **5. revoluta**
b.	Leaves hairless or almost so 10
10a.	Leaves sticky when young, strongly scented **8. illinita**
b.	Leaves not sticky when young, not strongly scented **10. tucumanensis**

1. E. pulverulenta (Ruiz & Pavon) Persoon. Figure 96(1), p. 168. Illustration: Sweet, British flower garden **7**: pl. 310 (1837).

Evergreen shrub to 4 m or more. Branches slightly angled, hairy, sticky when young. Leaves 5–10 × 2–4 cm, oblong or elliptic, apex rounded, base abruptly narrowed to a short stalk. Young leaves with dense, soft hairs, persisting beneath, margin with fine sharp teeth. Inflorescence a narrow terminal raceme, 10–20 × 2–3 cm. Flower-stalks 2–4 mm, hairy. Receptacle 1–2 mm. Calyx-lobes 1–1.5 mm, narrowly trianglular. Petals 3–5 × *c.* 2 mm, obovate, erect or spreading, white. Stamens and style 2–2.5 mm. Disk flat, the margin 5-lobed. Capsule 4–6 mm, obovoid. *Chile.* H5. Summer.

2. E. rosea Grisebach (*E. pterocladon* Hooker; *E. montana* Philippi). Figure 96(2), p. 168. Illustration: *Botanical Magazine*, 4827 (1855); Phillips & Rix, Shrubs, 222 (1989); Krüssman, Manual of cultivated broadleaved trees & shrubs **2**: f. 26c (1986).

Evergreen shrub to 3 m. Branchlets downy, reddish, distinctly angled with sinuous wings. Leaves 8–25 × 4–8 mm, narrowly obovate, acute, tapering to a short stalk, hairless except for the midrib above, glossy green, margin toothed, somewhat glandular. Inflorescence a stiff axillary raceme of 10–15 fragrant flowers, each borne in the axil of a leaf, or, towards the tip, a leafy bract. Flower-stalks short, reddish, with 2 narrow glandular-toothed bracteoles. Receptacle 1.5 mm. Calyx-tube 1–2 mm, lobes 1.5 mm, triangular, glandular-toothed. Petals 2.5–3 × *c.* 3 mm, erect; claw 5–6 × 1–1.5 mm, ovate, white, sometimes tinged with red, reflexed. Stamens 6–7 mm. Disk cylindric, slightly lobed, 2–2.5 mm. Style *c.* 8 mm, stigma

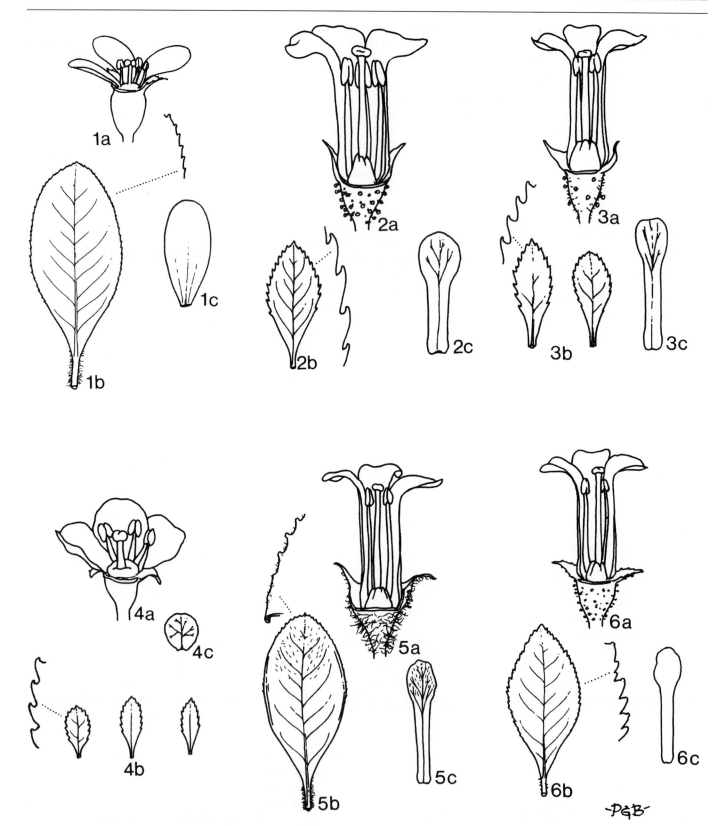

Figure 96. Diagnostic details of *Escallonia* species (a, flower; b, leaf; c, petal). 1a, b, c, *E. pulverulenta*. 2a, b, c, *E. rosea*. 3a, b, c, *E. alpina*. 4a, b, c, *E. virgata*. 5a, b, c, *E. revoluta*. 6a, b, c, *E. rubra*.

capitate to peltate, 1–1.5 mm across. *S Chile.* H4. Summer.

3. E. alpina de Candolle (*E. fonkii* Philippi). Figure 96(3), p. 168. Illustration: *Botanical Magazine*, n.s., 642 (1973).

Evergreen shrub 1–3 m, hairless or clothed with short, erect hairs. Shoots reddish at first, sparsely glandular, becoming grey, bark flaking. Leaves to 2.7 × 1.2 cm, spathulate to obovate, tip rounded with a sharp point, base wedge-shaped, stalkless, glossy green above with minute down on the midrib towards the base, paler beneath, sharply toothed towards the tip, each tooth gland-tipped. Inflorescence an axillary raceme of 4–15 flowers, leafy at the base and with leaf-like bracts towards the tip. Receptacle *c.* 2 mm. Calyx-tube *c.* 1.5 mm, lobes 2–3 mm, triangular, somewhat hairy and glandular. Petals erect, red; claw 6–7 × 1.5 mm, reflexed. Stamens *c.* 8 mm. Disk conical, slightly lobed. Style 7–8 mm, stigma capitate, *c.* 1.5 mm across. Capsule conical to spherical, 4–5 mm. *Chile, Argentina.* H4. Summer.

4. E. virgata (Ruiz and Pavon) Persoon (*E. philippiana* (A. Engler) Masters. Figure 96(4), p. 168. Illustration: Bean, Trees & shrubs hardy in the British Isles, edn 8, **2**: 125 (1973); Krüssmann, Manual of cultivated broad-leaved trees & shrubs **2**: f. 26a & e (1986); Phillips & Rix, Shrubs, 223 (1989).

Deciduous shrub to 2 m. Branches erect, lateral shoots short, angled, with very short hairs. Leaves 1–2 × 0.4–0.6 cm, obovate, hairless, apex acute or obtuse, almost stalkless, margin somewhat toothed. Flowers solitary or few in the upper leaf axils and terminal on short shoots, forming a leafy raceme. Receptacle 1.5–2.5 mm. Calyx-lobes *c.* 2 mm, triangular-ovate, with glandular teeth. Petals 3–6 × 2–3 mm, rounded, spreading, white or pale pink. Stamens and style 1.5–2 mm. Capsule obovoid, 4–6 mm. Disk flat. *Chile & S Argentina.* H4. Summer.

Hybrids with *E. rubra* are named **E. × langleyensis** Veitch. Figure 97(6), p. 170. The numerous cultivars vary considerably between the parental extremes. All are evergreen. The cultivar 'Apple Blossom' is typical. Illustration: Hay & Beckett, Reader's Digest encyclopedia of garden plants and flowers, edn 4, 260 (1987); Phillips & Rix, Shrubs, 222 (1989).

5. E. revoluta (Ruiz & Pavon) Persoon. Figure 96(5), p. 168. Illustration: *Botanical Magazine*, 6949 (1887).

Evergreen shrub to 6 m. Branchlets downy to densely hairy. Leaves obovate, 2–5 × 1.2–3 cm, acute to obtuse, narrowed to a short stalk, more or less downy on both surfaces, margin sharply toothed towards tip, somewhat curved under. Inflorescence a panicle or raceme, 4–10 cm, terminating a side-branch. Flower-stalks short, each with a small bracteole at the base. Receptacle 2–3 mm. Calyx 2–3 mm, lobes *c.* 2 mm, hairy, narrow, acute. Petals white, erect, claw 1.1–1.2 cm × 0.2–0.4 mm, tip ovate, *c.* 3 × 2 mm. Stamens *c.* 9 mm. Disk 1–2 mm, conical. Style *c.* 1.2 cm, stigma peltate to capitate. Capsule 4–5 mm, conical, hairy. *Chile.* H4. Late summer–autumn.

6. E. rubra (Ruiz & Pavon) Persoon (*E. punctata* de Candolle; *E. rubra* var. *punctata* de Candolle) Hooker. Figure 96(6), p. 168. Illustration: *Botanical Magazine*, 6599 (1881); Krüssmann, Manual of cultivated broad-leaved trees & shrubs **2**: f. 26d (1986); Phillips & Rix, Shrubs, 222 (1989).

Evergreen shrub 2–4 m. Branches slightly hairy, with conspicuous stalked glands. Leaves 2–8 × 1.5–4 cm, ovate to obovate, blunt or short acuminate, narrowed into a short stalk, glossy above, glandular-spotted beneath, margin coarsely toothed. Flowers in loose terminal and lateral panicles. Flower-stalks slender, 5–8 mm. Receptacle *c.* 4 mm. Calyx-lobes 2–3 mm, with stalked glands. Petals 1–1.2 cm, including a long claw *c.* 2 mm wide, erect except for the broader spreading tip, deep pink to deep red. Stamens and style *c.* 9 mm, disk conical, 5-lobed at tip. Capsule 7–8 mm, obovoid. *Argentina, Chile.* H4–5. Summer.

Var. **macrantha** (Hooker & Arnott) Reiche. Leaves 3–8 × 2–4 cm, narrowly to broadly elliptic, deeply toothed. Petals 1.5–1.8 cm, claw *c.* 4 mm wide. Stamens and style 1.2–1.4 mm, capsule 6–7 mm. *Chile.* H5. Summer–autumn.

7. E. leucantha Rémy (*E. bellidifolia* Philippi). Figure 97(1), p. 170.

Evergreen shrub to 5 m. Shoots downy. Leaves 1.2–2.5 × 0.8–1.2 cm, obovate to oblanceolate, acute or obtuse, tapered to a short stalk, hairless except on the midrib beneath, margin finely toothed. Inflorescence a large leafy panicle to 30 × 15 cm, composed of short terminal

racemes on side-shoots. Flower-stalks short, downy. Receptacle *c.* 1 mm. Calyx-lobes short, triangular, finely hairy. Petals *c.* 7 × 3 mm, spathulate, spreading, white. Stamens *c.* 5 mm. Style 6–7 mm, stigma peltate. Disk flat, slightly lobed. Capsule 3–4 mm, conical to bell-shaped. *Chile.* H4. Summer.

8. E. illinita Presl (*E. grahamiana* Hooker & Arnott). Figure 97(2), p. 170. Illustration: *Edwards's Botanical Register* **22**: pl. 1099 (1836).

Evergreen shrub to 3 m or more. Young branches densely glandular. Leaves 2–6 × 1–3 cm, obovate or elliptic, glossy above, sticky when young and strongly scented, narrowed to a long stalk 3–6 mm. Inflorescence a panicle, 1–1.2 cm, branches slightly hairy and glandular, bracts leaf-like. Receptacle *c.* 3 mm. Calyx-lobes 2.5 mm, awl-shaped. Petals white, erect, claw 9–11 × *c.* 2 mm, tip *c.* 3 mm wide, spreading. *Chile.* H4–5. Summer.

9. E. laevis (Vellozo) Sleumer (*E. organensis* Gardner). Figure 97(3), p. 170. Illustration: *Botanical Magazine*, 4274 (1846); Graf, Exotica, series 4, edn 12, 2042 (1985).

Evergreen shrub to 2 m. Branchlets stout, angled, glandular. Leaves to 7.5 × 2.5 cm, obovate to oblong, obtuse, tapering to a short reddish stalk, hairless, margin toothed towards the tip, often reddish. Inflorescence a short, dense terminal panicle. Receptacle 2–3 mm. Calyx-tube *c.* 1 mm, lobes *c.* 3 mm, narrowly triangular, hairless, sometimes glandular. Petals erect, red, claw *c.* 10 × 2 mm, tip *c.* 5 × 4 mm, ovate to rounded, reflexed at right angles to the claw. Stamens *c.* 8 mm. Style 9–10 mm, stigma capitate, somewhat bilobed. Disk flat. Capsule 4–5 mm, spherical. *Brazil.* H4–5. Autumn.

10. E. tucumanensis Hosseus. Figure 97 (4), p. 170. Illustration: *Botanical Magazine*, n.s., 565 (1970); Phillips & Rix, Shrubs, 223 (1989).

Evergreen shrub to 6 m. Young shoots reddish, hairless or slightly downy. Leaves 5–8 × 1.5–2 cm, oblong to elliptic, thin, shortly acuminate, base wedge-shaped, stalk short. Lower surface of leaf dotted with blackish glands, midrib sometimes downy, upper surface glossy, margin finely toothed. Inflorescence an axillary panicle of 4–15 flowers, leafy at the base and with leafy bracts to 15 × 4 mm subtending the

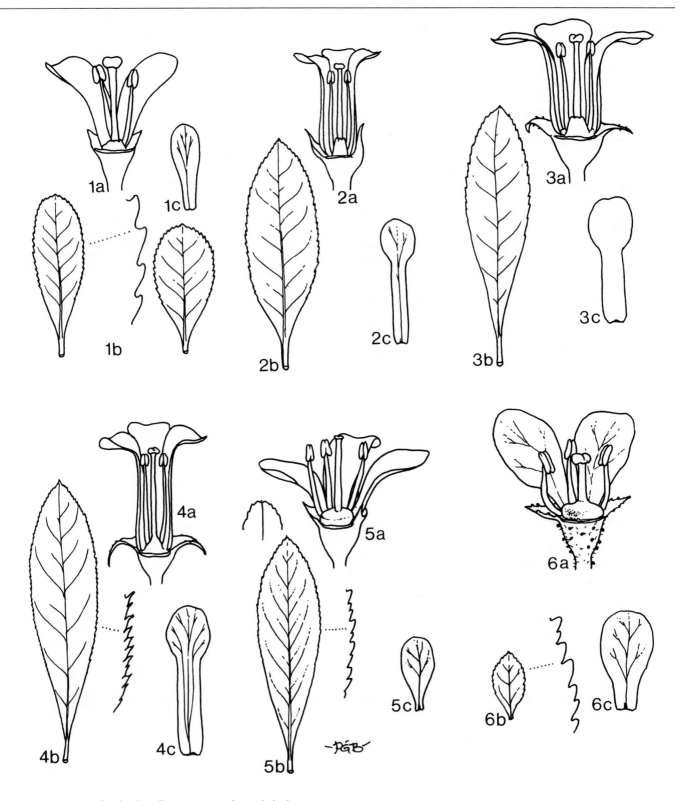

Figure 97. Diagnostic details of *Escallonia* species (a, flower; b, leaf; c, petal). 1a, b, c, *E. leucantha.* 2a, b, c, *E. illinita.* 3a, b, c, *E. laevis.* 4a, b, c, *E. tucumanensis.* 5a, b, c, *E. bifida.* 6a, b, c, *E.* × *langleyensis.*

flowers on slender stalks to 1.2 cm. Receptacle *c.* 3 mm. Calyx-tube 1–2 mm, lobes 3–5 × 1–1.5 mm, linear, toothed and glandular. Petals white, erect, claw 1–1.4 cm × 2–3 mm, tip 3–5 × 3–4 mm, reflexed. Stamens *c.* 1 cm. Disk thin, papillose, slightly conical. Style 9–11 mm, stigma capitate, *c.* 2 mm across. Capsule 5–6 mm, approximately spherical. *Argentina.* H4. Summer.

11. E. bifida Link & Otto (*E. montivedensis* (Chamisso & Schlechtendal) de Candolle; *E. floribunda* misapplied). Figure 97(5), p. 170. Illustration: *Botanical Magazine,* 6404 (1879); *Journal of the Royal Horticultural Society* **86**: f. 34 (1961); Krüssmann, Manual of cultivated broad-leaved trees & shrubs **2**: f. 26b (1986); Brickell (ed.), RHS A–Z encyclopedia of garden plants, 429 (2003).

Evergreen shrub to 3 m or more. Young shoots bluntly angled, hairy, becoming smooth. Leaves 3–8 × 1–1.5 cm, oblanceolate, blunt or rounded, base tapered to a short stalk. Margin finely toothed. Inflorescence a broad terminal panicle to 8 cm, becoming loose with age. Flower-stalks to 7 mm; receptacle *c.* 2 mm. Calyx-lobes 1.5–2 mm, triangular, with minute glandular teeth. Petals *c.* 7 × 3 mm, obovate, white, somewhat spreading. Stamens and style 6–7 mm, stigma capitate; disk low, flat-topped. Capsule 3–4 mm, broadly obovate to spherical. *E South America.* H4–5. Autumn.

7. ANOPTERUS Labillardière
A.C. Whiteley
Evergreen shrubs or small trees. Leaves alternate, simple, leathery, obovate to oblanceolate, tapered to both ends and with large rounded teeth, each gland-tipped. Flowers in upright terminal racemes, cup-shaped, white or pink-tinged, petals 6–9 concave, obovate; calyx small, with 6–9 toothed, triangular lobes. Stamens as many as petals, with flattened tapering filaments. Ovary superior, 1-celled with numerous ovules. Fruit a cylindric capsule with 2 recurving valves. Seeds winged.

A genus of 2 species from Australia. They require an acid or neutral soil, rich in humus and prefer some shade. Propagate by seed or cuttings.

1a. Leaves to 12 cm; flowers with 6
  calyx-lobes, petals and stamens
      1. glandulosus
  b. Leaves to 30 cm; flowers with up to
  9 calyx-lobes, petals and stamens
      2. macleayanus

1. A. glandulosus Labillardière. Illustration: *Botanical Magazine,* 4377 (1848); *Revue Horticole,* 310 (1868); Galbraith, Collins field guide to the wild flowers of south-east Australia, pl. 27 (1977); Krüssmann, Manual of cultivated broad-leaved trees & shrubs **1**: 159 (1984).

Tall shrub or tree to 10 m, hairless. Leaves 5–12 × 4–5 cm, dark glossy green, leaf-stalk 6–20 mm. Flowers *c.* 1.6 cm across. Calyx-lobes, petals and stamens 6. Capsule *c.* 1.2 cm. *Australia (Tasmania).* H5. Spring.

2. A. macleayanus F.J. Mueller.
Similar but leaves to 30 cm and flowers with up to 9 calyx-lobes, petals and stamens. *SE Australia.* H5.

8. COROKIA Cunningham
P.G. Barnes
Evergreen shrubs with alternate leaves. White or silvery T-shaped hairs present on stems, on leaf under-sides, flower-stalks, backs of petals and calyx-lobes. Flowers yellow, in a terminal or axillary panicle, raceme or cluster, or solitary. Receptacle top-shaped. Sepals, petals and stamens 5; the petals usually with a small incised scale at the base. Ovary inferior, 1- or 2-celled; style slender, with a somewhat 2-lobed stigma. Fruit a spherical or oblong berry with persistent sepals.

A genus of 3 species, native to New Zealand. They are easily cultivated but require a warm, sunny position and a well-drained soil. Propagation of cultivars is by cuttings; species by seed or cuttings. In edition 1 this genus was included in Cornaceae.

1a. Leaves less than 5 cm   2
  b. Leaves 4–15 cm   3
2a. Leaves rounded, 5–20 mm, abruptly
  narrowed to a winged stalk
      1. cotoneaster
  b. Leaves narrowly oblong,
  oblanceolate or spathulate, 1–4 cm,
  base wedge-shaped   **4. × virgata**
3a. Leaves narrowly lanceolate; flowers
  in panicles; fruit less than 8 mm
      2. buddleioides
  b. Leaves narrowly elliptic-oblong or
  oblanceolate; flowers in racemes;
  fruit *c.* 10 mm   **3. macrocarpa**

1. C. cotoneaster Raoul. Figure 98(4), p. 172. Illustration: *Botanical Magazine,* 8425 (1912); Hunt (ed.), Marshall Cavendish encyclopedia of gardening, 367 (1968); Bean, Trees & shrubs hardy in the British Isles, edn 8, **1**: 712 (1976); Brickell (ed.), RHS A–Z encyclopedia of garden plants, 308 (2003).

Shrub to 2 m or more, very variable in habit, with many densely intertwined branches, hairy when young, becoming hairless and dark with age. Leaves with rounded blade 5–20 mm × 3–5 mm (to 1.5 cm on shaded or juvenile shoots), abruptly narrowed into a winged stalk 5–20 mm. Flowers 6–8 mm, solitary or 2 or 3 together in upper leaf-axils or at ends of shoots. Petals 4–5 mm, oblong. Fruits 5–8 mm, red or yellow. *New Zealand.* H5. Summer.

'Little Prince' has an erect habit and leaves to 4 mm wide.

2. C. buddleioides Cunningham. Figure 98(1), p. 172. Illustration: *Botanical Magazine,* 9019 (1923); Metcalf, Cultivation of New Zealand trees and shrubs, pl. 12 (1987).

Erect shrub to 3 mm. Leaves 4–15 × 1–2 cm, narrowly lanceolate, acute, glossy above; stalk 5–10 mm. Flowers 7–10 mm, in slender terminal and axillary panicles up to 5 cm; petals 5–6 mm, pale or bright yellow, lanceolate, acute. Fruits 6–7 mm, blackish red. *New Zealand (North Island).* H5–G1. Summer.

3. C. macrocarpa Kirk. Figure 98(2), p. 172. Illustration: *Botanical Magazine,* 9168 (1929); Metcalf, Cultivation of New Zealand trees and shrubs, pl. 14, 15 (1987).

Shrub to 6 m, occasionally tree-like. Leaves 5–10 × 1–3 cm, narrowly elliptic-oblong or narrowly obovate, somewhat blunt; stalk to 1 cm. Flowers 8–10 mm, in axillary racemes to 4 cm; petals narrowly oblong, acute. Fruit *c.* 10 mm, broadly oblong, red or orange. *New Zealand (Chatham Island).* G1. Summer.

4. C. × virgata Turrill; *C. buddleioides* × *C. cotoneaster.* Figure 98(3), p. 172. Illustration: *Botanical Magazine,* 8466 (1912); Metcalf, Cultivation of New Zealand trees and shrubs, pl. 14, 15 (1987); Brickell (ed.), RHS A–Z encyclopedia of garden plants, 308 (2003).

Figure 98. Diagnostic details of *Corokia* species. 1, *C. buddleioides* (a, leafy branch with inflorescence; b. fruit). 2, *C. macrocarpa*. 3, *C. × virgata*, fruiting branch (a, flower from side). 4. *C. cot9oneaster*, leafy branch.

Dense shrub 2–3 m. Leaves 1–4 cm × 5–15 mm, sometimes purplish green, narrowly oblong, oblanceolate or spathulate, blunt. Flowers 1–1.2 cm wide, few together in upper leaf-axils or in short terminal panicles; petals narrowly oblong, acute. Fruit 5–7 mm, spherical, red or yellow. *New Zealand (North Island)*. H5–G1. Summer.

A natural hybrid. Several cultivars have been named, distinguished by the colour of the fruits or leaves, which vary from green to bronze-purple.

144. CUNONIACEAE

Woody. Leaves evergreen, opposite or whorled, trifoliolate or usually pinnate, usually stipulate; ptyxis conduplicate or supervolute. Flowers in racemes or heads, unisexual or bisexual, actinomorphic. Calyx, corolla and stamens perigynous or epigynous. Calyx of 4–5 (rarely to10) free sepals. Corolla of 4–5 petals (rarely to 10 or absent). Stamens 4–10, anthers sometimes opening by pore-like slits. Ovary of 2–3 united carpels; ovules many, axile; styles 2–3. Fruit a capsule or rarely a drupe or nut.

There are 25 genera with about 350 species. Species of about 8 genera are occasionally cultivated. *Bauera*, which is from Australia and has trifoliate, exstipulate leaves and floral parts up to 10, is sometimes separated off as the family Baueraceae.

1a. Leaves in opposite pairs, stalkless, of 3 leaflets, appearing as whorls of 6; sepals and petals 4–10; stamens numerous **1. Bauera**
 b. Combination of characters not as above **2**
2a. Leaves in whorls of 3 **5. Calycomis**
 b. Leaves in opposite pairs **3**
3a. Stipules persistent, with forward pointing teeth **4. Caldcluvia**
 b. Stipules falling early **4**
4a. Leaves winged between the leaflets **3. Weinmannia**
 b. Leaves not winged between the leaflets **5**
5a. Leaves with 3 leaflets or pinnately divided, with glandular forward-pointing teeth **2. Cunonia**
 b. Leaves simple or with 3 leaflets, not glandular **6**
6a. Leaves simple; flowers in spherical heads **8. Callicoma**
 b. Leaves simple or with 3 leaflets; flowers in panicles or racemes **7**
7a. Flowers in loose panicles; sepals lengthening, becoming reddish, rarely white after flowering **7. Ceratopetalum**
 b. Flowers in dense racemes; sepals inconspicuous **6. Geissois**

1. BAUERA Andrews
P.G. Barnes
Low evergreen shrubs, leaves opposite, with 3 leaflets, stalkless, appearing whorled. Flowers regular, solitary in upper leaf axils, long-stalked or nearly stalkless.

Sepals 4–10; petals 4–10, pink or white. Stamens numerous, on a conspicuous nectar-secreting disk. Ovary almost superior, 2-celled, styles 2, recurved. Fruit a 2-celled capsule.

A genus of 3 species in E Australia. Slightly tender shrubs requiring a sheltered situation in full sun. In temperate regions some winter protection is required although plants originating at high altitudes may be hardier. The smaller variants are well-suited to cultivation in pans in an alpine house. Propagate by cuttings in late summer or seed.

1a. Flowers stalked; leaflets less than 1.5 cm **1. rubioides**
 b. Flowers stalkless; leaflets usually more than 1.5 cm **2. sessiliflora**

1. B. rubioides Andrews. Illustration: *Botanical Magazine*, 715 (1804); Curtis, Student's flora of Tasmania **1**: f. 42 (1956); Galbraith, Collins field guide to the wild flowers of south-east Australia, pl. 14, (1977).

Shrub of variable habit, prostrate, scrambling or erect, to 1 m. Leaflets 5–15 mm, lanceolate, slightly toothed, glossy dark green. Flowers 5–18 mm across, on stalks usually longer than the leaves. Petals obovate, white, pale pink or occasionally bicoloured. *Australia*. H5. Summer.

Var. **microphylla** (Seringe) Bentham. Leaflets 5–7 mm; flowers 7–8 mm wide, mostly with 6 petals, white, on stalks mostly shorter than the leaves.

2. B. sessiliflora von Mueller. Illustration: Galbraith, Collins field guide to wild flowers of south-east Australia, pl. 14 (1977).

Erect shrub to 2 m. Leaflets 1.2–3.5 cm, entire, hairy. Flowers stalkless, pink. *Australia*. H5–G1. Summer.

2. CUNONIA Linnaeus
M.F. Gardner
Trees or shrubs. Leaves opposite, with 3 leaflets or pinnate; leaflets with glandular forward-pointing teeth; stipules large, falling early. Flowers white or cream, bisexual, small, borne in dense spike-like axillary racemes, sepals and petals 5, overlapping; stamens 10, flattened at the base. Ovary superior, 2-celled; styles 2, persistent. Capsules 2-valved, seeds numerous.

A genus of 17 species, 1 from South Africa and the rest from New Caledonia. Easily grown in a well-drained, peat-

enriched soil, outdoors in mild areas or a cool glasshouse elsewhere. Propagate from half-ripened cuttings taken in August and rooted under glass with bottom heat.

1. C. capensis Linnaeus. Illustration: *Loddiges' Botanical Cabinet*, 826 (1824); *Botanical Magazine*, 8504 (1913); Palgrave, Trees of Southern Africa, 205 (1977); Everett, New York Botanical Gardens illustrated encyclopedia of horticulture **2**: 942 (1981).

A large shrub or small tree to 15 m with pale flaky bark when young. Leaves with 3–5 pairs and one terminal leaflet; leaflets 5–10 cm, lanceolate to ovate; margins with forward-pointing teeth. Racemes to 14 cm, opposite; flowers white or cream. Fruit a 2-horned capsule, brownish. *South Africa (Cape Province & Natal)*. H5–G1. Summer.

3. WEINMANNIA Linnaeus
M.F. Gardner
Trees or shrubs. Leaves opposite, simple, with 3 leaflets or pinnate with wings between the leaflets; stipules falling early. Flowers small, clustered in terminal or axillary racemes; sepals and petals 4 or 5; stamens 8 or 10; ovary superior, 2-celled; styles 2. Fruit a dry capsule; seeds often hairy, rarely slightly winged.

A genus of about 190 species widely distributed throughout Central and South America, the Pacific Islands, New Zealand, Malaysia and Madagascar. Easily grown in a well-drained, loamy soil, in the open in milder areas or in a cool glasshouse elsewhere. Propagated by seed sown freshly under glass or autumn cuttings rooted under a mist unit with bottom heat.

1a. Leaves pinnately divided **1. trichosperma**
 b. Leaves simple **2. racemosa**

1. W. trichosperma Linnaeus. Illustration: Dimitri, La region de los bosques Andino-Patagonicos **2**: 35 (1972); Hoffmann, Flora silvestre de Chile zona austral, 101 (1982); Rodriguez et al., Flora arborea de Chile, 337, 338 (1983); Krüssmann, Manual of cultivated broad-leaved trees & shrubs **3**: 457 (1986).

Tree or shrub to 20 m. Leaves pinnate, 8–10 cm with triangular wings between the leaflets; leaflets 11–13, broadly elliptic, deep glossy green, with 3 coarse teeth on each side. Flowers white, crowded in erect 3–5 cm racemes, fragrant. Fruit glossy red when young. *Chile*. H5–G1. Spring.

2. W. racemosa Linnaeus filius. Illustration: Salmon, The native trees of New Zealand, 186, 187 (1980); Wilson, Stewart Island plants, 73 (1982); Salmon, A field guide to the native trees of New Zealand, 100 (1986); Evans, New Zealand in flower, 64 (1987).

Tree to 20 m. Leaves *c.* 10 cm, simple, oblong-lanceolate to elliptic, coarsely toothed, leaves of younger plants varying from simple to 3–5-parted, or 3-lobed. Flowers white to light pink borne in narrow, 10–12 cm racemes. Fruit reddish brown. *New Zealand.* H5–G1. Summer.

4. CALDCLUVIA D. Don
M.F. Gardner
Trees with opposite branchlets. Leaves opposite, simple, or pinnate, with glandular forward-pointing teeth, finely nerved; stipules leaf-like, sickle-shaped, with forward-pointing teeth, persistent. Flowers borne in axillary panicles; calyx 4 or 5, sepals deciduous; petals 4 or 5; stamens 8–10; styles 2. Capsules 2-beaked, 2-celled, many-seeded.

A genus of 11 species from southern Chile, New Zealand, tropical Australia, Malesia and New Guinea. Cultivation as for *Weinmannia.*

Literature: Hoogland, R. D., Studies in Cunoniaceae II: the genera Caldcluvia, Pullea, Acsmithia and Spiraeanthum, *Blumea* **25**: 481–505 (1979).

1a. Leaves simple, to 12 cm
1. paniculata
　b. Leaves pinnately lobed, to 25 cm
2. rosifolia

1. C. paniculata (Cavanilles) D. Don. Illustration: Dimitri, La region de los bosques Andino-Patagonicos **2**: 35 (1982); Hoffmann, Flora silvestre de Chile zona austral, 99 (1982); Rodriguez et al., Flora arborea de Chile, 96, 97 (1983); *Flora Patagonica* **4b**: 43 (1984).

Tree to 6 m. Leaves simple, 5–12 × 4–5 cm, oblong-lanceolate, shiny above, downy beneath, with glandular forward-pointing teeth, shortly stalked. Stipules 1.2–1.7 cm, toothed. Flowers 5–6 cm across, cream, borne in axillary panicles; sepals 2.5–3 cm, brownish, petals 4–4.5 cm; stamens free, 4–5.5 cm. Fruit a many-seeded leathery capsule, softly hairy, 7.5–8 cm. *S Chile.* H5–G1. Spring–summer.

2. C. rosifolia (A. Cunningham) Hoogland (*Ackama rosaefolia* A. Cunningham, *Weinmannia rosaefolia* (A. Cunningham)

A. Gray. Illustration: Kirk, Forest flora of New Zealand, 63 (1889); Cheeseman, Illustrations of the New Zealand flora, pl. 42 (1914); Eagle, Eagle's 100 trees of New Zealand, pl. 54 (1978); Salmon, The native trees of New Zealand, 188–189 (1980).

Tree to 12 m, covered with short brownish hairs. Leaves pinnately lobed with a terminal leaflet, to 25 cm; lobes 6–20, unequal, with sharp forward-pointing teeth, almost stalkless; stipules leaf-like. Flowers *c.* 3 mm across, borne in many-flowered branched panicles, to 15 cm; sepals ovate, *c.* 1 mm, persistent. Styles persistent. Capsules ovoid, 3–4 mm, softly hairy. *New Zealand (North Island).* G1. Spring–summer.

5. CALYCOMIS D. Don
M.F. Gardner
Shrub, 1–2 × 1–1.5 m. Leaves dark shiny green, 4–10 × 2–4.5 cm, in whorls of 3, ovate, stalkless, pointed at apex, rounded at base, with regular forward-pointing teeth, prominently veined beneath. Flowers white tinged pink, 5–6 mm across, borne at the ends of the branches in dense axillary clusters, to 8 cm. Capsules 2-celled, seeds covered with glands.

A genus of a single species from Australia. This is a very decorative plant grown for both its showy flowers and handsome foliage. Easily cultivated in a humus-rich, well-drained soil in a cool shady or semi-shady position. Propagate from seed or cuttings.

1. C. australis (A. Cunningham) Hoogland (*Acrophyllum venosum* A. Cunningham). Illustration: *The Botanist* **2**: pl. 95 (1839); *Botanical Magazine*, 4050 (1844) – as *Acrophyllum verticillatum.*
Australia (New South Wales). H5–G1. Summer.

6. GEISSOIS Labillardière
M.F. Gardner
Trees or shrubs. Leaves opposite, simple or with 3 leaflets (in ours), with forward-pointing teeth; stipules falling early, basal stipules conspicuous. Flowers bisexual, borne in racemes, axillary or clustered near ends of branches; calyx deeply divided into 4 or 5 lobes; petals absent, stamens 10, exserted; ovary 2-celled; styles 2. Fruit a slender capsule, seeds numerous, flattened, winged in upper part.

A genus of 18 species from Australia, New Caledonia, Vanuatu and Fiji.

Although *G. racemosa* Labillardière was introduced into cultivation in 1851, it is now rarely seen in gardens. These handsome plants are grown for their colourful young foliage and decorative leaves. They are easily cultivated in a deep organically rich, loamy soil which is moisture retentive.

1. G. benthamii F. Mueller. Illustration: Maiden, The forest flora of New South Wales **4**: pl. 221 (1916); Wrigley & Fagg, Australian native plants, edn 3, 512 (1979); Elliot & Jones, Encyclopaedia of Australian plants **4**: 352 (1986).

Tree, 10–18 × 5–12 m; trunk with large buttresses; bark grey and wrinkled. Leaves 20–30 cm, leaflets 3, leaf-stalks 2–7 cm, leaflets 8–18 × 3–5 cm, leathery, dark green and shiny, ovate to elliptic, veins prominent, coarsely toothed. Flowers cream to yellowish, borne in dense, axillary racemes, 10–15 cm; capsules 1.5–1.8 cm, cylindric, hairy. *Australia (New South Wales).* G1. Summer.

7. CERATOPETALUM Smith
M.F. Gardner
Trees or shrubs. Leaves opposite, simple or with 3 leaflets, dark green; stipules falling early. Flowers borne in branched axillary or terminal panicles; sepals 5, spreading after flowering; petals 5 or absent, linear, 3-lobed; stamens 10. Ovary semi-superior to almost superior, 2-celled, styles 2. Fruit 1-seeded.

A genus of 5 species from New Guinea and eastern Australia. Although the flowers are not very showy, the sepals enlarge after fertilisation and become very colourful. All species are easily cultivated in a well-drained soil in a sunny or semi-shady site. Propagate by seed or cuttings from half-ripened wood.

1a. Leaves simple　**1. apetalum**
　b. Leaves with 3 leaflets
2. gummiferum

1. C. apetalum D. Don. Illustration: Maiden, The forest flora of New South Wales, pl. 21 (1904); Maiden, Some principal common trees of New South Wales, 209 (1917); Holliday & Hill, Field guide to Australian trees, 75 (1969); Elliot & Jones, Encyclopaedia of Australian plants **3**: 10 (1989).

Tree 10–20 × 5–8 m; young shoots with 4 conspicuous ridges. Leaves simple, 3–25 × 1–7 cm, ovate to oblong-lanceolate, tapered at both ends, margins with shallow forward-pointing teeth; leaf-stalks 6–12 mm.

Flowers borne in axillary and terminal panicles 7–10 cm across. Sepals *c.* 6 mm, white, lengthening to 1.8 cm and becoming crimson or purple; petals usually absent. *Australia (New South Wales).* G1. Spring.

2. C. gummiferum J. Smith. Illustration: Maiden, Flowering plants and ferns of New South Wales **7**: pl. 25 (1898); *Botanical Magazine*, n.s., 312 (1958); *Hortus* III, 344 (1976); Elliot & Jones, Encyclopaedia of Australian plants **3**: 10, 11 (1989).

Shrub or tree 3–10 × 2–6 m; young shoots rounded. Leaves with 3 leaflets; leaflets 3–7 cm × 6–14 mm, narrowly oblong, blunt at the apex, dark green above, paler beneath; margins with shallow forward-pointing teeth; leaf-stalks 1.2–3.6 cm. Flowers *c.* 6 mm across, white, borne in terminal panicles. Sepals 3–4 mm lengthening to 1.2 cm and becoming bright red, rarely white; petals 2–3 mm. *Australia (New South Wales).* G1. Spring.

8. CALLICOMA Andrews
M.F. Gardner
Tree or shrub, 3–10 × 4–6 m, with long and slender branches. Leaves opposite, simple, 5–12 × 2–4 cm, elliptic to lanceolate, apex pointed, tapered or rounded at base, shiny above, whitish or rust-downy beneath, coarsely toothed; stipules falling early. Flowers stalkless, borne in many-flowered spherical heads, 1–2 cm across, petals absent; stamens 8–10, long-exserted, anthers pale yellow; ovary 2- or 3-celled, styles 2 or 3. Fruit a capsule, enclosed by calyx-lobes.

A genus of a single species native to Australia. Successful cultivation depends on a cool protected position in a moisture-retentive soil. Propagate from seed or cuttings.

1. C. serratifolia Andrews. Illustration: Maiden & Campbell, Flowering plants and ferns of New South Wales **2**: pl. 6 (1895); Blombery, A guide to native plants Australian plants, 107 (1967); Wrigley & Fagg, Australian native plants, edn 3, 227 (1979); Elliot & Jones, Encyclopaedia of Australian plants **2**: 409 (1982).

Australia (New South Wales). H5–G1. Spring–summer.

145. DAVIDSONIACEAE

Trees with irritant hairs. Leaves alternate, pinnate, large, with conspicuous, broad,

palmately veined stipules. Flowers in racemes or panicles, bisexual, actinomorphic. Perianth and stamens hypogynous. Perianth of 5 united segments. Stamens 10, arising from a disc. Ovary of 2 united carpels; ovules several per cell, axile. Fruit a drupe. .

A family of a single species and genus (*Davidsonia pruriens*), very occasionally grown in European gardens.

1. DAVIDSONIA F.J. Mueller
M.F. Gardner
Evergreen shrubs or trees, branches densely hairy. Leaves spirally arranged, pinnate, winged between the leaflets, margins irregularly toothed, covered with bright red irritant hairs when young; stipules conspicuous, palmately veined. Flowers borne in drooping racemes; bracts with glandular teeth; calyx 4- or 5-lobed; petals absent; stamens 8–10; ovary with 2 cells. Fruit a drupe-like berry; seeds flattened, fibrous.

A genus of 1 or 2 species native to Australia. Best cultivated in dappled shade in a well-drained, acid soil in mild areas, possibly under glass or as a house plant elsewhere. Propagate from fresh seed.

1. D. pruriens F.J. Mueller. Illustration: *Gardener's Chronicle*, 819 (1877); Elliot & Jones, Encyclopaedia of Australian plants suitable for cultivation **3**: 199 (1984); Jones, Ornamental rainforest plants in Australia, 216 (1986).

Slender, unbranched shrub or tree, 4–8 × 1–2.5 m; bark brown, corky; Leaves dark green, 30–80 cm, leaflets 5–25 × 3–9 cm, ovate-lanceolate, opposite, densely hairy. Flowers brown or reddish brown, *c.* 6 lmm across. Berries purple to blue-black, 3–5 lcm, hairy, ovoid; seeds flat with lacerated margins. *Australia (Queensland & New South Wales).* G1. Summer.

146. PITTOSPORACEAE

Woody, sometimes climbing. Leaves alternate or opposite, simple, exstipulate, often evergreen; ptyxis variable. Inflorescence of clusters or flowers solitary. Flowers bisexual, actinomorphic. Calyx, corolla and stamens hypogynous. Calyx of 5 free sepals. Corolla of 5 free petals. Stamens 5 anthers sometimes united. Ovary of 2–5 united carpels; ovules many,

axile; style 1, short, stigmas 2–5. Fruit a capsule or berry.

A family of about 9 genera and 300 species, concentrated in the southern hemisphere and Pacific region.

Literature: Bennett, E.M., New taxa and new combinations in Australian Pittosporaceae. *Nuytsia* **2**(4): 184–199 (1978); Bennett, E.M., Pittosporaceae. *Australian Plants* **15**: 20–36 (1988) [all Australian species except Billardiera].

1a. Robust shrubs; anthers ovate; fruit dehiscent   2
  b. Twining plants with weak stems, woody at base; anthers linear or ovate; fruit indehiscent berries   4
2a. Branches becoming spiny; flowers numerous; capsule not woody, stalked; seeds flattened   **2. Bursaria**
  b. Branches without spines, or flowers more or less solitary; capsule leathery or woody   3
3a. Petals less than 2.5 cm; seeds not winged   **4. Pittosporum**
  b. Petals more than 3 cm; seeds winged   **3. Hymenosporum**
4a. Flowers blue; anthers linear, fused around style   **5. Sollya**
  b. Flowers not blue; anthers ovate, free   **1. Billardiera**

1. BILLARDIERA Smith
E.C. Nelson
Vines or shrubs. Leaves entire. Flower solitary, or a few in clusters, pendent, bell-shaped, axillary or terminal. Sepals 5, petals 5, anthers 5, free. Fruit a berry.

A genus of about 30 species from Australia.

1a. Young shoots hairy; flowers yellow; fruit yellow-green   **2. scandens**
  b. Young shoots hairless; flowers green-yellow; fruit white, red, blue or purple(never green)   **1. longiflora**

1. B. longiflora Labillardière. Illustration: *Botanical Magazine*, 1507 (1812); Morley & Toelken, Flowering plants in Australia, 139 (1983); Brickell, The RHS gardeners' encyclopedia of plants & flowers, 176 (1989); Macoboy, What shrub is that? 57 (1989).

Vine, young shoots hairless. Leaves 2–5 cm × 4–6 mm, linear-lanceolate, hairless, dark green, sometimes tinged with purple. Sepals 2–7 mm. Petals 2–3 cm, green-yellow, solitary, terminal on short branchlets, urceolate (broadest towards mouth). Fruit a berry, *c.* 2 × 1 cm, shining purple. *Australia (New South Wales, Victoria, Tasmania).* G1–H5. Summer.

Variants with red, blue or white berries are cultivated, and the best of these may be perpetuated by cuttings, as seedlings will not always be true.

2. B. scandens Smith. Illustration: *Botanical Magazine*, 801 (1805), 1313 (1810); Cochrane et al., Flowers and plants of Victoria, t. 368 (1973).

Weak vine (rarely a shrub), young stems hairy. Leaves 1–5 cm × 4–6 mm, ovate lanceolate, margin wavy, usually hairy on lower surface. Sepals 6–9 mm; petals 1.5–2.5 cm, pale yellow. Berry 2–3 cm, elliptic, green-yellow, often downy. *E Australia.* G1–H5. Summer (most of year).

2. BURSARIA Cavanilles
E.C. Nelson
Trees or shrubs; branches often with spines. Leaves entire, almost stalkless. Inflorescence terminal, pyramidal. Capsule flattened, walls thin and parchment-like, with 2 carpels, each with 1 or 2 seeds.

A genus of 6 species, endemic to Australia. Only one species is currently listed from European gardens, but given the polymorphic nature of the species, more than one may be present.

Literature: Bennett, E.M., New taxa and new combinations in Australian Pittosporaceae. *Nuytsia* **2**(4): 184–199 (1978); Elliott, W.R. & Jones, D., Encyclopaedia of Australian plants suitable for cultivation **2**: 392–394 (1982).

1a. Leaves hairless; sepals persistent
1. spinosa
  b. Leaves hairy on lower surface; sepals deciduous   **2. lasiophylla**

1. B. spinosa Cavanilles. Illustration: *Botanical Magazine*, 1767 (1815); Macoboy, What shrub is that?, 66 (1989).

Shrub or rarely a small tree to 10 m; branches hairless, spiny. Leaves 1–5 cm × 5–7 mm, obovate, shining green above, hairless. Panicles large and showy; sepals 1–2 mm; corolla to 1 cm across, cream to white, fragrant. Capsule heart-shaped. *Australia (all states except Northern Territory).* G1–H5. Summer.

An extremely variable species sometimes split into 6 varieties.

2. B. lasiophylla E.M. Bennett.
Similar to *B. spinosa* but with leaves hairy on lower surface and sepals deciduous. *Australia (New South Wales,*

Victoria, South Australia). G1–H5. Spring–summer.

3. HYMENOSPORUM Mueller
E.C. Nelson
Evergreen tree or shrub, to 20 m. Leaves alternate, dark glossy green. Flowers in terminal clusters, to 5 cm, yellow turning pale cream or white, fragrant. Petals united into tube, silky. Capsule with 2 cells, seeds numerous, with a membranous wing.

A genus of a single species, confined to rainforests in Australasia.

1. H. flavum (R. Brown) Mueller. Illustration: Botanical Magazine, 4799 (1854); Morley & Toelken, Flowering plants in Australia, 139 (1983).

Australia (Queensland, New South Wales), New Guinea. G1. Summer.

4. PITTOSPORUM Gaertner
E.C. Nelson
Trees or shrubs, usually evergreen. Leaves usually simple, often clustered towards tips of shoots. Flowers terminal or axillary, usually solitary. Sepals free or occasionally fused at base.

A genus of about 200 species from the Pacific region and eastern Asia. Those best known in European gardens are from Australasia. They are cultivated principally as foliage plants. Propagation by cuttings is easy, but species may also be raised from seeds; cultivars must be perpetuated by vegetative means.

Literature: Allan, H.H., Flora of New Zealand, 1: 305–318 (1961); Metcalf, L.J., The cultivation of New Zealand trees and shrubs (revised edn) (1987).

1a. Branching widely spreading; leaves less than 1 cm × 2 mm; flowers *c.* 4 mm across   **1. anomalum**
  b. Plant not as above   2
2a. Leaves entirely hairless   3
  b. Leaves at least felted beneath   7
3a. Leaf-margins toothed (rarely entire)   **4. dallii**
  b. Leaf-margins entire   4
4a. Flowers dark red   **7. tenuifolium**
  b. Flowers cream to yellow-green   5
5a. Leaf-margins wavy, not curved downwards and inwards; flowers to 1 cm across   6
  b. Leaf-margins not wavy, often curved downwards and inwards; flowers large, to *c.* 2.5 cm across **8. tobira**
6a. Leaves lemon-scented when crushed   **5. eugenioides**

  b. Leaves not as above **9. undulatum**
7a. Leaves to 1 cm across, dark green, margins recurved; flowers dark red and cream   **2. bicolor**
  b. Leaves more than 2 cm across; flowers uniformly dark maroon   8
8a. Leaves elliptic, base rounded, not tapering; flowers bisexual **6. ralphii**
  b. Leaves elliptic to obovate, base tapering; flowers functionally unisexual   **3. crassifolium**

1. P. anomalum Laing & Gourlay. Illustration: Metcalf, The cultivation of New Zealand trees and shrubs, 260 (1987).

Shrub to about 1 m; branches widely diverging and so intricately interlaced, densely felted with grey-white hairs when young, soon becoming black. Foliage sparse. Adult leaves to 1 cm × 2 mm, linear-oblong, pinnatifid or prominently lobed. Flowers *c.* 4 mm across, pale cream, fragrant, almost stalkless, terminal or occasionally in leaf-axils, petals to 3 mm. *New Zealand.* H5. Spring–early summer.

An astonishing shrub, with pale grey interlaced branchlets. It is a veritable botanical curiosity and is hardly recognisable as a species of *Pittosporum*.

2. P. bicolor Hooker. Illustration: Cochrane et al., Flowers and plants of Victoria, 445 (1968).

Shrub to 5 m. Leaves to 3 cm × 5 mm, oblong, leathery, margins often recurved, dull dark green above, hairy and paler beneath. Flowers on long, slender stalks, petals recurved, red shading to almost translucent cream. Ovary hairy. Fruits more than 1 cm across, valves wrinkled on inside. *Australia (New South Wales, Tasmania, Victoria).* H5. Spring.

3. P. crassifolium A. Cunningham. Illustration: Salmon, The native trees of New Zealand, 144–145 (1980); Brickell, The RHS gardeners' encyclopedia of plants & flowers, 71 (1989).

Tree or shrub to 9 m. Leaves 5–10 × 2.4–2.6 cm, elliptic to obovate, margins recurved slightly, upper surface glossy, crackled. Flowers functionally unisexual; male flowers in umbels of 5–10; female flowers in pairs or solitary. Fruits to 3 cm, sparsely hairy; seeds black in golden yellow glutin. *New Zealand (North & South Islands).* H5. Summer.

'Variegatum' is often grown.

4. P. dallii Cheeseman. Illustration: Salmon, The native trees of New Zealand,

135 (1980); Brickell, The RHS gardeners' encyclopedia of plants & flowers, 72 (1989).

Tree or shrub to 6 m, spreading. Leaves 5–10 × 1–4 cm, alternate or almost whorled, crowded at tips of branches; blades thick, leathery, dark green and somewhat glossy above, paler beneath, margins regular toothed (rarely entire); leaf-stalks to 2 cm. Flowers to 1.5 cm across, in dense, terminal compound umbels, yellow-green (or white), fragrant. Sepals to 1 cm. Petals to 2 cm. *New Zealand (South Island)*. H5. Summer.

A rare species both in the wild and in cultivation, but reputed to be the most hardy under European conditions.

5. P. eugenioides A. Cunningham. Illustration: Salmon, The native trees of New Zealand, 132 (1980); Brickell, The RHS gardeners' encyclopedia of plants & flowers, 71 (1989).

Tree or small shrub to 12 m, shoots hairless. Leaves 5–12.5 × 2.5–4 cm, alternate, elliptic, margin wavy, glossy with pallid undersurface, aromatic (lemon-scented when crushed); leaf-stalks slender to 2.5 cm long. Flowers green-yellow, fragrant, sometimes unisexual (shrubs dioecious), to 1 cm diameter, in compound, terminal umbels, flower-stalks silky. Sepals to *c.* 2 mm. Petals less than 7 mm, spreading or recurved. Fruits black, wrinkled outside. *New Zealand*. H5. Summer.

Variegated plants are most commonly found in cultivation.

6. P. ralphii Kirk. Illustration: Salmon, The native trees of New Zealand, 142 (1980); Brickell (ed.), RHS A–Z encyclopedia of garden plants, 825 (2003).

Tree or shrub to 4 m; shoots, flowers and leaf-stalks densely felted. Leaves 7–12.5 × 2–2.5 cm, alternate, elliptic, margin sometimes wavy, glossy green above, densely felted below. Flowers dark maroon, in terminal clusters. Fruits hairy outside; seeds black. *New Zealand (North Island)*. H5. Summer.

7. P. tenuifolium Gaertner (*P. colensoi* Hooker; *P. nigricans* invalid; *P. mayi* invalid). Illustration: Salmon, The native trees of New Zealand, 140–141, 146–147 (1980); Walsh, An Irish florilegium, 149–150 (1984); Brickell, The RHS gardeners' encyclopedia of plants & flowers, 95, 145 (1989); Brickell (ed.), RHS A–Z encyclopedia of garden plants, 825 (2003).

Monoecious or dioecious tree or shrub, sometimes reaching over 10 m; bark usually very dark, almost black. Leaves to 6 × 2.5 cm, alternate elliptic, green and shining above, paler beneath, margin entire, sinuously wavy. Flowers *c.* 1 cm across, unisexual or rarely bisexual, solitary or in clusters, axillary, dark crimson, fragrant, almost stalkless. Sepals silky when young. Petals *c.* 1.2 cm, recurved, paler towards the base. Fruit grey-black at maturity; seeds black. *New Zealand*. H5. Spring.

P. tenuifolium is now most frequently seen in the less specialised gardens in its purple-leaved (e.g. 'Purpureum' and 'Tom Thumb') and variegated variants (particularly as 'Silver Queen' with white margins and grey-green blade). A similar plant, with grey green leaves, irregularly variegated with creamy white and invariably spotted with crimson is 'Garnettii' ('Sandersii'), a hybrid of uncertain parentage.

8. P. tobira Aiton. Illustration: Kurata, Woody plants of Japan, 234 (1985); Macoboy, What shrub is that? 274 (1989).

Shrub to 10 m; branches spreading. Leaves 3–10 × 2–4 cm, obovate, leathery, dark green shining above, ribs paler. Flowers to 2.5 cm across, in terminal cluster, cream-white becoming yellow when mature. Fruit yellow-green; seeds red. *Japan, China, Korea*. G1–H5. Spring.

9. P. undulatum Ventenat. Illustration: Cochrane et al., Flowers and plants of Victoria, 567 & 568 (1968).

Tree or shrub to 15 m. Leaves 7–15 × 2.5–5 cm, obovate, margin wavy, leathery, dark green and shining above, pale beneath, hairless. Flowers in terminal umbels, creamy white, fragrant. Fruits to 1 cm across, smooth; seeds red-brown. *E Australia*. H5. Spring–early summer.

5. SOLLYA Lindley
E.C. Nelson

Subshrubs with thin twining stems, usually hairless. Leaves stalkless, entire. Flowers in pendent clusters, or solitary, on slender stalks, blue. Anthers longer than filaments, joined to form cone around style; pollen shed inwards. Ovary with 2 cells. Fruit a juicy berry; seeds embedded in pulp.

A genus of 2 species native to SW Western Australia, only one of which is widely cultivated in Europe. It is tolerant of a considerable range of conditions but

seems to prefer a moist humus-rich soil. Propagation is by seed.

Literature: Burtt, B.L., The correct names of the Australian bluebell creepers, *Kew Bulletin* 74–76 (1948).

1. S. heterophylla Lindley (*Billardiera fusiformis* Labillardière; *Sollya fusiformis* (Labillardiere) Briquet). Illustration: *Edwards's Botanical Register*, 1466 (1832); Morley & Toelken, Flowering plants in Australia 139 (1983); Bennett & Dundas, The bushland plants of Kings Park, Western Australia, 58 (1988); Phillips & Rix, Shrubs, 221 (1989).

Weak-stemmed vine. Leaves 3–5 cm × 5–20 mm, oblong. Sepals less than 3 mm, acute, slightly pouched at base. Petals *c.* 1 cm, ovate, acute, overlapping to form an elegant bell. Filaments enlarged at base, anthers yellow. Ovary downy. Fruit a blue berry to 2.5 cm, cylindric, succulent. *SW Western Australia (naturalised in S Australia, Tasmania)*. G1–H5. Summer–autumn.

The fruits are edible when blue. White- and pink-flowered cultivars are reported in Australia but are very rarely seen in Europe.

---

147. BYBLIDACEAE

Herbs with insect-trapping glandular hairs on leaves. Leaves alternate, linear, spirally coiled when young, exstipulate. Flowers solitary, bisexual, actinomorphic. Calyx, corolla and stamens hypogynous. Calyx of 5 free sepals. Corolla of 5 free petals. Stamens 5, anthers opening by pores. Ovary of 2 united carpels; ovules many, axile. Fruit a capsule.

A single genus with 2 species, increasingly cultivated by insectivorous-plant enthusiasts, from Australia, superficially resembling the *Droseraceae*, although not related. It is closely allied to the *Roridulaceae*, with which it is sometimes combined.

1. BYBLIS Salisbury
E.C. Nelson

Description as for family.

A genus of 2 species. They do not release pollen unless stamens are vibrated; in natural habitats, insects' wing beats effect release. Propagation is from seed or (*B. gigantea*) by root cuttings in a peat: sand: perlite or peat: sand mixture, which is kept continually moist but not sodden, in a well-lit, heated glasshouse.

1a. Leaves to 30 cm; flowers 2–4 cm
    across   **1. gigantea**
 b. Leaves to 5 cm; flowers *c.* 1 cm
    across   **2. linifolia**

1. B. gigantea Lindley. Illustration: *Botanical Magazine*, 7846 (1902); Erickson et al., Flowers and plants of Western Australia, 33 (1973); Morley & Toelken, Flowering plants in Australia, 141 (1983).

Perennial herb or subshrub with woody rootstock, to 60 cm. Leaves to 30 cm. Flowers 2–4 cm across, iridescent; sepals 0.8–2 cm, narrow, ovate; petals lilac-pink, blue or purple. Seeds numerous. *SW Western Australia.* G1. Summer.

2. B. linifolia Salisbury. Illustration: Jessop, Flora of central Australia 103 (1981).

Annual herb (rarely perennial), to 30 cm. Leaves to 5 cm. Flowers *c.* 1 cm across; petals pink to pale mauve. *New Guinea, N tropical Australia (Western Australia, Northern Territory, Queensland).* G2. Summer.

148. RORIDULACEAE

Small shrubs. Leaves with insect-trapping glandular hairs, alternate but clustered at the ends of branches, exstipulate, entire or lobed. Flowers solitary or in racemes, bisexual, actinomorphic. Calyx, corolla and stamens hypogynous. Calyx of 5 free sepals. Corolla of 5 free petals. Stamens 5, anthers opening by pores. Ovary of 3 united carpels; ovules 1–4 per cell, axile; style 1, stigma capitate. Fruit a capsule.

This distinctive family contains a single genus, sometimes included in the Byblidaceae. The glandular hairs suggest that they are insectivorous though this has not been proved. They require well-ventilated conditions in a well-drained but moist soil. Propagation is from seed.

Literature: Obermeyer, A.A., Roridulaceae, *in Flora of Southern Africa* **13**: 201–204 (1970).

1. RORIDULA Linnaeus
E.C. Nelson
Leaves yellow-green, with gland-tipped hairs ("tentacles") and non-glandular hairs on the upper surface; lower side hairless or nearly so. Flowers pink to magenta. Sepals with glandular hairs, more or less equalling petals.

A genus of 2 species from the Cape Province of South Africa. They have explosive anthers which when touched suddenly reverse their position showering a pollinating insect with pollen.

1a. Leaves with prominent linear-
    lanceolate lobes   **1. dentata**
 b. Leaves entire   **2. gorgonias**

1. R. dentata Linnaeus (*R. muscicapa* Gaertner). Illustration: *Gardeners' Chronicle* **10**: 367 (1891); *Flora of Southern Africa* **13**: 203 (1970); Kondo & Kondo, Carnivorous plants of the world in colour, 133 (1983).

Shrub to 2 m. Leaves to 5 cm × 3 mm, linear-lanceolate with opposite, linear-lanceolate marginal lobes. Petals persistent, *c.* 1.2 cm. Ovary hairy; style with scarcely broadened stigma; carpels with 1 ovule per cell. *South Africa (SW Cape Province).* G1–H5. Late winter–spring.

2. R. gorgonias Planchon (*R. crinita* Gandoger). Illustration: Flora of Ssouthern Africa **13**: 203 (1970); Kondo & Kondo, Carnivorous plants of the world in colour, 134 (1983).

Low shrub to 50 cm. Leaves entire, linear-lanceolate, to 12 cm × 5 mm. Petals deciduous, *c.* 1.5 cm. Ovary hairless; style with broad stigma; carpels with 2–4 ovules per cell. *South Africa (SW Cape Province).* G1–H5. Spring.

149. ROSACEAE

Herbs, shrubs or trees, rarely climbing. Leaves usually alternate, rarely opposite or whorled, usually stipulate, simple or divided, evergreen or deciduous; ptyxis very variable. Inflorescences various. Flowers usually actinomorphic and bisexual. Calyx, corolla and stamens, or perianth and stamens perigynous (perigynous zone sometimes very small) or epigynous. Calyx of 4–6 free sepals or 4–6-lobed, epicalyx sometimes present. Corolla of 4–6 free petals (rarely absent). Stamens 4–many. Ovary of 1–many free carpels, or 2–5 united carpels; ovules 1–many per cell, axile or marginal; styles as many as the carpels, free or almost so or united for more than half their length. Fruits variable: follicles, achenes, 'berries' of drupelets, pomes.

A large and variable family with 115 genera and about 3200 species. Many genera are native to Europe and/or North America, and very many are cultivated as ornamental herbs, shrubs or trees. Many produce edible fruit, notably *Rubus*, *Malus* (apple), *Pyrus* (pear) and *Prunus* (cherry, plum, almond, apricot, peach, nectarine). Includes Quillajaceae.

1a. Leaves simple or shallowly lobed; if
    deeply lobed then at least part of
    the blade visible on either side of
    midrib   2
 b. Leaves pinnately or palmately
    divided to mibrib; leaflets stalked or
    stalkless   62
2a. Leaves evergreen, needle-like,
    solitary or clustered
                               33. Adenostoma
 b. Leaves deciduous, or if evergreen
    then not needle-like   3
3a. Deciduous perennial herbs, or
    plants with woody stock and
    annual leafy and flowering shoots
                                                  4
 b. Evergreen or deciduous trees and
    shrubs   5
4a. Leaves palmately divided; sepals
    yellowish green; petals absent
                                47. Alchemilla
 b. Leaves simple or 3–5-lobed; sepals
    reddish green; petals white
                                    23. Rubus
5a. Dwarf or prostrate shrublets; non-
    flowering stems not exceeding
    10 cm   6
 b. Shrubs or trees; or leafy shoots at
    least 20 cm   9
6a. Leaves deeply 3–5-lobed in upper
    part, resembling a mossy saxifrage
                                  16. Luetkea
 b. Leaves entire or margins scalloped
    or toothed   7
7a. Dense, cushion-forming tufted
    shrublet; leaves 2.5–4 mm; flowers
    almost stalkless, pink to purple
                                  15. Kelseya
 b. Loose shrublets, stems not tufted;
    leaves 5 mm or more; flowers
    borne on long stalks, creamy
    white, white or rarely yellow   8
8a. Leaves stalkless, light green above
    and beneath; flowers borne in
    spike-like racemes
                              14. Petrophytum
 b. Leaves stalked; margins scalloped
    or toothed, dark green above,
    white-felted beneath; flowers
    solitary on long stalks   **39. Dryas**

9a. Leaves evergreen   10
 b. Leaves deciduous   27
10a. Leaves often deeply lobed to more than halfway towards midrib   11
 b. Leaves entire or shallowly toothed   13
11a. Leaves 6 mm–1.5 cm, deeply divided into 3–9 very narrow lobes   **37. Cowania**
 b. Leaves 2 cm or more, lobes broad   12
12a. Flowers 2–5, in umbels; stamens 30–50   **50. Docynia**
 b. Flowers many, in corymbs or panicles, very rarely solitary; stamens 5–25   **71. Crataegus**
13a. Ovaries or ovary superior   14
 b. Ovary inferior   19
14a. Petals absent; style very long, conspicuously plumose; fruit a 1-seeded achene   **34. Cercocarpus**
 b. Petals present; other characters not as above   15
15a. Stamens numerous   16
 b. Stamens 10–20   17
16a. Fruit a hip   **24. Rosa**
 b. Fruit a drupe   **76. Prunus**
17a. Stamens 10   **1. Quillaja**
 b. Stamens 15–20   18
18a. Male flowers in racemes; female flowers solitary; borne on separate plants; fruit of 5 follicles united at base   **2. Kageneckia**
 b. Male and female flowers borne on the same plant; fruit a dry woody capsule   **4. Lindleya**
19a. Leaf-margin entire   20
 b. Leaf-margin toothed   24
20a. Flowers borne in erect racemes or panicles; flower-stalks fleshy   **63. Rhaphiolepis**
 b. Flowers solitary or in cymes at the end of lateral spurs   21
21a. Sepals with 2 bracteoles at base; fruit a dry capsule surrounded by calyx-lobes   **68. Dichotomanthes**
 b. Sepals without bracteoles; fruit fleshy, crowned by persistent calyx-lobes   22
22a. Stamens 10   **66. Heteromeles**
 b. Stamens *c.* 20   23
23a. Branches conspicuously thorny   **69. Pyracantha**
 b. Branches thornless   **67. Cotoneaster**
24a. Fruits 1–4 cm, usually pear-shaped, yellow, fragrant   **59. Eriobotrya**
 b. Fruits 1 cm or less, orange, red, dark red or bluish black, rarely yellow   25

25a. Fruits bluish black   **63. Rhaphiolepis**
 b. Fruits orange, red, dark red or yellow   26
26a. Branches conspicuously thorny; fruits usually compressed laterally   **69. Pyracantha**
 b. Branches usually thornless; fruits rounded or egg-shaped   **65. Photinia**
27a. Ovary or ovaries superior   28
 b. Ovaries inferior   44
28a. Leaves opposite; sepals and petals 4   **20. Rhodotypos**
 b. Leaves alternate; sepals and petals 5 or more   29
29a. Stipules absent   30
 b. Stipules present, but sometimes falling early   31
30a. Leaves lobed and toothed, widest below the middle; fruit of 5 hairy achenes enclosed in a persistent calyx   **18. Holodiscus**
 b. Leaves entire, widest above the middle; fruit of 5, 2-seeded, follicles joined at the base   **13. Sibiraea**
31a. Petals absent   **22. Neviusia**
 b. Petals present   32
32a. Petals yellow   33
 b. Petals white, pink, purple or red   35
33a. Branches thornless; fruit a group of 5–8 achenes surrounded by a persistent calyx   **21. Kerria**
 b. Branches thorny or with prickles; fruit a drupe or a hip   34
34a. Pith of young branches channelled, bark flaking; fruit a 1-seeded drupe   **77. Prinsepia**
 b. Pith of young branches entire, bark smooth; fruit a hip   **24. Rosa**
35a. Fruit a 1-seeded drupe   36
 b. Fruit not as above   38
36a. Sepals 10, petals reddish brown   **78. Maddenia**
 b. Sepals 5, petals white, pink or purplish   37
37a. Pith of young branches channelled, each flower-stalk with 2 bracts   **79. Oemleria**
 b. Pith of young branches entire, flower-stalks without bracts   **76. Prunus**
38a. Leaves entire or lobed, or if divided then not more than halfway towards midrib   39
 b. Leaves divided almost to midrib   42
39a. Leaf-margins entire or toothed (rarely lobed)   40
 b. Leaf-margins lobed and toothed   41

40a. Leaf-margins entire; flowers at least 2 cm across; fruit a woody capsule   **3. Exochorda**
 b. Leaf-margins toothed or lobed or, if entire, then flowers not more than 8 mm across; fruit a dehiscent follicle   **12. Spiraea**
41a. Flowers in rounded, umbel-like racemes; partial fruit bladder-like, opening by two slits   **9. Physocarpus**
 b. Flowers in long conical racemes or panicles; fruit not bladder-like   **10. Neillia**
42a. Leaves hairless or with few hairs, widest below the middle, ovate, doubly toothed and lobed; flowers in panicles; fruit an irregularly shaped follicle   **11. Stephanandra**
 b. Leaves densely hairy, especially beneath, widest above the middle, obovate, deeply divided into 3–7 lobes; flowers solitary; fruit an achene   43
43a. Flowers stalkless, *c.* 1 cm across; fruit an achene with persistent style   **36. Purshia**
 b. Flowers borne on long stalks, 2–3 cm across; fruit a head of hairy achenes   **38. Fallugia**
44a. Leaves palmately or pinnately lobed   45
 b. Leaves entire but often toothed   51
45a. Leaves rounded and entire in the upper half, partly pinnate or lobed below the middle   **62. × Amelosorbus** & **58. × Sorbaronia**
 b. Leaves lobed above the middle, palmately lobed or even lobed throughout   46
46a. Leaves lobed on upper half only   **72. × Crataemespilus** & **73. Crataegomespilus**
 b. Leaves palmately or evenly lobed throughout   47
47a. Fruit with 1–5 bony nutlets   **71. Crataegus**
 b. Fruit with 1–5 cells, each containing 1 or more seeds   48
48a. Styles free   **53. Pyrus**
 b. Styles joined at the base   49
49a. Flowers not more than 1 cm across   **58. × Sorbaronia**
 b. Flowers at least 1.5 cm across   50
50a. Leaves palmately lobed   **56. Eriolobus**
 b. Leaves never palmately lobed   **55. Malus**

51a. Stamens not more than 25   52
  b. Stamens numerous   54
52a. Stipules persistent, kidney-shaped
       49. Chaenomeles
  b. Stipules deciduous, triangular   53
53a. Petals not more than 1 cm; fruit
    not more than 1 cm, red to black
       64. Aronia
  b. Petals 1.5–2 cm; fruit 2.5 cm
    across, green at first becoming
    brown   **70. Mespilus**
54a. Stipules persistent, conspicuous,
    large and kidney-shaped
       49. Chaenomeles
  b. Stipules deciduous, usually
    inconspicuous   55
55a. Petals 2–3 cm   **48. Cydonia**
  b. Petals 6–18 mm   56
56a. Fruit with 1–5 bony nutlets
       67. Cotoneaster
  b. Fruit with 1–5 cells, each
    containing 1 or more seeds   57
57a. Flowers in racemes
       61. Amelanchier
  b. Flowers in umbels, corymbs,
    panicles or solitary   58
58a. Fruits 6–8 cm   **51. × Pyronia** &
       52. Pyrocydonia
  b. Fruit 4–40 mm   59
59a. Leaves narrowly oblong, stalkless
    or very shortly stalked, clustered at
    the ends of short shoots
       60. Peraphyllum
  b. Leaves not as above   60
60a. Calyx-lobes deciduous **57. Sorbus**
      & 54. × Sorbopyrus
  b. Calyx-lobes persistent
    on fruit   61
61a. Fruit 5–12 mm   **65. Photinia**
  b. Fruit 2.5–4 cm **54. × Sorbopyrus**
62a. Deciduous perennial herbs or
    plants with woody stock and
    annual flowering- shoots   63
  b. Evergreen or deciduous trees or
    shrubs   79
63a. Leaves with 3 leaflets   64
  b. Leaves pinnately divided or
    palmately divided with more than
    3 leaflets   71
64a. Petals sometimes absent
       44. Sibbaldia
  b. Petals present, usually 5–7   65
65a. Plants with spreading stolons or
    rhizomes   66
  b. Plants without stolons or rhizomes
              68
66a. Flowers white or pink; fruiting
    receptacle juicy   **45. Fragaria**
  b. Flowers yellow or creamy-yellow;
    fruiting receptacle dry   67

67a. Flowers solitary   **46. Duchesnea**
  b. Flowers in corymbs of 3–8
       41. Waldsteinia
68a. Styles lengthening in fruit, often
    with feathery hairs, or jointed
       40. Geum
  b. Styles not lengthening in fruit   69
69a. Calyx bell-shaped with 5 teeth
       5. Gillenia
  b. Calyx 5–lobed, with 5 epicalyx
    segments between the lobes   70
70a. Stamens 4 or 5   **44. Sibbaldia**
  b. Stamens 10–30   **43. Potentilla**
71a. Petals absent   72
  b. Petals present   74
72a. Leaves palmately divided or lobed;
    sepals 4, with 4 alternating
    epicalyx segments **47. Alchemilla**
  b. Leaves pinnately divided; sepals
    3–6 or, if 4, epicalyx absent   73
73a. Stipules forming a sheath around
    leaf-stalk, with a leafy lobed apex
    on each side; receptacle usually
    with barbed spines   **31. Acaena**
  b. Stipules leaf-like, crescent-shaped,
    joined to leaf-stalk; receptacle
    without barbed spines
       27. Sanguisorba
74a. Epicalyx segments present   75
  b. Epicalyx absent   77
75a. Leaves pinnate, with terminal leaflet
    much larger than lateral leaflets 76
  b. Leaves palmate or, if pinnate, then
    all leaflets similar in size
       43. Potentilla
76a. Styles persistent as awns on the
    fruit   **40. Geum**
  b. Styles deciduous   **42. Coluria**
77a. Flowers yellow   **25. Agrimonia**
  b. Flowers cream, white, pink or
    purplish   78
78a. Flowers unisexual in spike-like
    racemes; fruit a group of follicles
       17. Aruncus
  b. Flowers bisexual in large panicles;
    fruit a group of achenes
       19. Filipendula
79a. Leaves evergreen   80
  b. Leaves deciduous   89
80a. Petals absent   81
  b. Petals present   82
81a. Fruit a white berry
       30. Margyricarpus
  b. Fruit an achene; receptacle with
    barbed spines   **31. Acaena**
82a. Epicalyx segments present between
    calyx-lobes   **43. Potentilla**
  b. Epicalyx segments absent   83
83a. Stamens numerous   84
  b. Stamens 1–20   85

84a. Fruit a hip   **24. Rosa**
  b. Fruit a 1-seeded achene terminated
    by a feathery style   **37. Cowania**
85a. Leaves 11–40 cm   86
  b. Leaves not more than 7 cm   87
86a. Leaves 11–17 cm, pinnately divided
    into 3–9 leaflets, each further
    divided to midrib
       8. Lyonothamnus
  b. Leaves 20–40 cm, with 10–16
    leaflets   **26. Hagenia**
87a. Leaves to 2 cm, deeply 3–lobed and
    further divided   **16. Luetkea**
  b. Leaves 2.5–7 cm   88
88a. Leaves 2.5–3.5 cm; stamens 12–15
      75. × Pyracomeles
  b. Leaves 3–7 cm; stamens 15–20
      74. Osteomeles
89a. Ovary inferior; fruit a pome   90
  b. Ovary or ovaries superior; fruit not
    a pome   92
90a. Leaves only pinnate in lower part
    62. × Amelasorbus &
    58. × Sorbaronia
  b. Leaves pinnate throughout   91
91a. Leaflets entire, 0.5–1.5 cm
      74. Osteomeles
  b. Leaflets toothed, 1.5–5 cm
      57. Sorbus
92a. Petals absent   93
  b. Petals present   96
93a. Stamens 2–7   **31. Acaena**
  b. Stamens 25 or more   94
94a. Leaves with 5–7 leaflets
      32. Polylepis
  b. Leaves with 7–25 leaflets   95
95a. Leaflets 1–8 mm
      28. Sarcopoterium
  b. Leaflets 2–2.5 cm **29. Marcetella**
96a. Leaves 3-pinnate
      35. Chamaebatia
  b. Leaves 1- or 2-pinnate or with 3
    leaflets   97
97a. Leaves 2-pinnate   98
  b. Leaves pinnate or with 3 leaflets
              99
98a. Leaves fern-like, primary leaflets
    not more than 6 mm; fruit of 5
    follicles   **7. Chamaebatiaria**
  b. Leaves not as above, primary
    leaflets 4–20 cm; fruit a berry of
    fleshy drupelets   **23. Rubus**
99a. Fruit a hip   **24. Rosa**
  b. Fruit not as above   100
100a. Fruit a berry of fleshy drupelets
      23. Rubus
  b. Fruit not as above   101
101a. Leaves with 15–25 leaflets; fruit a
    capsule of united carpels
      6. Sorbaria

b. Leaves usually with 3–9 leaflets;
   fruit a head of achenes   102
102a. Style persistent, feathery
                              38. Fallugia
   b. Style deciduous, simple
                              43. Potentilla

1. QUILLAJA Molina
J. Cullen

Evergreen trees or shrubs. Leaves alternate,
leathery, toothed, wavy or almost entire,
shortly stalked; stipules small, soon falling.
Flowers few, in short racemes, often
functionally unisexual. Perigynous zone
small. Sepals 5, edge-to-edge in bud. Petals
5, narrowly obovate, greenish white.
Stamens 10: 5 borne at the apices of the
lobes of a 5-lobed nectar-secreting disc (the
lobes on the same radii as the sepals) and
5 borne in the centre of the flower on the
same radii as the petals. Ovary of 5 carpels,
free or almost so. Fruit a group of 5
follicles, each many-seeded and opening
widely, united at their bases.

A small genus from temperate and
subtropical S America, with one
occasionally grown species. Now
considered by many to be better placed in
its own family, *Quillajaceae*. Propagation is
usually by seed.

1. Q. saponaria Molina. Figure 99(7),
p. 183. Illustration: *Botanical Magazine*,
7568 (1897); Rodriguez, Matthei &
Quezada, Flora arbórea de Chile, pl. 73
(1983).

Shrub to 15 m. Young shoots densely
downy. Leaves elliptic to oblong, obtuse,
usually toothed, to 6 cm or more. Flowers
inconspicuous, to 1 cm across. Sepals
densely downy outside and inside. Petals
much narrower than sepals. Disc greenish
yellow. Fruit to 2 cm across, velvety when
young. *Chile.* H5–G1. Summer.

The bark contains saponins and extracts
from it have been used as soap substitutes.

2. KAGENECKIA Ruiz & Pavon
M.F. Gardner

Evergreen trees or shrubs. Leaves alternate,
margins with forward-pointing teeth;
stipules very small, falling early. Flowers
unisexual, terminal, borne separately on
different plants; male flowers in racemes or
corymbs, female flowers solitary.
Perigynous zone bell- or cone-shaped,
sepals 5, overlapping, petals 5; stamens
16–20, inserted at the mouth of the calyx;
carpels 5, free, pouch-like at base, styles 5.
Ovary superior. Fruit with 5 follicles,
united at the base, covered with spreading

stellate hairs; seeds numerous, broadly
winged in the upper part.

A genus of 3 species from Chile that
prefer a well-drained soil and a very
protected, sunny position. Propagated
under glass from seed sown in the spring.

1. K. oblonga Ruiz & Pavon. Illustration:
Edwards's Botanical Register, 1836 (1836) –
as *K. crataegoides* Don; Rodriguez, Matthei
& Quezada, Flora arbórea de Chile, 178
(1983); Hoffmann, Flora silvestre de Chile
zona austral, 175 (1982).

Tree or shrub, 4–8 m, shoots hairless.
Leaves 3–10 × 1–5 cm, leathery, ovate-
lanceolate, tapering at the base, rounded
or pointed at the tip. Female flowers *c.*
2 cm across, petals pure white, rounded, in
corymbs of 6–9; male flowers with 16–20
stamens. Fruit 2–3 cm across, light brown.
C Chile. H5–G1. Summer.

3. EXOCHORDA Lindley
E.C. Nelson

Deciduous shrubs. Leaves alternate, simple,
stalked. Flowers frequently unisexual, in
terminal racemes on previous year's
shoots. Sepals 5. Petals 5, white.
Perigynous zone broadly top-shaped with a
large disc. Stamens 15–30, inserted on the
margin of the disc. Styles 5, fused towards
base. Ovary superior. Fruit with 5 wings or
5-angled, at maturity separating into
5 hard capsules; 1 or 2 seeds in each
capsule, flattened, winged.

A genus of about 4 poorly differentiated
species from E Asia. For best results
selectively prune weakest young shoots
immediately after flowering. Propagation
by cutting for selected cultivars, or from
seed.

1a. Petal margins not overlapping
                              1. giraldii
   b. Petal margins overlapping   2
2a. Petals broadly obovate; stamens in
    groups of 3–5   **2. × macrantha**
   b. Petals circular; stamens in groups of
    3 or 5   **3. racemosa**

1. E. giraldii Hesse var. **wilsonii** Rehder.
Illustration: *Kew Magazine* **3**: 156 (1986);
Phillips & Rix, Shrubs, 86 (1989); Brickell
(ed.), RHS A–Z encyclopedia of garden
plants, 442 (2003).

Shrub to 4 m; branches erect. Leaves
ovate to elliptic, margins usually entire,
leaf-stalk 1–2.5 cm, slender, green. Flowers
to 5 cm across, 6–8 in racemes, uppermost
stalkless. Petals elliptic, long-clawed,
margins not overlapping, occasionally with

2–4 apical teeth. Stamens 20–25 in
weakly defined clusters. Fruit 1–1.5 cm.
China. H3. Spring.

E. serratifolia Moore differs in having
sharp, forward-pointing teeth. Rare in
gardens.

2. E. × macrantha (Lemoine) Schneider.
Illustration: Phillips & Rix, Shrubs, 82
(1989).

Shrub to 3 m. Flowers to 3 cm across;
racemes terminal on leafy axillary shoots,
to 10 cm, with 6–10 flowers. Petals
broadly obovate, short-clawed, margins
overlapping. Stamens 15–25, in groups of
3–5. H3. *Garden origin.* Spring.

A hybrid between *E. korolkowii*
× *E. racemosa*. It differs from the parents in
having stamens in unequal clusters. Most
commonly found as 'The Bride';
illustration: Brickell (ed.), RHS A–Z
encyclopedia of garden plants, 442
(2003).

3. E. racemosa (Lindley) Rehder
(*E. grandiflora* (Hooker) Lindley; *Spiraea
grandiflora* Hooker). Illustration: *Botanical
Magazine*, 4795 (1854).

Shrub to 3 m. Leaves elliptic to obovate,
apex obtuse or with a short point; leaves to
7.5 × 2.5 cm, on flowering-shoots usually
entire, those on sterile shoots weakly
toothed towards apex; leaf-stalk *c.* 1.2 cm.
Flowers 3–5 cm across; racemes erect, to
10 cm, with 6–10 flowers. Petals circular,
abruptly forming short basal claw, margins
overlapping. Stamens 15–25, in clusters of
3 or 5. Fruit *c.* 1 cm. *N China.* H3. Early
summer.

E. korolkowii Lavallée (*E. albertii*
Regel). Flowers smaller; stamens 5 in
clusters of 5. *Former USSR (Turkestan).* H3.
Spring.

4. LINDLEYA Humbolt, Bonpland &
Kunth
M.F. Gardner

Evergreen trees with leaves scattered;
stipules present. Flowers bisexual, solitary,
borne in leaf-axils towards the end of
branches; flower-talks with 2 bracts;
perigynous zone persistent, sepals 5,
overlapping; petals 5, spherical, stalkless;
stamens 15–20, inserted in mouth of
calyx, filaments unequal, flattened at base,
free, anthers recurved; ovary superior,
5-celled; styles 5, free, terminal, erect. Fruit
an oblong, 5-sided, woody capsule; seeds
c. 10, thinly winged.

A genus of 2 species native to Mexico.

1. L. mespiloides (Humbolt, Bonpland & Kunth) Rydberg. Illustration: *Revue Horticole*, 81 (1854); Schneider, Illustriertes Handbuch der Laubholzkunde **1**: 496 (1906).

Tree to 3 m, shoots grey, hairless and warty. Leaves 1.2–1.8 cm, oblong to lance-shaped, pointed, with minutely glandular teeth. Flowers terminal, mostly solitary, *c.* 1.8 cm across; calyx tubular, sepals 5; petals 5, white; stamens 20. Fruit roundish to ovate, *c.* 8 mm, 5-angled. *Mexico.* H5–G1. Summer.

5. GILLENIA Moench

S.G. Knees

Perennial herbs with erect branches to 1.2 m. Leaves with 3 lanceolate leaflets, margins irregularly and sharply toothed. Stipules paired, small, inconspicuous or larger and leaf-like. Flowers bisexual in long-talked terminal panicles; petals 5, oblong-ovate, white or pinkish. Perigynous zone bell-shaped. Sepals 5. Stamens 10–20; styles 5; carpels with 2–4 ovules. Fruit of 5 leathery follicles, each with 1–4 seeds, testa leathery, endosperm striped.

A genus of just 2 species native to E North America. Both are suitable for a shady position in the larger rock garden or herbaceous border. Propagation is either by division of the rootstock in spring or by seed.

1a. Stipules ovate-lanceolate, coarsely toothed, persistent; petals mainly white   **1. stipulata**
  b. Stipules narrowly linear, inconspicuous, margin smooth or slightly toothed, falling early; petals pink   **2. trifoliata**

1. G. stipulata (Mühlenberg) Baillon (*Spiraea stipulata* Mühlenberg). Illustration: Steyermark, Flora of Missouri, pl. 94 (1963); Strausbaugh & Core, Flora of West Virginia, edn 2, 467 (1978).

Stems to 1.2 m, sparsely hairy. Leaves 5–8 cm, those near the base often with deeply divided leaflets. Stipules 4–6 cm, ovate, leaf-like, persistent. Leaves and stipules gland-dotted beneath. Petals 1–1.3 cm, narrowly oblanceolate, white (rarely pink-tinged). Stamens 10–20. Fruit 6–7 mm. *E North America (New York to Illinois, S to Georgia, Kansas & Texas).* H2. Early–mid summer.

2. G. trifoliata (Linnaeus) Moench (*Spiraea trifoliata* Linnaeus). Illustration:

Baillon, Histoire des plantes **1**: f. 442, 389 (1869); Strausbaugh & Core, Flora of West Virginia, edn 2, 467 (1978); Brickell (ed.), RHS A–Z encyclopedia of garden plants, 481 (2003).

Stems 1–1.2 m, sparsely hairy. Leaves 5–7 cm, leaflets oblong-ovate, margins toothed, lower leaves not markedly different from upper leaves, leaf-stalks *c.* 8 mm. Stipules inconspicuous, narrowly linear, falling early. Lower surface of leaves and stipules without glands. Petals 1.2–2.2 cm, narrowly lanceolate, pink. Stamens 10–20. Fruit 5–6 mm. *E North America (New York, Ontario, Michigan, S to Georgia & Alabama).*

6. SORBARIA (de Candolle) Braun.

J. Cullen

Deciduous shrubs, frequently with stellate hairs. Leaves large, pinnate with up to 25 toothed leaflets, alternate, with stipules. Inflorescence a large, terminal panicle. Flowers with sepals, petals and stamens perigynous, the perigynous zone cup-shaped, lined by a nectar-secreting disc. Sepals 5. Petals 5. Stamens 20–50. Ovary of 5 (rarely 4) carpels, which are united towards the base, each with several ovules. Fruit a 'capsule' formed from the more or less united carpels, the follicles opening lengthwise along their outer margins. Seeds long, pale brown. Figure 99(12), p. 183.

A genus of about 10 species, mostly from Asia, grown for their handsome pinnate leaves and large panicles of white or cream flowers produced rather late in the year. Propagation is by cuttings.

Literature: Rahn, K., A survey of the genus Sorbaria, *Nordic Journal of Botany* **8**: 557–63 (1989).

1a. Panicle-branches spreading very widely, borne at more or less 90° to the main axis   2
  b. Panicle-branches erect or curving upwards, borne at an acute angle to the main axis   4
2a. Leaflets less than 1.5 cm wide, simply toothed   **3. aitchisonii**
  b. Leaflets more than 1.5 cm wide, doubly toothed   3
3a. Young leaves usually with some stellate hairs beneath; panicle and often the lower leaf surface with prominent stalked glands; stamens exceeding petals   **1. arborea**

  b. Young leaves with simple or clustered hairs beneath; stalked glands absent or very inconspicuous; stamens about as long as petals   **2. tomentosa**
4a. Stamens *c.* 20; leaflets with *c.* 25 pairs of veins   **6. assurgens**
  b. Stamens 40–50; leaflets with up to 20 pairs of veins   5
5a. Panicle lacking long glandular hairs   **4. sorbifolia**
  b. Panicle with long glandular hairs   **5. grandiflora**

1. S. arborea Schneider (*Spiraea arborea* (Schneider) Bean). Illustration: *Horticulture* **18**: 489 (1913) & **19**: 393 (1919); Huxley, Deciduous garden trees and shrubs, pl. 174 (1979).

Shrub 1.3–8 m. Young shoots thinly downy, sometimes with a few stellate hairs as well, more rarely hairless. Leaves with 13–17 (rarely as few as 9) leaflets, which are lanceolate, acuminate, doubly toothed and more than 1.5 cm wide; usually with stellate hairs beneath at least when young, often with stalked brown glands as well. Panicle-branches widely spreading; axis densely downy or with stellate hairs and with dark, stalked glands. Sepals 2.5–3 mm, reflexed. Petals white or cream, almost circular, shortly clawed, 2.5–3 mm. Stamens clearly exceeding petals. Capsule hairless, 3–4 mm. *W China.* H4. Late summer–autumn.

The type and density of hairs is very variable in this species. The 3 varieties formerly recognized (var. *arborea*, leaves with thin stellate hairs beneath; var. *glabrata* Rehder, with the leaves more or less hairless beneath; and var. *subtomentosa* Rehder, leaves with dense stellate hairs beneath) are merely different selections from the range of variation available in the wild.

2. S. tomentosa (Lindley) Rehder (*S. lindleyana* Maximowicz; *Spiraea lindleyana* Loudon).

Very similar to *S. arborea* but leaflets to 23, very long acuminate and with simple or somewhat clustered hairs beneath and stamens about as long as the petals. *Himalaya (E Afghanistan to W Nepal).* H4. Late summer–autumn.

3. S. aitchisonii (Hemsley) Rehder (*Spiraea aitchisonii* Hemsley).

Shrub to 3 m. Young shoots hairless, usually reddish. Leaflets 15–21, narrowly lanceolate, acuminate, less than 1.5 cm

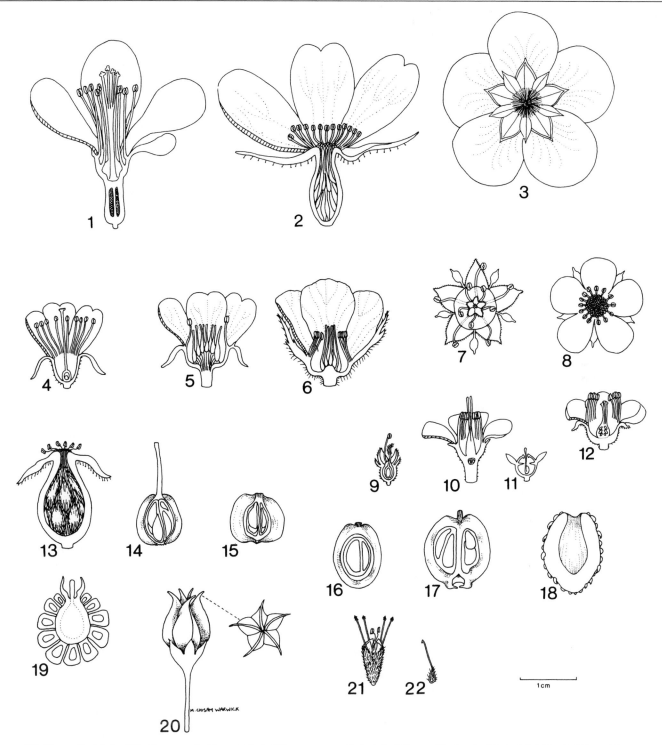

Figure 99. Diagnostic details of the Rosaceae. 1, *Chaenomeles* half-flower. 2, *Rosa* half-flower. 3, *Potentilla fruticosa* flower from below, showing calyx and epicalyx. 4, *Prunus* half-flower. 5, *Geum urbanum* half-flower. 6, *Rubus tricolor* half-flower. 7, *Quillaja saponaria* flower from above. 8, *Fragaria* from above, showing epicalyx. 9, *Acaena caesiiglauca* half-flower. 10, *Raphiolepis umbellata* half-flower. 11, *Alchemilla* half-flower. 12, *Sorbaria* half-flower. 13, *Rosa* half-fruit. 14, *Malus* half-fruit. 15, *Cotoneaster roseus* half-fruit. 16, *Prunus domestica* half-fruit. 17, *Sorbus cashmiriana* half-fruit. 18, *Fragaria alpina* half-fruit. 19, *Rubus* half-fruit. 20, *Physocarpus amurensis* whole fruit from side and above. 21, *Acaena caesiiglauca* whole fruit from side. 22, Geum urbanum whole achene from side.

wide, simply toothed, hairless. Panicle-branches widely spreading, axis and branches hairless. Sepals *c.* 2 mm. Petals white, *c.* 3.5 mm, circular, shortly clawed. Stamens exceeding petals. Capsule hairless. *W Himalaya.* H5. Late summer–autumn.

4. S. sorbifolia (Linnaeus) Braun (*Spiraea sorbifolia* Linnaeus). Illustration: Kitamura & Okamoto, Coloured illustrations of trees and shrubs of Japan, f. 214 (1977); Brickell (ed.), RHS A–Z encyclopedia of garden plants, 994 (2003).

Shrub to 2 m. Young shoots hairless or downy. Leaves with 13–23 very narrowly elliptic to oblong-lanceolate, acuminate, doubly toothed leaflets, each with *c.* 20 pairs of veins, hairless or with stellate hairs beneath. Panicle-branches erect, curving upwards, downy and with dark, stalked glands. Sepals *c.* 2 mm; petals white, *c.* 3.5 mm. Stamens 40–50, considerably exceeding petals. Ovary hairless or with sparse stellate hairs. Capsule to *c.* 5 mm, hairless or sparsely hairy. *N Asia, including Japan.* H3. Late summer–autumn.

A variant with stellate hairs on leaves and carpels has been called var. **stellipila** Maximowicz.

5. S. grandiflora (Sweet) Maximowicz (*S. pallasii* (G. Don) Pojárkova; *Spiraea grandiflora* Sweet).

Similar to *S. sorbifolia* but petals 4–5 mm, panicle narrow, more or less corymb-like, with long, glandular bristles. *E former USSR.* H3. Late summer–autumn.

6. S. assurgens Vilmorin & Bois (*Spiraea assurgens* Vilmorin & Bois).

Very similar to *S. sorbifolia* but leaflets with 25 or more pairs of veins, stamens 20, scarcely exceeding the petals. *?W China* H5. Late summer–autumn.

The origin of this species is uncertain. It was named from cultivated plants thought to have come from China.

7. CHAMAEBATIARIA (Brewer & Watson) Maximowicz
J. Cullen
Aromatic, deciduous, upright shrub to 1.5 m, with dense stellate hairs and stalked glands, at least on the young growth. Leaves alternate (sometimes appearing almost whorled on condensed lateral shoots), to 3 cm, ovate-lanceolate, bipinnate with very numerous small leaflets; stipules small, linear-lanceolate. Flowers to 1 cm across, in dense terminal panicles, perigynous zone small. Sepals 5,

densely hairy, often greyish or whitish. Petals 5, circular or almost so, spreading, white. Stamens numerous, borne on the margins of the entire nectar-secreting disc. Carpels 5, hairy, united only at their extreme bases. Fruit a group of 5 follicles, each few-seeded, united only at their extreme bases.

A genus of a single species, a handsome small shrub with fern-like foliage. It requires a sunny, well-drained position and is propagated by seed or cuttings.

1. C. millefolium (Torrey) Maximowicz (*Spiraea millefolia* Torrey). Illustration: *Journal of the Royal Horticultural Society* **95**: f. 125 (1970); Lenz & Dourley, California native trees and shrubs, 74 (1981).

W USA. H5. Spring–summer.

8. LYONOTHAMNUS A. Gray
J. Cullen
Slender trees to 16 m; bark reddish, scaling in narrow strips. Leaves evergreen, opposite, simple and entire to bipinnate (only variants with bipinnate leaves cultivated); leaflets 3–9, divided into triangular segments with a curved lower side and more or less straight upper side, close packed, not exactly opposite each other; all dark green above, whitish, greyish or greenish beneath, with soft hairs. Inflorescence a many-flowered panicle borne on a lateral shoot. Flowers bisexual, radially symmetric with a small perigynous zone. Sepals 5, densely white, cotton-hairy outside. Petals 5, white, spreading, 2–3 mm. Stamens *c.* 15. Carpels 2, free, glandular; style solitary. Fruit of 2 glandular follicles borne in the persistent perigynous zone; seeds 4 in each follicle.

A genus of a single species, from islands off the coast of California. It requires a sunny site, and is propagated by seed, cuttings or root suckers.

1. L. floribundus Gray subsp. **asplenifolius** (Greene) Raven (*L. asplenifolius* Greene). Illustration: Abrams, Illustrated flora of the Pacific states **2**: 407 (1944).

W USA (islands off the coast of California). H5–G1.

Subsp. *asplenifolius* has bipinnate leaves and rarely flowers in cultivation.

9. PHYSOCARPUS Maximowicz
J. Cullen & H.S. Maxwell
Deciduous shrubs with bark peeling in thin strips; hairs, when present, mostly stellate. Leaves alternate, with small, often toothed

stipules, stalked, usually with 3–5 palmately arranged main veins (other venation pinnate), margins toothed or scalloped and often 3- or 5-lobed; buds in the leaf-axils several, one above the other. Flowers in corymb- or umbel-like racemes; bracts usually small and falling early. Sepals, petals and stamens perigynous, perigynous zone cup-shaped. Sepals 5, generally hairy on both surfaces. Petals 5, usually almost circular, sometimes irregularly notched or toothed. Stamens 20–40, anthers usually reddish. Ovary superior of 2–5 (rarely 1) carpels, united at least at the base, each containing 2–5 ovules. Fruit a group of usually bladder-like follicles opening along both sutures. Seeds usually 2 per follicle, yellowish and shining, each with an appendage (caruncle).

A genus of about 10 rather similar species, 2 from E Asia, the rest from N America. They are easily grown, succeeding best in moist soils. Propagation is by seed or by cuttings.

1a. Carpels 3–5, united only towards the base   2
  b. Carpels 2 (rarely 1), united for at least half their length   6
2a. Carpels and follicles completely hairless   3
  b. Carpels densely hairy; follicles hairy (sometimes only along the sutures) 5
3a. Sepals hairy only around their margins; flower-stalks hairless or almost so   **3. ribesifolius**
  b. Sepals hairy on both surfaces; flower-stalks usually hairy   4
4a. Follicles about twice as long as sepals; leaves mostly longer than broad; racemes rather open   **1. opulifolius**
  b. Follicles scarcely exceeding sepals; leaves generally as broad as long; racemes condensed   **2. capitatus**
5a. Largest leaves 6 cm or more, leaves hairy beneath   **4. amurensis**
  b. Largest leaves to 6 cm, leaves hairless beneath, or with a few hairs along the veins only   **5. intermedius**
6a. Follicles bladdery, not flattened; styles more or less spreading   **6. monogynus**
  b. Follicles flattened and keeled towards the apex; styles erect   **7. malvaceus**

1. P. opulifolius (Linnaeus) Maximowicz (*Spiraea opulifolia* Linnaeus). Illustration: Schneider, Illustriertes Handbuch der

Laubholzkunde **2**: 443, 445, 447 (1912); Gleason, Illustrated flora of the north-eastern United States and adjacent Canada **2**: 284 (1952); *Botanical Magazine*, n.s., 459 (1964); Justice & Bell, Wild flowers of North Carolina, 94 (1968).

Much-branched shrub to 3 m; branches hairless or almost so. Leaves mostly ovate, longer than broad, 6–10 cm, usually 3-lobed (occasionally 5-lobed), irregularly toothed, mostly hairless, sometimes with a few hairs on the veins beneath. Racemes on short leafy shoots, many-flowered, rather open; flower-stalks usually with stellate hairs. Flowers 1–1.3 cm across. Sepals usually hairy on both surfaces. Petals white. Carpels usually 5, hairless or almost so. Follicles united only towards the base, hairless, shining, about twice as long as the persistent sepals. *E & C North America.* H1. Summer.

'Luteus' (var. *luteus* Zabel): young leaves bright yellow.

2. P. capitatus (Pursh) Kuntze (*Spiraea capitata* Pursh). Illustration: Schneider, Illustriertes Handbuch der Laubholzkunde **2**: 445, 447 (1912); Clark, Wild flowers of British Columbia, 246 (1973); Krüssmann, Manual of cultivated broad-leaved trees and shrubs **2**: 403 (1986).

Very similar to *P. opulifolius* but racemes many-flowered, condensed and rather head-like, follicles scarcely exceeding the sepals. *W North America.* H4. Summer.

3. P. ribesifolius Komarov (*P. amurensis* misapplied).

Shrub to 3 m; branches hairless. Leaves mostly broadly ovate, usually 3-lobed, the largest 6–10 cm or more, irregularly doubly toothed, hairless beneath or with a few hairs along the veins. Racemes on short leafy shoots, many-flowered, rather open; flower-stalks hairless. Flowers to 1 cm across. Sepals hairy only on margins. Petals white. Carpels 5, hairless. Follicles united only towards the base, hairless, somewhat longer than the persistent sepals. *E former USSR.* H1. Summer.

Much confused in gardens and in the literature with *P. amurensis*.

4. P. amurensis (Maximowicz) Maximowicz (*Spiraea amurensis* Maximowicz). Figure 99(20), p. 183. Illustration: Schneider, Illustriertes Handbuch der Laubholzkunde **2**: 443, 445, 447 (1912); Krüssmann, Manual of cultivated broad-leaved trees and shrubs **2**: 403 (1986).

Shrub to 3 m; branches hairless or somewhat hairy when young. Leaves broadly ovate, 3-lobed, the largest more than 6 cm, irregularly double-toothed, pale and densely hairy beneath with stellate hairs. Racemes on short leafy shoots, few- to many-flowered, open; flower-stalks usually densely stellate-hairy. Flowers to 1.5 cm across. Sepals hairy on both surfaces. Petals white. Carpels usually 5, with dense stellate hairs. Follicles united only towards the base, rather sparsely hairy, sometimes with hairs only near the sutures, somewhat longer than the sepals. *E Asia (former USSR, China, Korea).* H1. Summer.

5. P. intermedius Schneider

Usually a small shrub to 1.5 m; branches usually erect, crowded, hairless or almost so. Leaves circular to broadly ovate, simple or shallowly 3-lobed, irregularly double-toothed or scalloped, to 6 cm, hairless or sparsely hairy on the veins beneath. Racemes on short leafy shoots, many-flowered, rather dense; flower-stalks hairy. Flowers to 1.5 cm across. Sepals hairy on both surfaces. Petals white. Carpels 3–4, densely hairy. Follicles united only towards the base, hairy, about twice as long as the sepals. *E North America.* H2. Summer.

6. P. monogynus (Torrey) Coulter (*Spiraea monogyna* Torrey). Illustration: Schneider, Illustriertes Handbuch der Laubholzkunde **2**: 445, 447 (1912); Krüssmann, Manual of cultivated broad-leaved trees and shrubs **2**: 403 (1986).

Small shrub to 1 m; branches hairless or with a few stellate hairs. Leaves rounded to almost kidney-shaped, rarely longer than broad, 1–3 cm, usually 3- or 5-lobed (rarely entire), coarsely and irregularly toothed, hairless or with a few hairs on the veins beneath. Racemes on short leafy shoots, many-flowered, rather dense; flower-stalks hairless or with a few stellate hairs. Flowers 8–13 mm across. Sepals hairy on both surfaces. Petals white or pinkish. Carpels usually 2 (rarely 1), densely hairy. Follicles united for up to half their length, not flattened, densely hairy; styles spreading. *C USA.* H3. Summer.

7. P. malvaceus (Greene) Nelson. Illustration: *Botanical Magazine*, 7758 (1901) – as *Neillia torreyi*.

Shrub to 2 m; branches often with stellate hairs. Leaves ovate, 3- or 5-lobed, doubly toothed or scalloped, with stellate hairs beneath, 2–6 cm. Racemes on short leafy shoots, many-flowered, dense; flower-stalks with stellate hairs. Flowers to 1.5 cm across. Sepals densely hairy on both surfaces. Petals white. Carpels 2, with dense stellate hairs. Follicles united for about half their length, flattened and keeled towards the apex, densely hairy; styles erect. *W North America.* H3. Summer.

10. NEILLIA D. Don

J. Cullen

Arching deciduous shrubs to 3 m with brown, shredding bark. Leaves with variably deciduous stipules, ovate to lanceolate, toothed, variably divided into 1–5 (usually 3) lobes, usually hairy, at least on the veins beneath. Axillary buds multiple and one above the other on the vegetative shoots, sometimes also on flowering-shoots. Inflorescence a terminal panicle or raceme, borne on short, leafy, axillary shoots of the previous year. Flowers with bracts, shortly stalked. Sepals, petals and stamens perigynous, the perigynous zone bell-shaped or tubular, persisting and enlarging in fruit and often becoming bristly-glandular. Calyx-lobes 5, acute, downy within. Petals 5, white, pink or red. Stamens 15–30 (rarely as few as 10) in 2 or 3 whorls, inflexed in bud. Ovary superior. Carpels free, usually 1, rarely 2–5, each with 2–10 ovules. Fruit a follicle or group of follicles borne within the persistent, enlarged perigynous zone.

A genus of about 10 species from the Himalaya, China and E Asia. A few are easily grown as ornamental shrubs. They are propagated by seed or by cuttings.

Literature: Vidal, J., Le genre Neillia (Rosaceae), *Adansonia* **3**: 142–66 (1963); Cullen, J., The genus Neillia in mainland Asia and in cultivation, *Journal of the Arnold Arboretum* **52**: 137–58 (1971).

1a. Leaves on flowering-shoots with multiple buds one above the other in their axils; perigynous zone with adpressed bristles in flower; inflorescence a terminal panicle
  1. thyrsiflora
  b. Leaves on flowering- (but not vegetative-) shoots with single buds in their axils; perigynous zone hairless, downy or with spreading bristles in flower; inflorescence a raceme borne on a short, leafy, axillary shoot　2
2a. Perigynous zone bell-shaped, as broad as or broader than long
  2. affinis

b. Perigynous zone cylindric or tubular, longer than broad **3**
3a. Axis of inflorescence with tubercule-based, stellate hairs; ovules 2 **3. uekii**
b. Axis of inflorescence hairless or hairy with simple hairs; ovules 5–9 **4**
4a. Perigynous zone hairless; racemes with up to 17 (rarely to 21) flowers **4. sinensis**
b. Perigynous zone with fine adpressed hairs; racemes with 23–60 flowers (rarely as few as 19) **5. thibetica**

1. N. thyrsiflora D. Don. Illustration: Schneider, Illustriertes Handbuch der Laubholzkunde **1**: 445, 447 (1906); *Journal of the Arnold Arboretum* **52**: 148 (1971).

Shrub to 2 m. All leaves (on vegetative shoots and beneath the inflorescence) with multiple buds, one above the other, in their axils. Inflorescence a terminal panicle. Perigynous zone with adpressed bristles in flower. Petals usually white. *Himalaya, N Burma, China (Yunnan, Kwangsi), Vietnam & Indonesia (Java, Sumatra).* H5. Summer.

2. N. affinis Hemsley. Illustration: *Journal of the Arnold Arboretum* **52**: 148 (1971).

Shrub to 2 m. Only leaves on vegetative shoots with multiple buds, one above the other, in their axils. Racemes borne on short, leafy, axillary shoots. Petals pink. Carpels often more than 1. Perigynous zone hairless, downy or with spreading bristles in flower, bell-shaped, as broad as or broader than long. *W China (Yunnan, Sichuan) & N Burma.* H5. Summer.

Plants cultivated as *N. ribesioides* (see below under *N. sinensis*) are often this species. Two varieties have occurred in cultivation: var. **affinis** has racemes with 10 or more flowers that are spaced along the axis; and var. **pauciflora** (Render) Vidal has racemes with fewer flowers clustered towards the ends of the flowering shoots.

3. N. uekii Nakai. Illustration: Nakai, Flora sylvatica Koreana **4**: t. 13 (1916).

Shrub to 2 m. Only leaves on vegetative shoots with multiple buds, one above the other, in their axils. Racemes borne on short leafy axillary shoots, the axis with tuberculately based stellate hairs (which are ultimately deciduous). Petals usually white. Perigynous zone cylindric or tubular, longer than broad; ovules 2. *Korea.* H3. Summer.

4. N. sinensis Oliver. Illustration: Schneider, Illustriertes Handbuch der Laubholzkunde **1**: 445, 447 (1906); *Journal of the Arnold Arboretum* **52**: 148 (1971); Everett, New York Botanical Gardens illustrated encyclopedia of horticulture **7**: 1270 (1981).

Shrub to 3 m. Leaves variable, ovate to oblong, often lobed. Only leaves on vegetative shoots with multiple buds, one above the other, in their axils. Racemes with 9–17 (rarely to 21) flowers, borne on short, leafy, axillary shoots. Flower-stalks 2–7 mm. Perigynous zone hairless (in cultivated varieties), cylindric or tubular, longer than broad, whitish or pink, 5–11 mm. Petals white or pink. Carpel 1. *China.* H5. Summer.

A variable species with *c*. 6 varieties in the wild, of which only 2 are known in cultivation: var. **sinensis** with perigynous zone 6.5–11 mm; and var. **ribesioides** (Rehder) Vidal (*N. ribesioides* Rehder) with shorter perigynous zone. The name *N. ribesioides* is often misapplied in gardens (see *N. affinis* above).

5. N. thibetica Bureau & Franchet (*N. longeracemosa* Hemsley). Illustration: *Botanical Magazine*, n.s., **3** (1968); *Journal of the Arnold Arboretum* **52**: 148 (1971); Brickell (ed.), RHS A–Z encyclopedia of garden plants, 716 (2003).

Shrub to 3 m. Leaves ovate or ovate-oblong, usually downy beneath, often lobed. Only leaves on vegetative shoots with multiple buds, one above the other, in their axils. Racemes borne on short, leafy, axillary shoots, with 23–60 (rarely as few as 19) flowers. Flower-stalks 0.5–2 mm. Perigynous zone cylindric or tubular, longer than broad, pink, finely adpressed-downy and with a few bristles developing near the base after fertilisation. Petals usually pink. *W China (Sichuan).* H4. Summer.

The most widely grown of the species, usually found under the name *N. longeracemosa*.

11. STEPHANANDRA Siebold & Zuccarini
J. Cullen
Arching shrubs to 2 m or more. Leaves deciduous, ovate to narrowly ovate, lobed and doubly toothed, with stipules and multiple buds, one above the other in their axils. Flowers in a panicle borne terminally on long or short shoots, sepals, petals and stamens perigynous. Perigynous zone

hemispherical, lined with a nectar-secreting disc. Sepals 5. Petals 5, white. Stamens 10–20. Ovary superior, hairy, of a single carpel that contains a single ovule. Fruit an irregularly shaped follicle. Seed 1, pale to dark brown.

A genus of 3 or 4 species from Japan and China. They are easily grown for their small but numerous white flowers and handsome leaves. Propagation is by seed or division or by cuttings.

1a. Stamens 10; leaves (at least near the base) lobed for more than one third of the distance from margin to midrib; sepals without a conspicuous apical tooth **2. incisa**
b. Stamens 15–20; leaves lobed at most to one third of the distance from margin to midrib; sepals each with a conspicuous apical tooth **1. tanakae**

1. S. tanakae (Franchet & Savatier) Franchet & Savatier. Illustration: Kitamura & Okamoto, Coloured illustrations of trees and shrubs of Japan, f. 213 (1977); Brickell (ed.), RHS A–Z encyclopedia of garden plants, 1006 (2003).

Shrub to 2 m. Leaves broadly ovate, weakly 3–5-lobed, the lobes extending for at most one third the distance from margin to midrib, hairless above, downy along the main veins beneath. Stipules large, lanceolate, toothed. Bracts conspicuous, entire or with a few gland-tipped teeth. Perigynous zone hairless. Sepals 2.5–3 mm, each with a conspicuous apical tooth. Petals rounded to the base, *c*. 1.5 mm. Follicle hairy. *Japan.* H4. Summer.

2. S. incisa (Thunberg) Zabel (*Spiraea incisa* Thunberg; *Stephanandra flexuosa* Siebold & Zuccarini). Illustration: Schneider, Illustriertes Handbuch der Laubholzkunde **1**: 445, 447, 448 (1906); *Gartenpraxis* 1987: 8.

Shrub to 2 m. Leaves ovate to narrowly ovate, with 3–7 or more deep lobes, with adpressed bristles above and beneath along the veins and stalk. Stipules long-lanceolate, more or less entire. Bracts inconspicuous. Perigynous zone with sparse adpressed bristles. Sepals 3–3.5 mm. Petals *c*. 1.5 mm, distinctly clawed. Follicle spreading-hairy. *China (Shandong), Japan & Korea.* H3. Summer.

12. SPIRAEA Linnaeus
H.S. Maxwell & S.G. Knees
Deciduous shrubs, buds small with 2–8 exposed scales. Leaves alternate, simple,

lobed, toothed or occasionally entire; usually with short stalks, stipules usually absent. Flowers bisexual or unisexual in umbel-like racemes, corymbs or panicles; perigynous zone cup-shaped; sepals 5, petals 5, stamens 15–60, inserted between the disc and the sepals. Ovary superior, carpels 5, styles 5. Fruit a group of dehiscent follicles, seeds 2–10, oblong.

A genus of 80–100 species from temperate Asia, Europe and North America as far south as Mexico. Some of the species can be propagated from suckers or cuttings of well-ripened wood. As fertile seed is freely produced and cross fertilisation is very common, seed-raised plants cannot be guaranteed to come true. However, some of the most widely grown spiraeas are hybrids.

Literature: Silverside, A.J., The nomenclature of some hybrids of the Spiraea salicifolia group naturalised in Britain, *Watsonia* **18**: 147–51 (1990); Businsky, R. & Businska, L., The genus Spiraea in cultivation in Bohemia, Moravia and Slovakia, *Acta Pruhoniciana*, 72 (2002).

1a. Flowers in terminal conical panicles that are longer than wide   2
  b. Flowers in lateral corymbs, or if terminal then not longer than wide   12
2a. Leaves toothed for two-thirds their length or more   3
  b. Leaves toothed only at apex, or not more than one third their length, occasionally entire   8
3a. Leaves felted beneath   4
  b. Leaves hairless beneath or with few scattered hairs   5
4a. Lower leaf surface completely covered with greyish brown felted hairs; margins with round teeth, occasionally doubly toothed   **1. tomentosa**
  b. Lower leaf surface partially covered with whitish felted hairs; margins with acute teeth, usually doubly toothed   **2. × sanssouciana**
5a. Leaves 2.5–4 cm wide   **3. menziesii**
  b. Leaves 1.5–2.5 cm wide   6
6a. Flowering-stems totally hairless, leaf-margins with regular rounded teeth   **4. latifolia**
  b. Flowering-stems hairy; leaf-margins irregular and doubly toothed   7
7a. Stems light yellowish brown, rounded, softly downy   **5. × semperflorens**

  b. Stems dark reddish brown, grooved, almost hairless   **6. salicifolia**
8a. Flowers in narrow conical panicles not more than 2.5 cm wide   9
  b. Flowers in broad conical panicles 3–7 cm wide   10
9a. Leaves often entire towards inflorescence; leaf midrib and veins beneath light greyish green   **7. douglasii**
  b. Leaves usually toothed, even towards inflorescence; leaf midrib and veins beneath light brown   **8. × pseudosalicifolia**
10a. Leaves 1–3.5 cm, densely hairy beneath   **9. × brachybotrys**
  b. Leaves 3.5–7 cm, with few hairs or hairless   11
11a. Flowers in loose panicles 7–17 cm at widest point   **10. alba**
  b. Flowers in tight panicles not more than 5 cm at widest point   **11. × pyramidata**
12a. Flowers in terminal panicles or corymbs   13
  b. Flowers in lateral corymbs   24
13a. Flowers in broadly cone-shaped panicles   14
  b. Flowers in flat-topped or spreading panicles or corymbs   17
14a. Leaves toothed in upper half or third   15
  b. Leaves toothed for two-thirds or more   16
15a. Medium shrub 1.5–1.8 m; leaves 5–8 cm   **12. × watsoniana**
  b. Small shrub 60–120 cm; leaves 1.5–4.5 cm   **13. densiflora**
16a. Petals half as long as stamens   **14. × conspicua**
  b. Petals as long as stamens   **15. × fontenaysii**
17a. Dwarf straggly branched shrub not more than 30 cm; leaves widest above the middle   **16. decumbens**
  b. Erect shrubs usually 50 cm or more; leaves widest at or below middle   18
18a. Leaves widest at the middle   19
  b. Leaves widest below the middle   21
19a. Dwarf shrub 20–50 cm; leaves 1.7–2.7 cm   **17. betulifolia**, see var. **aemiliana**
  b. Erect shrub to 1 m; leaves 2–8 cm   20
20a. Leaf-margins scalloped   **17. betulifolia**
  b. Leaf-margins doubly toothed   **18. × foxii**

21a. Dwarf shrub to 40 cm; leaves 1–3 cm, thick, conspicuously puckered, leathery   **19. bullata**
  b. Shrub usually 1–2 m; leaves 5–15 cm, thin   22
22a. Fruit hairless   **20. japonica**
  b. Fruit hairy   23
23a. Corymbs to 13 cm across, yellow, downy; flowers 5–7 mm   **21. × revirescens**
  b. Corymbs to 15 cm across, covered with brown felted hairs; flowers 4–5 m   **22. micrantha**
24a. Leaves mostly 3 cm or less   25
  b. Leaves mostly 3 cm or more   38
25a. Leaves 1.5–3 cm   26
  b. Leaves commonly less than 1.5 cm   34
26a. Leaves conspicuously 3–5-lobed and toothed towards the apex   **23. trilobata**
  b. Leaves entire, with 1–3 shallow notches towards apex or shallowly toothed   27
27a. Leaves broadly ovate-rounded, less than twice as long as wide   28
  b. Leaves narrowly ovate-elliptic, more than twice as long as wide   29
28a. Leaves often entire, rounded; petals longer than the stamens   **24. nipponica** var. **rotundifolia**
  b. Leaves ovate-rounded, petals as long as stamens   **25. blumei**
29a. Leaves ovate-elliptic, mostly widest at or below the middle   30
  b. Leaves linear-lanceolate or obovate to ovate, mostly widest above the middle   32
30a. Petals shorter than stamens   **26. canescens**
  b. Petals as long as or longer than stamens   31
31a. Petals rounded, equalling stamens   **27. cana**
  b. Petals oblong, longer than stamens   **28. × cinerea**
32a. Leaves linear to lanceolate, at least 4 times longer than wide   **29. thunbergii**
  b. Leaves obovate to ovate, mostly less than twice as long as wide   33
33a. Fruit almost hairless; petals longer than or as long as stamens   **30. hypericifolia**
  b. Fruit covered with fine down; petals shorter than stamens   **31. crenata**
34a. At least some leaves 3-lobed or conspicuously toothed at apex   35

b. Leaves mostly entire or with 1–3 small teeth or indentations at apex 36

35a. Leaves downy above, grey- or white-felted beneath; fruit downy
32. yunnanensis

b. Leaves hairless, bright green above and below; fruit hairless
33. gemmata

36a. Leaves mostly 4–7 mm; flowers in groups of 6–8 **34. calcicola**

b. Leaves 8–15 mm; flowers in groups of 10 or more 37

37a. Corymbs hairless but with a bluish grey bloom **35. baldschuanica**

b. Corymbs downy, not bloomed
36. arcuata

38a. Leaves toothed at apex or only for upper third 39

b. Leaves toothed for most of their length or at least two-thirds 51

39a. Fruit hairless 40

b. Fruit downy or with longer hairs 46

40a. Leaves hairless 41

b. Leaves hairy 43

41a. Petals longer than stamens
37. × schinabeckii

b. Petals equalling or shorter than stamens 42

42a. Shrub to about 4 m, branches downy
38. veitchii

b. Shrub to about 1 m, branches hairless **39. virginiana**

43a. Leaves deeply toothed to 3-lobed
40. pubescens

b. Leaves finely toothed or entire 44

44a. Leaves entire **38. veitchii**

b. Leaves finely toothed 45

45a. Leaves 4–15 mm across, widest at or above the middle **41. sargentiana**

b. Leaves 1.5 cm or more across, widest at or below the middle
39. virginiana

46a. Leaves hairless **42. trichocarpa**

b. Leaves hairy 47

47a. Leaves 2–3 cm wide 48

b. Leaves 1.5 cm wide or less 49

48a. Petals shorter than stamens
43. media, see subsp. **polonica**

b. Petals longer than stamens
44. henryi

49a. Leaves almost entire or with a few very shallow teeth **45. wilsonii**

b. Leaves conspicuously toothed 50

50a. Leaves obovate; margins with conspicuous white hairs **43. media**

b. Leaves ovate-elliptic, margins without conspicuous hairs
46. mollifolia

51a. Leaf-margin with fine even teeth 52

b. Leaf-margin irregularly and often doubly toothed or lobed 59

52a. Leaves narrowly ovate-elliptic to lanceolate or if broadly ovate then very finely toothed 53

b. Leaves broadly ovate, distinctly wider at the base, teeth conspicuous 55

53a. Flowers borne in leafless, almost stalkless umbels; leaves ovate with very short, even teeth
47. prunifolia

b. Flowers borne in leafy, stalked umbels; leaves narrowly ovate to elliptic-lanceolate 54

54a. Flowers borne in downy clusters
48. bella

b. Flowers borne in hairless clusters
49. cantoniensis

55a. Fruit almost hairless
50. longigemmis

b. Fruit hairy, often densely so 56

56a. Flowers borne in hairless clusters 57

b. Flowers borne in hairy clusters 58

57a. Flowers perfect; leaves wedge-shaped at base **51. miyabei**, see var. **glabrata**

b. Male and female flowers borne separate; leaves truncate at base
52. amoena

58a. Flowers c. 8 mm, in corymbs 3–6 cm
51. miyabei

b. Flowers c. 6 mm, in corymbs 5–9 cm
53. rosthornii

59a. Petals twice as long as stamens 60

b. Petals half as long or as long as stamens 61

60a. Leaves densely felted beneath
54. × blanda

b. Leaves hairless beneath
55. × vanhouttei

61a. Underside of leaves and young shoots covered with yellowish brown, felted hairs **56. chinensis**

b. Underside of leaves hairless or with few scattered hairs
57. chamaedryfolia

1. S. tomentosa Linnaeus. Illustration: Gleason, Illustrated flora of the north-eastern United States and adjacent Canada **2**: 286 (1952); House, Wild flowers, pl. 94A (1961); Justice & Bell, Wild flowers of North Carolina, 93 (1968); Krüssmann, Manual of cultivated broad-leaved trees & shrubs **3**: 347 (1986).

Upright shrub, to 1.2 m, suckering, strong-growing, shoots with dense brown felted hairs when young, erect, angled. Leaves 4–8 × 2–4 cm, oblong-ovate, acute, coarsely and often doubly toothed, almost to base, wrinkled, dark green and almost hairless above, densely felted greyish brown hairs below, stalks 1–4 mm. Flowers purple-pink, in narrow, erect, dense brown felted cone-shaped panicles, 8–20 × 4–6.5 cm. Petals obovate, shorter than the stamens. Fruit downy; sepals reflexed. *E North America*. H2. Late summer–autumn.

Var. **alba** Weston. Differs in its white flowers.

2. S. × sanssouciana Koch (*S. nobleana* Hooker; *S. japonica* 'Paniculata'). Illustration: *Botanical Magazine*, 5169 (1860); Dippel, Handbuch der Laubholzkunde **3**: 496 (1893).

Shrub to 1.5 m, shoots with fine downy greyish hairs when young, brownish, angular, erect. Leaves 5–10 × 2–3 cm, oblong-lanceolate, doubly toothed on the apical half to two-thirds, light green, densely whitish felted below. Flowers pink to bright rose in terminal panicles, 10–17 × 3–7 cm, made up of several multi-topped cymes; petals half as long as stamens. Fruit hairless, slender, longer than the reflexed calyx; style spreading; sepals recurved. *Garden origin*. H2. Summer.

A hybrid between *S. douglasii* and *S. japonica*.

3. S. menziesii Hooker (*S. douglasii* var. *menziesii* (Hooker) Presl). Illustration: Schneider, Illustriertes Handbuch der Laubholzkunde **1**: 474, 482 (1906).

Shrub 1–1.5 m, young shoots finely downy, striped. Leaves 5–11 × 2.5–4 cm, oblong-lanceolate, usually hairless but sometimes with hairs on the veins, paler green below, stalks 3–5 mm. Flowers pink, in terminal conical panicles, which are longer than wide; stamens more than twice as long as the petals. *W North America (Alaska to Oregon)*. H2. Summer.

4. S. latifolia (Aiton) Borkhausen (*S. bethlehemensis* misapplied; *S. carpinifolia* Willdenow; *S. salicifolia* Linnaeus var. *latifolia* Aiton; *S. canadensis* misapplied). Illustration: Bailey, Standard cyclopedia of horticulture **3**: 3213 (1935); Gleason, Illustrated flora of the north-eastern United States and adjacent Canada **2**: 286 (1952); House, Wild flowers, pl. 93A (1961); Krüssmann, Manual of cultivated broad-leaved trees & shrubs **3**: 347 (1986).

Shrub 1.5–1.8 m, young shoots hairless, brownish red, erect, angular. Leaves

3–7 cm, broadly elliptic to obovate or oblong, acute at both ends, bright green above, bluish green below; margins with regular rounded teeth. Flowers white to pale pink, in hairless, broad conical-shaped panicles, longer than 20 cm, often with branches spreading horizontally at the end of the current year's growth. Petals shorter than the stamens, disc usually pink. Fruit hairless; style spreading. *N America*. H2. Summer–early autumn.

5. S. × semperflorens Zabel (*S. spicata* Dippel).

Shrub to 1.5 m, shoots downy, yellowish brown, round, finely striped. Leaves 6–10 cm, oblong-lanceolate, warty and sharply doubly toothed almost to base, acuminate, glaucous and almost hairless below. Flowers rose-pink, large, in well-branched, finely downy, narrowly cone-shaped panicles 17–20 × 3–5 cm. Petals half as long as the stamens. Fruit hairless. *Garden origin*. H2. Summer–early autumn.

A hybrid between *S. japonica* and *S. salicifotia*. 'Syringaeflora' is a spreading shrub 90–120 cm, branches slightly angled; leaves *c.* 7.5 × 2 cm, lanceolate to oblong, acuminate, almost hairless beneath, margins toothed above the middle; flowers pink in densely felted conical panicles; stamens longer than petals. H2. Summer–early autumn.

6. S. salicifolia Linnaeus. Illustration: Reichenbach, Icones florae Germanicae et Helveticae **24**: pl. 152 (1909); Csapody & Tóth, Colour atlas of flowering trees and shrubs, 96 (1982); Pignatti, Flora d' Italia **1**: 539 (1982); Daykin, Pictorial guide to shrubs and climbing plants, 96 (1986).

Narrowly erect shrub, to 2 m, suckering and strong growing, shoots finely downy when young, reddish brown, slightly angular. Leaves 4–8 × 1.5–2.5 cm, elliptic to oblong-lanceolate, acute at both ends, scabrous, sometimes doubly toothed, hairless on both sides, darker green above, short-stalked. Flowers pink to whitish, in slim, downy, erect pyramid-shaped terminal panicles *c.* 20 × 8 cm; petals half as long as stamens; calyx downy. Fruit hairy, style recurved. *SE Europe to NE Asia & Japan (naturalised in many parts of Europe)*. H2. Summer.

A hybrid, **S. × rosalba** Dippel (*S. × rubella* Dippel; *S. salicifolia × S. alba*), differs from *S. salicifolia* in its narrowly cone-shaped panicle and very pale pink petals.

7. S. douglasii Hooker. Figure 100(1), p. 190. Illustration: *Botanical Magazine*, 5151 (1859); Schneider, Illustriertes Handbuch der Laubholzkunde **1**: 474, 482 (1906); Clark, Wild flowers of British Columbia, 282 (1973); Krüssmann, Manual of cultivated broad-leaved trees & shrubs **3**: 347 (1986).

Erect shrub 1.5–2.5 m, shoots producing suckering stolons, brown to reddish, finely downy when young, slender. Leaves 4–10 × 1.5–2.5 cm, oblong, rounded at both ends, toothed towards the apex, leaves below inflorescence usually entire, dark green above, white densely felted hairs below, stalks 2–4 mm. Flowers purple-pink, in erect terminal panicles, to 20 cm, narrowly cone-shaped. Axis of flower with white densely felted hairs, stalk and calyx downy grey; stamens pink, longer than the petals. Fruit shiny, hairless, with erect or spreading styles; sepals recurved. *W North America (British Columbia to N California, naturalised in parts of N & C Europe)*. H2. Summer.

S. × illardii Hérincq; *S. alba × S. douglasii*. Intermediate between the parents in leaf shape, hairiness, panicle shape and flower colour but usually closer to *S. douglasii*.

8. S. × pseudosalicifolia Silverside (*S. × billardii* misapplied not Herincq). Illustration: Krüssmann, Manual of cultivated broad-leaved trees & shrubs **3**: 347 (1986).

Shrub to 2 m, shoots brown, finely downy. Leaves 5–8 cm, oblong-lanceolate, acute at both ends, coarsely toothed in the apical two-thirds, greyish green with densely felted light brown hairs on midrib and veins beneath when young. Flowers pink, *c.* 5 mm, in narrow densely felted 10–20 cm panicles. Stamens twice as long as petals. Fruit hairless; styles spreading. *Garden origin*. H2. Summer.

A hybrid between *S. douglasii* and *S. salicifolia*.

9. S. × brachybotrys Lange (*S. pruinosa* Zabel; *S. luxuriosa* Zabel).

Shrub to 2.5 m, shoots finely downy, striped, angular, more or less erect. Leaves 1–3.5 × 1–2.5 cm, narrowly elliptic-oblong, toothed towards the apex, or sometimes with only a few teeth at the tip, dull dark green, slightly downy above, paler with grey felted hairs beneath. Flowers in dense, usually leafy terminal panicles 3–10 cm, covered with felt-like hairs and with short side-shoots. Flower-stalks and calyx hairy;

petals pale pink; sepals reflexed. *Garden origin*. H2. Summer.

A hybrid between *S. canescens* and *S. douglasii*. Similar to *S. fontenaysii* but leaves with densely felted hairs beneath and reflexed sepals.

10. S. alba Duroi (*S. lanceolata* Borkhausen; *S. salicifolia* Linnaeus var. *paniculata* Aiton). Illustration: *Gardeners' Chronicle* **11**: 753 (1879); Schneider, Illustriertes Handbuch der Laubholzkunde **1**: 482 (1906); Gleason, Illustrated flora of the north-eastern United States and adjacent Canada **2**: 286 (1952); Stace, New flora of the British Isles, 398 (1991).

Suckering shrub, similar to *S. salicifolia*, to 2 m, branches slightly downy when young, outspread, reddish brown, angled. Leaves oblong-oblancelate, 3.5–5 × 1.3–3 cm, acute, sharply toothed, veins hairless or downy beneath, stalk *c.* 5 mm. Flowers white, in finely hairy, broadly conical panicles 15–30.5 cm. Sepals erect; stamens white, as long or longer than petals. Fruit hairless. *E USA to Canada*. H2. Summer.

11. S. × pyramidata Greene. Illustration: Clark, Wild flowers of British Columbia, 271 (1973).

Shrub to 1 m, spreading by suckers, branches hairless, reddish brown, round, upright. Leaves 3.5–7 cm, elliptic-oblong, obtuse to acute, coarsely and occasionally doubly toothed towards the apex, hairless beneath, stalks 2–5 mm. Flowers many, white to pinkish, in pyramidal or rounded, downy or hairless panicles. *NW USA*. H2. Summer.

A hybrid between *S. betulifolia* var. *lucida* and *S. menziesii*, which does sometimes occur in the wild.

12. S. × watsoniana Zabel (*S. nobleanum* Zabel not J.D. Hooker; *S. subvillosa* Rydberg). Figure 100(2), p. 190.

Shrub 1.5–1.8 m with erect shoots. Leaves 5–8 cm, elliptic-oblong to almost obtuse, toothed only towards the apex, base usually rounded, with densely grey felted hairs below. Flowers pink, many, in finely downy pyramidal terminal panicles. Fruit small. *WN USA (Oregon); has occurred in cultivation and in the wild with parents.* H2. Summer.

A hybrid between *S. douglasii* and *S. densiflora* subsp. *splendens*. Similar to *S. × sanssouciana*.

13. S. densiflora Rydberg. Illustration: Munz, California mountain wild flowers,

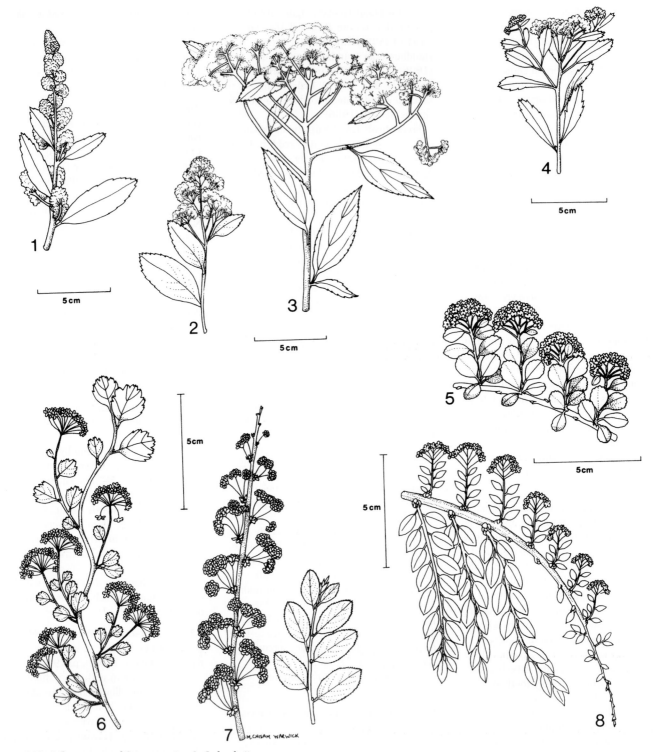

Figure 100. Inflorescences of *Spiraea* species. 1, *S. douglasii*.
2, *S.* × *watsoniana*. 3, *S. micrantha*. 4, *S. decumbens*. 5, *S. blumei*. 6, *S. trilobata*. 7, *S. prunifolia*. 8, *S. calcicola*.

pl. 16 (1963); Fries, Wild flowers of Mount Rainier and the Cascades, 130 (1970).

Shrub to 1.2 m, young shoots round, hairless, reddish brown. Leaves 1.5–4.5 cm, elliptic, rounded at both ends, toothed or scalloped towards the apex, hairless on both surfaces, deep green above, paler beneath, stalks 2 mm or less. Flowers pink, in dense 1.5–4 cm, hairless, dome-shaped erect panicles on current year's growth; petals shorter than stamens. Fruit hairless; sepals obtuse, ovate, erect. *W North*

America (British Columbia to Oregon). H2. Summer.

Subsp. **splendens** (K. Koch) Abrams (*S. splendens* K. Koch; *S. arbuscula* Greene). Illustration: Schneider, Illustriertes Handbuch der Laubholzkunde **1**: 468, 474 (1906). Shrub to 1.2 m, young shoots finely downy. Leaves ovate to elliptic-lanceolate, sometimes acute, toothed or doubly toothed towards the apex, with acuminate teeth. Flowers in finely downy panicles; sepals acute, triangular. *USA (Oregon to California).* H2. Summer.

'Nobleana': leaves mostly tapered, sometimes rounded at base, downy on the veins above, covered with a dull greyish down below. Flowers many, in broad corymbose panicles, 7.5–25.5 cm, at the end of the new growth. Flower-stalks grey-felted. H2. Summer.

14. S. × conspicua Zabel.

Upright shrub, *c.* 1 m, young shoots downy, dark brown, slightly angled. Leaves 3–6 cm, elliptic-oblong, acute at both ends, toothed or doubly toothed, almost hairless. Flowers pale pink to white, in finely downy, broad pyramidal panicles on long erect shoots; petals half as long as the stamens. *Garden origin.* H2. Summer–early autumn.

A hybrid between *S. japonica* 'Albiflora' and *S. alba*.

15. S. × fontenaysii Lebas (*S. fontenaysii* forma *alba* Zabel; *S. fontenaysiensis* Dippel).

Erect slender-branched shrub to 2 m, shoots downy when young, angular. Leaves 2–5 cm, elliptic-oblong, obtuse at both ends, scallop-toothed towards the apex, blue-green, almost hairless beneath, short-stalked. Flowers white, in downy pyramid-shaped panicles, 3–8 × 3–8 cm, on short side-shoots at the end of last year's growth; petals as long as the stamens. Fruit almost hairless; styles spreading. *Garden origin.* H2. Summer.

A hybrid between *S. canescens* and *S. salicifolia*.

16. S. decumbens W. Koch. Figure 100 (4), p. 190. Illustration: Schneider, Illustriertes Handbuch der Laubholzkunde **1**: 459, 474 (1906); Reichenbach, Icones florae Germanicae et Helveticae **24**: pl. 151 (1909); Pignatti, Flora d' Italia **1**: 540 (1982); Krüssmann, Manual of cultivated broad-leaved trees & shrubs **3**: 352 (1986).

Dwarf shrub, often prostrate, *c.* 25 cm, shoots slender, erect, buds obtuse, small with few scales. Leaves

1.3–3.8 × 6–13 mm, elliptic-oblong, widest above the middle, acute at both ends, hairless on both surfaces, toothed or doubly toothed, stalks *c.* 3 mm. Flowers white, *c.* 6 mm, in small, terminal flat-topped panicles or corymbs, 3–5 cm, on the current year's growth. Flowers not quite dioecious; petals as long as stamens. Fruit hairless, style erect; sepals spreading or recurved. *SE Europe (S Germany to S Tyrol).* H2. Summer.

Subsp. **tomentosa** Poech (*S. hacquetii* Fenzl & Koch; *S. lancifolia* Hoffmannsegg). Illustration: Dippel, Handbuch der Laubholzkunde **3**: 476 (1893); Schneider, Illustriertes Handbuch der Laubholzkunde **1**: 459, 474 (1906); Reichenbach, Icones florae Germanicae et Helveticae **24**: pl. 151 (1909). Dwarf shrub, often prostrate, 20–30 cm, shoots grey-downy. Leaves elliptic, 1.5–2.5 cm, toothed at apex, with grey densely felted hairs beneath. Flowers white, in flat-topped panicles, 2.5–3.5 cm, not projecting above the apical leaves. Sepals half-spreading in fruit. *Austria, Italy (Tyrol).* H2. Early summer.

17. S. betulifolia Pallas. Illustration: Schneider, Illustriertes Handbuch der Laubholzkunde **1**: 468, 474 (1906); Gleason, Illustrated flora of the north-eastern United States and adjacent Canada **2**: 285 (1952); Clark, Wild flowers of British Columbia, 279 (1973); Krüssmann, Manual of cultivated broad-leaved trees & shrubs **3**: 352 (1986).

Densely bushy shrub of rounded habit, 50–100 cm. Branches reddish brown, striped, hairless, slightly angled. Leaves 2–4 cm, broadly ovate to elliptic, usually rounded at apex, rounded to tapered at base, doubly or simply scalloped, sometimes only in the upper half, dark green above, grey-green with net–like venation beneath, usually hairless, sometimes downy beneath, stalks hairless, 1–6 mm. Flowers white or very occasionally pinkish, hairless, in dense terminal corymbs, 2.5–9 cm; stamens twice as long as petals; sepals reflexed in fruit. Fruit hairless; style erect. *NE Asia to C Japan.* H2. Summer.

Var. **corymbosa** (Rafinesque-Schmaltz) Maximowicz (*S. corymbosa* Rafinesque not Muhlenberg; *S. betulifolia* misapplied, not Pallas). Illustration: Schneider, Illustriertes Handbuch der Laubholzkunde **1**: 474 (1906). Leaves 3–8 cm, ovate to roundish, broadly tapered at base, scalloped or coarsely and doubly toothed towards the

apex, glaucous beneath with a few hairs, stalks 3–8 mm. Flowers white, to 5 mm, in rounded densely packed 5–10 cm corymbs; stalks downy 3–8 mm. Sepals erect in fruit. Fruit erect, hairless, glossy. *E USA.* H2. Summer.

Var. **lucida** (Douglas) C.L. Hutchinson (*S. lucida* J.D. Hooker; *S. corymbosa* var. *lucida* (Douglas) Zabel). Illustration: Schneider, Illustriertes Handbuch der Laubholzkunde **1**: 468, 474 (1906). Shoots brown to yellowish. Leaves 2–6 cm, broadly ovate to oblong, acute or sometimes rounded at the apex, rounded or broadly tapered at the base, scalloped or doubly and deeply toothed on the apical half to two-thirds, shiny above, paler beneath, stalks 3–6 mm. Flowers white, in densely packed hairless 3–10 cm corymbs. *NW USA.* H2. Summer.

Var. **aemiliana** (Schneider) Koidzume (*S. beauverdiana* Schneider). Dwarf shrub; 20–50 cm, leaves 1.7–2.7 cm, broadly rounded, scalloped, with distinct net-like venation. Flowers 4–5 mm, on a inflorescence 2–2.5 cm. *Japan.* H2. Summer.

18. S. × foxii (Vos) Zabel.

Compact shrub 0.5–1 m, shoots brown, almost hairless, flexuous. Leaves 5–8 cm, elliptic, doubly toothed on the upper two-thirds only, dull green above often with brown markings, light green beneath, hairless on both sides. Flowers whitish, sometimes tinged bluish pink, in finely downy, widely branched, flattish corymbs 10–20 cm across. Fruit obtuse at apex; styles spreading. *Garden origin.* H2. Summer.

A hybrid between *S. betulifolia* var. *corymbosa* and *S. japonica*, similar to *S. × margaritae*.

19. S. bullata Maximowicz (*S. crispifolia* misapplied). Illustration: Schneider, Illustriertes Handbuch der Laubholzkunde **1**: 468, 474 (1906); *Garden Answers*, 55 (1987).

Dwarf shrub to 40 cm, branches with long white hairs when young, eventually brown and hairless. Leaves 1–3 cm, rounded, ovate, conspicuously puckered, leathery, deeply toothed, greyish green beneath. Flowers deep rosy pink, in small dense corymbs forming a terminal panicle 4–8 cm across. Stamens reddish, slightly longer than petals. Fruit with style remnants stunted. *Japan, only known in cultivation.* H2. Summer.

20. S. japonica Linnaeus filius
(*S. × bumalda* Burvénich; *S. callosa*
Thunberg; *S. pumila* Zabel). Illustration:
Gleason, Illustrated flora of the north-
eastern United States and adjacent Canada
2: 286 (1952); Csapody & Tóth, Colour
atlas of flowering trees and shrubs, pl. 176
(1982); Pignatti, Flora d' Italia **1**: 540
(1982); Krüssmann, Manual of cultivated
broad-leaved trees & shrubs **3**: 352, pl.
123 (1986).

Shrub to 2 m, usually only slightly
branched, shoots erect, downy when
young, striped or almost circular. Leaves
5–12 cm, short-stalked, ovate-oblong to
narrowly elliptic, acute, deeply doubly
toothed, hairless on both surfaces or grey-
green and downy on the veins beneath.
Flowers light to dark pink-red, occasionally
white, *c.* 6 mm, in flat terminal downy
15–20 cm corymbs, leafy at the base.
Petals shorter than the stamens; calyx-
lobes erect at flowering, later spreading.
Fruit hairless. *Himalaya to China, Korea &
Japan.* H1. Summer.

Var. **acuminata** Franchet. Leaves
oblong-ovate to lanceolate, acuminate,
green, downy on the venation. Flowers
pink, in corymbs 10–14 cm. *C & W China.*
H2. Summer.

Var. **fortunei** (Planchon) Rehder
(*S. fortunei* Planchon; *S. callosa* Lindley not
Thunberg). Taller than 1.5 m, shoots
downy when young, rounded. Leaves
5–10 cm, oblong-lanceolate, sharply and
doubly toothed with hardened tips,
wrinkled above, hairless and glaucous
beneath. Inflorescence downy, many
branched; flowers pink. *E & C China.* H2.
Summer.

'Albiflora' ('Alba'; *S. albiflora* (Miquel)
Zabel). Illustration: Bailey, Standard
cyclopedia of horticulture **3**: 3212 (1935);
Krüssmann, Manual of cultivated broad-
leaved trees & shrubs **3**: 348 (1986).
Dwarf shrub 30–60 cm.

'Anthony Waterer' (*S. × bumalda*
'Anthony Waterer'). Illustration: Hillier
colour dictionary of trees & shrubs, 242
(1981); Csapody & Tóth, Colour atlas of
flowering trees and shrubs, pl. 177 (1982);
Amateur Gardening **105**: 5443 (1989).
Compact shrub, 1.2–1.5 m, with narrow
often variegated leaves and bright crimson
flowers.

'Little Princess'. Illustration: *Gardening
from 'Which'* 1988 (May 4): 149. Dwarf
shrub to 50 cm, wider than taller, erect,
low, well-branched. Leaves ovate, *c.*
2.5 × 1.2 cm, dull green above, often

wrinkled, hairless beneath. Flowers many,
pale lilac-pink, in *c.* 4 cm downy cymes,
short-stalked.

S. × margaritae Zabel. Illustration
Krüssmann, Manual of cultivated broad-
leaved trees & shrubs **3**: 355 (1986).
Shrub to 1.5 m, branches dark reddish
brown, finely downy, round, finely striped.
Leaves 5–8 × 3–4 cm, ovate-elliptic, acute
to wedge shaped, coarsely to doubly
toothed, dark green above, paler beneath
with a few hairs on the veins, short-
stalked. Flowers rose-pink becoming
lighter, 7–8 mm, in downy, loose, flat
corymbs, to 15 cm. Petals half as long as
the stamens; calyx downy. Fruit small,
hairless; styles usually erect. *Garden origin.*
H2. Summer–autumn.

A hybrid between *S. japonica* and
S. × superba (*S. albiflora × S. corymbosa*).
Foliage brightly coloured in autumn.

21. S. × revirescens Zabel.
Shrub, to 1.2 m, branches brown, downy
when young, slightly angled. Leaves
5–10 cm, ovate-oblong, acuminate, tapered
at base, deeply or doubly toothed, light
green, blue-green beneath, veins downy
yellow. Inflorescence downy, corymbs
10–13 cm, terminal; flowers rose-pink,
5–7 mm; petals shorter than the stamens.
Fruit downy. *Garden origin.* H2.
Summer–autumn.

A hybrid between *S. amoena* and
S. japonica.

22. S. micrantha J.D. Hooker (*S. japonica*
Linnaeus var. *himalaica* Kitamura). Figure
100(3), p. 190.
Small to medium-sized shrub, *c.* 1.5 m,
branches hairy. Leaves 7.5–15 cm, ovate-
lanceolate, toothed to doubly toothed,
warty, tapering to a long point at the apex,
almost rounded at the base, downy
beneath. Flowers mostly bisexual, 4–5 mm,
pale pink, in loose leafy corymbs to 15 cm,
with dense brown felted hairs. Fruit
densely hairy. *E Himalaya.* H2. Summer.

23. S. trilobata Linnaeus (*S. aquilegiifolia*
misapplied; *S. rotundifolia* misapplied;
S. grossulariifolia vera misapplied). Figure
100(6), p. 190. Illustration: Schneider,
Illustriertes Handbuch der Laubholzkunde
1: 455, 459 (1906); Bailey, Standard
cyclopedia of horticulture **3**: 3210 (1935);
Krüssmann, Manual of cultivated broad-
leaved trees & shrubs **3**: 351 (1986).
Shrub, 1–1.5 m, dense, compact, broad,
shoots hairless, slender, spreading, often
flexuous. Leaves 1.5–3 cm rarely to 4 cm,

rounded, scallop-toothed, usually 3–5
lobed, base rounded to broadly tapering,
bluish green above, stronger in colour
beneath, stalks 5–8 mm. Flowers white,
small, in many-flowered corymbs, 2–4 cm
on the end of short leafy twigs on last
year's growth, flower-stalks 6–19 mm,
hairless, slender. Petals longer than the
stamens. Fruit somewhat spreading; sepals
erect. *N Asia (China to Siberia & Turkestan).*
H2. Summer.

Similar to *S. vanhouttei* but smaller in all
parts.

24. S. nipponica Maximowicz var.
rotundifolia (Nicholson) Makino
(*S. bracteata* Zabel not Rafinesque).
Illustration: *Botanical Magazine*, 7429
(1895); Bean, Trees & shrubs hardy in the
British Isles, edn 8, **4**: 489 (1980);
Krüssmann, Manual of cultivated broad-
leaved trees & shrubs **3**: 351 (1986);
Daykin, Pictorial guide to shrubs and
climbing plants, 96 (1986).

Upright bushy shrub, branches hairless,
reddish brown, rounded, angular, long,
arching. Leaves 1.5–3 cm, ovate to
rounded, scalloped, apex rounded, usually
notched, sometimes entire, broadly
tapering at base, hairless, dark green
above, blue-green beneath, stalks to 4 mm.
Flowers *c.* 8 mm, white, many, in
semicircular corymbs; stalks leafy. Petals
roundish, longer than the stamens. Fruit
erect, slightly hairy; styles usually
spreading. *Japan.* H2. Summer.

Var. **tosaensis** (Yatabe) Makino. Differs
in having smaller flowers in dense umbel-
like corymbs and is probably one of the
commonest spiraeas in cultivation.

25. S. blumei D. Don (*S. chamaedryfolia*
Blume not Linnaeus; *S. obtusa* Nakai).
Figure 100(5), p. 190. Illustration:
Schneider, Illustriertes Handbuch der
Laubholzkunde **1**: 455, 459 (1906);
Krüssmann, Manual of cultivated broad-
leaved trees & shrubs **3**: 351 (1986).

Shrub to 1.5 m, branches spreading,
arching, rounded, hairless. Leaves
2–3.5 cm, ovate-rounded, obtuse, sharply
scalloped-toothed, sometimes almost 3–5-
lobed, distinctly veined beneath, bluish
green; stalks 6–8 mm. Flowers white, both
unisexual and bisexual, in many small
corymbs. Petals rounded-obovate, equal in
length to the stamens. Styles spreading,
longer than or as long as the stamens.
Japan, Korea. H2. Summer.

A variable species often confused with
S. trilobata. The main difference is the

shape of the leaves, i.e. more diamond-shaped and pinnately veined in *S. blumei*.

26. S. canescens D. Don (*S. vaccinifolia* misapplied, not Don; *S. flagelliformis* misapplied; *S. laxiflora* Lindley). Illustration: Schneider, Illustriertes Handbuch der Laubholzkunde **1**: 468, 474 (1906); Krüssmann, Manual of cultivated broad-leaved trees & shrubs **3**: 352 (1986); Brickell (ed.), RHS A–Z encyclopedia of garden plants, 1000 (2003).

Dense shrub to 2–3.5 m, branches downy, angular, ribbed, arching. Leaves 1–2.5 cm, elliptic to obovate, toothed at the apex, tapering to a short stalk, downy on the margins above, with greyish green hairs beneath. Flowers many, white, in semi-circular corymbs 3–5 cm, on the upper side of the branches; petals shorter than the stamens. Flower-stalk and calyx with greyish felted hairs. Fruit with long hairs; styles angled. *Himalaya*. H2. Summer.

27. S. cana Waldstein & Kitaibel. Illustration: Schneider, Illustriertes Handbuch der Laubholzkunde **1**: 451, 455 (1906); Reichenbach, Icones florae Germanicae et Helveticae **24**: pl. 148 (1909).

Dense shrub, to 1 m, rarely 2.5 m, branches downy, slender, round, buds small. Leaves 1–2.5 cm, elliptic-oblong, acute, mucronate, entire or sometimes toothed at the apex, downy grey on both sides, thicker beneath, silky when young. Flowers white, *c.* 6 mm, in downy, dense umbel-like corymbs on leafy stalks. Petals circular, as long as the stamens. Fruit downy; styles spreading; sepals reflexed. *SE Europe*. H2. Early summer.

28. S. × cinerea Zabel.
Densely branched shrub, 1.5 m, sometimes more, downy, brown, striped, angular. Leaves 2.5–3.5 cm, oblong, acuminate, entire or with 1 or 2 teeth at the apex with short recurved leathery tips, grey-downy when young becoming grey-green above, paler beneath. Flowers white, *c.* 6 mm, in many-flowered small, stalkless corymbs at the apex, or on leafy stalks to *c.* 2 cm towards the base of the shoots; petals oblong rounded, longer than the stamens; styles slanting to erect, usually falling before the fruit ripens. *Garden origin*. H2. Early summer.

A hybrid between *S. cana* and *S. hypericifolia*.

29. S. thunbergii Siebold. Illustration: Gleason, Illustrated flora of the northeastern United States and adjacent Canada **2**: 284 (1952); Csapody & Tóth, Colour atlas of flowering trees and shrubs, 96 (1982); Krüssmann, Manual of cultivated broad-leaved trees & shrubs **3**: 351 (1986); *Gardening from 'Which'* (March): 67 (1987) – as *S. thunbergii* 'Kew'.

Shrub, 1–1.5 m, young shoots downy at first, eventually hairless, bright green. Leaves 2–3.5 cm × 3–6 mm, linear-lanceolate, acute, warty and finely toothed towards the apex, hairless, often yellow, orange-red in the autumn. Flowers 6–8 mm, white, in stalkless corymbs, which are 3–5-flowered. Perigynous zone shallow, smooth; petals obovate, longer than the stamens; flower-stalks *c.* 8 mm. Styles spreading. *China, Japan*. H2. Spring–early summer.

'Arguta' (*S. × arguta* Zabel). Illustration: Schneider, Illustriertes Handbuch der Laubholzkunde **1**: 450 (1906); Huxley, Deciduous garden trees and shrubs, pl. 180 (1979); Hellyer, Collingridge illustrated encyclopedia of gardening, pl. 533 (1983); *Gardening Now* (August): 12 (1987). Shrub, 1–2.5 m, branches slender, nodding, downy. Leaves 2–4 cm × 5–15 mm, bright green, oblong-obovate or lanceolate, often doubly toothed, downy at first, becoming hairless. Flowers *c.* 8 mm, white, many, in corymbs. H2. Spring.

A hybrid between *S. thunbergii* and *S. multiflora*. Similar in habit to *S. thunbergii* but taller and more vigorous.

30. S. hypericifolia Linnaeus. Illustration: Schneider, Illustriertes Handbuch der Laubholzkunde **1**: 450, 451 (1906); Pignatti, Flora d' Italia **1**: 540 (1982); Krüssmann, Manual of cultivated broad-leaved trees & shrubs **3**: 351 (1986).

Shrub, 1–1.8 m, shoots finely downy, brown, rounded, arching. Leaves 1.5–3.5 × 1–1.4 cm, obtuse, obovate to lanceolate, entire or slightly scalloped towards the apex, grey-green above, lighter beneath, 3-veined at the base, slightly downy to hairless, almost stalkless. Flowers white, 3–6, in almost stalkless corymbs, if stalked then downy; petals circular, longer than the stamens. Fruit almost hairless; styles erect-recurved. *SE Europe to Siberia & C Asia*. H2. Late spring.

Subsp. **obovata** (Waldstein & Kitaibel) Huber. (*S. obovata* Waldstein & Kitaibel; *S. hypericifolia* var. *obovata* (Waldstein & Kitaibel) Maximowicz). Illustration:

Schneider, Illustriertes Handbuch der Laubholzkunde **1**: 451 (1906). Leaves narrowly obovate, rounded, entire or with 3–5-scalloped teeth towards the apex. Flowers *c.* 5 mm; sepals as long as perigynous zone; petals *c.* 3 mm, as long as the stamens. *SE Europe*. H2. Spring.

S. × multiflora Zabel. Shrub, *c.* 1.5 m, branches finely downy when young, brownish, slender. Leaves 2–3 × 1–1.5 cm, obovate, greyish green, paler and downy beneath when young, entire towards the long tapering base, veins 3–5. Flowers white, many, in usually stalkless corymbs, the lower ones usually on short stalks, leafy at the base.

A hybrid between *S. crenata* and *S. hypericifolia*.

31. S. crenata Linnaeus (*S. crenifolia* C.A. Meyer; *S. vaccinifolia* misapplied). Illustration: Schneider, Illustriertes Handbuch der Laubholzkunde **1**: 451, 455 (1906); Reichenbach, Icones florae Germanicae et Helveticae **24**: pl. 147 (1909); Krüssmann, Manual of cultivated broad-leaved trees & shrubs **3**: 351 (1986).

Densely bushy shrub to 1.5 m, shoots downy at first becoming hairless, reddish brown, striped, slender, round, slightly angled. Leaves 2–3.5 cm × 5–30 mm, obovate-oblong, acute or rounded, tapered at base, entire or with a few scalloped teeth towards the apex, 3-veined, greyish green, downy at first. Flowers white, *c.* 5 mm, in dense, semi-circular umbel-like corymbs to 2.5 cm, on small leafy stalks. Petals circular, obovate, usually shorter than the stamens. Fruit finely downy, styles erect, enclosed by sepals. *SE Europe to Caucasus & Altai Mts*. H2. Summer.

Similar to *S. hypericifolia*, which has 3-veined leaves from base but differs in its almost stalkless inflorescence.

32. S. yunnanensis Franchet (*S. sinobrahuica* W.W. Smith).

Shrub *c.* 2 m, shoots yellowish brown with dense felted hairs when young. Leaves 8–20 mm, ovate-rounded to obovate, doubly toothed to shallowly lobed or sometimes entire, rounded at apex, tapered towards the base, dull green and downy above, grey or white felted hairs below, stalks 2–5 mm. Flowers 1–2 cm, white, 10–20 in densely hairy corymbs on short leafy stalks. Calyx and flower-stalk downy. Petals roundish, almost twice as long as the 20 stamens. Fruit downy. *W China (Yunnan)*. H2. Summer.

33. S. gemmata Zabel (*S. mongolica* Koehne not Maximowicz). Illustration: Schneider, Illustriertes Handbuch der Laubholzkunde **1**: 451, 455 (1906); Krüssmann, Manual of cultivated broadleaved trees & shrubs **3**: 351 (1986).

Shrub 2–3 m, almost hairless, shoots reddish, slender, angled; winter buds slender, to 5 mm, pointed, longer than the leaf-stalk. Leaves 1–2 cm × 3–8 mm, narrowly elliptic to oblong, entire, sometimes with 3 teeth at the apex, hairless, bright green above and below, stalk short to 2.5 mm. Flowers 6–8 mm, white, 2–6 in very short or stalkless corymbs *c.* 2.5 cm; petals almost round, longer than the stamens. Fruit hairless; sepals spreading; style erect or spreading. *NW China, Mongolia.* H1. Summer.

34. S. calcicola W.W. Smith. Figure 100 (8), p. 190.

Shrub 70–150 cm, shoots reddish, hairless, slender, arching. Leaves 4–7 × 3–4 mm, ovate to elliptic, obtuse, entire, rounded at apex, tapered at base, hairless (leaves of strong shoots to 1.3 cm, semicircular, deeply 3-lobed, lobes coarsely and doubly scallop-toothed, truncate at base). Weaker shoots obovate-tapered, deeply toothed at apex. Flowers white tinged reddish on the outside, 6–8 in corymbs, stalks 2–5 mm, forming an inflorescence of 10–13 cm. Calyx-lobes hairless. Fruit hairless with erect styles. *W China (Yunnan).* H2. Summer.

35. S. baldschuanica B. Fedtschenko.

Dwarf shrub of rounded, compact habit, densely branched, thin, hairless. Leaves 8–150 mm, obovate, bluish green, toothed at the apex. Flowers white in small erect hairless corymbs; flower-stalks often with a bluish grey bloom. *SE former USSR.* H3. Summer.

36. S. arcuata J.D. Hooker. Illustration: Schneider, Illustriertes Handbuch der Laubholzkunde **1**: 462 (1906); Krüssmann, Manual of cultivated broadleaved trees & shrubs **3**: 351 (1986).

Shrub with short arching branches; stems reddish brown, strongly ribbed, downy when young, eventually almost hairless. Leaves 8–150 mm, oblong-elliptic, slightly downy, entire or occasionally with a few small teeth towards apex. Flowers *c.* 6 mm, white occasionally pink, 10–15 in downy 2–4 cm umbel-like corymbs at the ends of short leafy stalks; stamens 18–25, slightly longer than petals. Fruit hairless.

Himalaya (Nepal) to SW China. H2. Early summer.

37. S. × schinabeckii Zabel.

Shrub 1.2–1.8 m, shoots with scattered hairs, brownish yellow, striped towards the apex, flexuous, base 5–sided. Leaves 4.5–5 cm, oblong to ovate, hairless, doubly to deeply toothed, dark green above, bluish beneath. Flowers white, large, in corymbs; petals rounded, longer than the stamens. Styles almost curving outwards. *Garden origin.* H2. Summer.

A hybrid between *S. chamaedryfolia* and *S. trilobata.* Similar to *S. chamaedryfolia* but flowers larger.

38. S. veitchii Hemsley. Illustration: *Botanical Magazine*, 8383 (1911); Schneider, Illustriertes Handbuch der Laubholzkunde **2**: 960 (1912).

Shrub to 4 m, strong-growing, branches slightly downy, striped, reddish, 50–100 cm, arching. Leaves 2–5 cm × 8–20 mm, elliptic to oblong, occasionally obovate, obtuse, entire, base broadly tapered, hairless, or downy and glaucous beneath, stalks *c.* 2 mm. Flowers 4–5 mm, white, in finely downy, densely packed corymbs, 3–6 cm; Petals shorter than the stamens; calyx and stalks finely downy. Fruit hairless. *C & W China.* H2. Summer.

39. S. virginiana Britton. Illustration: Gleason, Illustrated flora of the northeastern United States and adjacent Canada **2**: 286(1952).

Shrub, 1 m or taller, hairless, with many branches. Leaves 2–5 × 1.5–1.8 cm, oblong-lanceolate, entire or with a few teeth towards the apex, acute, rounded to tapered at base, paler to glaucous beneath, hairless or finely hairy on veins beneath. Flowers white, many in slightly downy or hairless corymbs to 5 cm. Fruit hairless. *E USA.* H2. Summer.

Similar to *S. betulifolia* var. *corymbosa* but with leaves entire or with a few teeth and glaucous below.

40. S. pubescens Turczaninow. Illustration: Schneider, Illustriertes Handbuch der Laubholzkunde **1**: 455, 459 (1906).

Shrub, 1–2 m, branches densely felted, downy hairy when young, arching, slender, rounded. Leaves 3–4 cm, ovate to diamond-shaped to elliptic, downy above with grey felted hairs beneath, deeply toothed to almost 3-lobed, stalks 2–3 mm. Flowers 6–8 mm, white, in hairless, almost

semicircular corymbs. Petals as long as the stamens. Fruit hairless, styles spreading. *N China.* H2. Summer.

41. S. sargentiana Rehder.

Shrub, to 2 m, shoots downy when young, arching, round, slender, spreading, buds ovoid to obtuse with several scales. Leaves 2–4 cm × 4–15 mm, narrowly elliptic to obovate, finely toothed at the apex, entire at base, dull green and finely downy above, shaggy with paler hairs beneath, stalks 1–3 mm. Flowers *c.* 6 mm, creamy white, in dense, shaggy, numerous corymbs 2.5–4 cm; petals as long as stamens. Fruit almost hairless; styles spreading. *W China.* H2. Summer.

42. S. trichocarpa Nakai. Illustration: Nakai, Flora sylvatica Koreana **4**: pl 12 (1916); Bean, Trees & shrubs hardy in the British Isles, edn 1, **3**: 464 (1933).

Shrub, 1–2 m, shoots hairless, slender, rigid, angular, spreading. Leaves 2.5–6 × 1–2.5 cm, oblong to lanceolate, hairless or with scattered hairs, green, almost acute, entire or toothed near the apex, tapered at the base, stalk to 6 mm. Flowers *c.* 8 mm, white, in downy corymbs, 2.5–5 cm, at the end of short leafy twigs on the previous year's growth, forming an inflorescence over 30 cm, with the lower branches having 2–7 flowers, and often a leafy bract, stalks downy. Petals rounded, notched. Fruit downy. *Korea.* H2. Summer.

43. S. media Schmidt (*S. confusa* Regel & Koernicke). Illustration: Bean, Trees & shrubs hardy in the British Isles, edn 8, **4**: 487 (1980); RHS dictionary of gardening 2002 (1981); Krüssmann, Manual of cultivated broad-leaved trees & shrubs **3**: 351 (1986); Aeschimann et al, Flora alpina **1**: 729 (2004).

Upright shrub, 1–1.5 m, smooth, round branches, downy when young, sometimes hairless, yellow to brownish. Leaves 3–5 cm × 8–20 mm, ovate-oblong, base wedge-shaped, deeply toothed towards the apex, occasionally entire, bright green on both surfaces, more or less downy beneath, margins with long white hairs, stalks to 4 mm. Flowers *c.* 8 mm, white, many, in hairless or almost hairless racemes at the ends of small leafy side-shoots or sometimes in groups of small panicles. Petals circular, shorter than the stamens. Fruit downy; sepals reflexed; styles spreading or reflexed. *E Europe to NE Asia.* H2. Spring–early summer.

Subsp. **polonica** (Blocki) Pawloski
(*S. polonica* Blocki). Leaves 5–6 × 2–3 cm;
inflorescence covered with soft down,
petals yellowish, margins fringed. *Poland,
former Czechoslovakia.* H2. Summer.

S. × pikoviensis Besser (*S. nicoudiertii*
misapplied). Illustration: Reichenbach,
Icones florae Germanicae et Helveticae **24**:
pl. 149 (1909). Closely related to *S. crenata*
but shoots hairless, yellowish brown,
round. Leaves 2.5–5 cm, oblong, entire or
with a few teeth at the apex, very slightly
downy beneath. Flowers white, many, in
almost hairless corymbs, stalks 1.5–3 cm.
Petals circular, shorter than the stamens.
Fruit downy; styles usually straight, erect.
W Ukraine, former USSR. H2. Late spring.

A hybrid between *S. crenata* and
S. media.

44. S. henryi Hemsley. Illustration:
Botanical Magazine, 8270 (1909); RHS
dictionary of gardening **4**: 2002 (1956);
Huxley, Deciduous garden trees and
shrubs, pl. 181 (1979); Bean, Trees &
shrubs hardy in the British Isles, edn 8, **4**:
480 (1980).

Shrub, 2–3 m, open spreading habit,
shoots roundish, reddish brown, downy
when young. Leaves 3–8 × 2–3 cm,
oblanceolate, acute or rounded, coarsely
toothed at apex, entire on the smaller
leaves, hairless or almost downy above,
densely hairy beneath, stalks 4–6 mm.
Flowers white, *c.* 6 mm, in rounded
corymbs *c.* 5 cm on the end of short leafy
shoots; petals circular, longer than the
stamens; stalk and calyx downy. Fruit
hairy, slightly spreading. *C & W China.* H2.
Summer.

45. S. wilsonii Duthie. Illustration:
Botanical Magazine, 8399 (1911);
Schneider, Illustriertes Handbuch der
Laubholzkunde 2: 960 (1912).

Shrub 2–2.5 m, shoots downy when
young, reddish purple, arching. Leaves
2–5.5 cm, elliptic-obovate, entire or with a
few coarse teeth at the apex, base tapered,
dull green and downy above, greyish green
with longer and denser hairs below
especially on the veins, very short stalked.
Flowers *c.* 6 mm, white, many, in hairless,
semi-circular, terminal corymbs 3–5 cm, on
short leafy side-shoots, stalks hairless.
Petals as long as stamens; calyx hairless.
Fruit spreading, downy. *C & W China.* H2.
Summer.

Differs from *S. henryi* in having hairless
ovaries and hairless or slightly silky

flower-stalks. Leaves of flowering-shoots
entire, downy above and duller green.

46. S. mollifolia Rehder. Illustration:
Krüssmann, Manual of cultivated broad-
leaved trees & shrubs 3: 356 (1986).

Shrub 1.5–2 m, branches reddish purple,
hairy when young, arching. Leaves
2–3 cm × 5–10 mm, elliptic-obovate,
tapered at both ends (usually more so at
the apex), usually entire but sometimes
with a few teeth towards the apex, with a
few grey silky hairs on both sides, light
greyish green beneath, short-stalked.
Flowers *c.* 8 mm, white, in downy corymbs
to 2.5 cm, corymbs with short leafy stalks;
stamens 20. Fruit downy; sepals spreading;
styles erect or spreading. *W China.* H2.
Summer.

47. S. prunifolia Siebold & Zuccarini
(*S. prunifolia* var. *plena* Schneider). Figure
100(7), p. 190. Illustration: Schneider,
Illustriertes Handbuch der Laubholzkunde
1: 450, 451 (1906); Gleason, Illustrated
flora of the north-eastern United States and
adjacent Canada 2: 284 (1952);
Krüssmann, Manual of cultivated broad-
leaved trees & shrubs 3: 356 (1986).

Upright dense shrub, to 2–3 m, branches
downy at first, arching, nodding, slender.
Leaves 2.5–4.5 × 1–2 cm, ovate to oblong-
elliptic, acute at both ends, finely toothed,
bright green, shiny above, soft downy grey
beneath especially when young, stalk to
5 mm, orange to reddish brown in the
autumn. Flower-stalks 1.5–2 cm, slender,
hairless, flowers *c.* 1 cm white, double with
numerous broadly-ovate petals, 3–6 in
stalkless corymbs. Petals longer than
stamens. Fruit hairless, spreading. *Japan,
China (in cultivation but also found wild in
Hubei).* H2. Spring.

Only the double variant is found in
cultivation. The single-flowered variant is
S. prunifolia forma *simpliciflora* Nakai.
Korea, China, Taiwan.

48. S. bella Sims (*S. expansa* Wallich;
S. fastigiata Wallich). Illustration: *Botanical
Magazine,* 2426 (1823); Schneider,
Illustriertes Handbuch der Laubholzkunde
1: 468, 474 (1906); Bean, Trees & shrubs
hardy in the British Isles, edn 8, **4**: 471
(1980); Krüssmann, Manual of cultivated
broad-leaved trees & shrubs 3: 348 (1986).

Shrub to 1 m, occasionally to 1.8 m,
branches slender, spreading, angular,
downy when young. Leaves 2.5–5.5 cm,
ovate-elliptic to lanceolate, acute, rounded
to broadly tapered at base, doubly toothed

towards the apex, hairless on both
surfaces, or sometimes downy on the veins
beneath, glaucous beneath, stalks to 6 mm.
Flowers unisexual, pink to white, *c.* 6 mm,
in downy, loose, terminal corymbs 2–4 cm
on previous year's growth. In male flowers
stamens longer than the petals, in female
flowers stamens shorter. Sepals reflexed;
styles spreading. *Himalaya.* H2. Summer.

49. S. cantoniensis Loureiro
(*S. reevesiana* Lindley). Illustration:
Schneider, Illustriertes Handbuch der
Laubholzkunde 1: 459, 464 (1906);
Csapody & Tóth, Colour atlas of flowering
trees and shrubs, 96 (1982) – as *S.
cantoniensis* 'Lanceata'; Brickell (ed.), RHS
A–Z encyclopedia of garden plants, 1000
(2003).

Shrub to 1.8 m, hairless with slender,
arching, round branches. Leaves
2.5–5.5 × 1.3–1.8 cm, diamond-shaped to
lanceolate, 3-lobed or finely toothed, dark
green above, glaucous beneath with
distinct net-like venation, stalks 6–8 mm.
Flowers white, *c.* 1 cm, many in semicircular 2.5–5 cm corymbs on leafy stalks;
petals rounded to elliptic, longer than the
stamens. Fruit with sepals erect; styles
spreading. *China, Japan.* H2. Summer.

50. S. longigemmis Maximowicz.
Illustration: Bailey, Standard cyclopedia of
horticulture 3: 3212 (1935); Krüssmann,
Manual of cultivated broad-leaved trees &
shrubs 3: 352 (1986).

Shrub 1–1.5 m, branches hairless when
young, thin, spreading, angular. Leaves
3–6 cm, broadly ovate, distinctly wider at
the base, acute, toothed to doubly toothed,
glandular-tipped, bright green, hairless,
sometimes hairy on the veins beneath.
Flowers *c.* 6 mm, white, in loose, downy
corymbs 5–7 cm; stalk and calyx hairy.
Petals shorter than the prominent stamens.
Fruit almost hairless; style spreading. *NW
China.* H2. Summer.

51. S. miyabei Koidzumi (*S. silvestris*
Nakai). Illustration: Nakai, Flora sylvatica
Koreana 4: t. 5 (1916).

Upright shrub, 1–1.2 m, erect branches,
downy when young, slightly angular; buds
ovoid, small with several scales. Leaves
3–6 cm, ovate to elliptic, acute, rounded or
tapering to rounded at base, doubly
toothed, green and hairless or almost so on
both sides, stalks 2–5 mm. Flowers *c.*
8 mm, white, in downy flat corymbs
3–6 cm across. Petals almost round;
stamens 2–3 times longer than the petals.

Fruit with dense felted hairs; styles spreading. *Japan.* H2. Summer.

Var. **glabrata** Rehder. Differs in its larger leaves, which are wedge-shaped at the base, and flowers borne in hairless clusters. *C China.* H2. Summer.

52. S. amoena Spae (*S. expansa* K. Koch). Illustration: Schneider, Illustriertes Handbuch der Laubholzkunde **1**: 474 (1906).

Upright shrub to 2 m, stems downy, rounded, slender, few branches; buds hairy. Leaves to 10 × 3 cm, broadly-ovate, truncate at base, acute to acuminate, rounded to tapered, with fine even teeth, both simply and doubly toothed, usually only on the upper half, dark green above, sometimes hairless, glaucous, downy and hairy on the veins beneath, stalks 5–8 mm. Flowers dioecious, white to slightly pinkish red, in flat compound corymbs 5–20.5 cm borne on the current year's growth. Fruit hairy. *NW Himalaya.* H2. Summer.

53. S. rosthornii Pritzel.

Shrub to 2 m, shoots downy at first, spreading. Leaves 3–8 × 2–4 cm, broadly-ovate, acuminate, tapered or almost rounded at base, deeply and doubly toothed to slightly lobed, bright green above, often with soft downy hairs on both sides, especially on the veins beneath, stalks 5–8 mm. Inflorescence downy, flowers *c.* 6 mm, white, on long-stalked corymbs 5–9 cm across, at the end of leafy shoots, petals shorter than the stamens. Fruit downy; styles spreading. *W China.* H2. Summer.

54. S. × blanda Zabel (*S. reevesiana* misapplied). Illustration: Schneider, Illustriertes Handbuch der Laubholzkunde **1**: 464 (1906).

Shrub, 1.5–2 m, branches finely downy, brown, striped, angular, almost straight. Leaves 2.5–6 cm, ovate-oblong, acute, scalloped to sharply toothed, hairless above, with grey, dense, felted hairs beneath. Flowers white, in downy corymbs; petals circular, longer than the stamens. Fruit hairless; styles reflexed. *Garden origin.* H2. Spring–summer.

A hybrid between *S. cantoniensis* and *S. chinensis.*

55. S. × vanhouttei (Briot) Zabel (*S. aquilegiifolia* Briot var. *vanhouttei*). Illustration: Schneider, Illustriertes Handbuch der Laubholzkunde **1**: 464 (1906); Gleason, Illustrated flora of the north-eastern United States and adjacent

Canada **2**: 285 (1952); Hellyer, Collingridge illustrated encyclopedia of gardening, pl. 535 (1983); *Gartenpraxis* **8**: 8 (1989).

Shrub, to 2 m, branches, hairless, rod-like, long, arching. Leaves 1.5–4 × 1.5–3 cm, diamond-shaped to obovate, dark green above, bluish below, hairless, scalloped or coarsely toothed and occasionally 3–5-lobed at the apex. Flowers *c.* 8 mm, white, many, in flat corymbs 2.5–5 cm across; petals circular, twice as long as the partly sterile stamens. Fruit spreading; styles half-erect; sepals spreading. *Garden origin.* H2. Summer.

A hybrid between *S. cantoniensis* and *S. trilobata.*

56. S. chinensis Maximowicz (*S. pubescens* Lindley not Turczaninow). Illustration: Schneider, Illustriertes Handbuch der Laubholzkunde **1**: 455, 459 (1906).

Dense shrub, to 1.5 m, shoots with yellowish felted hairs when young, stems nodding, angular. Leaves 2.5–5 × 1.5–3.5 cm, ovate-rhombic to obovate, acute to rounded, rounded or broadly tapered at base, deeply toothed, glandular-tipped, occasionally 3-lobed, dark green and finely downy above, yellowish densely felted hairs below; stalks 2–8 mm. Flowers white, *c.* 1 cm, in densely hairy, many-flowered corymbs 2.5–5 cm. Sepals ovate-lanceolate, spreading at first, becoming erect. Stamens shorter or as long as petals. Fruit downy; styles spreading. *NE China.* H2. Summer.

57. S. chamaedryfolia Linnaeus (*S. flexuosa* Fischer). Illustration: Schneider, Illustriertes Handbuch der Laubholzkunde **1**: 455, 459 (1906); Gleason, Illustrated flora of the north-eastern United States and adjacent Canada **2**: 285 (1952); Pignatti, Flora d' Italia **1**: 540 (1982); Krüssmann, Manual of cultivated broad-leaved trees & shrubs **3**: 351, 357 (1986).

Shrub to 2 m, with stolons; branches hairless, yellowish, angled, flexuous. Leaves 2–7 × 2–3 cm, ovate-elliptic to oblong-lanceolate, acute, tapered, deeply toothed, bright green, usually hairless, occasionally with few scattered hairs below, stalk 5–10 mm. Flowers white, *c.* 1 cm, in corymbs of *c.* 4 cm, borne at the ends of short leafy shoots, occasionally stalkless on the upper shoots. Sepals reflexed, petals circular, shorter than the stamens. Fruit slightly downy, outspread, styles spreading. *NE Asia.* H2. Summer.

Var. **ulmifolia** (Scopoli) Maximowicz (*S. ulmifolia* Scopoli). Illustration: Schneider, Illustriertes Handbuch der Laubholzkunde **1**: 455 (1906); Reichenbach, Icones florae Germanicae et Helveticae **24**: pl. 150 (1909). Taller shrub, shoots less spreading, not as stiff. Leaves ovate, usually rounded at base, doubly toothed. Many-flowered semi-circular inflorescence *c.* 8 cm with longer stalks. Fruit with upright styles. *SE Europe to NE Asia & Japan.* H2. Summer.

Var. **flexuosa** (Fischer) Maximimowicz (*S. flexuosa* Fischer). Illustration: Schneider, Illustriertes Handbuch der Laubholzkunde **1**: 459 (1906). Smaller shrub, stems more angled or with wings; leaves smaller and narrower, simply toothed towards the apex or sometimes entire. Fewer flowers. *Siberia.* H2. Summer.

S. × gieseleriana Zabel. Illustration: Krüssmann, Manual of cultivated broad-leaved trees & shrubs **3**: pl. 129 (1986). Shrub, 1.5–3 m, downy grey, shoots obtuse, angular, somewhat flexuous. Leaves 3–4 × 1–2 cm, ovate, acute, downy, sharply toothed on the upper two-thirds, sometimes entire. Flowers *c.* 8 mm; petals as long as or shorter than the stamens. H2. Summer.

A hybrid between *S. cana* and *S. chamaedryfolia.*

S. × nudiflora Zabel (*S. hookeri* Zabel). Differs in having inflorescences to 8 cm with flowers in compound or sometimes simple, downy corymbs and fruits with reflexed sepals.

A hybrid between *S. chamaedryfolia* var. *ulmifolia* and *S. bella.*

13. SIBIRAEA Maximowicz
J. Cullen

Deciduous, prostrate to erect shrubs to 1 m; young shoots dark reddish, hairless. Leaves alternate (sometimes appearing whorled on condensed lateral shoots), linear-oblong to narrowly obovate, shortly stalked, obtuse but slightly mucronate, 3–10 × 0.6–2 cm, slightly hairy beneath and on the margins. Stipules absent. Flowers in racemes clustered at the tips of the shoots, to 6 mm across, mostly functionally unisexual (plants effectively dioecious). Perigynous zone small. Sepals 5, to 1 mm, triangular, erect. Petals white to greenish, 2–2.5 mm, almost circular but shortly clawed. Stamens numerous, borne on the margins of a cup-shaped, indented nectar-secreting disc. Ovary superior. Carpels 5, joined only at their extreme bases. Fruit a group of 5, 2-seeded follicles.

A genus of a single species with a remarkable distribution, centred in E Asia but with disjunct localities in SE Europe. The plant is easily grown and is propagated by seed or cuttings.

1. S. laevigata (Linnaeus) Maximowicz (*S. altaiensis* (Laxmann) Schneider; *Spiraea laevigata* Linnaeus; *S. altaiensis* Laxmann). Illustration: Csapody & Tóth, Colour atlas of flowering trees and shrubs, 99 (1982).

Former USSR (Siberia), China, former Yugoslavia. H3. Summer.

The correct name of the species is uncertain, both names cited above having been originally published within a very short time of each other. Variants with narrow, oblong-linear leaves have been described as var. **angustata** Rehder (*E Asia*). The European populations have been treated as var. **croatica** (Degen) Schneider, but are not in any way distinctive.

14. PETROPHYTUM (Torrey & Gray) Rydberg
J. Cullen
Small, low, hummock-forming shrubs with dense mats of evergreen basal leaves. Stems bearing reduced, bract-like leaves. Inflorescence a very dense, spike-like raceme. Flowers with bracts, sepals, petals and stamens perigynous, perigynous zone top-shaped to hemispherical, lined with a nectar-secreting disc. Sepals 5. Petals 5, creamy white. Stamens 20–40. Ovary superior of usually 5 (more rarely 3 or 7) free carpels, each containing 2–4 ovules, style long. Fruit a group of few-seeded follicles.

A genus of 3 species from N America. They are small shrubs, which grow best in crevices of rock or walls, as they are prone to injury from too much water. Propagation is by seed.

1a. Leaves 1-veined   **1. caespitosum**
  b. Leaves 3-veined   2
2a. Leaves ash-grey; follicles *c.* 3 mm   **3. cinerascens**
  b. Leaves green; follicles to 2 mm   **2. hendersonii**

1. P. caespitosum (Nuttall) Rydberg (*Spiraea caespitosa* Nuttall). Illustration: Schneider, Illustriertes Handbuch der Laubholzkunde **1**: 487 (1906); Rickett, Wild flowers of the United States **5**: pl. 47 (1971).

Basal leaves oblanceolate, tapering slightly towards the base, 7–15 × 1.5–4 mm, with a single vein visible, densely silvery silky. Stems to 20 cm; stem-leaves narrowly oblanceolate. Racemes 1.5–4 cm, dense. Sepals erect, densely hairy, *c.* 2 mm. Petals oblong-oblanceolate, 2–2.5 mm. Stamens 20, exceeding the petals. Carpels densely hairy. Follicles *c.* 2 mm. *W USA, extending E to Arizona and Texas.* H4. Summer.

2. P. hendersonii (Canby) Rydberg. Illustration: *The Rock Garden* **20**(79): 190 (1987); Brickell (ed.), RHS A–Z encyclopedia of garden plants, 795 (2003).

Basal leaves oblanceolate to obovate, 1–2.5 cm × 2–6 mm, with 3 veins visible beneath, hairy or hairless. Stems to 20 cm, stem-leaves narrowly obovate. Spikes dense, 2–4 cm. Sepals with marginal hairs, reflexed. Petals oblong to oblong-obovate, 2–2.5 mm. Stamens 35–40, as long as or slightly longer than petals. Carpels hairy. Follicles to 2 mm. *NW USA (Washington, Olympic Mts).* H5. Summer.

3. P. cinerascens (Piper) Rydberg.
Dense shrublet; shoots short, stout. Leaves to 2.5 cm, oblanceolate, obtuse or almost acute, thick, leathery, ash-grey with 3 veins, loosely hairy. Inflorescence-stem to 15 cm. Flower-stalks to 4 mm. Sepals 2 mm, lanceolate, ash-grey. Petals *c.* 2 mm, spathulate or oblanceolate, obtuse. Follicles *c.* 3 mm, sparsely hairy. *NW USA (Washington).* H5. Summer–autumn.

15. KELSEYA (Watson) Rydberg
M.F. Gardner
Evergreen, cushion-forming, tufted shrublets, 5–8 × 7–8 cm. Leaves entire, 2.5–4 mm, leathery, densely overlapping, covered with fine silky hairs. Flowers solitary, pinkish to purplish, fading to brown, almost stalkless; perigynous zone small; sepals 5; petals oblong-elliptic, 2–3 mm; stamens 10, longer than petals, reddish purple; styles 5. Fruit a follicle, *c.* 3 mm, splitting completely at 2 seams.

A genus of a single species from W North America, best grown in the alpine house, but also thriving out of doors in a trough or sink. It thrives if grown jammed between two pieces of rock in a well-drained gritty soil. Propagation is usually from seed.

1. K. uniflora (Watson) Rydberg. Illustration: *Bulletin of the Alpine Garden Society* **31**: 358 (1963); Everett, New York Botanical Gardens illustrated encyclopedia of horticulture **6**: 1893 (1981); Hitchcock et al., Flora of the Pacific northwest, 214 (1987).

W North America. H4. Spring–summer.

16. LUETKEA Bongard
J. Cullen
Low, evergreen shrubs with creeping woody stems and erect, herbaceous, sparsely hairy flowering-shoots, to 20 cm. Leaves to 2 cm, mostly towards the bases of the flowering-stems, alternate, stalked, usually divided into 3 narrow segments, which are themselves deeply 3-lobed, the ultimate segments more or less linear; leaves on the flowering stems smaller and less divided. Flowers in terminal racemes, 5–8 mm across; stalks short. Perigynous zone small. Sepals 5, triangular, acute. Petals white, *c.* 4 mm, exceeding the sepals. Stamens 20, somewhat united towards their bases, attached to the margins of a 10-lobed, fleshy nectar-secreting disc. Ovary superior, of 5 free carpels. Fruit a group of 5 follicles, each containing several seeds.

A genus of a single species from W North America. It may be grown in a sunny, well-drained position and is propagated by seed.

1. L. pectinata (Torrey & Gray) Kunze (*Spiraea pectinata* Torrey & Gray). Illustration: Rickett, Wild flowers of the United States **5**: pl. 47 (1971).

W North America. H4. Spring–summer.

17. ARUNCUS Linnaeus
J. Cullen
Large herbs with rather thick, woody rhizomes, dioecious or almost so. Leaves alternate, arising from swollen nodes, divided into 3 (rarely 4 or 5) leaflets, which themselves are similarly divided or more commonly pinnate, the ultimate segments irregularly doubly toothed. Flowers unisexual, in spike-like racemes that form a panicle (rarely a simple raceme). Sepals 5, small. Petals white or cream, slightly exceeding the sepals. Stamens numerous, small and vestigial in female flowers, attached at the margins of a cup-shaped, entire perigynous zone, which bears a nectar-secreting disc on the inner surface. Carpels 3 (rarely more), free, absent in male flowers. Fruit a group of follicles.

A genus of a few species from the northern hemisphere. They are similar in general appearance to species of *Astilbe* (Saxifragaceae, p. 95) and the two genera are sometimes confused; they can be most easily distinguished by the flowers of *Aruncus* having many stamens, those of *Astilbe* 4–10. They are easily grown, requiring rather damp, rich soil to do well, and are propagated by seed or by division of the rhizome.

1a. Flowers with conspicuous bracts
    3. aethusifolius
 b. Flowers without bracts   2
2a. Plant to 2 m or more; racemes to
    8 cm   **1. dioicus**
 b. Plant to 30 cm; racemes to 3 cm
    2. parvulus

1. A. dioicus (Walter) Fernald
(*A. sylvestris* Kosteletzky; *A. vulgaris* Rafinesque; *Spiraea aruncus* Linnaeus). Illustration: Justice & Bell, Wild flowers of North Carolina, 93 (1968); Polunin, Flowers of Europe, pl. 44 (1969); Hess et al., Flora der Schweiz **2**: 313 (1970); Aeschimann et al., Flora alpina **1**: 731 (2004).

Stems erect, unbranched, to 2 m or more, hairless. Leaves long-stalked, divided into 3 leaflets, which are themselves pinnate with 5 or more segments (the lowermost of which may also be somewhat pinnately divided); ultimate segments ovate-oblong, acuminate, irregularly doubly toothed, bright green, sparsely hairy on the veins beneath. Racemes spreading more or less at right angles to the panicle-axis, to 8 cm, axis with dense, crisped hairs. Flowers without bracts, to 5 mm across. Carpels usually 3. *Northern hemisphere*. H1. Summer.

A very variable species which has been divided into several varieties (sometimes recognised as species); the most commonly seen, apart from the species as described above are var. **triternata** (Maximowicz) Hara, with the leaflets divided into 3 segments (not pinnate) and 'Kneiffii', which has the ultimate segments deeply cut.

2. A. parvulus Komarov.
Similar to *A. dioicus*, but smaller, 15–30 cm; with shorter racemes to 3 cm. *E former USSR*. H1. Summer.

3. A. aethusifolius (Léveillé) Nakai.
Smaller, leaf-segments themselves deeply lobed, dark green above, racemes upright, flowers with conspicuous bracts. *Korea*. H3. Summer.

Very *Astilbe*-like, and doubtfully in cultivation.

18. HOLODISCUS Maximowicz
J. Cullen
Shrubs or small trees to 8 m. Leaves alternate, deciduous, simple, toothed and/or lobed, usually whitish or greyish hairy at least beneath. Stipules absent. Inflorescence a large terminal panicle with many flowers, lateral branches spreading or drooping; axis, branches and calyces densely hairy. Flowers small, shortly stalked, each subtended by a bract. Perigynous zone small. Sepals 5, triangular. Petals white, almost circular, shortly clawed, spreading. Stamens 20, borne on the margins of an entire, cup-shaped disc. Ovary superior, of 5 free carpels, usually hairy, each containing 2 ovules. Fruit a group of 5 indehiscent, 1-seeded achenes.

A genus of 8 species from W America, from Canada to Mexico. They grow well in sunny, open positions and may be propagated by seed, semi-ripe hardwood cuttings or by layering.

Literature: Ley, A., A taxonomic revision of the genus Holodiscus (Rosaceae), *Bulletin of the Torrey Botanical Club* **70**: 275–88 (1943).

1a. Leaves with deep major teeth or even lobes that are themselves toothed; blade of leaf truncate, rounded or somewhat tapered to base, distinct from the obvious stalk, which is 5–25 mm   **1. discolor**
 b. Leaves simply toothed; blade very gradually tapered to base, stalk indistinct, less than 2 mm   2
2a. Leaves toothed only at the rounded apex   **3. microphyllus**
 b. Leaves toothed almost to the base
    2. dumosus

1. H. discolor (Pursh) Maximowicz (*Spiraea discolor* Pursh). Illustration: *Bulletin of the Torrey Botanical Club* **70**: 287 (1943); Muller, Wild flowers of Santa Barbara, unnumbered pages (1958); Thomas, Flora of the Santa Cruz mountains, 198 (1961); Csapody & Tóth, Colour atlas of flowering trees and shrubs, 99 (1982).

Arching or spreading shrub to 6 m; bark reddish brown to grey or almost black, peeling; young shoots downy. Leaves ovate to elliptic, deeply toothed or even lobed, the lobes or major teeth themselves with teeth; base rounded, truncate or somewhat tapered, but clearly distinct from the stalk; blades 3–9 × 2–8 cm, sparsely hairy to hairless above, whitish hairy beneath; stalk

5–25 mm. Flowers to 5 mm across; petals white, to 2 mm, hairy towards the base outside. *W North America, from British Columbia to Baja California*. H3. Summer.

2. H. dumosus (Nuttall) Heller (*H. discolor* var. *dumosus* (Nuttall) Dippel). Illustration: *Bulletin of the Torrey Botanical Club* **70**: 287 (1943).

Widely spreading shrub to 3 m; bark of older shoots reddish, becoming grey and peeling; young shoots pale, very downy. Leaves 1–3 × 5–20 mm, ovate to spathulate, densely whitish hairy beneath, toothed to beyond half their length from the apex, teeth rarely themselves toothed, hairy above, base tapered very gradually; stalk absent or inconspicuous. Flowers to 5 mm across. Petals white, hairy towards the base outside. *W USA & NW Mexico*. H5. Summer.

Variable and not always clearly distinguishable from *H. discolor*.

3. H. microphyllus Rydberg. Illustration: *Bulletin of the Torrey Botanical Club* **70**: 287 (1943).

Generally a small spreading shrub, occasionally to 2 m; bark of older shoots reddish becoming grey or black and peeling; young shoots pale, hairy, sometimes glandular. Leaves obovate to spathulate, 5–18 × 3–10 mm, densely greyish downy on both surfaces, toothed only near the rounded apex, long-tapered at the base, without a distinct stalk. Petals 1.5–2 mm, white with a few hairs towards the base outside. *W USA, Mexico (Baja California)*. H5. Summer.

19. FILIPENDULA Miller
J. Cullen
Perennial herbs (often large) with tuberous roots or with rhizomes. Basal and lower stem-leaves pinnate with 3–5 or more major leaflets, often with smaller leaflets or lobes between them; terminal leaflet usually larger than the laterals, 3–5-lobed, the laterals lobed or not, all toothed. Inflorescence a large panicle with many flowers, branches arching erect or spreading. Flower-buds more or less spherical; flowers without bracts. Perigynous zone small. Sepals 5 or 6, small, erect in flower, persistent and reflexed in fruit. Petals 5 or 6, white, cream, pink or purplish red, spreading. Stamens 20–40, attached at the margin of a small, cup-shaped nectar-secreting disc. Carpels 5–10, free, sometimes compressed, 2-seeded, usually attached by their bases

but in 1 species attached on their inner faces, with part of the carpel projecting below the point of attachment. Fruit a group of achenes, erect and separate, or (in the species with laterally attached carpels) those of each flower spirally coiling together.

A genus of about 10 species of herbs, some of them very large, from north temperate regions, though with a concentration of species in E Asia. The name 'Spiraea palmata' has been used by different authors for several of the species, and material under this name (or under F. palmata) generally requires re-identification. The species are generally easily grown, requiring moist, rich soil, and are propagated by seed or by division.

Literature: Barnes, P.G., Confusion in cultivated meadowsweets (Filipendula Miller), The New Plantsman 5: 145–53 (1998); Lee, S. & Lee, J., A taxonomic study of Filipendula on the basis of fruit morphology, Korean Journal of Plant Taxonomy 28: 1–24 (1998).

Roots. Tuberous: 7.
Basal and lower stem-leaves. Pinnate with many, more or less equal, pinnately lobed or toothed leaflets: 7; pinnate with the terminal leaflet larger than the 3–5-lobed lateral leaflets: 5, 6; pinnate with the terminal leaflet larger than the unlobed lateral leaflets: 1, 2, 3, 4. Usually white- or greyish hairy beneath: 1, 2, 4, 5.
Sepals. With a small patch of sparse hairs inside: 1, 3, 6.
Petals. White or cream: 1, 2, 3 (an uncommon cultivar), 4, 5, 7; pink or purplish red: 3, 6. Blade tapering gradually into an indistinct claw: 1, 2, 7; blade abruptly tapered or truncate into a distinct claw: 4, 5, 6; blade deeply truncate or even auriculate at junction of blade and distinct claw: 3.
Achenes. Flattened, hairy on the keels: 1, 2, 3; not flattened, hairy all over: 7; neither flattened nor hairy: 4, 5, 6. Attached above the base, all those of each flower coiling spirally as they mature: 4.

1a. Leaves with numerous leaflets all more or less similar in size; carpels hairy all over the surface, not flattened; roots tuberous   **7. vulgaris**
  b. Leaves with rarely more than 5 leaflets, the terminal larger than the laterals; carpels compressed, with long hairs on the keels only, or neither compressed nor hairy   2

2a. Lateral leaflets of the basal and lower stem-leaves 3–5-lobed   3
  b. Lateral leaflets of the basal and lower stem-leaves not lobed   4
3a. Petals pink or purplish red; leaves green beneath, hairs only on the veins   **6. rubra**
  b. Petals white or cream; leaves densely white-hairy over the whole surface beneath   **5. palmata**
4a. Carpels hairless, not compressed, attached above the base, achenes of each flower coiling spirally as they mature   **4. ulmaria**
  b. Carpels compressed with hairs along the keels, attached at their bases, not coiling spirally   5
5a. Stems hairless; petals pink or pinkish red (white in an uncommon cultivar), the blade deeply and abruptly truncate or even auriculate at the junction with the distinct claw   **3. purpurea**
  b. Stems variably hairy; petals white or cream, the blade gradually tapered into the indistinct claw   6
6a. Plant 1.5–3.5 m; sepals with a small patch of sparse hairs inside   **1. kamtschatica**
  b. Plant to 80 cm; sepals entirely hairless   **2. vestita**

1. F. kamtschatica (Pallas) Maximowicz (*Spiraea kamtschatica* Pallas; *S. palmata* Miquel). Illustration: Komarov (ed.), Flora SSSR **10**: 287 (1941).

Very large rhizomatous perennial herb, stems 1.5–3.5 m, shortly hairy. Basal and lower stem-leaves pinnate with 3–5 leaflets, the terminal leaflet deeply 3–5-lobed, cordate at base, much larger than the unlobed lateral leaflets, all dark green and hairless above, densely or sparsely whitish hairy beneath, all irregularly doubly toothed. Stipules large, narrowly oblong, toothed. Inflorescence rather spreading, branches hairy. Sepals usually 5, hairless outside, with a small patch of sparse hairs inside. Petals usually 5, white, 3–4 mm, the blade gradually tapered into an indistinct claw. Carpels usually 5, attached at their bases, compressed with long hairs on the keels. Achenes 1–5, erect, with long hairs on the keels. *Japan. N China, E former USSR.* H3. Summer.

The largest and most spectacular of the species.

2. F. vestita (G. Don) Maximowicz.
Perennial herb to 80 cm. Stem with rather fine adpressed hairs of variable density. Basal and lower stem-leaves pinnate with 3–5 leaflets, terminal leaflet 3–5-lobed, cordate at base, much larger than the unlobed lateral leaflets, all leaflets irregularly doubly toothed, dark green and hairless above, densely grey or white-hairy beneath. Stipules broadly oblong, toothed. Inflorescence-branches erect arching, finely hairy. Sepals usually 5, hairless. Petals usually 5, c. 3 mm, white or very pale yellowish cream, gradually tapered to an indistinct claw. Carpels 5–10, attached at their bases, compressed, the keels with long hairs. *Himalaya from Afghanistan to Nepal; W China.* H4. Summer.

3. F. purpurea Maximowicz (*Spiraea palmata* Thunberg). Illustration: Botanical Magazine, 5726 (1868); Kitamura et al., Coloured illustrations of herbaceous plants of Japan **2**: pl. 29 (1978); Brickell (ed.), RHS A–Z encyclopedia of garden plants, 450 (2003).

Perennial rhizomatous herb to 1.3 m. Stems hairless. Basal and lower stem-leaves pinnate with 3–5 leaflets, terminal leaflet deeply 3-lobed, truncate at base, much larger than the unlobed lateral leaflets, all dark green and hairless above, with short hairs on the veins beneath, all irregularly doubly toothed. Stipules rather narrow, lanceolate, little-toothed. Inflorescence spreading, branches hairless or hairy. Sepals usually 5, pinkish, hairless outside, with a small patch of sparse hairs inside. Petals usually 5, 2–3.5 mm, pink or purplish red, the blade abruptly truncate or even auriculate where it joins the short but distinct claw. Carpels usually 5, attached by their bases, compressed, with long hairs along the keels. Follicles very compressed, with long hairs along the keels. *Originated in gardens in Japan, not known for certain in the wild.* H4. Summer.

Some plants in cultivation as F. purpurea 'Alba' are similar but have white petals; some of these, at least, have hairless carpels (and follicles?) and may be of hybrid origin.

4. F. ulmaria (Linnaeus) Maximowicz (*Spiraea ulmaria* Linnaeus). Illustration: Gleason, Illustrated flora of the north-eastern United States and adjacent Canada **2**: 300 (1952); Ary & Gregory, The Oxford book of wild flowers, 80 (1962); Polunin, Flowers of Europe, pl. 44 (1969); Garrard & Streeter, Wild flowers of the British Isles, pl. 35 (1983).

Upright perennial herb to 1.2 m, with pinkish rhizomes. Basal and lower stem-leaves pinnate with 2–5 pairs of leaflets, terminal leaflet larger than lateral leaflets, deeply divided into 3 lobes, dark green and hairless above, usually densely white-downy beneath, irregularly doubly toothed. Stipules rounded, toothed. Inflorescence-branches usually arching, with dense or sparse short hairs. Flowers fragrant. Sepals usually 5, yellowish, shortly hairy outside, hairless inside. Petals usually 5, cream-white, 3–5 mm, rather abruptly tapered into a distinct claw. Carpels 6–10, hairless, attached by part of their inner angle, projecting a little below the point of attachment. Achenes of each flower coiling spirally together as they mature. *Europe, W Asia, naturalised in E North America.* H1. Summer.

A double-flowered variant is known as 'Plena' and a variant with the leaves variegated is known as 'Aurea' or 'Aureovariegata' (var. *aureovariegata* Voss).

5. F. palmata (Pallas) Maximowicz (*Spiraea palmata* Pallas not other authors; *S. digitata* Willdenow). Illustration: Komarov (ed.), Flora SSSR **10**: 287 (1941).

Rhizomatous perennial herb to 1 m. Basal and lower stem-leaves pinnate with 3–5 leaflets, all leaflets deeply 3–5-lobed, the terminal leaflet considerably larger than the laterals, all doubly toothed, dark green and hairless above, densely white-downy beneath. Stipules oblong-rounded, toothed. Inflorescence rather spreading, branches hairless or with short hairs. Sepals usually 5, hairless. Petals usually 5, white, 2–3 mm, rather abruptly tapered into a distinct claw. Carpels 5–8, attached at their bases, hairless. Achenes erect. *E former USSR.* H1. Summer.

Uncommon in cultivation, often confused with other species (especially *F. purpurea*).

6. F. rubra (Hill) Robinson (*Spiraea lobata* Gronovius; *S. palmata* Murray). Illustration: Gleason, Illustrated flora of the north-eastern United States and adjacent Canada **2**: 300 (1952); Rickett, Wild flowers of the United States **1**: pl. 41 (1966); Brickell (ed.), RHS A–Z encyclopedia of garden plants, 450 (2003).

Upright rhizomatous perennial herb to 3 m. Basal and lower stem-leaves pinnate, all leaflets 3-lobed, the terminal leaflet larger than the laterals, all deeply doubly toothed, dark green and hairless above, green and hairy only on the veins beneath. Stipules oblong, toothed.

Inflorescence-branches arching upright, branches hairless. Sepals usually 5, hairless outside, with a small patch of sparse hairs inside. Petals usually 5, pink or (in 'Venusta') purplish red, rather abruptly tapered or truncate into a distinct claw. Carpels 6–10 attached at their bases, hairless. Achenes 6–10, erect, hairless. *E & C USA.* H3. Summer.

7. F. vulgaris Moench (*Spiraea filipendula* Linnaeus; *Ulmaria filipendula* (Linnaeus) Hill; *F. hexapetala* Gilibert). Illustration: Bonnier, Flore complète **3**: pl. 169 (1914); Polunin, Flowers of Europe, pl. 44 (1969); Garrard & Streeter, Wild flowers of the British Isles, pl. 35 (1983).

Perennial herb to 80 cm, with swollen, tuberous roots; vegetative parts hairless or almost so. Leaves pinnate with numerous leaflets all more or less the same size (diminishing slightly downwards), pinnately lobed or toothed. Stipules large, ovate-oblong, toothed. Inflorescence branches erect arching or spreading. Flowers fragrant. Sepals usually 6, whitish, hairless. Petals usually 6, creamy white, obovate, 6–7 mm, tapering gradually into an indistinct claw. Carpels 6–12, attached by their bases, hairy all over, not compressed. Achenes erect, hairy all over. *Europe, N Africa, W Asia.* H1. Summer.

'Flore Pleno' has double flowers and 'Grandiflora' has slightly larger cream-coloured flowers.

20. RHODOTYPOS Siebold & Zuccarini
J. Cullen

Deciduous shrub to 2 m or more; branches greyish, hairless. Leaves opposite, shortly stalked, 4–10 cm ovate to ovate-oblong, apex acuminate, base rounded, irregularly doubly toothed, with long whitish hairs on both surfaces. Stipules linear or thread-like. Flowers terminating short, leafy, lateral shoots, 2.5–4 cm across. Sepals 4, toothed, alternating with 4 additional small lobes. Petals 4, white, almost circular. Stamens numerous. Ovary superior. Carpels 4, free. Fruit a group of up to 4 shining, black, almost dry drupes surrounded by the persistent and enlarged calyx.

A genus of a single species from China, Japan and Korea, grown for its rose-like flowers. It is easily grown and propagated by softwood cuttings taken in summer.

1. R. scandens (Thunberg) Makino (*R. kerrioides* Siebold & Zuccarini). Illustration: *Botanical Magazine*, 5805 (1869); *Addisonia* **23**: pl. 759 (1954–59); Bean, Trees &

shrubs hardy in the British Isles, edn 8, **3**: 942 (1976); Csapody & Tóth, Colour atlas of flowering trees and shrubs, pl. 52 (1982).

China, Japan, Korea. H1. Late spring–summer.

21. KERRIA de Candolle
J. Cullen

Deciduous shrub to 3 m; branches green, hairless. Leaves simple, alternate, stalks short, 2.5–11 cm, lanceolate to ovate, apex acuminate, base cordate to truncate, doubly toothed, hairless or with adpressed bristles above, sparsely white-hairy beneath. Stipules narrow, falling early. Flowers solitary, terminating short, leafy lateral shoots, 2–5 cm across. Sepals 5. Petals 5 (many in 'Pleniflora', see below), yellow or orange-yellow. Stamens numerous. Ovary superior; carpels 5–8, free. Fruit a group of up to 8 achenes surrounded by the persistent calyx.

A genus of a single species probably originally from China, but widely cultivated there and elsewhere in the Far East for a long period, often as the double-flowered variant. It is easily grown and can be propagated by softwood cuttings taken in summer.

1. K. japonica (Linnaeus) de Candolle. Illustration: Perry, Flowers of the world, 262 (1972); Kitamura et al., Coloured illustrations of trees and shrubs of Japan **2**: 216 (1977); Everett, New York Botanical Gardens illustrated encyclopedia of horticulture 1895 (1981); *Il Giardino Fiorito* **54**: 31 (1988).

China, Japan & E Asia. H1. Late spring–summer.

The double-flowered 'Pleniflora' is most commonly grown. Variants with variegated leaves ('Variegata', 'Aureo-variegata') and with branches striped green and yellow ('Aureo-vittata') are also grown.

22. NEVIUSIA A. Gray
J. Cullen

Deciduous shrub, 1–2 m. Stems brownish, hairless. Leaves alternate, shortly stalked, ovate to ovate-oblong, 3–7 cm, acute or acuminate at the apex, rounded to slightly cordate at the base, irregularly doubly toothed, with adpressed, white bristles on both surfaces. Stipules very small, thread-like. Flowers in cymes or solitary, terminating short, leafy, lateral shoots, to 1.5 cm across. Perigynous zone small. Sepals 5, spreading, toothed. Petals absent.

Stamens numerous, yellowish white, exceeding the sepals. Carpels 2–4, free. Fruit a group of 2–4 achenes surrounded by the persistent perigynous zone and calyx.

A genus of a single species from the SE USA, grown occasionally for its attractive stamens. It is easily grown and propagated by softwood cuttings in summer.

1. N. alabamensis Gray. Illustration: *Botanical Magazine*, 6806 (1885); Dean et al., Wild flowers of Alabama, 81 (1973); *Arnoldia* **36**: 59 (1976); Csapody & Tóth, Colour atlas of flowering trees and shrubs, pl. 52 (1982).

SE USA (Alabama). H3. Summer.

23. RUBUS Linnaeus
J. Cullen

Erect, scrambling, trailing or prostrate shrubs or low shrubs, shoots rarely herbaceous and dying down in winter, often with prickles on the stems, leaf- and leaflet-stalks and inflorescences. Leaves deciduous or evergreen, entire, lobed or divided into 3–many leaflets, which are usually toothed and may themselves be lobed; stipules always present, often conspicuous, sometimes falling early. Flowers in clusters, racemes or panicles, sometimes solitary, usually bisexual. Sepals 4 or 5 or rarely more, spreading, erect or reflexed. Petals 4 or 5 or rarely more, very variable in size, white, pink, red or purple, spreading or erect and adpressed to the stamens. Stamens numerous; anthers sometimes hairy. Ovary superior; carpels 5–many, free, borne on a usually cylindric or conical receptacle. Fruit a 'berry' composed of 5–many variably coherent fleshy drupelets, which may separate from the receptacle as a (hollow) unit, or may fall from the plant by abscission of the flower-stalk while still attached to the receptacle. Figure 99(19), p. 183.

A large genus of an uncertain number of species from most parts of the world. Many of the species produce edible fruit, and some are grown commercially for this purpose; the fruit of other wild species is also collected and eaten.

The identification of the species is in general reasonably straightforward except for those of Subgenus *Rubus* (see p. 212). In many of the species the shoots (stems, canes) are biennial, bearing leaves during their first year and shoots bearing leaves and inflorescences arising from the axils of the first-year leaves in the second. The leaves of the first and second year may differ considerably in degree of division and/or lobing. In this account it is the leaves of the flowering-shoots that are described unless otherwise indicated. In some plants, shoots borne low down on the second-year stems do not flower, but grow rapidly and mimic first-year shoots; the leaves on these shoots are often aberrant.

Different species of the genus show a remarkable range of hair types on stems, leaf-stalks and leaves. Many species have long, soft, sometimes felted hairs, which may be mixed with stalked or stalkless glands, bristles (long, straight, parallel-sided hairs) or prickles (hard, straight or variously hooked small spines).

In a few species the flowers are functionally unisexual; stamens and carpels are both present, but the members of one or other set are infertile. This characteristic is sometimes difficult to judge.

Literature: Focke, W.O., Species Ruborum, *Bibliotheca Botanica* **72** (1910–11); Bailey, L.H., Certain cultivated Rubi, *Gentes Herbarum* **1**: 139–200 (1925); Newton, A. & Edees, E.S., Brambles of the British Isles (1988).

1a. Plants herbaceous, stems dying down each winter 2
  b. Plants with persistent, aerial, usually woody stems 3
2a. Leaves divided into 3 distinct leaflets; flowers bisexual **1. arcticus**
  b. Leaves simple, sometimes 3-lobed; flowers unisexual **2. chamaemorus**
3a. Aerial woody stems thin, creeping close to the soil surface, giving rise to short, erect, herbaceous flowering-shoots 4
  b. Aerial woody stems stout, prostrate, trailing, arching, scrambling or erect, not as above 6
4a. Stipules deeply divided into narrow segments; anthers with sparse, long hairs; leaves persistent, leathery and conspicuously wrinkled above **3. pentalobus**
  b. Stipules entire or toothed; anthers hairless; leaves not persistent, leathery and wrinkled as above 5
5a. Leaves simple, 3-lobed **4. calycinus**
  b. Leaves divided into 3 leaflets **5. nepalensis**
6a. Stipules broad, attached directly to shoot, free from leaf-stalks, and/or sometimes falling early 7
  b. Stipules narrow, lanceolate or linear, attached directly to the leaf-stalks 16
7a. Stems without prickles; stipules persistent; leaves always simple **6. tricolor**
  b. Stems with at least some prickles; stipules usually falling early; leaves simple or compound 8
8a. Leaves divided into leaflets, or lobed for more than one third of the distance from margin to stalk-attachment 9
  b. Leaves simple, not lobed or more shallowly lobed 11
9a. Leaves divided into distinct, shortly-stalked leaflets, which have many, conspicuous, parallel, lateral veins that end in many, even, hair-tipped teeth **8. lineatus**
  b. Leaves lobed, but not to the base, venation not as above 10
10a. Leaves densely white- or pale brown-hairy beneath; flowers in racemes longer than broad, borne in leaf-axils **7. henryi**
  b. Leaves densely brown-hairy beneath; flowers in more or less stalkless clusters in leaf-axils **12. reflexus**
11a. Inflorescence a large, widely spreading terminal panicle 12
  b. Inflorescences forming condensed clusters in leaf-axils 14
12a. Leaves rather leathery, lanceolate (rarely lanceolate-ovate), usually hairless or very sparsely hairy beneath; margins with small, distant teeth **9. ichangensis**
  b. Leaves rather thin, ovate-lanceolate, usually hairy, at least on the veins beneath; margins with regular, close, conspicuous teeth 13
13a. Flowering-shoots densely hairy; branches of inflorescence with conspicuous stalked glands **10. parkeri**
  b. Flowering-shoots hairless; branches of inflorescence without glands **11. lambertianus**
14a. Leaves as broad as long or broader, rounded to the short, abruptly acute apex **15. irenaeus**
  b. Leaves longer than broad, tapering gradually to the pointed apex 15
15a. Leaves not, or only extremely shallowly, lobed **13. flagelliflorus**
  b. Leaves distinctly 3-, 5- or 7-lobed **14. kumaonensis**

16a. Flowers all unisexual, in large, usually widely spreading panicles; leaflets usually short compared to their very long, prickly stalks   17

b. Flowers usually bisexual in inflorescences of various kinds, not as above; at least the lateral leaflets longer than their stalks   19

17a. Leaves simple (reduced to a terminal leaflet)   **20. parvus**

b. Leaves with 3–5 leaflets (the blades sometimes very reduced, but leaflet-stalks evident)   18

18a. Leaflets ovate or ovate-oblong, acutely toothed; fruits yellowish   **21. australis**

b. Leaflets lanceolate or oblong-lanceolate, with few, rather deep, triangular lobes; fruits red to orange   **22. squarrosus**

19a. Stems without prickles, upright, with flaky bark; leaves simple, lobed   20

b. Stems with prickles, upright, arching, scrambling, trailing or prostrate, bark not as above; leaves usually compound   23

20a. Flowers solitary, 5 cm or more across   21

b. Flowers 2 or more together in racemes or panicles, less than 5 cm across   22

21a. Sepals in fruit brownish outside, reddish inside; leaf-lobes acute   **17. trilobus**

b. Sepals in fruit greenish or brownish inside and out; leaf-lobes rather rounded   **16. deliciosus**

22a. Petals purple; flowers many, in panicles   **18. odoratus**

b. Petals white; flowers few, in racemes or rarely solitary   **19. parviflorus**

23a. Fruit separating from cylindric or conical receptacle, therefore hollow   24

b. Fruit not separating from receptacle, falling from parent plant by abscission of fruit-stalks   43

24a. Leaves on flowering-shoots simple, though often lobed   25

b. Leaves on flowering-shoots with 3 or more distinct leaflets   26

25a. Leaves to 3.5 cm; flowers solitary   **23. microphyllus**

b. Leaves 4 cm or more; flowers in racemes or clusters   **24. crataegifolius**

26a. Most flowers on a plant solitary, occasionally 2 or 3 together   27

b. Most flowers on a plant in racemes, panicles or clusters, occasionally a few solitary   30

27a. Leaflets mostly 3; petals pink   **25. spectabilis**

b. Leaflets always 5 or more; petals white   28

28a. Leaves, at least beneath, with scattered, stalkless, glistening yellow glands   **27. rosifolius**

b. Leaves without stalkless, glistening yellow glands   29

29a. Lateral leaflets 1–3 cm; sepals usually hairy and glandular outside, sparingly hairy inside towards the apex   **28. amabilis**

b. Lateral leaflets 3.5–9 cm; sepals usually hairless outside, densely hairy inside towards the apex   **26. illecebrosus**

30a. Leaflets on leaves of flowering-shoots mostly 5 or more, if 3, then terminal leaflet tending to be deeply 3-lobed   31

b. Leaflets on leaves of flowering-shoots always 3, the terminal not deeply 3-lobed   34

31a. Terminal leaflet very broadly obovate to diamond-shaped, broadest near the tip   **33. coreanus**

b. Terminal leaflet ovate, lanceolate or elliptic, broadest near the middle or below   32

32a. Leaflets 7–15, the terminal lanceolate, pinnately lobed   **31. thibetanus**

b. Leaflets 5 or 7 (rarely 3), the terminal not as above   33

33a. Inflorescence raceme-like, long, not flat-topped; leaflets scarcely lobed; stems conspicuously bloomed   **32. cockburnianus**

b. Inflorescence corymb-like, flat-topped; leaflets conspicuously lobed; stems not bloomed   **40. mesogaeus**

34a. Plant with numerous bristles as well as soft hairs and prickles   35

b. Plant without bristles   37

35a. Leaflets broadly elliptic, rounded at the tip, very evenly toothed   **35. ellipticus**

b. Leaflets ovate to narrowly elliptic, acute, irregularly toothed or shallowly lobed   36

36a. Terminal leaflet much larger than the lateral leaflets; flowers 1.5–2.5 cm across; bristles reddish or purplish   **34. phoenicolasius**

b. Terminal leaflet little larger than the laterals; flowers 1–1.8 cm across; bristles brownish or pale   **41. idaeus**

37a. Calyx hairless outside, though occasionally with small prickles near the base   38

b. Calyx variously hairy outside   39

38a. Carpels and styles hairy, the fruit very densely hairy; flowers 2.5–3.5 cm across   **30. lasiostylus**

b. Carpels, styles and fruits not as above; flowers 1–2 cm across   **29. biflorus**

39a. Inflorescence an open raceme, not flat-topped; prickles mostly straight or almost so; fruit usually red   **41. idaeus**

b. Inflorescence a dense, flat-topped cluster; prickles often hooked; fruit usually purplish or black   40

40a. Plant with gland-tipped hairs at least on flower-stalks and/or calyx   **39. glaucifolius**

b. Plant entirely without gland-tipped hairs   41

41a. Leaves rather finely and regularly toothed   **38. glaucus**

b. Leaves coarsely and irregularly toothed   42

42a. Mature fruit black; prickles on flower-stalk thin, not conspicuously broadened at base, little hooked   **36. occidentalis**

b. Mature fruit purplish; prickles on flower-stalks conspicuously broadened at base, hooked   **37. leucodermis**

43a. Stems prostrate or trailing on the ground, the flowering-shoots erect, arising from them   44

b. Stems upright, arching or scrambling, sometimes arching over and rooting at the tip   50

44a. Stems conspicuously bloomed   45

b. Stems not conspicuously bloomed   48

45a. Flowers bisexual; prickles broadly based   **55. caesius**

b. Flowers unisexual; prickles narrowly based, needle-like   46

46a. Most parts of the plant with conspicuous, stalked glands   **58. macropetalus**

b. Stalked glands absent or few, inconspicuous   47

47a. Leaves bright green, thinly hairy to hairless beneath   **56. vitifolius**

b. Leaves dull or grey-green, densely hairy beneath   **57. ursinus**

48a. At least some of the leaves of the first-year stems leathery and persistent; bristles usually present **52. trivialis**

b. Leaves all deciduous; bristles usually absent 49

49a. Flowers 2–3 cm across **53. flagellaris**

b. Flowers 3–4 cm across **54. roribaccus**

50a. All leaves with a whitish undersurface produced by dense hairs beneath 51

b. All leaves more or less green beneath, though often hairy 56

51a. Inflorescence a few-flowered cluster, not obviously longer than broad, without a distinct, long axis **51. cuneifolius**

b. Inflorescence generally many-flowered, raceme-like, longer than broad, with a distinct, long axis 52

52a. Stems conspicuously bloomed **47. ulmifolius**

b. Stems usually not bloomed 53

53a. Each leaf with the lowermost leaflets unstalked **49. linkianus**

b. Each leaf with the lowermost leaflets with short, but distinct stalks 54

54a. Prickles in the inflorescence mostly straight 55

b. Prickles in the inflorescence hooked **50. procerus**

55a. Petals white or pale pink **48. bifrons**

b. Petals deep pink **43. elegantispinosus**

56a. Inflorescence, leaf-stalks and new shoots with conspicuous, stalked glands **44. allegheniensis**

b. Stalked glands absent or few and inconspicuous 57

57a. Leaflets deeply lobed **42. laciniatus**

b. Leaflets toothed but not deeply lobed 58

58a. Flower-clusters raceme-like, much longer than broad **46. bellobatus**

b. Flower-clusters corymb-like, little longer than broad **45. frondosus**

Subgenus **Cylactis** (Rafinesque) Focke. Plants herbaceous, stems dying down each winter. Leaves divided into 3 distinct leaflets; stipules broad or narrow, generally free from the leaf-stalk. Flowers bisexual. Fruit consisting of few, large, loosely coherent drupelets.

1. R. arcticus Linnaeus. Figure 101(1), p. 204. Illustration: *Botanical Magazine,*

132 (1797); Polunin, Flowers of Europe, pl. 43 (1969).

Stems to 30 cm, borne on a persistent rhizome, dying down annually, softly hairy or hairless, without prickles. Leaves long-stalked, divided into 3 distinct leaflets (very rarely deeply 3-lobed only); leaflets broadly elliptic to almost diamond-shaped, 1.2–4.5 × 1.5–3.5 cm, shortly stalked, margins unevenly toothed, lateral leaflets often oblique at the base, usually softly hairy on both surfaces. Flowers 1–3 together, long-stalked, 1.5–2.5 cm across. Sepals and petals 5–7 or more, petals longer than sepals, pink. Carpels downy. Fruit made up of about 20 drupelets, red. *N Europe, N Asia, N America.* H1. Summer.

Subsp. **acaulis** (Michaux) Focke (*R. acaulis* Michaux) is smaller, has strictly 1-flowered stems, and the flowers are somewhat larger (to 3 cm across). *N North America.*

Subgenus **Chamaemorus** Focke. Plants herbaceous, stems dying down each winter, without prickles. Leaves simple or 3-lobed; stipules broad, leaf-like. Flowers solitary, terminal, unisexual. Fruits of rather few coherent drupelets.

2. R. chamaemorus Linnaeus. Figure 101(2), p. 204. Illustration: Ary & Gregory, The Oxford book of wild flowers, 80 (1962); Rickett, Wild flowers of the United States **1**: pl. 141 (1966); Polunin, Flowers of Europe, pl. 44 (1969); Keble Martin, The concise British flora in colour, edn 2, pl. 29 (1969).

Stems 5–20 cm, borne on a persistent rhizome, dying down each winter, with downwardly directed, soft hairs, some stalked glands and persistent stipules. Leaves 2–6.5 × 3–10.5 cm, almost circular to kidney-shaped, cordate at base, simple, usually shallowly 3–5-lobed, margins irregularly toothed, upper surface rough, lower surface with soft, white hairs and sometimes stalked glands. Flower solitary, 1.5–2.5 cm across, terminal; flower-stalks softly hairy and glandular. Sepals erect-spreading, ovate, reddish. Petals 5 or more, white, exceeding the sepals. Fruit golden orange when ripe. *N Europe, N Asia, N America.* H1. Summer.

Subgenus **Chamaebatus** Focke. Plants with woody, persistent aerial stems lying close to the soil surface and giving rise to erect, leafy flowering shoots; small prickles present. Leaves

simple, sometimes 3-lobed; stipules broad, ovate, sometimes divided, free from the leaf-stalks. Flowers solitary or 1–3 together, bisexual. Fruit with a few rather large drupelets.

3. R. pentalobus Hayata (*R. calycinoides* Hayata not Kuntze; *R. calcinoides* var. *macrophyllus* Li; *R. fockeanus* misapplied). Figure 101(3), p. 204. Illustration: *Botanical Magazine,* 9644 (1944); Li et al., Flora of Taiwan **3**: 110 (1977).

Prostrate small shrub, sometimes with arching branches. Stems creeping and rooting, hairy and with a few small prickles. Flowering-shoots hairy and with a few prickles. Leaves leathery, persistent, conspicuously wrinkled above, simple, rounded at the apex, base cordate, shallowly 3–5-lobed, margins unevenly and coarsely toothed, hairy above when young, persistently brownish hairy beneath. Stipules 7–10 mm, deeply divided into narrow, tapering segments. Flowers solitary, shortly stalked, to 2 cm across. Sepals ovate, irregularly toothed at the apex or entire. Petals white, shorter than to almost as long as sepals. Anthers with a few hairs. Fruit scarlet. *Taiwan.* H4. Summer.

This species was placed in Subgenus *Chamaebatus* by its original author. However, Sealy in the notes to the *Botanical Magazine* plate cited above, suggested it should be placed in Subgenus *Malachobatus*. It is anomalous in both subgenera, but apparently less so in *Chamaebatus.*

Its naming is very complex. It was first described as *R. calycinoides* (a name under which it is frequently found in gardens at present) by Hayata, who unfortunately overlooked an earlier use of this name, by Kuntze, for a Himalayan species of Subgenus *Malachobatus*, which is perhaps not cultivated (see below). This oversight was missed by Sealy and by the authors of the Flora of Taiwan, who use the name *calycinoides*; in the Flora a variety, *macrophyllus* Li, was recognised, under which *R. pentalobus* Hayata was cited as a synonym. The difference between the varieties was based mainly on size, and seems insignificant; the name *R. pentalobus* is therefore the earliest that can legitimately be used for the species.

The plant was introduced in the 1930s under the name *R. fockeanus*; genuine *R. fockeanus* Kurz is a species from Subgenus *Dalibarda* Focke (which has no

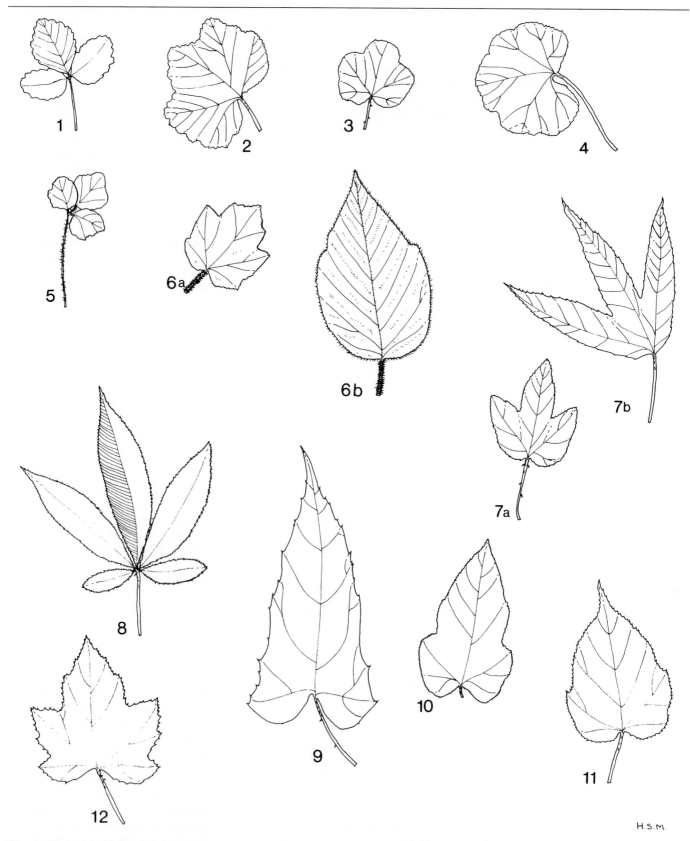

Figure 101. Leaves of *Rubus* species (× 0.5). 1, *R. articus*. 2, *R. chamaemorus*. 3, *R. pentalobus*. 4, *R. calycinus*. 5, *R. nepalensis*. 6a, b, *R. tricolor*. 7a, b, *R. henryi*. 8, *R. lineatus*. 9, *R. ichangensis*. 10, *R. parkeri*. 11, *R. lambertianus*. 12, *R. reflexus*.

other cultivated representatives), differing from *R. pentalobus* in its deciduous leaves with 3 distinct leaflets, slightly toothed stipules and lack of prickles; it occurs in the Himalaya and SW China.

4. R. calycinus Wallich. Figure 101(4), p. 204.

Aerial stems woody, thin, creeping, rooting, giving rise to short, leafy flowering-stems, which bear backwardly directed prickles and some long hairs. Leaves simple, slightly 3-lobed, margins finely toothed, 2–4.5 × 3.5–6.5 cm, almost circular to kidney-shaped, base cordate, with long, fine hairs above and prickles on the veins beneath; stalks long and prickly; stipules entire and toothed. Flowers terminal, solitary or 2 together, 1.5–3 cm across. Sepals ovate, toothed at the apex. Petals white, about as long as sepals. Anthers hairless. Fruit reddish, of numerous drupelets. *Himalaya.* H3. Spring.

Subgenus **Dalibardastrum** Focke. Plants with trailing or scrambling woody aerial stems, without prickles. Leaves simple or divided into 3 leaflets; stipules attached to the stem, free from the leaf-stalks, large and broad, persistent in our species. Flowers bisexual.

5. R. nepalensis (J.D. Hooker) Kuntze (*R. nutans* G. Don). Figure 101(5), p. 204. Illustration: *Botanical Magazine*, 5023 (1857).

Prostrate, small shrub; stems rooting, thin, creeping, covered with soft, curled hairs and irregularly spreading bristles. Leaves divided into 3 distinct leaflets; leaflets ovate-oblong to diamond-shaped, evergreen, shortly stalked, margins irregularly and finely toothed, softly hairy above and beneath; leaf-stalks hairy like the stems. Stipules ovate, entire or sparsely toothed, hairy. Flowers solitary, terminal, 2–3 cm across. Sepals narrowly triangular, acute or acuminate, entire or somewhat toothed at the tip; perigynous zone densely bristly. Petals as long as to a little longer than the sepals, white. Fruit red, somewhat concealed by the persistent sepals. *W Himalaya, from NW India to Nepal and perhaps Sikkim.* H3. Summer.

R. fockeanus Kurz is very similar (see under *R. pentalobus*).

6. R. tricolor Focke. Figures 99(6), p. 183 & 101(6), p. 204. Illustration: *Botanical Magazine*, 9534 (1938); Bean, Trees & shrubs hardy in the British Isles, edn 8, **4**: 237 (1980).

Large, deciduous, trailing or scrambling shrub to 2 m high or wide, with short stems covered with dense, spreading, brownish bristles. Flowering-shoots arching or prostrate, many-leaved, with similar bristles. Leaves simple, scarcely lobed or with 3–5 or more small lobes, ovate or broadly ovate, base cordate, apex acute or acuminate, 4.5–11.5 × 4–9 cm, margins finely toothed, dark green and hairless or with stiff, adpressed bristles along the veins above, densely white-hairy and with dense bristles along the veins beneath. Stipules large, persistent, ovate, toothed. Flowers few in terminal racemes, sometimes with solitary flowers in a few leaf-axils below the raceme, to 3 cm across. Sepals white-hairy and bristly outside, long-tapered, toothed. Petals white, somewhat shorter than the sepals. Anthers hairless. Fruit bright red. *SW China (Yunnan, Sichuan).* H5. Summer.

Subgenus **Malachobatus** Focke. Mostly shrubs with woody, upright, arching, scrambling or somewhat prostrate aerial branches, with prickles. Leaves simple or divided into several leaflets, when simple, often lobed; stipules large, attached directly to the stems, often early deciduous.

7. R. henryi Hemsley & Kuntze. Figure 101(7), p. 204. Illustration: *Botanical Magazine*, n.s., 33 (1948).

Evergreen, scrambling shrub to 6 m. Stems white-hairy, at least when young, with few prickles. Leaves rather deeply 3- or 5-lobed, the lobes extending to more (usually much more) than one third of the distance from margin to the point of stalk-attachment, rarely a few leaves almost unlobed; lobes elliptic to narrowly elliptic, to 12 cm, long-tapered at the apex, toothed, dark green above, densely white- or pale brown-hairy beneath. Stipules narrowly elliptic, entire, deciduous. Flowers borne in long racemes in the leaf-axils, 1–2 cm across. Sepals triangular, densely hairy, reflexed. Petals pink, more or less erect, somewhat shorter than the sepals, soon falling. Anthers with sparse, long hairs. Fruit black and shining, to 1.5 cm across. *C & W China.* H3. Spring–summer.

Var. **bambusarum** (Focke) Rehder (*R. bambusarum* Focke) – illustration: Brickell (ed.), RHS A–Z encyclopedia of garden plants, 939, 2003 – has narrow leaf-lobes, and is the most commonly cultivated variant.

8. R. lineatus Blume. Figure 101(8), p. 204. Illustration: Bean, Trees & shrubs hardy in the British Isles, edn 8, **4**: 228 (1980).

Deciduous or evergreen, scrambling shrub to 3 m; stems with close, parallel, adpressed white hairs at least when young, with a few short prickles or prickles virtually absent. Leaves divided into 3 or 5 shortly stalked leaflets; leaflets elliptic or elliptic-oblanceolate, 9–15 × 1.5–4 cm, tapering to base and apex, dark green above, greenish or silvery with dense, parallel, adpressed hairs beneath, lateral veins very conspicuous, numerous, parallel, ending in close, even, hair-tipped teeth at the margins. Stipules ovate, entire. Flowers few in short axillary clusters. Sepals ovate to ovate-lanceolate, acuminate, silky hairy outside, to 1 cm. Petals greenish white, 4–5 mm. Fruit red. *Himalaya & W China to Indonesia (E to Borneo).* H5. Spring–summer.

The leaflets of this species are very striking with numerous, parallel lateral veins, numerous regular teeth and, usually, shining, silvery undersurfaces.

R. splendidissimus Hara. Similar, but leaves usually with 3 leaflets, the hair on all parts more woolly, less regularly arranged and including some gland-tipped bristles. *Nepal, Bhutan.* H5. Spring–summer.

9. R. ichangensis Henry & Kuntze. Figure 101(9), p. 204.

Deciduous, scrambling or more or less prostrate shrub; stems with stalked glands and a few hooked prickles. Leaves thinly leathery, lanceolate or more rarely ovate-lanceolate, margins sometimes somewhat shallowly and sinuously lobed, with small, distant teeth, hairless or very sparsely hairy beneath and on the main veins above; stalks prickly. Stipules narrowly lanceolate, toothed, quickly deciduous. Flowers in a large, spreading, terminal panicle to 25 cm, 7–10 mm across. Sepals erect, ovate-triangular, acute, with dense short hairs inside and out, outside also with adpressed bristles and spreading, gland-tipped bristles. Petals white, about as long as sepals. Fruit of 12–20 drupelets, red. *C & W China.* H3. Summer.

10. R. parkeri Hance. Figure 101(10), p. 204.

Like *R. ichangensis* but shoots densely hairy and prickly, leaves thinner, broader and rather more conspicuously but still shallowly lobed, rather densely hairy;

panicle conspicuously glandular; fruit of fewer drupelets, black. *C China*. H3. Early summer.

11. R. lambertianus Seringe. Figure 101 (11), p. 204.

Like *R. ichangensis* but leaves broadly ovate, thinner but half-evergreen, conspicuously toothed, abruptly acuminate, stipules divided into narrow segments; petals shorter than sepals; fruit of 15–20 drupelets, red. *C & S China, Japan, Taiwan*. H3. Summer.

12. R. reflexus Ker Gawler. Figure 101 (12), p. 204. Illustration: *Botanical Magazine*, 7716 (1900); *Gartenpraxis* 1981 (Jan), 28.

Deciduous scrambling shrub; stems with very dense and close brown hairs and sparse, minute prickles. Leaves 6–13.5 × 6–15 cm, rather deeply 3- or 5-lobed, the lobes triangular or narrowly triangular, acute, finely to coarsely toothed, with brown hairs along the veins above and densely and closely covering the whole surface beneath. Stipules large, persistent, deeply divided into narrow, brown-hairy lobes. Flowers to 1.5 cm across, almost stalkless in condensed axillary clusters. Sepals ovate, densely hairy within, very densely brown-hairy outside. Petals about as long as sepals. Anthers hairless or with a few long hairs. Fruit spherical, purple-black. *S China*. H5. Early summer.

Plants cultivated as *R. moluccanus* are often this species. The name *R. moluccanus* Linnaeus applies to a species from Indonesia not in general cultivation.

13. R. flagelliflorus Focke.

Evergreen scrambling or trailing shrub. Stems hairy, at least when young, and with small prickles. Leaves ovate to ovate-lanceolate, not lobed, acute, cordate at base, 8–16 cm, margins toothed, hairy above when young, densely hairy with yellowish hairs beneath. Stipules deeply divided. Flowers in condensed axillary clusters. Sepals ovate, acute, densely hairy outside, hairless and purple inside, reflexed. Petals white, about as long as sepals, quickly deciduous. Fruit hemispherical, purple-black. *C & W China*. H4. Summer.

14. R. kumaonensis Balakrishnan (*R. reticulatus* J.D. Hooker).

Large scrambling shrub. Stems whitish hairy, with a few short, straight or weakly curved prickles. Leaves ovate to almost circular, with 3, 5 or 7 conspicuous but shallow, acute lobes, 10–20 × 10–20 cm, margins finely toothed, base cordate, upper surface with a conspicuous network of veins and with long spreading hairs, lower surface whitish with short, curled hairs and longer, somewhat spreading hairs. Stipules to 1 cm, lanceolate, deeply divided into narrow segments. Flowers to 1.5 cm across, in axillary clusters. Sepals ovate-acuminate, whitish hairy. Petals white, shorter than sepals. *N India*. H3. Summer.

Part of a complex of rather similar Himalayan species, including **R. rugosus** Smith with smaller leaves with rounded lobes and dense flower-clusters, and the genuine *R. calycinoides* Kuntze (see p. 203), which is larger with prickles on the main veins and stalks of the leaves and larger stipules. The name *R. moluccanus* Linnaeus has been applied to all of these species from time to time (see above).

15. R. irenaeus Focke. Figure 102(1), p. 207.

Trailing or almost prostrate shrub; stems densely hairy with short, curled hairs and scattered, longer, adpressed hairs, sometimes with a few prickles. Leaves very broadly ovate to kidney-shaped, 7–10 × 9.5–12 cm, very shallowly 3-lobed, margins regularly toothed, cordate at base, rounded to the abruptly acute apex, dark green above, whitish or pale brown beneath with dense, curled, whitish or brownish hairs and longer, spreading hairs. Stipules ovate-oblong, deeply toothed, to 3 cm. Flowers 1.5–2 cm across, in short, condensed clusters in the leaf axils. Sepals ovate, the apex acuminate-tailed, densely hairy. Petals white, slightly exceeding the sepals. Anthers with long hairs. Fruit red. *C China*. H3. Summer.

Subgenus **Anoplobatus** Focke. Arching or upright shrubs without prickles; leaves simple, lobed; stipules narrow, attached directly to the leaf-stalks. Flowers bisexual, usually large.

16. R. deliciosus Torrey. Figure 102(2), p. 207. Illustration: *Botanical Magazine*, 6062 (1873); Perry, Flowers of the world, 264 (1972); Bean, Trees & shrubs hardy in the British Isles, edn 8, **4**: 220 (1980).

Deciduous shrub to 3 m, with arching and spreading branches; bark brownish, peeling, without prickles, but with soft, white hairs while young. Flowering-shoots softly hairy. Leaves 2.5–6.5 × 2–8.5 cm, semi-circular to ovate in outline, with 3 or 5 rather rounded, coarsely and irregularly toothed lobes, base truncate to cordate, softly hairy above and beneath. Stipules oblong, entire. Flower solitary, 5–6.5 cm across. Sepals ovate, tailed, hairy and with scattered glands outside, remaining greenish or becoming completely brown in fruit. Petals exceeding the sepals. Fruit hemispherical, dark purple. *W USA*. H5. Summer.

17. R. trilobus Seringe. Figure 102(3), p. 207. Illustration: *Botanical Magazine*, n.s., 452 (1964).

Very similar to *R. deliciosus* but leaves 4–7.5 × 4–7 cm, darker green and with more acute lobes, sepals less glandular outside, becoming brown outside and red inside in fruit. *S Mexico*. H5–G1. Spring–summer.

Very difficult to distinguish from *R. deliciosus*, though from a different area. In gardens, the hybrid between them, 'Benenden', often known as R. × 'Tridel', makes identification even more difficult.

18. R. odoratus Linnaeus. Figure 102(4), p. 207. Illustration: *Botanical Magazine*, 323 (1801); House, Wild flowers, pl. 101 (1961); Justice & Bell, Wild flowers of North Carolina, pl. 89 (1968).

Upright or scrambling deciduous shrub to 3 m; bark peeling, pale brown; shoots without prickles, hairy when young, flowering-shoots with soft hairs and conspicuous stalked glands. Leaves broadly triangular, 7–17 × 8–24 cm, cordate at base, with 3 or 5 triangular, acute, toothed lobes, sparsely hairy above, hairy along the veins beneath. Stipules lanceolate. Flowers 3.3–5 cm across, usually numerous in spreading panicles, fragrant; branches of the panicle conspicuously glandular to the naked eye. Sepals ovate, tailed, glandular-hairy. Petals exceeding the sepals, bright purple. Fruit red. *N America, N Mexico*. H2. Late spring–summer.

19. R. parviflorus Nuttall. Figure 102(5), p. 207. Illustration: *The Garden* **102**: 224 (1977); Stace, New flora of the British Isles, fig. 405 (1991).

Upright or rather low and tangled deciduous shrub to 2 m; bark peeling, shoots without prickles but with soft hairs and shortly stalked glands when young, flowering-shoots with soft hairs and similar glands. Leaves almost circular in outline, 4–19 × 5–22 cm, with 3 or 5 (rarely 7) triangular, acute, coarsely toothed lobes, cordate at base, with soft spreading hairs and shortly stalked glands above and

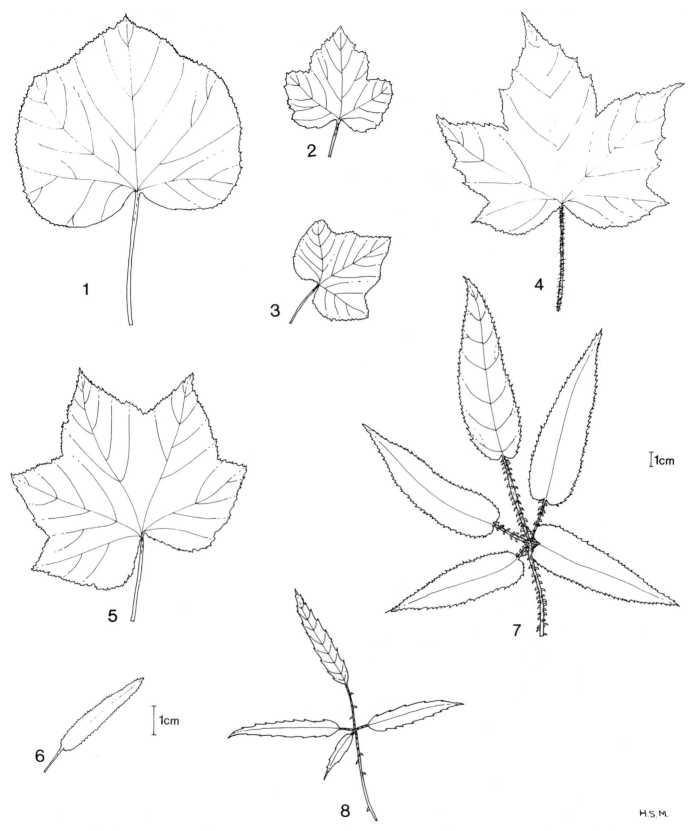

Figure 102. Leaves of *Rubus* species (× 0.5). 1, *R. irenaeus*.
2, *R. deliciosus*. 3, *R. trilobus*. 4, *R. odoratus*. 5, *R. parviflorus*. 6, *R. parvus*.
7, *R. australis*. 8, *R. squarrosus*.

beneath. Stipules lanceolate. Flowers 2–10 in corymb-like racemes (rarely solitary), 2.5–4 cm across; branches of racemes softly hairy and with shortly stalked glands (not conspicuous to the naked eye). Sepals ovate, tailed, hairy and glandular. Petals white, exceeding the sepals. Fruit broadly convex, red. *N America, N Mexico.* H2. Summer.

Subgenus **Lampobatus** Focke.
Scrambling evergreen shrubs with leathery leaves, usually compound, the leaflets often reduced and much shorter than their prickly stalks. Flowers small, unisexual, in usually widely spreading panicles.

20. R. parvus Buchanan. Figure 102(6), p. 207. Illustration: Salmon, Field guide to the alpine plants of New Zealand, pl. 204, 205 (1968).

Low-growing shrub with the main stems creeping and rooting, without prickles when mature. Leaves simple (reduced to a single, terminal leaflet), leathery, linear to linear-oblong, acute, shallowly cordate or rounded at base, margins regularly and finely toothed, borne on long, prickly stalks. Flowers in panicles to 2 cm across. Sepals ovate-acuminate, hairy. Petals white. Fruit red. *New Zealand (South Island).* H5–G1.

21. R. australis Forster. Figure 102(7), p. 207.

Young plants creeping and rooting. Shoots slender, hairy, usually with numerous slender prickles. Leaves evergreen, divided into 3–5 leaflets with long, prickly stalks; leaflets thin, oblong, toothed, acute, 1–3 × 1–2 cm. Adult plants forming mounds to 10 m or more. Adult leaves similar to the juvenile but relatively broader, leathery, rounded to the base, 3–5 × 1–3.5 cm; stalks and leaflet-stalks very prickly. Flowers in panicles 6–12 cm across. Sepals ovate, obtuse, hairy. Petals white, somewhat exceeding sepals. Fruit yellowish. *New Zealand.* H5–G1.

Normally seen only in the juvenile phase in cultivation.

22. R. squarrosus Fritsch. Figure 102(8), p. 207.

Similar to *R. australis* but with yellowish prickles, leaflets lanceolate or oblong-lanceolate, with few, rather triangular large teeth or lobes, to 7 × 3 cm, rounded at the base; the blades of the leaflets are often very reduced or absent; petals

yellowish; fruit orange-red. *New Zealand.* H5–G1.

The most frequently cultivated of the New Zealand species (often called *R. australis* in gardens). When the leaflet-blades are reduced, as is often the case, the plant resembles a mass of barbed wire, consisting mostly of the prickly stems, leaf- and leaflet-stalks.

Subgenus **Idaeobatus** Focke.
Most plants with biennial stems, with or rarely without prickles. Leaves simple or mostly divided into several leaflets. Stipules narrow, attached to the leaf-stalks. Flowers bisexual. Fruit separating from the cylindric or conical receptacle when ripe, and therefore hollow.

23. R. microphyllus Linnaeus filius (*R. incisus* Thunberg). Figure 103(1), p. 209.

Low, trailing shrub; stems reddish, white-bloomed, with a few prickles. Leaves simple, ovate or broadly ovate in outline, 1.5–3.5 × 1.5–3.2 cm, entire or shallowly 3-lobed, margins finely toothed, hairless above or sparsely hairy along the veins towards the base, hairy along the veins beneath, undersurface glaucous; base truncate or cordate, apex acute. Flowers 1.5–2.2 cm across, erect, solitary in the leaf-axils. Sepals lanceolate, hairless outside, densely hairy inside. Petals white or white flushed with pink, considerably longer than the sepals. *Japan, China.* H3. Spring.

24. R. crataegifolius Bunge. Figure 103 (2), p. 209.

Upright shrub to 3 m; stems prickly, reddish, not bloomed. Leaves simple, palmately 3- or 5-lobed, 4–16 × 3.5–19 cm, broadly ovate in outline, the middle lobe often narrowed to its base, base truncate or cordate, apex acute, margins irregularly doubly toothed; hairless or sparsely hairy on the veins above, more densely so on the veins beneath. Flowers 1.5–2 cm across, few in racemes or clusters. Sepals lanceolate, hairless or sparsely hairy outside, densely hairy inside. Petals white, exceeding the sepals. Fruit red. *Japan, China, Korea.* H3. Summer.

25. R. spectabilis Pursh. Figure 103(3), p. 209. Illustration: *Edwards's Botanical Register* **17**: pl. 1425 (1831).

Upright shrub to 2 m, spreading rapidly by rhizomes; stems with numerous fine prickles below. Leaves divided into usually

3 leaflets, hairless or with long hairs along the veins above, with rather dense long hairs along the veins beneath; leaflets deeply toothed or somewhat lobed, terminal leaflet ovate or broadly ovate, 4.5–9 × 2.5–7 cm, lateral leaflets 2.5–7 × 1.5–5 cm. Flowers solitary (occasionally 2 or 3 together) in the leaf-axils, long-stalked, drooping, fragrant, 2–3.5 cm across. Sepals lanceolate, acuminate, hairy inside and out. Petals pink or pinkish purple, considerably exceeding the sepals. Fruit orange. *W North America.* H3. Late spring.

This can become a very invasive weed in some situations; once established, it is very difficult to eradicate. A double-flowered variant ('Flore Pleno') is sometimes grown.

26. R. illecebrosus Focke. Figure 103(4), p. 209. Illustration: *Botanical Magazine*, 8704(1917).

Small shrub with creeping stems and annual, upright flowering-shoots to 1 m; stems prickly. Leaves mostly with 5–7 lanceolate, doubly toothed leaflets with conspicuous pinnate veins, hairless or sparsely bristly above, hairless beneath; terminal leaflet 4–10.5 × 1.5–2.5 cm, lateral leaflets 3.5–9 × 1–2.5 cm. Flowers mostly solitary in the leaf axils, or 2–3 together forming a flat-topped cluster, 2.5–4.5 cm wide. Sepals usually hairless outside, very densely hairy towards the apex inside. Petals white, exceeding the sepals. Fruit large, scarlet. *Japan.* H3. Late summer.

27. R. rosifolius Smith. Figure 103(5), p. 209. Illustration: *Botanical Magazine*, 6970(1887).

Erect or sometimes scrambling shrub to 2.5 m (though often less), stems softly hairy and prickly. Leaves with 5–7 lanceolate or more rarely ovate, doubly toothed leaflets, sparsely hairy above with adpressed bristles, beneath with similar bristles along the veins and with glistening, yellow stalkless glands on the surface; terminal leaflet 3–8 × 1.5–2.5 cm, lateral leaflets 2–7 × 1–2 cm. Flowers 2–3.5 cm across, solitary. Sepals lanceolate, long-acuminate, with fine spreading hairs and yellow glands outside, and dense, velvety hairs inside. Petals as long as to slightly longer than the sepals, white. Fruit red. *Himalaya to Japan & Indonesia (Sumatra).* H5. Summer.

A double-flowered variant, var. **coronarius** Sweet is sometimes grown.

Figure 103. Leaves of *Rubus* species (× 0.5). 1, *R. microphyllus*.
2, *R. crataegifolius*. 3, *R. spectabilis*. 4, *R. illecebrosus*. 5, *R. rosifolius*.
6, *R. amabilis*. 7, *R. biflorus*. 8, *R. lasiostylus*. 9, *R. thibetanus*.
10, *R. cockburnianus*.

28. R. amabilis Focke. Figure 103(6), p. 209.

Upright shrub to 2 m or more, stems with small prickles. Leaves with 7–11 ovate to lanceolate, conspicuously doubly toothed leaflets, hairless above, with long fine hairs and a few small prickles along the veins beneath; terminal leaflet 2.5–6 × 1.5–3.5 cm, lateral leaflets 1–3 cm × 5–20 mm. Flowers 3–5 cm across, solitary. Sepals lanceolate, hairy and glandular outside, sparingly hairy inside towards the apex. Petals white, longer than the sepals. Fruit red. *W China*. H5. Summer.

29. R. biflorus Smith. Figure 103(7), p. 209. Illustration: *Botanical Magazine*, 4678 (1852); Brickell (ed.), RHS A–Z encyclopedia of garden plants, 939 (2003).

Upright shrub to 3 m; stems very glaucous or whitish bloomed, with numerous prickles. Leaves with mostly 3 leaflets, sparsely hairy above, densely white-hairy beneath, margin with irregular, rounded teeth, sometimes somewhat lobed. Flowers 1–3 together in the leaf-axils, to 2 cm across. Sepals ovate, hairless outside, hairy inside towards the acuminate tip. Petals white, as long as or slightly longer than the sepals. Fruit yellow. *Himalaya*. H3. Spring–summer.

Var. **quinqueflorus** Focke, with conspicuously white stems and more flowers in the cluster, is usually grown.

30. R. lasiostylus Focke. Figure 103(8), p. 209. Illustration: *Botanical Magazine*, 7426 (1895).

Upright shrub to 2 m; stems glaucous or whitish bloomed, with numerous prickles. Leaves with mostly 3 (rarely 5) leaflets, sparsely bristly above, densely white-hairy beneath; terminal leaflet broadly ovate to almost circular, base somewhat cordate, 7.5–11 × 6.5–9 cm, considerably larger than the lateral leaflets (which are 6–9 × 3.5–6 cm), all irregularly coarsely doubly toothed or somewhat lobed. Flowers 1–few, drooping, in axillary clusters, 2.6–3.4 cm across. Sepals ovate-lanceolate, acuminate, hairless outside, densely white-hairy within. Petals reddish, shorter than the sepals. Styles hairy. Fruit red but densely covered in long, whitish hairs, which obscure the fruit colour. *W China*. H5. Summer.

31. R. thibetanus Franchet. Figure 103 (9), p. 209.

Shrub to 2 m with very bloomed, prickly stems. Leaves divided into 7 or more leaflets, lanceolate overall, tapering to the apex, rather densely greyish hairy above, very densely white-hairy beneath, toothed; terminal leaflet lanceolate, pinnately lobed towards its base, toothed towards its apex, 3–4.5 × 1–2 cm, lateral leaflets 9–20 × 4–10 mm. Flowers in racemes terminating short leafy shoots (sometimes some solitary in axils below the raceme), 7–9 mm across. Sepals densely hairy inside and out. Petals red-purple, hairy outside especially towards the base. Fruit black, bloomed. *W China*. H5. Spring–summer.

32. R. cockburnianus Hemsley (*R. giraldianus* Focke). Figure 103(10), p. 209. Illustration: *Journal of the Royal Horticultural Society* **94**: 514 (1969).

Upright shrub to 3 m; stems whitish bloomed, with scattered prickles. Leaves with 5–7 (rarely to 9) leaflets, dark green and sparsely hairy at least along the veins above, with whitish hairs beneath giving a greenish white colour to the surface, all irregularly doubly toothed; terminal leaflet ovate, sometimes lobed, to 10 × 5 cm; lateral leaflets narrower, 6–7.5 × 2.5–3 cm. Flowers numerous, in terminal racemes. Petals pinkish purple. Fruit black. *China*. H3. Spring–summer.

33. R. coreanus Miquel. Figure 104(1), p. 211.

Upright or arching shrub to 3 m; stems with conspicuous white bloom and numerous prickles. Leaves with usually 5 (rarely 7) leaflets, hairless above except along the veins, usually densely white-hairy beneath, margins irregularly toothed; terminal leaflet usually wedge-shaped or obovate, broadest towards the top, 2–5.5 × 1.5–4.5 cm; lateral leaflets ovate, 1.5–5 × 1–3.5 cm. Flowers 1–1.7 cm across, in flat-topped terminal racemes. Sepals lanceolate, acuminate, rather loosely hairy outside, more densely so inside. Petals pink, shorter than the sepals. Fruit red or blackish. *China, Japan, Korea*. H3. Spring–summer.

34. R. phoenicolasius Maximowicz. Figure 104(2), p. 211. Illustration: *Botanical Magazine*, 6479 (1889); *Gardening from 'Which'* (February): 28 (1988).

Upright shrub to 3 m or more; stems hairy and with numerous reddish purple bristles and few fine, almost straight prickles. Leaves with 3 broadly ovate to ovate, acute, shallowly lobed and toothed

leaflets, hairless or sparsely hairy above, densely white-hairy beneath, often also with a few bristles or prickles along the veins; terminal leaflet 4.5–11 × 2–11 cm, lateral leaflets 3.5–8.5 × 1.5–6 cm. Flowers 1.5–2.5 cm across, few, in short, rather congested racemes, stalks very bristly. Sepals lanceolate, acuminate, very bristly. Petals pink, shorter than the sepals. Fruit bright red. *Japan, Korea, N China*. H3. Summer.

35. R. ellipticus Smith. Figure 104(3), p. 211.

Erect shrub to 3 m; stems softly hairy and with numerous bristles and scattered, downwardly hooked prickles. Leaves with 3 distinct broadly elliptic to almost circular, rounded, evenly toothed leaflets, dark green and hairless above, densely white-hairy beneath, with bristles and a few prickles on the main veins; terminal leaflet 4–10 × 2.5–7.5 cm, lateral leaflets 2.5–6.5 × 1.5–4.5 cm. Flowers 8–14 mm across, numerous in rather dense panicles, the panicle- and flower-stalks bristly. Sepals oblong, acute, softly hairy on the lobes, softly hairy and bristly on the perigynous zone. Petals white, about as long as the sepals. Fruit yellow. *Himalaya to SW China*. H5. Spring.

Very distinctive with its bristly stems and broad leaflets with rounded tips.

36. R. occidentalis Linnaeus. Figure 104 (4), p. 211. Illustration: *Gentes Herbarum* **5**: 881 (1945).

Upright shrub to 2.5 m, the first-year stems ultimately arching and rooting at their tips, usually strongly white-bloomed. Leaves with 3 leaflets, which are coarsely and irregularly toothed, dark green above, densely white-hairy beneath; terminal leaflet 3–12 × 1.5–8 cm, long-stalked, often somewhat cordate at the base, lateral leaflets 2.5–10 × 1–6 cm. Flowers 1.2–1.7 cm across, mostly in a close, flat-topped cluster; flower-stalks with soft white hairs and thin, scarcely hooked, rather narrowly based prickles. Sepals triangular, acuminate, downy. Petals white, almost as long as sepals. Fruit black, bloomed. *E North America, extending west to Colorado*. H3. Summer.

37. R. leucodermis Douglas. Figure 104 (5), p. 211. Illustration: *Gentes Herbarum* **5**: 885 (1945).

Very similar to *R. occidentalis* but leaflets smaller (terminal 2–7.5 × 1.5–4.5 cm, laterals 1–6 × 1–3.5 cm), prickles on the

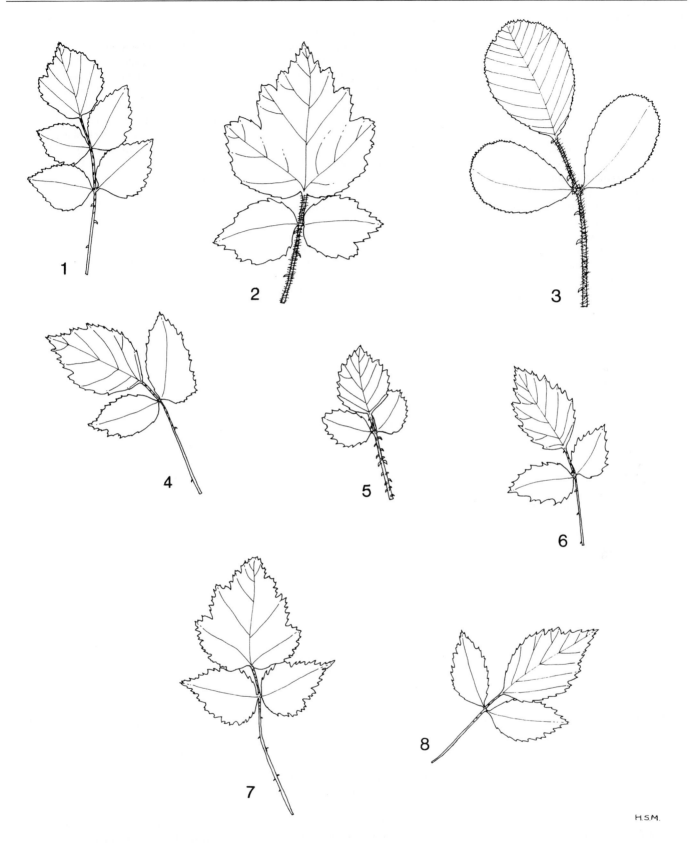

Figure 104. Leaves of *Rubus* species (× 0.5). 1, *R. coreanus.*
2, *R. phoenicolasius.* 3, *R. ellipticus.* 4, *R. occidentalis.* 5, *R. leucodermis.*
6, *R. glaucifolius.* 7, *R. mesogaeus.* 8, *R. idaeus.*

flower-stalks usually broadly based and hooked; fruit purplish. *W North America.* H4. Summer.

38. R. glaucus Bentham.

Similar to *R. occidentalis* but leaflets finely toothed, hairless above, fruit reddish black. *C & S America.* G1.

Doubtfully in cultivation in Europe.

39. R. glaucifolius Kellogg. Figure 104 (6), p. 211.

Upright arching or trailing shrub, the main stems not rooting at the tip. Leaves with 3 leaflets, which are coarsely and irregularly toothed, finely downy to hairless above, densely white-hairy beneath; terminal leaflet 3–5 × 2–3.5 cm, long-stalked, often more or less 3-lobed, lateral leaflets 2–4 × 1–2.5 cm. Flowers 1–1.5 cm across, mostly in close, flat-topped clusters; flower-stalks with stalked glands and dense hairs, prickles few to many, rather broadly based, mostly straight. Sepals upright, triangular, downy and usually with stalked glands. Petals white, somewhat shorter to slightly longer than the sepals. Fruit red to purple. *Western USA (California).* H5. Summer.

40. R. mesogaeus Focke. Figure 104(7), p. 211.

Stems scrambling to 3 m or more, hairy and with few broadly based prickles. Leaves with 5 leaflets, or if with 3 then the terminal leaflet tending to be 3-lobed; leaflets ovate, acute, lobed and toothed, dark green and sparsely hairy above, whitish or greyish with dense hairs beneath and with a few small prickles along the midrib; terminal leaflet 3–12 × 2–8 cm, lateral leaflets 1.5–9.5 × 1–5 cm. Flowers 9–16 mm across in short, flat-topped, corymbose racemes. Sepals very narrowly lanceolate, acuminate, densely white-hairy. Petals pink or white, shorter than sepals. Fruits black. *C & W China.* H3. Summer.

41. R. idaeus Linnaeus. Figure 104(8), p. 211. Illustration: Ary & Gregory, The Oxford book of wild flowers, 78 (1962); Keble Martin, The concise British flora in colour, edn 2, pl. 28 (1969).

Very variable, widely cultivated and often escaping. Stems to 2.5 m, first-year stems often glaucous or bloomed, with or without prickles. Leaves with 3 leaflets (up to 7 leaflets on the leaves of first-year stems), hairy or hairless above, densely white-hairy beneath, margins toothed; leaflets very variable in shape and size, the terminal stalked, 3–13 × 1.5–12.5 cm, laterals 2.5–10.5 × 1–6.5 cm. Flowers mostly in racemes, which are not close and flat-topped, 1–1.7 cm across, stalks with or without prickles. Sepals triangular, acuminate, usually downy. Petals white, shorter than the sepals. Fruit red (rarely yellowish). *Europe, N Asia, introduced elsewhere.* H1. Early summer.

The European raspberry, widely cultivated for fruit, less commonly for ornament, but spreading rapidly in gardens and orchards by both suckers and seed.

R. strigosus Michaux (*R. idaeus* var. *strigosus* (Michaux) Maximowicz). Illustration: *Gentes Herbarum* **5**: 869 (1945). Very like *R. idaeus* but often with glandular hairs and numerous bristles on stems, leaf- and flower-stalks. *N America.* H4. Summer.

Subgenus **Rubus**. Like Subgenus *Idaeobatus* but fruit not separating from the receptacle, dispersed with it by abscission of the fruit-stalk.

The species of this Subgenus are particularly difficult to identify. Because of their tendency to produce seed without fertilisation (apomixis), their ability, when fertile, to hybridise easily (the resultant hybrids often persisting by apomixis), and the occurrence of polyploidy, a vast complex of variants is found in Europe and N America. Many hundreds of species (microspecies) have been named, and their identification in the wild is difficult, even for experts. Though only a few species are reputedly grown for ornament, their relationship to the wild species is uncertain and identification is thus even more difficult. Fortunately, the plants are not particularly ornamental and very few are deliberately grown for this purpose, though wild plants (from seed distributed by birds) may well appear and persist in wild gardens or neglected areas.

The account below covers those named species that are reported to have occurred in cultivation (whether they actually have been or still are cultivated remains a matter of considerable doubt). It is intended for guidance only; for further information the copious literature on the group should be studied.

Elaborate hierarchies of sections, subsections and series have been devised to group these microspecies; unfortunately, work in Europe and North America has not been well coordinated, and it is not possible to reconcile the various systems here, as only a small number of the microspecies are included. The literature listed below should be consulted for details.

Literature: Sudre, H., Rubi Europae (1908–13); Bailey, L.H., Rubus in North America, *Gentes Herbarum* **5** (1941–45); Watson, W.C.R., Handbook of Rubi of Great Britain and Ireland (1958); Weber, H.E., Die Gattung Rubus L. (Rosaceae) in nordwestlichen Europa (1972); Weber, H.E., Rubi Westfalici (1985); Newton, A. & Edees, E.S., Brambles of the British Isles (1988).

42. R. laciniatus Willdenow.

Robust plant, stems armed with numerous, equal, hooked prickles. Leaves with 5 leaflets, which are deeply lobed, greenish and hairy or not beneath. Inflorescence widely spreading, with numerous, hooked prickles. Sepals deflexed. Petals white or pink, toothed at the apex. *Origin unknown, widely cultivated and naturalised.* H3. Summer.

A prickleless variant is also cultivated.

43. R. elegantispinosus (Schumann) Weber.

Like *R. laciniatus* but stems sometimes bloomed, prickles straight or deflexed, usually purple; leaflets 5, not lobed, with felted white hairs beneath; petals deep pink. *W Europe.* H3. Summer.

44. R. allegheniensis Bailey.

Robust, upright shrub to 3 m, yellowish green overall; stems with strong, scattered prickles. Leaves of 3 or 5 leaflets, toothed, blunt, the terminal cordate, greenish and hairy beneath, usually with stalked glands. Inflorescence raceme-like, with conspicuous stalked glands. Petals white, narrow. *E North America.* H3. Summer.

R. rosa Bailey. Illustration: *Gentes Herbarum* **5**: 541, 543 (1944). Similar, but inflorescence shorter and more compact. *E USA.*

45. R. frondosus Bigelow. Illustration: *Gentes Herbarum* **5**: 733 (1945).

Upright, to 2.5 m, branches sometimes arching and rooting at the tip, with few, straight or hooked prickles. Leaves usually with 3 toothed leaflets, green and hairy beneath. Flower-clusters corymb-like, flat, without an long axis. Petals white, rather narrow. *E USA.* H3. Summer.

46. R. bellobatus Bailey. Illustration: *Gentes Herbarum* **5**: 665, 667, 668 (1945).

Like *R. frondosus* but prickles straighter, flower-clusters raceme-like, longer than broad, with an long axis, and larger fruits. *E USA.* H3. Summer.

47. R. ulmifolius Schott. Illustration: Keble Martin, The concise British flora in colour, edn 2, pl. 28 (1969).

Stems robust, arching, conspicuously bloomed, with broadly based, straight or hooked prickles. Leaves divided into 3–5 leaflets, which are densely white-hairy beneath. Inflorescence raceme-like. Sepals deflexed. Petals pink or white, crumpled. *S, W & C Europe.* H3. Summer.

A variant with double flowers ('Bellidiflorus') is sometimes grown, as is one with variegated leaves ('Variegatus').

48. R. bifrons Trattinick.

Similar to *R. ulmifolius* but stems not bloomed, lowermost leaflets with short but distinct stalks, prickles in the inflorescence mostly straight; petals white or pale pink. *W & C Europe.* H3. Summer.

49. R. linkianus Seringe. Illustration: *Addisonia* **22**: pl. 722 (1943–46).

Similar to *R. ulmifolius* but stems not bloomed; lowermost leaflets without stalks; prickles in the inflorescence hooked; flowers always double. *Origin unknown.* H3. Summer.

50. R. procerus Boulay (*R. discolor* misapplied; *R. armeniacus* Focke).

Similar to *R. ulmifolius* but stem not bloomed, lowermost leaflets with short, distinct stalks, prickles in the inflorescence mostly hooked, petals white or pale pink; anthers hairy. *S, W & C Europe.* H3. Summer.

'Himalayan Giant' ('Theodore Reimers') has larger leaves and flowers and is generally grown for its fruit; it is naturalised in various parts of Europe (see Kent, D.H., Rubus procerus 'Himalayan Giant', *Kew Magazine* **5**: 32–36, 1988).

51. R. cuneifolius Pursh. Illustration: *Gentes Herbarum* **5**: 424, 429 (1943).

Small, stiffly erect, ash-grey plant, stems with numerous hooked prickles. Leaves usually with 3 leaflets, these rather obovate and obtuse, whitish beneath with dense, felted hairs. Flower-clusters with 3–5 flowers, short and not raceme-like, without a distinct axis. Petals white, rather broad. *E USA.* H3. Summer.

52. R. trivialis Michaux. Illustration: *Gentes Herbarum* **5**: 200, 204 (1941).

Variable trailing plant, the tips of the branches rooting; stem bearing numerous, hooked prickles and bristles. Leaves of first-year shoots leathery, persisting until flowering, usually with 5 leaflets; those on the flowering shoots similar but smaller and often with 3 leaflets. Flowers solitary or 2 or 3 together, on long, prickly and glandular stalks. Petals white to pinkish, rather broad, often overlappping. *Widely distributed in S & E USA.* H3. Summer.

53. R. flagellaris Willdenow. Illustration: *Gentes Herbarum* **5**: 245, 248 (1943).

Trailing plant, rooting at the tips of the branches and sometimes also at the nodes; stems with sparse, hooked prickles. Leaves all deciduous, those on the flowering-shoots generally with 3 leaflets, which are usually obovate and then abruptly tapered, sometimes shouldered. Flowers in clusters of 1–3 on long stalks, 2–3 cm across. Petals white, obtuse, not overlapping. *E North America.* H3. Summer.

54. R. roribaccus Rydberg. Illustration: *Gentes Herbarum* **5**: 255 (1943).

Similar to *R. flagellaris* but larger and more vigorous; leaflets ovate, gradually tapered to the apex; flowers 3–4 cm across. *SE USA.* H3. Summer.

Two other generally very similar species may also be found: **R. velox** Bailey with more oblong leaflets, and **R. almus** (Bailey) Bailey with elliptic leaflets softly hairy beneath, both from Texas.

55. R. caesius Linnaeus. Illustration: Polunin, Flowers of Europe, pl. 44 (1969).

Stem trailing, hairless, bloomed, with weak, broadly based, straight or hooked prickles. Leaves with 3 leaflets, downy beneath, the lateral leaflets often somewhat lobed. Inflorescence corymb-like with 2–5 flowers. Flowers 2–2.5 cm across, bisexual. Sepals erect. Petals white, broad. Stamens greenish. Fruit black, bloomed, the individual drupelets easily separable. *Most of Europe, introduced & naturalised elsewhere.* H1. Summer.

56. R. vitifolius Chamisso & Schlechtendahl. Illustration: *Gentes Herbarum* **5**: 49 (1941).

Plant trailing; stems with straight, narrowly based prickles (almost narrow enough to be considered bristles). Leaves on first-year shoots with 3 leaflets, or deeply 3-lobed, those on flowering-shoots 3-lobed or scarcely lobed at all, ovate, bright green, thinly hairy when young. Flowers few to numerous, unisexual

(stamens or carpels sterile), in large or small, raceme-like clusters. Petals white, oblong or ovate, not overlapping. Fruit black, oblong, hairy. *W USA (California).* H5. Summer.

57. R. ursinus Chamisso & Schlechtendahl. Illustration: *Gentes Herbarum* **5**: 55 (1941).

Like *R. vitifolius* but all parts, especially the leaves, more densely greyish hairy, the leaves dull green; leaves on first-year shoots with 3 or 5 leaflets, those on flowering-shoots with 3 leaflets or 3-lobed. Fruit oblong to conical. *W USA (Oregon, California).* H5. Summer.

R. loganobaccus L.H. Bailey. Like *R. ursinus*, but with stems to 4 m. Leaves with dense down beneath. Fruit to 4 cm, blackish, glossy. *California.*

The loganberry of commerce.

58. R. macropetalus Douglas. Illustration: *Gentes Herbarum* **5**: 63 (1941).

Like *R. vitifolius* and *R. ursinus*, but plant with numerous, conspicuous, stalked glands; all leaves with 3 or 5 leaflets; petals white, very variable in size. Fruit oblong, not hairy. *W USA.* H5. Summer.

24. ROSA Linnaeus
V.A. Mathews edited by V.A. Mathews
Deciduous, or sometimes evergreen, shrubs with erect, arching or scrambling, occasionally trailing stems, usually armed with prickles and/or bristles. Stipules usually present, persistent, usually joined to leaf-stalk for most of their length. Leaves alternate, usually odd-pinnate, rarely with 3 leaflets or simple; leaflets toothed. Perigynous zone spherical to urn-shaped. Flowers solitary or in corymbs, usually borne at the end of short branches, single to double. Sepals 5 (rarely 4), entire, or the 2 outer and half the third one with lateral lobes, and the inner 2 and half the third one entire, the tips acute to attenuate or broadened and leafy. Petals 5 (rarely 4) in single flowers, usually obovate, tip often notched, white, cream, pink, red, purplish, orange or yellow. Stamens (in single flowers) 30–200, in several whorls. Carpels many, each with 1 ovule, free, carried on the base and/or sides of the enclosing receptacle (perigynous zone). Styles free or united into a column, protruding from mouth of receptacle or not. Fruit ('hip' or 'hep') containing many achenes, usually red or orange, sometimes blackish or green, enclosed by the fleshy perigynous

zone. Achenes 1-seeded, bony, often hairy. Figure 99(2, 13), p. 183.

A genus of over 100 species (possibly 150), distributed through temperate and subtropical zones of the northern hemisphere. Problems of classification and naming are rife in this genus, many species of which have been cultivated and hybridised for centuries, and some of which are very polymorphic. The last complete account of the genus was by Lindley (Rosarum monographia) in 1820: a new revision is urgently needed to incorporate investigations into the complex cytology and the problems produced by hybridisation. Thus, it must be stressed that this account is somewhat provisional.

Literature: Willmott, E., The genus Rosa, 2 vols (1910–1914); Thomas, G.S., Shrub roses of today (1962); Krüssmann, G., The complete book of roses (1981), English translation of Rosen, Rosen, Rosen, 1974; Thomas, G.S., The old shrub roses (1955, revised edn 1983); Thomas, G.S., Climbing roses old and new (1983); Modern roses **9**. The international checklist of roses (1987); Austin, D., Botanica's roses: the encyclopedia of roses (2001); Quest-Ritson, C., American Rose Society encyclopedia of roses (2003); Austin, D., The rose (2008).

1a. Leaves simple, lacking stipules; if leaves pinnate and/or stipules present, then petals with a basal crimson blotch   2
  b. Leaves pinnate with 3–13 leaflets, stipules present   3
2a. Leaves simple, grey-green; fruit very bristly   **1. persica**
  b. Leaves simple or with up to 7 leaflets, dark green; fruit with a few bristles   **2. × hardii**
3a. Flowers yellow   4
  b. Flowers white, cream, green, orange, pink, red or purplish   13
4a. Margin of leaflets with simple teeth   5
  b. Margin of leaflets with compound teeth   10
5a. Leaflets 11–19   **6. xanthina**
  b. Leaflets 3–9   6
6a. Flowers 2–3 cm across   7
  b. Flowers 3.8–8 cm across   8
7a. Flowers single or double, borne in many-flowered clusters; sepals reflexed and falling after flowering   **104. banksiae**
  b. Flowers single, solitary; sepals erect and persistent after flowering   **7. ecae**

8a. Sepals reflexed and falling after flowering   **103. × odorata**
  b. Sepals erect and persistent after flowering   9
9a. Flowers double; prickles decurved   **12. hemisphaerica**
  b. Flowers single or semi-double; prickles straight or slightly curved, flattened, often reddish   **6. xanthina**
10a. Flowers 2.5–4.5 cm across   **9. primula**
  b. Flowers 5–7.5 cm across   11
11a. Flowers double   **11. × harisonii**
  b. Flowers single   12
12a. Leaflets 8–10 mm; tips of sepals not expanded; flowers scentless   **8. kokanica**
  b. Leaflets 1.5–4 cm; tips of sepals expanded; flowers smelling unpleasant   **10. foetida**
13a. Bracts absent   14
  b. Bracts present, sometimes very small and falling early   44
14a. Flowers coppery red inside, yellow-buff outside; margin of leaflets with double teeth   **10. foetida**
  b. Flowers not as above; margin of leaflets with simple teeth; if teeth double then flowers white, ageing to pinkish   15
15a. Flowers solitary or 2–4 together   16
  b. Flowers in clusters of 5–100 or more   24
16a. Leaves evergreen   17
  b. Leaves deciduous   19
17a. Petals white or creamy white   18
  b. Petals pale pink with deeper veins   **107. × anemonoides**
18a. Flowers in clusters of 3–20   **98. luciae**
  b. Flowers solitary   **106. laevigata**
19a. Receptacle and fruit prickly   **111. stellata**
  b. Receptacle and fruit lacking prickles   20
20a. Fruit dark brown, becoming blackish   **3. pimpinellifolia**
  b. Fruit yellow, orange, red, deep crimson or purplish red   21
21a. Fruit purplish red, drooping   **4. × reversa**
  b. Fruit neither purplish red nor drooping   22
22a. Sepals falling after flowering   **100. arvensis**
  b. Sepals persistent after flowering   23
23a. Sepals glandular on the back; petals white tinged with pink   **5. koreana**

  b. Sepals hairless or silky on the back; petals white, cream or rose-pink   **13. sericea**
24a. Flowers double   25
  b. Flowers single   26
25a. Stigmas hairless; leaflets 7–9 (rarely 5–11)   **82. multiflora**
  b. Stigmas hairy; leaflets 3–5   **85. × beanii**
26a. Flowers pale to deep pink   27
  b. Flowers white or cream   31
27a. Leaflets linear-lanceolate   **83. watsoniana**
  b. Leaflets narrowly elliptic to almost round   28
28a. Flowers 5–7.5 cm across   29
  b. Flowers 1.5–5 cm across   30
29a. Leaflets with compound teeth; sepals persistent after flowering; styles united   **90. moschata**
  b. Leaflets with simple teeth; sepals falling after flowering; styles free   **84. setigera**
30a. Prickles more or less curved; sepals smooth or sparsely glandular on the back; fruit 6–25 mm   **100. arvensis**
  b. Prickles curved downwards; sepals glandular-bristly on the back; fruit 6–7 mm   **82. multiflora**
31a. Leaflets 13–21 cm long   32
  b. Leaflets 1–10 cm long   33
32a. Leaves evergreen or semi-evergreen, grey-green to blue-green; flower-stalks downy and sparsely glandular-downy   **89. brunonii**
  b. Leaves deciduous, green; flower-stalks sparsely downy and densely glandular-downy   **91. longicuspis**
33a. Stigmas hairless   34
  b. Stigmas hairy   37
34a. Stipules laciniate   **82. multiflora**
  b. Stipules not laciniate   35
35a. Flowers 5–7.5 cm across; sepals downy and glandular-bristly on the back   **84. setigera**
  b. Flowers 2.5–5 cm across; sepals smooth or sparsely glandular, or downy and glandular on the back   36
36a. Leaflets 1–4 cm long; fruit usually hairless   **100. arvensis**
  a. Leaflets 5–10 cm long; fruit downy   **92. cerasocarpa**
37a. Leaves evergreen or semi-evergreen   38
  b. Leaves deciduous   39
38a. Leaflets acute to acuminate   **98. luciae**

b. Leaflets obtuse   **97. wichuraiana**
39a. Leaves 3–8 cm; stigmas with sparse hairs   **96. soulieana**
  b. Leaves 8–17 cm; stigmas hairy or woolly   40
40a. Sepals persistent after flowering   **90. moschata**
  b. Sepals falling after flowering   41
41a. Sepals with lobes   42
  b. Sepals entire (rarely with a few lateral lobes)   43
42a. Flower stalks 1.5–2 cm   **86. helenae**
  b. Flower stalks 2.5–3.5 cm   **88. mulliganii**
43a. Leaflets (3–)5; flowers in clusters of 10–25   **87. rubus**
  b. Leaflets (5–)7–9; flowers in clusters of up to 100 or more   **94. filipes**
44a. Leaves evergreen or semi-evergreen   45
  b. Leaves deciduous   52
45a. Flowers *c.* 1.5 cm across   **105. cymosa**
  b. Flowers at least 2 cm across   46
46a. Flowers 10–15 cm across   **102. gigantea**
  b. Flowers 2–10 cm across   47
47a. Bracts very small, falling quickly   48
  b. Bracts larger, persistent   49
48a. Styles not protruding; fruit *c.* 7 mm   **104. banksiae**
  b. Styles protruding; fruit 1–1.6 cm   **99. sempervirens**
49a. Petals pale pink to scarlet or crimson, or green, or yellow with an orange back   50
  b. Petals white; leaflets 5–11, obtuse   51
50a. Fruit 1.5–2 cm; leaflets 3–5, downy on midrib beneath **101. chinensis**
  b. Fruit 2–3 cm; leaflets 5–7, completely hairless beneath   **103. × odorata**
51a. Leaflets obtuse; bracts large, deeply toothed; sepals brown hairy on the back   **108. bracteata**
  b. Leaflets acuminate; bracts neither large nor deeply toothed; sepals smooth on the back   **103. × odorata**
52a. Sepals entire   53
  b. At least some sepals with lateral lobes   141
53a. Flowers white   54
  b. Flowers pale to deep pink, crimson, mauve-pink or purplish pink   64

54a. Flowers 6–7.5 (rarely to 9) cm across; leaflets conspicuously wrinkled above   **45. rugosa**
  b. Flowers 1.2–6.5 cm across; leaflets not wrinkled above   55
55a. Flowers 8 or more in a cluster; fruit red, becoming purplish   **81. beggeriana**
  b. Flowers solitary or up to 7 in a cluster; fruit red to orange-red   56
56a. Sepals hairy inside   57
  b. Sepals not hairy inside   58
57a. Sepals glandular-bristly outside; leaflets broadly elliptic to obovate   **74. fedtschenkoana**
  b. Sepals smooth outside; leaflets narrowly ovate to elliptic   **76. elegantula**
58a. Margin of leaflets with compound teeth   59
  b. Margin of leaflets with simple teeth   61
59a. Receptacle and fruit glandular-bristly   **67. wardii**
  b. Receptacle and fruit smooth   60
60a. Stipules with glandular teeth   **58. nutkana**
  b. Stipules lacking glands   **54. laxa**
61a. Leaflets 9–15   **68. murielae**
  b. Leaflets 5–9   62
62a. Margin of leaflets toothed in upper two-thirds only   **71. webbiana**
  b. Margin of leaflets toothed for whole length   63
63a. Flower-stalks smooth, hairless   **56. woodsii**
  b. Flower-stalks glandular   **54. laxa**
64a. Sepals falling after flowering   65
  b. Sepals persisting after flowering   72
65a. Leaflets acute   66
  b. Leaflets obtuse   70
66a. Mature leaflets green   67
  b. Mature leaflets grey-green or tinged with brown or purple, covered with a dull waxy bloom   **39. glauca**
67a. Leaflets dull above   **40. palustris**
  b. Leaflets shiny above   68
68a. Sepals with expanded leafy tips; fruit 1–1.5 mm wide   **41. virginiana**
  b. Tips of sepals not leafy and expanded; fruit 7–10 mm wide   69
69a. Stems with sparse prickles and dense bristles; flower-stalks 1–3 cm   **43. nitida**
  b. Stems with sparse prickles, bristles rarely present; flower-stalks 5–10 mm   **44. foliolosa**

70a. Flowers 5–7 cm across; receptacle glandular-bristly   **16. × francofurtana**
  b. Flowers 2.5–3.8 cm across; receptacle smooth   71
71a. Flowers rose-pink; stems with prickles intermixed with bristles   **80. gymnocarpa**
  b. Flowers purplish pink; stems with prickles only   **79. willmottiae**
72a. Margin of leaflets with compound teeth   73
  b. Margin of leaflets with simple teeth   94
73a. Styles protruding   74
  b. Styles not protruding   75
74a. Leaflets 3–5, acute; flowers 1.9–2.5 cm across   **59. corymbulosa**
  b. Leaflets 7–9 (rarely 5), obtuse; flowers 2.5–3.8 cm across   **77. multibracteata**
75a. Receptacle and fruit smooth   76
  b. Receptacle and fruit glandular-bristly   88
76a. Leaflets acute to acuminate   77
  b. Leaflets obtuse   83
77a. Sepals smooth on the back   78
  b. Sepals glandular on the back   80
78a. Stems usually unarmed, occasionally with slender prickles and bristles   **50. pendulina**
  b. Stems with large, straight prickles in pairs at the nodes   79
79a. Stems 1–1.5 m; fruit 1–1.3 cm   **53. davurica**
  b. Stems 1.5–3 m; fruit 1.5–2 cm   **58. nutkana**
80a. Fruit usually drooping   81
  b. Fruit erect   82
81a. Stem to 2 m; receptacle smooth or glandular-bristly   **50. pendulina**
  b. Stem 2–4 m; receptacle smooth   **51. oxyodon**
82a. Flowers 5–6.5 (rarely 4) cm across; fruit 1.5–2 cm   **58. nutkana**
  b. Flowers 1.2–2.5 cm across; fruit rarely more than 1 cm **73. giraldii**
83a. Flowers 1.2–2.5 cm across   **73. giraldii**
  b. Flowers 3.7–6.5 cm across   84
84a. Petals pale pink   **54. laxa**
  b. Petals deep pink to purplish pink or red   85
85a. Fruit 7–15 mm   **57. californica**
  b. Fruit 1.5–3 cm   86
86a. Fruit drooping, usually with a neck at the top   87
  b. Fruit erect, lacking a neck at the top   **58. nutkana**

87a. Stems up to 2 m; receptacle smooth or glandular-bristly **50. pendulina**
  b. Stems 2–4 m; receptacle smooth **51. oxyodon**
88a. Flowers solitary or in clusters of 2–6; petals hairless on the back 89
  b. Flowers in clusters of 5–20; petals slightly downy on the back **64. setipoda**
89a. Fruit 8–20 mm, if more, then drooping 90
  b. Fruit 2.5–7.5 cm, erect 92
90a. Fruit drooping **50. pendulina**
  b. Fruit erect 91
91a. Flowers 1.2–2.5 cm across; fruit rarely more than 1 cm **73. giraldii**
  b. Flowers (4–)5–6.5 cm across; fruit 1.5–2 cm **58. nutkana**
92a. Sepals hairy inside **69. moyesii**
  b. Sepals not hairy inside 93
93a. Flowers 3.5–5 cm across; fruit *c.* 2.5 cm **66. sweginzowii**
  b. Flowers 5–7.5 cm across; fruit 2.5–4 (rarely to 7.5) cm **63. macrophylla**
94a. Styles protruding 95
  b. Styles not protruding 96
95a. Leaflets acute; flowers 3.8–5 cm across; fruit 1.5–2.5 cm **60. davidii**
  b. Leaflets obtuse; flowers 2–3.8 cm across; fruit to 1.3 cm **78. forrestiana**
96a. Leaflets acute to acuminate 97
  b. Leaflets obtuse 121
97a. Flowers 1.5–3.8 cm across 98
  b. Flowers 3–9 cm across 103
98a. Sepals smooth and hairless on the back, or hairy on the margin 99
  b. Sepals hairy or glandular-bristly on the back 101
99a. Fruit 7–15 mm 100
  b. Fruit 2–2.5 cm **62. banksiopsis**
100a. Sepals hairy inside; leaflets glaucous green above **76. elegantula**
  b. Sepals hairless inside; leaflets shiny green above **47. arkansana**
101a. Leaflets usually 5–9, downy beneath **55. pisocarpa**
  b. Leaflets usually 9–15, hairless beneath or downy only on the veins 102
102a. Stems 50–120 cm; leaflets 2–6 cm; flowers 2.5–4 cm across; sepals smooth or glandular-bristly on the back; fruit 1 cm or more **47. arkansana**
  b. Stems 1.2–2.5 m; leaflets 6–20 mm; flowers 2–2.5 cm across; sepals

downy on both sides; fruit 1 cm or less **75. prattii**
103a. Fruit smooth 104
  b. Fruit glandular-bristly, sometimes only sparsely so 111
104a. Fruit 7–15 mm 105
  b. Fruit 1.5–2.5 cm 109
105a. Leaflets usually 5–7 (rarely 9) 106
  b. Leaflets usually 7–11 108
106a. Stems with pairs of hooked prickles at the nodes **52. majalis**
  b. Stems unarmed, or with straight or slightly curved prickles 107
107a. Flowers 3–4.5 cm across; sepals hairless or downy on the back **56. woodsii**
  b. Flowers 4.5–6.5 cm; sepals downy and glandular on the back **49. blanda**
108a. Leaflets shiny above; sepals hairless inside **47. arkansana**
  b. Leaflets glaucous on both sides; sepals hairy inside **76. elegantula**
109a. Leaflets conspicuously wrinkled above, margin toothed in the upper two-thirds only **45. rugosa**
  b. Leaflets not wrinkled above, margin toothed for the whole length 110
110a. Fruit dark red; leaflets 6–25 mm **72. sertata**
  b. Fruit scarlet or mid-red; leaflets 1.5–6 cm **46. acicularis**
111a. Petals pale pink or lilac-pink; sepals hairy inside 112
  b. Petals mid-pink to deep purplish pink or red; sepals hairy or not inside 113
112a. Leaflets elliptic to narrowly ovate; flowers 2–3.8 cm across **76. elegantula**
  b. Leaflets broadly elliptic to obovate; flowers *c.* 5 cm across **74. fedtschenkoana**
113a. Fruit 1–2.5 cm 114
  b. Fruit 2.5–7.5 cm 119
114a. Flowers 2–4 cm across; fruit 1–1.5 cm 115
  b. Flowers 3.5–6.5 cm across; fruit 1.5–2.5 cm 116
115a. Leaflets shiny above; sepals hairless inside **47. arkansana**
  b. Leaflets glaucous on both sides; sepals hairy inside **76. elegantula**
116a. Leaflets with short stalks **70. bella**
  b. Leaflets stalkless 117
117a. Fruit lacking a neck at the top **58. nutkana**
  b. Fruit with a neck at the top 118
118a. Fruit dark red **72. sertata**

  b. Fruit orange-red **61. caudata**
119a. Flowers solitary or in clusters of 2–5; petals hairless on the back 120
  b. Flowers in clusters of 5–20; petals downy on the back **64. setipoda**
120a. Leaflets elliptic to narrowly ovate; sepals hairless inside **63. macrophylla**
  b. Leaflets broadly elliptic to ovate; sepals downy inside **69. moyesii**
121a. Leaflets usually 5–7, rarely 9 122
  b. Leaflets usually 7–13, rarely less 133
122a. Sepals smooth or downy on the back 123
  b. Sepals with glands on the back, hairy or not 128
123a. Flower-stalks smooth, hairless 124
  b. Flower-stalks glandular 126
124a. Flowers 3–4.5 cm across; stipules narrow, flat 125
  b. Flowers *c.* 5 cm across; stipules broad, those on strong shoots usually inrolled **52. majalis**
125a. Prickles straight or slightly curved; flowers mid-pink **56. woodsii**
  b. Prickles recurved; flowers deep pink to crimson **57. californica**
126a. Flowers pale pink **54. laxa**
  b. Flowers bright or deep pink to crimson or purple-pink 127
127a. Flowers 3.7–4 cm across; prickles recurved; fruit with a neck at the top **57. californica**
  b. Flowers (4–)5–6.5 cm across; prickles usually straight; fruit lacking a neck at the top **58. nutkana**
128a. Fruit 8–13 mm 129
  b. Fruit 1.5–2.5 cm 130
129a. Flowers 2.5–3 cm across **55. pisocarpa**
  b. Flowers 4.5–6.5 cm across **49. blanda**
130a. Flowers pale pink or lilac-pink 131
  b. Flowers deep pink to purplish pink or bright red 132
131a. Prickles recurved, hooked or straight, intermixed with bristles **54. laxa**
  b. Prickles straight, bristles absent **71. webbiana**
132a. Leaflets 5–25 mm; stems lacking bristles **72. sertata**
  b. Leaflets 1.9–5 cm; young stems with bristles **58. nutkana**
133a. Fruit to 2.5 cm 134
  b. Fruit 2.5–6 cm 140

134a. Leaflets dull blue-green or
somewhat glaucous 135
b. Leaflets bright green, sometimes
shiny above 138
135a. Sepals hairy inside and on the
margin, not glandular on the back;
fruit 7–13 mm **76. elegantula**
b. Sepals not hairy inside,
glandular on the back; fruit
1.5–2.5 cm 136
136a. Stems with dense bristles
46. acicularis
b. Stems lacking dense bristles 137
137a. Flowers pale pink or lilac-pink,
borne on flower-stalks 1–1.3 cm
71. webbiana
b. Flowers deep pink to purplish pink,
borne on flower-stalks 1.5–3 cm
72. sertata
138a. Flowers pale pink, 4–5 cm across
54. laxa
b. Flowers mid-pink to deep pink,
2.5–4 cm across 139
139a. Stems to 50 cm; leaflets downy
beneath **48. suffulta**
b. Stems 50–120 cm; leaflets usually
hairless beneath, sometimes downy
on the veins **47. arkansana**
140a. Flower-stalks densely glandular-
bristly; fruit dark red;
flowers smelling of unripe apples;
petals downy on the back
64. setipoda
b. Flower-stalks moderately
glandular-bristly or smooth; fruit
orange-red; flowers scentless; pet
als hairless on the back
69. moyesii
141a. Sepals persistent after flowering,
spreading to erect 142
b. Sepals falling after flowering,
usually reflexed, sometimes
spreading 160
142a. Flowers double 143
b. Flowers single or semi-double 144
143a. Leaflets 5–7, broadly ovate to more
or less round; fruit red, not prickly
17. × centifolia
b. Leaflets usually 9–15, narrowly
ovate to obovate; fruit yellow-
green, prickly **109. roxburghii**
144a. Flowers white, pale pink or pale
lilac-pink 145
b. Flowers mid-pink to deep pink or
purple-pink 148
145a. Receptacle prickly or glandular-
bristly 146
b. Receptacle smooth **34. coriifolia**
146a. Margin of leaflets with compound
teeth, glandular and downy

beneath, aromatic when crushed;
flowers 2.5–5 cm across
26. rubiginosa
b. Margin of leaflets with simple
teeth, downy or hairless beneath,
not aromatic; flowers 5–10
(rarely 3) cm across 147
147a. Leaflets 1–2.5 cm; fruit prickly
110. hirtula
b. Leaflets 2.5–3.5 cm; fruit slightly
glandular-bristly **36. britzensis**
148a. Fruit yellow-green, prickly
109. roxburghii
b. Fruit orange-red to dark red or
blackish, smooth or glandular-
bristly 149
149a. Fruit with a neck at the top 150
b. Fruit lacking a neck at the top 151
150a. Stems to 2 m **65. hemsleyana**
b. Stems 2.4–3 m **64. setipoda**
151a. Prickles straight or slightly curved
152
b. Prickles hooked 158
152a. Margin of leaflets with simple teeth
153
b. Margin of leaflets with compound
teeth 154
153a. Stems 50–120 cm; leaflets usually
hairless beneath, sometimes downy
on the veins **47. arkansana**
b. Stems to 50 cm; leaflets downy
beneath **48. suffulta**
154a. Stems with prickles intermixed
with bristles 155
b. Stems with prickles only 156
155a. Margin of leaflets with 3–10 teeth
on each side; flower-stalks 2–5 mm,
usually smooth **27. sicula**
b. Margin of leaflets with 8–20 teeth
on each side; flower-stalks 2–15
(rarely to 25) mm, usually
glandular-bristly **28. pulverulenta**
156a. Young stems not bloomed
22. villosa
b. Young stems bloomed 157
157a. Leaflets more or less circular
23. mollis
b. Leaflets elliptic to broadly ovate
24. sherardii
158a. Stems 30–70 cm; leaflets 7–15 mm
28. pulverulenta
b. Stems 1–3 m; leaflets 1–3.5 cm 159
159a. Leaflets dark green above,
glandular and downy beneath
26. rubiginosa
b. Leaflets often bloomed above,
hairless or downy beneath, but not
glandular **37. dumalis**
160a. Styles united 161
b. Styles free 165

161a. Flowers 5–7.5 cm across; stigmas
woolly; fruit 1.5–2.5 cm
38. jundzillii
b. Flowers 1.5–5 cm across; stigmas
hairless or sparsely hairy; fruit to
1.2 cm 162
162a. Bracts usually absent; flowers fruit-
scented 163
b. Bracts present; flowers not scented
164
163a. Stigmas hairless; fruit red, 6–7 mm
82. multiflora
b. Stigmas sparsely hairy; fruit pale to
dark orange, c. 1 cm
96. soulieana
164a. Young stems with bristles as well
as prickles; leaflets 7–9, hairless on
both sides **93. maximowicziana**
b. Young stems bearing a few
prickles, but no bristles; leaflets 5
(rarely 3–7), more or less downy
on both sides, especially beneath
95. phoenicia
165a. Flowers semi-double or double 166
b. Flowers single 179
166a. Prickles hooked 167
b. Prickles straight or curved 170
167a. Margin of leaflets with compound
teeth **14. gallica**
b. Margin of leaflets with simple teeth
168
168a. Flowers c. 5 cm across; fruit
orange-red **21. × collina**
b. Flowers 6–8 cm across; fruit red
169
169a. Styles protruding **20. × alba**
b. Styles not protruding
17. × centifolia
170a. Margin of leaflets with compound
teeth 171
b. Margin of leaflets with simple teeth
172
171a. Flowers pale pink, fading to
whitish; sepals hairy inside and on
margin **15. × macrantha**
b. Flowers rose-pink or crimson, or
striped white, pink and red; sepals
neither hairy inside nor on margin
14. gallica
172a. Leaflets obtuse 173
b. Leaflets acute to acuminate 176
173a. Styles protruding; flowers semi-
double **19. × damascena**
b. Styles not protruding; flowers
double 174
174a. Bracts narrow 175
b. Bracts very large
16. × francofurtana
175a. Flowers 6–8 cm across; stems to
2 m **17. × centifolia**

b. Flowers 3.8–5 cm across; stems to
50 cm   **42. carolina**
176a. Flowers double   177
b. Flowers semi-double   178
177a. Flowers 6–8 cm across; stems to
2 m   **17. × centifolia**
b. Flowers 3.8–5 cm across; stems to
50 cm   **42. carolina**
178a. Styles protruding; leaflets downy
beneath   **19. × damascena**
b. Styles not or only slightly
protruding; leaflets hairless beneath
but with glands and tiny prickles
on the midrib   **15. × macrantha**
179a. Margin of leaflets with simple teeth
180
b. Margin of leaflets with compound
teeth   189
180a. Sepals hairy inside and on the
margin   **15. × macrantha**
b. Sepals hairless inside   181
181a. Petals deep pink, white at the base;
mature leaflets grey-green, often
tinged with brown or purple
39. glauca
b. Petals white or pale pink to mid-
pink, if deeper then not white at
the base; mature leaflets green 182
182a. Leaflets acute to acuminate   183
b. Leaflets obtuse   187
183a. Flowers 2.5–5 cm across   184
b. Flowers 5–7.5 cm across   186
184a. Prickles straight; receptacle
glandular-bristly   **42. carolina**
b. Prickles hooked; receptacle smooth
185
185a. Styles protruding   **35. canina**
b. Styles not protruding or only
slightly so   **32. stylosa**
186a. Leaflets 3–5; styles protruding
18. × richardii
b. Leaflets 5–9; styles not protruding
41. virginiana
187a. Stems to 1.5 m bearing prickles
intermixed with bristles
42. carolina
b. Stems 1.5–5 m bearing prickles but
no bristles   188
188a. Leaflets narrowly elliptic to ovate,
usually hairless on both sides,
although sometimes downy on the
veins beneath   **35. canina**
b. Leaflets broadly ovate to roundish,
downy on both sides
33. corymbifera
189a. Flowers rose-pink or crimson, or
striped white, pink and red
14. gallica
b. Flowers white to pale pink   190
190a. Receptacle smooth   191

b. Receptacle glandular-bristly   194
191a. Leaflets with glands beneath   192
b. Leaflets lacking glands beneath
35. canina
192a. Prickles unequal   **31. serafinii**
b. Prickles equal   193
193a. Sepals glandular on the back
29. micrantha
b. Sepals not glandular on the back
35. canina
194a. Leaflets 3–5 (rarely to 7); flowers
c. 7.5 cm across   **15. × macrantha**
b. Leaflets 5–7; flowers 2.5–5.5 cm
across   195
195a. Prickles equal; styles more or less
protruding   196
b. Prickles unequal; styles not
protruding   **30. horrida**
196a. Flowers 2.5–3 cm across
29. micrantha
b. Flowers 3.8–5.5 cm across
25. tomentosa

1. R. persica A.L. Jussieu (*Hulthemia
persica* (A.L. Jussieu) Bornmüller;
H. berberifolia (Pallas) Dumortier; *Rosa
berberifolia* Pallas; *R. simplicifolia* Salisbury).
Illustration: Harkness, Roses, pl. 1 (1978);
Thomas, A garden of roses, 127 (1987);
Austin, The heritage of the rose, 352
(1988); Phillips & Rix, Roses, 19 (1988).

Straggling, often decumbent, suckering
shrub. Stems yellowish brown, 40–90 cm,
hairless or downy, bearing straight or
curved, slender, yellow or reddish prickles.
Stipules absent. Leaves deciduous, simple,
stalkless, grey-green, 1.3–3.2 cm, broadly
elliptic to obovate, acute, toothed towards
apex, usually downy. Receptacle bristly.
Flowers solitary, single, 2.5–3 cm across.
Sepals lanceolate, downy and more or less
prickly on the back. Petals yellow with a
basal crimson spot. Fruit spherical,
sometimes flattened, blackish, very prickly.
*Iran, Afghanistan, former USSR (Central
Asia, SW Siberia).* H5–G1. Early summer.

2. R. × hardii Cels (× *Hulthemosa hardii*
(Cels) Rowley). Illustration: *Paxton's
Magazine of Botany* **10**: 195 (1843);
Harkness, Roses, pl. 2 (1978); Thomas,
A garden of roses, 93 (1987); Phillips &
Rix, Roses, 18 (1988).

A hybrid between *R. persica* and
R. clinophylla. Stems to 2 m. Stipules
present. Leaves simple or with up to 7,
dark green, hairless leaflets. Flowers *c.*
5 cm across. Sepals with some lateral lobes.
Petals deep yellow with a basal crimson
spot. Fruit yellowish orange, shortly hairy

with a few bristles. *Garden origin.* H5–G1.
Early summer.

3. R. pimpinellifolia Linnaeus
(*R. spinosissima* Linnaeus (1771) not
Linnaeus (1753). Illustration: Ross-Craig,
Drawings of British plants **9**: pl. 15 (1956);
Phillips & Rix, Roses, 20 (1988); Blamey &
Grey-Wilson, Illustrated flora of Britain and
northern Europe, 179 (1989).

Much-branched, suckering shrub. Stems
purple-brown, 90–200 cm, bearing dense
straight, slender prickles and stiff bristles,
especially dense at the stem-base. Stipules
narrow. Leaves deciduous; leaflets 7–9
(rarely 7–11), broadly elliptic to broadly
obovate or more or less circular, 6–20 mm,
obtuse, hairless although mid-vein
sometimes downy beneath, margin with
simple glandular teeth. Bracts absent.
Receptacle usually hairless. Flowers
solitary, single or double, 3.8–6 cm across.
Sepals entire, narrowly lanceolate,
acuminate, much shorter than petals,
hairless on the back but with woolly
margin, erect and persistent in fruit. Petals
creamy white. Fruit spherical or more or
less so, dark brown becoming blackish,
7–15 mm, smooth and hairless. *W & S
Europe, SW & C Asia east to NE China and
Korea.* H2. Summer.

'Grandiflora' ('Altaica'; *R. pimpinellifolia*
var. *altaica* (Willdenow) Thory;
R. pimpinellifolia var. *grandiflora* Ledebour).
Illustration: Beales, Classic roses, 139
(1985); Thomas, A garden of roses, 131
(1987); Phillips & Rix, Roses, 20 (1988).
Stems less bristly than the species and
flowers 5–7.5 cm across. 'Hispida'
(*R. pimpinellifolia* var. *hispida* (Sims) Boom;
R. spinosissima var. *hispida* (Sims) Koehne).
Illustration: Harkness, The rose, pl. 103
(1979). Leaflets 1.9–3.2 cm; flowers
5–7.5 cm across, opening pale yellow and
fade to cream. 'Andrewsii' (*R. spinosissima*
var. *andrewsii* Willmott). Flowers double,
rose-pink. 'Nana' (*R. spinosissima* var. *nana*
Andrews). Flowers semi-double, white.

R. × hibernica Templeton. Illustration:
Gault & Synge, The dictionary of roses in
colour, pl. 22 (1971); Thomas, A garden of
roses, 97 (1987); Walsh & Nelson, An
Irish florilegium **2**: 19 (1983); Austin, The
heritage of the rose, 367 (1988). A hybrid
between *R. pimpinellifolia* and *R. canina*
that occurred naturally in Ireland although
it is now thought to exist only in gardens.
Stems 4 m or more. Leaflets 5–7, sparsely
hairy beneath, especially on the veins.
Flowers single. Sepals with lateral lobes

and the apex expanded. Petals pale pink, paler towards the base. Fruit dark red, obovoid.

4. R. × reversa Waldstein & Kitaibel. Illustration: Phillips & Rix, Roses, 16 (1988).

A hybrid between *R. pimpinellifolia* and *R. pendulina*. Leaflets densely glandular. Petals carmine. Fruit spherical-ovoid, deep purplish red, drooping, *c.* 2 cm. *S Europe north to Switzerland and S France*. H4. Summer.

5. R. koreana Komarov.

Dwarf, densely branched shrub. Stems dark red, to 1 m, bearing dense bristles. Stipules narrow. Leaves deciduous; leaflets 7–11 (rarely to 15), elliptic to elliptic-obovate, 1–2 cm, more or less obtuse, hairless, sometimes slightly downy beneath, margins with simple glandular teeth. Bracts absent. Flowers solitary, single, 4–6 cm across, scented. Sepals entire, lanceolate, glandular on the back, erect and persistent in fruit. Petals white, tinged with pink. Fruit ovoid to oblong, orange-red, 1–1.5 cm. *N Korea, NE China.* H4. Summer.

6. R. xanthina Lindley.

Erect shrub. Stems brown when young, ageing to grey-brown, 1.5–3.5 m, bearing straight or slightly curved, broad-based prickles, which are sometimes much flattened on non-flowering shoots. Stipules narrow, with margins curved downwards and inwards. Leaves deciduous; leaflets 7–13, broadly elliptic to obovate or spherical, 8–20 mm, obtuse, usually hairless above and hairy beneath, margins with simple teeth. Bracts absent. Receptacle smooth. Flowers solitary, rarely 2, semi-double, 3.8–5 cm across. Sepals entire, lanceolate, acuminate, leafy and toothed at tip, hairless or sparsely hairy, erect and persistent in fruit. Petals bright yellow. Fruit spherical or broadly ellipsoid, brown-red or maroon, 1.2–1.5 cm, smooth and hairless. *N China, Korea.* H4. Early summer.

Forma **hugonis** (Hemsley) Roberts (*R. hugonis* Hemsley). Illustration: Bois & Trechslin, Roses, 15 (1962); Gibson, Shrub roses for every garden, pl. 2 (1973); Thomas, A garden of roses, 99 (1987); Phillips & Rix, Roses, 17 (1988). Differs in having leaflets elliptic to obovate and hairless flower-stalks bearing single flowers 4–6 cm across. *C China.* H4.

Forma **spontanea** Rehder (*R. xanthina* forma *normalis* Rehder & Wilson, in part). Illustration: Austin, The heritage of the rose, 353 (1988); Phillips & Rix, Roses, 20 (1988). Differs from *R. xanthina* itself in having single flowers 5–6 cm across. It is one of the parents of the splendid 'Canary Bird', the other parent being forma *hugonis*.

7. R. ecae Aitchison (*R. xanthina* var. *ecae* (Aitchison) Boulenger). Illustration: Gault & Synge, The dictionary of roses in colour, pl. 7 (1971); Harkness, The rose, pl. 101 (1979); Thomas, A garden of roses, 69 (1987); Phillips & Rix, Roses, 16 (1988).

Much-branched, erect, suckering shrub. Stems to 1.5 m, bearing dense, straight, flattened, reddish prickles. Stipules very narrow. Leaves deciduous, aromatic; leaflets 5–9, broadly elliptic to obovate or spherical, 4–8 mm, obtuse, glandular beneath, margins with simple, often glandular teeth. Bracts absent. Flowers solitary, single, 2–3 cm across. Sepals entire, hairless, spreading or deflexed in fruit, persistent. Petals deep yellow. Fruit spherical, shiny red-brown, 0.5–1 cm, smooth and hairless. *NE Afghanistan, NW Pakistan and adjacent former USSR, N China.* H4. Summer.

A parent (with 'Canary Bird') of 'Golden Chersonese' (illustration: Phillips & Rix, Roses, 15, 1988) which grows to 2 m and has inherited the aromatic leaves and deep yellow flowers of *R. ecae*. 'Helen Knight' (illustration: Phillips & Rix, Roses, 16, 1988) is a hybrid between *R. ecae* and probably *R. pimpinellifolia* 'Grandiflora' and has large yellow flowers and red-brown stems, which will reach 3 m against a wall.

8. R. kokanica (Regel) Juzepczuk (*R. xanthina* var. *kokanica* (Juzepczuk) Boulenger). Illustration: Phillips & Rix, Roses, 14 (1988).

Erect suckering shrub. Stems to 2 m, reddish brown when young. Stipules narrow. Leaves deciduous, aromatic; leaflets 5–7, broadly elliptic to obovate or more or less circular, glandular beneath, margins with compound glandular teeth. Bracts absent. Flowers solitary, single, *c.* 5 cm across. Sepals entire, hairless, erect and persistent in fruit. Petals bright yellow. Fruit spherical, brownish. *Iran, Afghanistan, Kazakhstan, Mongolia, NW China (Xinjiang).* H4. Early summer.

9. R. primula Boulenger (*R. ecae* subsp. *primula* (Boulenger) Roberts). Illustration: Gault & Synge, The dictionary of roses in

colour, pl. 39 (1971); Gibson, The book of the rose, pl. 15 (1980); Beales, Classic roses, 89 (1985); Phillips & Rix, Roses, 15 (1988).

Erect shrub. Stems slender, to 3 m, reddish brown when young, bearing stout, straight, somewhat compressed, broad-based prickles. Stipules narrow. Leaves deciduous, very aromatic; leaflets 9 (rarely 7–13), elliptic to obovate or oblanceolate, 6–20 mm, acute or obtuse, hairless above, beneath with large glands, margins with compound glandular teeth. Bracts absent. Receptacle smooth. Flowers solitary, single, 2.5–4.5 cm across. Sepals entire, hairless, erect and persistent in fruit. Petals primrose-yellow. Fruit spherical to obconical, brownish red to maroon, 1–1.5 cm, smooth. *Former USSR (Central Asia) to N & C China.* H4. Early summer.

10. R. foetida Herrmann (*R. lutea* Miller; *R. eglanteria* Linnaeus, in part). Illustration: Harkness, The rose, pl. 100 (1979); Thomas, A garden of roses, 75 (1987); Phillips & Rix, Roses, 15 (1988); Blamey & Grey-Wilson, Illustrated flora of Britain and northern Europe, 179 (1989).

Erect shrub. Stems erect or arching, 1–3 m, at first dark brown, later greyish, bearing few, straight or curved prickles of unequal length, abruptly broadened at the base, usually mixed with bristles. Stipules very narrow. Leaves deciduous; leaflets 5–9, elliptic to obovate, 1.5–4 cm, obtuse to almost acute, bright green and more or less hairless above, dull green, downy and more or less glandular beneath, margins with rather few, compound glandular teeth. Bracts absent. Receptacle smooth or bristly. Flowers solitary or sometimes 2–4, single, 5–7.5 cm across, with an unpleasant smell. Sepals entire or with a few lateral lobes, lanceolate, apex expanded, hairless or glandular-bristly, erect and persistent in fruit. Petals deep yellow. Stigmas hairy. Fruit spherical, red, 0.8–1 cm, smooth or bristly. *SW & WC Asia.* H4. Summer.

'Bicolor' (*R. bicolor* Jacquin; *R. foetida* var. *bicolor* (Jacquin) Willmott). Illustration: Thomas, A garden of roses, 77 (1987); Austin, The heritage of the rose, 356 (1988); Phillips & Rix, Roses, 15 (1988). Possibly a very old hybrid between *R. kokanica* and *R. hemisphaerica*. Petals coppery red inside and yellow-buff on the back. Branches occasionally revert to the yellow-flowered form. 'Persiana' (*R. foetida* var. *persiana* (Lemaire) Rehder).

Illustration: Austin, The heritage of the rose, 372 (1988); Phillips & Rix, Roses, 15 (1988); Brickell, The RHS gardeners' encyclopedia of plants and flowers, 152 (1989). Flowers very double, yellow, freely produced, smaller than those of *R. foetida* itself.

11. R. × harisonii Rivers. Illustration: Gault & Synge, The dictionary of roses in colour, pl. 20 (1971); Gibson, The book of the rose, pl. I (1980); Beales, Classic roses, 140 (1985).

A hybrid between *R. pimpinellifolia* and *R. foetida*. Erect shrub with occasional suckers. Stems 50–200 cm. Leaves deciduous; leaflets 5–9, elliptic, glandular beneath, margins with compound, somewhat glandular teeth. Bracts absent. Receptacle bristly. Flowers solitary, loosely double, 5–6 cm across, with an unpleasant smell. Sepals lobed at apex, slightly hairy on the back, margin glandular, erect and persistent in fruit. Petals sulphur-yellow. Fruit almost black, vertically compressed, bristly. *Garden origin.* H4. Summer.

'Lutea maxima' (*R. spinosissima* Linnaeus var. *lutea* Bean) has the same parentage as *R. × harisonii* but shows more influence of *R. foetida* in that it has buttercup-yellow flowers, the sepals sometimes with lateral lobes, and a smooth fruit.

12. R. hemisphaerica Herrmann (*R. rapinii* Boissier & Balansa). Illustration: Anderson, The complete book of 169 Redouté roses, 29 (1979); Thomas, A garden of roses, 95 (1987); Phillips & Rix, Roses, 14 (1988).

Stiffly erect, much-branched shrub. Stems to 2 m, bearing scattered, slender decurved prickles with broad bases, on the young shoots mixed with bristles. Stipules narrow. Leaves deciduous, slightly aromatic; leaflets 5–9, obovate, obtuse to slightly notched, grey-green and hairless above, glaucous and downy on the veins beneath, margins with simple teeth in upper two-thirds. Bracts absent. Receptacle sometimes glandular. Flowers solitary, double, 4–5 cm across, often sweetly scented. Sepals broadly lanceolate, toothed at the apex, erect and persistent in fruit. Petals sulphur-yellow. Stigmas woolly. Fruit spherical, dark red, 1.2–1.5 cm, smooth. *Gardens of SW Asia.* H4. Summer.

Uncommon in cultivation. The double-flowered plant, 'Flore Pleno' (*R. sulphurea* Aiton), is more usually cultivated. Illustration: Phillips & Rix, Roses, 15 (1988).

13. R. sericea Lindley. Illustration: Phillips & Rix, Roses, 16 (1988) – as subsp. *omeiensis.*

Upright or somewhat spreading shrub. Stems 2–4 m, grey or brown, bearing straight or curved, often upward-pointing, reddish prickles with broad bases, often mixed with slender bristles. Stipules narrow. Leaves deciduous; leaflets 7–11, elliptic, oblong or obovate, 6–30 mm, obtuse or more or less acute, hairy or not beneath, margins with simple teeth often confined to upper half. Bracts absent. Receptacle hairless. Flowers solitary, on short lateral shoots, single, 2.5–6 cm across. Sepals entire, hairless or silky on the back, erect or spreading and persistent in fruit. Petals usually 5, white or cream, notched at the apex. Fruit spherical to pear-shaped with a narrow stalk, dark crimson, scarlet, orange or yellow, 8–15 mm, smooth. *India (Sikkim), Bhutan to Burma and SW China* H4. Early summer.

Subsp. **omeiensis** (Rolfe) Roberts. (*R. omeiensis* Rolfe; *R. sericea* var. *omeiensis* (Rolfe) Rowley). Illustration: Thomas, A garden of roses, 134 (1987); Phillips & Rix, Roses, 17 (1988) – as *R. sericea*. Leaflets 11–19, silky beneath, petals usually 4, white. Fruit with a fleshy stalk. SW & C China .

Subsp. **omeiensis** forma **pteracantha** Franchet (*R. sericea* var. *pteracantha* (Franchet) Boulenger; *R. omeiensis* forma *pteracantha* (Franchet) Rehder & Wilson). Illustration: Gibson, The book of the rose, pl. 1 (1980); Austin, The heritage of the rose, 353 (1988) – as *R. sericea* var. *pteracantha*; Welch, Roses, 62 (1988); Phillips & Rix, Roses, 17 (1988). The stems bear flattened prickles, which can be up to 3.8 cm wide at the base and 1.2–2 cm deep, abruptly and sharply pointed, forming interrupted wings down the stems; prickles on young stems are red, turning grey and woody in the second year. *W China.*

14. R. gallica Linnaeus (*R. provincialis* Herrmann; *R. rubra* Lamarck). Illustration: Phillips & Rix, Roses, 28 (1988).

Low, erect shrub, suckering. Stems 50–200 cm, green to dull red, bearing slender, unequal, curved or sometimes hooked prickles, mixed with glandular bristles. Stipules narrow with spreading pointed tips. Leaves deciduous; leaflets 3–5 (rarely to 7), leathery, broadly elliptic to almost circular, acute or obtuse, dark green and hairless above, hairy and glandular beneath, at least on the veins,

margins with compound, usually glandular teeth. Bracts present. Receptacle with stalked or stalkless glands. Flowers solitary, or sometimes 2–4, single or semi-double, fragrant, 4–8 cm across. Sepals with some lateral lobes, glandular on the back, reflexed and falling after flowering. Petals rose-pink or crimson. Styles free, not protruding; stigmas woolly. Fruit spherical to ellipsoid, brick-red, *c.* 1.3 cm, densely glandular-bristly. *S & C Europe, E to Ukraine, Turkey, Iraq and the Caucasus (naturalised in E North America).* H4. Summer.

'Officinalis' (*R. gallica* var. *officinalis* Thory). Illustration: Anderson, The complete book of 169 Redouté roses, 43 (1979); Welch, Roses, 106 (1988); Phillips & Rix, Roses, 49 (1988). Shrub rarely more than 70 cm, flowers semi-double, crimson, very fragrant. 'Versicolor' (*R. gallica* var. *versicolor* Linnaeus; 'Rosa Mundi'). Illustration: Gibson, The rose gardens of England, 22 (1988); Le Rougetel, A heritage of roses, 11 (1988); Phillips & Rix, Roses, 40 (1988). A sport of 'Officinalis' with stems 1–2 m and semi-double flowers 7–9 cm across and striped white, pink and red. It often reverts to the non-striped type. It is frequently confused with *R. damascena* 'Versicolor'.

15. R. × macrantha misapplied, not Desportes. (*R.* 'Macrantha'). Illustration: Edwards, Wild and old garden roses, 34 (1975); Beales, Classic roses, 195 (1985); Austin, The heritage of the rose, 352 (1988); Phillips & Rix, Roses, 109 (1988).

Vigorous shrub. Stems 1.5–2 m, arching or spreading, green, bearing sparse, scattered, straight or slightly curved prickles, mixed with small straight bristles and stalked glands. Stipules narrow. Leaves deciduous; leaflets 3–5 (rarely to 7), dull green, ovate to oblong-ovate, acute to acuminate, hairless on both sides but glandular beneath and with tiny prickles on the midrib, margins with simple or compound teeth. Bracts present. Receptacle slightly glandular below. Flowers 2–5 in clusters, single to semi-double, fragrant, *c.* 7.5 cm across. Sepals with many lateral lobes, hairy inside and on margin, slightly glandular on the back, reflexed and falling after flowering. Petals pale pink, fading to whitish. Styles free, not or only slightly protruding. Fruit almost spherical, red, to 1.5 cm wide. *Garden origin.* H4. Summer.

A hybrid between *R. gallica* and possibly *R. canina*.

16. R. × francofurtana Muenchhausen
(*R. turbinata* Aiton).

Shrub with erect stems to 2 m, grey-green, the flowering-stems unarmed or bearing a few straight or curved prickles and bristles. Stipules broad. Leaves deciduous; leaflets 5–7, grey-green, broadly ovate to more or less round, obtuse, downy on the veins beneath, margins with coarse simple teeth. Bracts very large. Receptacle glandular-bristly. Flowers solitary or 2–6, usually double, slightly fragrant, 5–7 cm across. Sepals entire or with a few lateral lobes, hairy and glandular on margin and back, erect or reflexed, falling after flowering. Petals purplish pink with darker veins. Styles free, not protruding. Fruit obconical, red. *Garden origin*. H4. Summer.

A hybrid, probably between *R. gallica* and *R. majalis*.

17. R. × centifolia Linnaeus
(*R. provincialis* Miller in part (1788) not Herrmann, 1762). Illustration: Phillips & Rix, Roses, 60 (1988).

Loosely branched shrub producing a few suckers. Stems to 2 m, bearing many small, almost straight, scattered prickles and larger hooked ones. Stipules narrow. Leaves deciduous; leaflets 5–7, dull green, broadly ovate to more or less round, acute to obtuse, usually hairless above and downy beneath, margins with simple glandular teeth. Bracts narrow. Receptacle bearing glands. Flowers 1–few, double, more or less spherical, very fragrant, 6–8 cm across. Sepals with lateral lobes, glandular on the back, spreading, somewhat persistent in fruit. Petals usually pink, more rarely white or dark red. Styles free, not protruding. Fruit ellipsoidal or spherical, red. *Garden origin*. H4. Summer.

A complex hybrid involving *R. gallica*, *R. moschata*, *R. canina* and *R. damascena*. There are several cultivars: 'Bullata'. Illustration: Anderson, The complete book of 169 Redouté roses, 51 (1979); Thomas, A garden of roses, 51 (1987); Phillips & Rix, Roses, 59 (1988). Leaflets crinkled, brownish above when young; flowers pink. 'Cristata' (*R. centifolia* var. *cristata* Prévost). Illustration: Thomas, A garden of roses, 53 (1987); Phillips & Rix, Roses, 60, 61 (1988). Lower-growing; sepals with many marginal segments, which provide each pink flower with a green fringe; flowers less globular. 'Muscosa' (*R. muscosa* Miller; *R. centifolia* var. *muscosa* (Miller) Seringe). Illustration: Bois & Trechslin, Roses, 13 (1962); Thomas, A garden of roses, 55

(1987); Phillips & Rix, Roses, 64, 65 (1988). Shrub *c.* 1 m. Calyx and flower-stalks bearing much-branched scented glands, which form the so-called 'moss', hence the common name of 'moss rose'. 'Parvifolia' (*R. parvifolia* Ehrhart (1791) not Pallas (1788); *R. centifolia* var. *parvifolia* (Ehrhart) Rehder). Illustration: Phillips & Rix, Roses, 58 (1988). Stems to 1 m; leaflets small, dark green; flowers fragrant, deep pink suffused with purple and with a paler centre. It is possible that it is not in fact related to *R. centifolia* at all.

18. R. × richardii Rehder (*R. sancta* Richard, not Andrews; *R. centifolia* var. *sancta* (Richard) Zabel). Illustration: Beales, Classic roses, 196 (1985); Austin, The heritage of the rose, 350 (1988); Phillips & Rix, Roses, 57 (1988).

Low, spreading shrub. Stems to 1.3 m, bearing small, scattered, hooked prickles of unequal size. Stipules broad. Leaves deciduous; leaflets 3–5, ovate to narrowly elliptic, acute, wrinkled above, downy beneath, margins with simple glandular teeth. Bracts narrow. Receptacle smooth. Flowers several in loose clusters, single, 5–7.5 cm across. Sepals with lateral lobes, the apex leafy, downy and glandular on the back, reflexed and falling after flowering. Petals pale pink. Styles free, protruding; stigmas woolly. *Garden origin*. H4. Summer.

A hybrid of *R. gallica* and possibly *R. phoenicia* or *R. arvensis*.

19. R. × damascena Miller. Illustration: Bois & Trechslin, Roses, 11 (1962).

Shrub with stems to 2.2 m, densely armed with stout, curved, equally sized prickles and stiff bristles. Stipules often with toothed margins. Leaves deciduous; leaflets 5 (rarely to 7), grey-green, ovate to elliptic, acute to obtuse, hairless above, downy beneath, margins with simple teeth. Bracts narrow. Receptacle glandular-bristly. Flowers in clusters of up to 12, semi-double, fragrant. Sepals with lateral lobes and slender tips, glandular and hairy on the back, reflexed and falling after flowering. Petals pink. Styles free, protruding. Fruit obconical, red, to 2.5 cm, bristly. *Turkey*. H4. Summer.

A hybrid between *R. gallica* and *R. moschata*, *R. × damascena* is the summer damask rose. The autumn damask rose is var. **semperflorens** (Loiseleur & Michel) Rowley (*R. bifera* (Poiret) Persoon). Illustration: Krüssmann, The complete book of roses, 259 (1981); Phillips & Rix,

Roses, 54 (1988) – as 'Quatre Saisons'. It cannot be distinguished from *R. × damascena* except by the production of autumn flowers. 'Versicolor' (*R. damascena* var. *versicolor* Weston). Illustration: Anderson, The complete book of 169 Redouté roses, 115 (1979); Phillips & Rix, Roses, 56 (1988). Known as the York and Lancaster rose, it has loosely double flowers with deep pink or very pale pink petals, or petals which are partly deep pink and partly pale pink. 'Trigintipetala' (*R. × damascena* var. *trigintipetala* (Dieck) Keller). Flowers semi-double, *c.* 8 cm across, with *c.* 30 red petals. 'Portlandica' has bright red semi-double flowers, faintly scented and produced from midsummer to autumn.

20. R. × alba Linnaeus.

Spreading shrub with stout, arching stems 1.8–2.5 m, bearing scattered, hooked, unequal prickles, often mixed with bristles. Stipules broad. Leaves deciduous; leaflets 5 (rarely to 7), dull green, ovate to more or less round, shortly acuminate to obtuse, hairless above and downy beneath especially on the veins, margins with simple teeth. Bracts narrow. Receptacle slightly glandular-bristly, at least below. Flowers 1–3, semi-double or double, 6–8 cm across, fragrant. Sepals with lateral lobes and leafy tips, glandular-bristly on the back, reflexed and falling after flowering. Petals white to pale pink. Styles free, protruding. Fruit more or less spherical, red, 2–2.5 cm. *Garden origin*. H4. Summer.

Parentage of the hybrid is disputed. It may be *R. gallica* × *R. arvensis* or *R. corymbifera* or even *R. canina* × *R. × damascena*.

'Maxima' (*R.* 'Alba Maxima'). Illustration: Phillips & Rix, Roses, 40 (1988). Double flowers open pink but fade to creamy white. 'Semiplena'. Illustration: Gibson, The book of the rose, pl. xxii (1980); Phillips & Rix, Roses, 41 (1988). Flowers semi-double, white. Is a sport of 'Maxima'. 'Incarnata' (*R. incarnata* Miller; *R. × alba* var. *incarnata* (Miller) Weston). Illustration: Phillips & Rix, Roses, 41 (1988). Stems with very few prickles but densely bristly on the flowering branches below the bracts. Leaflets usually 7. Flowers double, pale pink.

21. R. × collina Jacquin.

Erect shrub with stems 1.5–2 m, bearing strong, hooked prickles, which are red when young. Leaves deciduous; leaflets 5 (rarely to 7), hairless above, downy on the veins beneath, margins with simple teeth. Bracts present. Receptacle often glandular. Flowers 1–3, more or less double, fragrant, *c.* 5 cm across. Sepals with lateral lobes and leafy tips, reflexed and falling after flowering. Petals pink. Fruit ovoid, orange-red, sometimes glandular. *Garden origin.* H4. Summer.

Possibly a hybrid between *R. canina* and *R. gallica.*

22. R. villosa Linnaeus (*R. pomifera* Herrmann, invalid). Illustration: Blamey & Grey-Wilson, Illustrated flora of Britain and northern Europe, 181 (1989).

Densely branched shrub, often suckering. Branches stiff, straight, 1–2.4 m, reddish when young, bearing scattered, slender, straight or slightly curved prickles of unequal size. Stipules broad. Leaves deciduous; leaflets 5–9, bluish green, elliptic, 3.2–6.5 cm, acute or obtuse, hairy on both sides, often with dense, apple-scented glands beneath, margins with compound glandular teeth. Bracts broad, often concealing receptacle. Receptacle densely glandular-bristly. Flowers 1–3 or more, single, slightly fragrant, 2.5–6.5 cm across. Sepals with a few lateral lobes, acuminate, glandular on the back, persistent in fruit. Petals deep pink. Styles free, not protruding; stigmas woolly. Fruit spherical to pear-shaped, dark red, 1–3 cm, glandular-bristly. *C & S Europe, Turkey, former USSR (Caucasus).* H4. Summer.

'Duplex' (*R. pomifera* forma *duplex* (Weston) Rehder). Illustration: Gault & Synge, The dictionary of roses in colour, pl. 38 (1971); Beales, Classic roses, 6 (1985). Leaflets grey-green; flowers semi-double, clear pink, *c.* 6.5 cm across; more free-flowering. Possibly a hybrid between *R. villosa* and a tetraploid rose of garden origin.

23. R. mollis Smith (*R. villosa* misapplied not Linnaeus). Illustration: Phillips & Rix, Roses, 29, 215 (1988); Blamey & Grey-Wilson, Illustrated flora of Britain and northern Europe, 181 (1989).

Similar to *R. villosa* and thought by some to be synonymous with it; differs in the young stems being bloomed, the smaller more or less circular leaflets, 1.2–3.5 cm, the less bristly receptacle and a smaller fruit, 1–1.5 cm. *Europe, W Asia.* H3. Summer.

24. R. sherardii Davies. Illustration: Ross-Craig, Drawings of British plants **9**: pl. 22 (1956): Blamey & Grey-Wilson, Illustrated

flora of Britain and northern Europe, 181 (1989).

Shrub with dense branches to 2 m and with a whitish waxy bloom when young, bearing straight or curved prickles. Leaves deciduous; leaflets 5–7 (rarely 3–7), bluish green, elliptic to broadly ovate, hairy on both surfaces, margins with compound teeth. Bracts present. Receptacle glandular-bristly. Flowers several in a cluster, single, 3–5 cm across. Sepals with lateral lobes, glandular-hairy on the back, erect and persistent in fruit. Petals deep pink. Styles free, not protruding: stigmas woolly. Fruit ovoid to obconical, red, 1.2–2 cm wide, glandular-bristly. *N & C Europe.* H3. Summer.

25. R. tomentosa Smith. Illustration: Ross-Craig, Drawings of British plants **9**: pl. 23 (1956): Blamey & Grey-Wilson, Illustrated flora of Britain and northern Europe, 181 (1989).

Shrub with arching, often zig-zag branches to 3 m, green when young and often bloomed, bearing sparse, straight or slightly curved prickles of equal size and often in pairs. Stipules with short free tips. Leaves deciduous, smelling resinous when crushed; leaflets 5–7, light- to grey-green, elliptic to ovate, obtuse, acute or acuminate, usually hairy on both sides and glandular beneath, margins with compound glandular teeth. Bracts present. Receptacle glandular-bristly. Flowers solitary or few, single, fragrant, 3.8–5.5 cm across. Sepals with lateral lobes and expanded tips, glandular-bristly on the back, reflexed and falling after flowering. Petals pale pink or white. Styles free, somewhat protruding; stigmas woolly or not. Fruit ovoid to spherical, orange-red, 1.8–2.5 cm with stalked glands. *Europe, Turkey, Caucasus (occasionally naturalised in USA).* H4. Summer.

26. R. rubiginosa Linnaeus (*R. eglanteria* Linnaeus, in part). Illustration: Krüssmann, The complete book of roses, 263 (1981); Phillips & Rix, Roses, 18, 106 (1988); Blamey & Grey-Wilson, Illustrated flora of Britain and northern Europe, 181 (1989); Brickell (ed.), The RHS gardeners' encyclopedia of plants and flowers, 149 (1989).

Dense, much-branched shrub with arching branches 2–3 m, bearing many, scattered, stout, hooked, unequal prickles and often with stiff bristles on the flowering branches. Stipules broad. Leaves deciduous, aromatic especially in damp

weather; leaflets 5–9, dark green, ovate to roundish, 1–3 cm, obtuse to acute, usually hairless above, glandular and downy beneath, margins with compound teeth. Bracts present. Receptacle glandular-bristly. Flowers 1–7, occasionally more, single, fragrant, 2.5–5 cm across. Sepals with lateral lobes, glandular-bristly on the back, erect and persistent in fruit. Petals pale to deep pink. Styles free, not protruding; stigmas woolly. Fruit ovoid to almost spherical, red or orange, 1–2.5 cm, smooth or glandular-bristly. *Europe, N Africa, Asia (naturalised in N America).* H4. Summer.

27. R. sicula Trattinick.

Densely branched, suckering shrub. Stems 50–150 cm, red when young, bearing sparse, slender, straight or slightly curved prickles of equal length, intermixed with glandular bristles. Leaves deciduous, slightly aromatic; leaflets 5–9, broadly ovate to roundish, 6–20 mm, hairless above, glandular and sometimes downy beneath, margins with glandular compound teeth. Bracts present. Receptacle with sparse glandular bristles. Flowers usually solitary, sometimes 2 or 3, single, 2.5–3.3 cm across. Sepals with a few lateral lobes, glandular on the back, margin hairy, erect or spreading and persistent in fruit. Petals bright pink. Styles free, not protruding; stigmas woolly. Fruit ovoid to almost spherical, red, eventually blackish, 1–1.3 cm, sparsely glandular-bristly. *Mediterranean area.* H4. Summer.

28. R. pulverulenta Bieberstein (*R. glutinosa* Smith; *R. dalmatica* Kerner; *R. glutinosa* var. *dalmatica* (Kerner) Keller).

Compact shrub with stems 30–70 cm, bearing numerous stiff, straight or decurved, whitish prickles intermixed with small glandular bristles. Stipules broad. Leaves deciduous; leaflets 5–7 (rarely 3 to 9), elliptic to obovate or roundish, 7–15 mm, both sides hairless or somewhat downy, glandular, margins with glandular compound teeth. Bracts present. Receptacle smooth or glandular-bristly. Flowers 1 or 2, single, 2.5–3.8 cm across. Sepals with a few gland-edged lateral lobes, the tips slightly expanded, erect and persistent in fruit. Petals rose-pink. Styles free, not protruding; stigmas woolly. Fruit ellipsoidal or almost spherical, dark red, to 2.5 cm, smooth or glandular-bristly. *S Europe to Turkey, Lebanon, Iran, Afghanistan, former USSR (Caucasus).* H4. Summer.

29. R. micrantha Smith. Illustration: Ross-Craig, Drawings of British plants **9**: pl. 25 (1956); Krüssmann, The complete book of roses, 258 (1981).

Much-branched shrub with arching stems to 3.5 m, bearing curved or hooked prickles of equal size; bristles usually absent. Stipules narrow. Leaves deciduous; leaflets 5–7, elliptic to broadly ovate, 1.5–4 cm, acute, hairless or hairy above, densely hairy and glandular beneath, margins with glandular compound teeth. Bracts present. Receptacle smooth or sparsely glandular-bristly. Flowers 1–4 (rarely to 8), single, 2.5–3 cm across. Sepals with lateral lobes and an expanded tip, glandular on the back, spreading or reflexed and falling after flowering. Petals pale pink or white. Styles free, more or less protruding; stigmas hairless. Fruit ovoid to almost spherical, red, 1.2–1.8 cm, smooth or sparsely glandular-bristly. *Europe except extreme north, NW Africa, SW Asia (naturalised in N America).* H3. Summer.

30. R. horrida Fischer (*R. biebersteinii* Lindley).

Low, stiffly branched shrub. Stems bearing short, stout, hooked prickles of unequal length, interspersed with glandular bristles. Leaves deciduous; leaflets 5–7, broadly elliptic to almost circular, 8–22 mm, more or less hairless, glandular beneath, margins with compound teeth. Bracts present. Receptacle glandular-bristly. Flowers solitary, single, 2.5–3 cm across. Sepals with lateral lobes, glandular on the back, reflexed and falling after flowering. Petals white. Styles free, not protruding; stigmas not woolly. Fruit ovoid to spherical, red, 1–1.6 cm, glandular-bristly. *SE Europe to Turkey and former USSR (Caucasus).* H4. Summer.

31. R. serafinii Viviani. Illustration: Phillips & Rix, Roses, 18 (1988).

Densely branched shrub with stems to 1.2 m, bearing stout, curved or hooked prickles of unequal length, often intermixed with bristles. Stipules narrow. Leaves deciduous, aromatic; leaflets 5–7 (rarely to 11), ovate to roundish, 5–12 mm, obtuse, glossy and hairless above, glandular beneath, margins with glandular compound teeth. Bracts present. Receptacle smooth. Flowers 1 (rarely to 3), single, 2.5–5 cm across. Sepals with lateral lobes, often glandular on the back, reflexed and falling after flowering. Petals pale pink. Styles free, not protruding; stigmas not woolly. Fruit obovoid to spherical, red,

0.8–1.2 cm wide, smooth. *Italy, Sicily, Corsica, Sardinia, Bulgaria & S former Yugoslavia.* H4. Early summer.

32. R. stylosa Desvaux.

Shrub with arching stems to 3 m, bearing stout, hooked, wide-based prickles. Stipules narrow. Leaves deciduous; leaflets 5–7, dark green, narrowly elliptic to ovate, 1.5–5 cm, acute or acuminate, glossy and hairless above, downy beneath, lacking glands, margins with simple teeth. Bracts narrow. Receptacle smooth. Flowers 1–8 or more, single, 3–5 cm across. Sepals with lateral lobes, often slightly glandular on the back, reflexed and falling after flowering. Petals usually white, sometimes pale pink. Styles free, not or slightly protruding; stigmas not woolly. Fruit ovoid to spherical, red, 1–1.5 cm, smooth. *W Europe, Bulgaria, W Asia.* H4. Summer.

33. R. corymbifera Borkhausen (*R. dumetorum* Thuillier).

Shrub with spreading or erect branches 1.5–3 m, bearing stout, hooked prickles. Leaves deciduous; leaflets 5–9, broadly ovate to roundish, 2.5–6 cm, obtuse, downy on both sides, margins with simple teeth. Bracts present. Flowers in clusters, single, 4–5 cm across. Sepals with some lateral lobes, usually hairless, sometimes slightly glandular, reflexed and falling after flowering. Petals white to pale pink. Styles free, not or slightly protruding. Fruit ovoid to almost spherical, orange-red, 1.2–2 cm. *Cooler parts of Europe and SW Asia, N Africa.* H3. Summer.

34. R. coriifolia Fries.

Densely branched shrub with erect or arching stems *c.* 2 m, bearing curved prickles of equal length. Leaves deciduous; leaflets 5–7, oblong to broadly elliptic, obtuse, hairless above, downy beneath at least on the veins, margins with simple or compound teeth. Bracts large. Receptacle hairless. Flowers 1–4, single. Sepals with lateral lobes, hairy or not on the back, erect to spreading and persistent in fruit. Petals white or pink. Styles free; stigmas woolly. Fruit ovoid to spherical, red, to 2.5 cm, smooth. *Europe, former USSR (Caucasus).* H3. Summer.

Very similar to *R. dumalis* and considered by some botanists to be the same. Alternatively, other botanists place it as a variety of *R. dumalis* subsp. *boissieri*, and due to the nomenclatural principle of priority its name would then become

R. dumalis Bechstein subsp. *boissieri* (Crépin) Nilsson var. *boissieri.*

35. R. canina Linnaeus. Illustration: Bois & Trechslin, Roses, 9 (1962); Gibson, The book of the rose, pl. 2 (1980); Phillips & Rix, Roses, 28 (1988); Blamey & Grey-Wilson, Illustrated flora of Britain and northern Europe, 179 (1989).

Vigorous shrub with arching, sometimes climbing stems 1.5–5.5 m, bearing scattered, strong, hooked, more or less equal prickles; bristles absent. Stipules narrow to broad. Leaves deciduous; leaflets 5–7, narrowly elliptic to ovate, 1.5–4 cm, acute or obtuse, usually hairless on both sides, sometimes glandular beneath or downy on veins, margins with simple or compound teeth. Bracts often broad. Receptacle smooth. Flowers solitary or 2–5, single, fragrant, 2.5–5 cm across. Sepals with lateral lobes, usually hairless on the back, reflexed and falling after flowering. Petals white or pale pink. Styles free, protruding; stigmas woolly or not. Fruit ovoid to almost spherical, red to orange, 1–3 cm, smooth. *Most of Europe & SW Asia, NW Africa (naturalised in North America).* H3. Summer.

36. R. britzensis Koehne. Illustration: Krüssmann, The complete book of roses, 267 (1981).

Erect shrub with stems 2–3 m, bearing sparse, small, scattered, slender prickles on the flowering-branches. Leaves deciduous; leaflets 7–11, grey-green, elliptic to ovate, 2.5–3.5 cm, hairless above, beneath with sparse glands on midrib, margins with simple teeth. Bracts present. Receptacle glandular-bristly. Flowers 1 or 2, single, 7–10 cm (rarely 3–10 cm) across. Sepals with lateral lobes, glandular-bristly on the back, erect and persistent in fruit. Petals pale pink fading to white, notched. Styles free; stigmas woolly. Fruit ovoid, dark red or brownish, 2.5–3 cm, slightly glandular-bristly. *S Turkey, N Iraq.* H4. Early summer.

37. R. dumalis Bechstein (*R. glauca* Loiseleur, not Pourret).

Shrub with more or less erect, often bloomed stems 1–2 m, bearing hooked, broad-based prickles. Stipules broad. Leaves deciduous; leaflets 5–7, broadly ovate to roundish, 1.2–3.5 cm, acute or obtuse, hairless and often bloomed above, hairless or downy beneath, margins with simple or compound teeth. Bracts broad. Receptacle smooth to glandular-bristly. Flowers 1–many, single. Sepals with lateral lobes, hairless or slightly glandular on the back, margin downy, erect and persistent in fruit. Petals pink. Styles free, scarcely protruding; stigmas woolly. Fruit ovoid to spherical, red, 1.5–2.2 cm, smooth or glandular-bristly. *Europe, Turkey.* H3. Summer.

38. R. jundzillii Besser (*R. marginata* misapplied, not Wallroth).

Erect or trailing, suckering shrub with stems 1–2.4 m, bearing a few slender, scattered, straight or decurved prickles, sometimes intermixed with bristles. Stipules large and broad. Leaves deciduous; leaflets 5–7, obovate to broadly elliptic, 2.5–4.5 cm, acute or acuminate, hairless above, more or less glandular and sometimes downy beneath, margins with compound glandular teeth. Bracts present. Receptacle often glandular-bristly. Flowers usually solitary, sometimes 2–8, single, slightly fragrant, 5–7.5 cm across. Sepals with lateral lobes, glandular on the back, spreading to reflexed and falling after flowering. Petals pale to rosy pink, fading with age. Styles united; stigmas woolly. Fruit ovoid to spherical, red, 1.5–2.5 cm, smooth or glandular-bristly. *W Europe to S former USSR (including Caucasus), Turkey.* H3. Summer.

39. R. glauca Pourret (*R. rubrifolia* Villars). Illustration: Gault & Synge, The dictionary of roses in colour, pl. 47, 48 (1971); Thomas, A garden of roses, 91 (1987); Phillips & Rix, Roses, 18 (1988); *Kew Magazine* **6**: pl. 115 (1989).

Erect, sparsely branched shrub with arching stems 1.5–3 m, dark red and bloomy when young, bearing sparse, straight or decurved, broad-based prickles, the strong shoots also with bristles. Stipules entire or with glandular teeth. Leaves deciduous; leaflets 5–9, grey-green, often tinged with brown or purple and covered with a dull waxy bloom, ovate to narrowly elliptic, 2–4.5 cm, acute, hairless on both sides, margins with simple teeth. Bracts present. Flowers solitary or 2–12, single, 2.5–4 cm across. Receptacle often with sparse glandular bristles. Sepals entire or with a few lateral lobes, smooth or glandular-bristly on the back, spreading and falling after flowering. Petals deep pink, white at the base. Styles free, protruding. Fruit ovoid to almost spherical, brownish red, 1.3–1.5 cm, smooth or with sparse glandular bristles. *Mountains of C & S Europe.* H3. Summer.

40. R. palustris Marshall (*R. corymbosa* Ehrhart). Illustration: Krüssmann, The complete book of roses, 268 (1981); Phillips & Rix, Roses, 21 (1988).

Erect, rather broadly spreading, suckering shrub with slender reddish or purplish brown stems to 2 m, bearing stout, more or less curved, broad-based prickles in pairs at the nodes, Stipules narrow, the halves of each pair rolled inwards to form a tube, Leaves deciduous; leaflets 5–7 (rarely to 9), dull green, oblong to elliptic, 2–6 cm, acute, hairless above, downy beneath at least on veins, margins with simple teeth. Bracts broad. Receptacle spherical, glandular-bristly. Flowers usually in clusters, rarely solitary, single, 4–5.5 cm across. Sepals entire, expanded at tips, glandular-bristly on the back, spreading and falling after flowering. Petals pink. Styles free, not protruding. Fruit spherical, red, *c.* 8 mm, smooth or glandular-bristly. *E North America.* H3. Summer.

41. R. virginiana Herrmann (*R. lucida* Ehrhart). Illustration: Niering & Olmstead, Audubon Society field guide to North American wildflowers – eastern region, pl. 581 (1979); Gibson, The book of the rose, pl. 15 (1980); Thomas, A garden of roses, 149 (1987); Phillips & Rix, Roses, 27 (1988).

Erect, often suckering shrub, with brownish red stems to 1.5 m, unarmed or bearing straight or recurved prickles in pairs at the nodes, and also bristles on the young stems. Stipules widening towards tip, glandular-toothed. Leaves deciduous; leaflets 5–9, obovate to oblong-elliptic, 2–6 cm, acute, glossy green and hairless above, beneath hairless, or downy on midrib, margins with coarse simple teeth except at base. Bracts present. Receptacle smooth or glandular-bristly. Flowers 1–8, single, 5–6.5 cm across, fragrant. Sepals entire or with a few lateral lobes, tips leafy, glandular and hairy on the back, spreading or reflexed and falling after flowering. Petals pale pink to bright pink. Styles free, not protruding; stigmas woolly. Fruit almost spherical, red, 1–1.5 cm wide, smooth or glandular-bristly. *E North America.* H3. Summer.

R. × mariae-graebneriae Ascherson & Graebner is a hybrid between *R. virginiana* and probably *R. palustris*. An erect shrub with a spherical shape and stems to 1.6 m, which bear slightly curved prickles and sometimes scattered bristles. Leaves

deciduous; leaflets glossy green, margins with simple teeth. Bracts present. Petals rose-pink. Fruit almost spherical, red. *Garden origin.* H3. Summer–early autumn.

R. × suionum Almquist is a hybrid (probably complex) whose origin is uncertain, but recent opinion suggests that *R. virginiana* may be involved. Leaves distinct, with 5 or 7 leaflets, and with a much smaller pair of leaflets below the terminal one. Flowers double, clear pink. *Garden origin.* H2. Summer.

42. R. carolina Linnaeus (*R. humilis* Marshall). Illustration: Peterson & McKenny, Field guide to wildflowers of north-eastern and north-central North America, 219, 257 (1968); Thomas, A garden of roses, 49 (1987); Phillips & Rix, Roses, 77 (1988).

Suckering shrub with slender stems 1–1.5 m, bearing scattered, straight, slender, unequal prickles in pairs at the nodes, and dense bristles especially on young stems. Stipules narrow, the halves of each pair often rolled inward to form a tube, entire to glandular-toothed. Leaves deciduous; leaflets 5–9, lanceolate to narrowly ovate or almost round, acute to obtuse, hairless on both sides or downy beneath, margins with simple teeth. Bracts present. Receptacle glandular-bristly. Flowers solitary or few in a cluster, single, 3.8–5 cm across. Sepals with lateral lobes, glandular-bristly on the back, spreading and falling after flowering. Petals pale to mid-pink. Styles free, not protruding. Fruit almost spherical, red, 7–9 mm, glandular-bristly. *E & C North America.* H3. Summer.

'Alba' (*R. virginiana* Herrmann var. *alba* Rafinesque). Illustration: Thomas, A garden of roses, 151 (1987). Leaflets always hairy beneath; flowers white.

'Plena'. Illustration: Thomas, Shrub roses of today, pl. 2 (1962). Dwarf shrub to 50 cm; flowers double, outer petals ageing almost to white.

43. R. nitida Willdenow. Illustration: Krüssmann, The complete book of roses, 268 (1981); Thomas, A garden of roses, 121 (1987); Phillips & Rix, Roses, 21, 214 (1988).

Erect, suckering shrub with often reddish stems 50–100 cm, bearing sparse, straight prickles and dense purplish brown or reddish bristles. Stipules broad, with glandular teeth. Leaves deciduous; leaflets 5–9, elliptic, 1–3 cm, acute, shiny green and hairless above, hairless or sparsely hairy beneath, margins with fine, simple teeth. Bracts present. Receptacle glandular-bristly. Flowers 1 (rarely to 3), single, fragrant, 4.5–6.5 cm across. Sepals entire, bristly or glandular on the back, spreading and falling after flowering. Petals deep pink. Styles free, not protruding. Fruit spherical or more or less so, dark scarlet, 8–10 mm, glandular-bristly. *E North America.* H3. Summer.

44. R. foliolosa Torrey & Gray. Illustration: Beales, Classic roses, 20, 212 (1985); Phillips & Rix, Roses, 27 (1988).

Shrub, suckering, with rather weak, reddish stems 50–100 cm, unarmed or bearing sparse slender, straight prickles, and only rarely with bristles. Stipules narrow. Leaves deciduous; leaflets 7–9 (rarely to 11), narrowly oblong, 1–5 cm, acute, glossy and hairless above, downy on midrib beneath, margins with fine simple teeth. Bracts present. Receptacle glandular-bristly. Flowers 1–5, single, fragrant, 5–6.5 cm across. Sepals entire, glandular-bristly on the back, spreading and falling after flowering. Petals white to rose-pink. Styles free, not protruding. Fruit almost spherical, red, *c.* 8 mm wide, glandular-bristly. *SE USA.* H4. Summer–early autumn.

45. R. rugosa Thunberg. Illustration: Niering & Olmstead, Audubon Society field guide to North American wildflowers – eastern region, pl. 582 (1979); Phillips & Rix, Roses, 100 (1988).

Erect shrub with stout stems 1.2–2.5 m, downy when young, bearing very dense prickles of unequal size and dense bristles. Stipules large. Leaves deciduous; leaflets 5–9, dark green, oblong to elliptic, 2.5–5 cm, usually acute, hairless and conspicuously wrinkled above, downy beneath with conspicuous veins, margins with shallow teeth except at base. Bracts large, enfolding the flower-stalks. Receptacle smooth. Flowers solitary or few, single, fragrant, 6–7.5 (rarely to 9) cm across. Sepals entire, expanded at tip, downy, erect and persistent after flowering. Petals purplish pink. Styles free, not protruding. Fruit almost spherical, with a neck at the top, red to orange-red, 2–2.5 cm, smooth. *Eastern former USSR, Korea, Japan, N China (naturalised in Britain & NE USA).* H2. Early summer–autumn.

Var. **rosea** Rehder has single, rose-pink flowers. Var. **albo-plena** Rehder has double, white flowers. 'Alba'. Illustration: Beales, Classic roses, 239 (1985); Austin, The heritage of the rose, 238 (1988);

Phillips & Rix, Roses, 101 (1988). Flowers solitary, white, opening from pale pink buds; very free fruiting.

R. × kamtchatica Ventenat is a hybrid involving *R. rugosa* and either *R. davurica* Pallas or *R. amblyotis* C.A. Meyer. Some authorities consider it to be a variety of *R. rugosa*, in which case its correct name is var. *ventenatiana* C.A. Meyer (*R. rugosa* var. *kamtchatica* (Ventenat) Regel). It differs from *R. rugosa* in the more slender stems, which bear fewer prickles, fewer wrinkled leaflets, and smaller flowers and fruit. *Former USSR (E Siberia, Kamtchatka).* H2. Summer.

R. rugosa has been crossed with *R. arvensis* to produce **R. × paulii** Rehder ('Paulii'). Illustration: Gault & Synge, The dictionary of roses in colour, pl. 34 (1971); Gibson, The book of the rose, pl. 36 (1980); Austin, The heritage of the rose, 356 (1988). Stems vigorous, trailing to 4 m, flowers in clusters, single, white, fragrant, to 6 cm across, with entire glandular sepals. 'Rosea'. Illustration: Gault & Synge, The dictionary of roses in colour, pl. 35, 1971; Austin, The heritage of the rose, 349 (1988); Phillips & Rix, Roses, 101 (1988). Flowers pink with paler centres.

R. × micrugosa Henkel (*R. vilmorinii* Bean). Illustration: Gault & Synge, The dictionary of roses in colour, pl. 28 (1971). A hybrid between *R. rugosa* and *R. roxburghii*, which is intermediate between the parents. Flowers pale pink, 10–12.5 (rarely 8–12.5) cm across. Fruit orange-red, almost spherical, with prickles. 'Alba'(illustration: Beales, Classic roses, 229, 1985) has white fragrant flowers produced over a period, and more erect growth.

46. R. acicularis Lindley. Illustration: Krüssmann, The complete book of roses, 272 (1981).

Loose shrub with stems 1 (rarely to 2.5) m, bearing straight or slightly curved, weak, slender, unequal prickles intermixed with dense, slender bristles. Stipules narrow. Leaves deciduous; leaflets 5–7 (rarely 3–9), ovate to elliptic, 1.5–6 cm, acute, blue-green and usually hairless above, greyish and downy beneath, margins with simple teeth. Bracts narrow, more or less equalling flower-stalks. Receptacle smooth. Flowers 1 (rarely to 3), single, fragrant, 3.8–6.2 cm across. Sepals entire, glandular on the back, erect and persistent after flowering. Petals rose-pink

to purplish pink. Styles free, not protruding; stigmas woolly. Fruit ellipsoidal or spherical to pear-shaped, with a neck at the top, red, 1.5–2.5 cm, smooth. *Circumpolar: N America, N Europe, Russia (Siberia), Kazakhstan, Mongolia, N China, Korea & Japan.* H1. Early summer.

Var. **nipponensis** (Crépin) Koehne. Illustration: Phillips & Rix, Roses, 24 (1988). Leaflets 7–9, oblong, obtuse; flowers bright pink. *C & S Japan.*

47. R. arkansana Porter. Illustration: Krüssmann, The complete book of roses, 273 (1981); Phillips & Rix, Roses, 20 (1988).

Erect, suckering shrub with stems 50–120 cm, bearing usually dense, straight, unequal prickles, and bristles. Stipules narrow. Leaves deciduous; leaflets 9–11 (rarely 3–7), obovate to elliptic, 2–6 cm, acute or obtuse, shiny and hairless above, sometimes downy on the veins beneath, margins with simple teeth. Bracts present. Receptacle smooth or slightly glandular. Flowers few to many in lateral clusters, single, 2.5–4 cm across. Sepals entire or with lateral lobes, smooth or glandular-bristly on the back, spreading to erect, and persistent after flowering. Petals deep pink. Styles free, not protruding. Fruit pear-shaped to almost spherical, red, 1–1.5 cm, smooth or slightly glandular. *C USA.* H3. Summer.

48. R. suffulta Greene (*R. arkansana* var. *suffulta* (Greene) Cockerell). Illustration: Beales, Classic roses, 249 (1985); Phillips & Rix, Roses, 2 (1988).

Low shrub with stems to 50 cm, bearing dense, straight prickles and bristles. Stipules broad. Leaves deciduous; leaflets 7–11, ovate-oblong to broadly elliptic, 1.5–4 cm, obtuse, bright green and hairless or downy above, downy beneath, margins with simple teeth. Bracts present. Receptacle smooth. Flowers in clusters, single, *c.* 3 cm across. Sepals entire or sometimes with lateral lobes, spreading to erect, and persistent after flowering. Petals mid-pink. Styles free, not protruding. Fruit spherical, red, *c.* 1 cm, smooth. *E & C North America.* H3. Summer.

Very similar to *R. arkansana* and considered to be conspecific with it by some authorities.

49. R. blanda Aiton. Illustration: Thomas, A garden of roses, 39 (1987).

Erect shrub 1–2 m, stems brown, unarmed or sometimes bearing a few, scattered, slender, straight prickles near the base. Stipules broadening towards top. Leaves deciduous; leaflets 5–7 (rarely to 9), elliptic to oblong-obovate, 2–6 cm, acute to obtuse, dull green and hairless above, usually downy beneath, margins with simple teeth. Bracts large, enfolding the flower-stalks. Receptacle smooth. Flowers solitary or 3–7, single, 4.5–6.5 cm across, fragrant. Sepals entire, downy and glandular on the back, erect and persistent in fruit. Petals rosy pink. Styles free, not protruding. Fruit ovoid to pear-shaped, red, *c*. 1 cm, smooth. *E & C North America*. H3. Early summer.

50. R. pendulina Linnaeus (*R. alpina* Linnaeus; *R. cinnamomea* Linnaeus (1753) not (1759); *R. pyrenaica* Gouan). Illustration: Gibson, The book of the rose, pl. 34 (1980); Rosetti, I fiori delle Alpi, pl. 267 (1980); Phillips & Rix, Roses, 24, 25 (1988).

Suckering shrub of variable habit, with green or red-brown stems 60–200 cm, usually unarmed but sometimes bearing slender prickles and bristles. Stipules broadening towards top. Leaves deciduous; leaflets 5–9 (rarely to 11), elliptic to broadly so, 2–6 cm, acute or obtuse, hairless or downy above, usually downy beneath and sometimes glandular, margins with compound glandular teeth. Bracts enfolding and equalling the flower-stalks, soon falling. Receptacle smooth or glandular-bristly. Flowers 1 (rarely to 5), single, 3.8–6.5 cm across. Sepals entire with an expanded tip, usually smooth, sometimes glandular on the back, erect and persistent in fruit. Petals deep pink or purplish pink. Styles free, not protruding; stigmas hairy. Fruit often pendent, almost spherical to ovoid, usually with a neck at the top, red, 1.5–3 cm, smooth or glandular-bristly. *Mts of S & C Europe*. H3. Early summer.

'Nana' is a dwarf, freely suckering selection to 30 cm.

R. × l'heriteriana Thory. Illustration: Thomas, A garden of roses, 107, (1987). Thought to be a hybrid between *R. pendulina* and *R. chinensis*. Shrub with reddish stems climbing to 4 m, unarmed or with a few prickles. Leaves deciduous; leaflets 3–7, ovate-oblong, with simple teeth, hairless on both sides. Bracts present. Receptacle smooth. Flowers many, in clusters, semi-double. Sepals entire. Petals purplish red, whitish at the base. Styles free, not protruding. Fruit almost

spherical, smooth. *Garden origin*. H4. Summer.

51. R. oxyodon Boissier (*R. alpina* var. *oxyodon* (Boissier) Boulenger; *R. pendulina* var. *oxyodon* (Boissier) Rehder).

Very similar to *R. pendulina* but stems 2–4 m; receptacle smooth, sepals usually glandular on the back; fruit smooth. *Former USSR (E Caucasus)*. H4. Summer.

52. R. majalis Herrmann (*R. cinnamomea* Linnaeus (1759) not (1753); *R. cinnamomea* var. *plena* Weston; *R. foecundissima* Muenchhausen; *R. majalis* 'Foecundissima'). Illustration: Thomas, A garden of roses, 115 (1987); Phillips & Rix, Roses, 26 (1988); Blamey & Grey-Wilson, Illustrated flora of Britain and northern Europe, 179 (1989).

Erect, suckering shrub with slender, red-brown stems 1.5–3 m, much-branched towards the top, bearing slender, hooked prickles in pairs at the nodes as well as scattered prickles on lower parts of stem, intermixed with bristles. Stipules broad, those on strong shoots often rolled inwards. Leaves deciduous; leaflets 5–7, elliptic to obovate, 1.5–5 cm, obtuse to acuminate, sometimes downy above, usually downy beneath, margins with simple teeth except at the base. Bracts large, equal to or longer than flower-stalks. Receptacle smooth. Flowers few, single or (in gardens) more usually double, *c*. 5 cm across. Sepals entire, with woolly margin and often hairy on the back, erect and persistent in fruit. Petals mid-pink to purplish pink. Styles free, not protruding; stigmas woolly. Fruit spherical or slightly lengthening, dark red, 1–1.5 cm, smooth. *N & C Europe, former USSR (Siberia)*. H2. Summer.

Herrmann based the species on the double-flowered form, which has been cultivated since the 16th century.

53. R. davurica Pallas.

Shrub with stems 1–1.5 m, bearing large, more or less straight, prickles in pairs at the nodes. Stipules narrow. Leaves deciduous; leaflets *c*. 7, oblong-lanceolate, 2.5–3.5 cm, acute, hairy and glandular beneath, margins with compound teeth. Bracts present. Receptacle smooth. Flowers 1–3, single. Sepals entire, margins hairy, erect and persistent in fruit. Petals pink. Styles free, not protruding. Fruit ovoid, red, 1–1.3 cm, smooth. *Russia (E Siberia), S Mongolia, NE China, Korea, Japan*. H2. Summer.

Very similar to *R. majalis*.

54. R. laxa Retzius. Illustration: Phillips & Rix, Roses, 21 (1988).

Shrub with slender, arching stems to 2.5 m, bearing few, large, recurved, hooked or straight prickles, intermixed with bristles. Stipules broad, not glandular. Leaves deciduous; leaflets 5–9, ovate to elliptic or oblong, 1.5–4.5 cm, obtuse, hairless above, usually hairy beneath, margins with simple or compound teeth. Bracts present. Receptacle smooth. Flowers 1–6, single, 4–5 cm across. Sepals entire, usually glandular on the back, erect and persistent after flowering. Petals white to pale pink, notched. Styles free, not protruding. Fruit ellipsoidal to spherical, red, *c*. 1.5 cm, smooth. *Former USSR (Siberia south to Tienshan & Pamir Alai), NW China*. H3. Summer.

55. R. pisocarpa A. Gray. Illustration: Krüssmann, The complete book of roses, 276 (1981); Phillips & Rix, Roses, 26 (1988).

Erect shrub with slender, arching stems 90–250 cm, unarmed or bearing few, straight, weak prickles in pairs at the nodes, and some bristles towards base. Stipules broad. Leaves deciduous; leaflets 5–7 (rarely to 9), elliptic to ovate, 1.3–4 cm, obtuse or somewhat acute, hairless above, downy beneath, margins with simple teeth. Bracts broad. Receptacle smooth. Flowers 4 or 5 (rarely 1–5), single, 2.5–3 cm across. Sepals entire, glandular-bristly on the back, erect and persistent in fruit. Petals rosy pink to purplish pink. Styles free, not protruding. Fruit ellipsoidal to spherical, purplish red to red or orange, to 1.3 cm, smooth. *W North America from British Columbia south to N California*. H3. Summer.

56. R. woodsii Lindley (*R. macounii* Greene).

Stiffly branched shrub with stems to 2 m, purplish or reddish brown, becoming grey, bearing many, straight or slightly curved prickles, sometimes intermixed with bristles. Stipules narrow. Leaves deciduous; leaflets 5–7 (rarely to 9), obovate to elliptic, 1–3 cm, acute to obtuse with simple teeth, hairless above, hairless or downy beneath. Bracts present. Receptacle smooth. Flowers 1–3 (rarely to 5), single, 3–4.5 cm across. Sepals entire, hairless or downy on the back, erect to spreading and persistent in fruit. Petals mid-pink, occasionally white. Styles free, not protruding. Fruit ovoid to spherical, usually with a neck at the top, 0.5–1.5 cm, smooth. *W & C North America*. H3. Summer.

Var. **fendleri** (Crépin) Rydberg (*R. fendleri* Crépin). Illustration; Phillips & Rix, Roses, 26 (1988). Differs in having glandular-bristly stipules and flower-stalks, leaflets that usually have compound teeth, smaller lilac-pink flowers and smaller fruit. *W North America to N Mexico.*

57. R. californica Chamisso & Schlechtendahl.

Erect shrub with stems 1.5–3 m, greenish when young, red-brown when older, bearing stout, broad-based, recurved prickles in pairs at the nodes, and often with bristles on the young shoots. Stipules narrow. Leaves deciduous; leaflets 5–7, ovate to broadly elliptic, 1–3.5 cm, obtuse, hairless or downy above, downy and often glandular beneath, margins usually with simple teeth. Bracts broad. Receptacle smooth or hairy when young. Flowers in clusters, single, fragrant, 3.7–4 cm across. Sepals entire, hairy on the back, erect and persistent in fruit. Petals deep pink to bright crimson. Styles free, not protruding. Fruit spherical or slightly lengthened, with a neck at the top, 7–15 mm, smooth. *W USA south to Mexico (Baja California).* H4. Summer.

'Plena' (forma *plena* Rehder). Illustration: Thomas, A garden of roses, 45 (1987); Austin, The heritage of the rose, 355 (1988); Phillips & Rix, Roses, 27 (1988). Stems to 2 m, flowers semi-double. The rose that is usually cultivated under this name is in fact *R. nutkana.*

58. R. nutkana C. Presl. Illustration: Beales, Classic roses, 237 (1985).

Robust shrub with stout, purplish brown stems 1.5–3 m, bearing stout, usually straight, broad-based prickles in pairs at the nodes and slender bristles on the young stems. Leaves deciduous; leaflets 5–7 (rarely to 9), dark green, ovate to elliptic, 1.9–5 cm, acute or obtuse, hairless above, glandular and sometimes downy beneath, margins with compound glandular teeth. Bracts present. Receptacle smooth. Flowers 1 (rarely to 3), single, 5–6.5 (rarely 4–6.5) cm across, with glandular-bristly stalks. Sepals entire, glandular-bristly or not, and more or less downy on the back, erect and persistent in fruit. Petals bright red to purple-pink, occasionally white. Styles free, not protruding. Fruit spherical to more or less so, red, 1.5–2 cm, smooth. *W North*

America from Alaska to N California. H3. Summer.

Frequently cultivated under the name *R. californica* 'Plena'.

Var. **hispida** Fernald (*R. spaldingii* Crépin). Illustration: Phillips & Rix, Roses, 20 (1988). Differs in having leaflets downy beneath and simple non-glandular teeth, hairless flower-stalks and glandular-bristly receptacles and fruit. W North America from *British Columbia to Utah (more easterly than R. nutkana itself).*

59. R. corymbulosa Rolfe. Illustration: *Botanical Magazine*, 8566 (1914).

Shrub with erect or sometimes prostrate or climbing stems to 2 m, unarmed or bearing a few straight, slender prickles. Leaves deciduous, turning purple in autumn; leaflets 3–5, elliptic to ovate-oblong, 1.3–5 cm, acute, dark green and sparsely downy above, glaucous and downy beneath, margins with compound teeth. Bracts present. Receptacle obovoid, glandular-bristly. Flowers in clusters of up to 12, single, 1.9–2.5 cm across. Sepals entire, downy and glandular-bristly on the back, erect and persistent in fruit. Petals deep pink, paler at the base. Styles free, slightly protruding; stigmas downy. Fruit ovoid-spherical to spherical, coral-red, 1–1.3 cm, glandular-bristly. *C China.* H4. Summer.

60. R. davidii Crépin. Illustration: Phillips & Rix, Roses, 24, 217 (1988).

Shrub with erect or arching stems, 1.8–4 m, bearing scattered, stout, straight or slightly curved, reddish, broad-based prickles. Leaves deciduous; leaflets 7–9 (rarely 5–11), dark green, ovate to elliptic, 6–50 mm, acute, wrinkled and hairless above, downy beneath especially on the veins, margins with simple, sometimes glandular teeth. Bracts present. Receptacle glandular-bristly, more or less downy. Flowers 3–12 or more, single, fragrant, 3.8–5 cm across. Sepals entire, often downy and glandular, erect and persistent in fruit. Petals rosy pink. Styles free, protruding; stigmas woolly. Fruit pendent, ovoid with a slender neck at the top, scarlet, 1.5–2 cm. *W & C China.* H4. Summer.

Var. **elongata** Rehder & Wilson, which has leaflets 5–7.5 cm that may be hairless beneath, fewer (3–7) flowers, and more elongated fruit to 2.5 cm, is now considered by many to represent part of the variation of *R. davidii.*

61. R. caudata Baker.

Erect shrub with reddish stems 1–4 m, bearing few, stout, scattered, broad-based prickles. Stipules broad. Leaves deciduous; leaflets 7–9, elliptic to ovate, 2.5–5 cm, acute, hairless on both sides except for midrib beneath, margins with simple teeth. Bracts present. Receptacle glandular-bristly. Flowers few in a tight cluster, single, 3.5–5 cm across. Sepals entire, glandular-bristly on the back, expanded at the tip, erect and persistent in fruit. Petals deep pink. Styles free, not protruding. Fruit ovoid-oblong with a neck at the top, orange-red, 2–2.5 cm, glandular-bristly. *C China.* H4. Summer.

Very similar to and possibly not distinct from *R. setipoda.*

62. R. banksiopsis Baker.

Similar to *R. caudata*: differs in the receptacle and sepals being smooth and the smaller flowers (2–3 cm across). *C China.* H4. Summer.

63. R. macrophylla Lindley. Illustration: Thomas, A garden of roses, 111 (1987); Phillips & Rix, Roses, 24, 25, 216 (1988).

Shrub with erect and arching, dark red to purple stems 2.5–4 (rarely to 5) m, unarmed or bearing a few stout, upward-pointing, straight prickles, which are often paired at the nodes. Stipules usually broad. Leaves deciduous; leaflets 9–11 (rarely 7–11), elliptic to narrowly ovate, 2.5–6.5 cm, acute or acuminate, hairless above, downy and sometimes glandular beneath, margins with simple or compound teeth. Bracts broad. Receptacle glandular-bristly. Flowers 1–5, single, 5–7.5 cm across. Sepals entire, bristly and more or less glandular on the back, expanded at the tip, erect and persistent in fruit. Petals mid-pink or deep pink to mauvish pink. Styles free, not protruding; stigmas woolly. Fruit almost spherical to bottle-shaped with a neck at the top, red, 2.5–4 (rarely to 7.5) cm, glandular-bristly. *Himalaya from NW India east to W China.* H4. Summer.

64. R. setipoda Hemsley & Wilson. Illustration: Gault & Synge, The dictionary of roses in colour, pl. 51 (1971); Phillips & Rix, Roses, 22 (1988).

Shrub with stout, reddish, erect or arching stems 2.4–3 m, bearing sparse, short, straight, broad-based prickles intermixed with bristles. Stipules large. Leaves deciduous, slightly aromatic; leaflets 7–9, elliptic to elliptic-ovate, 3–6 cm, acute to obtuse, mid-green and hairless above, greyish, glandular and downy on the veins beneath, margins with simple or compound teeth. Bracts large. Receptacle narrowly ellipsoidal, glandular-bristly, purplish. Flowers in loose clusters of up to 20, single, smelling of unripe apples, 3–5 cm across. Sepals usually entire, expanded at the tip, margin glandular, erect and persistent in fruit. Petals mid-pink to deep purplish pink, shading to white at the base, slightly downy on the back. Styles free, not protruding. Fruit bottle-shaped with a neck at the top, dark red, 2.5–5 cm, glandular-bristly. *C China.* H4. Summer.

65. R. hemsleyana Täckholm.

Very similar to, and possibly better regarded as a form of *R. setipoda*; differs in its smaller size (to 2 m) and sepals with lateral lobes. *C China.* H4. Summer.

66. R. sweginzowii Koehne (*R. moyesii* misapplied, not Hemsley & Wilson). Illustration: Gault & Synge, The dictionary of roses in colour, pl. 58, 59 (1971); Phillips & Rix, Roses, 24 (1988).

Shrub with reddish, spreading stems to 3.5 (rarely to 5) m, sometimes unarmed or bearing dense, 3-angled prickles, intermixed with bristles. Stipules broad. Leaves deciduous; leaflets 7–11, elliptic to broadly so, 2.5–5 cm, acute, hairless above, downy on veins beneath, margins with compound teeth. Bracts broad. Receptacle glandular-bristly. Flowers 1–3 (rarely to 6), single, 3.5–5 cm across. Sepals entire, glandular-bristly on the back, erect and persistent in fruit. Petals bright pink. Styles free, not protruding; stigmas woolly. Fruit bottle-shaped, glossy or orange-red, *c.* 2.5 cm, glandular-bristly. *C China.* H4. Summer.

67. R. wardii Mulligan. Illustration: Phillips & Rix, Roses, 24 (1988).

Similar to *R. sweginzowii*, differing in generally lacking prickles and bristles, in the smaller leaflets (1.3–1.9 cm), and in the white flowers. *China (SE Xizang).* H4. Summer.

68. R. murielae Rehder & Wilson.

Shrub with reddish, erect or arching stems 1.5–3 m, unarmed or bearing a few slender, straight prickles intermixed with pinkish bristles, dense on young shoots. Leaves deciduous; leaflets 9–15, elliptic to oblong, 1–4 cm, more or less acute, hairless on both sides, but downy on midrib beneath, margins with simple

glandular teeth. Bracts present. Receptacle smooth. Flowers 3–7, single, 2–2.5 cm across. Sepals entire, expanded at the tip, hairy or not on the back, erect and persistent in fruit. Petals white. Styles free, not protruding. Fruit bottle-shaped, orange-red, 1–2 cm, smooth. *SW China.* H4. Summer.

69. R. moyesii Hemsley & Wilson. Illustration: Thomas, A garden of roses, 117 (1987); Austin, The heritage of the rose, 368 (1988); Le Rougetel, A heritage of roses, 39 (1988); Phillips & Rix, Roses, 22, 217 (1988).

Erect shrub with stout, red-brown stems 2–3.5 m, bearing pale, scattered, stout, broad-based prickles especially on non-flowering-shoots, the lower parts of the stems also with bristles. Leaves deciduous; leaflets 7–13, broadly elliptic to ovate, 1–4 cm, acute, dark green above, somewhat glaucous beneath, both sides hairless except for midrib beneath, which is downy and sometimes prickly, margins usually with simple teeth. Bracts with glandular margins. Receptacle glandular-bristly. Flowers 1–2 (rarely to 4), single, 4–6.5 cm across. Sepals entire, expanded at the tip, downy inside, hairy and sparsely glandular on the back, erect and persistent in fruit. Petals pink to blood-red. Styles free, not protruding. Fruit bottle-shaped, with a neck at the top, orange-red, 3.8–6 cm, glandular-bristly at least towards the base. *W China.* H4. Summer.

'Fargesii' (*R. moyesii* var. *fargesii* Rolfe; *R. fargesii* misapplied, not Boulenger). Illustration: Beales, Classic roses, 232 (1985) – as *R. fargesii*; Phillips & Rix, Roses, 22 (1988). Flowers pink to rose-red; leaflets shorter (1–2 cm) and wider, obtuse. 'Geranium' Illustration: Phillips & Rix, Roses, 22, (1988); Brickell, The RHS gardeners' encyclopedia of plants and flowers, 151 (1989). Compact, growing only to 2.5 m; leaflets paler green; flowers scarlet; fruit broader, crimson, with a shorter neck. It is the cultivar most often grown. 'Sealing Wax'. Illustration: Phillips & Rix, Roses, 216 (1988). Fruits large, bright red; flowers pink.

Forma **rosea** Rehder & Wilson, with pink flowers and coarsely toothed leaflets, may not in fact be related to *R. moyesii*, but appears to be closer to *R. davidii*, which comes from the same area. However, some botanists give it as a synonym of *R. holodonta*. There is confusion about what

rose this name should be attached to, and investigation is needed.

R. holodonta Stapf is closely related to *R. moyesii* but has leaflets to 5 cm. It may not be worthy of recognition at species level. *W China.*

A hybrid between *R. moyesii* and *R. setipoda* is called **R. × wintoniensis** Hillier. Differs from *R. moyesii* in having clusters of 7–10 flowers with crimson petals that are white at the base, and aromatic leaves.

R. × highdownensis Hillier. Illustration: Gault & Synge, The dictionary of roses in colour, pl. 23 (1971); Beales, Classic roses, 232 (1985); Austin, The heritage of the rose, 371 (1988). A hybrid of *R. moyesii* with clusters of mid-red to deep pink flowers.

70. R. bella Rehder & Wilson. Illustration: Phillips & Rix, Roses, 2 (1988).

Shrub with spreading, purplish branches to 3 m, bearing a few slender, straight prickles, intermixed with bristles on the lower part of the stems. Stipules broad. Leaves deciduous; leaflets 7–9, shortly stalked, elliptic to broadly ovate, 1–2.5 cm, more or less acute, hairless on both sides, sometimes glandular on the veins beneath, margins with simple teeth. Bracts glandular. Receptacle glandular-bristly. Flowers 3, single, 4–5 cm across. Sepals entire, expanded at the tip, glandular on the back, erect and persistent in fruit. Petals bright pink. Styles free, not protruding. Fruit ovoid or ellipsoid, orange-red, 1.5–2.5 cm, glandular-bristly. *NE China.* H4. Summer.

71. R. webbiana Royle. Illustration: Edwards, Wild and old garden roses, 33 (1975); Thomas, A garden of roses, 153 (1987); Phillips & Rix, Roses, 22 (1988).

Shrub with slender stems 1–2 m, purplish brown, often bloomed when young, bearing few straight, slender, yellowish, broad-based prickles. Leaves deciduous; leaflets 5–9, obovate to broadly elliptic or roundish, 6–25 mm, obtuse, somewhat glaucous and hairless above, often slightly downy beneath, margins with simple teeth except in the lower third. Bracts present. Receptacle smooth or glandular-bristly. Flowers 1 (rarely to 3), single, faintly fragrant, 3.8–5 cm across. Sepals entire, glandular and often hairy on the back, erect and persistent in fruit. Petals white to pale pink or lilac pink, sometimes pink with a whitish base. Styles free, not protruding. Fruit spherical to broadly bottle-shaped, with a neck at the top, shiny red, 1.5–2.5 cm, smooth or glandular-bristly. *W & C Himalaya, China (Xizang), Afghanistan, former USSR (Central Asia)*. H4. Early Summer.

Var. **microphylla** Crépin (*R. nanothamnus* Boulenger). Illustration: Phillips & Rix, Roses, 23 (1988) – as *R. nanothamnus*. Stems only to 50 cm, leaflets smaller (to 1.6 cm), and flowers smaller (2.5–3.8 cm across). *Former USSR (Central Asia), Afghanistan, Kashmir.* More common in cultivation than the species.

72. R. sertata Rolfe.

Similar to *R. webbiana* but leaflets obtuse to acute, flowers deep pink to purplish pink on longer stalks (1.5–3 cm rather than 1–1.3 cm), and fruit dark red. *SW, C & EC China.* H4. Summer.

73. R. giraldii Crépin. Illustration: Phillips & Rix, Roses, 18 (1988).

Differs from *R. webbiana* in having acute or obtuse leaflets with compound teeth, very short flower-stalks often concealed by large bracts, smaller pink flowers 1.2–2.5 cm across with a white centre, and a smaller fruit, rarely exceeding 1 cm. *C China.* H4. Summer.

74. R. fedtschenkoana Regel. Illustration: Gault & Synge, The dictionary of roses in colour, pl. 15 (1971); Beales, Classic roses, 222 (1985); Thomas, A garden of roses, 71 (1987); Phillips & Rix, Roses, 22, 214 (1988).

Suckering shrub with vigorous erect stems 1–2.5 (rarely to 3) m, bearing slender, straight or curved prickles in pairs at the nodes, pinkish when young, sometimes reduced to bristles. Stipules acuminate. Leaves deciduous; leaflets 5–9, glaucous green, broadly elliptic to obovate, acute, hairless above, downy beneath, margins with simple teeth. Bracts present. Receptacle glandular-bristly. Flowers 1–4, single, smelling slightly unpleasant, *c.* 5 cm across. Sepals entire, glandular-bristly on the back, margin and inner surface downy, erect and persistent in fruit. Petals usually white, rarely pink. Styles free, not protruding. Fruit pear-shaped, red to orange-red, 1.5–2.5 cm, glandular-bristly. *Kazakhstan, NW China.* H4. Summer–early autumn.

75. R. prattii Hemsley. Illustration: Phillips & Rix, Roses, 18 (1988).

Shrub with erect purplish or reddish stems 1.2–2.5 m, unarmed or bearing a few, straight, yellow or pale brown prickles. Stipules entire, acute. Leaves deciduous: leaflets usually 11–15 (rarely 7–9), obovate to narrowly ovate or elliptic, 6–20 mm, acute, hairless above, downy on the veins beneath, margins with obscure teeth. Bracts present. Receptacle glandular-bristly. Flowers 3–7 (rarely 1–7), single, 2–2.5 cm across. Sepals entire, downy on both sides, erect and persistent in fruit. Petals pale to more usually deep pink. Styles free, not protruding. Fruit ovoid to bottle-shaped, scarlet to orange-red, 6–10 cm, glandular-bristly. *W China.* H5. Summer.

76. R. elegantula Rolfe (*R. farreri* Stearn).

Shrub producing densely prickly suckers, and stems 1–2 m, unarmed or bearing a few prickles and dense, red bristles, particularly in the lower part. Stipules narrow. Leaves deciduous, glaucous green, turning purple to crimson in the autumn; leaflets 7–11, narrowly ovate to elliptic, 1–2.5 cm, acuminate or obtuse, downy on midrib beneath, margins with simple teeth. Bracts present. Receptacle smooth or glandular. Flowers solitary to few, single, 2–3.8 cm across. Sepals entire, abruptly acuminate, hairy inside and on margin but hairless on the back, spreading and persistent in fruit. Petals white to pale pink or rose-pink. Styles free, not protruding. Fruit ovoid to top-shaped, red, 1.2–1.3 cm, smooth or glandular. *NW & W Central China.* H4. Early summer.

'Persetosa' (*R. elegantula* forma *persetosa* Stapf; *R. farreri* 'Persetosa'). Illustration: Harkness, The rose, pl. 98 (1979); Gibson, The book of the rose, pl. 1 (1980). More prickles on the stems and smaller leaflets (to 1.3 cm), smaller whitish to salmon-pink flowers to 1.2–2 cm across, opening from coral-pink buds, and orange-red fruit *c.* 7 mm. *NW China.*

77. R. multibracteata Hemsley & Wilson (*R. reducta* Baker). Illustration: Phillips & Rix, Roses, 25, 214 (1988).

Shrub with vigorous, arching stems 2–4 m, green when young, turning red-brown, bearing slender, straight prickles usually in pairs at the nodes. Stipules glandular. Leaves deciduous; leaflets 7–9 (rarely 5–9), obovate to elliptic or more or less circular, 6–15 mm, obtuse, dark green and hairless above, greyish green beneath and downy on the midrib, margins with compound teeth. Bracts broad. Receptacle glandular-bristly. Flowers solitary to many in terminal clusters, single, smelling slightly unpleasant, 2.5–3.8 cm across. Sepals entire, downy inside, glandular on the back, erect and persistent in fruit. Petals bright pink to lilac pink. Styles free, protruding; stigmas woolly. Fruit spherical to ovoid or bottle-shaped, orange-red, 1–1.5 cm, with a few glandular bristles. *SE China.* H4. Summer.

78. R. forrestiana Boulenger. Illustration: Gibson, Shrub roses for every garden, pl. 1 (1973); Austin, The heritage of the rose, 350 (1988); Phillips & Rix, Roses, 23, 214 (1988).

Shrub with vigorous, erect or spreading stems 2 m, bearing straight or forward-pointing, brown prickles in pairs at the nodes. Stipules broadly ovate. Leaves deciduous; leaflets 5–7 (rarely to 9), elliptic to round, 1–1.3 cm, obtuse, hairless above, hairless or hairy to glandular on veins beneath, margins with simple teeth. Bracts very broad. Receptacle smooth or glandular-bristly. Flowers 1–5, single, fragrant, 2–3.8 cm across. Sepals entire, often glandular on the back, erect and persistent in fruit. Petals pale to bright pink. Styles free, protruding. Fruit ovoid with a distinct neck, red, to 1.3 cm, smooth or glandular-bristly. *SW China.* H5. Summer.

R. latibracteata Boulenger. Similar but with larger leaflets to 2.5 cm; styles not protruding. *W China (Yunnan).* H5. Summer.

79. R. willmottiae Hemsley. Illustration: Thomas, A garden of roses, 157 (1987).

Densely branched shrub with erect or arching stems to 3 m, glaucous-bloomed when young, becoming red-brown, bearing slender, straight prickles, mostly in pairs at the nodes, bristles absent. Stipules glandular-hairy on the margins. Leaves deciduous; leaflets 7–9, obovate to oblong or almost round, 6–15 mm, obtuse, hairless on both sides, margins with simple or compound teeth. Bracts present. Receptacle smooth. Flowers usually solitary, single, slightly fragrant, 2.5–3.8 cm across. Sepals entire, hairless on the back, falling after flowering. Petals purplish pink. Styles free, not protruding. Fruit ovoid, pear-shaped or more or less spherical, orange-red, 1–1.8 cm. *W & NW China.* H4. Early summer.

'Wisley'. Illustration: Phillips & Rix, Roses, 214 (1988). Flowers deeper pink with narrower petals and sepals persistent in fruit.

R. × pruhoniciana Schneider (illustration: Gault & Synge, The dictionary of roses in colour, pl. 40, 1971) is a hybrid between *R. moyesii* and *R. willmottiae* or *R. multibracteata*. It is very similar to *R. willmottiae*, differing in having maroon-crimson flowers, and fruit that persists after the leaves have fallen. *Garden origin.*

80. R. gymnocarpa Torrey & Gray. Illustration: Phillips & Rix, Roses, 26 (1988).

Shrub with erect, slender stems 1–3 m, more or less unarmed or bearing slender, straight prickles, intermixed with bristles. Stipules narrow. Leaves deciduous; leaflets 5–9, elliptic to ovate or roundish, 1–4 cm, obtuse, usually hairless on both sides, sometimes with glands beneath, margins with compound, often glandular teeth. Bracts present. Receptacle smooth. Flowers 1–4, single, 2.5–3.8 cm across. Sepals entire, hairless and usually without glands, falling after flowering. Petals rose-pink. Styles free, not protruding. Fruit spherical, ellipsoid or pear-shaped, red, 6–10 mm, smooth. *W North America.* H3. Summer.

81. R. beggeriana Fischer & Meyer. Illustration: Phillips & Rix, Roses, 29 (1988).

Shrub with erect, sometimes climbing stems 1.8–3 m, reddish when young, bearing pale, more or less flattened, hooked prickles usually in pairs at the nodes. Stipules narrow. Leaves deciduous, aromatic; leaflets 5–9, narrowly elliptic to obovate, 7–30 mm, obtuse, grey-green and hairless above, glandular and sometimes downy beneath, margins with simple teeth. Bracts narrowly ovate. Receptacle smooth. Flowers in clusters of 8 or more at the ends of new shoots, single, smelling slightly unpleasant, 2–3.8 cm across. Sepals entire, hairy or not on the back, glandular or not, erect after flowering but eventually falling. Petals white. Styles free, not protruding. Fruit spherical, red, turning purplish, 6–10 mm, smooth. *Afghanistan, Kazakhstan, Mongolia, NW China.* H4. Summer.

82. R. multiflora Murray. Illustration: Niering & Olmstead, Audubon Society field guide to North American wildflowers – eastern region, pl. 211 (1979); Beales, Classic roses, 66 (1985); Thomas, A garden of roses, 119 (1987); Phillips & Rix, Roses, 38 (1988).

Strong-growing shrub with arching, trailing or sometimes climbing stems 3–5 m, bearing many small, stout, decurved prickles. Stipules with laciniate margins, usually glandular-bristly. Leaves deciduous; leaflets 7–9 (rarely 5–11), obovate or elliptic, 1.5–5 cm, acute, acuminate or obtuse, hairless above, hairless or downy beneath, margins with simple teeth. Bracts usually absent. Receptacle hairy. Flowers few to many in branched clusters, single, fruit-scented, 1.5–3 cm across. Sepals with lateral lobes, glandular-bristly on the back, shorter than petals even in bud, reflexed and falling after flowering. Petals cream, fading to white, occasionally pink. Styles united, protruding; stigmas hairless. Fruit ovoid to spherical, red, 6–7 mm. *S China, Taiwan, Japan, Korea (naturalised in USA).* H3. Summer.

The lower branches root where they touch the soil. Much used as a stock for grafting, especially for ramblers.

Var. **cathayensis** Rehder & Wilson (*R. cathayensis* (Rehder & Wilson) Bailey; *R. gentiliana* Léveillé & Vaniot). Illustration: Phillips & Rix, Roses, 35 (1988). Single, rosy pink flowers to 4 cm across, in flattish clusters. *China (Hubei, Gansu, Sichuan, Yunnan).*

Var. **nana** misapplied is a dwarf variety with pale pink or cream single, fragrant flowers. 'Grevillei' (*R. multiflora* var. *platyphylla* Thory; 'Platyphylla'). Illustration: Anderson, The complete book of 169 Redouté roses, 59 (1979). Leaflets large, wrinkled; flowers in clusters of 25–30 (rarely to 50) usually double flowers, deep pinkish purple fading to whitish. 'Carnea' (*R. multiflora* var. *carnea* Thory) has double, flesh-pink flowers. 'Wilsonii' has single white flowers *c.* 5 cm across.

R. × iwara Regel is a hybrid between *R. multiflora* and *R. rugosa*. It differs from *R. multiflora* in the stems bearing large, hooked prickles, and the leaflets being grey-downy beneath. The flowers are white and single. *Japan.*

83. R. watsoniana Crépin (*R. multiflora* var. *watsoniana* (Crépin) Matsumura).

Shrub with arching, trailing or climbing stems to 1 m, bearing small, scattered prickles. Stipules very narrow, entire. Leaves deciduous; leaflets 3–5, linear-lanceolate, 2.5–6.5 cm, hairless above and mottled with yellow or grey near the midrib, downy beneath, margins wavy and bearing simple teeth. Bracts absent. Flowers in clusters, single, 1–1.7 cm across. Sepals entire, hairy on the back, reflexed after flowering, finally falling. Petals pale pink, occasionally white. Styles united, protruding. Fruit spherical, red, 6–7 mm. *Japan, garden origin.* H3. Summer.

The fruits are usually sterile. Any seedlings that are produced are normal *R. multiflora* suggesting that *R. watsoniana* may be a mutant.

84. R. setigera Michaux. Illustration: Thomas, A garden of roses, 141 (1987); Phillips & Rix, Roses, 34 (1988).

Shrub with slender, spreading, trailing or rambling stems 2–5 m, bearing stout, more or less straight, broad-based, scattered prickles. Stipules narrow. Leaves deciduous; leaflets 3–5, ovate to ovate-oblong, 3–8 cm, acute to acuminate, deep green and hairless above, pale green and downy on the veins beneath, margins with coarse, simple teeth. Bracts absent. Receptacle glandular-bristly. Flowers 5–15 in loose clusters, single, sometimes fragrant, 5–7.5 cm across. Sepals with lateral lobes, downy and glandular-bristly on the back, reflexed after flowering, then falling. Petals deep pink, fading to pale pink or almost white. Styles united, protruding; stigmas hairless. Fruit spherical, red to greenish brown, *c.* 8 mm, glandular-bristly. *E & C North America.* H3. Summer.

85. R. × beanii Heath (*R. anemoniflora* Lindley; *R. triphylla* Roxburgh). Illustration: Thomas, A garden of roses, 31 (1987).

Shrub with spreading stems, bearing few, small, scattered slender, hooked prickles. Leaves deciduous; leaflets 3–5, ovate to narrowly ovate, 3.8–7.5 cm, acute or acuminate, hairless on both sides, margins with simple teeth. Bracts absent. Flowers in loose clusters, double, 2.5–4.5 cm across. Sepals with lateral lobes. Petals pale pink, the inner ones narrow and rather ragged. Styles united, protruding; stigmas hairy. *E China, garden origin.* H4. Summer.

There is enormous confusion about the origin of this rose: some botanists think it is a hybrid between *R. multiflora* and either *R. laevigata* or *R. banksiae*, others that *R. banksiae* and *R. moschata* are the parents.

86. R. helenae Rehder & Wilson. Illustration: Le Rougetel, A heritage of roses, 146 (1988); Phillips & Rix, Roses, 36 (1988).

Shrub with rambling stems 5–6 m, purplish brown when young, bearing short, stout, hooked prickles. Leaves deciduous; leaflets 7–9 (rarely 5–9), ovate to elliptic or obovate, 1.9–6 cm, acute, bright green and hairless above, greyish beneath and downy on the veins, margins with simple teeth. Bracts absent. Receptacle densely glandular. Flowers many in flattish clusters, single, fragrant, 2–4 cm across. Sepals with lateral lobes, glandular on the back, reflexed after flowering, eventually falling. Petals white. Styles united into a hairy column, protruding; stigmas hairy. Fruit ovoid, ellipsoid or pear-shaped, scarlet or orange-red, 1–1.5 cm, glandular. *C China, Vietnam, Thailand.* H4. Summer.

87. R. rubus Léveillé & Vaniot (*R. ernestii* Bean; *R. ernestii* forma *nudescens* Stapf and forma *velutescens* Stapf).

Vigorous shrub with spreading or semi-climbing, often purplish stems 2.4–5 m, bearing few, short, hooked prickles (young shoots hairy or more or less hairless). Stipules narrow. Leaves deciduous; leaflets 5 (rarely 3 to 5), elliptic ovate to oblong-obovate, 3–9 cm, acute, glossy and hairless above, greyish and usually downy beneath, often purple-tinged when young, margins with simple teeth. Bracts absent. Receptacle glandular-bristly. Flowers 1 or 2 in tight clusters, single, fragrant, 2.5–3.8 cm across, borne on stalks 1–2.5 cm. Sepals entire or with lateral lobes, downy and glandular on the back, reflexed after flowering and falling. Petals white, yellowish at the base. Styles united into a shortly protruding downy column; stigmas woolly. Fruit spherical or ovoid, dark red, 9–15 mm, glandular-bristly. *W & C China.* H4. Late summer.

88. R. mulliganii Boulenger. Illustration: Austin, The heritage of the rose, 370 (1988); Phillips & Rix, Roses, 38 (1988).

Similar to *R. rubus*, but stems 6 m or more, leaflets 5–7, flowers 4.5–5.5 cm across on stalks 2.5–3.5 cm, carried in looser clusters; the sepals always have lateral lobes. *W China (Yunnan).* H5. Summer.

89. R. brunonii Lindley (*R. moschata* Herrmann var. *napaulensis* Lindley). Illustration: Thomas, A garden of roses, 43 (1987); Phillips & Rix, Roses, 36, 39 (1988).

Vigorous shrub with arching or climbing stems 5–12 m, bearing short, stout, hooked prickles. Stipules narrow with spreading tips. Leaves deciduous, drooping, 17–21 cm; leaflets 5–7 (rarely to 9), narrowly ovate to elliptic or oblong-elliptic, 3–6 cm, acute or acuminate, grey-green or bluish green and somewhat hairy above,

downy beneath at least on veins, sometimes glandular, margins with simple teeth. Bracts absent. Receptacle hairy and usually glandular. Flowers in clusters, often with several clusters combined into a large compound inflorescence, single, fragrant, 2.5–5 cm across. Sepals with lateral lobes, hairy and slightly glandular on the back, reflexed at flowering time, falling after flowering. Petals white. Styles united, protruding; stigmas downy. Fruit almost spherical to obovoid, reddish brown, 7–18 mm, smooth. *Himalaya, from Afghanistan to SW China & Burma.* H4. Summer.

Many of the plants grown in gardens as *R. moschata* are in fact *R. brunonii*, having been distributed under the wrong name by nurseries.

'La Mortola' (illustration: Thomas, Climbing roses old and new, pl. 1, 1983) is a very strong-growing selection.

90. R. moschata Herrmann. Illustration: Thomas, Climbing roses old and new, pl. 3 (1983); Beales, Classic roses, 259 (1985); Phillips & Rix, Roses, 34 (1988).

Robust shrub with arching or semi-climbing, purplish or reddish stems 3–10 m, bearing few, scattered, straight or slightly curved prickles. Stipules very narrow. Leaves deciduous; leaflets 5–7, broadly ovate to broadly elliptic, 3–7 cm, acute or acuminate, shiny and hairless above, downy or not on veins beneath, margins with simple teeth. Bracts absent. Receptacle finely adpressed-hairy, rarely slightly glandular. Flowers in few-flowered loose clusters, single, with a musky scent, 3–5.5 cm across. Sepals entire or with lateral lobes, hairy on the back, reflexed and persistent after flowering. Petals white or cream, reflexing with age. Styles united, protruding; stigmas woolly. Fruit almost spherical to ovoid, orange-red, 1–1.5 cm, usually downy and sometimes glandular. *Unknown in the wild, cultivated in S Europe. Mediterranean & SW Asia (naturalised in N America).* H4. Late summer–autumn.

Var. **nastarana** Christ (illustration: Wilson & Bell, The fragrant year, 156, 1971) has smaller leaflets always hairless beneath, and larger, more numerous, pink-tinged flowers.

'Dupontii' (*R.* × *dupontii* Déséglise, in part; *R. nivea* Lindley, not de Candolle). Illustration: Gault & Synge, The dictionary of roses in colour, pl. 6 (1971); Beales, Classic roses, 255 (1985); Thomas, A garden of roses, 67 (1987); Austin, The

heritage of the rose, 351 (1988). Was thought for some time to be a hybrid between *R. moschata* and *R. gallica.* but this seems to be unlikely. It differs from *R. moschata* in its smaller size (2–3 m), leaflets with compound teeth, single creamy pink flowers 6–7.5 cm across, sepals glandular on the back, and free styles. *Garden origin.* H3. Summer.

91. R. longicuspis Bertoloni (*R. lucens* Rolfe; *R. yunnanensis* (Crepin) Boulenger). Illustration: Thomas, Climbing roses old and new, pl. 2 (1983); Phillips & Rix, Roses, 39 (1988).

Vigorous shrub with scrambling and climbing stems to 6 m or more, reddish when young, bearing few, short, curved or hooked prickles, which may be absent on flowering-branches. Stipules narrow. Leaves evergreen or semi-evergreen, to 20 cm, reddish when young; leaflets 5–7 (rarely 3–7), narrowly ovate to elliptic, leathery, 5–10 cm, acuminate, hairless on both sides although occasionally downy on midrib beneath, margins with simple teeth. Bracts absent. Receptacle hairy and usually glandular. Flowers to 15 in a loose cluster, single, smelling of bananas, *c.* 5 cm across, opening from narrowly ovoid buds. Sepals with lateral lobes, hairy and glandular on the back, reflexed after flowering and eventually falling. Petals white, silky on the back. Styles united, protruding. Fruit broadly ellipsoid to spherical, red to orange, 1.5–2 cm, often hairy and glandular. *NE India, W China.* H4. Early–mid summer.

Var. **sinowilsonii** (Hemsley) Yu & Ku. (*R. sinowilsonii* Hemsley). Illustration: Phillips & Rix, Roses, 39 (1988). Leaves to 30 cm, leaflets slightly downy beneath, flower-buds broadly ovoid, sepals hairless or almost so on the back. *SW China.* H4. Summer.

92. R. cerasocarpa Rolfe (*R. gentiliana* of Rehder & Wilson in part, not Léveillé & Vaniot).

Shrub with climbing or semi-climbing stems to 4.5 m, bearing few, stout, scattered, recurved prickles. Stipules narrow. Leaves deciduous, 17–20 cm; leaflets 5 (rarely 3–5), narrowly ovate to elliptic, leathery, 5–10 cm, acute to acuminate, hairless or more or less so on both sides, margin with simple teeth. Bracts absent. Receptacle downy and glandular. Flowers many in clusters, single, fragrant, 2.5–3.5 cm across, opening from abruptly pointed buds. Sepals usually with lateral lobes, downy and glandular on the back, reflexed and falling after flowering. Petals white. Styles united. Fruit spherical, deep red, 1–1.3 cm, downy. *W & C China.* H4. Summer.

93. R. maximowicziana Regel.

Shrub with arching and climbing stems bearing few, small, scattered, straight and hooked prickles, with bristles on the young branches. Stipules very narrow. Leaves deciduous; leaflets 7–9, ovate-elliptic to oblong, 2.5–5 cm, acute to acuminate, hairless on both sides, margins with simple teeth. Bracts present. Receptacle smooth. Flowers many in small clusters, single, 2.5–3.5 cm across. Sepals with lateral lobes, reflexed after flowering and finally falling. Petals white. Styles united; stigmas hairless. Fruit ovoid, red, 1–1.2 cm, smooth. E Russia, NE *China, Korea.* H3. Summer.

94. R. filipes Rehder & Wilson.
Illustration: Phillips & Rix, Roses, 38 (1988).

Shrub with arching and climbing stems to 9 m, purple when young, bearing few, small, hooked prickles. Stipules narrow. Leaves deciduous, coppery when young; leaflets 5–7, narrowly ovate to narrowly elliptic, 3.5–8 cm, acuminate, hairless above and beneath or sometimes downy on veins beneath, margins with simple teeth. Bracts absent. Flowers to 100 or more in large clusters to 30 cm or more wide, single, fragrant, 2–2.5 cm across. Sepals with a few lateral lobes, glandular and slightly downy or hairless on the back, reflexed and falling after flowering. Petals creamy white, sometimes downy on the back. Styles united, protruding; stigmas woolly. Fruit spherical to broadly ellipsoid, orange becoming crimson-scarlet, 8–15 mm. *W China.* H4. Summer.

In cultivation it is most commonly found as the cultivar 'Kiftsgate' (illustration: Phillips & Rix, Roses, 36, 1988; Brickell, The RHS gardeners' encyclopedia of plants and flowers, 169, 1989), in which the inflorescence can be to 45 cm across.

95. R. phoenicia Boissier.
Shrub with slender, climbing stems 3–5 m, bearing few, short, curved or hooked, broad-based, more or less equal prickles. Stipules glandular-toothed. Leaves deciduous; leaflets 5 (rarely 3–7), elliptic to roundish, 2–4.5 cm, acute to obtuse, more or less downy on both sides, more so beneath, margins with simple or

compound teeth. Bracts usually hairy. Receptacle hairless or slightly hairy. Flowers 10–14 in clusters, single, 4–5 cm across, opening from ovoid, rounded buds. Sepals with lateral lobes, often downy on the back, reflexed and falling after flowering. Petals white. Styles united, protruding; stigmas hairless. Fruit ovoid, red, 1–1.2 cm, smooth. *NE Greece, Cyprus, Turkey, Syria, Lebanon.* H5. Summer.

96. R. soulieana Crepin. Illustration: Thomas, A garden of roses, 143 (1987).

Robust shrub with erect, spreading or semi-climbing stems 3–4 m, bearing scattered, stout, decurved, compressed, broad-based prickles. Stipules narrow. Leaves deciduous, 6.5–10 cm; leaflets 7–9 (rarely 5–9), grey-green, obovate to elliptic, 1–3 cm, acute to obtuse, hairless on both sides although more or less downy on midrib beneath, margins with simple teeth. Bracts absent. Receptacle glandular. Flowers in many-flowered branched clusters 10–15 cm across; flowers single, with a fruity scent, 2.5–3.8 cm across, opening from yellow buds. Sepals entire or with a few lateral lobes, hairless or downy and often glandular on the back, reflexed and falling after flowering. Petals white. Styles united, protruding; stigmas sparsely hairy. Fruit ovoid to almost spherical, pale to dark orange, *c.* 1 cm, glandular. *W China.* H5. Summer.

97. R. wichuraiana Crepin. Illustration: Thomas, Climbing roses old and new, pl. 5 (1983); Thomas, A garden of roses, 155 (1987); Phillips & Rix, Roses, 35 (1988).

Shrub with prostrate, trailing or climbing stems 3–6 m, bearing strong, curved prickles. Stipules broad, toothed. Leaves evergreen or semi-evergreen; leaflets 5–9, dark green, elliptic to broadly ovate or roundish, obtuse, hairless on both sides except for midrib beneath, margins with simple teeth. Bracts absent. Receptacle sometimes glandular. Flowers 6–10 in loose clusters, single, fragrant, 2.5–5 cm across. Sepals entire or with a few lateral lobes, often downy or slightly glandular on the back, reflexed and falling after flowering. Petals white. Styles united, protruding; stigmas woolly. Fruit avoid or spherical, orange-red to dark red, 1–1.5 cm. *Japan, Korea, E China, Taiwan (naturalised in N America).* H4. Summer–early autumn.

'Variegata' has leaflets that are cream with pink tips when young, turning green, marked with cream.

R. × jacksonii Baker (illustration: Thomas, A garden of roses, 101, 1987) is a hybrid between *R. wichuraiana* and *R. rugosa.* It differs from *R. wichuraiana* by its larger, wrinkled leaflets and bright crimson flowers. *Garden origin.*

R. × kordesii Wulff was produced by Wilhelm Kordes who selfed 'Max Graf'. Like *R. × jacksonii* it is a hybrid between *R. wichuraiana* and *R. rugosa. R. × kordesii* is tetraploid and has been used as a parent for a number of repeat-flowering climbers. It differs from *R. wichuraiana* in having deep pink to red, recurrent flowers, which are single to semi-double and 7–8 cm across. *Garden origin.*

98. R. luciae Crépin.

Shrub with prostrate or climbing stems to 3.5 m, bearing small, scattered, pale brown, rather flattened, slightly hooked prickles. Stipules very thin. Leaves evergreen or semi-evergreen; leaflets 5 (rarely to 7), rather thin, 2–4.5 cm, the terminal one longer than the others, acute to acuminate, both sides hairless or more or less so, margins with simple teeth. Bracts absent. Receptacle smooth or glandular. Flowers 3–20 in clusters, single, fragrant, 2–3 cm across, opening from short, rounded buds. Sepals reflexed, falling after flowering. Petals white. Styles united, protruding; stigmas woolly. Fruit ovoid to spherical, red to purplish, *c.* 7 mm, smooth or glandular. *SE China, Japan (Ryukyu Islands), Korea, Philippines.* H4. Early summer and again in late summer.

Very similar to *R. wichuraiana* and considered by some botanists to be the same: if so, then the name *R. luciae* has priority.

99. R. sempervirens Linnaeus.
Illustration: Thomas, A garden of roses, 137 (1987); Phillips & Rix, Roses, 34 (1988); Blamey & Grey-Wilson, Illustrated flora of Britain and northern Europe, 179 (1989).

Shrub with prostrate, trailing or scrambling stems 6–10 m, unarmed or bearing straight or hooked, broad-based prickles. Stipules narrow. Leaves evergreen or semi-evergreen; leaflets 5–7 (rarely 3–7), narrowly ovate to elliptic, 1.5–6 cm, the terminal usually larger than the upper laterals, acuminate, hairless on both sides except for midrib, which may be downy beneath, margins with simple teeth. Bracts short, entire, more or less hairless, falling early. Receptacle often glandular-bristly. Flowers 3–10 (rarely 1–10) in clusters, single, slightly fragrant, 2.5–4.5 cm across, opening from blunt, ovoid buds. Sepals entire, glandular-bristly on the back, reflexed to spreading and falling after flowering. Petals white. Styles united, protruding; stigmas usually woolly. Fruit ovoid or spherical, orange-red, 1–1.6 cm, often glandular-bristly. *S Europe, NW Africa, Turkey*. H5. Summer.

100. R. arvensis Hudson. Illustration: Ross-Craig, Drawings of British plants **9**: pl. 13 (1956); Beales, Classic roses, 252 (1985); Thomas, A garden of roses, 33 (1987); Phillips & Rix, Roses, 35 (1988).

Shrub with trailing or climbing stems 1–2 m, bearing sparse, scattered, stout, more or less curved, equal prickles. Stipules narrow. Leaves deciduous; leaflets 5–7 (rarely 3–7), deep green, elliptic to broadly ovate or roundish, 1–4 cm, more or less acute, hairless on both sides, although sometimes downy on veins beneath, margins with simple teeth. Bracts absent. Receptacle smooth or somewhat glandular. Flowers 1–8, single, usually fragrant, 2.5–5 cm across. Sepals with lateral lobes, smooth or sparsely glandular on the back, reflexed and falling after flowering. Petals white to pink. Styles united, protruding; stigmas hairless. Fruit spherical to ovoid, red, 0.6–2.5 cm, usually smooth. *S, W & C Europe, S Turkey*. H2. Summer.

R. × polliniana Sprengel is a hybrid between *R. arvensis* and *R. gallica*, which differs from *R. arvensis* in its rather leathery leaflets that are slightly downy beneath, and flowers 5–6 cm across. *N Italy*.

101. R. chinensis Jacquin (*R. indica* Loureiro, not Linnaeus).

Shrub, varying in habit from dwarf to semi-climbing, to 6 m. Stems more or less unarmed or bearing scattered, more or less hooked, somewhat flattened prickles. Stipules narrow. Leaves evergreen; leaflets 3–5, lanceolate to broadly ovate, 2.5–6 cm, acuminate, glossy and hairless above, hairless beneath except for downy midrib, margins with simple teeth. Bracts narrow. Receptacle smooth or glandular. Flowers solitary or in clusters, single or semi-double, *c.* 5 cm across, often fragrant. Sepals entire or with a few lateral lobes, smooth or glandular on the back, reflexed and falling after flowering. Petals pale pink to scarlet or crimson. Styles free, somewhat protruding. Fruit ovoid to pear-shaped, greenish brown to scarlet, 1.5–2 cm. *China, garden origin*. H5. Summer–early autumn.

A repeat-flowering rose which contributed this feature to European roses when it was introduced around 1800.

Var. **spontanea** (Rehder & Wilson) Yu & Ku (*R. chinensis* forma *spontanea* Rehder & Wilson). Illustration: Phillips & Rix, Roses, 32 (1988). Climber or bush, usually 1–2.5 m. Leaflets lanceolate. Flowers 1–3, single, pink, turning red, 5–6 cm across. Sepals entire. Fruit orange. *W China*.

It is the wild type of cultivated *R. chinensis*.

'Minima' (*R. chinensis* var. *minima* (Sims) Voss). Stems 20–50 cm; flowers rose-red, single or double, usually solitary, *c.* 3 cm across, with pointed petals. 'Rouletii' (*R. roulettii* Correvon), which has stems 10–25 cm, and double, rosy pink flowers 1.9–2.5 cm across, is thought by some botanists to be synonymous with 'Minima'. 'Mutabilis' (*R. mutabilis* Correvon; *R. chinensis* forma *mutabilis* (Correvon) Rehder; 'Tipo Ideale'). Illustration: Thomas, Shrub roses of today, pl. VI (1962); Phillips & Rix, Roses, 68, 69 (1988). Stems 1–1.7 m, leaflets purplish or coppery when young; flowers single, fragrant, 4.5–6 cm across, petals yellow with an orange back, turning coppery salmon-pink and eventually deep pink. 'Pallida' has stems to 1 m or more, and clusters of semi-double, fragrant, blush-pink flowers. 'Semperflorens' (*R. chinensis* var. *semperflorens* (Curtis) Koehne) has stems 1–1.5 m, and semi-double, deep pink or crimson-scarlet, delicately scented flowers. 'Viridiflora' (*R. chinensis* var. *viridiflora* (Lavallee) Dippel). Illustration: Phillips & Rix, Roses, 69 (1988); Welch, Roses, 114 (1988). A monstrous variant probably derived from 'Pallida', in which the petals are green streaked with crimson or purplish and the stamens and pistils become leafy, narrow, with toothed segments.

102. R. gigantea Crepin (*R. × odorata* var. *gigantea* (Crepin) Rehder & Wilson; *R. odorata* 'Gigantea'). Illustration: Thomas, A garden of roses, 89 (1987); Phillips & Rix, Roses, 33 (1988).

Shrub with climbing stems 8–12 (rarely to 30) m, bearing stout, scattered, uniform, hooked prickles. Stipules narrow. Leaves evergreen or semi-evergreen; leaflets 5–7, elliptic to ovate, 3.8–9 cm, acuminate, glossy above, hairless on both sides, margins with simple, often glandular teeth. Bracts present. Receptacle smooth. Flowers solitary (rarely to 3), single, fragrant,

10–15 cm across, opening from slender, pale yellow buds. Sepals entire, smooth on the back, reflexed and falling after flowering. Petals white or cream. Styles free; stigmas downy. Fruit spherical or pear-shaped, red, or yellow flushed with red, 2–3 cm. *NE India, Burma, China (Yunnan)*. H5. Early summer.

103. R. × odorata (Andrews) Sweet.

Thought to be a hybrid between *R. chinensis* and *R. gigantea*. It differs from *R. gigantea* in having single or double flowers 5–8 cm across, with white, pale pink or yellowish petals. *SW China, garden origin*. H5. Mid–late summer.

104. R. banksiae Aiton filius.

Shrub with strong climbing stems to 12 m, unarmed or bearing very sparse, hooked prickles. Stipules very narrow, soon falling. Leaves evergreen; leaflets 3–7, oblong-lanceolate to elliptic-ovate, 2–6.5 cm, acute or obtuse, glossy and hairless above, sometimes downy beneath on the midrib, margin wavy, with simple teeth. Bracts very small, soon falling. Flowers many in umbels, single or double, fragrant, 2.5–3 cm across. Sepals entire, reflexed and falling after flowering. Petals white or yellow. Styles free, not protruding. Fruit spherical, dull red, *c.* 7 mm. *W & C China*. H4. Early summer.

Var. **banksiae** (*R. banksiae* var. *alboplena* Rehder; *R. banksiae* 'Banksiae'; *R. banksiae* 'Albo-Plena' or 'Alba Plena'). Illustration: Beales, Classic roses, 399 (1985); Phillips & Rix, Roses, 30 (1988). Flowers double, white, violet-scented.

Var. **normalis** Regel. Illustration: Phillips & Rix, Roses, 31 (1988). Flowers single, white, very fragrant, and stems usually bearing hooked prickles.

'Lutea' (*R. banksiae* var. *lutea* Lindley). Illustration: Harkness, Roses, pl. 4 (1978); Harkness, The rose, pl. 1 (1979); Thomas, A garden of roses, 35 (1987); Phillips & Rix, Roses, 30 (1988) – as 'Lutescens'. Flowers double, yellow, slightly fragrant; leaflets usually 5; stems generally unarmed. The hardiest taxon and the most floriferous. 'Lutescens' (*R. banksiae* forma *lutescens* Voss). Illustration: Austin, The heritage of the rose, 299, 330 (1988); Phillips & Rix, Roses, 30 (1988) – as 'Lutea'. Has single, yellow, highly scented flowers.

R. × fortuneana Lindley. Illustration: Beales, Classic roses, 400 (1985); Thomas, A garden of roses, 79 (1987). Thought to be a hybrid between *R. banksiae* and

R. laevigata. It differs from *R. banksiae* in having 3–5 rather thin leaflets, and solitary, double, creamy white flowers 5–10 cm across. *China, garden origin*. H5. Summer.

105. R. cymosa Trattinick (*R. microcarpa* Lindley, not Besser nor Retzius). Illustration: Phillips & Rix, Roses, 31 (1988).

Differs from *R. banksiae* in having more prickly stems, larger, branched inflorescences, and smaller, always single, white flowers (*c*. 1.5 cm across) with sepals bearing lateral lobes. *C & S China, Taiwan, Laos, Vietnam*. H4. Early summer.

Specimens have been found that are intermediate between *R. cymosa* and *R. banksiae*: it is likely that future work will prove that the two are the same.

106. R. laevigata Michaux. Illustration: Anderson, The complete book of 169 Redouté roses, 89 (1979); Krüssmann, The complete book of roses, 293 (1981); Thomas, A garden of roses, 105 (1987); Phillips & Rix, Roses, 33 (1988).

Vigorous shrub with climbing, green stems to 10 m or more, bearing scattered, stout, red-brown, hooked prickles. Stipules united at the base, soon falling. Leaves evergreen; leaflets 3 (rarely to 5), lanceolate to elliptic or ovate, rather leathery, 3–6 (rarely to 9) cm, acute or acuminate, glossy above, hairless on both sides, midrib sometimes prickly beneath, margins with simple teeth. Bracts absent. Receptacle very bristly. Flowers solitary, single, fragrant, 5–10 cm across. Sepals entire, bristly on the back, erect and persistent in fruit. Petals white or creamy white. Styles free, not protruding. Fruit pear-shaped, orange-red to red, 3.5–4 cm, bristly. *S China, Taiwan, Vietnam, Laos, Cambodia, Burma (naturalised in S USA)*. H5. Early summer.

'Cooperi'. Illustration: Gault & Synge, The dictionary of roses in colour, pl. 5 (1971); Phillips & Rix, Roses, 32 (1988). Some leaves with 5 or 7 leaflets and red, rather than green stems. As they age the petals become pink-spotted. It is possible that it is a hybrid between *R. laevigata* and *R. gigantea*.

107. R. × anemonoides Rehder (*R. 'Anemone'*). Illustration: Thomas, A garden of roses, 29 (1987); Phillips & Rix, Roses, 32 (1988).

Is thought to be a hybrid between *R. laevigata* and *R. × odorata*. Less vigorous

than *R. laevigata*, from which it differs in the stipules being more united and the petals pale pink with deeper veins. *Garden origin*. H5. Early summer.

108. R. bracteata Wendland. Illustration: Thomas, A garden of roses, 41 (1987); Phillips & Rix, Roses, 32 (1988).

Shrub with prostrate or climbing, brown-downy stems 3–6 m, bearing stout, broad-based, hooked prickles in pairs at the nodes, and numerous glandular bristles. Stipules united at the base, fringed. Leaves evergreen; leaflets 5–11, dark green, obovate to elliptic or oblong, 1.5–5 cm, obtuse, glossy and hairless above, downy beneath at least on midrib, margins with simple teeth. Bracts large, deeply toothed, downy. Receptacle hairy. Flowers usually solitary, single, smelling of fruit, 5–8 cm across. Sepals entire, brown-hairy on the back, reflexed and falling after flowering. Petals white. Styles free, not protruding. Fruit spherical, orange-red, 2.5–3.8 cm, hairy. *S & SE China, Taiwan, S Japan (naturalised in USA from Virginia to Texas & Florida)*. H5. Summer–late autumn.

109. R. roxburghii Trattinick (*R. 'Roxburghii'; R. roxburghii 'Plena'*). Illustration: Thomas, A garden of roses, 135 (1987); Phillips & Rix, Roses, 19 (1988).

Flowers double, pink, darker in the centre. It was originally introduced from Chinese gardens and is rarely grown.

Forma **normalis** Rehder & Wilson (*R. microphylla* Lindley). Illustration: Gault & Synge, The dictionary of roses in colour, pl. 45, 46 (1971); Harkness, Roses, pl. 3 (1978); Gibson, The book of the rose, pl. 34 (1980); Phillips & Rix, Roses, 18, 19 (1988). Spreading shrub with rather stiff stems to 5 m or more, with peeling grey or pale brown bark, and bearing few, straight, hooked prickles in pairs at the nodes. Stipules narrow, united. Leaves deciduous, 5–10 cm; leaflets usually 9–15 (rarely 7 or 17–19), narrowly ovate to obovate, 1–2.5 cm, acute or obtuse, hairless on both sides, margins with simple teeth. Bracts falling early. Receptacle prickly. Flowers usually solitary, single, fragrant, 5–7.5 cm across. Sepals with lateral lobes, downy and prickly on the back, erect and persistent in fruit. Petals mid-pink to deep pink. Styles free, not protruding. Fruit flattened-spherical, yellow-green, 3–4 cm, prickly. *S China, Japan*. H4. Summer.

R. × coryana Hurst. Illustration: Phillips & Rix, Roses, 19, 1988. Like *R. roxburghii*

but stems shorter, less prickly (to 2 m); flowers bright carmine. *Garden origin*.

A hybrid between *R. roxburghii* and probably *R. macrophylla*.

110. R. hirtula (Regel) Nakai (*R. roxburghii* var. *hirtula* (Regel) Rehder & Wilson). Illustration: Phillips & Rix, Roses, 18 (1988).

Like *R. roxburghii* but leaflets elliptic to oblong-elliptic, downy beneath; flowers single, pale pink or lilac-pink. Japan.

Similar to *R. roxburghii*: some botanists agree with Rehder & Wilson that it would be better treated as a variety.

111. R. stellata Wooton (*Hesperhodos stellatus* (Wooton) Boulenger). Illustration: Thomas, A garden of roses, 147 (1987).

Shrub of loose habit with erect, slender, stellate-hairy stems 60–120 cm, bearing dense, straight, slender, pale yellow, often paired prickles intermixed with glandular bristles. Stipules united. Leaves deciduous; leaflets 3–5, wedge-shaped to obovate, 5–12 mm, stellate-hairy on both sides or only beneath, margins with simple teeth except towards the base. Bracts absent. Receptacle with pale spines. Flowers solitary, single, 3.5–6 cm across. Sepals with lateral lobes, glandular and spiny on the back, margin woolly, erect and persistent in fruit. Petals soft pink to bright rose or dark purplish red. Stamens to 160 or more. Styles free, not protruding; stigmas woolly. Fruit hemispherical, flat-topped, not fleshy, dull red to brown-red, to 2 cm, prickly. *SW USA (New Mexico, Texas, Arizona)*. H4. Summer.

Var. **mirifica** (Greene) Cockerell (*R. mirifica* Greene; *R. stellata* subsp. *mirifica* (Greene) Lewis; *Hesperhodos mirijicus* (Greene) Boulenger). Illustration: Thomas, Shrub roses of today, pl. 4 (1962); Phillips & Rix, Roses, 18 (1988). Differs in lacking stellate hairs and having flowers with fewer (to 150) stamens. *SW USA (New Mexico, S Texas)*. H4. Summer.

25. AGRIMONIA Linnaeus
J. Cullen

Erect perennial herbs with rhizomes. Stems with short glandular hairs borne almost on the surface and longer, spreading or deflexed, simple hairs. Leaves sometimes forming a rosette at the base of the stem, pinnate with 7–13 large leaflets alternating in opposite pairs with much smaller leaflets; all deeply toothed. Stipules large, stem-clasping, deeply toothed. Flowers many in a raceme, each subtended by a bract and 2 bracteoles (all of which may be 3-lobed or entire), shortly stalked. Perigynous zone cylindric, bell-shaped or top-shaped, its upper part with numerous spreading or deflexed, hard, hooked bristles outside and below the calyx. Sepals 5, spreading in flower, persistent and more or less erect in fruit. Petals 5, yellow, golden yellow or orange-yellow, rarely whitish. Stamens 10–20, borne on the edge of a yellowish disc that roofs the perigynous zone. Carpels 2 (rarely more), free, their styles projecting through a small central hole in the roof of the perigynous zone. Fruit a group of 1 or 2 (rarely more) achenes enclosed in the hardened, often grooved, perigynous zone surmounted by the bristles and the persistent calyx.

A genus of a small number of species, all rather difficult to distinguish, mainly from the north temperate areas, but 1 species in South Africa and C Africa. They have a remarkable floral structure and their fruits are dispersed by attachment to animal fur by means of the hooked bristles. They are easily grown, responding best to fairly moist conditions, and are propagated by seed or by division.

Literature: Skalicky, V., Ein Beitrag zur Erkenntniss der Europaischen Arten der Gattung Agrimonia L., *Acta Horti Botanici Pragensis*, 87–108 (1962).

1a. Non-glandular hairs on the stem of 2 sizes, longer and spreading, shorter and finer, usually somewhat to very deflexed; glands on the lower leaf surface not easily visible, obscured by hairs   **1. eupatoria**
  b. Non-glandular hairs on the stem all more or less similar, spreading; glands on the lower leaf surface easily visible   2
2a. Lowermost bristles on the perigynous zone sharply deflexed, even in flower, clearly so in fruit; fruit-stalks 4–10 mm   **3. repens**
  b. Lowermost bristles on the perigynous zone spreading in flower, sometimes somewhat deflexed in fruit; fruit-stalks usually 1–3 mm   **2. procera**

1. A. eupatoria Linnaeus. Illustration: Coste, Flore de la France **2**: 58 (1903); Bonnier, Flore complète **4**: pl. 183 (1914); Ary & Gregory, The Oxford book of wild flowers, 16 (1980); Pignatti, Flora d'Italia **1**: 566 (1982).

Stems to 1.5 m with spreading non-glandular hairs and shorter and finer,

somewhat to very deflexed, non-glandular hairs as well as almost stalkless glands. Glands on the lower leaf surface not easily visible, obscured by hairs. Bracts 3-lobed. Flower-stalks very short, extending to 1–3 mm in fruit. Petals 3–6 mm, yellow, not notched. Perigynous zone tapered to the base, deeply grooved for most of its length, the lowermost bristles spreading. *Europe, N Africa, W & C Asia*. H1. Summer.

Very variable especially in size and density of hairs. 'Alba' has whitish petals.

2. A. procera Wallroth (*A. odorata* misapplied). Illustration: Coste, Flore de la France **2**: 58 (1903); Pignatti, Flora d'Italia **1**: 566 (1982).

Like *A. eupatoria* but non-glandular hairs on the stem all more or less the same in length and thickness, spreading; glands on the lower leaf surface easily visible; petals golden yellow, 5–8 mm, often notched; perigynous zone rather rounded to the base, scarcely grooved or with shallow grooves for up to half its length, outermost bristles of perigynous zone ultimately slightly deflexed. *Most of Europe, N Africa, SW Asia*. H1. Summer.

3. A. repens Linnaeus (*A. odorata* Miller).

Like *A. eupatoria* but non-glandular hairs on the stem all more or less the same in length and thickness, spreading; glands on the lower leaf surface easily visible; petals golden yellow, 5–7 mm, not notched; perigynous zone rounded to the base, deeply grooved over most of its length, the lowermost bristles conspicuously deflexed, even in flower. *Turkey, N Iraq, naturalised in parts of Europe*. H3. Summer.

26. HAGENIA Gmelin
J. Cullen

Dioecious trees to 18 m (in the wild) with open, umbrella-shaped crowns. Branches with dense, golden hairs. Leaves evergreen, 10–40 cm, pinnate with 5–8 pairs of leaflets, silvery silky beneath. Stipules large, sheathing the stems. Flowers numerous in panicles, those of the female flowers broader and more substantial than the male; each flower subtended by 2 bracts. Sepals 4 or 5 alternating with 4 or 5 epicalyx segments. Male flowers with 4 or 5 white or orange petals (rarely absent), 10–20 stamens with sparsely hairy anthers, borne on the edge of a nectar-secreting disc, and a pistillode. Female flowers reddish; petals not known; staminodes to 20, carpels 1 or 2, free,

surrounded by the perigynous zone. Fruit an achene or 2 achenes surrounded by the persistent perigynous zone surmounted by the calyx and epicalyx.

A genus of a single species from tropical E Africa. Though spectacular, it is rarely grown. It is propagated by seed, which is relatively short-lived.

1. H. abyssinica (Bruce) Gmelin. Illustration: Menninger, Flowering trees of the world for tropics and warm climates, pl. 326 (1962); Flora Zambesiaca **4**: 21 (1978).
Tropical E Africa from Ethiopia to Zambia. G2.

27. SANGUISORBA Linnaeus
J. Cullen

Herbs, often large, with thick, woody stocks. Leaves pinnate, often with numerous, shortly stalked, toothed leaflets. Stipules leaf-like, often crescent-shaped. Flowers in dense heads (spikes), which are spherical to narrowly cylindric, flowers female or bisexual, opening from the top of the head downwards or the bottom upwards; each flower with a bract and 2 bracteoles, which are often hairy. Perianth-segments 4, greenish, whitish or reddish, spreading. Stamens 4–30, filaments thread-like or dilated and flattened above, often projecting well beyond the perianth-segments. Carpels 1–3, free, borne within the rounded or 4-sided perigynous zone, the style(s) long or short, stigma a mop-like group of papillae. Fruit an achene or a group of achenes surrounded by the persistent perigynous zone, which is often winged or ornamented, surmounted by the remains of the perianth.

A genus of about 10 species from north temperate areas, including species formerly placed in *Poterium*. There are some woody species in the Canary Islands but these are not in general cultivation. The plants are easily grown and propagated by seed.

Literature: Nordborg, G., Sanguisorba L, Sarcopoterium Spach and Bencomia Webb et Berth., *Opera Botanica* **11**(2) (1966) and The genus Sanguisorba section Poterium, *Opera Botanica* **16** (1967).

1a. Head with female flowers at the top and bisexual flowers lower down; stamens in the lower bisexual flowers 20–30   **4. minor**
  b. Head containing only bisexual flowers; stamens in all flowers usually 4, rarely up to 12   2

2a. Stamens and styles scarcely projecting beyond the perianth-segments, filaments not dilated and flattened above; heads dark red, little longer than broad   **3. officinalis**

  b. Stamens and styles conspicuously projecting beyond the perianth-segments, filaments dilated and flattened above; heads whitish, pinkish or magenta, much longer than broad   3

3a. Flowers opening from the bottom of the head upwards; filaments white   **1. canadensis**

  b. Flowers opening from the top of the head downwards; filaments usually pink or pale magenta   **2. obtusa**

1. S. canadensis Linnaeus. Illustration: Rickett, Wild flowers of the United States **1**: pl. 37 (1966); Justice & Bell, Wild flowers of North Carolina, 91 (1968); Brickell (ed.), RHS A–Z encyclopedia of garden plants, 954 (2003).

Stems to 2 m, hairy or hairless below. Leaflets *c.* 13, oblong to ovate-oblong, base cordate, apex rounded, regularly and rather deeply toothed, midrib often with sparse hairs beneath. Inflorescences cylindric, 6–14 cm × 5–7 mm (excluding the stamens). Flowers white, opening from the bottom of the head upwards. Perianth-segments 2–3 mm. Stamens 4, projecting well beyond the perianth-segments, filaments white, dilated and flattened above. Carpel usually 1; style projecting beyond the perianth. *N America & E Asia.* H1. Summer.

Plants with particularly broad leaves are sometimes separated as **S. sitchensis** Meyer (illustration: Rickett, Wild flowers of the United States **5**: pl. 48, 1971).

2. S. obtusa Maximowicz (*S. magnifica* Schischkin). Illustration: *Botanical Magazine*, 8690 (1916); Huxley, Garden perennials and water plants, pl. 25 (1970); Kitamura et al., Coloured illustrations of herbaceous plants of Japan **1**: pl. 30 (1977).

Large herb to 2.5 m or more. Stems usually hairy below. Leaflets *c.* 13, rather close and sometimes overlapping, ovate-oblong, base cordate, apex very broadly rounded, coarsely toothed, usually with some hairs on the midrib and veins beneath. Inflorescences cylindric, 5–9 × *c.* 1 cm (excluding the stamens), flowers opening from the top downwards, white or pink. Perianth-segments *c.* 3 mm, greenish or pink. Stamens 4, projecting well beyond

the perianth-segments, filaments dilated and flattened above, pink or white. Carpel usually 1; style projecting well beyond the perianth. *Japan, E former USSR.* H2. Summer.

White-flowered variants have been called var. **albiflora** Makino.

S. hokusanensis Makino. Illustration: Takeda, Alpine flora of Japan in colour, pl. 41 (1959); Kitamura et al., Coloured illustrations of herbaceous plants of Japan **1**: pl. 30 (1977). Very similar, but stamens variable in number (to 12), filaments magenta. *Japan.* H5. Summer.

3. S. officinalis Linnaeus. Illustration: Bonnier, Flore complète **4**: pl. 183 (1914); Ary & Gregory, The Oxford book of wild flowers, pl. 4 (1980); Polunin, Flowers of Europe, pl. 45 (1969); Garrard & Streeter, Wild flowers of the British Isles, pl. 39 (1983).

Herb to 1 m or more, hairless or rarely sparsely hairy below. Leaflets 7–15, very variable in shape, ovate, ovate-oblong to linear-oblong, base usually truncate or somewhat cordate (abruptly tapered when leaflets long and narrow), apex rounded to bluntly acute, coarsely toothed. Inflorescences ovoid to broadly cylindric, 1.5–5 cm × 5–12 mm, opening from the top downwards, dark red. Perianth-segments *c.* 3 mm. Stamens 4, not or scarcely projecting beyond the perianth, filaments thread-like. Carpel usually 1; style scarcely projecting beyond the perianth. *North temperate areas.* H1. Summer.

4. S. minor Scopoli (*Poterium sanguisorba* Linnaeus). Illustration: Bonnier, Flore complète **4**: pl. 183 (1914); Garrard & Streeter, Wild flowers of the British Isles, pl. 39 (1983).

Herb smelling of cucumber when crushed. Stems to 50 cm or more, with long, wavy hairs below. Lower leaves with up to 25 ovate to almost circular leaflets, shortly stalked, base and apex rounded, deeply toothed, the terminal tooth smaller than those adjacent; upper leaves with fewer, narrower leaflets. Inflorescences oblong to spherical, greenish or tinged with purple, female flowers above, bisexual flowers below. Perianth-segments *c.* 3 mm. Stamens 20–30 in the lower bisexual flowers, fewer in those above, projecting beyond the perianth. Carpels usually 2, styles projecting beyond the perianth or almost so. *Europe, N Africa, W & C Asia.* H1. Summer.

28. SARCOPOTERIUM Spach
J. Cullen
Mound-forming shrub to 75 cm, the ends of the branches persisting as spines, the outer bark often silvery, peeling in strips to reveal the brown under-bark. Leaves pinnate with 9–25 very small, oblong to ovate, entire to 3–5-toothed leaflets, 1–8 mm, downy above, whitish hairy beneath, margins turned under. Flowers in leaf-opposed or terminal spikes, unisexual, spikes often with male flowers above, female below. Perianth-segments 4, greenish and often white-margined, *c.* 2 mm. Stamens numerous. Carpels 2, free, enclosed in the perigynous zone. Fruit berry-like, the perigynous zone becoming fleshy, 3–5 mm across, red to yellowish brown.

A genus of a single species from the Mediterranean area where it is often a dominant shrub in certain communities (phrygana). It is not very frequently grown and is propagated by seed.

1. S. spinosum (Linnaeus) Spach (*Poterium spinosum* Linnaeus). Illustration: Polunin & Huxley, Flowers of the Mediterranean, pl. 48 (1965); Polunin, Flowers of Europe, pl. 46 (1969); Pignatti, Flora d'Italia **1**: 568 (1982); *Il Giardino Fiorito* (June): 19 (1987).

Mediterranean area (Sardinia & Tunisia eastwards). H5. Spring.

29. MARCETELLA Sventenius
S.G. Knees
Dioecious trees or shrubs, with more or less erect branches. Leaves alternate, pinnate with a terminal leaflet; leaflets toothed. Flowers very small in tight racemes; sepals 5; petals absent; stamens 25–30, on male flowers; carpels 5. Fruit a dry samara.

A genus of 2 species from Macronesia, only one is sufficiently ornamental to be cultivated regularly.

1. M. moquiniana (Webb & Berthelot) Sventenius (*Bencomia moquiniana* Webb & Berthelot). Illustration: Webb & Berthelot, Histoire naturelle des Iles Canaries **3**: pl. 39 (1844); Bramwell & Bramwell, Wild flowers of the Canary Islands, pl. 163 (1974); Kunkel & Kunkel, Flora de Gran Canaria **1**: pl. 16 (1971); Kunkel Die Kanarischen Inseln und ihre Pflanzenwelt, 170 (1987).

Small tree or shrub to 4 m. Leaves 15–20 cm, leaflets 2–2.5 cm, toothed, glaucous green above, pale beneath; new leaves reddish, glossy. Sepals 5, concave, 1–2 mm; female flower spikes reddish purple; male flower spikes yellowish green. Fruit a dry samara; wings unequal. *Canary Islands (Gran Canaria, Tenerife).* H5–G1. Summer–autumn.

30. MARGYRICARPUS Ruiz & Pavon
S.G. Knees

Creeping or prostrate, evergreen dwarf shrub; branches ascending towards apex. Stems straw-coloured, internodes covered for most of their distance by clasping, papery stipules. Margins of stipules with white silky hairs. Leaves 1.4–1.8 cm × 8–12 mm, broadly ovate, pinnate with 3–5 pairs of glossy, linear lobes; margins strongly curved downwards and inwards, apex of lobes shortly pointed. Flowers solitary, stalkless, borne in leaf axils, but not on current year's growth. Sepals 5, persistent; corolla absent; stamens 1–3; carpel 1. Fruit 5–6 × 7–9 mm, an almost spherical, soft white berry, often tinged pale pink. Seeds ovate, *c.* 4 mm, reddish brown.

A genus of just one species, from the Andes. The evergreen foliage and pearly berries are the most attractive features of this plant, which can be grown in poor or sandy soils, provided drainage is good. Plants are easily raised from seed, or from cuttings rooted in late summer. Alternatively, plants can be raised from layers.

1. M. pinnatus (Lamarck) O. Kuntze (*M. setosus* Ruiz & Pavon; *Empetrum pinnatum* Lamarck). Illustration: Ruiz & Pavon, Flora Peruviana, t. 8 (1798–1802); Hunt (ed.), Marshall Cavendish encyclopedia of gardening, 1198 (1968).

South America (Colombia to S Chile, S Brazil, Uruguay to Argentina). H2. Summer.

31. ACAENA Linnaeus
P.F. Yeo

Perennial evergreen herbs or undershrubs. Stems usually prostrate, often rooting at nodes. Leaves alternate, pinnate with a terminal leaflet; stipules forming a sheath, often with an entire or divided leafy lobe on either side; leaflets usually toothed, asymmetric at the base. Plants bisexual or female and bisexual. Flowers borne in stalked spikes or dense ovoid or spherical heads. Cleistogamous flowers sometimes present in axils at base of flowering-stem. Perigynous zone hollow, almost closed at the mouth, usually bearing barbed spines.

Sepals 3–6, usually 4. Petals absent. Stamens 2–7, with dorsifixed anthers. Carpels 1 or 2, occasionally to 5, concealed in the receptacle except for the curved feathery stigmas. Fruit an achene, dispersed within the perigynous zone.

A genus of about 100 species in the south temperate area, California and the Hawaiian Islands. The natural variation pattern presents great difficulty for classification, especially in South America. The treatment here might not be sustainable if many more introductions from that region were made. The species are useful for ground cover and present a diverse array of leaf-form and leaf-colour; the flowers are inconspicuous but sometimes the spines are decorative. Propagation is by cuttings, division or seed. Hybridisation is frequent in gardens.

Literature: Bitter, G., Die Gattung Acaena, *Bibliotheca Botanica* **17**(74) (1910–1911); Bitter, G., Weitere Untersuchungen Uber die Gattung Acaena, *Feddes Repertorium* **10**: 489–501 (1912); Dawson, J.W., Natural Acaena hybrids in the vicinity of Wellington, *Transactions of the Royal Society of New Zealand* **88**: 13–27 (1960); Grondona, E., Las especies argentinas del genero 'Acaena' ('Rosaceae'), *Darwiniana* **13**: 308–42 (1964); Walton, D.W. & Greene, S.W., The South Georgian species of Acaena and their probable hybrid, *British Antarctic Survey Bulletin* **25**: 29–44 (1971); Yeo, P.F., Acaena (pp. 51–55) & The species of Acaena with spherical heads cultivated and naturalized in the British Isles (pp 193–221), in P.S. Green (ed.), Plants: Wild and Cultivated (1973); Yeo, P.F., Acaenas, *The Garden* **107**: 326–8 (1982).

Leaves. Feathery: **1**. Bluish glaucous and densely hairy: **12**.
Leaflet colour. Silvery on both sides: **3**. Silvery beneath and green above: **2, 9**. Green above, tinged with brown: **7, 11, 13**. Bright green above, not silvery beneath: **1, 4, 8, 10**. Bluish or greenish glaucous above: **4, 5, 6, 12, 14, 15**.
Leaflet length. In most leaves not more than 11 mm: **4, 6, 7, 11, 13, 15**. In some leaves more than 11 mm: **1, 3, 4, 5, 8, 9, 10, 12**.
Inflorescences. Hidden beneath leaves: **15**.
Spines on fruiting receptacle. All over surface: **1, 2, 3**. At apex only: **4, 5, 6, 8, 9, 10, 11, 12, 13, 14, 15**. Absent: **7, 14**. Not rigid: **13, 14**.

1a. Fruiting receptacle with spines not confined to apex   2
 b. Fruiting receptacle with spines confined to apex, occasionally absent   4
2a. Leaves feathery   **1. myriophylla**
 b. Leaves not feathery   3
3a. Leaflets silvery silky on both sides, with 8–24 teeth   **3. splendens**
 b. Leaflets silvery silky beneath, green above though with some silky hairs, with 2–7 teeth   **2. sericea**
4a. Leaflets near tip of leaf at least 1.3 times as long as broad, often 1.5–2.5 times as long as broad   5
 b. Leaflets near tip of leaf at most 1.5 times as long as broad   8
5a. Leaflets near tip of leaf usually with 17 or more teeth   6
 b. Leaflets near tip of leaf with not more than 13 teeth   7
6a. Leaflets not leathery, green beneath; spines on fruiting receptacle 2   **8. ovalifolia**
 b. Leaflets leathery, silvery silky beneath; spines on fruiting receptacle 4   **9. argentea**
7a. Leaflets bright glossy green, those near tip of leaf 2–2.5 times as long as broad   **10. novaezelandiae**
 b. Leaflets near tip of leaf light matt green, edged and veined with brown, those near base and stems, brown; leaflets near tip of leaf not more than twice as long as broad   **11. anserinifolia**
8a. Leaves not glaucous above   9
 b. Leaves glaucous above   11
9a. Largest leaves 3 cm; spines of fruiting receptacle 6–13 mm, soft, unbarbed, bright red   **13. microphylla**
 b. Largest leaves 5–9.5 cm; spines of fruiting receptacle not as above, sometimes absent   10
10a. Leaflets cut no more than one third of the way to the midrib into 7–11 teeth; spines of fruiting receptacle barbed at tip   **4. magellanica**, see subsp. **laevigata**
 b. Leaflets cut halfway to the midrib into 5–8 teeth; spines of fruiting receptacle not barbed, sometimes absent   **7. glabra**
11a. Leaflets 7–9   **12. caesiiglauca**
 b. Leaflets 11 or more in the majority of leaves   12
12a. Leaflets near tip of leaf mostly more than 10 mm   **5. affinis**
 b. Leaflets near tip of leaf mostly less than 10 mm   13

13a. Leaf-sheaths with an entire leafy lobe on each side or none
6. saccaticupula

b. Leafy lobes of at least some of the leaf-sheaths with 2–6 teeth   14

14a. Inflorescences hidden beneath leaves
15. buchananii

b. Inflorescences not hidden beneath leaves   15

15a. Leaflets near tip of leaf longer than wide; inflorescences 7–11 mm
4. magellanica, see subsp. **magellanica**

b. Leaflets near tip of leaf as broad as or broader than long; inflorescences less than 7 mm   **14. inermis**

1. A. myriophylla Lindley (*A. hieronymi* Kuntze). Figure 105(1), p. 239. Illustration: *Bibliotheca Botanica* **17**(74): 74, 77, t. 5, 7 (1910); *Darwiniana* **13**: 255–7 (1964); Cabrera, Flora de la provincia de Buenos Aires **3**: 392 (1967).

Perennial silky-hairy herb with trailing leafy stems to 75 cm. Leaves to 12 × 3 cm, stalk very short, with a short sheath, its lobes either absent or looking like the leaflets; leaflets 13–23, sometimes more, pinnatisect into linear-lanceolate acute segments. Flowering-stems 20–30 cm, leafy. Flowers in dense cylindric spikes, becoming interrupted below in fruit. Stamens longer than sepals; filaments white; anthers dark red. Stigmas green, longer than sepals. Fruiting perigynous zone with a dense covering of slender straw-yellow spines. *C & S Argentina.* H4. Summer.

2. A. sericea Jacquin filius. Figure 105(2), p. 239. Illustration: *Darwiniana* **13**: 333–5 (1964); Moore, Flora of Tierra del Fuego, 134 (1983).

Dwarf shrub with ascending stems, terminated by clusters of erect leaves, not rooting. Leaves to 11 × 2 cm, covered in silky hairs; stalk 0.7–1 times as long as blade, with a long sheath; leaflets 7–11, bright green above, whitish beneath, those near tip of leaf mostly 6–15 × 2.5–6 mm, one and a third to 3 times as long as wide, obovate to wedge-shaped, with 2–7 teeth. Flowering-stems to 30 cm, usually with one leaf near the middle subtending a few flowers. Most flowers in 1–3 dense ovoid or spherical heads. Stamens dark red, shorter than sepals. Stigmas flesh-pink, becoming dark red, shorter than sepals. Fruiting perigynous zone with short, conical spines throughout. *Patagonia.* H1. Summer.

3. A. splendens Hooker & Arnott. Illustration: *Darwiniana* **13**: 338 (1964).

Dwarf shrub with ascending stems, terminated by clusters of erect leaves, not rooting, forming a cushion. Leaves 4–22 cm, covered in silky hairs; stalk one third to two-thirds as long as blade, with a long sheath that usually lacks leafy lobes; leaflets usually 7–11, silvery white on both surfaces, those near tip of leaf 1.5–3.5 cm, 2.25–4.4 times as long as broad, elliptic or obovate, with 8–24 shallow teeth. Flowering-stems *c.* 40 cm. Flowers in a spherical head, or solitary, or clustered along the stalk. Stamens shorter than sepals; anthers purple or blackish. Fruiting perigynous zone with barbed spines throughout. *Andes of Chile & Argentina.* H3. Summer.

4. A. magellanica (Lamarck) Vahl.

Dwarf shrub with short or long trailing stems; rosettes in leaf-axils. Leaves to 9.5 cm, stalk short, sheath with or without an entire or 2-toothed leafy lobe. Leaflets 9–15, with 3–11 teeth. Flowering-stems *c.* 17 cm in fruit. Flowers in dense spherical or slightly lengthening heads; much smaller subsidiary heads sometimes present. Stamens and stigma dark red, as long as or slightly longer than sepals. Fruiting perigynous zone with 2–4 barbed spines at the apex; spines 2–6.5 mm.

Subsp. **magellanica** (*A. magellanica* subsp. *venulosa* (Grisebach) Bitter; *A. glaucophylla* Bitter). Figure 105(4), p. 239. Illustration: Green, Plants: wild and cultivated, 197 (1973); *Bibliotheca Botanica* **17**(74): 155, 156, t. 16a (1910). Leaves 2–4.5 cm. Leaflets hairless, light glaucous grey-green above, paler beneath, those near tip of leaf mostly 3.5–8 mm, fan-shaped, cut one third to halfway towards midrib into 3–8 blunt teeth. Flower-stems grey-green, flushed with purplish red. Main inflorescences 1.8–2.4 cm in fruit. *S Patagonia (Chile, Argentina).* H1. Summer.

Subsp. **laevigata** (Aiton filius) Bitter (*A. adscendens* Vahl; *A. laevigata* Aiton filius). Figure 105(3), p. 239. Illustration: *Bibliotheca Botanica* **17**(74): 171 (1910); Green, Plants: wild and cultivated, 199 (1973). Leaves mostly 3–9.5 cm, very sparingly glandular-hairy. Leaflets thickish, upper surface dark green, with a slight removable bloom, hairless; lower surface glaucous green with conspicuously netted veins, hairless or nearly so, those near tip of leaf 5.5–16 mm, oblong or nearly square, cut one quarter to one third of the

way towards midrib into 7–9, sometimes 11 teeth. Flower-stems green, sometimes flushed with red. Main inflorescences 2.1–2.4 cm in fruit. *Falkland Islands.* H1. Summer.

5. A. affinis J.D. Hooker (*A. distichophylla* Bitter; *A. adscendens* misapplied). Figure 105(5), p. 239. Illustration: Hooker, Flora Antarctica **2**: t. 96 (1845); *Bibliotheca Botanica* **17**(74): 178, 209, t. 20 (1910); Vallentin & Cotton, Illustrations of the flowering plants and ferns of the Falkland Islands, t. 17 (1921); Green, Plants: wild and cultivated, 201 (1973).

Stems trailing *c.* 1 m, herbaceous parts bright pink. Leaves mostly 6–12 cm, sometimes with subsidiary leaflets; stalk short; sheath *c.* 1.3 cm, with 2 entire or toothed leafy lobes. Leaflets 9–13, light bluish or greyish glaucous green and hairless above, or hairy on margins, grey-green with conspicuous veins and finely hairy beneath, those near tip of leaf mostly 9–18 mm, 1.5 times as long as broad, oblong or almost circular, with the lower edge running down the midrib for up to 4 mm, with 9–14 teeth. Flowering-stems red, 11–21 cm in fruit. Inflorescences spherical, 2–2.5 cm in fruit. Stamens and stigmas dark red, longer than the sepals. Spines 4, to 8 mm, barbed; additional minute spines sometimes present. *Magellan area of S America, Antarctic Islands.* H3–4. Summer.

6. A. saccaticupula Bitter. Figure 106(1), p. 240. Illustration: Green, Plants: wild and cultivated, 206 (1973); *The Garden* **107**: 327 (1982).

Stems trailing to 50 cm, coppery red. Leaves 2–7 cm; stalk *c.* one quarter as long as blade, sheath to 6 mm, with 1–6 leafy lobes on each side. Leaflets 11–13, upper surface glaucous grey or grey-blue, tinged with purple, red or brown, margins finely hairy, lower surface more strongly glaucous, sometimes with red veins; leaflets near leaf-tip 4.5–11 × 3.5–9.5 mm, fan-shaped or broadly oblong, with 5–9 teeth. Flowering-stems brownish red or deep red, 13–19 cm in fruit. Inflorescences spherical, 1.5–1.8 cm in fruit. Stamens slightly longer than sepals, filaments white, anthers red. Stigmas about as long as sepals, blackish red. Spines 4, 3–4 mm, barbed, thickened at base, pinkish red. *New Zealand.* H3. Summer.

This plant remained unidentified for some time after its introduction and was then named 'Blue Haze'. Illustration:

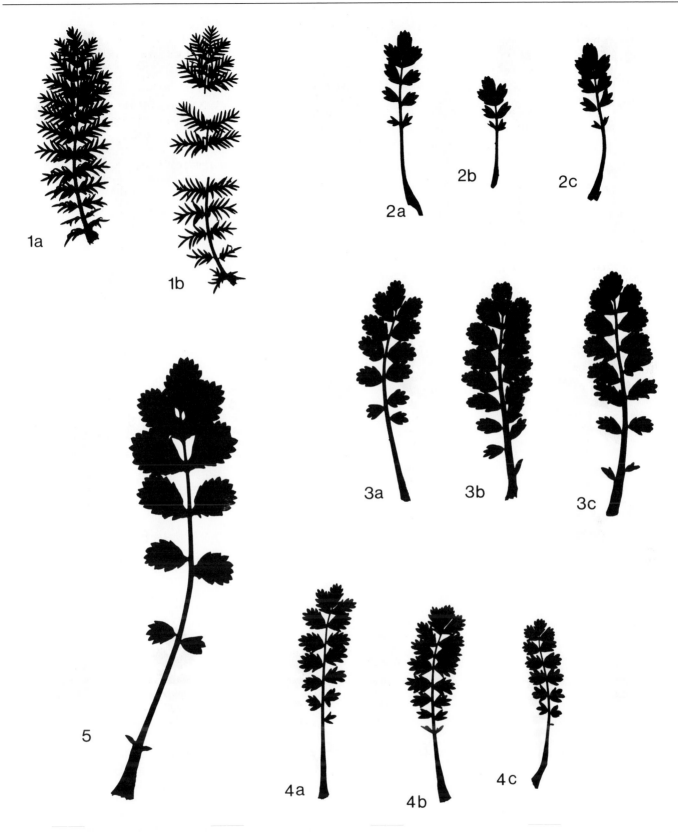

Figure 105. Leaf silhouettes of *Acaena* species (× 0.75). 1a, b,
A. myriophylla. 2, *A. sericea* (a, b, upper, c, under). 3a, b, c, *A. magellanica*
subsp. *laevigata*. 4a, b, c, *A. magellanica* subsp. *magellanica*. 5, *A. affinis*.

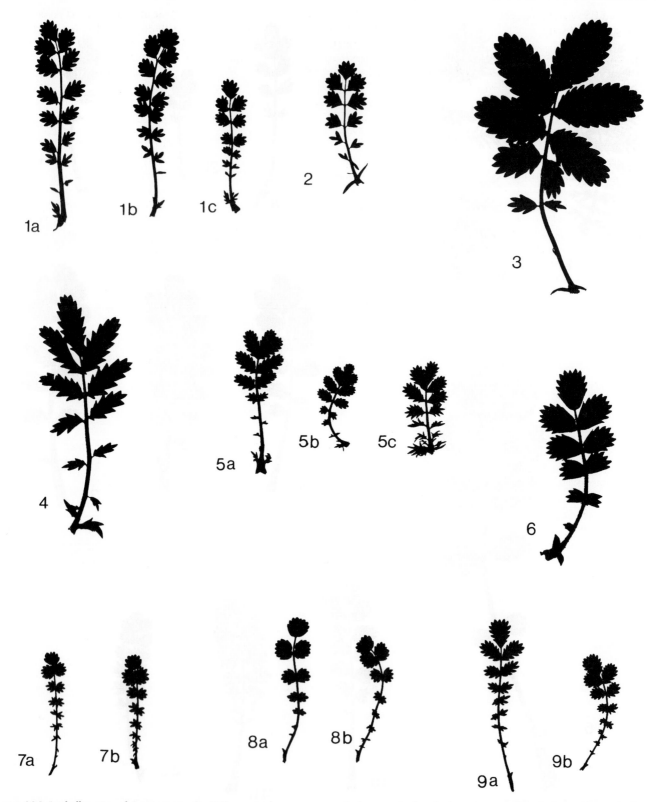

Figure 106. Leaf silhouettes of *Acaena* species (× 0.75 except where noted). 1a, b, c, *A. saccaticupula*. 2, *A. glabra*. 3, *A. ovalifolia*. 4, *A. novaezelandiae*. 5a, b, c, *A. anserinifolia*. 6, *A. caesiiglauca*. 7a, b, *A. microphylla* (× 1). 8a, b, *A. inermis*. 9a, b, *A. buchananii*.

Brickell (ed.), RHS A–Z encyclopedia of garden plants, 61 (2003).

7. A. glabra Buchanan. Figure 106(2), p. 240. Illustration: *Bibliotheca Botanica* **17** (74): 279, t. 34 (1911); Salmon, Field guide to alpine plants of New Zealand, 127 (1968); Green, Plants: wild and cultivated, 206 (1973).

Stems trailing to 30 cm; greenish straw-coloured. Leaves 3–5 cm; stalk not longer than interval between first two pairs of leaflets; sheath to 6 mm, usually with two entire leafy lobes. Leaflets 9–11, shortly stalked, rather thick and firm, hairless, upper surface slightly glossy, yellowish green, brown-tinged at margins, lower surface glaucous grey-green; leaflets near tip of leaf 5.5–11 mm, 1.3–1.5 times as long as broad, broadly obovate with wedge-shaped base, with 5–8 teeth. Flowering-stems brownish, 6–10 cm in fruit. Inflorescences spherical, *c.* 11 mm in fruit. Stamens white, just exceeding the sepals. Stigma white, shorter than sepals. Spines 4, to 2 mm or sometimes vestigial, not barbed. *New Zealand (Southern Alps).* H3. Summer.

8. A. ovalifolia Ruiz & Pavon. Figure 106(3), p. 240. Illustration: *Darwiniana* **13**: 230–3 (1964); Cabrera, Flora de la provincia de Buenos Aires **3**: 392 (1967); Green, Plants: wild and cultivated, 209 (1973); *The Garden* **107**: 327 (1982).

Stems trailing to 1 m; green, sometimes red-flushed. Leaves 5–12 mm; stalk to half as long as blade; sheath to 5 mm, leafy lobes with 2–4 teeth. Leaflets 7–9, bright green, finely wrinkled and hairless above, slightly glaucous and finely hairy beneath, silky beneath when young, at least on veins, those near tip of leaf 1.5–3 cm, 1.75–2 times as long as broad, elliptic or oblong, with 12–23 teeth. Flowering-stems green, 6–12 cm in fruit. Inflorescences spherical, 1.8–3 cm in fruit. Stamens and stigmas white, slightly longer than sepals. Spines 2, 8–10 mm, barbed, red. *S America (Magellan area to Colombian Andes).* H3. Summer.

9. A. argentea Ruiz & Pavon. Illustration: *Darwiniana* **13**: 236–8 (1964).

Like *A. ovalifolia* but with leaf-sheaths to 1.8 cm, their leafy lobes sometimes entire, leaflets leathery, green above, white with silky hairs beneath, anthers purple, spines of fruit 4, often unequal. *S Argentina.* H3. Summer.

10. A. novaezelandiae Kirk (*A. sanguisorbae* misapplied; *A. anserinifolia* misapplied). Figure 106(4), p. 240. Illustration: Cheeseman, Illustrations of the New Zealand flora **1**: t. 39 (1914); Salmon, Field guide to alpine plants of New Zealand, 85 (1968); Cochrane et al., Flowers and plants of Victoria, 120 (1973); Green, Plants: wild and cultivated, 211 (1973).

Stems trailing to 1 m; flushed with pinkish red. Leaves 3–10 cm; stalk not longer than interval between first 2 pairs of leaflets; sheath to 5 mm, with leafy-lobes, teeth 1–4 or absent; leaflets usually 9–13, bright glossy green, finely wrinkled, hairless or nearly so above, glaucous green and sparsely to densely silky-hairy beneath, those near tip of leaf 7–22 mm, 2–2.5 times as long as broad, oblong or obovate-lanceolate, with 8–15 teeth. Flowering stem pale green, sometimes flushed with red, to 11 cm, sometimes to 25 cm in fruit. Inflorescences spherical, 1.5–3 cm in fruit. Stamens white, longer than sepals. Stigmas white with a purplish tip, longer than sepals. Spines 4, 6–9.5 mm, barbed, red. *New Zealand, Australia, Tasmania.* H3–4. Summer.

11. A. anserinifolia (J.R. & G. Forster) Druce (*A. sanguisorbae* invalid). Figure 106(5), p. 240. Illustration: *Transactions of the Royal Society of New Zealand* **88**: 14, 15 (1960); Green, Plants: wild and cultivated, 212 (1973); *The Garden* **107**: 328 (1982).

Stems trailing to 30 cm, brown, finely hairy. Leaves 2–5 cm; stalk shorter than interval between first 2 pairs of leaflets, sheath to 3 mm with a pinnatisect leafy lobe on each side; leaflets 9–13; upper surface light matt green, more or less flushed with brown, sparsely silky-hairy, lower surface slightly glaucous green, sometimes tinged with purple, sparsely to densely silky-hairy; leaflets near tip of leaf 3–8 mm, mostly 1.3–2 times as long as broad, oblong or oblong-obovate, the lower edge often running down midrib for *c.* 1 mm, with 10–13 brush-tipped teeth. Flowering-stems brown, silky-hairy, 3–7.5 cm in fruit. Inflorescences spherical, 1.2–1.6 cm in fruit. Stamens and stigmas white, longer than sepals. Spines 4, 3.5–6 mm, barbed, red. *New Zealand.* H4. Summer.

Hybrids between this species, *A. novaezelandiae* and *A. inermis* are probably the most common hybrids in gardens.

12. A. caesiiglauca (Bitter) Bergmans. Figures 99(9, 21), p. 183 & 106(6), p. 240. Illustration: *Bibliotheca Botanica* **17** (74): t. 27 (1911); Green, Plants: wild and cultivated, 215 (1973); *The Garden* **107**: 327 (1982).

Stems trailing to 60 cm, pale brown, densely woolly. Leaves mostly 4–8 cm; stalk 1–2 times as long as interval between first 2 pairs of leaflets; sheath to 3 mm, lobes with 1–3 teeth or entire; leaflets 7–9; upper surface blue-grey glaucous, with sparse fine adpressed hairs; lower surface slightly paler and more glaucous, silky-hairy; both surfaces slightly purple-tinged in age; leaflets near tip of leaf 6–14 mm, 1.3–1.5 times as long as broad, oblong or broadly obovate, with 6–10 brush-tipped teeth. Flowering-stems pale brown, woolly, 10–14 cm in fruit. Flower-heads spherical, to 2.3 cm in fruit. Stamens white, much longer than sepals. Stigmas white, sometimes tinged with pink. Spines 4, 4.5–7 mm, barbed, olive-brown. *New Zealand.* H3. Summer.

13. A. microphylla J.D. Hooker. Figure 106(7), p. 240. Illustration: Wooster, Alpine plants, 29 (1872); Salmon, New Zealand flowers and plants in colour, edn 2, 128 (1976); Salmon, Field guide to alpine plants of New Zealand, 85 (1968); Green, Plants: wild and cultivated, 216 (1973); *The Garden* **107**: 328 (1982).

Stems trailing to 30 cm; light brown or brownish green, quickly rooting. Leaves 1–3 cm; stalk very short; sheath to 2 mm, with 2 entire leafy lobes. Leaflets 11–13, yellowish green above, strongly flushed with brown at margins, becoming entirely dull purplish brown, hairless, similar beneath but slightly glaucous and with fine adpressed hairs on the veins, those at tip of leaf 2–4.5 mm, with 3–7 brush-tipped teeth. Flowering stems brown, 1–4 cm in fruit. Inflorescences spherical, to 3 cm in fruit. Stamens and stigmas longer than sepals; stamens white; stigmas white with pink midline. Spines 4, to 1.3 cm, soft, thick, not barbed, bright red. *New Zealand (North Island).* H3. Summer.

14. A. inermis J.D. Hooker (*A. microphylla* misapplied). Figure 106(8), p. 240. Illustration: Salmon, Field guide to alpine plants of New Zealand, 84 (1968); Green, Plants: wild and cultivated, 217 (1973); *The Garden* **107**: 328 (1982).

Stems trailing to 30 cm; flesh-coloured to brownish glaucous, quickly rooting. Leaves 2–4 cm, sometimes to 6 cm; stalk one eighth to one fifth as long as leaf; sheath to 2.5 mm, with an entire leafy lobe on each side. Leaflets 11–13; upper surface dull bluish grey or brownish grey, slightly tinged with green (appearing marbled under a lens), becoming orange to straw-coloured in age, hairless; lower surface pale grey-green glaucous to purplish glaucous, with fine adpressed hairs on veins; leaflets at tip of leaf usually 2–5 mm, fan-shaped, square or transversely oblong, with 5–10 brush-tipped teeth. Flowering-stems brownish, 1–6 cm in fruit. Inflorescences spherical, to 1.6 cm in fruit if spiny. Stamens white, much longer than sepals. Stigmas white, about as long as sepals. Spines absent or to 6 mm, often unequal, soft, thick, not barbed, bright red. *New Zealand (South Island).* H3. Summer.

15. A. buchananii J.D. Hooker. Figure 106(9), p. 240. Illustration: *Bibliotheca Botanica* **17**(74): 290, t. 27, 29 (1911); Green, Plants: wild and cultivated, 219 (1973).

Stems trailing to 30 cm; pale green to pale brown. Leaves 1.5–5.5 cm; stalk one fifth to one quarter as long as blade; sheath 3–8 mm, with 2 entire leafy lobes. Leaflets 11–17, pale glaucous green above, hairless or with hairs near edge, about the same colour and hairy beneath, those at tip of leaf 3–9 mm, 1.25–1.5 times as long as broad, oblong or broadly ovate, with 6–12 brush-tipped teeth. Flowering-stems pale green, 3–11 mm. Inflorescences hidden beneath the leaves, shorter than broad, *c.* 3 cm wide in fruit, with conspicuous bracts. Stamens 6–7 mm, white. Stigmas much longer than sepals, white, becoming pink. Spines 4, 7–13 mm, greenish yellow, with a tuft of reflexed hairs at the tip. *New Zealand (South Island).* H3. Summer.

32. POLYLEPIS Ruiz & Pavon
M.F. Gardner
Small trees or shrubs; branches twisted, covered with scars of fallen leaves. Leaves alternate, with 3 leaflets or pinnate; leaflets in few pairs, leathery; leaf-stalks broadly membranous, sheathing at base. Racemes slender, flowers mainly pendent, bracts present, sepals 3–5, persistent, forming a calyx with a constricted throat, usually with 3 or 4 wings; petals absent; stamens numerous, inserted in calyx throat. Fruit a leathery achene enclosed in a hardened angular spiny or winged calyx tube.

A genus of 15 species from the Andes. Only occasionally cultivated but will thrive in a well-drained soil in a sunny position. Propagate from seed.

1. P. australis Bitter. Illustration: *Botanische Jahrbücher* **14**: 620 (1911); Krüssmann, Manual of cultivated broad-leaved trees & shrubs **2**: 423 (1986).

A broad, deciduous shrub to 3 m; branches contorted in a zig-zag manner, with long lengths of brown bark flaking away, shoots with long internodes. Leaves unevenly pinnate: the first leaf has only 1 leaflet, the second 5 leaflets and the third 5–7, with only the latter 2 having axillary shoots; leaflets bluish green, 2–4 × 1 cm, elliptic, apex notched, base rounded, margin scalloped. Fruit 3-winged. *Argentina.* H4.

33. ADENOSTOMA Hooker & Arnott
J. Cullen
Evergreen, resinous shrubs. Leaves needle-like, borne in bundles or singly and alternately; stipules present or absent. Flowers small, bisexual, numerous, in erect or spreading terminal panicles. Sepals 5, translucent, erect. Petals 5, white, longer than the sepals. Stamens 10–15 borne at the mouth of the perigynous zone above some nectar-secreting glands. Perigynous zone cylindric, somewhat tapering to the base, 10-grooved. Carpel 1, containing a single ovule and with an oblique or lateral style. Fruit an achene surrounded by the persistent perigynous zone, which hardens and becomes contracted towards the apex, the whole ellipsoid, black and hard.

A genus of 2 species from USA (California) and adjacent Mexico. Uncommon in gardens; propagated by seed or cuttings taken in spring.

1a. Leaves mostly borne in bundles, each bundle borne on an expanded base, which bears 2 small stipules; flower subtended by a bract and 2 bracteoles, each of which is green and 3-lobed with a narrowly triangular central lobe and much smaller lateral lobes
 1. fasciculatum
 b. Leaves all borne singly, without stipules; flower subtended by a single green, simple bract with, above it, several broader, mostly translucent bracts **2. sparsifolium**

1. A. fasciculatum Hooker & Arnott. Illustration: Baillon, Natural history of

plants, English translation, **1**: 373 (1871); Ornduff, Introduction to California plant life, pl. 4 (1974); Munz, Flora of southern California, 790 (1974); Wiggins, Flora of Baja California, 790 (1980).

Shrub to 3.5 m (often with a swollen, tuber-like base in the wild). Leaves mostly borne in bundles, each bundle borne on a hairy, swollen base which bears 2 hairy, narrowly triangular stipules. Leaves needle-like, somewhat swollen above, 4–12 mm. Flowers subtended by a bract and 2 bracteoles, each 3-lobed with a narrowly triangular terminal lobe and shorter lateral lobes. Petals *c.* 1.5 mm. Perigynous zone hairy or hairless. *USA (California), Mexico (Baja California).* H5–G1. Summer.

Occurs in the wild in areas subject to fires, which it survives by means of its woody tuber.

2. A. sparsifolium Torrey.
Erect, tree-like shrub to 6 m. Shoots and leaves with scattered, scale-like, brown, resin-secreting glands. Leaves borne singly, alternate, 6–15 mm, narrowly linear to thread-like, without stipules. Flowers subtended by a single, simple, narrowly triangular, green bract and several broader, mostly translucent bracts borne above it. Petals 2 mm or more. Perigynous zone with hairs in the grooves. *USA (S California), Mexico (Baja California).* H5–G1. Summer.

34. CERCOCARPUS Kunth
J. Cullen
Shrubs or small trees. Leaves more or less evergreen, alternate or clustered on short shoots, simple, entire or toothed, shortly stalked. Stipules rather small. Flowers solitary or in groups of 3 (rarely these groups somewhat aggregated), axillary or terminal on short lateral shoots. Perigynous zone hairy, narrow and tubular, broadening abruptly above into a 5-lobed perianth that ultimately opens very widely, the lobes and upper part of the widened tube reflexed. Petals absent. Stamens 10–25 (in our species), inserted in 2 or more series on the upper (reflexed) part of the expanded perianth; anthers hairy or hairless. Ovary superior, carpel 1, borne in the scarcely swollen base of the perigynous zone; ovule solitary; style terminal. Fruit an achene surmounted by the long, persistent, plumed-hairy style, surrounded at the base by the base of the perigynous zone, which becomes dry and brownish.

A genus of about 6 species from the south and west USA and adjacent Mexico. They are infrequently grown but can be quite ornamental in flower and fruit. They require a sunny, well-drained position and can be propagated by seed or by cuttings.

Literature: Martin, F.L., A revision of Cercocarpus, *Brittonia* **7**: 91–111 (1950).

1a. Anthers hairless; leaves with entire, downwardly rolled margins, acute at the apex   **2. ledifolius**
  b. Anthers hairy; leaves usually toothed, at least towards the rounded apex, margins not rolled downwards   **1. montanus**

1. C. montanus Rafinesque.

Shrub or small tree to 8 m; bark grey or brown, fissured on old trunks. Leaves firm, dark green above, paler beneath, variable in shape, mostly oblanceolate or elliptic, usually conspicuously tapered to the base, usually with at least a few teeth towards the rounded apex, 1–7 cm × 3–35 mm, veins conspicuous beneath, forming a pattern of near-squares, which are more densely hairy between than on the veins. Stipules to 8 mm. Flowers 3–6 mm wide. Perigynous zone 3–9 mm, densely hairy. Stamens 20–25, anthers hairy. Style 2–9 cm in fruit, densely hairy. *W USA, adjacent Mexico (Baja California)*. H3. Spring–summer.

Var. **montanus**. Generally a shrub to 4 m; leaves with rather coarse, ovate teeth which are not apiculate. *Rocky Mts.*

Var. **glaber** (Watson) Martin (*C. betuloides* Torrey & Gray; *C. betulaefolius* Nuttall). Illustration: *Hooker's Icones Plantarum* **4**: t. 322 (1841); Elias, The complete trees of North America, 559 (1980). Shrub or small tree to 8 m; leaves with few, small, apiculate teeth, or entire. *USA (Oregon, California, Arizona), Mexico (Baja California)*.

Var. **blancheae** (Schneider) Martin. Leaves very broadly elliptic, very leathery, coarsely toothed. *Islands off California.*

The naming of this species is in some doubt, and it is usually found in the literature as *C. betuloides*; however, Martin's revision has been followed here as the most general review of the genus.

2. C. ledifolius Torrey & Gray. Illustration: *Hooker's Icones Plantarum* **4**: t. 324 (1841); Elias, The complete trees of North America, 556 (1980); Krüssmann, Manual of cultivated broad-leaved trees and shrubs **1**: 314 (1986).

Shrub or small tree to 8 m; bark red-brown, furrowed on old trunks, the younger branches whitish. Leaves leathery, narrowly elliptic to linear, 1–4 cm, tapered to the base and the acute apex, margins entire, rolled downwards, ultimately hairless and shining dark green above, shortly brown-hairy beneath. Stipules to 2 mm. Flowers solitary or in pairs, 2–6 mm wide. Perigynous zone 3–6 mm, densely hairy. Stamens 10–25, anthers hairless. Style in fruit to 7 cm, often twisted or curved, densely hairy. *W USA, adjacent Mexico (Baja California)*. H5. Spring–summer.

35. CHAMAEBATIA Bentham
J. Cullen

Small, aromatic shrubs, most parts with dense, short, somewhat crisped, simple, whitish hairs and much longer, stalked glands. Leaves partially evergreen, 3-pinnate with very small ultimate segments. Stipules small, lanceolate or narrowly triangular. Flowers rather few in a flat-topped terminal panicle, each subtended by a bract and 2 alternate bracteoles. Perigynous zone bell-shaped, densely hairy and glandular outside, densely hairy inside towards the base. Sepals 5, spreading, ultimately somewhat reflexed, white-hairy inside, white-hairy and glandular outside. Petals 5, white, spreading. Stamens numerous. Carpel 1, hairy, with a short style. Fruit an achene enclosed in the persistent perigynous zone.

A genus of 2 species from western USA and adjacent Mexico, superficially very similar to *Chamaebatiaria millefolium* (p. 184), differing in its lack of stellate hairs, 3-pinnate leaves, more open, fewer-flowered panicle, somewhat larger flowers and single carpel forming a 1-seeded achene. It requires a sunny position and well-drained soil and can be propagated by seed or semi-ripe hardwood cuttings in summer.

1. C. foliolosa Bentham. Illustration: *Botanical Magazine*, 5171 (1860); Jepson, Flora of California **2**: 213 (1936).

Shrub to 1 m. Bark of older shoots dark, often bloomed. Leaves more or less stalkless. Flowers 1.5–2 cm across. *W USA (California)*. H5–G1. Summer.

36. PURSHIA de Candolle
J. Cullen

Shrubs to 3 m; shoots hairy when young. Leaves deciduous alternate, mostly borne in clusters on short lateral shoots, obovate, long-tapered to the base, rounded to the apex, which is divided into 3 oblong, blunt lobes, slightly hairy above, densely white-hairy beneath, margins rolled under. Stipules small, narrowly triangular. Flowers bisexual, solitary at the ends of the short lateral shoots, stalkless or almost so. Perigynous zone more or less funnel-shaped, white-hairy and glandular. Sepals 5, oblong, blunt, reflexed in flower, yellowish and hairless inside. Petals creamy white, spoon-shaped. Stamens c. 25. Ovary superior. Carpels 1 (rarely 2), style short. Achene(s) ovoid, hairy, c. 1 cm, projecting from the perigynous zone beyond the persistent sepals, tipped by the short, persistent style.

A genus of 2 species from western North America. The single cultivated species requires a well-drained soil in a sunny position, and is propagated by layering.

1. P. tridentata (Pursh) de Candolle. Illustration: *Edwards's Botanical Register* **17**: t. 1446 (1831); Abrams, Illustrated flora of the Pacific states **2**: 452 (1944); Clark, Wild flowers of the Pacific northwest, 254 (1976).

Sprawling shrub to 3 m. Leaves 5–30 mm, deciduous or mostly so. Flowers to 1 cm across. *W USA (Oregon, California to Wyoming & New Mexico)*. H5. Summer.

37. COWANIA D. Don
M.F. Gardner

Evergreen shrubs or small trees. Leaves alternate, lobed or pinnatifid, leathery, dotted with glands on upper surface, densely felted with white hairs beneath, margins rolled inwards; stipules joined to the leaf stalk. Flowers white, pale yellow or rich rose, solitary, very shortly stalked, terminal, on short leafy twigs; perigynous zone short; sepals 5, overlapping; petals 5, obovate, spreading; stamens numerous, inserted at the mouth of the calyx in 2 rows; styles 1–12, stalkless, covered with long hairs, lengthening in fruit; ovary superior. Fruit a one-seeded achene, terminating with a long feathery style.

A genus of 5 species from SW North America to central Mexico, occasionally seen in cultivation. These plants prefer a well-drained, limey soil in a warm sunny, protected position where low winter temperatures occur infrequently. Propagate from seed sown under glass in spring.

1a. Leaves with 5–9 lobes; flowers rich rose   **1. plicata**

b. Leaves with 3–5 lobes, flowers white
   to pale yellow   **2. stansburiana**

1. C. plicata D. Don. Illustration: *Botanical Magazine*, 8889 (1921); *The New Flora and Silva* **5**(1): fig. 21 (1932); RHS dictionary of gardening 564 (1956).

A rigid, well-branched shrub to 1.5 m, with peeling bark; young shoots reddish, very glandular, covered with white woolly hairs that soon fall away. Leaves 8–15 mm, obovate, pinnatisect, with 5–9 lobes. Flowers rich rose, *c.* 3 cm wide, sepals swept back, glandular; petals rounded obovate; anthers yellow; fruiting style *c.* 3 cm. *N Mexico.* H4–5. Summer.

Usually considered to be the most garden-worthy species of the genus.

2. C. stansburiana Torrey. Illustration: McMinn, Manual of the Californian shrubs, 226–7 (1939); Everett, New York Botanical Gardens illustrated encyclopedia of horticulture **2**: 901 (1981); Martin & Hutchins, Summer wildflowers of New Mexico, 117 (1986).

A freely branched, aromatic shrub to 3 m, bark dark and shredding, twigs reddish brown, glandular. Leaves 6–15 mm, obovate, pinnatisect, with 3–5 lobes; leaf-stalks 3–8 mm. Flowers tube-shaped, fragrant; flower-stalks glandular, *c.* 5 mm; sepals broadly obovate, 4–6 mm; forming a funnel-shaped calyx covered with glands; petals white or yellow; styles 5–10, 3–5 cm in fruit. *SW North America to N Mexico.* H4–5. Summer.

Some authorities consider this species to be a variety of *C. mexicana* D. Don.

38. FALLUGIA Endlicher
J. Cullen

Deciduous, upright shrub to 1.5 m; bark ultimately greyish or whitish. Leaves alternate, 8–15 mm, often with short leafy shoots in their axils, pinnate, lobes 3–7, oblong-linear, blunt, variably hairy above, densely white- or brown-hairy beneath between the raised main veins and the downwardly rolled margins. Flowers solitary or in few-flowered racemes terminating the branches, each with a bract and 2 alternate bracteoles, bisexual or female, 2–3 cm across. Perigynous zone shallowly cup-shaped, hairy outside, very densely white-hairy inside. Sepals 5, ovate, with 5 narrow epicalyx-segments alternating with them. Petals 5, white, obovate to almost circular. Stamens numerous. Carpels numerous, hairy, borne on a conical elevation within the

perigynous zone, each with a single ovule and a long style. Fruit a head of hairy achenes, each surmounted by the long, persistent, reddish, feathery-hairy style.

A genus of a single species from southern USA and adjacent Mexico. It has attractive flowers and fruits and requires a well-drained soil in a very sunny position. It is propagated by seed.

1. F. paradoxa (D. Don) Endlicher. Illustration: *Botanical Magazine*, 6660 (1882); Abrams & Ferris, Illustrated flora of the Pacific states **2**: 452 (1944); Lenz & Dourley, California native trees and shrubs, 99 (1981); Martin & Hutchins, Summer wildflowers of New Mexico, 115 & unnumbered colour plate (1986). *S USA & adjacent Mexico.* H5–G1. Summer.

39. DRYAS Linnaeus
J. Cullen

Low shrubs with creeping, branched, woody stems giving rise to tufts or rosettes of leaves. Leaves evergreen, shortly stalked, toothed or scalloped, margins usually rolled under to some extent, usually dark green and conspicuously wrinkled above, densely white hairy beneath with cobwebby hairs, often also with dark purplish or blackish stalked glands and/or long, bristle-like processes, which themselves bear white hairs, sometimes also with stalkless, resin-secreting glands. Stipules lanceolate. Flowers solitary on long stalks that rise above the leaves, sometimes functionally unisexual. Perigynous zone short, cup-like; sepals 7–10, often densely covered with dark stalked glands as well as cobwebby hairs. Petals 7–20, often 8, yellowish or white, forwardly directed or spreading. Stamens numerous, filaments sometimes hairy, deciduous. Ovary superior, carpels numerous. Fruit a head of achenes each with a long, persistent, plumed (silky-hairy) style, surrounded at the base by the persistent perigynous zone and sepals; the styles often twist together spirally while the fruits are ripening.

A genus whose classification is very confused; many authors recognise only a single species (*D. octopetala*), which may be divided into subspecies, varieties, etc., whereas others recognise up to 20 or more species. In this account the treatment proposed by Hultén (reference below) is followed; Hultén recognises 4 species (3 of them cultivated), some of which are further divided.

The plants are easily grown in a sunny position and may be propagated by seed, by cuttings taken in early autumn, or by layering.

Literature: Hultén, E., Studies in the genus Dryas, *Svensk Botanisk Tidskrift* **53**: 507–42 (1959).

1a. Flowers pendent or horizontal at flowering, not fully opening, the yellow petals directed forwards; filaments with sparse long hairs
                          1. drummondii
  b. Flowers erect, opening widely, the white, cream or faintly yellowish petals spreading; filaments hairless **2**
2a. Midrib of the leaves beneath with hairy processes and/or stalked glands
                          2. octopetala
  b. Midrib of the leaves beneath with white hairs only (or rarely almost hairless)   **3. integrifolia**

1. D. drummondii Richardson (*D. octopetala* Linnaeus var. *drummondii* (Richardson) Watson). Illustration: *Botanical Magazine*, 2972 (1830); Porsild, Rocky Mountain wild flowers, 233 (1974); Clark, Wild flowers of the Pacific northwest, 250 (1976).

Leaves elliptic to narrowly obovate, 1–4 cm × 5–15 mm, distinctly toothed, densely white-hairy beneath with hairy processes on the midribs. Flowering-stems 5–25 cm, flowers pendent or horizontal. Sepals covered with dark purple stalked glands. Petals yellow, to 1.5 cm, not opening widely, forwardly directed. Fruit erect. *N America.* H1. Late spring–summer.

2. D. octopetala Linnaeus. Illustration: Hay & Synge, Dictionary of garden plants in colour, pl. 60 (1969); Everard & Morley, Wild flowers of the world, pl. 20 (1970); Rickett, Wild flowers of the United States **6**: pl. 65 (1973); Garrard & Streeter, Wild flowers of the British Isles, pl. 35 (1983).

Leaves oblong to ovate or somewhat obovate, 5–50 × 2–25 mm, clearly toothed or scalloped, densely white-hairy beneath and with either or both stalked glands and hairy processes on the midrib. Flowering-stems to 50 cm. Sepals with dark purple stalked glands. Flowers erect. Petals white, 0.7–1.7 cm, spreading. *N temperate regions.* H1. Late spring–summer.

A variable species. Subsp. **octopetala** has hairy processes on the midribs of the leaves beneath and occurs throughout the range of the species; minor variants of it include forma **argentea** (Blytt) Hultén

(*D. lanata* Stein; *D. octopetala* var. *lanata* of catalogues), which has the leaves densely hairy above; a small, low-growing variant has been called var. **minor** Hooker; var. **pilosa** Babington (*D. babingtoniana* Porsild) has leaves some of which lack hairy processes on the midrib beneath.

Subsp. **hookeriana** (Juzepczuk) Hultén (*D. hookeriana* Juzepczuk). Illustration: Porsild, Rocky Mountain wild flowers, 235 (1974). This has only stalked glands on the midribs of the leaves beneath, and is from western North America.

The hybrid with *D. drummondii* has been named as **D. × suendermannii** Sündermann. It is very like subsp. *octopetala*, with ovate leaves, erect flowers with spreading petals and hairless filaments, but the petals are yellowish in bud, fading gradually to white as the flower matures. It is widely grown.

3. D. integrifolia Vahl (*D. tenella* Pursh; *D. octopetala* var. *integrifolia* (Vahl) Hooker). Illustration: Rickett, Wild flowers of the United States **6**: pl. 65 (1973).

Like *D. octopetala* but smaller and more compact. Leaves often almost entire or with the margins folded under so that the teeth (which are only towards the base of the leaf) are not visible; midribs of the leaves with only white cobwebby hairs (sometimes completely without hairs). *Greenland, N America, extreme eastern part of former USSR.* H1. Summer.

Variable and divided into 2 subspecies by Hultén; it is uncertain which of these is cultivated. Hybrids with *D. octopetala* are known throughout the range of the species, and may well also be cultivated.

40. GEUM Linnaeus
D.A.H. Rae
Perennial herbs with 2 types of leaves. Basal leaves usually unequally pinnate with the terminal leaflet often distinctly larger than the laterals. Stem-leaves much smaller. Flowers solitary or in corymbs on simple or branched stems, bisexual, yellow, orange, white or red. Perigynous zone short. Sepals 5 with 5 smaller lobes between. Petals 5 in a saucer or bell-shaped arrangement. Stamens many. Carpels many on a conical or cylindric receptacle. Fruit a group of achenes with long persistent styles; often feathery or jointed.

A genus of about 40 species with a wide distribution, mostly in temperate and cold regions. They are easily grown in rock gardens or borders in any reasonable soil. Propagated by seed and division, they hybridise easily in the garden giving rise to much confusion in their identification.

Literature: Smedmark, J.E.E., A re-circumscription of Geum, *Botanische Jahrbücher* **126**: 409–17 (2006).

1a. Terminal leaflet of basal leaves less than twice size of lateral leaflets   2
 b. Terminal leaflet of basal leaves at least twice size of lateral leaflets   5
2a. Plants with creeping non-flowering-stems (stolons); flowers solitary
                                1. reptans
 b. Plants without creeping non-flowering-stems; flowers solitary or to 7 per stem   3
3a. Leaflets lobed to at least half their length   4
 b. Leaflets shallowly lobed   **2. elatum**
4a. Flowers yellow; calyx green **3. rossii**
 b. Flowers cream to purple, never bright yellow; calyx often purple
                                4. triflorum
5a. Plants with creeping non-flowering-stems; flowers always solitary
                                6. uniflorum
 b. Plants without creeping non-flowering-stems; flowers solitary or to 7 per stem   6
6a. Petals dark orange-red to buff-pink; flowers nodding   **5. rivale**
 b. Petals yellow, white, cream, bright orange or scarlet; flowers erect   7
7a. Lateral leaflets of basal leaves usually less than 5 mm   **7. parviflorum**
 b. Lateral leaflets of basal leaves more than 5 mm   8
8a. Stems to 30 cm   9
 b. Stems more than 30 cm   10
9a. Basal leaves to 10 cm; terminal leaflet *c.* 2.5 cm across
                                8. montanum
 b. Basal leaves to 20 cm; terminal leaflet to 7.5 cm across
                                12. coccineum
10a. Leaf-stalks shaggy, with hairs to 5 mm   **10. bulgaricum**
 b. Leaves with short hairs or hairless   11
11a. Petals yellow   **9. urbanum**
 b. Petals scarlet, sometimes copper-coloured   **11. chiloense**

1. G. reptans Linnaeus. Illustration: Huxley, Mountain flowers of Europe, pl. 294 (1986); Parish, Flowers in the wild, 129 (1983); Brickell (ed.), RHS A–Z encyclopedia of garden plants, 479 (2003); Aeschimann et al., Flora alpina **1**: 759 (2004).

Stems to 20 cm, also with creeping non-flowering-stems to 40 cm. Basal leaves pinnate to 15 cm. Terminal leaflet only slightly larger than laterals, 2.5 cm across, heart- to kidney-shaped, deeply irregularly lobed. Lateral leaflets in 4–7 pairs, usually with 3–5 deeply cut wedge-shaped lobes. Stem-leaves few, slender to 2 cm, 3–5 deeply divided lobes. Flowers erect, solitary, *c.* 3 cm across. Petals yellow, scarcely longer than sepals. *European Alps.* H3. Summer.

G. × rhaeticum Brugger; *G. montanum × G. reptans*. Stems to 20 cm, solitary to numerous. Basal leaves pinnate to 25 cm. Terminal leaflet only slightly larger than laterals, heart to wedge-shaped, unevenly lobed. Lateral leaflets in 3–7 pairs, entire or shallowly 3-lobed. Stems-leaves few, small, slender, shallowly lobed. Flowers erect, *c.* 2.5 cm across. Petals golden yellow. *European Alps.* H3. Summer.

2. G. elatum Wallich. Illustration: *Botanical Magazine*, 6568 (1881); Polunin & Stainton, Flowers of the Himalaya, pl. 36, 38 (1985).

Stems slender and branched, to 35 cm. Basal leaves pinnate to 20 cm. Terminal leaflet usually smaller than largest lateral and similarly shaped, roundish and shallowly lobed. Lateral leaflets in 8–12 pairs, often with small and large pairs arranged alternately. Stem-leaves small with stipules. Flowers erect, 1–3 per stem, to 3 cm across. Petals golden yellow, often 2-lobed. Inflorescence stalks long and slender. *Himalaya.* H3. Summer.

3. G. rossii (R. Brown) Seringe. Illustration: Steere, Wild flowers of the United States **1**: 175 (1966).

Stems to 15 cm, slightly downy. Basal leaves pinnate to 10 cm, minutely downy. All leaflets, including terminal, to 1.5 cm, in 6–10 unequally shaped pairs, linear to wedge-shaped, deeply divided into 3 or more lobes. Stems-leaves similar to leaflets but longer, usually 3–5. Flowers erect, 3–5 per stem, to 2 cm across. Petals round, yellow. Sepals to same length as petals. *Alaska.* H3. Summer.

4. G. triflorum Pursh. Illustration: *Botanical Magazine*, 2858 (1828); Steere, Wild flowers of the United States **1**: 135 (1966).

Stems erect to 45 cm, hairy and with fine glandular hairs. Basal leaves irregularly pinnate to 15 cm with 7–15 principal pairs of leaflets each to 2.5 cm. Terminal leaflet not significantly different from laterals. Terminal and lateral leaflets narrowly wedge-shaped, margins very irregular; leaflets deeply divided, almost to base, glandular. Stem-leaves few, similar to basal leaflets, slender, to 2.5 cm. Flowers on long stems to about twice the length of the leaves, up to 3 flowers per stem, 2.5 cm across. Petals cream to purple, sometimes with a purplish red margin, oblong, not spreading, about as long as calyx. Calyx often tinged purplish to dark red. *N America (east of the Cascade Mts)*. H3. Summer.

5. G. rivale Linnaeus. Illustration: Bonnier, Flore complète **3**: pl. 170 (1914); Huxley, Mountain flowers of Europe, pl. 302 (1986); Steere, Wild flowers of the United States **1**: 135 (1966); Aeschimann et al., Flora alpina **1**: 761 (2004).

Stems 20–80 cm, glandular-downy. Basal leaves pinnate to 15 cm. Terminal and (often) top 2 laterals to 5 cm across, obovate or wedge-shaped, coarsely double-toothed. Other lateral leaflets in 1–5 pairs, to 1.5 cm, with deep forward-pointing teeth. Stem-leaves few, often 3–lobed and slender to 3 cm. Flowers nodding, 3–5 per stem, *c.* 2.5 cm across. Petals dark orange-red to buff-pinkish, scarcely longer than calyx-lobes. *Europe, Asia & N America*. H3. Early summer.

G. × tirolense Kerner; *G. montanum × G. rivale*. Usually almost identical to *G. rivale*.

6. G. uniflorum Buchanan. Illustration: Evans, New Zealand in flower, 107 (1987); Mark & Adams, New Zealand alpine plants, pl. 25 (1973); Salmon, Field guide to the alpine plants of New Zealand, 249 (1968).

Stems to 1.5 cm, creeping rootstock, forming broad patches. Basal leaves pinnate to 7.5 cm, hairy, especially on margin. Terminal leaflet *c.* 2.5 cm wide, heart to broadly kidney-shaped, slightly lobed to deeply scalloped. Lateral leaflets in 1 or 2 pairs, very small (to 3 cm), deeply lobed. Stem-leaves narrow and slender, to 5 mm. Flowers solitary on 10–15 cm slender stems, 2–3 cm across. Petals white or cream, roundish. *New Zealand (South Island)*. H4–5. Summer.

7. G. parviflorum Smith. Illustration: Mark & Adams, New Zealand alpine plants,

pl. 25 (1973); Moore & Irwin, The Oxford book of New Zealand plants, 31 (1978).

Stems erect, 10–50 cm across, hairy. Basal leaves pinnate, to 15 cm. Terminal leaflet to 5 cm across, kidney-shaped, faintly 3–5 lobed, scalloped, hairy on both sides. Lateral leaflets in 4–8 pairs, small, to 5 mm, margins deeply cut or with forward-pointing teeth. Stem leaves unevenly lobed, slender, to 1 cm, hairy. Flowers *c.* 1.2 cm across, in loose, few-flowered panicles, with long slender flower-stalks. Petals white or cream. *New Zealand & S America*. H4–5. Summer.

8. G. montanum Linnaeus. Illustration: Huxley, Mountain flowers of Europe, pl. 295 (1986); Parish, Flowers in the wild, 129 (1983); Brickell (ed.), RHS A–Z encyclopedia of garden plants, 479 (2003); Aeschimann et al., Flora alpina **1**: 759 (2004).

Stems to 30 cm (usually less than 10 cm in the wild), hairy. Basal leaves pinnate, to 10 cm. Terminal leaflet *c.* 2.5 cm across, heart-shaped to kidney-shaped or rounded, irregularly and deeply toothed. Lateral leaflets in 3–6 pairs, to 1 cm, toothed. Stem-leaves *c.* 1.5 cm, deeply cut. Flowers erect, usually solitary, 2–3 cm across. Petals golden yellow, almost circular. Sepals almost to length of petals. *S Europe*. H3. Summer.

9. G. urbanum Linnaeus. Figure 99(5, 22), p. 183. Illustration: Bonnier, Flore complète **3**: pl. 170 (1914); Polunin, Flowers of Europe, pl. 45 (1969); Aeschimann et al., Flora alpina **1**: 761 (2004).

Stems erect to 60 cm, slightly downy. Basal leaves pinnate, to 1.5 cm, hairy. Terminal leaflet to 8 cm across, roundish, scalloped. Lateral leaflets in 2 or 3 pairs, often unequal in size, usually 5–10 mm. Upper pair of lateral leaflets often similar in size to terminal leaflet. Stem-leaves to 2 cm, wedge-shaped to linear, unevenly lobed and with forward-pointing teeth. Stipules 1–3 cm, leaf-like. Flowers to 1.5 cm, erect, 1–3 per stem, in open cymes to 2 cm across. Petals yellow, spreading, about same size as sepals, obovate or oblong. *N Europe*. H3. Summer.

G. × intermedium Ehrhart; *G. urbanum × G. rivale*. Variable in all aspects between the 2 parents.

10. G. bulgaricum Pančič. Illustration: Polunin, Flowers of Greece and the Balkans, pl. 16 (1980).

Stems to 60 cm with soft and glandular hairs. Basal leaves pinnate, to 30 cm, with dense soft hairs to 5 mm. Terminal leaflet to 10 cm across, heart to kidney-shaped with uneven jagged and rounded forward-pointing teeth; lateral leaflets variable in size and number, but usually in 5–7 pairs, *c.* 1.5 cm. Stem-leaves clasping stem, to 2.5 cm, slender, deeply lobed. Flowers 2.5–3.5 cm across, 3–7. Petals bright yellow or orange, triangular. Styles with fine hairs, *c.* 1.2 cm. *Bulgaria*. H3. Summer.

G. × borisii Keller; *G. bulgaricum × G. reptans*. Illustration: Brickell, RHS gardeners' encyclopedia, 249 (1989). Variable, but usually more similar to *G. bulgaricum*.

11. G. chiloense Balbis (*G. quellyon* Sweet). Illustration: Brickell, RHS gardeners' encyclopedia of plants and flowers, 247 (1989).

Stems to 60 cm, hairy and glandular. Basal leaves pinnate, 10–30 cm, hairy. Terminal leaflet to 75 cm across, heart- to kidney-shaped with uneven jagged and rounded forward-pointing teeth; lateral leaflets *c.* 1.5 cm, variable in size and number, usually in 3 pairs. Stem-leaves clasping stem, to 2.5 cm, unevenly and deeply lobed. Flowers erect, 1–5, either solitary or in a corymb, *c.* 2.5 cm across. Petals scarlet, sometimes copper-coloured. Filaments red. *Chile*. H3. Summer.

Most of the frequently grown cultivars are derived from *G. chiloense*, e.g. 'Mrs Bradshaw', red, double; 'Lady Stratheden', yellow, double; 'Red Wings', red, semi-double.

12. G. coccineum Sibthorp & Smith. Illustration: Brickell (ed.), RHS A–Z encyclopedia of garden plants, 479 (2003).

Stems to 30 cm with soft dense hairs. Basal leaves pinnate, to 20 cm, with soft hairs. Terminal leaflet 3–4 times larger than lateral leaflets, to 7.5 cm across, rounded or heart-shaped, deeply lobed, (sometimes 3-lobed) or unevenly sharply toothed. Lateral leaflets to 2 cm, in 1–3 pairs, deeply lobed, variable. Stem-leaves few, to 2.5 cm, unequally lobed. Flowers *c.* 2.5 mm across, erect, 1–3 per stem. Petals yellow or orange, almost rounded. Epicalyx (lobe between sepals) distinctly different to sepals, *c.* 1 × 2 mm. *S Europe, Greece*. H3–4. Summer–autumn.

G. × jankae G. Beck; *G. coccineum × G. rivale*. Similar to *G. coccineum* but leaves more deeply toothed and less hairy.

41. WALDSTEINIA Willdenow
S.G. Knees

Rhizomatous perennial herbs, often with stolons. Leaves mostly basal, ternate or lobed; lobes 3–7. Flowers terminal on slender branched stems. Bracteoles 5; perigynous zone short; sepals 5, narrowly lanceolate, apex acute; petals 5, rounded, yellow; stamens numerous; style lengthening, soon deciduous. Ovary. Fruits of 2–6 achenes.

These low spreading perennials are valued as good ground-cover plants with their ability to quickly colonise otherwise difficult areas, such as shady corners, or north-facing borders. Propagation is usually by division, although plants may be grown from seed.

1a. Flowers 1–1.5 cm across; leaves with 5–7 lobes **1. geoides**
  b. Flowers 1.5–2 cm across; leaves usually ternate 2
2a. Leaflets 1.2–3 cm **2. ternata**
  b. Leaflets 3.5–5 cm **3. fragarioides**

1. W. geoides Willdenow. Illustration: *Botanical Magazine*, 2595 (1825); Reichenbach, Icones florae Germanicae et Helveticae **25**: t. 65 (1912).

Rhizome erect or shortly creeping. Leaves broadly cordate to kidney-shaped, lobes 5–7, coarsely toothed. Flowering-stems 15–25 cm with 3–7 flowers, bracts leafy, flowers 1–1.5 cm across, petals acuminate at base. *EC Europe (Bulgaria to Ukraine)*. H1. Spring

2. W. ternata (Stephan) Fritsch (*W. trifolia* Rochel; *W. sibirica* Trattinick). Illustration: Reichenbach, Icones florae Germanicae et Helveticae **25**: t. 66 (1912); Brickell (ed.), RHS A–Z encyclopedia of garden plants, 1084 (2003); Aeschimann et al., Flora alpina **1**: 763 (2004).

Plants with creeping, branched rhizomes and rooting stolons. Leaves ternate, leaflets 1.2–3 cm, stalkless, cuneate at base, shallowly lobed, margins with forward-pointing teeth. Flowering stems 10–15 cm with 3–7 flowers, bracts inconspicuous; flowers 1.5–2 cm across, petals rounded. *EC & S Europe, E Siberia, Japan*. H1. Spring.

3. W. fragarioides (Michaux) Trattinick. Illustration: *Revue Horticole*, 510 (1890); Gleason, Illustrated flora of north-eastern United States and adjacent Canada **2**: 291 (1952).

Plants with rhizomes and stolons. Leaves usually ternate, 9–18 cm, leaflets 3.5–5 cm, cuneate at base, toothed towards apex.

Flowering-stems 8–20 cm with 3–8 flowers; flowers 1.5–2 cm across; calyx 3–5 mm, silky, bracts 5–10 mm. Petals obovate-broadly elliptic, 8–10 mm, greatly exceeding sepals. *E North America*. H1. Spring.

42. COLURIA R. Brown
J. Cullen

Herbs with rhizomes. Stems erect. Leaves mostly basal, pinnate with 7 or more large leaflets with much smaller leaflets in between, terminal leaflet largest, often lobed, all toothed; stem-leaves few, reduced; stipules lanceolate. Flowers solitary or few in racemes. Perigynous zone cylindric-bell-shaped to funnel-shaped, conspicuously 10-veined. Sepals 5, alternating with 5 smaller, persistent epicalyx segments. Petals 5–7, almost circular, very shortly clawed. Stamens numerous, filaments hardened and persistent in fruit. Carpels rather few. Styles long, hairy towards the base and constricted there, deciduous. Fruit a group of achenes in the persistent perigynous zone surmounted by the persistent filaments.

A genus of about 5 species from eastern Asia, very rarely cultivated. Little is known of the requirements of the single species treated below, but it is propagated by seed and perhaps by division of the rhizome.

Literature: Evans, W.E., The genus Coluria, *Notes from the Royal Botanic Garden, Edinburgh* **15**: 48–54 (1925).

1. C. geoides (Pallas) Ledebour. Illustration: Komarov (ed.), Flora USSR **10**: 245 (1941).

Stems to 35 cm, densely hairy like the leaves. Basal leaves in outline gradually tapered to the base, terminal leaflet truncate at its base. Stem-leaves alternate, small, borne all along the stem. Flowers *c*. 2 cm across, bright yellow. Achenes covered in translucent papillae. *E Former USSR (Siberia), Mongolia*. H1. Spring.

43. POTENTILLA Linnaeus
A.C. Leslie & S.M. Walters

Perennial, rarely annual or biennial herbs or small shrubs. Leaves pinnate, palmate, with narrow leaflets or divided into 3 leaflets. Flowers solitary or in cymes, sepals and petals usually 5; epicalyx present. Perigynous zone small, more or less flat, receptacle dry or spongy. Petals yellow, white, pink, red, orange or bicoloured. Stamens 10–30. Carpels usually 10–80,

rarely as few as 4. Fruit a head of achenes. Style usually deciduous.

A genus of about 500 species, chiefly in the northern temperate and arctic regions. The flower is superficially like that of *Ranunculus* (volume 2, p. 382), and most easily distinguished by the presence of the epicalyx. Most species are easily grown in a well-drained soil in a sunny position, though a few may require alpine-house treatment. Herbaceous species may be propagated by seed, division or cuttings, woody species by seed or cuttings.

1a. Shrub 2
  b. Perennial herb, more or less woody at the base 6
2a. Flowering shoots annual, herbaceous; leaves toothed **5. salesoviana**
  b. Flowers borne on woody perennial twigs; leaves not toothed 3
3a. Leaves large, to 2 cm wide, leaflets broadly elliptic; stipules ovate, brownish, conspicuous on the twigs **4. arbuscula**
  b. Leaves smaller, leaflets oblong, elliptic or linear; stipules lanceolate, pale, not conspicuous on the twigs 4
4a. Petals white; epicalyx-segments broad, obtuse **3. davurica**
  b. Petals yellow; epicalyx-segments narrow, acute 5
5a. Flowers not more than 1.5 cm, borne on long, slender stalks; leaflets 5–9, usually 7, linear, small **2. parvifolia**
  b. Flowers usually more than 1.5 cm, on relatively short stalks; leaflets 5 or 7, oblong-lanceolate to elliptic **1. f ruticosa**
6a. Leaves pinnate 7
  b. Leaves palmate with narrow leaflets, or made up of 3 leaflets 13
7a. Petals purple **6. palustris**
  b. Petals yellow or white 8
8a. Petals white; leaflets ovate to almost circular **11. rupestris**
  b. Petals yellow; leaflets various 9
9a. Leaflets deeply pinnatisect **12. multifida**
  b. Leaflets entire or toothed 10
10a. Leaflets entire or toothed only at the apex **10. bifurca**
  b. Leaflets toothed along the margins 11
11a. Flowers solitary **7. anserina**
  b. Flowers in cymes 12
12a. Plant with stems producing stolons **8. lineatus**

b. Plant producing rhizomes
                    9. peduncularis
13a. Leaflets 3–7   14
b. Leaflets 3   32
14a. Petals yellow (sometimes with an
     orange spot at the base)   15
b. Petals white, pink, red or bicoloured
     yellow and pink/red   26
15a. Plant stoloniferous, freely rooting at
     the nodes   **32. reptans**
b. Plant not stoloniferous (though
     sometimes rooting from more or less
     prostrate woody stems)   16
16a. Leaves densely white-hairy beneath
                                               17
b. Leaves not densely white-hairy
     beneath   20
17a. Stems prostrate or ascending   18
b. Stems usually erect; petals usually
     more than 1 cm   **20. gracilis**
18a. Stems always prostrate; inflorescence
     densely white-hairy   **14. calabra**
b. Stems sometimes ascending;
     inflorescence loosely covered with
     white hairs   19
19a. Petals 4–5 mm; styles conical
                              13. argentea
b. Petals 3–4 mm; styles club-shaped
     and often distorted   **15. collina**
20a. Flowering-stems erect, robust, to 70
     cm   **16. recta**
b. Flowering-stems usually ascending,
     less than 40 cm   21
21a. Petals at least 1 cm   22
b. Petals less than 1 cm   23
22a. Flowering-stems to 40 cm; terminal
     tooth of leaflets as long as laterals
                              24. pyrenaica
b. Flowering-stems less than 25 cm;
     terminal tooth of leaflet shorter than
     laterals   **28. aurea**
23a. Leaves densely stellate-hairy beneath
                              31. cinerea
b. Leaves not densely stellate-hairy
     beneath   24
24a. Plant mat-forming, with more or less
     prostrate stems rooting at the nodes
                          30. neumanniana
b. Plant clump-forming, not rooting at
     the nodes   25
25a. Leaves more or less silvery-hairy
     beneath; petals without an orange
     basal spot   **19. nevadensis**
b. Leaves not silvery-hairy; petals
     usually with an orange basal spot
                              29. crantzii
26a. Petals usually white (rarely pale
     pink)   27
b. Petals deep pink, red or bicoloured
                                               30

27a. Filaments hairy, at least below
                          37. caulescens
b. Filaments hairless   28
28a. Stems usually shorter than leaves
                              42. alba
b. Stems usually exceeding leaves   29
29a. Stems 5–10 cm; leaflets often 5;
     flowers solitary or in groups of up to
     4   **38. clusiana**
b. Stems 10–30 cm; leaflets often 7;
     flowers in groups of more than 4
                          40. alchemilloides
30a. Gland-tipped hairs present on the
     stem   **21. thurberi**
b. Gland-tipped hairs absent   31
31a. Stems erect   **23. nepalensis**
b. Stems prostrate or ascending
                              33. × tonguei
32a. Leaves stellate-hairy beneath
                              31. cinerea
b. Leaves not stellate-hairy beneath   33
33a. Leaves densely white hairy beneath
                                               34
b. Leaves not densely white hairy
     beneath   35
34a. Leaflets obovate to circular,
     irregularly lobed   **17. villosa**
b. Leaflets elliptic to obovate, sharply
     toothed   **22. atrosanguinea**
35a. Petals more than 9 mm   36
b. Petals to 9 mm   41
36a. Flowers solitary (rarely
     paired)   37
b. Flowers 3–7 per inflorescence (rarely
     paired)   39
37a. Petals pink, rarely white; sepals
     purplish   **39. nitida**
b. Petals yellow; sepals not purplish   38
38a. Stock with persistent, hairy leaf-
     bases   **36. eriocarpa**
b. Stock lacking persistent, hairy leaf-
     bases   **35. cuneata**
39a. Epicalyx-segments obtuse
                              34. megalantha
b. Epicalyx-segments acute   40
40a. Stem with patent or somewhat
     adpressed hairs; petals yellow
                              18. grandiflora
b. Stem with adpressed hairs; petals
     deep orange
                          28. aurea, see subsp.
                                      chrysocraspeda
41a. Petals white, cream, pink or orange
                                               42
b. Petals yellow   46
42a. Stems 30–60 cm
                          22. atrosanguinea
b. Stems to 30 cm   43
43a. Leaves usually shiny green and
     hairless above   **45. tridentata**

b. Leaves not shiny green and hairy
     above   44
44a. Petals 6–9 mm   45
b. Petals *c.* 5 mm   **44. sterilis**
45a. Clump-forming herbaceous perennial
     to 10 cm (rarely more) **41. speciosa**
b. Rhizomatous or stoloniferous
     herbaceous perennial to 25 cm
                              43. montana
46a. Stems 30–60 cm
                          22. atrosanguinea
b. Stems to 30 cm   47
47a. Petals not notched at apex
                              41. speciosa
b. Petals notched at apex   48
48a. Plant lacking glands; leaves hairless
     above   **25. brauniana**
b. Plant more or less glandular hairy;
     leaves at least sparsely hairy above
                                               49
49a. Petals usually 1.5 times sepals;
     leaves densely hairy above; teeth on
     leaflets of basal leaves elliptic
     or oblanceolate, very obtuse
                              26. frigida
b. Petals at least 1.5 times sepals;
     leaves usually only sparsely hairy
     above; teeth on leaflets of basal
     leaves triangular-oblong, acute, or
     almost obtuse   **27. hyparctica**

1–4. Potentilla fruticosa Group

Deciduous shrubs to *c.* 1.5 m; much-branched; twigs thin. Leaves pinnate with 5–9 linear-lanceolate to broadly ovate-elliptic, entire leaflets (more rarely with 3 leaflets), more or less silky-hairy (rarely hairless). Stipules persistent, entire, acute or obtuse. Calyx with 5, more or less ovate, acute lobes. Epicalyx-lobes 5, of very varying size and shape, from linear to ovate and sometimes 2-fid or 3-fid. Corolla of 5 (rarely more in semi-double variants) large, obovate to circular, free petals, commonly yellow or white. Stamens numerous. Carpels few, free, each with a club-shaped style attached near the base. Fruit a head of small achenes, each with a ring of basal hairs. Plants sometimes dioecious; male flowers (often larger) with numerous well-developed stamens, carpels reduced to a clump of hairs; female flowers with sterile staminodes (often well developed), carpels well developed.
N temperate to Arctic areas of Asia & N America, extending to the Caucasus, Himalaya & China; rare and local in N Europe and mountains of S Europe. H1. Late spring–summer.

The classification of the group is complex, and there is yet little agreement on an appropriate treatment; the group includes the first four species listed below. As a whole they are tolerant of lime, preferring a well-drained soil, and are intolerant of shade. They are easily propagated by cuttings, but open-pollinated seed germinates readily and, since cross-pollination is frequent in the bisexual taxa and obligatory in the dioecious ones, new variants continually arise, especially those with unusual petal-colour; good examples are the red-flowered cultivars 'Red Ace' and 'Royal Flush', the pink-flowered semi-double 'Princess' and the orange-flowered 'Tangerine'.

Literature: Bean, W.J., Trees and shrubs hardy in the British Isles, edn 8, **3**: 328–43 (1976); Klackenberg, J., The holarctic complex Potentilla fruticosa (Rosaceae), *Nordic Journal of Botany* **3**: 181–91 (1983); Brearley, C, The shrubby potentillas, *The Plantsman* **9**: 90–109 (1987).

1. P. fruticosa Linnaeus. Figure 99(3), p. 183. Illustration: Raven & Walters, Mountain flowers, pl. 10 (1956); Rickett, Wild flowers of the United States **1**: 129 (1966); Garrard & Streeter, Wild flowers of the British Isles, 87 (1983).

Habit variable, usually more or less erect and loosely branched. Leaflets 5 or 7, 1–2 cm, oblong-lanceolate to elliptic, more or less hairy with long, white hairs; stipules lanceolate, acute, pale. Flowers usually in cyme-like groups, more rarely solitary. Petals 6–16 mm across, yellow, obovate to circular. *Widespread.*

May be divided into 2 subspecies, differing mainly in chromosome number and sexuality.

Subsp. **fruticosa**: dioecious, a rare plant of open calcareous rocks and riverside habitats. *N Europe.*

Subsp. **floribunda** (Pursh) Elkington (*P. floribunda* Pursh) is bisexual. *N America and very locally in the Pyrenees, W Alps and Bulgaria. Distribution in Asia uncertain.*

The European plant was cultivated early, both in Britain and Scandinavia, but is rarely seen in modern gardens, having largely been replaced by cultivars. Two modern cultivars, 'Goldfinger' and 'Jackman's Variety', are, however, thought to be selections from *P. fruticosa*, differing most obviously in their larger flowers.

2. P. parvifolia Lehmann (*P. fruticosa* 'Farreri'). Illustration: Krüssmann, Handbuch der Laubgehölze **2**: 462 (1976) – as *P. fruticosa* 'Parvifolia'.

Like *P. fruticosa* but leaves usually not exceeding 2 cm, with 5–9, usually 7, linear or linear-lanceolate leaflets, often densely hairy, grey-green above and white beneath, and small, bisexual, yellow flowers on long, slender stalks. *Former USSR (Siberia, C Asia), Mongolia, Himalaya*

Represented in gardens by several cultivars, e.g. 'Buttercup', 'Gold Drop' and 'Klondyke', which have compact habit and long flowering period. The commonly grown, pale yellow-flowered 'Katherine Dykes' is probably a hybrid of *P. parvifolia*.

3. P. davurica Nestler (*P. glabra* Loddiges; *P. mandshurica* Maximowicz; *P. veitchii* Wilson). Illustration: Krüssmann, Handbuch der Laubgehölze **2**: 462 (1976) – as *P. fruticosa* var. *davurica*; Brickell, RHS gardeners' encyclopedia of plants and flowers, 126 (1989).

Like *P. fruticosa* but more compact, hairless; flowers white; epicalyx segments broad and obtuse. *Former USSR (E Siberia to Pacific), Japan, China (N&W Xizang).*

Var. **mandshurica** (Maximowicz) Wolf and 'Manchu' differ in having a thick, adpressed silky hair-covering. Represented in gardens mainly by several white-flowered cultivars, especially 'Abbotswood', with dark green foliage and profuse flowering over a long period, and 'Farrer's White', a very vigorous shrub with larger flowers than 'Abbotswood'. Some yellow-flowered cultivars (e.g. 'Elizabeth') may be hybrids involving *P. davurica*.

4. P. arbuscula D. Don (*P. rigida* Lehmann). Illustration: The Plantsman **9**: 93 (1987).

A compact shrub to 60 cm with stouter twigs than *P. fruticosa*, large, ovate leaflets to 2 cm across and broadly ovate, brownish stipules, conspicuous on the twigs. Flowers large (usually 2–3 cm across); petals usually yellow; epicalyx-segments often 2-fid or 3-fid. *Himalaya, China (W & N Xizang).*

Typically the leaves have 5 leaflets, but plants with leaves with 3 leaflets, to which the name *P. rigida* is usually applied, are also in cultivation.

This is represented in gardens by several variants, e.g. 'Beesii' (*P. fruticosa* 'Nana Argentea'), a dense, leafy shrub with silvery-hairy leaflets and large, golden-yellow, male flowers. There is much confusion as to the specific limits of *P. arbuscula*, partly because flower-colour is

not preserved in herbarium material and not always stated by the original collector.

5. P. salesoviana Stephan. Illustration: *Botanical Magazine*, 7258 (1892); Krüssmann, Manual of cultivated broad-leaved trees & shrubs **2**: 439 (1986).

Upright shrub 30–100 cm, woody below, with annual, erect, flowering-shoots. Leaves pinnate; leaflets 7–13, sharply toothed, the terminal 2–4 cm, oblong, dark green and hairless above, white-hairy beneath. Flowers 3–7 together, 3–3.5 cm across. Petals obovate, apex not notched, white, sometimes tinged pink. Receptacle almost spherical, densely hairy. Achenes densely hairy. *Former USSR (Siberia, C Asia), Himalaya, China (Xizang).* H1. Summer.

6. P. palustris (Linnaeus) Scopoli (*Comarum palustre* Linnaeus). Illustration: Ross-Craig, Drawings of British plants **8**: t. 39 (1955); Polunin, Flowers of Europe, t. 46 (1969); Garrard & Streeter, Wild flowers of the British Isles, 82 (1983); Aeschimann et al., Flora alpina **1**: 763 (2004).

Herbaceous perennial with long, creeping, woody rhizomes. Stems decumbent, 15–45 cm, sparsely hairy. Leaves pinnate; leaflets 5–7, 2.6–6 × 1–2 cm, oblong, coarsely toothed, slightly greyish green and hairless above, greyish green and almost hairless beneath. Flowers several, in loose cymes, to 3 cm across. Petals 5–8 mm, ovate-lanceolate, acuminate, deep purple, persistent, half as long as the purplish sepals, which enlarge in fruit. Receptacle spongy. *Europe, Asia, N America.* H1. Summer.

Suitable for peaty places in a water garden.

7. P. anserina Linnaeus. Illustration: Ross-Craig, Drawings of British plants **8**: t. 35 (1955); Pignatti, Flora d'Italia **1**: 575 (1982); Garrard & Streeter, Wild flowers of the British Isles, 36 (1983); Aeschimann et al., Flora alpina **1**: 763 (2004).

Patch-forming herbaceous perennial. Stock short, ending in a rosette of leaves and producing prostrate, stoloniferous stems to 80 cm. Leaves 5–40 cm × 5–15 mm, pinnate, with 3–12 pairs of large leaflets alternating with smaller ones; large leaflets 1–6 cm, ovate or oblong, toothed, silvery silky-hairy above and beneath or only beneath, rarely green and sparsely hairy or hairless on both surfaces. Flowers solitary, 1–2 cm across. Petals 7–10 cm, obovate, apex not notched, yellow, much longer than the sepals. *Most of Europe, N & C Asia, N & S America, Australasia.* H1. Summer.

Attractive in its silver-leaved variants, but generally too invasive for all but the wildest parts of the garden.

8. P. lineatus Treviranus (*P. fulgens* Hooker) Illustration: *Botanical Magazine*, 2700 (1826) – as *P. splendens*; Iconographia cormophytorum Sinicorum **2**: 290 (1972).

Similar, but with a leafy stem and cymose inflorescence, and producing stolons. *Himalaya, China.* H1. Summer.

9. P. peduncularis D. Don. Illustration: Polunin & Stainton, Flowers of the Himalaya, t. 37 (1985).

Also similar, but rhizomatous and larger in all its parts and lacking the alternating, smaller leaflets; habit much more stiffly erect, the flowers 2–5, in cymes. *W Nepal to SW China.* H1. Summer.

10. P. bifurca Linnaeus. Illustration: Komarov (ed.), Flora SSSR **10**: t. 7 (1941).

Rhizomatous perennial. Stems 10–30 cm, woody at the base, silky-hairy to almost hairless. Leaves pinnate; leaflets 5–15, 8–20 × 3–8 mm, oblong-ovate, entire or with 2 or 3 apical teeth. Flowers in a loose cyme. Petals 4–8 mm, yellow, longer than the sepals. Achenes hairy at the base when young. *SE Europe, Asia.* H1. Summer.

11. P. rupestris Linnaeus. Illustration: Ross-Craig, Drawings of British plants **8**: t. 36 (1955); Grey-Wilson & Blamey, Alpine flowers of Britain and Europe, 99 (1979); Pignatti, Flora d'Italia **1**: 575 (1982); Aeschimann et al., Flora alpina **1**: 765 (2004).

Clump-forming herbaceous perennial. Stems 20–60 cm, erect, with gland-tipped hairs above, often reddish at the base. Leaves 7–15 cm, pinnate; leaflets in 2–4 distant pairs, decreasing in size below, 1–6 cm × 5–35 mm, ovate to almost circular, coarsely toothed, green above and beneath, hairy on both surfaces. Flowers in a loose cyme, 1.5–2.5 cm wide. Petals obovate, entire, 8–14 mm, white, longer than the sepals. *W & C Europe, N Africa, W & C Asia, N America.* H1. Spring–summer.

Dwarf variants from Corsica and Sardinia, with stems only 4–10 cm, fewer leaflets and 1 or 2 smaller flowers have been distinguished as var. **pygmaea** Duby ('Nana').

12. P. multifida Linnaeus. Illustration: Huxley, Mountain flowers, 54 (1967); Grey-Wilson & Blamey, Alpine flowers of Britain and Europe, 99 (1979); Pignatti, Flora d'Italia **1**: 575 (1982); Aeschimann et al., Flora alpina **1**: 765 (2004).

Clump-forming herbaceous perennial. Stems 5–40 cm, erect or ascending, sparsely to densely hairy. Leaves pinnate, leaflets 5–9, often crowded so appearing almost palmate, 5–40 × 3–20 mm, deeply pinnatisect with up to 5 linear ultimate segments, green above, silvery silky-hairy beneath. Flowers few to many, 1–1.5 cm across. Petals obovate, 5–7 mm, apex notched, yellow, equalling or slightly exceeding sepals. *Europe to Asia.* H1. Summer.

13. P. argentea Linnaeus. Illustration: Ross-Craig, Drawings of British plants **8**: t. 37 (1955); Keble Martin, The concise British flora in colour, t. 27 (1965); Garrard & Streeter, Wild flowers of the British Isles, pl. 82 (1983); Aeschimann et al., Flora alpina **1**: 767 (2004).

Mat-forming herbaceous perennial. Stems 15–50 cm, prostrate or ascending, usually downy. Leaves palmate with narrow leaflets; leaflets 5, 1–3 cm × 5–15 mm, wedge-shaped to obovate, with 2–7 obtuse teeth or lobes, green and hairless above, densely white-hairy beneath. Flowers numerous, 1–1.5 cm across. Petals 4–5 mm, obovate, yellow, only slightly longer than the sepals. Styles conical, tapering to the apex. *Throughout Europe, W&C Asia, N America.* H1. Summer.

14. P. calabra Tenore.

Similar to *P. argentea*, but always with prostrate stems. Leaves smaller, greyish green to white-hairy above and the inflorescence more densely white-hairy. *C & S Italy, Sicily, western part of Balkan Peninsula.* H1. Summer.

15. P. collina Wibel (*P. alpicola* Fauconnet).

Similar to *P. argentea* but with club-shaped styles that are often distorted. *C Europe.* H1. Summer.

16. P. recta Linnaeus. Illustration: Butcher, A new illustrated British flora **1**: 667 (1961); Pignatti, Flora d'Italia **1**: 578 (1982); Brickell, The RHS gardeners' encyclopedia of plants and flowers, 246 (1989); Aeschimann et al., Flora alpina **1**: 769 (2004).

Clump-forming herbaceous perennial. Stems 10–70 cm, erect, with both short, gland-tipped hairs and longer, white, glandless hairs. Leaves palmate; leaflets 5–7, 5–10 cm × 5–35 mm, oblong to obovate, coarsely toothed or shallowly lobed, green or greyish above, hairy above and beneath. Flowers numerous in loose cymes, 2–2.5 cm across. Petals 6–12 mm, obovate, apex deeply notched, yellow, longer than sepals. *C, E & S Europe, N Africa, W & C Asia, naturalised elsewhere.* H1. Summer

A variable species. Commonly grown variants include 'Sulphurea' ('Pallida') with pale yellow flowers and 'Warrenii' ('Macrantha') with larger, bright yellow flowers.

17. P. villosa Pursh. Illustration: Abrams, Illustrated flora of the Pacific states **2**: 436 (1944); Clark, Wild flowers of British Columbia, 259 (1973); Clark, Wild flowers of the Pacific northwest, 262 (1976).

Cushion-forming herbaceous perennial. Stems ascending, 10–30 cm, densely white or yellowish hairy. Leaves with 3 leaflets; leaflets 1.5–4 cm, obovate to almost circular, thick, irregularly lobed around the margins, green and slightly hairy above, densely white-hairy beneath, with impressed veins. Flowers few, 2–3 cm across. Petals 6–12 mm, broadly obovate, deeply notched at apex, golden yellow, much longer than sepals. *W North America, NE Asia.* H1. Early summer.

18. P. grandiflora Linnaeus. Illustration: Huxley, Mountain flowers, 55 (1967); Grey-Wilson & Blamey, Alpine flowers of Britain and Europe, 101 (1979); Pignatti, Flora d'Italia **1**: 579 (1982); Aeschimann et al., Flora alpina **1**: 771 (2004).

Clump-forming herbaceous perennial. Stems 10–40 cm, ascending, with dense, spreading hairs. Leaves with 3 leaflets; leaflets 1.5–4 × 1–3 cm, obovate to almost circular, coarsely and bluntly toothed, green above and beneath, sparsely hairy above, more densely so beneath. Flowers 2–5, 1.5–3.5 cm across. Epicalyx-segments acute. Petals broadly obovate, 1–1.5 cm × 9 mm, apex notched, yellow, much longer than sepals. *C & E Pyrenees, Alps.* H1. Late summer.

19. P. nevadensis Boissier. Illustration: Boissier, Voyage botanique dans le midi d'Espagne **1**: t. 59 (1838–45).

Clump-forming herbaceous perennial. Stems 15–30 cm with long, spreading hairs. Leaves palmate; leaflets 5, 4–20 × 3–15 mm, obovate or oblanceolate, scalloped-toothed, green above, silvery-hairy beneath. Flowers 1–4 together, 1–1.5 cm across. Petals 4–7 mm, obovate, apex notched, yellow, longer than the sepals. *S Spain (Sierra Nevada)*. H1. Spring–summer.

Var. **condensata** Boissier is smaller, more densely tufted and has the leaves silvery-silky on both surfaces.

20. P. gracilis J.D. Hooker. Illustration: Jepson, Manual of the plants of California, 489 (1925); Abrams, Illustrated flora of the Pacific states **2**: 433 (1944); Clark, Wild flowers of the Pacific northwest, 254 (1976).

Clump-forming herbaceous perennial. Stems 40–70 cm, usually erect, with dense, spreading hairs. Leaves palmate: leaflets 5–7 (or more), 2–6 cm, obovate or oblanceolate, conspicuously toothed along their whole length, green but with some silky hairs above, densely white-hairy beneath. Flowers numerous in loose cymes, 1–2 cm across. Petals obovate, usually more than 1 cm, apex notched, yellow, longer than sepals. *W North America*. H1. Summer.

Very variable. The description above is of var. **gracilis**. Plants with the leaves more sparsely hairy and often glandular beneath are distinguished as subsp. **nuttallii** (Lehmann) Keck (*P. nuttallii* Lehmann).

21. P. thurberi Lehmann. Illustration: Rickett, Wild flowers of the United States **4**: t. 53 (1970).

Clump-forming herbaceous perennial. Stems 30–60 cm, erect, with both gland-tipped and glandless hairs. Leaves palmate; leaflets 5–7, 2.5–5 cm, broadly oblanceolate, coarsely toothed, green above, hairy beneath. Flowers in loose cymes, 1.5–2 cm across. Petals dark brownish red, slightly longer than sepals. *SW USA*. H1. Summer

Similar to the Himalayan *P. nepalensis* (see below), but differing in having gland-tipped hairs.

22. P. atrosanguinea D. Don. Illustration: Coventry, Wild flowers of Kashmir **1**: t. 18 (1923); Blatter, Beautiful flowers of Kashmir **1**: t. 21 (1927); Stainton, Flowers of the Himalaya, a supplement, t. 137 (1988); Brickell (ed.),

RHS A–Z encyclopedia of garden plants, 844 (2003).

Clump-forming herbaceous perennial. Stems 30–60 cm, hairy. Leaves usually of 3 leaflets; leaflets 2–5 × 1–3 cm, elliptic, ovate or obovate, sharply toothed, dark green and hairless or grey silky-hairy above, densely white-hairy beneath. Flowers in loose cymes, 2–3 cm across. Petals 9–11 cm, obovate, apex notched, yellow, orange or red, longer than sepals. *W & C Himalaya*. H1. Summer–autumn.

Var. **atrosanguinea** has red flowers; yellow-flowered plants have been distinguished as var. **argyrophylla** (Lehmann) Grierson & Long (*P. argyrophylla* Lehmann). Var. **cataclina** (Lehmann) Wolf and *P. argyrophylla* var. *leucochroa* J.D. Hooker represent smaller variants of these two varieties.

Garden hybrids and selections of these varieties (and hybrids with *P. nepalensis*) are widely grown. Many are larger plants with more than 3 leaflets, and often have larger, double flowers; several are bicoloured, e.g. 'Yellow Queen', single to double yellow; 'Gibson's Scarlet', single, bright red; and 'Monsieur Rouillard', double, mahogany red with yellow blotches.

23. P. nepalensis J.D. Hooker. Illustration: *Botanical Magazine*, 9182 (1929); Perry, Collins guide to border plants, edn 2, t. 35 (1966); Brickell, RHS gardeners' encyclopedia of plants and flowers, 237 (1989).

Clump-forming herbaceous perennial. Stems 30–75 cm, hairy, often tinged red. Leaves palmate; leaflets 5, 2–8 × 1–3 cm, narrowly elliptic or oblong-ovate, coarsely toothed, sparsely hairy, green on both surfaces. Flowers in loose cymes, 1.3–3 cm across. Petals 7–14 mm, obovate, apex notched, typically pinkish red with darker bases, longer than the sepals. *Pakistan to C Nepal*. H1. Summer–autumn.

Variable in flower-colour: 'Miss Wilmott' is a crushed strawberry pink with a darker eye; 'Roxana' is yellowish pink with a darker eye and almost yellow rim. Both come largely true from seed.

P. × hopwoodiana Sweet; *P. nepalensis × P. recta*. Similar but the basal leaves often have 6 leaflets and the petals are pale yellow in the lower half (with a deep rose basal spot) and bright rose in the upper part, though with a paler margin.

24. P. pyrenaica de Candolle. Illustration: Coste, Flore de la France **2**: 22 (1903).

Clump-forming herbaceous perennial. Stems to 50 cm, ascending from a curving base, sparsely to densely hairy with somewhat adpressed hairs. Leaves with long stalks, palmate with 5 oblong, adpressed-hairy leaflets, toothed in the upper two-thirds, with terminal tooth equalling the others; stipules short, blunt. Flowers few, in condensed cymes. Petals large, yellow, obovate, apex notched, usually twice as long as the sepals. Epicalyx-segments shorter or narrower than the sepals. Style thread-like at apex. *Pyrenees, mts of N & C Spain*. H1. Summer.

25. P. brauniana Hoppe (*P. minima* Haller). Illustration: Coste, Flore de la France **2**: 25 (1903); Aeschimann et al., Flora alpina **1**: 773 (2004).

Dwarf plant, woody at base, with sparsely hairy stems not exceeding 5 cm, and small leaves with 3 obovate leaflets, hairless on the upper surface, toothed in the upper half. Flowers small, solitary. Petals yellow, obovate, apex notched, hardly exceeding sepals. Epicalyx-segments obtuse, more or less equalling sepals. *S Europe (Pyrenees to E Alps)*. H1. Late spring–summer.

26. P. frigida Villars. Illustration: Coste, Flore de la France **2**: 24 (1903); Fenaroli, Flora della Alpi, 175 (1955); Aeschimann et al., Flora alpina **1**: 773 (2004).

Like *P. brauniana* but stems to 10 cm and densely hairy throughout, with some glandular hairs mixed with long, spreading, glandless hairs. *S Europe (Pyrenees to E Alps)*. H1. Late spring–summer.

27. P. hyparctica Malte. Illustration: Polunin, Arctic flora, 271 (1959).

Like *P. brauniana* but often strongly glandular-hairy, with larger flowers with petals about 1.5 times as long as sepals. *Circumpolar, Arctic areas*. H1. Summer.

These last 3 species are best cultivated in a sink garden. *P. frigida* is said by some alpine gardeners to be the most attractive.

28. P. aurea Linnaeus. Illustration: Coste, Flore de la France **2**: 25 (1903); Fenaroli, Flora della Alpi, opposite 96 (1955); Blamey & Grey-Wilson, Illustrated flora of Britain and northern Europe, 187 (1989); Brickell (ed.), RHS A–Z encyclopedia of garden plants, 844 (2003); Aeschimann et al., Flora alpina **1**: 773 (2004).

Mat-forming perennial, woody at the base, with silkily hairy stems to 25 cm. Leaves palmate with 3 or 5 leaflets, silkily adpressed hairy on the edge and the veins beneath, toothed in the upper half with the terminal tooth obviously shorter than the others. Stipules narrow, pointed. Cyme with few relatively large flowers. Petals orange-yellow, darker at base, broadly obovate, apex shallowly notched, longer than the sepals. Epicalyx-segments narrow, pointed, more or less equalling sepals. *S & C Europe.* H1. Late spring–summer.

'Rathboneana' has semi-double flowers.

Subsp. **chrysocraspeda** (Lehmann) Nyman, with 3 leaflets only, occurs in the eastern part of the range. It is sometimes grown as the variant 'Aurantiaca' with deep orange flowers.

29. P. crantzii (Crantz) Fritsch (*P. alpestris* Haller; *P. salisburgensis* Haenke). Illustration: Garrard & Streeter, Wild flowers of the British Isles, 83 (1983); Blamey & Grey-Wilson, Illustrated flora of Britain and northern Europe, 187 (1989); Aeschimann et al., Flora alpina **1**: 773 (2004).

Clump-forming herbaceous perennial, woody at base, with hairy, ascending flowering-stems to 30 cm. Leaves palmate with 5 (rarely 3) obovate leaflets with few broad, blunt teeth, the terminal almost equalling the others, with spreading hairs beneath and on the margins; stipules rather broad, more or less acute. Flowers 1–10, in a loose cyme. Petals deep yellow, often with an orange basal spot, obovate, apex rather deeply notched, longer than sepals. *N, C & S Europe.* H1. Late spring–summer.

A variable plant in the wild. Variants with well-developed basal spots on the petals make very attractive rock-garden plants.

30. P. neumanniana Reichenbach (*P. verna* Linnaeus; *P. tabernaemontani* Ascherson). Illustration: Garrard & Streeter, Wild flowers of the British Isles, 83 (1983); Blamey & Grey-Wilson, Illustrated flora of Britain and northern Europe, 189 (1989); Aeschimann et al., Flora alpina **1**: 775 (2004).

Like *P. crantzii* but mat-forming with more or less prostrate, somewhat woody stems rooting at the nodes, leaves often with 7 leaflets, and narrow, pointed stipules. The flowers are smaller and the petals lack the orange basal spot. *N, W & C Europe.* H1. Late spring–summer.

A very variable plant of limited value in gardens. 'Nana' is a very dwarf variant with flowering-stems less than 5 cm.

31. P. cinerea Villars (*P. arenaria* Borckhausen; *P. tommasiniana* Schulz). Illustration: Coste, Flore de la France **2**: 26 (1903); Blamey & Grey-Wilson, Illustrated flora of Britain and northern Europe, 189 (1989); Aeschimann et al., Flora alpina **1**: 777 (2004).

Plant with the habit of *P. verna* but easily recognised by its grey colour, due to a dense covering of stellate hairs mixed with long, simple hairs and sometimes also glandular hairs. Leaves with 3 or 5 obovate leaflets, toothed in the upper half, with the terminal tooth shorter than the others. The flowers resemble those of *P. verna*. *C, S & E Europe.* H1. Late spring–summer.

A very variable plant, prettier than *P. verna* in rock gardens and occasionally grown.

32. P. reptans Linnaeus. Illustration: Ross-Craig, Drawings of British plants **8**: t. 34 (1955); Pignatti, Flora d'Italia **2**: 582 (1982); Garrard & Streeter, Wild flowers of the British Isles, 83 (1983); Aeschimann et al., Flora alpina **1**: 777 (2004).

Patch-forming herbaceous perennial. Stock short, ending in a rosette of leaves and producing prostrate, stoloniferous stems 20–100 cm, rooting at the nodes. Leaves palmate; leaflets usually 5, 5–70 × 3–25 mm, obovate or oblong-obovate, toothed, green above and beneath, hairless or sparsely hairy. Flowers solitary, 1.7–2.5 cm across, on long stalks. Petals 7–12 mm, obovate, apex notched, much longer than sepals. *Widespread in Europe, N Africa & Asia, naturalised elsewhere.* H1. Summer.

Generally regarded as a weed, but a double-flowered variant, forma **pleniflora** Bergmans is sometimes cultivated.

33. P. × tonguei Baxter (*P. tormentilla-formosa* Maund, invalid). Illustration: Maund, Botanic Garden **7**: t. 585 (1837); *Gentes Herbarum* **4**: 313 (1941).

Clump-forming herbaceous perennial. Stems prostrate or ascending, to 35 cm, hairy, not rooting at the nodes. Leaflets 3–5, 5–25 × 3–15 mm, narrowly obovate to obovate, coarsely toothed, often tinged bronze, hairy above and beneath. Flowers solitary or in few-flowered cymes, *c.* 1.5 cm across. Petals broadly obovate, apex notched, yellow with a red base, longer than sepals. *Garden origin.* H1. Summer–autumn.

Originally described as being '*Tormentilla reptans* × *Potentilla formosa*' (i.e.

P. reptans × *P. nepalensis*) but later authors have given *P. anglica* or *P. aurea* as the first parent.

34. P. megalantha Takeda (*P. fragariformis* Schlechtendahl). Illustration: *Kew Bulletin* **1911**: 252; Brickell, RHS gardeners' encyclopedia of plants and flowers, 247 (1989); Brickell (ed.), RHS A–Z encyclopedia of garden plants, 844 (2003).

Clump-forming, softly hairy herbaceous perennial. Stems to 30 cm, ascending. Leaves with 3 leaflets; leaflets 3–8 × 3–8 cm, thick, broadly elliptic to obovate, with few obtuse teeth, densely hairy but green above and beneath. Flowers 3–7, 3–4 cm across. Epicalyx-segments obtuse. Petals obovate, apex slightly notched, golden yellow, longer than sepals. *NE Asia.* H1. Summer.

A variant with double flowers, 'Perfecta Plena' is occasionally offered.

35. P. cuneata Lehmann (*P. ambigua* Cambessedes). Illustration: Polunin & Stainton, Flowers of the Himalaya, t. 39 (1985).

Mat-forming herbaceous perennial, the woody stock without persistent leaf-bases. Stems 2–10 cm, erect. Leaves all basal, of 3 leaflets; leaflets 6–13 × 4–10 mm, obovate, shallowly 3-toothed or lobed at the apex, green and hairy or hairless above, adpressed-hairy beneath. Flowers solitary, 1.3–2.5 cm across. Petals broadly obovate, 9–10 × 7–8 mm, yellow, longer than the sepals. Achenes covered with long silky hairs. *India (Kashmir) to SW China.* H1. Summer–autumn.

36. P. eriocarpa Lehmann. Illustration: Coventry, Wild flowers of Kashmir **3**: t. 26 (1930); Polunin & Stainton, Flowers of the Himalaya, t. 38 (1985); Brickell (ed.), RHS A–Z encyclopedia of garden plants, 844 (2003).

Mat- or cushion-forming herbaceous perennial, the woody stock with persistent, hairy leaf-bases. Stems 2–18 cm. Leaves of 3 leaflets; leaflets 1.2–3 cm × 5–15 mm, wedge-shaped to obovate, long-stalked, deeply toothed or 3–5-lobed in the upper half, bright green above, adpressed-hairy beneath. Flowers usually solitary, 2–3 cm across. Petals broadly obovate, apex deeply notched, yellow, longer than sepals. Achenes hairy. *Pakistan to China.* H1. Summer.

37. P. caulescens Linnaeus. Illustration: Huxley, Mountain flowers, 149 (1967);

Grey-Wilson & Blamey, Alpine flowers of Britain and Europe, 101 (1979); Pignatti, Flora d'Italia **1**: 583 (1982); Aeschimann et al., Flora alpina **1**: 777 (2004).

Clump-forming herbaceous perennial. Stems 5–30 cm, ascending, with adpressed or slightly spreading hairs. Leaves palmate; leaflets 5–7, 1–3 cm × 5–8 mm, oblong or oblong-obovate, with a few teeth at the apex, green above, silvery silky-hairy beneath. Flowers numerous in loose cymes, to 2 cm across. Petals obovate, 6–10 × 2.5–5 mm, apex sometimes notched, white (rarely pinkish), slightly longer than the sepals. Filaments hairy at base. *S Europe*. H1. Summer.

38. P. clusiana Jacquin. Illustration: Hegi, Illustrierte Flora von Mittel-Europa **4**: 820 (1922); Grey-Wilson & Blamey, Alpine flowers of Britain and Europe, 101 (1979); Pignatti, Flora d'Italia **1**: 583 (1982); Aeschimann et al., Flora alpina **1**: 779 (2004).

Clump-forming herbaceous perennial. Stems 5–10 cm, with hairs that are more or less adpressed. Leaves palmate; leaflets usually 5, 7–13 × 3–5 mm, obovate, 3–5-toothed at the apex, silky-hairy beneath. Flowers solitary or in few-flowered cymes. Petals broadly obovate, 9–10 × 6–8 mm, apex notched, white, much longer than sepals. Filaments hairless. *E Alps, W former Yugoslavia, Albania*. H1. Summer.

39. P. nitida Linnaeus. Illustration: Polunin, Flowers of Europe, t. 46 (1969); Grey-Wilson & Blamey, Alpine flowers of Britain and Europe, 103 (1979); Pignatti, Flora d'Italia **1**: 583 (1982); Aeschimann et al., Flora alpina **1**: 779 (2004).

Densely tufted perennial, with a woody stock. Stems 2–10 cm, densely silvery silky-hairy. Leaflets usually 3 (rarely 4 or 5), 5–10 mm, obovate or oblanceolate, entire or with a few teeth at the apex, silvery silky-hairy above and beneath. Flowers 1–2, *c.* 2.5 cm across. Petals 1–1.2 cm × 7–10 mm, broadly obovate, apex notched, pink, rarely white (forma **albiflora** Sauter), longer than the purplish sepals. *SW & SE Alps, N Apennines*. H1. Summer.

Variable in petal colour. 'Rubra' and 'Lissadell' are selections with darker pink flowers.

40. P. alchemilloides Lapeyrouse. Illustration: Coste, Flore de la France **2**: 18 (1903); Huxley, Mountain flowers, 149 (1967).

Clump-forming herbaceous perennial. Stems 10–30 cm, silky-hairy, exceeding the leaves. Leaves palmate; leaflets 5–7, 1–2.5 cm, oblong-elliptic, usually with 3 small teeth at the apex, green and hairless above, silvery silky-hairy beneath. Petals 8–10 mm, obovate, apex notched, white, longer than the sepals. Achenes hairy. *Pyrenees*. H1. Summer.

41. P. speciosa Willdenow. Illustration: Wehrhahn, Die Gartenstauden **2**: 625 (1931); Polunin, Flowers of Greece and the Balkans, 114 (1980).

Clump-forming herbaceous perennial. Stems 2–10 cm (rarely more), densely grey- or white-hairy. Leaves of 3 leaflets; leaflets 1.5–3 × 1–2 cm, broadly obovate to broadly ovate, thick, scalloped or scalloped-toothed at least in the upper two-thirds, green and almost hairless or densely hairy above, white-hairy beneath. Flowers several, 1–2 cm across. Petals 6–10 mm, obovate, white, rarely pale yellow, slightly longer than the sepals. *W & S parts of Balkan Peninsula, Crete, Syria, Iraq*. H2. Summer.

42. P. alba Linnaeus. Illustration: Polunin, Flowers of Europe, 161 (1969); Pignatti, Flora d'Italia **1**: 584 (1982); Brickell, RHS gardeners' encyclopedia of plants and flowers, 313 (1989); Aeschimann et al., Flora alpina **1**: 781 (2004).

Clump-forming herbaceous perennial with adpressed or slightly spreading hairs. Stems 5–20 cm, decumbent or ascending, usually shorter than the leaves. Leaves palmate; leaflets 5, 2–4 cm, oblong to obovate-lanceolate, with a few teeth near the tip, hairless and green above, silvery silky-hairy beneath. Flowers 1–5, 1.5–2.5 cm wide. Petals 7–10 mm, obovate, white, longer than sepals. Achenes smooth, with a few long hairs at the base. *C & E Europe*. H1. Spring–summer.

43. P. montana Brotero (*P. splendens* de Candolle). Illustration: Coste, Flore de la France **2**: 19 (1903); Bonnier, Flore complète **3**: t. 172 (1914).

Rhizomatous or stoloniferous herbaceous perennial. Stems 5–25 cm, downy. Leaflets usually 3, rarely 4 or 5, 1–3 cm, oblong to obovate, bluntly toothed in the upper half or only at the tip, hairy and green above, grey silky-hairy beneath. Flowers 1–4, 1.5–2.5 cm across. Petals 6–9 mm, obovate, tip notched, white, much longer

than sepals. Achenes smooth. *N Iberian Peninsula, W & C France*. H1. Spring–summer.

44. P. sterilis (Linnaeus) Garcke (*P. fragariastrum* Persoon; *Fragaria sterilis* Linnaeus). Illustration: Ross-Craig, Drawings of British plants **8**: t. 29 (1955); Pignatti, Flora d'Italia **1**: 585 (1982); Garrard & Streeter, Wild flowers of the British Isles, 82 (1983); Aeschimann et al., Flora alpina **1**: 781 (2004).

Mat-forming herbaceous perennial. Stems decumbent, 5–15 cm, with spreading hairs; stock usually also producing stolons. Leaves of 3 leaflets; leaflets 8–25 × 8–20 mm, broadly obovate, scalloped-toothed, sparsely hairy and slightly bluish green above, with more dense, spreading, grey hairs beneath. Flowers 1–5, 1–1.5 cm across. Petals *c.* 5 mm, obovate, apex notched, white or very pale pink, equalling or slightly longer than the sepals, widely separated. *W, C & S Europe*. H1. Spring.

45. P. tridentata Solander (*Sibbaldiopsis tridentata* (Solander) Rydberg). Illustration: Britton & Brown, Illustrated flora of the northern United States & Canada, edn 2, **2**: 262 (1913); Fernald, Gray's manual of botany, edn 8, 807 (1950).

Mat-forming perennial, with extensively creeping rhizomes and prostrate, woody bases. Flowering-stems ascending, 2.5–30 cm, adpressed hairy. Leaves of 3 leaflets; leaflets oblong to wedge-shaped, with usually 3 teeth at the apex, usually shiny green and hairless above, adpressed hairy beneath, leathery, evergreen. Flowers 1–6, 6–15 mm across. Petals ovate or obovate, apex entire, white or pale pink (forma **aurora** Graustein), longer than the sepals. *NE North America, Greenland*. H1. Summer.

44. SIBBALDIA Linnaeus
H.S. Maxwell
Tufted perennial herbs, often woody at base. Leaves deciduous, alternate, with 3 leaflets, hairy; stipules papery, attached at the base of leaf-stalks. Flowers bisexual, in cymes; perigynous zone short; sepals 5, epicalyx present; petals 5, sometimes absent; stamens 5, rarely 4 or 10; carpels 5–12. Fruit a cluster of 5–10 achenes, styles deciduous in fruit.

A genus of about 8 species from the colder parts of the northern hemisphere, only one of which is widely grown.
S. parviflora Willdenow (*S. cuneata*

Hornemann) is occasionally listed in specialist catalogues but only appears to differ from the species described below in being more hairy. *Sibbaldia* is very closely related to *Potentilla*. Propagation is usually from seed or by division of the rootstock.

1. S. procumbens Linnaeus (*Potentilla sibbaldi* Haller filius). Illustration: Takeda, Alpine flora of Japan in colour, pl. 42 (1963); Fitter, Wild flowers of Britain & northern Europe, 113 (1974); *The Living Countryside* **8**: 1893 (1988); Blamey & Grey-Wilson, Illustrated flora of Britain & northern Europe, 193 (1989).

Prostrate or tufted herb to 10 cm. Leaflets 5–25 mm, obovate to narrowly wedge-shaped, with 3–6 teeth or small lobes at apex. Flowering-stems 1–30 cm, with 3–7 flowers in dense terminal clusters. Sepals 1–4 mm; petals 1.5–4 mm, rarely absent, narrowly obovate, yellow; style lateral. Fruit 1–3 mm, shiny. *N America, Eurasia.* H1. Summer.

45. FRAGARIA Linnaeus
A.C. Leslie

Perennial herbs with stolons. Leaves all basal, divided into 3 leaflets. Flowers usually in cymes (rarely solitary) on axillary scapes. Perianth parts usually in 5 s, additional petals sometimes present. Perigynous zone short, flat. Epicalyx present. Petals usually white, rarely partly or entirely pink. Stamens and carpels numerous, but one or other reduced or poorly formed in functionally unisexual flowers. Perigynous zone and receptacle becoming fleshy and usually brightly coloured in fruit, either bearing achenes on its surface or sunk in pits. Figure 99(8), p. 183.

A genus of 12–15 species from north temperate and subtropical areas and South America. They grow best in rich loamy soils, but are tolerant of a wide variety of conditions, providing the soil is not waterlogged. They are all tolerant of shade but flower and fruit best in open, sunny sites. *F. vesca* is especially tolerant of chalky soils. Propagation of the species is generally by seed, but that of named cultivars must be by division or from stolons.

Literature: Staudt, G., Taxonomic studies in the genus Fragaria, *Canadian Journal of Botany* **40**: 869–86 (1962); Darrow, G.M., The strawberry (1966).

1a. Calyx-lobes adpressed to the young
fruit   2

b. Calyx-lobes spreading or reflexed in
young fruit   5
2a. Achenes sunk in deep pits in the
receptacle   **3. virginiana**
b. Achenes in shallow pits or almost
superficial   3
3a. Flowers 1–2, *c.* 1.5 cm wide
6. daltoniana
b. Flowers more than 2, more than
2 cm wide   4
4a. Leaflets thick, dark green, with a
conspicuous network of veins, silky-
hairy beneath; flowers usually
functionally unisexual **4. chiloensis**
b. Leaflets not as above; flowers usually
functionally bisexual **5. × ananassa**
5a. Achenes sunk in deep pits in the
receptacle   **3. virginiana**
b. Achenes sunk in shallow pits or
superficial   6
6a. Uppermost flower-stalks with
adpressed hairs   7
b. Uppermost flower-stalks with
spreading or reflexed hairs   8
7a. Flowers less than 2 cm wide; leaflets
hairy above; lateral leaflets stalkless
or very shortly stalked   **1. vesca**
b. Flowers more than 2 cm wide;
leaflets hairless or very sparsely
hairy above; lateral leaflets distinctly
stalked   **5. × ananassa**
8a. Terminal leaflets pale or yellowish
green, often more or less diamond-
shaped with rounded angles; flowers
often functionally unisexual
2. moschata
b. Terminal leaflets usually dark or
bluish green, often obovate or
obovate to diamond-shaped; flowers
usually bisexual   **5. × ananassa**

1. F. vesca Linnaeus. Illustration: Ross-Craig, Drawings of British plants **8**: t. 28 (1955); Keble Martin, The concise British flora in colour, 27 (1965); Grey-Wilson & Blamey, Alpine flowers of Britain and Europe, 103 (1979); Aeschimann et al., Flora alpina **1**: 783 (2004).

Scapes 5–30 cm. Leaflets 1–10 × 1.2–3.8 cm, ovate, obovate or diamond-shaped; lateral leaflets stalkless or very shortly stalked; all bright green and hairy above, with adpressed silky hairs beneath. At least the upper flower-stalks with adpressed hairs. Flowers 1.2–1.8 cm wide, usually bisexual. Calyx-lobes spreading or reflexed in fruit. Fruiting receptacle 1–2 cm, ovoid or almost spherical, rarely ovoid-conical, usually red; achenes projecting from the surface,

sometimes lacking at the base of the fruit. *Most of Europe, eastern N America; naturalised elsewhere.* H1. Spring-autumn.

Variants that, under suitable conditions, flower and fruit throughout the year have long been cultivated as Alpine strawberries (forma **semperflorens** (Duchesne) Staudt). Figure 99(18), p. 183. Other variants sometimes encountered are: forma **roseiflora** (Boulay) Staudt, with pinkish petals; forma **alba** (Duchesne) Staudt, with whitish fruits; forma **eflagellis** (Duchesne) Staudt, which lacks stolons (a character now incorporated in some Alpine strawberry cultivars); 'Monophylla' has leaves of a single leaflet; 'Multiplex' ('Flore Pleno', 'Bowles Double') has double flowers; and 'Muricata' (Plymouth strawberry) has leaf-like petals and stamens and enlarged achenes, each with a long, spine-like tip.

2. F. moschata Duchesne (*F. elatior* Ehrhart). Illustration: Coste, Flore de la France **2**: 27 (1903); Roles, Illustrations to Clapham, Tutin & Warburg, Flora of the British Isles **2**: t. 578 (1960); Pignatti, Flora d'Italia **1**: 586 (1982); Aeschimann et al., Flora alpina **1**: 785 (2004).

Scapes 10–50 cm. Leaflets 1.6–12.5 × 2.2–9.2 cm, the terminal often diamond-shaped with rounded angles, all distinctly stalked, pale or yellowish green and hairy above, with loosely adpressed hairs beneath. All flower-stalks with spreading or reflexed hairs. Flowers 1.5–3 cm wide, often functionally unisexual, the male flowers larger than the female. Calyx-lobes spreading or reflexed in fruit. Fruiting receptacle *c.* 2 cm, ovate, obovate or almost spherical, dark red, with a musky flavour; achenes projecting from the surface but absent from the contracted basal neck. *C Europe, naturalised elsewhere.* H1. Spring–summer.

3. F. virginiana Duchesne. Illustration: Bailey, The evolution of our native fruits, 427–9 (1898); Gleason, Illustrated flora of the north-eastern United States and adjacent Canada **2**: 289 (1952); Darrow, The strawberry, pl. 8–1 (1966).

Scapes 5–25 cm. Leaflets 1.5–7.5 cm × 9–45 mm, broadly ovate to obovate, all distinctly stalked, pale to mid green and sparsely hairy above, thinly to densely hairy beneath. Upper flower-stalks usually with adpressed, rarely spreading hairs. Flowers 6–25 mm wide, usually functionally unisexual, the male flowers larger than the female. Calyx-lobes adpressed to the young fruit. Fruiting receptacles 5–20 mm wide, spherical or ovoid, red. Achenes sunk in deep pits. *E North America, naturalised elsewhere*. H1. Spring–summer.

4. F. chiloensis Duchesne. Illustration: Nicholson, Illustrated dictionary of gardening **3**: 21 (1885); Abrams, Illustrated flora of the Pacific states **2**: 444 (1944); Darrow, The strawberry, pl. 5–2 & 9–1 (1966).

Scapes 3–25 cm. Leaflets 1.8–5 × 1.2–4 cm, obovate to broadly obovate, all distinctly stalked, dark glossy green and almost hairless above, with a strong network of veins and densely adpressed hairy beneath. Flower-stalks with adpressed or spreading hairs. Flowers 2–5.2 cm wide, usually functionally unisexual. Calyx-lobes adpressed to the young fruit. Fruiting receptacle 1.5–2 cm wide, spherical or ovoid, red. Achenes more or less superficial or in shallow pits. *West coast of N America (Alaska to California), west coast of S America (Peru to Argentina), Hawaii; naturalised elsewhere.* H1. Spring–summer.

5. F. × ananassa Duchesne. Illustration: Roles, Illustrations to Clapham, Tutin & Warburg, Flora of the British Isles **2**: t. 579 (1960); Darrow, The strawberry, pl. 6–1 to 6–4 (1966).

Scapes 15–50 cm. Leaflets very variable in size and shape, all distinctly stalked, rather bluish green and almost hairless above, with adpressed hairs beneath. Flower-stalks usually with adpressed hairs, rarely with spreading hairs in a few cultivars. Flowers 2–3.5 cm wide, usually bisexual. Calyx-lobes reflexed or adpressed to the young fruit. Fruiting receptable very variable in size and shape, to 5 cm wide, red. Achenes more or less superficial or in shallow pits. *Garden origin.* H1. Spring–summer.

A hybrid between *F. chiloensis* and *F. virginiana* and now the most commonly cultivated strawberry. Very variable in all its characters, some variants closely resembling one or other parent. There are many named cultivars, two of which are of ornamental value: 'Variegata', with leaves boldly blotched creamy white, and 'Serenata' which has pink flowers. The latter derives from a cross with *Potentilla palustris* (p. 249).

6. F. daltoniana Gay. Illustration: Polunin & Stainton, Flowers of the Himalaya, t. 38 (1985).

Scapes 1.5–5 cm, often with a solitary flower. Leaflets 7–20 × 5–15 mm, elliptic to obovate, all shortly stalked, dark glossy green and sparsely hairy above, with adpressed hairs beneath. Flower-stalks with adpressed hairs. Flowers to 1.5 cm wide. Calyx-lobes adpressed to young fruit; epicalyx-lobes toothed. Fruiting receptacle 1.2–2.5 × 1–1.3 cm, long-ovoid, red to white. Achenes sunk in shallow pits. *Himalaya (Uttar Pradesh to Sikkim)*, H1. Spring–summer.

46. DUCHESNEA Smith
P.S. Green
Like *Fragaria* but the flowers yellow and the fruit not juicy.

A genus of 6 species from India and SE Asia.

1. D. indica (Andrews) Focke (*Fragaria indica* Andrews). Illustration: *Edwards's Botanical Register* **1**: t. 61 (1815); Hegi, Illustriertes Flora von Mittel-Europa, (edn 2) **4**: f. 188 (1964); Everett, New York Botanical Gardens illustrated encyclopedia of horticulture **4**: 1148 (1981); Brickell (ed.), RHS A–Z encyclopedia of garden plants, 392 (2003).

Perennial herb, stems short, leaves in rosettes and on long runners that root at the nodes and tips to form new rosettes. Leaves with 3 leaflets, hairy, leaflets obovate to angular-ovate, margins coarsely toothed, terminal leaflet 1–3 cm × 7–25 mm, lateral leaflets slightly smaller. Flower-stalks 2.5–10 cm, slightly longer than the subtending leaves. Epicalyx and sepals hairy, persistent in fruit. Petals bright yellow. Fruit with achenes spread over a bright red, swollen, insipid receptacle to *c.* 2 cm. *China & India, widely introduced into warm climates.* H5–G1. Summer.

47. ALCHEMILLA Linnaeus
S.M. Walters
Deciduous perennials with a more or less developed woody stock and annual flowering-shoots. Basal leaves palmately 5–9 (rarely to 11) lobed, sometimes compound with separate leaflets. Stem-leaves with fewer lobes and relatively large toothed or lobed stipules. Inflorescence cymose, much-branched, with numerous small flowers on short flower-stalks. Perigynous zone cup-shaped. Perianth

greenish, of 4 outer epicalyx segments and 4 alternating sepals. Petals absent. Stamens 4, often with poorly developed anthers. Carpel 1, with basal style and head-like stigma, more or less protruding from a cup-like perigynous zone (receptacular cup, hypanthium), surrounded by a nectar-secreting disk (see Figure 99(11), p. 183). Fruit an achene enclosed in the dry perigynous zone.

Some 300 species have been described, mostly from the mountains of Europe and the Caucasus; many of these are known to be apomictic and high polyploids. Widespread in Europe and Asia, particularly in the north and on mountains; rare (and mostly introduced) in N America; also on mountains in Africa and the Andes. Most species are easily identified on well-grown material, but late-season second growth can be misleading, and should not be used for identification. The cultivated species of *Alchemilla* may be placed into three groups based upon their growth habit. These are: medium or tall, often robust plants; shorter, softly hairy plants; and low growing alpines. Several species, especially *A. mollis*, reproduce freely from seed, and all can be propagated vegetatively.

In descriptions of the leaves, the depth of lobing may be indicated by a fraction (e.g. on third), which represents the depth of the lobe as a proportion of the distance to the centre of the leaf. If the number of leaf-teeth is given, it refers to one side of the middle lobe.

1a. Low-growing with creeping, sometimes prostrate, annual stems freely rooting at the nodes   **2**
  b. Habit various, rhizomatous, not freely rooting   **4**
2a. Plant hairless or more or less so; epicalyx-segments represented by very small teeth   **23. pentaphyllea**
  b. Plant hairy; epicalyx-segments well developed, at least half as long as sepals   **3**
3a. Mat-forming; leaves 5-lobed (rarely 3-lobed), with a few small, blunt teeth   **24. ellenbeckii**
  b. Loosely spreading; leaves 7-lobed (rarely 5-lobed), with long, acute teeth   **25. abyssinica**
4a. Flowering-stems less than 2 cm   **5. faeroensis**, see var. **pumila**
  b. Flowering-stems at least 5 cm   **5**
5a. Leaves compound, or lobed halfway to centre or more   **6**

b. Leaves lobed less than halfway to centre   10

6a. Leaves with 5–7 free leaflets (sometimes outermost joined basally)   7

b. Leaves with 7–9 (rarely 5) very deep lobes (sometimes the middle one free)   8

7a. Leaf hairless above; teeth of leaflets *c.* 1 mm   **1. alpina**

b. Upper leaf surface thinly adpressed-hairy; teeth of leaflets at least 2 mm   **2. sericea**

8a. Middle leaf-lobe free to base or more or less so   **3. plicatula**

b. All leaf-lobes obviously joined at base   9

9a. Leaves circular in outline, with basal lobes touching or overlapping; leaf-teeth not conspicuous   **4. conjuncta**

b. Leaves kidney-shaped in outline, with widely spaced basal lobes; leaf-teeth conspicuous   **5. faeroensis**

10a. Hairs on flowering-stems and leaf-stalks, if present, more or less adpressed, often silky   11

b. Hairs on flowering-stems and leaf-stalks spreading, at an angle of 45° or more   17

11a. Plant hairless except for lower leaf surface and lowest part of stem   **18. glabra**

b. Plant with hairs on stems up to the inflorescence branches   12

12a. Upper surface of leaves hairless or slightly hairy along the folds only   13

b. Upper surface of leaves evenly hairy   15

13a. Epicalyx segments equalling sepals   **21. venosa**

b. Epicalyx segments shorter than sepals   14

14a. Upper surface of leaves hairless; leaf-outline kidney-shaped   **5. faeroensis**

b. Upper surface of leaves usually slightly hairy along folds; leaf-outline circular   **13. vetteri**

15a. Robust plant with flowering-stems to 30 cm and leaves thick in texture   **6. fulgens**

b. Small plants with slender flowering-stems to 20 cm and leaves thin in texture   16

16a. Basal lobes of mature summer leaves overlapping; leaf-teeth 5 or 6, acute   **10. sericata**

b. Basal lobes of leaves not overlapping; leaf-teeth 4 or 5, often obtuse   **11. rigida**

17a. Robust plants with flowering-stems to 60 cm; epicalyx-segments at least as long as sepals   18

b. Small or medium-sized plants with flowering-stems rarely more than 50 cm; epicalyx-segments shorter than sepals   20

18a. Upper leaf surface hairless   **20. epipsila**

b. Upper leaf surface hairy   19

19a. Leaf-lobes shallow, not more than one fifth; flower-stalks hairless   **19. mollis**

b. Leaf-lobes relatively deep, to two-fifths; flower-stalks hairy   **22. speciosa**

20a. Upper leaf surface hairless   **17. xanthochlora**

b. Upper leaf surface hairy   21

21a. Whole plant, including individual flower-stalks, hairy   22

b. At least some individual flower-stalks hairless   24

22a. Hairs on stems and leaf-stalks ascending   **9. elisabethae**

b. Hairs on stems and leaf-stalks spreading or slightly deflexed   23

23a. Leaf-lobes truncate, separated by deep, toothless incisions   **8. erythropoda**

b. Leaf-lobes rounded, separated by shallow incisions   **7. glaucescens**

24a. Flowers less than 2.5 mm across; lower part of stem and leaf-stalks densely hairy with some downwardly directed hairs   **16. tytthantha**

b. Flowers at least 2.5 mm across; lower part of stem and leaf-stalks hairy but without downwardly directed hairs   25

25a. Hairs on stem and leaf-stalks ascending, sometimes almost silky   **12. lapeyrousii**

b. Hairs on stems and leaf-stalks spreading at right angles   26

26a. Leaf-lobes of summer leaves almost triangular   **15. acutiloba**

b. Leaf-lobes rounded   **14. monticola**

Section **Alpinae** Camus. Leaves palmate or deeply palmately lobed, more or less silvery-hairy on the lower surface; epicalyx-segments less than half as long as sepals.

1. A. alpina Linnaeus. Figure 107(1), p. 257. Illustration: Coste, Flore de la France **2**: 63 (1903); Ross-Craig, Drawings of British plants **9**: pl. 16 (1956); Garrard & Streeter, Wild flowers of the British Isles,

pl. 37 (1983); Brickell (ed.), RHS A–Z encyclopedia of garden plants, 94 (2003).

Mat-forming, with creeping woody rootstock and slender, erect, densely adpressed-hairy flowering stems to 15 cm. Basal leaves compound with 5 (rarely 7) lanceolate-obovate leaflets with acute teeth towards the apex, dark green and hairless above, silvery-hairy beneath with closely adpressed hairs. *W & WC Europe.* H1. Summer.

Prefers non-calcareous soil. Most plants sold under this name are either *A. plicatula* or *A. conjuncta,* which are both more attractive and easier to cultivate.

2. A. sericea Willdenow. Figure 107(2), p. 259. Illustration: Komarov (ed.), Flora SSSR **10**: pl. 22 (1941).

Like *A. alpina* but with much more conspicuous and irregular teeth in the upper half of the leaflets, and with thin, adpressed hairs on the upper surface of the leaves. *Former USSR (Caucasus), Turkey.* H1. Summer.

Rarely cultivated and apparently rather difficult to grow.

3. A. plicatula Gandoger (*A. asterophylla* Buser; *A. hoppeana* misapplied). Figures 107 (3), p. 257 & 108(1), p. 258. Illustration: Coste, Flore de France **2**: 63 (1903).

Like *A. alpina* but more robust, with stems to 20 cm; leaves with 7 (rarely 9) segments, the middle one free to the base, the others joined near the base, and leaf outline circular, with somewhat overlapping basal segments. *Europe (Pyrenees to Balkan Peninsula), Turkey.* H1. Summer.

4. A. conjuncta Babington (*A. alpina* misapplied). Figure 107(4), p. 257. Illustration: Coste, Flore de France **2**: 63 (1903); Ross-Craig, Drawings of British plants **9**: pl. 5 (1956); Brickell (ed.), RHS A–Z encyclopedia of garden plants, 94 (2003); Aeschimann et al., Flora alpina **1**: 787 (2004).

Robust plant with flowering-stems to 30 cm, silky-hairy throughout. Leaves relatively thick, hairless above and densely silky-hairy beneath, more or less circular in outline, with 7 (rarely 9) broadly lanceolate to elliptic segments all distinctly joined at the base for up to one fifth of their length and with indistinct teeth almost hidden by the silky hair-covering. *Jura & SW Alps; naturalised in Scotland.* H1. Summer.

Often sold as *A. alpina.*

Figure 107. Leaf silhouettes of *Alchemilla* species (× 0.5). 1, *A. alpina.*
2, *A. sericea.* 3, *A. plicatula.* 4, *A. conjuncta.* 5a, *A. faeroensis,*
b, *A. faeroensis* var. *pumila.* 6, *A. fulgens.* 7, *A. glaucescens.*
8, *A. erythropoda.* 9, *A. elisabethae.* 10, *A. sericata.* 11, *A. rigida.*

12, *A. lapeyrousii.* 13, *A. vetteri.* 14, *A. monticola.* 15, *A. acutiloba.*
16, *A. tytthantha.* 17, *A. xanthochlora.* 18, *A. glabra.* 19, *A. mollis.*
20, *A. epipsila.* 21, *A. venosa.* 22, *A. speciosa.* 23, *A. pentaphyllea.*
24, *A. ellenbeckii.* 25, *A. abyssinica.*

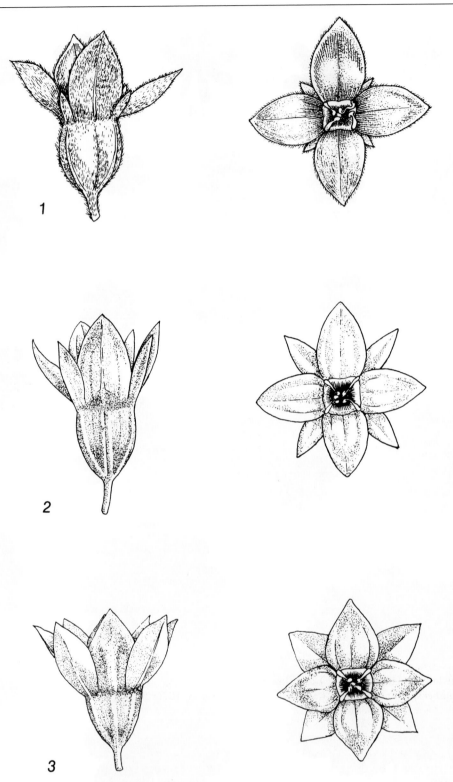

Figure 108. Flowers of *Alchemilla* species, side and front view.
1, *A. plicatula* (Section Alpinae) (× 10). 2, *A. tytthantha* (Section
Alchemilla) (× 10). 3, *A. mollis* (Section Alchemilla, Series Elatae) (× 10).

Section **Alchemilla**. Leaves palmately lobed, rarely to more than halfway; epicalyx-segments more than half as long as sepals.

5. A. faeroensis (Lange) Buser. Figure 107(5), p. 257.

Like *A. conjuncta* but leaves kidney-shaped in outline, usually 7-lobed, with deep incisions to about halfway to the centre, and with rather large, acute teeth around the upper edge of the lobes. *E Iceland, Faeroe Islands.* H1. Summer.

Both the typical plant and the dwarf var. **pumila** (Rostrup) Simmons, which does not exceed 5 cm, are easily cultivated.

6. A. fulgens Buser (*A. splendens* misapplied). Figure 107(6), p. 257. Illustration: Coste, Flore de la France 2: 64 (1903).

Like *A. conjuncta* but leaf-lobes broader and shorter, with acute teeth, separated by toothless, V-shaped incisions to less than halfway; upper surface of leaf more or less adpressed hairy. *Pyrenees.* H1. Summer.

7. A. glaucescens Wallroth (*A. minor* misapplied; *A. hybrida* misapplied; *A. pubescens* misapplied). Figure 107(7), p. 257. Illustration: Coste, Flore de la France 2: 64 (1903); Ross-Craig, Drawings of British plants 9: pl. 1 (1956); Garrard & Streeter, Wild flowers of the British Isles, pl. 37 (1983).

Plant to 20 cm, covered with densely spreading hairs throughout. Leaves circular in outline, 7–9-lobed, basal lobes usually overlapping; all lobes shallow, rounded, with 4–6 teeth: incision between lobes shallow or absent. Inflorescence with rather dense clusters of flowers. *Europe.* H1. Summer.

Neat and easily grown, but infrequently cultivated.

8. A. erythropoda Juzepczuk (*A. erythropodioides* Pawlowski). Figure 107 (8), p. 257. Illustration: Beckett, Concise encyclopedia of garden plants, 17 (1983).

Like *A. glaucescens* but grey or blue-green with deep incisions to two-fifths between the truncate leaf-lobes; dense hairs on leaf-stalks and lower part of flowering-stems slightly downwardly directed. Often developing a reddish colour on stems in sun. *Balkan Peninsula, W Carpathians, Turkey, Former USSR (Caucasus).* H1. Summer.

Plants referable to the closely allied **A. caucasica** Buser, differing mainly in lacking downwardly directed hairs, are

apparently also in cultivation under the name *A. erythropoda*.

9. A. elisabethae Juzepczuk. Figure 107 (9), p. 257.

Dwarf, very blue-green plant with flowering stems to 15 cm, thickly clothed with ascending hairs throughout. Leaves 7-lobed, with deep incisions (to two-fifths) between the flattened lobes; teeth on lobes 3 or 4, broad and blunt. *Former USSR (Caucasus).* H1. Summer.

A very attractive rock-garden plant, beginning to be popular.

10. A. sericata Reichenbach. Figure 107 (10), p. 257. Illustration: Reichenbach, Iconographia botanica exotica 1: t. 4 (1827).

Small, slender plant with thin flowering-stems to 20 cm, thinly silky-hairy throughout. Leaves 7-lobed, circular in outline, with overlapping basal lobes, with deep incisions to halfway between the semicircular lobes; teeth on lobes 5 or 6, acute; upper leaf surface light green, sparsely adpressed hairy, lower surface pale green with a denser, rather silky hair-covering. Flowering-stems and leaf-stalks quickly turning reddish in sun. *Former USSR (Caucasus).* H1. Summer.

11. A. rigida Buser. Figure 107(11), p. 257.

Like *A. sericata* but hair-covering less closely adpressed (a difference especially obvious on the underside of the leaves, which have a less silky appearance), and leaves a deeper green, kidney-shaped in outline, with basal lobes not or scarcely overlapping, with fewer (4 or 5) broader teeth on the lobes. *Turkey, Former USSR (Caucasus).* H1. Summer.

An attractive rock-garden plant.

12. A. lapeyrousii Buser (*A. hybrida* (Linnaeus) Linnaeus). Figure 107(12), p. 257. Illustration: *Watsonia* 5: 259 & pl. 12b (1963).

Medium-sized, rather robust plant with flowering-stem to 25 cm, clothed with forwardly directed hairs throughout except the flower-stalks, which are often hairless (a lens is needed to see this). Leaves 7–9-lobed, ascending-hairy on both surfaces and on leaf-stalks, kidney-shaped in outline, with more or less triangular lobes, with short incisions (to one third) between lobes; teeth on lobes 5 or 6, acute. *Pyrenees & SC France.* H1. Summer.

13. A. vetteri Buser. Figure 107(13), p. 257.

Small plant with flowering-stems to 20 cm, clothed throughout with more or less adpressed, silky hairs. Leaves hairless or only thinly hairy along the folds above, silky-hairy beneath, circular in outline, with 7–9 rather truncate lobes separated by long incisions to two-fifths; teeth on lobes 5 or 6, acute. *SW Europe, Spain to France (Maritime Alps).* H1. Summer.

An attractive rock-garden plant, rarely seen in cultivation.

Species **14–18**, which could be grouped together under the aggregate name *A. vulgaris*, have a limited value as garden plants, mainly because their flowers are relatively small (usually less than 3 mm across) and are greenish rather than yellowish.

14. A. monticola Opiz (*A. pastoralis* Buser). Figure 107(14), p. 257. Illustration: Garrard & Streeter, Wild flowers of the British Isles, pl. 37 (1983).

Medium-sized plant with flowering-stems to 50 cm, mature summer leaves to 8 cm across, densely spreading-hairy throughout except for the flower-stalks, which are usually hairless (use a lens). Leaves more or less circular in outline, 9–11-lobed; lobes rounded, separated by short incisions; leaf-teeth 7–9, rather regular, acute. *Europe.* H1. Summer.

One of the commonest 'Vulgaris' alchemillas throughout much of Europe. Occasionally brought into cultivation from the wild.

15. A. acutiloba Opiz (*A. acutangula* Buser). Figure 107(15), p. 257. Illustration: Garrard & Streeter, Wild flowers of the British Isles, pl. 37 (1983).

Robust plant to 65 cm, with unmistakable mature summer leaves to 10 cm across, with 9–13 almost triangular lobes with straight sides, and very unequal, acute teeth, the largest in the middle of each side, spreading hairy except for the inflorescence, which is almost hairless, and the upper leaf surface, which is only sparsely hairy. *Europe.* H1. Summer.

16. A. tytthantha Juzepczuk (*A. multiflora* Rothmaler). Figures 107(16), p. 257 & 108(2), p. 258. Illustration: *Watsonia* 4: 282 & pl. 17(1961).

Medium-sized plant with relatively slender flowering-stems to 50 cm and almost circular leaves with dense spreading hairs on both surfaces, with 9 (rarely to

11) shallow lobes and 6–8 small, acute teeth. Leaf-stalks and lower part of flowering-stems densely hairy with some downwardly pointing hairs. Flowers hairless, small (less than 2.5 mm). *Former USSR (Crimea)*. H1. Summer.

Naturalised in several places in S Scotland, presumably from botanic gardens. Tolerates light shade.

17. A. xanthochlora Rothmaler (*A. pratensis* misapplied). Figure 107(17), p. 257. Illustration: Ross-Craig, Drawings of British plants **9**: pl. 3 (1956); Garrard & Streeter, Wild flowers of the British Isles, pl. 37 (1983); Aeschimann et al., Flora alpina **1**: 789 (2004).

Medium-sized plant, often yellow-green, with robust flowering-stems to 50 cm and more or less kidney-shaped leaves, hairless or more or less so on the upper surface, with dense spreading hairs below. Leaf-lobes 9–11, rounded, with 7–9 wide, acute, almost equal teeth. Lower part of flowering-stems and leaf-stalks with spreading hairs, often somewhat forwardly directed; upper part of inflorescence, including flower-stalks and flowers, usually hairless. *N, W & C Europe, extending to Sweden & Greece. Summer.* H1.

One of the commonest 'Vulgaris' alchemillas in the wild. Not uncommonly brought into cultivation from the wild, but not recommended for gardens.

18. A. glabra Neygenfind (*A. alpestris* misapplied). Figure 107(18), p. 257. Illustration: Ross-Craig, Drawings of British plants **9**: pl. 4 (1956); Garrard & Streeter, Wild flowers of the British Isles, pl. 37 (1983); Aeschimann et al., Flora alpina **1**: 789 (2004).

Medium-sized plant with flowering-stems to 60 cm, easily distinguished from all other species likely to be seen in gardens by being almost hairless. Leaves more or less kidney-shaped, with 9–11 triangular-ovate lobes and no incisions between; leaf-teeth 7–9, rather wide but acute, and very unequal. Adpressed hairs are present on lowest stem internodes and on the leaf-stalks and on the outer half of veins on the lower leaf surface. *N & C Europe, extending to the Pyrenees & Balkans.* H1. Summer.

Sometimes cultivated from the wild.

Species **19–22** belong to the Series **Elatae** Rothmaler, and have larger flowers and a generally more decorative, yellower inflorescence than the other species.

19. A. mollis (Buser) Rothmaler (*A. grandiflora* Anon.). Figures 107(19), p. 257 & 108(3), p. 258. Illustration: *Journal of the Royal Horticultural Society* **73**: 309, f. 106 (1948); Beckett, Concise encyclopedia of garden plants, 17 (1983); Brickell (ed.), RHS A–Z encyclopedia of garden plants, 94 (2003); Aeschimann et al., Flora alpina **1**: 791 (2004).

Large, robust plant with flowering-stems to 80 cm, densely spreading-hairy throughout except for the flower-stalks (use a lens). Leaves to 15 cm, circular in outline with overlapping basal lobes; lobes 9–11, very shallow, semicircular, with 7–9 wide, ovate, unequal teeth. Inflorescence showy, yellowish; flowers relatively large, to 5 mm across, with epicalyx-segments at least as long as sepals, and giving the appearance of an 8-pointed star. *E Carpathians, Turkey, former USSR (Caucasus)*. H1. Summer.

In recent years this species has become by far the commonest alchemilla in cultivation in the British Isles. It flowers relatively late (summer rather than late spring), and is much used in flower-arrangements. 'Variegata', with variably developed yellowish leaf-markings, is sometimes grown. Plants grown as 'Robusta', and said to be more vigorous than the type, have proved on comparative cultivation to be indistinguishable. 'Mr Poland's Variety' is *A. venosa* (see below).

20. A. epipsila Juzepczuk (*A. indivisa* Rothmaler). Figure 107(20), p. 257.

Like *A. mollis* in stature and showy, yellowish inflorescence, but easily distinguished by the hairless upper leaf surface. The leaf-lobes are less shallow than in *A. mollis. Balkan Peninsula*. H1. Summer.

In cultivation in some gardens in Britain, and probably more widespread than presently recognised.

21. A. venosa Juzepczuk. Figure 107(21), p. 257.

Medium-sized plant, with rather slender flowering-stems to 40 cm, clothed up to the main inflorescence branches with more or less adpressed hairs. Inflorescence and flowers resembling *A. mollis*. Leaves somewhat thin and papery, circular in outline, with overlapping basal lobes; lobes 9–11, semi–circular, with clear incisions to one third; leaf-teeth 6–8, unequal, acute. Upper leaf surface hairless, lower with more or less adpressed hairs; leaf-stalks

more or less adpressed-hairy. *Former USSR (Caucasus)*. H1. Late summer.

Widespread in gardens in Britain, but not usually distinguished from *A. mollis*, which it resembles in flower. Distinguished by its more delicate stature, more or less adpressed hairs and earlier flowering time.

22. A. speciosa Buser. Figure 107(22), p. 257. Illustration: *Botanical Magazine*, n.s., 546 (2006).

Rather large plant, with the general appearance and floral characters of *A. mollis*, but differing most obviously in its deeply lobed leaves (to two-fifths), with narrower, acute teeth, its ascending hairs and hairy flower-stalks. *Former USSR (Caucasus)*. H1. Summer.

Rare in cultivation.

Section **Pentaphyllea** Camus. Stoloniferous, rooting at the nodes. Leaves palmate. Epicalyx-segments reduced to 4 small teeth.

23. A. pentaphyllea Linnaeus. Figure 107(23), p. 257. Illustration: Coste, Flore de la France **2**: 62 (1903); Guinochet & Vilmorin, Flore de France **5**: t. 1674 (1984); Aeschimann et al., Flora alpina **1**: 785 (2004).

Dwarf, often prostrate plant, hairless or nearly so, with short and slender flowering-shoots rooting at the nodes. Leaves to 2 cm, palmate with 5 leaflets deeply cut into long, narrow segments in the upper half. Flowers few; epicalyx-segments represented by very small teeth, much shorter than the ovate, blunt sepals. *Alps*. H1. Summer.

Very different from other European species and reproducing sexually. Rare in cultivation. It needs careful alpine house or trough cultivation.

Section **Longicaules** Rothmaler (including Section Parvifoliae Rothmaler). Wide-creeping plants freely rooting at the nodes. Epicalyx-segments at least half as long as sepals. The classification of this and other African groups is as yet relatively little understood, and there is some evidence that they are at least partly sexual, with consequent free hybridisation.

For further information, see Hedberg, O., Afro-alpine vascular plants, *Symbolae Botanicae Upsalienses* **15**(1): 102–17 & 281–92 (1957).

24. A. ellenbeckii Engler. Figure 107 (24), p. 257.

A small, prostrate, sparsely hairy, wide-creeping plant freely rooting at the nodes of slender stolons, and producing short, erect, few-flowered lateral shoots. Leaves not more than 2 cm across, deeply 5-lobed; lobes wedge-shaped, with few small blunt teeth. *Mountains, from Ethiopia to Kenya.* H4. Summer.

Very variable, but the clones now popularly on sale in garden nurseries are very uniform dwarf carpeting plants suitable for damp ground, and easily reproduced vegetatively. It will not, however, survive more than a little frost and needs protection in winter.

25. A. abyssinica Fresenius. Figure 107 (25), p. 257. Illustration: *Senckenbergia Biologia (Frankfurt)* **61**: 130, 131 (1980).

Wide-creeping plant with prostrate stems rooting at the nodes, and short, erect, few-flowered branches. Leaves to 4 cm, kidney-shaped in outline, mostly 7-lobed; lobes shortly wedge-shaped with incisions to almost halfway, with 3–5 long, acute, curved teeth. Leaf surfaces sparsely hairy; leaf-stalks and stems obviously hairy with long, spreading hairs. *Ethiopia to Kenya.* H4. Summer.

Another African species, **A. pedata** A. Richard, is in cultivation in Britain; it may prove to be hardier than the others, but is not yet widely available.

48. CYDONIA Miller
E.C. Nelson
Deciduous shrubs or trees, without thorns, to 8 m; young shoots with grey-white felt. Leaves entire, ovate to elliptic, to 10 × 5 cm, densely felted below; stipules falling early, hairy, glandular. Flowers solitary, white to pink. Sepals 5, persistent, hairy outside, toothed. Petals 5. Stamens 15–25. Ovary inferior. Carpels 5, walls cartilaginous in fruit; ovules numerous. Styles 5, free. Fruit a pome, closed at top with persistent sepals, pear-shaped, aromatic, golden yellow, covered with felt of short hair.

A genus of a single species, the quince, has been cultivated for many centuries and was employed as an aphrodisiac in classical times. Plants are used as stock for grafting pears. It can be propagated by cuttings and from seed.

1. C. oblonga Miller (*Pyrus cydonia* Linnaeus; *C. vulgaris* Persoon; *C. maliformis* Miller). Illustration: Bianchini et al., Fruits of the earth, 135 (1975); Graf, Tropica, 849 (1978).

Perhaps W Asia, naturalised in S Europe. H3. Spring.

As it has been so long in cultivation, numerous minor variants are reported including 'Lusitanica' (*Cydonia lusitanica* Miller), which is more vigorous, produces larger, pink flowers, but is not as hardy as the normal variants; 'Vranja' (illustration; Brickell (ed.), RHS A–Z encyclopedia of garden plants, 337, 2003) has very fragrant pale green fruit.

49. CHAENOMELES Lindley
E.C. Nelson
Deciduous shrubs or trees, sometimes with thorns. Leaves simple, toothed; stipules persistent, large, kidney-shaped. Flowers in axillary clusters. Sepals 5, hairless, deciduous, erect, entire. Petals 5 (numerous in some 'double-flowered' cultivars). Stamens more than 20. Ovary inferior. Carpels 5, ovules numerous; styles 5, free. Fruit a pome, closed at top, stalkless. Figure 99(1), p. 183.

Cultivars must be propagated vegetatively by cuttings (half-ripened wood, in June and July); species may be raised from seed, but variation should be expected and hybrids may be formed when other species or cultivars are grown nearby. Commonly treated as wall plants; prune after flowering, and in late summer train and remove surplus shoots.

Literature: Weber, C., Cultivars in the genus Chaenomeles, Arnoldia **23**: 17–75 (1963); Weber, C., The genus Chaenomeles (Rosaceae), *Journal of the Arnold Arboretum* **45**: 302–45 (1964).

1a. Tree or tall shrub at least 2 m; shoots stout, erect; young leaves with dense adpressed hairs beneath **1. cathayensis**
  b. Shrub with spreading shoots, rarely over 1 m (except when artificially trained); leaves hairless or at most sparsely hairy beneath 2
2a. Leaves coarsely toothed; young shoots felted becoming hairless, distinctly warty; petals orange-red to pink **2. japonica**
  b. Leaves finely toothed; young shoots hairless, never warty; petals pink rarely white **3. speciosa**

1. C. cathayensis (Hemsley) Schneider (*Cydonia cathayensis* Hemsley; *C. mallardii* invalid; *Chaenomeles lagenaria* var. *cathayensis* (Hemsley) Rehder; *C. lagenaria* var. *wilsonii* Rehder). Illustration: *Hooker's Icones Plantarum*, 2657 & 2658 (1901).

Shrub to small bushy tree, to 3 m or more; branches few, stout, stiffly erect, with numerous short, blunt spurs; shoots hairy when young, becoming hairless. Leaves elliptic to lanceolate, 5–8 × 2–3.5 cm, apex acute, margins toothed, teeth forward-pointing, with awn-like tips, felted underneath when young; stipules pointing forwards, to 1.5 × 2 cm. Flowers white to pink. Fruits to 15 cm, occasionally to 20 cm, apple-like, green, yellow or red. *China, Burma, Bhutan.* H4. Spring.

This produces the largest fruits, which have long been used medicinally in China.

C. × vilmoriniana Weber. A hybrid between *C. cathayensis* and *C. speciosa.* Similar to the former in general habit and leaf-shape, young leaves felted beneath; flowers white, flushed pink. Differs in its shoots, which are more slender, spurred and sharply spined; fruits few, c. 8 cm. *Garden origin.* H4. Spring.

Selections include: 'Mount Everest', flowers large, white changing to pink with yellow and lavender; 'Vedrariensis' (*C. hybrida* var. *vedrariensis* invalid), flowers white flushed pink.

C. × californica Weber. A hybrid between *C. cathayensis* and *C. × superba.* Similar to the former in having stiff, erect branches, differs in being more slender and sharply spurred; young shoots hairy. Leaves lanceolate, felted beneath when young. Flowers large, white (Vilmoriniana group), pink to red or bicoloured. Fruits apple-shaped. *Garden origin.* H4. Spring.

C. × clarkeana Weber. A hybrid between *C. cathayensis* and *C. japonica.* Low, spreading shrub, spines numerous; leaves intermediate in shape and character between the 2 parents; flowers large, pink to red.

2. C. japonica (Thunberg) Spach (*Pyrus japonica* Thunberg; *P. maulei* Masters; *Cydonia japonica* (Thunberg) Persoon).

Shrub to 1 m; branches loose, spreading, with short felt when young, becoming hairless and distinctly warty. Leaves ovate to spathulate, apex often notched, margins coarsely toothed, to 5 × 3 cm, hairless even when young; stipules toothed, to 1 × 2 cm. Flowers orange-red to pink. Stamens 40–60. Fruits c. 4 cm, aromatic. *Japan.* H2. Spring.

Often used to flavour confectionery and jams. A cultivar, 'Maulei', with single, orange to salmon-pink flowers is also grown.

Var. **pygmaea** Maximowicz has subterranean branches and lacks spines.

Var. **alpina** Maximowicz is smaller in all parts, and more branched than var. *japonica*.

3. C. speciosa (Sweet) Nakai (*Cydonia speciosa* Sweet; *C. japonica* (Loiseleur) Persoon; *Chaenomeles lagenaria* (Loiseleur) Koidzumi). Illustration: *Botanical Magazine*, 692 (1803).

Shrub 2–4 m; shoots erect or spreading, spined, hairless or occasionally hairy when young. Leaves to 10 × 4 cm, ovate to oblong, hairless, rarely sparsely hairy on veins below when young, sharply toothed; stipules pointing forward, to 1 × 2 cm. Flowers red, rarely white to pink. Fruit variable in shape and size. *China, Burma*. H3. Spring.

C. × superba (Frahm) Rehder (*Cydonia maulei* T. Moore var. *superba* Frahm). A hybrid between *C. japonica* and *C. speciosa*. Shrub to 1.7 m, branches spreading, spined; young shoots with short, rough hairs, becoming warty. Leaves resembling those of *C. japonica*. Flowers white to orange to red. Fruit apple-shaped. *Garden origin*. H3. Spring.

Numerous cultivars have been raised that belong to this hybrid, among the best is 'Rowallane Seedling' with single, red flowers.

50. DOCYNIA Decaisne
M.F. Gardner
Evergreen or semi-evergreen, trees or shrubs. Leaves ovate-elliptic to lance-shaped, entire, lobed, or margins with forward-pointing teeth; stipules present. Flowers 2–5, borne in stalkless umbels; calyx densely felted, sepal lobes lance-shaped; petals 5, broadly ovate, narrowed towards the base; stamens 30–50; ovary inferior, 5-celled with 3–10 ovules, styles 5, joined, with long hairs at base. Fruit fleshy, egg-shaped to pear-shaped, calyx persistent.

A genus of 5 species native to China, the East Indies and Vietnam. The name *Docynia* is an anagram of *Cydonia*. Plants of *D. rufifolia* (Léveillé) Rehder may be confused with *D. indica* and *D. delavayi*, particularly the latter.

1a. Leaves entire   **1. delavayi**
  b. Leaves divided or lobed, sometimes entire, but with finely toothed margins   **2. indica**

1. D. delavayi (Franchet) Schneider (*Pyrus delavayi* Franchet). Illustration: *Revue Horticole*, 45–7 (1918); Flora Republicae Popularis Sinicae **36**: 349 (1974); Wu, Wild flowers of Yunnan, **2**: 109 (1986).

Evergreen spreading shrub or tree to 7 m; young shoots hairy, becoming chocolate-brown and hairless, later black, sometimes developing spine-tipped short shoots. Leaves 3–7 × 2.3–2.8 cm, ovate to lance-shaped, pointed, entire, felted beneath; leaf-stalks, 1–1.5 cm, downy. Flowers pink in bud, very fragrant; calyx felted, petals white. Fruit yellow, *c.* 3 cm across, egg-shaped, downy. *China (Yunnan)*. H4. Spring.

2. D. indica Decaisne. Illustration: Brandis, Indian trees, fig. 124 (1906); Schneider, Illustriertes Handbuch der Laubholzkunde **1**: 726, 727 (1906).

Evergreen tree, 3–5 m, branches short, with thorns; young branches densely white-woolly, later almost hairless. Leaves 5–7 × 2.5–2.8 cm, ovate to oblong-ovate, sharp- pointed or tapering to a point, rounded towards the base, finely toothed, sometime entire, woolly hairy beneath, later hairless; leaf-stalks 1–1.5 cm; juvenile leaves lobed. Flowers borne in axillary umbels; calyx grey-yellow, hairy; petals white. Fruit yellow, 4–5 cm, egg-shaped to rounded. *India (Assam), Upper Burma, China (Yunnan)*. H4. Spring.

51. × PYRONIA Veitch
S.G. Knees & P.G. Barnes
Shrub or small tree to 8 m, resembling *Cydonia* in habit. Shoots brown, densely dotted with lenticels. Leaves deciduous, 5–8 × 2–4 cm, elliptic, downy beneath; margins entire or finely scalloped; leaf-stalk 1.5–2 cm. Flowers borne in groups of 3 at the shoot-tips; sepals triangular, hairy on both surfaces, with glandular teeth; corolla *c.* 1.7 × 1.5 cm, rounded with notch at apex, white with pink-tinged margins, especially in bud; stamens 20, anthers deep pink-violet; ovary inferior; styles more or less free at base, very hairy. Fruit 7–8 cm, egg-shaped, greenish yellow, spotted red; flesh white, lacking stone or grit cells.

A genus of sexual hybrids between *Cydonia* and *Pyrus* originally developed around 1895 in the Veitch nursery.

1. × P. veitchii (Trabut) Guillaumin; *Cydonia oblonga × Pyrus communis*. Illustration: *Journal of Heredity* **7**: 416 (1916).

Originated in cultivation. H4. Spring–early summer.

'Luxemburgiana' is more like *Pyrus* in habit with greenish yellow, pear-shaped fruits. Illustration: *Bulletin de la Société Dendrologique de France* **56**: 68 (1925).

52. × PYROCYDONIA Daniel
S.G. Knees & P.G. Barnes
Deciduous tree to 5 m. Leaves 3.5–7.5 × 2.5–4 cm, ovate to elliptic, rounded at base, acute at apex, more or less downy; margin toothed; leaf-stalk to 6 mm, downy. Sepals triangular, hairless except for inner-surface, margin toothed with glandular hairs; petals *c.* 1 × 1.2 cm, broadly ovate, white; stamens 20; ovary inferior; styles free, hairless at base. Fruits 6–8 cm, brownish with paler spotting.

This is a result of a graft chimaera, forming a tree that resembles quince.

1. P. daniellii Daniel *Cydonia oblonga × Pyrus communis*. Illustration: *Revue Horticole*, 28–9 (1914).

'Winkleri' has elliptic to ovate leaves to 4.5 cm, which are more downy.

53. PYRUS Linnaeus
S.G. Knees
Deciduous trees, sometimes with thorny branches. Buds with overlapping scales. Leaves alternate, inrolled in bud, stalked, entire or, very rarely, lobed; margins often with forward-pointing teeth. Flowers in umbel-like racemes, appearing with or before the leaves. Petals 5, white, or rarely pinkish, clawed, rounded to broadly oblong; stamens 18–30, anthers usually reddish; ovary inferior; styles 2–5, free, closely constricted at base by nectariferous disk; ovules 2 per cell. Fruit a spherical or pear-shaped pome; flesh with grit cells; seeds black or brownish black.

A genus of about 25 species, from Europe to E Asia, south to N Africa. Ornamental trees with conspicuous white flowers and foliage that often turns a good reddish colour in autumn. Propagation of the species is from seed or by grafting on to stock of *P. communis*, while most of the cultivars are grafted. A sunny position and not too moist soil will suit most, and several of the Mediterranean species can tolerate long dry periods.

Literature: Browicz, K., Conspectus and chorology of the genus Pyrus, *Arboretum Kornickie* **38**: 17–38 (1993).

1a. At least some branches terminating in thorns or spines   2

b. Branches usually thornless   14
2a. Leaves remaining hairy or
   downy, even when mature
                 15. eleagrifolia
b. Leaves hairless when young, or
   becoming so with maturity   3
3a. Leaf-blade hairless when young, very
   glossy   4
b. Leaf-blade downy or hairy when
   young, eventually hairless   8
4a. Leaf-stalk 5–6 cm   5
b. Leaf-stalk 2–5 cm   7
5a. Leaves 2.5–5 cm   **11. fauriei**
b. Leaves at least 7 cm   6
6a. Leaf-margin scalloped
                 9. calleryana
b. Leaf-margin deeply 3–5 lobed
                 10. dimorphophylla
7a. Leaf-blades ovate, wedge-shaped,
   rounded, or cordate at base; stalk
   often longer than blade   **2. cordata**
b. Leaf-blades ovate to oblong-
   lanceolate; stalk as long or shorter
   than blade   **4. syriaca**
8a. All leaves entire   9
b. At least some leaves pinnately
   divided or lobed   10
9a. Petals 6–12 mm   11
b. Petals 1.5–1.7 cm   **1. communis**
10a. Petals circular   **7. pashia**
b. Petals ovate to oblong   **19. regelii**
11a. Fruit with deciduous sepals
                 3. cossonii
b. Fruit with persistent sepals   12
12a. Leaf-blade 5–9 mm wide   13
b. Leaf-blade 1–2 cm wide
                 12. amygdaliformis
13a. Branches pendent   **5. salicifolia**
b. Branches spreading   **20. salvifolia**
14a. Leaf-blade with acute tips   15
b. Leaf-blade acuminate or obtuse   16
15a. Leaves 3–6 cm, leaf-stalk hairy
       15. elaegrifolia, see var.
                 kotschyana
b. Leaves 5–8 cm, leaf-stalk hairless
                 14. nivalis
16a. Styles 2 or 3   17
b. Styles 4 or 5   18
17a. Leaf-margin with broad, spreading
   teeth; leaf-stalk 2.5–6 cm
                 17. phaeocarpa
b. Leaf-margin with saw-like teeth; leaf-
   stalk 2–3.5 cm   **8. betulifolia**
18a. Young shoots hairless or almost so
                 19
b. Young shoots covered with felted
   hairs   20
19a. Shoots of second year purplish
   brown; leaf-blade long acuminate
                 13. bretschneideri

b. Shoots of second year yellowish
   brown; leaf-blade shortly acuminate
                 6. ussuriensis
20a. Petals 7–9 mm   **16. × michauxii**
b. Petals 1.5–1.7 cm   21
21a. Leaf base truncate   **6. ussuriensis**
b. Leaf base rounded, cordate or
   broadly wedge-shaped   **18. pyrifolia**

1. P. communis Linnaeus. Illustration:
Ross-Craig, Drawings of British plants **9**:
pl. 27 (1956); Polunin & Everard, Trees
and bushes of Europe, 75 (1976); Phillips,
Trees in Britain, Europe and North
America, 179, 219 (1978); Krüssmann,
Manual of cultivated broad-leaved trees &
shrubs **3**: 77 (1986).

Tree to 15 m, with broad crown and
occasionally thorny branches. Leaf-blade
4–5 × 2.5–3.5 cm, ovate to elliptic, almost
hairless, acuminate, margins with forward-
pointing teeth; leaf-stalks reddish, c.
1.5 cm. Flowers in groups of 7–9, often
tinged pink in bud. Calyx-lobes 7–9 mm,
lanceolate, covered with light brown hairs.
Petals 1.5–1.7 × 1–1.2 cm, ovate-oblong,
free. Stamens 18–20, anthers deep pinkish
red. Styles 3–5. Fruit pear-shaped to
spherical, 2–5 cm, yellowish green.
S Europe, SW Asia. H1. Spring.
   Numerous cultivated variants
exist.

2. P. cordata Desvaux. Illustration:
Decaisne, Jardin fruitier du Muséum **1**:
pl. 3 (1858); Ross-Craig, Drawings of
British plants **9**: pl. 28 (1956).
   Like *P. communis* but smaller in all its
parts, shrubby, rarely exceeding 4 m.
Leaves hairless; blades 1–4 × 2–3 cm,
ovate, or rounded, often cordate or wedge-
shaped at base; leaf-stalks 3–5 cm. Petals
8–10 × 5–9 mm. Fruit spherical to obovoid,
1–1.8 cm across, not tapering towards the
stalk, brown with white spots, eventually
turning reddish. *SW England, SW Europe
(France, Iberian Peninsula)*. H3. Spring.

3. P. cossonii Rehder (*P. longipes* Cosson
& Durieu; *P. communis* Linnaeus var.
longipes (Cosson & Durieu) Henry).
Illustration: Decaisne, Jardin fruitier du
Muséum **1**: pl. 4 (1858).
   Closely related to *P. cordata* but differing
in its more elliptic leaves. Leaf-blades
2.5–5 × 1–3 cm, white woolly beneath,
eventually hairless; leaf-stalks 2.5–5 cm,
reddish. Petals 1.2–1.5 cm × 9–11 mm.
Fruits spherical, c. 1.5 cm across, brown;
calyx-lobes deciduous. *N Africa (Algeria)*.
H3. Spring.

4. P. syriaca Boissier. Illustration:
Decaisne, Jardin fruitier du Museum **1**: pl.
9 (1858); Meikle, Flora of Cyprus **1**: 632
(1977); Krüssmann, Manual of cultivated
broad-leaved trees & shrubs **3**: 77, pl. 26
(1986).
   Similar to *P. cossonii*. Tree to 6 m, with
erect branches and spreading twigs. Leaves
hairless; blades 2–5 cm, oblong-lanceolate;
margin finely scalloped; leaf-stalk 2–5 cm,
green. Flowers in groups of 6–10 in dense
hemispherical corymbs. Petals
7–9 × 5–6 mm, oblong-ovate. Fruits
spherical, 2–2.5 cm across, greenish
yellow, on stalks to 5 cm. *Cyprus, SW Asia*.
H4. Spring.

5. P. salicifolia Pallas. Illustration:
Decaisne, Jardin fruitier du Museum **1**:
pl. 12 (1858); Phillips, Trees in Britain,
Europe and North America, 180 (1978);
Krüssmann, Manual of cultivated broad-
leaved trees & shrubs, **3**: pl. 27, 28 (1986);
Brickell (ed.), RHS A–Z encyclopedia of
garden plants, 872 (2003).
   Tree 5–8 m, branches usually pendent.
Young shoots very grey, densely covered in
forward-pointing silky hairs; short shoots
ending in a spine. Leaves
3–5 cm × 7–9 mm, lanceolate to narrowly
elliptic, wedge-shaped at base, pale greyish
green above, whitish green beneath,
darkening and becoming hairless with age,
stalkless. Flowers in groups of 5–7. Calyx-
lobes c. 3 mm, broadly triangular, obtuse.
Petals 9–12 × 8–9 mm, broadly ovate, free.
Stamens 20–22, anthers deep pinkish red.
Styles 5. Fruit 2–3 cm, greenish, on short
stalks. *W Asia (Caucasus to Turkey)*. H1.
Spring.
   Most commonly grown as the cultivar
'Pendula', which is said to be a more
pendent selection than the wild species.

6. P. ussuriensis Maximowicz.
Illustration: Decaisne, Jardin fruitier du
Muséum **1**: pl. 5 (1858); Krüssmann,
Manual of cultivated broad-leaved trees &
shrubs, **3**: 74, 77, pl. 27 (1986).
   Tree to 15 m. Young twigs yellowish
brown, hairless. Leaves eventually hairless,
very glossy on both surfaces; stalks
2.5–4.5 cm, green; blade
5.5–8 × 3.5–4 cm, broadly ovate to elliptic,
acuminate, base truncate, margin with
forward-pointing teeth. Flowers in groups
of 5–8. Calyx-lobes 4–6 mm, ovate, margin
with red teeth. Petals 1.5–1.9 ×
1.3–1.7 cm, sometimes 3-lobed towards
apex, overlapping. Stamens c. 20, anthers
deep reddish pink; styles 5. Fruit spherical,

greenish yellow, 3–4 cm across, on short thick stalks. *E Asia (E former USSR, Korea to Japan)*. H1. Spring.

Var. **hondoensis** (Kikuchi & Nakai) Rehder differs in its rusty brown, hairy young leaves, which are slightly narrower than in var. *ussuriensis*. *C Japan*.

Var. **ovoidea** (Rehder) Rehder. Illustration: Krüssmann, Manual of cultivated broad-leaved trees & shrubs **3**: 77, pl. 27 (1986). Differs in its more ovate leaves and distinctly ovoid fruits with yellow flesh. *N China, Korea.*

7. P. pashia D. Don. Illustration: Decaisne, Jardin fruitier du Muséum **1**: pl. 7 (1858); Polunin, Flowers of the Himalaya, pl. 35 (1985); Krüssmann, Manual of cultivated broad-leaved trees & shrubs, **3**: 73, pl. 27 (1986).

Tree 10–12 m, usually with thorns. Branches hairy when young, soon becoming hairless. Leaves hairy when young, stalks reddish, 3.5–4 cm; blade 7–9 × 3.7–4.5 cm, broadly ovate, apex acuminate, base rounded, margin with forward-pointing teeth. Flowers 6–8, flower-stalks hairy. Calyx-lobes *c.* 3 mm, rounded, concave. Petals 1–1.2 cm × 8–10 mm, circular. Stamens *c.* 20, anthers pale reddish pink. Styles 5. Fruit almost spherical, *c.* 2 cm across, brown, warty, on stalks 2–3 cm. *Himalaya to W China*. H1. Spring.

8. P. betulifolia Bunge. Illustration: Decaisne, Jardin fruitier du Muséum **1**: pl. 20 (1858); Bailey, Standard cyclopedia of horticulture **3**: 2869 (1935); Duke & Ayensu, Medicinal plants of China **1**: 552 (1985); Krüssmann, Manual of cultivated broad-leaved trees & shrubs, **3**: 73, 74, pl. 27 (1986).

Slender tree, 5–10 m; young branches thickly covered with persistent greyish felt, hairless in second year. Leaves hairy when young, with down persisting on veins beneath throughout the year. Blades ovate to oblong or diamond-shaped, acuminate, tapered or rounded at base, margin with coarse forward-pointing teeth, glossy dark green above, stalks 2–3.5 cm. Flowers in groups of 8–12; calyx-teeth 2–3 mm, triangular; petals 7–9 mm, ovate to oblong; styles 2 or 3. Fruit spherical, 1–1.5 cm across, brown with whitish lenticels. *N China*. H2. Spring.

9. P. calleryana Decaisne. Illustration: Krüssmann, Manual of cultivated

broad-leaved trees & shrubs, **3**: 71, 73, pl. 27 (1986).

Tree to 16 m, branches hairless, often very thorny. Leaf-blades 7–10 × 4.5–5.5 cm, ovate to broadly ovate, shortly acuminate, hairless; margin scalloped, stalks 5–6 cm. Flowers in groups of 6–12; calyx-lobes shaggy, 4–5 mm; petals 1.1–1.4 cm × 9–11 mm, oblong to oblong to ovate, rounded, often notched at apex; styles 2 or 3. Fruit spherical, *c.* 1 cm across, brownish; calyx-lobes deciduous. *C & S China*. H3. Early spring.

Two selections commonly grown are: 'Bradford', a thornless, fast-growing cultivar with good autumn colour and rusty brown fruits; and 'Chanticleer' (illustration: Brickell, RHS gardeners' encyclopedia of plants and flowers, 48, 1989; Brickell (ed.), RHS A–Z encyclopedia of garden plants, 872, 2003) with a narrowly conical crown and reddish purple leaves in autumn.

10. P. dimorphophylla Makino.
Similar to *P. calleryana* but with deeply 3–5-lobed young leaves. *Japan*. H3. Spring.

11. P. fauriei Schneider.
Similar to *P. calleryana* but smaller. Leaves 2.5–5 cm, elliptic, narrowly wedge-shaped at base; leaf-stalk 2–2.5 cm, downy. Flowers in clusters of 2–8. Fruit *c.* 1.3 cm. *Korea*.

12. P. amygdaliformis Vilmorin. Illustration: Bean, Trees & shrubs hardy in the British Isles, edn 8, **3**: 446 (1976); Phillips, Trees in Britain, Europe and North America, 179 (1978); Polunin, Flowers of Greece and the Balkans, pl. 18 (1980); Krüssmann, Manual of cultivated broad-leaved trees & shrubs **3**: 72, pl. 27 (1986).

Shrub or tree to 16 m, usually with thorny branches. Young shoots with long thin hairs. Leaves variable; blades 2.5–7 × 1–2 cm, ovate, oblong or obovate, acute or obtuse, eventually hairless; leaf-stalk 1–3 cm. Flowers in groups of 8–12; calyx-lobes 4–6 mm, lanceolate; petals 1–1.2 × 1–1.1 cm, broadly ovate; stamens *c.* 20, anthers dark red. Fruits spherical, 2–3 cm across, yellowish brown, on stalks 2–3 cm. *S Europe*. H3. Spring.

13. P. bretschneideri Rehder. Illustration: Bailey, Standard cyclopedia of horticulture **3**: 2869 (1935); Krüssmann, Manual of cultivated broad-leaved trees & shrubs, **3**: 76 (1986).

Tree to 15 m, with almost hairless branches, purplish brown in second year.

Leaf-blades 5–11 cm, ovate to elliptic, eventually hairless; apex long acuminate, margins with bristly, forward-pointing teeth; leaf-stalks 2.5–7 cm. Flowers in groups of 7–9. Petals 8–10 mm, ovate to oblong. Stamens *c.* 20, styles 4 or 5. Fruit 2.5–3 cm, with white flesh, spherical to ovoid, on stalks 3–4.5 cm. *N China*. H2. Spring.

14. P. nivalis Jacquin. Illustration: Polunin & Everard, Trees and bushes of Europe, 77 (1976); Krüssmann, Manual of cultivated broad-leaved trees & shrubs **3**: 72, pl. 27 (1986).

Small thornless tree to *c.* 13 m. Young shoots thickly covered with white woolly hairs. Leaves downy when young, eventually almost hairless above; blade 5–8 × 2–4 cm, elliptic to obovate, acute, wedge-shaped at base; margin entire or very shallowly scalloped towards apex; leaf-stalk 1–3 cm, whitish green. Flowers in groups of 6–9; calyx-lobes 3–4 mm, triangular, covered with very silky hairs; petals 1–1.2 × 0.7–0.8 cm, ovate; stamens *c.* 20; styles 5. Fruit 3–5 cm across, yellowish green, spherical, on stalks 3–6 cm. *C & SE Europe*. H2. Spring.

15. P. eleagrifolia Pallas. Illustration: Decaisne, Jardin fruitier du Muséum **1**: pl. 17 (1858); Polunin & Everard, Trees and bushes of Europe, 76 (1976); Polunin, Flowers of Greece and the Balkans, pl. 18 (1980); Krüssmann, Manual of cultivated broad-leaved trees & shrubs **3**: 72, pl. 27 (1986).

Small tree or shrub to 7 m, similar to *P. nivalis* but with thorny branches, and smaller in all its parts. Leaf-blades 4–7 × 1.2–2.5 cm, usually covered on both surfaces with white or greyish woolly hairs. Fruit 2–2.5 cm across, greenish. *Turkey*. H4. Spring.

Var. **kotschyana** (Decaisne) Boissier (illustration: Decaisne, Jardin fruitier du Muséum **1**: pl. 18, 1858) is more shrubby, usually thornless, and has shorter, wider leaves.

16. P. × michauxii Poiret; *P. amygdaliformis × P. nivalis*. Illustration: Decaisne, Jardin fruitier du Muséum **1**: pl. 16 (1858).

Small thornless tree to 9 m, with rounded crown. All parts covered with white woolly hairs when young. Leaf-blade 3–7 × 2–3.5 cm, ovate to oblong-elliptic, obtuse or abruptly acuminate, eventually glossy and hairless above; margin entire;

leaf-stalks 2–4 cm. Flowers in groups of 8 or 9. Calyx-lobes 3–4 mm, narrowly triangular; petals 7–9 × 5–6 mm; stamens 18–21, anthers dark red; styles 4 or 5. Fruit *c.* 3 × 2 cm, spherical or top-shaped, yellowish green, spotted brown. *Origin uncertain, possibly Middle East.* H3. Spring.

17. P. phaeocarpa Rehder. Illustration: *Moeller's Deutsche Gaertnerzeitung* **31**: 112 (1916); Bailey, Standard cyclopedia of horticulture **3**: 2870 (1935).

Tree to 15 m, young branches covered in downy hairs, eventually reddish brown and hairless. Leaf-blades 6–10 cm, ovate-elliptic to oblong-ovate, with long hairs when young, eventually hairless; margin with spreading teeth, apex long acuminate, base broadly wedge-shaped; leaf-stalks 2–6 cm. Flowers in groups of 3 or 4; petals 8–10 mm, ovate-oblong; styles 2 or 3 (rarely 4). Fruit 2–2.5 cm, pear-shaped, brown with paler spots. *N China.* H2. Spring.

18. P. pyrifolia (Burman) Nakai (*P. serotina* Rehder). Illustration: *Botanical Magazine*, 8226 (1908); RHS dictionary of gardening 1722 (1974); Krüssmann, Manual of cultivated broad-leaved trees & shrubs **3**: 76, pl. 26, 27 (1986); Hyashi, Woody plants of Japan, 324, 325 (1988).

Tree 5–12 m, young shoots with tufts of woolly hairs, hairless in second year. Leaf-blades 6.5–11 × 3.5–6 cm, oblong-ovate to ovate, apex acuminate, base rounded, cordate, or broadly wedge-shaped; eventually hairless; margin with bristly forward-pointing teeth; leaf-stalk 3–4.5 cm. Flowers in groups of 6–9, calyx-teeth 4–5 mm, lanceolate; petals 1.5–1.7 × 1.1–1.3 cm, ovate-elliptic; stamens 18–21; styles 4 or 5. Fruit 2.5–3.5 cm across, spherical, brownish, dotted with white lenticels. *C & W China.* H3. Spring.

Var. **culta** (Makino) Nakai has longer, wider leaves, to 15 cm and larger, apple-shaped, yellow or brown fruit.

Var. **stapfiana** (Rehder) Rehder has longer pear-shaped fruit to 6 cm.

19. P. regelii Rehder (*P. heterophylla* Regel & Schmalhausen). Illustration: Bean, Trees & shrubs hardy in the British Isles, edn 8, **3**: 452 (1976); Krüssmann, Manual of cultivated broad-leaved trees & shrubs **3**: 72, pl. 27 (1986).

Shrub or small tree, 5–9 m, often with thorny twigs. Leaves very variable, covered with soft downy hairs when young; blade 2–6 cm, entire or with 1–7 shallow or deep lobes; margins coarsely and irregularly toothed; leaf-stalks 2–6 cm. Flowers in groups of 5–7. Petals 6–9 mm, ovate to oblong. Fruit 2–3 cm across, pear-shaped or spherical. *Turkestan.* H3. Spring.

20. P. salvifolia de Candolle. Illustration: *Edwards's Botanical Register* **18**: t. 1482 (1832); Decaisne, Jardin fruitier du Muséum **1**: pl. 21 (1858); Schneider, Illustriertes Handbuch der Laubholzkunde **1**: f. 361 (1906).

Very thorny tree to 15 m, closely allied to *P. nivalis* and believed to be a hybrid between it and *P. communis.* Leaf-blades *c.* 5 cm, leaf-stalks 4–6 cm. Fruit *c.* 2.5 cm, pear-shaped, on stalks 2–3 cm. *S Europe (France eastward to Hungary).* H3. Spring.

54. × SORBOPYRUS Schneider
S.G. Knees & P.G. Barnes
Tree to 15 m; branches and buds covered in short whitish hairs. Leaves 6–10 cm, broadly elliptic, base rounded, apex acute, irregularly and coarsely toothed, hairy beneath. Flowers in 5–many-flowered corymbs, on slender stalks. Sepals 5, *c.* 5 mm, triangular, downy; petals 5, *c.* 9 × 6 mm, creamy white; stamens *c.* 20, anthers pale pink; ovary inferior; styles 2–5. Fruit *c.* 2.5 cm across, pear-shaped, downy, green or reddish yellow.

1. × S. auricularis (Knoop) Schneider; *Pyrus communis × Sorbus aria.* Illustration: *Edward's Botanical Register*, t. 1437 (1831).
Garden origin. H4. Spring–summer.

Var. **bulbiformis** (Tatar) Schneider is closer to *Pyrus* but has puckered leaves with coarse forward-pointing teeth; sepals irregularly recurved, to 6 mm; petals *c.* 1.3 × 1.1 cm; *c.* 25 deep pink anthers and fruits to 4 cm across.

55. MALUS Miller
H.S. Maxwell, S.G. Knees, & M.F. Gardner
Deciduous trees or shrubs, occasionally semi-evergreen; lateral shoots often with thorns. Leaves alternate, folded or rolled in bud, toothed or lobed, green to reddish purple. Flowers bisexual, stalked, borne in umbel-like clusters; calyx of 5 persistent lobes; petals 5, semi-circular to broadly ovate, white, rose or crimson; stamens 15–50, anthers rounded, ovary inferior, 3–5-celled, styles 2–5, always united at base, basal part shaggy with long hairs; fruit fleshy, with or without grit cells, almost spherical, calyx mostly persistent;

seeds 1 or 2 per cell, brownish or black. Figure 99(14), p. 183.

A genus of 25–30 species occurring throughout temperate parts of the northern hemisphere. As all the species freely hybridise with each other seed cannot be relied on as an effective means of propagation. Although some can be rooted from cuttings taken from leafless shoots during the winter, most are increased by grafting.

Literature: Huckins, C.A., Flower and fruit keys to the ornamental crab apples cultivated in the United States, *Baileya* **15**: 129–64 (1967); Fiala, J.L., Flowering crabapples: the genus Malus (1994); Robinson, J.P., Harris, S.A. & Juniper, B.E., Taxonomy of the genus Malus Mill., with emphasis on the cultivated apple, Malus domestica Borckh., *Plant Systematics and Evolution* **226**: 35–58 (2001).

1a. Leaf-margins irregularly and deeply toothed or often lobed 2
 b. Leaf-margins shallowly or evenly toothed but never lobed 15
2a. Flowers borne in umbel-like clusters; leaves usually with 2 lobes towards the base 3
 b. Flowers in corymbs; leaves variously lobed 5
3a. Fruit 1.6–3 cm, yellow **1. spectabilis**
 b. Fruit 6–15 mm, yellowish red to brown 4
4a. Umbels almost stalkless; fruits 1–1.5 cm, yellowish red **2. toringoides**
 b. Umbels stalked; fruits 6–8 mm, red or yellowish brown **3. sieboldii**
5a. Leaves 2–3 cm, deeply cut almost to midrib **4. transitoria**
 b. Leaves 3 cm or more, leaves lobed but not deeply cut 6
6a. Flowers 1–2 cm across 7
 b. Flowers 2.5–4 cm across 11
7a. Leaves with conspicuous densely felted hairs beneath, margins lobed for most of their length; styles 5 8
 b. Leaves almost hairless or with scattered hairs below; margins irregularly lobed; styles 3 or 4 9
8a. Leaves 5–7 cm; lobes pointed, often for half the distance to midrib or more **5. florentina**
 b. Leaves 6–12 cm; lobes rounded, not more than half the distance to midrib or less **6. yunnanesis**
9a. Leaves narrowly ovate-elliptic or lanceolate; flower-stalks and calyx covered with dense hairs **7. fusca**

- b. Leaves broadly ovate to rounded; flower-stalks and calyx hairless **10**
- 10a. Leaves with 1–2 pairs of lobes, usually towards apices; flowers *c.* 1.5 cm across; fruit yellow or purple **8. kansuensis**
- b. Leaves with 2 or more pairs of lobes; flowers *c.* 2 cm across; fruit yellow or green **9. honanensis**
- 11a. Flowers 2.5–3 cm across; fruit 1.5–2.5 cm **12**
- b. Flowers 3–4 cm across; fruit 3–4 cm **13**
- 12a. Leaves 3–7 cm, almost evergreen, hairless when mature **10. angustifolia**
- b. Leaves 7–12 cm, deciduous, remaining hairy, especially below **11. tschonoskii**
- 13a. Flower-stalks and outer calyx-lobes covered with matted hairs **12. ioensis**
- b. Flower-stalks and outer calyx-lobes hairless or with few scattered hairs **14**
- 14a. Leaves on flowering-shoots distinctly lobed; fruit yellow, fragrant **13. glaucescens**
- b. Leaves on flowering-shoots unlobed or only slightly lobed; fruit green, ribbed at apex **14. coronaria**
- 15a. Leaves 3–4 times as long as wide; blade narrowly ovate-elliptic to oblong **16**
- b. Leaves usually not more than twice as long as wide; blade broadly ovate to rounded **22**
- 16a. Fruits 3–4 cm, green, even when mature **17**
- b. Fruits 6–15 mm, red or brownish **18**
- 17a. Flower-stalks and calyx-lobes hairy **12. ioensis**
- b. Flower-stalks and calyx-lobes hairless **14. coronaria**
- 18a. Flower-stalks 5–15 mm **16. brevipes**
- b. Flower-stalks 2–5 cm **19**
- 19a. Flowers 2.5–3 cm across **20**
- b. Flowers 3–4 cm across **21**
- 20a. Leaf-margin with jagged teeth; fruit 8–10 mm **15. floribunda**
- b. Leaf-margin smooth or with fine teeth; fruit 1.3–1.5 cm **17. sikkimensis**
- 21a. Flowers usually deep pink; fruit 6–8 mm, red or brownish **18. halliana**
- b. Flowers white, rarely pink-tinged; fruit 8–10 mm, red or yellow **19. baccata**

- 22a. Fruit 1–1.5 cm **23**
- b. Fruit 2–6 cm **24**
- 23a. Flowers *c.* 2 cm across, white **20. prattii**
- b. Flowers 3.5–4 cm across, pink at first, later white **21. hupehensis**
- 24a. Flowers *c.* 3 cm across **25**
- b. Flowers 4–5 cm across **26**
- 25a. Leaves densely covered with matted hairs beneath; fruit *c.* 2.5 2.5 cm, brownish yellow, flushed purple **11. tschonoskii**
- b. Leaves hairless or with few scattered hairs beneath; fruit *c.* 2 cm, yellowish green or red **22. prunifolia**
- 26a. Leaves always green; young branches usually thorny; outer calyx-lobes hairless; petals white, often tinged pink **23. sylvestris**
- b. Leaves often bronze or purplish; young branches thornless; outer calyx-lobes hairy; petals white or deep pink **24. pumila**

1. M. spectabilis (Aiton) Borkhausen (*M. spectabilis* var. *plena* Bean; *Pyrus spectabilis* Aiton). Illustration: *Botanical Magazine*, 267 (1794); Bailey, Standard cyclopedia of horticulture **2**: 3292(1950) *Journal of the Royal Horticultural Society* **86**: 42 (1961); Phillips, Trees in Britain, Europe and North America, 139 (1978).

Tall shrub or small tree to 8 m; young branches downy, eventually reddish brown. Leaves 5–8 cm, elliptic to oblong, dark green above, shiny, paler below, downy; apex shortly acuminate, margins with adpressed teeth and occasionally with 2 lobes towards the base. Flowers borne in umbel-like clusters, dark pink in bud, eventually whitish; stalks 2–3 cm; calyx hairless to almost downy, lobes triangular to ovate; corolla 4–5 cm across, occasionally semi-double. Fruit 1.6–3 cm, yellow. *Believed to be from China but not known in the wild.* H2. Spring.

2. M. toringoides (Rehder) Hughes (*M. transitoria* var. *toringoides* Rehder; *Pyrus transitoria* var. *toringoides* Bailey; *P. toringoides* Osborne; *Sinomalus toringoides* Koidzumi). Illustration: *Botanical Magazine*, 8948 (1922); Krüssmann, Manual of cultivated broad-leaved trees & shrubs **2**: pl. 121 (1986).

Shrub or tree to 8 m; branches hairy at first eventually hairless. Leaves 3–8 cm, ovate, usually with 2 scalloped lobes on either side, rarely entire, eventually hairless except on the veins beneath.

Flowers 3–6, in almost stalkless umbel like clusters; calyx with densely felted hairs, corolla *c.* 2 cm across, white, slightly fragrant. Fruit 1–1.5 cm, spherical to almost pear-shaped, yellow or red, persistent. *W China.* H2. Late spring.

3. M. sieboldii (Regel) Rehder (*M. toringo* Nakai; *Pyrus sieboldii* Regel; *P. toringo* Siebold). Illustration: Bailey, Standard cyclopedia of horticulture **2**: 3295 (1915); Makino, Illustrated flora of Japan, 1407 (1956); *Massachusetts Horticultural Society, Horticulture* **34**(10): 506 (1956); Krüssmann, Manual of cultivated broad-leaved trees & shrubs **2**: pl. 121 (1986).

Low shrub or small tree to 4 m; branches blackish brown, nodding. Leaves 3–6 cm, ovate-elliptic, coarsely toothed, with 3–5 lobes, dark green above paler below, downy on both sides. Flowers in umbel-like clusters of 3–6, stalks 2–2.5 cm, corolla *c.* 2 cm, pale pink fading to white. Fruit 6–8 mm, spherical, red to yellowish brown, persistent. *Japan.* H2. Late spring.

Var. **sargentii** Rehder. Low-growing shrub to 2 m, with very thorny dense, spreading branches and larger flowers to 2.5 cm across, fruit dark red. Sometimes treated as a separate species.

M. × sublobata (Dippel) Rehder (*M. ringo* forma *sublobata* Dippel). Illustration: *Massachusetts Horticultural Society, Horticulture* **44**(30): 20 (1966). Similar but with leaves 4–8 cm and 1 or 2 lobes. Flowers to 4 cm across; fruits to 1.5 cm yellow. *Known only in cultivation.* H2. Late spring.

A hybrid between *M. prunifolia* and *M. sieboldii*.

M. × zumi (Matsumura) Rehder (*Pyrus zumi* Matsumura). Illustration: Krüssmann, Manual of cultivated broad-leaved trees & shrubs **2**: 287 (1986). Pyramidal tree; leaves 5–9 cm; flowers *c.* 3 cm across; fruits *c.* 1 cm, red.

A hybrid between *M. sieboldii* and *M. baccata* var. *mandshurica*.

4. M. transitoria (Batalin) Schneider (*Pyrus transitoria* Batalin; *Sinomalus transitoria* Koidzumi). Illustration: Icones cormophytorum Sinicorum, **2**: 2207 (1972); Krüssmann, Manual of cultivated broad-leaved trees & shrubs **2**: pl. 121 (1986).

Very similar to *M. toringoides* but smaller and branches densely felted when young. Leaves 2–3 cm, with deeper narrower lobes, very rarely entire. Flowers borne in corymbs; fruit usually less than 1 cm, bright red. *NW China.* H2. Late spring.

5. M. florentina (Zuccagni) Schneider (*Pyrus crataegifolia* Savi; *Crateagus florentina* Zuccagni; *Eriolobus florentina* (Zuccagni) Stapf). Illustration: *Botanical Magazine*, 7423 (1895); RHS dictionary of gardening 1239 (1956); Phillips, Trees in Britain, Europe and North America, 137 (1978); Krüssmann, Manual of cultivated broad-leaved trees & shrubs **2**: pl. 121 (1986).

Deciduous tree *c.* 8 × 6 m, pyramidal when young, becoming rounded; branches slender, dark brown, twigs shaggy with long hairs when young. Leaves 3–7 cm, broadly ovate, irregularly sharply toothed, wedge-shaped or heart-shaped at base, dull green above, grey-yellow felted hairs beneath, leaf-stalk 5–20 mm, downy, reddish. Flowers 1.5–2 cm across, in loose clusters of 2–6, flower-stalks slender, *c.* 2.5 cm, downy, pinkish; calyx very woolly, lobes narrow, pointed, falling early; petals white. Fruit *c.* 1 cm, round-ovate, stone-cells present, yellowish turning red. *Italy, former Yugoslavia & Greece.* H2. Summer.

Considered by some authorities to be × *Malosorbus florentina* (Zuccagni) Browicz, a bigeneric hybrid between *Malus sylvestris* and *Sorbus torminalis*.

6. M. yunnanesis (Franchet) Schneider (*Pyrus yunnanensis* Franchet). Illustration: *Botanical Magazine*, 8629 (1915); Icones cormophytorum Sinicorum, **2**: 2210 (1972); Krüssmann, Manual of cultivated broad-leaved trees & shrubs **2**: pl. 119 (1986).

Tree to 10 m, with narrow upright habit; shoots with densely felted hairs when young. Leaves 6–12 cm, broadly ovate, base rounded to cordate, margin double-toothed, sometimes with 3–5 pairs of lobes, densely felted beneath. Flowers borne in corymbs, 4–5 cm across; calyx shaggy; corolla *c.* 1.5 cm across, white, styles 5. Fruit 1–1.5 cm, deep red, often dotted. *W China.* H2. Late spring.

Var. **veitchii** Rehder. Differs in having distinctly cordate leaf-bases and flowers less than 1.2 cm across. Fruits to 1.3 cm with white dots. *C China.* H2. Late spring.

7. M. fusca (Raflnesque) Schneider (*M. rivularis* (Hooker) Roemer; *Pyrus fusca* Raflnesque; *P. rivularis* Hooker). Illustration: Sargent, Silva of North America **4**: 170 (1892); *Botanical Magazine*, 8798 (1919); Elias, The complete trees of North America, 604 (1980); Krüssmann, Manual of cultivated broad-leaved trees & shrubs **2**: pl. 117 (1986).

Deciduous shrub or tree, 6–8 m; crown rounded, spreading; bark with large flattened scales, reddish brown; shoots hairy when young. Leaves 3–10 × 1.5–4 cm, lanceolate, sharply toothed, 3-lobed on long shoots, leaf-stalks hairy. Flowers 1.5–2.2 cm across, in clusters of 6–12, flower-stalks hairy; calyx hairy, falling early; petals white to pinkish white; styles 3 or 4. Fruit 1.5 cm, broadest near the tip, yellow to red. *W North America.* H2. Spring–summer.

8. M. kansuensis (Batalin) Schneider (*Pyrus kansuensis* Batalin; *Eriolobus kansuensis* (Batalin) Schneider. Illustration: Icones cormophytorum Sinicorum, **2**: 2206 (1972); Krüssmann, Manual of cultivated broad-leaved trees & shrubs **2**: pl. 121 (1986).

Shrub or small tree to 5 m; young branches downy at first eventually hairless and reddish brown. Leaves 5–8 cm, broadly ovate with 3–5 lobes, usually towards the apex, margin finely toothed, dark green above, paler and hairy below. Flowers in corymbs of 4–10; calyx shaggy, corolla *c.* 1.5 cm across, creamy-white, styles 3. Fruit *c.* 1 cm, yellow to purple with paler dots. *NW China.* H2. Late spring.

9. M. honanensis Rehder (*Sinomalus honanensis* (Rehder) Koidzumi). Illustration: Icones cormophytorum Sinicorum, **2**: 2209 (1972).

Very like *M. kansuensis* but slower growing and with thinner twigs. Leaves 6–8 cm, broadly ovate, rarely ovate-oblong, with 2–5 pairs of ovate toothed lobes, downy beneath. Flowers to 10, in hairless corymbs; corolla *c.* 2 cm across, white; styles 3 or 4. Fruit *c.* 1 cm, yellow to green. *NE China.* H2. Late spring.

10. M. angustifolia (Aiton) Michaux (*Pyrus angustifolia* Aiton). Illustration: Sargent, Silva of North America **4**: 169 (1892); Britton & Brown, Illustrated flora of the northern United States & Canada, edn 2, **2**: 234 (1913); Elias, The complete trees of North America, 603 (1980); Krüssmann, Manual of cultivated broad-leaved trees & shrubs **2**: pl. 121 (1986).

Semi-evergreen tree or shrub, 5–7 m, broad, rounded; bark reddish brown, scaly, narrowly ridged, shoots hairy when young. Leaves 3–7 × 1.5–4 cm, lanceolate-oblong to ovate-oblong, tip rounded or pointed, tapering towards base, entire to coarsely toothed especially towards tip, sometimes lobed, shining green above, paler beneath,

hairless when mature. Flowers 2–3 cm across, borne in few-flowered corymbs on short lateral shoots; calyx teeth white, woolly inside; petals rose or white, fragrant; style 1. Fruit 1.5–2.5 cm across, yellowish green, fragrant. *SE North America.* H2. Spring–summer.

11. M. tschonoskii (Maximowicz) Schneider (*Pyrus tschonoskii* Maximowicz; *Eriolobus tschonoskii* Rehder). Illustration: *Botanical Magazine*, 8179 (1908); Bailey, Standard cyclopedia of horticulture **2**: 3294 (1915); Makino, Illustrated flora of Japan, 1411 (1956); Krüssmann, Manual of cultivated broad-leaved trees & shrubs **2**: pl. 119 (1986).

Tree to 12 m; branches erect, pyramidal, shoots with densely felted white hairs. Leaves 7–12 cm, ovate-elliptic to oblong, irregularly toothed and often shallowly lobed, with densely felted white hairs especially below. Flowers in corymbs of 2–5, corolla *c.* 3 cm across, white tinged pink. Fruit 2–3 cm, yellowish green with a red cheek. *Japan.* H2. Late spring.

Foliage especially good for autumn colour, turning red, yellow or purple.

12. M. ioensis (Wood) Britton (*Pyrus coronaria* var. *ioensis* Wood; *P. ioensis* (Wood) Bailey). Illustration: Britton & Brown, Illustrated flora of the northern United States and Canada **2**: 235 (1897); *Botanical Magazine*, 8488 (1913); Sargent, Manual of the trees of North America, 278 (1922); Krüssmann, Manual of cultivated broad-leaved trees & shrubs **2**: pl. 121 (1986).

Small tree with open habit, developing ornamental peeling bark with age; branches with densely felted hairs at first, becoming reddish brown, hairless. Leaves 5–10 cm, oblong-ovate, coarsely to sharply toothed, often shallowly lobed, dark green above, yellowish green below with densely felted hairs. Flowers in corymbs of 4–6, calyx-lobes covered with matted hairs; corolla *c.* 4 cm across, white or pale pink, fragrant. Fruit *c.* 3 cm, waxy, green, sometimes angled, calyx persisent. *C USA.* H2. Summer.

Several naturally occuring varieties are listed in the American literature including var. **bushii** Rehder, var. **creniserrata** Rehder, var. **palmeri** Rehder and var. **texana** Rehder; however, these are not thought to be widely available in Europe.

M. × denboerii Krüssmann. Differs from *M. ioensis* in its purple to bronze leaves and reddish pink flowers *c.* 2 cm across; fruits to 2.5 cm, reddish. *Cultivated origin.* H2. Summer.

A hybrid involving *M. ioensis* as the female parent; the other parent is uncertain but may be *M. × purpurea.*

M. × soulardii (Bailey) Britton. Illustration: Britton & Brown, Illustrated flora of the northern United States and Canada **2**: 235 (1897). Very like *M. ioensis* but with broader leaves to 8 cm; fruits to 5 cm, flattened, yellowish green. *USA.* H2. Summer.

13. M. glaucescens Rehder (*P. glaucescens* (Rehder) L.H. Bailey). Illustration: Sargent, Trees & shrubs **2**: 139 (1913); Sargent, Manual of the trees of North America, 381 (1922); Bailey, Standard cyclopedia of horticulture **2** 3297–9 (1915).

Shrub or small tree to 8 m; branches hairless, with few sharp thorns. Leaves 5–8 cm, ovate to triangular, with short triangular lobes, dark green above, bluish below, densely hairy at first. Flowers in corymbs of 5–7, calyx slightly shaggy on the outside, densely felted inside, corolla *c.* 3.5 cm, white to pink, fragrant. Fruit 3–4 cm, flattened, greenish yellow, indented at both ends. *North America.* H2. Late spring.

14. M. coronaria (Linnaeus) Miller (*M. bracteata* Rehder in Sargent; *Pyrus coronaria* Linnaeus). Illustration: *Botanical Magazine*, 2009 (1818); Britton & Brown, Illustrated flora of the northern United States and Canada **2**: 235 (1897); Elias, The complete trees of North America, 604 (1980); Krüssmann, Manual of cultivated broad-leaved trees & shrubs **2**: pl. 119 (1986).

Similar to *M. angustifolia* except shoots white woolly at first. Leaves 5–10 cm, ovate-elliptic to ovate-lanceolate, rounded to heart-shaped at base, irregularly toothed and lobed, hairy at first beneath. Flowers in corymbs of 4–6, flower-stalks and calyx-lobes hairless, corolla 3–4 cm across, pale pink, fragrant. Fruit 3–4 cm across, slightly flattened, greenish, ribbed toward apex. *E North America.* H2. Summer.

Var. **dasycalyx** Rehder. Differs in its smaller flowers to 3 cm wide, calyx often hairy; fruits yellowish green. *C North America.* H2. Summer.

Var. **lancifolia** (Rehder) Fernald (*Pyrus lancifolia* (Rehder) L.H. Bailey). Has more pointed leaves 3–11 cm, lanceolate. *E North America.* H2. Summer.

'Charlottae' (*Pyrus charlottae* misapplied) and 'Nieuwlandiana' (*P. nieuwlandiana* misapplied) both have larger semi-double flowers.

15. M. floribunda Van Houtte (*M. pulcherrima* (Ascherson & Graebner) K.R. Boynton; *Pyrus floribunda* Kirchner not Lindley). Illustration: *Revue Horticole*, pl. 591 (1871); *Flore des serres* **15**: pl. 1585 (1862–65); Phillips, Trees in Britain, Europe and North America, 137 (1978); Krüssmann, Manual of cultivated broad-leaved trees & shrubs **2**: pl. 75, pl. 117 (1986).

Shrub or small tree 4–8 m; branches long arching, shoots downy at first, eventually hairless. Leaves 4–8 cm, ovate-acuminate, sharply toothed, especially on the stronger shoots, dark green and hairless above, paler and downy beneath. Flowers in clusters of 4–7, stalks 2.5–4 cm, corolla 2.5–3 cm across, rose in bud, fading to pink or white; styles 4. Fruit 8–10 mm, red or yellow, calyx deciduous. *Japan.* H2. Spring.

16. M. brevipes (Rehder) Rehder (*M. floribunda* var. *brevipes* Rehder; *Pyrus brevipes* (Rehder) L.H. Bailey).

Resembling *M. floribunda* but low densely branched shrub; leaves 5–7 cm, ovate, acuminate, finely toothed. Flower-stalks 5–15 mm, corolla *c.* 3 cm across, white. Fruit 1.5 cm across, bright red, slightly ribbed. *Origin unknown.* H2. Spring.

17. M. sikkimensis (Wenzig) Schneider (*M. pashia* Wenzig var. *sikkimensis* Wenzig; *Pyrus sikkimensis* J.D. Hooker). Illustration: Bailey, Standard cyclopedia of horticulture **2**: 7430 (1950); Bean, Trees & shrubs hardy in the British Isles, edn 8, **2**: 709 (1973); Krüssmann, Manual of cultivated broad-leaved trees & shrubs **2**: pl. 119 (1986).

Tree 5–7 m; young branches covered in woolly hairs. Leaves 5–7 cm, ovate to oblong, margins entire or fine-toothed. Flowers in clusters of 4–9, calyx woolly, corolla *c.* 2.5 cm across, white. Fruit 1.3–1.5 cm across, almost pear-shaped, red or yellow, dotted. *North India.* H2. Late spring.

18. M. halliana Koehne. Illustration: Bailey, Standard cyclopedia of horticulture **2**: 3289 (1950); Makino, Illustrated flora of Japan, 1408 (1956); Csapody & Tóth, Colour atlas of flowering trees and shrubs, pl. 203 (1982); Krüssmann, Manual of cultivated broad-leaved trees & shrubs **2**: 286 (1986).

Shrub 2–4 m, rarely a small tree; young shoots soon becoming hairless. Leaves 4–8 cm, oblong-ovate, leathery, hairless. Flowers in clusters of 4–7, corolla 3–4 cm across, dark pink; styles 4. Fruit 6–8 mm, reddish brown. *Japan, China.* H2. Late spring.

Perhaps best known in cultivation as 'Parkmanii', which has paler pink, double flowers.

Var. **spontanea** (Makino) Koidzumi (*M. floribunda* var. *spontanea* Makino). This differs in its vase-shaped habit, almost white flowers and shorter stalks. Fruit yellowish green.

M. × atrosanguinea (Späth) Schneider (*M. floribunda atrosanguinea* misapplied; *Pyrus atrosanguinea* Spath). Illustration: *Massachusetts Horticultural Society, Horticulture* **34**(10): 506 (1956) & **35**(2): 76 (1957); *American Horticulturist* **56**(2): 23 (1977).

Leaves dark green; flowers simple, buds deep rosy red, not fading. Fruit *c.* 1 cm across, red or yellow.

A hybrid between *M. halliana* and *M. sieboldii.*

19. M. baccata (Linnaeus) Borkhausen (*M. sibirica* Borkhausen). Illustration: Icones cormophytorum Sinicorum, **2**: 2198 (1972); Bean, Trees & shrubs hardy in the British Isles, edn 8, **2**: 694 (1973); Philips, Trees in Britain, Europe and North America, 136 (1978); Krüssmann, Manual of cultivated broad-leaved trees & shrub **2**: pl. 121 (1986).

Tree or shrub to 5 m; young shoots hairless; leaves 3–8 cm, ovate acuminate, margin with long fine teeth, glossy above, hairless. Flowers borne in small clusters, calyx hairless, lobes acuminate, corolla 3–3.5 cm across, creamy white, fragrant. Fruit *c.* 1 cm, bright red or yellow. *NE Asia to N China.* H2. Spring.

Var. **mandshurica** (Maximowicz) Schneider (*Pyrus baccata* Linnaeus var. *mandshurica* Maximowicz). Illustration: *Botanical Magazine*, 6112 (1874). Differs in its leaves, which are often entire towards the base, and have small distant teeth towards the apex, downy below when young. Calyx-lobes often downy; corolla *c.* 4 cm across, pure white. Fruit 1–1.2 cm, bright red. *C Japan, C China.* H2. Early spring.

M. × hartwigii Koehne. Illustration: Krüssmann, Manual of cultivated broad-leaved trees & shrubs **2**: pl. 120 (1986). Differs in its larger leaves, usually 6–9 cm;

flowers slightly double, dark pink at first; fruit pear-shaped, yellowish green.

A hybrid between *M. baccata* and *M. halliana*.

20. M. prattii (Hemsley) Schneider (*Pyrus prattii* Hemsley; *Docynopsis prattii* (Hemsley) Koidzumi). Illustration: Lee, Forest botany of China, 170 (1935); Icones cormophytorum Sinicorum, **2**: 2208 (1972); Krüssmann, Manual of cultivated broad-leaved trees & shrubs **2**: pl. 119 (1986).

Upright shrub or tree to 7 m; young shoots hairy at first, soon becoming hairless. Leaves 6–15 cm, ovate-elliptic to oblong, base rounded, slightly hairy beneath. Flowers in clusters of 7–10, corolla *c.* 2 cm, white; styles 5. Fruit 1–1.5 cm across, red or yellow, dotted, calyx persistent. *China (Hubei, Sichuan).* H3. Spring.

21. M. hupehensis (Pampanini) Rehder (*M. theifera* Rehder). Illustration: *Botanical Magazine*, 9667 (1943–48); Icones cormophytorum Sinicorum, **2**: 2200 (1972); Phillips, Trees in Britain, Europe and North America, 138 (1978); Krüssmann, Manual of cultivated broad-leaved trees & shrubs **2**: pl. 121 (1986).

Shrub or tree 5–7 m; twigs stiff, outspread, hairy at first soon becoming hairless. Leaves 5–10 cm, ovate to oblong, sharply toothed, base rounded or heart-shaped. Flowers in clusters of 3–7; calyx-lobes hairy, reddish; corolla 3.5–4 cm across, pink at first eventually white. Fruit *c.* 1 cm, greenish yellow, with reddish cheeks, calyx deciduous. *China, India.* H2. Early spring.

22. M. prunifolia (Willdenow) Borkhausen (*Pyrus prunifolia* Willdenow). Illustration: *Botanical Magazine*, 6158 (1875); *Il Giardino Fiorito* **47**: 557 (1981); Icones cormophytorum Sinicorum, **2**: 2203 (1972); Krüssmann, Manual of cultivated broad-leaved trees & shrubs **2**: 286 (1986).

Small tree 5–10 m; twigs softly hairy when young. Leaves 5–10 cm, elliptic or ovate, sparsely hairy beneath, margin toothed. Flowers in clusters of 6–10, buds pink; calyx covered with white matted hairs; corolla *c.* 3 cm across, opening pure white. Fruit *c.* 2 cm across, yellow-green or red, calyx persistent. *Probably NE Asia (origin uncertain).* H2. Spring.

Var. **rinki** (Koidzumi) Rehder (*M. pumila* var. *rinki* Koidzumi; *M. ringo* Siebold; *M. asiatica* Nakai; *Pyrus ringo* Wenzig).

Illustration: *Botanical Magazine*, 8265 (1909). Differs only in being slightly more downy, with shorter flower-stalks and usually pinkish flowers. *Cultivated in China.* H2. Spring.

M. × robusta (Carrière) Rehder (*Pyrus microcarpa* C. Koch var. *robusta* Carrière). Illustration: Hay & Synge, Dictionary of garden plants in colour with house and greenhouse plants, pl. 1703 (1986). Vigorous shrub or small tree; branches spreading, arching. Leaves 7–10 cm, narrowly elliptic; flowers 2.5–4 cm across, mostly white. Fruit 2–2.5 cm, purplish red or yellow.

A hybrid between *M. baccata* and *M. prunifolia*.

'Persicifolia' (*M. robusta* var. *persicifolia* Rehder). Differs in its oblong-lanceolate, peach-like leaves.

23. M. sylvestris (Linnaeus) Miller (*Pyrus malus* Linnaeus var. *sylvestris* Linnaeus). Illustrations: Reichenbach, Icones florae Germanicae et Helveticae **25**: pl. 111 (1912); Ross-Craig, Drawings of British plants **9**: pl. 35 (1952); Phillips, Trees in Britain, Europe and North America, 139 (1978); Blamey & Grey-Wilson, Illustrated flora of Britain and northern Europe 193 (1989).

Deciduous tree or shrub, *c.* 7 m, shoots more or less spiny, short, bark brown, fissured. Leaves 3–11 × 2.5–5.5 cm, ovate to rounded, margin with fine irregular teeth, apex shortly pointed, base broadly wedge-shaped to rounded; leaf-stalk 1.5–3 cm. Flowers 3–4 cm across, borne on short shoots, flower-stalks hairless; sepals 3–7 mm, densely hairy on inside; petals white flushed with pink; anthers yellow; styles hairless or with a few long hairs at base. Fruit 2.5–4 cm, globe-shaped, yellowish green flushed with red. *Europe* H1. Spring–summer.

Subsp. **mitis** (Wallroth) Mansfeld (*M. domestica* Borkhausen). Shoots densely hairy, not spiny; leaves 4–13 × 3–7 cm, egg-shaped to elliptic with a rounded base, slightly hairy above, densely hairy beneath. Fruits *c.* 5 cm, variable in colour. *Europe.* H1. Spring–early summer.

This is the cultivated apple, which is widely naturalised and the parent of numerous orchard hybrids.

24. M. pumila Miller (*M. communis* Poiret; *M. domestica* Borkhausen; *M. pumila* var. *paradisiaca* (Linnaeus) Schneider; *M. sylvestris* var. *paradisiaca* (Linnaeus) Bailey; *M. dasyphylla* Borkhausen; *Pyrus*

pumila (Miller) Tausch; *P. malus* var. *paradisiaca* Linnaeus; *P. malus* var. *pumila* (Miller) Henry)). Illustration: Schneider, Illustriertes Handbuch der Laubholzkunde **1**: 718 (1906); Reichenbach, Icones florae Germanicae et Helveticae **25**: pl. 112 (1912); Edlin & Nimmo, The world of trees, 57 (1974); Krüssmann, Manual of cultivated broad-leaved trees & shrubs **2**: 286 (1986).

Small tree, 5–7 m, habit open, twigs usually thornless, young shoots covered with soft down. Leaves 4–10 cm, broadly ovate to rounded, hairy at first, eventually hairless above. Flowers 4–5 cm across; calyx hairy, lobes longer than the cup; corolla white, turning pink. Fruit 2–6 cm, green, indented at each end. *Europe, SE Asia.* H2. Spring.

M. × adstringens Rehder. Flowers usually pink; fruit shortly stalked, 4–5 cm across, red, yellow or green. This is a source of many garden cultivars. A hybrid between *M. baccata* and *M. pumila*.

M. × astracanica Dumont de Courset. Differs from *M. pumila* in its coarser leaves with saw-like teeth, bright red flowers and bloomed fruits. A hybrid between *M. prunifolia* and *M. pumila*.

M. × heterophylla Spach. Differs in its wider leaves, flowers pink in bud, opening white, *c.* 4 cm. Fruits green *c.* 6 cm across.

'Niedzwetzkyana' (*Malus niedwetzkyana* Dieck). Illustration: *Botanical Magazine*, 7975 (1904). Shrub to 4 m, branches upright, spreading, young shoots bright red, wood purplish. Leaves bronze-brown; flowers in clusters of 4–7, corolla *c.* 4.5 cm, dark red. Fruit 5–6 cm across, dark red, pulp red. Much used in hybridisation and the origin of many of the dark leaved and flowered forms.

M. × purpurea (Barbier) Rehder (*Malus floribunda* var. *purpurea* Barbier; *Pyrus purpurea* of gardens). Illustration: Phillips, Trees in Britain, Europe and North America 139 (1978); Csapody & Tóth, Colour atlas of flowering trees and shrubs, pl. 208 (1982). Wood purplish; leaves 8–9 cm, brownish red at first, eventually dark green; flowers 3–4 cm across, purplish red at first, but soon fading. Fruit 1.5–2.5 cm across, purplish red. A hybrid between *M. pumila* 'Niedzwetzkyana' and *M. × atrosanguinea*.

M. × moerlandsii Doorenbos. Abundantly flowering; fruit 1–1.5 cm across, purple. A hybrid between *Malus × purpurea* 'Lemoinei' and *M. sieboldii*.

56. ERIOLOBUS (de Candolle) Roemer

S.G. Knees

Deciduous trees. Leaves alternate, palmately lobed with long stalks; stipules falling early. Flowers bisexual in terminal stalkless or shortly stalked umbels. Calyx of 5 reflexed lobes, united at the base; corolla 5, white; stamens *c.* 20; ovary inferior, 5-celled; styles 5, united at the base. Fruit spherical to pear-shaped with grit cells, not sunken at calyx; seed 1 or 2, not compressed.

A genus of about 6 species from SE Europe and SW Asia. Propagation is usually from seed or grafting onto *Sorbus* species.

1. E. trilobatus (Poiret) Roemer

(*Crataegus trilobata* Poiret; *Pyrus trilobate* (Poiret) de Candolle; *Sorbus trilobata* (Poiret) Heynold; *Malus trilobata* (Poiret) Schneider). Illustration: Schneider, Illustriertes Handbuch der Laubholzkunde **1**: 726, 727 (1906); *Botanical Magazine*, 9305 (1933); Davis (ed.), Flora of Turkey and the east Aegean Islands **4**: 89 (1972).

Tree to 6 m; branches upright, young shoots densely covered with white hairs at first, eventually hairless and brownish red. Leaves 1.5–8 × 2–11 cm, usually 3-lobed, with each lateral lobe further divided, lobes acute, margins with saw-like teeth; leaf-stalks about as long as blade. Flowers in clusters of 3–10; sepals 8–12 × 2–3 mm, triangular; corolla *c.* 4 cm across. Fruit *c.* 3 cm across, yellowish green or reddish, calyx persistent; seeds *c.* 10 × 5 mm, brown. *SE Europe, SW Asia & Middle East.* H3. Summer.

57. SORBUS Linnaeus

H. McAllister & N.P. Taylor

Deciduous trees to rhizomatous shrubs. Buds ovoid to conic. Leaves alternate, simple and toothed to pinnately lobed or divided. Flowers bisexual in flat, rarely pyramidal clusters, white or more rarely pink to crimson; sepals and petals 5; stamens 10–20; carpels 2–5, free or more or less united, each with 2 ovules; ovaries inferior or semi-inferior; styles free or united at the base. Fruit usually a small pome, spherical or ovoid-spherical, brown to orange-red to crimson to white; cells 2–5, mostly with cartilaginous walls, each potentially with 2 seeds, though many fruits contain only a single viable seed.

A genus of about 100 sexual species (and an indefinite number of named and un-named apomictic microspecies) from the northern hemisphere. The apomictic species listed are either single clones reproducing clonally from seed (e.g. *S. pseudohupehensis*, *S. glabriuscula*, *S. vilmorinii*) or aggregates of several similar microspecies (e.g. *S. decora*, *S. graeca*, *S. hybrida*, *S. intermedia*). Most are suitable for gardens or landscape planting and are grown for one or more of the following features: flowers, decorative fruits, attractive habit, foliage. Most species are light-demanding, relatively small, exposure-tolerant, mountain trees or shrubs. Apomictic species should be grown from seed and, especially with the smaller shrubby species, allowed to form multi-stemmed bushes. Sexual species may need to be grafted as many are self-incompatible and only one clone may be available.

Measurements are mostly given as an upper limit for a healthy vigorous plant because size is so susceptible to environmental factors. Drought greatly reduces the size of many parts, especially leaves and fruit. Leaf characters given are those of leaves on fruiting or vegetative spur shoots; leaves on long vegetative shoots are often very different in form. Identification is usually easiest in fruit, though flower colour is important for the identification of some species. Non-fruiting specimens, especially of the apomicts, may be unidentifiable.

Extensive hybridisation between tetraploid apomicts closely related to *S. aria* and diploid *S. aria*, *S. torminalis* and *S. aucuparia* have given rise to a large number of apomictic microspecies, which are treated under Section Aria. Hybrids are also known between *S. aria* and *S. alnifolia* and between species of Sections Sorbus and Cotoneaster and Pyracantha. Recently hybrids have arisen between white-fruited species of subgenus *Sorbus* and *S. aria*, giving rise to *S. arranensis*-like plants with white fruit.

Literature: Kalkman, C., The Malesian species of the subfamily Maloideae (Rosaceae), *Blumea* **21**(2): 413–42 (1973); Rohrer, J.R., Robertson, K.R. & Phipps, J.B., Variation in structure among fruits of Maloideae (Rosaceae), *American Journal of Botany* **78**(12): 1617–35 (1991); Aldasoro, J.J., Aedo, C., Muñoz Garmendia, F., Pando de la Hoz, F. & Navarro, C., Revision of Sorbus subgenera Aria and Torminaria (Rosaceae-Maloideae), *Systematic Botany Monographs* **69**: 1–148 (2004); McAllister, H., The genus Sorbus: mountain ash and other rowans (2005).

Key 1 (key to subgenera).

1a. Leaves compound with 2–26 pairs of free leaflets and a similar terminal leaflet; flesh of fruit homogeneous **2**

  b. Leaves simple, lobed or, if with one or more pairs of free leaflets at base, then upper part of leaf-blade lobed or merely toothed and not resembling the lateral leaflets **3**

2a. Trees; ripe fruit greenish, often flushed red where exposed to sun or russetted, large, more than 20 mm long; bud green, with more than 4 bud-scales lacking apical leaf scars visible on winter bud
  32. domestica (subgenus **Cormus**)

  b. Trees to dwarf shrubs, ripe fruit less than 20 mm, orange-red to crimson to white, never with any green colouration when ripe; bud rarely green, mostly red to brown to black with fewer than 4 bud-scales lacking apical leaf scars visible on winter bud **Key 1** (subgenus **Sorbus**)

3a. Leaves green, hairless above, glabrescent beneath, with triangular acute lobes extending almost halfway to midrib, and 4–6 pairs of veins; fruit top-shaped, brown, flesh of fruit homogeneous
  53. torminalis (subgenus **Torminaria**)

  b. Leaves not as above; fruit flesh heterogeneous breaking into uniform sized cell masses when squashed **Key 2** (subgenus **Aria**) - see p. 276

Subgenus Sorbus.

Deciduous trees to rhizomatous shrubs. Buds ovoid-conic, (greenish) to reddish to black, with usually only 2 or 3 bud-scales lacking apical leaf scars visible.

A subgenus of about 40 sexual species and an indefinite number of named and un-named apomictic microspecies. Most are small, exposure-tolerant, light-demanding, shallow-rooted mountain trees or shrubs preferring light soils and cool moist summers. Only species such as *S. decora*, and *S. cashmiriana* and *S. pseudohupehensis* and their close relatives, thrive in warmer drier areas. *S. poteriifolia* is best grown in a peat garden.

Key 1. Subgenus Sorbus (pinnate-leafed rowans)

1a. Fruit orange to vermilion, rarely yellow **2**

b. Fruit crimson to white, rarely yellowish or creamy-white   14

2a. Fruit yellow   3

b. Fruit orange to vermilion   5

3a. Buds ovoid-conic, reddish with rust-coloured hairs; leaflets without papillae beneath at ×100 magnification   **9. 'Joseph Rock'**

b. Buds ovoid-spherical, dark or blackish red with predominantly white hairs; leaflets papillose beneath at ×100 magnification   4

4a. Fruit broader than long; stipules large, leafy, persistent, even in inflorescence   **2. esserteauiana**

b. Fruit about as broad as long; stipules small, often soon falling, especially in inflorescence   **1. aucuparia**

5a. Buds very sticky, ovoid, chestnut-coloured; twigs very thick   **5. sargentiana**

b. Buds sticky or not, if sticky then conic, reddish or blackish   6

6a. Buds never sticky, ovoid to spherical with predominantly white hairs; leaflets with papillose beneath at ×100 magnification   7

b. Buds sticky or not, conic with rust-coloured hairs; leaflets without papillae beneath at ×100 magnification   9

7a. Stipules small, often soon falling, even in inflorescence; leaflets not glossy or leathery   **1. aucuparia**

b. Stipules large, persistent, leafy, even in inflorescence; leaflets glossy, leathery   8

8a. Leaflets fewer than 15, large and broad to 9 × 3 cm; fruit broader than long   **2. esserteauiana**

b. Up to 29 leaflets, each long and narrow to 4.9 × 1 cm, fruit about as broad as long   **3. scalaris**

9a. Shrub with delicate thin twigs; stipules large, leafy, especially in inflorescence; leaves kite-shaped in outline with leaflet size increasing from base of leaf   **4. gracilis**

b. Combination of characters not as above   10

10a. Leaflets more than 3 times as long as broad, apex drawn out into a fine point   11

b. Leaflets less than 3 times as long as broad, apex more abruptly acute   13

11a. Buds usually sticky; leaflets c. 15 per leaf; gaps visible between calyx-lobes   **6. americana**

b. Buds more or less sticky or not sticky; leaflets 11–15 per leaf; calyx-lobes overlapping   12

12a. Tree or few-stemmed shrub; leaves deep green   **8. commixta**

b. Many-stemmed shrub; leaves light green and very glossy   **7. scopulina**

13a. Small tree or large shrub; buds large, conic, blackish, usually sticky; 12–17 leaflets per leaf   **10. decora**

b. Shrub; buds ovoid, greenish red, more or less sticky; 7–11 leaflets per leaf   **11. californica**

14a. Usually rhizomatous shrubs less than 1 m; leaves less than 10 cm, bearing fewer than 14 leaflets, which are less than 2 cm; fruit crimson, becoming pinkish white to white   15

b. Larger shrub or tree; not rhizomatous though very occasionally with root suckers; leaflets larger and/or more numerous; fruit white to crimson   16

15a. Shrub more than 20 cm; flowers white   **12. reducta**

b. Shrub less than 20 cm; flowers pink   **13. poteriifolia**

16a. Fruit uniformly crimson, at least at first, becoming almost white; flower white (if flower pink, see **27. S. microphylla** aggregate)   17

b. Fruit never uniformly crimson, white more or less flecked with crimson; flower white to crimson   21

17a. Fruit very firm, late ripening (November) and slowly becoming paler   **23. carmesina**

b. Fruit relatively soft, ripening earlier   18

18a. Leaflets oblong, more than 5 cm, with papillae beneath at a magnification of ×100; fruit ellipsoid but truncate and flat at calyx, crimson   **17. 'Ghose'**

b. Leaflets lanceolate, less than 5 cm, without papillae beneath; fruit apple-shaped, crimson becoming white   19

19a. Leaves with fewer than 23 leaflets each over 3 cm; fruit hairless, flushed pink, styles 3   **15. 'Pearly King'**

b. Leaves with 23 or more leaflets; fruit with scattered rust-coloured hairs, especially around stalks; styles 3–5   20

20a. Leaves with more than 25 leaflets each less than 3 cm long; fruit flecked with crimson; styles 3 or 4   **14. vilmorinii**

b. Leaves with usually 23–25 leaflets c. 3 cm; fruit flushed pink; styles 5 rarely 4   **16. bissetii**

21a. Buds conic, pointed, more or less hairless except at tip and scale margins; up to 17 (rarely to 21) leaflets per leaf; inflorescence pyramidal or corymb-like; styles less than 2.25 mm, separate at base   22

b. Bud ovoid, hairless or hairy; mostly 17 or more leaflets per leaf; inflorescence corymb-like; styles more than 2.25 mm, contiguous at base   29

22a. Leaves with sheathing base; inflorescence corymb-like; fruit small, to 7 × 6 mm, spindle-shaped but truncate and flat at calyx, fruit crimson to creamy-pearly white   23

b. Leaves without sheathing base; inflorescence corymb-like or pyramidal; fruit more than 7 × 6 mm, more or less apple-shaped, mostly porcelain-white sometimes more or less flushed or flecked pink   25

23a. Leaflets not leathery, margin toothed, recurved only at extreme base; buds more or less covered with rust-coloured hairs; fruit crimson   **17. 'Ghose'**

b. Leaflets leathery, margin toothed in upper half or only towards apex, recurved at least in lower half; buds usually more or less hairless   24

24a. Leaflets 9 or more, less than 10 cm; fruit crimson to pink   **18. insignis**

b. Leaflets fewer than 9, more than 10 cm; fruit creamy-pearly white   **19. harrowiana**

25a. Inflorescence corymb-like; fruit often initially crimson becoming creamy to pearly white, soft; styles 3 rarely to 5, 2.25–2.5 mm; calyx-lobes fleshy only in fruit   **14. vilmorinii**

b. Inflorescence pyramidal; fruit becoming more crimson on ageing, white sometimes flushed or flecked crimson, firm; styles 4 or 5 rarely 3, c. 2 mm; calyx-lobes very fleshy   26

26a. Fruit almost pure white except around calyx   27

b. Fruit white flushed pink distant from as well as around calyx, especially where exposed to direct sunlight   28

27a. Leaflets less than 4 cm; fruit mostly more than 8 × 8.5 mm; styles 3 or 4, rarely 5   **20. forrestii**

b. Leaflets more than 4 cm; fruit smaller than 8 × 8.5 mm; styles 4 or 5 **21. glabriuscula**
28a. Leaves blue-green, kite-shaped with leaflets decreasing in size towards base from topmost or second topmost pair; fruit very hard, to 7 × 8 mm **22. pseudohupehensis**
b. Leaves green with most leaflets more or less equal in size; fruit less hard, some more than 7 × 8 mm **24. muliensis**
29a. Fruit very soft, often more than 12 mm across, with fleshy calyx-lobes protruding somewhat at fruit apex giving fruit an inverted pear shape 30
b. Fruit firmer, most less than 10 mm across, with green to blackish (depending on maturity of fruit), largely non-fleshy calyx-lobes in depression at fruit apex (i.e. fruit apple-shaped) or forming a protruding but non-fleshy crown-like structure 31
30a. Fruit white **25. cashmiriana**
b. Fruit pink **26. rosea**
31a. Flowers pink **27. microphylla**
b. Flowers white 32
32a. Leaflets papillose beneath at × 100 magnification, oblong, ovate, blunt or mucronate 33
b. Leaflets without papillae beneath, ovate to lanceolate, acute 34
33a. Mostly more than 25 leaflets per leaf; leaflets ovate, less than 18 × 7 mm; fruit less than 7.5 × 9.5 mm; style over 3 mm **28. munda**
b. Up to 25 leaflets per leaf; leaflets oblong, more than 18 × 7 mm; fruit often more than 7.5 × 9.5 mm; style c. 2.5 mm **29. foliolosa**
34a. Multi-stemmed shrub usually less than 3 m; buds blackish with whitish hairs **30. frutescens**
b. Fewer-stemmed larger shrub; buds reddish to reddish black with rust-coloured hairs **31. koehneana**

1. S. aucuparia Linnaeus (*S. pohuashanensis* (Hance) Hedlund; *S. fastigiata* (Loudon) Hartweg & Rümpler; *S. rossica* Späth; *S. moravica* Dippel). Illustration: Hay & Synge, Dictionary of garden plants in colour, 239 (1969); Brickell, RHS gardeners' encyclopedia of plants & flowers, 54 (1989); McAllister, The genus Sorbus: mountain ash and other rowans, pl. 1 (2005).

Tree to 20 m but usually much smaller; bark smooth, grey with laterally elongated lenticels; buds ovoid, blackish with white, rarely a few rust-coloured hairs. Leaves with up to 15, rarely to 19, coarsely toothed leaflets, 6–9 cm, with papillae beneath. Flowers white; fruit red, rarely yellow, late summer, to 9–12 × 9–14 mm, like minute apples in shape, with calyx in depression at fruit apex. Carpels 3 or 4, apices hairy, not fused, forming a conical structure within calyx-lobes. *Eurasia & NW Africa.* H1. Spring–summer.

The populations in several regions have been named as separate species but differ morphologically only in minor characteristics. *S. pohuashanensis* of supposedly N Chinese origin is unlike known-origin plants from that region, having large leaflets and large clusters of flowers and fruit. It may be of SE European origin. Many cultivars of *S. aucuparia* are grown, including: 'Dickeana' with yellow fruit; 'Beissneri' with orange bark and fastigiate habit; 'Aspleniifolia' with deeply cut leaflets.

2. S. esserteauiana Koehne (*S. conradinae* Koehne). Illustration: *Botanical Magazine*, 9403 (1935); Krüssmann, Manual of cultivated broad-leaved trees & shrubs **3**: pl. 117 (1986).

Similar to *S. aucuparia* but with thicker twigs, more ovoid-spherical, lighter coloured, redder buds; larger, leathery leaflets, which are felty white-hairy beneath; large leafy persistent stipules; to 6.5–7 × 8.5–9.5 mm, hard, late-ripening fruits that are broader than long. *China (W Sichuan).* H2.

A yellow-fruited variant, 'Flava', occurs. Uncommon in cultivation. Many trees grown under this name are *S. esserteauiana* × *S. aucuparia*, presumably raised from seed of the self-incompatible *S. esserteauiana*.

3. S. scalaris Koehne. Illustration: Hay & Synge, Dictionary of garden plants in colour, 239 (1969).

Very closely related to *S. esserteauiana* but with more numerous (to 33) narrow leaflets; fruits about as broad as long. *China (W Sichuan).* H2.

4. S. gracilis (Siebold & Zuccarini) K. Koch. Illustration: Krüssmann, Manual of cultivated broad-leaved trees & shrubs **3**: pl. 117 (1986).

A delicate shrubby species with leaves kite-shaped in outline with up to 11 leaflets. Stipules large, leafy, persistent. Flowers and fruit in small clusters; flowers greenish white; fruit red, elongated and somewhat pear-shaped in the most commonly cultivated clone. *C Japan.* H2.

Very uncommon in cultivation. Requires an acid soil.

5. S. sargentiana Koehne. Illustration: Hay & Synge, Dictionary of garden plants in colour, 239 (1969); Krüssmann, Manual of cultivated broad-leaved trees & shrubs **3**: pl. 115 (1986).

A slow-growing small tree or large bush with thick twigs and conspicuous, large, ovoid, chestnut-coloured, very sticky buds. Stipules large, leafy, persistent. Leaves to 30 cm with up to 13 leaflets each to 13 cm, with veins impressed above. Flowers and fruit in large corymbs, fruit small, usually broader than long, to 8 × 9 mm, with calyx-lobes separated, not overlapping. Styles 3 or 4, to 2 mm, inserted on fused flat carpel tops. *W China.* H2.

The very closely related *S. wilsoniana* has been reintroduced recently and differs in its variable leaflet number and leaflets without impressed veins.

6. S. americana Marsh. Illustration: Krüssmann, Manual of cultivated broad-leaved trees & shrubs **3**: pl. 116 (1986).

Small tree or shrub very similar to *S. commixta* but with stouter twigs, often stickier, darker buds, longer narrower leaflets; fruits with calyx-lobes separated, not overlapping. Fruit to 7.5 × 8 mm, orange-red, hard, glossy, usually borne in large clusters. *C & E USA.*

7. S. scopulina Greene.

Shrub to 2 m, but otherwise differing from *S. americana* only in its usually greener buds and lighter green leaf with fewer glossier leaflets. *NW America (Rocky Mountains north to Alaska).* H1.

Very uncommon in cultivation.

8. S. commixta Hedlund (*S. serotina* Koehne; *S. matsumurana* misapplied; *S. rufoferruginea* (Schneider) Schneider; *S. wilfordii* Koehne; *S. 'Embley'; S. 'Jermyns'; S. discolor* (Maximowicz) Hedlund). Illustration: *Botanical Magazine*, n.s., 166 (1951); Krüssmann, Manual of cultivated broad-leaved trees & shrubs **3**: pl. 117 (1986); Brickell, RHS gardeners' encyclopedia of plants & flowers, 54 (1989).

Tree or shrub; buds conic, greenish red to red, to dark red, sticky or not, more or less hairless except at tip and scale margins where there are rust-coloured hairs. Leaves at first covered with rust-coloured hairs, soon more or less hairless except along veins beneath; leaflets to 17, oblong-lanceolate, finely and evenly toothed, tapered to a fine point, colouring brilliantly in autumn. Flower white, fruit orange-red to red, often small, hard, shiny, spherical with calyx not in depression at fruit apex. *Japan, Korea.* H1.

The commonest clone in cultivation is a fast-growing, tall, fastigiate tree with rather sparse clusters of small hard fruits, but the species is very variable in the wild with respect to stature, leaflet size and breadth, and fruit size but is constant in the characters mentioned in the description. The very closely related **S. randaiensis** (Hayata) Koidzumi, which has only recently been introduced, differs in having more numerous, more glossy leaflets and fruit with 4 or 5 carpels.

9. S. 'Joseph Rock' Illustration: *Botanical Magazine*, n.s., 554 (1969); Hay & Synge, *Dictionary of garden plants in colour,* 239 (1969); Brickell, *RHS gardeners' encyclopedia of plants & flowers,* 55 (1989); McAllister, *The genus Sorbus: mountain ash and other rowans,* 220, pl. 18, 19 (2005).

Tall fastigiate tree with reddish ovate buds bearing rust-coloured hairs. Leaves with up to 21 neat lanceolate leaflets, almost all of much the same size to 4.4 × 1.6 cm, developing brilliant autumn colouration. Flowers white. Fruit to 10 × 11 mm, yellow with red around calyx becoming very pale yellow. Styles 2.5 mm, mostly 3, distantly inserted on the more or less fused carpel tops. Almost certainly *S. commixta × S. monbeigii.* H2.

10. S. decora (Sargent) Schneider. Illustration: Hosie, *Native trees of Canada,* 207 (1979).

Small tree or shrub. Twigs stout bearing large conical blackish, usually very sticky buds. Leaves large with up to 15 leaflets, to 7.5 cm, each less than 2.5 times as long as broad. Fruits to 1.2 × 1.2 cm, in large more or less drooping clusters somewhat hidden by leaves, very attractive to birds. Apomictic agg. *E North America north to Greenland.* H1.

11. S. californica G.N. Jones (*S. cascadensis* G.N. Jones; *S. sitchensis* misapplied).

Shrub to 2 m with more or less sticky ovoid buds; flowering buds often incompletely covered by the bud-scales. Leaves with up to 11 leaflets, each less than 2.5 times as long as broad. Fruits borne in showy erect corymbs, very attractive to birds. Apomictic agg. *W North America (Cascade Mts).* H1.

Uncommon in cultivation.

S. sitchensis Roemer. Illustration: McAllister, *The genus Sorbus: mountain ash and other rowans,* pl. 3 (2005). An apomictic complex, which has only recently been introduced to cultivation in Europe and differs in its non-sticky buds that have a glaucous bloom and bear rust-coloured hairs, bluish green matt leaves and cerise-pink, not orange-red fruits. *W North America (Rocky Mts).*

12. S. reducta Diels. Illustration: Krüssmann, *Manual of cultivated broad-leaved trees & shrubs* **3**: pl. 118 (1986); Brickell, *RHS gardeners' encyclopedia of plants & flowers,* 300 (1989).

Shrub, rhizomatous or not, to 50 cm. Buds reddish with rust-coloured hair. Leaves to 10 cm with up to 15 ovate, glossy leaflets, which colour orange-red in autumn. Flowers white. Fruit to 9.5 × 11 mm, at first dull crimson, becoming more white with age. Style *c.* 2 mm. Apomictic. *N Burma to W China (Yunnan).* H2.

13. S. poteriifolia Handel-Mazzetti (*S. pygmaea* misapplied). Illustration: McAllister, *The genus Sorbus: mountain ash and other rowans,* pl. 17 (2005).

Strongly rhizomatous low shrub to 10 cm; buds reddish. Leaves glossy to 8 cm, with up to 15 ovate leaflets. Flowers pinkish with each petal having a deep pink zone in the centre. Fruit to 8.25 × 9.5 mm first dull crimson, ripening to almost pure white. Carpel apices fused; styles 3–5, *c.* 2 mm, Apomictic microspecies. *N Burma to W China (Yunnan).* H2.

Uncommon in cultivation.

14. S. vilmorinii Schneider. Illustration: *Botanical Magazine,* 8241 (1889); Hay & Synge, *Dictionary of garden plants in colour,* 239 (1969); Brickell, *RHS gardeners' encyclopedia of plants & flowers,* 66 (1989).

Shrub with slender twigs bearing ovoid reddish buds more or less covered with rust-coloured hairs. Leaves to 14 cm with up to 29 lanceolate leaflets to

2.3 cm × 7 mm. Flowers white. Fruit *c.* 9.5 × 9 mm, crimson becoming white flecked with crimson; carpel tops fused, more or less hairless, sunk in depression at fruit apex; styles *c.* 2 mm, more or less wedge-shaped at base; seeds usually pear-shaped and 1 per fruit, dark brown, to 5 × 2.5 mm. Apomictic microspecies. *W China.* H2.

The closely related sexual diploid **S. pseudovilmorinii** McAllister is very much commoner than the above in the wild and very variable with fruit colour ranging from crimson to pure white. Some provenances are more drought-tolerant and easier to grow than *S. vilmorinii.* Seeds usually less elongated and several per fruit. It is becoming more common in cultivation.

15. S. 'Pearly King' (*S. pluripinnata* (Schneider) Koehne, misapplied; *S. vilmorinii* 'Robusta'). Illustration: McAllister, *The genus Sorbus: mountain ash and other rowans,* 37 (2005).

Small spreading tree with reddish ovate buds bearing rust-coloured hairs. Leaves with up to 21 neat lanceolate leaflets almost all of much the same size to 4.4 × 1.6 cm. Flowers white. Fruit to 10 × 11 mm, dull crimson becoming pink. Apomictic (microspecies?). *Probably China (Sichuan, Gongga Shan).* H2.

16. S. bissetii Koehne. Illustration: McAllister, *The genus Sorbus: mountain ash and other rowans,* 165 (2005).

A thick-twigged shrub similar to *S. vilmorinii* but thicker, larger and stiffer in all its parts. Leaves with up to 25 leaflets. Fruit at first dull crimson becoming almost white, to 10.5 × 12 mm; styles *c.* 2.75 mm. Apomictic microspecies. *China (Yunnan, Sichuan).* H2.

Uncommon in cultivation.

17. S. 'Ghose' Illustration: McAllister, *The genus Sorbus: mountain ash and other rowans,* pl. 20 (2005).

Tree to 30 m with thick stiff twigs bearing large (to 3 cm) ovoid-conic, dark red buds, more or less covered with rust-coloured hairs, which are particularly dense at the apex; leaf-bases somewhat sheathing, leaves to 32 cm, with 13–15 leaflets, each to 10 × 2.7 cm, acute and toothed to base, with numerous rust-coloured hairs beneath. Flowers white in large corymbs. Fruit in large corymbs, small, to 7.5 × 7.75 mm, crimson, broadly ellipsoid but truncate at calyx; styles 3 or

4, c. 2.25 mm, more or less distantly inserted in the more or less fused hairy carpel apices. Seed to 3.5 × 2.25 mm, pale yellow brown. *Probably of hybrid origin but said to have come from near Darjeeling in N India.*

18. S. insignis J.D. Hooker. Illustration: Krüssmann, Manual of cultivated broad-leaved trees & shrubs **3**: pl. 115 (1986); *The Plantsman* **13**(4): 231 (1992).

Large tree with thick stiff twigs bearing large, ovoid-conic greenish to reddish more or less hairless buds. Leaf-base sheathing. Leaves to 33 cm with up to 17 leathery leaflets, each to 9 × 1.9 cm, toothed only in upper part and with margin recurved throughout, more or less hairless beneath. Flowers white in large corymbs. Fruit pinkish white, ovoid, to 6.75 × 5.5 mm; styles 3 or 4, c. 2 mm, more or less distantly inserted on the more or less fused hairy carpel apices. *C& W Himalaya.* H3.

Uncommon in cultivation.

19. S. harrowiana (Balfour & W.W. Smith) Rehder. Illustration: Krüssmann, Manual of cultivated broad-leaved trees & shrubs **3**: pl. 115 (1986); McAllister, The genus Sorbus: mountain ash and other rowans, pl. 5 (2005).

Very similar to *S. insignis* but with fewer (to 5), even larger (often over 10 cm) leaflets. *N Burma to W China (Yunnan).* H4.

Uncommon in cultivation.

20. S. forrestii McAllister & Gillham. Illustration: *Botanical Magazine*, n.s., 792 (1981); McAllister, The genus Sorbus: mountain ash and other rowans, 126, 127 (2005).

Small tree to 5 m. Buds ovoid-conic, acute, reddish brown, more or less hairless except for usually rust-coloured hairs at scale margins and especially at apex. Leaves to 22 cm with up to 19 distantly inserted, oblong-elliptic leaflets to 4.5 × 1.7 cm, all of much the same size and toothed in the upper half to two-thirds. Flowers white. Fruit at first green, becoming almost pure white except for reddish colour around calyx, to 9.5 × 9 mm; styles 4–5, c. 2.5 mm, distantly inserted on the fused almost hairless carpel apices. Seed brown. Apomictic microspecies. *W China (NW Yunnan).* H2.

Uncommon in cultivation.

21. S. glabriuscula (Cardot) McAllister (*S. hupehensis* Schneider misapplied; *S. wilsoniana* Schneider misapplied).

Illustration: Hay & Synge, Dictionary of garden plants in colour, 239 (1969).

Small tree to 8 m. Twigs stiff. Buds ovoid-conic, acute, greenish red, more or less hairless except for whitish hairs at scale margins and especially at apex. Leaves with up to 17 oblong leaflets, each to 6 × 2.2 cm, mucronate, all of much the same size. Flowers in pyramidal panicles; petals white, becoming reflexed. Fruit green, becoming white, hard, to 7.5 × 8 mm, calyx-lobes fleshy; styles 4 or 5, c. 2.25 mm, distantly inserted on fused more or less hairless carpel apices; seed brown. Apomictic microspecies. *W China (NW Yunnan).* H2.

22. S. pseudohupehensis Schneider (*S. hupehensis* Schneider var. *rosea*; 'Pagoda Pink'; 'November Pink'). Illustration: *Botanical Magazine*, n.s. 167 (1950); Krüssmann, Manual of cultivated broad-leaved trees & shrubs **3**: 341 (1986); Brickell, RHS gardeners' encyclopedia of plants & flowers, 54 (1989).

Small tree to 8 m. Buds ovoid-conic, acute, reddish, more or less hairless except for whitish hairs at scale margins and especially at apex. Leaf bluish green with up to 15 ovate leaflets to 5 × 2.2 cm but often, especially on the weaker shoots, decreasing in size from the apex towards the base giving the leaves a kite-shaped outline. Flowers in pyramidal panicles, white, petals becoming reflexed. Fruit green flushed crimson, becoming white flushed crimson, especially around the calyx, very hard, to 7 × 8 mm; calyx-lobes fleshy; styles 5, c. 2.25 mm, distantly inserted on fused, more or less hairless, carpel apices; seed brown. Apomictic microspecies. *China (probably Yunnan).* H3.

In addition to the following, several other similar apomictic microspecies have been introduced recently including **S. olivacea** McAllister, with olive tree-like leaflets and pink fruit from Sichuan, and **S. rushforthii** McAllister from Tibet.

23. S. carmesina McAllister. Illustration: McAllister, The genus Sorbus: mountain ash and other rowans, 114 (2005).

Fruit very firm, late-ripening (November) and slowly becoming paler. *China (NW Yunnan, N Cangshan).*

24. S. muliensis McAllister. Illustration: McAllister, The genus Sorbus: mountain ash and other rowans, 114 (2005).

Very closely related to *S. pseudohupehensis* but with leaflets longer,

narrower, more distantly inserted, more equally sized and fruits larger, to 8.5 × 9 mm, with 4–5 styles. Apomictic microspecies. *China (Sichuan).* H2.

Very uncommon in cultivation.

25. S. cashmiriana Hedlund. Figure 99 (17), p. 183. Illustration: Hay & Synge, Dictionary of garden plants in colour, 239 (1969); Brickell, RHS gardeners' encyclopedia of plants & flowers, 66 (1989).

Small tree or large shrub. Buds ovoid-conic, acute, reddish, more or less hairless except for rust-coloured hairs at scale margins and especially at apex. Leaves to 20 cm, with 17–21 lanceolate leaflets each to 3.7 × 1.5 cm. Flowers in corymbs, pink, more than 1 cm across. Fruit to 1.5 × 1.3 cm green, becoming almost pure white with white fleshy calyx-lobes protruding; styles 4–5, c. 3 mm, closely inserted on the fused carpel apices. Apomictic microspecies. *W Himalaya.* H2.

Related introductions (e.g. *S. ovalis*) from China (Sichuan) differ in having thinner twigs, smaller ovoid leaflets, usually white flowers, which are not so large as in *S. cashmiriana*; but the flowers and fruits are still larger and the protruding fleshy calyx more prominent that in any other white-fruited *Sorbus*.

26. S. rosea McAllister. Illustration: McAllister, The genus Sorbus: mountain ash and other rowans, pl. 13 & 14 (2005).

Similar to *S. cashmiriana* in overall form but very different in its pink fruit. *S. rosea* is smaller in stature than *S. cashmiriana*, has deeper pink flowers as well as fruit, and generally more reddish pigmentation of its twigs and leaves. The fruit is initially green, becoming a uniform soft pink colour before turning paler with darker reddish patches. Apomictic microspecies. *NW Pakistan.* H2.

27. S. microphylla (Wallich) Decaisne. Illustration: *Botanical Magazine*, n.s., 879 (1983); McAllister, The genus Sorbus: mountain ash and other rowans, 171 (2005).

Small tree or shrub. Buds ovoid. Leaves very variable to 17 cm with up to 33 leaflets. Flowers pale pink to almost crimson. Fruit white, sometimes crimson at first; calyx never fleshy, style bases inserted close together on the fused carpel apices. , *Himalaya.* H2. A very variable aggregate of numerous microspecies, of which **S. khumbuensis** McAllister is possibly the only one at all widespread in cultivation. Members of the aggregate are distinguished from all other species with numerous small leaflets by the combination of the pink flowers and initially crimson fruit that becomes pink to almost white on ripening.

28. S. munda Koehne. Illustration: *Botanical Magazine*, 9460 (1936); Hay & Synge, Dictionary of garden plants in colour, 239 (1969).

Shrub with greyish twigs. Buds ovoid, dark-red brown, covered with rusty-grey coloured hairs. Leaves to 14 cm with up to 31 ovate, blunt leaflets to 3 × 1 cm, with papillae beneath. Flowers white. Fruit at first green, becoming almost pure white, to 7.5 × 9.5 mm; calyx partially fleshy; styles 5, *c.* 3.5 mm, closely inserted on the more or less hairless fused carpel apices. Apomictic microspecies. *China (Sichuan).* H2.

29. S. foliolosa (Wallich) Spach (*S. ursina* (Wenzig) Decaisne). Illustration: *The Plantsman* **13**(4): 233 (1992).

Tree or shrub with stiff twigs bearing conic dark red buds with rust-coloured hairs confined to scale margins and bud apex. Leaves to 21 cm bearing up to 25 oblong, blunt or mucronate leaflets all of much the same size, to 4.7 × 1.5 cm, and with papillae beneath. Fruit at first green, becoming white, to 1 × 1.1 cm, often broader than long, conspicuously black at calyx; styles 5, *c.* 2.5 mm, closely inserted on the more or less hairless fused carpel apices. Apomictic microspecies. *C Himalaya.* H2.

Uncommon in cultivation.

30. S. frutescens McAllister (*S. koehneana* misapplied; *S. fruticosa* invalid). Illustration: McAllister, The genus Sorbus: mountain ash and other rowans, pl. 6 & p. 148 (2005).

Shrub to 2 m. Twigs dark brown with ovate, pointed blackish buds, hairless except for whitish hairs at scale margins and apex. Leaves dark green with up to 29 deeply toothed leaflets to 2.2 cm × 8 mm, without papillae beneath. Fruit green becoming pure white, to 9.5 × 12 mm; styles 5, *c.* 2.75 mm, closely inserted on the fused slightly hairy carpel tops; seed initially reddish becoming dark brown. Apomictic microspecies. *China (SW Gansu).* H1.

31. S. koehneana Schneider. Illustration: Krüssmann, Manual of cultivated broad-leaved trees & shrubs **3**: pl. 118 (1986).

Small tree or shrub. Buds ovoid-conic, dark-red, hairless except for rust-coloured hairs at scale margins and apex. Leaves with up to 25 leaflets, without papillae beneath. Flowers white. Fruit green, becoming white, to 7.5 × 10 mm, but most much less broad. Styles 5, *c.* 2.5 mm, more or less closely inserted on the fused carpel tops. Seed initially reddish, becoming brown. *N & W China.* H1.

Uncommon in cultivation. This diploid sexual species is widespread in China but no introduction to date has proved popular. Likely to prove more popular are very closely related shrubs from Sichuan with very small glossy leaflets which have been named **S. setschwanensis** Koehne and the apomictic tetraploid microspecies **S. eburneus** McAllister with up to 33 leaflets.

Subgenus Cormus (Spach) Boissier. Now sometimes treated as a separate genus. Leaves pinnate, carpels 5, fully united; ovary inferior. Fruits green to brown.

32. S. domestica Linnaeus. Illustration: *The Plantsman* **7**(2): frontispiece (1985).

Tree to 20 m; bark deeply furrowed; branches becoming hairless; buds sticky and shiny. Leaflets 11–21, narrowly oblong, 3–8 cm, with coarse, forward-pointing teeth, base symmetric, hairy beneath. Flowers *c.* 1.5 cm, in conical corymbs to 10 cm. Fruit to 3 cm, apple or pear-shaped, yellow-green, ripening to red. *C & S Europe, N Africa, Turkey.* H2. Spring–early summer.

'Signalman'; *S. domestica* × *S. scopulina*. A low narrowly pyramidal tree; leaves small; fruit large, pale orange.

Forma **pomifera** (Hayne) Rehder. Fruit apple-shaped, 2–3 cm. Forma **pyriformis** (Hayne) Rehder. Fruit pear-shaped, 3–4 cm.

Subgenus Aria Persoon.

Species of subgenus *Aria* vary from tall trees to large shrubs with one or a few trunks. These are characterised by ovoid-conic buds with many scales evident, looking superficially like those of *Acer pseudoplatanus* or a *Prunus*, and simple, toothed, or more or less lobed leaves, which often have a persistent white or greyish tomentum beneath. The fruits are often red but some species have yellowish, greenish or russet-coloured fruits. In the species with persistent calyx-lobes these are never wholly fleshy. The fruit flesh is heterogeneous giving a dryish granular texture.

They occur continuously across Eurasia from western Europe through the mountains of the middle east, the Himlaya, China to Korea and Japan, but are absent as native trees from N. America.

Subgenus *Aria* as interpreted here includes species usually placed in subgenus *Micromeles*. Rohrer et al. (1991) argue that the heterogeneous flesh of the fruits in all species and lack of consistency in the distribution of other characters makes it impossible to maintain the separation of *Micromeles*. The most natural division of the subgenus would seem to be into three groups:

Section **Aria** (Persoon) Dumortier. Species related to *S. aria*, with fruits that are wholly red when ripe and seeds that are round in cross-section. The embryos of these seeds must contain germination inhibitors as they do not germinate on excision. Almost all of these have leaves that are whitish tomentose beneath. This group occurs from western Europe to the Caucasus, is the only group found in this area, and does not appear to overlap in distribution with the other groups. Diploids (only *S. aria, S. chamaemespilus,* and possibly *?S. umbellata*) are sexual, and polyploids apomictic. Apomicts have hybridised with *S. aria, S. aucuparia, S. torminalis* and *S. chamaemespilus* to produce a vast range (100 named in Flora Europaea) of apomictic microspecies, which are often (as here) grouped with the pure 'aria' types under *Sorbus* subgenus *aria* or treated as the 'arranensis' and 'latifolia' aggregates. The species and most of the derived apomicts are light-demanding, exposure-tolerant, deep-rooted and relatively drought-tolerant mountain trees, often naturally found on limestone, but tolerant of a wide pH range in cultivation.

Section **Micromeles** (Decaisne) Rehder. Himalayan and east Asiatic species with usually more or less spherical fruits that never become uniformly red or orange, the fruit skin often containing much chlorophyll when ripe and may be densely covered with lenticels (russeted). The seeds are always laterally flattened and radially elongated as in subgenus *Cormus*, and the embryos germinate on excision.

The concept of this grouping is new and it may be divisible into two groups: the species with white tomentose leaf undersides mostly have persistent calyces and free carpel apices (formerly referred to subgenus *Aria*) and seem to form a natural grouping from *S. lanata* in the western Himalaya through *S. vestita*, *S. hedlundii*, *S. thibetica*, *S. pallescens* to *S. hemsleyi* and **S. dunnii** Rehder in eastern China, and are grown primarily for their foliage. Species with more or less glabrous leaf undersides mostly have deciduous calyces and fused carpel apices (formerly referred to *Micromeles*). Species of this last group are often more shade-tolerant than those of other sections but, coming from areas affected by monsoons, less tolerant of drought. The fruits are rarely attractive. They are usually grown for their flowers (*S. meliosmifolia*) or foliage (*S. caloneura*).

All species are diploid and sexual.

Section **Alnifoliae** (T.T. Yü) Aldasoro, Aedo & C. Navarro (*S. alnifolia*, *S. folgneri*, *S. zahlbruckneri*, *S. yuana*, *S. japonica*). East Asiatic species (China, Japan) with oblong fruits that become a distinctive translucent red or orange on ripening and whose seeds are pear-shaped, round in cross-section, and whose embryos contain germination inhibitors and do not germinate on excision. Leaf venation with main lateral veins running into teeth. The distribution of this group overlaps with that of section *Micromeles* but no hybrids have been reported.

Species of this section have attractive fruit and often autumn foliage colouration.

All species are diploid and sexual.

Key 2. Subgenus **Aria**

1a. Ripe fruit wholly and evenly coloured red to orange at maturity, usually with scattered lenticels, seeds tear-drop shaped, more or less circular in cross-section   2

b. Ripe fruit green to russet sometimes with red colouration over the green on the side exposed to the sun, occasionally orange, never evenly and brightly coloured, usually with large, conspicuous lenticels, seeds about as broad as long, flattened and therefore elliptic in cross-section (section **Micromeles**)   7

2a. Ripe fruit a translucent red or orange, sometimes with white waxy bloom, usually conspicuously longer then broad, calyx and top of receptacle deciduous (except *S. zahlbruckneri*, *S. yuana*) by time fruit is mature or calyx small and wholly dried up when fruit is ripe;

bud brown and glabrous resembling that of a *Prunus* (section **Alnifoliae**)   3

b. Fruit colour opaque without waxy bloom, globose or slightly longer than broad, calyx-lobes nearly always persistent in ripe fruit, fleshy at base; bud greenish to red-brown, scales usually somewhat hairy at margins (section **Aria**)   18

3a. Leaves conspicuously lobed (to about one fifth distance to midrib)   **36. japonica**

b. Leaves not conspicuously lobed   4

4a. Leaf white-tomentose beneath, or, if thinly cobwebby, then calyx persistent   5

b. Leaf glabrescent beneath   6

5a. Leaf densely white tomentose beneath, calyx and top of receptacle deciduous in ripe fruit   **35. folgneri**

b. Leaf thinly white cobwebby beneath, calyx persistent   **38. zahlbruckneri**

6a. Calyx persistent, leaf to 12.7×9.4 cm; fruit orange to 16×13 mm   **39. yuana**

b. Calyx deciduous, leaf and fruit smaller; fruit red   **37. alnifolia**

7a. Leaves densely white tomentose beneath; calyx persistent except in *S. hemsleyi* (*S. thibetica* group)   8

b. Leaves not densely white tomentose beneath; calyx deciduous except in *S. megalocarpa* ('*Micromeles*')   13

8a. Calyx deciduous; fruit *c.* 10 mm, brown   **49. hemsleyi**

b. Calyx persistent; fruit green to orange to brownish-orange   9

9a. Leaves deeply (*c.* one fifth of distance to midrib) lobed, fruit large (>10 mm)   **45. lanata**

b. Leaves not as deeply lobed; fruit often smaller (large in *S. vestita*, *S. megalocarpa*)   10

10a. Fruit longer than broad, somewhat flattened, green with small lenticels and red flush where exposed to sun   **48. pallescens**

b. Fruit broader than long, not flattened, lenticels large and numerous   11

11a. Leaves less than 10 cm long; fruit orange when ripe   **47. wardii**

b. Leaves longer than 10 cm, fruit greenish to brownish orange   12

12a. Leaves with 11 or more pairs of lateral veins; styles 2 or 3 (4)   **46. thibetica**

b. Leaves with 11 or fewer pairs of lateral veins; styles 3–5   **44. vestita**

13a. Fruits large (more than 10 mm), longer than broad, russet with dense covering of lenticels; flowering before coming into leaf   **43. megalocarpa**

b. Fruit smaller to *c.* 10 mm; flowering as young leaves expand   14

14a. Fruit small, less than 7 mm long, about as long as broad, green   **40. epidendron**

b. Fruit *c.* 10 mm long, at least as broad as long   15

15a. Leaves smooth, obovate, shiny; main lateral veins not reaching margins; fruit green   16

b. Leaves more or less bullate (veins impressed above), elliptic, craspedodromous (main lateral veins running into teeth); fruit russet   17

16a. Petiole *c.* 5 mm; leaves broadly obovate   **41. keissleri**

b. Petiole 5–10 mm; leaves elliptic-obovate   **42. aronioides**

17a. Leaves with 10–18 pairs of veins; young leaves chocolate-coloured, white cobwebby-tomentose on both sides   **33. caloneura**

b. Leaves with 16–34 pairs of veins; young leaves nether conspicuously chocolate-coloured nor cobwebby tomentose   **34. meliosmifolia**

18a. Petals erect, pink; leaves green beneath   **62. chamaemespilus**

b. Petals white; leaves white- or grey-tomentose beneath   19

19a. Leaves entire or lobed but without free leaflets at base   20

b. Leaves with one or more pairs of free leaflets at base   27

20a. Leaves not conspicuously lobed (lobes extending less than one fifth of way to midrib)   21

b. Leaves distinctly lobed (lobes extending more than one fifth of way to midrib)   23

21a. Fruit orange; leaves grey tomentose beneath   **57. bristoliensis**

b. Fruit red; leaves white tomentose beneath   22

22a. Leaves mostly ovate-elliptic with 10 or more pairs of veins but variable; fruit longer than broad; usually a tall tree   **52. aria**

b. Leaves widest above middle or more or less circular with fewer than 11 pairs of veins; fruit globose; multi-stemmed shrub to tree   **51. graeca**

23a. Fruit yellowish to brownish orange   24

b. Fruit orange-red to red; leaf-lobes more rounded, often more or less parallel-sided   25

24a. Leaves greenish to grey beneath, leaf-lobes more or less sharply triangular   **54. latifolia**

b. Leaves white beneath, leaf-lobes more rounded   **50. umbellata**

25a. Fruit orange-red, conspicuously longer than broad; leaves oblong with 3–4 pairs of relatively similar rounded lobes   **55. intermedia**

b. Fruit red   26

26a. Fruit globose; leaves oblong with 3 or 4 pairs of relatively similar rounded lobes   **58. mougeotii**

b. Fruit usually longer than broad; leaves elliptic with lobes less similar in size   **56. arranensis**

27a. Leaves with mostly more than 3 pairs of free leaflets   **61. teodori**

b. Leaves with 1–3 pairs of free leaflets   28

28a. Leaves relatively long and narrow, mostly more than 1.75 times as long as broad   **59. × thuriangiaca**

b. Leaves relatively broader, mostly less than 1.75 times as long as broad   **60. hybrida**

Section **Micromeles** (Decaisne) Rehder. Distinguished by its fully inferior ovaries, and receptacle apex and calyx usually (except *S. megalocarpa*) deciduous as a unit, leaving a clean scar at the apex of the mature fruit.

33. S. caloneura (Stapf) Rehder (*Micromeles caloneura* Stapf). Illustration: *Botanical Magazine*, 8335 (1910); Krüssmann, Manual of cultivated broad-leaved trees & shrubs **3**: 333 (1986).

Tree to 10 m. Stems and winter buds hairless. Leaves simple, ovate-elliptic to oblong, mostly to 11 × 5 cm, tapered at both ends, lateral veins in 10–15 pairs, which are not impressed above, more or less hairless except in the vein axils, margins double-toothed; leaf-stalk 6–12 mm. Flowers white, *c.* 1.2 cm across in dense, downy corymbs 5–10 cm across; anthers pinkish brown. Fruit spherical to pear-shaped, *c.* 1 cm, brown to bronze-coloured, covered with lenticels. *China*. H1.

34. S. meliosmifolia Rehder. Leaves with 19–24 pairs of impressed veins and showy white flowers and, as a living tree, quite different in appearance from *S. caloneura*. *Western China*.

35. S. folgneri (Schneider) Rehder (*Micromeles folgneri* Schneider). Illustration: Bean, Trees & shrubs hardy in the British Isles, edn 8, **4**: 427 (1980); Krüssmann, Manual of cultivated broad-leaved trees & shrubs **3**: 333 (1986).

Tree to 9 m, often with drooping branches. Stems white-felted at first; winter buds hairless. Leaves simple, narrowly ovate to lanceolate, to 10 × 4 cm, tapered at both ends, lateral veins in 8–10 pairs, with dense white felt beneath, margins finely toothed; leaf-stalk *c.* 12 mm. Flowers white, 8–12 mm across, numerous in dense, woolly corymbs to 10 cm across. Fruit ovoid, to 1.2 cm across, red at least on one side (yellow in 'Lemon Drop'). *China*. H5.

36. S. japonica (Decaisne) Hedlund (*Micromeles japonica* (Decaisne) Koehne; *S. japonica* var. *calocarpa* Rehder). Tree to 20 m, with larger, shallowly lobed leaves and smaller flowers. *Japan & Korea*. H5.

37. S. alnifolia (Siebold & Zuccarini) K. Koch (*Micromeles alnifolia* (Siebold & Zuccarini) Koehne; *S. alnifolia* var. *submollis* Rehder). Illustration: *Botanical Magazine*, 7773 (1901).

Tree to 20 m. Leaves sometimes shallowly lobed, with 6–12 pairs of lateral veins, more or less hairy beneath at first but never white-felted, margins double-toothed. Fruit bright red to deep pink. *E Asia*. H1.

38. S. zahlbruckneri C.K. Schneider. Similar to *S. alnifolia* but calyx persistent and typically with narrower leaves. *C China*.

39. S. yuana Spongberg. Similar to *S. alnifolia* but with calyx persistent and larger leaves and fruit and fruit orange rather than red. *China (W Hubei)*.

40. S. epidendron Handel-Mazzetti. Tree or shrub to 15 m. Stems and foliage loosely hairy at first, winter buds hairless. Leaves simple, elliptic to obovate, to 15 × 7 cm, but often much smaller, tapered at both ends, lateral veins 10–12 pairs, margins simply toothed; leaf-stalk of variable length. Flowers white, *c.* 6–7 mm across. Fruit spherical, truncate at apex, only 6–8 mm across, covered with lenticels, in dense clusters. *S to C China & adjacent Burma*. H2.

Very similar is **S. rhamnoides** (Decaisne) Rehder, distinguished by its smaller leaves with 12–14 pairs of veins, flowers to *c.* 1 cm across and fruit lacking lenticels. *E Himalaya & Burma*. H5.

41. S. keissleri (Schneider) Rehder (*Micromeles keissleri* Schneider).

Tree to 12 m. Stems densely covered with lenticels, hairy at first. Leaves simple, ovate to obovate, leathery, to 7.5 × 6 cm, soon hairless, tapering very gradually at base, abruptly at apex, lateral veins in 7–10 pairs, margins simply toothed; leaf-stalk ill-defined, to 5 mm. Flowers white, to 1 cm across, petals soon falling. Fruit greenish, covered with lenticels, to 1.8 cm across. *C to SW China, N Burma*. H1.

42. S. aronioides Rehder. Similar (as is the possibly synonymous **S. thomsonii** (King ex J.D. Hooker) Rehder) to *S. keissleri* but leaves with longer leaf-stalks and less distinctly obovate leaves.

43. S. megalocarpa Rehder. Illustration: *Botanical Magazine*, n.s., 259 (1955); Krüssmann, Manual of cultivated broad-leaved trees & shrubs **3**: pl. 118 (1986).

Shrub or tree to 8 m, with large, glossy winter buds to 1.8 cm resembling those of a horse-chestnut (but not sticky). Stems stout, to 5 mm thick the first year, hairless, dark but marked with numerous lenticels. Leaves simple, ovate to obovate, rather variable in size but sometimes to 20 × 11 cm, gradually tapered to rounded at base, lateral veins to 20 pairs, more or less hairless on both sides except for small tufts in the vein axils, margins finely simple to double-toothed; leaf-stalks 2–5 cm. Flowers expanding before the leaves, white, *c.* 1.8 cm across, in dense more or less globular clusters to 15 cm across. Fruit egg-shaped, to 3 × 2 cm, brown. *S to C China (Sichuan)*. H1.

The more commonly cultivated var. **cuneata** Rehder has shorter leaf-stalks and smaller fruits.

44. S. vestita (G. Don) Loddiges (*S. cuspidata* (Spach) Hedlund). Illustration: *Botanical Magazine*, 8259 (1909); Krüssmann, Manual of cultivated broad-leaved trees & shrubs **3**: pl. 119 (1986).

Tree to 23 m. Stems white-woolly at first, stout (*c.* 5 mm across at 1 year old), with obtuse, almost hairless winter buds. Leaves simple, variably oblong-ovate to broadly elliptic, to 22 × 12 cm, obtuse to acuminate at apex, rounded to tapered at base, lateral veins in 8–16 pairs, closely and finely white-felted beneath, margin single or irregularly double-toothed;

leaf-stalk 2.5–5 cm. Flowers white, 1.5–2.5 cm across, in very woolly corymbs to 7.5 cm across; styles 4 or 5, rarely 3. Fruit more or less spherical, *c.* 1.5 cm across or larger, yellow-brown sometimes tinged red, covered with lenticels. *Himalaya & N Burma.*

S. hedlundii Schneider differs mainly in its leaves with brownish hairs on the midrib and lateral veins beneath. *E Nepal to W Bhutan.* H3.

45. S. lanata (D. Don) Schauer.
Similar to *S. vestita* but with distinctly lobed leaves and with leaves less persistently white-woolly beneath, lateral veins in 9 or more pairs and fruit to 2.5 cm across or larger. *W Himalaya.* H5.

46. S. thibetica (Cardot) Handel-Mazzetti (*S. mitchellii* invalid). Illustration: Brickell, RHS gardeners' encyclopedia of plants & flowers, 52 (1989).
Similar to *S. vestita* but with more slender stems and leaves often smaller (but not in 'Mitchellii'), sometimes almost circular, more loosely hairy and less woolly beneath. Flowers smaller; styles 1–3. Fruit brown, yellow tinged reddish or green-orange. *E Himalaya to SC China.*

47. S.wardii Merrill.
Similar to *S. thibetica* but typically with smaller leaves and fruit. Doubtfully distinct but trees in cultivation of distinct appearance and occurring to east of *S. thibetica.*

48. S. pallescens Rehder.
Similar to *S. thibetica* but more slender in appearance with 2–5 styles and green to yellow somewhat laterally flattened fruit only *c.* 9 mm across. *China.* H1.

49. S. hemsleyi (C.K. Schneider) Rehder.
Similar to *S. pallescens* but the calyx is deciduous in the fruits, which are densely covered with lenticels like those of *S. caloneura.*

50. S. umbellata (Desfontaines) Fritsch. Illustration: Krüssmann, Manual of cultivated broad-leaved trees & shrubs **3**: pl. 120 (1986).
Small tree or shrub to 6 m. Leaves simple, varying from fan-shaped to rounded, 2.5–7 × 3–6.5 cm, tapered at base, lateral veins in 5–8 pairs, brilliant white-felted beneath, shallowly lobed or with large single to double toothing at margin; leaf-stalk 6–18 mm. Flowers white, 1–1.5 cm across, in woolly corymbs to 7.5 cm across; styles 2. Fruit spherical,

c. 1.2 cm. *SE Europe and adjacent parts of Asia.* H1.

51. S. graeca (Spach) Kotschy.
Like *S. umbellata*, but leaves to 9 cm, toothed rather than lobed, lateral veins 9–11 pairs, the teeth spreading and symmetrical. Fruit almost spherical, usually less than 1.2 cm, crimson, with few large lenticels. An aggregate of many apomictic microspecies. Apomictic agg. *Mediterranean area, EC Europe & Iraq.* H3.

S. rupicola (Syme) Hedlund. Illustration: Roles, Illustrations to Flora of the British Isles, edn 2, fig. 46E (1962).
Leaves to 14.5 cm, obovate, lateral veins mostly 7–9 pairs, marginal teeth forward-pointing, curved on their outer edges. Fruit with numerous lenticels. Apomictic agg. *NW Europe.* H3.

Rarely cultivated is **S. lancastriensis** E.F. Warburg, with leaves and fruits somewhat intermediate between the above. Apomictic. *NW England.*

52. S. aria (Linnaeus) Crantz (*S. majestica* invalid). Illustration: *Botanical Magazine*, 8184 (1908); Roles, Illustrations to Flora of the British Isles, edn 2, fig. 45J (1962); Krüssmann, Manual of cultivated broad-leaved trees & shrubs **3**: pl. 116 (1986); Brickell, RHS gardeners' encyclopedia of plants & flowers, 51 (1989) – as 'Lutescens'.
Tree 10–25 m, rarely a shrub. Leaves simple, variable in shape, 5–12 cm, lateral veins mostly 10–14 pairs, densely white-felted beneath, margins double-toothed to almost lobed towards apex; leaf-stalk 7–20 mm. Flowers white, *c.* 1.2 cm across, in woolly corymbs to 7.5 cm across. Fruit usually longer than broad, bright red, with lenticels. *W, C & S Europe.* H1.

53. S. torminalis (Linnaeus) Crantz. Illustration: Bean, Trees & shrubs hardy in the British Isles, edn 8, **4**: pl. 55, dust jacket (1980); Krüssmann, Manual of cultivated broad-leaved trees & shrubs **3**: pl. 115 (1986).
Tree 10–22 m. Young parts loosely woolly at first, eventually hairless. Leaves simple, broadly ovate-triangular, *c.* 13 × 13 cm, with 3 or 4 pairs of large, acute lobes, which grade into teeth towards leaf-apex, margins double-toothed; leaf-stalk 2.5–5 cm. Flowers white, 1–1.5 cm across, in loose, woolly inflorescences; styles 2. Fruit spherical to ovoid, *c.* 1.2 cm, brownish, densely covered

with lenticels. Suckering freely from roots. *Europe, SW Asia, N Africa.* H2.

The following (**54**, **55**) are aggregates of, or single, apomictic microspecies derived from *S. aucuparia* and *S. torminalis* and tetraploids derived from *S. aria.*

54. S. latifolia (Lamarck) Persoon. Illustration: Bean, Trees & shrubs hardy in the British Isles, edn 8, **4**: pl. 55 (1980); Krüssmann, Manual of cultivated broad-leaved trees & shrubs **3**: pl. 114 (1986).
Tree 10–20 m. Stems downy at first. Leaves broadly elliptic, to 10 cm, almost as wide or narrower, apex pointed, base rounded to broadly tapered, lateral veins and lobes in 7–9 pairs, persistently greyish felted beneath, margins irregularly single-to double-toothed; leaf-stalk 1.2–2.5 cm, hairy. Flowers white, 1.5–2 cm across, in woolly corymbs to 7.5 cm across. Fruit spherical, dull brownish red, with lenticels. Apomictic agg. *Portugal to SW Germany.* H1.
Believed to be of hybrid origin, involving *S. torminalis* and one of the *S. aria* group. The following are similar (see also *S. intermedia*):

S. devoniensis Warburg. Similar to *S. torminalis* but with more shallowly lobed leaves persistently greyish-green felted beneath and orange fruit.

S. croceocarpa P.D. Sell (*S.* 'Theophrasta'; *S. theophrasta* invalid). Illustration: Ross-Craig, Drawings of British plants **9**: 33 (1956); *Dendroflora* **3**: 62, fig. 1 (1966). Leaves 7.5 cm, with 8–11 pairs of lateral veins, double-toothed, not lobed. Apomictic. *Naturalised in S Britain.*

Rarely cultivated are **S. karpatii** Boros, a shrub with lobed leaves *c.* 7.5 × 6 cm, lateral veins 7–9 pairs, and **S. pseudovertesensis** Boros, with obscurely lobed leaves *c.* 8.5 × 5 cm, lateral veins 9–11 pairs. *Hungary.*

55. S. intermedia (Ehrhart) Persoon (*S. scandica* (Linnaeus) Fries). Illustration: Roles, Illustrations to Flora of the British Isles, edn 2, fig. 45G (1962); Krüssmann, Manual of cultivated broad-leaved trees & shrubs **3**: pl. 114 (1986).
Tree to 13 m. Leaves elliptic to oblong-elliptic, 7–12 cm, often rounded at base, lobed, lobe sinuses extending from one sixth to one third of the way to the midrib (on sucker shoots sometimes with a free leaflet at base), lateral veins in 7–9 pairs, yellowish grey-woolly beneath, margins single to double-toothed; leaf-stalk

1–1.5 cm. Flowers white, *c.* 1.8 cm across, in densely woolly corymbs to 12.5 cm across; anthers cream; styles 2. Fruit oblong, 1.2–1.5 cm, orange-scarlet, sparsely covered with lenticels. *W Europe (Scandinavia & Baltic region).* H1.

Apomictic and of hybrid origin, like the following, which are similar in most details:

S. anglica Hedlund. Illustration: Roles, Illustrations to Flora of the British Isles, edn 2, fig. 45 H (1962). Shrub to 2 m. Leaves more or less obovate, tapered at base, whitish grey-woolly beneath. Anthers pink tinged. Fruit depressed-spherical, 7–12 mm, crimson. *W Britain & Ireland.* H3.

S. minima (A. Ley) Hedlund. Illustration: Roles, Illustrations to Flora of the British Isles, edn 2, fig. 45F (1962). Shrub to 3 m. Leaves 6–8 cm, narrower, 1.8–2.2 times as long as broad. Flowers to 9 mm across. Fruit depressed-spherical, 6–8 mm. *Wales (Brecon).* H1.

56. S. arranensis Hedlund. Illustration: Roles, Illustrations to Flora of the British Isles, edn 2, fig. 45D (1962).

Leaves with lobe sinuses extending to half or more of the way to the midrib. Flowers *c.* 9 mm across; anthers cream or pink. Fruit 8–10 mm. *Scotland (Isle of Arran).* H2.

57. S. bristoliensis Wilmott. Illustration: Roles, Illustrations to Flora of the British Isles, edn 2, fig. 46J (1962).

Leaves more or less obovate, very shallowly lobed; leaf-stalk 1.2–2 cm. Flowers *c.* 1.2 cm across; anthers pink. Fruit 9–11 mm, bright reddish orange. *England (Avon Gorge).* H3.

S. bakonyensis (Jávorka) Kárpáti is similar to the preceding. *Hungary.* H3.

58. S. mougeotii Soyer-Willemet & Godron.

Leaves similar to those of *S. intermedia* but red fruits that are about as broad as long. *W Alps, Pyrenees.* H2.

S. × hostii (Jacquin) K. Koch is a hybrid between *S. mougeotii* and *S. chamaemespilus*, which forms a shrub to 4 m, with lobed leaves and flowers white tinged pink.

59. S. × thuringiaca (Use) Fritsch. (*S. thuringiaca*, misapplied in part; *S. quercifolia* invalid; *S. decurrens* invalid; *S. lanuginosa* invalid). Illustration: Roles, Illustrations to Flora of the British Isles, edn 2, fig. 45B (1962).

Tree to 15 m, with woolly winter buds. Leaves oblong-lanceolate to narrowly elliptic, 7.5–15 × 3.5–7.5 cm, dull grey-woolly beneath, with 1 or 2 pairs of free leaflets at base on extension growths (to 5 or more pairs in 'Decurrens' (*S. lanuginosa*)), or only lobed to the midrib on old spur shoots, principal lateral veins in 10–12 pairs, margins double-toothed; leaf-stalk 1.5–3 cm. Flowers white, *c.* 1.2 cm across, in corymbs 7.5–12 cm across; styles 2 or 3. Fruit depressed-spherical to ellipsoid, *c.* 9 mm across, bright red. *Europe.*

A hybrid, the unnamed parent being either *S. aria* or *S. intermedia.* Sometimes confused with the next species, *S. hybrida,* which is a tetraploid apomict (possibly derived from *S. aucuparia × S. rupicola).*

60. S. hybrida Linnaeus (*S. fennica* (Kalm) Fries; *S. meinichii* misapplied).

Smaller. Leaves to 10.5 cm, grey-white-woolly beneath. Flowers *c.* 1.5 cm across; styles 3. Fruit spherical, 1–1.5 cm across. Apomictic. *Scandinavia.* H1.

61. S. teodorii Liljefors.

Is of similar origin to *S. hybrida* but rarely cultivated. It is notable for its leaves with 4 pairs of free leaflets at the base. *Scandinavia.* H1.

S. meinichii (Hartmann) Hedlund has 5 pairs of free leaflets at the base.

Section **Chamaemespilus** (Medikus) Schauer. Sometimes now treated as a separate genus. Petals pink, erect; sepal erect.

62. S. chamaemespilus (Linnaeus) Crantz. Illustration: Krüssmann, Manual of cultivated broad-leaved trees & shrubs **3**: pl. 120 (1986).

Dwarf shrub to 2 m, rarely more. Leaves simple, ovate to obovate, 3–7.5 × 1.5–3.7 cm, green and almost hairless above and below, finely toothed; leaf-stalks 3–8 mm. Flowers pinkish red, in compact corymbs composed of small umbel-like clusters, petals remaining erect. Fruit 8–12 mm, scarlet. *C & S Europe.* H2.

58. × SORBARONIA Schneider
S.G. Knees

Deciduous shrubs or small trees, with slender, sometimes pendent branches; young shoots usually covered with white hairs. Leaves simple or lobed, margins toothed. Flowers in small dense clusters. Sepals 5; petals 5, white or pale pink; ovary inferior; styles 3 or 4; fruit red or almost black.

Hybrids between *Sorbus* and *Aronia* and intermediate between the two. Several other hybrids have been named but do not appear to be widely cultivated.

1a. Leaves pinnately divided or lobed, especially towards the base; margins without glandular hairs **1. hybrida**
  b. Leaves simple; margins with glandular hairs **2. alpina**

1. × **S. hybrida** (Moench) Schneider; *Sorbus aucuparia × Aronia arbutifolia.* Illustration: *Edwards's Botanical Register* **14**: 1196 (1828).

Leaves 3–8 cm, ovate to oblong, with 2 or 3 pairs of lobes towards the base, scalloped towards the apex, downy beneath, margins with forward-pointing teeth. Flowers in clusters 2–3 cm across, white or pinkish white. Fruit spherical to pear-shaped. 8–10 mm, dark purple. *Garden origin, before 1785.* H3. Spring–summer.

2. × **S. alpina** (Willdenow) Schneider; *Sorbus aria × Aronia arbutifolia.* Illustration: Schneider, Illustriertes Handbuch der Laubgeholzkunde **1**: 699 (1906).

Shrub to 8 m, similar to *Sorbus aria.* Leaves simple, margins glandular with forward-pointing teeth. Flowers in clusters 4–7 cm across, creamy white; styles 3–4. Fruit 7–9 mm across, red or brownish red. *Garden origin, before 1809.* H4. Spring–summer.

59. ERIOBOTRYA Lindley
C.M. Mitchem

Evergreen tree or large shrub. Leaves spirally arranged, stalked or almost stalkless, simple, leathery, with prominent veins extending to leaf-margin; stipules present. Inflorescence a terminal panicle, to 15 cm. Calyx-tube 5-lobed, fused to receptacle. Petals 5, usually white; stamens *c.* 20. Ovary inferior, with 2–5 chambers, each containing 2 ovules; styles 2–5, fused at base. Fruit a fleshy pome containing 1 or 2 large seeds; calyx-lobes usually persistent, except 2.

A genus of 20 species from the Himalaya to Japan and SE Asia, grown for fruit, timber or ornament. Propagation is by seed in spring or autumn; selected cultivars by budding or grafting on quince or seedling loquat. They require a deep, rich, sandy loam, in a sunny position. Fruit only ripens reliably in S Europe.

Literature: Sargent, C.S., Plantae Wilsonianae **1**: 194 (1912); Popenoe, W., Manual of tropical & subtropical fruit, 20

(1960); Vidal, J.E., Notes sur quelques Rosacéaes Asiatiques (3): Révision du genre Eriobotrya (Pomoideae), *Adansonia* **5**: 537–80 (1965); Kalkman, C, The Malesian species of the subfamily Maloideae (Rosaceae), *Blumea* **21**: 413–42 (1973).

1a. Leaf-stalk less than 1 cm 2
  b. Leaf-stalk greater than 1 cm 3
2a. Underside of leaves covered with densely felted hairs; leaf-stalk 1–5 mm; styles 5 **1. japonica**
  b. Underside of leaves sparingly covered with felted hairs; leaf-stalk 5–10 mm; styles 2 **2. hookeriana**
3a. Underside of leaves covered with densely felted hairs; leaf-stalk 1.5–2 cm **3. prinoides**
  b. Underside of leaves almost hairless; leaf-stalk 2–5 cm **4. deflexa**

1. E. japonica (Thunberg) Lindley (*Crataegus bibas* Loureiro; *Mespilus japonica* Thunberg; *Photinia japonica* Franchet & Savatier). Illustration: Hay & Synge, Dictionary garden plants in colour, pl. 1591 (1969); Masefield et al., The Oxford book of food plants, 105 (1969); Polunin & Everard, Trees & bushes of Europe, 84 (1976).

Large shrub to 6 m. Leaves elliptic, 12–30 × 3.5–10 cm, upper half of leaf-margin smooth or toothed; hairless above, woolly beneath; 12–22 pairs of veins. Leaf-stalk less than 5 mm; stipules subulate and persistent, to 1.5 cm. Inflorescence to 12 cm, flowers stalkless. Calyx-tube to 5 mm, woolly, lobes to 3 mm. Petals obovate, *c.* 10 × 6 mm, white; stamens *c.* 20, styles 5. Fruit ovoid to spherical, to 3 cm across, yellow. *China & Japan.* H5. Autumn.

2. E. hookeriana Decaisne (*E. dubia* (Lindley) Decaisne).

Slender tree to 10 m. Leaves elliptic to oblanceolate, 15–25 × 4–10 cm, margin with coarse forward-pointing teeth; hairless above, sparsely felty beneath; 15–30 pairs of veins. Leaf-stalk 5–10 mm; stipules semi-lunar, to 11 mm. Inflorescence a spreading panicle, to 15 cm, flowers almost stalkless. Calyx-tube 2–3 mm, lobes to 3 mm, white or pinkish; styles 2. Fruit an ellipsoid pome, 1.2–1.8 cm × 2–8 mm, yellow. *E Himalaya (Sikkim, Bhutan).* H5. Autumn.

3. E. prinoides Rehder & Wilson (*E. bengalensis* (Hooker) Dunn; *E. dubia* (Decaisne) Franchet). Illustration: *Adansonia* **5**: 564 (1965).

Tree to 12 m. Leaves elliptic, 8–15 × 3–7.5 cm, upper two-thirds of margin toothed; hairless above, densely felted hairs beneath; 10–12 pairs of veins. Leaf-stalk 1.5–2 cm. Inflorescence to 10 cm; calyx-tube to 3 mm, woolly, lobes to 2 mm, falling early. Petals white, oval, to 5 mm. Styles 2 or 3, woolly at base. Fruit ovoid, to 10 × 7 mm. *China (Sichuan, Yunnan).* H5. Autumn.

Var. **laotica** Vidal has almost circular leaves to 7 × 5 cm. Var. **prinoides** is most commonly cultivated.

4. E. deflexa (Hemsley) Nakai (*Photinia deflexa* Hemsley).

Large tree, to 12 m. Leaves oblanceolate or elliptic, 6–15 × 2.5–5 cm, coarsely lobed, hairless above, with a few hairs beneath; 10–15 pairs of veins; leaf-stalk 2–5 cm. Inflorescence a spreading panicle, to 15 cm. Calyx-tube to 4 mm, woolly, lobes to 4 mm. Petals white, almost circular; stamens *c.* 20; styles 2 or 3. Fruit ovoid or spherical, to 1.5 cm across, yellow. *Taiwan, S Vietnam.* H5. Autumn.

60. PERAPHYLLUM Torrey & Gray
C.M. Mitchem

Deciduous shrubs to 2 m, stems widely divergent, much-branched, bark grey. Leaves 2.5–4 cm × 4–10 mm, narrowly lanceolate to obovate, mucronate; margin smooth or with fine forward-pointing teeth. Stipules falling early. Inflorescence an erect corymb of 2 or 3 flowers. Flowers white, bisexual, to 2 cm across, subtended by 2 small bracts. Calyx persistent, to 1 cm, covered with silky hairs, fused to bell-shaped receptacle. Petals 8–10 mm, circular. Stamens 20, as long as petals, arranged around rose-coloured dish; anthers yellow. Ovary inferior, 2–4-celled; styles 2 or 3, not exceeding stamens; stigma red, capitate. Fruit a yellow, fleshy, spherical drupe to 2 cm, containing 1 seed.

A genus of one species from W North America. Prefers hot, dry summers and sandy or loamy soil. Propagation by seed or layering.

1. P. ramosissimum Torrey & Gray. Illustration: *Botanical Magazine*, 7420 (1895).

W North America (Oregon to California & Colorado). H1. Late spring–early summer.

61. AMELANCHIER Medikus filius
J.E. Richardson

Deciduous trees or shrubs. Leaves alternate, simple, entire or with forward-pointing teeth. Stipules deciduous. Flowers in terminal racemes, rarely solitary. Calyx-tube bell-shaped, lobes 5, narrow, reflexed, persistent. Petals 5, white, rarely pink. Stamens 10–20, inserted on the throat of the calyx; filaments awl-shaped. Styles 2–5. Ovary inferior, apex usually hairy, cells becoming double the number of styles. Fruit a small berry-like pome with 1–5 cells each containing 1 or more seeds.

A genus of about 30 species from N America, N and C Europe, Japan, China and Korea. Propagate from seed, by layering or by division of the suckers. Slow-growing and sun-loving.

1a. Leaves overlapping in bud; flowers 1–3 together **14. bartramiana**
  b. Leaves folded in bud not overlapping; flowers several to many, racemose 2
2a. Leaf-margin coarsely toothed (fewer than 6 teeth per cm) 3
  b. Leaf-margin finely toothed (more than 6 teeth per cm) 7
3a. Plant stoloniferous 4
  b. Plant not stoloniferous 5
4a. Petals 7–10 mm; apex of ovary hairy **4. humilis**
  b. Petals 1–1.5 cm; apex of ovary hairless **6. ovalis**
5a. Leaf apex usually truncate **1. alnifolia**
  b. Leaf apex acute to rounded to obtuse 6
6a. Shrub or small tree; petals 0.6–0.8 cm **2. utahensis**
  b. Straggling or arching slender shrub; petals 1–1.5 cm **3. sanguinea**
7a. Stoloniferous shrub 8
  b. Non-stoloniferous tree or shrub 9
8a. Tree or shrub to 8 m; leaves hairless except for midrib **7. canadensis**
  b. Shrub to 2 m; leaves densely white-hairy beneath when young **5. stolonifera**
9a. Apex of ovary hairless 10
  b. Apex of ovary hairy 13
10a. Racemes more than 8 cm 11
  b. Racemes less than 8 cm 12
11a. Leaves hairless when young; fruit to 1.8 cm across **10. laevis**
  b. Leaves densely white-hairy when young; fruit more than 1.8 cm across **9. × grandiflora**
12a. Tree or shrub to 20 m; leaf apex acute; fruit red-purple **8. arborea**
  b. Tree or shrub to 10 m; leaf apex obtuse to almost acute or bluntly shortly acuminate; fruit purple-black **11. lamarckii**

13a. Shrub or tree to 12 m; racemes
pendent				**12. asiatica**
  b. Shrub to 4 m; racemes erect
					13. spicata

1. A. alnifolia (Nuttall) Nuttall. (*Aronia alnifolia* Nuttall; *Amelanchier canadensis* var. *alnifolia* Torrey & Gray; *A. canadensis* var. *florida* Schneider; *A. oreophila* A. Nelson in part). Illustration: Britton & Brown, Illustrated flora of the northern United States and Canada **2**: 239 (1897).

Shrub or small tree to 12 m, becoming hairless and somewhat glaucous. Leaves 2–5.5 cm, thick, broadly elliptic or almost circular, apex usually truncate (rarely acutish), base rounded or almost cordate, coarsely toothed above the middle, vein pairs 7–12. Racemes to 4 cm, erect, with 5–15 flowers. Petals 8–16 mm, oblanceolate, tapered to the base, creamy white. Fruit *c*. 1 cm across, dark blue, juicy and edible. *W & C North America*. H1. Late spring–early summer.

Var. **cusickii** (Fernald) C. Hitchcock (*A. cusickii* Fernald). Shrub to 3 m; young branches red and glossy, becoming grey; leaves to 5.5 cm, hairless.

Var. **semiintegrifolia** (Hooker) C. Hitchcock. (*A. oxyodon* Koehne; *A. florida* Lindley). Shrub or small tree to 12 m, young branches rusty hairy, becoming hairless; leaves to 4 cm, hairy.

2. A. utahensis Koehne (*A. crenata* Greene; *A. mormonica* Schneider; *A. oreophila* A. Nelson in part; *A. prunifolia* Greene; *A. purpusii* Koehne).

Shrub or small tree to 5 m. Leaves to 3 cm, rounded to ovate, apex rounded, notched, base rounded to wedge-shaped, finely hairy, coarsely toothed, vein pairs 11–13. Racemes to 3 cm, erect or ascending, 3–6-flowered. Petals 6–8 mm, linear. Fruit to 1 cm across, spherical, purple-black. *W North America*. H1. Spring.

3. A. sanguinea (Pursh) de Candolle (*A. amabalis* Wiegand; *A. rotundifolia* Roemer). Illustration: Britton, North American trees, 439 (1908).

Straggling or arching slender shrub to 3 m. Young branches red or grey. Leaves 2.5–7 cm, ovate to almost rounded, apex acute to obtuse, base rounded to slightly wedge-shaped, densely white-yellow-hairy beneath, becoming hairless, coarsely toothed, almost reaching base, vein pairs 11–13. Racemes to 8 cm, erect to pendent, with 4–10 flowers. Petals 1–1.5 cm, white or light pink. Fruit 6–9 mm across,

spherical, purplish black, sweet. *E North America, S Canada*. H1. Spring.

4. A. humilis Wiegand.
Shrub to 1.25 m, stoloniferous, patch-forming. Leaves to 5 × 3 cm, elliptic to elliptic-oblong, base rounded or almost cordate, with coarse forward-pointing teeth to below the middle, very densely white-hairy beneath. Racemes erect, many-flowered. Petals 7–10 mm, oblong obovate. Ovary hairy at apex. Fruit almost black, sweet. *E North America*. H1. Spring.

5. A. stolonifera Wiegand (*A. spicata* misapplied). Illustration: Taylor's guide to shrubs, pl. 86 (1987).

Shrub to 2 m, stoloniferous, patch-forming. Leaves to 5 × 3 cm, oblong to almost circular, base rounded or almost cordate, with fine forward-pointing teeth in the upper two-thirds, densely white-hairy beneath when young, vein pairs 7–10. Racemes erect. Petals to 9.5 mm, obovate oblong. Ovary hairy at apex. Fruit blue-black, juicy. *E North America*. H1. Spring.

6. A. ovalis Medikus filius (*A. vulgaris* Moench; *A. rotundifolia* (Lamarck) Dumont de Courset). Illustration: *Botanical Magazine*, 2430 (1823).

Erect or spreading stoloniferous shrub to 3 m. Young twigs woolly. Leaves 2.5–5 cm, ovate to obovate, apex rounded or notched and mucronate, with coarse forward-pointing teeth, white-hairy beneath when young, becoming hairless. Racemes to 4 cm, erect, with 3–8 flowers, white hairy. Petals 1–1.5 cm, ovate oblong. Apex of ovary hairless. Styles 5, free. Fruit 6 mm across, spherical, dark blue to black. *C & S Europe*. H1. Spring.

7. A. canadensis (Linnaeus) Medikus filius (*A. botryapium* (Linnaeus filius) Borkhausen; *A. intermedia* Spach; *A. oblongifolia* (Torrey & Gray) M.J. Roemer). Illustration: Britton, North American trees, 237 (1908); *Botanical Magazine*, 8611 (1915).

Shrub to 8 m, stoloniferous. Leaves to 5 × 2.5 cm, elliptic, apex acute or rounded, base usually rounded, with fine, shallow, sharp teeth, hairless except for midrib and stalk, vein pairs 9–13. Racemes to 6 cm, erect, initially white-hairy. Petals to 9 × 3 mm, ovate to oblanceolate, obtuse, white. Fruit *c*. 1 cm across, spherical, purple-black. *E North America*. H1. Spring.

'Micropetala': tall, erect, leaves richly coloured in autumn; flowers small. 'Prince William': to 3.5 m, fruits purple, edible, abundant. 'Springtyme': to 3.5 m, ovate,

erect, compact; leaves yellow-orange in autumn. 'Tradition': to 7.5 m, ovate, erect, early flowering.

8. A. arborea (Michaux filius) Fernald (*A. canadensis* Siebold & Zuccarini). Illustration: Preston, North American trees, 250 (1969); Petrides, A field guide to eastern United States trees, pl. 42 (1988); Foote & Jones, Native shrubs and woody vines of the southeast, pl. 18 (1989).

Tree or shrub to 20 m. Bark light brown, tinged with red. Leaves to 10 cm, with fine forward-pointing teeth, oblong-ovate to ovate or ovate, apex acute, dark green and hairless above and pale below, vein pairs 11–17. Racemes to 7.5 cm, pendent, with 4–10 flowers. Petals 1–1.5 cm, spathulate to strap-shaped, pure white. Fruit *c*. 1 cm across, red-purple, dry and tasteless. *E North America*. H1. Spring.

9. A. × grandiflora Rehder (*A. botryapium* var. *lanceolata* misapplied; *A. canadensis* var. *grandiflora* Zabel). Illustration: Taylor's guide to trees, pl. 128, 129 (1988).

Differs from *A. arborea* in the larger flowers, longer, more slender, less hairy racemes and leaves purple with densely matted woolly hairs to densely white-hairy. Differs from *A. laevis* in the densely white-hairy young leaves, more flowers, shorter flower-stalks and larger fruit. *Garden origin*. H1. Spring–summer.

'Autumn Brilliance': bark light grey; leaves brilliant red in autumn. 'Ballerina': flowers pure white in large heads. 'Coles Select': leaves with bright autumn colours. 'Cumulus': erect, leaves leathery, orange-red in autumn; flowers abundant; fruit red, becoming purple. 'Princess Diana': small tree, gracefully spreading, leaves bright red in autumn; fruit purple. 'Robin Hill': narrowly erect; flowers pale pink, fading to white, nodding; fruit small, red, juicy. 'Rubescens': flowers bluish pink. 'Strata': branches horizontally.

10. A. laevis Wiegand. Illustration: Voss, Michigan flora, fig. 187 (1985); Taylor's guide to trees, pl. 128, 129 (1988).

Shrub or tree to 13 m. Leaves 4–6 cm, ovate, apex acute, base rounded, finely toothed almost to the base, hairless, vein pairs 12–17. Racemes to 12 cm, pendent, flowers numerous. Petals 1.2–2.2 cm. Fruit to 1.8 cm across, spherical, purple to almost black, sweet. *C North America*. H1. Spring.

'Prince Charles': somewhat rounded habit, flowers abundant; leaves orange and red in autumn; fruit blue, edible.

11. A. lamarckii Schröder (*A. botryapium* (Linnaeus) de Candolle; *A. canadensis* K. Koch; *A. canadensis* var. *botryapium* Koehne; *A. confusa* misapplied; *A. × grandiflora* Franco; *A. laevis* Clapham, Tutin & Warburg; *A. laevis* forma *villosa* Pelkwijk). Illustration: Bird, Flowering trees and shrubs, 15 (1989).

Tree or shrub to 10 m. Young branches bristly, becoming hairless. Leaves 8.5 × 5 cm, elliptic, oblong-elliptic or oblong-obovate, apex obtuse to acute or bluntly shortly acuminate, base rounded to wedge-shaped, with fine, sharp forward-pointing teeth, all the way around, copper-red and bristly when flowering, becoming hairless. Racemes to 7.5 cm, loose, with 6–10 flowers. Petals *c.* 14 × 5 mm, oblanceolate to elliptic. Ovary hairless at apex. Fruit *c.* 9.5 mm across, spherical, purple-black. *E Canada.* H1. Spring.

12. A. asiatica (Siebold & Zuccarini) Walpers (*A. canadensis* var. *japonica* Miquel). Illustration: Kurata, Illustrated important trees of Japan **4**: pl. 13 (1973).

Shrub or tree to 12 m, with slender spreading branches. Leaves to 7 cm, ovate to elliptic oblong, apex acute, base rounded or almost cordate, with fine forward-pointing teeth almost to the base, densely white or yellow hairy beneath, becoming hairless. Racemes pendent, densely flowered, woolly. Flowers *c.* 3 cm across. Petals *c.* 1.5 cm, ovate oblong. Ovary woolly at apex. Fruit blue-black. *Japan, China, Korea.* H1. Spring.

13. A. spicata (Lamarck) K. Koch (*A. ovalis* Borkhausen). Illustration: *Botanical Magazine*, 7619 (1898); Foote & Jones, Native shrubs and vines of the southeast, pl. 18 (1989).

Shrub to 4 m. Leaves to 5 cm, finely toothed, densely white-hairy when young, becoming hairless, vein pairs 7–9. Racemes to 4 cm, erect, with 4–10 flowers, woolly. Petals 4–10 mm, oblanceolate, white or pink. Styles united at base. Ovary apex hairy. Fruit to 8 mm across, spherical, purple-black. *NE North America.* H1. Spring.

14. A. bartramiana (Tausch) M. Roemer (*A. oligocarpa* M. Roemer). Illustration: *Botanical Magazine*, 8499 (1913).

Shrub to 2.5 m. Leaves 3–5 cm, elliptic to elliptic-oblong, apex acute or rounded, base wedge-shaped, with sharp and fine forward-pointing teeth to below the middle or nearly to the base, hairless when young, overlapping when in bud, vein pairs 10–16. Flowers 1–3, to 2.5 cm across. Petals *c.* 8 mm, obovate. Ovary woolly at apex. Fruit to 1.5 cm across, purplish black. *C to E North America.* H1. Spring.

62. × AMELASORBUS Rehder
J.E. Richardson
Hybrid between *Amelanchier* and *Sorbus*. Differs from *Amelanchier* in the partly pinnate leaves and paniculate inflorescences, and from *Sorbus* in the partly simple leaves, partly lobed or incompletely pinnate leaves and paniculate flowers with 5 pistils.

Cultivation as for *Amelanchier*.

1. × A. jackii Rehder.
Deciduous shrub to 2 m. Leaves to 10 cm, ovate to elliptic, toothed, margin slightly sinuous, irregularly lobed or entire, hairy becoming hairless. Inflorescence a panicle to 5 cm, petals *c.* 2 cm, oblong, white. Fruit small, almost spherical, dark red, blue bloomy. *NW USA.* H1. Spring–summer.

63. RHAPHIOLEPIS Lindley
C.M. Mitchem
Small evergreen trees or shrubs. Leaves leathery, simple, stalked; young leaves densely covered in felty hairs. Inflorescence a terminal panicle, felted. Floral parts in 5 s, subtended by bracts. Stamens 15–20. Ovary inferior, 2-celled; styles 2. Fruit a dry ovoid or spherical berry with a distinct scar at apex, with 1 or 2 large seeds.

There are 3–5 species all from the subtropical and warm temperate regions of south-east and east Asia. Two species and one hybrid are in cultivation. They prefer a warm, sunny position in a well-drained fertile soil. Plants are slow-growing and resent disturbance; usually grown as wall plants or as a low hedge. Propagation is by seed, autumn cuttings, layering or grafting.

Literature: Kalkman, C, The Malesian species of the Subfamily Maloideae (Rosaceae), *Blumea* **21**: 413–42 (1973); Kitamura, S., Short reports of Japanese plants, *Acta Phytotaxonomica et Geobotanica* **26**: 1–2 (1974); Ohashi, H., Rhaphiolepis (Rosaceae) of Japan, *Journal of Japanese Botany* **63**(1): 1–7 (1988).

1a. Leaves thin, to 3 cm across, deeply toothed **1. indica**
  b. Leaves thick and leathery, more than 3 cm across, shallowly toothed **2**
2a. Panicles erect; flowers fragrant **2. umbellata**
  b. Panicles ascending; flowers scentless **3. × delacourii**

1. R. indica (Linnaeus) Lindley (*R. fragrans* Geddes; *R. japonica* Siebold & Zuccarini; *R. major* Cardas; *R. salicifolius* Lindley; *Crateagus indica* Linnaeus). Illustration: *Botanical Magazine*, 1726 (1815).

Small tree or shrub, to 1 m. Leaves 7–11 × 2–3 cm, thin, dark glossy green above, paler beneath, narrowly-elliptic to oblanceolate, margin sharply toothed. Inflorescence a loose panicle. Bracts to 5 mm, narrowly lanceolate. Petals to 8 mm, pinkish white. Fruit to 1 cm, spherical, blue-black. *Japan.* H5. Spring–summer.

2. R. umbellata (Thunberg) Makino (*R. japonica* misapplied; *R. japonica* Siebold & Zuccarini var. *integerrima* Hooker & Arnott; *R. ovata* Briot). Figure 99(10), p. 183. Illustration: *Botanical Magazine*, 5510 (1865); *Addisonia* **2**: t. 70 (1917).

Sturdy shrub to 2 m. Leaves 3–9 × 3.5–6, thick and leathery, ovate to oblanceolate, dark green and hairless above, paler and covered with felted hairs beneath, margin with fine forward-pointing teeth in upper half. Inflorescence an erect terminal panicle; flowers fragrant. Bracts to 1 cm, awl-shaped. Petals to 1 cm, white, tinged rose. Fruit a single-seeded dry berry, to 1 cm, blue-black with slight bloom. *Japan & Korea.* H4. Spring–summer.

3. R. × delacourii André. Illustration: *Botanical Magazine*, n.s., 362 (1960); Miller Gault, The dictionary of trees and shrubs in colour, pl. 341 (1976); Krüssmann, Manual of cultivated broad-leaved trees & shrubs **2**: pl. 5 (1986).

Dome-shaped shrub to 2 m. Leaves 6–9.5 × 3–4.5 cm, leathery, broadly obovate to oblanceolate, margin finely toothed in upper half. Inflorescence an ascending panicle; flowers scentless. Bracts to 7 mm, narrowly lanceolate. Sepals to 5 mm, lanceolate. Petals to 1 cm, rose pink, narrowly obovate. *Garden origin.* H5. Early spring–summer.

A hybrid between *R. indica* and an unknown cultivar of *R. umbellata*, with a long flowering period.

Two cultivars are grown: 'Coates Crimson' and 'Spring Song'.

64. ARONIA Medikus filius

S.G. Knees

Deciduous shrubs; branches with closely adpressed, slender pointed buds. Leaves alternate, stalked, simple with toothed margins and blackish glands on midrib above. Stipules small, falling early. Flowers white or pale pink in small corymbs. Calyx-lobes 5, joined at base. Petals 5, spreading; stamens numerous, anthers purplish pink; ovary inferior, 5-celled, with woolly hairs towards the apex; styles 5, joined at their bases, carpels partly free. Fruit apple-like, with persistent remains of calyx.

Three closely related species from North America, cultivated for their attractive flowers and fruit; in addition, the leaves colour well in autumn. The fruits are inedible and probably gave rise to the widely used common name of chokeberry. Aronias have a confused history and have been included in *Pyrus*, *Sorbus* and *Mespilus*, but are probably most closely related to *Photinia*. All species are hardy and easily propagated from seed or by cuttings.

Literature: Hardin, J.W., The enigmatic chokeberries (Aronia, Rosaceae), *Bulletin of the Torrey Botanical Club* **100**: 178–84 (1973); Wiegers, J., Aronia Medik. in the Netherlands 1. Distribution and taxonomy, *Acta Botanica Neerlandica* **32**(5/6): 481–8 (1983).

1a. Leaves and inflorescence usually hairless; fruit purplish black

1. melanocarpa

  b. Leaves and inflorescence covered with whitish hairs; fruit red or dark brownish purple 2

2a. Fruit red **2. arbutifolia**

  b. Fruit dark brownish purple

3. × prunifolia

1. A. melanocarpa (Michaux) Elliott (*Mespilus arbutifolia* Michaux var. *erythrocarpa* Michaux; *Pyrus melanocarpa* (Michaux) Willdenow). Illustration: Britton & Brown, Illustrated flora of the northern States & Canada, edn 2, **2**: 291 (1913); *Botanical Magazine*, 9052 (1924); Brown & Brown, Woody plants of Maryland, 123 (1972); Krüssmann, Manual of cultivated broad-leaved trees and shrubs **1**: 173 (1984).

Shrub 50–350 cm, with hairless branches. Leaves 2–7 × 1–6 cm, elliptic or obovate to oblong-lanceolate, apex acuminate or almost obtuse; margins finely toothed. Inflorescence a hairless, loosely branched corymb with 4–11 flowers.

Petals white, *c.* 7 mm, almost round and abruptly narrowed into a stalk-like base. Fruit 6–8 mm across, almost spherical or slightly flattened, glossy purplish black. *E North America.* H2. Spring.

Three varieties have been described and plants are sometimes grown under these names. Var. **elata** Rehder is the largest with measurements for all its parts at the top of the ranges given above. Var. **grandifolia** (Lindley) Schneider (*Pyrus grandifolius* Lindley) has larger leaves, to 6 cm wide, but is otherwise similar to var. *elata*. A third variant with softly hairy, young leaves is var. **subpubescens** (Lindley) Schneider.

2. A. arbutifolia (Linnaeus) Elliott (*Pyrus arbutifolia* Linnaeus). Illustration: *Botanical Magazine*, 3668 (1838); Britton & Brown, Illustrated flora of the northern United States & Canada, edn 2, **2**: 290 (1913); Brown & Brown, Woody plants of Maryland, 133 (1972); Krüssmann, Manual of cultivated broad-leaved trees and shrubs **1**: 173 (1984).

Shrub usually to 3 m, more rarely reaching 6 m; branches covered with densely felted hairs. Leaves 4–7 × 2–4.5 cm, elliptic to oblong or obovate, apex acute or abruptly acuminate, hairless above except for the midrib, greyish with felted hairs beneath. Inflorescence 3–4 cm across, a dense corymb of 9–20 flowers. Calyx with glandular hairs; petals white or reddish pink, 6–8 mm. Fruit spherical or slightly pear-shaped, 5–7 mm across (8–10 mm in some varieties), bright scarlet or very dark red. *E USA.* H2 Spring.

Amongst the varieties sometimes grown are: var. **leiocalyx** Rehder, with hairless sepals and flower-stalks; var. **macrophylla** (Hooker) Rehder, which forms a large shrub or small tree to 6 m; var. **macrocarpa** Zabel has fruits 8–10 mm across. Var. **pumila** (Schmidt) Rehder (*Mespilus pumila* Schmidt) is a dwarf shrub with small leaves and dark red fruits. Like *A. melanocarpa*, these varieties are probably just extreme variants within the natural range of the species and may not merit formal recognition. However they are included here as they are sometimes encountered in the trade.

3. A. × prunifolia (Marshall) Rehder (*Pyrus floribunda* Lindley; *A. atropurpurea* Britton). Illustration: Britton & Brown, Illustrated flora of the northern United States & Canada, edn 2, **2**: 291 (1913);

Brown & Brown, Woody plants of Maryland, 133 (1972).

Shrub to 4 m, very similar to *A. arbutifolia* but with looser corymbs of 10–20 flowers. Calyx less hairy and usually lacking glandular hairs. Fruit almost spherical, 8–10 mm across, dark brownish purple. *E North America.* H2. Spring–early summer.

Considered by most recent authorities to be a hybrid between the two preceding species. A selection with good autumn colour and dark red fruits is 'Brilliant'.

65. PHOTINIA Lindley

M.F. Gardner & S.G. Knees

Evergreen or deciduous trees and shrubs. Leaves alternate, simple, usually with fine forward-pointing teeth or entire, with short stalks, stipules sometimes almost leaf-like, free. Flowers, normally white, woolly in bud, borne in terminal or axillary umbel-like panicles; petals 5 with a distinct claw, hairless or hairy at the base, sepals 5, persisting in fruit; stamens 15–25. Ovary semi-inferior, 2–5-celled, styles 2, rarely 3–5. Fruit more or less fleshy, seeds 1 or 2 to each cell.

A genus of about 40 species from the Himalaya to Japan and Sumatra. Most are easily cultivated in freely drained soils; some evergreen species are noted for their tolerance of chalky soils, but the deciduous ones require a soil that is neutral to acid. Many of the evergreen species require a sheltered position. Most are grown for their outstanding autumn colour and attractive berries, which last long into the winter; others are valued for their attractive reddish young leaves. Propagate from seed sown in early spring or by cuttings of half-ripened wood taken in early autumn and rooted under mist with bottom heat.

Literature: Kalkman, C, The Malesian species of the subfamily Maloideae (Rosaceae), *Blumea* **21**: 413–42 (1973).

Leaves. Evergreen: **1, 2, 3, 4, 5, 6, 7, 8, 9, 10**; deciduous: **11, 12, 13, 14**.
Leaf-margins. Entire: **7, 8, 10**; spiny: **1**; toothed: **2, 3, 4, 5, 6, 9, 11, 12, 13, 14**; gland-tipped: **11, 14**.
Fruits. Globe-shaped: **1, 2, 3, 4, 6, 7, 8, 9, 10**; egg-shaped: **5, 11, 12, 13, 14**.

1a. Leaf-margins entire 2
  b. Leaf-margins toothed 4
2a. Fruit red to blackish purple, not more than 5 mm across

7. integrifolia

b. Fruit bright red, pinkish red or yellow, 8 mm or more across   3

3a. Upper leaf surface shiny   **10. niitakayamensis**

b. Upper leaf surface dull, often reddish   **8. davidiana**

4a. Leaf-margin conspicuously spiny   **1. prionophylla**

b. Leaf-margin shallowly toothed, occasionally gland-tipped   5

5a. Non-flowering or internal shoots usually spiny   **5. davidsoniae**

b. All shoots without spines   6

6a. Leaves evergreen; fruit globe-shaped   7

b. Leaves deciduous; fruit egg-shaped   11

7a. Leaves 10–18 cm, very glossy above   **4. serratifolia**

b. Leaves 3–10 cm, dull or slightly shiny   8

8a. Petals white with a pinkish tinge; fruit black   **2. glabra**

b. Petals white; fruit yellowish, orange or red   9

9a. Shoots with conspicuous lenticels; fruit yellowish orange-red   **6. lasiogyna**

b. Lenticels inconspicuous; fruit red   10

10a. Flowers 6–8 mm across, in umbels 10–12 cm wide   **3. × fraseri**

b. Flowers 1.1–1.3 cm across, in umbels 5–10 cm wide   **9. nussia**

11a. Leaf-margin with gland-tipped teeth   12

b. Marginal teeth without glands, spine-tipped   13

12a. Leaves 12–18 × 5–6 cm, veins hairy beneath   **14. glomerata**

b. Leaves 5–13 × 2–4 cm, veins hairless   **11. beauverdiana**

13a. Leaves persistently shaggy beneath; flower-stalks hairy   **12. villosa**

b. Leaves and flower-stalks hairless   **13. parvifolia**

1. P. prionophylla (Franchet) Schneider (*Eriobotrya prionophylla* Franchet). Illustration: *Botanical Magazine*, 9134 (1927); Flora Reipublicae Popularis Sinicae **36**: 235 (1974); Wu, Wild flowers of Yunnan **1**: 133 (1986).

Evergreen shrub to 2 m, growth habit rigid; young shoots covered with a greyish down. Leaves 3.5–9 × 2.5–5 cm, very hard, leathery, obovate to ovate, wedge-shaped at the base, rounded and tipped with a short point, spiny-toothed, persistently downy and strongly veined beneath, finely downy above when young, soon becoming

hairless and dark green, leaf-stalks *c.* 1.2 cm. Flowers *c.* 8 mm across, white, borne in crowded, flattish umbels 5–7.5 cm across; petals obovate, incurved, stamens 20, yellow; calyx-tube woolly, sepals short, triangular, downy or hairless towards their tips. Fruit crimson, *c.* 6 mm across, globe-shaped, woolly at the apex. *China (Yunnan).* H5–G1. Summer.

Requires the protection of a warm wall in cooler districts.

2. P. glabra (Thunberg) Maximowicz (*Crataegus glabra* Thunberg). Illustration: Makino, Illustrated flora of Japan, **25**: 1405 (1956); Kurata, Illustrated important forest trees of Japan **3**: 57 (1971); Hayashi, Woody plants of Japan, 337 (1985).

Evergreen shrub to 4.5 m. Leaves 5–8 × 2–4 cm, bronzy when young, elliptic to oblong–obovate, pointed, tapered at base, with shallow forward-pointing teeth; stalks 1–1.5 cm, hairless. Flowers white tinged pink, hawthorn-scented, borne in loose terminal, many-branched umbels, 5–10 cm across, petals narrowly ovate, hairy at base on the inside. Fruit 4–6 mm wide, red changing to black, globe-shaped. *Japan & China.* H4–5. Summer.

The young growth is often susceptible to late spring frosts. Mainly represented in cultivation by the clones 'Rubens', which has been selected for its bronzy red young leaves, and 'Parfait' ('Variegata'), which has creamy white leaf-margins.

3. P. × fraseri Dress; *P. glabra × P. serratifolia.* Illustration: Brickell, RHS gardeners' encyclopedia of plants & flowers, 85 (1989).

Large evergreen shrub, 3–5 m, intermediate between the parents. Leaves 7–9 cm, glossy green above, lighter beneath, elliptic to more obovate, with fine forward-pointing teeth, abruptly short-pointed, broadly wedge-shaped at base, leaf-stalk 2.8–5.6 cm, hairy above when young. Flowers 6–8 mm across, in umbels 10–12 cm across, petals white, hairy on inside at base. *Garden origin.* H3–4. Summer.

'Birmingham', Red Robin' and 'Robusta' have been selected for their attractive bright coppery or red young leaves, which are susceptible to late spring frosts.

4. P. serratifolia (Desfontaines) Kalkman (*Crataegus serratifolia* Desfontaines; *P. serrulata* Lindley). Illustration: *Botanical Magazine*, 2105 (1819); Icones

cormophytorum Sinicorum, **2**: 2145 (1972); Guo-zi Hsu, The wild woody plants of Taiwan, 131 (1984); Krüssmann, Manual of cultivated broad-leaved trees & shrubs **2**: pl. 159, 397 (1986).

Evergreen shrub or tree 8–12 m, with sturdy branches. Leaves 10–18 × 5–8 cm, leathery, reddish when young, oblong, rounded or tapering at base, dark glossy green above, yellowish beneath, midrib hairless, margins shallowly toothed; leaf-stalks 2.5–4 cm, covered with whitish hairs. Flowers in umbels 10–18 cm across, petals hairless. Fruit 5–6 mm across, globe-shaped, red. *China.* H4–5. Spring–summer.

Sometimes used as a windbreak or hedge in less exposed sites as young growth is often susceptible to late spring frosts. The clone 'Rotundifolia' is lower growing and has leaves that are smaller and rounded.

5. P. davidsoniae Rehder & Wilson. Illustration: *Gardeners' Chronicle* **71**: 199 (1922); Icones cormophytorum Sinicorum, **2**: 2146 (1972); Flora of Jiangsu **2**: 283 (1982); Krüssmann, Manual of cultivated broad-leaved trees & shrubs **2**: pl. 159 (1986).

Evergreen tree or shrub, 6–10 m; inner branches often with conspicuous straight spines *c.* 3 cm; young shoots reddish, downy; buds extremely small. Leaves 5–15 × 2–4.5 cm, dark glossy green above, pale beneath, oblanceolate to narrowly elliptic, tapering at both ends, leaf-stalk 7–14 mm. Flowers many, *c.* 1 cm across, borne in terminal umbels 8–10 cm across, flower-stalks downy, calyx-tube funnel-shaped, lobes triangular, downy, persistent in fruit. Fruit egg-shaped, orange-red, *c.* 8 mm. *China (W Hubei).* H4–5. Summer.

6. P. lasiogyna (Franchet) Franchet (*Eriobotrya lasiogyna* Franchet). Illustration: Icones cormophytorum Sinicorum, **2**: 2149 (1972); Wu, Wild flowers of Yunnan **2**: 115 (1986).

Evergreen shrub or small tree, 3–6 m; branches slender, bark covered with lenticels. Leaves 5–9 × 2.5–4 cm, obovate, leathery, shiny above; apex acute or rounded, margins shallowly toothed; stalk 1.2–1.9 cm. Flowers in corymbs 8–10 cm across, fragrant; calyx-lobes short, broadly triangular; petals 5–7 mm, rounded, creamy white; styles 2–4. Fruit 5–8 mm across, globe-shaped, reddish or orange-yellow, woolly at apex. *China (Yunnan).* H4–5. Summer.

7. P. integrifolia Lindley. Illustration:
Flora Reipublicae Popularis Sinicae **36**:
223 (1974); Polunin & Stainton, Flowers
of the Himalaya, 466 (1985); Wu, Wild
flowers of Yunnan **2**: 113 (1986);
Stainton, Flowers of the Himalaya, a
supplement, pl. 29 (1988).

Evergreen tree or shrub, young branches
hairless. Leaves 8–10 × 3–4 cm, leathery,
oblong or ovate oblong, wedge-shaped or
rounded at base, tapered at apex, entire,
hairless, glossy above paler beneath; leaf-
stalk 1–1.5 cm, stipules 2–4 mm, falling.
Flowers borne in terminal, spreading
panicles, hairless to slightly hairy, flower-
stalks 1–3 mm; sepals hairless. Fruit
reddish to black-purple globe-shaped,
c. 5 mm across. *C & E Himalaya, S China &
N Vietnam, Malaysia.* H5. Summer.

8. P. davidiana (Decaisne) Kalkman
(*Stranvaesia davidiana* Decaisne).
Illustration: *Botanical Magazine*, 9008
(1923); Krüssmann, Manual of cultivated
broad-leaved trees & shrubs **3**: pl. 130
(1986); Wu, Wild flowers of Yunnan **1**:
143 (1986); Brickell, RHS gardeners'
encyclopedia of plants & flowers, 67
(1989).

Evergreen, sparsely branched tree or
shrub, 8–9 m; young shoots silky hairy,
soon becoming hairless. Leaves
6–12 × 2–3 cm, often reddish, dull, entire,
oblong-lanceolate, narrowed at the apex,
sharply tipped, hairless, except on veins
beneath; leaf-stalks 1–2 cm, often reddish,
covered with silky hairs; stipules awl-
shaped, soon falling. Flowers arranged in
loose, hairy, flat umbels 5–8 cm across,
anthers red or pinkish; styles 5. Fruit
bright red, globe-shaped, 7–9 mm wide.
China. H3. Summer.

'Fructuluteo' has very attractive yellow
fruit. 'Prostrata', is often used for ground
cover on the account of its prostrate stems,
although erect branches often occur.
'Palette' is a slow-growing variant with
leaves blotched creamy white, sometimes
tinged pink. 'Redstart' (*P. davidiana*
'Fructuluteo' × *P. × fraseri* 'Robusta'). This
seedling has red young growths and
flowers very freely. Fruit red, tipped with
yellow. *Garden origin.* H4. Summer.
'Winchester' is similar to 'Redstart' and
was raised from the same cross but has
thinner elliptic leaves to 13.5 cm and fruit
orange-red, flushed yellow.

9. P. nussia (D. Don) Kalkman
(*Stranvaesia nussia* (D. Don) Decaisne; *S.
glaucescens* Lindley). Illustration: *Edwards's
Botanical Register* **23**: 1956 (1837);
Polunin & Stainton, Flowers of the
Himalaya, 466 (1985).

Evergreen shrub to 5 m, young
branchlets covered with whitish down,
becoming hairless. Leaves 6–10 × 3–5 cm,
tough and leathery, dark glossy green
above, paler beneath with a downy midrib,
lanceolate to obovate, finely toothed.
Flowers 1.1–1.3 mm wide, borne in flat
almost nodding, terminal umbels, 5–10 cm
across, flower-stalks usually woolly,
occasionally hairless. Fruit orange, globe-
shaped, c. 8 mm, hairy when young, later
becoming red and hairless. *E Himalaya,
S China to Philippines.* H5. Summer.

Requires the warmth of a south-facing
wall in colder districts.

10. P. niitakayamensis Hayata.
Illustration: *Kew Magazine* **5**: 148 (1988);
Li, Woody flora of Taiwan, 327 (1963); Li,
Flora of Taiwan **3**: 142 (1977).

Evergreen shrub to 3 m, densely
branched. Leaves 5–10 × 2–3 cm, entire,
oblanceolate, tapered at the apex, rounded
to wedge-shaped at the base, dark shiny
green above, pale green almost dull
beneath with up to 15 pairs of veins, mid-
vein and margins hairy, leaf-stalk
0.7–1.7 cm, curved, grooved, densely hairy
above; stipules c. 5 mm, joined to leaf-stalk
at base. Flowers, 9–10 mm across, borne in
hemispherical, terminal umbels,
c. 4 × 8 cm, anthers lilac-pink. Fruit
c. 8 × 9 mm, bright pinkish red, globe-
shaped. *Taiwan.* H4–5. Summer.

11. P. beauverdiana Schneider.
Illustration: Icones cormophytorum
Sinicorum, **2**: 2153 (1972); Flora of
Jiangsu **2**: 283 (1982); Wu, Wild
flowers of Yunnan **2**: 111 (1986);
Krüssmann, Manual of cultivated
broad-leaved trees & shrubs **2**: pl. 159,
397 (1986).

Deciduous shrub or slender tree to 9 m,
bark downy when young, soon hairless.
Leaves 5–13 × 2–4 cm, leathery, narrowly
obovate to lanceolate, narrowly wedged-
shaped at base, long, pointed at tip, veins
in 8–14 pairs, conspicuous; margin with
stiff teeth, each bearing a small black
gland, leaf-stalk c. 1 cm. Flowers
numerous, 6–8 mm wide, borne in large,
flattish, terminal or axillary umbels c. 5 cm
wide. Fruit c. 6 mm, deep red, egg-shaped.
China (W Hubei) H2. Summer.

Var. **notabilis** (Schneider) Rehder &
Wilson. Illustration: Liu, Illustrations of
native & introduced ligneous plants of
Taiwan **1**: 435 (1960); Guo-zi Hsu, The
wild and woody plants of Taiwan, 128
(1984); Krüssmann, Manual of cultivated
broad-leaved trees & shrubs **2**: pl. 159
(1986). Leaves broad oblong-elliptic,
7–12 cm, with 12 pairs of veins.
Inflorescences more loose, 8–10 cm wide.
Fruit orange-red. In fruit this variety is
considered to be far superior to the species.

12. P. villosa (Thunberg) de Candolle.
Illustration: *Botanical Magazine*, 9275
(1932); Icones cormophytorum Sinicorum,
2: 2159 (1972); Flora of Jiangsu **2**: 284
(1982); Krüssmann, Manual of cultivated
broad-leaved trees & shrubs **2**: 397 (1986).

Deciduous shrub or small tree to 5 m,
young shoots downy, later hairless. Leaves
3–8 × 1.5–3 cm, obovate or ovate-
lanceolate, tapered at the base with a long
fine point at the apex, very tough, dark
green above, pale yellow-green beneath
with shaggy indumentum, finely toothed,
each tooth spine-tipped. Flowers borne in
hairy umbels, c. 5 cm wide, stalks warty.
Fruit c. 8 mm, bright red, egg-shaped.
Japan, China & Korea. H2. Summer.

Very variable in the amount of down
present on the leaves, young shoots and
flower-stalks. Much valued for its red fruits
and brilliant autumn colour.

Var. **laevis** (Thunberg) Dippel.
Illustration: *Botanical Magazine*, 9275
(1929); Krüssmann, Manual of cultivated
broad-leaved trees & shrubs **2**: 397
(1986). Leaves longer, pointed; branchlets
and flowers only slightly hairy to hairless;
fruit to 1.5 cm.

Forma **maximowicziana** (Léveillé)
Rehder. Leaves almost stalkless, distinctly
puckered, rounded and abruptly pointed at
the tip, base wedge-shaped, veins
conspicuously sunken, autumn colour very
striking. According to some authorities this
variant is better recognised as a seperate
species.

Var. **sinica** Rehder & Wilson. Slender
tree to 10 m, young shoots downy. Leaves
not tough, mostly elliptic, bright green,
soon hairless above, paler beneath,
especially downy on the midrib and veins,
hairless later. Flowers borne in racemes,
2.5–5 cm across. Fruit c. 1.2 cm, orange-
scarlet, conspicuously warted.

13. P. parvifolia (Pritzel) Schneider.
Illustration: Icones cormophytorum
Sinicorum, **2**: 2158 (1972); Flora of
Jiangsu **2**: 285 (1982); Guo-zi Hsu, The
wild woody plants of Taiwan, 130 (1984);

Krüssmann, Manual of cultivated broad-leaved trees & shrubs **2**: 397 (1986).

Deciduous shrub, 2–3 m, young shoots dark red, hairless. Leaves 3–6 cm, ovate to obovate, slenderly pointed, broadly wedge-shaped at the base, sharply toothed, deep green above, lighter beneath. Flowers 5 or 6 in terminal umbels, 5 or 6 cm across; flower-stalks 1.2–2.5 cm, hairless. Fruit dullish to orange-red, egg-shaped, *c.* 8 mm, crowned by persistent sepals. *China (Hubei)*. H2. Spring–summer.

14. P. glomerata Rehder & Wilson. Illustration: Icones cormophytorum Sinicorum, **2**: 2148 (1972); Flora Reipublicae Popularis Sinicae **36**: 223 (1974); Wu, Wild flowers of Yunnan **2**: 112 (1986).

Deciduous tree or shrub, shoots reddish when young, shaggy with long hairs. Leaves 12–18 × 5–6 cm, thin, leathery, narrow oblong to oblanceolate, tapered towards the base, margins with fine, glandular teeth, slightly rolled inwards, yellowish green above, lighter beneath with 6–9 pairs of veins, shaggy with long deciduous hairs. Flowers fragrant, nearly stalkless, in umbels 6–10 cm across, densely shaggy. Fruit 5–7 mm, egg-shaped, red. *China (Yunnan)*. H4–5. Summer.

66. HETEROMELES M.J. Roemer
M.F. Gardner
Evergreen shrubs, 2–10 m; bark grey, branchlets slightly hairy. Leaves 5–10 cm, elliptic or lance-shaped to oblong, leathery, sharply toothed, pale beneath; leaf-stalks 1–2 cm, downy. Flowers many, borne in flattish, corymb-like panicles; sepals triangular, 1–1.5 mm; petals white, *c.* 3 mm; stamens 10; styles 2 or 3, separate. Ovary inferior. Fruit bright red, 5–6 mm, persistent throughout the winter.

A genus with a single species from North America, grown for its attractive fruit and handsome foliage. Although easily cultivated in the open in milder areas, in colder districts it is best grown against a wall for protection. Propagation as for *Photinia*.

1. H. arbutifolia M. Roemer. Illustration: *Edwards's Botanical Register* **6**: 491 (1820) – as *Photinia arbutifolia*; Sargent, Manual of the trees of North America 392 (1922); Munz, Flora of southern California, 744 (1974); Lenz & Dourley, California native trees & shrubs, 107 (1981).

USA (California). H4–5. Summer.

67. COTONEASTER Medikus
J. Fryer & B. Hylmö
Evergreen or deciduous shrubs and small trees. Leaves alternate, simple, entire. Flowers in cymes, small clusters, or solitary; stamens 10–20; ovary inferior; carpels 1–5, usually with free styles. Fruit (pomes) somewhat succulent, containing 1–5 one-seeded, hard-walled nutlets.

The genus is distributed throughout Eurasia and North Africa. A marked concentration of species is found in the Himalaya and western China.

As with several other genera of Rosaceae, (e.g. *Crataegus*, *Rubus* and *Sorbus*) species with apomictic breeding systems are common. This means that plants raised from seed are usually genetically identical to their mother. The morphological differences between species with such breeding systems are often small, making their identification difficult.

The species show a great variation in growth habit: from tall trees 15–20 m (*C. frigidus*) to prostrate subalpine shrubs (*C. radicans*). Species from colder areas are deciduous, whereas those from warmer zones are mostly evergreen or semi-evergreen. In central and northern Europe most cultivated species are deciduous. One species (*C. lucidus*), native to the shores of Lake Baikal, Siberia, is one of the most hardy of all garden shrubs, and is frequently planted in Russia and Scandinavia (even north of the Arctic Circle). In those parts of Europe with milder climates, such as the British Isles, evergreen species are more common in cultivation. Many of these are cultivars selected from sexually reproducing species such as *C. salicifolius*, *C. dammeri* and *C. conspicuus*.

The fruits are much liked by birds, and in many areas several species have become naturalised following their escape from gardens. Bees also visit the shrubs in great numbers.

Shrubs thrive in any adequately drained soil and are easily propagated from cuttings, taken in late summer; or seed, which needs a period of chilling before it will germinate. Apomictic species breed true from seed while outbreeding species will give a degree of variation. Most cultivars should be increased by cuttings only.

Fireblight can sometimes be a problem, especially with some evergreen species; it can be controlled by spraying an antibiotic (streptomycin) or copper salts, and cutting out affected branches 30 cm into the green

unaffected wood. It is important to burn removed branches immediately.

Literature: Klotz, G., Uebersicht üeber die in Kultur befindlichen Cotoneaster, Arten und Formen, *Wissenschaftliche Zeitschrift der Martin-Luther-Universitat Halle-Wittenburg* **6**: 945–82 (1957); Flinck, K.E. & Hylmö, B., A list of series & species in the genus Cotoneaster, *Botaniska Notiser* **119**: 445–63 (1966); Hurusawa, I. & Kawakami, *Informationes Annuales Hortorum Botanicorum Facultatis Scientiarum Universitatis Tokyoensis* (1967); Klotz, G., Synopsis der Gattung Cotoneaster Medic. *1-Beiträge zur Phytotaxonomie* **10**: 7–81 (1982); Phipps, J.B. et al., A checklist of the subfamily Maloideae (Rosaceae), *Canadian Journal of Botany* **68**: 2209–69 (1990).

Key to Series

1a. Flowers in each group opening in sequence (or flowers solitary); petals mostly obovate, usually erect, (rarely spreading), purple, red or pink rarely white; filaments usually pink or red   2
  b. Flowers in each group opening more or less simultaneously (or flowers solitary); petals nearly circular, spreading, usually white or cream; filaments usually white   14
2a. Stamens 10–15   3
  b. Stamens *c.* 20   6
3a. Branches warted; leaves to 2 cm, thick, shiny, edges wavy; flowers solitary, pendent; fruit red, nearly spherical; nutlets 2
                                          2. Verruculosi
  b. Branches not warted; leaves to 5 cm, thin to moderately thick, dull or shiny; edges flat or wavy; flowers usually 2 or 3 per group (rarely 4–6), if solitary, then spreading or erect; fruit red or black, cylindric to obovate or rarely spherical or almost so; nutlets 2 or 3 (rarely 4)   4
4a. Petals spreading, plain red, falling early; filaments dark red; fruit red; nutlets 2   **3. Sanguinei**
  b. Petals erect, multicoloured, purple, red, pink, white, not falling early; filaments dark red, red or pale pink; fruit red or black; nutlets 2 or 3 (rarely 4)   5
5a. Leaves initially with sparse adpressed bristles beneath, margins flat or wavy; filaments dark red or pink; fruit orange to red   **4. Adpressi**

b. Leaves initially downy or shaggy beneath, margins flat; filaments pink (rarely red); fruit black **5. Nitentes**

6a. Leaves with adpressed bristles beneath 7

b. Leaves shaggy or with densely felted, or sparse hairs beneath 9

7a. Leaves less than 2 cm; flowers solitary or 2–3 per cyme, pendent; calyx hairless or very sparsely hairy or bristly; fruit orange or red; nutlets 3 **1. Distichi**

b. Leaves usually more than 2 cm; flowers 1–11 per cyme, usually almost erect; calyx with adpressed bristles; fruit red or black; nutlets 2–5 8

8a. Flowers very wide or narrowly bell-shaped; fruit obovoid or nearly spherical, 8–12 mm, red; nutlets 2–4 **9. Acuminati**

b. Flowers small, spreading, not bell-shaped; fruit sperical, to 7 mm, red or black; nutlets mostly 5 **10. Glomerulati**

9a. Surface of leaves blistered or wrinkled; fruit red, rarely orange (or in one species purple-black) 10

b. Surface of leaves flat, with the veins only slightly impressed; fruit black, red or maroon 11

10a. Leaves 1–3.5 cm (rarely 6 cm), wrinkled, often dull, leaf-margins sometimes recurved; anthers pale purple or white; fruit red; nutlets 2–4 **12. Franchetioides**

b. Leaves 1–15 cm, blistered, shiny, leaf-margins not recurved; anthers white; fruit red or purple-black; nutlets 3–5 **11. Bullati**

11a. Lower surface of leaves shaggy or with short erect hairs, often sparse; fruit black **6. Lucidi**

b. Lower surface of leaves with densely felted hairs; fruit red, maroon or black 12

12a. Petals erect or semi-spreading; fruit maroon or black **7. Melanocarpi**

b. Petals erect; fruit red 13

13a. Nutlets 3–5 **8. Cotoneaster**

b. Nutlets 2 **13. Zabelioides**

14a. Anthers white 15

b. Anthers purple or black 18

15a. Nutlets apparently 1 (= 2 nutlets joined) **15. Multiflori**

b. Nutlets 1 (not joined) or 2 (rarely 3) 16

16a. Fruit black, (in one species red-purple); nutlets 1 or 2 **17. Insignes**

b. Fruit red; nutlets 2 (rarely 3) 17

17a. Leaves hairy beneath; petals and filaments pink, or leaves hairless beneath; petals and filaments white; nutlets 2 (rarely 3) **14. Megalocarpi**

b. Leaves hairy beneath; petals and filaments white; nutlets 2 **16. Racemiflori**

18a. Fruit black **17. Insignes**

b. Fruit red, orange or salmon-pink 19

19a. Branches growing at right angles; nutlets 1 (rarely 2) per fruit **18. Hebephylli**

b. Branches growing at acute angles; nutlets 2–5 20

20a. Leaf-edges not recurved; flowers many per cyme; nutlets 2 21

b. Leaf-edges usually recurved; flowers solitary, 2–7 or many per cyme; nutlets 2–5 22

21a. Leaves thin, blade 10–15 cm **19. Chaenopetalum**

b. Leaves thick, blade usually less than 5 cm (rarely to 9 cm) **21. Pannosi**

22a. Leaves to 10 cm, upper surface wrinkled or with veins slightly impressed; flowers solitary, 2–4 or many per cyme; nutlets 2–5 **20. Salicifolii**

b. Leaves to 3 cm, upper surface flat; flowers solitary or 2–7 per cyme; nutlets 2 **22. Microphylli**

In each series the main species is fully described, comparison is then made with other species in that series.

1. Series **Distichi** Yü.

1a. Flowers solitary; leaves nearly circular **1. nitidus**

b. Flowers solitary or 2 or 3 per cyme; leaves elliptic **2. marquandii**

1. C. nitidus Jacques (*C. distichus* Lange; *C. rupestris* Charlton). Illustration: Saunders, Refugium botanicum **1**: t. 54 (1869) – as *C. rotundifolius*).

Shrub, erect, *c.* 3 m. Branches straggly, initially with adpressed bristles, stipules *c.* 4 mm, more or less persistent. Flowers solitary, pendent; stalks *c.* 5 mm. Calyx hairless; lobes obtuse. Petals erect, red and off-white. Stamens 20; anthers white, filaments red. Fruit 9–11 mm, pendent, obovate to almost spherical, orange, finally turning red, shiny; nutlets 3. Apomictic. *China (Yunnan).* H3. Spring.

2. C. marquandii Klotz.

Branches fastigiate. Leaves deciduous; blade elliptic to obovate-elliptic, apex

acuminate. Flowers solitary, or 2 or 3 per cyme; calyx with sparse adpressed bristles. Fruit 7–8 mm; bright orange. *Himalaya (Burma, Bhutan).* H3. Spring.

2. Series **Verruculosi** Klotz.

1a. Young branches densely warted, stiffly erect; leaves to 1.5 cm, rarely notched at apex **3. verruculosus**

b. Young branches sparsely warted, spreading, erect; leaves to 1 cm, notched at apex **4. cavei**

3. C. verruculosus Diels.

Shrub, erect, 60–150 cm. Branches erect, densely warted, with adpressed bristles. Leaves deciduous or semi-evergreen; blade 8–14 mm, moderately thick, circular to broadly elliptic or broadly obovate, margins slightly wavy; apex obtuse, mucronate, upper surface shiny, mid-green, with a few adpressed bristles, lower surface hairless; stalk 2–5 mm, slender, stipules red, oblong, some with hairy margins, persistent. Flowers solitary, pendent, almost stalkless. Calyx hairless; lobes usually obtuse. Petals erect, red, pink-edged. Stamens 12–15; anthers white, filaments pale pink. Fruit 8–10 mm, pendent, almost spherical, red, shiny; nutlets 2. Apomictic. *China (Yunnan).* H3. Spring.

4. C. cavei Klotz.

Shrub *c.* 80 cm; branches not so noticeably warted. Leaf-blade to 2 cm, apex notched, upper surface dark-green; stalk 1–3 mm. Petals deep-pink to red. Fruit 7–9 mm, spherical. *Himalaya (Nepal, Sikkim).* H3. Spring.

3. Series **Sanguinei** Klotz.

1a. Leaves with hairs only on the veins beneath **5. sanguineus**

b. Leaves with felted hairs beneath **6. rubens**

5. C. sanguineus Yu. Illustration: *Bulletin British Museum, Botany* **1**: 5, pl. 4 (1954).

Shrub, erect, 1.5–2.5 m. Branches dense, straggly, initially with adpressed yellowish bristles. Leaves deciduous, blade 1–3 cm, moderately thick, broadly ovate, apex acute to obtuse, sometimes mucronate, upper surface shiny, sparsely hairy, veins impressed, lower surface with only a few hairs on the veins; leaf-stalk 3–6 mm with adpressed bristles; stipules to 2 mm, lanceolate, margin shaggy, hairs not persistent. Flowers *c.* 1 cm across, solitary or in pairs, semi-erect, almost stalkless. Calyx hairless, wide bell-shaped, red, lobes shortly triangular, margins shaggy. Petals spreading, dark-red, falling early. Stamens 10 (rarely to 15); anthers white, filaments dark red. Fruit 9–11 mm, pendent, broadly obovoid to nearly spherical, light red; nutlets 2. Apomictic. *Himalaya.* H3. Spring.

6. C. rubens W.W. Smith.

Shrub 50–150 cm. Leaf-blade 2–4 cm, circular or wide elliptic, apex obtuse, mucronate, lower surface with felted hairs; leaf-stalk *c.* 1 mm, with felted hairs. Flowers *c.* 8 mm across. Calyx densely downy. *China (Yunnan).* H3. Spring.

4. Series Adpressi Hurusawa.

1a. Branching irregular, rooting at nodes; leaves thin, usually wavy-edged; fruit spherical, succulent   2
 b. Branching regular, seldom rooting at nodes; leaves moderately thick, usually not wavy-edged; fruit ellipsoid or cylindric, rarely spherical, not succulent   5
2a. Shrub to 30 cm   **7. adpressus**
 b. Shrub 50–100 cm   3
3a. Leaves broadly obovate   **12. atropurpureus**
 b. Leaves spherical   4
4a. Leaves very wavy; flower 9–10 mm; nutlets usually 2   **9. nanshan**
 b. Leaves not or only slightly wavy; flower 6–7 mm; nutlets usually more than 2   **8. apiculatus**
5a. Fruit orange-red   6
 b. Fruit red to rich dark red   9
6a. Leaves to 2 cm, concave   **13. hjelmqvistii**
 b. Leaves less than 1.5 cm, flat, or only slightly wavy   7
7a. Leaves broadly obovate, thin to moderately thick, slightly wavy   **12. atropurpureus**
 b. Leaves almost circular to broadly elliptic, thick, flat.   8
8a. Leaves to 1.2 cm; fruit 5–6 mm   **10. horizontalis**
 b. Leaves to 8 mm; fruit 4–5 mm   **11. perpusillus**
9a. Branching regular; leaves moderately thick, to 1.5 cm   **14. ascendens**
 b. Branching irregular, straggly; leaves thin, to 2.5 cm   **15. divaricatus**

7. C. adpressus Bois (*C. horizontalis* var. *adpressus* Schneider). Illustration: Tarouca & Schneider, Freiland-Laubgehölze, 164 (1922).

Shrub, prostrate, to 30 cm. Branches rigid, irregular, initially downy, rooting at nodes, internodes short. Leaves deciduous,

falling early; blade 5–15 mm, thin, broadly ovate to obovate, margin wavy, initially hairy, apex obtuse to nearly acute, mucronate, upper surface dull green, hairless, lower surface initially with adpressed hairs; stalk 1–2 mm. Flowers including calyx 6–7 mm, erect, solitary or in pairs, nearly stalkless. Calyx slightly downy, lobes triangular, sometimes mucronate, margin hairy. Petals erect, red and pink. Stamens 10 (rarely to 13); anthers white, filaments pink to red. Fruit 6–7 mm, erect, spherical, bright red, succulent; nutlets 2. Diploid, variable. *W China.* H3. Spring.

'Little Gem' is smaller, forming a compact cushion; it sometimes flowers but rarely sets fruit.

8. C. apiculatus Rehder & Wilson.

Prostrate, ends of branches ascending, 50–100 cm. Leaf-blade 1–2 cm, somewhat circular, margin usually flat, (rarely slightly wavy), apex terminating abruptly to a small point, upper surface shiny; stalks becoming purple. Fruit 1–1.2 cm, succulent; nutlets usually 3. Apomictic. *China (Sichuan).* H2. Spring.

9. C. nanshan Mottet (*C. adpressus* var. *praecox* Bois & Berthault).

Prostrate, ends of branches ascending, *c.* 1 m. Leaf-blade 1.2–2.5 cm, circular, apex somewhat acute or acuminate, margins extremely wavy. Flowers (including calyx) 9–10 mm. Petals red and pink, fringed off-white. Fruit 1–1.2 cm, nearly spherical, very succulent; nutlets usually 2. Apomictic. *China (Sichuan).* H2. Spring.

10. C. horizontalis Decaisne.

Height *c.* 1 m (3 m if supported). Branches regular (herringbone), not rooting at nodes. Leaves not falling early; blade to 1.2 cm across, shiny, almost circular to broadly elliptic, margins usually flat. Filaments dark red. Fruit 5–6 mm orange-red; nutlets 3. Apomictic. *W China.* H2. Spring.

11. C. perpusillus Flinck & Hylmö (*C. horizontalis* var. *perpusillus* Schneider; *C. horizontalis* 'Saxatilis').

Branches regular (herringbone), not rooting at nodes. Leaves not falling early; blade thick, shiny, margins usually flat. Fruit 4–5 mm, oblong to nearly spherical, orange-red. Apomictic. *China (Hubei).* H3. Spring.

12. C. atropurpureus Flinck & Hylmö (*C. horizontalis* var. *prostratus* misapplied).

Prostrate or ascending shrub, 50–100 cm (3 m supported). Branching regular. Leaf-blade thin to moderately thick, broadly obovate, margins slightly wavy; apex obtuse to truncate; upper surface somewhat shiny. Petals red with black bases. Filaments dark red. Fruit orange-red; nutlets 2 or 3. Apomictic. *China (Hubei).* H2. Spring.

'Variegatus' is a shrub to 50 cm; leaves edged white, tinged pink in autumn, sometimes wavy.

13. C. hjelmqvistii Flinck & Hylmö (*C. horizontalis* var. *robustus* Grootendorst).

Prostrate and ascending to 1.5 m (4 m if supported). Branching regular, not rooting at nodes. Leaves with intense autumn colour; blade to 2 cm, moderately thick, ovate to circular, concave, upper surface shiny, light-green. Fruit orange-red. Apomictic. *China (Sichuan).* H3. Spring.

Sometimes grown as *C. horizontalis* 'Darts Splendid' or 'Coralle'.

14. C. ascendens Flinck & Hylmö (*C. horizontalis* var. *wilsonii* Wilson; *C. horizontalis* var. *fructo-sanguineus* invalid).

Ascending; 1–2 m. Branches regular, not rooting at nodes. Leaf-blade moderately thick, elliptic, margins only slightly wavy, upper surface shiny light green. Flowers solitary or 2 or 3 per cyme. Petals pink and off-white. Fruit ellipsoid to almost spherical, red, nutlets 2 or 3. Apomictic. *China (Hubei).* H2. Spring.

15. C. divaricatus Rehder & Wilson.

Erect; 1–2 m. Branches wide-spreading, not rooting at nodes. Leaf-blade 8–25 mm, thin, ovate to elliptic, margins rarely wavy, upper surface shiny dark green. Flowers solitary or 2–4 per cyme. Stamens 12–15. Fruit 7–9 mm, ellipsoid to cylindric, very dark red. Apomictic. *China (Hubei).* H2. Spring.

'Gracia'. Prostrate and ascending, to 1 m; branches somewhat regular, wide spreading, not rooting at nodes; leaves with intense autumn colour; blade to 2 cm, thick, apex sometimes obtuse, upper surface with veins impressed; petals pink; Fruit light red, not numerous. H3. Spring.

'Valkenburg'. Prostrate and ascending, to 1.5 m; branching dense, spreading not rooting at nodes; leaf-blade to 2 cm, elliptic, moderately thick, upper surface with veins slightly impressed; petals pink; rarely setting fruit. H3. Spring.

5. Series **Nitentes** Flinck & Hylmö.

1a. Leaves 2–5 cm; flowers 3–6 per
cyme **18. tenuipes**
  b. Leaves to 2.5 cm; flowers 1–3 per
cyme 2
2a. Shrub to 3 m; stamens 10, filaments
pink **16. nitens**
  b. Shrub to 1.5 m; stamens usually 12,
filaments red **17. harrysmithii**

16. C. nitens Rehder & Wilson.
Illustration: *Botaniska Notiser* **115**: 32
(1962).

Shrub, erect, 2–3 m. Branches arched,
initially light red-brown with sparse
yellow-grey hairs. Leaves deciduous, blade
1–2.5 cm, thin, circular to ovate, apex
obtuse, upper surface hairless, shiny green,
lower surface initially sparsely hairy; stalk
2–3 mm, sparsely hairy, Flowers erect,
solitary or 2 or 3 per cyme; stalks 2–4 mm,
soon hairless. Calyx sparsely hairy, lobes
broadly triangular. Petals erect, purple-red
and off-white. Stamens 10; anthers white;
filaments pink. Fruit 7–8 mm, spreading,
obovoid, purple-black; nutlets 2. Apomictic.
China (Sichuan). H2. Spring.

17. C. harrysmithii Flinck & Hylmö.
Prostrate, ascending, to 1.5 m. Branches
regular, somewhat horizontal. Leaf-blade
1.5–2.5 cm, elliptic to ovate, apex acute to
acuminate, sparsely hairy above, the newly
unfolded leaves papillose, sparsely hairy
beneath. Flower-stalks 1–3 mm, sparsely
hairy. Stamens usually 12, filaments red.
Fruit 6–7 mm. *China (Sichuan).* H2. Spring.

18. C. tenuipes Rehder & Wilson.
Branches slender, graceful, pendent,
initially shiny, light purple-brown. Leaf-
blade 2–5 cm, narrowly elliptic, lower
surface shaggy; leaf-stalk slender. Flowers
3–6 per cyme. Petals pink and off-white.
Fruit 6–10 mm, cylindric. *China (Sichuan).*
H2. Spring.

6. Series **Lucidi** Pojárkova.

1a. Fruit red **20. acutifolius**
  b. Fruit black 2
2a. Leaves falling very early; nutlets
usually 3 **19. lucidus**
  b. Leaves falling medium to late;
nutlets usually 2 3
3a. Leaves thin, flat; upper surface not
blistered **23. laetevirens**
  b. Leaves moderately thick; upper
surface slightly blistered, shaggy 4
4a. Leaves to 8 cm (rarely 10 cm)
21. villosulus
  b. Leaves to 5 cm **22. ambiguus**

19. C. lucidus Schlechtendal. Illustration:
Komarov (ed.), Flora SSSR **9**: 252 (1939).
Shrub, erect, 1–3 m. Branches fairly
dense, initially with adpressed downy hairs.
Leaves deciduous, moderately thick, intense
autumn colour, falling early, blade 2–7 cm,
ovate-elliptic, apex acute, upper surface
shiny, flat, soon hairless; lower surface
initially shaggy with yellowish hairs, later
with sparse adpressed down; leaf-stalk
2–6 mm, hairy. Flowers 5–15 per loose
spreading cyme; stalks with yellowish
adpressed hairs. Calyx nearly hairless; lobes
broadly triangular, margins hairy. Petals
erect, pink, green and off-white. Stamens
20; anthers white, filaments pink. Fruit
8–10 mm, pendent, obovoid to spherical,
black, shiny; nutlets 3 (rarely 2). Apomictic.
Siberia (Lake Baikal). H1. Early spring.

20. C. acutifolius Turczaninov.
Similar to *C. lucidus*, but fruit red. *China
(Nei Mongol).*
Probably not in cultivation; plants
grown under this name are usually either
C. lucidus or *C. villosulus*, both of which
have black fruit.

21. C. villosulus Flinck & Hylmö
(*C. acutifolius* var. *villosulus* Rehder &
Wilson).
Height 2–4 m. Branching vigorous.
Leaves falling medium to late; blade
3–10 cm, ovate to oblong-ovate, apex
acute to acuminate, upper surface
somewhat blistered, lower surface and
calyx densely shaggy. Fruit obovoid slightly
downy; nutlets 2 (rarely 3). *China (Hubei).*
H2. Early spring.

22. C. ambiguus Rehder & Wilson.
Leaves falling medium to late; blade
2.5–5 cm, apex finely acuminate, upper
surface slightly blistered. Fruit nearly
spherical; nutlets 2. *C China.*
Very rarely grown.

23. C. laetevirens Klotz (*C. acutifolius* var.
laetevirens Rehder & Wilson).
Shrub 2–3 m. Branching open and
graceful. Leaf margins hairy, upper surface
light green, lower surface sparsely hairy;
stalk 3–5 mm. Flowers 3–7 per cyme.
Calyx-lobes acute or acuminate, mucronate.
Fruit oblong to nearly spherical; nutlets
usually 2. *China (Sichuan).* H2. Early spring.
Many cotoneasters labelled *C. ambiguus*
are this species.

7. Series **Melanocarpi** Pojárkova.

1a. Petals erect; fruit black or bluish
black 2

  b. Petals semi-spreading; fruit maroon
4
2a. Shrub 1.5–2 m; flowers to 15 per
cyme **24. melanocarpus**
  b. Shrub 2–4 m; flowers more than 15
to 40 per cyme 3
3a. Flower-stalks shaggy; late flowering
and ripening of fruit
25. polyanthemus
  b. Flower-stalks with only a few hairs;
earlier flowering and ripening of fruit
26. laxiflorus
4a. Leaves to 8 cm; cymes loose
27. ignavus
  b. Leaves to 3.5 cm; cymes compact
28. zeravschanicus

24. C. melanocarpus Loddiges.
Illustration: *Loddiges' Botanical Cabinet* **154**:
1531(1830).
Shrub, erect, 1.5–2 m. Branches loosely
erect, initially shiny red-brown, thickly
downy. Leaves deciduous, blade
2–6 cm, somewhat thin, ovate to
elliptic, apex acute, mucronate, upper
surface dull dark green, initially with
sparse hairs, lower surface with greyish
felted hairs; stalks 4–7 mm, felted.
Flowers 3–15 per cyme, spreading to
pendent; flower-stalks sparsely hairy, the
axils more so. Calyx almost hairless, lobes
obtuse. Petals erect, pink and off-white.
Stamens 20; anthers white, filaments pink.
Fruit 7–9 mm, pendent, obovoid to
spherical, black, with a waxy bloom;
nutlets 2 or 3. Apomictic. *Russia.* H1. Early
spring.

25. C. polyanthemus Wolf.
Height to 4 m. Upper surface of leaves
shiny. Flowers 15–40 per cyme, flowering
much later than *C. melanocarpus*; stalks
shaggy. Calyx sparsely hairy. Fruit obovoid
to pear-shaped with only a very weak
waxy bloom, ripening irregularly. *C Asia
(Tianshan Mts).* H1. Spring.

26. C. laxiflorus Lindley.
Height 2–2.5 m. Branches initially
brownish purple with a grey peeling
cuticle. Flowers 20–40 per cyme, loose,
pendent, stalks almost hairless. Fruit
almost spherical. *Origin unknown.* H2. Early
spring.

27. C. ignavus Wolf.
Branches stiffly erect. Leaf-blade thin,
nearly circular, obovate to broadly elliptic.
Flower-stalks noticeably dichotomously
branched. Petals semi-spreading. Fruit
nearly ellipsoid, maroon. *Turkmenistan.* H2.
Early spring.

28. C. zeravschanicus Pojárkova.

Leaf-blade 1.5–3.5 cm, almost circular, upper surface a medium green. Flowers 4–8 per cyme, compact, erect; stalks usually purple. Petals semi-spreading, very pale pink to off-white. Fruit 5–8 mm, dull dark maroon, with the dried remains of the calyx persisting as a coronet. *Tadzhikstan.* H2. Early spring.

8. Series **Cotoneaster**.

1a. Leaves broadly ovate, apex acute, upper surface nearly hairless
29. integerrimus
 b. Leaves nearly circular to broadly elliptic, apex obtuse, upper surface persistently hairy **30. tomentosus**

29. C. integerrimus Medikus. Illustration: Lindman (ed.), Nordens Flora **2**: 331 (1964).

Shrub, prostrate to erect, *c.* 1.5 m. Branches arched, initially with felted hairs. Leaves deciduous, blade 1–5 cm, somewhat thin, usually broadly ovate, apex acute (rarely obtuse), usually mucronate, upper surface dull, initially very sparsely hairy, lower surface with felted greyish hairs; stalks 4–7 mm with felted hairs. Flowers 3–7 per cyme, spreading to pendent, axils thinly felted; stalks with sparse hairs. Calyx hairless, lobes obtuse, margins sometimes with downy hairs. Petals erect-incurved, short, scarcely longer than calyx-lobes, pale pink to off-white. Stamens 20; anthers white, filaments very pale pink. Fruit spreading and pendent, nearly spherical, 8–11 mm, red; nutlets 3–5, (rarely 2). Apomictic. *Europe.* H1. Early spring.

30. C. tomentosus Lindley (*C. nebrodensis* Koch).

Shrub to 2 m. Leaf-blade 3–6 cm, nearly circular to broadly elliptic, apex obtuse, upper surface persistently hairy, lower surface with dense felted hairs. Flowers 5–12 per cyme, stalks and calyx with felted hairs. Nutlets 4 or 5. *C Europe (Alps).* H1. Early spring.

9. Series **Acuminati** Yü.

1a. Leaves thin, somewhat wavy-edged, elliptic-ovate; nutlets 2
31. acuminatus
 b. Leaves moderately thick, flat; nutlets 2–4   2
2a. Leaves circular-ovate, to 3 cm; nutlets 3 or 4 **33. simonsii**
 b. Leaves ovate-lanceolate, to 5 cm; nutlets 2 or 3   3

3a. Leaves acuminate, upper surface and calyx-lobes persistently hairy
32. mucronatus
 b. Leaves acute, upper surface and calyx-lobes becoming hairless
34. newryensis

31. C. acuminatus Lindley. Illustration: *Transactions of the Linnean Society of London* **101**: 13 (1821) *Loddiges' Botanical Cabinet* **41**: 919 (1824).

Shrub, erect, 2–4 m. Branches stiffly erect, vigorous, initially dark greenish brown, with adpressed bristles. Leaves deciduous, blade 3–6 cm, thin, elliptic-ovate, margin somewhat wavy, apex acuminate with a small point, upper surface shiny light green, with adpressed bristles, lower surface at first thickly covered with adpressed bristles. Leaf-stalk 2–5 mm with adpressed bristles. Flowers solitary or 2–4 per cyme, spreading to pendent; stalks usually 5–10 mm with adpressed bristles. Calyx wide bell-shaped, with adpressed bristles; lobes broadly triangular, obtuse or acute. Petals wide-erect, red-pink, green and off-white, stamens 20; anthers white, filaments pale pink. Fruit 8–12 mm, mostly pendent, broadly obovoid, bright red, shiny, hairy near apex; nutlets 2 (rarely 3). Diploid, variable. *Himalaya.* H3. Spring.

32. C. mucronatus Franchet.

Leaf-blade to 5 cm, thicker and flatter than *C. acuminatus*, ovate-lanceolate, apex acute to acuminate, mucronate, upper surface dark green, lower surface with dense adpressed bristles. Calyx bell-shaped, lobes acuminate. Nutlets 2 or 3. Apomictic. *China (Yunnan).* H3. Spring.

33. C. simonsii Baker.

Height to 4 m. Branches wide-spreading. Leaf-blade 1.5–3 cm, thicker and flatter than *C. acuminatus*, broadly ovate, apex acute to somewhat acuminate, lower surface with adpressed bristles. Calyx-lobes acuminate. Fruit 8–10 mm, obovoid; nutlets 3 or 4. Apomictic. *Himalaya (Sikkim, Bhutan).* H3. Spring.

Commonly used for hedging.

34. C. newryensis Barbier.

Shrub to 5 m. Branches spreading, initially with downy hairs. Leaf-blade moderately thick, 2–3 cm, elliptic, edges flat, apex acute, upper surface mid-green, becoming hairless, lower surface sparsely hairy; stalks shaggy. Flowers 5–11 per cyme, spreading. Calyx not bell-shaped, lobes obtuse or acuminate, hairless, margin

woolly. Petals pink with very little white. Fruit 8–10 mm spreading, obovoid to nearly spherical, light red; nutlets 2 or 3, usually 3. *China (Sichuan, Yunnan).* H3. Spring.

10. Series **Glomerulati** Flinck & Hylmö.

1a. Leaves shiny above; fruit red
35. glomerulatus
 b. Leaves dull above; fruit black
36. foveolatus

35. C. glomerulatus W.W. Smith (*C. nitidifolius* Marquand). Illustration: *Hooker's Icones Plantarum*, **23**: t. 3145 (1930).

Shrub, erect, very leafy, with intense autumn colour, 1–3 m. Branches spreading, initially golden brown, with sparse yellowish grey-felted hairs. Leaves deciduous; blade 4–6 cm, moderately thick, lanceolate-ovate, apex acuminate, upper surface shiny, pale to mid-green, lower surface with sparse adpressed bristles; stalks 2–3 mm, with adpressed bristles. Flowers small, *c.* 4 mm across, 3–9 per cyme, spreading; stalks with adpressed bristles. Calyx with adpressed bristles, lobes triangular, acute or acuminate. Petals erect, pink to off-white. Stamens 20 (rarely 16); anthers white, filaments pink. Fruit *c.* 5 mm, spreading or pendent, nearly spherical, red, shiny; nutlets 3–5, usually 5. Diploid, variable. *China (Yunnan).* H3. Spring.

36. C. foveolatus Rehder & Wilson.

Leaf-blade 3–8 cm, thin, broadly-elliptic to ovate, edges with adpressed bristles, apex acuminate, upper surface dull, mid-green, lower surface with adpressed bristles. Flowers 3–7 per cyme. Fruit black. Apomictic. *China (Hubei).* H2. Spring.

Rare in cultivation.

11. Series **Bullati** Flinck & Hylmö.

1a. Fruit orange-red or red   2
 b. Fruit dark maroon or purplish black   5
2a. Fruit orange-red   3
 b. Fruit red   4
3a. Leaves thin, usually 1.5 times as long as wide, lower surface becoming thinly hairy; nutlets 3 or 4 (rarely 5) **39. boisianus**
 b. Leaves thick or moderately thick, usually twice as long as wide, lower surface persistently felted or shaggy; nutlets 4 or 5 (rarely 3)
42. sikangensis
4a. Leaves to 7 cm **37. bullatus**
 b. Leaves to 15 cm **38. rehderi**

5a. Leaves to 5 cm; fruit maroon
40. obscurus
b. Leaves to 12 cm; fruit purplish black
41. moupinensis

37. C. bullatus Bois (*C. bullatus* var.
floribundus Rehder & Wilson). Illustration:
Botanical Magazine, 8284 (1909) – as *C. moupinensis* forma *floribunda*).

Shrub or small tree, erect, 2.5–4 m.
Branches with a spreading open habit,
initially light brown, with greyish yellow
down, older bark becoming blackish.
Leaves deciduous, blade 3–7 cm,
moderately thick, ovate to oblong-elliptic,
apex acute, upper surface shiny, dark
green, blistered, slightly downy, lower
surface shaggy with greyish green hairs,
densely on the veins; stalks 3–6 mm.
Flowers 12–30 per cyme, spreading; stalks
with greyish downy hairs. Calyx with
sparse adpressed bristles, lobes short,
triangular. Petals erect, red, green and off-
white. Stamens 20; anthers white. Fruit
6–8 mm, spreading, obovoid to nearly
spherical, red, somewhat shiny; nutlets 5
(rarely 4). Apomictic. *China (Sichuan)*. H2.
Early summer.

'Firebird': Leaf-blade somewhat convex,
lower surface with whitish down. Cymes
compact. Fruit 8–10 mm obovoid, orange-
red. H2. Early summer.

38. C. rehderi Pojarkova (*C. bullatus* var.
macrophyllus Rehder & Wilson).
Shrub 4–5 m. Branches with older bark
not so dark as *C. bullatus*. Leaf-blade
5–15 cm, apex acuminate, lower surface
with only a few hairs; stalks 1–4 mm.
Calyx nearly hairless, lobes with hairy
edges. Fruit 8–11 mm, rich red. *China
(Sichuan)*. H2. Early Summer.

39. C. boisianus Klotz.
Shrub 2.5–3 m. Branching habit more
dense. Leaf-blade 3–6 cm, thin, elliptic-
ovate, lower surface becoming thinly hairy;
stalks 2–3 mm. Flowers 9–18 per cyme.
Calyx with adpressed bristles. Fruit bright
orange-red; nutlets 3–4 (rarely 5). *China
(Sichuan)*. H3. Early summer.

40. C. obscurus Rehder & Wilson.
Shrub 1.5–2 m. Leaf-blade 3–5 cm,
ovate, apex acute or acuminate, upper
surface dull. Fruit maroon, nutlets 3 or 4.
China (Sichuan). H2. Early summer.

41. C. moupinensis Franchet.
Shrub to 3 m. Leaf-blade to 12 cm, apex
acuminate. Fruit purple-black; nutlets 5.
China (Sichuan). H2. Early summer.

42. C. sikangensis Flinck & Hylmö.
Height 2–3 m. Branches initially with
adpressed bristles, older bark greyish. Leaf-
blade 2–5 cm, thick, ovate to elliptic-ovate,
apex acuminate or rarely acute, lower
surface shaggy to felted. Flowers 6–15 per
cyme. Fruit *c.* 1 cm, obovoid, reddish
orange; nutlets 4 or 5 (rarely 3). *China
(Sichuan)*. H2. Early summer.

12. Series **Franchetioides** Flinck &
Hylmö.

1a. Leaves evergreen or semi-evergreen,
 lower surface with silvery white
 felted hairs; very late flowering and
 fruiting 2
 b. Leaves deciduous, lower surface with
 greyish felted hairs; medium to early
 flowering and fruiting 5
2a. Shrub 1–1.5 m; leaves to 2 cm; fruit
 spherical, red **46. amoenus**
 b. Shrub 1.5–3 m; leaves 2–6 cm; fruit
 obovoid or spherical, orange-red 3
3a. Leaves to 6 cm; anthers white
 44. sternianus
 b. Leaves to 3.5 cm; anthers pale
 purple 4
4a. Calyx-lobes triangular-acuminate;
 nutlets 3 (rarely 2) **43. franchetii**
 b. Calyx-lobes drawn out into a long
 mucro; nutlets 2 **45. wardii**
5a. Shrub to 3 m; fruit red
 47. dielsianus
 b. Shrub *c.* 1 m (rarely 1.5 m); fruit
 orange-red **48. splendens**

43. C. franchetii Bois. Illustration: *Revue
Horticole*, 379–81 (1902); *Botanical
Magazine*, 8571 (1914).
Shrub, erect, 1.5–3 m. Branches slender,
spreading, arched, initially with yellowish
brown felted hairs. Leaves semi-evergreen;
blade 2–3.5 cm, thick, slightly wrinkled,
elliptic to ovate, apex acute or acuminate,
upper surface shiny, sparsely hairy, lower
surface with silvery white felted hairs;
stalks 1–3 mm with felted hairs. Flowers
5–15 per long-stalked cyme, erect and
spreading; stalks with downy hairs. Calyx
with long silky hairs, lobes acuminate,
terminating in a long narrow point.
Petals erect, red and off-white. Stamens
20; anthers pale purple; filaments
white. Fruit 6–9 mm, spreading, obovoid,
orange to light red, downy; nutlets 3
(rarely 2). Apomictic. *China (Yunnan)*. H3.
Summer.
Often confused with *C. sternianus*, which
now seems to be the more commonly
grown of the two species.

44. C. sternianus Boom (*C. franchetii* var.
sternianus Turrill). Illustration: Stace, New
flora of the British Isles, 459 (1991).
Branches stiffly erect. Leaf-blade
2.5–6 cm, very thick. Cymes shorter
stalked, more erect. Calyx with woolly
hairs, lobes triangular, sometimes
mucronate. Petals fringed. Anthers white.
Fruit spherical, 8–11 mm, dark orange-red;
nutlets 3 (rarely 4). *China (Yunnan)*. H3.
Summer.

45. C. wardii W.W. Smith (*C.* 'Gloire de
Versailles').
Shrub 1.5–2.5 m. Branches thin and
very graceful. Leaves thin. Calyx-lobes
often drawn out into long points. Nutlets
2. *China (Xizang)*. H3. Summer.
Plants grown as *C. wardii* are usually
C. sternianus. *C. wardii* is very rare in
cultivation.

46. C. amoenus Wilson.
Shrub 1–1.5 m. Branching dense. Leaves
evergreen, blade 1–2 cm; stalks 2–5 mm.
Flowers 6–10 per cyme, compact, more
erect. Petals semi-spreading, white with
pink-red. Anthers pink-purple. Fruit
5–6 mm, spherical, red; nutlets 2 or 3
(rarely 4). *China (Yunnan)*. H3. Summer.

47. C. dielsianus Pritzel.
Shrub 2–3 m. Branches erect, very long
and thin, often with fan-like branching at
tips. Leaves deciduous, blade 1.5–2.5 cm,
ovate to obovate, lower surface with felted
yellowish grey hairs. Flowers 3–7 per
cyme. Calyx downy, lobes acuminate with
long mucros. Anthers white. Fruit *c.* 6 mm,
almost spherical, rich red; nutlets 3–4.
China. H2. Early summer.
'Rubens' (*C. rubens* misapplied, not W.W.
Smith). Height 1–1.5 m. Branches erect
and spreading. Leaves deciduous; blade
elliptic, apex acute to acuminate, lower
surface with grey felted hairs. Fruit
obovoid, shiny, rich red; nutlets 3. *Origin
unknown*. H2. Early summer.

48. C. splendens Flinck & Hylmö
(*C.* × 'Sabrina').
Shrub spreading and ascending to
1–2 m. Leaves deciduous; blade 1–2 cm,
thin, broadly elliptic to nearly circular,
upper surface somewhat shiny, bright
green, sparsely hairy, lower surface with
yellowish grey hairs. Flowers 3 (rarely
2–7) per cyme. Anthers white. Fruit
9–11 mm, nearly spherical, shiny, bright
orange-red; nutlets usually 4. *China
(Sichuan)*. H2. Early summer.

13. Series **Zabelioides** Flinck & Hylmö

1a. Leaves to 4 cm, elliptic to ovate, apex acute; flowers off-white to pale pink; fruit obovoid, dark red to nearly maroon **49. zabelii**

b. Leaves to 3 cm, broadly ovate, apex obtuse; flowers pink and red; fruit spherical, red **50. fangianus**

49. C. zabelii Schneider. Illustration: *Die Gartenwelt* **25**: 429(1931).

Shrub, erect, 1.5–2 m. Branches slender, spreading initially with densely, downy hairs. Leaves deciduous, blade 1.5–4 cm thin, ovate to elliptic, apex acute (sometimes obtuse), upper surface dull light green, sparsely hairy, lower surface shaggy with yellowish grey hairs; stalks to 3 mm, thickly hairy. Flowers 4–12 per cyme, loose, spreading, stalks thickly hairy. Calyx shaggy, lobes broadly triangular, obtuse. Petals erect, off-white to pale pink. Stamens 20 (rarely 18); anthers white; filaments pale pink. Fruit 7–8 mm pendent, obovoid, with downy hairs, dark red to nearly maroon; nutlets 2. Apomictic. *China*. H2. Early summer.

50. C. fangianus Yü.

Shrub to 3 m. Leaf-blade 1–3 cm, broadly ovate, apex usually obtuse. Flowers 5–15 per cyme, more compact. Petals pink and red. Fruit spherical, rich-red. *China (Hubei)*. H2. Early summer.

14. Series **Megalocarpi** Pojárkova.

1a. Leaves with lower surface persistently hairy; flowers white **51. megalocarpus**

b. Leaves with lower surface hairless; flowers pink **52. roseus**

51. C. megalocarpus Popov. Illustration: Flora of Kazakhstana **4**: 399 (1961).

Shrub, erect, 1.5–2.5 m. Branches thin, spreading, initially dark purple-brown, downy. Leaves deciduous, blade 2–5 cm, thin, ovate, apex acute, mucronate, upper surface shiny light green, initially sparsely hairy, lower surface somewhat shaggy; stalk 3–5 mm with shaggy hairs. Flowers 3–20 per cyme, erect, stalks with sparse downy hairs. Calyx nearly hairless, lobes acute. Petals semi-spreading, white. Stamens 20; anthers and filaments white. Fruit *c*. 1 cm, erect, spherical, red, shiny; nutlets 2 (rarely 3). Apomictic. *C Asia (Tianshan Mts)*. H2. Early spring.

52. C. roseus Edgeworth. Figure 99(15), p. 183.

Leaf-blade ovate-elliptic, both surfaces hairless. Cymes flat-topped. Calyx hairless; lobes acute to obtuse. Petals bright, clear pink. Nutlets 2. *Himalaya (Kashmir, Afghanistan)*. H2. Spring.

15. Series **Multiflori** Yü.

1a. Fruit carmine or red 2

b. Fruit maroon 3

2a. Leaves initially sparsely hairy, soon hairless; leaf-stalk and calyx hairless; flowers to 20 per cyme **53. multiflorus**

b. Leaves with lower surface persistently sparsely hairy; leaf-stalk and calyx initially hairy; flowers to 10 per cyme **54. hupehensis**

3a. Leaves thick, dark green, lower surface, leaf-stalk and calyx with felted hairs; very late flowering and leaf fall **56. veitchii**

b. Leaves thin, green or light green, lower surface, leaf-stalk and calyx persistently sparsely hairy or hairless; early or medium to early flowering and leaf fall 4

4a. Leaf-stalk and calyx hairless; flowers and petals flat, all white; early flowering **55. calocarpus**

b. Leaf-stalk and calyx sparsely hairy; flowers wide bell-shaped, petals concave, white, bases pale purple; medium-late flowering **57. przewalskii**

53. C. multiflorus Bunge. Illustration: Ledebour, Icones plantarum novarum vel imperfecti cognitarum, floram rossicam, **3**: t. 274 (1831).

Shrub or small tree, erect, 3–5 m. Branches slender, arched, initially shiny purple-brown, slightly downy. Leaves deciduous; blade 2–5 cm, thin, broadly ovate to nearly circular, apex usually obtuse, upper surface dull, yellowish green, hairless, lower surface initially sparsely hairy; stalk 3–10 mm, hairless. Flowers 8–10 mm across, erect and spreading, 10–20 per cyme; stalks nearly hairless. Calyx hairless, lobes triangular, acute or obtuse. Petals spreading, white. Stamens 20; anthers and filaments white. Fruit 1–1.2 cm, spreading, spherical, carmine, shiny, succulent; nutlets apparently 1 (2 nutlets joined). Apomictic. *C Asia (Kazakhstan)*. H2. Spring.

54. C. hupehensis Rehder & Wilson.

Shrub 1.5–3 m. Leaf-blade 15–35 mm, elliptic to ovate, apex usually acute, upper surface green, lower surface sparsely hairy.

Flowers *c*. 1.3 cm across, 5–10 per cyme, stalks sparsely hairy. Fruit carmine to red. *China (Hubei)*. H3. Spring.

55. C. calocarpus Flinck & Hylmö (*C. multiflorus* var. *calocarpus* Rehder & Wilson).

Leaf-blade broadly elliptic, apex acute; upper surface initially sparsely hairy, lower surface initially shaggy, becoming sparsely hairy. Flowers to 1.5 cm across; petals flat, sometimes reflexed. Fruit dark red to maroon. *China (Sichuan)*. H2. Early spring.

56. C. veitchii Klotz (*C. racemiflorus* var. *veitchii* Rehder & Wilson).

Shrub 3–4 m. Branches wide-spreading. Leaf-blade thick, broadly ovate, sometimes elliptic, apex acute, upper surface dark green, initially shaggy, becoming sparsely so, lower surface and leaf-stalks with felted hairs. Flowers 1.2–1.5 cm across, 5–15 per cyme. Calyx with felted hairs. Petals somewhat concave, pink in bud. Fruit 1.2–1.3 cm, dark red to maroon, frequently splitting. *China (Hubei)*. H2. Early summer.

57. C. przewalskii Pojárkova.

Shrub 2–5 m. Leaf-blade to 6 cm, broadly ovate, apex acute to acuminate; stalks sparsely hairy. Flowers *c*. 1.3 cm across. Calyx campanulate, sparsely hairy. Petals concave, bases pale purple. *China (Gansu)*. H2. Spring.

16. Series **Racemiflori** Klotz.

1a. Shrub usually less than 1 m; leaves circular, to 2 cm (rarely 3 cm) **59. nummularius**

b. Shrub to 3.5 m; leaves ovate to elliptic, to 5 cm 2

2a. Leaf-apex acuminate or acute **58. racemiflorus**

b. Leaf-apex acute or obtuse **60. tauricus**

58. C. racemiflorus Koch.

Shrub, erect, 1.5–3 m. Branches initially shiny dark purple, shaggy with grey hairs. Leaves deciduous, thin, blade 2–5 cm, broadly ovate to elliptic, apex acute or acuminate, mucronate, upper surface dull, hairless, lower surface with greyish felted hairs; stalks 4–8 mm, shaggy. Flowers *c*. 8 mm across, 6–15 per cyme, compact, erect, flower-stalks and calyx with felted hairs; lobes triangular, acute. Petals spreading, white. Stamens 20; anthers white; filaments white. Fruit 8–10 mm, spreading, nearly spherical, red, dull; nutlets 2. Apomictic. *Caucasus*. H2. Early summer.

59. C. nummularius Fischer & Meyer.

Shrub to 1 m. Leaf-blade 2 cm (rarely to 3 cm), circular (rarely broadly elliptic), apex mucronate, upper surface sparsely hairy. Flowers 3–7 per cyme, densely compact. Petals red in bud. Fruit 7–8 mm. *W Asia*. H3. Early summer.

60. C. tauricus Pojárkova.

Leaf-blade 1.5–4 cm, ovate or elliptic, apex obtuse or acute, upper surface hairless. Flower-stalks and calyx with sparse hairs. Petals red in bud. Fruit 7–8 mm. *Ukraine (Crimea)*. H2. Early summer.

17. Series **Insignes** Pojárkova.

1a.	Leaves circular	2
b.	Leaves elliptic or obovate	3
2a.	Shrub or small tree 3–6 m; leaves to 6 cm	**61. ellipticus**
b.	Shrub 1.5–2 m; leaves to 2.5 cm	**62. hissaricus**
3a.	Leaves obovate, hairless beneath, except midrib sparsely hairy; nutlets spherical, 2	**65. bacillaris**
b.	Leaves lanceolate to elliptic, at least initially hairy beneath; nutlets obovoid, 1 or 2	4
4a.	Leaves lanceolate-elliptic, lower surface initially sparsely hairy; nutlets 1 or 2	**66. transens**
b.	Leaves elliptic, lower surface persistently hairy; nutlets 2	5
5a.	Cymes compact; fruit black	**63. affinis**
b.	Cymes loose; fruit red to purple	**64. gamblei**

61. C. ellipticus Loudon (*C. lindleyi* Schneider; *C. insignis* Pojarkova).

Shrub or small tree, erect, 3–6 m. Branches wide-spreading, initially red-brown to dark brown, downy. Leaves deciduous, moderately thick, blade 2.5–6 cm, circular (rarely broadly elliptic to obovate), apex mucronate, upper surface hairless, or sparsely hairy on midrib, lower surface with greyish white hairs; leaf-stalk to 8 mm, sparsely hairy. Flowers *c.* 1 cm across, 5–20 per cyme, erect, stalk and calyx with felted hairs, lobes broadly triangular, acute or obtuse. Petals spreading, white, with tufts of hair at base, pink in bud. Stamens 20; anthers and filaments white. Fruit 7–9 mm, spreading, spherical, black with a bluish waxy bloom, apex open; nutlets 1 or 2. Apomictic. *W Himalaya*. H3. Early summer.

62. C. hissaricus Pojárkova.

Shrub 1.5–2 m, densely branched. Leaf-blade 1.5–2.5 cm (rarely longer), circular. Flowers *c.* 8 mm across, 5–12 per cyme. Petals usually white in bud. Fruit 6–7 mm. *C Asia (Tadzhikistan)*. H3. Early summer.

63. C. affinis Lindley.

Shrub or small tree to 8 m. Leaf-blade 4–10 cm, thin, elliptic. Flowers to 15 per cyme, compact. Anthers mauve. Fruit 1–1.2 cm, cylindric, purple becoming black, succulent; nutlets 2, obovate. *Himalaya*. H3. Early summer.

64. C. gamblei Klotz.

Very similar to *C. affinis* but with cymes loose, and the fruit remaining red to purple. *Himalaya (Sikkim, Bhutan)*. H3. Early summer.

65. C. bacillaris Lindley.

Branches suckering, arched and pendent, graceful. Leaf-blade 4.5–10 cm, obovate, lower surface hairless, except for sparsely hairy midrib. Calyx sparsely hairy. Petals with large tufts of hairs at base. Anthers mauve. Fruit 8 mm; nutlets 2, spherical. *Himalaya (Kashmir to Bhutan)*. H3. Early summer.

66. C. transens Klotz.

Small tree, to 5 m. Leaf-blade 5–10 cm, reddish when young, lanceolate to elliptic, lower surface initially sparsely hairy. Anthers dark mauve. Fruit *c.* 1 cm, maroon to purple-black; nutlets 1 or 2. *China (Yunnan)*. H3. Early summer.

Probably more closely related to series *Hebephylli*. Frequently grown as *C. affinis* or *C. bacillaris*.

18. Series **Hebephylli** Klotz.

1a.	Shrub to 2 m, densely branched; leaves elliptic-lanceolate, apex acute	**69. ludlowii**
b.	Shrub to 3 m, branches slender, arched; leaves broadly elliptic to almost circular, apex obtuse rarely acute	2
2a.	Fruit red	**67. hebephyllus**
b.	Fruit dark maroon	**68. monopyrenus**

67. C. hebephyllus Diels (*C. tibeticus* Klotz). Illustration: *Wissenschaftliche Zeitschrift der Friedrich-Schiller-Universitat, Jena*, 334 (1968).

Shrub, erect, 1.8–3 m. Branches spreading at right angles, graceful, initially shiny purple, downy. Leaves deciduous; blade 1.5–4 cm, thin, broadly obovate, apex obtuse, shortly mucronate, upper surface dull, initially reddish brown, later light green, hairless, lower surface initially sparsely hairy; stalks to 5 mm, sparsely hairy. Flowers *c.* 7 mm across, 6–20 per cyme, erect; stalks slender, sparsely hairy. Calyx initially sparsely hairy, lobes broadly triangular, acute. Petals spreading, white, sometimes with small tufts of hair at base. Stamens 20; anthers purple to black, filaments white. Fruit 6–7 mm, pendent, depressed spherical, carmine, becoming red, dull; nutlets 1 per fruit. Apomictic. *China (Yunnan, Xizang)*. H3. Early summer.

Plants grown as *C. hebephyllus* are usually *C. monopyrenus*.

68. C. monopyrenus Flinck & Hylmö (*C. hebephyllus* var. *monopyrenus* W.W. Smith).

Branches initially dark brown. Leaf-blade moderately thick, 2.5–5 cm, broadly elliptic, apex obtuse, sometimes acute, mucronate, upper surface dark green, reddish at fruiting, lower surface initially felted. Flowers *c.* 1 cm across. Fruit *c.* 1 cm, cylindric, becoming dark maroon. *China (Yunnan)*. H3. Early summer.

69. C. ludlowii Klotz.

Shrub to 2 m; densely branched. Leaves semi-evergreen; blade 1.7–3 cm, elliptic to elliptic-lanceolate, apex acute, upper surface rich green; stalks to 4 mm. Flowers *c.* 1 cm across, 4–12 per cyme, stalks thickly hairy. Fruit 8–9 mm, salmon-pink and orange. *Himalaya (Bhutan, Nepal, Xizang)*. H3. Early summer.

19. Series **Chaenopetalum** Koehne.

1a.	Leaves deciduous, thin, surface flat	**70. frigidus**
b.	Leaves evergreen or semi-evergreen, moderately thin, surface wrinkled	**71. × watereri**

70. C. frigidus Lindley. Illustration: *Edwards's Botanical Register* **15**: 1229 (1829).

Large shrub or tree, erect; 5–15 m. Branches spreading, widely arched, initially dark brown, downy. Leaves deciduous; blade 6–15 cm, thin, narrowly elliptic to obovate, apex acute, upper surface flat, dull green, hairless except for sparsely hairy midrib, lower surface initially shaggy; stalks 5–8 mm, downy. Flowers *c.* 7 mm across, 20–60 per cyme, erect and spreading; stalks and calyx felted, lobes triangular, acute. Petals spreading, white (also white in bud). Stamens 20; anthers purple; filaments white. Fruit *c.* 6 mm, spreading, nearly spherical, red; nutlets 2.

Diploid, variable. *Himalaya.* H3. Early summer.

The largest species in the genus. Many varieties cultivated. 'Fructo-Luteo' has creamy-yellow fruit. 'Anne Cornwallis' has thinner, more wavy leaf-blades than *C. frigidus*; fruit pear-shaped, orange.

71. C. × watereri Exell. Illustration: Stace, New flora of the British Isles, 459 (1991).

Shrub or small tree, to 6 m; branches vigorous, pendent, initially dark grey-brown. Leaves evergreen or semi-evergreen; blade 5–9 cm, moderately thick, narrowly elliptic, apex acute-acuminate, upper surface somewhat shiny, dark green, wrinkled, lower surface initially downy; stalks 5–7 mm, densely downy. Flowers 30–50 per cyme; stalks densely downy. Calyx densely downy, lobes broad, triangular. Fruit 8–9 mm; nutlets 2–4 (rarely 5). *Garden origin.* H3. Early summer.

The putative hybrid between *C. frigidus* and *C. salicifolius* to which the type name 'John Waterer' is given. 'Salmon Spray' has fruit 6–7 mm, pinkish orange. 'Coral Bunch' is similar to 'Salmon Spray', but with fewer, larger fruit of a lighter shade. 'Cornubia' has the fruit red, very abundant. 'Aldenhamensis' is a narrowly erect shrub to 4 m; fruit 5–7 mm, rich-red. 'St Monica' is a shrub, almost evergreen; branches initially very shiny purple-brown; fruit in large pendent bunches.

'Vicarii' has leaves to 15 cm; fruit dark-red, 5–7 mm.

20. Series **Salicifolii** Yü.

1a. Prostrate, creeping shrub with rooting branches; leaves less than 3 cm   **2**
 b. Erect shrub, to 5 m; leaves to 10 cm   **3**
2a. Leaves to 3 cm; nutlets usually 5   **77. dammeri**
 b. Leaves less than 1.5 cm; nutlets usually 4   **78. radicans**
3a. Leaves strongly wrinkled; nutlets 2–5   **4**
 b. Leaves only slightly wrinkled; nutlets 2 (rarely 3)   **6**
4a. Leaves broadly elliptic; nutlets 2   **74. rugosus**
 b. Leaves elliptic-lanceolate; nutlets 3–5   **5**
5a. Leaves elliptic-lanceolate, lower surface with felted hairs to somewhat thinly woolly   **72. salicifolius**
 b. Leaves narrowly lanceolate, lower surface extremely woolly   **76. floccosus**
6a. Leaves hairless, lower surface glaucous   **73. glabratus**
 b. Leaves with felted hairs beneath   **75. henryanus**

72. C. salicifolius Franchet. Illustration: Havens Planteleksikon, 169 (1987).

Shrub, erect, 2–5 m. Branches erect and spreading, initially red purple to greenish, thinly downy. Leaves evergreen; blade 3–10 cm, thick, elliptic-lanceolate, edges recurved, apex acute to acuminate, upper surface shiny, dark green, wrinkled, hairless, lower surface with felted to thinly woolly hairs, sometimes hairless by autumn. Midrib often reddish; stalks 5–8 mm, often reddish with felted hairs. Flowers *c.* 6 mm across, 30–100 per cyme, erect-spreading, dense; stalks with felted hairs. Calyx with downy hairs, lobes shortly triangular, acute. Petals spreading, white. Stamens 20, anthers purple to black, filaments white. Fruits *c.* 5 mm, spreading, nearly spherical, bright-red, shiny; nutlets 3–5. Many varieties cultivated. Diploid, variable. *China (Sichuan).* H3. Early summer.

'Exburyensis' has the leaf-blade 8–12 cm, upper surface mid-green; fruit golden-yellow.

'Rothschildianus' has the leaf-blade upper surface pale-green; fruit cream to pale yellow. 'Pink Champagne' is similar to the above two cultivars; fruit yellow, becoming pinkish. 'Repens' is prostrate to 1.5 m; branches spreading; leaf-blade 2.5–3.5 cm; stalk to 4 mm; flowers 3–10 per cyme. 'Herbstfeuer' ('Autumn Fire') is prostrate, 20–25 cm; branches spreading; leaf-blade 4–6 cm, colouring well in autumn; flowers 5–12 per cyme. 'Gnom' is also prostrate, to 15 cm; branches spreading; leaves evergreen, blade to 2.5 cm, upper surface very dark green; fruit 4–6 mm. 'Parkteppich' ('Park Carpet') is a shrub to 1 m; branches spreading; leaf-blade 2–3 cm; fruit light red. 'Hybridus Pendulus' is prostrate, 20–25 cm; branches spreading; as with the previous four cultivars it can be trained (or top-grafted onto a *C. × watereri* cultivar) to form a pendent bush or small tree. May also be trained against a wall; it is probably a hybrid between *C. salicifolius* and *C. dammeri*.

73. C. glabratus Rehder & Wilson.

Branching dense. Leaf-blade 4–7 cm, upper surface hairless, lower surface glaucous, hairs only on midrib; stalks sparsely hairy. Flower-stalks initially hairy. Calyx sparsely hairy; lobes broadly triangular, obtuse. Fruit 4–5 mm, bright orange-red; nutlets 2. *China (Sichuan).* H3. Early summer.

74. C. rugosus Pritzel.

Leaf-blade very thick, broadly elliptic, apex acute, upper surface very strongly wrinkled, lower surface densely shaggy. Flower-stalk and calyx with shaggy hairs. Nutlets 2. *C China.* H3. Early summer.

Possibly not in cultivation. Plants grown as *C. rugosus* are usually *C. salicifolius* or other species.

75. C. henryanus Rehder & Wilson.

Shrub 2–4 m; robustly branched. Leaves semi-evergreen, blade 8–12 cm, oblong-elliptic to oblong-lanceolate, upper surface very sparsely hairy, lower surface remaining felted and woolly, leaf- and flower-stalks shaggy. Calyx densely shaggy, lobes triangular, acute. Fruit 6–7 mm, dark red; nutlets 2 or 3. *China (Hubei).* H3. Early summer.

Rare in cultivation. Plants grown as this species are usually *C. salicifolius.*

76. C. floccosus Flinck & Hylmö (*C. salicifolius* var. *floccosus* Rehder & Wilson).

Shrub 2–4 m. Leaf-blade narrowly lanceolate, lower surface with a dense covering of woolly hairs, which fall away in tufts. Apomictic. *China (Sichuan).* H3. Early summer.

Uncommon in cultivation. Plants grown as this species are usually *C. salicifolius.*

77. C. dammeri Schneider (*C. humifusus* Veitch).

Prostrate. Branches long, creeping, rooting at nodes. Leaf-blade 1.5–3 cm, broadly obovate to elliptic, apex usually obtuse, upper surface with veins slightly impressed, lower surface sparsely hairy, veins somewhat prominent; stalks sparsely hairy. Flowers *c.* 9 mm across, solitary or 2–4 per cyme; stalks to 1.2 cm, initially sparsely hairy. Calyx initially sparsely hairy. Fruit 5–7 mm, shy-fruiting; nutlets usually 5. *China (Hubei).* H3. Early summer.

'Major' is more robust; leaf-blade 2.5–3.5 cm, almost circular. Often found as *C. dammeri* var. *radicans.*

78. C. radicans Klotz (*C. dammeri* var. *radicans* Schneider).

Prostrate. Branches long, creeping, rooting at nodes. Leaf-blade 1–1.5 cm, obovate to elliptic, upper surface mid-green, initially with a few long hairs, somewhat dull, veins slightly impressed, lower surface initially sparsely hairy; stalks 2–4 mm, initially sparsely hairy. Flowers *c.* 9 mm across, solitary or in pairs; stalks sparsely hairy. Calyx sparsely hairy, lobe apex acute to acuminate. Fruit red; nutlets 3 or 4, usually 4. *China (Sichuan).* H3. Early summer.

Often grown as *C. dammeri* 'Eichholz' or 'Oakwood'.

C. × suecicus Klotz. Prostrate, 40–60 cm. Branches arched, rooting at nodes. Leaves semi-evergreen; blade 1–2 cm, obovate to oblong, apex obtuse, upper surface not wrinkled, lower surface somewhat glaucous, initially sparsely hairy; stalk and calyx sparsely hairy. Flowers *c.* 8 mm across, to 6 per cyme. Fruit red, uncommon. H3. Early summer.

This is probably the hybrid between *C. conspicuus* and *C. dammeri*, to which the commonly grown 'Skogholm' belongs. 'Coral Beauty' is similar to 'Skogholm' but more leafy, and with more abundant coral-red fruit. 'Jürgl' is prostrate, to 50 cm, branches widely arched; leaf-blade to 1.5 cm, ovate, upper surface slightly wrinkled; flowers solitary or in pairs; fruit to 9 mm, red; nutlets 2 or 3.

21. Series Pannosi Flinck & Hylmö.

1a.	Leaves hairless; calyx-lobes hairless	2
b.	Leaves persistently hairy beneath; calyx-lobes hairy	3
2a.	Leaves to 8 cm	**80. glaucophyllus**
b.	Leaves to 4 cm	**86. meiophyllus**
3a.	Leaves less than 3 cm; flowers to 12 per cyme	**79. pannosus**
b.	Leaves more than 3 cm; flowers more than 12 per cyme	4
4a.	Leaves narrowly elliptic or narrowly obovate, more than twice as long as wide	**82. harrovianus**
b.	Leaves broadly elliptic or broadly obovate, less than twice as long as wide	5
5a.	Calyx only sparsely hairy with long silky hairs	**85. serotinus**
b.	Calyx with felted hairs	6
6a.	Leaves thick, obovate or broadly elliptic, upper surface with impressed and lower surface with raised veins	**83. lacteus**
b.	Leaves thin or moderately thick, elliptic, surfaces flat	7

7a.	Flowers to 30 per cyme; calyx-lobes woolly	**81. vestitus**
b.	Flowers to 75 per cyme; calyx-lobes felted with silky hairs	**84. turbinatus**

79. C. pannosus Franchet. Illustration: *Acta Phytotaxonomica et Geobotanica* **13**: 225–37(1967).

Shrub, erect, 2–3 m. Branches long and slender, initially with whitish felted hairs. Leaves semi-evergreen; blade 1–3 cm, thick, elliptic to ovate, apex acute, mucronate, upper surface dull, greyish green, initially sparsely hairy, lower surface with whitish felted hairs; stalks 3–7 mm with felted hairs. Flowers to 8 mm, 6–12 per cyme, erect to spreading; stalk and calyx plus lobes with felted hairs; lobes triangular acuminate, mucronate. Petals spreading, white. Stamens 20; anthers purple-black; filaments white. Fruit 6–8 mm, spreading, nearly spherical, red, shiny, apex persistently hairy; nutlets 2. Apomictic. *China (Yunnan).* H3. Summer. Late flowering, winter fruiting.

80. C. glaucophyllus Franchet.

Shrub 2–4 m. Leaf-blade 3–8 cm, ovate or elliptic, apex acute, upper surface green, hairless, lower surface glaucous, hairless; stalks 5–10 mm. Flowers *c.* 6 mm across, 15–40 per cyme; stalks with sparse downy hairs. Calyx-lobes hairless. Fruit 6–7 mm, obovoid, orange, hairless. *China (Yunnan).* H3. Summer.

81. C. vestitus Flinck & Hylmö (*C. glaucophyllus* var. *vestitus* W.W. Smith).

Shrub to 4 m. Leaf-blade to 6 cm, elliptic, upper surface green. Flowers to 30 per cyme. Fruit with woolly hairs. *China (Yunnan).* H3. Summer.

82. C. harrovianus Wilson.

Shrub to 4 m. Leaf-blade 3–5 cm, elliptic or obovate, apex acute with a long point, upper surface somewhat shiny, lower surface of older leaves finally hairless; stalks 4–5 mm, with silky hairs. Flowers *c.* 6 mm across, 20 to 60 per cyme; stalk and calyx with silky hairs. Fruit 4–5 mm, hairy. *China (Yunnan).* H3. Summer.

83. C. lacteus W.W. Smith.

Shrub or small tree, to 5 m. Branches initially reddish brown, shaggy to felted with yellowish hairs. Leaf-blade 3.5–9 cm, obovate or broadly elliptic, apex acute or obtuse, upper surface somewhat shiny, with veins impressed, lower surface with felted yellowish hairs, becoming thinner with age. Flowers *c.* 100 per cyme. Petals

milky white. Fruit to 6 mm, obovoid. *China (Yunnan).* H3. Summer.

84. C. turbinatus Craib.

Shrub 3–5 m. Leaf-blade 3.5–6 cm, green. Flowers *c.* 6 mm across, to *c.* 75 per cyme, stalk and calyx, including lobes, thickly covered with silky hairs. Fruit 4–5 mm, obconical, with silky hairs. *China (Yunnan).* H3. Summer.

85. C. serotinus Hutchinson.

Shrub or small tree, 3–5 m. Leaf-blade 4–7 cm, elliptic or obovate-elliptic, apex acute, with a long fine point, upper surface somewhat shiny, green, hairless, lower surface becoming almost hairless; stalks to 1 cm, with silky hairs. Flowers 5–6 mm, to 40 per cyme, loose; stalk and calyx, including lobes, with silky hairs. Fruit obovoid to nearly spherical, with sparse silky hairs. *China (Yunnan).* H3. Summer.

86. C. meiophyllus Klotz (*C. glaucophyllus* var. *meiophyllus* W.W. Smith).

Branching dense. Leaves evergreen; blade 2–4 cm, initially reddish, becoming green, broadly elliptic, apex obtuse (rarely acute), both surfaces hairless. Flowers 5–6 mm, 15–40 per cyme. Calyx-lobes hairless. Fruit orange-red, hairless. *China (Yunnan).* H3. Summer.

22. Series Microphylli Yü.

1a.	Flowers usually 2–7 per cyme	2
b.	Flowers usually solitary	6
2a.	Calyx with dense adpressed bristles	**93. buxifolius**
b.	Calyx with sparse adpressed bristles	3
3a.	Leaves to 3 cm	**94. marginatus**
b.	Leaves less than 1.5 cm	4
4a.	Leaves narrowly elliptic, sometimes ovate, apex acute to acuminate with a long mucro	**97. lidjiangensis**
b.	Leaves elliptic, broadly-elliptic, to broadly obovate, apex obtuse or acute, with a short mucro	5
5a.	Leaves 5–8 mm	**87. microphyllus**
b.	Leaves 8–15 mm, apex acute	**92. prostratus**, see 'Ruby'
6a.	Upper surface of leaves pale green, dull	**95. congestus**
b.	Upper surface of leaves medium to dark green, shiny, (or only slighty dull), hairless, hairy or initially sparsely hairy or bristly	7
7a.	Leaf-blade narrow, 2–5 times as long as wide	8
b.	Leaf-blade broad, less than twice as long as wide (usually 1.5)	10

8a. Leaves narrowly elliptic, upper surface only slightly shiny, papillose **88. conspicuus**

b. Leaves oblong to oblanceolate or narrowly obovate, upper surface very shiny, not papillose   9

9a. Leaves to 1.5 cm; fruit 8–10 mm **98. integrifolius**

b. Leaves less than 8 mm; fruit 4–5 mm **99. linearifolius**

10a. Shrub to 2 m, robust, erect or prostrate; leaves to 2 cm **89. rotundifolius**

b. Shrub prostrate to creeping; leaves less than 1.5 cm   11

11a. Flowers to 1 cm, often pink in bud; fruit 8–10 mm   12

b. Flowers to 9 mm, pink or white in bud; fruit 4–6 mm   13

12a. Shrub to 20 cm, branches flat to the ground **90. cochleatus**

b. Shrub to 50 cm, branches somewhat raised above ground **92. prostratus**

13a. Leaves elliptic, upper surface shiny, midrib initially with long adpressed hairs; flower buds often pinkish **91. cashmiriensis**

b. Leaves broadly obovate, upper surface somewhat dull, hairless; flower buds white **96. procumbens**

87. C. microphyllus Lindley. Illustration: *Edwards's Botanical Register* **13**: t. 114 (1827).

Shrub, prostrate or ascending, 50–100 cm. Branches dense, initially black-purple, with adpressed whitish bristles. Leaves evergreen, blade 5–8 × 3–5 mm, thick; broadly elliptic to broadly obovate, edges somewhat recurved, apex obtuse or acute, upper surface shiny, dark green, initially midrib sparsely hairy, lower surface with sparse adpressed bristles, glaucous, with a prominent vein pattern; stalks 1–2 mm, with adpressed bristles. Flowers *c.* 6 mm across, solitary or 2–5 per cyme, erect; stalks 1–2 mm with adpressed bristles.Calyx with adpressed bristles, lobes triangular, obtuse. Petals spreading, white, often pink in bud. Stamens 20; anthers purple to black; filaments white. Fruit 5–8 mm, depressed spherical, carmine, dull; nutlets 2. *Himalaya.* H3. Spring.

Several other species may be grown as *C. microphyllus.*

'Donard Gem' has the leaf-blade very dark-green, margin hairy; upper surface slightly shiny, sparsely hairy, lower surface

with dense adpressed hairs; fruit carmine to red, abundant.

88. C. conspicuus Marquand (*C. conspicuus* var. *decorus* Russell; *C. nanus* Klotz; *C. permutatus* Klotz; *C. pluviflorus* Klotz).

Shrub to 1.5 m. Leaf-blade 5–20 × 2–8 mm, green, narrowly elliptic, upper surface slightly shiny, initially sparsely hairy, papillose, lower surface without vein pattern, stalks 1–1.5 mm. Flowers *c.* 1 cm across, usually solitary, except at ends of branches, and then up to 5 per cyme. Calyx sparsely hairy, lobes acute or acuminate. Fruit 7–9 mm, red, shiny. Diploid, variable. *China (Xizang).* H3. Spring.

Many varieties are cultivated. 'Red Glory' is erect; to 2 m; leaf-blade 1–2 cm. 'Highlight' has the fruit light orange; nutlets 1 or 2. This is probably identical to *C. sherriffii* Klotz.

89. C. rotundifolius Lindley (*C. microphyllus* var. *uva-ursi* Lindley).

Shrub 50–200 cm. Branches more erect. Leaf-blade very thick, 7–20 × 5 × 11 mm. Flowers 1–1.3 cm across, always solitary. Calyx with sparse hairs, lobes acute. Fruit 7–10 mm, nearly spherical. Apomictic. *Himalaya.* H3. Spring.

90. C. cochleatus Klotz (*C. buxifolius* var. *cochleatus* Franchet).

Prostrate, to 20 cm. Branches downward curved, creeping, rooting at nodes. Leaf-blade 5–14 × 3–9 mm, broadly elliptic, apex obtuse. Flowers 8–10 mm across, always pink in bud. Fruit carmine-red. Apomictic. *W China (Yunnan).* H3. Spring.

Plants grown as *C. cochleatus* in the British Isles are usually *C. cashmiriensis.*

91. C. cashmiriensis Klotz.

Prostrate, to 30 cm. Branches creeping, often rooting at nodes. Leaf-blade 4–11 × 2–6 mm, elliptic, upper surface midrib initially with long adpressed hairs; stalks 1–3 mm. Flowers solitary. Fruit nearly spherical, red. Apomictic. *Himalaya (Kashmir).* H3. Spring.

92. C. prostratus Baker.

Prostrate, 50 cm. Branches long and thin, creeping. Leaf-blade 6–13 × 4–8 mm, apex not notched; stalks 2–4 mm. Flowers *c.* 1 cm across, solitary or sometimes in pairs. Fruit 9–10 mm, nearly spherical, red. *Himalaya.* H3. Spring.

'Ruby' (*C. rubens* misapplied not W.W. Smith). Branches stiffly projecting,

becoming wide-spreading. Leaf-blade 8–15 × 4–7 mm, elliptic; stalks reddish. Flowers *c.* 8 mm across, 3–9 per cyme. Fruit *c.* 8 mm, carmine. Often found as *C. sikkimensis* misapplied.

93. C. buxifolius Lindley.

Shrub to 1.5 m. Leaf-blade 1–1.8 × 5–7 mm, elliptic, upper surface with long adpressed hairs, lower surface with dense adpressed bristles; stalks 2–4 mm. Flowers 3–5 per cyme; calyx felted, lobes triangular, acute. Fruit nearly spherical, hairy. *S India (Nilgiri Hills).* H3. Spring.

The true species is rare in cultivation.

94. C. marginatus Schlechtendal.

Prostrate to erect, 2–3 m. Leaf-blade 1–2.5 cm × 5–11 mm, obovate or elliptic, apex obtuse, sometimes acute, margin with hairs. Stalk 3–5 mm. Flowers 8–10 mm across, 2–7 per cyme, white in bud. Fruit *c.* 8 mm, carmine to red; nutlets 2 (rarely 3). *Himalaya.* H3. Spring.

Often grown as *C. prostratus* var. *lanatus* of gardens.

'Eastleigh' is a shrub to 4 m; branches erect and spreading, robust; leaves to 3 cm; fruit to 1.3 cm, deep red, succulent.

95. C. congestus Baker (*C. pyrenaicus* Chancerel).

Shrub to 70 cm; densely branched, congested, often rooting at nodes. Leaf-blade 5–14 × 3–8 mm, thin, obovate, upper surface dull, pale-green, hairless, lower surface very sparsely hairy; stalks 3–5 mm, thin, often reddish. Flowers 7–9 mm across, solitary; stalk and calyx sparsely hairy. Fruit 6–10 mm, red; nutlets 2 (rarely 3). Diploid, variable. *Himalaya.* H3. Spring.

'Nanus' is mat-forming, hugging the ground; fruit rarely produced.

96. C. procumbens Klotz.

Prostrate, to 20 cm. Branches often very dense. Leaf-blade 6–12 × 4–11 mm, broadly obovate, upper surface somewhat dull, hairless, purple when very young; stalks 4–6 mm, thin, becoming hairless. Flowers 8–9 mm across, solitary, white in bud. Fruit 5–6 mm, spherical. *China.* H3. Spring.

'Streibs Findling' is probably this species.

97. C. lidjiangensis Klotz (*C. exellens* Marquand, invalid).

Shrub to 1.5 m. Branches slender, spreading and erect, initially with felted hairs. Leaf-blade 6–12 × 3–6 mm, elliptic, sometimes ovate, margin hairy, apex acute or acuminate, not notched, with a long point, upper surface with sparse adpressed bristles, lower surface with whitish densely felted hairs; stalks 2–3 mm, with felted hairs. Flowers 8–9 mm across, solitary or 2–4 (rarely to 8) per cyme; stalks 1–3 mm. Calyx felted, lobes acute or acuminate, often mucronate. Fruit 5–6 mm, obovoid or nearly spherical, carmine to red, shiny; calyx-lobes persisting as a coronet. Apomictic. *China (Yunnan)*. H3. Early summer.

Often found as *C. buxifolius*.

98. C. integrifolius (Roxburgh) Klotz
(*C. thymifolius* Baker).

Leaf-blade 7–15 × 3–6 mm, oblanceolate to oblong or obovate, apex notched, upper surface very dark green, initially with sparse adpressed bristles, lower surface with dense adpressed bristles; stalks 2–4 mm. Flowers *c.* 1.1 cm, usually solitary. Calyx-lobes triangular-acute. Fruit 8–10 mm. Diploid, variable. *Himalaya*. H3. Early summer.

99. C. linearifolius Klotz.

Shrub to 60 cm. Branches initially prostrate, dense. Leaf-blade 4–7 × 1–3 mm, oblong to oblanceolate, apex notched, upper surface very dark green, lower surface glaucous, without prominent veins; stalks 0.5–1.5 mm. Flowers solitary, *c.* 5 mm across, pink in bud. Fruit 4–5 mm. Diploid, variable. *Himalaya (Nepal)*. H3. Early summer.

68. DICHOTOMANTHES Kurz
M.F. Gardner
Evergreen trees or shrubs, young branches covered with dense white woolly hairs. Leaves alternate, entire, ovate, pointed, tapered towards the base, 3–10 × 1.5–3 cm, dark green, hairless above, glossy with pale silky hairs beneath; stipules very small, thread-like, falling early. Flowers borne in terminal corymbs *c.* 5 cm across; sepals woolly outside with 2 bracteoles at the base; petals 2–3 mm, white; stamens 15–20; ovary inferior. Fruit a dry, oblong capsule 6 mm, almost entirely surrounded by the calyx, which becomes fleshy with maturity.

A genus of a single species closely resembling *Cotoneaster*. Young plants are frost susceptible and are therefore best given the protection of a wall. Propagate by seed or late summer cuttings rooted under mist with gentle bottom heat.

1. D. tristaniicarpa Kurz. Illustration:
Journal of Botany **11**: pl. 133 (1873); *Hooker's, Icones Plantarum*, 2653 (1900); Icones cormophytorum Sinicorum, **2**: 2362 (1972); Wu, Wild flowers of Yunnan **2**: 109 (1986).

China (Yunnan). H4–5. Summer.

69. PYRACANTHA M.J. Roemer
S.G. Knees
Evergreen shrubs; branches usually thorny; buds small, softly hairy. Leaves alternate, shortly stalked, margins entire, shallowly scalloped or with 5 forward-pointing teeth; stipules minute, falling early. Flowers hermaphrodite in compound corymbs. Calyx of 5 short triangular teeth; petals white, rounded, often with a notched apex and sometimes narrowed into a short basal stalk; stamens *c.* 20, anthers yellow. Ovary half-inferior, carpels 5, free on central axis; styles 5. Fruit apple-like with persistent calyx, red, orange or yellowish, often slightly compressed laterally. Pyrenes 5.

A genus of 6 ornamental species from south-eastern Europe, Himalaya and China, often grown as wall shrubs. Pruning is not necessary but branches should be tied to supports if grown against walls. Many of the species and cultivars are susceptible to attack by fungal pathogens, particularly scab and fireblight, but resistant clones may become available following research and hybridisation programmes.

Literature: Egolf, D.R. & Drechsler, R.F., Chromosome numbers of Pyracantha (Rosaceae), *Baileya* **15**: 82–8 (1967); Egolf, D.R., Pyracantha 'Mohave', a new cultivar (Rosaceae), *Baileya* **17**: 79–82 (1970).

1a. Leaf-margin not scalloped or only finely toothed   2
   b. Leaf-margin scalloped or conspicuously toothed   3
2a. Leaves 3–7 cm, elliptic to lanceolate, underside densely covered with hairs   **4. atalantioides**
   b. Leaves 2.5–4.5 cm, oblanceolate to narrowly obovate, underside with few hairs or hairless   **7. koidzumii**
3a. Calyx and flower-stalks hairless   4
   b. Calyx and flower-stalks covered with hairs   5
4a. Young shoots with persistent rusty brown hairs   **2. crenulata**
   b. Young shoots with greyish hairs, eventually hairless and dark reddish brown   **6. rogersiana**
5a. Underside of leaves densely grey hairy   **5. angustifolia**

   b. Underside of leaves hairless or sparsely hairy when young   6
6a. Leaves elliptic to ovate-oblong, widest above the middle, underside hairless   **3. crenatoserrata**
   b. Leaves narrowly ovate-lanceolate, widest below the middle, underside hairy when young   **1. coccinea**

1. P. coccinea M.J. Roemer. Illustration:
Hegi, Illustrierte Flora von Mittel-Europa **4**: 688 (1922); Bean, Trees & shrubs hardy in the British Isles, edn 8, **3**: pl. 55 (1976); Krüssmann, Manual of cultivated broad-leaved trees & shrubs **3**: pl. 26 (1986); Phillips & Rix, Shrubs, 262 (1989).

Erect shrub to 6 m; young shoots grey downy; thorns 1.2–1.8 cm. Leaves 2–4 cm × 6–18 mm, narrowly ovate-lanceolate, acute; margin slightly scalloped, hairless except along margin near the base when young. Flowers-stalks slightly downy; flowers *c.* 8 mm across, creamy white; calyx-lobes broadly triangular, slightly downy. Fruit *c.* 10 × 8 mm, orange, in dense clusters. *SE Europe, east to the Caucasus*. H2. Summer.

Most widely grown as the selection 'Lalandei', which is slightly larger in all its parts and more colourful in fruit than the species.

2. P. crenulata (D. Don) M.J. Roemer
(*Mespilus crenulata* D. Don; *Crataegus crenulata* (D. Don) Lindley). Illustration: *Edwards's Botanical Register* **30**: 52 (1844); Polunin & Stainton, Flowers of the Himalaya, pl. 34 (1985); Krüssmann, Manual of cultivated broad-leaved trees & shrubs **3**: pl. 26 (1986).

Spiny shrub to 2 m; leaves crowded along the branches, lateral branches often terminating in spines; young shoots with rusty brown hairs. Leaves 1.5–3 cm, narrowly oblong, blunt, shiny, leathery; margin scalloped; leaf-stalk covered with rusty hairs when young. Flowers arranged in axillary clusters along the branches. Calyx with 5 blunt lobes, hairless. Petals 4–5 mm, rounded, white. Fruit spherical, *c.* 6 mm, orange-red with nuts protruding from the persistent calyx. *India, Burma & SW China*. H3. Spring–summer.

3. P. crenatoserrata (Hance) Rehder
(*P. yunnanensis* Chittenden; *P. gibbsii* var. *yunnanensis* Osborn; *P. fortuneana* (Maximowicz) Li; *Photinia crenato-serrata* Hance). Illustration: *Revue Horticole*, pl. 204 (1913); *Gardeners' Chronicle* **65**: 265,

266 (1919); *Botanical Magazine*, 9099 (1929).

Shrub 4–6 m, similar to *P. atalantioides* and *P. rogersiana*. Young shoots rust brown, hairy; leaves 2.5–7 × 1–2.5 cm, elliptic to ovate-oblong, broadest in upper third of blade, apex rounded, base long tapered; margin coarsely toothed, hairless beneath. Flowering-stems downy; calyx downy; petals 4–5 mm, white. Fruit c. 7 mm across, coral-red. *NW China.* H3–4. Summer.

4. P. atalantioides (Hance) Stapf (*P. gibbsii* A.B. Jackson; *Sportella atalantioides* Hance). Illustration: *Gardeners' Chronicle* **60**: 310 (1916); *Botanical Magazine*, 9099 (1925); Krüssmann, Manual of cultivated broad-leaved trees & shrubs **3**: pl. 26 (1986); Phillips & Rix, Shrubs, 262 (1989).

Shrub to 6 m. Leaves 3–7 × 1.2–3 cm, oblong-elliptic to lanceolate, hairless above, densely covered with greyish hairs beneath, eventually becoming blue-green; margin entire. Flowers 8–9 mm across, white to cream, in dense downy corymbs. Fruit 6–7 mm across, orange-red. *W China.* H3–4. Spring.

5. P. angustifolia (Franchet) Schneider (*Cotoneaster angustifolia* Franchet). Illustration: *Botanical Magazine*, 8345 (1910); RHS dictionary of gardening 1719 (1956); Krüssmann, Manual of cultivated broad-leaved trees & shrubs **3**: pl. 25, 26 (1986); Wu, Wild flowers of Yunnan **2**: 116 (1986).

Densely bushy shrub 3–4 m; stems rigid, horizontal, spine-tipped, covered in grey down during the first year. Leaves 1.2–3 × 0.3–1.2 mm, narrowly oblong or slightly obovate, apex rounded, often with a slight notch and a few stiff dark teeth on either side, base rounded or tapered, dark green and hairless above, grey felted beneath. Flowers 2–4 cm across, white, in downy corymbs to 5 cm across. Fruit c. 8 mm across, grey downy when young, eventually hairless and brilliant orange-yellow. *W China.* H2. Summer.

6. P. rogersiana (Jackson) Chittenden (*P. crenulata* var. *rogersiana* A.B. Jackson). Illustration: *Botanical Magazine*, n.s., 74 (1949); Phillips & Rix, Shrubs, 173, 262 (1989).

Shrub to 4 m. Leaves 1.5–3.5 cm × 5–10 mm, oblanceolate or narrowly obovate, tapered towards stalk; margin shallowly toothed. Flowers c. 7 mm across,

calyx-tube hairless, lobes triangular; petals creamy white; stamens with white filaments. Fruit c. 6 mm across, orange-red to yellow. *W China.* H1. Spring.

'Flava'. Illustration: Phillips & Rix, Shrubs, 262 (1989). A commonly grown, pale yellow fruiting form.

7. P. koidzumii (Hayata) Rehder (*Cotoneaster koidzumii* Hayata). Illustration: *Botanical Magazine*, n.s., 205 (1952); Li, Woody flora of Taiwan, 289 (1963); Li, Flora of Taiwan **3**: 91 (1977); Krüssmann, Manual of cultivated broad-leaved trees & shrubs **3**: 68 (1986).

Similar to *P. rogersiana*, but leaves usually entire, hairless beneath; calyx-tube and inflorescence with soft hairs. *Taiwan.* H4. Spring–early summer.

The cultivar 'Mohave' is a selection from the hybrid between *P. koidzumii* and *P. coccinea* 'Wyatt', raised in Washington in 1963. It is resistant to fireblight and *Fuscicladium* attack and bears large orange-red fruits from autumn to winter.

70. MESPILUS Linnaeus
S.G. Knees
Shrub or small tree, 2–5 m; branches occasionally spiny, young shoots woolly. Leaves alternate, deciduous, entire; blade 6–12 × 3–6 cm, oblong lanceolate; margin toothed, glandular; stalks 2–5 mm. Stipules deciduous. Flowers solitary, terminal on short shoots. Calyx softly hairy, with 2 bracteoles at the base, lobes 5, up to 4 times as long as tube. Petals 5, rounded, 1.5–2 cm, white; stamens 30–40, anthers almost joined. Ovary inferior. Carpel solitary at the base of the calyx-tube, 1-celled; style lateral; stigma capitate; ovules 2. Fruit dry, often 1-seeded, 2–5 cm across, slightly exerted from the fleshy calyx, which becomes enlarged in fruit, green at first, becoming brown.

A genus of a single species. Several cultivars have been selected for larger or seedless fruits, which are usually eaten once frosted, when they become sweet and palatable. Propagation is from seed, which may be slow to germinate, or by grafting or budding onto pear (*Pyrus*), quince (*Cydonia*) or hawthorn (*Crataegus*). A fertile or clay-rich soil is ideal for the medlar, which does not thrive in very dry soils. A position in full sun is preferable to semi-shade.

1. M. germanica Linnaeus. Illustration: Reichenbach, Icones florae Germanicae et Helveticae **25**: t. 100 (1914); Bean, Trees

& shrubs hardy in the British Isles, edn 8, **2**: pl. 101 (1973); Krüssmann, Manual of cultivated broad-leaved trees & shrubs **2**: pl. 124 (1986).

SE Europe east to W Asia & Iran. H1. Spring.

71. CRATAEGUS Linnaeus
S.G. Knees & M.C. Warwick
Deciduous, or rarely evergreen trees or shrubs, often with spiny stems and branches. Leaves alternate with entire or lobed margins, stipules present. Flowers bisexual, solitary or in corymbs. Sepals 5, petals 5, stamens 5–25, ovules solitary or 2–5, joined at base, free at apex. Fruit a drupe with 1–5 hard nutlets, each containing a single seed.

An important ornamental genus in which over 1000 species have been described, though this number has been reduced by recent authors. About 100 species are found in the old world, but the vast majority are from North America. Propagation is normally from seed but hybrids and selected clones are grafted or reproduced by cuttings.

Literature: Sargent, C.S., Manual of the trees of North America (1905).

Fruits. Bluish green, green or yellowish green: **14, 31, 36**; yellow: **5, 7, 14, 22, 31**; orange-yellow: **2, 4, 16, 21**; orange-red: **3, 7, 12, 13, 16, 33**; reddish brown or brown: **17, 23, 35, 37**; red or pinkish red: **5, 8, 9, 10, 11, 15, 18, 19, 20, 24, 25, 26, 27, 28, 29, 32, 34, 35, 36, 38, 39, 40, 42, 43, 44, 45, 46, 47**; bluish red: **11, 41, 47**; dark reddish purple to black: **1, 6, 23, 30, 36, 46, 47**; blackish brown: **23**.

1a. Inflorescence and young shoots covered with densely matted white hairs   2
 b. Inflorescence and young shoots hairless, or with long straight hairs   17
2a. Leaves deeply cut or lobed   3
 b. Leaves more or less entire, but margins with forward-pointing teeth or very shallowly lobed   7
3a. Leaf-stalk 1.5 cm or longer; fruit blackish purple   **1. pentagyna**
 b. Leaf-stalk not more than 1.2 cm; fruit yellow, orange or reddish   4
4a. Leaf-surface thickly covered with soft felted hairs   5
 b. Leaf-surface thinly covered with short hairs, but not felty to the touch   6

5a. Flowers 2–2.5 cm across; fruit
   yellow-orange   **2. tanacetifolia**
b. Flowers 1.6–1.9 cm across; fruit
   orange-red   **3. orientalis**
6a. Leaf-tips 3–5-lobed, surface almost
   hairless; flowers not more than 1.2
   cm across   **4. azarolus**
b. Leaves 7–9-lobed, surface with
   closely adpressed, white hairs;
   flowers 1.8 cm or more across
    5. × dippeliana
7a. Leaves widest at or below the
   middle, broadly ovate, normally
   5–10 cm wide; margins doubly saw-
   toothed, mostly truncate at base   8
b. Leaves usually widest at or above
   the middle, usually not more than
   2–5 cm wide, narrowly ovate to
   oblanceolate   13
8a. Flowers 6–12 mm across   9
b. Flowers 2–3 cm across   10
9a. Leaves truncate at base; fruit black
    6. chlorosarca
b. Leaves wedge-shaped at base; fruit
   yellowish orange or reddish
    7. calpodendron
10a. Stipules persistent after fruiting;
   anthers rose-pink **8. ellwangeriana**
b. Stipules falling early; anthers yellow
   or white   11
11a. Fruit pear-shaped, c. 1 cm across,
   light red; seeds 5   **9. submollis**
b. Fruit spherical, 1.2–1.8 cm across,
   light bluish red or red   12
12a. Stamens 20; fruit red, seeds 4 or 5
    10. mollis
b. Stamens 10; fruit light bluish red,
   seeds 3 or 4   **11. arnoldiana**
13a. Leaves longer than wide, ovate to
   lanceolate   14
b. Leaves more or less as wide as long,
   ovate to rounded   15
14a. Stamens 15–20; fruit c. 2.5 cm
   across, yellowish to dull orange-red,
   with conspicuous lenticels
    12. pubescens
b. Stamens usually 5–10; fruit
   1.6–1.9 cm across, orange-red
    13. × lavallei
15a. Leaves not more than 4 cm; thorns
   very slender; flowers solitary or in
   groups of 2 or 3   **14. uniflora**
b. Leaves 4–10 cm; thorns stout or
   absent; flowers numerous   16
16a. Anthers yellow; fruit not more than
   1 cm across, dull red; seeds 5
    15. collina
b. Anthers pink, fruit c. 1.8 cm across,
   yellowish red; seeds 2–5
    16. punctata

17a. Inflorescence and young shoots
   hairless   18
b. Inflorescence and young shoots with
   few straight hairs that sometimes
   persist   31
18a. Fruit 5–8 mm across   19
b. Fruit at least 10 mm across   22
19a. Leaves often deeply lobed, especially
   towards the base; fruit light brown
    17. dahurica
b. Leaves shallowly lobed; fruit red   20
20a. Leaves broadly triangular, truncate
   at base; fruit c. 8 mm
    18. phaenopyrum
b. Leaves ovate to lance-shaped or
   oblong, wedge-shaped at base; fruit
   5–8 mm   21
21a. Thorns 2.5–3 cm, stout; flowers
   1–1.2 cm across   **19. canbyi**
b. Thorns 3–4 cm, slender; flowers c.
   1.8 cm across   **20. viridis**
22a. Fruit brownish black, orange or
   yellowish   23
b. Fruit bright red to very dark red   25
23a. Leaves deeply lobed   24
b. Leaves shallowly lobed
    21. wattiana
24a. Fruit golden yellow   **22. altaica**
b. Fruit brownish black
    23. dsungarica
25a. Stamens 10   **24. holmesiana**
b. Stamens 15–20   26
26a. Anthers purple   **25. sanguinea**
b. Anthers pink or red   27
27a. Anthers red   28
b. Anthers pink   29
28a. Leaves always lobed; stamens 20;
   styles and seeds 5
    26. × durobrivensis
b. Leaves entire and lobed on the same
   plant; stamens 15–20; style and seed
   solitary   **27. heterophylla**
29a. Leaves truncate, very rarely wedge-
   shaped   **28. coccinioides**
b. Leaves wedge-shaped at base   30
30a. Leaves with margins very sharply
   lobed and doubly toothed; fruit
   rounded, red   **29. × flabellata**
b. Leaves with margins slightly lobed
   and toothed; fruit egg-shaped, red,
   eventually blackish   **30. douglasii**
31a. Calyx and inflorescence branches
   more or less hairless   32
b. Calyx and inflorescence branches
   covered with long straight
   hairs   36
32a. Leaves broadly ovate to rounded or
   diamond-shaped   33
b. Leaves ovate-elliptic to lanceolate   35
33a. Fruit yellowish green   **31. flava**

   b. Fruit red or orange-red   34
34a. Leaves 6–8 cm; fruit bright red
    32. wilsonii
b. Leaves 2–3.5 cm; fruit orange-red
    33. aprica
35a. Flowers 1.8–2 cm across **34. nitida**
b. Flowers 1.2–1.5 cm across
    35. crus-galli
36a. Leaves irregularly and deeply
   lobed for most of their length
   with at least some divisions almost
   to midrib, others dissecting
   each side of leaf-blade to at least
   halfway   46
b. Leaves almost entire or with shallow
   regular lobes, particularly in upper
   half of leaf   37
37a. Stamens 20, anthers red
    36. pruinosa
b. Stamens 5–10 or, if 20, anthers
   white, pale yellow or pink   38
38a. Flowers few, in clusters of
   3–10   39
b. Flowers many, in clusters of 12 or
   more   40
39a. Leaves truncate at base; flowers in
   groups of 3–7; fruit dull reddish
   brown   **37. intricata**
b. Leaves wedge-shaped at base; flowers
   in groups of 6–10; fruit red
    38. jackii
40a. Thorns 7–12 cm   **39. succulenta**,
   see var. **macrantha**
b. Thorns less than 7 cm   41
41a. Leaves narrowly ovate-elliptic,
   almost evergreen   **41. persistens**
b. Leaves broadly ovate to rounded,
   deciduous   42
42a. Leaf-margin toothed but rarely lobed
    40. persimilis
b. Leaf-margin lobed and irregularly
   toothed   43
43a. Stamens 20   **39. succulenta**
b. Stamens 5–10   44
44a. Flowers c. 2.5 cm across
    42. jonesiae
b. Flowers 2 cm or less   45
45a. Stamens 5–10; flowers c. 2 cm; fruit
   spherical, orange-red
    43. chrysocarpa
b. Stamens 10; flowers 1.5–2 cm
   across; fruit pear-shaped, glossy red
    44. pedicellata
46a. Leaves 5–10 cm, broadly angular to
   ovate; fruit c. 1.5 cm
    45. pinnatifida
b. Leaves less than 6 cm, broadly
   diamond-shaped, deeply lobed; fruit
   6–13 mm; flowers often double or
   pink   47

47a. Flowers 1.5–1.8 cm across; fruit
0.6–1.3 cm across; seeds 2 or 3
46. laevigata

b. Flowers 0.8–1.5 cm across; fruit
6–8 mm, seed solitary
47. monogyna

1. C. pentagyna Willdenow. Figure 109
(10), p. 301. Illustration: *Revue Horticole*,
310 (1901); Reichenbach, Icones florae
Germanicae et Helveticae **25**: pl. 681
(1912); Davis (ed.), Flora of Turkey and
the east Aegean Islands **4**: 143 (1972);
Krüssmann, Manual of cultivated broad-
leaved trees & shrubs **1**: pl. 153 (1984).

Tree to 5 m, young branches covered
with soft hairs and few thorns to 1 cm.
Leaves 2–6 cm, diamond-shaped to broadly
ovate, with short hairs beneath; lobes 3–7,
margins with forward-pointing teeth.
Flowers borne in loose shaggy panicles
4–7 cm across; corolla *c.* 1.5 cm across;
stamens 20, anthers red. Fruit 1.1–1.3 cm,
egg-shaped, blackish purple. *E Europe,
Caucasus to SW Asia, Iran.* H5. Late
spring–early summer.

2. C. tanacetifolia (Lamarck) Persoon.
Illustration: *Edwards's Botanical Register*
22: 1884 (1836); Schneider, Illustriertes
Handbuch der Laubholzkunde **1**: 786–7
(1906); Davis (ed.), Flora of Turkey and
the east Aegean Islands **4**: 143 (1972);
Krüssmann, Manual of cultivated broad-
leaved trees & shrubs **1**: pl. 153 (1984).

Shrub or small tree to 12 m, branches
upright, covered with short hairs at first.
Leaves *c.* 2.5 cm, ovate to diamond-shaped,
both sides shaggy; margins with forward-
pointing teeth, lobes 5–7, narrow. Stipules
conspicuous. Flowers in hairy, umbel-like
groups of 4–8; corolla 2–2.5 cm across.
Fruit 2–2.5 cm across, orange-yellow.
W Asia. H4. Spring.

3. C. orientalis Pallas (*Mespilus
odoratissima* Andrews). Figure 109(11),
p. 301. Illustration: *Botanical Magazine*,
2314 (1822); Phillips, Trees in Britain,
Europe and North America, 107 (1978);
RHS Dictionary of gardening 570 (1981);
Krüssmann, Manual of cultivated broad-
leaved trees & shrubs **1**: pl. 154 (1984).

Tree to 7 m, straggly, with short hairy
branches, eventually becoming thorny.
Leaves 4–5 cm, triangular or diamond-
shaped, deeply cut into 5–9 lobes, with
short hairs above and dense matted hairs
beneath. Flowers in white-hairy clusters of
7–10; corolla 1.6–1.9 cm across, white;
stamens 20. Fruit *c.* 1.5 cm across,

spherical, orange-red, hairy. *SE Europe,
W Asia.* H3. Early summer.

4. C. azarolus Linnaeus. Figure 109(12),
p. 301. Illustration: *Revue Horticole*, 441
(1856); Schneider, Illustriertes Handbuch
der Laubholzkunde **1**: 789, 790 (1906);
Phillips, Trees in Britain, Europe and North
America, 106 (1978); Krüssmann, Manual
of cultivated broad-leaved trees & shrubs **1**:
pl. 152 (1984).

Small tree or shrub, with thornless
branches, often covered with soft hairs.
Leaves 3–7 cm, ovate to diamond-shaped,
with 3–5 deep lobes, both sides hairy at
first, eventually glossy green above, greyish
beneath. Flowers few, in hairy umbel-like
clusters. Fruit *c.* 2 cm across, spherical,
yellowish orange. *S Europe, N Africa &
W Asia.* H3. Spring.

5. C. × dippeliana Lange. Illustration:
Schneider, Illustriertes Handbuch der
Laubholzkunde **1**: 797 (1906).

Shrub or small tree resembling
C. tanacetifolia but with leaves lobed only to
the middle of the blade. Lobes acute, with
forward-pointing teeth, hairy on both
surfaces. Flowers in dense clusters; calyx
and corolla hairy; stamens *c.* 20, anthers
red. Fruit *c.* 1.5 cm across, yellow or dull
red. *Garden origin (England).* H3. Spring.

Raised in cultivation *c.* 1830 and
thought to be a hybrid between *C. punctata*
and *C. tanacetifolia.*

6. C. chlorosarca Maximowicz. Figure
109(16), p. 301. Illustration: Dippel,
Handbuch der Laubholzkunde **3**: 450
(1893); Schneider, Illustriertes Handbuch
der Laubholzkunde **1**: 772, 774 (1906);
Krüssmann, Manual of cultivated broad-
leaved trees & shrubs **1**: 396, pl. 153
(1984).

Small tree of pyramidal habit; young
branches covered with fine hairs when
young; thorns 1–1.2 cm, straight. Leaves
5–9 cm, broadly triangular, hairy on both
surfaces, eventually hairless above; lobes
3–5 pairs, margins with forward-pointing
teeth. Flowers in umbel-like panicles
4–7 cm across; petals *c.* 9 mm. Fruit black,
flesh green. *China (Manchuria).* H2.
Spring–summer.

7. C. calpodendron (Ehrhart) Medikus
(*C. tomentosa* Duroi). Figure 109(1),
p. 301. Illustration: Britton & Brown
Illustrated flora of the northern United
States & Canada **2**: 2002 (1898); Bailey,
Standard cyclopedia of horticulture 885
(1935); Gleason, Illustrated flora of the

north-eastern United States and adjacent
Canada **2**: 375 (1952); Krüssmann,
Manual of cultivated broad-leaved trees &
shrubs **1**: 396, pl. 153 (1984).

Small tree to 6 m; branches erect,
thornless or with short thorns. Leaves
5–12 cm, broadly ovate, often lobed, hairy
beneath, dull green and hairless above,
margins with forward-pointing teeth.
Flowers in umbel-like panicles 6–12 cm
across. Fruit *c.* 1 cm, elliptic, yellowish or
orange-red, fleshy. *Canada (Ontario), south
to C USA.* H2. Summer.

8. C. ellwangeriana Sargent (*C. coccinea*
Linnaeus, in part). Illustration: *Botanical
Magazine*, 3432 (1835); Sargent, Silva of
North America **13**: pl. 671 (1902); Hough,
Handbook of the trees of the northern
states and Canada, 250, 251 (1947);
Botanical Magazine, n.s., 105 (1950).

Small tree very like *C. pedicellata* but
with leaves and flowers downy. *E USA.*
H3. Spring–summer.

9. C. submollis Sargent. Illustration:
Sargent, Silva of North America **4**: pl. 182
(1892); Schneider, Illustriertes Handbuch
der Laubholzkunde **1**: 797, 799 (1906);
Gleason, Illustrated flora of the north-
eastern United States and adjacent Canada
2: 371 (1952).

Large shrub or tree, 8–10 m, branches
thorny, softly hairy at first. Leaves 4–8 cm,
ovate, abruptly wedge-shaped at base,
occasionally rounded or heart-shaped,
hairy above and below; margins with
forward-pointing teeth, with 4 or 5 lobes
on each side. Flowers many in loose hairy
corymbs; calyx hairy, sepals with red-
stalked glands; corolla *c.* 2 cm across;
stamens 10, anthers white to yellow. Fruit
c. 1 cm across, pear-shaped, light red, seeds
5. *NE North America.* H2. Spring.

10. C. mollis (Torrey & Gray) Scheele.
Illustration: Sargent, Silva of North
America **13**: pl. 659 (1902); Gleason,
Illustrated flora of the north-eastern United
States and Canada **2**: 371 (1952); Phillips,
Trees in Britain, Europe and North
America 108 (1978); Krüssmann, Manual
of cultivated broad-leaved trees & shrubs **1**:
pl. 153 (1984).

Tree to 11 m, branches with or without
thorns. Leaves 6–10 cm, broadly ovate,
covered with dense hairs beneath,
eventually only on the veins; margins
doubly toothed, with 4 or 5 pairs of lobes.
Flowers in densely hairy corymbs; corolla
c. 2.5 cm across; stamens 20, anthers pale

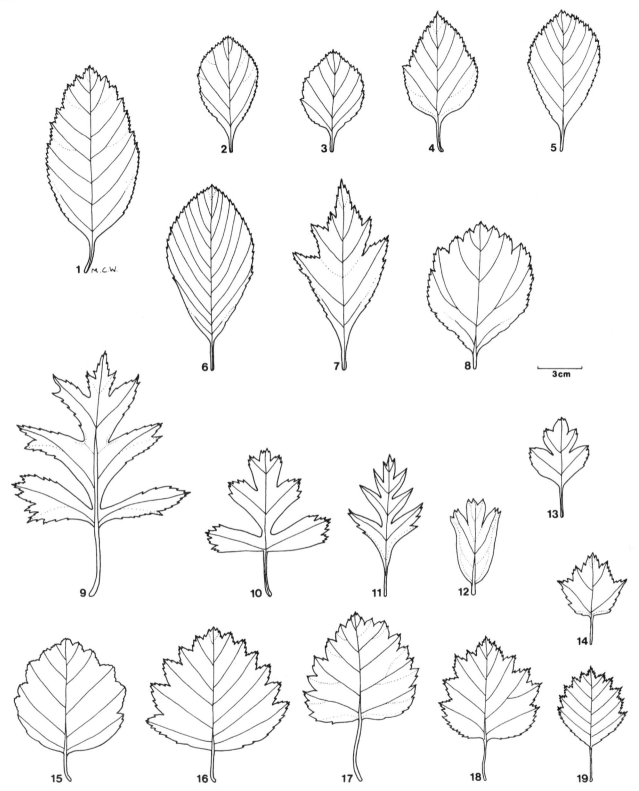

Figure 109. Leaves of *Crataegus* species. 1, *C. calpodendron*.
2, *C.* × *prunifolia*. 3, *C. collina*. 4, *C. viridis*. 5, *C. crus-galli*. 6, *C.* × *lavallei*.
7, *C. dahurica*. 8. *C. douglasii*. 9, *C. pinnatifida*. 10, *C. pentagyna*.

11, *C. orientalis*. 12, *C. azarolus*. 13, *C. laevigata*. 14, *C. phaenopyrum*.
15, *C. wilsonii*. 16, *C. chlorosarca*. 17, *C. wattiana*. 18, *C. pedicellata*.
19, *C. flabellata*.

yellow. Fruit 1.2–1.8 cm across, almost spherical, red, downy; seeds 4–5. *North America.* H2. Spring–early summer.

Plants sometimes offered as **C. arkansana** Sargent (illustration: Sargent, Silva of North America **13**: pl. 660, 1902) with slightly longer fruits are now considered to represent part of the natural range of the species.

11. C. arnoldiana Sargent. Illustration: Sargent, Silva of North America **13**: pl. 668 (1902); Bailey, Standard cyclopedia of horticulture **1**: 822 (1935); Gleason, Illustrated flora of the north-eastern United States and adjacent Canada **2**: 373 (1952).

Small tree 7–10 m, with stiff thorny branches, shaggy with long hairs at first. Leaves 4–5 cm, with 3–5 pointed, saw-toothed lobes on each side, ovate with rounded or truncated base; the leaves on long shoots much larger than those on short shoots. Flowers *c.* 3 cm across, numerous in many loose-flowered corymbs; stamens 10, anthers light yellow. Fruit *c.* 1.5 cm across, light bluish red, rarely red, spherical, shaggy at each end; seeds 3 or 4. *NE USA.* H4. Spring.

12. C. pubescens (Humbolt, Bonpland & Kunth) Steudel forma **stipulacea** (Loudon) Stapf (*C. stipulacea* Loudon; *C. mexicana* de Candolle). Illustration: Dippel, Handbuch der Laubholzkunde **3**: 427 (1893); *Botanical Magazine*, 8589 (1914).

Shrub to 3 m, branches with few thorns. Leaves 4–8 cm, wedge-shaped to elliptic or oblong-lanceolate, dark green above, paler with tufted hairs beneath; margins with scalloped teeth. Flowers in clusters of 6–12; corolla *c.* 2 cm wide. Fruit yellowish to dull orange-red. *Mexico.* H5. Early spring.

13. C. × lavallei Hérincq *C. pubescens* forma *stipulacea × C. crus-galli*. Figure 109 (6), p. 301. Illustration: Phillips, Trees in Britain, Europe and North America, 108 (1978); RHS dictionary of gardening **2**: 569 (1981).

Tree 5–7 m, young branches downy, thorns to 4 cm, few. Leaves 4–11 cm, oblong to elliptic, acute; margins unevenly toothed. Flowers borne in erect corymbs up to 7 cm across; corolla *c.* 2 cm across; stamens 5–20, anthers pink or yellowish. Fruit 1.6–1.9 cm across, spherical, orange-red, speckled, persisting into winter. *Raised in cultivation (France).* H4. Spring–early summer.

Most commonly grown as 'Carrieri', which has only 5–10 pink anthers.

14. C. uniflora Muenchhausen (*C. champlainensis* Sargent). Illustration: Sargent, Silva of North America **4**: pl. 191 (1892) & **13**: 669 (1902); Gleason, Illustrated flora of the north-eastern United States and adjacent Canada **2**: 345 (1952); Krüssmann, Manual of cultivated broad-leaved trees & shrubs **1**: 400 (1984).

Straggly shrub to 3 m, branches with thin thorns to 4 cm. Leaves to 4 cm, ovate, wedge-shaped at base, covered with short hairs at first, eventually hairless; margins finely toothed. Flowers solitary or in clusters of 2–3, almost stalkless; corolla *c.* 1.5 cm across, creamy white; stamens 20, anthers whitish. Fruit *c.* 1.2 cm, rounded to pear-shaped, yellow or greenish. *E North America.* H4. Spring–early summer.

15. C. collina Chapman. Figure 109(3), p. 301. Illustration: Sargent, Silva of North America **13**: 654 (1902); Britton & Brown, Illustrated flora of the northern United States & Canada, edn 2, **2**: 301 (1913).

Shrub or small tree to 8 m; branches spreading, with stout thorns. Leaves broadly ovate to elliptic, about as long as wide, margins doubly toothed; anthers yellow. Fruit *c.* 1 cm, spherical, dull red; pulp yellowish, mealy; seeds 5. *C & E USA.* H2. Spring.

16. C. punctata Jacquin. Illustration: Sargent, Silva of North America **4**: pl. 184 (1892); Hough, Handbook of the trees of the northern states and Canada, 246–7 (1947); Gleason, Illustrated flora of the north-eastern United States and adjacent Canada **2**: 349 (1952); Elias, The complete trees of North America, 615 (1980).

Tree 6–10 m, with a rounded crown and many horizontal branches, usually with stout thorns 5–8 cm, or occasionally thornless. Leaves 5–10 cm, ovate to obovate, shaggy beneath; margins irregularly toothed. Flowers numerous; corolla 1.5–2 cm across; stamens 20, anthers pink. Fruit rounded, *c.* 1.8 cm across, yellowish red, lightly spotted; seeds 2–5. *E North America.* H3. Spring.

17. C. dahurica Koehne. Figure 109(7), p. 301. Illustration: Schneider, Illustriertes Handbuch der Laubholzkunde **1**: 772, 774 (1906); Komarov (ed.), Flora SSSR **9**: 324 (1939); Krüssmann, Manual of cultivated broad-leaved trees & shrubs **1**: pl. 153 (1984).

Small tree or shrub, branches dark brown, with thorns to 4 cm. Leaves 2–5 cm, ovate to elliptic or diamond-shaped, acutely lobed, dark green, appearing early in spring, turning bright red in autumn. Flowers borne in loose hairless panicles; corolla *c.* 1.5 cm across; stamens 20, anthers purple. Fruit 6–8 mm across, light brown. *SE Siberia.* H1. Spring–early summer.

18. C. phaenopyrum (Linnaeus films) Medikus. Figure 109(14), p. 301. Illustration: Gleason, Illustrated flora of the north-eastern United States and adjacent Canada **2**: 345 (1952); Elias, The complete trees of North America, 619 (1980); Everett, New York Botanical Gardens illustrated encyclopedia of horticulture **3**: 909 (1981); Krüssmann, Manual of cultivated broad-leaved trees & shrubs **1**: pl. 153 (1984).

Tree to 10 m, branches with thorns to 7 cm. Leaves 3–7 cm, almost triangular; lobes 3–5, margins with sharp forward-pointing teeth. Flowers many in umbel-like panicles; corolla *c.* 1.2 cm across; stamens 20, anthers pink. Fruit *c.* 8 mm across, a glossy red, flattened sphere; seeds 2–5. *NE North America (Virginia to Alabama).* H4. Late spring–early summer.

19. C. canbyi Sargent. Illustration: Sargent, Silva of North America **13**: pl. 638 (1902); Britton, North American trees, 450 (1908); Britton & Brown, Illustrated flora of the northern United States & Canada, edn 2, **2**: 299 (1913).

Large shrub or small tree to 6 m, with a broad open crown, branches with thorns 2.5–3 cm, stout; resembling *C. crus-galli* but with broader leaves to 7 cm, widest near the middle. Flowers in hairless corymbs, corolla 1–1.2 cm across; stamens 10–20, anthers pink. Fruit 5–7 mm across, glossy and bright red with red pulp. *E North America.* H3. Spring.

20. C. viridis Linnaeus. Figure 109(4), p. 301. Illustration: Sargent, Silva of North America **4**: pl. 187 (1892); Britton & Brown Illustrated flora of the northern United States & Canada **1**: 1996 (1897); Gleason, Illustrated flora of the north-eastern United States and adjacent Canada **2**: 345 (1952); Elias, The complete trees of North America, 607, 613 (1980).

Tree to 12 m, with spreading branches and thin thorns to 4 cm. Leaves 2–6 cm, ovate to oblong, always wedge-shaped at the base; margins with forward-pointing teeth. Flowers borne in clusters 4–5 cm across; corolla *c.* 1.8 cm, white. Fruit

6–8 mm across, almost spherical, bright red, persistent into winter. *E USA*. H4. Spring–early summer.

21. C. wattiana Hemsley & Lace (*C. korolkowii* misapplied). Figure 109(17), p. 301. Illustration: *Revue Horticole*, 308 (1901); *Botanical Magazine*, 8818 (1919).

Small tree resembling *C. sanguinea*; young branches red-brown, thornless or with short thorns. Leaves 5–9 cm, ovate, acute, hairless; lobes 3–5 pairs. Stamens 15–20, anthers whitish. Fruit *c.* 1 cm across, rounded, orange-yellow, fleshy. *C Asia (Altai Mts to SW Pakistan)*. H2. Summer.

22. C. altaica (Loudon) Lange. Illustration: Schneider, Illustriertes Handbuch der Laubholzkunde **1**: 772, 774 (1906); Komarov (ed.), Flora SSSR **9**: 324 (1939); Icones cormophytorum Sinicorum **2**: 2142 (1972); Krüssmann, Manual of cultivated broad-leaved trees & shrubs **1**: pl. 153 (1984).

Small straggly tree to 4 m. Branches with few thorns 2–3 cm. Leaves ovate, deeply lobed with warty toothed margins, bright green. Flowers borne in loose umbel-like panicles; corolla *c.* 1 cm across. Fruit 1–1.2 cm across, spherical, golden yellow. *C Asia*. H1. Spring.

23. C. dsungarica Zabel. Illustration: Schneider, Illustriertes Handbuch der Laubholzkunde **1**: 772, 774 (1906).

Like *C. altaica* but with leaves 3–8 cm, hairless, ovate to diamond-shaped to broadly ovate, with 2–4 pairs of lobes. Flowers in hairless clusters and fruit blackish brown. *E Siberia, N China*. H1. Spring.

24. C. holmesiana Ashe. Illustration: Sargent, Silva of North America **13**: pl. 676 (1902); Hough, Handbook of the trees of the northern states and Canada, 252, 253 (1947); Gleason, Illustrated flora of the north-eastern United States and adjacent Canada **2**: 367 (1952); Krüssmann, Manual of cultivated broad-leaved trees & shrubs **1**: pl. 153 (1984).

Tree-like shrub occasionally reaching 10 m, with a loose conical crown of ascending branches; thorns slender. Leaves 6–8 cm, ovate to oblong-ovate, with 4–6 pairs of saw-toothed lobes. Flowers many, in hairless corymbs; stamens 10, anthers red. Fruit *c.* 1 cm across, light red; seeds 3, pulp dry or mealy. *E North America*. H4. Spring.

25. C. sanguinea Pallas. Illustration: Nuttall, The North American sylva, pl. 44 (1865); Schneider, Illustriertes Handbuch der Laubholzkunde **1**: 772, 774 (1906).

Tree to 7 m, with glossy, purplish brown branches and few stout thorns to 3 cm. Leaves 5–8 cm, diamond-shaped to broadly ovate, dark green above, with short hairs and paler beneath; margins with 2 or 3 pairs of rough, doubly toothed lobes. Flowers borne in small umbel-like panicles; anthers purple. Fruit 1–1.2 cm across, bright red, almost translucent. *E Siberia*. H1. Spring.

26. C. × durobrivensis Sargent. Illustration: Sargent, Trees & shrubs **1**: pl. 2 (1902).

Shrub 3–5 m, with hairless young shoots and thorns 3–5 cm. Leaves broadly ovate, wedge-shaped at base, hairless, divided into 2–4 triangular lobes. Flowers in hairless clusters; corolla *c.* 2 cm across; stamens 20, anthers red. Fruit *c.* 1.1 cm across, rounded, shiny crimson red. *N America*. H3. Spring.

Often described as one of the most ornamental species with among the largest flowers in the genus. It is thought by some to be a hybrid between *C. pruinosa* and *C. suborbiculata* Sargent.

27. C. heterophylla Flugge. Illustration: *Edwards's Botanical Register* **14**: pl. 1161 (1828) & **22**: pl. 84 (1836); Loudon, Arboretum et fruticetum Britanicum **2**: 864 & **4**: pl. 31 (1838); Schneider, Illustriertes Handbuch der Laubholzkunde **1**: 780, 782 (1906).

Shrub or small tree, 1.5–3 m; shoots hairless, with thorns 1–1.5 cm. Leaves 3–4.5 cm, shallowly 3–7-lobed or entire, rounded or broadly ovate, wedge-shaped at base. Flowers in groups of 8–15. Fruit 1.5–1.8 cm. *C & SW Asia*. H2. Summer.

28. C. coccinioides Ashe. Illustration: Sargent, Silva of North America **13**: pl. 674 (1902); Bailey, Standard cyclopedia of horticulture 1099 (1950); Gleason, Illustrated flora of the north-eastern United States and adjacent Canada **2**: 373 (1952); Bean, Trees & shrubs hardy in the British Isles, edn 8, **1**: 768 (1970).

Small tree, thorns 3–5 cm, straight. Leaves dull red on new growth, brownish beneath, truncate at base, very rarely wedge-shaped, orange-red in autumn. Flowers in clusters of 5–7, corolla *c.* 1.2 cm, stamens 20, anthers pink. Fruit

rounded, dark reddish pink. *C USA*. H2. Spring.

29. C. flabellata (Bosc) Koch. Figure 109 (19), p. 301. Illustration: Gleason, Illustrated flora of the north-eastern United States and adjacent Canada **2**: 361 (1952); Krüssmann, Manual of cultivated broad-leaved trees & shrubs **1**: pl. 153 (1984); Duncan & Duncan, Trees of the southeastern United States, 189 (1988).

Shrub to 6 m, branches with curved thorns, 4–10 cm. Leaves 3–7 cm, broadly ovate, sparsely shaggy above and on veins beneath at first; margins very sharply lobed and doubly toothed. Flowers borne in downy corymbs; calyx-lobes with forward-pointing teeth; corolla 1.5–2 cm across; stamens 15–20. Fruit 1–1.1 cm across, rounded, red. *E North America*. H3. spring.

30. C. douglasii Lindley (*C. rivularis* Nuttall). Figure 109(8), p. 301. Illustration: Sargent, Silva of North America **4**: 175 (1892); Gleason, Illustrated flora of the north-eastern United States and adjacent Canada **2**: 375 (1952); Phillips, Trees in Britain, Europe and North America, 106 (1978); Krüssmann, Manual of cultivated broad-leaved trees & shrubs **1**: 397 (1984).

Tree to 12 m with slender drooping branches, thorns to 3 cm, few. Leaves 3–8 cm, broadly oblong to ovate, hairless, except on midrib beneath; margins toothed and slightly lobed. Flowers in hairless panicles of 10–20. Corolla *c.* 1 cm wide; stamens 20. Fruit *c.* 1.2 cm, egg-shaped, wine-red, eventually reddish black. *W North America*. H2. Spring.

31. C. flava Aiton. Illustration: Sargent, Silva of North America **13**: pl. 693 (1902); Gleason, Illustrated flora of the north-eastern United States and adjacent Canada **2**: 351 (1952); Krüssmann, Manual of cultivated broad-leaved trees & shrubs **1**: 297 (1984); Brickell, RHS gardeners' encyclopedia of plants and flowers, 63 (1989).

Slow-growing, wide-spreading shrub or small tree to 6 m, branches with thorns to 2.5 cm. Leaves *c.* 2.5 cm, broadly ovate to rounded or diamond-shaped; stipules rounded, conspicuous. Flowers borne in small clusters of 3–7; corolla *c.* 1 cm across; stamens 10–20, anthers purple. Fruit *c.* 1 cm across, rounded to pear-shaped, greenish yellow. *E North America*. H4. Spring.

32. C. wilsonii Sargent. Figure 109(15), p. 301. Illustration: Icones cormophytorum Sinicorum **1**: 2140 (1972).

Tree to 6 m, resembling *C. calpodendron* but with leaves broadly ovate to rounded, glossy above, shaggy beneath. Fruit to 1 cm, elliptic, bright red. *C China.* H4. Spring.

33. C. aprica Beadle. Illustration: Sargent, Silva of North America **13**: pl. 698 (1902); Schneider, Illustriertes Handbuch der Laubholzkunde **1**: 792, 793 (1906).

Shrub or small tree to 6 m with outspread, undulating branches and straight, slender thorns 2–3.5 cm. Leaves 2–3.5 cm, broadly ovate to rounded, apex acute to rounded, hairy when young; margins with saw-like, gland-tipped teeth and shallow lobes towards the apex. Flowers borne in groups of 3–6, in small corymbs; sepals with gland-tipped, saw-like teeth; stamens 10; anthers yellow. Fruit *c.* 12 mm across, spherical, dull orange-red; seeds 3–5. *SE USA.* Spring. H4–5.

34. C. nitida (Britton & Brown) Sargent (*C. viridis* var. *nitida* Britton & Brown). Illustration: Gleason, Illustrated flora of the north-eastern United States and adjacent Canada **2**: 349 (1952); Sargent, Silva of North America **13**: 703 (1902); Schneider, Illustriertes Handbuch der Laubholzkunde **1**: 797, 799 (1906).

Tree to 7 m, like *C. viridis* but branches covered with short matted hairs at first; eventually hairless or with few long hairs; thornless or with few short thorns. Leaves 2–8 cm, oblong to elliptic, slightly lobed. Flowers 1.8–2 cm across. Fruit *c.* 1.5 cm, egg-shaped to spherical, dull red with a glaucous bloom. *S USA.* H5. Spring–early summer.

35. C. crus-galli Linnaeus. Figure 109(5), p. 301. Illustration: Sargent, Silva of North America **4**: pl. 178 (1892); Gleason, Illustrated flora of the north-eastern United States and adjacent Canada **2**: 345 (1952); Phillips, Trees in Britain, Europe and North America, 106 (1978); Krüssmann, Manual of cultivated broad-leaved trees & shrubs **1**: pl. 158 (1984).

Tree to 10 m, with a flat spreading crown and vase-shaped branching pattern; thorns *c.* 8 cm, slender, straight. Leaves 2–8 cm, oblong to ovate, wedge-shaped at base, rounded at tip, leathery; margins smooth. Flowers to 1.5 cm across; petals white; stamens 10, anthers pink. Fruit often persistent until spring, dullish red, pulp red. *E USA.*

Often grown as 'Pyracanthifolia' (illustration: Sargent, Silva of North America **13**: pl. 637, 1902), with wide-spreading, thorny branches.

C. × grignonensis Mouillefert; *C. crus-galli* × *C. pubescens.* Illustration: Krüssmann, Manual of cultivated broad-leaved trees & shrubs **1**: 398 (1984). Shrub to *c.* 5 m, branches vase-shaped, almost thornless. Leaves ovate, base wedge-shaped, covered with long and short hairs, persisting for some time; margins scalloped, with 2–4 pairs of lobes. Flowers in small, umbel-like clusters. Fruit to 1.5 cm, brownish red with grey spots.

Raised in Frankfurt *c.* 1873.

36. C. pruinosa (Wendland) K. Koch. Illustration: Sargent, Silva of North America **13**: 648 (1902); Schneider, Illustriertes Handbuch der Laubholzkunde **1**: 797, 799 (1906); Gleason, Illustrated flora of the north-eastern United States and adjacent Canada **2**: 365 (1952); Elias, The complete trees of North America, 618 (1980).

Tree to 6 m, branches hairless, with stout thorns 2.5–4 cm. Leaves 3–5 cm, broadly elliptic, wedge-shaped at base, red at first, eventually blue-green, margins irregularly toothed. Flowers borne in loose corymbs; corolla 2–2.5 cm across; stamens 20, anthers red. Fruit *c.* 1 cm across, rounded, blue-green at first, eventually dark red, pulp yellowish. *NE North America.* H2. Spring.

37. C. intricata Lange. Illustration: Schneider, Illustriertes Handbuch der Laubholzkunde **1**: 800, 801 (1906); Gleason, Illustrated flora of the north-eastern United States and adjacent Canada **2**: 355 (1952); Everett, New York Botanical Gardens illustrated encyclopedia of horticulture **3**: 911 (1981); Krüssmann, Manual of cultivated broad-leaved trees & shrubs **1**: pl. 152 (1984).

Shrub 1–3 m, branches with curved thorns 2–4 cm. Leaves ovate-elliptic, acutely lobed, bright green above, hairless; margins doubly toothed. Flowers borne in corymbs of 3–7; corolla 1.2–1.5 cm across; stamens 10, anthers yellow or pale pink. Fruit to 1.3 cm across, rounded, dull reddish brown; seeds 3–5. *E North America.* H3. Spring.

38. C. jackii Sargent. Illustration: Britton & Brown, Illustrated flora of the northern United States & Canada, edn 2, **2**: 306 (1913).

Small tree-like shrub, to 3 m. Leaves 2–3 cm, ovate-elliptic to broadly ovate, broadly wedge-shaped at base, rarely truncate; margins with shallow indistinct lobes and rough teeth. Flowers borne in groups of 6–10; anthers yellow. Fruit *c.* 1.2 cm, red, egg-shaped; seeds 2 or 3. *Canada (Quebec).* H2. Summer.

39. C. succulenta Link. Illustration: Sargent, Silva of North America **4**: pl. 181 (1892); Schneider, Illustriertes Handbuch der Laubholzkunde **1**: 775, 776 (1906); Gleason, Illustrated flora of the north-eastern United States and adjacent Canada **2**: 375 (1952); Elias, The complete trees of North America, 611 (1980).

Tree or shrub to 5 m, branches red-brown, straggly, with stout thorns to 7 cm. Leaves 5–8 cm, broadly ovate; margins doubly toothed. Flowers many, in hairy, umbel-like panicles; corolla *c.* 1.8 cm across, white; stamens 20, anthers pink, rarely white. Fruit *c.* 1 cm across, rounded, glossy red. *E North America.* H3. Early summer.

Var. **macracantha** (Loddiges) Eggleston. Illustration: Hough, Handbook of the trees of the northern states and Canada, 258, 259 (1947). Differs in its more numerous thorns, 7–12 cm, and having only 10 stamens with white or pale yellow anthers.

40. C. persimilis Sargent (*C.* × *prunifolia* (Poiret) Persoon). Illustration: *Edwards's Botanical Register* **22**: pl. 1808 (1836); Phillips, Trees in Britain, Europe and North America, 109 (1978); Everett, New York Botanical Gardens illustrated encyclopedia of horticulture **3**: 910 (1981); Krüssmann, Manual of cultivated broad-leaved trees & shrubs **1**: pl. 153 (1984).

Rounded deciduous tree 7–9 m. Branches with few slender thorns to 6 cm. Leaves 4–6 cm, ovate to oblong-ovate, toothed or shallowly lobed, glossy above. Flowers borne in corymbs 1.5–2 cm across. Petals white, anthers pale yellow. Fruit spherical, bright red, persisting until late autumn or early winter. *North America (Ontario to New York).* H2. Late spring–early summer.

Usually grown as the cultivar 'MacLeod' ('Prunifolia') and 'Prunifolia Splendens' (illustration: Lord, Flora, 433, 2003), which is more vigorous and bears larger leaves and flower corymbs.

41. C. persistens Sargent. Illustration: Sargent, Trees and shrubs **2**: 190 (1913).

Tall shrub or small tree, 3–4 m, with wide-spreading branches, many with thick thorns to 5 cm. Leaves lanceolate to oblong to ovate, coarsely toothed towards apex, almost evergreen, persisting well into the winter. Corolla *c.* 2 cm across; stamens 20; anthers white. Fruit *c.* 1.5 cm across, egg-shaped, dull bluish red, persistent; seeds 2 or 3. *Garden origin.* H3. Spring.

42. C. jonesiae Sargent. Illustration: Sargent, Silva of North America **13**: pl. 684 (1902); Britton & Brown, Illustrated flora of the northern United States & Canada, edn 2, **2**: 300 (1913).

Tall shrub or small tree to 6 m, young branches with long and short hairs, eventually hairless, glossy brown; thorns 5–7 cm. Leaves 7–10 cm, broadly ovate to rounded, wedge-shaped at base; margins with rough teeth and acute lobes above the middle, glossy dark green above, hairy beneath when young. Corolla *c.* 2.5 cm across; stamens 10, anthers pink. Fruit *c.* 1.5 cm across, bright red; seeds 2 or 3. *NE North America.* H2. Summer.

43. C. chrysocarpa Ashe (*C. rotundifolia* Moench). Illustration: Gleason, Illustrated flora of the north-eastern United States and adjacent Canada **2**: 355 (1952); *Canadian Journal of Botany* **65**: 2648 (1987); Looman & Best, Budd's flora of the Canadian prairie provinces, 441 (1987).

Tree to 6 m, branches dense, thorns slender. Leaves 3–5 cm, broadly ovate to rounded; lobes 3 or 4 pairs, doubly toothed. Flowers *c.* 2 cm across; sepals with glandular hairs. Fruit almost spherical, red, pulp yellow. *NE North America.* H2. Spring.

44. C. pedicellata Sargent (*C. coccinea* Linnaeus, in part). Figure 109(18), p. 301. Illustration: Sargent, Silva of North America **13**: pl. 677 (1902); Gleason, Illustrated flora of the north-eastern United States and adjacent Canada **2**: 367 (1952); Phillips, Trees in Britain, Europe and North America, 108 (1978); Krüssmann, Manual of cultivated broad-leaved trees & shrubs **1**: pl. 152 (1984).

Tree to 7 m, mature branches hairless, with straight or slightly curved thorns to 5 cm. Leaves 7–10 cm, broadly ovate, margins doubly toothed with 4 or 5 pairs of lobes. Flowers borne in loose downy corymbs; corolla 1.5–2 cm across; stamens 10, anthers pink. Fruit 1–1.8 cm,

pear-shaped, glossy red. *E USA.* H3. Spring–summer.

45. C. pinnatifida Bunge. Figure 109(9), p. 301. Illustration: Schneider, Illustriertes Handbuch der Laubholzkunde **1**: 769, 770 (1906); Bailey, Standard cyclopedia of horticulture **1**: 888 (1935); Icones cormophytorum Sinicorum **1**: 2137 (1972); Krüssmann, Manual of cultivated broad-leaved trees & shrubs **1**: 400 (1984).

Tree to 6 m, branches hairless, with few thorns to 1 cm. Leaves 5–10 cm, angular to ovate, lobes 5–9, glossy on both sides, soon falling. Flowers few in loose umbel-like panicles; corolla *c.* 1.5 cm; stamens 20. Fruit *c.* 1.5 cm across, red, minutely dotted; seeds 3 or 4. *N China.* H2. Late spring–early summer.

Var. **major** N.E. Brown is the variant most commonly seen in cultivation; it differs in its larger, thicker leaves and fruit to 2.5 cm across.

46. C. laevigata (Poiret) de Candolle (*Mespilus laevigata* Poiret; *C. oxyacantha* misapplied, not Linnaeus; *C. oxyacanthoides* Thuill). Figure 109(13), p. 301. Illustration: Hough, Handbook of the trees of the northern states and Canada, 260–1 (1947); Komarov (ed.), Flora SSSR **9**: 350 (1939); Phillips, Trees in Britain, Europe and North America, 107 (1978); Everett, New York Botanical Gardens illustrated encyclopedia of horticulture **3**: 909 (1981).

Shrub or small tree, 2–5 m, branches eventually hairless, with few short thorns. Leaves 4–5 cm, ovate with 3–5 rounded lobes, margins with forward-pointing teeth; stipules acuminate. Flowers in umbel-like panicles of up to 10; corolla 1.5–1.8 cm across, white or pink; stamens 20, anthers red. Fruit 6–13 mm, rounded to egg-shaped, deep red; seeds 2 or 3. *Europe, N Africa, W Asia.* H2. Spring–early summer.

Among the many cultivars of this species are: 'Paul's Scarlet', with bright red flowers; 'Francois Rigaud', with yellow fruits; and 'Rubra Plena' with double pink flowers.

C. × mordenensis Boom. Like *C. laevigata* but with larger leaves and flowers and differing from *C. succulenta* in its shorter thorns and more deeply cut leaves, hairless inflorescence and glandless sepals.

A hybrid between *C. laevigata* and *C. succulenta*.

47. C. monogyna Jacquin. Illustration: Schneider, Illustriertes Handbuch der Laubholzkunde **1**: 780, 782 (1906); Davis

(ed.), Flora of Turkey and the east Aegean Islands **4**: 143 (1972); Phillips, Trees in Britain, Europe and North America, 108 (1978); Everett, New York Botanical Gardens illustrated encyclopedia of horticulture **3**: 909 (1981).

Shrub or tree 2–10 m, branches thorny, hairless or with few short hairs, thorns 2–2.5 cm. Leaves 3–5 cm, with 3–7 deeply cut lobes, whitish green beneath, hairy; stipules entire. Flowers borne in umbellate panicles; calyx-lobes oblong; corolla 8–15 mm across, white; stamens 20, anthers red. Fruit 6–8 mm across, rounded to egg-shaped, bright red; seed solitary. *Europe, N Africa, Asia.* H2. Late spring.

72. × CRATAEMESPILUS G. Camus
S.G. Knees
Tall shrubs or small trees with densely leaved crowns; branches covered with soft hairs. Leaves deciduous, ovate to obovate, usually lobed; margins smooth or with forward-pointing teeth, hairy beneath. Flowers solitary or in groups of 2 or 3. Sepals and petals 5; stamens 14–28; styles 2 or 3. Fruit with 2 or 3 nutlets, almost spherical, covered with soft hairs.

Hybrids between the genera *Crataegus* and *Mespilus* that originated through normal fruiting rather than by grafting, vegetative means or chimaeras. Both hybrids are of French origin and are very intermediate between their parents.

1a. Leaves lobed towards tip, margin
 with forward-pointing uneven teeth
 1. grandiflora
 b. Leaves lobed, with smooth margin
 2. gillotii

1. × C. grandiflora G. Camus; *Crataegus laevigata × Mespilus germanica.* Illustration: *Botanical Magazine*, 3442 (1845) – as *Mespilus lobata*; *Revue Horticole*, 80 (1869); Reichenbach, Icones florae Germanicae et Helveticae **25**: pl. 107 (1914); Krüssmann, Manual of cultivated broad-leaved trees & shrubs **1**: pl. 152 (1986).

Shrub or small tree to 5 m. Leaves 3–7 cm, ovate to obovate, lobed in upper third, margin with uneven forward-pointing teeth; becoming yellow-brown in autumn. Flowers 2–2.5 cm across; sepals hairy; petals white. Fruit 1–1.6 cm across, spherical to egg-shaped, brown. *France.* H4. Early summer.

2. × C. gillotii Beck & Reichenbach; *Crataegus monogyna × Mespilus germanica.* Illustration: Reichenbach, Icones florae

Germanicae et Helveticae **25**: pl. 107 (1914).

Shrub or small tree to 4 m. Leaves 3–6 cm, ovate to obovate, lobed, margin smooth. Flowers *c.* 2 cm across; sepals hairy; petals white; styles 2. Fruit 1–1.5 cm. *France.* H4. Early summer.

73. × CRATAEGOMESPILUS Simon-Louis
S.G. Knees
Deciduous shrub or small tree to *c.* 5 m, with thorny twigs. Leaves 7–15 cm, narrowly oblong to elliptic, dark green above, lighter green and woolly beneath; margins with fine forward-pointing teeth. Flowers in groups of 5–8; sepals 5, erect, united in a woolly calyx; petals 5, white, 5–8 mm, rounded; stamens 15–20; styles 1–3. Fruit 1.5–2 cm across with persistent calyx-lobes; seeds 1–3, sterile.

A graft hybrid in which new shoots commonly revert back to the habit and form of their original parents.

1. C. dardarii Simon-Louis *Crataegus monogyna+Mespilus germanica.* Illustration: Schneider, Illustriertes Handbuch der Laubholzkunde **1**: 765, 766 (1906); *Kew Bulletin*, pl. 268 (1911); Reichenbach, Icones florae Germanicae et Helveticae **25**: 108 (1914); Krüssmann, Manual of cultivated broad-leaved trees & shrubs **1**: 394 (1986).

Garden graft hybrid that originated in France. H4. Spring–summer.

Two cultivars are occasionally grown. 'Jules d'Asniers' ('Asnieresii') is closer to *Crataegus* with flowers in groups of 3–12, and glossy brown fruit. 'Jouinii' is similar but flowers earlier and is totally sterile.

74. OSTEOMELES Lindley
S.G. Knees
Deciduous or semi-evergreen shrubs or small trees. Leaves alternate, pinnately divided; stipules present. Flowers white in terminal nodding corymbs. Calyx-teeth 5, acute; petals 5, ovate to oblong; stamens 15–20; ovary inferior; styles 5. Fruit with persistent remains of calyx and 5 seeds.

A genus of about 10 species from eastern Asia, Polynesia and New Zealand. All species can be propagated from semi-hardwood cuttings, or seed, although the latter is not commonly produced in cultivation.

1a. Leaves with 15–31 obovate or oblong to elliptic leaflets; apices straight, acute   **1. schweriniae**
  b. Leaves with 9–29, obovate or heart-shaped leaflets; apices very obtuse and usually recurved   2
2a. Leaflets 4–8 mm, rounded, obovate or heart-shaped   **2. subrotunda**
  b. Leaflets 8–15 mm, obovate-oblong   **3. anthyllidifolia**

1. O. schweriniae Schneider (*O. anthyllidifolia* Lindley, misapplied). Illustration: *Botanical Magazine*, 7354 (1894); Krüssmann, Manual of cultivated broad-leaved trees and shrubs **2**: 338 (1986); Brickell, The RHS gardeners' encyclopedia of plants & flowers, 106 (1989); Phillips & Rix, Shrubs, 34 (1989).

Deciduous or semi-evergreen shrub to 3 m. Branches and leaf-stalks with greyish hairs; leaves 3–7 cm, hairy beneath, with 15–31 leaflets, each 4–12 mm, obovate or oblong to elliptic; apices acute, straight, ending with a short point. Flowers *c.* 1.5 cm across. Sepals lanceolate-ovate, hairless on inner face, hairy on outer face. Petals ovate-oblong, white; stamens 15–18; styles hairy. Fruit spherical or ovoid, 6–8 mm, red at first, eventually bluish black, smooth. *W China.* H5. Spring–summer.

Var. **microphylla** Rehder & Wilson. Leaves with fewer leaflets, 3–5 mm, often hairless; inflorescence more compact. Thought to be more hardy in cultivation.

2. O. subrotunda K. Koch. Illustration: *Hooker's Icones Plantarum* **27**: pl. 2644 (1900); *Gardeners' Chronicle* **103**: 125 (1938); Makino, Illustrated flora of Japan, 1389 (1956).

Slow-growing, rigid, evergreen or semi-deciduous shrub, rarely exceeding 1 m. Branches twisted, hairy when young. Leaflets 9–17, 4–8 mm, rounded or obovate, sometimes heart-shaped, hairs beneath sparse and closely adpressed, apex obtuse, recurved, margin hairy. Flowers *c.* 1 cm across, in corymbs 2–3 cm wide; styles smooth. *SE China.* H5–G1. Summer.

Perhaps best grown in containers and overwintered in a cool glasshouse.

3. O. anthyllidifolia (Smith) Lindley (*Pyrus anthyllidifolia* Smith). Illustration: Krüssmann, Manual of cultivated broad-leaved trees and shrubs **2**: 338, pl. 145 (1986); Wagner et al., Manual of the flowering plants of Hawaii **2**: 1101 (1990).

Semi-evergreen shrub to 2 m; shoots densely hairy. Leaves glossy, sparsely hairy above, densely silky hairy beneath; leaflets 13–29, 8–15 mm, obovate-oblong, apex

obtuse. Flowers similar to *O. subrotunda* but with styles shaggy. Fruit spherical, to 8 mm, red and hairy at first, eventually bluish black and completely smooth. *Hawaii, Polynesia, Bonin Isles.* H5–G1. Summer.

75. × PYRACOMELES Guillaumin
S.G. Knees
Evergreen shrub to 2 m, with slender young shoots, greyish downy at first, soon hairless. Leaves 2.5–3.2 cm, pinnate at base, pinnately lobed towards apex; leaflets 5–9, ovate, rounded and toothed at apex, hairless, or nearly so. Flowers *c.* 1 cm across, numerous in terminal corymbs. Sepals 5; petals white; stamens 12–15. Fruit *c.* 4 mm across, spherical, coral-red with 4 or 5 nutlets.

A genus of hybrids between *Pyracantha* and *Osteomeles*, said to come true from seed.

1. × P. vilmorinii Rehder; *Pyracantha atalantioides × Osteomeles subrotunda.* Illustration: *Gardeners' Chronicle* **103**: 125 (1938).

Raised in cultivation. H4. Spring.

76. PRUNUS Linnaeus
J.C.M. Alexander, assisted by J. Cullen, C.J. King & P.P. Yeo
Deciduous or evergreen trees or shrubs of very variable habit, sometimes spiny. Leaves alternate, usually toothed, generally with 1 or more conspicuous glands (extra-floral nectaries) on the stalk; stipules present, sometimes falling early. Flowers solitary or borne in racemes, umbels, corymbs or clusters, with distinct campanulate or cylindric perigynous zones. Sepals 5, often toothed, sometimes reddish. Petals 5 (more in variants with semi-double or double flowers), white, cream, or pink to red, usually spreading. Stamens numerous. Ovary of a single carpel borne in the base of the perigynous zone; ovules 2. Fruit a drupe, generally fleshy, mostly 1-seeded, sometimes large. Stone pitted, ridged or smooth, sometimes keeled. Figure 99(4), p. 183.

A genus of more than 200 species widespread in the north temperate area, a few on mountains in the tropics (Andes, SE Asia). Several of the species produce economically important fruits (cherries, peaches, plums, damsons, nectarines, almonds, etc.) and many more are grown for their flowers, which are often very abundantly produced, frequently before the leaves have opened. Many are grown as street trees and some are used for hedging.

Most are easily grown in good soil and succeed in full sun. Propagation of most species is by seed (which requires stratification), though seed produced in gardens may well be of hybrid origin. Many may also be propagated by semi-ripe hardwood cuttings. Most of the cultivars require budding or grafting, with species such as *P. avium*, *P. mahaleb* and *P. padus* as stocks. Several of the evergreen species may be propagated by layering. The plants are susceptible to a wide range of fungal and insect pests and may require careful management.

A large number of species is grown for ornament, and many cultivars have been selected and raised (both from the species and from numerous hybrids between them). The classification and identification of the plants is therefore somewhat difficult. There is a hierarchy of subgenera and sections, based mainly on fruit, seed and bud characters. As these are not helpful in identification, they are not included here. The key below avoids fruit characters and is based on the flowers and fully expanded foliage leaves. For those species that flower well before the leaves open, specimens of flowers and mature leaves will be required.

Some authors raise some of the subgenera to the rank of genus. These are: *Prunus* in the strict sense (species related to *P. spinosa*), *Cerasus* Miller (species related to *P. avium*), *Armeniaca* Scopoli (*P. armeniaca*), *Persica* Miller (*P. persica*), *Amygdalus* Linnaeus (species related to *P. dulcis*), *Padus* Miller (species related to *P. padus*) and *Laurocerasus* Duhamel (species related to *P. laurocerasus*). Names based on these genera may be found occasionally in the botanical literature; they are not cited here as synonyms.

In the descriptions below, the perigynous zone should be assumed to be campanulate (that is, widening gradually from the rounded base for at least half its length) unless it is stated to be cylindric (that is, widening abruptly at the base and parallel-sided for usually more than half its length). A few species are intermediate and are described as tubular-campanulate. It should be noted that the account in the 1991 edition of the Royal Horticultural Society's dictionary of gardening uses these terms in a different sense. It should also be assumed that the flowers of the deciduous-leaved species are borne before the leaves, unless otherwise stated.

Literature: Miyoshi, M., Japanese mountain cherries, their wild forms and cultivars, *Journal of the College of Science of the Imperial University, Tokyo* **24**: 1–175 (1916); Wilson, E.H., The cherries of Japan (1916); Koehne, E., Die Kirschenarten Japans, *Mitteilungen der Deutsche Dendrologische Gesellschaft* **1917**: 1–65 (1917); Ingram, C., Ornamental cherries (1948); Sano, T., Flowering cherries (1961); Chadbund, G., Flowering cherries (1972); Ohwi, J. & Ohta, Y., Flowering cherries of Japan (1973); Flower Society of Japan, Manual of Japanese flowering cherries (1982); Kuitert, W., Japanese flowering cherries (1999).

1a. Flowers usually 10 or more in narrow, cylindric racemes, if fewer then very dense and bark shiny brownish yellow　2
　b. Flowers solitary or 2–10 in clusters or short, broad, open racemes　18
2a. Evergreen trees or shrubs　3
　b. Deciduous trees or shrubs　9
3a. Racemes to 4 cm　4
　b. Racemes more than 5 cm　5
4a. Leaves sparsely toothed
　　　88. caroliniana
　b. Leaves spiny　**86. ilicifolia**
5a. Racemes leafy towards base　6
　b. Racemes not leafy towards base　7
6a. Leaves with narrow acute tips
　　　74. salicifolia
　b. Leaves with obtuse or rounded tips
　　　73. virens
7a. Racemes erect　**85. laurocerasus**
　b. Racemes ascending, spreading or hanging　8
8a. Leaves coarsely toothed
　　　84. lusitanica
　b. Leaves entire to finely and distantly toothed　**87. lyonii**
9a. Racemes not leafy towards base　10
　b. Racemes leafy towards base　11
10a. Leaves gland-dotted beneath; bark shiny yellowish brown
　　　82. maackii
　b. Leaves not gland-dotted beneath
　　　83. buergeriana
11a. Margins of mature leaves scalloped or with incurved teeth at least in apical part　12
　b. Margins of mature leaves with straight projecting teeth　13
12a. Main stalk of inflorescence densely hairy; leaf-tips acute-acuminate
　　　76. vaniotii

　b. Main stalk of inflorescence hairless or sparsely hairy; leaf-tips rounded-acuminate　**72. serotina**
13a. Mature leaves with narrow tips 1 cm long or more　14
　b. Mature leaves with short abrupt tips less than 1 cm　15
14a. Leaf-bases rounded; styles projecting　**81. grayana**
　b. Leaf-bases cordate; styles not projecting　**80. ssiori**
15a. Petals more than 6 mm　**75. padus**
　b. Petals 5 mm or less　16
16a. Main stalk of inflorescence densely hairy　**78. napaulensis**
　b. Main stalk of inflorescence hairless or sparsely hairy　17
17a. Petals 2–3 mm; leaves shiny above, broadest towards tip, mature blades usually 10 cm or less
　　　79. virginiana
　b. Petals 4–5 mm; leaves dull above, broadest near middle, mature blades usually more than 10 cm
　　　77. cornuta
18a. Flowers stalkless or stalks to 9 mm
　　　19
　b. Flower-stalks more than 9 mm　64
19a. Style or ovary (and fruit) hairy　20
　b. Style, ovary and fruit hairless　39
20a. Leaves folded in bud　21
　b. Leaves rolled in bud　36
21a. Mature leaves to 2 cm wide　22
　b. Most mature leaves 2 cm wide or more　28
22a. Leaves densely white-felted beneath, of 2 types　23
　b. Leaves not white-felted beneath, all similar　24
23a. Flowers hairless or sparsely hairy outside, stalkless　**63. incana**
　b. Flowers densely hairy outside, stalks *c.* 3 mm　**65. bifrons**
24a. Young wood with a greyish or whitish bloom　25
　b. Young wood greenish, reddish, or purplish, not bloomed　27
25a. Often thorny; petals overlapping; sepals reddish, not or scarcely toothed　26
　b. Not thorny; petals not overlapping; sepals greenish, densely toothed
　　　33. tenella
26a. Leaves pale greyish green, widest near base; stalks reddish; flowers white or pale pink
　　　26. fenzliana
　b. Leaves dark greyish green, widest near middle; stalks green; flowers pink or red　**31. tangutica**

27a. Leaves narrowly elliptic, acute, entire to finely toothed, hairless above **28. kansuensis**

b. Leaves broadly elliptic, acuminate, coarsely toothed, bristly above **37. canescens**

28a. Leaves broadly elliptic or ovate, sometimes 3-lobed, with large broad triangular teeth 29

b. Leaves narrowly elliptic or lanceolate, scalloped or finely toothed 32

29a. Stipules narrowly linear, fringed, sometimes falling early 30

b. Stipules broadly elliptic to circular, persistent **37. canescens**

30a. Leaves hairless beneath except on midrib and veins; stipules falling by midsummer **42. subhirtella**

b. Leaves hairy beneath; stipules persistent 31

31a. Flowers pink, 1.8–3.5 cm wide, often double **32. triloba**

b. Flowers white with pinkish centre, *c.* 1.5 cm wide, single **62. tomentosa**

32a. Leaves with acute, forward-pointing teeth, 2 mm or more apart at widest part of leaf **29. davidiana**

b. Leaves scalloped or more closely toothed 33

33a. Leaves tapered at base, widest near the middle 34

b. Leaves rounded at base, widest below the middle 35

34a. Sepals hairy; angle between side-veins and midrib of leaf obtuse **27. persica**

b. Sepals hairless; angle between side-veins and midrib acute **67. glandulosa**

35a. Leaves hairy beside midrib beneath; stone smooth **30. mira**

b. Leaves hairless; stone pitted **25. dulcis**

36a. Leaves fringed with dense, long, narrow teeth to 2 mm or more **20. mandshurica**

b. Leaves scalloped, or with broad triangular teeth to 1 mm 37

37a. Leaves wedge-shaped to narrowly rounded at base; blade length *c.* 2 times the width **23. mume**

b. Leaves broadly rounded to truncate at base; blade length *c.* 1.5 times the width 38

38a. Most leaves with a gradual point *c.* 1.5 cm **21. sibirica**

b. Leaves with an abrupt point to 1 cm **22. armeniaca**

39a. Leaves rolled in bud 40

b. Leaves folded in bud 51

40a. Flowers 1.8 cm or more across 41

b. Flowers 1.5 cm or less across 44

41a. Leaves truncate or slightly cordate at base; flowers almost stalkless **19. brigantina**

b. Leaves wedge-shaped to rounded at base; flower-stalks 5 mm or more 42

42a. Leaves upright, narrow, with few veins diverging at a very acute angle **7. simonii**

b. Leaves horizontal or drooping, veins diverging at an angle of *c.* 45° 43

43a. First-year twigs shiny, green (purple in some cultivars); leaves hairless beneath except on midrib and veins **3. cerasifera**

b. First-year twigs dull, brown to grey; leaves hairy beneath, especially when young **2. domestica**

44a. Mature leaves boat-shaped, often curved along midrib 45

b. Mature leaves not boat-shaped 46

45a. Calyx-lobes hairless, margin without glands **16. angustifolia**

b. Calyx-lobes hairy inside, margin glandular **17. reverchonii**

46a. Leaves acute or acuminate at tip 47

b. Leaves rounded or obtuse at tip 49

47a. Flowers becoming pink with age 48

b. Flowers remaining white **15. munsoniana**

48a. Leaves rounded at base; sepals *c.* 1.6 cm **10. alleghaniensis**

b. Leaves tapered at base; sepals 2.2–2.5 cm **18. × orthosepala**

49a. Flowers becoming pink with age, stalks hairy **8. subcordata**

b. Flowers remaining white, stalks hairless 50

50a. Leaf-stalks hairless **4. cocomilia**

b. Leaf-stalks with short hairs **1. spinosa**

51a. Leaves purplish at first, remaining reddish brown when mature **70. × cistena**

b. Leaves green when mature 52

52a. Leaves long-acuminate; tips acute or obtuse 53

b. Leaves not acuminate, sometimes abruptly short-pointed 58

53a. Leaves with acute teeth standing out from margin 54

b. Leaves scalloped or with obtuse teeth, if teeth acute then not standing out from margin 56

54a. Sepals with gland-tipped marginal teeth **66. japonica**

b. Sepals without marginal teeth 55

55a. Mature leaves with 8–10 teeth per cm; flowers deep or mid-pink **44. cerasoides**

b. Mature leaves with 4–6 teeth per cm; flowers white to pale pink **47. hirtipes**

56a. Flowers 2 cm across or more **27. persica**, see var. **nucipersica**

b. Flowers less than 2 cm across 57

57a. Sepals hairy; leaf-bases rounded **10. alleghaniensis**

b. Sepals hairless; leaf-bases tapered **6. consociiflora**

58a. Leaves greyish glaucous beneath, not toothed or scalloped near the base 59

b. Leaves not glaucous beneath, toothed or scalloped all round 60

59a. Leaves tapered equally at tip and base, ascending **69. pumila**

b. Leaves tapered narrowly at base, abruptly at tip, spreading **71. besseyi**

60a. Perigynous zone tubular; flowers stalkless or very shortly stalked **64. jacquemontii**

b. Perigynous zone cup-shaped; flower-stalks 5 mm or more 61

61a. Leaves broadly elliptic, length less than 2 times the width **9. maritima**

b. Leaves narrowly ovate or elliptic, length *c.* 2.5 times the width or more 62

62a. Tree to 8 m; leaves to 10 × 4 cm **15. munsoniana**

b. Shrub to 2 m; leaves to 7 × 3 cm 63

63a. Leaves to 4 × 2 cm, boat-shaped, often curved along midrib **68. humilis**

b. Leaves to 7 × 3 cm, flat **67. glandulosa**

64a. Leaves rolled in bud 65

b. Leaves folded in bud 69

65a. Ovary and base of style (and fruit) hairy **24. × dasycarpa**

b. Ovary and style hairless 66

66a. Leaf-tips obtuse to rounded **8. subcordata**

b. Leaf-tips acute or acuminate 67

67a. Leaf-margin scalloped or with rounded teeth **5. salicina**

b. Leaf-margin with obtuse to acute teeth 68

68a. Flowers usually solitary, to 2.5 cm wide; leaves often purple **3. cerasifera**

b. Flowers 2–4 together, to 1.2 cm wide **10. alleghaniensis**

69a. Sepals becoming reflexed 70

b. Sepals not becoming reflexed 88

70a. Flower-clusters lacking a distinct stalk; flower-stalks all arising from the same point in each cluster (stalkless umbels) 71

b. Flower-clusters clearly stalked; flower-stalks arising at one or more points (stalked umbels or corymbs) 80

71a. Leaves purplish at first, becoming reddish brown when mature **70. × cistena**

b. Leaves green when mature 72

72a. Leaves with acute or acuminate teeth 73

b. Leaves with obtuse or rounded teeth 75

73a. Leaves hairy beneath **12. mexicana**

b. Leaves hairless or almost so beneath 74

74a. Leaf-tips strongly acuminate **11. americana**

b. Leaf-tips evenly tapered **58. pensylvanica**

75a. Flowers 2 cm wide or more 76

b. Flowers less than 2 cm wide 78

76a. Flower-clusters with leaf-like bracts **35. cerasus**

b. Flower-clusters without leaf-like bracts (large bud-scales may be present at base of cluster) 77

77a. Leaves with coarse, narrow, projecting rounded teeth; flower-stalks longer than 2 cm **34. avium**

b. Leaves scalloped or with short, broad rounded teeth; flower-stalks to 2 cm **13. nigra**

78a. Mature leaves 7 cm or more, rounded or broadly tapered at base **14. hortulana**

b. Mature leaves less than 6 cm, narrowly tapered at base 79

79a. Twigs grey or black; leaves scalloped, with dark marginal glands **36. fruticosa**

b. Twigs reddish brown; leaves indistinctly toothed **69. pumila**

80a. Flowers 3 cm or more across 81

b. Flowers 2.5 cm or less across 82

81a. Leaf-stalks and perigynous zone hairless **56. cyclamina**

b. Leaf-stalks and perigynous zone hairy **55. dielsiana**

82a. Teeth standing out from leaf-margin, at least 2 mm long 83

b. Teeth 1 mm or less, or leaves finely scalloped 85

83a. Teeth on leaf-margin with hair-like tips **59. pilosiuscula**

b. Teeth on leaf-margin rounded or obtuse 84

84a. Teeth on leaf-margin rounded; flowers c. 2.5 cm across; stalks arising close together **34. avium**

b. Teeth on leaf-margin obtuse; flowers c. 1.5 cm across; stalks arising from different points **61. maximowiczii**

85a. Mature leaves 6 cm or less, finely scalloped, with short abrupt tips **57. mahaleb**

b. Mature leaves more than 7 cm, toothed, with drawn out tips 86

86a. Sepals and perigynous zone hairy **54. pseudocerasus**

b. Sepals and perigynous zone hairless 87

87a. Flowers white; stalks of flower-clusters c. 6 mm **60. litigiosa**

b. Flowers pink; stalks of flower-clusters 2.5 mm or more **43. campanulata**

88a. Stalks of mature leaves at most very sparsely hairy 89

b. Stalks of mature leaves densely downy or hairy 98

89a. Mature leaf-blades 5 cm or less 90

b. Mature leaf-blades usually more than 5 cm 91

90a. Leaf-tips acuminate; midribs ginger hairy beneath **45. rufa**

b. Leaf-tips round or obtuse, rarely acute; midribs not ginger hairy **40. mugus**

91a. Teeth on mature leaves very fine and regular, at least 5 major teeth per cm 92

b. Teeth on mature leaves coarse, fewer than 4 per cm 93

92a. Flowers white; bark glossy red, peeling in strips **46. serrula**

b. Flowers pink; bark not glossy **44. cerasoides**

93a. Teeth on mature leaves acuminate or drawn out into hair-like tips 94

b. Teeth on mature leaves acute or obtuse, not hair-like 97

94a. Stems and leaves hairless or very slightly hairy 95

b. Stems, undersides of leaves or perigynous zone distinctly hairy 96

95a. Flowers in umbels; leaf-teeth acuminate **49. sargentii**

b. Flowers in racemes; leaf-teeth drawn out into hair-like tips **51. serrulata**

96a. Flowers white **52. speciosa**

b. Flowers pink **50. juddii**

97a. Flowers produced before the leaves from January–March; bracts falling as flowers open **47. hirtipes**

b. Flowers produced with the leaves in April and May; bracts persistent **41. nipponica**

98a. Teeth on mature leaves fine, less than 2 mm 99

b. Teeth on mature leaves coarse, at least 3 mm 100

99a. Flowers 1 cm wide or more; perigynous zone with dense brown hairs **45. rufa**, see var. **tricantha**

b. Flowers less than 1 cm wide; perigynous zone hairless or sparsely downy **10. alleghaniensis**

100a. Mature leaves 6 cm or less **38. incisa**

b. Mature leaves more than 6 cm 101

101a. Bracts on fully-developed inflorescence 1 cm or more 102

b. Bracts of fully developed inflorescence less than 1 cm 103

102a. Petals falling early; flowers solitary or paired **39. apetala**

b. Petals not falling early; flowers 5 or 6 in racemes **48. × yedoensis**

103a. Flowers double or semi-double, usually more than 3 cm wide **53. sieboldii**

b. Flowers single, usually less than 3 cm wide **51. serrulata**, see var. **pubescens**

1. P. spinosa Linnaeus. Illustration: Ross-Craig, Drawings of British plants **8**: t. 1 (1955); Masefield et al., The Oxford book of food plants, 67 (1969); Phillips, Trees in Britain, Europe and North America, 176 (1978); Krüssmann, Manual of cultivated broad-leaved trees and shrubs **3**: pl. 9 & f. 25 (1986).

Deciduous shrub or small tree to 4 m, often suckering; young shoots with thorns, downy when young. Leaves mostly 1–3 cm, rolled in bud, broadest above the middle, obtuse or very shortly acute, wedge-shaped at the base, more or less hairy; stalks 2–10 mm, shortly hairy. Flowers to 1.5 cm across, solitary or 2 together, white, on hairless stalks to 6 mm.

Style hairless. Fruit 1–1.5 cm across, spherical, blue-black, bloomed, with bitter green flesh. Stone nearly spherical. *Europe to W Siberia, Mediterranean area.* H1. Early spring.

Used for hedging, and, in double- and pink-flowered, purple-leaved variants ('Plena', 'Purpurea') for ornament. For making sloe gin the fruit is gathered in the wild.

2. P. × domestica Linnaeus. Figure 99 (16), p. 183.

Deciduous shrub or small tree; young shoots downy or hairless, not shiny; thorns few or none. Leaves 4–10 cm, rolled in bud, broadest above the middle, usually acute, wedge-shaped at the base, hairy beneath; borne horizontally or drooping; veins diverging at an angle of *c.* 45°; stalks 5–25 mm. Flowers white, 1.5–2 cm across, appearing with the leaves, solitary or 2–3 together on downy or hairless stalks 5–20 mm. Fruit 2–8 cm. *Europe, W Asia.* H1. Spring.

Probably a hybrid between *P. spinosa* and *P. cerasifera* var. *divaricata*; we follow Sell (*Nature in Cambridgeshire* **33**: 29–39, 1991 & **34**: 59–60, 1992) in taking a narrow view of *P. spinosa* and *P. cerasifera* and including deviating variants in *P. × domestica.*

Three subspecies are recognised, but there are many intermediates between them.

Subsp. **domestica**. Illustration: Boswell, English botany, edn 3, **3**: t. 410 (1886); Hegi, Illustrierte Flora von Mittel-Europa **4**: f. 1275k (1923); Masefield et al., The Oxford book of food plants, 69, f. 2–6 (1969). Thornless tree to 12 m, with the leaves hairy on both sides, the fruits 4–8 cm, almost spherical to oblong-ovoid, sweet, red or purple and the stone strongly flattened, sharply angled and free from the flesh.

The plum, widely cultivated as a fruit crop.

Subsp. **institia** (Linnaeus) Bonnier & Layens (*P. institia* Linnaeus; *P. × damascena* Dierbach). Illustration: Boswell, English botany, edn 3, **3**: t. 409 (1886); Taylor, Plums of England, 33 (1949); Masefield et al., The Oxford book of food plants, 67, f. 5, 6 (1969). Shrub or tree to 6 m, often thorny, leaves hairy on both sides, the fruits usually 2–3 cm, spherical or shortly ovoid, sweet or acid, yellow, red or purple, the stone not strongly flattened but angled and adhering to the flesh.

The damson or bullace (fruit flesh acid) or mirabelle (fruit flesh sweet).

Subsp. **italica** (Borkhausen) Gams & Hegi (*P. × italica* Borkhausen). Illustration: Hegi, Illustrierte Flora von Mittel-Europa **4**: f. 12751 & 1 (1923); Taylor, Plums of England, 73,129 (1949); Masefield et al., The Oxford book of food plants, 69, f. 5, 6 (1969). Small tree, intermediate in size between subsp. *domestica* and subsp. *institia*, with the leaves hairless above and the fruit 3–5 cm, almost spherical, sweet, green, the stone neither strongly flattened nor sharply angled, adhering to the flesh.

The greengage, probably of garden origin.

3. P. cerasifera Ehrhart (*P. myrobalana* Loiseleur). Illustration: *Botanical Magazine*, 6519 (1880); Taylor, Plums of England, 104 (1949); Masefield et al., The Oxford book of food plants, 67, f. 1 (1969); Phillips, Trees in Britain, Europe and North America, 171 (1978).

Deciduous shrub or small tree to 8 m, without suckers; young shoots hairless, remaining green into the second year (purple in some cultivars), occasionally with spines. Leaves 3–7 cm, rolled in bud, broadest below the middle, acute or acuminate, base rounded or wedge-shaped, hairless above, hairy only on the midrib and veins beneath; borne horizontally or drooping; veins diverging at an angle of *c.* 45°; stalks 5–10 mm, hairy. Flowers 2–2.5 cm across appearing with the leaves, usually solitary, on stalks 0.5–2.5 mm. Fruit spherical, yellow or red, sweet or insipid, spherical, 2–2.5 cm across; stone almost circular in outline and slightly flattened. *SE Europe, SW Asia.* H2. Spring.

Var. **divaricata** (Ledebour) Bailey (*P. monticola* Koch) is the species as found in the wild; it has a more slender habit, with looser branching and the leaves are rounded at the base; the flowers are small, white and the fruits are yellow, to 2 cm across. It is rarely cultivated.

The cultivated plant is usually found as dark-leaved cultivars grown as small ornamental trees or for hedging. These are: 'Pissardii' (var. *atropurpurea* Jaeger). Illustration: Nicholson & Clapham, The Oxford book of trees, 180 (1975); Mitchell, The complete guide to trees of Britain and northern Europe, 79 (1985). This has pink buds, pinkish white petals and blackish purple leaves. 'Nigra'. Illustration: Nicholson & Clapham, The Oxford book of trees, 180 (1975); Mitchell, The complete

guide to trees of Britain and northern Europe, 79 (1985). Petals pink, leaves dark red.

P. × blireana André; *P. cerasifera × P. mume*. Illustration: *Revue Horticole* 1905: 392 (1905); Macoboy, What tree is that?, 177 (1979). Like *P. cerasifera* but with broader, purple, acuminate leaves, double pink flowers on stalks 1–1.5 cm, and a downy ovary. *Garden origin.*

4. P. cocomilia Tenore (*P. pseudoarmeniaca* Heldreich & Sartorelli). Illustration: Polunin, Flowers of Greece and the Balkans, pl. 19 (1980); Pignatti, Flora d'Italia **1**: 616 (1982).

Like *P. cerasifera* but plant spiny, hairless, with leaves to 4 cm, rounded or obtuse, flowers usually 2 together, to 1 cm across on stalks no longer than the perigynous zones, fruit one third longer than wide, yellow. *S Italy, Sicily, S Balkan Peninsula.* H5. Spring.

5. P. salicina Lindley (*P. triflora* Roxburgh). Illustration: *Revue Horticole* **1895**: 160 (1895); Krüssmann, Manual of cultivated broad-leaved trees and shrubs **3**: f. 24, 25 (1986).

Deciduous tree to 10 m; young shoots hairless, becoming red-brown and shiny. Leaves 6–10 cm, rolled in bud, oblong-obovate to elliptic-obovate, acute or acuminate, wedge-shaped at the base, glossy, doubly round-toothed; stalks 1–2 cm. Flowers 1.5–2 cm across, solitary to 3 together on hairless stalks 1–1.5 cm. Sepals slightly toothed, hairless. Petals white. Style hairless. Fruit 5–7 cm, almost spherical, though with a depression at the attachment, green, yellow or red, with a deep furrow on the side. *China, Korea.* H2. Spring.

The Japanese plum, grown commercially for its fruit outside Europe.

6. P. consociiflora Schneider. Illustration: *Botanical Magazine*, n.s., 805 (1980).

Like *P. salicina* but leaves to 7 cm, folded in bud, with glandular teeth; flowers to 1.2 cm across, 2–5 together. *C China.* H2. Spring.

7. P. simonii Carrière. Illustration: *Revue Horticole* **1872**: 110 (1872); *The Garden* **70**: 225 (1906); Krüssmann, Manual of cultivated broad-leaved trees and shrubs **3**: f. 23 (1986).

Small deciduous tree with erect branches; young shoots hairless. Leaves 7–10 cm, rolled in bud, narrowly ovate to oblong-obovate, acuminate, wedge-shaped or rounded at the base, finely bluntly toothed, hairless; borne upright; veins few, diverging at a very acute angle. Flowers 2–2.5 cm across, white, solitary or to 3 together on stalks less than 1 cm. Style hairless. Fruit to 3 × 5 cm, with a deep furrow on the side, dark red with sweet, fragrant, yellow flesh. Stone small, almost circular in outline, rough, adhering to the flesh. *N China*. H2. Spring.

The apricot plum, cultivated for its fruit in areas free from frost.

8. P. subcordata Bentham. Illustration: Sargent, Silva of North America **4**: 154 (1892); Hitchcock et al., Vascular plants of the Pacific northwest **3**: 163 (1971); Krüssmann, Manual of cultivated broad-leaved trees and shrubs **3**: f. 11 (1986).

Deciduous shrub or small tree to 8 m, branches spreading; young shoots hairy or hairless, at first bright red, later greyish brown. Leaves 3–7 cm, rolled in bud, broadly ovate to almost circular, sometimes weakly cordate, obtuse, sharply and often doubly toothed, hairy beneath at least when young; stalks 1–2 cm. Flowers to 1.5 cm across, 2–4 together on stalks to 1.2 cm. Sepals downy on both sides. Petals white becoming pink. Style hairless. Fruit 1.5–3 cm, ellipsoid, dark red, bloomed, variably acid. *W USA*. H4. Spring.

The Pacific or Oregon plum, cultivated for its fruit in California.

9. P. maritima Marsh (*P. acuminata* Michaux; *P. pubescens* Pursh). Illustration: *Botanical Magazine*, 8289 (1909); Gleason, Illustrated flora of the north-eastern United States and adjacent Canada **2**: 332 (1952).

Straggling shrub to 2 m, lower branches often arching downwards; young shoots hairy. Leaves 4–7.5 cm, folded in bud, ovate, elliptic or obovate, acute, broadly wedge-shaped at base, finely saw-toothed, hairless above, downy beneath; stalks 4–6 mm, downy. Flowers 1.2–1.5 cm across, in umbels of 2–10, on stalks 5–7 mm. Sepals often toothed, downy on both sides. Petals white. Style hairless. Fruit 1.5–2 cm, spherical, red to purple, bloomed, with a deep furrow on the side; stone ovoid in outline, flattened. *E USA*. H2. Spring.

The Beach plum, grown for its fruit in North America and for ornament in Europe.

P. × dunbarii Rehder; *P. americana* × *P. maritima*. Similar, but shoots ultimately hairless, leaves larger, more coarsely toothed, acuminate, less downy beneath; fruits larger, purple, the stone more compressed. *Garden origin*. H2. Spring.

10. P. alleghaniensis Porter. Illustration: Sargent, Silva of North America **3**: t. 153 (1892); Gleason, Illustrated flora of the north-eastern United States and adjacent Canada **2**: 333 (1952).

Straggling shrub or tree to 6 m; shoots becoming reddish brown, sometimes downy at first, sometimes with thorns. Leaves rolled or folded in bud, 6–9 cm, narrowly elliptic to lanceolate, acuminate, wedge-shaped at the base, finely saw-toothed, hairy at least when young; stalks 7–12 mm, downy, rarely with glands. Flowers 1–1.2 cm across, appearing with the leaves, 2–5 together on stalks 5–16 mm. Sepals more or less hairy. Petals white becoming pinkish. Style hairless. Fruit 8–16 mm across, almost spherical, dark purple, bloomed; stone 6–12 mm, ovoid. *E USA*. H2. Spring.

11. P. americana Marsh (*P. lanata* (Sudworth) Mackenzie & Bush). Illustration: Sargent, Silva of North America **3**: t. 150 (1892); Sargent, Manual of the trees of North America, edn 2, 562, 563 (1922); Gleason, Illustrated flora of the north-eastern United States and adjacent Canada **2**: 332 (1952).

Deciduous small shrub or tree to 8 m, suckering and forming thickets; trunk divided 1–2 m above the ground, branches pendulous at tips; twigs with thorns. Leaves 5–12 cm, folded in bud, narrowly to broadly ovate or obovate, abruptly or gradually acuminate, broadly tapered or rounded at the base, finely and acutely saw-toothed, hairless or with hairs along the midrib beneath; stalks usually without glands. Flowers 2–3 cm across, 2–5 together on stalks 8–16 mm. Sepals sometimes hairy, often toothed but almost without glands on the teeth, reflexed at the end of flowering. Petals white. Style hairless. Fruit 2–3 cm, spherical or almost so, red or yellowish; stone not or slightly flattened. *E & C USA*. H2. Spring.

12. P. mexicana Watson (*P. arkansana* Sargent). Illustration: Sargent, Manual of the trees of North America, edn 2, 565 (1922); Steyermark, Flora of Missouri, 859 (1963).

Like *P. americana* but a non-suckering tree to 12 m, leaf-stalks with glands, leaves slightly cordate at base, downy beneath when mature, flowers 1.5–2 cm across; fruit purplish. *SC USA & N Mexico*. H3. Spring.

13. P. nigra Aiton (*P. americana* Marsh var. *nigra* Wanghenhein; *P. borealis* Poiret; *P. emarginata* (J.D. Hooker) Eaton). Illustration: *Botanical Magazine*, 1117 (1808); Sargent, Silva of North America **3**: t. 149 (1892); Sargent, Manual of the trees of North America, edn 2, 561 (1922); Rosendahl, Trees and shrubs of the upper Midwest, edn 2, 253 (1955).

Like *P. americana* but narrow-crowned, with upright branches; leaf-stalks 1.2–1.5 cm, with glands near the top, leaf-teeth obtuse and gland-tipped, flower-stalks, perigynous zone and sepals reddish, sepals gland-toothed, petals sometimes becoming pink; fruit more ovoid, stone flattened. *E Canada, E & NC USA*. H3. Spring.

14. P. hortulana Bailey. Illustration: Sargent, Silva of North America **3**: t. 151 (1892); Sargent, Manual of the trees of North America, edn 2, 568 (1922); Steyermark, Flora of Missouri, 859 (1963).

Tree to 10 m, not suckering; young shoots hairless. Leaves 7–11 cm, folded in bud, rather thick, ovate-lanceolate to oblong-obovate, gradually acuminate, broadly tapered or rounded at the base, shallowly saw-toothed with incurved, gland-tipped teeth, downy beneath when young; stalks 1.5–3 cm with several glands. Flowers 1.2–1.5 cm across, appearing with the leaves, 2–4 together, on stalks c. 1.2 cm. Sepals glandular-hairy, downy inside and sometimes outside, usually finally reflexed. Petals white, Style hairless. Fruit spherical, 1.8–3 cm, red or yellow; stone not much flattened, pointed at both ends. *C USA*. H4. Spring.

15. P. munsoniana Wight & Hedrick. Illustration: Sargent, Manual of the trees of North America, edn 2, 568 (1922); Gleason, Illustrated flora of the north-eastern United States and adjacent Canada **2**: 333 (1952); Braun, The woody plants of Ohio, 218 (1961); Steyermark, Flora of Missouri, 859 (1963).

Like *P. hortulana* but suckering and forming thickets; leaves rolled (or folded?) in bud, leaf-teeth each with a gland on the side, flowers borne on short spurs, sometimes appearing before the leaves, fruit 1.5–2 cm, stone truncate or obliquely truncate at the base. *SC USA*. H2. Spring.

There is some confusion in the literature as to whether the young leaves of this species are rolled or folded in bud.

16. P. angustifolia Marsh. Illustration: Sargent, Silva of North America **3**: 152 (1892); Sargent, Manual of the trees of North America, edn 2, 570 (1922); Gleason, Illustrated flora of the northeastern United states and adjacent Canada **2**: 333 (1952).

Deciduous shrub or small tree to 7 m, suckering and forming thickets; twigs flexuous, hairless, reddish, spiny. Leaves 2–5 cm × 5–20 mm, rolled in bud, lanceolate to oblong-lanceolate, acute, tapered to the base, hairless beneath except on the midrib, margins more or less incurved to give a boat-shape; stalk 6–12 mm, red, usually with glands at the top. Flowers 8–9 mm across, 2–4 together on stalks 3–12 mm. Sepals hairy inside at the base, without glands, finally reflexed. Petals white. Style hairless. Fruit to 1.2 cm, almost spherical, red or yellow; stone ovoid. *SE USA.* H4. Spring.

Var. **angustifolia** is described above.

Var. **watsonii** (Sargent) Bailey (illustration: *Garden & Forest* **7**: 135, 1894) grows only to 4 m and has smaller leaves and flowers, finer leaf-teeth and a thicker-skinned fruit. *SC USA.* H4. Spring.

Var. *angustifolia* has not been grown successfully in most of Europe, whereas var. *watsonii* is sometimes seen.

17. P. reverchonii Sargent.

Like *P. angustifolia* but a shrub to 2 m with leaf-stalks to 1.2 cm, leaves bluntly toothed, to 7 cm, sepals glandular-hairy, fruit to 2 cm, and the stone pointed at both ends. *USA (Oklahoma & Texas).* H2. Spring.

18. P. × orthosepala Koehne; *P. americana × P. angustifolia* var. *watsonii.* Illustration: *Garden & Forest* **7**: 187 (1894).

Like *P. angustifolia* but leaf-stalks to 1.8 cm, leaves to 7.5 × 3.5 cm, saw-toothed with incurved, thickened but rarely glandular teeth, acuminate, not boat-shaped; flowers to 1.6 cm across, sepals densely hairy inside, petals becoming pink, fruit to 2.5 cm, dark blue, bloomed, stone flattened. *Garden origin.* H2. Spring.

19. P. brigantina Villars. Illustration: Krüssmann, Manual of cultivated broad-leaved trees and shrubs **3**: f. 5 (1986).

Shrub or small tree to 6 m; young shoots hairless. Leaves rolled in bud, ovate to elliptic, shortly acuminate, truncate or slightly cordate at base, doubly toothed, hairy beneath at least on the veins. Flowers almost stalkless. Flowers white or pale pink, to 2 cm across, 2–5 in clusters. Fruit rounded, yellow, smooth. *SE France.* H5. Spring.

Long cultivated for the scented oil obtained from the seeds.

20. P. mandshurica (Maximowicz) Koehne. Illustration: Krüssmann, Manual of cultivated broad-leaved trees and shrubs **3**: pl. 6 (1986).

Small wide shrub to 6 m (perhaps a much larger tree in the wild); young shoots dark red-brown. Leaves rolled in bud, broadly elliptic to ovate, abruptly acuminate, fringed with dense, long, narrow teeth to 2 mm or more, hairy in the vein-axils beneath. Flowers solitary, pale pink, to 3 cm across, stalks to 5 mm. Style or ovary (and fruit) hairy. Fruit spherical to wider than long, to 2.5 cm across, yellow, sour; stone small, smooth. *N China, Korea.* H1. Spring.

21. P. sibirica Linnaeus. Illustration: Krüssmann, Manual of cultivated broad-leaved trees and shrubs **3**: pl. 6 & f. 5 (1986).

Small upright tree or shrub to 5 m. Leaves rolled in bud, ovate, blade length about 1.5 times the width, long-acuminate, point *c.* 1.5 cm long, broadly rounded to truncate at base, simply and finely toothed, reddish when young, hairless. Flowers solitary, almost stalkless, pink or white, to 3 cm across. Style or ovary (and fruit) hairy. Fruit almost spherical, 1.5–2 cm across, yellow with reddish patches; stone smooth, winged or angular. *E former USSR, N China.* H1. Spring.

Sometimes regarded as a variety of *P. armeniaca.*

22. P. armeniaca Linnaeus. Illustration: Gleason, Illustrated flora of the north-eastern United States and adjacent Canada **2**: 331 (1952); Krüssmann, Manual of cultivated broad-leaved trees and shrubs **3**: pl. 6 & f. 5, 24 (1986).

Round-crowned tree to 10 m or a smaller shrub; young shoots reddish brown, glossy, hairless. Leaves rolled in bud, broadly ovate or ovate-circular, broadly rounded to truncate at base; blade length about 1.5 times the width, abruptly acuminate, point to 1 cm, closely and obtusely toothed, with teeth to 1 mm. Flowers solitary or 2 together, pink, to 2.5 cm across, stalkless or stalks to 9 mm. Style and ovary hairy. Fruits yellow, often with a red patch, shortly hairy, to 3 cm across; stone large, smooth, with a furrowed margin. *N China.* H3. Spring.

The apricot, long cultivated for its fruit. Variegated ('Variegata') and pendulous-branched ('Pendula') variants are grown for ornament.

23. P. mume Siebold & Zuccarini. Illustration: *Revue Horticole* **1885**: t. 564 (1885); Krüssmann, Manual of cultivated broad-leaved trees and shrubs **3**: pl. 6 & f. 5, 16, 23 (1986).

Similar to *P. armeniaca* but young shoots green, leaves ovate to elliptic, long-acuminate, blade length about 2 times the width, wedge-shaped to narrowly rounded at base. Flowers stalkless, white to dark pink, to 3 cm across, fragrant; fruit sour to bitter, not separating from the pitted stone. *S Japan.* H3. Spring.

24. P. × dasycarpa Ehrhart; *P. armeniaca × P. cerasifera.* Illustration: Krüssmann, Manual of cultivated broad-leaved trees and shrubs **3**: f. 5 (1986).

Like *P. armeniaca* but leaves hairy on the veins beneath, flower-stalks more than 9 mm, hairy, ovary and base of style hairy; fruit red or blackish purple, bloomed, sour. *Origin uncertain, long cultivated in W & C Asia.* H4. Spring.

25. P. dulcis (Miller) D.A. Webb (*P. amygdalus* Batsch; *P. communis* Arcangeli). Illustration: Bonnier, Flore complète **3**: pl. 167 (1914); Lancaster, Trees for your garden, 106 (1974); Phillips, Trees in Britain, Europe and North America, 171 (1978); Edlin, The tree key, 145 (1978).

Upright tree to 10 m; young shoots hairless. Leaves folded in bud, narrowly elliptic or lanceolate, scalloped or finely toothed, rounded at base, widest below the middle, hairless. Flowers solitary or paired, stalkless or almost so, white or pale pink, 3–5 cm across. Style, ovary and fruit hairy. Fruit ovoid, flattened; stone pitted. *SW Asia, N Africa.* H4. Early spring.

The almond, long cultivated for its fruits (the kernels of the stones). Three varieties occur in the wild and in commercial cultivation, but they are not important in gardens, where several cultivars are grown.

P. × persicoides Ascherson & Graebner (*P. × amygdalopersica* (Weston) Rehder, invalid) is the hybrid between *P. dulcis* and *P. persica.* Illustration: Macoboy, What tree is that?, 176 (1979). It has occasional thorns, more finely toothed leaves and flowers 4–5 cm across, pink with a darker centre. Its naming is complicated and the name used above may prove not to be correct.

26. P. fenzliana Fritsch. Illustration: Krüssmann, Manual of cultivated broad-leaved trees and shrubs **3**: pl. 12 & f. 10 (1986).

Like *P. dulcis* but often thorny and with the young wood with a greyish or whitish bloom; leaves widest near the base, greyish or bluish green. *Former USSR (Caucasus).* H1. Spring.

27. P. persica (Linnaeus) Batsch. Illustration: Bonnier, Flore complète **3**: pl. 166 (1914); Edlin, The tree key, 145 (1978); Everett, New York Botanical Gardens illustrated encyclopedia of horticulture **8**: 2822 (1981); Lunardi (ed.) Macdonald encyclopedia of shrubs and trees, 127 (1988).

Tree to 8 m; young shoots hairless. Leaves folded in bud, narrowly elliptic or lanceolate, finely toothed, tapered at base, widest near the middle; angle between side-veins and midrib of leaf obtuse. Flowers solitary, stalkless or almost so, 2.5–3.5 cm across, pink or red. Sepals, style, ovary and fruit hairy. Fruit almost spherical, to 7 cm across. *N & C China.* H1. Late spring.

The peach, widely cultivated for its fruit since ancient times. Var. **nucipersica** Schneider, the nectarine, also long-cultivated, has hairless fruits; it is unknown in the wild.

28. P. kansuensis Rehder. Illustration: *Journal of the Royal Horticultural Society* **84**: 165 (1959).

Like *P. persica* but young shoots greenish, yellowish or purplish; flowering earlier. *NW China.* H1. Early spring.

29. P. davidiana (Carrière) Franchet. Illustration: Krüssmann, Manual of cultivated broad-leaved trees and shrubs **3**: pl. 6 & f. 10 (1986); *The Garden* **112**: 228 (1987).

Tree to 10 m; young shoots hairless. Leaves folded in bud, narrowly elliptic or lanceolate, finely toothed with acute, forward-pointing teeth 2 mm or more apart. Flowers solitary, stalkless or almost so, pale pink, to 2.5 cm across. Style and ovary hairy. Fruit hairy, yellowish, spherical, to 3 cm across; stone pitted. *NW China.* H1. Spring.

30. P. mira Koehne. Illustration: *Botanical Magazine*, 9548 (1939).

Tree to 10 m; young shoots hairless. Leaves folded in bud, narrowly elliptic or lanceolate, sparsely scalloped, rounded at base, widest below the middle, hairy beside midrib beneath. Flowers solitary or 2 together, stalkless or almost so, white, 2–2.5 cm across. Style and ovary hairy. Fruit hairy, yellowish, spherical, to 3 cm across; stone smooth. *W China.* H1. Spring.

31. P. tangutica (Batalin) Koehne. Illustration: *Botanical Magazine*, 9239 (1931).

Dwarf shrub to 4 m; young wood with a greyish or whitish bloom; often thorny. Leaves folded in bud, to 2 cm wide, dark greyish green, widest near middle, not white-felted beneath; stalks green. Flowers pink or red, solitary, stalkless, to 2.5 cm across. Sepals reddish, not or scarcely toothed. Petals overlapping. Style or ovary hairy. Fruit hairy, to 2 cm, across, with thin flesh, keeled. *W China (Sichuan).* H1. Spring.

32. P. triloba Lindley. Illustration: *Botanical Magazine*, 8061 (1906); Hillier colour dictionary of trees and shrubs, 172 (1981); Everett, New York Botanical Gardens illustrated encyclopedia of horticulture **8**: 2821 (1981).

Shrub or small tree to 5 m; young shoots hairless or downy. Leaves folded in bud, broadly elliptic or ovate, sometimes 3-lobed, with large, broad, triangular teeth, hairy beneath. Stipules narrowly linear, fringed, persistent or sometimes falling early. Flowers pink, 1.8–3.5 cm wide, often double, solitary or 2 together, stalkless or stalks to 9 mm. Style, ovary (and fruit) hairy. Fruit almost spherical, to 1.5 cm across, red. *China.* H1. Spring.

Cultivated material generally has double flowers; the single-flowered variant (var. **simplex** (Bunge) Rehder) is rarely seen.

33. P. tenella Batsch. Illustration: *Botanical Magazine*, 161 (1791); Polunin, Flowers of Greece and the Balkans, pl. 19 (1980); Hay & Beckett (eds), Reader's Digest encyclopedia of garden plants and flowers, edn ?, 556 (1985).

Branched, upright shrub to 1.5 m; young wood with a greyish or whitish bloom, not thorny. Leaves folded in bud, obovate to oblanceolate, to 2 cm wide, not white-felted beneath, all similar. Flowers solitary or to 3 together, stalkless, pink to red, to 3 cm wide. Sepals greenish, densely toothed. Petals not overlapping. Style and ovary hairy. Fruit hairy, yellowish, to 2 cm; stone

rough. *C Europe to former USSR (Siberia).* H1. Spring.

A few cultivars are available.

34. P. avium Linnaeus. Illustration: Bonnier, Flore complète **3**: pl. 167 (1914); Phillips, Trees in Britain, Europe and North America, 170 (1978); Edlin, The tree key, 147 (1978); Lunardi (ed.) Macdonald encyclopedia of shrubs and trees, 126 (1988).

Tree to 20 m (rarely more); young shoots hairless. Leaves folded in bud, ovate-oblong, toothed, with teeth coarse, arrow, rounded or obtuse, standing out from leaf-margin and at least 2 mm long. Flower-clusters clearly stalked or stalkless, without leaf-like bracts (large bud-scales may be present at base of cluster); flower-stalks arising at one or more points (stalked umbels or corymbs), or all arising from the same point in each cluster (stalkless umbels). Flowers 2 cm wide or more, white. Sepals becoming reflexed. Fruit dark red. *Europe, N Africa, SW Asia, former USSR (W Siberia).* H1. Spring.

'Plena' is a widely cultivated variant with double flowers.

P. × schmittii Rehder is the hybrid *P. avium × P. canescens*; it is intermediate between the parents but has glossy, brown bark with conspicuous lenticels.

P. × fontanesiana (Spach) Schneider is *P. avium × P. mahaleb*; similar to *P. avium* but with softly hairy young shoots.

35. P. cerasus Linnaeus. Illustration: Gleason, Illustrated flora of the north-eastern United States and adjacent Canada **2**: 330 (1956); Vedel & Lange, Trees and bushes in wood and hedgerow, 83 (1960); Brosse, Arbres de France et d'Europe occidental, 161 (1977).

Tree to 10 m, suckering, hairless or with a few hairs on the veins of the leaves beneath. Leaves folded in bud, elliptic to ovate, green when mature, with obtuse or rounded teeth. Flowers in dense clusters borne with the leaves, stalks 1–3.5 cm; clusters lacking a distinct stalk but with leaf-like bracts; flower-stalks all arising from the same point in each cluster (stalkless umbels). Flowers 2 cm wide or more. Sepals becoming reflexed. Fruit depressed-spherical, reddish black. *S Europe to N India.* H3. Spring.

A very variable species, with numerous varieties and cultivars available.

P. × eminens Beck (*P. reflexa* misapplied) is the hybrid *P. cerasus*

× *P. fruticosa*, which is intermediate between its parents.

36. P. fruticosa Pallas. Illustration: Hegi, Illustrierte Flora von Mittel-Europa **4**: t. 157 (1923); Jávorka & Csapody, Iconographia florae Hungaricae, 263 (1931); Krüssmann, Manual of cultivated broad-leaved trees and shrubs **3**: pl. 6 & f. 8 (1986).

Spreading shrub to 1 m; young shoots grey or black, hairless. Leaves folded in bud, scalloped, with obtuse or rounded teeth with dark marginal glands, green when mature, elliptic to obovate, less than 6 cm, narrowly tapered at base. Flowers 2–4 in stalkless umbels, white, to 1.5 cm across; flower-stalks 1.5–2.5 cm. Sepals becoming reflexed. Fruit dark red, more or less spherical. *C Europe to W Siberia*. H1. Spring.

37. P. canescens Bois. Illustration: Krüssmann, Manual of cultivated broad-leaved trees and shrubs **3**: pl. 7 (1986).

Shrub to 2 m, branchlets hairy, young wood greenish, reddish, or purplish, not bloomed. Leaves folded in bud, broadly elliptic or ovate, acuminate, coarsely toothed, with large, broad triangular teeth, sometimes 3-lobed, bristly above, not white-felted beneath, all similar. Stipules broadly elliptic to circular, persistent. Flowers 2–5 in corymbs, pale pink, to 1.2 cm across, petals not opening widely, hairy towards the base; bracts leaf-like; flowers stalkless or stalks to 9 mm. Perigynous zone cylindric. Style, ovary (and fruit) hairy. Fruit spherical, to 1 cm across, red. *W China (Sichuan, Hubei)*. H2. Late spring.

P. × dawyckensis Sealy. Illustration: *Botanical Magazine*, 9519 (1938); Krüssmann, Manual of cultivated broad-leaved trees and shrubs **3**: pl. 7 (1986). Similar to *P. canescens* but much hairier, petals notched, fruits ellipsoid, yellow-red. *China*.

Thought by Rehder and others to be the natural hybrid *P. canescens* × *P. dielsiana* but accepted as a species by Krüssmann.

38. P. incisa Thunberg. Illustration: *Botanical Magazine*, 8954 (1923); Bean, Trees and shrubs hardy in the British Isles, edn 8, **3**: t. 46 (1976); Flower Society of Japan, Manual of Japanese flowering cherries, 131, 132 (1982); Brickell, RHS gardeners' encyclopedia of plants and flowers, 59 (1989).

Usually a rounded shrub, more rarely tree-like and to 10 m. Leaves folded in bud, purplish when young, 6 cm or less, ovate to obovate, acute, teeth coarse, at least 3 mm long; stalks of mature leaves densely downy or hairy. Flowers solitary or to 3 in clusters, white to pale pink, to 2.5 cm across, perigynous zone cylindric; flower-stalks more than 9 mm. Sepals not becoming reflexed. Petals soon falling. Fruit ovoid, black-purple, to 8 mm across. *Japan*. H2. Spring.

P. × hillieri Hillier is the hybrid between *P. incisa* and *P. sargentii*. It is intermediate between the parents, forming a small, densely branched tree, which gives good autumn colour.

39. P. apetala Franchet & Savatier. Illustration: Flower Society of Japan, Manual of Japanese flowering cherries, 121 (1982); Krüssmann, Manual of cultivated broad-leaved trees and shrubs 3: pl. 7 & f. 26 (1986).

Like *P. incisa* but mature leaves more than 6 cm. long, flowers solitary or in pairs, bracts on fully developed inflorescence 1 cm or more. *Japan*. H2. Spring.

40. P. mugus Handel-Mazzetti. Illustration: Krüssmann, Manual of cultivated broad-leaved trees and shrubs **3**: pl. 7 (1986).

Like *P. incisa* but stalks of mature leaves at most very sparsely hairy; leaf-tips round or obtuse, rarely acute. Perigynous zone tubular-campanulate. *W China (Xizang)*. H1. Spring.

41. P. nipponica Matsumura (*P. nikkoensis* Koehne). Illustration: Makino, New illustrated flora of Japan, t. 1160, 1161 (1963); Kurata, Illustrated important forest trees of Japan **3**: 47 (1971); Flower Society of Japan, Manual of Japanese flowering cherries, 311 (1982).

Tree to 6 m; young shoots hairless. Leaves ovate or rarely obovate, folded in bud, teeth coarse, fewer than 4 per cm, acute or obtuse, not hair-like; stalks of mature leaves at most very sparsely hairy. Flowers produced with the leaves in April and May; bracts persistent. Flowers solitary or to 3 together, white or pale pink, to 2.5 cm across, perigynous zone tubular-campanulate; flower-stalks more than 9 mm. Sepals not becoming reflexed. Petals falling early. Fruit spherical, black, to 1 cm across. *Japan*. H1. Late spring.

P. kurilensis Miyabe (*P. nipponica* var. *kurilensis* (Miyabe) Wilson). Illustration: Flower Society of Japan, Manual of Japanese flowering cherries, 311 (1982). This has somewhat larger leaves and hairy leaf- and flower-stalks. *Japan (Kurile Islands)*.

42. P. subhirtella Miquel. Illustration: Phillips, Trees in Britain, Europe and North America, 175 (1978); Krüssmann, Manual of cultivated broad-leaved trees and shrubs **3**: pl. 8, 17 & f. 26 (1986); Lunardi (ed) Macdonald encyclopedia of shrubs and trees, 129 (1988).

Tree to 10 m; young shoots hairy. Leaves folded in bud, broadly elliptic or ovate, sometimes 3-lobed, with large, broad triangular teeth, hairless beneath except on midrib and veins. Stipules narrowly linear, fringed, falling by midsummer. Flowers 2–5 together in clusters, white or pink, to 2 cm across, perigynous zone cylindric, stalkless or stalks to 9 mm. Style, ovary (and fruit) hairy. Fruit to 1 cm across, black. *Japan*. H1. Autumn–spring.

'Autumnalis' is a selection with widely angled branches and double white flowers, which flowers from early winter to spring.

43. P. campanulata Maximowicz. Illustration: *Botanical Magazine*, 9575 (1939)– as *P. cerasoides* var. *campanulata*; Bean, Trees and shrubs hardy in the British Isles, edn 8, **3**: 357 (1976); Macoboy, What tree is that?, 177 (1979); Flower Society of Japan, Manual of Japanese flowering cherries, 122 (1982).

Shrub or small tree to 8 m; young shoots hairless. Leaves folded in bud, elliptic-ovate to ovate-oblong, teeth 1 mm or less. Flowers 2–5 in stalked umbels, pink or red, borne before or with the leaves; perigynous zone cylindric; stalks of flower-clusters 2.5 mm or more; stalks of flowers more than 9 mm. Sepals becoming reflexed. Sepals and perigynous zone hairless. Fruit ovoid, red, to 1.5 cm, stone smooth. *Japan, Taiwan*. H4. Spring.

44. P. cerasoides D. Don. Illustration: Polunin & Stainton, Flowers of the Himalaya, pl. 33 (1985); Flower Society of Japan, Manual of Japanese flowering cherries, 122 (1982); Krüssmann, Manual of cultivated broad-leaved trees and shrubs **3**: pl. 7 & f. 22 (1986); *The Garden* **112**: 225 (1987).

Like *P. campanulata* but leaves more leathery, ovate-rounded, more sharply toothed; fruit pointed. *Himalaya*. H4. Spring.

Possibly no longer in cultivation (though the name may be found misapplied in catalogues).

45. P. rufa J.D. Hooker. Illustration: *Journal of the Royal Horticultural Society* **102**: 354 (1977); Polunin & Stainton, Flowers of the Himalaya, pl. 33 (1985); Krüssmann, Manual of cultivated broad-leaved trees and shrubs **3**: pl. 7 & f. 26 (1986).

Small tree to 7 m; young shoots reddish hairy. Leaves narrowly ovate to obovate-lanceolate, folded in bud, to 5 cm, toothed, acuminate, midribs ginger-hairy beneath, stalks of mature leaves at most very sparsely hairy. Flowers usually solitary, white to pale pink, to 1.5 cm across; stalks more than 9 mm. Sepals not becoming reflexed, perigynous zone tubular, not hairy. Fruit broadly ellipsoid, dark red. *Himalaya*. H4. Spring.

Var. **tricantha** (Koehne) Hara. Inflorescence (flower-stalks, perigynous zone) hairy. *N India*.

46. P. serrula Franchet. Illustration: Bean, Trees and shrubs hardy in the British Isles, edn 8, **3**: t. 47 (1976); Phillips, Trees in Britain, Europe and North America, 174 (1978); Hillier colour dictionary of trees and shrubs, 175 (1981); Krüssmann, Manual of cultivated broad-leaved trees and shrubs **3**: pl. 17 (1986).

Tree to 10 m, bark glossy red, peeling in strips, in old plants scarred with rough, horizontal lenticels; young shoots downy. Leaves folded in bud, lanceolate, acuminate, teeth on mature leaves very fine and regular, at least 5 major teeth per cm; stalks of mature leaves at most very sparsely hairy. Flowers solitary or to 3 together, white, to 2 cm across, borne with the leaves; perigynous zone cylindric; stalks more than 9 mm. Sepals not becoming reflexed. Style hairy below. Fruits ovoid, to 7 mm. *W China*. H2. Late spring.

47. P. hirtipes Hemsley (*P. conradinae* Koehne; *P. helenae* Koehne). Illustration: *The Garden* **87**: 97 (1923); Krüssmann, Manual of cultivated broad-leaved trees and shrubs **3**: pl. 7 (1986).

Tree to 8 m, or smaller shrub. Leaves folded in bud, ovate to oblong-obovate, long-acuminate, green when mature; teeth 4–6 teeth per cm or fewer than 4 per cm, acute, coarse, standing out from the margin, not hair-like; stalks of mature leaves at most very sparsely hairy. Flowers solitary or to 4 together, white to pale pink, borne before the leaves from January–March; perigynous zone cylindric; bracts falling as flowers open. Sepals not becoming reflexed. Fruits ovoid, red. *C China*. H4. Winter–early spring.

48. P. × yedoensis Matsumura; *?P. subhirtella × P. speciosa*. Illustration: Phillips, Trees in Britain, Europe and North America, 175 (1978); Everett, New York Botanical Gardens illustrated encyclopedia of horticulture **8**: 2823 (1981); Krüssmann, Manual of cultivated broad-leaved trees and shrubs **3**: pl. 1, 6, 21 (1986); Brickell, RHS gardeners' encyclopedia of plants and flowers, 60 (1989).

Tree to 15 m; young shoots downy. Leaves folded in bud, elliptic, acuminate, more than 6 cm, teeth on mature leaves coarse, at least 3 mm long; stalks of mature leaves densely downy or hairy. Flowers 5 or 6 in stalked racemes, pinkish at first, becoming white, to 3.5 cm across; perigynous zone cylindric. Sepals not becoming reflexed. Petals not falling early. Fruits spherical, to 8 mm across, black. *Garden origin*. H2. Spring.

49. P. sargentii Rehder. Illustration: Phillips, Trees in Britain, Europe and North America, 172 (1978); Hillier colour dictionary of trees and shrubs, 175 (1981); Krüssmann, Manual of broad-leaved trees and shrubs **3**: pl. 2, 18 & f. 20 (1986); Brickell, RHS gardeners' encyclopedia of plants and flowers, 39 (1989).

Tree to 20 m; young shoots hairless or sparsely hairy. Leaves folded in bud, oblong-elliptic to obovate-oblong, teeth on mature leaves coarse, fewer than 4 per cm, acuminate or drawn out into hair-like tips; stalks of mature leaves at most very sparsely hairy; stems and leaves hairless or very slightly hairy. Flowers 2–4 in stalkless umbels, pink, to 4 cm across; perigynous zone tubular-campanulate; stalks more than 9 mm. Sepals not becoming reflexed. Petals notched at the apex. Fruits ovoid or oblong-ovoid, dark red. *Japan, former USSR (Sakhalin), Korea*. H1. Spring.

50. P. × juddii Anderson; *P. sargentii × P. × yedoensis*.

Like *P. sargentii* but young leaves copper-coloured, flowers larger, deeper pink, fragrant. Garden origin.

51. P. serrulata Lindley. Illustration: *Botanical Magazine*, 8012 (1905); Wilson, The cherries of Japan, pl. 6 (1916); Chadbund, Flowering cherries, pl. 2 (1972).

Tree to 10 m, though usually much smaller in cultivation; young shoots hairy or hairless. Leaves folded in bud, narrowly ovate, glossy, long acuminate; teeth on mature leaves coarse, fewer than 4 per cm, acuminate or drawn out into hair-like tips; stalks of mature leaves at most very sparsely hairy. Stems and leaves hairless or very slightly hairy. Flowers 3–5 together in racemes, white or pink, single or double; perigynous zone cylindric; stalks more than 9 mm. Sepals not becoming reflexed. Fruits ovoid, black, shining. *China, Japan, Korea*. H1. Early–late spring.

A highly variable species, generally considered to consist of 3 varieties (though it is uncertain which of these are genuinely in cultivation) and a large group of cultivars, long-cultivated in Japan and now widely grown throughout the temperate world.

Var. **spontanea** (Maximowicz) Wilson is a tall tree with grey-brown bark with large persistent lenticels, the young leaves copper to brown, hairless, flowers white, 2.5–3.5 cm across in shortly stalked corymbose clusters.

Var. **hupehensis** Ingram is similar to var. *spontanea* but the flowers appear earlier than the leaves, the young leaves are bronze and the flowers are in almost stalkless clusters.

Var. **pubescens** Wilson is similar to var. *spontanea* but the leaf-stalks and undersurfaces are hairy, the young leaves are green or slightly bronze and the flowers are in long-stalked corymbs.

The Japanese flowering cherries or Sato-Zakura Group (*P. lannesiana* (Carrière) Wilson; *P. donarium* Siebold) are usually considered under this species. The group is of complex hybrid origin and no attempt is made here to list or describe the many cultivars. They vary greatly in habit (strictly upright, vase-shaped, round-headed or weeping), colour of the young leaves (green to dark red), type of inflorescence, degree of doubleness of the flowers (petals 5–50 or more), petal colour (from cream through white, pinkish white, pink and dark pink) and the presence of leaf- or petal-like bodies replacing the ovary.

The following key, translated and adapted from an original by Arie Peeters (Wageningen, The Netherlands), covers 42 of the most commonly encountered cultivars and the varieties described above. If in doubt at any stage, the user is recommended to follow both alternative leads. Other keys can be found in The RHS dictionary of gardening **2**: 1086 (edn 1, 1951) and in Krüssmann, Manual of cultivated broad-leaved trees & shrubs **3**: 47–8 (1986). Reference should also be made to Ingram, Flowering cherries (1948) and the Japanese Flower Society's Manual of Japanese flowering cherries (1982).

1a. Petals 20 or fewer   2
  b. Petals more than 20   41
2a. Petals 5–10   3
  b. Petals 10–20   25
3a. Petals 5 (occasionally 6–8)   4
  b. Petals 6–10 (occasionally 5)   20
4a. Flowers pure white when fully open
                                                                      5
  b. Flowers pale pink or darker when fully open   14
5a. Flowers 4.8 cm wide or less   6
  b. Flowers more than 4.8 cm wide   13
6a. Flowers less than 4 cm wide   7
  b. Flowers 4–4.8 cm wide   10
7a. Leaf-stalks hairy   8
  b. Leaf-stalks hairless   9
8a. Flowers 2.5–3 cm wide, opening in May; inflorescence-stalks 1.5 cm or more; young leaves green or bronze-green   var. **pubescens**
  b. Flowers 3–3.7 cm wide, opening mid to late April; inflorescence-stalks usually 1 cm or less; young leaves coppery brown   **'Fudanzakura'**
9a. Young leaves coppery red to brown; flower clusters shortly stalked   var. **spontanea**
  b. Young leaves bronze; flower clusters almost stalkless   var. **hupehensis**
10a. Petaloids (small extra deformed petals) present in some flowers   11
  b. Petaloids absent   12
11a. Conical tree with flattened top; flowers fragrant; petals 12–13 mm wide   **'Jo-nioi'**
  b. Broadly columnar tree; flowers not very fragrant; petals 16–19 mm wide   **'Hatazakura'**
12a. Petals often ragged   **'Washino-o'**
  b. Petals entire or slightly notched   **'Taki-nioi'**
13a. Flowers 5–5.5 cm wide, sometimes with more than 5 petals; young leaves bronze-green   **'Ojochin'**
  b. Flowers 5.5–6 cm wide, rarely with more than 5 petals; young leaves brown   **'Taihaku'**
14a. Leaf-stalks hairy   15

  b. Leaf-stalks hairless   16
15a. Flowers 2.5–3 cm wide, open in May; inflorescence-stalks 1.5 cm or more; young leaves green or bronze-green   var. **pubescens**
  b. Flowers 3–3.7 cm wide, open mid to late April; inflorescence-stalks usually 1 cm or less; young leaves coppery brown   **'Fudanzakura'**
16a. Flowers less than 4.7 cm wide   17
  b. Flowers 4.7 cm wide or more   19
17a. Flowers sparse, 2 or 3 per cluster, open mid April; petals *c.* 16 × 12 mm; growth rapid   **'Benden'**
  b. Flowers abundant, 3–6 per cluster, open late April to early May; petals 1.8 × 1.3 cm or more; growth slow to medium   18
18a. Flowers 4–6 per cluster, open late April; petals *c.* 18 × 13 mm; growth slow   **'Taguiarashi'** (**'Ruiratt'**)
  b. Flowers 3 or 4 per cluster, open late April to early May; petals *c.* 19 × 16 mm; growth medium **'Hizakura'** (name often misapplied to 'Kanzan')
19a. Flowers on short shoots; petals smooth; clusters less than 5 cm **'Kirigayatsu'** (**'Mikuruma-gaeshi'**)
  b. Flowers not on short shoots; petals folded and wavy; clusters often more than 5 cm   **'Ojochin'**
20a. Flowers creamy white to yellowish green   **'Ukon'**
  b. Flowers white or pink   21
21a. Columnar tree with more or less erect inflorescences   **'Amanogawa'**
  b. Conical or broadly conical tree with spreading or hanging inflorescences
                                                                     22
22a. Flower-stalks and calyx purplish red; leaf-base wedge-shaped   **'Taoyame'**
  b. Not as above   23
23a. Flowers barely fragrant, often with only 5 petals; flower-stalk not firm
                                                           'Ojochin'
  b. Flowers very fragrant, rarely with 5 petals; flower-stalks firm   24
24a. Conical tree; flowers often slightly pink when fully open; style shorter than anthers; calyx-lobes clearly toothed   **'Ariake'**
  b. Spreading tree; flowers pure white when fully open; style as long as anthers; calyx-lobes occasionally toothed   **'Shirotae'**
25a. Leaves covered in woolly hairs
                                                         'Takasago'
  b. Leaves not covered in woolly hairs
                                                                     26

26a. Columnar tree with more or less erect inflorescences   **'Amanogawa'**
  b. Not as above   27
27a. Flowers cream-coloured to yellowish green   **'Ukon'**
  b. Flowers white or pink   28
28a. Flowers-stalks and calyx purplish to brownish red   29
  b. Not as above   30
29a. Flowers usually with 1 style; petals 5–15, somewhat notched; leaves broadest towards the tip; growth rapid   **'Taoyame'**
  b. Flowers sometimes with 1–14 extra styles among the anthers; petals 14 or 15, unnotched; leaves narrow, broadest towards the base; growth slow   **'Horinji'**
30a. Flowers almost always with only one perfect style and ovary   31
  b. More than 50% of flowers either with 2 perfect styles and ovaries, or with 1 or 2 ovaries which are leaf-like at base (sample at least 10 flowers)   38
31a. Flowers 5–5.5 cm wide   32
  b. Flowers 3.5–4.8 cm wide   34
32a. Flowers with more than 13 petals, barely fragrant, pale pink when fully open   **'Sumizome'**
  b. Flowers with fewer than 13 petals, fragrant, white or very slightly pink when fully open   33
33a. Conical tree; flowers often slightly pink when fully open; style shorter than anthers; calyx-lobes clearly toothed   **'Ariake'**
  b. Spreading tree; flowers pure white when fully open; style as long as anthers; calyx-lobes occasionally toothed   **'Shirotae'**
34a. Flowers pure white when fully open; young leaves green   **'Albo Plena'**
  b. Flowers pale pink or pink when fully open; young leaves bronze-green to brownish bronze   35
35a. Petals 1.8–2.1 cm wide; calyx-lobes obtuse, narrowed at base
                                     'Uzuzakura' (**'Hokusai'**)
  b. Petals 1–1.7 cm wide; calyx-lobes acute, not narrowed at base   36
36a. Petals scalloped, soft purplish pink (like *P. sargentii*), 1–1.3 cm wide
                                                      'Yae-marasaki'
  b. Petals slightly notched, very pale pink, 1.3–1.7 cm wide   37
37a. Inflorescence-stalks longer than 2.3 mm; calyx-lobes 8–9 mm; leaves very finely toothed, the teeth hardly visible in young leaves   **'Shujaku'**

b. Inflorescence-stalks less than 2.3 mm; calyx-lobes 6–7 mm; leaves more coarsely toothed, the teeth visible in young leaves **'Edozakura'**

38a. More than 50% of flowers with 2 perfect styles and ovaries, if only 1 then never leaf-like at base (sample at least 10 flowers) **'Imose'**

b. More than 50% of flowers with 1 or 2 imperfect styles and ovaries that are leaf-like at base 39

39a. Mature flower-buds conical to cylindric, lacking stripes on petal backs; flowers saucer-shaped when fully open; leaf-like ovary bases not hidden **'Ichiyo'**

b. Mature flower-buds ovoid, with dark pink stripes on petal backs; flowers cup-shaped when fully open; leaf-like ovary bases usually hidden 40

40a. Flowers 4–4.5 cm wide, often with unopened petals held in calyx-tube **'Itokukuri'**

b. Flowers 4.8–5.3 cm wide, rarely with unopened petals in calyx-tube **'Okikuzakura'**

41a. Petals 20–50 42
b. Petals more than 50 54

42a. Leaf-stalks hairy **'Taizanfukan'**
b. Leaf-stalks hairless 43

43a. Flowers pure white when fully open, less than 4 cm wide, with 1 perfect style and ovary **'Albo Plena'**
b. Not as above 44

44a. More than 50% of flowers with two perfect styles and ovaries, if only one then never leaf-like at base (sample at least 10 flowers) **'Imose'**
b. Not as above 45

45a. More than 50% of flowers with only one style and ovary, the latter leaf-like only at base (sample at least 10 flowers) 46
b. More than 50% of flowers with at least 2 leaf-like ovaries 48

46a. Mature flower-buds conical to cylindric, lacking stripes on petal backs; flowers saucer-shaped when fully open; leaf-like ovary-bases not hidden **'Ichiyo'**

b. Mature flower-buds ovoid, with dark pink stripes on petal backs; flowers cup-shaped when fully open; leaf-like ovary-bases usually hidden 47

47a. Flowers 4–4.5 cm wide, often with unopened petals held in calyx-tube **'Itokukuri'**

b. Flowers 4.8–5.3 cm wide, rarely with unopened petals in calyx-tube **'Okikuzakura'**

48a. Flowers less than 4.4 cm wide **'Fugenzo'**
b. Flowers more than 4.4 cm wide 49

49a. Flowering middle to end of May (very late, after 'Kanzan'); flowers almost white when fully open (sometimes turning purplish pink before falling); flower-clusters often more than 10 cm 50
b. Not as above 51

50a. Young leaves green; flowers remaining white **'Okumiyaku'** (**'Shimidsu'**)
b. Young leaves brown to bronze-green; flowers often turning purplish pink before falling **'Shirofugen'**

51a. Flowers almost white when fully open **'Ichiyo'**
b. Flowers distinctly pink when fully open (beware of 'Shirofugen', see 50) 52

52a. Flowers with 2 (rarely 3) leaf-like ovaries (sample at least 10 flowers) **'Sekiyama'** (**'Kanzan'**)
b. Some flowers with 4 leaf-like ovaries 53

53a. Many thin bare branches on crown; calyx-lobes clearly toothed on mature buds; anther connectives elongated **'Pink Perfection'**
b. No thin bare branches on crown; calyx-tubes not toothed on mature buds; anther connectives not elongated **'Daikoku'**

54a. Habit weeping **'Kiku-shidare'**
b. Habit ascending 55

55a. Flowering late April to early May; flowers less than 4 cm wide; inflorescence-stalk less than 2 cm **'Geraldinae'** (**'Asano'**)
b. Flowering early to mid May; flowers more than 4 cm wide; inflorescence-stalk more than 2 cm **'Hiyodora'**

52. P. speciosa (Koidzumi) Ingram (*P. lannesiana* (Carrière) Wilson forma *albida* Wilson). Illustration: Krüssmann, Manual of cultivated broad-leaved trees and shrubs **3**: pl. 13 (1986).

Tree to 12 m, bark pale grey-brown. Stems, undersides of leaves or perigynous zones distinctly hairy. Leaves folded in bud, elliptic-obovate to obovate, teeth on mature leaves coarse, fewer than 4 per cm, acuminate or drawn out into hair-like tips; stalks of mature leaves at most very sparsely hairy. Flowers 3 or 4, borne with the leaves in loose, rather long-stalked corymbs, white, to 1.5 cm across; perigynous zone cylindric; stalks more than

9 mm. Sepals not becoming reflexed. Fruit ovoid, black and shining. *Japan.* H2. Spring.

Sometimes considered as part of the Sako-Zakura group of *P. serrulata.*

53. P. × sieboldii (Carrière) Wittmack; *P. speciosa × P. apetala.* Illustration: Wilson, The cherries of Japan, pl. 8 (1916); Flower Society of Japan, Manual of Japanese flowering cherries, 322 (1982);

Tree to 8 m, with smooth, grey bark. Leaves folded in bud, elliptic to ovate, more than 6 cm, teeth coarse, at least 3 mm long; stalks of mature leaves densely downy or hairy. Flowers 3 or 4 in corymbs, double or semi-double, usually more than 3 cm wide; perigynous zone cylindric; stalks more than 9 mm, hairy. Sepals not becoming reflexed. *Garden origin.* H2. Spring.

54. P. pseudocerasus Lindley (*P. involucrata* Koehne). Illustration: *Edwards's Botanical Register* **10**: t. 800 (1824); Krüssmann, Manual of cultivated broad-leaved trees and shrubs **3**: pl. 7, 15 & f. 22 (1986).

Tree to 8 m; young shoots hairless or slightly hairy. Leaves folded in bud, broadly ovate to oblong-ovate; teeth 1 mm or less. Flowers 3–6 in stalked umbels or corymbs, before or with the leaves, pink in bud, white when open, to 2 cm across; stalks more than 9 mm. Sepals becoming reflexed. Sepals, perigynous zone and style hairy. Fruit ovoid, yellow to red, to 1.5 cm. *China (Hubei).* H2. Spring.

P. cantabrigiensis Stapf (*P. pseudocerasus* var. *cantabrigiensis* (Stapf) Ingram). Illustration: *Botanical Magazine,* 9129 (1928). This has longer, more acuminate leaves, a more dense inflorescence and petals deep pink in bud, opening paler. See Yeo, *Baileya* **20**: 11–17 (1976).

55. P. dielsiana Schneider. Illustration: Krüssmann, Manual of cultivated broad-leaved trees and shrubs **3**: pl. 7 (1986).

Shrub or tree to 20 m; young shoots hairless. Leaves folded in bud, oblong-obovate to oblong; stalks hairy. Flowers 3–5 in stalked umbels or corymbs, 3 cm or more across borne before the leaves; stalks more than 9 mm. Perigynous zone hairy. Sepals becoming reflexed. Fruit spherical, red, to 1 cm across. *C China.* H2. Spring.

56. P. cyclamina Koehne. Illustration: *Botanical Magazine,* n.s., 338 (1959); Krüssmann, Manual of cultivated broad-leaved trees and shrubs **3**: pl. 7 (1986).

Very like *P. dielsiana* but leaf-stalks and perigynous zone not hairy. *C China.* H2. Spring.

57. P. mahaleb Linnaeus. Illustration: Bonnier, Flore complète **3**: pl. 167 (1914); Phillips, Trees in Britain, Europe and North America, 172 (1978); Krüssmann, Manual of cultivated broad-leaved trees and shrubs **3**: pl. 17 & f. 15 (1986); Brickell, RHS gardeners' encyclopedia of plants & flowers, 59 (1989).

Tree to 10 m; young shoots downy. Leaves folded in bud, almost circular to broadly ovate, downy along the midrib beneath, 6 cm or less, finely scalloped, with short abrupt tips 1 mm or less. Flowers 6–10 in stalked umbels or corymbs, white, fragrant, to 1.5 cm across; stalks more than 9 mm. Sepals becoming reflexed. Fruit *c.* 6 mm across, black. *E Europe to C Asia.* H1. Late spring.

58. P. pensylvanica Linnaeus filius. Illustration: *Botanical Magazine*, 8486 (1913); Phillips, Trees in Britain, Europe and North America, 172 (1978); Krüssmann, Manual of cultivated broad-leaved trees and shrubs **3**: f. 15, 19 (1986).

Shrub or tree to 10 m; branches hairless. Leaves folded in bud, ovate to oblong-lanceolate, green when mature, with acute or acuminate teeth, hairless or almost so beneath; tips evenly tapered. Flowers 3–6 before or with the leaves in stalkless umbels, white, to 1.5 cm across; stalks more than 9 mm. Sepals becoming reflexed. Fruit red, spherical, to 6 mm across. *N America.* H1. Spring.

59. P. pilosiuscula Koehne. Illustration: *Botanical Magazine*, 9192 (1930).

Tree to 12 m; young shoots downy or hairless. Leaves folded in bud, oblong to oblong-obovate, teeth standing out from leaf-margin, at least 2 mm long, with hair-like tips. Flowers 2 or 3, borne with the leaves in stalked corymbs, white or pale pink, to 2 cm across; stalks more than 9 mm. Sepals becoming reflexed. Fruit red, ellipsoid, to 1 cm across. *C & W China.* H1. Late spring.

60. P. litigiosa Schneider. Illustration: Krüssmann, Manual of cultivated broad-leaved trees and shrubs **3**: pl. 8 (1986).

Small tree to 6 m; young shoots hairless. Leaves folded in bud, narrowly obovate to oblong-obovate, teeth 1 mm or less. Flowers 2 or 3 in stalked umbels, white, to 2.5 cm across; stalks of flower-clusters *c.* 6 mm; flower-stalks more than 9 mm. Sepals and perigynous zone hairless. Sepals becoming reflexed. Style hairy at the base. Fruit ellipsoid, red, to 1 cm. *W China (Hubei).* H2. Spring.

61. P. maximowiczii Ruprecht. *Botanical Magazine*, 8641 (1915); Krüssmann, Manual of cultivated broad-leaved trees and shrubs **3**: pl. 8 & f. 15 (1986).

Tree to 8 m in cultivation (larger in the wild); young shoots downy. Leaves folded in bud, obovate, teeth rounded or obtuse standing out from leaf-margin, at least 2 mm long. Flowers 5–10, borne after the leaves in stalked corymbs, *c.* 1.5 cm across, cream-white; stalks more than 9 mm. Sepals becoming reflexed. Fruit spherical, black. *E Asia.* H1. Late spring.

62. P. tomentosa Thunberg. Illustration: *Botanical Magazine*, 8196 (1908); Everett, New York Botanical Gardens illustrated encyclopedia of horticulture **8**: 2825 (1981); Krüssmann, Manual of cultivated broad-leaved trees and shrubs **3**: pl. 8, 20 & f. 28 (1986).

Shrub to 3 m, or rarely a small tree; young shoots downy. Leaves folded in bud, broadly elliptic or ovate, sometimes 3-lobed, with large, broad triangular teeth, hairy beneath. Stipules narrowly linear, fringed, sometimes falling early. Flowers usually 2 together, white or pinkish, 1.5–2 cm across, appearing with the leaves; stalkless or stalks to 9 mm. Style and ovary hairy. Fruit spherical, red, ultimately hairless or slightly hairy. *Himalaya to W China, Japan.* H1. Late spring.

63. P. incana (Pallas) Batsch. Illustration: *Edwards's Botanical Register* **25**: t. 28 (1939); Krüssmann, Manual of cultivated broad-leaved trees and shrubs **3**: f. 14 (1986).

Open shrub to 2 m; young shoots finely downy. Leaves folded in bud, oblong-ovate to obovate, to 2 cm wide, densely white-felted beneath, of two types. Flowers stalkless, solitary or 2 or 3 in stalkless clusters, borne with the leaves, pink, to 1 cm across, hairless or sparsely hairy outside. Perigynous zone cylindric. Style and ovary hairy. Fruit hairy, ovoid, red, 6–8 mm. *SE Europe, Turkey, Caucasus, Iran.* H2. Late spring.

64. P. jacquemontii J.D. Hooker. Illustration: *Botanical Magazine*, 6976 (1888); Krüssmann, Manual of cultivated broad-leaved trees and shrubs **3**: pl. 8 & f. 14 (1986).

Spreading shrub to 3 m; young shoots hairy or hairless. Leaves folded in bud, elliptic to obovate-oblong, green when mature, not acuminate, sometimes abruptly short-pointed; not glaucous beneath, toothed all round. Flowers solitary or 2 together, pink, to 2 cm across, stalkless or very shortly stalked. Perigynous zone cylindric. Style, ovary and fruit hairless. Fruit spherical, to 1.5 cm across, red. *W Himalaya.* H4. Spring.

65. P. bifrons Fritsch. Illustration: Krüssmann, Manual of cultivated broad-leaved trees and shrubs **3**: f. 14 (1986).

Like *P. jacquemontii* but smaller (to 1.5 m), flowers borne with the leaves, petals often downy towards the base inside, style, ovary and fruit hairy. *W Himalaya.* H2. Spring.

66. P. japonica Thunberg. Illustration: Makino, New illustrated flora of Japan, t. 1145 (1963); Krüssmann, Manual of cultivated broad-leaved trees and shrubs **3**: pl. 8 & f. 13 (1986).

Shrub to 1.5 m; young shoots hairless. Leaves folded in bud, ovate-lanceolate, ovate or broadly ovate, green when mature, long-acuminate, with acute teeth standing out from margin. Flowers borne with the leaves, 2 or 3 in clusters, pink, to 2 cm or more across, shortly stalked. Sepals with gland-tipped marginal teeth. Fruit spherical to ellipsoid, dark red, to 1 cm across. *C China, Korea, introduced early into Japan.* H1. Late spring.

67. P. glandulosa Thunberg (*P. japonica* misapplied). Illustration: *Botanical Magazine*, 8260 (1909); Hillier colour dictionary of trees and shrubs, 172 (1981); Krüssmann, Manual of cultivated broad-leaved trees and shrubs **3**: f. 13 (1986).

Shrub to 2 m, young shoots usually hairless. Leaves folded in bud, narrowly ovate or elliptic, tapered at base, widest near the middle, length *c.* 2.5 times the width or more, to 7 × 3 cm, flat, green when mature, not acuminate, sometimes abruptly short-pointed, not glaucous beneath. Flowers solitary or 2 together in clusters, white to pale pink, to 1.2 cm across, usually stalked. Sepals hairless. Style and ovary hairy. Fruit hairy, spherical, to 1.2 cm across, red. *C & N China, Japan.*

68. P. humilis Bunge (*P. bungei* Walpers). Illustration: *Botanical Magazine*, 7335

(1894); Krüssmann, Manual of cultivated broad-leaved trees and shrubs **3**: f. 13 (1986).

Shrub to 2 m, young shoots downy. Leaves folded in bud, to 4 × 2 cm, narrowly ovate or elliptic, length *c.* 2.5 times the width or more, boat-shaped, often curved along midrib, green when mature, not acuminate, sometimes abruptly short-pointed; not glaucous beneath. Flowers borne with the leaves, solitary or 2 together in clusters, pale pink, to 1.5 cm across, stalks 5 mm or more. Fruit almost spherical, bright red, to 1.5 cm across. *N China.* H1. Spring.

69. P. pumila Linnaeus. Illustration: Gleason, Illustrated flora of the north-eastern United States and adjacent Canada **2**: 330 (1952); Krüssmann, Manual of cultivated broad-leaved trees and shrubs **3**: pl. 8, 16 & f. 13, 14 (1986).

Small shrub to 1.5 m; young shoots reddish brown, hairless. Leaves folded in bud, indistinctly toothed, oblanceolate to narrowly obovate, green when mature, less than 6 cm, narrowly tapered at base, not acuminate, sometimes abruptly short-pointed, greyish glaucous beneath; not toothed or scalloped near the base. Flowers 2–4 together, white, to 1.5 cm across, in stalkless umbels. Sepals becoming reflexed. Fruit more or less spherical, shiny purple-black. *E North America.* H1. Late spring.

70. P. × cistena (Hansen) Koehne; *P. cerasifera* 'Atropurpurea' × *P. pumila*. Illustration: Hillier colour dictionary of trees and shrubs, 172 (1981); Everett, New York Botanical Gardens illustrated encyclopedia of horticulture **8**: 2825 (1981); Krüssmann, Manual of cultivated broad-leaved trees and shrubs **3**: pl. 2 (1986); Brickell, RHS gardeners' encyclopedia of plants and flowers, 123 (1989).

Shrub to 2 m at most. Leaves folded in bud, lanceolate-obovate, hairy beneath, purplish at first, becoming reddish brown when mature. Flowers solitary or 2 together in stalkless umbels, white. Sepals becoming reflexed. Fruit blackish purple. *Garden origin.* H1. Spring.

71. P. besseyi Bailey. Illustration: *Botanical Magazine*, 8156 (1907); Krüssmann, Manual of cultivated broad-leaved trees and shrubs **3**: pl. 8 (1986).

Prostrate shrub. Leaves folded in bud, elliptic to elliptic-lanceolate, green when mature, abruptly tapered, sometimes abruptly short-pointed, greyish glaucous beneath; not toothed or scalloped near the narrowly tapered base. Flowers 2–4 in clusters, white, to 1.5 cm across; flowers stalkless or stalks to 9 mm. Fruit spherical, purple-black, to 1.5 cm across. *W USA.* H2. Spring.

72. P. serotina Ehrhart. Illustration: Sargent, Silva of North America **4**: t. 159 (1892); Phillips, Trees in Britain, Europe and North America, 174 (1978); Bärtels, Gartengehölze, 357 (1981).

Deciduous trees to 35 m or shrubs; bark brown, inner aromatic; young shoots hairless. Leaves oblong, acuminate, margins scalloped or with incurved teeth at least in apical part, glossy above, pale beneath and hairy along the veins. Flowers white, 1–1.4 cm across, usually 10 or more in narrow, cylindrical racemes, leafy towards base. Fruit ovoid, 8–10 mm, dark purple. *N America.* H1. Late spring.

73. P. virens (Wooton & Standley) Shreve. Illustration: Sargent, Manual of the trees of North America, edn 2, 578 (1922).

Like *P. serotina* but smaller, some leaves tending to be evergreen, usually elliptic and acute, more finely toothed, the flowers smaller, in shorter racemes. *USA (Texas to Arizona), Mexico.* H4. Late spring.

74. P. salicifolia Kunth (*P. capuli* Sprengel). Illustration: *Revue Horticole* 1893: unnumbered plate (1893) – as *Cerasus capuli*; Popenoe, Manual of tropical and subtropical fruits, pl. 13 (1960); Krüssmann, Manual of cultivated broad-leaved trees and shrubs **3**: f. 31 (1986).

Evergreen trees or shrubs to 12 m; young shoots hairless. Leaves lanceolate with narrow acute tips, finely toothed, completely hairless or slightly hairy beneath. Flowers white *c.* 1 cm across, usually 10 or more in loose, cylindric racemes more than 5 cm, leafy towards base. Fruit spherical, purple-red, to 1.7 cm across. *Mexico to Peru.* H4. Late spring.

75. P. padus Linnaeus. Illustration: Phillips, Trees in Britain, Europe and North America, 173 (1978); Krüssmann, Manual of cultivated broad-leaved trees and shrubs **3**: f. 18 (1986); Brickell, RHS gardeners' encyclopedia of plants and flowers, 49 (1989).

Deciduous trees or shrubs to 15 m; young twigs at first downy. Leaves elliptic, with short abrupt tips less than 1 cm, margins with straight projecting teeth, hairless, bluish green beneath. Flowers white, fragrant, to 1.5 cm across, 10 or more in narrow, cylindric racemes, leafy towards base. Perigynous zone hairy within. Petals more than 6 mm. Fruit spherical, to 8 mm across, black. *Europe & W Asia to C Japan.* H1. Spring.

76. P. vaniotii Léveillé. Illustration: Krüssmann, Manual of cultivated broad-leaved trees and shrubs **3**: pl. 5 (1986).

Deciduous trees or shrubs to 15 m; young shoots hairless. Leaves oblong to obovate, margins scalloped or with incurved teeth at least in apical part, acute-acuminate. Flowers white, to 8 mm wide, 10 or more in narrow, loose, cylindric racemes, leafy towards base, main stalk of raceme densely hairy. Fruit spherical, orange-red to brown-red, to 8 mm across. *W China to Taiwan.* H2. Spring.

77. P. cornuta (Royle) Steudel. Illustration: *Botanical Magazine*, 9423 (1935); Polunin & Stainton, Flowers of the Himalaya, pl. 33 (1985); Krüssmann, Manual of cultivated broad-leaved trees and shrubs **3**: pl. 5 & f. 18 (1986).

Deciduous trees or shrubs to 5 m (more in the wild); young shoots hairless. Leaves elliptic to obovate, with short abrupt tips less than 1 cm, margins with straight projecting teeth, dull above, broadest near middle, mature blades usually more than 10 cm. Flowers white, 6–10 mm across, 10 or more in narrow, cylindric racemes, main stalk of raceme hairless or sparsely hairy, leafy towards base. Fruit ovoid, hairy, to 2 cm. *Afghanistan, Himalaya to W China.* H2. Spring.

78. P. napaulensis (Seringe) Steudel. Illustration: Stainton, Flowers of the Himalaya, a supplement, 30 (1988)

Like *P. cornuta* but leaves hairy beneath, stalks of the racemes densely hairy. *Himalaya, W China.* H5. Spring.

Similar to *P. sericea* (Batalin) Koehne, which is rare in cultivation; the two are sometimes regarded as a single species.

79. P. virginiana Linnaeus. Illustration: Sargent, Silva of North America **4**: t. 158 (1892); Phillips, Trees in Britain, Europe and North America, 176 (1978); Krüssmann, Manual of cultivated broad-leaved trees and shrubs **3**: pl. 8, 20 & f. 31 (1986).

Deciduous trees or shrubs, somewhat stoloniferous; young shoots hairless. Leaves broadly obovate to broadly elliptic, with short abrupt tips less than 1 cm, broadest towards tip, shiny above, mature blades usually 10 cm or less; margins with straight projecting teeth. Flowers usually 10 or more in narrow, cylindric racemes; stalk of raceme hairless or sparsely hairy, leafy towards base. Petals 2–3 mm. Fruit spherical, at first red, later black, to 1 cm across. *N America*. H1. Late spring.

80. P. ssiori Schmidt. Illustration: Shirasawa, Iconographie des essences forestières du Japon **2**: t. 28 (1908); Krüssmann, Manual of cultivated broad-leaved trees and shrubs **3**: pl. 5 & f. 18 (1986).

Deciduous trees or shrubs to 25 m in the wild; young shoots hairless. Leaves obovate-elliptic to oblong, with narrow tips 1 cm long or more, margins with straight projecting teeth, base cordate; dark green and hairless above, hairy in the vein-axils beneath. Flowers white, to 1 cm across, many in narrow, cylindric racemes, which are leafy towards the base. Styles not projecting. Fruit more or less spherical, black, to 1 cm across. *Japan*. H2. Spring.

81. P. grayana Maximowicz. Illustration: Shirasawa, Iconographie des essences forestières du Japon **1**: t. 46 (1900); Krüssmann, Manual of cultivated broad-leaved trees and shrubs **3**: pl. 8 & f. 18 (1986).

Small deciduous trees or shrubs, to 7 m; young shoots hairy or hairless. Leaves oblong-ovate, with narrow tips 1 cm long or more, margins with straight projecting teeth, bases rounded. Flowers white, to 1 cm across, many, in narrow, cylindric racemes, which are leafy towards the base. Styles projecting. Fruit spherical but pointed, ultimately black, to 8 mm across. *Japan*. H4. Late spring.

82. P. maackii Ruprecht. Illustration: Phillips, Trees in Britain, Europe and North America, 172 (1978).

Deciduous trees or shrubs to 10 m; bark shiny brownish yellow; young shoots downy. Leaves ovate-oblong, finely toothed, with dotted glands (and some hairs on the veins) beneath. Flowers white, to 1 cm across, 6–10 in narrow, irregular, cylindric racemes on the old wood, not leafy towards the base. Styles longer than stamens, hairy below. Fruit spherical, black, to 5 mm across. *N China, Korea*. H1. Spring.

83. P. buergeriana Miquel. Illustration: Shirasawa, Iconographie des essences

forestières du Japon **1**: to 46 (1900); Krüssmann, Manual of cultivated broad-leaves trees and shrubs **3**: f. 18 (1986).

Deciduous trees or shrubs to 10 m; young shoots hairy or not. Leaves elliptic or oblong-elliptic, acuminate, toothed, not gland-dotted beneath, but hairy in the vein axils. Flowers white, to 7 mm across, usually many in narrow, cylindric racemes, which are not leafy towards the base. Styles very short. Fruit spherical, black, borne on the persistent perigynous zone. *Japan, Korea*. H2. Spring.

84. P. lusitanica Linnaeus. Illustration: Phillips, Trees in Britain, Europe and North America, 172 (1978); Hillier colour dictionary of trees and shrubs, 173 (1981); Mitchell, The complete guide to trees of Britain and northern Europe, 83 (1985); Hay & Beckett (eds), Reader's Digest field guide to the trees and shrubs of Britain, 97 (1988).

Evergreen trees or shrubs to 20 m; young shoots hairless, red. Leaves oblong-ovate, coarsely toothed, dark green and glossy above, paler beneath. Flowers white, to 1 cm across, usually 10 or more in narrow, cylindric racemes, ascending, spreading or hanging, more than 5 cm, not leafy towards base. Fruit ovoid, to 8 mm, dark red. *Spain, Portugal, Azores*. H2. Late spring.

85. P. laurocerasus Linnaeus. Illustration: Nicholson & Clapham, The Oxford book of trees, 140 (1975); Phillips, Trees in Britain, Europe and North America, 172 (1978); Mitchell, The complete guide to trees of Britain and northern Europe, 83 (1985); Hay & Beckett (eds), Reader's Digest field guide to the trees and shrubs of Britain, 96 (1988).

Evergreen trees or more usually shrubs to 8 m; young shoots hairless, green. Leaves oblong to obovate or elliptic, entire or toothed, dark green above, paler beneath. Flowers white, to 8 mm across, many in narrow, erect, cylindric racemes, more than 5 cm, not leafy towards base. Fruit conical, dark red, to 8 mm. *E Europe, SW Asia*. E2. Spring.

Many selections have been made and treated as cultivars. See Krüssmann, Manual of cultivated broad-leaved trees and shrubs **3**: 36–8 (1986) for descriptions and a key.

86. P. ilicifolia (Nuttall) Walpers. Illustration: Sargent, Silva of North America **4**: t. 162 (1892); Phillips, Trees

in Britain, Europe and North America, 171 (1978); Krüssmann, Manual of cultivated broad-leaved trees and shrubs **3**: pl. 5 & f. 12 (1968); Hickman, The Jepson manual, 977 (1993).

Evergreen trees or shrubs to 9 m, holly-like. Leaves spiny, ovate to broadly lanceolate, 5–7 cm, acute, entire to finely and distantly toothed, hairless. Flowers white, to 8 mm across, usually 10 or more in narrow, ascending, spreading or hanging, cylindric racemes, more than 5 cm, not leafy towards base. Fruit almost spherical, purple-black, to 1.5 cm across. *USA (California)*. H5. Summer.

87. P. lyonii (Eastwood) Sargent (*P. integrifolia* Sargent not Walpers). Illustration: Abrams, Illustrated flora of the Pacific states **2**: 468 (1944); Krüssmann, Manual of cultivated broad-leaves trees and shrubs **3**: pl. 5 (1986); Hickman, The Jepson manual, 977 (1993).

Evergreen trees or shrubs to 14 m. Leaves ovate to ovate-lanceolate, acute, entire or finely and sparsely toothed, hairless. Flowers white, to 8 mm across, usually many in narrow, cylindric racemes, more than 5 cm, not leafy towards base. Fruit spherical, black, 1–2.5 cm across. *USA (California)*. H5. Spring.

88. P. caroliniana (Miller) Aiton. Illustration: Sargent, Silva of North America **4**: t. 160 (1892); Krüssmann, Manual of cultivated broad-leaved trees and shrubs, f. 7 (1986).

Evergreen trees or shrubs to 12 m. Leaves entire or sparsely toothed. Flowers cream-white, to 1 cm across, usually 10 or more in short, dense, cylindric racemes, to 4 cm. *SE USA*. H5. Spring.

77. PRINSEPIA Royle
H.S. Maxwell
Deciduous, arching shrubs with axillary spines and flaking bark; branches with chambered pith; buds small, naked and/or covered with a few small hairy scales. Leaves alternate, often clustered, simple, membranous to leathery, entire or toothed. Stipules, if present, small, persistent. Flowers stalked, in axillary bracteate racemes or 1–8 (rarely to 13) in the axils of previous years shoots. Perigynous zone tubular, cup-shaped; calyx 5-lobed, lobes equal or unequal, broad, short. Petals 5, distinct, equal, almost circular, clawed, white or yellow, spreading. Stamens 10 or many. Ovary superior. Fruit an oblique

1-seeded drupe, red or purple, juicy and edible.

A genus of 4 species from China, Taiwan and the Himalaya. Only 3 species are of any horticultural importance. *P. utilis* is cultivated as a hedging plant, *P. uniflora* and *P. sinensis* as ornamentals. They grow well in ordinary, well-drained soil and are best in full sun, although they can stand slight shade. No special care is required except a little thinning of the branches in late winter or early spring. They generally do not fruit very well in cultivation. Propagation is by seed, cuttings or layering.

1a. Leaves mucronate **2. uniflora**
  b. Leaves long acuminate **2**
2a. Flowers white; calyx-lobes irregular; stamens numerous; spines 1–5 cm, with leaves, buds or scales **3. utilis**
  b. Flowers yellow; calyx-lobes equal; stamens 10; spines 3–12 mm, without leaves, buds or scales **1. sinensis**

1. P. sinensis (Oliver) Bean (*Plagiospermum sinense* Oliver). Illustration: *Botanical Magazine*, 8711 (1917); Bean, Trees and shrubs hardy in the British Isles, edn 8, **4**: 344 (1980); Krüssmann, Manual of cultivated broad-leaved trees and shrubs **3**: 442 (1986).

Shrub with rather loose spreading habit, 2–3 m; bark yellowish, hairless or sometimes only with tufts of hair on the nodes and base of stipules. Spines 3–12 mm, curved, without leaves, buds or scales. Leaves oblong-lanceolate to lanceolate, apex long acuminate, base obtuse to rounded, entire or sparingly toothed, margin hairy. Leaves on fertile branches 2.5–5.6 cm × 6–16 mm; stalks 3–14 mm; sterile branches 3–9 cm × 8–24 mm; stalks 8–24 mm. Flowers yellow, single, in clusters of 1–8, borne in leaf-axils, on very short spurs, 1.3–2 cm across; calyx-lobes equal; flower-stalks hairless, 8–21 mm. Stamens 10. Fruit red or purple, 1.1–1.3 cm, ripening in August. *E Asia (Manchuria, Ussuri-land & N Korea).* H3. Spring.

2. P. uniflora Batalin. Illustration: Everett, New York Botanical Gardens illustrated encyclopedia of horticulture **8**: 2799 (1981); Brickell, The RHS gardeners' encyclopedia of plants and flowers, 105 (1989).

Loose spreading shrub to 1.6 m; bark reddish brown; stipules present; young shoots hairless. Spines 3–16 mm, curved, without leaves, buds or scales. Leaves membranous, dark glossy green, linear-oblong to narrow-oblong, acute or obtuse, base tapering, apex mucronate, entire or toothed. Leaves on fertile branches 8–44 × 3–8 mm; stalks 1–6 mm; on sterile branches 4.7–8.2 cm × 7–14 mm; stalks 2–4 mm. Flowers white; calyx-lobes equal with marginal hairs, 1–8 among the clustered leaves on very short spurs, *c.* 1.4 cm; stalks hairless, 3–7 mm. Petals *c.* 5 mm, obovate; stamens 10; anthers yellow. Fruit 1–1.5 cm, purple or dark red with a slight bloom. *Inner Mongolia, NW & C China.* H1. Early spring.

3. P. utilis Royle. Illustration: *Botanical Magazine*, n.s., 194 (1952); Polunin & Stainton, Flowers of the Himalaya, 31 (1985).

Vigorous spiny shrub, 1.5 m, occasionally to 3.6 m, young shoots green, downy at first, becoming hairless. Spines numerous, produced in every leaf-axil, 1–5 cm, straight, often leafy or with buds or scales. Leaves 1.1–10.5 cm × 5–20 cm, very variable, thin, leathery, more or less persistent, elliptic to elliptic-lanceolate, apex usually long acuminate, base narrowly tapered, usually toothed, marginal teeth glandular, stipules absent, dull green, hairless. Leaf-stalks 4–6 mm, base densely hairy, glandular, especially on young shoots. Inflorescence 1.5–6 cm, with hairless or finely downy stalks to 1.3 cm, flowers in racemes, 1–13 (usually 5–7), 1.3–2 cm across, creamy white, fragrant; bracts occasionally leaf-like; calyx irregular, with 2 small and 3 large calyx-lobes, margin entire or irregularly toothed. Stamens numerous. Fruit oblong-obovoid, 1–1.5 cm, dark purple, bloomed. *W Pakistan, India, Nepal, Bhutan & China.* H3. Late autumn–early winter.

78. MADDENIA J.D. Hooker & Thomson
M.F. Gardner
Deciduous, dioecious trees or shrubs. Leaves alternate, with glandular forward-pointing teeth; stipules large, glandular-toothed. Flowers unisexual, on short stalks, borne in short terminal dense racemes; sepals 10, small, some long and petal-like; stamens 25–40, inserted in calyx mouth in 2 more or less distinct rows. Male flowers with a solitary carpel, female flowers with 2. Ovary superior. Fruit a single-seeded drupe; stone ovoid, 3-keeled on one side.

A genus of 4 species from the Himalaya and China, which are easily cultivated in most garden soils.

1. M. hypoleuca Koehne. Illustration: *Deutsche Baumschule* **11**: 71 (1959); Krüssmann, Manual of cultivated broad-leaved trees & shrubs **2**: 265 (1986); Phillips & Rix, Shrubs, 33 (1989).

Shrub or small tree to 6 m; young twigs dark brown, hairless. Leaves ovate-oblong, 4–7 cm, double-toothed, tapering towards the end, rounded at base, dark green above, blue-white beneath, with 14–18 pairs of veins. Flowers red-brown at first, later green, borne in racemes, 3–5 cm; stamens 20–30, anthers yellow. Fruit *c.* 8 mm, ovoid, black. *C & W China.* H4. Winter–spring.

79. OEMLERIA Reichenbach
C.M. Mitchem
Deciduous shrubs to 3 m, which form dense thickets. Bark purple-brown; lenticels prominent on branchlets. Leaves 2.5–9 cm × 8–30 mm, obovate to lanceolate, undersides paler. Inflorescence a raceme of up to 13 flowers each subtended by 2 bracteoles. Flowers to 1 cm across, white. Calyx to 2 cm. Petals ovate and upright. In male flowers: numerous stamens around perigynous zone, to 4 mm, incurved; smaller and non-functional in female flowers. Ovary superior. Carpels 5; stigma bilobed. Fruit a 1-seeded, thinly fleshy drupe, bitter-tasting.

Literature: Allen, G.A., Flowering pattern and fruit production in the dioecious shrub, Oemleria cerasiformis (Rosaceae), *Canadian Journal of Botany* **64**(6): 1216–20 (1986); Anon., Osmaronia cerasiformis (Torrey & Gray) Greene, *Davidsonia* **5**: 12–15 (1974).

A genus of one species, grown mainly for its early, fragrant flowers and decorative fruits.

1. O. cerasiformis (W.J. Hooker & Arnott) Landon (*Osmaronia cerasiformis* W.J. Hooker & Arnott; *Nuttalia cerasiformis* (W.J. Hooker & Arnott) Greene; *O. davidiana* Baillon; *Exochorda davidiana* Baillon). Illustration: *Botanical Magazine*, 582 (1970); Krüssmann, Manual of cultivated broad-leaved trees and shrubs **2**: pl. 146 (1986).

Female plants are less commonly planted, as they are coarser and not as free-flowering. Prefers semi-shade and moist, humus-rich soils. Propagation by seed or layering. *W North America (British Columbia to California).* H1. Early spring.

150. CHRYSOBALANACEAE

Woody. Leaves alternate, stipulate, often leathery. Flowers in racemes, panicles or cymes, usually bisexual, actinomorphic or zygomorphic. Calyx, corolla and stamens perigynous. Calyx of 5 free sepals. Corolla of 4–5 free petals. Stamens 3–many. Ovary of 2–3 united carpels often 1 or 2 of them sterile, often asymmetrically placed in the tubular or cup-shaped perigynous zone; ovules 2, basal; styles free. Fruit a drupe.

There are 500 species in 17 genera distributed through the tropics of both hemispheres. A few species are occasionally cultivated as glasshouse plants.

Literature: Prance, G.T. & White, F., The genera of *Chrysobalanaceae*: A study in practical and theoretical taxonomy and its relevance to evolutionary biology, *Philosophical Transactions of the Royal Society of London* **320**: 1–184 (1988).

1a. Flowers radially symmetric; ovary inserted at or near base of perigynous zone   2
  b. Flowers bilaterally symmetric; ovary inserted at or near mouth of perigynous zone   3
2a. Stamens projecting, joined in groups; filaments hairy   **1. Chrysobalanus**
  b. Stamens included or projecting; filaments smooth, free   **2. Licania**
3a Stamens included   **3. Parinari**
 b.   Stamens projecting   4
4a. Inflorescence of little-branched panicles or racemes   **5. Atuna**
  b. Inflorescence of much branched corymbose panicles   **4. Maranthes**

1. CHRYSOBALANUS Linnaeus
E.H. Hamlet

Small trees and shrubs. Leaves smooth or with adpressed hairs beneath. Bracts and bracteoles small, without glands. Inflorescence terminal or axillary. Flowers bisexual. Calyx-lobes 5, acute; petals 5, longer than calyx-lobes. Stamens 12–26; filaments hairy, united at base. Ovary densely hairy, inserted at base of perigynous zone. Style downy. Fruit a fleshy drupe.

A genus of 2 species from tropical America, tropical Africa & West Indies. They can be grown in ordinary soil and need a frost-free climate. Propagation usually by seed, also by cuttings.

1. C. icaco Linnaeus. Illustration: van Steenis & de Wilde, Flora Malesiana, series 1, **18**(4): 644 (1989).

Small tree or shrub, to 5 m, branches smooth with lenticels. Leaf-stalks 2–5 mm; stipules 1–3 mm, deciduous. Leaves 2–8 ´ 1.2–6 cm, alternate, leathery, rounded to ovate-elliptic, slightly notched, rounded to obtuse, slightly tapered at base. Inflorescences terminal and axillary panicles or cymes with grey-brown matted hairs. Perigynous zone cup-shaped with densely felted hairs. Calyx-lobes 4 or 5, *c.* 2.5 mm, rounded to acute; petals 4 or 5, *c.* 5 mm, white, smooth. Stamens 12–16, longer than petals; filaments joined in groups, densely hairy. Ovary hairy. Fruit ovate to obovate, 1.5–5 cm. *Tropical W Africa (Guinea to Angola), C America to S Brazil, W Indies.* G2.

2. LICANIA Aublet
E.H. Hamlet

Trees or shrubs. Leaf-stalks smooth or with 2 or more stalkless glands. Leaves entire. Inflorescence a simple or branched racemose panicle, less often a panicle of cymes or spikes. Flowers bisexual. Calyx-lobes 5, acute; petals 4 or 5, or absent. Stamens 3–40, unilateral or forming a circle; filaments rarely joined and usually smooth. Fruit a fleshy drupe.

A genus of about 190 species from Africa, tropical Asia & America.

1a. Stamens 5–7   **3. incana**
  b. Stamens 8–14   2
2a. Inflorescence with grey downy hairs   **2. rigida**
  b. Inflorescence brown to rusty brown   **1. arborea**

1. L. arborea Seemann. Illustration: *Contributions from the United States National Herbarium* **27**: t. 29 (1928); Pennington & Sarukhan, Arboles Tropicales de Mexico, 161 (1968).

Tree to 60m. Leaves ovate-rounded to oblong, 5–12 ´ 2.5–8 cm, on fertile branches, larger on sterile branches. Leaf-stalks 5–12 mm, terete with densely felted hairs when young, later hairless. Inflorescences axillary or terminal racemose panicles, main axis and branches with densely felted brown to rusty brown hairs. Flowers 2.5–3 mm, solitary and densely clustered. Calyx-lobes acute with densely felted hairs; petals 5, oblong, downy. Stamens 8–12, inserted in a circle; filaments as long as calyx-lobes, joined for half their length. Fruit oblong, to

3 cm. *C America & W South America to Peru.* G2.

2. L. rigida Bentham.

Small tree to 15m. Leaves 6–16 × 2.8–6.5 cm, oblong to elliptic, leathery, rounded to slightly notched at apex, rounded to cordate at base. Inflorescences in racemose panicles; main axis and branches with densely felted grey hairs. Flowers 2.5–3.5 mm, in small groups. Calyx-lobes acute; petals 5, densely downy. Stamens *c.* 14; filaments densely downy. Fruit elliptic, 4–5.5 cm. *NE Brazil.* G2.

3. L. incana Aublet (*L crassifolia* Bentham). Illustration: Martius, Flora Brasiliensis **14**(2): t. 3 (1867).

Shrub or rarely small tree. Leaves 2.5–8.5 × 1.3–5.5 cm, ovate to oblong, acute to acuminate at apex, with point to 1 cm long, rounded at base, rarely slightly wedge-shaped. Inflorescences terminal and axillary spikes. *c.* 2 mm, in small clusters along the stem. Calyx-lobes acute; petals absent. Stamens 5–7, unilateral; filaments shorter than calyx-lobes, smooth. Style equalling filaments. Fruit ovoid to spherical, *c.* 1.6 cm. *S America (Colombia, Venezuela & Brazil).* G2.

3. PARINARI Aublet
E.H. Hamlet

Trees or shrubs, occasionally woody at base with herbaceous branches. Leaves alternate, entire. Leaf-stalks usually with 2 stalkless glands. Inflorescence in terminal or axillary much-branched panicles. Flowers 4–11 mm, bisexual. Perigynous zone cone to bell-shaped, slightly swollen at one side. Calyx-lobes 5, acute; petals 5. Stamens 6–8; filaments not exceeding calyx-lobes, unilateral with staminodes inserted opposite them. Ovary inserted laterally at the mouth of the receptacle. Style thread-like. Fruit a fleshy drupe.

A genus of about 40 species from tropical Africa, tropical Asia, tropical America and the Pacific Islands.

1a. Stipules persistent, partly clasping stem   **1. campestris**
  b. Stipules falling early, not clasping stem   **2. excelsa**

1. P. campestris Aublet. Illustration: *Flora of the Guianas* **85**: 102 (1986).

Tree to 25 m, young branches downy, becoming hairless and greyish with age. Leaf-stalks 2–7 mm, downy when young, terete; blades 6–13 × 3–6.5 cm, ovate, acuminate at apex, rounded to cordate at base, hairless above, with densely felted hairs beneath. Stipules broad, to 3 cm, acute at apex, persistent, partly clasping the stem. Inflorescences axillary and terminal panicles. Perigynous zone slightly bell-shaped to obconical. Petals 5, white, shorter than calyx-lobes. Stamens 7, fertile, unilateral, with 7 or 8 staminodes opposite them. Ovary and lower part of style densely hairy. Fruit 4–6 × 2–3 cm, oblong. *Tropical America (Venezuela, Trinidad, Guianas to Brazil).* G2.

2. P. excelsa Sabine. Illustration: Hutchinson & Dalziel, Flora of West Tropical Africa **1**: 429 (1958).

Tree to 40 m. Leaf-stalks 3–7 mm, downy when young, terete; blades 3–9 × 1.5–5 cm, ovate to oblong-elliptic, rounded to wedge-shaped at base, acuminate at apex. Stipules *c.* 1 mm, falling early. Inflorescences terminal, in loose panicles. Flower-stalks 1–2 mm. Perigynous zone slightly bell-shaped to obconical. Petals 5, white, ovate, falling early, shorter than calyx-lobes. Stamens 7, fertile, unilateral with 7 or 8 short thread-like staminodes opposite them. Ovary and base of style hairy. Fruit 2.5–4 × 1.8–2.5 cm, ellipsoid. *Tropical America & tropical Africa.* G2.

4. MARANTHES Blume
E.H. Hamlet
Trees with leaf-undersides hairless or with dense woolly hairs. Leaf-stalks usually with 2 glands, rarely without. Bracts and bracteoles without glands, not enclosing the young flowers in groups. Perigynous zone shape various, but narrowed to the base, hairless inside at base. Calyx-lobes rounded. Stamens 18–60, projecting beyond calyx-lobes, often forming a complète circle, or slightly unilateral; filaments hairless. Ovary inserted laterally at the mouth of the receptacle. Fruit a fleshy drupe.

A genus of 11 species from tropical Africa, except 1 from Malaysia to New Guinea, NE Australia and W Polynesia.

1. M. corymbosa Blume (*Parinari corymbosa* Miquel). Illustration: Prance, Flora Neotropica **9**: 203 (1972); *Candollea* **20**: 143–5 (1965); Van Steenis & de Wilde, Flora Malesiana, series 1, **10**(4): 672 (1989).

Tree to 40 m, young branches shortly downy or hairless. Leaves 6.5–15 ´ 2–8 cm, oblong-elliptic to lanceolate, leathery, slightly wedge-shaped at base,

apex acuminate. Leaf-stalks 4–6 mm, terete with 2 glands near base of blade. Inflorescences in terminal corymb-like panicles. Flowers 5–6 mm, obconical, bell-shaped; flower-stalks 2–3 mm. Calyx-lobes ovate-elliptic; petals 5 with marginal hairs. Stamens inserted in a semi-circle. Fruit oblong to pear-shaped. *SE Asia, Pacific Islands, tropical Australia & New Guinea.* G2.

5. ATUNA Rafinesque
E.H. Hamlet
Trees with complicated widely divergent branching. Leaves almost hairless; stipules large. Inflorescence a raceme or sparsely branched, contracted panicle. Flowers bisexual. Calyx-lobes 5, broadly ovate to lanceolate; petals 5, hairless, exceeding calyx-lobes. Stamens 10–25; filaments free, exserted. Fruit densely warty.

A genus of about 11 species from southern India, Thailand, Malaysia, Indonesia, Pacific Islands and New Guinea.

1. A. racemosa Rafinesque. Illustration: Smith, Flora Vitiensis Nova **3**: 49 (1985); Van Steenis & de Wilde, Flora Malesiana, series 1, **10**(4): 668 (1989).

Tree with young branches hairless or having adpressed bristles. Leaves 4.5–35 × 2–11 cm, ovate, elliptic, oblong or lanceolate, papery to leathery. Flowers in axillary racemes or little branched with up to 3 or more racemose branches on short main inflorescence-stalk. Calyx-lobes 4–7 mm, ovate to ovate-oblong. Petals to 10 mm, ovate or oblong, blue or white. *Thailand, Malaysia, Indonesia, Philippines, Pacific Islands & New Guinea.* G2.

Subsp. **racemosa** (*Parinari scabra* Hasskarl) has leaves 10–35 cm, usually elliptic, oblong or lanceolate but sometimes ovate, papery or thickly leathery, apex long, finely acuminate, 0.6–2.5 cm; leaf-stalks thick; flowers 1–1.7 cm.

Subsp. **excelsa** (Jack) Prance (*Parinari laurina* A. Gray; *P. macrophylla* Teijsmann & Binnendijk) with leaves 4.5–12 cm, usually ovate or oblong-ovate, leathery, apex bluntly acuminate, 3–10 mm; leaf-stalks thin; flowers 8–11 mm.

151. MIMOSACEAE
(LEGUMINOSAE-MIMOSOIDEAE)

Mostly shrubs or trees, rarely herbaceous. Leaves usually bipinnate, less often pinnate or reduced to expanded stalks and rachises

(phyllodes). Flowers in spikes, racemes or heads, these often aggregated into more elaborate compound inflorescences. Calyx, corolla and stamens hypogynous. Calyx usually 5-lobed, sometimes 3- or 4-lobed, or lobes altogether lacking. Corolla radially symmetric, of 5 (less often 4 or 3) free or united petals which are valvate in bud. Stamens 3–10–many, often conspicuous, anthers opening by slits, pollen often in masses of 4–332 grains clumped together. Ovary usually of a single carpel; ovules 1–many, marginal. Fruit a legume; seeds with a U-shaped line (pleurogram) on each face.

There are about 82 genera and 3,270 species. Many are grown as ornamentals in Europe, especially species of *Acacia*, *Mimosa* and *Albizia*. The family was included in the Leguminosae in edn 1 (as subfam. *Mimosoideae*). The editors are grateful to Dr G. Lewis and Dr B. Schrire of Kew, and Ms S. Andrews for their helpful comments on the accounts of this family.

Literature: Lewis, G., et al. (eds.), Legumes of the World (2005).

1a.	Stamens more than 10	2
b.	Stamens 10 or fewer, usually free, sometimes united at the base	7
2a.	Stamens free or nearly so	**7. Acacia**
b.	Stamens united into a tube	3
3a.	Fruits splitting longitudinally along 1 or 2 margins	4
b.	Fruits not splitting or breaking longitudinally into 1-seeded units	5
4a.	Seeds with arils	**13. Pithecellobium**
b.	Seeds without arils	5
5a.	Seeds black, standing out against the orange-red colour of the interior of the open fruit; stipules small, falling early	**14. Archidendron**
b.	Combination of characters not as above	6
6a.	Flowers of the same part-inflorescence uniform; valves of pod elastically dehiscing from apex, recurving	7
b.	Flowers of the same part-inflorescence heteromorphic; valves of pod not recurving from the apex or, indehiscent	8
7a.	Inflorescence terminal, branched, subtended by obvious, stipule-like bracts;stamens without any red coloration	**12. Zapoteca**
b.	Inflorescence not as above; at least the anthers red	**11. Calliandra**

8a. Inflorescence a spike or raceme
10. Paraserianthes
 b. Inflorescence a head   **9. Albizia**
6a. Leaflets bipinnate   **15. Samanea**
 b. Leaflets once-pinnate   **8. Inga**
7a. Fruits splitting longitudinally along 1
 or 2 margins   8
 b. Fruits not splitting or fruits breaking
 transversely into individual 1-seeded
 units   10
8a. Anthers hairy   **5. Leucaena**
 b. Anthers hairless   9
9a. Trees; seeds red, or red and black
1. Adenanthera
 b. Aquatic or terrestrial herbs; seeds
 usually brown   **6. Neptunia**
10a. Plants spineless or very rarely
 sparsely spiny; fruit splitting
 transversely into 1-seeded units   11
 b. Plants usually spiny; fruit
 indehiscent   **3. Prosopis**
11a. Trees, shrubs or herbs; fruits to 3 cm
4. Mimosa
 b. Trees, shrubs or woody climbers;
 fruit usually more than 30 cm
2. Entada

1. ADENANTHERA Linnaeus
D. Fränz

Evergreen, unarmed trees. Leaves bipinnate, leaflets opposite with several to many pairs of alternate segments. Racemes spike-like, axillary, solitary or sometimes paired, often aggregated at the tips of the shoots. Flowers bisexual, bracts minute. Calyx short, bell-shaped, 5-lobed. Petals 5, united below, not overlapping. Stamens 10, alternately long and short, each anther tipped with an early-falling gland. Ovary stalkless, with many ovules. Style thread-like, stigma small, terminal. Pods linear, curved or spirally twisted, opening longitudinally into 2 leathery or somewhat leathery valves which do not separate from the sutures nor lose their outer layer. Seeds thick, with a hard, shining, red or red and black seed coat.

A genus of 13 species from tropical Asia, Africa and Australia. They are grown in warm, humid conditions in a compost of peat and loam. Propagation is easy by cuttings taken at a node and placed in sand in a closed frame, or by seeds which require soaking in hot water before sowing.

1. A. pavonina Linnaeus. Illustration: Graf, Exotica, edn 12, 1370 (1985); Lewis et al. (eds.), Legumes of the world, 167 (2005).

Tree 4–20 m; young shoots usually hairless. Leaves to 40 cm, with 3–5 pairs of leaflets each with 5–9 oblong or ovate segments, 1.5–4.5 × 1.2–2.3 cm, rounded at the apex, minutely downy beneath. Racemes 9–26 cm, hairless or slightly downy; flower-stalks 2–3.5 cm. Flowers yellowish or white and yellow in the same racemes. Calyx 0.75–1 mm, usually hairless. Petals 3–4.5 cm. Filaments 2.5–4 mm. Pods 18–22 × 1.3–1.7 cm, brown, much coiled after opening. Seed lens-shaped, red, 8–10 × 7–9 mm. *India, Burma, SE Asia; often cultivated in the tropics elsewhere.* G1. Early summer.

2. ENTADA Adanson
D. Fränz

Trees, shrubs or woody climbers. Leaves bipinnate, each leaflet with 1–many segments. Flowers in spikes or spike-like racemes, these solitary or clustered. Flowers white or yellow, sometimes unisexual. Calyx with 5 lobes. Petals 5, free or nearly so, borne on a short perigynous zone. Stamens 10, fertile; anthers each with an early-falling apical gland. Pods straight or curved, flat or rarely spirally twisted, sometimes to 90 cm, woody, the outer layer usually peeling off, the inner portions consisting of many woody envelopes which break away from the persistent sutures as 1-seeded units. Seeds large, circular, compressed, deep brown, smooth.

A genus of about 30 species, widespread in the tropics, whose large seeds are frequently washed up on beaches in western Europe. The 1 species cultivated in Europe requires warm glasshouse treatment.

1. E. phaseoloides (Linnaeus) Merrill (*Lens phaseoloides* Linnaeus; *E. scandens* (Roxburgh) Bentham; *E. gigas* (Linnaeus) Fawcett & Rendle; *Mimosa scandens* Roxburgh; *E. schefferi* Ridley). Illustration: Polunin & Stainton, Flowers of the Himalayas, pl. 24 (1985).

Large woody climber to 25 m, unarmed; young shoots more or less hairless or downy. Leaves with 2 (rarely 1) pairs of leaflets, the rhachis ending in a forked tendril. Leaflets stalked, segments oblong or obovate, 2.5–5 cm, leathery. Flowers small, cream to greenish or yellowish, on distinct, slender flower-stalks 1–2 mm. Calyx hairless or shortly downy, 1–1.25 mm. Petals 2.5–3 mm. Filaments hairless, 3.5–6 mm. Pods spirally twisted, 60 or more × 7.5–10 cm, outer layer falling away to reveal the thick, papery, somewhat flexible inner layer. Seeds hard, dark brown or purplish, 4–5.5 cm across. *West Indies, tropical Africa & Asia, Pacific islands.* G2.

3. PROSOPIS Linnaeus
D. Fränz

Trees or shrubs, with or without axillary, solitary or paired spines, or sometimes the stipules spine-like. Leaves bipinnate with 1 or 2 pairs of leaflets; segments usually numerous, small, entire. Flowers small, greenish, in cylindric or spherical axillary spikes. Calyx bell-shaped, teeth very short and not overlapping. Petals 5, not overlapping, united below the middle or ultimately free, woolly on the inner surface. Stamens 10, free, projecting; each anther with an early-falling gland at the apex. Ovary covered with shaggy hairs; style thread-like. Pod sickle-shaped or spirally coiled, leathery and indehiscent, usually pulpy within. Seeds many, ovate, compressed.

A genus of 44 species from the tropics, subtropics and W North America, differing from *Acacia* in having fewer stamens and indehiscent pods. A few species are grown in cool glasshouse conditions. Propagation is by cuttings of firm young shoots taken close to the stem, rooted in sand under gentle heat. Some species are used as forage crops in desert regions.

1a. Tree with few spines   **1. chilensis**
 b. Tree with numerous spines   2
2a. Leaflet-segments close, *c.* 2 cm; pods
 beaked   **2. juliflora**
 b. Leaflet-segments distant, *c.* 1 mm;
 pods not beaked   **3. cineraria**

1. P. chilensis

Large trees with few spines. Leaves with 1 or 2 pairs of leaflets, each with 13–20 segments which are 1.5–4 cm, 10 or more times as long as broad. Pods sickle-shaped. *Chile, Argentina.* G1.

Fast-growing in dry conditions.

2. P. juliflora (Swartz) de Candolle (*P. cumanensis* Kunth; *P. pallida* Kunth). Illustration: Lewis et al., (eds.), Legumes of the world, 172 (2005).

Tree to 15 m, sometimes to 1 m across, or a shrub, with many spines 1.2–5 cm. Bark grey. Leaflet-segments in 11–19 pairs, very close (*c.* 6 mm apart), oblong or linear-oblong, thin in texture, to 2 cm, less than 5 times as long as broad, hairless or slightly hairy on the margin. Flowers fragrant, in cylindric spikes 5–10 × 1–1.5 cm, yellow (mainly from the projecting stamens). Pods almost straight, beaked, hairless, 10–20 cm. *Coastal Central & Northern South America, West Indies, Caribbean.* G1.

3. P. cineraria (Linnaeus) Druce (*P. spicigera* Linnaeus). Illustration: Pickering & Patzelt, Wild plants of Oman, 241 (2008).

Small to medium tree with many broadly based spines, and with thick, fibrous grey, rough bark and purplish brown, hard wood. Leaves with distant leaflet-segments *c.* 12.5 mm. Pods cylindric, constricted between the seeds and appearing necklace like, pulpy. *Oman & Iran to India & SE Asia.* G1.

4. MIMOSA Linnaeus
D. Fränz
Trees, shrubs or herbs of varying habit (rarely woody climbers), usually with spines. Leaves bipinnate (sometimes superficially palmate because of the short axis), often sensitive to touch, occasionally reduced to phyllodes; leaflets with few to many pairs of segments. Flowers small, stalkless, in spherical heads or cylindric spikes. Calyx minute or rudimentary. Petals 4 or 5, united. Stamens 4–10, projecting; pollen granular. Pods flat, straight to much curved, usually splitting into 1-seeded segments; outer layer not separating; sutures persistent.

A genus of between 400 and 500 species, widely distributed in the tropics and subtropics, but mostly found in S America. They require cool glasshouse treatment in Europe and grow well in a compost of equal parts of loam and peat to which a little sand should be added. They are propagated by seed or by cuttings in a sandy compost with bottom heat.

1a. Annual herb   **4. pudica**
  b. Shrubs   2
2a. Leaflet-segments ovate **1. sensitiva**
  b. Leaflet-segments linear-oblong 3
3a. Stems and leaves with recurved spines; flowers in spikes
    2. spegazzinii
  b. Stems and leaves with straight spines; flowers in heads **3. pigra**

1. M. sensitiva Linnaeus (*M. floribunda* Bentham).
Somewhat climbing, prickly evergreen shrub to 1.8 m. Leaves with 2 unequal leaflets; leaflet-segments ovate, acute, with adpressed hairs beneath, hairless above;

leaf-stalks prickly. Flowers purple. *Tropical America.* G2.

Less sensitive to touch than *M. pudica.*

2. M. spegazzinii Pirotta. Illustration: Graf, Exotica, edn 12, 1385 (1985).

Spiny, much branched, somewhat climbing shrub, with recurved spines at the bases of the leaf-stalks. Leaves with 2 leaflets 5–7.6 cm; segments very numerous, stalkless and close together, oblong or linear-oblong, acute, 3-nerved. Flowers rose-purple in spherical heads to 3.8 cm across borne in terminal racemes. Filaments rose-purple, anthers yellow. Pods 3- or 4-seeded, prickly, to 2.5 cm, linear. *Argentina, naturalised in USA (California).* G2.

3. M. pigra Linnaeus (*M. asperata* Linnaeus).
Shrub to 4.5 m, sometimes climbing or scrambling. Stems with broadly based spines to 7 mm, usually more or less adpressed-bristly. Leaf-stalks 3–15 mm, axis with erect or forwardly pointing slender prickles; leaflet-segments 6–16, sometimes with prickles between the segment-pairs, linear-oblong, 3.8–12.5 × 0.5–2 mm, margins often minutely bristly. Flowers mauve or pink in almost spherical, stalked heads to 1 cm across, borne 1–3 together in the upper axils. Stamens 8. Pods clustered, densely bristly all over, 3–8 cm × 9–14 mm. *Widespread in tropical Africa & America; introduced in tropical Asia.* G2.

4. M. pudica Linnaeus. Illustration: Rucker, Die Pflanzen im Haus, 291 (1982); Graf, Exotica, edn 12, 1387 (1985); Graf, Tropica, edn 3, 546, 562 (1986); Encke, Kalt- und Warmhauspflanzen, 302 (1987).

Grown as an annual, but probably perennial, sometimes woody below, to 50 cm, often prostrate or straggling; stems sparsely spiny, spines 2.5–5 mm, also bristly or almost hairless. Leaves not spiny, stalk 1.5–5.5 cm; axis short, the usually 2 leaflets appearing almost palmate; leaflet-segments 10–26 pairs, linear-oblong, 6–15 × 1.2–3 mm. Flowers lilac or pink, in shortly ovoid, stalked heads 1–1.3 cm × 6–10 mm, borne 1–5 together in the leaf-axils. Stamens 4. Pods clustered, densely prickly on the margins, 1–1.8 cm × 3–5 mm. *Originally from Brazil but now widely naturalised in warm countries.* G1.

5. LEUCAENA Bentham
D. Fränz
Trees or shrubs, without spines. Leaves evergreen, bipinnate, with a gland often present on the axis above, between the lowest pair of leaflets, sometimes glands present elsewhere on the axis; leaflets with 1–several pairs of segments. Flowers in rounded, stalked, axillary heads, borne 1–3 together. Flowers bisexual, stalkless. Calyx with 5 teeth. Petals 5, free, downy or hairless outside. Stamens 10, fertile, anthers without glands at their tips. Ovary stalked, downy or hairless. Pods oblong or linear-oblong, compressed, usually thinly leathery, splitting into 2 non-recurving lobes. Seeds lying more or less transversely in the pod, glossy, unwinged.

A genus of about 50 species from the SE USA, C America and S America; some species are grown as fodder and are naturalised in Texas, California and Hawaii. Cultivation as for *Acacia.*

Literature: Hughes, C., Monograph of *Leucaena, Systematic Botany Monographs:* **55** (1998).

1. L. leucocephala (Lamarck) de Wit (*Mimosa glauca* Linnaeus; *L. glauca* (Linnaeus) Bentham; *Acacia glauca* (Linnaeus) Moench; *A. leucocephala* Link; *Mimosa leucocephala* (Link) Lamarck; *Acacia frondosa* Willdenow). Illustration: Graf, Exotica, edn 12, 1385 (1985); Graf, Tropica, edn 3, 556 (1986).

Shrub or small tree to 9 m; young shoots densely and shortly grey-downy. Leaves with stalk 2.4–7 cm, often with a gland above at the junction of the lowest pair of leaflets, glands usually otherwise absent; leaflets 4–8 pairs, opposite; segments 10–20 pairs, 7–8 × 1.5–5 mm, glaucous beneath. Flower-heads spherical; stalks 2–5 cm. Calyx 2–3.5 mm, pale green, very shortly downy outside. Filaments 6.5–7.5 mm; anthers hairy. Pods 8–18 × 1.8–2.1 cm, on a stalk to 3 cm. Seeds elliptic to obovate, 7.5–9 × 4–5 mm. *Tropical America, extending to USA (Florida, Texas).* G2.

This species is widely naturalised in the Old World from cultivation as a crop.

6. NEPTUNIA Loureiro
D. Fränz
Aquatic or terrestrial, annual or perennial herbs, erect, prostrate or floating, without spines, branches often compressed or angled. Leaves bipinnate; leaflets with 8–15 pairs of segments. Flowers in

spherical or ellipsoid heads which are solitary and axillary. Flowers small, stalkless, the upper bisexual, the lower male and the lowest sterile with flattened staminodes. Calyx bell-shaped, small, 5-toothed. Petals 5, free or united at the base, not overlapping. Stamens 5 or 10, free; anthers each with an apical gland. Ovary stalked, with many ovules; style thread-like, stigma minute, terminal, concave. Pod flat, membranous, oblong, not contorted or spiralled, dehiscent. Seeds transverse, compressed, oblong-ellipsoid to obovoid, smooth.

A genus of 12 or more species widely distributed, mainly in the tropics. They are not easy to cultivate. The leaves are sensitive to touch, like those of species of *Mimosa*.

1a. Leaf without a gland on the axis between the lowest pair of leaflets   **1. oleracea**
 b. Leaf with a gland on the axis between the lowest pair of leaflets   **2. plena**

1. N. oleracea Loureiro (*Mimosa prostrata* Lamarck; *N. prostrata* (Lamarck) Baillon; *Mimosa natans* Roxburgh; *Desmanthus natans* (Roxburgh) Wight & Arnott). Illustration: Graf, Exotica, edn 12, 1385 (1985); Graf, Tropica, edn 3, 560 (1986); Lewis et al., (eds.), Legumes of the world, 174 (2005).

Annual aquatic herb with stems usually floating or creeping, rooting especially at the nodes, hairless or shortly downy when young; branches zig-zag. Leaves with stalk 2.5–9 cm; leaflets in 2–4 pairs, each divided into 7–22 pairs of oblong segments, 5–20 × 1.5–4 mm, hairless or with a few hairs on the margins. Flowers yellow, in heads 1.5–2.5 cm on stalks 6.5–30 cm. Calyx 1–3 mm. Petals 3–4 mm. Stamens 10, anthers completely without glands at their apices; staminodes 1.1.7–2.1 cm. Pods bent at an angle to the short stalk, shortly oblong, 1.3–3.8 × 1–1.2 cm. Seeds 5–5.5 × 3–3.5 mm. *Tropics.* G2.

2. N. plena (Linnaeus) Bentham (*Mimosa plena* Linnaeus; *Desmanthus plenus* Willdenow). Illustration: Bruggeman, Tropical plants, pl. 176 (1957); Graf, Exotica, edn 12, 1385, 1387 (1985).

Plant perennial, somewhat woody. Leaves with 2–4 pairs of leaflets, each with 10–30 pairs of segments. Flowers mostly pale yellow, male flowers brown. Heads ovoid, *c.* 3.8 × 2.5 cm on stalks to 15 cm.

The sterile flowers consist of a mass of reflexed, narrowly lanceolate staminodes. *American & Asian tropics.* G2.

7. ACACIA Miller
E.C. Nelson

Trees and shrubs, evergreen, Leaves compound, bipinnate and persistent in mature plants, or bipinnate leaves entirely absent after the seedling stage and replaced by phyllodes. Inflorescence fluffy, spherical or cylindric, solitary or in clusters, axillary; flowers yellow, hermaphrodite, Sepals 3–5, minute. Petals 3–5, fused into bell-shaped corolla. Stamens numerous, much longer than corolla, thus conspicuous. Ovary stalkless; ovules numerous. Legumes usually dehiscent, with 2 valves; seeds usually with a fleshy aril.

A huge genus of at least 900 (possibly 1200) species concentrated in Australia (Western Australia possesses more than 500 species), but also distributed in tropical and subtropical regions of Africa, Asia and America; especially abundant in arid and semi-arid habitats.

Acacia is undergoing intensive taxonomic studies in Australia at present and numerous new species are likely to be named. There is a proposal to split *Acacia* (cf. Pedley 1986) into three genera – most taxa included here would be assigned to the controversial genus *Racosperma* (cf. Maslin 1989).

Because of these taxonomic problems, and because in European gardens only a few of the hundreds of species are in general cultivation (although botanical gardens may harbour many more species), I have taken a very restricted view of *Acacia* in this treatment. The genus will receive comprehensive treatment in a forthcoming volume of Flora of Australia. Several species are sold as 'mimosa'.

Literature: Pedley, L., Derivation and dispersal of *Acacia* (Leguminosae) with particular reference to Australia, and the recognition of *Senegalia* and *Racosperma*. *Botanical Journal of the Linnean Society* **92**: 219–254 (1986); Maslin, B.R., Wattle become of *Acacia*. *Australian Systematic Botany Society Newsletter* **58**: 1–13 (1989).

1a. Mature plants with true pinnate or bipinnate leaves   2
 b. Mature plants with phyllodes (not pinnate or bipinnate leaves)   3
2a. Leaves with glaucous bloom; *c.* 5 cm, with 2–4 pairs of pinnae   **1. baileyana**

 b. Leaves not glaucous, much more than 5 cm, with more than 4 pairs of pinnae   **2. dealbata**
3a. Phyllodes triangular   **5. pravissima**
 b. Phyllodes linear or lanceolate, not triangular   4
4a. Phyllodes linear, much less than 5 mm across   5
 b. Phyllodes not linear, more than 5 mm across   6
5a. Inflorescence a loose spike, bright yellow   **7. riceana**
 b. Inflorescence cylindric, pale yellow   **9. verticillata**
6a. Inflorescence cylindric   **3. longifolia**
 b. Inflorescence spherical   7
7a. Phyllodes with several parallel veins prominent, to 15 × 3 cm dark green to grey green; inflorescence pale creamy yellow   **4. melanoxylon**
 b. Phyllodes with only mid-vein prominent; inflorescences bright yellow   8
8a. Phyllodes blue-green   **6. retinodes**
 b. Phyllodes green   **8. saligna**

1. A. baileyana Mueller. Illustration: *Botanical Magazine*, 9309 (1933); Elliot & Jones, Encyclopaedia of Australian plants suitable for cultivation **2**: 20, 21 (1982); Phillips & Rix, Shrubs, 18 (1989); Brickell, The RHS gardeners' encyclopedia of plants and flowers, 69 (1989).

Tree to 10 m. Leaves bipinnate, glaucous blue-green to grey; pinnules *c.* 5 × 2 mm, pinnae oblong, 2–4 pairs per leaf, 2–5 cm × 5–10 mm. Inflorescences spherical in a compound spike. *Australia (S New South Wales).* G1–H5. Summer.

Widely cultivated in southern Europe, but intolerant of damp winters and thus not common in northwestern gardens. A prostrate cultivar, and others with purple or golden foliage are reported in Australian gardens.

2. A. dealbata A. Cunningham. Illustration: Phillips & Rix, Shrubs, 18 (1989); Brickell, The RHS gardeners' encyclopedia of plants and flowers, 56 (1989); Brickell (ed.), RHS A–Z encyclopedia of garden plants, 60 (2003).

Tree to 30 m. Leaves bipinnate, usually green; pinnules linear, 2–5 mm, coarsely hairy; pinnae 10–20 pairs per leaf, *c.* 3 cm. Inflorescences spherical in spike. *Australia (New South Wales, Victoria, Tasmania).* G1–H5. Summer.

Probably the most tolerant of the species cultivated in Europe, and commonly sold as 'mimosa'; fast-growing in suitable conditions and not particular about soil conditions.

3. A. longifolia (Andrews) Willdenow. Illustration: *Botanical Magazine*, 1827 (1816), 2166 (1822).

Tree or shrub to 10 m. Phyllodes curved, lanceolate-elliptic with numerous veins, 7–15 × 1–2.5 cm, green. Inflorescences cylindric, to 5 cm. *SE Australia*.

Exceptionally tolerant of lime.

4. A. melanoxylon R. Brown (*Racosperma melanoxylon* (R. Brown) L. Pedley). Illustration: *Botanical Magazine*, 1659 (1814); Elliot & Jones, Encyclopaedia of Australian plants suitable for cultivation **2**: 82 (1982).

Tree to 20 m or bushy shrub. Bipinnate foliage only on juvenile plants, soon replaced; phyllodes lanceolate to elliptic, dark green to grey-green, with longitudinal veins, to 15 × 3 cm. Inflorescences spherical, pale cream-yellow, in spikes which are shorter than phyllodes. *E Australia*. G1–H5. Spring–early summer.

Possibly the most abundant of the species with phyllodes cultivated in western Europe. Plants will produce suckers.

5. A. pravissima Mueller. Illustration: Elliot & Jones, Encyclopaedia of Australian plants suitable for cultivation **2**: 99 (1982); Brickell, The RHS gardeners' encyclopedia of plants and flowers, 69 (1989).

Shrub *c.* 1.5 m, with long, sparsely branching arching shoots. Phyllodes approximately triangular, closely set along stems, 5–20 × *c.* 10 mm, with 2 or 3 conspicuous veins and gland on margin near branch. Inflorescences spherical, in spikes *c.* 10 cm, bright yellow. *Australia (New South Wales, Victoria)*. G1–H5. Spring.

A. cultriformis G. Don. Illustration: *Botanical Magazine*, n.s. 322 (1958–59); Brickell, The RHS gardners' encyclopedia of plants and flowers, 103 (1989); Brickell (ed.), RHS A–Z encyclopedia of garden plants, 60 (2003). Similar, but phyllodes silver-grey.

6. A. retinodes Schlechtendal. Illustration: *Botanical Magazine*, 9177 (1929–30); Elliot & Jones, Encyclopaedia of Australian plants suitable for cultivation **2**: 106 (1982).

Shrub, rarely a small tree to 6 m, with pendent branches. Phyllodes linear-lanceolate, often slightly curved, with prominent mid-vein, 5–20 cm × 5–20 mm, apex pointed, blue-green. Inflorescences spherical, in spikes shorter than phyllodes. *Australia*. H5–G1. Summer.

Tolerant of salt and thus suitable for cultivation in coastal gardens.

7. A. riceana Henslow. Illustration: *Botanical Magazine*, 5835 (1870); Curtis & Stones, Endemic flora of Tasmania **2**: 42 (1969).

Shrub to 6 m with pendent branches. Phyllodes arranged in whorls, linear, sharply pointed, 1–5 cm × *c.* 2 mm, dark green. Inflorescence a loose cylindric spike, lemon-yellow, *c.* 5 cm (longer than phyllodes). *Australia (Tasmania)*. G1–H5. Spring.

8. A. saligna (Labillardière) Wendland (*A. cyanophylla* Lindley). Illustration: Elliot & Jones, Encyclopaedia of Australian plants suitable for cultivation **2**: 110 (1982); Bennett & Dundas, The bushland plants of Kings Park, Western Australia, 17 (1989).

Shrub, sometimes with pendent branches. Phyllodes variable, usually curved slightly, otherwise linear-lanceolate, 8–30 × 1–8 cm, midrib prominent, green. Inflorescences spherical, in axillary spikes. *Australia (Western Australia)*. G1–H5. Spring.

9. A. verticillata (L'Héritier) Willdenow. Illustration: *Botanical Magazine*, 110 (1797).

Shrub to 5 m. Phyllodes linear, dark green, sharply pointed, 1–2 cm. Inflorescences cylindric, solitary or in groups of 2–3, to 1.5 cm, pale yellow. *SE Australia*. G1–H5. Spring.

This is reputed to be tolerant of lime and wet soils, but it is not common in European gardens.

8. INGA Miller
D. Fränz
Trees or shrubs without spines. Leaves pinnate, without a terminal leaflet but with 2–5 pairs of rather large lateral leaflets, usually with glands on the axis between the leaflet-pairs. Flowers white or yellowish, in heads, spikes, racemes or umbels, borne in the leaf-axils in clusters of 1–5. Stamens many, united below. Corolla small, tubular or bell-shaped. Pods narrow, more or less indehiscent, often thickened at the sutures, often 4-angled, with a white fleshy pulp around the seeds.

A genus of about 200 species from tropical America, extending to USA (Florida, S California). A few are grown in tropical house conditions, in a peat and loam compost, with abundant watering during the summer, but scarcely any in winter. The genus is similar to both *Calliandra* and *Acacia*, differing from the latter in its united stamens.

1a. Corolla white   **1. feuillei**
  b. Corolla brown, stamens white
                   2. edulis

1. I. feuillei de Candolle (*I. dulcis* Willdenow; *I. anomala* misapplied). Illustration: Lewis et al., (eds.), Legumes of the world, 200 (2005).

Tree to 9 m. Leaves simply pinnate, leaflets in 3 or 4 pairs, ovate-oblong, acute at both ends, hairless; leaf-stalks winged. Corolla white. Pods 30–60 cm, linear, flat, hairless, white inside with sweet, edible pulp. *Peru*. G2.

2. I. edulis Martius. Illustration: Graf, Exotica, edn 12, 1384 (1985); Graf, Tropical, edn 3, 563 (1986).

Tree to 15 m, with broad crown and grey bark. Leaves simply pinnate, glossy dark green, the leaflets separated by the winged axis. Corolla brown, hairy, contrasting with the white stamens. Pods with thickened, furrowed margins, 4-angled, containing edible, sweet pulp. *C & S America*. G2.

9. ALBIZIA Durazzini
D. Fränz
Trees or shrubs without spines, rarely climbing. Leaves bipinnate, leaflets each with 1–many pairs of variably sized ultimate segments; stipules usually small but sometimes large and leaf-like. Flowers in spherical heads, stalked, axillary and solitary or clustered. Flowers bisexual or occasionally some male only; 1 or 2 flowers in each head often larger and different in form from the others, apparently male. Calyx with 5 teeth or lobes. Corolla funnel-shaped or bell-shaped with 5 lobes. Stamens 19–50, fertile, their filaments united below into a tube which may project from the corolla. Ovary stalkless or shortly stalked, with many ovules; style thread-like, stigma minute. Pods oblong, straight, flat, usually dehiscent, without cross-walls, papery or leathery but not thickened or fleshy. Seeds ovate or circular, compressed.

A genus of 100–150 species mostly found throughout the tropics, but extending into the subtropics. It is closely related to *Acacia*, but has united filaments. The species require glasshouse conditions in much of Europe, and are usually propagated by seed.

1a. Ultimate segments of leaves more than 20 pairs 2
 b. Ultimate segments of leaves fewer than 20 pairs 3
2a. Stipules conspicuous; flowers mostly cream, sometimes tinged with purple **2. chinensis**
 b. Stipules inconspicuous, soon falling; flowers light pink **1. julibrissin**
3a. Shrub **3. basaltica**
 b. Tree 4
4a. Leaflets 8–24 mm wide, hairless beneath **4. lebbeck**
 b. Leaflets 4–11 mm wide, very downy beneath **5. adianthifolia**

1. A. julibrissin (Willdenow) Durazzini (*Acacia julibrissin* Willdenow; *Mimosa japonica* Thunberg). Illustration: Everett, New York Botanical Gardens illustrated encyclopedia, 23 (1980); Graf, Exotica, edn 12, 1368 (1985); Brickell (ed.), RHS A–Z encyclopedia of garden plants, 93 (2003).

Tree to 12 m, with broad, spreading crown; young shoots angular, hairless. Leaves bipinnate with 6–12 pairs of leaflets each divided into 20–30 pairs of ultimate segments, the entire leaf 23–46 cm long and half as wide; ultimate segments 8–13 × 2–3 mm, oblong, oblique, sometimes downy on the midrib beneath. Flowers light pink in terminal clusters of dense heads each terminating in a stalk 2.5–5 cm. Stamens numerous, thread-like, 2.5 cm or more. Pods flat, to 15 × 2 cm, constricted between the seeds. *Ethiopia, Iran to Japan, C China*. H4.

Though this is the hardiest of the species, young plants require glasshouse protection during their first year. The leaflets fold up in a sleeping position during the night.

Forma **rosea** (Carrière) Mouillefent (*A. rosea* Carrière; *A. nemu* invalid). Illustration: Noailles & Lancaster, Plantes de jardins mediterranee, 38 (1977). More dwarf and more bushy than the genuine species, flowers bright pink.

2. A. chinensis (Osbeck) Merrill (*Mimosa chinensis* Osbeck; *A. stipulata* Roxburgh). Illustration: Polunin & Stainton, Flowers of the Himalayas, pl. 24 (1984).

Large deciduous tree to 40 m, trunk to 1.5 m across, with a broad, flat-topped crown and smooth grey bark. Leaves like those of *A. julibrissin* but with conspicuous stipules and ultimate segments scarcely 3 mm wide; midrib of ultimate segments running along or near the upper margin. Flowers mostly cream, though sometimes tinged with purple. Pods 10–18 cm, light brown. *SE Asia, from Pakistan to Indonesia (Java)*. G2.

3. A. basaltica Bentham.

Shrub with cylindric branchlets, minutely glandular rusty-downy. Leaves with 1 or 2 pairs of leaflets, the leaf-stalk at most 1.3 cm; ultimate segments in 5–10 pairs, oblong or ovate, very obtuse, mostly 4–6 mm, leathery, minutely hoary-downy. Flowers in dense spherical heads of about 20–30 in upper axils, inflorescence-stalks scarcely exceeding the subtending leaves. Flowers to 3 mm. Calyx downy, shortly lobed, about ²/₃ as long as the corolla. Staminal tube nearly as long as corolla, free parts of the filaments much longer. Pod to 7.5 × 8–10 mm, leathery, very flat, with somewhat thickened margins. Seeds flat, circular. *Australia (Queensland)*. G1.

4. A. lebbeck (Linnaeus) Bentham (*Mimosa lebbeck* Linnaeus; *Acacia lebbeck* (Linnaeus) Willdenow; *Mimosa sirissa* Roxburgh). Illustration: Everett, New York Botanical Gardens illustrated encyclopedia of horticulture, 177 (1960); Graf, Exotica, edn 12, 1371 (1985); Graf, Tropica, edn 3, 539, 540 (1986).

Deciduous tree to 15 m, bark grey, young shoots downy. Leaves with 2–4 pairs of leaflets divided into 3–11 pairs of ultimate segments 1.5–4.8 cm × 8–24 mm, oblong or elliptic-oblong, somewhat asymmetric with midrib near upper margin, rounded at apex, hairless or rarely thinly downy above. Flowers yellowish white on stalks 1.5–4.5 mm. Calyx 3.5–5 mm, not slit on one side, sometimes shortly downy. Corolla 5.5–9 mm, outside of lobes finely downy. Staminal tube not or scarcely projecting beyond corolla; filaments 1.5–3 cm, pale green or greenish yellow above, white below. Pods oblong, 15–33 × 3–5.5 cm, hairless or almost so, leathery, glossy, more or less veined, straw-coloured. Seeds 7–11.5 × 7–9 mm, flattened. *India to Australia, widely naturalised elsewhere*. G1.

5. A. adianthifolia (Schumacher) Wight (*Mimosa adianthifolia* Schumacher; *A. fastigiata* (Meyer) Oliver).

Tree to 40 m with flattened crown; bark grey to yellowish, rough; young shoots coarsely and persistently rusty-or brown-downy. Leaves with 5–8 pairs of leaflets each divided into 9–17 pairs of obliquely diamond-shaped or oblong ultimate segments mostly 7–20 × 4–11 mm, thinly downy above, very downy beneath. Flowers on stalks 0.5–1 mm. Calyx 2.5–4 mm, downy. Corolla 6–11 mm, white or greenish white, downy. Staminal tube projecting 1.3–2 cm beyond corolla, red to greenish or pink. Pod oblong, flat or slightly transversely folded, densely and persistently downy, not glossy, prominently veined, usually pale brown. Seeds 7–9.5 × 6.5–8.5 mm. *E & S Africa*. G1.

10. PARASERIANTHES I.C. Nielsen
D. Fränz, edited by J. Cullen
Trees or shrubs. Leaves with inconspicuous, quickly-falling stipules, bearing extra-floral nectaries; bipinnate, leaflets with 5–22 ultimate segments, which are opposite. Inflorescence a spike or raceme. Flowers 5-parted; calyx-5-lobed, corolla 5-lobed, lobes edge-to-edge in bud. Stamens numerous, united at the base, anthers eglandular. Pod flat, straight, dehiscent along both sutures. Seeds flat or biconvex.

A genus of 4 species, formerly included (as in edn 1) in *Albizia*, but differing essentially in the inflorescence. Only one species is cultivated.

1. P. lophantha (Willdenow) I.C. Nielsen (*Acacia lophantha* Willdenow; *Albizia lophantha* (Willdenow) Bentham; *Albizia distachya* (Ventenat) McBride). Illustration: Parey's Blumengärtnerei, 854 (1958); Blombery, What flower is that?, 42 & pl. 60 (1972); Encke, Kalt-und Warmhauspflanzen, 300 (1987).

Shrub or small tree to 6 m; branches, leaf-stalks and inflorescence-stalks usually velvety-downy. Leaves with 14–24 leaflets, each with 40–60 ultimate segments which are to 1 cm, linear, silky-downy beneath. Flowers stalked in usually paired spikes to 5 cm, yellowish or white. *SW Australia*. H5–G1. Summer.

11. CALLIANDRA Bentham
D. Fränz
Shrubs or small trees. Leaves bipinnate. Flowers in large, rounded or spherical heads, bracts inconspicuous. Calyx 5-toothed or -lobed. Petals 5, united from the base to about the middle into a 5-lobed, funnel-shaped or bell-shaped corolla. Stamens numerous, long-projecting,

their coloured filaments providing most of the colour of the inflorescence, at least then anthers red. Ovary with many ovules; style thread-like. Pods linear, usually narrowed towards the base, flat, straight or almost so, not pulpy within, opening elastically from the apex, the flaps leathery with raised margins. Seeds circular to obovate, compressed.

A genus of about 200 species from tropical America, Asia and Madagascar. Several species are grown in an open compost of loam and peat. Most require warm greenhouse conditions and abundant sunshine. Propagation is by seed or by cuttings, which require bottom heat to root successfully.

1a. Leaves with more than 2 pairs of
    leaflets   2
 b. Leaves with 1 or 2 pairs of leaflets   4
2a. Leaves with 7–12 pairs of leaflets
                            1. houstoniana
 b. Leaves with 3–7 pairs of leaflets   3
3a. Inflorescence to 35 cm wide
                            2. eriophylla
 b. Inflorescence to 7.5 cm wide
                            3. tweedii
4a. Leaves 1-pinnate   **4. surinamensis**
 b. Leaves 2-pinnate   5
5a. Inflorescence spherical   **5. falcata**
 b. Inflorescence hemispherical
                            6. haematocephala

1. C. houstoniana (Miller) Standley
(*C. inermis* (Linnaeus) Druce; *Mimosa houstonii* L'Heritier; *Inga houstonii* (L'Héritier) de Candolle; *C. houstonii* (L'Héritier) Bentham).

Slender shrub to 6 m. Leaves with 7–12 pairs of leaflets; ultimate segments in 10–40 pairs, oblong-linear, somewhat curved, to 6 mm, acute. Flower-heads clustered in a terminal, raceme-like inflorescence to 36 cm. Flowers to 5 cm; corolla hairy, anthers purple-red. Pods to 12 cm, with dense, brown, stiff hairs. *C America.* G2.

2. C. eriophylla Bentham. Illustration: Orr, Wildflowers of western America, pl. 181 (1974); Graf, Exotica, edn 12, 1375 (1985).

Low, much-branched shrub to 30 cm, older branches grey, young shoots with broad ridges bearing downwardly pointing hairs. Stipules bristle-like. Leaflets 1–7 pairs; ultimate segments usually 5–8 pairs, oblong, 3–4 mm, obtuse or somewhat acute, bearing more or less flattened bristles. Heads few-flowered, in racemes or axillary. Calyx 1–1.5 mm. Corolla 4–6 mm.

Stamens to 2 cm, united at the base into a short tube, reddish purple. Pods 3–6 cm × *c.* 5 mm, tapering below the middle, densely downy with downwardly pointing hairs. *USA (Texas to Arizona & California), Mexico.* G1.

3. C. tweedii Bentham (*Inga pulcherrima* Sweet), Illustration: Graf, Exotica, edn 12, 1374 (1985); Tropica, edn 3, 548 (1986).

Shrub or small tree to 2 m, sparsely hairy. Leaves bipinnate with 3–6 or more pairs of leaflets each 3.8–6.3 cm; ultimate segments very numerous, overlapping, narrowly oblong, 6–8 mm, silky-hairy when young. Flower-heads hemi-spherical, 5–7.5 cm wide, each axillary on a downy stalk to 5 cm. Calyx and corolla yellowish green, shaggy-hairy, lobes erect. Stamens red, 2.5–3.8 cm, numerous. Pods to 5 cm, shaggy-hairy. *Brazil.* G1–2. Late winter–autumn.

4. C. surinamensis Bentham. Illustration: Milne & Milne, Living plants of the world, 107 (1967); Chin, Malaysian flowers in colour, 51 (1977); Graf, Exotica, edn 12, 1374 (1985); Graf, Tropica, edn 3, 548 (1986).

Small tree, spreading shrub or woody climber. Leaves with 1 pair of leaflets; ultimate segments in 8–12 pairs, oblong-lanceolate, to 1.3 cm. Flowers in axillary heads. Stamens united below into a projecting column, filaments red apically, white in the lower part. *Brazil, Surinam.* G2.

5. C. falcata Bentham (*C. fulgens* J.D. Hooker).

Evergreen shrub or small tree. Leaves with 2 leaflets each divided into 3 pairs of ultimate segments; leaf-stalk slender, downy; ultimate segments narrowly oblong, 2.5–6.5 cm × 8–16 mm. Flowers in spherical, shortly stalked heads to 6.5 cm wide. Corolla bright pink, to 1.3 cm. Stamens with scarlet filaments and crimson anthers. *Mexico.* G2. Spring.

6. C. haematocephala Hasskarl (*C. inaequilatera* Rusby). Illustration: Graf, Exotica, edn 12, 1374 (1985); Graf, Tropica, edn 3, 547 (1986); Brickell (ed.), RHS A–Z encyclopedia of garden plants, 208 (2003).

Loose shrub or small tree, to 7.5 m, spreading. Leaves with 2 pairs of leaflets, each divided into 4–10 pairs of ultimate segments; terminal ultimate segments to 9.2 cm, others 2–3.8 cm, obliquely oblong-lanceolate. Flowers in hemi-spherical heads 5–7.5 cm wide, bisexual or functionally

male. Corolla small, pink. Stamens *c.* 25, 2.5–3.1 cm, filaments white towards the base, red apically, united at the base into a tube as long as or longer than the corolla, with an irregular internal fringe; anthers black. Pods densely downy. *Bolivia, Brazil.* G2. Late winter.

12. ZAPOTECA H.M. Hern
D. Fränz, edited by J. Cullen
Like *Calliandra* but inflorescence terminal, branched, subtended by obvious, stipule-like bracts; stamens without any red coloration.

A genus of 6 species from South America, included in *Calliandra* in edn 1. Cultivation as for *Calliandra*.

1a. Inflorescence spherical; flowers white
                            1. portoricensis
 b. Inflorescence hemispherical; flowers
    yellowish   **2. formosa**

1. Z. portoricensis (Jacquin) Bentham (*Mimosa portoricensis* Jacquin; *Acacia portoricensis* (Jacquin) Willdenow; *Calliandra portoricensis* (Jacquin) Bentham). Illustration: Bruggeman, Tropical plants, pl. 230 (1957); Graf, Exotica, edn 12, 1375 (1985); Graf, Tropica, edn 3, 548 (1986).

Shrub or small tree to 8 m, with grooved, downy young shoots. Leaves bipinnate with 2–7 pairs of leaflets 3.8–7.5 cm, ultimate segments linear to narrowly oblong, overlapping, densely set; stalks and margins downy. Flowers in spherical axillary heads to 5 cm across. Corolla to 6 mm, green. Stamens many, white, *c.* 1.9 cm. Pods to 12.5 cm. *West Indies, C America, W Africa.* G2. Summer.

2. Z. formosa (Kunth) H. Hern (*C. gracilis* Baker; *C. moritziana* Cárdenas).

Evergreen shrub, much branched with hairy shoots. Leaves bipinnate with 1–3 pairs of leaflets, each divided into 4–6 pairs of ultimate segments 1–3.2 cm, obliquely ovate-oblong, downy on both surfaces. Flowers in hemi-spherical heads on a stalk 5–10 cm. Corolla yellowish, to 3 mm. Stamens 30–40, 2–2.5 cm, pale cream. *C America.* G2.

13. PITHECELLOBIUM Martius
D. Fränz
Erect, broad trees or shrubs, with or without axillary, stipular spines. Leaves bipinnate, the leaflets either small and themselves pinnately divided into numerous segments, or large and with 1–3 pairs of segments, rarely reduced to a single segment; glands

usually present on the leaf-stalk; stipules persistent. Flowers in head-like spikes, bisexual or variously unisexual, calyx and corolla 5- or rarely 6-lobed. Calyx bell-shaped, shortly toothed. Corolla tubular or funnel-shaped. Stamens few to many, much projecting, united into a tube, at least at the base. Ovary stalkless or stalked, with many ovules, style thread-like, stigma terminal, small or head-like. Pods compressed, curved, sickle-shaped, twisted or rarely straight, leathery, thick or somewhat fleshy, 2-lobed at the apex, not divided between the seeds. Seeds pulpy, ovate to circular, compressed, often dark-coloured, with variously expanded fleshy arils.

A genus of about 200 species from tropical America. Cultivation as for *Inga*. Only one is generally cultivated.

1. P. dulce (Roxburgh) Bentham (*Mimosa dulcis* Roxburgh; *Inga dulcis* (Roxburgh) Willdenow). Illustration: Lewis et al. (eds.), Legumes of the world, 209 (2005).

Very spiny tree to 20 m. Leaves bipinnate; ultimate segments obovate to oblong, obtuse, to 2.5 cm. Flowers in shortly stalked heads in a raceme-like panicle, white, downy. Calyx 1–1.5 mm. Corolla 3–4.5 mm. Pods twisted, 12.5–15.5 cm, red. Seeds black, glossy, 9–10 × 7–8 mm; aril pulpy. *C America, N South America; introduced in the Philippine Islands & elsewhere.* G2.

14. ARCHIDENDRON F. Mueller
D. Fränz, edited by J. Cullen
Trees or shrubs, differing (ours) from *Pithecellobium* in the small, early-falling stipules and the black seeds which have no arils, though they are borne against the red or orange-red interior of the pods.

A genus of 94 species from eastern Asia. Only one is cultivated; it was included in *Pithecellobium* in edn 1.

1. A. bigeminum (Linnaeus) I.C. Nielsen (*Pithecellobium bigeminum* (Linnaeus) Martius; *Inga bigemina* (Linnaeus) Willdenow).

A small tree, like *Pithecellobium dulce* but with white flowers and broad, flat pods which are curved into a ring, orange-red inside and containing black seeds which have no arils. *Himalayas, SE Asia to the Philippine Islands.* G2.

15. SAMANEA (Bentham) Merrill
D. Fränz
Spineless or rarely spiny trees or shrubs. Leaves several to many times pinnate, the leaflets 1–many times pinnate. Flowers in spherical heads, calyx and corolla 5-lobed. Stamens many, united into a tube at the base. Pods straight or somewhat curved, rigid, more or less constricted between the seeds, flat or cylindric, usually indehiscent, with cross-walls between the seeds.

A genus of about 30 species from C & S America, though introduced and cultivated elsewhere.

1. S. saman (Jacquin) Merrill (*Mimosa saman* Jacquin; *Pithecellobium saman* (Jacquin) Bentham; *Enterolobium saman* (Jacquin) Prain). Illustration: Graf, Exotica, edn 12, 1390, 1391 (1985); Graf, Tropica, edn 3, 560, 563, 564 (1986); Lewis et al., (eds.), Legumes of the world, 210 (2005).

Large tree to 35 m, trunk reaching 1 m or more across; branches widely spreading, young shoots velvety-downy. Leaves 2–4-pinnate, shining above, downy beneath. Flowers shortly stalked, in heads which are borne on stalks 10–12.5 cm. Calyx to 7 mm, downy. Corolla to 1.3 cm, yellowish, silky with shaggy hairs. Stamens 20, light crimson, only shortly united at the base. Pods stalkless, straight, thick-margined, leathery-fleshy, hairless, 15–20 × 1.3–2.5 cm, flattened or almost cylindric. *West Indies, C America.* G2.

Fast-growing, widely cultivated in the tropics as an ornamental shade-tree.

152. CAESALPINIACEAE (LEGUMINOSAE-CAESALPINIOIDEAE)

Trees, shrubs or herbs. Leaves usually pinnate, rarely bipinnate, never reduced to phyllodes, stipulate. Flowers usually bisexual and zygomorphic. Calyx, corolla and stamens hypogynous or perigynous. Calyx usually of 5 sepals or 5-lobed. Corolla of 3–5 petals which are imbricate so that the uppermost is within the laterals, which themselves are overlapped by the 2 lower. Stamens 3–10, often deflexed downwards, anthers often opening by pores, pollen granular. Ovary of a single carpel; ovules 1–many, marginal. Fruit a legume; seeds usually without a lateral line.

There are 162 genera and about 2,000 species. A few genera are native to Europe and/or North America, and species of various genera are grown as ornamentals (especially *Cercis*, *Gleditsia* which are hardy in Europe), *Delonix* and *Amherstia* which are grown in warmer regions. In edn 1 this family was included in Leguminosae (as subfam. *Caesalpinioideae*). The editors are grateful to Dr G. Lewis and Dr B. Schrire of Kew and Ms S. Andrews for their help in the preparation of these accounts.

Literature: Lewis, G. et al. (eds.), Legumes of the world (2005).

1a. Leaves simple, consisting of a single leaflet (which may be deeply lobed at the apex)   2
  b. Leaves pinnate or bipinnate or consisting of 2 or more completely distinct leaflets   3
2a. Flowers in clusters borne on the old wood, to 2 cm   **13. Cercis**
  b. Flowers not in clusters on the old wood, much longer than 2 cm   **14. Bauhinia**
3a. Leaves 1-pinnate (or apparently so), sometimes of only 2 leaflets   4
  b. Leaves clearly 2- or 3-pinnate   13
4a. Leaflets 2   **14. Bauhinia**
  b. Leaflets more than 2   5
5a. Petals and sepals similar, both small, or petals absent; flowers in catkin-like inflorescences, usually functionally unisexual   6
  b. Petals and sepals clearly differing; flowers not in catkin-like inflorescences, usually bisexual   7
6a. Petals absent; plants not spiny 10. Ceratonia
  b. Petals present; plants usually spiny   **1. Gleditsia**
7a. Fertile stamens 3   **18. Tamarindus**
  b. Fertile stamens 7 or more   8
8a. Flowers bright red or orange-red, borne in large heads directly on the main branches; young leaves limp, brightly coloured, at first enclosed in a tube of scale-leaves   **16. Brownea**
  b. Flowers yellow, or if red then not in large pendent heads; young leaves not as above, though sometimes limp   9
9a. Leaves divided into 2 leaflets at the extreme base, each easily mistaken for a 1-pinnate leaf; leaflet-axes flattened, ultimate segments more than 10 pairs   **9. Parkinsonia**
  b. Leaves strictly 1-pinnate; axis not flattened; leaflets to 8 pairs   10

10a. Anthers opening by terminal or
     basal pores; petals yellow or orange-
     yellow, rarely pink or whitish   11
  b. Anthers opening by slits; flowers red
     or pink   12
11a. The 3 lowers stamens with long,
     curved filaments much more than
     twice as long as the anthers which
     are short and hairy; large trees
                              11. Cassia
  b. The 2 or 3 lower stamens often
     longer than the others, but filaments
     at most twice as long as anthers;
     anthers hairless; small trees, shrubs
     or herbs   **12. Senna**
12a. Petals 5, 3 very large (to 5 cm) and
     2 very small; flowers in hanging
     racemes   **17. Amherstia**
  b. Petals 5, more or less equal, or some
     reduced to threads, the largest to
     1.8 cm; flowers in erect panicles
                            15. Schotia
13a. Petals and sepals similar, both small;
     flowers usually functionally
     unisexual   14
  b. Petals clearly differing from the sepals,
     often large; flowers bisexual   15
14a. Some leaves simply pinnate, others
     2-pinnate; plant usually spiny
                           1. Gleditsia
  b. All leaves 2-pinnate; plant not spiny
                        2. Gymnocladus
15a. Stigma broad, peltate; pod flattened,
     winged, indehiscent, elliptic, 1- or 2-
     seeded   **3. Peltophorum**
  b. Stigma not as above; pod variable,
     not as above   16
16a. Leaves with 2 leaflets, each with a
     flattened axis and ultimate segments
     conspicuously small for the size of
     the leaf   **9. Parkinsonia**
  b. Leaves with more than 2 leaflets, not
     as above   17
17a. Ultimate segments 15–25 pairs per
     pinna; petals usually red or scarlet,
     rarely yellow   **4. Delonix**
  b. Ultimate segments to 12 pairs; petals
     yellow or orange-yellow, sometimes
     red-spotted   18
18a. Lower sepal with toothed margins,
     comb-like; plant spiny   **8. Tara**
  b. Lower sepal not as above; plants not
     spiny   19
19a. Leaves terminating in a single leaflet
                          7. Libidibia
  b. Leaves terminating in a pair of
     leaflets   20
20a. Glands conspicuous on young stems
                      6. Erythrostemon
  b. Glands absent   **5. Caesalpinia**

1. GLEDITSIA Linnaeus
C.J. King

Deciduous trees, with trunks and branches
usually armed with simple or branched
spines. Leaves alternate, pinnate or
bipinnate, with numerous, slightly
scalloped leaflets, the terminal often absent.
Flowers small, radially symmetric,
greenish, unisexual and bisexual on the
same plant, in racemes or more rarely
panicles. Sepals 3–5. Petals 3–5, similar to
the sepals. Stamens 6–10. Style short with
a large terminal stigma. Fruit a flattened
pot, indehiscent or very late dehiscent.
Seeds 1–many, flattened, ovate to almost
circular.

A genus of about 14 species, mainly
from the Caspian region of E Asia, but with
2 or 3 in E North America and one in
warm temperate S America. Three species,
one of these having several cultivars, are
generally cultivated in Europe, usually for
their ornamental foliage and conspicuous
spines. They are rather tender when
young, thriving best in a sunny position in
loamy soils in the warmer parts of Europe.
Propagation is normally by seed, though
cultivars of *G. triacanthos* are increased by
grafting or budding on seedlings of the
type.

1a. Spines terete; leaves usually once
    pinnate; leaflets rarely more than
    14; pods usually less than 25 cm,
    almost straight   **1. sinensis**
  b. Spines flattened, at least at the base,
     rarely absent; leaves pinnate or
     bipinnate; leaflets often more than
     14; pods twisted   2
2a. Young shoots hairless, dark purplish
    brown; leaflets 24 or fewer; pods
    usually less than 30 cm   **2. japonica**
  b. Young shoots slightly downy at
     base, green; leaflets 14–22; pods
     usually more than 30 cm
                          3. triacanthos

1. G. sinensis Lamarck (*G. horrida*
Willdenow). Illustration: Krüssmann,
Handbuch der Laubgehölze, edn 2, **2**: t. 42
(1977); Bean, Trees & shrubs hardy in the
British Isles, edn 8, **2**: t. 41 (1978);
Csapody & Tœth, Colour atlas of flowering
trees and shrubs, t. 61 (1982).

Trees to 15 m, with stout, conical, often
branched spines. Young shoots hairless or
soon becoming so. Leaves usually once
pinnate, 12–18 cm, downy on the axis;
leaflets 8–14 or sometimes more, 3–8 ×
1.5–2.8 cm, ovate or ovate-lanceolate,
obtuse or acute, obliquely tapered at the
base, margins slightly wavy, more or less
downy on midrib and stalk. Flowers in
slender, pendent, downy racemes 6–8 cm.
Pods 12–25 × 2–3 cm, almost straight,
dark purplish brown, the walls dotted with
minute pits. *China*. H3.

2. G. japonica Miquel (*G. horrida*
(Thunberg) Makino; *Fagara horrida*
Thunberg). Illustration: Krüssmann,
Handbuch der Laubgehölze, edn 2, **2**:
t. 35 (1977); Lewis et al. (eds.), Legumes
of the world, 130 (2005).

Tree to 20 m, with stout, branched
spines, flattened at least at the base. Young
shoots dark purplish brown, hairless and
shining. Leaves pinnate or bipinnate,
20–30 cm, usually downy on the axis;
leaflets 14–24, 2–4 cm × 6–12 mm, ovate
to lanceolate, obtuse or acute, sometimes
matched, margins entire or shallowly
scalloped and slightly wavy, more or less
downy on midrib and stalk. Flowers in
slender racemes. Pods 20–30 × 2.5–3.5 cm,
curved and twisted. *Japan*. H3.

The description above refers to var.
japonica, the only one in cultivation. In
small cultivated trees the leaflets are only
1–1.5 cm, giving the plants a very different
appearance.

3. G. triacanthos Linnaeus. Illustration:
Edlin, The tree key, 153 (1978); Edlin &
Nimmo, Illustrated encyclopaedia of trees
of the world, 183 (1974); Phillips, Trees in
Britain, Europe and North America 121
(1978), Everett, New York Botanical
Gardens illustrated encyclopedia of
horticulture **5**: 1493 (1981).

Tree to 25 m in cultivation, with stout,
often 3-branched spines, somewhat
flattened at least at the base. Young shoots
slightly downy at the base. Leaves pinnate
or bipinnate, 10–20 cm, downy on the
axis; leaflets 14–32, 1.5–3.5 cm ×
5–15 mm, oblong-lanceolate, obtuse or
acute, margins wavy or shallowly
scalloped, downy at first above and
beneath. Flowers in slender, downy
racemes 5–7 cm. Pods 30–45 × 2.5–4 cm,
more or less curved and twisted, shining
dark brown. *E North America*. H3.

There are several different growth forms,
including a pendent variant, 'Bujotii'
(Pendula), 'Elegantissima', a cultivar with
shrubby habit, and a spineless variant,
forma **inermis** de Candolle. The foliage of
this species turns golden yellow in the
autumn, and in the spineless cultivar
'Sunburst' the young growth is bright
yellow (illustration: Lancaster, Trees for

your garden, edn 2, 71, 1979; Hillier, The Hillier colour dictionary of trees and shrubs, 114, 1981).

2. GYMNOCLADUS Lamarck
J. Cullen

Trees, without spines. Leaves deciduous, bipinnate, stipules present or absent, stipels present or absent (ours). Inflorescence a terminal or axillary raceme or panicle. Flowers radially symmetric, functionally unisexual, the males with a rudimentary ovary, the females with abortive or sterile anthers; both sexes on the same plant. Calyx tubular, 5-toothed with equal teeth, the calyx not completely covering the corolla in bud. Petals 4 or usually 5, similar to the sepals, hairy, cream or marked with purple outside. Stamens 10 in 2 whorls, those of 1 whorl longer than those of the other, filaments free. Ovary stalkless, with 4 or more ovules. Pod oblong, often curved, woody, pulpy inside, containing 1 or more seeds, opening along the upper suture.

A genus of 4 species from E North America and E China. Only the American species is cultivated; it lacks the stipels which are found in the other 3 species. It requires a deep, rich soil, but grows slowly and does not flower regularly; propagation is by seed.

Literature: Lee, Y.T., The genus *Gymnocladus* and its tropical affinity, *Journal of the Arnold Arboretum* **57**: 91–112 (1976).

1. G. dioica (Linnaeus) Koch (*G. canadensis* Lamarck). Illustration: MacDonald encyclopedia of trees, 123 (1978); Elias, The complete trees of North America, 652 (1980); Csapody & Tœth, A colour atlas of flowering trees and shrubs, 139 (1982); Taylor's guide to trees, 156, 157 (1988).

Tree to 10 m, bark dark grey, fissured. Leaflets ovate, 3–7 × 2–4 cm, abruptly acuminate, broadly rounded to truncate at the base, hairless except for a few scattered hairs on the midrib beneath; in autumn the leaflets fall individually, leaving the bare axes. Racemes or panicles of female flowers longer than those of male flowers, branches of all white-hairy; flowers fragrant. Calyx-tube 6–12 mm, hairy, teeth 4–6 mm. Petals 5–7 mm, hairy on both surfaces, greenish white. Pod 15–25 cm. *E North America.* H3. Summer.

3. PELTOPHORUM (Vogel) Bentham
J. Cullen

Round-topped deciduous trees. Young shoots, leaf-stalks, axes and calyces covered in red-brown or greyish hairs. Leaves bipinnate, ultimate segments with short mucros at their tips. Stipules soon falling. Racemes terminal. Calyx with a short tube and 5 more or less equal lobes which overlap and completely enclose the corolla in bud. Corolla bright yellow; petals more or less equal, each with a long claw and broader blade, margins wavy. Stamens 10, filaments free, unequal, hairy at the base. Ovary flattened, shortly hairy; style hairless, twisted near the base and again towards the apex; stigma broad, peltate. Pods pendent, flattened, indehiscent, elliptic, 1- or 2-seeded.

A genus of about 9 species from the tropics. Only one is grown; it will tolerate a small degree of frost, but generally requires glasshouse treatment in most of Europe. Propagation is by seed.

1. P. africanum Sonder. Illustration: *Flowering Plants of Africa* **36**: t. 1434 (1964); Palgrave, Trees of southern Africa, t. 90 (1977); Palmer, A field guide to the trees of southern Africa, t. 11 (1977).

Tree to 6 m or more. Leaves to 10 cm, densely or sparsely adpressed hairy on both surfaces. Calyx-tube bell-shaped, *c.* 2 mm, lobes longer, green, spreading. Petals to 2 cm, bases of the claws hairy. Pod to 6 × 1.7 cm. *C Tropical Africa, South Africa.* G1.

The leaves are touch-sensitive, the leaflets folding upwards together on stimulation.

4. DELONIX Rafinesque
J. Cullen

Widely branched, rather flat-topped trees, without spines. Leaves bipinnate, with very numerous ultimate segments; stipules absent. Flowers very showy, in terminal or axillary racemes. Calyx of 5 sepals, all equal, free almost to the base, attached to a short perigynous zone, edge-to-edge in bud, usually reddish. Petals 5, red, scarlet or yellow, long-clawed, with abruptly broadened blades which are rounded, often broader than long and wavy on their margins; the uppermost is more gradually tapered than the rest and is sometimes marked with a contrasting colour. Stamens 10, widely spreading, filaments free, anthers dorsifixed. Ovary stalkless with many ovules. Pod woody, compressed, containing many seeds, more or less solid between them.

A genus of about 10 species, 2 from Africa, the rest from Madagascar. One of the Madagascan species (below) is now widely cultivated in the tropics and subtropics world-wide, though it is not easily grown well in Europe, requiring warm glasshouse conditions and considerable space. It is tolerant of most soils and is fast-growing in suitable sites. Propagation is by seed.

1. D. regia (Hooker) Rafinesque (*Poinciana regia* J.D. Hooker). Illustration: *Botanical Magazine*, 2884 (1829); Pertchik & Pertchik, Flowering trees of the Caribbean, 95 (1951); Menninger, Flowering trees of the world, t. 102, 108 (1962); Leathart, Trees of the world, 167 (1977).

Tree to 15 m. Leaves to 60 cm, with very numerous ultimate segments, bright green above, paler beneath. Flowers 7–10 cm across, red or scarlet (rarely yellow), the upper petal marked with white or yellow; claws of petals hairy outside. Filaments with swollen, hairy bases. Pod 60 × 5–7 cm, many-seeded. *Originally from Madagascar, now widespread in the tropics & subtropics.* G2. Summer.

The Flamboyant, is one of the most spectacular and widely cultivated tropical trees. A variant with completely yellow flowers is sometimes grown.

5. CAESALPINIA Linnaeus
J. Cullen

Trees or shrubs (ours), sometimes climbing or scrambling, often with thorn-like prickles. Leaves bipinnate with up to 12 pairs of leaflets, terminating in a pair of leaflets; stipules deciduous or persistent. Flowers in terminal or axillary racemes, panicles or clusters, bracts usually evident. Calyx of 5 almost completely free, overlapping sepals, the lowest outermost, often larger than and partially or completely enclosing the others, sometimes toothed. Petals 5, the uppermost often slightly smaller than the others and more distinctly clawed, all yellow or reddish. Stamens 10, filaments free, usually hairy, shorter than to much longer than the petals. Ovary stalkless, containing few ovules. Pod variably shaped, dehiscent (sometimes very tardily) or indehiscent, woody, few-seeded.

A genus in its strict sense (as recognised here; a broad sense of the genus was employed in edn 1) of no more than 25 species, mainly from the tropics but extending into temperate areas in N & S America and southern Africa. They require a dryish, warm soil in a sunny position, and, when established, grow rapidly. Propagation is by seed.

Literature: Isely, D., Leguminosae of the United States: II. Subfamily *Caesalpinioideae*, *Memoirs of the New York Botanical Garden* **25**(2): 33–51 (1975).

1a. Trees; young shoots covered with rusty brown hairs; pod prickly
        1. echinata
  b. Trees or shrubs, sometimes scrambling; young shoots not as above; pods not prickly   2
2a. Stamens 5–8 cm, considerably exceeding the petals
        4. pulcherrima
  b. Stamens at most to 3 cm, not or scarcely exceeding the petals   3
3a. Leaflets asymmetric, oblong to diamond-shaped, attached by 1 corner; flower-stalks 1–1.5 cm; upright shrub   **2. sappan**
  b. Leaflets not as above; flower-stalks more than 1.5 cm; plant scrambling or climbing   **3. decapetala**

1. C. echinata Lamarck. Illustration: Die Natürlichen Pflanzenfamilien **3**(3): 175 (1894).

Tree, branches densely prickly, the young shoots covered with rusty brown hairs. Leaves without prickles, with 2–4 pairs of leaflets, each divided into 7–10 pairs of ultimate segments which are oblong to diamond-shaped, rounded at the apex, somewhat tapered to the base, 1.2–2 cm. Racemes axillary and terminal; flower-stalks *c*. 2 cm. Sepals hairy, the lowest not toothed. Petals yellow. Stamens shorter than petals. Pod oblong, to 7 cm, prickly. *Brazil*. G2.

Doubtfully in cultivation in Europe, but reported in some lists.

2. C. sappan Linnaeus. Illustration: *Flora of Thailand* **4**(l): 66 (1984).

Shrub or tree, with prickles, hairy or not. Leaves with 7–10 pairs of leaflets, these divided into 10–20 pairs of ultimate segments which are asymmetric, oblong to diamond-shaped, attached at one corner, 1.4–2.5 cm × 7–10 mm, apex rounded, with or without black dots beneath. Racemes (rarely panicles) terminal, brown hairy at first; flower-stalks 1–1.5 cm. Sepals unequal, 8–12 mm, the lowest somewhat hooded in bud. Petals 1–1.4 cm, yellow. Filaments slightly longer than petals,

densely hairy. Pod elliptic to oblong, 6–10 × 3–4 cm, eventually dehiscent. *Malaysia, Indonesia, Indo-China, S China*. G2.

3. C. decapetala (Roth) Alston. Illustration: *Botanical Magazine*, 8207 (1908); Hay & Synge, Dictionary of garden plants in colour, pl. 1471 (1969); *Gartenpraxis* (September) 1976: 422; Noailles & Lancaster, Mediterranean plants and gardens, 50 (1977).

Scrambling or climbing shrub; stems and leaf-stalks with few to many prickles. Leaves with 6–10 pairs of leaflets, these divided into 7–12 pairs of ultimate segments which are obovate to elliptic, rounded at both ends, very shortly stalked, 1–2.5 cm × 5–15 mm, veins conspicuous, silky-hairy on both surfaces. Racemes axillary and terminal, many-flowered, flowers often hanging; flower-stalks 1.5–3 cm. Sepals to 1 cm, unequal, the lowest not toothed. Petals pale yellow, sometimes some or all of them red-spotted, 1–1.3 cm. Stamens slightly longer than petals. Pod 7–10 × 2–3 cm, flattened, eventually opening along the upper suture. *Tropical & subtropical Asia, widely introduced elsewhere*.

Var. **decapetala**, from the more tropical parts of Asia has rather distant flowers on stalks 2–3.5 cm. G2.

Var. **japonica** (Siebold & Zuccarini) Isely (*C. japonica* Siebold & Zuccarini; *C. sepiaria* Roxburgh var. *japonica* (Siebold & Zuccarini) Makino) is from China and Japan, and has the flowers closely packed, on stalks 1.5–2.5 cm. H5–G1.

4. C. pulcherrima (Linnaeus) Swartz (*Poinciana pulcherrima* Linnaeus). Illustration: *Botanical Magazine*, 995 (1807); Bruggeman, Tropical plants, pl. 227 (1957); Perry, Flowers of the world, 159 (1970); *Der Palmengarten*, **No. 2**: 67 (1987); Lewis et al., (eds.), Legumes of the world, 140 (2005).

Shrub or small tree to 5 m or more, with or without prickles, hairless, often somewhat glaucous. Leaves divided into 5–10 pairs of leaflets, these divided into 6–10 pairs of ultimate segments which are elliptic to obovate, 1–2.5 cm × 5–8 mm, rounded to the apex, somewhat tapered to the base. Flowers in large, terminal racemes; flower-stalks 4–6 cm. Sepals unequal, orange-red, the lowest broader than the others and enclosing them in bud. Petals 1.5–2 cm, obovate, with wavy margins, orange-yellow to yellow. Stamens with filaments 5 cm or more, much longer

than the petals. Pod flattened, 6–12 × 1.5–2 cm, dehiscent. *West Indies, widely introduced in the tropics*. G1.

6. ERYTHROSTEMON Klotzsch
J. Cullen
Very like *Caesalpinia* but young growth with conspicuous glands.

A genus of 12 or 13 species, mostly from C & S America, but extending into Mexico and the southern USA. Only one species is grown. Cultivation as for *Caesalpinia*.

1. E. gilliesii (Hooker) Klotzsch (*Caesalpinia gilliesii* (Hooker) Dietrich. Illustration: *Botanical Magazine*, 4006 (1843); *American Horticulturist* **56**: 23 (1977); *Der Palmengarten*, **No. 2**: 67 (1987); Lewis et al. (eds.), Legumes of the world, 141 (2005).

Shrub or tree to 5 m, without prickles, young shoots shortly hairy and glandular. Leaves divided into 8–12 leaflets, terminating in a pair of leaflets; leaflets divided into 7–11 pairs of ultimate segments which are elliptic to elliptic-oblong, 3–8 × 1–5 mm, rounded to the apex, slightly tapered to the base, with black dots along the margins beneath. Racemes terminal, glandular and sticky, and with hairs without glands; flower-stalks 1.5–3 cm or more. Sepals 1.5–2 cm, the lowest broader than the others. Petals bright yellow with orange markings, 2–3.5 cm. Stamens with red filaments 5–8 cm, much exceeding the petals. Pod flattened, 6–12 × 1.5–2 cm, oblong, often curved, dehiscent, hairy at least when young. *Southern tropical & temperate S America, naturalised in several countries*. G1.

7. LIBIDIBIA (de Candolle)
Schlechtendahl.
J. Cullen
Very like *Caesalpinia* but differing as indicated in the key to genera, above.

A genus of 6–8 species from the new world tropics. Only a single species is cultivated; cultivation as for *Caesalpinia*.

1. L. coriaria (Jacquin) Willdenow.
Small tree or shrub, without prickles. Leaves with 4–8 pairs of leaflets, terminating in a single leaflet, each leaflet divided into 15–25 ultimate segments which are 4–10 × 1–2 mm, stalkless, oblong, hairless, rounded at apex and base, often with dark dots beneath. Flowers in axillary or apparently terminal panicles or clusters; flower-stalks 1–2 mm. Sepals

3–4 cm, the lowermost not toothed. Petals yellow or pale yellow, 4–6 cm. Stamens scarcely exceeding the petals. Pod 3–5 cm × 8–20 mm, ovoid to oblong, often becoming twisted, indehiscent. *C & N South America, West Indies.* G2.

8. TARA Molina.
J. Cullen
Very like *Caesalpinia*, but differing in the characters indicated in the key to genera, above.

A genus of 3 species from C and northern S America. Only a single one is occasionally grown; cultivation as for *Caesalpinia*.

1. T. spinosa (Molina) Britton & Rose (*Caesalpinia spinosa* (Molina)Kuntze; *C. tinctoria* Humboldt & Kunth). Illustration: Lewis et al. (eds.), Legumes of the world, 138 (2005).

Shrub or small tree; young shoots brown-hairy. Prickles usually present on stems and leaves. Leaves divided into 2–5 leaflets, terminating in a pair of leaflets; leaflets divided into 5–7 pairs of ultimate segments which are 1.5–4.5 cm × 8–20 mm, rounded to notched at the apex, slightly tapered towards the base, with conspicuous veins and, usually, dark dots beneath. Racemes clustered, terminal, many-flowered; flower-stalks 5–10 mm. Sepals unequal, 5–6 mm, usually deciduous as the flower opens, the lowest deeply toothed, comb-like. Petals yellow and reddish, 6–7 mm. Stamens very unequal, the longest equalling or slightly exceeding the petals. Pod oblong, 6–10 × 1–2.5 cm, indehiscent. *S America.* G2.

9. PARKINSONIA Linnaeus
J. Cullen
Trees or shrubs, usually with a pair of spines at each node. Leaves 2–3-pinnate, the 1–3 leaflets divided into rather small, numerous, ultimate segments. Flowers in axillary or terminal racemes, almost radially symmetric; flower-stalks jointed. Calyx divided almost to the base into 5 segments which are reflexed or deciduous after flowering. Corolla yellow, sometimes red-spotted, petals clawed. Stamens 10, filaments free and usually hairy below, almost equal, about as long as the petals. Pod leathery or woody, indehiscent or very slowly dehiscent. Seeds few.

A genus of about 10 species from southern USA, C America and Africa, some of them (not ours) sometimes separated as

the genus *Cercidium* Tulasne. Only 1 is commonly grown, requiring a warm, well-drained site. Propagation is by seed.

1. P. aculeata Linnaeus. Illustration: Menninger, Flowering trees of the world, pl. 320 (1970); *Gartenpraxis* (January) 1978: 47; Everett, New York Botanical Gardens illustrated encyclopedia of horticulture, 2494 (1981); *Il Giardino Fiorito* (April) 1987: 4; Lewis et al., (eds.), Legumes of the world, 153 (2005).

Shrub or small tree to 10 m, shoots conspicuously green. Leaves 2-pinnate, stalkless, so that each leaflet appears to be a 1-pinnate leaf; axes of the leaflets flattened, to 30 cm, bearing numerous, shortly stalked ultimate segments which are ovate to oblong and 2–5 mm, often quickly deciduous. Calyx 6–7 mm. Corolla yellow with some red spots, to 2 cm across. Pod 2–10 cm × 5–6 mm, mostly rounded in section but some parts flattened. *S USA & adjacent Mexico, widely introduced in other parts of the world.* H5–G1. Spring.

The flattened leaflet-axes bearing small ultimate segments render this species easily identifiable among the cultivated woody legumes.

10. CERATONIA Linnaeus
J. Cullen
Evergreen shrubs or small trees to 10 m. Leaves pinnate, with or without a terminal leaflet; leaflets oblong to almost circular, rounded to the base and the apex, 3–5 × 3–4 cm, finely veined; stipules small, deciduous. Flowers small, numerous, borne in catkin-like racemes on spurs from the old wood, often unisexual, plants monoecious or dioecious or rarely with mixed unisexual and bisexual flowers in each inflorescence. Perigynous zone disc-like. Sepals 5, soon falling. Petals absent. Stamens 5, filaments free, anthers somewhat versatile. Ovary with several ovules; stigma peltate. Pod oblong, 10–20 × 1.5–2 cm, woody, dark brown, indehiscent.

A genus of a single species from the Mediterranean area, remarkable for its very catkin-like inflorescences. It is easily grown in well-drained soil in a sunny site and is usually propagated by seed, though cuttings will root if given bottom heat.

1. C. siliqua Linnaeus. Illustration: Bonnier, Flore complÒte **3**: pl. 165 (1914); Polunin, Flowers of Europe, pl. 49 (1969); Polunin & Everard, Trees and bushes of Europe, 100 (1976); Edlin, The illustrated

encyclopaedia of trees, 183 (1978); Lewis et al., (eds.), Legumes of the world, 133 (2005).

Mediterranean area. H5–G1. Autumn.

The edible fruits yield carob and are processed into various health-food products, and also fed to stock often in cattle cake.

11. CASSIA Linnaeus
J. Cullen
Trees, rarely shrubs. Leaves pinnate, without a terminal leaflet, often with 1 or more glands on the leaf-stalk or axis. Stipules variable, usually deciduous. Flowers usually in axillary or terminal racemes or panicles, more rarely solitary and axillary. Perigynous zone absent. Calyx of 5 almost completely free sepals, often unequal in size. Corolla usually yellow or orange-yellow, more rarely pink or whitish, the petals somewhat unequal in size, the uppermost usually the smallest. Stamens 10, usually the 1–3 upper small and sterile; of the fertile stamens 2 or 3 are usually longer than the others and may be borne on curved, swollen filaments which are more than twice as long as the anthers; anthers usually hairy, usually somewhat arrow head-shaped at the base, opening by terminal, or more rarely basal pores. Pod often large, variable in shape, usually divided into 1-seeded chambers, dehiscent or indehiscent.

There are about 30 species, mostly from the tropics but with extensions into the subtropics and temperate areas in many regions. Only a small number is cultivated for ornament, but many more have been cultivated for their medicinal properties.

Most of the species require glasshouse protection in much of Europe. The woody species may be propagated by seed or by cuttings, and the perennial herbaceous species by division.

Literature: Irwin, H.S. & Barneby, R.C., The American *Cassiinae*. A synoptical revision of Leguminosae Tribe *Cassieae* Subtribe *Cassiinae* in the New World, *Memoirs of the New York Botanical Garden* **35**: 1–918 (1982); Andrews, S. & Knees, S.G., Confusing *Cassias*, *The Kew Magazine* **5**(2): 76–81 (1988); Randell, B.R., *Cassia* and *Senna*, a review of the groups of plants formerly known as *Cassia*, *Australian Plants* **20**: 238–249 (2000).

1a. Flowers pink or pinkish orange, whitish outside; leaves with 8–20 pairs of leaflets **1. grandis**

b. Flowers yellow; leaves with up to 7 (rarely to 8) pairs of leaflets
2. fistula

1. C. grandis Linnaeus filius (*C. renigera* invalid).

Tree to 30 m in the wild, trunk sometimes spiny. Leaves with 8–15 pairs of leaflets which are oblong, rounded and very shortly stalked at the base, bluntly acute at the apex, shortly spreading hairy on both surfaces. Stipules deciduous. Flowers numerous in erect racemes. Sepals more or less equal. Petals pink inside, whitish outside, 1.5–2.5 cm. Fertile stamens 7, the central 4 with short, straight filaments and long anthers opening by apical pores, the lower 3 with very long, curved, swollen filaments and short anthers opening by basal pores; anthers covered with spreading hairs. Pod pendent, woody, 40–100 × 4–6 cm, with thickened sutures. *Tropical America, widely introduced in Asia.* G2.

2. C. fistula Linnaeus. Illustration: Menninger, Flowering trees of the world, pl. 120 (1970); Everard & Morley, Wild flowers of the world, pl. 112 (1970); Perry, Flowers of the world, pl. 161 (1972); Polunin & Stainton, Concise flowers of the Himalayas, pl. 24 (1985).

Tree to 16 m, with widely spreading branches. Leaves with 3–8 pairs of leaflets which are shortly stalked, ovate to lanceolate, 6–16 × 3–6 cm, acute or somewhat rounded at the apex, closely veined, finely downy at least when young. Stipules deciduous. Flowers numerous in long, open, hanging racemes. Sepals unequal. Petals yellow, veined, 2.5–5 cm. Fertile stamens 7, the upper 4 with short, straight filaments, the lower 3 with long, curved, swollen filaments many times longer than the anthers; all anthers finely adpressed hairy, opening by terminal pores. Pod to 50 × 1.7 cm, terete, cylindric, dark brown to black. *India, SE Asia, widely introduced elsewhere.* G2.

12. SENNA Miller
J. Cullen
Differing from *Cassia* in being small trees, shrubs or herbs, with stamens 10 (rarely 7), the lower 3 with filaments longer than those of the other stamens, but not more than twice as long as the anthers; anthers hairless.

A genus of about 300 species, mostly tropical. Several are grown as ornamentals. Literature: as for *Cassia*, see above.

1a. Herbs, sometimes woody at the base, but without persistent aerial branches   2
  b. Shrubs or small trees, with persistent aerial branches   4
2a. Leaflets obovate, rounded at apex; pods cylindric, very narrow, 10–20 × 3–4 cm   **11. obtusifolia**
  b. Leaflets ovate, ovate-lanceolate or elliptic, acute; pods shorter and broader   3
3a. Leaves with 4 or 5 pairs of leaflets which are acuminate
10. occidentalis
  b. Leaves with 6–8 pairs of leaflets, acute but not acuminate
9. hebecarpa
4a. Flower-buds concealed in greenish black bracts; anthers bent outwards at the top where the pores are; stipules persistent, conspicuous
1. didymobotrya
  b. Flower-buds not as above; anthers not bent at the top; stipules deciduous or persistent but not conspicuous   5
5a. Leaflets narrowly linear, terete, grooved   **8. artemisioides**
  b. Leaflets broader, flat, not grooved   6
6a. Plant completely covered with short, dense hairs   **5. multiglandulosa**
  b. Plant hairless or sparsely hairy   7
7a. Leaves with a gland between the lowest pair of leaflets or between each pair of leaflets   8
  b. Leaves completely without glands   9
8a. Gland present only between the lowermost pair of leaflets; raceme corymb-like   **2. corymbosa**
  b. Glands present between each pair of leaflets; racemes not corymb-like
7. septentrionalis
9a. Small tree; flower-buds conspicuously spherical, hard; pods linear-oblong, woody, to 30 cm
4. siamea
  b. Shrubs; flower-buds not as above; pods papery, much shorter   10
10a. Leaflets oblong-obovate, rounded at the apex   **5. italica**
  b. Leaflets lanceolate or narrowly elliptic, acute   **6. alexandrina**

1. S. didymobotrya (Fresenius) Irwin & Barneby (*Cassia didymobotrys* Fresenius; *C. myrsifolia* misapplied; *C. nairobensis*

Bailey & Bailey). Illustration: Vieira, Flores de Madeira, t. 86 (1986).

Shrub or small tree to 3 m. Leaves with 7–15 pairs of leaflets which are elliptic or elliptic-obovate, rounded but mucronate at the apex, finely hairy on both surfaces, 1.5–4 cm × 7–15 mm; stipules persistent, conspicuous, to 1 cm. Flowers numerous in erect racemes, each concealed by a large, greenish black bract. Sepals more or less equal. Petals golden yellow, veined, 1.5–2.5 cm. Fertile stamens 7, the lower 2 longer, anthers sharply bent at the top where the pores are. Pod stalked, oblong, flat, 8–10 × 1.5–2 cm. *Tropical Africa, introduced elsewhere.* G2.

2. S. corymbosa (Lamarck) Irwin & Barneby (*Cassia corymbosa* Lamarck; *Adipera corymbosa* (Lamarck) Britton & Rose; *C. floribunda* invalid). Illustration: *Botanical Magazine*, 633 (1803); *Gartenpraxis*, (April) 1986: 41.

Hairless shrub or small tree to 4 m or more. Leaves with 2 or 3 pairs of leaflets which are oblong-lanceolate, acute; a shortly stalked gland present between the lowest pair of leaflets above. Stipules deciduous. Flowers few in corymbs at the branch apices. Sepals unequal. Petals golden yellow, 1–1.5 cm. Fertile stamens 7, the lower 3 longer; anthers slightly arrow head-shaped, opening by apical pores. Pod cylindric, finally membranous, 4–10 cm × 7–10 mm. *Temperate S America.* G1.

The plant known as **S. × floribunda** (*Cassia floribunda* Cavanilles; *C. corymbosa* (Lamarck) var. *plurijuga* Bentham) has up to 6 pairs of leaflets and is a hybrid between *S. multiglandulosa* (Jacquin) Irwin & Barneby and *S. septentrionalis* (Viviani) Irwin & Barneby.

3. S. multiglandulosa (Jacquin) Irwin & Barneby (*Cassia tomentosa* Linnaeus filius; *Adipera tomentosa* (Linnaeus filius) Britton & Rose).

Densely downy shrub or small tree to 6 m. Leaves evergreen with 5–9 pairs of leaflets which are elliptic to oblong or slightly obovate, 1–4 cm × 8–12 mm, rounded or acute at the apex, yellowish or brownish hairy when young; inconspicuous and deciduous glands present between some of the lower leaflet-pairs. Stipules deciduous. Flowers in axillary racemes. Sepals unequal, 2 narrower than the others and more hairy. Petals yellow, *c.* 1.5 cm. Fertile stamens 7, 2 longer; anthers arrow head-shaped, opening by terminal pores. Pods oblong,

somewhat compressed, 8–12 cm ×
8–10 mm, hairy at first. *Tropical America.*
G1.

4. S. siamea (Lamarck) Irwin & Barneby
(*Cassia siamea* Lamarck; *C. floribunda*
misapplied; *Sciacassia siamea* (Lamarck)
Britton & Rose).

Hairless tree to 10 m. Leaves with 6–12
pairs of leathery leaflets which are elliptic-
oblong, rounded or slightly notched at the
apex, 3–8 × 1–2 cm. Stipules deciduous.
Flowers in dense, erect, corymb-like
panicles, the buds hard and spherical, the
axes robust. Sepals unequal. Petals yellow,
1–1.5 cm. Fertile stamens to 9, sterile 1–3,
2 or 3 of the fertile longer; anthers slightly
arrow head-shaped, opening by terminal
pores. Pod narrowly oblong, flat, to
30 × 1.5 cm, woody, sutures thickened. *SE
Asia, introduced elsewhere.* G2.

5. S. italica Miller (*Cassia italica* (Miller)
Sprengel; *C. obovata* Colladon).

Perennial herb, somewhat woody at the
base. Leaves with 5 or 6 pairs of leaflets
which are oblong-obovate, rounded at the
apex, 2–3.5 × 1–1.5 cm, somewhat
glaucous; stipules deciduous. Racemes
axillary. Sepals more or less equal. Petals
yellow, veined, *c.* 1.5 cm. Fertile stamens 7,
3 longer; anthers arrowhead-shaped, dark
brown with a yellowish stripe on each side,
opening by terminal pores. Pods papery,
very flattened, usually curved, rounded at
both ends, brownish to black, to
5 × 1.5 cm. *Tropical Africa to India.* G1.

6. S. alexandrina Miller (*Cassia senna*
Linnaeus; *C. angustifolia* Vahl; *C. acutifolia*
Delile; *Senna angustifolia* (Vahl) Batka).
Illustration: Bentley & Trimen, Medicinal
plants **2**: t. 89–91 (1877); *Gartenpraxis*,
April 1986: 39.

Perennial herb, somewhat woody below,
with sparse, adpressed, bristle-like hairs.
Leaves with 4–7 (rarely more) pairs of
leaflets which are lanceolate or narrowly
elliptic, acute at the apex, 2–5 cm ×
5–12 mm; stipules deciduous. Racemes
axillary. Sepals more or less equal. Petals
yellow, veined, 1–1.5 cm. Fertile stamens
7, 2 longer; anthers arrowhead-shaped,
opening by terminal pores. Pods papery,
very flattened, somewhat curved, rounded
at both ends, brownish to black, to
5 × 1.5 cm. *N & E Africa to India.* G1.

7. S. septentrionalis (Viviani) Irwin &
Barneby (*C. laevigata* Willdenow; *Adipera
laevigata* (Willdenow) Britton & Rose;

C. floribunda misapplied). Illustration:
Gartenflora **3**: t. 77 (1854).

Hairless shrub to 3 m. Leaves with 3–5
pairs of leaflets which are elliptic-ovate to
lanceolate, tapered to the acute apex,
3–7 × 1.5–2 cm, with a shortly stalked
gland between each leaflet-pair above.
Stipules deciduous. Racemes axillary, few-
flowered. Sepals unequal. Petals orange-
yellow, veined, *c.* 1.5 cm. Fertile stamens 7,
2 or 3 longer; anthers slightly arrowhead-
shaped, opening by terminal pores. Pod
cylindric-oblong, 5–10 × 1–1.5 cm, sutures
thickened. *Tropical America.* G1.

8. S. artemisioides (de Candolle) Radnell
(*Cassia artemisioides* de Candolle;
C. chatelainiana misapplied). Illustration:
Everard & Morley, Wild flowers of the
world, pl. 130 (1970); *Botanical Magazine*,
n.s., 599 (1972); Rotherham et al.,
Flowers and plants of New South Wales
and southern Queensland, pl. 484 (1975).

Shrub to 2 m, covered in ashy-white
hairs. Leaves with 3–8 pairs of leaflets
which are linear, terete and grooved,
1–2.5 cm × 1–2 mm; stipules deciduous.
Racemes axillary, clustered towards the
tips of the branches. Sepals more or less
equal, the 3 outer hairy. Petals yellow,
7–16 mm. Stamens 10, the 2 lower a little
longer; anthers slightly arrowhead-shaped,
opening by terminal pores. Pods oblong,
flat, 4–8 cm × 8–10 mm. *E Australia.* G1.

9. S. hebecarpa (Fernald) Irwin &
Barneby (*Cassia hebecarpa* Fernald;
C. hebecarpa var. *longipila* Baun;
C. marilandica misapplied; *Ditremexa
marilandica* mispplied). Illustration: Rickett,
Wild flowers of the United States **1**: pl. 61
(1966).

Perennial herb to 1 m or more, usually
conpsicuously downy. Leaves with 6–8
pairs of leaflets which are elliptic to elliptic-
lanceolate, often with marginal hairs,
acute, 2–5 cm × 8–20 mm; leaf-stalk with a
conspicuous gland near the base; stipules
deciduous. Flowers in axillary and terminal
racemes, hanging in bud. Sepals unequal.
Petals yellow, veined, *c.* 1 cm. Fertile
stamens 7, 3 longer; anthers slightly
arrowhead-shaped, dark, opening by
terminal pores. Pod oblong, flattened,
7–10 × 5–7 cm, hairy at least when young,
black, distinctly segmented, the segments
almost square. *E & C USA.* H3. Summer.

Much confused and intergrading with
C. marilandica Linnaeus, which is of more
southerly distribution, is less hairy, and has
narrower pod-segments.

10. S. occidentalis (Linnaeus) Link
(*Cassia occidentalis* Linnaeus; *Ditremexa
occidentalis* (Linnaeus) Britton & Rose).
Illustration: *Edwards's Botanical Register* **1**:
t. 83 (1816); *Revue Horticole* **1897**: 156.

Small or large hairless annual herb.
Leaves with 4 or 5 pairs of leaflets which
are ovate to ovate-lanceolate, long-tapered
to the acuminate apex; leaf-stalks with 1
or rarely 2 glands near the base; stipules
deciduous. Flowers few, in axillary
racemes. Sepals unequal. Petals yellow,
1.2–1.5 cm. Fertile stamens 7, 2 or 3
longer; anthers slightly arrowhead-shaped,
opening by terminal pores. Pods narrowly
oblong, flattened, 8–14 cm × 6–8 mm,
sutures thickened, pale. *Tropical &
subtropical America, introduced elsewhere.* G1.

11. S. obtusifolia (Linnaeus) Irwin &
Barneby (*Cassia obtusifolia* Linnaeus; *C. tora*
misapplied; *Emelista tora* misapplied).
Illustration: Rickett, Wild flowers of the
United States **2**: pl. 116 (1967); Polunin &
Stainton, Concise flowers of the Himalayas,
pl. 24 (1987).

Small to large annual herb, hairless or
downy. Leaves with usually 3 pairs of
leaflets which are obovate or broadly
obovate, rounded to the ultimately finely
pointed apex, 2–8 × 1–2.5 cm; an
inconspicuous gland present on the leaf-
stalk just below the lowest leaflet-pair;
stipules somewhat persistent. Flowers
axillary, mostly solitary. Sepals unequal.
Petals yellow, veined, *c.* 1 cm. Fertile
stamens 7, 2 or 3 longer; anthers opening
by terminal pores. Pod linear-cylindric,
4-angled, curved downwards from near
the base, 10–20 cm × 3–4 mm. *Tropical &
temperate America north to Pennsylvania &
Michigan; introduced elsewhere.* H4.
Summer.

13. CERCIS Linnaeus
J. Cullen

Shrubs or small trees, flowering before the
leaves expand. Leaves simple, palmately
veined, ovate, circular or kidney-shaped;
stipules deciduous. Flowers usually in
umbel-like clusters (strictly short racemes
but with very short axes) borne on short
shoots on the previous year's growth, more
rarely in pendent racemes. Calyx obliquely
and widely bell-shaped, shallowly
5-toothed. Corolla usually rose-pink to
purple, rarely white; petals clawed, keel
longer than standard. Stamens 10,
filaments free, curved downwards and
included between the keel petals. Ovary

several-seeded. Pods oblong, flat, with a narrow wing along the lower suture, several-seeded, eventually dehiscent.

A genus of about 6 species from N America, the Mediterranean area and E Asia. In spite of their wide and scattered distribution, the species (apart from *C. racemosa*) are very similar and difficult to distinguish. This difficulty is compounded by the fact that the plants flower before their leaves are expanded.

They require a rich soil in a sunny site. Propagation is mainly by seed, but some can be grafted on to stocks of *C. siliquastrum* or *C. canadensis*.

Literature: Hopkins, M., *Cercis* in North America, *Rhodora* **44**: 193–211 (1942); Isely, D., Leguminosae of the United States: 2. Subfamily *Caesalpinioideae*, *Memoirs of the New York Botanical Garden* **25**: 134–150 (1975).

1a. Flowers in distinct, long, pendent racemes   **1. racemosa**
 b. Flowers in umbel-like clusters   2
2a. Flowers 1.5 cm or more; all leaves rounded or notched at the apex, 6–12 cm   **3. siliquastrum**
 b. Flowers to 1.5 cm; usually at least some of the leaves acute or shortly acuminate at the apex, if all rounded or notched then leaves 2.5–7 cm   3
3a. Buds of next season's inflorescences with black, overlapping scales clearly visible in the leaf-axils by midsummer; flowers 1.1–1.5 cm, stalks 1–2 cm   **2. chinensis**
 b. Buds of next season's inflorescences small, without obvious black, overlapping scales, not conspicuous; flowers 1–1.5 cm, stalks 8–15 mm   **4. canadensis**

1. C. racemosa Oliver. Illustration: *Hooker's Icones Plantarum* **19**: t. 1894 (1889); Schneider, Illustriertes Handbuch der Laubholzkunde **2**: 5, 6 (1907); *Botanical Magazine*, 9316 (1933).

Small tree to 10 m, with black bark. Leaves broadly ovate, 6–10 × 6–10 cm, somewhat cordate or truncate at the base, pointed at the apex, pale and densely crisped hairy beneath, especially on or near the veins; stalks 2–4 cm. Racemes pendent, sometimes clustered; axes and flower-stalks densely crisped hairy; flower-stalks 8–12 mm in flower, 1.5–2 cm in fruit. Calyx 4–5 mm, margins of teeth with sparse fine hairs. Corolla rose-pink, 1–1.3 cm. Pods oblong, tapered at both

ends, reddish to dark maroon, 6–11 × 1–1.6 cm, wing 1.5–3 mm. *W China*. H4. Early summer.

2. C. chinensis Bunge. Illustration: Schneider, Illustriertes Handbuch der Laubholzkunde **2**: 5, 6 (1907); Bailey, Standard cyclopedia of horticulture **1**: 721 (1935).

Shrub or small tree to 10 m, bark variable, pale brown to grey or almost black. Leaves ovate to almost circular, with narrow, translucent margins, 7–13 × 8–12 cm, base cordate (sometimes widely so), apex usually shortly acute or acuminate, more rarely some leaves with rounded or notched apices, usually with tufts of hair in the vein-axils beneath; stalks 2.5–4.5 cm. Buds of next year's inflorescences conspicuous in the leaf-axils by midsummer, with overlapping, black scales. Flower-clusters with axis *c*. 2 mm. Calyx 3–4 mm, margins finely hairy; flower-stalks 5–10 mm in flower, 1.5–2 cm in fruit. Corolla bright pinkish purple, 1.2–1.5 cm. Pods oblong, tapered at both ends, flat, 8–13 × 1–1.7 cm, reddish when young, wing 2–2.5 mm. *China*. H3. Spring.

3. C. siliquastrum Linnaeus. Illustration: *Botanical Magazine*, 1138 (1808); Bonnier, Flore complète **3**: pl. 165 (1914); Polunin & Huxley, Flowers of the Mediterranean, pl. 54 (1965); Polunin, Flowers of Europe, pl. 50 (1969); Lewis et al., (eds.), Legumes of the world, 59 (2005).

Small tree or shrub to 8 m; bark greyish, young shoots dark. Leaves broadly ovate to almost circular, 6–10 × 8–14 cm, base cordate with a wide sinus, apex smoothly rounded or notched, hairless beneath even when young; stalks 3–4 cm. Flower clusters each with a hairy axis to 8 mm; flower-stalks 1–2 cm in flower, 1.5–2.5 cm in fruit. Calyx 4–6 mm, each tooth with a patch of short hairs at and just below the apex. Corolla bright purple or purplish red, 1.5–2 cm. Pods oblong, 8–12 × 1.4–2 cm, tapered at both ends, reddish when young, wing 1.5–2 mm. *Mediterranean area*. H4. Spring.

A variant with hairy flower-stalks, calyces and pods, subsp. **hebecarpa** (Bornmuller) Yaltirik, is native in the eastern Mediterranean area and may be grown there. 'Alba' (forma *alba* Rehder) has white flowers.

4. C. canadensis Linnaeus.
Shrub or small tree to 10 m; bark blackish, smooth or rough. Leaves ovate to

almost circular or kidney-shaped, 3.5–12 × 5–14 cm, with translucent margin very narrow or absent, cordate to truncate at the base, acute, shortly acuminate or rounded at the apex, hairless or finely hairy along the veins beneath. Flower-clusters without an axis or the axis very short, hairy. Calyx 2.5–3.5 cm, often reddish purple, margins finely hairy. Corolla bright purple-pink or rarely white, 9–14 mm. Pods oblong, flattened, 4–10 × 8–20 mm, tapered at both ends, sometimes glaucous. *N America*.

The above description covers a bewildering array of variants from N America. Following Isely (reference above) there are 2 species, *C. canadensis* in the east, *C. occidentalis* in the west. In intervening areas, especially Texas, various intermediate forms are found, rendering discrimination of the species extremely difficult. Isely remarks that 'taxa ... are most conveniently identifiable on the basis of where they come from'. This is of no help to the gardener, who will probably know nothing of the ultimate wild origins of his material.

Isely treats *C. canadensis* as consisting of 3 varieties (2 of them possibly cultivated) and *C. occidentalis* as a simple species. Brief diagnoses are given below, but it must be stressed that the possibility of making accurate identifications is not great.

Var. **canadensis**. Illustration: Gleason, Illustrated flora of the north-eastern United States and adjacent Canada **2**: 383 (1952); *Horticulture* **35**: 76 (1957); Justice & Bell, Wild flowers of North Carolina, 98 (1968). Leaves mostly with acute or shortly acuminate apices (if apices rounded then leaves 7 cm or more), thin; flowers 9–12.5 mm. *E North America from Canada to Florida*. H3. Spring.

'Alba' (var. *alba* Anon.), illustration: Brickell (ed.), RHS A–Z encyclopedia of garden plants, 252 (2003), has white flowers.

Var. **texensis** (Watson) Hopkins (*C. reniformis* Watson). Leaves leathery, usually rounded at the apex, if acute, to 8 cm at most, flowers 9–13 mm. *SE USA (Texas, Oklahoma)*. H5. Spring.

A white-flowered variant is known in the USA as 'Texas White' (illustration: *American Nurseryman* **167**: 31, 1988).

C. occidentalis Gray. Illustration: Abrams, Illustrated flora of the Pacific States **2**: 477 (1944); Munz, California spring wild flowers, 78 & pl. 58 (19) (1959). Leaves thick, usually rounded at the apex, flowers 1.2–1.4 cm. *W USA (California, Arizona, Utah)*. H5. Spring.

14. BAUHINIA Linnaeus
J. Cullen

Trees, shrubs or woody climbers, some with tendrils, a few spiny (not ours). Leaves of 2 distinct leaflets or more commonly of a single leaflet which is notched or lobed at the apex; stipules variable, usually deciduous. Flowers in racemes, terminal, axillary or terminating lateral shoots. Calyx either of 5 sepals, edge-to-edge in bud, or united in bud and splitting along 1 or more lines as the flower opens, often spathe-like and reflexed. Petals 5, slightly unequal, spreading, usually conspicuously clawed. Fertile stamens 10, 5, 3 or 1, filaments free, often hairy. Ovary stalked with few to many ovules. Pod dehiscent or not, stalked, leathery or woody.

There are about 250 species from the tropics, of which only a few are in general cultivation. They succeed in a well-drained soil in a sunny position. Propagation is normally by seed, but propagation by suckers (in those species that produce them) is also possible; cuttings are sometimes used, but they are difficult to root.

Literature: Isely, D., Leguminosae of the United States: 2. *Caesalpinioideae, Memoirs of the New York Botanical Garden* **25**(2): 15–31 (1975).

Leaves. Of 2 distinct leaflets: **1**; of a single leaflet variably 2-lobed at the apex: **2, 3, 4, 5, 6.** To 6 cm: **1** (leaflets), **2, 6;** 7 cm or more: **3, 4, 5.**
Tendrils. Present: **1.**
Calyx. Of 5 sepals: **1;** splitting once or more, but not into 5 sepals: **2, 3, 4, 5, 6.**
Petals. White or cream: **2, 3, 4;** pink, violet, violet-blue or deep violet: **1, 4, 5;** salmon pink to orange: **6.** Without distinct claws: **2, 3;** with distinct claws: **1, 4, 5, 6.** Fertile stamens: Ten: **2, 3;** five: **4;** three or one: **1, 5, 6.**

1a. Leaves made up of 2 distinct leaflets; tendrils present   **1. yunnanensis**
  b. Leaves of a single leaflet variably notched or lobed at the apex; tendrils absent   2
2a. Petals less than 2 cm   **2. racemosa**
  b. Petals more than 2 cm   3
3a. Fertile stamens 10; buds narrowly ovoid, very long-acuminate, terminating in 4 or 5 thread-like points   **3. acuminata**
  b. Fertile stamens fewer than 10; buds not as above   4
4a. Fertile stamens 5   **4. variegata**
  b. Fertile stamens 1 or 3   5
5a. Petals salmon-pink to orange-red, 2.5–4.5 cm, obviously clawed   **6. galpinii**
  b. Petals whitish pink or purple-mauve, 4–5 cm, scarcely clawed   **5. purpurea**

1. B. yunnanensis Franchet. Illustration: *Botanical Magazine*, 7814 (1902).

Prostrate or erect woody shrub to 3 m or more often a woody climber with paired, flattened, usually tightly coiled tendrils in the leaf-axils. Leaves made up of 2 distinct leaflets with a short projection between them; leaflets asymmetric, ovate, 2.5–5 cm, hairless. Flowers in racemes terminating lateral branches; stalks 2–4 cm. Buds pear-shaped. Sepals 5, becoming free and reflexed or spreading, 5–7 mm. Petals pink or pinkish violet, 1–1.5 cm, obovate, tapered gradually into narrow claws, some hairy outside and on the margins. Fertile stamens 3. Pod stalked, oblong, to 12 × 2 cm, dehiscent. *SW China (Yunnan).* G1.

2. B. racemosa Lamarck. Illustration: *Hooker's Icones Plantarum* **2**: t. 141 (1837).

Shrub or small tree to 6 m, branches pendent, without tendrils. Leaves broader than long, 3–6 × 4.5–7 cm, truncate or slightly cordate at the base, bilobed to about one-third, lobes rounded, downy or hairless, greyish beneath. Flowers in long, hairy racemes; stalks to 4 mm. Buds obliquely obovoid, pointed, hairy. Calyx spathe-like. Petals white or pale yellow, *c.* 1 cm, spreading. Fertile stamens 10. Pod stalked, straight or curved, to 20 × 2 cm. *Burma, NE India.* G2.

3. B. acuminata Linnaeus. Illustration: *Botanical Magazine*, 7866 (1902).

Shrub to 3 m, without tendrils. Leaves ovate to almost circular, 7–15 × 7–15 cm, cordate at the base, lobed to one-third or less at the apex, the lobes usually acute, with short, crisped hairs on the conspicuous veins beneath. Racemes short, few-flowered. Buds narrowly ovoid, very long-acuminate, terminating in 4 or 5 thread-like, erect or spreading points. Calyx splitting once, spathe-like, 2.5–4 cm. Petals white or cream, obovate or elliptic, scarcely clawed, 2.5–4.5 cm. Fertile stamens 10. Pod stalked, oblong, flattened, 5–15 × 1–2 cm. *India, Malaysia, S China.* G2.

4. B. variegata Linnaeus. Illustration: *Botanical Magazine*, 6818 (1885); Menninger, Flowering trees of the world, pl. 78 (1962); *Journal of the Royal Horticultural Society* **98**: f. 21 (1973); *Gartenpraxis* **1977**: 586.

Shrub or small tree to 10 m, without tendrils. Leaves broadly circular in outline, truncate or cordate at the base, lobed to about one-third at the apex with rounded or obtuse lobes, 6–20 × 7–18 cm, hairless. Flowers few, in short, finely hairy racemes; flower-stalks 2–3 mm, perigynous zone long and stalk-like. Buds appearing stalked, obovoid, blunt at the apex. Calyx 1–2 cm, spathe-like, notched at the apex. Petals white to pinkish, reddish or bluish purple, often mottled and with pronounced, coloured veins, 4–6 cm, obovate, clawed. Fertile stamens 5. Pod stalked, to 30 × 2.5 cm, flattened. *Himalaya, China.* G2.

The white-flowered variant, 'Candida' (var. *candida* Hamilton) is often grown (illustration: *Botanical Magazine*, 7312, 1893).

B. × blakeana Dunn. Illustration: Menninger, Flowering trees of the world, pl. 79 (1962). Supposedly the hybrid between *B. variegata* and *B. purpurea*. It has more numerous, larger flowers than *B. variegata* and the petals are deep rose-lavender changing to dark red as the flower matures.

5. B. purpurea Linnaeus. Illustration: Menninger, Flowering trees of the world, pl. 80 (1962); Perry, Flowers of the world, 159 (1970); Polunin & Stainton, Concise flowers of the Himalayas, t. 206 (1987); Lewis et al. (eds.) Legumes of the world, 61 (2005).

Tree or shrub to 7 m. Leaves more or less circular or broadly elliptic, truncate or cordate at the base, lobed to one-third or half with rounded or pointed lobes, 7–15 × 7.5–12 cm, hairless or with fine adpressed hairs beneath. Racemes many-flowered, hairy; flower-stalks 2–10 mm. Buds narrowly pear-shaped, pointed, ribbed. Calyx 2–4 cm, splitting on more than 1 line. Petals whitish, pink or purple-mauve, 4–5 cm, distinctly clawed, obovate. Fertile stamens 3. Pod stalked, flattened, to 25 × 2.5 cm. *Himalaya, Burma, S China.* G2.

6. B. galpinii N.E. Brown. Illustration: *Botanical Magazine*, 7494 (1896); Palgrave, Trees of Central Africa, 75 (1956); Menninger, Flowering vines of the world,

f. 87 (1970); Brickell (ed.), RHS A–Z encyclopedia of garden plants, 163 (2003).

Shrub to 4 m, usually scrambling or climbing, without tendrils. Leaves broadly circular to kidney-shaped, rounded to slightly cordate at the base, lobed at most to one-third at the apex with rounded lobes, finely hairy beneath, 2.5–5 × 2–5.5 cm. Racemes few-flowered, rusty-brown hairy; flower-stalks 2–3 cm. Buds narrowly obovoid, pointed, hairy. Calyx spathe-like, 2–5 cm. Petals salmon-pink to orange, 2.5–4.5 cm, each with a more or less diamond-shaped, pointed blade narrowing abruptly into the long, conspicuous claw. Fertile stamens 3. Ovary hairy. Pod to 10 × 2.5 cm, flattened, dehiscent. *Zimbabwe, South Africa.* G2.

15. SCHOTIA Jacquin
J. Cullen

Trees or shrubs. Leaves pinnate, without a terminal leaflet, leaflets leathery; stipules small, deciduous. Flowers numerous in panicles borne terminally or sometimes on the older wood. Perigynous zone bell-shaped or funnel-shaped. Calyx of 4 unequal lobes, overlapping in bud. Petals 5, occasionally some or all of them reduced to linear threads. Stamens 10, alternately longer and shorter, filaments entirely free or fused at the base. Ovary containing several ovules, stalked, the stalk united with the perigynous zone. Pod oblong, compressed, woody, tapered to the apex, eventually dehiscent, several-seeded. Seeds with or without arils.

A genus of about 20 species from Africa south of the Zambesi. Only a small number is grown. They require greenhouse protection in most of Europe (though they can be grown out of doors in the Mediterranean area), with a well-drained soil and sunny site. They are propagated by seed.

1a. Stamens free to the base; leaves with 6–18 pairs of leaflets **1. afra**
 b. Stamens united at the base; leaves with 3–8 pairs of leaflets 2
2a. Petals 5, normally developed; flower-stalks very short, to 2 mm **2. latifolia**
 b. All or some of the petals reduced to linear threads; flower-stalks at least 5 mm **3. brachypetala**

1. S. afra (Linnaeus) Thunberg (*S. speciosa* Jacquin). Illustration: *Botanical Magazine*, 1153 (1809) – as *S. tamarindifolia*;

Flowering Plants of Africa **42**: p. 1665 (1973); Palgrave, Trees of southern Africa, t. 76 (1977); Palmer, A field guide to the trees of southern Africa, pl. 10 (1977).

Shrub or small tree to 7 m. Leaves with 6–18 pairs of leaflets which are hairless or finely downy, linear to oblong or elliptic, 5–20 × 1–10 mm, obtuse and mucronate at the apex. Flowers in almost spherical panicles; flower-stalks 3–9 mm. Calyx leathery, red. Petals 5, red to pink, 1–1.8 cm, downy within. Stamens exceeding the petals, filaments free to the base. Pods 3–15 × 2–5 cm. Seed without an aril or aril very small. *South Africa (Cape Province), Namibia.* H5–G1.

Variable in the number, size and shape of the leaflets.

2. S. latifolia Jacquin. Illustration: *Flora of Southern Africa* **16**(2): 29 (1977).

Tree to 10 m or more with rounded crown. Leaves with 3–5 pairs of leaflets which are hairless or downy, elliptic-oblong to obovate, 1.5–6.5 × 1–3.5 cm, tapered or rounded at the base, rounded or acute but scarcely mucronate at the apex. Flowers in rather open, terminal panicles; flower-stalks very short (to 2 mm). Calyx reddish brown. Petals 5, pink to pale pink, 9–11 mm. Stamens slightly exceeding the petals, united at the base with the tube split along 1 side. Pods 5–14 × 3–4.5 cm. Seeds each with a large, yellow aril. *South Africa (Cape Province, Transvaal).* H5–G1.

3. S. brachypetala Sonder. Illustration: *Flowering Plants of South Africa* **20**: pl. 777 (1940); *Flora of Southern Africa* **16**(2): 29 (1977); Palgrave, Trees of southern Africa, t. 77 (1977); Palmer, A field guide to the trees of southern Africa, pl. 10 (1977); Lewis et al., (eds.) Legumes of the world, 73 (2005).

Tree to 16 m with rounded crown. Leaves with 4–8 pairs of leaflets which are elliptic to oblong or obovate, 2.5–8.5 × 1–4.5 cm, hairless or shortly downy, rounded to the apex. Flowers in congested, more or less spherical panicles; flower-stalks 5–12 mm. Calyx leathery, red. Petals 5, all or some of them reduced to linear threads, those fully developed red, 1.3–1.8 cm. Stamens with filaments united at the base, tube split at 1 side or not. Pods 5–17 × 3.5–5 cm. Seeds each with a large, yellow aril. *Mozambique, Zimbabwe, South Africa.* H5–G1.

16. BROWNEA Jacquin
J. Cullen

Small trees with spreading crowns, branches ultimately pendent. Leaves evergreen, pinnate with 2–18 pairs of leaflets; young leaves brightly coloured, enclosed at first in a tube of scale-leaves, later limp and hanging for a few days, ultimately expanding; stipules long and thread-like, quickly deciduous. Flowers in pendent heads terminating the branches (rarely borne on the old wood); heads enclosed in large bracts which fall as the flowers open; each flower subtended by a large bract and by 2 bracteoles which are united into a 2-lobed tube around the flower. Sepals 5, red or orange-red. Petals 5, free, more or less equal, clawed, red or orange-red. Stamens 10–11 (rarely more), filaments united at least at the extreme base (sometimes into 2 or 3 bundles). Ovary hairy, with few ovules. Pod oblong, woody, often contorted, few-seeded.

A genus of about 30 species from tropical S America and the Caribbean, whose classification is very confused. They require glasshouse treatment in Europe and are prized not only for their brightly coloured, spectacular inflorescences, but also for the interesting, coloured, limp young leaves. They require a rich soil and high temperatures, and can be damaged by overwatering. Propagation is by hardwood cuttings.

1a. Flower-heads to 5 cm across, funnel-shaped when open; young leaves purplish, becoming pinkish brown; stipules 1–2 cm **3. coccinea**
 b. Flower-heads more than 5 cm across, spherical when open; young leaves mottled or pinkish buff becoming pinkish brown; stipules more than 10 cm 2
2a. Shoots persistently hairy; stipules to 30 cm; petals exceeding stamens **1. grandiceps**
 b. Shoots soon hairless, warty; stipules to 12 cm; stamens exceeding petals **2. ariza**

1. B. grandiceps Jacquin. Illustration: *Botanical Magazine*, 4839 (1855); Corner, Wayside trees of Malaya, pl. 80 (1940); Everard & Morley, Wild flowers of the world, pl. 170 (1970); Lewis et al. (eds.), Legumes of the world, 98, 109 (2005).

Tree to 5 m or more. Shoots with dense, persistent, pale brown hairs. Leaves with 11 or more pairs of leaflets (3–5 pairs in those near the inflorescences), stalk and axis densely hairy with pale brown hairs; leaflets increasing in size upwards, oblong-obovate or elliptic-oblong, the longest to 15 × 4 cm, unequal and rounded at the base (cordate in the lowermost leaflets), acuminate at the apex, densely and persistently hairy on the midrib beneath; young leaves brownish green mottled with pink and white, later pinkish brown, the stalk and axes whitish. Stipules to 30 cm. Flower-heads spherical, *c.* 15 cm across, many-flowered; bracteoles 2.5–3 cm, tube to 1.5 cm, densely hairy; sepals red, darker than petals, 4–4.5 cm; petals long-clawed, red-scarlet, 7–8 cm. Stamens usually 11, to 5 cm, filaments united at the base for *c.* 8 mm. *N South America.* G2.

2. B. ariza Bentham. Illustration: *Botanical Magazine*, 6469 (1880); Brickell (ed.), RHS A–Z encyclopedia of garden plants, 196 (2003).

Similar to *B. grandiceps*, but bark warty, shoots hairless or soon so if slightly hairy when young, leaflets hairless, the young leaves uniformly pinkish buff becoming pinkish brown, then green, stipules to 12 cm; flower-heads somewhat smaller, petals *c.* 6.5 cm, stamens exceeding petals. *N South America (Venezuela only?).* G2.

3. B. coccinea Jacquin. Illustration: *Botanical Magazine*, 3964 (1843); Corner, Wayside trees of Malaya, pl. 79 (1940).

Tree to 10 m or more. Shoots hairless, shining, warty. Leaflets 3–9 pairs, leathery, lanceolate to obovate, the largest 13–17 × 5–7 cm, unequal and rounded (cordate in lowest leaflets) at the base, acuminate at the apex, greyish green, hairless. Young leaves purplish, becoming pinkish brown. Stipules 1–2 cm. Flower-heads spherical in bud, funnel-shaped when open, *c.* 5 cm across. Bracteoles *c.* 3 cm, tube *c.* 1.5 cm, covered with adpressed, bristle-like hairs. Sepals bright red, *c.* 4 cm. Petals long-clawed, *c.* 5.5 cm, vermilion. Stamens exceeding petals, filaments united into a tube for *c.* 3 cm. *Northern South America, Jamaica?* G2.

Material grown in the tropics as *B. latifolia* Jacquin and *B. capitella* Jacquin (*B. speciosa* Reichenbach) is very similar; further research is needed on these plants to establish their distinctions (if any) and the correct names.

17. AMHERSTIA Wallich
J. Cullen
Evergreen trees to 20 m or more. Leaves pinnate with 4–7 pairs of leaflets, without a terminal leaflet; leaflets oblong, oblong-lanceolate or obovate, abruptly acuminate, rounded at the base, 10–30 × 2–8 cm, whitish or pale beneath; young leaves bronze, limp and hanging; stipules 2.5–4 cm, soon deciduous. Flowers in hanging racemes to 45 cm; flower-stalks long, each bearing 2 large, persistent bracteoles. Sepals 4, coiling when the flower is open. Petals 5, 3 large and 2 very small, the uppermost to 5 cm long and wide, with a whitish, pink-striped and -spotted claw, the blade deep pink with a yellow blotch at the apex, the other 2 large petals pink with the apices yellow-blotched. Stamens 9 or 10, 5 longer and 4 or 5 shorter, the filaments united into a tube for about half their length. Ovary stalked, with few ovules, yellow, the stalk and style red. Pods to 15 × 4 cm, containing 4–6 seeds.

A genus of a single species from Burma, widely grown as a tropical ornamental. In Europe it requires high temperatures in a glasshouse, rich, well-drained soil and constantly moist air to succeed, but will flower when quite small if well grown. Propagation is by seed or by hardwood cuttings.

1. A. nobilis Wallich. Illustration: *Botanical Magazine*, 4453 (1849); Corner, Wayside trees of Malaya, pl. 78 (1940); Menninger, Flowering trees of the world, pl. 100 (1962); Everard & Morley, Wild flowers of the world, pl. 113 (1970); Lewis et al. (eds.), Legumes of the world, 90 (2005).

Burma, widely cultivated elsewhere. G2.

18. TAMARINDUS Linnaeus
J. Cullen
Tree to 20 m or more with a rounded crown and ultimately drooping branches. Leaves pinnate without a terminal leaflet; leaflets opposite, in 9–16 pairs, oblong, symmetric, very shortly stalked, rounded at the base, rounded or slightly notched and shortly mucronate at the apex, 8–20 × 4–8 mm, hairless. Flowers 6–10 in short racemes borne on the branches, each flower initially enclosed in a large, quickly deciduous bract. Perigynous zone narrowly funnel-shaped. Sepals 4, 8–9 mm, unequal, reflexed and soon falling after the flower opens. Petals 5, the 2 lower minute, the upper 3 larger, 9–13 mm, reddish-veined on a pale yellow background, margins wavy. Fertile stamens 3, filaments united below, the top of the fused zone with a number of small teeth. Pod indehiscent, oblong-cylindric though somewhat 4-cornered, often irregular due to abortion of some ovules, 7–12 × 1–3 cm, succulent and pulpy when young, later dry, indehiscent. Seeds few.

A genus of a single species from Africa or India, widely cultivated in tropical regions, especially E Asia and mainly for the sake of the sour pulp of the immature pods, which is used in cookery, for flavouring drinks and in folk-medicine. It will grow well in most soils, preferring moist conditions but surviving some degree of aridity. Propagation is by seed (seedlings very prone to damping off) or by budding.

1. T. indica Linnaeus. Illustration: Bentley & Trimen, Medicinal plants **2**: t. 92 (1876); Corner, Wayside trees of Malaya, pl. 119, 120 (1940); Palgrave, Trees of southern Africa, t. 79 (1977); Matthew, Illustrations on the flora of the Tamil Nadu Carnatic, 238 (1983); Lewis et al., (eds.), Legumes of the world, 90 (2005).

Widespread in the tropics, origin probably Africa or India. G2.

153. FABACEAE
(LEGUMINOSAE-PAPILIONOIDEAE)

Herbs or woody or climbing plants. Leaves usually alternate, stipulate, pinnately compound, sometimes 2–3-pinnate or of a single or 2–3 leaflets, rarely absent; ptyxis of leaflets almost aways conduplicate, but supervolute in some species of *Lathyrus*. Inflorescence various, usually a raceme. Flowers usually bisexual and zygomorphic. Calyx, corolla and stamens hypogynous or perigynous. Calyx usually of 5 free sepals or forming a 5-lobed tube, Corolla of 5 petals, all free or sometimes the lower 2 united towards the apex, imbricate in bud such that the uppermost petal is outermost, overlapping the others, and often larger than them. Stamens usually 10 united into a tube or 9 united into a tube and 1 at least partly free; anthers opening by slits. Ovary of a single carpel (rarely more); ovules 1–many, marginal; style usually 1. Fruit usually a legume (sometimes indehiscent or breaking into 1-seeded segments); seeds usually without a lateral line.

There are about 450 genera, and over 11,000 species. Many genera are native to Europe and/or North America, and many are grown as ornamentals. Many of the genera produce crops of economic importance (peas, beans, etc.). In edn 1 this family was included in the Leguminosae (as subfam. *Papilionoideae*).

In almost all genera of this subfamily the 5 petals are diversified. The uppermost petal, which overlaps all the others in bud is generally large, with a long claw and the upright or reflexed blade; this is known as the standard. Within this are 2 lateral petals, often rather narrow, known as the wings. Finally, within these are 2 larger petals which are usually united at least along their lower margins; these, which are known as the keel, generally enclose the stamens, ovary and style. In some cases the wings are coupled to the keel by means of a fold or tuck, and their decoupling is sometimes important in pollination. In a few cases (*Amorpha*), the wings and keel are absent and in *Clitoria* and *Centrosema* the flowers are reversed so that they are borne on the plant with the standard lowermost.

The way in which the filaments of the stamens are united is important in the identification of the genera. The filaments may be entirely free (or just minutely joined at the extreme base), or may be all united (monadelphous stamens), either into a tube around the ovary and style, or into a sheath which is open along the top or 9 of them may be fused into a sheath while that of the uppermost is free (diadelphous stamens); in this latter case the uppermost filament may be entirely free from the rest, or may be free at the base, but united with the others higher up. Also, in the case of a few genera with all 10 filaments united into a tube, the uppermost may progressively become free from the apex of the tube as the flower ages. This character often causes confusion, as it is often difficult to discern in herbarium specimens; however, in living flowers it is generally easy to see and to understand.

The editors are grateful to Dr G. Lewis and Dr B. Schrire of Kew, and Ms S. Andrews for their help with these accounts.

Literature: Lewis, G., et al. (eds.), Legumes of the world (2005).

1a. Trees or shrubs with flowers of 3 types   **93. Laburnocytisus**
  b. Woody or herbaceous plants with flowers of 1 type   2

2a. Filaments of all 10 stamens completely free for at least 90% of their length (sometimes all slightly fused at the extreme base)   3
  b. Filaments of all or most (usually 9) of the stamens united for much of their length, often the uppermost completely free at the base (sometimes united to the others higher up), rarely the stamens variably united in groups, or rarely the stamens 9 only, filaments all united   25
3a. Most leaves pinnately divided, with more than 3 leaflets   4
  b. Most leaves simple, or with 3 leaflets, or leaves with blades absent   10
4a. Plant sticky with stalkless glands; leaflets fleshy and irregularly toothed; pod glandular, breaking into 1-seeded segments   **36. Adesmia**
  b. Plant without the above combination of characters   5
5a. Leaflets narrow, heather-like (ericoid)   **74. Burtonia**
  b. Leaflets expanded, not narrow and heather-like   6
6a. Lower 4 petals similar to each other, not differentiated into wings and keel; pod woody, inflated, containing 3–5 chestnut-like seeds   **1. Castanospermum**
  b. Lower 4 petals differentiated into wings and keel; pods and seeds not as above   7
7a. Bracts borne on the flower-stalks; corolla dark blue   **2. Bolusanthus**
  b. Bracts borne at the bases of the flower-stalks; corolla variously coloured, not dark blue   8
8a. Base of the leaf-stalk swollen, enclosing the axillary bud   **4. Cladrastis**
  b. Base of the leaf-stalk not swollen as above, axillary bud not enclosed   9
9a. Pods flattened, linear-oblong, dehiscent; standard reflexed   **3. Maackia**
  b. Pods terete, constricted between the seeds, often winged, not dehiscent but breaking into 1-seeded segments; standard reflexed or forwardly-directed   **5. Sophora**
10a. Leaf-blades absent   11
  b. Leaf-blades present, simple or of 3 leaflets   12

11a. Mature leaves reduced to thread-like stalks; flowers 1 per bract   **76. Viminaria**
  b. Leaves entirely absent from mature plants; flowers 2 per bract   **75. Sphaerolobium**
12a. Leaves simple (sometimes formed from a single leaflet)   13
  b. Leaves of 3 leaflets (sometimes these side-by-side, without a common stalk)   20
13a. Calyx umbilicate at base   **84. Podalyria**
  b. Calyx not umbilicate at base   14
14a. Leaves opposite or in whorls   15
  b. Leaves alternate   16
15a. Leaves opposite; stipules present; calyx 1–1.5 cm   **70. Brachysema**
  b. Leaves in whorls; stipules absent; calyx 6–7 mm   **71. Oxylobium**
16a. Ovules 4 or more   17
  b. Ovules 2   18
17a. Leaves ovate, cordate at base, usually with the margins sharply toothed   **72. Chorizema**
  b. Leaves linear to linear-lanceolate, not cordate at the base, margins not toothed   **71. Oxylobium**
18a. Leaves not heather-like, margins not rolled under, or leaves terete; bracteoles persistent, borne close beneath and overlapping the calyx   **78. Pultenaea**
  b. Leaves heather-like, margins rolled under; bracteoles absent or distant from and not overlapping the calyx   19
19a. Pod flattened, triangular in outline; calyx 2.5–3 mm   **77. Daviesia**
  b. Pod ovoid or almost spherical, not flattened; calyx 3–6 mm   **79. Dillwynia**
20a. Leaflets side-by-side, without a common stalk, heather-like   **74. Burtonia**
  b. Leaflets not as above, borne on a common leaf-stalk, not heather-like   21
21a. Plant herbaceous, shoots dying back to the root in winter   22
  b. Plant woody, aerial shoots persistent through the winter   23
22a. Plants hairless or almost so, often glaucous; corolla violet-blue, yellow or white; pod inflated   **90. Baptisia**
  b. Plants usually conspicuously hairy, not glaucous; corolla deep purple or yellow; pod flat   **89. Thermopsis**

23a. Calyx very deeply 5-lobed, lobes edge-to-edge in bud, tube very short, black outside; keel deep, fringed along the suture; leaves often almost opposite **73. Gompholobium**
  b. Combination of characters not as above 24
24a. Standard well developed, about as long as wings; corolla more than 2.5 cm **88. Piptanthus**
  b. Standard much shorter than wings; corolla 1.8–2.5 cm **87. Anagyris**
25a. Stamens 10 with all their filaments united for a great proportion of their length, certainly all united in the lower part, forming a tube or sheath 26
  b. Stamens 10 with the filament of the uppermost free at the base and for some distance above it, sometimes united with those of the other 9 well above the free base, or filaments variably united in groups or stamens 9, filaments all united 60
26a. Most leaves with more than 3 leaflets, mostly pinnate, occasionally palmate 27
  b. All leaves simple or of 3 leaflets only 36
27a. Leaves palmate **91. Lupinus**
  b. Leaves pinnate 28
28a. Corolla reduced to the standard only, which envelops the stamens and style **34. Amorpha**
  b. Corolla with wings and keel as well as standard 29
29a. Leaves with conspicuous, brown, stalkless glands beneath; flowers *c.* 3 cm **35. Amicia**
  b. Leaves without conspicuous, brown stalkless glands beneath; flowers to 2.5 cm, often much smaller 30
30a. Stipules and stipels persistent; corolla blue to purple **27. Hardenbergia**
  b. Stipules present but usually falling early, or absent, stipels absent; corolla variously coloured, not usually blue to purple 31
31a. Evergreen woody climber; pod indehiscent, compressed, hairy, few-seeded, somewhat winged along the sutures **8. Derris**
  b. Usually deciduous trees, shrubs, subshrubs or herbs; pod various, not as above 32
32a. Trees or upright shrubs 33
  b. Low shrublets or herbs 34

33a. Corolla violet or pink; the 2 upper calyx-teeth reflexed **83. Virgilia**
  b. Corolla golden yellow to orange; the 2 upper calyx-teeth forwardly-directed **6. Pterocarpus**
34a. Stems and leaves sticky, glandular-hairy; stipules united to the leaf-stalk **64. Ononis**
  b. Stems and leaves not glandular and sticky; stipules absent or free from the leaf-stalk 35
35a. Calyx-teeth about as long as tube; stipules persistent; corolla pale violet to white **47. Galega**
  b. Calyx-teeth shorter than the tube; stipules absent; corolla yellow or red, rarely creamy white **53. Anthyllis**
36a. Mature leaves consisting of a single leaflet or mature plants without leaves 37
  b. Mature leaves all or mostly with 3 leaflets 48
37a. Plants spiny, either the ends of the branches hardened and spiny, or leaves replaced by spiny phyllodes 38
  b. Plants not at all spiny 40
38a. Corolla blue **96. Erinacea**
  b. Corolla yellow 39
39a. Leaves absent from mature plants, replaced by spine-tipped phyllodes **104. Ulex**
  b. Leaves present on mature plants, spine-tipped phyllodes absent **102. Genista**
40a. Stamens united into a sheath which is open on the upper side 41
  b. Stamens united into a tube closed on the upper side 44
41a. Corolla purple **81. Hovea**
  b. Corolla yellow or red or yellow and red 42
42a. Stamens all of the same length; standard to 1.5 cm **82. Bossiaea**
  b. Stamens alternately longer and shorter; standard more than 1.5 cm 43
43a. Keel beaked; wings puckered or sculptured between the veins outside; corolla yellow **86. Crotalaria**
  b. Keel not beaked; wings not puckered or sculptured; corolla red **80. Templetonia**
44a. Calyx divided to the base on the upper side 45
  b. Calyx not divided to the base on the upper side 46

45a. Young stems broadly winged **99. Chamaespartium**
  b. Young stems not broadly winged **98. Spartium**
46a. Seeds appendaged; upper lip of calyx with 2 very short teeth **95. Cytisus**
  b. Seeds not appendaged; upper lip of calyx deeply toothed or notched 47
47a. Pod dehiscent, not inflated **102. Genista**
  b. Pod indehiscent, inflated **100. Retama**
48a. Leaflets toothed **64. Ononis**
  b. Leaflets not toothed 49
49a. Plant spiny 50
  b. Plant not spiny 51
50a. Corolla blue **96. Erinacea**
  b. Corolla yellow **102. Genista**
51a. Leaves dotted with dark glands, often with a resinous or bituminous smell 52
  b. Leaves not dotted with dark glands, not smelling as above 53
52a. Foetid biennials to perennial herbs; leaves with 3 leaflets **32. Bituminaria**
  b. Scented herbs or shrubs; leaves pinnate with a terminal leaflet **33. Psoralea**
53a. Stamens united into a sheath open at the top; pods inflated and hard, the seeds ultimately loose and rattling within them **86. Crotalaria**
  b. Stamens united into a tube closed at the top; pods various, not usually as above 54
54a. Seed with an appendage along a long side; corolla purple with a yellow spot at the base of the standard; leaflets usually folded upwards along the midrib **85. Hypocalyptus**
  b. Seed without an appendage, or with a small appendage along a short side; corolla yellow, pink or purple; leaflets not usually as above 55
55a. Pod covered with glandular warts **103. Adenocarpus**
  b. Pod not covered with glandular warts 56
56a. Small tree with flowers in long, hanging racemes **92. Laburnum**
  b. Shrubs (rarely small trees) with flowers in erect inflorescences 57
57a. Upper lip of calyx with 2 very short teeth; seeds appendaged **95. Cytisus**

b. Upper lip of calyx deeply 2-toothed; seeds not appendaged   58

58a. Tall shrubs; leaf-stalk 1.5–5 cm; pod 3.5–5 cm   **94. Petteria**

b. Small shrubs; leaf-stalk to 1.5 cm; pod less than 3.5 cm   59

59a. Leaves and branches mostly alternate; calyx not inflated   **102. Genista**

b. Leaves and branches mostly opposite; calyx somewhat inflated   **97. Echinospartium**

60a. Leaves of 3 leaflets, or of 5 leaflets and palmate, rarely simple or absent(sometimes replaced by a tendril)   61

b. Leaves pinnate with more than 3 leaflets   88

61a. Leaves simple or absent from mature plants   62

b. Leaves of 3 or 5 leaflets, present on mature plants   68

62a. Leaves simple   63

b. Leaves absent from mature plants   65

63a. Spineless annual; flowers yellow, several, in heads   **57. Coronilla**

b. Spiny perennial or small shrub; flowers yellow, pink or red, solitary or in pairs   64

64a. Corolla yellow   **53. Anthyllis**

b. Corolla pink or red   **46. Alhagi**

65a. Leaves replaced by a tendril subtended by 2 enlarged leaf-like stipules   **60. Lathyrus**

b. Tendrils absent; stipules minute or absent   66

66a. Pods linear, many-seeded, somewhat compressed between seeds   **101. Notospartium**

b. Pods ovoid to oblong, 1-few-seeded, not compressed between seeds   67

67a. Flower-stalk woolly; raceme with *c.* 20 flowers; pods indehiscent   **49. Chordospartium**

b. Flower-stalk hairless or finely downy; racemes usually with fewer flowers, if with *c.* 20, then pods dehiscent   **50. Carmichaelia**

68a. Leaflets with the veins running to the margin, which is usually toothed   69

b. Leaflets with veins joining and looping near the margin, which is not toothed   73

69a. Standard and keel deep blue, wings pink; stipules free from the leaf-stalk   **65. Parochetus**

b. Flowers not of the above colours; stipules joined to leaf-stalk to some extent   70

70a. At least 5 of the filaments dilated below the anthers; pod usually 1-or 2-seeded, usually enclosed in the persistent corolla and calyx   **69. Trifolium**

b. Filaments not dilated below the anthers; pods usually several-seeded, not enclosed as above, corolla usually falling   71

71a. Pods coiled or sickle-shaped, often spiny   **68. Medicago**

b. Pods not as above   72

72a. Pods nutlet-like, with 1–few seeds; plants smelling of new-mown hay (coumarin)   **66. Melilotus**

b. Pods straight or curved, long, many-seeded; plants smelling variously, but not of new-mown hay   **67. Trigonella**

73a. Standard with a projecting spur on the back   **28. Centrosema**

b. Standard without a spur on the back   74

74a. Stipels absent   75

b. Stipels present (rarely soon falling or gland-like)   78

75a. Shrub; flowers in axillary racemes congested into panicles at the tips of the branches   76

b. Herb or small, spiny shrub; flowers solitary or in pairs   77

76a. Keel somewhat curved, acute; bracts each subtending 1 flower; bracteoles quickly deciduous   **17. Campylotropis**

b. Keel straight, blunt; bracts each subtending 2 flowers; bracteoles persistent   **18. Lespedeza**

77a. Small, spiny shrub   **53. Anthyllis**

b. Herb, without spines   **54. Tetragonolobus**

78a. Plant a shrub with spines an the stems and often also on the leaves   **19. Erythrina**

b. Plant a shrub, woody or herbaceous climber or herb, without spines   79

79a. Flowers reversed, with the standard lowermost   **29. Clitoria**

b. Flowers not reversed, standard uppermost   80

80a. Style coiled through 2 or 3 coils   **31. Phaseolus**

b. Style straight or curved but not coiled as above   81

81a. Bracteoles absent   82

b. Bracteoles present   83

82a. Corolla blue-purple or whitish   **27. Hardenbergia**

b. Corolla reddish or yellow and black   **26. Kennedia**

83a. Trees or woody climbers with flowers in pendent racemes   84

b. Combination of characters not as above   85

84a. Pod not flattened, often bristly or velvety (causing irritation)   **20. Mucuna**

b. Pod flattened, wing-like, initially hairy, becoming beige with age   **22. Butea**

85a. Calyx obviously 2-lipped, the upper lip larger, somewhat lobed, the lower consisting of 3 small teeth   **24. Canavalia**

b. Calyx not 2-lipped, 5-toothed, teeth not as above   86

86a. Style flattened and bearded   **30. Lablab**

b. Style not flattened and bearded   87

87a. Stipules attached to the leaf-base above their bases, part extending below the point of attachment   **25. Pueraria**

b. Stipules attached to the leaf-stalks by their bases, often early deciduous   **16. Desmodium**

88a. Most leaves terminating in a short or minute point, a spine or a tendril, without a terminal leaflet   89

b. Most leaves with a terminal leaflet   99

89a. At least some leaves terminating in tendrils   90

b. Tendrils absent   94

90a. Leaflets with parallel veins and/or stems winged   **60. Lathyrus**

b. Leaflets with pinnate veins; stems not winged   91

91a. Calyx-teeth all equal, all at least twice as long as the tube   **61. Lens**

b. Calyx-teeth unequal, at least 2 of them less than twice as long as the tube   92

92a. Calyx-teeth more or less leaf-like; stipules to 1 cm   **62. Pisum**

b. Calyx-teeth not at all leaf-like; stipules much smaller   93

93a. Style hairy all round, or on the lower side, or entirely hairless   **59. Vicia**

b. Style hairy on the upper side only   **60. Lathyrus**

94a. Stamens 9, filaments all united   **7. Abrus**

b. Stamens 10, filaments of 9 united, that of the uppermost free at least at the base   95

95a. Leaves ending in a spine which hardens and persists after the leaflets have fallen   96

b. Leaves ending in a short or minute soft point which does not harden and persist   98

96a. Creeping or mat-forming shrubs or herbs with shoots not clearly differentiated into long and short shoots   **44. Astragalus**

b. Mostly upright or spreading shrubs with shoots clearly differentiated into short (flowering and leafing) shoots and long, extension shoots   97

97a. Flowers solitary or clustered on the short shoots on jointed stalks or rarely in a few-flowered umbel; corolla usually yellow, more rarely orange or pink   **42. Caragana**

b. Flowers 2–5 in a raceme; corolla purplish pink to magenta   **41. Halimodendron**

98a. Flowers in axillary racemes; stipules soon falling, never spine-like   **14. Sesbania**

b. Flowers solitary or clustered on jointed stalks; stipules persistent, often hardened and spine-like   **42. Caragana**

99a. Stipels present and persistent, though sometimes small   100

b. Stipels entirely absent   103

100a. Flowers reversed, the standard lowermost   **29. Clitoria**

b. Flowers not reversed, standard uppermost   101

101a. Large woody climber with numerous flowers in long, pendent racemes   **12. Wisteria**

b. Trees or shrubs; inflorescences various, not usually as above   102

102a. Hairs mostly medifixed; stipules not spine-like; shrubs   **15. Indigofera**

b. Hairs basifixed or absent; stipules usually persisting as spines; trees or shrubs   **13. Robinia**

103a. Veins of leaflets running out to the toothed margin   **63. Cicer**

b. Veins of leaflets looping and joining within the entire margin   104

104a. Stipules absent or represented by minute glandular points, the lowermost pair of leaflets sometimes mimicking stipules 105

b. Stipules present, developed, not represented by minute glandular points   108

105a. Leaves more or less stalked, with a variable number of leaflets (from 1 to 27), but generally some with more than 5 leaflets   106

b. Leaves almost stalkless, all with 5 leaflets, the lowermost pair sometimes mimicking stipules 107

106a. At least some flowers in terminal or apparently terminal inflorescences   **53. Anthyllis**

b. All flowers axillary or in axillary inflorescences   **43. Calophaca**

107a. Keel beaked   **55. Lotus**

b. Keel not beaked   **56. Dorycnium**

108a. Pods inflated, translucent, papery   109

b. Pods not as above   111

109a. Flowers yellow; leaflets 2–6 pairs   **40. Colutea**

b. Flowers not yellow; leaflets usually more numerous   110

110a. Corolla scarlet; keel not coiled upwards   **39. Sutherlandia**

b. Corolla purplish blue, pink, reddish brown or rarely white; keel sometimes coiled upwards   **38. Swainsona**

111a. Pod indehiscent, breaking transversely into 1-seeded segments   112

b. Pod dehiscent or indehiscent, not breaking transversely into 1-seeded segments, sometimes the pod itself 1-seeded   114

112a. Flowers in racemes, the flower-stalks arising serially from the axis   **51. Hedysarum**

b. Flowers in umbel-like clusters or heads, all the flower-stalks arising from more or less the same point   113

113a. Segments of the pod crescent-shaped, horseshoe-shaped or rectangular with an arched sinus   **58. Hippocrepis**

b. Segments of the pod linear to oblong, straight or slightly curved   **57. Coronilla**

114a. Back of the standard and/or other parts of the corolla hairy   115

b. Corolla hairless   117

115a. Flowers in axillary panicles; pod 4-winged   **10. Piscidia**

b. Flowers in terminal or leaf-opposed racemes; pod not 4-winged   116

116a. Corolla bluish purple   **9. Mundulea**

b. Corolla yellow, pinkish, cream or white   **11. Tephrosia**

117a. Twiner, roots bearing strings of tubers; keel coiled upwards within the standard   **23. Apios**

b. Combination of characters not as above   118

118a. Shrubs; anthers consisting of a single sac; pod small, opening by the falling of the sides from a persistent framework to which the seeds are attached   **50. Carmichaelia**

b. Shrubs or herbs; anthers consisting of 2 sacs; pod not as above   119

119a. Standard reflexed, acuminate, keel and wings deflexed, narrow like a parrot's beak, all scarlet or rarely pink or white   **37. Clianthus**

b. Flowers not as above   120

120a. Leaves dotted with orange-brown glands   **48. Glycyrrhiza**

b. Leaves not dotted with orange-brown glands   121

121a. Pod indehiscent, more or less circular in outline, margins usually toothed, the sides often with toothed veins or pitted   **52. Onobrychis**

b. Pod not as above   122

122a. Keel toothed on the upper side; leaflets oblique at the base, or if narrow, then curved   **45. Oxytropis**

b. Keel toothed on the lower side or not toothed; leaflets symmetric at the base   **44. Astragalus**

1. CASTANOSPERMUM Cunningham
J. Cullen

Trees to 30 m in the wild, with smooth, grey to brown bark and spreading branches. Leaves odd-pinnate with 8–17 leaflets. Flowers long-stalked in racemes arising on the old (leafless) shoots. Calyx cup-shaped, very weakly and equally 5-toothed. Standard reflexed, the other 4 petals similar to each other. Stamens 10, free; anthers attached at their middles to filaments. Fruit a large pod containing 3–5 large, rounded, chestnut-like seeds.

A genus of a few species from N & E Australia, New Caledonia, the New Hebrides, etc. The one cultivated species is easily grown in rich, moist soils, but good drainage is essential. Propagation is by seed, which germinates rapidly.

1. C. australe Hooker. Illustration: Elliot & Jones, Encyclopaedia of Australian plants **2**: 478 (1982); Morley & Toelken,

Flowering plants in Australia, 154 (1983); Lewis et al., (eds.), Legumes of the world, 229 (2005).

Tree to 8 m in cultivation. Leaves dark green and glossy above. Racemes 5–15 cm, flower-stalks *c.* 2.5 cm. Flowers 3–4 cm, reddish or yellowish (var. **brevivexillum** F.M. Bailey). Pods 10–25 × 4–6 cm, woody, inflated. *Australia (Queensland, northern New South Wales).* G2.

Seedlings are grown as house plants in Australia.

2. BOLUSANTHUS Harms
J. Cullen

Tree to 15 m in the wild, often much smaller in cultivation. Bark grey. Branches arching, branchlets with brownish silky hairs. Leaves deciduous, pinnate with 9–15 leaflets including a terminal leaflet; leaflets slightly unequal-sided and sickle-shaped, shortly stalked, brownish silky hairy above and beneath, especially when young. Inflorescence a many-flowered raceme 14–20 cm; flower-stalks to 2 cm, silky hairy, each bearing a bract towards the base. Calyx cup-shaped, deeply 5-toothed, densely silky-hairy. Corolla dark blue, the standard to 1.5 cm, sometimes with a pale patch towards the base of the blade. Stamens 10, filaments free. Pod linear-oblong, flattened, to 7 × 1.2 cm, indehiscent, containing several seeds.

A genus of a single species from S & E Africa, requiring a warm, sunny site. Propagation is by seed (though this is slow).

1. B. speciosus (Bolus) Harms. Illustration: *Flowering Plants of South Africa* 1: pl. 23 (1921); Eliovson, South African wild flowers for the garden, edn 4, opposite 195 (1965); Wild flowers of South Africa, 67 (1980); Fabian & Germishuizen, Transvaal wild flowers, pl. 62 (1982).

South Africa (Transvaal, Natal, Swaziland), Botswana, Mozambique, Zimbabwe. G1–2.

3. MAACKIA Ruprecht & Maximowicz
J. Cullen

Deciduous trees. Leaves pinnate with up to 17 opposite leaflets and a terminal leaflet; axillary buds exposed. Flowers numerous, rather small, in terminal racemes or panicles; bracts at the bases of the flower-stalks. Calyx cylindric to bell-shaped, obliquely 5-toothed. Corolla whitish, blade of the standard reflexed upright. Stamens 10, their filaments united at the extreme

base only. Pods compressed, linear-oblong, dehiscent, containing 1–5 seeds.

A genus of about 6 species from E Asia. They are not widely grown, being less ornamental than *Cladrastis*, but are of easy cultivation, requiring good soil and a sunny position. Propagation is normally by seed, but root cuttings have also been used.

Literature: Takeda, H., Maackia and Cladrastis, *Notes from the Royal Botanic Garden Edinburgh* **8**: 95–104 (1913); Andrews, S., Trees of the year, Cladrastis and Maackia, International Dendrology Society Yearbook for 1996: 12–26 (1997).

1. M. amurensis Ruprecht (*Cladrastis amurensis* (Ruprecht) Koch). Illustration: *Botanical Magazine*, 6551 (1881); Csapody & Tóth, A colour atlas of flowering trees and shrubs, 131 (1982); Brickell (ed.), RHS A–Z encyclopedia of garden plants, 659 (2003).

Tree to 20 m in the wild, usually much smaller and shrubby in cultivation; bark peeling; young shoots minutely hairy at first. Leaves to 30 cm, with 7–11 ovate to elliptic leaflets, rounded to the base, tapered, sometimes rather bluntly, to the apex, hairy beneath when young, later often hairless. Flowers in stiffly erect racemes 10–15 cm. Petals whitish; keel and wings to 1.1 cm. Pods 3.5–5 cm, compressed, slightly winged. *E former USSR, N China, Korea, Taiwan.* H1. Summer.

Var. **buergeri** (Maximowicz) Schneider (*Cladrastis amurensis* var. *buergeri* Maximowicz). Illustration: Addisonia **3**: pl. 87 (1918). Plant generally smaller; leaflets persistently and densely hairy beneath. *Japan.* H3? Summer.

M. chinensis Takeda (*M. hupehensis* Takeda). Very similar to *M. amurensis* var. *buergeri* but leaflets 7–17, shiny silvery-hairy beneath when young, hairs persistent, apex rather obtuse, flowers in panicles 15–20 cm. *C China.* H3? Summer.

4. CLADRASTIS Rafinesque
J. Cullen

Trees or shrubs. Leaves pinnate with 7–13 alternate, shortly stalked leaflets including a terminal leaflet, the base of the leaf-stalk swollen and enclosing the axillary bud; stipels sometimes present. Flowers usually many in erect or pendent panicles; bracts at the bases of the flower-stalks. Calyx usually bell-shaped to cylindric with 5 short, equal teeth or lobes. Corolla white or

pinkish, sometimes marked with yellow. Stamens 10, filaments almost completely free. Pod membranous, flattened, sometimes winged, often somewhat constricted between the few seeds, indehiscent.

A genus of about 6 species, 1 from N America, the rest from E Asia. They require a rich soil and a sunny position, and may be propagated by seed or by root cuttings.

Literature: Takeda, H., Cladrastis and Maackia, *Notes from the Royal Botanic Garden Edinburgh* **8**: 95–104 (1913); Andrews, S., Trees of the year, Cladrastis and Maackia, International Dendrology Society Yearbook for 1996: 12–26 (1997).

1a. Leaflets green beneath, with small, linear stipels; pod conspicuously winged   **3. platycarpa**
  b. Leaflets grey-green beneath, without stipels; pod not conspicuously winged   2
2a. Panicles pendent; leaflets 7–9, 4–6 cm or more wide; calyx 7–8 mm   **1. kentukea**
  b. Panicles upright; leaflets 9–13, to 3.5 cm wide; calyx 4–5 mm   **2. sinensis**

1. C. kentukea (Dumont de Courset) Rudd (*C. lutea* (Michaux) Koch; *C. tinctoria* Rafinesque). Illustration: *Botanical Magazine*, 7767 (1901); Justice & Bell, Wild flowers of North Carolina, 99 (1968); Bean, Trees & shrubs hardy in the British Isles, edn 8 **1**: 631 (1970); *Il Giardino Fiorito* **48**: 564 (1982).

Tree to 20 m; bark smooth, wood yellow. Leaves with 7–9 ovate to broadly obovate leaflets 8–15 × 4–6 cm or more, apex rather abruptly tapered or acuminate, base rounded to tapered, dark green above, greyish green beneath, almost hairless. Panicles pendent, branches hairless or with sparse curled hairs. Flowers white, fragrant. Calyx cylindric to cup-shaped, 7–8 mm, with 5 rounded, equal, hairy teeth. Standard 2–2.5 cm. Pod 5–10 × 1–1.5 cm, not winged. *S & C USA.* H4. Early summer.

2. C. sinensis Hemsley. Illustration: *Notes from the Royal Botanic Garden Edinburgh* **8**: pl. 26 (1913); *Botanical Magazine*, 9043 (1924); Lewis et al., (eds.), Legumes of the world, 233 (2005).

Tree to 25 m or shrubby, bark brown, smooth. Leaves with 9–13 oblong to oblong-lanceolate leaflets

6–11 × 1.5–3.5 cm, long-tapered to the apex, rounded to the base, dark green and hairless above, greyish green and usually hairy, at least along the midrib, beneath. Panicles erect, branches hairy. Flowers white or pinkish. Calyx bell-shaped, 4–5 mm, equally 5-lobed with obtuse lobes, hairy all over, brownish. Standard 1–1.5 cm. Pod 5–10 cm × 7–11 mm, not winged. *W & C China*. H5. Summer.

3. C. platycarpa (Maximowicz) Makino (*Platyosprion platycarpum* (Maximowicz) Maximowicz). Illustration: Kurata, Illustrated important forest trees of Japan **1**: pl. 76 (1968); Kitamura & Murata, Coloured illustrations of woody plants of Japan **1**: pl. 71 (1977); Krüssmann, Manual of cultivated broad-leaved trees and shrubs **1**: 337 (1986).

Tree to 20 m, bark smooth, greyish. Leaves with 7–9 narrowly ovate to ovate-oblong leaflets 7.5–10 × 2.5–4 cm, tapered to acuminate at the apex, tapered to rounded at the base, green above, paler beneath, generally with some crisped hairs on stalks, margins and main veins, each leaflet with 2 linear stipels at the base. Panicles upright, many-flowered, densely hairy. Calyx bell-shaped, *c.* 6 mm, with 5 equal, obtuse teeth, densely hairy all over, brownish. Petals white, the standard *c.* 1.5 cm, with a yellow spot at the base. Pod *c.* 8 × 1.5 cm, conspicuously winged. *Japan*. H5. Summer.

5. SOPHORA Linnaeus
J. Cullen

Trees or shrubs. Leaves pinnate with up to 43 leaflets (including a terminal leaflet) which are generally opposite; stipules small; axillary buds small, exposed. Flowers bisexual, in terminal or axillary racemes or panicles, pea-flower-like (with blade of standard reflexed upright) or with all the petals forwardly directed; bracts borne at the bases of the flower-stalks. Calyx cup-shaped, truncate, usually obliquely so, with 5 small teeth. Stamens 10, filaments slightly united at the base. Pod stalked, terete, constricted between the seeds, sometimes longitudinally winged, often not dehiscing but breaking into 1-seeded segments.

A genus of about 50 species, widespread (though more than half of the cultivated species are from the southern hemisphere). It is very variable and has in the past been divided into several genera. Only a small number are generally cultivated. They require warm, sunny sites and rich soil, and may need the protection of a wall or glasshouse in northern parts. Propagation is generally by seed, but some species will root from cuttings of young wood which have a heel of older wood.

Literature: Heenan, P.B., de Lange, P.J. & Wilton, A.D., *Sophora* in New Zealand, *New Zealand Journal of Botany* **39**: 17–53 (2001).

Habit Very large trees: **1**; shrubs with short branches which terminate in spines: **3**; small shrubs with branches diverging at very wide angles: **7**.
Leaflets More than 21: **6**; 2.5–5 cm: **1**, **2**; less than 2.5 cm: **3**, **4**, **5**, **6**, **7**.
Inflorescence. Panicle: **1**; raceme with 4–10 flowers: **2**, **3**, **4**, **5**, **6**; raceme with 2–3 flowers: **7**.
Flowers. Pea-flower-like, with blade of standard reflexed upright: **1**, **2**, **3**; with all petals forwardly directed: **4**, **5**, **6**, **7**; white, cream, pinkish or bluish: **1**, **2**, **3**; yellow: **4**, **5**, **6**, **7**.
Standard: To 1.5 cm: **1**, **2**, **3**; more than 2.5 cm: **4**, **5**, **6**, **7**.
Pods. With up to 6 seeds: **1**, **2**, **3**, **4**; with 6 or more seeds: **5**, **6**, **7**; longitudinally winged: **5**, **6**, **7**.

1a. Flowers to 1.5 cm, pea-flower-like, the blade of the standard reflexed upright, white, cream, pinkish or bluish　　2
　b. Flowers 3 cm or more, all the petals forwardly directed　　4
2a. Plants with spines formed from the ends of short lateral shoots　　**3. davidii**
　b. Plants without spines　　3
3a. Large trees; inflorescence a panicle　　**1. japonica**
　b. Small trees to 6 m; inflorescence a raceme　　**2. affinis**
4a. Leaflets 1.8 cm or more; pod not winged　　**4. macrocarpa**
　b. Leaflets to 1.8 cm; pod longitudinally winged　　5
5a. Small shrub with branches diverging at very wide angles; racemes with 2–3 flowers　　**7. prostrata**
　b. Small trees or shrubs with branches diverging at narrow angles; racemes with 4–10 flowers　　6
6a. Leaflets 11–21; wings exceeding the standard　　**5. tetraptera**
　b. Leaflets 23–43; wings about as long as standard　　**6. microphylla**

1. S. japonica Linnaeus (*Styphnolobium japonicum* (Linnaeus) Schott). Illustration: *Botanical Magazine*, 8764 (1918); Polunin & Everard, Trees and bushes of Europe, 105 (1976); Csapody & Tóth, Colour atlas of flowering trees and shrubs, 135 (1982); Everett, New York Botanical Gardens illustrated encyclopedia of horticulture **9**: 3183 (1982).

Tree to 30 m with rounded crown and often contorted and twisted main branches; bark greyish and corrugated; young shoots green, downy. Leaves 15–25 cm with 7–17 lanceolate to ovate leaflets 2.5–5 cm, broadly tapered to rounded at the base, acute at apex, shortly stalked, dark green and hairless above, paler or glaucous and downy beneath. Flowers 1–1.5 cm in loose terminal panicles to 30 cm. Calyx 3–4 mm. Petals cream, white or marked with pinkish purple 'Violacea', the standard with blade reflexed upright. Pods 5–8 cm, shortly stalked, terete, hairless, constricted between the 1–6 seeds. *China, Korea*. H4. Late summer.

A widely planted tree often used to line avenues in the warmer parts of Europe. It is seldom grown further north, where it rarely flowers. Var. **pubescens** (Tausch) Bosse has narrower leaflets and is more densely hairy. Variants with drooping branches ('Pendula') or variegated leaves ('Variegata') are sometimes grown. A survey of currently available cultivars is provided by Schalk, P.H., *Sophora japonica*, *Dendroflora* **22**: 69–27 (1985).

Robinia pseudoacacia (p. 350) is often planted with *S. japonica* and is very similar to it when not in flower and fruit; however, the *Robinia* has dark reddish brown young shoots, leaflets with more rounded tips and, usually, paired, short spines at the base of each bud.

2. S. affinis Torrey & Gray (*Styphnolobium affinis* (Torrey & Gray) Walpers). Illustration: Elias, The complete trees of North America, 662 (1980).

Tree to 6 m, with rounded crown. Leaves 10–20 cm with 13–19 elliptic leaflets 2–4 cm, tapered to the base, somewhat rounded at the apex, hairless or slightly hairy beneath. Flowers 1–1.5 cm in axillary racemes to 15 cm. Calyx bell-shaped, downy. Petals white tinged with pink, the standard with blade reflexed upright. Pods 3–8 cm, black, downy, constricted between the few seeds. *S USA (Arkansas, Texas)*. H4. Early summer.

This and the previous species are now often separated off from *Sophora* into the genus *Styphnolobium* Schott, characterised by the white or cream corollas with the standard with a reflexed blade.

3. S. davidii (Franchet) Skeels (*S. viciifolia* Hance). Illustration: *Botanical Magazine*, 7883 (1903); Csapody & Tóth, Colour atlas of flowering trees and shrubs, 135 (1982); Brickell (ed.), RHS A–Z encyclopedia of garden plants, 993 (2003).

Rounded shrub to 3 m; branchlets brownish, downy, the ends of many short branches persisting as spines. Leaves 3–9 cm, of 11–17 leaflets; leaflets elliptic-oblong, 5–12 mm, rounded to the base, rounded or somewhat tapered to the mucronate, often somewhat notched apex, hairy when young. Flowers 1–1.5 cm in racemes terminating short branches. Calyx downy, cylindric to bell-shaped, 4–7 mm, shortly 5-toothed, often violet. Petals bluish white, standard with blade reflexed upright. Pods hairy, 3–6 cm × 3–5 mm, deeply but gradually constricted between the 1–4 distant seeds. *C & W China*. H4. Spring–summer.

4. S. macrocarpa J.E. Smith. Illustration: *Botanical Magazine*, 8647 (1916); Perry, Flowers of the world, 165 (1972).

Shrub or small tree; young brances densely downy. Leaves 7–10 cm, divided into 11–17 leaflets; leaflets oblong or lanceolate-oblong 1.8–2.5 (rarely more?) cm × 7–12 mm, broadly rounded at the base and apex, finely downy above and beneath. Flowers in short axillary racemes. Calyx downy, cup-shaped, obliquely truncate with 5 very small teeth, 1–1.4 cm. Petals yellow, all forwardly directed, standard 2.8–3.5 cm, wings and keel 2.5–3 cm. Pods downy, 7–12 cm, long-stalked, very deeply constricted between the 1–4 seeds, not winged. *Chile*. H5.

5. S. tetraptera Miller. Illustration: *Botanical Magazine*, 167 (1796); Everard & Morley, Flowers of the world, pl. 129 (1970); Johnson, The international book of trees, 211 (1973); Salmon, The native trees of New Zealand, 197–199 (1980).

Large shrub or small-tree to 12 m; bark greyish, young shoots hairy. Leaves 8–15 cm, with 11–21 leaflets; leaflets ovate to elliptic-oblong, tapered to the base, rounded or notched at the apex, 7–18 mm, silky hairy on both surfaces. Flowers bright to golden yellow in racemes of 4–10. Calyx bell-shaped, obliquely truncate with 5

small teeth, silky hairy, 1–1.5 cm. Petals all forwardly pointing; standard 3–3.5 cm, wings 3.7–4 cm, keel exceeding wings. Pod to 20 cm, stalked, very deeply constricted between the 6 or more seeds, with 4 wings extending the whole length of the pod. *New Zealand*. H5–G1. Spring.

A variant which flowers as a low shrub is cultivated in New Zealand (and perhaps elsewhere) as 'Gnome'.

6. S. microphylla Aiton. Illustration: *Botanical Magazine*, 1442 (1812); *Journal of the Royal Horticultural Society* **96**: f. 247 (1971); Salmon, The native trees of New Zealand, 200, 201 (1980); Lewis et al., (eds.), Legumes of the world, 226, 243 (2005).

Like *S. tetraptera* but a smaller tree with a distinct juvenile phase as a shrub with slender branches which diverge at narrow angles; leaves tending to droop, with 23–43 leaflets which are generally smaller (in var. **longicarinata** (Simpson) Allan the leaflets are numerous, small and broadly elliptic); standard 3–4 cm, wings about as long as the standard, keel longer. *New Zealand*. H5–G1. Spring.

Var. **fulvida** Allan. More densely hairy with brown hairs.

A very closely related plant occurs in temperate South America; it has been named **S. macnabiana** Graham or **S. microphylla** subsp. **macnabiana** (Graham) Yakovlev. Illustration: *Botanical Magazine*, 3735 (1840). Standard somewhat shorter than the wings. It may be in cultivation.

7. S. prostrata Buchanan. Illustration: Salmon, The native trees of New Zealand, 201 (1980).

Like *S. tetraptera* and *S. microphylla* but a low shrub to 2 m with branches diverging at very wide angles; leaves to 2.5 cm with *c.* 17 leaflets; flowers in racemes of 2 or 3, the keel to 3 cm, longer than the wings which are themselves longer than the standard. *New Zealand*. H5–G1. Spring.

6. PTEROCARPUS Jacquin
J. Cullen

Evergreen or deciduous trees; sap often red. Leaves pinnate, the leaflets several, opposite or not, terminal leaflets present; stipules often falling quickly, stipels absent. Flowers in racemes or panicles. Calyx funnel- or bell-shaped, shortly 5-lobed, the upper 2 lobes united for much of their length. Corolla yellow, orange, whitish (ours) or violet; wings longer than keel.

Stamens 10, filaments all united. Ovary sometimes shortly stalked, with few ovules. Pod compressed, indehiscent, the central part hardened and seed-bearing, winged. Seeds with small arils.

A tropical genus of about 20 species; only 1 is occasionally grown, and little is known of its requirements. Propagation is by seed.

1. P. angolensis de Candolle. Illustration: Palgrave, Trees of Central Africa, 330, 331 (1956); *Flora of Tropical East Africa, Leguminosae* **3**: 90 (1971); Palmer, A field guide to the trees of southern Africa, pl. 13 (1977); Lewis et al., (eds.), Legumes of the world, 322 (2005).

Deciduous tree to 20 m, with open crown (rarely a shrub); bark grey, fissured. Young shoots grey- or silver-hairy. Leaves to 35 cm with 11–19 lanceolate to elliptic, shortly acuminate leaflets, generally persistently hairy beneath. Racemes often appearing before the leaves; flowers scented. Calyx 8–10 mm, hairy. Corolla 1.6–2 cm, golden yellow to orange. Fruit almost circular, 1–1.5 cm, stalked, with feathery bristles on the central, seed-bearing part. *Tropical Africa*. G2.

7. ABRUS Adanson
J. Cullen

Woody climbers, scramblers or low shrubs. Leaves pinnate without a terminal leaflet, with numerous, opposite lateral leaflets; stipules small, persistent; stipels small, linear, usually present. Flowers aggregated on short, wart-like outgrowths in racemes or spikes; bracts and bracteoles small, often falling early. Calyx funnel-shaped, truncate with 5 small teeth. Corolla pale purple to yellowish; standard notched at the apex, keel longer than wings. Stamens 9, the filaments all united and united at the base to the standard. Ovary with many ovules, downy; style curved, persistent. Pods oblong to linear, irregularly swollen or flattened, beaked. Seeds with arils.

A genus of 4 (Breteler) or more than 15 (Verdcourt) species from tropical regions. Only 1 is grown, mainly for the sake of its seeds, which, though extremely poisonous, have often been used as beads. It is easily grown and propagated by seed or by cuttings.

Literature: Breteler, F.J., Revision of Abrus Adanson with special reference to Africa, *Blumea* **10**: 607–624 (1960); Verdcourt, B., Studies in the Leguminosae-Papilionoideae for the Flora of East Tropical

Africa 2: a reappraisal of the species of the genus Abrus Adanson, *Kew Bulletin* **24**: 235–253 (1970).

1. A. precatorius Linnaeus. Illustration: Bentley & Trimen, Medicinal plants **2**: pl. 77 (1877); Marloth, The flora of South Africa **2**: pl. 29 (1925); *Blumea* **10**: 618 (1960); Lewis et al., (eds.), Legumes of the world, 388, 390, 391 (2005).

Leaves with 8–17 pairs of leaflets; leaflets oblong to slightly obovate, rounded to the base and the mucronate apex, 6–25 × 3–10 mm, with sparse adpressed whitish hairs. Calyx *c.* 3 mm. Corolla *c.* 1.5 cm, reddish, purplish or yellowish. Pod oblong, compressed, hairy, sometimes warty, 2–5 × 1–1.5 cm. Seeds 5–7 × 4–5 mm, scarlet with a black blotch around the hilum, glossy. *Tropics.* G2.

Verdcourt divides the species into 2, subsp. **precatorius** with hairy pods and subsp. **africanus** Verdcourt, with pods both hairy and warty (mainly African); both may be in cultivation.

8. DERRIS Loureiro

J. Cullen

Evergreen (ours) or deciduous woody climbers. Leaves odd-pinnate with more or less opposite lateral leaflets and a terminal leaflet; stipules falling early, stipels absent. Flowers numerous in racemes or panicles. Calyx broadly bell-shaped, truncate, with 5 short teeth. Corolla white, pink or yellowish. Stamens 10, filaments all united into a tube (in the cultivated species). Ovary hairy, several-seeded. Pod indehiscent, compressed, hairy, few-seeded, somewhat winged along both sutures.

A genus of about 60 species from the tropics; only a few are grown, mostly as a source of the insecticide 'Derris' (rotenone).

1. D. elliptica (Wallich) Bentham. Illustration: *Botanical Magazine*, 8530 (1913) – as *D. oligosperma*.

Large, vigorous, evergreen climber with blackish stems and hairy young shoots. Leaves 20–40 cm, with 9–13 leaflets; leaflets obovate-oblong, acute, persistently brown-hairy beneath. Racemes to 25 cm, many-flowered, axis and branches hairy. Calyx to 8 mm, hairy. Corolla pink or rarely white, to 1.2 cm; standard with 2 thickenings at the base of the blade, hairy on the back; keel hairy towards the tip. Pod hairy with 1–4 seeds. *Burma to Malaysia, cultivated elsewhere.* G2.

D. malaccensis Prain. Illustration: Duke, Handbook of legumes of world

economic importance, 76 (1981). Very similar but leaves and petals hairless. *Malaysia to New Guinea.* G2.

Both species are grown commercially in the tropics as a source of rotenone.

9. MUNDULEA (de Candolle) Bentham

J. Cullen

Small trees or shrubs. Leaves pinnate, leaflets opposite or almost so, terminal leaflet present; stipules small, triangular, stipels absent. Flowers in terminal racemes. Calyx bell-shaped, 5-toothed. Corolla bluish purple, standard silky-hairy outside; wings and keel downy on the margins near the base. Stamens 10, filaments of 9 united into a tube from the base, the uppermost free at the base, united to the others higher up. Ovary with several ovules, hairy. Pod densely hairy, several-seeded, indehiscent, often constricted between the seeds. Seeds without arils.

A genus of about 15 species from Africa to Malaysia, the rest from Madagascar, of which only the non-Madagascan species is grown. Little is known of its requirements in Europe. Propagation is by seed.

1. M. sericea (Willdenow) Chevalier. Illustration: *Flora of Tropical East Africa*, Leguminosae Part **3**: 156 (1971); Palmer & Pitman, Trees of southern Africa **2**: 920, 921 (1972); Palgrave, Trees of southern Africa, pl. 100 (1977).

Tree or shrub to 7 m; bark smooth or fissured. Leaves to 10 cm, leaflets 13–17, ovate to lanceolate, hairy at least beneath. Flowers in pairs in hairy racemes. Calyx *c.* 6 mm, the upper 2 teeth united for most of their length. Standard to 2 cm. Pod yellowish brown, velvety. *S & tropical Africa, India, Malaysia.* G2.

10. PISCIDIA Linnaeus

J. Cullen

Deciduous trees. Leaves pinnate with 5–9 opposite leaflets and a terminal leaflet; stipules large, falling early, stipels absent. Flowers in dense, hairy, axillary panicles. Calyx 5-lobed, the upper 2 lobes united for a greater part of their length than the others. Corolla pink, white (sometimes striped with pink or red) or red, standard hairy on the back. Stamens 10, 9 of the filaments united at the base and for most of their length, the uppermost free at the base, united to the others higher up; anthers versatile. Pod 4-winged, indehiscent, containing 3–7 seeds.

A genus of about 10 species in subtropical and tropical America. Only 1 is grown; it can be propagated by seed or by cuttings.

1. P. piscipula (Linnaeus) Sargent. Illustration: Fawcett & Rendle, Flora of Jamaica **4**(2): 83 (1920); Correll & Correll, Flora of the Bahama Archipelago, 676 (1982).

Tree to 10 m. Leaves to 25 cm, with elliptic-ovate, acute or obtuse leaflets, usually hairless above, minutely hairy beneath. Flowers appearing before the leaves. Corolla to 1.2 cm, hairy, standard 1.3–1.5 cm, more or less circular, notched. Pod stalked, 3–8 cm, wings lobed and crisped. *USA (Florida), West Indies.* G2.

11. TEPHROSIA Persoon

J. Cullen

Perennial herbs or shrubs, usually conspicuously hairy. Leaves pinnate with 5–31 leaflets (in the cultivated species) including a terminal leaflet; stipules small, deciduous or persistent, stipels absent. Flowers in terminal or leaf-opposed racemes. Calyx 5-lobed, the upper lobes united for more of their length than the others. Corolla small to large, standard hairy on the back, wings and keel hairy or not. Stamens 10, with 9 of the filaments united at the base and for most of their length, the uppermost filament usually free at the base but united to the others above. Ovary with 4–16 ovules, usually hairy; style hairy or hairless. Pod linear to oblong, usually compressed, hairy at least on the sutures, the walls often cracking obliquely with age. Seeds without arils.

A genus of about 400 species mostly from the tropics, a few extending into temperate areas. Only a few are grown as ornamentals (though others are grown as green manures or for fish poisons in some areas), are of easy culture and propagated by seed.

1a. Style hairy throughout its length; standard 1.5 cm or more   2
  b. Style hairless (rarely hairy just at the extreme base); standard to 1 cm   3
2a. Pods and ovary hairy only along the sutures; standard pink inside, orange outside   **1. grandiflora**
  b. Pods and ovary hairy all over the surface; standard yellow to cream outside, cream to white inside   **2. virginiana**
3a. Leaves white-hairy beneath; pods with short, erect hairs   **3. purpurea**

b. Leaves hairy but green beneath; pods with adpressed hairs

4. capensis

1. T. grandiflora (Aiton) Persoon. Illustration: Batten & Bokelmann, Wild flowers of the Eastern Cape Province, pl. 70 (1966); Fabian & Germishuizen, Transvaal wild flowers, pl. 61b (1982).

Shrub 50–200 cm; young shoots with adpressed hairs. Leaves with 7–17 leaflets; leaflets narrowly elliptic to narrowly obovate, tapered to the base, abruptly rounded to the blunt or notched, mucronate apex, adpressed hairy on both surfaces, densely so, giving a whitish appearance beneath. Calyx deeply 5-lobed, densely adpressed hairy, tube 3–5 mm, lobes 4–7 mm. Corolla pink, standard orange outside, densely hairy, 1.8–2.5 cm. Style hairy throughout its length. Pod compressed, oblong, hairy only on the sutures. Seeds 9–16. *South Africa, naturalised elsewhere.* H5–G1.

2. T. virginiana (Linnaeus) Persoon. Illustration: Gleason, Illustrated flora of the north-eastern United States and adjacent Canada **2**: 413 (1952); Rickett, Wild flowers of the United States **1**: 281 (1966); Justice & Bell, Wild flowers of North Carolina, 102 (1968).

Low shrub; young shoots with whitish spreading hairs. Leaves with 15–37 leaflets; leaflets narrowly elliptic to obovate, tapered to the base, tapered or somewhat rounded to the mucronate apex, hairless above (except along the margins), with spreading, often somewhat curled hairs beneath. Calyx deeply 5-lobed, densely covered with whitish spreading hairs, tube 2–3 mm, lobes 3–6 mm. Standard 1.5–2 cm, lemon-yellow to cream and densely hairy outside, cream to white inside; wings and keel usually pink, hairy. Style hairy throughout its length. Pods hairy all over the surface. Seeds 6–11. *E & C North America.* H3. Spring–summer.

T. vogelii J.D. Hooker. Illustration: *Transactions of the Linnean Society of London* **29**: t. 31 (1873); Lewis et al., (eds.), Legumes of the world, 386 (2005). Similar, but leaflets 11–31; upper calyx-lobes united for most of their length; standard *c.* 2.5 cm; pod 10–12 × 1–1.3 cm. *Tropical Africa.* G2.

T. candida de Candolle. Similar to *T. vogelii* but standard, wings and keel white, pod 6–9 cm × 7–9 mm. *Tropics.* G2.

3. T. purpurea (Linnaeus) Persoon. Illustration: *Flora of East Tropical Africa, Leguminosae*, **3**: 187 (1971).

Many-stemmed shrub to 1.3 m; young shoots whitish with silky, adpressed hairs. Leaves with 7–15 leaflets; leaflets obovate, gradually tapered to the base, abruptly rounded to the blunt, truncate or notched, mucronate apex, adpressed silky hairy, densely so and whitish beneath. Calyx deeply 5-lobed, tube 1.5–3 mm, teeth 2–4 mm, adpressed hairy. Corolla purplish red; standard 5–10 mm, densely hairy outside, wings and keel densely hairy. Style hairless. Pods somewhat flattened, with erect, short hairs all over the surface. Seeds up to 6. *Tropics.* G2.

A variable species, divided into several subspecies by some authors; it is uncertain which of these are in cultivation.

4. T. capensis Persoon. Illustration: Kidd, Wild flowers of the Cape peninsula, 107 (1983); Burman & Bean, Hottentots Holland to Hermanus, 117 (1985).

Many-stemmed slender shrub; young shoots with adpressed hairs. Leaves with 9–15 leaflets; leaflets very narrowly elliptic to narrowly obovate, long-tapered to the base, rather abruptly rounded to the blunt, shortly mucronate apex, hairless above, adpressed hairy beneath. Calyx deeply 5-lobed, tube 2–2.5 mm, lobes 2–2.5 mm, all adpressed hairy. Corolla bright pink or purplish; standard 6–10 mm, adpressed hairy outside; wings and keel not hairy. Style hairless except at the extreme base. Pods flattened, with adpressed hairs all over the surface, containing *c.* 6 seeds. *South Africa, Mozambique.* G1.

12. WISTERIA Nuttall
J. Cullen

Large, deciduous woody climbers. Leaves pinnate with up to 19 opposite lateral leaflets and a terminal leaflet; stipules falling early, narrow stipels present; leaflet-stalks somewhat swollen. Flowers numerous in often large, pendent racemes. Calyx bell-shaped or somewhat cylindric, obliquely 5-lobed, the 2 upper lobes often united for most of their length. Corolla blue, purple, lilac, pink or white; standard large, usually with 2 humps at the base of the blade. Stamens 10, 9 of them with their filaments united, that of the uppermost free. Ovary with several ovules. Pod large, compressed, with several seeds.

A genus of 6 species from N America and E Asia, most of them valued as ornamental climbers for growing against walls, on arches or in arbours. Though climbing, some of them can be grown as standards by careful pruning. They are generally easily grown, requiring a sunny site and appropriate support. Propagation by seed is possible but not recommended, as many inferior variants will be produced; selected variants are best propagated by grafting.

Literature: McMillan-Browse, P., Some notes on members of the genus *Wisteria*, *The Plantsman* **6**: 109–122 (1984), which gives much detail on selected clones and on propagation techniques; Valder, P., Wisterias: a comprehensive guide (1995).

1a. Flower-stalks more than 1 cm, usually flexuous; pods hairy   2
  b. Flower-stalks up to 1 cm, usually stiff, ascending; pods hairless   4
2a. Leaflets persistently silky-hairy on both surfaces   **3. venusta**
  b. Leaflets becoming hairless or almost so on both surfaces   3
3a. Stems twining clockwise; leaflets 13–19; flowers opening serially from the base of the raceme, usually when the leaves are fully expanded   **1. floribunda**
  b. Stems twining anticlockwise; leaflets 7–13, usually 9–11; flowers in the raceme all opening at more or less the same time, before the leaves are fully expanded   **2. sinensis**
4a. Racemes 4–10 cm, without glands   **4. frutescens**
  b. Racemes 20–35 cm, with glands on axis and flower-stalks   **5. macrostachya**

1. W. floribunda (Willdenow) de Candolle (*W. brachybotrys* Siebold & Zuccarini). Illustration: *Botanical Magazine*, 7522 (1897); Rose, Climbers and wall plants, t. 47 (1982); *The Plantsman* **6**: 117 (1984); Lewis et al., (eds.), Legumes of the world, 366 (2005).

Large woody climbers to 8 m or more, twining clockwise. Leaflets 13–19, narrowly ovate or elliptic to lanceolate, rounded to the base, acuminate towards the ultimately blunt apex, hairless above, hairless or very sparsely hairy beneath (except when young); leaves usually fully expanded as flowering begins. Racemes 20–50 cm or more (to 1.5 m in some cultivars), with many fragrant flowers. Flower-stalks 1–3 cm, flexuous and spreading; flowers opening serially from the base of the raceme. Calyx

4–6 mm, silky-hairy. Corolla white, blue, violet or pink, standard 1.4–2 cm, broadly oval. Pod 10–15 cm, narrowed towards the base, finely downy. *Japan.* H3. Late spring–early summer.

Introduced in the form of various clones differing in flower-colour and size of raceme; new clones have been regularly produced since. The most striking are: 'Alba' (flowers white), 'Multijuga' (racemes very long), 'Rosea' (flowers pink), 'Violacea Plena' (flowers violet, double) and 'Issai' (flowers opening before the leaves are fully expanded) the latter perhaps of hybrid origin.

2. W. sinensis (Sims) Sweet. Illustration: *Botanical Magazine*, 2083 (1819) – as *Glycine sinensis*; Menninger, Flowering vines of the world, f. 107 (1970); Perry, Flowers of the world, 162 (1972); *The Plantsman* **6**: 117 (1984).

Very similar to *W. floribunda* but less vigorous, climbing anticlockwise; leaflets 7–13 (usually 9–11), densely hairy at first, later more or less hairless; flowers in the raceme opening more or less simultaneously before the leaves are fully expanded. *China.* H4. Spring.

Variable, with numerous selected clones available. The species name is often misspelled 'chinensis'.

W. × formosa Rehder is the hybrid between *W. sinensis* and *W. floribunda*; it is generally intermediate between its parents, but all the flowers in the raceme open more or less simultaneously.

3. W. venusta Rehder & Wilson. Illustration: *The Plantsman* **6**: 113 (1984).

Climbing to about 10 m. Leaflets 9–13, ovate to elliptic or oblong to oblong-lanceolate, persistently hairy on both surfaces, the hairs producing a silvery sheen. Racemes downy, 10–15 cm; flower-stalks more than 1 cm, flexuous, spreading, downy. Flowers white or pinkish (purple in 'Violacea'), of similar size to those of *W. floribunda*. Pod 15–20 cm, hairy. *Japan.* H3. Summer.

First described from cultivated, white-flowered plants from Japan; the common wild-type there has purple flowers. A variant with double, white flowers ('Plena') has been grown.

4. W. frutescens (Linnaeus) Poiret. Illustration: *Botanical Magazine*, 2103 (1819) – as *Glycine frutescens*; Dean, Mason & Thomas, Wild flowers of Alabama, 89 (1983).

Climbing to 12 m; young shoots hairless or almost so. Leaves with 9–15 leaflets which are narrowly ovate or lanceolate to elliptic, tapered or somewhat rounded at the base, acuminate at the apex, sparsely hairy along the main veins above, hairy and pale beneath, margins more densely hairy. Racemes 4–10 cm, dense, axis and flower-stalks hairy; flower-stalks to 1 cm, stiff, ascending. Calyx cylindric to bell-shaped, densely hairy, 6–9 mm, the upper teeth very short. Corolla lilac-purple, often with a yellowish spot at the base of the standard; standard 1.5–2 cm. Pod hairless, compressed, 5–10 cm. *SE USA.* H4.

A white-flowered variant ('Nivea') has been grown.

5. W. macrostachya (Torrey & Gray) Torrey & Gray. Illustration: Loughmiller & Loughmiller, Texas wild flowers, 140 (1984).

Slender climber to 8 m; young shoots hairless. Leaflets usually 9, ovate to elliptic or lanceolate, hairy beneath when young, rounded to almost cordate at the base, acuminate at the apex. Racemes dense, 20–35 cm, axis and flower-stalks with glandular hairs; flower-stalks to 1 cm, stiff, ascending. Corolla lilac-purple. Pod hairless, somewhat constricted between the seeds. *SE USA.* H4. Summer.

13. ROBINIA Linnaeus
J. Cullen

Trees or shrubs. Leaves pinnate with opposite or somewhat alternate leaflets and a terminal leaflet; stipules often persistent as paired spines; stipels present, small, needle-like. Flowers in erect or pendent racemes; bracts present, soon falling. Calyx bell-shaped, unequally 5-toothed. Corolla white to pink or pale purple; standard broad, reflexed upwards, keel-petals united to each other at the base. Stamens 10, 9 of them with united filaments, that of the uppermost free or partly so. Pod flattened, bristly or not, rarely winged along the upper suture, containing several seeds.

A genus of 4 species (see below) from eastern and southern USA. Traditionally many more species have been recognised, but recent work (cited below) indicates that only 4 have any real structural basis, and even these hybridise (in the wild) to some extent.

The plants are easy to grow, particularly *R. pseudoacacia*, and will tolerate quite poor soils (*R. pseudoacacia* tends to grow very rapidly and coarsely in good soils). They

are propagated by seed, by suckers (root division), or, in the case of cultivars and those species or varieties that rarely form ripe pods, by grafting on to *R. pseudoacacia* stocks.

Literature: Isely D. & Peabody, F.J., *Robinia* (Leguminosae: Papilionoideae), *Castanea* **49**(4): 187–202 (1984).

1a. Large trees; flowers usually white (pink in some cultivars), standard 1.5–2 cm; pods hairless; racemes without glands   **1. pseudoacacia**
  b. Small trees or shrubs; flowers usually pink to purple; pods sparsely to densely bristly; racemes usually with glands on axis and flower-stalks   **2**
2a. Leaves persistently hairy on both surfaces   **2. neomexicana**
  b. Leaves ultimately hairless at least above   **3**
3a. Leaflets mostly 13–21, persistently hairy beneath; plant not conspicuously bristly   **4. viscosa**
  b. Leaflets 9–13, becoming hairless beneath; plant often conpsicuously bristly   **3. hispida**

1. R. pseudoacacia Linnaeus. Illustration: Justice & Bell, Wild flowers of North Carolina, 101 (1968); Edlin & Nimmo, The world of trees, 60 (1974); Polunin & Everard, Trees and bushes of Europe, pl. Ill, 112 (1976); *Il Giardino Fiorito* **48**: 281 (1982).

Usually large trees with dark brown, fissured bark; young shoots hairless or sparsely hairy. Leaves with 11–23 leaflets (reduced to a large terminal leaflet and 0–2 pairs of small lateral leaflets in 'Unifoliola' also known as 'Monophylla'); leaflets lanceolate, ovate, elliptic or oblong-elliptic, rounded at the apex, tapering to rounded at the base, shortly stalked, hairless or with some fine soft hairs beneath at maturity; stipules persisting as triangular spines (absent in 'Inermis'). Flowers numerous, white, fragrant, in long, pendent racemes. Calyx to 8 mm. Standard 1.5–2 cm, yellow-blotched at the base of the blade. Pod oblong, flattened, hairless, narrowly winged along the upper suture. *E USA.* H2. Spring.

A very variable tree in which numerous cultivars (Varieties) have been named, based on crown-shape, leaf-coloration, shape and texture and habit. It is very widely planted in C & S Europe, and naturalised there, and is used (often together with the superficially very similar

Sophora japonica, p. 346) as a street or avenue tree.

'Decaisneana' (*R. pseudoacacia* var. *decaisneana* Carriere). Illustration: *Addisonia* **19**: pl. 624 (1935–36). Stipule-spines reduced and pink flowers; widely grown, as is 'Bella Rosea' with glandular branches and pink flowers. Hybridises with other species of the genus, and these hybrids are sometimes grown:

R. × slavinii Rehder (*R. margarettae* Ashe; *R. pseudoacacia* × *R. hispida*.) Shrub or small tree with leaflets rather variable in number, bristly inflorescences and large, pink flowers and warty pods. It occurs sporadically in the wild as well as in gardens.

R. × holdtii Beissner. Thought to be *R. pseudoacacia* × *R. neomexicana* and possibly including back-crosses to *R. pseudoacacia*. Like *R. pseudoacacia* but with keel and wings white, the standard pinkish, pods glandular-bristly.

R. × ambigua Poiret; (*R. pseudoacacia* × *R. viscosa*). Shrub with glandular twigs and whitish or deep pink corolla.

2. R. neomexicana Gray (*R. luxurians* (Dieck) Silva-Taroucana & Schneider). Illustration: Martin & Hutchins, Summer wild flowers of New Mexico, 145 (1986).

Shrub or small tree to 10 m; young shoots downy at first. Leaves with 13–25 leaflets; leaflets lanceolate, narrowly ovate or oblong, rounded at the base and apex, permanently downy on both surfaces. Stipule-spines straight, slender, to 1 cm. Calyx 8.5–10 mm, downy and glandular. Corolla pink, standard 2–2.7 cm. Pod covered with glandular bristles. *S USA (New Mexico, Arizona).* H4? Summer.

3. R. hispida Linnaeus.

Shrub to 3 m, suckering widely, branches usually with spreading bristles to 1.5 mm, these glandular when young, and fine, soft white hairs. Leaflets 9–13, ovate to broadly elliptic to oblong, rounded to the base, rounded or somewhat acute at the apex, hairless on both surfaces when mature. Racemes rather few-flowered, axis and flower-stalks with glandular bristles and soft, whitish hairs. Calyx 0.9–1.4 cm. Standard 2.8–4 cm, pink to rose purple, as are wings and keel. Pods rarely formed (see below), densely bristly. *E USA.* H2. Summer.

According to Isely & Peabody, this species consists mainly of triploid, usually sterile clones, which rarely produce fruit,

but spread by suckering. Because of this, it is very variable and divided into several varieties, of which all but var. *kelseyi* are known in the wild.

Var. **hispida**. Illustration: Justice & Bell, Wild flowers of North Carolina, 101 (1968); Hay & Synge, Dictionary of garden plants in colour, pl. 1836 (1969); Everard & Morley, Wild flowers of the world, pl. 153 (1970); *Gartenpraxis* **1987**: 10. With conspicuous, spreading bristles on young shoots and leaves as well as on the raceme-axis and flower-stalks; fruit very rarely produced.

Var. **nana** (Elliott) de Candolle (*R. ellottii* (Chapman) Small). Low shrubs to 1 m, with spreading bristles only on the raceme-axis and flower-stalks; leaflets usually less than 1.5 cm; stipular spines usually conspicuous; fruit very rarely formed.

Var. **rosea** Pursh (*R. boyntonii* Ashe). Shrub or small tree more than 1 m, with spreading bristles only on raceme-axis and flower-stalks; leaflets usually 1.5–3 cm; stipular spines reduced or often absent; fruit very rarely produced.

Var. **fertilis** (Ashe) Clausen (*R. fertilis* Ashe). Shrub to 3 m with dense, spreading bristles on young shoots and leaves; leaflets mostly 1.2–1.8 times as long as broad; fruit regularly produced.

Var. **kelseyi** (Hutchinson) Isely. Illustration: *Botanical Magazine*, 8213 (1908); *Journal of the Royal Horticultural Society* **36**: f. 134 (1911); *Addisonia* **1**: t. 3 (1916). Shrub to 3 m, with bristles only on raceme-axis and flower-stalks; leaflets mostly 1.8–2.4 times as long as broad; fruit regularly produced.

4. R. viscosa Ventenat.

Shrubs or small trees to 12 m; young shoots with stalked or stalkless glands and soft, white, sometimes stellate hairs. Leaves with 15–21 leaflets; leaflets lanceolate-oblong to ovate-oblong, rounded to the base and apex, persistently hairy beneath. Racemes glandular and sticky, with 10–25 flowers. Calyx 7–10 mm; corolla pink to pinkish purple, standard 2–2.5 cm. Pods infrequent, glandular-bristly. *SE USA.* H3. Summer.

Var. **viscosa**. Illustration: Gleason, Illustrated flora of the north-eastern United States and adjacent Canada **2**: 414 (1952). Young shoots and racemes with flat, stalkless or almost stalkless glands (stalked glands present in the raceme).

Var. **hartwigii** (Koehne) Ashe (*R. hartwigii* Koehne). Illustration: *Gardeners's*

Chronicle **90**: 389 (1931). Young shoots with conspicuously stalked glands; racemes with stalked glands. Pods more frequently formed than in var. *viscosa*.

14. SESBANIA Scopoli

H.S. Maxwell & J. Cullen

Short-lived shrubs or small trees. Leaves pinnate, terminating in a short extension of the axis (rarely with a terminal leaflet), leaflets often numerous; stipules present, soon falling; stipels small or absent. Flowers usually in axillary racemes. Calyx broadly bell-shaped, truncate or oblique, usually weakly toothed. Petals long-clawed; standard spreading or reflexed, often notched at the apex, often with 2 appendages on the inner surface towards the base. Stamens 10, 9 of them with their filaments united, that of the uppermost free. Pod long, thin, sometimes winged. Seeds numerous, without arils.

A genus of about 50 species from the tropics and subtropics, often found in seasonally damp places. Only a small number are grown. Propagation is by seed.

1a. Calyx 1.7–2.5 cm; standard 7.5–10 cm **1. grandiflora**
  b. Calyx and standard much smaller **2**
2a. Leaflets 14–20; pod 5–10 cm with 4 conspicuous wings **2. punicea**
  b. Leaflets 20–40; pod 6.5–25 cm, thin, unwinged **3. sesban**

1. S. grandiflora (Linnaeus) Poiret.

Small, soft-wooded tree to 10 m. Leaflets 30 or more, oblong or slightly obovate-oblong, tapered or rounded to the base, rounded to the sometimes notched apex; axis of leaf hairless, leaflet-stalks and often the undersurface of the leaflets hairy. Flowers few. Calyx broadly bell-shaped, oblique, scarcely toothed, hairless, 1.7–2.5 cm. Corolla white or red; standard reflexed, ovate-oblong, 7.5–10 cm; keel parrot's beak-shaped, 9.5–13 cm. Pod 30 cm or more, flattened, ridged on the corners, very slightly constricted between the numerous seeds. *Areas around the Indian Ocean, from Mauritius to India, Sri Lanka and further east.* G2. Summer.

2. S. punicea (Cavanilles) Bentham (*S. tripetii* invalid). Illustration: *Botanical Magazine*, 7353 (1894); Herter, Flora illustrada del Uruguay, t. 1607 (1954); Menninger, Flowering trees of the world, t. 284, 285 (1962) as – *Daubentonia punicea* and *D. tripetii*; Cabrera, Flora de la provincia de Buenos Aires **3**: 499 (1967).

Shrub or small tree. Leaflets 14–20, oblong or slightly obovate-oblong, rounded or tapered to the base, rounded to the mucronate apex, very slightly hairy beneath. Flowers *c.* 10 in axillary racemes. Calyx bell-shaped, truncate, wavy-toothed, hairless, 5–6 mm. Corolla bright reddish purple; standard reflexed, 1.5–2 cm; keel parrot's beak-shaped, 2–2.5 cm. Pod 5–10 cm, conspicuously and broadly 4-winged, slightly constricted betweeen the seeds. *S Brazil, Uruguay, Argentina.* G2.

3. S. sesban (Linnaeus) Merrill
(*S. aegyptiaca* Poiret).

Much–branched shrub to 3 m. Leaflets 20–40, oblong, parallel-sided, rounded to the base and apex, sparsely hairy to almost hairless on both surfaces, glaucous. Flowers few to many. Calyx broadly bell-shaped, truncate, weakly 5-toothed, 4–7 mm. Standard 1–1.6 cm, notched at apex, yellow or partly or completely suffused with purple; wings and keel mostly yellow, keel 1.3–1.8 cm. Pod 6.5–25 cm, linear-cylindric, thin, twisted, often somewhat to deeply constricted between the seeds, not winged. *Old World tropics.* G2. Summer.

Very variable in flower colour and pod size and degree of constriction. Many varieties have been named, but it is uncertain which of these have been cultivated.

15. INDIGOFERA Linnaeus
B.D. Schrire & S. Andrews
Shrubs (ours). Hairs usually medifixed (i.e. attached at a point along their length), often at the middle, and mostly adpressed, or with unequal branches, when usually spreading. Leaves with a terminal leaflet and 7–27 opposite lateral leaflets (ours); stipules small to large, persistent or soon falling; stipels usually present, small. Racemes axillary, few- to many-flowered, usually borne on the current year's growth, more rarely borne on older wood; bracts usually small. Calyx bell-shaped, deeply 5-toothed with the lowermost tooth usually the longest, more rarely truncate and scarcely toothed. Corolla usually pink to purplish, rarely white or yellowish. Standard generally obovate, reflexed upwards, variously hairy outside. Wings forming a horizontal plane plane above the keel, variously hairy towards the base. Keel with a projection of spur on each side, usually hairy at least along the suture. Stamens 10, 9 of them with the filament sunited, the uppermost free (ours); anthers

appendaged at their tips, sometimes hairy. Ovary several-seeded. Pod variable, usually linear-cylindric, scarcely compressed, mor rarely compressed or constricted between the seeds, always with thin partitions between the individual seeds.

A genus of about 750 species from the tropics and subtropics. Only a small number is grown. The flowers have an explosive polllination mechanism, involving the decoupling of the wings from the 2 backwardly directed outgrowths or spurs of the keel, which causes a discharge of the pollen; following this, the keel tends to reflex downwards and backwards and then falls, together with the wings, so that most herbarium specimens lack these organs.

The species are generally easily grown and valued for their late, pink to purple flowers. They are propagated by seed or by cuttings of semi-ripe wood rooted ini a heated case.

Literature: Schrire, B. D. Tribe Indigofereae. In G. Lewis et al., (eds.), Legumes of the world, 361–365 (2005).

1a. Racemes spreading, descending or
   pendulous   2
 b. Racemes erect or ascending   6
2a. Corolla 1.3–1.8 cm; stamens
   1,1–1.4 cm   3
 b. Corolla 8.5–12 mm; stamens
   6.5–10 mm   4
3a. Standard hairless on the back except
   for the downy margins; terminal
   leaflet broadly obovate to rhombic-
   elliptic, apex rounded to obtuse
                                  1. kirilowii
 b. Standard hairy on the back; terminal
   leaflet ovate-lanceolate, acute
                                     2. decora
4a. Calyx-lobes shorter than the tube;
   leaflets, ovary and pods hairless;
   terminal leaflet 8–30 mm wide;
   racemes loosely 10–30-flowered
                                    3. fortunei
 b. Calyx-lobes more or less equalling to
   up to twice the length of the tube;
   leaflets, ovary and pods hairy;
   terminal leaflet 3–10 mm wide;
   racemes densely 30–100-flowered   5
5a. Racemes pendulous, to 30 cm;
   standard densely white-downy on
   the back,bluish pink; leaflets 13–27
                                     4. pendula
5b. Racemes spreading more or less
   horizontally with tips ascending;
   standard thinly bristly-downy, bright
   reddish pink; leaflets 7–13 (rarely to
   17)   **5. howellii**

6a. Standard dark crimson to rust-
   coloured, bronze or maroon   7
 b. Standard pink, rose or violet   8
7a. Bracts less than 1 mm wide,
   lanceolate-subulate, not as below;
   terminal leaflet 4–10 mm wide;
   leaflet-pairs 5–15 mm apart; leaf-
   stalk 5–15 (rarely to 20) mm; ovary
   and pod hairy   **6. szechuensis**
 b. Bracts 1.5–3 mm wide, cuspidate at
   the apex, boat-shaped and entirely
   enclosing young flower-buds;
   terminal leaflet 1.5–4.5 cm wide;
   leaflet-pairs more than 2 cm apart;
   leaf-stalk 2.5–5 cm; ovary and pod
   hairless   **7. hebepetala**
8a. Ovary and pod hairless   9
 b. Ovary and pod with adpressed hairs
                                            13
9a. Standard hairless on the back (or if
   bristly only so at the tips or
   margins);flower-stalks 1–3 mm   10
 b. Standard bristly-downy on the back;
   flower-stalks 2–9 mm   11
10a. Racemes 6–35 cm, stalkless with
   sterile bracts at the base; bracts
   3–10 × 1.5–3.5 mm wide, boat-
   shaped or broadly ovate; leaflets
   11–21 (rarely as few as 7)
                                    8. cassioides
 b. Racemes 4–6 (rarely to 8) cm, stalk
   5–15 mm, without sterile bracts at
   the base; bracts 1.5–2 cm, narrowly
   triangular; leaflets 5–11
                                  9. himalayensis
11a. Flowers contrasting pink and white;
   racemes dense, spike-like, 20–50-
   flowered; calyx 1–2 mm, lobes
   triangular, shorter than tube
                                    10. jucunda
 b. Flowers uniformly pink or white;
   racemes relatively loosely 10–13-
   flowered   12
12a. Leaflets 7–11 (rarely to 13);
   terminal leaflet 1–3 cm wide
   (rarely as little as8 mm); leaflet-pairs
   mostly 1–2 cm apart; leaf-stalk
   1.5–5 cm; calyx 2.5–3.5 mm, lobes
   triangular, shorter than tube
                                    3. fortunei
 b. Leaflets 13 –27 (rarely as few as 9);
   terminal leaflet 5–9 mm wide; leaflet-
   pairs mostly less than 1 cm apart;
   stalk 7–15 mm; calyx 1.5–2 mm,
   more or less truncate or lobes
   broadly triangular, much shorter
   than tube   **11. australis**
13a. Flower-stalks 4.5–6 mm; leaf-rhachis
   stlightly winged or flattened
                                     12. henryi

b. Flower-stalks 1–3 mm; leaf-rhachis
   not as above   14
14a. Standard with spreading hairs
    outside   **15. mairei**
b. Standard minutely bristly-downy
   outside, hairs not spreading as above
                                               15
15a. Racemes more or less stalkess (stalk
    0–0.5 mm); flowers 5–7 mm,
    stamens 4–6 mm   16
b. Raceme-stalk 5–20 mm (if less then
   flowers 9–10 mm, stamens 7–9 mm)
                                               17
16a. Standard hooded apically; terminal
    leaflet narrowly ovate to elliptic,
    larger than laterals, 2–5 cm; leaf-
    stalk 2–5 cm; stamens 5–6 mm
                              13. amblyantha
b. Standard not hooded apically;
   terminal leaflet obovate to elliptic,
   5–15 (rarely to 25) mm; leaf-stalk
   6–18 mm; stamens 4–5 mm
                              14. bungeana
17a. Leaflets 5–13; terminal leaflet
    2–5 mm wide; leaflet-pairs often less
    than0.7 mm apart   **16. rigioclada**
b. Leaflets 9–19 (rarely to 21); terminal
   leaflet 4–15 mm wide; leaflet-
   pairs5–15 mm apart   18
18a. Young racemes in bud greyish
    green; racemes usually ascending to
    erect;leaf-stalk 1–8 mm; stamens
    8–11.5 mm   **17. heterantha**
b. Young racemes in bud dark brown;
   racemes stiffly erect; leaf-stalk
   8–20 mm; stamens 6–7 mm
                              18. hancockii

1. I. kirilowii Palibin. Illustration:
Schneider, Illustriertes Handbuch der
Laubholzkunde **2**: 65 (1907–1912);
Botanical Magazine 8580 (1914); Bailey,
Standard cyclopedia of horticulture **2**:
1646 (1935).

Low bushy shrubs to *c.* 1 m. Leaflets
usually 7–11 (rarely 5), leaflet-pairs
1.5–2.5 cm or more apart; terminal leaflet
2–5 × 1–3 cm, broadly obovate to rhombic-
elliptic, apex rounded to obtuse; stalk
1–3 cm. Racemes spreading to descending
or pendulous, loosely 10–30-flowered.
Flowers pink, 1.5–1.8 cm. Calyx 2.5–4 mm,
lobes broadly triangular, shorter than to
more or less equalling tube. Standard
hairless outside except for the finely downy
margins. Stamens 1.2–1.4 cm. Ovary and
pods hairless. *NE China, Korea, Japan.* H2.
Summer.

2. I. decora Lindley. Illustration: *Edwards's
Botanical Register* **32**: t. 22 (1846);

Botanical Magazine, 5063 (1858); Bailey,
Standard cyclopedia of horticulture **2**:
1645 (1935); Brickell, RHS A–Z
encyclopedia of garden plants, 567 (2003).

Low bushy shrubs to 50–100 cm.
Leaflets usually 7–11 (rarely as few as 5 or
as many as 13), leaflet-pairs mostly 1–2 cm
or more apart, terminal leaflet
2.5–10 × 1–3.5 cm, ovate-lanceolate,
acute; leaf-stalk 5–26 mm. Racemes
spreading to descending or pendulous,
loosely 10–30-flowered. Flowers pink or
white, 1.3–1.8 cm. Calyx 2.5–3 mm, lobes
triangular, shorter than tube. Standard
white bristly-downy outside. Ovary and
pods hairless. *E. China, Japan.* H5. Late
summer.

The above description applies to var.
decora, which is not reliably hardy; it needs
a sheltered position outside and is often
killed to the root in winter.

3. I. fortunei Craib (*I. alba* A. Goualt; *I.
subnuda* Craib). Illustration: *Revue
Horticulture* (Paris), ser. IV, **3**: 361, t. 19
(1854).

Low bushy shrub 30–60 cm. Leaflets
7–11 (rarely to 13), hairless, leaflet-pairs
mostly 1–2 cm or more apart; terminal
leaflet 1.5–5.5 cm × 8–30 mm, elliptic; leaf-
stalk 1.5–5 cm. Racemes erect to spreading
or descending, loosely 10–30-flowered.
Flowers pink or white, 8.5–12 mm. Calyx
2.5–3.5 mm, lobes shorter than tube.
Standard white bristly-downy outside.
Stamens 7–10 mm. Ovary and pods
hairless. *E. China.* H5.

Not reliably hardy in much of Europe;
requires a sheltered position outside.

4. I. pendula Franchet. Illustration:
Schneider, Illustriertes Handbuch der
Laubholzkunde **2**: 65 (1907); *Botanical
Magazine* 8745 (1918); Lewis et al., (eds.),
Legumes of the world, 360 (2005).

Erect to scrambling shrub with 1–many
stems, to 5 m. Leaflets 13–23 (rarely as
few as 11 or as many as 27); leaflet-pairs
1.2–2 cm or more apart; terminal leaflet
1–3.5 cm × 5–15 mm, obovate to elliptic;
leaf-stalk 1–6 cm. Racemes pendulous, to
30 cm, densely 30–100-flowered. Flowers
pink, 9–11 mm. Calyx 2.5–3.5 mm, lobes
more or less equalling tube. Standard blue-
pink, densely white-downy outside. Ovary
and pods hairy. *W China.* H5. Summer.

5. I. howellii Craib & W.W. Smith
(*I. verticillata* Gagnepain; *I. potaninii*
misapplied).

Erect to spreading, 1–many-stemmed
shrub to 4 m. Leaflets usually 9–13 (rarely
as few as 7 or as many as 17); leaflet-pairs
1.2–2 cm or more apart; terminal leaflet
8–35 × 3–20 mm, obovate to oblanceolate
or ellliptic; leaf-stalk 1–4 cm (rarely as little
as 5 mm). Racemes spreading more or less
horizontally with tips ascending, densely
many-flowered. Flowers bright reddish
pink, 8.5–10.5 mm. Calyx 2–3 (rarely to
4) mm, lobes more or less equalling to
twice as long as the tube. Standard thinly
bristly-downy outside. Stamens 6.5–8 mm.
Ovary and pods hairy. *W China.* H5.
Summer.

6. I. szechuensis Craib (*I. potaninii* Craib).
Spreading to bushy shrub 1–2 m.
Leaflets 7–13 (rarely as few as 5), leaflet-
pairs 5–15 mm apart; terminal leaflet
7–16 × 4–10 mm, obovate to elliptic; leaf-
stalk 5–20 mm. Racemes erect or
ascending, densely many-flowered, bracts
less than 1 mm wide, lanceolate-subulate.
Flowers dark reddish pink, 9–9.5 mm.
Calyx 2–3.5 mm, lobes more or less
equalling the tube. Standard dark crimson,
with white bristles outside. Ovary and pods
hairy. *W China.* H5. Summer.

7. I. hebepetala Baker. Illustration:
Botanical Magazine, 8208 (1908).

Erect to spreading, 1–many-stemmed
shrub to 4 m. Leaflets 7–9 (rarely 5); leaflet
pairs more than 2 cm apart; terminal
leaflet 2.5–8 × 1.5–4.5 cm, elliptic-oblong,
wider than the laterals; leaf-stalk
2.5–5.5 cm. Racemes erect or ascending,
densely many-flowered; bracts 1.5–3 mm
wide, cuspidate at the apex, boat-shaped
and entirely enclosing the young flower-
buds. Flowers dark reddish pink, 8–11 mm.
Calyx 2–2.5 mm, lobes shorter than tube.
Standard dark crimson to rust-coloured,
bronze or maroon, sparsely downy outside.
Stamens 8–10 mm. Ovary and pods
hairless. *Himalaya to NW China.* H5.
Summer.

Plants with leaflets hairless above have
been named var. *glabra* Ali.

8. I. cassioides De Candolle (*I. pulchella*
Roxburgh; *I. elliptica* Roxburgh; *I. arborea*
Roxburgh; *I. purpurascens* Roxburgh).
Illustration: *Botanical Magazine*, 3348
(1834).

Erect to spreading, 1–many-stemmed
shrubs to 3 m. Leaflets 7–11 (rarely to 21),
leaflet-pairs mostly 1–2 cm apart; terminal
leaflet 1–3 cm × 5–22 mm, obovate to
elliptic; leaf-stalk 5–50 mm. Racemes erect

or ascending, 6–35 cm, densely many-flowered, stalkless with sterile bracts at the base; bracts 3–10 × 1.5–3.5 mm, boat-shaped or broadly ovate. Flowers pink to purplish pink, 5–12 mm. Flower-stalks 1..2.5 mm. Calyx 2.5–3.5 (rarely to 5) mm, lobes shorter than to more or less equalling tube. Standard hairless outside (or with hairs on the margin only). Stamens 7–10 mm (rarely as little as 4 mm). Ovary and pods hairless. *India to SW China.* H5. Summer.

Rather tender, but will survive outside in sheltered areas.

9. I. himalayensis Ali.

Erect, 1–many-stemmed shrub to 3 m. Leaflets 7–11 (rarely 5), leaflet-pairs less than 1.5 cm apart; terminal leaflet 9–25 × 4–11 mm, elliptic to oblong; leaf-stalk 5–15 mm. Racemes erect or ascending, 4–8 cm, more or less loosely 10–30-flowered; inflorescence-stalk 5–15 mm, without sterile bracts at the base. Flowers bright pink, 9–11 mm, stalks 2–3 mm. Calyx 2.5–3 mm, lobes more or less equalling to twice as long as the tube. Standard hairless outside, bristly only at the tip.

Ovary and pods hairless. *Himalaya.* H5. Early to late summer.

10. I. jucunda Schrire. Illustration: *Botanical Magazine,* n.s., 216 (1997).

Erect, 1–many-stemmed shrubs to 3 m. Leaflets 9–17 (rarely 7), leaflet-pairs 5–15 mm apart; terminal leaflet 7–30 × 6–20 mm, obovate to oblanceolate or elliptic; leaf-stalk 6–24 mm. Racemes erect or ascending, dense, more or less spike-like, 20–50-flowered. Flowers scented, contrasting bright pink and white, 7–9 mm; stalks 2.5–9 mm. Calyx 1–2 mm, lobes broadly triangular, shorter than the tube. Standard densely downy outside. Stamens 5–6.5 mm. Ovary and pods hairless. *South Africa.* G1.

A conservatory plant in Britain.

11. I. australis Willdenow. Illustration: *Edwards's Botanical Register* **5**: t. 386 (1819); Cochrane et al., Flowers and plants of Victoria, f. 384 (1973); King & Burns, Wild flowers of Tasmania, 33 (1969); Rotherham et al., Flowers and plants of New South Wales, pl. 244 (1975).

Erect or bushy, 1–many-stemmed shrubs to 2 m. Leaflets 13–27 (rarely as few as 9), bluish green, leaflet-pairs mostly less than 1 cm apart; terminal leaflet

5–35 × 5–9 mm, ovate to oblong, longer than laterals; leaf-stalk 7–15 mm. Racemes erect or ascending, relatively loosely 10–30-flowered. Flowers uniformly pink, 6–9 mm; stalks 1.5–5 mm. Calyx 1.5–2.5 mm, lobes, if present broad, shorter than tube. Standard softly bristly-downy outside. Stamens 4–7 mm. Ovary and pods hairless. *Australia (New South Wales, Victoria, South Australia, Tasmania).* H5–G1.

Tender, but will survive outside in sheltered areas.

12. I. henryi Craib.

Scrambling, 1–many-stemmed shrubs 1–2 m. Leaflets 5–15 (rarely 17), leaflet-pairs less than 1.5 cm apart; terminal leaflet 1.5–2.3 cm × 5–10 mm, obovate to elliptic, rhachis flattened, slightly winged; leaf-stalk 5–20 mm. Racemes erect or ascending, loosely-flowered. Flowers pink, 9–10 mm. Calyx *c.* 3 mm, lobes more or les equalling the tube. Standard bristly-downy outside. Stamens 7–9 mm. Ovary and pods hairy. *W China.* H5? Summer.

13. I. amblyantha Craib (*I. potaninii* misapplied). Illustration: *Dendroflora* **22**: 67 (1985); Brickell, RHS A–Z encyclopedia of garden plants, 567 (2003).

Erect, 1–many-stemmed shrub to 4 m. Leaflets usually 7–11, more rarely 5 or 13, leaflet-pairs mostly 1–2 cm apart; terminal leaflet 1.5–5 cm × 8–20 mm, narrowly ovate to elliptic; leaf-stalk 6–20 mm. Racemes erect, spike-like, 5–15 cm, densely many-flowered; inflorescence-stalk 0–5 mm. Flowers pink to salmon pink, 6–7 mm; flower-stalks 1–2.5 mm. Calyx 2–3.5 mm, lobes shorter than to more or less equalling tube. Standard hooded, bristly-downy on the outside. Stamens 5–6 mm. Ovary and pods hairy. *C to E China.* H3. Late summer.

14. I. bungeana Walpers (*I. pseudotinctoria* Matsumura).

Erect shrubs 40–100 cm. Leaflets 5–9 (rarely to 11), leaflet-pairs less than 1 cm apart; terminal leaflet 5–25 × 3–15 mm, obovate to elliptic; leaf-stalk 6–20 mm. Racemes erect, spike-like, 4–10 cm, densely many-flowered; inflorescence-stalk 1–5 mm. Flowers pink to purple, 5–6 mm, stalks 1–2 mm. Calyx 2–2.5 mm, lobes more or less equalling to twice the length of the tube. Standard bristly-downy outside. Stamens 4–5 mm. Ovary and pods hairy. *China, Korea, Japan.* H3? Late summer.

15. I. mairei Pampanini (*I. monbeigii* Craib).

Erect, 1–many-stemmed shrub to 3 m. Leaflets 5–13 (rarely to 17), leaflet-pairs 1–1.5 cm apart; terminal leaflet 5–35 × 3–20 mm, obovate to elliptic; leaf-stalk 5–20 mm. Racemes erect or ascending, 3.5–8 cm, more or less dense, many-flowered; inflorescence-stalk 1–10 mm. Flowers pink, 7–9 mm, stalks 1–2 mm. Calyx 2.5–3 mm, lobes shorter than to as long as tube. Standard with conspicuous spreading hairs outside. Stamens 6–8 mm. Ovary and pods hairy. *W China.* H5. Summer.

16. I. rigioclada Craib.

Shrubs to 1 m. Leaflets 5–13, leaf-pairs usually less than 7 mm apart; terminal leaflet 3–15 × 2–5 mm, obovate to elliptic; leaf-stalk 4–10 mm. Racemes erect or ascending, 3–8 cm, more or less dense, many-flowered; inflorescence-stalk 1–10 or rarely to 20 mm. Flowers pink, 9–10 mm, flower-stalks 1–2 mm. Calyx 2.5–3 mm, lobes more or less equalling tube. Standard with adpressed bristles outside. Stamens 7–9 mm. Ovary and pods hairy. *W China.* H5? Summer.

17. I. heterantha Brandis (*I. gerardiana* Baker var. *heterantha* Baker; *I. heterantha* Brandis var. *gerardiana* (Baker) Ali; *I. quadrangularis* Graham; *I. jacquemontii* misapplied; *I. divaricata* misapplied; *I. splendens* misapplied; *I. floribunda* misapplied; *I. canescens* misapplied). Illustration: *Edwards's Botanical Register* **28**: t. 57 (1842) – as *I. dosua*; Brickell (ed.) RHS A–Z encyclopedia of garden plants, 568 (2003); Lewis et al., (eds.), Legumes of the world, 365 (2005).

Erect to spreading, 1–many-stemmed shrub to 3 m. Leaflets usually 13–19 (rarely 9 or to 21), leaflet-pairs less than 1 cm apart; terminal leaflet 3–25 × 2–15 mm, obovate to oblanceolate or elliptic; leaf-stalk 1–9 mm. Racemes erect or ascending, 3–8 cm, more or less dense, 10–50-flowered, young inflorescences greyish green in bud with few brownish hairs; inflorescence-stalk 5–30 cm. Flowers pink to purple, 9–13 mm; flower-stalks 1.5–3 mm. Calyx 2.5–4 mm, lobes equalling to more or less twice as long as the tube. Standard densely bristly-hairy outside. Ovary and pods hairy. *Himalaya, W China.* H3.

18. I. hancockii Craib (*I. forrestii* Craib).

Erect to spreading, 1–many-stemmed shrubs to 3 m. Leaflets 9–19, leaflet-pairs

5–15 mm apart; terminal leaflet 6–18 × 4–10 mm, obovate to elliptic; leaf-stalk 8–20 mm. Racemes stiffly erect, 3–12 cm, more or less dense, many-flowered, young inflorescences in bud dark brown densely covered with brownish hairs; inflorescence-stalk 4–15 mm. Flowers pink to purple, 7–9 mm, stalks 1–2 mm. Calyx 2–2.5 mm, lobes more or less equalling the tube. Standard densely bristly hairy on the outside. Stmanes 6–7 mm. Ovary and pods hairy. *W China*. H4. Summer.

16. DESMODIUM Desvaux

J. Cullen & H. S. Maxwell

Herbs with woody bases, or shrubs. Leaves usually with 3 leaflets, occasionally reduced to the terminal leaflets only; stipules present, attached by their bases, often early deciduous, stipels present, small. Flowers in axillary and/or terminal racemes or panicles with more than 1 flower to each major bract; bracteoles present. Calyx funnel-shaped, deeply lobed. Standard large, reflexed upwards. Stamens 10, filaments of 9 of them united, that of the upper stamen free or free at the base but united to the others above the base. Pod generally segmented and breaking into 1-seeded segments, more rarely opening along the lower suture. Seeds with or without arils.

A genus of about 300 species (split into several genera by recent authors), mostly from the tropics and subtropics, where several are grown as fodder plants. Only a few are grown for ornament in Europe, where they require a warm, sunny site or the protection of a glasshouse. Propagation is by seed or by division of the roots.

Literature: Ohashi, H., A monograph of the subgenus *Dollinera* of the genus *Desmodium* (Leguminosae), *Bulletin of the University Museum, Tokyo* **2**: 259–320 (1971); Ohashi, H., The Asiatic species of *Desmodium* and its allied genera (Leguminosae), *Ginkgoana* **l** (1973).

1a. Pods not jointed, opening along the lower suture; leaflets usually 3, occasionally 1, the terminal leaflet 4–6 times longer than the lateral leaflets   **1. motorium**
 b. Pods jointed, breaking into 1-seeded segments; leaflets 1 only, or, if 3, then the terminal 1–2 times longer than the laterals   2

2a. Leaflet almost always 1, very large and densely white woolly hairy beneath and with a conspicuous network of lateral veins
  4. yunnanense subsp. **praestans**
 b. Leaflets almost always 3, not as above   3
3a. Filament of upper stamen free to the base; leaflets much longer than broad, usually lanceolate to oblong
  2. canadense
 b. Filament of the upper stamen united to the others for about half its length; leaflets almost as broad as long, broadly ovate to almost circular   **3. elegans**

1. D. motorium (Houttuyn) Merrill

(*D. gyrans* (Linnaeus filius) de Candolle; *Codariocalyx motorium* (Houttuyn) Ohashi).

Erect, branched shrub to 2 m; young shoots hairless. Leaves mostly with 3 leaflets, occasionally reduced to the terminal leaflet; terminal leaflet ovate-elliptic or narrowly so, 4–6 times as long as the small lateral leaflets; all leaflets hairy above when young, persistently whitish hairy beneath. Racemes or panicles terminal and axillary, axes and flower-stalks with spreading or hooked hairs. Calyx 2–5 mm. Corolla mauve, lilac or orange, standard 7–12 mm. Pod curved, flattened, hairy, not segmented but opening along the lower suture, scarcely constricted between the seeds on the upper suture, more deeply so on the lower; seeds 3–11, with arils. *Himalaya & China, SE Asia to N Australia.* G1. Autumn.

Not particularly attractive but of interest because of the rotatory movements performed by the lateral leaflets when exposed to bright sunlight. The species is placed by many authors in the small genus *Codariocalyx* Hasskarl on account of its dehiscent pods containing seeds with conspicuous arils.

2. D. canadense (Linnaeus) de Candolle.

Illustration: Gleason, Illustrated flora of the north-eastern United States and adjacent Canada **2**: 429 (1952); Rickett, Wild flowers of the United States **1**: pl. 79 (1966); *Horticulture* **51**: 26 (1973).

Large, erect, perennial herb to 2 m; young shoots with sparse, spreading or ascending, somewhat kinked or hooked, bristle-like hairs. Leaves all of 3 leaflets, the terminal 1–1.4 times as long as the laterals; leaflets oblong-lanceolate to oblong, sparsely hairy above with bristle-like hairs, pale and somewhat more

densely bristly beneath. Panicles upright, terminal and axillary. Calyx 2–5 mm, densely hairy. Corolla purple, standard 9–18 mm. Upper stamen with filament free to the base. Pod with 3–5 joints, scarcely notched along the upper suture, deeply so along the lower, covered with short, spreading, hooked, bristle-like hairs. *E North America.* H1. Summer.

3. D. elegans de Candolle subsp. elegans

(*D. tiliifolium* (D. Don) Wallich). Illustration: *Revue Horticole* **1902**: 458, 459; Schneider, Illustriertes Handbuch der Laubholzkunde **2**: 109 (1907); Polunin & Stainton, Flowers of the Himalayas, pl. 26 (1985).

Shrub to 4 m; young shoots densely adpressed hairy with curled hairs. Leaves with 3 leaflets, the terminal 1.3–1.7 times longer than the laterals; all leaflets broadly ovate to almost circular, adpressed bristly above, whitish or pale and more densely bristly beneath, often markedly silvery hairy when young. Panicles spreading, terminal and axillary, spreading hairy. Calyx 4–6 mm. Corolla pinkish purple, standard 9–15 mm. Upper stamen with its filament united to the others for about half its length. Pod with up to 10 joints, notched between the seeds along the upper suture, very deeply notched along the lower, adpressed bristly. *Himalaya, China.* H5. Summer.

D. elegans is a very variable species, divided into several subspecies and varieties by Ohashi (references above); the cultivated variant, commonly known as *D. tiliifolium*, belongs to subsp. *elegans* var. *elegans*.

4. D. yunnanense Franchet subsp. praestans (Forrest) Ohashi (*D. praestans* Forrest). Illustration: *Journal of the Royal Horticultural Society* **68**: f. 36 (1943); *Botanical Magazine*, n.s., 407 (1962).

Shrub to 4 m; young shoots with very densely white-woolly hairs. Leaves usually of a single terminal leaflet (its stalk jointed or with tiny stipels indicating the position of the suppressed lateral leaflets); leaflet mostly 10–25 cm (rarely as little as 5.5 cm) × 5–17.5 cm, broadly ovate to almost circular, greyish above, with dense white-woolly hairs and with a conspicuous network of veins beneath. Panicles terminal, much branched and widely spreading, with white-woolly hairs. Calyx 3–6 mm. Corolla pinkish purple, standard 9–15 mm. Upper stamen with its filament united to the others for about half its length. Pod with 4–7 (more?) joints, deeply constricted along both sutures, hairy with silky hairs. *W China*. H5. Summer.

17. CAMPYLOTROPIS Bunge
J. Cullen

Deciduous shrubs. Leaves made up of 3 leaflets, the terminal leaflet usually conspicuously stalked; stipules present, stipels absent. Flowers in axillary racemes crowded into panicles at the tips of the branches; bracts each subtending a single flower, bracteoles soon deciduous. Calyx bell-shaped, 5-toothed, the 2 upper teeth united for most of their length; flower-stalk jointed just beneath the calyx. Corolla purple; standard reflexed upright; keel curved, sickle-shaped, acute. Stamens 10, 9 of them with united filaments, that of the uppermost free. Pod short, 1-seeded, ovoid, ellipsoid or almost spherical, compressed, not opening.

A genus of about 65 species from Asia. Only 1 is grown in Europe, where it requires a well-drained soil and a sunny, open position. It is propagated by seed or by division.

1. C. macrocarpa (Bunge) Rehder
(*Lespedeza macrocarpa* Bunge). Illustration: Schneider, Illustriertes Handbuch der Laubholzkunde **2**: 111, 112 (1907).

Shrub to 1 m; young shoots angled, with adpressed or somewhat spreading hairs. Leaflets elliptic to oblong, 2–5 cm, rounded to the base and the conspicuously mucronate (sometimes also notched) apex, hairless above, pale and adpressed hairy beneath; stipules narrowly lanceolate. Racemes dense. Calyx to 4 mm. Corolla purple, standard 1–1.2 cm. Pod ellipsoid, flattened, 1.2–1.5 cm. *N & C China*. H1. Late summer.

18. LESPEDEZA Michaux
J. Cullen

Annual or perennial herbs or shrubs. Leaves borne in spirals or in 2 distinct ranks; leaflets 3, more or less equal; terminal leaflet stalked or not; stipules persistent, stipels absent; winter buds with spiral scales or scales in 2 ranks. Flowers axillary or in axillary racemes, each bract subtending 2 flowers; in several species corolla-less, cleistogamous flowers are produced as well as normal, opening flowers; bracteoles 2, closely overlapping the base of the calyx, persistent. Calyx funnel- or bell-shaped, 5-toothed, the upper teeth often united for much of their length.

Standard clawed or not; keel straight, blunt. Stamens 10, filaments of 9 united, that of the uppermost usually free. Pod 1-seeded, not dehiscent.

A genus of about 40 species from temperate and subtropical parts of N America, Asia and Australia. All of the cultivated species are known from Japan, where they have been extensively studied. They are easily grown in a warm, sunny site and are propagated by seed.

Literature: Akiyama, S., A revision of the genus *Lespedeza* section *Microlespedeza* (Leguminosae), *Bulletin of the University Museum, Tokyo* **33** (1988) – contains numerous diagnostic illustrations which are not further cited here.

1a. Plant annual   2
  b. Plant a shrub or long-persisting herb   3
2a. Stems with downwardly pointing hairs; calyx hairy; pod slightly longer than the calyx   **7. striata**
  b. Stems with upwardly pointing hairs; calyx hairless; pod twice as long as the calyx   **8. stipulacea**
3a. Terminal leaflet not or very shortly stalked; cleistogamous flowers present with normal flowers   **6. cuneata**
  b. Terminal leaflet conspicuously stalked; cleistogamous flowers absent   4
4a. Leaves and scales of the winter-buds in 2 ranks   5
  b. Leaves and scales of the winter-buds spirally arranged   6
5a. Standard and keel pale yellow   **4. buergeri**
  b. Standard and keel reddish purple   **5. maximowiczii**
6a. Standard conspicuously and abruptly clawed at the base   **3. formosa**
  b. Standard gradually tapered to the base   7
7a. Racemes congested, usually shorter than the subtending leaves; wings longer than keel   **2. cyrtobotrya**
  b. Racemes open, much longer than the subtending leaves; wings shorter than keel   **1. bicolor**

1. L. bicolor Turczaninow. Illustration:
Gartenflora **1860**: t. 299 ; Huxley, Deciduous garden trees and shrubs, pl. 99 (1979).

Shrub to 2 m with persistent aerial branches; young shoots adpressed hairy. Leaves and scales of winter-buds spirally arranged. Terminal leaflet stalked,

2–6 × 1–3.5 cm, elliptic to ovate or obovate, slightly longer than the laterals; all hairless or almost so above, adpressed hairy beneath. Racemes exceeding the subtending leaves, 2–10 cm, with 4–12 loosely arranged flowers; bracteoles narrow, scarcely ribbed, hairy outside. Calyx 3.1–4.7 mm, teeth acute. Standard 9–12 mm, gradually tapered to the base, red-purple inside; keel shorter than the standard but longer than the wings. Pod 5–7 × 4–6 mm, hairy or not, scarcely stalked. *Japan, Korea, N China, E former USSR*. H3. Autumn.

2. L. cyrtobotrya Miquel. Illustration:
Kitamura & Murata, Coloured illustrations of herbaceous plants of Japan, pl. 23 (1981).

Similar to *L. bicolor* but to 2.5 m; racemes congested, 1–2 cm, shorter than the subtending leaves; calyx 4.5–6.2 mm, teeth acuminate; pods slightly smaller, always hairy. *Japan, Korea, N China, E former USSR*. H3. Autumn.

3. L. formosa (Vogel) Koehne (*L. sieboldii* Miquel). Illustration: Thrower, Plants of Hong Kong, 106 (1971); Lewis et al., (eds.), Legumes of the world, 432 (2005).

Shrub to 2 m, many aerial parts dying in winter; young shoots adpressed hairy. Leaves and scales of winter-buds spirally arranged. Terminal leaflet conspicuously stalked, 2–6 × 1.5–3.5 cm, slightly larger than the laterals, all densely hairy beneath, hairy or not above. Racemes 2–15 cm, exceeding the subtending leaves, open, with 4–14 flowers; bracteoles narrow, scarcely ribbed, hairy. Calyx 3.5–6 mm, teeth acute. Standard 9.5–13.5 mm, abruptly and conspicuously narrowed into a distinct claw, red-purple; keel longer than the red-purple wings. Pod compressed, 7–12 × 4–5 mm, hairy, stalked. *Himalaya to Japan*. H3. Late summer.

A very variable species, particularly in the shape and size of the calyx–teeth (see Akiyama, S. & Ohba, H., Taxonomy of *Lespedeza formosa* (Vogel) Koehne, *Bulletin of the University Museum, Tokyo* **31**: 217–229, 1988). Related to it are 2 species known with certainty only from cultivation (ancient cultivation in Japan) which have been confused among themselves and with *L. formosa* and are probably of hybrid origin:

L. thunbergii (de Candolle) Nakai (*Desmodium penduliflorum* Oudemans; *L. penduliflora* (Oudemans) Nakai).

Illustration: Kitamura & Murata, Coloured illustrations of herbaceous plants of Japan, pl. 23 (1981); Brickell (ed.), RHS A–Z encyclopedia of garden plants, 620 (2003); Lewis et al., (eds.), Legumes of the world, 435 (2005); *Amateur Gardening* September 1987: 21. Similar to the above but leaflets 3.5–9 × 2.5–5 cm, hairless above except along the midrib; racemes 10–15 cm; calyx 5.3–5.5 mm; standard 1.2–1.5 cm, shortly clawed, red-purple; wings deeper red-purple. H3. Summer.

Possibly a hybrid of *L. patens* Nakai, a species very similar to *L. formosa*.

L. japonica Bailey. Illustration: Kitamura &, Murata, Coloured illustrations of herbaceous plants of Japan, pl. 23 (1981). Very similar to both *L. formosa* and *L. thunbergii*, but leaflets usually persistently hairy above, and flowers white ('Japonica') or at least partially white (other cultivars). H3. Late summer.

The name *L. japonica* covers a range of variants widely cultivated in Japan and thought to be hybrids of *L. formosa* with various other species.

4. L. buergeri Miquel. Illustration: Schneider, Illustriertes Handbuch der Laubholzkunde 2: 112 (1907); Kitamura & Murata, Coloured illustrations of herbaceous plants of Japan, pl. 23 (1981).

Shrub to 3 m; young shoots adpressed hairy. Leaves and scales of winter-buds arranged in 2 ranks. Terminal leaflet stalked, 1–5 cm × 5–30 mm, narrowly to broadly ovate or elliptic, usually hairless above, adpressed hairy beneath. Racemes 2–7 cm, usually exceeding the subtending leaves, open, with 4–10 flowers; bracteoles brownish, broad, several-ribbed, hairy. Calyx 1.8–3 mm, teeth acute. Standard 7–9 mm, pale yellow with purple patches inside, distinctly clawed; wings about as long as standard, purple or reddish purple; keel pale yellow, exceeding the standard. Pod compressed, 1–1.5 cm × 5–5.5 mm, hairy or not, stalked. *Japan, E China.* H3. Summer.

5. L. maximowiczii Schneider. Illustration: Lewis et al., (eds.), Legumes of the world, 435 (2005).

Like *L. buergeri* but inflorescences often shorter, bracteoles *c.* 5-ribbed, standard 9–10 mm, red-purple, wings deeper red-purple, shorter than the standard, keel paler than wings, exceeding the standard; pod almost stalkless. *Japan, Korea, S China.* H5. Summer.

6. L. cuneata G. Don (*L. sericea* Miquel). Illustration: Schneider, Illustriertes Handbuch der Laubholzkunde 2: 112, 114 (1907); Gleason, Illustrated flora of the north-eastern United States and adjacent Canada 2: 436 (1952); Kitamura & Murata, Coloured illustrations of herbaceous plants of Japan, pl. 23 (1981).

Shrub to 1 m with ascending-upright, densely leafy branches; young shoots with hairs which are spreading at their bases and then curve upwards. Terminal leaflet not or very shortly stalked, 7–20 mm, long-tapered to the base, truncate or notched at the apex, hairless above, sparsely to densely hairy beneath. Racemes dense, shorter than the subtending leaves; bracteoles narrowly triangular, hairy. Cleistogamous flowers present as well as normal flowers. Normal flowers with calyx to 4 mm, teeth acuminate, corolla whitish, standard 6–7 mm, keel longer than standard. Pod 1.5–2 mm, not stalked, hairy. *E & SE Asia to N Australia; introduced in E & S USA.* H5. Late summer.

7. L. striata Murray (*Kummerowia striata* (Murray) Schindler). Illustration: Gleason, Illustrated flora of the north-eastern United States and adjacent Canada 2: 436 (1952); Kitamura & Murata, Coloured illustrations of herbaceous plants of Japan, pl. 22 (1981).

Erect annual herb, much branched; stems with downwardly pointing whitish hairs. Stipules broadly lanceolate, brown, obliquely inserted. Terminal leaflet 1–1.5 cm × 5–8 mm, not notched, slightly larger than the laterals, margin and main vein beneath hairy. Flowers axillary, normal and cleistogamous. Normal flowers *c.* 5 mm, reddish purple; calyx 3–3.5 mm, hairy. Cleistogamous flowers without corollas. Pod flat, almost circular, abruptly acute, *c.* 3.5 mm, slightly exceeding calyx. *Japan, China, Korea, Taiwan, introduced in E & S USA.* H5. Flowering almost throughout the year.

8. L. stipulacea Maximowicz (*Kummerowia stipulacea* (Maximowicz) Makino). Illustration: Gleason, Illustrated flora of the north-eastern United States & adjacent Canada 2: 436 (1952).

Similar to *L. striata* but stems with adpressed, upwardly pointing hairs, leaflets usually notched, calyx hairless, pod about twice as long as the calyx, abruptly rounded at the apex. *Japan, China, Korea, E former USSR, introduced in E & S USA.* H5. Flowering almost throughout the year.

This and the previous species are now generally placed in the genus *Kummerowia*, all of whose species are annual.

19. ERYTHRINA Linnaeus

U. Oster & J. Cullen

Trees, shrubs or rarely herbs, usually deciduous and spiny. Leaves of 3 leaflets; stipules persistent, glandular stipels usually present at the bases of the stalks of the lateral leaflets. Flowers in terminal racemes, produced with the leaves or when the plant is leafless. Calyx mostly cup-shaped, truncate. Corolla usually red, standard large, folded over and concealing the much shorter wings and keel in bud, remaining so, or spreading and reflexing to reveal the other petals. Stamens 10, filaments of 9 united, that of the uppermost free. Pods woody, flat or cylindric, usually constricted between the seeds. Seed often red to scarlet and with a black blotch.

A genus of more than 100 species of tropical and warm temperate regions of the world, of which only a few are grown in Europe (many more are cultivated in the United States). In most areas they require glasshouse protection. They are propagated by seed or by cuttings.

1a. Standard at last erect and widely spreading at full-flowering, revealing the wings and keel **1. crista-galli**
  b. Standard constantly folded over and concealing the wings and keel 2
2a. Standard at most to 9 mm wide; leaflets mostly shallowly 3-lobed **4. herbacea**
  b. Standard more than 1 cm wide; leaflets not 3-lobed 3
3a. Leaf-stalks hairless, usually spiny; calyx truncate, without obvious teeth **2. corallodendron**
  b. Leaf-stalks downy, not spiny; calyx with 5 small teeth **3. caffra**

1. E. crista-galli Linnaeus. Illustration: Perry, Flowers of the world, 165 (1970); Noailles & Lancaster, Mediterranean plants and gardens, 78 (1977); *Gartenpraxis* July 1977: 324; *The Garden* **112**: 392 (1987) & **113**: 577 (1988).

Shrub to small or large tree, dependent on growing conditions; stems with strong spines. Leaves with spines on the stalk and midribs; leaflets ovate, ovate-lanceolate or almost circular, rounded to the base, abruptly pointed at the apex; glandular stipels conspicuous. Flowers appearing with the leaves, in terminal racemes. Standard bright red, to 5 cm, at least 1 cm wide, ultimately spreading and revealing the much shorter wings and keel, which are greenish with red tips. Pod woody, to 40 cm. Seeds black with brownish markings. *S Brazil, Uruguay, Paraguay, N Argentina.* G1. Summer.

Copious nectar is produced by the flowers and often drips from them.

2. E. corallodendron Linnaeus.

Tree to 8 m, usually spiny. Leaves spiny or not; leaflets ovate to diamond-shaped, the terminal often larger and more acuminate than the laterals; stalks not hairy; glandular stipels not conspicuous. Flowers few, in racemes, produced when the plant is leafless, deep red-crimson. Standard to 7.5 cm, at least 1 cm wide, constantly folded over the much shorter wings and keel. Pod to 10 cm, beaked. Seeds scarlet with a black spot. *S USA, Mexico.* H5–G1.

3. E. caffra Thunberg. Illustration: de Wit, Plants of the world **2**: pl. 275 (1963–5); *Flowering Plants of Africa* **43**: pl. 1709 (1976).

Spiny tree to 8 m. Leaves without spines, or with very small spines on the downy stalks; leaflets triangular-ovate, rounded at the base, long-acuminate at the apex; glandular stipels conspicuous. Flowers in dense racemes produced when the plant is more or less leafless, brilliant scarlet. Calyx hairy, distinctly 2-lipped, toothed. Standard 3.5–5 cm, at least 1 cm wide, constantly folded over the much shorter wings and keel. Pod to 12.5 cm. Seeds red with a black spot. *E & S Africa.* H5–G1.

4. E. herbacea Linnaeus (*E. arborea* (Chapman) Small). Illustration: *Botanical Magazine*, 877 (1805); Dean et al., Wild flowers of Alabama and adjoining states, 87 (1983).

Perennial herb, shrub or small tree, spiny. Leaves with or without spines on the stalk or midrib; leaflets triangular to arrowhead-shaped, usually with 2 rounded lateral lobes and a narrow, acuminate terminal lobe; glandular stipels conspicuous. Flowers scarlet, in terminal racemes, produced with the leaves. Calyx truncate, without obvious teeth. Standard to 5 cm × 9 mm, constantly folded over the much shorter wings and keel. Pods to 20 cm, shortly beaked. Seeds bright scarlet with a black blotch. *S USA, adjacent Mexico.* H5–G1. Spring–summer.

20. MUCUNA Adanson
S.G. Knees
Woody climbers, climbing herbs and erect shrubs. Leaves with 3 stalked leaflets; stipules falling early, stipels often present. Flowers large, in axillary clusters or racemes, often long-stalked and hanging, very showy, purple, red, greenish, yellow or white; bracts and bracteoles deciduous. Calyx-lobes 4 or 5, 2-lipped above; standard rounded, shorter than other petals; keel hardened at apex, sharply beaked; ovaries few–several, ovules *c.* 12; style fine, stigma very small; stamens 10, 9 united, 1 free, anthers in 2 alternate rings of 5, the inner ring shorter. Fruit a pod, often bristly or velvety; seeds oblong to spherical, hilum present or absent.

A genus of about 100 species from tropical Asia, Africa and the Americas. All can be cultivated from seed and will need support from a wall or strong trellis.

Literature: Verdcourt, B., A manual of New Guinea legumes (1979); Wilmot-Dear, M., A revision of *Mucuna* in China & Japan, *Kew Bulletin* **39**(1): 23–65 (1984); Wilmot-Dear, M., A revision of *Mucuna* in the Pacific, *Kew Bulletin* **45**(1): 1–35 (1990).

1a. Semi-woody, climbing herb, short-lived perennial or annual to 4 m; leaves conspicuously hairy beneath
　　　　　　　　　　　1. pruriens
　b. Vigorous woody climbers, 12–30 m; leaves usually hairless beneath, or if hairy then only when young　2
2a. Evergreen climber to 12 m; flowers blackish purple, unpleasantly scented
　　　　　　　　　　2. sempervirens
　b. Climber 20–30 m; flowers bright orange-red　　　　　　　　　3
3a. Leaflets 5–7.5 cm across
　　　　　　　　　　　3. bennettii
　b. Leaflets 8.5–13.5 cm across
　　　　　　　　4. novaguineensis

1. M. pruriens (Linnaeus) de Candolle (*Marcanthus cochinchinensis* Loureiro; *Mucuna nivea* (Roxburgh) Wight & Arnott; *M. aterrima* (Piper & Tracy) Holland; *M. deeringiana* (Bort) Merrill). Illustration: *Botanical Magazine*, 4945 (1856); *Kew Bulletin*, **39**(1): 32 (1984).

Semi-woody climbing herb, short-lived perennial or annual to 4 m. Stems rough, covered with long bristly hairs when young, eventually hairless. Terminal leaflet 9–16 × 5–10 cm, ovate, obovoid or elliptic, leaf-stalks 2–30 cm. Racemes to 30 cm; calyx covered with pale brown hairs; corolla 2–4 cm, deep blackish purple to

lilac or white. Fruit 5–9 × 1–2 cm, oblong, covered with irritant, orange-brown hairs; seeds 1.3–1.7 cm across. *Asia (widely naturalised in tropical Africa & America).* G2.

Var. **utilis** (Wight) Burck, differs in having stems without long hairs and light brown hairs on the fruit.

2. M. sempervirens Hemsley. Illustration: *Botanical Magazine*, 7978 (1904); *Kew Bulletin* **39**(1): 27 (1984); *Kew Magazine* **3**: 139 (1986); Phillips & Rix, *Shrubs*, 155 (1989).

Vigorous evergreen climber to 12 m, old stems to 30 cm across. Terminal leaflets 8–16 × 3.5–9 cm, narrowly elliptic to ovate, leaf-stalks 7–16 cm. Flower-clusters 10–36 cm; calyx with red and brown bristles; corolla 2–4 cm, dark purple or reddish. Fruit 30–50 × 3–3.5 cm, covered with red-brown velvety hairs and bristles; seed 2–3 cm across. *India, Sikkim, Bhutan, Burma, W China.* H5–G1.

3. M. bennettii F.J. Mueller. Illustration: Graf, Exotica, series 4, **2**: 1388 (1985); Graf, Tropica, edn 3, 533 (1986); *Kew Bulletin* **45**(1): 33 (1990); Brickell (ed.), RHS A–Z encyclopedia of garden plants, 701 (2003).

Woody climber to 20 m; stems rough, mostly hairless. Terminal leaflets 11–15 × 5–7.5 cm, narrowly elliptic, leaf-stalks 10–14 cm. Flower clusters 2–10 cm; calyx to 1.5 cm, with sparse bristles; corolla 3–6 cm; bright red. Fruit unknown. *New Guinea.* G2.

The name of this species is often misapplied to *M. novaguineensis*.

4. M. novaguineensis R. Scheffer (*M. bennettii*, misapplied). Illustration: Verdcourt, A manual of New Guinea legumes, 449, 456 (1979); Lewis et al., (eds.), Legumes of the world, 405 (2005).

Woody climber to 30 m; stems to 5 cm across, covered with hairs at first, eventually hairless. Terminal leaflets 10–19 × 8.5–13.5 cm, elliptic, apex acute, base rounded, hairless. Flower clusters 7–60 cm; calyx 6–13 mm; corolla 5–8 cm, brilliant orange-red. Fruit 16–27 × 4–6 cm; seeds *c.* 4 cm across. *New Guinea.* G2.

21. STRONGYLODON Vogel
M.C. Warwick
Evergreen, woody stemmed twining climbers and shrubs. Leaves pinnate, with 3 leaflets; stipules falling early; terminal leaflets lanceolate to circular, apex acuminate, rounded at base; lateral leaflets

slightly smaller in size, oblique at base. Flowers axillary in many-flowered drooping racemes. Calyx bell-shaped, 5-lobed. Corolla orange, red, blue or bluish green, standard reflexed, acute at apex with 2 appendages above the claw, wing half as long as standard, attached to the keel, petals at base, keel petals united, as long as standard. Stamens 10, 9 united and 1 free, filaments hairless. Ovary with 1–12 ovules; pods wrinkled; seeds black or brown, smooth or wrinkled.

A genus of about 20 species native to Old World tropics, only one of which is commonly grown. It is frost-tender (min. 18 °C) and requires a humus-rich, moist but well-drained soil and partial shade in summer. After 3 years flowers are produced on new or old wood and stems should be trained along wires to allow flowers to hang down.

Seeds have short viability and must be planted soon after pod opens. Overnight soaking and an incision in the seed coat will assist germination. Cuttings of half-ripened wood can be rooted in a few weeks in a misty, warm greenhouse. Light pruning may be required if plant becomes too rampant.

1. S. macrobotrys Gray. Illustration: *Botanical Magazine*, n.s., 627 (1972); Perry, Flowers of the world, 164 (1972); *Kew Magazine* **1**: 189 (1984); Brickell, RHS gardeners' encyclopedia of plants and flowers, 164 (1989).

Woody climber to 20 m, stems to 3 cm thick. Leaves 12–15 × 5–7 cm, pinnate with leaflets oblong to elliptic; stipules 3–4 mm, ovate to triangular. Racemes pendent, to 3 m, bracteoles *c.* 1.5 mm, ovate, with short hairs at margin. Flower-stalk 1.8–4 cm, hairless. Calyx 8–14 mm, 5-lobed. Flowers 4–6 cm, bluish green, standard 3.7–4.8 cm, ovate, reflexed with claw 3–50 mm; with 2 appendages 0.7–1 cm, above the claw; wings 2–2.4 cm, oblong to elliptic, with claw *c.* 1.1 cm; keel 1.1–1.3 cm with claw tapered into a gently curved beak. Stamens 4.8–7.2 cm, filaments hairless. Style 3.8–5.5 cm, hairless, stigma terminal; ovary *c.* 6 mm, covered with adpressed hairs. Pods 8.5–15 × 5–7 cm, elliptic, inflated and wrinkled with 6–12 smooth, black seeds with a long hilum and thin seed coat. *Philippine Islands.* G2. Winter–spring.

S. macrobotrys with its pale flowers in long racemes away from the foliage suggests bat pollination whereas other species are bird pollinated.

22. BUTEA Willdenow
U. Oster & J. Cullen
Trees or woody climbers (ours), rarely herbs. Leaves divided into 3 leaflets; stipules deciduous, stipels present. Flowers in dense terminal racemes or panicles; bracteoles present. Calyx bell-shaped, 5-toothed, the upper 2 teeth partly united and often larger than the others. Corolla large, standard often reflexed. Stamens 10, the filaments of 9 united, that of the uppermost free. Ovary with 1 ovule. Pod flattened, wing-like, the single seed borne at the uppermost end.

A genus of 7 species from tropical Asia, rarely grown and requiring glasshouse protection. Propagation is by seed.

1. B. superba Roxburgh. Illustration: Menninger, Flowering vines of the world, t. 110 (1970).

Large woody climber. Leaflets leathery, broadly ovate, widely tapered to the base, rounded to the apex, to 28 × 20 cm. Racemes long, hanging, axis densely brown hairy. Calyx bell-shaped, 5-toothed, brown-hairy, *c.* 2 cm. Corolla reddish (drying yellowish), all the petals densely to loosely white hairy outside. Standard reflexed upwards, *c.* 5 cm, somewhat shorter than the curved, pointed keel. Pods to 15 × 4 cm. *Burma, Thailand.* G2.

Other species of *Butea* are cultivated in the Asian tropics and perhaps in N America, e.g. **B. monosperma** (Lamarck) Taubert (*B. frondosa* Willdenow), which is a tall tree with hairy leaflets and pods. *India, Burma.*

23. APIOS Medikus
U. Oster & J. Cullen
Twining perennial herbs, some (ours) with tuberous roots. Leaves alternate, pinnate with several leaflets, including a terminal leaflet; stipules subulate, persistent; stipels absent. Flowers in short racemes, sometimes leaf-opposed. Calyx cup-like, scarcely toothed. Corolla with a broad standard, the keel much longer but coiled upwards within the standard, blunt and notched at the apex. Stamens 10, 9 with their filaments united, that of the uppermost free. Ovary with coiled style. Pod flattened, several-seeded, the sides spiralling after dehiscence.

A genus of about 10 species from America and Asia. Only 1 is occasionally grown, propagated by seed and by tubers.

1. A. americana Medikus (*A. tuberosa* Moench). Illustration: *Botanical Magazine*,

1198 (1809); Everard & Morley, Wild flowers of the world, pl. 153 (1970); Dean et al., Wild flowers of Alabama and adjoining states, 89 (1983).

Twiner to 3 m, roots bearing strings of tubers. Leaflets ovate or narrowly ovate, acuminate, rounded to cordate at the base. Racemes dense, generally leaf-opposed; flowers fragrant, brown, calyx *c.* 5 mm, standard *c.* 1 cm. Pod to 10 cm. *E North America (New Brunswick to Texas).* H3. Summer.

The tubers are edible.

24. CANAVALIA Adanson
J. Cullen
Climbers, often large, but sometimes cultivated as bushy annuals. Leaves long-stalked, made up of 3 leaflets; stipules soon falling; terminal leaflet with 2 deciduous or occasionally persistent stipels. Flowers in axillary racemes or panicles; bracteoles 2, soon falling. Calyx cylindric, 2-lipped, the upper lip very large, somewhat 2-lobed, the lower lip of 3 small teeth. Corolla showy, usually purplish or violet, more rarely reddish, bluish or almost white; blade of standard reverse heart-shaped. Stamens 10, filaments of 9 united, that of the uppermost free, at least at the base. Pods large, somewhat flattened. Seeds large, each with a long, linear hilum.

A genus of about 50 species from the tropics and subtropics, some grown in these areas for food (young leaves, young pods) or as fodder. They are rarely grown as ornamentals, though 3 species are recorded. They require rich soil in a warm situation and are propagated by seed.

Literature: Sauer, J., Revision of *Canavalia*, Brittonia **16**: 106–181 (1964).

1a. Leaflets acute but not acuminate **1. ensiformis**
  b. Leaflets gradually or abruptly acuminate 2
2a. Leaflets gradually acuminate to an acute tip; calyx at least 1.5 cm; hilum more than 1.2 cm **2. gladiata**
  b. Leaflets abruptly acuminate to an obtuse tip; calyx less than 1.5 cm; hilum less than 1.2 cm **3. virosa**

1. C. ensiformis (Linnaeus) de Candolle. Illustration: *Botanical Magazine*, 4027 (1843); Masefield et al., The Oxford book of food plants, pl. 41 (1969); Duke, Handbook of legumes of world economic importance, 39 (1981).

Leaflets to 20 cm, ovate or broadly ovate, base rounded to truncate, apex acute but not acuminate, sparsely hairy along the veins above and beneath. Bracteoles obtuse, to 2 mm. Flower-stalk to 2 mm. Calyx to 1.4 cm, sparsely hairy, upper lip as long as tube, its apex abruptly constricted behind the pointed tip; lowest tooth to 2.5 mm, acute, exceeding the acute lateral teeth. Standard to 2.7 cm. Pod to 30 × 3.5 cm, somewhat compressed, the sides rolling spirally on dehiscence. Seeds to 2 cm, oblong, somewhat compressed, cream or white with an inconspicuous mark near the hilum which is *c.* 9 mm. *SW USA, C America, West Indies, S America, introduced elsewhere.* G2.

2. C. gladiata (Jacquin) de Candolle. Illustration: Duke, Handbook of legumes of world economic importance, 41 (1981).

Similar to *C. ensiformis* but leaflets thinner, conspicuously but gradually acuminate; calyx to 1.6 cm, almost hairless, upper lip constricted just behind the tip which is not pointed; standard *c.* 3.5 cm; pod to 40 × 5 cm; seeds to 3.5 cm, reddish brown (rarely whitish), hilum *c.* 2 cm. *Origin uncertain, probably E Asia, naturalised in many parts of the tropics & subtropics.* G2.

3. C. virosa (Roxburgh) Wight & Arnott (*C. polystachya* (Forsskål) Schweinfurth).

Like *C. ensiformis* and *C. gladiata* but leaflets to 15 cm, abruptly acuminate to an obtuse apex, more densely hairy on both surfaces; calyx to 1.2 cm, hairy, upper lip shorter than tube; standard *c.* 3 cm; pod to 17 × 3 cm; seeds *c.* 2 cm, brown or reddish brown with black marbling; hilum *c.* 1 cm. *Tropical & S Africa to India.* G2.

25. PUERARIA de Candolle
J. Cullen

Woody climbers (ours), often with large underground tubers. Leaves of 3 leaflets, the lateral leaflets conspicuously unequal-sided, tending to be lobed; stipules attached near the middle, projecting above and below the point of attachment (ours); stipels narrow. Racemes axillary, dense, with 2 or more flowers at each node. Bracteoles 2, borne at the base of the calyx. Calyx bell-shaped, deeply 5-toothed, the lowest tooth the longest. Corolla white, purplish or bluish. Stamens with the filaments united into a tube, the uppermost free at the base, becoming more so with age. Pod oblong, flattened, more or less straight, containing 5–20 seeds.

A genus of 17 species from tropical and temperate eastern Asia and the Pacific. Only a single species is grown and is propagated by seed; this species is widely grown in Asia and introduced elsewhere for the sake of starch produced from the tuber, as a fodder plant and for erosion control.

1. P. montana (Loureiro) var. **lobata** (Willdenow) Maesen & Almeida (*P. thunbergiana* (Siebold & Zuccarini) Bentham; *P. lobata* (Willdenow) Ohwi).

Woody climber (to 30 m in the wild) with large tubers. Terminal leaflet ovate to almost circular, acute or abruptly acuminate, densely hairy above and beneath, 8–20 × 5–22 cm; lateral leaflets smaller. Racemes dense, very hairy. Flowers 1–2.5 cm. Pods 4–13 cm, densely shaggy-hairy. *E Asia, N to Japan, Pacific Islands; introduced elsewhere.* G1.

26. KENNEDIA Ventenat
J. Cullen & H.S. Maxwell

Woody scramblers or prostrate, scrambling herbs. Leaves with 3 (rarely more) leaflets; stipules sometimes falling early; stipels usually present, linear. Flowers in loose or dense axillary racemes (rarely solitary); bracteoles absent. Calyx tubular, deeply 5-toothed. Corolla reddish or black and yellow. Stamens with 9 of the filaments united, that of the uppermost free. Pods flattened, containing several seeds with spongy partitions between them. Seeds appendaged.

A genus of about 11 species from Australia; only a few are in cultivation. Propagation is by seed or cuttings taken in spring or early summer. The name is often spelled 'Kennedya'.

Literature: Silsbury, J.H. & Brittan, N.H., Distribution and ecology of the genus *Kennedya* Vent. Western Australia, Australian Journal of Botany **3**: 113–135 (1955).

1a. Flowers in pairs in loose racemes   2
 b. Flowers not in pairs, in dense umbels or racemes or rarely solitary   3

2a. Flowers black and yellow; leaflets ovate to almost circular
   4. nigricans
 b. Flowers red or reddish purple; leaflets narrower   **5. rubicunda**
3a. Flowers numerous, almost stalkless in dense, umbel-like racemes
   1. coccinea
 b. Flowers few, clearly stalked, in racemes or rarely solitary   4

4a. Stipels persistent; bracts persistent, conspicuous   **2. prostrata**
 b. Stipels soon falling; bracts absent or soon falling   **3. eximia**

1. K. coccinea Ventenat. Illustration: Farrall, West Australian native plants in cultivation, 172 (1970).

Prostrate or scrambling, somewhat woody. Leaflets usually 3 (rarely more), leathery, adpressed bristly above and beneath, elliptic or ovate to narrowly lanceolate, rounded or acute at the apex, 3–6 cm × 5–32 mm. Flowers almost stalkless, numerous in dense umbel-like racemes which are long-stalked. Calyx densely hairy. Corolla red, 1.3–1.5 cm, keel shorter than wings. Pod compressed, hairy. *Australia (Western Australia).* H5–G1. Spring.

2. K. prostrata R. Brown. Illustration: Morcombe, Australia's western wild flowers, 92 (1968); Cochrane et al., Flowers and plants of Victoria, f. 43 (1969); Galbraith, Collins' fieldguide to wild flowers of south-east Australia, pl. 15 (1977).

Prostrate, scarcely woody. Leaflets 3, lanceolate to almost circular, 1.5–2.5 × 1.2–1.6 cm, rounded though mucronate at the apex, hairless above, adpressed-bristly beneath; stipules large, persistent, stipels linear, persistent. Flowers 1–6 in loose racemes, bracts conspicuous, inflorescence-stalk longer than flower-stalks. Calyx with spreading bristles. Corolla *c.* 2 cm, scarlet-pink. Pod compressed, hairy or not. *Australia.* H5–G1. Spring.

3. K. eximia Lindley. Illustration: *Paxton's Magazine of Botany* **16**: 36 (1849).

Like *K. prostrata* but stipels soon falling, bracts absent or soon falling. *Australia (Western Australia).* H5–G1. Spring.

4. K. nigricans Lindley. Illustration: *Edwards's Botanical Register* **20**: t. 1715 (1835); Erickson et al., Flowers and plants of Western Australia, pl. 238 (1973); Lewis et al., (eds.), Legumes of the world, 406 (2005).

Woody climber. Leaflets 3, leathery, 2.5–12 cm, broadly ovate to almost circular, rounded at the apex, hairy on both surfaces, less so above. Flowers in pairs in loose axillary racemes. Calyx densely hairy. Corolla 3–4 cm, black with a patch of yellow on the standard. Pod compressed. *Australia (Western Australia).* H5–G1. Spring.

5. K. rubicunda (Schneevoogt) Ventenat.
Illustration: Cochrane et al., Flowers and
plants of Victoria, f. 479 (1968);
Rotherham et al., Flowers and plants of
New South Wales and southern
Queensland, pl. 237 (1975); Galbraith,
Collins' fieldguide to the wild flowers of
south-east Australia, pl. 15 (1977); Brickell
(ed.), RHS A–Z encyclopedia of garden
plants, 597 (2003).

Woody climber. Leaflets 3, leathery,
elliptic, narrowly ovate or lanceolate, acute
or acuminate at the apex, hairless above,
adpressed hairy beneath, 4–9 × 2–4.5 cm.
Flowers in pairs in loose axillary racemes.
Calyx densely adpressed hairy. Corolla
2.5–3.5 cm, red or reddish purple. Pod
compressed, hairy, to 8 cm. *Australia.*
H5–G1. Spring.

27. HARDENBERGIA Bentham
J. Cullen & H.S. Maxwell
Shrubs or woody twiners. Leaves of 3–5
leaflets, when the terminal leaflet stalked
and the laterals opposite or whorled, or
reduced to a single leaflet; stipules and
stipels persistent (even when leaf reduced
to a single leaflet). Flowers in pairs in
axillary racemes. Bracteoles absent. Calyx
cylindric to bell-shaped, 5-toothed, the
upper teeth united for most of their length.
Corolla blue or purple, rarely white.
Stamens 10 with all their filaments united
into a tube or the uppermost at least
partially free. Pod flattened or not, with or
without spongy partitions between the
seeds.

A genus of a small number (perhaps
only 2) of species from Australia. They are
easily grown and propagated by seed or by
cuttings taken in late summer.

1a. Leaves reduced to a single leaflet
                      2. violacea
  b. Leaves of 3–5 leaflets
                  1. comptoniana

1. H. comptoniana (Andrews) Bentham.
Illustration: *Botanical Magazine*, 8992
(1924); Gardner, Wild flowers of
Western Australia, 73 (1959);
Marchant et al., Flora of the Perth
region **1**: 267 (1987); Brickell (ed.), RHS
A–Z encyclopedia of garden plants, 502
(2003).

Shrub or woody climber. Leaves of 3–5
leaflets, the 2–4 lateral leaflets opposite or
whorled, shortly stalked, the terminal
leaflet with a longer stalk; leaflets leathery,
3–14.5 × 1.3–5 cm, lanceolate to ovate,
rounded to the base, long-tapered to the

ultimately blunt though mucronate apex,
hairless or almost so (except for their
stalks); stipels small, persistent. Racemes
very long, axillary, flower-stalks hairy.
Calyx bell-shaped, hairless. Corolla blue to
purple, rarely white, 7–11 mm. Pod
cylindric, swollen, 3.5–4 cm. *Australia
(Western Australia).* H5. Winter–spring.

2. H. violacea (Schneevoogt) Stearn.
Illustration: Rotherham et al., Flowers and
plants of New South Wales and southern
Queensland, pl. 160 (1975); Cunningham
et al., Plants of western New South Wales,
396 (1981); Lewis et al., (eds.), Legumes of
the world, 406 (2005).

Shrub or woody climber. Leaves of a
single, leathery, ovate to lanceolate leaflet,
3.5–8 cm × 7–37 mm, obtuse at the apex,
rounded to truncate at the base, hairless or
almost so. Racemes sometimes branched at
the base. Calyx bell-shaped, hairless,
purple. Corolla 7–9 mm, deep purple with
yellow spots at the base of the standard.
Pods flattened, *c.* 3 cm. *Australia.* H5.
Winter–spring.

28. CENTROSEMA (de Candolle)
Bentham
J. Cullen
Herbaceous climber. Leaves with 3 leaflets,
stipules and stipels persistent. Flowers 1–5,
in axillary racemes with swollen nodes.
Bracteoles 2, conspicuous. Flowers borne
upside-down (standard below). Calyx
5-toothed. Standard large, with a small
projecting spur towards the base. Stamens
10, filaments of 9 united that of the
uppermost free, anthers alternately longer
and shorter. Stigma hairy. Pods long-
beaked, with thickened sutures.

A genus of about 45 species mostly from
the American tropics; only 1 species, from
both temperate, subtropical and tropical
America is grown. It is propagated by seed.

1. C. virginianum (de Candolle)
Bentham. Illustration: Rickett, Wild flowers
of the United States **2**: pl. 118 (1966);
Dean et al., Wild flowers of Alabama, 91
(1973).

Stems thin from a woody base, twining,
finely hairy with hooked hairs. Terminal
leaflet stalked, *c.* 4–5 cm, laterals similar in
size, very shortly stalked, hairy on both
surfaces. Bracteoles large, ribbed, enclosing
the calyx in bud. Corolla 2.5–4 cm, bluish
violet blotched with yellow or white. Pod
6–12 cm, valves twisting when open. *E
USA, N South America.* H5. Late summer.

29. CLITORIA Linnaeus
J. Cullen
Herbaceous or somewhat woody climbers.
Leaves pinnate with 3–9 leaflets including
a terminal leaflet; stipules and stipels
persistent. Flowers large, axillary, solitary
or in racemes, borne upside-down
(standard below); bracteoles large, leaf-like
but much shorter than the calyx. Calyx
tubular to funnel-shaped, 5-toothed, the
upper teeth more or less united. Standard
large, without a spur. Stamens 10, 9 of
them with their filaments united, that of
the uppermost at least partially free. Style
hairy. Pod flattened, containing several
seeds.

A genus of about 70 species from the
tropics. Only 1 is grown, in rich soil in a
warm, sunny, humid situation. It is
propagated by seed or by cuttings taken in
early spring.

Literature: *Clitoria ternatea*: a select
bibliography, *Medicinal and Aromatic Plants
Abstracts*, **10**(2): 163–166 (1988).

1. C. ternatea Linnaeus. Illustration:
Botanical Magazine, 1542 (1813);
Reichenbach, Flora exotica **4**: t. 226
(1835); Bruggemann, Tropical plants, t. 10
(1957); Menninger, Flowering vines of the
world, pl. 150 (1970).

Climber with annual stems from a more
or less persistent stock. Leaves with 5–9
leaflets with smaller and larger, rather
sparse, hooked hairs. Flowers solitary.
Calyx green, 2–2.5 cm. Standard 4–5 cm,
blue with paler markings and white centre,
or entirely white ('Album'). Pod flattened,
7–12 cm, hairy with hooked hairs. *Tropics,
precise origin unknown.* G2. Summer.

30. LABLAB Adanson
J. Cullen
Annual to perennial herbs or climbers.
Hairs, when present, not hooked. Leaves
with 3 leaflets; stipules and stipels
persistent. Flowers in groups of 2–4 in
racemes; bracteoles present. Calyx broadly
bell-shaped, 5-toothed. Corolla large,
standard broad and with auricles at the
base; wings joined to the keel. Stamens 10,
9 with their filaments united, that of the
uppermost free, at least at the base. Style
flattened, bearded. Pod broad containing
several black or white seeds.

A genus of about 60 species from the
Old World tropics, the single cultivated
species often separated into the genus
Lablab Adanson (of which it is the sole
species); it is widely grown in the tropics as

a vegetable and many cultivars are known. Only occasionally grown as an ornamental, its propagation is by seed.

1. L. purpureus (Linnaeus) Sweet (*Dolichos lablab* Linnaeus; *Lablab niger* Medikus; *L. vulgaris* Savi). Illustration: Masefield et al., The Oxford book of food plants, pl. 45 (1969); Smitinand, Wild flowers of Thailand, pl. 101 (1975); *Gartenpraxis* April 1988: 10.

Perennial but usually grown as an annual, twining or rarely erect, to 10 m, hairy or hairless. Leaflets ovate, 4–15 × 4–10 cm, truncate at base, abruptly acuminate at apex. Flowers purple or white, 1.2–2.5 cm. Pod papery, hairy or not, the upper margin straight, the lower curved. *Origin uncertain, now found throughout the tropics.* H5–G1. Summer.

In edn 1 this was treated as a species of the genus *Dolichos* Linnaeus.

31. PHASEOLUS Linnaeus
J. Cullen
Like *Dolichos* but plants with hooked hairs, corolla usually smaller, often red, keel and style coiled through a number of revolutions, stigma terete and bearded longitudinally, pods usually narrower and more or less cylindric.

A genus of 60 species from the New World, mainly from the tropics. Many of its species (together with those of the allied genus *Vigna* Savi, some species having names in both genera) are widely cultivated as vegetables (beans of various kinds). Few of the species are cultivated purely as ornamentals; those that are so cultivated are generally grown as annuals and propagated by seed.

Literature: Marechal, R., Mascherpa, J.-M. & Stainer, F., Étude taxonomique des genres *Phaseolus* et *Vigna*, Boissiera **28** (1978).

1. P. coccineus Linnaeus (*P. multiflorus* Lamarck). Illustration: *Horticulture* **52**: 38 (1974); *Gartenpraxis* March 1976: 111 & April 1988: 12; Lewis et al., (eds.), Legumes of the world, 428 (2005).

Perennial grown as an annual, climbing. Leaflets ovate, to 12 cm. Flowers bright scarlet. Style with 1–2 revolutions. Pods to 30 cm, linear-cylindric, hairy or not. Seeds broadly oblong. *Widely cultivated; origin uncertain, perhaps Mexico.* H2.

The Scarlet runner bean; very variable and with many cultivars of economic importance.

P. caracalla Linnaeus (*Vigna caracalla* (Linnaeus) Verdcourt). Illustration: *Horticulture* **40**: 203 (1962); *Il Giardino Fiorito* **54**: 35 (1988, July/August). Perennial; flowers with reflexed and contorted, pink, white or yellow standard and pink to violet wings; style with 3–5 spirals. *Tropical S America.* G2.

32. BITUMINARIA Fabricius
J.R. Akeroyd
Foetid biennial to perennial herb, somewhat woody at the base, shortly and stiffly hairy; stems 30–120 cm, erect but often rather straggling, branched below. Stipules small, narrow. Leaves with 3 leaflets on long stalks, dotted with glands; leaflets 1–6 cm, narrowly lanceolate to ovate or almost circular, entire. Flowers in dense heads on axillary stalks longer than leaves, subtended by paired bracts. Calyx with 5 unequal teeth. Corolla 1.5–2 cm, bluish violet, sometimes pink or white. Stamens united for much of their length. Fruit indehiscent, ovoid, flattened, 1-seeded, hairy, with long, curved beak; seed 5–6 mm.

A genus of three species from the Mediterranean area, one of which is cultivated. Propagation is by seed.

1. B. bituminosum (Linnaeus) Stirton (*Psoralea bituminosa* Linnaeus; *Aspalthium bituminosum* (Linnaeus) Fourreau). Illustration: Polunin, Flowers of Europe, pl. 54 (1969); Zohary, Flora Palaestina **2**, pl. 66 (1972); Aeschimann et al., Flora alpina **1**: 859 (2004); Lewis et al., (eds.), Legumes of the world, 450 (2005).
Mediterranean area. H1. Summer.
A variable species. The whole plant has a distinctive smell of pitch, and the sap will blister the skin in sunlight.

33. PSORALEA Linnaeus
S.G. Knees
Scented herbs or shrubs. Leaves alternate, pinnate with a terminal leaflet, rarely simple, with translucent dots. Flowers solitary, in heads, racemes, spikes or sometimes clustered. Calyx-lobes 5, nearly equal; corolla wings about as long as keel; standard ovate or rounded; stamens 10, all united or 1 free. Fruit a short, 1-seeded, indehiscent legume.

1. P. pinnata Linnaeus. Illustration: Andrews, The botanist's repository **7**: 474 (1807); *Journal of Horticulture*, series 3, **7**: 281 (1884); Fryson, Flora of Nilgiri & Pulney hill-tops **2**: 85 (1915); Gibson,

Wild flowers of Natal (Coastal Region), pl. 43 (1975).

Compact shrub to 4 m. Leaves crowded along erect stems, with 5–11, needle-like leaflets, 2.5–3 cm, linear to linear-lanceolate, acute. Flowers solitary or clustered; calyx *c.* 9 mm; corolla *c.* 1.3 cm, blue, striped white; stamens 9 united, 1 free; style *c.* 1.1 cm, curved. *S Africa, naturalised elsewhere.* H5–G1. Summer–autumn.

34. AMORPHA Linnaeus
J. Cullen
Shrubs. Leaves pinnate with up to 45 leaflets, including a terminal leaflet, leaflets usually with brownish glands on the lower surface; stipules soon falling, stipels thread-like. Flowers very numerous in dense, spike-like racemes which may be clustered to form an apparent panicle, flower-stalks short. Calyx funnel-shaped, 5-toothed, sometimes with glands near the top. Corolla reduced to the standard (wings and keel absent) which is purplish, clawed and envelops the stamens and style. Stamens 10, their filaments all united into a tube below; anthers yellow, brown or purplish. Pod 1-seeded, indehiscent, glandular.

A genus of 15 very similar species from N America and adjacent Mexico. They are best grown in a sunny site and are propagated by division or by cuttings.

Literature: Wilbur, R.L., A revision of the north American genus *Amorpha* (Leguminosae-Psoraleae), *Rhodora* **77**: 337–409 (1975).

1a. Shrub of 1 m or more; leaf-stalks obvious, longer than the width of the lowermost leaflet; pod 5–9 mm
 3. fruticosa
 b. Shrub to 1 m, often much less; leaf-stalks very short or absent; pods 3–5.5 mm 2
2a. Racemes clustered; plants often ash-grey, hairy; glands on lower leaf surface small and inconspicuous
 1. canescens
 b. Racemes solitary; plants not hairy as above; glands on the lower leaf surface conspicuous **2. nana**

1. A. canescens Pursh. Illustration: *Botanical Magazine*, 6618 (1882); Rickett, Wild flowers of the United States **1**: pl. 81 (1966); *Gartenpraxis* September 1976: 425.

Shrub to 80 cm, rarely more. Leaves spreading, to 12 cm, stalk very short or almost absent, with 27–47 leaflets; leaflets

1–2.5 cm, narrowly elliptic, tapered to the base and the acute or somewhat rounded, mucronate apex, shortly stalked, variably (often densely) with grey-white hairs on both surfaces and small, inconspicuous brownish glands beneath. Racemes clustered, to 20 cm or more. Calyx 3–4 mm, densely hairy. Standard 4.5–6 mm, bright violet, abruptly tapered into the claw. Anthers yellow to brownish. Pod 3–5 mm, glandular and usually hairy. *E North America (Manitoba to Texas).* H3. Summer.

2. A. nana Nuttall (*A. microphylla* Pursh). Illustration: *Botanical Magazine*, 2112 (1820); Rickett, Wild flowers of the United States **4**: pl. 122 (1970); Lewis et al., (eds.), Legumes of the world, 302 (2005).

Small shrub to 60 cm. Leaves spreading, to 7 cm, stalk very short or almost absent, with 7–22 leaflets; leaflets 6–12 mm, narrowly elliptic to oblong, broadly tapered to the base, rather rounded to the mucronate apex, sparsely hairy on the margins and main vein beneath or completely hairless, with a few, large, conspicuous brownish glands. Racemes solitary, to 9 cm. Calyx *c.* 3 mm, usually hairless except on the margins of the teeth. Standard 4.5–6 mm, abruptly tapered into the claw, dark purple. Anthers purplish. Pods 4.5–5.5 mm, glandular, not hairy. *C North America (Manitoba & Saskatchewan to New Mexico).* H3. Summer.

3. A. fruticosa Linnaeus. Illustration: Rickett, Wild flowers of the United States **1**: pl. 81 (1966); Justice & Bell, Wild flowers of North Carolina, 100 (1968); *Botanical Magazine*, n.s., 604 (1970–72); Polunin & Everard, Trees and bushes of Europe, 113 (1976); Aeschimann et al., Flora alpina **1**: 859 (2004).

Shrub 1–4 m. Leaves spreading, to 25 cm, stalk obvious, at least as long as the width of the lowermost leaflet; leaflets 9–21 (rarely more), 2–5 cm, oblong or elliptic, tapering or rounded to the base and the mucronate apex, variably hairy beneath and with small, inconspicuous glands or glandless. Racemes clustered, to 20 cm. Calyx 3–4 mm, sparsely hairy and with a few glands. Standard 5–6 mm, gradually tapering into the claw, dark reddish purple. Anthers yellow. Pod 5–9 mm, glandular, hairy or not, *N America (Quebec to Florida, California, Wyoming & N Mexico).* H3. Summer.

35. AMICIA Humboldt, Bonpland & Kunth
J. Cullen

Softly wooded shrubs. Stems upright. Leaves pinnate, with 4 or 5 leaflets including a terminal leaflet, all with orange-brown glands on the lower surface; stipules large, almost circular, united at the base, enclosing the young growth, soon falling; stipels represented by dense tufts of hair. Flowers axillary, solitary or in few-flowered racemes; bracts similar to stipules but smaller. Calyx with a large, almost circular (stipule-like) upper tooth, the lateral teeth very small, the lower teeth narrow. Corolla large, yellow, keel blunt, standard glandular outside. Stamens 10, their filaments all united, the tube so formed split above. Pod several-seeded, indehiscent, breaking into 1-seeded segments.

A genus of 7 species in C & S America. The cultivated species requires a sunny site with protection from cold and wind; it may be killed to the root in hard winters. Propagation is by cuttings taken in summer.

1. A. zygomeris de Candolle. Illustration: *Botanical Magazine*, 4008 (1843); *Paxton's Magazine of Botany* **13**: 173 (1847); McVaugh, Flora Novo-Galiciana **5**: 277 & frontispiece (1987); Lewis et al., (eds.), Legumes of the world, 312 (2005).

Shrub to 2 m; stems with spreading hairs of varying lengths. Leaflets green above, glaucous beneath, oblong, rounded to the base, deeply and widely notched at the apex, to 6 × 5 cm. Upper calyx-tooth to 1.5 cm. Standard to 3 × 2.5 cm, yellow or orange-yellow, keel yellow with a brownish or purplish blotch towards the apex. *Mexico.* H5.

A handsome flowering shrub and striking foliage plant, whose leaflets show sleep-movements.

36. ADESMIA de Candolle
J. Cullen

Herbs or shrubs, usually with stalkless or stalked glands, often sticky. Leaves pinnate (ours) with many leaflets, including a terminal leaflet; stipules small, persistent, stipels absent. Flowers in axillary racemes (ours). Calyx bell-shaped, 5-toothed. Corolla yellow or reddish. Stamens 10, their filaments free. Pod indehiscent, breaking transversely into 1-seeded segments, often glandular.

A genus of 230 species in S South America (Peru and Brazil to Tierra del Fuego), of which 1 is occasionally grown as a curiosity. Propagation is by seed when available.

Literature: Burkart, A., Sinopsis del genero sudamericano de Leguminosas, *Adesmia* DC., *Darwiniana* **14**: 463–568 (1967).

1. A. boronioides Reiche. Illustration: *Botanical Magazine*, 7748 (1900).

Low shrub, very glandular and sticky, glands stalkless. Leaves with many, irregularly toothed, notched, fleshy leaflets. Flowers yellow (standard sometimes with red lines), 8–10 mm. Pod with few segments, the upper suture straight, the lower deeply indented, the segments rounded, very glandular. *Chile, Argentina.* H5–G1.

37. CLIANTHUS Lindley
J. Cullen

Herbs or soft-wooded shrubs. Leaves pinnate with numerous lateral leaflets and a terminal leaflet; stipules large, persistent, stipels absent. Flowers very showy in erect or pendent axillary racemes, each with a bract and 2 bracteoles. Calyx bell- or cup-shaped, 5-toothed. Corolla scarlet or scarlet and black, pink or white (a cultivar). Standard reflexed, narrow, acuminate, wings deflexed, shorter than the keel; keel large, acuminate, curved like a parrot's beak. Stamens 10, filaments of 9 of them united, that of the uppermost free. Ovary shortly stalked; style bearded. Pod firm, turgid and beaked, containing several seeds.

A genus of 2 species, 1 from Australia, the other from New Zealand. Both are cultivated and are among the most spectacular of the smaller legumes. Their cultivation is not entirely easy in Europe, though they are more easily grown in the southern hemisphere, where they are often treated as annuals. They require a sunny, open site and well-drained, deep, neutral to alkaline soil. Propagation is by seed (which requires scarification) or cuttings; young plantlets are very susceptible to damping-off. Seedlings of *C. formosus* are sometimes grafted on to seedling stocks of *C. puniceus* or species of *Colutea*.

1. C. puniceus (G. Don) Lindley. Illustration: *Botanical Magazine*, 3584 (1837); *Journal of the Royal Horticultural Society* **84**: f. 38 (1959); Moore & Irwin, The Oxford book of New Zealand plants,

162 (1978); Moggi & Guignolini, Fiori da balcone e da giardino, 86 (1982).

Soft-wooded shrub with branches to 2 m, sparsely adpressed hairy throughout. Leaves with up to 15 pairs of leaflets; leaflets narrowly lanceolate to linear-oblong, rounded to the blunt apex. Racemes with up to 15 flowers, pendent; bracteoles small, distant from the calyx. Calyx broadly bell-shaped, tube 7–10 mm, teeth 1.5 mm. Corolla scarlet, reddish or rarely white; standard to 6 cm, erect or somewhat reflexed, without a swelling at the base of the blade; wings much shorter than keel; keel curved forwards, to 8 cm, acuminate and beaked. Pods to 8 cm. *New Zealand (North Island)*. H5–G1.

In 'Albus' the flowers are white and in 'Roseus', reddish.

38. SWAINSONA Salisbury
J. Cullen

Herbs or shrubs, covered generally with short, adpressed hairs. Leaves pinnate, with a terminal leaflet; stipules conspicuous, persistent, stipels absent. Flowers in axillary racemes: bracts present, bracteoles present, rarely very small or absent. Calyx bell-shaped, with 5 nearly equal teeth. Standard upright, blade more or less circular; wings curved or twisted; keel broad, curving or somewhat coiled upwards, obtuse (ours). Stamens 10, 9 of them with filaments united, that of the uppermost free. Ovary stalked or not, with several ovules; style longitudinally bearded. Pod ovoid, membranous, translucent and inflated (ours).

A genus of about 50 species from Australia; only 3 are in general cultivation. They are easily grown in a dry, sunny site, and are propagated by seed or by cuttings.

1a. Corolla clear scarlet **1. formosa**
  b. Corolla purplish blue, pink, red, reddish brown or rarely white **2. galegifolia**

1. S. formosa (G. Don) J. Thompson (*Clianthus dampieri* R. Brown; *C. formosus* (G. Don) Ford & Vickery). Illustration: *Botanical Magazine*, 5051 (1858); Erickson et al., Flowers and plants of Western Australia, 143, 147 (1973); Rotherham et al., Flowers and plants of New South Wales and southern Queensland, 162 (1975); Elliot & Jones, Encyclopaedia of Australian plants **3**: 50, 51 (1984).

Sprawling annual or perennial herb, densely woolly throughout. Leaves with up to 10 pairs of leaflets; leaflets elliptic to oblong, acute or rounded at the apex, greyish with dense woolly hairs, at least beneath. Racemes erect, few-flowered; bracteoles close to and overlapping the calyx. Calyx tubular to bell-shaped, tube 6–7 mm, teeth 5–7 mm, all greyish with woolly hairs. Corolla clear scarlet; standard erect, vertical, with a usually black swelling at the base of the blade; wings shorter than keel; keel deflexed downwards vertically but curving outwards, acuminate. Pods to 6 cm, hairy. *Australia (Queensland, New South Wales, Western Australia, South Australia, Northern Territories)*. H5–G1.

In edn. 1 this species was included in *Clianthus*.

2. S. galegifolia R. Brown. Illustration: *Botanical Magazine*, 792 (1805) – as *Colutea galegifolia* & 1725 (1815) – as *S. coronillifolia*; Rotherham et al., Flowers and plants of New South Wales and southern Queensland, 63 (1975); Galbraith, Collins' fieldguide to the wild flowers of south-east Australia, pl. 15 (1977); Cunningham et al., Plants of western New South Wales, 413 (1981).

Shrub to 1 m or more. Leaves with 5–10 pairs of oblong, obtuse or notched leaflets. Calyx 3–6 mm. Corolla purplish blue, pink, red, reddish brown or rarely white; standard to 1.6 cm wide, with a spot of paler colour at the base, and with 2 oblique plate-like swellings at the base; keel broad, obtuse. Pod inflated, membranous, 2.5–5 cm, on a short stalk. *Australia (Queensland, New South Wales)*. H5–G1.

S. procumbens Mueller. Illustration: Cochrane et al., Flowers and plants of Victoria, 89 (1968); Hodgson & Paine, Field guide to Australian wild flowers **1**: 183 (1971–77). Similar, but leaflets narrow and distant, corolla mauve-blue, the standard broader than long, deeply notched, keel coiled upwards through somewhat more than 1 revolution. *Australia (South & Western Australia, New South Wales)*. H5–G1. Summer.

39. SUTHERLANDIA R. Brown
H.S. Maxwell

Evergreen erect or prostrate shrubs, 60–200 cm. Leaves alternate, pinnate with a terminal leaflet. Leaflets green with greyish hairs. Stipules small, usually linear, broader at base. Inflorescence a slender, axillary raceme. Flowers showy, rose to bright red or purple. Calyx bell-shaped, lobes 5, triangular, all segments pointing forwards; standard folded, veined, shorter than the keel. Stamens 10, 9 united and 1 separate. Flowers followed by 'puffy' pale green somewhat translucent pods.

A genus of 5 species native to South Africa, only 1 species is in general cultivation. They are easily grown in a greenhouse or in full sun outdoors in a well-drained leaf-rich soil and can be tolerant of a few degrees of frost. Propagation is easy from seed or from cuttings; they should be pruned in late winter.

When the pods of *S. frutescens* float on water they are seen as being like ducks hence the name 'Duck Plant'; it was also at one time thought to be a cancer cure.

1. S. frutescens (Linnaeus) R. Brown (*Colutea frutescens* Linnaeus). Illustration: Everett, The New York Botanical Gardens illustrated encyclopedia of horticulture **10**: 3263 (1981); Phillips & Rix, Shrubs 26 (1989); Brickell, The RHS gardeners' encyclopedia of plants & flowers, 133 (1989); Macoboy, What shrub is that? 322 (1989).

Shrub 60–160 cm. Leaves 4.5–9 cm, with 13–21 narrowly oblong to linear-elliptic leaflets, 5–20 mm, sometimes slightly mucronate, usually hairy beneath, hairless above or almost so. Inflorescence a 6–10 flowered raceme, 2.5–8 cm, flower-stalks to 8 mm, with stiff hairs; bracts *c.* 2 mm, ovate. Flowers bright red, 2.5–5 cm; keel 2.5–3.5 cm, claw 8–15 mm. Ovary stalked. Pods papery, translucent, inflated, broadly elliptic to almost spherical, 3.5–5.5 × 2–4 cm. *S Africa, Namibia*. Late spring–summer.

40. COLUTEA Linnaeus
J. Cullen

Shrubs or rarely small trees. Leaves pinnate, with 2–6 pairs of leaflets and a terminal leaflet, often somewhat glaucous; stipules persistent, stipels absent. Flowers rather few in axillary racemes, with bracts. Calyx usually broadly bell-shaped, with 5 short teeth. Corolla yellow to orange-red, the claws of the petals usually projecting from the calyx. Standard upright, with 2 small swellings above the claw; wings shorter than to longer than the keel, each with a distinct auricle on the upper margin and sometimes a spur on the lower; keel rounded or beaked at the apex. Stamens 10, 9 with their filaments united, that of the uppermost free. Ovary stalked. Pod inflated, papery, and translucent, indehiscent or splitting towards the apex, stalked, the stalk often projecting from the persistent calyx; seeds several.

A genus of 26 rather similar species from Eurasia and E Africa (Ethiopia). The species have been frequently confused, both in the wild and in gardens and the most commonly seen cultivated *Colutea* is the hybrid *C. × media*. They are easily grown in a sunny position in well-drained soil and are propagated by seed or by cuttings taken in autumn.

Literature: Browicz, K., A revision of the genus *Colutea*, *Monographiae Botanicae* **16** (1963).

1a. Ovary and pod hairless   2
 b. Ovary hairy, pod often hairy   4
2a. Keel terminating in a distinct beak; flowers orange-red   **4. orientalis**
 b. Keel blunt, not beaked; flowers yellow   3
3a. Wings shorter than to as long as keel, rarely longer, without a spur on the lower margin   **1. arborescens**
 b. Wings longer than keel, with a distinct spur on the lower margin   **2. cilicica**
4a. Older shoots grey-brown, matt; leaflets usually more than 1.5 cm   **1. arborescens**
 b. Older shoots red- or purple-brown, shining; leaflets to 1.5 cm   5
5a. Keel distinctly beaked; corolla orange-yellow   **5. buhsei**
 b. Keel blunt, not beaked; corolla yellow   **3. gracilis**

1. C. arborescens Linnaeus. Illustration: Polunin & Everard, Trees and bushes of Europe, 112 (1976); Krüssmann, Manual of cultivated broad-leaved trees and shrubs **1**: 361 (1986); *Il Giardino Fiorito*, June 1987: 55; Aeschimann et al., Flora alpina **1**: 839 (2004); Lewis et al., (eds.), Legumes of the world, 482 (2005).

Shrub to 5 m; young shoots hairy at first, older shoots grey-brown, peeling, matt. Leaves with 3–6 pairs of leaflets; leaflets ovate to broadly elliptic or rarely obovate, to 3 × 2 cm, rounded to the base and to the obtuse or notched apex, hairless above, sparsely adpressed hairy beneath. Racemes with 3–8 flowers. Calyx 6–8 mm, covered with adpressed hairs which are white, dark brown to black or a mixture of the two. Corolla yellow, standard with reddish veins; wings shorter than to as long as keel, rarely longer, without a spur on the lower margin; keel 1.6–2 cm, blunt at apex. Ovary hairless or hairy. Pod

5–8 cm, usually hairless. *C, S & E Europe.* H2. Summer.

Often recorded from east of Europe, but there mistaken for *C. cilicica.* Two subspecies occur: subsp. **arborescens** from C & E Europe, with ovary hairless, and subsp. **gallica** Browicz, from SW and part of C Europe, with ovary hairy. Both are apparently in cultivation. Browicz mentions 2 cultivars, 'Crispa' with the leaflet-margins wavy, and 'Bullata' with the surface wrinkled.

C. × media Willdenow is the hybrid between this and *C. orientalis*, and is more widely cultivated than either of its parents. It is intermediate between the 2 in most characters, with brownish red or orange flowers with the keel 1.5–1.6 cm, scarcely beaked. Illustration: Seabrook, Shrubs for your garden, 45 (1975); Csapody & Tóth, Colour atlas of flowering trees and shrubs, pl. 67 (1982); Brickell (ed.), RHS A–Z encyclopedia of garden plants, 298 (2003).

2. C. cilicica Boissier & Balansa. Illustration: Flora SSSR **11**: pl. 22 (1945).

Similar to *C. arborescens* but leaflets usually smaller (mostly to 2 × 1.4 cm), calyx broadly bell-shaped, 7–9 mm, wings always longer than the keel and with a spur on the lower margin, keel 2–2.2 cm, ovary always hairless. *Greece, Turkey, Syria, Lebanon, Israel, former USSR (Caucasus, Crimea & adjacent areas).* H4. Summer.

Difficult to distinguish from *C. arborescens*; intermediates occur in E Europe and may also be in cultivation.

3. C. gracilis Freyn & Sintenis. Illustration: Flora SSSR **11**: pl. 22 (1945).

Like *C. arborescens* but bark on older shoots brownish red or brownish violet, shining; leaflets 3–7 × 2–6 mm, rounded or very slightly notched at apex, corolla pale yellow, wings longer than keel, without a spur on the lower margin, keel 1.4–1.6 cm; ovary always hairy, pod 3–4.5 cm, sparsely hairy. *Iran/former USSR border area (Kopet Dağ).* H3? Summer.

4. C. orientalis Miller. Illustration: Flora SSSR **11**: pl. 22 (1945); Krüssmann, Manual of cultivated broad-leaved trees and shrubs **1**: 361 (1986).

Shrub to 3 m, young shoots hairless, older shoots brownish grey, matt. Leaves with 2–4 pairs of leaflets; leaflets often bluish green, broadly obovate, tapered to the base, rounded or notched at the apex, to 1.8–1.5 cm. Racemes with 3–5 flowers. Calyx 5–6 mm, covered in mostly black

hairs. Corolla orange-red, standard with darker veins and a pale yellow spot near the base of the blade; wings shorter than keel; keel 1.1–1.3 cm, beaked. Ovary hairless. Pod to 5 × 2 cm, stalk short or absent. *Former USSR (Caucasus).* H3. Summer.

5. C. buhsei (Boissier) Shaparenko (*C. persica* misapplied var. *buhsei* Boissier). Illustration: Flora SSSR **11**: pl. 22 (1945).

Shrub to 3 m; young shoots hairy at first, older shoots brownish red and shining. Leaves with 3 or 4 pairs of leaflets; leaflets to 1.5 × 1.2 cm, broadly obovate, rounded or somewhat notched at the apex, hairless above, with adpressed hairs beneath. Raceme with 2–5 flowers. Calyx 7–8 mm, sparsely hairy with mixed white and dark brown to black hairs. Corolla orange-yellow, wings longer than keel, keel 2–2.2 cm, beaked. Ovary hairy. Pod 6–7.5 cm, hairy, shortly stalked. *N Iran & adjacent former USSR.* H2. Summer.

41. HALIMODENDRON de Candolle
J. Cullen

Round-headed shrub with pale brown, striped bark, with weakly distinguished long and short shoots. Leaves on the extension shoots persisting as stout spines to 6 cm, each with 2 much smaller stipular spines at the base. Foliage leaves with usually 4 leaflets, pinnately arranged, the axis tipped with a short spine; leaflets glaucous, oblong-obovate, tapered to the base, rounded to the apex, to 4 cm, hairy or hairless; stipules of the short-shoot leaves membranous. Racemes terminating the (axillary) short shoots, with 2–5 flowers each with a bract and 2 bracteoles. Calyx 5–7 mm, bell-shaped, finely hairy, truncate, with 5 small teeth. Corolla 1.5–2 cm, pale purplish pink or magenta, standard erect with its sides reflexed. Stamens 10, filaments of 9 of them united, that of the uppermost free. Ovary shortly stalked, with several ovules. Pod stalked, inflated, obovoid or obovoid-oblong, leathery, beaked, yellowish brown, containing several seeds.

A genus of a single species occurring from E Europe (Russia) to C & E Asia, usually found on saline soils. It is often grown grafted on to stocks of *Caragana arborescens* (p. 365) but can also be grown on its own roots. It requires a sunny position on well-drained soil. Propagation is by seed or by grafting.

1. H. halodendron (Pallas) Voss
(*H. argenteum* (Lamarck) de Candolle).
Illustration: Flora SSSR **11**: t. 20 (1945);
Schischkin, Botanicheskii atlas, t. 33
(1963).

E Europe to C & E Asia. H3. Late spring.

42. CARAGANA Lamarck
J. Cullen

Shrubs or small trees, often with widely
spreading or pendent, little-branched
branches. Shoots generally of 2 kinds: long,
extension shoots on which the leaves are
borne alternately, their axes often
persisting as spines, their stipules also often
spiny and persistent; and short shoots
borne in the axils of the extension-shoot
leaves, very long-lived, terminated by
usually membranous stipules and bearing
clusters of leaves and the flowers. Leaves
with 4 or more leaflets, mostly pinnate but
some with the leaflets borne in a close
cluster (palmate), the terminal leaflet
replaced with a spine. Stipules present,
joined at the base, divergent at the apex,
membranous at first, persistent, some with
thickened midribs which persist as spines
as the membranous tissue decays. Flowers
solitary or in clusters, each with its own
jointed stalk arising from the short shoot,
rarely in few-flowered, stalked umbels;
bracteoles sometimes persisting at the joint.
Calyx bell-shaped or cylindric, truncate or
oblique at the mouth, 5-toothed. Corolla
usually yellow, more rarely orange or pink,
the petals with long claws projecting from
the calyx; standard erect, its margins
reflexed, wings usually each with a
backwardly projecting auricle. Stamens 10,
filaments of 9 united, that of the
uppermost free. Ovary containing several
ovules. Pod several-seeded, hairless or
hairy inside.

A genus of perhaps 65 species from Asia
(just extending into Europe in the western
former USSR). They are often of
picturesque form and old plants can make
striking specimens. Their shoot
architecture, with strongly differentiated
long and short shoots is remarkable; it
is very difficult to identify material
consisting only of extension shoots. The
classification of the genus is very confused
and in need of considerable revision; the
identification of cultivated material is
not easy.

They are in general easily grown in a
very sunny position in well-drained soil.
Propagation is best by seed, though
cuttings and layering can also be used.

Literature: Komarov, V.L., Monografiya
roda *Caragana*, Acta Horti Petropolitani **29**:
178–362 (1908); Sanchi, C., Sistema
roda *Caragana* Lam, Novosti Sistematiki
Vysshich Rastenii **32**: 76–90 (2000).

Leaf-spines. Present on long shoots: **3, 4, 5,
6, 7, 8, 9, 10, 11, 12, 13, 14**; absent:
1, 2, 3, 4, 5.
Leaves. Mostly pinnate with 4 leaflets in 2
pairs, or more: **1, 2, 3, 4, 5, 6, 7, 8, 9,
10**; leaflets 4 in a cluster, not pinnately
arranged: **11, 12, 13, 14**. *Flowers.* In
an umbel: **4**. With persistent bracteoles:
4, 7.
Calyx. Truncate at the apex with well-
separated teeth: **1**; oblique at the apex
with adjacent teeth: **2, 3, 4, 5, 6, 7, 8,
9, 10, 11, 12, 13, 14**. Hairy on the
surface: **1, 2, 4, 6, 7, 8, 9**; hairless on
the surface (though hairy on the
margins of the teeth): **3, 5, 10, 11, 12,
13, 14**.
Corolla. Orange: **14**; white flushed with
pink, or pink: **9**.

1a. Leaves composed of 4 leaflets which
    form a cluster (i.e. are not pinnately
    arranged) usually below the tip of
    the leaf-axis   2
  b. Leaves composed of 4 or more
    leaflets which are pinnately arranged
    5
2a. Leaves of the short shoots stalked
    (sometimes shortly so), those on
    the long shoots also shortly stalked
    (leaving a single scar on the
    spine)   3
  b. Leaves on the short shoots stalkless
    (those on the long shoots shortly
    stalked)   4
3a. Leaflets obovate to reversed-
    triangular, broadest almost at the
    apex which is rounded or truncate,
    completely hairless; flower-stalk
    jointed above the middle **12. frutex**
  b. Leaflets very narrowly obovate or
    linear-obovate, broadest $1/4$–$1/3$ of the
    length from the apex, which is
    tapered, with flaky scales beneath;
    flower-stalk jointed at the middle or
    below   **11. densa**
4a. Calyx to 5 mm; corolla orange or
    pinkish yellow; ovary hairless
    14. aurantiaca
  b. Calyx 7–10 mm; corolla clear yellow;
    ovary downy **13. pygmaea**
5a. Leaf-stalks and axes of the leaves on
    the long shoots all deciduous, not
    persisting as spines   13

  b. Leaf-stalks and axes of the leaves
    on the long shoots persisting as
    spines, or at least some of them so
    doing   6
6a. Flowers borne in an umbel of 3–5
    on a stalk which arises from the
    short shoot **4. brevispina**
  b. Flowers borne singly or in clusters,
    each with its own stalk arising from
    the short shoot   7
7a. Most leaves with 2 pairs of
    leaflets (occasionally a few with 3
    pairs)   8
  b. Most leaves with 3 or more pairs of
    leaflets (occasionally a few with 2
    pairs)   11
8a. Surface of calyx hairy, ribbed
    6. maximowicziana
  b. Surface of calyx hairless (margins of
    teeth usually hairy)   9
9a. Calyx 1 cm or more; leaflets broadly
    obovate-elliptic, 1–1.5 cm × 5–7 mm,
    smoothly rounded to the apex which
    has a point to 1 mm, hairless
    10. sinica
  b. Calyx to 1 cm; leaflets various,
    not as above, narrower (to 5 mm)
    or if broader then with the point
    at the apex at least 2 mm and
    with sparse silvery hairs on both
    surfaces   10
10a. Leaves dark green, hairless above,
    clearly stalked, leaflets oblong-
    obovate, in 2 very close pairs (some
    almost with the 4 leaflets clustered);
    point at the apex to 1 mm
    5. spinosa
  b. Leaves pale grey-green, with sparse
    silky hairs on both surfaces,
    indistinctly stalked, leaflets obovate,
    clearly pinnately arranged, point at
    the apex 1–2.5 mm **3. decorticans**
11a. Corolla 2.5–3 cm, cream flushed
    with pink or entirely pale to deep
    pink; spines very numerous, thin,
    flexible; whole plant covered with
    long, spreading hairs **9. jubata**
  b. Corolla to 2.5 cm, yellow (standard
    sometimes orange); spines rather
    few, stout, not flexible; plant
    variously hairy, not as above 12
12a. Mature leaflets densely silky-hairy on
    both surfaces; flower-stalks jointed in
    the lower half, without persistent
    bracteoles **8. gerardiana**
  b. Mature leaflets with only the
    margins hairy; flower-stalks jointed
    in the upper half with 3 persistent
    bracteoles at the joint
    7. franchetiana

13a. Leaflets to 8 mm at most, veins diverging from midrib at a very acute angle; calyx cylindric, mouth oblique, teeth adjacent
  2. microphylla

  b. Leaflets more than 8 mm, veins diverging from the midrib at a wide angle; calyx truncate, teeth distant
  1. arborescens

1. C. arborescens Lamarck. Illustration: Huxley, Deciduous garden trees and shrubs in colour, pl. 28 (1979); Csapody & Tóth, Colour atlas of trees and shrubs, 147 (1982); Davis, Gardener's illustrated encyclopaedia of trees and shrubs, 21, 22 (1987); *Gartenpraxis* May 1987: 56 August 1988: 24.

Shrub or small tree to 6 m, bark grey-brown. Short shoots well developed. Leaves of main shoots deciduous, the leaflets falling first, the axes sometimes shortly retained but not persistent and forming spines. Leaves pinnate with 4–6 pairs of leaflets; leaflets oblong or oblong-elliptic, rounded to the base and minutely pointed apex, to 2.5 × 1 cm, veins diverging from the midrib at a wide angle, hairless above, silky-hairy beneath; stipules membranous, not spiny. Flower-stalks to 2.5 cm, hairy, jointed above the middle. Calyx bell-shaped, 6–8 mm, truncate with small, distant teeth, hairy all over the surface. Corolla yellow, to 1.8 cm. Pod 2.5–3 cm, hairless. *E former USSR, Mongolia, NE China.* H2. Summer.

A widespread and variable species; 'Pendula' has pendent branches and 'Nana' is rather low-growing. Several species related to *C. arborescens* are reputedly in cultivation; these are scarcely distinguishable from *arborescens* or from each other.

C. boisii Schneider (*C. arborescens* var. *crasseaculeata* Bois). Smaller, leaves dark but clear green above, paler beneath, stipules of the main shoot leaves spiny, calyx-teeth somewhat larger and more conspicuous. *W China.* H5. Summer.

C. fruticosa Besser. Like *C. boisii* with spiny stipules, but leaves bright green above and beneath. *E former USSR, Korea.* H2. Summer.

C. × sophorifolia Tausch is reputedly the hybrid between *C. arborescens* and *C. microphylla*. It is very like *arborescens*, showing little influence of its other parent, but is lower growing, with smaller leaflets and spiny stipules on the main shoots. *Garden origin.* H2. Summer.

2. C. microphylla Lamarck.

Shrub to 3 m, but often much smaller; bark pale grey-brown. Short shoots well developed, covered in membranous stipules. Leaves of the main shoots not persistent as spines, but stipules forming short spines. Leaves with 4 or more pairs of leaflets; leaflets mostly to 5 × 3 mm (rarely to 8 mm), obovate-elliptic, rounded to the base and to the conspicuously mucronate apex, grey-green, densely silvery-hairy on both surfaces, veins diverging from the midrib at a very acute angle. Flower-stalk to 1.2 cm, densely hairy, jointed at about the middle. Calyx 1 cm or more, hairy all over the surface, cylindric, ribbed, mouth oblique, teeth *c.* 1 mm, triangular, adjacent. Corolla yellow, to 2.5 cm. Pod 2.5–3 cm. *E former USSR, Mongolia.* H2. Summer.

3. C. decorticans Hemsley. Illustration: *Acta Horti Petropolitani* **29**: t. 14 (1908).

Shrub or small tree, bark shining yellow-brown. Short shoots well developed, covered in membranous stipules. Axes of leaves of long shoots persistent or not as spines; stipules of these leaves spiny, persistent. Leaves with usually 2 pairs (rarely 3 pairs) of leaflets, clearly pinnately arranged; leaflets *c.* 1 cm × 5 mm, grey-green, obovate, tapered to the base, abruptly narrowed to a long point 1–2.5 mm, veins diverging from the midrib at an acute angle, persistently and sparsely silvery-hairy on both surfaces. Flower-stalks to 1.8 cm, shortly hairy, jointed above the middle. Calyx cylindric, to 6 mm, mouth oblique, hairless, teeth *c.* 1.5 mm, triangular, adjacent. Corolla yellow, to 2 cm. Pod to 4 cm, dark brown, hairless. *Afghanistan.* H3. Summer.

4. C. brevispina Royle. Illustration: Brickell (ed.), RHS A–Z encyclopedia of garden plants, 229 (2003).

Shrub to 3 m, bark grey-brown, striped. Short shoots small, covered by membranous stipules. Leaves of long shoots persisting as spines to 5 cm, each with 2 much smaller spiny stipules at the base. Leaves pinnate with 5 or more pairs of leaflets; leaflets oblong to elliptic, rounded to the base and to a short point at the apex, hairless, veins conspicuous, spreading at a wide angle from the midrib, dark green above, pale greyish green beneath. Flowers 3–5 in a stalked umbel; flower-stalks hairy; bracteoles persistent. Calyx cylindric to bell-shaped, to 1 cm, finely hairy throughout, mouth oblique, teeth

triangular, 2–3 mm, adjacent. Corolla yellow, sometimes flushed with orange, to 2 cm. Pod to 5 cm, hairy outside and inside. *W Himalaya.* H4. Summer.

5. C. spinosa de Candolle (*C. ferox* Lamarck).

Shrub to 2 m with long arching branches; bark grey-brown, striped. Short shoots well developed, covered in membranous stipules. Axes of long shoot leaves persistent as spines to 3 cm; stipules of these leaves membranous, their midribs ultimately becoming spiny. Leaves with 2 pairs of leaflets close but usually pinnately arranged, rarely some with the 4 leaflets clustered; leaflets to 1.8 cm × 4 mm, oblong-obovate, tapered to the base, rounded to the ultimately pointed apex, hairless when mature (silky-hairy when young), veins at an acute angle to the midrib. Flower-stalks hairy, very short, mostly hidden within the short shoot. Calyx cylindric, mouth oblique, hairless except for the margins of the teeth, to 8 mm, teeth to 1.5 mm, adjacent. Corolla yellow to 1.8 cm. Pod to 2 cm, hairless. *E former USSR, N China.* H2. Summer.

6. C. maximowicziana Komarov. Illustration: *Acta Horti Petropolitani* **29**: t. 11 (1908).

Much-branched, intricate shrub to 2 m; bark brown, shining. Short shoots well developed, covered in spiny stipules and short, somewhat persistent leaf-axes. Leaves on long shoots persistent as spines 1.5–4 cm, with rather broad stipular spines at the base. Leaves of 2 or rarely 3 pairs of pinnately arranged leaflets; leaflets narrowly obovate, gradually tapered to the base, rounded to the ultimately mucronate apex, to 10 × 3 mm, hairy. Flower-stalks short, hairy, mostly hidden within the short shoot. Calyx cylindric, ribbed, to 1 cm, hairy all over the surface, mouth oblique, teeth *c.* 1 mm, adjacent. Corolla yellow, *c.* 2 cm. Pod to 2 cm, hairy. *W China.* H4. Summer.

7. C. franchetiana Komarov. Illustration: *Acta Horti Petropolitani* **29**: t. 13 (1908).

Spreading or compact shrub to 2 m; bark grey-brown, young shoots hairless. Short shoots well developed, covered with membranous stipules. Leaves of the main shoots persistent as distinct, stout spines to 5 cm; stipules of these leaves membranous, not spiny. Leaves with 4–6 pairs of leaflets, pinnately arranged; leaflets to 1.2 cm × 4 mm, obovate, tapered to the base, rather abruptly tapered to a long fine point, densely silky-hairy when young, only hairy on the margins when mature. Flower-stalks hairy, to 1.5 cm, jointed above the middle and bearing 3 persistent bracteoles at the joint. Calyx cylindric, finely hairy throughout, to 1.5 cm, mouth oblique, teeth *c.* 3 mm, triangular, adjacent. Corolla yellow, the standard sometimes with orange markings, to 2.5 cm. Pod to 5 cm, hairy. *W China.* H4. Summer.

8. C. gerardiana (Graham) Bentham. Illustration: *Acta Horti Petropolitani* **29**: t. 13 (1908).

Like *C. franchetiana* but old leaf-spines dense, and shoots and leaves densely hairy; flower-stalks very short, without persistent bracteoles at the joint; corolla pale yellow to almost white; pod to 2.5 cm. *W Himalaya.* H3. Summer.

9. C. jubata (Pallas) Poiret. Illustration: Csapody & Tóth, Colour atlas of trees and shrubs, 147 (1982); Bärtels, Gardening with dwarf trees and shrubs, 115 (1986).

Shrub to 5 m, though usually much less; branches covered with spines and dense hair, the bark scarcely visible, but brown and striped. Short shoots well developed, covered with slender spines from old leaves. Leaves on the main shoots persistent as dense spines which are thin, flexible and hairy; stipules membranous. Leaves pinnate with 4–7 pairs of leaflets, the whole densely covered with long, spreading hairs; leaflets oblong, rounded to the base, shortly tapered to an acute but scarcely mucronate apex. Flower-stalks very short, hidden within the short shoot. Calyx reddish, broadly cylindric, to 1.2 cm, mouth oblique, teeth triangular, to 2 mm, adjacent, the whole densely hairy. Corolla cream streaked with pink or entirely pale to deep pink, to 3 cm. Pod to 2 cm, hairy. *C & E former USSR, N China.* H3. Spring–summer.

Old plants have a bizarre appearance due to their long branches densely covered with spines.

10. C. sinica (Buc'hoz) Rehder (*C. chamlagu* Lamarck).

Shrub to 2 m, bark grey-brown. Short shoots small, covered with membranous stipules. Leaves of the main shoots persistent as spines which are rather distant, to 2 cm, each with a pair of shorter spiny stipules at the base. Leaves with 2 pairs of leaflets, pinnately arranged;

leaflets to 1.5 cm × 7 mm, broadly obovate-elliptic, tapered to the base, rounded to the apex, which is shortly pointed, hairless. Flower-stalks hairless, to 1.2 cm, jointed above the middle. Calyx broadly cylindric, 1.1–1.4 cm, mouth oblique, teeth triangular, *c.* 2 mm, adjacent, the whole hairless except for the margins of the teeth. Corolla yellow, 2.5–3 cm. Pod to 3.5 cm, hairless. *N & W China.* H3. Summer.

11. C. densa Komarov. Illustration: *Acta Horti Petropolitani* **29**: t. 7 (1908).

Shrub to 2 m, bark glossy, brown. Short shoots rather small, covered by spiny stipules and leaf-stalks. Leaves of long shoots persisting as distant spines to 1 cm, each with 2 much smaller stipular spines at the base. Leaves with 4 leaflets, clustered (not pinnately arranged) in the upper third of the spine; leaflets glaucous, to 1.2 cm × 3 mm, very narrowly obovate or linear-obovate, broadest 1/4–1/3 below the tapered apex, veins at an acute angle to the midrib, all with flaky scales beneath. Flower-stalks to 1.8 cm, jointed somewhat below the middle, finely crisped hairy. Calyx cylindric, *c.* 6 mm, mouth oblique, teeth triangular, *c.* 2 mm, adjacent, the whole hairless except for the margins of the teeth. Corolla yellow, 1.8–2 cm. Pod to 2 cm, hairless. *W China.* H4. Summer.

12. C. frutex (Linnaeus) Koch (*C. frutescens* de Candolle). Illustration: *Acta Horti Petropolitani* **29**: t. 5 (1908); Schischkin, Botanicheskii atlas, t. 35 (1963); Csapody & Tóth, Colour atlas of trees and shrubs, 147 (1982); *Gartenpraxis* August 1988: 24.

Shrub to 3 m or more, with long branches; bark glossy brown, striped. Short shoots well developed, covered in stipules which are mostly membranous. Long shoot leaves persisting as spines to 6 cm, each with 2 small stipular spines at the base. Leaves of 4 clustered leaflets (not pinnately arranged) borne together in the upper third of the spine; leaflets obovate-reversed triangular, to 1.3 cm × 5 mm, tapered to the base from the widest part which is almost at the truncate or broadly rounded, finely pointed apex, all glaucous and hairless, veins at an acute angle to the midrib. Flower-stalks *c.* 1.2 cm, jointed above the middle. Calyx cylindric, *c.* 7 mm, mouth oblique, teeth triangular, *c.* 1.5 mm, adjacent, the whole hairless except for the margins of the teeth. Corolla yellow,

c. 1.8 cm. Pod to 4 cm, hairless. *European & Asiatic former USSR.* H3. Summer.

A variable species. Plants grown under this name in the early part of the century are not the same as the wild material. They have no obvious spines (though the material available is all of young extension shoots) and the leaves are long-stalked, with 4 long, broad, rounded leaflets borne in a cluster at the top of the stalk; in some leaves there is a small soft, pointed tip between the leaflets, in others this is absent. The calyx is small, rather truncate, with small, distant teeth, and the corolla is larger. These specimens have not been matched with any known wild species, and may perhaps be of hybrid origin (the lack of spines, broad leaflets, truncate calyx and large corolla all suggest *C. arborescens*). It is uncertain whether or not any such material still survives in the living state.

13. C. pygmaea de Candolle. Illustration: Bärtels, Gartengeholze, 122 (1981); Davis, The gardener's encyclopaedia of trees and shrubs, 92 (1987); *Gartenpraxis* August 1988, 24; Lewis et al., (eds.), Legumes of the world, 491 (2005).

Upright to prostrate shrub to 1 m, with long, slender, sometimes trailing branches; bark grey-brown, striped. Short shoots rather small, covered in membranous stipules. Leaves of the long shoots persisting as spines to 8 mm, each with 2 somewhat shorter stipular spines at the base. Leaves with 4 leaflets borne in a cluster (not pinnately arranged) almost at the base of the spine; leaflets very narrowly linear-obovate, to 13 × 2 mm, tapered to the base and to the pointed apex, veins diverging at an acute angle from the midrib, all sparsely hairy. Flower-stalks to 1.7 cm, jointed just above the middle, at least the upper part sparsely hairy. Calyx cylindric, 7–10 mm, mouth oblique, teeth triangular, 2–4 mm, adjacent, all hairless except for the teeth margins. Corolla pale yellow, 1.8–2 cm. Ovary downy, pod to 3 cm, hairless when mature. *E former USSR, adjacent China.* H3. Summer.

14. C. aurantiaca Koehne. Illustration: Bean, Manual of trees & shrubs hardy in the British Isles, edn 8, **1**: 493 (1970).

Very similar to *C. pygmaea* but flower-stalks completely hairless, calyx to 5 mm, corolla orange-yellow or pinkish-yellow, 1.5–2 cm, ovary completely hairless. *Afghanistan, adjacent former USSR.* H4. Summer.

43. CALOPHACA de Candolle
S.G. Knees

Deciduous shrubs or low perennial herbs with alternate leaves. Leaves pinnate with 3–13 pairs of leaflets and a terminal leaflet. Flowers solitary or in racemes; calyx tubular with 5 slender teeth; corolla violet or yellow; stamens 10, with 9 united and 1 free. Fruit a cylindric legume.

A genus of about 10 species, all native to Asia, often grown in mixed borders or on rocky slopes. They require good drainage and can be propagated from seed or by grafting onto *Laburnum* species.

1a. Leaves with 5–8 pairs of leaflets; flowers 4–9 in racemes 7–10 cm **1. wolgarica**
  b. Leaves with 8–12 pairs of leaflets; flowers up to 12 in racemes 13–20 cm **2. grandiflora**

1. C. wolgarica (Linnaeus filius) de Candolle. Illustration: Loudon, Arboretum & Fruticeum Britannicum **2**: 635 (1838); Schneider, Illustriertes Handbuch der Laubholzkunde **2**: 104 (1907); Bailey, Standard cyclopedia of horticulture, 636 (1914); Lewis et al., (eds.), Legumes of the world, 491 (2005).

Shrub 90–120 cm. Leaves 5–8 cm, downy beneath; leaflets 6–12 mm, rounded–ovate, bristle-tipped, in 5–8 pairs. Flowers 4–9 in racemes 7–10 cm; calyx *c.* 8 mm, shaggy; corolla *c.* 2.5 cm, bright yellow. Pod to 3 cm, with 1–2 seeds. *S Former USSR (Turkestan).* H2. Summer.

2. C. grandiflora Regel. Illustration: *Gartenflora*, pl. 1231 (1886); Dippel, Handbuch der Laubholzkunde **3**: 717 (1893); Schneider, Illustriertes Handbuch der Laubholzkunde **2**: 104 (1907).

Like *C. wolgarica*, but less downy and larger in all its parts. Leaves 4–8 cm; leaflets 2–2.5 × 1.1–1.3 cm, ovate, pointed, in 8–12 pairs. Flowers up to 12 in racemes 13–20 cm; corolla 2.5–2.8 cm, bright yellow. Pod oblong, with 1–2 seeds. *S Former USSR (Turkestan).* H2. Summer.

44. ASTRAGALUS Linnaeus
J. Cullen

Herbs or small shrubs. Leaves pinnate, either with a terminal leaflet or the axis continuing and persistent as a hardened spine; leaflets symmetric at base; stipules persistent, often conspicuous; stipels absent. Hairs, when present, basifixed or medifixed, grey, white or dark brown to black, colours often mixed. Flowers in racemes or spikes, axillary or arising directly from the basal rosette on a scape or scape absent. Bracts usually present. Calyx bell-shaped to tubular, 5-toothed, sometimes very deeply so. Corolla with wings and keel usually shorter than the variably-shaped standard. Keel sometimes toothed on the lower side. Stamens 10, the filaments of 9 united, that of the uppermost free. Pod variously shaped, usually several-seeded, divided along its length by a septum developed from an infolding of the lower suture.

One of the largest genera of flowering plants, with perhaps 2000 or more species, mainly from W & C Asia, but extending across the northern hemisphere and scattered in the southern hemisphere; in W & C Asia they are characteristic components of steppe vegetation. Fortunately for the present purpose, very few of them are grown as ornamentals. An elaborate hierarchy of subgenera and sections exists, but is of little help with the cultivated species, which are scattered through it; this hierarchy is ignored here, and the cultivated species fall into 3 groups: a) the annuals (no. **1**); b) the perennial, non-spiny species (nos. **2, 3, 4, 5, 6, 9 & 11**) and the spiny ('tragacanthoid') species (nos. **7, 8, 10, 12 & 13**).

Two characters require a little explanation. The hairs are either basifixed or medifixed, and this difference is important in distinguishing the species. It is sometimes difficult to decide whether hairs are of one type or the other; in general, however, under a magnification of 15 times or more, medifixed hairs can be rotated somewhat on their attachment when one end is gently pushed with a needle. Hairs attached closer to one end than the other occur in the genus as a whole, but are not important in the recognition of the cultivated species. The shape of the standard is sometimes important in the distinction of the spiny species; in most, the blade is truncate or hastate, abruptly contracted into a narrow claw, while in others the claw is broadened upwards, becoming almost as wide as the blade, from which it is demarcated by a quite small constriction; the terms 'stenonychioid' and 'platonychioid' are used to describe these conditions in some of the literature

All the species are easily grown in a sunny site in well-drained soil, and are propagated by seed.

Literature: There is a very extensive literature on the genus, but this is of little help with the cultivated species. All the species included here are covered in either *Flora Europaea* **2**: 108–124 (1968) or Davis (ed), Flora of Turkey **3**: 49–254 (1970), which should be referred to for further information.

Habit. Annual herb: **1**; perennial herb: **2, 3, 4, 5, 6, 9, 10, 11**; small shrub: **7, 8, 12, 13**. Stemless, with scape: **4, 11**; stemless without scape: **6**.

Leaves. With a terminal leaflet: **1,– 2, 3, 4, 5, 6, 9, 11**, with a terminal spine: **7, 8, 10, 12, 13**. Leaflets fewer than 10 pairs: **2, 3, 4, 5, 7, 8, 10, 11, 12, 13**; leaflets 10–20 pairs: **1, 2, 3, 4, 6, 11, 12, 13**; leaflets more than 20 pairs: **9**.

Hairs. All basifixed: **1, 2, 3, 4, 5, 6, 7, 8, 9, 10**; all medifixed: **11, 12, 13**.

Calyx. Deeply divided (almost to the base), the teeth obscured by long, white hairs: **7, 8**; with a distinct tube, bell-shaped or tubular, variously hairy but not obscured by long, white hairs: **1, 2, 3, 4, 5, 6, 9, 10, 11, 12, 13**. Usually less than 1 cm: **1, 2, 3, 4, 5, 7, 8, 10, 12, 13**; usually more than 1 cm: **6, 9, 10, 11**.

Corolla. Yellowish (sometimes buff or flushed with pink): **1, 2, 5, 6, 7, 8, 9, 12**; whitish: **1, 4, 10, 11, 12, 13**; bluish or purple: **3, 4, 10, 11, 12**.

1a. Leaves terminating in a leaflet, the axis not continuing as a persistent, hardened spine 2
  b. Leaves terminating in a persistent, hardened spine, without a terminal leaflet 9
2a. Plant annual **1. boeticus**
  b. Plant perennial 3
3a. Most hairs clearly and distinctly medifixed **11. monspessulanus**
  b. Most hairs clearly and distinctly basifixed 4
4a. Stems absent, the inflorescence-stalk either absent (the inflorescence borne between the basal leaves) or present 5
  b. Stems present, the inflorescences borne in the axils of the stem-leaves 6
5a. Leaves (including stalks and axes) with spreading, brown hairs; spikes stalkless or almost so, borne between the rosette leaves; calyx 1.2–1.5 cm; corolla yellow **6. exscapus**

b. Leaves with adpressed, white hairs; spikes borne on a distinct, though sometimes short stalk; calyx 4–5 mm; corolla white or bluish purple **4. depressus**

6a. Corolla purplish or bluish **3. danicus**

b. Corolla yellow or whitish **7**

7a. Calyx 1.4–2 cm; leaves with 20 or more pairs of leaflets **9. centralalpinus**

b. Calyx 5–10 mm; leaves with up to 15 pairs of leaflets **8**

8a. Leaves with 3–7 pairs of leaflets; stipules free **5. glycyphyllos**

b. Leaves with 8–15 pairs of leaflets; at least the stipules of the upper leaves united at the base **2. cicer**

9a. Most hairs on the plant medifixed **10**

b. Most hairs on the plant basifixed **11**

10a. Calyx-teeth ¼–½ as long as tube; corolla pinkish, purple, yellow or white **12. angustifolius**

b. Calyx-teeth up to ¼ as long as the tube; corolla always whitish **13. massiliense**

11a. Calyx with distinct and visible tube, hairy, but not as below; corolla white to purple, 1–2.2 cm **10. sempervirens**

b. Calyx with tube very short, the teeth and tube plumed and obscured by dense, long, white hairs; corolla whitish, yellow, buff or pinkish, sometimes with purple veins **12**

12a. Leaves densely greyish white-hairy; standard with blade truncate or hastate, abruptly contracted into the narrow claw **8. microcephalus**

b. Leaves green, sparsely hairy or hairless; standard with claw broadened towards the top, demarcated from the blade by a slight constriction **7. gummifer**

1. A. boeticus Linnaeus. Illustration: Bonnier, Flore complète **3**: pl. 144 (1914).

Annual herb. Stems erect or sprawling, to 60 cm. Leaves with 10–15 pairs of leaflets, with a terminal leaflet; leaflets 8–18 mm, oblong or oblong-obovate, tapered to the base, truncate or notched at the apex, rather sparsely adpressed hairy beneath with basifixed hairs. Stipules free, triangular. Flowers 5–15 in dense spikes. Calyx 5–8 mm, tubular, the teeth as long as the tube, covered in dark brown to black and white hairs. Corolla yellow or whitish, 1–1.4 cm. Pod 2–4 cm. oblong, triangular in section, grooved beneath,

hooked at apex, shortly hairy. *Canary Islands. Mediterranean area east to Iran.* H1. Spring.

2. A. cicer Linnaeus. Illustration: Bonnier, Flore complète **3**: pl. 144 (1914); Polunin, Flowers of Europe, pl. 54 (1969); Aeschimann et al., Flora alpina **1**: 841 (2004).

Erect or spreading perennial to 50 cm. Leaves with 8–15 pairs of leaflets, with a terminal leaflet; leaflets 1.5–3.5 cm, lanceolate or ovate-lanceolate to oblong, tapered to base and apex, with short, adpressed, rather sparse basifixed hairs on both surfaces; stipules narrowly lanceolate, united at the base. Flowers 10–25 in dense spikes. Calyx 7–10 mm, tubular, teeth about half as long as tube, covered with adpressed, basifixed hairs, some white, most dark brown to black. Corolla yellow, 1.4–1.7 cm. Pods ovoid-spherical to spherical, inflated, hooked at apex, covered with long, somewhat spreading white hairs and shorter, adpressed. dark brown to black hairs. *Most of Europe. Turkey.* H1. Summer.

3. A. danicus Retzius (*A. hypoglottis* misapplied). Illustration: Keble Martin, The concise British flora in colour, pl. 24 (1965); Garrard & Streeter, The wild flowers of the British Isles, pl. 32 (1983); Aeschimann et al., Flora alpina **1**: 841 (2004).

Perennial herb, stems prostrate to ascending, to 30 cm. Leaves with 6–13 pairs of leaflets, with a terminal leaflet; leaflets 5–16 mm, oblong-elliptic to ovate-oblong, tapered to base and apex (rarely somewhat notched at apex), with rather sparse, long, white, basifixed hairs on both surfaces; stipules united round the stem for at least one-third of their length. Flowers numerous in dense, oblong to almost spherical spikes. Calyx tubular. 6–8 mm, teeth half or more as long as tube, covered in dense, adpressed, basifixed, dark brown hairs. Corolla purplish or bluish, 1.5–1.8 cm. Pod ovoid, inflated, with long, dense, spreading, somewhat swollen-based, white, basifixed hairs. *Most of Europe. W Asia.* H3. Summer.

4. A. depressus Linnaeus. Illustration: Bonnier, Flore complète **3**: pl. 144 (1914); Aeschimann et al., Flora alpina **1**: 843 (2004).

More or less stemless perennial with woody base. Leaves in rosettes, with 6–14 pairs of leaflets, with a terminal leaflet; leaflets 5–9 mm, obovate, tapered to the

base, truncate or notched at the apex, with adpressed, basifixed (rarely the hairs attached close to, but not quite at the end) hairs on at least the lower surface; stipules free. Spikes borne on a scape, with 7–30 flowers, cylindric or spherical. Calyx 4–5 mm, tubular, teeth short, with sparse, mostly dark brown hairs. Corolla white or bluish purple, 1–1.4 cm. Pod cylindric, slightly curved, hairless or with sparse white hairs. *C & S Europe, Cyprus, Turkey.* H3. Summer.

5. A. glycyphyllos Linnaeus. Illustration: Bonnier, Flore complète **3**: pl. 143 (1914); Ary & Gregory, The Oxford book of wild flowers, pl. 3 (1962); Keble Martin, The concise British Flora in colour, pl. 24 (1969); Polunin, Flowers of Europe, pl. 54 (1969); Aeschimann et al., Flora alpina **1**: 845 (2004).

Perennial herb with trailing or ascending stems to 2 m. Leaves with 3–7 pairs of leaflets, with a terminal leaflet; leaflets ovate to elliptic or broadly elliptic, 2.5–4.5 cm, rounded to the base and the sometimes shortly mucronate apex, hairless above, sparsely adpressed hairy with white, basifixed hairs beneath; stipules free, triangular. Racemes (flowers slightly stalked) dense, cylindric, with 12–30 or more flowers. Calyx bell-shaped to tubular, 5–7 mm, hairless or with dark brown or black hairs on the teeth. Corolla pale yellow or cream, 1–1.7 cm. Pod cylindric, curved upwards, hairless or with white, adpressed hairs. *Eurasia.* H1. Summer.

A variant with adpressed, dark brown to black hairs on the calyx is known as var. **glycyphylloides** (de Candolle) Matthews (*A. glycyphylloides* de Candolle).

6. A. exscapus Linnaeus. Illustration: Bonnier, Flore complète **3**: pl. 143 (1914).

Stemless perennial with stout, woody stock. Leaves mostly densely hairy (including stalk and axis) with brownish, spreading, basifixed hairs; leaflets in 12–19 pairs, terminal leaflet present, 1–2.5 cm, elliptic-oblong, rounded at base and apex, upper surface sometimes sparsely hairy or almost hairless; stipules free, narrowly triangular. Spikes stalkless or almost so, borne between the rosette leaves, with 3–10 flowers. Calyx tubular, 1.2–1.5 cm, with dense, brown, spreading hairs. Corolla yellow, 2–3 cm. Pod oblong, with dense, brown, spreading hairs. *Spain to C & E Europe.* H3. Late spring–summer.

Part of a complex of species which are difficult to distinguish.

7. A. gummifer Labillardière.

Shrub to 30 cm. Hairs all basifixed. Leaflets 4–7 pairs, the leaf terminating in a hardened, persistent spine; leaflets elliptic, 5–10 mm, usually mucronate, sparsely hairy or hairless; stipules triangular ovate. Flowers in groups of 2 or 3 in the leaf-axils, forming ovoid or cylindric inflorescences. Calyx 5–7 mm, divided into teeth almost to the base, the whole covered with long, spreading, white hairs. Corolla buff or yellow (sometimes flushed with pink), 1–1.2 cm, persistent in fruit; standard with the claw broadened above, then slightly constricted into the blade. Pods 1- or 2-seeded. *Turkey, Lebanon.* H5. Summer.

A source of gum tragacanth.

8. A. microcephalus Willdenow.

Like *A. gummifer* but cushion-forming, leaves (including axis and spine) greyish white with dense, spreading, basifixed hairs; stipules narrowly lanceolate; corolla yellow with purple veins, standard with the blade truncate or hastate, abruptly contracted into the narrow claw. *Turkey, former USSR (Caucasus), Iran.* H5. Summer.

Belonging to a different section of the genus from *A. gummifer*, but superficially very similar.

9. A. centralalpinus Braun-Blanquet (*A. alopecuroides* misapplied). Illustration: *Botanical Magazine*, 3193 (1832); Bonnier, Flore complète **3**: pl. 144 (1914).

Perennial herb with stout, erect stems to 1 m, covered with spreading, basifixed hairs. Leaves with 20–30 pairs of leaflets, with a terminal leaflet; leaflets 1–3 cm, elliptic, ovate or lanceolate, rounded to the base, tapered to the apex, hairless above, densely spreading-hairy with basifixed hairs beneath; stipules conspicuous, lanceolate to narrowly triangular. Flowers numerous in stalkless, ovoid to cylindric spikes. Calyx 1.4–2 cm, cylindric, densely brown-hairy, the teeth as long as or slightly shorter than the tube. Corolla yellow, 1.5–2 cm. Pod ovoid, compressed, with long, spreading, brown hairs. *W Alps (France, Italy), Bulgaria.* H4. Summer.

For many years confused with **A. alopecuroides** Linnaeus from Spain and extreme SW France. Illustration: Bonnier, Flore complète **3**: pl. 144, 1914, – as *A. narbonensis*. This is much less hairy, has spherical, somewhat stalked inflorescences, calyx-teeth longer than the tube and the corolla 2.2–2.7 cm.

10. A. sempervirens Lamarck (*A. aristatus* L'Héritier). Illustration: *Loddiges' Botanical Cabinet* **13**: t. 1278 (1827); Bonnier, Flore complète **3**: pl. 142 (1914); Brickell (ed.), RHS A–Z encyclopedia of garden plants, 154 (2003); Aeschimann et al., Flora alpina **1**: 847 (2004).

Somewhat tufted perennial herb with woody stock; stems to 40 cm, trailing or ascending. Leaves with 4–10 pairs of leaflets, the axis persisting as a weak spine; leaflets narrowly oblong, narrowly oblanceolate or linear, covered, like the axis, in whitish, spreading, basifixed hairs. Stipules joined to the leaf-stalks for about half their length. Spikes with up to 10 flowers. Calyx 7–15 mm, slightly inflated, covered with dense, greyish, somewhat spreading hairs. Corolla white to purple, 1–2.2 cm. Pod ovoid, densely hairy. *S Europe, N Africa.* H4. Summer.

11. A. monspessulanus Linnaeus. Illustration: *Botanical Magazine*, 373 (1797); Bonnier, Flore complète **3**: pl. 143 (1914); Polunin & Smythies, Flowers of southwest Europe, pl. 19 (1973); Aeschimann et al., Flora alpina **1**: 849 (2004); Lewis et al., (ed.), Legumes of the world, 481 (2005).

Erect, stemless perennial herb, scapes to 30 cm. Leaves with 7–20 pairs of leaflets, with a terminal leaflet; leaflets 5–15 mm, ovate or oblong to almost circular, rounded to the base and apex, hairless above, with sparse, whitish, medifixed hairs beneath. Spikes ovoid or oblong, often rather loose, with up to 30 flowers. Calyx 9–16 mm, covered with mostly brown, medifixed hairs. Corolla purplish (rarely whitish), 2–3 cm. Pod cylindric, sparsely hairy with medifixed hairs (hairs soon falling), somewhat curved. *S Europe, N Africa.* H5. Summer.

12. A. angustifolius Lamarck. Illustration: Brickell (ed.), RHS A–Z encyclopedia of garden plants, 154 (2003).

Dwarf, cushion-forming shrublet. Leaves with 5–12 pairs of leaflets, the axis continuing as a persistent spine which is sparsely hairy at first; leaflets 3–7 mm, narrowly elliptic to obovate, rounded to the sometimes mucronate apex, densely ashy grey-hairy on both surfaces with medifixed hairs; stipules lanceolate, united for about half their length. Flowers 3–14 in racemes. Calyx tubular, 6–10 mm, ribbed, with a mixture of white and dark brown to black medifixed hairs; teeth one quarter to half as long as tube. Corolla

pinkish, purple, yellow or whitish, 1.5–1.8 cm. Pod oblong-cylindric, usually hairy. *Balkan Peninsula to Turkey.* H5. Summer.

13. A. massiliensis (Miller) Lamarck (*A. tragacantha* Linnaeus, in part). Illustration: Bonnier, Flore complète **3**: pl. 142 (1914); Polunin & Smythies, Flowers of southwest Europe, pl. 19 (1973).

Rather open shrub to 30 cm. Leaves with 6–12 pairs of leaflets, the axis continuing as a stout spine; leaflets 4–6 mm, oblong to elliptic, rounded or truncate to the obtuse apex, densely silvery-silky above and beneath with medifixed hairs. Racemes with 3–8 flowers. Calyx 5–7 mm, tubular, the teeth to one-quarter as long as the tube, covered with dark brown to black and white, medifixed hairs. Corolla whitish, 1.3–1.7 cm. Pod oblong, densely hairy. *SW Europe.* H5. Summer.

45. OXYTROPIS de Candolle
J.R. Akeroyd

Herbaceous perennials or subshrubs. Leaves pinnate; leaflets entire. Flowers in axillary spikes or racemes, similar to those of *Astragalus* but the keel with an acute beak at the apex.

A genus of about 100 species from the arctic and mountainous regions of Eurasia and North America. Propagation is by seed or division in the spring. The species in cultivation in Europe are compact perennials, suitable for the rock garden or alpine glasshouse; most of them prefer calcareous soils.

1a. Plant glandular and foetid **3. foetida**
 b. Plant neither glandular nor fetid   2
2a. Flowering stems leafy   3
 b. Leaves all basal   5
3a. Leaflets obtuse   **5. jacquinii**
 b. Leaflets acute   4
4a. Corolla bluish   **4. lapponica**
 b. Corolla pale yellow   **6. pilosa**
5a. Corolla yellow to cream, sometimes tinged violet   6
 b. Corolla blue, purple, pink or violet   7
6a. Raceme ovoid; calyx-hairs whitish   **1. campestris**
 b. Raceme almost spherical; calyx-hairs blackish   **8. ochroleuca**
7a. Leaflets not distant or in whorls   8
 b. Leaflets distant or in whorls   9
8a. Leaves silky-hairy, usually with fewer than 25 leaflets   **2. halleri**
 b. Leaves downy, with at least 25 leaflets   **7. pyrenaica**

9a. Leaflets not in whorls; corolla
    c. 2 cm, distinctly longer than calyx
    9. lambertii

b. Leaflets in whorls of 3–4; corolla
    1.2–1.5 cm, slightly longer than
    calyx **10. splendens**

1. O. campestris (Linnaeus) de Candolle. Illustration: Keble Martin, Concise British flora in colour, pl. 23 (1967); Barneby, European alpine flowers in colour, pl. 42 (1967); Lunardi (ed.), Macdonald encyclopedia of alpine flowers, pl. 173 (1985); Blamey & Grey-Wilson, Illustrated flora of Britain and northern Europe, 207 (1989).

Tufted, downy perennial, with stout rootstock. Leaves all basal; leaflets 17–25, elliptic or lanceolate, acute, hairy. Flowering-stems 8–20 cm, about as long as leaves, hairy. Raceme ovoid, dense. Corolla 1.5–2 cm, pale yellow or cream, the wings and keel often tinged violet. Fruit 1.4–1.8 cm, ovoid-oblong, hairy. *Mountains of Europe.* H1. Summer.

2. O. halleri Koch (*O. sericea* (de Candolle) Simonkai; *O. uralensis* misapplied). Illustration: Keble Martin, Concise British flora in colour, pl. 23 (1965); Blamey & Grey-Wilson, Illustrated flora of Britain and northern Europe, 207 (1989); Aeschimann et al., Flora alpina **1**: 857 (2004); Lewis et al., (eds.), Legumes of the world, 480 (2005).

Similar to *O. campestris*, but silky-hairy; leaflets 11–28; flowering-stems longer than leaves; corolla 1.5–2 cm, blue to purple; fruit 1.5–2 cm, ovoid, more densely hairy. *Mountains of Europe.* H1. Summer.

3. O. foetida (Villars) de Candolle. Illustration: Aeschimann et al., Flora alpina **1**: 857 (2004).

Similar to *O. campestris*, but glandular-hairy and foetid; leaflets 21–51, linear-lanceolate to narrowly oblong, the margins down-curled; flowering-stems longer than leaves; corolla *c.* 2 cm, pale yellow; fruit narrower. *W Alps.* H1. Summer.

4. O. lapponica (Wahlenberg) Gay. Illustration: Polunin & Stainton, Flowers of the Himalayas, pl. 30 (1985); Polunin, Collins photoguide to wild flowers, 274 (1988); Blamey & Grey-Wilson, Illustrated flora of Britain and northern Europe, 207 (1989); Aeschimann et al., Flora alpina **1**: 851 (2004).

Loosely tufted perennial. Leaflets 17–29, lanceolate to oblong-lanceolate, acute, adpressed-hairy. Flowering-stems 5–25 cm,

decumbent to ascending, leafy, hairy. Racemes almost spherical, rather loose. Corolla 8–12 mm, bluish violet. Fruit 8–15 mm, narrowly ellipsoid, pendent, densely appressed-hairy with dark hairs. *Mts of Eurasia, Scandinavia.* H1. Summer. Grows best on lime-poor soils.

5. O. jacquinii Bunge (*O. montana* misapplied).

Similar to *O. lapponica*, but leaflets 25–41, lanceolate to narrowly ovate, obtuse, sparsely hairy; corolla violet to purple; fruit ovoid, somewhat swollen, not pendent. *Alps, French Jura.* H1. Summer.

6. O. pilosa (Linnaeus) de Candolle (*Astragalus pilosus* Linnaeus). Illustration: Barneby, European alpine flowers in colour, pl. 42 (1967); Blamey & Grey-Wilson, Illustrated flora of Britain and northern Europe, 207 (1989); Aeschimann et al., Flora alpina **1**: 857 (2004).

Perennial with long silvery hairs. Leaflets 19–27, oblong to linear-oblong, acute. Flowering-stems 15–40 cm, erect, leafy. Racemes ovate-oblong, compact. Corolla 1.2–1.4 cm, pale yellow. Fruit 1.5–2 cm, ovoid to cylindric, densely hairy. *Mountains of Europe & W Asia.* H1. Summer.

7. O. pyrenaica Godron & Grenier (*O. montana* subsp. *samnitica* (Arcangeli) Hayek). Illustration: Polunin & Smythies, Flowers of southwest Europe, 148 (1973).

Tufted, downy perennial. Leaflets 25–41, lanceolate to narrowly elliptical or oblong, acute, silky-hairy. Flowering-stems 5–20 cm. Raceme compact, but extending in fruit. Corolla bluish-violet or purplish. Fruit 1.5–2 cm, oblong to narrowly ovoid. *Pyrenees, Alps & mountains of N part of Balkan Peninsula.* H2. Summer.

8. O. ochroleuca Bunge.

Leaflets 21–31, oblong-lanceolate or elliptic. Flowering-stems 5–25 cm, ascending, longer than leaves. Raceme almost spherical, dense, many-flowered. Corolla small, pale yellowish. Calyx with blackish hairs. Fruit oblong-ovoid, pendent, finely hairy. *Mountains of C Asia.* H1. Summer.

9. O. lambertii Pursh. Illustration: *Botanical Magazine*, 2147 (1820); *Edwards's Botanical Register*, **13**: 1054 (1827).

Tufted, silky-hairy perennial. Leaflets 11–23, 1–4 cm, rather distant, linear-lanceolate, acute. Flowering-stems 15–30 cm. Racemes to 15 cm. Corolla *c.* 2 cm, pinkish purple. Fruit 2–3 cm,

distinctly longer than calyx, cylindric, shortly hairy. *N America.* H1. Summer.

10. O. splendens Douglas.

Tufted, densely hairy perennial. Leaflets in whorls of 3 or 4, lanceolate, acute, silky-hairy. Flowering-stems 10–30 cm, longer than leaves, with spreading hairs. Raceme up to 15 cm, loose. Calyx white-hairy. Corolla 1.2–1.5 cm, blue to purplish. Fruit *c.* 1 cm, slightly longer than calyx, ovoid, hairy. *NW North America.* H1. Summer.

46. ALHAGI Gagnepain

J. Cullen

Bushy, somewhat woody perennial herbs with rigid spines on the lower branches and with the upper, short, lateral flowering branches terminating in spines. Leaves simple, stalkless; stipules minute. Flowers borne singly or in pairs in the axils of minute bracts; flower-stalks very short or absent. Calyx bell-shaped, shallowly 5-toothed. Corolla pink to red, standard and keel longer than the wings, the keel obtuse. Stamens 10, 9 of them with their filaments united, that of the uppermost free. Ovary containing several ovules. Pod indehiscent, constricted between the 1–5 seeds.

A genus of a single species from arid habitats from Europe and North Africa to Central Asia, occasionally grown as curiosities. They require winter-protection in most of Europe, a well-drained soil and a sunny site. Propagation is by seed or cuttings.

Literature: Keller, B.A. & Shaparenko, K.K., Materiali k sistematiko-ekologicheskogo monografii roda *Alhagi* Tournefort ex Adanson, *Sovetskaya Botanika* **3–4**: 151–185 (1933).

1. A. maurorum Medikus subsp. **maurorum** (*A. pseudalhagi* (Bieberstein) Desvaux; *A. camelorum* Fischer). Illustration: Lewis et al., (eds.), Legumes of the world, 492 (2005).

To 1 m, hairless; branches mostly spreading at an acute angle. Leaves 1–2 cm × 3–4 mm, lanceolate to oblanceolate. Calyx *c.* 2 mm, shallowly toothed, the sinuses between the teeth shallowly U-shaped. Corolla pink, 7–10 mm. Pod 8–30 × 2–3 mm, hairless, dark brown, strongly contracted between the seeds. *Eastern Europe to C Asia.* H4. Summer.

Subsp. **graecorum** (Boissier) Awmack & Lock (*A. graecorum* Boissier; *A. mannifera*

Desvaux). Illustration: Lewis et al., (eds.), Legumes of the world, 492 (2005). Very similar but stems, calyx, ovary and pods hairy, the branches at a wider angle, and calyx-teeth with V-shaped sinuses between them. *East Mediterranean area, Arabia to east Asia.* H4. Summer.

47. GALEGA Linnaeus
J.R. Akeroyd

Erect herbaceous perennials with a bushy habit. Leaves irregularly pinnate; leaflets obtuse or acute. Stipules conspicuous. Flowers many in stalked racemes longer than leaves. Stamens 10, 9 of them with filaments united. Fruit cylindric, slender, constricted between seeds, beaked; seeds many.

A genus of 6 species in Europe, SW Asia and East Africa. Propagation is by seed or division of plants during winter. The two species grown in European gardens are also grown as fodder crops and are locally naturalised in W & N Europe.

1a. Stipules arrow-shaped; fruits spreading to somewhat erect
    1. officinalis
  b. Stipules ovate; fruits deflexed
    2. orientalis

1. G. officinalis Linnaeus (*G. bicolor* Regel, *G. persica* Persoon, *G. tricolor* Hooker). Illustration: Hay & Synge, Dictionary of garden plants in colour, t. 1140, 1141 (1969); Polunin, Flowers of Europe, pl. 53 (1969); Aeschimann et al., Flora alpina **1**: 837 (2004); Lewis et al., (eds.), Legumes of the world, 487 (2005).

Plant 30–160 cm, hairless to sparsely hairy. Leaflets 9–17, 1.5–5 cm × 4–15 mm. elliptic to lanceolate, mucronate. Flowers 30–50 in racemes. Calyx-teeth about as long as tube. Corolla 1–1.5 cm, lilac, purple or white. Fruit 2–5 cm, spreading to somewhat erect; seeds 2–10. *C & S Europe to W Pakistan.* H1. Summer–autumn.

2. G. orientalis Lamarck. Illustration: Brickell, RHS gardeners' encyclopedia of plants and flowers, 211 (1989); Brickell (ed.), RHS A–Z encyclopedia of garden plants, 466 (2003); Aeschimann et al., Flora alpina **1**: 837 (2004).

Similar to *G. officinalis* but leaflets 3–6 × 1–2.5 cm, acuminate; stipules ovate to broadly ovate; calyx hairy, the teeth shorter than tube; corolla bluish violet; fruit deflexed. *Caucasus.* H1. Summer.

A variable species, especially in flower colour, and numerous varieties and cultivars have been recognised. Several of these are derived from the hybrid **G. × hartlandii** Clarke (*G. officinalis* × *G. orientalis*). Illustration: Brickell, RHS gardeners' encyclopedia of plants and flowers, 190 (1989).

48. GLYCYRRHIZA Linnaeus
J. Cullen

Glandular, rhizomatous perennial herbs. Leaves pinnate, with a terminal leaflet, the leaflets dotted with orange-brown glands; stipules lanceolate, minute; stipels absent. Flowers numerous in axillary racemes or spikes which are loose or dense and head-like. Calyx 5-toothed, 2-lipped, the tube narrowly funnel-shaped, the upper teeth short, the lower as long as or longer than the tube. Corolla white, yellow, mauve, violet or bluish; keel obtuse or acute. Stamens 10, 9 of them with their filaments united, that of the uppermost free. Ovary with few ovules. Pod compressed, dehiscent with the sides contorting, seeds 1–several.

A genus of about 20 species from Eurasia, North America and temperate South America. The 2 species generally grown are both European and require a rich, deep soil. Propagation is by division.

1a. Racemes or spikes congested, head-like, 2–5 cm; corolla 5–7 mm; pod covered with spine-like bristles
    2. echinata
  b. Racemes or spikes loose, not head-like, 5 cm or more; corolla 0.9–1.8 cm; pod not covered with bristles
    1. glabra

1. G. glabra Linnaeus. Illustration: Bonnier, Flore complète **3**: t. 148 (1914); Masefield et al., The Oxford book of food plants, 199 (1969); Brickell (ed.), RHS A–Z encyclopedia of garden plants, 489 (2003); Aeschimann et al., Flora alpina **1**: 859 (2004).

Perennial to 60 cm; stems sparsely hairy, glandular above. Leaves with 5–9 pairs of leaflets which are ovate to elliptic, rounded to the base, pointed or rounded at the apex. Racemes loose, 5 cm or more; flower-stalks short. Calyx-teeth to 6 mm. Corolla 9–15 mm, blue or violet. Pod oblong, 1.5–2.5 cm × 4–5 mm, red-brown, with thickened sutures, laterally compressed, with 1–6 seeds, hairless or hairy but without bristles. *S Europe, N Africa, SW Asia.* H3. Summer.

The species is extensively grown as a crop in some areas for its rhizomes which yield liquorice.

2. G. echinata Linnaeus. Illustration: *Botanical Magazine*, 2154 (1820); Jordanov, Flora Reipublicae popularis Bulgaricae **6**: 183 (1976).

Erect perennial to 1 m or more; stems sparsely hairy above. Leaves with 4–8 pairs of leaflets which are narrowly elliptic, acute at both ends, 1–3 cm × 5–15 mm. Racemes or spikes dense, head-like, 2–5 cm; flower-stalks not visible. Calyx *c.* 2 mm. Corolla whitish to violet, 5–7 mm. Pod obovoid, flattened, reddish brown, 1–1.6 cm × 6–8 mm, covered in glands and spine-like bristles, 1–3–seeded. *E Mediterranean area, SW Asia.* H5.

49. CHORDOSPARTIUM Cheeseman
J. Cullen

Shrub to 8 m, often with a trunk to 3 cm across in the wild. Branches drooping, leafless except for small, adpressed, triangular, brownish scales at the nodes, terete, grooved. Flowers usually numerous in cylindric racemes, borne singly or in groups of 2–5 at the nodes; inflorescence-stalk hairy, flower-stalks woolly. Flowers to 1 cm, pale lavender or whitish, the standard with purple veins. Calyx 3–4 mm, cup-shaped, hairy, with 5 minute teeth. Stamens 10, 9 of them with their filaments united, that of the uppermost free. Ovary hairy, containing 1–5 ovules; styles long, incurved, hairy on the upper side towards the apex. Pods hairy, indehiscent, swollen, *c.* 5 mm, usually 1–seeded.

A genus of a single species from New Zealand, notable for its drooping, leafless branches. It is relatively easily grown in a warm site and is propagated by seed.

1. C. stevinsonii Cheeseman (*Carmicaehlia stevinsonii* Cheesman). Illustration: *Botanical Magazine*, 9654 (1943); Metcalf, The cultivation of New Zealand trees and shrubs, pl. 4 (1972); Moore & Irwin, The Oxford book of New Zealand plants, 165 (1978); Salmon, The native trees of New Zealand, 202, 203 (1980).

New Zealand (South Island). H5–G1. Early summer.

Very threatened in its native habitat, but established in cultivation. Now usually included in *Carmichaelia*.

50. CARMICHAELIA R. Brown
J. Cullen

Shrubs or small trees; habit diverse, ranging from small, compact, patch-forming plants to larger, erect shrubs or trees. Leaves pinnate with 3–7, often notched leaflets, usually absent from mature plants, their axils marked by small notches in the flattened or terete branchlets. Flowers solitary or in racemes borne in the notches, sometimes more than 1 raceme from each notch; flower-stalks hairy or finely downy. Calyx bell-shaped, 5-toothed. Petals distinctly clawed. Stamens 10, the filaments of 9 united, that of the uppermost free; anthers each with only a single pollen-sac. Pod dry with thickened margins which may project into the cell, hard, indehiscent or dehiscent only at the apex or by one or both sides separating from the persistent, thickened margins. Seeds 1–few.

A genus of 39 species from New Zealand and Lord Howe Island. A considerable number of names can be found in seed catalogues and plant lists, but it is uncertain how many are in general cultivation in Europe.

Most of the species are leafless when mature and have a broom-like appearance. Their classification is very difficult and is based largely on the pods and the seeds, which may not be produced in cultivation. The account below covers most of the species likely to be found in gardens, and is based largely on Allan's Flora of New Zealand 1: 373–397 (1961). The key given here avoids the use of pod and seed characters, and should therefore be used with caution.

Most of the species are doubtfully hardy in Europe; they are normally propagated by seed.

Literature: Simpson, G., A revision of the genus *Carmichaelia*, Transactions of the Royal Society of New Zealand **75**: 231–287 (1945); Slade, B.F., Cladode anatomy and leaf trace systems in New Zealand brooms, ibid., **80**: 81–96 (1952); Heenan, P.B., A taxonomic revision of Carmichaelia in New Zealand, *New Zealand Journal of Botany* **33**: 455–475 (1995), **34**: 157–177 (1996).

1a. Mature plant with leaves present during spring and summer 2
  b. Mature plant leafless or almost so 3
2a. Erect or spreading shrub; racemes tight, with 10–40 flowers
     1. angustata
  b. Scrambling shrub; racemes loose, with 2–6 flowers **6. kirkii**
3a. Erect or spreading shrubs or trees, 50 cm or more in height 4

  b. Prostrate, sprawling or patch-forming plants, rarely higher than 15 cm 7
4a. Branchlets 1–2 mm wide 5
  b. Branchlets 3 mm or more wide 6
5a. Branchlets very compressed; racemes with a conspicuous stalk to 1 cm
     4. arborea
  b. Branchlets scarcely compressed; racemes stalkless or stalk very short
     3. flagelliformis
6a. Branchlets compressed, 8–12 mm wide; flowers 2–3 cm **2. williamsii**
  b. Branchlets terete, 3–4 mm across; flowers at most 7 mm **5. petriei**
7a. Branchlets terete, not compressed
     10. curta
  b. Branchlets conspicuously compressed 8
8a. Branches arising from rhizomes; flowers solitary or in pairs at each notch **8. uniflora**
  b. Branches arising from a taproot; flowers mostly in racemes of 3 or more arising from each notch 9
9a. Branches and calyx hairless
     7. enysii
  b. Branches and calyx hairy 10
10a. Flowers *c.* 3 × 2 mm **11. nigrans**
  b. Flowers *c.* 10 × 5–7 mm **9. astonii**

1. C. angustata Kirk.

Erect or spreading shrub to 2 m; branches hairless or almost so, slightly compressed, to 2 mm across, grooved, borne at wide angles and drooping. Leaves present during spring and summer, pinnate with 3–7 leaflets. Racemes usually solitary, tight, with 10–40 flowers; inflorescence-stalk hairy, *c.* 2 cm. Flowers *c.* 4 × 3 mm. Calyx *c.* 2 × 2 mm, hairy, teeth very short. Petals whitish, veins purple, standard with a purple blotch at the base, or all flushed purplish. Pods brownish, 7–8 × 3–4 mm, ovate-oblong, opening at the apex only; beak *c.* 2 mm. Seeds 2–4, pale to dark brown. *New Zealand (South Island)*. H5–G1.

C. odorata J.D. Hooker. Illustration: *Botanical Magazine*, 9479 (1937). Similar, but racemes with not more than 12 flowers, calyx shorter, branches usually hairy, pod broader. *New Zealand (North Island)*. G1.

2. C. williamsii Kirk. Illustration: *Botanical Magazine*, n.s., 70 (1949); Moore & Irwin, The Oxford book of New Zealand plants, 165 (1978); Salmon, The native trees of New Zealand, 206 (1980).

Shrub to 4 m. Branchlets very compressed, 8–12 mm wide, grooved and

hairless. Leaves present only in young plants. Racemes 1 or 2 per notch, each with 1–5 flowers, inflorescence-stalks hairy, *c.* 1 cm. Flowers 2–3 cm. Calyx 5–6 × 4–5 mm, teeth narrowly triangular. Corolla yellowish, veins purple. Pod 2–3 cm × 6–7 mm, opening along its whole length, elliptic-oblong, dark brown to black. Seeds 6–15, red or red mottled with black. *New Zealand (North Island)*. H5–G1.

C. aligera Simpson (*C. australis* misapplied). Illustration: Allan, Flora of New Zealand **1**: 374 (1961); Salmon, The native trees of New Zealand, 206, 207 (1980). Similar, but racemes with 8–12 flowers; flowers much smaller (*c.* 4 × 3 mm, calyx 1.5–2 × 1.5–2 mm), pods 8–10 × 4–5 mm, black; seeds orange-red mottled with black. *New Zealand (North Island)*. H5–G1.

3. C. flagelliformis J.D. Hooker. Illustration: Allan, Flora of New Zealand **1**: 374 (1961).

Shrub to 3 m. Branchlets somewhat compressed, grooved, *c.* 1 mm across. Leaves borne only on young plants. Racemes 2–4 per notch, each with 3–7 flowers disposed in an almost umbellate manner; inflorescence-stalk very short or almost absent. Flowers *c.* 4 × 3 mm. Calyx *c.* 1 mm, with minute teeth. Corolla whitish, veined and flushed with purple. Pods compressed, ovoid, opening along their whole length, dark brown to black, with straight, stout beaks. Seeds 1 or 2, red mottled with black. *New Zealand (North Island)*. H5–G1.

C. arenaria Simpson. Similar, but a shrub to 50 cm, racemes with short stalks, flowers slightly larger, pods longer, with 2–4 seeds. *New Zealand (South Island)*. H5–G1.

4. C. arborea (Forster) Druce (*C. australis* misapplied). Illustration: Salmon, The native trees of New Zealand, 206 (1980).

Shrub or small tree to 5 m. Branchlets compressed, hairless, mostly 1–2 mm wide. Leaves present only in young plants. Racemes 1–3 to each notch, each with 3–5 flowers disposed in an almost umbellate manner; inflorescence-stalk to 1 cm. Flowers *c.* 5 × 4 mm. Calyx *c.* 2 × 2 mm, with minute teeth. Corolla whitish, the standard with purple centre and veins, the keel greenish at the base. Pods oblong, compressed, 8–10 × 4–5 mm, opening along their whole length, dark brown to black; beak sharp, oblique. Seeds 2–4, pale yellow-green mottled with black. *New Zealand (South Island)*. H5–G1.

C. ovata Simpson. Similar but pods shorter, dark brown, the beak also shorter. *New Zealand (South Island).* H5–G1.

C. violacea Kirk. Also similar, but a smaller shrub (to 1 m), flowers in racemes of 3–8, pods slightly hairy, smaller. *New Zealand (South Island).* H5–G1.

5. C. petriei Kirk.

Shrub to 2 m. Branchlets terete, 3–4 mm across. Leaves borne usually only on young plants. Racemes 1–3 per notch, *c.* 6 × 5 mm. Calyx *c.* 3 × 2 mm, with minute, acute teeth. Corolla greenish white flushed with purple and with purple veins. Pods oblong, 8–10 × *c.* 4 mm, turgid, dark brown. Seeds greenish yellow mottled with black. *New Zealand (South Island).* H5–G1.

6. C. kirkii J.D. Hooker. Illustration: Allan, Flora of New Zealand **1**: 374 (1961); Moore & Irwin, The Oxford book of New Zealand plants, 165 (1978).

Woody scrambler to 2 m or more. Branchlets more or less terete, striped, silky-hairy or hairless, *c.* 1 mm across. Leaves borne on adult plants during spring and summer, pinnate with 3–5 leaflets; leaflets hairless, stalk and axis silky hairy. Racemes with 2–6 flowers. Flowers 6–10 × 4–8 mm. Calyx *c.* 4 × 2 mm, with narrowly triangular teeth. Corolla white to cream with purple veins and blotches. Pods 8–15 × 4–6 mm, turgid, ellipsoid, opening from the base to the apex. Seeds 2–4, white, sometimes tinged with pale blue or mottled with black. *New Zealand (South Island).* H5–G1.

7. C. enysii Kirk. Illustration: Allan, Flora of New Zealand **1**: 374 (1961); Salmon, Collins' guide to the alpine plants of New Zealand, 165 (1985); Brickell (ed.), RHS A–Z encyclopedia of garden plants, 233 (2003).

Dwarf shrub to 5 cm, forming dense patches to 10 cm across, arising from a stout taproot. Adult plants leafless. Branchlets compressed, 1–2 mm wide. Racemes 1 or 2 from each notch, each with 1–3 flowers. Flowers *c.* 5 × 4 mm. Calyx *c.* 1 mm with short acute or blunt teeth. Corolla greenish to purplish, veins dark purple. Pods 6–8 × 4–5 mm, pale, more or less circular in profile, one side falling completely, the other persistent. Seeds 1 or rarely more, dull black. *New Zealand (South Island).* H5–G1.

C. orbiculata Colenso (*C. enysii* var. *orbiculata* (Colenso) Kirk). Very similar, but branchlets 2–3 mm wide, pods shorter, seeds dull olive-green mottled with black. *New Zealand (North Island).* H5–G1.

8. C. uniflora Kirk.

Small shrub with underground runners (rhizomes), forming compact or open patches, to 6 cm. Branchlets compressed, 1–1.5 mm across. Leaves absent in mature plants. Flowers 6–10 × 4–6 mm, solitary or in pairs in the notches. Calyx *c.* 3 × 2 mm, with short, acute teeth. Corolla whitish, standard purplish towards the base, and with purple veins; keel purple. Pods elliptic-oblong, *c.* 10 × 4 mm, somewhat compressed, dark brown to black. Seeds 4–6, bluish black. *New Zealand (South Island).* H5–G1.

9. C. astonii Simpson. Illustration: Allan, Flora of New Zealand **1**: 374 (1961).

Dwarf shrub arising from a woody root, forming hard patches to 15 cm, and 20 cm across. Branchlets very compressed, grooved, hairy, 4–8 mm wide. Leaves absent from adult plants. Racemes with 3–7 flowers. Flowers *c.* 10 × 5–7 mm. Calyx *c.* 5 × 3 mm, hairy with teeth 2–3 mm. Corolla whitish, veined with purple, standard with a purple basal blotch. Pods oblong, indehiscent, 1.2–2 cm × 4–7 mm, dark brown to black, beak to 3 mm, oblique. Seeds 4–8, pale green with sparse black spots, ultimately brownish. *New Zealand (South Island).* H5–G1.

C. monroi J.D. Hooker. Illustration: Allan, Flora of New Zealand **1**: 374 (1961); Abrams, New Zealand alpine plants, pl. 26 (1973). Similar but branchlets 2–4 mm wide, pods *c.* 1.5 cm × 4 mm, hooked; seeds 8–12. *New Zealand (South Island).* H5–G1.

10. C. curta Petrie.

Sprawling shrub, rather little branched, branches to 1 m, terete, 1–2 mm across, grooved, hairy at least when young. Racemes tight with 8–10 flowers. Flowers *c.* 4 × 3 mm. Calyx *c.* 1 × 1 mm, hairy, with minute teeth. Corolla creamy yellow striped with purple; standard with 2 bands of purple along the blade. Ovary hairy. Pod hairless when mature, 3–5 × 2–3 mm, turgid, brown; beak *c.* 1 mm, curved. Seeds 1–3, usually 2, pale green mottled with black. *New Zealand (South Island).* H5–G1.

11. C. nigrans Simpson. Illustration: Allan, Flora of New Zealand **1**: 374 (1961).

More or less prostrate shrub, branches to 1 m. Branchlets ascending, compressed, *c.* 1 cm wide, sparsely hairy. Racemes 1–3 at each notch, each with 5–10 flowers. Flowers *c.* 3 × 2 mm. Calyx short, sparsely hairy, with very short teeth. Corolla white veined with purple, standard with a purple blotch at the base. Pods *c.* 4 × 1.5 mm, oblong, somewhat compressed, black; beak curved, oblique. Seeds 2–4, pale brown. *New Zealand (South Island).* H5–G1.

51. HEDYSARUM Linnaeus

S.G. Knees & M.C. Warwick

Annual or perennial herbs or low shrubs. Stems several, grooved, often covered with adpressed hairs. Leaves alternate, pinnate with a terminal leaflet, the leaflets almost stalkless; stipules united or free, papery. Flowers erect or nodding, borne in stalked axillary racemes; calyx with 5 teeth, corolla with 5 petals, standard longer than keel, pink, purple, yellow or white, ovary with 4–8 ovules. Fruit a stalked segmented pod, with or without wings; each segment containing 1 seed, segments flattened or rounded; seeds 2–6.

A genus of about 100 species from much of the north temperate world. Most species are best grown from seed, but some can be propagated by cuttings. The shrubby species can be cut back to near ground level each spring to encourage fresh vigorous growth.

Literature: Rollins, R., The genus *Hedysarum* in North America, *Rhodora* **42**: 217–239 (1940).

Flowers. Yellowish or creamy white: **4**; red or purple: **1, 2, 3, 5, 6, 7, 8**.
Pods. Spiny: **1, 6**; smooth **2, 3, 4, 5, 7,8**.
Stipules. Free: **1**; joined **2, 3, 4, 5, 6, 7, 8**.

1a.	Stipules free	**1. coronarium**
b.	Stipules joined	2
2a.	Leaflets 21–27	**6. multijugum**
b.	Leaflets usually less than 21	3
3a.	Flowers creamy white	**4. boutignyanum**
b.	.Flowers pink, red or purplish	4
4a.	Pods 7–12 mm wide	**7. occidentale**
b.	Pods less than 7 mm wide	5
5a.	Stems 1–1.5 m	**5. microcalyx**
b.	Stems 10–70 cm	6
6a.	Leaflets narrowly elliptic to lanceolate, not more than 5 mm wide	**8. boreale**
b.	Leaflets broadly ovate, mostly *c.* 1 cm or more wide	7
7a.	Bracts 6–15 mm	**2. hedysaroides**
b.	Bracts 1–4 mm	**3. alpinum**

1. H. coronarium Linnaeus (*Sulla coronaria* (Linnaeus) Medikus). Illustration: *Flore des serres* **3**: 1382 (1858); Robinson, The English flower garden, pl. 128 (1883); Reichenbach, Icones Florae Germanicae et Helveticae **22**: 195 (1903); Bonnier, Flore complète **3**: pl. 164 (1914).

Perennial or biennial to 50 cm; stipules free. Leaflets 7–15, elliptic, 3–4 cm. Flowers in crowded racemes of 10–35, fragrant, on long stalks; calyx-teeth as long as tube, sparsely hairy; corolla *c.* 1.8 cm, deep red. Pod minutely roughened, becoming spiny; seeds 2–4. *Europe*. H3. Summer.

'Album'. Illustration: Waldstein & Kitaibel, Plantarum Rariorum Hungaricae, **2**: 3 (1805). Flowers white, sometimes cultivated. The species is now frequently separated from *Hedysarum* and placed in the gensu *Sulla* Medikus.

2. H. hedysaroides (Linnaeus) Schinz & Thellung (*Astragalus hedysaroides* Linnaeus; *H. obscurum* Linnaeus; *H. caucasicum* Bieberstein). Illustration: Bonnier, Flore complète **3**: pl. 164 (1914); Jávorka & Csapody, Iconographia florae Hungaricae, 291 (1932); Hess, Landholt & Hizel, Flora der Schweiz **2**: 572 (1970); Jelitto & Schacht, Hardy herbaceous perennials **1**: 277 (1990).

Perennial with mostly unbranched stems 10–60 cm, hairless or with few hairs; stipules 1–2 cm, joined, brown. Leaflets 7–21, ovate to elliptic, 1–2.5 cm. Racemes 3–8 cm, with 15–20 flowers; bracts 6–15 mm; calyx-teeth 1.5–3 mm, triangular; corolla 1.3–2.5 cm, reddish violet or white. Pod 3–5 mm wide, with 2–5 seeds. *Arctic Russia to SC Europe*. H1. Summer.

3. H. alpinum Linnaeus (*H. sibiricum* Ledebour). Illustration: *Botanical Magazine*, 3316 (1821); *Edwards's Botanical Register* **10**: 808 (1824); Hitchcock & Cronquist, Flora of the Pacific northwest, 261 (1973); Budd et al., Flora of the Canadian prairie provinces, 478 (1971).

Multi-stemmed herbaceous perennial, stems 20–70 cm, grooved; stipules 1.5–2 cm, joined, brown; leaves with 17–21 leaflets, narrowly elliptic, 2.5–3.5 cm. Bracts 1–4 mm; calyx-teeth almost equal, narrowly triangular; corolla 1.1–1.5 cm, red or purplish carmine. Pod 3.5–6 mm across, wings *c.* 0.5 mm; seeds 1–5. *Circumpolar*. H5. Spring–summer.

Var. **grandiflorum** Rollins is very similar but lower growing with larger flowers, 1.4–1.8 cm.

4. H. boutignyanum Alleizette. Illustration: *Bulletin de la Société Botanique de France* **75**: 41 (1928).

Hairless perennial, 50–60 cm. Leaflets 7–11, ovate to elliptic, 2–3 cm; stipules 2–3 cm, joined. Racemes 5–20 cm, bracts 1–1.5 cm; calyx-teeth shorter than tube; corolla 1.2–1.6 cm, cream or white, sometimes with bluish veins. Pod *c.* 5 mm across, with 2–5 seeds, margins hairy. *France (SW Alps)*. H3. Summer

5. H. microcalyx Baker. Illustration: *Botanical Magazine*, 6931 (1887).

Almost hairless perennial with slender stems 1–1.5 m upper leaves whorled; stipules 1.5–4 cm, elliptic, conspicuous. Leaflets 3–4 cm, hairless. Corolla 1.6–1.9 cm, bluish purple, wings shorter than keel. Pod 5–7 mm across, margins narrowly winged, seeds 2 or 3. *Himalaya (Kashmir to Uttar Pradesh)*. H4. Summer–autumn.

6. H. multijugum Maximowicz. Illustration: Dippel, Handbuch der Laubholzkunde **3**: 718 (1893); Schneider, Illustriertes Handbuch der Laubholzkunde **2**: 108 (1907); Bean, Trees & shrubs hardy in the British Isles, edn 7, **2**: 93 (1951); Hegi, Illustrierte Flora von Mittel-Europa, edn 2, **4**: 1484 (1975).

Woody based perennial with stems to 1.5 m. Leaflets 21–27, ovate, 1–1.2 cm; stipules 2–3 mm. Calyx 4–6 mm; corolla 1.9–2.3 cm, crimson-purple. Pods *c.* 5 mm across, with conspicuous spines, seeds 2–4. *Mongolia*. H2. Summer–autumn.

Var. **apiculatum** Sprague, illustration: *Botanical Magazine*, 8091 (1906), with loose erect panicles of flowers, is probably the most commonly cultivated variant of this species.

7. H. occidentale Greene. Illustration: Rickett, Wild Flowers of the United States **5**(2): 361 (1971); Hitchcock & Cronquist, Flora of the Pacific northwest, 261 (1973); Stewart, Wildflowers of the Olympic cascades, pl. 127 (1988).

Stems woody, 40–80 cm, branched above. Leaves hairy above, with 13–19 leaflets, each 1–3 cm, ovate to lanceolate; stipules 2–3 cm, joined. Flowers in racemes of 20–80; calyx-lobes 4–5 mm, corolla 1.6–2.2 cm, reddish purple. Pod 7–12 mm across, margins winged for 1–2 mm, seeds

1–4. *W USA & Canada*. H2. Summer–autumn.

8. H. boreale Nuttall. Illustration: Hitchcock & Cronquist, Flora of the Pacific northwest, 261 (1973); Kuijt, A flora of Waterton Lakes National Park, 376 (1982); Scotter & Flygare, Wildflowers of the Canadian Rockies, 139 (1986).

Herbaceous perennial, with many stems to 45 cm, arising from a woody rootstock. Leaves 4–8 cm, with 9–13 leaflets, each 2–5 mm wide, hairy; stipules *c.* 1 cm, joined. Flowers numerous, erect; upper calyx-lobes slender; corolla 1.2–1.9 cm, red. Pod 5–7 mm wide, with 2–6 seeds. *N America (Saskatchewan to Oklahoma and Arizona)*. H3–4. Spring–summer.

Some of the material offered as this species may be *H. alpinum*.

Var. **mackenzii** (Richardson) C.L. Hitchcock. Illustration: *Botanical Magazine*, 6386 (1878). Leaves hairless or very sparsely hairy above, adpressed hairs beneath, with 13–17 leaflets and darker flowers. *North America (Canada to N USA), Arctic, E Siberia*. H2. Spring.

52. ONOBRYCHIS Miller

J.R. Akeroyd

Herbaceous or subshrubby perennials, or annuals. Leaves irregularly pinnate; leaflets entire. Flowers in elongate, stalked axillary racemes. Calyx bell-shaped, with 5 equal linear teeth. Stamens united except for the uppermost which is free. Fruit indehiscent, flattened, almost circular, usually spiny on veins and margins; seed solitary.

A genus of about 100 species in Eurasia and N Africa. Propagation is by seed sown in the spring.

1a. Annual; flowers 2–8 in racemes
　　　　　　　　　　　　7. caput-galli
　b. Perennial; flowers more than 8 in
　　　racemes　　　　　　　　　　　　2
2a. Standard hairy on back; margin of
　　　fruit entire or with 2 rows of teeth　3
　b. Standard hairless; margin of fruit
　　　with 1 row of teeth　　　　　　　4
3a. Leaflets greyish and densely hairy
　　　beneath　　　　　　**1. hypargyrea**
　b. Leaflets green and sparsely hairy
　　　beneath　　　　　　　**2. radiata**
4a. Fruit 4–6 mm, with teeth on sides
　　　and margins　　　　　**4. arenaria**
　b. Fruit usually at least 6 mm, with
　　　teeth on margin only　　　　　　5

5a. Calyx-teeth 1–2 times as long as
tube **5. montana**

b. Calyx-teeth 2–4 times as long as
tube 6

6a. Hairs of calyx whitish; margin of
fruit with 6–8 teeth to 1 mm
3. viciifolia

b. Hairs of calyx brownish; margin of
fruit with 3–5 teeth to 1.5 mm
6. alba

1. O. hypargyrea Boissier (*O. tournefortii*
misapplied).

Velvety-hairy perennial; stems erect,
30–100 cm. Leaflets 9–15, 2.5–5 cm,
ovate-oblong, acute, hairless above, densely
greyish-hairy beneath. Calyx with shaggy
hairs, the teeth slightly longer than tube.
Corolla 1.5–2 cm, pale yellow with pink
veins. Fruit 1.4–1.8 cm, with shaggy hairs,
the margin entire or shortly toothed. *C &
W Turkey, former Yugoslavia (Macedonia),
N Greece*. H1. Summer.

2. O. radiata (Desfontaines) Bieberstein.
Illustration: *Edwards's Botanical Register* **32**:
37 (1815).

Softly-hairy perennial; stems 30–60 cm,
erect. Leaflets 11–21, 1–3 cm, ovate-
oblong, obtuse or almost acute, hairless
above, green and sparsely hairy beneath.
Calyx with shaggy hairs, the teeth 2–3
times as long as tube. Corolla 1.5–2 cm,
pale yellow with red veins. Fruit
1.1–1.8 cm, shaggy-hairy to hairless, the
margin with small teeth to 1.5 mm.
E Turkey, Caucasus. H1. Summer.

3. O. viciifolia Scopoli (*O. sativa*
Lamarck). Illustration: Keble Martin,
Concise British flora in colour, pl. 24
(1965).

Robust, hairy perennial; stems
30–80 cm, usually erect. Leaflets 7–25,
1–3.5 cm, narrowly ovate to oblong,
mucronate. Calyx with rather shaggy
hairs, the teeth 2–3 times as long as tube.
Corolla 1–1.4 cm, the standard about as
long as keel, bright pink with purplish
veins. Fruit 5–8 mm, strongly net-veined,
with teeth to 1 mm on lower margin.
C & W Europe. H1. Summer.

Mostly grown as a fodder crop and in
amenity seed mixtures.

4. O. arenaria (Kitaibel) de Candolle.
Illustration: Polunin, Flowers of Greece and
the Balkans, 310 (1980); Aeschimann
et al., Flora alpina **1**: 955 (2004).

Similar to *O. viciifolia* but smaller and
less erect; corolla 0.8–1.2 cm; fruit

4–6 mm, softly-hairy, toothed on sides and
margin. *C & SE Europe to C Asia*. H1.
Summer.

5. O. montana de Candolle (*O. viciifolia*
var. *montana* (de Candolle) Lamarck & de
Candolle). Illustration: Barneby, European
alpine flowers in colour, pl. 43 (1967);
Macdonald encyclopedia of alpine flowers,
pl. 172 (1985); Aeschimann et al., Flora
alpina **1**: 955 (2004).

Similar to *O. viciifolia*, but smaller and
more woody at base; stems procumbent to
almost erect; leaflets 11–17, lanceolate to
oblong; raceme denser; corolla purplish
pink, the standard shorter than keel;
calyx–teeth 1–2 times as long as tube; fruit
7–12 mm, shortly hairy, toothed on upper
margin. *Mountains of C & SE Europe*. H1.
Summer.

6. O. alba (Waldstein & Kitanov) Desvaux.
Perennial, rather woody at base; stems
10–30 cm, procumbent to ascending.
Leaflets 13–19, linear to narrowly elliptic
or oblong, shortly appressed-hairy. Raceme
dense. Calyx hairy, the teeth 2–2.5 times
as long as tube. Corolla 0.7–1.3 cm, pink
to pinkish purple. Fruit 4–7 mm, with
shaggy hairs, net-veined, toothed on sides
and margin, the marginal teeth to 1.5 mm.
Balkan Peninsula, C & S Italy. H1. Summer.

A variable species; the plant grown in
gardens is subsp. **laconica** (Boissier) Hayek
(*O. laconica* Boissier).

7. O. caput-galli (Linnaeus) Lamarck.
Illustration: Polunin & Huxley, Flowers of
the Mediterranean, 33 (1965).

Ascending to erect annual, 20–40 cm.
Leaflets 11–15, narrowly oblong to
obovate, greyish-hairy. Flowers 2–8 in
raceme. Calyx hairy, the teeth 1–2 times
as long as tube. Corolla 5–6 mm, pinkish.
Fruits *c.* 8 mm, strongly net-veined,
toothed, especially on margin.
Mediterranean area & S Europe. H2.
Summer.

53. ANTHYLLIS Linnaeus
J. Cullen
Annual or perennial herbs or low shrubs.
Leaves usually pinnate with a terminal
leaflet, rarely reduced to 3 leaflets or to a
solitary (terminal) leaflet. Stipules small,
falling early. Flowers usually in dense
heads, each subtended by 2 usually large
bracts, more rarely in clusters or solitary in
the axils of single bracts. Calyx tubular,
bell-shaped or unequally swollen,

sometimes constricted towards the mouth,
with equal or unequal teeth. Corolla
usually yellow, more rarely cream or red.
Stamens with the filaments all united or
the filament of the uppermost stamen free
for up to half its length. Pod shortly stalked
or stalkless, often indehiscent, usually
enclosed in the persistent, papery calyx,
containing 1–many seeds.

A genus of about 20 species mainly from
the Mediterranean area but extending
north to Finland and Iceland, west to
Madeira, east to the Caucasus and perhaps
beyond, and south to Ethiopia. The plants
are easily grown and will tolerate poor
soils. They are best propagated by seed
which gives faster germination if scarified
before sowing.

Literature: Cullen, J., The *Anthyllis
vulneraria* complex – a resumé, *Notes from
the Royal Botanic Garden Edinburgh* **35**:
1–38 (1976).

Habit. Shrubs: **1**, **2**; perennial herbs: **3**, **4**;
annual herbs: **4**; (rarely), **5**.
Leaves. All simple or with 3 leaflets: **1**, **5**;
at least some with 5 or more leaflets: **2**,
4, **5**; most with 17–41 leaflets: **3**.
Flowers. In heads: **2**, **3**, **4**; in clusters or
solitary: **1**, **5**.
Calyx. Tubular: **1**, **2**, **3**; bell-shaped: **2**:
unequally swollen and constricted
towards the apex: **4**, **5**. Teeth as long as
tube, fringed with long hairs: **3**.
Standard. Much exceeding the other petals:
3.

1a. Low shrubs with woody branches 2
b. Herbs, occasionally woody at the
extreme base 3
2a. Flowers in heads; bracts palmately
lobed **2. barba-jovis**
b. Flowers in clusters or solitary; bracts
simple **1. hermanniae**
3a. Calyx tubular, not swollen, not
constricted towards the mouth;
leaflets 17–41 **3. montana**
b. Calyx unequally swollen, constricted
towards the mouth; leaflets 1–17,
rarely more 4
4a. Calyx-teeth unequal, mouth of calyx
oblique; legume 1-seeded
4. vulneraria
b. Calyx-teeth equal, mouth of calyx
straight; legume 2-seeded,
constricted between the seeds
5. tetraphylla

1. A. hermanniae Linnaeus. Illustration:
Botanical Magazine, 2576 (1825); Huxley &

Taylor, Flowers of Greece and the Aegean, pl. 117 (1977).

Shrub to 50 cm, branches woody and contorted, their tips becoming spine-like. Leaves simple or with 3 leaflets, silky hairy. Flowers axillary, solitary or in clusters; bracts simple. Calyx tubular, 3–5 mm, teeth shorter than tube, mouth straight. Corolla yellow. Pod with 1 seed. *Mediterranean area.* H5. Summer.

2. A. barba-jovis Linnaeus. Illustration: Botanical Magazine, 1927 (1817); Bonnier, Flore complète **3**: pl. 126 (1914).

Shrub to 1 m with woody branches. Leaves with 9–19 more or less equal leaflets, silky hairy above and beneath. Flowers in terminal heads, bracts palmately lobed. Calyx 4–6 mm, tubular to bell-shaped, teeth shorter than tube, mouth straight. Corolla rather pale yellow. Pod with 1 seed. *S Europe.* H5. Summer.

3. A. montana Linnaeus. Illustration: Bonnier, Flore complète **3**: pl. 126 (1914); *Botanical Magazine*, n.s., 333 (1959); Polunin, Flowers of Europe, pl. 621 (1969); Ceballos et al., Plantas silvestres de la peninsula Iberica, 162 (1980); Aeschimann et al., Flora alpina **1**: 941 (2004).

Perennial herb, often woody at the base. Leaves mostly near the base of the flowering-stems, with 17–41 more or less equal leaflets, downy on both surfaces. Flowers in heads; bracts deeply palmately lobed. Calyx tubular, teeth as long as tube with long hairs on their margins, mouth of calyx straight. Corolla red to purple, the standard much exceeding the other petals. *Alps & mountains of S Europe.* H3. Summer.

4. A. vulneraria Linnaeus. Illustration: Bonnier, Flore complète **3**: pl. 127 (1914); Polunin, Flowers of Europe, pl. 622 (1969); Aeschimann et al., Flora alpina **1**: 943, 944 (2004); Lewis et al., Legumes of the world, 459 (2005).

Perennial (rarely annual) herbs. Leaves with 3–15 (rarely to 19) leaflets or the lowermost reduced to 3 leaflets or a solitary terminal leaflet, when more than 1 leaflet present the terminal leaflet sometimes larger than the others, all variably silky-hairy. Flowers in heads; bracts palmately lobed. Calyx unequally inflated, with 5 short unequal teeth, mouth oblique. Corolla cream, yellow, pink or red. Pod 1-seeded (rarely 2-seeded, when not constricted between the seeds).

Europe, N Africa, SW Asia. H1. Spring–summer.

A complex species divided into more than 30 subspecies (some treated as separate species by some authors) across its range. It is uncertain which of the subspecies are in cultivation, though the following have been seen:

Subsp. **vulneraria**. Stem-leaves with more or less equal leaflets; bract-lobes acute; calyx to 5 mm wide; corolla yellow. *NW Europe.* H1. Spring–summer.

Var. **langei** (Sagorski) Jalas is very similar but is more frequently branched, very silver-silky and occurs only in coastal areas.

Subsp. **polyphylla** (de Candolle) Nyman. Similar to subsp. *vulneraria* but leaves with more numerous leaflets and stems shaggy-hairy in the lower part. *C & E Europe.* H1. Summer.

Subsp. **argyrophylla** (Rothmaler) Cullen. Small plants; lower leaves with 1–3 leaflets, the terminal much larger than the laterals, stem-leaves small, leaflets more or less equal, all very densely silver-silky; bract-lobes acute; corolla red. *S Spain.* H5. Summer.

Sometimes grown as *A. webbiana* Hooker, a name of uncertain application.

Subsp. **alpestris** (Schultes) Ascherson & Graebner (*A. alpestris* Schultes). Illustration: Rasetti, I fiori dell Alpi, f. 314 (1980). Small plants; leaves with few leaflets, stem-leaves with the terminal leaflet larger than the laterals; bract-lobes obtuse; calyx more than 5 mm wide, grey-hairy; corolla pale yellow. *European mts, from N Spain to the Balkans.* H3. Summer.

Subsp. **pyrenaica** (Beck) Cullen (*A. pyrenaica* Beck). Illustration: Bonnier, Flore complète **3**: pl. 127 (1914) – as *A. dillenii*. Like subsp. *alpestris* but calyx white-hairy, corolla pinkish red. *Pyrenees, mts of N Spain.* H3. Summer.

5. A. tetraphylla Linnaeus. Illustration: *Botanical Magazine*, 108 (1788); Bonnier, Flore complète **3**: pl. 127 (1914); Huxley & Taylor, Flowers of Greece and the Aegean, pl. 115 (1977).

Prostrate annual herb. Leaves with 3–5 leaflets, the terminal much larger than the others, all spreading, hairy. Flowers in axillary clusters. Calyx unequally inflated at flowering, with equal, short teeth, mouth straight. Corolla yellow, keel often red at apex. Pod 2-seeded, constricted between the seeds. *Mediterranean area.* H3. Spring.

54. TETRAGONOLOBUS Scopoli
J. Cullen
Plants similar to *Lotus* but leaves with 3 leaflets, genuine stipules better-developed and leaf-like. Flowers solitary or 2 together. Calyx-teeth equal. Corolla yellow or dark red. Pod almost square in section, the angles winged.

A small genus, often included in *Lotus*, from Europe and the Mediterranean area, easily grown and propagated by seed.

1a. Perennial; corolla yellow
1. maritimus
b. Annual; corolla crimson or dark red
2. purpureus

1. T. maritimus (Linnaeus) Roth (*Lotus siliquosus* Linnaeus; *T. siliquosus* (Linnaeus) Roth). Illustration: *Botanical Magazine*, 151 (1791); Hegi, Illustrierte Flora von Mittel-Europa **4**: 1375–6 (1923); Polunin & Smythies, Flowers of southwest Europe, pl. 22 (1973).

Perennial to 40 cm, hairy or not. Leaflets oblanceolate to obovate, to 3 cm; stipules ovate. Flowers solitary, borne on stalks considerably longer than the leaves. Corolla pale yellow, 2.5–3 cm. Style with a membranous wing on one side. Pod 3–6 cm, wings to 1 mm wide. *Europe, N Africa, SW Asia.* H3. Summer.

2. T. purpureus Moench (*Lotus tetragonolobus* Linnaeus; *T. edulis* Link). Illustration: Bonnier, Flore complète **3**: pl. 144 (1914); Fenaroli, Flora Mediterranea, 122 (1962); Polunin & Huxley, Flowers of the Mediterranean, pl. 72 (1965); Huxley & Taylor, Flowers of Greece and the Aegean, pl. 144 (1977).

Annual herb to 40 cm, downy. Leaflets to 4 cm, obovate to almost diamond-shaped. Stipules ovate. Flowers solitary, stalks shorter than or equalling leaves. Corolla crimson or dark red. Style without a membranous wing at one side. Pod 3–9 cm, wings 2–4 mm wide. *S Europe, Mediterranean area.* H3. Summer.

55. LOTUS Linnaeus
J. Cullen
Annual or perennial herbs, sometimes woody below. Leaves pinnate with 5 leaflets (ours) including a terminal leaflet, axis often short, the 2 lowermost leaflets often resembling stipules. Genuine stipules very small, glandular. Flowers in heads or solitary. Calyx bell-shaped to almost

tubular, with 5 equal or unequal teeth. Corolla scarlet, crimson to dark purplish brown or yellow, the keel with a conspicuous beak. Stamens 10, filaments of 9 united, that of the uppermost free. Pod cylindric, sometimes broadly so, straight or curved. Seeds numerous.

A genus of perhaps 150 species from temperate regions of which only a small number is grown. They will tolerate a wide range of soils and are propagated by seed or cuttings.

Literature: Arrambari, A.M., A cladistic analysis of *Lotus* of the Old World, *Canadian Journal of Botany* **78**: 351–360; A cladistic analysis of *Lotus* of the New World, *Cladistics* **16**: 283–297 (both 2000).

Habit. Annual: **4**; perennial: **1, 2, 3, 5**.
Leaflets. Very narrow, linear to narrowly elliptic: **1, 2**; relatively broader: **3, 4, 5**. Grey-or whitish-hairy: **1, 5**; green, hairy or not: **2, 3, 4**.
Flowers. Red or purplish brown: **1, 2**; yellow, sometimes tinged with red: **3, 4, 5**. Solitary or 2 together: **1, 4**; in heads of 3 or more: **2, 3, 5**. Keel 3–4 cm: **1**. Standard reflexed, with a broad black line down the centre: **1**.
Calyx-teeth. Clearly unequal: **1, 5**; equal or almost so: **2, 3, 4**.
Pod. 4–8 cm across, grooved above: **4**; narrower, not obviously grooved above: **1, 2, 3, 5**.

1a. Flowers scarlet, the keel to 4 cm, curved, the standard narrow, reflexed, with a broad black line down its centre   **1. berthelotii**
  b. Flowers yellow or brownish, keel much smaller, standard not as above
                                                                           2
2a. Leaflets very narrow, linear to very narrowly elliptic, much longer than wide; flowers brownish purple   **2. jacobaeus**
  b. Leaflets relatively broader; flowers yellow, sometimes tinged with red
                                                                           3
3a. Calyx-teeth unequal, the 2 upper longer, curved upwards; leaves and stems,grey- or whitish-hairy
                                                                 5. creticus
  b. Calyx-teeth more or less equal, not as above; leaves and stems green, hairy or not
                                                                           4
4a. Flowers solitary or 2 together; pod short, curved, 4–8 mm in diameter, grooved above; annual   **4. edulis**

  b. Flowers 3 or more together in heads; pod straight, 2–3 mm in diameter, scarcely grooved above; perennial
                                                                 3. corniculatus

1. L. berthelotii Masferrer. Illustration: *Botanical Magazine*, 6733 (1884) – as *L. peliorhynchus*; Menninger, Flowering vines of the world, pl. 93 (1970); Bramwell & Bramwell, Wild flowers of the Canary Islands, 64 (1974); Gartenpraxis 1981: 464–465.

Sprawling or scrambling perennial. Leaves and stems conspicuously white- or grey-hairy. Leaves of 5 leaflets all attached at the same point, leaflets to 1.5 cm, linear; axillary shoots usually present in the leaf-axils. Flowers solitary or paired in the leaf-axils. Calyx unequally toothed. Corolla bright red, the standard narrow, reflexed, black-centred, the keel 3–4 cm, curved, claw-like. Fruit rarely formed. *Canary Islands, Cape Verde Islands*. H5–G1. Summer.

Very spectacular with its red keel looking like a parrot's beak. Now very rare and threatened in the wild, but well established in cultivation, propagated by cuttings.

2. L. jacobaeus Linnaeus. Illustration: *Botanical Magazine*, 79 (1787); *Kew Magazine* **4**: 87 (1987).

Perennial. Leaves of 5 leaflets, the upper 3 usually separated from the lower 2 by a stalk; leaflets linear to very narrowly elliptic or very narrowly obovate, to 4 cm. Flowers 3 or more together in axillary heads. Calyx-teeth more or less equal. Corolla purplish brown, keel to 1 cm. Pod linear-cylindric. *Cape Verde Islands*. H5–G1. Summer.

3. L. corniculatus Linnaeus. Illustration: Bonnier, Flore complète **3**: pl. 142 (1914); Keble Martin, The concise British flora in colour, pl. 23 (1965); Hay & Synge, Dictionary of garden plants in colour, 156 (1969); Aeschimann et al., Flora alpina **1**: 939 (2004); Lewis et al., (eds.), Legumes of the world, 463 (2005).

Sprawling to erect perennial, hairy or not. Leaflets 5, the upper 3 separated from the lower 2 by a short stalk, all obovate to almost circular. Flowers 3 or more together in heads. Calyx-teeth more or less equal. Corolla bright yellow, occasionally tinged with red, keel to 1.5 cm. Pod linear-cylindric, to 3 mm across, not conspicuously grooved above. *Eurasia, N Africa*. H1. Spring–summer.

A very widespread, common and variable plant in Europe. A variant with double flowers ('Pleno') is sometimes grown.

4. L. edulis Linnaeus. Illustration: Bonnier, Flore complète **3**: pl. 142 (1914); Pignatti, Flora d'Italia **1**: 746 (1982).

Annual herb, somewhat downy. Leaves with a short stalk between the 3 upper and 2 lower leaflets; leaflets obovate. Flowers solitary or paired, on a long (common) stalk. Calyx-teeth more or less equal. Corolla yellow, keel to 1.5 cm. Pod curved, widely cylindric, 4–8 mm across, deeply grooved above. *Mediterranean area*. G1–H5. Summer.

5. L. creticus Linnaeus. Illustration: Bonnier, Flore complète **3**: pl. 141 (1914); Polunin & Smythies, Plants of southwest Europe, pl. 22 (1973).

Rather upright perennial herb. Leaves and stems grey- or whitish-hairy. Leaves with 5 leaflets all arising from more or less the same point, obovate. Flowers 3 or more together in heads. Calyx-teeth unequal, the 2 upper longer than the others and curved upwards. Corolla yellow, keel to 1.5 cm, with a purple beak. Pod linear-cylindric, 2–3 mm across, not conspicuously grooved above. *Mediterranean area*. H5. Spring.

56. DORYCNIUM Miller
J. Cullen
Perennial herbs or small shrubs. Leaves almost stalkless, pinnate with 5 leaflets, the lower pair often appearing superficially like stipules; true stipules minute. Flowers in axillary heads. Calyx bell-shaped with 5 equal or unequal teeth. Corolla whitish, often with red or pink spots or lines; keel obtuse, dark red. Stamens 10, 9 with the filaments united, that of the uppermost free. Pod oblong, ovoid or narrowly cylindric. Seeds 1–many.

A genus of about 8 species, mainly from the Mediterranean area, often included in *Lotus*. They are not especially attractive and are rarely grown. They are, however, easy to cultivate; propagation is by seed.

Literature: Rikli, M., Die Gattung *Dorycnium* Vill, *Botanische Jahrbücher* **31**: 314–404(1901).

1a. Leaflets all very shortly stalked, without an obvious longer stalk to the upper 3 leaflets   **1. hirsutum**

b. Leaves with a very obvious stalk between the upper 3 and lower 2 leaflets   **2. broussonetii**

1. D. hirsutum (Linnaeus) Seringe (*Bonjeania hirsuta* (Linnaeus) Reichenbach). Illustration: Bonnier, Flore complète **3**: pl. 140 (1914); Polunin & Walters, Guide to the vegetation of Britain and Europe, 135 (1985); Aeschimann et al., Flora alpina **1**: 935 (2004).

Shrub to 50 cm. Leaflets to 1.8 cm, all equally and shortly stalked, elliptic to narrowly obovate, with long, dense, spreading hairs. Flowers in heads of 4–10 subtended closely by bracts which are similar to the leaves. Calyx-teeth slightly unequal. Corolla 1–2 cm, white or pinkish. Pod 6–12 mm. *Mediterranean area.* H4. Spring–summer.

D. pentaphyllum Scopoli (*D. suffruticosum* Villars). Illustration: Aeschimann et al., Flora alpina **1**: 935 (2004); Lewis et al., (eds.), Legumes of the world, 464 (2005). A small shrub less hairy than the above, with narrower leaflets and corollas 3–7 mm. *C & E Europe, Mediterranean area.* H3. Spring–summer.

May occasionally be grown.

2. D. broussonetii (Choisy) Webb & Berthelot. Illustration: Bramwell & Bramwell, Flores silvestres de las Islas Canarias, 67 (1974).

Shrub to 1.5 m. Leaves with a conspicuous stalk between the lower 2 leaflets, which are borne at the junction with the stem, and the upper 3. Leaflets elliptic to obovate, 2–4 cm. Flowers in heads of 5–8. Calyx-teeth slightly unequal. Corolla white with pink spots or lines. Pod much exceeding the calyx. *Canary Islands.* H5. Spring.

57. CORONILLA Linnaeus
J.M. Lees
Annual or perennial herbs or low shrubs. Leaves pinnate with a terminal leaflet, rarely simple or with only 3 leaflets; stipules variable, free or united to each other. Flowers in heads in the leaf-axils. Calyx bell-shaped, more or less 2-lipped. Petals yellow or pinkish, keel acute. Stamens 10, filaments of 9 united, that of the uppermost free. Pod breaking up into 1-seeded linear to oblong segments, which are not constricted between them, round in section or ridged or angled.

A genus of about 25 species from Europe, western and central Asia and the Canary Islands. They are easily grown in most soils. Propagation is by seed.

Literature: Uhrova, A., Revision der Gattung *Coronilla* L., *Beihefte Botanischer Centralblatt* **53** (13): 1–174 (1935); Jahn, A., Beiträge zur Kentniss der Sippenstruktur einiger Arten der Gattung *Coronilla* L., *Feddes Repertorium* **85**: 455–532 (1974).

1a. Lower leaves simple or with 3 leaflets, the terminal leaflet much larger than the laterals; annual   **7. scorpioides**
  b. Lower leaves pinnate, the leaflets more or less equal; perennial or shrubby   2
2a. Corolla white, pink or purple   **6. varia**
  b. Corolla yellow   3
3a. Claw of standard 2–3 times as long as calyx   **1. emerus**
  b. Claw of standard equalling or only slightly longer than calyx   4
4a. Corolla 1.2–1.8 cm; leaves glaucous   **5. orientalis**
  b. Corolla 5–12 mm; leaves not glaucous   5
5a. Perennial herb; heads with 12–22 flowers   **4. coronata**
  b. Small low shrub; heads with 4–15 flowers   6
6a. Stipules wedge-shaped to obovate, deciduous, 2–10 mm; leaflets without scarious margins   **2. valentina**
  b. Stipules united, persistent, *c.* 1 mm; leaflets with scarious margins   **3. minima**

1. C. emerus Linnaeus. Illustration: Bonnier, Flore complète **3**: pl. 162 (1914); Aeschimann et al., Flora alpina **1**: 949 (2004).

Shrub to 2 m. Leaves with 2–4 pairs of leaflets which are 1–2 cm, obovate, mucronate, glaucous. Stipules 1–2 mm, free, membranous. Corolla 1.4–2 cm, pale yellow; claw of standard 2–3 times as long as calyx. Pod 5–11 cm, with 3–12 segments which are 8–10 mm. *C & SE Europe, also locally extending to S Norway, the Pyrenees & E Spain.* H2. Summer.

This species is now frequently included in *Hippocrepis*, as *H. emerus* (Linnaeus) Lassen.

2. C. valentina Linnaeus. Illustration: Bonnier, Flore complète **3**: pl. 162 (1914).

Shrub to 1 m. Leaves with 2–6 pairs of leaflets which are obovate and notched. Stipules wedge-shaped to obovate, free, deciduous. Heads with 4–12 flowers. Corolla 7–12 mm, yellow. Legume 1–5 cm, with 1–10 segments which are 5–7 mm, spindle-shaped, somewhat compressed and with 2 obtuse angles. *Mediterranean area, S Portugal.* H5. Summer.

C. glauca Linnaeus (*C. valentina* subsp. *glauca* (Linnaeus) Battandier). Illustration: Bonnier, Flore complète **3**: pl. 162 (1914); Polunin & Smythies, Flowers of southwest Europe, pl. 23 (1973); Brickell (ed.), RHS A–Z encyclopedia of garden plants, 308 (2003); Lewis et al., (eds.), Legumes of the world, 458 (2005). Similar but with 2 or 3 pairs of leaflets, stipules ovate or lanceolate, membranous, 2–6 mm; pod with 1–4 (rarely to 10) segments. *Mediterranean area.* H5. Summer.

3. C. minima Linnaeus. Illustration: Bonnier, Flore complète **3**: pl. 161 (1914) Aeschimann et al., Flora alpina **1**: 947 (2004).

Small shrub to 50 cm. Leaves with 2–6 pairs of leaflets which are elliptic, obovate or almost circular, 2–15 mm, stalkless and with translucent margins. Stipules *c.* 1 mm, united to each other, membranous, persistent. Heads with up to 15 flowers, flower-stalks 2–4 mm. Corolla 5–12 mm, yellow. Pod 1–3.5 cm, with 1–7 segments which are 4.5–5.5 mm, oblong, 4-angled. *W Europe extending to Switzerland & Italy.* H3. Summer.

4. C. coronata Linnaeus (*C. montana* Scopoli). Illustration: Bonnier, Flore complète **3**: pl. 161 (1914); Aeschimann et al., Flora alpina **1**: 947 (2004).

Perennial herb, 10–70 cm. Leaves with 3–7 pairs of leaflets which are elliptic to obovate, 1.5–4 cm, shortly stalked and with narrow translucent margins. Stipules 3–5 mm, united to each other, membranous, deciduous. Heads with 12–22 flowers, flower-stalks 4–6 mm. Corolla 7–11 mm, yellow. Pod 1.5–3 cm, segments 1–9, 6–7.6 mm, ovoid-oblong, obtusely angled. *C Europe, extending E to the Crimea.* H4. Summer.

5. C. orientalis Miller (*C. iberica* Bieberstein; *C. cappadocica* Willdenow).

More or less prostrate perennial herb with a woody base. Stems to 50 cm. Leaves with 3–5 pairs of leaflets which are broadly wedge-shaped, glaucous and more or less notched. Stipules 3–5 mm, wedge-shaped to obovate. Heads with 3–9 flowers. Flowers 1.2–1.8 cm, yellow, claws of petals about as long as the sepals. Pod 2–4.5 mm, curved, with 2–11 segments. *SW Asia*. H5.

6. C. varia Linnaeus (*Securigera varia* (Linnaeus) Lassen). Illustration: Bonnier, Flore complète **3**: pl. 161 (1914); Huxley & Taylor, Flowers of Greece and the Aegean, t. 113 (1977); Aeschimann et al., Flora alpina **1**: 949 (2004).

Perennial herb to 1.2 m. Leaves with 5–12 pairs of leaflets which are oblong or elliptic, to 3 cm, with narrow, translucent margins. Stipules 1–6 mm, free, membranous. Heads with 10–20 (rarely as few as 5) flowers. Corolla 8–15 mm, white, pink or purple. Pod 1.5–8 cm, with 3–12 segments which are 4–6 mm, oblong, 4-angled. *C & S Europe to C Russia*. H2. Summer.

7. C. scorpioides (Linnaeus) Koch. Illustration: Aeschimann et al., Flora alpina **1**: 949 (2004).

Annual herb to 40 cm. Leaves simple or with 3 leaflets, the terminal leaflet to 4 cm, elliptic to almost circular, much larger than the kidney-shaped lateral leaflets. Stipules 1–2 mm, united to each other, membranous. Heads with 2–5 flowers. Corolla 4–8 mm, yellow. Pod 2–6 cm, curved, segments 2–11, oblong, obtusely 4-angled. *S Europe*. H5. Summer.

58. HIPPOCREPIS Linnaeus
J.R. Akeroyd

Herbaceous or shrubby perennials (cultivated species). Leaves irregularly pinnate; leaflets entire. Stipules small, linear to lanceolate. Flowers in umbels on long axillary stems. Calyx elongately bell-shaped with 5 teeth. Corolla yellow. Stamens united except for the uppermost which is free. Fruit flattened, indehiscent, of a single segment or segmented and breaking up when ripe, the segments with deep sinuses.

A genus of some 15 species in Europe and the Mediterranean area. Propagation is by seed or division of plants. The plant described above as *Coronilla emerus* is now frequently included here.

1a. Leaflets obtuse; fruit covered with reddish or brown papillae
                         1. comosa
  b. Leaflets acute; fruit smooth
                        2. balearica

1. H. comosa Linnaeus. Illustration: Polunin, Collins photoguide to wild flowers, p. 216 (1988); Brickell, RHS gardeners' encyclopedia of plants and flowers, 326 (1989); Aeschimann et al., Flora alpina **1**: 951 (2004).

Stems 10–40 cm, prostrate to ascending, somewhat woody at base. Leaflets 7–17, 5–15 × 2–4 mm, obovate to linear, obtuse, usually hairy beneath. Flowers 5–12 in umbel. Corolla 6–12 mm, yellow. Fruit 1.5–3 cm, covered with reddish or brown papillae; sinuses of segments semi-circular to almost circular. *S & W Europe & Mediterranean area*. H1. Summer.

A variable species; a variant with pale yellow corollas is often grown in gardens.

2. H. balearica Jacquin. Illustration: *Botanical Magazine*, 427 (1798); Polunin & Smythies, Flowers of southwest Europe, pl. 22 (1973).

Subshrub; stems 20–60 cm, erect, woody below. Leaflets 11–21, 4–12 × 1–4 mm, linear or oblong, acute. Flowers 2–10 in umbel. Corolla 1–1.5 cm. Fruit 1.5–4.5 cm, smooth; sinuses of segments semicircular to circular. *Balearic Islands*. H5. Summer.

59. VICIA Linnaeus
F.K. Hibberd

Perennial and annual herbs; stems angular in cross-section but never winged. Leaves usually with many pairs of leaflets, the axis ending in a tendril or short point; leaflets folded in bud (*V. faba* exceptionally sometimes has rolled leaflets). Flowers blue, purple, yellow, orange or whitish. Wing petals attached to keel. Style cylindric or flattened, hairy all round or with a tuft or hairs on the outer face. Pod diamond-shaped or flattened-cylindric. Germination hypogeal.

A genus of about 150 species distributed throughout the temperate northern hemisphere, extending into tropical Africa and south America. Propagation is by seed.

Literature: Kupicha, F.K. The infrageneric structure of *Vicia*. *Notes from the Royal Botanic Garden Edinburgh* **34**: 287–326 (1976); Kenicer G. & Norton S., 632. Vicia americana Leguminosae. *Botanical Magazine* **25**: 342–349 (2008).

Leaves. With a tendril: **3, 5**: without a tendril: **1, 2, 4, 6**. Stipules with a nectary (a dark spot, visited by ants): **4, 5, 6**; stipules without nectaries: **1, 2, 3**.

Flowers. Inflorescence distinctly stalked: **1, 2, 3**; flowers more or less stalkless in leaf axils: **4, 5, 6**.

1a. Leaf-axis ending in a short point   2
  b. Leaf-axis ending in a tendril   5
2a. Racemes distinctly stalked   4
  b. Racemes with very short stalk, or flowers stalkless in the leaf axil   3
3a. Annual, with large edible seeds 2 cm or more across   **6. faba**
  b. Perennial, with small inedible seeds   **4. oroboides**
4a. Leaves with 3–6 pairs of leaflets; flowers orange-yellow   **1. crocea**
  b. Leaves with 1 pair of leaflets, flowers purple   **2. unijuga**
5a. Flowers in a dense stalked raceme   **3. cracca**
  b. Flowers 1–2, stalkless in leaf-axils   **5. sativa**

1. V. crocea (Desfontaines) Fedtschenko (*Orobus croceus* Desfontaines).

Softly-hairy perennial with erect stems 40–70 cm, the pair of stipules at each node very unequal in shape and size. Leaf-axis ending in a short point; leaflets 3–6-paired, ovate-acuminate, 3.5–6 × 1.5–3 cm. Racemes many-flowered, arranged in a panicle; flowers 1.5–2 cm, corolla orange-yellow. *Turkey, N Iran, Caucasia*. H2. Early summer.

Very rare in cultivation. *V. crocea* is remarkably similar to *Lathyrus aureus*, but the two can be distinguished as follows:

V. crocea has folded young leaflets, unequal stipules, the end of the style hairy all round, and a short pod narrowed to a slender stalk, whereas *L. aureus* has rolled leaflets, equal stipules, the style hairy only on the inner face, and a long linear pod.

2. V. unijuga A. Braun (*Orobus lathyroides* Linnaeus). Illustration: *Botanical Magazine*, 2098 (1819); Phillips & Rix, Perennials **2**: 61 (1990).

Erect perennial 30–40 cm. Leaves with one pair of leaflets, the axis ending in a short point; leaflets ovate-acuminate, with hairy margins. Flowers 3–18 in a loose raceme; corolla purple. *Mongolia, Siberia, China, Korea & Japan*. H1. Early summer.

3. V. cracca Linnaeus. Illustration: Hegi, Illustrierte Flora von Mittel-Europa **4**: t. 170, fig. 1 (1923); Keble Martin, Concise British flora in colour, pl. 24 (1965); Polunin, Flowers of Europe, 55 (1969); Grey-Wilson & Blamey, Alpine flowers of

Britain & Europe, 119 (1979); Aeschimann et al., Flora alpina **1**: 863 (2004).

Downy perennial with stems scrambling or climbing to 2 m. Leaves with many pairs of leaflets and a branched tendril; leaflets linear-lanceolate, 1–2.5 cm. Racemes dense, many-flowered, on stalks 2–10 cm. Flowers 1–1.2 cm, purplish blue. *Widespread in N Europe & Asia.* H1. Summer.

4. V. oroboides Wulfen (*Orobus lathyroides* Sibthorp & Smith, not Linnaeus). Illustration: Hegi, Illustrierte Flora von Mittel-Europa **4**: 1549 (1923); Aeschimann et al., Flora alpina **1**: 869 (2004).

A hairless or downy perennial with erect stems 5–60 cm. Leaves with 1–3 pairs of leaflets, the leaf axis ending in a short point; leaflets ovate to oblong, acute, up to 8 × 4 cm. Racemes 3–8 flowered, inflorescence stalks very short; flowers *c.* 1.8 cm, whitish or pale yellow. *Italy, Austria, Hungary & former Yugoslavia.* H2. Summer.

Rare in cultivation.

5. V. sativa Linnaeus. Illustration: Keble Martin, Concise British flora in colour, pl. 25 (1965); Polunin, Flowers of Europe, 55 (1969); Aeschimann et al., Flora alpina **1**: 873 (2004); Lewis et al., Legumes of the world, 506 (2005).

Annual, with stems to 1 m. Leaves with 4–8 pairs of leaflets and a tendril; leaflets linear to obovate, notched at apex, 1–2 cm, sparsely hairy. Flowers 1 or 2, stalkless at a node, 1–3 cm, purple. *Europe, Asia, N Africa.* H2. Summer.

6. V. faba Linnaeus. Illustration: Bonnier, Name this flower, pl. 16 (1917); Hegi, Illustrierte Flora von Mittel-Europa **4**: t. 170, fig. 4 & p. 1558 (1923); Masefield et al., Oxford book of food plants, 41 (1969); Lewis et al., (eds.), Legumes of the world, 504 (2005).

Stout erect hairless annual with angular stems from 30–200 depending on the cultivar. Leaves with 2–4 pairs of leaflets, the axis ending in a point; leaflets elliptic, ovate or obovate, 5–10 cm. Flowers 1–several, almost stalkless in leaf-axils, 1.6–3 cm, often white with blackish wings, sometimes red. Pods large, 5–20 × 1–2 cm, containing edible seeds in a 'woolly' lining. *Unknown in the wild, but widely cultivated as a crop (broad bean, tic bean or horse bean).* H2. Early summer.

60. LATHYRUS Linnaeus
F.K. Hibberd

Herbaceous perennials and annuals with often branched or unbranched climbing or sprawling stems which are often winged. Leaves with 2 or more leaflets, the axis ending in a tendril or a short point; leaflets rolled up in bud. Racemes axillary, 1 to many-flowered. Wing petals attached to keel. Style flattened, with a brush of hairs on the inner face, often twisted through 90 degrees. Germination hypogeal.

A genus of about 150 species distributed throughout the temperate northern hemisphere and extending into tropical east Africa and South America. Propagation is by seed or division. Species with erect stems and leaves with more than a pair of leaflets are woodland, while the rest (including *L. japonicus*) grow best in sun and well-drained soil. Despite occasional references to hybrids in horticultural literature, hybridisation between Eurasian species has been proved virtually impossible (Davies, 1958). Lathyrus is a rare example of a genus with flowers in the three primary colours, red, blue and yellow.

Literature: Davies, A.J.S., A cytotaxonomic study in the genus *Lathyrus*. Ph.D. Thesis, Manchester University (1958); Bässler, M., Revision der eurasiatischen Arten von Lathrys L. Sect. Orobus (L.) Grenier & Godron. *Feddes Repertorium* **84**: 329–477 (1973). Kupicha, F.K., The infrageneric structure of *Lathyrus*. *Notes from the Royal Botanic Garden Edinburgh* **41**: 209–244 (1983); Norton, S., Some observations on *Lathyrus rotundifolius*. *New Plantsman* **1**(2): 78–83 (1994); See also Kenicer, G. & Norton, S., *Botanical Magazine* **25** (4) (2006).

Stems. Clearly winged: **7, 8, 9, 10, 11, 12, 13**; angular in cross-section but not winged: **1, 2, 3, 4, 5, 6, 14, 15**.
Leaves. With 4 or more leaflets: **1, 2, 3, 4**; with a pair of leaflets: **5, 6, 7, 8, 9, 10, 11, 12, 13, 14, 15**. Tendril present: **1, 4, 5, 6, 7, 8, 9, 10, 11, 12, 13, 15**: absent: **2, 3, 14**. Leaf-stalk absent: **15**.
Flowers. Corolla orange or yellow: **1, 2, 12**; pink or crimson: **3, 5, 6, 7, 8, 9, 10, 11**; blue or purple: **3, 4, 11, 13, 14, 15**. (Several species have white-flowered forms.) Flowers sweetly scented:

6, 11. Style twisted through 90°: **5, 6, 7, 8, 9, 10, 11, 12, 13**.

1a. Leaves with 4 or more leaflets   2
  b. Leaves with a pair of leaflets   5
2a. Leaf-axis ending in a branched tendril   3
  b. Leaf-axis ending in a short point   4
3a. Stems erect; flowers orange
          1. davidii
  b. Stems prostrate; flowers pale purple
          4. japonicus
4a. Leaflets elliptic; flowers orange
          2. aureus
  b. Leaflets ovate-acuminate; flowers blue-purple, pink or white **3. vernus**
5a. Plants hairy   6
  b. Plants hairless   8
6a. Stems winged; tall annuals with tendrils   7
  b. Stems unwinged; prostrate perennial without tendrils **14. laxiflorus**
7a. Flowers blue, purple, pink, red or white, sweetly scented **11. odoratus**
  b. Flowers yellow, unscented
          12. chloranthus
8a. Stems winged   9
  b. Stems not winged   13
9a. Flowers blue; annual, with stems to 45 cm; pod oblong, *c.* 3 cm
          13. sativus
  b. Flowers crimson, pink, purplish or white; perennial, with stems climbing or sprawling to 2 m or more; pod linear, 5 cm or more   10
10a. Racemes with 1 or 2 flowers; standard 3.5 cm **10. tingitanus**
  b. Racemes with 3–11 flowers; standard 2 cm or less   11
11a. Leaflets net-veined, 4.5 cm or less
          7. rotundifolius
  b. Leaflets parallel-veined, 5 cm or more   12
12a. Stipules more than half as wide as stem; lowest calyx-tooth 8 mm
          8. latifolius
  b. Stipules less than half as wide as stem; lowest calyx-tooth 4 mm
          9. sylvestris
13a. Leaf-stalk absent, the two leaflets and two large stipules forming a cluster at each node **15. nervosus**
  b. Leaf-stalk present, stipules much smaller than leaflets   14
14a. Tendrils branched; leaflets elliptic (widest at the middle); standard *c.* 4 cm **5. grandiflorus**
  b. Tendrils usually unbranched; leaflets often broadest above the middle; standard *c.* 1 cm **6. tuberosus**

1. L. davidii Hance. Illustration: *Gartenflora* **32**: t. 1127 (1883).

Hairless perennial, with erect unbranched stems 80 cm; stems not winged. Leaves ending with a branched tendril, and having 2–4 pairs of leaflets, these ovate, 5–9 × 2.5–5.5 cm, yellowish green, soft-textured; stipules large, conspicuous. Racemes with 8–20 flowers; inflorescence-stalk 2–6 cm. Flower-stalk 17–20 cm. Calyx-teeth represented by mere points around the mouth of the tube. Corolla orange. *Siberia, China, Korea & Japan.* H4. Summer.

2. L. aureus (Steven) Brandza (*L. luteus* misapplied). Illustration: Stojanoff & Stefanoff, Flora of Bulgaria, edn **2**: 655 (1933); Flora Republicii Populare Romine **5**: t. 78 (1957).

Perennial with erect unbranched stems up to 80 cm; stems not winged; stems and leaf-stalks bearing dark glandular hairs (hand-lens needed), and plants also sparsely downy with curly hairs. Leaves with 3–5 pairs of leaflets, the leaf-axis ending in a short point 3–6 mm; leaflets elliptic, acute to slightly acuminate, 3.5–5 × 2–2.5 cm. Racemes dense, many-flowered, with inflorescence-stalk 7–11 cm. Flower 1.7–1.9 cm. Calyx-teeth unequal, the lowest one longest, *c.* 3 mm. Corolla orange. *Native to countries bordering the Black Sea.* H2. Early summer.

L. aureus belongs to a complex of closely related species differentiated by leaflet size and shape, flower size, presence or absence of glandular hairs and relative lengths of calyx-teeth (Bässler, 1973). Other members of this group, apart from *L. aureus*, may possibly be in cultivation.

3. L. vernus (Linnaeus) Bernhardi (*Orobus vernus* Linnaeus; *L. cyaneus* misapplied). Figure 110(1), p. 384. Illustration: *Botanical Magazine* **15**: 521 (1801); Hegi, Illustrierte Flora von Mittel-Europa **4**: t. 172, fig. 3 & p. 1576 (1923); Grey-Wilson & Blamey, Alpine flora of Britain and Europe, 121 (1979); Phillips & Rix, Perennials **1**: 60 (1990); Aeschimannet al., Flora alpina **1**: 879 (2004).

Hairless perennial with unwinged stems 25 cm. Leaves with 2 or 3 pairs of leaflets, the leaf-axis ending in a point 5–13 mm; leaflets ovate-acuminate, 3–8 × 1.2–4 cm. Racemes with 3–6 flowers; inflorescence-stalks 3–5 cm. Flowers 1.3–1.7 cm. Calyx-teeth unequal, the lowest longest, 2.5–3 mm. Corolla purplish blue, vivid blue, pure white or with pale pink standard and keel and white wings. *Continental Europe to E Siberia.* H1. Spring and early summer.

4. L. japonicus Willdenow (*L. maritimus* (Linnaeus) Bigelow). Figure 110(2), p. 384. Illustration: Hegi, Illustrierte Flora von Mittel-Europa **4**: t. 172 & p. 1586 (1923); Keble Martin, Concise British flora in colour, pl. 25 (1965); Polunin, Flowers of Europe, 56 (1969); Phillips & Rix, Perennials **2**: 58 (1990).

Hairless or hairy perennial, unbranched or with prostrate branches to 1 m; stems not winged. Leaves ending in a short, branched tendril and having 2–5 pairs of leaflets, these elliptic, 1.5–5 cm × 8–25 mm, leathery and glaucous; stipules *c.* 2 cm; resembling leaflets in size and texture. Racemes with 3–10 flowers, inflorescence-stalks 3.5–5 cm. Flowers 1–2.5 cm. Calyx-teeth unequal, the lowest tooth 4–5 mm. Corolla bicoloured: standard pale purple, wings and keel white tinged with purple. *Circumpolar, on shingle & sandy beaches* . H1. Summer.

Cultivated plants of *L. japonicus* do not have such leathery leaflets as those grown in the wild.

Subsp. **japonicus.** Flowers up to 2.5 cm with pods over 1 cm wide. *E Asia, arctic Europe & North America.*

Subsp. **maritimus** (Linnaeus) Ball. Flowers up to 1.6 cm and pods less than 1 cm wide. *Baltic area, British coasts, Alaska, California & the Great Lakes area of N America.*

5. L. grandiflorus *Sibthorp & Smith.* Figure 110(3), p. 384. Illustration: *Botanical Magazine*, 1938 (1817); Polunin, Flowers of Greece & the Balkans, pl. 23 (1980); Huxley & Taylor, Flowers of Greece & the Aegean, fig. 99 (1977); Phillips & Rix, Perennials **2**: 58 (1990).

A far-running hairless perennial, with slender but strongly climbing stems to 2 m or more; stems not winged. Leaves with leaf-stalk 1.5–2.5 cm, one pair of leaflets and a branched tendril; leaflets elliptic, 23–37 × 1.2–2.5 cm, slightly wavy-margined. Flowers 1 or 2, on inflorescence-stalks 3–4 cm. Flowers *c.* 3 cm, standing 4 cm when flattened; calyx-teeth unequal, the lowest tooth 3 mm, standard deep pink, wings crimson, keel pale pink. *Italy, Sicily, former Yugoslavia, Albania & Bulgaria.* H3. Early summer.

6. L. tuberosus Linnaeus. Illustration: *Botanical Magazine* **4**: 111 (1790); Hegi, Illustrierte Flora von Mittel-Europa **4**: t. 171, fig. 4 (1923); Polunin, Flowers of Europe, 56 (1969); Phillips & Rix, Perennials **2**: 59 (1990); Aeschimann et al., Flora alpina **1**: 885 (2004).

Hairless perennial with slender rhizomes bearing ovoid tubers 3–4.5 cm, these edible and nutty-flavoured; stems branched, scrambling to 1 m, unwinged. Leaves with leaf-stalk 5–8 mm, one pair of leaflets and a simple or 3-fid tendril; leaflets elliptic or narrowly obovate, 2–4.3 cm × 7–18 mm. Racemes with 3–9 flowers; inflorescence-stalk 4.5–5 cm. Flowers sweetly scented, *c.* 9 mm; calyx–teeth unequal, the lowest 3–5 mm; corolla bright pink. *Most of Europe east to C Asia.* H1. Summer.

7. L. rotundifolius Willdenow (*L. heterophyllus*, misapplied). Illustration: *Botanical Magazine*, 6522 (1880); Phillips & Rix, Perennials **2**: 59 (1990); *The New Plantsman* **1**(2), pl. 79 (1994).

Hairless perennial with stems to 2 m or more; stems winged. Leaves with leaf-stalk 1–2.5 cm, one pair of leaflets and a branched tendril; leaflets elliptic, 3–4.5 × 1.5–3.5 cm. Racemes 4–11-flowered; inflorescence-stalk 4–8 cm. Flowers 1.4–1.7 cm, calyx-teeth unequal, the lowest 3 mm; corolla either bright purplish pink or a soft brownish red. *Crimea, Caucasia, E. Turkey & neighbouring Iraq & Iran.* H2. Summer.

L. undulatus Boissier, a close relative of *L. rotundifolius* distinguished by its wavy-edged leaflets, is probably not in cultivation.

8. L. latifolius Linnaeus. Figure 110(4), p. 384. Illustration: Hegi, Illustrierte Flora von Mittel-Europa **4**: 1597 (1923); Grey-Wilson & Blamey, Alpine flowers of Britain and Europe, 121 (1979); Phillips & Rix, Perennials **2**: 58 (1990); Brickell (ed.), RHS A–Z encyclopedia of garden plants, 611 (2003); Aeschimann et al., Flora alpina **1**: 887 (2004).

Glaucous, hairless perennial with robust, strongly winged stems climbing or sprawling to form dense mounds. Leaves with leaf-stalk 3.5–5 cm, one pair of leaflets and a branched tendril; leaflets oblong-elliptic, 8–11 × 2.5–4 cm. Racemes with 6–11 flowers; inflorescence-stalks 12–14 cm. Flowers *c.* 1.8 cm; calyx-teeth unequal, the lowest *c.* 8 mm. Corolla of uniform coloration, varying from deep pink to white. *S Europe.* H2. Summer.

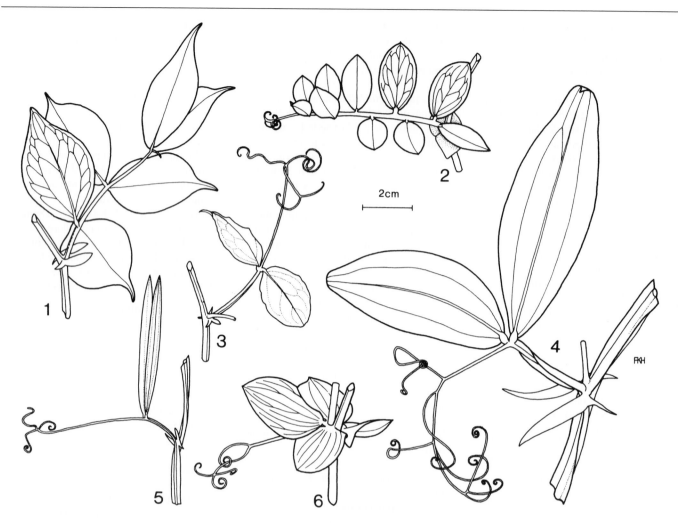

Figure 110. Leaves of Lathyrus species. 1, *L. vernus*. 2, *L. japonicus*.
3, *L. grandiflorus*. 4, *L. latifolius*. 5, *L. sativus*. 6, *L. nervosus*.

9. L. sylvestris Linnaeus. Illustration: Ross-Craig, Drawings of British plants **7**: t. 73 (1954); Keble Martin, Concise British flora in colour, pl. 25 (1965); Phillips & Rix, Perennials **2**: 59 (1990); Aeschimann et al., Flora alpina **1**: 887 (2004).

Bright green, hairless perennial with stems to 2 m or more; stems winged. Leaves with leaf-stalk 2, 5–4 cm, one pair of leaflets and a branched tendril; leaflets linear-elliptic, 5–15 cm × 5–40 mm. Racemes with 3–8 flowers on inflorescence-stalks 10–20 cm. Flowers *c.* 1.4 cm; calyx-teeth unequal, the lowest *c.* 4 mm; wing petals purplish, standard pink, the colour fading towards the margins and much paler on the reverse. *Europe, NW Africa, Caucasia.* H1. Summer.

10. L. tingitanus Linnaeus. Illustration: *Botanical Magazine*, 100 (1789); Hegi,

Illustrierte Flora von Mittel-Europa **4**: 1565 (1923); Polunin, Flowers of Europe, 56 (1969) – as *L. latifolius*; Polunin & Smythies, Flowers of southwest Europe, 20 (1973).

Hairless perennial with winged stems climbing strongly to 3 m or more. Leaves with leaf-stalk 2–3 cm, one pair of leaflets and a branched tendril; leaflets linear-elliptic, narrowly elliptic or narrowly oblong-elliptic, 4–7 cm × 5–10 mm. Racemes with 1 or 2 flowers on inflorescence-stalk 4–9 cm. Flowers 1.8–2.4 cm; calyx-teeth subequal, *c.* 3 mm; corolla pale pink or crimson-magenta, *c.* 3.5 cm when flattened. *Iberian Peninsula, Sardinia, Morocco, Algeria, Canary Islands.* H3. Summer.

11. L. odoratus Linnaeus. Illustration: *Botanical Magazine*, 60 (1788); Everard &

Morley, Wild flowers of the world, pl. 29 (1970); Perry, Flowers of the world, 160 (1972); Brickell (ed.), RHS A–Z encyclopedia of garden plants, 611 (2003).

Hairy annual with erect, strongly winged stems climbing to 2 m or more. Leaves with leaf-stalk 3–4.5 cm, one pair of leaflets and a branched tendril; leaflets ovate-elliptic, 5–6.5 × 3.5–4 cm. Racemes with 2–4 flowers; inflorescence-stalk *c.* 18 cm. Flowers strongly scented, 3 cm; calyx-teeth equal, 7–8 mm; standard wine-red, keel and wings purple. *Italy, Sicily.* H3. Summer.

The wild species is described above. Modern 'sweet peas' are descended from the cultivars 'Countess Spencer' and 'Gladys Unwin', which both had wavy standards. After 1900 this 'Spencer' type ousted the old-fashioned varieties with stiff upright or hooded standards, and there are now innumerable cultivars in every colour except yellow.

12. L. chloranthus Boissier & Balansa. Illustration: Kayagin (ed.), Flora Azerbaijan **5**: t. 49 (1954).

Hairy annual with winged stems climbing to 70 cm. Leaves with 1 pair of leaflets and a branched tendril; leaflets elliptic, 2–6 × 7–20 mm. Flowers 1 or 2 on inflorescence-stalk *c*. 6 cm. Calyx-teeth subequal, *c*. 7 mm. Flowers *c*. 2 cm, bright greenish yellow, standard downy on the back, sometimes with a red blotch. *C & E Turkey, Armenia, N Iraq, Iran*. H3. Summer.

13. L. sativus Linnaeus. Figure 110(5). p. 370. Illustration: *Botanical Magazine*, 115 (1790); Hegi, Illustrierte Flora von Mittel-Europa **4**: t. 171, fig. 6 (1923); Polunin, Flowers of Europe, 56 (1969); Aeschimann et al., Flora alpina **1**: 891 (2004).

Hairless annual with winged stems climbing weakly to 45 cm or more. Leaves with leaf-stalk 1 or 2 cm, one pair of leaflets and a simple or branched tendril; leaflets narrowly elliptic, 4–6 cm × 5–8 mm. Flowers solitary; inflorescence-stalk 1.8–3.2 cm. Flowers *c*. 1.3 cm, calyx-teeth subequal, 3–4 mm; corolla blue, blue and white or white, fading towards the base and sometimes with pink veins. *S & C Europe, N Africa & SW Asia*. H3. Early summer.

14. L. laxiflorus (Desfontaines) O. Kuntze (*Orobus hirsutus* Linnaeus). Illustration: *Botanical Magazine*, 2345 (1822); Stojanoff & Stefanoff, Flora of Bulgaria, edn **2**: 654 (1933); Phillips & Rix, Perennials **2**: 58 (1990).

Softly downy perennial, with unwinged stems arching and spreading, 13–18 cm. Leaves with leaf-stalk 1–1.5 cm and one pair of leaflets; leaf-axis ending in a short point 3–5 mm; leaflets broadly elliptic, 2–3 × 1.5–2 cm. Racemes with 3 or 4 flowers; inflorescence-stalk 2.5–4 cm. Flowers 1.3–1.5 cm; calyx-teeth equal and slender, 6–7 mm; standard purple, wings and keel bright blue. *E Mediterranean area, Caucasia, Iran & Syria*. H3. Early summer.

15. L. nervosus Lamarck. (*L. magellanicus* misapplied). Figure 110(6). p. 384. Illustration: *Botanical Magazine*, 3987 (1842); Phillips & Rix, Perennials **1**: 60 (1990); Brickell (ed.), RHS A–Z encyclopedia of garden plants, 611 (2003).

Hairless perennial with unwinged stems trailing or weakly climbing to 1 m,

remaining evergeen in winter. Leaves very thick and leathery, glaucous with a white bloom; leaf-stalk absent, leaf-axis ending in a stiff, branched tendril; stipules as conspicuous as the pair of leaflets; leaflets elliptic, 3.5–4 × 2.5–3 cm. Racemes with *c*. 10 flowers borne in whorl-like clusters of 3 or 4; inflorescence-stalk 7–8.5 cm. Flowers 1.5–1.8 cm; calyx-teeth unequal, the lowest *c*. 5 mm; corolla purplish blue, paler on keel and wings. *S America*. H5. Summer.

61. LENS Miller
F.K. Hibberd
Low, hairy annuals, with stems angular but not winged. Leaves with a few pairs of small leaflets and tendrils; leaflets fold in bud. Racemes few-flowered; flowers small, pale and inconspicuous; calyx-teeth equal, as long as the corolla, wing petals adhering to the keel; style flattened, hairy on the inner surface. Pod stalked, oblong, 1–3-seeded; seeds lens-shaped. Germination hypogeal.

A genus of 5 species, distributed around the Mediterranean, in south-west Asia and tropical Africa. Propagation is by seed.

Literature: Davis, P.H. and Plitmann, U. in Davis, Flora of Turkey **3**: 325–328 (1970).

1. L. culinaris Medikus (*L. esculenta* Moench). Illustration: Reichenbach, Icones Florae Germanicae **22**: t. 265 (1903); Hegi, Illustrierte Flora von Mittel-Europa **4**: t. 171, fig. 1 (1923); Masefield, et al., Oxford book of food plants, 43 (1969); Lewis et al., (eds.), Legumes of the world, 506 (2005).

Slender hairy annual to 40 cm. Leaflets elliptic, 7–15 × 1.5–3.5 mm. Flowers pale mauve or white, *c*. 8 mm. *Unknown in the wild*. Summer.

Widely cultivated in Europe and Asia for its seeds (lentils).

62. PISUM Linnaeus
F.K. Hibberd
Hairless annuals with stems circular in cross-section. Leaves with large leafy stipules, 1–4 pairs of leaflets and a strong, branched tendril; leaflets fold in bud. Racemes few-flowered, flowers showy, calyx-teeth subequal, wing petals united with the keel; style with retroflexed margins, hairy on the inner face. Pod cylindric. Germination hypogeal.

A genus of 2 species, occurring in the Mediterranean area and south-west Asia. Propagation is by seed.

1. P. sativum Linneaus subsp. **sativum**
Stems 10–200 cm. Inflorescence-stalk up to twice as long as stipules, 1–3-flowered. Flowers 1.6–3 cm.

Var. **sativum**. Illustration: Reichenbach, Icones Florae Germanicae **22**: t. 270 (1903); Bonnier, Name this flower, pl. 17 (1917); Masefield et al., Oxford book of food plants, 43 (1969). Flowers white. Seeds spherical, green, yellow or white, smooth or wrinkled, sweet. *The garden pea is unknown in the wild*. H2. Summer.

Var. **arvense** (Linnaeus) Poiret. Illustration: Hegi, Illustrierte Flora von Mittel-Europa **4**: t. 170, fig. 5 (1923); Polunin, Flowers of Europe, 56 (1969); Lewis et al., Legumes of the world, 508 (2005). Flowers bicoloured, with pink standard and purple wings. Seeds dark, angular. *Europe & SW Asia*. H2. Summer.

Var. *arvense*, the field pea, is grown for fodder and silage.

63. CICER Linnaeus
F.K. Hibberd
Perennial and annual herbs, often spiny, conspicuously glandular-hairy. Leaves ending in a leaflet, spine or tendril; leaflets 3–many, toothed, the veins terminating in the teeth. Racemes 1–few-flowered. Wing petals free from the keel. Style hairless. Fruit inflated, seeds beaked. Germination hypogeal.

A genus of 40 species distributed around the Mediterranean and east to the Himalayas and central Asia. Propagation is by seed.

Literature: van der Maesen, L.J.F., *Cicer L.*, a monograph of the genus, with special reference to the chickpea (*Cicer arietinum* L.), its ecology and cultivation. *Mededelingen Landbouwhogeschool Wageningen* **72**: 10 (1972).

1. C. arietinum Linnaeus. Illustration: *Botanical Magazine*, 2274 (1821); Hegi, Illustrierte Flora von Mittel-Europa **4**: 1500 (1923); Purseglove, Tropical crops, Dicotyledons **1**: t. 37 (1968); Masefield et al., The Oxford book of food plants, 39 (1969); Aeschimann et al., Flora alpina **1**: 861 (2004).

Sturdy annual, *c*. 30 cm. Leaves with 5–7 pairs plus a terminal leaflet. Flowers solitary, 1–1.2 cm thick, mauve or white. Pod 1.7–3 cm, containing 1–3 seeds. *Unknown in the wild*.

Extensively cultivated in S Europe, N Africa and Asia for the seeds (chick pea).

64. ONONIS Linnaeus
J.M. Lees

Annual or perennial herbs or dwarf shrubs, usually sticky, glandular-hairy. Leaves usually with 3 leaflets, rarely reduced to a single leaflet or pinnate with a terminal leaflet; leaflets usually toothed. Stipules united to the leaf-stalk. Flowers in spikes, racemes or panicles. Calyx bell-shaped or tubular. Corolla yellow, pink or purple, rarely almost white; keel more or less beaked. Stamens 10, their filaments all united. Pod oblong or ovate. Seeds 1–many.

A genus of about 75 species, mostly from the Mediterranean area, but also occurring in the Canary Islands and east to NW India and Mongolia. They are easily grown. Propagation is usually from seed.

Literature: Sirjaev, G., Generis *Ononis* L., Revisio critica, *Beihefte Botanischer Centralblatt* **49**(2): 381–665 (1932).

1a. Flowers in spikes or racemes, sometimes with several flowers at each node; flower-stalks not jointed, or with a joint not more than 1.5 mm from the base; pod ovate or diamond-shaped   **5. arvensis**
  b. Flowers in panicles, sometimes condensed and with the primary branches 1-flowered; flower-stalks distinctly jointed, the joint more than 1.5 mm from the base; pod linear to oblong   2
2a. Corolla yellow, sometimes with red or violet veins   **4. natrix**
  b. Corolla pink or purple, occasionally white   3
3a. Terminal leaflet with a long stalk   **1. rotundifolia**
  b. Terminal leaflet stalkless or with a very short stalk   4
4a. Stem 25–100 cm, erect, usually many-flowered; most bracts with a single leaflet   **2. fruticosa**
  b. Stem 5–35 cm, prostrate, 1–6-flowered; bracts with 1 or 3 leaflets   **3. cristata**

1. O. rotundifolia Linnaeus. Illustration: Bonnier, Flore complète **3**: pl. 124(1914); Polunin & Smythies, Flowers of southwest Europe, pl. 21 (1973); Ceballos et al., Plantas silvestres de la peninsula Iberica, 169 (1980); Aeschimann et al., Flora alpina **1**: 895 (2004).

Erect, branched dwarf shrub, 10–50 cm. Stem densely glandular-hairy. Leaves with 3 leaflets which are *c.* 2.5 cm, elliptic to almost circular, obtuse, coarsely toothed,

sparingly glandular; terminal leaflet with a long stalk. Primary branches of the inflorescence *c.* 3 cm, increasing to 6 cm in fruit, ending in a spine. Flower-stalks 3–6 (rarely to 20) mm. Corolla 1.6–2 cm, pink or whitish. Pod 2–3 cm. Seeds 10–20, *c.* 3 mm, minutely warty. *SE Spain to E Austria & C Italy.* H4. Summer.

2. O. fruticosa Linnaeus. Illustration: Bonnier, Flore complète **3**: pl. 124 (1914); Polunin & Smythies, Flowers of southwest Europe, pl. 21 (1973); Brickell (ed.), RHS A–Z encyclopedia of garden plants, 741 (2003); Aeschimann et al., Flora alpina **1**: 895 (2004).

Erect dwarf shrub to 1 m. Stem downy towards the apex, the young stems downy throughout. Leaves mostly with 3 leaflets, more or less stalkless; leaflets 7–25 mm, oblong-oblanceolate, somewhat leathery, hairless, toothed. Primary branches of the inflorescence 1–3 cm, in the axils of translucent bracts. Flower-stalks 2–10 mm. Corolla 1–2 cm, pink. Pod 1.8–3 cm. Seeds *c.* 4, to 2.5 mm, minutely warty. *S Europe.* H4. Summer.

3. O. cristata Miller (*O. cenisia* Linnaeus). Illustration: Bonnier, Flore complète **3**: pl. 124 (1914); Ceballos et al., Plantas silvestres de la peninsula Iberica, 170 (1980); Aeschimann et al., Flora alpina **1**: 895 (2004).

Creeping perennial, with rhizomes, 5–35 cm. Stems shortly glandular-hairy. Leaves with 3 leaflets which are 5–10 mm, oblong or oblanceolate, somewhat leathery. Primary branches of the inflorescence 6–30 mm, shortly pointed. Corolla 1–1.5 cm, pink. Pod 9–12 mm. Seeds *c.* 5, to 2.5 mm, warty. *Mountains of S Europe.* H3. Summer.

4. O. natrix Linnaeus. Illustration: Bonnier, Flore complète **3**: pl. 126 (1914); Taylor, Wild flowers of Spain and Portugal, 55 (1972); Ceballos et al., Plantas silvestres de la peninsula Iberica, 167 (1980); Aeschimann et al., Flora alpina **1**: 897 (2004); Lewis et al., Legumes of the world, 502 (2005).

Erect, much-branched dwarf shrub, 15–60 cm. Stem densely glandular-hairy. Leaves with 3 leaflets or the lower leaves pinnate with more leaflets; leaflets variable, ovate to linear. Flowers in loose, leafy panicles; primary branches 1-flowered. Corolla 6–20 mm, yellow, frequently with red or violet veins. Pod 1–2.5 cm. Seeds

4–10, *c.* 2 mm, smooth or minutely warty. *SW Europe.* H3. Summer.

5. O. arvensis Linnaeus (*O. hircina* Jacquin; *O. altissima* Lamarck; *O. spinosa* Linnaeus subsp. *hircina* (Jacquin) Gams). Illustration: Aeschimann et al., Flora alpina **1**: 901 (2004).

Shrubby perennial herb, 30–100 cm. Stems erect, variably hairy. Leaves mostly with 3 leaflets; leaflets 6–30 mm, elliptic to ovate. Flowers stalked, borne in pairs at each node in dense spikes at the ends of the branches; bracts with 1–3 leaflets; flower-stalk not jointed or joint not more than 1.5 mm from the base. Corolla 1–2 cm, pink. Pod 5–9 mm, about equalling the calyx. Seeds 1–3, *c.* 2.5 mm, dark brown, warty. *Eurasia.* H1. Summer.

O. spinosa Linnaeus. Similar but with flowers borne singly; stem downy with hairs in a single line, pod equalling or slightly exceeding the calyx. *W, C & S Europe.* H3. Summer.

65. PAROCHETUS D. Don
E.H. Hamlet

Creeping herbaceous perennial. Leaves palmate with 3 heart-shaped, finely toothed leaflets, veins running out to margins. Stipules free from the leaf-stalk. Flowers blue or pale purple, 1–4, stalked, in umbels, those in the lower leaf axils very small and not opening with pods ripening on or below the soil surface. Calyx bell-shaped, deeply cleft; lobes 5, acute, almost equal. Petals free, obovate, short-clawed; standard deep blue (rarely purple) wings pinkish, long-clawed, oblong-obovate; keel shorter than wings, turned inwards. Stamens 10, filaments united except for that of the uppermost, which is free. Style bent; stigma small. Pod linear, acutely beaked; seeds 8–20.

A genus of 1 species from Asia, tropical E and southern Africa. It can be grown outside preferring a wel-drained soil and partially shaded position or in a cool greenhouse. Suitable for rock gardens or hanging baskets. Propagation is by cuttings.

1. P. communis D. Don. Illustration: *Flora of Tropical East Africa*, **4**(2): 1015 (1971); Polunin & Stainton, Flowers of the Himalayas, pl. 28 (1984); *The New Plantsman* **2**(2), pl. 107 (1995); Brickell (ed.), RHS A–Z encyclopedia of garden plants, 770 (2003).

E tropical & S Africa to E & SE Asia.

66. MELILOTUS Miller
E.H. Hamlet

Annual or biennial herbs, smelling of newly mown hay. Leaves with 3 pinnately veined leaflets; leaflets usually linear to elliptic-oblong, short-stalked, veins running out to the margin; stipules lanceolate to awl-shaped, fused to the leaf-stalk to some extent. Flowers small, yellow, white or white tipped with blue, in erect, axillary, spike-like racemes. Calyx shortly bell-shaped; lobes 5, almost equal, awl-shaped to lanceolate, acute to acuminate, shorter than the tube. Petals falling early; standard obovate-oblong, narrow at base, nearly stalkless; wings oblong, auricled at the base. Keel blunt, clawed, obtuse. Stamens 10, filaments of 9 united, that of the uppermost free. Style thread-like; stigma small. Pod nutlet-like, straight, beaked, spherical or obovoid. Seeds 1–few.

A genus of 20 species from Mediterranean Europe, SW Asia and adjacent Africa. They are moderately hardy, drought-resistant and prefer alkaline soils. Propagation is by seed.

1a. Flowers white **1. alba**
  b. Flowers yellow 2
2a. Standard and wings equal, longer than keel; pod transversely wrinkled **3. officinalis**
  b. Standard, wings and keel equal; pod net-veined **2. altissima**

1. M. alba Medikus. (*M. officinalis* (Linnaeus) Lamarck subsp. *alba* (Medikus) Ohashi & Tateishi). Illustration: Zohary (ed.), Flora Palaestina 2: pl. 224 (1972); Cunningham et al., Plants of western New South Wales, 405 (1981); Keble Martin, The new concise British flora in colour, pl. 22 (1982); Pignatti, Flora d'Italia 1: 706 (1982).

Erect, branched annual or perennial, 30–150 cm. Leaves with 3 pinnately veined leaflets; leaflets 1–3 cm × 5–15 mm, narrowly oblong-obovate to almost circular, with teeth pointing forwards. Stipules bristle-like, entire. Racemes loose and slender, many-flowered. Petals 4–5 mm, white; wings and keel nearly equal, shorter than standard. Stamens united for about half their length. Pod 3–5 mm, obovoid, mucronate, net-veined, hairless, greyish brown when ripe. *Eurasia.* H1. Summer–autumn.

2. M. altissimus Thuillier. Illustration: Polunin, Flowers of Europe, pl. 57 (1969); Keble Martin, The new concise British flora

in colour, pl. 22 (1982); Reader's Digest field guide to the wild flowers of Britain, 113 (1985).

Erect branched biennial or short-lived perennial, 60–150 cm. Leaflets oblong-ovate or wedge-shaped, obtuse, with teeth pointing forwards. Stipules subulate to bristle-like, entire. Racemes 2–5 cm, many-flowered, lengthening in fruit. Flowers 5–7 mm, yellow; wings, standard and keel equal. Pod 5–6 mm, obovoid, acute, net-veined, downy, black when ripe, usually 2-seeded. Style long and persistent. *Europe & Mediterranean area.* H1. Summer.

3. M. officinalis (Linnaeus) Lamarck. Illustration: Phillips, Wild flowers of Britain, 73 (1977); Chinery, Field guide to the plant life of Britain & Europe, 49 (1987); Ceballos et al., Plantas silvestres de la peninsula Iberica, 165 (1980); Aeschimann et al., Flora alpina 1: 903 (2004).

Trailing or erect, branched biennial, 40–250 cm. Leaflets of lower leaves obovate to ovate, those of the upper ovate-lanceolate, all with teeth pointing forwards. Racemes loose and slender, many-flowered. Flowers 4–7 mm, yellow; wings and standard equal, longer than keel. Pod 3–5 mm, transversely wrinkled, mucronate, hairless, brown when ripe, usually 1-seeded. Style often deciduous. *Eurasia, naturalised in N America.* H1. Summer.

67. TRIGONELLA Linnaeus
E.H. Hamlet

Erect annual herbs, often strongly scented. Leaves with 3 toothed leaflets, veins running out to the leaflet margins; stipules joined to the leaf-stalk. Flowers yellow, blue or white in heads, umbels or short dense racemes in leaf-axils, rarely solitary. Calyx 5-toothed, teeth ovate, acuminate. Petals free, standard obovate or oblong; wings oblong, auricled; keel oblong, shorter than wings, obtuse. Stamens free from petals, filaments of 9 united that of the uppermost free. Fruit variable, oblong or oblong-linear, compressed or terete, with 1-many seeds.

A genus of 80 species from Eurasia, Southern Africa and Australia. They prefer a well-drained soil and sunny position. Propagation is by seed.

1a. Flowers 1.2 cm or more **3. foenum-graecum**
  b. Flowers less than 1 cm 2

2a. Stems hollow; pod abruptly contracted into a beak **1. caerulea**
  b. Stems solid; pod tapering into a beak **2. procumbens**

1. T. caerulea Seringe. Illustration: Polunin, Flowers of Europe, pl. 57 (1969); Hegi, Illustrierte Flora von Mittel-Europa, 4: 1234 (1975); Everett, The New York Botanical Gardens illustrated encyclopedia of horticulture, 3401 (1982); Pignatti, Flora d'Italia 1: 710 (1982); Aeschimann et al., Flora alpina 1: 907 (2004).

Stems 20–100 cm, erect, sparsely hairy, hollow. Leaflets 2–5 cm × 5–20 mm, ovate to oblong, notched, finely toothed. Racemes spherical, dense, many-flowered. Inflorescence-stalk 2–5 cm. Calyx *c.* 3 mm. Flowers *c.* 6 mm, blue or white. Pod 4–5 × 3 mm, erect or almost horizontal, diamond-shaped to obovate, abruptly contracted into a beak, *c.* 2 mm. Seeds ovoid, brown, finely warty. *Cultivated for fodder in Europe & widely naturalised as a weed.* Summer.

2. T. procumbens (Besser) Reichenbach. Illustration: Polunin & Walters, A guide to the vegetation of Britain & Europe, 152 (1985).

Stems 20–50 cm, decumbent or erect, solid. Leaflets 1–3 cm × 3–10 mm, oblong to linear-oblong, finely toothed. Inflorescence stalk 2–6 cm. Flowers numerous in dense ovoid heads. Calyx 3–3.5 mm, bell-shaped, teeth lanceolate. Flowers 5.5–7 mm, lilac-blue. Pod 1–3-seeded, oblong, somewhat compressed, tapering into a beak. Seeds oblong, smooth. *Europe & SW Asia.*

3. T. foenum-graecum Linnaeus. Illustration: Polunin, Flowers of Europe, pl. 58 (1969); Zohary, Flora Palaestina 2: pl. 199 (1972); Hegi, Illustrierte Flora von Mittel-Europa 4: 1232 (1975).

Stems 10–50 cm, erect. Leaflets 1–3 cm × 5–15 mm, obovate to oblanceolate, toothed or sometimes incised. Flowers 1 or 2 in leaf-axils. Calyx 7–8 mm, tubular; teeth lanceolate-linear, as long as calyx-tube. Flowers 1.2–1.8 cm, yellowish white, sometimes tinged with lilac. Pods terete or somewhat compressed, linear, straight or somewhat curved, net-veined, gradually tapering into a beak. Seeds 10–20. *Mediterranean area, probably as an escape from cultivation. Of doubtful origin.* Summer.

Widely cultivated as a spice plant (fenugreek).

68. MEDICAGO Linnaeus
E.H. Hamlet

Annual or perennial herbs or shrubs. Leaves with 3 pinnate leaflets; leaflets obovate, margins toothed, veins running out to margins. Stipules fused to leaf-stalk to some extent. Flowers small, yellow or violet, rarely variegated, in axillary spikes or small heads. Calyx bell-shaped; lobes 5, almost equal. Petals free from staminal tube, standard obovate or oblong, narrowed at the base; wings oblong, auricled, clawed, longer than the obtuse keel. Stamens 10, filaments of 9 united that of the uppermost free. Style awl-shaped, smooth; stigma almost capitate, oblique. Pod spirally coiled, or sickle-shaped, often covered with spines, with 1–several seeds.

A genus of about 50 species from temperate Eurasia, the Mediterranean area and Africa. They prefer a well-drained alkaline soil. Propagation is by seed.

Literature: Lesins, K.A. & Lesins, I., Genus Medicago (Leguminosae), A Taxogenetic Study (1979).

1a. Flowers violet, lavender, pink or
    white   **2. sativa**
  b. Flowers yellow   2
2a. Herb; flowers less than 5 mm
                               3. lupulina
  b. Shrub; flowers more than 5 mm
                                 1. arborea

1. M. arborea Linnaeus. Illustration: Polunin & Everard, Trees & bushes of Europe, 113 (1976); Polunin, Flowers of Greece & the Balkans, 97 (1980); Lewis et al., (eds.), Legumes of the world, 503 (2005).

Shrub 1–4 m, densely covered with adpressed silky hairs. Stipules triangular, entire. Leaflets variable in size, 1–2 cm × 8–18 mm, obovate to ovate, wedge-shaped at base, entire or finely toothed at apex. Inflorescence short, with 4–8 flowers. Calyx-teeth shorter than tube. Petals 1–1.5 cm, yellow, standard elliptic. Pods greyish yellow, 1.2–1.5 cm across, net-veined. *Mediterranean & Canary Islands, S Europe to SW Asia.* Summer.

2. M. sativa Linnaeus. Illustration: Zohary, Flora Palaestina **2**: pl. 204 (1972); Phillips, Wild flowers of Britain, 84 (1977); Cunningham et al., Plants of western New South Wales, 404 (1981); Reader's Digest field guide to the wild flowers of Britain, 110(1985).

Perennial with stems 30–120 cm, prostrate to erect, arising from the crown. Leaflets 8–28 × 3–15 mm, obovate at lower nodes, wedge-shaped or linear-oblanceolate at upper nodes. Inflorescence with 7–35 flowers; petals 6–12 mm. Calyx half the length of the petal. Corolla violet, lavender, rarely pink or white; standard twice or more as long as wide; wings longer than keel. Pod yellow-brown. *Eurasia, naturalised in N America and probably elsewhere.* Summer–autumn.

3. M. lupulina Linnaeus. Illustration: Zohary, Flora Palaestina **2**: pl. 203 (1972); Phillips, Wild flowers of Britain, 53 (1977); Reader's Digest field guide to the wild flowers of Britain, 112 (1985); Aeschimann et al., Flora alpina **1**: 907 (2004).

Perennial, biennial or annual herbs; stems decumbent, 20–80 cm, branching from the base. Leaflets 1.1–1.4 cm × 6–17 mm, broadly ovate to obovate. Inflorescence with 14–24 flowers, petals 2.5–3.5 mm. Calyx 1.5–2.3 mm. Corolla yellow; standard rounded to broadly ovate; wings shorter than keel. Pod an ash-grey to black nutlet with a coiled tip, 1-seeded. *Eurasia, N Africa, naturalised in N America.* Summer–autumn.

69. TRIFOLIUM Linnaeus
U. Oster

Annual, biennial or perennial herbs, occasionally somewhat woody. Leaves usually with 3 leaflets, rarely palmately divided with 5–8 leaflets, the leaflets usually toothed, the veins running out to the margins; stipules large, persistent, joined to the leaf-stalk to some extent. Flowers in heads or short spikes, rarely solitary. Calyx tubular, 5-toothed, teeth equal or unequal. Petals persistent or deciduous, attached to each other and to the staminal tube. Stamens 10, 9 with their filaments united, that of the uppermost free, with all or 5 of the filaments swollen towards the apex. Pod enclosed in the persistent calyx or shortly protruding, indehiscent or dehiscent by the inner suture or by a hardened lid; seeds 1–4, rarely to 10.

A genus of about 300 species, mostly from north temperate regions, a few from the southern hemisphere. Many are important as fodder or green manure crops, but only a small number is grown for ornament. They are easily grown in good soil and are propagated by seed

(which may require scarification) or, in the case of the perennial species, by division.

1a. Flowers solitary   **5. uniflorum**
  b. Flowers in heads   2
2a. Leaflets 5 or more   **1. lupinaster**
  b. Leaflets 3   3
3a. Calyx with only 5 distinct veins
                               4. hybridum
  b. Calyx with 10–20 distinct veins   4
4a. Calyx hairless   5
  b. Calyx hairy   6
5a. Stipules attached to the leaf-stalk for
    at least 6/7 of their length; seeds 1
    or 2   **2. alpinum**
  b. Stipules attached to the leaf-stalk for
    up to 1/2 their length; seeds 3 or 4
                                 3. repens
6a. Calyx with 20 distinct veins
                                10. rubens
  b. Calyx with 10 distinct veins   7
7a. Flowers stalked; fruiting calyx
    inflated   8
  b. Flowers not stalked; fruiting calyx
    not inflated   9
8a. Flowers reversed so that the
    standard is lowermost
                             7. resupinatum
  b. Flowers not reversed, standard
    uppermost   **6. fragiferum**
9a. Inflorescence terminal
                             11. pannonicum
  b. Inflorescence axillary   10
10a. Inflorescence long egg-shaped to
    cylindric   **8. incarnatum**
  b. Inflorescence spherical, egg-shaped
    or oblong-conical   11
11a. Lowest tooth of calyx 3-veined, at
    least at the base   **12. alexandrinum**
  b. Lowest tooth of the calyx 1-veined
                                9. pratense

1. T. lupinaster Linnaeus. Illustration: *Botanical Magazine*, 879 (1805); Hegi, Illustrierte Flora von Mittel-Europa, edn 2, **4**: 1313 (1975).

Herbaceous perennial, stems 15–50 cm, erect or ascending, hairless or with a few adpressed hairs. Leaf-stalks to 1 cm, shorter than stipules, and united to them; leaflets 5–8 (only 3 in the first-developed leaves), lanceolate-oblong, toothed. Inflorescence-stalk 10–30 cm, with axillary heads forming an umbel; each head with up to 20 flowers; flower-stalks 1–2 cm, deflexed after flowering. Calyx-teeth about as long as tube. Corolla 1–2 cm, crimson red, or rarely yellowish white. Pod twice as long as the calyx, with 1–9 seeds. *E Europe to E Asia.* H2. Early summer.

2. T. alpinum Linnaeus. Illustration: Polunin, Flowers of Europe, pl. 60 (1969); Hegi, Illustrierte Flora von Mittel-Europa, edn 2, **4**: 1313 & t. 163 (1975); Aeschimann et al., Flora alpina **1**: 915 (2004).

Hairless herbaceous perennial to 18 cm, with a tap-root. Leaf-bases covering the stems; leaf-stalks 2–12 cm, stipules 4–9 cm, attached to the leaf-stalks; leaflets 3, 1–5 cm, lanceolate to linear. Inflorescence-stalk 5–15 cm; heads with 3–12 flowers on stalks to 2 mm. Calyx-teeth *c.* 6 mm, the lower to 1 cm. Corolla 1.8–2.5 cm, pink, purple or rarely cream. Pod stalked, obovoid, containing 1 or 2 seeds. *S Europe, from Spain to the Alps.* H3. Summer.

A high-altitude plant whose cultivation in the lowlands is difficult.

3. T. repens Linnaeus. Illustration: Jávorka & Csapody, Iconographia Florae Hungaricae, t. 2010 (1929); Keble Martin, The new concise British flora in colour, pl. 22 (1969); Hegi, Illustrierte Flora von Mittel-Europa, edn 2, **4**: t. 164 (1975); Lewis et al., (eds.), Legumes of the world, 501 (2005).

More or less hairless perennial with a tap-root, stems creeping and rooting at the nodes. Leaf-stalks to 2 cm; leaflets 3, 5–30 mm, broadly obovate, base tapered, with pale marks above and translucent veins, margins with fine forwardly pointing teeth; stipules membranous, sheathing, with subulate tips, veins green or red. Flower-heads spherical; flowers stalked, scented. Calyx-tube whitish, teeth lanceolate, green, the 2 upper slightly longer than the rest and separated by a narrow, acute sinus. Corolla 8–13 mm, cream to pink, deflexed and brown after flowering. Pod linear, flat, constricted between the 3 or 4 seeds. *Europe, W Asia, N Africa, widely introduced elsewhere.* H1. Late spring to autumn.

A variable species, widely cultivated. 'Atropurpureum' (forma *atropurpureum* Anon.), with dark purple-red leaflets, is often cultivated in gardens.

4. T. hybridum Linnaeus. Illustration: Jávorka & Csapody, Iconographia Florae Hungaricae, f. 2009 (1929); Polunin, Flowers of Europe, pl. 59 (1969); Keble Martin, The new concise British flora in colour, pl. 22 (1969); Hegi, Illustrierte Flora von Mittel-Europa, edn 2, **4**: t. 164 (1975).

Hairless biennial or perennial herb, usually to 40 cm, sometimes to 90 cm; stems erect or ascending, if prostrate, not rooting at the nodes. Leaves with 3 leaflets, leaf-stalks to 10 cm; leaflets 1–2 cm × 5–15 mm, obovate or inverted-heart-shaped, rarely elliptic; stipules greenish, broadly ovate, gradually narrowing to the tip. Flower-heads spherical, apparently terminal; inflorescence-stalk about twice as long as the subtending leaf. Flowers stalked, stalks to twice as long as calyx-tube, deflexed after flowering. Calyx-tube 1–1.5 mm, with 5 distinct veins and 5 more obscure veins; teeth 2–3 mm, the upper 2 slightly longer than the lower 3 and separated by a broad sinus. Corolla 5–10 mm, at first whitish, later pink and finally turning brown. Pod ellipsoid, protruding from the calyx, with 2–4 seeds. *Now found in most of Europe & elsewhere, but probably native only to W Europe.* H1. Late spring–summer.

5. T. uniflorum Linnaeus. Illustration: Sweet, British flower garden, ser. 2, **2**: 200 (1838); Brickell, RHS dictionary of gardening, 2146 (1956); Polunin & Huxley, Flowers of the Mediterranean, pl. 63 (1965).

Perennial with woody tap-root. Stems prostrate, tufted, with very short internodes. Leaf-stalks 1–8 cm, leaves with 3 leaflets; leaflets 4–15 mm, circular, obovate or diamond-shaped, acute or obtuse, with a short, sharp, flexible point, strongly veined, hairless or with adpressed hairs; stipules broadly triangular, membranous, long-acuminate, overlapping. Flowers solitary, axillary, stalks 1–7 mm, usually shorter than calyx-tube, deflexed and sometimes thickened in fruit. Calyx-tube 5–7 mm, hairless or downy; teeth almost equal, narrowly lanceolate, half as long as tube or less. Corolla 1.2–2.5 cm, cream, purple or parti-coloured. Pod linear, acute, hairy above. *Mediterranean area from Sicily eastwards.* H4. Late spring–summer.

6. T. fragiferum Linnaeus. Illustration: Keble Martin, The new concise British flora in colour, pl. 23 (1969); Hegi, Illustrierte Flora von Mittel-Europa, edn 2, **4**: t. 164 (1975); Aeschimann et al., Flora alpina **1**: 921 (2004).

Perennial (rarely biennial) herb, hairy or hairless. Stems 2–40 cm, usually prostrate and often rooting at the nodes, sometimes tufted. Leaf-stalks to 15 cm; leaflets 3, obovate, elliptic or inverted-heart-shaped, 3–20 mm; stipules *c.* 1 cm, membranous, lanceolate. Inflorescence-stalk 5–20 cm, hairless or with scattered hairs; head hemi-spherical and 1–1.5 cm wide in flower, later somewhat ellipsoid. Calyx shaggy, with glandular hairs, at least in the upper part, the upper lip greatly inflated and helmet-like in fruit. Corolla pale pink or rarely white. Pod not projecting from the calyx, with 1 or 2 seeds. *Most of Europe.* H2. Summer.

Tolerates somewhat brackish soils.

7. T. resupinatum Linnaeus. Illustration: Jávorka & Csapody, Iconographia Florae Hungaricae, t. 1996 (1929); Hegi, Illustrierte Flora von Mittel-Europa, edn 2, **4**: 1320, 1321 (1975); Aeschimann et al., Flora alpina **1**: 921 (2004).

Hairless annual herb 10–60 cm, stems ascending, prostrate or erect. Lower leaves congested and rosette-like, long-stalked, the upper shorter, almost stalkless. Leaflets 3, 0.5–2.5 cm, obovate, rarely elliptic, tapered to the base; margins with fine, forwardly pointing teeth; stipules lanceolate, membranous. Flower-heads axillary, spherical, 8–20 mm. Flowers inverted (standard downwards). Calyx 5–10 mm, pear-shaped in fruit, hairy on the upper side, with clearly visible veins, upper 2 teeth divergent. Corolla 2–8 mm, pink to purple. Pod spherical to ovoid, membranous, containing 1 or 2 seeds. *Mediterranean area, introduced elsewhere.* H4. Spring–early summer.

8. T. incarnatum Linnaeus. Illustration: *Botanical Magazine*, 328 (1796); Jávorka & Csapody, Iconographia Florae Hungaricae, t. 2026 (1929); Polunin, Flowers of Europe, pl. 59 (1969); Hegi, Illustrierte Flora von Mittel-Europa, edn 2, **4**: 1329 (1975).

Annual herb. Stems 20–50 cm, simple or branched from the base, erect or ascending; stems and leaves with adpressed or spreading hairs. Leaflets 3, obovate to almost circular, 8–35 mm, finely toothed towards the tip; stipules ovate, blunt, membranous or green, tips obscurely toothed, often reddish. Inflorescence long-stalked, elongate-egg-shaped to cylindric, 10–50 cm, flowers stalkless. Calyx-teeth about as long as the tube, acute, linear, spreading stellately in fruit. Corolla blood-red, pink, cream or white. Pod ovoid. *S Europe.* H4. Late spring–early summer.

Sensitive to frost and to drought during spring.

9. T. pratense Linnaeus. Illustration: Jávorka & Csapody, Iconographia Florae Hungaricae, t. 2021 (1929); Hegi, Illustrierte Flora von Mittel-Europa, edn 2,

4: 1331 & t. 162 (1975); *The Living Countryside* **No. 7**: 1523 (1977).

Tufted perennial herbs (often short-lived), more or less adpressed hairy throughout. Stems to 1 m. Leaflets 3, obovate or oblong-lanceolate to almost circular, 1–4 cm × 5–20 mm, often hairy only beneath, often with a lighter green or reddish stripe above; stipules ovate-lanceolate, abruptly contracted into bristle-like tips. Heads axillary, 2–4 cm, spherical to ovoid, enclosed at the base by stipules of reduced leaves; flowers stalkless. Calyx-tube adpressed hairy, 10-veined; teeth subulate, the lowest 1-veined, twice as long as the tube. Corolla 1.2–1.5 cm, reddish purple or pink, rarely white. Pod ovoid, 1-seeded. *Europe, introduced elsewhere*. H1. Late spring–summer.

Widely grown as a fodder crop.

10. T. rubens Linnaeus. Illustration: Jávorka & Csapody, Iconographia Florae Hungaricae, t. 2013 (1929); Polunin, Flowers of Europe, pl. 59 (1969); Hegi, Illustrierte Flora von Mittel-Europa, edn 2, **4**: 1347 & t. 162 (1975); Lewis et al., Legumes of the world, 501 (2005); Aeschimann et al., Flora alpina **1**: 931 (2004).

Usually hairless perennial with a creeping rhizome. Stems erect, 20–60 cm. Leaflets 3, oblong-lanceolate or rarely elliptic, to 6–7 × 1–1.5 cm, toothed; stipules to 7 cm, lanceolate, attached to the stems for more than half their length. Inflorescence axillary. Heads cylindric, to 8 × 2.5 cm, stalk to 4 cm. Calyx-tube with 20 veins; teeth subulate, hairy, the lowest much longer than the others. Corolla *c.* 1.5 cm, purple or rarely white. Pod ovoid, 1-seeded. *C & S Europe*. H4. Early summer.

11. T. pannonicum Jacquin. Illustration: Jávorka & Csapody, Iconographia Florae Hungaricae, t. 2016 (1929); Hegi, Illustrierte Flora von Mittel-Europa, edn 2, **4**: 1350 (1975); Aeschimann et al., Flora alpina **1**: 933 (2004).

Perennial with long tap-root and short rhizomes. Stems erect, 20–50 cm (rarely more), hairy. Leaflets 3, oblong-lanceolate, 3–6 cm × 8–18 mm, tips pointed or rounded; free tips of stipules linear, herbaceous, to 3 cm. Inflorescence-stalk to 8 cm. Heads terminal, ovoid or cylindric, 5–8 cm. Calyx-teeth linear to subulate, the lowest twice as long as the others, all hairy. Corolla 2–2.5 cm, yellowish white. Pod 1-seeded. *E,C & S Europe*. H4. Early summer.

12. T. alexandrinum Linnaeus. Illustration: Hegi, Illustrierte Flora von Mittel-Europa, edn 2, **4**: 1282 (1975).

Annual herb to 70 cm, erect, branched, sparsely hairy. Leaflets 3, elliptic or oblong, 1.3–2.5 cm × 5–10 mm; leaf-stalks 2–5 cm, stipules 7–14 mm, free part subulate, the upper stipules dilated at the base. Inflorescence-stalk to 3 cm; heads ovoid or oblong-conical, 1.5–2 cm. Calyx hairy, tube tapered towards the base, teeth unequal, spine-like, the lowest 3-veined at least at the base, about as long as the tube, the others shorter and 1-veined. Corolla 8–10 mm, cream. Pod slightly protruding from the calyx. *E Mediterranean area*. H4. Summer.

70. BRACHYSEMA R. Brown
J. Cullen

Shrubs, sometimes prostrate, sometimes scrambling or climbing. Leaves opposite (ours) or alternate, simple (reduced to a single leaflet), leathery, occasionally reduced to small scales on flattened stems (not ours). Stipules present (ours) or absent. Flowers axillary, solitary or in clusters. Calyx 5-toothed. Corolla usually red; standard recurved, narrow, shorter than the wings and keel. Stamens 10, filaments free. Ovary with several ovules. Pods swollen, leathery, containing several seeds with arils.

A genus of 7 species from Western Australia. Only a single species is generally grown, though others have been cultivated in the past. It is propagated by seed or by cuttings.

1. B. celsianum Lemaire (*B. lanceolatum* Meissner). Illustration: *Botanical Magazine*, 4652 (1852); Erickson et al., Flowers and plants of Western Australia, 127 (1973).

Shrub, sometimes scrambling, to 2 m and 3 m wide; young shoots covered with silver or white hairs. Leaves opposite, ovate, entire, 2–5 cm, dark green and more or less hairless above, densely silvery-hairy beneath; stipules narrow, subulate. Flowers 1–3 in the leaf-axils. Calyx to 1.5 cm, funnel-shaped, deeply 5-toothed, densely covered with silver hairs. Corolla to 2.5 cm, red. Pods hairy, *c.* 1.5 cm. *Western Australia*. H5–G1.

71. OXYLOBIUM Andrews
J. Cullen

Shrubs, sometimes somewhat scrambling. Leaves simple (reduced to a single leaflet), leathery, alternate, opposite or in whorls (these sometimes rather irregular), margins often recurved. Stipules absent or subulate. Flowers in terminal and axillary racemes or panicles which are sometimes dense and head-like, yellow, red or orange. Calyx funnel-shaped, not umbilicate at base, deeply 5-toothed. Standard broad. Stamens 10, filaments free. Ovary stalked or not, ovules 4 or more, usually densely hairy, with several seeds; seeds without appendages.

A genus of about 30 species from Australasia. Several have been cultivated, but only 2 seem to be generally available. They are easily grown in a warm, rather dry site, and may be propagated by cuttings.

1a. Leaves mostly in whorls of 3–4, mostly more than 4 mm wide; stipules apparently absent **1. ellipticum**
  b. Leaves alternate, mostly less than 4 mm wide; stipules present, subulate **2. linearifolium**

1. O. ellipticum R. Brown. Illustration: *Botanical Magazine*, 3249 (1833).

Erect shrub to 2.5 m; young shoots with dense adpressed or somewhat spreading hairs. Leaves crowded, usually in whorls of 3 or 4 but the whorls sometimes irregular, elliptic to linear, 1–3 cm × 3–11 mm (rarely longer and proportionately narrower), apex sharply mucronate, margins recurved, hairless above, densely brown-hairy beneath; stipules apparently absent. Flowers in dense, head-like racemes or panicles at the ends of the branches. Calyx 6–7 mm, with adpressed or somewhat spreading hairs. Corolla yellow or orange-yellow. Standard broad, 1–1.2 cm, longer than wings and keel. Pod shortly cylindric, *c.* 8 mm, densely spreading-hairy. *Australia (Tasmania, Victoria, New South Wales)*. H5–G1.

2. O. linearifolium (G. Don) Domin (*O. callistachys* misapplied).

Shrub; young shoots with adpressed hairs. Leaves sparse, alternate, very narrowly linear or linear-lanceolate, to 8 cm × 4 mm, apex sharply mucronate, margins recurved, hairless above, brown-hairy beneath; stipules small, subulate. Flowers sparse in long-stalked racemes. Calyx 8–10 mm, densely hairy. Standard yellowish, broad, *c.* 1.5 cm; wings and keel reddish, about as long as the standard. *Western Australia*. H5–G1.

72. CHORIZEMA Labillardière

J. Cullen

Shrubs, sometimes sprawling or scrambling. Leaves alternate, leathery, simple (reduced to a single leaflet), margins usually toothed (ours); stipules small or absent. Flowers in terminal or rarely axillary racemes. Calyx 5-toothed. Corolla red, orange or pinkish purple, standard longer than the wings. Stamens 10, filaments free. Ovary with 8–many ovules. Pod ovoid or compressed, rather soft, several-seeded. Seeds without appendages.

A genus of 18 species from Australia (mostly Western Australia), a few cultivated for their bright flowers and interesting, often holly-like foliage. They require a moist but well-drained soil in a partially shaded but sunny site and are propagated by seed, which requires scarification.

1. C. cordatum Lindley. Illustration: *Edwards's Botanical Register* **24**: pl. 10 (1838); *Botanical Magazine*, n.s., 237 (1954); Morcombe, Australia's western wild flowers, 101 (1968); *Growing Native Plants* **11**: 269 (1981).

Shrub, often sprawling or climbing through other vegetation. Leaves to 5 cm, ovate, acute at apex, cordate at base, margin usually with coarse, broadly triangular, sharply pointed teeth, but occasionally entire or almost so, more or less hairless above and beneath. Racemes to 15 cm. Standard scarlet or orange, yellow towards the base of the blade, wings and keel pinkish purple. Pods inflated. *Western Australia.* H5–G1.

There are 2 very similar (probably not distinct) species whose names sometimes appear in the literature:

C. varium Lindley. Very similar but young shoots and the undersides of the leaves hairy. *Western Australia.* H5–G1.

C. ilicifolium Labillardiere. Illustration: *Botanical Magazine*, 1029 (1807); Erickson et al., Flowers and plants of Western Australia, 58 (1973); Brickell (ed.), RHS A–Z encyclopedia of garden plants, 266 (2003); Lewis et al., (eds.), Legumes of the world, 350 (2005). A very small shrub with longer, narrower leaves which are somewhat hairy beneath. *Western Australia.* H5–G1.

73. GOMPHOLOBIUM J.E. Smith

J. Cullen

Shrubs. Leaves alternate to almost opposite, usually compound with 3 (ours) or more leaflets; stipules minute or absent. Flowers axillary, rather large and showy; buds long-stalked. Calyx very deeply 5-lobed, the tube very short, black outside, the lobes edge-to-edge in bud. Petals shortly clawed; standard large, broad, longer than the wings and keel; keel very deep. Stamens 10, their filaments free. Ovary with usually 8 ovules. Pod inflated, ovoid to spherical, several-seeded.

A genus of 26 species from Australasia. They are striking plants which caused great interest when they were introduced to Europe in the 19th century, but only 1 is currently available. It is showy, with narrow, greyish leaflets, black, caper-like flower-buds and yellow flowers. It is not particularly easy to grow, requiring good drainage and a sheltered site. Propagation is by seed or by cuttings taken from current growth when just firm, but young plants are difficult to establish.

1. G. latifolium J.E. Smith. Illustration: *Botanical Magazine*, 4171 (1845); Rotherham et al., Flowers and plants of New South Wales and southern Queensland, 59 (1975); Elliot & Jones, Encyclopaedia of Australian plants **4**: 380 (1986).

Upright shrub to 2 m; shoots somewhat hairy or rough, at least at the nodes. Leaves alternate or almost opposite, shortly stalked, divided into 3 leaflets; leaflets 3–6 cm × 3–6 mm, linear-obovate, tapered to the base, abruptly acute or truncate (though with a short mucronate point) at the apex, hairless. Calyx 1–1.5 cm, black outside, the lobes pale inside, sometimes with glands there, margins densely fringed. Corolla bright to dull yellow, standard broad, 2–2.5 cm long; keel deep, fringed along the suture between the petals. Pods ovoid, stalked, to 1.8 cm. *Australia (Victoria, New South Wales, Queensland).* H5–G1.

74. BURTONIA R. Brown

J. Cullen

Small shrubs; young shoots hairy or hairless. Leaves alternate, pinnate or made up of 3 leaflets borne side-by-side, without a common leaf-stalk; leaflets narrow, heather-like, with downwardly curved margins in ours; stipules minute or absent. Flowers solitary in the upper leaf-axils but numerous, forming a leafy terminal raceme. Calyx deeply 5-lobed with very short tube, the lobes edge-to-edge in bud. Corolla purple, red, orange or yellow; standard with broad blade, longer than wings and keel; keel deep, wider than wings. Stamens 10, filaments free. Ovary shortly stalked or stalkless, containing 2 ovules. Pods more or less spherical. Seeds without appendages.

A genus of about 12 species from Australia. Like *Gompholobium*, several species have been introduced, but only 2 are available today. Both of these have narrow, heather-like (ericoid) leaflets. They require moist but well-drained soil, a sheltered site and are very susceptible to frost damage. Propagation is by seed, which requires scarification; young plants are difficult to establish.

1a. Young branches hairy; standard orange-red with a yellow blotch, wings and keel brownish

         2. hendersonii

  b. Young branches hairless; standard pinkish purple, wings and keel reddish   **1. scabra**

1. B. scabra (Smith) R. Brown. Illustration: *Botanical Magazine*, 4392 (1848) – as *B. pulchella*, & 5000 (1857); Gardner, Wildflowers of Western Australia, edn 1, 61 (1959); Erickson et al., Flowers and plants of Western Australia, 73 (1973); Elliot & Jones, Encyclopaedia of Australian plants **2**: 395 (1982).

Shrub with hairless young shoots. Leaves dense, usually longer than the internodes and flower-stalks, to 1.5 cm × 1 mm, greyish, rough. Calyx black, 5–7 mm, the teeth with fringed margins. Standard pinkish purple, 1.5–1.7 cm; wings and keel reddish. *Australia (Western Australia).* H5–G1.

2. B. hendersonii (Paxton) Bentham. Illustration: Erickson et al., Flowers and plants of Western Australia, 119 (1973).

Like *B. scabra* but young shoots hairy, leaves not rough, usually shorter than the internodes and flower-stalks, standard orange-red with a yellow blotch at the base, wings and keel brownish. *Australia (Western Australia).* H5–G1.

75. SPHAEROLOBIUM J.E. Smith

J. Cullen

Hairless perennial herbs or shrubs, leafless (ours). Flowers in a raceme or in lateral clusters, small, 2 per bract. Calyx 5-lobed, the 2 upper lobes broad and with diverging tips, united for most of their length, the 3 lower narrowly triangular to lanceolate, more deeply divided. Standard with a broad blade, longer than wings; keel longer or shorter than wings. Stamens 10, filaments free. Ovary with 2 ovules; style incurved, usually with a membrane or ring of hairs just below the stigma. Pod spherical or compressed, 1–2-seeded. Seeds without appendages.

A genus of about 15 species from Australia. Only 1 is grown for its curious appearance; little is known of its cultivation requirements, but these are presumably similar to those of *Burtonia* (p. 391).

1. S. vimineum J.E. Smith. Illustration: *Botanical Magazine*, 969 (1807); Rotherham et al., Flowers and plants of New South Wales and southern Queensland, 37 (1975); Galbraith, Collins' field guide to the wild flowers of south-east Australia, pl. 38 (1977).

Perennial herb or low shrub, prostrate, sprawling or erect, to 70 cm; stems wiry, branched, leafless. Flowers small, in pairs (often with a projection between them) in racemes, each pair subtended by a scale-like bract; flower-stalks recurved, jointed just beneath the calyx. Calyx dark or black, 2–3 mm, lobes with fringed margins. Standard 4–6 mm, yellow with a red flush or blotch in the centre; wings yellow, keel greenish. *Australia (New South Wales, Victoria, South Australia, Tasmania).* H5–G1.

76. VIMINARIA J.E. Smith
J. Cullen
Shrub to 3 m or more with branched shoots, sometimes with a trunk to 1.5 m; young shoots minutely hairy. Leaves alternate, usually reduced to thread-like stalks, leaflets absent; stipules lanceolate, fringed or minutely toothed. Flowers stalked in long terminal racemes, 1 per bract; bracts similar to the stipules. Calyx green, membranous, cylindric to funnel-shaped, 3–4 mm, with 5 short equal teeth which are sparsely fringed. Standard yellow with a dark blotch, lines or flush towards the base, broad, 8–10 mm, longer than wings; wings and keel reddish brown. Stamens 10, filaments free. Ovary short, containing 2 ovules. Pod ovoid-oblong, indehiscent, containing 1 seed which has a small appendage.

A genus of a single species widespread in Australia. Cultivation as for *Sphaerolobium* (p. 391).

1. V. juncea (Schrader) Hoffmannsegg (*V. denudata* J.E. Smith). Illustration: *Botanical Magazine*, 1190 (1809) – leaves

unusual; Cochrane et al., Flowers and plants of Victoria, 25 (1968); Galbraith, Collins' fieldguide to the wild flowers of south-east Australia, pl. 36 (1977); Lewis et al., (eds.), Legumes of the world, 342 (2005).
Australia. H5–G1.

77. DAVIESIA J.E. Smith
J. Cullen
Shrubs, sometimes spiny. Leaves alternate, simple, needle- or heather-like, sometimes ending in spines, or absent (see comment under species 2). Stipules usually absent. Flowers solitary or in clusters in the leaf-axils, usually concentrated towards the tips of the branches. Bracteoles not close beneath and overlapping the calyx. Calyx deeply or shallowly 5-toothed, not umbilicate at base. Corolla yellowish, often marked or blotched with red or other colours; standard with a broad blade, about as long as the wings and keel. Stamens 10, filaments free. Ovules 2. Pod flattened, triangular in outline, containing 2 seeds.

A genus of about 75 species from Australia. Only a few are grown, requiring a well-drained soil and a sunny position. They are propagated by seed, which requires scarification.

Literature: Crisp, M.D. et al., Contributions towards a revision of *Daviesia, Australian Systematic Botany* **4**(2): 229–298 (1991), **8**(6): 1155–1249 (1995), **10**(1): 31–48 (1997), **10**(3): 321–329 (1997).

la. Leaves of 2 forms, most of them needle-like but some laterally compressed and with a hump towards the apex on the upper edge; stems hairless   **2. incrassata**
  b. Leaves all of the same form, heather-like; stems with crisped hairs   **1. acicularis**

1. D. acicularis J.E. Smith. Illustration: *Botanical Magazine*, 2679 (1826); Cunningham et al., Plants of western New South Wales, 388 (1981).

Shrub to 60 cm or more; shoots rough with short, crisped hairs. Leaves crowded, ascending-erect, heather-like with margins curved downwards and a prominent midrib beneath, not spine-tipped though acute, hairy or not. Flowers solitary or in small clusters in the leaf-axils, much shorter than the subtending leaves. Calyx *c.* 3 mm, deeply 5-toothed. Standard yellow, reddish-blotched towards the base; keel reddish.

Pod *c.* 1 cm. *Australia (New South Wales, Queensland, South Australia).* H5–G1.

2. D. incrassata J.E. Smith. Illustration: *Botanical Magazine*, 4244 (1846); Gardner, Wildflowers of Western Australia, edn 1, 66 (1959); Lewis et al., (eds.), Legumes of the world, 353 (2005).

Glaucous shrub to 1.5 m, shoots hairless. Leaves (see comment below) mostly terete, grooved, ending in a sharp spine; some laterally compressed, widening upwards and with a hump on the upper side towards the hard, spiny apex; all leaves distant, spreading, internodes clearly visible. Flowers in axillary clusters of 3–8 in the leaf-axils, exceeding the subtending leaves. Calyx shallowly 5-toothed, *c.* 2.5 mm. Standard reddish orange or yellow, sometimes marked with green or black; keel reddish, sometimes black-tipped. Pod slightly inflated, to 1.4 cm. *Australia (Western Australia).* H5–G1.

Some authors regard this species as leafless, considering the organs described above as leaves to be modified shoots.

78. PULTENAEA J.E. Smith
J. Cullen
Shrubs. Leaves usually alternate, simple, flat or terete (though grooved above); stipules usually present, small, lanceolate (ours). Flowers in clusters in the leaf-axils towards the tips of the shoots or in terminal heads with individual flowers very shortly stalked. Bracteoles persistent, borne close below and overlapping the calyx. Calyx bell-shaped, not umbilicate at base, shallowly to deeply toothed with 5 more or less equal teeth. Corolla yellow or pink; standard with a broad, notched blade, somewhat longer than the wings and keel. Stamens 10, filaments free. Ovary containing 2 ovules. Pod ovate, compressed, exceeding the persistent calyx. Seeds with arils.

A genus of about 120 species from Australia. Only a small number is available in Europe today, though many more were introduced in the past. They require a well-drained soil in a sunny position and may be propagated by seed or by cuttings.

Literature: Woolcock, D.T. & C.E., Pultenaea, with reference to related genera of the pea-flower family, *Australian Plants* **12**: 95–103, 116–121 (1982), 304–309 (1984); **13**: 72–83 (1984), 250–257, 324–329 (1986); **14**: 130–137, 179–185 (1987) and continuing. de Kok,

R.P.J. & West, J.E., A revision of Pultenaea, 1–2, *Australian Systematic Botany* **15**(1): 81–113 (2002), **16**(2): 229–273 (2003). Orthia, L.A., de Kok, R.P.J. & Crisp, M.D., A revision of Pultenaea 3–4, *Australian Systematic Botany* **17**(3): 273–326 (2004), **18**: 149–206 (2005).

1a. Leaves terete, warty; flowers pink
3. subalpina
b. Leaves flat, not warty; flowers mostly yellow **2**
2a. Leaves narrowly ovate, 4–7 × 2–3 mm **1. gunnii**
b. Leaves narrowly obovate or rarely narrowly oblong, 1–2 cm × 2–3 mm **2. flexilis**

1. P. gunnii Bentham. Illustration: Cochrane et al., Flowers and plants of Victoria, 33 (1968); Galbraith, Collins' field guide to the wild flowers of south-east Australia, pl. 39 (1977); Australian Plants **12**: 121 (1982).

Wiry shrub to 1 m; young shoots hairy. Leaves alternate, 4–7 × 2–3 mm, narrowly ovate, convex and hairless above, hairy beneath; stipules pointed. Flowers 3–8 in terminal heads. Calyx *c.* 4 mm, very deeply 5-toothed with lanceolate teeth, hairy. Corolla mainly yellow, keel dark red. Pod *c.* 5 mm, somewhat swollen, hairy. *Australia (Victoria, Tasmania).* H5–G1.

2. P. flexilis J.E. Smith. Illustration: *Australian Plants* **12**: 306 (1984).

Erect shrub to 4 m; shoots hairy at least when young. Leaves alternate, 1–2 cm × 2–3 mm, usually narrowly obovate, occasionally narrowly oblong, hairless or very sparsely hairy. Flowers in clusters in the leaf-axils towards the tips of the branches. Calyx *c.* 3 mm, with 5 rather broad, shallow, more or less equal teeth, hairless or the tooth-margins hairy. Corolla yellow. Pod *c.* 6 mm. *Australia (New South Wales, S Queensland).* H5–G1.

3. P. subalpina (Mueller) Druce (*P. rosea* Mueller). Illustration: *Botanical Magazine*, 6941 (1887).

Erect shrub; shoots hairy with dark, spreading hairs. Leaves alternate, to 1.5 cm, terete, grooved above, warty, hardened into a short point at the apex. Flowers in stalkless terminal heads. Calyx 4–5 mm, densely hairy, deeply 5-toothed with lanceolate teeth. Corolla pink. Pod 4–5 mm. *Australia (Victoria).* H5–G1.

79. DILLWYNIA J.E. Smith

J. Cullen

Heath-like shrubs; young shoots usually hairy. Leaves alternate, simple, needle-like or heather-like, often spine-tipped. Flowers axillary, solitary, paired or in small clusters, or in short terminal racemes; bracteoles not close beneath and overlapping the calyx. Calyx 5-toothed, the 2 upper teeth larger and more united than the 3 smaller, lower teeth; not umbilicate at base. Petals yellow, yellow-orange or reddish, standard broader than long, wings narrow, keel straight or incurved. Stamens 10, filaments free. Ovary containing 2 ovules, style hooked near the apex. Pod ovoid to almost spherical, short.

A genus of 22 species from Australia. They require a well-drained, sandy soil and a sunny position and can be propagated by seed or by cuttings.

1a. Leaves warty, blunt at the apex
1. floribunda
b. Leaves not warty, spine-tipped
2. juniperina

1. D. floribunda J.E. Smith. Illustration: *Botanical Magazine*, 1545 (1813); King & Burns, Wild flowers of Tasmania, 31 (1969); Rotherham et al., Flowers and plants of New South Wales and southern Queensland, 36 (1975); Elliot & Jones, Encyclopaedia of Australian plants **3**: 277 (1984).

Erect shrub, 15–200 cm; shoots hairy, sometimes densely so. Leaves needle-like, more or less terete, 1–1.5 cm, spreading or ascending, warty, blunt at the apex. Flowers in pairs in the leaf-axils, numerous and forming long, many-flowered, leafy racemes, the shoot often continuing vegetative growth above. Calyx top-shaped (narrowed to the base), 5–6 mm, sparsely hairy with long, wavy hairs. Corolla yellow or yellow-orange, petals falling after flowering. Pods to 7 mm, scarcely exceeding calyx. *Australia (New South Wales, Queensland).* H5–G1.

2. D. juniperina Loddiges. Illustration: Elliot & Jones, Encyclopaedia of Australian plants **3**: 278 (1984).

Erect shrub; shoots densely adpressed hairy. Leaves needle-like, narrow, more or less triangular in section, 1–1.5 cm, widely spreading, not warty, ending in hardened spines. Flowers in short, few-flowered, head-like racemes terminating the shoots. Calyx bell-shaped (rounded at base), 3–4 mm, adpressed hairy. Petals yellow or

yellow-orange, persistent. Pod *c.* 6 mm. *Australia (Queensland, New South Wales, Victoria).* H5–G1.

80. TEMPLETONIA R. Brown

J. Cullen

Hairless shrubs with angular or grooved branches. Leaves alternate, simple, entire (absent in some non-cultivated species); stipules minute or present as small spines. Flowers axillary, solitary or 2 or 3 together, red in ours. Calyx 5-toothed, the 2 upper teeth almost completely united, the 2 lateral teeth shorter, the lowermost somewhat longer. Standard reflexed. Stamens with filaments all united into a sheath open on upper side; anthers alternately longer and shorter, the longer erect, the shorter versatile. Ovary with several ovules; style incurved, thread-like. Pod very flattened, often oblique, the sides completely separating. Seeds with appendages.

A small genus from Australia. Only 1 species is grown, requiring well-drained soil and a sunny site.

1. T. retusa R. Brown. Illustration: Botanical Magazine, 2088 (1819) – as *T. glauca*, & 2334 (1822); *Journal of the Royal Horticultural Society* **91**: f. 211 (1966); Fairall, West Australian native plants in cultivation, 227 (1970); Erickson et al., Flowers and plants of Western Australia, 20 (1973).

Hairless shrub to 2 m; young branches deeply grooved. Leaves thick, leathery, pale green, oblong-obovate, shortly stalked, rounded at the base, rounded, truncate or notched at the apex, but often with a small point. Calyx bell-shaped, 7–9 mm, teeth shallow, margins hairy. Corolla red; standard rather narrow, strongly reflexed, 4–4.5 cm; wings and keel about as long as standard. Pod stalked, flat, oblong, tapering more gradually at base than apex, 3–4.5 cm × 6–11 mm, containing several seeds. *Australia (Western Australia, South Australia).* H5–G1.

81. HOVEA R. Brown

J. Cullen

Shrubs, usually hairy. Leaves alternate, simple, entire or prickly-toothed. Stipules thread-like or absent. Flowers blue or purple in axillary clusters or short racemes, rarely solitary. Calyx with a large upper lip formed from the 2 upper teeth which are almost completely united and truncate; lower 3 teeth much smaller. Petals purple, clawed; standard broad, notched, longer than the wings and keel; keel short, incurved. Stamens 10, with filaments all united into a sheath split on the upper side, sometimes also on the lower side, some stamens sometimes almost entirely free; anthers alternately longer and shorter, the longer erect, the shorter versatile. Ovary containing 2 or more ovules; style incurved, rather thick. Pod swollen, spherical or ovoid. Seeds with appendages.

A genus of perhaps 20 species from Australia, of which only a few are cultivated. They require a well-drained soil and a sunny site and are propagated mainly by seed.

Literature: Ross, J.H., Notes on *Hovea*, *Muelleria* **6**: 425–428 (1988), **7**: 21–38, 135–139 (1998), 203–206 (1990), 349–359 (1991), **9**: 15–28 (1996).

1a. Leaves heather-like, very narrowly lanceolate, margins curved downwards alongside the prominent midrib, sharply pointed **3. pungens**
  b. Leaves broad, not heather-like, sharply pointed only if toothed **2**
2a. Leaves entire **1. elliptica**
  b. Leaves with toothed margins **2. chorizemifolia**

1. H. elliptica (Smith) de Candolle (*H. celsii* Bonpland). Illustration: *Botanical Magazine*, 2005 (1818); *Edwards's Botanical Register* **4**: pl. 280 (1818); Erickson et al., Flowers and plants of Western Australia, 59 (1973).

Shrub to 3 m; young branches slightly grooved, adpressed hairy. Leaves elliptic, entire, to 7 × 3 cm on main shoots, usually smaller on lateral branches, tapered to the base and to the blunt though slightly mucronate apex, hairless above, with brown or blackish hairs beneath. Flowers in clusters. Calyx to 5 mm, densely brown-hairy. Corolla bright purple, standard broad, to 1.2 cm. Pod ovoid, to 1 cm. *Australia (Western Australia)*. H5–G1.

2. H. chorizemifolia de Candolle. Illustration: Gardner, Wild flowers of Western Australia, edn. 1, 70 (1959); Lewis et al., (eds.), Legumes of the world, 257 (2005).

Shrub to 2 m, with thickened roots; young shoots slightly grooved, adpressed hairy. Leaves narrowly to broadly oblong–elliptic, 3.5–6 × 1–2.5 cm, with coarse, spine-like teeth on margins, tip spine-like, hairless on both surfaces.

Flowers in shortly stalked clusters. Calyx funnel-shaped, 3.5–5 mm, adpressed hairy. Corolla purple, standard to 8 mm. Pods ovoid, often broader than long. *Australia (Western Australia)*. H5–G1.

3. H. pungens Bentham. Illustration: Erickson et al., Flowers and plants of Western Australia, 42 (1973); Lewis et al., (eds.), Legumes of the world, 252 (2005).

Shrub to 1 m; young shoots slightly grooved, adpressed hairy. Leaves heather- or needle-like, with margins folded downwards alongside the conspicuous midrib, very narrowly lanceolate, 2.5–3.5 cm × 2–3 mm, hard, widely spreading, sharply pointed, hairless above, with distant, long hairs on the midrib beneath. Flowers solitary or in shortly stalked clusters in the axils. Calyx 5–6 mm, adpressed hairy. Corolla bright purple, standard broad, to 1 cm. Pod very turgid, shiny, broader than long. *Australia (Western Australia)*. H5–G1.

82. BOSSIAEA Ventenat
J. Cullen

Shrubs; branches terete, angular, winged or flattened. Leaves alternate (ours) or opposite, simple, entire or rarely toothed; stipules small, brown, lanceolate or thread like. Flowers axillary or in clusters of 2 or 3 in the leaf-axils, yellow, orange, red or brownish. Calyx 5-toothed with the 2 upper teeth much larger than the lower 3. Petals yellow or red, clawed; standard broad, reflexed; wings narrow, longer than keel. Stamens 10, with filaments united into a sheath open along the upper side; anthers all of the same length, versatile. Ovary stalked or not, with several ovules; style incurved. Pods flat, not winged, the sides completely separating, sutures thickened. Seeds with appendages.

A genus of about 20 species from Australia. Only a few are grown, requiring a well-drained but constantly moist soil and a sunny site. Propagation is by seed or by cuttings of just hardened wood.

Literature: Woolcock, D.T., *Bossiaea* and related genera, *Australian Plants* **11**: 95–128 (1981); Ross, J.H., A conspectus of the western Australian *Bossiaea* species, *Muelleria* **23**: 15–142 (2006).

1a. Stems conspicuously flattened; leaves linear to elliptic or obovate on the one plant **4. heterophylla**
  b. Stems terete or somewhat winged; leaves various, but all of similar form on the one plant **2**

2a. Leaves not in 2 distinct ranks; young shoots adpressed hairy **1. cinerea**
  b. Leaves in 2 distinct ranks; young shoots with some spreading hairs or hairless or almost so **3**
3a. Branches spreading at very wide angles; leaves well spaced; standard 9–12 mm **3. disticha**
  b. Branches ascending at narrow angles; leaves crowded; standard to 8 mm **2. linophylla**

1. B. cinerea R. Brown. Illustration: *Botanical Magazine*, 3895 (1842) – as *B. tenuicaule*; Cochrane et al., Flowers and plants of Victoria, 30 (1968); Galbraith, Collins' fieldguide to wild flowers of south-east Australia, pl. 40 (1977); Elliot & Jones, Encyclopaedia of Australian plants **2**: 358 (1982).

Shrub to 2 m; young shoots terete or angular, adpressed hairy, sometimes whitish. Leaves borne in several ranks, narrowly ovate, to 2 cm × 8 mm, rounded to the base, tapered to the acute, mucronate apex, sparsely hairy above, more densely so beneath with adpressed hairs. Flowers solitary in the leaf-axils, long-stalked. Calyx *c.* 4 mm, with shallow teeth, hairless except for the tooth-margins. Corolla yellow, standard reddish on the back, to 1.5 cm. Pod to 2 cm × 5 mm. *Australia (Victoria, South Australia, New South Wales, Tasmania)*. H5–G1.

2. B. linophylla R. Brown. Illustration: *Botanical Magazine*, 2491 (1824); Elliot & Jones, Encyclopaedia of Australian plants **2**: 359 (1982).

Shrub to 3 m; shoots terete or somewhat winged, hairless or sparsely hairy. Leaves in 2 ranks, crowded, linear, 1.5–2 cm × 1–2 mm, acute, sparsely hairy above and beneath, margins somewhat curved downwards. Flowers in small, stalked racemes in the leaf-axils. Calyx 3–4 mm, sparsely adpressed hairy. Corolla mostly yellowish, keel red; standard to 8 mm. Pods stalked, flattened, elliptic to narrowly elliptic, 1–2 cm × 5–6 mm, hairless. *Australia (Western Australia)*. H5–G1.

3. B. disticha Lindley. Illustration: *Edwards's Botanical Register* **4**: pl. 55 (1841).

Shrub to 1.5 m; branches terete, long and very widely spreading, rather spreading-hairy. Leaves in 2 ranks, very narrowly ovate to oblong, 6–15 × 2–4 mm, rounded at the base, gradually then ultimately abruptly tapered to the mucronate apex, sparsely adpressed hairy above, more densely so beneath. Flowers long-stalked, solitary in the leaf-axils. Calyx 4–5 mm, teeth rather deep and narrow, all adpressed hairy. Corolla mainly yellow, keel red; standard 9–12 mm. Pod stalked, oblong, abruptly tapered at each end, c. 2 cm × 8 mm, hairy along the upper suture, at least at first. *Australia (Western Australia).* H5–G1.

4. B. heterophylla Ventenat. Illustration: *Botanical Magazine*, 1144 (1808) – as *B. lanceolata*; Rotherham et al., Flowers and plants of New South Wales and southern Queensland, 48 (1975).

Shrub to 2 m; shoots conspicuously flattened, hairless, sparsely hairy or with tufts of minute hairs in the leaf-axils. Leaves in 2 ranks, linear to elliptic or obovate, very variable in shape on the one plant, 7–15 × 2–10 mm, tapered to the acute apex, tapered or rounded to the base, hairless. Flowers solitary in the leaf-axils, stalked. Calyx 4–6 mm, hairless except for the fringed tooth-margins. Corolla yellow; standard 1–1.5 cm, reddish on the back. Pod stalked, oblong, flat, abruptly tapered at apex and base, to 3 cm × 6 mm, hairless. *Australia (New South Wales, Victoria, Queensland).* H5–G1.

83. VIRGILIA Poiret
J. Cullen

Shrubs or small trees to 15 m or more in the wild. Leaves pinnate with a terminal leaflet and 6–12 pairs of lateral leaflets (rarely more or fewer); stipules linear, acute; stipels absent. Flowers in racemes or occasionally panicles, axillary and terminal; bracts persistent or falling early. Flowers violet to pink, rarely whitish. Calyx bell-shaped, umbilicate at the base, 2-lipped, upper lip of 2 reflexed teeth, the lower of 3 straight teeth. Standard broad, reflexed, wings curved, keel curved and beaked. Stamens 10, filaments united into a tube; anthers versatile. Ovary with 5–8 ovules; style curved, stigma with a fringe of hairs. Seeds with small appendages.

A genus of 2 species from the Cape Province of South Africa. They require a light, sandy, well-drained soil and a sunny site. Propagation is by seed.

Literature: Van Wyk, B.E., A revision of the genus *Virgilia* (Fabaceae), *South African Journal of Botany* **52**: 347–353 (1986).

1. V. oroboides (Bergius) Salter (*V. capensis* Linnaeus). Illustration: *Botanical Magazine*, 1590 (1813); *Flowering plants of South Africa* **8**: pl. 305 (1928); *Journal of the Royal Horticultural Society* **99**: f. 194 (1974); Palgrave, Trees of southern Africa, t. 96 (1977).

Shrub or small tree; bark smooth or rough. Leaflets with dense adpressed hairs beneath; stipules 3–12 mm. Flowers in a raceme or panicle; bracts persistent, 7–15 × 4–10 mm; bracteoles present but minute. Calyx 6–10 mm, densely hairy. Corolla pale pink to violet; standard to 2 cm; beak of keel pink, yellowish green or dark purple. Pods oblong, gradually tapered to the base, abruptly tapered to the apex, to 4 × 1 cm, densely hairy. *South Africa (Cape Province).* H5–G1. Spring–summer.

Van Wyk recognises 2 subspecies: subsp. **oroboides** with pale pink flowers, the keel-beak pink or yellowish green, bark rough; and subsp. **ferruginea** van Wyk, with violet to violet-purple flowers, the keel-beak dark purple, bark usually smooth. Both subspecies are likely to have been introduced into cultivation in Europe.

V. divaricata Adamson. Illustration: Menninger, Flowering trees of the world, pl. 308 (1962); Palmer & Pitman, Trees of southern Africa **2**: 903 (1973). Similar, but leaflets hairless beneath or hairy only on the midrib, bracts 2–5 × 1–3 mm, falling early, bracteoles absent; bark smooth. *South Africa (Cape Province).* H5–G1.

This species has also been introduced to cultivation, but it is uncertain whether or not it is still to be found. Van Wyk mentions that intermediates between the 2 species are cultivated in South Africa, and some of these have certainly been grown in Europe.

84. PODALYRIA Lamarck
J. Cullen

Small to large, usually densely hairy shrubs. Leaves simple, very shortly stalked, somewhat leathery; stipules usually falling early. Flowers solitary in the leaf-axils or in few-flowered axillary racemes which are sometimes concentrated at the shoot tips; bracteoles usually present. Calyx 5-toothed, the 2 upper teeth united further than the 3 lower, the base umbilicate when the flower is open. Standard reflexed, broad, notched, longer than the keel and wings. Stamens 10, filaments free or very slightly united at the extreme base. Ovary with several ovules. Pod hard, woody, containing several seeds.

A genus of 25 species from South Africa, mainly from the Cape Province. They need a well-drained soil and a sunny site; propagation is best by seed, though cuttings of well-ripened wood have also been used.

1a. Standard 2.5 cm wide or more; bracteoles united and forming a hood over the flower-bud, deciduous as the flower opens **3. calyptrata**
 b. Standard to 2 cm wide; bracteoles not as above **2**
2a. Inflorescence-stalks 2–4-flowered, exceeding the leaves; standard 1.5–2 cm **1. biflora**
 b. Inflorescence-stalks mostly 1-flowered, not exceeding the leaves; standard to 1 cm **2. sericea**

1. P. biflora (Retzius) Lamarck (*P. argentea* Salisbury). Illustration: *Botanical Magazine*, 753 (1804); Rice & Compton, Wild flowers of the Cape of Good Hope, t. 34 (1951); Jackson, Wild flowers of Table Mountain, 26 (1977).

Densely branched shrub to 60 cm; young shoots densely hairy. Leaves ovate to elliptic or almost circular, 1–2.5 cm, tapered to the base, rounded or obtusely tapered to the apex which is mucronate and bent downwards, densely silky-hairy on both surfaces and with longer hairs on the margins, at least when young. Flowers in axillary racemes of 2–4; inflorescence-stalk exceeding the subtending leaf; bracteoles small. Calyx 7–10 mm, densely hairy, teeth usually longer than the tube. Corolla pink or white, standard 1.5–2 × 1.5–2 cm, longer than the whitish keel and wings. Pod cylindric, to 4 cm, densely hairy. *South Africa (Cape Province).* H5–G1.

2. P. sericea R. Brown. Illustration: *Botanical Magazine*, 1923 (1817); Batten & Bokelmann, Wild flowers of the eastern Cape Province, pl. **67**(5) (1966); Mason, Western Cape sandveld flowers, 129 (1972).

Like *P. biflora* but the dense hairs giving the plant a silvery sheen; flowers solitary in the axils, on short stalks; standard pink with a purple blotch at the base, to 1 cm; pod to 3 cm. *South Africa (Cape Province).* H5–G1.

3. P. calyptrata Willdenow. Illustration: *Botanical Magazine*, 1580 (1813) – as *P. styraciflora*; Menninger, Flowering trees of the world, 307 (1962); Palgrave, Trees of southern Africa, t. 98 (1977); Lewis et al., (eds.), Legumes of the world, 269 (2005).

Large shrub to 3 m or more, young shoots hairy. Leaves elliptic to elliptic-obovate, sometimes broadly so, 2–5 cm, tapered or rounded to the base, rounded to the apex where there is a recurved mucro, finely hairy on both surfaces. Flowers few, fragrant, in stalked axillary racemes; bracteoles forming a cup-shaped hood over the flower bud, deciduous as the bud opens. Calyx to 1.3 cm, silky. Corolla pink, standard 2 × 2.5 cm or more. Pod densely shaggy hairy, to 4 cm. *South Africa (Cape Province)*. H5–G1.

85. HYPOCALYPTUS Thunberg
J. Cullen

Shrubs; shoots often dark reddish. Leaves made up of 3 leaflets; stipules triangular to linear. Flowers numerous in a terminal raceme; bracts narrow, usually falling early; bracteoles present. Calyx bell-shaped to cylindric, ultimately umbilicate at the base, 5-toothed, the 2 upper teeth united for most of their length, the lower 3 narrowly triangular. Corolla violet; standard with broad, reflexed blade; keel curved, scarcely beaked. Stamens 10, their filaments united into a tube; anthers alternately longer and shorter, the longer erect, the shorter versatile. Ovary stalked, containing 3–30 ovules. Pod stalked, linear-oblong (ours) or inflated, containing several seeds. Seeds more or less oblong, each with an appendage along a long side.

A genus of 3 species from the Cape Province of South Africa. They require a well-drained soil and a sunny site, and are propagated by seed.

Literature: Dahlgren, R., The genus *Hypocalyptus* Thunb. (Fabaceae), *Botaniska Notiser* **125**: 102–125 (1972).

1. H. sophoroides (Bergius) Baillon. Illustration: *Botanical Magazine*, 3894 (1842); *Botaniska Notiser* **125**: 107 (1972); Moriarty, Outeniqua, Tsitsikamma and eastern Little Karroo, 119 (1981).

Much branched shrub to 3 m. Leaflets oblanceolate to obovate, 1–3 cm × 6–20 mm, tapered to the base, obtuse to notched, though ultimately pointed, at the apex, hairless above, hairless or very sparsely hairy beneath, usually folded upwards along the midrib so that the upper surface is scarcely exposed; stipules 1–4 mm, often soon falling. Raceme dense, usually with 30 or more flowers; axis hairy. Calyx hairless, 6–8 mm.

Corolla purple; standard 1 cm or more, with a yellow spot at the base of the blade. Pod oblong-linear, abruptly tapered at each end, 3.5–6.5 cm × 4–6 mm. *South Africa (Cape Province)*. H5–G1. Summer.

86. CROTALARIA Linnaeus
J. Cullen

Annual or perennial herbs or shrubs. Leaves simple (reduced to a single leaflet) or compound with 3 (ours) or more leaflets; stipules usually falling early; stipels absent. Flowers in racemes which are often long; bracts and bracteoles usually present. Calyx deeply 5-toothed, variable, sometimes 2-lipped. Flowers yellow (ours); standard with a large, reflexed blade with a fold or groove down the middle; wings often conspicuously puckered between the veins outside; keel usually very curved, beaked in cultivated species, often fringed above and on the suture. Stamens 10, filaments all united into a sheath open at the top, anthers alternately longer and versatile, shorter and erect. Ovary with several ovules; style hairless or with 1 or 2 lines of hairs. Pods turgid, hard, the seeds ultimately loose and rattling within them.

A genus of about 600 species, mainly from tropical regions and the southern hemisphere, only a few cultivated. They are easily grown and propagated by seed or, in the case of the shrubby species, by cuttings.

1a. Leaves stalked, of 3 leaflets　　2
　b. Leaves stalkless, of a single leaflet　　3
2a. Leaves hairy above; stipules very narrowly lanceolate, inconspicuous
　　　　　　　　　　　1. micans
　b. Leaves hairless above; stipules obovate, leaflet-like, conspicuous but often falling　　**2. capensis**
3a. Shrub; leaves hairy above; calyx-tube less than 3 mm　　**3. juncea**
　b. Tall annual herb; leaves hairless above; calyx-tube 4–6 mm
　　　　　　　　　　　4. retusa

1. C. micans Link (*C. anagyroides* Humboldt, Bonpland & Kunth). Illustration: Bernal, Flora de Colombia **4**: 35 (1986); Lewis et al., (eds.), Legumes of the world, 272 (2005).

Erect or spreading shrub to 1 m; shoots densely hairy. Leaves stalked, of 3 leaflets, the terminal larger than the laterals, all obovate to elliptic, tapered to the base, rounded or tapered to the mucronate apex, hairy above and beneath; stipules very narrowly lanceolate. Racemes rather short. Calyx bell-shaped, tube 3.5–5 mm, lowest

tooth 4–6 mm, all sparsely hairy. Corolla yellow; standard 1.5–2 cm; wings as long as or slightly longer than keel; keel densely fringed along its upper margin, bases of the suture also fringed. Pod stalked, broadly cylindric, 2.5–4 × 1–1.5 cm, finely hairy. *N South America*. G1.

2. C. capensis Jacquin. Illustration: *Botanical Magazine*, 7950 (1904); *Flowering Plants of South Africa* **10**: 386 (1930); Rice & Compton, Wild flowers of the Cape of Good Hope, t. 41 (1950); Kidd, Cape Peninsula, 99 (1973).

Erect shrub to 2 m; shoots densely hairy. Leaves stalked, of 3 leaflets, the terminal somewhat larger than the laterals, all obovate to elliptic, tapered to the base, usually rounded at the apex, hairless above, adpressed hairy beneath, margins fringed; stipules obovate, leaflet-like, soon falling. Racemes rather short. Calyx bell-shaped, tube 5–6 mm, lowest tooth 6–8 mm, all sparsely hairy. Corolla mostly yellow striped with red brown, fading orange; standard 2–2.5 cm; wings shorter than keel; keel whitish, fringed along the upper margin and also along the suture. Pod stalked, broadly cylindric, 4–6 cm × 5–10 mm, finely adpressed hairy, bluish green. *South Africa (Cape Province, Natal)*. H5–G1.

3. C. juncea Linnaeus. Illustration: *Botanical Magazine*, 1933 (1815); Bernal, Flora de Colombia **4**: 61 (1986).

Spreading shrub to 3 m; young shoots ridged, hairy. Leaves reduced to a single leaflet, stalkless, linear-oblong to oblong-elliptic, tapered or rarely rounded to the acute apex, hairy on both surfaces; stipules minute or absent. Racemes to 30 cm, many-flowered. Calyx very obviously 2-lipped, with a very short tube, densely spreading hairy, and with a pronounced bulge at the base beneath; lower teeth 1–1.5 cm. Corolla bright yellow; standard 1.5–2 cm, with brown hairs outside; wings shorter than keel; keel fringed on the upper surface towards the base and along the suture, twisted towards the apex. Pod stalked, ovoid-cylindric, blunt at both ends, to 3.5 × 1.5 cm, hairy, often blackish. *Origin unknown, long cultivated in many areas, especially India*. G1. Late summer.

Cultivated as a fibre plant in many areas.

4. C. retusa Linnaeus. Illustration: *Edwards's Botanical Register* **3**: t. 253

(1818); *Botanical Magazine*, 2561 (1825); Bernal, Flora de Colombia **4**: 88 (1986).

Annual herb to 1 m; shoots adpressed hairy. Leaves reduced to a single leaflet, stalkless, obovate to elliptic-obovate, tapered to the base, abruptly rounded to the usually notched apex, hairless above, adpressed hairy beneath, often wavy-margined towards the apex; stipules inconspicuous or absent. Racemes long, many-flowered. Calyx bell-shaped, sparsely hairy or hairless, tube 4–6 mm, upper lip much larger than lower. Corolla yellow; standard 1.5–2 cm; wings somewhat shorter than keel, keel fringed along the suture, twisted towards the apex. Pod stalked, cylindric, hairless, 3–4 cm × 8–10 mm. *Tropical Africa & Asia.* G1. Late summer.

87. ANAGYRIS Linnaeus
J.R. Akeroyd

Deciduous shrub 1–3 m, foetid and poisonous. Twigs green. Leaves with 3 leaflets 3–8 × 1–3 cm, elliptic, more or less obtuse, hairy beneath. Stipules minute, lanceolate. Flowers up to 20 in short axillary racemes on previous year's wood. Calyx bell-shaped, with 5 triangular teeth. Corolla 1.8–2.5 cm, yellow, the standard about half as long as the other petals, usually with a blackish spot; petals of keel free. Stamens free. Fruit 10–18 cm, shortly stalked, pendent, flat, constricted between seeds, hairless; seeds few, large.

A genus of a single species. Propagation is by seed or cuttings taken in summer.

1. A. foetida Linnaeus (*A. neapolitana* Tenore). Illustration: Sibthorp & Smith, Flora Graeca **4**: t. 366 (1823); Polunin & Everard, Trees and bushes of Europe, 106 (1976); Lewis et al., (eds.), Legumes of the world, 264 (2005).

Mediterranean area & SW Asia. H3. Spring.

88. PIPTANTHUS Sweet
J. Cullen

Shrubs or small trees to 4 m, branches with wide pith. Leaves stalked, of 3 leaflets; stipules conspicuous, fused for two-thirds or more of their length. Flowers in a loose or dense stalked terminal raceme, 3 arising from each bract. Calyx bell-shaped, 5-toothed, the upper 2 teeth fused for most of their length. Corolla yellow, standard reflexed, as long as the wings, with a conspicuous claw, blade notched at the apex. Stamens 10, their filaments free.

Ovary hairless or hairy, with 3–10 ovules. Pod oblong, flattened, leathery, with up to 10 seeds.

A genus of 2 species from the Himalaya and SW China. They are relatively easy to grow in most soils and are best propagated by seed.

Literature: Turner, B.L., Revision of the genus *Piptanthus* (Fabaceae: Thermopsideae), *Brittonia* **32**: 281–285 (1980).

1. P. nepalensis (J.D. Hooker) D. Don (*P. bicolor* Craib; *P. concolor* Craib; *P. forrestii* Craib; *P. concolor* subsp. *harrowii* Stapf; *P. concolor* subsp. *yunnanensis* Stapf; *P. laburnifolius* (D. Don) Stapf; *P. laburnifolius* forma *sikkimensis* Stapf and forma *nepalensis* Stapf). Illustration: Sweet, British flower garden **3**: t. 264 (1828); Schneider, Illustriertes Handbuch der Laubholzkunde **2**: 20, 22 (1907); *Botanical Magazine*, 9234 (1931); Polunin & Stainton, Concise flowers of the Himalayas, pl. 26 (1985).

Shrub to 4 m, with brittle stems. Leaflets generally lanceolate, acute, the terminal 2–15 cm, somewhat larger than the laterals; all hairless to finely hairy and bright green above, generally downy and paler beneath, veins conspicuous but not raised. Calyx 1.3–1.6 cm, downy. Corolla bright yellow, sometimes with brownish markings towards the base of the blade of the standard; keel 2.8–3.2 cm. Pod shortly stalked, hairy or hairless, 3–22 cm × 8–20 mm. *Himalaya, SW China.* H5. Early summer.

The species, subspecies and forms described by Craib (*Gardeners' Chronicle* **60**: 228, 1916) and Stapf (in the notes to the Botanical Magazine plate cited above) are merely selections from the total range of variability of the species.

P. tomentosus Franchet. Leaves densely hairy on both surfaces, veins beneath prominently raised. Flowers somewhat smaller (keel 1.4–1.6 cm). *SW China.*

89. THERMOPSIS R. Brown
C. Fraile

Tall, usually hairy perennial herbs with woody rhizomes. Leaves with 3 leaflets which vary from linear to elliptic, oblanceolate or obovate; stipules conspicuous, persistent, often leaf-like. Flowers in terminal or axillary, compact or loose racemes. Calyx 5-toothed, sometimes 2-lipped with the upper lip truncate or notched. Corolla deep purple or yellow.

Stamens 10, filaments free. Ovary more or less stalkless, containing numerous ovules. Pod flat, straight or recurved, many-seeded.

A genus of about 20 species from N America and Asia. They require deep well-drained soil and are propagated by seed or by division of established plants.

Literature: Larisey, M.M., A revision of the North American species of the genus *Thermopsis*, Annals of the Missouri Botanical Garden **27**: 245–258 (1940); St John, H., New and noteworthy Northwestern plants, part 9, notes on North American *Thermopsis*, Torreya **41**: 112–115 (1941); Chen, C.J., Mendenhall, M.G. & Turner, B.L., Taxonomy of *Thermopsis* in North America, *Annals of the Missouri Botanical Garden* **81**: 714–742 (1994).

1a. Corolla deep purple **1. barbata**
 b. Corolla yellow 2
2a. Calyx to 7 mm 3
 b. Calyx more than 7 mm 6
3a. Leaflets to 2 cm wide; fruit strongly recurved **6. rhombifolia**
 b. Leaflets more than 2 cm wide; fruit spreading or erect 4
4a. Leaflets hairless above, finely downy beneath **2. caroliniana**
 b. Leaflets hairy on both surfaces 5
5a. Stem and leaflets with densely felted hairs throughout; pods erect **7. macrophylla**
 b. Stem with adpressed bristles, leaflets adpressed hairy; pods spreading **5. gracilis**
6a. Leaflets more than 3 cm wide **8. fabacea**
 b. Leaflets less than 3 cm wide 7
7a. Leaves adpressed hairy on both surfaces; pods spreading **4. mollis**
 b. Leaves hairless above, slightly downy beneath; pods erect or recurved 8
8a. Leaves almost stalkless; pods recurved **3. lanceolata**
 b. Leaves clearly stalked; pods erect, closely adpressed to the axis **8. fabacea**

1. T. barbata Royle. Illustration: *Botanical Magazine*, 4868 (1855); Hara, Photoalbum of plants of eastern Himalaya, pl. 213 (1968); The Garden **100**: 353 (1975); Lewis et al., (eds), Legumes of the world, 262 (2005).

Stems erect, to 45 cm, the whole plant (stems, leaf-stalks, bracts, flower-stalks and calyces) covered with shaggy hairs. Leaves stalkless, opposite, the leaflets also stalkless, forming whorls on the stem, 1–3 cm × 5–10 mm, lanceolate, acuminate or acute, hairy on both surfaces. Flowers in short racemes, opposite. Calyx with 5 free teeth. Corolla deep purple. Pod broadly oblong, 2.5–3.5 cm, contracted to a mucronate apex, shaggy. *Himalaya*. H3. Summer.

2. T. caroliniana Curtis. Illustration: *The Green Scene* **5**(1): pl. 1 (1971); Brickell (ed.), RHS A–Z encyclopedia of garden plants, 1033 (2003).

Stems erect, sparingly downy to hairless. Leaves alternate, leaflets 5–7.5 × 2.5–4 cm, ovate to obovate, hairless above and finely downy beneath; stipules broadly ovate, leaf-like. Flowers in a compact terminal raceme. Calyx to 7 mm. Corolla yellow. Pod to 4.5 cm, erect, covered with dense, shaggy hairs. *E USA (N Carolina to Georgia)*. H5. Summer.

3. T. lanceolata R. Brown.

Stems erect, covered with shaggy hairs. Leaves almost stalkless, leaflets oblong-lanceolate, hairless above, silky-hairy beneath. Flowers apparently whorled in the raceme, yellow. Calyx more than 7 mm. Pod 4–5 cm, recurved, covered with shaggy hairs. *Former USSR (Siberia)*. H3. Summer.

4. T. mollis Curtis.

Stems erect, hairy. Leaves shortly stalked; leaflets 2–4 × 1–2 cm, elliptic-ovate, with tawny hairs along the veins both above and beneath. Flowers in compact racemes, yellow. Calyx more than 7 mm. Pod 3–4 cm, spreading, downy. *SE USA*. H5. Summer.

5. T. gracilis Howell. Illustration: Abrams, Illustrated flora of the Pacific States **2**: 488 (1944).

Stem erect, covered with adpressed bristles, somewhat angular. Leaflets 3.5–5.5 × 2–3 cm, ovate or elliptic, with adpressed hairs on both surfaces. Flowers in loose racemes, yellow. Calyx to 7 mm, with broadly triangular teeth, covered in shaggy hairs. Pods 4–5 cm, shaggy-hairy, spreading. *W North America*. H4. Summer.

6. T. rhombifolia Richardson. Illustration: Rickett, Wild flowers of the United States **6**: pl. 137(1973); Brickell (ed.), RHS A–Z encyclopedia of garden plants, 1033 (2003); Lewis et al., (eds.), Legumes of the world, 265 (2005).

Stem erect, adpressed downy throughout. Leaflets 2–3 × 1–2 cm, ovate, tapered to the base, hairless above, adpressed hairy beneath and with a

prominent central vein; stipules exceeding the leaf-stalks. Flowers in loose racemes, yellow. Calyx less than 7 mm. Pod strongly recurved. *W North America*. H4. Summer.

7. T. macrophylla Hooker & Arnott. Illustration: Abrams, Illustrated Flora of the Pacific States **2**: 488 (1944); Rickett, Wild flowers of the United States **5**: pl. 109 (1971).

Stem erect; stem, leaves and pods woolly with densely felted hairs. Leaflets 4–7 × 3–4 cm, obovate or oblanceolate, tapered to the base; stipules longer than the leaf-stalks, leaf-like. Calyx to 7 mm. Flowers yellow. Pod 6–8 cm, erect. *W USA*. H5. Summer.

8. T. fabacea (Pallas) de Candolle. Illustration: *Edwards's Botanical Register* **15**: t. 1272 (1829).

Stems erect, shaggy-hairy to hairless. Leaflets 3.5–7.5 × 1.5–3.5 cm, elliptic to obovate, hairless above, slightly downy beneath; stipules longer than the leaf-stalks, leaf-like. Flowers yellow, alternate or in pairs in erect racemes. Calyx 7 mm or more. Pod 4.5–7.5 cm, erect and adpressed to the axis, downy. *W USA*. H5. Summer.

The western North American **T. montana** Torrey & Gray (including *T. ovata* (Robinson) Rydberg) may sometimes be misidentified as *T. fabacea*; it usually has narrower, linear to obovate leaflets.

90. BAPTISIA Ventenat
J. Cullen

Perennial herbs with woody rhizomes, often glaucous, stems somewhat woody towards the base when old. Leaves with 3 leaflets (ours), shortly stalked; stipules minute, thread-like and deciduous or large, leaf-like and persistent. Flowers in racemes. Calyx bell-shaped, 2-lipped, the upper lip of 2 almost completely fused teeth, the lower of 3 distinct teeth. Corolla white, cream, yellow or blue; standard reflexed. Stamens 10, their filaments free. Ovary with numerous ovules. Pod inflated, variably shaped, many-seeded.

A genus of about 50 species from E & C USA; many hybrids occur in the wild and some may have been introduced. The 3 species generally found in cultivation are easily grown and can be propagated by seed or by division of the rhizome.

Literature: Larisey, M.M., A monograph of the genus *Baptisia*, Annals of the Missouri Botanical Garden **27**: 119–244 (1940); Turner, B.L., Overview of the

genus *Baptisia*, *Phytologia* **88**: 253–268 (2006).

1a. Corolla blue or violet-blue
                3. australis
 b. Corolla white, cream or yellow   2
2a. Corolla white, keel 2–2.5 cm; leaflets
   2.5–7 cm   **1. leucantha**
 b. Corolla cream or yellow, keel
   1–1.3 cm; leaflets 1–1.5 cm
               2. tinctoria

1. B. leucantha Torrey & Gray. Illustration: *Botanical Magazine*, 1177 (1809) – as *Podalyria alba*; Rickett, Wild flowers of the United States **1**: pl. 76 (1966).

Plant to 2 m, hairless almost throughout, glaucous. Leaflets 2.5–7 × 1.5–3 cm, obtuse, sometimes notched but mucronate at apex; stipules lanceolate, almost as long as to as long as leaf-stalks. Racemes long, flowers numerous, stalks 3–10 mm. Calyx densely white-hairy inside and on the margins. Corolla white, the standard sometimes purple-blotched; wings and keel 2–2.5 cm. Pod black, glaucous, shortly stalked, ovoid to oblong. *E & C USA*. H3. Spring–summer.

2. B. tinctoria (Linnaeus) Ventenat. Illustration: *Botanical Magazine*, 1099 (1808) – as *Podalyria tinctoria*; Rickett, Wild flowers of the United States **2**: pl. 121 (1967).

Plant to 1 m, almost completely hairless. Leaves shortly stalked, leaflets 1–1.5 cm × 6–10 mm, obovate, rounded or slightly notched at apex, tapered to the base; stipules minute, thread-like, deciduous. Racemes terminating lateral branches, short; flower-stalks 4–5 mm. Calyx rather sparsely hairy inside and on the margins. Corolla cream or yellow; wings and keel 1–1.3 cm. Pod black, glaucous, wrinkled, ovoid to spherical. *E USA*. H2. Spring–summer.

3. B. australis (Linnaeus) R. Brown (*B. exaltata* Sweet). Illustration: *Botanical Magazine*, 509 (1801); Rickett, Wild flowers of the United States, **1**: pl. 77 (1966); Hay & Synge, Dictionary of garden plants in colour, pl. 1001 (1969); Taylor's guide to perennials, 232 (1986).

Plant to 1.5 m, more or less hairless, glaucous. Leaves with stalks to 1 cm, leaflets obovate to oblanceolate, usually rounded at the apex, long-tapered to the base, 4–8 × 1.5–3 cm; stipules ovate-lanceolate to lanceolate, usually persistent, longer than the leaf-stalk. Racemes loose, terminal. Calyx sparsely hairy inside and on the margins. Corolla violet-blue to blue, wings and keel 2–2.7 cm. Pod greyish to brownish black, wrinkled, oblong-ellipsoid. *E USA.* H2. Spring–summer.

91. LUPINUS Linnaeus
D.R. McKean

Annual, biennial or perennial herbs or woody low shrubs. Leaves palmate, stipules attached to the base of the stalk. Flowers in racemes or spikes, in whorls or alternate. Calyx 2-lipped, each lip either deeply cleft to about the mid-point or each with 2 or 3 tiny teeth (almost entire) or entire; bracteoles attached to the calyx. Stamens 10, filaments of all joined together in a tube. Fruit a flat hairy pod.

A genus of about 200 species, widely distributed but absent from Australasia and South Africa. Most are from North, Central and South America. They prefer poor, well-drained, gravelly, acid soils, in full sun. In cultivation they can be forage crops, ornamentals or human food. The annuals are generally disease-free, but the perennials can suffer from root and crown rot. The leaves may be attacked by powdery mildew, or cucumber mosaic virus (mottled leaves). Honey fungus can also kill plants. Particular strains of the edible species are selected to avoid the toxic alkaloids which they all contain. Enthusiasts may be able to obtain some dwarf alpine lupins from America, e.g. *L. alopecuroides*, *L. confertus*, *L. lepidus*, *L. lyallii* and *L. ornatus*, but they are not commonly available and all need the protection of an alpine house.

Literature: Agardh, J. G., Synopsis Generis Lupini 1–43 (1835); Gladstones, J.S., Lupines of the Mediterranean Region and North Africa, *Western Australian Department of Agriculture Technical Bulletin* **26**: 1–48 (1974); Kurlovich, B.S., (ed), Lupins: geography, classification and genetic resources (2002).

Habit. Woody low shrub: **1,2**, **3**; perennial: **4, 5, 6, 7, 9**; annual: **8, 9, 10, 11, 12, 13, 14, 15, 16, 17, 18**. Shaggy hairy: **4, 6, 8, 13, 16**; shortly or silky hairy: **1, 2, 3, 5, 7, 9, 10, 11, 12, 14, 15, 17, 18**; hairless or nearly so: **9**.

Flowers. Mainly whorled: **2, 5, 6, 7, 11, 12, 13, 16, 17**; lower flowers (at least) alternate **1, 3, 5, 7, 8, 10, 13, 14, 18**.

Upper calyx-lip. Deeply divided: **2, 3, 8, 10, 11, 15, 16, 17, 18**; entire or almost entire: **1, 4, 5, 6, 7, 9, 12, 13, 14**.
Lower calyx-lip. Deeply divided: **8**; entire or almost entire: **1, 2, 3, 4, 5, 6, 7, 9, 10, 11, 12, 13, 14, 15, 16, 17, 18**.
Corolla. Yellow: **1, 11, (10, 12**, rarely); blue to pink, rarely white: **2, 3, 4, 5, 6, 7, 8, 10, 13, 15, 16, 17, 18**; white: **7, 12, 14**.

1a. Woody low shrub   2
  b. Herbaceous plant   4
2a. Flowers yellow (rarely lilac or blue)   **1. arboreus**
  b. Flowers mainly blue   3
3a. Lower leaf-stalks longer than leaflets; keel hairy on margins   **2. albifrons**
  b. Leaf-stalks about equalling the leaflets; keel hairless on margins   **3. chamissonis**
4a. Perennial   5
  b. Annual   8
5a. Upper surface of leaves hairy   **4. nootkatensis**
  b. Upper surface of leaves hairless   6
6a. Leaflets 9–17, 5–12 cm   **5. polyphyllus**
  b. Leaflets usually fewer and shorter   7
7a. Prostrate plant   **6. littoralis**
  b. Erect plant   **7. perennis**
8a. Lower calyx-lip deeply cut   **8. micranthus**
  b. Lower calyx-lip entire or with shallow notches   9
9a. Stems almost hairless   **9. mutabilis**
  b. Stems hairy   10
10a. Plant less than 30 cm; flowers yellow   **10. subcarnosus**
  b. Plant more than 30 cm or flowers yellow or not   11
11a. Flowers yellow   12
  b. Flowers not yellow   13
12a. Upper leaf surface hairy   **11. luteus**
  b. Upper leaf surface hairless   **12. densiflorus**
13a. Upper calyx-lip deeply cut to about mid-point   17
  b. Upper calyx-lip with a shallow notch or entire   14
14a. Stems fairly densely long-hairy; flowers dark blue   15
  b. Stems much less hairy; flowers white or tinged blue   16
15a. Leaflets oblanceolate; flowers white, tinged or veined pink or violet, purple or blue   **12. densiflorus**
  b. Leaflets oblong-obtuse; flowers blue, standard with a pink centre   **13. mexicanus**
16a. Keel hairy on upper margin; pod *c.* 2 × 1 cm   **12. densiflorus**
  b. Keel hairless on upper margin; pod 6–10 × 1.1–2 cm   **14. albus**
17a. All flowers alternate   18
  b. At least the upper flowers in whorls   19
18a. Leaves slightly fleshy and acute; lower calyx-lip simple   **15. texensis**
  b. Leaves not fleshy, obtuse; lower calyx-lip minutely 3-toothed   **10. subcarnosus**
19a. Stems densely shaggy-hairy   **16. pilosus**
  b. Stems shortly or sparsely hairy   20
20a. Flowers distinctly whorled   **17. nanus**
  b. Lower flowers alternate   **18. angustifolius**

1. L. arboreus Sims. Illustration: *Botanical Magazine*, 682 (1803); Dunn & Gillett, The lupines of Canada and Alaska, f. 53, 55 (1966); Hay & Beckett, Reader's Digest encyclopaedia of garden plants and flowers, 421 (1978); Rickett, Wild flowers of the United States **4**(2): pl. 126 (1970); Rose, The wild flower key, 179 f. 1 (1981).

Much-branched shrub to 3 m. Stems minutely hairy. Leaves with 5–12 leaflets, 2–6 cm × 5–10 mm, obovate-oblong, mucronate, silky-hairy beneath, hairless above; stipules subulate. Raceme-stalk to 10 cm, racemes to 30 cm. Flowers 1.4–1.7 × 1.5 cm, alternate or in whorls, golden yellow, rarely lilac or blue, scented. Upper calyx-lip notched or almost entire; lower entire. *USA (California), but naturalised near the sea in Britain & Ireland.* H5. Summer.

'Golden Spire' is a cultivar with golden yellow flowers.

2. L. albifrons Bentham. Illustration: *Edwards's Botanical Register* **19**: t. 1642 (1834); Rickett, Wild flowers of the United States **4**(2): pl. **131**(1970); Brickell (ed.), RHS A–Z encyclopedia of garden plants, 652 (2003).

Shrub 75–150 cm. Leaves with 7–10 leaflets, to 3 cm × 7 mm, spathulate-obovate, densely silvery-silky, leaf-stalk to 10 cm. Raceme-stalk 5–13 cm, racemes 8–30 cm. Flowers *c.* 1.4 cm across, mainly in whorls, petals blue or reddish purple, standard hairy on the back. Calyx upper lip deeply divided, lower lip entire. Pods 3–5 cm × *c.* 8 mm, 5–9 seeded. *USA (California).* H5. Summer.

The name *L. excubitus* Jones sometimes appears in the horticultural literature, but it may not be distinct from *L. albifrons*.

3. L. chamissonis Eschscholtz. Illustration: *Botanical Magazine*, 8657 (1916); Abrams, Illustrated flora of the Pacific States **2**: f. 2606 (1944); Rickett, Wild flowers of the United States **4**(2): pl. 127 (1970).

Low, dense, bushy shrub, 75–150 cm, minutely spreading hairy. Leaves with 6–9 leaflets, to 2.5 cm × 4–6 mm, minutely soft-hairy on both sides. Racemes 6–15 cm; flowers alternate or in whorls, blue or lavender, standard with a central yellow spot, slightly hairy on reverse, 1.2–1.6 cm across, keel hairy on upper margin, stalks *c.* 7 mm, spreading-hairy; bracts deciduous. Calyx upper lip deeply cleft, lower lip entire. Pods dull yellow, 3–5 cm × 7–8 mm. *USA (California)*. H5. Spring–summer.

Succeeds best in well-drained soil by a south-facing wall.

4. L. nootkatensis Don. Illustration: *Botanical Magazine*, 1311 (1810); *Loddiges' Botanical Cabinet* **9**: t. 897 (1824); Dunn & Gillett, The lupines of Canada and Alaska, f. 70, 71 (1966).

Stout perennial to 70 cm, stem and upper side of leaves shaggy-hairy. Leaves with 7 or 8 leaflets, 2–6 × 1–1.5 cm, oblanceolate, mucronate. Calyx upper lip notched, lower lip entire. Racemes *c.* 10 cm; flowers purple-blue, pink, white or parti-coloured, 1.2–1.6 cm across. Bracts deciduous. Pods *c.* 5 cm, brown and shaggy-hairy. *W USA, NE Asia*. H3. Early summer.

Naturalised in shingle and sand in Norway, Ireland and Scotland. Possibly best regarded as a subspecies of *L. perennis*.

5. L. polyphyllus Lindley. Illustration: *Edwards's Botanical Register* **13**: t. 1096 (1827); Reichenbach, Flora Exotica **3**: t.176 (1835); Dunn & Gillett, The lupines of Canada and Alaska, f. 73–75 (1966).

Herbaceous, mainly unbranched perennial, adpressed short hairy, 50–150 cm. Leaves with 9–17 leaflets, elliptic-oblanceolate, 5–12 × 1–2 cm, hairless above, sparsely hairy beneath. Raceme stalk 4–14 cm, raceme 18–40 cm. Flowers mainly in whorls, sometimes alternate, blue, purple or reddish. Calyx silvery-hairy, lips more or less equal, each entire or minutely toothed, 3.6–7.5 mm. Pod curved, 2.8–5 cm × 8–10 mm, densely

woolly. *W North America (California to Alaska)*. H5. Summer–autumn.

The following cultivars are known: 'Moerheimii', with pink and white flowers, 'Albiflorus' and 'Albus' with white flowers, 'Atroviolaceus', with dark purple flowers, 'Caeruleus' with blue, 'Carmineus' with red and 'Roseus' with pink flowers.

6. L. littoralis Douglas. Illustration: *Edwards's Botanical Register* **14**: t. 1198 (1828); *Botanical Magazine*, 2952 (1829); Dunn & Gillett, The lupines of Canada and Alaska, f. 57–59(1966).

Prostrate perennial forming mats, usually shaggy-hairy. Leaves with 6–8 leaflets, linear-oblanceolate, obtuse to rounded, mucronate, hairy, becoming hairless above, 1.3–2.5 cm × 3.5–6 mm, leaf-stalks 2–4 cm, bristly; stipules subulate. Raceme-stalk 3.5–5.5 cm, raceme to 10 cm. Flowers mainly in whorls, *c.* 1.2 cm across, standard pale blue, purplish to white, wings bright blue to pale purple, keel finely hairy on inside margin; bracts long-tapering, *c.* 1.5 cm, flower-stalks *c.* 4 mm. Calyx upper lip notched, lower lip entire, soft silvery-hairy, *c.* 5 mm. Pods 3–4 cm × 6 mm, sparsely rough-hairy. *W North America (California to British Columbia)*. H5. Summer.

7. L. perennis Linnaeus. Illustration: *Botanical Magazine*, 202 (1792); Barton, Flora of North America **2**: t. 38 (1822); Strong, American flora **3**: 128 (1849); Dunn & Gillett, The lupines of Canada and Alaska, f. 60, 61, (1966).

Stout perennial to 70 cm, stems minutely hairy. Leaves with 7–11 leaflets, 3–3.8 cm × 8–12 mm, oblanceolate, hairless above, hairy beneath, leaf-stalks to 9 cm, stipules subulate. Racemes to 20 cm, loose. Flowers alternate or whorled, purple-blue, pink, white, or parti-coloured, stalks *c.* 6 mm. Calyx upper lip notched, 4–6 mm, lower lip almost entire, *c.* 6 mm. Pod 3–5 cm × 8–9 mm, hairy. *E North America*. H2. Summer.

8. L. micranthus Gussone (*L. hirsutus* misapplied). Illustration: Fiori et al., Flora analytica d'Italia **2**: f. 1866 (1965); Zohary, Flora Palaestina **2**: pl. 56 (1972).

Annual to 50 cm, branching from base, stems and leaf-stalks coarsely hairy, stipules linear, pointed, attached to the leaf-stalk for about half of their length, leaves with 5–9 leaflets, 1–5 cm × 6–20 mm, obovate, tapered at base, mucronate, coarsely hairy above and

beneath. Raceme to 12 cm, raceme-stalk to *c.* 2 cm. Lower flowers alternate, upper ones whorled, mainly 1–1.4 cm, but very variable in size, blue with a white basal spot on standard; bracts subulate, persistent; bracteoles linear. Calyx upper lip deeply 2-toothed, lower lip about twice as long as upper, slightly 3-toothed. Pods 3–5 cm × 9–12 mm, coarsely hairy. *Mediterranean area*. H5. Summer.

9. L. mutabilis Sweet (*L. cruckshanksii* Hooker). Illustration: Sweet, British flower garden **2**: pl. 130 (1825); *Botanical Magazine*, 2682 (1826); *Edwards's Botanical Register* **18**: t. 1539 (1832); Lewis et al. (eds.), Legumes of the world, 288 (2005).

Annual or short-lived perennial to 1.5 m, almost hairless, glaucous. Leaves with 7–9 leaflets, oblanceolate, spathulate, 6 × 1.2 cm, rarely shortly hairy beneath, leaf stalks 4–8 cm. Raceme-stalk *c.* 10 cm, raceme 10–20 cm, bracts deciduous. Flowers *c.* 2 cm across, stalks 5–14 mm, hairless or minutely hairy; petals, blue or white, standard yellow centered, wings very broad, keel hairy on the margin. Calyx silvery-hairy, upper lip slightly notched, lower lip entire. Pods *c.* 8 × 1.6 cm, adpressed hairy. *Colombia, Peru*. H5–G1. Summer.

10. L. subcarnosus Hooker. Illustration: *Botanical Magazine*, 3467 (1836); Rickett, Wild flowers of the United States, **3**(1): pl. 81 and title page (1969).

Annual to 25 cm, silky-hairy. Leaves with 5–7 leaflets *c.* 3.5 cm × 6 mm, obovate-lanceolate, slightly fleshy, becoming hairless above, silky beneath, stipules subulate, bristle-pointed. Raceme-stalk to 10 cm, raceme *c.* 15 cm. Flowers alternate, dark blue, rarely white or yellow, with a white spot in centre of standard. Calyx silky, upper lip shorter than lower, deeply 2-toothed, lower lip almost entire, minutely 3-toothed. Pods *c.* 4 cm × 6 mm, densely long-hairy. *USA (Texas)*. H5. Spring–early summer.

11. L. luteus Linnaeus. Illustration: *Botanical Magazine*, 140 (1791); Reichenbach, Icones Flora Germanicae et Helveticae **22**: t. 6 (1903); Hegi, Illustrierte Flora von Mittel-Europa, edn 2, **4**: f. 1310 (1975).

Annual to 80 cm, branched from the base, hairy. Leaves with 7–11 leaflets, 3–6 cm × 8–15 mm, obovate, oblong, long-hairy above, less densely so beneath. Raceme-stalk to 12 cm, raceme to 25 cm.

Flowers golden yellow, sweetly scented, in well-spaced whorls, on stalks *c.* 2 mm. Bracts obovate, soon falling; bracteoles linear. Calyx upper lip deeply 2-toothed, lower lip with 3 shallow teeth. Pod 4–6 × 1–1.4 cm, densely long hairy. *Mediterranean area.* H5. Summer.

12. L. densiflorus Bentham. Illustration: *Edwards's Botanical Register* **20**: t. 1689 (1834); Abrams, Illustrated flora of the Pacific States **2**: f. 2563 (1944).

Erect annual to 50 cm, adpressed hairy, single-stemmed or branched towards the top. Leaves with 7–9 leaflets, oblanceolate, 3–5 cm, long-hairy beneath. Racemes to 25 cm. Flowers white, tinged or veined pink or violet, purple or blue, rarely yellow, 1.4–1.8 cm, in close or distant whorls, standard elliptic, apex rounded, keel hairy on upper margin, stalks *c.* 2 mm, bracts reflexing. Calyx upper lip with 2 short teeth, lower lip with 3 larger teeth. Pod 2 × 1 cm, long spreading hairy. *USA (California).* H4. Summer.

13. L. mexicanus Lagasca. (*L. hartwegii* Lindley; *L. bilineatus* Bentham). Illustration: *Edwards's Botanical Register* **25**: t. 31 (1839); Sweet, Ornamental flower garden **1**: t. 28 (1854); Hay & Beckett, Reader's Digest encyclopaedia of garden plants and flowers, 421 (1972).

Erect annual to 1 m with long spreading shaggy hairs, and shorter-hairy beneath the long hairs. Leaves with 7–9 leaflets, oblong-obtuse, to 4 cm × 7 mm, long adpressed-hairy beneath, hairless above; stalks to 9 cm; stipules linear. Raceme-stalk *c.* 10 cm, raceme to 20 cm. Flowers in whorls or alternate, 1.4–1.6 cm, blue, standard with a pink centre. Bracts feathery, deciduous. Calyx upper lip slightly notched, lower lip entire, densely hairy; bracteoles present. Pod densely felted, to 3.8 cm × 6 mm, about 9-seeded. *Mexico.* H4. Summer.

14. L. albus Linnaeus. Illustration: Reichenbach, Icones Florae Germanicae et Helveticae **22**: t. 10 (1903); Polunin, Flowers of Europe, pl. 521 (1969).

Annual to 1.2 m, stems and leaf-stalks sparsely silky. Leaves with 5–9 leaflets, 2–6 cm × 1–2 mm, oblong-obovate, finely mucronate, almost hairless above, shaggy beneath, margins hairy; stipules finely pointed. Raceme to 50 cm, raceme-stalk to *c.* 2 cm. Lower flowers alternate, upper ones whorled, *c.* 1.5 cm across, white, sometimes tinged blue or violet with a

white central spot on standard, on short stalks to 2 mm. Bracts soon falling; bracteoles tiny or absent. Calyx-lips more or less equal, upper lip entire, lower lip entire or minutely 3-toothed. Pods 7–15 × 1–2 cm, shaggy at first, longitudinally wrinkled on drying. *Greece to W Turkey & Crete.* H5. Summer.

15. L. texensis Hooker. Illustration: *Botanical Magazine*, 3492 (1836); Rickett, Wild flowers of the United States **3**(1): pl. 79 (1969).

Annual to 25 cm, branching at ground level. Leaves with 5 leaflets, lanceolate, acute, to 3 cm × 8 mm, hairless above, silky beneath; stipules subulate. Raceme-stalk to 8 cm, raceme to 8 cm. Flowers alternate, *c.* 1 cm across, deep blue, standard with a white, yellow or reddish central blotch, stalks *c.* 5 mm. Calyx silky, upper lip shorter than lower, deeply toothed, lower lip entire. Pod 2.5 cm × 4 mm. *USA (California, Texas).* H4. Early summer.

This species is very similar to *L. subcarnosus* but with more acute and less fleshy leaves and an entire lower calyx lip.

16. L. pilosus Murray (*L. hirsutus* Linnaeus; *L. varius* Linnaeus). Illustration: Sibthorp & Smith, Flora Graeca **7**: t. 684 (1830); Lewis et al. (eds.), Legumes of the world, 287 (2005).

Erect, little-branched annual to 80 cm, long shaggy-hairy. Leaves with 7–11 leaflets, oblong-obovate, 2.5–6 × 1–1.8 cm, with long soft hairs on both surfaces. Raceme-stalk to 8 cm, raceme to 20 cm. Flowers 1.5–2 cm across, in whorls, deep blue, rarely pink with a white spot in centre of standard, base of keel white, darkening with age; stalks *c.* 5 mm. Calyx upper lip deeply divided, lower lip entire. Pod 8–2.5 cm, densely woolly. *Greece & Crete to Israel.* H4. Spring.

17. L. nanus Bentham. Illustration: *Edwards's Botanical Register* **20**: t. 1705 (1835); Sweet, British flower garden **3**: pl. 257 (182426); Abrams & Ferris, Illustrated flora of the Pacific States **2**: t. 2579 (1944); Brickell (ed.), RHS A–Z encyclopedia of garden plants, 653 (2003).

Erect annual 50 cm, branching at base or simple-stemmed, minutely adpressed- or spreading-hairy. Leaves with 5–7 leaflets, 1.5–3 cm × 4 mm, linear-lanceolate, channelled, adpressed-hairy above; leaf-stalks 4–8 cm. Raceme-stalk to 8 cm, raceme to 20 cm. Flowers in whorls, *c.* 1.4 cm across, wings azure blue, keel

white with dark purple tip, standard azure with yellowish or white spot, fragrant. Calyx upper lip deeply toothed, lower lip with 3 tiny teeth. Pods silky at first, to 3 cm × 4 mm, up to 7-seeded. *USA (California).* H4. Spring–early summer.

18. L. angustifolius Linnaeus. Illustration: Reichenbach, Icones Flora Germanicae et Helveticae **22**: t. 10 (1903); Polunin, Flowers of Europe, pl. 520 (1969).

Annual to 1.5 m, stems sparsely silky. Leaves with 5–9 leaflets, to 5 cm × 6 mm, linear to linear-spathulate, hairless above, sparsely silky beneath. Flowers alternate, upper ones in whorls, *c.* 1–1.5 cm across, light to dark blue, tinged purple, rarely pink or white. Bracts oblanceolate-obovate, deciduous; bracteoles short. Calyx upper lip deeply 2-toothed, lower lip entire or with 2 or 3 tiny teeth. Pods 3.5–6 cm × 7–15 mm. *W France to Morocco & Mediterranean area.* H5. Summer.

92. LABURNUM Fabricius
H. Ern

Small trees. Leaves made up of 3 leaflets. Flowers in simple, axillary or apparently terminal, leafless racemes, pendent while flowering. Calyx bell-shaped, slightly 2-lipped, lips undivided or shortly toothed. Corolla yellow. Stamens 10, all filaments united into a tube. Pod dehiscent, flattened, slightly constricted between the seeds. Seeds numerous, compressed.

A genus of 2 species native to the mountains of SC and SE Europe, both of which (and their hybrid) are widely used as ornamentals. The pods of all are very poisonous.

1a. Racemes more than 40 cm
　　　　　　　3. × watereri
　b. Racemes at most 40 cm　　　2
2a. Twigs adpressed-downy, greyish green; pod adpressed-downy, the hairs more or less persisting till maturity　**1. anagyroides**
　b. Twigs almost hairless, green; pod hairless　　　**2. alpinum**

1. L. anagyroides Medikus (*Cytisus laburnum* Linnaeus; *L. vulgare* Berchtold & Presl). Illustration: Polunin & Everard, Trees and bushes of Europe, 106 (1976); Aeschimann et al., Flora alpina **1**: 817 (2004); Lewis et al. (eds.), Legumes of the world, 291, 297 (2005).

Small tree to 7 m. Twigs greyish green, adpressed-downy. Leaflets 3–8 cm, elliptic to elliptic-obovate, usually obtuse and shortly mucronate, greyish green, adpressed-downy beneath when young. Racemes 10–30 cm, loose. Corolla *c.* 2 cm, golden yellow. Pod 4–6 cm, adpressed-downy when young, almost hairless when mature but some hairs usually persisting, upper suture unwinged. Seeds black. *SC Europe.* H1. Late spring.

2. L. alpinum (Miller) Berchtold & Presl (*Cytisus alpinus* Miller). Illustration: *Botanical Magazine*, 176 (1791); Aeschimann et al., Flora alpina **1**: 819 (2004).

Shrub or tree to 5 m or to 10 m in cultivation. Twigs hairless (sometimes hairy when young), green. Leaflets 3–8 cm, light green beneath. Racemes 15–30 cm, rather dense. Corolla 1.5 cm, yellow. Pod 4–5 cm, hairless, upper suture with wing 1–2 mm wide. Seeds brown. *SC & SE Europe.* H1. Early summer.

3. L. × watereri (Kirchner) Dippel. Illustration: Hora, The Oxford encyclopaedia of trees of the world, 209 (1981); Brickell (ed.), RHS A–Z encyclopedia of garden plants, 603 (2003); Lewis et al. (eds.), Legumes of the world, 291 (2005).

The hybrid between *L. anagyroides* and *L. alpinum*, intermediate in all characters except for the very long racemes. *SC Europe & of garden origin.*

Very widely cultivated; currently it has almost replaced the wild species in most parks and gardens. 'Vossii' has even longer, very many-flowered racemes.

93. × LABURNOCYTISUS C.K. Schneider
S.G. Knees
Deciduous spreading tree to 8 × 6 m, with greyish bark. Leaves with three elliptic-ovate leaflets, each 2–6 cm, dark green, adpressed-downy, especially beneath. Flowers of three types: yellowish *Laburnum*-like, purplish *Cytisus*-like and yellow and pink flowers in *Laburnum*-like racemes.

A graft chimera, originally raised in 1826. Needs full sun and will grow in most soil types, except those in waterlogged situations. Propagation is by grafting onto *Laburnum anagyroides* in late summer.

1. L. × adamii (Poiteau) Schneider (*Cytisus adamii* Poiteau; *Laburnum adamii* (Poiteau) Kirchner). Illustration: *La Belgique Horticole* **21**: 16–18 (1871); Phillips, Trees in Britain, Europe and North

America, 50, 128 (1978); Krüssmann, Manual of cultivated broad-leaved trees & shrubs **2**: pl. 38 (1986).

Graft chimera raised in gardens. H3. Spring–early summer.

The graft-hybrid between *Cytisus purpureus* and *Laburnum anagyroides*.

94. PETTERIA C. Presl
H. Ern
Non-spiny shrubs. Leaves made up of 3 leaflets. Flowers in terminal, erect, leafless racemes. Calyx bell-shaped to tubular, two lipped; upper lip divided to about two-thirds, lower 3-toothed. Corolla yellow; Pod linear-oblong, dehiscent, straight or slightly curved, somewhat inflated. Seeds without appendages.

A genus of one species in the west Balkan Peninsula, resembling *Laburnum*, but less attractive.

1. P. ramentacea (Sieber) C. Presl (*Cytisus ramentaceus* Sieber). Illustration: Polunin, Flowers of Greece and the Balkans, 288 (1980).

Erect shrub to 3 m. Branches terete or obscurely angled, hairless; twigs loosely adpressed-hairy when young, later more or less hairless. Leaves with stalks 1.5–5 cm; leaflets 2–7 × 1.2–3 cm, elliptic to obovate, rounded, occasionally slightly notched; dull green on both surfaces; hairless above, with adpressed hairs along the mid-vein and margins beneath. Flowers borne in terminal racemes 4–7 cm of 10–20. Calyx adpressed-hairy. Standard 1.6–2 × 1.4–1.5 cm, pentagonal, notched. Pod 3.5–5 cm × 8–10 mm, beaked, light brown, hairless, margins slightly thickened. Seeds 5–9, orange-brown. *W former Yugoslavia, N Albania.* H2. Late spring.

95. CYTISUS Linnaeus
F.T. de Vries
Shrubs or small trees without spines, from 10 cm to more than 6 m. Leaves with 1 or 3 leaflets, alternate, sometimes crowded, often soon falling, the branches and branchlets performing most of the photosynthesis. Flowers axillary, forming leafy or leafless, terminal or lateral racemes. Calyx 2-lipped, upper lip with 2 usually short teeth (rarely deeply cleft). Corolla white, yellow, purple or dark-brown; keel more or less sickle-shaped. Stamens 10, filaments all united into a tube. Stigma curved upwards, or rarely rolled up. Pods linear or oblong, with

numerous seeds which bear appendages (strophioles).

A genus of about 100 species from Europe and N Africa, closely related to *Ulex* and *Genista*, but differing in the appendaged seeds and lateral racemes. It is often divided into several genera, but a wide circumscription is appropriate here. They require full sun and a dry, neutral to acid, poor sandy soil. They are propagated by seed, cuttings or grafting, and should not be moved once established.

Literature: Van de Laar, H.J., *Cytisus* en *Genista*, *Dendroflora* **8**: 3–18 (1971).

1a. Calyx upper lip with 2 short teeth; keel sickle-shaped   2
  b. Calyx upper lip deeply cleft; keel oblong   35
2a. Calyx bell-shaped; leaves of 3 leaflets, simple, or both on the same plant; style curved or rolled up   3
  b. Calyx tubular; leaves of 3 leaflets; style rolled up   22
3a. Leaves all simple   4
  b. At least some leaves of 3 leaflets   8
4a. Branches grooved   5
  b. Branches angled, sometimes terete when older   6
5a. Leaf-stalks to 1 cm; leaves small, to 1.5 cm; procumbent shrub to *c.* 40 cm; flowers yellow   **9. × beanii**
  b. Leaves stalkless, to 2 cm; shrub to *c.* 3 m; flowers white, yellow, pink or red   **14. × praecox**
6a. Calyx and pod hairless   **20. diffusus**
  b. Calyx and pod downy   7
7a. Shrub to *c.* 30 cm; angles unwinged; flower-stalk 1–2.5 cm; bracteoles 2 or 3   **19. decumbens**
  b. Shrub to *c.* 60 cm; angles winged; flower-stalk to *c.* 1 cm; bracteole 1   **18. procumbens**
8a. Leaves all of 3 leaflets   9
  b. Leaves on lower branches of 3 leaflets, on younger or flowering branches simple, stalkless   15
9a. Inflorescence a leafless raceme   10
  b. Inflorescence a raceme consisting of leafy clusters or forming an umbel-like head   11
10a. All leaves stalked; calyx downy   **2. nigricans**
  b. Leaves on flowering branches stalkless; calyx hairless   **1. sessilifolius**
11a. Flowers white or pinkish-white   12
  b. Flowers yellow, sometimes red-streaked   13

12a. Branches angled; calyx more or less hairless **6. filipes**

b. Branches grooved; calyx downy **5. supranubius**

13a. Inflorescence an umbel-like head, sometimes several combined to a raceme; leaf-stalk to *c.* 4 mm **4. fontanesii**

b. Inflorescence a raceme, consisting of leafy clusters; leaf-stalk to *c.* 3 cm 14

14a. Branches grooved; leaves downy on both, sides; style rolled up **7. ardoinii**

b. Branches angled; leaves hairless above; style curved up **3. emeriflorus**

15a. Style rolled up; flowers usually solitary or paired in leafy bundles; branches angled; calyx usually hairless 16

b. Style curved up; flowers to 5 in leafy bundles; branches grooved or angled; calyx usually downy 19

16a. Compound leaves stalkless or nearly so; pod black, white-haired 17

b. Stalk of compound leaves to 1.5 cm; pod hairless, downy only on the margins 18

17a. Calyx 5–7 mm; compound leaves each with a short stalk to *c.* 5 mm **17. cantabricus**

b. Calyx 4–5 mm; compound leaves stalkless **16. grandiflorus**

18a. Stalk of compound leaves not more than 1.5 cm; calyx hairless **15. scoparius**

b. Stalk of compound leaves to *c.* 2.5 cm; calyx somewhat downy **13. × dallimorei**

19a. Branches grooved 20

b. Branches angled 21

20a. Procumbent shrub to *c.* 30 cm; stalk of compound leaves 0.5–1 cm **8. × kewensis**

b. Upright shrub to *c.* 1 m; compound leaves stalkless **10. purgans**

21a. Leaves hairless above; flower-stalk 0.7–1 cm **11. ingramii**

b. Leaves downy on both sides; flower-stalk to *c.* 0.5 cm sides **12. multiflorus**

22a. Leaves 4–11 cm; leaf-stalk to 3.5 cm; flowers borne singly in leafless racemes **32. battandieri**

b. Leaves to *c.* 5 cm; leaf-stalk to *c.* 2.5 cm; flowers in clusters of more than 1, in leafy bundles or umbel-like heads 23

23a. Flowers 1–7 in several leafy clusters, forming a leafy raceme 24

b. Flowers many in umbel-like heads, terminal and lateral 31

24a. Leaf-stalk not more than 1.5 cm 25

b. Leaf-stalk to *c.* 2.5 cm 28

25a. Flowers white or yellowish 26

b. Flowers yellow with reddish brown markings 27

26a. Leaves softly-downy **27. palmensis**

b. Leaves silky-downy **25. proliferus**

27a. Leaves 1–3.5 cm, downy on both sides; keel yellow **28. supinus**

b. Leaves not more than 2 cm, hairless above; keel reddish brown **26. demissus**

28a. Flowers yellow with reddish brown markings; branches terete, at least when older 29

b. Flowers purple-pink, or standard white, wings yellow and keel lilac-pink; branches angled 30

29a. Leaves more or less hairless when older; terminal flower with 1 bracteole **22. glaber**

b. Leaves on both sides downy; bracteoles absent **23. hirsutus**

0a. Calyx hairless except on margins; flowers purple-pink **21. purpureus**

b. Calyx downy; standard white, wings yellow, keel lilac-pink **24. × versicolor**

31a. Flowers white or yellowish 32

b. Flowers deep yellow; standard sometimes with a dark brown spot in the centre 33

32a. Calyx dark brown; flower-stalk 5–15 mm **25. proliferus**

b. Calyx whitish; flower-stalk 2–6 mm **30. albus**

33a. Branches long, erect, downy; standard with a dark brown spot in the cent **28. supinus**

b. Branches densely grey-green downy; standard yellow 34

34a. Standard narrowly obovate, rounded **31. tommasinii**

b. Standard broad, slightly notched **29. austriacus**

35a. Leaves of 3 leaflets **28. supinus**

b. Leaves simple 36

36a. Standard uniformly downy; leaves to *c.* 6 cm, or stalkless **36. linifolia**

b. Standard hairless or nearly so, or downy on the midrib, especially near the tip; leaves not more than 4.5 cm, always stalked 37

37a. Stipules 2.5–6 mm, persistent, prominent, giving a scaly appearance to the older twigs; mucro of leaflet 0.4–0.6 mm **35. maderensis**

b. Stipules 0.5–2.5 mm, not persisting or prominent; mucro of leaflet to 0.3 mm, or leaf apex obtuse 38

38a. Inflorescences indeterminate, flowers borne in congested racemes or heads on short axillary branches; standard hairless **37. monspessulanus**

b. Inflorescences determinate, flowers borne in short to long terminal racemes; standard downy or hairless on midrib 39

39a. Leaf-stalk 5–25 mm; most leaflets at least 10 mm, narrowly elliptic-obovate to oblanceolate **34. stenopetalus**

b. Leaf-stalk 1–6 mm; most leaflets 3–8 mm, obovate-oblong to oblanceolate **33. canariensis**

1. C. sessilifolius Linnaeus. Figure 111 (4), p. 404. Illustration: *Botanical Magazine*, 255 (1794); Vicioso, Genisteas Españolas, 43 (1953); Csapody & Tóth, Colour atlas of flowering trees and shrubs, 141 (1982).

Upright bushy shrub to *c.* 2 m, entirely hairless. Branches long, green, angled when young. Leaves of 3 leaflets, usually reduced to bracts and stalkless on flowering branches, otherwise stalk to 2.5 cm. Leaflets 6–18 × 5–14 mm, obovate to broadly elliptic, rounded or mucronate, light green. Flowers 4–12 in loose, leafless racemes, light yellow; stalk 4–7 mm, with 2 or 3 persistent bracteoles. Calyx 2–3 mm, bell-shaped, hairless. Standard 1–1.2 cm, rounded to notched; keel beaked. Pods 1.5–4 cm × 4–9 mm, oblong to linear, curved at the base, smooth; seeds 4–10. *S Europe, N Africa.* H3. Late spring.

2. C. nigricans Linnaeus (*Lembotropis nigricans* (Linnaeus) Grisebach. Figure 111 (1), p. 404. Illustration: *Botanical Magazine*, 8479 (1913); Bean, Trees & shrubs hardy in the British Isles, edn 8, 822 (1970); Reader's Digest encyclopaedia of garden plants and flowers, 205 (1972); Brickell (ed.), RHS A–Z encyclopedia of garden plants, 345 (2003).

Upright deciduous shrub to *c.* 2 m, entirely adpressed silky-downy; branches rod-like. Leaves of 3 leaflets, stalk 2–20 mm. Leaflets 6–30 × 5–16 mm, obovate-oblong to linear-elliptic, nearly hairless above, adpressed downy beneath. Racemes to 20 cm, leafless; flowers many, solitary, yellow, becoming darker. Flower-stalk 4–8 mm, with a long, persistent bracteole. Calyx 2–4 mm, bell-shaped, downy. Standard 9–12 × 6–9 mm,

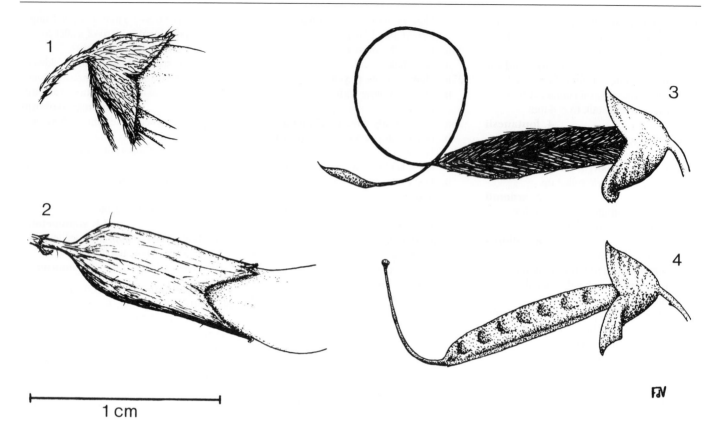

Figure 111. Diagnostic details of Cytisus species. 1, Adpressed downy bell-shaped calyx, with one linear bracteole (*C. nigricans*). 2, Nearly hairless tubular calyx, with three minute bracteoles of which two have fallen off (*C. purpureus*). 3, Rolled up style; pod with long hairs on the margin (*C. scoparius*). 4, Curved up style; pod hairless (*C. sessilifolius*).

obovate-oblong, slightly notched; wings shorter. Pods 2–4 cm × 3–7 mm, linear-oblong, adpressed silky-downy; seeds 2–8. *C & SE Europe & C Russia*. H3. Early summer.

Now often referred to the genus *Lembotropis*.

3. C. emeriflorus Reichenbach
(*C. glabrescens* Sartorius not Schrank). Illustration: *Botanical Magazine*, 8201 (1908); Bean, Trees & shrubs hardy in the British Isles, edn 8, 817 (1970).

Procumbent to erect shrub to *c.* 70 cm, rounded; branches rigid, angled, shaggy-hairy when young. Leaves of 3 leaflets; stalk 1.5–3 cm, slightly downy; leaflets 1–2 cm × 3–4 mm, obovate to elliptic, hairless above, adpressed downy beneath. Racemes consisting of leafy bundles, loose; flowers 1–5 per bundle, golden yellow, sometimes red-streaked; flower-stalk 1–2.5 cm, with 1 long, persistent bracteole. Calyx 2–3 mm, bell-shaped, downy. Standard 1–1.5 cm, broadly obovate, slightly notched. Pods 2.5–4 cm × 6–8 mm,

oblong to narrowly obovate, somewhat bowed, pointed, smooth black; seeds to 6. *SE Europe*. H3. Late spring.

4. C. fontanesii Bell (*Spartium biflorum* Desfontaines; *Chronanthus biflorus* (Desfontaines) Frodin & Heywood. Illustration: Vicioso, Genisteas Españolas, 44 (1953–1955).

Upright or also ascending shrub, to *c.* 50 cm; branches angled, downy only when young. Leaves of 3 leaflets; stalk 2–4 mm; leaflets 4–10 × *c.* 1 mm, linear to lanceolate, with some erect hairs. Flowers solitary or paired, or to 8 in umbel-like heads, sometimes combined to leafy racemes, golden yellow; flower-stalk 2–4 mm, white-downy; bracteole 1. Calyx 3–4 mm, bell-shaped, with bristle-like white hairs. Standard 1–1.7 cm, notched; keel about as long. Pods 1–1.5 cm × 6–8 mm, broadly obovate, pointed, enveloped by the withered standard, with translucent valves, brown-black, few-seeded. *E & S Spain, Balearic Islands*. H4. Late spring.

5. C. supranubius (Linnaeus filius) Kuntze (*C. fragrans* Lamarck). Illustration: *Botanical Magazine*, 8509 (1913); Roles, Illustrations to Clapham, Tutin & Warburg Flora of the British Isles, 172 (1957–65).

Upright shrub, to *c.* 3 m; branches thick, grooved, blue-green bloomed, downy only when young. Leaves of 3 leaflets, soon falling; stalk 5–12 mm, adpressed-downy; leaflets 4–8 × 1–2 mm, linear to narrowly lanceolate, adpressed-downy on both sides. Racemes consisting of leafy clusters; flowers 1–4 or many per cluster, milk-white, pink-toned, fragrant. Flower-stalk 2–4 mm, with 2 small bracts. Calyx 2–3 mm, bell-shaped, downy, soon falling; upper lip reduced, hardly toothed. Standard 9–15 × 6–8 mm, slightly notched; keel shorter than wings. Pods 1.5–2.5 cm × 3–5 mm, elongate, tapered at base and tip, black, hairless; seeds 2–7. *Canary Islands (Tenerife)*. H5. *Spring*.

6. C. filipes (Webb & Berthelot) Masferrer (*Spartocytisus filipes* Webb). Illustration:

Roles, Illustrations to Clapham, Tutin & Warburg, Flora of the British Isles, 171 (1957–65).

Small shrub, almost entirely hairless; branches densely arranged, thin, striped to angled, green. Leaves of 3 leaflets, soon falling; stalk 3–10 mm, thread-like; leaflets 5–12 × 1–4 mm, oblanceolate to narrowly obovate, thin, somewhat downy beneath. Racemes to c. 40 cm, consisting of leafy clusters; flowers 1–4 per cluster, pure white, fragrant. Flower-stalk 3–6 mm, with 3 bracteoles. Calyx 2–3 mm, bell-shaped, membranous, soon falling. Standard 8–12 × 4–6 mm, narrowly obovate, rounded; keel not very curved. Pods 2–3 cm × 5–6 mm, oblong, curved at the base, hairless, black, few-seeded. *Canary Islands*. H5. Late winter–spring.

7. C. ardoinii Fournier. Illustration: Hay & Synge, The colour dictionary of flowering plants, pl. 45 (1971); Brickell (ed.), RHS A–Z encyclopedia of garden plants, 344 (2003); Aeschimann et al., Flora alpina **1**: 821 (2004).

Low shrub to c. 60 cm, flat to domed; branches grooved, downy only when young. Leaves of 3 leaflets; stalk 6–12 mm, with persistent bud-scales; leaflets 4–10 × 1–3 mm, narrowly oblong to obovate-oblong, rounded, both sides long-adpressed-downy. Racemes consisting of leafy clusters; flowers 1–4 per cluster, golden yellow. Flower-stalk 4–7 mm. Calyx 2–4 mm, bell-shaped, long-downy. Standard c. 1 cm, rounded to slightly notched; keel hardly beaked; style rolled up. Pods 1–2.5 cm × 4–7 mm, linear-oblong, curved, adpressed-hairy, few-seeded. *S France*. H3. Spring.

8. C. × kewensis Bean. Illustration: *Botanical Magazine*, n.s., 299 (1956–57); Millar Gault, Dictionary of shrubs in colour, pl. 133 (1976); *Gardening from Which*, July 1987: 235.

Prostrate shrub, to c. 30 cm; branches stiff, grooved, hairless. Leaves usually of 3 leaflets; stalk 5–10 mm, adpressed-downy; leaflets 7–15 × 1.5–3 mm, oblong to obovate, adpressed-downy beneath, nearly hairless above. Racemes to c. 40 cm, consisting of leafy clusters; flowers 1–3 per cluster, cream-white to sulphur-yellow. Flower-stalk 5–8 mm, downy; bracteoles 3, small, linear; calyx 2–4 mm, bell-shaped, somewhat downy. Standard 1.2–1.5 cm, notched, hairless. Pods 1.5–2 cm, narrowly oblong, straight, densely downy. *Garden origin*. H2. Late spring.

The hybrid between *C. ardoinii* and *C. multiflorus*, developed at Kew Gardens around 1900.

9. C. × beanii Nicholson. Illustration: Hay & Synge, Colour dictionary of flowering plants, pl. 46 (1971); Millar Gault, Dictionary of shrubs in colour, pl. 132 (1976); *Gardening from Which*, May 1988: 151; Brickell (ed.), RHS A–Z encyclopedia of garden plants, 344 (2003).

Procumbent shrub, to c. 40 cm and to twice as wide; branches grooved, greenish brown, hairless. Leaves simple, 8–15 × 2–5 mm, elliptic to narrowly obovate, both sides slightly downy; stalk to c. 1 cm, downy. Racemes leafy; flowers 1–3 per axil, deep yellow. Flower-stalk 3–8 mm, slightly downy, with 3 small bracteoles. Calyx 1–2 mm, bell-shaped, downy. Standard 8–14 mm, rounded. Pods black with long white hairs when young. *Garden origin*. H2. Spring.

The hybrid between *C. ardoini* and *C. purgans*, developed at Kew Gardens in 1892.

10. C. purgans (Linnaeus) Boissier. Illustration: *Botanical Magazine*, 7618 (1898); Vicioso, Genisteas Espanolas, 46 (1953–1955); Krüssmann, Handbuch der Laubgehölze, **1**: pl. 157 (1976).

Upright shrub, to c. 1 m and as wide or wider; branches stiff, grooved to striped, downy when young. Leaves stalkless, on young and flowering twigs simple, otherwise of 3 leaflets, usually falling soon. Leaflets 8–12 × 2–3 mm, linear-lanceolate to narrowly obovate, hairless above, silky-downy beneath. Racemes from compact, head-like to elongate and c. 20 cm, consisting of leafy clusters; flowers 1–4 per cluster, fragrant, golden yellow. Flower-stalk 4–9 mm, with 2 small bracteoles. Calyx 2–3 mm, bell-shaped, downy. Standard 1–1.2 cm × 8–11 mm, broadly obovate, slightly notched; all petals equal. Pods 1–3 cm × 5–8 mm, obovate-oblong, black; seeds 3–6. *SE Europe & N Africa*. H3. Late spring–early summer.

11. C. ingramii Blakelock. Illustration: *Botanical Magazine*, n.s., 211 (1952–1953).

Densely leafy upright shrub, to c. 2 m; branches angled, hairless. Leaves simple, stalkless on flowering and younger branches, otherwise of 3 leaflets; leaflets 1–3 cm × 6–15 mm, elliptic to oblong, silvery-downy beneath. Racemes leafy, few-flowered; flowers 1 or 2 per axil, cream-white to yellow. Flower-stalk 7–10 mm,

adpressed-downy, with 3 minute bracteoles. Calyx 4–7 mm, bell-shaped, adpressed-downy. Standard 1.5–2.5 cm, slightly notched, cream-white. Pods 3–3.5 cm × 7–10 mm, elliptic-oblong, straight, brownish black, with long hairs; seeds to 8. *N Spain*. H4. Late spring–early summer.

12. C. multiflorus (L'Héritier) Sweet (*C. albus* Link not Hacquet). Illustration: *Botanical Magazine*, 8693 (1917); Polunin & Smythies, Flowers of southwest Europe, pl. 16 (1973); Polunin & Everard, Trees and bushes of Europe, 108 (1976); Davis, The gardener's illustrated encyclopedia of trees & shrubs, 111 (1987).

Tall shrub to c. 3 m; branches rod-like, angled, downy only when young. Leaves simple on flowering and younger branches, otherwise of 3 leaflets, soon falling; stalk to 3 mm; leaflets 7–10 × 0.5–2 mm, linear-lanceolate to oblong, silky-downy. Racemes long, consisting of many leafy clusters; flowers 1–3 per cluster, pure white to yellowish. Flower-stalk 2–5 mm, silky-downy, with 1 minute bract. Calyx c. 5 mm, bell-shaped, silky-downy. Standard c. 1 cm, rounded; keel often shorter than wings. Pods 1.5–2.5 cm × 5–10 mm, oblong, pointed, adpressed-downy; seeds 2–6, with large appendages. *Spain & N Africa*. H3. Late spring.

This species has several cultivars, that are hardier (e.g. 'Durus') or have slightly pinkish-blushed (e.g. 'Incarnatus') or earlier appearing flowers ('Toome's Variety').

13. C. × dallimorei Rolfe. Illustration: *Gardeners' Chronicle* **51**: 198 (1912); *Botanical Magazine*, 8482 (1913).

Upright shrub, like *C. scoparius*, to c. 2 m; branches angled, hairless. Leaves of 3 leaflets, sometimes also simple on flowering branches; stalk 5–25 mm, simple leaves stalkless; leaflets like those of *C. scoparius*. Raceme leafy, with solitary or paired, lilac-pink, rosy purple or yellow flowers. Flower-stalk 5–10 cm, sometimes downy, with 3 minute bracteoles. Calyx 3–7 mm, bell-shaped, sometimes downy. Standard 1–2 cm, notched; keel long, beaked; wings carmine-red; style rolled up. Pods c. 2.5 cm, with long hairs on the margin. *Garden origin*. H3. Late spring.

The hybrid between *C. multiflorus* and *C. scoparius* 'Andreanus', developed at Kew Gardens about 1900.

14. C. × praecox Bean. Illustration: Millar Gault, Dictionary of shrubs in colour, pl. 135 (1976); *Gardening from Which*, March 1987: 66; *Amateur Gardening*, September 1987: 19; Taylor's guide to shrubs 176, 177 (1987).

Bushy shrub, to *c.* 3 m; branches long, slightly grooved, grey-green. Leaves usually simple, stalkless, 1–2 cm × *c.* 2 mm, lanceolate to linear-spathulate, silky-downy on both sides, soon falling. Raceme long, leafy, many-flowered; flowers solitary or paired, showy cream-white, yellow, pink or red, with unpleasant fragrance. Flower-stalk 3–8 mm, silky-downy, with 3 bracteoles. calyx *c.* 2 mm, bell-shaped, slightly downy. Standard 1–1.5 cm × 5–9 mm, obovate, rounded. Pods 1.5–2.5 cm × 4–5 mm, ovate, tapered gradually, blackish, patent hairy, few-seeded. *Garden origin.* H3. Spring.

The hybrid between *C. multiflorus* and *C. purgans*, developed in Warminster (England). The species has many cultivars, that have variations in flower-colour, time of flowering or hardiness.

15. C. scoparius (Linnaeus) Link (*Sarothamnus scoparius* (Linnaeus) Wimmer). Figure 111(3), p. 404. Illustration: Polunin & Everard, Trees & bushes of Europe, 108 (1976); Millar Gault, Dictionary of shrubs in colour, pl. 142 (1976); Bärtels, Gardening with dwarf trees and shrubs, 126 (1983); Aeschimann et al., Flora alpina **1**: 823 (2004).

Erect shrub, to *c.* 2.5 m; branches rod-like, angled, deep green, downy only when young. Leaves simple on flowering and younger branches, otherwise of 3 leaflets; stalk 5–15 mm, simple leaves stalkless; leaflets 5–20 × 2–9 mm, elliptic-oblong to obovate. Racemes leafy, many-flowered; flowers 1 or 2 per cluster, golden yellow. Flower-stalk 5–20 mm, with 2 minute bracteoles. Calyx 4–7 mm, bell-shaped, hairless. Standard 2–3 × 1–1.6 cm, notched; keel beaked; style rolled up. Pods 2.5–7 cm × 4–12 mm, oblong, black, with white or brownish hairs on the margins; seeds 6–15. *W, SE & C Europe, introduced elsewhere.* H2. Spring.

From this species very many varieties and hybrids with many different species have been developed, varying in flower-colour (deep-brown and yellow, e.g. 'Andreanus') and time of flowering (very long, e.g. 'Dukaat').

Subsp. **maritimus** (Rouy) Heywood, from the coasts of NW Europe, has most of its branches prostrate; it is occasionally grown.

16. C. grandiflorus (Brotero) de Candolle (*Sarothamnus grandiflorus* (de Candolle) Webb). Illustration: Vicioso, Genisteas Españolas, 50 (1953–1955); Polunin & Everard, Trees & bushes of Europe, 108 (1976).

Erect shrub, to *c.* 3 m; branches angled and silvery-downy only when young. Leaves stalkless, simple on flowering and younger branches, otherwise of 3 leaflets; leaflets 6–12 mm, awl-shaped to elliptic-obovate, slightly silvery-downy, especially on margins. Raceme leafy, many-flowered; flowers 1 or 2 per axil, golden yellow. Flower-stalk 6–9 mm, with 3 minute bracteoles. Calyx *c.* 4 mm, bell-shaped, hairless. Standard 1.5–2.5 × 1.5–2.5 cm, rounded to notched; all petals equal; style rolled up. Pods 2–4.5 × *c.* 1 cm, straight or somewhat bowed, black with long, white hairs; seeds 2–11. *S Spain, S & W Portugal.* H4. Spring.

17. C. cantabricus (Willkomm) Reichenbach (*Sarothamnus cantabricus* Willkomm).

Prostrate, compact shrub to *c.* 2 m; branches angled, hairless. Leaves simple on flowering and younger branches, otherwise of 3 leaflets; stalk 2–5 mm, simple leaves stalkless; leaflets 7–11 × *c.* 5 mm, obovate to lanceolate, somewhat downy above, more densely so beneath. Racemes leafy, with solitary, golden yellow flowers. Flower-stalk *c.* 1 cm, with 2 or 3 small bracteoles. Calyx 5–7 mm, bell-shaped, hairless. Standard 2–2.5 cm, deeply notched; style rolled up. Pods 3–5 cm × 6–8 mm, pointed, black, covered with long white hairs; seeds 3–10. *Spain & SW France.* H5. Late spring–early summer.

18. C. procumbens (Willdenow) Sprengel. Illustration: *Edwards's Botanical Register* **47**: 1150 (1817); Hegi, Illustrierte Flora von Mittel-Europa **4**: 1207 (1923); Krüssmann, Handbuch der Laubgehölze, **1**: 157(1976).

Procumbent, leafy shrub to *c.* 60 cm, entirely adpressed-downy; branches angled when young, terete and striped when older. Leaves simple, stalkless, 1.5–2.5 cm × 2–4 mm, oblong-obovate to lanceolate, less downy above. Raceme to *c.* 15 cm, consisting of leafy clusters; flowers 1–5 per cluster, yellow. Flower-stalk 7.5–10 mm, with one small bracteole. Calyx 2–5 mm, bell-shaped. Standard 1–1.5 cm, rounded; all petals hairless. Pods 3–3.5 cm × *c.* 5 mm, narrowly oblong, adpressed-downy; seeds 5–10. *SE Europe.* H3. Late spring–early summer.

19. C. decumbens (Durande) Spach (*C. prostratus* Simonkai). Illustration: *Botanical Magazine*, 8230 (1908); Bärtels, Das grosse Buch der Gartengehölze, 142 (1981); Krüssmann, Handbuch der Laubgehölze, **1**: pl. 157 (1976).

Like *C. procumbens*, but to *c.* 30 cm, downy only when young; branches angled. Leaves simple, stalkless, 8–20 × 3–6 mm, oblong obovate to oblanceolate, adpressed-downy above, more densely so beneath. Racemes consisting of leafy clusters; flowers 1–3 per cluster, golden yellow. Flower-stalk 1–2.5 cm, with 2 or 3 minute bracteoles. Calyx *c.* 5 mm, bell-shaped, with spreading hairs. Standard 1–1.5 cm, rounded; all petals hairless. Pods 2–3.5 cm × *c.* 6 mm, elliptic, black, erect downy, with 3–9 seeds. *S Europe.* H3. Late spring.

20. C. diffusus (Willdenow) Visiani. Illustration: Schlechtendahl et al., Flora von Deutschland, edn 5, 2295 (1880–1887).

Procumbent or ascending shrub, to *c.* 40 cm, almost entirely hairless; branches thin, angled. Leaves simple, stalkless, 1–2 cm, oblong to lanceolate, punctate. Racemes consisting of leafy clusters; flowers 1–3 per cluster, yellow. Flower-stalk 0.6–1 cm, slightly erect-downy, with 2 or 3 minute bracteoles. Calyx *c.* 4 mm, bell-shaped. Standard 1–1.2 cm, rounded; all petals equal. Pods 1.5–2.5 cm × 4–5 mm, elliptic to oblong, curved, black and hairless; seeds 2–6. *W Germany to Austria.* H2. Late spring.

21. C. purpureus Scopoli (*Chamaecytisus purpureus* (Scopoli) Link). Figure 111(2), p. 404. Illustration: *Botanical Magazine*, 1176 (1809); Bean, Trees & shrubs hardy in the British Isles, edn 8, 825 (1970); Bärtels, Das grosse Buch der Gartengehölze, 141 (1973); Millar Gault, Dictionary of shrubs in colour, pl. 136 (1976).

Procumbent or ascending shrub to *c.* 60 cm, entirely hairless or with occasional long hairs; branches rod-like, slightly angled. Leaves of 3 leaflets; stalk 1–2.5 cm; leaflets 1–2.5 cm × 4–7 mm, obovate-elliptic to oblong, pointed, dark green above. Racemes to *c.* 35 cm, consisting of leafy clusters; flowers 1–3 per cluster, purple-red to rose-pink. Flower-stalk

2–5 mm, with 2 or 3 minute bracteoles. Calyx 7–11 mm, tubular; lip-margins woolly hairy. Standard 1.5–2.5 cm, notched, with a dark patch in the centre. Pods 2.5–4 cm × 3–5 mm, narrowly oblong to obovate, curved, black, bloomed; seeds 4–7. *C & SE Europe*. H2. Late spring–early summer.

This species has a few cultivars: with darker purple (e.g. 'Atropurpureus'), lighter red (e.g. 'Albocarneus') to even whitish (e.g. 'Albus') flowers.

22. C. glaber Linnaeus filius (*C. biflorus* L'Héritier; *C. ratisbonensis* Schäffer; *C. elongatus* Waldstein & Kitaibel; *Chamaecytisus glaber* (Linnaeus filius) Rothmaler). Illustration: *Edwards's Botanical Register* **47**: 1191 (1815); *Botanical Magazine*, 8661 (1916); Hegi, Illustrierte Flora von Mittel-Europa, **4**: pl. 159(1923).

Variable, prostrate to erect shrub to *c.* 2 m; branches terete, rough grey or black downy when young. Leaves of 3 leaflets; stalks 5–25 mm, downy; leaflets 1–2.5 cm × 5–12 mm, obovate to oblanceolate, erect downy when young, more or less hairless when older. Racemes consisting of leafy clusters; flowers 2–4 per cluster, yellow with red-brown markings. Flower-stalk 2–7 mm, downy, without bracteoles, or with 1, linear, on terminal flowers. Calyx *c.* 1 cm, tubular, erect downy. Standard 1.5–3 cm, notched; petals sometimes downy. Pods 2–3.5 cm × 4–5 mm, narrowly ovate to oblong, densely downy; seeds 4–8. *C Europe, former USSR (Caucasus to Siberia)*. H2. Late spring.

23. C. hirsutus Linnaeus (*Chamaecytisus hirsutus* (Linnaeus) Link). Illustration: *Botanical Magazine*, 6819 (1885); Hegi, Illustrierte Flora von Mittel-Europa **4**: 1177 (1923); Polunin, Flowers of Europe, 51 (1969).

Variable, prostrate to upright shrub, covered with erect, long hairs; branches terete. Leaves of 3 leaflets; stalk 1–2.5 cm; leaflets 1–3 cm × 4–18 mm, obovate to elliptic, downy on both sides, especially on the margins. Raceme consisting of 1–many dense, leafy clusters; flowers 1–4 per cluster, golden yellow. Flower-stalk 1–7 mm, without bracteoles. Calyx 1–1.5 cm, tubular. Standard 2.5–3 cm, rounded or slightly notched, with a red-brown central patch, hairless; keel with a central row of long hairs. Pods 2.5–4 cm × 5–8 mm, oblong to obovate-

oblong, curved, with rough hairs, especially on the margins. *S & SE Europe to former USSR (Caucasus)*. H3. Late spring–early summer.

24. C. × versicolor (Kirchner) Dippel. Illustration: Krüssmann, Handbuch der Laubgehölze, **1**: pl. 157 (1976); Aeschimann et al., Flora alpina **1**: 823 (2004).

Broad or upright shrub to *c.* 70 cm; branches rod-like, slightly angled, shaggy downy when young. Leaves of 3 leaflets; stalk 1–2.5 cm, adpressed downy; leaflets 1–3 cm × 4–10 mm, obovate to elliptic, densely downy beneath, nearly hairless above. Raceme to *c.* 35 cm, consisting of leafy clusters; flowers 1–4 per cluster, multicoloured. Flower-stalk 3–7 mm, densely downy, with 2 or 3 minute bracteoles. Calyx 7–13 mm, tubular, downy. Standard 1.8–2.2 cm, rounded, whitish; wings yellow; keel lilac-pink. Pods downy. *Garden origin*. H3. Late spring.

A hybrid between *C. hirsutus* and *C. purpureus*.

25. C. proliferus Linnaeus filius. Illustration: Botanical Magazine, 1908 (1817); Roles, Illustrations to Clapham, Tutin & Warburg Flora of the British Isles, 28 (1957-1965); Kunkel & Kunkel, Flora de Gran Canaria **1**: 21 (1974).

Evergreen, open shrub to *c.* 5 m, entirely silky-downy; branches flexible, terete, dark brown. Leaves of 3 leaflets, stalk 5–10 mm, winged; leaflets 2–3 cm × 3–10 mm, elliptic to elongate-oblong, pointed, shortly stalked. Racemes from compact and umbel-like to *c.* 75 cm, consisting of leafy, shortly stalked clusters; flowers 4–7 per cluster, white. Flower-stalk 5–15 mm; bracteole 1, long, soon falling. Calyx 1–1.2 cm, tubular, dark brown, downy. Standard 1.5–2.5 cm, notched, downy on the outside, like the keel. Pods 3–5.5 cm × 7–11 mm, oblong, pointed, black-brown, downy; seeds 3–9. *Canary Islands (Tenerife, Gomera)*. H5. Spring–summer.

26. C. demissus Boissier (*C. hirsutus* Linnaeus var. *demissus* (Boissier) Halácsy). Illustration: Krüssmann, Handbuch der Laubgeholze, **1**: 126 (1976).

Low, wide shrub, to *c.* 10 cm; branches thin, grooved, grey-downy. Leaves of 3 leaflets, stalk 4–8 mm, erect downy; leaflets 1–2 cm × 3–6 mm, oblong-obovate to oblanceolate, more or less hairless above, with spreading hairs beneath. Racemes consisting of leafy clusters; flowers yellow,

standard and keel reddish brown. Flower-stalk 2–5 mm, dark, erect downy, without bracteoles. Calyx 1.2–1.5 cm, tubular, downy. Standard 2–3 × 1.2–1.5 cm, notched; keel much shorter. *N & E Turkey, Greece*. H3. Late spring–early summer.

27. C. palmensis (Christ) Hutchinson (*C. proliferus* var. *palmensis* Christ).

Evergreen shrub to 2.5 m; branches long, grooved only when young, yellowish downy. Leaves of 3 leaflets; stalk 0.7–1.5 cm, winged, downy; leaflets 1–3.5 × 4–10 mm, narrowly elliptic to oblanceolate, pointed, shortly stalked, finely and softly downy. Raceme leafy, consisting of clusters; flowers 2–5 per cluster, white or yellowish. Flower-stalk 5–10 mm, densely downy, with 3 linear bracteoles. Calyx 7–10 mm, tubular, densely downy, lips deeply cut. Standard 1.5–2 × *c.* 1 cm, rounded; all petals equal. Pods 2.5–4.5 cm × *c.* 5 mm, oblong, curved, brown-yellow-downy; seeds 6–12, glossy black. *Canary Islands (La Palma)*. H5. Late winter–spring.

28. C. supinus Linnaeus (*C. rochelli* Wierzbicki; *Chamaecytisus supinus* (Linnaeus) Link). Illustration: Schlechtendahl et al., Flora von Deutschland, edn 5, 2313 (1880–1897); Hegi, Illustrierte Flora von Mittel-Europa **4**: 1174 (1923); Vicioso, Genisteas Españolas, 42 (1953–1955).

Rounded shrub to *c.* 1.2 m and as wide; branches terete, with long erect hairs. Leaves of 3 leaflets, stalk 3–15 mm, adpressed downy; leaflets 1–3.5 cm × 5–15 mm, oblong-elliptic to broad obovate, more downy beneath. Flowers 2–10 in umbel-like heads with bracts at the base, sometimes combined to form leafy racemes; flower-stalk 2–5 mm, downy, with 1 bracteole. Calyx 8–14 mm, tubular, upper lip deeply cut, erect downy. Standard 1.7–2.5 cm, rounded, dull yellow with brown spots in the center, silky outside. Pods 1.5–3.5 cm × 4–6 mm, narrowly obovate to oblong, adpressed or erect downy; seeds 3–6. *C & S Europe*. H3. Summer.

The name *C. rochelii* Wierzbicki is sometimes applied to plants of this species with conspicuously adpressed-downy pods.

29. C. austriacus Linnaeus (*C. canescens* Presley; *Chamaecytisus austriacus* (Linnaeus) Link). Illustration: Schlechtendahl et al., Flora von Deutschland, edn 5, 2312

(1880–1887); Hegi, Illustrierte Flora von Mittel-Europa **4**: 1174(1923).

Upright or procumbent shrub to *c.* 1 m; branches thin, terete, or very slightly grooved, dense grey-downy. Leaves of 3 leaflets; stalk 2–12 mm, downy; leaflets 1–3 cm × 3–10 mm, narrowly elliptic to obovate, downy, especially on margins. Flowers 4 or more, in umbel-like heads, yellow; flower-stalk 2–5 mm, downy, with 3 linear bracteoles. Calyx 1.2–1.5 cm, tubular, downy. Standard 1.5–2 cm, slightly notched, downy outside; keel and wings much shorter. Pods 2–4 cm × 5–7 mm, narrowly oblong to obovate, downy, yellowish when young; seeds to 10. *C & SE Europe.* H2. Summer.

30. C. albus Hacquet (*Chamaecytisus albus* (Hacquet) Rothmaler). Illustration: *Botanical Magazine*, 1438 (1812).

Upright shrub to *c.* 80 cm; branches slender, terete, downy. Leaves of 3 leaflets; stalk 4–12 mm, downy; leaflets 1–3 cm × 7–13 mm, narrowly elliptic to oblong-obovate, downy on both sides. Flowers 3–many in umbel-like heads, white or yellowish; flower-stalk 2–6 mm, with 2 or 3 bracteoles. Calyx *c.* 1 cm, tubular, whitish downy. Standard 1.6–2 cm, silky outside. Pods 2–3.5 cm × 5–6 mm, rough-haired, with numerous seeds. *Hungary & W Germany.* H2. Late spring–early summer.

31. C. tommasinii Visiani (*Chamaecytisus tommasinii* (Visiani) Rothmaler).

Rounded deciduous shrub to *c.* 50 cm, entirely downy; branches terete, greyish green. Leaves of 3 leaflets; stalk 5–15 mm, winged; leaflets 1.2–2.5 cm × 2–8 mm, elliptic-lanceolate to obovate, both sides adpressed downy. Flowers many in umbel-like heads, yellow; flower-stalk 2–5 mm, downy; bracteoles 2 or 3, linear. Calyx 1–1.3 cm, tubular, deeply cut. Standard 1.5–2.2 cm × *c.* 8 mm, narrowly obovate, rounded. Pods 1.5–2.5 cm × 4–5 mm, oblong, somewhat bowed, yellowish brown, downy; seeds 3–8. *W former Yugoslavia & N Albania.* H3. Late spring.

32. C. battandieri Maire (*Argyrocytisus battandieri* (Maire) Reynaud). Illustration: *Botanical Magazine*, 9528 (1938–1939); Millar Gault, Dictionary of shrubs in colour, pl. 131 (1976); de Bray, Manual of old-fashioned shrubs, pl. 33 (1986); *Gardening from Which*, November 1986: 340.

Strong-growing, upright, semi-deciduous shrub to *c.* 5 m and as wide, entirely densely silver-grey-downy. Leaves of 3 leaflets; leaf-stalk 1–3.5 cm; leaflets 4–11 × 2.5–5 cm, oval-elliptic, mucronate, light green. Racemes 2–15 cm, erect, leafless; flowers many, solitary, golden yellow, smelling of pineapple. Flower-stalk 2–3 mm, with 3 deciduous bracteoles. Calyx tubular, with lower teeth deeply cut, soon falling. Standard 1–2 cm, rounded; all petals equal, silvery-grey-downy. Pods 3–5 cm × 6–8 mm, narrowly oblong, bowed at the base, yellow-brown, densely silvery-grey-downy; seeds 5–10, bulging. *N Africa.* H4. Early summer.

Now often found under the generic name *Argyrocytisus*.

33. C. canariensis (Linnaeus) Kuntze (*Cytisus candicans* Lamarck; *C. ramosissimus* Poiret; *Spartium albicans* Cavanilles; *Teline canariensis* (Linnaeus) Webb & Berthelot). Illustration: Webb & Berthelot, Phytographia Canariensis **3**(2): t. 41 (1842); *Boletim da Sociedade Broteriana* **65**: 277 (1971); Kunkel, Flora de Gran Canaria **1**: pl. 23 (1974); Bramwell & Bramwell, Wild flowers of the Canary Islands, pl. 29 & 168 (1974).

Erect or spreading, much-branched shrub, to *c.* 3 m; branches downy when young. Leaf-stalk 1–6 mm; stipules 0.5–2 mm. Leaflets 3–13 × 1.5–4.5 mm, obovate-oblong to oblanceolate, almost hairless above, densely downy beneath; tip mucronate; stalk 0.2–0.6 mm. Racemes 1–6 cm, terminal, with 4–20 flowers; bracts of the lower 3–10 flowers leaf-like with 3 leaflets, the rest simple. Flower-stalk 1–5 mm; calyx 2–6 mm, lower teeth distinct. Standard 1–1.3 cm, notched, with V-shaped downy area on the midrib towards tip. Pod 1.5–3 cm × 3–5 mm, somewhat bowed, shortly downy, with 5–8 seeds. *Canary Islands (Tenerife, Gran Canaria).* H5. Early spring.

34. C. stenopetalus (Webb & Berthelot) Christ (*Teline stenopetala* (Webb & Berthelot) Webb & Berthelot). Illustration: *Botanical Magazine*, n.s., 327 (1958–1959); Bramwell & Bramwell, Wild flowers of the Canary Islands, pl. 30 (1974).

Shrub or small tree, to *c.* 6 m; branches downy when young. Leaf-stalk 5–25 mm; stipules 1–2.5 mm. Leaflets 8–45 × 3–18 mm, narrowly elliptic-obovate to oblanceolate, very sparsely to densely downy above, adpressed-downy beneath, stalk 1–2.5 mm; tip acute or mucronate.

Racemes 5–13 cm, terminal, with 6–26 flowers; bract of lowest flower sometimes leaf-like with 3 leaflets, but usually simple like the rest. Flower-stalk 2–4 mm; calyx 3–5 mm, upper teeth acute, lower teeth minute. Standard 1–1.6 cm, notched, hairless or downy on midrib. Pods 1.5–4 cm × 3–7 mm, densely shaggy-downy, with 2–5 seeds. *Madeira.* H5. Spring.

Var. **sericeus** Pitard & Proust (*Cytisus stenopetalus* var. *magnofoliosus* Kuntze). Illustration: Guerra, Vegetacion y Flora de la Palma, 202 (1983). Leaves sparsely to densely downy above, leaflets 2–4.5 cm.

Var. **microphyllus** Pitard & Proust (*Cytisus stenopetalus* var. *gomerae* Pitard & Proust). Leaves hairless above, leaflets 8–25 mm.

35. C. maderensis Masferrer (*Cytisus candicans* Holle; *Teline maderensis* Webb & Berthelot). Illustration: *Boletim da Sociedade Broteriana* **65**: 292 (1971); Vieria, Flores da Madeira, pl. 4 (1986).

Erect shrub to *c.* 2 m; branches downy when young, stiff. Leaf-stalk 7–11 mm; stipules 2.5–6 mm. Leaflets 6–15 × 2–10 mm, oblong-obovate to oblanceolate, sparsely downy above, densely so beneath; stalk 0.5–1 mm; tip acute, with a mucro 0.4–0.6 mm. Raceme 1–4 cm, terminal, rather compact, with 5–15 flowers. Bracts of lower 1 or 2 flowers leaf-like with 3 leaflets, those of the rest simple, reduced. Flower-stalk 2.5–4 mm; calyx 5–7 mm, lower teeth distinct. Standard 1.1–1.5 cm, hairless. Pods 2–3.5 cm × 4–8 mm, densely downy. *Madeira.* H5. Late spring–summer.

36. C. linifolius (Linnaeus) Lamarck (*Teline linifolia* (Linnaeus) Webb & Berthelot). Illustration: Bramwell & Bramwell, Wild flowers of the Canary Islands, pl. 169 (1974).

Erect shrub to *c.* 3 m, branches adpressed downy when young. Leaf-stalk absent or to 6.5 mm; stipules 0.2–6 mm. Leaflets 1–6 cm × 2–10 mm, narrowly oblanceolate to elliptic, sparsely white-adpressed-downy above, densely so beneath; tip obtuse to acute, mucronate. Raceme 1–3 cm, terminal, dense with 4–20 flowers; bracts of lower flowers leaf-like with 3 leaflets, those of the rest simple, reduced. Flower-stalk 2.5–7 mm; calyx 5.5–15 mm, lower teeth distinct, linear. Standard 9–18 mm, densely downy. Pods 1.5–3.5 cm × 4–6 mm, adpressed to spreading hairs. *W Mediterranean (excluding*

Corsica, Sicily, Sardinia), Canary Islands, Moroccan coastlands. H5. Late spring–early summer.

Subsp. **linifolius**. Illustration: *Botanical Magazine*, 442 (1799); *Boletim da Sociedade Broteriana* **65**: 295 (1971). Leaves stalkless, without stipules; leaflets 1.7–2.5 cm × 2–4.5 mm, narrowly oblong, margins markedly curved upwards and inwards; stalk 1–2.5 mm. Flowers in racemes of 6–16; bracteoles 0.6–2.3 mm; calyx 6–10 mm; standard 1–1.4 cm.

Subsp. **rosmarinifolia** (Webb & Berthelot) Gibbs & Dingwall. Illustration: Webb & Berthelot, Phytographia canariensis, **3**(2): t. 44 (1836). Leaves stalkless, without stipules; leaflets 8–10 × *c.* 2 mm, linear-oblong, margins markedly curved downwards and inwards; base stalked *c.* 0.3 mm; tip acute. Raceme 1 cm, with usually 4 flowers; flower-stalk to 2.5 mm; bracteoles *c.* 4 mm; calyx *c.* 5.5 mm; standard *c.* 1 cm. Pod *c.* 1.7 cm × 5 mm, downy.

Subsp. **pallida** (Poiret) Gibbs & Dingwall (*Genista splendens* Webb & Berthelot). Illustration: Webb & Berthelot, Phytographia canariensis **3**(2): t. 43 (1836); *Boletim da Sociedade Broteriana* **65**: 298 (1971). Leaf-stalk to 2.5 mm and stipules to 6 mm; leaflets 3–6 cm × 4–11 mm, narrowly elliptic, margins slightly curved downwards and inwards; base stalked 1.5–2.2 mm; tip acute. Flower-stalk 4–9 mm; bracteoles 7–13 mm; calyx 9–15 mm; standard 1.4–1.7 cm, densely downy. Pods *c.* 3–3.5 cm × 4–5 mm, densely downy.

37. C. monspessulanus Linnaeus (*Genista candicans* Linnaeus; *Teline medicagoides* Medicus; *Cytisus pubescens* Moench; *T. monspessulana* (Linnaeus) Koch; *Genista monspessulanus* (Linnaeus) L.A.S. Johnson). Illustration: *Botanical Magazine*, 8685 (1916); Polunin & Smythies, Flowers of southwest Europe, pl. 18 (1973); Polunin & Everard, Trees & bushes of Europe, 109 (1976); Lamarda & Valsecchi, Alberi e Arbusti spontanei delle Sardegna, 280–283 (1982).

Erect, much branched shrub, to 3 m, branches densely downy when young; leaf-stalk 0.8–5 mm; stipules 0.5–1.5 mm. Leaflets 5–13 × 2–9 mm, broadly obovate to oblanceolate, sparsely to densely appressed or spreading-downy on both surfaces; stalk 0.3–2 mm; tip mucronate or obtuse. Raceme short, indeterminate, with dense clusters of 4–7 flowers at the ends of short lateral branches; bracts of the lower 2 or 3 flowers like the leaves, the rest simple, reduced. Flower-stalk 1.5–3 mm; calyx 4–7 mm, lower teeth minute. Standard 1–1.3 cm, hairless. Pods 1.5–2.5 cm × 3–5 mm, densely woolly, with 3–6 seeds. *Mediterranean area.* H5. Late spring.

Now often found in teh genus Genista. Nearly all of species Nos. **33–37** have spontaneously or in cultivation crossed with nearly all other species to form attractive hybrids. It is often difficult to trace back to the parents of these hybrids.

96. ERINACEA Adanson
H. Ern

Spiny shrubs with opposite or alternate branches. Leaves simple, sometimes made up of 3 leaflets; shortly stalked. Flowers 1–3, in axillary or more or less terminal clusters. Calyx inflated, bell-shaped, 2-lipped; upper lip with 2 teeth; lower with 3 teeth, the teeth one-third as long as the tube. Corolla blue-violet. Pod narrowly oblong, dehiscent. Seeds without appendages.

One species in SW Europe and N Africa. An interesting plant for the sunny rock-garden, but growing very slowly and only flowering for a short period.

1. E. anthyllis Link (*E. pungens* Boissier). Illustration: *Botanical Magazine*, 676 (1803); Vicioso, Genisteas Españolas **2**: 162 (1955); Polunin & Smythies, Flowers of southwest Europe, pl. 18 (1973); Lewis et al., Legumes of the world, 294 (2005).

Hummock-forming shrub, 10–30 cm; branches with stout spines. Leaves *c.* 5 mm, narrowly oblanceolate, falling early. Corolla 1.6–1.8 cm. Pod 1.2–2 cm, glandular and with shaggy hairs. Seeds 4–6. *Mountains of Spain, just extending to France in E Pyrenees; NW Africa.* H2. Late spring.

97. ECHINOSPARTIUM (Spach) Rothmaler
H. Ern

Small shrubs with opposite branches. Leaves made up of 3 leaflets, shortly stalked or stalkless. Calyx somewhat inflated, bell-shaped, 2-lipped, upper lip deeply 2-fid; lower with 3 prominent teeth; all teeth as long as or longer than the tube; corolla yellow. Pod dehiscent. Seeds without appendages.

A genus of 3–5 species in the Pyrenees and the Iberian Peninsula. Only 1 species is in cultivation.

1. E. horridum (Vahl) Rothmaler (*Cytisanthus horridus* (Vahl) Gams; *Genista horrida* (Vahl) de Candolle; *Spartium horridum* Vahl). Illustration: Vicioso, Genisteas Españolas **1**: 28 (1953); Polunin & Smythies, Flowers of southwest Europe, 227 (1973).

A spiny shrub up to 40 cm. Leaflets 4–9 mm, narrowly oblanceolate, silky-hairy beneath, nearly hairless above; pulvini prominent. Flowers usually 2 on each branch, opposite. Calyx 7–12 mm, sparsely silky-hairy; teeth acute. Standard 1.2–1.6 cm, nearly hairless to sparsely silky-hairy. Pod 9–14 × 4–5 mm, with shaggy hairs. Seeds 1–3, ovoid, brown. *Pyrenees & SC France.* H2. Early summer.

98. SPARTIUM Linnaeus
J.R. Akeroyd

Deciduous, unarmed shrub 1–4 mm. Branches many, erect, rush-like, grooved, green, hairless. Leaves few, falling early, 1–3 cm, linear–oblong to narrowly elliptic or lanceolate, with appressed–silky hairs beneath. Stipules absent. Flowers in loose terminal racemes, fragrant. Calyx sheath–like, split above, 1–lipped, with 5 small teeth. Corolla 2–3 cm, bright yellow. Stamens united. Fruits 3–8 cm, linear-oblong, flat, silky–hairy when young. Seeds many.

A genus of a single species. Propagation is by seed. Plants require a well–drained, preferably calcareous, soil and are drought–tolerant.

1. S. junceum Linnaeus (*Genista odorata* Moench). Illustration: *Botanical Magazine*, 85 (1789) – as *Genista juncea*; Hay & Synge, RHS dictionary of garden plants in colour, t. 1914 (1969); Polunin, Flowers of Europe, pl. 52 (1969); Brickell, RHS gardeners' encyclopedia of plants and flowers, 115 (1989).

Mediterranean area, but now naturalised elsewhere. H1. Summer–early autumn.

'Ochroleucum' is a cultivar with pale yellow corollas.

99. CHAMAESPARTIUM Adanson
H. Ern

Dwarf shrubs without spines, the young stems distinctly winged and flattened. Leaves simple or absent. Flowers in dense, terminal racemes. Calyx tubular, 2-lipped; upper lip deeply 2-fid, lower with 3 distinct teeth; corolla yellow, the standard broadly ovate, equalling the wings and keel. Pod dehiscent. Seeds with or without appendages.

A small genus of 2–4 species in C Europe and the Mediterranean area, systematically placed well in between *Cytisus* and *Genista*. Only 1 species is in horticultural use.

1. C. sagittale (Linnaeus) P. Gibbs (*Genista sagittalis* Linnaeus; *Genistella sagittalis* (Linnaeus) Gams; *Pterospartum sagittale* (Linnaeus) Willkomm). Illustration: Vicioso, Genisteas Españolas **1**: 138 (1953); Botanical Magazine, n.s., 332 (1959); Aeschimann et al., Flora alpina **1**: 831 (2004).

Dwarf shrub with procumbent, woody, mat-forming stems and usually erect, herbaceous, simple or little-branched flowering-stems 10–50 cm; wings constricted at the nodes, without teeth or lobes. Leaves 5–20 × 4–7 mm, elliptic, hairless or almost so above, downy beneath. Calyx 5–8 mm; corolla 1–1.2 cm, the standard usually hairless. Pod 1.4–2 cm × 4–5 mm, downy. Seeds 2–5, without appendages. *Continental & Mediterranean Europe*. H1. Early summer.

A nice plant for the sunny rock-garden. The W Mediterranean subspecies **delphinense** (Verlot) Soják from S France and **undulatum** (Ern) Soják from Andalucía may prove to be even more attractive (but less hardy) than the species, due to their silvery indumentum and their prostrate growth. The species is now generally treated as part of *Genista*.

100. RETAMA Rafinesque
H. Ern
Shrubs without spines. Leaves simple; falling early. Flowers in racemes. Calyx urn-shaped, bell-shaped or obconical, 2-lipped. Corolla white to yellow. Pod ovoid to spherical, indehiscent or finally incompletely dehiscent along ventral suture. Seeds without appendages.

A genus of 4 species in the Canary Islands, Mediterranean area and W Asia. Only 1 species is in cultivation.

1. R. monosperma (Linnaeus) Boissier (*Genista monosperma* Lamarck; *Lygos monosperma* (Linnaeus) Heywood). Illustration: *Botanical Magazine*, 683 (1803); Polunin & Smythies, Flowers of southwest Europe: 44 (1973); Bramwell &

Bramwell, Wild Flowers of the Canary Islands, f. 173 (1974); Maire, Flore de l'Afrique du nord **16**: 199 (1987).

Erect shrub up to 4 m. Branches pendent, silky-hairy when young. Leaves linear-lanceolate, silky-hairy, falling early. Flowers fragrant, in loose racemes. Calyx *c.* 3.5 mm, without hairs, splitting along a line around the circumference and falling after flowering; upper lip with 2 triangular teeth, lower with 3 linear-subulate teeth, teeth often hairy on the margin. Corolla 1–1.2 cm, white; standard rhombic-ovate, hairy; wings oblong, obtuse, as long as or shorter than the keel. Pod 1.4–1.8 cm, obovoid with a short beak; wrinkled when mature. Seeds 1 or 2, blackish. *S Portugal, SW Spain, Canary Islands, N Africa*. H5. Early spring.

101. NOTOSPARTIUM J.D. Hooker
S.G. Knees
Shrubs or small trees with slender, grooved and drooping, leafless branches. Flowers borne in racemes; calyx bell-shaped; corolla with shortly reflexed standard, keel obtuse, wings less than keel; stamens 10, 9 united, 1 free; ovary almost stalkless, style curved. Pods flattened, jointed between the seeds; seeds 6–15.

A genus of three species all native to South Island, New Zealand, requiring a sunny site with freely drained, sandy soil. They are easily propagated by seeds sown in autumn or from semi-ripe cuttings taken in summer. Some may need staking if not grown against a wall. This genus is often now considered to be part of *Carmichaelia* (genus No. 50 in this account).

1a. Branchlets not more than 2 mm across; flowers with narrow standard
 1. torulosum
 b. Branchlets 2.5–3.5 mm across; flowers with broad standard 2
2a. Flowers *c.* 0.8 cm, pink-veined, densely packed on racemes; fruits to 2.5 mm across **2. carmichaeliae**
 b. Flowers *c.* 1.2 cm, purple-veined, loosely arranged; fruits to 4 mm across **3. glabrescens**

1. N. torulosum Kirk.
Shrub to 4 m; branchlets *c.* 2 mm across. Racemes slender, to 5 cm; flowers *c.* 8 mm, loosely arranged, lilac, flushed purple; standard narrow. Pods 1.5–2.5 cm, not more than 2 mm across; seeds *c.* 1 mm, dark brown, up to 15 per pod. *New Zealand*. H5–G1. Summer.

2. N. carmichaeliae J.D. Hooker.
Illustration: *Botanical Magazine*, 6741 (1884); RHS dictionary of gardening **3**: 1383 (1981); Everett, New York Botanical Gardens encyclopedia of horticulture, 2340 (1981).

Shrub to 4 m; branchlets compressed, to 3.5 mm across. Racemes slender, densely flowered, to 5 cm; flowers *c.* 8 mm, pink-flushed and -veined; standard broad. Pods 8–17 mm, to 2.5 mm across with 6–10 seeds per pod. *New Zealand*. H5–G1. Summer.

3. N. glabrescens Petrie. Illustration: *Botanical Magazine*, 9530 (1939); Salmon, The native trees of New Zealand, 204, 205 (1980); Everett, New York Botanical Gardens encyclopedia of horticulture, 2340 (1981); Brickell (ed.), RHS A–Z encyclopedia of garden plants, 728 (2003).

Shrub or small tree to 10 m, with slender drooping branches; branchlets compressed, 2.5–3.5 mm across. Racemes slender, open flowered, to 5 cm; flowers *c.* 1.2 cm, purple flushed and veined; standard broad. Pods 8–25 mm, up to 4 mm across; seeds *c.* 2.5 mm, reddish yellow, mottled black, about 6 per pod. *New Zealand*. H5–G1. Summer.

102. GENISTA Linnaeus
H. Ern
Spiny or non-spiny shrubs with alternate or opposite branching. Leaves mostly deciduous, sometimes very early so, shortly stalked, simple or made up of 3 leaflets, alternate or opposite. Flowers bisexual, alternate or opposite in racemes or axillary clusters or heads. Calyx tubular, usually with prominent upper and lower lips, the upper lip 2-fid, the lower with 3 distinct teeth. Corolla yellow. Standard broadly ovate or triangular, acute, hairless or downy, as long as or shorter than the keel. Keel narrowly oblong, hairless or downy. Wings as long as standard, hairless. Stamens 10, their filaments all united into a tube. Pod either narrowly oblong and compressed or sickle- or diamond-shaped and more or less inflated, several-seeded, more rarely ovoid acuminate and 1- or 2-seeded, hairless or downy. Seeds without appendages.

A genus of about 100 species in Europe (most numerous in the Iberian Peninsula), SW Asia, N Africa, the Canary Islands and Madeira; introduced elsewhere. Genistas vary from dwarf and prostrate shrubs to more than 6 m. In many species the leaves fall very early. All prefer a sunny position and a well–drained, light, loamy or even stony soil. Whereas some species (e.g. *G. cinerea*, *G. fasselata*, *G. hispanica*, *G. januensis*, *G. pulchella*, *G. radiata* and *G. sylvestris*) tolerate or even prefer calcareous soils, others, like *G. anglica*, *G. berberidea*, *G. falcata* or *G. florida* will not thrive under such conditions since they inhabit acid soils in nature. The smaller species do well in rock-gardens: *G. anglica*, *G. berberidea* and *G. falcata* can give a distinctive note to the heather garden, and the taller species like *G. aetnensis*, *G. florida* and *G. tenera* make fine solitary plants or may be used on sunny edges of tree-plantings. Most are relatively short-lived. All are preferably propagated by seed. There seem not to be any hybrids in cultivation, though there are some sterile clones on the market, some of them with double flowers.

Literature: Balls, E.K., Two months' collecting in Morocco, *Journal of the Royal Horticultural Society* **69**: 357–362 (1944); Vicioso, C., Genisteas Españolas I: Genista-Genistella, 18–135 (1953); Gibbs, P.E., A revision of the genus *Genista*, *Notes from the Royal Botanic Garden Edinburgh* **27**: 11–99 (1966); Gibbs, P.E., Taxonomic notes on some Canary Island and North African species of *Cytisus* and *Genista*, Lagascalia **4**: 33–41 (1974); Canto, P. & Jesus Sanchez, M., Revision del agregado *Genista* cinerea (Leguminosae), *Candollea* **43**: 73–92 (1988).

Habit. Prostrate or low-growing: **1, 7, 9, 10, 14, 15, 16, 19**; large shrubs more than 2 m: **11, 12, 13**. Spiny shrubs: **1, 2, 3, 4, 5, 6, 7, 8, 9, 10**; non-spiny: **1, 11, 12, 13, 14, 15, 16, 17, 18, 19, 20, 21**.
Leaves. Of 3 leaflets: **3, 4, 20, 21**; simple: **1, 2, 3,5, 6, 7, 8, 9, 10, 11, 12, 13, 14, 15, 16, 17, 18, 19**.
Flowers. Congested in heads: **7, 20**.
Pods. Ovoid-acuminate, 1–2-seeded: **3, 4, 7, 10**.

1a. Plant more or less spiny 2
 b. Plant not spiny 11
2a. Pod narrowly oblong and compressed; seeds 1–12; standard broadly ovate, as long as wings and keel 3
 b. Pod ovoid-acuminate, usually 1-seeded, or diamond- or sickle-shaped and inflated, 2–8-seeded; standard triangular or ovate with a rounded or acute apex, usually shorter than the keel 4
3a. Plant with weak spines; bracteoles absent **1. pulchella**
 b. Plant with stout spines; bracteoles present **2. scorpius**
4a. Flowers or flowering branches borne directly on the axillary spines **3. fasselata**
 b. Flowers or flowering branches not borne directly on the axillary spines 5
5a. Most leaves of 3 leaflets **4. triacanthos**
 b. All leaves simple 6
6a. Leaves with spine-like stipules **5. berberidea**
 b. Leaves without spine-like stipules 7
7a. Bracts, at least of the lowermost flowers, 1 mm or less, or absent; bracteoles minute 8
 b. Bracts, at least of the lowermost flowers more than 1 mm; bracteoles usually conspicuous 10
8a. Calyx hairless or very sparsely silky; pod sickle-shaped; seeds up to 18 **6. falcata**
 b. Calyx densely hairy; pod ovoid-acuminate; seeds 1 or 2 9
9a. Flowers congested in heads; standard about as long as the keel **7. hispanica**
 b. Flowers in loose racemes; standard ½–⅓ as long as the keel **8. germanica**
10a. Leaves hairless; pod slightly sickle-shaped; seeds 4–6 **9. anglica**
 b. Leaves downy beneath; pod ovoid-acuminate; seeds 1 or 2 **10. sylvestris**
11a. Leaves simple, but sometimes withering early and plant appearing leafless 12
 b. At least the lower leaves of 3 leaflets 21
12a. Large shrubs, 1–4 m or more 13
 b. Prostrate or erect shrubs 10–100 cm 15
13a. Branches practically without leaves when flowering; pod 1–1.5 cm **11. aetnensis**
 b. Leaves present at flowering and persisting later; pod 1.5–2.5 cm 14
14a. Flowers in loose racemes which are 7–18 cm **12. florida**
 b. Flowers in rather dense racemes which are 3–5 cm **13. tenera**
15a. Keel and standard hairless 16
 b. Keel and usually the standard silky-hairy 18
16a. Stems usually 3-winged; leaves with a narrow, translucent, finely toothed margin **14. januensis**
 b. Stems not winged; leaves without a translucent, toothed margin 17
17a. Leaves on the main stem 9–50 mm, ovate, lanceolate, elliptic, oblong or oblanceolate, downy, or hairless and hairy on margin and midrib beneath **15. tinctoria**
 b. Leaves on the main stem 3–10 mm, linear-oblanceolate, almost hairless **16. lydia**
18a. Plant erect with long, flexuous branches **17. cinerea**
 b. Plant decumbent, much-branched 19
19a. Upper surface of leaves downy **1. pulchella**
 b. Upper surface of the leaves hairless or nearly so 20
20a. Flowers in dense, terminal racemes; bracteoles present; leaves almost stalkless **18. sericea**
 b. Flowers usually in long racemes on ascending branches; bracteoles absent; leaves usually shortly stalked **19. pilosa**
21a. Flowers in 2–12-flowered terminal heads **20. radiata**
 b. Flowers in usually long and loose racemes **21. ephedroides**

1. G. pulchella Visiani (*G. humifusa* Villars; *G. villarsii* Clementi). Illustration: Coste, Flore de la France **1**: 301 (1901).

Spreading, non-spiny or weakly spiny shrub, 10–30 cm. Young branches and surfaces of leaves with dense, long, silky, adpressed or spreading hairs. Leaves 2–9 × 1.5–3 mm, stalkless, narrowly elliptic. Flowers borne singly in the axil of each bract in a congested raceme; bracteoles absent; flower-stalks 2–5 mm. Calyx *c.* 4 mm. Standard 7–10 mm, ovate, with dense silky hairs. Pod 1.2–1.5 cm × 5–6 mm, felted. Seeds 2–4. *Mountains of SE France, Albania, W former Yugoslavia.* H4. Spring.

2. G. scorpius (Linnaeus) de Candolle. Illustration: Bonnier, Flore complète **2**: pl. 119 (1912); Vicioso, Genisteas Españolas **1**: 79 (1953).

Erect, rarely spreading, intricately branched shrub, 50–200 cm, with stout axillary spines. Leaves 3–11 × 1.5–2 mm, simple, sparsely hairy beneath, almost hairless above. Flowers borne on short branches arising from the spines, or directly on the spines. Flower-stalks

2–5 mm. Calyx 3–5 mm, hairless or almost so, lips shorter than tube. Standard 7–12 mm. Pod 1.5–4 cm, hairless. Seeds 2–7, compressed, ovoid, olive-green. *Spain, S France, Morocco.* H4. Spring.

3. G. fasselata Decaisne (*G. sphacelata* Spach). Illustration: Zohary, Flora Palaestina **2**: pl. 61 (1972).

Spiny shrub, 1–2 m, hairy, with dark scale-like leaf-bases (pulvini). At least some leaves with 3 leaflets; leaflets 3–15 × 1–3 mm, narrowly oblanceolate, silky, falling early. Flowers borne singly or in loose clusters on spines or unarmed branches; bracts leaf-like, bracteoles minute. Calyx 4–5 mm, hairless, lips *c.* one-third as long as tube. Standard 6–7 mm. Pod 8–10 × 5 mm, ovoid, obliquely beaked, wrinkled. Seeds 2. *S Aegean area, Crete, Cyprus, Israel.* H5. Summer.

4. G. triacanthos Brotero (*G. scorpioides* Spach). Illustration: Vicioso, Genisteas Españolas **1**: 49, 52 (1953).

Erect shrub, 50–100 cm, with axillary spines. Leaves of 3 leaflets; leaflets 3–8 × 1–2 mm, oblanceolate, almost hairless, without spiny stipules. Flowers in loose racemes; lowest bracts *c.* 2 mm, simple, uppermost reduced. Calyx 2.5–4 mm, hairless. Standard *c.* 6 mm, triangular, hairless, shorter than the keel. Pod ovate to diamond-shaped, 8–10 mm, blackish, becoming hairless at maturity. Seeds 1–2. *W Iberian Peninsula, Morocco.* H5. Spring.

5. G. berberidea Lange. Illustration: Vicioso, Genisteas Españolas **1**: 77 (1953).

Erect shrub 50–100 cm; young branches with dense spreading hairs. Leaves lanceolate, *c.* 5 × 3 mm, almost hairless. Stipules spine-like. Flowers 1–4 together at the ends of short branchlets; bracteoles *c.* 1 mm. Calyx 5–6 mm, covered with dense spreading hairs, lower lip twice as long as the tube. Standard 8–10 mm. Pod 5–11 mm, with sparse spreading hairs along the sutures, and with 4–6 dark, olive-green, shining seeds. *NW Spain, N Portugal.* H4. Spring.

6. G. falcata Brotero. Illustration: Vicioso, Genisteas Españolas **1**: 74 (1953).

Erect spiny shrub, 50–100 cm. Young branches and lower surfaces of the leaves sparsely silky. Leaves ovate to oblong-lanceolate, obtuse or acute, 4–12 × 2–6 mm. Flowers in short lateral racemes; bracts and bracteoles minute or absent. Calyx *c.* 5 mm, hairless to sparsely

downy, lips as long as the tube. Standard *c.* 9 mm. Pod hairless, 1–2.5 cm, with 10–18 shining yellow seeds. *Western part of the Iberian Peninsula.* H4. Spring.

7. G. hispanica Linnaeus. Illustration: Bonnier, Flore complète **2**: pl. 118 (1912); Vicioso, Genisteas Españolas **1**: 65 (1953); Polunin, Flowers of Europe, pl. 52 (1969); Brickell (ed.), RHS A–Z encyclopedia of garden plants, 470 (2003); Aeschimann et al., Flora alpina **1**: 831 (2004).

Decumbent to erect shrub, 10–50 cm, with axillary spines. Leaves simple, stalkless, 6–10 × 3–5 mm, lanceolate to oblanceolate, lower surface with dense adpressed or spreading hairs. Flowers in dense, terminal, almost head-like racemes; bracts *c.* 1 mm; bracteoles absent. Standard hairless, broadly ovate, slightly notched, about equalling wings and keel. Pod ovate to diamond-shaped, 6–9 mm, becoming hairless at maturity. Seeds 1 or 2, brown, somewhat shining. *N Spain, S France.* H3. Early summer.

Subsp. **hispanica**, from S France westwards to the E Pyrenees and E Spain, has branches and leaves with spreading hairs and standard 6–8 mm, whereas subsp. **occidentals** Rouy (*G. occidentalis* (Rouy) Coste) occurs in the western Pyrenees and northern Spain and has branches and leaves with adpressed hairs and the standard 8–11 mm.

8. G. germanica Linnaeus. Illustration: Coste, Flore de la France **1**: 298 (1901); Bonnier, Flore complète **2**: pl. 118 (1912); Polunin & Smythies, Flowers of southwest Europe, pl. 17 (1973); Pignatti, Flora d'Italia **1**: 641 (1982); Aeschimann et al., Flora alpina **1**: 831 (2004).

Erect shrub 30–60 cm, usually with axillary spines. Leaves simple, 8–20 × 4–5 mm, elliptic or lanceolate, lower surface with long, more or less spreading hairs. Flowers in loose racemes; bracts *c.* 1 mm; bracteoles absent. Calyx *c.* 5 mm, silky. Standard *c.* 8 mm, ovate with an acute apex, about two-thirds as long as the keel. Pod 8–10 × 5–6 mm, hairy. Seeds 2–4, ovoid, brown. *Most of Europe except the western fringe.* H1. Early summer.

9. G. anglica Linnaeus. Illustration: Bonnier, Flore complète **2**: pl. 119 (1912); Vicioso, Genisteas Españolas **1**: 71 (1953); Garrard & Streeter, Wild flowers of the British Isles, pl. 29 (1983).

Decumbent to erect shrub, 40–100 cm, with hairless or hairy branches and

axillary spines. Leaves 4–10 × 2–3 mm, lanceolate or elliptic, hairless. Stipules not spine-like. Flowers in short racemes; bracts leaf-like, in groups; bracteoles less than 1 mm. Calyx 3–4 mm, hairless, lips longer than tube. Standard 6–8 mm. Pod hairless, 1.4–2 cm × 5–6 mm, slightly inflated and sickle-shaped, with 4–6 shining black seeds. *W Europe, extending eastwards to S Sweden, N Germany & SW Italy, Morocco.* H3. Spring.

10. G. sylvestris Scopoli. Illustration: *Botanical Magazine*, 8075 (1906); Pignatti, Flora d'Italia **1**: 641 (1982).

Decumbent shrub, 20–50 cm, with weak axillary spines; young branches with silky hairs. Leaves 1–2 cm × 1–3 mm, simple, narrowly oblong or elliptic, lower surface with sparse hairs. Flowers in long, loose terminal racemes. Calyx 5–7 mm, with sparse silky hairs, upper lip about as long as lower, the lower teeth equal. Standard 7–8 mm, triangular, heart-shaped, claw less than 2 mm. Pod hairless at maturity, *c.* 8 × 4 mm, ovoid, beaked. Seeds 1 or 2, blackish. *Albania, former Yugoslavia, C & S Italy.* H3. Early summer.

11. G. aetnensis (Bivonia-Bernardi) de Candolle. Illustration: *Botanical Magazine*, 2674 (1826); Polunin & Everard, Trees and bushes of Europe, 109 (1976); Pignatti, Flora d'Italia **1**: 643 (1982); *Gardening from Which*, December 1986: 404.

Shrub to 6 m, without spines. Leaves simple, withering early, as do the leaf-like bracts. Flowers in loose racemes. Calyx *c.* 3 mm, almost hairless; upper teeth obtuse and about one-third as long as the tube, lower teeth minute. Standard more or less hairless. Pod 1–1.5 cm × 5 mm, hairless, sickle-shaped. Seeds 2–4. *Italy (Sardinia, Sicily).* H3. Summer.

12. G. florida Linnaeus. Illustration: Vicioso, Genisteas Espanolas **1**: 104 (1953); Polunin & Smythies, Flowers of southwest Europe, pl. 17 (1973).

Erect shrub without spines, 1–3 m. Leaves simple, 5–25 × 2–5 mm, oblanceolate, sometimes sparingly hairy above, shortly stalked to almost stalkless. Flowers borne singly in the axils of each bract in long, loose racemes; lowermost bracts leaf-like, upper reduced. Calyx 4–6 mm. Standard broadly ovate, almost hairless. Pod flattened, oblong to oblong-lanceolate, pointed, 1.5–2.5 cm, greyish silky. Seeds 2–4, black and shining. *Iberian*

Peninsula, W Pyrenees, Morocco. H3. Early summer.

There are several variants (or even species) belonging to this somewhat complex species. Subsp. **leptoclada** (Spach) Coutinho. *N Spain, just extending in to the French Pyrenees.* H3. In Berlin it flowers 4–6 weeks earlier than subsp. *florida.* Plants from Morocco (e.g. those cultivated as Balls 1944) show a beautiful contrast between the silvery leaves and the golden yellow, fragrant flowers (H4).

13. G. tenera (Murray) Kuntze (G. *virgata* (Aiton) de Candolle; *Cytisus tenera* Jacquin; *Spartium virgatum* Aiton). Illustration: *Botanical Magazine,* 2265 (1821).

Shrub without spines to 2 m or rarely to 4 m. Leaves simple, grey-green, more or less stalkless, *c.* 1.2 cm × 3 mm, silky beneath. Flowers in racemes 3–5 cm, terminating the short shoots of the current year. Calyx with silky-hairs. Flowers bright yellow, standard roundish, *c.* 1.2 cm across. Pods *c.* 2.5 cm, densely silky-hairy. Seeds 3–5. *Madeira.* H4. Summer.

A sterile variant of G. *tenera* with a more graceful habit and producing a profusion of rich golden yellow, fragrant flowers has been known for a long time as G. *cinerea* Hort. and is now being distributed under the cultivar name 'Golden Shower'.

14. G. januensis Viviani (G. *lydia* Boissier var. *spathulata* (Spach) Hayek; G. *scariosa* Viviani; G. *triangularis* Willdenow; G. *triquetra* Waldstein & Kitaibel). Illustration: *Botanical Magazine,* 9574 (1939); Pignatti, Flora d'Italia **1**: 637 (1982); Aeschimann et al., Flora alpina **1**: 827 (2004).

Prostrate to erect shrub, 10–50 cm; stems and branches usually 3-winged. Leaves of the flowering branches 5–12 × 2–4 mm, elliptic to obovate; leaves on the non-flowering branches 5–40 × 3–7 mm, elliptic to lanceolate; all hairless and with narrow, translucent, finely toothed margins. Flowers in short racemes on ascending lateral branches. Calyx 3.5–4 mm, almost hairless, lips shorter than the tube. Standard 0.9–1 cm, broadly ovate. Pod 2 cm × 4 mm, hairless. Seeds 5–8. *Italy, NW Balkan Peninsula.* H3. Spring.

15. G. tinctoria Linnaeus. Illustration: Vicioso, Genisteas Españolas **1**: 115 (1953); Garrard & Streeter, Wild flowers of the British Isles, pl. 29 (1983); Brickell (ed.), RHS A–Z encyclopedia of garden plants, 471 (2003); Aeschimann et al.,

Flora alpina **1**: 827 (2004); Lewis et al. (eds.), Legumes of the world, 295 (2005).

Prostrate to erect shrub, 10–200 cm, without spines, very variable in habit, leaf-shape and degree of hairiness. Leaves 9–50 × 2.5–15 mm, simple, hairless or densely silky-hairy. Flowers borne singly in the axils of bracts in short racemes towards the ends of the branches or in long simple or compound racemes; bracts leaf-like; bracteoles *c.* 1 mm; flower-stalk 1–2 mm. Calyx 3–7 mm, hairless to densely silky hairy. Corolla hairless; standard 8–15 mm, broadly ovate. Pod linear-lanceolate, straight or slightly curved, 2–3 cm × 2–3 mm, generally becoming hairless at maturity. Seeds 6–12, greenish. *Europe, W Asia.* H1. Summer.

'Plena' is a garden variety which grows as a dwarf, semi-prostrate shrub with double flowers; owing to the more numerous petals it is very brilliant in colour and is regarded as one of the best of all dwarf yellow-flowered shrubs.

16. G. lydia Boissier (G. *antiochia* Boissier; G. *rhodopea* Velenovský; G. *rumelica* Velenovský). Illustration: *Botanical Magazine,* n.s., 292 (1957); Brickell (ed.), RHS A–Z encyclopedia of garden plants, 471 (2003).

Prostrate or erect shrub, without spines, 10–200 cm. Stems not winged. Leaves 3–16 × 1–3 mm, simple, linear-oblanceolate or linear-oblong, almost hairless, entire and without a translucent margin. Flowers in short racemes of 2–8 on the lateral branches. Calyx 3.5–5 mm, hairless, lips almost as long as the tube. Standard 1–1.2 cm, broadly ovate. Pod hairless, 2–3 cm × 3–5 mm, with 2–6 seeds. *E Mediterranean area.* H3. Late spring.

17. G. cinerea (Villars) de Candolle. Illustration: *Botanical Magazine,* 8086 (1906); Vicioso, Genisteas Españolas **1**: 99 (1953); Ceballos et al., Plantas silvestres de la peninsula Iberica, 157 (1980); Aeschimann et al., Flora alpina **1**: 827 (2004).

Erect shrub without spines, 50–150 cm. Leaves simple, 5–10 × 2–3 mm, variable in shape, silky-hairy beneath. Flowers mostly paired and borne directly on the main branches; bracts in tufts. Standard hairless or with a central ridge of hairs, rarely uniformly silky hairy. Pod 1.2–1.9 cm × 4–5 mm, oblong or lanceolate, straight, shortly pointed, greyish silky or with shaggy hairs. Seeds

2–5, kidney-shaped to ovoid, brown, shining. *W Mediterranean area.* H3. Summer.

Out of the 5 subspecies recognised by Canto & Jesus Sanchez (reference above), subsp. **cinerea** and subsp. **ausetana** Bolòs & Vigo are hardy in the milder parts of Europe. The more Mediterranean subspecies, namely subsp. **murcica** (Cosson) Cantó & Sanchez, subsp. **speciosa** Rivas Martínez et al., (G. *ramosissima* misapplied) and subsp. **valentina** (Sprengel) Rivas Martínez (G. *valentina* (Sprengel) Steudel) may be recommended for zones H4 and H5.

18. G. sericea Wulfen. Illustration: Pignatti, Flora d'Italia **1**: 638 (1982); Aeschimann et al., Flora alpina **1**: 827 (2004).

Much-branched shrub without spines, 80–120 cm. Leaves 5–25 × 2–5 mm, almost stalkless, narrowly elliptic, oblanceolate or obovate, shortly mucronate, silky beneath, hairless or almost so above. Flowers borne singly in the bract-axils, in terminal clusters of 2–5; flowering branches often slender and flexuous. Bracteoles *c.* 1 mm, borne at the middle of the flower-stalk, which is 2–3 mm. Standard 1–1.4 cm, broadly ovate, tapered to the base, silky. Pod 1–1.5 cm × 5–6 mm, downy. Seeds 1–4. *Mountains of the W part of the Balkan Peninsula, NE Italy.* H3. Early summer.

19. G. pilosa Linnaeus. Illustration: Vicioso, Genisteas Españolas **1**: 120 (1953); Brickell (ed.), RHS A–Z encyclopedia of garden plants, 471 (2003); Aeschimann et al., Flora alpina **1**: 829 (2004).

Prostrate to more or less erect shrub without spines, 20–50 cm. Leaves 5–12 mm, simple, usually oblanceolate, shortly stalked to almost stalkless, adpressed silky beneath, hairless above. Flowers borne singly or in pairs in the bract-axils, in loose racemes on ascending branches. Bracteoles absent. Calyx 4–5 mm. Standard 8–10 mm, broadly ovate with sparse, adpressed, silky hairs. Pod 1.8–2.8 cm, oblong, compressed, with shaggy hairs. Seeds 3–8, ovoid, greenish, shining. *W & C Europe.* H2. Early summer.

20. G. radiata (Linnaeus) Scopoli (*Cytisanthus radiatus* (Linnaeus) Lang; *Spartium radiatum* Linnaeus). Illustration: *Botanical Magazine,* 2260 (1821); Coste, Flore de la France **1**: 299 (1901); Hegi,

Illustrierte Flora von Mittel-Europa, edn 2, **4**: t. 159 (1975); Pignatti, Flora d'Italia **1**: 642 (1982).

Erect shrub without spines, 30–100 cm. Branches opposite at almost every node. Leaves of 3 leaflets, opposite; leaflets 5–20 × 2–4 mm, oblanceolate, silky beneath, almost hairless above. Flowers almost opposite and almost stalkless in terminal clusters of 4–12; lowermost flowers with simple, usually shortly 3-fid, translucent bracts much shorter than the flowers; bracteoles 1–3 mm. Calyx 4–6 mm, lips about as long as the tube. Standard 8–14 mm, broadly ovate, as long as or slightly longer than the keel, hairless or with a central ridge of silky hairs. Pod *c.* 5 × 3 mm, with a short curved beak, silky hairy. Seeds 2 or 3. *S Alps to C Italy & W former Yugoslavia, very locally to SW Romania & C Greece.* H3. Summer.

21. G. ephedroides de Candolle (*Cytisanthus ephedroides* (de Candolle) Gams). Illustration: Pignatti, Flora d'Italia **1**: 642 (1982).

Erect shrub without spines, 50–100 cm. Leaves (at least the lower) of 3 leaflets, falling early; leaflets 4–15 × 2–3 mm, silky on both surfaces. Flowers alternate or almost opposite in loose racemes; lowermost bracts of 3 leaflets, the uppermost simple. Calyx 3–6 mm, silky-hairy, lips and upper teeth about as long as tube. Standard 7–10 mm, broadly ovate or diamond-shaped, *c.* two-thirds as long as the keel, sparsely silky. Pod *c.* 1 cm × 5 mm, with a curved beak, silky-hairy. Seeds 2–3. *S Italy, Sardinia, Sicily.* H5. Early summer.

103. ADENOCARPUS de Candolle
H. Ern

Unarmed shrubs with alternate branches. Leaves divided into 3 leaflets, the leaves sometimes on short shoots. Flowers in terminal racemes or clusters. Calyx tubular, 2-lipped, sometimes with glandular tubercles, the upper and lower lips prominent, the upper deeply 2-fid, the lower with 3 distinct teeth. Corolla orange-yellow. Stamens 10 with their filaments united into a tube. Pod oblong, dehiscent, covered with glandular warts. Seeds numerous, oblong, each with a small appendage along one of the shorter sides.

A genus of about 15 species, most of them from mountain areas of N Africa and southern Spain; however, one species

(*A. mannii*) is widely distributed in the high mountains of tropical Africa, and 3 are restricted to the Canary Islands.

Most *Adenocarpus* are very showy and can be highly recommended as ornamental shrubs, both for the deep yellow flowers and for their interesting foliage. In spite of this, only a few species are in horticultural use. They all need well-drained soil and full exposure to sunlight.

Literature: Rivas-Goday, S. & Fernandez-Galiano, E., *Adenocarpus hispanicus* (Lamk.) DC. como planta ornamental, *Anales del Instituto Botanico A.J. Cavanilles* **12** (2): 1–7 (1954); Vicioso, C., Genisteas Españolas **2**: 232–252 (1955); Gibbs, P.E., A revision of the genus *Adenocarpus*, *Boletim Sociedade Broteriana* **41**: 67–121 (1967).

1a. Leaflets 1.5–3.5 × 1–1.5 cm, broadly ovate or elliptic   **1. anagyrifolius**
 b. Leaflets less than 1.5 cm, or if more then very narrowly elliptic or very narrowly oblanceolate   2
2a. Leaf-stalks 4–15 mm   3
 b. Leaf-stalks 1–4 mm   5
3a. Leaflets 1–2 mm wide, markedly rolled upwards at the margins and appearing almost linear   **2. decorticans**
 b. Leaflets 3–7 mm wide, broadly elliptic to oblanceolate, margins only slightly rolled upwards   4
4a. Flower-stalks 1–7 mm; standard 1–1.5 cm   **3. complicatus**
 b. Flower-stalks 7–15 mm; standard 1.5–2.5 cm   **4. hispanicus**
5a. Calyx with glandular papillae; standard more or less hairless   **5. viscosus**
 b. Calyx without glandular papillae; standard covered with silky hairs   **6. foliolosus**

1. A. anagyrifolius Cosson & Balansa. Illustration: *Boletim Sociedade Broteriana* **41**: 114, map 6 (1967).

Erect shrub to 2 m. Leaf-stalks 1–2 cm; leaflets 1.5–3.5 × 1–1.5 cm, broadly ovate or elliptic, apex with a short mucro, lower surface with sparse adpressed hairs, upper surface more or less hairless. Flowers in loose racemes; flower-stalks 2–4 mm. Calyx 5–7 mm, covered with adpressed to somewhat spreading hairs. Standard *c.* 1.2 cm, broadly ovate, covered with short adpressed hairs. Pods 3–4.5 cm × 4–6 mm, covered with dense glandular papillae. *Morocco (Haut & Moyen Atlas).* H5. Early summer.

2. A. decorticans Boissier (*A. boissieri* Webb; *A. speciosus* Pomel). Illustration: *Botanical Magazine*, n.s., 48 (1949); *Boletim Sociedade Broteriana* **41**: 107, map 4 (1965); Brickell (ed.), RHS A–Z encyclopedia of garden plants, 77 (2003).

Erect shrub to 3 m, sometimes becoming tree-like. Leaflets 9–18 × 1–2 mm, very narrowly elliptic and with margins markedly rolled upwards, silky-hairy on both surfaces. Flowers in racemes; stalks 4–8 mm. Calyx *c.* 8 mm, densely silky-hairy, not glandular. Standard *c.* 1.5 cm, broadly ovate, silky-hairy. Pod 2–6 cm × 8–10 mm, oblong, densely covered with glandular papillae. *S Spain, Morocco, Algeria.* H5. Early summer.

3. A. complicatus (Linnaeus) Grenier & Godron (*A. bivonae* (Presl) Presl; *A. divaricatus* Sweet; *A. graecus* Grisebach; *A. intermedius* de Candolle; *A. parvifolius* (Lamarck) de Candolle). Illustration: *Botanical Magazine*, 1387 (1811); Lewis et al. (eds.), Legumes of the world, 289 (2005).

Erect shrub to 2 m or more; twigs and leaves sparsely to densely silky-hairy or with spreading hairs. Leaflets 5–25 × 2–7 mm, oblanceolate. Flower-stalks 1–7 mm. Calyx 5–7 mm, with or without glandular papillae, more or less hairless or silky-hairy. Standard 1–1.5 cm, silky hairy. Pod 1.5–4.5 cm × 4–6 mm, narrowly oblong, covered with glandular papillae. *Mediterranean area, extending to NW France.* H3. Early summer.

A variable and widespread species divided into several subspecies, though it is uncertain which of these is in cultivation. Hardiness depends very much on the origin of the plants or seeds; specimens from higher elevations of C Spain (province of Segovia) are hardy enough to produce seed in Berlin.

4. A. hispanicus (Lamarck) de Candolle. Illustration: Vicioso, Genisteas Españolas **2**: 234 (1955); *Boletim Sociedade Broteriana* **41**: 107, map 4 (1965).

Erect shrub to 2 m. Leaflets 1.5–3 cm × 3–8 mm, oblanceolate, acuminate, silky on both surfaces. Flowers in congested racemes; stalks 7–15 mm. Calyx 8–12 mm, with somewhat spreading hairs and glandular papillae, especially on the teeth. Standard 1.5–2.5 cm, broadly ovate, silky-hairy. Pod 2–5 cm × 8–10 mm, oblong, densely covered with glandular papillae. *Western half of the Iberian*

Peninsula & one small area in the coastal ranges of Morocco. H4. Summer.

Subsp. **argyrophyllus** (Rivas Goday) Rivas Goday, which is found in the Moroccan locality cited above, has leaves densely silvery white-hairy above, the calyx with few or no glandular papillae and the standard densely downy. This is a highly decorative plant which may prove somewhat more tender than subsp. **hispanicus**.

5. A. viscosus (Willdenow) Webb & Berthelot. Illustration: Bramwell & Bramwell, Wild flowers of the Canary Islands, f. 164 (1974).

Erect, densely leafy and somewhat sticky shrub. Leaves on short shoots, stalks 2–5 mm, leaflets 3–7 × 1.5–2.5 mm, narrowly elliptic, margins conspicuously rolled upwards, both surfaces with spreading hairs, the lower densely so. Flowers in terminal racemes. Calyx 7–8 mm, with long hairs and glandular papillae. Standard *c.* 1.2 cm, broadly ovate, more or less hairless or with sparse hairs towards the apex. Pod 2–3.5 cm × 5–6 mm, with glandular papillae and sparse hairs. *Canary Islands (La Palma, Tenerife).* G1. Spring.

6. A. foliolosus (Aiton) de Candolle. Illustration: Bramwell & Bramwell, Wild flowers of the Canary Islands, f. 165 (1974).

Erect, densely leafy shrub. Leaves usually dense on short shoots, stalks 1–3 mm; leaflets 3–6 × 1.5–2.5 mm, obovate or narrowly lanceolate, margins somewhat rolled upwards, lower surface with adpressed or spreading hairs, upper hairless. Flowers in terminal racemes; stalks 4–12 mm. Calyx 7–8 mm, usually with dense, long hairs. Standard *c.* 1.2 cm, broadly ovate, with dense silky hairs. Pod 1.2–4 cm × 4–5 mm, narrowly oblong, with glandular papillae (sometimes very sparse) and sparse hairs, containing 3–5 seeds. *Canary Islands (Gomera, Gran Canaria & Tenerife).* G1. Spring.

104. ULEX Linnaeus
H.S. Maxwell
Spiny, dense shrub. Leaves usually alternate, to 1.5 cm, ternate, linear, sharply pointed when young, reduced to green spines or scales when mature; stipules absent. Flowers fragrant, 1.5–2.5 cm, solitary or few in small axillary clusters or racemes, produced in the leaf axis of the previous years growth,

shortly stalked, golden-yellow; bracteoles 2, small, below flower. Calyx persistent, 2-lipped, lower lip with 3 small teeth, upper lip with 2 small teeth. Corolla persistent, standard ovate, wing and keel obtuse. Stamens 10, all united, alternating in 2 lengths; style slightly curved. Fruit a dehiscent, small, hairy, broadly-ovate to linear-oblong, explosive pod containing 1–6 seeds.

A genus of about 20 species from Europe and N Africa but now naturalised in the mid-United States. Only 3 species are in general cultivation in Europe. The plants grow best in poor, sandy, acidic soil in full sun, on grassland or scrub. They are often used as windbreaks in coastal areas or as ground cover for dry banks. Propagation can be from cuttings or seed but they do tend to hybridise in the wild. They can be grown in pots but once planted out they do not transfer well; pruning should be done after flowering.

1a. Bracteoles 2–5 mm; spines rigid, deeply grooved; flowers 2–2.5 cm **1. europaeus**
 b. Bracteoles 0.5–0.8 mm; spines softer, slightly grooved; flowers to 1.6 cm **2**
2a. Calyx 1–1.5 cm; standard 1.2–2 cm; fruit 0.8–1.5 cm with 1–2 seeds **2. gallii**
 b. Calyx 5–10 mm; standard 7–15 mm; fruit 5–8.5 mm with 3–6 seeds **3. minor**

1. U. europaeus Linnaeus (*U. hibernicus* misapplied; *U. strictus* (Mackay) Webb). Illustration: Ross-Craig, Drawings of British plants **7**: pl. 5 (1954); Fitter & Fitter, The wild flowers of Britain and Northern Europe, 121, 1 (1974); Csapody & Tóth, A colour atlas of flowering trees and shrubs, pl. 64, 140 (1982); Keble Martin, The new concise British flora, pl. 21 (1982).

Densely spiny glaucous shrub to 2.5 m; stems erect or ascending, sparsely hairy to densely felted with reddish-brown to black hairs. Spines to 2.5 cm, rigid, deeply grooved. Flowers 2–2.5 cm, coconut smelling, stalks to 6 mm, with adpressed hairs; bracteoles 2–5 mm; calyx persistent, 1–2 cm, with spreading hairs, shorter than the corolla; wing petals straight, longer than keel; standard even longer, 1.2–1.8 mm. Fruit to 1–2 cm, black with dense shaggy grey or brownish black hairs; seeds 2 or 3, dehiscing in summer. *W & C Europe.* H2. Throughout year.

'Flore Pleno' (*U. europaeus* 'Plenus'; *U. europaeus* var. *flore-pleno* G. Don;

U. europaeus forma *plenus* Schneider). Illustration: Bean, Trees & shrubs hardy in the British Isles **4**: pl. 92 (1980). Has a more compact habit and longer lasting double flowers. Propagation has to be from cuttings, as no seed is set.

2. U. gallii Planchon. Illustration: Ross-Craig, Drawings of British plants **7**: pl. 6 (1954); Fitter, Fitter & Blamey, The wild flowers of Britain and Northern Europe, 121, 1b (1974); Keble Martin, The new concise British flora, pl. 21 (1982).

Densely spiny, dark green shrub, 1.5–2 m; stems ascending but often prostrate in coastal areas, branches hairy. Spines slightly grooved, to 2.5 cm. Flowers to 1.6 cm; stalks 3–5 mm, with adpressed hairs, bracteoles 0.5–0.8 mm; calyx finely downy, 1–1.5 cm, smaller than standard and keel; wing petals curved, longer than keel; standard 1.2–2 cm. Fruit 8–15 mm; seeds 1 or 2, dehiscing in spring. *W Europe.* H3. Summer–autumn.

3. U. minor Roth (*U. nanus* Symons). Illustration: Ross-Craig, Drawings of British plants **7**: pl. 7 (1954); Fitter, Fitter & Blamey, The wild flowers of Britain and Northern Europe, 121, 1a (1974); Keble Martin, The new concise British flora, pl. 21 (1982).

Dense dwarf shrub 5–150 cm, with close habit, usually prostrate; stems with brownish hairs, young twigs and spines not glaucous. Spines slender, straight or slightly curved, soft, to 1.5 cm, very slightly grooved, with shaggy hairs at base. Flowers to 1.5 cm, paler yellow than *U. europaeus*; stalks 3–5 mm, with adpressed hairs; bracteoles 0.6–0.8 mm; calyx 5–10 mm, slightly downy when young, becoming hairless; wing petals straight, as long as or longer than keel; standard 7–15 mm, longer than calyx. Fruit 5–8.5 mm, shaggy haired, seeds 3–6, dehiscing in spring. *W Europe.* H3. Late summer–autumn.

---

154. LIMNANTHACEAE

Herbs. Leaves alternate or basal, divided, exstipulate; ptyxis conduplicate. Flowers solitary, bisexual, actinomorphic. Calyx, corolla and stamens hypogynous. Calyx of 3–5 free sepals. Corolla of 3–5 free petals. Stamens 6–10. Ovary of 3–5 carpels, bodies of carpels free; style 1, divided above into as many stigmas as there are carpels;

ovules 1 per carpel, ascending. Fruit a group of nutlets.

Two genera with 6 species, native to western North America. Species of *Limnanthes* are widely cultivated.

1. LIMNANTHES R. Brown

C.J. King

Sepals usually 5, sometimes 4 or 6; petals usually 5, sometimes 4 or 6, with a U-shaped band of hairs on the claw. Ovary of 5 carpels; stigmas 5, capitate.

A genus of 7 species from western North America. Some recent authors include this family in the Geraniaceae. Only one species is in general cultivation, and this grows best in a moist, moderately fertile soil in an open, sunny situation. Propagation is by seed sown where the plants are to flower, either in spring, or in autumn to flower the following year.

Literature: Clifton, R., Geraniales species checklist vol. 5 part 2, Geraniaceae Tribes Limnantheae and Oxalideae (excluding Oxalis) (2004).

1. L. douglasii R. Brown. Illustration: *Botanical Magazine*, 3554 (1837); Hay & Synge, Dictionary of garden plants in colour, 41 (1969); Hay & Beckett, Reader's Digest encyclopaedia of garden plants and flowers, 409 (1978); Brickell (ed.), RHS gardeners' encyclopaedia of plants & flowers, 280 (1989); Brickell (ed.), RHS A–Z encyclopedia of garden plants, 635 (2003).

Stem to 40 cm, branched from the base. Leaves 5–12 cm, hairless, with 3–11 pinnate divisions, each division lobed or toothed. Flower-stalks 5–10 cm. Flowers 2–3 cm across, fragrant. Sepals 6–10 mm, lanceolate, hairless. Petals 8–16 mm, wedge-shaped, deeply notched at apex, yellow tipped with white. Stamens 5–8 mm. Nutlets 2.5–4 mm, dark. *USA (California, Oregon).* H1. Summer.

A large-flowered variant is cultivated as 'Grandiflora', and several colour variants are treated as varieties. These include var. **sulphurea** Mason, petals yellow; var. **nivea** Mason, petals white, and var. **rosea** (Hartweg) Mason, petals white with pink veins.

155. OXALIDACEAE

Herbs, shrubs or trees. Leaves alternate or all basal, pinnate, palmate or trifoliolate, exstipulate. Inflorescence various. Flowers bisexual, actinomorphic. Calyx, corolla and stamens hypogynous. Calyx of 5 free or united sepals. Corolla of 5 free petals which are contorted in bud. Stamens 10, filaments sometimes united into a tube. Ovary of 5 united carpels; ovules 1–more, axile; styles 5, free. Fruit a capsule, often explosive.

A family of 3 genera and about 900 species, widely spread throughout the world, mainly tropical or subtropical. Some recent authors include this family in the Geraniaceae. Several species are cultivated: *Averrhoa* mainly for its edible fruit, *Oxalis* and *Biophytum* as ornamentals.

Literature: Knuth, R., Oxalidaceae, in Engler, A. (ed) Das Pflanzenreich **95** (1930); Clifton, R., Geraniaceae species checklist, **5**(2), Geraniaceae Tribes Limnantheae and Oxalideae (excluding Oxalis) (2004).

1a. Leaves palmate **3. Oxalis**
  b. Leaves pinnate 2
2a. Annual or perennial herbs; fruit a dry, dehiscent capsule **1. Biophytum**
  b. Evergreen trees; fruit a fleshy berry **2. Averrhoa**

1. BIOPHYTUM de Candolle

C.J. King

Annual or perennial herbs, with simple or branched stems, sometimes woody at the base. Leaves in whorls or clusters, even-pinnate, the axis ending in a bristle. Leaflets many, opposite or nearly so, oblong to circular, stalkless or with only very short stalks. Flowers in umbels or spherical clusters. Bracts small, persistent, forming an involucre at the base of the flower-stalks. Corolla white, pink, yellow or orange. Styles notched or bifid at apex. Capsule ovoid or oblong, the carpels separating at maturity.

A tropical genus of about 50 species. Some have leaflets sensitive to touch, one of these, *B. sensitivum*, being commonly grown as a novelty for this reason. The species need to be grown in a mixture of loam and peat in a greenhouse with a minimum temperature of 15 °C. The soil should be kept fairly moist and the plant should be protected from strong sunlight. Abundant seeds ejected from the capsules germinate readily in the surrounding soil.

1a. Stem simple; inflorescence many-flowered **1. sensitivum**
  b. Stem branched; flowers solitary or in pairs **2. proliferum**

1. B. sensitivum (Linnaeus) de Candolle (*Oxalis sensitiva* Linnaeus). Illustration: *Edwards's Botanical Register* **31**: t. 68 (1845); Heywood, Flowering plants of the world, 209 (1978); Everett, New York Botanical Gardens illustrated encyclopedia of horticulture **2**: 424 (1981); Graf, Exotica, edn. 11, 1845, 1848 (1982).

Plant more or less hairy; stem erect or ascending to 25 cm. Leaves to 16 cm, clustered at ends of stems; leaflets in 6–21 pairs, 4–18 × 2.5–8 mm. Inflorescences 3–12 cm. Bracts 2–4 mm, subulate, rigid. Flower-stalks to 2.5 mm. Sepals 4–8 mm, linear-lanceolate, rigid, with 3–5 veins. Petals 5–7 mm, oblong-obovate, yellow, sometimes with pale purple markings. Capsule *c.* 3 mm, seeds 2–4 per cell. *Tropical Africa & Asia to the Philippines.* G2. Summer.

Leaflets react rapidly to stimulation in warm conditions.

2. B. proliferum (Arnott) Wight (*Oxalis prolifera* Arnott). Illustration: Bond, Wild flowers of the Ceylon hills, 27 (1953).

Plant more or less hairy; stem prostrate and rooting at nodes or ascending to 25 cm. Leaves to 2.5 cm, in whorls at ends of stems and at nodes; leaflets in 5–10 pairs, *c.* 4 × 2 mm. Inflorescences *c.* 5 mm. Bracts *c.* 1 mm, subulate. Flower-stalks *c.* 1 cm. Sepals to 4 mm, lanceolate. Petals *c.* 6 mm, yellow. Capsule shorter than the sepals. *Sri Lanka.* G2.

2. AVERRHOA Linnaeus

M.F. Watson

Evergreen trees. Leaves alternate, pinnate, each with a terminal leaflet, somewhat sensitive to light; leaflets many, alternate to almost opposite, entire. Flowers very small, in short clusters, sometimes on naked branches or beneath leaves. Petals white, red or purple. Fruit a long, fleshy berry, with 5 deep ridges, star-shaped in cross-section.

A genus of 2 species, native to tropical Asia, and now introduced into the Americas. In Europe they are grown as hot-house plants; both species require humid atmospheric conditions and shading from direct sunlight. A growing medium of equal parts peat and loam gives best results. Propagate by softwood cuttings, budding, grafting or layering.

Literature: Lourteig, A., Flora of Panama, Part 4, Oxalidaceae. *Annals of the*

Missouri Botanical Garden **67**: 823–850 (1980).

1a. Bark grey-brown; leaflets 3–8, broadly ovate **1. bilimbi**
  b. Bark red-brown; leaflets 7–20, narrowly elliptic **2. carambola**

1. A. bilimbi Linnaeus. Illustration: Das Pflanzenreich **95**: 419 (1930).

Tree to 15 m with rather erect branches and grey-brown bark; new branches and leaf-axes with dense yellow or red-brown hairs. Leaves to 30 cm, with 7–20 pairs of leaflets; leaf-stalks 3–17 cm; leaflets narrowly elliptic, asymmetric, somewhat hairy, 2.5–15 × 1–5 cm. Flower clusters *c.* 8 × 4 cm, usually solitary in leaf-axils. Flowers purple, to 1.5 cm across, *c.* 20 per cluster. Fruit to 7.5 × 2.5 cm, ridges rounded. *S Asia*. G2. Summer.

2. A. carambola Linnaeus. Illustration: Graf, Exotica, edn. 8, 1336 (1979); *Annals of the Missouri Botanical Garden* **67**: 825 (1980).

Tree to 25 m, with spreading branches and red-brown bark. Leaves 16–20 cm, with 3–8 pairs of leaflets; leaflets broadly ovate, asymmetric, mainly hairless, 1.5–3 × 1–2 cm. Flowers to 1 cm across, in clusters of 15–25; clusters *c.* 4 × 2 cm, 1–3 per leaf-axil; petals violet with a white border. Fruit yellow, 5–12.5 × 5–6 cm, ridges acute. *S Asia*. G2. Summer.

Frequently grown for its fruit in the tropics.

3. OXALIS Linnaeus
M.F. Watson

Annual or perennial, stemmed or stemless herbs and shrubs, often with tubers or bulbs; very rarely aquatic. Leaves palmate; leaflets 3–20 or more (rarely 1), often folding down at night. Flowers with differing style and stamen lengths, axillary, often in cymes or umbel-like cymes, sometimes solitary; petals white, pink, red or yellow; filaments fused into a tube. Fruit a capsule; seed enclosed in a fleshy aril which springs the seed from the capsule at maturity.

A genus of up to 800 species centred in southern Africa and South America, but spread worldwide. Several species are cultivated as ornamentals in Europe and some have become common weeds. *O. tuberosa* is the only species commonly grown for its edible tubers; other species are eaten occasionally. Most will thrive in sun or partial shade in a well-drained compost or soil with added peat or leafmould. Several of the bulbous species are particularly suited to the rock garden. Some species are susceptible to fungal rust. Propagation is by bulbs, seed or occasionally cuttings.

Literature: Jacquin, N.J., Oxalis, (1794); Salter, T.M., The genus *Oxalis* in South Africa, *Journal of South African Botany Supplementary* **1**: 1–355 (1944); Young, D.P., Oxalis in the British Isles, *Watsonia* **4**: 51–69 (1958); Denton, M.F., A monograph of Oxalis, section *Ionoxalis* (Oxalidaceae) in North America, *Michigan State University Biological Series* **4**(10): 455–615 (1973); Lourteig, A. Oxalidaceae, *Flora illustrada Catarinense* (1983); Clifton, R., Geraniales species checklist, vol. 5 part 1, Oxalis checklist of species with some descriptions and cultural notes, edn 2 (2004).

1a. Stem visible, or a very shallowly buried, non-tuberous, creeping rhizome 2
  b. Stem not visible, leaves arising below soil level directly from bulb or tuberous woody rhizome 24
2a. Stem visible for less than 3 cm, or a slender rhizome 3
  b. Stem visible for more than 5 cm 11
3a. Plant with a non-tuberous rhizome 4
  b. Plant with a bulb or tuberous woody rhizome 6
4a. Plant less than 5 cm; leaflets bronze-green, hairless, less than 5 × 5 mm **15. magellanica**
  b. Plant more than 5 cm; leaflets green, somewhat hairy, more than 7 × 7 mm 5
5a. Plant to 20 cm, robust; leaflets 2.5 × 2.7 cm or more; flowers pale lilac or darker, rarely white **17. oregana**
  b. Plant less than 12 cm, slender; leaflets less than 1.5 × 2 cm; flowers white, with pale pink veins, occasionally pink **16. acetosella**
6a. Plant with a tuberous, woody rhizome 7
  b. Plant with a bulb 8
7a. Rhizomes globular, in tight clusters; leaflets dark-spotted, particularly around notch **12. rubra**
  b. Rhizomes globular to cylindric with constrictions, not in tight clusters; leaflets with usually only marginal dots **13. articulata**
8a. Plant less than 12 cm; flowers solitary; sepals without apical warts 9

  b. Plant usually more than 12 cm; flowers in umbels of 2 or more; sepals with 2 orange apical warts 10
9a. Leaflets triangular to ovate, not notched, usually hairless; flowers pink to violet **35. depressa**
  b. Leaflets heart-shaped, densely hairy; flowers yellow **33. melanosticta**
10a. Leaf-stalks 2–30 cm; leaflets larger than 1.5 × 2 cm; flowers in umbels of 3–10; corolla purple-violet **31. purpurata**
  b. Leaf-stalks 2–5 cm; leaflets to 1 × 1.5 cm; flowers in umbels of 2–4; corolla pale violet, rarely white **32. caprina**
11a. Plant bulbous; flowers solitary 12
  b. Plant not bulbous, annual or perennial arising from tubers or taproot; flowers solitary or not 14
12a. Leaves almost stalkless; leaflets strap-shaped; whole plant hairy **39. hirta**
  b. Leaf-stalks at least 1 cm; leaflets not strap-shaped; plant mostly hairless 13
13a. Leaflets deeply divided to at least the middle, lobes narrow; flowers purple-red; sepals streaked with purple-black warts **37. bifida**
  b. Leaflets heart-shaped, rounded; flowers white or very pale lilac; sepals with converging orange apical warts **36. incarnata**
14a. Stems succulent, fleshy, at least 5 mm thick, becoming decumbent with age; flowers yellow 15
  b. Stems not as above; flowers yellow or pink 18
15a. Plant with swollen tubers (4 × 3 cm) at or below soil surface, densely hairy throughout, somewhat purple-tinged **1. tuberosa**
  b. Plant without swollen tubers, usually hairless, purple colouring only on senescent parts 16
16a. Leaves succulent, shiny above, frosted beneath with watery papillae; outer sepals triangular with hastate bases **10. megalorrhiza**
  b. Leaves fleshy, but not as above; sepals oblong, tapering 17
17a. Leaf-stalks succulent, dilated in centre and tapering towards the ends; leaflets 5–10 × 4–7 mm, reddening with age and falling early, leaving stalks; flowers in cymes of 5–7 on stalks to 10 cm **5. herrerae**

b. Leaf-stalks somewhat succulent but not tapering; leaflets larger, to 12 × 9 mm; flowers in cymes of 9–16 on stalks to 30 cm	**11. peduncularis**

18a. Plant erect	19

b. Plant prostrate or ascending	22

19a. Bushy perennial with persistent red stems	**3. vulcanicola**

b. Annual or single-stemmed, sparingly branched perennial	20

20a. Hairy perennial; leaflets heart-shaped with triangular lobes	**2. ortgiesii**

b. Hairless annual; leaflets heart-shaped with round lobes	21

21a. Stem 4–6 cm, unbranched; flowers yellow in umbel-like cymes of 4–14	**4. valdiviensis**

b. Stems 10–35 cm, branched; flowers pink (rarely white) in cymes of 1–3	**6. rosea**

22a. Leaflets 5–15 × 8–20 mm or more, green or purple; flowers in umbels of 2–6; capsules deflexed	**7. corniculata**

b. Leaflets usually less than 7 × 7 mm, green; flowers solitary	23

23a. Stems rather woody; leaf-stalks 2–4 cm; leaves hairy; capsule cylindric, *c.* 11 mm, not deflexed	**9. chrysantha**

b. Stems not woody; leaf-stalks less than 2 cm; leaves hairless above; capsule spherical, to 6 mm, deflexed	**8. exilis**

24a. Leaflets 5 or more	25

b. Leaflets 3 or 4	30

25a. Leaflets narrow, finger-like, 5–12, rarely less; flowers yellow	**38. flava**

b. Plant not as above	26

26a. Plant 15–30 cm; leaflets wedge- or strap-shaped; flowers in umbels of 9 or more, sepals with apical warts	**27. lasiandra**

b. Plant less than 15 cm; leaflets heart-shaped; flowers solitary or in umbels of 2 or 3 flowers; sepals without warts	27

27a. Leaflets 8–12, glaucous green with purple, crinkled margins; flowers sweet scented, solitary	**21. laciniata**

b. Leaflets 9–22, glaucous blue to green without crinkled margins; flowers unscented, in umbels of 1–3 flowers	28

28a. Plant with large tuberous base; leaf-stalks red-brown, leaflets silver-grey, hairless; umbels with 1–3 flowers	**18. adenophylla**

b. Plant with slender creeping rhizome; leaves not as above; flowers solitary	29

29a. Rhizome covered in white scales; leaflets 9–20, not divided past the centre, glaucous blue with short hairs	**20. enneaphylla**

b. Rhizome covered in orange scales; leaflets 10–14, divided to the base, glaucous grey, densely hairy	**19. patagonica**

30a. Leaflets 4 (rarely 3), often with a purple band or sector at the base	**28. tetraphylla**

b. Leaflets 3	31

31a. Flowers yellow	32

b. Flowers not yellow	36

32a. Leaflets strap-shaped, finger-like; flowers solitary; sepals with indistinct apical warts	**38. flava**

b. Leaflets heart-shaped to diamond-shaped; if sepals have warts then flowers in umbel-like cymes	33

33a. Plant 20–40 cm; leaflets cordate, flat, hairless; flowers in umbel-like cymes of 3–20 flowers	**29. pes-caprae**

b. Plant to 18 cm; leaflets not as above; flowers solitary	34

34a. Leaflets diamond-shaped to circular, dark green above, purplish beneath	**34. purpurea**

b. Leaflets heart-shaped to rounded, usually notched, green	35

35a. Plant very hairy; leaflets heart-shaped to rounded, rather thick with dense, white-hairy margins	**33. melanosticta**

b. Plant mainly hairless; leaflets heart-shaped, thin, sparsely hairy	**22. lobata**

36a. Flowers all solitary	37

b. At least some inflorescences with 2 or more flowers	39

37a. Leaflets strap-shaped, finger-like; sepals with indistinct warts	**38. flava**

b. Leaflets rounded to diamond-shaped or obovate; sepals without warts	38

38a. Leaf-stalks 2–8 cm, with white hairs; leaflets diamond-shaped to ovate, dark green above, purplish beneath	**34. purpurea**

b. Leaf-stalks less than 2 cm, hairless; leaflets rounded to triangular-ovate, grey-green	**35. depressa**

39a. Leaflets strap-shaped, reversed triangular or widely triangular in outline, sometimes deeply divided	40

b. Leaflets rounded, heart-shaped, rarely divided below the middle	43

40a. Leaflets strap-shaped to reversed triangular, entire or shallowly notched, each usually with a purple band or sector at base	**28. tetraphylla**

b. Leaflets not as above	41

41a. Leaflets lobed to about the middle, lobes obovate; flowers pale violet, rarely white, in umbels of 2–4	**32. caprina**

b. Leaflets not lobed to middle or if so, lobes linear; flowers generally darker pink-violet, in umbels of 3–32	42

42a. Bulb-scales with 5–11 veins; leaflets broadly triangular, rarely divided below the middle, to 7 × 7 cm; flowers in umbels of 6–32	**25. latifolia**

b. Bulb-scales with 3 veins; leaflets triangular in outline, deeply divided to at least the middle, less than 3 × 3 cm; flowers in umbels of 3–10	**26. drummondii**

43a. Plants with woody rhizomes	44

b. Plant with scaly bulbs	46

44a. Leaflets glaucous green, not spotted, 2.5–4 cm; flowers white to pale violet; sepals without apical warts	**14. trilliifolia**

b. Leaflets green, usually with dark spots, 1.8–2.5 cm; flowers red-mauve, rarely white; sepals with apical warts	45

45a. Rhizome globular in tight clusters; leaves spreading, dark-spotted, particularly around notch; inflorescence an umbel-like cyme	**12. rubra**

b. Rhizome globular to cylindric with constrictions, not tightly clustered; leaves erect with only marginal dots; inflorescence usually cyme-like	**13. articulata**

46a. Bulb large, rounded to oblong, at least 8 mm across, producing few bulbils, or several on underground runners	47

b. Bulb small, round, usually less than 8 mm across, producing numerous stalkless bulbils	49

47a. Bulb rounded, small, *c.* 1 cm across; plant hairless; orange spots around leaf-notch; flowers lavender-pink or paler	**24. violacea**

b. Bulb oblong, more than 1 cm across; plant somewhat hairy; leaves unspotted; flowers purple-red	48

48a. Bulb without contractile root, fleshy underground runners producing apical bulbils; leaves thin, dark green above, dark purple beneath (variable) **31. purpurata**

b. Bulb with fleshy white contractile root and no runners; leaves rather leathery, bright green, occasionally purple beneath **30. bowiei**

49a. Bulbils produced on horizontal underground runners, sometimes very short; bulb-scales with 5–11 veins; leaflets broadly triangular to rounded; inflorescence umbel-like on hairless stalks **25. latifolia**

b. Bulbils stalkless, bulb-scales with 3 veins; leaflets broadly heart-shaped, rounded; inflorescence an irregular cyme on spreading, hairy stalks **23. corymbosa**

1. O. tuberosa G.F. Molina (*O. crenata* Jacquin). Illustration: Masefield et al., The Oxford book of food plants, 179 (1969).

Erect to decumbent, succulent-stemmed perennial to 25 cm; rhizome branched, tips swollen into tubers *c.* 4 × 3 cm, covered in small triangular scales; stems fleshy, with dense adpressed hairs, green-purple, leafy in the upper part, to 30 × 1 cm. Leaf-stalks 7–10 cm, hairless or sparsely hairy; leaflets 3, broadly heart-shaped, green or purple-tinged (particularly beneath), with dense adpressed hairs, rather thick, to 2.5 × 2.2 cm. Inflorescences 15–17 cm; flowers to 2 cm across, in umbels of 5–8, arising from upper leaf-axils; sepals without warts; petals yellow. *Colombia.* H2. Summer.

A long-cultivated root crop, Oca, of the high Andes. Three colours of tuber have been selected; yellow, white and red. The red and yellow types have lost the ability to flower.

2. O. ortgiesii Regel. Figure 112(6), p. 420. Illustration: Everard & Morley, Wild flowers of the world, pl. 171 (1970); Graf, Exotica, edn. 8, 1332 (1976); Beckett, The RHS encyclopaedia of house plants, 372 (1987).

Persistent-stemmed, erect perennial to 45 cm; rhizome horizontal, brown-red, to 20 cm × 5 mm, producing a single, sparsely branched, purplish, hairy stem 2–40 cm. Leaves crowded near stem apex; stalks erect to spreading, densely hairy, 4–8 mm; leaflets 3, heart-shaped, deeply divided into 2 large triangular lobes at tips, olive-green above, red-purple beneath, both surfaces hairy, to 6 × 3.5 cm. Flowers borne in

many-flowered cymes from the upper leaf-axils on stems to 30 cm; flowers stalkless and falling early, leaving tentacle-like stalks; sepals without warts, purple-edged; corolla *c.* 2.5 cm across, lemon-yellow with darker veins. *Peru.* G2. Spring–winter.

3. O. vulcanicola D. Smith (*O. siliquosa* Anon.). Illustration: Graf, Exotica, edn. 8, 1336 (1976); Everett, New York Botanical Gardens illustrated encyclopedia of horticulture 7: 2454 (1981).

Bushy perennial with loose, succulent, persistent stems, or spreading and mat-forming with age; stems 20–70 cm, much branched, hairy above, 2–3 mm thick, reddish. Leaves generally concentrated from the middle of the stem upwards, very dense at apex; stalks 3–7 cm; leaflets 3, heart-shaped, green, flushed red above, magenta beneath, almost hairless above, hairy beneath, to 4 × 2 cm. Inflorescences 4–8 cm, flowers 1–1.5 cm across, in umbel-like cymes of 4–7; sepals without warts; petals yellow with purple-red veins. *El Salvador to Panama.* G1. Summer–autumn.

4. O. valdiviensis Barnéoud (*O. valdiviana* Anon.).

Compact, erect-stemmed annual, 10–25 cm, hairless with thick fleshy taproots. Stems densely leafy, unbranched. Leaf-stalks erect, 4–14 cm; leaflets 3, broadly heart-shaped with a narrow notch, 1.2–2 × 1–2 cm, pale green, thin. Flowers in forked cymes of 4–14, borne on stems 8–18 cm, from upper leaf-axils; sepals often with red margins and without apical warts; corolla yellow, usually with brown veining, 1–1.5 cm across. Capsule *c.* 6 mm, rounded, drooping. *Chile.* H2. Summer.

Readily sets seed.

5. O. herrerae R. Kunth (*O. peduncularis* misapplied; *O. succulenta* misapplied). Illustration: Graf, Exotica, edn. 8, 1332, 1334 (1976).

Perennial shrublet, 10–30 cm; branches to 8 mm thick, reddish, becoming browner with age, hairless. Leaf-stalks 2–5 cm, to 3 mm thick at widest point, spreading, fleshy, dilated in the middle, tapering to the ends; leaflets 3, broadly heart-shaped, 5–10 × 4–7 mm, green, reddening with age, falling early, leaving leaf-stalks. Cymes on erect stems to 10 cm, with 5–7 flowers; flower-stalks to 5 mm, unequal; sepals without warts, reddened at edges; petals 1–1.5 cm across, yellow with red veins. *Peru.* G1. Summer.

6. O. rosea Jacquin (*O. racemosa* Savigny; *O. floribunda* misapplied; *O. rubra* A. Saint-Hilaire 'Delicata' misapplied). Illustration: *Botanical Magazine*, 2415 (1823), 2830 (1830).

Erect annual, 20–40 cm; stems hairless, often reddish at base, much branched and leafy throughout. Leaf-stalks to 3 cm, spreading; leaflets 3, heart-shaped, to 11 × 11 mm, pale green, occasionally reddened beneath. Inflorescences to 10 cm; flowers numerous, in loose cymes of 1–3; sepals red-tipped but without apical warts; corolla 1–1.5 cm across, pink with darker veins and a white throat, rarely entirely white. Capsules rounded. *Chile.* G1. Spring.

Seed sets readily.

7. O. corniculata Linnaeus. Illustration: Jacquin, Oxalis, t. 54 (1794); Keble Martin, The concise British flora in colour, **20** (1965); Aeschimann et al., Flora alpina **1**: 1093 (2004).

Creeping, much-branched, mat-forming, short-lived perennial; stems 10–30 cm, slender, prostrate to ascending. Leaves numerous along stems; stipules fused, rectangular; leaflets 3, heart-shaped, 5–15 × 8–20 mm, green, usually hairless above and hairy beneath. Flowers in umbels of 2–6 or borne on axillary stems 1–10 cm; sepals without warts; corolla *c.* 1 cm across, light yellow, sometimes with a red throat. Capsules 1.2–1.5 cm, cylindric, erect on deflexed stalks. *Origin unknown.* H3. Spring–autumn.

A cosmopolitan weed causing particular problems in glasshouses. Selfed seed is set very readily, and fired from explosive capsules.

Var. **atropurpurea** Planchon (*O. corniculata* Linnaeus var. *purpurata* Parlatore; *O. tropaeoloides* Schlechter) differs in having purple-bronze foliage, and all parts tinged purple. Also a cosmopolitan weed, but can grow in more exposed areas, e.g. cracks in paths.

8. O. exilis A. Cunningham (*O. corniculata* Linnaeus var. *microphylla* J.D. Hooker). Illustration: Garrard & Streeter, Wild flowers of the British Isles, 66 (1983).

Small, wiry, mat-forming, short-lived perennial to 4 cm; stem slender, creeping to 15 cm. Leaf-stalks 6–13 mm, erect, thin; leaflets 3, heart-shaped, sometimes deeply notched, green, 3–5 × 3–7 mm. Flowers solitary on erect stems to 1.8 cm, yellow, to 8 mm across; sepals without warts. Capsules spherical to cylindric, sparsely

Figure 112. Leaves of Oxalis species. 1, *O. acetosella*. 2, *O. drummondii*. 3, *O. tetraphylla*. 4, *O. caprina*. 5, *O. hirta*. 6, *O. ortgiesii*. 7, *O. purpurea*. 8, *O. corymbosa*. 9, *O. latifolia*. 10, *O. laciniata*. 11, *O. depressa*. 12, *O. lasiandra*. 13, *O. adenophylla*. 14, *O. flava*. Scale approximately × 0.5 cm.

hairy, on deflexed stalks, 5–8 × 2–3 mm. *Australia, New Zealand*. H3. Spring–summer.

Naturalised from gardens in Britain and the Channel Islands.

9. O. chrysantha Progel. Illustration: *Quarterly Bulletin of the Alpine Garden Society* **33**: 238 (1965); *Journal of the Royal Horticultural Society* **102**: 394–395 (1977).

Mat-forming, hairy perennial with creeping aerial stems to 20 cm, slender, rather woody, rooting and producing buds and leaf-rosettes at basal nodes. Leaves numerous along stem, stalks 2–4 cm;

leaflets 3, triangular to heart-shaped, *c.* 7 × 7 mm, green, whitish hairy. Flowers solitary on stems *c.* 3.5 cm; sepals without warts; corolla *c.* 1.5 cm across, golden yellow with red markings at mouth of throat. Capsule *c.* 11 mm, cylindric, hairy. *Brazil.* H5. Summer–autumn.

10. O. megalorrhiza Jacquin (*O. carnosa* Molina). Illustration: *Botanical Magazine*, 2866 (1828); Graf, Exotica, edn. 8, 1331, 1334 (1976); Everett, New York Botanical Gardens illustrated encyclopedia of horticulture **7**: 2451 (1981).

Perennial 30–40 cm, rhizome horizontal, woody, usually unbranched. Stems little-branched, 1–2 cm thick, hairless, fleshy, erect to sprawling, base semi-woody and covered with scales, bearing leaves and flowers towards the ends. Leaf-stalks erect, fleshy, to 8 cm; leaflets 3, heart-shaped, 1.1–1.8 × 1–2 cm, succulent, green, shiny above, with watery papillae appearing crystalline beneath. Flowers in umbels of 2–5, on stalks as long as leaves; sepals of 2 distinct shapes, 3 outer broadly triangular with hastate bases, 2 inner smaller and narrower, without warts; corolla to 2.5 cm across, bright yellow. Capsules *c.* 6 × 4 mm, hairless, rounded, enclosed within sepals until mature. *Chile.* G1. Summer–autumn.

Readily sets self-seeded and may be a pest in greenhouses.

11. O. peduncularis Humboldt, Bonpland & Kunth (*O. herrerae* misapplied). Illustration: Graf, Exotica, edn. 8, 1332 (1976); Everett, New York Botanical Gardens illustrated encyclopedia of horticulture **7**: 2451 (1981); Beckett, The RHS encyclopaedia of house plants, 373 (1987).

Erect to ascending, leafy-stemmed perennial to 60 cm; stems reddish, succulent, 5–7 mm thick, hairless. Leaves crowded towards stem-apex, stalks somewhat thickened, hairless; leaflets 3, obovate, *c.* 12 × 9 mm, quite thick, bright or pale green with purple margins, hairless above, with short adpressed hairs beneath. Flowers in cymes of 9–16 on erect stalks, held well above the leaves; sepals hairless with purple margins and without apical warts; corolla *c.* 1.5 cm across, yellow with red veins. *Peru, Ecuador.* H5. Spring–summer.

12. O. rubra A. Saint-Hilaire (*O. floribunda* misapplied; *O. rosea* misapplied). Illustration: Graf, Exotica, edn. 8, 1331, 1334 (1976); Everett, New York Botanical

Gardens illustrated encyclopedia of horticulture **7**: 2453 (1981).

Clump-forming, stemless perennial to 40 cm, arising from a semi-woody tuberous crown; tubers round to cylindric, covered with reddened scales, branching and forming tight clusters in old plants, which may rise above the soil surface. Leaves to 30 cm, numerous, spreading to erect, long-stalked; leaflets 3, heart-shaped, *c.* 1.8 × 1.5 cm, green, with dark spots particularly around notch, hairless above, hairy beneath and along margins. Flowers in umbel-like cymes of 6–12, on erect stems; sepals with several apical warts; corolla 1–1.5 cm across, red, pink or white. *S Brazil to Argentina.* H3. Summer.

This species is very similar to *O. articulata.* Cultivated plants may have reddened foliage. 'Alba' is a selected variant with white flowers.

13. O. articulata Savigny (*O. floribunda* Lehmann). Illustration: *Botanical Magazine*, 6748 (1884); *Gardeners' Chronicle* **159**: 535 (1966).

Rhizomatous perennial, 10–40 cm; rhizome to 14 × 2 cm, dark brown, tuberous, semi-woody, spherical, becoming cylindric with constrictions, little branched. Leaves 6–25 cm, basal, numerous, stalks erect to sprawling, hairless or with adpressed hairs; leaflets 3, heart-shaped, 1–2.5 × 2–3.5 cm, green usually with orange margins, oblong spots beneath, hairless above, hairy beneath. Flowers in cymes of 5–10, held well above the leaves on erect-spreading stems to 40 cm; sepals each with 2 orange apical warts; corolla to 2 cm across, bright mauve-pink, rarely white. *Paraguay.* H3. Summer–autumn.

Var. **hirsuta** Progel. Differs in leaves hairy above, and all parts generally more hairy. European cultivated material is usually this variety. Specimens without basal parts can be distinguished from the superficially similar *O. corymbosa* by the lack of spreading hairs on the leaf-stalks and stems.

14. O. trilliifolia J.D. Hooker.
Stemless perennial; rhizome stout, brown, vertical, 4–5 cm × 2–3 mm, covered in leaf-bases. Leaf-stalks 9–25 cm × 1–2 mm, erect, rather succulent; leaflets 3, heart-shaped, 2.5–4 cm, glaucous green, hairless above, hairy beneath and along margins. Flowers in umbel-like cymes of 2–8, on stems 10–25 cm; sepals without warts; corolla

c. 2 cm across, white to pale violet. *Pacific North America.* H4. Summer.

15. O. magellanica G. Forster (*O. lactea* J.D. Hooker). Illustration: Salmon, Collins guide to the alpine plants of New Zealand, 187 (1968); Mark & Adams, New Zealand alpine plants, 57 (1973); Galbraith, Collins field guide to the wild flowers of south-east Australia, pl. 66 (1977); *Gartenpraxis* **3**: 137 (1978).

Prostrate, stoloniferous, carpet-forming perennial to 4 cm, arising from a slender scaly rhizome. Leaf-stalks 2–4 cm, erect, sparsely hairy; leaflets 3, heart-shaped, *c.* 5 × 5 mm, bronze-green, hairless. Flowers solitary on short, erect stems 1.5–3.5 cm; sepals without warts; corolla *c.* 1 cm across, pure white. *South America & Australasia.* H4. Late spring–summer.

Plant reminiscent of a small *O. acetosella*; grows best in well-drained stony areas.

16. O. acetosella Linnaeus. Figure 112 (1), p. 420. Illustration: *Quarterly Bulletin of the Alpine Garden Society* **33**: 237 (1965); Keble Martin, The concise British flora in colour, 20 (1965); Hay & Beckett, Reader's Digest encyclopaedia of garden plants and flowers, 489 (1971); Graf, Exotica, edn. 8, 1332 (1976).

Creeping perennial, 3–12 cm; rhizome 1–2 mm thick, slender, pale green, scaly. Leaf-stalks erect, to 8 cm; leaflets 3, heart-shaped, to 1.5 × 2 cm, pale green, sparsely hairy. Flowers solitary, borne slightly above the leaves; sepals without warts; petals 1.5–2 cm across, pearl-white, veined pinkish purple. Capsule 3–4 mm, ovoid-spherical. *N temperate America & Europe.* H1. Spring.

Seed sets readily.

Var. **purpurascens** C. Martius. Flowers rose with purple veining. Less common in nature, but frequently grown as 'Rosea'. Both are suitable for use as ground-cover in shady places.

17. O. oregana Nuttall. Illustration: Graf, Exotica, edn. 8, 1335 (1976).

Creeping perennial, 6–20 cm. Rhizome horizontal, 2–2.5 mm thick, brown, quite robust, scaly with persistent stipules. Leaf-stalks 3–20 cm, erect with spreading hairs; leaflets 3, widely heart-shaped, 2.5–3.5 × 2.7–4.5 cm, green, hairless above, margins and undersurfaces with long hairs. Flowers solitary on erect, hairy stems, level with or just above leaves. Sepals hairy, without warts; corolla 2–2.5 cm across, pale lilac or darker, occasionally white. Capsules 7–8 mm, spherical. *W North America.* H2. Spring–autumn.

A very vigorous plant often used for ground-cover in cool, shaded positions. It is reminiscent of a robust *O. acetosella.*

18. O. adenophylla Gillies. Figure 112 (13), p. 420. Illustration: *Journal of the Royal Horticultural Society* **75**: f. 86 (1950); *Quarterly Bulletin of the Alpine Garden Society* **43**: 187 (1975); Graf, Exotica, edn. 8, 1334, 1336 (1976); Everett, New York Botanical Gardens illustrated encyclopedia of horticulture **7**: 2450 (1981).

Stemless perennial, 10–15 cm, with a large, brown, scale-covered, tuberous base. Leaf-stalks 5–12 cm, erect to spreading, reddish brown; leaflets 9–22, heart-shaped, *c.* 6 × 6 mm, silver-grey, hairless. Flowers in umbels of 1–3, level with the foliage; sepals without warts; corolla *c.* 2.5 cm across, lilac-pink to violet, darker veined, with 5 purple spots in the white centre. *Chile.* H2. Spring–early summer.

A common plant of rock gardens.

19. O. patagonica Spegazzini (*O. enneaphylla* Cavanilles var. *patagonica* (Spegazzini) Skottsberg).

Stemless perennial to 5 cm; rhizome semi-jointed, creeping, covered beneath with rounded orange scales, congested at apex. Leaf-stalks 2–5 cm, erect; leaflets 10–14, heart-shaped, 7–8 × 3.3–5 mm, divided to base and folded lengthwise, glaucous grey, densely hairy. Flowers solitary on glaucous grey, hairy stalks; sepals without warts, hairy; corolla to 2.5 cm across, red, pink to pale blue, hairy. *Chile, Argentina.* H2. Late spring–summer.

20. O. enneaphylla Cavanilles. Illustration: *Botanical Magazine*, 6256 (1876); Hay & Beckett, Reader's Digest encyclopaedia of garden plants and flowers, 490 (1971); Rix & Phillips, The bulb book, 155 (1981); Heywood & Chant, Popular encyclopaedia of plants, 251 (1982).

Stemless perennial to 14 cm; rhizome slender, creeping, horizontal, *c.* 5 × 2 cm, covered in thick white scales with bulbils in their axils. Leaf-stalks 1.5–8 cm, erect, occasionally hairy; leaflets 9–20, heart-shaped, 4–12 × 2–8 mm, partially folded, somewhat fleshy, glaucous blue, shortly hairy. Flowers solitary, held just above the leaves; sepals without warts; corolla *c.* 2 cm across, pale pink or blue to white, fragrant. *Chile & Argentina to Falkland Islands.* H1. Spring–autumn.

'Alba', 'Rosea' and 'Rubra' are named selections with white, rose-pink and red flowers. 'Minutifolia' is a dwarf form less than 5 cm, with small leaves and pale rose-pink flowers.

'Ione Hecker'. *O. enneaphylla* × *O. laciniata.* Illustration: *Quarterly Bulletin of the Alpine Garden Society* **47**: 44 (1979). Leaves as above, but with rather narrower segments and a deeper green. Flowers solitary; corolla to 3 cm across, vivid blue at edge, deepening to dark purple at centre. Generally larger and flowering more profusely than *O. laciniata.*

21. O. laciniata Cavanilles (*O. squarroso-radicosa* Steudel). Figure 112(10), p. 420. Illustration: *Quarterly Bulletin of the Alpine Garden Society* **28**: 345 (1960); **35**: 107 (1967); *Journal of the Royal Horticultural Society* **87**: f. 150 (1962); Das Pflanzenreich **130**: f. 18 (1930); Rix & Phillips, The bulb book, 155 (1981).

Spreading, stemless perennial 5–10 cm; rhizome branching freely just beneath ground level, forming a long chain of tiny, linked, scaly bulbils. Leaves arising from apex of rhizome on erect pinkish stalks 2.5–7 cm; leaflets 8–12, heart-shaped, to 2 cm, folded lengthwise, glaucous green with purple, wavy margins, hairless to softly hairy. Flowers solitary, on stems just above the foliage; sepals without warts, tips reddened; corolla *c.* 2.5 cm across, violet-crimson to lilac, or paler, all with darker veining and a green throat, sweetly scented. *Chile, Argentina.* H1. Late spring–summer.

This plant readily sets seed and has been hybridised with *O. enneaphylla.*

22. O. lobata Sims. Illustration: *Botanical Magazine*, 2386 (1823); *Quarterly Bulletin of the Alpine Garden Society* **3**: 84 (1935); **33**: 237 (1965); Rix & Phillips, The bulb book, 183 (1981); Brickell (ed.), RHS A–Z encyclopedia of garden plants, 755 (2003).

Stemless, bulbous perennial with tuberous roots, 8–10 cm; bulbs round, to 2.5 cm across, densely covered in brown woolly scales. Leaf-stalks 4–5 cm, erect; leaflets 3, heart-shaped, *c.* 5 × 6 mm, the laterals rather folded lengthwise, light green, usually spotted above, hairless or with long hairs, *c.* 5 × 6 mm. Flowers solitary, stalks almost twice as high as foliage; sepals with apical warts; corolla *c.* 1.5 cm across, golden yellow, dotted and lined with red. *Chile.* H2. Late summer–autumn.

23. O. corymbosa de Candolle (*O. bipunctata* Graham; *O. martiana* Zuccarini). Figure 112(8), p. 420. Illustration: *Botanical Magazine*, 2781 (1827); 3938 (1842); Graf, Exotica, edn. 8, 1335 (1976); Everett, New York Botanical Gardens illustrated encyclopedia of horticulture **7**: 2453 (1981).

Bulbous, stemless perennial 15–40 cm, producing numerous loosely scaly, stalkless bulbils with 3 veins per scale. Leaves erect to spreading, 10–35 cm; leaf-stalks and stems with long, white, spreading simple hairs, 0.5–2.5 mm; leaflets 3, broadly heart-shaped, 2.5–4.5 × 3–6.2 cm, rounded, green with dark spots on the undersurface, sparsely hairy. Flowers in irregularly branched cymes of 8–15; sepals with 2 orange apical warts; corolla *c.* 1.5 cm across, red to purple with darker veins and a white throat. *Brazil, Argentina.* Spring–early summer.

'Aureo-reticulata' is identical to small specimens of the above, but the leaves have light yellow veining, which is probably virus-induced. It does not set seed in Europe, but is liable to spread by bulbils in milder climates. The leaves have been used as a tamarind substitute. This species is particularly susceptible to the rust *Puccinea oxalidis.* For specimens without basal parts see *O. articulata.*

24. O. violacea Linnaeus. Illustration: *Botanical Magazine*, 2215 (1821); Everett, New York Botanical Gardens illustrated encyclopedia of horticulture **7**: 2450 (1981).

Stemless, bulbous, usually hairless perennial to 20 cm or more; bulbs brown, scaly, rounded, *c.* 1 cm across; scales 3-veined. Leaf-stalks erect, 7–13 cm; leaflets 3, heart-shaped, 8–20 mm × 1–2.8 cm, green with orange dots around the apical notch area only. Flowers held well above the leaves, in umbels of 2–16; sepals without apical warts; corolla 1.5–2 cm across, lavender-pink or paler (rarely white), with a green throat; styles 2. *North America.* H2. Spring–autumn.

25. O. latifolia Humboldt, Bonpland & Kunth (*O. intermedia* misapplied; *O. vespertilionis* Zuccarini). Figure 112(9), p. 420. Illustration: Bichard & McClintock, Wild flowers of the Channel Islands, pl. 88 (1975); Garrard & Streeter, Wild flowers of the British Isles, 67 (1983).

Bulbous, stemless perennial, 7–25 cm, producing bulbils on numerous short underground runners; bulbils scaly, scales with 5–11 veins. Leaf-stalks 8–25 cm, erect to spreading; leaflets 3, broadly triangular to heart-shaped, to 7 × 7 cm, dark green, hairless. Flowers in umbels of 6–32; sepals with 2 orange-brown apical warts; corolla 1.5–2 cm across, violet-pink or paler, with a green throat. *Mexico to Peru.* H4. Summer–autumn.

Widely naturalised in mild climates where it can become a serious weed. A 'Cornwall type' has been described, which has much rounder leaflets and very short runners. This can be confused with *O. corymbosa*, but is differentiated on the number of bulb-scale veins.

26. O. drummondii A. Gray
(*O. vespertilionis* Torrey & Gray; *O. vespertilionis* misapplied). Figure 112(2), p. 420.

Bulbous perennial to 20 cm or more; bulb composed of quite open, papery, 3-veined scales. Leaf-stalks 5–16 cm, erect to spreading, hairless; leaflets 3, V-shaped, lobed to ¼ or so of their length, lobes to 3 cm × 5 mm. Flowers in umbels of 3–10, held well above the leaves; sepals with fused red apical warts; corolla *c.* 2 cm across, purple; styles 2. *Mexico.* H5. Spring–summer.

Distinguished from *O. latifolia* by the 3-veined bulb-scales, flowers with 2 styles and generally deeper leaf lobing.

27. O. lasiandra Zuccarini (*O. floribunda* misapplied). Figure 112(12), p. 420. Illustration: *Botanical Magazine*, 3896 (1842); Graf, Exotica, edn. 8, 1331 (1976).

Bulbous, stemless perennial, 15–30 cm or more, with a thick taproot, densely covered with scaly bulbils at apex; scales with 15–25 veins. Leaves to 15 cm, basal, erect; stalks reddish green with sparse to abundant, spreading, jointed hairs *c.* 2 mm; leaflets 5–10, narrowly wedge-shaped to strap-like, to 5 × 2 cm, apex rounded, usually shallowly notched, green, hairless. Flowers in umbels of 9–26, on succulent stems with long, spreading, jointed hairs; stems twice as high as leaves; sepals and bracts with red apical warts; corolla *c.* 2 cm across, crimson to violet with a yellow throat. *Mexico.* G1. Summer–autumn.

28. O. tetraphylla Cavanilles (*O. deppei* Loddiges). Figure 112(3), p. 420. Illustration: Graf, Exotica, edn. 8, 1332, 1333, 1335 (1976); Beckett, The RHS encyclopaedia of house plants, 372 (1987);

Bulbous, stemless perennial; bulbs large, covered in hairy scales, 1.5–3.5 cm. Leaf-stalks 10–40 cm, erect to spreading, sparsely to moderately hairy; leaflets 4 (rarely 3), strap-shaped to reversed triangular, entire or shallowly notched, 2–6.5 × 2–3 cm, green, usually with a V-shaped purple band near the base pointing to the apex, usually hairless above, somewhat hairy beneath. Flowers in umbels of 5–12, on erect stalks 15–50 cm; sepals with red apical warts; corolla 1–2 cm across, red to lilac-pink, rarely white, throat green-yellow. *Mexico.* G1. Summer.

'Iron Cross' (illustration: Brickell (ed.), RHS A–Z encyclopedia of garden plants, 756, 2003), differs from the above in the segment formed by the coloured band on the leaf being entirely purple. This showy plant has been cultivated in Europe for its edible bulbs.

29. O. pes-caprae Linnaeus (*O. cernua* Thunberg; *O. libyca* Viviani). Illustration: Jacquin, Oxalis, t. 6 (1794); *Botanical Magazine*, n.s., 237 (1973); Graf, Exotica, edn. 8, 1331, 1332 (1976); Everett, New York Botanical Gardens illustrated encyclopedia of horticulture **7**: 2452 (1981).

Bulbous, stemless perennial, 20–40 cm; bulb with a white fleshy contractile root and a vertical stem to soil surface; stem and crown bear numerous stalkless bulbils. Leaves numerous, erect to spreading, slightly succulent; stalks 3–12 cm; leaflets 3, heart-shaped, 1.6–2 × 2.3–3.2 cm, bright green, often with dark purple spots, hairless. Flowers in cymes of 3–20, stalks somewhat nodding, stems erect, twice as high as foliage; sepals with 2 orange apical warts; corolla 2–3 cm across, deep golden yellow. *South Africa (Cape Province).* H5. Spring–early summer.

Widely naturalised in milder climates, where it can be a serious weed of cultivation. Reproduces by bulbils in Europe because only the short-styled variant was introduced and fertile seed is not set. A double-flowered variant, 'Flore Pleno', exists.

30. O. bowiei Lindley (*O. bowieana* Loddiges; *O. purpurata* Jacquin var. *bowiei* (Lindley) Sonder; *O. floribunda* misapplied). Illustration: *Quarterly Bulletin of the Alpine Garden Society* **3**: 83 (1935); Graf, Exotica, edn. 8, 1332, 1333 (1976); Everett, New York Botanical Gardens illustrated

encyclopedia of horticulture **7**: 2457 (1981).

Bulbous, stemless perennial, 20–30 cm, with glandular hairs on leaf-stalks, stems and calyx. Bulb long with a smooth brown tunic covering the pale brown flesh inside, with a long, fleshy, white contractile root. Leaf-stalks stout, erect, 5–15 cm; leaflets 3, rounded to broadly heart-shaped, shallowly notched, rather leathery, green, occasionally purple beneath, hairy along margin, to 5 × 5 cm. Flowers in umbels of 3–12, on erect stems 10–30 cm, bright rose-red to pink, with a greenish throat, 3–4 cm across; calyx without orange apical warts. *South Africa.* G1. Summer–autumn.

31. O. purpurata Jacquin.

Bulbous perennial, 10–30 cm; bulb without contractile root, rhizomatous with vertical stem to soil surface and crown, and fleshy underground runners producing apical bulbils or occasionally leaves. Leaves numerous on long, hairless to hairy, erect to spreading stalks to 30 cm; leaflets 3, heart-shaped, 1.5–5 × 2.3–7 cm, rounded, dark green and hairless above, dark purple (variable) and hairy beneath, margins hairy. Flowers in umbel-like groups of 3–10, on sparsely hairy stems; sepals with 2 orange apical warts; corolla *c.* 2.5 cm across, purple to violet with a yellow throat. *South Africa.* H5. Summer.

A very variable species.

32. O. caprina Linnaeus. Figure 112(4), p. 420. Illustration: *Botanical Magazine*, 237 (1793); Jacquin, Oxalis, t. 76 (1794).

Stemless, or very short-stemmed, almost hairless perennial, 15–20 cm, with vertical rhizome from scaly bulbs. Leaves basal or at stem-apex; leaf-stalks erect to sprawling, 2–5 cm; leaflets 3, widely triangular in outline, deeply divided to about the middle, lobes obovate, 5–10 × 5–15 mm. Flowers in umbels of 2–4, on rather weak stems to 20 cm; sepals with 2 orange apical warts; corolla pale violet, rarely white, with a pale green throat. *South Africa (Cape Province).* H5. Spring–early summer.

33. O. melanosticta Sonder. Illustration: Graf, Exotica, edn. 8, 133, 1335, 1336 (1976).

Almost stemless, bulbous perennial to 2.5 cm; bulbs 6.5–8.5 mm, small, scaly. Leaf-stalks erect to spreading, very hairy, 1.5–2.5 cm; leaflets 3, heart-shaped to rounded, rather thick, green with orange spots which blacken on drying, usually hairy on both surfaces, margins with dense white hairs. Flowers borne on erect stems the same height as leaves or shorter; sepals densely hairy without apical warts; corolla to 2 cm across, yellow. *South Africa*. H5. Late spring–summer.

34. O. purpurea Linnaeus (*O. variabilis* Jacquin; *O. grandiflora* Jacquin; *O. speciosa* Jacquin). Figure 112(7), p. 420. Illustration: *Botanical Magazine*, 1683, 1712 (1815); Graf, Exotica, edn. 8, 1331 (1976); Everett, New York Botanical Garden illustrated encyclopedia of horticulture **7**: 2452, 2453 (1981); *Quarterly Bulletin of the Alpine Garden Society* **50**: 274 (1982).

Bulbous perennial to 15 cm, arising from a smooth, rounded, blackish bulb to 1.7 cm across. Leaves often spreading to prostrate; leaf-stalks 2–8 cm, covered in white hairs; leaflets 3, diamond-shaped to circular or widely obovate, 4–40 × 4–30 mm, dark green above, spotted purple or deep purple beneath, hairless, but margins with dense hairs. Flowers solitary, equal or shorter than leaves; sepals without warts; corolla 3–5 cm across, rose-purple, deep rose to violet and pale violet, yellow, cream or white with a yellow throat. *South Africa (Cape Province)*. H5. Autumn–winter.

A very variable species, particularly in leaf shape and flower size and colour. Large-flowered variants have been recognised as cultivars, such as 'Grand Duchess' with flowers *c.* 5 cm across.

35. O. depressa Ecklon & Zeyher (*O. convescula* Jacquin; *O. inops* Ecklon & Zeyher). Figure 112(11), p. 420. Illustration: Jacquin, Oxalis, t. 55 (1794); *Quarterly Bulletin of the Alpine Garden Society* **30**: 291 (1962); Hay & Synge, Dictionary of garden plants in colour, 16 (1973); Rix & Phillips, The bulb book, 155 (1981).

Bulbous, almost stemless perennial, 4–12 cm, with a slender underground vertical rhizome, 5 cm or more from the bulb to the soil surface. Leaf-stalks 8–20 mm, erect to spreading, in a crown, hairless; leaflets 3, rounded to triangular or ovate, grey-green, sometimes dark-spotted, hairless or with a few marginal hairs, 3–10 × 5–16 mm. Flowers solitary on erect stems above leaves; sepals without warts; corolla 1.5–2 cm across, bright pink to rose-violet with a yellow throat. *South Africa*. H3. Summer.

An attractive, mat-forming plant, which may become invasive as abundant offsets are readily produced. Purple and white-flowered variants are known.

36. O. incarnata Linnaeus. Illustration: Jacquin, Oxalis, t. 71 (1794); Graf, Exotica, edn. 8, 1331 (1976); Everett, New York Botanical Gardens illustrated encyclopedia of horticulture **7**: 2451 (1981).

Bulbous perennial, hairless, 10–50 cm, with erect to sprawling, thin, branching stems; bulb rounded to 2 cm across. Leaves in whorls up the stem, stalks 2–6 cm; leaflets 3, heart-shaped, 8–20 × 5–15 mm, translucent green with dark spots under the margins; leaf-axils may bear reddish brown bulbils. Flowers solitary on erect stems 3–7 cm; sepals with several converging apical warts; corolla *c.* 2 cm across, white or very pale lilac, with darker veins and a yellow throat. *Namibia*. H5. Autumn.

37. O. bifida Thunberg (*O. filicaulis* Jacquin). Illustration: Jacquin, Oxalis, t. 79 (1794).

Perennial with a weak, erect or procumbent branched stem to 30 cm, arising from a scaly bulb. Leaves congested at the stem-apices; leaf-stalks 1.5–4 cm, slender; leaflets 3, narrowly heart-shaped, divided to the middle or beyond, 5–6 × 5–6 mm, green with long black marginal spots, hairless above, slightly hairy beneath. Flowers solitary on stems 2.5–8 cm; calyx with purple-black apical warts; corolla *c.* 1.2 cm across, purple-red with a greenish throat. *South Africa (Cape Province)*. H5. Spring–summer.

38. O. flava Linnaeus (*O. pectinata* Jacquin). Figure 112(14), p. 420. Illustration: Jacquin, Oxalis, t. 73, t. 75, t. 78(2) (1794); Graf, Exotica, edn. 8, 1331, 1334. (1976).

Robust, stemless perennial, 3–25 cm, with a brown rhizome *c.* 10 cm × 3 mm, covered with membranous scales. Leafstalks 2–6 cm, erect to spreading, dilated beneath a basal articulation; leaflets 2–12, narrow, finger-like, 7.5–20 mm, oblong with an apical notch, green, hairless. Flowers solitary on stems held slightly above the foliage; sepals with several indistinct, orange apical warts; corolla *c.* 2.5 cm across, bright yellow, white or very pale rose-violet, with a yellow throat. *South Africa (Cape Province)*. H5. Spring–early summer.

A very variable species.

39. O. hirta Linnaeus (*O. fulgida* Lindley; *O. hirtella* Jacquin; *O. multiflora* Jacquin; *O. rubella* Jacquin). Figure 112(5), p. 420.

Illustration: *Botanical Magazine*, 103 (1807), 1698 (1814); *Quarterly Bulletin of the Alpine Garden Society* **53**: 372 (1985); Beckett, The RHS encyclopaedia of house plants, 372 (1987); *Kew Magazine* **10**(2): pl. 217(1993).

Erect, decumbent or trailing, bulbous perennial with stems to 30 cm, branching above; bulbs to 1.5 cm across, small, round, each producing one stem which dies down each year; all parts of plant hairy. Leaves almost stalkless; leaflets 3, 1–1.5 cm × 1–3 mm, linear to oblong, with an apical notch, green, variably hairy. Flowers solitary on stems 1–5 cm in upper leaf-axils; sepals without warts; corolla *c.* 2.5 cm across, red-violet, purple or paler to white, (rarely yellow) with a yellow throat. *South Africa (Cape Province)*. G1. Autumn. Very variable.

---

156. GERANIACEAE

Herbs or shrubs, sometimes succulent. Leaves with stipules, alternate or opposite, simple, pinnate or palmate, usually toothed. Flowers in umbels, sometimes composed of only 2 or 3 flowers and then referred to as cymules (see *Geranium*), occasionally solitary. Flowers radially or bilaterally symmetric. Sepals 5. Petals 5 or sometimes fewer. Nectaries usually 5, at bases of stamens or, in *Pelargonium*, one at the apex of a spur. Stamens 10 or 15, shortly united at the base, some occasionally with staminodes. Ovary superior, of 5 carpels, each containing 2 ovules and united to a central column; style simple, divided at apex into 5 linear lobes covered on the inward-facing side with stigmatic papillae and more or less recurved when receptive. Fruit with the central column greatly elongated. Mature carpels consisting of a carpel-body containing 1 seed (rarely 2) and a ribbon-like piece, the awn; they separate from the axis of the column on drying out. Splitting involves 3 stages: (1) the carpel bodies gently separate from the column, (2) an interlude, the 'pre-explosive interval', (3) the peeling away of the awns from below upwards, which takes place with explosive force. Seeds are either expelled from the carpels or dispersed within them. In the former case the tip of the awn sometimes remains attached to the tip of the column; in the latter the awn may or may not remain attached to the carpel-body.

A family now usually restricted to 5 genera with about 700 species. They are plants of temperate climates in all continents.

Literature: Knuth, R., Geraniaceae, in A. Engler (ed.), Das Pflanzenreich **53** (1912)

1a. Sepal on the upper side of the flower with a slender spur produced at its base and attached to the flower-stalk

           4. Pelargonium

  b. Sepals all the same, without spurs 2

2a. Stamens 15, all fertile

           3. Sarcocaulon

  b. Stamens 10, either all fertile or alternate ones sterile (devoid of anthers)   3

3a. Stamens (in our species) all with anthers   **1. Geranium**

  b. Alternate stamens without anthers

           2. Erodium

1. GERANIUM Linnaeus

P.F. Yeo

Cultivated species herbaceous, sometimes woody at the base; annual, biennial or perennial. Vegetative parts, sepals and fruits usually hairy. Stem-leaves usually paired but the lowest and uppermost sometimes alternate. Leaves palmately divided or palmate, usually with the divisions lobed and the lobes toothed. Flowers usually in pairs on a Y-shaped structure (the cymule), sometimes solitary or in umbels, bisexual or apparently so, radially symmetric. Petals 5. Stamens 10, almost free, usually about as long as the sepals. Seeds expelled from the carpels or dispersed within them; in the latter case the awn may or may not remain attached to the carpel-body (further details below). Awn curved or coiled, not twisted more than half a turn.

A genus of about 300 species in temperate climates (including tropical mountains), of which about 80 have been grown for ornament. They are suitable for varying exposures to sun and shade according to species, and for various uses according to their stature; the largest may be used in formal herbaceous plantings and the smallest in the rock garden or in the alpine house. Propagation is by division or seed, rarely by cuttings (for woody species and those that root at the nodes).

In the following account, the leaves described are the main ones, which are usually basal or borne on the lower part of the stem. It should be noted that the term 'lobe' here applies to the segments into which the primary leaf-divisions are cut, and that the descriptions of the teeth apply also to the tips of the lobes. The stem of the 'Y' below each pair of flowers is the axis, its branches are the flower-stalks. Flower-stalk and sepal-length are measured at flowering time. Sepal-length, in both measurement and comparison, excludes the mucro. Stigmas are best measured just after petal-fall. The fruit is measured from the bases of the carpel-bodies to the bases of the stigmatic branches. It usually has a slender tip that corresponds with the style at the flowering stage; this is always included in the measurement of length but is only mentioned in descriptions when it is absent, distinctively long or different in closely related species.

Literature: Knuth, R., Geraniaceae, in A. Engler (ed.), Das Pflanzenreich **53**: 43–221, 575–83 (1912); Jones, G.N. & Jones, F.J., A revision of the perennial species of Geranium of the United States and Canada, *Rhodora* **45**: 5–53 (1943); Carolin, R.C., The genus Geranium L. in the south western Pacific area, *Proceedings of the Linnean Society of New South Wales* **89**: 326–61 (1964); Davis, P.H., Geranium sect. Tuberosa, revision and evolutionary interpretation, *Israel Journal of Botany* **19**: 91–113 (1970); Tokarski, M., Morphological and taxonomical analysis of fruits and seeds of the European and Caucasian species of the genus Geranium L., *Monographiae Botanicae* **36** (1972); Forty, J., A survey of hardy geraniums in cultivation, *The Plantsman* **2**: 67–78 (1980); Forty, J., Further notes on hardy geraniums, *The Plantsman* **3**: 127–128 (1981); Hilliard, O.M. & Burtt, B.L., A revision of Geranium in Africa south of the Limpopo, *Notes from the Royal Botanic Garden, Edinburgh* **42**: 171–225 (1985); Yeo, P.F., Fruit-discharge type in Geranium (Geraniaceae): its use in classification and its evolutionary implications, *Botanical Journal of the Linnean Society* **89**: 1–36 (1984); Yeo, P.F., Hardy geraniums (1985); Yeo, P.F., Geranium – Freiland-Geranien für Garten und Park, M. Zerbst (transl.) (1988); Clifton, R., Geraniales species check-list series **1**(2), edn 4, (2004).

The species of *Geranium* vary according to their type of fruit. When ripe or nearly ripe fruit is available, the differences can provide a useful check on identifications. A classification of fruit types is given below.

Fruit. Does not break up until the calyx rots: **49**; breaks up as soon as ripe: **1, 2, 3, 4, 5, 6, 7, 8, 9, 10, 11, 12, 13, 14, 15, 16, 17, 18, 19, 20, 21, 22, 23, 24, 25, 26, 27, 28, 29, 30, 31, 32, 33, 34, 35, 36, 37, 38, 39, 40, 41, 42, 43, 44, 45, 46, 47, 48, 50, 51, 52, 53, 54, 55, 56, 57, 58, 59, 60, 61, 62, 63, 64**.

Carpel-bodies. Remaining attached to tip of column after the release of seeds, with an orifice through which the seed is expelled during splitting of the fruit: **1, 2, 3, 4, 5, 6, 7, 8, 9, 10, 11, 12, 13, 14, 15, 16, 17, 18, 19, 20, 21, 22, 23, 24, 25, 26, 27, 28, 29, 31, 32, 33, 34, 35, 36, 37, 38**; falling from the column when the fruit splits, the column afterwards thus naked, seeds released or retained: **39, 40, 41, 43, 44, 45, 46, 47, 50, 51, 52, 53, 54, 55, 56, 57, 58, 59, 60, 61, 62, 63, 64**.

Carpel-body orifice (where carpel-body remains attached to column). Partly screened by bristles borne on a projecting horny knob at the lower end: **1, 2, 3, 4, 5, 6, 7, 8, 9, 10, 11, 12, 13, 14, 15, 16, 17, 18, 19, 20, 21, 22, 23, 24, 25, 26, 27, 28, 29, 31, 32, 33, 34, 35, 36, 37**; with the edge drawn out at the base into a prong that partly blocks the orifice or is bent inwards: **38**.

Carpel-body orifice (where carpel-body separates from column). Open when fruit splits allowing expulsion of seed: **39, 40, 41, 43, 44, 45, 46, 47**; sufficiently closed up to prevent emergence of seed when fruit splits: **48, 50, 51, 52, 53, 54, 55, 56, 57, 58, 59, 60, 61, 62, 63, 64**.

Awn (when carpel-body is sufficiently closed to prevent emergence of seed). Falling separately from carpel-body: **48, 50, 51, 52, 53, 54, 55, 56, 57, 58, 59**; remaining attached to carpel-body and becoming coiled: **60, 61, 62, 63, 64**.

1a. Stamens twice as long as sepals or more   2

  b. Stamens not more than 1.5 times as long as sepals   5

2a. Leaves cut almost to the base on either side of the middle division   3

  b. Leaves not cut to the base   4

3a. Petals 6–9 mm wide; petal-width less than half the length when the claw is included; sepals not adpressed to petal-bases; middle division of leaf-blade wedge-shaped, not distinctly stalked **57. canariense**

b. Petals 1.3–1.8 cm wide, *c.* two thirds as wide as long; sepals adpressed to petal-bases; middle division of leaf-blade clearly stalked **58. palmatum**

4a. Leaves usually 10–15 cm wide, their divisions palmato-pinnately lobed; sepals 7–9 mm **52. macrorrhizum**

b. Leaves not more than *c.* 5 cm wide, their divisions with 3 lobes at the apex; sepals not more than 6.5 mm **53. dalmaticum**

5a. Leaves cut or almost cut to the base on either side of the middle division 6

b. Leaves not cut to the base 17

6a. Leaves silky on both surfaces **63. argenteum**

b. Leaves not silky on both surfaces 7

7a. Leaves silky beneath 8

b. Leaves not silky beneath 9

8a. Leaf-segments thread-like; flowers usually deep pink **32. incanum** var. **multifidum**

b. Leaf-segments wider; flowers violet with a pale centre **34. robustum**

9a. Flower-stalks reflexed after flowering, usually making an angle of less than 50° with the axis; immature fruits reflexed **7. pratense**

b. Flower-stalks erect after flowering, or making an angle of not less than 50° with the axis; immature fruits usually erect 10

10a. Leaves forming rosettes in winter; plant without tubers; petals hairless; seed retained in carpels 11

b. Leaves not forming rosettes in winter; plant with underground tubers; petals with hairs at the base; seed released from carpels 14

11a. Plant in flower usually more than 80 cm; petals 1.3–1.8 cm wide **59. maderense**

b. Plant in flower usually less than 80 cm; petals not more than 1.3 cm wide 12

12a. Petal-claw as long as blade **55. robertianum**

b. Petal-claw half as long as blade 13

13a. Flowers 1.5–2 cm wide; sepal-mucro *c.* 0.5 mm **54. cataractarum**

b. Flowers 2.2–3.3 cm wide; sepal-mucro *c.* 2 mm **56. rubescens**

14a. Flowers widely trumpet-shaped; petals rounded or faintly notched; carpel with a blunt, horny, bristle-bearing projection at base, visible after the fruit has split **25. pylzowianum**

b. Flowers flat or saucer-shaped; petals distinctly notched; carpel without a horny projection at base 15

15a. Flowers 3.5–4.5 cm wide; slender tip of style 3–5 mm **41. malviflorum**

b. Flowers 2–3.5 cm wide; slender tip of style not more than 2.5 mm 16

16a. Plant without glandular hairs; style-tip not more than 1 mm, indistinct **39. tuberosum**

b. Plant with glandular hairs above; style-tip 1.5–2.5 mm **40. macrostylum**

17a. Flowers solitary 18

b. Flowers in pairs or umbels 20

18a. Leaf-divisions further divided into lanceolate or elliptic lobes with mostly lanceolate teeth **27. sanguineum**

b. Leaf-divisions shallowly lobed at apex 19

19a. Plant densely grey-hairy; flowers 2.5–3 cm wide **29. traversii**

b. Plant not densely hairy; foliage green or dark brown; flowers *c.* 1 cm wide **31. sessiliflorum** subsp. **novaezelandiae**

20a. Flower-stalks of immature fruits always erect, or plant completely sterile 21

b. Flower-stalks of immature fruits spreading or reflexed 44

21a. Plant trailing and flowers scattered **30. × riversleaianum**

b. Plant erect or flowers arranged in definite inflorescences 22

22a. Flowers facing sideways or nodding 23

b. Flowers upwardly inclined or erect 28

23a. Stamens closely adpressed to style; petals entire or with several irregular notches 24

b. Stamens not closely adpressed to style; petals each with a more or less distinct central notch 26

24a. Evergreen perennial **60. phaeum**

b. Deciduous perennial 25

25a. Hairs of the lower stem spreading; leaves divided half to two-thirds of

the way to the stalk into 5 or 7, the divisions not overlapping **10. platyanthum**

b. Hairs of the lower stem adpressed; leaves divided two-thirds to four-fifths of the way to the stalk into 7 or 9, the divisions overlapping **11. erianthum**

26a. Petals, apart from dark veins, white to medium violet; leaves grey-green, closely wrinkled, with very shallowly lobed, non-overlapping divisions **44. renardii**

b. Petals deep violet to violet-purple; leaves green, not very finely wrinkled, the divisions lobed for a quarter of their length or more, often overlapping 27

27a. Fruits not producing ripe seed **42. × magnificum**

b. Fruits developing normally **43. platypetalum**

28a. Flowers more or less funnel-shaped or trumpet-shaped 29

b. Flowers flat or saucer-shaped 35

29a. Petals white **5. rivulare**

b. Petals strongly coloured or with dark net-veining 30

30a. Leaves circular or kidney-shaped, the divisions broadest at the apex; petals not notched **48. polyanthes**

b. Leaves angular, the divisions broadest below the apex; petals notched 31

31a. Petals strongly net-veined all over **2. versicolor**

b. Petals not net-veined 32

32a. Leaf-divisions clearly lobed as well as toothed 33

b. Leaf-divisions toothed but not distinctly lobed 34

33a. Lobes of leaf-divisions with many teeth **1. endressii**

b. Lobes of leaf-divisions with only 1 or 2 teeth each **30. × riversleaianum**

34a. Plant without glandular hairs; carpel with a blunt, horny, bristle-bearing projection at base, visible after the fruit has split **3. nodosum**

b. Plant with glandular hairs above; carpel without a horny projection at base **47. gracile**

35a. Flowers dull purple with a blackish centre **15. procurrens**

b. Flowers not so coloured 36

36a. Plant without glandular hairs **45. libani**

b. Plant with glandular hairs on upper parts 37

37a. Flowers *c.* 1.5 cm wide
28. thunbergii

b. Flowers more than 2 cm wide 38

38a. Petals distinctly notched and with strongly feathered darker veins 39

b. Petals not or scarcely notched, with inconspicuous, unfeathered, darker veins or none 40

39a. Leaves with sides of divisions lobed and toothed nearly to the base; plant not dormant in summer
42. × magnificum

b. Leaves with sides of divisions not lobed and toothed below the widest part; plant dormant in summer after fruiting
46. peloponnesiacum

40a. Divisions of leaves profusely, finely lobed and toothed on most of their length; fruit not more than 2.5 cm
4. sylvaticum

b. Divisions of leaves sparingly lobed and toothed near the apex; fruit more than 2.5 cm 41

41a. Some flowers in umbels; sepal-mucro 2–3.5 mm **35. maculatum**

b. Flowers in pairs; sepal-mucro 0.5–1.5 mm 42

42a. Sepals *c.* 6 mm; petals white or nearly so; fruit not more than 2.1 cm; stigmas 3–4 mm
37. richardsonii

b. Sepals 7–11 mm; petals deep reddish pink; fruit 2.6–4 cm; stigmas 4–6 mm 43

43a. Flower-stalk 3.5–6 times as long as sepals; mucro 1–1.5 mm
12. swatense

b. Flower-stalk 1–2 times as long as sepals; mucro 0.5–1.25 mm
36. nervosum

44a. Flowers facing sideways or nodding 45

b. Flowers erect or upwardly inclined 55

45a. Stigmas 5–6 mm **14. lambertii**

b. Stigmas not more than 4 mm 46

46a. Flowers distinctly nodding; petals more or less strongly reflexed 47

b. Flowers facing sideways or very slightly nodding; petals not or slightly reflexed 50

47a. Sepal-mucro 4–7 mm
62. aristatum

b. Sepal-mucro not more than 2 mm 48

48a. Petals blackish red **18. sinense**

b. Petals pink to dull purple or bluish 49

49a. Stigmas 2.5–4 mm; carpels not with 4 or 5 ribs at apex, releasing the seed **19. pogonanthum**

b. Stigmas not more than 2 mm long; carpels with 4 or 5 ribs round the upper end, not releasing the seed
61. reflexum

50a. Plant to 10 cm; leaves to 5 cm wide
21. farreri

b. Plant normally 30 cm or more; leaves mostly more than 5 cm wide 51

51a. Flowers to 2.5 cm wide; petals usually greyish violet to dark brownish red; carpels with 4 or 5 ribs round the upper end, not releasing the seed **60. phaeum**

b. Flowers normally 3.5 cm wide or more; petals usually blue; carpels not with 4 or 5 ribs at apex, releasing the seed 52

52a. Stamens enlarged at base; fruit more than 3 cm 53

b. Stamens not enlarged at base; fruit less than 3 cm 54

53a. Plant not creeping underground, usually 75–130 cm; leaves cut six-sevenths to nine-tenths of the way to the stalk, all segments narrow and acute or, if relatively broad, not turned outwards **7. pratense**

b. Plant creeping by underground rhizomes, usually not more than 50 cm; leaves cut three-quarters to four-fifths of the way to the stalk, segments shorter, broader and sometimes blunt, often turned outwards **9. himalayense**

54a. Sepals 6–10 mm; some glandular hairs on upper parts
19. pogonanthum

b. Sepals 1–1.2 cm; glandular hairs absent or confined to underside of leaves **20. yunnanense**

55a. Leaves rounded or kidney-shaped in outline, their divisions broadest at the apex and palmately lobed 56

b. Leaves angular, their divisions broadest below the apex and usually pinnately or palmato-pinnately lobed 61

56a. Carpels hairless; plant annual
51. brutium

b. Carpels usually hairy; plant perennial 57

57a. Flowers 1.2–1.8 cm wide
50. pyrenaicum

b. Flowers at least 2.2 cm wide 58

58a. Beak of fruit not thickened; carpels with bristly teeth at apex, neither

shed at maturity nor releasing the seed **49. albanum**

b. Beak of fruit thickened in the normal way; carpels not with bristly teeth at apex, either shed at maturity or releasing the seed 59

59a. Flowers funnel-shaped; plant with very small underground tubers linked by slender rhizomes
25. pylzowianum

b. Flowers not funnel-shaped; rootstock compact, not tuberous and rhizomatous 60

60a. Petals 1.5–2.5 times as long as broad, not notched; carpels releasing the seed at maturity, themselves remaining attached to the beak of the fruit
38. asphodeloides

b. Petals less than 1.5 times as long as broad, often notched; carpels not releasing the seed, carpel-body and awn separating from the beak of the fruit **64. cinereum**

61a. Immature fruits erect, flower-stalk sharply bent near the calyx 62

b. Immature fruits not erect, held approximately in line with the flower-stalk 69

62a. Flowers not more than *c.* 1.5 cm wide **28. thunbergii**

b. Flowers 2 cm wide or more 63

63a. Leaves cut into linear segments 2–5 mm wide **33. caffrum**

b. Leaves with the segments essentially lanceolate and/or wider than 5 mm 64

64a. Plant with glandular hairs above
6. psilostemon

b. Plant without glandular hairs 65

65a. Plant downy or velvety; sepals 9–12 mm; petal-veining strong, feathered **23. wlassovianum**

b. Plant hairy but not downy or velvety; sepals 6–10 mm; petal-veining rather weak, not feathered 66

66a. Petals deep purplish red; stigmas 2.5–3 mm 67

b. Petals white to pink; stigmas 3.5–6 mm 68

67a. Nectaries topped with hair tufts
13. wallichianum

b. Nectaries not topped with hair tufts
22. palustre

68a. Leaves boldly marbled with yellow-green; lobes of leaf-blade divisions more or less obtuse, ovate or oblong; plant rhizomatous and with numerous small underground tubers   **26. orientalitibeticum**

b. Leaves green or only faintly marbled; lobes of leaf-blade divisions acute, lanceolate: underground parts not as above
24. yesoense

69a. Leaves divided not more than three-quarters of the way to the stalk; slender tip of style, if any, not more than 3.5 mm   70

b. Leaves divided six-sevenths of the way to the stalk or more; slender tip of style 7–8 mm   72

70a. Stamens longer than sepals; stigmas and slender tip of style 2.5–3.5 mm   **3. nodosum**

b. Stamens not more than 2/3 as long as sepals; stigmas c. 4.5 mm; slender tip of style, if any, not more than 1.5 mm   71

71a. Plant not spreading by rhizomes; stems few, 6–12 mm thick
17. rubifolium

b. Plant spreading by underground rhizomes; stems numerous, to 6 mm thick   **16. kishtvariense**

72a. Plant not creeping underground; leaf-divisions with rather freely toothed lobes; one or both of the flower-stalks in each pair shorter than the sepals; immature fruits and their stalks sharply reflexed
7. pratense

b. Plant creeping underground; leaf-divisions with the lobes entire or with one or two teeth; both of the flower-stalks in each pair as long as the sepals or longer; immature fruits slightly reflexed on spreading or slightly reflexed stalks   **8. clarkei**

1. G. endressii Gay. Figure 113(1), p. 429. Illustration: Bonnier, Flore complète **2**: 101 (1913); Thomas, Plants for ground cover, t. 17 (1970); Brickell (ed.), RHS A–Z encyclopedia of garden plants, 474 (2003).

Evergreen perennial to 45 cm with long surface rhizomes. Glandular hairs present on sepals and sometimes axes and flower-stalks. Leaves 5–10 cm wide, wrinkled, divided for four-fifths to five-sixths into 5; divisions diamond-shaped, palmato-pinnately lobed about halfway to midrib; lobes acutely toothed. Flowers 3–3.7 cm wide, erect, funnel-shaped, borne just above the leaves, on stalks to 2.5 times as long as the sepals. Sepals 7–9 mm. Petals 1.6–1.8 cm, about twice as long as wide,

entire or slightly notched, deep pink, with hairs at the base extending across front surface. Stamens c. two-thirds as long as sepals. Stigmas 2.5–3.5 mm. Immature fruits and stalks erect. Fruit 2–2.5 cm. *Western Pyrenees (France, Spain).* H3. Summer.

Fertile hybrids between *G. endressii* and *G. versicolor* (**G.** × **oxonianum** Yeo) are usually distinguishable by the moderately notched pink petals with more or less conspicuous net-veining, but if this is weak they are not easy to recognise unless direct comparison with *G. endressii* is possible; plants may reach 80 cm and have petals to 2.6 cm. *Naturalised in SW England & NW France.*

2. G. versicolor Linnaeus (*G. striatum* Linnaeus). Figure 113(2), p. 429. Illustration: *Botanical Magazine*, 55 (1788); Ross-Craig, Drawings of British plants **6**: 29 (1952); Yeo, Hardy geraniums, 69, t. 4 (1985).

Similar to *G. endressii* but rhizomes shorter; leaves not divided beyond four-fifths, teeth blunter; flowers 2.5–3 cm wide, on stalks to 1.5 times as long as the sepals; petals white with purple net-veining, distinctly notched and stamens slightly longer than sepals. *S Europe (Sicily to Greece).* H3. Summer.

3. G. nodosum Linnaeus. Figure 114(2), p. 430. Illustration: Hegi, Illustrierte Flora von Mittel-Europa **4**(3): 1707 (1924); Polunin, Flowers of Europe, t. 63 (1969); Phillips, Wild flowers of Britain, 123 (1977); Yeo, Hardy geraniums, 72, 73 (1985).

Deciduous perennial to 50 cm with long rhizomes on or below the surface; glandular hairs absent. Leaves 5–20 cm wide, glossy beneath; divided for two-thirds into 5; divisions broadly to narrowly elliptic, scarcely lobed, toothed except at base; blades of upper leaves usually with 3 lanceolate divisions. Flowers 2.5–3 cm wide, erect, funnel-shaped, scattered, on stalks usually shorter than sepals. Sepals 7–9 mm. Petals 1.5–2.2 cm, at least twice as long as wide, deeply notched, purplish pink or violet with darker veins, with hairs at the base extending across front surface. Stamens longer than sepals but not more than 1.5 times as long. Stigmas 2.5–3.5 mm. Immature fruits and their stalks erect, horizontal or nodding. Fruit 2.2–3 cm with slender tip 2.5–3.5 mm. *C France to Pyrenees, C Italy, C Yugoslavia.* H3. Summer.

4. G. sylvaticum Linnaeus. Figure 122(4), p. 447. Illustration: Hegi, Illustrierte Flora von Mittel-Europa **4**(3): t. 174 (1924) – as *G. pratense*; Hess et al., Flora der Schweiz **2**: 626 (1970); Aeschimann et al., Flora alpina **1**: 1055 (2004).

Deciduous tufted perennial to 70 cm, with glandular hairs above. Leaves 10–22 cm wide, divided for four-fifths or more into 7 or 9; divisions broadest above the middle, palmato-pinnately cut about halfway to midrib into oblong, many-toothed lobes; teeth more or less acute. Flowers 2.2–3 cm wide, erect, saucer-shaped, numerous, crowded, on stalks mostly 1–2 times as long as the sepals. Sepals 5–7 mm. Petals c. 1.5 cm, not more than 1.5 times as long as wide, entire or slightly notched, purplish violet with a white base, less often pink or white, with hairs at the base extending across front surface. Stamens much longer than sepals. Stigmas 2–3 mm. Immature fruits and stalks erect. Fruit 2–2.5 cm. *Europe, N Turkey, Caucasus.* H1. Early summer.

'Mayflower' (illustration: Brickell (ed.), RHS A–Z encyclopedia of garden plants, 477, 2003) has larger flowers to 3.5 cm across.

5. G. rivulare Villars (*G. aconitifolium* L'Héritier). Figure 115(1), p. 431. Illustration: Bonnier, Flore complète **2**: 101 (1913); Hess et al., Flora der Schweiz **2**: 626 (1970); Yeo, Hardy geraniums, 75, t. 5 (1985); Aeschimann et al., Flora alpina **1**: 1055 (2004).

Deciduous tufted perennial to 45 cm, without glandular hairs except on sepals and fruit. Leaves 10–15 cm wide, divided nearly to base into 7 or 9; divisions broadest at or above the middle, cut about four-fifths of the way to midrib into more or less lanceolate lobes with few to many teeth, giving a broken outline; teeth narrow, acute. Flowers 1.5–2.5 cm wide, erect, funnel-shaped, rather numerous and crowded, on stalks 1.5–4 times as long as the sepals. Sepals 5–7 mm. Petals c. 1.5 cm, more than 1.5 times as long as wide, entire, white, finely veined with violet, hairy at the base extending across front surface. Stigmas c. 2 mm. Immature fruits and stalks erect. Fruit 2.5–2.8 cm. *W & C European Alps.* H1. Early summer.

6. G. psilostemon Ledebour (*G. armenum* Boissier; *G. backhouseanum* Regel). Figure 122(5), p. 447. Illustration: *Gartenflora* **22**: t. 778 (1873); *The Garden* **11**: 478 (1877); Hay & Beckett, Reader's Digest

Figure 113. Leaves of *Geranium* species (× 0.5). 1, *G. endressii*.
2a, b, *G. versicolor*. 3a, b, *G. pratense* var. *stewartianum*, c, *G. pratense*
(Nepalese), d, *G. pratense* (European). 4, *G. transbaicalicum*.

Figure 114. Leaves of *Geranium* species (× 0.5). 1a, b, *G. himalayense*.
2, *G. nodosum*. 3, *G. platyanthum*. 4, *G. clarkei*. 5, *G. rubifolium*.
6, *G. sinense*.

Figure 115. Leaves of *Geranium* species (× 0.75). 1, *G. rivulare*.
2, *G. swatense*. 3, *G. wallichianum*. 4, *G. lambertii*. 5, *G. procurrens*.
6, *G. kishtvariense*. 7, *G. pogonanthum*. 8, *G. yunnanense*.

encyclopaedia of garden plants and flowers, edn 2, 300 (1978); Yeo, Hardy geraniums, 77, t. 6 (1985); Brickell (ed.), RHS A–Z encyclopedia of garden plants, 477 (2003).

Deciduous tufted perennial to 1.2 m, with glandular hairs above. Stipules 5–12 mm, narrowly triangular to lanceolate with a long fine tip. Leaves 15–30 cm wide, divided for c. four-fifths into 7; divisions broadest above the middle, lobed about halfway to midrib in upper two-thirds; lobes oblong, many-toothed; teeth acute or nearly so. Flowers c. 3.5 cm wide, not crowded, shallowly bowl-shaped, on stalks 2–3 times as long as the sepals. Sepals 8–9 mm. Petals c. 1.8 cm, nearly as wide, entire or slightly notched, brilliant reddish purple with base and veins black, with a few basal hairs on the edges. Stigmas 2.5–3 mm, black or nearly so. Immature fruits erect on more or less deflexed stalks. Fruit 3.2–3.5 cm. *NE Turkey, SW Caucasus.* H2. Summer.

7. G. pratense Linnaeus. Figure 113(3c, d), p. 429. Illustration: Ross-Craig, Drawings of British plants **6**: t. 32 (1952); Grey-Wilson & Blamey, Alpine flowers of Britain and Europe, 131 (1979); Yeo, Hardy geraniums, 81, 82 (1985); Aeschimann et al., Flora alpina **1**: 1055 (2004).

Deciduous tufted perennial, 75–130 cm, with glandular hairs on upper parts. Leaves 10–25 cm wide, divided for six-sevenths to nine-tenths into 7 or 9; divisions broadest above the middle, deeply pinnately lobed; lobes narrowly oblong with several teeth; teeth mostly lanceolate, acute or nearly so. Flowers 3.5–4.5 cm wide, saucer-shaped, facing sideways, numerous and crowded, on stalks usually not longer than sepals, but held well above leaves on axes to 10 cm. Sepals 7–12 mm with mucro 1.5–3.5 mm; calyx inflated after flowering. Petals 1.6–2.4 × 1.3–2 cm, entire, blue, varying to white or rarely pink, with hairs at base forming a tuft on each edge but not extending all across front surface. Filaments with base strongly enlarged; anthers usually dark blue. Stigmas 2–2.5 mm. Immature fruits and stalks directed downwards. Fruit 2.8–3.4 cm with slender tip 7–8 mm. *Europe, C Asia, Himalaya, W China.* H1. Summer.

The above description applies to plants from Europe and much of northern Asia. There are several named cultivars, including some with small double flowers.

Himalayan plants show much variation; a variant found in Nepal and some areas further west in the Himalaya (illustration: Yeo, Hardy geraniums, 84, 1985) has leaves less than 20 cm, divisions broadest at middle, more shortly tapered to either end, and with shorter, broader, more obtuse lobes and teeth; its petals often have dark veins.

Also in the W Himalaya is the May-flowering var. **stewartianum** Nasir (Figure 113(3a, b), p. 429), which is lower and bushier than European plants; its leaves are less finely cut though not such short and broad segments as in the variant just described; the flowers are upwardly inclined on longer stalks, and have violet to pink petals; fruits similar to those of *G. clarkei.*

G. transbaicalicum Sergievskaya. Figure 113(4), p. 429. Illustration: Yeo, Hardy geraniums, t. 8, 83, (1985). From around Lake Baikal in Siberia; only c. 40 cm, and slender-stemmed, with relatively small, very finely cut leaves and much brownish purple pigmentation.

A sterile plant intermediate between the European variant of *G. pratense* and *G. himalayense*, and doubtless a hybrid between them, is frequently cultivated under the name 'Johnson's Blue'.

8. G. clarkei Yeo. Figure 114(4), p. 430. Illustration: Yeo, Hardy geraniums, t. 11, 86 (1985).

Differs from *G. pratense* as follows: rhizomatous, not more than 50 cm; leaves 5–15 cm wide, divided nearly to base, with narrower, entire or few-toothed lobes; flowers larger, 4.2–4.8 cm wide, tilted upwards, less crowded, on stalks 0.6–1.6 times as long as sepals; sepals larger, with a relatively shorter mucro, not forming inflated calyx after flowering; petals purplish or white with bluish-pink veins; immature fruits and stalks slightly nodding to uptilted. *W Himalaya.* H1. Summer.

'Kashmir White' (illustration: Brickell (ed.), RHS A–Z encyclopedia of garden plants, 474, 2003) has white flowers.

9. G. himalayense Klotzsch (*G. grandiflorum* Edgeworth not Linnaeus). Figure 114(1), p. 430. Illustration: *Flora and Sylva* **1**: 54 (1903); Hay & Beckett, Reader's Digest encyclopaedia of garden plants and flowers, edn 2, 299 (1978); Graf, Exotica, edn 11, 1177 (1982); Yeo, Hardy geraniums, 87, 88, t. 12 (1985).

Deciduous carpet-forming rhizomatous perennial to 50 cm, with glandular hairs

on upper parts. Leaves 6–12 cm wide, divided for three-quarters to four-fifths into 7; divisions broadly diamond-shaped, almost palmately 3-lobed; lobes about as wide as long, often curved outwards, with several teeth; teeth obtuse or acute. Flowers 4–6 cm wide, saucer-shaped, facing sideways, not crowded, rather numerous, held well above the leaves on long axes, although one stalk in each pair is usually shorter than the calyx. Sepals 8–12 mm with mucro 0.5–1.5 mm. Calyx not inflated after flowering. Petals 2–3 × 1.8–2.5 cm, entire, blue or pale blue, often purplish towards base, with hairs at base forming a tuft on each edge but not extending across front surface. Filaments much enlarged at base; anthers dark blue. Stigmas 2–3.5 mm. Immature fruits and stalks directed downwards. Fruit 3.2–3.5 cm with slender tip 7–10 mm. *Himalaya (NE Afghanistan to C Nepal); Pamir area of former USSR.* H1. Summer.

There is a double-flowered cultivar; it is less vigorous in growth than the natural form and has flowers of much smaller diameter.

10. G. platyanthum Duthie (*G. eriostemon* de Candolle not Poiret). Figure 114(3), p. 430. Illustration: Sweet, Geraniaceae **2**: 197 (1822–24); Shimizu, New alpine flora of Japan in colour **1**: 199 (1982).

Deciduous tufted perennial, 30–50 cm, with glandular hairs throughout, or only on upper parts. Simple hairs on lower parts of plant spreading. Leaves 5–25 cm wide, divided for half to two-thirds into 5 or 7; divisions broadly ovate or oblong, not overlapping, shallowly 3-lobed at apex and shallowly toothed, or not lobed, merely unevenly toothed. Teeth more or less acute. Flowers 2.5–3.2 cm wide, flat, facing sideways or nodding, numerous, very crowded, on stalks 1–1.5 times as long as sepals, many of them in umbels. Sepals 7–9.5 mm. Petals to 1.6 × 1.6 cm, entire or irregularly notched, light violet-blue, with hairs at base forming a tuft on each edge but not extending all across front surface. Stamens adpressed to style; filaments blackish purple except at base, basal half covered with long spreading hairs; anthers dull blue. Stigmas 1.5–3 mm. Immature fruits and stalks erect, held well above the flowers. Fruit 2.6–3.2 cm, with slender tip 5–6 mm. *E Siberia to W China, Korea, Japan.* H1. Early summer.

Japanese and NE Chinese plants usually have the leaves similar to those of

G. erianthum and are distinguished from that by the spreading hairs of the lower parts.

11. G. erianthum de Candolle. Figure 123, p. 449. Illustration: Makino, Illustrated flora of Japan, enlarged edn, 398 (1941); Shimizu, New alpine flora of Japan in colour **1**: 199 (1982); Yeo, Hardy geraniums, 90, 91, t. 13 & 14 (1985); Brickell (ed.), RHS A–Z encyclopedia of garden plants, 474 (2003).

Differs from *G. platyanthum* as follows: simple hairs on lower parts of plant adpressed; leaves divided for two-thirds to four-fifths into 7 or 9, the divisions overlapping their neighbours, more elongated, the lobes more freely and deeply toothed; flowers slightly larger, 2.7–3.7 cm wide, facing sideways; petals entire, light to deep violet-blue, darkly veined; slender tip of fruit 6–7 mm. *E Siberia, Japan, Kuril Islands, Sakhalin, USA (Alaska), NW Canada.* H1. Early summer.

12. G. swatense Schönbeck-Temesy. Figure 115(2), p. 431. Illustration: Rechinger (ed.), Flora Iranica **69**: t. 1 (1970).

Deciduous perennial with a swollen taproot, to 25 cm, with glandular hairs above. Basal leaves few. Stem leafy, at first erect, soon trailing, with pairs of unequal leaves. Leaves 2–8 cm wide, marbled, divided for three-quarters to four-fifths into 5; divisions diamond-shaped, 3-lobed in the apical third; lobes with obtuse to acute teeth. Flowers 2.5–4 cm wide, saucer-shaped, more or less erect, rising above the leaves at each node, on stalks 3.5–6 times as long as sepals. Sepals 8–11 mm with mucro 1–1.5 mm, flushed with purple, sometimes recurved. Petals 1.6–1.8 × 1.1–1.7 cm, entire, pinkish violet to bright pink with fine darker veins, with hairs at the base forming a tuft on the edges and across front surface. Stamens shorter than sepals; filaments hairy in basal half, much enlarged at base; anthers dull blue or yellow. Stigmas *c.* 4.5 mm, red. Immature fruits erect on spreading or reflexed stalks. Fruit 2.2–3.2 cm, with slender tip *c.* 3 mm. *NE Afghanistan, N Pakistan.* H1.

13. G. wallichianum D. Don. Figure 115 (3), p. 431. Illustration: *Botanical Magazine*, 2377 (1823); Hay & Beckett, Reader's Digest encyclopaedia of garden plants and flowers, edn 2, 301 (1978); Polunin & Stainton, Flowers of the Himalaya, t. 21

(1985); Yeo, Hardy geraniums, 92, t. 15 (1985).

Deciduous perennial with compact rootstock and usually trailing stems, without glandular hairs. Plants of flowering size without basal leaves. The 4 stipules of a leaf-pair joined to make 2, these 6–18 × 4–11 mm, at first encasing the tip of the shoot. Leaves 3.5–15 cm wide, divided for two-thirds to four-fifths into 5 or sometimes 3; divisions wide, shortly tapered above and below the middle, shallowly lobed and acutely toothed. Flowers 2.7–3.6 cm wide, saucer-shaped, upwardly inclined, scattered, on stalks 2–9 times as long as the sepals. Sepals 6–9 mm with mucro 1.5–4 mm, Petals 1.6–1.7 × 1.3–1.6 cm. rounded or notched, purple to blue with darker veins; petal-base sometimes white, with a tuft of hairs on each edge and usually with hairs all across front surface. Filaments curving outwards, darkly coloured; anthers black. Stigmas 5.5–7 mm, black or dark red. Immature fruits erect on reflexed stalks. Fruit 2.6–4 cm. *W Himalaya (NE Afghanistan to Kashmir).* H1. Summer.

Conspicuously variable in overall hairiness and in colour of leaves and flowers.

14. G. lambertii Sweet (*G. candicans* misapplied; *G. grevilleanum* Wallich). Figure 115(4), p. 431. Illustration: Sweet, Geraniaceae **4**: t. 338 (182628); Ohashi, Flora of the eastern Himalaya, Third report, t. 4 (1975); *Journal of the Royal Horticultural Society* **109**: 37 (1984); Stainton, Flowers of the Himalaya, a Supplement, t. 17 (1988).

Deciduous perennial with compact rootstock and trailing stems; glandular hairs present on floral parts only. Basal leaves few or none. Leaves 5–15 cm wide, divided for half to two-thirds into 5; divisions wide, shortly tapered to apex; lobes shortly oblong; teeth acute or obtuse and mucronate. Flowers 3.5–5.5 cm wide, shallowly bowl-shaped or saucer-shaped, nodding or facing sideways, scattered, on stalks 1–3 times as long as sepals. Sepals 9–11 mm with mucro 1–2 mm. Petals 2–2.9 × 1.2–2 cm, entire, pale pink or white and then often red-stained at base; base with a tuft of hairs on each edge and hairs extending all across front surface. Filaments densely hairy, curving outwards, blackish or dark purple; anthers black. Stigmas 5–6 mm, black. Immature fruits

erect on reflexed stalks. Fruit 3.1–3.5 cm. *Himalaya (C Nepal to Bhutan & adjacent Xizang).* H2. Summer.

15. G. procurrens Yeo. Figure 115(5), p. 431. Illustration: *Botanical Magazine*, n.s., 644 (1973); Yeo, Hardy geraniums, 94, 95 (1985); Stainton, Flowers of the Himalaya, a supplement, t. 17 (1988).

Deciduous perennial with compact rootstock and trailing red stems that root at the nodes; glandular hairs present on upper parts. Basal leaves few. Some of the stipules joined in pairs. Leaves 5–10 cm wide, divided for two-thirds to five-sixths into 5 or rarely 7; divisions diamond-shaped, pinnately lobed or 3-lobed at the apex; lobes few-toothed; teeth obtuse or acute. Flowers 2.5–3.5 cm wide, saucer-shaped, upwardly inclined, scattered, on stalks 3–5 times as long as the sepals. Sepals 6–9 mm with mucro not more than 1 mm. Petals 1.6–1.8 cm × 8–12 mm, entire, deep dull pinkish purple with V-shaped black mark at base and black veins; base with a tuft of hairs on each edge and hairs all across front surface. Filaments curving outwards, black; anthers black. Stigmas 4–4.5 mm, black. Immature fruits erect on erect stalks. Fruit *c.* 2.2 cm. *Himalaya (E Nepal, Sikkim).* H2. Late summer–early autumn.

16. G. kishtvariense Knuth. Figure 115 (6), p. 431. Yeo, Hardy geraniums, 96, t. 17 (1985).

Deciduous perennial to 40 cm, spreading by underground rhizomes, with glandular hairs above. Stems with swollen nodes, more or less scrambling. Basal leaves few or none. Stipules to 6 mm, narrow, some of them joined in pairs. Leaves mostly 4–9 cm wide, bright green, wrinkled, divided for two-thirds to three-quarters into 3 or 5; divisions more or less diamond-shaped, shallowly and sometimes indistinctly 3-lobed, with many narrow acute teeth. Flowers *c.* 4 cm wide, saucer-shaped, more or less erect, scattered, on stalks 1.5–4 times as long as the sepals. Sepals 7–9 mm with mucro 2–3 mm. Petals 1.7–2.1 cm, entire, deep or purplish pink with a white V-shaped area at base and fine purple veins; hairs at base forming a tuft on each edge and extending all across front surface and some way up the edges. Stamens only two-thirds as long as sepals; filaments with base strongly enlarged and white, tips divergent; anthers blackish. Stigmas *c.* 4.5 mm, blackish red. Immature fruits and stalks reflexed. Fruit 2–2.2 cm, without a slender tip. *W Himalaya (Kashmir)*. H2. Summer.

17. G. rubifolium Lindley. Figure 114(5), p. 430. Illustration: *Edwards's Botanical Register* **26**: t. 67 (1840); *Plants and Gardens* 1989 (Summer): 73.

Deciduous perennial to 1.2 m with thick compact rootstock, glandular above. Stems few, thick, erect, copiously branched above. Basal leaves few or none. Stipules 6–15 mm, narrow, some of them joined in pairs. Leaves mostly 5–17 cm wide, wrinkled above, divided to two-thirds into 5 or 7; divisions broadly diamond-shaped to narrowly ovate, shallowly and indistinctly lobed, unevenly saw-edged. Flowers very numerous, similar to *G. kishtvariense*, but with petals sometimes slightly notched. Fruit to 2.7 cm, with a slender tip 1–1.5 mm. *Himalaya (W Kashmir)*. H2.

18. G. sinense Knuth (*G. platypetalum* Franchet, not Fischer & Meyer; *G. delavayi* misapplied). Figure 114(6), p. 430. Illustration: Yeo, Hardy geraniums, 98, t. 18 (1985).

Deciduous tufted perennial to 40 cm, with a thick knobbly rootstock and glandular hairs on upper parts. Some of the stipules joined in pairs. Leaves 5–20 cm wide, divided for three-quarters to seven-eighths into 5 or 7; divisions diamond-shaped with narrow base, lobed a third to halfway to midrib or merely unevenly toothed; teeth acute or nearly so. Flowers *c.* 2 cm wide, protruding in the centre, nodding, moderately numerous, more or less crowded. Sepals 5.5–7.5 mm, reflexed, with mucro 0.75–1.5 mm. Petals 9–10 × 6–7 mm, entire or irregularly lobed, reflexed, blackish red with a dull pink base, with hairs present on edges and front surface at base but very few. Nectary solitary, encircling the stamens. Stamens longer than sepals, adpressed to the style, with few or no hairs; filaments dark red; anthers nearly black. Stigmas 1.5–2 mm. Immature fruits directed downwards, on reflexed or spreading stalks. Fruit 2.2–2.6 cm. *SW China*. H2. Summer.

19. G. pogonanthum Franchet (*G. yunnanense* misapplied). Figure 115(7), p. 431. Illustration: *Notes from the Royal Botanic Garden, Edinburgh* **34**: 198 (1975); *Journal of the Royal Horticultural Society* **109**: 37 (1984); Yeo, Hardy geraniums, 99 (1985).

Deciduous tufted perennial to 60 cm, with a thick knobbly rootstock and sometimes glandular hairs on upper parts. Some of the stipules joined in pairs. Leaves 5–13 cm wide, divided for two-thirds to nine-tenths into 5 or 7; divisions diamond-shaped, palmato-pinnately lobed about halfway to midrib; lobes oblong, toothed; teeth acute. Flowers 2.5–3.5 cm wide, inversely bowl-shaped or saucer-shaped, nodding, scattered. Sepals 6–10 mm with mucro 1–2 mm. Petals 1–2 cm, about half as wide, entire, adpressed to the sepals and, if long enough, recurved beyond them, pink; base with a dense tuft of hairs on each edge and hairs extending all across front surface. Stamens longer than sepals; filaments adpressed to the style, curved outwards at tips, purplish red to pink, covered with hairs below; anthers blue-black. Stigmas 2.5–4 mm. Immature fruits directed downwards, on reflexed stalks. Fruit 2.2–2.8 cm, with slender tip 2–6 mm. *SW China, W & N Burma*. H2. Summer.

20. G. yunnanense Franchet. Figure 115 (8), p. 431. Illustration: *Journal of the Royal Horticultural Society* **109**: 37 (1984).

Like *G. pogonanthum* but glandular hairs absent or confined to undersides of leaves, leaves divided only to four-fifths to six-sevenths, with fewer, shallower, blunter lobes and teeth; with longer sepals, 1–1.2 cm, broader petals, 1.2–1.3 cm wide, forming bowl-shaped flowers, more extensively hairy and more divergent stamens and smaller fruit, 1.8–2.7 cm, with slender tip 2–3 mm. *SW China, N Burma*. H2. Summer.

21. G. farreri Stapf (*G. napuligerum* misapplied). Figure 119(1), p. 443. Illustration: *Gardeners' Chronicle* **75**: 333 (1924) & **87**: 241 (1930); *Botanical Magazine*, 9092 (1926); Hay & Beckett, Reader's Digest encyclopaedia of garden plants and flowers, edn 2, 301 (1978); Graf, Exotica, edn 11, 1177 (1982).

Deciduous perennial to 12 cm, with tufted rootstock, without glandular hairs. Leaves not more than 5 cm wide, divided for three-quarters to four-fifths into 7; divisions broadest near the apex, 3-lobed at apex for one quarter to two-thirds their length; teeth few, obtuse to acute, or none. Flowers 3–3.5 cm wide, flat or saucer-shaped, facing sideways, sometimes crowded, on stalks *c.* 2.5 times as long as the sepals. Sepals 7–9 mm with mucro 0.5–1 mm. Petals 1.3–1.5 × 1–1.5 cm,

entire, with wavy margins, very pale pink to white, with hairs at the base forming a tuft on each edge and another in the middle of the front surface. Stamens longer than sepals; anthers blue-black. Stigmas *c.* 1.5 mm. Immature fruits spreading horizontally on reflexed stalks. Fruit 2.6–2.9 mm, with slender tip 7–9 mm. *W China*. H1. Early summer.

22. G. palustre Linnaeus. Figure 117(2), p. 438. Illustration: Sweet, Geraniaceae **1**: t. 3 (1820); Coste, Flore de la France **1**: 249 (1901); Hegi, Illustrierte Flora von Mittel-Europa, edn 1 **4**(3): 1653 (1924); Fitter et al., Wild flowers of Britain and northern Europe, 135 (1974).

Deciduous bushy perennial to 60 cm with compact rootstock; without glandular hairs. Leaves 5–12 cm wide, divided for two-thirds to five-sixths into 5 or 7; divisions diamond-shaped, coarsely pinnately lobed halfway to the midrib or more; lobes oblong or tapered, few-toothed; teeth obtuse to acute. Flowers 3–3.5 cm wide, widely trumpet-shaped, upwardly inclined, scattered, on stalks 2.5–4 times as long as the sepals. Sepals 6.5–8 mm with mucro *c.* 1 mm. Petals 1.6–1.8 × *c.* 1 cm, entire or slightly notched, deep purplish red, base with hairs on margins, not forming a tuft, and extending across front surface. Stamens longer than sepals. Stigmas 2.5–3 mm. Immature fruits erect on deflexed stalks. Fruit *c.* 2 cm. *E & C Europe*. H1. Summer.

23. G. wlassovianum Link. Figure 117 (3), p. 438. Illustration: Sweet, Geraniaceae **3**: t. 228 (1824–26); Iconographia cormophytorum Sinicorum **2**: 528 (1972).

Deciduous bushy perennial to 40 cm with compact rootstock, without glandular hairs. Some of the stipules joined in pairs. Leaves 5–15 cm wide, divided for two-thirds to five-sixths into 5 or 7, downy; divisions slightly tapered below, shortly tapered above, palmato-pinnately lobed for *c.* one quarter their length; lobes shortly oblong, sometimes curved outwards, few-toothed; teeth acute. Flowers 3–4 cm wide, saucer-shaped to widely funnel-shaped, upwardly inclined, scattered, on stalks 1.5–3.5 times as long as the sepals. Sepals 9–12 mm, with mucro 1–2 mm. Petals 1.7–2.2 cm × 9–13 mm, *c.* 1.5 times as long as broad, entire, pale pink to dusky purple, strongly dark-veined, with hairs at base forming a single tuft on front surface and edges. Stamens longer than sepals, curving outwards. Stigmas 3.5–4.5 mm. Immature fruits erect on spreading or reflexed stalks. Fruit 2.4–2.9 mm. *E Siberia, Mongolia, Pacific area of former USSR, N China.* H1. Summer.

24. G. yesoense Franchet & Savatier. Figure 116(1), p. 436. Illustration: Takeda, Alpine flora of Japan in colour, 34 (1959); Kitamura et al., Coloured illustrations of herbaceous plants of Japan (Choripetalae), 90 (1972); Shimizu, New alpine flora of Japan in colour **1**: 199 (1982).

Deciduous bushy perennial to 40 cm with compact rootstock, without glandular hairs. Leaves 5–10 cm wide, divided for three-quarters to five-sixths or more into 7; divisions pinnately or palmato-pinnately cut into lanceolate lobes; lobes with a few short or long and slender acute teeth. Flowers 2.2–3 cm wide, saucer-shaped to shallowly funnel-shaped, upwardly inclined, scattered, on stalks 3–6 times as long as sepals. Sepals 6–10 mm with mucro 1–2 mm. Petals 1.4–2 cm × 6–13 mm, entire, pink with darker veins or white, hairs at base extending all across front surface. Stigmas 3.54.5 mm. Immature fruits erect on spreading or reflexed stalks. Fruit 1.7–2.4 cm. *C & N Japan, Kuril Islands.* H1. Summer.

25. G. pylzowianum Maximowicz. Figure 116(2), p. 436. Illustration: *The Garden* **85**: 345 (1921) – as *G. stapfianum*; *Kew Magazine* **1**: t. 18 (1984) – as *G. orientalitibeticum*.

Deciduous perennial 15–25 cm, without glandular hairs, forming small underground tubers, 5 × 3 mm, linked by slender rhizomes. Basal leaves arising directly from the tubers. Leaves 2–6 cm wide, kidney-shaped, semicircular or pentagonal, divided nearly to the base into 5 or 7; divisions broadest near apex, palmato-pinnately cut into lanceolate or tongue-shaped lobes with a few acute or obtuse teeth. Flowers 2.7–3.2 cm wide, widely trumpet-shaped, erect, few, on stalks 3.5–6 times as long as sepals. Sepals 6–9 mm with mucro 0.5–1 mm. Petals 1.2–2.3 × 1.1–1.4 cm, entire or slightly notched, deep pink, graduated to white at base, with fine darker veins, hairs at base extending all across front surface. Filaments curved outwards at the tips. Stigmas 2–3 mm. Immature fruits erect on more or less reflexed stalks. Fruit 1.8–2.1 cm. *W China (Gansu to Yunnan).* H1. Early summer.

26. G. orientalitibeticum Knuth (*G. stapfianum* misapplied; *G. roseum* invalid). Figure 116(3), p. 436. Illustration: *Kew Magazine* **1**: t. 17 (1984) – as *G. pylzowianum*; Yeo, Hardy geraniums, 112, 113, t. 19 (1985).

Like *G. pylzowianum* in habit but differing as follows: to 35 cm; tubers 5–10 × 4–6 mm; leaves mostly 5-sided, 2.5–7 cm wide, strongly marbled in two shades of green; leaf-lobes ovate or oblong, teeth more or less obtuse; flowers 2.3–2.7 cm wide, flat at edges but with shallowly cup-shaped centre; petals purplish pink with basal third white; stigmas 4–6 mm; fruit 2.2–2.5 cm. *W China (Sichuan).* H1. Summer.

27. G. sanguineum Linnaeus. Figure 116(4), p. 436. Illustration: Ross-Craig, Drawings of British plants **6**: 28 (1952); Hess et al., Flora der Schweiz **2**: 623 (1970); Fitter et al., Wild flowers of Britain and northern Europe, 135 (1974); Grey-Wilson & Blamey, Alpine flowers of Britain and Europe, 131 (1979).

Deciduous bushy perennial to 30 cm with underground rhizomes, without glandular hairs. Basal leaves few. Leaves 2–8 cm wide, divided for three-quarters to seven-eighths into 5 or 7; divisions pinnately or palmato-pinnately cut into lanceolate or elliptic lobes; lobes entire or with 1 or 2 teeth, the lower curved outwards; teeth mostly lanceolate, acute. Flowers 2.5–4.2 cm wide, saucer-shaped, erect or upwardly inclined, scattered, solitary, on stalks 2–5 times as long as sepals. Sepals 6.5–10 mm with mucro 1.5–3.5 mm. Petals 1.4–2.1 × 1.3–1.7 cm, entire, shallowly or deeply notched, deep purplish red, pink or white, usually with darker veins, with hairs at base in a dense tuft on either edge and usually extending all across front surface. Stamens shorter than sepals. Stigmas 3.5–4.5 mm. Immature fruits and stalks erect. Fruit 2.6–3.6 cm. *Europe, Caucasus, N Turkey.* H1. Summer.

Var. **striatum** Weston (*G. lancastrense* (Miller) Nicholson). Illustration: Graf, Tropica, 490 (1978) – as *G. grandiflorum*; Yeo, Hardy geraniums, t. 20 (1985); Brickell (ed.), RHS A–Z encyclopedia of garden plants, 477 (2003). Low-growing with flowers appearing pale flesh-pink, an effect produced by diffuse pink veining on a nearly white ground-colour. *NW coast of England.*

28. G. thunbergii Lindley & Paxton (*G. nepalense* Sweet var. *thunbergii* (Lindley & Paxton) Kudo). Figure 116(5), p. 436. Illustration: *Paxton's flower garden* **1**: 186 (1851) & and revised edn, **1**: 191 (1882); Makino, Illustrated flora of Japan, enlarged edn, 396 (1941); Yeo, Hardy geraniums, 117 (1985).

Evergreen sprawling perennial with a weak rootstock, to 25 cm, sometimes rooting at the nodes, with glandular hairs on upper parts. Winter leaves usually purple-blotched in the sinuses. Leaves 3–10 cm wide, divided for two-thirds to four-fifths into 5; divisions broadest near the apex or diamond-shaped or elliptic, coarsely lobed at the apex for a quarter to a third their length; lobes short and broad, shallowly toothed; teeth obtuse and mucronate or acute. Flowers 1.1–1.5 cm wide, upwardly inclined, on stalks 1–2.5 times as long as sepals. Sepals 5–6 mm with mucro 1 mm or less. Petals 7–10.5 × 4–6 mm, entire or slightly notched, white to deep purplish pink. Stamens much shorter than sepals. Stigmas 1–2 mm. Immature fruits erect on spreading, erect or rarely reflexed stalks. Fruit 1.5–2.1 cm, without a slender tip. *N China, Taiwan, Japan & neighbouring archipelagoes.* H2. Summer.

G. nepalense Sweet, from the Himalaya, S India and S China, is sometimes confused with *G. thunbergii*, partly for nomenclatural reasons; it lacks glandular hairs and is more slender and even smaller-flowered.

29. G. traversii J.D. Hooker. Figure 116 (6), p. 436. Illustration: Cheeseman, Illustrations of the New Zealand flora **1**: t. 25 (1914); Salmon, New Zealand flowers and plants in colour, edn 2, 164 (1967); Moore & Irwin, The Oxford book of New Zealand plants, 54 (1978).

Evergreen perennial with compact rootstock and shortly trailing white-felted stems to 20 cm, without glandular hairs. Leaves 5–10 cm wide, divided for *c.* three-fifths into 7, hoary grey-green; divisions broadly wedge-shaped, widest at the apex and there 3-lobed for a quarter to a third their length; lobes broader than long, occasionally toothed, obtuse and weakly mucronate. Flowers 2.5–3 cm wide, saucer-shaped, upwardly inclined, scattered, solitary, on stalks 6–10 times as long as sepals. Sepals 6.5–8 mm with mucro 0.5–1 mm. Petals 1.2–1.3 cm, nearly as wide as long, entire, white or pink with fine darker veins, with hairs at

Figure 116. Leaves of *Geranium* species (× 0.75). 1, *G. yesoense*. 2, *G. pylzowianum*. 3, *G. orientalitibeticum*. 4, *G. sanguineum*. 5a, b, *G. thunbergii*. 6, *G. traversii*. 7, *G. incanum* var. *multifidum*. 8, *G. caffrum*. 9, *G. robustum*.

base on edges and sparsely across front surface. Stamens shorter than sepals. Stigmas *c.* 1.5 mm. Immature fruit and stalks erect. Fruit 1.1–1.6 cm, without a slender tip. *New Zealand (Chatham Islands).* H5. Summer.

Very variable in nature but those cultivated at present are uniform and pink-flowered.

30. G. × riversleaianum Yeo. Figure 117 (5), p. 438. Illustration: Yeo, Hardy geraniums, 68, t. 3 (1985).

Similar to *G. traversii* but with much longer trailing stems and less dense hair-covering. Leaves 2.5–10 cm wide, downy, divided for two-thirds to four-fifths into 5 or 7; divisions broadly diamond-shaped to wedge-shaped, palmately or palmato-pinnately lobed; teeth very obtuse to acute. Flowers shortly funnel-shaped to saucer-shaped, in pairs, on stalks 1.5–4 times as long as sepals. Sepals to 6 mm. Petals pink to deep pinkish purple, slightly notched. Stigmas 1.5–3.5 mm. Fruit not developing. *Garden origin.* H4. Summer.

This plant is a putative hybrid between *G. endressii* and *G. traversii*. A number of clones has arisen, of which the oldest and best-known is 'Russell Prichard' (illustration: Brickell (ed.), RHS A–Z encyclopedia of garden plants, 477, 2003), which has intensely coloured flowers.

31. G. sessiliflorum Cavanilles subsp. **novaezelandiae** Carolin. Figure 119(2), p. 443. Illustration: Salmon, Field guide to the alpine plants of New Zealand, 160 (1968); Yeo, Hardy geraniums, 120 (1985).

Evergreen perennial with compact rootstock, without glandular hairs. Stems *c.* 15 cm, spreading. Leaves 1.5–4.5 cm wide, rounded, divided for half to three-fifths into 5 or 7, green or dark brown ('Nigricans'); divisions widest at apex and there shallowly 3-lobed; lobes occasionally toothed, more or less obtuse. Flowers *c.* 1 cm wide, funnel-shaped, erect, not or scarcely raised above the foliage, solitary on stalks 1–4 times as long as sepals. Sepals 3.5–6 mm, with long spreading hairs, mucro 1–1.5 mm. Petals 6.5–7.5 × 3 mm, white. Immature fruit erect on spreading or reflexed stalks. Fruit *c.* 1 cm. *New Zealand.* H3. Early summer–early autumn.

32. G. incanum Burman var. **multifidum** Hilliard & Burtt (*G. multifidum* Sweet not Andrews). Figure 116(7),

p. 436. Illustration: Sweet, Geraniaceae **3**: t. 245 (1824–26); Batten & Bokelman, Wild flowers of the eastern Cape Province, t. 74, f. 2 (1966).

Evergreen perennial with compact rootstock and bushy habit, with slightly woody stems to 50 cm, without glandular hairs. Leaves mostly alternate. Leaves *c.* 5 cm wide, divided to the base into 5; divisions pinnately lobed and some lobes toothed; segments of all orders linear, *c.* 1 mm wide, green above, silky-hairy beneath. Flowers 2.5–3.5 cm wide, funnel-shaped in the centre, more or less erect, on stalks 3–8 times as long as sepals. Sepals 5–6 mm, silvery-silky, with mucro *c.* 1 mm. Petals 1.3–1.8 cm × 9–14 mm, notched, usually deep pink; hairs at base forming a dense tuft on either edge and a small separate tuft on middle of front surface. Stigmas 3–3.5 mm. Immature fruits erect on more or less reflexed stalks. Fruit *c.* 2.7 cm. *South Africa.* H5. Early summer–early autumn.

Var. **incanum** has smaller flowers with white petals. Illustration: *Notes from the Royal Botanic Garden, Edinburgh* **42**: 196 (1985).

33. G. caffrum Ecklon & Zeyher. Figure 116(8), p. 436. Illustration: Saunders, Refugium botanicum **3**: t. 147 (1870); *Notes from the Royal Botanic Garden, Edinburgh* **42**: 174 (1985).

Evergreen perennial with compact rootstock and bushy habit, with slightly woody stems to 50 cm, with or without glandular hairs in the upper parts. Leaves 5–7.5 cm wide, divided to the base or nearly so into 5 or 7; divisions pinnatisect three-quarters to five-sixths the way to the midrib; lobes 2–5 mm wide, neglecting the teeth, and 2–6 times as long; teeth few, shallow and ovate to deep and linear, acute. Flowers 2–3 cm wide, shallowly funnel-shaped, more or less erect, scattered, on stalks 3–4 times as long as sepals. Sepals 5–7 mm with mucro to 1 mm. Petals 9–14 × 4–10 mm, entire or notched, white to pink with darker veins, with hairs at base in a weak tuft on each edge, not extending across front surface. Stigmas 3–3.5 mm. Immature fruits erect on erect or spreading stalks. Fruit 2.3–2.5 cm with slender tip to 4 mm. *South Africa.* H3. Summer.

Variable as a result of hybridisation in nature.

34. G. robustum Kuntze. Figure 116(9), p. 436. Illustration: *Notes from the Royal Botanic Garden, Edinburgh* **42**: 173 (1985).

Evergreen perennial with compact rootstock and initially erect stems to 1.75 m, woody at base; glandular hairs present on upper parts. Leaves 4–12 cm wide, with a felt of grey hairs beneath, divided to the base into 5; divisions broadest below the middle, pinnatisect to *c.* five-sixths of their width into oblong, many-toothed lobes; teeth elliptic, obtuse or acute. Flowers numerous but scattered, 3–3.5 cm wide, widely funnel-shaped, erect, on stalks 6–15 times as long as sepals. Sepals 6–9 mm, densely hairy, with mucro *c.* 0.75 mm. Petals 1.5–2 × 1.1–1.3 cm, notched, bluish violet with white base; base with hairs in a dense tuft on each edge but not extending across front surface. Stamens longer than sepals. Stigmas 4–4.5 mm. Immature fruit erect on more or less spreading stalks. Fruit *c.* 2.7 cm with slender tip 3–4 mm. *South Africa.* H4. Summer.

35. G. maculatum Linnaeus. Figure 118 (1), p. 440. Illustration: Sweet, Geraniaceae **4**: t. 332 (1826–28); Rickett, Wild flowers of the United States **1**: 235 (1966); Thomas, Perennial garden plants, edn 2, t. 14 (1982); Voss, Michigan flora **2**: 500 (1985); Brickell (ed.), RHS A–Z encyclopedia of garden plants, 476 (2003).

Deciduous perennial to 70 cm, with short thick rhizomes, sometimes with small-headed glandular hairs on upper parts. Stem-leaves widely spaced. Leaves 5–20 cm or more wide, divided for seven-eighths to nine-tenths into 5 or 7; divisions rather narrow, broadest above the middle, with a long entire wedge-shaped base and a palmato-pinnately lobed or merely toothed apex; lobes and teeth very acute. Inflorescence umbel-like. Flowers *c.* 3 cm wide, shallowly bowl-shaped, upwardly inclined, stalks 1.5–2 times as long as sepals. Sepals 8–10 mm with mucro 2–3.5 mm. Petals 1.8–1.9 × 1–1.3 cm, entire or shallowly notched, deep to pale pink with a white base, or white, with hairs at base forming a tuft on each edge but not extending across front surface. Stamens divergent. Stigmas 2–3 mm. Immature fruits erect on erect stalks. Fruit 2.7–2.9 cm with slender tip 4–6 mm. *E North America.* H1. Summer.

36. G. nervosum Rydberg. Figure 117(6), p. 438. Illustration: *Edwards's Botanical Register* **28**: t. 52 (1842) – as *G. erianthum*;

Figure 117. Leaves of *Geranium* species (× 0.5 except where noted).
1, *G. richardsonii*. 2, *G. palustre*. 3, *G. wlassovianum*. 4a, b, *G.* × *magnificum*
(x 1.5). 5, *G.* × *riversleaianum*. 6, *G. nervosum*.

Abrams, Illustrated flora of the Pacific States **3**: 5 (1951); Yeo, Hardy geraniums, 125 (1985).

Deciduous perennial to 70 cm, with compact woody rootstock; upper parts with glandular hairs. Leaves 5–20 cm wide, divided for *c.* four-fifths into 5 or 7; divisions broadest above the middle, palmato-pinnately cut for a quarter to half their length into 3–5 lobes; lobes asymmetric with several teeth on outer edge and none or 1 on inner; teeth acute. Flowers 3.5–4 cm wide, flat, erect, rather numerous and densely arranged, on stalks 1–2 times as long as sepals. Sepals 7–10 mm with mucro 0.5–1.25 mm. Petals 1.3–2.2 cm × 7–16 mm, usually notched, pale to deep pink with darker veins, with hairs at base forming a tuft on the edges and front surface and less densely distributed over basal third to half of front surface. Filaments curving outwards. Stigmas 4–6 mm. Immature fruits erect; stalks erect, spreading or reflexed. *W North America.* H1. Summer.

37. G. richardsonii Fischer & Trautvetter. Figure 117(1), p. 438. Illustration: *Botanical Magazine*, 3124 (1832); Hitchcock et al., Vascular plants of the Pacific northwest **3**: 386 (1969); Rickett, Wild flowers of the United States **5**: 291 (1971) & **6**: 347 (1973); Yeo, Hardy geraniums, 128, t. 23 (1985).

Deciduous perennial to 60 cm with thick compact rootstock; upper parts with glandular hairs. Leaves 5–12 cm wide, divided for two-thirds to five-sixths into 5 or 7; divisions gradually or abruptly tapered both ways from above the middle, 3-lobed at the apex for *c.* one third their length; lobes ovate, with few coarse acute or obtuse and mucronate teeth. Flowers 2.5–2.8 cm wide, flat, erect, scattered, on stalks 1.2–4 times as long as sepals. Sepals *c.* 6 mm, with mucro *c.* 1.5 mm. Petals *c.* 1.2–1.9 cm × 6–15 mm, entire, white or pale pink, sometimes with dark veins, front surface strongly hairy in basal half to three-quarters. Stamens longer than sepals, curving outwards. Stigmas 3–4 mm, green to yellowish. Immature fruits erect; stalks erect, spreading or reflexed. Fruit 1.7–2.1 cm with slender tip 1–2 mm. *W North America.* H1. Summer.

38. G. asphodeloides Burman.

Evergreen biennial or perennial of usually bushy habit to 40 cm, with a compact much-branched rootstock and thickened roots; glandular hairs present, at least on upper parts. Leaves 4–12 cm wide, divided for two-thirds to five-sixths into 5 or 7; divisions broadly wedge-shaped or almost oblong with truncate or shortly tapered apex, 3-lobed for a fifth to a quarter their length; lobes rather freely toothed; teeth short, broad, obtuse to acute. Flowers 2.3–3.5 cm wide, usually flat, upwardly inclined, loosely to densely arranged, on stalks 1.5–4 times as long as sepals. Sepals 6–8 mm with mucro from less than 1–2 mm. Petals 1–1.5 cm × 4–7 mm, 1.5–2.5 times as long as wide, entire or nearly so, white to rather deep pink or purple, often with darker veins; hairs at base forming a dense tuft on either edge but not extending all across front surface. Stamens shorter than sepals. Stigmas not more than 1 mm. Immature fruits erect on spreading or reflexed stalks. Fruit 1.6–2.2 cm with slender tip 4–5 mm. *S Europe (Sicily to Crimea & Caucasus), Turkey, N Iran, Syria, Lebanon.* H4. Summer.

Subsp. **asphodeloides**. Figure 118(2c), p. 440. Illustration: Fiori & Paoletti, Iconographia florae Italicae, edn 3, 302 (1933); Savulescu, Flora Republicii Populare Romane **6**: 113 (1958); Yeo, Hardy geraniums, 130, t. 24 (1985). Has main leaves 4–8 cm wide, few basal leaves, red-tipped glandular hairs, very numerous white to pink flowers usually with narrow petals. *Widespread.*

Subsp. **sintenisii** (Freyn) Davis. Figure 118(2b), p. 440. Similar but tends to be biennial and has more basal leaves to 12 cm wide and petals either pale pink or deep purple. *Turkey.*

Subsp. **crenophilum** (Boissier) Bornmüller. Figure 118(2a), p. 440. Basal leaves *c.* 9 cm wide, stem-leaves scarcely differing from the basal, numerous colourless glandular hairs, more erect but eventually scrambling stems, and relatively broad and deeply coloured petals. *Syria, Lebanon.*

39. G. tuberosum Linnaeus. Figure 118(3), p. 440. Illustration: Reichenbach, Icones florae Germanicae et Helveticae **5**: t. 194 (1841); Fiori & Paoletti, Iconographia florae Italicae, edn 3, 303 (1933); Bonnier, Flore complète **2**: t. 102 (1913); Huxley & Taylor, Flowers of Greece and the Aegean, t. 125 (1977).

Tuberous perennial, without glandular hairs. Tubers 6–15 mm wide, short. Stem to 30 cm, erect, with 1 or 2 pairs of well-developed leaves. Leaves 5–10 cm wide, with 7 divisions, of which at least the central one is free to the base; divisions broadest about the middle, pinnatisect almost to the midrib, largest lobes with narrow acute teeth; all segments 3 mm wide or less. Flowers 2–3.5 cm wide, flat, more or less erect, crowded, mostly in umbels. Sepals 5–8 mm with mucro to 1 mm. Petals 9–17 × 7–13 mm, deeply notched, purplish pink with darker veins, hairs at base forming a tuft on each edge but not extending across front surface. Stamens shorter than sepals. Stigmas *c.* 2.5 mm. Immature fruits and stalks erect. Fruit *c.* 2 cm with slender tip to 1 mm. *Mediterranean area eastwards to W Iran.* H1. Late spring.

Dormant from summer to early spring.

40. G. macrostylum Boissier. Figure 118(4), p. 440.

Like *G. tuberosum* but leaves appearing in autumn, upper parts of plant with glandular hairs, petals varying to pale pink with darker base, and fruit with slender tip 1.5–2.5 mm. *SE Europe, C & SW Turkey.* H3. Late spring.

41. G. malviflorum Boissier & Reuter (*G. atlanticum* misapplied). Figure 118(6), p. 440. Illustration: *Botanical Magazine*, 6452 (1879); Polunin & Smythies, Flowers of south-west Europe, 247 (1973); Yeo, Hardy geraniums, 135, t. 25 (1985).

Tuberous, summer-deciduous perennial; without glandular hairs. Tubers to 6 × 1.5 cm. Stem to 50 cm, with only 1 pair of well-developed leaves. Basal leaves appearing in autumn or early spring. Leaves mostly 10–15 cm wide, divided into 7 with the middle division free to the base. Divisions broadest about the middle, pinnatisect to within 1.5–3 mm of the midrib into several simple, pinnatisect or toothed lobes; teeth more or less acute. Flowers 3.57–4.5 cm wide, erect, at first saucer-shaped, crowded, mostly in umbels. Sepals 7–9 mm with mucro *c.* 1.5 mm. Petals 1.6–2.2 × 1.2–1.7 cm, notched, violet-blue or violet with darker veins, hairs at base forming a tuft on each edge but not extending all across front surface. Stigmas *c.* 1.5 mm. Immature fruits and stalks erect. Fruit 2.3–2.5 cm with slender tip 3–5 mm. *S Spain, Morocco, Algeria.* H4. Late spring.

42. G. × magnificum Hylander (*G. ibericum* misapplied; *G. platypetalum* misapplied). Figure 117(4), p. 438. Illustration: Hay & Beckett, Reader's Digest

Figure 118. Leaves of *Geranium* species (× 1.5). 1, *G. maculatum.* 2a, *G. asphodeloides* subsp. *crenophilum.* 2b, *G. asphodeloides* subsp. *sintenisii.* 2c, *G. asphodeloides* subsp. *asphodeloides.* 3, *G. tuberosum.* 4, *G. macrostylum.* 5, *G. renardii.* 6, *G. malviflorum.*

encyclopaedia of garden plants and flowers, edn 2, 300 (1978); Yeo, Hardy geraniums, 139–141, t. 27 (1985); Brickell (ed.), RHS A–Z encyclopaedia of garden plants, 476 (2003).

Deciduous perennial to 70 cm with compact rootstock, strongly hairy and with glandular hairs above. Leaves 10–20 cm wide, divided for two-thirds to three-quarters into 7, 9 or 11; divisions broadest about the middle, overlapping, palmato-pinnately lobed for a quarter to a third their length; lobes shortly oblong, often with 2 orders of teeth. Flowers 4–5 cm wide, saucer-shaped, facing sideways or upwardly inclined, crowded, mostly in umbels. Sepals 9–12 mm with mucro 2–3.5 mm. Petals 2.2–2.4 × 1.6–2.2 cm, notched, deep violet or blue-violet, with darker veins, with hairs at base forming a tuft on each edge but not extending across front surface. Stamens shorter than sepals, divergent, becoming recurved. Stigmas *c.* 2.5 mm. Immature fruits and stalks erect. Fruit not maturing but sometimes partially developed and with a slender tip 4–7 mm. *Garden origin.* H1. Summer.

This is a garden hybrid between *G. platypetalum* × *G. ibericum* Cavanilles, which are very similar to each other but differ in chromosome number. Though it is sterile at least 3 clones exist and they are often grown under the name of one or other parent species. It is best distinguished from these by its sterility.

43. G. platypetalum Fischer & Meyer. Figure 121(1), p. 445. Illustration: Yeo, Hardy geraniums, 136, 137 (1985).

Like *G.* × *magnificum* but to 40 cm, with leaf-divisions not overlapping, flowers 3–4.5 cm wide, flat or saucer-shaped, facing sideways, fruit developing normally, 2.9–3.5 cm with slender tip 4–5 mm. The glandular hairs are variable in length, the longest nearly as long as the non-glandular hairs (in *G.* × *magnificum* they are rather uniform in length and less than half as long as the non-glandular hairs). *Caucasus, NE Turkey, NW Iran.* H1. Summer.

44. G. renardii Trautvetter. Figure 118 (5), p. 440. Illustration: Hay & Beckett, Reader's Digest encyclopaedia of garden plants and flowers, edn 2, 301 (1978); Yeo, Hardy geraniums, 142, t. 28 (1985); Brickell (ed.), RHS A–Z encyclopaedia of garden plants, 477 (2003).

Deciduous perennial with thick woody rootstock trailing on the surface, with or without glandular hairs above. Leaves to 10 cm wide, grey-green, felted, the felt of the upper surface deciduous, finely wrinkled, divided halfway to the leaf-stalk into 5 or 7; divisions as broad or broader than long, not overlapping, rounded, with 3 or 5 shallow lobes at apex; lobes obtusely toothed. Flowers 3–1 cm wide, flat, facing sideways, crowded. Sepals 7–9 mm with mucro 0.5–1 mm. Petals 1.5–1.8 cm × 9–10 mm, notched, white or pale violet, with a strong pattern of darker veins, with hairs at base forming a tuft on each edge but not extending across front surface. Stamens longer than sepals. Stigmas *c.* 2 mm. Immature fruits and stalks erect. Fruit *c.* 3 cm with slender tip 4–5 mm. *Caucasus.* H2. Early summer.

45. G. libani Davis. Figure 120(1), p. 444.

Summer-deciduous perennial with thick compact rootstock, to 50 cm, without glandular hairs. Leaves 10–20 cm wide, glossy above, divided for two-thirds to three-quarters into 5 or 7; divisions pinnatifid about halfway to the midrib in the upper two-thirds, not overlapping; lobes oblong, obtusely toothed. Stems with only 1 pair of well-developed leaves. Flowers 2.8–3.2 cm wide, saucer-shaped, erect, rather crowded, mostly in umbels, on stalks about twice as long as sepals. Sepals 7–9 mm with mucro 1–1.5 mm. Petals 1.7–2 × 1.3–1.6 cm, notched, violet-blue or light violet, with darker veins, with hairs at base forming a tuft on each edge but not extending all across front surface. Stigmas *c.* 1 mm. Immature fruits and stalks erect. Fruit 3–3.7 cm. *Lebanon, W Syria, central S Turkey.* H4. Late spring.

46. G. peloponnesiacum Boissier. Figure 120(2), p. 444. Illustration: Yeo, Hardy geraniums, 144, t. 29 (1985).

Similar to *G. libani* but stems with 2 or 3 pairs of well-developed leaves, with glandular hairs above; leaves matt, more or less marbled and sometimes marked with brown in the sinuses above; flower-stalks 3–5 times as long as sepals, 3.8–4.3 cm wide, with sepals 9–12 mm; petals to 2.5 cm, occasionally with hairs on front surface at base. *Greece.* H2. Late spring.

47. G. gracile Nordmann. Figure 122(1), p. 447.

Deciduous perennial to 70 cm with compact rootstock; upper parts with glandular hairs. A few sparsely and coarsely toothed leaves produced first; leaves otherwise 10–20 cm wide, divided for two-thirds to three-quarters into 5 or 7; divisions diamond-shaped to ovate, not distinctly lobed but with uneven forwardly pointing teeth, the larger obtuse, the smaller acute. Flowers *c.* 2.5 cm wide, funnel-shaped, erect or upwardly inclined, scattered on stalks 1.5–2.5 times as long as sepals. Sepals 6–7 mm with mucro 1.5–2 mm. Petals *c.* 2.1 × 1 cm, notched, pale to deep pink with darker veins and white basal third; hairs at base not forming tufts and not extending all across front surface. Stamens longer than sepals. Stigmas *c.* 2.5 mm. Immature fruits and stalks erect. Fruit 2.7–3.4 cm with a slender but not well delimited tip. *NE Turkey, S Caucasus, N Iran.* H1. Summer–autumn.

48. G. polyanthes Edgeworth & Hooker. Figure 121(2), p. 445. Illustration: Polunin & Stainton, Flowers of the Himalaya, t. 20 (1985); Yeo, Hardy geraniums, 147, t. 30 (1985).

Deciduous perennial 15–45 cm, with a very thick compact rootstock; upper parts with glandular hairs. Leaves to 5 cm wide, rounded, smooth-textured and slightly fleshy, divided for two-thirds to four-fifths into 7 or 9; divisions broadest at apex where they are 3-lobed for two-thirds to four-fifths their length; lobes shortly toothed; teeth obtuse and mucronate or acute. Flowers *c.* 2.5 cm wide, funnel-shaped, erect, in umbels. Sepals 6–7 mm, mucro to 0.75 mm. Petals 1.2–1.5 cm × 8–10 mm, entire or slightly notched, bright deep purplish pink with darker veins; hairs at base forming a tuft on each edge but not extending all across front surface. Stamens shorter than sepals. Stigmas *c.* 1.5 mm. Immature fruits and stalks erect. Fruit 1.5–1.8 cm with no slender tip. Carpels with netted raised veins, minutely hairy. *Nepal to SW China.* H2. Summer.

49. G. albanum Bieberstein. Figure 120 (4), p. 444. Illustration: *Botanical Magazine*, 3732 (1839); Yeo, Hardy geraniums, 149, t. 32 (1985).

Evergreen perennial with compact rootstock and sprawling or scrambling stems to 1.75 m, with or without glandular hairs. Leaves 10–20 cm wide, rounded, divided for half to three-quarters into 5, 7 or 9; divisions broadest near apex where they are 3-lobed for a quarter to a third their length; lobes freely toothed; teeth obtuse or acute. Flowers 2.2–2.5 cm wide, saucer-shaped, upwardly inclined, scattered, on stalks 3–6.5 times as long as the sepals. Sepals 6–7.5 mm with mucro to 0.5 mm. Petals 1–1.4 × 1–1.1 cm, slightly notched, bright pink with paler shades towards base, darker veins; hairs at base forming a small tuft on each edge and scattered across front surface, Stigmas *c.* 1.5 mm. Immature fruits erect on spreading or reflexed stalks. Fruit 1–1.3 cm, with unthickened beak. Carpels ribbed on the sides and knobbed on the midrib at the top, bristly, not falling until the calyx decays. *SE Caucasus, N Iran.* H2. Summer.

50. G. pyrenaicum Burman. Figure 119 (3), p. 443. Illustration: Ross-Craig, Drawings of British plants **6**: 33 (1952); Phillips, Wild flowers of Britain, 37 (1977); Fitter et al., Wild flowers of Britain and northern Europe, 135 (1974); Aeschimann et al., Flora alpina **1**: 1059 (2004).

Evergreen perennial to 40 cm or scrambling higher, with compact rootstock, with many small glandular hairs among the glandless ones. Leaves 5–10 cm wide, rounded, divided for half to two-thirds into 7, sometimes 9; divisions broadest at apex where they are lobed for *c.* one third their length; lobes not longer than broad, usually toothed; teeth obtuse. Flowers 1.2–1.8 cm wide, widely funnel-shaped, upwardly inclined, scattered, on stalks 2–5 times as long as the sepals. Sepals *c.* 4.5 mm with insignificant mucro. Petals 8–10 × 6–7 mm, bifid, bluish pink, rarely purple or almost white, with slightly darker veins and sometimes a pale zone at base; hairs at base forming a small tuft near each edge but not extending all across front surface. Immature fruits erect on reflexed stalks. Stigmas *c.* 1.5 mm. Fruit 1.4–1.6 cm without a slender tip. Carpels covered with adpressed hairs, rarely hairless, inconspicuously ribbed near apex. *SW & W Europe to Turkey & the Caucasus, naturalised northwards.* H2. Late spring–autumn.

51. G. brutium Gasparrini. Figure 120(3), p. 444. Illustration: Polunin & Huxley, Flowers of the Mediterranean, t. 82 (1965) – as *G. molle grandiflorum*; Yeo, Hardy geraniums, 151 (1985).

Annual similar to *G. pyrenaicum* but with stem-leaves alternate, flowers 2–2.5 cm wide, petals 1.2–1.5 cm × 7–11 mm when plants are in their prime, much smaller in older plants; petals less deeply notched, bright pink; carpels hairless, densely covered with slanting ribs and scattered knobs. *Sicily, S Italy, Balkan Peninsula, Turkey.* H4. Spring–summer.

The carpels are very similar to those of **G. molle** Linnaeus, a widespread weedy species with very small flowers.

52. G. macrorrhizum Linnaeus. Figure 120(5), p. 444. Illustration: *Botanical Magazine*, 2420 (1823); Hay & Synge, Dictionary of garden plants in colour, 144 (1969); Hess et al., Flora der Schweiz **2**: 627 (1970); Phillips, Wild flowers of Britain, 44 (1977).

Evergreen perennial to 50 cm, carpet-forming, with thick underground and superficial rhizomes. Plant densely covered with aromatic glandular hairs. Leaves 10–15 cm wide, divided for two-thirds to three-quarters into 7; divisions broadest at or above the middle, palmato-pinnately lobed; lobes very short with obtuse and mucronate or acute teeth. Stem with 1–3 pairs of well-developed leaves. Flowers 2–2.5 cm wide, facing sideways, crowded, mostly in umbels on stalks about as long as sepals. Sepals 7–9 mm with mucro 2.5–3.5 mm, often red, erect, forming an inflated calyx. Petals 1.5–1.8 cm × 7–11 mm, entire, each with a ridged claw as long as the blade, deep purplish red to white, with hairs at base on edges and sometimes on ridges. Stamens *c.* 2.5 times as long as sepals. Immature fruits and stalks erect. Fruit 3.3–3.7 cm with slender tip to 1.8 cm. Carpels with wavy horizontal ribs. *Central S Europe, Balkan Peninsula, Carpathian Mountains.* H2. Early summer.

G. × cantabrigiense Yeo; *G. dalmaticum × G. macrorrhizum.* Figure 120(6), p. 444. Illustration: Yeo, Hardy geraniums, 154, t. 34 (1985); Brickell (ed.), RHS A–Z encyclopedia of garden plants, 474 (2003). Hybrids between these 2 species will probably key out here. They are intermediate between the parents in flowering time and morphology, with green, smooth and slightly shiny leaves with abruptly enlarged divisions and rounded lobes. The petals are pink or, in 'Biokovo', white with a pink streak at the base of the blade.

53. G. dalmaticum (Beck) Rechinger. Figure 119(4), p. 443. Illustration: Yeo, Hardy geraniums, 154, 155, t. 35 (1985); Brickell (ed.), RHS A–Z encyclopedia of garden plants, 474 (2003).

Almost deciduous perennial to 25 cm, carpet-forming with underground and superficial rhizomes, aromatic, with glandular hairs on upper parts. Flowering stems usually with 1 pair of well-developed leaves. Leaves to 5 cm wide, hairless except on veins, glossy, divided for four-fifths to six-sevenths into 5 or 7; divisions broadest near apex, where they are 3-lobed for a quarter to a third their length; lobes with or without a few coarse teeth; teeth ovate, obtuse and mucronate or acute. Flowers like those of *G. macrorrhizum* but sepals only to 6.5 mm, petals pink or white, claw only about half as long as blade; slender tip of fruit *c.* 1.4 cm. *SW former Yugoslavia, Albania.* H3. Summer.

54. G. cataractarum Cosson. Figure 120(7), p. 444. Illustration: Yeo, Hardy geraniums, 156, t. 36 (1985); Taylor, Collecting garden plants, 210 (1988).

Evergreen aromatic perennial to 30 cm, covered with very small-headed glandular hairs; rootstock compact, thick. Leaves 5–10 cm wide, with 5 divisions, the central one free to the base and stalked, the laterals on each side joined to a short common stalk; divisions pinnately cut one-fifth to five-sixths the way to the midrib into oblong, obtusely or acutely toothed lobes. Flowers 1.5–2 cm wide, funnel-shaped, upwardly inclined, crowded on stalks 0.75–1.75 times as long as sepals. Sepals 5–6 mm with mucro not more than 0.5 mm, erect, forming an inflated calyx. Petals 1.5–1.6 cm × 6–8 mm, entire, each with a claw half as long as the blade, bright pink, hairless. Stamens longer than sepals. Stigmas *c.* 1 mm. Immature fruits and stalks erect. Fruit *c.* 2 cm with slender tip 6–7 mm, hairless. Carpels with a network of ribs. *S Spain, Morocco.* H4. Early summer–autumn.

55. G. robertianum Linnaeus. Figure 121(3), p. 445. Illustration: Ross-Craig, Drawings of British plants **6**: t. 40 (1952); Fitter et al., Wild flowers of Britain and northern Europe, 134 (1974); Phillips, Wild flowers of Britain, 32 (1977); Brickell (ed.), RHS A–Z encyclopedia of garden plants, 477 (2003).

Foetid annual to 30 cm or scrambling higher, usually overwintering as a rosette. Glandular hairs present. Stems succulent and brittle. Leaves 4–11 cm wide, with 5 divisions, the central one free to the base and long-stalked, the laterals on each side joined to a short common stalk; divisions pinnatisect or pinnate to bipinnate;

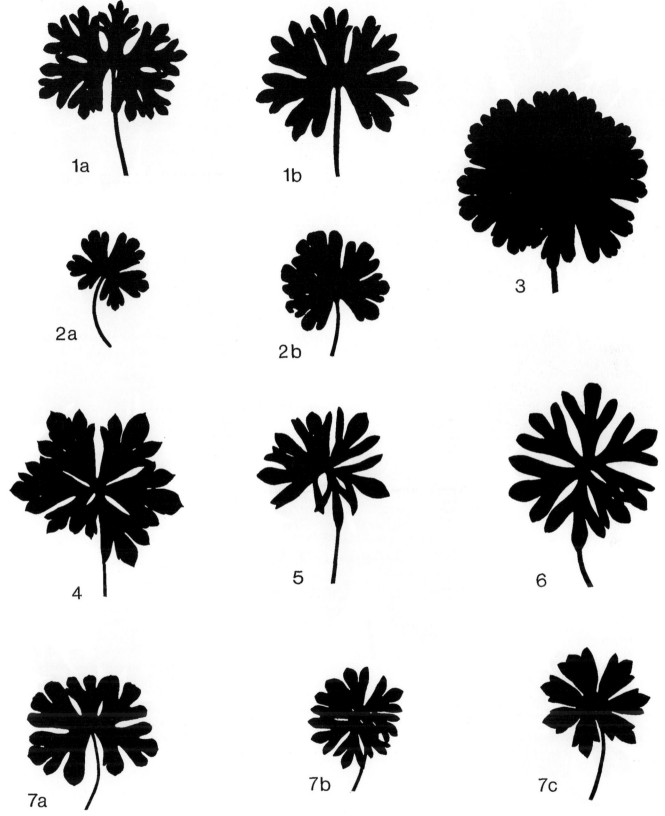

Figure 119. Leaves of *Geranium* species (× 1). 1a, b, *G. farreri*. 2a, b, *G. sessiliflorum* subsp. *novaezelandiae*. 3, *G. pyrenaicum*. 4, *G. dalmaticum*. 5, *G. argenteum*. 6, *G.* × *lindavicum*. 7a, b, *G. cinereum* var. *subcaulescens*, c, *G. cinereum* var. *cinereum*.

Figure 120. Leaves of *Geranium* species (× 0.75). 1, *G. libani*.
2, *G. peloponnesiacum*. 3, *G. brutium*. 4, *G. albanum*. 5, *G. macrorrhizum*.
6, *G. × cantabrigiense*. 7, *G. cataractarum*.

Figure 121. Leaves of *Geranium* species (×0.5). 1, *G. platypetalum*.
2, *G. polyanthes*. 3, *G. robertianum*. 4a, b, *G. phaeum*. 5, *G. reflexum*.
6, *G. × monacense*.

ultimate lobes obtuse, often mucronate. Flowers 1.1–1.6 cm wide, funnel-shaped, upwardly inclined, scattered, on stalks 0.5–1.5 times as long as sepals. Sepals 4–6 mm with mucro 0.75–1.25 mm, erect. Petals 1–1.4 cm × 3–6 mm, entire, with a ridged claw as long as the blade, bright pink with paler streaks or sometimes white. Stamens longer than sepals. Stigmas 0.5–1.5 mm. Immature fruits and stalks erect. Fruit 1.4–1.8 cm, hairless except for slender tip 4–5 mm. Carpels with a network of ribs on sides and 1–3 collars round apex, hairless or finely hairy, with a plume of very long hairs at apex. *Europe, NW Africa, Canary Islands, W Asia, Himalaya, SW China, E North America (perhaps introduced)*. H3. Late spring–autumn.

56. G. rubescens Yeo (not *G. rubescens* Andrews). Figure 122(8), p. 447. Illustration: *Journal of the Royal Horticultural Society* **95**: 399 (1970); *Botanical Journal of the Linnean Society* **67**: 299, 333 (1973).

Overwintering annual to 60 cm, like a large version of *G. robertianum* with deep red stems and leaf-stalks; leaves mostly 10–23 cm wide; flowers 2.2–3.3 cm wide with dark throat; sepal-mucro *c.* 2 mm; petals 1.8–2.2 cm × 7–13 mm; fruit 2.4–2.8 cm with slender tip 6–7 mm; carpels hairless and without a plume. *Madeira*. H4. Late spring–autumn.

Crosses with *G. canariense*, producing fertile hybrids. The name *G. rubescens* was first used for a different species, now regarded as a *Pelargonium*. In the interests of stability, the species described here has not been given a new name.

57. G. canariense Reuter (not *G. canariense* Poiret). Figure 122(2), p. 447. Illustration: *Journal of the Royal Horticultural Society* **95**: 399 (1970); *Botanical Journal of the Linnean Society* **67**: 301, 330 (1973).

Aromatic evergreen short-lived perennial with a rosette on a stalk usually 5–15 cm; plant to 60 cm in flower, branching after flowering, glandular-hairy. Leaves 8–25 cm wide, succulent and brittle, with 5 divisions, the central one free, wedge-shaped at the base and almost stalkless, laterals free nearly to the base; divisions pinnatisect into lanceolate pinnatifid lobes, the ultimate lobes with a few small teeth. Flowering shoots axillary or terminal. Flowers 2.3–3.6 cm wide, more or less star-shaped, facing sideways, crowded, on stalks

0.5–1.5 times as long as sepals. Sepals 8–10 mm with mucro *c.* 1 mm, divergent. Petals 1.7–2.4 cm × 6–9 mm, entire or slightly notched, with a ridged claw to half as long as blade; blade oblong or lanceolate, deep pink, sometimes with a dark base. Stamens 2–3 times as long as sepals. Stigmas *c.* 2 mm. Immature fruits slightly nodding on spreading or slightly nodding stalks. Fruit *c.* 3.5 cm with slender tip 1.6–1.8 cm. Carpels with a network of ribs, with or without fine hairs. *Canary Islands*. H5. Late spring–summer.

The name *G. canariense* was first used for a different species, now regarded as a *Pelargonium*. In the interests of stability, the species described here has not been given a new name.

58. G. palmatum Cavanilles (*G. anemonifolium* L'Héritier). Figure 122 (7), p. 447. Illustration: *Botanical Magazine*, 206 (1792); Sweet, Geraniaceae **3**: t. 244 (1824–26); *Journal of the Royal Horticultural Society* **95**: 399 (1970); *Botanical Journal of the Linnean Society* **67**: 302, 326 (1973).

Like *G. canariense* but with little or no evident stem below the rosette; plant in flower to 1.2 m; leaves to 35 cm wide with long-stalked middle division; flowering shoots axillary, flower-stalks 1.5–5 times as long as sepals, flowers slightly bilaterally symmetric; sepals adpressed to petals; petals 2–3 × 1.3–1.8 cm, dark at base, with claw to only a third as long as blade; stigmas 3–4 mm; fruit *c.* 3 cm, with slender tip 1.2–1.4 cm. *Madeira*. H5–G1. Summer.

59. G. maderense Yeo. Figure 122(6), p. 447. Illustration: *Journal of the Royal Horticultural Society* **95**: 399 (1970); *Botanical Journal of the Linnean Society* **67**: 303, 323 (1973); Yeo, Hardy geraniums, 163, t. 37 (1985); Brickell (ed.), RHS A–Z encyclopedia of garden plants, 475 (2003).

Like *G. canariense* but with stem below rosette to 60 cm and plant in flower to 1.5 m; leaves to 60 cm wide, divided to the base into 5, the middle division long-stalked; leaf and inflorescence-stalks dull purplish brown (not green or red-flushed); flowering shoots terminal, flower-stalks 3–20 times as long as sepals; sepals adpressed to petal-bases; petals 1.9–2.1 × 1.3–1.8 cm, dull purplish pink with blackish purple base and claw *c.* an eighth as long as blade; stamens not longer than sepals; stigmas *c.* 2.5 mm. Fruit *c.* 2.5 cm with slender tip to 7 mm. *Madeira*. H5–G1. Spring–early summer.

60. G. phaeum Linnaeus. Figure 121(4), p. 445.

Evergreen tufted perennial to 80 cm with short stout superficial rhizomes; glandular hairs very small. Stems often purple-dotted, oblique, usually once-forked, hooked at the tip, bearing pairs of flowers turned to one side. Leaves 10–20 cm wide, divided for a fifth to four-fifths into 7 or 9; divisions broadest at or above the middle, palmato-pinnately cut about halfway to the midrib into oblong or ovate, usually many-toothed, lobes; teeth obtuse to acute. Flowers 2.2–2.5 cm wide, flat or convex in the middle, facing sideways, on stalks 0.5–4 times as long as sepals. Sepals 6–11 mm with mucro not more than 0.5 mm. Petals 1.1–1.4 cm, about as wide, nearly black to medium greyish purple, paler at base, occasionally white; hairs at base moderately tufted on edges and scattered across front surface. Stamens longer than sepals, adpressed to the style except at the tips, with long hairs in the lower half. Stigmas 1.25–2 mm. Immature fruits upwardly inclined on upwardly inclined or horizontal stalks. Fruit 1.9–2.6 cm with slender tip 3–4 mm. Carpels with 4 or 5 ridges round the apex. *Pyrenees to Alps & E Europe to W former USSR*. H1. Late spring–early summer.

Var. **phaeum**. Illustration: Reichenbach, Icones florae Germanicae et Helveticae **5**: t. 197 (1841); Ross-Craig, Drawings of British plants **6**: 30 (1952); Polunin, Flowers of Europe, t. 646 (1969); Yeo, Hardy geraniums, 164, 166, t. 38 (1985). Leaves often marked with brown in the sinuses; petals with edge ruffled and/or with small lobes or with a central triangular point, dark dull red or purplish to nearly black, white at base.

Var. **lividum** (L'Héritier) Persoon. Illustration: Reichenbach, Icones florae Germanicae et Helveticae **5**: t. 197 (1841); Yeo, Hardy geraniums, 165, t. 38 (1985). Leaves unmarked; petals entire, medium greyish purple with a darker blue zone above the white base.

Plants with dark blue-violet petals are perhaps hybrids between these 2 varieties.

61. G. reflexum Linnaeus. Figure 121(5), p. 445. Illustration: Reichenbach, Icones florae Germanicae et Helveticae **5**: t. 196 (1841); Pignatti, Flora d'Italia **2**: 8 (1982); Yeo, Hardy geraniums, 168, t. 38 (1985).

Differs from *G. phaeum* in its more nodding flowers, only 1.3–1.6 cm wide, with strongly reflexed petals that are only

Figure 122. Leaves of *Geranium* species (× 0.33). 1, *G. gracile*.
2, *G. canariense*. 3, *G. aristatum*. 4, *G. sylvaticum*. 5, *G. psilostemon*.
6, *G. maderense*. 7, *G. palmatum*. 8, *G. rubescens*.

half as wide as long; petal colour bright pink to dull violet with a white base and a bluish band above this; hairs at base of petal forming a strong tuft on each edge but not scattered across front surface; stamens with only very small fine hairs at the base; immature fruits and stalks horizontal or nodding. *C Italy, S former Yugoslavia, N Greece.* H3. Summer.

G. × monacense Yeo; *G. phaeum × G. reflexum*. Figure 121(6), p. 445. Has intermediate petal-shape (best seen when the petals are detached and flattened) and petal colouring; colour variants exist reflecting the influence of one or other of the varieties of *G. phaeum*.

62. G. aristatum Freyn. Figure 122(3), p. 447. Illustration: *Hooker's Icones Plantarum* **33**: t. 3276 (1935); Yeo, Hardy geraniums, 169, t. 39 (1985).

Deciduous tufted perennial to 75 cm, strongly hairy, with glandular hairs above. Leaves 15–20 cm wide, divided for three-quarters into 5, 7 or 9; divisions palmato-pinnately lobed above the middle; teeth acute. Flowers 2.2–2.8 cm wide, obconical, sharply nodding. Sepals 7–8 mm, reflexed, with mucro 4–7 mm. Petals 1.3–1.6 cm, half as wide as long, entire or lobed, strongly reflexed, greyish violet or nearly white, with darker veins and base; hairs at base forming a tuft on each edge. Stamens adpressed to the style except at the tips, with long hairs on the lower half. Stigmas *c.* 1.5 mm. Immature fruits erect on reflexed stalks. Fruit *c.* 3.5 cm. Carpels with 4 or 5 ridges round the apex. *S former Yugoslavia, S Albania, NW Greece.* H2. Summer.

63. G. argenteum Linnaeus. Figure 119 (5), p. 443. Illustration: *Botanical Magazine*, 504 (1821); Bonnier, Flore complète **2**: t. 102 (1913); Hegi, Illustrierte Flora von Mittel-Europa **4**(3): 1675 (1924); Hess et al. , Flora der Schweiz, edn 1, **2**: 629 (1970).

Evergreen perennial to 15 cm with compact rootstock, densely covered with adpressed hairs, without glandular hairs. Leaves to 5 cm wide, silvery, divided nearly to the base into 7; divisions broadest near the apex, usually 3-lobed to beyond the middle; lobes linear to oblanceolate, entire, crowded, not lying flat. Flowers 2–2.5 cm wide, saucer-shaped, more or less erect, few, on stalks 2.5–4 times as long as sepals. Sepals 5.5–8.5 mm with mucro to 1 mm. Petals 1.2–1.4 cm, more or less

notched, pale pink to white, often with rather weak slightly netted darker veins, with hairs at base in a dense tuft on each edge and extending all across front surface. Stamens shorter than sepals, pale in colour. Stigmas 1–1.5 mm. Immature fruits erect on more or less reflexed stalks. Carpels with 1–3 ribs round apex. *SE France, N Italy, N former Yugoslavia.* H1. Summer.

64. G. cinereum Cavanilles.

Evergreen perennial to 25 cm with condensed or long, trailing, above-ground, perennial stems; glandular hairs absent. Leaves to 7.5 cm wide, divided for two-thirds to four-fifths into 5 or 7; divisions broadest at apex, lobed at apex for one fifth to half their length; lobes entire or few-toothed, obtuse to acute. Leaves on the stem few and reduced or none. Flowers 2.5–3 cm wide, saucer-shaped, more or less erect, rather few, on stalks 2.5–7 times as long as sepals. Sepals 6.5–11 mm with mucro to 1 mm. Petals 1.2–1.8 × 1–1.6 cm, entire or notched, with hairs forming a weak to strong tuft on each edge, sometimes extending across front surface. Stigmas to 2 mm. Immature fruits erect on more or less reflexed stalks. Fruit 2–4.3 cm. Carpels with 1–3 ribs round apex. *NW Africa, S Europe, Turkey, W Syria, Lebanon.* H2. Summer.

Var. **cinereum**. Figure 119(7c), p. 443. Illustration: L'Héritier, Geraniologia, t. 37 (1792, reprint 1978); Grey-Wilson & Blamey, Alpine flowers of Britain and Europe, 131 (1979); Yeo, Hardy geraniums, 171, 173, t. 41 (1985). Flowering shoots prostrate. Leaves greyish green with wedge-shaped divisions lobed for a fifth to a third their length; lobes triangular. Petals notched, net-veined in purplish red on a white or pale violet-pink ground. Filaments pale. *Pyrenees.*

Var. **subcaulescens** (de Candolle) Knuth (*G. subcaulescens* de Candolle). Figure 119(7a, b), p. 443. Illustration: Huxley & Taylor, Flowers of Greece and the Aegean, 122 (1977); Graf, Exotica, edn 11, 1177 (1982); Yeo, Hardy geraniums, 171, t. 43 (1985). Rootstock compact or trailing. Flowering shoots erect. Leaves dark greyish green with divisions lobed for a third to half their length; lobes 1–2 times as long as broad, sometimes more, broadly obtuse to acute; petals entire or notched, deep purplish red with darker veins, often

black at base; stamens blackish red. *?Italy, Balkan peninsula, Turkey.*

There are other varieties, some in cultivation, and some distinctive garden hybrids between var. *cinereum* and var. *subcaulescens*, a well-known cultivar of this parentage being 'Ballerina'.

G. × lindavicum Knuth; *G. cinereum × G. argenteum*. Figure 119(6), p. 443.

Illustration: Yeo, Hardy geraniums, 171, t. 40 (1985). Hybrids between the parents are intermediate and vary considerably according to which variety of *G. cinereum* is involved.

2. ERODIUM L'Héritier
P.F. Yeo
Annual, biennial or perennial herbs, usually evergreen. Leaves either simple and toothed, usually also lobed, or pinnate with variously dissected leaflets. Flowers in umbels or occasionally solitary on a jointed stalk, bisexual or unisexual (plants then dioecious). Petals 5, frequently distinctly unequal, the 2 upper being shorter and broader than the 3 lower and sometimes bearing dark blotches. Stamens 10, united into a tube at the base, alternately fertile (with an anther) and sterile (with staminodes, with no anther). Seeds dispersed within the carpels. Awn remaining attached to carpel-body and in our species divided into a lower corkscrew-like part and an upper tail-like part bent to one side.

A genus of about 100 species in temperate areas, especially the Mediterranean area and west Asia. About 30 are used in horticulture and they hybridise readily in gardens. Erodiums are mostly suitable for the alpine house and rock garden, only a few of the largest being suitable for the front of a border. They need a deep, well-drained soil and usually a sunny situation.

In the descriptions, measurements of the flower-stalk and sepals apply to flowering time; sepal-length excludes the mucro, but unless otherwise stated the mucro does not add significantly to the length. Descriptions of the carpels apply to the ripe fruit. Glandular hairs in the pits at the apices of the carpels are often spherical, without an evident stalk. Petal blotches are often elaborately composed, for example, by a peppering of black on a translucent ground.

Literature: Knuth, R., Geraniaceae, in A. Engler (ed.), Das Pflanzenreich **53**: 221–90, 583–7 (1912); Guittonneau, G.G.,

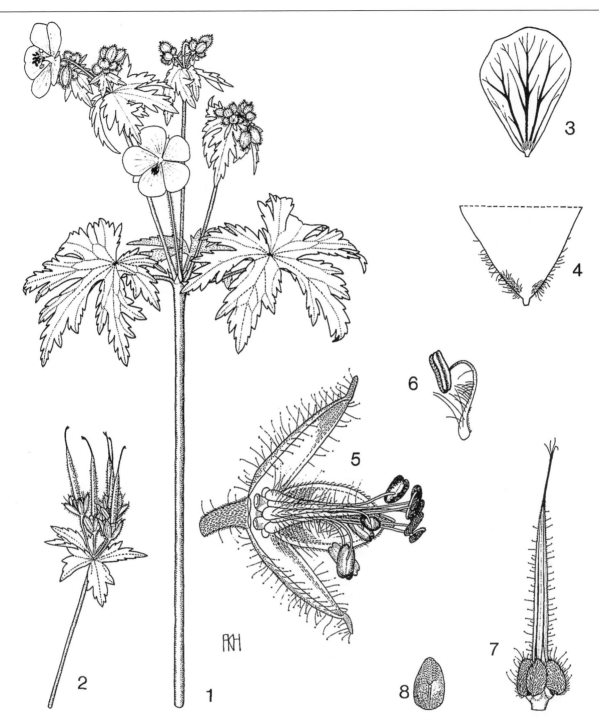

Figure 123. Diagnostic details of *Geranium erianthum*. 1, Shoot (× 1). 2, Fruiting branch (× 1). 3, Petal (× 3). 4, Petal-base from front (× 6). 5, Flower in male stage with 2 sepals, petals and 2 stamens removed (× 6). 6, Stamen with filament seen from back (× 6). 7, Fruit in pre-explosive interval (× 3). 8, Seed (× 6).

Contribution à l'étude biosystématique du genre Erodium L'Hér. dans le bassin méditérranéen occidental, *Boissiera* **20**: 154 (1972); Bacon, L., Some erodiums for the rock garden, *Bulletin of the Alpine Garden Society* **41**: 233–9 (1973); Leslie, A.C., The hybrid of Erodium corsicum Lém. with E. reichardii (Murray) DC., *The Plantsman* **2**: 117–26 (1980); Clifton, R.T. F., Geranium family species check list, Part 1, Erodium, edn 4, (1988); El-Oqlah, A., A revision of the genus Erodium L'Héritier in the Middle East, *Feddes Repertorium* **100**: 97–118 (1989).

Stem. Evident above rosettes: **1, 3, 4, 5, 6, 7, 8, 9, 10, 16, 17**; little or none above rosettes: **2, 3, 11, 12, 13, 14, 15, 18, 19, 20, 21**.

Leaves. Simple: **1, 2, 3, 4, 5, 6**; compound: **4, 5, 7, 8, 9, 10, 11, 12, 13, 14, 15, 16, 17, 18, 19, 20, 21**.

Pinnate leaves. Without subsidiary leaflets between the main ones: **4, 16, 17, 18, 19, 20, 21**; with subsidiary leaflets: **7, 8, 9, 10, 11, 12, 13, 14, 15**.

Petals. Unblotched: **1, 2, 3, 4, 7, 8, 9, 10, 13, 14, 15, 16, 18**; blotched: **5, 6, 11, 12, 13, 16, 17, 19, 20, 21**.

Carpels. With furrow around each apical pit: **1, 4, 5, 6, 16, 17, 18, 21**; without furrow around each apical pit: **2, 3, 7, 8, 9, 10, 11, 12, 13, 14, 15, 16, 19, 20**.

Apical pits of carpels. With glands: **1, 2, 3, 5, 6, 8, 9, 10, 11, 12, 13, 15**; without glands: **4, 7, 13, 14, 16, 17, 18, 19, 20, 21**.

1a. Leaves simple or with the blade divided to the base into 3, not pinnate   2
b. Leaves pinnate or pinnatisect, sometimes more than once   7
2a. Sepal mucro 2–7 mm; petals soon falling   **4. gruinum**
b. Sepal mucro not more than 2 mm; petals not usually falling quickly   3
3a. Inflorescence-stalks with 4–11 flowers; sepals 5.5 mm or more   4
b. Inflorescence-stalks with 3 or fewer flowers; sepals not more than 4 mm   5
4a. Sepal mucro not more than 0.5 mm; flower-stalks after flowering reflexed but the immature fruits erect   **5. trifolium**
b. Sepal mucro *c.* 2 mm; flower-stalks after flowering reflexed and the immature fruits usually reflexed also   **6. pelargoniiflorum**
5a. Plant stemless, without glandular hairs on leaves; leaves not more than 15 mm, sparsely hairy, toothed but scarcely lobed; petals white or pale pink with simple or feathered darker veins   **2. reichardii**
b. Plant with evident stems and with glandular hairs on leaves; leaves 5–24 mm, downy, unlobed or lobed and toothed; petals usually with pink net-veining   6
6a. Apical pits of carpel ringed by a furrow; leaves grey-green   **1. corsicum**
b. Apical pits of carpel not ringed by a furrow; leaves dark or pale green to greyish green   **3. × variabile**
7a. Leaves with subsidiary leaflets between the main ones   8
b. Leaves without subsidiary leaflets between the main ones   16

8a. Plant with evident leafy stem above the basal leaves   9
b. Plant without evident stems above the basal leaves (or sometimes with a stem to 2 cm)   12
9a. Petals yellow   **8. chrysanthum**
b. Petals white, pink or purple   10
10a. Most of the leaf-segments less than 2 mm wide or, if wider, petals white   **9. absinthoides**
b. Most of the leaf-segments more than 2 mm wide; petals not white   11
11a. Stems, leaf- and inflorescence-stalks with adpressed hairs; leaves silvery silky   **7. guicciardii**
b. Stems, leaf- and inflorescence-stalks with spreading hairs; leaves bright green, though with some adpressed hairs   **10. alpinum**
12a. Upper 2 petals with bold blotches   13
b. Upper 2 petals unblotched or with only small faint blotches   14
13a. Blotches on upper petals one-third–two-thirds the length of the petal, drawn out into one or more long points; ground colour of petals often white; leaves often grey-green to whitish   **11. cheilanthifolium**
b. Blotches on upper petals *c.* $^1\!/_2$–$^4\!/_5$ the length of the petal, rounded in outline, with only small teeth on margins; ground colour of petals pink; leaves green   **12. glandulosum**
14a. Leaves whitish silvery   **15. rupestre**
b. Leaves green or greyish green   15
15a. Plant hairy; sepals *c.* 7.5 mm; petals 1.2–1.3 cm   **13. foetidum**
b. Plant hairless or nearly so; sepals 9–11 mm; petals 1.5–2 cm   **14. rodiei**
16a. Plant with evident leafy stem above the basal leaves   17
b. Plant without evident stems above the basal leaves   19
17a. Inflorescence-stalks 3–7 cm; sepals *c.* 4 mm; fruit 3–4 cm   **16. cicutarium**
b. Inflorescence-stalks 5–14 cm; sepals 6 mm or more; fruit 4 cm or more   18
18a. Leaves with 7 or fewer leaflets; sepals 7–9 mm with mucro to 1 mm; the upper 2 petals blotched   **17. recoderi**
b. Leaves with 11 or more leaflets; sepals *c.* 5 mm with mucro less than 1 mm; petals unblotched   **18. acaule**
19a. Petals unblotched; carpel-body 5–6 mm   **18. acaule**
b. Petals usually blotched; carpel-body 7 mm or more   20

20a. All the leaf-segments linear or linear-oblanceolate   **19. carvifolium**
b. Leaf-segments wider or only the ultimate ones linear   21
21a. Sepals less than 1 cm   **20. castellanum**
b. Sepals 1–1.5 cm   **21. manescavi**

1. E. corsicum Léman. Figure 124(1), p. 451. Illustration: *Botanical Magazine*, 8888 (1938); *The Plantsman* **2**: 122 (1980); Everett, New York Botanical Garden illustrated encyclopedia of horticulture **4**: 1252 (1981); *Bulletin of the Alpine Garden Society* **55**: 125 (1987).

Perennial, much branched, not producing rosettes. Plant with glandular and non-glandular hairs. Stems to 25 cm, somewhat twiggy. Leaves 2.5–7 cm; stalks hairy; blades 8–24 mm, simple, greyish green-downy, obtusely lobed. Inflorescence-stalks to 5 cm, with 1–3 flowers. Flower-stalks to 1.7 cm. Sepals *c.* 3 mm. Petals 7–10 mm, equal, nearly as wide as long, overlapping, bright pink with darker veining, usually white towards the base, unblotched. Fruit 1.5–2 cm. Carpel-body *c.* 3.5 mm, its apical pits glandular, ringed by a furrow. *Corsica, Sardinia.* H3.

There is an albino-flowered cultivar.

2. E. reichardii (Murray) de Candolle (*E. chamaedryoides* (Cavanilles) L'Héritier). Figure 124(2), p. 451. Illustration: *Botanical Magazine*, 18 (1787); *The Plantsman* **2**: 122 (1980); Everett, New York Botanical Garden illustrated encyclopedia of horticulture **4**: 1251 (1981); *Bulletin of the Alpine Garden Society* **41**: 125 (1987).

Perennial without evident stems, forming rosettes on the ground. Plant with non-glandular hairs. Leaves to 4 cm; blades to 1.5 cm, dark green, sparsely hairy, simple, circular to elliptic, heart-shaped, scarcely lobed. Inflorescence-stalks to 3 cm, with 1 flower. Flower-stalks to 2 cm, very slender. Sepals 3.5–4 mm. Petals *c.* 1 cm, slightly unequal, white or pink, with fine purple veins, unblotched. Fruit *c.* 1.5 cm. Carpel-body *c.* 3.5 mm, its apical pits glandular, not ringed by a furrow. *Balearic Islands.* H3.

The cultivars 'Roseum' and 'Flore Pleno', customarily assigned to this species, belong in fact to *E. × variabile*.

3. E. × variabile Leslie. Illustration: *The Plantsman* **2**: 122 (1980); *Bulletin of the Alpine Garden Society* **58**: 79 (1990).

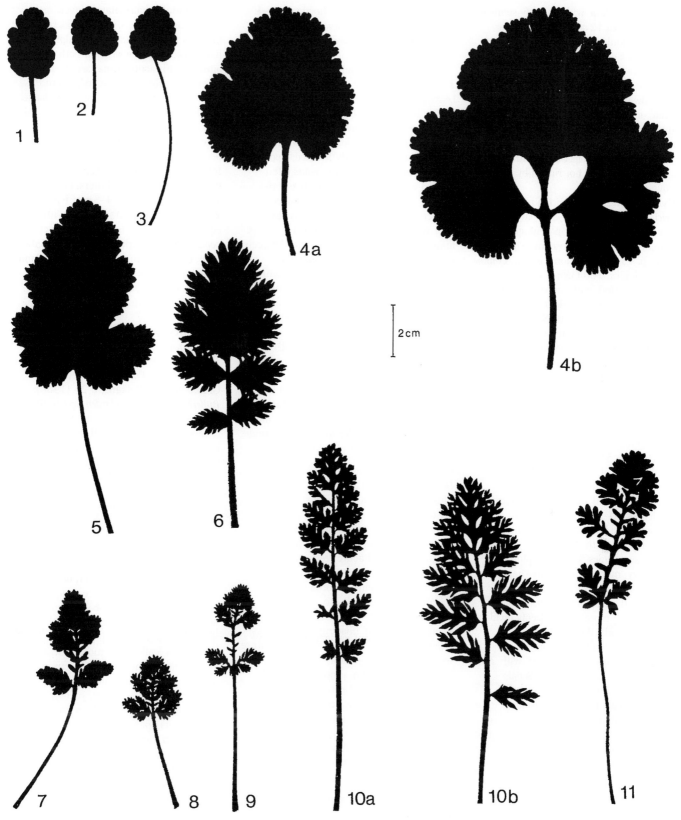

Figure 124. Leaf silhouettes of *Erodium* species. 1, *E. corsicum*.
2, *E. reichardii*. 3, *E.* × *variabile* 'Roseum'. 4a, b, *E. trifolium*.
5, *E. pelargoniiflorum*. 6, *E. recoderi*. 7, *E. foetidum*. 8, *E. glandulosum*.

9. *E. cheilanthifolium* stem-leaf. 10, *E. cicutarium*, a, basal leaf, b, stem-leaf.
11, *E. chrysanthum*.

Perennial, with or without leafy stems to 20 cm. Leaves to 12 cm, with glandular and non-glandular hairs; stalks to 10 cm; blades 5–23 × 4–26 mm, dark or greyish green, heart-shaped, lobed or unlobed. Inflorescence-stalks usually 1.5–3 cm, with 1 or 2 flowers. Petals 7–11 × 4–8 mm, obovate or broadly obovate, sometimes slightly notched, reddish purple with darker veins or white with reddish purple veins, unblotched. Carpel-body with the apical pits glandular, not ringed by a furrow. *Garden origin*.

Hybrids between *E. corsicum* and *E. reichardii*; very variable and often sold under the name of either parent species. The stemless cultivars 'Roseum', Figure 124(3), p. 451 (illustration: Griffith, Collins guide to alpines, 160 (1964) – as *E. chamaedryoides*), with more or less deep pink petals, and the double-flowered 'Flore Pleno' (illustration: Everett, New York Botanical Garden illustrated encyclopedia of horticulture 4: 1251, 1981), have also been incorrectly assigned to *E. reichardii*.

4. E. gruinum (Linnaeus) L'Héritier. Illustration: Hegi, Illustrierte Flora von Mittel-Europa 4(3): 1718 (1924); Polunin, Flowers of Europe, edn 1, t. 64 (1969); *Bulletin of the Alpine Garden Society* **41**: 235 (1973) & **58**: 70 (1990); Pignatti, Flora d'Italia **2**: 15 (1982).

Annual or biennial with leaves in a rosette and evident stems to 20–50 cm. Basal leaves 10–20 cm; stalks with scattered hairs; blades about as long as stalk or longer, with 3 leaflets or simple, but then sometimes 3-lobed or pinnately lobed, hairy, more or less heart-shaped to ovate, toothed. Stem-leaves with shorter stalks than the basal but blades to 12 × 7 cm and more or less deeply divided into 3, the middle division largest and pinnately lobed. Inflorescence-stalks 4–20 cm, the lowest the longest, with 2–6 flowers. Flower-stalks to 5 cm. Sepals 1–1.8 cm with mucro 2–7 mm. Petals 1.5–2.5 cm, pale to deep violet-blue, unblotched, falling very early in the day. Fruit 7.5–11.5 cm. Carpel usually 2-seeded, body *c.* 14 mm, its apical pits ringed by a furrow. *Mediterranean coasts except the NW.* H2.

5. E. trifolium (Cavanilles) Cavanilles (*E. hymenodes* L'Héritier). Figure 124(4), p. 451. Illustration: *Botanical Magazine*, 1174 (1809); Sweet, Geraniaceae **1**: t. 23 (1820); Das Pflanzenreich **53**: 585 (1912); Everett, New York Botanical Garden

illustrated encyclopedia of horticulture 4: 1252(1981) – as *E. pelargoniiflorum*; *Bulletin of the Alpine Garden Society* **58**: 79, 80, 82 (1990).

Biennial or perennial with leafy stems to 35 cm. Plant densely glandular-hairy. Leaves to 22 cm; stalks to twice as long as blades; blades ovate to heart-shaped, some of the lower leaves divided to the base into 3, margins shallowly lobed; lobes truncate, rather finely and bluntly toothed; upper surface finely wrinkled, the veins often stained with brown. Inflorescence-stalks usually 2.5–10 cm, with 4–11 flowers. Flower-stalks usually 1.5–2 cm, reflexed after flowering but carrying erect immature fruits. Sepals 5.5–7 mm, scarcely mucronate. Petals 8–14 × 5–6 mm, obovate, clawed at the base, white with feathered pinkish purple veins, the upper 2 each with a blotch above the claw formed by brownish red spots lying between the veins. Fruit 3.5–4 cm. Carpel-body *c.* 7 mm, its apical pits longer than wide, with stalkless glands, each ringed by a furrow. *Tunisia, Algeria, Morocco.* H4.

E. montanum Cosson & Durand is similar but is much smaller in all parts; the leaf-blades are less wrinkled and less hairy and have more angular lobes; the ground colour of the petals is pink. *Morocco.* H4.

6. E. pelargoniiflorum Boissier & Heldreich. Figure 124(5), p. 451. Illustration: *Botanical Magazine*, 5206 (1860).

Similar to *E. trifolium* but all leaves simple, the upper sometimes with broadly pointed lobes and acute teeth, their upper surface pale green, neither wrinkled nor with darkened veins; inflorescence-stalks to 12 cm and flower-stalks to 2.8 cm, the flowers thus more elevated above the foliage but slightly nodding; flower-stalks reflexed after flowering, with the immature fruits usually downwardly directed; sepals with mucro *c.* 2 mm; ground-colour of petals white; upper 2 petals with a transversely 2- or 3-zoned blotch; apical pits of carpel broader than long with stalked glands. *SW Turkey.* H4.

Less common in cultivation than the preceding species, which is often mistaken for it.

7. E. guicciardii Boissier. Figure 125(3), p. 453. Illustration: *Bulletin of the Alpine Garden Society* **41**: 237 (1973) & **58**: 79 (1990).

Dioecious perennial with evident leafy stems to 40 cm. Plant with glandular and

non-glandular hairs. Basal leaves numerous, to 27 × 5.5 cm; stalks to 1.5 times as long as blades, with adpressed hairs; blades silvery silky on both surfaces, pinnate with acute pinnatisect leaflets, or bipinnate; subsidiary leaflets present between the main ones; subsidiary leaflets and lobes of the main ones 1–2 mm wide, or to 3.5 mm in cultivation, linear-lanceolate or oblanceolate, acute, occasionally with a few teeth. Inflorescence-stalks to 10 cm with up to 8 flowers, with adpressed hairs. Flower-stalks 1.25–2.75 cm. Sepals 4–5 mm with mucro 2–3 mm. Petals 8–10 × 5–8 mm, about equal, pink to purple with darker veins coalescent as a dark spot at the extreme base. Fruit 5–7 cm. Carpel-body 8–10 mm, its apical pits not ringed by a furrow. *Albania, Greek & former Yugoslavian Macedonia.* H2.

8. E. chrysanthum de Candolle. Figure 124(11), p. 451. Illustration: Huxley & Taylor, Flowers of Greece and the Aegean, t. 131 (1977); Polunin, Flowers of Greece and the Balkans, t. 26 (1980); *Bulletin of the Alpine Garden Society* **58**: 70 (1990).

Dioecious perennial with large loose rosettes and evident leafy stems to 12 cm. Plant with glandular and non-glandular hairs. Basal leaves to 15 × 2.5 cm; blades about as long as stalks, clothed with dense, coarse, curled hairs, grey-green or, at certain angles of view, silvery, and with a silvery edge, pinnate with pinnatisect leaflets or bipinnate; subsidiary leaflets present between the main ones; subsidiary leaflets and lobes of the main ones obovate or obovate-lanceolate, more or less obtuse, sometimes themselves lobed. Stem-leaves with relatively shorter stalks and with sparser hairs than basal, pinnate with pinnatisect leaflets and more or less acute lobes. Inflorescence-stalks 5–7 cm, with 4–6 flowers. Flower-stalks 1.5–2 cm. Sepals 5–6 mm. Petals 7–12 mm, yellow, unblotched. Fruit 4.5–7.5 cm. Carpel-body 7–9 mm, its apical pits glandular, not ringed by a furrow. *Greece.* H4.

Some plants in cultivation under this name are probably hybrids with other dioecious species. Hybrids between this species and **E. amanum** Boissier & Kotschy (*E. absinthoides* var. *amanum* (Boissier & Kotschy) Brumhard) have been named **E. × lindavicum** Knuth and described as having red stems and pale yellow petals, smaller than those of

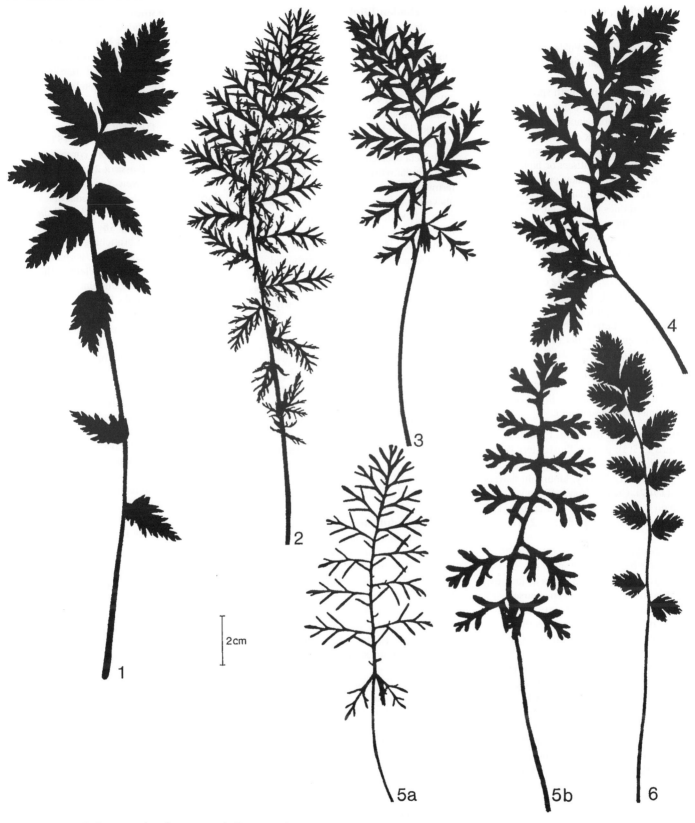

Figure 125. Leaf silhouettes of *Erodium* species. 1, *E. manescavi*.
2, *E. carvifolium*. 3, *E. guicciardii*. 4, *E. alpinum*. 5, *E. absinthoides*,
a, (Turkey), b, (Greece). 6, *E. castellanum*.

E. chrysanthum, but there is also a white-petalled cultivar.

9. E. absinthoides Willdenow. Figure 125(5), p. 453. *Bulletin of the Alpine Garden Society* **58**: 68 (1990).

Dioecious perennial with sprawling leafy stems to 30 cm. Plant with glandular and non-glandular hairs. Basal leaves to 10 or sometimes 20 cm; stalks 6–9 cm; blades grey-green, bipinnate with pinnatisect or bipinnatisect leaflets; subsidiary leaflets present between the main ones; subsidiary leaflets and lobes of the main ones linear to linear-spathulate, the wider ones 1–2 mm wide. Upper leaves smaller with relatively shorter stalks and more sparsely lobed leaflets. Inflorescence-stalks 3–5 cm with 2–8 flowers. Flower-stalks 1.5–2 cm. Sepals 7–8 mm. Petals 1–1.5 cm × 6–8 mm, obovate, violet, pink or white, unblotched. Fruit 4–7 cm. Carpel-body *c.* 9 mm, its apical pits glandular, not ringed by a furrow. *C Balkan Peninsula, Turkey, former Soviet Armenia.* H4.

White-petalled plants with leaf-lobes more than 2 mm wide are E. × lindavicum (see under *E. chrysanthum*).

10. E. alpinum (Burman) L'Héritier. Figure 125(4), p. 453. Illustration: Fiori & Paoletti, Iconographia florae Italicae, edn 3, 305 (1933); Huxley, Mountain flowers, 65 (1967); Grey-Wilson & Blamey, Alpine flowers of Britain and Europe, 133 (1979); Pignatti, Flora d'Italia **2**: 16 (1982).

Perennial with stems to 40 cm, with glandular hairs above. Leaves to 25 × 8 cm; stalks to 2.5 times as long as blades, shaggy; blades green, sparsely adpressed-hairy, bipinnatisect; pinnules entire or with up to 7 acute teeth, or themselves pinnatisect with ovate or oblong acutely toothed lobes; subsidiary leaflets present between the main ones, obtuse to acute. Inflorescence-stalks 8–35 cm, with 2–11 flowers, with spreading hairs. Flower-stalks 2–4 cm. Sepals 6.5–8 mm with mucro 2–3 mm. Petals 1–1.4 cm, about equal in size, obovate or broadly obovate, purplish pink with darker veins, unblotched. Fruit 5.5–6.5 cm. Carpel-body *c.* 10 mm, its apical pits glandular, not ringed by a furrow. *C Italy.* H4.

Hybridises with **E. ciconium** (Linnaeus & Juslen) L'Héritier to produce perennial offspring, some of which have garden value.

11. E. cheilanthifolium Boissier (*E. guttatum* misapplied; *E. leucanthum* misapplied; *E. petraeum* subsp. *crispum* misapplied; *E. trichomanifolium* misapplied). Figure 124(9), p. 451. Illustration: Das Pflanzenreich **53**: 279 (1912); *Boissiera* **20**: 94, t. 2 (1972); Grey-Wilson & Blamey, Alpine flowers of Britain and Europe, 133 (1979); Valdes, Talavera & Fernández-Galiano, Flora vascular de Andalucía occidental **2**: 279 (1987).

Perennial with branching rootstock terminated by leaf-rosettes. Leafy stems above the rosettes absent or not more than 2 cm. Glandular hairs present or absent. Leaves to 6 × 2 cm; stalks slightly longer than blades; blades green, grey-hairy or white-felted, bipinnate with pinnatifid pinnules; subsidiary leaflets present between the main ones; subsidiary leaflets and lobes of the main ones narrowly ovate, obtuse. Inflorescence-stalks usually 5–7 cm, with 2–4 flowers. Flower-stalks 1.5–2.2 cm. Sepals 5–6 mm. Petals *c.* 1 cm, unequal, white, pink or bluish pink with or without darker veins, the upper 2 each with a speckled blackish blotch ending in 1 or more points. Fruit 3–4 cm. Carpel-body *c.* 4.5 mm, its apical pits glandular, not ringed by a furrow. *S Spain.* H4.

The name **E. × willkommianum** Knuth applies to a hybrid between the present species and *E. glandulosum* in which the leaves are thinly hairy.

12. E. glandulosum (Cavanilles) Willdenow (*E. macradenum* L'Héritier; *E. petraeum* (Gouan) Willdenow subsp. *lucidum* (Lapeyrouse) Webb & Chater). Figure 124(8), p. 451. Illustration: *Botanical Magazine*, 5665 (1867); Boissiera **20**: 94, t. 2 & 5 (1972); Polunin & Smythies, Flowers of south-west Europe, t. 23 (1973); Jelitto, Schacht & Fessler, Die Freiland Schmuckstauden, edn 3, 231 (1985).

Similar to *E. cheilanthifolium*. Plant always glandular. Leaves to 10 × 3 cm; stalks longer than blades; blades green, with curled hairs, bipinnate with pinnatisect pinnules or 3 times pinnate; subsidiary leaflets and lobes of the main ones narrowly linear or spathulate. Inflorescence-stalks to 12 cm, with up to 5 flowers. Flower-stalks *c.* 2.5 cm. Petals pink, with slightly darker feathered veins, the blotches on the upper 2 usually large with translucent spots in the centre, rounded, with small teeth on the margins. *Pyrenees & adjacent N Spain.* H4.

E. × kolbianum Knuth. Hybrids between *E. glandulosum* and *E. rupestre* have grey leaves, but less so than in *E. rupestre*, and pale pink to nearly white petals, less intensely veined than in *E. rupestre*, the 2 upper blotched.

13. E. foetidum (Linnaeus & Nathorst) L'Héritier (*E. petraeum* (Gouan) Willdenow). Figure 124(7), p. 451. Illustration: Bonnier, Flore complète **2**: t. 103 (1913); *Boissiera* **20**: 94, t. 2 & 5 (1972); Everett, New York Botanical Garden illustrated encyclopedia of horticulture **2**: 1252 (1981); *Bulletin of the Alpine Garden Society* **58**: 77, 79 (1990).

Perennial with branching rootstock terminated by leaf-rosettes, without leafy stems above the rosettes. Plant with glandular and non-glandular hairs. Leaves *c.* 12 × 3.5 cm; stalks longer than blades; blades slightly greyish green, bipinnate with pinnatisect pinnules; subsidiary leaflets present between the main ones; subsidiary leaflets and lobes of the main ones linear-lanceolate. Inflorescence-stalks to 14 cm, with up to 7 flowers. Flower-stalks to 3.5 cm. Sepals *c.* 7.5 mm. Petals 1.2–1.3 cm, almost equal in size, about as broad as long, overlapping, pink with darker veins, unblotched or the 2 upper with small weak blotches. Fruit 2.5–3.5 cm. Carpel-body 6–7 mm, its apical pits usually glandular, not ringed by a furrow. *S France.* H4.

Plants grown in Britain as 'Merstham Pink' and 'County Park' appear to represent this species.

14. E. rodiei (Braun-Blanquet) Poirion. Illustration: *Boissiera* **20**: 94, t. 2 & 5 (1972).

Similar to *E. foetidum* but with leaves hairless or nearly so, with no glandular hairs, leaflets only once pinnatisect, inflorescence-stalks with up to 18 flowers, floral parts larger, sepals 9–11 mm; petals 1.5–2 cm; fruit 3–4.5 cm; carpels 7–10 mm. *SE France (near Grasse).* H4.

15. E. rupestre (Pourret) Marcet (*E. supracanum* L'Héritier). Illustration: *Boissiera* **20**: 94 (1972).

Perennial with branching rootstock terminated by rosettes of leaves, without leafy stems above the rosettes. Plant without glandular hairs. Leaves to 10 × 3 cm; stalks to 1.5 times as long as blades; blades whitish silvery above with adpressed hairs, green and nearly hairless beneath and on the edges, bipinnate or bipinnatisect, the pinnules entire or sometimes lobed; subsidiary leaflets present between the main ones; subsidiary leaflets and lobes of the main ones 1–1.5 mm wide, linear to ovate, more or less acute, wider and less spreading than in the 3 preceding species. Inflorescence-stalks to 6.5 cm, with up to 4 flowers. Flower-stalks to 1.5 cm. Sepals *c.* 5 mm. Petals 8–10 × 5–7 mm, nearly equal, overlapping, pale pink with darker veins, unblotched. Fruit 2.2–2.5 cm. Carpel-body *c.* 5 mm, its apical pits glandular, not ringed by a furrow. *NE Spain.* H4.

'Sara Francesca', with more finely and profusely cut leaves with spreading segments, is presumably a hybrid of *E. rupestre* (also grown under the misapplied name *E. hybridum*). Plants with the leaves whitish silvery above and green beneath and with veiny petals, but upper 2 with solid dark purple untoothed and unspeckled blotches on a pale pink ground, are frequently cultivated. They may also be hybrids of this species, particularly 'Natasha' (sometimes, perhaps incorrectly, also called 'Bidderi') in which the leaf segments are directed forwards as in *E. rupestre* (see also *E.* × *kolbianum* under *E. glandulosum*).

16. E. cicutarium (Linnaeus) L'Héritier. Figure 124(10), p. 451. Illustration: Coste, Flore de la France **1**: 254 (1901); Bonnier, Flore complète **2**: t. 104 (1913); Hegi, Illustrierte Flora von Mittel-Europa **4**(3): 1721, t. 174 (1924); Ross-Craig, Drawings of British plants, edn 1, **6**: 42 (1952).

Annual or biennial with leaf-rosettes pressed to the ground and erect or decumbent branching stems to 60 cm. Plant with glandular and non-glandular hairs. First-formed basal leaves *c.* 12 × 2.5 cm; stalks slightly longer than blades; blades pinnate with circular to broadly ovate and unevenly scalloped leaflets; without subsidiary leaflets between the main ones. Later-formed basal and lower stem-leaves with the blades usually longer than the stalks, its leaflets unevenly pinnatifid and with oblong-ovate lobes, or pinnatisect with pinnatifid to pinnatisect lobes, the ultimate segments then being lanceolate or linear-lanceolate and acute. Upper stem-leaves to 3.5 cm wide, sometimes stalkless. Inflorescence-stalks 3–7 cm, with 2–10 flowers. Flower-stalks 1–2.5 cm. Sepals *c.* 4 mm. Petals 4–10 mm, obovate, white to pink, usually unequal, upper 2 sometimes with moderate-sized

blotches. Fruit 3–4 cm. Carpel-body *c.* 6 mm, its apical pits non-glandular, usually ringed by a furrow. *Most of Europe.* H4.

Garden merit is sometimes claimed for this very variable weed, perhaps depending on the variant grown. The petals usually fall in the middle of the day.

17. E. recoderi Auriault & Guittonneau. Figure 124(6), p. 451. Illustration: *Lagascalia* **11**: 83 (1983).

Annual; stem erect at first, to 6 mm thick. Plant densely glandular-hairy. Blades of first-formed leaves simple, pinnatifid, ovate to ovate-oblong, of later ones to 4.5 cm, about as long as stalks, with 1 or 2, rarely 3, pairs of leaflets and no subsidiary leaflets; lobes or leaflets pinnatifid into toothed lobes. Inflorescence-stalks 5–8 cm with 4–6, occasionally to 10, flowers. Flower-stalks 2–2.5 cm. Flowers *c.* 2.6 cm wide. Sepals 7–9 mm; mucro *c.* 1 mm. Petals spreading horizontally or recurved, the upper 9–12 × 10 mm, the lower 1.3–1.4 cm × 6–8 mm, all pink, upper 2 with small rounded blotches. Fruit *c.* 5.5 cm. Carpel-body 9–10 mm, its apical pits non-glandular, ringed by a shallow furrow. *S Spain.* H5.

18. E. acaule (Linnaeus) Becherer & Thellung (*E. romanum* (Burman) L'Héritier). Illustration: *Botanical Magazine*, 377 (1797); Coste, Flore de la France **1**: 254 (1901); Pignatti, Flora d'Italia **2**: 17 (1982); Aeschimann et al., Flora alpina **1**: 1067 (2004).

Perennial usually without a leafy stem, without glandular hairs. Leaves to 25 × 4 cm; stalks not more than half as long as blades; blades green, pinnate, without subsidiary leaflets between the main ones; leaflets 11 or more, pinnatifid with more or less ovate, entire or acutely toothed lobes. Inflorescence-stalks 5–20 cm, with 4–9 flowers. Flower-stalks 1–2.5 cm. Sepals *c.* 5 mm with mucro less than 1 mm. Petals to 1.1 cm, about equal in size, obovate, bright pink, unblotched. Fruit 4–5.5 cm. Carpel-body 5–6 mm, its apical pits non-glandular, ringed by a furrow. *Mediterranean coastal area, Portugal.* H4.

19. E. carvifolium Boissier & Reuter. Figure 125(2), p. 453. Illustration: *Bulletin of the Alpine Garden Society* **41**: 236 (1973) & **58**: 73, 79 (1990).

Perennial with no evident stem. Plant with glandular and non-glandular hairs. Leaves 10–30 × 1.5–7 cm; stalks much

shorter than blades; blades green, pinnate with pinnatisect leaflets, without subsidiary leaflets between the main ones; pinnules more or less spreading, linear to oblanceolate, 3–8 times as long as wide, entire or pinnatifid with forwardly directed lobes. Inflorescence-stalks 10–30 cm, with 4–10 flowers. Flower-stalks 1.5–2.5 cm. Sepals 5.5–7 mm. Petals 7–20 mm, obovate to spathulate, purplish red, the 2 upper with blotches half or more as long as the petals. Fruit 4–5 cm. Carpel-body 7–10 mm, its apical pits non-glandular, not ringed by a furrow. *C Spain.* H4.

Hybrids with *E. castellanum* occur.

20. E. castellanum (Pau) Guittonneau (*E. daucoides* misapplied). Figure 125(6), p. 453. Illustration: *Boissiera* **20**: 114, t. 3 & 6 (1972); *Bulletin of the Alpine Garden Society* **58**: 69, 79 (1990).

Perennial with no evident stem. Plant with abundant glandular hairs. Leaves to 28 × 6 cm, more or less prostrate; stalks shorter than blades; blades bright green, pinnate, without subsidiary leaflets between the main ones; leaflets crowded, oblong, pinnatisect at base, sometimes merely pinnatifid towards apex, the lobes toothed to pinnatifid. Inflorescence-stalks 12–30 cm, with 8–20 flowers. Flower-stalks 2–3 cm. Sepals 5–7 mm with mucro 1–1.7 mm. Petals 1.2–1.6 cm, upper 2 varying from as broad as long to obovate, bright pink to pinkish red with large or small pale greyish to dark red blotches, the 3 lower narrower, each with a weak blotch. Fruit 4.5–5.5 cm. Carpel-body 7–10 mm, its apical pits non-glandular, not ringed by a furrow or with merely a depression next to the midrib. *N Spain.* H4.

Hybridises with *E. carvifolium* and *E. manescavi*.

21. E. manescavi Cosson. Figure 125(1), p. 453. Illustration: Coste, Flore de la France **1**: 253 (1901); Bonnier, Flore complète **2**: t. 103 (1913); Grey-Wilson & Blamey, Alpine flowers of Britain and Europe, 133 (1979); *Bulletin of the Alpine Garden Society* **58**: 79, 80 (1990).

Perennial, without evident leafy stems. Plant hairy, with glandular hairs above. Leaves to 30 × 10 cm, not prostrate; stalks much shorter than blades; blades bright green, pinnate with pinnatifid leaflets; without subsidiary leaflets between the main ones; lobes of leaflets toothed. Inflorescence-stalks to 50 cm, with 5–20 flowers. Flower-stalks to 6 cm. Sepals 1–1.5 cm; mucro 2–3 mm. Petals

1.4–2.2 cm × 7–16 mm, broadly obovate, deep purplish pink with netted darker veins, the 2 upper faintly to strongly blotched. Fruit 4.5–7.5 cm. Carpel-body 1–1.2 cm, its apical pits non-glandular, ringed by a rather weak furrow. *W Pyrenees.* H4.

E. × hybridum Knuth is similar but smaller, with more divided leaves and smaller, less intensely coloured petals. The parentage is given as *E. daucoides × E. manescavi* but as *E. castellanum* is grown as *E. daucoides* and can hybridise with *E. manescavi* this must be doubtful.

3. SARCOCAULON (de Candolle) Sweet
C. Innes

Perennial, succulent, mainly low-growing, rigid, short-branched low shrubs, often with swollen rootstock. Branches with hard resinous bark. Leaves usually of 2 kinds: primary, which have long stalks that become spines; and secondary, with shorter stalks that may persist as blunt stumps, and are situated in the axils of the primary, often forming clusters. Leaves simple or pinnatisect. Flowers bisexual, axillary, solitary on a jointed stalk, the lower part being the inflorescence-stalk, the upper the flower-stalk, radially symmetric. Petals rounded or somewhat truncate at apex, sometimes hairy. Stamens 15, all fertile, in groups of 3, in which the central one is the longest. Seeds dispersed within the carpels. Awn remaining attached to carpel-body, not twisted, plumed. The flowers frequently remain somewhat cup-shaped but widths given here apply to fully opened flowers.

A genus of possibly 14 species native to South Africa (Cape Province), Namibia and Angola. They are peculiarly attractive plants with something of a 'bonsai' appeal, though only a few are cultivated. They are leafless for much of the year and capable of surviving long periods of extreme drought, and are frequently difficult to identify. They require glasshouse or similar protection, and a bright, sunny and warm location is essential to successful culture. A truly porous, sandy, slightly acid soil is necessary. The short growing season coincides with the European winter and early spring months, so a temperature minimum of 15 °C must be maintained and only at this time should any water be given. Propagation is by seed or cuttings.

Literature: Rehm, S., Die Gattung Sarcocaulon (DC.) Sweet, *Botanische Jahrbucher* **67**: 664–75 (1935); Stiles, P.,

Sarcocaulon lorrei, *Bulletin of the African Succulent Plant Society* **7**: 36 (1972); Moffett, R.O., The genus Sarcocaulon, *Bothalia* **12**: 581–613 (1979).

1a. Leaves simple 2
 b. Leaves 1–3 pinnatisect 5
2a. Leaves without indentations apart from the apical notch 3
 b. Leaves lobed or toothed 4
3a. Sepals 8–14 mm with mucro 2–4 mm; petals yellow **1. l'heritieri**
 b. Sepals 5–10 mm with mucro less than 1 mm; petals pale pink to white **2. vanderietiae**
4a. Leaves mostly obovate, bluish and leathery, margins sinuous to toothed, slightly wavy; petals yellow to pale yellow **3. crassicaule**
 b. Leaves broadly ovate to circular, green and membranous, pleated and finely toothed; petals pink **4. mossamedense**
5a. Plant spineless or with short blunt spines to 6 mm; petals pink with dark patch near base **5. multifidum**
 b. Plant with spines *c.* 2.5 cm; petals whitish with lemon-yellow base **6. herrei**

1. S. l'heritieri (de Candolle) Sweet (*S. canariense* invalid). Illustration: L'Héritier, Geraniologia, t. 42 (1792; reprint 1978); Jacobsen, Handbook of succulent plants **2**: 757 (1960); *Bothalia* **12**: 586, t. 2 (1979).

Plant 60–75 cm, slender, erect, with rather few spreading branches to 30 cm, *c.* 1 cm thick, with pale greenish to grey waxy-bloomed bark. Spines *c.* 7 mm, whitish. Leaf-blades 2–3 cm, broadly obovate or those of the long-stalked leaves circular, all notched at apex, glaucous. Flower-stalks *c.* 2 cm, rarely shorter. Flowers 2–3 cm wide. Sepals 8–14 × 4–5 mm, hairless. Petals *c.* 25 × 8 mm, obovate, lemon-yellow. *South Africa (Namaqualand in Cape Province).* G1. Winter–spring.

2. S. vanderietiae Bolus (*S. burmannii* misapplied). Illustration: *Botanical Magazine*, 5729 (1868); Batten & Bokelmann, Wild flowers of the eastern Cape Province, t. 74, f. 5 (1966).

Plant 12–15 × 25–30 cm, compact; spiny, with stems *c.* 1 cm thick. Branches arising close to ground-level, olive-green to dark greyish, with rows of thin, straight spines 2 cm or more. Leaves 8–10 × 8–10 mm, obovate to heart-shaped. Flowers to 3 cm across or slightly more.

Sepals 5–10 mm, hairless or minutely hairy. Petals 1.5–2 × 1.4–1.6 cm, obovate, pale pink to white. *South Africa (E Cape Province).* G1. Autumn–spring.

3. S. crassicaule Rehm (*S. burmannii* misapplied; *S. spinosum* misapplied). Illustration: Lamb & Lamb, The illustrated reference on cacti and other succulents **2**: 346 (1959); *Bothalia* **12**: t. 3a (1979); Webb, The Pelargonium family, 41 (1984).

Plant 20–25 × 40–50 cm, branching repeatedly from 3–5 cm above the ground. Branches 1–1.5 cm thick, with somewhat torn, brownish grey or greyish yellow bark, with rows of thick, whitish grey, recurved spines *c.* 3 cm. Leaves 1.5–1.9 × 1.4–1.9 cm, slightly wavy-edged, bluish and leathery, finely hairy or hairless, with sinuous to toothed margins and occasionally finely toothed in addition. Blades of long-stalked leaves usually obovate. Short-stalked leaves with stalks *c.* 1 mm and blades with wedge-shaped base. Flower-stalks to 4 cm. Flowers *c.* 5 cm wide. Sepals 7–13 × 3–5 mm, more or less finely hairy. Petals *c.* 2.5 × 2.2 cm, pale yellow to white. *South Africa (Karoo), Namibia.* G1. Early–late spring.

4. S. mossamedense (Oliver) Hiern. Illustration: *Bothalia* **12**: t. 4a (1979).

Plant to 30 × 40 cm. Stems *c.* 3 cm thick at the base with several branches of the same thickness, all with thin greyish skin-like bark. Tips of branches with rows of short blunt spines. Leaves 2–3 × 1.2–2.9 cm, mainly ovate to circular, green and membranous, pleated and with fine teeth, minutely and finely hairy, rarely hairless. Stalks of long-stalked leaves 2–4 cm. Short-stalked leaves solitary or to 3 in a cluster, with stalks *c.* 1 cm. Flower-stalks 2 cm or more. Flowers *c.* 3.5 cm wide. Sepals *c.* 10 × 4 mm. Petals 1.7–2.5 × 1.2–1.8 cm, bright pink with paler, sometimes almost white, base. *Namibia, Angola.* G1. Early–late spring.

5. S. multifidum Knuth (*S. herrei* misapplied). Illustration: Paterson, A narrative of four journeys, edn 2, 113, t. 2 (1790); Jacobsen, Handbook of succulent plants **2**: 758 (1960); *Bothalia* **12**: 608, t. 4c (1979).

Plant *c.* 20 × 25 cm. Branches 8–10 × 1–2 cm, few, horizontally spreading, greyish brown or greenish white, more or less spineless. Leaves in 2 rows on upper sides of branches, all of one kind. Leafstalks *c.* 2 cm; blades *c.* 2 × 3 cm, erect, 2 or 3 times pinnatisect, finely hairy. Flower-stalks 1.2–3.2 cm. Flowers 3.5–4 cm wide. Petals *c.* 1.9 × 1.7 cm, pale pink, rarely white, with a deep red spot at the base giving the flower a dark throat. *South Africa (Cape Province), Namibia.* H1. Early–late spring.

6. S. herrei Bolus (*S. lorrei invalid*). Illustration: *Bothalia* **12**: t. 4b (1979).

Plant stiff, semi-erect with slightly ascending branches to 25 × 1 cm, with yellowish grey bark. Spines in straight rows. All herbaceous parts densely silky-hairy. Leaves *c.* 1.8 × 1.2 cm, broadly ovate to triangular, 2 or 3 times pinnatisect. Stalks of long-stalked leaves to 2.5 cm, forming spines to 3 mm thick at the base. Short-stalked leaves in clusters of 2–7. Flower-stalks *c.* 2 cm. Flowers *c.* 2.2 cm wide. Sepals *c.* 1 cm. Petals to 2 × 1.5 cm, whitish with lemon-yellow base, with a few scattered hairs on front surface. *South Africa (Cape Province).* H1. Winter.

4. PELARGONIUM L'Héritier
D.M. Miller

Shrubs, herbaceous perennials, sometimes woody at base, or annuals, some with tuberous roots. Stems erect or spreading, sometimes succulent. Leaves alternate, simple or pinnate or palmate, sometimes fleshy or aromatic. Flowers in umbels or pairs, bisexual, bilaterally symmetric, sometimes fragrant. Sepals 5, the upper with a slender spur at its base that is joined to the flower-stalk throughout its length and is swollen at its extremity where there is a nectar gland. Petals 5, sometimes 4 or 2, occasionally absent, clawed, upper 2 often larger than lower 3. Stamens 10, of which no more than 7 bear fertile anthers. Seeds dispersed within the carpels. Awn remaining attached to carpel-body, twisted, plumed.

A genus of over 200 species, mostly native to South Africa with a few from tropical Africa, Australia and the Middle East. Some species have been in cultivation for over 300 years and as a result the nomenclature has been sometimes confused and a very large number of hybrids have been raised for decorative use. All but the Middle Eastern species should be grown in frost-free conditions, in bright light and in a well-drained compost. Most species should be kept fairly dry in the winter, especially those that have a dormant period. Excessive humidity

encourages grey mould and stem rot. Most may be propagated by cuttings and the species by seed. Some species have tubers that may sometimes be divided. One main pest is white-fly.

The genus has been divided into about 14 sections mainly on morphological characteristics, which in some cases correlate with climatic conditions and geographical location. The sizes of plants can vary considerably depending on cultural conditions. Most of the following account follows the classification of Van der Walt et al. cited below.

Literature: Sweet, R., Geraniaceae **1–5** (1820–30); Harvey, W.H. & Sonder, O.W., Flora Capensis **1**: 259–308 (1860); Knuth, R., Pelargonium in A. Engler (ed.), Das Pflanzenreich **4**: 316–545 (1912); Moore, H.E., Pelargoniums in cultivation, Parts I & II, *Baileya* **3**: 5–26, 41–6, 71–9 (1955); Clifford, D., Pelargoniums (1970); Van der Walt, J.J.A., Pelargoniums of southern Africa **1** (1977); Van der Walt, J.J.A. & Vorster, P.J., Pelargoniums of southern Africa **2** (1981) & **3** (1988); Clifton, R., Geraniales species checklist, 1(4), edn 4, Pelargonium (2004); James, C.M., Gibby, M., & Barratt, J.A., Molecular studies in Geraniaceae: a Taxonomic appraisal of Sect. Ciconium and the origins of the zonal and ivy-leaved cultivars, *Plant Systematics and Evolution* **243**: 131–46 (2004)

Tubers or thickened rhizomes. Present: **2, 4, 5, 10, 26**.
Leaves. Pinnately veined: **2, 3, 5, 6, 7, 8, 14, 38, 39**. Palmately veined: **1, 10, 11, 12, 13, 15, 16, 17, 18, 19, 20, 21, 22, 23, 24, 25, 26, 28, 29, 30, 32, 33, 34, 41, 42, 43, 44, 45, 46**. Compound: **3, 5, 6, 7, 8, 11**. Simply lobed: **1, 4, 10, 12, 13, 14, 15, 16, 17, 18, 19, 20, 21, 22, 23, 24, 25, 26, 28, 29, 30, 31, 32, 33, 34, 39, 41, 42, 45, 46**. Fleshy: **6, 15, 24, 25**. Aromatic: **7, 13, 21, 27, 29, 30, 31, 32, 33, 34, 35, 36, 37, 38, 39, 40, 41, 42, 43, 44, 45**.
Stem. Very short or absent: **2, 4, 5, 9**.
Succulent: **1, 6, 10, 26**. Bearing persistent remains of stipules or leaf-stalks: **1, 6, 8, 26, 27, 28, 29, 30, 31**.
Flowers. Nearly radially symmetric: **1, 8, 13, 17**. Fragrant: **3, 4, 5**. White to cream: **1, 6, 7, 10, 11, 12, 20, 21, 22, 26, 27, 29, 30, 31, 43, 44, 45**. Pink to purple: **2, 8, 9, 10, 11, 12, 13, 14, 15, 16, 18, 24, 25, 27, 28, 32, 33, 34, 35, 36, 37, 38, 39, 40, 41,**

42, 44, 46. Red: **7, 17, 19**. Yellowish green to purplish brown: **3, 4, 5, 23**.
Fertile stamens. Five: **1, 2, 6, 14, 15, 16**. Six: **4**. Seven: **3, 5, 7, 8, 9, 10, 13, 17, 18, 19, 20, 21, 22, 23, 24, 27, 29, 30, 31, 32, 33, 34, 35, 37, 38, 39, 40, 41, 42, 43, 44, 45, 46**. Variable: **11, 12, 26, 28**.
Infertile stamens. **25, 36**.

1a. Plant with thickened stem, simple heart-shaped leaves and more or less radially symmetric white flowers without markings **1. cotyledonis**
  b. Plant not as above **2**
2a. Scrambling plant, stem succulent, 3- or 4-angled; flowers normally with 4 petals **10. tetragonum**
  b. Plant not as above **3**
3a. Flowers scented at night, never pink or red **4**
  b. Flowers not scented at night, usually in shades of white, pink or red **6**
4a. Leaves finely pinnately divided several times; flowers yellow to purple **5. triste**
  b. Leaves not as above **5**
5a. Stem-nodes not swollen; leaves large, lobed; flowers purplish black **4. lobatum**
  b. Stem with swollen nodes; leaves more or less fleshy, glaucous, pinnately divided into 3–5 leaflets; flowers greenish yellow **3. gibbosum**
6a. Plant with swollen knobbly stems; leaves fleshy and pinnately divided: flowers white **6. carnosum**
  b. Plant not as above **7**
7a. Stems bearing persistent but not membranous stipules or leaf-stalks, which sometimes become spiny **8**
  b. Stems without persistent stipules or leaf-stalks **14**
8a. Stems with short stout spines **26. echinatum**
  b. Stems with slender remains of leaf-stalks but not spiny **9**
9a. Leaves strongly aromatic; flowers usually white **10**
  b. Leaves not or faintly aromatic; flowers not white **13**
10a. Leaves divided into linear segments; flowers sometimes pink **27. abrotanifolium**
  b. Leaves simple; flowers white **11**
11a. Leaves bright green, apple-scented **30. odoratissimum**
  b. Leaves green or apple-scented **12**
12a. Leaves dark green, turpentine-scented **29. dichondrifolium**

b. Leaves glaucous with a spicy nutmeg scent   **31. fragrans**

13a. Leaves finely divided; flowers pink   **8. hirtum**

b. Leaves simple: flowers pink to dark purplish pink   **28. reniforme**

14a. Plant stemless with leaves in a rosette   15

b. Plant stemmed; stems sometimes short   16

15a. Plant with tubers; leaves pinnately lobed   **2. incrassatum**

b. Plant with thickened rhizomes; leaves palmately lobed   **9. endlicherianum**

16a. Leaves compound   17

b. Leaves simple but if very deeply divided not cut completely to the base   18

17a. Flowers red   **7. fulgidum**

b. Flowers pale pink to white   **11. myrrhifolium** var. **coriandrifolium**

18a. Flowers with short spur less than twice sepal length   19

b. Flowers with spur at least twice sepal length   40

19a. Leaves grey-green; stamens 3–5; upper petals dark purplish pink, lower petals white   **14. 'Splendide'**

b. Plant not as above   20

20a. Leaves oblong, deeply pinnately divided   **11. myrrhifolium** var. **myrrhifolium**

b. Leaves not as above, if deeply divided not oblong in shape   21

21a. Flowers less than 1 cm across   22

b. Flowers more than 1 cm across   23

22a. Leaves scalloped, aromatic; flowers usually deep purplish red   **13. grossularioides**

b. Leaves, not aromatic; flowers pale pink or white   **25. 'Saxifragoides'**

23a. Lower 3 petals linear, upper 2 petals greater than 3 times width of lower petals   24

b. Lower petals not linear, upper 2 petals less than 3 times width of lower petals   26

24a. Flowers white; leaves peppermint-scented   **45. tomentosum**

b. Flowers pink; leaves not peppermint-scented   25

25a. Leaves with shallow rounded lobes; upper petals pink with a dark purple central mark and a white basal blotch   **42. papilionaceum**

b. Leaves deeply cut, lobes with acute apices; upper petals with purple marks only   **41. hispidum**

26a. Leaves strongly balsam-scented   27

b. Leaves not balsam-scented   30

27a. Leaves sometimes slightly sticky, soft, hairy, lobes always obtuse   **39. panduriforme**

b. Leaves sticky, lobes usually acute   28

28a. Leaves not rough   **38. glutinosum**

b. Leaves rough or harsh   29

29a. Leaves divided almost to midrib into linear segments to 3 mm wide, with fine sharp teeth; petals often less than 1.5 cm   **35. denticulatum**

b. Leaves not so deeply divided, segments over 4 mm wide with irregular coarse teeth; petals more than 1.5 cm   **40. quercifolium**

30a. Leaves deeply divided almost to midrib   31

b. Leaves usually with 3–5 lobes   33

31a. Leaves diamond-shaped   **44. scabrum**

b. Leaves triangular in shape   32

32a. Leaf-segments less than 3 mm wide with margins strongly rolled under, not rose-scented   **37. radens**

b. Leaf-segments at least 3 mm wide with margins somewhat curved under, rose-scented   **36. 'Graveolens'**

33a. Leaves less than 1 cm, always lemon-scented   **33. crispum**

b. Leaves more than 1 cm, sometimes lemon-scented   34

34a. Leaves often cup-shaped; flowers more than 2.5 cm across   **34. cucullatum**

b. Leaves never cup-shaped; flowers less than 2.5 cm across   35

35a. Leaf-bases wedge-shaped; leaves often lemon-scented   **44. scabrum**

b. Leaf-bases truncate to cordate; leaves not lemon-scented   36

36a. Flowers white, pink or mauve, in tight heads; flower-stalks 1–2 mm   37

b. Flowers white or cream; inflorescence more open; flower-stalks more than 2 mm   39

37a. Plant upright; leaves somewhat rough, margins not crisped   **46. vitifolium**

b. Plant decumbent to weakly erect; leaves usually softly hairy   38

38a. Plant woody at base; leaves usually strongly rose-scented; inflorescence usually with 10–20 flowers; flowers pink to pale purple   **32. capitatum**

b. Plant sometimes swollen but not woody at base; leaves faintly aromatic; inflorescence usually with 5–10 flowers; flowers usually white   **12. australe**

39a. Plant a low-growing herb with trailing branches; leaves often zoned   **21. mutans**

b. Plant erect, woody at base; leaves never zoned   **43. ribifolium**

40a. Leaves usually peltate; plant trailing   **24. peltatum**

b. Leaves never peltate; plants not trailing   41

41a. Leaf-bases wedge-shaped, leaves glaucous and somewhat succulent; flowers pale salmon-pink   **15. acetosum**

b. Plant not as above   42

42a. Flowers yellowish green or greyish blue-green   **23. quinquelobatum**

b. Flowers not yellowish green or greyish blue-green   43

43a. Flowers usually pink or red, sometimes cream; petals more than 1 cm; stipules ovate or broadly ovate   44

b. Flowers cream or white; petals less than 1 cm; stipules lanceolate   **22. elongatum**

44a. Flowers rarely red, irregular; petals narrow   45

b. Flowers bright red, almost regular; petals rounded   47

45a. Leaf-bases shallowly cordate to truncate; flowers salmon-pink   **16. salmoneum**

b. Leaf-bases cordate; flowers not salmon-pink   46

46a. Inflorescence usually with more than 20 flowers; petals exceeding 2 cm; plant with short hairs   **18. zonale**

b. Inflorescence rarely more than 6-flowered; petals not exceeding 2 cm; plant with long hairs clearly visible   **20. alchemilloides**

47a. Plant erect; leaves almost circular, shallowly lobed   **17. inquinans**

b. Plant decumbent to semi-erect; leaves with triangular lobes   **19. tongaense**

Section **Isopetalum** (Sweet) de Candolle. Stems succulent. Leaves simple. Flowers more or less radially symmetric.

1. P. cotyledonis (Linnaeus) L'Héritier. Figure 127(10), p. 462. Illustration: Sweet, Geraniaceae **2**: 126 (1822–24): Van der Walt & Vorster, Pelargoniums of southern Africa **3**: 33 (1988).

Plant with short, slightly branched, succulent stems to 5 cm across, with scaly bark, rough with persistent stipules, becoming brownish with age; leaves clustered at ends of stems, often falling in dry season. Leaves 2–5 cm across, rounded with cordate bases to almost peltate, leathery, glossy dark green above with conspicuous impressed veins, densely grey hairy beneath; leaf-stalks to 8 cm. Inflorescence branched with several umbels on long stems, each with up to 15 flowers; flower-stalks to 8 mm; flowers to 1 cm across, more or less regular; sepals to 6 × 2 mm; petals *c.* 10 × 6 mm, elliptic, white without markings; spur *c.* 1 mm; fertile stamens 5. *St. Helena.* G1. Late spring–early summer.

Section **Hoarea** (Sweet) de Candolle (including Section *Seymouria* (Sweet) Harvey). Plants usually dormant in dry season; foliage and flowers usually appearing separately. Stemless perennial plants with one or more tubers; petals 5 or 2; fertile stamens 5. Often difficult to propagate.

2. P. incrassatum (Andrews) Sims (*P. roseum* (Andrews) de Candolle). Illustration: *Botanical Magazine*, 761 (1804); Sweet, Geraniaceae **3**: 262 (1824–26); Clifford, Pelargoniums, f. 79 (1970); Van der Walt & Vorster, Pelargoniums of southern Africa **2**: 79 (1981).

Plant with large underground tuber. Leaves basal, 3–6 × 2–5 cm, deeply pinnately lobed, with soft greyish white hairs; leaf-stalks as long as blades. Inflorescence to 30 cm, with 20–40 flowers; flower-stalks to 5 mm; flowers bright magenta; sepals *c.* 8 mm, narrow; upper 2 petals *c.* 20 × 8 mm, spathulate; lower 3 petals much smaller, *c.* 10 × 4 mm, narrowly spathulate with inrolled edges; spur to 4 cm. *South Africa (W Cape).* G1. Spring.

Section **Polyactium** de Candolle. Plants dormant in dry season, usually with 1 or more underground tubers; foliage and flowers appearing together. Leaves often pinnately veined; inflorescence with many flowers; flowers with a long spur and almost equal petals, often scented at night.

3. P. gibbosum (Linnaeus) L'Héritier. Figure 126(8), p. 460. Illustration: Sweet, Geraniaceae **1**: 61 (1820–22); Clifford, Pelargoniums, f. 48 (1970); Van der Walt

& Forster, Pelargoniums of southern Africa **1**: 17 (1977).

Spreading to scrambling plant, stems succulent at first, becoming woody with swollen nodes. Leaves *c.* 12 × 7 cm, pinnate with 1 or 2 pairs of leaflets, somewhat succulent, glaucous; leaf-stalks about equal to blades. Inflorescence with up to 15 flowers; flower-stalks very short; flowers 1.5–2 cm across, greenish yellow, scented at night; sepals *c.* 6 mm, strongly reflexed; petals obovate, 10–12 × 5–6 mm, slightly reflexed; spur *c.* 3 cm; fertile stamens 7. *South Africa (W coast).* G1. Summer.

4. P. lobatum (Burman) L'Héritier. Illustration: *Botanical Magazine*, 1986 (1818); Sweet, Geraniaceae **1**: 51 (1820–22); Van der Walt, Pelargoniums of southern Africa **1**: 24 (1977).

Plant with large irregular tuber below ground and very short stem above ground. Leaves variable, to 30 cm but usually much less in cultivation, with 3 or more lobes, softly hairy; stalks to 15 cm. Inflorescence to 30 cm or more, branched with clusters of 5–20 flowers; flowers *c.* 2 cm across, fragrant at night; flower-stalks very short; sepals *c.* 1 cm, lanceolate, strongly reflexed; petals *c.* 1.2 cm × 6 mm, rounded, very dark purple, edged yellowish green; spur to 3 cm; fertile stamens 6. *South Africa (S & SW Cape).* G1. Spring.

5. P. triste (Linnaeus) L'Héritier. Figure 126(1), p. 460. Illustration: *Botanical Magazine*, 1641 (1814); Sweet, Geraniaceae **1**: 85 (1820–22); Clifford, Pelargoniums, 38 (1970); Van der Walt, Pelargoniums of southern Africa **1**: f. 46 (1977).

Plant with large tuber and very short succulent stem. Basal leaves to 45 × 5–15 cm, usually less in cultivation, hairy; stalks to 20 cm with large heart-shaped stipules. Inflorescence to 30 cm or more, with smaller leaves, umbels with 5–20 flowers; flowers *c.* 1.5 cm across, scented at night, stalks very short; sepals 6–8 mm; petals obovate, slightly reflexed, usually brownish purple with broad yellowish margins but sometimes yellow or brown; upper 2 petals *c.* 15 × 7 mm, lower 3 slightly smaller; spur to 3 cm; fertile stamens 7. *South Africa (NW to S Cape).* G1. Spring–early summer.

Section **Otidia** (Sweet) Don. Leaves sometimes falling in the dry season. Plants with succulent stems, fleshy, pinnate,

compound leaves and very small stipules, sometimes with tubers. Flowers with almost equal petals; fertile stamens 5.

6. P. carnosum (Linnaeus) L'Héritier. Figure 127(19), p. 462. Illustration: *Baileya* **3**: f. 32 (1955); Clifford, Pelargoniums, f. 38 (1970); Van der Walt, Pelargoniums of southern Africa **1**: 8 (1977); Webb, The Pelargonium family, pl. 10 (1984).

Plant to more than 30 cm, with short thickened, branched succulent stems with remains of persistent leaf-stalks; rootstock thickened. Leaves very variable, to 15 × 5 cm, usually less in cultivation, oblong, more or less succulent, often deeply pinnately divided, grey-green to green; stalks 2–10 cm, thickened at base. Inflorescence often branched with several 2–8-flowered umbels; flowers less than 1 cm across; flower-stalks very short; 2 upper petals *c.* 7 × 5 mm, ovate, white to greenish tinged, marked with reddish purple lines; 3 lower petals slightly smaller, unmarked; spur *c.* 6 mm. *SW Africa.* G1. Summer–autumn.

Section **Ligularia** (Sweet) Harvey. Plants with branching stems often woody at base. Leaves usually pinnately divided; stems sometimes covered with remains of leaf-stalks or stipules; fertile stamens 7.

7. P. fulgidum (Linnaeus) L'Héritier. Figure 126(10), p. 460. Illustration: Sweet, Geraniaceae **1**: 69 (1820–22); Clifford, Pelargoniums, f. 72 (1970); Van der Walt, Pelargoniums of southern Africa **1**: 16 (1977); Webb, The Pelargonium family, pl. 22 (1984).

Plant spreading or scrambling, stems somewhat succulent, softly hairy, with fairly persistent membranous stipules. Leaves to 10 × 7 cm, oblong, with silver sheen; stalks *c.* 5 cm. Inflorescence branched, heads with 4–9 flowers; flower-stalks *c.* 8 mm; flowers nearly 2 cm across, scarlet; 2 upper petals to 2 × 1 cm, with darker veins; 3 lower petals slightly smaller; spur to 1.8 cm. *South Africa (west coast).* G1. Spring–early summer.

P. × schottii J.D. Hooker; probably *P. fulgidum* × *P. lobatum*. Illustration: *Botanical Magazine*, 5777 (1869); Webb, The Pelargonium family, pl. 49 (1984). Leaves deeply pinnately divided with long soft hairs, and large, deep purplish red flowers marked with purplish black blotches.

P. × ardens Loddiges. Figure 126(13), p. 460. Illustration: Sweet, Geraniaceae

Figure 126. Leaf silhouettes of *Pelargonium* species. 1, *P. triste*.
2, *P. radens*. 3, *P. denticulatum*. 4, *P.* 'Filicifolium'. 5, *P. myrrhifolium* var.
corandrifolium. 6, *P. hirtum*. 7, *P. abrotanifolium*. 8, *P. gibbosum*.

9, *P. quinquelobatum*. 10, *P. fulgidum*. 11, *P. ribifolium*. 12, *P. scabrum*.
13, *P.* × *ardens*. 14, *P.* 'Lady Plymouth'. 15, *P.* 'Graveolens'.
16, *P. glutinosum*. 17, *P. hispidum*.

1: 45 (1820–22); Webb, The Pelargonium family, pl. 22 (1984). A hybrid presumed to be of the same parentage with deeply pinnately lobed leaves and slightly smaller, bright but dark red flowers marked with dark brownish black.

8. P. hirtum (Burman) Jacquin. Figure 126(6), p. 460. Illustration: Sweet, Geraniaceae **2**: 113 (1822–24); Van der Walt, Pelargoniums of southern Africa **1**: 22 (1977).

Low-growing, hairy plant with erect or spreading, thickened branches, brownish with age and with persistent leaf-stalks. Leaves to 5.5×2 cm, like those of the carrot, faintly aromatic; leaf-stalks to 4 cm; stipules lanceolate, acute, attached to leaf-stalks for at least half their length. Inflorescence leafy with several clusters of 2–6 flowers; flower-stalks to 4 mm; flowers *c.* 1.5 cm across, bright to purplish pink, almost regular; sepals *c.* 5 mm, reflexed; petals spreading, upper 2, *c.* 1.1 cm $\times$ 6 mm, obovate, with darker spot at base, lower 3 slightly narrower without darker spots; spur *c.* 5 mm. *South Africa (Cape Peninsula).* G1. Late winter–spring.

Section **Jenkinsonia** (Sweet) Harvey. Leaves palmately lobed; petals 4 or 5, lower 3 very small; fertile stamens 7.

9. P. endlicherianum Fenzl. Illustration: *Botanical Magazine*, 4946 (1856); Clifford, Pelargoniums, f. 62 (1970).

Herbaceous, hairy, perennial plant with thickened rhizomes, to 30 cm or more. Basal leaves to 6 cm across, stem-leaves smaller; leaf-stalks to 7 cm. Inflorescence with 5–15 flowers; stems to 20 cm, fairly stout; flowers bright purplish pink; flower-stalks *c.* 5 mm; upper 2 petals to 3×1.5 cm, recurved; lower 3 petals minute or absent; spur to 2 cm. *Turkey.* H4. Summer.

10. P. tetragonum (Linnaeus) L'Héritier. Figure 127(12), p. 462. Illustration: *Botanical Magazine*, 136 (1791); Sweet, Geraniaceae **1**: 99 (1820–22); Van der Walt, Pelargoniums of southern Africa **1**: 45 (1977); Webb, The Pelargonium family, pl. 41 (1984).

Sprawling plant with green, 3- or 4-angled, succulent stems jointed at nodes, internodes long. Leaves few, 2–3×3.5–4 cm, usually with some hairs, often soon deciduous; leaf-stalks *c.* 3 cm. Inflorescence of 2 flowers; flower-stalks very short; flowers normally with 4 petals, cream to pale pink; sepals linear

1.1–1.5 cm; upper 2 petals to 4 b $\times$ 1.5 cm, spathulate, veined with dark red; lower 2 petals half size of upper; filaments bent in centre; spur to 4 cm or more. *S South Africa.* G1. Spring.

Section **Myrrhidium** de Candolle. Herbaceous, scrambling perennials, sometimes woody at base, often dying back in the dry season, or annuals. Leaves pinnately divided; petals 4 or 5, upper 2 often much larger than lower; fertile stamens 5 or 7.

11. P. myrrhifolium (Linnaeus) L'Héritier var. **coriandrifolium** (Linnaeus) Harvey (*P. coriandrifolium* Jacquin). Figure 126(5), p. 460. Illustration: Sweet, Geraniaceae **1**: 34 (1820–22); Van der Walt & Vorster, Pelargoniums of southern Africa **3**: 99 (1988); Webb, The Pelargonium family, pl. 31 (1984).

Low-growing, short-lived herb, sometimes woody at base. Leaves to 7×4 cm or more, compound, hairy or hairless; stalks to 7 cm. Inflorescence branched, each cluster with 2–5 flowers; stem *c.* 10 cm; flowers to 2 cm across, white or pink, almost stemless; petals 4 or 5, upper 2 petals to 3×1 cm, obovate, clawed, distinctly lined with red; lower 2 or 3 petals to 1.5 cm, unmarked; spur to 1 cm; fertile stamens 5, occasionally 7. *South Africa (W & SW Cape).* G1. Spring–summer.

The species is very variable in the wild but is usually represented in cultivation by the variety described above, which has the largest flowers and most finely divided leaves.

Var. **myrrhifolium** is sometimes grown but is less showy with smaller flowers and coarser, less divided leaves.

Section **Peristera** de Candolle. Small, short-lived perennials or annuals. Leaves palmate or pinnate; flowers small; fertile stamens 5 or 7.

12. P. australe Willdenow (*P. erodioides* J.D. Hooker). Figure 127(23), p. 462. Illustration: Clifford, Pelargoniums, f. 42 (1970); Van der Walt & Vorster, Pelargoniums of southern Africa **3**: 13 (1988).

Straggling, hairy, herbaceous perennial to 30 cm. Leaves to 10 cm across, usually much less, with 5–7 shallow lobes, slightly aromatic; leaf-stalks to 20 cm. Inflorescence with 5–10 or more flowers, compact; stem to 10 cm; flower-stalks short; flowers to 1.5 cm across, almost

regular, pink to white; sepals *c.* 5×3 mm, triangular; upper 2 petals *c.* 1 cm $\times$ 4 mm, apex often notched, veins deep pink; lower 3 petals very slightly smaller, clawed, sometimes with faint veins; spur 2–5 mm; fertile stamens usually 7. *SE Australia, Tasmania.* G1. Spring–summer.

This is a variable species and very distinct plants may be found in cultivation, such as one collected in Tasmania with a rosette of dark green leaves, a reddish leaf-stalk and smaller, white flowers. Others have pink flowers and leaves which are purplish red below. The species has been confused in cultivation with *Erodium trifolium* (see p. 452) from N Africa, which has larger white flowers veined with red in a looser inflorescence and leaves with a strong unpleasant smell.

13. P. grossularioides (Linnaeus) L'Héritier. Figure 127(21), p. 462. Illustration: Van der Walt, Pelargoniums of southern Africa **1**: 20 (1977).

Spreading, short-lived, almost hairless, herbaceous plant, often with red-tinged stems, sparingly branched and with long internodes. Leaves to 4×5 cm, aromatic; leaf-stalks to 10 cm or more. Inflorescence with many flowers, compact; stems to 15 cm; flower-stalks to 5 mm; flowers *c.* 8 mm across, almost regular, usually deep purplish pink, sometimes paler; upper 2 petals to 6×1–2 mm, oblong, marked with deeper pink; lower 3 petals unmarked; spur *c.* 3 mm; fertile stamens 7. *S & SE Africa.* G1. Spring–summer.

Plants in cultivation under this name with variegated foliage and larger pink flowers are hybrids, probably of *P. crispum*, and do not belong to this species.

Section **Campylia** (Sweet) de Candolle. Plants with short stems but extensive root systems. Leaves simple, ovate to linear; leaf-stalks long; fertile stamens 5.

There are probably no species of this section widely cultivated in Europe. The plant often grown as *P. tricolor* Curtis, *P. violareum* Jacquin or *P. tricolor* var. *arborea* invalid, appears to be a hybrid, which should correctly be named 'Splendide'.

14. P. 'Splendide'; probably *P. tricolor* × *P. ovale* (Burman) L'Héritier. Figure 127(22), p. 462. Illustration: Clifford, Pelargoniums, f. 80 (1970); Webb, The Pelargonium family, pl. 50 (1984); Brickell (ed.), RHS A–Z encyclopedia of garden plants, 783 (2003).

Figure 127. Leaf silhouettes of *Pelargonium* species. 1, *P. panduriforme*. 2, *P. papilionaceum*. 3, *P. cucullatum*. 4, *P. tomentosum*. 5, *P. alchemilloides*. 6, *P. peltatum*. 7, *P. tongaense*. 8, *P. zonale*. 9, *P. inquinans*. 10, *P. cotyledonis*. 11, *P. acetosum*. 12, *P. tetragonum*. 13, *P. salmoneum*. 14, *P. crispum*. 15, *P. crispum* 'Variegatum'. 16, *P. echinatum*. 17, *P. reniforme*. 18, *P.* 'Saxifragoides'. 19, *P. carnosum*. 20, *P. odoratissimum*. 21, *P. grossularioides*. 22, *P.* 'Splendide'. 23a, *P. australe*, b, Tasmanian variant. 24, *P. dichondrifolium*. 25a, *P. fragrans* 'Variegatum', b, *P. fragrans*. 26, *P. vitifolium*. 27, *P. capitatum*.

Plant to 30 cm, with short, branching, somewhat woody stems. Leaves to 4 × 2 cm, grey-green, hairy; leaf-stalks to 7 cm. Inflorescence branched, each cluster with 2 or 3 flowers; stems to 6 cm; flowers to 3 cm across; flower-stalks *c.* 1.5 cm; upper 2 petals to 1.8 × 1.5 cm, almost circular, dark red with blackish base; lower 3 petals slightly smaller, white; spur to 5 mm. *Garden origin.*

Section **Ciconium** (Sweet) Harvey (including Section Eumorpha (Ecklon & Zeyher) Harvey and Section Dibrachya (Sweet) Harvey). Shrublets or herbaceous plants often with thick, more or less fleshy stems. Leaves simple, palmately veined; fertile stamens usually 7; spurs often long.

15. P. acetosum (Linnaeus) L'Héritier. Figure 127(11), p. 462. Illustration: *Botanical Magazine*, 103 (1790); Van der Walt, Pelargoniums of southern Africa 1: 2 (1977); Webb, The Pelargonium family, pl. 12 (1984); Brickell (ed.), RHS A–Z encyclopedia of garden plants, 778 (2003).

Branching, more or less hairless plant to 60 cm in cultivation with brittle stems, sometimes woody at base. Leaves few, 2–6 × 1–2 cm, wedge-shaped at base, fleshy, glaucous, sometimes with reddish margins; leaf-stalks *c.* 1 cm. Inflorescence with 2–7 flowers; stems to 10 cm; flowers 4–5 cm across, salmon-pink; flower-stalks *c.* 5 mm; sepals narrow, *c.* 1.2 cm: upper 2 petals *c.* 2.5 cm × 5 mm, very narrow, erect, marked with darker veins; lower 3 petals similar but unmarked and slightly broader; spur 2.5–3 cm; fertile stamens 5. *South Africa (E Cape).* G1. Spring–summer.

16. P. salmoneum R.A. Dyer. Figure 127 (13), p. 462. Illustration: *Botanical Magazine*, 9357 (1934).

Erect branched shrublet to 1 m with short glandular hairs. Leaves to 4.5 × 5 cm, thick, green to somewhat glaucous, unmarked; leaf-stalks to 4 cm. Inflorescence with 4–14 flowers, with reflexed buds but erect flowers; stem to 15 cm; flowers to 4 cm across, salmon-pink; flower-stalks to 5 mm; petals almost equal, *c.* 2.5 × 1.5 cm, obovate, upper 2 petals veined darker; spur to 2.5 cm; fertile stamens 5. *S South Africa.* G1. Spring–summer.

This plant was for many years considered a distinct species but is sometimes considered to be a hybrid, possibly involving *P. acetosum*.

17. P. inquinans (Linnaeus) L'Héritier. Figure 127(9), p. 462. Illustration: Clifford, Pelargoniums, f. 53 (1970); Van der Walt, Pelargoniums of southern Africa 1: 23 (1977).

Erect branching, velvety hairy plant to 2 m, woody at base. Leaves to 8 cm across, almost circular, unmarked; leaf-stalks to more than 6 cm. Inflorescence usually with 10–20 upright flowers with reflexed buds; stems to 8 cm; flowers to 4 cm across, almost bilaterally symmetric, bright red; flower-stalks very short; petals *c.* 2 × 1.2 cm, rounded, upper 2 very slightly smaller; stamens and style barely protruding, free section of filaments very short; spur to 2.5 cm. *South Africa (E Cape).* G1. Spring–autumn.

This species and *P. zonale* are generally considered to be the main parents of the modern zonal pelargoniums, **P. × hortorum** Bailey.

P. × kewense R.A. Dyer; probably *P. zonale × P. inquinans.* Illustration: *Journal of the Royal Horticultural Society* 59: f. 80 (1934). Plants usually with unmarked leaves but sometimes with a zone when young, leaf-bases shallowly cordate; inflorescence with up to 27 bright red flowers with narrowly oblong to obovate petals.

P. scandens misapplied (not *P. scandens* Ehrhart). Illustration: Webb, The Pelargonium family, pl. 42 (1984). This plant is very similar and probably of similar parentage but the leaves have a very distinct and consistent zone. The flowers are bright red with narrow petals. Although this plant is commonly grown its nomenclature is subject to question and under investigation.

18. P. zonale (Linnaeus) L'Héritier. Figure 127(8), p. 462. Illustration: Das Pflanzenreich 4: f. 57 (1912); Clifford, Pelargoniums, f. 52 (1970); Van der Walt, Pelargoniums of southern Africa 1: 50 (1977).

Erect or scrambling plant to 1 m, usually hairy, woody at base. Leaves 5–8 cm across, usually with a darker brownish purple, horseshoe-shaped mark; leaf-stalks to 5 cm. Inflorescence with *c.* 50 flowers, usually fewer, buds reflexed but flowers upright; stems to 20 cm; flowers to 3 cm across, usually pale pink, sometimes white or red; flower-stalks very short; sepals reflexed; petals more or less equal in size, *c.* 20 × 7 mm, narrow, marked with darker veins, upper 2 erect, lower 3 spreading;

spur to more than 2.5 cm. *South Africa.* G1. Spring–autumn.

19. P. tongaense Vorster. Figure 127(7), p. 462. Illustration: Van der Walt & Vorster, Pelargoniums of southern Africa 3: 141 (1988).

Semi-erect to decumbent, hairy, herbaceous plant. Leaves 4–7 cm across, thick to somewhat fleshy with triangular lobes; leaf-stalks to 10 cm. Inflorescence with up to 10 flowers; stems to more than 20 cm; flowers to 2.5 cm across, bright red; flower-stalks very short; upper 2 petals *c.* 1.7 cm × 7 mm, ovate; lower 3 petals slightly broader; stamens and style barely protruding, free section of filaments very short; spur to 4 cm. *South Africa (NE Natal).* G1. Summer.

20. P. alchemilloides (Linnaeus) L'Héritier. (*P. alchemillifolium* Salisbury; *P. malvaefolium* Jacquin). Figure 127(5), p. 462. Illustration: Van der Walt, Pelargoniums of southern Africa 1: 3 (1977).

Herbaceous, decumbent, hairy, perennial plant. Leaves to 10 × 12 cm, often less, sometimes with a dark reddish brown horseshoe-shaped zone; leaf-stalks to 10 cm; stipules broadly ovate. Inflorescence with *c.* 5 flowers, sometimes more; stems to 20 cm; flowers to nearly 2 cm across, cream, white or pale pink; flower-stalks *c.* 5 mm, strongly reflexed after flowering; upper petals 1–2 cm × 2–10 mm, spathulate, sometimes with darker markings; lower petals smaller; spur to *c.* 8 mm. *S & E Africa.* G1. Spring–summer.

This species is very variable in flower and foliage characteristics and has been divided into several varieties. It may be split into 2 or more distinct species.

21. P. mutans Vorster (*P. grandiflorum* misapplied; *P. alchemilloides* misapplied). Illustration: *Flowering plants of Africa* 52: 2026 (1992).

Herbaceous sprawling plant with rather thick stems. Leaves to 10 cm across, aromatic, bright green, normally with a purple zone about halfway to margin, circular, 5-lobed; leaf-stalks 3–4 cm; stipules broadly ovate. Inflorescence with *c.* 8 flowers; stems to 25 cm; flowers to 2 cm across, white to cream; flower-stalks to 1 cm; upper petals *c.* 20 × 5 mm, linear, reflexed, faintly lined; lower petals 2 or 3, *c.* 15 × 3 mm, straight; stamens almost equal, often curved upwards; spur 1–2 cm. *South Africa (Natal).* G1. Summer.

Most plants grown as *P. grandiflorum* belong to this species. The true **P. grandiflorum** (Andrews) Willdenow (illustration: Van der Walt, Pelargoniums of southern Africa **1**: 19, 1977), with very large pink flowers and glaucous, palmately lobed, often zoned leaves, belongs to the Section *Glaucophyllum* and is probably not in general cultivation.

22. P. elongatum (Cavanilles) Salisbury (*P. tabulare* (Linnaeus) L'Héritier). Illustration: Das Pflanzenreich **4**: f. 55 (1912); Van der Walt, Pelargoniums of southern Africa **1**: 44 (1977).

Straggling herbaceous plant with long and short glandular hairs. Leaves 1.5–4 × 2–5 cm, usually with a dark brown to purplish horseshoe-shaped mark; leaf-stalks to 7 cm or more; stipules lanceolate. Inflorescence with 1–5 flowers; stems to 15 cm; flowers to 1 cm across, usually white to cream; flower-stalks very short; upper 2 petals *c.* 1 cm × 4 mm, narrow, usually veined with red; lower 3 petals slightly smaller; spur 1–2 cm. *South Africa (SW Cape)*. G1. Spring–late summer.

This plant is often found in cultivation incorrectly as *P. tabulare* **(**Burman filius) L'Héritier, a distinct species with pink flowers.

23. P. quinquelobatum Hochstetter Figure 126(9), p. 460. Illustration: Clifford, Pelargoniums, f. 31 (1970); Webb, The Pelargonium family, pl. 37 (1984).

Herbaceous, hairy, perennial plant with straggling, semi-prostrate stems. Leaves to 10 × 8 cm, deeply palmately lobed, dull green; leaf-stalks 8–12 cm. Inflorescence with *c.* 5 flowers; stems to 30 cm; flowers to 1.5 cm across, pale yellowish green or greyish blue-green; flower-stalks to 5 mm; upper 2 petals to 1.5 × 5 mm, oblong, lined with pink; lower petals slightly smaller, unmarked; spur to 3 cm. *E Africa*. G1. Summer.

24. P. peltatum (Linnaeus) L'Héritier. Figure 127(6), p. 462. Illustration: *Botanical Magazine*, 20 (1787); Van der Walt, Pelargoniums of southern Africa **1**: 33 (1977).

Trailing or climbing, more or less hairless, perennial. Leaves 3–7 cm across, usually peltate, somewhat fleshy, hairless or with short soft hairs, often with darker circular zone; leaf-stalks to 5 cm. Inflorescence with 2–9 flowers; stems to 8 cm; flowers to 4 cm across, white or pink to pale purplish pink; flower-stalks very

short; upper 2 petals 2.5 × 1 cm, obovate, marked with darker veins; lower 3 petals shorter but broader, unmarked; spur 3–4 cm. *South Africa*. G1. Spring–summer.

This species is the main parent of the ivy-leaved pelargoniums of gardens. In the wild it is variable in leaf characteristics and flower colour and in the past has been divided into several different species or varieties that are not usually recognised today. Plants with non-peltate leaves have been described as *P. lateripes* misapplied, which is a name of uncertain status and may refer to an early hybrid.

25. P. 'Saxifragoides'. Figure 127(18), p. 462. Illustration: Clifford, Pelargoniums, f. 28 (1970).

Low-growing, spreading, hairless plant. Leaves to 2 × 2.5 cm, ivy-shaped, thick and fleshy; leaf-stalks to 2.5 cm or more. Inflorescence with 2–5 flowers; stems 5–10 cm; flowers less than 1 cm across, pink to white; flower-stalks to 4 mm; upper 2 petals *c.* 5 × 3 mm, oblong, reflexed, with darker markings; lower 3 petals narrower, spreading, unmarked; spur *c.* 2 mm. *Garden origin*. G1. Spring–summer.

This plant was originally described as a species, but is now thought to be a hybrid or an extreme variant of *P. peltatum*.

Section **Cortusina** (de Candolle) Harvey. Plants often woody at base, with short thick stems covered with persistent stipules of leaf-stalks. Leaves simple, palmate, deciduous; leaf-stalks long; inflorescence unbranched; fertile stamens 6 or 7.

26. P. echinatum Curtis. Figure 127(16), p. 462. Illustration: *Botanical Magazine*, 309 (1795); Sweet, Geraniaceae **1**: 54 (1820–22); Van der Walt, Pelargoniums of southern Africa **1**: 13 (1977).

Erect, woody at base, stems thick spiny, roots tuberous. Leaves to 6 cm, across, grey-green, paler and hairy below; leaf-stalks to 10 cm; stipules awl-shaped, becoming hard and spiny. Inflorescence with 3–8 flowers; stems 3–4 cm; flowers 1.5–2 cm across, usually white in cultivation but also pink to purple in wild; flower-stalks 3–4 mm; upper 2 petals *c.* 2 × 1 cm, ovate with dark red blotches; lower 3 petals smaller, usually unmarked; spur 3–4 cm; fertile stamens 6 or 7. *W South Africa*. G1. Spring.

'Miss Stapleton' (illustration: Sweet, Geraniaceae, **3**: 212, 1824–26) is similar but has bright purplish pink flowers.

Section **Reniformia** (Knuth) Dreyer. Woody-based plants, often covered with persistent stipules or leaf-stalks. Leaves simple, palmate, evergreen, often aromatic; inflorescence branched; fertile stamens 7.

27. P. abrotanifolium (Linnaeus) Jacquin. Figure 126(7), p. 460. Illustration: Sweet, Geraniaceae **4**: 351 (1826–28); Clifford, Pelargoniums, f. 32 (1970); Van der Walt, Pelargoniums of southern Africa **1**: 1 (1977).

Branching, erect plant to 50 cm or more; stems becoming woody with age and bearing remains of leaf-stalks. Leaves 5–15 × 5–20 mm, finely divided, aromatic, grey-green; leaf-stalks about equal to leaf-blades. Inflorescence with 1–5 flowers, not branched; flowers *c.* 1.5 cm across, often white in cultivation but pink-flowered plants may also be found; flower-stalks very short; sepals *c.* 5 mm, reflexed; petals *c.* 1.5 × 3 mm, narrowly obovate, upper 2 veined reddish purple, lower 3 slightly wider, unmarked or with faint markings; spur *c.* 1.5 cm. *South Africa*. G1. Spring–summer.

28. P. reniforme Curtis. Figure 127(17), p. 462. Illustration: Van der Walt, Pelargoniums of southern Africa **1**: 40 (1977).

Erect or trailing plant to 1 m, woody at base, with small tuberous roots. Leaves 2–3 cm or more across; grey-green with velvety texture, silvery below, somewhat aromatic; leaf-stalks to 5 cm or more, persistent. Inflorescence branched, with 3–12 flowers; stems 3–10 cm; flowers 1–2 cm across, bright pink to deep purplish pink; flower-stalks to 1 cm; sepals reflexed; upper 2 petals 1–1.2 cm × 3–4 mm, narrowly oblong, erect, spotted and veined darker pink; lower 3 petals slightly broader, spreading, unmarked; spur 1–2 cm; fertile stamens 6 or 7. *South Africa (E Cape)*. G1. Spring–summer.

29. P. dichondrifolium de Candolle (*P. cradockense* Knuth; *P. relinquifolium* Brown; *P. burchellii* Knuth). Figure 127 (24), p. 462. Illustration: Das Pflanzenreich **4**: f. 58 (1912); Van der Walt & Vorster, Pelargoniums of southern Africa **3**: 47 (1988).

Perennial with short stems becoming woody, to 20 cm or more. Leaves crowded, 1–2 cm across, dark grey-green, aromatic, turpentine-scented; leaf-stalks to 10 cm, persistent; stipules very small. Inflorescence with 2–5 flowers, branched; flowers to

2 cm across, white; flower-stalks very short; upper 2 petals *c.* 1 cm, oblong, marked red; lower petals broader, unmarked; spur 2–3 cm or more; fertile stamens 7. *South Africa (E Cape)*. G1. Summer.

30. P. odoratissimum (Linnaeus) L'Héritier. Figure 127(20), p. 462. Illustration: Sweet, Geraniaceae **3**: 299 (1824–26); Clifford, Pelargoniums, f. 71 (1970); Van der Walt, Pelargoniums of southern Africa **1**: 30 (1977).

Low-growing herbaceous plant with short main stem but spreading to trailing flowering stems. Leaves to 4 cm across, sometimes much larger, light green, apple-scented; leaf-stalks to 5 cm, persistent; stipules small. Inflorescence branched, with 3–10 flowers; stems to 5 cm or more; flowers to 1.2 cm across, white; flower-stalks to 1 cm, green; calyx green; 2 upper petals to 1 cm × 3 mm, oblong, marked with red; lower petals slightly smaller, unmarked; spur 5–7 mm; fertile stamens 7. *South Africa (E & S Cape)*. G1. Spring–summer.

31. P. fragrans Willdenow. Figure 127 (25b), p. 462. Illustration: Sweet, Geraniaceae **2**: 172 (1822–24).

Erect, branched plant, to 30 cm, woody at base. Leaves to 2.5 × 3 cm, with soft, velvety texture, glaucous, with sharp spicy scent; leaf-stalks to more than 4 cm. Inflorescence branched, each cluster with 4–8 flowers; stems *c.* 4 cm; flowers to 1.5 cm across, white; flower-stalks 4–6 mm, often brownish red; calyx brownish red; upper 2 petals erect, to 8 × 3 mm, oblong, marked with red; lower petals spreading, unmarked; spur to 6 mm; fertile stamens 7. *South Africa*. G1. Spring–summer.

For many years this plant was considered to be a hybrid, (*P. odoratissimum × P. exstipulatum*) but is now treated as a species by many botanists. A cultivar with leaves edged creamy yellow is named 'Variegatum' (Figure 127(25a), p. 462).

Section **Pelargonium** (de Candolle) Harvey. Plants with much-branched stems, shrubs or woody at base; leaves usually palmately lobed or divided, often aromatic; upper petals distinctly larger than lower; fertile stamens 7.

32. P. capitatum (Linnaeus) L'Héritier. Figure 127(27), p. 462. Illustration: Van

der Walt, Pelargoniums of southern Africa **1**: 7 (1977).

Decumbent, somewhat spreading, or weakly erect, softly hairy plant to 1 m; stems becoming more or less woody at base. Leaves 2–8 cm across, rose-scented, with long soft hairs; leaf-stalks to 4 cm. Inflorescence very compact, with 10–20 flowers; stems to 10 cm; flowers to 1.5 cm across, usually mauve-pink; flower-stalks to 1 mm; upper 2 petals *c.* 1.8 cm × 5 mm, narrowly obovate, veined darker pink; lower petals slightly smaller, less marked; spur to 3 mm. *South Africa*. G1. Spring–summer.

This is one of the species cultivated for the extraction of geranium oil. It appears to be represented in gardens by the cultivar 'Attar of Roses' with a more upright habit and rougher, strongly aromatic leaves, which is probably a hybrid of *P. capitatum*.

33. P. crispum (Bergius) L'Héritier. Figure 127(14), p. 462. Illustration: Sweet, Geraniaceae **4**: 383 (1826–28); Van der Walt & Vorster, Pelargoniums of southern Africa **3**: 37 (1988).

Erect plant to 70 cm, with stems becoming woody at base. Leaves to 1 × 1.5 cm, margins crisped, rough, strongly lemon-scented; leaf-stalks *c.* 5 mm; stipules conspicuous. Inflorescence usually with 1 or 2 flowers; stems to 1 cm; flowers to 2 cm across, usually pink; flower-stalks *c.* 5 mm; upper 2 petals to 2 × 1 cm, spathulate, with darker markings, apex notched; lower petals narrow oblong, unmarked; spur to 8 mm. *South Africa (SW Cape)*. G1. Spring–summer.

Garden plants show a wide range of leaf size. Cultivars with distinctly larger leaves are grown as 'Major' and those with leaves edged with creamy white are named 'Variegatum' (Figure 127(15), p. 462). Illustration: Hay & Synge, Dictionary of garden plants, f. 586, 1969).

34. P. cucullatum (Linnaeus) L'Héritier (*P. acerifolium* L'Héritier; *P. angulosum* (Miller) L'Héritier). Figure 127(3), p. 462. Illustration: Van der Walt, Pelargoniums of southern Africa **1**: 12 (1977); Van der Walt & Vorster Pelargoniums of southern Africa **2**: 45 (1981) & **3**: 40 (1988); Brickell (ed.), RHS A–Z encyclopedia of garden plants, 780 (2003).

Erect, branched, hairy shrub to 2 m. Leaves usually *c.* 4.5 × 5 cm, rounded to triangular, toothed and sometimes shallowly lobed, often hood- or cup-shaped with long hairs, often with a thin reddish

margin; leaf-stalks *c.* 2 cm. Inflorescence branched, each head with *c.* 5 flowers; stems *c.* 5 cm; flowers to more than 3 cm across, bright purplish pink; flower-stalks to 1 cm; upper 2 petals to 2.5 × 1.5 cm, broadly obovate, veined deeper pink; lower petals slightly smaller, narrower, unmarked; spur 5–12 mm. *SW Africa*. G1. Spring–summer.

Cultivars with semi-double flowers are found in gardens as 'Flore Plenum'. This very variable species has been divided into 3 subspecies, mainly on the leaf characteristics. It is the principal parent of the Regal pelargoniums, **P. × domesticum** Bailey.

35. P. denticulatum Jacquin. Figure 126 (3), p. 460. Illustration: Sweet, Geraniaceae **2**: 109 (1822–24); Van der Walt & Vorster, Pelargoniums of southern Africa **2**: 51 (1981).

Erect, branched shrub to more than 1.5 m. Leaves 6–8 × 7–9 cm, divided almost to midrib, strongly balsam-scented, rough, very sticky, margins irregularly and finely toothed; leaf-stalks *c.* 5 cm. Inflorescence with *c.* 6 flowers; stems to 5 cm; flowers to 2 cm across, purplish pink; flower-stalks to 2 mm; upper 2 petals *c.* 1.8 cm × 6 mm, narrowly spathulate, veined darker pink, apex usually notched; lower petals slightly smaller, unmarked; spur 8–9 mm. *South Africa (S Cape)*. G1. Spring–summer.

A cultivar with very finely divided leaves and with the upper 2 petals very deeply 2-fid, is grown as 'Filicifolium' ('Fernaefohum'). Figure 126(4), p. 460.

36. P. 'Graveolens'. Figure 126(15), p. 460. Illustration: L'Héritier, Geraniaceae, t. 17 (1792, reprint 1978);); Brickell (ed.), RHS A–Z encyclopedia of garden plants, 780 (2003).

Erect, branching plant to more than 1 m, woody at base. Leaves *c.* 4 × 6 cm, with margins of segments somewhat curved under, rough, green, strongly rose- or mint-scented; leaf-stalks *c.* 3 cm. Inflorescence with 5–10 flowers; stems to 4 cm; flowers to 1.5 cm across, pink to pale pink; flower-stalks to 3 mm; upper 2 petals 1–1.5 cm × 5–6 mm, narrowly obovate, with rounded or notched apex, veined dark purplish pink; lower petals smaller, unmarked; spur 5–10 mm; anthers usually stunted and rarely bearing fertile pollen. *Garden origin*. G1. Spring–summer.

This plant has been cultivated as a source of geranium oil and many cultivars with similar characteristics are grown such as 'Lady Plymouth' (Figure 126(14), p. 460) with cream leaf-margins and 'Little Gem' with a more compact habit and rose-pink flowers. The plant illustrated in Van der Walt & Vorster, Pelargoniums of southern Africa **3**: 63 (1988) as *P. graveolens* misapplied, not L'Héritier, is a species only seen in specialist collections and has white flowers and deeply divided, softly hairy, somewhat mint-scented leaves. It is not the same as the plant illustrated by L'Héritier.

37. P. radens H.E. Moore (*P. radula* (Cavanilles) L'Héritier). Figure 126(2), p. 460. Illustration: Van der Walt, Pelargoniums of southern Africa **1**: p. 38 (1977); Webb, The Pelargonium family, pl. 38 (1984).

Erect, branched plant to almost 1 m, woody at base. Leaves to 5×6 cm, very rough, margins rolled under, strongly aromatic; leaf-stalks to 5 cm. Inflorescence with *c.* 5 flowers; stems to 3 cm; flowers *c.* 1.5 cm across, pale or pinkish purple; flower-stalks to 1 cm; upper 2 petals *c.* 1.7 cm × 6 mm, narrowly obovate, with darker margins; lower petals slightly narrower, unmarked; spur *c.* 6 mm. *South Africa (S & E Cape)*. G1. Spring–summer.

Plants grown as 'Radula' are cultivars closely related to *P.* 'Graveolens'.

38. P. glutinosum (Jacquin) L'Héritier. Figure 126(16), p. 460. Illustration: *Botanical Magazine*, 143 (1791); Van der Walt & Vorster, Pelargoniums of southern Africa **3**: 57 (1988).

Erect, branching, sticky shrub to well over 1 m, strongly balsam-scented. Leaves variable in shape, usually *c.* 5×5.5 cm, sometimes dark green; leaf-stalks *c.* 3 cm. Inflorescence with 1–8 flowers, stems to 8 cm; flowers *c* 1.5 cm across, pale to dark pink; flower-stalks very short; upper 2 petals usually *c.* 1.3 cm × 6 mm but variable in size, spathulate with darker markings, lower petals unmarked; spur *c.* 1 cm. *S South Africa*. G1. Spring.

39. P. panduriforme Ecklon & Zeyher. Figure 127(1), p. 462. Illustration: Van der Walt & Vorster, Pelargoniums of southern Africa **3**: 109 (1988).

Erect, branched shrub, to more than 1.5 m, strongly balsam-scented, sometimes slightly sticky and with long hairs. Leaves usually *c.* 3.5×2.5 cm, soft with long hairs; leaf-stalks to 2 cm or more. Inflorescence with 2–20 flowers; stems to 8 cm; flowers *c.* 3 cm across, pink; flower-stalks very short; upper 2 petals to 3.5×1.3 cm, spathulate, with darker markings; lower petals smaller with faint markings; spur to 1 cm. *S South Africa*. G1. Spring–summer.

40. P. quercifolium (Linnaeus) L'Héritier. Illustration: Van der Walt & Vorster, Pelargoniums of southern Africa **3**: 117 (1988).

Erect, branched, sticky shrub to more than 1.5 m, strongly balsam-scented. Leaves usually to 5×5 cm, rough with long glandular hairs; leaf-stalks 4–5 cm. Inflorescence with 2–6 flowers; stems *c.* 5 cm; flowers *c.* 1.5 cm across, pink; flower-stalks 1–2 mm; upper 2 petals to 2 cm × 7 mm, spathulate, apex notched, with darker purplish pink markings; lower petals smaller, unmarked; spur *c.* 1 cm. *S South Africa*. G1. Spring–summer.

This species is variable in the wild and cultivars referred to this species may be geographical variants of the species or hybrids of *P. quercifolium* × *P. panduriforme*.

41. P. hispidum (Linnaeus) Willdenow. Figure 126(17), p. 460. Illustration: Van der Walt & Vorster, Pelargoniums of southern Africa **2**: 73 (1981).

Erect, branched, aromatic plant to more than 2 m, woody at base, covered with rough hairs and bristles. Leaves *c.* 9×10 cm, with conspicuous veins; leaf-stalks *c.* 8 cm. Inflorescence branched, each cluster with 6–12 flowers, stems *c.* 4 cm; flowers to 1.5 cm across, pale to dark pink; flower-stalks to 7 mm; upper 2 petals *c.* 1.2 cm × 7 mm, obovate, with purple marks; lower petals much smaller, to 8 mm, narrow; spur *c.* 4 mm. *South Africa (SW & S Cape)*. G1. Summer.

42. P. papilionaceum (Linnaeus) L'Héritier. Figure 127(2), p. 462, Illustration: Sweet, Geraniaceae **1**: 27 (1820–22); Van der Walt, Pelargoniums of southern Africa **1**: 32 (1977).

Erect, hairy, branched plant to more than 2 m with a strong unpleasant scent, woody at base. Leaves to 7×10 cm, with long hairs; leaf-stalks to 8 cm. Inflorescence branched, each cluster with 5–10 flowers; stems to 6 cm; flowers *c.* 2 cm across, light to dark pink; flower-stalks *c.* 1 cm; upper 2 petals *c.* 20 × 8 mm, obovate, pink with a dark purple central mark and a white basal blotch; lower petals *c.* 7 × 2 mm, very narrow, paler; spur to 5 mm. *S South Africa*. G1. Spring–summer.

P. cordifolium (Cavanilles) Curtis (*P. cordatum* L'Héritier). Illustration: *Botanical Magazine*, 165 (1791); Van der Walt, Pelargoniums of southern Africa **1**: 9 (1977). Similar to *P. papilionaceum* but apparently no longer widely cultivated. The leaves of *P. cordifolium* are smaller, paler and hairy below without an unpleasant scent, and the flowers larger but with very narrow lower petals, much paler in colour than the spathulate pink to purple upper petals. *South Africa (S & E Cape)*. G1. Summer.

43. P. ribifolium Jacquin (*P. populifolium* Ecklon & Zeyher; *P. trilobatum* Ecklon & Zeyher, not Schrader). Figure 126(11), p. 460. Illustration: Van der Walt & Vorster, Pelargoniums of southern Africa **2**: 121 (1981).

Erect, branching, aromatic plant to nearly 2 m, woody at base. Leaves 4–5 × 5–6 cm, rough with glandular hairs, light green, leaf-stalks *c.* 4 cm. Inflorescence branched, each cluster with 6–12 flowers; stems to 4 cm; flowers *c.* 1.2 cm across, white, flower-stalks to 1 cm; upper 2 petals 1.7 × 1.1 cm, obovate, with dark red lines; lower petals *c.* 1.1 cm × 3 mm, narrower, sometimes marked with red lines; spur to 8 mm with a very conspicuous nectar-gland at base. *South Africa (E Cape)*. G1. Spring.

44. P. scabrum (Burman filius) L'Héritier. Figure 126(12), p. 460. Illustration: Das Pflanzenreich **4**: f. 59 (1912); Clifford, Pelargoniums, f. 40 (1970); Van der Walt, Pelargoniums of southern Africa **1**: 42 (1977).

Erect, branched herb to more than 1 m, often woody at base, strongly lemon-scented, covered with rough stiff hairs. Leaves variable, *c.* 4×4 cm, diamond-shaped; leaf-stalks *c.* 2 cm. Inflorescence sometimes branched with clusters of up to 6 flowers; stems to 2 cm; flowers *c.* 1.5 cm across, almost white to dark or purplish pink; flower-stalks to 1 cm; upper petals *c.* 1.5 cm × 5 mm, narrowly spathulate, with darker markings; lower petals to 10 × 3 mm; spur *c.* 5 mm. *W & SW South Africa*. G1. Early spring–summer.

The leaves of this species are very variable and very narrow-leaved plants have been named var. **balsameum** (Jacquin) Harvey (*P. balsameum* Jacquin), although in the wild there is a continuous gradation between the variants.

45. P. tomentosum Jacquin. Figure 127 (4), p. 462. Illustration: *Botanical Magazine*, 518 (1801); Sweet, Geraniaceae **2**: 168 (1822–24); Van der Walt & Vorster, Pelargoniums of southern Africa **2**: 145 (1981); Brickell (ed.), RHS A–Z encyclopedia of garden plants, 783 (2003).

Low-growing, but wide-spreading, branched herb to 50 cm, woody at base, strongly peppermint-scented, with long, soft, velvety hairs. Leaves 4–6 × 5–7 cm, very soft; leaf-stalks often twice the length of blades. Inflorescence branched, each cluster with up to 15 flowers; stems to 15 cm; flowers *c.* 1.5 cm across, white; flower-stalks to 2 cm; upper 2 petals *c.* 9 × 5 mm, obovate, with purple lines; lower petals to 1.1 cm but very narrow; spur to 2 mm. *South Africa (SW Cape)*. G1. Spring–summer.

This species grows well in semi-shaded positions.

46. P. vitifolium (Linnaeus) L'Héritier. Figure 127(26), p. 462. Illustration: Van der Walt, Pelargoniums of southern Africa **1**: 49 (1977).

Erect, hairy, strongly aromatic plant to nearly 1 m, woody at base. Leaves *c.* 6 × 8 cm, hairy and somewhat rough to touch; leaf-stalks *c.* 5 cm. Inflorescence very compact, with *c.* 10 flowers; stems to 10 cm; flowers *c.* 1.5 cm across, pink to pale purplish pink; flower-stalks very short; upper petals *c.* 1.5 cm × 4 mm, obovate, with darker markings; lower petals slightly narrower, unmarked; spur to 5 mm. *South Africa (SW & S Cape)*. G1. Spring–summer.

157. TROPAEOLACEAE

Herbs. Leaves alternate or opposite, simple or divided, stipulate or exstipulate; ptyxis flat or conduplicate. Flowers bisexual, zygomorphic, solitary, axillary, Calyx and corolla partly perigynous, stamens hypogynous. Calyx of 5 free or united sepals, 1 spurred. Corolla of 2 or 5 free petals. Stamens 8. Ovary of 3 united carpels; ovules 1 per cell, axile; style 1, lobed at the apex. Fruit a schizocarp.

There are 3 genera from Central and South America, of which only *Tropaeolum* is in cultivation.

Literature: Sparre, B. & Andersson, A., A taxonomic revision of the Tropaeolaceae, *Opera Botanica* **108**: 1–139 (1991); Clifton, R., Order Geraniales species checklist series, Tropaeolaceae (1999).

1. TROPAEOLUM Linnaeus
V.A. Matthews

Annual or perennial, herbaceous, somewhat succulent climbers, sometimes procumbent; perennials grow from a tuber. Leaves with twining stalks usually much longer than blades. Stipules usually very small and falling early, often present only in seedlings. Flowers usually borne on pendant stalks. Bracteoles absent (except in *T. ciliatum*). Sepals 5, 1 spurred, the 2 lower often larger than the 3 upper. Petals usually 5, sometimes 2, equal or the 2 upper different from the 3 lower. Fruit a schizocarp with 3 fleshy indehiscent carpels.

A genus of 86 species endemic to the New World, occurring from Mexico to temperate South America, mainly in mountainous areas. The climbing species attach themselves by twisting their stalks around any convenient support. Propagation is by seed (which, if not fresh, can take more than a year to germinate), tubers, or stem cuttings taken in spring. Most species are frost-tender. The plants dislike waterlogged soil. All species thrive in a sunny position. Annuals (and perennials treated as annuals) will produce lush leaves at the expense of flowers if grown in rich soil. If grown outdoors, the tubers of perennial species should be lifted and stored over the winter in dry, frost-free conditions (*T. speciosum* and *T. polyphyllum* are less tender but can be left in situ as their tubers pull themselves deep below the soil surface).

Leaves. Undivided: **14, 15, 16;** palmately lobed: **1, 13, 17;** palmately divided into leaflets: **2, 3, 4, 5, 6, 7, 8, 9, 10, 11, 12.**
Leaflets. Stalked: **12.**
Stipules. Large, kidney-shaped, glandular-hairy: **2.**
Leaf-stalks. More or less absent or up to 3.5 mm: **7.**

Petals. Two: **12;** three lower ones linear: **17;** three lower each with a spot: **15, 16;** three lower ones 3–9 mm: **1, 2, 8, 9, 10, 11, 12, 17;** three lower ones 1–2 cm: **3, 4, 5, 6, 7, 8, 11, 13, 14, 16, 17;** three lower ones 3–4 cm: **15.**

1a. Petals 2; leaves palmately divided into stalked leaflets
      12. pentaphyllum
  b. Petals 5; leaves palmately lobed or divided into stalkless leaflets   2
2a. Stipules large, persistent   3
  b. Stipules very small and falling early, or absent   4
3a. Stipules deeply divided into lobes, not glandular-hairy; bracteoles absent; petals bright red, 1.2–1.6 cm
      1. speciosum
  b. Stipules kidney-shaped, glandular-hairy; bracteoles present; petals yellow or orange, 4–7 mm
      2. ciliatum
4a. Leaves almost round or kidney-shaped   5
  b. Leaves palmately lobed or divided into leaflets   7
5a. Leaves almost entire; flowers 3–6 cm across   **15. majus**
  b. Leaves with each vein-ending projecting in a point; flowers to 3.5 cm across   6
6a. Petals yellow to orange, the lower 3 each with a dark or light red central spot   **16. minus**
  b. Petals orange-red or purplish red, not spotted   **14. peltophorum**
7a. Leaves stalkless or with stalks to 3.5 mm   **7. sessilifolium**
  b. Leaves with stalks 6–110 mm   8
8a. Lower 3 petals linear, with long marginal hairs; spur 8–12 mm, hooked at tip   **17. peregrinum**
  b. Lower 3 petals not as above; spur not hooked at tip, or if so, then 1.5–2.2 cm   9
9a. Petals pale to dark violet-blue   **11. azureum**
  b. Petals whitish, yellow, orange, red or reddish pink   10
10a. Sepals orange-red, tipped with black, or sometimes bluish, red or yellow with greenish margins
      10. tricolorum
  b. Sepals shades of green, dusky red, pinkish or brownish purple, rarely yellowish   11
11a. Spur 3–5 mm   12
  b. Spur 1–2.5 cm   13
12a. Flowers *c.* 4 cm across, borne in apparent clusters; petals 8–10 mm
      8. hookerianum
  b. Flowers to 1.3 cm across, solitary; petals 6–8 mm   **9. brachyceras**
13a. Leaves palmately lobed
      13. tuberosum
  b. Leaves palmately divided into leaflets   14
14a. Leaflets entire, or sometimes the central one lobed   15

b. All leaflets lobed 16
15a. Spur 1–1.5 cm **3. polyphyllum**
b. Spur 1.8–2 cm **6. leptophyllum**
16a. Leaflets linear to lanceolate, acute
4. myriophyllum
b. Leaflets obovate, obtuse **5. incisum**

1. T. speciosum Poeppig & Endlicher.
Illustration: Hay & Synge, Dictionary of
garden plants in colour, f. 1994 (1969);
Grey-Wilson & Matthews, Gardening on
walls, pl. 22 (1983); Brickell, RHS
gardeners' encyclopedia of plants & flowers,
168 (1989); Phillips & Rix, Bulbs, 229
(1989).

Perennial with hairless, climbing stems
to 3 m or more, growing from a more or
less fleshy, creeping rhizome. Stipules
deeply divided into lobes. Leaf-stalks
shorter than the flower-stalks. Leaves
palmately lobed into 5–7 obovate to
spathulate leaflets, which are downy
beneath. Flowers to 3.5 cm, each borne on
a 3–13 cm stalk. Sepals reddish, the 3
upper 6–7 mm, the 2 lower 1–1.2 cm. Spur
2–3 cm, narrow, curved or almost straight.
Petals bright red, 1.2–1.6 cm, notched, the
lower 3 with long claws. Fruit blue, held in
the persistent calyx, which turns deep red.
Chile. H4. Summer.

2. T. ciliatum Ruiz & Pavón. Illustration:
Phillips & Rix, Bulbs, 228 (1989).

Hairless perennial with slender stems
climbing to 3 m, growing from a fleshy
rhizome. Stipules kidney-shaped, glandular-
hairy. Leaves palmately divided into 5 or 6
leaflets. Bracteoles similar to stipules but
smaller. Flowers to 4 cm across. Sepals
green or purplish, 5–10 mm, triangular,
acute; spur 5–12 mm, stout, straight.
Petals yellow or orange, with purplish
veins, 4–7 mm, the lower 3 long-clawed.
Chile. H4. Summer.

3. T. polyphyllum Cavanilles. Illustration:
Hay & Synge, Dictionary of garden plants
in colour, f. 216 (1969); Wright, Complete
handbook of garden plants, 250 (1984);
Brickell, RHS gardeners' encyclopedia of
plants & flowers, 248 (1989); Phillips &
Rix, Bulbs, 229 (1989).

Tuberous, hairless perennial. Stems
stout, glaucous, prostrate, 25–50 cm,
simple or branched at the base. Stipules
absent. Leaf-stalks 2–4 cm. Leaves light to
bluish green, palmately divided almost to
the base into 5–9 obovate leaflets, which
are entire or sometimes the central one is
lobed. Flowers borne on stalks to 6 cm.
Sepals yellow-green to purplish,

1.1–1.5 cm, triangular; spur 1–1.5 cm,
narrowly conical. Petals lemon-yellow to
brownish orange (most commonly deep to
mid-yellow), 1–1.5 cm, clawed, the 2 upper
broad, blunt or notched, often with
purplish veins, the 3 lower narrower,
notched. Fruit brownish, smooth. *Chile,
Argentina*. H4. Summer.

4. T. myriophyllum (Poeppig &
Endlicher) Sparre (*T. polyphyllum* var.
myriophyllum Poeppig & Endlicher).

Differs from *T. polyphyllum* in having
more slender, somewhat climbing stems,
and lanceolate to linear, lobed, acute
leaflets. *Chile*. H4. Summer.

5. T. incisum (Spegazzini) Sparre
(*T. polyphyllum* var. *incisum* Spegazzini).

Tuberous, hairless perennial. Stems
prostrate, 45–75 cm (to 1.2 m in
cultivation). Stipules absent. Leaf-stalks
1–4 cm. Leaves 2–3.5 cm across, palmately
divided into 5–7 obovate, obtuse, lobed
leaflets. Flower-stalks 4.5–6 cm. Sepals
greenish pink or yellowish, 8–12 mm,
triangular; spur 1.2–1.7 cm, narrow. Petals
yellowish, orange or reddish pink, often
with purple veins, clawed, the 2 upper
obtuse, 1.2–1.4 cm, the 3 lower notched,
or all petals with more or less frilled
margins, 1–1.3 cm. Fruit blackish. *Chile,
Argentina*. H5. Summer.

6. T. leptophyllum G. Don.
Hairless perennial with large tubers.
Stems slender, climbing to 1 m. Stipules
absent. Leaf-stalks 3–4 cm. Leaves
palmately divided into 6–9 linear to
lanceolate leaflets. Flowers *c*. 3.5 cm, each
borne on a 1–1.2 cm stalk. Sepals
1–1.2 cm, triangular, acute; spur 1.8–2 cm,
narrowly conical, straight. Petals orange,
yellow, or white tinged with pink, entire or
notched, the 2 upper 1.2–1.4 cm, the 3
lower 1.5–1.8 cm, clawed. Fruit dark green
to blackish. *Chile*. H5. Summer.

7. T. sessilifolium Poeppig & Endlicher.
Illustration: *The Garden* **104**: 149 (1979).
Tuberous, hairless perennial. Stems
prostrate or sometimes climbing, simple or
sometimes branched. Stipules absent. Leaf-
stalks almost absent or up to 3.5 mm.
Leaves green or glaucous, palmately
divided into 3–5 leaflets. Flower-stalks
2–2.5 cm. Sepals green or yellow-green,
c. 8 mm, triangular, acute; spur often
purplish, *c*. 1.5 cm, straight. Petals dirty
yellow to whitish, often flushed with violet,
obovate, clawed, often notched, 1–1.2 mm
the 2 upper broader than the 3 lower and

often with red veins. Fruit blackish. *Chile*.
H5. Summer.

8. T. hookerianum Barnéoud (*T.
brachyceras* var. *hookerianum* (Barnéoud)
Buchenau). Illustration: *The Garden* **104**:
149 (1979).

Tuberous perennial. Stems hairless or
downy, climbing to 3 m, branched at the
base. Stipules absent. Leaf-stalks 6–10 mm.
Leaves palmately divided into 5–7 linear-
lanceolate to obovate, blunt or mucronate,
hairless leaflets. Flowers *c*. 4 cm, apparently
in clusters of 2–8, each borne on a 4–5 cm
stalk. Sepals greenish yellow, 6–8 mm,
ovate, blunt; spur 3–5 mm, conical. Petals
yellow, 8–10 mm, almost equal, clawed,
obtuse or slightly notched, the upper 2
sometimes with brown veins. Fruit
brownish. *Chile*. H5. Summer.

9. T. brachyceras Hooker & Arnott.
Illustration: *Botanical Magazine*, 3851
(1841).

Hairless perennial with small tubers.
Stems slender, climbing to 1.5 m. Stipules
absent. Leaf-stalks 1–1.5 cm. Leaves to
3 cm wide, palmately divided into 5–7,
obovate to linear-lanceolate, obtuse leaflets.
Flowers to 1.3 cm. Sepals green or rarely
pale bluish green, 5–7 mm, obtuse; spur
3–5 mm, conical, blunt. Petals yellow,
6–8 mm, more or less notched, clawed, the
2 upper with purplish lines. Fruit dark
brown. *Chile*. H5. Summer.

10. T. tricolorum Sweet (*T. tricolor*
Lindley). Illustration: Hay & Synge,
Dictionary of garden plants in colour,
f. 890 (1969); Pereire, The flower garden,
293 (1988); Brickell, RHS gardeners'
encyclopedia of plants & flowers, 163
(1989); Jellito & Schacht, Hardy
herbaceous perennials, 670 (1990).

Tuberous, hairless perennial. Stems
slender, climbing to 4.5 m. Stipules absent.
Leaf-stalks 1.5–3 cm. Leaves to 3 cm wide,
palmately divided into 5–7 linear to
obovate leaflets. Flower-stalks 4–5 cm.
Sepals usually orange-red, tipped with
black, but sometimes bluish, red or yellow
with a greenish margin, 5–8 mm, obtuse;
spur purplish, red or yellow with a dark
blue or blackish green tip, 1.5–2.5 cm,
narrow, straight or slightly curved. Petals
orange or yellow, 4–5 mm, entire, obtuse,
with short claws. Fruit blackish. *Chile*. G1.
Late spring–summer.

11. T. azureum Colla (*T. violaeflorum*
Dietrich). Illustration: *Botanical Magazine*,
3985 (1843); Pizarro, Flores silvestres de

Chile, pl. 21 (1966); Phillips & Rix, Bulbs, 229 (1989).

Tuberous, hairless perennial. Stems slender, climbing to 2.5 m, much-branched. Stipules absent. Leaf-stalks to 2 cm. Leaves to 5 cm wide, palmately divided into 4–6 leaflets; leaflets linear-lanceolate (lower leaves) to obovate (upper leaves). Flower-stalks 1–4 cm, longer than leaves. Flowers *c.* 2 cm across. Sepals green, 8–10 mm, ovate-lanceolate; spur 2–4 mm, conical, straight. Petals pale to dark violet-blue, whitish or yellowish towards the base, 8–12 mm, almost equal, obovate, clawed, blunt or notched. Fruit dark brown. *Chile.* G1. Summer–autumn.

Plants in cultivation have violet-blue flowers. Variants with white or yellow flowers are reported from the wild.

12. T. pentaphyllum Lamarck (*T. quinatum* Hellenius). Illustration: *Botanical Magazine*, 3190 (1832).

Tuberous, hairless perennial. Stems climbing to 3 m or more. Stipules absent. Leaf-stalks 4–6 cm. Leaves palmately divided into 3–7 stalked, elliptic, usually acute leaflets. Flowers 2–3.5 cm, each borne on stalks 6–9 cm. Sepals green, spotted with red, 6–10 mm, triangular-ovate, acute; spur red or pink and green, 2–2.5 cm, stout, conical with inflated tip. Petals 2, deep red to purplish, 3–5 mm, obovate, blunt. Fruit blue-black, fleshy. *Bolivia, Brazil, Argentina, Uruguay.* G1. Summer.

13. T. tuberosum Ruiz & Pavon. Illustration: Hay & Synge, Dictionary of garden plants in colour, f. 891 (1969); Brickell, RHS gardeners' encyclopedia of plants & flowers, 176 (1989); Phillips & Rix, Bulbs, 229 (1989); Davis, The gardener's illustrated encyclopedia of climbers and wall shrubs, 155 (1990).

Hairless perennial with large yellowish tubers streaked with dark red. Stems climbing to 3 m. Stipules tiny, falling early. Leaf-stalks to 10 cm. Leaves grey-green, paler beneath, 5–7 cm wide, with 5 or 6 mucronate lobes, which are directed upwards or sideways. Flowers *c.* 3.5 cm, each borne on a 12–20 cm stalk. Sepals dusky red, the 3 upper 8–10 mm, the 2 lower 1.2–1.4 cm; spur dusky red, 1.5–2.2 cm, straight, often hooked at the tip. Petals orange-yellow or orange-red, streaked with brown towards the base, entire, the 2 upper 6–9 mm, shortly clawed, the 3 lower 1–1.5 cm, long-clawed. Fruit 4–5 mm, dark brown to

blackish, wrinkled. *Venezuela, Colombia, Ecuador, Peru, Bolivia.* H5. Summer–autumn.

Subsp. **silvestre** Sparre. Illustration: Phillips & Rix, Bulbs, 228 (1989). Has fleshy rhizomes rather than tubers and smaller flowers 2.5–3 cm. *Colombia, Ecuador, Peru, Bolivia, Argentina.*

Subsp. **tuberosum** is cultivated as a crop in the Andes from S Venezuela to C Bolivia: the tubers are cooked and eaten. There are two sorts of subsp. *tuberosum* in cultivation: a short-day plant that only begins to flower in September and a day-neutral variant that flowers from May until the first frosts. The latter is sometimes sold under the name 'Sidney'. 'Ken Aslet' (illustration: Brickell, RHS gardeners' encyclopedia of plants & flowers, 174, 1989) has orange flowers and rather large tubers.

14. T. peltophorum Bentham (*T. lobbianum* Hooker). Illustration: *Botanical Magazine*, 4097 (1844).

Downy annual with climbing stems to 2 m, from a long, spindle-shaped root. Stipules tiny, thread-like, falling early. Leaf-stalks 5–7 cm. Leaves almost round, 2–9 cm wide, each vein-ending projecting in a point, hairy beneath, margins wavy. Flowers *c.* 2.5 cm, stalks to 15 cm. Sepals greenish purple, *c.* 1 cm, lanceolate to ovate, blunt; spur 2.5–3 cm, curved, with the basal half inflated. Petals orange or purplish red, the 2 upper 1.5–1.6 cm, rounded, the 3 lower 1.1–2.2 cm, long-clawed and with fringed margins. Fruit 8–10 mm, brown, wrinkled. *Colombia, Ecuador, Peru.* H5. Summer.

'Spitfire' has reddish stems and orange flowers, which are more tubular and have red markings in the throat.

15. T. majus Linnaeus. Illustration: Hay & Synge, Dictionary of garden plants in colour, f. 387 (1969); Pereire, The flower garden, 292 (1988).

Hairless annual, usually with stems climbing to 3 m. Stipules tiny, falling early. Leaf-stalks to 20 cm. Leaves 5–17 cm wide, almost round or sometimes kidney-shaped, usually entire. Flowers 3–6 cm across, each borne on a stalk equal to or shorter than leaf-stalks. Sepals green or yellow-green, 1.5–1.8 cm; spur 2–3.5 cm, inflated in the upper half. Petals cream, yellow (sometimes with a basal red blotch or stripes), orange, red or brownish maroon, the 2 upper 3–4 cm, rounded, clawed, the 3 lower 2.7–3.5 cm, rounded or toothed,

long-clawed. Fruit to 1 cm, ribbed, wrinkled, brownish. *Garden origin.* H3. Summer–autumn.

Not known in the wild. Sparre considers it to be of hybrid origin, probably from a cross between *T. minus* and *T. ferreyrae* Sparre. *T. peltophorum* was probably involved in the later hybrids which appeared in the 17th and 18th centuries. Cultivars with single, semi-double and double flowers are available, as are dwarf selections and cultivars with semi-trailing and non-climbing stems. 'Alaska' is dwarf and has leaves irregularly splashed with creamy white.

16. T. minus Linnaeus. Illustration: *Botanical Magazine*, 98 (1790); Pereire, The flower garden, 293 (1988).

Smaller than *T. majus*, with stems which do not climb but sometimes scramble. Leaves downy beneath, each vein-ending projecting in a point. Flowers to 3.5 cm across. Sepals greenish yellow, 1–1.3 cm, broadly lanceolate; spur 2.5–3 cm, cylindric, curved. Petals yellow to orange, the 2 upper 1.7–1.8 cm, shortly clawed, the 3 lower 1.5–1.6 cm, each with a dark or light red central spot, long-clawed. Fruit 5–6 mm, brown, wrinkled. *Peru, Ecuador.* H3. Summer.

17. T. peregrinum Linnaeus (*T. canariense* invalid). Illustration: Grey-Wilson & Matthews, Gardening on walls, pl. 22 (1983); Wright, Complete handbook of garden plants, 527 (1984); Davis, The gardener's illustrated encyclopedia of climbers and wall shrubs, 154 (1990).

Annual with hairless stems, climbing to 2 m or more. Stipules absent. Leaf-stalks 4–11 cm. Leaves pale green, 3.5–8 cm wide, palmately 5-lobed. Flowers to 2.5 cm across, stalks 6–12 cm. Sepals green, the 3 upper smaller than the 2 lower; spur green, 8–12 mm, stout, hooked at tip. Petals bright yellow, the 2 upper 1.5–2 cm, obovate, fringed, often spotted with red or purple at the base, long-clawed, the claw recurved with a tooth at the bend, the 3 lower 8–10 mm, linear, with long marginal hairs. Fruit 1–1.4 cm, dark brown or blackish, stalked. *Peru.* H5. Summer.

---

158. ZYGOPHYLLACEAE

Herbs or shrubs. Leaves usually opposite, usually compound, stipulate, often fleshy; ptyxis variable. Inflorescence cymose or flowers solitary. Flowers usually bisexual and actinomorphic. Calyx, corolla and stamens hypogynous, disc usually present. Calyx of 4–5 free sepals. Corolla of 4–5 free petals. Stamens 5–15. Ovary of 2–5 united carpels; ovules many, axile; style 1 or stigmas sessile. Fruit a capsule, drupe-like or a schizocarp.

A family of some 250 species in 27 genera, largely from the tropics and warm temperate areas, particularly in arid and saline habitats. *Guaiacum* Linnaeus is a small genus of 5 or 6 species that supply lignum-vitae, the hardest commercial timber. Reference to these species in horticultural literature probably relates to their original introduction, when nearly all plants of potential commercial interest were cultivated. *Larrea* Cavanilles is another small genus containing about 4 species of desert shrubs with aromatic foliage (often referred to as creosote bushes). Although cultivated in North America they do not seem to be grown in Europe outside institutional collections. A few species of the family with limited ornamental value are sometimes cultivated as curiosities.

1a. All leaves alternate　　**1. Peganum**
　b. Leaves mostly opposite　　2
2a. Leaves simple or with 1 pair of leaflets　　**2. Zygophyllum**
　b. Leaves with 5 or more pairs of leaflets　　3
3a. Leaves with 5–7 pairs of leaflets; sepals and petals 5　　**3. Tribulus**
　b. Leaves with 7–9 pairs of leaflets; sepals and petals 4 or 5　　**4. Porlieria**

1. PEGANUM Linnaeus
S.G. Knees

Branched, hairless, perennial herbs. Leaves alternate, irregularly divided, stipules narrowly pointed. Flowers solitary, leaf-opposed or lateral. Sepals 4 or 5, persistent; petals 4 or 5; disc a ring; stamens 12–15; ovary spherical. Fruit a capsule; seeds with endosperm.

A genus of 5 or 6 species from the Mediterranean area to Mongolia and S North America.

1. P. harmala Linnaeus. Illustration: Zohary, Flora Palestina **2**: pl. 352 (1972); Polunin & Smythies, Flowers of south-west Europe, 254 (1973); Täckholm, Student's flora of Egypt, edn 2, 310 (1974); Valdés et al., Flora vascular de Andalucía occidental **2**: 266 (1987).

Much-branched, erect perennial, 20–100 cm; rootstock tough, woody. Leaves 2–8 cm, fleshy; leaflets linear-lanceolate; stipules small. Flowers 1–2 cm, stalked. Sepals linear, sometimes toothed towards the base, persistent. Petals greenish white. Fruit 7–10 mm, spherical, stalked. *SE Europe, Mediterranean area.* H5–G1. Spring.

2. ZYGOPHYLLUM Linnaeus
S.G. Knees

Small shrubs or branched perennial herbs with stout rootstocks. Leaves opposite, simple or with 2 leaflets. Sepals and petals 4 or 5; stamens 8–10. Fruit an inflated 4- or 5-angled capsule; seeds with endosperm.

A genus of 80 species, usually found in arid areas or deserts from the Mediterranean area to Central Asia, South Africa and Australia. Although several species have been briefly cultivated in Europe, only a few are now thought to be in regular cultivation.

1a. Leaves simple; flowers in crowded panicles　　**3. prismatocarpum**
　b. Leaves divided; flowers solitary　　2
2a. Hairless herb with jointed stems; petals 5–7 mm　　**1. fabago**
　b. Shrub with white hairs and smooth stems; petals 3–4 mm　　**2. album**

1. Z. fabago Linnaeus. Illustration: Fiori & Paoletti, Iconographia florae Italicae, f. 2521 (1901); Zohary, Flora Palestina **2**: pl. 364 (1972); Polunin & Smythies, Flowers of south-west Europe, 254 (1973); Townsend & Guest, Flora of Iraq **4**(1): 300 (1980).

Hairless perennial herb with jointed stems 20–75 cm, sometimes becoming woody towards the base. Leaves dark green, divided into 2 leaflets; leaf-stalks winged, 5–20 mm; leaflets 4–6 cm, but often only *c.* 1 cm towards the top of the plant, obovate to ovate, obtuse at tips. Stipules 1–2 mm, lanceolate. Flowers bilaterally symmetric, solitary in the axils of upper leaves. Sepals 5–8 mm, oblong to obovate, green with transparent margins; petals 5–7 mm, white or pale yellow, sometimes orange towards the base; stamens 8–10 mm, filaments orange or pinkish, anthers yellowish. Fruit 1.5–3.5 cm, cylindric, 5-angled. Seeds 3–4 mm, flattened. *Middle East, SE Europe, naturalised in W Mediterranean area.* H5. Spring–summer.

2. Z. album Linnaeus filius. Illustration: Sibthorp & Smith, Flora Graeca **4**: t. 371 (1823); Zohary, Flora Palestina **2**: 366 (1972).

Small shrub with ascending or decumbent stems to 1.5 m, covered with white hairs. Leaves fleshy, bluish green; leaflets 4–8 mm, oblong or elliptic. Flowers solitary in leaf-axils: Sepals 2–3 mm; petals 3–4 mm, spathulate, white; stamens 3.5–4 mm. Fruit star-shaped, 8–10 mm, 5-lobed; seeds brown, ovoid, *c.* 2.5 mm. *N Africa, Middle East & Mediterranean area.* G1. Summer–autumn.

3. Z. prismatocarpum E. Meyer.

Small shrub with grey bark and green, spreading branches. Leaves simple, almost clasping the stem, rounded to ovate, widest above the middle; stipules minute or absent. Flowers in crowded, pyramidal panicles 5–8 cm. Sepals 2–3 mm, ovate; petals 5–6 mm, cream; stamens like petals. Fruit 8–10 mm, oblong with 5 furrows and slightly winged, brownish purple. *South Africa.* H5–G1. Summer.

3. TRIBULUS Linnaeus
S.G. Knees

Much-branched, prostrate or sprawling herbs. Leaves opposite in unequal pairs, sometimes alternate, pinnate, stalked with stipules at the base. Flowers solitary, stalked, borne in leaf-axils. Sepals 5; petals 5, falling early; disc 5-lobed; stamens 10 (rarely 5), inserted below the disc margin, filaments unequal; ovary stalkless, 5-celled, ovules 3–5 per cell; style simple, 5-angled; stigma 5-angled. Fruit splitting into 5 at maturity.

A genus of 25 species, widely distributed in tropical and warm temperate areas, particularly in arid climates. Propagation is always from seed, which may be sown in a sandy loam with bottom heat.

1. T. terrestris Linnaeus. Illustration: Reichenbach, Icones florae Germanicae et Helveticae **5**: t. 161 (1841); Polunin, Flowers of Europe, pl. 64 (1969); Blundell, Wild flowers of east Africa (Collins Guide), 245 (1987); Valdés et al., Flora vascular de Andalucía occidental **2**: 267 (1987).

Hairy annual or biennial; stems to 50 cm, often woody at base. Leaves oblong, 1.5–5 cm, with 5–7 pairs of leaflets, silvery green, 3–15 mm. Stipules lanceolate to acuminate, 1.5–3 mm. Flower-stalks 5–15 mm; sepals lanceolate, 4–5 mm; petals narrowly obovate, 4.5–6 mm, yellow; stamens 2.5–3 mm; stigma 5-rayed;

ovules 3 per cell. Fruit rounded, *c.* 1 cm across, each mericarp with 2 or 4 stout spines and several slender spines. Seeds 3 per mericarp. *Tropics & subtropics.* G1–2. Spring–autumn.

Although considered weedy in much of its range, this interesting annual is sometimes cultivated in Europe as an ornamental curiosity.

4. PORLIERIA Ruiz & Pavon
S.G. Knees

Shrubs with spreading stiff branches. Leaves opposite, pinnately divided, the leaflets almost opposite, entire. Flowers solitary or paired, borne in leaf-axils; sepals 4 or 5, rounded and often unequal, falling early; petals 4 or 5, narrowed into a claw towards the base.

A genus of 6 species from Central and South America, only one of which appears to be in cultivation. Propagation is by cuttings of ripe wood, rooted in sand with extra light and bottom heat. Plants thrive in a well-drained mixture of peat and loam.

1. P. hygrometrica Ruiz & Pavón. Illustration: Ruiz & Pavón, Florae Peruvianae, pl. 9 (1794); Ruiz & Pavón, Florae Peruviana **4**: pl. 343 (1802); *Nuovo Giornale Botanico Italiano* **24**: pl. 2 (1892).

Shrub to 80 cm with light grey bark; the young shoots brown and covered with soft down. Leaves with 7–9 pairs of leaflets; leaflets 2–10 × 1.5–2 mm, oblong-elliptic; leaf-stalks absent or to 1 mm. Calyx deeply 4-lobed, sepals 2–3 mm, broadly ovate; petals 4, 3–4 mm, whitish; stamens exceeding petals, yellowish. Fruit a capsule of 4 united segments, 7–8 mm across. *South America (Chile, Bolivia, Peru).* G1. Spring.

One of the most interesting features of this plant is the sensitivity of its leaflets, which open wide during the day or in good weather and close together at night or when weather conditions deteriorate.

159. LINACEAE

Herbs or shrubs. Leaves alternate or opposite, entire, stipulate or exstipulate; ptyxis flat or conduplicate. Inflorescence a cyme. Flowers bisexual, actinomorphic. Calyx, corolla and stamens hypogynous. Calyx of 4–5 free or united sepals. Corolla of 3–5 free petals. Stamens 4–5, 10 or 15 sometimes united at base. Ovary of 3–5

united carpels usually becoming 6–10-celled by the growth of secondary septa; ovules 1–2 per cell, axile; styles 3–5, free. Fruit a capsule or drupe.

A cosmopolitan family of 15 genera with some 300 species. The two cultivated genera are herbs or shrubs.

1a. Leaves elliptic-obovate to oblong-obovate, stalked, pinnately veined **1. Reinwardtia**
  b. Leaves narrow, sometimes spathulate, stalkless, single or parallel-veined **2. Linum**

1. REINWARDTIA Dumortier
S.L. Jury

Hairless shrub or woody-based perennial to 1 m. Leaves alternate, stalked, elliptic-obovate to oblong-obovate, often toothed, pinnately veined. Stipules minute, falling early. Flowers solitary in leaf-axils or in dense corymbs at the ends of branches. Sepals 5. Petals 5, falling early, yellow, twisted, basally united into a tube. Stamens 5, joined at the base, alternating with 5 staminodes. Styles 3–5. Fruit a capsule with 3–5 cells.

A genus of a single species from central and east Asia, which is cultivated for its beautiful flax-like flowers.

1. R. indica Dumortier (*R. trigyna* (Roxburgh) Planchon; *R. tetragyna* Planchon). Illustration: Brickell, RHS gardeners' encyclopedia of plants & flowers, 138 (1989); Brickell (ed.), RHS A–Z encyclopedia of garden plants, 882 (2003).

E India to China & Malaya. G1. Summer & occasionally throughout the year.

2. LINUM Linnaeus
T. Upson

Annual to perennial herbs or small shrubs. Leaves stalkless, often narrow with single or parallel veins, entire. Stipules sometimes present, modified to paired glands at base of leaves. Flowers short-lived, but new ones open daily, parts in 5s. Petals free, rarely united at base, blue, yellow, red, pink or white. Petals alternate with 5 tooth-like staminodes. Filaments united at base, heterostylous in many species. Capsule usually spherical to almost so with a beak dehiscing into 10 valves. Seeds brown or black.

A genus of about 200 species, principally distributed throughout the temperate areas of the world, rare in the tropics. Plants are suitable for the rock garden, dry garden or perennial border; flowering mid-spring to

late summer; best in a sunny position on well-drained soil. Propagation is by seed for annuals, perennials and shrubs; shoot-cuttings can be taken from spring to summer. Plants are usually short-lived so rejuvenation is frequently required.

1a. Flowers yellow 2
  b. Flowers blue, red, pink or white 7
2a. Shrub to 1 m, with woody branches; leaves thick **1. arboreum**
  b. Plants with a variably developed woody stem to 60 cm; leaves usually not especially thick 3
3a. Inflorescence almost head-like **2. capitatum**
  b. Inflorescence cyme-like or flowers solitary 4
4a. Petals 2.5–3.5 cm, rarely only 2.2 cm, gradually narrowed into a claw; beak of capsule *c.* 2 mm **3. campanulatum**
  b. Petals 1–2.5 cm, rarely to 3 cm, usually abruptly narrowed into a claw; beak of capsule usually *c.* 1 mm 5
5a. Inflorescence with 1–9 flowers **4. elegans**
  b. Inflorescence with more than 10 flowers 6
6a. Stems not more than 20 cm; inflorescence with 10–15 flowers **5. 'Gemmell's Hybrid'**
  b. Stems to 60 cm; inflorescence with 25–40 flowers **6. flavum**
7a. Flowers red or reddish pink **7. grandiflorum**
  b. Flowers blue, pink or white 8
8a. Leaves with glandular, sticky hairs **8. viscosum**
  b. Leaves not glandular or sticky 9
9a. Annual, erect and unbranched without a woody base **9. usitatissimum**
  b. Perennial, usually branched, with at least a woody rootstock 10
10a. Stigma globular; sepals unequal, the outer narrow, inner with broad papery margins **10. perenne**
  b. Stigma club-shaped; sepals equal, without broad papery margins 11
11a. Sepal-margin glandular-hairy 12
  b. Sepal-margin not glandular-hair 13
12a. Stamens and stigma at same height in flower; sepals 5–8 mm; petals 1–2 cm **11. tenuifolium**
  b. Stamens and stigma at different heights in flower; sepals 4–6 mm; petals 1.5–3.5 cm **12. suffruticosum**

13a. Petals 2. 5–4 cm, bright blue;
     capsule 7–9 mm   **13. narbonense**
  b. Petals 1–2.5 cm, white, sometimes
     tinged blue; capsule 4–5 mm
                          14. monogynum

1. L. arboreum Linnaeus. Illustration:
Huxley & Taylor, Flowers of Greece and
the Aegean, pl. 132 (1977); Brickell, RHS
gardeners' encyclopedia of plants & flowers,
298 (1989); Brickell (ed.), RHS A–Z
encyclopedia of garden plants, 638
(2003).

Hairless shrub 30–50 cm, rarely to 1 m.
Leaves thick and persistent, spathulate,
5–20 × 3–10 mm, borne in rosettes at the
ends of woody branches. Inflorescence a
compact cyme. Sepals 5–8 mm, lanceolate,
acuminate. Petals 1.2–1.8 cm, golden
yellow. *S Aegean area.* H4. Summer.

Needs a warm sunny spot, even in
favourable areas, or an alpine house.
Grows best on limestone.

2. L. capitatum Schultes.

Herbaceous perennial growing from a
woody rootstock with well-developed
rhizomes, often terminating in leaf rosettes.
Flowering stems 10–40 cm. Rosette leaves
oblong to spathulate, obtuse; stem-leaves
linear to lanceolate, acute. Inflorescence a
cyme with 5–15 flowers. Sepals 5–6 mm,
oblong-lanceolate, acuminate. Petals
1.5–2 cm, yellow. *Balkan Peninsula, C & S
Italy.* H4. Summer.

3. L. campanulatum Linnaeus.
Illustration: Aeschimann et al., Flora alpina
1: 1025 (2004).

Hairless perennial 20–30 cm, woody at
base. Lower leaves spathulate, upper stem-
leaves oblanceolate with narrow
transparent margins and minute glands on
each side at the base. Flowers in loose
cymes; petals pale yellow, often veined
orange, 2–3 cm across with long claws.
S Europe. H4. Summer.

4. L. elegans Boissier (*L. iberidifolium*
Aucher).

Dwarf plant with woody branched stem
and compact basal leaf rosettes. Flowering
stems to 15 cm. Lower leaves obovate to
spathulate, 3-veined, thick with a
conspicuous transparent margin.
Inflorescence with 3–7 flowers; flowers
2–3 cm wide; petals yellow with distinct
veining. *Balkan Peninsula.* H5. Late spring.

5. L. 'Gemmell's Hybrid'. Illustration:
Griffith, Collins guide to alpine & rock
garden plants, pl. 19 (1985).

Woody perennial 12–15 cm, spreading
30–40 cm. Stems at ground level bearing
numerous stalkless glaucous-green leaves;
basal leaves spathulate and in whorls; stem
leaves oblong and more scattered.
Inflorescence a corymb of up to 15 flowers,
cup-shaped and opening flat, 2–3 cm
across. Petals obovate, deep yellow, veined
pale orange, *c.* 15 × 5 mm. Sepals edged
with glandular hairs. *Garden origin.* H2.
Summer.

A vigorous and free-flowering hybrid
between *L. campanulatum* and *L. elegans*
raised from a chance seedling at the
nursery of the Gemmell Brothers near
Glasgow, *c.* 1940.

6. L. flavum Linnaeus. Illustration:
Polunin, Flowers of Greece and the
Balkans, pl. 26 (1980); Brickell, RHS
gardeners' encyclopedia of plants & flowers,
325 (1989) – as var. *compactum*;
Aeschimann et al., Flora alpina **1**: 1025
(2004).

Robust, erect and hairless woody-based
perennial to 60 cm with relatively few or
no flowering rosettes. Leaves
2–3.5 cm × 3–12 mm, 3–5-veined, the
lower spathulate, upper lanceolate.
Inflorescence branched with 25–40
flowers. Petals *c.* 2 cm, golden yellow.
Sepals 5–8, lanceolate with marginal
glandular hairs. *C & SE Europe to C Russia.*
H2. Summer.

7. L. grandiflorum Desfontaines.
Illustration: Hay & Beckett (eds), Reader's
Digest encyclopedia of garden plants &
flowers, edn 2, 412 (1978).

Erect, branching, leafy hairless annual,
30–40 cm. Leaves linear to ovate-
lanceolate, acuminate and greyish green.
Flowers 3–4 cm, red or reddish pink, borne
in loose panicles. Sepals lanceolate-
acuminate with glandular margins,
c. 1 cm. *N Africa.* H5. Summer.

8. L. viscosum Linnaeus. Illustration:
Barneby, European alpine flowers in
colour, pl. 48 (1967); Aeschimann et al.,
Flora alpina **1**: 1029 (2004).

Perennial, stems erect to 60 cm, covered
with soft hairs. Leaves glandular and
sticky, 3–11 mm wide, lanceolate or
ovate-lanceolate, 3–5-veined. Flowers
borne in erect terminal corymbs.
Sepals lanceolate, 6–9 mm with glandular
hairs. Petals 3–5 times as long as sepals,
pink or rarely blue. *S & SC Europe.* H2.
Summer.

9. L. usitatissimum Linnaeus (*L. crepitans*
Dumortier). Illustration: Aeschimann et al.,
Flora alpina **1**: 1029 (2004).

Annual with an unbranched erect stem
to 50 cm. Leaves linear or lanceolate,
1.5–3 mm wide, 3-veined. Flowers in
corymb-like panicles. Sepals 6–9 mm.
Petals blue. *Probably of Asian origin;
recorded throughout Europe.* H3. Summer.

Anciently and extensively cultivated for
the fibre (flax) and oil from the seed
(linseed). Many commercial varieties have
been selected for value as fibre or the oil
content of their seeds.

10. L. perenne Linnaeus (*L. sibiricum* de
Candolle; *L. album* Boissier; *L. alpinum*
Jacquin; *L. lewisii* Pursh). Illustration: Keble
Martin, The concise British flora in colour,
pl. 19 (1965); Brickell, RHS gardeners'
encyclopedia of plants & flowers, 296
(1989).

Hairless perennial, sometimes woody at
base; stems to 60 cm, erect, ascending or
decumbent. Leaves 1–4 mm wide, linear-
lanceolate, entire with 1–3 veins. Petals
1–1.5 cm, 3–4 times as long as sepals, pale
blue. Sepals ovate, unequal, outer narrow,
inner with broad papery margins. *C & S
Europe to N America.* H2. Summer.

'Album', flowers white; Lewisii', flowers
pink; 'Caerulea', flowers sky blue.

Subsp. **alpinum** (Jacquin) Ockendon
differs in having dark blue petals.

11. L. tenuifolium Linnaeus. Illustration:
Barneby, European alpine flowers in
colour, pl. 48 (1967); Polunin, Flowers of
Greece & the Balkans, pl. 663 (1980);
Aeschimann et al., Flora alpina **1**: 1031
(2004).

Perennial, usually hairless, 20–45 cm,
with few sterile stems, erect, ascending or
decumbent. Leaves 0.5–2 mm wide, linear,
1-veined; margins with minute saw-like
teeth. Inflorescence a loose cyme. Sepals
5–8 mm with glandular margins. Petals
c. 2 cm, pale pink almost white. *C & S
Europe.* H3. Summer.

12. L. suffruticosum Linnaeus
(*L. salsoloides* Lamarck). Illustration:
Griffith, Collins guide to alpine & rock
garden plants, pl. 19 (1985).

Hairless perennial to 50 cm, sometimes
woody at base with many short, non-
flowering shoots. Stems procumbent with
ascending lateral shoots. Leaves 0.2–1 mm
wide, linear or bristle-like with rough,
minutely toothed, inrolled margins. Sepals
4–6 mm, ovate-acuminate. Petals white

with a pink to violet claw, 3–4 times as long as sepals. *SW Europe, from C Spain to NW Italy extending to NC France.* H3. Summer.

13. L. narbonense Linnaeus (*L. paniculatum* Moench). Illustration: Hay & Beckett (eds), Reader's Digest encyclopedia of garden plants & flowers, edn 2, 412 (1978); Brickell, RHS gardeners' encyclopedia of plants & flowers, 243 (1989); Brickell (ed.), RHS A–Z encyclopedia of garden plants, 639 (2003).

Hairless perennial with erect or ascending stems to 50 cm. Leaves linear or lanceolate, long-acuminate. Petals *c.* 2.5 cm, 2.5–3 times as long as sepals, bright blue. Capsule 7–9 mm, almost spherical with a narrow beak. *W & C Mediterranean area, N Spain & NE Portugal* H4. Late spring–summer.

14. L. monogynum Forster (*L diffusum* Schultes).

Stout perennial herb to 60 cm. Stems woody at base, erect, simple or branching. Leaves alternate, linear to elliptic, 5–30 mm. Inflorescence with few to many flowers in simple or compound corymbs. Sepals elliptic, acute to acuminate, 5–15 mm. Petals 1–2.5 cm, white sometimes tinged blue. *New Zealand.* H5. Summer.

Needs a sunny, sheltered position with light, well-drained soil.

---

160. EUPHORBIACEAE

Plants woody or herbaceous, sometimes succulents, milky sap often present, frequently acrid and poisonous. Leaves usually alternate and stipulate, simple or compound, rarely absent, sometimes replaced by cladodes; ptyxis very variable. Inflorescences various, sometimes a cup of bracts with glandular margins (cyathium). Flowers unisexual, actinomorphic. Perianth and stamens or calyx, corolla and stamens hypogynous. Perianth of 2–6 free or united segments, more rarely calyx of 5 or 10 sepals, corolla of 5 free or united petals. Stamens 1–many, free or united. Ovary of 2–4 (usually 3) or rarely more united carpels; ovules 1–2 per cell, axile; styles 2–4 or rarely more, usually 3, free or slightly joined at the base, often divided above. Fruit usually schizocarpic; seeds often carunculate.

A large and variable family of 326 genera and 7750 species. Seven genera are native to Europe, 47 to North America. Species of about 20 are cultivated for ornament; the most important is *Euphorbia*, which contains both normal herbaceous and shrubby species and succulent, more or less leafless shrubs. Includes Phyllanthaceae.

The inflorescences of plants of the Tribe Euphorbieae (genera included here are *Euphorbia*, *Monadenium*, *Pedilanthus* and *Synadenium*) are remarkable structures, which can superficially be mistaken for individual flowers. Each ultimate inflorescence (cyathium) consists of a cup-like structure that has an irregular or 5-toothed margin, and bears around the top a number of coloured, nectar-secreting glands (these may be united and not distinguishable into separate glands). The simplest case is found in *Euphorbia*, in which there are 1–5 yellow or orange-red glands that spread in a remarkable petal-like formation. In the centre of the cup is a single, stalked female flower, consisting of a naked ovary, which, when mature, projects from the cup, leaning to one side. Attached to the inside of the cup are several male flowers, each consisting of a solitary, stalked stamen, often subtended by a small, feather-like bract (bracteole); the stamen is jointed to the stalk, and this joint is clearly visible at a magnification of × 15 (in some tropical, non-cultivated genera related to *Euphorbia*, a small perianth is borne at the joint). In fruit, the cyathium remains, with the ripe schizocarp projecting from it. In *Monadenium* and *Synadenium* the glands are united: in *Synadenium* they are united into a continuous ring, whereas in *Monadenium* the whole cup is split down one side. In *Pedilanthus* the cyathium is much modified into a remarkable, bilaterally symmetric structure, which is borne at almost a right-angle to the cyathium-stalk. The cup has a swollen base that narrows to a 5-lobed, 2-lipped apex; the whole cyathium is often brightly coloured, and the glands are small.

In all these genera the cyathia (the primary inflorescences) may be grouped into more complex inflorescences, which are often umbel-like, though there is usually a solitary cyathium at the centre of each pseudo-umbel.

1a. Flowers borne in inflorescences of various kinds, but not in cyathia   2
 b. Flowers borne in cyathia   26

2a. Leaves reduced to tiny scales, their function taken over by flattened, leaf-like cladodes, which bear flowers around their margins
  4. Phyllanthus
 b. Leaves normal, not reduced to scales; flowers in inflorescences of various kinds, but not as above   3
3a. Flowers (at least the male) with a perianth consisting of sepals and petals   4
 b. Flowers with a perianth of a single whorl, petals absent   8
4a. Leaves palmately veined, usually palmately lobed   5
 b. Leaves pinnately veined, entire, toothed or irregularly lobed   6
5a. Petals yellow or red; sepals 5, overlapping in bud   **16. Jatropha**
 b. Petals white; sepals 2, edge-to-edge in bud   **8. Aleurites**
6a. Annual herbs with stellate hairs
  7. Chrozophora
 b. Perennial herbs, shrubs or trees without stellate hairs   7
7a. Leaves evergreen, usually coloured, often irregularly lobed; male flowers in long axillary racemes, stamens 20–30   **15. Codiaeum**
 b. Leaves deciduous, green, usually not lobed; male flowers solitary or in axillary clusters, stamens 5–6
  5. Andrachne
8a. Leaves palmately veined and lobed; stamens much branched
  14. Ricinus
 b. Leaves not palmately veined and lobed; stamens not as above   9
9a. Leaves mostly opposite   10
 b. Leaves alternate   11
10a. Herbs without milky sap; carpels 2; stamens 8–15   **10. Mercurialis**
 b. Shrubs with milky sap; carpels 3; stamens 3   **20. Excoecaria**
11a. Inflorescence surrounded by 2 brightly coloured bracts; sterile flowers numerous, forming a cushion   **24. Dalechampia**
 b. Inflorescence not as above   12
12a. Plant densely covered with stinging hairs   **17. Cnidoscolus**
 b. Plant lacking stinging hairs   13
13a. Ovary with 2 ovules in each cell   14
 b. Ovary with 1 ovule in each cell   18
14a. Plant to 15 m; flowers in long, catkin-like or branched inflorescences; fruit a small drupe
  1. Antidesma

b. Plant to 4 m; flowers in axillary clusters or solitary; fruit not a small drupe 15
15a. Pistillode clearly present in the male flowers **3. Securinega**
b. Pistillode absent from male flowers 16
16a. Disc conspicuous in the female flowers **4. Phyllanthus**
b. Disc absent from female flowers 17
17a. Styles united, stigmas 3, spreading, 2-lobed; leaves white or white to pink-blotched **2. Breynia**
b. Styles 3, free, distant, arising on the margin of the ovary; leaves green **6. Sauropus**
18a. Plant with clear sap 19
b. Plant with milky sap 21
19a. Stamens 30 or more, sometimes as many as 100 **9. Mallotus**
b. Stamens usually 8, rarely as many as 16 20
20a. Bark with conspicuous triangular leaf-scars; ovary 2-celled **11. Macaranga**
b. Bark without conspicuous leaf-scars; ovary 3-celled **13. Acalypha**
21a. Stamens with filaments united into a column; ovary with 3 or more cells 22
b. Stamens with free filaments; ovary with 2 or 3 cells 24
22a. Stamens 2 **23. Hippomane**
b. Stamens many 23
23a. Stigmas much divided **19. Hura**
b. Stigmas undivided **12. Hevea**
24a. Stamens 5 or more **18. Homalanthus**
b. Stamens 2 or 3 25
25a. Styles united, only the stigmas free; seed covered by a white, oily appendage (aril) **22. Sapium**
b. Styles free; seed not covered as above **21. Stillingia**
26a. Cyathium very strongly bilaterally symmetric, 2-lipped, broad at base, tapered to the apex, borne at an angle to the cyathium-stalk **28. Pedilanthus**
b. Cyathium more or less radially symmetric, though sometimes split down one side, not 2-lipped, not shaped as above, continuing the line of the cyathium-stalk 27
27a. Glands of the cyathium 1–5, distinct **25. Euphorbia**
b. Glands of the cyathium united 28
28a. Glands of the cyathium united into a continuous ring **26. Synadenium**

b. Glands of the cyathium united into a ring that is split at one side **27. Monadenium**

1. ANTIDESMA Linnaeus
J Cullen

Evergreen trees or shrubs; bark greyish: Leaves alternate, pinnately veined, entire, shortly stalked; stipules present though sometimes falling early. Flowers small, unisexual, both sexes on the same plant, in axillary or terminal spikes, racemes or rarely panicles; both male and female flowers 1 to each bract. Male flowers with a cup-like or 3–5-lobed perianth, with an irregularly lobed nectar-secreting disc outside the 3 or 4 stamens (ours). Female flowers with a shortly 3- or 4-lobed perianth, a cup-like nectar-secreting disc and a 3-celled ovary with 2 ovules in each cell; styles 3, very short, sometimes 2-lobed at the apex. Fruit a small drupe, usually containing a single fertile seed, which has no appendages.

There are about 160 species mainly from the Himalaya to south-east Asia, a few in Africa and Madagascar. The single cultivated species requires warm glasshouse conditions, in which it will flower as a shrub, before eventually becoming a small tree. It is propagated by seed.

1. A. bunius (Linnaeus) Sprengel. Illustration: *American Horticulturist* **63**: 26–7 (1984).

Hairy shrub or small tree to 15 m. Leaves 8–25 × 3. 5–10 cm, lanceolate-oblong to lanceolate, rarely somewhat ovate or obovate, acuminate but ultimately obtuse at the apex, rounded to the base, sometimes hairy in the vein-axils beneath; stipules soon falling. Male flowers in hairy spikes 6–17 cm, unpleasantly scented; stamens long-projecting, anthers purple. Female flowers shortly stalked in racemes, perianth densely hairy within towards the base. Drupes spherical, 5–10 mm across, at first reddish, later dark purple. *Himalaya & Sri Lanka to N Australia*. G2.

2. BREYNIA Forster & Forster
J. Cullen

Evergreen shrubs. Leaves alternate, pinnately veined, entire, shortly stalked, with small, narrowly triangular stipules. Flowers minute, unisexual, both sexes on the same plant, axillary, solitary or in clusters. Male flowers with a top-shaped perianth, tapered below, with 6 small, incurved lobes at the top; stamens 3,

filaments united into a column. Female flowers with a leathery, shallowly cup-shaped, 6-lobed perianth enlarging in fruit; ovary 3-celled, each cell with 2 ovules; styles 1, united below: stigmas 2, lobed. Fruit fleshy and indehiscent (ours); seeds 4–6.

A genus of about 15 species from tropical Africa, Asia and Polynesia. Only one is grown as a foliage plant in glasshouses and conservatories; it is easily grown, producing its brightly coloured young leaves best in a sunny position and following heavy pruning. It is propagated by root- or stem-cuttings.

1. B. disticha Forster & Forster forma **nivosa** (Bull) Radcliffe-Smith (*B. nivosa* (Bull) Small). Illustration: *Floral Magazine*, n.s., **30**: t. 120 (1874); Clay & Hubbard, The Hawaii garden, tropical exotics, 69 (1977); Brickell (ed.), RHS A–Z encyclopedia of garden plants, 193 (2003).

Shrub to 4 m but often much less. Leaves broadly elliptic to almost circular, rounded at base and apex (sometimes notched), strikingly white or pinkish when young. Flowers minute, long-stalked. Fruit fleshy, indehiscent, to 8 mm across, depressed spherical, tapered to the base. *Polynesia, introduced elsewhere*. G1.

'Roseotincta'. Illustration: Encke, Kalt- und Warmhauspflanzen, 238 (1987). Older leaves variegated dark green and cream, young leaves suffused with reddish purple; tips of branches bright pink.

3. SECURINEGA Comerson
R.F.L. Hamilton

Deciduous shrubs or small trees, often spiny. Normally bisexual or unisexual. Leaves often small, alternate, simple, entire, short-stalked, with stipules. Flowers small, greenish, without petals, axillary. Male flowers in clusters; sepals 5 or 6, greenish white; stamens 5 or 6. Female flowers sometimes solitary, with 5 or 6 sepals; ovary 3-celled with 3 free or shortly joined and bifid styles. Fruit a capsule containing 3–6 seeds.

A genus of 20–25 species in temperate and subtropical parts of S Europe, Africa, Asia, and South America, largely of scientific interest. *S. suffruticosa*, a graceful shrub with bright foliage, is occasionally cultivated. It grows well in any rich loamy soil and is propagated from seed or by cuttings of greenwood, under glass. The mature wood is extremely hard, hence the generic name, from Latin *securis*, a hatchet, and *nego*, I refuse.

1. S. suffruticosa (Pallas) Rehder (*S. fluggeoides* Muller; *S. ramiflora* Müller; *S. japonica* Miquel, in part). Illustration: Tarouca et al., Freiland-Laubgehölze, edn 2, 384, 446 (1922); Everett, New York Botanical Gardens illustrated encyclopedia of horticulture **9**: 3099 (1981).

Deciduous shrub to 2 m with hairless, brown or yellowish green, slender, spreading branches. Leaves elliptic to lanceolate-ovate, acute or obtuse, wedge-shaped at base, to 6 cm, mostly bright green above, paler and slightly glaucous beneath. Flowers bright yellowish green. Male flowers *c.* 3 mm across, up to 10 per cluster. Female flowers solitary, upright. Capsule more or less spherical, *c.* 5 mm across, on a long, slender stalk. *Mongolia, China*. H4. Summer.

4. PHYLLANTHUS Linnaeus
M. C. Tebbitt
Monoecious or rarely dioecious herbs, shrubs and trees, often with leaf-like flattened branches (cladodes). Leaves very variable in size and sometimes reduced to scales, alternate (very rarely opposite), entire, in 2 vertical rows; stalks shorter than blades. Flowers often borne on the sides of the cladodes; male flowers with 4, 5 or 6 sepals, stamens 2–5, free or fused; no rudimentary ovary; female flowers with sepals like those of the male. Petals absent in both sexes.

A genus of about 650 species native to tropical and subtropical areas. In Europe the cultivated species are best grown in pots in warm glasshouses where they are of interest for their graceful appearance and unusual flowering habit. They require a fertile, porous, peaty soil and do best when their roots are slightly crowded. The plants require a moderate degree of atmospheric humidity and shading from bright sunlight. The winter night temperature should not be allowed to fall below 16 °C.

la. Annual herb **1. amarus**
  b. Shrub or tree 2
2a. Stamens 3, fused 3
  b. Stamens not as above 6
3a. Cladodes inconspicuous, deciduous branches to 25 cm **2. emblica**
  b. Cladodes conspicuous, deciduous branches commonly longer than 25 cm 4

4a. Main stem with terminal crown of leafy branches, apex reddish brown with densely felted hairs; sepals 5 **3. mimosoides**
  b. Main stem with branches along length, apex not as above; sepals 6 5
5a. Flowers borne along sides of cladodes; cladodes *c.* 20 cm wide **4. angustifolius**
  b. Flowers clustered towards apex of cladodes; cladodes *c.* 10 cm wide **5. arbuscula**
6a. Flower-stalks 5–10 mm; stamens 2, united **6. pulcher**
  b. Flower-stalks less than 5 mm; stamens 4, free **7. acidus**

1. P. amarus (Linnaeus) Skeels (*P. niruri* Linnaeus). Illustration: Graf, Exotica, Series 4, edn 12, 1054 (1985).
Annual herb to 50 cm, glaucous; cladodes present, deciduous branches to 10 cm, appearing pinnate; leaves 1–1. 8 cm in the middle of the branch, shorter above and below, oblong-elliptic or ovate, rounded-blunt, membranous, hairless. Sepals 5 or 6, unequal; in female flowers *c.* 1 mm, elliptic-ovate; in male flowers 0.5–0.8 mm, ovate, stamens 3, united for three-quarters of their length. *Tropical W Africa*. G2.

2. P. emblica Linnaeus.
Tree *c.* 8 m, much branched. Cladodes present but inconspicuous, deciduous branches *c.* 25 cm, appearing pinnate. Leaves 1.2–2 cm × 2–5 mm, linear-oblong, obtuse to acute; stalks 0.3–0.7 mm. Flower-stalks, 1–2.5 mm; 5 or 6 sepals, oblong-obovate or spathulate; male flowers with 3 united stamens. *Tropical Asia*. G2.

3. P. mimosoides Swartz.
Shrub to 5 m, stems with terminal crown of leafy branches, apex reddish brown with densely felted hairs; cladodes present, deciduous branchlets to 70 cm, appearing pinnate; leaves 5–11 × 2.4–2.5 mm, asymmetrically oblong, mucronate, almost stalkless. Flower-stalks 1–3.5 mm, obtuse to almost truncate; calyx-lobes 5; stamens 3, fused in a column. *Lesser Antilles*. G2.

4. P. angustifolius (Swartz) Swartz. Illustration: Graf, Exotica, Series 4, edn 12, 1053 & 1056A (1985).
Shrub, 60–300 cm; cladodes 50–100 × 20–25 cm, linear-lanceolate to lanceolate, appearing pinnate along deciduous branches; scale-leaves *c.* 3 mm, linear, acuminate. Flowers borne along

sides of cladodes, flower-stalks 2–4 mm; sepals in both sexes 6, *c.* 1 mm, outer elliptic, inner rhomboid or broadly elliptic, reddish cream; stamens 3, fused. *Jamaica*. G2.

5. P. arbuscula (Swartz) J. F. Gmelin (*P. speciosa* Jacquin).
Variable shrub or small tree, 2–7 m; cladodes 40–75 × 10–12 cm, lanceolate, appearing pinnate along smooth deciduous branches; leaves 4–8 mm, lanceolate to linear, acuminate, membranous. Flowers clustered towards apices of cladodes; sepals 6, unequal; in male flowers, outer *c.* 1.4 mm, inner to 2.2 mm, elliptic to obovate-elliptic; in female flowers outer *c.* 1.5 mm, elliptic, inner *c.* 2 mm, broadly obovate-elliptic or roundish, cream or greenish to scarlet; stamens 3, fused. *Jamaica*. G2.

6. P. pulcher (Baillon) Mueller.
Shrub to 1.5 m, cladodes present, deciduous branches to 20 cm, appearing pinnate, covered with reddish brown hairs. Basal leaves scale-like, 1.8–2.8 cm × 8–14 mm, oblong to elliptic, abruptly acute and scarious-mucronate; stalks 0.8–1 mm. Flower-stalks 5–10 mm; calyx-lobes 4, ovate; stamens 2, united; petals yellow. *Java, Sumatra, Borneo*. G2.

7. P. acidus (Linnaeus) Skeels (*P. distichus* (Linnaeus) Mueller).
Tree to 10 m, sparsely branched, hairless, cladodes to 50 cm, appearing pinnate, deciduous branchlets 25–52 cm; leaves 5–9 × 2.5–4.5 cm, ovate to ovate-lanceolate, acute; stalks 2.5–4 mm. Flower-stalks to 5 mm, both sexes with 4 calyx-lobes; stamens 4, free; flowers pink to red. *S Asia*. G2.

5. ANDRACHNE Linnaeus
R. F L. Hamilton
Low, deciduous shrubs or perennial herbs. Leaves alternate, 2-ranked, usually simple and entire, with or without stipules. Flowers bisexual or incompletely unisexual, on long stalks. Male flowers in axillary clusters; sepals 5, rarely 6, larger than the petals, free or shortly joined; petals 5, rarely 6, free or joined, with disc glands; stamens shorter than and equal in number to the sepals, free or joined around a rudimentary ovary. Female flowers solitary; petals minute or absent; sepals 5, rarely 6, free or joined; styles 3, free or shortly joined, stigmas more or less deeply bifid. Ovary 3-celled. Fruit more or less spherical,

6-seeded, separating into 3 carpels, each opening into 2 parts.

A genus of about 20 species distributed through southern Europe, Africa, Asia, and North and South America; mainly of scientific interest, rarely found outside botanic gardens. A sunny situation is required, but otherwise the species are easily grown. They can be propagated from seed or from greenwood cuttings under glass; some species sucker freely.

1a. Branches ascending, hairless; leaves
 ovate **1. colchica**
 b. Branches angled, downy;
 leaves elliptic to obovate
 2. phyllanthoides

1. A. colchica Fischer & Meyer. Illustration: Everett, New York Botanical Gardens illustrated encyclopedia of horticulture **1**: 157 (1980).

Dense shrub to 1 m, with slender, simple, ascending branches. Leaves 6–18 mm, ovate, blunt, rounded at the base, entire with thickened margins, thin and hairless; stalks 8–10 mm. Sepals green, edged with white. Petals of male flowers yellowish, spathulate, with 2 lobes near the base. Female flowers with hair-like petals, much shorter than the sepals. Fruit *c.* 5 mm. *Former USSR (Caucasus)*. H3. Summer.

2. A. phyllanthoides (Nuttall) Muller (*A. roemeriana* Müller). Illustration: Britton & Brown, Illustrated flora of the northern United States and Canada **3**: 518 (1898).

Shrub to 1 m. densely branched, with downy, angled shoots. Leaves elliptic to obovate, 8–18 mm, blunt, hairless or downy beneath; stalks 4–5 mm. Flowers *c.* 5 mm, yellowish green. Male flowers in clusters, female flowers solitary. *Central USA*. H3. Summer.

6. SAUROPUS Blume
J. Cullen
Evergreen shrubs. Leaves alternate, shortly stalked, pinnately veined, with stipules. Flowers unisexual, both sexes on the same plant, axillary, solitary or in small clusters. Male flowers with a disc-shaped, flattened perianth with 6 short, inflexed lobes; stamens 3, their filaments united into a short column. Female flowers with 6 broad perianth-segments united at the base, sometimes enlarging in fruit; ovary 3-celled, each cell with 2 ovules; styles 3, free, borne on the outer margins of the ovary. Fruit a fleshy or leathery capsule

borne on the persistent perianth, containing 6 seeds.

A genus of about 30 species from subtropical and tropical Asia. Only one is grown as a foliage plant. Cultivation as for *Breynia*.

1. S. androgynus (Linnaeus) Merrill (*S. albicans* Blume). Illustration: Das Pflanzenreich **147**: 217 (1922); Die natürlichen Pflanzenfamilien, edn 2, **19C**: 57, 71 (1931); Ochse & Rarkhuizen van den Brink, Vegetables of the Dutch East Indies, 291 (1931).

Shrub to 2 m, hairless. Leaves ovate to elliptic or ovate-oblong, broadly rounded to the base, tapered to the ultimately blunt apex. Flowers minute, 1–5 in each axil, the male flowers towards the base of the shoot. Fruit white, almost spherical, 1–1.5 cm across. *Sri Lanka, India, Himalaya & SW China to south-east Asia*. G2.

The young shoots (including the fruits) are eaten as a vegetable in SE Asia.

7. CHROZOPHORA A. Jussieu
R.F.L. Hamilton
Annual herbs more or less densely covered in stellate hairs. Leaves alternate. Male flowers in terminal spike-like racemes or axillary clusters; sepals and petals 5, stamens 5–15, joined at the base; rudimentary ovary absent. Female flowers usually solitary, occasionally 2 or 3 at the base of the male flowers; sepals 10; petals minute or absent; styles 3, bifid; ovary 3-celled, separating into 3 carpels, each opening into 2 parts.

A genus of 6 species, from the Mediterranean area, Asia and tropical Africa.

1. C. tinctoria (Linnaeus) A. Jussieu. Illustration: Reichenbach, Icones florae Germanicae et Helveticae **5**: 152 (1841); Bonnier, Flore complète **10**: pl. 550 (1929).

Plant green to greyish green, thinly hairy. Stems to 50 cm, more or less branched. Leaves alternate, ovate to diamond-shaped, with entire or wavy-toothed margins, rounded at apex, tapered to base; stalks as long as or up to twice the length of the blades. Male flowers with whitish yellow petals; stamens 9–11. Female flowers on long stalks, curling after fertilisation; petals sepal-like. Capsule shortly pointed, covered with mushroom-shaped hairs. *Europe*. H4. Summer.

The source of a dye called turn-sole, long used for colouring food-stuffs and linen.

8. ALEURITES Forster & Forster
J. Cullen
Evergreen or deciduous trees with milky sap; hairs sometimes stellate. Leaves large, long-stalked, generally with 2 prominent glands at the stalk apex, entire or somewhat lobed; main veins 3–7, palmate. Flowers unisexual in large cyme-like panicles, males and females generally in the same inflorescence. Calyx cup-shaped, 2-lobed. Corolla disc-shaped, of 5 (rarely to 8) spreading petals. Stamens 8–20, in 3 series, usually some of them sterile (staminodes) in male flowers, all sterile in female flowers. Ovary 2–5-celled, each cell containing 1 ovule; styles bifid; pistillode absent from male flowers. Fruit a large nut, somewhat fleshy outside.

A genus of about 15 species from south-east Asia, several of them cultivated in the tropics for oil (tung oil, chinawood oil) produced by their fruits. Few are grown as ornamentals in Europe; they will tolerate most soils and are propagated by seed or by hardwood cuttings.

1a. All hairs stellate; petals 7–9 mm;
 leaves evergreen **1. moluccana**
 b. Most hairs simple; petals at least
 1.5 cm; leaves deciduous 2
2a. Plant with a trunk; male flowers
 hairy inside; fruit warty or wrinkled
 3
 b. Plant lacking a trunk; male flowers
 hairless inside; fruit smooth
 2. fordii
3a. Fruit warty **3. cordata**
 b. Fruit wrinkled **4. montana**

1. A. moluccana (Linnaeus) Willdenow. Illustration: Das Pflanzenreich **42**: 130 (1910).

Tree to 10 m with a well-developed trunk. Leaves persistent, alternate in clusters towards the tips of the branches, ovate to narrowly ovate, base rounded to cordate, apex acute to acuminate, entire or 3–5-lobed, to 20 cm, dark green, with stellate hairs beneath. Flowers unisexual in large cymes, which are mixed or male only. Calyx 3–5 mm. Petals usually 5, whitish, those of the male flowers hairy within towards the base. Fruit ovoid, 3-ridged. *Tropical E Asia*. G2.

2. A. fordii Hemsley. Illustration: *Hooker's Icones Plantarum* **29**: t. 2801, 2802 (1906); Menninger, Flowering trees of the world for tropics and warm climates, pl. 191 (1962); Everett, New York Botanical Gardens illustrated encyclopedia of horticulture, 104 (1980).

Flat-topped tree to 12 m, without a well-developed trunk. Leaves deciduous, ovate to heart-shaped, entire or slightly lobed, cordate at base, acute at apex, to 22 cm, glossy, with simple hairs beneath. Flowers in cymose panicles appearing before the leaves are fully developed, each panicle with mostly male flowers. Calyx to 1 cm. Petals 5 (rarely to 8), whitish, 1.8–3 cm, those of the male flowers tinted with red and yellow at the base, hairless inside. Fruit more or less spherical, somewhat pointed at the apex, smooth. *C & W China.* G2.

3. A. cordata (Thunberg) Steudel (*Verniciacordata* (Thunberg) Airy Shaw).

Like *A. fordii* but with a trunk; calyx 7–11 mm; petals whitish, *c.* 2 cm, those in the male flowers hairy within towards the base; fruit ovoid, warty. *SE Asia.* G1.

Rare in cultivation.

4. A. montana (Loureiro) Wilson. Illustration: Menninger, Flowering trees of the world for tropics and warm climates, pl. 191 (1962).

Like *A. fordii* but with a trunk; inflorescences appearing after the leaves have expanded; male and female flowers often in separate inflorescences; petals 5, 1.8–2.3 cm, those of the male flowers tinged yellowish and hairy inside towards the base; fruit ovoid, 3-ridged, wrinkled. *SE China, Burma.* G1.

Rare in cultivation.

9. MALLOTUS Loureiro
J. Cullen
Trees or shrubs without milky latex, usually with stellate hairs. Leaves alternate or opposite, stalked, simple or lobed, usually with dot-like brown glands on the lower surface; stipules present, small, subulate. Flowers small, unisexual, male and female on the same plant, in terminal or axillary racemes or rarely panicles; male flowers 1–3 to each bract, female flowers 1 to each bract. Male flowers with a perianth of 3 or 4 segments and numerous stamens with free filaments. Female flowers with a 3–5-lobed perianth and a 2–4-celled ovary with a single ovule in each cell. Fruit a schizocarp splitting into 2–4 (usually 3) mericarps, which eventually separate from the persistent central axis. Seeds without appendages.

A genus of about 100 species from the Old World tropics, mostly from Asia. Only one is occasionally grown, and little is known of its cultivation requirements.

1. M. philippinensis (Lamarck) Mueller (*Croton philippinensis* Lamarck) Illustration: Bentley & Trimen, Medicinal plants **4**: t. 236 (1875); Das Pflanzenreich **63**: 185 (1914).

Tree; twigs with dense brownish hair-covering, mainly of small stellate hairs forming a felt, with longer, simple hairs projecting. Leaves long-stalked, alternate (rarely almost opposite), elliptic or elliptic-oblong, 5–25 × 2–10 cm, tapered to the base, acuminate at the apex, pinnately veined, with a dense felt of stellate hairs beneath, among which are found brown, dot-like glands. Racemes terminal and axillary; male flowers with 40 or more stamens; ovary usually 3-celled. Fruit deeply 3-lobed, bright red when ripe. Seeds black. *Himalaya, SE Asia.* H5-G1.

10. MERCURIALIS Linnaeus
R.F.L Hamilton
Annual or perennial herbs with watery latex. Leaves opposite, with small stipules. Flowers usually unisexual; petals absent and sepals 3 in both sexes. Male flowers in long axillary spikes; stamens 8–15. Female flowers axillary, solitary, in stalkless clusters, or in spikes; 2 or 3 sterile stamens sometimes present; styles 2; ovary of 2 cells each with 1 seed.

A genus of 8 species in Europe, temperate Asia and the Mediterranean area.

1. M. perennis Linnaeus. Illustration: Keble Martin, The new concise British flora in colour, 75 (1982); Garrard & Streeter, Wild flowers of the British Isles, 109 (1983).

Roughly hairy perennial with extensive rhizomes. Stems unbranched, erect, 15–40 cm. Leaves 2–8 cm, smaller on the lower part of the stem, ovate-lanceolate to elliptic, with toothed margins; stalks usually 5–10 mm. Flowers 4–5 mm across. Male flowers in axillary spikes. Female flowers 1–3, axillary, their stalks lengthening in fruit. Fruit roughly hairy, 4–8 mm wide. *Europe, N Africa & SW Asia.* H2. Summer.

Occasionally grown for ground cover.

11. MACARANGA Thouars
S.G. Knees
Shrubs or trees, usually dioecious. Leaves alternate, simple, entire or lobed; palmately

or pinnately veined, sometimes peltate. Flowers in axillary panicles; calyx 2–4-lobed, corolla absent; stamens usually 8 (ours), ovary 2–6-celled; ovules 1 per cell. Fruit a capsule.

A genus of about 240 species, native to tropical Africa, south-east Asia, Australia and Melanesia. The few cultivated species are grown primarily as foliage plants. Cultivation and propagation as for *Codiaeum*.

1. M. grandiflora (Blanco) Merrill (*M. porteana* Andre). Illustration: *Revue Horticole*, 176 (1888); *Gardeners' Chronicle* **16**: 284 (1894); *Botanical Magazine*, 7407 (1895); Graf, Tropica, edn 3, 429 (1986).

Shrub or small tree to 5 m or more, bark with conspicuous triangular leaf-scars. Leaves 60–100 cm, ovate to rounded, peltate, prominently veined, dark green above, rosy beneath; stalks to 60 cm; stipules 7–15 cm. Flowers in axillary, pyramidal spikes or racemes, inconspicuous, pale red; mostly male, but a few female towards the tip of the flower-stalk; ovary 2-celled. Capsules 8–10 mm. *Philippines.* G1. Spring.

A striking foliage plant.

12. HEVEA Aublet
M.C. Warwick
Tropical monoecious trees with milky latex. Leaves alternate, on long slender stalks; leaflets 3, entire. Inflorescence a paniculate cyme, flowers small. Sepals 5; corolla absent; stamens 5–10, filaments joined to form a column; ovary 3-celled. Fruit a 3-seeded capsule.

A genus of about 12 species from the Amazon basin in South America, and an important source of rubber. Seedlings were raised at Kew from seeds collected in Brazil by Sir Henry Wickham, and then introduced to Malaya and Indonesia. These first cultivated rubber trees were the ancestors of the plantations in the Far East. It is grown as an ornamental in glasshouses and often included in collections of plants important to man. Strictly tropical, it requires a minimum night temperature of 16 °C and high humidity with a coarse porous soil.

1. H. brasiliensis (A. Jussieu) J. Mueller (*Siphonia brasiliensis* A. Jussieu). Illustration: *Hooker's Icones Plantarum* **26**: pl. 2573, 2574 (1899); Leathart, Trees of the world, 174 (1977); Graf, Exotica, Series 4, edn 12, 1052–4 (1985); Graf, Tropica, edn 3, 429 (1986).

Deciduous tree to 30 m; bark grey and smooth. Leaflets 30–60 cm, elliptic, thick, leathery, purplish bronze when young, turning to brown or red before falling; leaf-stalks to 30 cm. Flowers fragrant, white, tiny, borne in clusters, with many male flowers surrounding a few females. Fruit 3-lobed, *c.* 2.2 cm, rupturing explosively to release seeds. *Brazil (Mato Grosso), cultivated in SE Asia & Sri Lanka.* G2.

13. ACALYPHA Linnaeus

A.C. Whiteley

Herbs or shrubs, monoecious or occasionally dioecious. Leaves alternate, simple, usually toothed. Flowers small, without petals, often in dense terminal or axillary spikes or racemes. Male flowers in axils of minute bracts. Calyx 4-lobed. Stamens 8–16. Female flower-spikes subtended by leafy bracts. Sepals 3–5. Ovary 3-celled, each cell containing 1 ovule. Styles 3, usually long and much branched, often brightly coloured. Fruit a capsule.

A genus of about 430 species distributed throughout the tropics and subtropics. A few are cultivated for flowers or foliage. Propagate by cuttings.

1a. Leaves usually variegated; female flower-spikes to 6 mm across
    1. wilkesiana
  b. Leaves green; female flower-spikes 1.2–2.5 cm across   **2**
2a. Leaves to 5 cm; female flower-spikes to 8 cm   **3. hispaniolae**
  b. Leaves 10–20 cm; female flower-spikes 15–50 cm   **2. hispida**

1. A. wilkesiana J. Mueller (*A. tricolor* Seeman). Illustration: Graf, Tropica, edn 2, 408, 409, 423 (1981); Macoboy, What shrub is that?, 21, 22 (1990); Brickell, RHS gardeners' encyclopedia of plants & flowers, 111 (1989); Brickell (ed.), RHS A–Z encyclopedia of garden plants, 62 (2003).

Evergreen shrub to 5 m. Leaves 10–20 × 6–12 cm, elliptic to ovate, shortly acuminate, base rounded or wedge-shaped, sharply toothed, bronze-green mottled with orange, red or purple. Female flower-spikes axillary, to 20 cm × 6 mm, reddish. *Pacific islands.* G1. Spring–autumn.

Many cultivars are grown, with variously coloured leaves, either mottled or margined. Some cultivars also have laciniate leaf-margins. Commonly grown are: 'Obovata', bronze-green with pink margins; 'Godseffiana', green with white margins; 'Macafeeana', marbled in shades of red and bronze.

2. A. hispida N.L. Burman (*A. sanderi* N.E. Brown; *A. sanderana* Schumacher). Illustration: *Botanical Magazine*, 7632 (1899); Hay & Synge, Dictionary of garden plants in colour, 51 (1969); Graf, Tropica, edn 2, 408, 423 (1981); Macoboy, What shrub is that?, 22 (1990).

Dioecious evergreen shrub to 5 m. Leaves 10–22 × 7–15 cm, broad-ovate, acuminate, base rounded, coarsely toothed, downy on the veins beneath. Female flower-spikes axillary, very dense, 15–50 × 1.2–2.5 cm. Sepals minute. Styles long, much-branched, red. *Malaysia.* G1. Spring–autumn.

'Alba' has female flowers with white styles and leaves narrowly margined with white.

3. A. hispaniolae Urbatsch.

Prostrate shrublet. Stems slender, sparsely hairy, to 50 cm. Leaves to 5 × 4 cm, broadly ovate, obtuse, with coarse rounded teeth; stalks 1.5–4 cm. Female flower-spikes terminal, very dense, 3–8 × 1.2–2 cm. Styles long, much branched, red. *Haiti, Hispaniola.* G1. All year.

This species may be a variant of *A. chamaedrifolia* (Lamarck) J. Mueller.

14. RICINUS Linnaeus

A.C. Whiteley

Annual herbs, shrubs or small, short-lived trees, 1–7 m. Stems green or reddish, often glaucous, becoming hollow. Branching occurs at the nodes immediately below an inflorescence. Leaves alternate, spherical, peltate, 10–75 cm across, palmately 5–11-lobed, the lobes up to half the length of the leaf. Stipules united, sheathing, 1–3 cm. Leaf-stalks 8–50 cm, with 2 nectaries at the base, 2 at the junction with the leaf-blade and 1 or more towards the base on the upper side. Leaves and stalks green or reddish, often glaucous; blades lobed, sharply toothed and acuminate. Flowers in narrow terminal panicles 10–40 cm. Male and female flowers separate within the panicle, male below, female above. Male flowers in 3–16-flowered cymes on stalks 5–15 mm; sepals 3–5, ovate, 5–7 mm; petals absent; stamens numerous, 5–10 mm, filaments much branched with many anthers. Female flowers in 1–7-flowered cymes on stalks 4–5 mm; sepals 3–5, united, soon falling; petals absent;

ovary superior, 3-celled; ovules 1 per cell; stigmas 3, style short. Ovary covered with fleshy spines, enlarging in fruit, finally rigid. Fruit a 3-lobed capsule, 1.5–2.5 cm wide, green or reddish, becoming brown and woody. Capsule splits into 3, often explosively. Seeds 5–15 mm, ovoid, pale brown to black, strongly mottled, with a yellowish white caruncle.

A genus of a single very variable species, an important crop plant in the tropics, the seeds yielding castor oil. Easily grown in well-drained soil and full sun. It is often grown as an exotic and architectural 'dot' plant in annual bedding schemes planted by parks departments and local authorities. Propagation is by seeds, which are poisonous.

1. R. communis Linnaeus. Illustration: *Botanical Magazine*, 2209 (1821); Purseglove, Tropical crops **1**: 183 (1968); Hay & Synge, Dictionary of garden plants in colour, 46 (1969); Graf, Tropica, edn 2, 428 (1981).

Widely naturalised throughout the tropics & subtropics. H5–G1. Summer–autumn.

Several ornamental cultivars are widely grown for tropical effect in temperate areas. 'Gibsonii', 'Sanguineus' and 'Zanzibariensis' all have reddish leaves, especially leaf-stalks.

15. CODIAEUM Jussieu

J. Cullen

Shrubs (ours) or rarely trees, hairless, sap somewhat milky. Leaves evergreen, leathery and often shiny, stalked, very variable in shape, usually brightly coloured or spotted with colour, sometimes with the margins wavy or the blades curled up spirally. Flowers unisexual in male or female axillary racemes (rarely a female flower present at the base of the male raceme), usually both sexes on the same plant. Male flowers often paired in long racemes, with 1 or 2 flowers at each node subtended by a whorl of small bracts, long-stalked, the stalks jointed below the middle; buds spherical; sepals 5, reflexed when the flower is open, often reddish; petals shorter than sepals, 5, white, free, often irregular in shape; stamens 20–30 (ours) borne on the domed receptacle, forming a spherical tuft; anthers small, flattened; pistillode absent. Female racemes solitary, nectar-secreting glands forming a cup; ovary 3-celled with 1 ovule in each cell. Fruit a spherical capsule.

A genus of about 6 species from south-east Asia to the Pacific. One species is ubiquitously grown as a house or glasshouse foliage plant, known informally as 'Croton', though this is the latin name of a distinct genus not cultivated for amenity. Its leaves are immensely variable in shape, size and colour (see below). It is relatively easily grown in good soil, but requires a temperature of not less than 20 °C and bright sunlight to do well. Propagation is by cuttings.

1. C. variegatum (Linnaeus) Blume.
Shrub to 2 m or more. Leaves variable. Male racemes to 30 cm, female racemes shorter. Sepals *c.* 5 mm. *SE Asia to Polynesia.* G2.

All the cultivated material belongs to var **pictum** Müller, which is unknown in the wild. Its leaves are variable in shape, size and colour, and many hundreds of cultivars have been selected, raised and named. These have been grouped into several *formae* (probably best thought of as cultivar groups) by Pax (Das Pflanzenreich **47**: 24, 1911) on the basis of leaf shape.

Platyphyllum group: leaves 2–4 times as long as wide, entire, usually obovate.

Ambiguum group: leaves 5 or more times longer than wide, entire, lanceolate or narrowly lanceolate.

Taeniosum group: leaves linear, to 1 cm wide.

Crispum group: leaves spirally twisted or with wavy margins.

Lobatum group: leaves more or less 3-lobed.

Cornutum group: leaves entire, the midrib projecting from below the apex as a thread or horn.

Appendiculatum group: leaf-blades interrupted, narrowed to the midrib in the middle.

Within each of these groups cultivars are distinguished by the precise shape of the leaves and their colour: this may be light or dark green with spots of various single or mixed colours, or may be variegated, or have the veins picked out in a contrasting colour, or may be almost 1-coloured (not green) or multicoloured in zones. Many of the existing cultivars have names of Latin form and may be found labelled as species in collections, producing an extensive pseudosynonymy that is not cited above; all cultivated codiaeums belong to the species here described. Because of the variety of cultivars, no illustrations are cited above.

16. JATROPHA Linnaeus
M.C. Tebbitt
Monoecious or dioecious perennial herbs, shrubs, or trees with milky or watery latex. Leaves often large in cultivated species, to 30 cm, alternate, stalked, palmately veined (ours), entire or sometimes palmately lobed or cut, sometimes peltate, usually with conspicuously branched thorn-like stipules. Inflorescence a many-branched, flat-topped cyme; flowers not aggregated within a common involucre, male flowers with 5 sepals and 5 petals, female flowers lack petals; petals yellow, purple or scarlet to deep orange-red.

A genus of about 125 species; native to tropical and subtropical North and South America, Africa and Asia. Easily grown from seed and cuttings of young branches that have been allowed to dry out before setting in a damp sandy soil. Mature plants require a well-drained and sunny position in a temperate house.

1a. Leaves shallowly lobed; flowers greenish yellow or yellowish white **3. curcas**
  b. Leaves deeply lobed; flowers coral-red or scarlet **2**
2a. Leaves peltate; shrub to 2.5 m **2. podagrica**
  b. Leaves not peltate; shrub or tree to 6 m **1. multifida**

1. J. multifida Linnaeus (*Adenoropium multifldum* Pohl). Illustration: Das Pflanzenreich **42**: 40 (1910); Graf, Exotica, Series 4, edn 12, 1051 & 1055 (1985).
Shrub or tree to 6 m. Leaves *c.* 30 cm wide, nearly round, deeply cleft into many lobes that may or may not be lobed again, whitish beneath, stalked. Flowers scarlet. *Tropical America.* G1.

2. J. podagrica J. D. Hooker. Illustration: Rowley, Illustrated encyclopedia of succulents, 239 (1978); Everett, New York Botanical Gardens illustrated encyclopedia of horticulture **6**: 1852 (1981); Rauh, Wonderful world of succulents, pl. 31, 4 (1984); Brickell (ed.), RHS A–Z encyclopedia of garden plants, 586 (2003).
Shrub to 2.5 m, with a rough, swollen main stem; deciduous. Leaves to 30 cm wide, rounded-ovate, with 3–5 deep, round lobes, peltate, whitish beneath; stalks to 10 cm. Flower-stalks long, red; flowers coral-red. *Guatemala, Panama.* G1.

3. J. curcas (*Curcas curcas* (Linnaeus) Britton & Millspaugh). Illustration: Das

Pflanzenreich **42**: 78 (1910); Everett, New York Botanical Gardens illustrated encyclopedia of horticulture **6**: 1852 (1981); Graf, Exotica, Series 4, edn 12, 1060 (1985).
Shrub or small tree, 2–6 m, deciduous. Leaves 25–30 × 6–15 cm, rounded-ovate, with 3–7 shallow lobes, lower surface whitish; stalks about as long as blades. Flowers greenish yellow or yellowish white. *Mexico to South America.* G1.

17. CNIDOSCOLUS Pohl
M.C. Tebbitt
Monoecious or rarely dioecious perennial herbs, shrubs or small trees, with milky latex; frequently evergreen. Leaves alternate, stalked, entire, lobed, toothed or divided, usually with stinging hairs. Flowers in cymes; calyx showy, white; petals absent.

A genus of about 75 species from temperate and tropical South America.

1. C. urens (Linnaeus) Arthur (*Jatropha wrens* Linnaeus). Illustration: *Botanical Journal of the Linnean Society* **85**: 217 (1982).
Monoecious shrub or large herb, densely covered with stinging hairs. Leaves 7–15 × 8–30 cm, ovate to almost rounded, with 3–5 lobes, margins toothed or entire; stalks as long as or longer than blades. Flowers white, sweet-scented; inflorescence-stalks 3.5–12 cm; calyx hairy. *S Mexico to Argentina.* G1. Year round.

18. HOMALANTHUS Jussieu
J. Cullen
Hairless trees or shrubs with milky latex. Leaves alternate, long-stalked; stipules present, somewhat translucent. Flowers unisexual, small, in terminal racemes; female flowers few, at the bases of the racemes only; bracts finely toothed, with 2 fleshy glands (ours) at the base. Male flowers with a perianth of 2 segments (ours), which are finely toothed, and 5–many stamens with very short filaments. Female flowers with a shallowly cup-shaped, slightly toothed perianth, not distinguishable into individual segments (ours); ovary 2-celled with a single ovule in each cell; styles free except at the extreme base, spreading. Fruit a schizocarp splitting eventually into 2 mericarps. Seeds each with a fleshy appendage.

A genus of about 30 species from south-east Asia and Australasia. Only one is occasionally grown, requiring warm glasshouse conditions. The generic name is sometimes misspelt 'Omalanthus'.

1. H. populifolius Graham. Illustration: *Botanical Magazine*, 2780 (1827); Das Pflanzenreich **52**: 42 (1912); Galbraith, Collins field guide to the wild flowers of south-east Australia, pl. 87 (1977).

Shrub or small tree. Leaves broadly ovate (often broader than long), 6–14 cm, dark green above, pale greyish green beneath, bronze when young, often bright red in the autumn, broadly tapered to the base, acute at apex, margins slightly wavy. Racemes 6–10 cm. Fruit to 10 mm. *Australia (Queensland, New South Wales, Victoria), New Guinea.* G2. Summer.

19. HURA Linnaeus
J. Cullen
Trees with usually spiny branches and irritant milky latex. Leaves alternate, long-stalked; stipules present. Flowers unisexual, both sexes borne on the same plant, the males in dense spikes, the females few at the bases of the male spikes or solitary in the upper leaf-axils. Male flowers small, with a cup-like, somewhat toothed perianth and numerous stamens with filaments united into a thick column in the centre, bearing the anthers in 2 or 3 whorls. Female flowers much larger, dull red; perianth cup-like, somewhat toothed; ovary several-celled with a single ovule in each cell; style long, somewhat flower-like, tubular below, expanded, spreading and divided at the apex into numerous stigmatic lobes. Fruit capsule-like, woody, dehiscing explosively when ripe.

There are 2 species from Central and South America and the West Indies, of which one is grown. It prefers a light, open soil and may be propagated by cuttings.

1. H. crepitans Linnaeus. Illustration: Das Pflanzenreich **52**: 272 (1912); Bailey, Standard cyclopedia of horticulture, **2**: 1615 (1935).

Tree to 30 m, but usually much smaller in cultivation. Leaves ovate or oblong-ovate, cordate at the base, acuminate at the apex, lateral veins conspicuous, parallel, hairy beneath on the veins, margins irregularly toothed to almost entire. Male spikes long-stalked, 2–5 cm. Perianth of female flowers to 8 mm; style 3–4 cm, the expanded stigmas to 2 cm across. Fruit to 3 cm across. *C & South America, West Indies.* G1.

20. EXCOECARIA Linnaeus
J. Cullen
Shrubs (ours) with milky sap. At least the lower leaves opposite (ours); leaves persistent, thick but membranous, stalked, margins often toothed; stipules minute. Flowers very small, unisexual, both sexes usually borne on the same plant; male flowers in axillary spikes, female in short axillary cymes (rarely a few female flowers present at the base of male spikes). Male flowers with 3 overlapping perianth-segments and 3 stamens that project beyond the perianth-segments and have almost spherical anthers. Female flowers with 3 perianth-segments (sometimes united at the base); ovary 3-celled with 1 ovule per cell; styles 3, united at the extreme base, the free parts undivided, curling downwards along the ovary, the tips ultimately coiling upwards. Fruit a schizocarp splitting into 3 mericarps.

A genus of perhaps 30 little-known species from tropical areas. Only one is grown as a foliage plant in warm glasshouses. It tolerates most soils and is propagated by cuttings.

1. E. cochinchinensis (Loureiro) Merrill (*E. bicolor* Hasskarl). Illustration: Lecomte, Flore generale de L'Indo-Chine **5**: 405 (1926).

Erect shrub to 1.5 m. Leaves mostly opposite, narrowly elliptic to obovate, rather rounded to the base, acuminate at the apex, to 15 cm, toothed, glossy dark green above, bright purplish red beneath. Female inflorescences small; bracts margined with glandular hairs, each with a swollen glandular hump on each side towards the base. *Vietnam, Thailand & adjacent areas.* G2.

Most cultivated material has leaves red or purplish red beneath and corresponds with the plant named as *E. bicolor* var. *purpurascens* Pax & Hoffman, whereas wild material generally has leaves green beneath and has been named *E. bicolor* var. *viridis* Pax & Hoffmann.

21. STILLINGIA Linnaeus
J. Cullen
Hairless shrubs or herbs (ours) with milky latex. Leaves alternate (ours), shortly stalked, margins obtusely toothed with gland-tipped teeth; stipules present, sometimes divided. Flowers unisexual, small, in spikes, the female flowers 1 per bract at the base of the spike, the males several per bract, occupying the rest of the

spike; bracts each with 2 conspicuous glands towards the base. Male flowers with a thin, cup-like, obscurely 2-lobed perianth and 2 or 3 stamens with free filaments and anthers projecting beyond the perianth. Female flowers with 3 free perianth-segments (ours); ovary 2- or 3-celled, with 1 ovule per cell; styles 2 or 3, mostly free, spreading. Fruit a capsule splitting into 2 or 3 single-seeded segments.

A genus of about 25 species mostly from tropical and subtropical America, some from Madagascar and Polynesia. Only one North American species is occasionally grown. Propagate by seed.

1. S. sylvatica Linnaeus. Illustration: Das Pflanzenreich **52**: 192 (1912); Rickett, Wild flowers of the United States **2**: pl. 53 (1967); Dean et al., Wild flowers of Alabama, 97 (1983).

Herbaceous perennial to 1 m. Leaves rather leathery, very shortly stalked, elliptic to narrowly obovate or lanceolate, obtuse at the apex, finely toothed. Spikes to 7 cm, at first dense, later lengthening and becoming loose; glands on the bracts conspicuous, cup-like. *E USA from Virginia southwards.* H5–G1. Summer.

22. SAPIUM P. Brown
J. Cullen
Hairless shrubs or trees with irritant milky latex. Leaves alternate, long-stalked, sometimes toothed; stipules present, small. Flowers unisexual, small, in terminal spikes with a few female flowers at the base, the rest male (rarely some spikes entirely male); bracts broad, subtending several flowers, each with 2 glands towards the base. Male flowers with a small, 2- or 3-lobed perianth; stamens 2 or 3, filaments free, anthers projecting from the perianth. Female flowers with a 2- or 3-lobed or -toothed perianth; ovary with 2 or 3 cells, a single ovule in each; styles united (ours), bearing 2 or 3 long, spreading stigmas at the apex. Fruit a capsule (ours) usually with 3 seeds. Seeds covered in a white, oily appendage (aril) in our species.

A genus of over 100 species from tropical areas. Only one is grown; it is easily cultivated and propagated by seed or by cuttings.

1. S. sebiferum (Linnaeus) Roxburgh. Illustration: Das Pflanzenreich **52**: 238 (1912); The Garden **111**: 569 (1986).

Small or large tree, hairless, with slender branches. Leaves broadly ovate, often broader than long, truncate or tapered at the base, acuminate at the apex, lateral veins pronounced, margins entire. Spikes to 10 cm. Fruits conspicuously 3-lobed, pointed, *c.* 1 cm across. *C, W & S China, Taiwan; introduced elsewhere.* G1.

Oil and wax are extracted from the seed-covering (aril).

23. HIPPOMANE Linnaeus
J. Cullen

Shrub or tree to 15 m, hairless, with copious, milky, irritant, poisonous latex. Leaves alternate, long-stalked, ovate to almost circular, 5–10 cm, truncate, rounded or slightly cordate at base, acuminate at apex, leathery, entire or slightly toothed or scalloped; stipules triangular, 4–8 mm. Flowers small, unisexual, in terminal spikes with thick axes; male flowers *c.* 8 to each glandular bract, occupying most of the spike, female flowers 1 to each bract, few at the base of the spike. Male flowers with a 2- or 3-lobed perianth and 2 stamens, filaments joined into a column at the base, the column projecting from the perianth; pistillode absent. Female flowers with the perianth deeply 3-lobed; ovary 6–9-celled, each cell containing 1 ovule; styles united at the extreme base, free above, spreading, undivided. Fruit a more or less spherical drupe to 3 cm across, borne on short stalks.

A genus of a single species from Central America and the West Indies. It is rarely grown, being tropical and not particularly ornamental. It may be propagated by cuttings.

1. H. mancinella Linnaeus. Illustration: Das Pflanzenreich **52**: 262 (1912); Fournet, Flore de Guadeloupe et de Martinique, 1556 (1978); Correll & Correll, Flora of the Bahama archipelago, 820 (1982).

C America, West Indies. G2.

24. DALECHAMPIA Linnaeus
J. Cullen

Low to medium shrubs, some scrambling; milky latex absent. Leaves alternate, simple (ours), lobed or divided; stipules conspicuous. Flowers unisexual, male and female in the same inflorescence (ours), subtended by 2 whorls of more or less opposite bracts, 2 small and green alternating with 2 much larger, unequal and coloured. Male flowers several, borne on jointed stalks among a coloured mass of sterile male flowers, which is surrounded

by whitish bracts; perianth-segments 4 or 5, hairy; stamens 5–many, attached by short stalks to a stout central column formed from the fused filaments; the fertile and sterile male flowers are massed on the side of the inflorescence adjacent to the larger of the 2 coloured bracts. Female flowers with 5 or more narrow perianth-segments with hairy margins and a 3-lobed, silky-hairy ovary with a terminal style; stigma simple; cells of the ovary 3, each containing 1 ovule; each female flower subtended by a small pinkish green bract. Fruit hairy, at least at first, a 3-lobed capsule dehiscing elastically. Seeds without appendages.

A genus of about 100 species, some with stinging hairs, from the tropics (most from South America). Only one is commonly grown for its remarkable inflorescences. These superficially consist of 2 pinkish red (rarely white), unequal, leaf-like bracts with a yellow mass of closely packed cylindric bodies (the apical parts of the sterile male flowers) between them, through which project slightly longer, whitish, fertile male flowers. This whole mass is enclosed by whitish bracts. At the lower side of these (subtended by the smaller of the 2 large bracts) are 3 female flowers whose only conspicuous feature is formed by the 3 long, white styles. The plant is easily grown in a warm glasshouse, preferring well-drained compost, and is propagated by cuttings.

1. D. spathulata (Scheidweiler) Baillon (*D. roezliana* Mueller). Illustration: *Botanical Magazine*, 5640 (1867); de Wit, Plants of the world **1**: pl. 151, 152 (1963); Everard & Morley, Wild flowers of the world, pl. 178 (1970); Encke, Kalt- und Warmhauspflanzen, 238 (1987).

Small shrub. Leaves somewhat leathery, spathulate to obovate, to 20 × 9 cm, margins irregularly and bluntly toothed and wavy, tapered to the ultimately abruptly rounded base, acuminate at the apex, dark green and finely hairy on the midrib above, paler and finely hairy beneath. Stipules ovate-triangular. Inflorescences axillary, much shorter than the leaves that subtend them. Large bracts ovate, dull reddish pink or rarely whitish, fading to green as the fruit develops. Mass of sterile male flowers to 1 cm wide, bright yellow. Fruit deeply 3-lobed, to 1.5 cm across, hairy. *C & N South America.* Flowering irregularly throughout the year.

The commonest variant has reddish pink bracts and is sometimes called var. **spathulata** (*D. roezliana* var. *rosea* Mueller); a white-bracted variant, 'Alba' or var. *alba* invalid, is sometimes grown.

25. EUPHORBIA Linnaeus
S. Carter (succulent species) & J. Cullen, revised by S.G. Knees

Trees, shrubs, perennial, biennial or annual herbs, often succulent and spiny or cactus-like. Copious milky, irritant latex present in all parts. Leaves on the stem usually alternate (often opposite or whorled in the inflorescence), more rarely opposite, variable in shape, occasionally lobed, often small and very short-lived in succulent species; stipules present or absent, when present often glandular or represented by spines. Ultimate inflorescences flower-like (cyathia), consisting of a cup formed from 5 fused bracts and containing a central, stalked female flower surrounded by several male flowers; bracteoles often present between the male flowers; margins of the cup 5-lobed, and bearing 1–5 (rarely more) nectaries, which may be flat or pouched and entire or horned, in some cases with petal-like extensions giving the cyathium the appearance of a flower with a corolla. Cyathia sometimes unisexual, with males and females usually on separate plants. Each male flower consists of a single stamen whose filament is jointed to a short stalk. The female flower consists of a 3-celled naked ovary with 1 axile ovule in each cell; styles 3, united at the base, often further divided above. The cyathia may be solitary or borne in more complex inflorescences (see below). Fruit a hard, brittle or spongy capsule opening along the septa, borne on the usually elongate female flower-stalk. Seeds usually with appendages (caruncles).

A genus of 1600 or more species, cosmopolitan. From the point of view of the gardener, the genus consists of 2 groups differing in overall structure and cultivation requirements: the succulent (cactus-like) species with thick fleshy stems, and the non-succulent. These are treated separately below (except for a single number-sequence).

1a. Stems and branches not succulent, or rarely slightly fleshy, not ridged or angled; rarely spiny (when spines present, formed from the hardened ends of the inflorescence-branches of the previous year);

leaves persistent through the growing season; roots never large and tuberous **Group A**

  b. Stems and branches succulent, or if only slightly fleshy then cylindric, pencil-like and apparently leafless, often longitudinally ridged or angled, often tuberculate and/or spiny (spines on horny pads called spine-shields, which are sometimes joined along the ridges, or the hardened remains of inflorescence-stalks); stems sometimes much abbreviated and the roots large and tuberous **Group B**

Group A

The non-succulent cultivated euphorbias cover a wide range of growth-forms, from tropical trees and shrubs to hardy, native, European herbs (many weedy). They are spread across a wide spectrum of the subgenera and sections into which the genus is divided; as these are not particularly helpful as regards identification of cultivated plants, they are not mentioned further in this account.

The inflorescences of many of these species are complex, and require some explanation. The cyathia are occasionally solitary, but are usually borne in a false umbel (referred to below simply as an umbel, for convenience; it is not a genuine umbel as there is usually a single cyathium terminating the main branch, in the centre). This umbel may have 1–many branches (rays) and is generally subtended by a whorl of leaves (umbellary leaves), which are generally similar to the stem-leaves, though often shorter and broader.

Each ray of the umbel may terminate in a cyathium or, more commonly, may be further branched. Each cyathium and branch is subtended by a pair of opposite leaves (more rarely a whorl of 3 leaves), which are frequently of a different shape and colour from the stem and umbellary leaves; these are called 'ray-leaves' in the descriptions below. Below the terminal umbel, axillary flowering branches may be present; these are termed 'axillary rays' in the descriptions below.

Most of the species included below are easily grown in good soil, some requiring partial shade, others tolerating open, sunny situations. They are propagated by seed, division (perennials) or by cuttings (shrubby species). Many of the herbaceous species, particularly the annuals, are

invasive weedy plants and care should be taken in introducing these to gardens, as they may well swamp other, more choice plants.

Literature: the literature on the non-succulent euphorbias is extensive and not particularly helpful as far as the identification of cultivated plants is concerned. Most useful are the accounts of the genus in Floras of temperate areas (especially Flora Europaea **2**: 213–26, 1968, and Davis (ed.), Flora of Turkey **7**: 571–630, 1982. There are also two linked papers on the cultivated species: Turner, R., A review of spurges for the garden, *The Plantsman* **5**: 129–56 (1984); and Radcliffe-Smith, A. & Turner, J., Key to the non-succulent hardy and half-hardy Euphorbia spp., ibid., 157–61.

1a. Perennial herb or shrub   2
  b. Annual or biennial herb   40
2a. Shrubs with persistent, woody aerial branches   3
  b. Herbs dying down to the base or basal rosette of leaves in winter   11
3a. Glands 1–3, pouch-like; leaves broad, ovate-elliptic, ovate or fiddle-shaped, the ray-leaves bright red or creamy white   **5. pulcherrima**
  b. Glands 4 or 5, not pouch-like; leaves various, not as above   4
4a. Low shrub, the ends of the previous year's shoots persisting as spines   **15. acanthothamnos**
  b. Taller shrub, not spiny   5
5a. Glands with red, orange, yellow or white petal-like extensions, the cyathium appearing to possess a corolla   6
  b. Glands without petal-like extensions   7
6a. Lower leaves often whorled, ray-leaves white; whole plant densely hairy; stipules present, glandular   **1. leucocephala**
  b. Lower leaves not whorled, ray-leaves green; plant mostly hairless; stipules absent   **4. fulgens**
7a. Inflorescence a terminal panicle of umbels, without distinct umbellary or ray-leaves, very strongly honey-scented   **10. mellifera**
  b. Inflorescence a terminal umbel or a solitary cyathium, umbellary and ray-leaves present, not strongly honey-scented   8
8a. Cyathia solitary, terminal   **11. balsamifera**
  b. Cyathia in terminal umbels   9

9a. Inflorescences completely suffused with dark red-purple; leaves obovate; ray-leaves united at base   **9. atropurpurea**
  b. Inflorescences yellowish green; leaves oblong, narrowly oblong or oblong-lanceolate; ray-leaves free   10
10a. Leaves truncate, notched or very rounded at the apex, which is not mucronate; glands conspicuously horned   **8. obtusifolia**
  b. Leaves tapered then somewhat rounded to the mucronate apex; glands not horned, sometimes irregularly toothed or lobed   **7. dendroides**
11a. Glands with white, petal-like extensions, which are hairy on the outside   **3. corollata**
  b. Glands without petal-like extensions   12
12a. Glands 8 to each cyathium   **25. capitulata**
  b. Glands 4 or 5 to each cyathium   13
13a. Ray-leaves united at their bases to form cups   14
  b. Ray-leaves free at their bases   17
14a. Capsules hairy   **43. characias**
  b. Capsules hairless   15
15a. Stem-leaves rounded, truncat or cordate at the base   **42. oblongifolia**
  b. Stem-leaves tapered to the base   16
16a. Ray-leaf cups 1–2.5 cm across; first year stem-leaves much larger than those of the second year   **41. amygdaloides**
  b. Ray-leaf cups 1–5 cm across; first year stem-leaves not or only slightly larger than those of the second year   **44. macrostegia**
17a. Plants conspicuously glaucous   18
  b. Plants bright or dark green, not conspicuously glaucous   24
18a. Glands horned, margins between the horns indented, often notched or somewhat lobed or toothed   19
  b. Glands not horned, margins smooth   21
19a. Hairy bracteoles present between the male flowers; capsules 3–4.5 mm   **32. nicaeensis**
  b. Bracteoles absent from between the male flowers; capsules at least 4.5 mm   20
20a. Leaves obovate or spathulate   **30. myrsinites**
  b. Leaves lanceolate   **31. rigida**
21a. Plant quite smooth, not at all papillose   **33. seguieriana**

b. Plants papillose, often conspicuously so, but sometimes the papillae easily visible only on the leaf-margins near the base   22

22a. Ray-leaves minutely toothed
                                35. pithyusa

b. Ray-leaves not minutely toothed, absolutely entire   23

23a. Plant very densely and conspicuously papillose; leaves acuminate   **36. decipiens**

b. Plant not conspicuously papillose; leaves rounded but sometimes mucronate at the apex
                                32. nicaeensis

24a. Glands horned, margins between the horns indented, often notched or somewhat lobed or toothed   25

b. Glands not horned, margins smooth, entire   29

25a. Capsules 5–6 × 6–7 mm; stem-leaves always shiny on the upper surface
                                40. lucida

b. Capsules not more than 4 × 5 mm; stem leaves dull on the upper surface
                                       26

26a. Seeds pitted   **34. portlandica**

b. Seeds smooth   27

27a. All stem-leaves narrowly linear, crowded, those of the axillary shoots densely clustered   **37. cyparissias**

b. Stem-leaves variable but not as above   28

28a. Stem-leaves oblanceolate, rounded at the apex   **39. esula**

b. Stem-leaves broadest below the middle, acute at the apex
                                38. virgata

29a. Capsule smooth (though sometimes hairy)   30

b. Capsule covered with warts   34

30a. Capsule hairy   31

b. Capsule hairless   32

31a. Stem-leaves linear-lanceolate; plant usually more than 1 m
                                24. altissima

b. Stem-leaves oblong or oblong-lanceolate; plants less than 1 m
                                16. corallioides

32a. Ray-leaves strikingly red-tinged; tufts of hair present on the stems below the leaf-insertions   **12. griffithii**

b. Ray-leaves yellow or greenish; stems completely hairless   33

33a. Capsule c. 7 × 8 mm; young growth green   **14. wallichii**

b. Capsule to 6 × 7 mm; young growth bright red   **13. sikkimensis**

34a. Warts on capsule small, hemispherical and sparse   35

b. Warts on capsule large, at least some cylindric (if all more or less hemispherical, then very dense)   37

35a. Umbel with 1—5 rays, stems weak, thin, at most to 7 mm thick at base
                                20. soongarica

b. Umbel with more than 5 rays; stems robust, often 1 cm thick or more at the base   36

36a. Capsule 2–3 mm, sparsely verruculose, glabrous **18. schillingii**

b. Capsule 4–6 mm, covered with sparse hemispherical warts
                                19. palustris

37a. Stems with reduced, scale-like leaves at the base   **23. dulcis**

b. Stems without scale-leaves at the base   38

38a. Stem-leaves finely toothed, linear, usually less than 2 cm
                                22. uliginosa

b. Stem-leaves entire or rarely with a few minute teeth towards the apex, usually more than 2 cm   39

39a. Seeds warty   **17. epithymoides**

b. Seeds smooth   **21. hyberna**

40a. Biennial herb with opposite leaves and spongy capsules   **45. lathyris**

b. Annual herbs with alternate leaves and hard or brittle capsules   41

41a. Umbellary leaves sometimes white-margined; glands with petal-like extensions, hairy on the back
                                2. marginata

b. Umbellary leaves not white-margined; glands without petal-like extensions   42

42a. Glands 1 or 2, each broad, 2-lipped; leaves varying from linear to fiddle-shaped on the same plant, ray-leaves red-blotched at the base
                                6. cyathophora

b. Glands 4 or 5, flat, not 2-lipped; leaves not conspicuously variable on the same plant, the uppermost not red-blotched at the base   43

43a. Leaves minutely toothed   44

b. Leaves absolutely entire   45

44a. Leaves rounded to the blunt apex; seeds pitted or roughened
                                26. helioscopia

b. Leaves acute, tapered; seeds smooth
                                27. stricta

45a. Leaves linear to oblong, not stalked
                                28. exigua

b. Leaves obovate, often broadly so, distinctly stalked   **29. peplus**

1. E. leucocephala Lotsy. Illustration: Clay & Hubbard, The Hawaii garden, tropical exotics, 75 (1977).

Shrub or small tree to 4 m; young parts crisply hairy. Leaves mostly whorled below, variable, oblong, obovate, elliptic or oblong-lanceolate, long-stalked; stipules glandular. Primary umbel with 3–6 rays of differing lengths, usually further branched. Ray-leaves few, white, obovate, with stalks as long as or longer than the blades. Glands 5, small, entire, with white, petal-like extensions, which are hairy outside. Capsule smooth, c. 7 × 8 mm. C America. G1. Spring.

Widely cultivated in the tropics.

2. E. marginata Pursh (*E. variegata* Sims). Illustration: *Journal of the Royal Horticultural Society* 99: pl. 205 (1974); *Gartenpraxis* 1986(March) 25.

Erect annual herb to 1 m. Stems hairless below, sparsely hairy above. Leaves narrowly elliptic, ovate, oblong or obovate, rounded or tapered to the base, acute at the apex, very shortly stalked, hairless or slightly hairy towards the base; ray-leaves and sometimes the umbellary leaves white-margined; stipules short, lanceolate, soon falling. Umbel usually with 3 rays. Glands 4, ovate, greenish, with rounded, white, petal-like extensions. Capsules densely hairy. C, E & S USA. H1. Summer.

In most of Europe, a half-hardy annual; its latex is extremely irritant and poisonous.

3. E. corollata Linnaeus. Illustration: Justice & Bell, Wild flowers of North Carolina, 108 (1968).

Slender perennial herb to 1 m, usually sparsely hairy throughout. Leaves narrowly elliptic to oblong, shortly stalked, tapered or rounded to the base, rounded at the apex. Main umbel with c. 5 rays, these many times branched, producing a large compound inflorescence. Glands 5, oblong, pouched, each with a white, petal-like extension. Capsule smooth, brown, c. 5 × 6 mm. *E North America, from Ontario to Florida.* H2. Summer–autumn.

4. E. fulgens Klotzsch (*E. jacquiniflora* J. D. Hooker). Illustration: Kaier, Indoor plants in colour, pl. 85 (1961); Hay et al., Dictionary of indoor plants in colour, pl. 223 (1974); Vieria, Flores de Madeira, pl. 108 (1986).

Spreading shrub to 2 m, with arching, spreading or drooping, mostly hairless branches. Leaves lanceolate, acute, tapered to the base and apex, very long-stalked, dark green above, paler beneath. Stipules absent. Cyathia in axillary cymes shorter than the subtending leaves. Glands 5, each with an orange-scarlet or yellow, petal-like appendage. Capsules rarely formed in cultivation. *Mexico*. H5–G1. Winter–spring.

Widely grown for the cut-flower trade.

5. E. pulcherrima Klotzsch (*Poinsettia pulcherrima* (Klotzsch) Graham). Illustration: Clay & Hubbard, The Hawaii garden, tropical exotics, 77 (1977); Vieria, Flores de Madeira, pl. 109 (1986); *The Garden* **112**: 554(1987).

Shrub to 3 m or more in suitable locations. Leaves variable, lanceolate to ovate-elliptic or fiddle-shaped, entire, toothed or lobed, very long-stalked; stipules glandular. Ray-leaves bright red or greenish cream. Umbel terminal with few rays. Cyathia with 1 (rarely 2), pouched, yellow-brown gland. Capsule rarely formed in cultivation. *C America*. H5–G1. Winter.

A very showy small tree, which can be easily propagated by cuttings; with suitable treatment the young plants can be induced to flower, and are sold as small house plants, especially around the Christmas period. The plant is variable in leaf-shape and the exact colour of the ray-leaves; numerous cultivars have been developed.

6. E. cyathophora Murray (*Poinsettia cyathophora* (Murray) Klotzsch & Garcke; *E. heterophylla* misapplied; *P. heterophylla* misapplied). Illustration: Duncan, Wild flowers of the southeastern United States, 97 (1974).

Annual herb to 1 m. Leaves mostly hairless, varying from linear to ovate or obovate and from entire to toothed or lobed (sometimes fiddle-shaped) on the same plant; all leaves shortly stalked, shiny on the upper surface, ray-leaves red-blotched towards the base. Stipules glandular. Umbels dense. Cyathia with 1 or 2 broad, 2-lipped, yellow-brown glands. Capsule smooth. *Mexico*. H5-G1. Summer.

Often found under the name *E. heterophylla* Linnaeus; this name, however, applies to another Mexican species not in cultivation, which has the upper leaves green or purple-spotted, without red blotches at the base.

7. E. dendroides Linnaeus. Illustration: Bonnier, Flore complète **10**: pl. 543 (1929); Polunin & Huxley, Flowers of the Mediterranean, pl. 90 (1965); Ceballos et al., Plantas silvestres de la peninsula Iberica, 188 (1980); Polunin, Concise guide to the flowers of Britain and Europe, pl. 66 (1987).

Shrub to 2 m. Leaves 2.5–6.5 cm × 3–8 mm, oblong-lanceolate, shortly stalked, tapered to the base, rounded to the mucronate apex, hairless. Axillary rays absent. Main umbel terminal, umbellary leaves yellowish, persistent; ray-leaves yellow, free. Cyathia with 5 rounded or somewhat lobed yellowish green glands. Capsule 5–6 mm, smooth or almost so. Seeds appendaged. *Mediterranean area*. H5–G1. Spring.

8. E. obtusifolia Poiret. Illustration: Bramwell & Bramwell, Wild flowers of the Canary Islands, t. 186 (1974); Kunkel & Kunkel, Flora de Gran Canaria **3**: pl. 136 (1978).

Like *E. dendroides* but leaves oblong, narrower, truncate, notched or very rounded to the apex, which is not mucronate; glands clearly horned. *Canary Islands*. H5–G1. Spring–summer.

In subsp. **obtusifolia** the umbellary leaves persist until fruiting, whereas in subsp. **regis-jubae** (Webb & Berthelot) Maire (*E. regis-jubae* Webb & Berthelot) they fall early.

9. E. atropurpurea Broussonet. Illustration: Bramwell & Bramwell, Wild flowers of the Canary Islands, f. 36 (1974).

Like *E. dendroides* but leaves glaucous, obovate, to 10 × 2.5 cm, rounded or notched at the apex, the whole inflorescence dark red-purple; ray-leaves united at the base to form cups; glands not horned or lobed. *Canary Islands*. H5–G1. Spring–summer.

10. E. mellifera Alton. Illustration: Bramwell & Bramwell, Wild flowers of the Canary Islands, pl. 189 (1974); *Journal of the Royal Horticultural Society* **100**: 608 (1975); Brickell (ed.), RHS A–Z encyclopedia of garden plants, 439 (2003).

Shrub to 3 m. Stem-leaves lanceolate or very narrowly elliptic, shortly stalked, to 16 × 2 cm or more, tapered to the base and to the mucronate apex, mostly hairless except along the midrib beneath; upper leaves with axillary rays. Inflorescence a panicle of umbels, without distinct umbellary or ray-leaves. Cyathia with 5 orange-red glands that become greenish brown, very strongly honey-scented. Capsules *c.* 4 × 6 mm, covered in large, cylindric warts. Seeds appendaged. *Canary Islands, Madeira*. H5–G1. Spring.

Surprisingly hardy in western Europe, though often cut to the base in severe winters.

11. E. balsamifera Aiton. Illustration: Bramwell & Bramwell, Wild flowers of the Canary Islands, pl. 184 (1974); Kunkel & Kunkel, Flora de Gran Canaria **3**: pl. 134 (1978); *Gartenpraxis* 1978 (Sept): 469 (1978); Brickell (ed.), RHS A–Z encyclopedia of garden plants, 437 (2003).

Like *E. mellifera* but shorter, leaves much smaller, to 6 × 1 cm, oblong-obovate; umbel unbranched, terminal; capsule to 6 × 9 mm, wrinkled but not warty, somewhat hairy, seeds not appendaged. *Canary Islands*. H5–G1. Spring.

12. E. griffithii J.D. Hooker. Illustration: Hara, Photo-album of plants of eastern Himalaya, pl. 91 (1968); Perry, Flowers of the world, 115 (1972); *The Plantsman* **5** (3): frontispiece (1984) – 'Dixter'; *Gardening from 'Which'* 1989 (March): 37.

Perennial herb. Young growth green. Stems erect, to 80 cm, branched above, hairless except for small tufts of hairs below the insertion of each leaf. Stem-leaves linear to lanceolate, 4–13 cm × 8–22 mm, acute, tapered to the base, very shortly stalked; umbellary and ray-leaves reddish. Umbel terminal with *c.* 5 rays. Cyathia with 4 or 5 orange or yellow, entire glands. Capsule smooth, hairless, *c.* 5 mm. *Bhutan, China (Xizang)*. H3. Spring–summer.

Variable in the precise coloration of the ray-leaves, In 'Fireglow' they are very bright red, whereas in 'Dixter' they are more orange-red. Some variants have leaves with whitish midribs.

13. E. sikkimensis Boissier. Illustration: *The Garden* **103**: 311 (1978); Brickell (ed.), RHS A–Z encyclopedia of garden plants, 440 (2003).

Very similar to *E. griffithii* but young growth reddish, stems completely hairless, ray-leaves yellowish green; capsule to 6 × 7 mm. *N India (Sikkim), Bhutan*. H3. Summer.

Probably not distinct from *E. griffithii*, with which it vicariates.

14. E. wallichii J. D. Hooker. Illustration: *Botanical Magazine*, n.s., 442 (1964); *The Garden* **102**: 449 (1977).

Like *E. griffithii* and *E. sikkimensis* but young growth green, stems completely hairless; ray-leaves yellow; capsule *c.* 7 × 8 mm. *W Himalaya*. H3. Summer.

15. E. acanthothamnos Boissier. Illustration: *Journal of the Royal Horticultural Society* **88**: 388 (1963); Polunin & Huxley, Flowers of the

Mediterranean, t. 89 (1965); Polunin, Concise guide to the flowers of Britain and Europe, pl. 66 (1987).

Intricately branched shrub to 30 cm, hairless, older inflorescence branches persistent as hardened spines in pairs and diverging at wide angles. Leaves elliptic to obovate, acute or rounded at the apex. Ray-leaves yellowish. Cyathium with 4 or 5 yellowish brown, entire glands. Capsule 3–4 mm, bearing conical warts. Seeds smooth. *E Mediterranean area.* H5. Spring.

16. E. corallioides Linnaeus. Illustration: Pignatti, Flora d'Italia **2**: 37 (1982); *The Plantsman* **5**: 138 (1984).

Upright perennial herb with rather few erect stems to 80 cm. Stem-leaves oblong to oblong-lanceolate or oblanceolate, almost stalkless, rounded to the apex, covered with long, soft, spreading, white hairs. Ray-leaves yellowish or somewhat red-tinged. Cyathia with 5 rounded, yellow-green glands. Capsule almost smooth, hairy, 3–4 mm. *Italy, Sicily.* H5. Spring.

17. E. epithymoides Linnaeus (*E. polychroma* Kerner). Illustration: *Journal of the Royal Horticultural Society* **91**: f. 137 (1966); Perry, Flowers of the world, 115 (1972); *Gardening from 'Which'* 1988 (Feb): 13; Aeschimann et al., Flora alpina **1**: 1001 (2004).

Perennial herbs with stout stocks, generally densely and softly downy. Stems erect, stout, 20–40 cm. Leaves narrowly elliptic, oblong or lanceolate, 3–6 × 1–2.5 cm, rounded at the base and usually rounded at the apex, sometimes very obscurely and finely toothed. Ray-leaves yellow, sometimes purplish-tinged. Cyathium with 4 or 5 small, entire, rounded, yellowish glands. Capsules 3–4 mm, covered in dense, cylindric warts. *S & CE Europe.* H2. Spring–summer.

Part of a complex of species from south and east Europe which are difficult to distinguish. Several cultivars are available.

18. E. schillingii. Radcliffe-Smith. Illustration: *Kew Magazine* **4**(3): 80 (1987); Brickell (ed.), RHS A–Z encyclopedia of garden plants, 440 (2003); Lord (ed.), Flora, 589 (2003).

Erect, herbaceous perennial 80–100 cm, spreading to 30 cm across. Leaves narrowly elliptic-oblong or oblong-lanceolate, 10–13 × 1–3 cm, deep green often with conspicuous whitish midribs. Inflorescence with bright greenish yellow

bracts. Umbel with 6–8 rays. Ray-leaves 2 or 3, rhombic-suborbicular, 2–2.5 × 1.8–2 cm. Cyathium with 5 rounded, yellow-ochre glands. Capsule 2–3 mm, sparingly verruculose, glabrous. *E. Nepal.* H4. Summer–autumn.

19. E. palustris Linnaeus. Illustration: Bonnier, Flore complète **10**: pl. 541 (1929); Pignatti, Flora d'Italia **2**: 38 (1982); Brickell (ed.), RHS A–Z encyclopedia of garden plants, 439 (2003); Aeschimann et al., Flora alpina **1**: 999 (2004).

Robust, tufted, hairless perennial herb. Stems erect, to 1.5 m, stout, 7–15 mm thick at the base. Leaves elliptic to linear-lanceolate, 2–8 cm × 3–15 mm, tapered or rounded to the base, rounded at the apex, margins minutely and obscurely toothed. Umbel with 5 or more rays. Ray-leaves almost circular, yellowish. Axillary rays usually present. Cyathium with 4 or 5 rounded, entire, yellowish glands. Capsule 4–6 mm, covered with sparse, hemispherical warts. Seeds smooth. *Europe, W Asia.* H1. Spring–summer.

20. E. soongarica Boissier.
Like *E. palustris* but generally shorter, stems less robust (to 7 mm thick at the base), leaves 2–11 cm × 5–22 mm, umbel with 1–5 rays. *E Europe, temperate Asia.* H1. Spring–summer.

21. E. hyberna Linnaeus. Illustration: Bonnier, Flore complète **10**: pl. 542 (1929); Pignatti, Flora d'Italia **2**: 38 (1982); Garrard & Streeter, Wild flowers of the British Isles, pl. 50 (1983); Aeschimann et al., Flora alpina **1**: 1001 (2004).

Perennial herb with a stout rhizome. Stems erect, to 60 cm, becoming leafless below. Leaves oblong to oblanceolate, somewhat tapered to the base, rounded or notched at the apex, hairless above, with sparse spreading hairs beneath. Ray-leaves yellow; rays usually 5. Cyathium with 4 or 5 rounded, entire, yellowish glands. Capsule 5–6 mm, with cylindric warts. Seeds smooth. *W & S Europe.* H3. Spring–summer.

Subsp. **hyberna** has the capsule clearly stalked, whereas subsp. canuti (Parlatore) Tutin (*E. canuti* Parlatore), from the Maritime Alps, has the capsule almost stalkless.

22. E. uliginosa Boissier.
Perennial with a stout woody stock, hairless or somewhat downy. Stems erect,

to 60 cm. Leaves linear-oblong, 5–20 × 1–3 cm, leathery, rounded at the apex, finely but distinctly toothed. Ray-leaves triangular, yellowish, rays 2–5. Cyathium with 4 or 5 entire, rounded, yellowish green glands. Capsule 2.5–3 mm, covered in cylindric to club-shaped warts. Seeds smooth. *Spain, Portugal.* H5. Spring–summer.

Occurs in the wild in seasonally damp places.

23. E. dulcis Linnaeus. Illustration: Bonnier, Flore complète **10**: pl. 541 (1929); Pignatti, Flora d'Italia **2**: 39 (1982); Garrard & Streeter, Wild flowers of the British Isles, pl. 50 (1983); Aeschimann et al., Flora alpina **1**: 1001 (2004).

Usually a somewhat hairy perennial herb with thick, jointed rhizomes. Stems 20–50 cm, slender, erect, bearing reduced but conspicuous scale-leaves at the base. Stem-leaves 2.5–7.5 cm × 8–15 mm, narrowly obovate to oblong-obovate, tapered to the base, shortly stalked, rounded to the apex. Ray-leaves triangular, somewhat cordate at the base, yellowish. Cyathium with 4 or 5 entire, rounded, yellowish glands. Capsule 3–1 mm, with 3 deep grooves, covered with cylindric warts, sometimes hairy. Seeds smooth, dark brown. *Most of Europe.* H1. Spring.

24. E. altissima Boissier.
Tufted perennial herb with a stout rootstock, usually softly downy throughout. Stems robust, to 2 m, erect. Leaves lanceolate or narrowly elliptic, 4–12 cm × 8–12 mm, shortly stalked, tapered at the base and the acute apex. Umbellary leaves much shorter and broader than stem-leaves; ray-leaves broadly ovate to almost circular, acute, yellowish green. Axillary rays numerous. Cyathia with 4 or 5 entire, rounded, yellowish glands. Capsule to 5 mm, smooth or with slightly granular (but not warty) surface. Seeds smooth. *Turkey, Syria, Lebanon, N Iraq.* H4. Summer.

25. E. capitulata Reichenbach.
Small, straggling, more or less hairless evergreen perennial herb; stems numerous, arising from slender rhizomes. Leaves close, overlapping, to 10 × 5 mm, obovate, tapered to the base, rounded to the apex. Ray-leaves yellow. Cyathium solitary, terminal, appearing deformed, with 8 purplish, rounded, entire glands. Capsule

grooved, warty. Seeds smooth. *Balkan Peninsula.* H3.

26. E. helioscopia Linnaeus. Illustration: Bonnier, Flore complète **10**: pl. 544 (1929); Garrard & Streeter, Wild flowers of the British Isles, pl. 50 (1983); Aeschimann et al., Flora alpina **1**: 1005 (2004).

Annual herb, stems to 40 cm, sparingly branched from the base, hairless or with sparse hairs. Stem-leaves obovate-spathulate, increasing in size upwards, shortly stalked, tapered to the base, rounded at the apex, toothed in the upper half. Umbel contracted, rather flat; ray-leaves obliquely obovate, broad, yellowish. Cyathia with rounded, entire, greenish glands. Capsule to 3.5 mm, 3-lobed, smooth. Seeds pitted or the surface roughened. *Europe, N Africa, SW Asia.* H4. Late winter–summer.

A frequent garden weed, occasionally tolerated as an ornamental.

27. E. stricta Linnaeus (*E. serrulata* Thuillier). Illustration: Bonnier, Flore complète **10**: pl. 544 (1929); Garrard & Streeter, Wild flowers of the British Isles, pl. 50 (1983); *The Plantsman* **5**: 148 (1984); Aeschimann et al., Flora alpina **1**: 1005 (2004).

Slender, upright annual herb, hairless. Stems often reddish, to 1 m, with very numerous axillary rays in the upper part, these rays spreading at wide angles. Leaves oblong-oblanceolate, tapered to the base, shortly stalked, somewhat tapered at the apex, finely toothed. Ray-leaves short and broad, yellowish. Cyathia with 5 entire, yellowish glands. Capsule 2.5–3 mm, 3-lobed, covered in cylindric warts. Seeds smooth. *Europe to W Asia.* H1. Spring–summer.

A delicate annual, occasionally grown, but spreading rapidly and very invasive.

28. E. exigua Linnaeus. Illustration: Bonnier, Flore complète **10**: pl. 548 (1929); Garrard & Streeter, Wild flowers of the British Isles, pl. 51 (1983); Aeschimann et al., Flora alpina **1**: 1007 (2004).

Bushy, hairless, annual herb, much branched from the base, stems to 20 cm. Stem and umbellary leaves linear, linear-oblong or linear-obovate, tapered to the base but stalkless, notched or 3-toothed at the apex. Rays 3–5, ray-leaves lanceolate, somewhat cordate at the base, yellowish. Cyathia with entire, rounded, reddish

glands. Capsule to 1.5 mm, smooth. Seeds finely warty. *Europe, N Africa, W Asia.* H1. Spring–summer.

An invasive species.

29. E. peplus Linnaeus. Illustration: Bonnier, Flore complète **10**: pl. 548 (1929); Garrard & Streeter, Wild flowers of the British Isles, pl. 51 (1983); Aeschimann et al., Flora alpina **1**: 1009 (2004).

Hairless annual herb, usually much branched, stems to 40 cm. Stem-leaves obovate, often broadly so, stalked, tapered to the base, rounded at the apex. Ray-leaves stalkless, ovate to diamond-shaped, yellow. Rays 3 or 4, much branched. Cyathia with yellow, half-moon-shaped, horned glands. Capsule 3-lobed, ridged, to 2 mm. Seeds pitted. *Europe, N Africa, SW Asia, widely introduced elsewhere.* H1. Late winter–summer.

A very invasive species.

30. E. myrsinites Linnaeus. Illustration: Polunin & Huxley, Flowers of the Mediterranean, pl. 94 (1965); Hay & Synge, Dictionary of flowering plants in colour, pl. 69 (1971); *Journal of the Royal Horticultural Society* **91**: f. 138 (1966) & **97**: f. 180 (1972); Aeschimann et al., Flora alpina **1**: 1005 (2004).

Glaucous, hairless perennial herb. Stems trailing to ascending, simple, robust, to 40 cm. Leaves rather thick, obovate-spathulate to almost circular, tapered then ultimately rounded to the base, acute and mucronate at the apex. Axillary rays present or not. Ray-leaves broader and shorter than stem-leaves, bright lime-green. Rays usually 5 or more, branched, variable in length. Cyathium with glands that are horned, the horns dilated towards the apex or weakly lobed, yellowish. Bracteoles absent from between the male flowers. Capsule 5–7 mm, smooth or minutely granular. Seeds with granular surface, greyish brown. *S Europe, N Africa, SW Asia.* H3. Spring–summer.

'Washfied' has a red tinge to the inflorescence.

31. E. rigida Bieberstein (*E biglandulosa* Desfontaines). Illustration: Polunin & Huxley, Flowers of the Mediterranean, pl. 92 (1965); Polunin, Flowers of Greece and the Balkans, pl. 28 (1980); *The Plantsman* **5**: 144 (1984); Brickell (ed.), RHS A–Z encyclopedia of garden plants, 440 (2003).

Similar to *E. myrsinites* but stems ascending, to 60 cm, leaves narrower,

lanceolate; ray-leaves almost circular; capsule 6–7 mm, sparsely papillose; seeds smooth, pale grey to white. *Mediterranean area east to former USSR (Caucasus).* H3. Summer.

32. E. nicaeensis Allioni. Illustration: Bonnier, Flore complète **10**: pl. 546 (1929); Pignatti, Flora d'Italia **2**: 46 (1982); Aeschimann et al., Flora alpina **1**: 1009 & 1010 (2004).

Glaucous perennial herb, often red-suffused; hairless, papillose though the papillae often inconspicuous and most easily seen on the leaf-margins towards their bases. Rootstock stout. Stems upright, to 80 cm. Leaves lanceolate to oblong, more or less entire, tapered to the base, stalkless, rounded though often mucronate at the apex. Axillary rays present or absent. Rays 5–18; ray-leaves ovate to kidney-shaped, yellowish. Cyathium with yellowish glands that are rounded and entire or somewhat horned. Capsule 3–4.5 × 3–5 mm, rough, sometimes hairy. Seeds smooth or almost so, pale grey. *C & E Europe, N Africa, former USSR (Caucasus).* H3. Summer.

A variable species. Subsp. **glareosa** (Bieberstein) Radcliffe-Smith has a smaller capsule and fewer rays in the umbel.

33. E. seguieriana Necker. Illustration: Brickell (ed.), RHS A–Z encyclopedia of garden plants, 440 (2003); Aeschimann et al., Flora alpina **1**: 1013 (2004).

Hairless, glaucous perennial herb, without any papillae, quite smooth. Stems erect, to 60 cm, often branched from the base. Leaves linear to elliptic-oblong, leathery, entire, tapered to the base and to the pointed apex. Rays 5–30; ray-leaves triangular to heart- or kidney-shaped, yellowish. Axillary rays usually numerous. Cyathium with entire, rounded, yellowish glands. Capsule 2–3 mm, roughened on the keels. Seeds smooth, pale grey. *Most of Europe.* H2. Summer.

A variable species. Subsp. **seguieriana** has 5–15 rays in the umbel; subsp. **niciciana** (Borbás) Rechinger (*E. niciciana* Novak), from the Balkan peninsula, has 15–30 rays.

34. E. portlandica Linnaeus. Illustration: Bonnier, Flore complète **10**: pl. 547 (1929); Garrard & Streeter, Wild flowers of the British Isles, pl. 51 (1983).

Hairless perennial herb with a woody rootstock, usually branched from the base. Stems ascending or erect, to 40 cm. Leaves evergreen, obovate to oblanceolate, tapered to the base, rather abruptly tapered to the acute apex. Ray-leaves triangular to diamond-shaped, yellow. Rays 3–6. Cyathium with yellow, distinctly horned glands. Capsule 2.5–3 × 3–3.2 mm, grooved, somewhat granular on the keels. Seeds pale grey, pitted. *W Europe, N Africa.* H3. Spring–summer.

Generally maritime in the wild.

35. E. pithyusa Linnaeus. Illustration: Bonnier, Flore complète **10**: pl. 543 (1929); Pignatti, Flora d'Italia **2**: 48 (1982).

Papillose, glaucous perennial, often woody at the base. Stems erect, to 60 cm, often much branched below. Leaves linear-lanceolate or ovate-lanceolate, those towards the base of the stem reflexed and overlapping, each tapered to a pointed apex. Umbellary leaves obtuse but mucronate; ray-leaves broadly ovate to almost circular, toothed. Axillary rays usually numerous. Cyathium with entire, rounded, yellowish glands. Capsule 2–3.5 × 2.5–3.5 mm, 3-grooved, roughened on the keels. Seeds dark grey to whitish, smooth or somewhat roughened. *W Mediterranean area.* H3. Summer.

36. E. decipiens Boissier & Buhse.

Glaucous, very densely and conspicuously papillose perennial herb, woody at the base, forming large clumps. Stems to 25 cm, ascending. Leaves narrowly oblanceolate, tapered to the base, acuminate at the apex. Ray-leaves small, elliptic, abruptly narrowed to a long point, yellowish. Axillary rays absent. Umbel with 5 or more rays, short, congested. Cyathia somewhat reddish with entire, oblong to rounded, yellow or purplish glands. Capsule *c.* 4 mm, 3-grooved, smooth, purplish black. *Iran.* H5. Spring–summer.

37. E. cyparissias Linnaeus. Illustration: Bonnier, Flore complète **10**: pl. 546 (1929); Strid, Wild Flowers of Mount Olympus, pl. 78 (1980); Garrard & Streeter, Wild flowers of the British Isles, pl. 51 (1983); *Gardening from 'Which'* **1987**(August): 245.

Hairless, rhizomatous perennial. Stems several from the crown, erect, to 50 cm, densely leafy; axillary shoots usually present, congested, densely leafy. Leaves linear, obtuse or rarely somewhat acute. Umbellary leaves linear-oblong; ray-leaves triangular, diamond-shaped or almost circular, yellow. Rays usually more than 8;

axillary rays sometimes present. Cyathia with the glands yellow, distinctly but shortly horned. Capsule to 3.5 mm, 3-grooved, surface granular. Seeds smooth, shiny, grey. *Most of Europe, introduced in North America.* H1. Summer.

This species spreads extensively by the growth of the rhizomes in winter, and also, to some extent, by seed. It is best confined to deep pots kept clear of the soil surface. A number of cultivars is available, some less invasive than others.

38. E. virgata Waldstein & Kitaibel (*E. waldsteinii* (Sojak) Radcliffe-Smith; *E. esula* subsp. *tommasiniana* (Bertoloni) Nyman). Illustration: Bonnier, Flore complète **10**: pl. 546 (1929); Aeschimann et al., Flora alpina **1**: 1015 (2004).

Hairless perennial herb. Stems several from the crown, erect, to 1 m, often branched. Leaves linear-oblong to oblong-lanceolate, broadest below the middle, tapered to the base and apex, dull on the upper surface. Ray-leaves free but superficially appearing united in pairs, broadly rounded-triangular to kidney-shaped, yellowish. Rays 6–15 (rarely more). Cyathium with yellow glands with distinct, long horns. Capsule to 4.5 mm, deeply 3-grooved, surface granular. Seeds smooth, grey. *W & C Europe to C Asia.* H2. Summer.

Somewhat invasive in suitable sites.

39. E. esula Linnaeus. Illustration: Bonnier, Flore complète **10**: pl. 546 (1929); Garrard & Streeter, Wild flowers of the British Isles, pl. 51 (1983); Aeschimann et al., Flora alpina **1**: 1015 (2004).

Variable perennial, similar to *E. virgata* but with leaves widest above the middle and rounded at the apex. *Most of Europe, introduced elsewhere.* H1. Summer.

Doubtfully cultivated. A widespread species which hybridises with *E. virgata* in eastern Europe; the hybrid,

E. × pseudovirgata (Schur) Soó (*E. uralensis* misapplied), has spread into western Europe, and may occasionally be cultivated.

40. E. lucida Waldstein & Kitaibel. Illustration: Pignatti, Flora d'Italia **2**: 48 (1982).

Rhizomatous perennial herb, similar to *E. virgata* but stem-leaves conspicuously shiny above, capsules 5–6 × 6–7 mm. *C & E Europe to former USSR (Siberia).* H3. Summer.

41. E. amygdaloides Linnaeus. Illustration: Ceballos et al., Plantas silvestres de la peninsula Iberica, 187 (1980); Garrard & Streeter, Wild flowers of the British Isles, pl. 51 (1983); *The Plantsman* **5**(3): frontispiece (1984); Polunin, Concise guide to the flowers of Britain and Europe, pl. 66 (1987).

Hairy, tufted or rhizomatous perennial herb, the shoots usually biennial, flowering on the growth of the second year, the first year's leaves persistent in rosettes at the tops of the first-year stems. First-year leaves stalked, obovate, tapered to the shortly stalked base, rounded to the obtuse apex. Second-year leaves smaller, more elliptic, scarcely stalked. Axillary rays usually present. Ray-leaves broadly ovate, united in pairs to form cups 1–2.5 cm across. Cyathia with horned, yellow glands. Capsule 3–4 × 2.5–4 mm, grooved, hairless. Seeds smooth, blackish. *Mediterranean area, C Europe, SW Asia.* H3. Summer.

Variable with a number of varieties in the wild and several cultivars and selections in gardens.

Var. **amygdaloides** is tufted, and has dull, membranous, usually hairy leaves. Var. **robbiae** (Turrill) Radcliffe-Smith, thought to have originated in Turkey, is widespread in cultivation, tolerating dry shade; it has spreading rhizomes and more shiny, leathery, hairless leaves. Illustration: *Botanical Magazine*, n.s., 208 (1952–53); Everard & Morley, Wild flowers of the world, pl. 35 (1970).

E. × martinii Rouy is the name applied to the natural hybrid between this species and *E. charasias.*

42. E. oblongifolia (C. Koch) C. Koch (*E. amygdaloides* Linnaeus var. *oblongifolia* C. Koch).

Like *E. amygdaloides* but first-year stems flowering, leaves elliptic, rounded, truncate or cordate at base; cups 2–3 cm across; capsule to 5 mm, seeds smooth, dark grey. *E Turkey, former USSR (Caucasus).* H3. Summer.

43. E. charasias Linnaeus. Illustration: Bonnier, Flore complète **10**: pl. 542 (1929); Polunin & Huxley, Flowers of the Mediterranean, pl. 93 (1965); Polunin, Concise guide to the flowers of Britain and Europe, pl. 66 (1987); Brickell (ed.), RHS A–Z encyclopedia of garden plants, 438 (2003).

Like *E. amygdaloides* but glaucous, much taller (to 2 m), first-year leaves linear to oblanceolate, dense, those of the second year reduced, much less dense; axillary rays numerous; ray-leaf cups 2–3 cm across; glands variable in shape; capsules 4–7 × 5–6 cm, densely hairy; seeds smooth, silver-grey. *Portugal, Mediterranean area.* H3. Spring–summer.

A variable species. Subsp. **characias**, from the western Mediterranean area has shorter stems (to 80 cm) and glands reddish brown, whereas subsp. **wulfenii** (Koch) Radcliffe-Smith (*E. veneta* Willdenow) from the eastern Mediterranean area is taller and has greenish yellow glands.

44. E. macrostegia Boissier.

Like *E. amygdaloides* but hairless, leaves leathery, glossy green, those of the second year a little smaller than those of the first; ray-leaf cups 2–5 cm across; glands with rather short horns; capsule *c.* 5 mm, smooth, not hairy; seeds smooth, pale grey. *Turkey, Syria, Lebanon, Iran.* H4. Summer.

Material in cultivation under this name may well be *E. amygdaloides*.

45. E. lathyris Linnaeus. Illustration: Pignatti, Flora d'Italia **2**: 43 (1982); Garrard & Streeter, Wild flowers of the British Isles, pl. 50 (1983); Aeschimann et al., Flora alpina **1**: 1007 (2004).

Hairless, glaucous, robust biennial herb, stems erect, to 1.5 m. Leaves opposite, linear-lanceolate to lanceolate, bluish green, shallowly cordate at the base, acuminate at the apex. Umbellary leaves oblong-lanceolate to ovate-lanceolate; ray-leaves triangular-ovate, pale green, acuminate. Cyathium with horned glands, which are yellowish green with purple blotches. Capsule smooth, spongy, 1.3–1.7 cm. Seeds grey or brown, surface somewhat rough. *C & E Mediterranean area, widely naturalised elsewhere.* H1. Summer.

Group B

The succulent species of *Euphorbia*, like the non-succulent ones, cover a wide range of growth forms, from fleshy herbs to trees and shrubs; they are spread among a number of subgenera and sections. Almost all originate from the African continent, especially South Africa, plus a number from Madagascar, the Arabian peninsula, India and Malaysia, with just a few from Central and South America and one from Australia.

The stems and usually also the branches, are thick and fleshy, acting as water storage organs, and the form each

species takes is usually characteristic of the section to which it belongs. As these sections can only be used in a very broad sense for the purposes of identification, they are not further mentioned in this account. Stems and branches can be smoothly cylindric and apparently leafless, or covered in tubercles, or longitudinally ridged. The tubercles are swellings at the bases of fallen leaves, sometimes with spines in their axils formed by hardened inflorescence-stalks. The ridges are often prominent and produced into angles, formed by longitudinal rows of tubercles, or stipules modified as spines or bristles, or longitudinal series of 'spine-shields'. These spine-shields are horny pads, which surround the leaf-scars and bear the stipules modified as small prickles, and a pair of large thorn-like spines.

Inflorescence characteristics are fairly uniform within groups (sections), as solitary cyathia, simple cymes (1 central cyathium with 2 lateral) or branched cymes in an umbel with just a few (2–5) rays, all on very short or fairly long stalks. The structure of the cyathium is identical to that in the non-succulents, with individual variation in ray-leaf size and ornamentation of the glands. The ray-leaves are usually much smaller and less leaf-like than in the non-succulents, and are called 'bracts' in the descriptions below.

All the succulent species require pot culture under glass; a few exceptions can withstand Mediterranean winters. A minimum of 5 °C is advisable, with 7–10 °C giving a better chance of success. All need a very well-drained compost, with plenty of added grit. Opinions vary as to the use of soil-less versus soil-based composts and plastic versus clay pots. Watering should be done with care, the compost never being allowed to remain wet. Unlike the treatment for cacti (Vol. 2, p. 210), it is not advisable to allow the compost to dry out for a long period during the winter, as roots will then die off and it can be extremely difficult to start growth again in the spring without causing rot to set in.

Propagation can be by seed or by cuttings. Seed is usually easy to set with South African species, as long as both male and female plants are available with unisexual species. Cuttings should be allowed to dry off, usually for at least a week, to allow a firm callus to form, otherwise rot will set in at once. Some species can be split up, but, again, any

damaged parts should be allowed to dry off thoroughly.

In many of the succulent species the latex is extremely caustic; it should not be allowed to enter a cut, or come in contact with delicate areas of the skin, the eyes, lips or nostrils, as severe inflammation can result. Normally any burning or aching sensation will wear off in an hour or so, but the eyes should be washed out with plenty of water. If much latex gets on to hands or clothes, a little petrol is the best solvent.

Literature: a very complete account of the South African species, with keys, is given in White, A., Dyer, R.H. & Sloane, B.L, The succulent Euphorbieae of southern Africa, 2 volumes (1941). Excellent photographs, descriptions and articles appear in the recently published volumes of *The Euphorbia Journal* (Strawberry Press, Mill Valley, California), volumes 1–7 (1983–91). A comprehensive list of species with brief descriptions and some photographs of plants in cultivation appears in Jacobsen, H., Handbook of succulent plants **1** (1960), but its accuracy is not always reliable.

1a. Stems and branches neither tuberculate nor with spines, or rarely with spines formed by the sharpened tips of branchlets　　2
　b. Stems and branches with spines, or if spineless then tuberculate　　16
2a. Stem covered in large leaf-scars; leaves with cyathia in a tuft at the stem-apex　　**108. millotii**
　b. Stem not as above　　3
3a. Dwarf herb, branching underground from a large tuberous root; leaves at ground level, to 10 cm　　**59. silenifolia**
　b. Shrubs or trees; leaves to 4 cm　　4
4a. Branchlets spine-tipped　　5
　b. Branchlets not spine-tipped　　6
5a. Small tree; all branchlets sharply spine-tipped, cyathia densely clustered, dark red　　**52. stenoclada**
　b. Compact shrub; branchlets becoming woody and spine-tipped; cyathia in umbels with 1–3 rays, yellow　　**58. lignosa**
6a. Stems flattened, to 2 cm wide　　**50. xylophylloides**
　b. Stems cylindric, to 10 cm thick　　7
7a. Stems with 4–6 longitudinal ridges　　8

b. Stems smooth, finely lined or with callused leaf-scars 9

8a. Stems 4–8 mm thick, with 3 obscure ridges below each leaf-scar **46. pteroneura**

b. Stems *c*.1 cm thick, with 4–5 rounded ridges separated by deep grooves **47. sipolisii**

9a. Nectary glands red, with conspicuous white petal-like appendages **48. antisyphilitica**

b. Nectary glands yellow, without appendages 10

10a. Stems marked with fine lines **49. tirucalli**

b. Stems without fine lines 11

11a. Branches smooth, without leaf-scars 12

b. Branches with callused leaf-scars 13

12a. Branches in whorls **53. aphylla**

b. Branches not in whorls **55. sarcostemmoides**

13a. Cyathia in dense clusters **51. alluaudii**

b. Cyathia in umbels of 3–7 cymes 14

14a. Stems *c*.1 cm thick **57. dregeana**

b. Stems 5–7 mm thick 15

15a. Cyme-bracts with marginal hairs; nectary glands slightly lobed **54. mauritanica**

b. Cyme-bracts hairless; glands entire **56. schimperi**

16a. Stems with longitudinal rows of bristly stipules, or with spines formed by hardened inflorescence-stalks, or tuberculate 17

b. Stems with thorn-like spines in pairs on horny pads 62

17a. Stems with longitudinal rows of bristly stipules 18

b. Stems not as above 20

18a. Cyathia supported by a pair of conspicuous white spreading bracts **105. lophogona**

b. Cyathia with clasping tubular-shaped bracts 19

19a. Bracts bright red **106. viguieri**

b. Bracts white **107. leuconeura**

20a. Leaves large, 7–15 cm, semi-persistent 21

b. Leaves tiny, deciduous 25

21a. Cyathia stalkless at the tubercle-bases **62. clandestina**

b. Cyathia on long stalks 22

22a. Tubercles shallow 23

b. Tubercles very prominent 24

23a. Cyathia in loosely branching cymes **60. bubalina**

b. Cyathia solitary **61. clava**

24a. Stem to 15 cm; cyathia solitary on long stalks **63. bupleurifolia**

b. Stem 30–60 cm; cyathia in branching cymes **64. monteiri**

25a. Main stem extremely short, with many branches radiating from the plant centre more or less at ground level 26

b. Plant not as above 36

26a. Main stem branching below ground and giving rise to several tufts of slender branches **67. muirii**

b. Main stem not branching below ground 27

27a. Cyathia with a conspicuous mass of white woolly bracteoles 28

b. Bracteoles not conspicuously woolly 30

28a. Nectary glands green, entire **69. esculenta**

b. Nectary glands yellow or white, with fringed processes 29

29a. Glands yellow, sometimes entirely white, often bifid, with short white processes **70. inermis**

b. Glands and processes entirely yellow **71. superans**

30a. Branches 30–90 × *c*. 2 cm; nectary glands green with white or creamy processes 31

b. Branches *c*. 20 cm × 5–15 mm, usually much shorter; glands red, yellow, or green, entire or toothed but without processes 33

31a. Glands with 3–6 long white processes branched at the tip **65. caput-medusae**

b. Glands with short white or creamy processes 32

32a. Gland-processes 3 or 4, very short, white **66. bergeri**

b. Gland-processes 4–6, creamy, branched at tips and recurved **68. tuberculatoides**

33a. Glands entire, dark green to brownish red; branches 1–2.5 cm **75. gorgonis**

b. Glands toothed, yellow; branches 2–20 cm 34

34a. Branches *c*. 20 × 1–1.5 cm; glands notched on the inner margin **74. woodii**

b. Branches *c*. 30 × 8 mm 35

35a. Cyathia produced on the short branches around the plant centre; glands greenish yellow **72. pugniformis**

b. Cyathia produced from the central tubercle-like branches; glands bright yellow **73. flanaganii**

36a. Stem unbranched, 30–50 cm, distinctly tuberculate, with persistent, hardened, rigid, often sterile inflorescence-stalks **76. schoenlandii**

b. Plant not as above 37

37a. Main stem to 30–50 cm, densely covered with short tuberculate branches and with a few persistent hardened inflorescence-stalks **77. multiceps**

b. Plant not as above 38

38a. Main stem short with very numerous tuberculate branches forming large clumps; nectary glands with several large finger-like processes 39

b. Main stem unbranched, or with relatively few thick branches, with tubercles inlongitudinal rows, glands without processes 43

39a. Plants with compact branching from a thick tuberous root 40

b. Plants spreading by means of rhizomes 41

40a. Branches spherical, 1–2 cm across, strung together like beads **78. globosa**

b. Branches short, very dense, forming a compact cushion **82. polycephala**

41a. Branches no more than 5 mm thick, with prominent tubercles **81. wilmaniae**

b. Branches *c*.10 mm thick, with rounded tubercles 42

42a. Cyathium-stalks *c*.4 mm **80. tridentata**

b. Cyathium-stalks 1–10 cm **79. ornithopus**

43a. Stem unbranched, or rarely branching, more or less spherical, with tubercles forming 8 ridges 44

b. Stem branching, usually cylindric, with tubercles forming 5–many ridges or angles 46

44a. Ridges on plant body scarcely raised; spines never present **100. obesa**

b. Ridges prominent, some spines (persistent inflorescence-stalks) usually present 45

45a. Stem to 12 cm; some spines usually present, mostly unbranched **101. meloformis**

b. Stem to 30 cm; spines always present, branching **102. valida**

46a. Spines absent 47

b. At least a few spines present (persistent inflorescence-stalks) 50

47a. Branches spherical, in compact clumps; tubercles very prominent, in 12 or more ridges **83. susannae**

b. Branches cylindric; tubercles small, slightly raised in 5–9 ridges or angles 48

48a. Branches to 5 cm thick, with 7–9 definite angles **97. anoplia**

b. Branches 1–2 cm thick, with 5 ridges 49

49a. Branches produced on thin stalks *c.* 1 cm **84. tubiglans**

b. Branches without a stalk **85. jansenvillensis**

50a. Stem with numerous branches from the base, forming a compact cushion, 10–30 cm 51

b. Stem branching at the base and above to form shrubs 50–300 cm 56

51a. Tubercles prominent along definite angles 52

b. Tubercles scarcely raised along definite angles 54

52a. Branches 4–6 cm thick, with up to 17 angles **88. mammillaris**

b. Branches 1.5–4 cm thick, with 7–12 angles 53

53a. Branches very numerous, 1.5–2.5 cm thick, forming tight clumps **86. submammillaris**

b. Branches few to numerous, 2–4 cm thick, often lengthening and sprawling **87. fimbriata**

54a. Branches usually with 7 angles **89. pulvinata**

b. Branches with 8–12 angles 55

55a. Branches 2–3 cm thick with 8–9 angles; spines to 8 mm **90. aggregata**

b. Branches 3–4.5 cm thick with 9–12 angles; spines 1.5–3 cm **91. ferox**

56a. Stems with 5 or rarely 6 angles **92. pentagona**

b. Stems with 7 or more angles 57

57a. Spines solitary along the angles 58

b. Spines mostly in groups of 2–5 60

58a. Spines with 3–5 spreading branches at the tips **99. stellispina**

b. Spines simple 59

59a. Stems with 7 or 8 angles; spines to 3 cm **93. heptagona**

b. Stems with usually 11 angles; spines to 1 cm **94. cereiformis**

60a. Stems 10–15 cm thick with *c.* 14 angles; spines very numerous, grouped 2–5 together, 1–4 cm **98. horrida**

b. Stems 1–4 cm thick; spines 1–3 together, to 1 cm 61

61a. Stems with 7–10 angles; cyathium-stalks 1–2 cm **95. inconstantia**

b. Stems with 12–20 angles; cyathium-stalks *c.*4 mm **96. polygona**

62a. Floriferous shrubs with cyathia cupped by a pair of conspicuous brightly coloured bracts (usually red) 63

b. Plants not very floriferous and cyathia without conspicuous bracts 64

63a. Stems *c.*8 mm thick or often more, with stout spines **103. milii**

b. Stems 3–5 mm thick, with needle-like spines **104. beharensis**

64a. Spine-shields (horny pads bearing the paired spines) always completely separate 65

b. All or most spine-shields joined along the branch-angles 83

65a. Trees or shrubs, with stems and branches erect to at least 15 cm and 1 cm thick 66

b. Dwarf plants, with branches spreading, or if erect for more than 15 cm then less than 1 cm thick 98

66a. Stems with 2 or 3 angles, usually winged 67

b. Stems with 4–8 angles, occasionally some also with 3 angles 73

67a. Stems with 2 winged angles **127. ramipressa**

b. Stems with 3 angles 68

68a. Stems 5–17 cm wide, with angles straight or wavy 69

b. Stems 1–5 cm wide, with angles lobed 70

69a. Stems 7–10 cm thick, angles fairly stout, scarcely winged; leaves to 6 cm **113. antiquorum**

b. Stems 5–17 cm wide, thinly winged; leaves to 20 cm **120. amplophylla**

70a. Stems strongly variegated 71

b. Stems uniformly green 72

71a. Stems pale green with a central yellowish band; branches spreading **112. lactea**

b. Stems dark glossy green with transverse light-green mottling; branches strictly erect **116. trigona**

72a. Stems 3–5 cm wide, with wing-margins deeply lobed; leaves often present, 3–5 cm **115. barnhartii**

b. Stems 1–2 cm wide, with wing-margins shallowly lobed; leaves absent **126. grandidens**

73a. Leaves persistent, 15–30 cm 74

b. Leaves absent, or if present then 1–6 cm 76

74a. Stems cylindric, with spirally arranged tubercles **109. neriifolia**

b. Stems with 4 or 5 longitudinal series of tubercles 75

75a. Leaf-margins straight **110. nivulia**

b. Leaf-margins wavy **111. undulatifolia**

76a. Stems with 5–7 angles; leaves often present on mature growth, 2–6 cm **114. royleana**

b. Stems with 3–8 angles; leaves absent, or occasionally on seedling growth, 1–5 cm 77

77a. Stems 2–3 cm thick with 4 angles, densely branching to form compact clumps **125. resinifera**

b. Stems and branching not as above 78

78a. Stems 1.5–2.5 cm wide with 4 or 5 angles **129. tetragona**

b. Stems 4–12 cm wide with 3–8 angles 79

79a. Stems with 3 or 4 angles; spine-shields 1–3 cm apart 80

b. Stems with 4–8 angles; spine-shields not more than 1 cm apart 81

80a. Stems 4–8 cm wide with 4 angles, not winged **119. ingens**

b. Stems *c.* 10 cm wide with 3 or 4 winged angles **121. deightonii**

81a. Stems *c.* 4 cm wide with 4–6 angles; cyathia deep red **122. canariensis**

b. Stems 4–12 cm wide with 4–8 angles; cyathia yellow 82

82a. Stems with 5–8 angles; spine-pairs diverging at *c.* 120°; capsule spherical, red **117. abyssinica**

b. Stems with 4 or 5 angles; spine-pairs diverging at *c.*180°; capsule 3-lobed, yellowish green **118. ammak**

83a. Stem-angles distinctly winged and deeply constricted into segments usually about as long as wide 84

b. Stem-angles not winged and segments, if any, much longer than wide 86

84a. Stems with 3 wings; spines to 7 cm **131. grandicornis**

b. Stems with 4–6 wings; spines not more than 2.5 cm 85

85a. Stems with 4–6 wings, uniformly green **130. cooperi**

b. Stems with 4 wings, strongly streaked with yellow **132. grandialata**

86a. Stems trailing, less than 1 cm thick, with 4 angles **143. saxorum**

b. Stems erect, 5–100 mm thick, with 3–10 angles **87**

87a. Stems with usually 3 angles only, 5–10 cm thick **128. triangularis**

b. Stems with 4–10 angles, rarely some present with 3 angles **88**

88a. Stems constricted into segments **89**

b. Stems not constricted into segments **95**

89a. Stems usually variegated with yellowish green streaks **133. pseudocactus**

b. Stems uniformly green **90**

90a. Some stems present with 3 angles **134. franckiana**

b. All stems with 4–8 angles **91**

91a. Plants rhizomatous **135. coerulescens**

b. Plants not rhizomatous **92**

92a. Stems 2–3 cm thick **138. polyacantha**

b. Stems 4–15 cm thick **93**

93a. Stems only slightly segmented; horny margins along the angles 2–3 mm wide **139. ledienii**

b. Stems obviously segmented; horny margins 5 mm wide or more **94**

94a. Stems with 5–8 angles; capsules spherical **136. virosa**

b. Stems with 5 angles; capsules 3-lobed **137. avasmontana**

95a. Cyathia red to dull red **96**

b. Cyathia yellow to dull yellow **97**

96a. Spine pairs conspicuously unequal **123. officinarum**

b. Spine pairs equal **124. echinus**

97a. Stems 1–1.5 cm thick with 4–6 angles, uniformly green **140. griseola**

b. Stems 1.5–2 cm thick with 4 angles, variegated **141. heterochroma**

98a. Branches 8–10 mm thick with 4 angles; spine-shields narrowly triangular; capsules stalkless **142. schinzii**

b. Branches 8–25 mm thick with 2–4 angles; spine-shields bluntly triangular; capsules on long stalks **99**

99a. Branches 8–12 mm thick with 3 or 4 angles **100**

b. Branches 1–2.5 cm thick with 2 or 3 angles **101**

100a. Branches to 20 cm, with 3 or 4 deeply lobed angles **144. knuthii**

b. Branches c. 10 cm, with 4 rounded angles **146. micracantha**

101a. Branches 1–2.5 cm wide, with usually 3 angles **145. squarrosa**

b. Branches 1–1.5 cm wide, with 2 angles **147. stellata**

46. E. pteroneura Berger. Figure 128(2), p. 492. Illustration: White et al., The succulent Euphorbieae (southern Africa) **1**: 12 (1941); Jacobsen, Handbook of succulent plants **1**: 467, 468 (1960); Jacobsen, Lexicon of succulent plants, pl. 79 (1974).

Shrub c. 50 cm, with erect branching. Branches cylindric, 5–8 mm thick, with 3 longitudinal ridges from each leaf-scar. Leaves on short stalks, ovate, to 4 × 2 cm, short-lived. Cyathia 1–4 at the branch-tips, yellow, with a pair of enfolding yellowish green bracts c. 1 cm across. Capsule 3-lobed, c. 5 mm across, on a curved stalk. *Mexico.* G1.

47. E. sipolisii N.E. Brown. Illustration: Jacobsen, Handbook of succulent plants **1**: 468 (1960); The Euphorbia Journal **1**: 112 (1983).

Like *E. pteroneura* but stems with 4 or 5 rounded ridges and grooved sides, c. 1 cm thick, constricted into segments c. 10 cm, greyish green. *Brazil.* G1.

Uncommon in cultivation except in specialist collections.

48. E. antisyphilitica Zuccarini. Illustration: Jacobsen, Handbook of succulent plants **1**: 407 (1960); The Euphorbia Journal **2**: 97 (1984); The new RHS dictionary of gardening **2**: 263 (1992).

Stoloniferous shrub to 50 cm. Branches erect, cylindric, c. 5 mm thick, somewhat woody. Leaves small and very short-lived. Cyathia solitary, 1–4 cm apart, c. 1 cm across including the red glands, which bear conspicuous, white, petal-like appendages. Capsule 3-lobed, to 5 mm across, on a curved stalk. *Mexico.* G1.

Generally seen only in specialist collections.

49. E. tirucalli Linnaeus. Illustration: White et al., The succulent Euphorbieae (southern Africa) **1**: 101 (1941); Jacobsen, Lexicon of succulent plants, pl. 82 (1974); The Euphorbia Journal **1**: 115 (1983)

& **3**: 72 (1985); Graf, Exotica, Series 4, edn 12, 1026 (1985).

Densely branched shrub to 4 m, or tree to 12 m if left undisturbed. Branches cylindric, 5–7 mm thick, with fine longitudinal lines. Leaves lanceolate, to 15 × 2 mm, very short-lived. Cymes condensed at the branch-tips, forming tight clusters of cyathia. Cyathia c. 4 mm across, yellow, unisexual, male and female cyathia on separate plants. Capsule almost spherical, c. 9 mm across, on stalk to 1 cm. *Widespread throughout tropical Africa; also naturalised in the Arabian peninsula, Madagascar, India & the Far East.* G1.

The most widespread of all the succulent species of *Euphorbia*, it has become naturalised through its use as a protective hedge around settlements, grown from branch-cuttings, which root extremely easily. The copious, irritant latex provides an efficient deterrent against marauding animals.

50. E. xylophylloides Lemaire (*E. enterophora* Drake). Illustration: Jacobsen, Handbook of succulent plants **1**: 601 (1960); Jacobsen, Lexicon of succulent plants, pl. 83 (1974); The Euphorbia Journal **2**: 105 (1984) & **3**: 25 (1985).

Tree to 20 m, with spreading main branches densely re-branching at the tips. Secondary branches flattened, to 2 cm wide. Leaves tiny, at the tips of the branches, very short-lived. Cyathia in tight clusters at the branch-tips, c. 4 mm across, unisexual, male and female cyathia on separate plants. Capsule 3-lobed, c. 5 mm across. *S Madagascar.* G1.

51. E. alluaudii Drake (*E. leucodendron* Drake). Illustration: Jacobsen, Handbook of succulent plants **1**: 461 (1960); Jacobsen, Lexicon of succulent plants, pl. 77 (1974); The Euphorbia Journal **3**: 26 (1985).

Similar to *E. xylophylloides* but a shrubby tree to 4 m.

Subsp. **alluaudii** has cylindric branches.

Subsp. **onoclada** (Drake) Friedman & Cremers (*E. onoclada* Drake) has branches constricted at intervals into oblong segments with prominent leaf-scars. *S Madagascar.* G1.

Uncommon in cultivation except in specialist collections.

52. E. stenoclada Baillon. Illustration: Jacobsen, Handbook of succulent plants **1**: 408 (1960); Jacobsen, Lexicon of succulent plants, pl. 81 (1974); Graf, Tropica, edn 1,

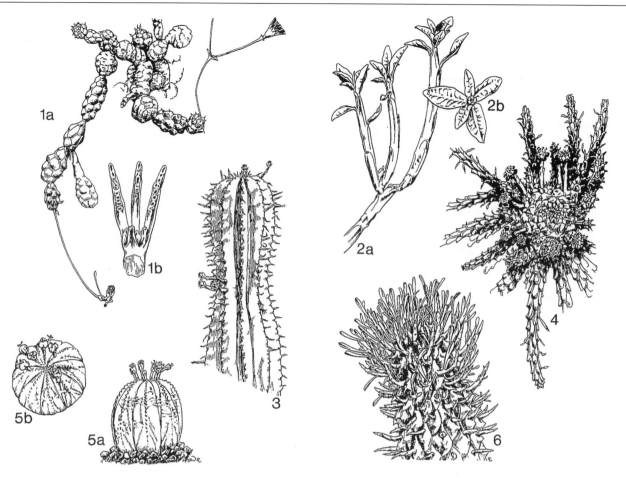

Figure 128. Diagnostic details of *Euphorbia* species. 1, *Euphorbia globosa* (a, branching habit (× 0.5); b, cyathium gland (× 4.5)). 2, *E. pteroneura* (a, branching habit (× 0.5); b, cyathium with surrounding bracts and leaves (× 0.75)). 3, *E. inconstantia*, stem apex, with sterile and fertile inflorescence-stalks (× 0.5). 4, *E. pungiformis*, plant with cyathia on short stalks (× 0.5). 5, *E. obesa* (a, plant with male flowers; b, plant with female flowers in fruit (× 0.5)). 6, *E. schoenlandii* stem apex with sterile inflorescence-stalks (× 0.5).

422, 425 (1978); *The Euphorbia Journal* **3**: 22 (1985).

Like *E. xylophylloides* but branches cylindric with numerous, short, spine-tipped branchlets, 2–5 cm, cyathia dark red. *Madagascar*. G1.

Uncommon in cultivation.

53. E. aphylla Willdenow. Illustration: Jacobsen, Handbook of succulent plants **1**: 463 (1960); Bramwell & Bramwell, Wild flowers of the Canary Isles, pl. 185 (1974).

Shrub to 1 m, with whorled branching. Branches cylindric, to 10 × 1 cm, broadening towards the tips. Leaves lanceolate, *c.* 5 mm, very short-lived. Cyathia clustered at the branch tips, *c.* 4 mm across, on stalks *c.* 5 mm. Capsule 3-lobed, *c.* 5 mm across, on a short, curved stalk. *Canary Islands.* H5–G1.

54. E. mauritanica Linnaeus. Illustration: White et al., The succulent Euphorbieae (southern Africa) **1**: 107 (1941); Jacobsen, Handbook of succulent plants **1**: 450 (1960); Jacobsen, Lexicon of succulent plants, pl. 75 (1974); *The Euphorbia Journal* **3**: 119 (1985) & **4**: 83 (1987).

Shrub 1–1.75 m, with dense, erect branching. Branches cylindric, *c.* 5 mm thick. Leaves lanceolate, to 1.5 cm, short-lived, leaving obvious scars. Cymes 5–7 in terminal umbels, with rays *c.* 1.5 cm; bracts with marginal hairs. Cyathia *c.* 7 mm across, with yellow glands slightly lobed on the outer margin. Capsule 3-lobed, *c.* 6 mm across, on a curved stalk. *Widespread in South Africa, especially Cape Province.* H5–G1.

There are several varieties based on the thickness and colour of the branches and cyme-rays.

55. E. sarcostemmoides Willis. Illustration: *The Euphorbia Journal* **4**: 123 (1987).

Similar to *E. mauritanica* but smaller, cymes branching repeatedly, with cyathia at the tips on very short, congested rays; cyathia *c.* 4 mm across, greenish yellow, unisexual; capsules on short stalks. *Australia (dry areas across the centre).* G1.

A recently described species, rare in cultivation.

56. E. schimperi Presl. Illustration: Graf, Exotica, Series 4, edn 12, 1046 (1985); *The Euphorbia Journal* **5**: 118 (1988).

Similar to *E. mauritanica* but often scrambling to 4 m, bracts hairless, cyathia larger with glands not lobed and capsule *c.* 1 cm across. *Arabian peninsula & possibly*

NE Africa in mountain areas from the Red Sea hills to N Somalia. G1.

Uncommon in general cultivation.

57. E. dregeana Boissier. Illustration: White et al., The succulent Euphorbieae (southern Africa) **1**: 127 (1941); *The Euphorbia Journal* **2**: 104 (1984).

Similar to *E. mauritanica* but branches *c.* 1 cm thick, cymes branching 1–3 times with very short rays, capsule spherical with 6 longitudinal ridges, *c.* 9 mm across. *South Africa (Cape Province), Namibia.* G2.

Uncommon in general cultivation.

58. E. lignosa Marloth. Illustration: White et al., The succulent Euphorbieae (southern Africa) **1**: 136 (1941); Jacobsen, Lexicon of succulent plants, pl. 78 (1974); *The Euphorbia Journal* **1**: 88 (1983) & **5**: 49 (1988); Graf, Exotica, Series 4, edn 12, 1058 (1985)

Compact shrub to 50 cm × 1 m, densely branching. Branches cylindric, with short branchlets at right angles, becoming woody and spine-tipped. Leaves lanceolate, *c.* 1 cm, very short-lived. Cymes 1–3 from the branch-tips, with very short rays. Cyathia *c.* 8 mm across, with yellow glands toothed on the outer margins. Capsule spherical, *c.* 5 mm across. *Namibia.* G2.

59. E. silenifolia (Haworth) Sweet. Illustration: White et al., The succulent Euphorbieae (southern Africa) **1**: 229 (1941); *The Euphorbia Journal* **2**: 136 (1984) & **8**: 102 (1992).

Main stem extremely short, merging into the large tuberous root, *c.* 5 cm thick, branching at the apex underground into several short branches. Leaves tufted from the branch-tips at ground-level, lanceolate, to 10 cm, tapering into a stalk 2–8 cm. Cymes usually branching once, 3–5 in umbels on stalks 3–12 cm, ray-leaves ovate, *c.* 10 × 6 mm. Cyathia *c.* 6 mm across, dark brown. Capsule spherical, *c.* 6 mm across. *South Africa (S Cape Province).* G1.

Rare in cultivation.

60. E. bubalina Boissier. Illustration: White et al., The succulent Euphorbieae (southern Africa) **1**: 252 (1941); *The Euphorbia Journal* **2**: 44, 45 (1984); Graf, Exotica, Series 4, edn 12, **1**: 1045 (1985); The new RHS dictionary of gardening **2**: 259 (1992).

Shrub *c.* 1 m with few branches. Branches 1–2 cm thick, with spiral rows of flattened tubercles. Leaves lanceolate, 7–10 cm, semi-persistent. Cymes loosely branching, in umbels with 2 or 3 rays on stalks to 12 cm, which harden and persist for some time. Ray-leaves rounded, often reddish. Cyathia *c.* 5 mm across, yellow. Capsule almost spherical, *c.* 8 mm across. *South Africa (Cape Province east to Natal).* G1.

61. E. clava Jacquin. Illustration: White et al., The succulent Euphorbieae (southern Africa) **1**: 256 (1941); *The Euphorbia Journal* **1**: 64 (1983) & **2**: 40 (1984); Graf, Exotica, Series 4, edn 12, **1**: 1049 (1985).

Shrub *c.* 1 m with few branches. Main stem *c.* 5 cm thick, branches *c.* 2.5 cm thick, with spiral rows of shallow tubercles. Leaves lanceolate, *c.* 13 cm. Cyathia *c.* 8 mm across, green, supported by 3 narrow, rounded, green bracts, *c.* 1 cm across, solitary on stalks 10–15 cm, with several tiny leaves, withering and persisting for some time. Capsule spherical, *c.* 8 mm across. *South Africa (SE Cape Province).* G1.

62. E. clandestina Jacquin. Illustration: White et al., The succulent Euphorbieae (southern Africa) **1**: 243 (1941); Jacobsen, Handbook of succulent plants **1**: 418 (1960); Jacobsen, Lexicon of succulent plants, pl. 68 (1974); *The Euphorbia Journal* **1**: 63 (1983); **2**: 43 (1984).

Similar to *E. clava* but branches to 5 cm thick, tubercles more prominent, cyathia at the bases of the tubercles on extremely short stalks that are not persistent. *South Africa (Cape Province).* G1.

63. E. bupleurifolia Jacquin. Illustration: White et al., The succulent Euphorbieae (southern Africa) **1**: 231 (1941) Jacobsen, Handbook of succulent plants **1**: 415 (1960); Jacobsen, Lexicon of succulent plants, pl. 67 (1974); The new RHS dictionary of gardening **2**: 266 (1992).

Stem to 15 cm, branching only occasionally, 5–8 cm thick, densely covered with spirally arranged, dark brown, shiny tubercles. Leaves in a tuft from the stem-apex, lanceolate, 1–1.5 cm. Cyathia *c.* 7 mm across, supported by a pair of rounded bracts, *c.* 1 cm across, on stalks 1–5 cm, unisexual, males and females on separate plants. Capsule spherical, *c.* 1 cm across, hairy. *South Africa (Cape Province to Natal).* G1.

Easily cultivated and, since its introduction in 1791, a favourite with growers of succulents.

64. E. monteiri J.D. Hooker. Illustration: White et al., The succulent Euphorbieae (southern Africa) **1**: 266 (1941); *The Euphorbia Journal* **1**: 95 (1983) & **7**: 115, 152 (1991); Graf, Exotica, Series 4, edn 12, 1046, 1057 (1985).

Stem usually unbranched, 30–60 × 5–10 cm, with prominent tubercles spirally arranged. Leaves in tufts from the stem-apex, lanceolate, to 15 × 3 cm. Cymes branching, 3–5 in an umbel on leafy, semi-persistent stalks 15–30 cm, with ray-leaves broadly ovate, 2–6 × 2–3 cm. Cyathia *c.* 1 cm across, including the yellow glands, which are fringed with 3–6 long processes. Capsule 3-lobed, *c.* 1 cm across. *S Angola, Namibia, Botswana, South Africa (N Cape Province).* G1.

Uncommon in cultivation.

65. E. caput-medusae Linnaeus. Illustration: White et al., The succulent Euphorbieae (southern Africa) **1**: 350 (1941); Jacobsen, Lexicon of succulent plants, pl. 68 (1974); *The Euphorbia Journal* **4**: 78 (1987); The new RHS dictionary of gardening **2**: 266 (1992).

Main stem very short, to 20 cm thick, with numerous radiating branches *c.* 90 × 2 cm, curving upwards, densely covered with very prominent spiralled tubercles. Leaves tiny, short-lived. Cyathia *c.* 1.5 cm across including the green glands, which bear 3–6 low white processes branched at the tips on their outer margins. Capsule almost spherical, *c.* 1 cm across. *South Africa (Cape Peninsula).* G1.

66. E. bergeri N.E. Brown. Illustration: White et al., The succulent Euphorbieae (southern Africa) **1**: 360 (1941); Jacobsen, Handbook of succulent plants **1**: 471 (1960); *The Euphorbia Journal* **3**: 89 (1985); Graf, Exotica, Series 4, edn 12, **1**: 1021 (1985).

Like *E. caput-medusae* but branches shorter and more slender, cyathia smaller and glands with 3–7 short, greenish white processes. *Known only in cultivation.* G1.

May be a hybrid with *E. caput-medusae* as one parent.

67. E. muirii N.E. Brown. Illustration: White et al., The succulent Euphorbieae (southern Africa) **1**: 342, 343 (1941); *The Euphorbia Journal* **1**: 96 (1983) & **2**: 69 (1984).

Like *E. caput-medusae* but branches to 18 × 1.5 cm, trailing and re-branching to form similar plant-bodies to the parent, cyathial glands yellow with 4–6 linear processes. *South Africa (Cape Province)*. G1.

68. E. tuberculatoides N.E. Brown. Illustration: White et al., The succulent Euphorbieae (southern Africa) **1**: 376 (1941); *The Euphorbia Journal* **7**: 90 (1985).

Like *E. caput-medusae* but with thicker branches forming clumps to 45 cm, cyathia on persistent stalks 1–3 cm, glands with 3–5 short, creamy-yellow processes on the outer margins, branched at the tips and recurved. *South Africa (Cape Province)*. G1.

Uncommon in cultivation and rare in the wild.

69. E. esculenta Marloth. Illustration: White et al., The succulent Euphorbieae (southern Africa) **1**: 380 (1941); Jacobsen, Handbook of succulent plants **1**: 429 (1960); Jacobsen, Lexicon of succulent plants, pl. 70 (1974); *The Euphorbia Journal* **3**: 88 (1985).

Like *E. caput-medusae* but branches to 20 cm, cyathia on stalks to 1 cm, with small green glands and filled with a mass of white, woolly bracteoles. *South Africa (Cape Province)*. G1.

70. E. inermis Miller. Illustration: White et al., The succulent Euphorbieae (southern Africa) **1**: 389 (1941); Jacobsen, Handbook of succulent plants **1**: 441 (1960); *The Euphorbia Journal* **3**: 89 (1985); Graf, Exotica, Series 4, edn 12, **1**: 1050 (1985).

Similar to *E. caput-medusae* but branches more numerous, to 35 × 1.5 cm, with some persistent inflorescence-stalks to 5 mm, cyathia filled with a mass of white, woolly bracteoles, glands irregular in shape, usually yellow and deeply 2-lobed with the lobes ending in white processes. *South Africa (Cape Province)*. G1.

In cultivation in a pot the branches elongate and the plant loses its compact shape. It does better planted out.

71. E. superans Nel (*E. supernans* misspelt). Illustration: *The Euphorbia Journal* **2**: 68 (1984).

Very similar to *E. inermis* but glands entirely yellow and processes minute. *South Africa (E Cape Province)*. G1.

72. E. pugniformis Boissier. Figure 128 (4), p. 492. Illustration: White et al., The succulent Euphorbieae (southern Africa) **1**: 335 (1941); Jacobsen, Handbook of succulent plants **1**: 468 (1960); Jacobsen, Lexicon of succulent plants, pl. 79 (1974);

Graf, Exotica, Series 4, edn 12, 1047 (1985).

Main stem extremely short at the apex of a tuberous root 5–8 cm thick, with very many, densely crowded, radiating branches, and spirally arranged tubercles at the plant centre. Outer branches lengthening to *c.* 3 cm, *c.* 8 mm thick, with small tubercles spirally arranged. Leaves oblong, *c.* 5 mm. Cyathia produced from the short central branches, *c.* 5 mm across, with yellow glands, which are slightly notched on the outer margin, sweet-scented. Capsule 3-lobed, *c.* 4 mm across, usually slightly hairy. *South Africa (Cape Province)*. G1.

73. E. flanaganii N.E. Brown. Illustration: *The Euphorbia Journal* **1**: 77 (1983) & **2**: 68 (1984); Graf, Exotica, Series 4, edn 12, **1**: 1059 (1985).

Very similar to *E. pugniformis* but the cyathia very numerous from the tubercle-like branches at the centre of the plant body. *South Africa (Cape Province)*. G1.

74. E. woodii N.E. Brown. Illustration: White et al., The succulent Euphorbieae (southern Africa) **1**: 326 (1941); Jacobsen, Handbook of succulent plants **1**: 488 (1960); Rowley, Illustrated encyclopedia of succulents, 226 (1978); *The Euphorbia Journal* **3**: 88 (1985); The new RHS dictionary of gardening **2**: 2266 (1992).

Similar to *E. pugniformis* but larger, with branches to 20 × 1–1.5 cm, glands obviously notched. *South Africa (Natal)*. G1.

75. E. gorgonis Berger. Illustration: White et al., The succulent Euphorbieae (southern Africa) **1**: 340 (1941); *The Euphorbia Journal* **2**: 68 (1984) & **3**: 90 (1985); The new RHS dictionary of gardening **2**: 266 (1992).

Main stem extremely short at the apex of a tuberous root 5–10 cm thick, with many crowded, radiating branches, and prominent pointed tubercles at the centre. Branches 8–25 × 5–10 mm, with spirally arranged, rounded tubercles. Cyathia *c.* 5 mm across, red. Capsule 3-lobed, *c.* 4 mm across, slightly hairy. *South Africa (Cape Province)*. G1.

76. E. schoenlandii Pax (*E. fasciculata* Thunberg). Figure 128(6), p. 492. Illustration: White et al., The succulent Euphorbieae (southern Africa) **1**: 289 (1941); Jacobsen, Handbook of succulent plants **1**: 474 (1960); *The Euphorbia Journal* **1**: 110 (1983) & **3**: 110 (1985).

Stem unbranched, 30–50 × 10–15 cm, with large, prominent, spirally arranged tubercles. Leaves tiny, short-lived. Inflorescence-stalks often sterile, persistent, rigid and curved upwards, 2.5–5 cm, with a few hard projections from the tiny fallen bracts. Cymes 1–3 in umbels with very short rays. Cyathia *c.* 8 mm across, including the greenish glands, which are fringed with 3–8 long processes. Capsule 3-lobed, *c.* 6 mm across. *South Africa (Cape Province)*. G1.

Once common in the wild, but now endangered owing to the destruction of its natural habitat.

77. E. multiceps Berger. Illustration: White et al., The succulent Euphorbieae (southern Africa) **1**: 460 (1941); *The Euphorbia Journal* **2**: 124 (1984); Graf, Exotica, Series 4, edn 12, **1**: 1047 (1985).

Like *E. schoenlandii* but main stem thicker, densely covered with spreading branches to 7 × 3 cm, progressively shorter towards the top, forming a pyramidal crown. *South Africa (Cape Province)*. G1.

Rare in cultivation and endangered in the wild.

78. E. globosa (Haworth) Sims. Figure 128(1), p. 492. Illustration: White et al., The succulent Euphorbieae (southern Africa) **2**: 495 (1941); Jacobsen, Handbook of succulent plants **1**: 434 (1960); Jacobsen, Lexicon of succulent plants, pl. 71 (1974); The new RHS dictionary of gardening **2**: 266 (1992).

Main stem extremely short at the apex of a tuberous root to 3 cm thick, with numerous spherical branches which re-branch, producing a cushion of marble-like branches 1–2 cm across, all with large, flattened tubercles, Flowering branches narrower, to 6 cm, with solitary cyathia on stalks *c.* 2.5 mm, or with branching cymes on stalks 2–8 cm. Cyathia *c.* 2 cm across, glands green and divided into 3 or 4 long, finger-like processes and with minute, white-margined pits on the surface. Capsule produced from the solitary cyathia, almost spherical, *c.* 8 mm across. *South Africa (Cape Province)*. G1.

79. E. ornithopus Jacquin. Illustration: White et al., The succulent Euphorbieae (southern Africa) **2**: 509 (1941); Graf, Exotica, Series 4, edn 12, **1**: 1046 (1985); *The Euphorbia Journal* **3**: 76 (1985) & **8**: 80 (1992).

Main stem extremely short at the apex of a cylindric, rhizomatous, tuberous root to 5 cm thick, branching at ground level. Branches 1–3 cm × c. 8 mm, with large, flattened tubercles. Leaves to 5 mm, short-lived. Cyathia solitary or in branching cymes on stalks 1–10 cm, bearing a few tiny bracts. Cyathia c. 1 cm across, glands with 3 or 4 finger-like processes and the surface covered with tiny, white-margined pits. Capsule almost spherical, c. 8 mm across. *South Africa (Cape Province).* G1.

80. E. tridentata Lamarck. Illustration: White et al., The succulent Euphorbieae (southern Africa) **2**: 502 (1941); Graf, Exotica, Series 4, edn 12, **1**: 1046 (1985); *The Euphorbia Journal* **3**: 76 (1985).

Main stem extremely short at the apex of a tuberous, rhizomatous root, with numerous branches of irregular size from ground level. Branches spreading or erect when flowering, to 15 × 1 cm, with slightly raised tubercles. Cyathia at the tips of the branches on stalks to 4 mm, bearing a few tiny bracts. Cyathia c. 1.5 cm across, glands with 3 or 4 white, wrinkled, finger-like processes. Capsule almost spherical, c. 8 mm across. *South Africa (Cape Province).* G1.

81. E. wilmaniae Marloth. Illustration: White et al., The succulent Euphorbieae (southern Africa) **2**: 514 (1941); *The Euphorbia Journal* **5**: 149 (1988).

Like *E. tridentata* but branches thinner with prominent tubercles, lengthening considerably underground or in shade, much shorter in full sun, producing large mats at ground level. *South Africa (Cape Province).* G1.

In cultivation in Europe the branches seldom remain short.

82. E. polycephala Marloth. Illustration: White et al., The succulent Euphorbieae (southern Africa) **2**: 525 (1941); *The Euphorbia Journal* **5**: 39 (1988).

Main stem extremely short at the apex of a large, tuberous root, densely branching to form a compact cushion to 1 m across at ground level or just above. Branches very short, c. 1.5 cm thick, with flattened tubercles. Cyathia c. 1 cm across, glands with 3 white-margined, finger-like processes. *South Africa (Cape Province).* G1.

Rare in cultivation and endangered in the wild.

83. E. susannae Marloth. Illustration: White et al., The succulent Euphorbieae (southern Africa) **2**: 529 (1941); Jacobsen, Handbook of succulent plants **1**: 479

(1960); *The Euphorbia Journal* **1**: 114 (1983).

Main stem c. 3 cm thick from a deep taproot, branching at the base and below ground level in the wild. Branches to 8 cm, with 12–16 longitudinal ridges of very prominent tubercles. Cyathia unisexual, on very short stalks, c. 3 mm across, glands greenish yellow; male and female flowers on separate plants. Capsule deeply 3-lobed, c. 4 mm across, purplish. *South Africa (W Cape Province).* G1.

A favourite for cultivation because of its compact growth and easy propagation.

84. E. tubiglans Marloth. Illustration: White et al., The succulent Euphorbieae (southern Africa) **2**: 542 (1941); Jacobsen, Handbook of succulent plants **1**: 408, 483 (1960); *The Euphorbia Journal* **2**: 139 (1984).

Main stem very short at the apex of a tuberous root to 5 cm thick, producing 2–5 branches to 8 × 2 cm, each on a stalk-like neck c. 1 cm long, and with 5 longitudinal ridges of very small tubercles. Cyathia on stalks c. 2 cm, 3 mm across, reddish, supported by 3 small bracts; unisexual, male and female cyathia on separate plants. Capsule 3-lobed, c. 3 mm across. *South Africa (Cape Province).* G1.

85. E. jansenvillensis Nel. Illustration: White et al., The succulent Euphorbieae (southern Africa) **2**: 545 (1941); *The Euphorbia Journal* **8**: 94 (1992).

Similar to *E. tubiglans* but branches 20–30 cm, not stalked, bracts below the cyathia larger (to 4 mm). *South Africa (Cape Province).* G1.

86. E. submammillaris (Berger) Berger. Illustration: White et al., The succulent Euphorbieae (southern Africa) **2**: 587 (1941); Graf, Exotica, Series 4, edn 12, **1**: 1024 (1985).

Stems densely branching, forming a tight clump 10–20 cm. Branches 1.5–2.5 cm thick, with 7–10 angles, deeply grooved. Inflorescence-stalks (sterile) persistent, spiny, 1–2 cm. Cyathia c. 5 mm across, solitary on stalks c. 4 mm, bearing a few, tiny, purple bracts, unisexual male and female cyathia on separate plants. *Known only in cultivation.* G1.

Described from a cultivated plant of unknown origin and possibly a hybrid.

87. E. fimbriata Scopoli. Illustration: White et al., The succulent Euphorbieae (southern Africa) **2**: 589 (1941); Jacobsen, Handbook of succulent plants **1**: 431

(1960); Jacobsen, Lexicon of succulent plants, pl. 70 (1974); *The Euphorbia Journal* **3**: 111 (1985).

Main stem short, with numerous branches from the base to form a clump to 30 cm, or branches lengthening in shaded positions to nearly 1 m. Branches 2–4 cm thick, with 7–12 longitudinal rows of tubercles forming prominent angles. Inflorescence-stalks (sterile) persistent, spiny, c. 1.5 cm, bearing a few tiny bracts. Cyathia c. 6 mm across, yellow to purplish, unisexual male and female cyathia on separate plants. Capsule 3-lobed, c. 5 mm across. *South Africa (Cape Province).* G1.

88. E. mammillaris Linnaeus. Illustration: White et al., The succulent Euphorbieae (southern Africa) **2**: 596 (1941); Jacobsen, Handbook of succulent plants **1**: 449 (1960); Jacobsen, Lexicon of succulent plants, pl. 74 (1974); *The Euphorbia Journal* **1**: 91 (1983).

Stems branching from the base to form compact clusters, 20 cm. Branches erect, 4–6 cm thick, with 7–17 angles and grooves. Inflorescence-stalks (sterile) persistent, spiny, to 1 cm. Cyathia c. 5 mm across, solitary on stalks to 2 mm, yellowish to purple, unisexual male and female cyathia on separate plants. Capsule 3-lobed, c. 6 mm across. *South Africa (Cape Province).* G1.

89. E. pulvinata Marloth. Illustration: White et al., The succulent Euphorbieae (southern Africa) **2**: 603 (1941); Jacobsen, Handbook of succulent plants **1**: 469 (1960); Jacobsen, Lexicon of succulent plants, pl. 79 (1974); *The Euphorbia Journal* **2**: 130 (1984).

Stems densely branching to form a compact cushion to 30 cm × 1 m across. Branches c. 3 cm thick, with 7–10 angles and grooves. Inflorescence-stalks (sterile) c. 1 cm. Cyathia on very short stalks, c. 5 mm across, purplish, unisexual male and female cyathia on separate plants. Capsule 3-lobed, c. 4.5 mm across. *South Africa (Cape Province, Transvaal).* G1.

90. E. aggregata Berger. Illustration: White et al., The succulent Euphorbieae (southern Africa) **2**: 612 (1941); *The Euphorbia Journal* **3**: 100 (1985).

Stems densely branching, forming large, compact cushions c. 10 cm. Branches 2–3 cm thick, with 8 or 9 angles and grooves. Inflorescence-stalks (sterile) persistent, spiny, 6–8 mm. Cyathia c. 3 mm across, purplish green, unisexual male and female cyathia on separate plants. Capsule spherical, *c.* 6 mm across. *South Africa (Cape Province).* G1.

Var. **alternicolor** (N.E. Brown) White, Dyer & Sloane has transverse bands of whitish green on the branches.

91. E. ferox Marloth. Illustration: White et al., The succulent Euphorbieae (southern Africa) **2**: 619 (1941); Jacobsen, Handbook of succulent plants **1**: 430 (1960); *The Euphorbia Journal* **3**: 110 (1985); Brickell (ed.), RHS A–Z encyclopedia of garden plants, 439 (2003).

Stems densely branching, forming large compact cushions *c.* 15 × 100 cm. Branches 3–4.5 cm thick, with 9–12 prominent angles and grooves. Inflorescence-stalks persistent, stout and spiny, 1.5–3 cm, solitary. Cyathia *c.* 3 mm across, purplish, unisexual male and female cyathia on separate plants. Capsule spherical, *c.* 6 mm across. *South Africa (Cape Province).* G1.

92. E. pentagona Haworth. Illustration: White et al., The succulent Euphorbieae (southern Africa) **2**: 626 (1941); *The Euphorbia Journal* **2**: 23 (1984).

Shrub to 3 m, with compact branching. Branches 1–4 cm thick, with 5 or 6 prominent angles. Sterile inflorescence-stalks persistent, spiny, 1–1.5 cm. Cyathia *c.* 4 mm across, purple, in unbranched cymes on short stalks, unisexual male and female cyathia on separate plants. Capsule 3-lobed, *c.* 5 mm across. *South Africa (Cape Province).* G1.

93. E. heptagona Linnaeus (*E. morinii* Berger; *E. enopla* misapplied). Illustration: White et al., The succulent Euphorbieae (southern Africa) **2**: 646 (1941); Jacobsen, Handbook of succulent plants **1**: 439 (1960); Graf, Exotica, Series 4, edn 12, 1025 (1985).

Shrub to 1 m, with erect branching. Branches 3–4 cm thick, with usually 7 or 8 angles. Sterile inflorescence-stalks persistent, spiny, to 3 cm. Cyathia *c.* 4 mm across, green, on stalks 1.5 cm, unisexual, male and female flowers on separate plants. *South Africa (Cape Province).* G1.

94. E. cereiformis Linnaeus. Illustration: White et al., The succulent Euphorbieae (southern Africa) **2**: 657, 659 (1941); Graf, Exotica, Series 4, edn 12, **1**: 1025 (1985).

Shrub to 1 m, with loose branching. Branches 2.5–5 cm thick, with usually 11 pronounced angles and grooves.

Inflorescence-stalks hardening into spines *c.* 1 cm. Male and female cyathia on separate plants. *Known only in cultivation.* G1.

An untidy plant. There is much controversy as to whether the plant cultivated under this name is the one originally described by Linnaeus.

95. E. inconstantia Dyer. Figure 128(3), p. 492. Illustration: White et al., The succulent Euphorbieae (southern Africa) **2**: 667 (1941); Jacobsen, Handbook of succulent plants **1**: 441 (1960).

Shrub *c.* 1 m, branching from the base. Branches 4–8 cm thick, with 7–10 angles and grooves. Sterile inflorescence-stalks persisting, spiny, 2 or 3 together, *c.* 1 cm. Cyathia *c.* 4 mm across, on stalks 1–2 cm, with several small bracts. Capsule 3-lobed, *c.* 6 mm across, purple, unisexual male and female cyathia on separate plants. *South Africa (Cape Province).* G1.

96. E. polygona Haworth. Illustration: White et al., The succulent Euphorbieae (southern Africa) **2**: 671, 672 (1941); Graf, Exotica, Series 4, edn 12, **1**: 1029 (1985).

Shrub to 1.5 m, branching from the base. Branches 7–10 cm thick, with 7 angles when young, later developing 12–20 angles, very deeply grooved. Sterile inflorescence-stalks persistent and spiny, 2 or 3 together, 5–10 mm. Cyathia *c.* 7 mm across, dark purple, on stalks *c.* 4 mm; unisexual, males and females on separate plants. Capsule spherical, *c.* 6 mm across. *South Africa (Cape Province).* G1.

97. E. anoplia Stapf. Illustration: White et al., The succulent Euphorbieae (southern Africa) **2**: 661, 662 (1941); *The Euphorbia Journal* **3**: 101 (1985).

Like *E. polygona* but branches *c.* 5 cm thick with 7–9 angles and deep grooves, spineless, cyathia on very short stalks, dark red, producing only male flowers. *Origin unknown.* G1.

Described from a cultivated plant of unknown origin and almost certainly a hybrid, probably with *E. polygona* as one parent.

98. E. horrida Boissier. Illustration: White et al., The succulent Euphorbieae (southern Africa) **2**: 678 (1941); Jacobsen, Lexicon of succulent plants, pl. 72 (1974); Rowley, Illustrated encyclopedia of succulents, 226 (1978); *The Euphorbia Journal* **1**: 85 (1983).

Shrub to 1 m, branching from the base. Branches 10–15 cm thick, with *c.* 14 angles, deeply grooved. Sterile inflorescence-stalks persistent, spiny, grouped 2–5 together, 1–4 cm, very numerous. Cyathia *c.* 4 mm across, green, on stalks 5–8 mm, unisexual, males and females on separate plants. *South Africa (Cape Province).* G1.

A favourite for cultivation, with several colour variants, ranging from green to grey, often described as varieties.

99. E. stellispina Haworth. Illustration: White et al., The succulent Euphorbieae (southern Africa) **2**: 712, 713 (1941); Jacobsen, Handbook of succulent plants **1**: 476 (1960); Jacobsen, Lexicon of succulent plants, pl. 80 (1974).

Shrub to 50 cm, branching densely from the base. Branches 3–7.5 cm thick, with 10–16 angles, deeply grooved. Sterile inflorescence-stalks persistent, spiny, 4–10 mm, with 3–5 spiny, spreading branches from the tips. Cyathia *c.* 4 mm across on stalks 2–10 mm (stalks sometimes branching at the tips); unisexual, males and females on separate plants. *South Africa (Cape Province).* G1.

100. E. obesa J.D. Hooker. Figure 128(5), p. 492. Illustration: White et al., The succulent Euphorbieae (southern Africa) **2**: 549 (1941); Jacobsen, Handbook of succulent plants **1**: 549 (1960); Jacobsen, Lexicon of succulent plants, pl. 77 (1974); Rowley, Illustrated encyclopedia of succulents, 226 (1978); The new RHS dictionary of gardening **2**: 266 (1992).

Stems unbranched, to 20 × 9 cm, with usually 8 shallow longitudinal ridges formed by minute tubercles, and with purple markings in transverse bands. Cyathia solitary or in unbranched cymes on very short stalks, *c.* 3 mm across, red, unisexual, males and females on separate plants. Capsule 3-lobed, *c.* 7 mm across. *South Africa (Cape Province).* G1.

A favourite plant in cultivation, easily propagated by seed and easily hybridised with other species; it is, nevertheless, very rare in the wild.

101. E. meloformis Alton. Illustration: White et al., The succulent Euphorbieae (southern Africa) **2**: 565 (1941); Jacobsen, Handbook of succulent plants **1**: 451 (1960); Jacobsen, Lexicon of succulent plants, pl. 74 (1974); *The Euphorbia Journal* **5**: 96 (1988); The new RHS dictionary of gardening **2**: 266 (1992).

Stems usually unbranched, 10–12 × 9–10 cm, with minute tubercles in 8 prominent ridges, and marked with transverse purple bands. Cyathia *c.* 4 mm across in usually branching cymes on stalks 1–6 cm, which occasionally harden and persist; unisexual, males and females on separate plants. Capsule 3-lobed, *c.* 6 mm across. *South Africa (Cape Province).* G1.

102. E. valida N.E. Brown. Illustration: White et al., The succulent Euphorbieae (southern Africa) **2**: 573 (1941); Jacobsen, Handbook of succulent plants **1**: 485 (1960); Jacobsen, Lexicon of succulent plants, pl. 81 (1974); *The Euphorbia Journal* **1**: 40, 118 (1983).

Stem unbranched, to 30 × 18 cm, with minute tubercles in 8 prominent ridges and marked with purple transverse bands. Cyathia *c.* 4 mm across, green, in branching cymes to 4 cm, stalks to 2 cm, the stalks and branches hardening and persisting after flowering. Cyathia unisexual, males and females on separate plants. *South Africa (Cape Province).* G1.

103. E. milii Des Moulins (*E. bojeri* J.D. Hooker). Illustration: Jacobsen, Lexicon of succulent plants, pl. 75 (1974); Rowley, Illustrated encyclopedia of succulents, 225 (1978); Graf, Exotica, Series 4, edn 12, 1026 (1985); The new RHS dictionary of gardening **2**: 259 (1992).

Shrub to 1.5 m, branches dark reddish brown, *c.* 8 mm thick, with *c.* 5 longitudinal ridges bearing spines to 1 cm on each side of the leaf-scars. Leaves obovate, *c.* 4 × 1.5 cm. Cymes branching several times, with slender stalks. Cyathia small, cupped by a pair of conspicuous, brilliant red bracts *c.* 8 mm across. Capsule 3-lobed, *c.* 4 mm across, dark red. *Madagascar.* G1.

An extremely variable species in which many varieties have been described, almost all from cultivated material of unknown wild origin. The best known is var **splendens** (Hooker) Urseh & Leandri, a larger plant to 2 m, with larger leaves, longer spines and bigger cyathial bracts. Var **hislopii** (N.E. Brown) Leandri is also sometimes cultivated, with branches to 3 cm thick, leaves to 18 × 5 cm. Other varieties occur within this wide range, many including variants with yellow bracts.

104. E. beharensis Leandri. Illustration: *The Euphorbia Journal* **1**: 58 (1983).

Like *E. milii* but branches very slender, 3–5 mm thick, spines 1–1.5 cm. *Madagascar.* G1.

Only likely to be found in specialist collections.

105. E. lophogona Lamarck. Figure 129 (6), p. 498. Illustration: Jacobsen, Handbook of succulent plants **1**: 448 (1960); Jacobsen, Lexicon of succulent plants, pl. 174 (1974); Graf, Tropica, edn 1, 420 (1978); *The Euphorbia Journal* **4**: 18 (1987).

Solitary stems to 50 cm, sparsely branched. Branches with 4 or 5 angles marked by longitudinal rows of closely packed, stiff, fringed stipules on either side of large leaf-scars. Leaves persistent, obovate, to 12 × 5 cm, glossy, bright green with paler veins, stalks pinkish *c.* 2 cm. Cymes on stalks 4–7 cm at the stem-apex, branching 2–4 times. Cyathia *c.* 5 mm across, subtended by a pair of conspicuous, white, rounded bracts *c.* 8 mm across. *S Madagascar.* G1.

106. E. viguieri Denis. Illustration: Jacobsen, Handbook of succulent plants **1**: 486 (1960); Jacobsen, Lexicon of succulent plants, pl. 82 (1974); *The Euphorbia Journal* **1**: 119 (1983), **3**: 18 (1985) & **4**: 15 (1987).

Like *E. lophogona* but *c.* 1 m, leaves often pinkish at the base, cyathia enveloped by a pair of bright red, clasping bracts giving a tubular appearance. *W Madagascar.* G1.

107. E. leuconeura Boissier. Illustration: Jacobsen, Handbook of succulent plants **1**: 447 (1960); Rowley, Illustrated encyclopedia of succulents, 229 (1978); *The Euphorbia Journal* **1**: 88 (1983) & **4**: 19 (1987).

Similar in habit to *E. lophogona* but stem with 4 angles, stipules bristle-like, cymes congested on short stalks from the leaf-axils, with cyathia enclosed by a pair of white, clasping bracts. *Origin unknown.* G1.

Known only in cultivation and possibly of hybrid origin.

108. E. millotii Ursch & Leandri. Illustration: *The Euphorbia Journal* **3**: 37 (1985).

Like *E. lophogona* but more branched, branches cylindric, without the ridges of spiny stipules, leaves in tufts at the stem-apex, lanceolate to 6 cm, cymes on shorter stalks, cyathia pendent and enveloped by a cup-shaped pair of reddish green bracts. *N Madagascar.* G1.

Rare in cultivation.

109. E. neriifolia Linnaeus. Illustration: Jacobsen, Handbook of succulent plants **1**: 457 (1960); Jacobsen, Lexicon of succulent plants, pl. 76 (1974); *The Euphorbia Journal* **5**: 140 (1988); The new RHS dictionary of gardening **2**: 263 (1992).

Shrub or tree to 6 m. Branches 1.5–2 cm thick, with prominent tubercles forming 4 or 5 obscure angles, but becoming rounded. Spine-shields small, rounded, with 2 stout spines to 1 cm. Leaves persistent, somewhat fleshy, obovate-oblong, 15–30 × *c.* 4 cm. Cymes branching once, with reddish cyathia each *c.* 5 mm across on stalks 5–10 mm. Capsule on slender stalks, deeply 3-lobed, *c.* 1 cm across. Seeds ovoid, smooth. *S India, Burma, Malaysia.* G1.

110. E. nivulia Buchanan-Hamilton. Illustration: Jacobsen, Lexicon of succulent plants, pl. 76 (1974); *The Euphorbia Journal* **1**: 97 (1983) & **3**: 134 (1985); Graf, Exotica, Series 4, edn 12, **1**: 1048 (1985).

Very similar to *E. neriifolia* but with cylindric branches and spirally arranged tubercles. *N India.* G1.

111. E. undulatifolia Janse. Illustration: Jacobsen, Handbook of succulent plants **1**: 484 (1960); *The Euphorbia Journal* **3**: 134 (1985); Graf, Exotica, Series 4, edn 12, **1**: 1058 (1985).

Like *E. neriifolia* but branches with obvious angles and leaves with wavy margins. *Origin unknown.*

Probably of hybrid origin, with *E. neriifolia* as one of the parents.

112. E. lactea Haworth. Illustration: Jacobsen, Handbook of succulent plants **1**: 445 (1960); Jacobsen, Lexicon of succulent plants, pl. 74 (1974); Graf, Tropica, edn 1, 417, 421, 422 (1978); *The Euphorbia Journal* **2**: 118 (1984).

Much-branched shrubby tree to 3 m. Branches with 3 wings, 3–5 cm wide, with a distinct, pale, longitudinal stripe between the wings. Spine-shields small, 2–3 cm apart on the wavy margins of the wings, with short, paired spines. Leaves *c.* 2 cm. Cyathia small, yellow. *Sri Lanka, Réunion, Mauritius.* G1.

113. E. antiquorum Linnaeus. Illustration: Jacobsen, Handbook of succulent plants **1**: 407 (1954); *The Euphorbia Journal* **1**: 54 (1983) & **7**: 98 (1991); The new RHS dictionary of gardening **2**: 263 (1992).

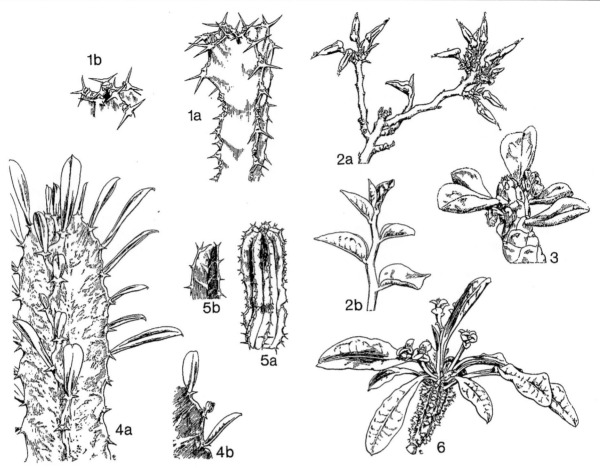

Figure 129. Diagnostic details of Euphorbiaceae species (continued). 1, *Euphorbia coerulescens* (a, branch (× 0.5); b, branch apex, showing spinescence (× 0.75)). 2, *Pedilanthus tithymaloides* (× 0.5) (a, flowering branch; b, branch in leaf). 3, *Monadenium lugardae* fruiting branch (× 0.5). 4, *Euphorbia ingens* (× 0.5) (a, apex of seedling with semi-persistent leaves; b, stem margin, showing spinescence). 5, *E. echinus* (a, branch (× 0.5); b, branch angles, showing spinescence (× 0.75)). 6, *E. lophogona* flowering branch (× 0.5).

Similar to *E. lactea* but to 10 m, with stouter, grey-green branches to 10 cm wide and with 3–5 wavy angles. Cyathia on stalks 1–3 cm. *India, Sri Lanka, Malaysia.* G1.

114. E. royleana Boissier. Illustration: Jacobsen, Handbook of succulent plants **1**: 472 (1960); *The Euphorbia Journal* **1**: 110 (1983); Graf, Exotica, Series 4, edn 12, **1**: 1056 (1985).

Similar to *E. lactea* but branches to 7 cm thick, with 5–7 angles forming straight ribs. *NE India, Burma.* G1.

115. E. barnhartii Croizat (*E. trigona* Roxburgh). Illustration: Jacobsen, Handbook of succulent plants **1**: 411 (1960); Jacobsen, Lexicon of succulent plants, pl. 2, 67 (1974); *The Euphorbia Journal* **3**: 102 (1985).

Like *E. lactea* but branches constricted into conical segments and leaves obovate, 3–5 × 1–2 cm. *NE India to SE Asia.* G1.

116. E. trigona Haworth (*E. hermentiana* Lemaire). Illustration: Jacobsen, Handbook of succulent plants **1**: 482 (1960); *The Euphorbia Journal* **1**: 117 (1983) & **7**: 14 (1991).

Shrub or small tree with closely packed, erect branches. Branches with 3 wings, 4–5 cm wide, dark green with conspicuous paler marbling. Spine-shields small, *c.* 2 cm apart on the wavy margins of the wings, with short, paired spines. Leaves obovate, 1–6 cm, semi-persistent. *Wild origin unknown, but presumed to be southern India.* G1.

Popular as a pot-plant in hotels, etc., as it thrives well in shady places. It has never been known to flower.

117. E. abyssinica Gmelin (*E. acrurensis* N.E. Brown; *E. candelabrum* Kotschy var. *erythraeae* Berger; *E. erythraeae* (Berger) N. E. Brown). Illustration: Jacobsen, Handbook of succulent plants **1**: 404 (1960); *The Euphorbia Journal* **1**: 50 (1983); Graf, Exotica, Series 4, edn 12, **1**: 1021 (1985); The new RHS dictionary of gardening **2**: 263 (1992).

Broad-crowned tree 5–10 m. Branches with 4–8 angles, constricted into rounded segments 4–12 cm wide. Spine-shields closely set, triangular, *c.* 7 mm, with 2 stout spines 2–10 mm. Leaves ovate-lanceolate, 2–5 cm, on young growth only. Cyathia *c.* 8 mm across, yellow. Capsule spherical, 2–2.5 cm across, fleshy and bright red but hardened, brown and 3-lobed before dehiscence. Seeds spherical, smooth. *NE Sudan (Red Sea hills), Ethiopia (Eritrea).* H5–G1.

118. E. ammak Schweinfurth. Illustration: Jacobsen, Handbook of succulent plants **1**: 405 (1960); *The Euphorbia Journal* **1**: 53 (1983) & **5**: 115 (1988).

Similar to *E. abyssinica* but branches with 4 or 5 angles, segments longer and less constricted, spines widely spreading, capsules 3-lobed, *c.* 1 cm across. *S Arabian peninsula, Yemen.* H5–G1.

119. E. ingens Boissier. Figure 129(4), p. 498. Illustration: White et al., The succulent Euphorbieae (southern Africa) **2**: 924 (1941); Jacobsen, Handbook of succulent plants **1**: 442 (1960); Graf, Tropica, edn 1, 425 (1978); *The Euphorbia Journal* **3**: 70 (1985) & **7**: 65 (1991).

Densely branched tree *c.* 10 m, with a short trunk and a broad crown. Branches with 4 angles, 4–8 cm wide, constricted into segments 8–15 cm. Spine-shields triangular, *c.* 5 mm, with 2 small spines. Leaves lanceolate, 1–2 cm, on young growth only. Cyathia to 1 cm across, yellow. Capsule spherical, *c.* 1 cm across, fleshy but hardened and 3-lobed before dehiscence. Seeds ovoid, smooth. *South Africa (Transvaal), northwards through Zimbabwe, Mozambique, Malawi, Zambia & possibly further north.* G1.

The stems and branches of seedlings are sometimes attractively variegated. In cultivation the branch-angles often become thinly fleshy and more wing-like.

120. E. amplophylla Pax (*E. obovalifolia* misapplied). Illustration: Jacobsen, Handbook of succulent plants **1**: 530 (1960); *The Euphorbia Journal* **7**: 64 (1991).

Similar to *E. ingens* but branches with 3 thinly fleshy wings 5–17 cm wide, leaves 10–20 × 2–6 cm, persistent for some time on the young growth. *Ethiopia southwards through E Africa to northern Zambia & Malawi.* G1.

This is the largest species of *Euphorbia*, occurring in the wild in mist forest at high altitudes (above 1800 m). Since the publication of the Flora of Tropical Africa, in 1912, the name *E. obovalifolia* has been misapplied to this tree; *E. obovalifolia* Richard, as originally described, was in fact a seedling plant of *E. abyssinica*.

121. E. deightonii Croizat. Illustration: *The Euphorbia Journal* **1**: 71 (1983) & **8**: 34 (1992).

Like *E. ingens* but a large shrub, branches with 3 or 4 wings, *c.* 10 cm wide, cyathia *c.* 5 mm across, green, capsules 3-lobed, *c.* 1 cm across, dull red. *W Africa (Sierra Leone, Ghana, Nigeria).* G1.

122. E. canariensis Linnaeus. Illustration: Jacobsen, Handbook of succulent plants **1**: 416 (1960); Bramwell & Bramwell, Wild flowers of the Canary Isles, pl. 181 (1974); *The Euphorbia Journal* **4**: 32 (1987); The new RHS dictionary of gardening **2**: 263 (1992).

Shrub to 3 m, with dense, erect branching, forming large clumps. Branches with 4–6 angles, *c.* 4 cm wide. Spine-shields ovate, closely set, with short, paired spines. Cyathia *c.* 6 mm across, deep red. Capsule deeply 3-lobed, *c.* 8 mm across, on short stalks. *Canary Islands.* H5–G1.

Seedlings of this species are particularly attractive, with a dark bronze-green, slightly wrinkled surface; the young branches are bright green, later turning grey-green.

123. E. officinarum Linnaeus. Illustration: Jacobsen, Handbook of succulent plants **1**: 460, 461 (1960); Jacobsen, Lexicon of succulent plants, pl. 77 (1974).

Shrub *c.* 1 m, with loose branching. Branches rounded, with 6–10 angles (or ridges), *c.* 3 cm thick. Spine-shields joined along the angles, with short, paired spines *c.* 5 mm, *c.* 5 mm apart. Cyathia *c.* 4 mm across, red. *W coast of Morocco.* H5–G1.

Var. **beaumieriana** (Hooker & Cosson) Maire (*E. beaumieriana* Hooker & Cosson) is a more sturdy shrub to 2 m, branches with 9 or 10 angles, to 5 cm thick, and spines of variable length (5–20 mm). This and the typical var. **officinarum** are possibly identical with the following species, which is more popular in cultivation. All 3 occur along the same coastal areas of Morocco.

124. E. echinus Hooker & Cosson. Figure 129(5), p. 498. Illustration: Jacobsen, Handbook of succulent plants **1**: 427, 428 (1960); Jacobsen, Lexicon of succulent plants, pl. 70 (1974); *The Euphorbia Journal* **1**: 74 (1983).

Shrub to 1 m. Branches rounded, with 5–8 angles, 4–5 cm thick. Spine-shields joined along the ribs, pale buff, with slender, paired spines 4–10 mm, *c.* 6 mm apart. Cyathia *c.* 4 mm across, dull red. Capsule on slender curved stalks, 3-lobed,

c. 5 mm across. *W coast of Morocco.* H5–G1.

125. E. resinifera Berg. Illustration: Jacobsen, Handbook of succulent plants **1**: 471 (1960); Jacobsen, Lexicon of succulent plants, pl. 80 (1974); Rowley, Illustrated encyclopedia of succulents, 227 (1978); *The Euphorbia Journal* **2**: 133 (1984).

Densely branched shrub forming large compact cushions to 75 cm × 5 m or more. Branches with 4 angles, 2–3 cm thick. Spine-shields small, 5–10 mm apart along the angles, with paired spines 5–6 mm. Cyathia small, red. Capsule on a long, slender stalk, 3-lobed, *c.* 5 mm across, red. *Morocco (Atlas Mountains).* H5–G1.

126. E. grandidens Haworth. Illustration: White et al., The succulent Euphorbieae (southern Africa) **2**: 899 (1941); Jacobsen, Handbook of succulent plants **1**: 435 (1960); *The Euphorbia Journal* **3**: 63 (1985); Graf, Exotica, Series 4, edn 12, **1**: 1027 (1985).

Tree to 20 m, with branches clustered at the tips of the naked trunk and main branches. Branches 1–2 cm across, sharply 3-angled with wavy margins. Spine-shields small, with 2 spines *c.* 5 mm and to 3 cm apart. Cyathia *c.* 5 mm across, yellow. Capsule on a curved stalk, 3-lobed, *c.* 8 mm across. *South Africa (Transvaal, Cape Province).* G1.

127. E. ramipressa Croizat (*E. alcicornis* Berger). Illustration: Jacobsen, Handbook of succulent plants **1**: 470 (1960); Jacobsen, Lexicon of succulent plants, pl. 79 (1974); Graf, Exotica, Series 4, edn 12, **1**: 1027 (1985).

A shrub or small tree very like *E. grandidens* but branches with 2 wings. It is not known to have ever flowered. *Origin unknown.*

Described from a cultivated plant of unknown origin.

128. E. triangularis Desfontaines. Illustration: White et al., The succulent Euphorbieae (southern Africa) **2**: 894 (1941); *The Euphorbia Journal* **3**: 66, 133 (1985).

Tree to 20 m, with branches clustered at the tips of the numerous main branches. Branches with usually 3 angles, constricted into segments 10–30 × 5–10 cm. Spine-shields separate or joined along the angles, with paired spines to 8 mm and 8–18 mm apart. Cyathia *c.* 5 mm across, yellow.

Capsule on a curved stalk, 3-lobed, *c.* 7 mm across. Seeds ovate, smooth. *South Africa (Cape Province, SE Transvaal).* G1.

129. E. tetragona Haworth. Illustration: White et al., The succulent Euphorbieae (southern Africa) **2**: 914 (1941); Jacobsen, Handbook of succulent plants **1**: 480 (1960); Jacobsen, Lexicon of succulent plants, pl. 81 (1974); *The Euphorbia Journal* **3**: 131 (1985).

Tree to 13 m, with main branches spreading and rebranching towards the tips. Branches 1.5–2.5 cm thick, with 4 or 5 angles. Spine-shields triangular, 5–8 mm, with 2 spines 5–10 mm, 1–1.8 cm apart. Cyathia *c.* 6 mm across, yellow. Capsule on a slender stalk, 3-lobed, *c.* 9 mm across. *South Africa (Cape Province).* G1.

130. E. cooperi Berger (*E. angularis* misapplied; *E. lemaireana* misapplied). Illustration: White et al., The succulent Euphorbieae (southern Africa) **2**: 872 (1941); Jacobsen, Handbook of succulent plants **1**: 421, 422 (1960); Jacobsen, Lexicon of succulent plants, pl. 68 (1974); *The Euphorbia Journal* **3**: 68 (1985).

Large shrub or small tree to 5 m. Branches seldom re-branched, with 4–6 stout wings, deeply constricted into oblong-conical segments 15–30 × 10–15 cm. Spine-shields joined along the branch-angles, with paired spines 5–10 mm and 1–2 cm apart. Cyathia *c.* 7 mm across, yellow. Capsule deeply 3-lobed, *c.* 1 cm across. Seeds spherical, smooth, grey. *South Africa (Natal, Transvaal), Zimbabwe, Mozambique, Malawi, Zambia, Tanzania.* G1.

In cultivation the branches are often etiolated and the shape of the segments distorted, with thin wings and weak spines. There are 2 varieties that are rarely cultivated in Europe, var. **calidicola** Leach (Mozambique and Malawi) and var. **ussanguensis** (N.E. Brown) Leach (Zambia and Tanzania).

The genuine *E. angularis* Klotzsch is a shrub which occurs on Goa Island, and Mozambique, and is not in cultivation. *E. lemaireana* Boissier was described from a cultivated plant; its origin and identity are uncertain and the name should not be used.

131. E. grandicornis N.E. Brown. Illustration: White et al., The succulent Euphorbieae (southern Africa) **2**: 861 (1941); Jacobsen, Handbook of succulent plants **1**: 435 (1960); Jacobsen, Lexicon of succulent plants, pl. 71 (1974); *The*

Euphorbia Journal **1**: 81 (1983) & **4**: 60 (1987).

Shrub to 2 m. Branches with 3 wings, deeply constricted into broad, lobed segments to 12 × 15 cm. Spine-shields joined along the wing-edges, with paired spines of irregular length, 1–7 cm, often with small prickles at their bases. Cymes grouped in horizontal lines 1–3 together, each with 3 cyathia. Cyathia *c.* 8 mm across, yellow. Capsule 3-lobed, *c.* 1.3 cm across, red. Seeds spherical, smooth. *South Africa (Transvaal), Mozambique.* G1.

E. breviarticulata Pax. Illustration: *The Euphorbia Journal* **1**: 59 (1983) & **4**: 97 (1987).

Often confused with *E. grandicornis* but is easily distinguished by its cymes crowded in groups of 1–9. *E Africa.*

132. E. grandialata Dyer. Illustration: White et al., The succulent Euphorbieae (southern Africa) **2**: 866 (1941); *The Euphorbia Journal* **1**: 80 (1983), **2**: 88 (1984) & **4**: 58 (1987).

Like *E. grandicornis* but branches with 4 wings, marked with yellowish radial lines, spines 1.5–2.5 cm. *South Africa (Transvaal).* G1.

A rare species, threatened with extinction in its native habitat, but protected by law.

133. E. pseudocactus Berger. Illustration: White et al., The succulent Euphorbieae (southern Africa) **2**: 835 (1941); Jacobsen, Handbook of succulent plants **1**: 467 (1960); Jacobsen, Lexicon of succulent plants, pl. 78 (1974); *The Euphorbia Journal* **1**: 106 (1983).

Spreading shrub to 1 m. Branches with 3–5 angles, irregularly constricted into segments to 15 × 5 cm, usually streaked radially with yellowish green. Spine-shields joined along the branch-angles, with paired spines *c.* 1 cm. Cyathia *c.* 7 mm across, yellow. Capsule 3-lobed, *c.* 1.5 cm across, red. Seeds spherical, smooth. *South Africa (Cape Province).* G1.

134. E. franckiana Berger. Illustration: White et al., The succulent Euphorbieae (southern Africa) **2**: 854, 855 (1941); Jacobsen, Handbook of succulent plants **1**: 432 (1960); *The Euphorbia Journal* **1**: 78 (1983) & **1**: 432 (1984).

Like *E. pseudocactus* but branches plain green with 3 or 4 angles, spine-shields sometimes separated. *Garden origin.* G1.

Described from a cultivated plant of unknown origin, this appears to be identical with unmarked plants of *E. pseudocactus.*

135. E. coerulescens Haworth. Figure 129(1), p. 498. Illustration: White et al., The succulent Euphorbieae (southern Africa) **2**: 839 (1941); Jacobsen, Handbook of succulent plants **1**: 420 (1960); Graf, Tropica, edn 1, 422 (1978); *The Euphorbia Journal* **2**: 101 (1984).

Shrub *c.* 1.5 m, spreading by underground rhizomes. Branches sparsely rebranched, glaucous, with 4–6 angles, constricted into rounded segments to 5 cm thick. Spine-shields joined into a broad horny margin, with paired spines *c.* 1 cm. Cyathia *c.* 6 mm across, bright yellow. Capsule 3-lobed, *c.* 6 mm across, red. *South Africa (Cape Province).* G1.

136. E. virosa Willdenow. Illustration: White et al., The succulent Euphorbieae (southern Africa) **2**: 791 (1941); Jacobsen, Handbook of succulent plants **1**: 487 (1960); Jacobsen, Lexicon of succulent plants, pl. 82 (1974); *The Euphorbia Journal* **1**: 120 (1983), **5**: 58 (1988) & **7**: 2 (1991).

Like *E. coerulescens* but not rhizomatous, branches with 5–8 angles, *c.* 6 cm thick, cyathia *c.* 1 cm across, capsules spherical, *c.* 1 cm across, fleshy but hardening before dehiscence. *South Africa (Namaqualand).* G2.

137. E. avasmontana Dinter. Illustration: White et al., The succulent Euphorbieae (southern Africa) **2**: 817 (1941); Jacobsen, Handbook of succulent plants **1**: 409 (1960); *The Euphorbia Journal* **1**: 56 (1983).

Like *E. coerulescens* but not rhizomatous, branches with 5 angles, *c.* 6 cm thick, slightly constricted into segments 5–15 cm long, cyathia *c.* 8 mm across, yellow. *Namibia.* G2.

138. E. polyacantha Boissier (*E. thi* Schweinfurth). Illustration: Jacobsen, Handbook of succulent plants **1**: 466 (1960); Graf, Tropica, edn 1, 424 (1978); Graf, Exotica, Series 4, edn 12, **1**: 1025 (1985); *The Euphorbia Journal* **1**: 105 (1983) & **7**: 155 (1991).

Like *E. coerulescens* but usually less than 1 m, not rhizomatous, branches 2–3 cm thick, spine-shields closely set with spines 2–5 mm long, cyathia *c.* 4 mm across, capsules blackish. *NE Ethiopia (including Eritrea), Sudan (Red Sea hills).* H5–G1.

139. E. ledienii Berger. Illustration: White et al., The succulent Euphorbieae (southern Africa) **2**: 851 (1941); *The Euphorbia Journal* **2**: 22 (1984); Graf, Exotica, Series 4, edn 12, **1**: 1058 (1985).

Shrub to 2 m. Branches with 4–7 angles, shallowly and irregularly constricted into oblong segments 4–6 cm thick. Spine-shields separate or joined in a horny margin, with paired spines to 6 mm. Cyathia *c.* 5 mm across, yellow. Capsule on a curved stalk, 3-lobed, *c.* 7 mm across. *South Africa (Cape Province).* G1.

140. E. griseola Pax. Illustration: White et al., The succulent Euphorbieae (southern Africa) **2**: 776, 777 (1941); *The Euphorbia Journal* **1**: 82 (1983), **2**: 22 (1984) & **4**: 50 (1987).

Like *E. ledienii* but 30–150 cm, branches with 4–6 angles, 1–1.5 cm thick, spines to 1 cm, cyathia *c.* 4 mm across, capsule on slender, curved stalk, 3-lobed, *c.* 5 mm across. *South Africa (Transvaal), Botswana, Zimbabwe, Zambia, Malawi.* G2.

Very variable in habit, thickness of branches and spine formation; several subspecies have been described.

141. E. heterochroma Pax. Illustration: White et al., The succulent Euphorbieae (southern Africa) **2**: 779 (1941); *The Euphorbia Journal* **1**: 84 (1983) & **7**: 117, 118, 137 (1991).

Like *E. ledienii* but branches with 4 angles, 1.5–2 cm thick, regularly marked with light and dark green, spines sometimes absent, capsules on slender, curved stalks, 3-lobed, *c.* 5 mm across. *Tanzania, Kenya.* G2.

There are several related species occurring in East Africa, with more branch-angles, stronger spines or red-coloured cyathia and capsules. Some of these may be found in specialist collections.

142. E. schinzii Pax. Illustration: White et al., The succulent Euphorbieae (southern Africa) **2**: 745 (1941); *The Euphorbia Journal* **3**: 128 (1985), **5**: 90 (1988) & **7**: 136 (1991).

Stems numerous and rhizomatous from a thick, fleshy root, giving rise to dense tufts of branches. Branches with 4 angles, sometimes lobed, 10–20 cm × 8–10 mm. Spine-shields triangular, separated, 5–10 mm, with paired spines *c.* 1 cm. Cyathia *c.* 3 mm across, yellow. Capsule 3-lobed, *c.* 4 mm across, stalkless. *South Africa (Transvaal), SE Botswana, Zimbabwe.* G2.

A variable species, with almost every population in the wild possessing its own characteristics.

143. E. saxorum Bally & Carter. Illustration: *The Euphorbia Journal* **7**: 56 (1991).

Like *E. schinzii* but stems longer, trailing and rooting, with 4 angles, 5–8 mm thick, dark green mottled with reddish green, spine-shields usually joined along the angles, cyathia crimson, *c.* 6 mm across, solitary, capsule dark red. *SE Kenya (on rocky outcrops in a very small area).* G2.

Likely to be found only in specialist collections.

144. E. knuthii Pax. Illustration: White et al., The succulent Euphorbieae (southern Africa) **2**: 733 (1941); Jacobsen, Lexicon of succulent plants, pl. 73 (1974); *The Euphorbia Journal* **2**: 118 (1984).

Like *E. schinzii* but branches straggly, with 3 or 4 angles, strongly marked with a longitudinal band of pale green, cyathia *c.* 5 mm across, greenish yellow, capsule on curved stalks, deeply 3-lobed, *c.* 5 mm across. *S Mozambique, SE Zimbabwe.* G2.

145. E. squarrosa Haworth. Illustration: White et al., The succulent Euphorbieae (southern Africa) **2**: 725 (1941); Jacobsen, Handbook of succulent plants **1**: 475 (1960); Jacobsen, Lexicon of succulent plants, pl. 81 (1974); *The Euphorbia Journal* **3**: 130 (1985).

Stem very short at the apex of a large tuberous root, producing numerous branches, rarely re-branching. Branches to 15 cm, with usually 3 angles, 1–2.5 cm thick. Spine-shields small, closely set along the deeply lobed branch-angles, with paired spines 1–6 mm. Cyathia *c.* 5 mm across, green. Capsule on long curved stalk, deeply 3-lobed, *c.* 5 mm across. *South Africa (Cape Province).* G2.

146. E. micracantha Boissier. Illustration: White et al., The succulent Euphorbieae (southern Africa) **2**: 729 (1941); *The Euphorbia Journal* **3**: 117 (1985).

Like *E. squarrosa* but branches with 4 very rounded angles, 8–12 mm thick. *South Africa (Cape Province).* G2.

147. E. stellata Willdenow. Illustration: White et al., The succulent Euphorbieae (southern Africa) **2**: 719 (1941); *The Euphorbia Journal* **1**: 113 (1983); The new RHS dictionary of gardening **2**: 266 (1992).

Stem very short at the apex of a large tuberous root, producing a tuft of spreading branches. Branches to 15 cm, with 2 angles only, 1–1.5 cm wide, usually marked longitudinally with a pale green band. Spine-shields small on the closely set lobes of the angles, with paired spines *c.* 4 mm. Cyathia *c.* 4 mm across, yellow. Capsule on a curved stalk, deeply 3-lobed, *c.* 5 mm across. Seeds ovoid, smooth. *South Africa (Transvaal).* G2.

26. SYNADENIUM Boissier

S. Carter

Shrubs or trees, with fleshy cylindric branches and copious milky latex. Leaves alternate, fleshy, hairless or hairy, obovate or rarely lanceolate, margins entire or finely toothed; stipules present as small dark brown glands. Cyathia in dichotomously branching axillary cymes; nectary glands joined into a continuous rim around the cup of the cyathium. Ovary 3-celled with 1 ovule in each cell. Fruit a 3-lobed capsule projecting from the cup on a short stalk and opening along the septa. Seeds with an appendage (caruncle), usually minute.

A genus confined to east and southern tropical Africa, with about 20 closely related species.

Literature: No complete account of the genus has been made, but those species most often in cultivation are dealt with in White, A., Dyer, R.H. & Sloane, B.L, *The succulent Euphorbieae of southern Africa* **2** (1941). The east African species are described in detail in the Flora of tropical east Africa, Euphorbiaceae **2** (1988).

1a. Leaves green, or sometimes pink-tinged; margins entire   **1. grantii**
   b. Leaves glossy green, or more usually purple or purple-streaked; margins minutely toothed   **2. compactum**

1. S. grantii J.D. Hooker. Illustration: *Botanical Magazine*, 5633 (1867).

Shrubby tree to 10 m. Leaves to 15 cm, obovate, margins entire or occasionally crinkled towards the tip, hairless except for a few long hairs towards the base, uniformly green or sometimes tinged pink beneath. Cyathia downy towards the base, *c.* 6 mm across, red with a dull crimson glandular rim. Capsule *c.* 7 mm across. Seeds ovoid, *c.* 2.5 mm. *Uganda, Kenya, Tanzania.* G1. Year round.

2. S. compactum N.E. Brown. Illustration: *The Euphorbia Journal* **1**: 127 (1983).

Like *S. grantii* but leaves hairless, glossy green, margins minutely toothed and cyathia densely and minutely hairy. *Kenya.* G1.

More usually seen in cultivation is var. **rubrum** Carter (*S. grantii var. rubrum* invalid), with leaves entirely purple, or heavily purple-streaked and flecked. This can be identified at once by its finely toothed leaf-margins. *Widely distributed in cultivation in tropical Africa, but apparently originating in the wild in Kenya.* G1. Year round.

27. MONADENIUM Pax
S. Carter

Small trees, shrubs or perennial herbs, sometimes with a tuberous root. Stem and branches fleshy or often succulent and tuberculate. Milky latex present. Leaves fleshy, entire or toothed; stipules absent, or present as small glands or spines. Cyathia in axillary cymes; bracts in pairs, persistent and often united along one edge and enveloping the cup of the cyathium. Nectary glands joined into a horseshoe-shaped rim, with the developing capsule protruding through the gap. Ovary 3-celled with 1 ovule in each cell. Fruit an oblong capsule on a curved stalk, 3-lobed, with fleshy ridges along the septa. Seeds oblong, with a cap-like caruncle.

A genus of over 50 species, mostly from east Africa, but extending northwards into Ethiopia and Somalia, and southwards as far as the Transvaal in South Africa. All are highly prized by succulent-plant enthusiasts and many can be found in specialist collections.

Literature: A beautifully illustrated book by Bally, P.R.O., The genus Monadenium (1961), is a monographic work containing details of all but a few recently described species.

1a. Rootstock tuberous; branches spreading, 1–2 cm thick　**1. stapelioides**
　b. Rootstock fleshy, not tuberous; branches erect, 2–5 cm thick　2
2a. Branches without prominent tubercles, spineless　**2. lugardiae**
　b. Branches with prominent tubercles tipped by clusters of small spines　**3. schubei**

1. M. stapelioides Pax (*M. succulentum* Schweikerdt). Illustration: *Flowering plants of South Africa* **20**: pl. 776 (1940); Bally, Genus Monadenium, pl. 12 (1961).

Rootstock tuberous, producing numerous stems, branching at the base. Branches to 30 × 2 cm, with prominent upward-pointing tubercles, crowned only rarely with minute prickles on young growth. Leaves in tufts at the branch-tips, 1–5 cm. Bracts joined, longer than the cyathium, greenish white flushed with pink. Cyathium *c.* 5 mm across, nectary rim dull red. Capsule *c.* 5 mm, 3-lobed, with a toothed ridge along each angle. Seeds *c.* 3 mm, grey, with a creamy caruncle. *Kenya, N Tanzania.* G2.

Found only in specialist collections.

2. M. lugardiae N.E. Brown. Figure 129 (3), p. 498. Illustration: White, Dyer & Sloane, Succulent Euphorbieae 2: 942 (1941); Bally, Genus Monadenium, pl. 11 (1961); Rauh, Die Grossartige Welt der Sukkulenten, 30, f. 1 (1967); Court, Succulent flora of southern Africa, 90 (1981).

Shrub to 50 cm. Branches 2–3 cm thick, covered with spirally arranged, slightly raised tubercles. Leaves at the branch-tips, 2–6 cm. Cymes branching once; bracts as long as the cyathium, green. Cyathium *c.* 6 mm across, with nectary rim yellow. Capsule *c.* 6 mm, with a toothed ridge along each angle. Seeds *c.* 3.5 mm, slightly warty with a creamy caruncle. *South Africa (Transvaal), Botswana.* G1.

3. M. schubei (Pax) N.E. Brown. Illustration: Bally, Genus Monadenium, pl. 20 (1961); Lamb & Lamb, Illustrated reference on cacti and other succulents **4**: 1101 (1975); Rauh, Die Grossartige Welt der Sukkulenten, 30, f. 4 (1967); *The Euphorbia Journal* **5**: 69 (1988).

Spreading perennial. Branches erect to 60 cm, rarely more, 3–5 cm thick, with very prominent spirally arranged tubercles. Leaves in tufts, 3–8 cm; stipules evident as a cluster of 5 prickles around the tip of each tubercle. Cymes branching once. Bracts as long as the cyathium, greyish green flushed with pink. Cyathium *c.* 5 mm across, nectary rim deep pink. Capsule *c.* 6 mm, 3-lobed, with a toothed ridge along each angle. Seeds *c.* 3.5 mm, with a creamy caruncle. *Tanzania.* G2.

Found only in specialist collections.

28. PEDILANTHUS Poiteau
S. Carter

Shrubs or small trees, with woody or fleshy branches and milky latex. Leaves fleshy, entire, with small stipules. Cyathia in dichotomous cymes, usually tightly clustered; bracts in pairs, small, persistent. Nectary glands 2, 4 or 6, enclosed in a spur-like extension of the cyathium, often brightly coloured. Ovary 3-celled with 1 ovule in each cell. Fruit a 3-lobed capsule. Seeds without a caruncle.

A genus of 14 species in Central America, the West Indies and South America.

Literature: A very detailed monographic account has been produced by Dressler, R. L., The genus Pedilanthus (Euphorbiaceae), *Contributions from the Gray Herbarium* **182**: 1–86 (1957).

1a. Nodes of the branches in a zig-zag arrangement; flowers *c.* 1.5 cm　**1. tithymaloides**
　b. Nodes of the branches straight; flowers *c.* 2.5–3 cm　2
2a. Leaves 1 cm　**2. macrocarpus**
　b. Leaves 5–10 cm　**3. bracteatus**

1. P. tithymaloides (Linnaeus) Poiteau. Figure 129(2), p. 498. Illustration: Rauh, Die Grossartige Welt der Sukkulenten, 31, f. 3 (1987); Macoboy, What shrub is that?, 260 (1990).

Succulent-stemmed shrub, 1–2 m. Branches with nodes 3–6 cm, usually arranged in a zig-zag pattern. Leaves 1–1.5 cm, ovate, deciduous. Cyathia enclosed by bright red, beak-shaped bract-like extensions, *c.* 1.5 cm in terminal and axillary clusters. Capsules deeply 3-lobed, *c.* 5 mm across. *C & N South America & West Indies.* G1. Spring.

There are several varieties distinguished by branch characters, leaf-size and colour, and size of the 'flower'. The one most often cultivated in Europe is var. **smallii** (Millspaugh) Dressler, with an obviously zig-zag branch pattern, and slightly larger leaves (1.5–3 cm), which are sometimes variegated with yellowish green or pink.

2. P. macrocarpus Bentham. Illustration: Lamb & Lamb, Illustrated reference on cacti and other succulents **1**: 225 (1955); *The Euphorbia Journal* **3**: 38 (1985).

Like *P. tithymaloides* but nodes of the branches straight, leaves *c.* 1 cm; flowers 2–3.5 cm, capsules much larger, *c.* 2 × 1.5 cm, with 6 short horns around the base. *Mexico.* G1.

3. P. bracteatus (Jacquin) Boissier (*P. pavonis* (Klotzsch & Garcke) Boissier).

Like *P. tithymaloides* but nodes of the branches straight, leaves 4.5–10 cm; flowers 2–3 cm, capsules shallowly lobed, 1–1.3 cm across. *Mexico.* G1.

161. DAPHNIPHYLLACEAE

Trees or shrubs. Leaves alternate, crowded, entire, exstipulate, usually evergreen; ptyxis flat. Flowers in axillary racemes, unisexual, actinomorphic. Perianth and stamens hypogynous. Male flowers: perianth of 3–8, free, imbricate segments, stamens 6–12. Female flowers: perianth absent, staminodes few, small or absent, ovary of 2, imperfectly united carpels; styles 1–2, persistent, undivided; ovules 2 per cell, pendulous. Fruit a drupe, 1-seeded.

A family of one genus with about 24 species native to eastern Asia and Malaysia. Only 2 species are commonly cultivated.

1. DAPHYNIPHYLLUM Blume
S.G. Knees
Description as for family.

Propagation is by cuttings of partially ripened wood taken in late summer and rooted with bottom heat, or perhaps more usually from seed, which is freely produced in cultivation in *D. macropodum*.

Literature: Engler, H.G.A., Daphniphyllaceae, Das Pflanzenreich **22** (1919).

1a. Shrub or small tree reaching 3–7 m in cultivation **1. macropodum**
  b. Low spreading shrub to 1 m, rarely exceeding 2m **2. humile**

1. D. macropodum Miquel. Illustration: Das Pflanzenreich **22**: 10 (1919); Bean, Trees & shrubs hardy in the British Isles, edn 8, **2**: 24 (1973); Brickell, RHS gardeners' encyclopedia of plants & flowers, 86 (1989); Macoboy, What shrub is that?, 122 (1990).

Shrub or small tree 3–7 m with compact habit. Branches stout, glaucous and red-tinged when young. Leaves 8–20 × 2–8 cm, ovate, obovate or oblong, dark glossy green above, glaucous whitish green beneath. Flowers pale green, inconspicuous, strongly scented; sepals *c.* 5 mm, triangular, falling early; stamens with short filaments, anthers to 5 mm; stigma 2–4-lobed. Fruit 6–9 mm, rounded. *Japan, Korea, China.* H2. Spring.

'Variegatum' has leaves with broad creamy white margins but is less hardy than non-variegated variants.

2. D. humile Maximowicz (*D. macropodum* var. *humile* (Maximowicz) Rosenthal;

D. jezoense Anon.). Illustration: Makino, Illustrated flora of Japan, 381 (1948).

Low spreading shrub, normally to 1 m, occasionally reaching 2 m. Leaves 5–12 × 1.5–5 cm, ovate; otherwise very similar to *D. macropodum*. Flowers rarely seen in cultivation. *Japan (Yezo).* H2–3.

162. RUTACEAE

Woody or herbaceous plants. Leaves alternate or opposite, simple or compound, exstipulate, usually aromatic, gland-dotted, often evergreen; ptyxis usually conduplicate, rarely flat. Inflorescences various. Flowers usually bisexual, usually actinomorphic. Calyx, corolla and stamens usually hypogynous, disc usually present. Calyx of 3–6 free or united sepals. Corolla of 3–6 free or united petals or rarely absent. Stamens 3–12 or rarely more, staminodes sometimes present. Ovary of 4–5 or, rarely, many united carpels, often free but united by single styles or otherwise styles rarely free, as many as carpels, ovary often borne on a short stalk; ovules 1–many, axile. Fruit fleshy, a capsule or a group of samaras.

A family of 150 genera consisting of about 1600 species found mainly in Africa, Australia and America. Some important ornamental plants belong to this family, but it is chiefly of note because of 'citrus fruits' (oranges, grapefruit, etc.).

1a. Herbaceous perennials, shrubs or trees, without spines; fruit a capsule, berry or samara 2
  b. Trees or shrubs, with thorns or spines on shoots; fruit usually fleshy or juicy 21
2a. Herbaceous perennials or small shrubs; leaves 1–3-pinnate; flowers in panicles, cymes or racemes 3
  b. Shrubs or trees; leaves simple, sometimes divided; flowers often solitary 5
3a. Leaves pinnate; petals yellow **1. Ruta**
  b. Leaves 1–3-pinnate; petals white or purplish 4
4a. Leaves 2- or 3-pinnate; flowers less than 1 cm across; sepals and petals 4 **2. Boenninghausenia**
  b. Leaves pinnate; flowers at least 2.5 cm across; sepals and petals 5 **3. Dictamnus**

5a. Leaves simple 6
  b. Leaves compound 15
6a. Leaves opposite; sepals fused into a cup; petals 4, fused to form a campanulate corolla; stamens 8 **11. Correa**
  b. Leaves alternate, rarely opposite; sepals 4 or 5, free; petals 4 or 5, free or fused only at base 7
7a. Stamens more than twice the number of petals; fruit a berry 8
  b. Stamens not more than twice the number of petals; fruit not a berry 10
8a. Petals not winged **19. Fortunella**
  b. Petals narrowly winged 9
9a. Stamens 20–40; ovary with 8–15 cells **23. Citrus**
  b. Stamens 16–20; ovary with 3–15 cells **20. × Citrofortunella**
10a. Deciduous shrub; flowers unisexual; male flowers in axillary racemes, stamens 4; female flowers solitary **12. Orixa**
  b. Evergreen shrub; flowers bisexual 11
11a. Stamens 4 or 5, staminodes absent **18. Skimmia**
  b. Stamens 10, or stamens 5 with 5 staminodes 12
12a. Stamens 5; staminodes 5 (sometimes enclosed in petal claw) 13
  b. Stamens 10 (sometimes 8), all fertile 14
13a. Staminodes free, with gland at apex; ovary with stalked glands **5. Adenandra**
  b. Staminodes enclosed or concealed by petal claw; ovary not glandular **4. Coleonema**
14a. Anthers hairless **9. Eriostemon**
  b. Anthers with hairy appendages **10. Crowea**
15a. Stamens equal in number to petals or absent 16
  b. Stamens twice the number of petals 19
16a. Evergreen shrubs or trees; leaves opposite, with 3 leaflets; flowers bisexual; sepals, petals and stamens 4 **6. Zieria**
  b. Deciduous trees or shrubs; leaves pinnate; leaflets usually more than 3; flowers frequently unisexual; sepals, petals and stamens often more than 4 17
17a. Leaves usually with 3 leaflets; fruit a winged samara **17. Ptelea**
  b. Leaves with more than 3 leaflets; fruit a berry or capsule, not winged 18

18a. Buds in leaf-axils exposed; fruit of 4 or 5 dehiscent carpels **13. Tetradium**

b. Buds in leaf-axils concealed by swollen leaf-stalk base; fruit an indehiscent drupe with 5 seeds **16. Phellodendron**

19a. Leaves with 3 leaflets; petals white with velvety hairs **8. Acradenia**

b. Leaves simple, with 3 leaflets or compound; petals without hairs, white, pink, brown or yellow-green 20

20a. Flowers white; petals and sepals usually 5; stamens 10; ovary hairy **14. Choisya**

b. Flowers pink-red, red-brown or yellow-green, rarely white; petals and sepals 4; stamens 8; ovary not hairy **7. Boronia**

21a. Flowers unisexual or bisexual; sepals, petals and stamens 3–5 **15. Zanthoxylum**

b. Flowers usually bisexual; stamens at least twice the number of petals 22

22a. Flowers red; stamens 10–12; fruit with hard white shell **25. Limonia**

b. Flowers white or at most tinged with red; stamens more than 20; fruit not white 23

23a. Leaves simple 24

b. Leaves with 3 leaflets 25

24a. Branches with long spines **22. × Citroncirus**

b. Branches with short spines **23. Citrus**

25a. Petals glandular; stamens 35–45; fruit with woody coat **24. Aegle**

b. Petals not glandular; stamens 20–60; fruit not woody 26

26a. Leaves prominently winged; fruit dull lemon-yellow **21. Poncirus**

b. Leaves narrowly winged; fruit orange or yellow **22. × Citroncirus**

1. RUTA Linnaeus

E.C. Nelson

Aromatic perennial herbs, shoots becoming woody towards base. Leaves spotted with glands, alternate, pinnately divided, segments linear to obovate. Flowers in a cyme. Sepals 4. Petals 4 (except central flower which has 5), dull, dark yellow, hooded, margins toothed or hairy (rarely entire). Stamens 8–10 with hairless filaments. Styles fused. Fruit a 4- or 5-celled capsule.

A genus of 5 species in Europe and western Asia, with 2 species in Macaronesia. Only *R. graveolens* is widely cultivated as a medicinal or ornamental herb. A closely related genus is *Haplophyllum* Jussieu, which includes *H. patavinum* (Linnaeus) G. Don (*Ruta patavinum* Linnaeus); this may be distinguished from *Ruta* by its rough-haired filaments, less divided leaves (simple or with 3 leaflets), and 5 petals in all flowers. Many plants grown under this name are *Ruta graveolens*. *Ruta* species may be propagated by cuttings or from seed; cultivars must be increased vegetatively.

1a. Petals toothed, without fringe of hairs; leaf-segments less than 1 cm across **1. graveolens**

b. Petals with a fringe of hairs; leaf-segments less than 6 mm across **2. chalepensis**

1. R. graveolens Linnaeus. Illustration: Heywood, Flowering plants of the world, 203 (1978); Stodola, Volak & Bunney, The illustrated book of herbs, f. 206 (1984); Brickell, RHS gardeners' encyclopedia of plants & flowers, 145 (1989).

Hairless herb to 50 cm; shoots and leaves with glaucous bloom. Leaves pinnately divided, occasionally pinnae also pinnatisect; segments less than 1 cm wide, obovate. Inflorescence loose; flower-stalks equalling or longer than fruit; bracts leaf-like, lanceolate. Sepals lanceolate, acute. Petals toothed, wavy. Fruit hairless; segments with obtuse apices. *SE Europe.* H4. Summer.

Grown for many centuries as a medicinal herb but dangerous if taken in large quantities; the sap can cause a severe rash especially when skin is exposed to bright sunlight. As an ornamental it is especially prized for its grey-blue foliage. Two cultivars are widely available, 'Jackman's Blue' (compact habit with grey-blue leaves) and 'Variegata' (leaf-margins white).

2. R. chalepensis Linnaeus. Illustration: Sibthorp & Smith, Flora Graeca **4**: t. 368 (1823); Polunin & Huxley, Flowers of the Mediterranean, pl. 80 (1965).

Like *R. graveolens* but ultimate segments of leaves narrower, 1–6 mm across. Inflorescence-bracts heart-shaped to ovate. Petals with a fringe of hair. Capsule-segments with acuminate apices. *S Europe.* H3. Summer.

2. BOENNINGHAUSENIA Meissner

E.C. Nelson

Herbaceous perennial or woody, low shrubs to 50 cm, hairless, stems and foliage glaucous; shoots terete. Leaves alternate, 2 or 3 times divided into 3; leaflets 1–2.5 cm × 7–20 mm, entire, obovate to elliptic, very glaucous (rarely white) beneath, with glandular spots; stalks 5–10 cm. Flowers in panicles, white, bisexual. Sepals 4, oblong, fused at base, to 1 mm, persistent in fruit. Petals 4, oblong, to 4 mm. Stamens 6–8, longer than petals. Disc urn-shaped. Ovary with 4 cells each with 6–8 ovules. Style short, deciduous. Fruits of 4 follicles, to 3 mm; seeds black, tubercled, kidney-shaped.

A genus of one species, widely distributed from India to Japan, and hardy in most European gardens. It can be easily raised from seed, and is tolerant of lime.

1. B. albiflora (Hooker) Meissner (*B. japonica* Nakai). Illustration: Die natürlichen Pflanzenfamilien **4**: f. 71B (1896); Iconographia cormophytorum Sinicorum **2**: 549, t. 2828 (1972); Krüssmann, Handbuch der Laubgehölze **1**: 160 (1976). *Temperate Asia (Himalaya to Japan).* H4. Summer–autumn.

3. DICTAMNUS Linnaeus

C. Gorman

Deciduous, perennial herb to 1 m, with woody rootstock. Leaves to 7 cm, alternate, pinnate, gland-dotted. Leaflets 9–11, ovate, 2.5–7.5 cm, with fine forwardly pointing teeth, dark green, lemon-scented. Flowers to 2.5 cm across, bilaterally symmetric, white to purple, in terminal racemes; stalks with bracts. Sepals 5, lanceolate, minute. Petals 5, pointed, narrow, the lowest bent downwards. Stamens 10, curved upwards. Fruit a 5-lobed capsule.

A genus of one variable and widespread species, notable for its propensity for exuding an ethereal, flammable oil from the leaves; this can be ignited and thus the vernacular name 'burning bush'.

1. D. albus Linnaeus (*D. caucasicus* (Fischer & Meyer) Grossheim; *D. fraxinella* Persoon; *D. hispanicus* Willkomm; *D. purpureus* invalid), Illustration: *Botanical Magazine*, 8961 (1922); Everett, New York Botanical Gardens illustrated encyclopedia of horticulture **4**: 1064 (1981); Brickell (ed.), RHS A–Z encyclopedia of garden plants, 375 (2003).

S Europe to Siberia & N China. H5. Summer.

4. COLEONEMA Bartling & Wendland

E.C. Nelson

Aromatic, evergreen shrubs resembling *Erica* species. Leaves linear, simple, alternate, gland-dotted. Flowers in clusters or solitary towards tips of branches, bisexual. Sepals 5, margins with hairs or membranous. Petals 5, white to red, narrowing into basal claws that are channelled. Disc cup-shaped or with 5 forked lobes. Stamens 5; staminodes 5, fused to or concealed within channel in petal claws. Fruit with 5 carpels.

A genus of about 5 species, mainly native in the south-western area of South Africa.

1a. Petals white, *c.* 4 mm   **1. album**
  b. Petals pink, *c.* 1 cm   **2. pulchrum**

1. C. album (Thunberg) Bartling & Wendland (*Diosma alba* Thunberg).

Shrub to 2 m. Leaves *c.* 1.5 cm, apex with sharp straight point. Flowers axillary, solitary. Sepals thin, ovate, with fine, marginal hairs. Petals white, *c.* 4 mm. *South Africa.* G1–H5. Spring.

2. C. pulchrum J.D. Hooker (*Diosma tenuifolium* Presl). Illustration: *Botanical Magazine*, 3340 (1834); Brickell (ed.), RHS A–Z encyclopedia of garden plants, 296 (2003).

Shrub 1–2 m; branches slender, arching. Leaves to 4 cm, linear, with a sharply pointed apex. Sepals with fine, marginal hairs. Petals pink, to 1 cm. *South Africa.* G1–H5. Spring.

5. ADENANDRA Willdenow

E.C. Nelson

Small shrubs. Leaves alternate or occasionally opposite, simple, entire, often with 2 glands at base of stalk. Flowers solitary or in terminal racemes, bisexual. Sepals 5. Petals 5, stalkless or sometimes with short claws, hairless. Disc cup-shaped with 5–10 lobes. Stamens 5, shorter than sepals; anthers with a stalked, spoon-shaped gland at apex; staminodes 5, with terminal gland. Ovary with 5 cells, each containing 2 ovules, covered with stalked glands. Style solitary; stigma 5-lobed, disc-shaped.

There are about 40 named species and most are native in the Cape Province of South Africa. Few are cultivated in Europe, none being widely grown.

1a. Leaf margins curved downwards and inwards; flowers solitary, almost stalkless   **1. uniflora**
  b. Leaf-margins not curved downwards and inwards; flowers in umbels, stalked   **2. fragrans**

1. A. uniflora (Linnaeus) Willdenow (*Diosma uniflora* Linnaeus). Illustration: *Botanical Magazine*, 273 (1794); *Loddiges' Botanical Cabinet*, t. 493 (1820).

Dwarf shrub, less than 50 cm. Leaves lanceolate, apex prolonged into a point, margins curved downwards and inwards. Flowers solitary on very short stalks. Petals white or pink inside, purple outside, twice as long as sepals. *South Africa (Cape Province).* G1–H5. Spring.

2. A. fragrans (Sims) Roemer & Schultes (*Diosma fragrans* Sims). Illustration: *Botanical Magazine*, 1519 (1813).

Shrub, 40–100 cm. Leaves linear to oblong, apex prolonged into a point, to 4 cm. Flowers *c.* 2 cm across, in umbels; stalks to 5 cm. Petals pink on outer side, white inside, 3 times as long as sepals. *South Africa (Cape Province).* G1–H5. Summer.

6. ZIERIA Smith

E.C. Nelson

Evergreen shrubs or small trees. Leaves with 3 leaflets (rarely simple), aromatic, opposite. Flowers white or pink, in cymes or solitary. Sepals and petals 4, deciduous. Stamens 4. Disc with 4 gland-like lobes opposite sepals. Carpels 4, each with 2 ovules; fruit dehiscent, dry, not inflated.

More than 20 species are currently recognised. All are native to eastern Australia. Propagation is by cuttings. In Australia they are recommended as plants for shaded places. It is very doubtful that there are more than 2 species cultivated in Europe.

1a. Leaflets more than 6 cm; branches and leaf-stalks with stellate hairs, not glandular   **1. arborescens**
  b. Leaflets less than 5 cm; branches and leaf-stalks usually hairless, glandular   **2. smithii**

1. Z. arborescens Sims (*Z. macrophylla* Bonpland; *Z. smithii* var. *macrophylla* (Bonpland) Bentham). Illustration: *Botanical Magazine*, 4451 (1849); Cochrane et al., Flowers and plants of Victoria, t. 460 (1973).

Tall shrub or small tree to 4 × 3 m; branches with short stellate hairs. Leaves with 3 leaflets, dark green, Leaflets to 10 × 3 cm, upper surface hairless and often shiny, lower surface with grey felt. Flowers

white to 1.2 cm across. *Australia (New South Wales, Tasmania, Victoria).* G1–H5. Reputed to tolerate most soils, and to be fast growing.

2. Z. smithii Andrews. Illustration: *Andrews' Botanical Repository* **9**: t. 606 (1810); *Botanical Magazine*, 1395 (1811).

Shrub to 2 × 2 m; branches hairless or nearly so, with warty glands. Leaves with 3 leaflets, very aromatic but not pleasantly so. Leaflets to 5 cm, hairless. Flowers white, less than 1 cm across. *Australia (New South Wales, Victoria, Queensland).* G1. Spring.

7. BORONIA Smith

E.C. Nelson

Shrubs or woody perennials. Leaves opposite, simple or compound. Flowers solitary, or in cymes and clusters. Sepals and petals 4. Stamens 8, in 2 whorls, those opposite sepals sometimes sterile; filaments often expanded towards apex. Carpels fused only at apex, each containing 2 ovules. Style solitary.

A genus of about 70 species endemic to Australia (1 species is reported from New Caledonia), where many are now frequently cultivated in gardens. In Europe few are in cultivation. In general they require lime-free soil, which is well-drained but not dry. Most of the species in European gardens tolerate dappled shade. The western Australian species may be propagated from seed, but cuttings are best. Use firm young growth, bottom heat of about 25 °C, and a rooting hormone.

1a. Flowers brown to red-brown or yellow-green   **1. megastigma**
  b. Flowers pink to pink-red, rarely white   2
2a. Shoots and leaves hairy, flowers *c.* 5 mm across   **2. molloyae**
  b. Shoots and leaves hairless or flowers more than 1 cm across   3
3a. Leaflets 5–9; flowers in cymes, cup-shaped, *c.* 2 cm across **3. floribunda**
  b. Not as above   4
4a. Flowers solitary, globular, *c.* 1 cm across   **4. heterophylla**
  b. Flowers in clusters, saucer-shaped, *c.* 2 cm across   5
5a. Leaflets 5–9, entire   **5. pinnata**
  b. Leaflets 7–15, toothed   **6. thujona**

1. B. megastigma Bartling. Illustration: *Botanical Magazine*, 6049 (1973); Elliot & Jones, Encyclopaedia of Australian plants **2**: 344 (1982); Brickell (ed.), RHS A–Z encyclopedia of garden plants, 187 (2003).

Shrub to 3 m. Leaves compound; leaflets 3–5, to 1.5 cm, linear, apex obtuse. Flowers fragrant, on short axillary stalks, solitary, pendent, globular, *c.* 1 cm across. Petals dark brown to red-brown or yellow-green outside, yellow inside. *Western Australia.* G1. Spring–summer.

The characteristic brown flowers are not especially attractive, but they are wonderfully fragrant (some people cannot smell the perfume). It is especially popular for conservatories and as a cut-flower. In Australia many different variants have been selected and named, including 'Lutea', with yellow flowers, and 'Chandleri' with dark brown flowers.

2. B. molloyae Drummond (*B. elatior* Bartley). Illustration: *Botanical Magazine,* 6285 (1877).

Shrub to 4 m with hairy branches. Leaves compound, pinnate, to 5 cm; leaflets 5–15, hairy, aromatic, elliptic, *c.* 10 × 2 mm. Flowers solitary on axillary shoots, pendent, *c.* 5 mm across, campanulate. Petals red-pink. *Western Australia.* G1–H5. Spring–summer.

Tolerant of a wide range of soil conditions, light frost, and dappled shade. The petals retain their vivid colour and persist for a long time after pollination. It is often used as a cut flower.

3. B. floribunda Sprengel.

Shrub to 1.5 m. Leaves compound, pinnate; leaflets 5–9, *c.* 2.5 cm. Flowers fragrant, in cymes, cup-shaped, *c.* 2 cm across. Petals pale pink, rarely white. *Australia (New South Wales).* G1–H5. Spring.

One of the most reliable species with large, fragrant flowers, and tolerance of light frosts. Prefers moist, well-drained, sandy soil.

4. B. heterophylla Mueller. Illustration: *Botanical Magazine,* 6845 (1885).

Shrub to 3 m. Leaves compound, pinnate, to 5 cm; leaflets linear-elliptic, *c.* 3 cm, or leaves simple. Flowers slightly fragrant, solitary on short axillary stalks, globular, *c.* 1 cm across. Petals pink-red. *Western Australia.* G1–H5. Spring.

Tolerant of shade, light frost and a variety of different soil conditions; fast-growing. Used as a cut flower.

5. B. pinnata Smith. Illustration: *Botanical Magazine,* 1763 (1815); Elliot & Jones,

Encyclopaedia of Australian plants **2**: 347 (1982).

Shrub to 2 m. Leaves pinnate, to 5 cm, aromatic (camphor-like when crushed); leaflets 5–9, ovate to lanceolate, 5–25 × 3–4 mm. Flowers in axillary clusters, saucer-shaped, *c.* 2 cm across. Petals pink. *Australia (New South Wales).* G1–H5. Spring.

Prefers dappled shade and is frost-tolerant.

6. B. thujona Penfold & Welch. Illustration: Elliot & Jones, Encyclopaedia of Australian plants **2**: 351 (1982).

Like *B. pinnata* but leaflets 7–15, toothed. *Australia (New South Wales).* G1–H5. Spring.

8. ACRADENIA Kippist
E.C. Nelson

Trees or shrubs with evergreen, opposite leaves; leaflets 3. Flowers borne in panicles, bisexual. Sepals 5 or 6, fused towards base, persistent in fruit. Petals 5 or 6, deciduous. Stamens 10–12, those alternating with the petals equalling the petals, those opposite the petals shorter; filaments tapering; anthers versatile, mucronate. Ovary with 5 cells, each with 2 ovules, fused at base, and joined on inner side at the middle by the style, each with prominent gland on upper side. Fruit composed of 1–5 one-seeded follicles, fused at base.

A genus of 2 species endemic in Australia, but only one is in general cultivation.

Literature: Hartley, T.G., A revision of the genus Acradenia (Rutaceae), *Journal of the Arnold Arboretum* **58**: 171–81 (1978).

1. A. frankliniae Kippist. Illustration: Curtis & Stones, Endemic flora of Tasmania **5**: 167 (1975); *Journal of the Arnold Arboretum* **58**: 171–81 (1978).

Shrub or small tree to 10 m; shoots hairy, becoming glaucous. Leaves with prominent wart-like glands; stalks hairy or hairless, to 8 mm. Leaflets leathery, dark green above, paler beneath, hairless or sparsely hairy, elliptic to obovate, to 6 × 2 cm, tapering towards base, with rounded teeth in apical portion. Inflorescence terminal; flowers *c.* 5 mm across. Petals white, with scattered velvety hairs and green glands, *c.* 5 mm. *Australia (Tasmania).* H5. Spring–early summer.

9. ERIOSTEMON Smith
E.C. Nelson

Aromatic shrubs or small trees; branches with prominent warty glands. Leaves

alternate, simple, without stipules. Inflorescences axillary or terminal, or flowers solitary. Flowers bisexual. Sepals 5, free, persistent. Petals 5, free, with glandular spots, white, blue or pink. Stamens 8–10, usually erect, those opposite sepals longest; anthers hairless; filaments usually hairy. Carpels 5 (rarely 3 or 4), free or fused at base, each with 2 ovules; style solitary; stigma 5-lobed. Fruit separating into 5; seed usually solitary.

A genus of more than 30 species, all native in Australia. Only one is found in general cultivation in European gardens, and even then it is quite uncommon. Propagation is by seed. *Eriostemon* will flourish in the same conditions as *Correa.*

Literature: Wilson, P.G., A taxonomic revision of the genera Crowea, Eriostemon and Phebalium (Rutaceae), *Nuytsia* **1**: 19–60 (1970).

1. E. myoporoides de Candolle. Illustration: *Botanical Magazine,* 3180 (1832); Elliot & Jones, Encyclopaedia of Australian plants **3**: 477, 478 (1984).

Shrub to 2 m; shoots hairless, glaucous, very warty to almost smooth. Leaves stalkless, oblong to obovate, 1.5–11 cm, with numerous warty glands. Flowers axillary in clusters of 4–6, occasionally solitary; stalks to 2 cm. Petals to 0.8 mm, white, hairless, keeled. *Eastern Australia.* G1–H5. Spring.

10. CROWEA Smith
E.C. Nelson

Evergreen shrubs. Leaves alternate, simple, with warty glands, hairless. Flowers axillary or terminal, on stalks. Bracteoles persistent and occasionally leaf-like. Sepals 5, free. Petals 5, free, persistent in fruit. Stamens 10, those opposite sepals slightly longer than those opposite petals, curving inwards over ovary before pollination; filaments with keeled or convex inner surfaces, and hairs along margins; anthers with hairy appendages, 0.2–0.3 mm. Ovary hairless with 5 free carpels, without sterile apex, each with 2 ovules. Fruit of 1–5 follicles.

A genus of 3 species endemic to Australia, formerly popular as conservatory plants, but now uncommon in European gardens.

Literature: Wilson, P.G., A taxonomic revision of the genera Crowea, Eriostemon and Phebalium (Rutaceae), *Nuytsia* **1**: 15–19 (1970).

1a. Leaves toothed, decurrent at base with 2 narrow, toothed wings along stalks **1. angustifolia**
  b. Leaves entire, not decurrent, stalks distinct 2
2a. Branches hairless, acutely angular or narrowly winged **2. exalata**
  b. Branches with minute hairs, obtusely angular or more or less flattened **3. saligna**

1. C. angustifolia Smith (*C. angustifolia* Turczaninov; *C. dentata* Bentham). Illustration: Elliot & Jones, Encyclopaedia of Australian plants **3**: 118 (1984).

Shrub to 3 m. Branches hairless, terete and with narrow, toothed wings decurrent from leaf-bases. Leaves hairless, stalkless, linear to elliptic to obovate, occasionally with toothed margins. Flowers usually solitary, axillary, on short stalks that thicken towards apex. Bracteoles 2–4, to 1.5 mm. Sepals to 2 mm, hairless except for small marginal hairs. Petals to 1.2 cm, white or pink darkening with age. Stamens *c.* 7 mm; filaments glandular towards apex. Disc purple. Ovary *c.* 0.7 mm. Style equalling stamens; stigma globular, about twice as wide as style. *Australia (Western Australia).* G1–H5. Spring.

Var. **angustifolia**. Leaves linear to oblong; flowers pink. Var. **dentata** (Bentham) Wilson. Leaves elliptic to obovate; flowers usually white.

2. C. exalata Mueller (*Eriostemon crowei* Mueller). Illustration: Elliot & Jones, Encyclopaedia of Australian plants **3**: 119 (1984); Brickell (ed.), RHS A–Z encyclopedia of garden plants, 328 (2003).

Shrub to 1 m. Shoots glandular, slightly angular or more or less terete, with minute hairs in sunken lines. Leaves entire, spathulate to elliptic, 1.1–5 cm × 1–6 mm, hairless. Flowers solitary, terminal, usually on short axillary shoots. Sepals *c.* 2 mm, with minute hairs. Petals overlapping, to 1.2 cm, pink to pale mauve. Stamens with pink flattened filaments *c.* 3 mm. Disc dark green. Style *c.* 5 mm, thick; stigma globular, *c.* 5 mm across. *Australia (New South Wales, Victoria).* G1–H5. Spring.

3. C. saligna Andrews. Illustration: *Andrews' Botanical Repository* **2**: t. 70 (1800); Ventenat, Jardin de la Malmaison **1**: t. 7 (1803); *Botanical Magazine*, 989 (1807); Heywood, Flowering plants of the world, 203 (1978).

Shrub to 1.5 m. Shoots hairless with narrow wings decurrent from leaf-bases. Leaves stalkless, hairless, elliptic, 3–6 cm × 4–13 mm. Flowers axillary, solitary; stalks terete, 5-grooved, 5–13 mm. Bracteoles 2, basal, *c.* 1 mm. Sepals to 3 × 3 mm, hairless except for minute marginal hairs. Petals 1.2–2 cm × 4–7 mm, pink to purple (very rarely white). Stamens to 5 mm, arranged in a pyramid, margins joining. Style short and thick, less than 1 mm; stigma globular, *c.* 1 mm across. *Australia (New South Wales).* G1–H5. Spring.

11. CORREA Andrews
E.C. Nelson

Shrubs (rarely trees) with stellate hairs on shoots, foliage and flowers. Leaves simple, opposite, usually entire. Flowers solitary or clustered in terminal cymes, hanging or erect. Calyx cup-shaped with an entire or lobed margin. Corolla of 4 fused petals, sometimes splitting to base. Stamens 8, inserted at base of 8-lobed disc; filaments linear, those opposite the petals usually shortest and with broadened bases, hairless; anthers projecting or included. Ovary hairy, with 4 cells each with 2 ovules. Style slender, almost equalling stamens. Stigma with 4 minute lobes.

A genus of 11 species, all endemic to Australia. Like many Australian endemic genera, they tend to be intolerant of lime, and should be cultivated in a well-drained, acid loam. In southern parts, species and cultivars may be grown outdoors without protection; some will succeed in sheltered gardens in north-western Europe but it is more usual to cultivate them in a frost-free glasshouse. Propagation may be by seed or cuttings; cultivars must be propagated vegetatively.

Literature: Wilson, P.G., A taxonomic revision of the genus Correa (Rutaceae), *Transactions of the Royal Society of South Australia* **85**: 21–53 (1961).

1a. Corolla usually deciduous after pollination; filaments expanding towards base 2
  b. Corolla persisting after pollination; filaments not markedly broadened towards base 10
2a. Flowers green; calyx with 4 prominent, deep, triangular lobes **1. calycina**
  b. Flowers red, pink, green or white; if flowers greenish, calyx not deeply

lobed; if lobed, lobes linear or rounded, or calyx toothed 3
3a. Flowers white or pale pink; corolla splitting to base into 4 petals, less than 1.5 cm **2. alba**
  b. Flowers pink, red, green or green-yellow; corolla not usually splitting to base, more than 1.5 cm 4
4a. Prostrate shrubs; flowers not pendent; calyx with teeth between lobes (appearing 8-lobed) **3. decumbens**
  b. Erect shrubs; flowers usually pendent; calyx not lobed or if lobed without distinct teeth between lobes 5
5a. Flowers red with green lobes or all green or yellow-green; flower-stalks less than 4 mm; calyx closing over fruit immediately after pollination 6
  b. Flowers cream, pink, pale red or green (not bicoloured); flower-stalks more than 4 mm; calyx remaining open 8
6a. Bracteoles persistent **4. reflexa**
  b. Bracteoles deciduous 7
7a. Flowers green **5. glabra**
  b. Flowers red with green lobes **6. schlechtendalii**
8a. Leaves light green; flower-stalks not felted, becoming thicker towards apex **7. pulchella**
  b. Leaves dark green; flower-stalks felted, not thickened towards apex 9
9a. Flowers yellow-green or pale cream; corolla more than 2 cm **8. backhousiana**
  b. Flowers yellow-green or red with green lobes; corolla less than 2 cm **4. reflexa**
10a. Calyx with distinctly protruding base **9. baeuerlenii**
  b. Calyx cup-shaped **10. lawrenciana**

1. C. calycina Black. Illustration: Phillips & Rix, Shrubs, 19 (1989).

Shrub to 3 m. Leaves ovate to oblong, 2–4 cm × 8–14 mm, hairless above, hairy beneath; stalks to 8 mm. Flowers solitary, green. Calyx square in cross-section, hairy outside and inside, with 4 prominent triangular lobes to 1.5 cm. Corolla cylindric to 3 cm, deciduous. Stamens protruding; filaments opposite petals slightly broadened at base. Ovary hairy. Style hairless. *South Australia.* H5–G1. Spring.

A rare species in the wild, distinguished by its distinctly lobed calyx, which is hairy inside.

2. C. alba Andrews (*C. rufa* (Labillardière) Ventenat). Illustration: *Andrews' Botanical Repository* **1**: t. 18 (1798); Ventenat, Jardin de Malmaison **1**: t. 13 (1803); *Edwards' s Botanical Register* **6**: t. 515 (1821); Elliot & Jones, Encyclopaedia of Australian plants **3**: 90 (1984).

Shrub to 1.5 m, or prostrate. Leaves ovate to obovate, 8–50 × 6–35 mm, hairless or hairy above, hairy beneath; stalks to 8 mm. Flowers solitary or in clusters, green-white or pale pink. Calyx 4-lobed, 0.2–0.4 mm deep, with cream or brown hairs, constricted after pollination; lobes linear or rounded. Corolla 1.1–1.3 cm; lobes becoming more or less free and spreading. Stamens not protruding; filaments distinctly broader at base. Ovary hairy; style hairless. *E Australia (New South Wales, Victoria, Tasmania)*. H5–G1. Spring–summer.

One of the hardiest species, long cultivated outdoors in Ireland in shelter of south-facing walls.

Var. **alba** has leaves 1.5–3.5 cm, and flower-stalks more than 2.5 mm. *New South Wales*.

3. C. decumbens Mueller. Illustration: Elliot & Jones, Encyclopaedia of Australian plants **3**: 93 (1984).

Prostrate shrub. Leaves narrow, elliptic to oblong, to 45 × 9 mm, hairless above, with light brown felt beneath, margins slightly recurved. Flowers erect, terminal on short axillary branches, red or pink with green corolla-lobes. Calyx *c.* 4 mm deep, with 8 lobes; 4 lobes linear, to 5 mm, 4 lobes triangular, less than 2 mm. Corolla cylindric, 1.8–2.5 cm. Stamens markedly projected; filaments opposite the petals slightly broader at base. Ovary with rusty brown hairs. Style hairless, stigma prominently 4-lobed. *South Australia*. H5. Spring–summer.

In cultivation this hardy species is usually prostrate, but erect, spreading shrubs are found in the wild.

4. C. reflexa (Labillardière) Ventenat (*C. rubra* Smith; *C. speciosa* Andrews; *C. virens* Smith). Illustration: *Andrews' Botanical Repository* **10**: t. 653 (1812); *Botanical Magazine*, 1746 (1815); Phillips & Rix, Shrubs, 18 (1989); Elliot & Jones, Encyclopaedia of Australian plants **3**: 94–6 (1984).

Spreading or erect shrub to 1.5 m. Leaves oblong to round, *c.* 5 × 3 cm, upper surface smooth to rough, hairless to hairy; lower surface hairy to felted; stalks to 5 mm. Bracteoles persistent. Flowers erect or pendent, solitary or 2 or 3 at ends of short branches, green-yellow to red with green lobes. Calyx to 0.6 mm deep, with cream-brown felt, entire or with 4 or 8 slight lobes, becoming folded to flattened after pollination. Corolla cylindric, sometimes swollen at middle, 1.5–4 cm, deciduous. Stamens slightly projecting; filaments opposite the petals broadened at base and convex. Style hairy at base. Ovary hairy. *Australia (Queensland, New South Wales, Victoria, Tasmania, South Australia, Western Australia)*. H5–G1. Spring–summer.

A widespread, variable species; Wilson (1961) gives keys to the 4 varieties recognised among wild populations: var. **cardinalis** (Hooker) Court has oblong leaves and large red flowers and may be of hybrid origin.

5. C. glabra Lindley.

Shrub to 3 m. Leaves elliptic, 1–4 cm × 5–17 mm, usually hairless above, hairless or felted beneath; stalks 2–4 mm. Flowers solitary, on terminal or axillary shoots, pale green. Calyx 3–10 mm deep, *c.* 5 mm across, minutely lobed, becoming flattened and closed after pollination. Corolla cylindric, 1.5–3 cm, lobes sometimes recurved. Stamens projecting to half length of corolla; filaments broadened at base. Ovary hairy. Style hairless or hairy below. Fruit white. *E Australia (New South Wales, Victoria, South Australia)*. H5–G1. Summer.

A variable species in the wild, resembling *C. schlechtendalii* in some of its variants.

6. C. schlechtendalii Behr. Illustration: Phillips & Rix, Shrubs, 19 (1989).

Shrub to 2 m. Leaves elliptic or oblong, 1.5–4.5 cm × 8–15 mm, hairless above, sometimes sparsely hairy beneath; stalks to 5 mm. Bracteoles deciduous. Flowers solitary, on terminal or lateral shoots, red with green lobes. Calyx to 0.6 mm deep, hairless or sparsely hairy, entire or with 4 small teeth, becoming flattened after pollination. Corolla cylindric, 1.6–3 cm, deciduous. Stamens projecting to half length of corolla; filaments broadened at base. *Australia (South Australia, Victoria)*. H5–G1. Summer.

7. C. pulchella Sweet. Illustration: Sweet, Flora Australasica, t. 1 (1827); *Edwards's Botanical Register* **15**: t. 1224 (1829); *Botanical Magazine*, 4029 (1843).

Low shrub, rarely more than 60 cm. Leaves pale green, linear to elliptic or ovate, 1–2 cm × 2–15 mm, usually hairless; stalks to 4 mm. Flowers solitary, pendent or held horizontally, pink-red. Calyx usually hairless, *c.* 4 mm deep, not lobed, remaining open after pollination. Corolla cylindric, 1.5–2.7 cm, deciduous. Stamens almost equalling corolla; filaments distinctly broadened at base. Ovary hairless to hairy. Style hairless or slightly hairy at base. Disc white to purple. *South Australia*. H5–G1. Summer.

C. × harrisii Paxton; *C. pulchella × C. reflexa*. Illustration: *Paxton's Magazine of Botany* **7**: 79 (1840); *The Garden* **113**: 48 (1988); Phillips & Rix, Shrubs, 18 (1989). Shrubs to 1 m. Flowers pendent, red. Corolla cylindric, lobes reflexing. Stamens protruding.

8. C. backhousiana J.D. Hooker (*C. speciosa* var. *backhousiana* (Hooker) Rodway). Illustration: *Hooker's Icones Plantarum* **1**: t. 2 (1837); *Botanical Magazine*, n.s., 289 (1957); Phillips & Rix, Shrubs, 18 (1989); Brickell (ed.), RHS A–Z encyclopedia of garden plants, 309 (2003).

Shrub or bushy tree to 5 m. Leaves ovate to elliptic, *c.* 3 × 2 cm, leathery, dark green, hairless above, with rusty brown felt beneath; stalks 4–7 mm. Flowers solitary or in clusters, green or pale cream. Calyx *c.* 5 mm deep, with rusty brown hairs outside, minutely lobed or wavy, remaining cup-shaped after pollination. Corolla cylindric, 2–3 cm. Stamens included or protruding; filaments opposite the petals distinctly broader at base. Ovary hairy; style hairless or with stellate hairs at base. *Australia (Tasmania)*. H5. Summer.

9. C. baeuerlenii Mueller. Illustration: *Hooker's Icones Plantarum* **23**: t. 2245 (1892); Elliot & Jones, Encyclopaedia of Australian plants **3**: 92 (1984).

Shrub. Leaves ovate to elliptic, 2–6.5 × 1–2.2 cm, sparsely hairy when young, soon hairless. Flowers solitary, terminal or axillary, green-yellow. Calyx to 7 mm deep, *c.* 6 mm across, sparsely hairy, wavy or irregularly toothed; base becoming saucer-shaped, *c.* 1 cm across, horizontal at first, later reflexed. Corolla cylindric, 2–3 cm. Stamens protruding; filaments slightly broadened at base. Ovary very hairy. Style hairless. *Australia (Clyde River valley, New South Wales)*. H5. Summer.

A rare and restricted species, remarkable for its chef's hat-shaped calyx, which is distinctive.

10. C. lawrenciana J.D. Hooker
(*C. ferruginea* Backhouse). Illustration:
Maund, Botanist **3**: t. 124 (1839); Elliot &
Jones, Encyclopaedia of Australian plants
3: 93 (1984).

Shrub or tree to 10 m, Leaves elliptic to
ovate, 2.5–8 × 7–60 mm, hairless above,
hairless to felted beneath. Flowers solitary
or 2–7 in a cyme, green-yellow to red.
Calyx 3–10 mm deep, *c.* 5 mm across,
variously lobed. Corolla cylindric,
1.5–3.2 cm, *c.* 7 mm across. Stamens
projecting to half length of corolla;
filaments usually not broadened at base.
Ovary hairy. Style hairless. *Australia (New
South Wales, Victoria, Tasmania)*. H5.
Spring–summer.

Withered corolla and filaments persist
even after fruit has dehisced.

12. ORIXA Thunberg
E.C. Nelson
Shrubs to 3 m, dioecious, deciduous;
branches without spines, with grey felt
when young. Leaves simple, alternate, dark
green (turning pale yellow in autumn),
aromatic, obovate to elliptic,
5–12 × 3–7 cm, entire, hairless except for
on young leaves. Flowers unisexual;
males in short (*c.* 3 cm), axillary
racemes, green; females solitary; flower-
stalks 1–2 cm. Sepals 4, fused at base.
Petals 4, spreading, *c.* 0.3 mm. Male
flowers with 4-lobed disc and 4 stamens.
Female flowers with deeply 4-lobed ovary;
single style; 4-lobed stigma. Fruit of 4
carpels; 1 seed per carpel, explosively
expelled when ripe.

A genus of 1 species. The aroma of the
leaves is described as spicy or fetid, and the
plant is used in Japan for hedging.

1. O. japonica Thunberg. Illustration:
Iconographia cormophytorum Sinicorum
2: 546, t. 2822 (1972); Hayashi, Woody
plants of Japan, 382 (1985). *Korea, China,
Japan*. H4. Summer.

'Variegata' has cream leaf-margins,
shading to grey-green.

13. TETRADIUM Louriero
E.C. Nelson
Deciduous trees; young shoots with pith,
lenticels prominent; axillary buds
prominent. Leaves compound, pinnate.
Inflorescence a terminal corymb. Flowers
unisexual; sepals 4 or 5; petals 4 or 5.
Stamens 4 or 5 or absent. Fruit with 4 or
5 follicles, each with 1 or 2 smooth, shiny,
black seeds.

These are distinguished from *Euodia* (in
the strict sense) by their pinnate leaves
and terminal inflorescences of unisexual
flowers; *Euodia* has leaves simple or with 3
leaflets, axillary inflorescences and bisexual
flowers. *Tetradium* includes all the hardy
species hitherto cultivated in Europe as
Euodia (often incorrectly spelled as *Evodia*).
Tetradium species can be raised easily from
seed.

Literature: Hartley, T.G., A revision of
the genus Tetradium (Rutaceae), *Gardens
Bulletin of Singapore* **34**: 91–131 (1981);
Flanagan, M., Notes on the genus
Tetradium, *Kew Magazine* **5**: 181–91
(1988).

1a. Leaflets with conspicuous oil-glands 2
 b. Leaflets without conspicuous oil-
glands 3
2a. Leaflets with regular forwardly
pointing teeth **3. fraxinifolium**
 b. Leaflets entire or irregularly toothed
4. ruticarpum
3a. Leaflets hairy on veins beneath
1. daniellii
 b. Leaflets hairless and glaucous
2. glabrifolium

1. T. daniellii (Bennett) Hartley (*Euodia
daniellii* (Bennett) Hemsley; *E. henryi* Dode;
E. hupehensis Dode; *E. sutchuenensis* Dode;
E. velutina Rehder & Wilson; *E. vestita*
Rehder & Wilson). Illustration:
Iconographia cormophytorum Sinicorum
2: 549, t. 2827 (1972); Phillips, Trees in
Britain, Europe and North America, 116
(1978).

Tree or shrub, 15–20 m. Leaves to
50 cm; leaflets 5–15, ovate to elliptic,
wedge-shaped or cordate at base, tapering
to point at apex, without oil-glands
beneath, hairy beneath at least on veins,
but not glaucous. Flowers small, white.
Korea, China. H3. Late summer.

Plants cultivated as *Euodia velutina* have
dense white felt on leaflet undersurfaces;
the upper surface is also downy; this
appears to represent an extreme variant of
the species. Plants grown as *E. hupehensis*
have dark green, almost leathery leaflets.

2. T. glabrifolium (Bentham) Hartley
(*Euodia glauca* Miquel; *E. fargesii* Dode).
Illustration: Iconographia cormophytorum
Sinicorum **2**: 548, t. 2825 (1972).

Distinguished from *T. daniellii* by its
hairless leaflets with glaucous
undersurfaces and pointed apices. Flowers
with 5 carpels, each with 2 ovules; fruit

with 5 one-seeded follicles. *E Asia*. H4.
Summer.

Very rare in cultivation.

3. T. fraxinifolium (Hooker) Hartley
(*Euodia fraxinifolia* (Hooker) Bentham).

Differs from *T. daniellii* as it possesses
conspicuous oil-glands on the lower
surface of the leaves (which may be up to
75 cm); leaflets with teeth pointing
forwards. *Nepal to SW China, N Thailand to
N Vietnam*. H5. Summer.

4. T. ruticarpum (Jussieu) Hartley (*Euodia
rutaecarpa* (Jussieu) Bentham; *E. baberi*
Rehder & Wilson).

Resembles *T. fraxinifolium* in possessing
conspicuous oil-glands, but has leaves to
40 cm; leaflets entire or irregularly toothed.
China, Taiwan. H4–5.

14. CHOISYA Kunth
E.C. Nelson
Shrubs with aromatic, opposite, palmately
divided leaves; leaflets 3–15. Flowers in
terminal or axillary panicles, or solitary.
Sepals usually 5, deciduous. Petals usually
5, white, hairless. Stamens usually 10,
inner whorl (opposite petals) shorter than
outer whorl. Ovary hairy, with 5 carpels
fused at base; carpels with hairless apical
horn joined by the centrally attached 5-
furrowed style; stigmas 5, capitate. Fruit of
2 carpels each with 1 or 2 seeds.

A genus of 5 species from south-western
USA and Mexico; only *C. ternata* is widely
cultivated in Europe, and it is represented
in cultivation by a sterile clone.

Literature: Muller, C.M., A revision of
Choisya, *American Midland Naturalist* **24**
(3): 729–42 (1940); Benson, L. & Darrow,
R.A., Trees and shrubs of the southwestern
deserts, edn 3, 131–2 (1981).

1a. Leaves with 3 leaflets; leaflets oblong
to obovate, entire, without
prominent glands; flowers in a
diffuse panicle **1. ternata**
 b. Leaves with more than 3 leaflets;
leaflets linear, with very prominent
warty glands; flowers solitary in leaf-
axils or rarely in pairs 2
2a. Leaflets 5–13 **2.
dumosa**, see var. **dumosa**
 b. Leaflets 3–7 3
3a. Petals 1–1.3 cm **2.
dumosa**, see var. **arizonica**
 b. Petals less than 1 cm **2.
dumosa**, see var. **mollis**

1. C. ternata Kunth. Illustration: *Botanical
Magazine*, n.s., 318 (1958); Phillips & Rix,

Shrubs, 35 (1989); Brickell (ed.), RHS A–Z encyclopedia of garden plants, 265 (2003).

Shrub to 2 m, hairless. Leaves with 3 leaflets, bright green, hairless and without warty glands; leaflets oblong to obovate, to 5 × 3 cm. Flowers in terminal panicles, *c.* 3 cm across, fragrant. *Mexico.* H5. Spring–early summer.

A useful evergreen that tolerates lime, and will often produce flowers out of season in late autumn. Recently a sickly yellow-green foliaged cultivar ('Sundance') has been introduced and is widely promoted; it is more tender than the parent clone, being intolerant of cold wind and frost.

'Aztec Pearl' (a selected seedling of the hybrid *C. dumosa* var. *arizonica* × *C. ternata*) is intermediate between parent species.

2. C. dumosa (Torrey) Gray (*Astrophyllum dumosum* Torrey; *C. arizonica* Standley; *C. mollis* Standley). Illustration: Die natürlichen Pflanzenfamilien **4**: f. 69 (1896).

Shrub to 2 m, shoots hairy. Leaves opposite, palmate, stalks hairy; leaflets 3–13, to 5 cm × 5 mm. Flowers solitary (rarely in pairs) in leaf-axils, to 2.5 cm across. Carpels 5, hairy, to 7 mm, warty or with prominent sharp points on apex and outer sides. *USA (New Mexico, Arizona, Texas).* H5–G1. Summer.

This variable species is now considered to consist of 3 varieties. At least 2 of these are in cultivation in Europe. Var. **arizonica** (Standley) Benson. Illustration: Benson & Darrow, Trees and shrubs of the southwestern deserts, 132 (1981); Phillips & Rix, Shrubs, 35 (1989). Distinguished by the densely felted shoots and leaf-stalks; leaflets 3–7, to 5 cm × 5 mm; flowers large; carpels with long prominent points. Var. **dumosa**. Illustration: Benson & Darrow, Trees and shrubs of the southwestern deserts, edn 3, 131 (1981). Shoots not densely felted, hairs spreading; leaflets bright green, 5–13, to 4 cm × 3 mm; flowers small; carpels with prominent glands. Var. **mollis** (Standley) Benson. Illustration: Phillips & Rix, Shrubs, 34 (1989). Like *C. dumosa* var. *dumosa* but with 3–5 leaflets, to 4.5 cm × 5 mm; flowers large.

15. ZANTHOXYLUM Linnaeus
A. Brady
Deciduous or semi-evergreen trees or shrubs, more or less spiny, bark aromatic. Leaves alternate, pinnate, dotted with glands. Flowers unisexual or bisexual, usually in cymes or panicles, yellow-green. Sepals 3–5. Petals absent or up to 5. Stamens 3–5. Fruit a capsule or follicle.

A genus of about 200 species, native to North and South America, Africa, Asia and Australia. Species described here are generally hardy and are best grown in loamy soil. Propagation is by seed, semi-ripe shoot cuttings, root cuttings or suckers. The generic name is sometimes found, especially in older literature, spelled *Xanthoxylum*, *Xanthoxylon* or *Zanthoxylon*. Until recently 9 species from eastern Asia were included in the genus *Euodia*.

Literature: Hartley, A revision of the Malesian species of Zanthoxylum (Rutaceae), *Journal of the Arnold Arboretum* **47**: 171–221 (1966).

1a. Leaflets downy beneath, aromatic; flowers in axillary cymes, appearing before leaves **1. americanum**
  b. Leaflets hairless or nearly so beneath; flowers terminal, rarely axillary 2
2a. Leaf-stalks with prominent wing; prickles in pairs at base of each leaf **2. armatum**
  b. Leaf-stalks not or scarcely winged; prickles not as above 3
3a. Leaves 25–70 cm 4
  b. Leaves less than 25 cm 5
4a. Leaflets 11–27, 7–13 cm **3. ailanthoides**
  b. Leaflets 5–19, to 5 cm **4. clava-herculis**
5a. Spines hooked **5. stenophyllum**
  b. Spines straight 6
6a. Spines in pairs; flowers in panicles **6. piperitum**
  b. Spines solitary; flowers in cymes 7
7a. Leaflets 13–23, lanceolate **7. schinifolium**
  b. Leaflets 7–1, ovate **8. simulans**

1. Z. americanum Miller (*Z. fraxineum* Willdenow). Illustration: Britton & Brown, Illustrated flora of the northern United States and Canada **2**: 353 (1897); Gleason, Illustrated flora of the north-eastern United States and adjacent Canada **2**: 467 (1952); Phillips, Trees in Britain, Europe and North America, 214 (1978).

Deciduous tree or shrub to 8 m. Shoots downy, prickles in pairs under leaf-buds. Leaves *c.* 30 cm; leaflets 5–11, ovate, 4–7 cm, downy beneath, dark green above, aromatic. Flowers yellow-green, in axillary cymes, appearing before leaves. Fruit black. *E North America.* H1. Early spring.

Called the toothache tree for its medicinal properties.

2. Z. armatum de Candolle (*Z. planispinum* Siebold & Zuccarini; *Z. alatum* Roxburgh invalid; *Z. alatum* var. *planispinum* (Siebold & Zuccarini) Rehder & Wilson). Illustration: *Botanical Magazine*, 8754 (1918); Hayashi, Woody plants of Japan, 380 (1985).

Dioecious, deciduous shrubs to 4 m. Shoots hairless; spines in pairs at base of each leaf, broad, flat, 6–20 mm. Leaves 13–25 cm; stalks 1.5 cm, winged; leaflets 3–10, ovate to lanceolate, 3–12 cm, toothed. Flowers yellow, in racemes 3–5 cm. Fruit red, warty, spherical, 4–6 mm across. *Pakistan to Japan, Philippines & Indonesia.* H1. Spring.

Z. planispinum is sometimes separated as it has fewer (3–5) leaflets, but the number of leaflets in this species is very variable.

3. Z. ailanthoides Siebold & Zuccarini. Illustration: Iconographia cormophytorum Sinicorum **2**: 542, t. 2814 (1972).

Deciduous tree to 20 m; shoots with dense, persistent prickles. Leaves pinnate, 25–70 cm; leaflets 11–27, ovate to lanceolate, 7–13 cm, hairless, toothed. Cymes to 13 cm; flowers white-green. Stamens yellow. Fruit red, shining, beaked, spherical, *c.* 5 mm across. *Japan, China, Taiwan.* H2. Spring.

4. Z. clava-herculis Linnaeus. Illustration: Sargent, Silva of North America **1**: t. 29 (1891); Gleason, Illustrated flora of the north-eastern United States and adjacent Canada **2**: 467 (1952).

Dioecious shrub or tree to 10 m; trunk and branches blue-grey, with stout prickles at apex of corky protrusions. Leaves to 30 cm; leaflets 5–19, lanceolate to ovate, to 5 cm, lustrous above; stalks and midrib prickly, slightly toothed. Flowers in terminal cymes. Sepals minute. Petals green. Fruit with single seed, wrinkled. *S USA.* H3. Spring.

5. Z. stenophyllum Hemsley. Illustration: Iconographia cormophytorum Sinicorum **2**: 544, t. 2818 (1972).

Deciduous, climbing shrub to 4 m. Shoots hairless, prickles short, hooked, stiff. Leaves hairless, main stalks prickly; leaflets 7–13, ovate, oblong to lanceolate, 4–10 cm, wedge-shaped at base, finely toothed, slender-pointed. Flowers in panicles 5–9 cm across. Fruits red, glossy *c.* 4 mm across. *W China.* H3. Spring.

6. Z. piperitum de Candolle. Illustration: Hegi, Illustrierte Flora von Mittel-Europa **5** (1): 47 (1925); Bean, Trees & shrubs hardy in the British Isles, edn 8, **4**: 773 (1980); Hayashi, Woody plants of Japan, 380 (1985); Brickell (ed.), RHS A–Z encyclopedia of garden plants, 1096 (2003).

Compact tree or shrub to 7 m. Shoots downy, prickles in pairs, 6–12, *c.* 1 cm. Leaves 10–15 cm, midrib downy, prickly; leaflets 11–23, ovate, 1.5–4 cm, hairless, toothed, apex of terminal leaflet notched. Flowers green, in panicles, 1.5–3 cm, terminal on short axillary shoots. Fruits red, seeds black. *N China, Korea, Japan.* H2.

Seeds are used as pepper in Japan.

7. Z. schinifolium Siebold & Zuccarini. Illustration: Iconographia cormophytorum Sinicorum **2**: 543, t. 2816 (1972).

Dioecious, deciduous shrub to 3 m. Shoots hairless; branch prickles solitary, to 1.2 cm. Leaves 8–20 cm, main stalks prickly; leaflets 13–23, lanceolate, 1.5–4 cm, sometimes minutely downy, bluntly toothed, apex notched. Flowers green, *c.* 1.2 cm across, in cymes 5–6 cm across. Seed blue-black. *China, Japan, Korea.* H1. Spring.

8. Z. simulans Hance. Illustration: Iconographia cormophytorum Sinicorum **2**: 539, t. 2807 (1972); Brickell (ed.), RHS A–Z encyclopedia of garden plants, 1096 (2003).

Spreading shrub or tree to 8 m. Shoot with stout prickles, *c.* 1.5 cm, flattened at base. Leaves to 12 cm, axis and midrib prickly; leaflets 7–11, ovate, 1.5–5 cm, hairless, glossy, bluntly toothed. Flowers in cymes *c.* 5 cm wide. Fruit red with darker spots. *China, Japan.* H2. Spring.

16. PHELLODENDRON Ruprecht

F.T. de Vries

Tall, deciduous trees; bark often thick and corky. Leaves to 40 cm, opposite, pinnate with a terminal leaflet; base of stalks swollen, hiding the bud completely. Flowers unisexual, inconspicuous, yellow-green, in terminal panicles. Petals 5–8; sepals 5–8, small. Male flowers with 5 or 6 stamens, alternating with and twice as long as sepals; ovary rudimentary. Female flowers with 5 or 6 staminodes; ovary 5-celled with short stout style and 5-lobed stigma. Fruit *c.* 1 cm, an orange-like, indehiscent drupe with 5 seeds and a tough black skin.

A genus of 10 species of aromatic trees of north-east Asia, mainly attractive because of their leaves and black fruits in autumn, and an often picturesque habit. They grow well on chalky, rich soil, and are propagated by seed, or by cuttings with a heel of older wood in summer, or root-cuttings in winter.

1a. Bark thick and corky when mature 2
 b. Bark thin, plated, sometimes finely channelled 3
2a. Young branches grey to orange-yellow; leaves glossy green above, bluish and hairless beneath **1. amurense**
 b. Young branches reddish brown; leaves dull green above, green and downy beneath **2. lavallei**
3a. Leaves bluish and hairless beneath, except on the midrib; panicles nearly hairless **3. sachalinense**
 b. Leaves light green and densely downy beneath; panicles very downy 4
4a. Bark deep brown; leaflets very asymmetric **4. japonicum**
 b. Bark grey-brown; leaflets symmetric **5. chinense**

1. P. amurense Ruprecht. Illustration: Sargent, Trees & shrubs **13**: 93 (1905); Köhne, Deutsche Dendrologie, 348 (1893); Makino, Illustrated flora of Japan, 1166 (1956).

Tree to 15 m, with a wide crown. Bark light grey, thick, corky, deeply fissured; young branches grey to orange-yellow, hairless; winter-buds coated with silvery hairs. Leaves 25–35 cm, with 5–13 leaflets, hairy only on margins and base of midribs, main stalks nearly hairless; leaflets 5–13 × 1.5–7 cm, ovate to ovate-lanceolate, with an obtuse or rounded base and a long-pointed tip, dark glossy green above, bluish and paler beneath. Panicle 7–15 × 3.5–8 cm, erect, terminal, corky when older; fruit 8–13 mm. *E former USSR, N China, Japan.* H1. Early summer.

2. P. lavallei Dode (*P. amurense* var. *lavallei* (Dode) Sprague). Illustration: *Botanical Magazine*, 8945 (1922).

Tree to 15 m, with a wide crown. Bark brown, thick, corky. Young branches purplish brown, downy at first; winter-buds reddish brown. Leaves 20–40 cm, with 5–13 leaflets, white hairy on margins, midribs and chief veins; main stalk softly downy; leaflets 5–16 × 2.5–4.5 cm, ovate-elliptic to oblong-lanceolate,

asymmetrically rounded or wedge-shaped at base, tip acuminate, dull yellow-green above, lighter green beneath. Panicle *c.* 12 × 6–8 cm, loose, downy. Ovary hairless. Fruit 8–11 mm, juicy. *Japan.* H3. Early summer.

3. P. sachalinense (F. Schmidt) Sargent (*P. amurense* var. *sachalinense* F. Schmidt). Illustration: Lee, Forest botany of China, 192 (1935); Sargent, Trees & shrubs **13**: 94 (1905).

Tree *c.* 15 m. Bark dark brown, thin, finely channelled, eventually plated. Young branches red-brown; winter-buds rusty-downy. Leaves 22–30 cm, with 5–11 leaflets, hairless or sometimes downy on midribs; main stalk downy; leaflets 6–12 × 2–5 cm, broadly ovate to oblong, bases rounded or tapered, tips acuminate, dull green above, bluish beneath. Panicle 8–15 × 6–8 cm, nearly hairless, densely flowered. Fruit *c.* 1 cm. *Japan, E former USSR (Sakhalin), Korea.* H1. Early summer.

4. P. japonicum Maximowicz (*P. amurense* var. *japonicum* (Maximowicz) Ohwi). Illustration: Sargent, Trees and shrubs **13**: 95 (1905); Csapody & Tóth, A colour atlas of flowering trees and shrubs, 155 (1982).

Tree *c.* 15 m, bushy headed, of stiff habit. Bark deep brown, thin, channelled, plate-like; young branches reddish brown, bloomed. Leaves 23–37 cm, with 7–15 leaflets, thick grey down on veins beneath; stalks densely downy; leaflets 5–11 × 2.5–4.5 cm, broadly ovate to oblong, with very asymmetrically rounded or truncate bases, and acuminate tips, dull green above, light green beneath. Panicle 6–10 × 4–5 cm, very downy, erect. Ovary hairless. Fruit 9–13 mm. *Japan.* H3. Early summer.

5. P. chinense Schneider. Illustration: Schneider, Illustriertes Handbuch der Laubhölzkunde **2**: 79c (1907–12); Brickell (ed.), RHS A–Z encyclopedia of garden plants, 799 (2003).

Tree *c.* 20 m. Bark thin, dark grey-brown, slightly fissured; young branches reddish brown, thinly downy. Leaves 20–37 cm, with 7–13 leaflets, pale downy beneath, especially on the midribs; stalks densely downy; leaflets 5–12 × 2.5–4.5 cm, oblong-ovate to oblong-lanceolate, margins nearly parallel, with rounded, broadly tapered bases, and pointed tips; dull dark yellow-green above, downy, light green beneath. Panicle 5–20 × 6–12 cm, very downy. Ovary downy. Fruit 9–11 cm. *C China*. H2. Early summer.

A slightly less downy variety of this species is often called var. **glabriusculum** Schneider.

17. PTELEA Linnaeus

C. Gorman & E.C. Nelson

Aromatic, spineless, deciduous shrubs and trees. Leaves alternate, with 3–5 stalkless leaflets. Flowers green or yellow-white, mostly unisexual, in terminal corymbs. Sepals 4 or 5, minute. Petals 4 or 5. Stamens 4 or 5; filaments hairy on inner surface. Fruit a 2-seeded samara with a distinct, thin, flat, broadly encircling wing.

This genus from North America is notoriously difficult taxonomically with species estimated from 3 to about 60. The fruits were once used as a substitute for hops in brewing.

Literature: Greene, E.L., The genus Ptelea in western and southwestern United States and Mexico, *Contributions from the US National Herbarium* **10**: 49–78 (1906).

1a.	Bark white when mature
	3. lutescens
b.	Bark brown or red-brown when mature	2
2a.	Leaflets 2–7 cm, leaf-stalks 2–5 cm
	1. crenulata
b.	Leaflets 6–12 cm, leaf-stalks 5–10 cm
	2. trifoliata

1. P. crenulata Greene (*P. baldwinii* Torrey var. *crenulata* (Greene) Jepson; *P. aptera* Parry). Illustration: Everett, New York Botanical Gardens illustrated encyclopedia of horticulture **8**: 2841 (1981).

Plant to 5 m, bark red-brown. Leaflets narrowly ovate to lanceolate, 2–7 cm, entire, toothed or scalloped, yellow-green, hairless above, more or less downy beneath; terminal leaflet longest; stalks 2–5 cm. Flowers *c*. 8 mm across; filaments downy towards base. Samara disc-shaped, to 2 cm across, wing narrow; style persistent. *USA (California)*. H3. Summer.

2. P. trifoliata Linnaeus (*P. isophylla* Greene; *P. serrata* Small). Illustration: Gleason, Illustrated flora of the north-eastern United States and adjacent Canada, 469 (1952); Everett, New York Botanical Gardens illustrated encyclopedia of horticulture **8**: 2841 (1981).

Tree with a rounded crown, to 8 m; bark brown. Leaflets 3 (rarely 1) ovate-elliptic, 6–12 × 2–3.5 cm, dark green, shiny above,

paler and downy (when young) beneath, margins sometimes vaguely scalloped; stalks 5–10 cm. Flowers *c*. 8 mm across, on slender downy stalks; corymbs 5–7 cm across. Petals densely hairy within. Ovary usually hairless. Samara disc-shaped, 1.5–2.5 cm across, straw to pale green, net-veined around seed. *E USA, S Canada*. H4. Summer.

A variable species. Cultivars are listed with fastigiate habit, with blue-green foliage and with 1 or 5 leaflets. Fruit superficially resembles that of *Ulmus* but rarely contains viable seed in British gardens.

3. P. lutescens Greene.

Differs from *P. trifoliata* by its young shoots yellow-grey maturing to shining pale grey, bark white; leaflets hairless, lanceolate 4.5–9 × 1–2 cm, faintly round-toothed, apex slenderly pointed. Fruit often 2 or 3 winged, wrinkled, 2–2.5 cm wide, glandular in the centre with a flattened ovate seed *USA (Grand Canyon, Arizona)*. H5. Summer.

18. SKIMMIA Thunberg

N.P. Taylor & H.S. Maxwell

Unisexual or bisexual, evergreen shrubs to trees, more or less aromatic and almost hairless. Basal half of shoots with oblong to lanceolate, acute bracts, falling early; lenticels inconspicuous. Young stems green, brownish or reddish becoming creamy grey to yellow. Leaves clustered near tips of each season's growth, alternate, simple, entire or slightly scalloped near tip, dotted with translucent glands, persisting for 1 or more years. Flowers many, small, in short terminal compound panicles, white or yellow, sometimes tinged with pink, normally unisexual, dioecious; female and bisexual inflorescences often with only 1–5 flowers; flower-parts usually in 4s or 5s. Carpels 2–5, fused. Style short or about equal to the ovary. Stigma irregular with 2–5 lobes. Fruit fleshy, red or black, more or less spherical, containing 1–4 (rarely 5) seeds.

A genus of 4 species from China, Japan and the Himalaya. They are easily propagated by seed or cuttings and require a slightly shady area, usually with some shelter during winter and a moist, not too alkaline soil.

1a.	Leaves slightly aromatic when rubbed, with 4–8, rarely 9–11, pairs of veins (not always visible), stalks green or reddish	**1. japonica**

b.	Leaves strongly aromatic when rubbed, with 7–20 pairs of veins (not always visible), stalks green or brownish	2
2a.	Flowers opening fully; petals spreading or reflexed, white or greenish white; fruit red or black	3
b.	Flowers not opening fully; petals *c*. 45° or erect, creamy white or yellow; fruit red	5
3a.	Low creeping shrub, usually less than 1 m; leaves to 10 × 3.5 cm; fruit black	**2.**
	laureola, see subsp. **laureola**
b.	Shrubs or trees eventually to 7 m; some or all leaves to 15 × 3.5 cm; fruit black or red	4
4a.	Flower parts always in 5s; plants male, female or bisexual; fruit black
	2. laureola, see subsp. **multinervia**
b.	Flower parts sometimes in 4s, plants always female; fruit red
	3. × confusa, see 'Isabella'
5a.	Inflorescence more or less spherical, to 5 cm; flowers unpleasantly scented, greenish yellow to yellow; plants male or female; fruit orange-red	**4. anquetilia**
b.	Inflorescence pyramidal, to 15 cm; flowers sweetly scented, creamy white; plants always male
	3. × confusa

1. S. japonica Thunberg. Illustration: *Botanical Magazine*, 8038 (1905); RHS dictionary of gardening 1962 (1951); Bean, Trees & shrubs hardy in the British Isles **4**: pl. 47 (1980); Everett, New York Botanical Gardens illustrated encyclopedia of horticulture **9**: 3162 (1982).

Creeping or erect shrub to 3 m, rarely more in cultivation, weakly aromatic. Young stems green, brownish or reddish. Leaves oblanceolate to obovate or elliptic, variable in size, dark to light or yellowish green; tips acuminate, acute, rounded or notched; bases tapering to rounded; blades papery to leathery, smooth or finely granular when dry, with 4–8, rarely 9–11, pairs of veins (not always visible). Leafstalks to 2.2 cm, green or reddish. Male inflorescences well-developed, broadest towards tips, with numerous ascending branches; female and bisexual inflorescences conspicuous, or small with 1–5 flowers. Flowers unisexual or bisexual, with parts in 4s or 5s, very sweetly scented. Petals spreading, white, sometimes tinged with pink or red. Anthers yellow. Fruit red. *E Asia*. H2. Spring.

Subsp. **japonica**. To 3 m, rarely taller. Leaves oblanceolate to obovate or oblong, sometimes elliptic, acute to rounded, less often acuminate or notched, leathery, with 6–8, rarely 9–11, pairs of veins. Flowers unisexual with parts mostly in 4s, but in 5s or 6s at tips of the main inflorescence branches. *E Asiatic Islands (Sakhalin, S Kurils, south to Taiwan).* Var. **japonica** (*S. oblata* Moore; *S. fragrans* Carrière; *S. rubella* Carrière). Shrub or small tree to 5 m, rarely with adventitious roots. Leaves 15 × 5 cm (rarely to 17 × 6 cm). *Japan, Ryukyu, Taiwan.* Var. **intermedia** Komatsu (*S. repens* Nakai; *S. japonica* var. *repens* (Nakai) Ohwi). Low, densely-branched shrub, often creeping and with adventitious roots, rarely exceeding 50 cm. Leaves 9 × 2.6 cm (rarely to 13 × 3.5 cm). *E Asiatic Islands (Sakhalin, Kuril), Japan.* Subsp. **reevesiana** (Fortune) N.P. Taylor & Airy Shaw (*S. reevesiana* (Fortune)). Illustration: *Kew Magazine* **4**: pl. 90 (1987) – as 'Chilan Choice'; *Garden Answers* (September): 62 (1990); Brickell (ed.), RHS A–Z encyclopedia of garden plants, 988 (2003). Low shrub or tree to 7 m, erect or rarely spreading, normally without adventitious roots. Leaves dark green, elliptic, 6.5–10 × 2–2.5 cm, acuminate, papery to leathery, with 4–6 (rarely 7 or 8) pairs of veins. Flowers bisexual with parts in 5s (rarely unisexual with parts in 4s in Taiwanese collections and some cultivated variants). *S & SE China, Taiwan, S Vietnam, Philippines.*

2. S. laureola (de Candolle) Walpers.

Creeping or erect shrubs or trees with well-defined trunk, to 13 m, strongly aromatic. Young stems 3–6 mm thick. Leaves oblanceolate to narrowly elliptic, very variable in size, entire, very dark green when mature, leathery, finely granular when dry, with 7–20 pairs of veins, acute to acuminate or sharp, rarely rounded; base gradually tapering. Stalks to 3.2 cm, usually shorter, light green or tinged with brown. Inflorescences spherical to pyramidal, often well-branched at base, female and bisexual inflorescences sometimes with less than 10 flowers. Flowers bisexual or unisexual, parts in 5s, sweetly scented at first. Petals spreading to reflexed, creamy or greenish white. Anthers orange-yellow. Fruit black. *Nepal, NE India, Burma & N Vietnam to China.* H2. Spring.

Subsp. **multinervia** (C.C. Huang) N.P. Taylor & Airy Shaw (*S. multinervia*

C.C. Huang). Illustration: *Kew Magazine* **4**: pl. 90 (1987). Stems arising from a common base or trunk. Leaves to 24 × 6 cm, mostly oblanceolate, with 10–20 (rarely fewer) pairs of more or less conspicuous veins in the largest leaves; stalks to 3.2 cm. Leaves sometimes similar to those of subsp. *laureola*. *Nepal, NE India, Bhutan, Burma, China, N Vietnam.* Subsp. **laureola**. Creeping shrub, seldom exceeding 1 m; young stems 3–4 mm thick. Leaves to 10 × 3.5 cm, usually with 7–10 pairs of veins (usually indistinct); stalks to 8 mm. *Nepal, NE India, Bhutan, Burma, China.* Subsp. lancasteri N.P. Taylor (*S. melanocarpa* Rehder & Wilson). Differs from subsp. *laureola* in its thinner, more gradually tapered leaves with finer points. *N & SC China.*

3. S. × confusa N.P. Taylor (*S. laureola* misapplied; *S. multinervia* misapplied). Illustration: *Kew Magazine* **6**: pl. 136 (1989).

Spreading shrub, 50–300 cm, strongly aromatic. Young stems 3–5 mm thick, green later with yellowish grey bark. Leaves oblanceolate to narrowly elliptic, 7.5–15 × *c.* 2.5 cm, entire, acute to acuminate, gradually tapering at the base, leathery, yellowish green or darker, finely granular above when dry, with 10–15 pairs of veins; stalks poorly defined, to *c.* 1.5 cm, green, rarely slightly reddish brown. Inflorescences large, male *c.* 15 × 15 cm, well-branched, pyramidal. Flowers unisexual, parts in 4s or 5s, sweetly scented; male flowers with petals remaining at *c.* 45° not fully expanded, creamy white, anthers orange; female flowers with petals fully expanded, white. Fruit occasionally developed, depressed and spherical, bright red, to 1 cm. *Garden origin.* H2. Spring.

A hybrid between *S. anquetilia* and *S. japonica* subsp. *japonica*, which is widely distributed in British gardens and frequently offered for sale wrongly identified as *S. japonica* or *S. laureola*. The most common clones in cultivation are 'Chelsea Physic' and 'Kew Green' (illustration: Brickell (ed.), RHS A–Z encyclopedia of garden plants, 987, 2003), both male, and 'Isabella', a female plant. 'Chelsea Physic' is the smaller of the 2 male plants with a lower more spreading habit, generally narrower leaves and smaller inflorescences.

4. S. anquetilia N.P. Taylor & Airy Shaw (*S. laureola* misapplied). Illustration: Everett,

New York Botanical Gardens illustrated encyclopedia of horticulture **9**: 3162 (1982); *Botanical Magazine*, n.s., 789 (1980); *The Plantsman* **1**(4): 226 (1980).

Creeping or erect shrub to 2 m, pungently aromatic throughout. Young stems to 6 mm thick, later with a waxy, yellowish and often smooth bark. Leaves narrowly oblanceolate to oblong elliptic, to 18 × 4.5 cm, entire, rounded to gradually acuminate, tapering towards base for *c.* two-thirds of length, leathery, often yellowish green, finely granular above when dry, with up to 15 pairs of veins; stalks indistinct, to *c.* 1.3 cm, yellow-green. Inflorescences very compact, more or less spherical, to 5 cm. Flowers unisexual, parts in 5s, unpleasantly scented. Petals remaining more or less erect, never spreading, markedly boat-shaped, greenish yellow to yellow. Anthers orange. Fruit orange to red. *Afghanistan, Pakistan, NW India, W Nepal* H2. Spring.

19. FORTUNELLA Swingle
E.C. Nelson
Evergreen trees or shrubs. Leaves alternate, simple, leathery, dotted with glands, dark green above, paler beneath. Flowers solitary or in clusters. Sepals 5. Petals 5, white. Stamens 16–20. Stigma capitate, hollow. Fruit an aromatic berry with thick skin, each cell with 2 ovules.

A genus of 4 or 5 species. The fruits are sold as kumquats (entire fruit edible). The shrubs are ornamental too. The genus is not clearly segregated from *Citrus* and intergeneric hybrids may occur (× *Citrofortunella*).

1a. Shrubs with spines; leaves to 7.5 cm; fruit round   **1. japonica**
  b. Shrubs almost spineless; leaves to 10 cm; fruit ovoid   **2. margarita**

1. F. japonica (Thunberg) Swingle (*Citrus japonica* Thunberg; *C. madurensis* Louriero). Illustration: Siebold, Flora Japonica, **1**: t. 15 (1836).

Shrub, branches usually with spines. Leaves elliptic, to 7.5 cm, apex obtuse. Fruit with 5 or 6 cells, round, to 3 cm across; rind sweet, deep orange. *Perhaps native in China.* G1–H5. Summer.

The sweet fruit is sold as round kumquat or marumi kumquat.

2. F. margarita (Louriero) Swingle. Illustration: Iconographia cormophytorum Sinicorum **2**: 556, t. 2841 (1972), reprinted in Krüssmann, Handbuch der Laubgehölze **2**: 53 (1976).

Like *F. japonica* but differs in having longer leaves (to 10 cm), and a persistent style; fruit ovoid, to 2.5 cm across, with 4 or 5 cells; rind sweet, orange-yellow. *Perhaps from S China.* G1–H5. Summer.

20. × CITROFORTUNELLA Ingram & Moore
E.C. Nelson

Evergreen shrubs or small trees resulting from cross-pollination of *Citrus* and *Fortunella* species. Leaves simple; stalks winged. Flowers and fruits intermediate between parental species.

Three species are generally listed; their fruits are sometimes of considerable commercial value. The name × *Citrofortunella* (*Baileya* **19**: 169, 1975) is properly formed despite a statement to the contrary in The Kew index of taxonomic literature, 280 (1975).

1a.	Fruit bright orange	**1. microcarpa**
b.	Fruit yellow	2
2a.	Leaves lanceolate	**2. swinglei**
b.	Leaves elliptic-ovate	**3. floridana**

1. C. microcarpa (Bunge) Wijnands; *Citrus reticulata* × *?Fortunella margarita* (*C. mitis* (Blanco) Ingram & Moore; *Citrus mitis* Blanco). Illustration: Brickell, RHS gardeners' encyclopedia of plants & flowers, 119 (1989); Everett, New York Botanical Gardens illustrated encylopedia of horticulture **3**: 773 (1981); Brickell (ed.), RHS A–Z encyclopedia of garden plants, 278 (2003).

Shrub or small tree; branches without spines or with short prickles. Leaves 5–10 cm; stalks with very narrow wings. Flowers solitary or in pairs, axillary. Fruit spherical or slightly compressed, to 4 cm across; skin loose, bright orange; pulp acid. G1–H5. Summer.

One of the hardiest citrus shrubs; the fruit can be used as a substitute for lemon or lime, and they remain on trees during winter.

2. C. swinglei Ingram & Moore; *Citrus aurantiifolia* × *Fortunella margarita*.

Differs from *C. floridana* in having lanceolate leaves; flowers with pink buds, fruits with 7 or 8 segments, 3–5 cm across, pulp yellow, skin bright yellow. G1–H5. Summer.

Also known as limequat.

3. C. floridana Ingram & Moore; *Citrus aurantiifolia* × *Fortunella japonica*.

Leaves dark green above, 5–8 cm, elliptic-ovate. Flowers white or streaked with pink. Fruit with 6–9 segments, ovoid or spherical, skin yellow. G1–H5. Summer.

'Eustis', white flowers, hardy, fruit 3–4 cm across. Fruit marketed as limequat.

21. PONCIRUS Rafinesque
C. Gorman

Deciduous shrubs to 7 m. Branches stiff, flattened, smooth, green, with green spines 2.5–5 cm. Leaves compound, often borne on old wood; stalks slightly winged; leaflets 3–5, elliptic to obovate, to 4 cm, gland-dotted. Flowers solitary or in pairs, axillary, borne before leaves on previous year's shoots, fragrant. Sepals 5. Petals 4 or 5, white, concave, obovate, narrowed towards base. Stamens 20–60, free, pink, irregular. Fruit with 6–8 cells, spherical, 3–5 cm across, yellow, densely downy, very fragrant, pulp thin, acidic, seeds numerous.

A genus of one species, which hybridises with *Citrus* species; some botanists still include it in the latter genus.

1. P. trifoliata (Linnaeus) Rafinesque (*Citrus trifoliatus* Linnaeus; *Aegle sepiaria* de Candolle). Illustration: *Botanical Magazine*, 6513 (1880); Gleason, Illustrated flora of the north-eastern United States and adjacent Canada, 467 (1952); Hayashi, Woody plants of Japan, 376 (1985); Phillips & Rix, Shrubs, 86 (1989). *Korea, N China.* H4. Late spring.

A robust, hardy shrub for which many exaggerated claims have been made, including the ability of a hedge of it to stop military tanks! The bigeneric hybrid *Poncirus trifoliata* × *Citrus sinensis*, commonly called citrange, is treated under × *Citroncirus webberi*.

22. × CITRONCIRUS Ingram & Moore
E.C. Nelson

Evergreen trees or shrubs resulting from cross-pollination of *Citrus* and *Poncirus* species. Branches with thorns. Leaves simple or with 3 leaflets. Flowers white. Fruit to 8 cm across, pulp acidic and bitter. H5. Summer.

A single species is known from this cross.

1. × C. webberi Ingram & Moore; *Citrus sinensis* × *Poncirus trifoliata*.

Usually known as citrange. Several cultivars have been named: 'Troyer' is widely used as stock for citrus crops.

23. CITRUS Linnaeus
C. Gorman & E.C. Nelson

Evergreen shrubs or small trees, occasionally with spines beside buds. Leaves alternate, simple, thick and leathery; stalks winged and usually conspicuously jointed at junction with blades. Flowers white to purple, in axillary cymes, or pairs, or solitary, usually fragrant, bisexual. Petals usually 5. Stamens at least 4 times petals. Style deciduous. Fruit a segmented, aromatic, gland-dotted berry, yellow-green to orange when ripe, with leathery skin containing juicy pulp.

A genus of about 16 species is generally recognised but many more have been named; they represent one of the most important genera in the fruit industries of subtropical and tropical countries, including Mediterranean areas of Europe.

Literature: Swingle, W.T., The botany of Citrus, *Proceeding of the Florida State Horticultural Society* (1943); Scora, R.W. & Nicolson, D., The correct name for the shaddock, Citrus maxima, not C. grandis (Rutaceae), *Taxon* **35**: 592–5 (1986).

1a.	Leaf-stalks not winged or narrowly winged	2
b.	Leaf-stalks with broad, prominent wing	7
2a.	Stamens more than 5 times number of petals	3
b.	Stamens usually 4 or 5 times number of petals	4
3a.	Leaf-stalks with distinct joint below blades; fruit with pronounced apical nipple, less than 12.5 cm, skin bright yellow	**1. limon**
b.	Leaf-stalks without joint below blades; fruit without nipple, over 15 cm across, skin pale yellow or green	**2. medica**
4a.	Fruit ovoid, yellow-green, sometimes with apical nipple	**3. aurantiifolia**
b.	Fruit spherical, green to yellow-orange, orange to dark orange, without nipple	5
5a.	Fruit not more than 8 cm across, skin thin, easily separating from segments	**4. reticulata**
b.	Fruit not as above, skin tightly adhering to segments	6
6a.	Spines less than 5 mm; leaf-stalks less than 1 cm; fruit with fewer than 10 segments	**5. tachibana**
b.	Spines over 1 cm; leaf-stalks 1–2.5 cm; fruit with 10–13 segments	**6. sinensis**

7a. Leaf-stalks similar in shape and size
   to blades; pulp of fruit with
   numerous droplets of acrid oil
                                      7. hystrix
  b. Leaf-stalks and blades dissimilar;
   pulp of fruit without oil droplets   8
8a. Leaf-midribs with sparse hairs; fruits
   more than 20 cm across   **8. maxima**
  b. Leaf-blades entirely hairless; fruits
   less than 15 cm across   9
9a. Leaves with wavy margins, to 10
   cm; fruit orange-red, less than
   10 cm across   **9. aurantium**
  b. Leaves with scalloped margins,
   10–15 cm; fruit yellow, 10–15 cm
   across   **10. × paradisi**

1. C. limon (Linnaeus) Burmann
(*C. limonium* Risso). Illustration:
Purseglove, Tropical crops **2**: 507 (1968);
Heywood, Flowering plants of the world,
203 (1978); Hayashi, Woody plants of
Japan, 375 (1985).

Small tree, 3–6 m. Young shoots angled,
hairless, becoming terete, with stout
axillary spines. Leaves pale green, broadly
elliptic, 5–10 × 3–6 cm, with forwardly
pointing teeth; stalks short, distinctly
jointed at apex, narrowly winged. Flowers
solitary or in short racemes, 3.5–5 cm
across, bisexual or male. Petals reflexed,
white inside, purple outside. Stamens
20–40, joined in groups. Fruit oblong-
ovoid with a terminal nipple, segments
8–10, 6.5–12.5 cm; skin rough to smooth;
pulp acidic. *SE Asia.* H5. Throughout the
year.

The commercial lemon, perhaps a hybrid
between *C. aurantiifolia* and *C. medica*.

C. limetta Risso. Fruits sold as sweet
lime; may be a mutant of *C. limon* but
has pure white flowers and sweet pulp.
C. × limonia Osbeck. A dwarf shrub,
perhaps a cultivar of a hybrid between
C. limon and *C. reticulata*. Frequently
grown as a pot-plant.

2. C. medica Linnaeus. Illustration:
Purseglove, Tropical Crops **2**: 509 (1968);
Krüssmann, Handbuch der Laubgehölze **1**:
361 (1976); Bonnier, Flore complète **2**:
111 (1913).

Tree to 3 m, of irregular growth. Young
twigs angular becoming terete, hairless,
with short axillary spines. Leaves elliptic-
ovate to ovate-lanceolate, 10–18 cm,
hairless, toothed, veins prominent; stalks
narrowly margined, not jointed at blades.
Flowers in panicles, racemes or clusters,
bisexual or male. Petals 5, white inside,
purple outside, 3–4 cm wide. Stamens
more

than 30, joined in groups of 4 or more.
Fruit 15–25 cm, segments 10–13, yellow,
fragrant; skin very thick, rough, warty;
pulp pale yellow or green, sparse, acidic to
sweet; seeds small, white. *India.* H5. Most
of year.

Fruit is sold as the citron.

3. C. aurantiifolia (Christmann) Swingle
(*C. lima* Lunan). Illustration: Purseglove,
Tropical crops **2**: 501 (1968).

Hairless, much-branched, spiny shrub or
tree. Leaves elliptic-ovate, apex blunt,
5–8 cm; stalks narrowly winged. Flowers
less than 2.5 cm across, white, in racemes.
Stamens 20–25. Fruit yellow-green, ovoid,
sometimes with a small terminal nipple, to
8 × 6 cm, skin thin, smooth, segments 10,
containing acidic pulp. *Southeast Asia.*
G1–H5. Summer.

Fruit is sold as the lime. The plant may
have originated as a hybrid between
C. medica and another unknown species.

4. C. reticulata Blanco (*C. deliciosa*
Tenore; *C. nobilis* Louriero). Illustration:
Purseglove, Tropical crops **2**: 513 (1968);
Hayashi, Woody plants of Japan, 372
(1985).

Small tree to 8 m, with a dense head;
branches slender, sometimes spiny. Leaves
narrowly ovate, elliptic or lanceolate,
4–8 × 1.5–4 cm, usually scalloped, dark
shining green above, yellow-green beneath;
stalks short, scarcely winged. Flowers
white, solitary or in clusters, axillary,
1.5–2.5 cm across. Stamens 20. Fruit
flattened-spherical, 5–8 cm across;
segments 10–14, separating easily; skin
shiny, thin, green to yellow-orange to dark
orange, easily separated from segments;
pulp sweet. *SE Asia, China.* G1–H5.
Summer.

The fruit is sold as the tangerine or
clementine (fruits deep orange-red),
mandarin (fruits yellow to pale orange),
and satsuma (the hardiest, widely
cultivated in Japan).

5. C. tachibana (Makino) Tanaka.
Illustration: Li, Woody flora of Taiwan,
364 (1963); Hayashi, Woody plants of
Japan, 370 (1985).

Shrub with slender branches, 3–4 m;
spines *c.* 3 mm. Leaves leathery, ovate-
elliptic, 8–9 × 3.5–4 cm, slightly toothed;
stalks *c.* 8 mm, slightly winged. Flowers
axillary, small, solitary. Sepals *c.* 1.5 cm.
Fruit spherical, depressed dorso-ventrally,
2.5 × 4.5 cm, segments 7–9; skin smooth,

orange-yellow. *Taiwan, Japan.* H5–G1.
Summer.

6. C. sinensis (Linnaeus) Osbeck.
Illustration: Purseglove, Tropical crops **2**:
515 (1968); Krüssmann, Handbuch der
Laubgehölze **1**: 361 (1976).

Tree with rounded crown, 6–12 m.
Young twigs angular, becoming terete,
with few slender spines 1.5–5 cm. Leaves
to 10 cm, elliptic to oblong-ovate, dark
green above, paler beneath, sometimes
slightly toothed; stalks 1–2.5 cm, narrowly
winged. Flowers small, white, fragrant,
solitary or 2–6 in short raceme. Petals 4 or
5. Stamens 20–25, joined in groups. Fruit
sweet, spherical and dorso-ventrally
flattened to ovoid; segments 10–13; skin
orange-yellow, nearly smooth, tightly
adhering to pulp; core solid. *China.* G1–H5.
Summer.

The fruit is sold as the (sweet) orange,
navel orange or blood orange. This is
perhaps a hybrid involving *C. maxima* and
C. reticulata, originally produced in China.

7. C. hystrix de Candolle. Illustration:
Duhamel, Trait des arbres et arbustes, t. 7,
39 (1852).

Tree or shrub to 12 m. Twigs angular
when young becoming terete with short
axillary spines, hairless. Leaves ovate-
oblong, slightly notched at apex, round at
base, *c.* 8 × 6 cm; stalks winged and same
length as blades. Flowers white, cream or
pink, axillary, fragrant. Petals 4. Stamens
24–30, free. Fruit ellipsoid, to 7 cm across;
skin green to yellow, glandular; pulp green,
containing numerous droplets of acrid oil,
acidic. *Malaysia.* G1. Summer.

8. C. maxima (Burmann) Merrill
(*C. grandis* (Linnaeus) Osbeck). Illustration:
Purseglove, Tropical crops **2**: 504 (1968);
Krüssmann, Handbuch der Laubgehölze **1**:
361 (1976).

Spreading tree to 6 m. Twigs and young
branches with or without slender spines,
downy. Leaves elliptic, 5–20 × 2–12 cm,
midribs sparsely downy; stalks broadly
winged appearing as secondary heart-
shaped blades. Flowers white to cream,
solitary or clustered, 3–7 cm across. Petals
4 or 5. Stamens 20–25. Fruit very large
(to 8 kg), spherical; skin yellow-orange,
very coarse, thick with scattered large oil-
glands; segments 11–16; pulp sweet or
sour; vesicles large. *Possibly native in
Malaysia, Polynesia.* G2. Summer.

Fruit sold as shaddock, pomelo,
pompelmous or pampelmousse.

9. C. aurantium Linnaeus. Illustration: Purseglove, Tropical crops **2**: 503 (1968); Krüssmann, Handbuch der Laubgehölze **1**: 361 (1976); Heywood, Flowering plants of the world, 203 (1978); Brickell (ed.), RHS A–Z encyclopedia of garden plants, 279 (2003).

Tree with rounded crown, to 10 m. Shoots angular when young, becoming terete, with long slender axillary spines. Leaves elliptic, wavy-edged, to 10 cm; stalks 2–3 cm, obovate in outline, broadly winged above, tapering to a wingless base. Flowers white, solitary or few in axils, fragrant. Stamens 20–24. Fruit orange-red, nearly spherical, to 7.5 cm across; skin thick; segments 10–12, containing acidic pulp, hollow when ripe. *South-east Asia.* H5–G1. Summer.

This fruit is sold as Seville or sour orange, and is possibly a hybrid between *C. maxima* and *C. reticulata*. This is used in manufacture of cointreau and curacao, and subsp. **bergamia** (Risso & Poitet) Wight & Arnott is a source of oil of bergamot. Bergamot orange is subsp. **aurantium** 'Bouquet'.

10. C. × paradisi Macfadyen (*C. grandis* var. *racemosa* (Roemer) Stone). Illustration: Purseglove, Tropical crops **2**: 511 (1968); Hayashi, Woody plants of Japan, 374 (1985).

Tree 10–15 m, with a rounded crown. Twigs angular and hairless, with spines. Leaves 10–15 cm, pale green when young, hairless, broadly elliptic, scalloped, rounded or sometimes cordate at base; stalks broadly winged forming heart-shaped secondary blades. Flowers white, 4 or 5 cm across, 2–20 in axillary clusters or terminal racemes. Petals 4 or 5, slightly reflexed. Stamens 20–25. Fruit spherical, flattened dorso-ventrally, 10–15 cm across, in clusters; skin thick, yellow; segments 11–14; pulp acid to sweet; vesicles coarse. *Garden origin.* G1. Summer.

Fruit is sold as the grapefruit, which arose in the West Indies in the 18th century; probably a complicated hybrid involving *C. maxima*. Usually with pale yellow pulp, but red-fleshed cultivars are grown, usually 'March' and 'Ruby'.

24. AEGLE Linnaeus
E.C. Nelson
Trees to 10 m. Branches with solitary and paired spines, 1–2 cm. Leaves alternate, usually with 3 leaflets, aromatic; stalks 2–4 cm, not winged, slightly swollen at base, jointed at apex; leaflets ovate to elliptic, central leaflet largest, to 7.5 × 5 cm, green or pale green, hairless, with minute glands. Flowers in axillary racemes to 5 cm. Sepals 4 or 5, less than 2 mm, covered with minute hairs, deciduous. Petals 4 or 5, white to green-white, to 15 × 8 mm, densely covered with glands. Stamens 35–45, unequal, filaments 4–7 mm, pollen white. Fruit with hard, woody shell; segments 8–20, each with 6–10 woolly seeds, pulp clear, sticky.

A genus of one species from southern Asia. Pulp from the fruits is edible and the tree is widely cultivated in tropical regions.

1. A. marmelos (Linnaeus) Correa de Serra. Illustration: *Paxton's Magazine of Botany* **16**: t. 238 (1849); Die natürlichen Pflanzenfamilien **4**: f. 113 (1896). *India, Burma, perhaps Indo-China.* G1. Summer.

25. LIMONIA Correa
C. Gorman
Spiny deciduous trees to 10 m. Leaves alternate, pinnate, with 1–4 pairs of leaflets, and a single terminal leaflet, obovate to elliptic, 3–4.5 × 1.5–3 cm, gland-dotted, sometimes toothed; stalks winged. Flowers unisexual or bisexual, in loose terminal and axillary panicles, dull red. Calyx with 5 teeth. Petals 5. Stamens 10–12. Fruit spherical or pear-shaped, shell white and hard, 5–6 cm across, initially with 4–6 segments, which join into 1; seeds hairy; pulp pink, aromatic, acidic.

A genus of one species from south-eastern Asia, yielding useful timber and gum. *Citrus* species may be grafted on to it.

1. L. acidissima Linnaeus (*Feronia limonia* (Linnaeus) Swingle; *F. elephantum* Correa). *India to Indonesia.* G2. Summer.

Both leaves and flowers have an anise fragrance. The pulp may be eaten after removing seeds. It is also a soap substitute.

163. CNEORACEAE

Shrubs, sometimes with medifixed hairs. Leaves alternate, simple, exstipulate; ptyxis conduplicate. Inflorescence a cyme. Flowers bisexual, actinomorphic. Calyx, corolla and stamens hypogynous, disc absent. Calyx of 3–4 free sepals. Corolla of 3–4 free petals. Stamens 3–4. Ovary of 3–4 united carpels, stalked; ovules 1–2 per cell, axile; styles 3–4, free. Fruit a schizocarp.

A family of 2 genera from Cuba (perhaps introduced there), the Canary Islands and the west Mediterranean area.

1. CNEORUM Linnaeus
J. Cullen
Evergreen shrubs to 1 m, usually rather upright. Leaves entire, leathery, greyish green, margins somewhat rolled under. Flowers small; petals yellow. Fruit usually of 3 mericarps which fall from a central axis; mericarps fleshy at first outside, very hard within.

A genus of 1 or 2 species from Cuba (see above) and the Mediterranean area. The one cultivated species is easily grown in pots or in well-drained soil in a sunny position, and is propagated by seeds or by cuttings.

1. C. tricoccon Linnaeus. Illustration: Bonnier, Flore complète **2**: pl. 115 (1913); Polunin & Smythies, Flowers of southwest Europe, pl. 23 (1973); Ceballos et al., Plantas silvestres de la peninsula Iberica, 190 (1980); Guittoneau & Huon, Connaître et reconnaître la flore et la vegetation méditeranéennes, 155 (1983).

Shrub, 30–100 cm; bark grey-brown. Shoots hairless or finely downy. Leaves narrowly oblong, tapered to stalkless bases, very rounded but mucronate at the apex, 1–3 cm × 3–8 mm. Sepals c. 1 mm. Petals yellow, to 8 mm. Mericarps reddish at first, ultimately black, warty, almost spherical, c. 5 mm across. *W Mediterranean area.* H5–G1. Spring–summer.

164. SIMAROUBACEAE

Woody. Leaves alternate, simple or compound, exstipulate; ptyxis conduplicate. Inflorescences various. Flowers usually unisexual, actinomorphic. Calyx, corolla and stamens hypogynous, disc present. Calyx 3–8-lobed. Corolla of 4–8 free petals or rarely absent. Stamens 4–14 or rarely more. Ovary of 2–5 united carpels; ovules 2 per cell, axile; styles 1–5, free. Fruits various, often samaras.

A family of 20 genera and about 120 species. Most are found in the tropics and subtropics and include several species of economic importance. The timber is valued in some species and others produce bark with useful glucosides. A few are grown purely for their ornamental value.

Although the genus *Picramnia* Swartz is mentioned in some of the literature it does not appear to be widely grown, even in southern Europe.

1a. Leaves with 3–7 leaflets; petals 4–5 cm, red   **4. Quassia**
 b. Leaves with 7–35 leaflets; petals 3–30 mm, white, greenish white, cream or greenish yellow   2
2a. Leaves with one or more pairs of glands at the base; fruit a winged samara, *c.* 3.5 cm   **1. Ailanthus**
 b. Leaves without glands; fruit a drupe or separating into 4 parts, 3–25 mm   3
3a. Leaves with 9–15 leaflets; flowers in loose corymbs 8–15 cm, greenish yellow; fruit of 1–3 fleshy or leathery drupes, 3–7 mm across   **2. Picrasma**
 b. Leaves with 7–20 leaflets; flowers in axillary racemes 10–25 cm, greenish white to cream; fruit 8–25 mm, separating into 4   **3. Kirkia**

1. AILANTHUS Desfontaines
S.G. Knees

Deciduous, dioecious tall trees, often with fissured grey bark. Leaves alternate, fetid, pinnate with a terminal leaflet, the basal lobes with conspicuous glands near their bases. Stipules minute, falling early. Flowers usually unisexual, in terminal panicles. Calyx of 5 sepals (rarely 6). Petals 5 (rarely 6), not overlapping; disc hemispherical. Stamens lacking appendages, 2–5 in bisexual flowers, 10 in male flowers. Ovary 5-celled; stigmas 5; ovule solitary, pendent. Fruits of 1–5 samaras, containing a single seed. Seed without endosperm.

A genus of 10 species, from temperate and tropical Asia and Australasia. Only one species is regularly cultivated in our area. Propagation is usually from the freely produced suckers or root-cuttings taken in winter, or from seed sown in autumn.

1a. Young branches smooth   **1. altissima**
 b. Young branches covered with soft spines   **2. vilmoriniana**

1. A. altissima (Miller) Swingle (*Toxicodendron altissimum* Miller; *A. glandulosa* Desfontaines; *A. japonica* Anon.). Illustration: Marshall Cavendish encyclopedia of gardening **1**: 24 (1968); Hutchinson, Evolution and phylogeny of flowering plants, 392 (1969); Hui-Lin Li, Flora of Taiwan

3: 539 (1977) – as var. *tanakai*; Phillips, Trees in Britain, 58, 82 (1978).

Narrow deciduous tree, 20–30 m, with yellowish grey, deeply fissured bark. Leaves 45–60 cm (often 1 m or more on new growth, produced after severe pruning). Leaflets 11–25, lanceolate, 10–17 × 2.5–3 cm, margins entire except for 2 or 3 coarse teeth near base, lowest pair gland-tipped. Stipules falling early, *c.* 0.5 mm. Calyx-lobes 1–2 mm; petals 3–3.5 mm, white. Stamens in 2 whorls. Carpels 5; styles joined towards the base. Samaras 3–4 cm × 7–9 mm, reddish brown when ripe. *China.* H2. Summer.

A species probably not distinct from *A. altissima* is referred to in the literature as *A. giraldii* Dode. The name *A. japonica* is sometimes seen in nursery catalogues but does not appear to be validated in the literature. It may refer to another variant of *A. altissima*.

2. A. vilmoriniana Dode (*A. glandulosa* Desfontaines 'Spinosa'). Illustration: Vilmorin, *Fruticeum Vilmorinianum*, 31 (1904); *Revue Horticole*, 445 (1904).

Like *A. altissima*, but all parts covered in soft hairs; young branches covered with soft spines. Leaves 60–120 cm, with 21–35 leaflets, central axis sometimes spiny. Leaflets broadly acuminate. Samaras *c.* 5 cm. *W China (Sichuan).* H3. Summer.

2. PICRASMA Blume
S.G. Knees

Trees or shrubs with very bitter constituents. Leaves alternate, pinnate with a terminal leaflet. Flowers inconspicuous, in axillary panicles. Calyx of 4 or 5 teeth, often enlarging in fruit; disc thickened. Stamens 4 or 5, hairy. Ovary with 3–5 cells, free. Styles distinct at apex and base, but joined in the middle. Ovules erect, solitary. Fruit of 1–3 fleshy or leathery drupes. Seeds erect, almost transparent.

A genus of about 6 species from eastern Asia and the Pacific islands. Only one is cultivated, and although the flowers and fruit are not particularly ornamental, the autumn foliage is outstanding, with hues of rich orange and red.

1. P. quassioides (D. Don) Bennett (*Simaba quassioides* D. Don; *P. ailanthoides* Planchon). Illustration: *Botanical Magazine*, n.s., 279 (1956); Hui-Lin Li, Flora of Taiwan **3**: 542 (1977); Phillips, Trees in Britain, 55, 152 (1978); Brickell, RHS gardeners' encyclopedia of plants & flowers, 69 (1989).

Small or medium-sized tree, rarely exceeding 12 m. Branches reddish brown with prominent leaf-scars, lenticels yellow. Leaves 25–37 cm; leaflets 9–15, ovate or oblong-ovate, 4–11 cm, acuminate at apex, broadly wedge-shaped at base, margins shallowly toothed. Flowers unisexual, axillary, greenish yellow, in loose corymbs 8–15 cm across. Drupes 5–7 mm across, spherical, blue, purplish or red, supported by persistent calices. *India, China, Korea, Japan.* H2. Summer.

3. KIRKIA Oliver
S.G. Knees

Trees or shrubs, branches often sticky towards the tips and clearly marked with old leaf-scars. Leaves spirally arranged, mostly crowded towards the ends of branches, pinnate with a terminal leaflet. Flowers unisexual or bisexual in axillary racemes. Sepals 4, erect, united at base. Petals 4, overlapping. Stamens 4, alternate with petals, as long as petals in male flowers, much shorter in female flowers; anthers dorsifixed. Disc fleshy. Carpels 4–8; ovules 1 per cell. Styles 4–8, stigmas head-like. Fruit woody, separating into four 3-sided mericarps on a slender central stalk. Seed slender, triangular in cross-section, slightly curved.

A small genus of 5 species. sometimes placed in its own family (Kirkiaceae). It is endemic to the African continent and contains economic and ornamentally important species. The 2 species mentioned below are propagated from seed or cuttings, will grow well in freely draining soils and are drought resistant. The leaves colour well in autumn.

Literature: Coates Palgrave, K., Trees of southern Africa, 353–355 (1977); Stannard, B., A revision of *Kirkia* (Simaroubaceae), *Kew Bulletin* **35**(4): 829–839 (1981).

1a. Leaves 20–35 cm; leaflets 3.5–7 cm; fruit 1–2.5 cm   **1. acuminata**
 b. Leaves 8–12 cm; leaflets 1–2.5 cm; fruit 8–10 mm   **2. wilmsii**

1. K. acuminata Oliver. Illustration: *Hooker's Icones Plantarum*, t. 1036 (1868); Palmer & Pitman, Trees of southern Africa **2**: 1002, 1004, 1005 (1972); Coates Palgrave, Trees of southern Africa, 354, pl. 117 (1977); *Kew Bulletin* **35**(4): 832 (1981).

Deciduous tree 15–20 m, with smooth silver bark, becoming fissured with age. Leaves 20–35 cm with 7–12 pairs of ovate to lanceolate leaflets. Leaflets 3.5–7 × 1–2.5 cm, softly hairy, especially on midribs, or almost hairless, margins scalloped or finely toothed; central axis unwinged. Inflorescences usually much shorter than leaves, 10–25 cm, with leaf-like bracts. Flowers greenish white to cream; sepals *c.* 2 mm; petals 4–6 mm; styles 4, *c.* 1.5 mm. Fruits 1–2.5 cm × 5–11 mm, oblong, 4-sided, with four 1-celled mericarps. *Tropical & southern Africa (Congo to South Africa).* G2–1. Spring–early summer.

2. K. wilmsii Engler. Illustration: Palmer & Pitman, Trees of southern Africa **2**: 1006 (1972); Coates Palgrave, Trees of southern Africa, 355 (1977); *Kew Bulletin* **35**(4): 832 (1981).

Deciduous tree to 8 m with grey bark. Leaves 8–12 cm with 10–20 pairs of leaflets. Leaflets 10–25 × 2.5–6 mm, bases often unequal, oblong to elliptic, hairy or sparsely hairy; margins coarsely toothed; central axis winged. Inflorescences as long as or longer than leaves, bracts *c.* I cm. Flowers greenish white to cream; sepals *c.* 1.5 mm; petals 2.5–3 cm; styles 4, united, *c.* 1 mm. Fruits 8–10 × 4–5 mm, oblong, 4-sided. *South Africa.* H5–G1. Spring–early summer.

4. QUASSIA Linnaeus
S.G. Knees

Trees or shrubs, bisexual or monoecious. Leaves simple or compound, alternate, stalked, usually with pitted glands on lower surfaces; stipules absent. Inflorescences axillary or terminal, clustered, racemes, panicles or umbels. Flowers with 4–6 floral parts. Calyx lobed; petals overlapping or twisted, longer than sepals. Stamens twice as many as petals, with appendages near the base. Disc cylindric or almost spherical. Carpels free or apparently joined basally; styles fused, stigmas stellate or capitate; ovule apical. Fruit a drupe, or sometimes woody.

A genus of about 40 species, throughout the tropics and subtropics. Only one species is regularly cultivated in temperate climates.

1. Q. amara Linnaeus. Illustration: *Botanical Magazine*, 497 (1800); Heywood, Flowering plants of the world, 199 (1978); Graf, Tropica, edn. 3, 895 (1986).

Small tree or shrub to 9 m. Leaves 23–27 cm, pinnate with a terminal leaflet, central axis conspicuously winged. Leaflets 3–7, elliptic, abruptly acuminate, wedge-shaped at base, 13–17 cm. Inflorescence a raceme, with jointed flower-stalks. Flowers bisexual; sepals 5; petals 5, red, 4–5 cm. Stamens 10, with basal appendages, softly hairy towards the bases. Ovary of 5, almost united carpels; style 1. Fruit of several separate drupes. *Mexico, C & N South America, West Indies.* H5–G2. Summer.

---

165. BURSERACEAE

Woody with aromatic resins. Leaves alternate, usually compound, exstipulate. Inflorescences panicles or flowers solitary. Flowers unisexual or bisexual, actinomorphic. Calyx, corolla and stamens hypogynous, with a disc. Calyx 3–5-lobed. Corolla of 3–5 free petals (rarely absent). Stamens 6–10. Ovary of 2–5 united carpels; ovules 2 per cell, axile; style 1 or stigma sessile. Fruit a drupe or capsule.

A family of 18 genera and 540 species from tropical areas.

1a. Inflorescences axillary; flowers without densely felted hairs **1. Bursera**
  b. Inflorescences terminal (in ours); flowers with densely felted hairs **2. Canarium**

1. BURSERA Linnaeus
E.H. Hamlet

Trees or shrubs, often with thin papery bark. Leaves deciduous, variously pinnate, usually crowded at end of branches; leaflets opposite, entire or toothed. Inflorescences axillary or almost terminal. Flowers small, unisexual and/or bisexual. Calyx and corolla with 3–5 lobes. Stamens 6–10, almost equal. Style short; stigma with 3–5 lobes. Capsule ovoid, 3-angled, with 1-several seeds.

A genus of about 50 species from America. They can be grown in ordinary soil. Propagation is by seed or cuttings.

1a. Leaves irregularly finely scalloped **1. fagaroides**
  b. Leaves entire **2. simaruba**

1. B. fagaroides Engler. Illustration: Die Natürlichen Pflanzenfamilien **3**(4): 250 (1897); Sanchez, La flora del Valle de Mexico, f. 176 (1969).

Shrubs or small trees; bark grey-brown. Leaves 5–11 cm, on short lateral twigs; leaflets 7–15, lanceolate to ovate or obovate, opposite, stalkless, 15–40 × 3–15 cm. Flowers solitary or in pairs, on short lateral twigs. Sepals deciduous. Flower-parts in 5s; petals 3–4 mm, linear-lanceolate. Fruit 7–9 mm, 3-angled. *USA (Arizona), Mexico.* G1.

2. B. simaruba Sargent. Illustration: Sargent, Silva of North America **1**: pl. 41, 42 (1891); Pennington & Sarukhan, Arboles Tropicales de Mexico, 235 (1968); Fournet, Flore illustrée des phanéeogames de Guadeloupe et de Martinique, 1075 (1978); Proctor, Flora of the Cayman Islands, 573 (1984).

Trees or shrubs to 15 m with old bark copper-red, peeling off in papery layers. Leaves 9–32 cm, alternate on ends of branches; leaflets 5–11, broadly ovate to elliptic-lanceolate, 2.7–11.5 × 1.5–5 cm, entire, acuminate, oblique at bases. Inflorescences many-flowered, 2.5–10 cm. Male flower parts in 5s, female flower parts in 3s. Petals ovate-elliptic. *S USA, Central America, West Indies to N South America.*

2. CANARIUM Linnaeus
E.H. Hamlet

Trees or shrubs. Leaves spirally arranged or rarely whorled in 3s, usually more or less in groups at ends of twigs, pinnate with a terminal leaflet. Inflorescences terminal (in ours). Flowers unisexual, parts in 3s. Petals white, free, nearly always oblong-obovoid, very rarely with a claw, fleshy and often rather thick, except for the margins. Stamens 6, in 1 whorl, free to entirely joined, in female flowers sterile and often less well developed. Stigma 3-lobed. Disc 6-lobed, in male flowers usually strongly developed, in female flowers often joined to staminodes, and if receptacle concave, almost entirely joined to it. Fruit ovoid-oblong.

A genus of 75 species from tropical Africa, Asia, north Australia and the Pacific islands. They can be grown in a peaty loam soil, in a warm glasshouse. Propagation is by cuttings.

1. C. indicum Linnaeus (*C. commune* Linnaeus in part). Illustration: Van Steemis, Flora Malesiana **5**: 267–269 (1956).

Tree with buttresses. Leaves 3–8, opposite; leaflets 5–35 × 2–16 cm, obovate to oblong-lanceolate, entire. Inflorescence 15–40 cm, terminal; flowers 1–1.5 cm with densely felted hairs. In male flowers calyx 5–7 mm; stamens free and disc cushion-like. In female flowers calyx 7–10 mm, staminodes joined to disc, and disc joined to receptacle. Fruit ovoid. *Indonesia (Moluccas), New Guinea, Solomon Islands, New Hebrides.* G2.

166. MELIACEAE

Trees, wood often scented. Leaves usually alternate, mostly pinnate, exstipulate; ptyxis conduplicate. Inflorescences cymose panicles. Flowers usually bisexual, actinomorphic or rarely slightly zygomorphic. Calyx, corolla and stamens hypogynous with disc. Calyx 4–6-lobed. Corolla of 4–6 or rarely 8 free petals. Stamens usually 8–12, filaments usually united into a tube. Ovary of 2–20 united carpels; ovules usually 2 or more per cell, axile; style 1, short. Fruit a berry, capsule or drupe.

A family of 52 genera and about 550 species, largely restricted to the tropics, where many are important timber trees (the mahoganies, species of **Swietenia** Jacquin). Few are hardy in Europe and only a very small number are generally grown.

Literature: Pennington, T.D. & Styles, B.T., A generic monograph of the Meliaceae, *Blumea* **22**: 419–540 (1975).

1a. Shrubs or small trees with simple
    leaves   **1. Turraea**
 b. Trees with pinnate leaves   2
2a. Leaves 2- or 3-pinnate; filaments
    united   **2. Melia**
 b. Leaves 1-pinnate; filaments free   3
3a. Flowers sweetly scented; leaflets
    toothed or entire, with oblique or
    obtuse bases   **3. Toona**
 b. Flowers smelling of rotten onions;
    leaflets entire, rounded or obtuse at
    base   **4. Cedrela**

1. TURRAEA Linnaeus
M. Cheek
Small trees or shrubs to 8 m. Leaves usually simple and entire. Calyx cup-shaped or 5-lobed. Petals 5 or 6, narrowly spathulate. Staminal-tube cylindric, anthers 8–10, inserted on the rim of the tube, interspersed with appendages. Ovary with 5–20 cells, each containing 2 ovules side by side. Style swollen below the stigma. Fruit a capsule. Seeds glossy black, red or orange, usually with contrasting arils.

A genus of about 65 species mainly from Africa and Madagascar. Very few are grown.

1a. Leaves with densely felted hairs,
    margins entire   **2. floribunda**
 b. Leaves hairless, often 3-lobed   2

2a. Leaves linear to obovate, rounded at
    apex   **1. obtusifolia**
 b. Leaves elliptic to oblong, 3-lobed at
    apex   **3. heterophylla**

1. T. obtusifolia Hochstetter. Illustration: Graf, Exotica, edn. 12, 1550 (1985).

Evergreen shrub to 3 m. Leaves linear to obovate, to 5 cm, slightly 3-lobed, rounded at apex. Flowers white, scentless. Petals 2.5–4 cm. Fruit (if developed) almost spherical, opening to reveal *c.* 10 brilliant orange seeds with tiny, concealed, white arils. *Southern Africa.* G1. Almost throughout the year.

2. T. floribunda Hochstetter (*T. heterophylla* Sonder). Illustration: Graf, Exotica, edn. 12, 1551 (1985).

Deciduous shrub or small tree to 4 m. Leaves elliptic, 8–15 cm, entire, covered with densely felted hairs. Flowers white to yellow, sweetly scented in the evening. Petals 4–10 cm × 3–5 mm. Fruit (if developed) very woody, opening to reveal black seeds, each partly enclosed in a dull orange aril. *Southern C & E Africa.* G1. Autumn, when leafless.

3. T. heterophylla J.E. Smith (*T. lobata* Lindley).

Evergreen shrub to 2 m. Leaves elliptic-oblong, 3–10 cm, 3-lobed at the apex, margins sometimes wavy, hairless, glossy, dark green. Flowers white, scentless. Petals 1.2–1.5 cm × 3–4 mm. Fruit developing without pollination, opening to reveal a bright orange interior with scattered black seeds, each partly enveloped in a bright orange aril. *W Africa.* G2. Throughout the year.

2. MELIA Linnaeus
M. Cheek
Trees with stellate hairs. Leaves 2- or 3-pinnate, stalks with paired glands at the base, leaflets toothed. Flowers bisexual or male. Calyx deeply 5-lobed. Petals 5 or 6. Staminal-tube cylindric, ribbed, anthers interspersed with appendages. Nectary ring-like. Ovary 4–8-celled, each cell with 2 ovules. Stigma slightly lobed. Fruit indehiscent, fleshy.

A genus of 3 species in Africa and southeast Asia. A single species is cultivated as an ornamental tree in the warmer parts of the world. Species of the very similar genus *Azadirachta* Jussieu, which is widely grown in the tropics, are sometimes confused with those of *Melia*;

they lack stellate hairs and the leaves are 1-pinnate.

Literature: Mabberley, D.J., A monograph of *Melia* in Asia and the Pacific, *Gardens Bulletin of Singapore* **37**(1): 49–64 (1984).

1. M. azedarach Linnaeus. Illustration: *Botanical Magazine*, 1066 (1808); Graf, Exotica, edn. 12, 1551 (1985); Brickell (ed.), RHS A–Z encyclopedia of garden plants, 684 (2003).

Deciduous tree 2–15 m. Leaves 10–45 cm, leaflets ovate-elliptic, 2–3.5 cm, deeply toothed, almost hairless. Flowers in clusters of 10–70. Petals 7–8 × 2–3 mm, pale mauve or white. Staminal-tube deep purple, flared at apex. Fruit 1–1.5 cm, yellowish, glossy. *India, China to Australia; locally naturalised in S Europe.* H5–G1. Late summer.

'Umbraculifera' has a dense, flattened head of radiating branches.

3. TOONA (Endlicher) M.J. Roemer
M. Cheek
Deciduous or semi-deciduous, small to medium, monoecious trees, ill-smelling when bruised. Branches with prominent lenticels and large leaf-scars. Leaves pinnate with more than 5 pairs of oblique or obtuse, sometimes toothed, stalked leaflets. Inflorescences hanging, with long branches and up to several hundred white, sweetly scented flowers, which appear bisexual but are unisexual. Calyx saucer-shaped, sometimes deeply 5-lobed. Petals 5, small. Stamens 5, filaments free. Ovary 5-celled, each with 6–10 ovules. Ovary and stamens borne on a short stalk. Fruit ellipsoid, woody; seeds winged at one or both ends.

A genus very similar to *Cedrela*, of 6 poorly defined species, from India to Northern Australia. The first species below is unlikely to be hardy in most of Europe; the second is used as a street tree in Paris. Measurements of the leaves refer to those from flowering branches; leaves from saplings are much larger.

1a. Leaflets leathery, entire, base obtuse
                             1. ciliata
 b. Leaflets papery, finely toothed, base
    oblique   **2. sinensis**

1. T. ciliata Roemer (*Cedrela toona* Roxburgh). Illustration: Graf, Exotica, edn. 12, 1551 (1985).

Semi-deciduous tree to 20 m. Leaves usually with 7 or 8 pairs of ovate-lanceolate leaflets 7–13 × 2.5–1 cm, acute or acuminate, obtuse at base, margins entire, leathery, hairless. Flowers white, sweetly scented; petals *c.* 3 mm. Seeds winged at both ends. *India to Australia.* G1–2.

2. T. sinensis (Jussieu) Roemer (*Cedrela sinensis* Jussieu; *Ailanthus flavescens* Carriére). Illustration: Krüssmann, Manual of cultivated broad-leaved trees & shrubs **1**: pl. 110 (1984); Macoboy, What tree is that?, 64, 65 (1979); Brickell (ed.), RHS A–Z encyclopedia of garden plants, 1042 (2003).

Similar to *T. ciliata*, but deciduous, leaflets papery, finely toothed, with oblique bases; petals 4–5 mm. *N China to Malaysia.* H5. Summer.

4. CEDRELA Browne
M. Cheek
Very similar to *Toona*, but flowers ill-scented (of rotting onions); calyx cup-shaped, often divided to the base; petals 3–7 mm; seeds winged at one end only.

A genus of about 8 species from tropical America. The one cultivated species is Spanish Cedar, from which cigar boxes are traditionally made, an important timber tree in the tropics.

1. C. odorata Linnaeus (*Cedrela mexicana* Roemer). Illustration: Macoboy, What tree is that?, 65 (1986).

Tree to 3 m, smelling strongly of garlic. Leaves with 7 or 8 pairs of leaflets. Flowers whitish, smelling of rotten onions. Petals 6–7 mm, downy outside. Seeds winged at one end. *Mexico to Ecuador.* G1–2. Unlikely to flower in Europe.

167. MALPIGHIACEAE

Woody, often with medifixed hairs. Leaves usually opposite, rarely alternate or whorled, simple or rarely lobed or divided, stipulate or exstipulate, usually evergreen; ptyxis flat or conduplicate. Inflorescences racemes, umbels or panicles. Flowers usually bisexual, slightly zygomorphic, rarely actinomorphic. Calyx, corolla and stamens hypogynous. Calyx of 5 free sepals, often some or all of them with 2 nectar-secreting glands on the outside. Corolla of 5 free petals, often fringed and/ or unequal, Stamens 10, filaments united, anthers sometimes opening by pores,

sometimes some sterile. Ovary of 3 united carpels; ovules 1 per cell, axile; styles 3, free or united. Fruits various, chiefly winged mericarps.

A family of about 70 genera and 1250 species from the tropics and subtropics, mainly concentrated in Central and South America. Only a few are in general cultivation. Both genera and species are difficult to identify and many in cultivation appear to be incorrectly named. The medifixed hairs are characteristic of the family; in some species they are easily detached and can cause skin irritation.

Literature: Niedenzu, F., in A. Engler (ed.) Das Pflanzenreich **141** (1928); Cuatrecasas, J., Prima Flora Colombiana 2, *Webbia* **13**: 343–664 (1958), **15**: 393–398 (1960); Arènes, J., Répartition géographique des Malpighiacées vivantes et fossiles, *Compte rendu sommaire des Séances de la Société de Biogéographie* **33** (290): 8l–108, suppléments 1–3 (1957); Robertson, K.R., The Malpighiaceae in the southeastern United States, *Journal of the Arnold Arboretum* **53**: 101–112 (1972); Anderson, W.R., The origin of the Malphigiaceae, *Memoirs of the New York Botanic Garden* **64**: 210–224 (1990).

1a. Carpels free or only partially united   2
  b. Carpels united for all or most of their length   4
2a. Carpels completely free
                              8. Banisteriopsis
  b. Carpels partially united   3
3a. Stamens all with fertile anthers
                              7. Heteropteris
  b. Some stamens with infertile and much-reduced anthers
                              9. Stigmaphyllon
4a. Anthers opening by pores   5
  b. Anthers opening by slits   6
5a. Carpels usually 2; leaves mostly alternate   **2. Acridocarpus**
  b. Carpels 3; leaves mostly opposite
                              3. Tristellateia
6a. Stipules borne on the base of the leaf-stalk   **1. Galphimia**
  b. Stipules absent or very small and borne on the stem at the base of the leaf-stalk   7
7a. Climber; sepal-glands absent
                              5. Sphedamnocarpus
  b. Usually trees or shrubs, rarely climbers; sepal-glands usually present   8

8a. Sepal-glands 1–10 and small; stamens free at the base **4. Hiptage**
  b. Sepal-glands 6–10, usually at least $^2/_3$ as long as the sepals; stamens united at base   **6. Malpighia**

1. GALPHIMIA Cavanilles
B. MacBryde
Dwarf shrubs, shrubs or rarely small trees. Leaves opposite, usually with 2 glands near the base of the blade; stipules small, borne on the leaf-stalk. Flowers in racemes or panicles terminating the branches. Sepals 5, glands absent or very small at the sinuses. Petals yellow, somewhat unequal. Stamens 10 in 2 whorls; anthers opening by slits. Carpels 3, completely united, usually hairless; styles 3, subulate, each with a tiny stigma. Fruit a capsule-like schizocarp, the mericarps late in separating.

A much-confused genus (see *Taxon* **16**: 76–77, 1967) of about 12 species from the southern USA to northern Argentina. The plants are easily grown in full sun in a glasshouse in well-drained soil. They should be pruned when not in flower or when flowering is at a minimum. Propagation is by seed or cuttings.

1. G. gracilis Battling (*Thryallis gracilis* (Battling) Kuntze; *T. glauca* misapplied; *G. nitida* Reasoner & Reasoner). Illustration: Maund, Botanist **1**: t. 18 (1837); Clay & Hubbard, The Hawai'i garden **2**: 57 (1977); *Annals of the Missouri Botanical Garden* **67**: 879 (1980); Kingsbury, 200 conspicuous, unusual or economically important plants of the Caribbean, (unnumbered) (1988) – as *G. glauca*.

Shrub to 4 m; branches with red-brown hairs at first. Leaf-blades usually hairless, 1–7.5 cm × 4–34 mm, elliptic-oblong to ovate-lanceolate. Flowers 1.4–2.2 cm across, numerous, fragrant, in racemes. Petals ribbed on the back, falling well before the fruit is mature, bright yellow, fading somewhat reddish. *C Mexico; naturalised in the Caribbean.* H5–G1. Flowering for most of the year, though mainly in spring–autumn.

Widely grown in the tropics and subtropics. Genuine **G. glauca** Cavanilles (*Malpighia glauca* (Cavanilles) Persoon; *Thryallis glauca* (Cavanilles) Kuntze) is from Brazil and is probably not in general cultivation; it has petals that are keeled on the back and which persist until the fruit is mature.

2. ACRIDOCARPUS Guillemin & Perrotet
B. MacBryde

Shrubs (often trailing or climbing) or trees. Leaves usually alternate, often glandular on the blade-base beneath. Stipules absent or very small and borne on the base of the leaf-stalk. Flowers in racemes, corymbs, umbels or panicles. Sepal-glands 1–10, very small or absent. Petals yellow or white, usually unequal. Stamens usually unequal, anthers opening by pores. Carpels usually 2 (ours), completely united, hairy; styles usually 2, persistent. Schizocarp of 2 samaras, the narrow wings thickened on their upper edges.

A mainly African genus of up to 35 species, of which only one is cultivated. It can be grown in a cool glasshouse in rich soil and should be watered plentifully. Pruning should be carried out after flowering. Propagation is by seed, cuttings or layers.

1. A. natalitius jussieu Illustration:
Botanical Magazine, 5738 (1868); van der Spuy, South African shrubs and trees for the garden, 65 (1971); Graf, Exotica, edn. 11, 1525 (1982); *Flora of Southern Africa* 18(3): 70 (1986).

Shrub to 3 m, strong climber or tree to 6 m; branchlets with red-brown to yellowish hairs at first. Leaves leathery, puckered, dark green and glossy, 3–15 × 1.3–5 cm (× 4–15 mm in var. *linearifolius* Launert), oblong to ovate or oblanceolate, with usually 2 glands at the base. Flowers many, 2–4 cm wide, in conical racemes. Sepals 5, 2 of them each with 1 or 2 glands. Petals golden yellow, margins lacerated. Samaras red-brown, wings triangular-ovate to oblong. *S Mozambique to E South Africa*. G1. Summer.

The cut wood has a faint peppery smell.

3. TRISTELLATEIA Thouars
B. MacBryde

Woody climbers, shrubs or small trees. Leaves usually opposite (rarely in whorls of 3 or 4), usually with 2 glands on the margins beneath at the base of the blade or on the stalk. Stipules very small, borne on the leaf-stalk. Flowers slightly bilaterally symmetric, in racemes or panicles. Sepals 5, glands absent, or 1–10 and very small. Petals yellow, equal or almost so. Stamens 10, anthers opening by apical pores. Carpels usually 3, united, hairy; style usually 1, slender. Schizocarp of 3 samaras.

A genus of 21 species mainly from Madagascar but extending to East Africa and Polynesia. The one cultivated species is easily grown in full sun in well-drained soil; it requires regular watering, particularly in summer. Propagation is by seed, cuttings or layers.

Literature: Arenes, J., Monographic du genre *Tristellateia*, Memoires du Museum national d'Histoire naturelle **21**: 275–330 (1947).

1. T. australasiae Richard (*T. australis* Richard; *T australasica* Jussieu). Illustration: *Botanical Magazine*, 8334 (1910); Li, Flora of Taiwan 3: 555 (1977); Polunin, Plants and flowers of Singapore, 53 (1987); Courtright, Tropicals, 132 (1988).

Climber to 10 m, adpressed white-hairy at first. Leaves shiny above, 3.5–14 × 2–7 cm, ovate to ovate-oblong, usually with 2 glands at or near the base. Flowers many, 2–2.5 cm across in open racemes terminating the shoots or on 2-leaved short shoots; slightly fruit-scented. Sepals without glands. Petals slightly unequal, yellow becoming red. Stamens 10. Carpels usually 3, completely united, papillose. Samaras brown. *Malaysia, Indochina to Japan, Fiji & N Australia*. G2. Summer.

4. HIPTAGE Gaertner
B. MacBryde

Shrubs, climbers or trees. Leaves opposite, blades with 2 glands at base and often smaller glands along the margins. Stipules absent or very small and glandular, borne beside the leaf-stalks. Flowers bilaterally symmetric, usually in racemes. Sepals 5, without glands or glands 1–10 and very small. Petals yellow, pink, rose or white, strongly reflexed and wavy-edged, toothed or fringed. Stamens 10, filaments curved, anthers opening by slits. Carpels 3, completely united, each somewhat 3-lobed; style 1. Schizocarp of 1–3 samaras, each with 3 side-wings.

A genus of up to 45 species, mostly in Southeast Asia. The one cultivated species is grown in sun or partial shade in rich soil; it should be watered regularly, though drier intervals may stimulate flowering. Propagation is by seeds, cuttings or layers.

1. H. benghalensis (Linnaeus) Kurz (*Banisteria tetraptera* Sonnerat; *H. madablota* Gaertner; *Gaertner a racemosa* (Cavanilles) Roxburgh). Illustration: Li, Flora of Taiwan 3: 553 (1977); Walden & Hu, Wild flowers of Hong Kong around the year, pl. 16

(1977); Mathew, Illustrations on the flora of the Tamil Nadu Carnatic, 92 (1983); McMakin, A field guide to the flowering plants of Thailand, t. 124 (1988).

Climber to 30 m, rarely a shrub; young branches with silvery to yellow, silky hairs at first. Leaves usually glossy above, 7–22 × 2–10 cm, ovate-lanceolate to elliptic or oblong, usually with 2 glands beneath at the base and more along the margins. Flowers numerous, 1–2.5 cm wide, fragrant, in densely hairy racemes. Sepals 5, usually hairy, 2 of them each with a linear-oblong convex gland. Petals whitish or pale pink, sometimes with some yellow, hairy outside. Style hairy at the base. Samaras dark reddish to shiny brown. *India to Japan, China, Southeast Asia*. G2. Spring–summer.

A frequent ornamental in India and Southeast Asia, but not often seen elsewhere.

5. SPHEDAMNOCARPUS Bentham
B. MacBryde

Woody climbers or rarely shrubs or herbs with only the rootstock woody, Leaves mostly opposite, occasionally deciduous, usually with 2–4 glands on the blade or stalk. Flowers in umbels grouped in panicles. Sepals equal or almost so, without any glands. Petals usually equal, yellow or white. Stamens 10 or rarely more, anthers opening by slits. Carpels 3 or rarely 4, completely united, hairy. Schizocarp of 3 (rarely 4) samaras.

A genus of about 14 species, mostly in Madagascar, but extending to South Africa and Angola. The single cultivated species needs full sun and a well-drained soil; if not supported it will generally form a shrub. Propagation is by seed or cuttings.

Literature: Launert, E., A revision of the continental African species of *Sphedamnocarpus* Planch., *Boletim da Sociedade Broteriana*, ser. 2, **35**: 33–47 (1961).

1. S. pruriens (Jussieu) Szyszylowicz. Illustration: *Botanical Magazine*, 7894 (1903); *The Flora of South Africa* 2: pl. 37 (1925); Plowes & Drummond, Wild flowers of Rhodesia, pl. 93 (1976); Heywood, Flowering plants of the world, 213 (1978).

Branches erect to twining or trailing, 2–4 m, often with silvery grey or yellow hairs when young. Leaves with short-tipped hairs at least beneath, 1.5–13.5 × 1–10 cm, almost circular to lanceolate or oblong; stipules absent or beside the leaf-base. Flowers 8–25 mm wide, in open umbels. Sepals usually hairy. Petals yellow to orange, crisped, entire or with fine teeth. Ovary densely hairy; styles divergent. Samaras purplish red to golden brown. *Mozambique & eastern South Africa to Angola & Namibia.* G1. Summer–autumn.

A variable species. The sharp-tipped hairs are easily detached and can cause irritation.

6. MALPIGHIA Linnaeus
B. MacBryde

Shrubs or trees (rarely climbing). Leaves opposite, with usually 2 (rarely to 6) glands near the midrib to the base. Stipules small, borne on the stem beside the leaf-base. Flowers usually bilaterally symmetric, 1–few in axillary corymbs or umbels or numerous in racemes. Sepals 5, glands 6–10, usually large. Petals pink, purple or white, blades often wavy or finely toothed. Stamens equal or unequal, united at the base. Carpels 3, completely united, hairless; styles equal or unequal. Fruit an angular or spherical drupe containing usually 3 pyrenes.

A genus of about 45 species from Mexico, the Caribbean and Central & South America. The species are difficult to distinguish and many in collections may be misnamed. The fruits are edible. They can be grown in full sun in a glasshouse (some outside in favoured areas) and require to be well watered in summer. Propagation is by cuttings, layers, grafts or seed.

Literature: Anderson, W.R., Notes on neotropical Malpighiaceae 2, *Contributions from the University of Michigan Herbarium* **16**: 55–108 (1987), **17**: 39–54 (1990).

1a. Leaves holly (*Ilex*)-like, to 1.5 times as long as broad, usually toothed, eachtooth ending in a bristle **4. coccigera**
 b. Leaves not as above, relatively longer, entire **2**
2a. Leaves hairy beneath **3. urens**
 b. Leaves hairless beneath **3**
3a. Leaves dispersed along the long shoots; leaf-tips acute or acuminate **1. glabra**
 b. Leaves usually crowded on short shoots; leaf-tips obtuse or notched **2. emarginata**

1. M. glabra Linnaeus (*M. punicifolia* Linnaeus; *M. nitida* Miller not Jacquin; *M. fallax* Salisbury; *M. biflora* Poiret; *M. oxycocca* Greisbach var. *biflora* (Poiret)

Niedenzu; *M. semeruco* Jussieu). Illustration: *Botanical Magazine*, 813 (1805); Liu, Illustrations of native and introduced ligneous plants of Taiwan **1**: 351 (1960); *Annals of the Missouri Botanical Garden* **67**: 902 (1980); Wasowski & Wasowski, Native Texas plants, 251 (1988).

Shrub or small tree to 3 m (rarely to 10 m, or scrambling); young branches and leaves with soft, sharp-pointed adpressed hairs at first. Leaves dispersed along the shoots, glossy to dull, 2–11 × 1–5.5 cm, ovate to elliptic-lanceolate or oblong, acute or acuminate at the tip. Flowers 1–2 cm wide, slightly fragrant, 2–8 in umbels or racemes. Sepals with usually 6 large glands. Petals pink, lilac or white. Styles straight, truncate. Fruit scarlet, usually unlobed, 5–15 mm wide. *USA (Texas), Mexico, Greater Antilles, Central & South America.* H5–G1. Spring–autumn (but can flower all the year).

Grown widely as an ornamental and also for the insipid acid fruit.

M. mexicana Jussieu (*M. tomentosa* Moricand; *M. oaxacana* Loesener; *M. guadalajarensis* (S. Watson) Rose; *M. edulis* J.D. Smith; *M. cordata* Small; *M. hintonii* Bullock). Illustration: *Bulletin de l'Herbier Boissier* **2**: pl. 20 (1894); Das Pflanzenreich **91**: 620 (1928). Similar, but leaves with persistent, medifixed, needle-like bristles beneath and obtuse or truncate tips, fruit 1.7–2.5 cm wide. *Mexico, Guatemala, Costa Rica.* G1. Summer.

2. M. emarginata de Candolle (*M. punicifolia* misapplied; *M. glabra* misapplied; *M. berteriana* Sprengel; *M. retusa* Bentham; *M. urens* var. *lanceolata* (Grisebach) Grisebach; *M. umbettata* Rose). Illustration: Hermann & Celhay, Plants and flowers of Tahiti, 110 (1974); Adams, Caribbean Flora, 33 (1976); *Annals of the Missouri Botanical Garden* **67**: 902 (1980); Graf, Tropica, edn. 3, 458 (1986).

Shrub or small tree to 8 m, erect, sometimes suckering; branches at first with small, often needle-like adpressed hairs. Leaves usually crowded on short shoots, glossy above, soon hairless, 1–10 × 1–5 cm, obovate to elliptic, tip obtuse or notched. Flowers 1–2.5 cm wide in umbels. Sepals densely hairy. Petals pink, lilac or white. Stamens unequal. Ovary hairless, lobed. Fruit scarlet, crimson, orange-red or pale purple, 1–3.2 cm wide. *Mexico to northern Central America.* G1. Spring–autumn.

Sometimes grown in the tropics for its edible fruit.

3. M. urens Linnaeus. Illustration: Cavanilles, Monadelphiae classis dissertationes No. **8**: t. 235 (1789); Lamarck, Tableau encyclopédique et méthodique, Botanique 1, **2**: pl. 381 (1793); Spach, Histoire naturelle des vegetaux, phanérogames **2**: pl. 20 (1834).

Shrub or small tree to 3 m; young branches greenish, usually with tubercle-based adpressed bristles, persisting as tubercles. Leaves becoming hairless above, bristles persisting beneath and on the margins, 6–18 × 2–6.5 cm, mostly lanceolate to elliptic-lanceolate; stipules 1.5–3.5 mm, dark red to almost black, persistent. Flowers 8–16 mm wide in umbels or corymbs. Sepals usually bristly. Petals purplish, usually fading to white, winged on the back. Ovary usually lobed. Styles straight or curved. Fruit red, 1–1.5 cm in diameter, spherical. *Hispaniola.* G2. Summer.

Much confused. The following 3 species may be grown as *M. urens*:

M. incana Miller (*M. campechiensis* Poiret). Shoots smooth, stipules at most to 0.7 mm, umbel hairy. *S Jamaica.* G2. Summer–autumn.

M. oxycocca Grisebach. Illustration: Hawkes & Sutton, Wild flowers of Jamaica, pl. 15 (1974). Shoots smooth or with lenticels but without tubercles; stipules to 1.5 mm; umbel hairless. *Jamaica.* G2. Spring-autumn (or all year if well grown).

M. martinicensis Jacquin. Illustration: *Flora of the Lesser Antilles* **4**: 601 (1988). Shoots with lenticels but not tuberculate; stipules 1.5–2 mm; umbel hairless. *Lesser Antilles.* G2. Summer–autumn.

4. M. coccigera Linnaeus (*M. coccifera* Cavanilles; *M. heteranthera* Wight; *M. heterantha* Morley; *M. variifolia* Turczaninow). Illustration: *Edwards's Botanical Register* **7**: t. 568 (1821); Bailey, Manual of cultivated plants, edn. 2, 614 (1949); Liu, Illustrations of native and introduced ligneous plants of Taiwan **1**: 350 (1960); Graf, Exotica, edn. 11, 1518 (1982).

Shrub to 2.5 m; branches hairless or with stiff adpressed hairs when young, the hairs sometimes persisting as tubercles. Leaves partly crowded on short shoots, holly (*Ilex*)-like, glossy, flat to wavy, 6–30 × 4–24 mm, almost circular to obovate, toothed with the tip of each tooth with a spiny bristle. Flowers 1 or 2 together, 1.2–2 cm wide. Petals pink, lilac or white. Stamens unequal. Ovary unequally 3-lobed, 1 style more slender and shorter than the others. Fruit orange-red. *Puerto Rico, Lesser Antilles.* G2. Spring–summer.

Grown worldwide, sometimes as a houseplant.

7. HETEROPTERIS Kunth
B. MacBryde

Woody, sometimes slender climbers to shrubs or small trees. Leaves usually opposite and with glands on the blade or stalk. Stipules absent or minute, by the side of the leaf-base. Flowers bilaterally symmetric, in umbels, corymbs or racemes, sometimes grouped into panicles. Sepal-glands usually 8–10, rarely absent. Petals equal or unequal, usually uniformly yellow, pink, white, orange, bronze, lilac or maroon. Stamens 10, all fertile, equal or unequal. Carpels 3, only partially united, hairy; styles 3, persistent, stout, alike or not. Schizocarp of 3 samaras.

A genus of over 125 species from South America, with one species in West Africa. They are cultivated as for *Malphighia*.

Literature: Anderson, W.R., Notes on neotropical Malpighiaceae 1–3, *Contributions from the University of Michigan Herbarium* **15**: 93–136 (1982); **16**: 55–108 (1987); **17**: 39–54 (1990).

1. H. beecheyana Jussieu (*Banisteria beecheyana* (Jussieu) C. Robinson). Illustration: Das Pflanzenreich **91**: 308 (1928); Menninger, Flowering vines of the world, 218, pl. 1ll (1970); Graf, Tropica, edn. 4, 620 (1992).

Climber to 3 m or shrub to 3.5 m; branchlets densely felted with rusty hairs. Leaf-blades often wrinkled, 1.5–10 cm × 9–65 mm, ovate to ovate-circular, oblong or obovate, obtuse or notched, usually with rusty or yellowish long spreading hairs beneath. Flowers 1–1.2 cm wide, in hairy umbels grouped in leafy, conical panicles. Sepal-glands 6–10. Petals unequal, pink to lilac, the shorter with glandular-hairy margins. Stamens unequal. Samaras 1–3, pink-rose to red-brown, adpressed-hairy. *Mexico to Bolivia.* G2. Summer–autumn.

A variable species. Three other species may occasionally be found in cultivation.

H. chrysophylla (Lamarck) de Candolle (*Banisteria chrysophylla* Lamarck). Illustration: *Botanical Magazine*, 3237 (1833); Morton, Exotic plants, 67 (1971);

Graf, Exotica, edn. 11, 1525 (1982). Leaf-blades smooth, hairy beneath; flowers 1–1.5 cm wide; petals yellow-orange; stamens unequal; samaras yellow-brown. *Bolivia & N Argentina to Brazil (& Uruguay).* G1–2. Summer.

H. angustifolia Grisebach. Illustration: Das Pflanzenreich **91**: 324 (1928); *Lilloa* **9**: 6 (1943); *Flora de la Provincia de Buenos Aires* **4**: 55 (1965). Leaf-blades smooth, thin, hairless beneath, linear-lanceolate; flowers 1.3–2 cm wide; stamens equal or almost so; samaras dark red. *Southern South America.* G1–2. Spring–summer.

H. umbellata Jussieu. Illustration: *Archives du Muséum d'Histoire naturelle* **3**: pl. 14 (1843); Boletim de Botanica (Sao Paulo) **9**: 167 (1987). Leaf-blades smooth, hairless beneath, ovate or elliptic; flowers to 1 cm wide; stamens unequal; samaras yellow-orange to red-brown. *Bolivia & Argentina to SE Brazil (& Uruguay?).* G1–2. Summer.

8. BANISTERIOPSIS C. Robinson
B. MacBryde

Woody climbers, shrubs, dwarf shrubs or small trees; stems sometimes annual from a woody rootstock. Leaves opposite or in whorls of 3, with glands on blade or stalk. Flowers bilaterally symmetric, in umbels or cymes, sometimes grouped into a panicle. Sepals with 8 glands (rarely absent). Petals yellow, pink or white, usually unequal. Stamens 10, usually unequal. Carpels 3, completely free from each other, hairy. Styles 3. Fruit of 3 samaras.

A genus of over 90 species, mainly from Brazil. The one cultivated species requires full sun or partial shade and a well-drained soil; it should be watered frequently, though less in winter; it is propagated by cuttings or layers.

1. B. caapi (Grisebach) C. Morton (*Banisteria caapi* Grisebach; *Banisteria quitensis* Niedenzu; *Banisteriopsis quitensis* (Niedenzu) Morton; *Banisteria inebrians* (C. Morton) MacBride). Illustration: *Webbia* **13**: t. 40, 41, 507, 509 (1958); Emboden, Narcotic plants, pl. 54 (1972); Schultes & Hofmann, Plants of the gods, 35, 120–122 (1979); Schultes & Raffauf, The healing forest, 276–279 (1990).

Robust climber, bark light chocolate-coloured; stems sometimes silky-hairy when young. Leaves shining and hairless above, 5–20 × 2.5–11 cm, ovate, acuminate at apex, sparsely hairy beneath with 2–5 pairs of glands near the margin.

Flowers few to many in 4-flowered woolly or velvety umbels grouped in panicles. Sepals usually with 8 glands. Petals yellow fading to white, fringed. Stamens unequal. Styles terete, slender. *Brazil (W Amazonia).* G2. Summer.

The natural range of this species is obscured by native cultivation, extending to Venezuela, Ecuador and Bolivia. Parts of the plant are used to produce a hallucinatory drink.

9. STIGMAPHYLLON Jussieu
B. MacBryde

Woody climbers, rarely shrubs or perennials with annual stems and a woody rootstock. Leaves usually opposite, blades entire to palmately lobed (very variable in a single population), margins sometimes glandular, with usually 2 large glands on the base of the blade or on the stalk. Stipules very small, borne beside the leaf-base. Flowers bilaterally symmetric in panicles, umbels, corymbs or racemes. Sepal-glands 8. Petals yellow or some orange or red. Stamens unequal, some sterile. Carpels 3, partially united, hairy; styles 3, stout, usually unequal. Schizocarp of 3 samaras

A genus of about 110 species in Central and South America, one in West Africa. The few in cultivation are often wrongly named. They require full or partial shade in a well-drained soil. Propagation by cuttings or layers; cuttings are often slow to establish.

Literature: Anderson, C., Stigmaphyllon (Malpighiaceae) in Mexico, Central America and the West Indies, *Contributions from the University of Michigan Herbarium* **16**: 1–48 (1987).

1a. Leaves heart-shaped, with overlapping auricles at the base, margins ciliate **3. ciliatum**
 b. Leaves circular to linear, base without auricles, margins not ciliate **2**
2a. Leaf-stalks 2–30 mm, blades hairless to sparsely silky hairy beneath **1. emarginatum**
 b. Leaf-stalks 1–6 cm, blades densely velvety to woolly beneath **2. littorale**

1. S. emarginatum (Cavanilles) Jussieu (*Banisteria fulgens* Linnaeus; *S. lingulatum* (Poiret) Small; *S. periplocifolium* (de Candolle) Jussieu; *Banisteria microphytta* Hamilton; *S. microphyllum* misapplied; *S. haitiense* Urban & Niedenzu;

S. rubinervium Alain). Illustration: Fawcett & Rendle, Flora of Jamaica **4**: 236 (1920); Das Pflanzenreich **91**: 14, 467 (1928); Menninger, Flowering vines of the world, 220 (1970); Courtright, Tropicals, 130 (1988).

Climber to 15 m; branches coarsely rusty-hairy when young. Leaves white silky-hairy when young, entire, 1–13 cm × 5–105 mm, almost circular to ovate, elliptic or linear, base rounded to cordate, apex obtuse or notched, usually with 2 stalkless glands on the stalk. Flowers 1.8–2.8 cm wide, usually in racemes. Petals golden yellow, entire to bluntly toothed. Stamens unequal, anthers sometimes all fertile. Samaras 1–3, often reddish. *West Indies*. G2. Spring–autumn.

Two other similar species may occasionally be grown:

S. diversifolium (Kunth) Jussieu (*S. ledifolium* (Kunth) Small; *S. sericeum* Grisebach; *S. cordifolium* Niedenzu). Illustration: Fournet, Fleurs et plantes des Antilles, 72 (1976); Sastre & Portecop, Plantes fabuleuses des Antilles illustrée, 79 (1985); Honychurch, Caribbean wild plants and their uses, edn. 2, 57 (1986). Leaves woolly beneath; umbels, racemes or panicles with 6–27 flowers; always some anthers infertile. *Cuba, Lesser Antilles*. G2. Summer.

S. ellipticum (Kunth) Jussieu. Illustration: Herklots, Flowering tropical climbers, 131 (1976); Cronquist, An integrated system of classification of flowering plants, 769 (1981). A slender climber; leaves usually finely woolly beneath; umbels 1–9-flowered; some anthers always sterile. *Cuba, Lesser Antilles*. G2. Summer.

2. S. littorale Jussieu (*Banisteria bonariensis* Hooker & Arnott). Illustration: *Botanical Magazine*, 6623 (1882); Battfreund, Flora Argentina **3**: 139 (1909); Mathias (ed.), Flowering plants in the landscape, 123 (1982).

Robust climber to 10 m, or dwarf shrub with thick, more or less woody tubers; branches yellow-hairy when young. Leaves hairless or very sparsely hairy, 5–10 × *c.* 4–12 cm, broadly ovate to almost circular or oblong, base truncate to cordate, tip acute to notched; stalks with 2 stalkless glands near the base. Flowers 1.5–3.5 cm wide, 10–20 in an umbel. Petals golden yellow, margins fringed. Stamens unequal. Styles reddish, sometimes hairless. Samaras

1–3, green. *S Brazil, Paraguay, Argentina*. G2. Summer–autumn.

3. S. ciliatum (Lamarck) Jussieu (*Banisteria ciliata* Lamarck). Illustration: *Paxton's Magazine of Botany* **15**: 77 (1849); *Botanical Magazine*, n.s., 526 (1968); Everard & Morley, Wild flowers of the world, pl. 178 (1970); Bor & Raizada, Some beautiful Indian climbers and shrubs, edn. 2, 204, pl. 22, 67, 68 (1982).

Slender woody climber to 9 m, branchlets silky-hairy when very young. Leaves hairless or almost so, entire, 2–10 × 2–7 cm, ovate to heart-shaped with usually long, overlapping auricles at the base, tip obtuse to acuminate, margins with cilia; glands 2, on the stalk. Flowers 3–4.5 cm wide, slightly scented, 3–9 in umbels. Petals bright yellow, edges wavy or fringed. Stamens unequal, the anthers of some reduced or absent. Styles hairless. *South America from Trinidad to Uruguay; naturalised on Barbados*. G2. Summer or flowering continuously.

168. POLYGALACEAE

Herbs or shrubs, rarely climbers. Leaves usually alternate, entire, exstipulate; ptyxis flat or supervolute. Inflorescences racemes, spikes or panicles. Flowers bisexual, zygomorphic. Calyx, corolla and stamens usually hypogynous or calyx hypogynous, corolla and stamens perigynous. Calyx usually of 5 free sepals, lateral pair petal-like. Corolla usually of 3 petals, often joined to staminal tube. Stamens 8–10 (rarely 4–5), filaments united, anthers opening by pores. Ovary usually of 2 united carpels; ovules usually 1 per cell, axile; style 1, stigmas as many as carpels. Fruit usually a capsule; seeds with arils.

A cosmopolitan family of 18 genera and about 950 species.

1a. Fruit fleshy; style 2-branched
          3. Nylandtia
  b. Fruit dry; style unbranched   2
2a. Herb or shrub; fruit a capsule
          1. Polygala
  b. Tree; fruit a winged samara
          2. Securidaca

1. POLYGALA Linnaeus
E.H. Hamlet

Annual or perennial herbs or shrubs, sometimes spiny. Leaves alternate, more

rarely opposite or whorled, ovate, lanceolate or linear, sometimes rudimentary. Flowers in a terminal or lateral raceme, spike or head. Sepals 5, falling early or persistent, unequal, the inner 2 petal-like, sometimes shortly clawed, all free, or the inner 2 united, stalkless. Petals 3 (rarely 5), the upper 2 basally joined to the staminal-tube, the lower boat-shaped, clawed, entire or bearing a crest of 2 lobes, both entire or divided into a number of appendages; lateral petals minute, simple, bifid or deeply 2-lobed, sometimes absent. Disc sometimes present. Stamens 8 (rarely 4, 5 or 9), sometimes only 6 fertile with 2 staminodes; filaments linear, joined into a split tube, sometimes with marginal hairs; anthers opening at apex. Fruit a membranous capsule, compressed, elliptic or obovate, often notched.

An almost cosmopolitan genus of about 500 species. They can be grown outside or in a cool glasshouse. Shrubs are propagated by cuttings inserted in sandy peat, in a frame in autumn or by suckers removed in autumn. Herbaceous plants are propagated by division in spring. All may be propagated by seed sown in a frame, in spring or autumn.

1a. Plant with leaf-rosettes   2
  b. Plant without leaf-rosettes   3
2a. Plant erect   **1. amara**
  b. Plant spreading   **2. calcarea**
3a. Leaves opposite   **5. fruticosa**
  b. Leaves alternate   4
4a. Leaves mostly in a terminal cluster
          8. paucifolia
  b. Leaves spaced along stem   5
5a. Flowers in clusters of more than 3, often many-flowered   6
  b. Flowers solitary or in clusters of up to 3   7
6a. Plant a shrub; flowers deep purple to pale lilac   **7. virgata**
  b. Plant a herb; flowers white or greenish   **9. senega**
7a. Leaves 2.5–5 cm, oblong to obovate
          6. myrtifolia
  b. Leaves 1.5–3 cm, ovate or linear to linear-lanceolate   8
8a. Plant 5–15 cm; capsule 6–8 mm
          3. chamaebuxus
  b. Plant to 5 cm; capsule 1–1.2 cm
          4. vayredae

1. P. amara Linnaeus. Illustration: Grey-Wilson & Blamey, The alpine flowers of Britain & Europe, 137 (1979); Aeschimann et al., Flora alpina **1**: 1039 (2004).

Erect perennial; stems 5–20 cm, numerous, growing from the centre of a basal rosette. Basal leaves 1.5–3.5 cm × 6–10 mm, elliptic to obovate, upper lanceolate to oblong, broadest near the middle, hairless, acute. Flowers in racemes of 8–25, white, pink, violet or blue. Petals 3.5–6.5 mm, distinctly jointed between tube and keel, and between keel and crest. Wings 4.5–8 mm in fruit, elliptic. Capsule 3.5–5.5 mm. *Mountains of eastern C Europe & N former Yugoslavia.* H2. Summer.

2. P. calcarea Schultz. Illustration: Polunin & Smythies, Flowers of southwest Europe, pl. 27 (1973); Brickell (ed.), RHS A–Z encyclopedia of garden plants, 836 (2003); Aeschimann et al., Flora alpina **1**: 1039 (2004).

Stems spreading, 5–20 cm, with decumbent, usually leafless stolons ending in leaf-rosettes. Rosette leaves 5–20 × c. 10 mm, spathulate to obovate, hairless to sparsely hairy. Flowering-stems arising from the rosettes, upper leaves smaller, linear-lanceolate. Racemes with 6–20 flowers; bracts linear-lanceolate. Flowers 6–7 mm, blue or white. Wings c. 5 mm, obovate to oblong-elliptic, veins joining at margin. Capsule 4–6 mm. *W Europe, northwards to S England.* H2. Spring–summer.

Requires a gritty, well-drained soil with lime or chalk and a sunny, open position.

3. P. chamaebuxus Linnaeus. Illustration: Tosco, Mountain flowers, 37 (1978); Grey-Wilson & Blamey, The alpine flowers of Britain & Europe, 137 (1979); Lippert, Fotoatlas der Alpénblumen, 68 (1981); Aeschimann et al., Flora alpina **1**: 1033 (2004).

Dwarf creeping shrub, 5–15 cm. Leaves alternate, spaced along stem, 1.5–3 cm × 3–12 mm, ovate to linear-lanceolate, leathery. Flowers solitary or in pairs in leaf-axils. Outer sepals unequal; larger wings spreading horizontally, white to yellow, ageing to reddish brown. Petals 1–1.4 cm, upper shorter than keel; keel bright yellow with very small 2–6-lobed crest; tube and upper petals similar in colour to wings. Capsule 6–8 mm, stalkless, surrounded by a wing less than 1 mm wide. *Most of Europe.* H1. Spring–summer.

Var. **grandiflora** Neilreich. Illustration: Brickell (ed.), RHS A–Z encyclopedia of garden plants, 836 (2003). Flowers purplish red and yellow. Prefers a peaty, lime-free soil and a partially shady position.

4. P. vayredae Costa. Illustration: *Botanical Magazine*, 9009 (1924); Grey-Wilson & Blamey, The alpine flowers of Britain & Europe, 137 (1979).

Similar to *P. chamaebuxus* but plant to 5 cm; leaves 1.5–2 cm × 2–6 mm, linear to linear-lanceolate, acute. Flowers solitary or in very short racemes of 2 or 3 flowers in the leaf-axils. Outer sepals 3–4 mm, membranous, pale green or purplish; wings to 1.4–1.7 cm × 5–7 mm, broadly and bluntly obovate, bright rose-purple. Upper 2 petals 1.1–1.2 cm, obliquely obovate-lanceolate; keel a spherical hood with a 5–7-lobed, yellow crest. Capsule winged, 1–1.2 × c. 1 cm. Seeds ellipsoid, c. 4 mm. *Spain.* H1. Spring.

Prefers a peaty, lime-free soil and a partially shady position.

5. P. fruticosa Bergius (*P. oppositifolia* Linnaeus; *P. oppositifolia* var. *latifolia* (Ker-Gawler) Harvey). Illustration: Jeppe, Natal wild flowers, pl. 39g (1975).

Erect branched shrub, 30–200 cm. Leaves 8–40 mm, opposite, lanceolate or ovate, mostly acute, with short stalks. Flowers reddish purple or greenish, in short terminal racemes. Outer sepals 3–5 mm, ovate, margin hairy; wings 8–15 mm, unequal, acute or slightly obtuse. Keel including crest 8–14 mm, claw shorter than blade; crest large. Capsule broadly obovate with obvious marginal rim. Seeds ellipsoid, downy. *South Africa (E Cape, Natal).* G1. Summer.

P. × dalmaisiana also keys out here, see below.

6. P. myrtifolia Linnaeus. Illustration: Perry, Flowers of the world, 235 (1972); Moriarty, South African wild flower guide 2, 129 (1982); Anon., Wild flowers of South Africa, Reprinted, 77 (1984).

Erect shrub 1–2.5 m. Leaves alternate, spaced along stem, 2.5–5 cm, oblong to obovate. Flowers few, on short racemes, appearing lateral; bracts persistent. Outer sepals 5–8 mm, ovate, concave, margins hairy; wings 1.2–1.8 cm, violet-purple, finely mucronate. Petals lilac or rarely white, shading to deep violet at apex of keel; upper petals short, 2-lobed, upper lobe reflexed. Keel including crest 1.2–2.2 cm, claw much shorter than wing. Capsule elliptic-spherical, notched, narrowly winged. Seeds ellipsoid, downy. *South Africa (E Cape, Natal), locally naturalised in the W Mediterranean area.* H5. Summer.

Var. **grandiflora** (Loddiges) Hooker. Illustration: *Botanical Magazine*, 3616

(1837). Has purple flowers for most of the year. Slightly tender and in cold areas should be pot-grown in a cool, moist, partially shaded and frost-free glasshouse.

P. × dalmaisiana Bailey (*P. myrtifoha* var. *dalmaisiana* Voss). Illustration: Brickell (ed.), RHS A–Z encyclopedia of garden plants, 836 (2003). Leaves resembling those of *P. myrtifolia* in shape but may be opposite and alternate on the same plant. Flowers to 3 cm, bright purple or rosy-red, keel and foot white. A hybrid between *P. myrtifolia* var. *grandiflora* and *P. fruticosa* var. *cordata* Harvey.

7. P. virgata Thunberg. Illustration: Letty, Wild flowers of the Transvaal, 187 (1962); Gibson, Wild flowers of Natal (coastal region), 52 (1975); Anon., Wild flowers of South Africa, 77 (1984).

Shrub 1.5–3 m, sparsely to fairly densely crisped-downy, eventually hairless. Leaves 2–9 cm × 3–20 mm, alternate, linear to narrowly elliptic, downy when young, becoming hairless, acute, tapered to the base. Inflorescence a terminal, many-flowered, loose raceme, 3–15 cm; bracts c. 3 mm, deciduous; bracteoles c. 2 mm. Flower-stalks to 7 mm, slender. Flowers deep purple to pale lilac; wings 1–1.7 cm × 9–13 mm, very broadly elliptic to almost spherical. Upper petals c. 7 × 3–3.5 mm, broadly and irregularly spathulate; keel c. 13 × 5 mm, margins minutely hairy; crest 4–5 mm, purple. Style 1–1.2 cm, curved; stamens 8. Capsule c. 10 × 8 mm, obovate, margins winged, c. 1 mm wide. *Tropical & southern Africa.* H5–G1. Summer.

Prefers a light soil and an open position.

8. P. paucifolia Willdenow. Illustration: House, Wild flowers, pl. 125B (1961); Rickett, Wild flowers of the United States **2**: 527 (1967); Gillett, The milkworts of Canada, 8 (1968); Zichmanis & Hodgins, Flowers of the wild, 115 (1982).

Perennial herb with rhizomes; stems erect, 5–15 cm. Leaves 1.5–5 × 1–2.5 cm, alternate, ovate to elliptic, 3–6 in a terminal cluster; lower smaller, 4–10 mm, bract-like, widely spaced along the stem. Flowers of 2 types; normal flowers 1–5, in an axillary but apparently terminal cluster, rose-purple, occasionally white. Sepals 5, 3 hood-like, wings more deeply coloured, 1.5–1.7 cm, obovate. Upper petals 1.2–1.8 cm, lower about equal in length, pouch c. 6 mm, fringed crest 3–3.5 × c. 6 mm. Stamens 6, fused to the lower petal, free at apex. Abnormal flowers small, non-opening, *c.* 2 mm on short stems with bracts, rarely leafy. Capsule 6–7 mm, obovate, slightly winged. *E North America.* H1. Spring–summer.

Prefers a peaty soil and shady position.

9. P. senega Linnaeus. Illustration: House, Wild flowers, pl. 125A (1961); Rickett, Wild flowers of the United States **2**: 527, 529 (1967); Gillett, The milkworts of Canada, 12 (1968); Strausbauch & Core, Flora of West Virginia, 593 (1977).

Perennial, downy, often with yellowish green stems; leaves crowded, arising from a woody root. Leaves 1.3–7 cm × 2–30 mm, alternate, lanceolate or ovate to elliptic, dark green above, pale beneath, becoming smaller and finally scale-like, purplish towards the base, acuminate, midrib prominent on lower surface, margins rough, often inwardly curved. Inflorescence a dense, terminal spike-like panicle, 1–4.5 cm. Flower-stalks 4–7 mm; flowers white or greenish; stamens 7. Capsule 2.5–3.5 × *c.* 3.5 mm. Seeds pear-shaped, curved. *N America.* H1. Summer.

2. SECURIDACA Linnaeus
E.H. Hamlet
Shrubs, small trees, or climbers, sometimes spiny. Leaves alternate, entire. Flowers in terminal and axillary racemes or panicles. Sepals 5, unequal, inner 2 larger, petal-like. Petals 3, free, central petal usually clawed, or sometimes 2 extra scale-like petals present. Stamens 8, filaments linear, joined into a split tube. Style borne obliquely on ovary, flattened above; stigma terminal. Fruit a samara; seeds spherical.

A genus of 80 species from Asia and America, with one species in Africa. Our species requires a warm, frost-free position in well-drained sandy soil. It is propagated from seed, soaked thoroughly before sowing; seedlings resent disturbance.

1. S. longepedunculata Fresenius. Illustration: Letty, Wild flowers of the Transvaal, 186 (1962); Drummond & Palgrave, Common trees of the Highveld, pl. 44 (1973); Drummond, Common trees of the central watershed woodlands of Zimbabwe, 109 (1981); Germishuizen, Transvaal wild flowers, 153 (1982).

Deciduous tree. Leaves sometimes in clusters, variable in size and shape, leathery, 1–5 cm × 5–20 mm, ovate-oblong to linear-lanceolate, apex obtuse or rounded, base narrowly tapered. Leaf-stalk

slender, 2–10 mm. Flowers bisexual, long-stalked, 1–1.8 cm, pink or lilac to purple, sweetly scented, in terminal and axillary racemes, appearing with the young leaves. Sepals 5, unequal, the inner 2 petal-like; petals 3, free, central petal clawed; stamens 8, joined to form a split tube. Style 8–10 mm; stigma 1.5–2 mm. Fruit a nut with veined wing, turning rosy-red or brown when mature. *Tropical & southern Africa.* H5. Summer.

3. NYLANDTIA Dumortier
E.H. Hamlet
Evergreen shrubs. Flowers purplish pink, sometimes white. Differs from *Polygala* in having a 2-branched style and fleshy fruit.

A genus of a single species from South Africa. Rare in cultivation.

1. N. spinosa (Linnaeus) Dumortier. *South Africa (Cape Province).* G1.

---

169. CORIARIACEAE

Shrubs, branches angular. Leaves opposite, entire, exstipulate, ptyxis flat. Flowers solitary or in racemes, usually bisexual, actinomorphic. Calyx, corolla and stamens hypogynous. Calyx of 5 free sepals. Corolla of 5 free petals, each keeled inside. Stamens 10. Ovary of 5–10 free carpels; ovules 1 per carpel, apical; styles as many as carpels. Fruit an achene surrounded by the fleshy corolla.

A family of a single genus with a wide distribution from central and western South America, Mediterranean area, Himalaya and eastern Asia to New Zealand.

1. CORIARIA Linnaeus
M.F. Gardner
Description as for the family.

A genus of 5 species occasionally grown for their racemes of colourful fruits and attractive autumn leaf colour. Although some species are said to have edible fruits, many are known to have poisonous leaves and fruits. All are easily grown in a well-drained, loamy soil; however most species will be killed by severe winters. Propagate by seed or cuttings taken from half-ripened shoots.

1a. Bases of lateral branches usually without bud-scales; inflorescence borne terminally, in the leaf-axils or on current year's growth   2

b. Bases of lateral branches surrounded by persistent bud-scales; inflorescence borne in leaf-axils of previous year's growth   3
2a. Stems herbaceous; fruit-stalks longer than 1 cm   **1. terminalis**
b. Stems woody; fruit-stalks less than 1 cm   **2. ruscifolia**
3a. Racemes finely to densely hairy
                                    3. napalensis
b. Racemes hairless   4
4a. Enlarged persistent petals bright coral-red, exceeding mature achenes and enclosing them   **4. japonica**
b. Enlarged persistent petals reddish brown, not exceeding mature achenes   **5. myrtifolia**

1. C. terminalis Hemsley. Illustration: *Revue Horticole*, pl. 160 (1907); *Botanical Magazine*, 8525 (1913); Heywood, Flowering plants of the world, 200 (1978); Brickell, RHS gardeners' encyclopedia of plants & flowers, 141 (1989).

Shrub to 1 m with herbaceous stems, spreading by underground rhizomes; bases of lateral branches usually without bud-scales. Leaves 2.5–7.5 cm, ovate, acute at tips, usually with 5–7 veins. Flowers bisexual, in terminal racemes 15–20 cm, borne on current year's growth; petals greenish. Fruit black or orange, stalks 1–1.5 cm. *Bhutan, China, India (Sikkim).* H5. Summer.

Some authorities consider the mature fruit to be orange, turning black when withered. The orange-fruited variants have been called var. **xanthocarpa** Rehder & Wilson. Illustration: Brickell (ed.), RHS A–Z encyclopedia of garden plants, 305 (2003).

2. C. ruscifolia Linnaeus. Illustration: Muñoz Schick, Flora del parque nacional Puyehue, 212 (1980); Hoffmann, Flora silvestre de Chile, 151 (1991).

Shrub to 2 m; bases of lateral branches usually without bud-scales. Leaves 1–7.5 cm × 8–32 mm, elliptic to lanceolate on lateral branches, rounded on main stem, acute to cordate at base, acute to mucronate at tip, veins 3–5. Flowers bisexual, in terminal or axillary racemes 10–25 cm, borne on current year's growth; sepals ovate, green to red. Fruit black, stalks 3–6 mm. *Chile, Argentina.* H5. Summer.

Subsp. **microphylla** (Poirret) Skog (*C. microphylla* Poirret; *C. thymifolia* Humboldt). Similar to above except leaves 5–31 × 2–16 mm. Flowers in terminal racemes, 10–15 cm. *Mexico, W tropical South America & New Zealand.*

3. C. napalensis Wallich (*C. sinica* Maximowicz). Illustration: Polunin & Stainton, Flowers of the Himalayas, 458 (1985), and supplement, pl. 20 (1988).

Shrub 1–2.5 m, bases of lateral branches surrounded by persistent bud-scales. Leaves 3.5–10 × 2–8 cm, elliptic to ovate, rounded or shallowly cordate at base, hairless, with 3–5 veins at base. Flowers bisexual, in racemes 2.5–10 cm, finely to densely hairy, borne in leaf-axils of previous year's growth; sepals ovate, *c.* 1 mm, rounded; petals reddish; anthers crimson. Fruits black. *N India, Nepal, W China.* H5. Spring.

4. C. japonica Gray. Illustration: *Botanical Magazine*, 7509 (1896); Bailey, Standard cyclopedia of horticulture, 846 (1935); Krüssmann, Manual of cultivated broad-leaved trees & shrubs **1**: pl. 133(1984); Takatori, Color atlas medicinal plants of Japan, f. 58 (1980).

Shrub to 1 m, bases of lateral branches surrounded by persistent bud-scales. Leaves 6–8 × 2–3.5 cm, ovate to lanceolate, rounded at base, acuminate at tip, 3-veined. Flowers unisexual, in racemes 4–6.5 cm, borne in the axils of previous year's growth, hairless; petals greenish; sepals *c.* 3.5 mm. Fruits with fleshy petals exceeding the mature achenes and enclosing them, bright coral-red. *Japan.* H4. Summer.

5. C. myrtifolia Linnaeus. Illustration: Krüssmann, Manual of cultivated broad-leaved trees & shrubs **1**: 366 (1976); Polunin, Trees and bushes of Britain and Europe, 118 (1976); Lopez Gonzalez, Guia de Incafo de los arboles y arbustos de la penninsula Iberica, 302 (1982); Valdés, Flora vascular de Andalucía occidental **1**: 128 (1987).

Shrub 1–3 m, bushy, bases of lateral branches surrounded by persistent bud-scales. Leaves 3–6 cm, elliptic to lanceolate, acute or cuspidate, 3-veined. Flowers unisexual or bisexual, in racemes 2–5 cm, in axils of previous year's growth, hairless; petals greenish. Fruits with fleshy, dark reddish brown petals not exceeding mature achenes. *SW Europe.* H4. Summer.

170. ANACARDIACEAE

Woody plants with resinous bark. Leaves alternate, simple or compound, exstipulate; ptyxis conduplicate or rarely flat.

Inflorescences racemes or panicles. Flowers bisexual or unisexual, actinomorphic. Calyx, corolla and stamens hypogynous, disc often present. Calyx of 3–5 free sepals or rarely of 5 united sepals. Corolla of 3–5 free petals, rarely absent. Stamens 1–10. Ovary of 3–5 united carpels; ovules 1 per cell, apical or basal; style always 1, sometimes divided above. Fruit a drupe, 1-seeded.

A family of about 73 genera mostly tropical or subtropical but with some genera from the Mediterranean area and temperate North America. The resin from many species is often allergenic. The family provides a number of ornamental and important fruit trees including the sumachs (*Rhus*), cashews (*Anacardium*), pistachios (*Pistacia*) and mangoes (*Mangifera*).

1a. Carpels 5 and free or carpel solitary 2
 b. Carpels united 3
2a. Fruit kidney-shaped or ovoid, fleshy **5. Mangifera**
 b. Fruit kidney-shaped, dry **6. Anacardium**
3a. Petals absent **3. Pistacia**
 b. Petals present 4
4a. Stamens 8–20 5
 b. Stamens 1–6 8
5a. Stamens 12–20 **11. Sclerocarya**
 b. Stamens 8–10 6
6a. Petals not overlapping **10. Spondias**
 b. Petals overlapping 7
7a. Ovary 1-celled **8. Schinus**
 b. Ovary with 4 or 5 cells **12. Harpephyllum**
8a. Style or stigma more or less at apex of ovary **1. Rhus**
 b. Style or stigma on the side of ovary 9
9a. Ovary sunk into a cup-shaped or tubular hollowed receptacle **9. Semecarpus**
 b. Ovary not sunk as above 10
10a. Fruiting panicles feathery **2. Cotinus**
 b. Fruiting panicles not feathery 11
11a. Terminal leaflet on distinctly winged stalk **7. Loxostylis**
 b. Terminal leaflet-stalk not winged **4. Schinopsis**

1. RHUS Linnaeus
M.C. Tebbitt
Deciduous or evergreen shrubs, climbers or trees, dioecious or with both unisexual and bisexual flowers. Leaves alternate, odd-pinnate, with 3 leaflets or entire. Inflorescence a panicle; flowers small. Calyx 5-lobed. Petals 5, longer than calyx-lobes. Stamens 5, reduced to staminodes in

female flowers. Stigmas 3, at apex of ovary. Fruit a drupe, often with thin red flesh.

A genus of about 150 species from temperate and subtropical North America, southern Africa, subtropical east Asia and northeast Australia. Commonly grown for their architectural beauty which in some species is particularly noticeable in the autumn when they have lost their leaves and are in fruit. The hardy species may be grown outside as specimen plants or as windbreaks, while the half-hardy species are best grown in a cool glasshouse in areas prone to frost. All species require full sun and a well-drained soil. Propagate by semi-ripe cuttings taken in the summer, autumn sown seed or by root cuttings taken during the winter.

Literature; Coombes, A.J., Cut-leaved sumacs, *The New Plantsman* **1**(2): 107–113 (1994).

1a. Leaves pinnate 2
 b. Leaves with 3 leaflets or simple 11
2a. Adult leaflets entire 3
 b. Adult leaflets with forward-pointing teeth 7
3a. Fruiting panicles pendent 4
 b. Fruiting panicles erect 5
4a. Branches hairless or almost so, leaves almost hairless beneath **10. potaninii**
 b. Branches hairy; leaves hairy on veins beneath **9. punjabensis** var. **sinica**
5a. Inflorescence an axillary panicle; fruit off-white **11. trichocarpa**
 b. Inflorescence a terminal panicle; fruit brownish and hairless or red and hairy 6
6a. Branches tinged red; flowers greenish; fruit red, hairy **7. copallina**
 b. Branches not tinged red; flowers brownish, hairy; fruit brownish yellow, hairless **12. sylvestris**
7a. Branchlets hairy 8
 b. Branchlets hairless 10
8a. Central axis of leaf not winged; leaf-stalk densely covered in rust-coloured hairs **6. typhina**
 b. Central axis of leaf winged; leaf-stalk not as above 9
9a. Inflorescence loose; leaves to 7 cm; fruit brownish purple **4. coriaria**
 b. Inflorescence dense; leaves to 10 cm; fruit red **7. copallina**
10a. Leaflets 3–7, with brown hairs beneath; flowers creamy white **8. chinensis**

b. Leaflets 9–15, hairless beneath; flowers green **5. glabra**
11a. Leaves simple 12
b. Leaves with 3 leaflets 13
12a. Inflorescence a short dense spike; flowers light yellow, tinged red **2. ovata**
b. Inflorescence a panicle; flowers white to pink **1. integrifolia**
13a. Plant evergreen, to 8 m; inflorescence a panicle; fruit shiny brown, thinly fleshy **3. lancea**
b. Plant deciduous, to 2.5 m; inflorescence a spike; fruit red, hairy 14
14a. Terminal leaflets 3-lobed, leaflets to 2 cm; flowers pale green **14. triloba**
b. Terminal leaflets entire, leaflets to 7.5 cm; flowers yellow **13. aromatica**

1. R. integrifolia (Nuttall) S. Watson.

Evergreen shrub or tree to 10 m. Leaves simple, to 5 cm, elliptic, entire or with forwardly pointing teeth. Inflorescence a hairy panicle. Flowers white to pink. Fruit dark red, hairy. *SW North America.* H5. Spring.

2. R. ovata S. Watson. Illustration: Elias, The complete trees of North America field guide and natural history, 801 (1980); Macoboy, What shrub is that?, 292 (1990).

Evergreen shrub to 3 m. Leaves simple, to 7 cm, ovate, often folded, usually entire. Inflorescence a short dense spike. Flowers light yellow, tinged red. Fruit deep red, hairy. *SW North America.* H5. Spring.

3. R. lancea Linnaeus. Illustration: Palmer & Pitman, Trees of southern Africa, 1254 (1972).

Evergreen tree to 8 m. Branches red. Leaves with 3 leaflets; leaflets to 18 cm, leathery, usually entire, stalkless. Inflorescence a dense terminal panicle. Flowers pale green-yellow. Fruit shiny brown, thinly fleshy. *SE southern Africa.* H5. Spring.

4. R. coriaria Linnaeus. Illustration: Krüssmann, Manual of cultivated broad-leaved trees & shrubs **3**: 199 (1986); Polunin & Smythies, Flowers of southwest Europe, pl. 25 (1973).

Deciduous shrub to 3 m. Branches densely hairy. Leaves pinnate, central axis winged; leaflets 9–15, to 7 cm, elliptic to oblong, with coarse forwardly pointing teeth, hairy beneath. Inflorescence a loose terminal panicle. Flowers greenish. Fruit brownish purple, hairy. *S Europe.* H5. Late summer.

5. R. glabra Linnaeus. Illustration: Elias, The complete trees of North America, field guide and natural history, 804 (1980); Brickell, The RHS gardeners' encyclopedia of plants & flowers, 111 (1989); *The New Plantsman* **1**(2): 109 (1994).

Shrub or small tree to 3 m. Branches hairless. Leaves pinnate; leaflets 9–15, to 7 cm, elliptic to oblong, with coarse forwardly pointing teeth, hairless beneath, turning red in autumn. Inflorescence a dense panicle. Flowers green. Fruit scarlet, hairy. *S Canada to NE USA.* H1. Late summer.

6. R. typhina Linnaeus (*R. hirta* Linnaeus). Illustration: Preston, North American trees, 288 (1969); Elias, The complete trees of North America, field guide and natural history, 803 (1980); *The New Plantsman* **1**(2): 109 (1994); Marshall Cavendish encyclopaedia of gardening, 1840 (1970).

Suckering deciduous tree or shrub to 10 m. branchlets densely hairy. Leaves compound; leaflets 11–13, to 11 cm, oblong-lanceolate, margins with forwardly pointing teeth, hairy, turning red in autumn; leaf-stalks densely covered in rust-coloured hairs. Inflorescence a rusty red, dense terminal panicle. Flowers green. Fruit red, hairy. *E North America.* H1. Summer.

'Dissecta' has the leaflets pinnately dissected. 'Laciniata' (illustration: Brickell, The RHS gardeners' encyclopedia of plants & flowers, 92 (1989); Macoboy, What shrub is that?, 293, 1990) has the leaflets and bracts laciniately toothed.

R. × pulvinata Greene (*R. hybrida* Rehder). Intermediate between the parents (*R. glabra × R. hirta*), branches and inflorescence with short fine hairs. Often found in gardens as *R. glabra*.

Hybrids between these parents with partly or deeply bipinnate leaves have been described as the Autumn Lace group, with a particular cultivar having been selected as 'Red Autumn Lace' (illustration: *The New Plantsman* **1**(2): 109, 111, 1994).

7. R. copallina Linnaeus (*Schmaltzia copallina* Small). Elias, The complete trees of North America, field guide and natural history, 805 (1980).

Deciduous shrub or small tree to 6 m. Branches tinged red. Leaves pinnate, central axis winged; leaflets 9–21, to 10 cm, oblong to obovate or ovate to lanceolate, usually entire, hairy beneath. Inflorescence a dense terminal panicle. Flowers greenish. Fruit red, hairy. *E North America.* H1. Late summer.

8. R. chinensis Miller (*R. javanica* Thunberg not Linnaeus; *R. rosbeckii* Deene; *R. semialata* Murray). Illustration: Thrower, Hong Kong trees **2**: 93 (1977); Polunin & Stainton, Concise flowers of the Himalayas, pl. 23 (1985).

Deciduous shrub or flat-headed tree to 8 m. Leaves pinnate, central axis winged; leaflets 3–7, to 11 cm, ovate-oblong, coarsely scalloped, with brown hairs beneath, almost stalkless. Inflorescence a large broad panicle. Flowers creamy white. Fruit red, densely hairy. *China, Japan.* H4. Late summer.

9. R. punjabensis Stewart var. **sinica** (Diels) Rehder & Wilson.

Tree to 12 m; branches hairy. Leaves pinnate, central axis winged; leaflets 7–11, more when young, to 12 cm, ovate to oblong, entire, stalkless, veins hairy below. Inflorescence a pendent hairy panicle. Flowers cream-white. Fruit red, densely hairy. *C & W China.* H2. Late summer.

10. R. potaninii Maximowicz.

Deciduous tree to 8 m. Leaves pinnate, central axis often winged; leaflets 7–11, ovate-oblong to ovate-lanceolate, adult leaves entire, almost stalkless. Inflorescence a pendent panicle. Flowers off-white. Fruit dark red, densely hairy. *C & W China.* H1. Early summer.

11. R. trichocarpa Miquel. Illustration: Brickell, The RHS gardeners' encyclopedia of plants & flowers, 68 (1989); Brickell (ed.), RHS A–Z encyclopedia of garden plants, 904 (2003).

Tree to 15 m, branchlets hairy when young. Leaflets 13–17, to 10 cm, ovate to oblong, entire, veins hairy beneath, turning red in autumn. Flowers in axillary panicles, green. Fruit off-white. *Temperate E Asia.* H3. Summer.

12. R. sylvestris Siebold & Zuccarini.

Tree to 10 m. Leaves compound; leaflets 7–13, to 10 cm, ovate. Inflorescence a panicle. Flowers brownish, hairy. Fruit brownish yellow, hairless. *Temperate E Asia.* H3. Summer.

13. R. aromatica Aiton (*R. canadensis* Marsh). Illustration: Krüssmann, Manual of cultivated broad-leaved trees & shrubs **3**:

200 (1978); Phillips & Rix, Shrubs, 36 (1989).

Shrub to 2.5 m, usually prostrate. Leaves aromatic, with 3 leaflets; leaflets to 7.5 cm, ovate, almost stalkless, hairy, coarsely toothed, with forwardly pointing teeth. Inflorescence a terminal spike, appearing before the leaves. Flowers yellow. Fruit red, hairy. *SE Canada to S & E North America.* H1. Spring.

'Laciniata' has narrow, deeply lobed leaves. Var. **serotina** (Greene) Rehder is an upright shrub with leaflets hairless to almost so, the terminal leaflets fan-shaped to obovate. *C North America.* H1. Spring.

14. R. triloba Nuttall. Illustration: Krüssmann, Manual of cultivated broad-leaved trees & shrubs **3**: 200 (1978).

Deciduous shrub to 2 m; young shoots hairy, becoming hairless. Leaves with 3 leaflets, terminal leaflet 3-lobed; leaflets to 2 cm, ovate, stalkless or almost so, usually with coarse forwardly pointing teeth. Inflorescence a spike. Flowers pale green. Fruit red, hairy. *W North America.* H1. Spring.

2. COTINUS Adanson
J.E. Richardson
Deciduous trees or shrubs with yellow wood. Leaf-blades simple, entire or slightly toothed, stipules absent. Inflorescences large, loose terminal panicles often with feathery divisions. Flowers small, rarely bisexual, often yellow. Sepals 5, overlapping. Petals 5, twice as long as sepals. Stamens 5. Ovary 1-celled with 3 lateral styles.

A genus of 3 species from southern Europe to northern China and the southeastern United States. Grown for their foliage, flower-heads and autumn colour. They require fertile but not over-rich soil. Propagate by softwood cuttings in summer or by seed in autumn.

1a. Leaves broadly elliptic to nearly rounded, entire, hairless; fruiting panicle very showy **1. coggygria**
　b. Leaves broadly elliptic to obovate, wedge-shaped, rounded to slightly notched at the tip, hairy beneath when young; fruiting panicle not showy **2. obovatus**

1. C. coggygria Scopoli (*Rhus cotinus* Linnaeus). Illustration: Polunin & Stainton, Flowers of the Himalayas, 458 (1985); Williams, Hamlyn guide to plant selection, 122 (1988).

Deciduous shrub or rarely small tree to 5 m. Leaves 3–12 cm, broadly elliptic, blunt to nearly rounded, entire, light green becoming yellow or red in autumn. Inflorescence 8–20 cm, loose, spreading or drooping, densely feathery giving the appearance of a smoky haze. Flowers *c.* 3 mm, yellowish to pale purple; stalks lengthening after flowering, covered in long, grey or purple silky hairs. Fruit 3–5 mm, unequally lobed. *Southern Europe, W Asia, Pakistan to C Nepal, N China.* H1. Spring–summer.

Dark-leaved variants 'Royal Purple' and 'Rubrifolius' provide good colour in the shrub border.

2. C. obovatus Rafinesque (*C. americanus* Nuttall; *Rhus cotinoides* Chapman). Illustration: Williams, Hamlyn guide to plant selection, 123 (1988).

Deciduous shrub or short tree to 12 m, usually dioecious. Leaves 5–15 cm, broadly elliptic to obovate, wedge-shaped, rounded to slightly notched at the tip, light purple and hairy beneath when young, maturing to mid-green, turning orange-red or purple in autumn. Inflorescence 10–15 cm, loose, spreading or drooping with few flowers. Flowers greenish yellow; stalks with purplish or brownish, rather inconspicuous hairs. Fruit *c.* 3 mm across, kidney-shaped. *USA.* H1. Spring.

3. PISTACIA Linnaeus
J.E. Richardson
Dioecious trees or shrubs. Leaves alternate, pinnate with or without a terminal leaflet, rarely of 3 leaflets or simple, membranous or leathery. Flowers minute, unisexual, in axillary, loose or dense panicles or clustered racemes. Bracts leaf-like in texture or membranous. Bracteoles often membranous, unequal in size and form. Petals absent. Stamens 3–8, anthers ovate or oblong; filaments very short, inserted on the disc. Ovary spherical or ovoid, 1-celled with a short style and 3 stigmas. Fruit a 1-seeded drupe, obovoid to spherical or rarely transversely ovoid or oblong.

A genus of 11 species from the Mediterranean area, Asia and Malaysia. Propagate by seed in spring or from semi-ripe cuttings in autumn.

Literature: Zohary, M., A monographical study of the genus *Pistacia*, *Palestine Journal of Botany* **5**: 187–227 (1950–52).

1a. Leaves with a terminal leaflet 2
　b. Leaves without a terminal leaflet 3

2a. Tree or shrub to 6 m; leaflets 3–5 cm **2. terebinthus**
　b. Tree to 10 m; leaflets 5–12 cm **3. vera**
3a. Tree 10–27 m; fruit *c.* 3 mm across **4. chinensis**
　b. Tree or shrub to 4 m; fruit *c.* 5 mm across **1. lentiscus**

1. P. lentiscus Linnaeus (*P. massiliensis* Miller; *P. lentiscus* var. *latifolia* Cosson; *P. lentiscus* var. *massiliensis* (Miller) Fiori; *P. lentiscus* var. *angustifolia* de Candolle; *Lentiscus vulgaris* Fourreau). Illustration: *Botanical Magazine*, 1967 (1818); Polunin, Flowers of Europe, pl. 68 (1969); Brickell (ed.), RHS A–Z encyclopedia of garden plants, 824 (2003).

Evergreen tree or shrub to 4 m. Leaves 3–10 cm, persistent, leathery, pinnate with 2–7 pairs of leaflets; terminal leaflet absent. Leaflets 1.5–4.5 cm × 6–16 mm, stalked, ovate, oblong-lanceolate or elliptic, obtuse, mucronate, hairless. Male flowers bright red at first. Female flowers brownish. Fruit 4–5 mm across, red turning black at maturity, spherical, oblique, crowned with a trifid style. *Mediterranean area.* H4. Spring–summer.

2. P. terebinthus Linnaeus (*P. terebinthus* subsp. *vulgaris* Engler; *P. terebinthus* var. *angustifolia* Lee & Lamarck; *Terebinthus vulgaris* Tournefort). Illustration: Polunin, Flowers of Europe, pl. 69 (1969).

Deciduous tree or shrub 2–6 m. Leaves 10–20 cm, pinnate with a terminal leaflet; stalks 3–8 cm; axis hairless. Leaflets 3–5 cm in 3–6 pairs, stalkless or with semi-terete to angular stalks, hairless, shiny above, pale beneath, ovate-oblong or oblong, rarely lanceolate, obtuse or acute, mucronate, rarely acuminate. Flowers bright reddish purple. Fruit 5–7 × 5–6 mm, red to purple-brown in autumn, obovoid, slightly oblique. *W Mediterranean area.* H4. Spring–summer.

3. P. vera Linnaeus (*P. trifolia* Linnaeus; *P. narbonensis* Linnaeus; *P. reticulata* Willdenow).

Deciduous tree 3–10 m. Leaves 10–20 cm, leathery, pinnate with a terminal leaflet, rarely simple; stalks 5–10 cm. Leaflets 5–12 × 5–6 cm, stalked or stalkless, in 1–3 pairs, hairless or sparsely hairy on midrib, shiny above, pale beneath, broadly lanceolate to ovate or circular, often abruptly shortly acuminate, acute, rarely obtuse or slightly notched, mucronate, rarely blunt. Fruit

10–20 × 6–12 mm, oblong-linear, almost obovoid or spherical, laterally compressed. *C Asia, Afghanistan, Iran.* H4. Spring.

4. P. chinensis Bunge. Illustration: Lee, Forest botany of China, 719 (1935).

Tree 10–27 m. Bark dark grey. Branches silky grey-hairy. Leaves pinnate without a terminal leaflet; leaflets *c.* 6 × 1.5 cm, oblong-elliptic, broadly acuminate, wedge-shaped at base, deep shiny green above, light green and slightly grey-hairy beneath. Flowers small, aromatic, bright red, in dense terminal and lateral panicles. Fruit *c.* 3 × 3 mm, nearly spherical, slightly oblique. *China.* H3.

4. SCHINOPSIS Engler
J.E. Richardson
Dioecious trees (sometimes with both unisexual and bisexual flowers). Young branches hairy at the tips. Shorter branches spiny. Leaves alternate, simple or compound. Inflorescence a lateral or terminal panicle. Flowers unisexual and/or bisexual. Sepals 5, ovate, slightly overlapping, persistent. Petals oblong-lanceolate, obtuse, overlapping, each with a prominent central vein. Stamens 5, alternating with and slightly shorter than petals. Fruit a samara, oblong. Seed oblique, nearly rhomboid.

A genus of 7 species from southern South America.

Literature: Meyer, T. & Barkley, F.A., Revision del genero *Schinopsis* (Anacardiaceae), *Lilloa* **33**: 207–257 (1973).

1a. Leaves simple, entire **1. balansae**
 b. Leaves compound or compound and simple leaves borne on the same plant **2. quebracho-colorado**

1. S. balansae Engler (Quebrachia morongii Britton; *S. balansae* var. *pendula* Tortorelli).

Tree to 25 m, dioecious or bearing both unisexual and bisexual flowers. Leaves 4–10 × 1.5–2.5 cm, simple, alternate, oblong to oblong-lanceolate, hairless; stalks 5–10 mm. Inflorescence an axillary or terminal panicle. Petals 2–2.5 × 1.3–1.4 mm, oblong to obovate, hairless. Samara red, soon drying and turning maroon. *Bolivia, Paraguay, NW Argentina.*

2. S. quebracho-colorado
(Schlechtendal) Barkley & Meyer. (*Aspidiosperma quebracho-colarado* Schlechtendal; *Loxopterigium lorentzii*

Grisebach; *Quebrachia lorentzii* (Grisebach) Engler, *Schinopsis lorentzii* (Grisebach) Engler; *S. heterophylla* Ragonese & Castiglioni).

Tree to 25 m. Leaves pinnate with a terminal leaflet, stalk 2–3 cm, finely hairy; leaflets 13–31, stalkless, 1.8–3 cm × 4–7 mm, slightly leathery, opposite to alternate, lanceolate to oblong-lanceolate, hairy on the main veins above, densely hairy beneath. Inflorescence usually a terminal panicle. Petals 5, *c.* 2 × 1–2 mm, oblong-obovate. Fruit turning from green to bright red to reddish brown on maturity. *Bolivia, Paraguay, Argentina.* Spring.

The young leaves are notable for their variability.

5. MANGIFERA Linnaeus filius
J.E. Richardson
Trees bearing both unisexual and bisexual flowers. Leaves simple, alternate, leathery or papery. Stipules absent. Inflorescence a terminal or axillary panicle. Flowers small; bracts deciduous. Sepals 4 or 5. Petals 4 or 5, overlapping, occasionally with glandular ridges. Stamens 1–12, usually 5, often only 1 fertile, staminodes sometimes present. Ovary hairless, 1-celled, oblique or spherical; style lateral. Fruit a drupe, kidney-shaped or ovoid, fleshy, resinous. Seed large, compressed.

A genus of about 40 species from Asia. Usually propagated by seeds but budding and grafting must be used for certain cultivars.

Literature: Mukherji, S., A monograph on the genus *Mangifera* Linnaeus filius, *Lloydia* **12**: 73–136 (1949).

1a. Trees to 30 m; panicles hairless or shortly downy; fruit with unpleasant odour 2
 b. Trees to more than 30 m; panicles hairy or with densely felted hairs; fruit without unpleasant odour 3
2a. Leaves to 50 cm, elliptic or obovate; flowers *c.* 1 cm; fruit green **2. foetida**
 b. Leaves to 30 cm, oblong-lanceolate; flowers *c.* 7 mm; fruit yellowish green with yellow spots **3. × odorata**
3a. Leaves leathery; petals to 8 mm; disc minute, stalk-like **4. caesia**
 b. Leaves thinly leathery or papery; petals to 5 mm; disc swollen, with 4 or 5 lobes, broader than ovary **1. indica**

1. M. indica Linnaeus (*Manga indica* Ray; *M. simiarum* Rumphius; *M. calappa* Rumphius; *M. dodol* Rumphius; *Mangas domestica* Hermann; *Mangifera arbor* Hermann; *M. amba* Forssk l; *M. domestica* Gaertner; *M. sativa* Römer & Schultes; *M. indica* var. *parrie* Blume). Illustration: *Botanical Magazine*, 4510 (1850); Simpson & Ogorzaly, Economic botany, plants in our world, 134 (1986).

Spreading tree 20–45 m. Leaves to 30 × 7 cm, thinly leathery or papery, lanceolate, acuminate or acute; stalks 1.5–6 cm. Panicles pyramidal, dense, 20–35 cm with densely felted hairs. Flowers yellowish or cream. Petals 5, 3–5 × 1–2 mm, oblong-ovate, free from disc. Stamens 5, 3 or 4 infertile, inserted inside the disc at its base; disc swollen, with 4 or 5 lobes, broader than ovary. Fruit 4–25 × 1–10 cm, fragrant, usually ovoid, pointed or heart-shaped, green, yellow, reddish or yellow and reddish when ripe, flesh thick with sweet juice, edible. *Burma, naturalised throughout tropical Asia.* G1. Flowering varies according to where the plant is grown.

2. M. foetida Loureiro (*Manga foetida* Rumphius; *Mangifera horsfeldii* Miquel; *M. indica* Blume not Linnaeus; *M. leschenaultii* Marchand; *M. foetida* var. *blumii* Pierre).

Tree 18–24 m. Leaves 15–50 × 5–17 cm, thickly leathery, elliptic-oblong to broadly elliptic, sometimes slightly obovate, notched or rounded at apex, base slightly wedge-shaped, upper surface raised between stout nerves; stalks 2–6 cm. Panicle *c.* 30 cm, terminal or axillary, stout, hairless, blood-red when fresh, black when dry, branches cymose with 4 or more flowers. Flowers *c.* 1 cm × 6 mm across, deep red, pink inside. Petals 8–9 × 2–2.5 mm, linear-oblong, acute, ridge long, 3-fid. Disc narrow, often reduced to form a stalk to the ovary, rarely obsolete. Stamens 5, united at the base with a short receptacle; 1 fertile, *c.* 8 mm, 4 sterile, to 5 mm, unequal. Drupe 8–10 × 6 × 4–6 cm, green, ovoid, not compressed, apex oblique, fetid-smelling. *Burma; widely naturalised throughout tropical Asia.* G1. Winter–summer.

3. M. × odorata Griffith (*M. foetida* var. *odorata* Pierre; *M. foetida* var. *cochinchinensis* Pierre; *M. foetida* var. *kawinii* Blume; *M. oblongifolia* J.D. Hooker).

Tree to 30 m. Leaves 15–30 × 5–10 cm, thickly leathery, oblong or elliptic-lanceolate, shortly acuminate, slightly

wavy at the margins; stalks 3–6 cm.
Panicle 12–50 cm, dark green, stout,
hairless or shortly downy. Flowers *c.*
7 × 5 mm across, odorous, flesh-coloured.
Petals *c.* 6 × 2 mm, greenish, suffused with
blood red, oblong-lanceolate, free or joined
at base, ridge elevated at base, branching
upwards into 3 short arms reaching up to
the middle. Disc narrow, often reduced to
form a stalk to ovary, rarely obsolete.
Stamens 5, 1 (rarely 2) fertile, *c.* 5 mm,
others sterile, 2–3 mm. Drupe
10–11 × 5–6 cm, oblong with a beak
below the apex, unpleasant-smelling,
yellowish green with yellow spots and a
central reddish disc. *Origin unknown;
naturalised in Malaysia.* Gl. Spring–autumn.

A hybrid between *M. indica* and
M. foetida.

4. M. caesia Jack (*M. foetida* Blume;
M. kemang Blume; *M. verticillata* Robinson).

Tree to 35 m. Leaves 12–30 × 5.5–9 cm,
whorled at apex, leathery, elliptic, obovate
to oblong or lanceolate, wedge-shaped,
shortly acuminate or obtuse; stalks 2–6 cm,
flattened. Inflorescence a terminal panicle,
15–75 cm, pyramidal, finely or very shortly
downy. Flowers *c.* 1 cm, pale lilac;
bracteole broadly elliptic, densely hairy
outside. Petals 5, attached to disc,
c. 8 × 1 mm, linear, ridge very conspicuous,
margins inrolled. Disc minute, stalk-like.
Stamens 5, 1 fertile, shorter than petals,
staminodes minute. Drupe
18–19 × 9–10 cm, rough, obovate-oblong,
right shoulder bulged above, slightly
broader towards base, brown to yellow-
brown or pale green, edible. *Sumatra,
Malay Peninsula; naturalised elsewhere in
Malaysia.* G1. Summer–autumn.

6. ANACARDIUM Linnaeus
J.E. Richardson
Trees or shrubs with grey or brown bark.
Leaves simple, entire, alternate, leathery.
Inflorescence terminal or axillary,
paniculate or corymbose. Calyx 5-lobed,
deciduous. Petals 5 or none, if present then
very narrow. Stamens 7–10, all or only a
few fertile. Ovary 1-celled, stalkless; style 1.
Fruit kidney- or heart-shaped, dry, borne
on a greatly enlarged fleshy stalk.

A genus of 8 tropical American species
some of which are naturalised throughout
the tropics. Propagate from seed or
cuttings, preferably in sandy or stony soils.

1. A. occidentale Linnaeus. Illustration:
Da Veiga Soares, Arvores nativas do Brasil,
37 (1990).

Tree to 12 m. Bark smooth, brown.
Leaves 4–22 × 2–15 cm, obovate to
broadly elliptic, leathery, hairless; stalks
5–20 mm. Inflorescence a panicle or
corymb to 26 cm, hairy. Bracts 5–10 mm,
ovate to oblong. Flower-stalks 2–5 cm.
Sepals to 5 mm, unequal, ovate-lanceolate.
Petals 7–15 mm, linear, yellowish pink or
pale green striped red, becoming red. Fruit
to 3 × 2 cm, kidney or heart-shaped.
Central & South America. H5.

The dry fruit or cashew nut may be
eaten. The swollen fruit-stalk, known as
the cashew apple may be eaten raw or
sliced and sweetened with sugar when
ripe.

7. LOXOSTYLIS Reichenbach
J.E. Richardson
Dioecious tree to 6 m. Bark grey, young
growth red. Leaves dark green when
mature, alternate, with 2–6 pairs of
opposite, stalkless leaflets and a terminal
leaflet borne on a broadly winged stalk.
Leaflets 1.3–6 cm × 6–13 mm, lanceolate,
sometimes with short thorn-like points.
Flowers in terminal racemes. Petals 5.
Sepals 5. Male flowers white; stamens 5.
Female flowers green with greenish white
sepals which become longer and red. Fruit
ovoid.

A genus of a single species. Seeds
germinate easily but seedlings transplant
badly.

1. L. alata Reichenbach. Illustration:
Palmer & Pitman, Trees of southern Africa,
1204 (1972). *South Africa.* H5–G1.

8. SCHINUS Linnaeus
J.E. Richardson
Evergreen, dioecious, resinous trees or
shrubs. Leaves alternate, simple or
compound; leaflets stalkless. Flowers small
in axillary or terminal panicles, with
bracts. Sepals 5. Petals 5, overlapping,
yellow or green. Stamens 10, inserted on a
disc. Ovary stalkless, 1-celled, styles 3.
Fruit a drupe.

A genus of about 28 species from South
America naturalised in North America, the
Canary Islands and China. Grown mainly
for foliage, in freely draining soil and full
light. Propagate by seed in spring or semi-
ripe cuttings in summer.

Literature: Barkley, F.A., Schinus
Linnaeus, *Brittonia* **5**: 160–198 (1944).

1a. Tree to 25 m; flowers greenish
　　yellow　　　　　　　　**1. molle**

　b. Tree or shrub to 6 m; flowers not
　　greenish yellow　　　　　　2
2a. Leaves with 3–13 leaflets
　　　　　　　　3. terebinthifolius
　b. Leaves simple　　　　　　3
3a. Flowers yellow; leaves stalkless
　　　　　　　　2. polygamus
　b. Flowers white; leaves stalked
　　　　　　　　4. patagonicus

1. S. molle Linnaeus. Illustration:
Rodriguez et al., Flora arborea de Chile,
324 (1983); Brickell (ed.), RHS A–Z
encyclopedia of garden plants, 965 (2003).

Fast-growing evergreen, weeping tree to
25 m, with graceful pendent branches.
Leaves to 30 cm, fern-like, divided into
many linear-lanceolate, glossy, rich green
leaflets 4–9 cm × 6–10 mm, often toothed.
Panicles 10–20 cm, much branched. Flowers
greenish yellow; sepals 5, *c.* 0.5 mm,
circular; petals 2–3 mm. Fruits 6–10 mm
across, pinkish-red in branched pendent
clusters. *Mexico, Chile, S Brazil Uruguay,
N Argentina.* H4. Late winter–summer.

This is the source of pink peppercorns
widely used in cookery.

2. S. polygamus (Cavanilles) Cabrera
(*S. dependens* Ortega). Illustration:
Rodriguez et al., Flora arborea de Chile,
329 (1983).

Shrub or small tree to 5 m. Leaves
simple, entire, usually stalkless. Flowers
yellow, in short axillary racemes; sepals 4
or 5, *c.* 0.7 mm, ovate. Fruit red to deep
purple. *W South America.* H5. Late spring.

3. S. terebinthifolius Raddi.
Evergreen shrub or tree to 3 m or more,
usually of bushy spreading habit. Branches
not pendent. Leaves with 3–13 (usually 7)
leaflets, 4–7 × 1.2–4 cm, oblong, very dark
green above, lighter beneath. Flowers
white. Fruits 6–10 mm, red produced only
if plants of both sexes are grown together.
Brazil. H4. Summer–autumn.

4. S. patagonicus (Philippi) Johnston.
Illustration: Correa, Flora Patagonica **5**:
101 (1969).

Tree to 5 m. Branches grey, hairless.
Leaves 1.6–6.5 × 1.1–3.5 cm, simple,
alternate, ovate-elliptic, hairless; stalks
2.5–18 mm. Inflorescences axillary, central
axis 6–28 mm, hairless or scarcely hairy.
Bracts 0.5–1 mm, triangular, hairy on the
margin. Flowers *c.* 3 mm; stalks 3.5–7 mm;
bracteoles *c.* 0.7 mm, ovate, hairless. Petals
2.5–3 mm, oblong, white. Fruits 5–7 mm
across, spherical, hairless, violet. *S South
America.* H4.

9. SEMECARPUS Linnaeus
J.E. Richardson

Trees. Leaves alternate, simple, entire, leathery; stipules absent. Flowers small, unisexual and bisexual, in terminal (rarely axillary) panicles. Sepals 3 or 5, deciduous. Petals 3 or 5, ovate or oblong-ovate, overlapping. Stamens 5 or 6; filaments thread-like, longer than petals in male flowers, short in bisexual flowers; anthers usually oblong. Ovary sunk into a cup-shaped or tubular hollow receptacle, absent or vestigial in male flowers, superior in bisexual flowers, 1-celled; styles 3, divergent; stigmas head-like or bilobed. Fruit a drupe, oblong or nearly spherical, oblique, on a fleshy receptacle. Seed pendent.

A genus of 38 species from tropical Asia and Australia, only one generally cultivated.

1. S. anacardium Linnaeus.
Deciduous tree of moderate size. Bark dark brown with acrid juice. Leaves 18–60 × 10–30 cm, crowded at the ends of branches, obovate-oblong, rounded at apex, sometimes shortly auricled at the base, entire, leathery, hairless above when mature, ash grey and more or less hairy beneath. Flowers 5–8 mm across, greenish yellow, stalked, in bundles arranged in large hairy terminal panicles; female panicles shorter than male. Petals to 5 mm, ovate. Fruits to 2.5 cm across, obliquely ovoid or oblong, smooth, shiny, black when ripe. *Himalaya*. H5. Spring–summer.

10. SPONDIAS Linnaeus
J.E. Richardson

Trees. Leaves usually clustered near the ends of the branchlets, alternate, odd-pinnate. Flowers small, short-stalked in racemes or panicles. Sepals 4 or 5, deciduous. Petals 4 or 5, not overlapping. Stamens 8–10. Ovary stalkless, with 4 or 5 cells; styles 4 or 5. Fruit a fleshy drupe.

A genus of about 12 species from Southeast Asia and tropical America, cultivated for their edible fruits.

1a. Leaf-veins branching near margins; flowers yellowish white; fruit 2.5–5 cm **1. mombin**
 b. Leaf-veins not branching, almost parallel; flowers greenish white; fruit to 8 cm **2. cytherea**

1. S. mombin Linnaeus (*S. lutea* Linnaeus).

Graceful tree to 20 m. Leaves 20–30 cm, pinnate, central axis hairy; leaflets 7–17, almost entire, 6.5–10 cm, ovate-lanceolate, acuminate, veins branching near margins. Inflorescence a terminal panicle, 15–32 cm. Flowers small, yellowish white. Fruit 2.5–5 cm, ovoid, yellow. *Tropics*. H5.

2. S. cytherea Sonnerat (*S. dulcis* Forster).
Spreading graceful tree to 20 m. Leaves 20–36 cm, pinnate; leaflets 11–23, entire or slightly scalloped, 6–9 cm, ovate-oblong, acuminate, veins almost parallel, not branching. Inflorescence a terminal panicle, 20–30 cm. Flowers small, greenish white. Fruit 7–8 × 2.5–8 cm, ovoid or obovoid, golden yellow. *Society Islands*. H5.

11. SCLEROCARYA Hochstetter
J.E. Richardson

Trees or shrubs. Bark grey. Leaves alternate, pinnate with a terminal leaflet, stalked; leaflets circular to broadly oblong, toothed to entire. Inflorescence a panicle; sepals 4 or 5, oblong or round, overlapping; petals 4 or 5, oblong or obovate, obtuse, overlapping; stamens 12–20; ovary with 2 or 3 cells; styles 2 or 3. Fruit a drupe.

A genus of 3 species from tropical and southern Africa and the Pacific islands. Propagated from stem-cuttings or seed.

1. S. birrea Hochstetter (*S. caffra* Sender).
Deciduous tree to 20 m, usually dioecious. Leaves alternate, hairless, in clusters at end of branches; leaflets 7–13 (rarely to 17) pairs, 3–10 × 1.5–4 cm, opposite to almost so, ovate to elliptic, dark green above, blue-green beneath, apex broadly tapering, finally abrupt, base broadly tapering to rounded, young leaves toothed. Sepals red. Petals yellow. Stamens or staminodes 15–25. Male flowers in few short lateral spikes, female flowers solitary or a few together. Fruit 3–5 cm across, fleshy, almost spherical, yellow when mature. *Tropical & South Africa*. H5. Autumn.

12. HARPEPHYLLUM Bernhardi
J.E. Richardson

Evergreen dioecious tree, 6–15 m. Leaves to 30 cm, alternate, dark green, leathery and shiny, with 4–8 pairs of leaflets; terminal leaflet 5–10 × 1.3–2.5 cm, lanceolate or broader; lateral leaflets slender, slightly sickle-shaped, sharply pointed and narrowed to the base with the midrib to one side. Flowers whitish, borne on lateral racemes. Stamens 10. Ovary with 4 or 5 cells. Fruits *c.* 2.5 × 1.3 cm, red when ripe, long, plum-shaped.

A genus of a single species from southern Africa. The leaves remain on the tree for about 2 years before turning red and falling. The tree is never bare and is characterised by having red leaves in the green shiny crown. The rather sour pulp of the fruit is edible and makes a good jelly. Grows easily from seed or stem-cuttings.

1. H. caffrum Bernhardi. Illustration: Palmer & Pitman, Trees of southern Africa, 1194–1196 (1972). *South Africa, Zimbabwe, Mozambique, St Helena*. H5.

171. ACERACEAE

Monoecious or dioecious trees and shrubs. Buds with or without 4–many scales. Leaves opposite, stalked, simple, with 3–13 lobes, palmate or pinnate, with 3 leaflets, entire or toothed, evergreen or deciduous. Inflorescences terminal or lateral, racemose, corymbose, paniculate or sometimes umbellate. Flowers unisexual, male and female in the same inflorescences or in separate inflorescences. Sepals 4 or 5. Petals 4 or 5, rarely united or absent, disc outside, inside or around the insertion of the filaments, usually in a ring, rarely lobed or absent. Stamens 4–12, mostly 8. Ovary superior, 2-celled, with 2 axile ovules in each cell; stigmas 2. Fruit composed of 2 single-seeded samaras. Germination epigeal, rarely hypogeal.

A family of 2 genera mainly from temperate and subtropical areas of the northern hemisphere.

1a. Buds with scales; leaves simple, lobed or pinnate with 1–5 pairs of leaflets; leaf-stalks, if pinnate, not containing latex; wing of samara on one side of nutlet **1. Acer**
 b. Buds without scales; leaves pinnate, 4–9 pairs of leaflets; leaf-stalks containing latex; nutlet encircled by a broad wing **2. Dipteronia**

1. ACER Linnaeus
P.C. de Jong

Trees or shrubs, monoecious, occasionally dioecious, deciduous, occasionally (semi-) evergreen. Bud scales 2- to many-paired, edge-to-edge or overlapping. Leaves usually palmately lobed with 3–13 (usually 5) lobes, sometimes simple or with 3 leaflets, rarely palmate or pinnate, stalks sometimes containing latex. Inflorescences terminal and/or axillaryi, racemose, corymbose, paniculate and sometimes umbellate. Flowers unisexual, insect-pollinated, rarely wind-pollinated. Sepals usually 5, less often 4. Petals usually 5, rarely 4, occasionally absent or united. Stamens mostly 8, less often 4 or 5, occasionally 10–12. Ovary superior, 2-celled, each cell with 2 ovules. Fruit mostly composed of 2, sometimes 3 or 1 single-seeded samaras. Germination epigeal, occasionally hypogeal.

A large and complex genus, including about 120 species, widely distributed in the temperate regions of the northern hemisphere and subtropical Southeast Asia. A few species are found in tropical Asia, one of them, *A. laurinum* crossing the equator in Indonesia.

Some species may produce large trees from 25 to 40 m in the wild, but their role for timber production is limited. *A. saccharum* is the most valuable species for timber production, as well as being the source for maple syrup. The wood of *A. pseudoplatanus* is used for making violins. Over 60% of the species are cultivated commercially or are found in collections. These species are mainly the hardy ones. Of the cultivated species 25% are common or rather common. Many cultivars have been selected, but for the greater part they belong to 3 species: *A. palmatum*, *A. platanoides* and *A. pseudoplatanus*.

Selections of *A. platanoides*, *A. pseudoplatanus*, *A. rubrum* and *A. saccharinum* are used extensively for specimen and street planting. *A. campestre*, *A. cappadocicum*, *A. negundo* and *A. saccharum* are more rarely used. Most maples are valuable for park planting, but several species are still little known or underestimated. *A. campestre* and *A. tataricum* subsp. *ginnala* are often utilised for hedges. The ornamental maples are the most important, especially the cultivars of *A. palmatum*. They show a wide variation in leaf form and colour and shrub habit. Some species of Section *Macrantha*, notably *A. capillipes*, *A. davidii* and *A. pensylvanicum* have very attractive white-striped branches. *A. griseum* has cinnamon-brown, flaking bark. Maples are highly valued for the brillant red and orange autumn colours of the leaves. Notable for autumn colour are *A. japonicum*, *A. palmatum*, *A. rubrum* and *A. saccharum*, also *A. capillipes*, *A. circinatum*, *A. davidii*, *A. micranthum*, *A. pseudosieboldianum*,

A. tataricum subsp. *ginnala*, *A. sieboldianum* and *A. triflorum*.

The cultivars are mostly propagated by grafting on rootstocks of the same or a closely related species. The species can be easily grown from seed, but there are still problems with several of the species that are rarely found in cultivation. A few, notably *A. pseudoplatanus* and *A. saccharinum* are suitable as rootstock for a number of unrelated species. Some of the species that are rare in cultivation have not been grafted successfully on any rootstock except those of the same species.

A few species are naturalised outside their original areas, notably *A. negundo* in central Europe, India and China and *A. pseudoplatanus* in western Europe.

A large number of species, including all dioecious ones, have a strong tendency to develop fruits which are perfectly formed, but which contain no seeds; the fruits have to be cut to examine the presence of seeds.

The genus has been studied intensively by various authors resulting in a large number of monographs, revisions and papers dealing with special subjects such as flowering and chemistry. This has resulted in a classification of the genus with a natural grouping in sections and series. For this classification the reproductive organs and the number of bud-scales are most important. The key presented below is based mainly on these characters.

Literature: Fang, W.P., A monograph of Chinese Aceraceae, Contributions to the Biological Laboratories of the Science Society China, *Nanking Botany Series* **11** (1–9): 1–346 (1939); Ogata, K., A dendrological study on the Japanese Aceraceae with special reference to the geographical distribution, *Bulletin of the Tokyo University Forest* **60**: 1–99 (1965); Murray, A.E., A monograph of Aceraceae, Pennsylvania State University (thesis), Ann Arbor, Michigan, (1970); de Jong, P.C., Flowering and sex expression in *Acer L.* A biosystematic study, *Mededelingen Landbouwhogeschool Wageningen* **76**(2): 1–202 (1976); Vertrees, J.D., Japanese maples (1978) & Japanische Ahorne (1993); Delendick, T.J., A survey of foliar flavonoids in the Aceraceae, *Memoirs of the New York Botanical Garden* **54**: 1–129 (1990); de Jong, P.C., Taxonomy and distribution of Acer, International Dendrology Society Yearbook 1990: 6–10 (1990); Ohashi, H., The nomenclature of *Acer pictum* Thunberg ex Murray and its

infraspecific taxa (Aceraceae), *Journal of Japanese Botany* 68: 315–325 (1993); van Gelderen, D.M., de Jong P.C. & Oterdoom, H.J., Maples of the World (1994).

1a.	Leaves compound	2
b.	Leaves simple	10
2a.	Leaves palmately lobed (Series **Pentaphylla**)	
	43. pentaphyllum	
b.	Leaves with 3 leaflets or pinnate	3
3a.	Leaves pinnate (Series **Negundo**)	
	29. negundo	
b.	Leaves with 3 or 5 leaflets	4
4a.	Bud-scales 11–15-paired (Section **Trifoliata**)	5
b.	Bud-scales 2–4-paired	8
5a.	Bark flaking, grey or rust-brown	6
b.	Bark smooth, dark grey or grey-brown	7
6a.	Bark rust-brown; leaflets with coarse, forwardly pointing teeth	
	53. griseum	
b.	Bark grey; leaflets entire or with coarse, forwardly pointing teeth	
	56. triflorum	
7a.	Leaves and fruits hairless	
	54. mandshuricum	
b.	Leaves and fruits downy	
	55. maximowiczianum	
8a.	Inflorescences terminal or partly axillary; flowers with parts in 5s; leaves partly simple	
	24. glabrum subsp. **douglasii**	
b.	Inflorescences axillary; flowers with parts in 4s (Series **Cissifolia**)	9
9a.	Leaflets with coarse, forwardly pointing teeth; flowers stalked; petals long, yellow-green	
	30. cissifolium	
b.	Leaflets entire or with coarse, forwardly pointing teeth; flowers nearly stalkless; petals white, small and early falling	**31. henryi**
10a.	Bud-scales 2–4-paired	11
b.	Bud-scales 4- or more paired	40
11a.	Bud-scales always 2-paired	12
b.	Bud-scales 2 4-paired	24
12a.	Inflorescence a large panicle with a coiled cyme; flowers 20–400 (Section **Parviflora**)	13
b.	Inflorescence corymbose or racemose; flowers 7–25, sometimes to 50 (Section **Macrantha**)	16
13a.	Leaves undivided (Series **Distyla**)	
	2. distylum	
b.	Leaves with 3–5 (sometimes to 7) lobes	14
14a.	Leaves broadly 3–5-lobed, finely toothed; inflorescences pendent;

flowers with small petals; nutlets spherical (Series **Parviflora**)
      1. nipponicum

b. Leaves coarsely toothed; inflorescences erect; flowers with longs white petals (Series **Caudata**)   15

15a. Leaves mostly 3-lobed, grey-downy or hairless beneath, bases cordate, coarsely toothed   **4. spicatum**

b. Leaves mostly 5-lobed, downy beneath, bases not cordate, more sharply toothed   **3. caudatum**

16a. Leaves with 5–7 lobes   17

b. Leaves to 5-lobed   18

17a. Flowers small, reddish; fruits purplish   **18. micranthum**

b. Flowers rather large, yellowish; fruits greenish   **23. tschonoskii**

18a. Leaves very finely toothed
    19. pectinatum

b. Leaves coarsely toothed   19

19a. Young shoots with rich blue-white bloom   20

b. Young shoots not or only slightly glaucous   21

20a. Rust-brown down on leaf undersides, stalks and flower-stalks; nutlets small, spherical
    21. rufinerve

b. Rust-brown down only in vein-axils of leaf undersides; nutlets flat
    22. tegmentosum

21a. Leaves broadly 3-lobed, rarely unlobed   22

b. Leaves partly broadly 3-lobed, most often shallowly 3-lobed or unlobed, sometimes 5-lobed   23

22a. Leaf-stalks red; young shoots red and somewhat glaucous; fruits to 2 cm, red   **15. capillipes**

b. Leaf-stalks and young shoots greenish; fruits to 3 cm, green
    20. pensylvanicum

23a. Young shoots purplish; leaves with purple-black margins; usually small shrubs   **16. crataegifolium**

b. Young shoots reddish or greenish red; leaves without dark margins; large shrub or tree   **17. davidii**

24a. Bud-scales 2–3-paired   25

b. Bud-scales always 4-paired (Section **Palmata**)   31

25a. Inflorescences large, erect, terminal panicles; nutlets slightly veined (Series **Caudata**)   26

b. Inflorescences corymbose, sometimes racemose, small; nutlets strongly veined (Section **Glabra**)   27

26a. Leaves mostly 3-lobed, grey-downy or hairless beneath, bases cordate, coarsely toothed   **4. spicatum**

b. Leaves mostly 5-lobed, downy beneath, bases not cordate, more sharply toothed   **3. caudatum**

27a. Inflorescences terminal and sometimes partly axillary; sepals and petals 5; branches red-brown (Series **Glabra**)
    24. glabrum

b. Male inflorescences axillary, female terminal on short shoots and axillary on long shoots; sepals and petals 4; branches green or reddish
    28

28a. Leaves undivided to shallowly 3-lobed   **28. stachyophyllum**

b. Leaves 3- or 5-lobed   29

29a. Leaves coarsely toothed
    27. barbinerve

b. Leaves somewhat finer doubly toothed   30

30a. Leaves 3–5-lobed; lobes long-acuminate   **25. acuminatum**

b. Leaves 5- rarely 7-lobed; lobes not long-acuminate   **26. argutum**

31a. Leaves undivided (Series **Sinensia**)
    14. laevigatum

b. Leaves with 3–13 lobes   32

32a. Leaves with 5–13 lobes; inflorescences corymbose (Series **Palmata**)   33

b. Leaves with 3–5 (rarely 7) lobes; inflorescences paniculate, rarely corymbose (Series **Sinensia**)   38

33a. Leaves with 5–7 lobes
    7. palmatum

b. Leaves with 7–13 lobes   34

34a. Leaves with 7–11 lobes; inflorescences erect   35

b. Leaves with 9–13 lobes; inflorescences erect or pendent   36

35a. Leaves with 7–9 lobes; young shoots sticky; flowers reddish; wings of fruit held horizontally
    5. circinatum

b. Leaves with 7–11 lobes; flowers greenish; wings of fruit held at an obtuse angle
    10. sieboldianum

36a. Inflorescences erect; leaves with 9–13 lobes   **9. shirasawanum**

b. Inflorescences pendent; leaves with 9–11 lobes   37

37a. Flowers deep purple; leaf-lobes acuminate; dried leaves not persisting on the tree in early winter   **6. japonicum**

b. Flowers reddish; leaf-lobes long acuminate; dried leaves persisting on the tree in early winter
    8. pseudosieboldianum

38a. Inflorescences corymbose
    13. oliverianum

b. Inflorescences paniculate   39

39a. Leaves with white dots beneath
    12. erianthum

b. Leaves without white dots beneath
    11. campbellii

40a. Inflorescences axillary from leafless buds   41

b. Inflorescences terminal and axillary   46

41a. Bud-scales 8–12-paired; inflorescences racemose (Series **Lithocarpa**)   42

b. Bud-scales 4–7-paired; inflorescences umbellate, in clusters (Section **Rubra**)   44

42a. Leaf-lobes rather regularly toothed
    60. sterculiaceum

b. Leaf-lobes lobed and irregularly toothed   43

43a. Leaf-stalks containing latex; perianth-segments free
    59. sinopurpurascens

b. Leaf-stalks not as above; perianth-segments of female flowers united
    58. diabolicum

44a. Flowers without petals; sepals united; fruits large, mostly single
    65. saccharinum

b. Flowers with free sepals and petals; fruits small   45

45a. Leaves small, shallowly 3-lobed, lobes slightly lobed
    63. pycnanthum

b. Leaves with 3–5 deeper lobes, lobes distinctly lobed
    64. rubrum

46a. Leaf-stalks containing latex; disc around the bases of the filaments
    47

b. Leaf-stalks not as above; disc not as above   55

47a. Inflorescences pendent; stamens 8–12; nutlets convex (Series **Macrophylla**)
    57. macrophyllum

b. Inflorescences upright; stamens 8; nutlets flat (Section **Platanoidea**)
    48

48a. Leaf-lobes lobed   49

b. Leaf-lobes entire   51

49a. Leaves large, hairless except in vein-axils   **51. platanoides**

b. Leaves small to moderate, hairy beneath   50

50a. Leaf-lobes pointed, margins mostly with 2 pairs of blunt teeth
49. miyabei

b. Leaf-lobes rounded, margins mostly with 1 pair of blunt teeth
46. campestre

51a. Bark deeply furrowed; leaf-base truncate; nutlets rather thick
52. truncatum

b. Bark not deeply furrowed; nutlets flat   52

52a. Branches finely white-striped at least when young
47. cappadocicum

b. Branches not striped   53

53a. Leaves to 8 × 10 cm; fruits to 3 cm
46. campestre

b. Leaves to 10 × 15 cm; fruits to 6 cm   54

54a. Leaf-lobes not long-acuminate; fruits to 4 cm   **50. pictum**

b. Leaf-lobes long-acuminate; fruits to 6 cm   **48. longipes**

55a. Plants dioecious; leaves undivided, with double, forwardly pointing teeth, shortly stalked; inflorescences racemose (Section **Indivisa**)   **32. carpinifolium**

b. Plants monoecious; leaves not as above; inflorescences corymbose   56

56a. Leaves 3-lobed; flowers with 5 stamens and disc around the bases of the filaments; nutlet with hole on one side (Section **Pubescentia**)
61. pentapomicum

b. Leaves entire to 7-lobed; flowers with 8 stamens and disc borne outside the bases of the filaments; nutlet without hole on one side   57

57a. Nutlets elliptic-convex, veined (Section **Ginnala**)   **62. tataricum**

b. Nutlets mostly keeled, convex and not veined   58

58a. Leaves undivided to 3-lobed, usually leathery and persistent; flowers whitish, petals longer than sepals, stamens not protruding (Series **Trifida**)   59

b. Leaves with 3–7 lobes, rarely unlobed, papery, sometimes leathery and (semi-) persistent; flowers green-yellow, petals absent or rather broad and as long as sepals; stamens protruding   60

59a. Leaves entire, sometimes shallowly 3-lobed, persistent   **45. oblongum**

b. Leaves 3-lobed, persistent or not
44. buergerianum

60a. Flowers without petals, sepals united on long stalks; nutlets

spherical, not keeled (Series **Saccharodendron**)
42. saccharum

b. Flowers with free sepals and petals; nutlets not as above   61

61a. Leaves large, with 5–7 lobes, papery; inflorescences large; flowers with rather short stalks (Series **Acer**)   62

b. Leaves small, with 3–5 lobes, leathery, if papery then flowers with long drooping stalks (Series **Monspessulana**)   65

62a. Inflorescences paniculate, pendent
35. pseudoplatanus

b. Inflorescences corymbose, erect   63

63a. Nutlets keeled, convex; young shoots and leaf undersides somewhat or very glaucous; buds long-pointed   **33. caesium**

b. Nutlets not keeled; shoots not glaucous; buds not as above   64

64a. Leaves deeply to rather deeply incised; buds ovoid, not sharply pointed   **34. heldreichii**

b. Leaves rather slightly incised; buds narrowly pointed   **36. velutinum**

65a. Leaves semi-persistent, leathery, undivided to 3-lobed; flower-stalks rather short   66

b. Leaves deciduous, leathery or papery; with 3–5 lobes; flowers with long drooping stalks   67

66a. Leaves to 2 × 4 cm, dark green
41. sempervirens

b. Leaves to 3 × 5 cm, grey-green
39. obtusifolium

67a. Leaves to 6 × 8 cm, broadly 3-lobed
38. monspessulanum

b. Leaves more than 6 × 8 cm, with 3–5 lobes   68

68a. Leaves relatively large; buds ovate, rather large   **40. opalus**

b. Leaves relatively small; buds elliptic, rather small   **37. hyrcanum**

Section **Parviflora** Koidzumi. Trees or shrubs, monoecious, deciduous. Bud-scales 2, sometimes partly 3-paired. Leaves with 3–7 lobes or undivided, papery, margins with forwardly pointing teeth. Inflorescences terminal, sometimes partly axillary, paniculate, with 35–400 flowers. Sepals and petals 5. Disc within the filaments or around the bases. Stamens 8. Fruits with a soft pericarp. Parthenocarpic tendency low.

Series **Parviflora**. Bud-scales 2-paired. Leaves large, with 3–5 lobes, margin with double, forwardly pointing teeth.

Inflorescences with 2 or 3 pairs of leaves, large, pendent, With reddish brown down, with 300–400 flowers. Flowers small, disc around the bases of the filaments. Nutlets spherical.

1. A. nipponicum Hara (*A. parviflorum* Franchet & Savatier not Ehrhard). Illustration: Kurata, Illustrated important forest trees of Japan **2**: pl. 54 (1968); *Mededelingen Landbouwhogeschool Wageningen* **76**(2): f. 23, 1–4 (1976); Satake, Wild flowers of Japan, woody plants **2**: pl. 19 (1985); The new RHS dictionary of gardening **1**: 33 (1992).

Tree to 15 m in the wild, to 10 m in cultivation. Bark grey. Shoots dark green, smooth and somewhat shiny. Buds red. Leaves to 15 × 20 cm, with 3–5 shallow lobes, lobes broadly ovate, acute, base deeply cordate, margins with double, forwardly pointing teeth, light green on upper surface, with reddish brown down beneath. Inflorescences large, sometimes with only male flowers. Flowers small; ovary rusty downy. Fruits to 5 cm, nutlets spherical, green, rusty downy at first, wings held at near right angles. *Japan (Honshu, Kyushu, Shikoku)*. H2. Late spring.

A distinct maple, rare in cultivation. The leaves resemble some species of Section *Macrantha*, but *A. nipponicum* does not have the typical snake bark of that section. Other species with similar leaves, such as *A. caesium* and *A. sterculiaceum*, have a higher number of bud-scales. In sunny positions the leaves often develop black margins.

Series **Distyla** (Ogata) Murray. Bud-scales 2-paired, not overlapping. Leaves entire. Inflorescences with 1 pair of leaves, upright, with 20–70 flowers. Flowers with hairy sepals; disc lobed with the stamens inserted between the lobes. Nutlets elliptic-convex.

2. A. distylum Siebold & Zuccarini. Figure 130(1), p. 536. Illustration: Kurata, Illustrated important forest trees of Japan **2**: pl. 50 (1968); Vertrees, Japanese maples, 151 (1978); Satake, Wild flowers of Japan, woody plants **2**: pl. 20 (1989); The new RHS dictionary of gardening **1**: 33 (1992).

Tree to 10 m in the wild, to 6 m in cultivation. Bark light grey. Shoots greenish brown, somewhat hairy. Buds brownish, downy. Leaves broadly heart-shaped, shortly acuminate, margins scalloped, stalks short. Inflorescences

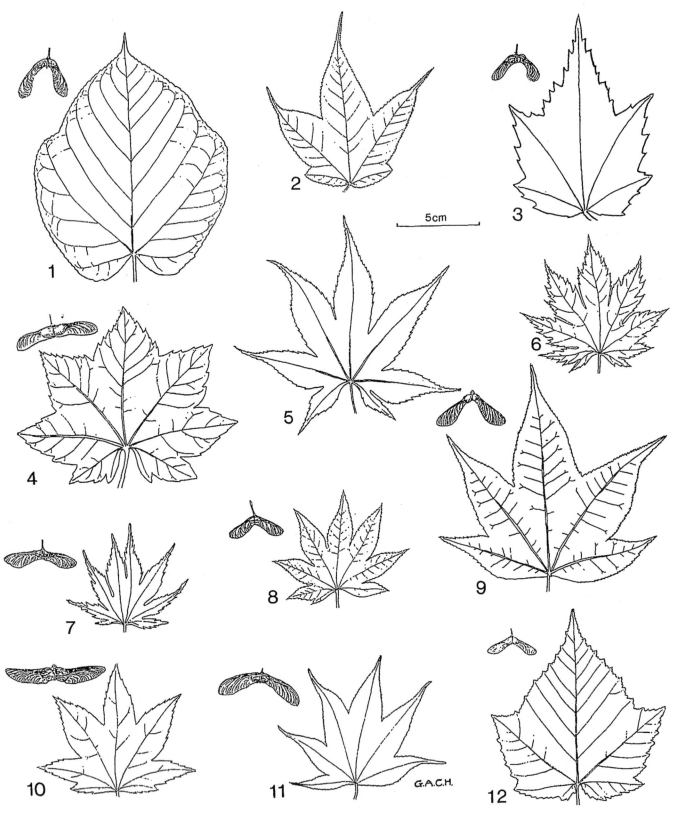

Figure 130. Leaves and fruits (top left) of Acer species. 1, *A. distylum.*
2, *A. caudatum* subsp. multoserratum. 3, *A. spicatum.* 4, *A. circinatum.*
5, *A. palmatum.* 6, *A. pseudosieboldianum.* 7, *A. shirasawanum.*
8, *A. sieboldianum.* 9, *A. campbellii.* 10, *A. erianthum.* 11, *A. oliverianum.*
12, *A. capillipes.*

paniculate, erect. Flowers yellowish. Fruits to 3 cm, nutlets almost spherical, covered with reddish hairs at first, wings spreading at an acute angle. *Japan.* H3. Late spring.

A distinct species, rare in cultivation. The leaves are somewhat similar to those of *A. davidii* and *A. sikkimense*, but those species have a striped bark.

Series **Caudata** Pax. Bud-scales 2-, sometimes 3-paired. Leaves with 3–5, sometimes 7 lobes. Inflorescences with 2 or 3 pairs of leaves, upright; 50–200 flowers, stalks of male flowers which have finished flowering partly dropped. Flowers with narrow petals, twice as long as the sepals; male flowers which have finished flowering drop stamens first, disc lobed, partly around the bases of the filaments, partly within them. Nutlets small, veined, one side convex, the other somewhat concave.

3. A. caudatum Wallich subsp. **caudatum** (*A. papilio* King). Illustration: Flora Reipublicae Popularis Sinicae **46**: f. 38 (1981).

Small tree or shrub to 10 m in the wild, to 5 m in cultivation. Bark light grey, falling off in thin flakes. Shoots red or reddish brown, downy when young. Buds downy, scales 2-, sometimes 3-paired. Leaves 5-lobed, somewhat circular, lobes triangular-ovate, acuminate, margins with coarse, double, forwardly pointing teeth, yellow-brown downy beneath, stalks red. Inflorescences large, terminal or axillary, upright panicles. Flowers greenish white, petals long and narrow. Fruits to 3 cm, nutlets flat and veined, wings held under an acute angle. *E Himalaya.* H2. Late spring.

Rare in cultivation. It differs from the Himalayan maples, *A. acuminatum* and *A. pectination*, in the reproductive organs, the grey bark and the down on the underside of the leaves.

Subsp. **multiserratum** (Maximowicz) Murray (*A. multiserratum* Maximowicz; *A. caudatum* var. *multiserratum* (Maximowicz) Rehder; *A. erosum* Pax; *A. caudatum* var. *prattii* Rehder; *A. caudatum* var. *georgei* Diels). Figure 130(2), p. 536. Illustration: Flora Reipublicae Popularis Sinicae **46**: pl. 39 (1981). A poorly known subspecies and probably not in cultivation. Rather intermediate between subsp. *caudatum* and subsp. *ukurunduense*. *C China to N Burma.*

Subsp. **ukurunduense** (Trautvetter & Meyer) Murray (*A. ukurunduense* Trautvetter & Meyer). Illustration: Kurata, Illustrated important forest trees of Japan **4**: pl. 25 (1973); Flora Reipublicae Popularis Sinicae **46**: pl. 39 (1981); Satake, Wild flowers of Japan, woody plants **2**: pl. 18 (1989), The new RHS dictionary of gardening **1**: 33 (1992). Small tree or shrub to 15 m in the wild, to 6 m in cultivation. Young shoots orange-grey to red, downy, later grey-yellow. Leaves deeply 5- sometimes 7-lobed, lobes tapered to the apex, ovate or triangular. *NE Asia, Japan (Hokkaido, Honshu).* H1. Late spring.

Rather rare in cultivation. It strongly resembles subsp. *caudatum*, especially in the reproductive organs. The young branches are more reddish and the leaves deeply lobed. The plants show a more upright growth.

4. A. spicatum Lamarck. Figure 130(3), p. 536. Illustration: Dippel, Handbuch der Laubholzkunde **2**: 423 (1892); Bean, Trees & shrubs hardy in the British Isles, edn. 8, **1**: 234 (1970).

Large shrub or small tree to 10 m, to 6 m in cultivation. Bark dark grey. Shoots grey downy and reddish when young. Buds reddish, scales 2-paired, not overlapping. Leaves 3- sometimes 5-lobed, base cordate, wrinkled downy on upper surface, grey downy or hairless beneath, margins with coarse, double, forwardly pointing teeth. Reproductive organs very similar to those of *A. caudatum*, slightly smaller. *E North America.* H1. Late spring.

More common in cultivation than the closely related *A. caudatum*. It differs from that species in the mostly 3-lobed and coarsely toothed leaves.

Section **Palmata** Pax. Trees and shrubs, monoecious, deciduous or (semi-) evergreen, terminal buds mostly abortive. Bud-scales always 4-paired. Leaves 3–many, lobed or simple. Inflorescences corymbose. Sepals 5, red or green-red; petals 5, white and rolled inwards, disc outside the filaments. Stamens 8. Nutlets elliptical-spherical. Parthenocarpic tendency low to moderate.

Series **Palmata**. Leaves deciduous, 5–7, sometimes to 13-lobed, margins with forwardly pointing teeth. Inflorescences terminal, sometimes lateral, 1 pair of leaves, 5–25 flowers, axis short, male flowers which have finished flowering completely fallen.

5. A. circinatum Pursh. Figure 130(4), p. 536. Illustration: Dippel, Handbuch der Laubholzkunde **2**: 462 (1892); *Mededelingen Landbouwhogeschool Wageningen* **76**(2): f. 15 (1976); Vertrees, Japanese maples, 148, 149 (1978); *The Plantsman* **5**(1): 48 (1983).

Small tree or shrub to 12 m, to 5 m in cultivation. Bark grey-brown, hairless. Shoots rather thick, hairless, green-brown and sticky when young. Leaves with 7–9 lobes, almost circular, pale green on upper surface, downy at first beneath, base cordate, lobes ovate and acute, margins with double, forwardly pointing teeth. Inflorescences small, upright, with 5–19 flowers. Flowers rather large, sepals reddish purple; petals white and small. Fruits to 4 cm, nutlets almost spherical, veined, wings red when young and held horizontally. *W North America.* H2. Spring.

Not rare in cultivation and often with deep red and orange autumn colours. Similar to *A. japonicum*, but differing in the sticky young shoots, the upright inflorescences and the somewhat smaller leaves.

6. A. japonicum Murray. Illustration: Kurata, Illustrated important forest trees of Japan **2**: pl. 43 (1968); Vertrees, Japanese maples, 139–141 (1978); Satake, Wild flowers of Japan, woody plants **2**: pl. 13 (1989); The new RHS dictionary of gardening **1**: 33 (1992).

Tree or shrub to 12 m, but to 6 m in cultivation. Bark grey, smooth. Shoots rather thick, hairless, green or reddish. Leaves to 15 cm wide, with 7–11 lobes, almost circular, lobes ovate to lanceolate, acuminate, with irregular, double, forwardly pointing teeth, bright green on upper surface, white-haired on the veins beneath. Inflorescences pendent, long-stalked, with 11–25 flowers. Flowers with sepals dark purple; petals rose-purple. Fruits to 3.5 cm, nutlets spherical, downy when young, veined, angle between wings very variable. *Japan (Hokkaido, Honshu).* H2. Spring.

This maple somewhat resembles *A. circinatum* and *A. pseudosieboldianum*. The latter has smaller leaves with 11–13 lobes. *A. sieboldianum* has leaves with 7–9 lobes. The leaves of *A. japonicum* turn brilliant red and orange during autumn. Rather common in cultivation. Cultivars grown are: 'Aconitifolium' with deeply lobed leaves and 'Vitifolium' with large leaves. Other cultivars include 'Ezonomomiji', 'Green Cascade' (dissected leaves) and 'Meigetsu'.

7. A. palmatum Murray subsp.
palmatum. Figure 130(5), p. 536.
Illustration: *Bulletin Tokyo University Forest*
60: f. 9. (1965); Kurata, Illustrated
important forest trees of Japan **2**: pl. 55
(1968); Vertrees, Japanese maples, (1978);
Salake, Wild flowers of Japan, woody
plants **2**; pl. 9 (1989).

Tree or shrub to 25 m in the wild, to
12 m in cultivation. Bark grey or greyish
light brown, smooth. Shoots rather thin,
hairless, green or reddish. Leaves with 5–7
lobes, to 5 × 4.5 cm, lobes lanceolate,
acuminate, green on upper surface, paler
beneath, hairless, margins with double,
forwardly pointing teeth. Inflorescences
upright, with 10–20 purple flowers. Fruits
to 3 cm, nutlets with thin pericarp, wings
held at an obtuse angle. *Japan (C & S
Honshu, Kyushu, Shikoku), Taiwan.* H1.
Spring.

This well-known species is characterised
in the wild by the rather small leaves, the
upright inflorescences and the thin-walled
nutlets.

Subsp. **amoenum** (Carrière) Hara
(*A. amoenum* Carrière). Illustration: Kurata,
Illustrated important forest trees of Japan
1: pl. 83 (1971); *Bulletin Tokyo University
Forest* **60**: f. 10, 11 (1965); Satake, Wild
flowers of Japan, woody plants **2**: pl. 10
(1989). Tree to 15 m, to 6 m in cultivation.
Bark grey to greyish light brown, smooth.
Shoots green, hairless. Leaves with 5–9
lobes, usually 7-lobed, to 8 × 7 cm, lobes
ovate, acuminate, margins with regular
forwardly pointing teeth, sometimes doubly
toothed. Inflorescences with 15–30
pendent flowers. Fruits to 4 cm, nutlets
with woody, veined pericarp, wings held at
a right or obtuse angle. *Japan, Korea.* H1.
Spring.

This subspecies is very variable but
differs from *A. palmatum* subsp. *palmatum*
with the larger, to 9-lobed leaves, the
pendent inflorescences and the woody
veined nutlets.

Subsp. **matsumurae** Koidzumi
(*A. matsumurae* (Koidzumi) Koidzumi).
Illustration: *Bulletin Tokyo University Forest*
60: f. 11 (1965). Tree to 10 m. Leaves
deeply 9-lobed, margins with double,
forwardly pointing teeth. *Japan (NW
Honshu).* H2. Spring.

A. palmatum is widely cultivated. Over
250 cultivars have been selected. Most of
them are small shrubs. Some of them may
produce bright red and orange autumn
colours. They are further classified into
several groups.

Palmatum Group: leaves with 5–7
lobes, e.g. 'Aureum', 'Bloodgood',
'Koreanum', 'Moonfire', 'Osakazuki'.

Dissectum Group: leaves deeply divided,
e.g. 'Dissectum', 'Dissectum Nigrum',
'Filigree', 'Garnet', 'Ornatum'.

Deeply Divided Group: leaves deeply
incised, e.g. 'Benikagami', 'Burgundy Lace',
'Inazuma', 'Nicholsoii', 'Sherwood Flame',
'Trompenburg'.

Elegans Group: leaves mostly 7-lobed, e.
g. 'Beni Kagami', 'Elegans', 'Katsura'.

Linearilobum Group: leaves long,
narrow lobed, e.g. 'Atrolineare',
'Linearilobum Atropurpureum', 'Red
Pygmee'.

Variegated Group: e.g. 'Butterfly',
'Higasayama', 'Versicolor'.

8. A. pseudosieboldianum (Pax)
Komarov. Figure 130(6), p. 536.
Illustration: *Acta Instituti Botanici
Academiae Scientarum URSS*, ser. 1, **1**: f. 25
(1933); Vertrees, Japanese maples, 156
(1978); Flora Reipublicae Popularis Sinicae
46: f. 30 (1981).

Small tree or shrub to 8 m, to 6 m in
cultivation. Bark grey-brown, smooth.
Shoots greenish or greenish purple. Leaves
with 9–11 deep lobes, almost circular, base
cordate, lobes ovate-lanceolate, acuminate,
margins with double, forwardly pointing
teeth, shiny green on upper surface, lightly
downy beneath, stalk downy.
Inflorescences pendent, downy, with
15–30 purple-red flowers. Fruits to 2.5 cm,
nutlets spherical, hairless, wings held near
horizontally. *China (Manchuria), Korea.* H1.
Spring.

This maple is rather rare in cultivation.
It often turns bright red and orange during
autumn. It is similar to *A. japonicum*, but
differs in the smaller flowers and the
longer, narrower leaf-lobes. It is distinct
from other species in this section by the
persistence of dried leaves on the trees
during winter.

Subsp. **takesimense** (Nakai) de Jong
(*A. takesimense* Nakai). Strongly resembles
the species, but has smaller leaves usually
with 13 lobes. *Korea (Ullung Do).*

9. A. shirasawanum Koidzumi. Figure
130(7), p. 536. Illustration: *Bulletin Tokyo
University Forest* **60**: f. 9 (1965); Kurata,
Illustrated important forest trees of Japan
3: pl. 45 (1971); Vertrees, Japanese
maples, 157, 158 (1978); Satake, *Wild
flowers of Japan, woody plants* **2**: pl. 12
(1989).

Tree or shrub to 20 m, to 6 m in
cultivation. Bark light greyish brown.
Shoots green, sometimes glaucous at first,
hairless. Leaves with 9–13 lobes, almost
circular, base cordate, lobes ovate,
acuminate, margins with fine, double,
forwardly pointing teeth, green on upper
surface, paler and thin downy on main
veins beneath. Inflorescences erect, with
9–20 flowers. Sepals red-purple; petals
yellow-white; disc outside the filaments.
Stamens 8. Fruits to 3 cm, nutlets
spherical, veined. *Japan (Honshu, Shikoku).*
H1. Spring.

Rather rare in cultivation, but 'Aureum',
a well-known selection of this species, was
until recently erroneously included under
A. japonicum. This maple is characterised
by the high number of leaf-lobes, upright
inflorescences, light bark and bright red,
long inner bud-scales when the leaves are
falling.

Var. **tenuifolium** Koidzumi
(*A. tenuifolium* (Koidzumi) Koidzumi).
Illustration: *Bulletin Tokyo University Forest*
60: f. 9 (1965); Kurata, Illustrated
important forest trees of Japan **3**: pl. 47
(1971); Satake, Wild flowers of Japan,
woody plants **2**: pl. 12, 13 (1989). Shrub
to 8 m. Leaves somewhat smaller and
thinner, usually 9-lobed. Wings of the
fruits smaller and more parallel instead of
horizontal. *Japan (Honshu, Kyushu,
Shikoku).* H2. Rare in cultivation.

10. A. sieboldianum Miquel. Figure 130
(8), p. 536. Illustration: *Bulletin Tokyo
University Forest* **60**: f. 8 (1965); Kurata,
Illustrated important forest trees of Japan
3: pl 46 (1971); Vertrees, Japanese maples,
158, 159 (1978); Satake, Wild flowers of
Japan, woody plants **2**: pl. 11 (1989).

Tree or shrub to 10 m, to 6 m in
cultivation. Bark light grey. Shoots
greenish brown, cobwebby and sometimes
with dense shaggy hairs when young.
Leaves with 7–11 ovate-oblong lobes,
mostly 9-lobed, circular, dark green on
upper surface, often with purplish margins,
veins downy beneath, margins with
double, forwardly pointing teeth.
Inflorescences erect, often densely hairy,
with 15–20 pale yellow flowers. Fruits to
2 cm, nutlets spherical, strongly veined,
wings held at an obtuse angle. *Japan.* H2.
Spring.

Relatively common in cultivation and
often with bright red and orange autumn
colours. It differs from *A. palmatum* in the
typical purplish margins of the leaves and

the pale yellow flowers. Cultivars include 'Kinugasayama', 'Miyaminishiki', 'Sodenouchi'.

Series **Sinensia** Pojarkova. Leaves deciduous, with 3–7 lobes, margins with forwardly pointing teeth, sometimes entire. Inflorescences mostly with a long axis, with 20–250 flowers, stalks of male flowers which have finished flowering mostly partly dropped. Flowers often with recurved calyx, petals often lobed, disc sometimes downy.

11. A. campbellii Hooker & Hiern subsp. **campbellii**. Figure 130(9), p. 536. Illustration: Flora Reipublicae Popularis Sinicae **46**: f. 42 (1981); The new RHS dictionary of gardening **1**: 33 (1992).

Tree to 30 m, to 6 m in cultivation. Bark grey-brown. Shoots reddish brown, hairless. Leaves with 5–7 lobes, to 12 × 15 cm, lobes shortly triangularly lobed and with shallow, forwardly pointing teeth, base truncate or cordate, vein-axils beneath tufted with whitish hairs. Inflorescences paniculate, axis to 15 cm, 30–60 flowers. Sepals yellow, petals white, disc and ovary with dense shaggy hairs. Fruits to 4 cm, nutlets spherical, hairless, wings held at an acute to right angle. *E Himalaya, SW China.* H4. Spring.

Rather rare in cultivation. In its present delimitation the species consist of a number of rather poorly known subspecies. Some of them are rather variable which makes identification difficult. All are rarely cultivated.

Subsp. **flabellatum** (Rehder) Murray (*A. flabellatum* Rehder). Illustration: Flora Reipublicae Popularis Sinicae **46**: f. 40 (1981). Leaves usually 7-lobed, lobes usually ovate-oblong, acuminate, margins with sharp, irregular forwardly pointing, adpressed, acute teeth. Inflorescences smaller, axis to 12 cm. Disc and ovary hairless. Wings of fruit spreading horizontally. *C China to N Burma.* H4.

Var. **yunnanense** Rehder from Yunnan is somewhat intermediate between subsp. *campbellii* and subsp. *flabellatum*.

Subsp. **sinense** (Pax) de Jong (*A. sinense* Pax). Illustration: Flora Reipublicae Popularis Sinicae **46**: f. 45, 46 (1981). Leaves 5-lobed, slightly leathery, lobes oblong-ovate or triangular-ovate, acuminate, usually somewhat glaucous beneath, margins with adpressed, forwardly pointing teeth. Inflorescence-axis 5–7 cm. Disc slightly downy, ovary white-hairy. Wings

of fruits held at a right angle to spreading horizontally. *E & C China.* H3.

Subsp. **wilsonii** (Rehder) de Jong (*A. wilsonii* Rehder). Illustration: Flora Reipublicae Popularis Sinicae **46**: f. 51 (1981). Leaves 3-lobed, sometimes 5-lobed, base rounded, rarely almost cordate or truncate, lobes ovate-oblong, tailed acuminate, margins entire, but sometimes with adpressed forwardly pointing teeth near the apex. Inflorescence-axis to 9 cm. Disc hairless; ovary downy. Fruit nutlets veined, wings spreading horizontally. *C China.* H3.

12. A. erianthum Schwerin. Figure 130 (10), p. 536. Illustration: Flora Reipublicae Popularis Sinicae **46**: f. 40 (1981).

Shrub or tree to 10 m in the wild, to 5 m in cultivation. Bark greenish or brownish grey. Shoots green or purplish green with white lenticels. Leaves 5- rarely 7-lobed, lobes ovate or triangular-ovate, acuminate, green on upper surface, light green, net-veined, slightly downy and with conspicuous white tufts of silky hairs beneath, margins with sharp, adpressed, forwardly pointing teeth. Inflorescences erect, axis to 10 cm. Sepals green, shaggy hairy. Petals white. Ovary densely shaggy-haired. Fruits to 3 cm, nutlet spherical, densely shaggy hairy when young, wings spreading horizontally, often bright red when young. *C China.* H3. Spring.

Rather rare in cultivation. It may be recognised by the long erect inflorescences, woolly young fruits and white tufts of hairs on the undersides of the leaves.

13. A. oliverianum Pax. Figure 130(11), p. 536. Illustration: Flora Reipublicae Popularis Sinicae **46**: f. 44 (1981).

Shrub or tree to 10 m in the wild, to 7 m in cultivation. Bark greenish or greenish brown. Shoots purplish green, hairless. Leaves 5-lobed, base almost cordate or truncate, lobes triangular-ovate or oblong-ovate, acuminate, dark green and glossy on upper surface, pale green and hairless beneath, vein-axils downy, margins with fine, forwardly pointing, adpressed teeth. Inflorescences corymbose, with 20–35 flowers. Sepals purplish green. Petals white. Ovary slightly shaggy hairy. Fruits to 3.5 cm, nutlets spherical, veined, wings spreading nearly horizontally. *C & S China, Taiwan.* H2. Spring.

Rather rare in cultivation and bearing a strong resemblance to *A. palmatum* but differ in its more papery leaves. In

A. palmatum the leaves, when 5-lobed, are rather small.

Series **Penninervia** Metcalf. Leaves (semi-)evergreen, always unlobed, margins entire or with forwardly pointing teeth. Otherwise as for Series *Sinensia*.

14. A. laevigatum Wallich. Illustration: Flora Reipublicae Popularis Sinicae **46**: f. 61 (1981): The new RHS dictionary of gardening **1**: 33 (1992).

Tree to 15 m. Bark dark green. Shoots olive-green, hairless, without terminal buds. Leaves leathery, 3-veined at base, lanceolate-oblong, acute, olive-green on upper surface, net-veined beneath, margins entire or with remote, forwardly pointing teeth; stalks *c*. 1 cm. Inflorescences erect, paniculate, axis to 9 cm. Sepals purplish green. Petals white. Disc purple, slightly downy. Fruits to 4 cm; nutlets spherical, wings held at an acute angle. *C & E China, E Himalaya.* H4. Spring.

Rare in cultivation. Differs from species of other sections with similar leaves by the lack of terminal buds.

A. cordatum Pax (*A. laevigatum* subsp. *cordatum* (Pax) Murray). Leaf-base cordate. *C China.* H4. Rarely cultivated and probably better distinguished as a subspecies of *A. laevigatum*.

A. fabri Hance (*A. fargesii* Rehder; *A. reticulatum* Champion). Illustration: Flora Reipublicae Popularis Sinicae **46**: f. 62 (1981). Leaf-margin entire or with finer, forwardly pointing teeth than those of *A. laevigatum*. *C & SE China.* H4. *Rarely cultivated. Closely related to* A. laevigatum *and strongly resembling it.*

Section **Macrantha** Pax. Trees and mostly shrubs, monoecious, but often all flowers male or female, deciduous. Axillary buds often stalked. Bud-scales always 2-paired, not overlapping. Inflorescences racemose, rarely partly corymbose, terminal and axillary, with 1 pair of leaves, with 10–25, sometimes to 40 flowers. Sepals and petals 5, disc within the bases of the filaments. Stamens 8. Nutlets flat, sometimes spherical. Parthenocarpic tendency moderate, but sometimes strong in cases when the flowers of a tree are not pollinated.

Because, in cultivation, species of this section often produce only female flowers, any resultant seed is usually hybrid. These hybrids are mostly very vigorous, often very similar to one of the parents.

15. A. capillipes Maximowicz. Figure 130 (12), p. 536. Illustration: Kurata, Illustrated important forest trees of Japan **2**: pl. 46 (1968); Bean, Trees & shrubs hardy in the British Isles, edn. 8, **1**: 190 (1970); Satake, Wild flowers of Japan, woody plants **2**: pl. 16 (1989).

Tree or shrub to 20 m in the wild, to 8 m in cultivation. Bark dark grey, fissured. Branches green with dark and light stripes lengthwise. Young shoots purplish or pinkish red, smooth and often glaucous. Leaves to 11 × 10 cm, 3-lobed, base almost cordate, terminal lobe large, triangular, side lobes small, dark green on upper surface, light green beneath, margins with somewhat irregular, forwardly pointing teeth; stalks red. Inflorescence small, with 15–25 greenish flowers. Fruits to 2 cm, red; nutlets spherical, small, wings held at a right to an obtuse angle. *Japan (C & S Honshu, Shikoku).* H2. Spring.

Rather common in cultivation.

A. rubescens Hayata (*A. morrissonense* Li not Hayata). Branches prominently white-striped. Leaves with 3–5 lobes, on adult trees also partly entire, dark green on upper surface, stalk red. *Taiwan.* H4. Rarely cultivated. Young plants strongly resemble *A. capillipes*.

16. A. crataegifolium Siebold & Zuccarini. Figure 131(1), p. 541. Illustration: Dippel, Handbuch der Laubholzkunde **2**: 417 (1892); Kurata, Illustrated important forest trees of Japan **2**: pl. 42 (1968); Vertrees, Japanese maples, 150 (1978); Satake, Wild flowers of Japan, woody plants **2**: pl. 14 (1989).

Shrub or small tree to 7 m in the wild, to 4 m in cultivation. Bark white-striped, hairless. Shoots purple-red with white stripes and white lenticels, buds small. Leaves to 4 × 6 cm, with 1–3 lobes, sometimes to 5, ovate, base almost cordate or truncate, dark green on upper surface and with purple-black margins, paler beneath, vein-axils downy when young, margins with irregular, forwardly pointing teeth. Inflorescences small, erect, with 9–12 small flowers. Sepals purplish. Petals greenish. Fruits to 3 cm, purplish, hairless; nutlets flat, wings held horizontally. *S Japan.* H2. Spring.

Rather rare in cultivation. 'Veitchii', leaves marked rose and white.

17. A. davidii Franchet. Figure 131(2), p. 541. Illustration: Flora Reipublicae Popularis Sinicae **46**: f. 65 (1981); *The Plantsman* **5**(1): 46 (1983); *Memoirs New York Botanical Garden* **54**: f. 18 (1990); The new RHS dictionary of gardening **1**: 31 (1992).

Shrub or tree to 15 m in the wild, to 10 m in cultivation. Bark green or red-purple when young and with light green to white stripes. Young shoots green or reddish and often somewhat glaucous, axillary buds of vigorous shoots stalked. Leaves to 15 × 8 cm, entire or with 2 obtuse basal lobes, base rounded or almost cordate, glossy green on upper surface, paler beneath, vein-axils beneath red-brown downy when young, margins with forwardly pointing, deep, adpressed teeth. Inflorescences to 15 cm, pendent, with 15–40 greenish yellow flowers. Fruits to 3.5 cm, hairless, nutlets cylindric to more spherical, with a hole on one side, wings held at an obtuse angle or spreading nearly horizontally. *China.* H2. Spring.

Rather common in cultivation. A variable species and very widely distributed in China. Hard to distinguish from subsp. *grosseri* and some related species when young. In young plants the leaves are mostly 3-lobed, adult trees have entire leaves except on vigorous shoots. There is an increasing number of cultivars described, among them 'Ernest Wilson', 'George Forrest', 'Karmen', 'Madeline Spitta', 'Rosalie' and 'Serpentine'.

Subsp. **grosseri** (Pax) de Jong (*A. grosseri* Pax; *A. hersii* Rehder; *A. tegmentosum* subsp. *grosseri* (Pax) Murray; *A. davidii* var. *horizontal* Pax). Illustration: *Mededelingen Landbouwhogeschool Wageningen* **76**(2): f. 18 (1976). Bark always green and white-striped. Leaves to 6 × 5 cm, on short shoots of adult trees, often entire, base cordate, on other shoots 3- to sometimes 5-lobed with abruptly pointed side-lobes, margins with forwardly pointing, adpressed, acute teeth. Inflorescences rather small, with 9–15 flowers. Fruits to 3 cm, nutlets rather flat. *N & C China.* H2. Spring. Rather common in cultivation. The leaves of the short shoots of the adult trees resemble those of the species, but on vigorous shoots they are wider and with the lobes slightly shorter than the central lobe. In the original introduction of Hers (*A. hersii*) from Henan the side-lobes are very distinct and tapered at the apex into a long point. Because of the large variation within this subspecies there is no need for a var. *hersii* (Rehder) Rehder.

A. sikkimense Miquel (*A. hookeri* Miquel). Illustration: Flora Reipublicae Popularis Sinicae **46**: f. 65 (1981). Bark green. Leaves entire, leathery, base cordate, 3-veined, elliptic-oblong, abruptly short-acuminate, margins entire to toothed with close, forwardly pointing teeth. Inflorescence pendent with 35–50 small, very shortly stalked flowers. Fruit to 2 cm, nutlets small. *E Himalaya, SW China.* H5–G1. A variable species, often epiphytic in its natural habitat. Rare in cultivation, usually named *A. hookeri*.

A. chienii Hu & Cheng. Illustration: Flora Reipublicae Popularis Sinicae **46**: f. 68 (1981). Leaves 3-lobed, base cordate, side-lobes small, pointed, when young densely reddish downy beneath, margin with fine, forwardly pointing teeth. Fruits small, shortly stalked. *SW China.* H5–G1. Spring. Rarely cultivated. The fine, forwardly pointing teeth on the leaves and the large racemes with small fruits possibly suggest a close relation with *A. sikkimense*.

18. A. micranthum Siebold & Zuccarini. Figure 131(3), p. 541. Illustration: *Bulletin Tokyo University Forest* **60**: f. 12 (1965); Kurata, Illustrated important forest trees of Japan **5**: pl. 42 (1976); *Mededelingen Landbouwhogeschool Wageningen* **76**(2): f. 21–23 (1976); Vertrees, Japanese maples, 152 (1978); Satake, Wild flowers of Japan, woody plants **2**: pl. 18 (1989).

Shrub or small tree to 10 m in the wild, to 6 m in cultivation. Bark grey. Branches greyish green and somewhat russet. Young shoots purplish brown, smooth. Leaves to 7 × 7 cm, 5-lobed (sometimes 7-lobed), base cordate, lobes ovate, long-acuminate, hairless except for some down at the base beneath, margins with double, forwardly pointing teeth. Inflorescence small, pendent, with 15–30 reddish green flowers *c.* 4 mm across. Fruits to 2 cm; nutlets purplish, spherical, wings held nearly horizontally. *Japan (Honshu, Kyushu, Shikoku).* H2. Late spring.

Rather rare in cultivation. It may develop red and orange autumn colours. Often confused with *A. tschonoskii*, but in that species the inflorescence is upright and partly corymbose, the flowers and fruits are larger and the wings of the fruits are at an acute to right angle. The latter also differs in the thicker young branches and the larger leaves, which turn yellow and orange in autumn.

19. A. pectinatum Wallich. Illustration: Flora Reipublicae Popularis Sinicae **46**: f. **67**(1981).

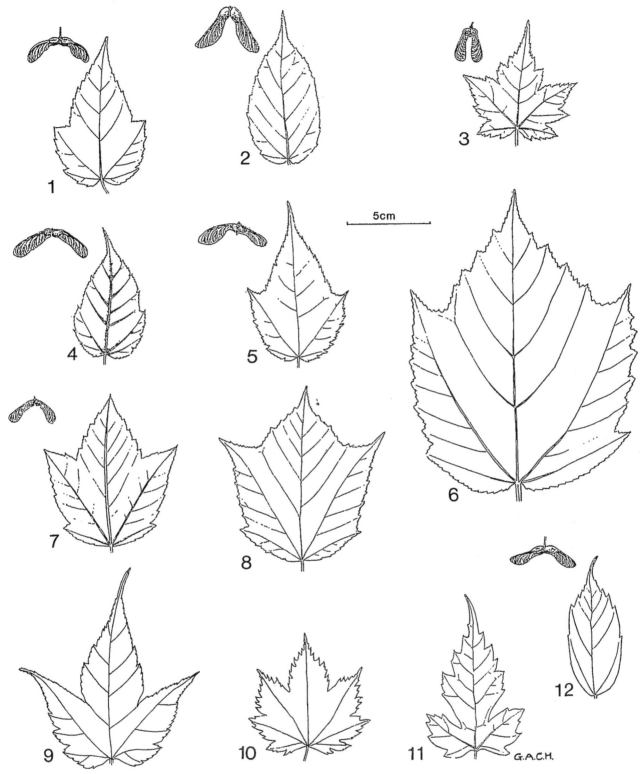

Figure 131. Leaves and fruits (top left) of Acer species.
1, *A. crataegifolium.* 2, *A. davidii.* 3, *A. micranthum.* 4, *A. pectinatum* subsp. laxiflorum. 5, *A. pectinatum* subsp. forrestii. 6, *A. pensylvanicum.*
7, *A. rufinerve.* 8, *A. tegmentosum.* 9, *A. acuminatum.* 10, *A. argutum.* 11, *A. barbinerve.* 12, *A. stachyophyllum.*

Tree to 18 m in the wild, shrub to 6 m in cultivation. Branches green, slightly white-striped, reddish or purplish when young. Leaves to 14 × 7 cm, with 3–5 lobes, central lobe relatively long, lobes triangular, acuminate to tailed, dark green on upper surface, veins beneath covered with rust coloured hairs when young, margins with sharp, fine, double, forwardly pointing teeth, stalks red. Inflorescences drooping. Flowers pale yellow. Fruits to 2.5 cm; nutlets flat, wings held horizontally. *Himalaya*. H4. Spring.

The description above is for subsp. **pectinatum**, which is rare in cultivation. *A. pectinatum* is a variable species with the following subspecies.

Subsp. **forrestii** (Diels) Murray (*A. forrestii* Diels). Figure 131(5), p. 541. Illustration: Brickell (ed.), RHS A–Z encyclopedia of garden plants, 67 (2003). Leaves to 12 × 9 cm, ovate, 3-lobed, lobes triangular-ovate, central lobe tailed-acuminate, lateral lobes half as long as the central lobe, acuminate, lower surface pale green, glaucescent, rarely with barbed hairs in the vein-axils. Flowers reddish green. Wings of fruits spreading at an obtuse angle to almost horizontal. *SW China*. H3. Spring. This subspecies is rather common in cultivation and is characterised by its almost hairless leaves.

Subsp. **laxiflorum** (Pax) Murray (*A. laxiflorum* Pax). Figure 131(4), p. 541. Leaves to 12 × 8 cm, slightly leathery, ovate, 3-lobed, sometimes almost entire, lobes triangular-ovate, central lobe long-acuminate, lateral lobes small, acute, dark green on upper surface, paler and reddish-downy on the veins while young, margins with sharp, forwardly pointing teeth, stalks reddish downy when young. Fruits purplish when young, wings spreading at right or obtuse angles. *C China*. H3. Spring. Rare in cultivation, differs from subsp. *forrestii* by the dense reddish down and somewhat thicker leaves.

Subsp. **maximowiczii** (Pax) Murray (*A. maximowiczii* Pax). Illustration: Flora Reipublicae Popularis Sinicae **46**: f. 66 (1981). Usually a small shrub. Leaves to 10 × 8 cm, with 3–5 lobes, ovate-oblong, central lobe long, acuminate, lateral lobes acute, lower surface pale green, reddish downy on the axils of the main veins, margins with sharp, double, forwardly pointing teeth. Fruits purplish when young, wings spreading at an obtuse angle. *C China*. H3. Spring. Rather rare in cultivation. The leaves are quite variable. It

differs from subsp. *forrestii* in the leaves which are mostly 5-lobed, with doubly toothed margins and teeth pointing forwards.

Subsp. **taronense** (Handel-Mazzetti) Murray (*A. taronense* Handel-Mazzetti). Illustration: Flora Reipublicae Popularis Sinicae **46**: f. 68 (1981). Leaves to 15 × 10 cm, 5-lobed, triangular-ovate, central lobe triangular-ovate, acuminate, lateral lobes acute, deep green on upper surface, densely reddish downy beneath on the veins when young, stalks bright red. Inflorescences with 30–40 nearly stalkless, small flowers. *SW China, N Burma*. H4. Spring. Rare in cultivation. Intermediate between subsp. *pectinatum* and subsp. *laxiflorum*, differing from both in the short, acuminate central lobe.

A. caudatifolium Hayata (*A. kawakamii* Koidzumi; *A. morrisonense* Hayata). Leaves ovate, shallowly lobed on young plants, entire on adult plant, apex long-tailed, dull green on upper surface, margins with sharp, double, forwardly pointing teeth. *Taiwan*. H4. Rarely cultivated. Similar to *A. pectinatum* subsp. *forrestii* but differing in its long, tapered, tailed apex.

20. A. pensylvanicum Linnaeus (*A. striatum* du Roi). Figure 131(6), p. 541. Illustration: *Mededelingen Landbouwhogeschool Wageningen* **76**(2): f. 17 (1976).

Tree or shrub to 12 m, to 8 m in cultivation. Bark red-brown. Branches green, white-striped, smooth. Leaves to 18 × 12 cm, 3-lobed, obovate, base almost cordate, lobes forwardly pointing, acuminate, bright green on upper surface, reddish downy on the veins when young. Inflorescences to 12 cm, pendent. Flowers pale yellow. Fruits to 3 cm, wings spreading at an obtuse angle. *NE USA to E Canada*. H2.

Rather common in cultivation. Leaves turning yellow in autumn. The branches of 'Erythrocladum' turn bright red in winter.

A. × conspicuum van Gelderen & Oterdoom; *A. davidii* × *A. pensylvanicum*. Leaves large, with 3–5 lobes, basal lobes underdeveloped when 5-lobed. Intermediate between the parent species. Cultivars include 'Phoenix' with bright red shoots in winter and 'Silver Vein' with conspicuously white-striped shoots.

21. A. rufinerve Siebold & Zuccarini. Figure 131(7), p. 541. Illustration: Kurata, Illustrated important forest trees of Japan **1**: pl. 84 (1964); Vertrees, Japanese

maples, 156 (1978); Satake, Wild flowers of Japan, woody plants **2**: pl. 14, 15 (1989).

Tree or shrub to 15 m. Bark grey. Branches green with lengthwise stripes, stripes becoming dark. Young shoots glaucous blue-white. Leaves to 12 × 10 cm, with 3–5 lobes, base truncate or cordate, central lobe triangular, larger than side-lobes, dark green on upper surface, paler and with reddish down along the veins when young, margins with double or irregular, forwardly pointing teeth. Inflorescences reddish, small, flowers 9–17, pale yellow. Fruits to 2 cm, nutlet spherical and without a hole at one side, wings held at an obtuse angle. *C & S Japan*. H3. Spring.

Rather common in cultivation. The small spherical nutlets without a hole at one side and the glaucous young shoots are very characteristic. The autumn colours vary from orange to deep purple.

22. A. tegmentosum Maximowicz. Figure 131(8), p. 541. Illustration: *Acta Instituti Botanici Academiae Scientarum URSS*, ser. 1, **1**: f. 26 (1933).

Tree to 10 m, often sparingly branched. Bark dark grey-striped. Branches finely striped. Young shoots glaucous white, especially during winter. Leaves to 15 × 10 cm, 3- to sometimes obscurely 5-lobed, base cordate, lobes ovate, acuminate, dull green on upper surface, when young paler and reddish downy along the veins beneath, margins with double, forwardly pointing teeth. Reproductive organs as for *A. pensylvanicum*. *China (Manchuria), Korea*. H1. Spring.

Rare in cultivation. Leaves come out very early in the spring. This species is closely related to *A. pensylvanicum*, but the latter has no glaucous young shoots. It resembles *A. rufinerve* for the glaucous young shoots, but differs in the white-striped older branches, the sparing branching and the flat nutlets of the fruit with a hole at one side. Autumn colour bright yellow.

23. A. tschonoskii Maximowicz subsp. **tschonoskii**. Illustration: *Bulletin Tokyo University Forest* **60**: f. 12 (1965); Kurata, Illustrated important forest trees of Japan **5**: pl. 16 (1976); Satake, Wild flowers of Japan, woody plants **2**: pl. 17, 19 (1989).

Tree or shrub to 10 m, in cultivation to 5 m. Bark grey. Branches light green and somewhat white-striped. Young shoots reddish brown, hairless. Leaves to 8 × 9 cm, deeply 5-lobed, sometimes to 7-lobed, circular-ovate, base cordate, green on upper surface, paler beneath when young, reddish downy along the veins beneath, margins with sharp, double, forwardly pointing teeth; stalks reddish green. Inflorescences corymbose or racemose, erect, with 7–12 yellowish flowers. Fruits to 3 cm, wings at an acute to right angle. *Japan.* H2. Spring.

Rather rare in cultivation. Autumn colour yellow-orange. It resembles *A. micranthum* and *A. pectinatum* subsp. *maximowiczii*. The latter has variable leaves with less lobes and pendent inflorescences.

Subsp. **komarovii** (Pojarkova) Urussov & Nedolushko (*A. komarovii* Pojarkova; *A. tschonoskii* var. *rubripes* Komarov; *A. tschonoskii* subsp. *koreanum* Murray). Illustration: Flora Reipublicae Popularis Sinicae **46**: f. 66 (1981). Trees, but sometimes shrubs to 12 m. Young shoots and leaf-stalks bright red. Leaves ovate, longer than wide, with shallow, forwardly pointing, adpressed, somewhat obtuse teeth. *Korea, China (Manchuria).* H1. Rare in cultivation.

Section **Glabra** Pax. Shrubs or small trees. Bud-scales 2–4-paired. Leaves 3-lobed, sometimes 5-lobed, undivided or (partly) with 3 leaflets, with forwardly pointing teeth. Inflorescences corymbose or racemose, terminal and sometimes partly axillary, with 1 pair of leaves or leafless. Sepals and petals in 4s or 5s; disc around the bases of the stamens or within them. Nutlets flat, hairless, strongly veined. Parthenocarpic tendency strong.

Series **Glabra**. Monoecious, but often functionally dioecious. Bud-scales 2–4-paired. Leaves 3-, sometimes 5-lobed or (partly) with 3 leaflets. Inflorescence corymbose or racemose, terminal and partly axillary, with 1 pair of leaves, axillary inflorescences sometimes leafless, with 3–11 flowers. Sepals and petals 5. Stamens 8.

24. A. glabrum Torrey. Illustration: *The Plantsman* **5**(2): 48 (1983); The new RHS dictionary of gardening **1**: 31 (1992).

Shrub or small tree, to 10 m in the wild, to 5 m in cultivation. Bark reddish grey. Branches red-brown, almost hairless. Leaves with 3–5 lobes, to 12 cm wide, almost cordate to wedge-shaped at base, hairless, dark green on upper surface, paler blue-green beneath, lobes acute to

acuminate, margins with coarse, double, forwardly pointing teeth. Inflorescences corymbose with 7–11 flowers or sometimes partly racemose with 3–5 flowers. Disc around the bases of the stamens or within them. Fruits *c.* 2.5 cm, nutlets veined, wings parallel. *W North America.* H2. Spring.

Rather rare in cultivation. *A. glabrum* is rather variable and is divided further into 4 subspecies.

Subsp. **douglasii** (Hooker) Wesmael. Vigorous shrub. Leaves 3-lobed, often deeply so, developing 3 leaflets on vigorous shoots, margins with forwardly pointing teeth. *NW North America.* H2.

Series **Arguta** (Render) Pojarkova. Dioecious. Bud-scales 2-paired, not overlapping. Inflorescences racemose, 7–15 flowers, pendent. Sepals and petals 4. Male flowers with 4–6 stamens and on inflorescences developed from leafless axillary buds. Female flowers without rudimentary stamens and on inflorescences developed from terminal buds of short shoots with 1 pair of leaves.

25. A. acuminatum D. Don. Figure 131 (9), p. 541. Illustration: Dippel, Handbuch der Laubholzkunde **2**: 424 (1892).

Small tree or shrub to 6 m. Bark smooth, green. Young stems smooth, purple-brown. Buds red, scales not overlapping. Leaves with 3–5 lobes, to 12 cm, base cordate, the basal lobes insignificant, lobes triangular, long-acuminate, dark green on upper surface, paler beneath, margins with double, forwardly pointing teeth. Inflorescences short racemes. The male flowers are born on racemes from leafless buds, the female from mixed buds with 1 pair of leaves or on vigorous shoots partly from leafless buds. Flowers yellow, disc around the bases of the filaments. Fruits 4–5 cm; nutlets hairless, strongly veined, wings held at a right angle. *W Himalaya.* H3. Spring.

This species is rare in cultivation. Within series *Arguta* it may be confused with the variable species *A. stachyophyllum*. The reproductive organs of *A. acuminatum* differ clearly from those of *A. caudatum* and *A. pectinatum* (also from W Himalaya).

26. A. argutum Maximowicz. Figure 131 (10), p. 541. Illustration: Kurata, Illustrated important forest trees of Japan **2**: pl. 41 (1968); Satake, Wild flowers of Japan, woody plants **2**: pl. 8 (1989).

Shrub or small tree to 8 m in the wild, to 5 m in cultivation. Bark grey or grey-brown. Young shoots purplish brown or scarlet, often with reddish hairs or downy. Leaves 5-lobed, noted for the net-like veins, lobes ovate, margins with double, forwardly pointing teeth, pale green and hairless on upper surface, paler and downy beneath, especially on the whitish veins. Reproductive organs as *A. acuminatum.* Fruits *c.* 2 cm. *Japan (CS Honshu, Shikoku).* H2. Spring.

This maple has an upright habit and often bright red young shoots. The leaves with net-like veins resemble those of *A. caudatum* subsp. *ukurunduense* and *A. spicatum*, but these species differ in their reproductive organs. The autumn colour of the leaves is yellow. Rarely cultivated.

27. A. barbinerve Maximowicz. Figure 131(11), p. 541. Illustration: *Acta Instituti Botanici Academiae Scientarum URSS*, ser. 1, **1**: f. 35 (1933); Flora Reipublicae Popularis Sinicae **46**: f. 71 (1981); The new RHS dictionary of gardening **1**: 31 (1992).

Shrub or small tree to 6 m. Bark grey-yellow and smooth. Young shoots green, rarely reddish, downy. Leaves 5-lobed, membranous, roundish ovate, base cordate or almost so, margins with coarse, forwardly pointing teeth, deep green on upper surface, paler and downy on the veins beneath. Reproductive organs as for *A. acuminatum.* Fruits 4–6 cm. *SE Siberia, NE China, Korea.* H1. Spring.

Rather rare in cultivation. It resembles *A. argutum* and *A. stachyophyllum*, but differs in having coarse, forwardly pointing teeth on the leaves and down.

28. A. stachyophyllum Hiern (*A. tetramerum* Pax). Figure 131(12), p. 541. Illustration: Flora Reipublicae Popularis Sinicae **46**: f. 71–73 (1981).

Small tree, but mostly a multi-stemmed shrub to 10 m, suckering. Bark greyish or yellowish brown, smooth. Shoots green or reddish green and smooth, often bright red when young. Leaves entire or shallowly 3-lobed, sometimes distinctly 3-lobed, ovate or obovate, apices acuminate, base truncate or rounded, bright green on upper surface, lighter and at first downy beneath, margins with double, forwardly pointing teeth. Reproductive organs as for *A. acuminatum.* Fruits *c.* 3 cm, sometimes to 4.5 cm, wings held at an acute angle. *E Himalaya to C China.* H1. Spring.

A variable species and distinct by its suckering habit (especially subsp.

betulifolium). The young branches are often slightly striped, similar to the so called snake-bark maples (Section *Macrantha*). Rarely cultivated.

Subsp. **betulifolium** (Maximowicz) de Jong. Multi-stemmed erect shrub to 5 m, suckering. Leaves entire or slightly 3-lobed, broadly wedge-shaped, usually 3-nerved at base, margins with rather coarse, forwardly pointing teeth. *C & W China*. H1. Spring. This subspecies has the most northern distribution and is distinguished by the small leaves and strong suckering habit.

Section **Negundo** Pax. Trees or shrubs, dioecious, deciduous. Bud-scales in 2 or 3 pairs. Leaves pinnate, with 3–7, sometimes to 9 leaflets. Inflorescences racemose, with 15–40 flowers. Sepals and petals 4. Stamens 4–6, absent in female flowers. Nutlets small, ellipsoid to rather flat, veined. Parthenocarpic tendency strong.

Series **Negundo**. Bud-scales 3-paired, sometimes 2-paired. Leaves pinnate, with 3–7, sometimes 9 leaflets, the lowest pair and the terminal leaflet sometimes with 3 lobes. Inflorescences racemose on female individuals and compound racemose on male individuals; flower-stalks long, drooping. Flowers without petals, greenish, disc absent.

29. A. negundo Linnaeus. Figure 132(1), p. 545. Illustration: *Mededelingen Landbouwhogeschool Wageningen* **76**(2): f. 31–32 (1976); Flora Reipublicae Popularis Sinicae **46**: f. 82 (1981); The new RHS dictionary of gardening **1**: 31 (1992).

Tree to 25 m, often with more than one trunk, crown widely spreading. Bark grey, smooth. Branches hairless, green or grey, sometimes glaucous. Leaves to 22 cm, pinnate, with 3–7, sometimes 9 leaflets, but mostly 5. The lowest pair and the terminal leaflet sometimes with 3 lobes. Leaflets ovate, coarsely toothed towards apex, bright green on upper surface, lighter beneath, slightly downy or hairless. Inflorescences lateral, the males in clusters of racemes with 15–50 flowers, the females in small, drooping racemes with 3–11 flowers and sometimes with 1 pair of small leaves. Flowers greenish yellow, the females with long white stigmas and without rudimentary stamens. Fruit to 4 cm, nutlets small, ellipsoid-spherical to rather flat, veined, wings at an acute angle. *Central & North America*. H1. Spring.

A vigorous tree and common in cultivation. Variable in down and the colour of young shoots. Naturalised in several countries of Asia and Europe. The selected cultivars mainly have variegated leaves, e.g. 'Auratum', 'Flamingo', 'Aureomarginatum', 'Variegatum'. The species is rather variable and several subspecies and varieties are distinguished, among them are:

Subsp. **californicum** (Torrey & Gray) Wesmael. Shoots and upper surface of leaves downy, underside with densely felted hairs. Leaves usually with 3 leaflets, margins often entire. *USA (California)*.

Var. **violaceum** (Kirchner) Jaeger. Branches becoming violet, young shoots covered with a glaucous bloom. Leaflets softly downy beneath. *Midwest USA*. Probably a selection rather than a real botanical variety.

Series **Cissifolia** (Koidzumi) Pojarkova. Bud-scales 2-paired, not overlapping. Leaves of 3 leaflets. Inflorescence racemose, with 15–40 flowers. Sepals and petals 4, very distinct. Stamens 4, sometimes 6. Disc around the bases of the filaments.

30. A. cissifolium (Siebold & Zuccarini) Koch. Figure 132(2), p. 545. Illustration: Kurata, Illustrated important forest trees of Japan **2**: pl. 48 (1968); Satake, Wild flowers of Japan, woody plants **2**: pl. 26 & 27 (1989); The new RHS dictionary of gardening **1**: 31 (1992).

Tree or large shrub to 20 m in the wild, to 12 m in cultivation. Bark grey, smooth. Shoots grey or greenish, downy when young. Leaves with 3 leaflets, occasionally partly palmate with 4 or 5 leaflets; leaflets to 9 cm, ovate, apices acute, with coarse, forwardly pointing teeth, dark green on upper surface, paler beneath, hairless except for some tufts at the vein-axils beneath. Leaf-stalks to 8 cm, leaflet-stalks very short. Inflorescences long, erect racemes, developed from axillary buds with a pair of leaves or leafless, downy, with 15–35 fragrant flowers. Petals yellow-green, twice the length of sepals; disc around the bases of the filaments; flower-stalks 3–5 mm. Fruit hairless; nutlets flat, veined, wings to 1.5 cm. *Japan*. H1. Spring.

This maple is rather commonly cultivated and has a typical diagonal mode of branching. It may develop rich orange-yellow autumn colours.

31. A. henryi Pax (*A. cissifolium* subsp. *henryi* (Pax) Murray). Figure 132(3),

p. 545. Illustration: Flora Reipublicae Popularis Sinicae **46**: f. 82 (1981); Brickell (ed.), RHS A–Z encyclopedia of garden plants, 65 (2003).

Tree or large shrub to 10 m tall and wide in the wild, to 5 m in cultivation. Bark grey to olive-green, smooth. Shoots green and downy, becoming olive-green and smooth. Leaves with 3 leaflets; leaflets to 10 cm, stalkless to almost so, elliptic, variably entire or with remote, forwardly pointing teeth, green on both sides, veins downy beneath, apices long-acuminate, stalks to 10 cm. Inflorescence long, erect to pendent racemes, downy, with 20–40 flowers. Flowers almost stalkless. Petals white, as long as sepals and shed very early during flowering. Fruit on drooping racemes 10–15 cm, hairless, almost stalkless; nutlets flat, veined, wings to 2 cm. *C China*. H2. Spring.

Rather rare in cultivation and may easily be confused with *A. cissifolium*, however, their flowers are clearly distinct. Those of the latter have short leaf-stalks and long yellow-green petals, while the flowers of *A. henryi* are almost stalkless and have small, white petals that are dropped early. Young plants of both species develop similar leaves, but those of *A. cissifolium* have a conspicuous yellowish orange colour in the young growth. *A. henryi* does not develop leaves with the typical entire margins before the trees are a considerable size. The leaves are also dark green, becoming purplish in the autumn.

Section **Indivisa** Pax. Trees or shrubs, deciduous, dioecious. Terminal bud usually lacking. Bud-scales 9–13 pairs, overlapping. Leaves simple, sharply toothed and with many pairs of veins. Inflorescences racemose, with 5–15 flowers. Nutlets narrowly ellipsoid, spherical to rather flat, hairless. Parthenocarpic tendency strong.

32. A. carpinifolium Siebold & Zuccarini. Figure 132(4), p. 545. Illustration: Kurata, Illustrated important forest trees of Japan **2**: pl. 47 (1968); Vertrees, Japanese maples, 147 (1978); Satake, Wild flowers of Japan, woody plants **2**: pl. 21 (1989); The new RHS dictionary of gardening **1**: 26 (1992).

Shrub or tree to 10 m. Bark dark grey. Young twigs hairless, brown. Leaves oblong-acuminate, pinnately nerved, margins with double, forwardly pointing teeth; stalks short. Inflorescences terminal and axillary, short drooping racemes,

Figure 132. Leaves and fruits (top left) of Acer species. 1, *A. negundo* (a, male; b, female). 2, *A. cissifolium*. 3, *A. henryi*. 4, *A. carpinifolium*. 5, *A. hyrcanum*. 6, *A. monspessulanum*. 7, *A. sempervirens*. 8, *A. opalus*. 9, *A. saccharum*. 10, *A. buergerianum*. 11, *A. campestre*. 12, *A. cappadocicum*. 13, *A. miyabei*. 14, *A. pictum*.

hairless, the male with 9–15 flowers, female with 5–9, with 1 pair of leaves. Sepals and petals 5, yellow-green. Disc around the bases of the filaments or within them. Stamens 4–10, mostly 6 in male flowers, variable in female flowers. *Japan (Honshu, Kyushu, Shikoku)*. H2. Spring.

Leaves very distinct and resembling those of *Carpinus* but opposite. Sepals and petals 5 in female flowers, petals 4 or absent in male flowers.

Section **Acer**. Trees or shrubs, monoecious, deciduous or semi-evergreen. Bud-scales in 5–13 pairs, overlapping. Leaves mostly 5-, sometimes 3-lobed, rarely unlobed. Inflorescences corymbose, terminal or axillary. Sepals and petals 5. Disc outside the bases of the filaments. Stamens mostly 8, with long protruding filaments in male flowers. Nutlets ovoid, often keeled. Parthenocarpic tendency strong or moderate.

Series **Acer**. Leaves mostly 5-lobed, large, with coarse, forwardly pointing teeth. Buds large with 5–10 pairs of scales. Inflorescences rather large, with 25–150 flowers. Nutlets ovoid-spherical, occasionally somewhat keeled. Parthenocarpic tendency moderate.

33. A. caesium Brandis.

Tree to 25 m in the wild, to 12 m in cultivation. Bark and branches grey-brown. Young shoots purplish green. Buds large, elliptic. Leaves to 14 × 20 cm, 5-lobed, almost leathery, base cordate, lobes broadly obtuse, basal lobes often obsolete, apex shortly acuminate, green on upper surface, blue-green and downy beneath, margins with double, forwardly pointing teeth. Inflorescences corymbose, with 25–40 flowers. Flowers greenish white, ovary hairless. Fruits to 6 cm; nutlets convex and ridged, wings spreading at a nearly erect or an acute angle. *Himalaya, SW China*. H3. Spring.

Rare in cultivation. It is easily distinguished by the somewhat glaucous leaf undersides and the narrow pointed buds.

Subsp. **giraldii** (Pax) Murray (*A. giraldii* Pax). Illustration: Flora Reipublicae Popularis Sinicae **46**: f. 35 (1981). Young branches glaucous white. Leaves to 11 × 15 cm, 3-lobed, but often with 2 more very small basal lobes, lobes broader and shorter than in subsp. *caesium*, undersides darker glaucous. *C & SW China*. H2. Spring.

Rather rare in cultivation and easily recognised by the white young branches.

34. A. heldreichii Boissier.

Trees to 25 m. Bark grey, peeling in flakes as in *A. pseudoplatanus* but starting at a greater age. Branches dark brown-red. Buds ovate, dark brown. Leaves to 10 × 14 cm, deeply 5-lobed, papery, lobes oblong-lanceolate, dark green on upper surface, paler and somewhat glaucous beneath with shaggy hairs on the veins when young, margins coarsely toothed. Inflorescences corymbose, erect, with 20–50 flowers. Flowers yellow-green; ovary downy. Fruits to 5 cm; nutlets brownish yellow, wings held at an acute angle, reddish when young. *SE Europe*. H2. Spring.

Rather rare in cultivation. Characterised by the dark brown buds, upright inflorescences and the usually deeply incised leaves.

Subsp. **trautvetteri** (Medvedev) Murray (*A. trautvetteri* Medvedev). Illustration: *Acta Institute Botanici Academiae Scientarum URSS*, ser. 1, **1**: f. 20 (1933); Brickell (ed.), RHS A–Z encyclopedia of garden plants, 65 (2003). Inner bud-scales bright red during leaf-break. Leaves deeply 5-lobed, somewhat less papery, margins coarsely and angularly toothed. Wings of fruits bright red when young. *N Turkey, W Caucasus*. H2. Spring. Rather rare in cultivation. Those from NW Turkey are intermediate with subsp. *heldreichii*.

35. A. pseudoplatanus Linnaeus. Illustration: *Mededelingen Landbouwhogeschool Wageningen* **76**(2): f. 6–10 (1976); The new RHS dictionary of gardening **1**: 26 (1992).

Trees to 40 m. Bark grey, peeling in flakes. Branches dark grey. Young shoots brownish green, hairless. Buds green, scales with black margins. Leaves to 12 × 16 cm, 3- to mostly 5-lobed, base cordate, lobes ovate, dark green on upper surface, grey-green and somewhat glaucous beneath with shaggy hairs on the veins when young, margins with coarse, forwardly pointing teeth. Inflorescences paniculate, pendent, with 25–60 yellowish green flowers. Stamens protruding. Ovary downy. Fruits to 5 cm, nutlets green, convex, wings held at an acute angle. *W & C Europe*. H2. Spring.

Very common. Characterised by the green buds, the scales with black margins and the large, drooping inflorescences. Various selections have variegated leaves

('Leopoldii', 'Nizettii', 'Simon Louis Frères', 'Variegatum'), red leaf-undersides ('Atropurpureum'), pink leaf-break ('Brilliantissimum', 'Prinz Handjery'), yellow leaf-break ('Flavomarginatum', 'Worleei'), or are suitable for street planting ('Erectum', 'Negenia', 'Rotterdam').

The species easily hybridises in cultivation with other species of Series *Acer*. The hybrids mostly show a strong resemblance to *A. pseudoplatanus*, especially the buds and inflorescences.

A. × hybridum Fischer & Meyer; *A. opalus* × *A. pseudoplatanus*. Tree to 15 m. Leaves 3-lobed, base cordate, lobes pointing forward.

36. A. velutinum Boissier (*A. insigne* Boissier & Buhse). Illustration: *Acta Instituti Botanici Academiae Scientarum URSS*, ser. 1, **1**: f. 18 (1933); Brickell (ed.), RHS A–Z encyclopedia of garden plants, 70 (2003).

Tree to 25 m. Bark grey, smooth. Branches dark greyish brown. Buds brown, narrowly pointed. Leaves to 22 × 20 cm, 5-lobed, base truncate or almost cordate, lobes broadly lanceolate, bright green on upper surface, somewhat glaucous and sparsely to very densely downy beneath, margins with coarse, forwardly pointing teeth. Inflorescences corymbose, erect, with 25–50 yellow-green flowers. Ovary downy. Fruits to 4 cm, nutlets downy when young, wings held at a right or obtuse angle. *Caucausus, N Iran*. H2. Spring.

Rather rare in cultivation. It resembles *A. pseudoplatanus* but differs in the upright inflorescences, the brown buds and shiny leaves. The hairiness of the leaf undersides is variable.

Var. **vanvolxemii** (Masters) Rehder (*A. vanvolxemii* Masters). Illustration: Dippel, Handbuch der Laubholzkunde **2**: 432 (1892). Leaves to 30 × 25 cm, not shiny on upper surface, hairiness limited to veins beneath. Wings of fruits held almost horizontally. *E Caucasus*. H2. Spring.

This variety is very vigorous and sometimes produces a hybrid with *A. pseudoplatanus* as the second parent. However, in contrast with other hybrids of the latter species it has the typical corymbose inflorescence and brown buds of *A. velutinum*. Rarely cultivated.

Series **Monspessulana** Pojarkova. Trees deciduous or semi-evergreen. Bud-scales 8–12-paired. Leaves mostly 3-, sometimes 5-lobed, rarely unlobed, often leathery, margins entire or with coarse, forwardly pointing teeth. Inflorescences with 10–50 flowers, mostly with long, drooping flowerstalks. Nutlets keeled-convex. Parthenocarpic tendency strong.

37. A. hyrcanum Fischer & Meyer (*A. stevenii* Pojarkova; *A. tauricolum* Boissier & Balansa). Figure 132(5), p. 545. Illustration: Dippel, Handbuch der Laubholzkunde **2**: 447 (1892).

Tree or shrub to 15 m in the wild and to 8 m in cultivation, deciduous. Bark grey to grey-brown. Shoots brownish, hairless. Leaves to 7 × 10 cm, 3- to mostly 5-lobed, papery, base cordate, lobes often lobed, dark green on upper surface, glaucous with downy veins beneath, margins entire to finely toothed. Inflorescences corymbose, with 15–25 pendent flowers. Flowers green-yellow. Fruits to 2.5 cm; nutlets keeled-convex, hairless, wings nearly parallel. *SE Europe, Turkey, Caucasus, Crimea.* H1. Spring.

A variable species, rather rare in cultivation. The leaves resemble those of *A. campestre* and *A. saccharum* subsp. *grandidentatum*. The former differs in all other organs, the latter has a united perianth. Several subspecies are recognised and these tend to grade into each other. Somewhat distinct is subsp. **reginae-amaliae** (Orphanides & Boissier) de Jong (*A. reginae-amaliae* Orphanides & Boissier; *A. opalus* subsp. *reginae-amaliae* (Orphanides) Aldèn) from Greece and W Turkey with very small leaves.

38. A. monspessulanum Linnaeus. Figure 132(6), p. 545. Illustration: The new RHS dictionary of gardening **1**: 26 (1992).

Tree to 12 m, deciduous. Bark grey. Branches brown, slender. Buds small, pointed. Leaves to 6 × 8 cm, 3-lobed, leathery, base almost cordate to cordate, lobes triangular-ovate, dark green on upper surface, paler beneath, margins entire or with remote, forwardly pointing teeth, lobed in the juvenile stage. Inflorescences corymbose, with 10–20 pendent flowers. Flowers green-yellow; ovary hairy. Fruits to 2.5 cm; nutlets keeled-convex, hairless, wings parallel to somewhat overlapping. *Mediterranean area to Iran & Turkestan.* H2. Spring.

Rather common in cultivation. The species is rather variable but remains easily recognisable by its (almost) entire, distinctly 3-lobed leaves. A number of subspecies are distinguished. They are rather poorly known and rare in

cultivation, among them subsp. **cinerascens** (Boissier) Yaltirik, subsp. **microphyllum** (Boissier) Bornmüller and subsp. **turcomanicum** (Pojarkova) Murray.

39. A. obtusifolium Sibthorp & Smith (*A. syriacum* Boissier & Gaillardot).

Shrub or small tree to 9 m, evergreen. Bark grey. Branches brownish green, hairless. Leaves to 3 × 5 cm, unlobed to shallowly 3-lobed, leathery, grey-green on upper surface, slightly paler beneath, margins with remote, forwardly pointing teeth to entire. Inflorescences corymbose, erect, with 10–20 flowers. Flowers yellow-green, ovary hairless. Fruits to 2.5 cm; nutlets keeled-convex, wings held at an acute to nearly right angle. *Cyprus, E Mediterranean.* H4. Spring.

Rare in cultivation. Differs from *A. sempervirens* in the larger, grey-green leaves.

40. A. opalus Miller. Figure 132(8), p. 545. Illustration: Dippel, Handbuch der Laubholzkunde **2**: 445 (1892).

Tree to 15 m, deciduous. Bark greybrown. Branches brown, hairless. Buds ovate, dark brown. Leaves to 8 × 10 cm, shallowly 5-lobed, leathery, base cordate to truncate, dark green on upper surface, paler and more or less downy along veins beneath, margins irregularly toothed. Inflorescences corymbose with 15–30 pendent flowers. Flowers yellow-green. Fruits to 4 cm; nutlets keeled-convex, hairless, wings held parallel at a right angle. *S Europe.* H2. Spring.

Rather common in cultivation. A variable species and further divided into a number of subspecies.

Subsp. **hispanicum** (Pourret) Murray (*A. hispanicum* Pourret). Leaves to 6 × 8 cm, downy along veins beneath. *N Spain.*

Subsp. **obtusatum** (Willdenow) Gams (*A. obtusatum* Wildenow). Leaves to 10 × 12 cm, 5–7-lobed, densely hairy beneath. *SE Europe.* H2. Spring. The most common subspecies in cultivation.

A. × coriaceum Tausch (*A. × martinii* Jordan; *A. × rotundilobum* Schwerin). Small tree growing wider than high. Leaves to 7 × 10 m, 3-lobed, leathery, base more or less cordate, margins scalloped. H2. Spring.

This was regarded as a hybrid of *A. monspessulanum* and *A. pseudoplatanus*, but the latter is most unlikely. It looks perfectly intermediate between *A. monspessulanum* and *A. opalus.* As a consequence

A. × rotundilobum and *A. × martinii* both with the same parentage, are synonymous. *A. × coriaceum* is rather rare in cultivation.

41. A. sempervirens Linnaeus (*A. orientale* Miller; *A. creticum* Linnaeus). Figure 132(7), p. 545. Illustration: Dippel, Handbuch der Laubholzkunde **2**: 433 (1892).

Shrub or small tree to 10 m, semievergreen. Bark grey. Branches greybrown, hairless. Leaves to 2 × 4 cm, indistinctly 3-lobed to unlobed, leathery, hairless, dark green on upper surface, olive-green beneath, margins entire or with a few small teeth. Inflorescences corymbose, erect, with 7–15 flowers. Flowers greenish yellow, ovary hairless. Fruits to 2 cm; nutlets keeled-convex, wings parallel or held at an acute angle. *Greece, Crete, S Aegean islands, W Turkey.* H4. Spring.

Rather rare in cultivation. Leaves variable, even on the same plant.

Series **Saccharodendron** (Rafinesque) Murray. Bud-scales 6–9-paired. Leaves 3–5-lobed, sometimes to 7-lobed, margins entire to coarsely toothed. Inflorescences terminal and rather frequently axillary, the latter partly from leafless buds. Flowers without petals, perianth-segments united. Fruits with spherical nutlets. Parthenocarpic tendency strong.

42. A. saccharum Marshall subsp. **saccharum**. Figure 132(9), p. 545. Illustration: Dippel, Handbuch der Laubholzkunde **2**: 438 (1892); The new RHS dictionary of gardening **1**: 26 (1992).

Trees to 40 m in the wild and to 15 m in cultivation. Bark grey, furrowed. Branches brown-grey, hairless. Buds heart-shaped, narrowly pointed. Leaves to 10 × 14 cm, 3–5-lobed, sometimes to 7-lobed, base cordate, lobes lobed, acuminate, green on upper surface, paler to glaucous and hairless to hairy beneath. Inflorescences corymbose with 20–40 pendent flowers. Flowers greenish yellow, perianth united, ovary downy. Fruits to 4 cm; nutlets dark brown, spherical, wings held at an acute angle. *E North America (Rocky Mountains).* H1. Spring.

Rather common in cultivation. Variable and now comprising all species of Series *Saccharodendron.* Well known for its maple syrup and for the orange and red autumn colours. The leaves resemble those of *A. platanoides,* but lack latex in the leaf-stalks.

Subsp. **floridanum** (Chapman) Wesmael (*A. barbatum* Michaux in part). Small tree. Bark chalky grey. Leaves to 6 × 8 cm, 3–5-lobed, base almost cordate, lobes obtuse, glaucous and hairy beneath. *Coastal plain, SE USA.* H2. Spring. Rare in cultivation.

Subsp. **grandidentatum** (Torrey & Gray) Desmarais (*A. grandidentatum* Torrey & Gray). Illustration: Dippel, Handbuch der Laubholzkunde **2**: 440 (1892). Shrub or small tree to 20 m. Bark dark brown. Leaves to 6 × 8 cm, 3-lobed, base cordate, lobes triangular, entire or with 3 secondary lobes. *W North America (W Montana to New Mexico).* H1. Rare in cultivation.

Subsp. **leucoderme** (Small) Desmarais (*A. leucoderme* Small). Small tree to 10 m. Bark chalky grey. Leaves to 6 × 8 cm, 3–5-lobed, base nearly cordate, lobes triangular, acute and with 2 large teeth. *USA (North Carolina to N Louisiana).* H1. Rare in cultivation.

Subsp. **nigrum** (Michaux) Desmarais (*A. nigrum* Michaux). Illustration: Dippel, Handbuch der Laubholzkunde **2**: 439 (1892); *The Plantsman* **5**(1): 48 (1983). Tree to 40 m in the wild and to 15 m in cultivation. Bark grey-black, deeply fissured. Leaves to 12 × 14 cm, 3- to sometimes 5-lobed, dark green on upper surface, yellow-green beneath, margins rather shallowly and roundedly toothed, leaf-stalks often with stipules. *E North America (Iowa to S Quebec & S Appalachians).* H1. Rather rare in cultivation and the only maple with stipules.

Section **Pentaphylla** (Hu & Cheng) de Jong. Monoecious. Leaves simple, 3-lobed or palmate, usually glaucous beneath, margins entire to lobed, with forwardly pointing teeth. Bud-scales 4–8-paired, grey-brown. Inflorescences corymbose, terminal and axillary. Flowers 5-parted, perianth yellow-white, petals longer than sepals. Stamens mostly 8. Disc outside the bases of the filaments. Fruits with keeled, convex nutlets. Parthenocarpic tendency strong.

Series **Pentaphylla** (Hu & Cheng) Murray. Leaves compound with 5, sometimes to 7, palmately arranged leaflets.

43. A. pentaphyllum Dieis. Illustration: Vertrees, Japanese maples, 155 (1978); Flora Reipublicae Popularis Sinicae **46**: f. 78 (1981); *The Plantsman* **5**(1): 46 (1983); *Memoirs New York Botanical Garden* **54**: f. 14 (1990).

Small tree to 10 m. Branches slender, spreading. Young shoots thin, brownish. Buds sharply pointed, blackish. Leaves palmate with 5, sometimes to 7, oblong-lanceolate leaflets, to 7 × 1.5 cm, light green on upper surface, glaucous beneath, margins almost entire to slightly toothed; leaf-stalks red, 4–7 cm. Inflorescences corymbose, terminal. Flowers with a yellowish perianth. Fruits with convex nutlets, samaras *c.* 2 cm. *C China.* H5. Spring.

A rare species. Plants in cultivation are vegetatively propagated or grown from seed from trees in the Strybing Arboretum, San Francisco, California (USA). The latter were grown from seed collected by J. Rock in 1929.

Series **Trifida** Pax. Leaves mostly persistent, simple or 3-lobed, margins entire to lobed, with forwardly pointing teeth, usually 3-veined.

44. A. buergerianum Miquel (*A. trifidum* Hooker & Arnott, not Hooker). Figure 132 (10), p. 545. Illustration: Vertrees, Japanese maples, 145–147 (1978) – cultivars; Flora Reipublicae Popularis Sinicae **46**: f. 55 (1981); The new RHS dictionary of gardening **1**: 26 (1992).

Tree to 20 m in the wild, usually a small tree or shrub in cultivation. Branches hairless. Leaves deciduous, 3-lobed, sometimes simple, 6–10 × 2.5–6 cm, dark green on upper surface, paler or glaucous beneath, downy when young, margins entire to scalloped; leaf-stalks 2.5–6 cm. Inflorescences corymbose, stalks downy. Flowers small, perianth whitish. Fruits hairless, to 2.5 cm, wings small, parallel or convergent. Parthenocarpic tendency strong. *SE China, Taiwan.* H3. Spring.

This maple shows a very distinct juvenile stage in which the leaves have margins with forwardly pointing teeth and narrow lobes. On adult trees the leaves are more leathery and shallowly lobed. The lobes are wider and almost entire, base rounded.

The species was described from Japanese material, but it is not endemic there. It is widely cultivated in China, Japan, Korea and South Africa and often shows brilliant scarlet to orange-yellow autumn colours. Rather rarely cultivated in Europe.

Subsp. **formosanum** (Hayata) Murray & Lauener. Leaves leathery, base cordate, probably evergreen. *Taiwan.* G1. Probably not cultivated.

Subsp. **ningpoense** (Hance) Murray. Leaves unlobed or with minor lobes, semi-evergreen. *C China.* G1. Rarely cultivated.

A. paxii Franchet. Leaves 3-lobed or entire, evergreen, leathery. *SW China.* G1. Closely related to *A. buergerianum* and also showing distinct juvenile leaves. Rarely cultivated.

45. A. oblongum de Candolle (*A. lanceolatum* Molliard). Illustration: Flora Reipublicae Popularis Sinicae **46**: f. 56 (1981).

Tree to 15 m, semi-evergreen. Bark grey. Branches green, hairless. Leaves to 12 × 4 cm, oblong to oblong-ovate, hairless, green on upper surface, glaucous beneath, margins entire. Inflorescences corymbose. Flowers whitish. Fruits to 4 cm, nutlets keeled-convex, wings held nearly horizontally. *Himalaya, SW China.* H5–G1. Spring.

Rarely cultivated. Juvenile trees have leaves which are shallowly 3-lobed and with forwardly pointing teeth. The remaining species of this series with entire leaves are very poorly known. Rarely cultivated are **A. albo-purpurascens** Hayata from Taiwan and **A. coriacaeifolium** Léveillé (*A. cinnamomifolium* Hayata) from S China.

Section **Platanoidea** Pax. Trees and shrubs, monoecious, deciduous. Buds rather large, scales 5–8-paired. Leaves with 3–5 lobes, sometimes 7, rarely undivided, margins entire, remotely toothed, leaf-stalks containing latex. Inflorescences corymbose, terminal and axillary, with 7–50 flowers. Flowers 5-parted. Disc around the bases of the filaments. Stamens 8. Nutlets flat. Parthenocarpic tendency moderate.

46. A. campestre Linnaeus. Figure 132 (11), p. 545. Illustration: *Mededelingen Landbouwhogeschool Wageningen* **76**(2): f. 14 (1976).

Tree to 25 m in the wild and to 12 m in cultivation. Bark grey, slightly fissured. Branches brown-grey, sometimes corky-ridged or downy. Buds small, greenish. Leaves to 8 × 10 cm, 3–5-lobed, base cordate, dull green on upper surface, downy beneath; lobes obtuse, entire or lobed. Inflorescences corymbose, with 9–20 flowers. Flowers yellow-green; sepals and petals slightly hairy; ovary hairy. Fruits to 3 cm, nutlets flat, hairy or hairless, wings held horizontally. *Europe, W Asia.* H1. Spring.

Common in cultivation. A variable species that has given rise to numerous and often rather obscure variants. Distinguishable subspecies, such as subsp. *leiocarpon* Wallroth with hairless nutlets and subsp. *marsicum* (Gussone) Hayek with nearly entire lobes have a poor geographic basis and need further study. Among the cultivars are: 'Nanum', a dwarf bush, 'Postelense', young leaves yellow, 'Schwerinii', leaves purplish, 'Royal Ruby' and 'Red Shine', bronze-red fruits, 'Elsrijk' and 'Evelyn', street trees.

47. A. cappadocicum Gleditsch subsp. **cappadocicum** (*A. colchicum* Gordon; *A. laetum* Meyer). Figure 132(12), p. 545. Illustration: Dippel, Handbuch der Laubholzkunde **2**: 456 (1892); Flora Reipublicae Popularis Sinicae **46**: f. 19 (1981).

Trees to 25 m in the wild and to 15 m in cultivation, suckering. Bark grey. Branches green-red and finely white-striped, often somewhat glaucous when young. Leaves to 10 × 15 cm, 5-sometimes 7-lobed, base cordate, lobes triangular-acuminate, bright green on upper surface, net-veined and vein-axils hairy beneath, margins entire; leaf-stalks long. Inflorescences corymbose with 15–25 flowers. Flowers green-yellow, sepals and petals slightly hairy. Fruits to 5 cm, nutlets flat, hairless, wings held at an obtuse angle. *N Turkey, Caucasus, N Iran, W Himalaya*. H1. Spring.

Rather common in cultivation and with autumn leaves deep yellow. A distinct character is the occasional production of suckers. A well-known selection is 'Aureum' with yellow leaves. The leaves of 'Rubrum' are reddish when young. The species is now comprised of a number of geographic elements, formerly recognised as separate species.

Subsp. **divergens** (Pax) Murray (*A. divergens* Pax; *A. quinquelobum* Koch, not Saporta). Illustration: *Acta Instituti Botanici Academiae Scientarum URSS*, ser. 1, **1**: f. 13 (1933). Shrub or small tree to 8 m. Bark grey. Branches brown-grey. Young shoots at first as described for the species, but only for 1 or 2 years. Leaves to 6 × 8 cm, 3–5-lobed, papery, base truncate, lobes ovate, acuminate, shiny green on upper surface, paler beneath, margins entire. *N Turkey, W Caucasus*.

Rather rare in cultivation. Leaves very variable in size and often very small.

Subsp. **lobelii** (Tenore) Murray (*A. lobelii* Tenore). Illustration: Brickell (ed.), RHS A–Z encyclopedia of garden plants, 64 (2003). Trees to 20 m. Young shoots mostly with blue-white bloom. Leaves to 15 × 15 cm, 5-lobed, lobes ovate, acuminate, slightly waved, margins entire. Fruits held nearly horizontally. *C & S Italy*. H2. Rather rare in cultivation and nearly always with an upright habit. This subspecies has white-striped branches similar to those of subsp. *cappadocicum*.

Subsp. **sinicum** (Rehder) Handel-Mazzetti. Illustration: Flora Reipublicae Popularis Sinicae **46**: f. 20 (1981). Trees to 20 m in the wild, to 9 m in cultivation. Young branches reddish brown. Leaves to 7 × 10 cm, 3–5-lobed, reddish when young. Fruits to 3.5 cm, dark red, wings held at an obtuse angle. *C & SW China*. H2. Rather rare in cultivation. Characterised by the pendent habit of the branches, the red summer leaves and the purple-red fruits. Bark and branches as in subsp. *cappodocicum*. The 3-lobed var. *tricaudatum* (Veitch) Rehder belongs to subsp. *sinicum*.

A. × zoesckease Pax (*A. neglectum* Lange, not Walpers); *A. campestre* × *A. cappadocicum*. Illustration: Dippel, Handbuch der Laubholzkunde **2**: 452 (1982). Shoots slightly white-striped. Leaves 3–5-lobed, usually dark green, margins of lobes entire to slightly lobed. Flowers and fruits similar to those of *A. campestre*. Rather common in cultivation, notably the cultivars 'Annae' and 'Elongatum'.

48. A. longipes Rehder (*A. fulvescens* Rehder). Illustration: Flora Reipublicae Popularis Sinicae **46**: 21–23 (1981); *The Plantsman* **5**(1): 48 (1983).

Tree to 20 m in the wild and to 10 m in cultivation. Bark grey, slightly fissured. Young branches reddish green. Leaves to 10 × 15 cm, usually 3- sometimes 5-lobed or undivided, lobes long-acuminate, often yellow-grey hairy beneath, margins entire. Inflorescences corymbose, loose, with 25–40 flowers. Flowers yellow-green. Fruits to 6 cm, nutlets flat, wings held at a right or obtuse angle. *C & SW China*. H2–3. Spring.

Rare in cultivation. A poorly known species related to *A. platanoides*.

Subsp. **amplum** (Rehder) de Jong (*A. amplum* Rehder). Branches purplish green, sometimes glaucous-white. Leaves to 20 × 18 cm, usually 5-lobed, sometimes 3-lobed or undivided, papery, lobes oblong-ovate, abruptly acuminate, often wavy, central lobe much longer than lateral lobes, glossy dark green on upper surface, paler and net-veined beneath. *C China*. Rare in cultivation.

A. tenellum Pax. Illustration: Flora Reipublicae Popularis Sinicae **46**: f. 18 (1981). Small tree. Bark grey. Leaves small, 3-lobed, sometimes undivided, triangular, base truncate or cordate, margins entire. *C China*. H2. Spring. Rarely cultivated and probably a small-leaved variant of *A. longipes*.

49. A. miyabei Maximowicz. Figure 132 (13), p. 545. Illustration: Kurata, Illustrated important forest trees of Japan **2**: pl. 51 (1968); Satake, Wild flowers of Japan, woody plants **2**: pl. 22 (1989); The new RHS dictionary of gardening **1**: 28 (1992).

Trees to 25 m in the wild, to 8 m in cultivation. Bark yellowish light grey, peeling in small flakes. Branches grey, somewhat corky and fissured, hairy when young. Leaves to 14 × 12 cm, 5-lobed, basal lobes small, base cordate, central lobes ovate, lobed with mostly 2 pairs of blunt teeth, apex pointed, dull green on upper surface, hairy beneath, leaf-Stalks hairy. Inflorescences corymbose, partly from leafless axillary buds, with 7–20 flowers. Flowers yellow, flower-stalks and ovary hairy. Fruits to 3 cm, nutlets flat, hairy, wings held horizontally. *Japan (Hokkaido, Honshu)*. H1. Spring.

Rather rare in cultivation and strongly resembling *A. campestre*. *A. miyabei* mostly has larger leaves with narrow pointed lobes and thicker, fissured branches.

A. × hillieri Lancaster. *A. miyabei* × *A. cappadocicum* 'Aureum'. Intermediate between parents. The seedling of the original cross has the cultivar name 'West Hill'; another seedling was named 'Summergold'. Rare in cultivation.

50. A. pictum Murray (*A. mono* Maximowicz). Figure 132(14), p. 545. Illustration: *Acta Instituti Botanici Academiae Scientarum URSS*, ser. 1, **1**: f. 8 (1933); Flora Reipublicae Popularis Sinicae **46**: f. 19 (1981).

Tree to 25 m in the wild and to 10 m in cultivation. Bark grey, somewhat fissured. Branches often with warty lenticels. Shoots yellowish brown, hairless or hairy when young. Buds dark purple or red. Leaves to 10 × 12 cm, 5–7-lobed, base cordate, lobes ovate-triangular to broad and obtuse, margins entire or sometimes with large teeth. Inflorescences corymbose, with

10–60 flowers. Flowers green-yellow, perianth hairless. Fruits to 4 cm, nutlets flat, wings held at various angles. *NE Asia to Japan & C China.* H1. Spring.

A very variable species and still poorly known. Rather common in cultivation, but without data on its area of origin it is difficult to identify to one of the several published varieties. *A. pictum* has light, often warty branches whereas *A. cappadocicum* has white-striped branches, *A. longipes* has brown branches and *A. truncatum* has furrowed branches.

Var. **ambiguum** (Pax) Dippel. Illustration: *Bulletin Tokyo University Forest* **60**: f. 7 (1965); Kurata, Illustrated important forest trees of Japan **2**: pl. 52 (1968); Satake, Wild flowers of Japan, woody plants **2**: pl. 23 (1989). Branches with warty lenticels. Young shoots yellowish brown, hairless. Leaves 5–7-lobed, circular to semicircular, base rounded to cordate, covered with short yellowish hairs beneath. *Japan, S Korea.* H1. Rather rare in cultivation.

Var. **mayrii** (Schwerin) Nakai (*A. mayrii* Schwerin). Illustration: *Bulletin Tokyo University Forest* **60**: f. 5 (1965); Kurata, Illustrated important forest trees of Japan **1**: pl. 82 (1971); Satake, Wild flowers of Japan, woody plants **2**: pl. 23 (1989). Young twigs purplish red, hairless and glaucous. Leaves mostly semicircular, 5-lobed, bright green on upper surface, hairless beneath, leaf-stalks red. Fruits large, light coloured. *Japan (Hokkaido, Honshu).* H1. Rather rare in cultivation. Shoots, leaves and fruits give this variety a bright appearance.

Var. **trichobasis** Nakai (*A. pictum* var. *savatieri* Pax, in part). Illustration: *Bulletin Tokyo University Forest* **60**: f. 6 (1965). Branches light grey. Young shoots light brown, hairless. Leaves 7–9-lobed, occasionally 5-lobed, circular, base cordate, hairless except axils of basal veins beneath. *Japan (C Honshu).* According to Ogata, Pax's var. *savatieri* was based on a mixture of this variety and for the major part on var. *glabrum* (Leveille & Vaniot) Kara. Rare in cultivation.

Subsp. **okomotoanum** (Nakai) de Jong (*A. okomotoanum* Nakai). Leaves to 14 × 12 cm, 7–9-lobed. Fruits to 6.5 cm, wings held at an acute angle. *Korea (Ullung Do).* H4. Recently introduced and still rare in cultivation.

51. A. platanoides Linnaeus. Figure 133 (1), p. 551. Illustration: *Acta Instituti*

Botanici Academiae Scientarum URSS, ser. 1, **1**: f. 12 (1933); *Mededelingen Landbouwhogeschool Wageningen* **76**(2): f. 11–13 (1976).

Tree to 30 m. Bark dark grey, slightly fissured. Branches green to brown, hairless. Leaves to 20 × 25 cm, 5-lobed, base cordate, lobes tailed, remotely toothed, dark green on upper surface, shining and with hairy vein-axils beneath. Inflorescences corymbose, with 20–60 flowers. Flowers greenish yellow, sometimes yellow. Fruits to 5 cm, nutlets green, flat, wings held horizontally. *N Europe to Turkey & Caucasus.* H1. Spring.

Very common in cultivation and often used for public planting. Several selections were made with typical shapes (e.g. 'Globosum', 'Columnare'), red leaves (e.g. 'Crimson King', 'Deborah', 'Faassen's Black', 'Goldsworth Purple', 'Schwedleri'), variegated leaves, (e.g. 'Drummondii'), dissected leaves (e.g. 'Palmatifidum'), lane trees (e.g. 'Cleveland', 'Emerald Queen').

Subsp. **turkestanicum** (Pax) de Jong (*A. turkestanicum* Pax; *A. cappadocicum* subsp. *turkestanicum* (Pax) Murray). Illustration: *Acta Instituti Botanici Academiae Scientarum URSS*, ser. 1, **1**: f. 10 (1933). Small tree to 8 m. Branches brown. Leaves to 10 × 12 cm, 5–7-lobed, base cordate to truncate, lobes ovate, acuminate, entire, the central lobes sometimes with 1 pair of teeth. *Turkestan, NE Afghanistan.* H2. Spring. Rare in cultivation. Reproductive organs as in the species, the leaves more like *A. cappadocicum*.

'Integrilobum' (*A. × dieckii* (Pax) Pax). Shoots with relatively short internodes. Leaves 5–7-lobed, margins entire. This cultivar was thought to be a hybrid of *A. cappadocicum* subsp. *lobelii* and *A. platanoides*, but it is in fact no more than an abnormal leaved *A. platanoides*. Rarely cultivated.

52. A. truncatum Bunge. Figure 133(2), p. 551. Illustration: Flora Reipublicae Popularis Sinicae **46**: f. 18 (1981).

Tree or shrub to 10 m. Bark deeply furrowed. Branches purplish or greyish brown, hairless, green when young. Leaves to 10 × 12 cm, 5- sometimes 7-lobed, base truncate, sometimes almost cordate, lobes triangular-ovate, acuminate, dark green on upper surface, paler and net-veined beneath, margins entire. Inflorescences corymbose, with 15–25 flowers. Flowers yellowish green. Fruits to 3.5 cm, nutlets

large, somewhat spherical, wings as large as nutlet and held at an obtuse angle. *N China.* H1. Spring.

Rather rare in cultivation, but grown as a street tree in Beijing (China). This species is distinguished from *A. pictum* by the deeply fissured bark, the thick nutlets and oily seeds.

Section **Trifoliata** Pax. Trees or shrubs, monoecious or dioecious, deciduous. Buds large, scales 11–15-paired, overlapping. Leaves with 3 leaflets, with coarse, forwardly pointing teeth or entire. Inflorescences 3-flowered, sometimes to 25 flowers. Flowers 5-parted. Disc outside the bases of the filaments, more rarely among them. Stamens 10–12, sometimes to 14. Nutlets large, strongly convex, with thick woody walls. Parthenocarpic tendency strong.

53. A. griseum (Franchet) Pax. Figure 133(3), p. 551. Illustration: *Mededelingen Landbouwhogeschool Wageningen* **76**(2): f. 24 (1976); Vertrees, Japanese maples, 152 (1978); Flora Reipublicae Popularis Sinicae **46**: f. 78 (1981); The new RHS dictionary of gardening **1**: 28 (1992).

Tree to 12 m, monoecious, but often functionally dioecious. Bark red-brown, hairless. Branches red-brown, peeling in thin flakes, young shoots hairy. Buds overlapping, scales 11–15-paired. Leaves with 3 leaflets, lateral leaflets smaller, all leaflets elliptic to oblong-ovate, margins with 3–5 pairs of coarse teeth, dark green on upper surface, blue-green beneath, veins and leaf-stalks hairy. Inflorescences 1–4, mostly 3-flowered. Flowers 5-parted, yellow. Fruits to 5 cm, nutlets downy, pericarp thick, wings held nearly at a right angle. *C China.* H2. Spring.

This maple is easily distinguished by its beautiful cinnamon-brown bark. The leaves may develop an orange-red autumn colour. Becoming more common in cultivation.

54. A. mandschuricum Maximowicz. Figure 133(4), p. 551. Illustration: *Acta Instituti Botanici Academiae Scientarum URSS*, ser. 1, **1**: f. 34 (1933); Flora Reipublicae Popularis Sinicae **46**: f. 79 (1981); *The Plantsman* **5**(1): 46 (1983); The new RHS dictionary of gardening **1**: 28 (1992).

Tree or shrub to 10 m, dioecious. Branches grey-brown, hairless. Buds dark, overlapping, scales 11–15-paired. Leaves with 3 leaflets, lateral leaflets smaller and

Figure 133. Leaves and fruits (top left) of Acer species. 1, *A. platanoides*. 2, *A. truncatum*. 3, *A. griseum*. 4, *A. mandschuricum*. 5, *A. triflorum*. 6, *A. diabolicum*. 7, *A. tataricum*. 8, *A. pycnanthum*. 9, *A. sterculiaceum*. 10, *A. rubrum*. 11, *A. saccharinum*.

almost stalkless, leaflets oblong to oblong-lanceolate, acuminate, dark green on upper surface, glaucous beneath, midribs sparsely hairy, margins with forwardly pointing teeth, leaf-stalk often longer than the largest leaflet. Inflorescences terminal and axillary, mostly 3-flowered, flower-stalks hairless. Flowers 5- or 6-parted, yellow-green. Fruits to 3.5 cm; nutlets hairless, wings at an acute or nearly right angle. *China (Manchuria), Korea*. H1. Spring.

Rather rare in cultivation and easy to distinguish from the other species of Section *Trifoliata* by its hairless leaves and shoots and smooth bark. The nutlets have two cavities, one with the embryo, the second empty.

55. A. maximowiczianum Miquel (*A. nikoense* Maximowicz). Illustration: Kurata, Illustrated important forest trees of Japan **2**: pl. 53 (1968); Satake, Wild flowers of Japan, woody plants **2**: pl. 26 (1989).

Tree to 25 m in the wild, monoecious, but often functionally dioecious. Bark dark grey-brown, smooth. Young shoots reddish brown, hairy. Buds large; highest pairs of axillary buds as large as terminal bud and often with secondary buds forming a cluster with 5–9 buds; bud-scales 11–15-paired. Leaves with 3 leaflets, leaflets to 15 × 6 cm, ovate to elliptic-oblong, the terminal leaflet almost stalkless, the laterals stalkless, dark green on upper surface, glaucous and hairy beneath, margins entire or sparsely toothed; stalks 2–5 cm, hairy. Inflorescences terminal and lateral, 3–5-flowered, but mostly 3; flower-stalks hairy. Flowers 5- or 6-parted, yellow. Stamens 8–12. Fruits to 5 cm, nutlets spherical, hairy, very woody, wings held at a right angle. *Japan, C China*. H1. Spring.

Rather rare in cultivation. Easy to distinguish from other species of Section *Trifoliata* by the smooth bark and large hairy leaves with almost entire margins. The buds accumulate at the ends of vigorous shoots. This maple may develop bright red and orange autumn colours.

56. A. triflorum Komarov. Figure 133(5), p. 551. Illustration: Brickell (ed.), RHS A–Z encyclopedia of garden plants, 70 (2003).

Tree or shrub to 12 m, dioecious. Bark light grey, peeling. Shoots warty and soon hairless. Bud-scales 11–15-paired. Leaves with 3 leaflets, leaflets to 9 × 3.5 cm, ovate to oblong-lanceolate, entire to with coarse, forwardly pointing teeth, deep green and

sparingly hairy on upper surface, paler green and with densely hairy veins beneath. Inflorescences mostly 3-flowered; flower-stalks sparingly hairy. Flowers mostly 5-parted, yellow. Fruits to 5 cm, nutlets almost spherical, densely hairy, wings spreading at a right angle. *Korea, N China*. H1.

This maple resembles *A. maximowiczianum*, but has a light grey, peeling bark and smaller, less hairy leaves. It may develop bright red and orange autumn colours.

Section **Lithocarpa** Pax. Trees, monoecious or dioecious, deciduous. Buds large, 5–8- or 8–12-paired. Leaves large, 3–5-lobed, margins lobed to entire, sometimes with forwardly pointing teeth; stalks sometimes containing latex. Inflorescences racemose or corymbose with long axis. Flowers large, 5- or 6-parted. Parthenocarpic tendency strong.

Series **Macrophylla** Pojarkova. Monoecious. Bud-scales 5–8-paired. Leaves deeply 5-lobed, margins lobed to entire; stalks containing latex. Inflorescences corymbose, axis long, drooping, terminal and axillary, with 30–80 flowers. Flowers 5- or 6-parted. Disc around the bases of the filaments. Stamens 8–12. Nutlet keeled-convex with stiff hairs.

57. A. macrophyllum Pursh. Illustration: *Mededelingen Landbouwhogeschool Wageningen* **76**(2): f. 29–30 (1976); The new RHS dictionary of gardening **1**: 28 (1992); Brickell (ed.), RHS A–Z encyclopedia of garden plants, 65 (2003).

Tree to 30 m. Bark rough, fissured. Branches hairless, green or reddish brown when young. Bud-scales 5–8-paired. Leaves to 30 cm wide, deeply 5-lobed, dark shining green on upper surface, paler and hairy beneath, lobes lobed to entire; stalks containing latex. Inflorescences long, pendent panicles. Flowers 5- or 6-parted. Disc around the bases of the filaments. Stamens 10–12. Fruits to 6 cm, nutlets keeled, convex with stiff hairs, wings held at a right angle. *W North America*. H2. Spring.

Rather rare in cultivation. It is easily recognised by the large leaves and the leaf-stalks containing latex.

Series **Lithocarpa**. Trees, dioecious, deciduous. Buds large, scales 8–12-paired, overlapping. Leaves large, 3–5-lobed, sometimes almost entire, margins remotely toothed or with forwardly pointing teeth,

sometimes entire; stalks sometimes containing latex. Inflorescences racemose, sometimes partly corymbose, axillary from leafless buds, with 5–11 flowers.

58. A. diabolicum Koch (*A. purpurascens* Franchet & Savatier). Figure 133(6), p. 551. Illustration: Kurata, Illustrated important forest trees of Japan **2**: pl. 49 (1968); Satake, Wild flowers of Japan, woody plants **2**: pl. 24 (1989); *Memoirs New York Botanical Garden* **54**: f. 22 (1990).

Tree to 15 m. Bark grey, smooth. Shoots yellowish brown, hairy when young. Buds large, scales 8–12-paired. Leaves 5-lobed, densely hairy on both sides when young, lobes broadly ovate and shortly acuminate, margins with remote, forwardly pointing teeth. Flowers 5-parted, yellow or purplish red, perianth of male flowers united. Disc around the bases of the filaments. Stamens mostly 8 and lacking in female flowers. Fruits to 4 cm, nutlets with stiff hairs and thick woody pericarp, wings parallel. *Japan*. H2. Spring.

Rare in cultivation. It differs from the related *A. sterculiaceum* in the toothing of the leaves. Most useful, however, for identification are the reproductive organs. The male flowers are often bright purple-red. The synonym *A. purpurascens* was based on a male tree with such flowers and purple young leaves.

59. A. sinopurpurascens Cheng (*A. diabolicum* subsp. *sinopurpurascens* (Cheng) Murray).

Small tree to 10 m, dioecious. Bark grey, nearly smooth. Shoots grey and fissured, greenish and slightly hairy when young. Leaves 5-lobed, basal lobes small, base cordate, lobes narrowly pointed, central lobe lobed; stalks containing latex. Flowers purplish, ovary hairy. Fruits to 4 cm, nutlets hairy, wings held at a right angle. *E China*. H2. Spring.

Rarely cultivated.

60. A. sterculiaceum Wallich subsp. **sterculiaceum** (*A. villosum* Wallich, not Presl & Presl; *A. schoenermarkiae* Pax). Figure 133(9), p. 551. Illustration: Flora Reipublicae Popularis Sinicae **46**: f. 76 (1981); The new RHS dictionary of gardening **1**: 28 (1992).

Trees to 15 m in the wild and to 8 m in cultivation. Bark dark brown. Branches stout, reddish or dark brown, with densely felted hairs. Buds ovate, large. Leaves to 16 × 20 cm, 3–5-lobed, papery, heart-shaped, deep green on upper surface, hairy and grey-green beneath, margins with coarse, forwardly pointing teeth. Flowers yellowish. Fruits to 7 cm, nutlets strongly veined, wings held at a right angle. *Himalaya*. H3. Spring.

Rather rarely cultivated.

Subsp. **franchetii** (Pax) Murray (*A. franchetii* Pax). Illustration: *Flora Reipublicae Popularis Sinicae* **46**: f. 75 (1981); *The Plantsman* **5**(1): 57 (1983). Branches with densely felted hairs or hairless. Leaves broadly 3-lobed, occasionally 5-lobed and with 2 small basal lobes. Fruits to 5 cm. *C & SW China*. H2–3. Spring. Rarely cultivated. Leaves smaller and less hairy than those of subsp. *sterculiaceum*. The plants from the most southern origins are intermediate in form.

Subsp. **thomsonii** (Miquel) Murray (*A. thomsonii* Miquel). Leaves to 25 × 30 cm, 3–5-lobed, leathery, lobes rather short, margins with remotely, forwardly pointing teeth. Fruits to 12 cm, wings held at an acute angle. *E Himalaya*. H4. Rarely cultivated.

Section **Pubescentia** (Pojarkova) Ogata. Small trees or shrubs, monoecious, deciduous. Bud-scales 6–10-paired. Leaves 3-lobed, margins with coarse, forwardly pointing teeth. Inflorescences terminal and axillary, the latter partly from leafless buds. Flowers 5-parted. Disc lobed, around the bases of the filaments. Stamens 5. Nutlets rather flat, one side convex, the other somewhat concave. Parthenocarpic tendency moderate.

61. A. pentapomicum Brandis (*A. pubescens* Franchet; *A. regelii* Pax). Illustration: *Acta Instituti Botanici Academiae Scientaram URSS*, ser. 1, **1**: f. 15 (1933).

Tree or shrub to 9 m. Bark grey, smooth. Branches brown-grey. Buds brownish black. Leaves to 10 × 15 cm, but mostly rather small, 3-lobed, leathery, lobes deeply incised, slightly tailed to truncate, bases almost cordate, margins with coarse, forwardly pointing teeth, grey-green beneath; stalks 3–8 cm. Inflorescences corymbose, developed from terminal, and axillary buds; axillary buds sometimes leafless. Flowers yellowish; stamens 5. Fruits to 2.5 cm, nutlets flat and with a hole at one side, wings held at an acute angle. *NW Himalaya, Turkestan*. H1. Spring.

Very rare in cultivation and when without flowers or fruits hard to distinguish from certain variants of *A. opalus* or *A. hyrcanum*.

Section **Ginnala** Nakai. Shrubs or small trees, monoecious, deciduous. Bud-scales 5–10-paired, overlapping. Leaves simple or 3-lobed. Inflorescences corymbose, terminal and axillary, erect, with 30–90 flowers. Flowers 5-parted, perianth greenish white, somewhat rolled inwards during flowering. Disc outside the filaments. Stamens 8. Nutlets rather flat, elliptic, strongly veined. Parthenocarpic tendency moderate.

62. A. tataricum Linnaeus. Figure 133 (7), p. 551. Illustration: *Acta Instituti Botanici Academiae Scientarum URSS*, ser. 1, **1**: f. 23 (1933); The new RHS dictionary of gardening **1**: 36 (1992).

Shrub or small tree to 10 m. Bark reddish brown to grey-brown, smooth. Buds rather small. Leaves simple or slightly 3-lobed, 5–10 cm, broadly ovate, acuminate, base rounded or almost cordate, margins with irregular, double, forwardly pointing teeth, bright green, veins hairy beneath. Fruits with red and mostly almost parallel wings; nutlets elliptic and strongly veined. *SE Europe to Iraq*. H1. Late spring.

A variable species. The description above is of subsp. **tataricum** in which the trees normally have entire leaves, but young plants mostly develop 3-lobed leaves and some variants have slightly 3-lobed leaves on adult plants. The autumn colour is yellow.

Subsp. **aidzuense** (Franchet) de Jong (*A. aidzuenze* (Franchet) Nakai; *A. ginnala* var. *aidzuense* (Frinchet) Ogata; *A. subintegrum* Fojarkova). Illustration: Kurata, Illustrated important forest trees of Japan **2**: pl. 45 (1968); Satake, Wild flowers of Japan, woody plants **2**: pl. 8, 9 (1989). Shrub to 7 m. The young shoots are rather thick. The leaves mostly have small, slightly acuminate lobes and turn yellow in autumn. The young fruits have very bright red wings. *Japan*. H1. This subspecies is somewhat intermediate between subsp. *tataricum* and subsp. *ginnala*.

Subsp. **ginnala** (Maximowicz) Wesmael (*A. ginnala* Maximowicz). Illustration: *Acta Instituti Botanici Academiae Scientarum URSS*, ser. 1, **1**: f. 21 (1933); *Flora Reipublicae Popularis Sinicae* **46**: f. 36 (1981); Brickell (ed.), RHS A–Z encyclopedia of garden plants, 70 (2003). A large shrub to 6 m, usually as tall as wide. The shoots are slightly slender and spreading. The leaves are mostly 3-lobed

with the central lobe markedly longer than outer lobes and often also 3-lobed. This subspecies mostly has a more spreading habit and often displays brilliant orange and red colours in autumn. Most common in cultivation. *NE Asia*. H1.

Subsp. **semenovii** (Regel & Herder) Murray (*A. semenovii* Regel & Herder). Illustration: *Flora Reipublicae Popularis Sinicae* **46**: f. 37 (1981). A densely branched shrub to 4 m. Shoots thin, grey-brown. Leaves to 4 cm, 3–5-lobed, autumn colour yellow. *Former USSR (Tienshan, Turkestan), N Afghanistan*. H1. This is the most distinct subspecies and rather rare in cultivation. The various subspecies easily hybridise, leading to intermediates.

Section **Rubra** Pax. Trees, monoecious or dioecious, deciduous; axillary buds forming clusters on the shoots. Bud-scales 4–7-paired, red. Inflorescences in clustered umbels, mostly 5-flowered, axillary, from leafless buds. Flowers 5-parted, perianth red or green-red. Petals and disc sometimes lacking and sepals united. Disc within the bases of the filaments. Stamens mostly 5. The flower-stalks of female flowers show enormous and rapid growth after pollination and reach 3–10 cm. Nutlets small, somewhat convex or large, obovoid, ripening during early summer. Parthenocarpic tendency low. Seeds germinate immediately after dropping.

63. A. pycnanthum Koch. Figure 133(8), p. 551. Illustration: Kurata, Illustrated important forest trees of Japan **4**: pl. 24 (1973); Satake, Wild flowers of Japan, woody plants **2**: pl. 7 (1989).

Tree to 25 m, dioecious. Branches reddish brown to grey brown. Leaves shallowly 3-lobed, almost circular to ovate, to 8 × 7 cm across, hairless, shining green on upper surface, glaucous beneath, margins with remote, forwardly pointing teeth, leaf-stalks to 6 cm. Inflorescences clustered, appearing before the leaves from leafless axillary buds. Flowers red or reddish. Fruits hairless, to 3 cm. *Japan (C Honshu)*. H2. Early spring.

A. pycnanthum is the Japanese counterpart of *A. rubrum* from N America. It is a rare species, both in the wild and in cultivation. The native stands are restricted to a small area west of Tokyo. It was only recently introduced into cultivation. Both species show a strong mutual resemblance. *A. pycnanthum* differs in the smaller and shallowly 3-lobed, sometimes entire leaves.

Young trees do not flower before they are 15 years old, while those of *A. rubrum* may produce flowers when 5–7 years old. As for *A. saccharinum* the inflorescences mainly appear in the upper part of the crown. The autumn colour of the leaves varies from yellow to orange-red.

64. A. rubrum Linnaeus. Figure 133(10), p. 551. Illustration: The new RHS dictionary of gardening **1**: 36 (1992).

Trees to 40 m in the wild, dioecious, or rarely monoecious. Bark smooth, grey. Branches brownish red to grey, red and hairless when young. Buds red or reddish and clustered at the end of the shoots. Leaves 3–5-lobed, to 10 cm, almost heart-shaped, lobes triangular-ovate, rather abruptly acuminate, dark green on upper surface, glaucous beneath, margins irregularly scalloped, with forwardly pointing teeth. Inflorescences bundled, mostly 5-flowered, axillary and often in clusters. Flowers red, sometimes green-red. Disc within the bases of the filaments. Stamens mostly 5. Fruits hairless on slender stalks, *c.* 2.5 cm, bright red when young, ripening at the end of May. *E & C North America.* H1. Early spring.

A. rubrum, the red maple, is rather variable in leaf-shape and hairiness of the leaf undersides. It shows brilliant red autumn colours. It may be confused with *A. saccharinum*. The leaves of *A. rubrum* are less deeply divided and the inflorescences are produced all over the crown and not just in the upper part as for *A. saccharinum*.

Var. **drummondii** (Nuttall) Sargent. Branches red, densely white felted hairy when young. Leaves densely white felted-hairy beneath. *SE USA.* H2. Early spring.

Some other varieties are often of garden origin and are replaced in cultivation by cultivars that were selected because of the shape of the crown (e.g. 'Bowhall', 'Gerling', 'Scanlon', 'Tilford'), or autumn colour (e.g. 'October Glory', 'Red Sunset', 'Schlesingeri').

65. A. saccharinum Linnaeus (*A. dasycarpum* Ehrhardt). Figure 133(11), p. 551. Illustration: Brickell (ed.), RHS A–Z encyclopedia of garden plants, 69 (2003).

Tree to 40 m, monoecious. Bark grey. Branches hairless, pendent. Leaves deeply 5-lobed, to 15 cm, margins with double, forwardly pointing teeth, light green and hairless on upper surface, glaucous and lightly hairy beneath. Inflorescences bundled in dense clusters in the upper part of the crown, mostly 5-flowered. Flowers

greenish. Petals absent. Sepals united. Disc absent. Stamens mostly 5, those of male flowers projecting, filaments white. Stigmas of female flowers projecting, red or reddish. Fruits mostly with 1 developed nutlet, to 5 cm and ripening early summer. *E North America.* H1. Late winter–early spring.

More vigorous than the other species of Section *Rubra*. It also differs in the pendent branches, especially in the lower part of the crown. The autumn colour is mostly yellow and sometimes orange-red. Common in cultivation and frequently used in public planting. Cultivars include 'Pyramidale' and 'Wieri'.

A. × freemanii Murray; *A. rubrum × A. saccharinum*. A natural hybrid with intermediate morphological characters. The flowers are sterile. Some clonal selections, such as 'Armstrong', 'Elegant' and 'Schwerinii', were formerly placed under one of the parent species.

2. DIPTERONIA Oliver
P.C. de Jong

Trees or shrubs, deciduous, monoecious. Buds naked. Leaves opposite, pinnate; stalks containing latex. Inflorescences terminal, large, paniculate. Flowers unisexual. Sepals and petals 5, white. Disc outside the filaments. Stamens 8; ovary downy. Fruits composed of 2 single-seeded samaras; nutlet flat, encircled by a broad wing.

A genus of 2 species in central and southwest China. Propagation from seed.

Literature: Fang, W.P., A monograph of Chinese Aceraceae, Contributions to the Biological Laboratories of the Science Society China, Nanking Botany Ser. 11, **1**: 14–18 (1939); de Jong, P.C., Sex expression in *Acer L.*, *Mededelingen Landbouwhogeschool Wageningen* 76(2): 97, 98 (1976); Delendick, T.J., A survey of foliar flavonoids in the Aceraceae, *Memoirs of the New York Botanical Garden* **54**: 1–128 (1990).

1. D. sinensis Oliver. Illustration: *Mededelingen Landbouwhogeschool Wageningen* 76(2): f. 33 (1976); *Memoirs New York Botanical Garden* **54**: f. 15 (1990).

Tree or shrub to 16 m in the wild, to 6 m in cultivation. Branches purple or purplish green. Buds naked. Leaves to 40 cm, pinnate, 7–11-paired, lowest pair and terminal leaflet sometimes with 3 leaflets; leaflets to 10 × 4 cm, oblong-ovate

to oblong-lanceolate, with remote, forwardly pointing teeth; stalks containing latex. Panicles large, erect, terminal and axillary, with 50–200 flowers. Fruits to 2.5 cm, hairless. China. H3. Late spring.

Rare in cultivation and often short-lived. It is easily grown from seed. It is easy to distinguish from other genera with pinnate leaves by the opposite leaves, and leaf-stalks which contain latex.

---

172. SAPINDACEAE

Trees, shrubs or woody climbers with tendrils in the inflorescences. Leaves usually alternate and compound, stipulate or exstipulate; ptyxis conduplicate. Inflorescences cymose, racemose or paniculate. Flowers unisexual (often superficially bisexual), actinomorphic or zygomorphic. Calyx, corolla and stamens hypogynous, disc outside stamens, often elaborate. Calyx of 4–5 free sepals often unequal. Corolla of 4–5 free petals or rarely absent. Stamens 5–8. Ovary usually of 3 united carpels; ovules 2 per cell, axile; style 1, short, sometimes lobed above. Fruits various; seeds with arils.

A family of 145 genera and about 1300 species, mainly from the tropics. Only a few genera and species are grown.

Literature: Radlkofer, L., Sapindaceae, in A. Engler, Das Pflanzenreich **98** (1934).

1a. Climbers with tendrils; stipules present, though small   2
  b. Upright trees or shrubs, without tendrils or stipules   4
2a. Herbaceous climber; fruit inflated, membranous   **8. Cardiospermum**
  b. Woody climbers; fruit neither inflated nor membranous   3
3a. Leaves of 3 leaflets which are themselves deeply 3-lobed   **6. Serjania**
  b. Leaves 2–4-pinnate   **7. Paullinia**
4a. Petals absent   **2. Dodonaea**
  b. Petals present   5
5a. Flowers bilaterally symmetric   6
  b. Flowers radially symmetric   7
6a. Flowers yellow, appearing after the leaves   **1. Koelreuteria**
  b. Flowers rose-pink, appearing before the leaves   **5. Ungnadia**
7a. Stamens 5; leaves without a terminal leaflet   **3. Filicium**
  b. Stamens 8; leaves with a terminal leaflet   **4. Xanthoceras**

1. KOELREUTERIA Laxmann
J. Cullen

Deciduous trees with thick, fissured bark. Leaves alternate, without stipules, pinnate or bipinnate, with alternate or opposite ultimate segments. Flowers bilaterally symmetric, in large, loose terminal panicles, unisexual, but male and female in the same inflorescence. Calyx 5-lobed, 3 of the lobes longer than the other 2, margins glandular or hairy. Petals 4 or 5, clawed, the base reflexed, the base of the blade bearing lobed outgrowths. Stamens mostly 8, filaments hairy. Ovary 3-celled, ovules 2 per cell, attached to the outer walls of the ovary (placentation parietal). Fruit an inflated capsule, papery. Seeds without arils.

A genus of 3 species from China and Taiwan to Fiji, of which 2 are grown. They require good soil and a sheltered position. Propagation is by seed or by root-cuttings.

Literature: Meyer, F.G., A revision of the Genus *Koelreuteria* (Sapindaceae), *Journal of the Arnold Arboretum* **57**: 129–166 (1976).

1a. Leaves pinnate, leaflets deeply toothed or irregularly lobed **1. paniculata**
 b. Leaves bipinnate, the leaflets finely and regularly toothed **2. bipinnata**

1. K. paniculata Laxmann (*K. apiculata* Rehder & Wilson). Illustration: *Journal of the Arnold Arboretum* **57**: f. 8, 9 (1976); Phillips, Trees in Britain, Europe and North America, 128 (1978); Kunkel, Flowering trees in subtropical gardens, 269 (1978); Macoboy, What tree is that?, 132 (1979).

Tree to 12 m (usually less in cultivation), with furrowed bark. Leaves pinnate, leaflets 11–18, coarsely toothed or lobed, acuminate, usually with a small point. Flowers numerous, slightly fragrant; petals 4, bright yellow, their appendages yellow at first, later orange-red, claw 1–2.5 mm, blade 4.5–7 mm. Filaments hairy, anthers lavender. Capsules 4–7 cm. *China*. H4. Early–late summer.

2. K. bipinnata Franchet. Illustration: Menninger, Flowering trees of the world, pl. 377 (1962); *Journal of the Arnold Arboretum* **57**: f. 10, 12 (1976).

Similar to *K. paniculate* but leaves bipinnate, leaflets usually 8–10 on each major division, entire or finely and uniformly toothed, petal-claw 2–3 mm, blade 5.5–9.5 mm, anthers grey to black. *China*. H4. Summer.

2. DODONAEA Miller
J. Cullen

Evergreen, dioecious shrubs or small trees. Leaves alternate, simple or pinnate; if pinnate, with or without a terminal leaflet. Flowers inconspicuous, solitary or in axillary or terminal cymes or panicles. Sepals 3–7, falling early. Petals absent. Stamens 6–16 (usually 8), absent or rudimentary in female flowers. Capsule with 2–6 angles or wings, usually membranous. Seeds with or without arils.

A genus of 68 species in the tropics and subtropics, most confined to Australia. They require warm conditions and a well-drained soil; propagation is by seed or by cuttings taken in summer.

Literature: West, J.G., A revision of *Dodonaea* (Sapindaceae) in Australia, *Brunonia* **7**: 1–194 (1984).

1a. Leaves simple or lobed at the apex 2
 b. Leaves pinnate 3
2a. Capsule-wings rounded, their longest axis that of the capsule; leaves not rigid **1. viscosa**
 b. Capsule-wings acute or oblique, their longest axis at right angles to the capsule; leaves rigid **4. stenophylla**
3a. Capsule unwinged; flowers solitary or few in a terminal cyme **5. humilis**
 b. Capsule winged; flowers many, in axillary panicles 4
4a. Leaflets 16–30 **2. multijuga**
 b. Leaflets 3–7 **3. adenophora**

1. D. viscosa Jacquin. Illustration: Graf, Tropica, 876 (1978); Graf, Exotica, edn. 11, 2038 (1982); Brickell, RHS gardeners' encyclopedia of plants & flowers, 119 (1989); Fairley & Moore, Native plants of the Sydney district, 226, 227 (1989).

Spreading shrub or small tree to 8 m. Leaves simple, linear to obovate or spathulate, 1–15 cm × 1–40 mm, entire or finely toothed, apex rarely notched or 3-lobed. Flowers in terminal panicles. Sepals 3 or 4, 1.3–3 mm. Stamens 6–10, usually 8. Ovary hairy or hairless. Capsule with 2–4 conspicuous wings with their longest axis that of the capsule, hairless. *Pantropical and also in the subtropics.* H5–G1.

Very variable in the wild in leaf and capsule shape. Often used as a hedge plant or a sand binder in the tropics and subtropics. 'Purpurea', with reddish purple leaves, is widely grown.

2. D. multijuga D. Don. Illustration: Flora of Australia **25**: 128 (1985); Fairley & Moore, Native plants of the Sydney district, 228 (1989).

Dioecious shrub to 1.5 m. Leaves 2–4.8 cm, pinnate with a terminal leaflet; leaflets 16–30, usually oblong to broadly obovate, entire, margins rolled downwards. Flowers in axillary panicles. Sepals 4 or 5, 3–4 mm. Stamens 8. Capsule 3-winged, sparsely hairy. *Australia (New South Wales).* G1.

3. D. adenophora Miquel. Illustration: Flora of Australia **25**: 129 (1985).

Like *D. multijuga* but a taller shrub (to 2.5m), leaflets 3–7, concave or folded upwards, sepals 4, 1–1.8 mm, stamens 6–8, capsule 4-winged. *Australia (Western Australia).* G1.

4. D. stenophylla Mueller. Illustration: Flora of Australia **25**: 129 (1985).

Dioecious shrub to 4 m. Leaves simple, erect, rigid, linear, acute, 3–11 cm hairless. Flowers few in axillary cymes or rarely in terminal panicles. Sepals 4, 1.2–1.8 mm. Stamens 8. Capsule usually 4-winged, the wings pointed or oblique, their longest axis at right-angles to that of the capsule, hairless. *Australia (Queensland, N New South Wales).* G1.

5. D. humilis Endlicher.

Spreading shrub. Leaves pinnate with a terminal leaflet, leaflets 3–15, obovate, 3–8 × 2–7 mm, hairless or sparsely hairy on the midrib beneath, Flowers solitary or few in terminal cymes. Sepals 4, 1.8–2.8 mm. Stamens 8. Capsule 4-lobed, not winged, glandular-hairy. *Australia (South Australia).* G1.

3. FILICIUM J.D. Hooker
J. Cullen

Small to large trees. Leaves alternate, pinnate, without a terminal leaflet, the lateral leaflets forming opposite pairs, all covered with glandular scales which secrete resin. Flowers small in axillary or terminal panicles. Sepals 5, persistent. Petals 5, not or little larger than the sepals. Stamens 5, inserted in grooves in a conspicuous disc which is covered with white hairs. Fruit an ellipsoid drupe containing 1 or 2 seeds.

A genus of 3 species from the Old World tropics, of which 1 is occasionally grown as a foliage plant in conservatories.

1. F. decipiens (Wight & Arnott) Thwaites. Illustration: Brandis, Indian trees, 188 (1906); McMillan, Tropical planting and gardening, edn. 6, 116 (1991).

Small tree with grey-hairy young shoots. Leaves congested at the tips of the branches, rigid, the stalk and axis winged, leaflets linear-oblong to oblong. Panicles rigid, erect. Drupe ovoid, shining, purple. *From SW Africa to India, Sri Lanka & Fiji.* G2.

4. XANTHOCERAS Bunge
J. Cullen

Deciduous tree or shrub to 8 m. Leaves stalkless, alternate, pinnate, with a terminal leaflet. Flowers appearing with the leaves, in racemes, those of the terminal raceme usually female, those of the lateral racemes usually male, though all appearing to have stamens and ovaries. Sepals 5. Petals 5, much larger than the sepals. Stamens 8, inserted on a disc which is 5-lobed, each lobe with an upright appendage half as long as the stamens. Ovary 3-celled, with a short, thick style. Capsule opening by 3 flaps, each cell containing several seeds.

A genus of a single species from northern China. It is generally hardy, but needs a long growing season to do well. It is propagated by seed or by root-cuttings.

1. X. sorbifolium Bunge. Illustration: *Botanical Magazine*, 6923 (1887); Bean, Trees and shrubs hardy in the British Isles, edn. 8, **4**: 759 (1980); Csapody & Tóth, A colour atlas of flowering trees and shrubs, 183 (1982); Krüssmann, Manual of cultivated broad-leaved trees & shrubs **3**: 489 (1986).

Leaves with 9–17 lanceolate, opposite or alternate leaflets, each sharply toothed with forwardly pointing teeth. Flowers fragrant. Sepals oblong, hairy, yellowish. Petals to 2 cm, white, yellowish or greenish veined at the base at first, the venation becoming intense red with age. *N China.* H2. Late spring.

5. UNGNADIA Endlicher
J. Cullen

Deciduous shrub or small tree, Leaves pinnate, the terminal leaflet stalked, the lateral leaflets stalkless. Flowers in clusters, appearing before the leaves, bilaterally symmetric. Calyx 5-lobed. Petals 4 or 5, with hairy claws and a tuft of hairs at the apex. Disc irregular, tongue-shaped. Stamens 7–10, unequal. Ovary stalked.

Fruit a leathery capsule opening by 3 flaps, each cell 1-seeded.

A single species from the southern USA and adjacent Mexico, very occasionally grown.

1. U. speciosa Endlicher. Illustration: *Flore des serres* **10**: t. 1059 (1854–55); *Gartenflora* **29**: 50 (1880); Martin & Hutchins, A flora of New Mexico **1**: 1219 (1980).

Shrub or tree to 10 m, with hairy shoots. Leaflets 5–9, oblong-ovate to ovate-lanceolate, toothed, hairy when young. Flowers to 2.5 cm across, petals rose-pink. Capsule to 5 cm across, borne on a long stalk. *USA (Texas, New Mexico), N Mexico.* G1. Spring.

6. SERJANIA Miller
J. Cullen

Woody climbers with tendrils in the inflorescences. Leaves pinnate or divided into 3 leaflets which may themselves be deeply divided; stipules small. Flowers small, bilaterally symmetric, in terminal or axillary racemes or panicles. Sepals 4 or 5, 2 of them united. Petals 4, each bearing a scale inside towards the base. Stamens 8. Disc with 4 glands. Ovary 3-celled. Fruit a schizocarp with 3 wings.

A genus of over 200 species from tropical and subtropical America. Only one is occasionally grown as a tropical house climber. Propagation is by seed or cuttings.

1. S. cuspidata Cambessedes. Illustration: *Gartenflora* **63**: 325 (1914).

Large, vigorous climber; stems triangular in section, the angles densely brown-hairy. Leaves of 3 leaflets which are themselves deeply 3-lobed and coarsely toothed. Flowers yellow, fragrant, sepals unequal. *Brazil.* G2.

7. PAULLINIA Linnaeus
J. Cullen

Like *Serjania*, but leaves 2–4-pinnate, stipules very small.

A genus of about 150 species from tropical America, one of which is cultivated for its fern-like foliage. Cultivation as for *Serjania*.

1. P. thalictrifolia Jussieu.

Large climber. Branches terete, velvety. Leaves mostly 3-pinnate, the axes slightly winged, the ultimate segments scalloped or toothed. Racemes usually short. Petals pale pink, each bearing a basal scale which is ciliate on one side. *Brazil.* G2.

8. CARDIOSPERMUM Linnaeus
J. Cullen

Herbaceous climbers. Leaves alternate, twice divided into 3 leaflets, the ultimate segments toothed. Flowers bilaterally symmetric, in axillary panicles which are long-stalked. Sepals 4, 2 larger and 2 smaller. Petals usually 4, obovate. Disc 1-sided, 4-lobed. Stamens 6–8. Ovary 3-celled with 1 ovule per cell. Fruit a membranous, inflated capsule.

A genus of 12 species, mainly from tropical America, 2 of them grown for their flowers and interesting fruits.

1a. Plants with pale brown hairs; flowers 8–12 mm; fruit 4–8 cm
  1. grandiflorum
  b. Plants hairless or with sparse, colourless hairs; flowers to 4 mm; fruit 1–3 cm **2. halicacabum**

l. C. grandiflorum Swartz. Illustration: Flora of Australia **25**: 9 (1985).

Plant covered with brown hairs. Panicle flat-topped, corymb-like. Flowers 8–12 mm, white; petals 9–11 mm, each with a crested scale inside. Fruit 4–8 cm, 6-ribbed, downy. *Tropical America; widely introduced elsewhere in the tropics.* G2.

2. C. halicacabum Linnaeus. Illustration: *Botanical Magazine*, 1049 (1807).

Slender climber, either hairless or with sparse, colourless hairs. Panicle umbel-like, with 3 or 4 branches. Flowers to 4 mm, white; petals to 3 mm, only 2 of them with an internal scale. Fruit 1–3 cm, downy. *Tropics, often introduced as a weed.* G2.

173. HIPPOCASTANACEAE

Woody. Leaves opposite, palmate, exstipulate; ptyxis conduplicate. Inflorescence racemose. Flowers usually bisexual, zygomorphic. Calyx and corolla perigynous, stamens usually hypogynous, disc present. Calyx of 4–5 united sepals. Corolla of 4–5 free petals. Stamens 5–9. Ovary of 3 united carpels; ovules 2 per cell, axile; style 1. Fruit a capsule containing large seeds.

A small family comprising 2 genera, *Aesculus* Linnaeus and *Billia* Peyritsch. *Aesculus* is native to temperate North America, the Sino-Himalayan area and a small area of southeast Europe (S Albania, N Greece). *Billia*, which is not found in cultivation in Europe, contains 2 evergreen species occuring in Central and South America.

Literature: Hardin, J.W., Studies in Hippocastanaceae, *Brittonia* **9**(2): 173–195 (1957); **12**(5): 26–38 (1960).

1. AESCULUS Linnaeus
J.R. Edmondson

Deciduous trees and shrubs with opposite, palmate leaves and prickly or smooth fruit. Leaves with 5–7 (rarely to 9) unequal leaflets borne on a long stalk. Flowers borne in large panicles occurring at the apex of branches formed during the current season's growth. Petals somewhat irregular, white or brightly coloured, free. Ovules 6, producing 1 (or rarely 2) seeds by abortion. Fruit with a fleshy coat containing a shiny brown seed.

A genus of around 23 species; some from North America are prone to hybridisation. They are easily cultivated in a wide range of preferably deep and well-drained soils. The mode of dispersal of the fruits is obscure, but the seeds of some are eaten by deer and bears.

1a. Fruits prickly; petals 4 or 5 2
 b. Fruits not prickly; petals 4 4
2a. Petals 4; panicles cylindric, to 18 cm **7. glabra**
 b. Petals usually 5; panicles conical, to 30 cm 3
3a. Petals white tinged with pink or yellow at base **12. hippocastanum**
 b. Petals pink to deep red **11. × carnea**
4a. Panicles pyramidal 5
 b. Panicles cylindric 8
5a. Shrub, with many stems arising from suckering base; leaflets scalloped **8. parviflora**
 b. Tree, with single erect trunk; leaflets toothed 6
6a. Leaves 35–55 cm (including stalk); fruits pear-shaped **6. turbinata**
 b. Leaves not more than 35 cm (including stalk); fruits more or less spherical 7
7a. Leaflet-stalks, 4–10 mm; hilum of seed occupying ½ of diameter **4. chinensis**
 b. Leaflet-stalks 10–15 mm; hilum of seed occupying ⅓ of diameter **5. wilsonii**
8a. Shrub; flowers red 9
 b. Tree; flowers white, pink or yellow 10
9a. Leaves hairless beneath or sparsely downy only in the vein-axils **10. pavia**

 b. Leaves greyish to brownish with densely felted hairs beneath **9. splendens**
10a. Flowers yellow **1. flava**
 b. Flowers white or pale pink 11
11a. Leaflets 5, 5–10 cm; small tree with fragrant flowers **2. californica**
 b. Leaflets 6 or 7, 20–30 cm; tall tree with odourless flowers **3. indica**

1. A. flava Solander (*A. octandra* Marshall). Illustration: Krüssmann, Manual of cultivated broad-leaved trees & shrubs **1**: pl. 25, f. 83, 85 (1984); Csapody & Tóth, A colour atlas of flowering trees and shrubs, 187 (1982); Brickell, The RHS gardeners' encyclopedia of plants & flowers, 55 (1989); Brickell (ed.), RHS A–Z encyclopedia of garden plants, 84 (2003).

Tree to 30 m, with dark brown bark and non-resinous winter buds. Leaves with 5 or 7 obovate to ovate leaflets, 7–17 × 3–5 cm, evenly toothed, with reddish brown hairs on lower surface. Panicles cylindric, erect, to 16 × 7 cm. Petals 4, yellow; stamens not projecting. Fruits almost spherical, oblique, smooth, 2-seeded. *E USA*. H2. Spring.

A. dallimorei Sealy. Arose as a chimaera from a graft of *A. flava* on a scion of *A. hippocastanum*. Its leaves resemble those of *A. hippocastanum* but bear the rusty red hairs characteristic of *A. flava*; its inflorescence resembles that of *A. hippocastanum* but the flowers have 4 petals in 2 dissimilar pairs, of variable colour ranging from white with red markings to greenish yellow. Potentially a valuable ornamental tree, but only recently taken extensively into cultivation.

A. × hybrida de Candolle; *A. flava* × *A. pavia* (*A. flava* var. *purpurascens* Gray). Resembles *A. flava* but differs in its reddish or purplish petals with glandular margins.

2. A. californica (Spach) Nuttall (*Pavia californica* (Spach) Hartweg). Illustration: *Botanical Magazine*, 5077 (1858); Bean, Trees & shrubs hardy in the British Isles, edn. 8, **1**: 251 (1970); Krüssmann, Manual of cultivated broad-leaved trees & shrubs **1**: pl. 26, f. 83, 85 (1976); Brickell, The RHS gardeners' encyclopedia of plants & flowers, 58 (1989).

Small tree to 13 m. Winter buds resinous. Leaves with 5 oblong to ovate leaflets 5–10 × 2–4 cm. Panicles cylindric, 10–15 × 5–7 cm. Petals 4, white or pale pink; stamens distinctly projecting. Fruits rough, without spines, 5–7.5 cm, with 1

shiny brown seed. *USA (California)*. H2. Summer.

The flowers have a distinctive fragrance.

3. A. indica Colebrooke (*Pavia indica* Cambessedes). Illustration: *Botanical Magazine*, 5117 (1859); Bean, Trees & shrubs hardy in the British Isles, edn. 8, **1**: 259 (1970); Krüssmann, Manual of cultivated broad-leaved trees & shrubs **1**: pl. 26, f. 83 (1976); The new RHS dictionary of gardening **1**: 76 (1992).

Tall tree to 35 m. Winter buds very resinous. Leaves with 6 or 7 irregularly toothed, lanceolate to obovate leaflets 20–30 × 7–10 cm. Panicles cylindric, 30–40 × 12–15 cm. Petals 4, upper pair white, tinged at base with yellow and red, lower pale rose-pink. Fruits rough, without spines, 5–8 cm, with 1 spiny brown seed. *NW Himalaya*. H2. Early summer.

'Sydney Pearce' (illustration: Brickell, The RHS gardeners' encyclopedia of plants & flowers, 52, 1989), with flowers more richly coloured and arranged in a denser spike, received an RHS Award of Merit in 1967.

4. A. chinensis Bunge (*A. sinensis* Anon.). Illustration: Krüssmann, Manual of cultivated broad-leaved trees & shrubs **1**: f. 83 (1976); Brickell, The RHS gardeners' encyclopedia of plants & flowers, 39 (1989); Brickell (ed.), RHS A–Z encyclopedia of garden plants, 84 (2003).

Tree to 30 m. Winter buds resinous. Leaves with 5 or more usually 7 narrowly oblong to obovate leaflets, dark green, rapidly becoming hairless when young. Panicles conical, 20–30 × 5–10 cm at base. Petals 4, white; stamens slightly projecting. Fruits rough, without spines, c. 5 cm, almost spherical; hilum covering about ½ of diameter of seed. *N China*, 1H4. Late spring.

This species was only introduced to cultivation in 1912, and mature specimens are therefore rare. It can in time be expected to grow at least as tall as *A. hippocastanum*.

5. A. wilsonii Rehder. Illustration: Krüssmann, Manual of cultivated broad-leaved trees & shrubs **1**: pl. 26 (1976); Flora Reipublicae Popularis Sinicae **46**: 280 t. 84(1981).

Tree to 25 m. Leaves with 7 oblong to ovate leaflets, the young leaflets downy beneath. Panicles conical, 25–40 × 8–12 cm at base. Petals 4, white. Fruits with a much thinner husk than

A. chinensis, the hilum covering *c.* ¹⁄₃ of diameter of seed. *C China.* H4. Early summer.

Rarely found in cultivation.

6. A. turbinata Blume (*A. dissimilis* Blume). Illustration: *Botanical Magazine*, 8713 (1917); Krüssmann, Manual of cultivated broad-leaved trees & shrubs **1**: pl. 24, (1976).

Tree to 35 m, young parts with long rust-brown hairs. Winter buds very resinous. Leaves with 5–7 very large narrowly obovate-oblong, stalkless leaflets 20–35 × 8–12 cm, margins evenly toothed. Panicles conical, 15–25 × 5–10 cm at base. Petals 4, creamy white to pale pink; stamens slightly projecting. Fruits rough, without spines, 5 × 3 cm, pear-shaped. *Japan.* H2. Early summer.

Much more slow-growing than *A. hippocastanum* and with much less showy inflorescences, this species is notable mainly for the size of its leaves.

7. A. glabra Willdenow (*A. pallida* Willdenow). Illustration: Krüssmann, Manual of cultivated broad-leaved trees & shrubs **1**: pl. 24, 25, f. 83–85 (1976); The new RHS dictionary of gardening **1**: 76 (1992).

Tree to 20 m, with resinous buds and fissured bark. Leaves with 5–7 obovate to ovate leaflets 8–15 × 5–8 cm, margins evenly toothed. Panicles oblong, 10–18 × 5–8 cm. Petals 4, greenish yellow; stamens distinctly projecting. Fruits broadly obovoid, with small prickles. *C & SE USA.* H2. Late spring.

Var. **sargentii** Rehder. Leaves with 6 or 7 longer acuminate leaflets with 2-toothed margins. Var. **leucodermis** Sargent with smooth whitish bark. *USA (Missouri).*

A. × neglecta Lindley. Illustration: Brickell, The RHS gardeners' encyclopedia of plants & flowers, 62, (1989); Brickell (ed.), RHS A–Z encyclopedia of garden plants, 84 (2003). A similar yellowish flowered tree of unknown origin with downy calyces and flower-stalks, may have originated in the SE United States and is perhaps of hybrid origin; the yellow or red-flowered var. *georgiana* (Sargent) Sargent, first described as *A. georgiana* Sargent (*A. silvatica* Bartram), is believed by some to represent one of its parents, *A. flava* being the other.

8. A. parviflora Walter (*A. macrostachya* Michaux; *Pavia macrostachya* (Michaux) Loiseleur). Illustration: Krüssmann, Manual

of cultivated broad-leaved trees & shrubs **1**: pl. 24, (1976); Csapody & Tóth, A colour atlas of flowering trees and shrubs, 183 (1982); Brickell, The RHS gardeners' encyclopedia of plants & flowers, 88 (1989); The new RHS dictionary of gardening **1**: 76 (1992).

Shrub 3–5 m, with spreading suckering growth. Leaves with 5 or 7 leaflets, 7–10 × 3–6 cm, obovate, shallowly scalloped, densely grey-felted beneath. Panicles cylindric, 20–30 × 6–9 cm. Petals 4, white; stamens long-projecting. Fruit almost spherical, smooth. *SE USA.* H4. Summer.

9. A. splendens Sargent.

Shrub 3–4 m; buds not resinous. Leaves with 5 lanceolate leaflets, 10–15 × 5–8 cm, margins evenly toothed, under surface densely greyish or rusty felted. Panicles narrowly pyramidal, 15–25 × 7–9 cm. Petals 4, scarlet; stamens projecting. Fruit obovoid to almost spherical, smooth. *SE USA.* H2. Spring.

10. A. pavia Linnaeus. Illustration: Krüssmann, Manual of cultivated broad-leaved trees & shrubs **1**: pl. 25, (1976); Csapody & Tóth, A colour atlas of flowering trees and shrubs, 187 (1982); The new RHS dictionary of gardening **1**: 76 (1992).

Shrub 3–4 m; buds not resinous. Leaves with 5 lanceolate to narrowly oblong leaflets, margins irregularly toothed, under surface hairless. Panicles oblong, 8–15 cm. Petals 4, reddish, margins glandular, often standards remaining closed at flowering period and hence colour inconspicuous. Fruits almost spherical, smooth. *S USA.* H2. Late spring.

Rare in cultivation except where grafted on to other species, often standards. 'Humilis' (*A. pavia* var. *humilis* Lindley), a prostrate shrub, was sometimes used in this way to produce the 'weeping' form of the tree known as *Pavia pendula* Anon.

11. A. × carnea Heyne. *A. hippocastanum × A. pavia.* Illustration: Krüssmann, Manual of cultivated broad-leaved trees & shrubs **1**: f. 83 (1976); Csapody & Tóth, A colour atlas of flowering trees and shrubs, 186 (1982); Brickell, The RHS gardeners' encyclopedia of plants & flowers, 39 (1989).

Tree 10–25 m; winter buds slightly resinous. Leaves with 5 or 7 irregularly toothed, obovate leaflets, the terminal one

the largest, very shortly stalked. Panicle narrowly conical, 15–20 × 8–10 cm. Petals 4 or 5, deep red; stamens projecting. Fruit spherical, slightly prickly, *c.* 3.5 cm across. *Garden origin.* H2. Spring.

A fertile polyploid hybrid. 'Briottii' has large, more brightly coloured panicles.

A. × plantierensis Andre, a hybrid whose parents are believed to be *A. hippocastanum* and *A. × carnea*, has the stalkless leaflets and conical panicles of the former and the pinkish petals of the latter.

12. A. hippocastanum Linnaeus (*A. septenata* Stokes). Illustration: Krüssmann, Manual of cultivated broad-leaved trees & shrubs **1**: f. 83, 85 (1976); Csapody & Tóth, A colour atlas of flowering trees and shrubs, 185 (1982); Brickell, The RHS gardeners' encyclopedia of plants & flowers, 38 (1989); The new RHS dictionary of gardening **1**: 76 (1992).

Tall tree to 30 m. Winter buds very resinous. Leaves with 5–7 irregularly toothed, obovate, stalkless leaflets, the terminal one largest. Panicles conical, 30 × 10 cm. Petals 4 or more often 5, white tinged with yellow (later red) at base. Fruits spiny, 5–7 cm, with 1 (rarely 2) shiny brown seeds. *Albania, N Greece.* H2. Spring.

Several cultivars are widely grown. 'Pyramidalis' has branches erect-horizontal, growing upwards at an angle of 45°. 'Baumannii' has double flowers and is sterile. 'Laciniata' has up to 9 leaflets reduced to an irregular jagged, linear blade.

174. MELIOSMACEAE

Trees or shrubs. Leaves evergreen or deciduous, simple or compound, alternate and without stipules. Flowers small, in panicles, bisexual, bilaterally symmetric. Sepals 3–5 or calyx with 3–5 lobes, or sepals apparently more than 5 (see below). Petals usually 5, free, usually 3 large and 2 smaller. Stamens 5, on the same radii as the petals, often only 2 of them fertile, these on the same radii as the 2 smaller petals. Disc usually present, lobed. Ovary superior, 2-celled, ovules 2 per cell, placentation axile; style short, stigma simple. Fruit a berry or drupe.

A family of 4 genera, of which only one is grown. They occur in Southeast Asia and tropical America. In edition 1, this family was included in the *Sabiaceae*.

1. MELIOSMA Blume

J. Cullen

Trees or shrubs with open branching and stout twigs. Leaves simple or pinnate. Flowers small, white or cream, fragrant, in large, usually terminal panicles. Sepals 3–5 but sometimes apparently more (to 13), the additional 'sepals' being formed from small, sterile bracts, sometimes the lowermost a little distant from the rest. Petals usually 5, the 3 outer unequal, much larger than the 2 inner which are notched or bifid. Stamens 5, 2 fertile, with strap-shaped filaments broadening above into a wide cup containing the anther which has a broad connective. Fruit a drupe.

This genus has been intensively studied by Beusekom (reference below), who treats it as consisting of about 25 species from South and Central America and eastern Asia. Only Asiatic species are cultivated. The structure of the small flowers is unusual and remarkable. Plants can be grown in full sun in good soil, and can be propagated by seed, layers or cuttings.

Literature: van Beusekom, C.F., Revision of Meliosma (Sabiaceae), section Lorenzanea excepted, living and fossil, geography and phylogeny, *Blumea* **19**: 355–539 (1971).

1a. Leaves pinnate   2
  b. Leaves simple   3
2a. Leaves terminating in 2 or 3 leaflets together; outer petals ovate or circular, not or little wider than long   **4. pinnata**
  b. Leaves terminating in a single leaflet; outer petals kidney-shaped, clearly wider than long   **5. veitchiorum**
3a. Leaf-veins 8–17 pairs, always some of them branching; leaves abruptly acuminate; sepals 5, equal   **1. parviflora**
  b. Leaf-veins 7–30 pairs, mostly unbranched; sepals apparently 5–13 (rarely fewer), usually unequal, the outer smaller and sometimes more or less remote from the others   4
4a. Leaves evergreen   **3. simplicifolia**
  b. Leaves deciduous   **2. dilleniifolia**

1. M. parviflora Lecomte.

Deciduous shrub or small tree to 10 m. Leaves obovate to almost spathulate, abruptly acuminate, irregularly toothed towards the apex, veins 8–17 pairs, several of the veins branching. Panicles terminal, erect, loose or dense, 10–35 cm; flower-stalks absent or to 2 mm. Sepals 5, 0.5–0.7 mm, equal. Outer petals hairless; inner petals 0.5–1 mm, bifid to halfway, with acute, fringed lobes. Fruit spherical, 5–6 mm across. *China (Sichuan, Hubei, Jiangxi, Zhejiang)*. H3. Late spring.

2. M. dilleniifolia (Wight & Arnott) Walpers.

Deciduous shrub or small tree to 15 m. Leaves obovate or elliptic to elliptic-oblong, acute, closely or remotely toothed, veins 8–30 pairs, straight, none or very few of them branched. Panicles terminal, erect or deflexed, 7–50 cm; flower-stalks 4–5 mm. Sepals 5, 0.75–1.75 mm, unequal, the outer 1 or 2 smaller and sometimes distant from the others. Outer petals hairless; inner petals 0.6–1.3 mm, bifid to halfway, with fringed lobes. Fruit spherical, 4–5 mm across. *Himalaya to Japan*. H3. Late spring.

A very variable species with a wide distribution; Beusekom recognises several subspecies, of which 3 are reputedly in cultivation.

Subsp. **cuneifolia** (Franchet) Beusekom (*M. cuneifolia* Franchet). Illustration: *Botanical Magazine*, 8357 (1911); *Gardeners' Chronicle* **86**: 343 (1929); Bean, Trees & shrubs hardy in the British Isles, edn. 8, **2**: pl. 99 (1973); Krüssmann, Manual of cultivated broad-leaved trees & shrubs **2**: 316 (1986). Leaves with conspicuous hairy patches (domatia) in the vein-axils beneath; panicles 10–50 cm, erect, with spreading branches, the main axis more or less straight. *W & C China*.

Subsp. **flexuosa** (Pampanini) Beusekom (*M. flexuosa* Pampanini; *M. pendens* Rehder & Wilson). Leaves with 12–21 pairs of veins; panicles 7–22 cm, erect or deflexed, main axis more or less zig-zag, lateral branches deflexed. *E China*.

Subsp. **tenuis** (Maximowicz) Beusekom (*M. tenuis* Maximowicz). Leaves with 6–14 pairs of veins; panicles erect to deflexed, 7–15 cm, with the main axis zig-zag and the lateral branches deflexed. *Japan*.

Subsp. **dilleniifolia**, from the Himalaya, is apparently not or rarely grown; it is like subsp. *cuneifolia* but the leaves have no domatia beneath.

M. meliantha Krüssmann is a name applied to material in cultivation which probably belongs to subsp. *cuneifolia* or subsp. *flexuosa*.

3. M. simplicifolia (Roxburgh) Walpers.

Evergreen shrub or tree to 20 m. Leaves elliptic to obovate-lanceolate, acute to acuminate, entire or with very few teeth (ours), veins 7–25 pairs, mostly straight, none or few of them branching. Panicles terminal, 10–60 cm, erect, loose or dense; flower-stalks usually absent, rarely to 3 mm. Sepals apparently 8–13 (ours), 0.75–2 mm, unequal, the outer smaller and often remote from the rest. Outer petals hairless; inner petals 0.5–1.5 mm, deeply bifid. Fruit spherical, 4–10 mm across. *E Asia, from India to Indonesia & Indochina*. H5–G2. Late spring.

A widespread and highly variable species, divided by Beusekom into 8 subspecies. Only subsp. *pungens* (Walpers) Beusekom (*M. pungens* Walpers), as described above, is grown. It occurs in India, Sri Lanka and Sumatra. Other subspecies have toothed leaves and fewer apparent sepals.

4. M. pinnata (Roxburgh) Walpers.

Illustration: *Gardeners' Chronicle* **86**: 345 (1929) – as *M. oldhamii*.

Small to large evergreen or deciduous tree. Leaves pinnate with 2–11 pairs of leaflets increasing in size upwards, and terminated by 2 or 3 leaflets together (ours); leaflets elliptic, obovate or ovate-oblong. Panicle terminal, erect or somewhat pendent, 10–70 cm; flower-stalks absent or to 3 mm. Sepals 4 or 5, 1–1.5 mm, unequal, the outer somewhat smaller than the inner. Outer petals hairy or hairless, ovate or circular, not or little wider than long; inner petals 0.6–1 mm, more or less deeply bifid, the lobes fringed. Fruit 3–11 mm across. *Most of tropical and subtropical Southeast Asia*. H5–G2.

Another highly variable species, divided by Beusekom into 9 subspecies, some of them further divided into varieties; only subsp. *arnottiana* (Walpers) Beusekom var. *oldhamii* (Maximowicz) Beusekom (*M. oldhamii* Maximowicz) is grown; it has deciduous leaves and a loose panicle. *C & E China, Korea, Japan*. H4.

5. M. veitchiorum Hemsley. Illustration: *Blumea* **19**: 521 (1971); Brickell (ed.), RHS A–Z encyclopedia of garden plants, 685 (2003).

Deciduous tree to 20 cm, with short internodes. Leaves pinnate with 3–6 pairs of lateral leaflets and a single terminal leaflet; leaflets elliptic to oblong. Panicles terminal, erect or pendent, 25–60 cm, loose or dense; flower-stalks 0.5–6 mm. Sepals usually 4, 1.5–2.5 mm, the outermost often much smaller than the rest. Outer petals kidney-shaped, considerably broader than long; inner petals *c.* 1 mm, notched to halfway. Fruit spherical, more than 1 cm across. *C China.* H4. Late spring.

175. MELIANTHACEAE

Herbaceous or woody plants. Leaves alternate, pinnate, stipulate, stipules between leaf-stalk and stem, often large; ptyxis conduplicate or conduplicate-plicate. Flowers in racemes. Flowers unisexual or bisexual, zygomorphic. Calyx and corolla perigynous, stamens hypogynous, with disc. Calyx of 4–5 free sepals. Corolla of 4–5 free petals. Stamens 4–5 or rarely 10, free or united. Ovary of 4–5 united carpels; ovules 1–many per cell, axile; style 1, stigma 4–5-lobed. Fruit a capsule.

A family of 2 genera and about 13 species found in southern and tropical Africa; only one of the genera is cultivated. In edition 1, this family included the Greyiaceae.

1. MELIANTHUS Linnaeus
J.D. Wann

Evergreen perennial herbs in cultivation, becoming woody, with unpleasant odour when bruised. Leaflets toothed, asymmetric, hairless or hairy. Flowers conspicuous, solitary or 2–4 together at nodes or in racemes, profusely nectariferous. Petals 4 or 5, upper ones partly fused into a hooded tube, the lower ones forming a short hairy spur. Stamens 4, protruding, free. Ovary 4-lobed. Style usually hairy at base. Fruit a papery capsule, 4-lobed or 4-winged. Seeds black, shining.

A genus of 5 species found in South Africa, grown primarily for their foliage. They require sun and fertile, well-drained soil. Long stems may be pruned in early spring. Propagate by seed in spring or by greenwood cuttings in summer. Red spider mite may be a problem.

1a. Leaves to 50 cm; leaflets 7–11, 5–13 cm, hairless; racemes to 80 cm **1. major**
  b. Leaves not more than 20 cm; leaflets 9–13, 4–5 cm; racemes not more than 38 cm **2. comosus**

1. M. major Linnaeus. Illustration: Marloth, Flora of South Africa **2**: 160, t. 54 (1925); Eliovson, Wild flowers of southern Africa, pl. 184 (1980); Phillips & Rix, Perennials **2**: 199 (1991).

Stems to 3 m, thick, sparingly branched, terete, glaucous, ridged to winged at internodes. Leaves to 50 cm, almost erect, then arching or spreading, hairless, glaucous blue; leaflets 7–11, ovate-oblong, 5–13 × 2–2.5 cm. Stipules to 10 cm. Racemes to 80 cm. Flowers rich brownish red. *South Africa.* H5. Summer.

2. M. comosus Vahl (*M. minor* Linnaeus). Illustration: *Botanical Magazine*, 301 (1795).

Stems to 1.5 m, downy, not glaucous. Leaves to 20 cm; leaflets 9–13, oblonglanceolate, 4–5 × 1–2 cm, grey-green and coarsely downy below. Racemes to 38 cm. Flowers brick-red. *South Africa.* H5. Summer.

176. GREYIACEAE

Soft-wooded trees. Leaves alternate, simple, exstipulate. Flowers in racemes, bisexual, actinomorphic. Calyx, corolla and stamens perigynous, disc present. Calyx of 5, free, persistent sepals. Corolla of 5 free petals. Stamens 10, borne on the disc which itself bears 10 staminode-like projections. Ovary of 5 united carpels, 1-celled but parietal placentas very intrusive; style 1; ovules many. Fruit a capsule.

A single genus with 3 species, one or two of which are quite frequently cultivated.

2. GREYIA Hooker & Harvey
E.C. Nelson

Shrubs or trees with alternate, simple leaves which persist for a year or more in cultivation. Flowers red, in terminal, branched racemes; sepals 5, persistent in fruit; petals 5, not fused, overlapping to form a campanulate corolla; stamens 10, in 2 whorls, longer than corolla; ovary composed of 5 fused carpels each with numerous, minute ovules. Fruit a schizocarp, almost cylindric, with 5 prominent grooves, dehiscent.

A genus of about 3 species restricted to South Africa. These shrubby trees can be propagated from seed sown in spring, or from cuttings (use young shoots). They are easy to grow in a frost-free glasshouse, or outdoors in southern Europe, being tolerant of most soil conditions and of drought. *G. sutherlandii* will tolerate being grown in pots and will even blossom when treated as bonsai.

Literature: Palgrave, W.C., Trees of southern Africa (1991); Schönland, S., Notes on the genus *Greyia* Hooker & Harvey, *Records of the Albany Museum* **3**: 40–51 (1914).

1a. Leaves retaining dense, woolly, white or yellow hairs beneath; flowers scarlet **1. radlkoferi**
  b. Leaves hairless, or with minute hairs only (not woolly) beneath; flowers red **2. sutherlandii**

1. G. radlkoferi Szyszylowicz. Illustration: *Records of the Albany Museum* **3**: pl. 1 (1914); Palgrave, Trees of southern Africa, 548 (1991).

Tree to 5 m, or shrub; bark on young shoots yellow-green. Leaf-stalk to 10 cm, with minute glandular hairs; blade ovate to circular, to 12 × 14 cm, with tapering base, margin with coarse, shallow lobes especially towards the blunt apex, lobes irregularly toothed, green and becoming hairless above, with persistent yellowish or whitish woolly hairs beneath. Flowers campanulate, stalks to 1.5 cm, in long heads; petals to 2 cm, scarlet; stamens only slightly longer than petals; blooming before or with new foliage. *E South Africa (Transvaal N Natal).* H5-G1. Late winter–early spring.

2. G. sutherlandii Hooker & Harvey. Illustration: *Flore des serres* **7**: t. 1739 (1867); *Botanical Magazine*, **2**: 6040 (1873); Eliovson, Wild flowers of southern Africa, t. 283, 285 (1980); Palgrave, Trees of southern Africa, t. 173 (1991).

Tree to 7 m, or shrub; bark on young shoots red-grey. Leaf-stalk to 7 cm, hairless; blade ovate to circular, cordate at base, margin with shallow lobes throughout, lobes irregularly toothed, green and hairless above, paler green and hairless or sometimes with minute, fine hairs beneath. Flowers campanulate, stalks to 1.8 cm, in dense, rounded heads; petals *c.* 2 cm, red; stamens distinctly longer than petals. *E South Africa (Natal S Transvaal).* H5–G1. Spring.

A white-flowered variant is reported from South Africa. In European gardens plants are grown that are intermediate between the two species in certain characters; these may represent natural variation within the genus, or perhaps are hybrids raised from seeds harvested in gardens.

177. BALSAMINACEAE

Annual or perennial herbs, rarely almost shrubby. Stems simple or branched, often thick and fleshy. Leaves alternate or whorled, usually stalked; blades simple with scalloped or saw-toothed margin, sometimes with stalkless or stalked glands at the base or running onto the leaf-stalk; stipules absent, but sometimes nectariferous glands in their place. Flowers solitary or in clusters or borne in racemes (sometimes umbel-like), usually lateral but occasionally terminal by suppression of the stem apex. Sepals 3 or 5, the lateral 1 or 2 pairs small and inconspicuous but the lower sepal larger, boat-shaped to deeply pouched, very variable in shape, with a short or long, swollen or thread-like spur. Petals 5, free or the lateral fused into 2 pairs in which the petals may be of similar or dissimilar size and shape; upper petal flat, concave or forming a hood or helmet, often crested. Stamens 5, closely united around the ovary and falling as a single unit, often before the stigmas are receptive. Ovary with a very short style; stigma generally 5-lobed. Fruit a berry or a 5-celled fleshy capsule which explodes elastically at maturity to expel the seeds forcibly from the plant.

A family of 2 genera found primarily in tropical and subtropical areas of Africa, Madagascar, India, Sri Lanka and Southeast Asia as far southeast as New Guinea and the Solomon Islands, but with many species in temperate areas of the Himalaya, China and Japan, with a few species in Europe and North America south to Mexico. *Hydrocera triflora* (Linnaeus) Wight & Arnott from Sri Lanka to Indonesia is not in cultivation.

1. IMPATIENS Linnaeus
C. Grey-Wilson
Description as for family but lateral petals always united into 2 pairs and fruit a capsule.

A large genus of 900 or more species. The annual species can be grown in the open garden where, provided they are pollinated, seed will be set in abundance. Perennial species are generally grown in conservatories or glasshouses with shading and high humidity during the summer months. *I. walleriana* and *I. hawkeri* are widely grown as houseplants and for summer bedding schemes. The tuberous-rooted *I. tinctoria* can be overwintered in

temperate areas provided the plant is protected by a deep layer of straw and other insulating material. Seed of most perennial species is shortly viable and needs to be sown as quickly as possible. Annual species will generally self-sow in the garden and require a winter dormancy before germination can take place.

Literature: Hooker, J.D., An epitome of the British Indian species of *Impatiens*, *Records of the Botanical Survey of India* **4**(1): 110 (1904), (2): 11–35 (1905), (3): 37–58 (1906); Grey-Wilson, C., Impatiens of Africa (1980); Grey-Wilson, C., A survey of the genus *Impatiens* in cultivation, *The Plantsman* **5**(2): 86–102 (1983).

1a. Flower-stalks absent or very short, certainly not more than 5 mm　2
 b. Flower-stalks present, at least 1.5 cm, but up to 30 cm　7
2a. Leaves in distinct whorls of 3–7　**6. hawkeri**
 b. Leaves alternate, rarely almost opposite　3
3a. Erect annuals with pendent, hairy capsules　**14. balsamina**
 b. Perennials with usually hairless, generally ascending capsules　4
4a. Flowers basically red, sometimes with some yellow; lower sepal deeply pouched, curved into a short recurved spur　**16. niamniamensis**
 b. Flowers yellow, mauve, pink or purple; lower sepal more or less abruptly constricted into a straight, curved or incurved spur　5
5a. Lateral sepals 4, in 2 pairs; flowers pink, mauve or purple　**15. arguta**
 b. Lateral sepals 2, in 1 pair; flowers basically yellow　6
6a. Trailing plant with reddish stems and small hairy leaves not more than 1.2 cm　**17. repens**
 b. Erect plant with green stems and hairless leaves at least 7 cm　**18. auricoma**
7a. Leaves in distinct whorls of 3 or more　8
 b. Leaves alternate, rarely some almost opposite or opposite on the same plant　10
8a. Stout annuals; flowers pouched with a short incurved spur　**10. glandulifera**
 b. Perennials; flowers more or less flat with a long thread-like spur　9
9a. Flowers deep red, 2–2.5 cm across; spur 2.8–4.5 cm　**1. verticillata**

 b. Flowers white or pale pink, 5–8 cm across; spur 5.5–9.8 cm　**2. sodenii**
10a. Flowers essentially flat with a long thread-like spur greatly exceeding the lower sepal　11
 b. Flowers deeply pouched with a short spur not longer than the pouched lower sepal　15
11a. Flowers 4.5–6.5 cm across, white with red or magenta markings　12
 b. Flowers 1.5–4.5 cm across, red, orange, pink, mauve or purple or if white then without markings　13
12a. Petals all conspicuous; lateral sepals 2, in 1 pair; plant without underground tubers　**9. grandis**
 b. Petals not all conspicuous, the upper lateral petals far smaller than the others; lateral sepals 4, in 2 pairs; plant with underground tubers　**7. tinctoria**
13a. Leaves thick and fleshy, matt green; flowers white with a pink spur and pink pollen　**5. gordonii**
 b. Leaves thin and somewhat fleshy, rather bright, often shiny green; flowers in shades of red, orange, pink and purple, if white then spur white and pollen white or cream　14
14a. Leaves hairless, with 5–8 pairs of lateral veins, margins scalloped　**3. walleriana**
 b. Leaves hairy at least beneath, with 8–14 pairs of lateral veins, margins saw-toothed　**4. usambarensis**
15a. Flowers yellow, cream or whitish with brown spots　16
 b. Flowers purple, pink, mauve or white　18
16a. Flowers not more than 1.8 cm; lower sepal shallowly boat-shaped; petals cream or pale yellow　**13. parviflora**
 b. Flowers at least 2 cm; lower sepal deeply boat-shaped to pouched; petals deep yellow, often spotted in part　17
17a. Upper petal prominently crested; spur abruptly curved through 180°　**20. cristata**
 b. Upper petal without a pronounced crest; spur gradually curved through less than 90°　**19. noli-tangere**
18a. Lateral sepals 4, in 2 pairs; tuberous-rooted perennial　**8. flanaganae**
 b. Lateral sepals 2, in 1 pair; annual, not tuberous-rooted　19
19a. Flowers borne in simple racemes; lower sepal deeply pouched with an incurved spur　**11. sulcata**

b. Flowers not borne in simple racemes, lowermost flowers on inflorescence aggregated into false whorls; lower sepal boat-shaped with a short, straight or somewhat curved spur
12. balfourii

1. I. verticillata Wight.

Bushy hairless perennial, to 70 cm. Leaves in whorls of 3–7, stalked; blades narrowly elliptic to lanceolate, 10–15 cm, with a scalloped or saw-toothed margin. Inflorescence an umbel-like raceme with 4–12 flowers. Flowers rather flat, 2–2.5 cm across. Lower sepal boat-shaped, with a slender curved spur 2.8–4.5 cm; petals bright scarlet, the lateral united petals equal in size, ovate, upper petal similar in size to the lateral, narrowly obovate. Capsule spindle-shaped, hairless. *SW India.* G2. Summer–autumn.

2. I. sodenii Engler (*I. elgonensis* Fries; *I. magniflca* Schulze; *I. olivieri* Watson; *I. thomsonu* Oliver; *I. uguenensis* Warburg). Figure 134(8), p. 563. Illustration: Grey-Wilson, Impatiens of Africa, 90, pl. 11, 12, 14, 15 (1980); *The Plantsman* **5**(2): 91 (1983).

Stout, bushy perennial with stiff, thick, fleshy stems, reaching 1.5 m, simple to moderately branched. Leaves in dense whorls of 6–10, stalkless or shortly stalked; blades oblanceolate, 5–18 × 1.8–4.8 cm, with a saw-toothed margin. Inflorescence with 1 or 2 flowers; flowers flat, 5–8 cm across. Lower sepal shallowly boat-shaped with a slender curved spur 5.5–9.8 cm. Petals white or pale pink, often with darker markings, occasionally pure white; the lateral petals equal in size, broadly oblong to almost circular, upper petal similar in size, obcordate. Capsule spindle-shaped, hairless. *S Kenya & N Tanzania.* G1. Summer–autumn.

3. I. walleriana J.D. Hooker (*I. holstii* Engler; *I. petersiana* Grignan; *I. sultanii* J.D. Hooker). Figure 134(10), p. 563. Illustration: Grey-Wilson, Impatiens of Africa, 80, pl. 3, 4, 5 (1980); *The Plantsman* **5**(2): 91 (1983); Heywood, Flowering plants of the world, 211 (1978); Brickell (ed.), RHS A–Z encyclopedia of garden plants, 566 (2003).

Spreading to upright bushy perennial, usually hairless. Leaves alternate, stalked; blades ovate to broadly elliptic, occasionally obovate, 2.5–13 × 2–5.5 cm, green, occasionally variegated, sometimes pink or reddish beneath, with a scalloped

margin which is glandular at the base. Inflorescence with 1 or 2 (occasionally 3–5) flowers; flowers flat, 2.4–4 cm across. Lower sepal shallowly boat-shaped, with a slender curved spur 2.8–4.5 cm. Petals white, pink, mauve, violet, purple, orange, red or crimson, rarely bicoloured, lateral petals equal, obovate, upper petal similar in size, obcordate. Capsule spindle-shaped, hairless. *E Africa (from S Kenya to Malawi & Mozambique, including Zanzibar).* G1. Spring–late autumn.

Widely grown as a pot-plant and for summer bedding; often treated as an annual in horticulture.

4. I. usambarensis Grey-Wilson. Illustration: Grey-Wilson, Impatiens of Africa, 19, 20, pl. 1, 2 (1980).

Similar to *I. walleriana*, but more robust, to 1 m. Leaves narrowly ovate to oblong-elliptic with 8–14 pairs of lateral veins, hairy on the veins beneath, margin with fine forwardly pointing teeth. *NE Tanzania (Usambara Mountains).* G1. Summer–autumn.

5. I. gordonii Baker (*I. thomasettii* J.D. Hooker). Illustration: *The Plantsman* **5**(2): 91 (1983).

Erect perennial with moderately branched stems to 60 cm, hairless. Leaves alternate, stalked; blades ovate-lanceolate to lanceolate-elliptic, 8–15 × 3.2–5.8 cm, matt green, with a scalloped or saw-toothed margin. Inflorescence generally with 2 or 3 flowers; flowers flattish, 4–4.5 cm across. Lower sepal shallowly boat-shaped with a slender, curved, flesh-pink spur, 6–7.2 cm. Petals white, lateral petals equal, obovate, upper petal obcordate; anthers pink. Capsule spindle-shaped, hairless. *Seychelles (Mahe Island).* Gl. Summer–autumn.

Rare and protected in the wild.

6. I. hawkeri Bull (*I. herzogii* Schumann; *I. mooreana* Schlechter; *I. petersiana* invalid; *I. schlecteri* Warburg). Figure 134(4), p. 563.

Bushy perennial to 1 m, though often less, hairless or somewhat hairy. Leaves in whorls of 3–7, stalked; blades variable from linear-lanceolate to elliptic or ovate, 4–24 cm × 5–60 mm, green to purplish, reddish, bronze, or occasionally variegated, with a saw-toothed margin. Flowers flat, usually solitary, 4–6 cm across. Lower sepal shallowly boat-shaped, with a slender curved spur 3–7.5 cm. Petals white, red, orange, mauve, lilac, pink or purple; all

petals obcordate, more or less equal in size. Capsule spindle-shaped, hairless. *New Guinea eastwards to the Solomon Islands.* G1. Spring–autumn.

Commonly grown as an annual pot-plant, generally under the collective name New Guinea Hybrids.

7. I. tinctoria Richard subsp. **tinctoria** (*I. prainiana* Gilg). Figure 134(9), p. 563. Illustration: Grey-Wilson, Impatiens of Africa, 154, 156 (1980).

Robust herbaceous perennial with a fleshy tuberous rootstock and stout erect stems, hairless. Leaves alternate, long-stalked; blades oblong to oblong-lanceolate, 7.5–23 × 2.6–9.8 cm, bright to deep green, with a saw-toothed margin. Inflorescence a many-flowered raceme with a stalk 8.5–30 cm; flowers somewhat cupped, 4.5–6.5 cm across. Lower sepal deeply boat-shaped with a slender, curved, white spur 8–13 cm. Petals white with pink or magenta markings towards the base, except for the upper petal; lateral petals very uneven, the upper of each pair ⅓ the size of the lower, oblong, the lower semicircular, overlapping and forming a prominent lip; upper petal hooded, half the size of the lateral united petals. Capsule club-shaped, hairless. *Ethiopia, S Sudan, E Zaire, W Uganda.* H5–G1. Summer–early autumn.

Subsp. **elegantissima** (Gilg) Grey-Wilson. Illustration: Grey-Wilson, Impatiens of Africa, 154, 157, pl. 31 (1980). Occasionally seen in cultivation, spur only 3–7 cm. *C & S Kenya, E Uganda.*

8. I. flanaganae Hemsley. Illustration: Grey-Wilson, Impatiens of Africa, 159 (1980).

Like *I. tinctoria*, but less robust and with smaller flowers 1.8–2.5 cm across; spur 3.2–4 cm; petals bright pink with a yellowish base. *South Africa (Natal).* H5–G1. Summer.

Rare in cultivation.

9. I. grandis Heyne (*I. hookeriana* Arnott). Figure 134(5), p. 563. Illustration: *The Plantsman* **5**(2): 99 (1983).

Robust perennial with stout fleshy stems to 1.5 m, becoming somewhat woody with age, hairless. Leaves alternate, stalked, bright green; blades broadly elliptic to ovate, 8.5–17 × 4.6–8 cm, with a pair of glands at the base and a scalloped margin. Inflorescence an umbel-like raceme with 3–6 flowers; inflorescence-stalk 9.5–15 cm; flowers rather flat, 4.5–6 cm across. Lower

Figure 134. Flowers and foliage of *Impatiens* species (× 0.6). 1, *I. balfourii*. 2, *I. balsamina*. 3, *I. glandulifera*. 4, *I. hawkeri* (a, b, c & d, four different leaves). 5, *I. grandis*. 6, *I. naimniamensis*. 7, *I. repens*. 8, *I. sodenii*. 9. *I. tinctoria* subsp. *tinctoria*. 10, *I. walleriana*.

sepal funnel-shaped, tapering into a curved spur 5.8–7.3 cm. Petals white with bold purple or magenta markings; lateral petals somewhat dissimilar in size, obovate to obcordate with rather frilled margins; upper petal broadly obcordate. Capsule club-shaped, hairless; seed rarely produced in cultivating *SW India, Sri Lanka*. G2. Winter–spring.

Rare in cultivation.

10. I. glandulifera Royle (*I. roylei* Walpers). Figure 134(3), p. 563. Illustration: Phillips, Wild flowers of Britain, 116 (1977); *The Plantsman* **5**(2): 99 (1983); Heywood, Flowering plants of the world, 211 (1985).

Robust hairless annual with a very stout, fleshy, hollow stem to 2.5 m. Leaves in whorls of 3–5, rarely opposite, stalked; blades narrowly lanceolate to lanceolate-elliptic, 6–23 × 1.5–7 cm, deep green, with a saw-toothed margin. Inflorescence a loose, often many-flowered, raceme; flowers pouched, 2.4–3.5 cm. Lower sepal pouched with a short incurved spur which is not more than 8 mm. Petals white to pink, mauve or purple, straw-coloured and spotted towards the base; lateral petals dissimilar, the upper of each pair ¹/₃ the size of the lower, the lower forming a prominent lip; upper petal hooded. Capsule club-shaped, hairless. *W Himalaya; widely naturalised along waterways in W Europe*; H2. Summer–early autumn.

11. I. sulcata Wallich (*I. gigantea* Edgeworth). Illustration: Grierson & Long, Flora of Bhutan **2**(1): 99 (1991).

Like *I. glandulifera* but less vigorous and with alternate or opposite leaves; blades broadly lanceolate to ovate with a scalloped margin. Flowers usually dark pink or purple and with a more prominent upper petal. *C & E Himalaya*. H2. Summer–autumn.

Often mistaken for *I. glandulifera* in cultivation, where it is far rarer.

12. I. balfourii J.D. Hooker. Figure 134 (1), p. 563. Illustration: *The Plantsman* **5** (2): 99 (1983).

Hairless annual with an erect, branched stem to 50 cm. Leaves alternate, stalked; blades ovate to ovate-oblong, 2–13 × 1.5 × 7 cm, bright mid-green, with a finely scalloped margin. Inflorescence with 3–9 flowers in a raceme, the lowermost in a false whorl; flowers pouched, 2.5–4 cm. Lower sepal white,

boat-shaped with a short straight or slightly curved or bent spur 7–12 mm. Petals white with the lower lateral petals bright pink; upper lateral petals ovate, ¹/₃ the size of the lower which are asymmetrically elliptic and 2-lobed; upper petal hooded. Capsule narrowly club-shaped, hairless. *W Himalaya*. H2. Midsummer–autumn.

13. I. parviflora de Candolle. Illustration: Phillips, Wild flowers of Britain, 116 (1977).

A slender hairless annual; stems erect 10–70 cm, simple or branched. Leaves alternate, stalked, pale green, the upper crowded, larger than the lower; blades ovate-elliptic, 4–20 × 2–9 cm, with a saw-toothed margin. Inflorescence a 3–10-flowered raceme; flowers small and somewhat pouched, 6–18 mm. Lower sepal shallowly boat-shaped with a short, almost straight spur 2–5 mm. Petals cream or pale yellow; lateral petals very dissimilar in size, the upper of each pair *c.* ¼ the size of the lower; upper petal hooded. Capsule narrowly club-shaped, hairless. *C Asia*. H2. Summer–autumn.

Occasionally cultivated but widely naturalised in Europe and often considered a weed.

14. I. balsamina Linnaeus. Figure 134(2), p. 563. Illustration: Grey-Wilson, Impatiens of Africa, 223 (1980); Heywood, Flowering plants of the world, 211 (1985); Brickell, RHS gardeners' encyclopedia of plants & flowers, 264 (1989); Grierson & Long, Flora of Bhutan **2**(1): 99 (1991).

A rather fleshy, hairy or almost hairless annual, with a thick fleshy erect stem to 60 cm, simple or branched. Leaves alternate, generally rather dense, with very short stalks; blades lanceolate to narrow-elliptic or oblanceolate, 2.5–9 × 1–2.6 cm, mid- to dark green with a saw-toothed margin. Flowers pouched, borne in axillary bundles of 2 or 3, or solitary, 3–5 cm across. Lower sepal boat-shaped with a slender curved spur, 1.6–2.1 cm. Petals white to pink, orange, red or crimson, 3–5 cm across; lateral petals obcordate, the upper of each pair *c.* ½ the size of the lower; upper petal hooded. Capsule spindle-shaped, pendent, densely hairy. *C & S India; widely naturalised in tropical and subtropical areas of the world*. H5. Summer–early autumn.

Cultivated variants include those with larger double and semi-double flowers.

15. I. arguta J.D. Hooker & Thomson (*I. leveillei* J.D. Hooker; *I. pseudoarguta* J.D. Hooker). Illustration: Grierson & Long, Flora of Bhutan **2**(1): 93 (1991).

Herbaceous perennial with decumbent to erect, rather slender stems, hairless. Leaves alternate, stalked; blades elliptic to ovate-elliptic, 3–12 × 1.2–4.5 cm, deep green, margin with fine forwardly pointing teeth. Flowers solitary or in clusters of 2 or 3 at the middle stem-nodes, deeply pouched, 2.8–3.5 cm. Lower sepal deeply pouched, whitish, pink or violet, constricted into a short incurved spur 3–5 mm. Petals pink, purple or violet with a whitish or yellowish base; lateral petals markedly dissimilar in size, the upper *c.* ¹/₃ the size of the lower; upper petal hooded. Capsule narrowly club-shaped, hairless. *C & E Himalaya (Nepal to Burma), SW China*. H5. Summer–autumn.

16. I. niamniamensis Gilg (*I. congolensis* invalid). Figure 134(6), p. 563. Illustration: Grey-Wilson, Impatiens of Africa, 207, pl. 47 (1980); *The Plantsman* **5**(2): 99 (1983); Brickell (ed.), RHS A–Z encyclopedia of garden plants, 566 (2003).

Erect perennial with a thick fleshy stem, simple or sparingly branched, often reddish, hairless. Leaves alternate, stalked, congested towards the top of the stem, broadly ovate to ovate-oblong or elliptic, 5.5–22 × 3–8.5 cm, deep green, with a coarsely scalloped margin. Flowers pouched, borne in bundles of 2–6 at the nodes, held in a more or less vertical plane, partly hidden below the leaves, 2.6–3.4 cm. Lower sepal deeply pouched, curved into a short recurved spur, pink to red or crimson, often yellow or greenish towards the top. Petals pink, red or yellow; lateral sepals small, scarcely protruding from the rim of the lower sepal; upper petal hooded. Capsule spindle-shaped, hairless. *W & C tropical Africa*. G2. Autumn–early summer.

Often grown under the cultivar name 'Congo Cockatoo'.

17. I. repens Moon. Figure 134(7), p. 563. Illustration: *The Plantsman* **5**(2): 99 (1983); Brickell, RHS gardeners' encyclopedia of plants & flowers, 248 (1989).

Trailing, somewhat hairy perennial with fleshy reddish stems, often rooting at the nodes. Leaves alternate, stalked, small and rather scattered; blades kidney-shaped, 7–12 mm × 1–1.7 cm, with a scalloped margin. Flowers pouched, solitary, 1.5–2 cm across, held just above the foliage. Lower sepal deeply boat-shaped, constricted into a short incurved spur, yellowish or whitish. Petals bright yellow, netted with reddish brown towards the base; lateral petals small, just exceeding the rim of the lower sepal; upper petal hooded. Capsule spindle-shaped, hairy. *C Sri Lanka*. G2. Winter–spring.

Rare and endangered in the wild.

18. I. auricoma Baillon. Illustration: *The Plantsman* **5**(2): 99 (1983).

Bushy hairless perennial, with erect, branched stems to 80 cm. Leaves alternate, stalked; blades obovate to broadly elliptic, 7–14 × 3.5–7 cm, mid-green, with a scalloped or saw-toothed margin. Flowers pouched, solitary or in clusters of 2 or 3, 1.2–1.5 cm across. Lower sepal deeply boat-shaped, with a short recurved swollen spur, partly concealed by the large pair of lateral sepals. Petals yellow; lateral petals small, scarcely exceeding the rim of the lower sepal; upper petal hooded. Capsule spindle-shaped, hairless. *Comoro Islands*. G2. Autumn–spring.

19. I. noli-tangere Linnaeus. Illustration: Keble Martin, Concise British flora in colour, pl. 20 (1965); *The Plantsman* **5**(2): 99 (1983).

Erect hairless annual, 20–120 cm; stems usually sparingly branched. Leaves alternate, shortly stalked; blades oblong to ovate-elliptic, 1.5–10 × 1–1.8 cm, bluish green, with a coarsely scalloped margin. Inflorescence raceme-like, with 2–6 flowers, stalk 8–40 mm; flowers pouched, 2–3.5 cm. Lower sepal deeply pouched, more or less funnel-shaped, constricted into a short recurved spur, yellow with small brown dots. Petals yellow, with or without brown marks; lateral petals markedly dissimilar in size the upper of each *c.* ½ the size of the lower; upper petal hooded. *Europe & Asia*. H1. Summer–autumn.

Occasionally cultivated but often a weed close to water.

20. I. cristata Wallich (*I. scabrida* invalid). Illustration: Grierson & Long, Flora of Bhutan **2**(1): 93 (1991).

Rather rough-hairy annual with erect to prostrate branched stems to 40–70 cm.

Leaves alternate, stalked, rather pale green; blades ovate to lanceolate, 3.8–11.8 × 1.5–8 cm, margin saw-toothed. Inflorescence a short 2- or 3-flowered raceme, with a stalk 8–15 mm; flowers deeply pouched, 2.5–4 cm, often with a somewhat twisted 'face'. Lower sepal deeply pouched and somewhat twisted, with a short incurved spur, yellowish. Petals yellow or whitish with brown spots towards the base; lateral petals dissimilar in size, the upper of each pair *c.* ½ the size of the lower; upper petal hooded with a pronounced pointed crest. Capsule linear-cylindric, hairy in the centre. *W & C Himalaya*. H1. Summer–autumn.

---

178. CYRILLACEAE

Woody. Leaves alternate, simple, exstipulate. Inflorescences racemose. Flowers bisexual, actinomorphic. Calyx, corolla and stamens hypogynous, disc absent. Calyx of 5 united sepals. Corolla of 5, free or united petals. Stamens 5 or 10. Ovary of 2–4 united carpels; ovules 1–2 per cell, axile; style 1, short, stigmas 2–4. Fruit dry, indehiscent.

A family of 3 genera from warm America, with several ornamental species grown for their fragrant flowers and good autumn colour. The two regularly cultivated species are important bee trees for commercial honey production.

Literature: J.L. Thomas, A monographic study of the Cyrillaceae, *Contributions from the Gray Herbarium of Harvard University* **186**: 3–114(1960).

1a. Calyx persistent, becoming larger and hard in fruit; stamens 5
1. Cyrilla
  b. Calyx deciduous, not enlarging after flowering; stamens 10 **2. Cliftonia**

1. CYRILLA Linnaeus
S.G. Knees
Evergreen or deciduous tree or shrub, 1–2 m (rarely to 6 m). Leaves 4–10 cm,

oblanceolate to obovate, glossy deep green above, paler beneath. Flowers borne in racemes 8–15 cm on previous year's wood. Bracts acute, *c.* 2 mm; flower-stalks 2–4 mm. Calyx persistent, enlarging and becoming hard after flowering. Petals *c.* 5 mm. Stamens 5; style thick, short, stigma with 2 or 3 lobes. Fruit *c.* 2 × 3.5 mm, green at first, eventually pinkish red. Seeds 2 or 3.

A genus of one species which is very variable. Evergreen variants do not grow well outside the tropics but deciduous ones originating from Texas and neighbouring states are relatively hardy. Propagation is easy from seed or cuttings and plants should be grown on in a slightly acid, peaty loam.

1. C. racemiflora Linnaeus. Illustration: *Botanical Magazine*, 2456 (1824); Heywood, Flowering plants of the world, 125 (1978); Macoboy, What shrub is that?, 117 (1990); Gentry, A field guide to the families and genera of woody plants of northwest South America, 386 (1993).

SE USA to N South America. H5–G1. Summer.

2. CLIFTONIA Gaertner
S.G. Knees
Evergreen shrub or small tree to 5 m. Leaves 4–5 cm, obtuse, wedge-shaped at base, dark green above, shortly stalked. Flowers fragrant, in racemes 3–6 cm; calyx deciduous; petals 5, white; stamens 10; style short, stigmas 3. Fruit *c.* 6 mm, wings 2–5, papery.

A genus of one species from southern USA.

1. C. monophylla (Lamarck) Britton & Sargent (*Ptelea monophylla* Lamarck). Illustration: *Botanical Magazine*, 1625 (1813); Sargent, Silva of North America **2**: t. 52 (1892); Heywood, Flowering plants of the world, 125 (1978).

USA (Georgia to Florida). H5–G1.

GLOSSARY

abscission-zone. A predetermined layer at which leaves or fern-fronds break off.

achene. A small, dry, indehiscent, 1-seeded fruit, in which the fruit-wall is of membranous consistency and free from the seed.

aciculus. see *Nomocharis* Vol. 1, p. 102.

actinomorphic. Radially symmetric

acuminate. With a long, slender point.

adnate. Attached to an organ of a different kind (e.g. stamens adnate to corolla-tube)

adpressed. Closely applied to a leaf or stem and lying parallel to its surface, but not adherent to it.

adventitious. (1) Of roots: arising from a stem or leaf, not from the primary root derived from the radicle of the seedling. (2) Of buds: arising somewhere other than in the axil of a leaf.

aggregate fruit. A collection of small fruits, each derived from a single free carpel, closely associated on a common receptacle, but not united. *Ranunculus* and *Rubus* provide familiar examples.

allopolyploidy. Polyploidy in a plant of hybrid origin.

alternate. Arising singly, 1 at each node; not opposite or whorled (figure G1(2), p. 568).

anastomosing. Describes veins of leaves which rejoin after branching from each other or from the main vein or midrib.

anatropous. Describes an ovule which turns through 180° in the course of development, so that the micropyle is near the base of the funicle (figure G4(2), p. 571).

annual. A plant which completes its life-cycle from seed to seed in less than 1 year.

anther. The uppermost part of a stamen, containing the pollen (figure G3(3 & 4), p. 570).

apetalous. Describes a flower without a corolla (petals).

apical. Describes the attachment of an ovule to the apex of a 1-celled ovary (figure G4 (8), p. 571).

apiculate. With a small point.

apomictic. Reproducing by asexual means, though often by the agency of seeds, which are produced without the usual sexual nuclear fusion.

arachnoid. Describes hairs which are soft, long and entangled, suggestive of cobwebs.

areole. See Cactaceae, Vol. 2, p. 209.

aril. An outgrowth from the region of the hilum, which partly or wholly envelops the seed; it is usually fleshy.

ascending. Prostrate for a short distance at the base, but then curving upwards so that the remainder is more or less erect; sometimes used less precisely to mean pointing obliquely upwards.

attenuate. Drawn out to a fine point.

auricle. A lobe, normally 1 of a pair, at the base of the blade of a leaf, bract, sepal or petal.

awn. A slender but stiff bristle on a sepal or fruit.

axil. The upper angle between a leaf-base or leaf-stalk and the stem that bears it (figure G1(1), p. 568).

axile. A form of placentation in which the cavity of the ovary is divided by septa into 2 or more cells, the placentas being situated on the central axis (figure G4 (10), p. 571).

axillary. Situated in or arising from an axil (figure G1(1), p. 568).

back-cross. A cross between a hybrid and a plant similar to one of its parents.

basal. (1) Of leaves: arising from the stem at or very close to its base. (2) Of placentation: describes the attachment of an ovule to the base of a 1-celled ovary (figure G4(6 & 7), p. 571).

basifixed. Attached to its stalk or supporting organ by its base, not by its back (figure G3(3), p. 570).

berry. A fleshy fruit containing 1 or more seeds embedded in pulp, as in the genera *Berberis*, *Ribes* and *Phoenix*. Many fruits (such as those of *Ilex*) which look like berries and are usually so called in popular speech, are, in fact, drupes.

biennial. A plant which completes its lifecycle from seed to seed in a period of more than 1 year but less than 2.

bifid. Forked; divided into 2 lobes or points at the tip.

bilaterally symmetric. Capable of division into similar halves along 1 plane and 1 only (figure G3(9), p. 570).

bipinnate. Of a leaf: with the blade divided pinnately into separate leaflets which are themselves pinnately divided (figure G1 (18), p. 568).

blade. A broadened part, furthest from the base, of a petal, corolla or similar organ, which has a relatively narrow basal part – the claw or tube (figure G3(1), p. 570).

bract. A leaf-like or chaffy organ bearing flower in its axil or forming part of an inflorescence, differing from a foliage-leaf in size, shape, consistency or colour (figure G2(2), p. 569).

bracteole. A small, bract-like organ which occurs on the flower-stalk, above the bract, in some plants.

bulb. A seasonally dormant underground bud, usually fairly large, consisting of a number of fleshy leaves or leaf-bases.

bulbil A small bulb, especially one borne in a leaf-axil or in an inflorescence.

bulblet. A small bulb developing from a larger one.

caducous. Falling early.

caespitose. Forming a tuft.

calyptra. Applied to any cap-like covering.

calyx. The sepals; the outer whorl of a perianth (figure G3(1), p. 570).

campanulate. Bell-shaped; a broad tube terminating in a flared limbs or lobes.

campylotropous. Describes an ovule which becomes curved during development and lies with its long axis at right angles to the funicle (figure G4(4), p. 571).

capitate. Compact and approximately spherical, head-like.

capitellate. Grouped into very small, dense or compact clusters.

capitulum. An inflorescence consisting of small flowers (florets), usually numerous, closely grouped together so as to form a 'head', and often provided with an involucre.

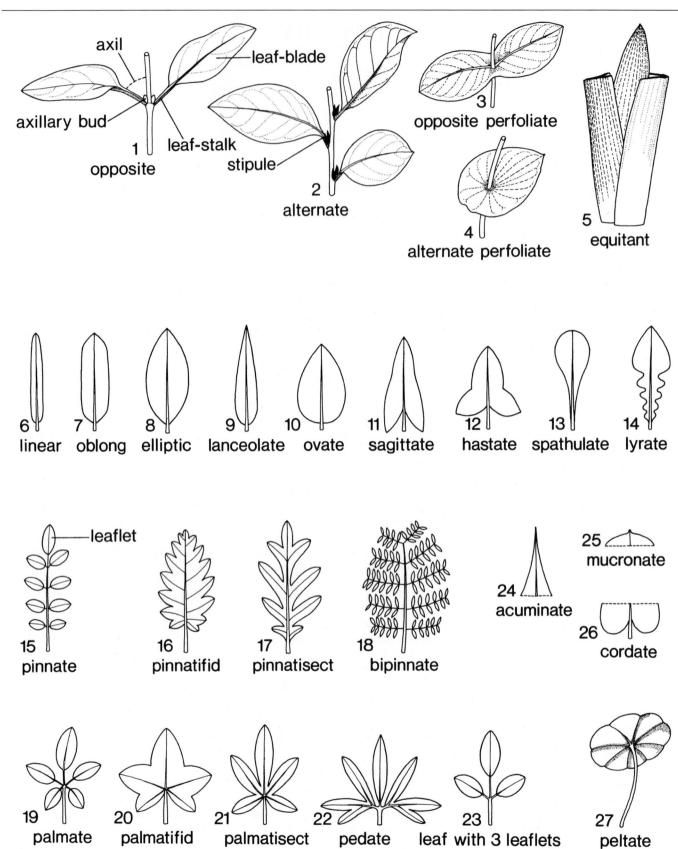

Figure G1. Leaves. 1–5, leaf insertion types. 6–14, leaf-blade outlines. 15–23, Leaf dissection types. 24–26, Leaf apex and base shapes. 27, Attachment of leaf-stalk to leaf-blade.

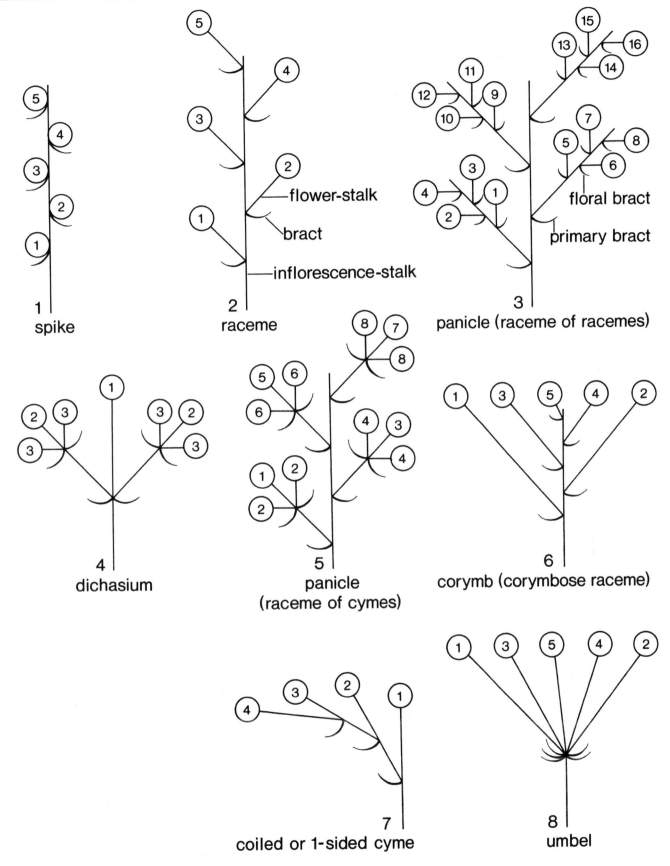

Figure G2. Inflorescences.

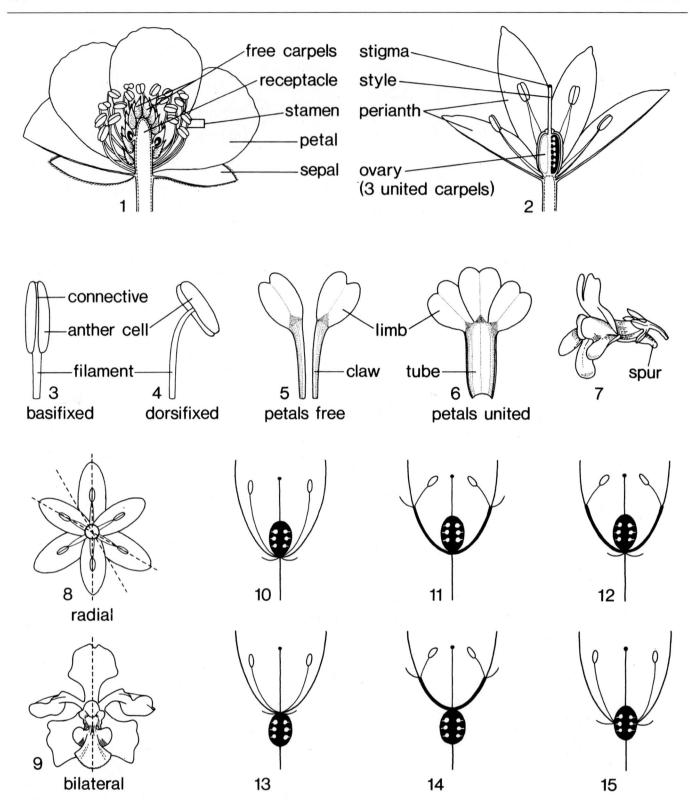

Figure G3. 1, 2, Two flowers illustrating floral parts. 3, 4, Two stamens showing alternative types of anther attachment. 5–7, Some terms relating to petals. 8, 9, Floral symmetry, planes of symmetry shown by broken lines. 10–15, Position of ovary. 10–12, Superior ovaries. 13, 14, Inferior ovaries. 15, Half-inferior ovary. 11, Perigynous zone bearing sepals, petals and stamens. 12, Perigynous zone bearing petals and stamens. 14, Epigynous zone bearing sepals, petals and stamens.

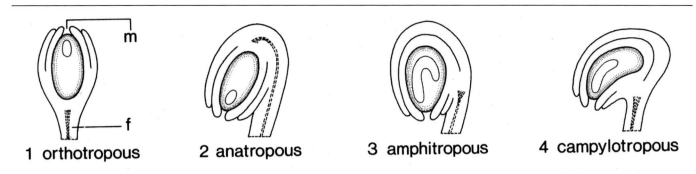

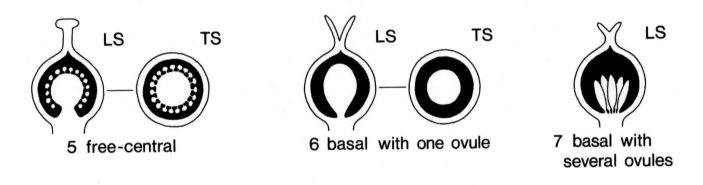

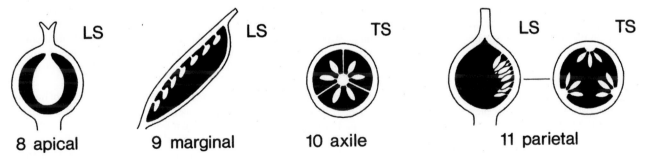

Figure G4. Ovules and placentation. 1–4, Ovule forms (f, funicle; m, micropyle). 5–11, Placentation types (LS, longitudinal section; TS, transverse section).

capsule. A dry, dehiscent fruit derived from 2 or more united carpels and usually containing numerous seeds.

carpel. One of the units (sometimes interpreted as modified leaves) situated in the centre of a flower and together constituting the gynaecium or female part of the flower. If more than 1, they may be free or united. They contain ovules and bear a stigma (figure G3(1 & 2), p. 570).

carpophore. See *Silene*, (Vol. 2, p. 184).

caruncle. A soft, usually oil-rich appendage attached to a seed near the hilum.

catkin. An inflorescence of unisexual flowers, made up of relatively conspicuous, usually overlapping bracts, each of which subtends a small apetalous flower or a group of such flowers; catkins are generally pendent, but some are erect.

cephalium. See Cactaceae, vol. 2, p. 209.

chromosome. One of the small, thread-like or rod-like bodies consisting of nucleic acid and containing the genes, which appear in a cell nucleus shortly before cell division.

cilate. Fringed on the margin with usually fine hairs (cilia).

circinate. Coiled at the tip, so as to resemble a crozier.

cladode. A branch which takes on the functions of a leaf (the leaves being usually vestigial). It may be flattened, as in *Ruscus* (vol. 1, p. 168), or needle-like, as in many species of *Asparagus* (vol. 1, p. 164).

clavate. Club-shaped, thickened towards the apex.

claw. The narrow base of a petal or sepal, which widens above into the limb or blade (figure G3(5), p. 570).

cleistogamous. Describes a flower with reduced corolla, which does not open but sets seed by self-pollination.

clone. The sum-total of the plants derived from the vegetative reproduction of an individual, all having the same genetic constitution.

column. (1) A solid structure in the centre of a flower, consisting of the style and stigmas united to the stamen or stamens. It is characteristic of the Orchidaceae (Vol 1. figures 22–23, p. 444, 445). (2) see Vol. 1, p. 339 (Graminae: *Stipa*).

compound. (1) Of a leaf: divided into separate leaflets. (2) Of an inflorescence: bearing secondary inflorescences in place of single flowers. (3) Of a fruit: derived from more than 1 flower.

compressed. Flattened from side to side.

comus. A tuft of infertile flowers or bracts sometimes found at the top of a dense inflorescence.

conduplicate. Of young leaves or leaflets in bud: folded upwards along the midrib.

cone. A compact, cylindrical or shortly conical inflorescence or fruiting inflorescence with closely overlapping bracts or scales.

connate. Attached to other organs of the same kind (e.g. petals connate).

connective. The tissue which separates the 2 lobes of an anther, and to which the filament is attached (figure G3(3), p. 570).

cordate. Describes the base of a leaf-blade which has a rounded lobe on either side of the central sinus (figure G1(26), p. 568).

corm. An underground, thickened stem-base, often surrounded by papery leaf-bases, and superficially resembling a bulb.

corolla. The petals; the inner whorl of the perianth (figure G3(1), p. 570).

corona. (1) A tubular or ring-like structure attached to the inside of the perianth (or perigynous or epigynous zone), either external to the stamens or united with their filaments; it is usually lobed or dissected (figure 163, Vol. 4, p. 497). (2) A ring-like structure or circlet of appendages on the outside of a tube formed by united filaments.

coronal scale. See *Silene*, vol. 2, p. 184

corymb. A broad, flat-topped inflorescence. In the strict sense the term indicates a raceme in which the lowest flowers have stalks long enough to bring them to the level of the upper ones (figure G2(6), p. 569), but the term *corymbose* is often used to indicate a flat-topped cymose inflorescence.

costapalmate. See Vol. 1, p. 361 (Palmae).

cotyledon. One of the leaves preformed in the seed.

crisped. (1) Of hairs: strongly curved, so that the tip lies near the point of attachment. (2) Of leaves, leaflets or petals: finely and complexly wavy.

cristate. With elevated, irregular ridges.

crownshaft. See Vol. 1, p. 361 (Palmae).

culm. See Vol. 1, p. 324 (Gramineae).

cultivar. A variant of horticultural interest or value, maintained in cultivation, and not conveniently equatable with an infraspecific category in botanical classification. A cultivar may arise in cultivation or may be brought in from the wild. Its distinguishing name can be

Latin in form, e.g. 'Alba', but is more usually in a modern language, e.g. 'Madame Lemoine', 'Frühlingsgold', 'Beauty of Bath'.

cuneate. Wedge-shaped.

cupule. A group of bracts, united at least at the base, surrounding the base of a fruit or a group of fruits.

cyme. An inflorescence in which the terminal flower opens first, other flowers being borne on branches arising below it (figure G2(5 & 7), p. 569).

cystolith. A concretion of calcium carbonate found within the cells of the leaf in some plants; they can sometimesbe seen when the leaf is viewed against the light, or felt as tiny hard lumps when the leaf is drawn between finger and thumb.

decumbent. More or less horizontal for most of its length, but erect or semi-erect near the tip.

decurrent. Continued down the stem below the point of attachment as a ridge or ridges.

dehiscent. Splitting, when ripe, along 1 or more predetermined lines of weakness.

dendroid. Branched, tree-like.

dichasial. Resembling a dichasium.

dichasium. A form of cyme in which each node bears 2 equal lateral branches (figure G2(4), p. 569).

dichotomous. Dividing into 2 equal branches; regularly forked.

dimorphic. Of two forms.

dioecious. With male and female flowers or cones on separate plants.

diploid. Possessing in its normal vegetative cells 2 similar sets of chromosomes.

disc. A variously contoured, ring-shaped or circular area (sometimes lobed) within a flower, from which nectar is secreted.

dissected. Deeply divided into lobes or segments.

distichous. Alternate, but arranged on opposite sides of the stem, not spiral.

distylic. Having the flowers of different plants either with long styles and shorter stamens or with long stamens and shorter styles.

domatia. Pockets or dense tufts of hairs in the axils of the main veins on the underside of a leaf.

dorsifixed. Attached to its stalk or supporting organ by its back, usually near the middle (figure G3(4), p. 570).

double. Of flowers: with petals much more numerous than in the normal wild state.

drupe. An indehiscent fruit in which the outer part of the wall is soft and usually fleshy, but the inner part stony. A drupe may be 1-seeded as in *Prunus* or *Juglans*, or may contain several seeds, as in *Ilex*. In the latter case each seed is enclosed in a separate stony endocarp and constitutes a pyrene.

drupelet. A miniature drupe forming part of an aggregate fruit.

ellipsoid. As *elliptic* but applied to a solid body.

elliptic. About twice as long as broad, tapering equally both to the tip and the base (figure G1(8), p. 568).

emarginate. Notched.

embryo. The part of a seed from which the new plant develops; it is distinct from the endosperm and seed-coat.

endocarp. The inner, often stony layer of a fruit-wall in those fruits in which the wall is distinctly 3-layered.

endosperm. A food-storage tissue found in many seeds, but not in all, distinct from the embryo and serving to nourish it and the young seedling during germination and establishment.

entire. With a smooth, uninterrupted margin; not lobed or toothed.

epicalyx. A group of bracts attached to the flower-stalk immediately below the calyx and sometimes partly united with it.

epigeal. The mode of germination in which the cotyledons appear above ground and carry on photosynthesis during the early stages of establishment.

epigynous. Describes a flower, or preferably the petals, sepals and stamens (or perianth and stamens) of a flower in which the ovary is inferior (figure G3(13 & 14), p. 570).

epigynous zone. A rim or cup of tissue on which the sepals, petals and stamens are borne in some flowers with inferior ovaries (figure G3(14), p. 570).

epiphyte. A plant which grows on another plant but does not derive any nutriment from it.

equitant. Used of leaves folded so that they are V-shaped in section at the base, the bases overlapping regularly, as in many Iridaceae (figure G1(5), p. 568).

exocarp. The outer, skin-like layer of a fruit-wall in those fruits in which the wall is distinctly 3-layered.

exstipulate. Without stipules.

fall. See Iridaceae (figure 10, Vol. 1, p. 243).

farina. The flour-like wax present on the stem and leaves of many species of Primula and of a few other plants.

fascicle. A bunch of leaves often enclosed at the base by a sheath.

fasciculate. A cluster or bundle, the unit of which are independent but appear to arise from a common point.

fastigiate. With all branches more or less erect, giving the plant a narrow tower-like outline.

filament. The stalk of a stamen, bearing the anther at its tip (figure G3(3 & 4) p. 570).

filius. Used with authority names to distinguish between parent and offspring when both have given names to species, e.g. Linnaeus (C. Linnaeus, 1707–1778), Linnaeus filius (C. Linnaeus, 1741–1783, son of the former).

fimbriate. Fringed.

flagellate. Bearing whip-like structures

floret. A small flower, aggregated with others into a compact inflorescence.

folioferous. Bearing leaf-like structures.

follicle. A dry dehiscent fruit derived from a single free carpel, and with a single line of dehiscence.

free. Not united to any other organ except by its basal attachment.

free-central. A form of placentation in which the ovules are attached to the central axis of a 1-celled ovary (figure G4(5), p. 571).

fruit. The structure into which the gynaecium is transformed during the ripening of the seeds; a *compound fruit* is derived from the gynaecia of more than one flower. The term 'fruit' is often extended to include structures which are derived in part from the receptacle (*Fragaria*), epigynous zone (*Malus*) or inflorescence-stalk (Ficus) as well as from the gynaecium.

funicle. The stalk of an ovule (figure G4(1), p. 571).

fusiform. Spindle-shaped; cylindric, but tapered gradually at both ends.

gland-dotted. With minute patches of secretory tissue usually appearing as pits on the surface, as translucent dots when held up to the light, or both.

glandular. (1) Of a hair: bearing at the tip a usually spherical knob of secretory tissue. (2) Of a tooth: similarly knobbed or swollen at the tip.

glaucous. Green strongly tinged with bluish grey; with a greyish waxy bloom.

glochid. See Cactaceae, Vol. 2, p. 209.

glume. A small bract in the inflorescence of a grass (Gramineae) or sedge (Cyperaceae); also used in a narrower sense to denote the 2 small bracts at the base of a grass-spikelet (figure 13, Vol. 1, p. 326).

graft-hybrid. A plant which, as a consequence of grafting, contains a mixture of tissues from 2 different species. Normally the tissues of 1 species are enclosed in a 'eskin' of tissue from the other species.

gynaecium. The female organs (carpels) of a single flower, considered collectively, whether they are free or united.

gynophore. The stalk which is present at the base of some pvaries (and the fruits developed from them).

half-inferior. Of an ovary: with its lower part inferior and its upper part superior (figure G3(15), p. 570).

haploid. Possessing in its normal vegetative cells only a single set of chromosomes.

hastate. With 2 acute, divergent lobes at the base, as in a mediaeval halberd (figure G1 (12), p. 568).

hastula. See Vol. 1, p. 361 (Palmae).

haustorium. The organ with which a parasitic plant penetrates its host and draws nutriment from it.

herb. A plant in which the stems do not become woody, or, if somewhat woody at the base, do not persist from year to year.

herbaceous. Of a plant: possessing the qualities of a herb as defined above.

heterostylic. Having flowers in which the length of the style relative to that of the stamens varies from one plant to another.

hilum. The scar-like mark on a seed indicating the point at which it was attached to the funicle.

hybrid. A plant produced by the crossing of parents belonging to 2 different named groups (e.g. genera, species, subspecies etc.). An F1 hybrid is the primary product of such a cross. An F2 hybrid is a plant arising from a cross between 2 F1 hybrids (or from the self-pollination of an F1 hybrid).

hydathode. A water-secreting gland immersed in the tissue of a leaf near its margin.

hypocotyl. That part of the stem of a seedling which lies between the top of the radicle and the attachment of the cotyledon(s).

hypogeal. The mode of germination in which the cotyledons remain in the seed-coat and play no part in photosynthesis.

hypogynous. Describes a flower, or, preferably the petals, sepals and stamens (or perianth and stamens) of a flower in which the ovary is superior and the petals, sepals and stamens (or perianth and stamens) arise as individual whorls on the receptacle (figure G3(10), p. 570).

incised. With deep, narrow spaces between the teeth or lobes.

included. Not projecting beyond the organs which enclose it.

indefinite. More than 12 and possibly variable in number.

indehiscent. Without preformed lines of splitting; opening, if at all, irregularly by decay.

indumentum. Any hairy covering.

induplicate. See Vol. 1, p. 361 (Palmae).

inferior. Of an ovary: borne beneath the sepals, petals and stamens (or perianth and stamens) so that these appear to arise from its top (figure G3(13 & 14), p. 570).

inflorescence. A number of flowers which are sufficiently closely grouped together to form a structural unit (figure G2, p. 569).

infraspecific. Denotes any category below species level, such as subspecies, variety and form. To be distinguished from *subspecific*, which means relating to subspecies only.

integument. The covering of an ovule, later developing into the seed-coat. Some ovules have a single integument, others 2.

internode. The part of a stem between 2 successive nodes.

interprimary vein. See Vol. 1, p. 373 (Araceae).

involucel. A whorl of united bracteoles borne below the flower in some Dipsacaceae.

involucre. A compact cluster of bracts around the stalk at or near the base of some flowers or inflorescences or around the base of a capitulum; sometimes reduced to a hair or ring of hairs.

keel. A narrow ridge, suggestive of the keel of a boat, developed along the midrib (or rarely other veins) of a leaf, petal, sepal, or glume.

labellum. See Orchidaceae, vol. 1, p. 439.

lacinidate. With the margin deeply and irregularly divided into narrow and unequal teeth.

lamellate. Made up of thin plates.

lanceolate. 3–4 times as long as wide and tapering more gradually towards the tip (figure G1(9), p. 568).

layer. To propagate by pegging down on the ground a branch from near the base of a shrub or tree, so as to induce the formation of adventitious roots.

leaflet. One of the leaf-like components of a compound leaf (figure G1(15), p. 568).

legume. The typical fruit of the Fabaceae, formed from a single carpel and opening down both sutures; legumes are, however, variable, some indehiscent, others breaking into 1-seeded segments, etc.

lemma. The lower and usually stouter of the 2 horny or membranous bracts which enclose the floret of a grass (Gramineae) (figure 13), Vol. 1, p. 326).

lenticel. A small, slightly raised interruption of the surface of the bark (or of the outer corky layer of a fruit) through which air can penetrate to the inner tissues.

lignotuber. A woody tuber, generally underground.

ligule. (1) A small membranous flap (more rarely a line or hairs) at the base of a leaf-blade. (figure 13, Vol. 1, p. 326. (2) A whitish area, superficially somewhat similar to the ligule of a grass at the base of the upper surface of the leaf-blade in members of the Cyperaceae (figure 12, Vol. 1, p. 303).

limb. A broadened part, furthest from the base, of a petal, corolla or similar organ, which has a relatively narrow basal part – the *claw* or tube (figure G3(5 & 6), p. 570).

linear. Parallel-sided and many times longer than broad (figure G1(6), p. 568).

lip. (1) That petal in an orchid flower which differs from the other 2; it usually occupies the lowest position in the flower and serves as an alighting place for insects (figure 11, Vol. 1, p. 271). (2) A staminode or petal, suggestive of the lip of an orchid, in the flower of some other monocotyledonous families (e.g. Zingiberaceae). (3) A major division of the apical part of a bilaterally symmetric calyx or corolla in which the petals or sepals are united; there is normally an upper and a lower lip, but either may be missing (e.g. Labiatae).

lodicule. Small, scale-like structures in the floret of most grasses (Gramineae) which occure between the lemma and the reproductive organs. By their swelling they cause the lemma and palea to diverge at flowering time (figure 13(8), Vol. 1, p. 326).

lyrate. Pinnatifid or pinnatisect, with a large terminal and small lateral lobes.

marginal. Of placentation: describing the placentation found in a free carpel of which contains more than 1 ovule.

medifixed. Of a hair: lying parallel to the surface on which it is borne and attached

to it by a stalk (usually short) at its mid-point.

membranous. Thin and semi-transparent.

mericarp. A carpel, usually 1-seeded, released by the break-up at maturity of a fruit formed from 2 or more joined carpels.

mesocarp. The central, often fleshy layer of a fruit-wall in those fruits in which the wall is distinctly 3-layered.

micropyle. A pore in the integument(s) of an ovule and the coat of a seed (figure G4 (1), p. 571)

monocarpic. Flowering and fruiting once and then dying.

monoecious. With separate male and female flowers or cones on the same plant; in flowering plants male flowers may contain non-functional carpels (and *vice versa*).

monopodial. A type of growth-pattern in which the terminal bud continues growth from year to year.

mucronate. Provided with a short narrow point at the apex (figure G1(25), p. 568).

mycorrhiza. A symbiotic association between the roots of a green plant and a fungus.

nectary. A nectar-secreting gland.

neuter. Without either functional male or female parts.

node. The point at which 1 or more leaves or flower parts are attached to an axis.

nut. A 1-seeded indehiscent fruit with a woody or bony wall.

nutlet. A small nut, often a component of an aggregate fruit.

obconical. Shaped like a cone, but attached at the narrow end.

obcordate. Inversely heart-shaped, the notch being apical.

oblanceolate. As *lanceolate*, but attached at the more gradually tapered end.

oblong. With more or less parallel sides and about 2–5 times as long as broad (figure G1(7), p. 568).

obovate. As *ovate*, but attached at the narrower end.

obovoid. As *ovoid*, but attached at the narrower end.

obsolete. Rudimentary: scarcely visible; reduced to insignificance.

ochrea. A sheath, made up of the stipules of each leaf-base, which surrounds the stem in many species of Polygonaceae; the ochrea is oftern scarious.

opposite. Describes 2 leaves, branches or flowers attached on opposite sides of the axis at the same node.

orthotropous. Describes an ovule which stands erect and straight (figure G4(1), p. 571).

ovary. The lower part of a carpel, containing the ovule(s) (i.e. excluding style and stigma); the lower, ovule-containing part of a gynaecium in which the carpels are united (figure G3(2), p. 570).

ovate. With approximately the outline of a hen's egg (though not necessarily blunt-tipped) and attached at the broader end (figure G1(10), p. 568).

ovoid. As *ovate*, but applied to a solid body,

ovule. The small body from which a seed develops after pollination (figure G4, p. 571).

palea. The upper, and usually smaller and thinner, of the 2 bracts enclosing the floret of grass (Gramineae) (figure13, Vol. 1, p. 326).

palate. A swelling in the mouth of the corolla-tube.

palmate. Describes a compound leaf composed of more than 3 leaflets, all arising from the same point, as in the leaf of *Aesculus* (figure G1(19), p. 568).

palmatifid. Lobed in a palmate manner, with the incisions pointing to the place of attachment, but not reaching much more than halfway to it (figure G1(20), p. 568).

palmatisect. Deeply lobed in a palmate manner, with the incisions almost reaching the base (figure G1(21), p. 568).

panicle. A compound raceme, or any freely branched inflorescence of similar appearance (figure G2(3 & 5), p. 569).

papillose. Covered with small blunt protuberances (papillae).

parietal. A form of placentation in which the placentas are borne on the inner surface of the walls of a 1-celled ovary (figure G4 (11), p. 571).

pectinate. With leaves or leaflets in 2 opposite, regular, eye-lash-like rows.

pedate. With a terminal lobe or leaflet, and on either side of it an axis curving outwards and backwards, bearing lobes or leaflets on the outer side of the curve (figure G1(22), p. 568).

peltate. Describes a leaf or other flat structure with a stalk attached other than at the margin (figure G1(27), p. 568).

pepo. A gourd fruit; a 1-celled, many-seeded, inferior frui, with parietal placentas and pulpy interior.

perennial. Persisting for more than 2 years.

perfoliate. Describes a pair of stalkless opposite leaves of which the bases are united, or a single leaf in which the auricles are united, so that the stem appears to pass through the leaf or leaves (figure G1(3 & 4), p. 568).

perianth. The calyx and corolla considered collectively, used especially when there is no clear differentiation between calyx and corolla; used also to denote a calyx or corolla when the other is absent (figure G3(2), p. 570).

pericarpel. See Cactaceae, Vol. 2, p. 211.

perigynous. Describing a flower, or, preferably the petals, sepals and stamens (or perianth and stamens) of a flower in which the ovary is superior and the petals, sepals and stamens (or perianth and stamens) are borne on the margins of a rim or cup which itself is borne on the receptacle below the ovary (it often appears as though the sepals, petals and stamens, or perianth and stamens, are united at their bases) (figure G3(11 & 12), p. 570).

perigynous zone. The rim or cup of tissue on which the sepals, petals and stamens (or perianth and stamens) are borne in a perigynous flower (figure G3(11 & 12), p. 570).

petal. A member of the inner perianth-whorl (corolla) used mainly when this is clearly differentiated from the calyx. The petals usually function in display and often provide an alighting place for pollinators (figure G3(1), p. 570).

petaloid. Like a petal in texture and colour.

phyllode. A leaf-stalk taking on the function and, to a variable extent, the form of a leaf-blade.

pinna. The primary division of a pinnate frond or leaf; it may be simple or itself divided.

pinnate. Describes a compound leaf or frond in which distinct leaflets or pinnae are arranged on either side of the axis or rachis (figure G1(15), p. 568). If these leaflets are themselves of a similar compound structure the leaf or frond is termed bipinnate (similarly, tripinnate, etc.).

pinnatifid. Lobed in a pinnate manner, with the incision reaching not much more than halfway to the axis or rachis (figure G1(16), p. 568).

pinnatisect. Deeply lobed in a pinnate manner, with the incisions almost reaching the axis or rachis (figure G1 (17), p. 568).

pistillode. A sterile ovary in a male flower.

placenta. A part of the ovary, often in the form of a cushion or ridge, to which ovules are attached.

placentation. The manner of arrangement of the placentas.

plicate. Of young leaves in bud, folded along the main veins.

pollen-sac. One of the cavities in an anther in which the pollen is formed; in flowering plants each anther normally contains 4 pollen-sacs, 2 on either side of the connective and separated by a partition which shrivels at maturity.

pollinium. Regularly shaped masses of pollen formed by a large number of pollen grains cohering (Orchidaceae: figure 23, Vol. 1, p. 445).

polyploid. Possessing in its normal vegetative cells more than 2 sets of chromosomes.

proliferous. Giving rise to plantlets or additional flowers on stems, leaves or in the inflorescence.

protandrous. With anthers beginning to shed their pollen before the stigmas in the same flower are receptive.

protogynous. With stigmas becoming receptive before the anthers in the same flower shed their pollen,

pseudobulb. A swollen, above-ground internode, green (at least when young), (figure 19, Vol. 1, p. 441).

ptyxis. The way in which a leaf is rolled or folded (or not) in bud.

pulvinus. A swollen region at the base of a leaflet, leaf-blade or leaf-stalk.

pustulate. With low blister-like projections.

pyrene. A small nut-like body enclosing a seed, 1 or more of which, surrounded by fleshy tissue, make up the fruit of, for example, *Ilex.*

raceme. An inflorescence consisting of stalked flowers arranged on a single axis, the lower opening first (figure G2(2), p. 569).

rachilla. See Vol. 1, p. 324 (Gramineae).

rachis. Of a compound fern frond or pinna: the central axis.

radially symmetric. Capable of division into 2 similar halves along 2 or more planes of symmetry (figure G3(8), p. 570).

radiate. (1) Spreading outwards from a common centre. (2) Possessing ray flowers as in Compositae.

radicle. The root preformed in the seed and normally the first visible root of a seedling.

ramiform. Bearing branches.

raphe. A perceptible ridge or stripe, at one end of which is the hilum, on some seeds.

receptacle. The tip of an axis to which the floral parts or perigynous zone (when present) are attached (figure G3(1), p. 570).

reduplicate. See Vol. 1, p. 361 (Palmae).

reflexed. Bent sharply backwards from the base.

resupinate. See Vol. 1, p. 439 (Orchidaceae).

rhizome. A horizontal stem, situated underground or on the surface, serving the purpose of food storage or vegetative reproduction or both; roots or stems arise from some or all of its nodes.

rootstock. The compact mass of tissue from which arise the new shoots of a herbaceous perennial. It usually consists mainly of stem tissue, but is more compact than is generally understood by rhizome.

rostellum. See Vol. 1, p. 439 (Orchidaceae).

ruminate. See Vol. 1, p. 361 (Palmae).

runner. A slender above-ground stolon with very long internodes.

sagittate. With a backwardly directed basal lobe on either side, like an arrow-head (figure G1(11), p. 568).

samara. A winged, dry, indehiscent fruit or mericarp.

saprophytic. Dependent for its nutrition on soluble organic compounds in the soil. Saprophytic plants do not photosynthesise and lack chlorophyll; some plants, however, are *partially saprophytic* and combine the two modes of nutrition.

scabrid. Roughly hairy.

scale-leaf. A reduced leaf, usually not photosynthetic.

scape. A leafless flower-stalk or inflorescence-stalk arising usually from ground level.

scarious. Dry and papery, often translucent.

schizocarp. A fruit which, at maturity, splits into its constituent mericarps.

scion. A branch cut from one plant to be grafted on the rooted stock of another.

seed. A reproductive body adapted for dispersal, developed from an ovule and consisting of a protective covering (the seed-coat), an embryo and, usually, a food reserve.

semi-parasite. A plant which obtains only part of its nourishment by parasitism.

sepal. A member of the outer perianth whorl (calyx) when 2 whorls are clearly differentiated as calyx and corolla, or when comparison with related plants shows that the corolla is absent. The sepals most often function in protection and support of other floral parts (figure G3(1), p. 570).

septum. An internal partition.

sheath. The part of a leaf or leaf-stalk which surrounds the stem, being either tubular or with free but overlapping edges.

shrub. A woody plant with several stems or branches from near the base, and of smaller stature than a tree.

simple. Not divided into separate parts.

sinus. The gap or indentation between 2 lobes or teeth.

spadix. A spike with numerous florets borne on a usually fleshy axis which may project beyond the topmost floret, often wholly or partly enclosed by a large bract (*spathe*) (figure 15, Vol. 1, p. 377).

spathe. A large bract at the base of a flower or inflorescence and wholly or partly enclosing it. The term is used in some families to denote collectively 2 or 3 such bracts.

spathulate. With a narrow basal part, which towards the apex is gradually expanded into a broad blunt blade.

spicate. Similar to a spike.

spike. An inflorescence or subdivision of an inflorescence consisting of stalkless flowers arranged on a single axis (figure G2(1), p. 569).

spikelet. A small spike forming one of the units of a complex inflorescence (figure 13, Vol. 1, p. 326).

spur. An appendage or prolongation, more or less conical or cylindric, often at the base of an organ. The spur of a corolla or single petal or sepal is usually hollow and often contains nectar (figure G3(7), p. 570)

stamen. The male organ, producing pollen, generally consisting of an anther borne on a filament (figure G3(1), p. 570).

staminode. An infertile stamen, often reduced or rudimentary or with a changed function.

standard. See Iridaceae (figure 10, Vol. 1, p. 243).

stellate. Star-like, particularly of branched hairs.

stigma. The part of a style to which pollen adheres, normally differing in texture from the rest of the style (figure G3(2), p. 570).

stipule. An appendage, usually 1 of a pair, found beside the base of the leaf-stalk in many flowering plants, sometimes falling early, leaving scars. In some cases the 2 stipules are united; in others they are partly united to the leaf-stalk.

stock. A rooted plant, often with the upper part removed, on to which a scion may be grafted.

stolon. A far-creeping, more or less slender, above-ground or underground rhizome giving rise to a new plant at its tip and sometimes at intermediate nodes.

stoma. A microscopic ventilating pore in the surface of a leaf or other herbaceous part of a plant.

strophide. An oily appendage on a seed.

style. The usually slender upper part of a carpel or gynaecium, bearing the stigma (figure G3(2), p. 570).

subtend(ed). Used of any structure (e.g. a flower) which occurs in the axil of another organ (e.g. a bract); in this case the bract subtends the flower.

subulate. Narrowly cylindric, and somewhat tapered to the tip.

sucker. An erect shoot originating from a bud on a root or a rhizome, sometimes at some distance from the stem of the parent plant.

superior. Of an ovary: borne at the morphological apex of the flower so that the petals, sepals and stamens (or perianth and stamens) arise on the receptacle below the ovary (figure G3 (10–12), p. 570).

supervolute. Of young leaves in bud: rolled vertically so that one margin is inside and one outside.

suture. A line marking an apparent junction of neighbouring organs.

sympodial. A type of growth pattern in which the terminal bud ceases growth (often by the production of an inflorescence), further growth being carried on by lateral buds (see Vol. 1, p. 439 (Orchidaceae)).

syncarp. A multiple fruit made up of the products of several ovaries.

tendril. A thread-like structure which by its coiling growth can attach a shoot to something else for support.

terete. Approximately circular in cross-section; not necessarily perfectly cylindric, but without grooves or ridges.

tessellated. With a chequered pattern of light and dark squares.

tetraploid. Possessing in its normal vegetative cells 4 similar sets of chromosomes.

throat. The part of a calyx or corolla transitional between tube and limb or lobes.

trichome. See Cactaceae, Vol. 2, p. 209.

trigonous. 3-angled.

triploid. Possessing in its normal vegetative cells 3 similar sets of chromosomes. Most triploid plants are highly sterile.

tristylic. Having the flowers of different plants with long, short, or intermediate-length styles; the stamens of each flower are of 2 lengths which are not the same as the style-length of that flower.

truncate. As though with the tip or base cut off at right angles.

tuber. A swollen underground stem or root used for food-storage.

tubercle. A small, blunt, wart-like protuberance.

tunic. The dead covering of a bulb or corm.

turion. A specialised perennating bud in some aquatic plants, consisting of a short shoot covered in closely packed leaves, which persists through the winter at the bottom of the water.

umbel. An inflorescence in which the flower-stalks arise together from the top of an inflorescence-stalk. This is a *simple umbel* (figure G2(8), p. 569); in a *compound umbel* the several stalks arising from the top of the inflorescence-stalk terminate not in flowers but in secondary umbels.

undivided. Without major divisions or incisions, though not necessarily entire.

urceolate. Shaped like a pitcher or urn, hollow and contracted at or just below the mouth.

valve. The part of the fruit, covering the seeds, that falls from the septum during dehiscence (Cruciferae).

vascular bundle. A strand of conducting tissue, usually surrounded by softer tissue.

vein. A vascular strand, usually in leaves or floral parts and visible externally.

venation. The pattern formed by the veins in a leaf.

ventricose. Swollen or inflated on one side.

versatile. Of an anther: flexibly attached to the filament by its approximate mid-point so that a rocking motion is possible.

vessel. A microscopic water-conducting tube formed by a sequence of cells not separated by end-walls.

viscidium. See Vol. 1, p. 439 (Orchidaceae).

viviparous. Bearing young plants, bulbils or leafy buds which can take root; they can occur anywhere on the plant and may be interspersed with, or wholly replace, the flowers in an inflorescence.

whorl. A group of more than 2 leaves or floral organs inserted at the same node.

wing. A thin, flat extension of a fruit, seed, sepal or other organ.

xerophytic. Drought tolerant. Can also describe the environment in which drought-tolerant plants live.

zygomorphic. Bilaterally symmetric.

INDEX